Geophysical Framework of the Continental United States

Edited by
L. C. Pakiser and
Walter D. Mooney

During the past half century, our knowledge of the continental crust and upper mantle of the United States has been significantly increased through geophysical observations. This volume presents a concise summary of the gross geophysical features of the crust and upper mantle, and highlights some of the historical developments that led to this knowledge.

Velocity models derived from seismic observations can be related to the structure, composition, and rheologic properties of the crust and upper mantle and to important solid-to-solid phase transformations.

Magnetic investigations, radioactive age dating, seismicity, and geological studies have revealed the worldwide patterns of sea-floor spreading, continental drift, and plate tectonics.

Gravity, electrical, heat-flow, and stress measurements have been applied increasingly to studies of the continental lithosphere.

Important regional similarities and differences have been discovered, and simple concepts of isostasy and crustal deformation have had to be modified or discarded.

This Memoir constitutes a comprehensive review and evaluation of these investigations.

The primary purpose of this volume is to provide, in a single publication, a comprehensive review and evaluation of our knowledge of the structure of the crust and upper mantle of the continental United States as determined from geophysical observations. The editors and authors believe that such an overview will be useful to geologists and geophysicists engaged in research on the structure, composition, and geologic evolution of the continental crust and upper mantle and in exploration for deposits of economic minerals and fossil fuels. The book should also be useful as a textbook for upper-division and graduate courses in geology and geophysics.

Although the emphasis is geophysical, the main goal of this volume is to demonstrate what geophysics has revealed and can reveal about the large-scale geologic and tectonic framework of the continent and what can be inferred from the geophysical observations about the composition of the Earth's crust and upper mantle.

The chapters in this volume constitute the most complete compilation of knowledge currently available on the geophysical framework of the continental United States. More than 100 authors, coauthors, and reviewers combined efforts to provide the latest scientific findings.

The Geological Society of America
Memoir 172

Geophysical Framework of the Continental United States

Edited by

L. C. Pakiser
U.S. Geological Survey
MS 966, Box 25046
Denver Federal Center
Denver, Colorado 80225

and

Walter D. Mooney
U.S. Geological Survey
MS 977, 345 Middlefield Road
Menlo Park, California 94025

1989

Published by The Geological Society of America, Inc.
3300 Penrose Place, P.O. Box 9140, Boulder, Colorado 80301

Printed in U.S.A.

GSA Books Science Editor Campbell Craddock

Library of Congress Cataloging-in-Publication Data

Geophysical framework of the continental United States / edited by
L. C. Pakiser, Walter D. Mooney.
p. cm. — (Memoir / Geological Society of America ; 172)
Includes bibliographical references.
ISBN 0-8137-1172-X
1. Geology—United States. 2. Geophysical—United States.
3. Earth—Crust. 4. Earth—Mantle. I. Pakiser, L. C. (Louis
Charles), 1919– . II. Mooney, Walter D. III. Series: Memoir
(Geological Society of America) ; 172.
QE77.G34 1989
557.3—dc20 89-26023
CIP

10 9 8 7 6 5 4 3 2

Contents

Preface

The scientific literature has become so vast, and the time available to read it so scarce, that editors should perhaps explain to prospective readers what justifies the volumes they have nudged through the publication process. The need for a single volume that summarizes the geophysical framework of the continental United States occurred independently to us in the course of our research at the U.S. Geological Survey, and while teaching at the University of New Orleans (L.C.P.) and Stanford University (W.D.M.). While existing journal articles, symposia volumes, and textbooks contain some of the information found in this volume, nowhere could we find a modern uniform treatment of the geophysics of the lower 48 states. Discussions with our colleagues convinced us that a summary volume would be of use to a great many people.

We began this volume with an outline of contents that we felt were essential, and immediately realized the impossibility of preparing the text by ourselves. We therefore invited knowledgeable scientists to write or collaborate on the individual chapters. Several chapters that were suggested to us were added, and it was decided to include a series of chapters on methods that would provide a concise summary of current practice in geophysics, together with up-to-date bibliographies. First drafts of these contributions were presented at an authors' meeting held at the Colorado School of Mines in Golden, Colorado, in March 1986, and most papers were substantially revised during the following eighteen months. Acceptance of individual manuscripts by the Editors preceded by as much as two years acceptance of the completed Memoir by the Society on October 31, 1988. The first manuscripts, by Rob Van der Voo and R. Ernest Anderson, were accepted in late 1986.

We have aimed toward a broad readership: toward geophysicists, geologists, and graduate students working or studying in the United States, and toward foreign scientists who have use for a single volume that summarizes geophysical investigations in our country. We and the authors hope that this volume succeeded in meeting the needs of this diverse group.

Louis C. Pakiser
Walter D. Mooney

Acknowledgments

This volume was in part made possible by the capable administration and support of several scientists in the U.S. Geological Survey who held the following positions during the 3-year-long period in which the manuscripts were prepared: Director, Dallas L. Peck; Chief Geologist, Robert M. Hamilton; Chief of the Office of Earthquakes, Volcanoes, and Engineering, John R. Filson; Chief of the Branch of Seismology, William E. Ellsworth; and Chief of the Branch of Geologic Risk Assessment, Albert M. Rogers. We are grateful to them for their help.

We are grateful to the following people who helped with informal review comments, lively discussions, and sage advice while this volume was being prepared: Elizabeth L. Ambos, R. Ernest Anderson, Harley Benz, Thomas M. Brocher, Rufus Catchings, Nikolas I. Christensen, Campbell Craddock, William H. Diment, Inci Ertan, Louis A. Fernandez, Ernst R. Flueh, Gary S. Fuis, Lee E. Gladish, James H. Leutgert, Jill McCarthy, Thomas V. McEvilly, Robert B. Smith, George A. Thompson, Allan W. Walter, and Craig S. Weaver.

The following U.S. Geological Survey employees assisted in a variety of ways: Louise Hobbs, general editoral assistant; Eleanor Omdahl, preparation of manuscripts; and Ralph Ricotta, drafting. Irene Munoz edited copy for the entire volume on contract with the Geological Society of America. We are indebted to them for their able and devoted work.

Formal reviews of the many chapters of this volume were made at the Editors' request by Clarence R. Allen, Myrl Beck, Jr., John C. Behrendt, Michael Berry, David Blackwell, Larry Braile, Larry Brown, Clark Burchfiel, Clement G. Chase, Campbell Craddock, William H. Diment, Charles L. Drake, Jerry P. Eaton, Robert Engdahl, Eric Frost, Andrew Griscom, Edwin L. Hamilton, Robert D. Hatcher, Donald Helmberger, Robert Herrmann, David P. Hill, Thomas Hildenbrand, Jack Hillhouse, William J. Hinze, David E. James, David L. Jones, Robert W. Kay, Charlotte G. Keen, G. Randy Keller, Thomas R. LaFehr, Frank McKeown, George McMechan, Brian J. Mitchell, William R. Muehlberger, K. Douglas Nelson, Keith Priestley, John Proffett, Douglas W. Rankin, Mitchell W. Reynolds, Kabir Roy-Chowdhury, Alan S. Ryall, Eli A. Silver, Paul K. Sims, Scott B. Smithson, Parke D. Snavely, Robert Speed, John H. Stewart, John D. Unger, Rob Van der Voo, Herbert Wang, Brian P. Wernicke, Hatten S. Yoder, Jr., Isidore Zietz, Mary Lou Zoback, and 15 anonymous reviewers. Additional reviews were made at the request of the authors and/or their institutions; these reviews are acknowledged in the individual chapters of this Memoir. All reviewers were helpful in improving the manuscripts and in maintaining the high standards of this volume. We are grateful to all of them. Our spouses, Helen M. Pakiser and Josephine A. Gandolfi, provided patient understanding and moral support while we were occupied in assembling this volume.

Finally, and most of all, we are grateful to the many authors of the chapters of this volume who have lent their creativity and hard work to making this vast body of information and ideas on the continental crust and upper mantle available to their colleagues in the Earth Sciences.

Louis C. Pakiser
Walter D. Mooney

Geological Society of America
Memoir 172
1989

Chapter 1

Introduction

L. C. Pakiser
U.S. Geological Survey, MS 966, Box 25046, Denver Federal Center, Denver, Colorado 80225
Walter D. Mooney
U.S. Geological Survey, MS 977, 345 Middlefield Road, Menlo Park, California 94025

ABSTRACT

During the past half century, our knowledge of the continental crust and upper mantle of the United States has been significantly increased through geophysical observations. We here present a concise summary of the gross geophysical features of the crust and upper mantle, and highlight some of the historical developments that led to this knowledge. Velocity models derived from seismic observations can be related to the structure, composition, and rheologic properties of the crust and upper mantle and to important solid-to-solid phase transformations. Magnetic investigations, radioactive age dating, seismicity, and geological studies have revealed the worldwide patterns of sea-floor spreading, continental drift, and plate tectonics. Gravity, electrical, heat-flow, and stress measurements have been applied increasingly to studies of the continental lithosphere. Important regional similarities and differences have been discovered, and simple concepts of isostasy and crustal deformation have had to be modified or discarded. This memoir constitutes a comprehensive review and evaluation of these investigations.

PURPOSES

The primary purpose of this volume is to provide, in a single publication, a comprehensive review and evaluation of our knowledge of the structure of the crust and upper mantle of the continental United States, exclusive of Alaska, as determined from geophysical observations. We believe that such an overview will be useful to geologists and geophysicists engaged in research on the structure, composition, and geologic evolution of the continental crust and upper mantle and in exploration for deposits of economic minerals and fossil fuels. The work should also be useful as a textbook for upper-division and graduate courses in geology and geophysics. Although the emphasis is geophysical, our main goal is to demonstrate what geophysics has revealed and can reveal about the large-scale geologic and tectonic framework of the continent and what can be inferred from geophysical observations about the composition of the Earth's crust and upper mantle.

An enormous amount of information on geophysical investigations of the crust and upper mantle of the United States has been published in hundreds of papers in many scientific journals. Some of this information has been summarized in monographs, commonly based on special conferences or symposia. This volume, however, represents the first systematic attempt to include in one resource the results of nearly all of the geophysical investigations of the crust and upper mantle of the continental United States, together with their geological significance.

The memoir is organized into four parts, each of which contains 3 to 12 chapters by authors from universities, government agencies, and research institutions. The topics progress from a review of the several geophysical methods of studying the Earth's crust and upper mantle (Part 1) to a region-by-region review of crustal and upper-mantle structure (Part 2), and from continental overviews of studies based on the different geophysical methods (Part 3) to geological and petrological syntheses based largely on the geophysical results (Part 4).

We conclude the volume with a discussion of major unsolved problems—the challenges of the future that will continue to make careers in geology and geophysics exciting and rewarding for many generations to come.

CRUST AND UPPER MANTLE

Detailed geophysical and compositional models for the crust and upper mantle are developed in later chapters. In this section we provide a working model, derived mainly from seismic and

Pakiser, L. C., and Mooney, W. D., 1989, Introduction, *in* Pakiser, L. C., and Mooney, W. D., Geophysical framework of the continental United States: Boulder, Colorado, Geological Society of America Memoir 172.

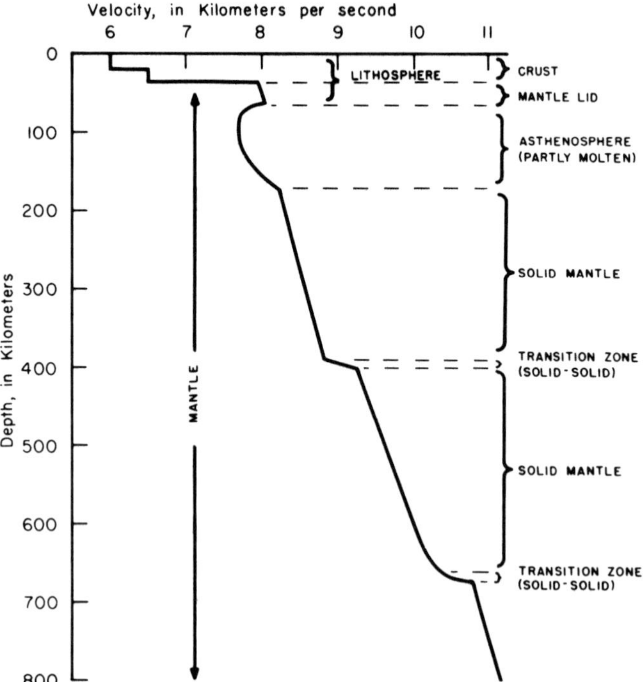

Figure 1. Compressional wave velocity model of the upper 800 km of the crust and upper mantle of the western United States. Velocity data are from Burdick and Helmberger (1978). Geologic interpretation has been added.

magnetic data, that is useful as background for subsequent discussions. For an overview of the velocity structure of the outer few hundred kilometers of the continent, we have adopted the upper-mantle compressional or P-wave velocity model of the western United States proposed by Burdick and Helmberger (1978). This model was derived by matching long-period body-wave forms with synthetic seismograms, except that the velocity distribution in the crust and uppermost mantle was based on direct arrivals from nuclear blasts (Carder and others, 1966) and various published models. The Burdick-Helmberger model is representative of the entire tectonically active western region, and it is broadly similar to the earlier Helmberger and Wiggins (1971) upper-mantle model of the midwestern United States.

The crust of the Burdick-Helmberger model (Fig. 1) consists of an upper layer 20 km thick, of velocity 6.0 km/sec, and a lower layer 15 km thick, of velocity 6.5 km/sec. The velocity of the upper mantle between the Moho and the base of the lithosphere is about 8.0 km/sec. The low-velocity asthenosphere, in which the velocity decreases gradually to 7.7 km/sec, extends from 65 km to a depth of about 170 km. At greater depths, the velocity increases downward from 8.2 km/sec at the base of the asthenosphere to a narrow transition zone at a depth of 400 km in which the velocity jumps abruptly by about 5 percent. Below the 400-km transition zone the velocity increases downward to a second transition zone at a depth of 670 km, in which the velocity jumps abruptly by about 4 percent. The velocity of the upper mantle at a depth of 800 km is 11.2 km/sec (Fig. 1).

We can relate the velocity model of the crust and upper mantle to rock composition in a general way. The upper crust is made up of heterogeneous rocks ranging from felsic to mafic (and uncommonly, ultramafic), but typically averages about granodiorite in bulk composition. The composition of the lower crust is more difficult to estimate. The velocities suggest that the average composition of the lower crust is probably that of granulite of intermediate composition approximating that of diorite (Christensen and Fountain, 1975), but the effects of magmatic and tectonic processes and geological observations require that the lower crust also must be highly heterogeneous.

Most earth scientists agree that the upper mantle is ultramafic and dominated by peridotite. Ringwood (1962, 1975, p. 176–205) has proposed a compositional model for the upper mantle that he termed "pyrolite," which has a composition equivalent to three parts of peridotite and one part of olivine tholeiite. The olivine of the peridotite and tholeiite would be the primary mineralogic constituent of the upper mantle in this model.

The seismogenic lithosphere consisting of the crust and mantle lid is solid and fairly brittle at fast strain rates. The asthenosphere is inferred to be soft, making feasible the sliding lithosphere of plate tectonics. The tholeiite fraction of the asthenosphere is probably molten.

The 400-km transition zone probably represents a solid-to-solid phase transformation from magnesium-rich olivine above the transition zone to spinel below. The pressure and the likely temperature of the transition zone are compatible with those of the olivine-spinel transformation (Akimoto and Fujisawa, 1968; Ringwood and Major, 1970; Ringwood, 1975, p. 397–404).

The nature of the 670-km transition zone is less understood. The discontinuity may represent another phase transformation from spinel to close-packed oxides of iron, magnesium, silicon, and aluminum, and/or compact silicates such as perovskite (Birch, 1952; Ringwood, 1975, p. 404–460; Knittle and Jeanloz, 1987), or a change in chemical composition, or both.

The two-layer velocity model of the crust from Burdick and Helmberger (1978) is an average for the western United States. In any given area, velocity variations with depth in the crust are much more complex, suggesting geologic complexity. For example, a proposed velocity-depth model of the upper 20 km of the crust in the Klamath Mountains of northern California contains four discontinuities and three low-velocity zones (Fig. 2). Fuis and others (1987) have derived a geologic model of the crust of the Klamath Mountains by comparing the velocity-depth model determined from seismic-refraction observations with laboratory measurements of rock velocities at pressures and temperatures appropriate for the depths of burial (Fig. 2). The upper 7 km of the crust is interpreted to consist of ultramafic and metasedimentary rocks associated with the Trinity ophiolite. The velocity-depth data below the Trinity ultramafic body can be interpreted in several ways, but the geologic model shown is consistent with

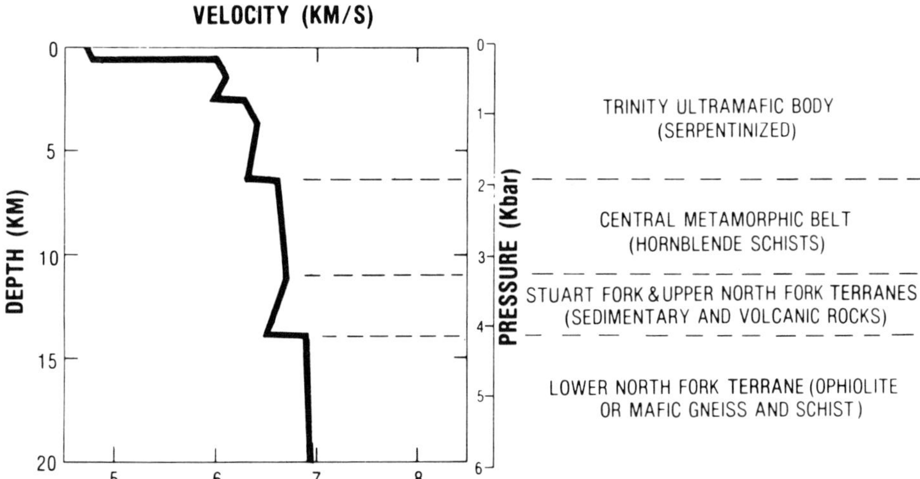

Figure 2. Comparison of velocity-depth model of Klamath Mountains with rock units suggested by the geology of the area. Rock compositions were inferred by comparing the velocity-depth model with laboratory measurements of rock velocities at appropriate pressures, and for depths below 15 km, appropriate temperatures. Modified from Figure 9 in Fuis and others (1987). The figure illustrates the complexity of crustal structure determined with modern seismic data.

the known geology of the area and laboratory measurements of rock velocities. The velocity-depth model of the Klamath Mountains is not uniquely required by the observations, but it is not likely to be significantly modified by alternate interpretations. Such models illustrate the detailed information that can be extracted from seismic observations and the structural and compositional complexity of the upper crust in the tectonically active western United States.

PLATE-TECTONICS FRAMEWORK

The North American plate (Fig. 3), which includes the continental United States, is bounded on the west by a system of transform faults and trenches, most notably the San Andreas fault of California and the Aleutian trench, and on the east by the Mid-Atlantic Ridge. The Pacific plate moves relatively northwest at a rate of about 6 cm/yr, sliding along the San Andreas and other transform faults, to be subducted beneath southern Alaska and the Aleutian chain. The North American plate moves relatively west from the Mid-Atlantic Ridge.

The ancient basement rocks of the North American craton range in age from about 1 to nearly 4 b.y. (Fig. 3). The basement rocks adjacent to the southeastern and western continental margins are younger. The crustal rocks of the western disturbed belt include younger accreted terranes, which are fault-bounded blocks of oceanic crust and island arcs that were transported atop the plates of oceanic lithosphere and jammed against the North American plate. The basement rocks of the oceanic crust are much younger than those of the continental craton (Fig. 3). They range in age from early Jurassic to Holocene; all are younger than 200 m.y. old. The basement rocks of the oceanic crust reflect the worldwide pattern of sea-floor spreading from oceanic ridges, such as the East Pacific Rise and the Mid-Atlantic Ridge. The rates of spreading and age distributions were determined from the patterns of magnetic anomalies caused by magnetization of cooling oceanic crust following reversals of the Earth's magnetic field and from the ages of rock samples taken from the sea floor. The dating of the magnetic-anomaly stripes makes it possible to map the isochrons separating rocks of different ages (Fig. 3).

The Atlantic and Gulf Coast margins of the United States are passive. Little crustal deformation is now taking place in the eastern two-thirds of the continent. The crust of the western third of the United States is being strongly deformed and in places injected with magma by the dynamic processes associated with the active western continental margin and the related tectonic and igneous activity of the Western Cordillera.

SOME HISTORICAL HIGHLIGHTS OF GEOPHYSICAL STUDIES IN THE U.S.

Almost a quarter century passed following the discovery by Mohorovičić (1910; see also Bonini and Bonini, 1979) of the seismic discontinuity that now bears his name before Gutenberg (1932) made the first determinations of the thicknesses of crustal layers from near-earthquake sources in southern California. Soon after, Lawson (1936) used average crustal densities and isostatic principles to estimate that the root of the Sierra Nevada, California, extends downward into the mantle to a depth of 68 km. Lawson's colleague Byerly (1937) demonstrated that seismic waves from earthquakes in northern and central California, as recorded at seismograph stations in Owens Valley just east of the Sierra Nevada, traveled slower than did seismic waves recorded at the same stations from earthquakes in Nevada and southern California. Byerly deduced that the crustal root of the Sierra Nevada is 40 to 70 km wide, and later concluded (Byerly, 1939) that it extends into the mantle to a depth of 71 km. The Sierran root is also identified by a pronounced low on the gravity map of the United States (Woollard and Joesting, 1964).

Gutenberg and others (1932) and Byerly and Dyk (1932) also used explosions in quarries to study velocity layering in the Earth's crust, but the first successful determinations of crustal thickness from explosions were made in New England by Leet (1936) and Slichter (1938, 1939).

Gutenberg (1955) postulated the existence of a low-velocity layer in the mantle at depths of about 100 km on the basis of severe attenuation of P-waves at epicentral distances of about 600 km from earthquakes in several regions.

In the late 1940s, the Carnegie Institution of Washington, under the leadership of Merle Tuve, launched a systematic explosion-seismic study of the continental crust of the United States. In a remarkable cross section (Fig. 4), Tuve and his co-workers (Tatel and Tuve, 1955, p. 49) summarized the relations of regional topography and crustal thickness from the East Coast region of Maryland and Virginia to Puget Sound in Washington. They demonstrated that variations in crustal structure are not simply related either to surface altitude, crustal thicknesses, or densities. In the caption to the cross section they concluded: "According to Pratt's hypothesis, all depths should be equal; according to Airy's hypothesis, the central columns should have boundaries at depths of 70 km; neither is true."

Following the pioneering work of Tuve and his co-workers, in 1956 the University of Wisconsin began a program of explosion studies of continental structure (Steinhart and Meyer, 1961). Wisconsin seismic teams determined crustal thicknesses in Arkansas and Missouri (1958), Wisconsin and upper Michigan (1958–1959), and Montana (1959). Contrary to expectations, but in support of the conclusions of Tatel and Tuve (1955), Meyer and others (1961) found that the crust actually thins in the northern Rocky Mountains when compared with the Great Plains to the east. Steinhart and Woollard (1961, p. 383) concluded from these results (see also Woollard, 1959) that "high elevations do not always mean a thick crust, and a low surface elevation is not always associated with a thin crust."

Woollard studied the relations among gravity, surface altitude, isostasy, and local and regional geology from 1936 to the end of his career. In a paper of fundamental importance, Wool-

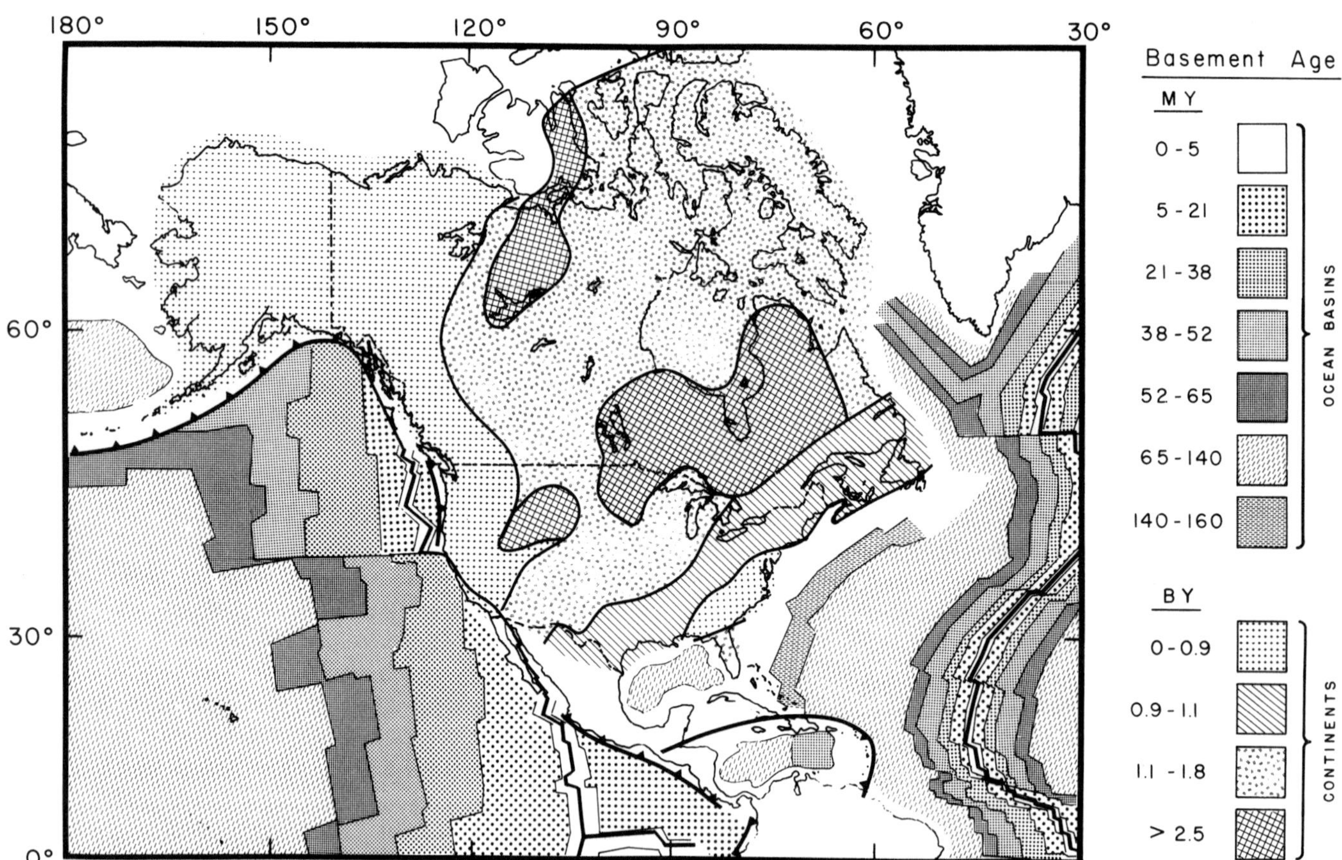

Figure 3. Age distribution of basement rocks of the North American continent and the Atlantic and Pacific Ocean basins. Adapted by Stuart and Paloma Nishenko from maps prepared by John Sclater and Linda Meinke (oceanic data) and published as Figure 18-21 in Press and Siever (1982) and by the Shell Oil Company (Precambrian data) and published in Cook and Bally, 1975. Trench and transform-fault systems along the Aleutian chain and western continental margin have been added from various sources. Locations of East Pacific Rise and Mid-Atlantic Ridge are shown by heavy lines in zones of basement 0 to 5 m.y. old.

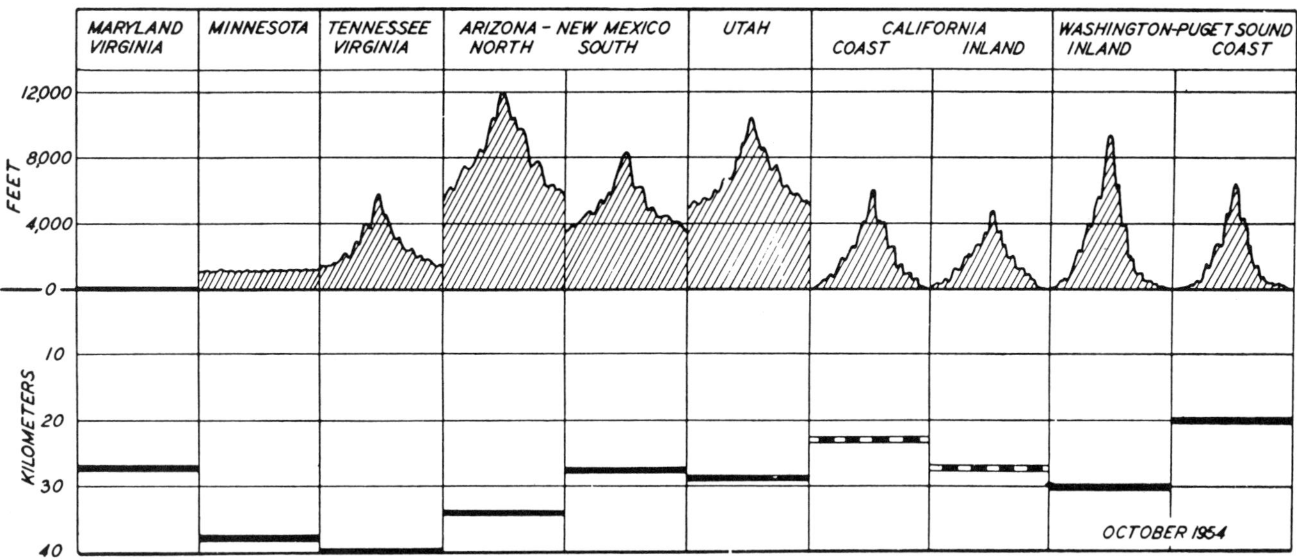

DEPTHS FROM CROSS-OVER. NO INCREASE OF VEL WITH DEPTH.

Figure 4. Regional topography and crustal thickness. After Tatel and Tuve (1955). Representation of topography is schematic. Crustal thickness in California is questioned. This figure is historically significant because it summarizes the first continentwide refraction reconnaissance of the crust and demonstrates for the first time that there is no simple model for isostatic compensation.

lard (1966) demonstrated that major basins in the North American craton, such as the Illinois and Michigan basins, are marked by positive gravity anomalies and uplifts such as the Cincinnati arch by negative anomalies, whereas in much of the tectonically active western United States—for example, the Basin and Range province—basins are commonly expressed by negative anomalies and uncompensated uplifts by positive anomalies. He concluded that these differences are related to changes in crustal density and tectonic activity. Woollard (1966) compared gravity anomalies and seismic determinations of crustal thickness and found that areas of thick crust (e.g., the Great Plains) are characterized by positive gravity anomalies, whereas areas of thin crust (e.g., the Basin and Range province) are characterized by negative anomalies. This seemingly contradictory result requires that isostatic compensation must be related, at least in part, to density variations in the upper mantle. Woollard (1966, p. 557) concluded, "Clearly, there is no single universal mechanism governing isostasy."

Publication of the Bouguer gravity anomaly map of the United States, under the leadership of Woollard and Joesting (1964), was a major achievement in exploration of the continental lithosphere. In his transcontinental geophysical profile, Woollard (1943) discovered a strong positive feature in northeastern Kansas (Fig. 5). Subsequently, Lyons (1950) compiled a gravity map of the United States. A striking feature of this map is the Midcontinent Gravity High extending from east-central Minnesota through Iowa and Nebraska into Kansas, which Lyons (1959) named the Greenleaf anomaly. Theil (1956) first showed its relation to the Keweenawan rocks near Lake Superior, later discussed in detail by Craddock and others (1963).

Over a span of 30 years, many institutions have studied the crust and upper mantle of the Basin and Range province, one of the most exhaustively studied regions of the continent. Tatel and Tuve (1955) expressed surprise in finding from seismic data that the crust is only 28 to 34 km thick in the Basin and Range to Colorado Plateau transition zone of Arizona, New Mexico, and Utah; from isostatic considerations, they had expected the crust of this elevated region to be about 70 km thick. Carder and Bailey (1958) obtained similar results in the Basin and Range province in the vicinity of the Nevada Test Site. Later, Press (1960), Berg and others (1960), and Diment and others (1961) discovered the characteristic low velocity of the upper mantle of the province, although they regarded the velocities they found at depths of 24 to 28 km—7.6 to 7.8 km/sec—as possibly too low for the upper mantle. Nevertheless, the unexpectedly thin crust and low-velocity upper mantle have been accepted by most investigators as characteristic features of the Basin and Range province since U.S. Geological Survey seismic teams demonstrated that the discontinuity below which the velocity is about 7.8 km/sec is the most prominent one in the province and should be regarded as the Mohorovičić discontinuity (Pakiser, 1963). Later investigators (see reviews by Smith, 1978; Pakiser, 1985) have added important details and new insights, but our knowledge of variations in crustal thickness and average velocities of the crust and upper mantle has changed little in the past 25 years. Recent seismic-reflection and detailed refraction studies, however, have greatly changed our ideas about the seismic structure of the crust and the nature of the lower-crust to upper-mantle transition zone.

One of the common but disconcerting experiences of scientific experiments is to develop a model for testing and then to

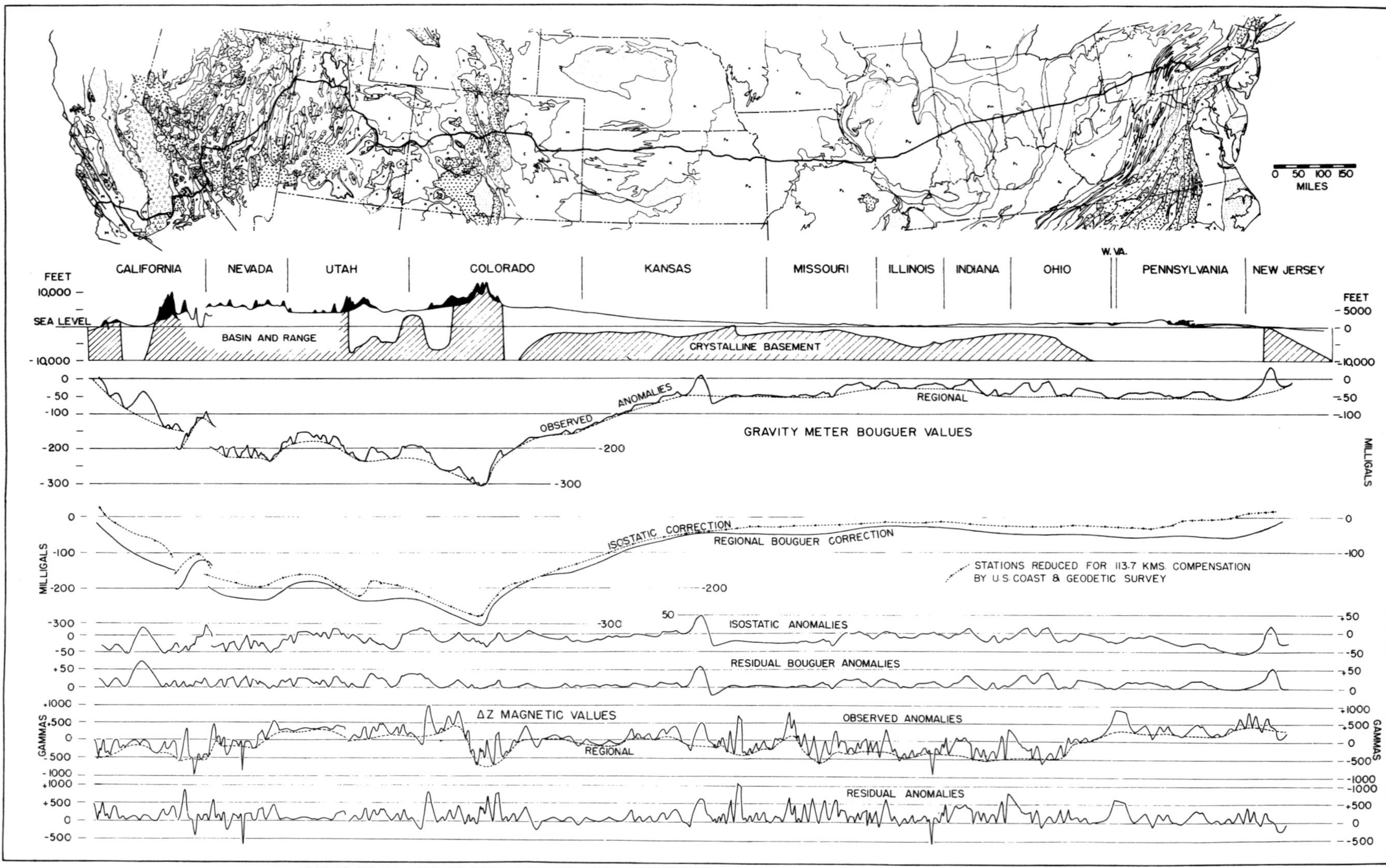

Figure 5. Transcontinental gravity and magnetic profile. After Woollard (1943). Notice the Midcontinent Gravity High in Kansas.

have it shattered by the experiment set up to support it. This happened in the Lake Superior seismic experiment of 1963, a cooperative effort of several U.S. and Canadian institutions led by the Carnegie Institution of Washington. It had been assumed that the crust of the Canadian Shield was simple and fairly uniform laterally, and the experiment was designed to determine the velocity structure of a "typical" shield area (Steinhart, 1964). A series of explosions of about 1,000 kg weight were detonated along the axis of Lake Superior, and travel times of seismic waves generated by these explosions were measured along many azimuths by stations in Canada and the United States. The crustal structure revealed was enormously complicated, with depths to the Mohorovičić discontinuity ranging from less than 30 km just west of Lake Superior to 55 km or more beneath the eastern half of the lake (Smith and others, 1966; O'Brien, 1968). Thus, maximum depths to Moho beneath this area of low surface altitude and relief rival those beneath the lofty Sierra Nevada (Eaton, 1963, 1966).

A second cooperative Lake Superior seismic experiment in 1966, Project Early Rise, was supported by the Defense Advanced Research Projects Agency (DARPA) and coordinated by the U.S. Geological Survey. This experiment led to development of several upper-mantle velocity models to depths of about 700 km in different areas of North America (Iyer and others, 1969). The P-wave upper-mantle low-velocity zone characteristic of the western United States was absent from models east of the Rocky Mountains derived from Project Early Rise data. The Midwestern model of Helmberger and Wiggins (1971) does contain a prominent upper-mantle low-velocity zone. The shear-wave velocity model for the Canadian Shield derived from surface waves exhibits only a subdued upper-mantle low-velocity zone (Brune and Dorman, 1963), whereas models for the Alps and oceanic areas exhibit pronounced S-wave low-velocity zones (Dorman, 1969).

The airborne magnetometer was developed during World War II, and much of the continental United States has since been covered by aeromagnetic surveys. Zietz (1969) demonstrated that many major basement provinces of the continent have characteristic magnetic-anomaly patterns and that important province boundaries are commonly revealed by distinct magnetic gradients. Through Zietz's leadership, a composite magnetic anomaly map of the United States was recently published (Zietz, 1982). Mayhew and others (1985) used satellite magnetometry (Magsat) results to study the magnetic properties of the crust. From estimates of the intensity of magnetization, they concluded that the Moho is the magnetic bottom of the continents except where the Curie isotherm (about 550°C for magnetite) is elevated within the crust. Curie depth models can be used to estimate regional heat flow and thus provide a constraint on crustal thermal models.

Electrical studies of the deep crust and upper mantle are a relatively recent development. Schmucker (1960, 1964) first recognized the low conductivity of the lower crust and upper mantle. Gough (1974) studied the relations among heat flow, crustal structure, and electrical conductivity in western North America. He demonstrated the existence of a zone of high conductivity in the lower crust and upper mantle, especially in areas of high heat flow, and recognized that transition zones from high heat flow to low heat flow are conductivity boundaries as well.

Seismic, gravity, magnetic, electrical, and other geophysical studies have provided much information about the structure and physical properties of the crust and upper mantle, but by themselves they tell us little about the composition of the rocks. Birch (1960, 1961) made laboratory measurements of seismic velocities of different rocks at pressures up to those appropriate for the lower crust and uppermost mantle. He related those measurements systematically to rock densities and compositions. From his, and later work by others (Christensen, 1982), reasonable inferences can now be made about the compositions and rheology of rocks of the crust and upper mantle from seismic velocities and densities.

Birch (1954) also pioneered the measurement of heat flow in the United States, and with two of his students discovered the important linear relation between heat flow and heat production in granitic rocks (Birch and others, 1968). As of 1962, only 11 reliable heat-flow determinations had been made in the United States (Lachenbruch and Sass, 1977). By the late 1960s, this number had been increased to about 350. At present, more than 1,000 heat-flow determinations are available for study. The 1977 heat-flow map of Lachenbruch and Sass (1977) revealed the characteristic difference between heat flow in the eastern and western areas of the United States. Heat flow is typically about 1 microcalorie/cm^2sec east of the Rocky Mountains and about 2 microcalories/cm^2sec in many areas to the west. Individual measurements exhibit a wide range of values in both regions, however. These results require that radioactivity must be largely concentrated in the upper 10 km or so of the crust, thus placing important constraints on the nature of the lower crust.

By the mid-1960s, nearly all the large-scale features of the velocity structure of the continental crust and upper mantle of the United States had been discovered (Herrin and Taggart, 1962; Pakiser and Steinhart, 1964; Herrin, 1969; Warren and Healy, 1974). Later studies, usually cooperative ones following the models of the Lake Superior experiments, have provided a rich body of knowledge and insights about the nature of the continental crust and upper mantle, especially those studies coinciding with, and later than, the plate-tectonics revolution. The main features of the earlier work have stood the test of time and have provided the framework for later research.

The revolution in knowledge of sea-floor spreading, continental drift, and plate tectonics could not have been made without the early recognition by Vacquier (1959) of magnetic evidence for large horizontal displacements along faults in the ocean floor, the recognition of linear magnetic-anomaly patterns off the west coast of North America by Mason and Raff (1961), the documentation of periodic reversals of the Earth's magnetic field by Cox and his co-workers (Cox and others, 1964), the recognition of the nature of transform faults by Wilson (1965), the recognition by Vine and Matthews (1963) that the magnetic anomalies over ocean ridges can be explained by reversals in the Earth's magnetic field and cooling of the spreading sea floor, and by correlation of the relations of seismicity to movement along transform faults and subduction zones (Isacks and others, 1968). This story, illustrated by Figure 3, has become one of the most familiar of 20th-century science.

Some of the most significant discoveries of the nature of the layering and tectonic features of the lower crust are now being made using the methods of reflection seismology. The exciting results of the last 10 years, reviewed elsewhere in this volume, were foreshadowed by early experiments in the 1950s. Junger (1951) demonstrated the existence of reflected energy from seismic waves generated by small explosions and returned from interfaces about 20 km deep in eastern Montana. In 1958 and 1959, Hasbrouck (1964) experimented with recording deep reflections in eastern Colorado. He was able to identify reflected energy from the lower crust and even from depths appropriate for the Moho. The recent advances in reflection seismology could not have been made without the development of the reflection common-depth-point field method by Mayne (1962) in about 1950 and the VIBROSEIS instrumental system by Crawford and his associates (Crawford and others, 1960) at the Continental Oil Company in the late 1950s.

The foregoing discussion has summarized our knowledge of the gross features of the structure and composition of the crust and upper mantle of the United States, and has highlighted the major historical developments that have led to this knowledge. We have not attempted to present here a complete history of research on the crust and upper mantle of the continental United States but have included what we consider to be some of the most significant developments. These and other developments provided the essential foundation on which later studies have been based. The chapters in this volume provide a comprehensive review and evaluation of these and later advances in knowledge and provide a modern view of the *Geophysical Framework of the Continental United States.*

ACKNOWLEDGMENTS

We thank Campbell Craddock, William Diment, Jerry Eaton, and Warren Hamilton for constructive reviews of this introduction.

REFERENCES CITED

Akimoto, S., and Fujisawa, H., 1968, Olivine-spinel solid solution equilibria in the system Mg_2SiO_4-Fe_2SiO_4: Journal of Geophysical Research, v. 73, p. 1467–1479.

Berg, J. W., Jr., Cook, K. L., Narans, H. D., Jr., and Dolan, W. D., 1960, Seismic investigation of crustal structure in the eastern part of the Basin and Range province: Seismological Society of America Bulletin, v. 50, p. 511–535.

Birch, F., 1952, Elasticity and constitution of the earth's interior: Journal of Geophysical Research, v. 57, p. 227–286.

—— , 1954, The present state of geothermal investigations: Geophysics, v. 19, p. 645–659.

—— , 1960, The velocity of compressional waves in rocks to 10 kilobars, Part 1: Journal of Geophysical Research, v. 65, p. 1083–1102.

—— , 1961, The velocity of compressional waves in rocks to 10 kilobars, Part 2: Journal of Geophysical Research, v. 66, p. 2199–2224.

Birch, F., Roy, R. F., and Decker, E. R., 1968, Heat flow and thermal history in New England and New York, *in* Zen, E., White, W. S., Hadley, J. B., and Thompson, J. B., Jr., eds., Studies of Appalachian geology; Northern and Maritime: New York, Interscience, p. 437–451.

Bonini, W. E., and Bonini, R. R., 1979, Andrija Mohorovičić; Seventy years ago an earthquake shook Zagreb: EOS Transactions of the American Geophysical Union, v. 60, p. 699–701.

Brune, J., and Dorman, J., 1963, Seismic waves and earth structure in the Canadian shield: Seismological Society of America Bulletin, v. 53, p. 167–210.

Burdick, L. J., and Helmberger, D. V., 1978, The upper mantle P velocity structure of the western United States: Journal of Geophysical Research, v. 83, p. 1699–1712.

Byerly, P., 1937, Comment on the Sierra Nevada in the light of isostasy, by Andrew C. Lawson: Geological Society of America Bulletin, v. 48, p. 2025–2031.

—— , 1939, Near earthquakes in central California: Seismological Society of America Bulletin, v. 29, p. 427–462.

Byerly, P., and Dyk, K., 1932, Richmond quarry blast of September 12, 1931, and the surface layering of the earth in the region of Berkeley: Seismological Society of America Bulletin, v. 22, p. 50–55.

Carder, D. S., and Bailey, L. F., 1958, Seismic wave traveltimes from nuclear explosions: Seismological Society of America Bulletin, v. 48, p. 377–398.

Carder, D. S., Gordon, D. W., and Jordan, J. N., 1966, Analysis of surface foci traveltimes: Seismological Society of America Bulletin, v. 56, p. 815–840.

Christensen, N. I., 1982, Seismic velocities, *in* Carmichael, R. S., ed., Handbook of physical properties of rocks, Volume 2: Boca Raton, Florida, CRC Press, p. 1–228.

Christensen, N. I., and Fountain, D. M., 1975, Constitution of the lower continental crust based on experimental studies of seismic velocities in granulite: Geological Society of America Bulletin, v. 86, p. 227–236.

Cook, T. D., and Bally, A. W., eds., 1975, Stratigraphic atlas of North and Central America: Princeton, New Jersey, Princeton University Press, 272 p.

Cox, A., Doell, R. R., and Dalrymple, G. B., 1964, Reversals in the earth's magnetic field: Science, v. 144, p. 1537–1543.

Craddock, C., Thiel, E. C., and Gross, B., 1963, A gravity investigation of the Precambrian of southeastern Minnesota and western Wisconsin: Journal of Geophysical Research, v. 68, p. 6015–6032.

Crawford, J. M., Doty, W., and Lee, M. R., 1960, Continuous signal seismograph: Geophysics, v. 25, p. 95–105.

Diment, W. H., Stewart, S. W., and Roller, J. C., 1961, Crustal structure from the Nevada Test Site to Kingman, Arizona, from seismic and gravity observations: Journal of Geophysical Research, v. 66, p. 201–214.

Dorman, J., 1969, Seismic surface-wave data on the upper mantle, *in* Hart, P. J., ed., The earth's crust and upper mantle: American Geophysical Union Geophysical Monograph 13, p. 257–265.

Eaton, J. P., 1963, Crustal structure from San Francisco, California, to Eureka, Nevada, from seismic-refraction measurements: Journal of Geophysical Research, v. 68, p. 5789–5806.

—— , 1966, Crustal structure in northern and central California from seismic evidence, *in* Bailey, E. H., ed., Geology of California: California Division of Mines and Geology Bulletin, v. 190, p. 419–426.

Fuis, G. S., Zucca, J. J., Mooney, W. D., and Milkereit, B., 1987, A geologic interpretation of seismic-refraction results in northern California: Geological Society of America Bulletin, v. 98, p. 53–65.

Gough, D. E., 1974, Electrical conductivity under western North America in relation to heat flow, seismology, and structure: Journal of Geomagnetism and Geoelectricity, v. 26, p. 105–123.

Gutenberg, B., 1932, Traveltime curves at small distances and wave velocities in southern California: Beitrage zur Geophysik, v. 35, p. 6–45.

——, 1955, Wave velocities in the earth's crust, *in* Poldervaart, A., ed., Crust of the earth: Geological Society of America Special Paper 62, p. 19–34.

Gutenberg, B., Wood, H., and Buwalda, J., 1932, Experiments testing seismographic methods for determining crustal structure: Seismological Society of America Bulletin, v. 22, p. 185–246.

Hasbrouck, W. P., 1964, A seismic crustal study in eastern Colorado [Ph.D. thesis]: Golden, Colorado School of Mines, 133 p.

Helmberger, D., and Wiggins, R. A., 1971, Upper mantle structure of midwestern United States: Journal of Geophysical Research, v. 76, p. 3229–3245.

Herrin, E., 1969, Regional variations of P-wave velocity in the upper mantle beneath North America, *in* Hart, P. J., ed., The earth's crust and upper mantle: American Geophysical Union Monograph 13, p. 242–246.

Herrin, E., and Taggart, J., 1962, Regional variations in P_n velocity and their effect on the location of epicenters: Seismological Society of America Bulletin, v. 52, p. 1037–1046.

Isacks, B., Oliver, J., and Sykes, L. R., 1968, Seismology and the new global tectonics: Journal of Geophysical Research, v. 73, p. 2565–2577.

Iyer, H. M., Pakiser, L. C., Stuart, D. J., and Warren, D. H., 1969, Project Early Rise; Seismic probing of the upper mantle: Journal of Geophysical Research, v. 74, p. 4409–4441.

Junger, A., 1951, Deep reflections in Big Horn County, Montana: Geophysics, v. 16, p. 499–505.

Knittle, E., and Jeanloz, R., 1987, Synthesis and equation of state of (Mg, Fe) SiO_3 perovskite to over 100 gigapascals: Science, v. 235, p. 665–670.

Lachenbruch, A. H., and Sass, J. H., 1977, Heat flow in the United States and the thermal regime of the crust, *in* Heacock, J. H., ed., The nature and physical properties of the earth's crust: American Geophysical Union Geophysical Monograph 20, p. 626–675.

Lawson, A. C., 1936, The Sierra Nevada in the light of isostasy: Geological Society of America Bulletin, v. 47, p. 1691–1712.

Leet, L. D., 1936, Seismological data on surface layers in New England: Seismological Society of America Bulletin, v. 26, p. 129–145.

Lyons, P. L., 1950, A gravity map of the United States: Tulsa Geological Society Digest, v. 18, p. 33–43.

——, 1959, The Greenleaf anomaly; A significant gravity feature: Kansas Geological Survey Bulletin, v. 137, p. 105–120.

Mason, R. G., and Raff, A. D., 1961, Magnetic survey off the west coast of North America: Geological Society of America Bulletin, v. 72, p. 1259–1265.

Mayhew, M. A., Johnson, B. D., and Wasilewski, P. J., 1985, A review of problems and progress in studies of satellite magnetic anomalies: Journal of Geophysical Research, v. 90, p. 2511–2522.

Mayne, W. H., 1962, Common reflection point horizontal data stacking techniques: Geophysics, v. 27, p. 927–938.

Meyer, R. P., Steinhart, J. S., and Bonini, W. E., 1961, Montana, 1959, *in* Steinhart, J. S., and Meyer, R. P., eds., Explosion studies of continental structure: Carnegie Institution of Washington Publication 622, p. 305–343.

Mohorovičić, A., 1910, Potres od 8 × 1909: Godishje Izvjesce Zagrebackog Meteoroloskog Observatorija za godinu 1909: Zagreb, Yugoslavia, Yearbook of Meteorological Observatories for 1909, 56 p.

O'Brien, P.N.S., 1968, Lake Superior; A reinterpretation of the 1963 seismic experiment: Journal of Geophysical Research, v. 72, p. 2669–2689.

Pakiser, L. C., 1963, Structure of the crust and upper mantle in the western United States: Journal of Geophysical Research, v. 68, p. 5747–5756.

——, 1985, Seismic exploration of the crust and upper mantle of the Basin and Range province, *in* Drake, E. T., and Jordan, W. M., eds., Geologists and ideas; A history of North American geology: Geological Society of America Centennial Special Volume 1, p. 453–469.

Pakiser, L. C., and Steinhart, J. S., 1964, Explosion seismology in the western hemisphere, *in* Odishaw, H., ed., Research in geophysics; Volume 2, Solid earth and interface phenomena: Cambridge, Massachusetts Institute of Technology Press, p. 123–147.

Press, F., 1960, Crustal structure in the California-Nevada region: Journal of Geophysical Research, v. 65, p. 1039–1051.

Press, F., and Siever, R., 1982, Earth, 3rd ed.: San Francisco, W. H. Freeman, 613 p.

Ringwood, A. E., 1962, A model for the upper mantle: Journal of Geophysical Research, v. 67, p. 857–866.

——, 1975, Composition and petrology of the earth's mantle: New York, McGraw-Hill, 618 p.

Ringwood, A. E., and Major, A., 1970, The system Mg_2SiO_4-Fe_2SiO_4 at high pressures and temperatures: Physics of the Earth and Planetary Interiors, v. 3, p. 89–108.

Schmucker, U., 1960, Deep anomalies in electrical conductivity: Scripps Institution of Oceanography Annual Progress Report, S10 Reference 61-13, 18 p.

——, 1964, Anomalies in the geomagnetic variations in the southwestern United States: Journal of Geomagnetism and Geoelectricity, v. 15, p. 193–221.

Slichter, L. B., 1938, Seismological investigations of the earth's crust using quarry blasts [abs.]: Geological Society of America Bulletin, v. 49, p. 1927.

——, 1939, Seismic studies of crustal structure in New England by means of quarry blasts [abs.]: Geological Society of America Bulletin, v. 50, p. 1934.

Smith, R. B., 1978, Seismicity, crustal structure, and intraplate tectonics of the interior of the Western Cordillera, *in* Smith, R. B., and Eaton, G. P., eds., Cenozoic tectonics and regional geophysics of the Western Cordillera: Geological Society of America Memoir 152, p. 111–144.

Smith, T. J., Steinhart, J. S., and Aldrich, L. T., 1966, Lake Superior crustal structure: Journal of Geophysical Research, v. 71, p. 1141–1172.

Steinhart, J. S., 1964, Lake Superior seismic experiment; Shots and travel times: Journal of Geophysical Research, v. 69, p. 5335–5352.

Steinhart, J. S., and Meyer, R. P., eds., 1961, Explosion studies of continental structure: Carnegie Institution of Washington Publication 622, 409 p.

Steinhart, J. S., and Woollard, G. P., 1961, Seismic evidence concerning continental crustal structure, *in* Steinhart, J. S., and Meyer, R. P., eds., Explosion studies of continental structure: Carnegie Institution of Washington Publication 622, p. 344–383.

Tatel, H. E., and Tuve, M. A., 1955, Seismic exploration of a continental crust, *in* Poldervaart, A., ed., Crust of the earth: Geological Society of America Special Paper 62, p. 35–50.

Theil, E. C., 1956, Correlation of gravity anomalies with the Keweenawan geology of Wisconsin and Minnesota: Geological Society of America Bulletin, v. 67, p. 1079–1100.

Vacquier, V., 1959, Measurement of horizontal displacement along faults in the ocean floor: Nature, v. 183, p. 452–453.

Vine, F. J., and Matthews, D. H., 1963, Magnetic anomalies over ocean ridges: Nature, v. 199, p. 947–949.

Warren, D. H., and Healy, J. H., 1974, Structure of the crust in the conterminous United States, *in* Mueller, S., ed., The structure of the earth's crust: Elsevier Scientific Publishing Company, Developments in Geotectonics 8, p. 203–213.

Wilson, J. T., 1965, A new class of faults and their bearing on continental drift: Nature, v. 207, p. 343–347.

Woollard, G. P., 1943, Transcontinental gravitational and magnetic profile of North America and its relation to geologic structure: Geological Society of America Bulletin, v. 44, p. 747–789.

——, 1959, Crustal structure from gravity and seismic measurements: Journal of Geophysical Research, v. 64, p. 1521–1544.

——, 1966, Regional isostatic relations in the United States, *in* Steinhart, J. S., and Smith, T. J., eds., The earth beneath the continents: American Geophysical Union Geophysical Monograph 10, p. 557–594.

Woollard, G. P., and Joesting, H. R., 1964, Bouguer gravity anomaly map of the United States: U.S. Geological Survey Special Map, scale 1:2,500,000.

Zietz, I., 1969, Aeromagnetic investigations of the earth's crust in the United States, *in* Hart, P. J., ed., The earth's crust and upper mantle: American Geophysical Union Geophysical Monograph 13, p. 404–415.

——, 1982, Composite magnetic anomaly map of the United States; Part A, Conterminous United States: U.S. Geological Survey Geophysical Investigations Map GP—954A, scale 1:2,500,000.

Manuscript Accepted by the Society October 31, 1988

Printed in U.S.A.

Geological Society of America
Memoir 172
1989

Chapter 2

Seismic methods for determining earthquake source parameters and lithospheric structure

Walter D. Mooney
U.S. Geological Survey, MS 977, 345 Middlefield Road, Menlo Park, California 94025

ABSTRACT

The seismologic methods most commonly used in studies of earthquakes and the structure of the continental lithosphere are reviewed in three main sections: earthquake source parameter determinations, the determination of earth structure using natural sources, and controlled-source seismology. The emphasis in each section is on a description of data, the principles behind the analysis techniques, and the assumptions and uncertainties in interpretation. Rather than focusing on future directions in seismology, the goal here is to summarize past and current practice as a companion to the review papers in this volume.

Reliable earthquake hypocenters and focal mechanisms require seismograph locations with a broad distribution in azimuth and distance from the earthquakes; a recording within one focal depth of the epicenter provides excellent hypocentral depth control. For earthquakes of magnitude greater than 4.5, waveform modeling methods may be used to determine source parameters. The seismic moment tensor provides the most complete and accurate measure of earthquake source parameters, and offers a dynamic picture of the faulting process.

Methods for determining the Earth's structure from natural sources exist for local, regional, and teleseismic sources. One-dimensional models of structure are obtained from body and surface waves using both forward and inverse modeling. Forward-modeling methods include consideration of seismic amplitudes and waveforms, but lack the formal resolution estimates obtained with inverse methods. Two- and three-dimensional lithospheric models are derived using various inverse methods, but at present most of these methods consider only traveltimes of body waves.

Controlled-source studies of the Earth's structure are generally divided by method into seismic refraction/wide-angle reflection and seismic reflection studies. Seismic refraction profiles are usually interpreted in terms of two-dimensional structure by forward modeling of traveltimes and amplitudes. The refraction method gives excellent estimates of seismic velocities, but relatively low resolution of structure. Formal resolution estimates are not possible for models derived from forward modeling, but informal estimates can be obtained by perturbing the best-fitting model. Inversion methods for seismic refraction data for one-dimensional models are well established, and two- and three-dimensional methods, including tomography, have recently been developed.

Seismic reflection data provide the highest resolution of crustal structure, and have provided many important geological insights in the past decade. The acquisition and processing of these data have been greatly advanced by the hydrocarbon exploration industry. However, reliable crustal velocity control is generally lacking, and the origin of deep crustal reflections remains unclear, resulting in nonunique interpretations. A new

Mooney, W. D., 1989, Seismic methods for determining earthquake source parameters and lithospheric structure, *in* Pakiser, L. C., and Mooney, W. D., Geophysical framework of the continental United States: Boulder, Colorado, Geological Society of America Memoir 172.

form of lithospheric seismology has recently emerged that combines the advantages of seismic refraction and seismic reflection profiles, and the distinction between the two methods is steadily diminishing.

Major challenges for future work will be the collection of data that are more densely sampled in space, and the development of interpretation methods that provide quantitative estimates of the uncertainties in the calculated models.

INTRODUCTION

A wide range of seismologic methods are used to study the structure of the lithosphere and the seismicity of the Earth. This chapter reviews seismic methods in three sections: the determination of earthquake source parameters; the determination of the Earth's structure using natural (earthquake) sources; and the determination of structure using manmade (controlled) sources. After a brief historical introduction, data-analysis techniques and assumptions and uncertainties associated with each method are discussed. Seismologic theory and practice are far too extensive to be completely covered in the available space; existing texts serving that function include Richter (1958), Grant and West (1965), Dobrin (1976), Claerbout (1976, 1985), Telford and others (1976), Červený and others (1977), Coffeen (1978), Garland (1979), Aki and Richards (1980), Ben-Menahem and Singh (1981), Lee and Stewart (1981), Sheriff and Geldart (1982, 1983), Kennett (1983), Bullen and Bolt (1985), Tarantola (1987), Waters (1987), and Yilmaz (1987). The purpose here is to provide a critical review of the methods used in other chapters of this volume, and particularly to discuss their uncertainties and limitations, and to indicate where more detailed information may be obtained.

Historical background

Wherever documents of human history are uncovered, hints of interest in earthquakes can be found. Chinese records of earthquakes date to the Chou dynasty (ca. 1122 to 221 B.C.), the Bible recounts the collapse of the walls of Jericho (ca. 1100 B.C.), and Aristotle (b. 384 B.C.) devoted his attention to the study and classification of earthquakes. Until the end of the 19th century, the cause of earthquakes was rarely associated with faults.

Fault scarps observed after two California earthquakes (Fort Tejon, 1857, and Owens Valley, 1872) played an important role in establishing the earthquake-fault association (Richter, 1958). The 1906 San Francisco earthquake was the most thoroughly studied of its time (Lawson, 1908), and provided data for the elastic rebound concept of Reid (1910), which is still the generally accepted theory of earthquake rupture and the origin of seismic waves (Howell, 1986).

The transition from a purely descriptive study of earthquakes to a quantitative study came with the development, in 1892, of a reliable seismograph by John Milne (1850–1913) in Japan. In 1897, Robert Oldham was the first to merge seismologic observations with theory when he identified compressional, shear, and surface waves on an actual seismogram. At the time of the 1906 San Francisco earthquake, there were already dozens of seismographs of various designs operating worldwide, most prominently in Germany, Italy, and Japan. Some 50 years later it had become clear that a more reliable and uniform determination of world seismicity required a network of standardized seismic stations. Thus, the World-Wide Standardized Seismographic Network (WWSSN), consisting of 120 continuously recording stations, began operation in the 1960s (Oliver and Murphy, 1971). It was recognized as well that regional networks generally provide much-improved accuracy and sensitivity for determining local seismicity compared with worldwide networks. By the 1970s, about 100 microearthquake networks were operating at various locations throughout the world (Lee and Stewart, 1981).

The history of early controlled-source seismic investigations of Earth's structure is summarized by DeGolyer (1935), Bullen and Bolt (1985), James and Steinhart (1966), and Sheriff and Geldart (1982). An Irish scientist, Robert Mallet, apparently was the first, in 1848, to report on the use of an explosive source in seismology (Aki and Richards, 1980, p. 268). One hundred years passed before a systematic program of controlled-source seismic profiling of the deep crust was begun. In the meantime, earthquake traveltime data were used to define the crust/mantle boundary (Mohorovičić, 1909) and seismic velocity layering within the crust (Conrad, 1925).

Economic incentives spurred the development of seismic refraction exploration methods in the 1920s. In 1924 an oil field was discovered in Texas using seismic refraction fan-shooting techniques. Marine seismic refraction investigations were initiated some 10 years later when Ewing and others (1937) recorded explosions at four stations located south of Woods Hole, Massachusetts. Subsequently, deep-water marine measurements were begun in the 1950s by Ewing and his colleagues. In Germany, seismic refraction measurements of the deep continental crust began in 1947, and in the United States the active pursuit of the method was begun shortly thereafter by the Carnegie Institution of Washington (Tuve, 1951; Tatel and Tuve, 1955). As of the early 1960s, seismic refraction profiles were collected with regularity both in the United States and worldwide.

The history of seismic reflection profiling is much more closely tied to hydrocarbon exploration activities (Sheriff and Geldart, 1982). The modern seismic reflection system as we know it today dates from the 1930s when field parties recorded 10 to 12 channels of data on photographic paper, using some signal processing and enhancement capabilities provided by vacuum tube circuits. The common depth point (CDP) recording method (Mayne, 1962, 1967) was invented in 1950 as a way of providing further signal enhancement, and it came into routine use in the mid-1960s. The VIBROSEIS method was developed in 1953 as an alternative to explosive sources; by the mid-1970s it accounted for about 60 percent of land recording and today is the most common source for on-land seismic reflection studies.

Investigations of the use of the seismic reflection method for studies of the deep crust began about the same time as seismic refraction studies, but required many more years to become a routine method (Mintrop, 1949). Much of the earliest work was done in Europe, but results in the United States were also reported by Junger (1951) and Dix (1965). Later, experiments on three continents—in Australia by the Bureau of Mineral Resources, in Canada (Kanasewich and Cumming, 1965; Clowes and others, 1968), and in Germany (Meissner, 1967; Dohr, 1970)—gave conclusive evidence for intracrustal and Moho reflections at near-vertical incidence. The Consortium for Continental Reflection Profiling (COCORP) undertook a pilot study of deep reflection seismology in northern Texas in 1975 (Oliver and others, 1976), and shortly thereafter a continuously operating seismic crew was engaged. To date, more than 10,000 km of data have been collected in the United States over a broad geographic area by COCORP and other institutions (Brown and others, 1986; Phinney and Roy-Chowshury, this volume; Smithson and Johnson, this volume). The high resolution of structure obtained with the seismic reflection method accounts for its extensive application.

Reflection profiling is also highly effective in the marine environment due to the excellent coupling of the source (air guns) with the water medium, the lack of surface waves, and the relatively uniform bottom conditions, which minimize static corrections. These advantages have been effectively exploited by scientists in the British Institutions for Reflection Profiling Syndicate (BIRPS; Matthews and Cheadle, 1986; Matthews and others, 1987). High-quality marine data exist for both the Atlantic and Pacific margins of the United States (Hutchinson and others, 1988; Trehu and others, this volume; Couch and Riddihough, this volume; Phinney and Roy-Chowdhury, this volume), and recently, deep reflection data were collected on the Great Lakes in a joint U.S.–Canadian project (Behrendt and others, 1988).

DETERMINATION OF EARTHQUAKE SOURCE PARAMETERS

Earthquake data

There are three basic sources of data used in earthquake studies: permanent regional seismic networks, temporary dense networks of portable seismographs, and global seismic observatories (broad-band seismic installations). Permanent regional networks are used where the level of seismic activity or potential seismic hazard is high enough to pose a significant public safety risk. Typically, each seismograph consists of a single vertical-component 1-Hz geophone, and high-gain analogue telemetry electronics, which are of limited use for extracting waveform information due to signal clipping; however, some networks use three-component digital recording and telemetry. There are some 50 seismic networks in the United States, each with an average of 30 stations spaced some 30 to 100 km apart. The largest permanent network is operated in California by the U.S. Geological Survey (USGS) and the California Institute of Technology; it consists of about 600 stations, with the densest coverage within 100 km of the San Andreas fault.

After a major earthquake occurs, a temporary network of portable seismographs may be installed in the epicentral region. Until recently, these networks, typically 10 to 20 seismographs spaced 5 to 10 km apart, consisted of vertical-component smoked-drum or pen-and-ink seismographs. Currently, these networks generally include three-component seismographs (e.g., Archuleta and others, 1982; Borcherdt and others, 1985) and accelerometers that provide more useful information for the determination of locations, magnitudes, and seismic moments than do vertical-component seismographs.

For larger earthquakes (magnitude 4.5 or greater), the study of broad-band waveforms (teleseisms) recorded at global seismic observatories provides additional information on the hypocentral depth and nature of faulting. Teleseismic data are particularly useful because they include waves emanating at a wide range of angles from the earthquake source, and usually include seismic information over a wider frequency band (e.g., Langston and Helmberger, 1975; Dziewonski and others, 1981).

Analysis of earthquake data

A three-component seismograph will register P- and S-wave (among other) arrivals from an earthquake; the differences in their respective arrival times are sufficient to calculate the distance from the seismograph to the earthquake, provided that the crustal velocity structure is approximately known. If three local seismographs with a broad azimuthal distribution record the earthquake, the hypocentral location can be determined accurately, if—again—the crustal velocity structure is well known. In practice, most micro-earthquakes are recorded by regional vertical-component seismographs, with only limited three-component recordings, and the local velocity structure is subject to numerous uncertainties. The impact of these limitations can be reduced by recording the earthquakes with many proximal stations. Therefore, one philosophy behind micro-earthquake network design is to operate as many vertical-component seismographs as possible, distributed evenly over the study area. An alternative philosophy is to operate three-component, broad-band, high–dynamic range seismographs distributed more broadly, and to use the complete waveforms to determine locations and source parameters. Richter (1958) and Lee and Stewart (1981) have discussed standard methods for locating earthquakes using network data.

The magnitude of an earthquake may be defined in various ways, depending on the portion and type of seismogram used in the measurement. Local earthquakes are generally assigned a magnitude in one of two ways. Local magnitude (M_L) is based on the maximum amplitude (normalized to a distance of 100 km) recorded on a standard Wood-Anderson seismograph (Richter, 1958). Coda magnitude (M_c), designed to carry M_L to values of less than 4.0, is derived from the duration of the recorded seismogram (the coda) after correction for the event-seismograph distance and regional attenuation variations (Lee and Stewart, 1981). Teleseismic earthquakes are often assigned a body-wave magnitude (m_b), which is determined from the amplitudes of short-period body waves, and/or a surface-wave magnitude (M_s), which is determined from the amplitude of surface waves with a period of 20 sec, a measurement requiring data from a long-period seismograph (Richter, 1958). However, deep earthquakes do not necessarily generate large surface waves, which makes M_s and M_L estimates unreliable. Moment magnitude (M_w) is a magnitude scale based directly on the radiated energy (Hanks and Kanamori, 1979) and is related to the seismic moment (M_o, measured in dyne-centimeters) by:

$$M_w = (2/3) \log M_o - 10.67 \quad (1)$$

and

$$M_o = \mu SA \quad (2)$$

where μ is the shear modulus in dyne/cm^2; S is the average fault slip in cm; A is the area of the fault surface in cm^2.

The earliest instrumental recordings of earthquakes showed a systematic variation in first ground motion with azimuth, and it was soon recognized that this first-motion variation could be used to infer the sense and nature of faulting at the source. A double-couple source (a pair of perpendicular coupled forces) generates either compressions or dilatations, depending on the azimuth with respect to the fault plane (Fig. 1A). Null records, or nodal phases, are observed where the motion changes from one sense to the other. The pressure axis (P-axis) is defined as the center of the dilatational quadrant, and the tensional axis (T-axis) is defined as the center of the compressional quadrant. These relations follow from basic principles of rock mechanics. As can be inferred from Figure 1A, the most important criterion for a reliable determination of an earthquake's focal mechanism is that it be recorded at many azimuths and distances, thereby giving uniform coverage of the focal sphere. Figure 1B illustrates three common fault types, and the sense of P-wave first motions (compressional or dilatational) in lower hemisphere projections, as is most commonly used. Focal mechanisms provide a powerful means of determining fault planes and slip directions, and consequently, crustal and lithospheric plate motions. Focal mechanisms for a variety of fault geometries and tectonic settings are presented by Dewey and others (this volume). For larger earthquakes it is now common to proceed in the analysis beyond first motions and find a model of fault slip that matches the observed regional and teleseismic waveforms and amplitudes (Helmberger, 1974; Langston and Helmberger, 1975; Nábělek, 1985). The determination of focal mechanisms from the analysis of long-period body waves has been discussed by Dziewonski and others (1981).

An important physical parameter associated with movement on a fault is the stress drop. Aki (1967) first suggested a striking similarity among all earthquakes, namely, a constant stress drop. Hanks (1977) concluded that, for 12 orders of magnitude in seismic moment, the stress drop is constant, with an uncertainty of ±1 order of magnitude. The stress drop ($\Delta\sigma$, measured in bars) associated with a locally recorded earthquake is generally calcu-

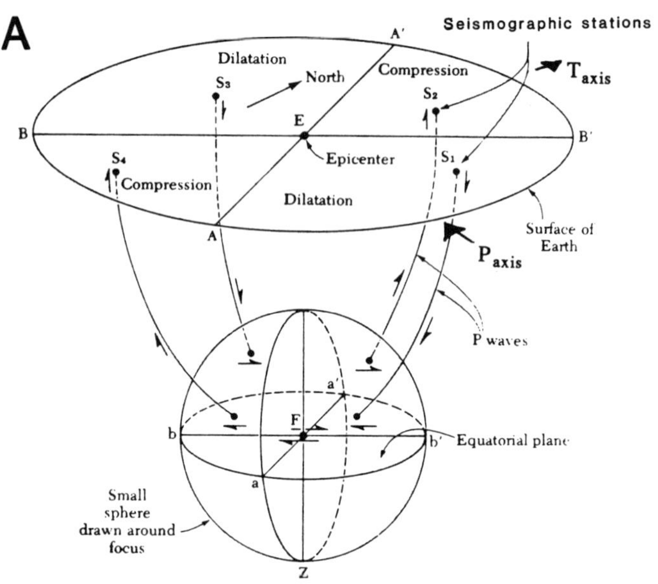

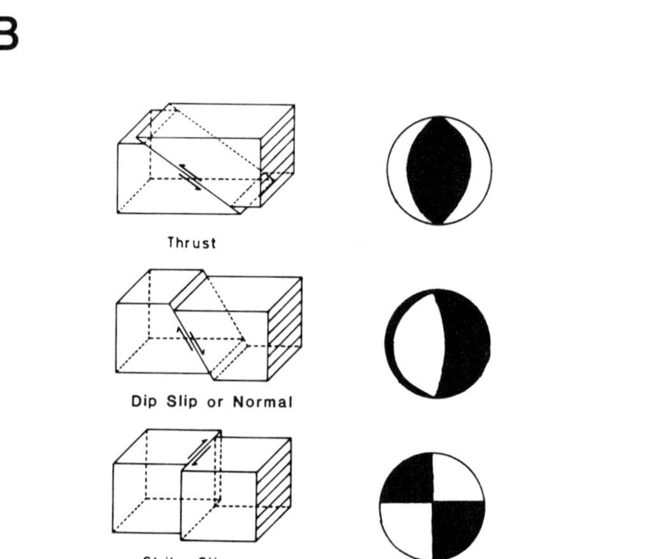

Figure 1. A, Upper-hemisphere compressional and dilatational quadrants in a three-dimensional view around a pure strike-slip earthquake focus. The focal mechanism of an earthquake can be determined if first-motion data are available covering all four quadrants. From Bolt (1982). B, Block models of three simple fault motions (thrust, normal, and strike slip), and the corresponding lower-hemisphere projection of first motions (black for compression, white for dilation).

A

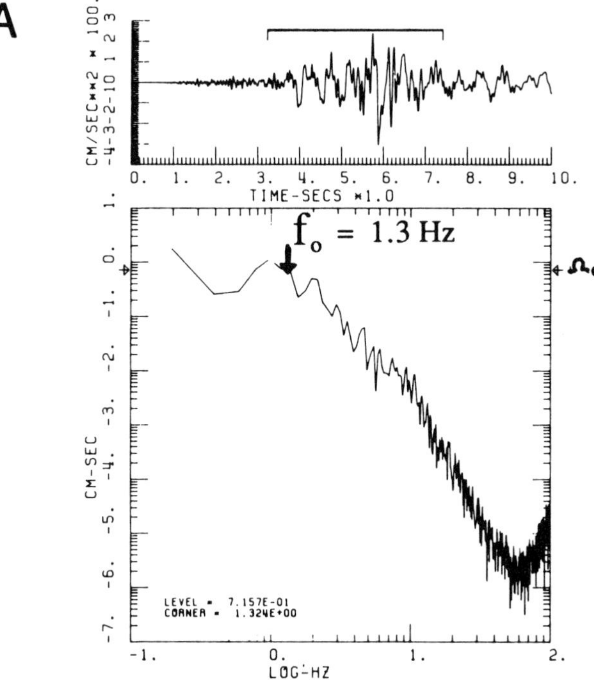

B

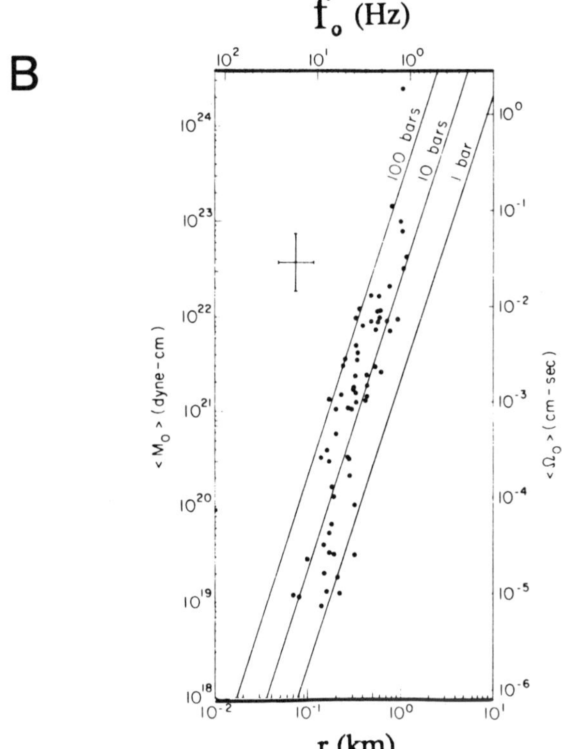

Figure 2. A, Seismogram and corresponding shear-wave frequency spectrum for a microearthquake. Source radius and stress drop are calculated from spectral values [equations (3) and (4)]. B, Seismic moment (M_0) versus source radius (r), low-frequency spectral level (Ω_0), and corner frequency (f_0) for microearthquakes in Mammoth Lakes, California. Stress drops of 1, 10, and 100 bars are indicated by sloping lines. Note error bar indicating uncertainty. From Archuleta and others (1982).

lated from the frequency spectrum of the shear wave. According to the model of Brune (1970):

$$r = 2.34V_s/(2f_0) \quad (3)$$
$$\Delta\sigma = 7M_0/16r^3) \quad (4)$$

where r is the source radius in km, V_S is the shear-wave velocity in km/s, f_0 is the corner frequency in the displacement spectrum (Fig. 2), and M_0 is the seismic moment in dyne-cm.

Stress drop is most reliably determined from these relationships using horizontal-component records that are used to calculate the S-wave spectrum (Fig. 2).

The seismic moment was defined above as the product of the fault area, average fault slip, and rock shear strength; the total seismic moment integrated over the source volume is the seismic moment tensor (Madariaga, 1983). The moment tensor measures the inelastic deformation at the source during the earthquake, and its value at the end of the rupture process measures the permanent inelastic strain produced by the event. For large earthquakes, the seismic moment and seismic moment tensor provide more accurate quantitative measurements of fault rupture than seismic magnitudes or stress drops (Aki and Richards, 1980). Mahdyiar (1987) has presented a nomograph that allows convenient visualization of the relationships between seismic source radius (r), moment magnitude (M_W), seismic moment (M_0), stress drop ($\Delta\sigma$), and the corner frequency (f_0) (Fig. 3).

Uncertainties in earthquake source parameter determinations

Errors in source parameters depend on the distribution and density of station coverage. For an earthquake within a regional network, epicentral determinations are more accurate than hypocentral determinations. As a general rule, epicentral locations are accurate to one-tenth the average station spacing, and hypocentral locations are two to three times less accurate. The most accurate focal depths are obtained when the nearest seismograph is within one focal depth of the epicenter. Earthquake locations outside the networks will have substantially larger errors. Likewise, the uncertainty in a focal mechanism depends on the azimuthal and distance coverage and on the data quality. For earthquakes with magnitude greater than 4.5 that have been recorded since the mid-1960s, a reliable focal mechanism can be obtained by waveform (synthetic seismogram) modeling (e.g., Langston and Helmberger, 1975; Nábělek, 1985). In many cases, the source properties of an earthquake of magnitude greater than 6 that occurred more than 30 years ago can be determined from the analysis of the surviving seismograms (e.g., Dozer, 1986).

Uncertainties in the determination of seismic moments and stress drops are, like location uncertainties, dependent on station coverage. The amplitude measurements on which magnitude calculations are based are affected by two major factors. The first, seismic attenuation (Q) along the propagation path, acts to reduce the measured amplitude. In general, attenuation effects are

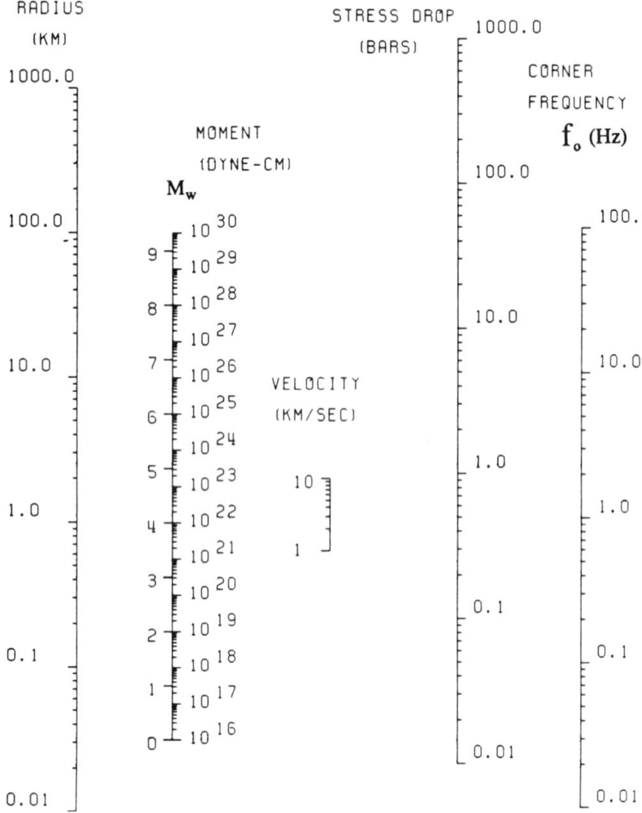

Figure 3. Composite nomograph of source parameters derived from equations (1), (3), and (4) in text. For the example in Figure 2A, the corner frequency (f_o) in the shear-wave spectrum is 1.3 Hz, and we may assume a crustal shear-wave velocity (V_s) of 3.0 km/sec. A line connecting these two values yields a source radius (r) of 0.8 km, a seismic moment (M_o) of 7.0×10^{21} dyne-cm, a moment magnitude (M_w) of 3.8, and a stress drop ($\Delta\sigma$) of 0.6 bars. From Mahdyiar (1987).

much larger for the western United States than the eastern. The second factor, the local station site response, may, in extreme cases, play the dominant role in determining seismic amplitudes. Uncertainties in seismic moments are estimated to be 50 percent or less; stress drop uncertainties are larger: half an order of magnitude or more (Archuleta and others, 1982; Fig. 2).

DETERMINING EARTH STRUCTURE FROM NATURAL SOURCES

Data

The use of earthquake sources to determine the Earth's structure has certain advantages over manmade sources: worldwide distribution (particularly the distribution with depth), strong seismic energy, and particularly high-amplitude shear and surface waves. By comparison, manmade sources with high yields (very large chemical explosions or nuclear sources) are limited in number and distribution, and are relatively weak in shear-wave energy. The disadvantages of natural sources arise in part from their unpredictability in time and space, and the uncertainties in hypocentral location and origin time.

The data used to determine earth structure from natural sources come from the same networks used to determine earthquake source parameters: permanent regional networks, temporary aftershock networks, and intermediate-period and broad-base seismic observatories such as WWSSN stations or other global networks. In addition, data are obtained from special temporary deployments of intermediate-period seismographs (as much as 20 sec) at broad spacings (25 to 100 km) that record teleseismic body- and surface-wave arrivals. Typically, these temporary deployments last 2 to 4 months, in order to record a sufficient number of earthquakes.

Methods of analysis

More than a dozen methods exist for the determination of Earth's structure using earthquake sources. Some of these methods can also be applied to manmade sources, particularly nuclear explosions.

One-dimensional methods: Body waves. The observed arrival times (or seismograms) from earthquake sources are commonly displayed in a traveltime plot (or record section), with distance from the source plotted against time. In such a display, the traveltime curves of refracted and reflected arrivals may be identified in the data. These curves can then be inverted for velocity-depth structure subject to several assumptions. The most important assumption is that the average structure between the source (or sources) and the recording array can be reasonably approximated by a one-dimensional velocity-depth function (i.e., assuming no lateral velocity variations). The questionable validity of this assumption is probably the largest source of error. A second source of error is the common assumption that no low-velocity zones occur within the one-dimensional velocity-depth function.

Once a P- or S-wave traveltime curve has been measured, the Herglotz-Wiechert integral (Bullen and Bolt, 1985) can be used to determine the one-dimensional velocity-depth function that matches an observed traveltime curve. This integral relates the velcoity-depth function to the slowness (reciprocal apparent velocity) versus distance curve:

$$Z(p_1) = \frac{1}{\pi}\int_0^{X_1} \cosh^{-1}(p_o/p_1)\,dx \qquad (5)$$

whre $dT/dX = p$, p_o is the apparent slowness at the surface, and p_1 is the apparent slowness at a distance X_1.

Once the traveltime curve has been inverted using this integral, the resultant velocity-depth model can be evaluated and modified by iterative forward modeling, wherein a series of small adjustments are made to the initial model to improve the overall traveltime fit. This is essentially the procedure used by Jeffreys and Bullen (1935, 1940) to develop the first models for the velocity structure of the Earth.

All traveltime observations are in some way incomplete or contain scatter due to near-surface effects, lateral velocity variations, and incomplete sampling of the wavefield. For this reason, it is often desirable to know the permissible bounds on the velocity-depth structure that a given traveltime curve provides rather than simply deriving one particular velocity-depth model. This question has been approached from various points of view, the most common one being "tau-*p* inversion" (Bessonova and others, 1974, 1976). In this approach, the observed time-distance curve is transformed into a delay time (tau)–apparent slowness (*p*) curve that is a single-valued function (i.e., triplications in the traveltime curve do not occur). This simple transformation allows the inversion process to conveniently include the extremal bounds (i.e., limits) on the velocity-depth structure. Bessonova and others (1976) illustrated the application of the method to earthquake traveltime data (Fig. 4). Walck and Clayton (1984) applied a variation of the tau-*p* method wherein an entire earthquake record section is directly transformed by means of slant stacking (e.g., McMechan and others, 1982; Waters, 1987) to yield the tau-*p* curve needed for inversion.

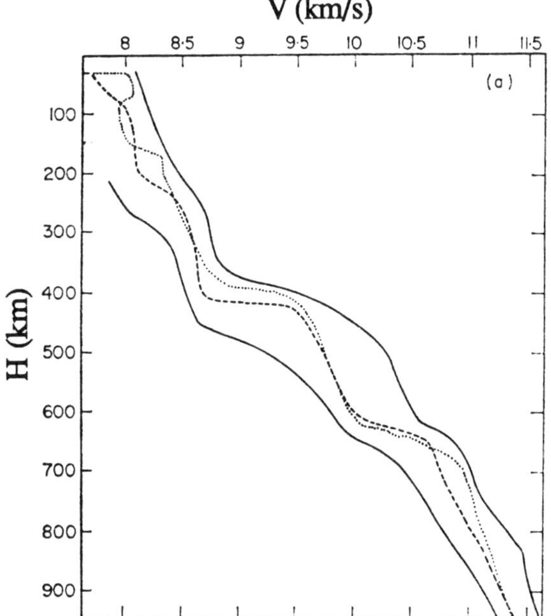

Figure 4. Application of "tau-*p*" inversion of earthquake traveltime for the determination of upper mantle structure. Shown are velocity-depth extremal bounds for 98 percent confidence level (solid lines), "best fitting" velocity model (dashed line), and comparison with previous model CIT 208 (dotted line). From Bessonova and others (1976).

An alternative approach to tau-*p* inversion is to search the parameter space for many models that fit the data. Because there is no direct means of finding all models that fit a given data set, an indirect means is used: a random search of the valid model space. In this so-called "hedgehog" method, many thousands of models (or even millions if sufficient computer time is available) are obtained by perturbing an originally satisfactory model and evaluating the new model for its fit to the observed data. Müller and Mueller (1979) and Knopoff (1972) effectively applied this method to seismic refraction and surface-wave data analysis, respectively.

Seismic amplitudes provide an important additional constraint on velocity structure because amplitudes are particularly sensitive to the details of velocity gradients and discontinuities. In fact, two velocity-depth models that are nearly indistinguishable in terms of their traveltimes will generally have important differences in amplitude-distance behavior. These differences can be determined by the calculation of synthetic seismograms (the numerical calculation of the response of an idealized medium to seismic wave propagation). Ideally, these seismograms should be calculated with the same frequency band as the observed seismograms and should include converted and multiple phases. Synthetic seismogram calculations are far more complicated than traveltime calculations and therefore require considerable computer resources. Synthetic seismogram modeling usually enters the process after a best-fitting velocity-depth model has been obtained from traveltimes.

Trial-and-error amplitude fitting is the most common method used to refine a velocity-depth model with synthetic seismograms. A series of adjustments are made to the initial model until a satisfactory amplitude and traveltime fit is obtained. A variety of methods exist for the calculation of synthetic seismograms for earthquake and explosive sources, including Cagniard–de Hoop (Helmberger and Wiggins, 1971), reflectivity (Fuchs and Müller, 1971; Kind, 1978), finite differences (Boore, 1972; Kelly, 1976), finite elements (Smith, 1975), the "WKBJ" method (Chapman, 1978), recursive methods (Kennett, 1974, 1983), and discrete wave-number methods (Bouchon, 1982; Luco and Apsel, 1983). Trial-and-error modeling using synthetic seismograms has been applied to the determination of upper mantle structure by Helmberger and Wiggins (1971), Grand and Helmberger (1984), and Burdick (1981). Aki and Richards (1980) and Spudich and Archuleta (1987) have given a comprehensive treatment of the theoretical basis for several of these methods.

Synthetic seismogram modeling can also be applied to determine the velocity structure beneath a single station that records nearly vertically incident teleseismic body waves. In this analysis, the complexity in the waveform is used to identify crustal and upper-mantle converted and multiply reflected phases (Langston, 1977; Owens and others, 1984). Based on the amplitudes and times of these phases, the depths and velocity contrasts of the major crustal and upper-mantle discontinuities can be calculated. Owens and others (1984) and Owens and Zandt (1985) have applied this method to crustal studies and obtained excellent agreement with neighboring seismic refraction data (Fig. 5).

Arrival times of local earthquakes provide sufficient information for the calculation of an average one-dimensional model of the velocity structure beneath a seismic network. Although such models are not optimal for interpreting tectonic structure, a one-dimensional velocity model with appropriate station corrections provides the easiest and most efficient means of obtaining accurate earthquake locations. To obtain this model, traveltime residuals are computed for network stations using an assumed initial velocity model. A linear set of equations is generated relating the traveltime residuals to a series of small perturbations in origin times, hypocentral coordinates, and the velocity of the model. This set of equations is solved using damped least squares, and the initial model and hypocentral parameters are adjusted accordingly. The final model is obtained after several iterations, and includes individual station terms which to some degree correct for lateral velocity heterogeneities and site-dependent variations (Crosson, 1976; Eberhart-Phillips and Oppenheimer, 1984).

One-dimensional methods: Surface waves. The use of seismic surface waves (Rayleigh and Love waves) for investigations of crustal and upper-mantle structure began in the 1950s with the advent of high-quality seismic observatories and digital computers. Surface-wave analysis is an effective complement to seismic-refraction and earthquake body-wave studies for several reasons. Surface-wave propagation is dependent mainly on the shear-wave velocity structure (Bullen and Bolt, 1985), whereas many seismic-refraction and body-wave studies are restricted only to the compressional-wave structure. Together the shear- and compressional-wave velocities can be used to determine Poisson's ratio, and infer composition and other physical properties. Also, surface-wave investigations can easily provide information regarding deep structure: periods between 10 and 100 sec are routinely analyzed, corresponding to depths ranging from the upper crust to about 200 km (Bullen and Bolt, 1985).

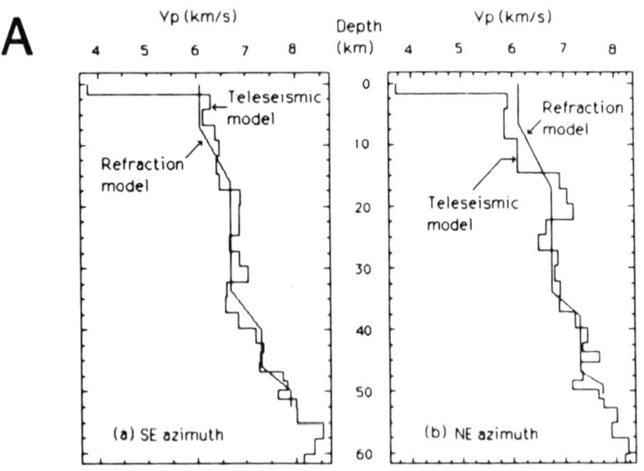

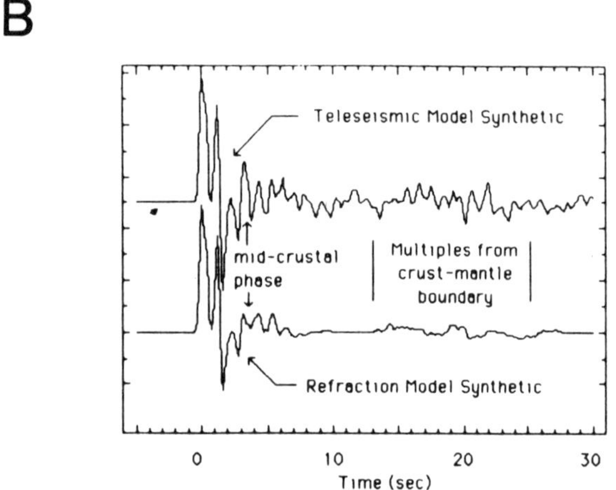

Figure 5. Teleseismic waveform modeling (also called receiver transfer function modeling) applied to data from the Cumberland Plateau, Tennessee (Zandt and Owens, 1986). A, Comparison of models derived from waveform modeling with nearby seismic refraction model. The two models agree well in average velocity structure. B, Example of two synthetic seismograms for teleseismic waveform model and refraction model showing more complex (and more realistic) synthetic for the teleseismic waveform model due to greater complexity in the velocity structure.

The most useful surface-wave data for crustal and upper-mantle studies are recorded by seismographs with a long-period response (100 sec or longer; Fig. 6). Stable temperature, barometric pressure, and tilt are required for these seismographs, which are generally operated at isolated observatories, such as the WWSSN locations. Although the number of stations is limited, the worldwide distribution of earthquakes provides an abundant source of surface-wave observations. Recently, digital recording of long-period seismographs has made possible computer processing of surface-wave recordings, thereby permitting the interpretation of seismic attenuation and anisotropy (Anderson and Hart, 1976; Liu and others, 1976; Mitchell and others, 1976; Mitchell and Herrmann, 1979; Tanimoto and Anderson, 1985; Yomogida and Aki, 1985).

Long-period surface-wave propagation depends on the entire crustal and upper-mantle velocity structure, but as a general rule, the shear-wave velocity structure at a depth approximately 0.4 times the Rayleigh wave wavelength (0.25 times the wavelength, and the near-surface velocity, for Love waves) has the greatest influence on the phase velocity for a wave with a given period (Fig. 6B). Surface waves are observed as dispersed seismograms, with the longer period waves arriving ahead of the shorter period waves because they sample greater depths (which generally have higher velocities; Fig. 6A). If a broad range of wavelengths is recorded, the lithospheric velocity structure can be estimated by working from the lowest phase velocity (shortest period and shallowest depth) to the highest phase velocity (longest period and greatest depth). Methods of measuring phase velocities of surface waves and inverting them for shear-wave structure have been discussed by Press (1956), Brune and others (1960), Brune (1969), and Kovach (1978).

Despite the differences in the equations of motion-governing surface waves and body waves, many of the analysis techniques previously discussed for body waves can be applied to surface-wave analysis after appropriate modifications. Trial-and-error fitting or least-squares inversion of the observed dispersion curves is used to find models that fit the data. This process can usually be constrained by converting an existing compressional-wave velocity structure to a shear-wave and density model and using it as a starting model. Generalized inverse theory (Backus and Gilbert, 1967; Wiggins, 1972; Aki and Richards, 1980; Menke, 1984; Tarantola, 1987) provides a method of evaluating uniqueness and

A

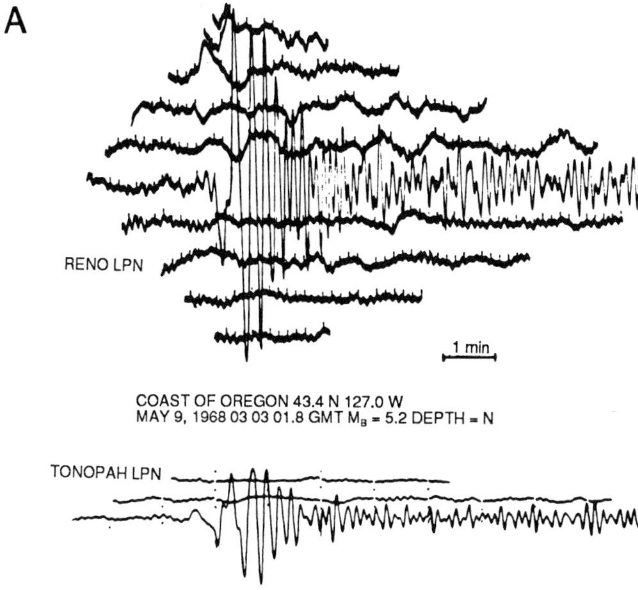

B

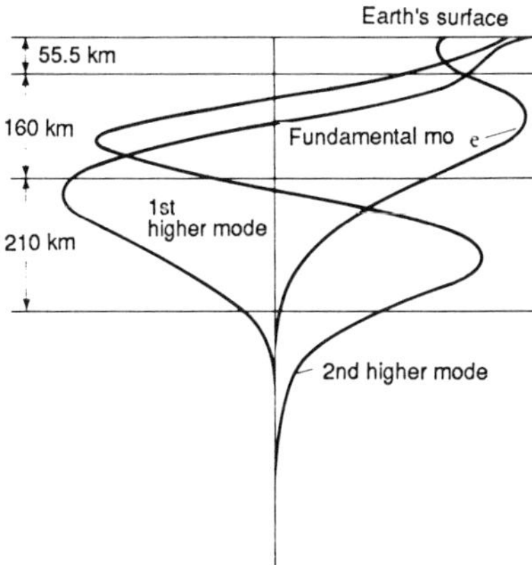

Figure 6. A, Typical surface-wave seismograms: Love waves recorded on long-period seismographs located at Reno and Tonopah, Nevada, for a magnitude 5.2 earthquake off the coast of Oregon. Well-dispersed waveforms are evident. From Priestley and Brune (1978). B, The patterns of displacement with depth for three modes of Love waves for a period of 30 sec. It is apparent that each mode will be sensitive to somewhat different portions of the Earth's structure. From Bolt (1982).

estimating errors in the derived models. Monte Carlo random model-selection techniques (Press, 1968) and the "hedgehog" procedure of varying the physical parameters (Knopoff, 1972) provide a means of evaluating the range of models that will satisfy the observed data.

One of the most important results of surface-wave analysis is the clear distinction in shear-wave structure between different tectonic provinces (Brune, 1969; Knopoff, 1972). Surface waves have also been highly effective in defining the base of the lithosphere, generally taken as the top of an upper-mantle shear-wave low-velocity zone. Measurements of seismic attenuation from surface waves indicate that attenuation is significantly higher in the upper mantle than in the crust, and a maximum in attenuation appears to correlate with the base of the lithosphere (Anderson and Hart, 1976; Mitchell and others, 1976), supporting theories of a more plastic asthenosphere.

Assumptions and uncertainties in one-dimensional analysis. The assumption that a one-dimensional velocity-depth model is a reasonable representation of Earth structure is the largest source of uncertainty in the methods discussed above. However, two points are worth noting. First, the one-dimensional assumption substantially reduces the degrees of freedom in the solution, thereby allowing for a well-constrained estimate of the average velocity structure. Second, a general stratification of the crust and upper mantle on the horizontal scale of 100 km and greater is evidenced by the excellent fit of one-dimensional synthetic seismograms to seismic-refraction and surface-wave data, and the gross results of seismic reflection profiling.

The presence of low-velocity zones within the crust will cause errors in the interpretation of seismic data when some one-dimensional inverse methods are used. Neither Herglotz-Wiechert integration nor tau-*p* inversion methods, as generally applied, allow for low-velocity zones. If a low-velocity zone is present, layers below the zone will appear shallower than their true depths. Even when a low-velocity zone is correctly identified, the velocity within the zone is subject to a large uncertainty.

Amplitude modeling using synthetic seismograms significantly reduces uncertainties in the one-dimensional interpretation of seismic data unless the one-dimensional assumption is seriously in error. Seismic amplitudes are particularly sensitive to velocity gradients and velocity discontinuities; thus a one-dimensional model derived from synthetic seismogram analysis will be better constrained and include more detail than a model derived only from traveltime analysis.

Zandt and Owens (1986) presented a comparison of teleseismic waveform modeling for crustal structure with a seismic refraction model for the same area. Their comparison shows that the two methods were in excellent agreement in one study area, with a mean difference of about 0.2 km/sec at any depth (Fig. 5A). Given this agreement, further investigations of the methodology, assumptions, and uncertainties in teleseismic waveform modeling are warranted.

For surface-wave modeling, consideration of the range of models that fit a given surface-wave data set indicates that while particular details within a model may not be required by the data, the gross shear-wave velocity structure will be well estimated (Kovach, 1978). Since the compressional- and shear-wave structure is closely correlated, the uncertainties of the model will be significantly reduced if seismic refraction data are available to constrain the crustal thickness and compressional-wave velocity structure.

Two- and three-dimensional methods of determining Earth's structure

It is often of greater interest to determine two- or three-dimensional velocity variations in the Earth than to characterize the average one-dimensional structure. Examples include studies of continental margins, the roots of mountain belts, detecting subducted slabs, and studies of calderas and geothermal areas. In these studies, a two-dimensional model is derived from a linear seismic array, and a three-dimensional model is derived from an areal array.

The time-term method (Willmore and Bancroft, 1960) is one of the oldest methods for determining two- and three-dimensional structure. The name is derived from the fact that the method separates the total traveltime of a particular source-receiver pair into three parts:

$$T_{i,j} = A_i + A_j + X_{i,j}/V_{ref} \quad (6)$$

where $T_{i,j}$ is the observed traveltime from source i to station j; A_i and A_j are the source and station time-terms, respectively; $X_{i,j}$ is the distance between source i and station j, and V_{ref} is the refractor velocity.

The advantage of expressing the traveltimes in this way is that it assigns a common time-term to each station and source, and fixes the refractor velocity for each arrival branch. Having done so, well-developed methods for the least-squares solution of a set of simultaneous equations may be used, if there are more equations than unknowns. The solution requires that a sparse N by N matrix be inverted, where N is the number of stations or sources, whichever is larger. The method is well suited to a variety of situations, such as where a large number of earthquakes have been recorded by a seismic network (Oppenheimer and Eaton, 1984; Hearn, 1984). Seismic anisotropy may also be included in the method simply by including an azimuthal dependence in the refractor velocity term (Bamford, 1977; Kohler and others, 1982; Walck and Minster, 1982).

The time-term method, as generally applied, assumes that the ray paths from source to receiver are along a series of nearly horizontal refractors with constant velocity. In areas of complex structure, delay-time methods are more effective than time-terms because delay times more accurately account for actual propagation paths (O'Brien, 1968).

In alternative methods, the Earth is modeled by a set of discrete blocks, each with constant velocity, or by a nonuniform three-dimensional grid, with velocity values specified at the intersections (grid points). Either method allows for an arbitrarily complex three-dimensional structure as the size of the blocks or the gridpoint spacing is decreased (assuming that sufficiently dense data are available to determine the velocity structure). In the case of the three-dimensional grid model, the velocity (and its spatial partial derivatives) at a particular point along a seismic ray path may be computed by linear interpolation between the surrounding eight gridpoints. This permits rapid calculation of approximate traveltimes to be used in an iterative simultaneous inversion for the three-dimensional velocity structure and hypocentral parameters using the traveltime residuals (Thurber, 1983). Parameter separation (Pavlis and Booker, 1980) may be used on the matrix of the hypocentral and velocity partial derivatives to separate the location problem from the velocity calculation, thereby reducing the size of the problem and increasing the amount of data that can be used. Eberhart-Phillips (1986) illustrates the effective use of this method in an inversion for crustal velocity structure in northern California using 200 earthquakes and 200 velocity gridpoints (Fig. 7).

In some applications of three-dimensional inversion, very large data sets are available, thereby requiring methods other than those using matrix inversion. For example, Hearn and Clayton (1986) considered about 45,000 earthquakes, each recorded by a subset of the southern California network (160 stations), yielding over 300,000 Pg arrival times. Such a large data set is not amenable to methods using matrix inversions. Instead, a form of tomography (Worthington, 1984; Humphreys and Clayton, 1988), a method widely used in medicine for imaging of the interior of the body, is used. Similar to time-term analysis, each traveltime is associated with terms corresponding to the source, receiver, and the connecting ray path, which is divided into N discrete calls. An initial reference model is derived from a least-squares fit to all data, and seismic ray paths from all sources to all stations are traced. The traveltime residuals calculated for all source-station pairs are then distributed into the ray-path cells and the station and source terms, thereby yielding a revised model. The final model is obtained when the traveltime residuals converge after several iterations. An advantage of this method is that the calculation is done on a ray-by-ray basis, so there is no limit on the amount of data that can be used.

A powerful method for determining two- and three-dimensional lithospheric models is based on the inversion of teleseismic body waves. The inversion procedure is similar to those previously described, except that there is no need to determine simultaneously the hypocentral location and origin time; the modeling is based on relative arrival times throughout the array. The earth structure in the volume to be inverted generally consists of a series of layers that are divided into a number of blocks (Fig. 8). Within each block, deviation from the initially assigned velocity is determined from the traveltime data. The minimum block size that makes physical sense for this method is that equal to one wavelength (e.g., 6 to 8 km for 1 Hz) of the seismic energy recorded. However, in many cases, the block size is larger than a wavelength due to the limited amount of data available for the inversion.

Aki and others (1977) presented the mathematical formation of the teleseismic inversion problem, including the consideration of the effects of errors in the standard earth model, the initial block model, and the source parameters for the events used. The block inversion of teleseismic data has been widely applied in the past 10 years. Aki (1982) presented a summary of these results,

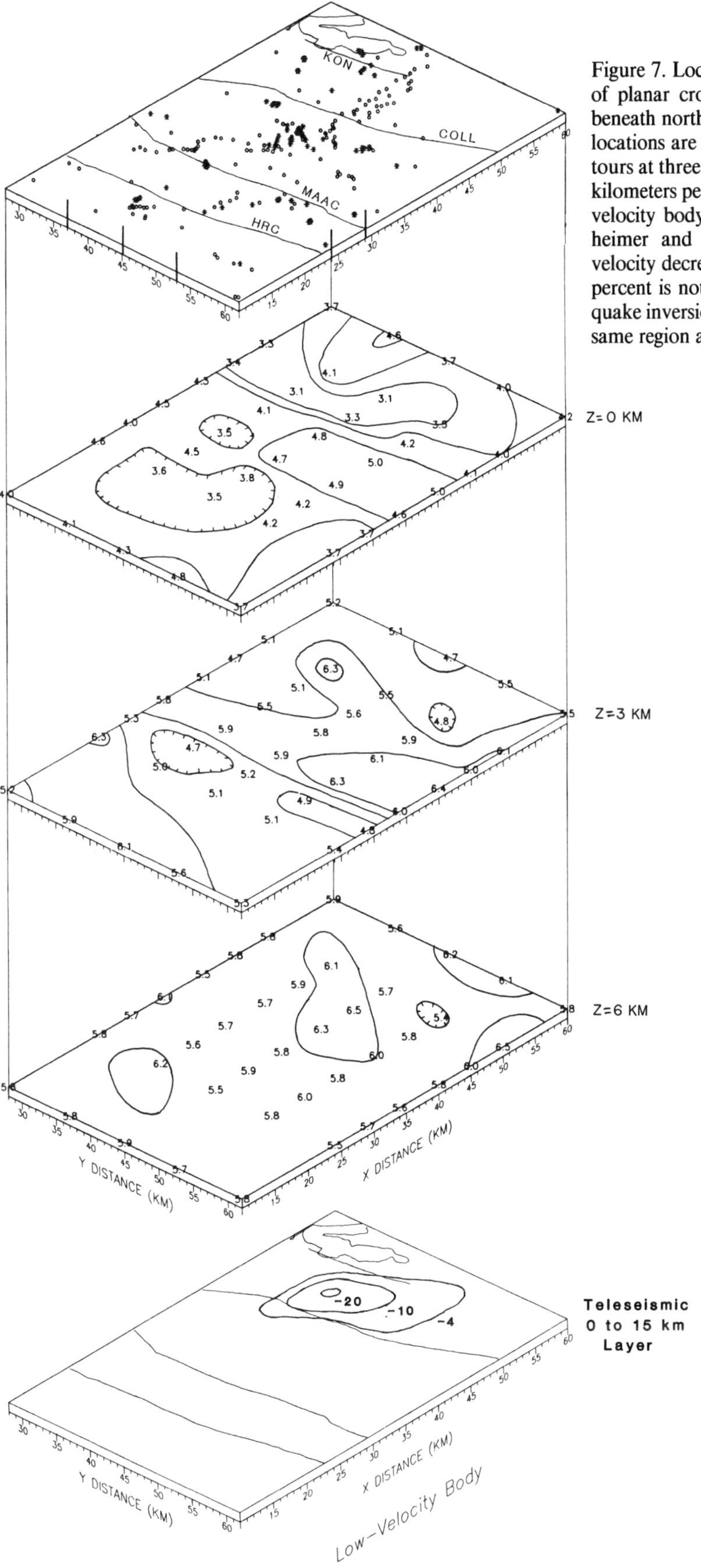

Figure 7. Local earthquake traveltime inversion: perspective plot of planar cross-sections of the upper crustal velocity structure beneath northern California (Eberhart-Phillips, 1986). Epicentral locations are shown in top (geographic) section. Isovelocity contours at three depths (0, 3, and 6 km) are shown; velocities are in kilometers per second. Bottom plane shows the upper-crustal low-velocity body determined from teleseismic residuals by Oppenheimer and Herkenoff (1981); contours indicate percentage velocity decrease. Teleseismic velocity decrease of as much as 20 percent is not seen in any of the three depths in the local earthquake inversion, which in fact shows high seismic velocities in this same region at a depth of 6 km.

A

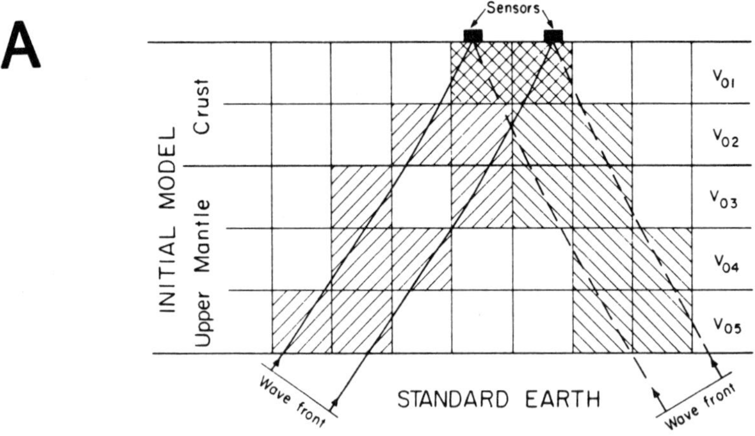

B

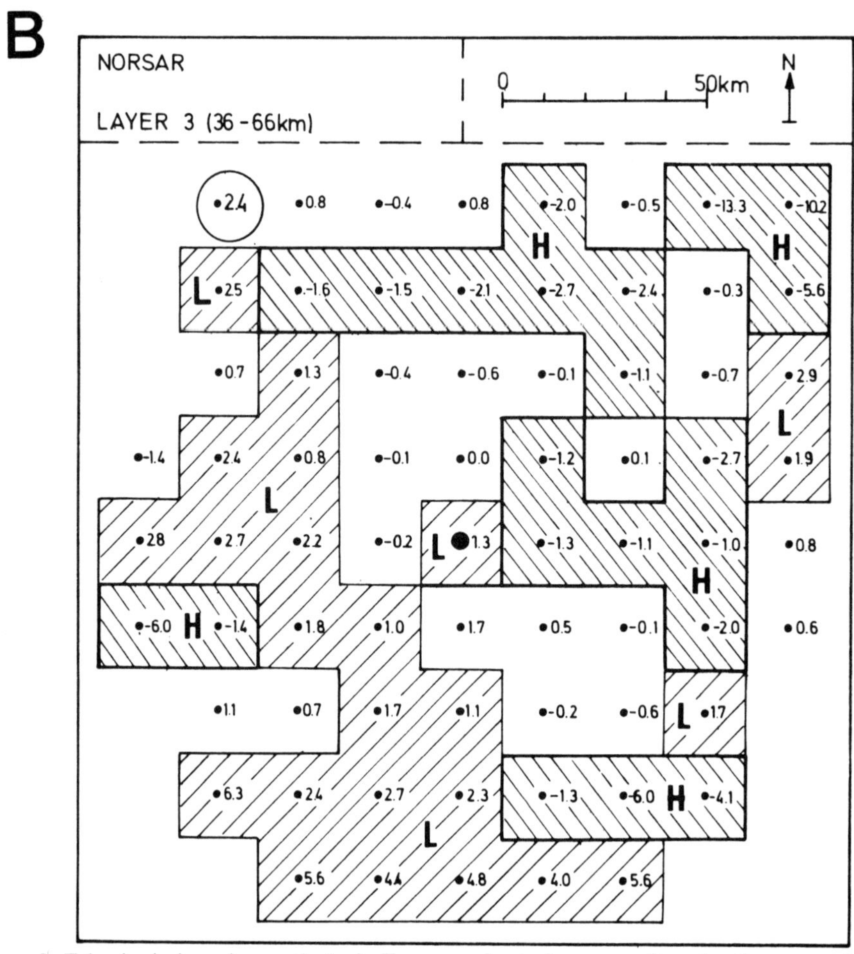

Figure 8. Teleseismic inversion method. A, Two wavefronts from opposite azimuths approaching surface sensors, traversing different blocks in the model, except for the two upper crustal blocks. B, Example from Norwegian "NORSAR" seismic array of a result from teleseismic inversion: high- and low-velocity regions (numbers in percentage of deviation from initial model) within blocks at a depth of 36 to 66 km. From Aki and others (1977).

and Iyer (1984a, b) discussed the application of the method to the study of calderas and magma chambers beneath volcanoes.

Assumptions and uncertainties in two- and three-dimensional methods. The time-term method is subject to several restrictive assumptions. The method will not work well if there is high relief on the refractor because the ray path from the refractor to a source or receiver will have an azimuthal dependence. Furthermore, the method does not account for lateral velocity variations or velocity gradients within the refractor. The result is that, if the refractor velocity is underestimated, the time-terms will be systematically underestimated as well (and vice versa).

Tomographic methods can be applied to laterally varying media, but often have blurred solutions because of the parameterization of the problem and because the solution is obtained iteratively rather than by a matrix inversion. The solution may be sharpened by using various processing techniques, although such processes may introduce artifacts.

Of the presently used three-dimensional methods, the matrix inversion method described by Thurber (1983) appears to result in the fewest artifacts, and a resolution matrix provides a measure of how well-constrained the velocity is at each grid point. Teleseismic inversion methods (Aki and others, 1977) also provide resolution matrices that indicate those portions of the block model with an insufficient number of transmitted rays to reliably constrain the velocity anomaly. It is important to note that error analysis in inverse modeling is parameter-specific, and that if the model parameters are not correct, the actual "answer" will be much different from the inverse "solution," even though the error analysis will not reflect this.

DETERMINING EARTH STRUCTURE FROM CONTROLLED SOURCES

Seismic refraction profiling

Data. The work of Tuve (1951) is the earliest reference to a crustal-scale seismic refraction profile in the compilation by Braile and others (this volume). In the intervening 35 years (1951 to 1986), some 150 additional profiles have been reported, at the rate of four to five a year. Compared with earlier profiles, the most significant advances in data acquisiton have come in the number and spacing of shot points and recorders that constitute modern seismic refraction profiling. Many pre-1970 profiles were of a reconnaissance nature and used shot-point spacings of 200 km or larger and recorder spacings of 10 to 20 km. These experiments succeeded in outlining the average compressional-wave velocity structure of the crust and upper mantle on regional scales. In some cases, large explosions were recorded to ranges well in excess of 1,000 km, yielding measurements of the seismic velocity of the upper mantle that remain unsurpassed in terms of quality (Iyer and Hitchcock, this volume). The refraction experiments of the 1960s provides sufficient information to allow for the first contour plots of crustal thickness in the United States (Pakiser and Steinhart, 1964; Warren and Healy, 1973). James and Steinhart (1966) provided an excellent critical review of both crustal reflection and refraction seismology up to the mid-1960s.

Since the mid-1970s, the need for more geologically realistic crustal models has led to considerably more interest in the two-dimensional details of the velocity structure of the crust. This need has necessitated profiles with closer spaced seismographs and shot points. The improvement in the density with which data are collected is about an order of magnitude: 15 to 25-km shot-point intervals and 0.3 to 2.0-km seismograph spacings are now common (Fig. 9). Airgun sources fired in lakes and over continental margins provide continental refraction data with 100-m shot-point intervals. Recently, these airgun profiles have been recorded by land arrays using 100- to 300-m recorder intervals, yielding data of unprecedented density. Prodehl (1984), Meissner (1986), and Mooney and Brocher (1987) have provided modern summaries of results from seismic refraction profiling.

Methods of analysis. For most seismic refraction profiles, the most prominent arrivals are refractions within "layers" in the crust, and wide-angle reflections from velocity discontinuities. In general, middle- and lower-crustal wide-angle reflections have higher amplitudes than the refractions, and the interpretation process emphasizes matching the arrival times and amplitudes of these reflections. In recognition of the importance of these wide-angle reflections, some authors prefer the term "refraction/wide-angle reflection" profiling to "refraction profiling"; here I use the briefer term for economy.

There have been steady improvements in methods for analyzing seismic refraction data since the first profiles were collected. Early investigators were often well aware of the potential for detailed traveltime and amplitude interpretation of two- and three-dimensional structure, but lacked both the requisite data density and the theoretical and computational capabilities necessary to make such interpretations. Prior to about 1970, most seismic refraction interpretations were based on analysis of refraction and wide-angle reflection arrival times, with the amplitudes of the most important phases generally interpreted in rather general terms (e.g., Healy, 1963; Hill and Pakiser, 1966). A variety of methods were applied in individual studies, in some cases with statistical analysis of errors in velocities and depths (e.g., Steinhart and Meyer, 1961). Two-dimensional models of crustal structure were derived from the interpretation of reversed profiles in terms of dipping layers, or by joining one-dimensional solutions for individual shot points.

The time-term method was often applied to determine depth variations on a given refractor (e.g., James and Steinhart, 1966). Smith and others (1966) gave a detailed mathematical treatment of the method as it was developed to that time. The Herglotz-Wiechert integral (equation 5) was also employed to derive one-dimensional models.

An alternative interpretational method was widely applied to North American refraction data by Prodehl (1970, 1979) and Prodehl and Pakiser (1980). This method is essentially a modification of the Herglotz-Wiechert integral to allow for low-velocity

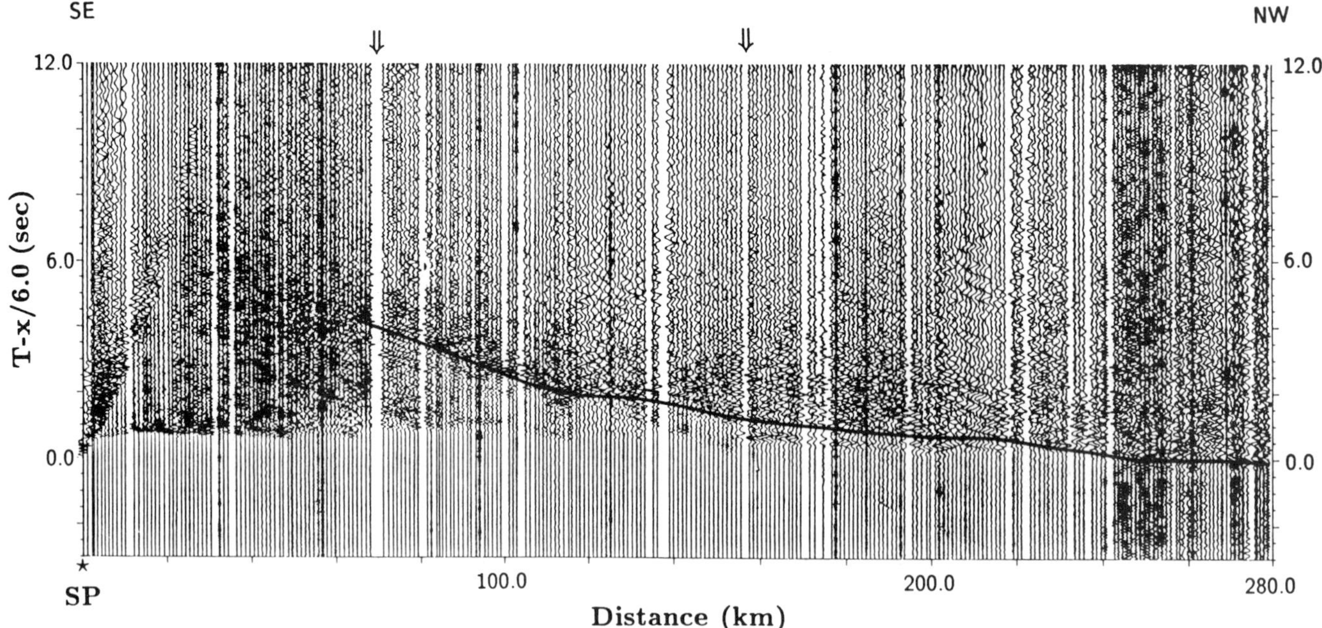

Figure 9. Modern seismic refraction/wide-angle reflection data recorded in central Alaska by the U.S. Geological Survey with an instrument spacing of 1 km. Crustal and upper-mantle phases are clearly recorded. Data are plotted with a reduction velocity of 6.0 km/sec. Moho reflection is indicated by solid line. Data of this quality and density are typical of those recorded over the past 10 years. More recently, data have been collected with recorded spacings of as little as 50 m, yielding data that can be processed similar to seismic reflection data (see text).

zones; it is well suited to the computation of one-dimensional velocity structures with either positive or negative velocity gradients and with transitional velocity discontinuities rather than sharp interfaces. The method and its application to real data are described by Prodehl (1979). Typically, the results from several shot points are presented in the form of isovelocity cross sections (c.f., Prodehl and Lipman, this volume), as opposed to layered cross sections.

Since the mid-1970s, major improvements in the analysis of crustal seismic refraction data have been made possible by two new capabilities: the processing of digital seismic traces and numerical modeling algorithms for two- and three-dimensional structures. The most important advances are:

1. Digital data processing allows for the routine application of methods that formerly were more cumbersome to apply to analogue data. These processes include frequency and velocity filtering, slant stacking of traces, and various modifications to the display of the traces (e.g., automatic gain control and variable-area shading).

2. Lateral variations in the velocity structure are accounted for with traveltime and amplitude calculation methods, as are the presence of seismic low-velocity zones within the crust and upper mantle (Mooney and Prodehl, 1984).

3. Ray theoretical and full-wave (complete) synthetic seismograms are calculated for the hypothesized crustal velocity model and are quantitatively compared with the observed data (Braile and Smith, 1974; Červený and others, 1977; McMechan and Mooney, 1980; Fig. 10).

4. The effects of apparent Q ("quality factor": inverse seismic attenuation) due to intrinsic attenuation and scattering may be accounted for.

5. Some quantitative estimates of model accuracy can be made when various inversion procedures are applied to the data.

In addition to these advances, seismic interpretation methods, which were previously applied only to seismic reflection data, are now applied to densely recorded refraction data. For example, McMechan and others (1982), Milkereit and others (1985) and McMechan and Fuis (1987) have applied automatic inversion methods based on slant stacking to refraction data, and Klemperer and Luetgert (1987) have applied normal-movement (NMO) velocity analysis and common midpoint (CMP) stacking methods. As denser seismic refraction/wide-angle reflection data are collected, the somewhat arbitrary line between reflection and refraction seismology will disappear.

Assumptions and uncertainties. There are important and restrictive assumptions made in many (but not all) seismic refraction interpretations published in the past 30 years: (1) that a planar, layered structure is a valid approximation; (2) that lateral velocity variations are smaller than vertical variations and occur on the same distance scale as the shot-point interval; and (3) that the principal crustal phases are correctly identified and have not been confused with multiples, phase conversions, or noise. This identification process is referred to as phase correlation.

The first assumption will not be valid in many situations, such as in highly folded and faulted areas or portions of the crust that have been heavily intruded. In such cases, known geologic

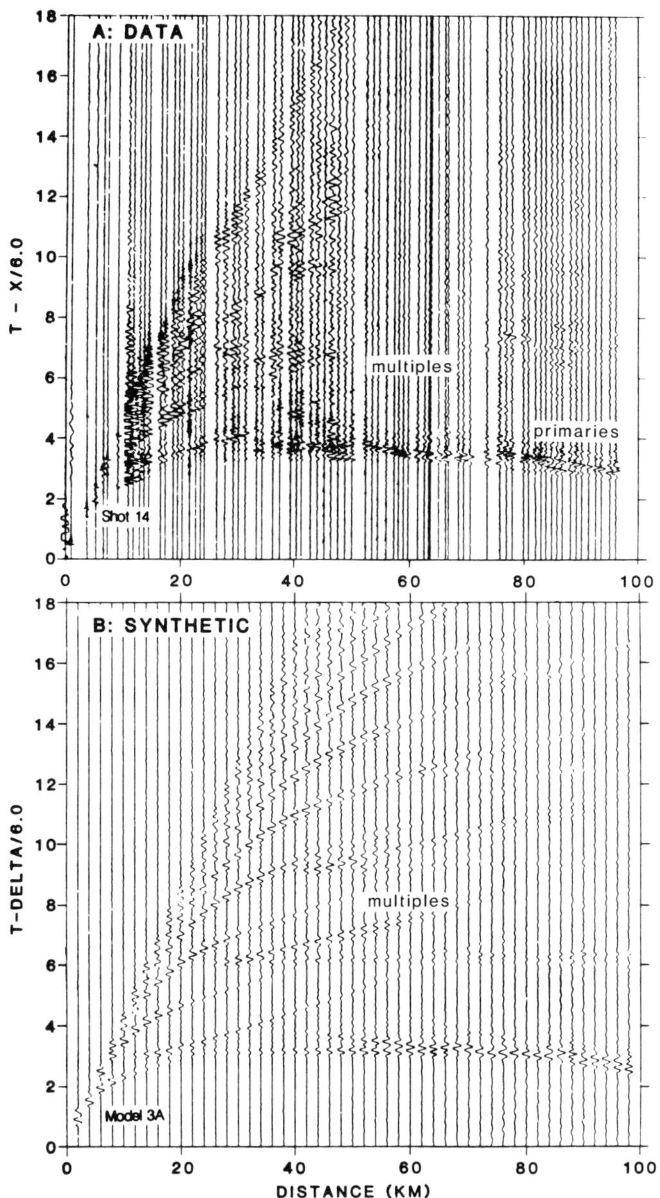

Figure 10. Synthetic seismogram modeling of seismic refraction data. A, Observed data from the Great Valley of California with prominent free-surface multiples. B, Synthetic seismograms calculated with the reflectivity method, including primary and multiple phases, and seismic attenuation. From Hwang and Mooney (1986).

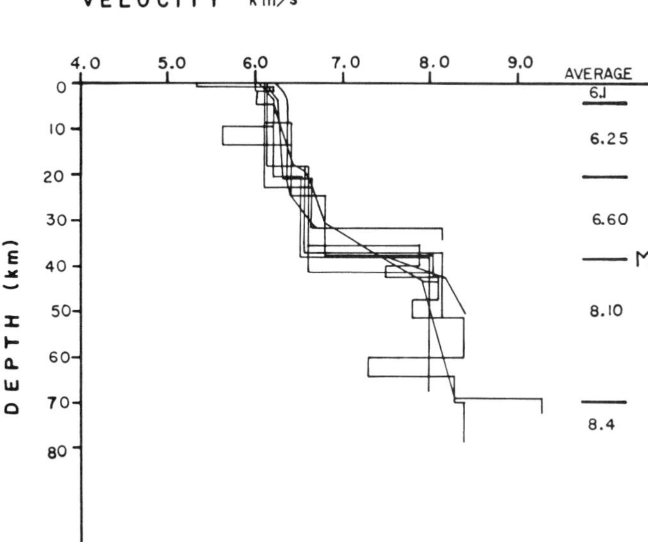

Figure 11. Comparison of seven separate traveltime interpretations of a seismic refraction profile recorded across the Saudi Arabian Shield in 1978. The average model is indicated in the column on the right-hand side of the figure. This comparison illustrates the uncertainty in seismic refraction data interpretation, particularly that arising from somewhat different phase correlations by different interpreters. Other comparisons (Walter and Mooney, 1987) show closer agreement when each interpretation includes amplitude modeling using synthetic seismograms. From Mooney and Prodehl (1984).

constraints should be incorporated into the model, or the results should be qualified by a statement that they represent only the average properties of the crustal structure. The second assumption is derived from the observation that there is a general horizontal stratification of seismic velocity due to increases in pressure, temperature, density, and metamorphic grade with depth. The third assumption, concerning seismic phase correlation, is continuously reevaluated during the interpretation process on the basis of reciprocal traveltimes, amplitude and frequency behavior of the phases, and subjective input based on experience. Detailed comparisons of interpretations of identical data sets have established that a final velocity model is more a function of the particular phase correlation used than of the interpretation method (Ansorge and others, 1982; Mooney and Prodehl, 1984; Finlayson and Ansorge, 1984; Walter and Mooney, 1987; Fig. 11).

Many of the existing refraction profiles in North America were collected in the 1960s, and Pakiser and Steinhart (1964) and James and Steinhart (1966) have summarized the uncertainties in the interpretation of refraction data as of that time. The determination of the seismic velocity of the upper crust has the lowest uncertainty, with an estimated error of 3 percent. This error estimate implies that a layer with a seismic velocity of 6.0 km/sec could actually have a velocity of 5.8 to 6.2 km/sec. Determinations of deeper crustal velocities and Moho depths have 10 percent estimated errors, but the upper-mantle velocity is uncertain to only 3 to 5 percent for the more detailed, reversed profiles due to the isolation of the Pn traveltime curve over large offsets. When considering uncertainties in crustal thickness and Pn velocity, the ratio of the horizontal-to-vertical scale should be kept in mind. For example, the measurement of a 40-km-thick crust with a 400-km-long refraction profile has a horizontal-to-vertical scale ratio of 10, which yields a very stable estimate in the vertical direction.

Since the 1970s, improved data quality and analysis methods have allowed more details of crustal velocity structure to be resolved, and uncertainties in layer depths, thicknesses, and velocities have been reduced approximately in half, particularly for the upper half of the crust. The assumption of planar layers is no longer necessary when two-dimensional ray-trace modeling is applied. Iterative forward modeling has been widely used to interpret this improved data. While forward modeling allows the inclusion in the velocity model of geologic and other constraints from geophysical data, it does not provide a quantitative estimate of errors (as does inverse modeling). Therefore, it is incumbent on the modeler to perturb the best-fitting model to assess the sensitivity of key model parameters to the fit of the data. In this way, some quantitative estimates of model resolution can be made. The statement by James and Steinhart (1966, p. 320) that "tradition has been strong in governing the types of velocity-depth functions fitted to the crust" remains true, and more formal means of evaluating errors and uncertainties in forward modeling remain to be developed.

Since seismic amplitudes are very sensitive to velocity gradients and discontinuities, these features can be refined through detailed synthetic seismogram modeling. The result is that velocity gradients (km/sec/km) are often cited along with average layer velocities. These velocity gradients are probably accurate to ±50 percent or better when they are determined from high-quality data. The nature of seismic discontinuities has received much recent attention, and it is not uncommon that velocity transition zones and laminated zones are modeled in the crust instead of first-order discontinuities (Meissner, 1973; Deichmann and Ansorge, 1983; Braile and Chiang, 1986; Sandmeier and Wenzel, 1986). Estimates of transition-zone thickness are dependent on the frequency and bandwidth of the seismic energy used to probe the discontinuity, but as a general rule, an error of as much as ±50 percent must be admitted.

Seismic reflection profiling

Data. Seismic reflection data provide the highest resolution information on the structural geometry of the crust. The application of reflection profiling to deep-crustal problems has led to great advances in the understanding of crustal structure, composition, and evolution. However, the resolution and data quality of reflection profiles is quite sensitive to the selection of data acquisition parameters. The selection process is often a difficult one; it involves a cost/benefit evaluation wherein factors such as the number of recording channels and shot points are balanced with the final cost per kilometer of profile. The earliest COCORP profiles were collected with a 48-channel recording system and 24-fold common depth point (CDP) coverage, 100-m geophone group intervals, and vibration points at every group. Four or five mechanical vibrator trucks were used as the seismic source array (Schilt and others, 1979; Oliver and others, 1983; Brown and others, 1986). Recent COCORP profiles were obtained with a 96- to 120-channel recording system, and the USGS has used an 800- to 1,000-channel sign-bit recording system, with 25-m group interval and 120-fold coverage (Stewart and others, 1986; Zoback and Wentworth, 1986). Field parameters vary for data collected by the University of Wyoming, Virginia Polytechnical Institute, the California Consortium for Crustal Studies (CALCRUST), and other groups; they have been discussed by Barazangi and Brown (1986a, b) and Matthews and Smith (1987).

For studies of the deep crust, it is often possible to reprocess exploration industry VIBROSEIS data to recover longer two-way times by recorrelating with the original sweep and allowing the sweep to extend off the end of the trace, thereby obtaining approximate correlations to greater times (Okaya, 1986; Trehu and Wheeler, 1987). A limitation to the application of this method is that the original recording parameters of industry data are normally optimized for shallow structure.

Methods of analysis. The acquisition and processing of seismic reflection data has been well advanced by the hydrocarbon exploration industry. These techniques are described in numerous textbooks that range from introductory to advanced (Dobrin, 1976; Telford and others, 1976; Claerbout, 1976, 1985; Coffeen, 1978; Sheriff and Geldart, 1983; Jenyon and Fitch, 1985; Waters, 1987; Yilmaz, 1987). However, it should be kept in mind that the application of reflection seismology to the deep crystalline crust presents special problems not encountered in the exploration of the sedimentary rocks in the upper crust. Here, I emphasize those aspects of seismic processing that are of particular importance to deep crustal reflection seismology. Additional perspectives can be found in Taner and others (1970), Mair and Lyons (1976), Smithson and others (1980, 1986), Johnson and Smithson (1986), Mayer and Brown (1986), Zhu and Brown (1986), and Matthews and Smith (1987). Bally (1983) presented numerous examples of deep crustal seismic reflection data with an emphasis on the reflection signature of various tectonic regimes.

The processing of land seismic reflection profiling begins with a consideration of near-surface effects. In general, there are variations in both elevation and near-surface velocities along a seismic profile. Static corrections (time shifts) are derived for these variations. Usually a reference datum is selected, and both shot points and geophones are corrected to that datum. The proper derivation and application of static corrections is one of the most difficult problems in reflection seismology because of the severity of lateral velocity variations in the near-surface (Sheriff and Geldart, 1983).

Virtually all continental seismic reflection data are recorded using the common depth point (CDP) method, which results in substantial signal enhancement except in the most geologically complex situations. The summing of the field traces with a common depth point requires the application of a normal-moveout correction (NMO) for the hyperbolic curvature of reflection traveltimes. Some knowledge of the crustal velocity structure is needed for the proper application of the NMO correction, and an optimal velocity-depth function is commonly chosen from a series of stacking velocity, NMO, and semblance plots (Jenyon and

Fitch, 1985). Standard seismic refraction analysis may also be applied to first arrivals, and the resulting shallow-velocity determinations may be used to supplement the analysis of the velocity panels. For the deeper crust, superior velocity estimates are obtained with longer array lengths (on the order of 10 to 20 km) because larger moveouts occur for larger offsets (Klemperer and Oliver, 1983). In theory, the root-mean-square velocity can be determined to depths of two or three times the maximum source-receiver offset (e.g., Bartelson and others, 1982; Hajnal, 1986). However, lateral velocity inhomogeneities and low signal-to-noise ratios degrade these velocity estimates, and in practice, few deep velocity estimates based on reflection data have been published. Coincident seismic refraction/wide-angle reflection measurements provide the most reliable deep seismic velocity estimates (Mooney and Brocher, 1987; Smithson and Johnson, this volume).

Many unwanted seismic signals that detract from the coherency of deep reflections are recorded during a seismic reflection survey. The most obvious unwanted signals are cultural noise; in many cases, noise can be removed by filtering or editing of the individual traces prior to stacking. Ground roll, shear waves, and other undesirable seismic energy can be largely eliminated from the unstacked data by a judicious combination of editing, muting, and velocity-filtering procedures (Sheriff and Geldart, 1983). Deconvolution is a process in which distortions of the wavefield due to multiples and filtering effects of the Earth are removed from the data (Sheriff and Geldart, 1983; Waters, 1987).

The CDP stacked section is a standard mode of presentation of seismic reflection data. Although further processing is often both desirable and possible, the CDP stacked section has the advantage of representing a standard level of processing, and is therefore the section with the fewest processing artifacts. The CDP section is often referred to as a "structural section" in recognition of the approximate image of the geologic structure it exhibits.

Lateral discontinuities and isolated velocity anomalies will produce diffractions in a CDP stacked section and may obscure or mimic deep reflections. True deep reflections may be well recorded, but may not appear in their correct subsurface positions because of the effects of dips and lateral velocity variations. Migration is a defocusing process that attempts to collapse diffraction hyperbolas and position reflectors at their actual locations in depth. Migration after CDP stacking is the most commonly applied method, and it is generally effective in structures with dips of less than about 30° and modest lateral velocity variations (Claerbout, 1976; Fig. 12). Other data artifacts can be reduced by (pre-stack) two-dimensional filters (fan, moveout, or pie-slice filters) that eliminate arrivals with particular apparent velocities as measured by the geophone array (Sheriff and Geldart, 1983).

The presentation of seismic reflection data at page size presents a special challenge since such data are normally interpreted at large scale, and processed sections reproduce poorly when reduced. Line drawings are commonly made of the data to display the principal reflections visible in the original section. However, because the construction of a line drawing is highly subjective, recently there have been efforts to produce page-size plots of seismic reflection data that share the visual advantages of line drawings, but which use an automated, and therefore more objective, selection of the reflections to be highlighted in the display. Kong and others (1985) have described a coherency method based on slant stacking of reflection panels that succeeds in producing page-size reflection sections that are as clear as line drawings (Fig. 13; Phinney and Roy-Chowdhury, this volume).

Forward modeling of seismic data makes is possible to calculate the theoretical response of the interpreted geologic structure and to compare this response with the observed reflection data. In the absence of deep drilling to constrain crustal interpretations, modeling is an important step in crustal interpretation and has been effectively applied to a number of data sets (Hale and Thompson, 1982; Wong and others, 1982; Fountain and others, 1984; Fountain, 1986; Peddy and others, 1986; Blundell and Raynaud, 1986; Reston, 1987).

Assumptions and uncertainties. Numerous factors involved in the acquisition and processing of seismic reflection data lead to uncertainties in interpretation.

1. Poor data quality may result from a variety of causes, particularly imperfect coupling of vibrators or geophones to the ground, high cultural noise, or low, uneven-fold data arising from skips or crooked-line recording.

2. Large static corrections due to inhomogeneous near-surface conditions may result in poor stacked data. Static corrections can be reliably determined if sufficiently small group intervals are used, but the need for long arrays (for deep velocity control) conflicts with the advantages of short group intervals.

3. Three-dimensional control on reflector geometry, through the use of off-line recording arrays, is generally not available because of logistical and budget limitations. Consequently, it is difficult to determine how much reflected energy is coming from out-of-the-plane sources, and most interpretations assume that all reflections come from within the vertical plane.

4. Horizontal resolution is limited in the deep crust. Seismic reflections are observable only from features with horizontal lengths on the order of one Fresnel zone, which may be expressed as:

$$R = (0.5\ LH) = 0.5\ V\ (2\ T/F) \qquad (7)$$

where R is the Fresnel zone radius, L is the wavelength at the dominant frequency, H is the depth, V is the average velocity, T is the arrival time, and F is the frequency (Sheriff and Geldart, 1982). This relationship yields a Fresnel zone radius of 3 km for a 20-Hz reflection from 30 km depth in a crust with an average velocity of 6 km/sec.

5. Migration of deep seismic data often does not result in significant improvements in the quality of the image of the reflectors. This is due to generally low signal levels, the incomplete recording of the wavefield of deep reflections with conventional recording arrays, and distortions to the wavefield by near-surface

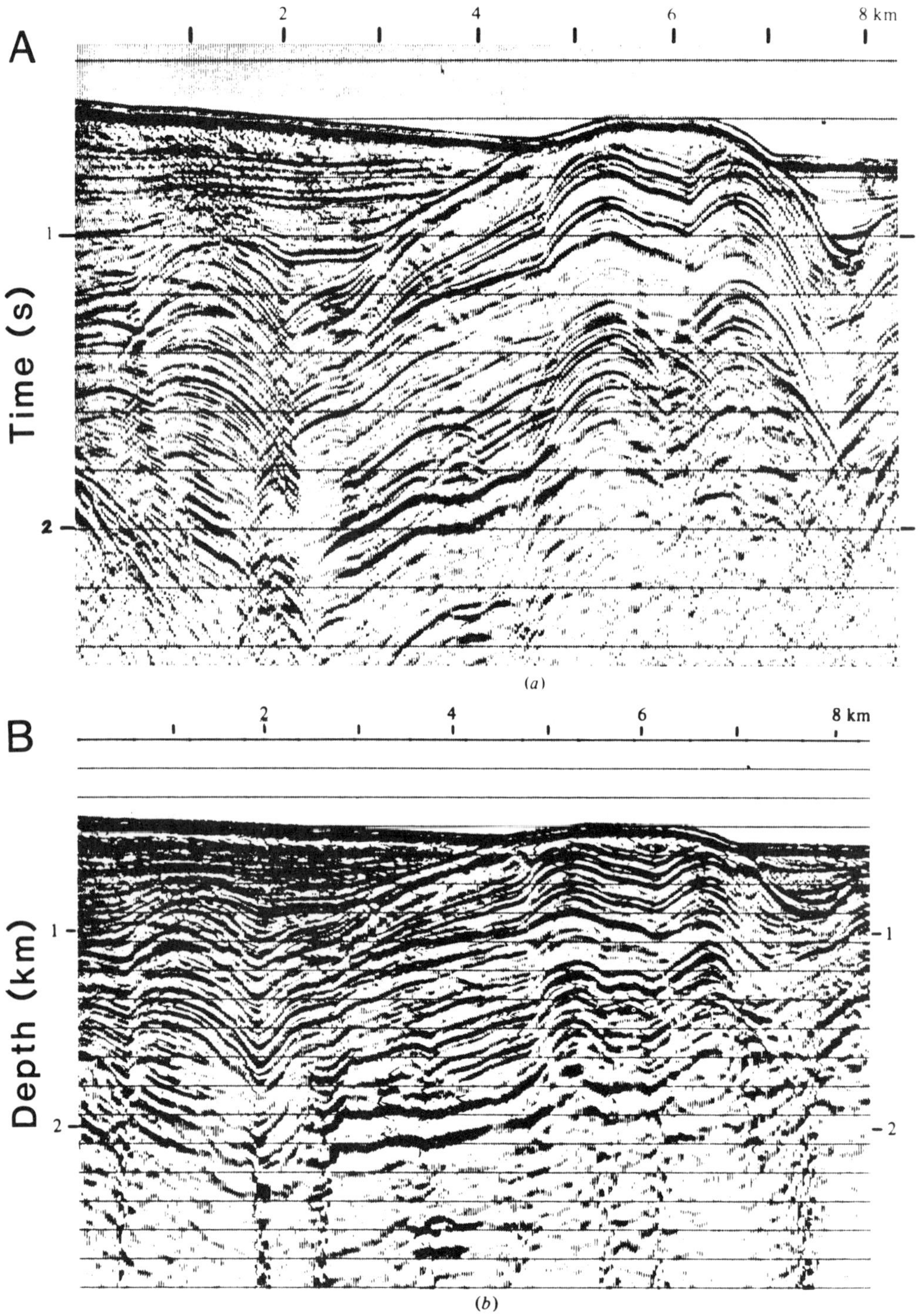

Figure 12. Example of depth migration of shallow seismic reflection data. A, Unmigrated time section. B, Finite-difference time migration with depth conversion. Note collapsed diffractions. From Hatton and others (1981) *in* Sheriff and Geldart (1983).

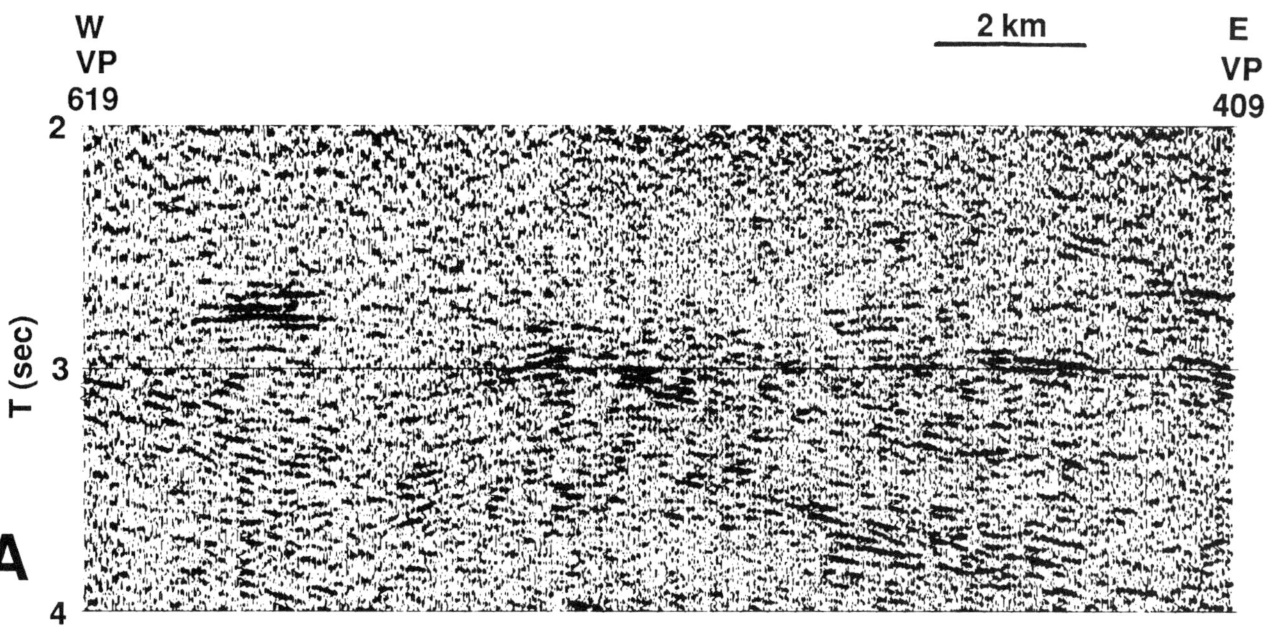

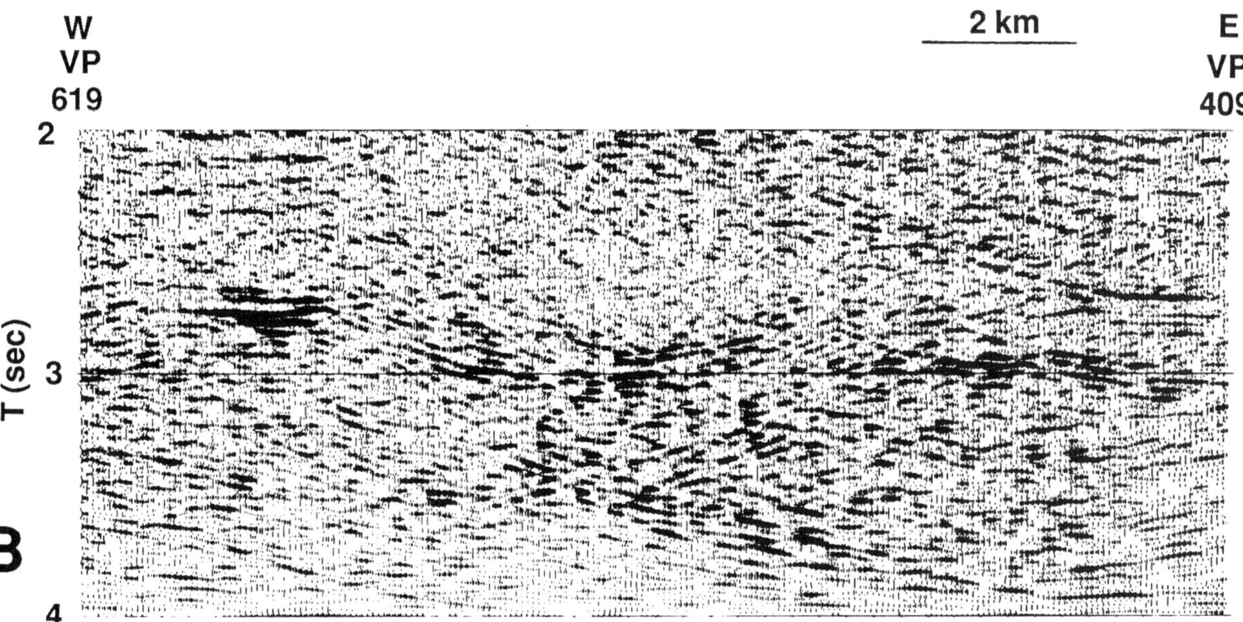

Figure 13. Seismic reflection data with coherency filtering applied to enhance reflections. Reproduction at small scale is significantly improved. A, Portion of the COCORP southern Appalachian line 4A stacked section from 2 to 4 sec, sampled at 4 msec, with 600 traces spaced at 33.5 m. B, Constant velocity (5.5 km/sec) time migration of the stacked section (A). From Kong and others (1985).

features (Warner, 1987). If dips greater than 30° are present, pre-stack migration (Claerbout, 1985) is desirable; however, this process requires substantial computer resources.

A vertical seismic profile (VSP) (Gal'perin, 1974; Balch and Lee, 1984) is recorded in a borehole using surface sources. Since the stratigraphy of the hole may be inferred from borehole geophysical logs, the VSP provides a direct measure of the reflection response of the layers, and may be used to calibrate other seismic surveys. To date, VSPs have had only limited application to crustal reflection profiling (Smithson and Johnson, this volume).

In addition to uncertainties arising from data acquisition and processing, there are other more general issues of seismic wave propagation in the Earth that raise questions regarding seismic reflection interpretation. Blundell and Raynaud (1986) and Reston (1987) have shown that layered reflections, as commonly observed in the lower crust, may be produced by out-of-plane reflections in a model that includes only a single corrugated layer, or by spatial interference effects. Thus, the origin of deep crustal reflections remains a critical question in seismology. Future 3-D surveys will provide important new insights.

CLOSING REMARKS

This chapter has emphasized the data, analysis techniques, assumptions, and uncertainties of seismic methods as they are used to obtain the results summarized in the other chapters of the volume. Rather than emphasizing future directions in seismology, I have focused on current and past practice. From this perspective, it is appropriate to comment briefly on future needs in seismologic research.

The common reliance on forward modeling for the interpretation of seismic data presents a problem for those who seek to quantify resolution and uncertainties in geophysical models. While inverse modeling of traveltimes includes formal resolution estimates, forward modeling shows that waveforms and amplitudes are critical in confirming and refining model details. Waveforms and amplitudes have yet to be fully considered in published inverse solutions for two- or three-dimensional structure. Similarly, the determination of seismic source parameters within complex Earth structures is a promising field. Three-dimensional waveform modeling methods, which are feasible only on the most powerful computers (except for very long-wavelength data), are now being developed as seismologists strive to develop realistic Earth models.

As important as these issues of data modeling are, the highest priority in seismology is high-quality data acquisition; seismology, like all of the earth sciences, is a data-driven science. This is evident from the timing of the periods of the most rapid growth and progress, which have coincided with new data initiatives, such as the installation of the WWSSN network, the Early Rise/Lake Superior seismic refraction investigation (both early 1960s), the installation of the large seismograph networks (beginning in the late 1960s), and the COCORP seismic reflection program (1976 to present). The greatest impact on the future of seismology will come from our willingness and resourcefulness in collecting ample high-quality data, and using acquisition techniques that capture as much as the wavefield as possible.

ACKNOWLEDGMENTS

Discussions over the past decade with many scientific colleagues have aided greatly in preparing the work of this chapter. Critical reviews by E. L. Ambos, W. L. Bakun, H. M. Benz, T. M. Brocher, R. H. Colburn, E. R. Engdahl, J. C. Lahr, S. Larkin, G. A. McMechan, T. V. McEvilly, D. H. Oppenheimer, L. C. Pakiser, K. Roy-Chowdhury, S. Schapper, and S. B. Smithson are greatly appreciated. Figure 9 was prepared by B. Beaudoin.

REFERENCES CITED

Aki, K., 1982, Three-dimensional seismic inhomogeneities in the lithosphere and asthenosphere; Evidence for decoupling in the lithosphere and flow in the asthenosphere: Reviews of Geophysics and Space Physics, v. 20, p. 161–170.

Aki, K., and Richards, P. G., 1980, Quantitative seismology; Theory and methods (2 volumes): San Francisco, W. H. Freeman, 932 p.

Aki, K., Christoffersson, A., and Husebye, E. S., 1977, Determination of the three-dimensional seismic structure of the lithosphere: Journal of Geophysical Research, v. 82, p. 277–296.

Anderson, D. L., and Hart, R. S., 1976, Absorption and the low-velocity zone: Nature, v. 263, p. 397–398.

Ansorge, J., Prodehl, C., and Bamford, D., 1982, Comparative interpretation of explosion seismic data: Journal of Geophysics, v. 51, p. 69–84.

Archuleta, R. J., Cranswick, E., Mueller, C., and Spudich, P., 1982, Source parameters of the 1980 Mammoth Lakes, California, earthquake sequence: Journal of Geophysical Research, v. 87, p. 4595–4607.

Backus, G., and Gilbert, F., 1967, Numerical application of a formalism for a geophysical inverse problems: Journal of Geophysics, v. 13, p. 247–276.

Balch, A. H., and Lee, M. W., eds., 1984, Vertical seismic profiles; Techniques, applications, and case histories: Boston, International Human Resources Development Corporation, 488 p.

Bally, A. W., 1983, Seismic expression of structural styles; A picture and work atlas: American Association of Petroleum Geologists Studies in Geology Series 15, volumes 1, 2, and 3.

Bamford, D., 1977, P_n-velocity anisotropy in a continental upper mantle: Geophysical Journal of the Royal Astronomical Society, v. 49, p. 29–48.

Barazangi, M., and Brown, L., 1986a, Reflection seismology; A global perspective: American Geophysical Union Geodynamics Series, v. 13, 311 p.

—— , eds., 1986b, Reflection seismology; The continental crust: American Geophysical Union Geodynamics Series, v. 14, 339 p.

Bartelson, H., Lueschen, E., Krey, Th., Meissner, R., Scholl, H., and Walther, Ch., 1982, The combined seismic reflection-refraction investigation of the Urach geothermal anomaly, *in* Haemel, R., ed., The Urach Geothermal Project: Stuttgart, Schweizerbart'sche Verlagsbuchhandlung, p. 247–262.

Behrendt, J. C., Green, A. G., Cannon, W. F., Hutchinson, D. R., Lee, M. W., Milkereit, B., Agena, W. F., Spencer, C., 1988, Crustal structure of the Midcontinent rift system; Results from GLIMPCE deep seismic reflection studies: Geology, v. 16, p. 81–85.

Ben-Menahem, A., and Singh, S. J., 1981, Seismic waves and sources: New York, Springer-Verlag, 1108 p.

Bessonova, E. N., Fishman, V. M., Ryaboy, V. Z., and Sitnikova, G. A., 1974,

The tau method for inversion of travel times; 1, Deep seismic sounding data: Geophysical Journal of the Royal Astronomical Society, v. 36, p. 377–398.

Bessonova, E. N., Fishman, V. M., Shrirman, M. G., Strikova, G. A., and Johnson, L. R., 1976, The tau method of inversion of travel times; 2, Earthquake data: Geophysical Journal of the Royal Astronomical Society, v. 46, p. 87–108.

Blundell, D. J., and Raynaud, B., 1986, Modeling lower crustal reflections observed on BIRPS profiles, *in* Barazangi, M., and Brown, L., eds., Reflection seismology; A global perspective: American Geophysical Union Geodynamics Series, v. 13, p. 287–295.

Bolt, B. A., 1982, Inside the Earth; Evidence from earthquakes: San Francisco, W. H. Freeman, 191 p.

Boore, D. M., 1972, Finite difference methods for seismic wave propagation in heterogeneous materials, *in* Bolt, B. A., ed., Seismology; Surface waves and earth oscillations; Methods in computational physics, v. 11: New York, Academic Press, p. 1–37.

Borcherdt, R. D., Fletcher, J. B., Jensen, E. G., Maxwell, G. L., Van Schaak, J. R., Warrick, R. E., Cranswick, E., Johnston, M.J.S., and McClearn, R., 1985, A general earthquake-observation system: Bulletin of the Seismological Society of America, v. 75, p. 1783–1825.

Bouchon, M., 1982, The complete synthesis of seismic crustal phases at regional distances: Journal of Geophysical Research, v. 87, p. 1735–1741.

Braile, L., and Smith, R. B., 1974, Guide to the interpretation of crustal refraction profiles: Geophysical Journal of the Royal Astronomical Society, v. 40, p. 145–176.

Braile, L. W., and Chiang, C. S., 1986, The continental Mohorovičić discontinuity; Results from near-vertical and wide-angle seismic refraction studies, *in* Barazangi, M., and Brown, L. D., eds., Reflection seismology; A global perspective, American Geophysical Union Geodynamics Series, v. 13, p. 257–272.

Brown, L., Barazangi, M., Kaufman, S., and Oliver, J., 1986, The first decade of COCORP; 1974–1984, *in* Barazangi, M., and Brown, L. D., eds., Reflection seismology; A global perspective: American Geophysical Union Geodynamics Series, v. 13, p. 107–120.

Brune, J. N., 1969, Surface waves and crustal structure, *in* Hart, P. J., ed., The earth's crust and upper mantle: American Geophysical Union Geophysical Monograph Series, v. 13, p. 230–242.

—— , 1970, Tectonic stress and the spectra of shear waves from earthquakes: Journal of Geophysical Research, v. 75, p. 4997–5009.

Brune, J. N., Nafe, J., and Oliver, J., 1960, A simplified method for the analysis and synthesis of dispersed wave trains: Journal of Geophysical Research, v. 65, p. 287–304.

Bullen, K. E., and Bolt, B. A., 1985, An introduction to the theory of seismology, 4th ed.: Cambridge, Cambridge University Press, 499 p.

Burdick, L. J., 1981, A comparison of the upper mantle structure beneath North America and Europe: Journal of Geophysical Research, v. 86, p. 5926–5936.

Červený, V., Molotkov, I. A., Pšenčík, I., 1977, Ray method in seismology: Prague, Czeckoslovakia, University of Karlova (Charles), 214 p.

Chapman, C. H., 1978, A new method for computing synthetic seismograms: Geophysical Journal of the Royal Astronomical Society, v. 54, p. 481–518.

Claerbout, J. F., 1976, Fundamentals of geophysical data processing: Palo Alto, California, Blackwell Scientific Publications, 274 p.

—— , 1985, Imaging the Earth's interior: Palo Alto, California, Blackwell Scientific Publications, 398 p.

Clowes, R. M., Kanasewich, E. R., and Cumming, G. L., 1968, Deep crustal seismic reflections at near vertical incidence: Geophysics, v. 33, p. 441–451.

Coffeen, J. A., 1978, Seismic exploration fundamentals: Tulsa, Petroleum Publishing Co, 261 p.

Conrad, V., 1925, Laufzeitkurven des Taueren Erdbeben vom, 28, November 1923: Mitteiler Erdbeban Kommission Wien, Akademische Wissenschaft, v. 59, p. 1–23.

Crosson, R. S., 1976, Crustal structure modeling of earthquake data; 1, Simultaneous least squares estimation of hypocenter and velocity parameters: Journal of Geophysical Research, v. 81, p. 3036–3046.

Deichmann, N., and Ansorge, J., 1983, Evidence for lamination in the lower continental crust beneath the Black Forest (southwestern Germany): Journal of Geophysics, v. 52, p. 109–118.

DeGolyer, E., 1935, Notes on the early history of applied geophysics in the petroleum industry: Society Petroleum Geophysicists Journal, v. 6, p. 1–10.

Dix, C. H., 1965, Reflection seismic crustal studies: Geophysics, v. 30, p. 1068–1084.

Dobrin, M. B., 1976, Introduction to geophysical prospecting, 3rd ed.: New York, McGraw-Hill, 619 p.

Dohr, G., 1970, Reflexionsseismische Messungen im Oberrheingraben mit digitaler Aufzeichnungstechnik und Bearbeitung, *in* Graben problems: Stuttgart, Schweizerbart Verlag, p. 207–218.

Dozer, D., 1986, Source parameters and faulting processes of the 1959 Hebgen Lake, Montana, earthquake sequence: Journal of Geophysical Research, v. 90, p. 4537–4555.

Dziewonski, A. M., Chow, T. A., and Woodhouse, J. H., 1981, Determination of earthquake source parameters from waveform data for studies of global and regional seismicity: Journal of Geophysical Research, v. 86, p. 2825–2852.

Eberhart-Phillips, D., 1986, Three-dimensional velocity structure in northern California Coast Ranges from inversion of local earthquake arrival times: Bulletin of the Seismological Society of America, v. 76, p. 1025–1052.

Eberhart-Phillips, D., and Oppenheimer, D., 1984, Induced seismicity in The Geysers Geothermal Area, California: Journal of Geophysical Research, v. 89, p. 1191–1207.

Ewing, M., Carry, A. P., and Rutherford, H. M., 1937, Geophysical investigations in the emerged and submerged Atlantic Coastal Plain; Part 1, Geological Society of America Bulletin, v. 48, p. 753–802.

Finlayson, D. M., and Ansorge, J., 1984, Workshop proceedings; Interpretation of seismic wave propagation in laterally heterogeneous structures: Canberra, Australia, Bureau of Mineral Resources, Geology, and Geophysics Report 258, 207 p.

Fountain, D. M., 1986, Implications of deep crustal evolution for seismic reflection interpretation, *in* Barazangi, M., and Brown, L., eds., Reflection seismology; The continental crust: American Geophysical Union Geodynamics Series, v. 14, p. 1–7.

Fountain, D. M., Hurich, C. A., and Smithson, S. B., 1984, Seismic reflectivity of mylonite zones in the crust: Geology, v. 12, p. 195–198.

Fuchs, K., and Müller, G., 1971, Computation of synthetic seismograms with the reflectivity method and comparison with observations: Geophysical Journal of the Royal Astronomical Society, v. 23, p. 417–433.

Gal'perin, E. I., 1974, Vertical seismic profiling: Society of Exploration Geophysicists Special Publication 12, 270 p.

Garland, G. D., 1979, Introduction to geophysics (mantle, core, and crust): Philadelphia, W. B. Saunders Company, 494 p.

Grand, S. P., and Helmberger, D. V., 1984, Upper mantle shear structure of North America: Geophysical Journal of the Royal Astronomical Society, v. 76, p. 399–438.

Grant, F. S., and West, G. F., 1965, Interpretation theory in applied geophysics: New York, McGraw-Hill, 584 p.

Hajnal, Z., 1986, Crustal reflection and refraction velocities, *in* Barazangi, M., and Brown, L. D., eds., Reflection seismology; A global perspective: American Geophysical Union Geodynamics Series, v. 13, p. 247–256.

Hale, L. D., and Thompson, G. A., 1982, The seismic reflection character of the continental Mohorovičić discontinuity: Journal of Geophysical Research, v. 87, p. 4625–4635.

Hanks, T. C., and Kanamori, H., 1979, A moment magnitude scale: Journal of Geophysical Research, v. 84, p. 2348–2350.

Hatton, L., Larner, K., and Gibson, B. S., 1981, Migration of seismic data from inhomogeneous media: Geophysics, v. 46, p. 751–767.

Healy, J. H., 1963, Crustal structure along the coast of California from seismic-refraction measurements: Journal of Geophysical Research, v. 68, p. 5777–5787.

Hearn, T. M., 1984, P_n travel times in southern California: Journal of Geophysical

Research, v. 89, p. 1843–1855.

Hearn, T. M., and R. W. Clayton, 1986, Lateral velocity variations in southern California; 1, Results for the upper crust from P-waves: Bulletin of the Seismological Society of America, v. 76, p. 495–509.

Helmberger, D., and Wiggins, R. A., 1971, Upper mantle structure in midwestern United States: Journal of Geophysical Research, v. 76, p. 3229–3245.

Helmberger, D. V., 1974, Generalized ray theory for shear dislocations: Bulletin of the Seismological Society of America, v. 64, p. 45–64.

Hill, D. P., and Pakiser, L. C., 1966, Crustal structure between the Nevada test site and Boise, Idaho, from seismic-refraction measurements, *in* Hart, P. J., ed., The earth beneath the continents: American Geophysical Union Geophysical Monograph 10, p. 391–419.

Howell, B. F., 1986, History of ideas on the cause of earthquakes: EOS Transactions of American Geophysical Union, v. 67, p. 1323–1326.

Humphreys, E., and Clayton, R. W., 1988, Adaption of back projection tomography to seismic travel time problems: Journal of Geophysical Research, v. 93, p. 1073–1085.

Hutchinson, D. R., Klitgord, K. D., Lee, M. W., and Trehu, A. M., 1988, U.S. Geological Survey deep seismic reflection profile across the Gulf of Maine: Geological Society of America Bulletin, v. 100, p. 172–184.

Hwang, L., and Mooney, W. D., 1986, Velocity and Q structure of the Great Valley, California, based on synthetic seismogram modelling of seismic refraction data: Bulletin of the Seismological Society of America, v. 76, p. 1053–1067.

Iyer, H. M., 1984a, A review of crust and upper mantle structure studies of the Snake River Plain–Yellowstone volcanic system; A major lithospheric anomaly in the western U.S.A.: Tectonophysics, v. 105, p. 291–308.

——, 1984b, Geophysical evidence for the locations, shapes and sizes, and internal structures of magma chambers beneath regions of Quaternary volcanism: Royal Society of London Philosophical Transactions, v. 310, p. 473–510.

James, D. E., and Steinhart, J. S., 1966, Structure beneath the continents; A critical review of explosion studies 1960–1965, *in* Hart, P. J., ed., The earth beneath the continents: American Geophysical Union Monograph 10, p. 293–312.

Jeffreys, H., and Bullen, K. E., 1935, Times of transmission of earthquake waves: Bureau of Central Seismology International, Series A, Fasc. v. 11, 202 p.

——, 1940, Seismological tables, British Association: London, Gray Milne Trust, 50 p.

Jenyon, M. K., and Fitch, A. A., 1985, Seismic reflection interpretation: Stuttgart, Lubrecht and Cramer, Geoexploration Monograph Series 1, no. 8, 318 p.

Johnson, R. A., and Smithson, S. B., 1986, Interpretive processing of crustal seismic reflection data; Examples from Laramie Range COCORP data, *in* Barazangi, M., and Brown, L., eds., Reflection seismology; The continental crust: American Geophysical Union Geodynamics Series, v. 14, p. 197–208.

Junger, A., 1951, Deep basement reflections in Big Horn County, Montana: Geophysics, v. 16, p. 449–505.

Kanasewich, E. R., and Cumming, G. L., 1965, Near vertical-incidence seismic reflections from the "Conrad" discontinuity: Journal of Geophysical Research, v. 70, p. 3441–3446.

Kelly, K. R., Ward, R. W., Treitel, S., and Alford, R. M., 1976, Synthetic seismograms; A finite difference approach: Geophysics, v. 41, p. 2–27.

Kennett, B.L.N., 1974, Reflections, rays, and reverberations: Bulletin of the Seismological Society of America, v. 64, p. 1685–1696.

——, 1983, Seismic wave propagation in stratified media: Cambridge, Cambridge University Press, 342 p.

Kind, R., 1978, The reflectivity method for a buried source: Journal of Geophysics, v. 44, p. 603–612.

Klemperer, S. L., and Luetgert, J. H., 1987, A comparison of reflection and refraction processing and interpretation methods applied to conventional refraction data from coastal Maine: Bulletin of the Seismological Society of America, v. 77, p. 614–630.

Klemperer, S. L., and Oliver, J. E., 1983, The advantage of length in deep crustal reflection profiles: First Break, April, p. 20–27.

Knopoff, L., 1972, Observation and inversion of surface-wave dispersion: Tectonophysics, v. 13, p. 497–519.

Kohler, W. M., Healy, J. H., and Wegener, S. S., 1982, Upper crustal structure of the Mount Hood, Oregon, region revealed by time term analysis: Journal of Geophysical Research, v. 87, p. 339–355.

Kong, S. M., Phinney, R. A., and Roy-Chowdhury, K., 1985, A non-linear signal detector for enhancement of noisy seismic record sections: Geophysics, v. 50, p. 539–550.

Kovach, R. L., 1978, Seismic surface waves and crustal and upper mantle structure: Reviews of Geophysics and Space Physics, v. 16, p. 1–13.

Langston, C. A., 1977, Corvallis, Oregon, crustal and upper mantle receiver structure from teleseismic P and S waves: Bulletin of the Seismological Society of America, v. 67, p. 713–725.

Langston, C. A., and Helmberger, D. V., 1975, A procedure for modelling shallow dislocation sources: Geophysical Journal of the Royal Astronomical Society, v. 42, p. 117–130.

Lawson, A. C., 1908, The California earthquake of April 19, 1906: Washington, D.C., Carnegie Institution of Washington, v. 1, 451 p.

Lee, W.H.K., and Stewart, S. W., 1981, Principles and applications of microearthquake networks: New York, Academic Press, Advances in Geophysics, v. 2, 293 p.

Liu, H. P., Anderson, D. L., and Kanamori, H., 1976, Velocity dispersion due to anelasticity; Implications for seismology and mantle composition: Geophysical Journal of the Royal Astronomical Society, v. 47, p. 41–58.

Luco, J. E., and Apsel, R. J., 1983, On the Green's functions for a layered half-space, Part 1: Bulletin of the Seismological Society of America, v. 73, p. 909–929.

Madariaga, R., 1983, Earthquake source theory; A review, *in* Kanamori, H., and Boschi, E., eds., Earthquakes; Observations, theory, and interpretation; Proceedings of the International School of Physics "Enrico Fermi," Course 85: Amsterdam, North-Holland Publishing Co., p. 1–44.

Mahdyiar, M., 1987, A momograph to calculate source radius and stress drop from corner frequency, shear velocity, and seismic moment: Bulletin of the Seismological Society of America, v. 77, p. 264–265.

Mair, J. A., and Lyons, J. A., 1976, Seismic reflection techniques for crustal structure studies: Geophysics, v. 41, p. 1272–1290.

Matthews, D. H., and Cheadle, M. J., 1986, Deep reflections from the Caledonides and the Variscides of Britain and comparison with the Himalayas, *in* Barazangi, M., and Brown, L. D., eds., Reflection seismology; A global perspective: American Geophysical Union Geodynamics Series, v. 13, p. 5–19.

Matthews, D. H., and Smith, C., eds., 1987, Deep seismic reflection profiling of the continental lithosphere: Geophysical Journal of the Royal Astronomical Society, v. 89, 447 p.

Matthews, D. H., and the BIRPS group, 1987, Some unsolved BIRPS problems: Geophysical Journal of the Royal Astronomical Society, v. 89, p. 209–216.

Mayer, J. R., and Brown, L. D., 1986, Signal penetration on the COCORP Basin and Range–Colorado Plateau survey: Geophysics, v. 51, p. 1050–1055.

Mayne, H., 1962, Common reflection point horizontal data stacking techniques: Geophysics, v. 27, p. 927–938.

——, 1967, Practical considerations in the use of common reflection point techniques: Geophysics, v. 32, p. 225–229.

McMechan, G. A., and Fuis, G. S., 1987, Ray equation migration of wide-angle reflections from southern Alaska: Journal of Geophysical Research, v. 92, p. 407–420.

McMechan, G. A., and Mooney, W. D., 1980, Asymptotic ray theory and synthetic seismograms for laterally varying structure; Theory and application to the Imperial Valley, California: Bulletin of the Seismological Society of America, v. 70, p. 2021–2035.

McMechan, G. A., Clayton, R., and Mooney, W. D., 1982, Application of wave field continuation to the inversion of refraction data: Journal of Geophysical Research, v. 87, p. 927–935.

Meissner, R., 1967, Exploring deep interfaces by seismic wide angle measurements: Geophysical Prospecting, v. 15, p. 598–617.

——, 1973, The "Moho" as a transition zone: Geophysical Surveys, v. 1, p. 195–216.

——, 1986, The continental crust; A geophysical approach: London, Academic Press, 426 p.

Menke, W., 1984, Geophysical data analysis; Discrete inverse theory: Orlando, Florida, Academic Press, 260 p.

Milkereit, B., Mooney, W. D., and Kohler, W. M., 1985, Inversion of seismic refraction data in planar dipping structure: Geophysical Journal of the Royal Astronomical Society, v. 82, p. 81–103.

Mintrop, L., 1949, On the stratification of the earth's crust according to seismic studies of a large explosion and earthquakes: Geophysics, v. 14, p. 321–336.

Mitchell, B. J., and Herrmann, R. B., 1979, Shear velocity structure in the eastern United States from the inversion of surface-wave group and phase velocities: Bulletin of the Seismological Society of America, v. 69, p. 1133–1148.

Mitchell, B. J., Leite, L.W.B., Yu, Y. K., and Herrmann, R. B., 1976, Attenuation of Love and Rayleigh waves across the Pacific at periods between 15 and 110 seconds: Bulletin of the Seismological Society of America, v. 66, p. 1189–1202.

Mohorovičić, A., 1909, Das Beben vom 8. X. 1909: Jahrbuch des Meterologisches Observatorium Zagreb, v. 9, p. 1–63.

Mooney, W. D., and Brocher, T. M., 1987, Coincident seismic reflection/refraction studies of the continental lithosphere; A global review: Reviews of Geophysics, v. 25, p. 723–742.

Mooney, W. D., and Prodehl, C., eds., 1984, Proceedings of the 1980 workshop of the IASPEI on the seismic modeling of laterally varying structures; Contributions based on data from th e1978 Saudi Arabian refraction profile: U.S. Geological Survey Circular 937, 158 p.

Müller, G., and Mueller, S., 1979, Travel-time and amplitude interpretation of crustal phases on the refraction profile Delta-W, Utah: Bulletin of the Seismological Society of America, v. 69, p. 1121–1132.

Nábělek, J., 1985, Geometry and mechanism of faulting of the 1980 El Asnam, Algeria, earthquake from inversion of teleseismic body waves and comparison with field observations: Journal of Geophysical Research, v. 90, p. 12713–12728.

O'Brien, P.N.S., 1968, Lake Superior crustal structure; A reinterpretation of the 1963 seismic experiment: Journal of Geophysical Research, v. 73, p. 2669–2689.

Okaya, D. A., 1986, Seismic profiling of the lower crust; Dixie Valley, Nevada, *in* Barazangi, M., and Brown, L., eds., Reflection seismology; The continental crust: American Geophysical Union Geodynamics Series, v. 14, p. 269–279.

Oliver, J., and Murphy, L., 1971, WWNSS; Seismology's global network of observing stations: Science, v. 174, p. 254–261.

Oliver, J. E., Dobrin, M., Kaufman, S., Meyer, R., and Phinney, R., 1976, Continuous seismic reflection profiling of the deep basement, Hardeman County, Texas: Geological Society of America Bulletin, v. 87, p. 1537–1546.

Oliver, J., Cook, F., and Brown, L., 1983, COCORP and the continental crust: Journal of Geophysical Research, v. 88, p. 3329–3347.

Oppenheimer, D. H., and Eaton, J. P., 1984, Moho orientation beneath central California from regional travel times: Journal of Geophysical Reesarch, v. 89, p. 267–282.

Oppenheimer, D. H., and Kerkenoff, K. E., 1981, Velocity-density properties of the lithosphere from three-dimensional modeling at The Geysers–Clear Lake region, California: Journal of Geophysical Research, v. 86, p. 6057–6065.

Owens, T. J., Zandt, G., Taylor, S. R., 1984, Seismic evidence for an ancient rift beneath the Cumberland Plateau, Tennessee; A detailed analysis of broadband teleseismic P waveforms: Journal of Geophysical Research, v. 89, p. 7783–7795.

Owens, T. J., and Zandt, G., 1985, The response of the continental crust-mantle boundary observed on broadband teleseismic receiver functions: Geophysical Research Letters, v. 12, p. 705–708.

Pakiser, L. C., and Steinhart, J. S., 1964, Explosion seismology in the western hemisphere, *in* Research in geophysics; v. 2, Solid earth and interface phenomena: Cambridge, Massachusetts Institute of Technology Press, p. 123–147.

Pavlis, L., and Booker, R., 1980, The mixed discrete-continuous inverse problem; Application to the simultaneous determination of earthquake hypocenters and velocity structure: Journal of Geophysical Research, v. 85, p. 4801–4810.

Peddy, C., Brown, L. D., and Klemperer, S. L., 1986, Interpreting the deep structure of rifts with synthetic seismic section, *in* Barazangi, M., and Brown, L., eds., Reflection seismology; A global perspective: American Geophysical Union Geodynamic Series, v. 13, p. 301–311.

Press, F., 1956, Determination of crustal structure from phase velocity of Rayleigh waves; 1, southern California: Geological Society of America Bulletin, v. 67, p. 1647–1658.

——, 1968, Earth models obtained by Monte Carlo inversion: Journal of Geophysical Research, v. 73, p. 5223–5234.

Priestley, K., and Brune, J., 1978, Surface waves and the structure of the Great Basin of Nevada and western Utah: Journal of Geophysical Research, v. 83, p. 2265–2272.

Prodehl, C., 1970, Seismic refraction study of crustal structure in the western United States: Geological Society of America Bulletin, v. 81, p. 2629–2645.

——, 1979, Crustal structure of the western United States: U.S. Geological Survey Professional Paper 1034, 74 p.

——, 1984, Structure of the earth's crust and upper mantle, *in* Landoldt Bornstein, n.s., v. 11A: Heidelberg, Springer, p. 97–206.

Prodehl, C., and Pakiser, L. C., 1980, Crustal structure of the southern Rocky Mountains from seismic measurements: Geological Society of America Bulletin, v. 91, p. 147–155.

Reid, H. F., 1910, Mechanics of the earthquake, *in* The California earthquake of 1906: Yearbook of the Carnegie Institution of Washington, v. 2, 192 p.

Reston, T. J., 1987, Spatial interference, reflection character, and the structure of the lower crust under extension; Results from 2-D seismic modelling: Annales Geophysicae, v. 5, p. 339–347.

Richter, C. F., 1958, Elementary seismology: San Francisco, W. H. Freeman, 768 p.

Sandmeier, K.-J., and Wenzel, F., 1986, Synthetic seismograms for a complex crustal model: Geophysical Research Letters, v. 13, p. 22–25.

Schilt, S., Oliver, J., Brown, L., Kaufman, S., Albaugh, D., Brewer, J., Cook, F., Jensen, L., Krumhansl, P., Long, G., and Steiner, D., 1979, The heterogeneity of the continental crust; Results from deep crustal seismic reflection profiling using the VIBROSEIS technique: Reviews of Geophysics and Space Physics, v. 17, p. 354–368.

Sheriff, R. E., and Geldart, L. P., 1982, Exploration seismology; Vol. 1, History, theory, and data acquisition: Cambridge, Cambridge University Press, 253 p.

——, 1983, Exploration seismology; Vol. 2, Data processing and interpretation: Cambridge, Cambridge University Press, 221 p.

Smith, T. J., Steinhart, J. S., and Aldrich, L. T., 1966, Lake Superior crustal structure: Journal of Geophysical Research, v. 71, p. 1141–1181.

Smith, W. D., 1975, The application of finite element analysis to body wave propagation problems: Geophysical Journal of the Royal Astronomical Society, v. 42, p. 747–768.

Smithson, S. B., Brewer, J. A., Kaufman, S., Oliver, J. E., and Zawislak, R. L., 1980, Complex Archean lower crustal structure revealed by COCORP crustal reflection profiling in the Wind River Range, Wyoming: Earth and Planetary Science Letters, v. 46, p. 295–305.

Smithson, S. B., Johnson, R. A., Hurich, C. A., Valasek, P. A., and Branch, C., 1987, Deep crustal structure and genesis from contrasting reflection patterns; An integrated approach: Geophysical Journal of the Royal Astronomical Society, v. 87, p. 67–72.

Spudich, P., and Archuleta, R. J., 1987, Techniques for earthquake ground-motion calculation with applications to source parameterization of finite faults, *in* Bolt, B.A., ed., Seismic strong motion synthetics: Orlando, Florida, Academic Press, p. 205–265.

Steinhart, J. S., and Meyer, R. P., 1961, Explosion studies of continental structure: Washington, D.C., Carnegie Institution Washington Publication 622, 409 p.

Stewart, D. B., Unger, J. D., Phillips, J. D., Goldsmith, R., Poole, W. H., Spencer, C. P., Green, A. G., Loiselle, M. C., and St-Julien, P., 1986, The

Quebec–Western Maine seismic reflection profile; Setting and first year results, *in* Barazangi, M., and Brown, L., eds., Reflection seismology; The continental crust: Washington, D.C., American Geophysical Union Geodynamic Series, v. 14, p. 189–199.

Taner, M. T., Cook, E. E., and Neidell, N. S., 1970, Limitations of the reflection seismic method: Lessons from computer simulations: Geophysics, v. 35, p. 55–573.

Tanimoto, T., and Anderson, D. L., 1985, Lateral heterogeneity and azimuthal anisotropy of the upper mantle; Love and Rayleigh waves 100-250 s: Journal of Geophysical Research, v. 90, 1842–1858.

Tarantola, A., 1987, Inverse problem theory; Methods for data fitting and model parameter estimation: Amsterdam, Elsevier, 613 p.

Tatel, H. E., and Tuve, M. A., 1955, Seismic exploration of a continental crust: Geological Society of America Special Paper 62, p. 35–50.

Telford, W. M., Geldart, L. P., Sheriff, R. E., and Keys, D. A., 1976, Applied geophysics: Cambridge, Cambridge University Press, 860 p.

Thurber, C. H., 1983, Earthquake locations and three-dimensional crustal structure in the Coyote Lake area, central California: Journal of Geophysical Research, v. 88, p. 8226–8236.

Trehu, A. M., and Wheeler, W. H., 1987, Possible evidence for subducted sediments beneath California: Geology, v. 15, p. 254–258.

Tuve, M. A., 1951, The earth's crust: Washington, D.C., Year Book of the Carnegie Institution of Washington, v. 50, p. 69–73.

Walck, M. C., and Clayton, R. W., 1984, Analysis of upper mantle structure using wave field continuation of P-waves: Bulletin of the Seismological Society of America, v. 74, 1703–1719.

Walck, M. C., and Minster, J. B., 1982, Relative array analysis of upper mantle lateral velocity variations in Southern California: Journal of Geophysical Research, v. 87, p. 1757–1772.

Walter, A. W., and Mooney, W.D., 1987, Interpretations of the SJ-6 seismic reflection/refraction profile, south-central California, USA: U.S. Geological Survey Open-File Report 87–73, 132 p.

Warner, M., 1987, Migration; Why doesn't it work for deep continental data?: Geophysical Journal of the Royal Astronomical Society, v. 89, p. 21–26.

Warren, D. H., and Healy, J. H., 1973, Structure of the crust in the conterminous United States: Tectonophysics, v. 20, p. 203–213.

Waters, K. H., 1987, Reflection seismology; A tool for energy resource exploration: New York, John Wiley & Sons, 537 p.

Wiggins, R., 1972, The general linear inverse problem; Implications of surface waves and free oscillations on earth structure: Reviews in Geophysics and Space Physics, v. 10, p. 251–285.

Willmore, P. L., and Bancroft, A. M., 1960, The time-term approach to refraction seismology: Geophysical Journal of the Royal Astronomical Society, v. 3, p. 419–432.

Wong, Y. K., Smithson, S. B., and Zawislak, R. L., 1982, The role of seismic modeling in deep crustal reflection interpretation; Part 1, Laramie: University of Wyoming Contributions to Geology, v. 20, p. 91–109.

Worthington, M. H., 1984, An introduction to geophysical tomography: First Break, November, p. 20–26.

Yilmaz, O., 1987, Seismic data processing: Society of Exploration Geophysists Investigations in Geophysics, v. 2, 526 p.

Yomogida, K., and Aki, K., 1985, Waveform synthesis of surface waves in a laterally heterogeneous earth by the Gaussian beam method: Journal of Geophysical Research, v. 90, p. 7665–7688.

Zandt, G., and Owens, T. J., 1986, Comparison of crustal velocity profiles determined by seismic refraction and teleseismic methods: Tectonophysics, v. 128, p. 155–161.

Zhu, T., and Brown, L. D., 1986, Signal penetration; Basin and Range to Colorado Plateau: Geophysics, v. 51, p. 1050–1054.

Zoback, M. D., and Wentworth, C. M., 1986, Crustal studies in central California using an 800-channel seismic reflection seismology; A global perspective: American Geophysical Union Geodynamics Series, v. 13, p. 183–196.

Manuscript Accepted by the Society October 31, 1988

Printed in U.S.A.

Geological Society of America
Memoir 172
1989

Chapter 3

Gravity methods in regional studies

Robert W. Simpson and Robert C. Jachens
U.S. Geological Survey, 345 Middlefield Road, Menlo Park, California 94025

ABSTRACT

The gravitational attractions of bodies in the crust and upper mantle offer important clues to the geologic evolution and anatomy of the continents. Regional gravity maps prepared from millions of observations provide a density and uniformity of coverage that can put other more isolated geophysical studies into a broader framework. Reliable and efficient instruments for collecting gravity observations have been available for many decades. Gravity reduction procedures are now fairly standardized, although misconceptions exist, and it is easy to lose sight of the end goal in the multiple reduction steps. In the last decade, computers have greatly facilitated the presentation and interpretation of large regional data sets. Numerous mathematical inversion techniques have been described in the literature, although for many purposes the best interpretive tool still remains the construction of simple, forward two-dimensional models, constrained by as much geological and geophysical information as can be obtained.

OVERVIEW

In this review, we focus on methods used to construct and interpret regional gravity maps of the sort that appear in other chapters in this volume. We begin, however, by mentioning briefly some other current lines of gravity research, because rapid advances in these areas will undoubtedly facilitate the study of regional gravity fields. A great impact has already been made by ingenious ways of observing the Earth's gravity field from artificial satellites orbiting the planet.

The low-order harmonics of the Earth's gravity field (wavelengths longer than about 2,000 km) can be inferred from perturbations in the orbits of satellites. Models of the Earth's global gravity field constructed in this way (e.g., Lerch and others, 1985) provide constraints on the mass distributions in the Earth's mantle and core. This information is valuable as a test for models of convective geometries in the mantle (e.g., Hager, 1984) and as an independent test of seismically inferred density contrasts in the interior of the Earth (e.g., Dziewonski, 1984). It has been proposed that anomalies in the Earth's gravity field could be observed to wavelengths as small as 100 km by measuring directly the changes in distance between two satellites, one of which follows the other at a distance of several hundred kilometers (Taylor and others, 1983; Keating and others, 1986). Such resolution would permit the study of upper-mantle and crustal sources (Wagner and Sandwell, 1984) and time variations in the gravity field (Wagner and McAdoo, 1986). The uniform coverage over most of the Earth's surface that is obtainable with satellite observations would also help to place the existing nonuniform surface coverage into context.

The Earth's gravity field can also be inferred by mapping variations in sea level with satellite radar altimeters (e.g., Marsh and others, 1986). Sea level responds to lateral changes in the gravity field—for example, extra water will be attracted by dense features in the subsurface such as seamounts—and if sea level can be measured with sufficient precision, and if tidal and other extraneous effects can be successfully removed, a map of the oceanic geoid can be calculated from which gravity anomalies can be inferred (Haxby and others, 1983; Haxby and Weissel, 1986). These efforts have rapidly increased our knowledge of the ocean floors and their underpinnings.

A different line of research involves the development of high-precision, portable, absolute gravity meters (Marson and Faller, 1986). Over the years most surface gravity readings have been obtained with instruments that measure only differences in gravitational attraction from place to place, so that absolute values of the field must be inferred by comparing surveys to a base network of known values. In the past, the base net was often

Simpson, R. W., and Jachens, R. C., 1989, Gravity methods in regional studies, *in* Pakiser, L. C., and Mooney, W. D., Geophysical framework of the continental United States: Boulder, Colorado, Geological Society of America Memoir 172.

established by relating observations to a single laboratory where an absolute value of gravity had been determined. Presently, portable absolute gravity instruments capable of making about one reading per day are routinely attaining accuracies on the order of 10 μmGal (e.g., Zumberge and others, 1986). Such instruments promise to serve useful roles in geodetic studies, and for establishing base-station networks from which relative surveys can emanate and meter calibrations can be obtained.

It is possible to monitor extremely small temporal changes in the gravity field using existing relative gravity meters in repeated surveys; this capability has proved useful in a number of geologically interesting areas. High-precision gravity observations have been used to monitor vertical tectonic movements (e.g., Jachens and others, 1983), mass changes in geothermal fields (Allis and Hunt, 1986), crustal loading (Lambert and others, 1986), deformation and magma movement at active volcanoes (Dzurisin and others, 1980), postglacial rebound (Becker and others, 1984), and groundwater variations (Lambert and Beaumont, 1977). Carefully conducted surveys are capable of detecting changes in elevation on the order of 1 cm, or equivalent subsurface changes in mass.

Another active line of research uses gravity observations to estimate the strength of the crust and lithosphere. Most topographic loads and density heterogeneities within the crust are supported isostatically as if the crust (and perhaps the upper mantle also) were floating on a fluid substratum. The buoyant support for such loads may occur locally if the crust is weak, but a strong crust can support loads over broad distances, and the effects should be reflected in the gravity anomalies. Examples of such studies in the conterminous United States include Banks and others (1977), Karner and Watts (1983), McNutt (1980, 1983), and Sheffels and McNutt (1986). Gravity observations, along with other geophysical data such as seismic information and heat-flow measurements, also provide useful constraints for modeling studies that attempt to explain the evolution of tectonic features (e.g., Lachenbruch and others, 1985).

The remainder of this chapter focuses on traditional onland and offshore gravity survey methods and interpretation techniques designed primarily to study crustal bodies of geological interest. We also discuss the sources of error, the implicit assumptions, and the inherent uncertainties in the interpretations so that the reader may be better equipped to evaluate gravity models. The references cited are those most familiar to us. Useful overviews are provided by LaFehr (1980), who has summarized the history of the gravity method as an exploration tool; by Paterson and Reeves (1985), who have summarized various interpretative approaches; and by Hinze (1985), who has edited a volume of regional potential field investigations that employ many state-of-the-art approaches. Available bibliographies (e.g., Colorado School of Mines, 1984) and annotated indices (e.g., Zwart, 1983) also offer a wealth of additional references. Appendix A (in pocket inside back cover) to this work is a bibliography of studies that contain quantitative models or in some cases more qualitative interpretations of the gravity field for regions in the conterminous United States.

INSTRUMENTS AND DATA COLLECTION

Over the years, a wide variety of portable devices for measuring variations in gravity over the Earth's surface have been invented. Many of the instruments in common use today measure changes in the attraction of gravity by allowing the operator to balance (compensate for) the gravitationally induced deflections of a small weight suspended from an arrangement of springs and beams. Such instruments can easily measure gravity differences between sites to better than one-tenth of a milligal (μGal), and with special procedures can be used to measure relative gravity to accuracies and precisions of a few microgal (μGal). (The unit of gravity is the Gal, after Galileo; 1 Gal is the strength of a field producing an acceleration of 1 cm/s^2. The Earth's field, measured at the surface, has a strength of approximately 980 Gal, whereas typical anomalies of interest produced by crustal sources are 1 to 100 mGal in amplitude.) A recent biography of Lucien LaCoste (Clark, 1984), who was one of the inventors of a widely used type of gravity meter, describes some of the challenging technical problems that had to be solved to measure differences in gravity of only a few parts per billion of the total gravity field. Meters of this sort have been used to collect a few million gravity readings over the conterminous United States, and nearly as many offshore, mostly over the past three or four decades. Such meters measure only differences in gravity, although with great precision, so that absolute values can be found only by direct or indirect comparison with locations where the absolute value of the gravity field is already known. Gravity values accurate to better than a few tenths of a milligal are adequate for most geologic investigations so that today the accuracy of measured gravity does not pose any limitation for most onland studies. Instruments designed for marine observations incorporate special features to minimize the effects caused by the tilts and accelerations of the ship (Dehlinger, 1978; LaCoste, 1983). Accuracies of 1 mGal in marine gravity observations have been achievable for two decades (LaFehr and Nettleton, 1967).

Much effort has also been directed toward creating gravity meters capable of operating below the Earth's surface in boreholes (Labo, 1986; LaFehr, 1983), and above the surface in aircraft (Hammer, 1983). Because neither type of measurement has yet contributed in a large way to the regional data, neither is discussed further here.

Determining accurate locations and elevations for the sites of gravity observations poses a greater problem for the overall accuracy of a gravity survey than the actual measurement of the gravity field. Traditional methods for onland positioning are based on topographic maps, surveying, or barometric leveling, and are either expensive and time-consuming, or yield values that are so uncertain that they place a serious limitation on the validity of the reduced gravity data. Offshore, not only locations but platform velocities and accelerations enter into the data reduction process, and must be adequately recorded or compensated for (Dehlinger, 1978). Geodetic Positioning Systems based on satellites (see e.g., Archinal and Mueller, 1986), as they become less expensive and faster to operate, promise to greatly facilitate the acquisition and reduction of gravity observations.

DATA REDUCTIONS

The reduction of gravity data is an excellent example of the enhancement of "signal" by the removal of predictable "noise." Gravitational variations that can be attributed to known causes—such as to the Earth's rotation, its gross ellipsoidal shape, or its topographic relief—can be removed by calculations based on realistic Earth models incorporating these causes. The anomalies remaining after the predicted effects have been removed are presumed to be caused by unknown sources, which it is the interpreter's task to identify, or by deficiencies in the Earth models themselves, which the interpreter needs to refine.

In the past few decades, the most important advance in the reduction of gravity data has been the widespread application of computers to certain of the more tedious reduction steps. In particular, with the help of digital topographic data sets, computer-assisted corrections for topography are routinely made (e.g., Plouff, 1977), at least for features at some distance from the observation site (i.e., in the "outer zones"), that are adequately represented by the resolution of the available digital topographic data. Isostatic anomalies also are readily calculated with the help of computers (Jachens and Roberts, 1981; Simpson and others, 1983). Both of these computations previously were extremely time-consuming. Hayford and Bowie (1912) estimated that it required 17 hr for a "computer" (theirs was one Miss Sarah Beall) to calculate the topographic and isostatic adjustments for a single gravity observation.

Descriptions of the gravity reduction process can be found in many standard references (e.g., Nettleton, 1976; Dobrin, 1976; Swick, 1942; Heiskanen and Moritz, 1967). The usual reductions attempt to predict and account for: (1) the total mass of the Earth, (2) the ellipsoidal shape of the Earth, and (3) the rotation of the Earth. These three effects are incorporated in a formula that yields *theoretical gravity* on the surface of the normal ellipsoid defined in some geodetic reference system (Chovitz, 1981). (Commonly, accelerations and deformations caused by lunar and solar tides are removed at an early stage; for marine observations, accelerations related to the motion of the ship must also be removed.) In addition, adjustments are commonly made for (4) the elevation of the observation above sea level (the *free-air* adjustment), and for (5) the attractions of nearby topographic or bathymetric relief (the *simple* or *complete Bouguer* adjustment, depending on the care taken to accurately represent the topography). For some purposes, such as modeling of very long profiles, it is necessary to account for (6) the fact that the two previous adjustments (4 and 5 above) used sea level (the surface of the geoid) as the reference elevation rather than the surface of the ellipsoid; this difference is called the *indirect effect* (Chapman and Bodine, 1979). Less commonly, (7) the attractions of compensating masses supporting topographic loads are predicted by using a model of isostatic compensation (the *isostatic* adjustment). A final (8) *geologic* correction that predicts the effects of known density distributions associated with geologic features such as sedimentary basins has been used in some applications (e.g., Woollard, 1966). (This final correction is also implicit in the modeling process in which the interpreter attempts to explain the last remaining gravity anomalies in terms of geologic bodies, thereby reducing the residual anomaly field at last to zero everywhere.)

Each of the major refinements to the basic theoretical gravity prediction provides an anomaly of the same name when the predicted result is subtracted from the observed gravity value (i.e., free-air, Bouguer, or isostatic anomaly, respectively). Examples of these anomalies over simple bodies approximating geologic features are shown in Figure 1. The best anomaly to use in any situation depends on the purpose of the investigation.

Bouguer gravity anomaly maps have traditionally been used onland because they remove the noise introduced by local topography. However, on continental scales, Bouguer gravity maps display broad anomalies inversely correlated with regional topography. These anomalies are caused by masses that isostatically support the topographic loads. Because these masses have not been accounted for in the Bouguer reduction steps, they reveal their presence on Bouguer gravity maps as broad gravity lows over mountain ranges and broad highs over oceanic depths. Free-air gravity anomaly maps are often used in offshore gravity investigations and are, in one point of view, the offshore equivalent of a Bouguer gravity map, because the sea-surface topography is necessarily at sea level so that no adjustment is required. Traditionally, however, in making Bouguer gravity maps, offshore topography has been taken to be bottom topography, so that the Bouguer correction includes an adjustment for water depths that conceptually replaces water with material of some suitable reduction density, often 2.67 g/cm^3.

Free-air gravity maps are probably better suited for study of deep ocean basins than for shallow shelf regions where bathymetric features are closer to the gravity meter and can obscure anomalies caused by buried sources. A Bouguer reduction can remove such bathymetric effects to first order. Isostatic residual gravity maps may be best suited, both onland and offshore, for enhancing anomalies caused by geologic features in the crust, because the isostatic adjustment removes to first order both the effects of topography and of the compensating masses to topography. The proper anomaly to use is best determined by the goal at hand; computerized data reduction and plotting make experimentation easy.

At present, gravity observations are commonly reduced using Gravity Formula 1967, which yields theoretical gravity values on the surface of the normal ellipsoid defined by the Geodetic Reference System 1967 (International Association of

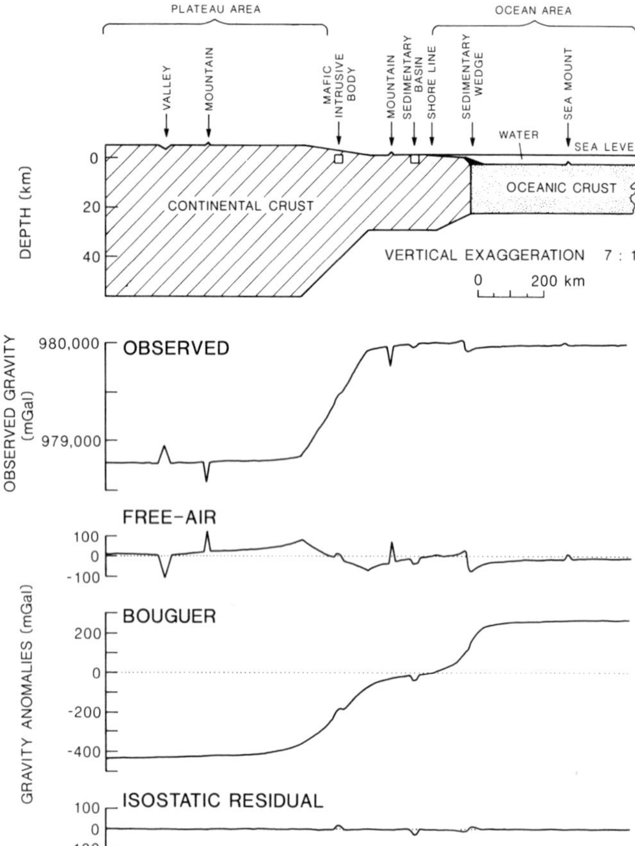

Figure 1. Observed gravity and various gravity anomalies calculated for a simple two-dimensional model. Although geometries are greatly simplified, the model does serve to demonstrate the intended effects of the various adjustments to observed gravity. The profile is assumed to extend in an east-west direction so that there is no latitudinal variation. The 980,000-mGal level for observed gravity values at sea level was chosen arbitrarily. The bottom of the model was constructed assuming local Airy compensation for the broader topographic features, and oceanic crust was assumed to be 0.2 g/cm^3 more dense than continental crust. The isostatic residual anomaly shown in the figure was calculated assuming equal densities for both types of crust; this incorrect assumption leads to the small low-high isostatic residual anomaly pair over the continental-oceanic transition. The high-low free-air anomaly pair in the same location is largely the result of change in water depth—the sedimentary wedge and the change in crustal densities contribute to the free-air anomaly at this location in only a minor way. Note that the vertical scale for the anomalies is twice that for the observed gravity.

Geodesy, 1971). In closed form, where g represents the normal force of attraction in Gal on the surface of the reference ellipsoid:

$$g = g_e \frac{1 + k \sin^2 \phi}{\sqrt{1 - e^2 \sin^2 \phi}},$$

where ϕ is the latitude, and

$$g_e = 978.031\ 845\ 58 \text{ Gal}$$
$$k = 0.001\ 931\ 663\ 383\ 21$$
$$e^2 = 0.006\ 694\ 605\ 328\ 56.$$

The companion International Gravity Standardization Net 1971 (Morelli, 1974) consists of a set of worldwide gravity base-station values that should be used in conjunction with Gravity Formula 1967. Chovitz (1981) has given a lucid summary of the history and role of the various geodetic reference systems in gravity studies.

Although the reduction of gravity data seems like a straightforward process, certain common misconceptions about the significance of the various reduction steps often lead to confusion at the interpretation stage. For example, the application of a free-air adjustment is often loosely referred to as a "reduction of the observation to sea level." This is not at all what has been accomplished, because the observation is necessarily fixed in space by its geometric relationship with neighboring massive bodies. A better way to view the free-air adjustment is as a refinement that allows the prediction of the attraction of gravity at the actual elevation above sea level at which the observation was made. The predicted and observed effects are differenced to produce the free-air anomaly.

In general we have found the best conceptual framework for thinking about the gravity reduction process to be the "prediction" approach (Simpson and others, 1987). Each step in the reduction process is regarded as an adjustment to the predicted gravity (starting with the theoretical gravity predicted on the reference ellipsoid) and as a corresponding refinement to a conceptual density model that approximates the true Earth. Gravity anomalies are the differences between the actual gravity values and the values calculated from our Earth model. The goal of gravity reduction and interpretation is to make the model as close to the real Earth as possible. If the model ever finally became correct in every detail, no anomalies would remain.

The alternate conceptual approach to the reduction process, which in our experience has often led us astray, is to regard each reduction step as a correction to the observed gravity. The result is then compared to the theoretical gravity on the reference ellipsoid. The difference between the predictive and corrective approaches seems to be nothing more than a rearrangement of terms to be added and subtracted. The real difference, however, lies in the utility of the predictive conceptual model in illuminating and motivating the reduction steps. Although a corresponding conceptual Earth model can be imagined for the corrective approach, it seems to us to be less in tune with the goals at hand—the explanation of gravity anomalies in terms of density contrasts within the Earth and the construction of an increasingly refined Earth model.

GRAVITY DATA BASES AND MAPS FOR THE UNITED STATES

Gravity data for the conterminous United States are available in several formats. The most up-to-date and comprehensive data set in map form is the *Gravity Anomaly Map of the United States* published by the Society of Exploration Geophysicists (1982). This colored contour map shows Bouguer gravity anomaly values onland and free-air gravity anomaly values offshore.

Wavelength-filtered versions of this data set are described in the chapter by Kane and Godson in this volume. An isostatic residual gravity map prepared from the same data is discussed by Jachens and others in this volume. Other recently published maps displaying continent-wide gravity data are the wavelength-filtered gravity maps of Hildenbrand and others (1982), and the combined gravity and topography maps of Guinness and others (1982) and Arvidson and others (1984). Gravity maps of many individual states have also been published at the same scale as state geologic maps, usually by state geological surveys or by the U.S. Geological Survey.

In addition to map presentations, digital data bases of gravity observations and of averaged topographic and bathymetric data are available. The National Geophysical Data Center (NGDC, National Oceanic and Atmospheric Administration, Code E/GC12, 325 Broadway, Boulder, Colorado 80303) maintains and distributes extensive files of both onland and offshore gravity observations. Topographic data, needed for terrain corrections and for defining isostatic models, are also available for various areas in 5-min, 3-min, 1-min, 30-sec, and 15-sec averages.

Gridded data sets, constructed in some instances partly from observations not yet released to the public, can also be obtained on magnetic tape. For example, the data grid with 4-km sample interval used to construct the Gravity Anomaly Map of the United States (Society of Exploration Geophysicists, 1982) is available (Godson and Scheibe, 1982a, b). Such data grids, although often not detailed enough for local studies, help to place local studies into regional context and serve as a starting point for further analyses and reductions of the gravity field on a continent-wide scale. The amount of editing effort that goes into the construction of such grids from the original point observations is still burdensome, although more automated and interactive approaches to this time-consuming task are emerging.

INTERPRETATION OF GRAVITY DATA

The interpretation of gravity data is best accomplished if constraints can be supplied from geological observations or from other kinds of geophysical techniques. Every contact that can be specified and every density contrast that can be bracketed reduces the inherent ambiguity in the gravity inversion process. The importance of collecting rock specimens for density determinations cannot be overemphasized. Data from borehole density logs and densities inferred from borehole gravity readings can also be invaluable. Sometimes, gravity profiles over topographic features allow more reliable bulk densities to be calculated than can be gotten from a nonrepresentative collection of lithologies exposed at the surface (Nettleton, 1939). When the rocks are inaccessible, the interpreter must turn to tabulations of densities (e.g., Johnson and Olhoeft, 1984) or to relations of seismic velocity to rock density (e.g., Birch, 1961; Bateman and Eaton, 1967; Nafe and Drake, 1968; Gardner and others, 1974; Hill, 1978), but densities determined from the rocks exposed in the study area are generally far more useful.

Interpretation usually consists of four phases: (1) identification and isolation of specific anomalies of interest, usually through appropriate enhancement and display techniques; (2) qualitative interpretation in which the appropriate geologic setting and generalized characteristics of the source bodies are reframed in terms of density distributions; (3) quantitative modeling to determine the numerical values of the source coordinates and densities; and (4) analysis of the geologic implications of the resulting model.

Phase 1

Advances in computer-driven plotting devices, such as inkjet and electrostatic plotters, have greatly increased the interpreter's options for viewing anomaly fields in the first place. It is often helpful to display a data set in a number of different ways before beginning a model. Contour maps, shaded-relief figures, color plots, or combinations of these used to display various data bases simultaneously, bring out different aspects of the problem. Numerous excellent examples of such display techniques can be found in Hinze (1985) and in Paterson and Reeves (1985). Many automatic algorithms for enhancing aspects of the gravity anomaly field have also proved useful, including wavelength-filtering, vertical derivatives, upward continuation, and gradient analyses designed to locate to first approximation the positions of structures and contacts (Cordell and Grauch, 1985; Blakely and Simpson, 1986). Density mapping offers yet another way to analyze large regional data sets (Gupta and Grant, 1985) in a relatively automatic way.

The isolation of an anomaly from other nearby anomalies or from broader regional anomalies constitutes the classical interpretation problem of *regional-residual separation.* The *residual* field refers to those anomalies, usually of restricted extent, that one is interested in understanding, whereas the *regional* field is the broader part of the gravity field, rich in longer wavelengths, that is commonly caused by sources that are deeper or laterally removed from the sources of interest. Most textbooks on gravity methods discuss a variety of approaches to this separation in some detail. Visual-graphical techniques (e.g., Dobrin, 1976) have been used successfully, and give the interpreter considerable control and a feeling for the ambiguities involved. A more automated approach uses a contouring program to construct a regional field by contouring the data set with the observations defining the anomalies of interest removed. Analytical techniques such as wavelength filtering and polynomial fitting can be more objective but may also introduce artifacts into the anomaly field (Ulrych, 1968). It may sometimes be possible to use the spectral characteristics of the gravity field (Syberg, 1972) to design a filter best suited to separate the anomalies from different sources. It may also be possible to infer from the spectrum some information as to how separable the set of anomalies may be. In many other cases, the application of an isostatic correction can neatly remove, at least to first order, regional fields caused by compensating masses to topographic loads.

Phase 2

Once an anomaly is identified and isolated, a qualitative understanding of the anomaly sources (phase 2) is a necessary prerequisite to quantitative modeling. This generally requires a thorough understanding of the geologic setting, the possible densities of causative bodies, and the constraints imposed by other geophysical data. The qualitative model usually includes some or all of the following factors: approximate locations of lateral boundaries, approximate attitude of the boundaries, estimates of maximum permissible depth of the sources, and approximate density contrasts between the sources and their surroundings. The qualitative model developed at this stage should be complete enough that the quantitative model in the next stage only sets numerical constraints on the allowable geometry and density. To attempt quantitative modeling without a well-considered qualitative model usually leads to unacceptable ambiguity in the final model.

Phase 3

The third phase of the interpretation process is the calculation of quantitative models of source bodies. Approaches using the equations for simple bodies and prepared nomographs can be found in the literature (see for example, Grant and West, 1965). Such tools, together with hand calculator programs that are now readily available (e.g., Society of Exploration Geophypsicists, 1981), can help in the construction of preliminary models either in the field or in the office. A useful guide for estimating the approximate thickness of a laterally extended source body is the formula for the vertical attraction g_z of an infinite slab of material of density ρ and thickness t:

$$g_z(mGal) \approx 0.04\ \rho(\mathrm{g/cm^3}) \cdot t\ (\mathrm{meters}).$$

For example, a negative anomaly of 10 mGal over a sediment-filled basin implies a sediment thickness of about 800 m if the density of the deposits relative to the surrounding basement is assumed to be –0.3 g/cm^3. Depth estimates for source bodies based on gradient steepness and basic anomaly dimensions (e.g., Grant and West, 1965; Bowin and others, 1986) also provide useful semi-quantitative insights before the construction of a quantitative model has actually begun.

Inversion by iterated forward modeling. The most widely used interpretive approach at present is probably the iterated use of two- or two-and-one-half dimensional algorithms that calculate the gravitational field from sources specified as horizontal prisms with simple polygonal cross sections (Talwani and others, 1959; Cady, 1980; Saltus and Blakely, 1983; Won and Bevis, 1987). (Two-and-one-half-dimensional methods allow the user to place finite ends on the horizontal prisms that would normally extend to infinity in two-dimensional models.) An alternate method for the forward calculation of gravity anomalies has been developed by Parker (1972). This approach involves the use of a Fourier transform and the summation of a rapidly convergent series. Its computational speed can greatly facilitate schemes requiring numerous iterations.

In the forward approach, a model is constructed, its field compared with the observed gravity field, adjustments are made by the interpreter, and the process repeated until a satisfactory level of agreement is attained. More automatic inversion schemes are available, but the trial-and-error use of forward modeling offers several distinct advantages: (1) it is usually simple to construct such forward models—especially in the two and two-and-one-half dimensional cases (three-dimensional forward models, e.g., Talwani and Ewing, 1960; Plouff, 1976, are also becoming easier to construct as interactive software for generating and editing models becomes more sophisticated, e.g., Chuchel, 1985); (2) it is possible to enter known geologic or geophysical constraints into the model in a straightforward manner; (3) the approach allows the interpreter to obtain a hands-on feel for the sensitivity of the fit to changes in various model parameters; and (4) it is possible for the modeler to inject intuition and common sense into the process.

Forward models can be constructed to match either free-air, Bouguer, or isostatic gravity anomalies. If free-air anomalies are used, topographic features above sea level should be included as bodies in the model using their correct densities rather than density contrasts. (Note that error may be introduced by representing a real three-dimensional topographic body by a two-dimensional model.) If Bouguer gravity anomalies are used, it is possible to correct for departures of topographic features from the chosen reduction density (usually 2.67 g/cm^3) by modeling the features as bodies with density contrast equal to the difference between their true density and the reduction density. Using Bouguer gravity anomalies commonly requires that the gravity values predicted from the model be datum-shifted, possibly by hundreds of milligal, if the geometry of the compensating masses has not been included in the model. At times, compensating masses at a considerable lateral distance from the model area can produce sloping regional fields even in areas of low topographic relief, as in the plains area east of the Rocky Mountains. In such cases, it may be a considerable simplification to model the isostatic anomalies, because topographically generated regional fields have already been approximately removed by a simple model of the compensating masses. Experience has shown that models of upper crustal sources based on isostatic anomalies frequently require no datum adjustment.

If one is attempting to include in the model certain density interfaces in the lower crust and upper mantle, such as the M-discontinuity, one is probably implicitly modeling compensating masses to topographic loads. In such situations, isostatic anomalies need to be modeled with proper allowance made for the compensating masses already implicit in the anomalies. Couch's approach (see the chapter in this volume) is to model free-air values using true densities (rather than density contrasts) down to a depth of 50 km and to difference the modeled free-air gravity values from 6,442 mGal, which is the value found for a standard oceanic crustal column that is presumably in isostatic equilibrium (Couch and Woodcock, 1981; Barday, 1974). Requiring a 50-km-thick model over the area of interest to conform to this free-air gravity value has the effect of isostatically balancing the mass in the model column, at least to first order. Because isostatic equilibrium prevails over most of the Earth's surface, it is quite a useful constraint to compare one's crustal section against a standard section that is in equilibrium. In Couch's approach the model must be long enough to include all of the important subsurface masses, some of which may be distant from the study area. For example, the compensating masses under the Rocky Mountains exert a profound effect for a considerable distance into the plains to the east; if their influence is not accounted for in the model, then incorrect inferences could be made about the densities of the bodies that actually do appear in one's model. Jachens and others (this volume) have proposed a way to model "partial" isostatic anomalies that attempts to incorporate information for the attractions of distant compensating masses, while at the same time allowing the modeler to specify all of the bodies and interfaces along the line of profile, including those that are providing compensation for topographic loads.

Takin and Talwani (1966) have provided examples that demonstrate the effects of ignoring the sphericity of the Earth or of assuming an infinite lateral extent for geologic bodies in the construction of long two-dimensional profile models. Such approximations built into most models must not be forgotten. Yet as a general philosophy in constructing forward models, we believe that the models should be kept as simple as possible, especially where geologic or other geophysical constraints are limited in number. Perfect correspondence between calculated and observed values is not necessarily desirable, because it tends to mislead people into thinking that the model actually shows something real. Keep models simple.

Automatic inverse methods. True inverse methods, as opposed to the iterated forward methods of modeling, start with the observed gravity anomaly as the input and calculate directly the geometries or densities of subsurface bodies that produce the anomalies. Because there are infinitely many source geometries that can produce a given anomaly, the automatic inversion usually makes some initial assumptions to constrain the possible models. The simplest approach specifies the geometry of the bodies and solves for the best-fitting set of densities (e.g., Cady, 1980). Others specify a density contrast and perhaps a top surface for the body (ground level) and solve for the geometry of the bottom surface (Bott, 1960; Cordell and Henderson, 1968). Many require that some reasonable quantity be minimized or constrained. For example, compactness may be required of the source bodies (Last and Kubik, 1983), or the ideal body satisfying bounds of density or depth may be solved for (Parker, 1975). Other approaches have been suggested by Goodacre (1980), Pedersen (1977), Oldenburg (1974), Al-Chalibi (1972), and Tanner (1967), to name just a few. Bibliographic reference sources mentioned above list many other papers on inversion techniques. A useful tutorial on least-squares inversion methods applied to geophysical problems is given by Lines and Treitel (1984).

One difficulty with such methods seems to be that it is not a simple matter to find constraints that are plausible or appropriate for all geologic contexts. A more fundamental problem occurs because the signal (the observed anomaly) is smoothed by distance because of the $1/r^2$ dependence in Newton's law. Thus, any noise in the observations is amplified by the inversion process, and a fundamental resolution problem results (Parker, 1977). For this reason, models should be kept as simple as possible consistent with the available constraints.

Other automatic inversion schemes allow the user interactively to fix certain boundaries and densities in the model and to vary selected other parameters until a best fit is found (e.g., Webring, 1985). Such approaches allow considerable operator intervention, but still require common sense and intuition in their use. The end result, for example, can depend on the order in which parameters such as body corners are allowed to vary. As before, the robustness of the final solution depends on the application of as many constraints as can be mustered from other sources.

Integrated modeling. One deficiency in most modeling schemes is the difficulty in simultaneously modeling multiple data sets, such as seismic and gravity, although some two-dimensional programs allow magnetic and gravity anomalies to be modeled in tandem (e.g., Menichetti and Guillen, 1983). It is also often difficult to blend in the known geologic constraints because of problems of map scale and projection. It seems that more integrated display and modeling approaches would be very powerful.

Phase 4

The fourth and final interpretative phase consists of the geological interpretation of the model. The inferred sources and their geometries must make sense geologically. It is perhaps a bit artificial to single this out as a discrete step, because most interpreters will have added their geological biases in all of the previous phases. It is most useful at this stage, however, to show the model to a geologist who knows something about the rocks in the area; usually both the interpreter and the geologist learn something from this exercise.

SOURCES OF ERROR AND UNCERTAINTY

The largest sources of error in most onland gravity surveys are uncertainties in the elevation of the observation and in the topographic correction. An error of 1 m in elevation translates into approximately a 0.2-mGal error in the Bouguer gravity anomaly. Location errors also contribute: at latitude 45°, an uncertainty of 100 m in the north-south direction produces a change of about 0.08 mGal in predicted gravity. Terrain corrections can reach values of tens of milligals in areas of rugged topography with associated errors probably on the order of 10 to 20 percent of the total terrain correction. Terrain effects of this magnitude are, fortunately, rare. Typical regional gravity studies of a reconnaissance nature, in which elevations are accurate to within 5 m and in which terrain corrections have been made with normal care, probably have anomaly values accurate to within 2 mGal approximately 90 percent of the time. The largest errors are usually in areas of high topographic relief. Marine surveys, if carried out with the aid of radio transponder location systems, can presently attain accuracies of better than 2 mGal. Many of the older marine surveys, however, have uncertainties of 10 mGal or worse.

A Bouguer reduction density of 2.67 g/cm^3 is often assumed in regional gravity studies. In areas where topographic features are composed of sedimentary or volcanic rocks with average densities substantially different from this value, gravity anomalies that correlate with topography may be generated. For example, using a reduction density of 2.67 g/cm^3 for a volcano with bulk density of 2.2 g/cm^3 will cause a difference in calculated anomalies of 2 mGal per 100 m of altitude. Nettleton (1939) has described a technique for inferring average terrain densities from gravity profiles over topographic features. In many cases, it is convenient to use a standard reduction density such as 2.67 g/cm^3 and to accommodate the true densities of topographic features at the interpretation stage by incorporating the topography as bodies into two- or three-dimensional models. These topographic bodies should be assigned density contrasts equal to the discrepancy between the reduction density and the true average density.

One source of uncertainty that is most difficult to evaluate occurs near the beginning of the modeling process with the extraction of a regional field designed to isolate the anomalies produced by the bodies of interest. The removal of a regional field is especially difficult when the interpreter is trying to model a significant thickness of the crust, including deeper interfaces that can produce long-wavelength anomalies.

A last source of uncertainty in the interpretation of gravity anomalies arises from the inherent nonuniqueness of the inverse problem for potential fields: an infinite number of source geometries and densities can be found that will satisfy the observed gravity anomalies. For this reason it is essential to constrain the geometries of gravity models using as many different kinds of geologic and geophysical observations as are available. Rock samples should be collected and their densities determined to constrain the models. If seismic velocity data are available, the interpreter should try to infer densities using standard velocity-density curves. It is unlikely that all such inferred densities will prove totally acceptable in the quantitative gravity models, but disagreements and discrepancies can be illuminating.

One of the great challenges in the interpretation of geophysical data is to facilitate the integration of the various separate interpretations so that they more readily constrain and limit each other. The available geological data also need to be more readily accessible in digital form at the interpretation stage, so that the constraints they impose can be more quickly applied. Such real-time integrations should greatly aid in reducing the uncertainties that exist in each method on its own.

ACKNOWLEDGMENTS

R.W.S. thanks Martin Kane and George Thompson, who first introduced him to gravity methods. Both of these teachers shared their hard-won insights with such enthusiasm and obvious love for the subject that their lessons will not soon be forgotten.

REFERENCES CITED

Al-Chalibi, M., 1972, Interpretation of gravity anomalies by non-linear optimisation: Geophysical Prospecting, v. 20, p. 1–16.

Allis, R. G., and Hunt, T. M., 1986, Analysis of exploitation-induced gravity changes at Wairakei geothermal field: Geophysics, v. 51, p. 1647–1660.

Archinal, B., and Mueller, I., 1986, GPS; An overview: Professional Surveyor, v. 6, no. 3, p. 7–11.

Arvidson, R. E., Bindschadler, D., Bowring, S., Eddy, M., Guinness, E., and Leff, C., 1984, Bouguer images of the North American craton and its structural evolution: Nature, v. 311, p. 241–243.

Banks, R. J., Parker, R. L., Huestis, S. P., 1977, Isostatic compensation on a continental scale; Local versus regional mechanisms: Geophysical Journal of the Royal Astronomical Society, v. 51, p. 431–452.

Barday, R. J., 1974, Structure of the Panama Basin from marine gravity data [M.S. thesis]: Corvallis, Oregon State University, 99 p.

Bateman, P. C., and Eaton, J. P., 1967, Sierra Nevada Batholith: Science, v. 158, p. 1407–1417.

Becker, M., Groten, E., Hausch, W., and Stock, B., 1984, Numerical models of Fennoscandian land uplift based on repeated gravity measurements and potential changes; Proceedings of the IAG-Symposium, IUGG General Assemblage, Hamburg, 1983, v. 1: Columbus, Ohio State University, p. 83-103.

Birch, F., 1961, The velocity of compressional waves in rocks to 10 kilobars, 2: Journal of Geophysical Research, v. 66, p. 2199–2224.

Blakely, R. J., and Simpson, R. W., 1986, Locating edges of source bodies from magnetic or gravity anomalies: Geophysics, v. 51, p. 1494–1498.

Bott, M.H.P., 1960, The use of rapid digital computing methods for direct gravity interpretation of sedimentary basins: Geophysical Journal, v. 3, p. 63–67.

Bowin, C., Scheer, E., and Smith, W., 1986, Depth estimates from ratios of gravity, geoid, and gravity gradient anomalies: Geophysics,v. 51, p. 123–136.

Cady, J. W., 1980, Calculation of gravity and magnetic anomalies of finite-length right polygonal prisms: Geophysics, v. 45, no. 10, p. 1507–1512.

Chapman, M. E., and Bodine, J. H., 1979, Considerations of the indirect effect in marine gravity modeling: Journal of Geophysical Research, v. 84, p. 3889–3896.

Chovitz, B. H., 1981, Modern geodetic earth reference models: EOS Transactions of the American Geophysical Union, v. 62, no. 7, p. 65–67.

Chuchel, B. A., 1985, POLYGON; An interactive program for constructing and editing the geometries of polygons using a color graphics terminal: U.S. Geological Survey Open-File Report 85–233–A, 38 p.

Clark, R. D., 1984, Lucien La Coste: Geophysics; The leading edge of exploration, v. 3, no. 12, p. 24–29.

Colorado School of Mines, 1984, Reduction, analysis, and interpretation of gravity and magnetic survey data; A bibliography: Golden, Colorado School of Mines Center for Potential Fields Studies, 71 p.

Cordell, L. E., and Grauch, V.J.S., 1985, Mapping basement magnetization zones from aeromagnetic data in the San Juan basin, New Mexico, *in* Hinze, W. J., ed., The utility of regional gravity and magnetic anomaly maps: Society of Exploration Geophysicists, p. 181–197.

Cordell, L. E., and Henderson, R. G., 1968, Iterative three-dimensional solution of gravity anomaly data using a digital computer: Geophysics, v. 33, p. 596–601.

Couch, R., and Woodcock, S., 1981, Gravity and structure of the continental margins of southwestern Mexico and northwestern Guatemala: Journal of Geophysical Research, v. 86, p. 1829–1840.

Dehlinger, P., 1978, Marine gravity: Amsterdam, Elsevier Scientific Publishing Co., 322 p.

Dobrin, M. B., 1976, Introduction to geophysical prospecting, 3rd ed.: New York, McGraw-Hill, 630 p.

Dziewonski, A. M., 1984, Mapping the lower mantle; Determinations of lateral heterogeneity in P velocity up to degree and order 6: Journal of Geophysical Research, v. 89, p. 5929–5952.

Dzurisin, D., Anderson, L. A., Eaton, G. P., Koyanagi, R. Y., Lipman, P. W., Lockwood, J. P., Okamura, R. T., Puniwai, G. S., Sako, M. K., and Yamashita, K. E., 1980, Geophysical observations of Kilauea volcano, Hawaii; 2, Constraints on the magma supply during November 1975-September 1977: Journal of Volcanology and Geothermal Research, v. 7, p. 241–269.

Gardner, G.H.F., Gardner, L. W., and Gregory, A. R., 1974, Formation velocity and density; The diagnostic basics for stratigraphic traps: Geophysics, v. 39, p. 770–780.

Godson, R. H., and Scheibe, D. M., 1982a, Description of magnetic tape containing conterminous U.S. gravity data in a gridded format: U.S. Department of Commerce National Technical Information Service PB82-254798 (magnetic tape with description), 5 p.

——, 1982b, Description of magnetic tape containing conterminous U.S. free-air gravity anomaly data in gridded format: U.S. Department of Commerce National Technical Information Service PB82-253378 (magnetic tape with description), 5 p.

Goodacre, A. K., 1980, Estimation of the minimum density contrast of a homogeneous body as an aid to the interpretation of gravity anomalies: Geophysical Prospecting, v. 28, p. 408–414.

Grant, F. S., and West, G. S., 1965, Interpretation theory in applied geophysics: New York, McGraw-Hall, 584 p.

Guinness, E. A., Arvidson, R. E., Strebeck, J. W., Schulz, K. J., Davies, G. F., and Leff, C. E., 1982, Identification of a Precambrian rift through Missouri by digital image processing of geophysical and geologic data: Journal of Geophysical Research, v. 87, p. 8529–8545.

Gupta, V. K., and Grant, F. S., 1985, Mineral-exploration aspects of gravity and aeromagnetic surveys in the Sudbury-Cobalt area, Ontario, *in* Hinze, W. J., ed., The utility of regional gravity and magnetic anomaly maps: Society of Exploration Geophysicists, p. 392–412.

Hager, B. H., 1984, Subducted slabs and the geoid; Constraints on mantle rheology and flow: Journal of Geophysical Research, v. 89, p. 6003–6015.

Hammer, S., 1983, Airborne gravity is here!: Geophysics, v. 48, p. 213–223.

Haxby, W. F., and Weissel, J. K., 1986, Evidence for small-scale mantle convection from Seasat altimeter data: Journal of Geophysical Research, v. 91, p. 3507–3520.

Haxby, W. F., Karner, G. D., LaBrecque, J. L., and Weissel, J. K., 1983, Digital images of combined oceanic and continental data sets and their use in tectonic studies: EOS Transactions of the American Geophysical Union, v. 64, p. 995–1004.

Hayford, J. F., and Bowie, W., 1912, The effect of topography and isostatic compensation upon the intensity of gravity: U.S. Coast and Geodetic Survey Special Publication 10, 132 p., 5 plates.

Heiskanen, W. A., and Moritz, H., 1967, Physical geodesy: San Francisco, W. H. Freeman, 364 p.

Hildenbrand, T. G., Simpson, R. W., Godson, R. H., and Kane, M. F., 1982, Digital colored residual and regional Bouguer gravity maps of the conterminous United States with cut-off wavelengths of 250 km and 1,000 km: U.S. Geological Survey Geophysical Investigations Map GP-953-A, scale 1:7,500,000.

Hill, D. P., 1978, Seismic evidence for the structure and Cenozoic tectonics of the Pacific Coast states, *in* Smith, R. B., and Eaton, G. P., eds., Cenozoic tectonics and regional geophysics of the western Cordillera: Geological Society of America Memoir 152, p. 145–174.

Hinze, W. J., ed., 1985, The utility of regional gravity and magnetic anomaly maps: Society of Exploration Geophysicists, 454 p.

International Association of Geodesy, 1971, Geodetic reference system 1967: International Association of Geodesy Special Publication 3, 116 p.

Jachens, R. C., and Roberts, C. W., 1981, Documentation of a FORTRAN program, "isocomp," for computing isostatic residual gravity: U.S. Geological Survey Open-File Report 81-574, 26 p.

Jachens, R. C., Thatcher, W., Roberts, C. W., and Stein, R., 1983, Correlation of changes in gravity, elevation, and strain in southern California: Science, v. 219, p. 1215–1217.

Johnson, G. R., and Olhoeft, G. R., 1984, Density of rocks and minerals, *in* Carmichael, R. S., ed., Handbook of physical properties of rocks: Boca Raton, Florida, CRC Press, v. 3, p. 1–38.

Karner, G. D., and Watts, A. B., 1983, Gravity anomalies and flexure of the lithosphere at mountain ranges: Journal of Geophysical Research, v. 88, p. 10449–10477.

Keating, T., Taylor, P., Kahn, W., and Lerch, F., 1986, Geopotential research mission; Science, engineering, and program summary: National Aeronautics and Space Administration Technical Memorandum 86240, 209 p.

Labo, J. A., 1986, A practical introduction to borehole geophysics: Society of Exploration Geophysicists Geophysical Reference Series, 330 p.

Lachenbruch, A. H., Sass, J. H., and Galanis, S. P., Jr., 1985, Heat flow in southernmost California and the origin of the Salton Trough: Journal of Geophysical Research, v. 90, p. 6709–6736.

LaCoste, L., 1983, LaCoste and Romberg straight-line gravity meter: Geophysics, v. 48, p. 606–610.

Lambert, A., and Beaumont, C., 1977, Nano variations in gravity due to seasonal groundwater movements; Implications for the gravitational detection of tectonic movements: Journal of Geophysical Research, v. 82, p. 297–306.

Lambert, A., Liard, J. O., and Mainville, A., 1986, Vertical movement and gravity change near the La Grande-2 Reservoir, Quebec: Journal of Geophysical Research, v. 91, p. 9150–9160.

Last, B. J., and Kubik, K., 1983, Compact gravity inversion: Geophysics, v. 48, p. 713–721.

LeFehr, T. R., 1980, History of geophysical exploration; Gravity method: Geophysics, .v 45, p. 1634–1639.

——, 1983, Rock density from borehole gravity surveys: Geophysics, v. 48, p. 341–356.

LaFehr, T. R., and Nettleton, L. L., 1967, Quantitative evaluation of a stabilized platform shipboard gravity meter: Geophysics, v. 32, p. 110–118.

Lerch, F. J., Klosko, S. M., Patel, G. B., and Wagner, C. A., 1985, A gravity model for crustal dynamics (GEM-L2): Journal of Geophysical Research, v. 90, p. 9301–9311.

Lines, L. R., and Treitel, S., 1984, Tutorial; A review of least-squares inversion and its application to geophysical problems: Geophysical Prospecting, v. 32, p. 159–186.

Marsh, J. G., Brenner, A. C., Beckely, B. D., and Martin, T. V., 1986, Global mean sea surface based upon the Seasat altimeter data: Journal of Geophysical Research, v. 91, p. 3501–3506.

Marson, I., and Faller, J. E., 1986—The acceleration of gravity; Its measurement and its importance: Journal of Physics E, Scientific Instrumentation, v. 19, p. 22–32.

McNutt, M. K., 1980, Implications of regional gravity for state of stress in the Earth's crust and upper mantle: Journal of Geophysical Research, v. 85, p. 6377–6396.

——, 1983, Influence of plate subduction on isostatic compensation in northern California: Tectonics, v. 2, 399–415.

Menichetti, V., and Guillen, A., 1983, Simultaneous interactive magnetic and gravity inversion: Geophysical Prospecting, v. 31, p. 929–944.

Morelli, C., ed., 1974, The International Gravity Standardization Net 1971: International Association of Geodesy Special Publication 4, 194 p.

Nafe, J. E., and Drake, C. L., 1968, Physical properties of rocks of basaltic composition, *in* Hess, H. H., and Poldevaart, A., eds., Basalts, the Poldevaart treatise on rocks of basaltic composition, vol. 2: New York, Wiley-Interscience, p. 483–502.

Nettleton, L. L., 1939, Determination of density for reduction of gravimeter observations: Geophysics, v. 4, p. 176–183.

—— , 1976, Gravity and magnetics in oil prospecting: New York, McGraw-Hill, 464 p.

Oldenburg, D. W., 1974, The inversion and interpretation of gravity anomalies: Geophysics, v. 39, p. 526–536.

Parker, R. L., 1972, The rapid calculation of potential anomalies: Geophysical Journal of the Royal Astronomical Society, v. 31, p. 447–455.

—— , 1975, The theory of ideal bodies for gravity interpretation: Geophysical Journal of the Royal Astronomical Society, v. 42, p. 315–334.

—— , 1977, Understanding inverse theory: Annual Review of Earth and Planetary Science, v. 5, p. 35–64.

Paterson, N. R., and Reeves, C. V., 1985, Applications of gravity and magnetic surveys; The state of the art in 1985: Geophysics, v. 50, p. 2558–2594.

Pedersen, L. B., 1977, Interpretation of potential field data; A generalized inverse approach: Geophysical Prospecting, v. 25, p. 199–230.

Plouff, D., 1976, Gravity and magnetic fields of polygonal prisms and application to magnetic terrain corrections: Geophysics, v. 41, p. 727–741.

—— , 1977, Preliminary documentation for a FORTRAN program to compute gravity terrain corrections based on topography digitized on a geographic grid: U.S. Geological Survey Open-File Report 77-535, 45 p.

Saltus, R. W., and Blakely, R. J., 1983, Hypermag; An interactive two-dimensional gravity and magnetic modeling program: U.S. Geological Survey Open-File Report 83-241, 91 p.

Sheffels, B., and McNutt, M., 1986, Role of subsurface loads and regional compensation in the isostatic balance of the Transverse Ranges, California: Evidence for intracontinental subduction: Journal of Geophysical Research, v. 91, p. 6419–6431.

Simpson, R. W., Jachens, R. C., Blakely, R. J., and Saltus, R. W., 1983, Airyroot; A FORTRAN program for calculating the gravitational attraction of an Airy isostatic root out to 166.7 km: U.S. Geological Survey Open-File Report 83-883, 66 p.

Simpson, R. W., Hildenbrand, T. G., Godson, R. H., and Kane, M. F., 1987, Digital colored Bouguer gravity, free-air gravity, station location, and terrain maps for the conterminous United States: U.S. Geological Survey Geophysical Investigations Map GP 953-B, scale 1:7,500,000.

Society of Exploration Geophysicists, 1981, Manual of geophysical hand-calculator programs: Tulsa, Oklahoma, Society of Exploration Geophysicists.

—— , 1982, Gravity anomaly map of the United States (exclusive of Alaska and Hawaii): Society of Exploration Geophysicists, scale 1:2,500,000.

Swick, C. H., 1942, Pendulum gravity measurements and isostatic reductions: U.S. Coast and Geodetic Survey Special Publication 232, 82 p.

Syberg, F.J.R., 1972, A Fourier method for the regional-residual problem of potential fields: Geophysical Prospecting, v. 20, p. 47–75.

Takin, M., and Talwani, M., 1966, Rapid computation of the gravitation attraction of topography on a spherical earth: Geophysical Prospecting, v. 14, p. 119–142.

Talwani, M., and Ewing, M., 1960, Rapid computation of gravitational attraction of three-dimensional bodies of arbitrary shape: Geophysics, v. 25, p. 203–225.

Talwani, M., Worzel, J. L., and Landisman, M., 1959, Rapid gravity computations for two-dimensional bodies with application to the Mendicino submarine fracture zone: Journal of Geophysical Research, v. 64, p. 49–59.

Tanner, J. G., 1967, An automated method of gravity interpretation: Geophysics, v. 13, p. 339–347.

Taylor, P. T., Keating, T., Kahn, W. D., Langel, R. A., Smith, D. E., and Schnetzler, C. C., 1983, GRM; Observing the terrestrial gravity and magnetic fields in the 1990's: EOS Transactions of the American Geophysical Union, v. 64, p. 601–611.

Ulrych, T. J., 1968, Effect of wavelength filtering on the shape of the residual anomaly: Geophysics, v. 33, p. 1015–1018.

Wagner, C. A., and McAdoo, D. C., 1986, Time variations in the Earth's gravity field detectable with geopotential research mission intersatellite tracking: Journal of Geophysical Research, v. 91, p. 8373–8386.

Wagner, C. A., and Sandwell, D. T., 1984, The GRAVSAT signal over tectonic features: Journal of Geophysical Research, v. 89, p. 4419–4426.

Webring, M., 1985, SAKI; A Fortran program for generalized linear inversion of gravity and magnetic profiles: U.S. Geological Survey Open-File Report 85-112, 108 p.

Won, I. J., and Bevis, M., 1987, Computing the gravitational and magnetic anomalies due to a polygon; Algorithms and Fortran subroutines: Geophysics, v. 52, p. 232–238.

Woollard, G. P., 1966, Regional isostatic relations in the United States, *in* The Earth beneath the continents: American Geophysical Union Geophysical Monograph 10, p. 557–594.

Zumberge, M. A., Sasagawa, G., and Kappus, M., 1986, Absolute gravity measurements in California: Journal of Geophysical Research, v. 91, p. 9135–9144.

Zwart, W. J., ed., 1983, Cumulative index of geophysics; 1936–1982: Geophysics, v. 48, no. 10A, 543 p.

Manuscript Accepted by the Society October 31, 1988

Printed in U.S.A.

Geological Society of America
Memoir 172
1989

Chapter 4

Crustal studies using magnetic data

Richard J. Blakely
U.S. Geological Survey, 345 Middlefield Road, Menlo Park, California 94025
Gerald G. Connard
Northwest Geophysical Associates, Inc., P.O. Box 1063, Corvallis, Oregon 97339

ABSTRACT

The magnetic method plays an important role in mineral, petroleum, and geothermal exploration. It also has made important contributions to geologic mapping, structural geology, and plate-tectonic theory. In particular, magnetic measurements using aircraft provide a relatively inexpensive way to trace magnetic rock units beneath covered areas, to reveal the shape of subsurface magnetic bodies, and to interpolate subsurface geologic information between widely spaced seismic data and other localized geophysical measurements. Computerized interpretation procedures currently fall into two categores: techniques designed to enhance the data, which include various display and filtering procedures, and modeling experiments, which may be either forward (trial-and-error) or inverse in nature.

INTRODUCTION AND HISTORY

Magnetism of rocks has been a source of curiosity since the time of Thales, a philosopher of ancient Greece (ca. 600 B.C.), who supposedly was the first to contemplate the attractive and repulsive forces of lodestones (Needham, 1962). A mining compass was used to locate iron ore bodies in Sweden as early as 1640 (Jakosky, 1940), making magnetic field interpretation one of the oldest geophysical exploration techniques (Heiland, 1963). The acquisition and interpretation of high-precision aeromagnetic data, however, is a relatively new science with origins during World War II.

Early measurements of crustal magnetic fields were motivated by the search for iron ores. Because oxides and sulfides of iron are commonly associated with other mineral commodities, magnetic measurements still play an important role in mineral exploration, but the magnetic method also contributes to geological mapping, petroleum exploration, structural geology, and plate-tectonic theory. The study of crustal magnetic fields can locate plutons, ophiolites, and other magnetic units beneath less magnetic rocks and thereby permits extrapolation of lithotectonic features from known outcrops. Magnetic studies can locate fault contacts and reveal their dip and configuration beneath the surface, describe the history of young volcanic terranes, provide constraints on the thickness and shape of sedimentary basins, and locate the depth to the Curie-temperature isotherm, an important parameter for geothermal exploration and petroleum maturation studies. Study of the magnetic field over oceans played an essential role in the plate-tectonic revolution, especially in defining sea-floor spreading and continental drift. On continents, crustal fields can help delineate suture zones and terrane boundaries.

Early measurements of the magnetic field for exploration purposes were made with land-based, balanced magnets similar in principle of operation to today's widely used gravimeters. It was recognized early that airborne measurements would provide superior uniform coverage compared to surface measurements because of the aircraft's ability to quickly cover remote and inaccessible areas. Edelmann made the first airborne measurement in 1910 via balloon (Heiland, 1935), although balanced-magnet instruments were not generally amenable to the accelerations associated with moving platforms. Military considerations spurred the development of a suitable magnetometer for aeromagnetic measurements. In 1941, Victor Vacquier, Gary Muffly, and R. D. Wyckoff, employees of Gulf Research and Development Company under contract with the U.S. government, modified 10-year-old fluxgate technology, combined it with suitable stabilizing equipment, and thereby developed a magnetometer for airborne detection of submarines. In 1944, J. R. Balsley and Homer

Blakely, R. J., and Connard, G. G., 1989, Crustal studies using magnetic data, *in* Pakiser, L. C., and Mooney, W. D., Geophysical framework of the continental United States: Boulder, Colorado, Geological Society of America Memoir 172.

Jensen, of the U.S. Geological Survey, used a magnetometer of similar design in the first modern airborne geophysical survey near Boyertown, Pennsylvania (Jensen, 1961).

A second major advance in instrumentation was the development of the proton-precession magnetometer by Varian Associates in 1955. Particles with magnetic moments precess in a magnetic field with a frequency proportional to the intensity of the field. The proton-precession magnetometer simply measures the precession frequency of protons in water or a liquid hydrocarbon and thereby measures the magnitude of the total magnetic field. The proton-precession magnetometer does not require elaborate stabilizing or orienting devices. Consequently, it is relatively inexpensive, very compact, easy to operate, and has revolutionized land-based, shipboard, and airborne measurements. Reford (1980) discussed magnetometers capable of considerably greater resolution, including the optically pumped and Overhauser-effect magnetometers. These high-sensitivity instruments are used increasingly, but the proton-precession magnetometer remains an important instrument for regional aeromagnetic surveys.

Magnetic fields are vector quantities. Special magnetometers are sometimes used in airborne surveys to measure the three orthogonal components of the field (e.g., Blakely and others, 1973), but the majority of airborne magnetometers measure the *total field,* the magnitude of the Earth's field in the direction of maximum field strength. Proton-precession magnetometers inherently sample the total field, and directional instruments, such as the fluxgate magnetometer, are usually aligned mechanically in the direction of maximum field strength in order to provide the total field.

Magnetic gradiometers have been constructed recently from high-sensitivity total-field magnetometers (Hood, 1980). In theory, gradient measurements offer important advantages: vertical-gradient measurements increase the definition of near-surface sources, lateral-gradient (across flightline) measurements reduce aliasing problems between flightlines, and longitudinal-gradient measurements can be used to remove temporal effects. In practice, however, gradiometer measurements are much more susceptible to noise caused by aircraft systems and motions. Errors that are acceptable in total-field surveys can greatly exceed the acceptable standard in gradiometer surveys, and special care is required (Hardwick, 1984).

Because of the simplicity of instrumentation, aeromagnetic surveys are among the cheapest sources of geophysical information. For example, a complete aeromagnetic survey (flightlines spaced 1 mi apart and tie-lines spaced 5 mi apart) of a 1° by 2° area costs roughly one-tenth the price of a single east-west seismic reflection profile across the same area. Moreover, the two-dimensional coverage provided by an aeromagnetic survey complements and ties together more localized types of measurements. Aeromagnetic data, therefore, provide a relatively inexpensive way to trace magnetic lithologic units beneath covered areas, to reveal the shape of subsurface magnetic bodies, and to interpolate subsurface geologic information between widely spaced seismic reflection profiles, refraction profiles, well logs, or other more localized geophysical measurements.

Several excellent reviews of the current state of aeromagnetic interpretation techniques have been published recently by Reford (1980), Paterson and Reeves (1985), and Hinze (1985). It is not our purpose to duplicate their efforts here, but instead to give a brief outline of the method with which the reader can understand aeromagnetic interpretations (and their inherent limitations) in subsequent chapters of this volume. In this brief chapter, we can only highlight what we consider to be significant historical and recent advances. The reader is encouraged to pursue the subject further with the reference list.

ROCK MAGNETIC PROPERTIES

In this review we use SI (Systeme Internationale) units for magnetic quantities, but most articles on geomagnetism and paleomagnetism published prior to 1980—and many published subsequently—used the venerable Gaussian (CGS) system of units. Tables 1 and 2 describe the magnetic quantities of importance in the following discussion. Considerable confusion still exists in the paleomagnetic and geomagnetic community regarding these two systems, and recent articles by Payne (1981) and Shive (1986) describe the history behind the confusion.

Magnetite (Fe_3O_4) and its solid solutions with ulvospinel (Fe_2TiO_4) are the most important magnetic minerals to aeromagnetic studies of crustal rocks. Other minerals, such as hematite, pyrrhotite, and alloys of iron, are important in certain geologic situations, but the volume percentage, size, shape, and history of magnetite grains are of greatest importance in most aeromagnetic surveys.

Magnetic grains can be represented as individual magnetic dipoles. From the perspective of anomaly interpretation, the magnetic dipoles within a rock unit coalesce into a vector distribution of dipole moment per unit volume $\mathbf{M}(x, y, z)$ which we call magnetization. The total magnetization is the vector sum of two components: induced magnetization $\mathbf{M}_i$, which is proportional in magnitude and generally parallel to the ambient field $\mathbf{H}$, and remanent magnetization $\mathbf{M}_r$, which has a direction and intensity dependent on the origin and tectonic history of the rock. Consequently,

$$\begin{aligned}\mathbf{M} &= \mathbf{M}_i + \mathbf{M}_r \\ &= \chi\mathbf{H} + \mathbf{M}_r\end{aligned}$$

where χ is the magnetic susceptibility of the rock. The relative importance of remanent and induced magnetization is sometimes expressed as the Koenigsberger ratio,

$$Q = \frac{|\mathbf{M}_r|}{\chi|\mathbf{H}|}.$$

Typical values of $\mathbf{M}_r$ and χ for representative rock types are provided by Clark (1966) and Carmichael (1982). Generally speaking, mafic rocks are more magnetic than silicic rocks.

TABLE 1. MAGNETIC QUANTITIES IN SI AND GAUSSIAN UNITS

Term	Symbol	SI Unit	Gaussian Unit	Conversion Factor
Magnetic induction	**B**	tesla (T)	gauss (G)	10^4
Magnetic field	**H**	$A \cdot m^{-1}$	oersted (Oe)	$4\pi \cdot 10^{-3}$
Magnetic polarization	**J**	tesla	$emu \cdot cm^{-3}$	$(4\pi)^{-1} \cdot 10^4$
Magnetic dipole moment per unit volume (magnetization)	**M**	$A \cdot m^{-1}$	$emu \cdot cm^{-3}$	10^{-3}
Susceptibility	χ	dimensionless	dimensionless	$(4\pi)^{-1}$

Notes:

The conversion factors (last column) are constants that covert quantities reported in SI units to Gaussian units; for example, SI magnetization of 1 $A \cdot m^{-1}$ equals a Gaussian magnetization of 0.001 $emu \cdot cm^{-3}$.

Magnetic anomalies due to crustal sources have typical amplitudes less than $2 \cdot 10^{-6}$T. It is convenient to report these small amplitudes in units of nanotesla (nT), where 1 nT = 10^{-9}T. The nanotesla is equal to the Gaussian unit, gamma, where 1 gamma = 10^{-5} G.

Hence, basalts are usually more magnetic than rhyolites, and gabbros more magnetic than granites. Also, extrusive rocks generally have a higher remanent magnetization and lower susceptibility than intrusive rocks with the same chemical composition. Sedimentary and metamorphic rocks often have low remanent magnetizations and susceptibilities. These statements and the compilations referenced above are only statistical guidelines with many exceptions. Values of $\mathbf{M}_r$ and χ often vary by several orders of magnitude within the same outcrop, for example. Interpretation of any magnetic survey should include investigation of the magnetic properties of representative rock samples from the area of study whenever feasible.

DATA ACQUISITION AND PROCESSING

Survey design

Thoughtful planning of a magnetic survey is a crucial initial step in the application of magnetic data to geological problems. In many ways, the geological interpretation actually begins with this step, because state-of-the-art interpretation techniques rarely can salvage a poorly designed survey.

Reid (1980) established several important theoretical limitations on flightline geometry. He showed that the horizontal spacing between flightlines Δx (and the separation of individual measurements along flightlines) should be no greater than twice the average depth $\bar{d}$ to the magnetic sources of interest. (For vertical gradient surveys, Δx should not exceed $\bar{d}$.) Wider spacings can cause aliasing errors: short-wavelength anomalies are inadequately sampled, appear longer in wavelength, and are distorted in resulting maps and interpretations.

To reduce aliasing errors perpendicular to flightline directions, aeromagnetic surveys are usually planned with most flightlines normal to the strike of dominant geologic features. Additional "tie-lines" are flown at right angles to the main flightlines and ideally have a spacing of no more than five times the flightline spacing. Tie-lines provide additional data between the main flightlines, serve as a check on the quality of the survey, are used to remove diurnal variations, and provide a means to correct for navigational and other systematic errors (Mittal, 1984).

The relationship between Δx and $\bar{d}$ shows that low-altitude surveys generally require more closely spaced measurements than high-altitude surveys. Ground-based measurements, for example, must be especially close together because of the proximity of near-surface geologic sources and the possibility of "cultural sources" (power lines, pipes, fences, etc.). Clearly, ground magnetic surveys are more appropriate for detailed investigations of shallow sources than for regional crustal studies.

Aeromagnetic surveys are usually flown at either constant barometric altitude (CBA) or at constant terrain clearance (draped), and the relative merits of each of these methods are the subject of a number of recent papers (Grauch and Campbell, 1984; Hansen, 1984a; Pearson and others, 1985; Paterson and Reeves, 1985). Draped surveys, by putting the magnetometer as close as feasible to the ground surface, provide an important advantage in learning about near-surface sources. However, draped surveys are more difficult to fly, process, and interpret than CBA surveys. Perhaps the most serious disadvantage of draped surveys is the difficulty in determining actual flight eleva-

TABLE 2. RELATIONSHIP BETWEEN VARIOUS MAGNETIC QUANTITIES

SI System	Gaussian System
$\mathbf{B} = \mu_o(\mathbf{H} + \mathbf{M})$	$\mathbf{B} = \mathbf{H} + 4\pi\mathbf{J}$
$\mathbf{J} = \mu_o\mathbf{M}$	$\mathbf{J} = \mathbf{M}$
$\mathbf{M} = \chi\mathbf{H}$	$\mathbf{M} = \chi\mathbf{H}$

Note: The constant μ_o is the permeability of free space ($\mu_o = 4\pi \cdot 10^{-7}$ $H \cdot m^{-1}$).

tion; the interpreter is rarely certain of the actual location of the sensor, except perhaps in the case of carefully controlled helicopter surveys.

Navigation

Navigational systems currently used for magnetic surveys include (roughly in order of increasing accuracy and cost) photography or videotape of the ground, combined with Doppler radar, LORAN C, inertial guidance, radio transponder beacons, and portable radar transponders.

The accuracy of the photographic method depends on identification of mapped geographic features, such as roads or power lines. This technique is obviously unsuitable for surveys over water, ice, and many wilderness areas. Inertial guidance navigation is used extensively for aeromagnetic surveys over these kinds of areas (e.g., Vogt and others, 1979). The accuracies possible with LORAN C navigation, where it is available, are quite variable. Under typical conditions, LORAN C positions are likely to be in error by at least 100 m (Spradley, 1985).

Portable radar transponders provide the most accurate horizontal positions (within a few meters) and can provide real-time information to assist the pilot in following predetermined flightlines (Connard, 1979; Petrick and others, 1984). However, these systems require line-of-sight contact between the aircraft and the ground-based radar transponders, which increases the logistical cost of this type of navigation, particularly over rugged terrain or for large surveys. Radio transponder beacons operate at longer ranges than, and can approach the accuracy of, radar transponders. However, radio transponders are more expensive and require more logistical support. The increased expense of using either type of transponder system is partially offset by the reduced cost of data reduction.

The determination of vertical altitude relies on barometric and radar altimeters. Barometric altimeters are subject to small-scale meteorological variations that are usually unknown, and radar altimeters have limited accuracy, particularly in rough topography where echoes may be nonvertical.

Data reduction

In aeromagnetic surveys, data reduction actually begins at the time of the measurement, with compensation for the magnetic effects of the aircraft (Hardwick, 1984). On completion of the survey, reduction of raw magnetic measurements consists of merging the magnetic measurements with navigational information (both functions of time) and removing all noncrustal components of the measurements, such as diurnal variations and the main field of the Earth. Both of these steps are subject to substantial errors. Jensen (1965) pointed out that errors in horizontal position as small as 30 m or in altitude as small as 10 m will cause magnetic anomaly errors larger than 1 nT, even in very low gradients.

Noncrustal components of magnetic measurements result from both temporal and spatial variations of the main magnetic field of the Earth (caused by sources in the core) and external fields (caused, for example, by electrical currents in the ionosphere). Long-wavelength (>2,000 km) anomalies and secular variations with periods greater than 6 m can be removed effectively by subtracting the appropriate International Geomagnetic Reference Field (IGRF) (Peddie, 1983; IAGA, 1986). Paterson and Reeves (1985) showed a striking example of how removal of the IGRF enhances a magnetic anomaly map.

Short-period temporal activity caused by external magnetic sources occurs with a broad range of periods, from 0.01 sec up to the entire period of the survey, and is loosely labeled "diurnal" activity or variation. There are three basic approaches for removing temporal variations from magnetic surveys. The most straightforward approach establishes one or more magnetic base stations at fixed locations and assumes that diurnal variations at the base stations are representative of the entire survey area. This technique works well when all parts of the survey are within about 80 km of a base station (Eggers and Thompson, 1984; Connard, 1979). However, Hartman and others (1971) and Whitman and Niblett (1961) demonstrated that significant errors can occur when attempting to extrapolate base-station information over larger distances, particularly in high magnetic latitudes. A traditional approach for removing diurnal effects from aeromagnetic surveys uses statistical analysis of magnetic anomaly discrepancies at line intersections (Foster and others, 1970; Yarger, 1978; Sander and Mrazak, 1982). Recently, both vertical and horizontal gradiometers have been used to remove temporal effects from both aeromagnetic and marine magnetic surveys (Hardwick, 1984; Eggers and Thompson, 1984).

Most methods of computer contouring and computer analysis (e.g., upward continuations discussed subsequently) of map data require that the data be interpolated onto a rectangular grid. Interpolation of magnetic data is neither simple nor unique, however. Several different techniques are currently in use, including minimum curvature (Briggs, 1974), bi-cubic spline (Bhattacharyya, 1969), weighted averaging (Pelto and others, 1968), and kriging (Olea, 1974). Errors in navigation or removal of temporal fields from the original data can cause spurious patterns in contour maps such as "herringbones," that are unrelated to geologic sources (e.g., Paterson and Reeves, 1985F). For these reasons, some interpreters prefer to analyze the original magnetic data in profile form.

Other types of surveys

Although the title of this memoir restricts us primarily to aeromagnetic studies of continental regions, a few comments should be included concerning two other important sources of magnetic data.

Marine magnetic measurements. Marine magnetic measurements are usually collected as a secondary priority to some other type of marine geophysical survey. Consequently, marine surveys usually are made along irregular tracklines with relatively long times between trackline intersections. Magnetic base stations are often difficult to establish near enough to the survey area to directly remove temporal variations. Eggers and Thompson (1984) and Hansen (1984b) described schemes for using horizontal-gradient measurements to deal with diurnal problems in marine surveys.

In recent years, deeply towed magnetometers have provided much information about magnetization of oceanic crust, reversals of the Earth's magnetic field, and the process of sea-floor spreading (e.g., Macdonald and others, 1983). The logistical problems associated with this type of survey are formidable, and the reader is referred to the literature for further information (e.g., Spiess and Mudie, 1970).

Satellite magnetic measurements. The seven-month flight of Magsat in 1979 and 1980 eclipsed all previous satellite missions in importance to crustal studies. Anomalies due to crustal sources are in the 5- to 10-nT range at typical Magsat altitudes (300 to 500 km) and are several times larger in amplitude than the estimated accuracy of Magsat total-field measurements. Consequently, analyses of Magsat anomalies are providing important information about large-scale structure of the crust (e.g., von Frese and others, 1986). Techniques for acquisition of satellite data are clearly beyond the scope of this chapter. The interested reader is referred to a special issue of the *Journal of Geophysical Research* (Langel, 1985) and to a recent review article by Mayhew and LaBrecque (1987).

Availability of existing magnetic data

A comprehensive index of published aeromagnetic data is available from the U.S. Geological Survey (Hill, 1986a–g) and provides a convenient way to determine the availability of public data from any specific area of the U.S. This index contains bibliographies and index maps for selected publications by the U.S. Geological Survey, other federal agencies, universities, states, and professional societies.

Magnetic data collected with public funds, such as aeromagnetic surveys flown or contracted by the U.S. Geological Survey, are available to the public from the National Geophysical Data Center (NGDC, National Oceanic and Atmospheric Administration, Code E/GC12, 325 Broadway, Boulder, Colorado 80303). Many of these surveys are somewhat localized in scope, but at least two surveys cover the entire conterminous United States at low altitude. The Project MAGNET Survey of the United States, which was flown by the U.S. Naval Oceanographic Office along 43 north-south flightlines spaced approximately 1° of longitude apart, is available from the NGDC. In addition, magnetic measurements were made as part of the National Uranium Resource Evaluation (NURE) along low-altitude flightlines spaced 1 to 6 mi apart, and are available from the U.S. Geological Survey as fiche or paper reproductions (USGS, Books and Open-File Reports, Denver Federal Center, MS 41, Box 25425, Denver, Colorado 80225) or in digital form (USGS, EROS Data Center, User Services, Sioux Falls, South Dakota 57198). The low altitude of, and wide spacing between, flightlines in both of these surveys make interpretation of magnetic sources between flightlines difficult (e.g., Sexton and others, 1982), but individual profiles are often useful for geologic interpretations.

INTERPRETATION TECHNIQUES: OVERVIEW FOR THE NON-SPECIALIST

This section briefly outlines the logical steps that often are used to develop a geologic interpretation from a new magnetic survey. The nonspecialist may wish to read this section and then skip the subsequent section, which describes the same techniques in more mathematical detail.

Phase I

Interpretation of magnetic data often consists of two phases, one basically qualitative in nature and the other more quantitative and dependent on computers. In many geological situations, a qualitative understanding of the sources of magnetic anomalies may be sufficient for the purposes of the study. In other situations, quantitative description of the sources (depth, thickness, shape, and magnetization) may be required. In any case, the qualitative initial phase should precede the more quantitative techniques of the second phase.

Phase I requires little more than contoured magnetic data (or profiles), a ruler, and a calculator. Plan-view shapes of magnetic bodies can be sketched from the location of high-gradient contours; comparison with published contour maps generated from simple source geometries (e.g., Vacquier and others, 1951) can help in this regard. Depths to certain isolated sources can be estimated from the distance over which gradients remain uniform (Peters, 1949). Grant and West (1965) have described several ways of using the shape and amplitude of anomalies to estimate thickness of magnetic bodies and the dip of their contacts with less magnetic units. Effects of topography in magnetic terranes can be assessed by comparing magnetic contour maps with topographic contour maps.

One product of a Phase I interpretation would probably be a sketch map that shows the boundaries between units with different magnetic properties and shows the geologic and lithotectonic interpretation of these boundaries. Magnetic and topographic profiles with cartoon cross sections would help show inferred subsurface configurations of the magnetic units.

Phase II

The arsenal of computerized interpretation techniques has flourished over the last 25 years. These techniques can be classed generally as either data enhancement methods or modeling methods.

Enhancement techniques. The likely first step on receiving a new magnetic survey is to display the data in line-contour, color-contour, shaded relief, or some other form (Cordell and

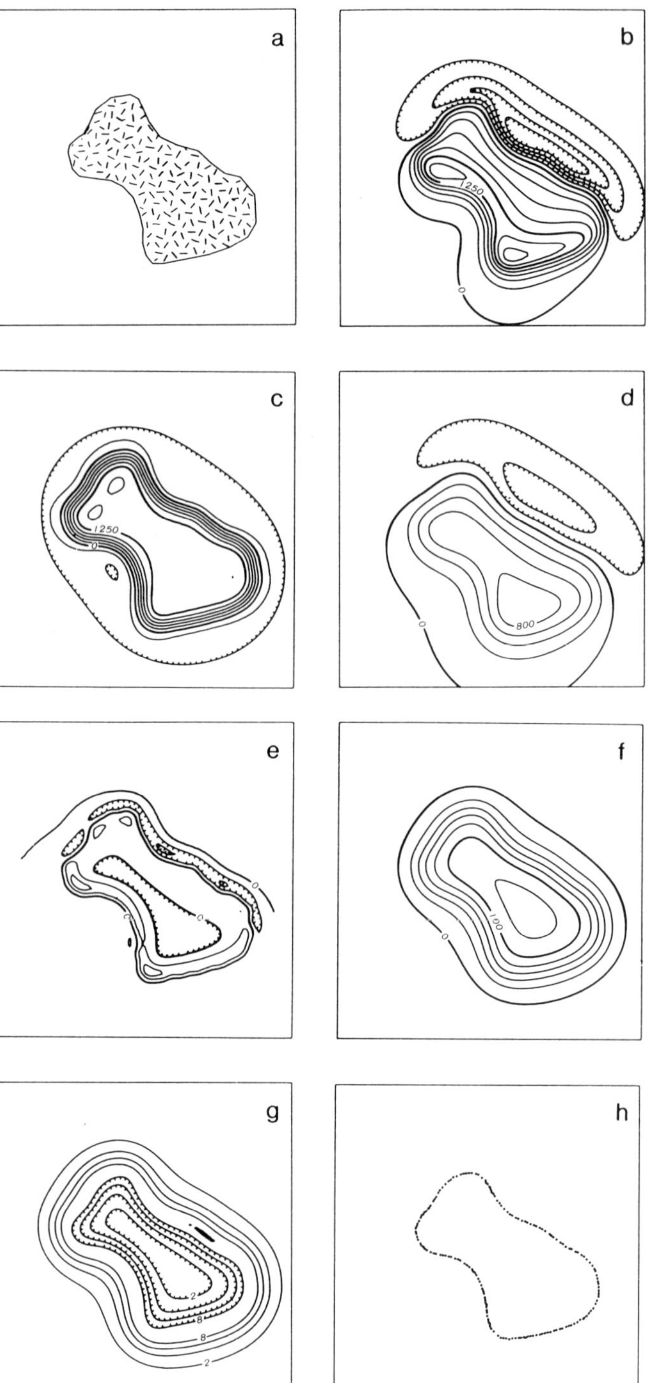

Figure 1. Ways to process aeromagnetic data. Hachures on contour maps indicate direction of closed minima. North direction is from bottom to top for each map. Boxes surrounding maps are 100 km in length. a, Hypothetical magnetic body used to generate sample aeromagnetic data. Stipple pattern indicates plan view of magnetic body with vertical sides, intensity of magnetization 10 A · m^{-1}, inclination of magnetization 60°, declination of magnetization 20°, and thickness 2 km. b, Total-field anomaly due to body shown in a. Anomaly is measured on horizontal surface 4 km above body and in a regional field with inclination 60° and declination 20°. Contour interval, 250 nT. c, Reduction to the pole of anomaly shown in b. Contour interval, 250 nT. d, Upward continuation. Anomaly shown in b continued upward 5 km from observation surface. Contour interval, 200 nT. e, Second vertical derivative of anomaly shown in b. Contour interval, 50 nT · km^{-2}. f, Pseudogravity transformation of anomaly shown in b, assuming magnetization of 1 A · m^{-1} and density contrast of 0.1 g · gm^{-3}. Contour interval, 20 "pseudo"-mGal. g, Magnitude of maximum horizontal gradient of pseudogravity transformation of anomaly shown in b. Contour interval, 2 "pseudo"-mGal · km^{-1}. h, Location of maxima of data shown in g. Note correspondence with true edges of magnetic source shown in a. Maxima located according to the method described by Blakely and Simpson (1986).

Knepper, 1987). The original data can be compared in this way with various filtered versions, some of which are discussed below and illustrated in Figure 1.

The shape of a magnetic anomaly depends not only on the shape of the source body, but also on the direction of magnetization and the direction of the ambient field. Consequently, identically shaped sources with nonparallel magnetizations can produce differently shaped anomalies. *Reduction to the pole* (Fig. 1c) is a filter operation that removes this ambiguity; it transforms magnetic anomalies into the anomalies that would be caused by identical magnetic sources but with vertical magnetization and with measurement in a vertical magnetic field. *Upward continuation* (Fig. 1d) transforms observed anomalies into the anomalies that would be observed at a higher altitude, and is useful for accentuating anomalies caused by deep sources at the expense of anomalies caused by shallow sources. *Vertical derivatives* (Fig. 1e) enhance anomalies due to shallow sources and help delineate their edges. The *pseudogravity transformation* (Fig. 1f) converts a magnetic anomaly into the gravity anomaly that would be caused by a density distribution exactly proportional to the magnetization distribution. It allows a one-to-one comparison of magnetic anomalies with observed gravity anomalies. The *horizontal gradient* of the pseudogravity anomaly (Fig. 1g) outlines edges of magnetic sources.

Modeling techniques. Modeling experiments help determine the subsurface configuration of magnetic units. The most common approach is *forward modeling:* an initial guess of the shape and magnetization is used to calculate a magnetic anomaly, the calculated anomaly is compared with the observed anomaly, and the shape and magnetization are modified in order to improve agreement between the two. These three steps (calculation, comparison, and adjustment) continue until the calculated and observed anomalies are in satisfactory agreement.

The modeling procedure can be automated by a variety of *inverse modeling* techniques, of which there are two general types: (1) if the bounding shape of the source is known, the distribution of magnetization within the source can be determined by least-squares (e.g., Bott, 1967) or Fourier-transform (e.g., Parker, 1973) techniques; and (2) if simplifying assumptions can be developed for the distribution of magnetization, the shape of magnetic sources can be automatically derived. For example, nonlinear least-squares approximations can be used to determine cross-sectional shapes of magnetic bodies (e.g., Johnson, 1969), or power-spectral analysis can be used to estimate the depth to the top and bottom of magnetic sources (e.g., Connard and others, 1983).

It is essential in any interpretation to appreciate the nonuniqueness property of possible causative bodies. For every observed magnetic anomaly, an infinite number of possible source configurations exist that will each produce the observed anomaly. The interpreter must strive to employ all other available information, such as surface geology, seismic reflection and refraction data, gravity anomalies, electrical measurements, and heat-flow data, in order to narrow the infinite set of possible solutions to a more manageable number. However, even in the most tightly constrained cases, source models are not unique, and slightly different constraining assumptions may result in quite different source configurations.

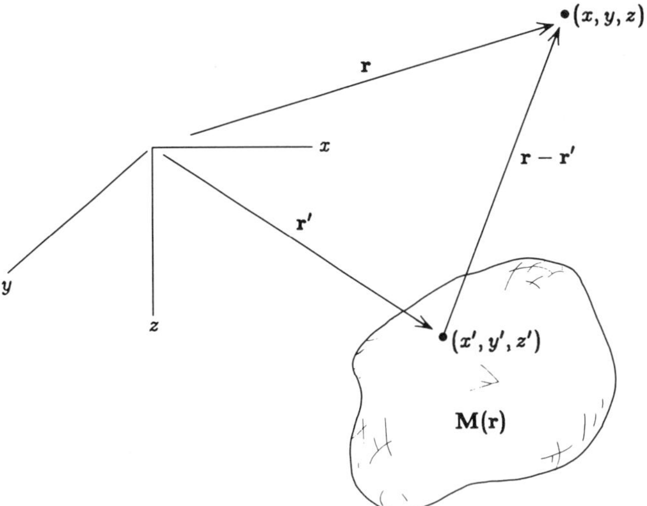

Figure 2. Coordinate system and vector notation used to describe the total-field anomaly at an external point due to a distribution of magnetic material $M(x, y, z)$.

INTERPRETATION TECHNIQUES: DETAILED DISCUSSION

The geophysicist is usually presented with a total-field anomaly measured on a nearly planar surface and given the task of describing certain parameters of the source; e.g., the distribution of magnetization, the depth to the top of the source, or the direction of magnetization. This task is made much more challenging by the nonuniqueness property of potential fields: an infinite variety of magnetic distributions all produce precisely the same total-field anomaly. In order to reduce the infinite number of mathematical answers to a manageable number of geological answers, it is essential that all geological and geophysical information and intuition be considered in the interpretation. Field or laboratory measurements (Hillhouse, 1987) of magnetic properties of rock samples from the Earth's surface often provide important constraints (e.g., Criss and Champion, 1984), but surface rocks usually have no or minimal relationship to underlying magnetic sources. Several recent studies have tried to ascertain magnetic properties of typical deep crustal sources from uplifted and eroded crustal sections, from xenoliths, and from plutonic and high-grade metamorphic terrains (Williams and others, 1985; Schlinger, 1985; Frost and Shive, 1986).

The starting equation in SI units for most magnetic interpretation techniques provides the total-field anomaly h at a point outside a distribution of magnetic material $M(x, y, z)$:

$$h(\mathbf{r}) = \frac{\mu_0}{4\pi}\,\hat{\mathbf{B}} \cdot \nabla \left\{ \int_R \mathbf{M}(\mathbf{r}') \cdot \nabla \frac{1}{|\mathbf{r} - \mathbf{r}'|}\, dx'\, dy'\, dz' \right\} \tag{1}$$

(Fig. 2). We have used vector notation in equation (1) to represent spatial coordinates, i.e., $\mathbf{r} = (x, y, z)$ and $\mathbf{r}' = (x', y', z')$. Integration is over region R, which includes all of the magnetic material (but not the field point $\mathbf{r}$); primed coordinates are the variables of integration. Unit vector $\hat{\mathbf{B}}$ is in the direction of the regional field, and μ_0 is the permeability of free space (Table 2). Several assumptions are implicit in equation (1). First, it is assumed that the regional field has a constant direction throughout the survey region and is large in magnitude with respect to the perturbing field. Second, the magnetization is assumed to be sufficiently well-behaved for the integral to exist.

Equation (1) has a complicated appearance. It can be more easily understood if it is rewritten as a filtering operation. If we let $\mathbf{M}(\mathbf{r}') = M(\mathbf{r}')\,\hat{\mathbf{M}}$, where $M(\mathbf{r}')$ is the intensity of magnetization and $\hat{\mathbf{M}}$ is a unit vector describing a fixed direction of magnetization, equation (1) becomes

$$h(\mathbf{r}) = \int_R M(\mathbf{r}')\, \phi(\mathbf{r} - \mathbf{r}')\, dx'\, dy'\, dz', \tag{2}$$

where ϕ is a Green's function dependent on coordinates $\mathbf{r}$ and $\mathbf{r}'$ and on unit vectors $\hat{\mathbf{B}}$ and $\hat{\mathbf{M}}$. Equation (2) shows that the total-field anomaly at any external point is simply the sum of all the infinitesimal elements of magnetization weighted by ϕ.

We have divided the computer-aided interpretation of aeromagnetic data into three general categories—forward modeling, inverse techniques, and data enhancement—and will discuss each of these separately in the following sections. Many of these approaches also are applicable to gravity data (Simpson and Jachens, 1987) and other forms of potential field data.

Forward modeling

Space domain. A *forward calculation* is the computation of $h(\mathbf{r})$ in equation (1) after making some simplifying assumptions about $\mathbf{M}(\mathbf{r})$. This procedure is mathematically unique in that only one total-field anomaly results from a given distribution of magnetization. A *forward model* is developed by trial-and-error adjustment of $\mathbf{M}(\mathbf{r})$, while considering all available geological and geophysical information, until the calculated anomaly matches the observed anomaly arbitrarily closely. Hence, the application of the forward method largely hinges on methods to approximate $\mathbf{M}(\mathbf{r})$ conveniently and realistically.

For example, geologic structures and magnetic anomalies often extend for long distances in one direction, so that magnetic sources may be suitably approximated by infinitely extended prisms lying parallel to the geologic structures. The cross-sectional shapes of the prisms are approximated by polygons (Talwani and Heirtzler, 1964; Saltus and Blakely, 1983), and it is the interpreter's task to determine the shape of the polygons by trial-and-error adjustment. Although this "two-dimensional" forward method is more than 20 years old, we believe that it is probably the most widely used algorithm in potential field interpretation today. A useful variation, known as the "2½-dimensional" calculation, allows the infinitely extended prisms to be replaced with prisms of finite length (Shuey and Pasquale, 1973; Rasmussen and Pedersen, 1979; Cady, 1980).

Three-dimensional sources (those that cannot be approximated as horizontal prisms) are usually approximated by one or more bodies each with uniform magnetization and simplified shape. For example, magnetic bodies can be represented by a stack of laminae (Talwani, 1965) or layers (Plouff, 1976), and each lamina (or layer) is in turn approximated in plan view by a polygon. This technique is especially amenable to topographic models because topographic contours provide at once the shape of individual laminae (or layers). Bott (1963), Barnett (1976), and Okabe (1979) described an alternative approach in which the surface of each body is approximated by a set of polygonal facets, reminiscent of a geodesic dome.

Fourier domain. More recently, a number of authors have used Fourier transforms to perform the forward calculation. In the following, we represent the Fourier transform of a two-dimensional function $f(x, y)$ by the notation $F[f(x, y)]$, where

$$F[f(x,y)] = \int_{-\infty}^{\infty}\int_{-\infty}^{\infty} f(x,y)\, e^{-i(k_x x + k_y y)} dx\, dy.$$

The independent variables of $F[f(x, y)]$ are wavenumbers k_x and k_y, which are inversely proportional to wavelengths λ_x and λ_y in the x and y directions, respectively, i.e., $k_x = 2\pi/\lambda_x$ and $k_y = 2\pi/\lambda_y$. Hence, the Fourier transform maps a function of x and y (space domain) into a function with coordinates related to wavelengths (Fourier domain). Algorithms that rapidly accomplish the Fourier transform have existed for many years (e.g., Singleton, 1969). It is the speed of these algorithms and the insights provided by the Fourier domain representation that have attracted recent interest.

In one approach, we let $h(\mathbf{r})$ be measured on a horizontal surface so that $h(\mathbf{r}) = h(x, y)$, let the magnetization be confined to a horizontal layer with top and bottom surface at depths z_t and z_b, respectively, and let the magnetization be constant within any vertical column of the layer. Then for this very simplified layer source, equation (2) can be written as:

$$h(x,y) = \int_{-\infty}^{\infty}\int_{-\infty}^{\infty} M(x',y')\, \phi_1(x-x', y-y')\, dx'\, dy', \tag{3}$$

where

$$\phi_1(x',y') = \int_z \phi(x',y',z')\, dz'.$$

Equation (3) is simply a two-dimensional convolution (Bracewell, 1965) of two functions $M(x, y)$ and $\phi(x, y)$. The Fourier transform of equation (3) becomes multiplication of the Fourier transforms of $M(x, y)$ and $\phi(x, y)$ (Bracewell, 1965):

$$F[h(x,y)] = F[M(x,y)] \cdot F[\phi_1(x,y)], \tag{4}$$

where

$$F[\phi_1(x,y)] = \theta(k_x, k_y)\,[e^{-kz_1} - e^{-kz_2}], \tag{5}$$

$$k = \sqrt{k_x^2 + k_y^2},$$

and $\theta(k_x, k_y)$ is a function that depends only on the direction of magnetization and direction of the regional field (Schouten, 1971). Certain assumptions are required about the behavior of $h(x, y)$ and $M(x, y)$ in order for their Fourier transforms to exist. For instance, $|h(x, y)|$ and $|M(x, y)|$ must approach zero as x and y approach the practical limits of integration, a criterion that is rarely met in geological applications because magnetic anomalies continue beyond the limits of the survey. One way to treat this particular problem is to first multiply $h(x, y)$ by a weighting function (e.g., a two-dimensional boxcar function) to force proper behavior of $h(x, y)$.

The function described by equation (5) was termed the *earth filter* by Schouten and McCamy (1972) because it mathematically converts magnetization of upper crustal rocks into total-field anomalies. Note that $\theta(k_x, k_y)$ in equation (5) contains all terms related to the direction of magnetization and direction of regional field, whereas the exponential terms contain all information about the location of the layer. Although the simplicity of the layer model is not applicable to many geologic situations, it has proved useful for magnetic anomalies over oceanic crust (e.g., Schouten and McCamy, 1972; Blakely and Cox, 1972) and is discussed with respect to depth-to-source calculations in a later section.

In the more general case where the magnetic distribution depends on depth, equation (2) cannot be written as a convolution and equation (4) does not exist. Parker (1973), however, described a way to calculate anomalies over more general distributions of magnetization using Fourier transforms. The magnetization is still assumed to be uniform in the vertical direction between top and bottom surfaces z_t and z_b, respectively, but the elevations of the top and bottom surfaces are functions of horizontal coordinates, i.e., $z_t = z_t(x, y)$ and $z_b = z_b(x, y)$. In the Fourier domain, Parker (1973) showed that:

$$F[h(x,y)] = \theta(k_x, k_y) \sum_{n=1}^{\infty} \frac{(-k)^n}{n!} F[M(x,y)\,(z_t^n(x,y) - z_b^n(x,y))]. \tag{6}$$

Hence, the calculation is performed by summing Fourier transforms of the magnetization multiplied by powers of the topography. Parker (1973) showed that convergence is most rapid if the top and bottom surfaces are treated separately with the origin placed at a level midway between the highest and lowest point of each surface. Computer programs implementing this method are readily available (e.g., Blakely, 1981).

Inverse techniques

Inverse methods involve the *direct* determination of certain parameters of the source from measured data (as opposed to the trial-and-error or *indirect* determination). Because of the infinite number of sources that will each cause an identical field, it is necessary to constrain the determination in some way. Constraints may be imposed mathematically (e.g., "magnetizations should have a minimum amount of variation from point to point"), geophysically (e.g., "magnetizations must not exceed reasonable values determined from sample measurements"), or geologically (e.g., "strike-slip faults cause near-vertical magnetic boundaries").

Inverse problems can be divided into two types: linear and nonlinear (Parker, 1977). On the one hand, equation (2) shows that the total-field anomaly is a linear function of the magnetization. For example, if the magnetization is doubled, the amplitude of the total-field anomaly is also doubled but its shape is unchanged. Determination of the magnetization from the total-field anomaly is, therefore, a *linear inverse* problem. On the other hand, information about the shape of the body is contained as nonlinear terms in the function ϕ. Determination of the shape of the body is, therefore, a *nonlinear inverse* problem.

Linear inverse techniques

A magnetic anomaly will always be "smoother" than the magnetic distribution that causes it because ϕ in equation (2) is a "smoothing" operator. Consequently, the determination of magnetization from observed measurements of h is an inherently unstable procedure (Parker, 1977). The slightest errors in measurements of h will tend to generate unrealistic solutions for M, solutions that nevertheless satisfy observed values of h.

Direction of magnetization. One simple linear inverse problem involves determining the best single magnetization for an entire body given the shape of the body and its magnetic field (Talwani, 1965; Parker, 1971). If we assume that the magnetization of the body is uniform, then equation (1) can be written in the form

$$h_i = \begin{pmatrix} V_{x,i} & V_{y,i} & V_{z,i} \end{pmatrix} \begin{pmatrix} M_x \\ M_y \\ M_z \end{pmatrix} \quad i = 1, 2, \ldots L, \tag{7}$$

where h_i is the calculated anomaly at a single point, M_x, M_y, and M_z are the three orthogonal components of magnetization, and $V_{x,i}$, $V_{y,i}$, and $V_{z,i}$ represent integral terms which depend on the geometry of the source. The integral terms can be approximated using methods such as those described by Talwani (1965) or Plouff (1976) discussed earlier. The objective is to find M_x, M_y, and M_z so that calculated values of h_i are as close as possible in a least-squares sense to observed field values h_i'. In other words, the three components of magnetization are required that minimize

$$E^2 = \sum_{i=1}^{L} (h_i - h_i')^2. \tag{8}$$

This method has had great success in studies of anomalies over seamounts. Each seamount is treated as a single paleomagnetic sample, so that its calculated direction of magnetization provides a virtual geomagnetic pole (VGP) position. Virtual geomagnetic poles calculated in this way for Pacific seamounts provided a Cretaceous polar-wander curve for the Pacific plate relative to the Earth's rotation axis and independent of data from other plates (Francheteau and others, 1970).

Matrix inversion. Bott (1967) described a simple extension to the above method and calculated the distribution of magnetization in a crustal section. He divided the layer into N polygonal cells and assumed that the magnetization throughout the kth cell was constant and equal to M_k. The total-field anomaly at point x,y is given by h_i. Then, the integral equation (2) becomes a matrix equation:

$$h_i = \sum_{k=1}^{N} M_k \phi_{ik} \qquad i = 1, 2, \ldots, L \qquad L > N, \tag{9}$$

where ϕ_{ik} is the field at point i due to cell k with unit magnetization. Bott's objective was to calculate a set of M_k such that h_i was as close as possible in a least-squares sense to observed anomaly values h_i'; i.e., N values of M_k are calculated so that equation (8) is minimized. Bott (1967) and Bott and Hutton (1970a) used this approach to determine magnetization of oceanic crust from ocean-surface anomalies. In this case, they could assume the cells had vertical sides and extended to infinite distances parallel to the linear sea-floor–spreading anomalies.

A series of papers (Emilia and Bodvarsson, 1969; Bott and Hutton, 1970b; Emilia and Bodvarsson, 1970; and Van den Akker and others, 1970) nicely describe the instabilities inherent in equation (9) and in all inverse methods. If the cells are made too small relative to their depths (i.e., if too much resolution is attempted), an instability in the calculated magnetization results: values of M_k vary unrealistically from cell to cell even though the calculated anomaly is in excellent agreement with the observed anomaly.

Divide by Earth filter. Equation (4) shows a simple way to calculate $M(x, y)$ from $h(x, y)$ if the magnetization is confined to a horizontal layer: (1) Fourier transform $h(x, y)$; (2) divide by $F[\phi(x, y)]$; and (3) inverse Fourier transform, i.e.,

$$M(x, y) = F^{-1} \left[\frac{F[h(x, y)]}{F[\phi(x, y)]} \right] \tag{10}$$

(Schouten and McCamy, 1972). In equation (10), we have used F^{-1} to denote the inverse Fourier transform. Equation (10) is very ill-behaved because $F[\phi(x, y)]$ is zero at $k_x = k_y = 0$ and also approaches zero for large wavenumbers. These problems can be treated by bandpass filtering $1/F[\phi(x, y)]$, but instabilities are still inherent (Blakely and Schouten, 1974).

Parker-Huestis method. The above approach is useful for situations where the magnetization is confined to a horizontal layer. Parker and Huestis (1974) described an iterative technique for the case where topography exists and where vertical thickness of the magnetic layer is uniform. By requiring that $t = z_t(x, y) - z_b(x, y)$ be a constant, equation (6) can be rewritten as:

$$F[M(x, y)] = \frac{F[h(x, y)]}{\theta\,(k_x, k_y)\,(1 - e^{-kt})} + \sum_{n=1}^{\infty} \frac{(-k)^n}{n!} F[M(x, y)\, z_t^n\,(x, y)]. \quad (11)$$

Equation (11) can be solved iteratively: an initial guess is made for $M(x, y)$ on the right side of equation (11) in order to determine a new $M(x, y)$. The new $M(x, y)$ is then used on the right side of equation (11) to find the next iteration. Iterations continue until the solution meets some convergence criterion. This inverse method has been very useful for deeply towed marine magnetic surveys where the magnetic effects of topography cannot be ignored (e.g., Macdonald and others, 1983).

Parker and Huestis (1974) also presented an iterative technique to calculate the *annihilator* for the above source geometry. The annihilator is a distribution of magnetization that, for the particular source geometry, produces no magnetic field. It therefore represents the nonuniqueness of the inverse problem: any amount of the annihilator can be added to the solution without changing the calculated anomaly. For example, an infinitely extended, uniformly magnetized layer produces no magnetic field. Hence, $M(r) = M_o$ is an annihilator for the magnetized layer, and techniques that determine the distribution of magnetization within a layer—such as equation (10)—can provide the magnetization only to within an additive constant.

Nonlinear inverse techniques

Potential fields are rather complicated functions (involving logarithms and arctangents) of the coordinates of the boundary of the sources. Hence, determination of the shape, size, or location of the source from observed anomaly values is a nonlinear inverse problem and more difficult to solve than the linear problem. In fact, nonlinear techniques often use assumptions that replace nonlinear equations with linearized approximations.

Shapes of isolated bodies. Least-squares and Taylor's series approximations can be used iteratively to determine the shape of magnetic sources from measured anomalies (Johnson, 1969; McGrath and Hood, 1973). The total-field anomaly h_i at points along a profile are assumed to be due to two-dimensional prisms, with polygonal cross sections and infinitely extended perpendicular to the profile. The anomaly at each observation point is a nonlinear function of the coordinates of the corners of the polygon,

$$h_i = h_i\,(x_1, z_1, x_2, z_2, \ldots x_L, z_L).$$

However, h_i can be approximated as a linear function by expanding in a Taylor's series and dropping high-order terms. This linear approximation is compared with observed values h_i' using least-squares techniques in order to determine best adjustments for corner coordinates. The corners are adjusted accordingly, and a new comparison with h_i' is made. This least-squares comparison is repeated until some convergence criterion is met. Algorithms for this type of solution are readily available (Webring, 1985). Interpretations using this approach should be cautious; solutions may depend on the initial guess at corner coordinates or on the presence of local minima in E^2.

Depth-to-source calculations. Calculations of depth to magnetic sources can be divided into two types: those that directly examine the shapes of magnetic anomalies (e.g., Naudy, 1971; O'Brien, 1972; Phillips, 1979), and those that analyze the statistical properties of ensembles of anomalies (e.g., Spector and Grant, 1970). In either case, simplifying assumptions for the distribution of magnetization must be made.

Several exceptionally simple techniques use the first approach. Peters (1949) showed, for example, that certain simple sources produce anomalies with gradients that are approximately uniform over horizontal distances that depend on the source depth, and Grant and West (1965) published a set of nomograms for determining the depth of various three-dimensional magnetic bodies. A quick depth-to-source estimate is possible, therefore, with simply a ruler and contoured magnetic data.

Other techniques require computer assistance. Profiles of total-field anomalies over very thin, semi-infinite sheets have a simple form given by:

$$h(x) = \frac{A(x - x_o) + Bz_o}{(x - x_o)^2 + z_o^2},$$

where x_o and z_o coordinates of the top of the sheet, and coefficients A and B depend on the dip of the sheet, the direction of magnetization, and the direction of the regional field. The above equation has four unknowns that can be determined from four or more points of $h(x)$. Hence, if thin dikes (or abrupt interfaces) are an appropriate model for the composition of the magnetic layer, then the above equation can be used to locate the top of the magnetic layer by examining sets of points along the profile. This approach, which was first discussed by Werner (1953), is commonly known as "Werner deconvolution." It has been further developed and automated by Hartman and others (1971) and by Ku and Sharp (1983).

An alternative approach uses the statistical properties of groups of anomalies as viewed in the Fourier domain. If magnetic sources are confined between depths z_t and z_b, then the power-density spectrum of the magnetic field is given by the product of the earth filter and the power-density spectrum of the magnetization:

$$\Phi_h\,(k_x, k_y) = \alpha\,(e^{-kz_t} - e^{-kz_b})^2\,\Phi_M\,(k_x, k_y),$$

where α is a constant dependent on the direction of magnetization and direction of regional field. The magnetization usually is assumed to have a high degree of randomness so that $\Phi_M(k_x, k_y)$ is constant. Then

$$\Phi_h(k_x, k_y) = A\,(e^{-kz_t} - e^{-kz_b})^2, \tag{12}$$

where A is a constant, and z_t and z_b can be estimated from the power-density spectrum of the total-field anomaly. This approach has been used to estimate depth to magnetic sources either from profile data (Treitel and others, 1971; Blakely and Hassanzadeh, 1981) or from gridded, two-dimensional data (Spector and Grant, 1970; Connard and others, 1983).

Depth to bottom of sources. In certain geologic situations, the depth to the bottom of magnetic sources corresponds to the depth to the Curie-temperature isotherm. Magnetic data, therefore, can help determine the thermal structure of upper crustal rocks (Bhattacharyya and Leu, 1975; Connard and others, 1983). This is an exceedingly difficult task because the bottom of magnetic sources contributes very low-amplitude and very long-wavelength components to the observed anomaly as compared to shallower sources.

Bhattacharyya and Leu (1977) determined depth to the bottom of individual sources by examination of isolated anomalies. They calculated the location of the "centroid" (center of mass) of each magnetic source from the moments of their respective anomalies. If the depth to the top of the source is also known, perhaps determined with the method of Spector and Grant (1970), the depth to the bottom of the source is easily calculated. This method depends on isolating individual anomalies from all surrounding anomalies and regional fields, a very difficult requirement.

Connard and others (1983) applied equation (12) to aeromagnetic data over the Cascade Range of Oregon in order to estimate depth to the bottom of magnetic sources. They divided the survey into overlapping cells and attempted to determine an average z_b for each cell. Plotted as position, the various values of z_b provided a crude map showing presumed depths to the Curie-temperature isotherm.

Extremal solutions. Rather than attempt to define the shape of magnetic sources, we might more safely find a lower bound on some parameter of the source. Huestis and Parker (1977), for example, found extremal solutions for the distribution of magnetization using techniques of linear programming. Such solutions minimize some mathematical measure of the magnetization, called a norm, and at the same time satisfy observed anomaly values. Magnetization distributions thus determined are not treated as "real"; they are used to estimate certain key parameters about the geometry of the source. For example, because a thin magnetic layer implies high magnetization, a minimum layer thickness can be determined by deciding on a geophysically reasonable range of values for the magnetization.

Data enhancement techniques

In many situations, much can be learned directly from gridded magnetic data without explicitly modeling the anomalies. The product of this class of techniques is usually a map, often contoured, that accentuates certain attributes of the source. Most of these techniques can be applied in either the space or Fourier domain. The latter approach is often more efficient and instructive, but Fourier transforms have important limitations: irregular survey boundaries and gaps in coverage are difficult to contend with, large anomalies or noise in one part of the survey tend to propagate to other parts of the survey, and the entire survey must be analyzed at once (Gibert and Galdeano, 1985).

Wavelength filters. The Fourier domain is a convenient place to apply filters that are designed to eliminate undesirable anomalies rich in particular wavelengths. Low-pass filters are sometimes applied to eliminate short-wavelength noise. Short-wavelength terrain effects, for example, could be eliminated with a low-pass filter. As is true for all wavelength filters, real anomalies with wavelengths similar to the noise will also be rejected by the filter. Besides the obvious application to eliminate noise, both high- and low-pass filters are sometimes used to distinguish anomalies based on their characteristic wavelengths.

Upward and downward continuation (Fig. 1d). Upward continuation is the process of transforming potential field data measured on one surface to some other higher surface. No assumptions are required about specific sources. Because the data are being "moved" away from the sources, this procedure is mathematically stable. Upward continuation, a type of filtering operation, can be performed either in the space domain as a convolution (Grant and West, 1965, p. 311) or in the Fourier domain as multiplication (Hildenbrand, 1983; Bhattacharyya and Chan, 1977). If the magnetic field is measured on a horizontal surface and is desired on a higher horizontal surface, the upward continuation field in the Fourier domain is given by:

$$F[h_U(x, y)] = F[h(x, y)]\,e^{-k\Delta z} \qquad \Delta z > 0, \tag{13}$$

where Δz is the upward continuation distance (Hildenbrand, 1983; Blakely, 1977). Equation (13) shows that upward continuation attenuates all wavelengths (except $k = 0$); the higher the wavenumber, the greater the attenuation. Hence, upward continuation suppresses the shortest wavelengths of a magnetic anomaly (Fig. 1d). Because shallow sources produce shorter wavelength anomalies than deeper ones, upward continuation can be considered a way to accentuate anomalies due to deeper sources, at the expense of shallower ones.

Equation (13) is not valid if the field is measured on an uneven surface and/or desired on an uneven surface. Grauch (1984) described a way to accomplish uneven surface continuation using a Taylor's series approximation where each element of the series includes a Fourier transform of the data. Parker and

Klitgord (1972) described a method, applicable only to profile data, which makes use of the Schwarz-Christoffel transformation.

It would seem from equation (13) that *downward* continuation could be achieved by simply changing the sign of the exponential; i.e.,

$$F[h_D(x, y)] = F[h(x, y)]\, e^{+k\Delta z} \qquad \Delta z > 0. \tag{14}$$

However, the exponential increase of the downward continuation filter amplifies short-wavelength noise as well as short-wavelength anomalies, and the procedure is highly unstable.

Reduction to the pole (Fig. 1c). The shape of a magnetic anomaly depends not only on the shape of the perturbing body, but also on the direction of magnetization and the direction of the regional field. Reduction to the pole transforms an observed anomaly into the anomaly that would be observed if the magnetization and regional field were vertical (as they would be for total-field measurements over a body with induced magnetization at the north magnetic pole). A symmetric body produces a symmetric anomaly if both of these vectors are vertical. Hence, reduction to the pole is a way to remove the asymmetries caused by nonvertical magnetization or regional field and produce a simpler set of anomalies to interpret. In particular, anomalies are more nearly centered over their respective causative bodies (Fig. 1c).

Reduction to the pole can be achieved either in the space domain by a convolution (Grant and West, 1965, p. 314–315) or in the Fourier domain according to:

$$F[h_p(x, y)] = F[h(x, y)]\, \frac{2\pi}{\theta\,(k_x, k_y)} \tag{15}$$

(Hildenbrand, 1983; Blakely, 1977), where $\theta\,(k_x, k_y)$ is the same function that appears in equations (5) and (6). The reduction-to-pole filter is ill-defined near $k_x = k_y = 0$ so certain instabilities can occur, especially if the original inclinations are shallow (Silva, 1986).

A symmetric body also produces a symmetric (albeit inverted) anomaly if both the direction of magnetization and regional field are horizontal (as they would be for induced magnetization at the equator). Hence, "reduction to the equator" may be useful for surveys measured at low latitudes where reduction to the pole may be mathematically unstable (Silva, 1986; Gibert and Galdeano, 1985).

Vertical derivatives (Fig. 1e). Vertical derivatives amplify short-wavelength information at the expense of long-wavelength information. Vertical-derivative maps, usually the second vertical derivative, show high gradients along edges of magnetic sources. Hence, they are sometimes used to locate edges of magnetic bodies. Vertical derivatives can be calculated either in the space domain (Grant and West, 1965, p. 221) or in the Fourier domain. The second vertical derivative, for example, is given by:

$$F[\frac{d^2}{dz^2}\, h(x, y)] = F[h(x, y)]\, k^2.$$

Pseudogravity transformation (Fig. 1f, g, h). Poisson's relation states that a body with uniform magnetization and density produces a magnetic field proportional to the derivative of the gravity field. It is a straightforward procedure, therefore, to calculate from magnetic data the gravitational attraction that would be observed if the magnetization distribution were replaced by a density distribution of exact proportions. This calculation, called the pseudogravity transform (Fig. 1f), can be applied either in the space domain (Baranov, 1957) or in the Fourier domain (Hildenbrand, 1983).

Thus Poisson's relation provides a way to compare magnetic data with gravity data in the case where magnetic sources are also expected to be mass concentrations (e.g., Cordell and Taylor, 1971; Chandler and others, 1981). Poisson's relation also provides a way to locate edges of magnetic bodies. In this case, the existence of an anomalous mass distribution is irrelevant; Poisson's relation is used to convert magnetic anomalies into hypothetical gravity anomalies only because certain properties of gravity anomalies are easier to analyze. In particular, the steepest gradient of a gravity anomaly usually is located approximately over the edges of gravity sources, as shown by Figure 1g, h (Heiland, 1963; Cordell and Grauch, 1982; Blakely and Simpson, 1986).

PITFALLS

We stressed earlier the difficulty presented by the nonuniqueness property of magnetic fields, but several other potential pitfalls should be avoided in magnetic interpretation. The temptation exists, for example, to consider contoured magnetic data as a rudimentary geology map; i.e., to interpret positive anomalies as magnetic rock units and negative anomalies as nonmagnetic or reversely magnetized units. Magnetization, however, is a vector quantity (unlike density, for example, which is a scalar quantity). Consequently, the shape of a magnetic anomaly depends on the direction of magnetization as well as on the shape of the source. In some situations, remanent magnetization can be assumed to be negligible ($Q = 0$) so that the total magnetization is parallel to the Earth's field, but remanent magnetization is often an appreciable or dominant component of the total magnetization, and its direction is often unknown.

Moreover, the vector nature of magnetization produces so-called edge-effect or polarization anomalies. At northern latitudes, for example, an isolated normally magnetized body will produce a positive anomaly located approximately over the center of the body and a negative anomaly located slightly to the north of the body (see Fig. 1b). Edge-effect anomalies should not be treated as separate isolated sources.

Removal of regional fields poses another difficult problem. The total magnetic field measured near the Earth's surface has its sources in the fluid core of the Earth and in that part of the crust shallower than the Curie-temperature isotherm (typically located 25 km below the Earth's surface in tectonically stable continental regions). We attempt to remove the core field by subtraction of the proper geomagnetic reference field, as discussed earlier, so that remaining magnetic anomalies represent sources in the crust only. It is frequently necessary, however, to isolate anomalies of particular interest from anomalies due to nearby or regional sources. For example, the anomaly caused by the Trinity ultramafic sheet in northern California is very broad and contaminates the anomaly caused by Mount Shasta, a nearby Quaternary volcano. Modeling experiments of the anomaly over Mount Shasta should be preceded by removal of that part of the anomaly caused by the Trinity ultramafic sheet. Removal of regional fields is clearly an interpretive process in itself.

Finally, topographic features in magnetic terranes (e.g., relatively young and unmetamorphosed volcanic regions) can cause significant anomalies in magnetic surveys that tend to obscure other anomalies of interest. Removal of effects of magnetic topography is more difficult than gravity terrain corrections; magnetizations vary by orders of magnitude (and by sign) from point to point and from rock type to rock type, whereas densities vary typically by only a few percent. Terrain effects have been studied by direct modeling (Plouff, 1976), by correlation with terrain models (Blakely and Grauch, 1983), by comparison with terrain models in the Fourier domain (Blakely, 1985), and by intelligent filters (Grauch, 1985).

FUTURE DIRECTIONS

At present, uncertainties in navigation and diurnal variations are the largest sources of error in magnetic anomaly measurements. Positioning systems based on satellites are expected to improve both the accuracy and cost of navigation.

Gradient measurements, which have the advantage of being independent of temporal variations, are gaining in popularity and may eventually replace total-field surveys as the standard for crustal studies (Donovan and others, 1984; Korhonen, 1984). Development of new airborne sensors, such as the SQUID magnetometer, promise more sensitive instruments and three-component gradiometers in small packages.

Several factors ensure a future for magnetic interpretations. First, digital magnetic, topographic, and bathymetric data are more available to the public than ever before through the National Oceanic and Atmospheric Administration and the U.S. Geological Survey. Second, several excellent regional compilations of magnetic data at statewide and continent-wide scales (e.g., Hinze and Zietz, 1985; U.S. Geological Survey and Society of Exploration Geophysicists, 1982; Hood and others, 1985; Dods and others, 1985; Bond and Zietz, 1986) have recently been published and will continue to appear. Third, interpretation algorithms usually suitable for microcomputers continue to be developed for rapid and accurate interpretation of large quantities of magnetic measurements. Future integration of magnetic interpretations with geologic and other geophysical information will promote a better understanding of the geology and tectonic history of the Earth's crust.

ACKNOWLEDGMENTS

We thank Robert Simpson, Robert Jachens, William Hinze, Tom Le Fehr, Marco Polo Pereria da Boa Hora, Richard Lu, and Richard Couch for critical reviews of this manuscript. We also thank Walter Mooney for his advice and comments.

REFERENCES CITED

Baranov, V., 1957, A new method for interpretation of aeromagnetic maps; Pseudogravimetric anomalies: Geophysics, v. 22, p. 359–383.

Barnett, C. T., 1976, Theoretical modeling of the magnetic and gravitational fields of an arbitrarily shaped three-dimensional body: Geophysics, v. 41, p. 1353–1364.

Bhattacharyya, B. K., 1969, Bicubic spline interpolation as a method for treatment of potential field data: Geophysics, v. 34, p. 402–423.

Bhattacharyya, B. K., and Chan, K. C., 1977, Reduction of magnetic and gravity data on an arbitrary surface acquired in a region of high topographic relief: Geophysics, v. 42, p. 1411–1430.

Bhattacharyya, B. K., and Leu, L.-K., 1975, Analysis of magnetic anomalies over Yellowstone National Park; Mapping of Curie point isothermal surface for geothermal reconnaissance: Journal of Geophysical Research, v. 80, no. 32, p. 4461–4465.

——, 1977, Spectral analysis of gravity and magnetic anomalies due to two-dimensional structure: Geophysics, v. 42, no. 1, p. 41–50.

Blakely, R. J., 1977, Documentation for subroutine REDUC3, an algorithm for the linear filtering of gridded magnetic data: U.S. Geological Survey Open-File Report 77-784, 27 p.

——, 1981, A program for rapidly computing the magnetic anomaly over digital topography: U.S. Geological Survey Open-File Report 81-298, 46 p.

——, 1985, The effect of topography on aeromagnetic data in the wavenumber domain; Proceedings of the International Meeting on Potential Fields in Rugged Topography, Bulletin No. 7; Lausanne, Switzerland, Institut de Géophysique de Université de Lausanne, p. 102–109.

Blakely, R. J., and Cox, A., 1972, Identification of short polarity events by transforming marine magnetic profiles to the pole: Journal of Geophysical Research, v. 77, no. 23, p. 4339–4349.

Blakely, R. J., and Grauch, V.J.S., 1983, Magnetic models of crystalline terrane; Accounting for the effect of topography: Geophysics, v. 48, p. 1551–1557.

Blakely, R. J., and Hassanzadeh, S., 1981, Estimation of depth to magnetic source using maximum entropy power spectra, with application to the Peru-Chile Trench, *in* Nazca plate; Crustal formation and Andean convergence: Geological Society of America Memoir 154, p. 667–682.

Blakely, R. J., and Schouten, H., 1974, Comments on paper by Hans Schouten and Keith McCamy, "Filtering marine magnetic anomalies": Journal of Geophysical Research, v. 79, no. 5, p. 773–774.

Blakely, R. J., and Simpson, R. W., 1986, Approximating edges of source bodies from magnetic or gravity anomalies: Geophysics, v. 51, no. 7, p. 1494–1498.

Blakely, R. J., Cox, A., and Iufer, E. J., 1973, Vector magnetic data for detecting short polarity intervals in marine magnetic profiles: Journal of Geophysical Research, v. 78, p. 6977–6983.

Bond, K. R., and Zietz, I., 1986, Composite magnetic anomaly map of the western United States west of 96° longitude: U.S. Geological Survey Geophysical Investigations Map GP-977.

Bott, M.H.P., 1963, Two methods applicable to computers for evaluating magnetic anomalies due to finite three dimensional bodies: Geophysical Prospecting, v. 11, no. 3, p. 292–299.

——, 1967, Solution of the linear inverse problem in magnetic interpretation with application to oceanic magnetic anomalies: Geophysical Journal of the Royal Astronomical Society, v. 13, p. 313–323.

Bott, M.H.P., and Hutton, 1970a, A matrix method for interpreting oceanic magnetic anomalies: Geophysical Journal of the Royal Astronomical Society, v. 20, p. 149.

——, 1970b, Limitations on the resolution possible in the direct interpretation of marine magnetic anomalies: Earth and Planetary Science Letters, v. 8, p. 317.

Bracewell, R., 1965, The Fourier transform and its application: New York, McGraw-Hill, 381 p.

Briggs, I. C., 1974, Machine contouring using minimum curvature: Geophysics, v. 39, p. 39–48.

Cady, J. W., 1980, Calculation of gravity and magnetic anomalies of finite-length right polygonal prisms: Geophysics, v. 45, no. 10, p. 1507–1512.

Carmichael, R. S., 1982, Handbook of physical properties of rocks: Boca Raton, Florida, CRC Press, v. 2, 345 p.

Chandler, V. W., Koski, J. S., Hinze, W. J., and Braile, L. W., 1981, Analysis of multisource gravity and magnetic anomaly data sets by moving-window application of Poisson's theorem: Geophysics, v. 46, p. 30–39.

Clark, S. P., 1966, Handbook of physical constants: Geological Society of America Memoir 97, 587 p.

Connard, G. G., 1979, Analysis of aeromagnetic measurements from the central Oregon Cascades [Master's thesis]: Corvallis, Oregon State University, 101 p.

Connard, G. G., Couch, R. W., and Gemperle, M., 1983, Analysis of aeromagnetic measurements from the Cascade Range in central Oregon: Geophysics, v. 48, p. 376–390.

Cordell, L., and Grauch, V.J.S., 1982, Mapping basement magnetization zones from aeromagnetic data in the San Juan basin, New Mexico, *in* W. J. Hinze, ed., The utility of regional gravity and magnetic anomaly maps: Society of Exploration Geophysicists, p. 181–197.

Cordell, L., and Knepper, D. H., 1987, Aeromagnetic images; Fresh insight to the buried basement, Rolla quadrangle, southeast Missouri: Geophysics, v. 52, p. 218–231.

Cordell, L., and Taylor, P. T., 1971, Investigation of magnetization and density of a North Atlantic seamount using Poisson's theorem: Geophysics, v. 36, p. 919–937.

Criss, R. E., and Champion, D. E., 1984, Magnetic properties of granitic rocks from the southern half of the Idaho Batholith; Influences of hydrothermal alteration and implications for aeromagnetic interpretations: Journal of Geophysical Research, v. 89, p. 7061–7076.

Dods, S. D., Teskey, D. J., and Hood, P. J., 1985, The new series of 1:1,000,000-scale magnetic anomaly maps of the Geological Survey of Canada; Compilation techniques and interpretation, *in* Hinze, W. J., ed., The utility of regional gravity and magnetic anomaly maps: Society of Exploration Geophysicists, p. 69–87.

Donovan, T. J., Hendricks, J. D., Roberts, A. A., and Eliason, P. T., 1984, Low-altitude aeromagnetic reconnaisance for petroleum in the Arctic National Wildlife Refuge, Alaska: Geophysics, v. 49, p. 1338–1353.

Eggers, D. E., and Thompson, D. T., 1984, An evaluation of the marine magnetic gradiometer: Geophysics, v. 49, p. 771–778.

Emilia, D. A., and Bodvarsson, G., 1969, Numerical methods in the direct interpretation of marine magnetic anomalies: Earth and Planetary Science Letters, v. 7, p. 194.

——, 1970, More on the direct interpretation of magnetic anomalies: Earth and Planetary Science Letters, v. 8, p. 320–321.

Foster, M. R., Jines, W. R., and Weg, K. V., 1970, Statistical estimation of systematic errors at intersections of lines of aeromagnetic survey data: Journal of Geophysical Research, v. 75, p. 1507–1511.

Francheteau, J., Harrison, C.G.A., Sclater, J. G., and Richards, M. L., 1970, Magnetization of Pacific seamounts; A preliminary polar curve for the northeastern Pacific seamounts; A preliminary polar curve for the northeastern Pacific: Journal of Geophysical Research, v. 15, p. 2035–2062.

Frost, B. R., and Shive, P. N., 1986, Magnetic mineralogy of the lower continental crust: Journal of Geophysical Research, v. 91, p. 6513–6521.

Gibert, D., and Galdeano, A., 1985, A computer program to perform transformations of gravimetric and aeromagnetic surveys: Computers and Geosciences, v. 11, p. 553–588.

Grant, F. S., and West, G. S., 1965, Interpretation theory in applied geophysics: New York, McGraw-Hill, 584 p.

Grauch, V.J.S., 1984, TAYLOR; A Fortran program using Taylor series expansion for level-surface or surface-level continuation of potential field data: U.S. Geological Survey Open-File Report 84-501, 31 p.

——, 1985, A new magnetic-terrain correction method for aeromagnetic mapping; Proceedings of the International Meeting on Potential Fields in Rugged Topography, Bulletin No. 7: Lausanne, Switzerland, Institut de Géophysique de Université de Lausanne, p. 110–113.

Grauch, V.J.S., and Campbell, D. L., 1984, Does draping aeromagnetic data reduce terrain-induced effects?: Geophysics, v. 49, p. 75–80.

Hansen, R. O., 1984a, Discussion on: "Does draping aeromagnetic data reduce terrain-induced effect?" by Grauch, V.J.S., and Campbell, D. L.: Geophysics, v. 49, p. 2070–2071.

——, 1984b, Two approaches to total field reconstruction from gradiometer data: Presented at 54th Annual Society of Exploration Geophysicists Meeting, Atlanta, Georgia, p. 245–247.

Hardwick, C. D., 1984, Important design considerations for inboard airborne magnetic gradiometers: Geophysics, v. 49, p. 2004–2018.

Hartman, R. R., Tesky, D. J., and Friedberg, L. L., 1971, A system for rapid digital aeromagnetic interpretation: Geophysics, v. 36, p. 891–918.

Heiland, C. A., 1935, Geophysical mapping from the air; Its possibilities and advantages: Engineering and Mining Journal, v. 136, p. 609–610.

——, 1963, Geophysical exploration: New York, Hafner Publishing Co., 1013 p.

Hildenbrand, T. G., 1983, FFTFIL; A filtering program based on two-dimensional Fourier analysis: U.S. Geological Survey Open-File Report 83-237, 61 p.

Hill, P. L., 1986a, Bibliographies and location maps of aeromagnetic and aeroradiometric publications for the states west of approximately 104° longitude (exclusive of Alaska and Hawaii): U.S. Geological Survey Open-File Report 86-525-A, 130 p.

——, 1986b, Bibliographies and location maps of aeromagnetic and aeroradiometric publications for the states west of the Mississippi River and east of approximately 104° longitude: U.S. Geological Survey Open-File Report 86-525-B, 48 p.

—— 1986c, Bibliographies and location maps of aeromagnetic and aeroradiometric publications for the states east of the Mississippi River and north of the Ohio and Potomac Rivers: U.S. Geological Survey Open-File Report 86-525-C, 90 p.

——, 1986d, Bibliographies and location maps of aeromagnetic and aeroradiometric publications for the states east of the Mississippi River and south of the Ohio and Potomac Rivers: U.S. Geological Survey Open-File Report 86-525-D, 57 p.

——, 1986e, Bibliographies and location maps of aeromagnetic and aeroradiometric publications for Alaska and Hawaii: U.S. Geological Survey Open-File Report 86-525-E, 130 p.

——, 1986f, Bibliographies and location maps of aeromagnetic and aeroradiometric publications for Puerto Rico and large areas of the conterminous United States: U.S. Geological Survey Open-File Report 86-525-F, 29 p.

——, 1986g, Lists and location maps of aeromagnetic and aeroradiometric publications from the Department of Energy NURE Program: U.S. Geological Survey Open-File Report 86-525-G.

Hinze, W. J., ed., 1985, The utility of regional gravity and magnetic anomaly maps: Society of Exploration Geophysicists, 454 p.

Hinze, W. J., and Zietz, I., 1985, The composite magnetic-anomaly map of the conterminous United States, *in* W. J. Hinze, ed., The utility of regional gravity and magnetic anomaly maps: Society of Exploration Geophysicists, p. 1–24.

Hood, P. J., 1980, Aeromagnetic gradiometry; A superior geological mapping tool for mineral exploration programs; Proceedings of the Workshop on SQUID Applications to Geophysics, June 1980: Los Alamos, New Mexico,

Los Alamos National Laboratory, 73 p.

Hood, P. J., McGrath, P. H., and Teskey, D. J., 1985, Evolution of Geological Survey of Canada magnetic-anomaly maps; A Canadian perspective, *in* W. J. Hinze, ed., The utility of regional gravity and magnetic anomaly maps: Society of Exploration Geophysicists, p. 62–68.

Huestis, S. P., and Parker, R. L., 1977, Bounding the thickness of the oceanic magnetized layer: Journal of Geophysical Research, v. 82, no. 33, p. 5293–5303.

IAGA Division I, Working Group 1, 1986, International Geomagnetic Reference Field revision 1985: EOS Transaction of the American Geophysical Union, v. 67, p. 523–524.

Jakosky, J. J., 1940, Exploration geophysics: Los Angeles, Trija Publishing Co., 1195 p.

Jensen, H., 1961, The airborne magnetometer: Scientific American, v. 204, p. 151–162.

——, 1965, Instrument details and applications of a new airborne magnetometer: Geophysics, v. 30, p. 875–882.

Johnson, W. W., 1969, A least-squares method of interpreting magnetic anomalies caused by two-dimensional structures: Geophysics, v. 34, no. 1, p. 65–74.

Korhonen, J. V., 1984, Experience in compiling low-altitude aeromagnetic surveys: Presented at the 54th Annual Society of Exploration Geophysicists Meeting, Atlanta, Georgia, p. 242–243.

Ku, C. C., and Sharp, J. A., 1983, Werner deconvolution for automated magnetic interpretation and its refinement using Marquardt's inverse modeling: Geophysics, v. 48, p. 754–774.

Langel, R. A., 1985, Introduction to the special issue; A perspective on MAGSAT results: Journal of Geophysical Research, v. 90, p. 2441–2444.

Macdonald, K. C., Miller, S. P., Luyendyk, B. P., Atwater, T. M., and Shure, L., 1983, Investigation of a Vine-Matthews magnetic lineation from a submersible; The source and character of marine magnetic anomalies: Journal of Geophysical Research, v. 88, p. 3403–3418.

Mayhew, M. A., and LaBrecque, J. L., 1987, Crustal geologic studies with Magsat and surface magnetic data: Reviews of Geophysics and Space Physics, p. 971–981.

McGrath, P. H., and Hood, P. J., 1973, An automated least-squares multimodel method for magnetic interpretation: Geophysics, v. 38, no. 2, p. 349–358.

Mittal, P. K., 1984, Algorithm for error adjustment of potential field data along a survey network: Geophysics, v. 49, p. 467–469.

Naudy, H., 1971, Automatic determination of depth on aeromagnetic profiles: Geophysics, v. 36, p. 717–722.

Needham, J., 1962, Science and civilisation in China; v. 4, Physics and physical technology; Part 1, Physics: Cambridge, Cambridge University Press, 434 p.

O'Brien, D. P., 1972, CompuDepth; A new method for depth-to-magnetic basement computation: Presented at the 42nd Annual Society of Exploration Geophysicists Meeting, Annaheim, California, p. 26–27.

Okabe, M., 1979, Analytical expressions for gravity anomalies due to homogeneous polyhedral bodies and translations into magnetic anomalies: Geophysics, v. 44, p. 730–741.

Olea, R. A., 1974, Optimal contour mapping using universal kriging: Journal of Geophysical Research, v. 79, p. 695–702.

Parker, R. L., 1971, The determination of seamount magnetism: Geophysical Journal of the Royal Astronomical Society, v. 24, no. 4, p. 321–324.

——, 1973, The rapid calculations of potential anomalies: Geophysical Journal of the Royal Astronomical Society, v. 31, no. 4, p. 447–455.

——, 1977, Understanding inverse theory: Annual Reviews of Earth and Planetary Sciences, v. 5, p. 35–64.

Parker, R. L., and Huestis, S. P., 1974, The inversion of magneitc anomalies in the presence of topography: Journal of Geophysical Research, v. 79, no. 11, p. 1587–1593.

Parker, R. L., and Klitgord, K. D., 1972, Magnetic upward continuation from an uneven track: Geophysics, v. 37, no. 4, p. 662–668.

Payne, M. A., 1981, SI and Gaussian CGS units, conversions and equations for use in geomagnetism: Physics of the Earth and Planetary Interiors, v. 26, p. 10–16.

Paterson, N. R., and Reeves, C. V., 1985, Applications of gravity and magnetic surveys; The state-of-the-art in 1985: Geophysics, v. 50, p. 2558–2594.

Pearson, W. C., Crosby, R. O., and Parker, R. L., 1985, Gravity/magnetics; 3, Techniques: Presented at the 55th Annual Society of Exploration Geophysicists Meeting, Washington, D.C., p. 212–215.

Peddie, N. W., 1983, International geomagnetic reference field and its evolution and the difference in total field intensity between new and old models for 1965–1980: Geophysics, v. 48, p. 1691–1696.

Pelto, C. R., Elkins, T. A., and Boyd, H. A., 1968, Automatic contouring of irregularly spaced data: Geophysics, v. 33, p. 424–430.

Peters, L. J., 1949, The direct approach to magnetic interpretation and its practical applications: Geophysics, v. 14, p. 290–320.

Petrick, W. R., Stodt, J. A., Olsen, S. L., and Sorenson, J. L., 1984, High-resolution aeromagnetics using ultralight aircraft: Presented at the 54th Annual Society of Exploration Geophysicists Meeeting, Atlanta, Georgia, p. 243–244.

Phillips, J. D., 1979, ADEPT; A program to estimate depth to magnetic basement from sampled magnetic profiles: U.S. Geological Survey Open-File Report 79-367, 35 p.

Plouff, D., 1976, Gravity and magnetic fields of polygonal prisms and application to magnetic terrain corrections: Geophysics, v. 41, p. 727–741.

Rasmussen, R., and Pedersen, L. B., 1979, End corrections in potential field modeling: Geophysical Prospecting, v. 27, p. 749–760.

Reford, M. S., 1980, Magnetic method: Geophysics, v. 45, p. 1640–1658.

Reid, A. B., 1980, Aeromagnetic survey design: Geophysics, v. 45, p. 973–976.

Saltus, R. W., and Blakely, R. J., 1983, Hypermag; An interactive, two-dimensional gravity, and magnetic modeling program: U.S. Geological Survey Open-File Report 83-241, 91 p.

Sander, E. L., and Mrazak, C. P., 1982, Regression technique to remove temporal variation from geomagnetic data: Geophysics, v. 47, p. 1437–1443.

Schlinger, C. M., 1985, Magnetization of lower crust and interpretation of regional crustal anomalies; Example from Lofoten and Vesterålen, Norway: Journal of Geophysical Research, v. 90, p. 11484–11504.

Schouten, H., and McCamy, K., 1972, Filtering marine magnetic anomalies: Journal of Geophysical Research, v. 77, no. 35, p. 7089–7099.

Schouten, J. A., 1971, A fundamental analysis of magnetic anomalies over oceanic ridges: Marine Geophysical Researches, v. 1, p. 111–114.

Sexton, J. L., Hinze, W. J., von Frese, R.R.B., and Braile, L. W., 1982, Long-wavelength aeromagnetic anomaly map of the conterminous United States: Geology, v. 10, p. 364–369.

Shive, P. N., 1986, Suggestions for the use of SI units in magnetism: EOS Transactions of the American Geophysical Union, v. 67, p. 25.

Shuey, R. T., and Pasquale, A. S., 1973, End corrections in magnetic profile interpretation: Geophysics, v. 38, p. 507–512.

Silva, J.B.C., 1986, Reduction to the pole as an inverse problem and its application to low-latitude anomalies: Geophysics, v. 51, p. 369–382.

Singleton, R. C., 1969, An algorithm for computing the mixed radix fast Fourier transform: Institute of Electrical and Electronic Engineers Transactions, Audio and Electroacoustics, v. AU-17, p. 93–103.

Spector, A., and Grant, F. S., 1970, Statistical models for interpreting aeromagnetic data: Geophysics, v. 35, p. 293–302.

Spiess, F. N., and Mudie, J. D., 1970, Small-scale topographic and magnetic features of the sea floor, *in* Maxwell, A. E., ed., The sea, v. 4, part 1: New York, John Wiley & Sons, p. 205–251.

Spradley, L. H., 1985, Surveying and navigation for geophysical exploration: Boston International Human Resources Development Corporation, 289 p.

Talwani, M., 1965, Computation with the help of a digital computer of magnetic anomalies caused by bodies of arbitrary shape: Geophysics, v. 30, no. 5, p. 797–817.

Talwani, M., and Heirtzler, J. R., 1964, Computation of magnetic anomalies caused by two-dimensional structures of arbitrary shape, *in* Computers in the mineral industries: Stanford, California, Stanford University Publications of the Geological Sciences, v. 9, pt. 1, p. 464–480.

Treitel, S., Clement, W. G., and Kaul, R. K., 1971, The spectral determination of depths to buried magnetic basement rocks: Geophysical Journal of the Royal Astronomical Society, v. 24, p. 415–429.

U.S. Geological Survey and Society of Exploration Geophysicists, 1982, Composite magnetic anomaly map of the United States; Part A, Conterminous United States: U.S. Geological Survey Geophysical Investigations Map GP–954–A, scale 1:1,250,000, 2 sheets.

Vacquier, V., Steenland, N. C., Henderson, R. G., and Zietz, I., 1951, Interpretation of aeromagnetic maps: Geological Society of American Memoir 47, 151 p.

Van den Akker, F. B., Harrison, C.G.A., and Mudie, J. D., 1970, Even more on the direct interpretation of magnetic anomalies: Earth and Planetary Science Letters, v. 9, p. 405.

Vogt, P. R., Taylor, P. T., Kovacs, L. C., and Johnson, G. L., 1979, Detailed aeromagnetic investigation of the Arctic Basin: Journal of Geophysical Research, v. 83, p. 1071–1089.

Von Frese, R.R.B., Hinze, W. J., Olivier, R., and Bentley, C. R., 1986, Regional magnetic anomaly constraints on continental breakup: Geology, v. 14, p. 68–71.

Webring, M., 1985, SAKI; A Fortran program for generalized linear inversion of gravity and magnetic profiles: U.S. Geological Survey Open-File Report 85-122, 28 p.

Werner, S., 1953, Interpretation of magnetic anomalies at sheet-like bodies: Sveriges Geologiska Undersok, Ser. C. C., Arsbok 43, N: 06.

Whitman, K., and Niblett, E. R., 1961, The diurnal problem in aeromagnetic surveying in Canada: Geophysics, v. 26, p. 221–228.

Williams, M. C., Shrive, P. N., Fountain, D. M., and Frost, B. R., 1985, Magnetic properties of exposed deep crustal rocks from the Superior Province of Manitoba: Earth and Planetary Science Letters, v. 76, p. 176–184.

Yarger, H. L., 1978, Diurnal drift removal from aeromagnetic data using least squares: Geophysics, v. 46, p. 1148–1156.

Manuscript Accepted by the Society October 31, 1988

Printed in U.S.A.

Geological Society of America
Memoir 172
1989

Chapter 5

Paleomagnetic methods

J. W. Hillhouse
U.S. Geological Survey, 345 Middlefield Road, Menlo Park, California 94025

ABSTRACT

This chapter briefly summarizes the basic principles and methods of paleomagnetic research that are germane to the geophysical framework of North America. Under the assumption that the geomagnetic field is approximated by a geocentric axial dipole source, paleomagnetic data are useful for determining paleolatitudes, displacements, and azimuthal rotations of crust during the distant past. As confirmed by paleomagnetic data, the time-averaged geomagnetic field has maintained the configuration of a geocentric dipole aligned with the Earth's rotation axis since 5 Ma. Paleomagnetism and paleoclimatic evidence from older rocks, when corrected for sea-floor spreading, strongly support the axial dipole hypothesis back to Jurassic time; evidence from Proterozoic rocks is also supportive. The major pitfalls in determining valid paleomagnetic poles are insufficient sampling to ensure complete time-averaging of the nondipole field, and incomplete removal of magnetic overprints that mask the primary depositional remanent magnetization. These problems can be avoided through careful design of the sampling scheme, thorough analysis of the components of magnetization, and application of conventional reliability tests.

INTRODUCTION

Paleomagnetism, the natural remanent magnetism of rocks, provides a record of the geomagnetic field throughout geologic time. The imprint of the ancient geomagnetic field is pertinent to studies of sea-floor spreading, drift of the continents and terranes, geologic structures, and stratigraphic correlation. In the context of the geophysical framework of the United States, this review emphasizes the application of paleomagnetic methods to the tectonic history of North American orogenic belts, providing background material for articles by M. E. Beck, Jr., and R. Van der Voo (this volume).

Paleomagnetism is a relatively young discipline, which developed as part of the plate tectonics revolution in earth science. Cox and Doell (1960) summarized the early development of paleomagnetic methods and described many of the techniques that are still used in modern laboratories. Techniques and instrumentation used in paleomagnetic research were most recently reviewed by Collinson (1983). Excellent references on the application of paleomagnetism, especially to plate tectonics, are by Irving (1964), McElhinny (1973), and Tarling (1971). An up-to-date reference about the geomagnetic dynamo, paleomagnetic observations, and planetary fields was written by Merrill and McElhinny (1983). Rock magnetism, the physical basis for paleomagnetic research, is the subject of texts by Stacey and Banerjee (1974) and O'Reilly (1984).

My review begins with a discussion of the geomagnetic field and how it changes with time. Properties of the time-averaged field, as determined from paleomagnetism, are then discussed in regard to the geocentric axial dipole hypothesis. The processes of magnetization in rocks and their reliability as recorders of the ancient field are briefly surveyed. Instrumentation, sources of errors in magnetic measurements, statistical treatment of paleomagnetic data, and reliability tests are also reviewed. As a prelude to the work by Beck and Van der Voo (this volume), the tectonic applications of paleomagnetism are introduced in the concluding section.

GEOMAGNETIC FIELD AND THE AXIAL DIPOLE HYPOTHESIS

Charts of the geomagnetic field are derived from direct measurements taken at observatories and from satellite surveys. Since 1838, when C. F. Gauss first represented the global field in

Hillhouse, J. W., 1989, Paleomagnetic methods, *in* Pakiser, L. C., and Mooney, W. D., Geophysical framework of the continental United States: Boulder, Colorado, Geological Society of America Memoir 172.

mathematical terms, the field has been described in two parts: (1) as an inclined, geocentric dipole that comprises 90 percent of the total field at the Earth's surface, and (2) as an irregular nondipole field that makes up the remaining 10 percent (Merrill and McElhinny, 1983, p. 15). Both parts of the field undergo secular variation, indicated by temporal changes in the magnetic inclination, declination, and intensity at the surface of the Earth. At a given observation point, most of the yearly change in magnetic direction can be attributed to variation of the nondipole field, except at points very close to the geomagnetic poles. Nondipole features, which are usually charted as anomalies in the field intensity, tend to drift and fluctuate in strength. The geomagnetic poles, which are defined as the intersection points of the inclined dipole axis with the Earth's surface, contribute to the yearly variation, although the poles move much more slowly than the nondipole features.

The study of paleomagnetism allows description of the past form of geomagnetic field throughout geologic time. We know that the field flip-flops at irregular intervals between normal and reversed polarity states, with the cumulative amounts of time being approximately equal for both states (Cox, 1969). The polarity transitions are completed very rapidly in terms of geologic time, so that transitional field directions are rarely found in rocks despite the great number of transitions. In general, paleomagnetic studies indicate that secular variation of the geomagnetic field is strongly influenced by the rotation of the Earth, so that the dipole axis tends to wobble irregularly about the rotation axis. This is true for both polarity states. Although the position of the geomagnetic pole might differ from the geographic pole by several tens of degrees at a given instant, the long-term average position of the geomagnetic pole is nearly coincident with the geographic pole.

An analogy to geomagnetic variation is the temperature variation recorded during the last century at a city such as San Francisco. Although the records show variations in the temperature over a broad range on a daily or monthly basis, the annual mean temperature remains constant at about 13.7°C. During the last century, stability of the annual mean temperature is largely due to the constancy of the Earth's orbit about the sun. Similarly, but on a time scale of thousands of years, the geomagnetic pole exhibits a large temporal variation in position, and yet the mean position of the pole is very stable due to the constancy of the Earth's rotation axis with respect to the solid body of the Earth.

The assumption of coincident geomagnetic and geographic poles, or geocentric axial dipole hypothesis, is the basis for plate tectonic interpretations of paleomagnetic data. Under this assumption, at a given site the time-averaged geomagnetic latitude is equal to the geographic latitude. How precisely has the time-averaged geomagnetic pole matched the geographic pole throughout geologic time? When paleomagnetic poles are averaged from a worldwide distribution of volcanic rocks 2 m.y. old and younger, the mean paleomagnetic pole is within 2° of the geographic pole (Wilson and McElhinny, 1974; McElhinny and Merrill, 1975). However, a distinction must be made between time averages derived from worldwide compilations and those from a limited region. If, for example, paleomagnetic poles are averaged from a thick stratigraphic sequence of lava flows at a particular site, there is a tendency for the time-averaged geomagnetic latitude to be slightly lower than the actual geographic latitude (Wilson and Ade-Hall, 1970; Wilson and McElhinny, 1974; Merrill and McElhinny, 1983, p. 184). The difference is about 3° to 5° in the region below 60° of latitude and decreases toward zero near the poles.

The discoverer of this effect suggested that the dipole source was possibly offset from the center of the Earth (Wilson, 1970, 1971). Cox (1975) subsequently suggested that large negative nondipole features tend to persist near the equator; as the features drifted past a given site, the magnetic field vector would swing more frequently in the direction of lower inclinations. Others believe that long-term quadrupole and octupole components of the field are responsible for the geomagnetic latitude anomaly (Coupland and Van der Voo, 1980). Due to symmetry of the anomaly, it averages to zero when paleomagnetic results are combined from sites evenly distributed in longitude around the Earth. The effect can introduce as much as 5° of error in time-averaged geomagnetic latitudes derived from a limited region, as determined from paleomagnetic data spanning the last 5 m.y.

Testing the geocentric axial dipole hypothesis for eras before 5 Ma is complicated by plate motions and possible movement of the entire lithosphere relative to axis of rotation. Reconstructions of Pangea, derived without the use of paleomagnetic pole positions, confirm that the field had a predominantly geocentric dipole source during Early Jurassic time. The dipolar nature of the field is demonstrated by the nearly exact coincidence of continental paleomagnetic poles when spreading in the Atlantic Ocean is backtracked according to the pattern of magnetic stripes on the sea floor (see, for example, Dalrymple and others, 1975). Confirmation of the dipole field being aligned with the rotation axis for pre-Miocene time requires paleoclimatic evidence of ancient latitude, such as the distribution of coral, limestone, and glacial deposits (Merrill and McElhinny, 1983, p. 175). Plots of various paleoclimatic indicators against paleomagnetically determined latitudes generally show that the axial dipole hypothesis is tenable throughout Phanerozoic time. Algal stromatolites of late Proterozoic age exhibit heliotropic structures that are consistent with the north-south axis derived from a preliminary paleomagnetic study of the deposits; this indicates the existence of an axial dipole field as early as 850 Ma (Vanyo and Awramik, 1982).

APPARENT AND TRUE POLAR WANDER

Studies during the past three decades have determined paleomagnetic poles from continental rocks of Archaean and younger age (McElhinny, 1973, p. 283–309; Irving and Irving, 1982). When plotted on the globe, the paleomagnetic poles from tectonically stable regions of a given continent define a path called the Apparent Polar Wander Path (APWP) for the continent. For a discussion of the North American APWP, see the chapter by Van der Voo in this volume. In the present geographic coordinate system, APWPs from the continents diverge from one another with increasing age due to past relative motions of the continents. Under the tenets of plate tectonics, it is movement of the continents relative to a stable rotation axis that causes the "apparent wander" of the paleomagnetic pole. In contrast, true polar wander is defined as movement of the lithosphere and lower mantle as a whole relative to the rotation axis (major axis of inertia) of the Earth (Goldreich and Toomre, 1969). As defined, true polar wander is the component of "apparent wander" common to all plates.

Geophysicists have looked for an absolute frame of reference to test the stability of the time-averaged geomagnetic reference frame and hence the validity of the geocentric axial dipole hypothesis. Of course, the system of celestial observations that is currently used to establish an absolute reference frame for the Earth cannot be applied to the geologic past. A promising substitute is an absolute reference frame based on the positions of fixed "hot spots" in the ocean basins. The geographic expressions of hot spots are long, linear chains of seamounts, such as the Hawaiian-Emperor chain. In this example, the volcanically active island of Hawaii marks the position of the hotspot. According to the hypothesis, as lithospheric plates drift over the hot spots, a linear chain of extinct volcanos is left behind (Morgan, 1972). The hot spots are believed to be deeply rooted in the mantle, so that they remain fixed relative to the sublithosphere and to one another. Although there is some disagreement concerning the fixity of hot spots and the ages of some seamount chains, the evidence generally supports the existence of a fixed distribution of hot spots since 100 Ma (Duncan, 1981). By using the hot-spot reference frame to correct paleomagnetic data for motion of the continents, it is possible to compare positions of the time-averaged geomagnetic pole with the present rotation axis, which defines the "hot-spot reference pole."

Such comparisons show a significant offset of the time-averaged geomagnetic pole from the "hot-spot reference pole" during Mesozoic time (Jurdy, 1981; Harrison and Lindh, 1982; Gordon, 1983; Andrews, 1985). The offset between the two reference frames is estimated at about 20° during Early Jurassic time. True polar wander, in the sense that the entire lithosphere or mantle, or both, have rolled relative to the spin axis of the Earth, is a possible explanation of the offset that does not require violation of the axial dipole assumption. If, on the other hand, the outer shells of the earth have not rotated relative to the spin axis, then the offset can be attributed to long-term wandering of the dipole axis away from the rotation axis. These issues have not been resolved, as there is disagreement concerning the validity of the assumption of fixed hot spots (Chase and Sprowl, 1984) and whether the entire lithosphere has rolled relative to the spin axis (Jurdy, 1981; Gordon and Cape, 1981). As discussed in the previous section, paleoclimatic indicators are generally consistent with the axial dipole assumption, but they are too imprecise to prove that true polar wander on the order of 20° has occurred.

NATURAL REMANENT MAGNETIZATION OF IGNEOUS AND SEDIMENTARY ROCKS

In rocks, acquisition and preservation of the ancient geomagnetic signal critically depends on lithology and mineralogy. Basalts and granites, for example, may acquire stable magnetic remanence, usually parallel to the ambient field, when titanomagnetite crystals cool below the magnetic blocking temperature, which depends on the size, shape, and titanium content of the crystals. For pure magnetite, the highest blocking temperature is 580°C, the Curie temperature above which ferromagnetic behavior vanishes. During deposition of sedimentary rocks, the magnetic moments of mineral grains such as magnetite, hematite, and pyrrhotite may be aligned with the ambient field and become locked in place after compaction. Red beds, an attractive target for paleomagnetic study, usually carry a magnetic remanence dominated by chemically precipitated hematite or by relict iron-bearing minerals that have been oxidized to hematite.

Current research regarding the sources of magnetism in rocks employs electron microscopy and microprobe techniques in addition to reflected light microscopy. Scanning electron microscopy and the related analytical methods are particularly useful for identifying iron oxides and their various crystal forms (McCabe and others, 1983; Hoblitt and Larson, 1975). Transmitted electron techniques, with their higher resolution, are necessary in studies of ultrafine magnetic grains, which typically carry the more stable fraction of remanence. The simplest carrier of remanence is the single-domain grain, which has a diameter of approximately 0.1 μm in the case of equant particles of magnetite. Larger grains tend to form two or more magnetic domains with opposite directions of magnetization in order to minimize the external magnetostatic energy. Multidomain structures in titanomagnetites and pyrrhotite have been observed through microscopic examination of magnetic colloids floated on polished grain mounts (Soffel, 1981; Halgedahl and Fuller, 1980). The technique reveals complex patterns of lamellae that are controlled by the crystallographic axes, structural defects, and shapes of the grains.

A fundamental question in paleomagnetism is whether the magnetization direction of a rock closely parallels the ambient field direction at the time of deposition. Early work on Hawaiian lava flows of historic age showed that the magnetization of subaerially deposited basalt was in accord with observatory records of the geomagnetic field direction (Doell and Cox, 1963). The magnetic fidelity of sedimentary rocks is more difficult to assess, because the detrital magnetization process depends so much on grain size or shape, magnetic mineralogy, and environment of deposition. The fidelity of deep-sea sediments is quite good, as demonstrated by Opdyke and Henry's (1969) compilation of paleomagnetic data from oceanic cores. Their data from deep-sea sites covering a broad range of latitude are consistent with the axial dipole hypothesis. Conversely, varved lake deposits exhibit a systematic shallowing in inclination compared to the inclination of the ambient field (Johnson and others, 1948).

Whether a sedimentary rock develops a systematic inclination error seems to depend on the movement of grains after settling from the water column (Verosub, 1977). King (1955) suggested that a shallow bias of the inclination could be generated if spherical magnetic grains roll into small intergrain depressions after touching down on the substrate. Alternatively, settling and compaction of elongate grains having magnetic moments parallel to their long axes might explain some cases of inclination error. In deep-sea sediments, the good fidelity is possibly due to extensive bioturbation on the sea floor, whereby jostling of the magnetic grains within the slurry of sediment and water allows proper post-depositional alignment of the grains with the ambient field.

The discovery of magnetite-producing bacteria in many depositional environments opens up the possibility that the presence of magnetotactic bacteria is more important than bioturbation in explaining the absence of inclination error in deep-sea sediments (Blakemore, 1975; Kirschvink and Chang, 1984). Such bacteria create single-domain crystals of magnetite that are of the proper size to carry very stable magnetic remanence. As the bacteria move about, chains of ultrafine magnetite particles within each cell remain aligned with the ambient field. After the bacteria die, the organic material that binds the magnetic particles into chains may also bind the chains to other particles of sediment, locking in a stable remanence.

All magnets tend to lose their remanence slowly due to thermal perturbations acting against the alignment of electron-spin moments within the ferromagnetic domains. In rocks, the original remanence decays and is replaced by magnetization parallel to the current ambient field, with the rate of acquisition of this viscous remanent magnetization being proportional to the Boltzmann energy, kT (O'Reilly, 1984, p. 94). The process occurs slowly when burial of the rock in a normal thermal gradient leads to mild heating and is accelerated by heating due to metamorphism. Fortunately for paleomagnetists, single-domain magnetite and hematite, which are common constituents of rocks, are stable recorders of the magnetic field. A major percentage of the original remanence in these materials can persist for billions of years at temperatures below 200°C (Tarling, 1971, p. 16).

PALEOMAGNETIC MEASUREMENTS AND ANALYSIS OF ERRORS

Paleomagnetic sampling involves the collection of oriented specimens, usually with a hand-held, gasoline-powered rock drill (Collinson, 1983, p. 192). When sample orientations are determined with a clinometer and solar compass, the error in orientation is usually no more than ±0.5°. The basic quantity measured in the laboratory is the average dipole moment of the specimen, a vector specified by the magnetic direction and total moment. The magnetization is calculated by dividing the total moment by the volume of the specimen. The ideal specimen would be a uniformly magnetized sphere, which is a true dipole source; however, the more easily obtained cylindrical specimen having a diameter roughly equal to its length is commonly used.

Three main types of magnetometers are routinely used in studies of rock magnetism: astatic, spinner, and cryogenic magnetometers (Collinson, 1983). The astatic magnetometer consists of an arrangement of magnets having a near-zero net moment that is suspended by a torsion fiber. When a specimen is placed near the magnets, the interacting magnetic fields twist the fiber suspension, creating a deflection that is proportional to the magnetic strength of the specimen. The deflection is amplified either electronically or optically to enhance the precision. In spinner magnetometers, the specimen is rotated near a detector, such as a coil or a flux-gate sensor (Foster, 1966). The sinusoidal electronic signal from the sensor is amplified and compared with a reference signal indexed to the rotating shaft. The magnetic moment is determined from the amplitude and phase angle of the signal from the sensor.

Cryogenic magnetometers take advantage of the quantum behavior of magnetic flux in superconducting metals (Goree and Fuller, 1976). Metals such as lead and niobium reach a level of near-zero resistance to electric currents, or superconductivity, when immersed in liquid helium (4°K). The rock specimen is inserted into a superconducting Helmholtz coil, causing a persistent current to be generated. The current then passes through a transfer coil near a "weak-link" detector. One such type of detector is a dielectric cylinder coated with a thin superconducting film that is scored longitudinally with a thin slit. The weak link is a small bridge that is left within the slit. As quanta of magnetic flux enter the film, the weak link is driven alternatively between superconducting and nonsuperconducting states, creating pulses in the output signal that are counted.

The sensor in each of the above magnetometers has a unidirectional axis of sensitivity, so it is necessary to place the specimen in the instrument in several orientations orthogonal to one another to determine the three-dimensional moment vector. A large amount of redundancy is usually built into the sequence of measurements to decrease the effects of electronic noise and to reduce possible errors due to inhomogeneous magnetization in the specimen. Since the development of sensitive flux-gate spinner magnetometers and cryogenic magnetometers, the magnetic directions of very weakly magnetized rocks can be measured with great accuracy. Errors due to lack of precision of the magnetic measurements are generally negligible compared to other uncertainties in paleomagnetic studies, such as the correction for tilt of the bedding.

Magnetic field vectors are corrected for tilt of the beds by rotating the vector about the strike by the angle of the dip. If the structure is known in enough detail, then an additional correction can be made to the magnetic declination, as would be necessary in the case of a plunging fold (Cox and Doell, 1960; MacDonald, 1980). Errors in declination sometimes cannot be avoided in structurally complex areas where undetected oblique faulting or multiple episodes of folding have occurred.

Given the assumption of a geocentric axial dipole field, the inclination (I) determines the paleolatitude (λ) by the simple dipole formula: $2 \tan \lambda = \tan I$. In most studies, the plane of the bedding, and hence the inclination, can be determined quite accurately. Because the calculation of paleolatitudes is not affected by errors in declination, complex local structures pose no problem provided the bedding plane is well defined. Conversely, the determination of large-scale tectonic rotations depends on the declination, which is prone to errors caused by locally complex structure.

When comparing paleomagnetic data sets from different regions, it is convenient to transform magnetic field vectors to virtual geomagnetic poles (VGP) to remove the latitudinal variation of the dipole field. The transformation from directions to VGPs goes from a spherical coordinate system, defined by vertical and azimuthal axes, to the global geographic coordinate system, defined by latitude and longitude. Again using the axial dipole assumption, the inclination sets the angular distance (P) from the collection site to the VGP by the formula, $\tan I = 2 \cot P$, and the declination is the azimuth from the site to the VGP (McElhinny, 1973, p. 23–24). The mean of VGP's in a given study is commonly termed the paleomagnetic pole and is assumed to coincide with the ancient geographic pole, provided the following criteria have been satisfied:

1. The sampled stratigraphic interval spans enough time to fully represent the natural secular variation of the dipole and nondipole fields. Short-term deviations from the axial dipole field should average to zero when the collection of VGPs covers the full range of secular variation, except for the small (5° or less) bias in locally determined time-averaged geomagnetic latitudes. Compilation of VGPs from rocks of Holocene age gives a mean paleomagnetic pole within a degree or two of the geographic pole, suggesting that about 10,000 yr is an adequate interval to fully represent geomagnetic secular variation.

2. The stable magnetization was acquired at the time of deposition, and all superimposed magnetic components due to viscous remanent magnetization, lightning, and chemical changes have been removed. Techniques for removing magnetic overprints are discussed in the following section, and tests concerning the age of magnetization are discussed in the section on "Reliability Tests.

The statistical model of Fisher (1953) is most commonly used to set confidence limits on paleomagnetic vectors, whether directions or poles are being considered. The model assumes that the unit vectors on a sphere are distributed with azimuthal symmetry about the mean and with a probability density (P) given by:

$$P = (\kappa/4\pi \sinh \kappa) \exp(\kappa \cos \theta),$$

where θ is the angle between individual vectors and the mean, and κ is the precision parameter. Given a set of N unit vectors, κ is estimated by $k = (N-1)/(N-R)$, where R is the vector sum. "k" approaches 0 for a uniform distribution of vectors on the surface of the sphere and increases as the cluster of vectors becomes more concentrated (McElhinny, 1973, p. 78–79). In practical terms, the Fisher distribution simulates a Gaussian normal distribution for points distributed on a spherical surface of unit radius.

Figure 1. Example of secular variation of virtual geomagnetic poles (VGP) from 36 Pliocene and younger basalts in New Mexico (square). VGPs are plotted on an equal area projection of the Northern Hemisphere centered on the geographic pole (NP). VGPs corresponding to normally magnetized basalts are plotted as solid circles; VGPs of reversed polarity (open circles) are transposed from the Southern Hemisphere. The paleomagnetic pole (X), or mean VGP, is encircled by the 95 percent confidence limit and the angular standard deviation (larger circle). Data supplied by E. A. Mankinen from U.S. Geological Survey files.

A statistic that describes the angular dispersion of vectors is the angular standard deviation (s) given by: $s = 81/\sqrt{k}$ in degrees. As a measure of directional scatter, s defines the spherical angle about the mean which encircles 63 percent of the data points on the unit sphere. It is important to note that s pertains to the range of variation about the sample mean, but not the confidence level of the mean. The circle of 95 percent confidence, or α_{95}, is given by:

$$\alpha_{95} = \cos^{-1}\left(1 - \frac{N-R}{R}\{P^{-1/(N-1)} - 1\}\right)$$

where $P = 0.05$, R is the unit vector sum, and N is the number of vectors. For small angles, the statistic can also be approximated by: $\alpha_{95} = 140/\sqrt{(kN)}$. In practical terms, the true mean of the sample population of vectors has a 95 percent probability of lying within α_{95} degrees of the sample mean. The α_{95} is commonly used to set confidence limits on paleomagnetic poles for tectonic studies.

A collection of VGPs from Pliocene and younger volcanic rocks from New Mexico illustrates a typical paleomagnetic data set and the use of these statistics (Fig. 1). Each VGP was averaged from 5 to 9 independently oriented cores from a basalt flow. The mean VGP from each flow provides an independent reading of the ancient field during normal and reversed polarity states. By selecting flows spanning a sufficient range of ages, several million years in this case, a full representation of the natural secular variation is obtained. Dipole wobble and the passage of nondipole features across the area created the near-circular distribution of VGPs seen in this example. The mean VGPs from all flows are then averaged together to obtain the paleomagnetic pole. Consistent with the axial dipole hypothesis, the mean paleomagnetic pole lies within 2° of the geographic pole. The angular standard deviation is about 18°, typical of the natural dispersion due to geomagnetic secular variation. The α_{95}, which decreases as the number of samples increases, is about 5.5°. So, despite a large natural range of variation, the mean paleomagnetic pole can be defined within a small circle of confidence when a sufficient number of samples have been collected.

Deciding what constitutes a sufficient number of samples depends on the inherent directional scatter of the paleomagnetic data, the latitude of the site, and the desired smallness of the α_{95}. The following analysis of errors is based on a hypothetical, but typical, study of volcanic rocks that have not been severely remagnetized since original cooling. In determining the magnetic direction of a particular flow, errors due to orienting the specimens and measuring magnetic vectors are negligible. The effects of lightning, even after magnetic cleaning methods are applied, and minor disturbance of the outcrop typically introduce a within-flow angular standard deviation, s_w, of about 3.5°. In tilted regions, errors introduced by the tilt corrections might add another component s_t of 5°. The total dispersion, s_f, in determining flow directions is given by $s_f^2 = s_w^2 + s_t^2$, which in this example equals 6.2°. The next step is to analyze the angular dispersion when directions from a large number of lava flows are grouped together.

By far the largest source of directional scatter in a collection of lava flows spanning several thousand years is geomagnetic secular variation, which varies with the latitude of the site. The dispersion of magnetic field directions due to secular variation, s_v, can be estimated from compilations of paleomagnetic data from young volcanic rocks in tectonically stable areas. Analyzing the data in terms of VGPs is more correct than using directions, because the distribution of VGPs best satisfies the assumption of circular symmetry made by the Fisher model. Hence, the remainder of this discussion of dispersion utilizes VGPs. As a consequence of the field-to-pole transformation function, angular dispersion of VGPs due to secular variation increases with increasing latitude of the sampling site (Cox, 1970). From compilation of many studies, McElhinny and Merrill (1975) showed that the angular standard deviation of VGPs, S_v, increases from 8° at the equator to 16° at the pole. The conversion of s_f into the VGP frame also introduces a latitude dependence that increases the dispersion at higher latitude, giving S_f for VGPs (Cox, 1970). When combined by the relation $S^2 = S_v^2 + S_f^2$, the overall dispersion of VGPs due to errors and the natural variation ranges from $S = 9°$ at the equator to $S = 20°$ at the pole.

The net effect of all angular dispersion on the α_{95} is shown as a function of latitude for sample collections of several different sizes (Fig. 2). Sites that acquired magnetization at a high paleolatitude require a greater number of samples to achieve a α_{95} comparable to the value that would be determined near the equator. For example, if an α_{95} of 5° about the paleomagnetic pole is desired, the minimum number of samples required would be 10 at the equator, while at least 45 samples would be required at 70° N. In both cases it is assumed that the ages of the samples span at least several thousand years to represent the full range of secular variation.

ANALYSIS OF MAGNETIC COMPONENTS

A serious problem in determining paleomagnetic poles from the orogenic belts is the removal of secondary magnetic components that mask the primary magnetization acquired during deposition. Metamorphism, magnetic viscosity, chemical alteration, and lightning are the main sources of secondary magnetizations. Usually, such magnetic overprints can be stripped away by progressive demagnetization in alternating magnetic fields or by progressive heating in a very low-field furnace (Thellier, 1966; Collinson, 1983, p. 308–359). Progressive demagnetization by immersion in acid removes secondary oxidation products from suitable sedimentary rocks, such as red beds. Changes in magnetic direction during such treatments are analyzed either graphically, for instance on an orthogonal vector diagram (As and Zijderveld, 1958; Zijderveld, 1967), or numerically by line-fitting methods (Kirschvink, 1980). These methods of magnetic vector analysis

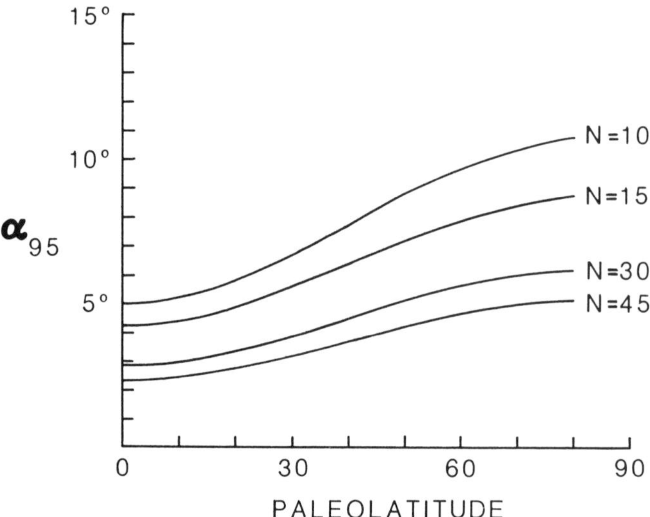

Figure 2. Dependence of 95 percent confidence limit, α_{95}, for virtual geomagnetic poles as a function of paleolatitude of the sampling site and the number (N) of samples collected. Model incorporates errors of measurements (see text) and natural secular variation according to compilations by McElhinny and Merrill (1975).

reveal the components that make up the total magnetization and ensure that a stable magnetic direction has been isolated in each rock specimen (Fig. 3).

In some cases when the primary and overprinting magnetizations have similar resistances to the demagnetization treatments, overprints cannot be removed completely. Such data can sometimes be salvaged by analyzing planes defined by the decay of magnetization during the demagnetization treatments (Fig. 4). The method of Halls (1976) fits great circles to the array of directions obtained by demagnetization of each specimen, then solves for the intersection of the circles to define the underlying primary component. Success of the method depends on having a near-random distribution of secondary components or a good variety of bedding corrections to ensure a strong intersection of the circles. A refined variant of the method uses three-dimensional least-squares analysis to define the demagnetization plane (Kirschvink, 1980) and a statistical technique based on the Bingham distribution to analyze the intersection of planes (Onstott, 1980). The Bingham distribution, which may be circularly symmetric, elliptical, or dispersed along a circular girdle, is ideally suited to the problem of finding the best intersection for a group of demagnetization planes, as the problem reduces to fitting a great circle to the girdle of poles corresponding to the demagnetization planes.

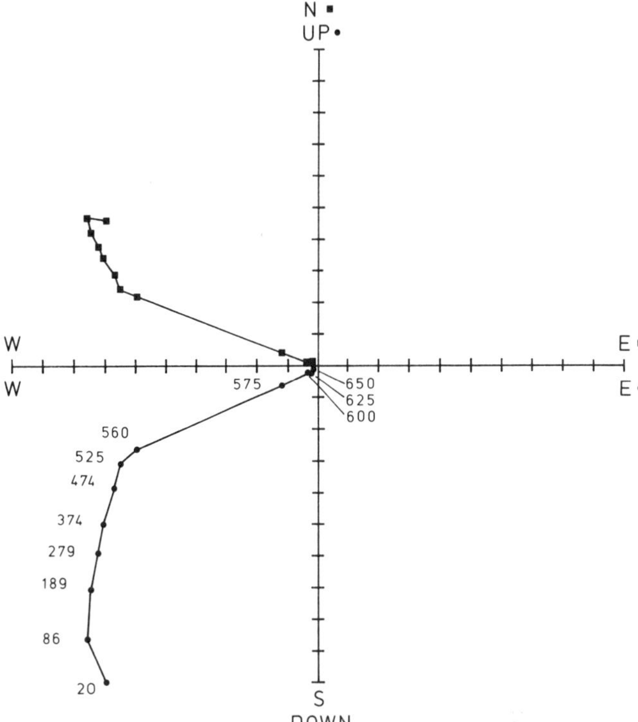

Figure 3. Example of an orthogonal vector diagram depicting thermal demagnetization of Triassic metabasalt from Hells Canyon, Oregon (Hillhouse and others, 1982). The magnetization vector is projected into the horizontal plane (squares) and the west-vertical plane (circles). Heating beyond 560°C isolates the characteristic component from the lower temperature secondary component, as indicated by linear decay of the magnetization to the origin.

RELIABILITY TESTS

Once the stable magnetic component has been identified, several tests help to determine whether the magnetization was acquired at the time of deposition or during original cooling of the rocks. The "fold test" of Graham (1949) establishes the necessary condition that the magnetization predates any folding of the strata. In applying the fold test, a comparison is made of the angular dispersion of magnetic directions before and after corrections are made for tilt of the bedding, preferably from coeval beds on opposing limbs of a fold (McElhinny, 1973, p. 89). The test is positive when application of the tilt corrections significantly decreases the angular dispersion of directions. The "conglomerate test" can be applied when cobbles of the sampled strata occur in an overlying conglomerate bed. If the lower bed gives a coherent magnetic direction while the cobbles give a random distribution of directions, then the possibility of complete remagnetization of the strata can be ruled out. Another check for remagnetization is the "baked contact test," which can be applied in contact aureoles around intrusions or in baked sediment below volcanic flows. If the magnetization is uniform within the igneous rock and the baked zone, but differs in the unbaked part of the host rock, then the test demonstrates magnetic stability of the igneous rock.

The "consistency of reversals test" uses the bipolar nature of the field to test for secondary components of magnetization. The mean directions of normal and reversed polarity specimens are compared. The test is positive when the two means are antipolar, at least within the confidence limits. Besides ensuring that secondary components have been removed, a successful reversal test indicates that a sufficient time span was sampled to represent the full range of secular variation. Other factors to consider in assessing the reliability of a paleomagnetic data set are the shape of the vector distribution and the amount of angular dispersion. The distribution of VGPs should be analogous to the expected distribution of secular variation, that is, having near-circular symmetry and an angular standard deviation ranging from 9° to 20°, depending on the paleolatitude. Elliptical or "streaked" distributions usually imply the presence of secondary magnetization. The shape of paleomagnetic distributions can be tested by several quantitative methods, such as point-density contouring (Robin and Jowett, 1986), analysis of modes (Van Alstine, 1980), and determination of elongation using the Bingham distribution (Onstott, 1980). When the amount of dispersion is abnormal, unusually low angular dispersion could be due to insufficient time sampling of the field or due to a completely uniform remagnetization of the strata. Unusually high angular dispersion also indicates the presence of secondary components of magnetization.

TECTONIC APPLICATIONS OF PALEOMAGNETISM

Under the assumption that paleomagnetic poles represent the geocentric axial dipole field, grids of paleolatitude can be

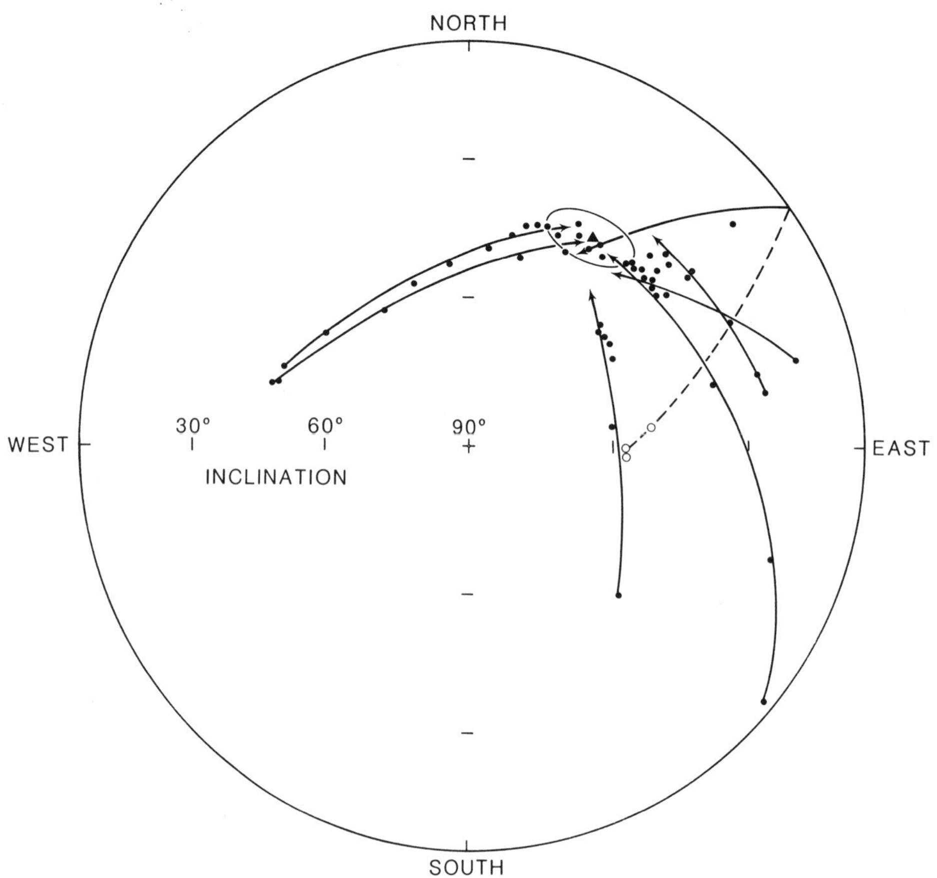

Figure 4. Example of planes analysis (Kirschvink, 1980; Onstott, 1980) to compensate for unremovable secondary magnetization after alternating-field demagnetization of six specimens from an ash-flow tuff, apparently struck by lightning. First, the path (dots) of directional change following progressive demagnetization of each specimen is fitted by a plane through the origin of the coordinate system. The intersection of each plane with the unit sphere is depicted by a great circle in this equal area plot; solid lines in the lower hemisphere and dashed line in the upper hemisphere. Then the best intersection of the planes is computed, giving an estimate of the primary direction (triangle) and 95 percent confidence oval.

developed for the geologic past. Grids that are derived from data collected within the tectonically stable continental interiors are useful for testing continental reconstructions. For post-Triassic time, such reconstructions are usually inferred without paleomagnetic data, using methods such as fitting coastlines together or by matching magnetic anomalies across sea-floor–spreading ridges. Van der Voo (this volume) discusses the development and application of paleomagnetic reference grids for North America. On a smaller geographic scale, paleolatitudes of orogenic belts are compared with the continental grid to detect displacements in latitude between terranes and the craton. Ancient longitudes are indeterminable from paleomagnetic data, so that displacements toward or away from the geographic pole are detectable, while east-west displacements are not. Therefore, paleomagnetic data yield minimum estimates of terrane movements.

The paleomagnetic azimuth (paleodeclination) is useful for determining rotations about a vertical axis. The method is best suited to rocks that have not undergone large displacements in latitude relative to the reference craton, because the pole of rotation may be assumed to be near the region of study. Paleomagnetism provides a wealth of information concerning deformation at plate margins and in intracontinental rifts. Combined with reconstruction of relative plate motions (e.g., Engebretson and others, 1985) and other geophysical data, paleomagnetic measurements are an integral part of studies pertaining to the crustal structure and tectonic history of plate margins. Beck (this volume) reviews paleomagnetic evidence for translation and rotations of terranes within the Cordillera of North America, and presents a detailed discussion of how such movements are calculated.

A major problem in the tectonic interpretation of paleomagnetic data is the problem of remagnetization. Remagnetization is common in the orogenic belts, and recent reevaluation of the reference poles has also shown that there are pervasive overprints in some established paleomagnetic poles from the North American craton (Van der Voo, this volume). Plutonism, underthrusting, and elevated thermal gradients in the crust impart secondary components of magnetization that must be recognized and removed. Therefore, in addition to the conventional reliability tests, magnetic studies have benefited from the use of thermal maturation indicators, such as vitrinite reflectance and conodont color alteration indices. The decay of magnetization in response to temperature level and the duration of heating are parallel to changes in these indicators, as determined from rock magnetic studies (Pullaiah and others, 1975; Middleton and Schmidt, 1982). The effect of tectonic strain on the magnetic fabric of rocks is not well known and may be an important factor in the remagnetization process. Chemical changes at low temperature during diagenesis, or weathering of sedimentary rocks, are other possible sources of remagnetization deserving further research.

To a large extent, the burden of proving tectonic interpretations has stimulated innovations in paleomagnetic techniques. For example, the need to understand the complex magnetic history of orogenic belts in Europe and North America led to better techniques for analyzing multi-component magnetizations, such as the methods for fitting lines and planes to demagnetization vectors. Greater emphasis is now placed on the determination of rock magnetic properties to determine the source of magnetization, and hence, to assess the reliability of the magnetic signal. Magnetic granulometry and domain structure are other important properties for estimating the relaxation time, and hence, longevity of the magnetization. Finally, the thermal response of magnetic remanence during laboratory experiments can be extrapolated to conditions of time and temperature in the natural environment to help distinguish primary and overprinted components of magnetization.

REFERENCES CITED

Andrews, J. A., 1985, True polar wander; An analysis of Cenozoic and Mesozoic paleomagnetic poles: Journal of Geophysical Research, v. 90, p. 7737–7750.

As, J. A., and Zijderveld, J.D.A., 1958, Magnetic cleaning of rocks in palaeomagnetic research: Geophysical Journal of the Royal Astronomical Society, v. 1, p. 308–319.

Blakemore, R. P., 1975, Magnetotactic bacteria: Science, v. 190, p. 377–379.

Chase, C. G., and Sprowl, D. R., 1984, Proper motion of hotspots; Pacific plate: EOS Transactions of the American Geophysical Union, v. 65, p. 1099.

Collinson, D. W., 1983, Methods in rock magnetism and palaeomagnetism: New York, Chapman and Hall, 503 p.

Coupland, D. H., and Van der Voo, R., 1980, Long-term nondipole components in the geomagnetic field during the last 130 m.y.: Journal of Geophysical Research, v. 85, p. 3529–3548.

Cox, A., 1969, Geomagnetic reversals: Science, v. 163, p. 237–245.

——, 1970, Latitude dependence of the angular dispersion of the geomagnetic field: Geophysical Journal of the Royal Astronomical Society, v. 20, p. 263–269.

——, 1975, The frequency of geomagnetic reversals and the symmetry of the nondipole field: Reviews of Geophysics and Space Physics, v. 13, no. 3, p. 35–51.

Cox, A., and Doell, R. R., 1960, Review of paleomagnetism: Geological Society of America Bulletin, v. 71, p. 645–768.

Dalrymple, G. B., Gromme, C. S., and White, R. W., 1975, Potassium-argon age and paleomagnetism of diabase dikes in Liberia; Initiation of central Atlantic rifting: Geological Society of America Bulletin, v. 86, p. 399–411.

Doell, R. R., and Cox, A., 1963, The accuracy of the paleomagnetic method as evaluated from historic Hawaiian lava flows: Journal of Geophysical Research, v. 68, p. 1997–2009.

Duncan, R. A., 1981, Hot spots in the southern oceans; An absolute frame of reference for motion of the Gondwana continents: Tectonophysics, v. 74, p. 29–42.

Engebretson, D. C., Cox, A., and Gordon, R. G., 1985, Relative motions between oceanic plates in the Pacific basin: Geological Society of America Special Paper 206, 59 p.

Fisher, R. A., 1953, Dispersion on a sphere: Proceedings of the Royal Society London, v. A217, p. 295–305.

Foster, J. H., 1966, A paleomagnetic spinner magnetometer using a fluxgate gradiometer: Earth and Planetary Science Letters, v. 1, p. 463–467.

Goldreich, P., and Toomre, A., 1969, Some remarks on polar wandering: Journal of Geophysical Research, v. 74, p. 2555–2567.

Gordon, R. G., 1983, Late Cretaceous apparent polar wander of the Pacific plate; Evidence for a rapid shift of the Pacific hot spots with respect to the spin axis: Geophysical Research Letters, v. 10, p. 709–712.

Gordon, R. G., and Cape, C. D., 1981, Cenozoic latitudinal shift of the Hawaiian hotspot and its implications for true polar wander: Earth and Planetary Science Letters, v. 55, p. 37–47.

Goree, W. S., and Fuller, M. D., 1976, Magnetometers using R.F.-driven SQUID's and their application in rock magnetism and paleomagnetism: Reviews of Geophysics and Space Physics, v. 14, p. 591–608.

Graham, J. W., 1949, The stability and significance of magnetism in sedimentary rocks: Journal of Geophysical Research, v. 54, p. 131–167.

Halgedahl, S., and Fuller, M., 1980, Magnetic domain observations of nucleation processes in fine particles of intermediate titanomagnetite: Nature, v. 288, p. 70–72.

Halls, H. C., 1976, A least-squares method to find a remanence direction from converging remagnetization circles: Geophysical Journal of the Royal Astronomical Society, v. 45, p. 297–304.

Harrison, C.G.A., and Lindh, T., 1982, Comparison between the hot spot and geomagnetic field reference frames: Nature, v. 300, p. 251–252.

Hillhouse, J. W., Gromme, C. S., and Vallier, T. L., 1982, Paleomagnetism and Mesozoic tectonics of the Seven Devils volcanic arc in northeastern Oregon: Journal of Geophysical Research, v. 87, p. 3777–3794.

Hoblitt, R. P., and Larson, E. E., 1975, New combination of techniques for determination of the ultrafine structure of magnetic minerals: Geology, v. 3, p. 723–726.

Irving, E., 1964, Palaeomagnetism and its application to geological and geophysical problems: New York, J. Wiley & Sons, 399 p.

Irving, E., and Irving, G. A., 1982, Apparent polar wander paths; Carboniferous through Cenozoic and the assembly of Gondwana: Geophysical Surveys, v. 5, p. 141–188.

Johnson, E. A., Murphy, T., and Torreson, O. W., 1948, Prehistory of the Earth's magnetic field: Terrestrial Magnetism and Atmospheric Electricity, v. 53, p. 349–372.

Jurdy, D. M., 1981, True polar wander: Tectonophysics, v. 74, p. 1–16.

King, R. F., 1955, Remanent magnetism of artificially deposited sediments: Monthly Notices of the Royal Astronomical Society, Geophysical Supplement, v. 7, p. 115–134.

Kirschvink, J. L., 1980, The least-squares line and plane and the analysis of palaeomagnetic data: Geophysical Journal of the Royal Astronomical Society, v. 62, p. 699–718.

Kirschvink, J. L., and Chang, S. R., 1984, Ultrafine-grained magnetite in deep-sea sediments; Possible bacterial magnetofossils: Geology, v. 12, p. 559–562.

MacDonald, W. D., 1980, Net tectonic rotation, apparent tectonic rotation, and the structural tilt correction in paleomagnetic studies: Journal of Geophysical Research, v. 85, p. 3659–3669.

McCabe, C., Van der Voo, R., Peacor, D. R., Scotese, C. R., and Freeman, R., 1983, Diagenetic magnetite carries ancient yet secondary remanence in some Paleozoic sedimentary carbonates: Geology, v. 11, p. 221–223.

McElhinny, M. W., 1973, Palaeomagnetism and plate tectonics: Cambridge, Cambridge University Press, 358 p.

McElhinny, M. W., and Merrill, R. T., 1975, Geomagnetic secular variation over the past 5 m.y.: Reviews of Geophysics and Space Physics, v. 13, p. 687–708.

Merrill, R. T., and McElhinny, M. W., 1983, The Earth's magnetic field; Its history, origin, and planetary perspective: New York, Academic Press, 401 p.

Middleton, M. F., and Schmidt, P. W., 1982, Paleothermometry of the Sydney Basin: Journal of Geophysical Research, v. 87, p. 5351–5359.

Morgan, W. J., 1972, Plate motions and deep mantle convection: Geological Society of America Memoir 132, p. 7–22.

Onstott, T. C., 1980, Application of the Bingham distribution function in paleomagnetic studies: Journal of Geophysical Research, v. 85, p. 1500–1510.

Opdyke, N. D., and Henry, K. W., 1969, A test of the dipole hypothesis: Earth and Planetary Science Letters, v. 6, p. 138–151.

O'Reilly, W., 1984, Rock and mineral magnetism: New York, Chapman and Hall, 220 p.

Pullaiah, G. E., Irving, E., Buchan, K. L., and Dunlop, D. J., 1975, Magnetization changes caused by burial and uplift: Earth and Planetary Science Letters, v. 28, 133–143.

Robin, P. F., and Jowett, E. C., 1986, Computerized density contouring and statistical evaluation of orientation data using counting circles and continuous weighting functions: Tectonophysics, v. 121, p. 207–223.

Soffel, H. C., 1981, Domain structure of natural fine-grained pyrrhotite in a rock matrix (diabase): Physics of the Earth and Planetary Interiors, v. 26, p. 98–106.

Stacey, F. D., and Banerjee, S. K., 1974, The physical principles of rock magnetism: New York, American Elsevier Publishing Company, 195 p.

Tarling, D. H., 1971, Principles and applications of palaeomagnetism: London, Chapman and Hall, 164 p.

Thellier, E., 1966, Methods and apparatus for alternating current and thermal demagnetization, *in* Runcorn, S. K., ed., Methods and techniques in geophysics, v. 2: New York, John Wiley & Sons, p. 205–247.

Van Alstine, D. R., 1980, Analysis of the modes of directional data with particular reference to paleomagnetism: Geophysical Journal of the Royal Astronomical Society, v. 61, p. 101–113.

Vanyo, J. P., and Awramik, S. M., 1982, Length of day and obliquity of the ecliptic 850 Ma ago; Preliminary results of a stromatolite growth model: Geophysical Research Letters, v. 9, p. 1125–1128.

Verosub, K. L., 1977, Depositional and postdepositional processes in the magnetization of sediments: Reviews of Geophysics and Space Physics, v. 15, p. 129–143.

Wilson, R. L., 1970, Permanent aspects of the Earth's non-dipole magnetic field over upper Tertiary times: Geophysical Journal of the Royal Astronomical Society, v. 19, p. 417–437.

——, 1971, Dipole offset; The time-average paleomagnetic field over the past 25 million years: Geophysical Journal of the Royal Astronomical Society, v. 22, p. 491–504.

Wilson, R. L., and Ade-Hall, J. M., 1970, Paleomagnetic indications of a permanent aspect of the non-dipole field, *in* Runcorn, S. K., ed., Palaeogeophysics: New York and London, Academic Press, p. 307.

Wilson, R. L., and McElhinny, M. W., 1974, Investigation of the large scale paleomagnetic field over the last 25 million years; Eastward shift of the Icelandic spreading ridge: Geophysical Journal of the Royal Astronomical Society, v. 39, p. 570–586.

Zijderveld, J.D.A., 1967, AC demagnetization of rocks; Analysis of results, *in* Collinson, D. W., Creer, K. M., and Runcorn, S. K., eds., Methods in paleomagnetism: New York, Elsevier, p. 254–286.

Manuscript Accepted by the Society October 31, 1988

Printed in U.S.A.

Geological Society of America
Memoir 172
1989

Chapter 6

Electrical structure of the crust and upper mantle beneath the United States; Part 1, Methods for determining the conductivity profile

George V. Keller
Department of Geophysics, Colorado School of Mines, Golden, Colorado 80401

ABSTRACT

The electrical conductivity of the Earth is a property that responds primarily to changes in the water content of a rock or to changes in temperature. At the temperatures present on the Earth's surface, conductivity depends most strongly on the amount and salinity of the ground water in a rock. At great depths, where the ambient temperature exceeds 500° to 600°C, conductivity depends on the motion of ions or electrons in the solid minerals of a rock, which in turn depends strongly on temperature. At intermediate depths, ranging from tens to hundreds of kilometers, conductivity is a function of both temperature and water content, with the relative importance varying with rock type.

Many methods have been used to study the conductivity of the Earth. Relatively direct measurements are made in wells and boreholes, to depths of about 8 km. Direct-current resistivity sounding has been used to explore the depths of sedimentary basins and the upper part of the crystalline basement. Most information about the crust and mantle has been obtained using the time-varying part of the Earth's magnetic field. In the magnetotelluric method, simultaneous observations of the magnetic and electric fields at a single site are used to determine a conductivity profile. In the geomagnetic deep-sounding method, the magnetic field is observed simultaneously at many stations, forming a network. The conductivity structure is determined from the spatial variation of the time-varying magnetic field strengths. Relatively simple methods of interpretation used in the past have provided ambiguous results, so that much of the information about the electrical structure of the Earth lacks the desired degree of reliability. More recently, methods of data acquisition and interpretation have been developed to make future use of the magnetotelluric and geomagnetic deep-sounding methods considerably more reliable.

Based on results from all these methods, the conductivity structure of the Earth appears to be grossly zoned. The zones, or shells, in order from the surface inward are: (1) a surface veneer of wet, conductive rocks, to depths ranging from a few hundred meters to a few tens of kilometers: (2) a zone of very low conductivity crystalline rocks that are relatively dry and cool, extending to depths of several tens or hundreds of kilometers; (3) a zone of gradually increasing conductivity, attributed to the development of thermally excited conduction mechanisms; (4) a zone beginning at a depth between 300 and 700 km, in which the conductivity rapidly increases by several orders of magnitude; and (5) a zone of gradually increasing conductivity to the base of the mantle. Sparse evidence indicates that the core is highly conductive, with a conductivity of 100,000 to 1 million Siemans per meter (S/m).

Keller, G. V., 1989, Electrical structure of the crust and upper mantle beneath the United States; Part 1, Methods for determining the conductivity profile, *in* Pakiser, L. C., and Mooney, W. D., Geophysical framework of the continental United States: Boulder, Colorado, Geological Society of America Memoir 172.

INTRODUCTION

For more than 100 years, it has been known that the interior of the Earth is electrically conductive. Gauss (1838) was apparently the first to recognize that part of the Earth's magnetic field is of internal origin and part of external origin; he devised a mathematical technique for separation of the two parts. In the ensuing years, many investigators have attempted to obtain better and better profiles of conductivity as a function of the depth in the Earth. Until fairly recently, the focus of such research has been on improving the knowledge of the conductivity profile so that the external part of the magnetic field can better be studied; only in recent years has it been recognized that the conductivity profile can in itself provide us with information about the petrologic and physical nature of the Earth's interior.

It is known that the main part of the Earth's magnetic field originates from currents flowing in a highly conductive core (Rikitake, 1966); at the surface of the Earth, this primary magnetic field has a size of 50,000 nanoTeslas (nT), more or less, and is nearly time invariant. Small, secular changes in the strength of the Earth's primary field have been observed over the centuries, but for time periods shorter than a few years, it is clear the time-varying portion of the Earth's magnetic field is of external origin.

A small part of the Earth's total magnetic field is contributed by electrical currents flowing high above the planet. These currents arise as electrically charged gases flowing outward from the Sun (the "solar wind") that interact with the primary magnetic field of the Earth. The solar wind varies in character with time, being not quite predictable, but yet somewhat structured.

Perhaps the first variable component of the external magnetic field to be recognized was the diurnal variation. This effect produces changes with a dominant period of 1 day, as well as harmonics, and has a strength of tens of nanoTeslas (about one part in a thousand of the internal field). As this time-varying field penetrates the Earth, it induces current flow in conductive rocks, thus converting its energy to heat. The diurnally varying field often penetrates to a depth of several hundred kilometers before its strength is reduced below observable values.

Another time-varying component of the magnetic field is provided by magnetic storms and substorms (Rostoker, 1966, 1972). Such storms, which occur sporadically, are periods of increased magnetic activity lasting for several days. Substorms are small, more localized events that last an hour or two. During storm periods, rapid changes of the magnetic field with amplitudes of several hundred nanoTeslas and more (of the order of 1 percent of the primary field) take place. Storm-time variations in the magnetic field can be used to study the conductivity profile at depths of 1,000 to 2,000 km. Substorms have a spectrum covering periods from 1,000 to 100,000 sec, which can be used to study the profile at depths from 100 to 1,000 km.

In addition to diurnal variations, storms, and substorms, rapid variations of the magnetic field with very small amplitudes occur continuously. These variations, called micropulsations, have amplitudes usually less than 1 nT, and periods ranging from 1 sec to several thousand seconds. Their amplitudes are only a part in 100,000 of the primary field. Micropulsations provide the basis for the magnetotelluric method (Tikhonov, 1950; Cagniard, 1953; Kaufman and Keller, 1980) of determining the conductivity profile, which is effective for depths ranging from a few hundred meters to a hundred kilometers or more.

The longest period variations in the magnetic field used in determining the conductivity profile are those with periodicities of a few years to a few tens of years. In this range, there is some disagreement as to whether the fields are of internal or external origin. Externally, an 11-yr cycle in the intensity of solar activity gives rise to periodicities of 11 yr and harmonics in the magnetic field seen at the surface of the Earth. Internally, it is believed that impulsive changes can take place in the primary magnetic field source in the Earth's core; these give rise to transient changes in the field at the Earth's surface with durations of a few years and longer. In either case, the strength of the long-term variation is of the order of 10 nT. Use of these long-period changes yields information about the conductivity in the inner part of the mantle, perhaps even down to the mantle/core boundary.

A summary of available information about the overall conductivity depth profile is shown in Figure 1. This collection is drawn from Lahiri and Price (1939), Banks (1967, 1972), Ducruix and others (1979), Currie (1968), Eckhardt and others

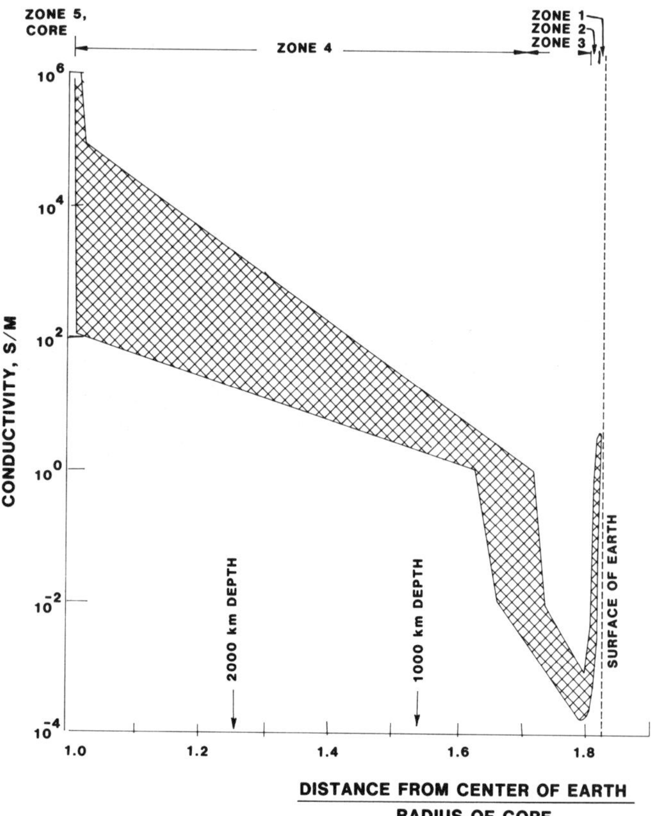

Figure 1. Summary of the estimates of the conductivity profile in the Earth, from the bottom of the mantle to the surface.

(1963), McDonald (1957), Yukutake (1959), Alldredge (1977), Runcorn (1955), Berdichevskiy and others (1972, 1974, 1976), Kolomijtseva (1972), Vanyan (1981), Courtillot and LeMouel (1979), Fainberg and Rotanova (1974), Feldman (1976), Golovkov and others (1971), Lahiri and Price (1939), Lubimova and Feldman (1970), Price (1967a,b, 1970), Sochelnikov (1968), Tozer (1969, 1970, 1981), Eckhardt and others (1963), and Hutton (1976).

The generalized conductivity depth profile appears to divide the Earth into six concentric zones, identified by characteristic ranges of electrical conductivity. In order from the surface inward, these zones are:

Zone 1: The surface veneer of wet rocks and the waters in the world's oceans are highly conductive. These materials have a thickness that ranges from a few hundred meters to a few tens of kilometers.

Zone 2: A zone of highly resistive crystalline rocks are the near-surface expression of the crust. These are relatively dry and relatively cool igneous and metamorphic rocks. The thickness of this resistant zone ranges from 10 km or less in regions of high heat flow or current tectonic activity to several hundred kilometers in cold, stable shield areas.

Zone 3: A zone of gradually increasing conductivity is present beneath the resistant zone. The increasing conductivity has been attributed to the development of thermally excited conduction mechanisms as the temperature reaches levels of a few hundred degrees (C) or higher.

Zone 4: At a depth below about 300 km but above 700 km, the conductivity rapidly increases from about 0.01 S/m to a value of approximately 1 S/m. The resolution with which this change can be located is poor, so the change may be abrupt or transitional over a depth interval of several hundred kilometers.

Zone 5: Beneath the transition at the base of zone 4, the conductivity probably continues to increase gradually with depth. There is some dispute about the conductivity at the lower boundary of the mantle, with some investigators arguing the conductivity in the bottom 500 km is no more than 100 S/m, while others insist the conductivity should be 100 to 100,000 S/m.

Zone 6: Evidence on the probable electrical conductivity of the core is indirect, based on consideration of other, better known, physical properties. Tozer (1959) argued the conductivity of the core is probably between 100,000 and 1 million S/m. Thus depending on the preferred interpretation of secular variations giving values for conductivity at the core/mantle boundary, the contrast in conductivity at that boundary may be minor (a factor of 10) or major (a factor of 10,000).

In addition to these six concentric shells, which are recognized uniformly around the Earth, there are many reports of an anomalously conductive region within zone 3 at many locations. Such anomalous regions have been attributed to high heat flow, causing at least partial melting at shallow depths in the mantle or even within the crust, or to the presence of significant amounts of highly mineralized water in the rock at these depths.

With the exception of the boundary between the surficial rocks and the crystalline basement, the other boundaries in the conductivity profile do not seem to be correlative with the major boundaries defined from seismic refraction, gravity, and other studies, such as the Moho discontinuity or the base of the lithosphere. While some efforts have been made to identify an electrical boundary at the same depth as the Moho, no significant data support such a correlation. If the Moho boundary is characterized by a change in the electrical properties of the rocks above and below, the effect must be of second order and not yet detectable with the survey methods in use.

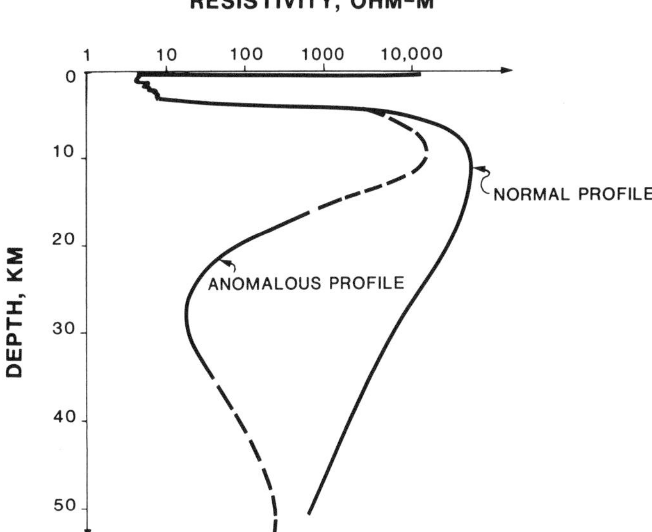

Figure 2. Expanded conductivity-depth profile in a normal crust and in one in which an anomalous region is present in Zone 2.

Lateral changes in the electrical properties of the Earth have been recognized only in the upper three zones listed above. If there are lateral changes in the lower three zones, their effect, as measured from the surface, is lost in an overwhelming dependence of conductivity on depth in the Earth. Thus, the focus in this compilation is on the upper three zones. Figure 2 shows schematically the conductivity profile in the outer three zones.

ELECTRICAL PROPERTIES OF ROCKS

Several mechanisms contribute to electrical conduction in rocks. The principal mechanisms are aqueous electrolytic conduction, solid electrolytic conduction, semiconduction, and metallic conduction. Conduction depends strongly on the composition of a rock, but with the range of temperatures that exist in the Earth over the complete conductivity-depth profile, temperature is the more important parameter (Adam, 1979; Alvarez and others, 1978; Haak, 1982; Tozer, 1970; Waff and Weill, 1975; Noritomi, 1961; Parkhomenko, 1967, 1982; Lebedev and Khitarov, 1964; Khitarov and others, 1970; Tyburczy and Waff, 1983; Volarovich and Parkhomenko, 1976; Keller, 1982).

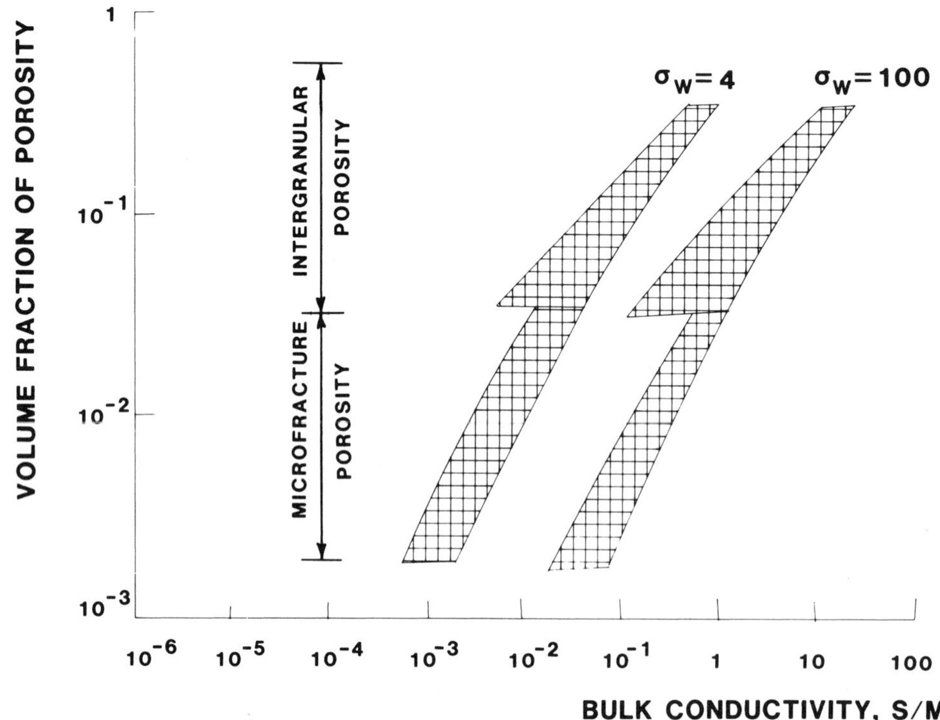

Figure 3. Correlation between rock conductivity and water content as a function of conductivity of the water in the rock. A conductivity of 4 Siemans/m corresponds to present-day sea water. A conductivity of 100 s/m can be approached by highly saline and hot water.

At the low temperatures characteristic of the upper part of the profile, almost all normal rock-forming minerals are nearly insulating semiconductors; the only significantly conductive rock component is the aqueous electrolyte that fills the pore spaces. The amount of conduction depends on the amount of water present, on its salinity, and on the geometrical distribution of the pore structures. An empirical expression known as Archie's law (Keller and Frischknecht, 1966) is widely used to describe this dependence:

$$\sigma_B = a\sigma_w W^m, \tag{1}$$

where σ_B is the bulk conductivity of the rock, σ_w is the conductivity of the electrolyte in the pore structure, W is the volume fraction of electrolyte-filled pore space, and a and m are empirically determined parameters. In sedimentary rocks, where the pore space is the volume left over between grains, reasonable values for these two parameters are $a = 0.6$ to 1.0 and $m = 1.6$ to 2.0. For crystalline rock where porosity is present primarily as joints and microfractures, appropriate values are $a = 1.4$ to 2.0 and $m = 1.3$ to 1.6. Figure 3 shows the relationship between bulk resistivity and porosity for a saturating solution like sea water at surface temperature (σ_w= 4. S/m) and for a saturating solution of high salinity at high temperature (σ_w= 100 S/m). A water conductivity of 100 S/m represents a highly concentrated electrolyte at high temperature.

At surface temperatures, not enough energy is provided to the solid minerals in a rock to excite charge carriers into conduction. At the temperatures present at depths of only a few hundred kilometers in the Earth, both solid electrolytic and electronic conduction mechanisms can be excited to a significant degree. Thermally excited conduction obeys the Stefan-Boltzman Law, in which the number of charge carriers available at any given instant is determined statistically by the relationship:

$$\sigma = \sigma_o e^{\frac{-U}{kT}}, \tag{2}$$

where σ_o is the conductivity at a reference temperature, U is an activation energy, k is Boltzmann's constant, and T is absolute temperature.

During the past two decades, numerous studies have been carried out on the electrical conductivity of dry rock-forming minerals and rocks (Parkhomenko, 1967, 1982; Volarovich and Parkhomenko, 1976; Shankland, 1975; Shankland and Waff, 1977; Shankland and others, 1981; Shankland and Ander, 1983; Olhoeft, 1976, 1981; Cermak and Lastovickova, 1986; Keller, 1963; Duba, 1972; Dvorak, 1973). A summary of these studies is shown in Figure 4.

As may be seen from the figure, at a given temperature, the conductivity of a dry rock can fall within a range covering two or three orders of magnitude. There is a tendency for rocks of more silicic composition to have a lower conductivity than rocks of more basic composition, but there is considerable overlap between the fields for dry rocks of various compositions. Overall, temperature is a more important parameter than composition in determining the conductivity of a dry rock (Karlya and Shankland, 1983).

If the temperature of a rock is sufficiently high to initiate melting, the conductivity is further increased. Conductivity in the molten phase depends on composition, and particularly on the gases in solution, but appears most commonly to fall in the range from 1 to 100 S/m in the case of basic and ultrabasic rocks (Shankland and Waff, 1977; Murase and McBirney, 1973; Presnall and others, 1972; Tyburczy and Waff, 1983). It should be noted that the range of conductivity values for molten minerals is the same as for hot saline water. With increasing pressure, the ionic bonding energy in a rock is likely to increase slightly, reducing the conductivity accordingly. On the other hand, transitions to more compact mineral forms at depth in the Earth may well lead to an increase in conductivity because the excitation energy required for electron semiconduction is reduced. It appears likely that the transition from zone 3 to zone 4 in the conductivity profile can be explained by a marked increase in electron semiconduction at the pressures and temperatures involved.

Because the rocks in zone 1, the surface layer of relatively conductive rocks, can have a profound effect on the results of many of the electrical sounding methods designed for deep penetration, it is important to consider their properties in particular. Much information on the surface layer is available as a data base.

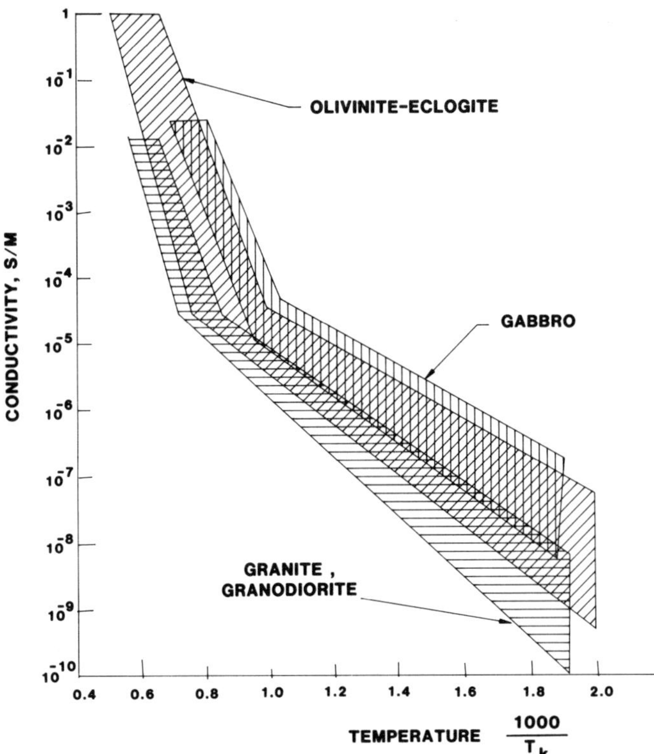

Figure 4. Conductivity of dry crystalline rocks as a function of temperature.

In recent years, 80,000 to 100,000 oil wells have been drilled each year in the United States, with accurate measurements of electrical conductivity of the surrounding rock being made in almost all cases. Libraries of such information exist, with more than 1.6 million well records on file.

Surveys of electrical conductivity in boreholes were first used in the 1920s, and by 1940 had become routinely used in evaluating boreholes drilled for hydrocarbon recovery in the United States. Over the years, the equipment used in such surveys was gradually improved, and as a consequence of the evolutionary development of equipment, the characteristics of the surveys ("logs") change with time.

Until the mid-1950s, the standard conductivity surveying tool was described as being an "Electrical Survey" (ES). In this method, an array of electrodes was lowered in the well, and apparent resistivity (the inverse of conductivity) was recorded continuously as the tool moved through the hole. From the mid-1950s to the mid-1970s, the standard choice of tool for determining conductivity was the Induction Electrical Survey (IES). In the induction log, the coupling between two coils, one a transmitter operating at about 30 kHz and the other a receiver, is measured and converted to conductivity. Since the mid-1970s, there has been further improvement in conductivity logging tools, notably the addition of the Spherically Focused Log (SFL). That log provides a detailed record of the electrical properties within a 20- to 30-in radius from the axis of the well bore (Dewan, 1983).

In exploration of the conductivity profile, the effect of a layer with a conductivity markedly greater than that in the layers above and below is equivalent to the effect of a vanishingly thin sheet with a conductance defined by:

$$S = \int_{z_l}^{z_l+h} \sigma(z)\, dz, \tag{3}$$

where $\sigma(z)$ is the conductivity as a function of depth, z, and h is the thickness of the conductive sequence (Hummel, 1929; Keller and Frischknecht, 1967).

In investigations of the electrical structure of the Earth at lithospheric and mantle depths, the entire sequence of conductive rocks and ocean water above the electrical basement can be treated as a thin conducting sheet, characterized by a conductance value that varies laterally over the sheet (Price, 1949).

The depths of the oceans and the conductivity of sea water are well known, and hence the conductance contributed by the oceans can easily be modeled. Conductivity of sea water lies between 3 and 4 S/m, depending on salinity and temperature. Beneath the thermocline, where temperature and salinity are relatively stable, the conductivity is close to 3.5 S/m. Conductance is about 3,000 S for each kilometer of water depth; in the deep oceans, the conductance can exceed 20,000 S.

Rocks, even those composing the conductive sequence at the surface of the Earth, are less conductive than the ocean. In an early study, it was observed that conductivity changed with geologic age in a reasonably predictable manner (Keller and Frischknecht, 1966).

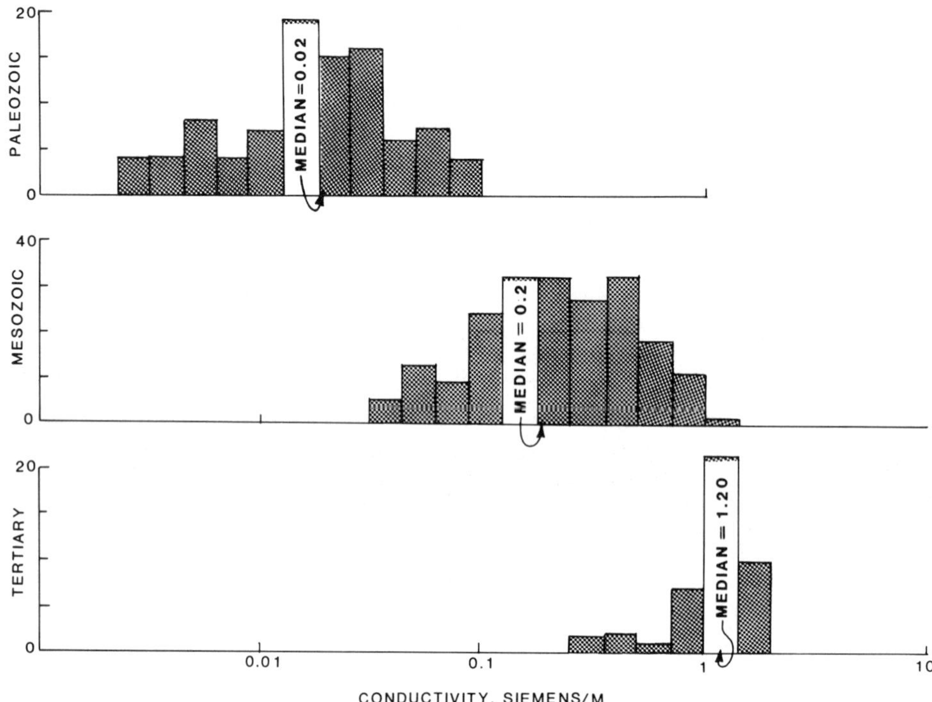

Figure 5. Histograms of average resistivities determined from well logs for intervals of Tertiary, Mesozoic, and Paleozoic age. Based on 126 well logs from the Gulf Coast basins, the Appalachian basins, and the Colorado Plateau.

The correlation between conductivity and age observed in the solid and weathered zone cannot be expected to be representative of rocks of similar ages at greater depths of burial. A few data are available from summaries of conductivity logs run in wells in basins of various ages in the United States (Keller, 1967). A plot of these conductivities and age is shown in Figure 5. The highest conductivities are observed for sedimentary rocks of Tertiary age, with progressively lower conductivities for rocks of Mesozoic and Paleozoic ages.

Figure 6 shows an integral conductance curve for the section penetrated by a deep well. Often the cumulative conductance curve will have a form of several nearly straight-line segments with differing slopes. This behavior implies that the interval over which the slope is nearly constant can be treated in large-scale soundings as though the entire interval so characterized was filled with rock of uniform conductivity.

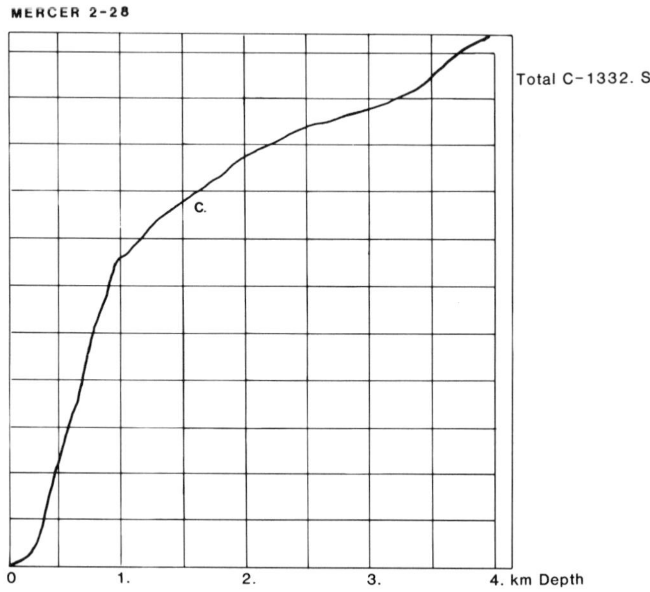

Figure 6. Cumulative conductance curve for the Mercer #2-28 well in the Imperial Valley, California. The vertical scale has been normalized to the maximum conductance.

CONTROLLED SOURCE METHODS FOR CONDUCTIVITY SOUNDING

A way of categorizing the many techniques for measuring earth resistivity useful for our purposes is on the basis of whether a controlled source is used to energize the Earth, or if naturally existing electromagnetic fields are used (Keller, 1971b). Information from the deeper part of the crust and beyond has only been obtained using the natural-field methods, but there has been some investigation of zone 2, the crystalline part of the crust and mantle, using controlled-source methods.

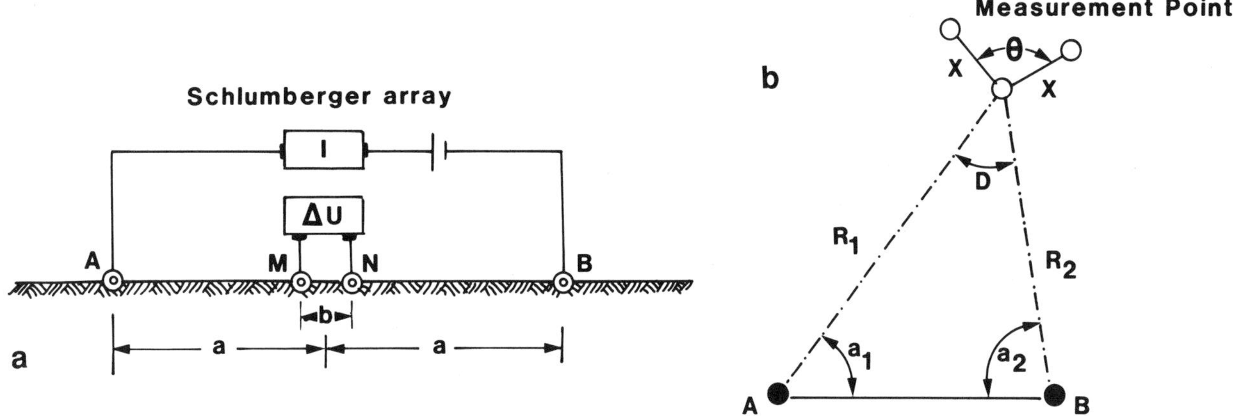

Figure 7. a, Layout of the Schlumberger array. b, Layout of the dipole mapping array.

Direct-current methods

Direct-current (DC) sounding methods (Keller and Frischknecht, 1966; Keller and others, 1966; Keller and Furgerson, 1977; Keller, 1966) are the best known and understood of the electrical probing methods, perhaps because they have been in use for many years, or perhaps because the results obtained with such surveys can be analyzed more easily than the results obtained with other sounding methods.

Many specific field techniques have been used in DC soundings, but of the crustal-scale surveys, only two have found favor, the Schlumberger technique and the dipole technique. In making a sounding with the Schlumberger array, four colinear electrode contacts with the Earth are used (see Fig. 7a). The two outer electrodes are moved progressively away from the center of the spread, while the inner two electrodes are used to measure the voltage drop developed by the current.

With the dipole array, five electrode contacts are used (see Fig. 7b). Current is supplied to the ground with one pair of relatively closely spaced electrodes, while two components of the electric field are mapped around the source using a triplet of electrodes. Because the electric field is mapped at many locations, perhaps hundreds, around a single source location, the method is sometimes called the roving dipole method.

The Schlumberger and dipole arrays have advantages and disadvantages relative to one another when they are used for crustal-scale resistivity soundings; thus, the choice of one or the other is not always clear cut. A major difficulty encountered with the Schlumberger array is that the span between current electrodes must be increased to hundreds of kilometers to provide information about the lower and upper mantle, even under favorable conditions (Van Zijl, 1969, 1975; Van Zijl and others, 1970). The voltage applied to these lines must be hundreds of volts. Extensive precautions must be taken so that such a length of current-carrying wire does not constitute a hazard to life. One solution is to make use of out-of-service power lines (Keller, 1968; Cantwell and Orange, 1965; Cantwell and others, 1964, 1965), but when this is done, the locations at which soundings can be made are severely limited.

Advocates of the use of dipole surveys for crustal-scale studies feel the operational ease of the method is the chief advantage over the Schlumberger array. With a dipole source length of 1 to 10 km, it is usually possible to make soundings even in densely inhabited areas. A serious disadvantage of the dipole array is its sensitivity to effects caused by relatively small lateral inhomogeneities in the near-surface resistivity.

With both the Schlumberger and dipole arrays, the depth to which conductivity can be determined from field measurements is crudely related to the maximum spacing between electrodes. In the case of identifying the depth to the base of zone 1 in the crustal conductivity profile, definition of the probing depth is straightforward. The sounding curve for this case has two asymptotes when plotted to logarithmic scales, as shown in Figure 8. The left-hand asymptote is approximately a straight line, yielding an average conductivity for the surface layers—those above the resistant basement surface. The right-hand asymptote is a line rising with unit slope, described by the equation:

$$a\rho_a = S, \tag{4}$$

where "a" is the electrode spacing factor, taken to be half the distance between current electrodes for the Schlumberger array, or the total distance between dipole centers in the dipole array, and ρ_a is the "apparent" or observed resistivity (the inverse of conductivity). The quantity $a(a)$ is the apparent conductivity computed from the observations as a function of spacing, and S is the integrated conductance of the surface layers, defined in equation (3).

Considering the curve in Figure 8, we can identify the depth capability of a conductivity sounding curve in terms of the intersection of these two asymptotes, which occurs for a Schlumberger spacing equal to the thickness of the conductive sequence.

The problem of detecting the thickness of a first layer that is relatively conductive is straightforward with the DC sounding

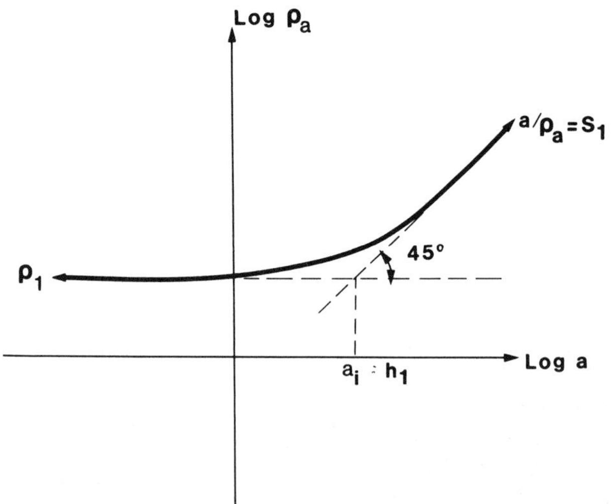

Figure 8. Asymptotes for a Schlumberger sounding curve measured over a two-layer sequence, with the lower layer being highly resistant.

methods; a problem in ambiguity arises when we try to probe through the second layer, the resistive part of the crust. The crust and upper mantle can be thought of grossly as representing a three-layer sequence, in which the middle layer has the highest resistivity. In all probability, the contrast in electrical properties between the middle layer and its neighbors is quite large, being measured in orders of magnitude. Theoretically calculated apparent resistivity curves have the form shown in Figure 9.

The apparent resistivity curve for a crustal sequence passes through a maximum value at spacings considerably greater than the thickness of the surface layer. For the Schlumberger array, the spacing at which the maximum value of apparent resistivity is observed is (Keller, 1971a, 1975):

$$a_{max} = \frac{(S_1 T_2)^{1/2}}{2}, \tag{5}$$

where S_1 is the integrated conductance in the surface layer, and T_2 is the transverse resistance of the resistant portion of the crust, defined by:

$$T = \int_{z_i}^{z_i+h} \rho(z)\, dz, \tag{6}$$

where $\rho(z)$ is the resistivity (inverse of conductivity) as a function of depth in zone 2.

The spacing needed to establish the maximum on the apparent resistivity sounding increases with both the conductance of the surface layers and the transverse resistance of the crust. The results of crustal-scale soundings that have been reported in the literature are reviewed in detail below, but generally, such surveys have yielded values for T_2 ranging from 10^6 to 10^9 ohms. The conductance of the rocks in zone 1, which will be recorded at that spacing, is given in Figure 10.

The spacing needed in order to establish the maximum on the apparent resistivity curve and thus identify the presence of zone 3 in the Earth's conductivity profile increases both with the conductance of the surface layers and with the transverse resistance of the crust. Considering Figure 10 for reasonable conductivity profiles, spacings of at least a few hundred kilometers and perhaps as great as 1,000 km are required. When measurements are made on this scale, it becomes doubtful that the electrical properties of the various zones are laterally uniform. Results of numerical analysis (unpublished) of a wide variety of models for lateral changes in conductivity of the zone 1 have shown that, in most cases, the effect of such changes is to create a false "rollover" or maximum on the apparent resistivity curve. In view of this, the value for T_2 derived from a resistivity sounding curve must be considered only to be a minimum value for zone 2.

Controlled-source electromagnetic methods

Electromagnetic methods, in which the Earth is probed with an artificial time-varying magnetic field, have long been used in shallow exploration for mineral deposits, as well as for other applications. A few efforts to use controlled-source electromagnetic methods for deeper exploration have been reported. In these, currents are driven into regions to be explored using an artificially generated time-varying magnetic field.

Electromagnetic methods differ from direct-current methods primarily in that excitation of currents in an area to be explored is not inhibited by the existence of the highly resistive rock in zone 2. On the other hand, with inductive methods, there is little capability for determining the slight induction that takes place in zone 2; such methods become insensitive to the electrical properties of zone 2. Several variants of controlled-source electromagnetic methods have been applied to crustal studies in the United States (Keller and Jacobson, 1983a,b).

An important distinction is that made between monopolar and bipolar systems (see Fig. 11). In a monopolar system, the energizing magnetic field is developed by passing a time-varying current waveform through one loop of wire, while the signal is detected with a second loop of wire lying on the ground concentric with the first (Kaufman and Keller, 1983). In order to obtain significant penetration of the magnetic field, it is necessary to provide a large moment to the source field, the moment being the product of current and area in the source. Experience has shown the radius of the transmitter loop must be of the order of the depth to be explored to achieve success.

In exploration to depths as great as 15 km, as is now being carried out, it has been inconvenient to lay out such large source loops; instead, a grounded wire that forms a secant to such a large loop is used (see Fig. 11). The principle of reciprocity can be invoked to argue that the grounded-wire system is equivalent to the concentric loop system in all except detail. Both the loop-on-loop and the grounded-wire/loop receiver systems constitute monopolar systems or systems in which the transmitter and receiver are centered on the same point.

A bipolar system involves the use of two loops. One is a transmitter energized with time-varying current, and the other is a

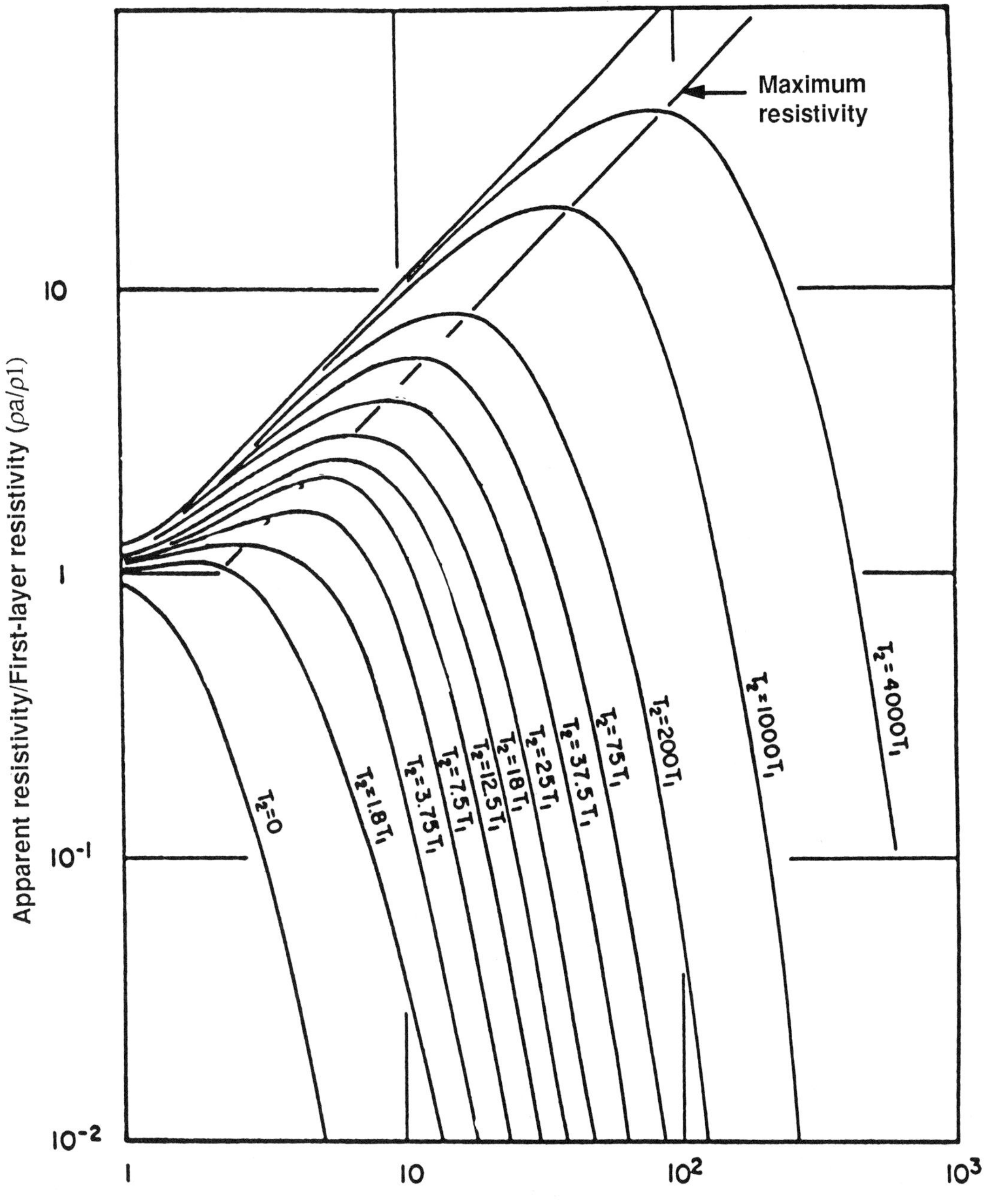

Figure 9. Apparent resistivity sounding curves obtained with the Schlumberger array over a three-layer sequence, with the middle layer being of high resistivity.

receiver loop; the centers of the two loops are separated by a significant distance, usually several times the depth to which penetration is desired. The advantage of the bipolar system is that the voltage received from the current flowing in the source loop is reduced to about the same level as the signal received from induction currents at the depth of investigation. With the monopolar system, the voltage received directly from the transmitter can be overwhelming in comparison to the signal returning from depth in the earth.

The monopolar systems are said to have a significant advantage over the bipolar system in that the response comes form a relatively more local region in the subsurface, and thus should be less complicated by the effects of lateral changes in the electrical properties of the Earth.

The electromagnetic and direct-current methods tend to be supplementary in their capabilities, in the case of a conductivity profile containing a low-conductivity interlayer. The direct-current method is limited to determination of the transverse

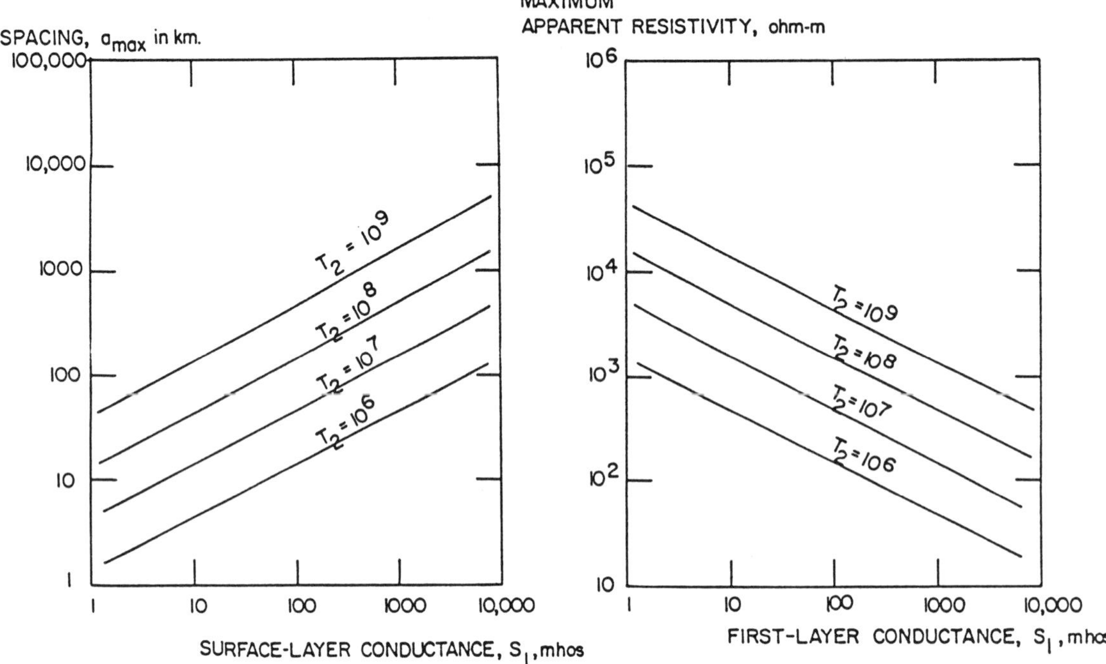

Figure 10. Maximum apparent resistivity and the spacing at which it is measured when the Schlumberger array is used over a three-layer sequences in which the middle layer is highly resistant.

resistance, *T,* of such a profile, while the electromagnetic methods are limited to the determination of only the thickness of such a layer. Therefore, measurements with both types of sounding would, in principle, permit complete definition of such a layer. Unfortunately, in the case of a highly conductive interlayer, both types of measurement are limited to the determination of conductance, *S;* therefore, resolution of conductivity and thickness in a relatively thin conductor does not appear to be feasible with any combination of measurement methods.

Finally, it must be recognized that, with controlled-source methods, the amount of energy required increases immensely as the depth of interest increases. The effort to reach depths of several tens of kilometers has been argued to be worthwhile only because controlled-source methods may be capable of providing somewhat better resolution than the natural-field methods discussed in the next section.

NATURAL-FIELD METHODS

Currently, the use of controlled-source methods is limited ultimately by our inability to provide sufficiently strong fields to obtain penetrations beyond 15 or 20 km, even under the best of conditions. The natural electromagnetic field of the Earth, generated by current systems of tens of thousands of amperes of strength flowing in the outer atmosphere and magnetosphere, provides an effective means for studying electrical structure of the Earth. Much of the information we have about the electrical properties of rocks in the deep interior comes from studies made with two related methods—the magnetotelluric method and the geomagnetic deep-sounding method.

The theory for the two methods has been treated by Schmucker and by Berdichevskiy and Zhdanov (1984). Details of the magnetotelluric method have been handled in papers and monographs by Cagniard (1953), Tikhonov (1950), Berdichevskiy and Dmitriev (1976), Clarke and Goldstein (1981), Filloux (1979), Madden (1971), Madden and Swift (1969), Vozoff (1972), and Word and others (1971). The geomagnetic deep-sounding method has been discussed by Banks (1973, 1979), Gough and Ingham (1983), Frazer (1974), and Gough (1973, 1974, 1983).

The approach of Berdichevskiy and Zhdanov (1984) clearly illustrates the close relationship of the magnetotelluric and geomagnetic deep-sounding methods. Consider the Earth to have a conductivity profile in which conductivity varies with depth, but not laterally (see Fig. 12). Assume that time-varying currents flow in a region above the earth, but separated from the earth by an insulating atmosphere so that we do not have to consider any current flow directly into the earth. Because of the very low frequencies present in the geomagnetic variational field, it is quite reasonable to ignore displacement currents and radiating fields, so that Maxwell's equations can be written as:

$$\text{curl}\,\vec{H} = \begin{cases} \vec{J}^{\,Q} & r \geqslant R \quad \text{(above the earth)} \\ \sigma_n \vec{E} & r \leqslant R \quad \text{(in the earth)} \end{cases}$$

where $\vec{H}$ is the magnetic field intensity, $\vec{J}^{\,Q}$ is the source current

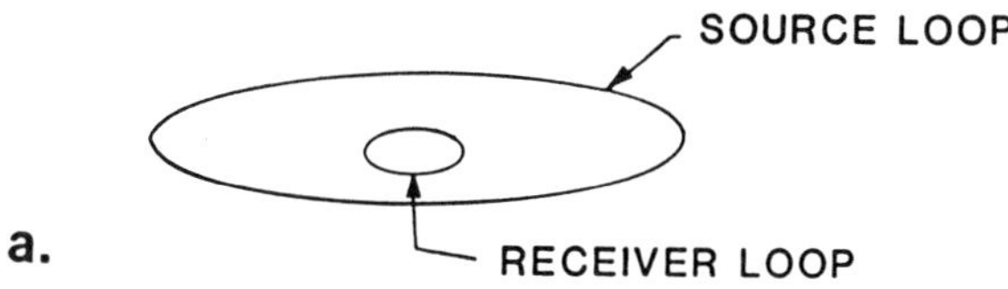

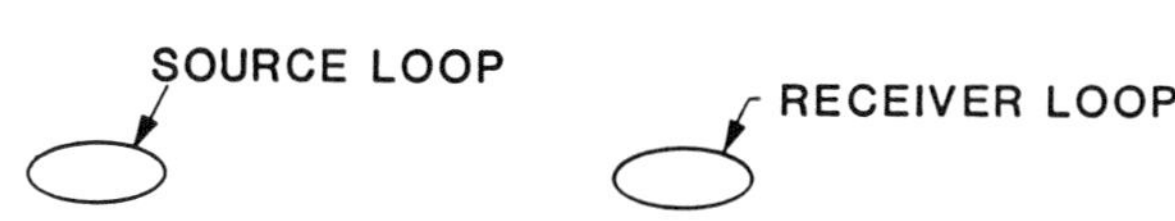

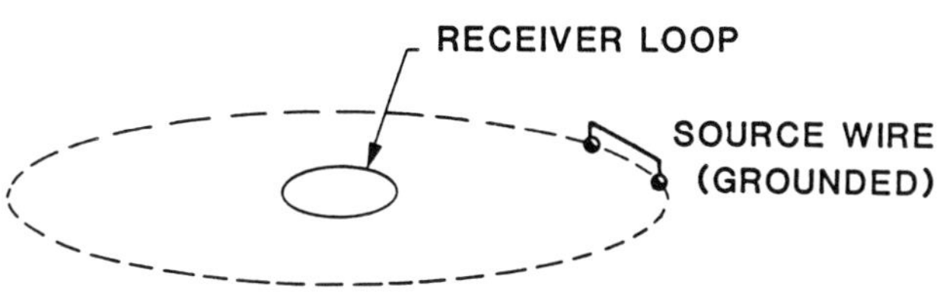

Figure 11. Source-receiver configurations used in electromagnetic soundings. a, Loop on loop or central loop induction method. b, Separated loop method. c, Grounded-wire method, as a modification of the loop on loop method.

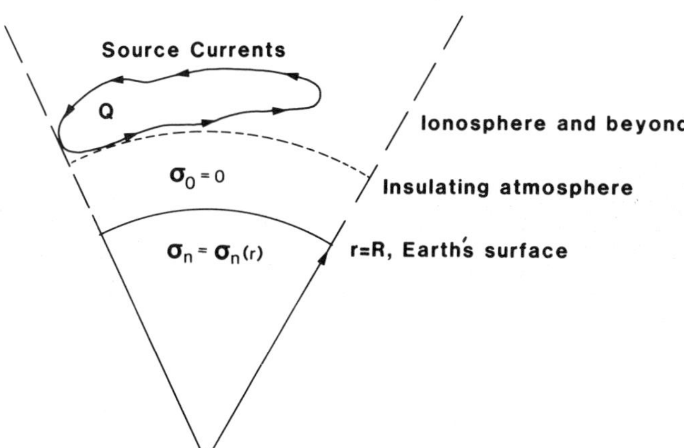

Figure 12. Earth model used in discussing the natural-field electromagnetic methods.

density, $\vec{E}$ is the electric field intensity, σ_n is the normal conductivity profile in the Earth, and R is the radius of the Earth.

The differential equations in (7) can be solved rather easily for values of the electric and magnetic field strengths at the surface of the Earth, where we can most readily observe them; see equations (6.7) through (6.11) in Berdichevskiy and Zhdanov (1984), for example. Each of the field components is linearly dependent on the source strength, while some of the components are also dependent on the conductivity profile in the Earth. Because the source distribution cannot easily be determined, it is not possible to determine the conductivity profile from observation of a single field component.

An ingeneous solution to this problem of indeterminancy is to consider the ratio of two components that depend differently on the conductivity profile, but for which the linear dependence on source strength will cancel. This ratio, known as an "impedance," will be independent of source strength but still carry information about the conductivity profile.

From the solution of Maxwell's equations, two useful impedances can be formed:

$$Z_m = \pm\, e_\theta / h_\phi$$

or

$$Z_m = \pm\, i\omega\mu m_o \frac{R}{m(m^{-1})} \frac{h_r}{h_\theta}, \tag{8}$$

where e_θ is a component of the electric field tangent to the Earth's surface and h_ϕ is the orthogonal magnetic field component, also tangent, h_r is the radial component of the magnetic field at the Earth's surface, and h_θ is a tangential component, ω is frequency in radians per second, μ_o is magnetic permeability, and m is the index for the spherical harmonics into which the field observations have been transformed. The use of lower-case letters for the field components indicates that the observed fields have been transformed to spatial spectrums by spherical harmonic analysis.

Note that the impedance can be computed from simultaneous observations of magnetic and electric field variations—the magnetotelluric method—or solely from magnetic field variations—the geomagnetic deep-sounding method. In either case, measurements must be made simultaneously at a large enough number of stations so that the spatial harmonic analysis can be carried out. Almost universally in the MT method, the Cagniard (1953) assumption that the source field is planar is made, so that only the zero-order term exists (P_o = 1), and observations at a single station can be used to compute impedance.

The concept of apparent conductivity can be introduced in electromagnetic methods. To arrive at a definition of apparent conductivity, we assume the conductivity profile to be uniform. Then, at limiting high frequencies,

$$\sigma_a = -\omega\mu_o / |Z_m|^2, \tag{9}$$

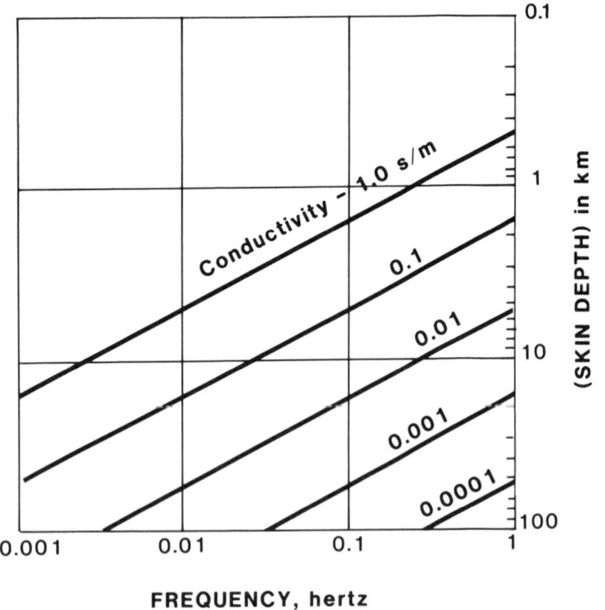

Figure 13. Skin depth as a function of frequency and conductivity for the case in which displacement currents can be neglected.

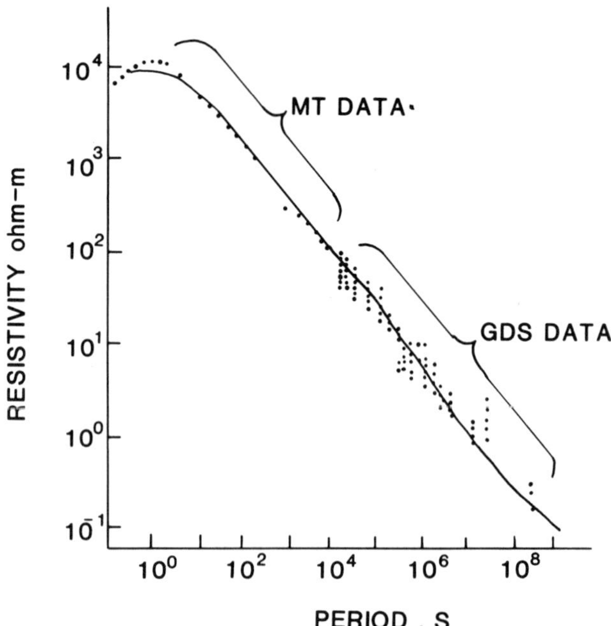

Figure 14. Apparent resistivities observed as a function of wave period over European shield areas, using the magnetotelluric method. References listed in the text.

which is the Cagniard formula for plane waves and $m = 0$. At limiting low frequencies (as in the GDS method),

$$\sigma_a = (m + 1)^2/\omega\mu_o R^2. \tag{10}$$

It should be clear that this definition cannot be carried all the way to the limit ($\omega = 0$), because in that case no induction would occur in the Earth.

The frequency content of magnetic variations is of importance, because the depth to which induction at detectable levels in the Earth can take place is controlled by skin depth, which is frequency-dependent:

$$\sigma = (2/\omega\mu\sigma)^{1/2}, \tag{11}$$

where σ is skin depth in meters.

In a medium with uniform conductivity, attenuation of electrical and magnetic field variations will follow a declining exponential behavior, so that at a distance of one skin depth, an attenuation of one neper (a reduction by the factor 0.73) will have taken place. Skin depth will also depend on conductivity, as shown by the curves in Figure 13.

Figure 14 illustrates practical results obtained with the MT and GDS methods, presented in terms of apparent conductivity as a function of frequency. The GDS data cover a range of frequencies from 10^{-9} to 10^{-3} Hz (taken from Fainberg and Rotanova, 1974; Berdichevskiy and others, 1976; Fainberg, 1981). The MT data were obtained in eastern Europe (Vanyan and Berdichevskiy, 1977; Vanyan, 1981). The combined data sets characterized an Earth with a conductivity profile dominated by increases of conductivity with depth.

The magnetotelluric (MT) sounding method:

Ideally, a magnetotelluric sounding curve would yield the conductivity profile beneath the observation site; in fact, because the Earth is not laterally uniform, the actual behavior of magnetotelluric sounding curves is far more complicated. First, the natural electromagnetic field has an instantaneous direction of polarization—that is, the direction of the magnetic field vector—which controls the direction of current flow in the Earth, and hence the direction in which the conductivity is measured. The polarization of the natural magnetic field usually changes with time, and over the course of a set of measurements, the polarization direction will probably vary over all possible directions. Also the conductivity structure of the Earth is likely to vary markedly with the direction in which current flow is induced, with the result that each direction of magnetic field polarization will yield a conductivity value appropriate for the direction of induced current flow. Instead of a specific, unique magnetotelluric sounding curve for a given observation station, one obtains a family of curves, bounded by some maximum curve and some minimum curve. The simplest product of a magnetotelluric sounding is a pair of limiting apparent resistivity curves, as shown in the example in Figure 15.

The volume of information provided by a single magnetotelluric sounding is generally viewed as being an advantage of the method. The difficulty that arises is in the use of one-dimensional modeling where it is assumed that the Earth consists of a sequence of horizontal layers, all laterally uniform. With a family of observed curves, the question arises as to which, if any, is the best representation of the conductivity profile beneath the observation site. A more positive approach is one in which the data from a single magnetotelluric sounding provide more opportunity for interpretation, in that the directional dependence of the curves can be used to elucidate information about the lateral changes in conductivity as well as the vertical changes (Hermance, 1982; Wannamaker and others, 1984; Stodt and others, 1981; Jupp and Vozoff, 1975, 1977; Losecke and Muller, 1975; Reddy and others, 1977; Swift, 1971; Weidelt, 1975).

One-dimensional interpretations are so simple that it is tempting to make such an interpretation even when it is clear from the spread between maximum and minimum apparent resistivity curves that the electrical structure is two- or three-dimensional. Modeling of such complicated structures poses several difficulties. First, the modeling is often numerically tedious, requiring considerable use of computer resources. Second, even more frustrating, is the realization that to obtain a meaningful two- or three-dimensional interpretation, it is usually necessry to have more complete field observations than are normally obtained.

An example of the behavior of magnetotelluric sounding curves over a relatively simple structure is shown in Figure 16. Here the structure is a three-layer sequence, but the surficial layer is interrupted by a strip with high conductivity. A series of sounding curves is shown in Figure 16 for observations made at locations along a profile crossing the conductive strip. The strip changes the shape of some of the sounding curves as though there were an additional conductive zone at depth. Each of the curves shown in Figure 16 was interpreted as if it were truly a one-

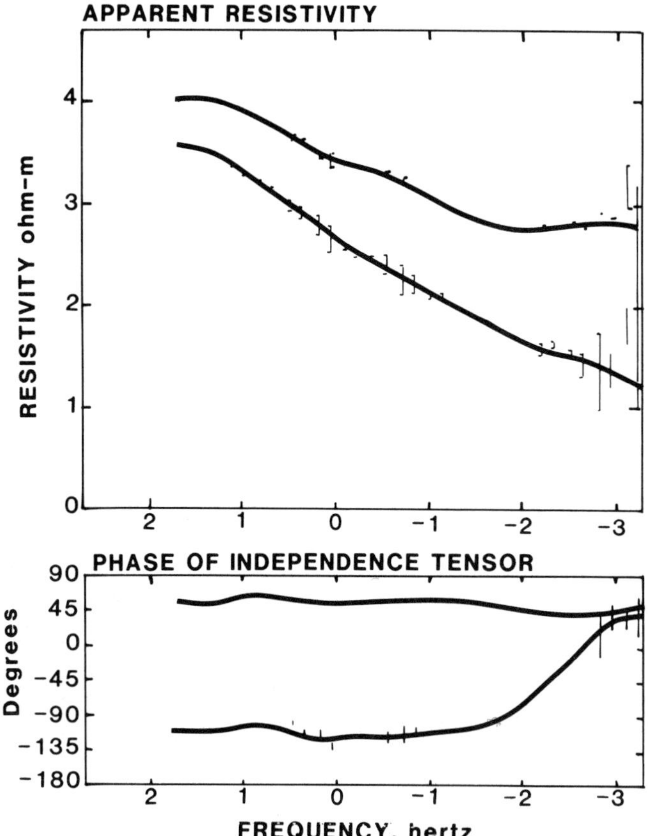

Figure 15. Output of a magnetotelluric sounding at a single station. The upper two curves are values of apparent resistivity after the observed tensor impedance has been rotated into principal directions, yielding maximum and minimum values for apparent resistivity. The lower two curves represent the phase shift between the orthogonal electric and magnetic field signals after the tensor impedance has been rotated. The frequency scale has been expressed as a base-10 logarithm.

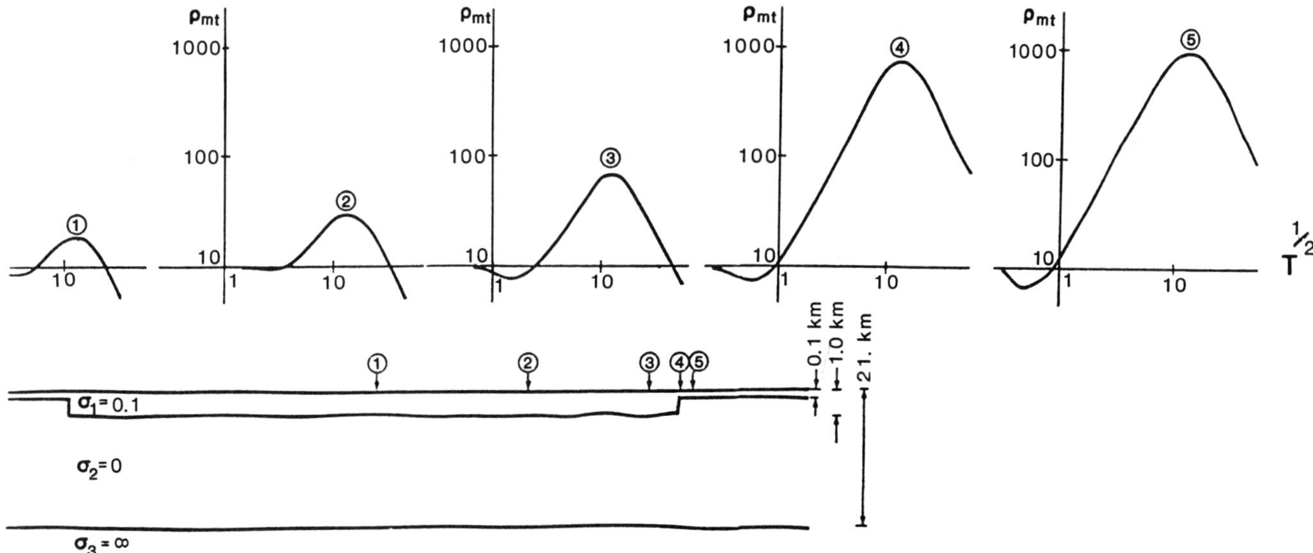

Figure 16. Profile of MT sounding curves (E-polarized transversely) over a graben-like structure in the surface of the resistant crust. Note that the MT curves have been compressed by plotting as a function of the square root of period ($T^{-1/2}$).

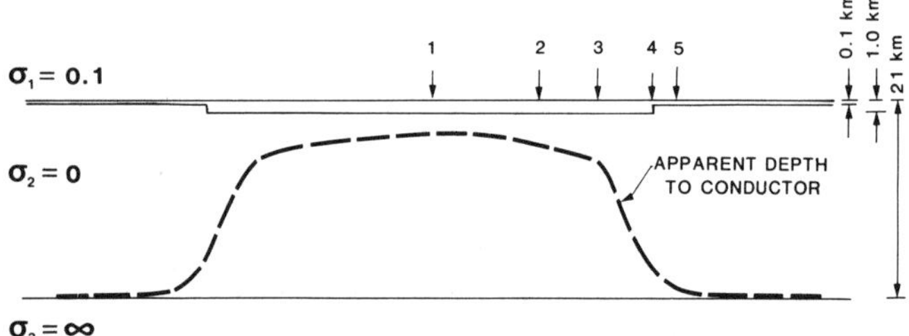

Figure 17. Interpreted depth to the surface of the conductive layer at depth for the sounding curves shown in Figure 16, using one-dimensional interpretation.

dimensional set of data. The resulting interpretations are shown as a stitched section (a cross section in which adjacent one-dimensional interpretations are plotted and correlated) in Figure 17. The introduction of false conductive layers is a serious problem when only one-dimensional interpretation procedures are applied to magnetotelluric sounding curves.

Another difficult problem arises when a three-dimensional conductivity structure affects the field data but is of too small a size to recognize. In acquiring a magnetotelluric sounding curve, magnetic field components and electric field components are recorded simultaneously at an observation site. The magnetic field is sensed with equipment that sums magnetic fields from all the currents flowing in a large volume of ground around the observation site. The electric field is sensed with pairs of electrodes separated by a short distance, usually of the order of 100 m. The electric field is not an average value for a large volume, as is the observed magnetic field, but is representative mainly of the ground immediately around and between the electrodes. This can lead to a "static shift" of the computed apparent resistivity when the small volume of ground around the electrodes has an electrical conductivity markedly different than that of the surface layer as a whole. For example, if the electrode contacts are placed in a pod of highly conductive soil, the electric field will be shorted (it will be reduced in amplitude from the value appropriate for the surface layer as a whole), and the magnetotelluric curve will be shifted down, as shown in the example in Figure 18 (Keller, 1964).

The problem of static shift in magnetotelluric sounding is a difficult one. The effect can often be recognized because the maximum and minimum curves will remain parallel to one another, while for larger three-dimensional structures, the behavior of the curves will be more complicated. Even if the presence of a static shift is recognized, there is no way of compensating for the effect using only the magnetotelluric observations. Recently it has been recognized that the static shift can be combatted by using another electrical method to determine the conductivity of the surface layer, so that the observed magnetotelluric sounding curves can be shifted up or down to match.

When large data sets are available, the static shift effect can be removed to some degree by averaging curves from nearby observation sites. For this to be fully successful, one must argue that the static shift is a process that statistically will have a zero average; that is, shifts in one direction are as common and of the same general size as shifts in the other direction. However, numerical evaluation of the static shifts caused by small inclusions of various shapes and conductivity contrasts shows that the downward shift caused by a local increase in conductivity is considerably stronger than the upward shift caused by a local decrease in conductivity. As the contrast becomes greater, for a resistive inclusion, the increase in electric field intensity quickly saturates, but for a conductive inclusion, the electric field intensity decreases without limit.

A third problem to be considered in evaluating the capabilities of magnetotelluric sounding is the screening effect that occurs when the surface layer is conductive. As noted earlier, the conductance in zone 1 may range from a few Siemans to 6,000 S. Even in the case of a laterally uniform Earth, the presence of this zone seriously affects the capability of the method for mapping the conductivity structure at depth. Consider a conductivity profile such as that shown in Figure 19. Here, the Earth is considered to have a relatively thin zone 1, in which the electrical properties are expressed as a conductance value, underlain by a crystalline basement (zone 2) of infinitely high resistivity, and underlain in turn by a lower crust and upper mantle in which the conductivity is assumed to increase continuously with depth. One-dimensional apparent resistivity curves for this conductivity profile are shown

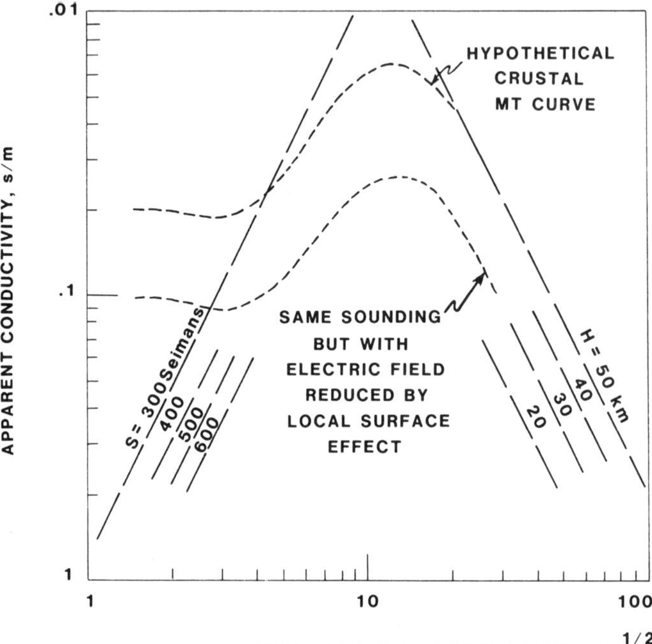

Figure 18. Example of the effect of "biasing," or the vertical shift of a magnetotelluric sounding curve. The shift occurs when the electric field is measured within a small anomalously conductive region at the surface. A downward shift causes an erroneous increase in the conductance of the surface layer, and an erroneous decrease in the depth to the third layer.

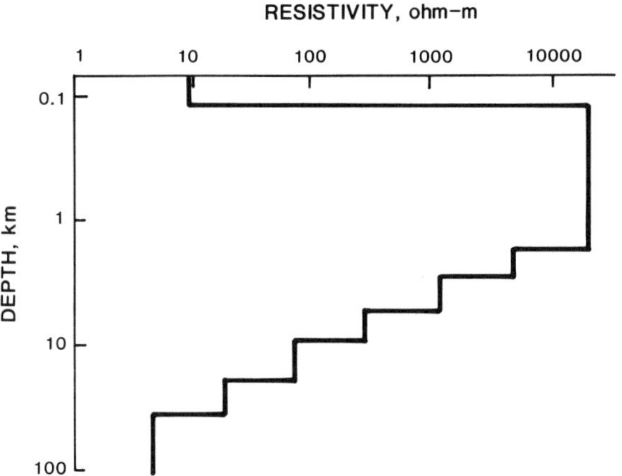

Figure 19. Stepwise approximation of a continuous resistivity profile through the crust. MT curves for this profile are shown in Figure 20.

in Figure 20 for various specified values of conductance in zone 1. The presence of the surface layer suppresses the effect of the insulating part of the crust on the sounding curves and shoves the expression of the deeper conductive layers to the right.

Geomagnetic deep sounding (GDS)

The use of Cagniard's assumption of plane-wave propagation of the natural electromagnetic field into the Earth is not the only approach to the quantitative use of that field. At micropulsation frequencies that provide penetrations more than 100 km or so into the Earth, the assumption that the field can be treated as a plane-wave field appears to be reasonable much of the time. For longer period fields, the assumption is not nearly as valid, and it becomes necessary to consider the structure of the incoming natural field as well as the structure of the Earth. This is done in the geomagnetic deep-sounding method, or the magneto-variational method, as it is sometimes called.

In geomagnetic deep sounding, the usual experimental approach is to record variations in the magnetic field simultaneously at many sites. Ordinarily, longer period variations are observed with the DGS method rather than the MTS method, with emphasis being placed on the observation of magnetic events such as bays and storms. These events have periods of hundreds of seconds and longer, providing easy penetration through all but the most conductive of surface layers.

Stations are often located on an approximate grid, with the distance between stations being several hundred kilometers. In geomagnetic deep sounding, only the three components of the magnetic variation field are measured, and not the electric field, thus avoiding problems with static shift.

The first step in GDS analysis is usually separation of the primary field arising from current flow outside the Earth from the secondary field caused by currents induced in the Earth. Several approaches can be used, including the classic approach of Gauss and Schmidt (Chapman and Bartels, 1940; Yanovskiy, 1978; Berdichevskiy and Zhdanov, 1984). In this method, spectral analysis in terms of spherical harmonics is carried out, and the inside and outside contributions are identified by forming the derivative with respect to the vertical. The formulas for doing this are called the Gauss-Schmidt formulas (equation (29.21), p. 190, Berdichevskiy and Zhdanov, 1984) for a spherical Earth. Similar formular were developed by Siebert (1958) and Berdichevskiy and Zhdanov (1984, p. 193, equation [29.33]).

These approaches require somewhat idealized conditions in that, in one case, data from a completely spherical Earth are required, and in the other, a perfectly flat Earth is needed. Useful information has been obtained for less ideal circumstances with GDS surveys that cover only a small fraction of the Earth's surface, but yet a large enough area that curvature must be considered.

A more modern approach based on work by Kertz (1954), Siebert and Kertz (1957), and Siebert (1962) permits a more realistic approach to modeling the Earth. This approach uses integration along closed paths, rather than spherical harmonic analysis, but is otherwise based on the same properties of electromagnetic field behavior. The expressions for the line integrals are given in equation 30.13, p. 199 of Berdichevskiy and Zhdanov (1984).

In many practical applications, the analysis formulas have been converted to a two-dimensional structure for either the source field or the Earth structure in order to reduce the amount of information needed from field observations. However, the line integral approach can be applied for a fully three-dimensional Earth (see, for example, Siebert, 1958; Scheube, 1958; Hartmann, 1963), but such analyses place high standards for recordings and for data acquisition, both in terms of the number of stations and their distribution. An ideal data set would consist of a uniform grid with several tens of stations in each direction, to avoid aliasing and truncation effects in spectral decomposition. No surveys have yet been completed with as many as 1,000 stations; most have made use of 50 to 100 sites.

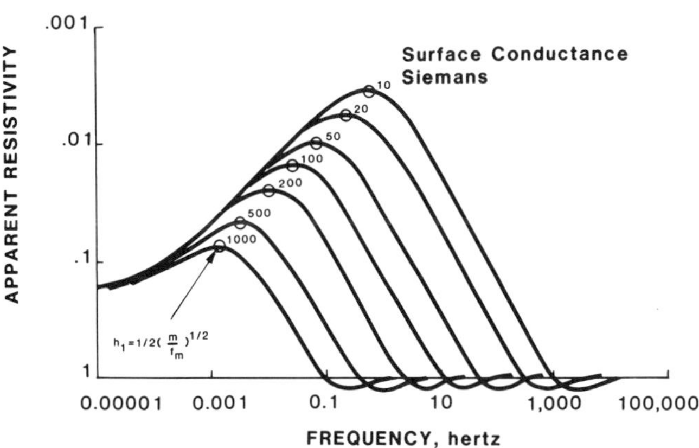

Figure 20. Magnetotelluric sounding curves computed for the gradational conductivity profile shown in Figure 19.

As with the MT method, a conductive layer may well mask the structure of deeper regions. In the case of the surface layer (the sedimentary sequence or the oceans), if its properties are reasonably well known, the effect often can be modeled as a thin conducting sheet with spatially varying conductance (Price, 1949). Then the effect of the surface sheet can be removed from the data, at least within the accuracy to which the surface conductance is known.

Even when no independent information is available about the conductance of the surface layer, some information is available directly from field observations. In order to identify changes in conductance of the surface layer, both the electric and magnetic fields must be measured simultaneously. Recognition of surface effects is based on the solution of two integral equations; this involves significant computational complexities. However, when only two-dimensional structures are considered, the mathematics is simplified markedly, leading to a pair of formulas known as the Schmucker formulas (see equations [47.11] and [47.12], p. 320, in Berdichevskiy and Zhdanov, 1984). Interpretation of deep anomalies after surface distortions to the secondary field have been suppressed can be accomplished by analytical continuation or by optimization of an assumed finite difference model.

The question of the probable uniqueness of an interpretation often arises. Three proofs of uniqueness exist for particular cases: a model in which conductivity varies only with depth (Tikhonov, 1965), a model in which resistivity varies both with depth and one horizontal direction—the so-called 2D model—but with the variation in conductivity being an analytic function (Weidelt, 1978), and a 2D model in which the variation in conductivity can be made up of piecewise continuous analytic functions (Gusarov, 1981). In all three cases, observations over an infinite band of frequencies are assumed. Intuitively, these proofs suggest that the time-frequency dependence of the fields provides determination of vertical changes in conductivity, while spatial frequency provides indications of lateral changes in conductivity. Because of this, unless sampling is pursued equally aggressively in both frequency domains, one or the other axis of conductivity determination will suffer. In MT surveying, time frequency is sampled more densely than space-frequency, so that MT surveys do not often offer desired resolution for lateral changes. In GDS surveys, the time-frequency bands and space-frequency bands are both usually undersampled, leading to a significant but comparable ambiguity in interpretations of all axes.

Berdichevskiy and Zhdanov (1984) suggest a two-stage approach to interpretation, which appears to work well. The first step consists of analytic continuation of the observed secondary magnetic fields downward into the Earth, so that the shape of the field surfaces collapses to a representation of the physical surface of a conductive zone. Analytic continuation of an electrodynamic field is quite similar to analytic continuation of static gravity and magnetic fields which has been done for many years. An example of the collapse of field onto a body of known shape (in this case, a cylindrical conducting mass) is shown in Figure 21.

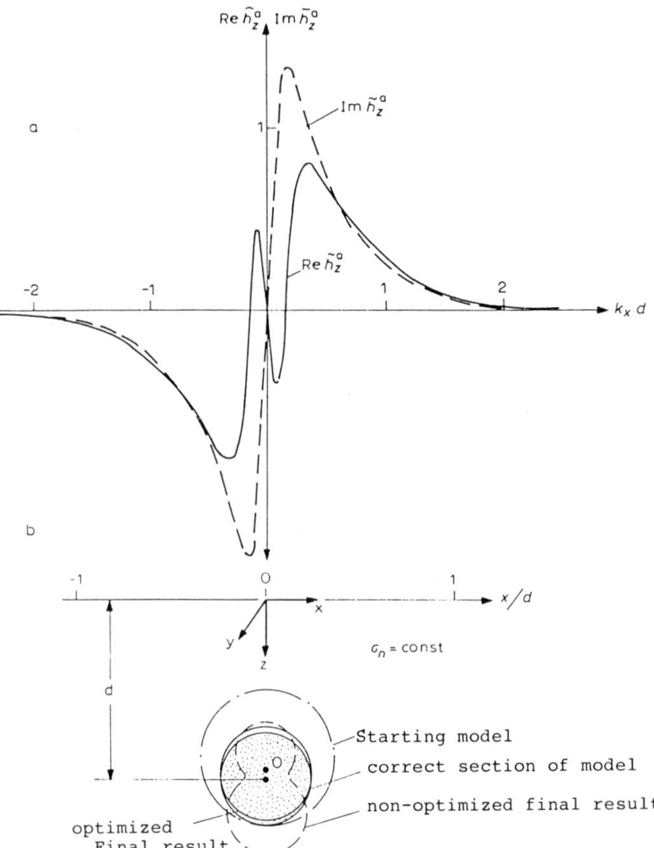

Figure 21. Example of the use of the method of tightening of contours (analytic continuation) to locate a conductive body from measurements of the induced magnetic field (from Berdichevskiy and Zhdanov, 1984). The upper curves are the imaginary and real parts of the induced vertical magnetic field, calculated for a cylindrical conductive body. The lower cross sections show successive estimates of the cross section compared with the actual model.

Once the rough form and location of a causative conductive mass has been established by continuation (effectively, by "tightening" of the field contours about the mass), the interpretation can be refined using an iterative optimization approach. Several finite differences and finite-element programs have been described in the literature for use in direct model calculations needed in optimization (e.g., see Hohmann, 1973; Hohmann and Ting, 1978; Oristiglio and Hohmann, 1984).

SUMMARY

Natural-field electromagnetic methods are capable of providing information about electrical properties at depths of tens and hundreds of kilometers. In early applications, limitations to our knowledge of the requirements for useful results led to surveys that yielded somewhat ambiguous results. For example, it was not generally recognized that the static shift effect could markedly change the interpretation of an MT sounding. It is now recognized that an auxilliary measurement of the near-surface part of the conductivity profile using some other electrical sounding method can essentially eliminate the ambiguity of static shift. In addition, it appears that if a spatial array of MT soundings is interpreted simultaneously, the effect of variations in surficial conductance changes can be removed. Thus, for data sets that are adequately sampled in both time and space, high-quality interpretations can be made. Unfortunately, such adequate sampling is rare in existing data, particularly in the case of older surveys.

REFERENCES CITED

Alldredge, L. R., 1977, Deep mantle conductivity: Journal of Geophysical Research, v. 82, no. 33, p. 5427–5431.

Adam, A., 1979, Connection between the electric conductivity increase due to phase transition and heat flow, *in* Schmucker, U., ed., Electromagnetic induction in the earth and moon: Dordrecht, D. Reidel Publishing Co., p. 115–123.

Alvarez, R., Reynoso, J. P., Alverez, L. J., and Martinez, M. L., 1978, Electrical conductivity of igneous rocks; Composition and temperature relations: Bulletin Volcanologique, v. 41, p. 317–327.

Ander, M. E., 1981, Magnetotelluric exploration in Arizona and New Mexico for hot dry rock geothermal energy using SQUID magnetometers, *in* Weinstock, H., and Overton, W. C., Jr., SQUID applications to geophysics: Society of Exploration Geologists, p. 61–65.

Banks, R. J., 1967, Geomagnetic variations and the electrical conductivity of the upper mantle: Geophysical Journal of the Royal Astronomical Society, v. 17, p. 457–487.

—— , 1972, The overall conductivity distribution in the earth: Journal of Geomagnetism and Geoelectricity, v. 24, p. 337–351.

—— , 1973, Data processing and interpretation in geomagnetic deep sounding data: Physics of the Earth and Planetary Interiors, v. 7, p. 339–348.

—— , 1979, The use of equivalent current systems in the investigation of geomagnetic deep sounding data: Geophysical Journal of the Royal Astronomical Society, v. 56, p. 139–157.

Berdichevskiy, M. N., and Dmitriev, V. I., 1976, Basic principles of interpretation of magnetotelluric sounding curves, *in* Adam, A., ed., Geoelectric and geothermal studies: Budapest, Akademai Kiado, p. 165–221.

Berdichevskiy, M. N., and Zhdanov, M. S., 1984, Advanced theory of deep electromagnetic sounding: Amsterdam, Elsevier, 408 p.

Berdichevskiy, M. N., Vanyan, L. L., Feldman, I. S., and Porstendorfer, G., 1972, Conducting layers in the Earth's crust and upper mantle: Gerlands Beitr. Geophys., v. 81, p. 187–196.

Berdichevskiy, M. N., Dmitriev, V. I., and Merschchnikova, N. A., 1974, The study of gradient media by deep electromagnetic sounding: Izvetiya AN SSSR, Fizika Zemli, v. 6, p. 61–72.

Berdichevskiy, M. N., Fainberg, E. B., Rotanova, N. M., Smirnova, J. B., and Vanjan, L. L., 1976, Deep electromagnetic investigations: Annals of Geophysics, v. 32, p. 143–155.

Berdichevskiy, M. N., Zhdanov, M. S., and Fainberg, E. B., 1976, Electrical conductivity functions in the magnetotelluric and magnetovariation methods: Annals of Geophysics, v. 32, no. 3, p. 301–318.

Cagniard, L., 1953, Basic theory of the magnetotelluric method of geophysical prospecting: Geophysics, v. 18, no. 3, p. 605–635.

Cantwell, T., and Orange, A. S., 1965, Further deep resistivity measurements in the Pacific Northwest: Journal of Geophysical Research, v. 70, no. 16, p. 4068–4072.

Cantwell, T., Galbraith, J. N., and Nelson, P., 1964, Deep resistivity results from New York and Virginia: Journal of Geophysical Research, v. 69, p. 4367–4376.

Cantwell, T., Nelson, P., Webb, J., and Orange, A. S., 1965, Deep resistivity measurements in the Pacific Northwest: Journal of Geophysical Research, v. 70, no. 8, p. 1931–1937.

Cermak, V., and Lastovickova, M., 1986, Temperature profiles in the earth of importance to deep electrical conductivity models: Physics of the Earth and Planetary Interiors (in press).

Chapman, S., and Bartels, J., 1940, Geomagnetism; Volume 2, Analysis of data and physical theories: Oxford, London, p. 543–1049.

Clarke, J., and Goldstein, N. E., 1981, Magnetotelluric measurements, *in* Weinstock, H., and Overton, W. C., Jr., SQUID applications to geophysics: Society of Economic Geologists, p. 49–59.

Courtillot, V., and LeMouel, J. L., 1979, Comments on "Deep mantle conductivity": Journal of Geophysical Research, v. 84, p. 4785–4790.

Currie, R. G., 1968, Geomagnetic spectrum of internal origin and lower mantle conductivity: Journal of Geophysical Research, v. 73, p. 2779–2786.

Dewan, J. T., 1983, Essentials of open-hole log interpretation: Tulsa, Oklahoma, PennWell Books, 361 p.

Ducruix, J., Courtillot, V., and Le Mouel, J. L., 1979, The late 1960s secular variation impulse, the eleven year magnetic variation, and the electrical conductivity of the deep mantle: Geophysical Journal of the Royal Astronomical Society, v. 61, p. 73–94.

Duba, A. S., 1972, Electrical conductivity of olivine: Journal of Geophysical Research, v. 77, p. 2483–2495.

Dvorak, Z., 1973, Electrical conductivity of several samples of olivinites, peridotites, and dunites as a function of pressure and temperature: Geophysics, v. 38, p. 14–24.

Echkardt, D., 1968, Theory and interpretation of electromagnetic impedance of the earth: Journal of Geophysical Research, v. 73, p. 5317–5326.

Eckhardt, D., Larner, K., and Madden, T., 1963, Long-period magnetic fluctuations and mantle electrical conductivity estimates: Journal of Geophysical Research, v. 68, no. 23, p. 6279–6289.

Fainberg, E. B., 1981, On the inverse problem of deep electromagnetic sounding of the earth: Geomagnetism and Aeronomy, v. 21, p. 715–719.

Fainberg, E. B., and Rotanova, N. M., 1974, Distribution of conductivity and temperature inside the earth by other data of deep electromagnetic sounding: Geomagnetism and Aeronomy: v. 10, p. 674–678.

Feldman, I. S., 1976, On the nature of conductive layers in the earth's crust and upper mantle, *in* Adam, A., ed., Geoelectric and geothermal studies: Budapest, Akademiai Kiado, p. 721–730.

Filloux, J. H., 1979, Magnetotelluric and related electromagnetic investigations in geophysics: Reviews of Geophysics and Space Physics, v. 17, no. 2, p. 282–294.

Frazer, M. C., 1974, Geomagnetic deep sounding with arrays of magnetometers: Reviews of Geophysics and Space Physics, v. 12, no. 3, p. 401–420.

Gauss, C. F., 1838, Allgemeine Theorie des Erdmagnetisms: Resultate magn. Verein 1838, Werk, v. 5, p. 121–193.

Golovkov, V. N., Kolomijtseva, G. I., Berdichevskiy, M. N., and Rotanova, N. M., 1971, On the earth's conductivity determination by the data on secular geomagnetic field variations: Geomagnetism and Aeronomy, v. 11, p. 1197–1229.

Gough, D. I., 1973, The interpretation of magnetometer array studies: Geophysical Journal of the Royal Astronomical Society, v. 35, p. 83–98.

—— , 1974, Electrical conductivity under western North America in relation to heat flow, seismology, and structure: Journal of Geomagnetism and Geoelectricity, v. 26, p. 105–123.

—— , 1983, Electromagnetic geophysics and global tectonics: Journal of Geophysical Research, v. 88, p. 3367–3377.

Gough, D. I., and Ingham, M. R., 1983, Interpretation methods for magnetometer arrays: Reviews of Geophysics and Space Physics, v. 21, no. 4, p. 805–827.

Gusarov, A. L., 1981, On uniqueness of solution of inverse magnetotelluric problem for two-dimensional media, *in* Demitriev, V. I., ed., Mathematical models in geophysics: Moscow University, p. 31–61.

Haak, V., 1982, Electrical conductivity of minerals and rocks at high temperatures and pressures, *in* Angenheister, G., ed., Physical properties of rocks: Berlin, Sprinter-Verlag, p. 291–307.

Hartmann, O., 1963, Behandlung lokaler erdmagetischer Felder als Randwertaufgabe der potentialtheorie: Abh Akad Wiss Goltingen Muth-phys. kl Beitr. Intl. Geophy. Jubar, 9.

Hermance, J. J., 1982, The asymptotic response of three-dimensional basin offsets to magnetotelluric fields at long periods; The effects of current channeling: Geophysics, v. 44, no. 11, p. 1562–1573.

Hohmann, G. W., 1973, Three-dimensional induced polarization and electrical geophysics, v. 40, no. 2, p. 309.

Hohmann, G. W., and Ting, S. C., 1978, Integral equation modelling of three-dimensional magnetotelluric response: Geophysics, v. 46, no. 2, p. 182–197.

Hummel, I., 1929, Der sheinbare spezifische Widerstand bei vier planparallelen schichter: Zeitschr. fur Geophysik, no. 5-6, pages?

Hutton, V.R.S., 1976, The electrical conductivity of the Earth and planets: Reports on Progress in Physics, v. 39, p. 487–572.

Jupp, D.L.B., and Vozoff, K., 1975, Stable iterative methods for the inversion of geophysical data: Geophysical Journal of the Royal Astronomical Society, v. 42, p. 957–976.

—— , 1977, Two-dimensional magnetotelluric investion: Geophysical Journal of the Royal Astronomical Society, v. 50, p. 333–352.

Karlya, K. A., and Shankland, T. J., 1983, Electrical conductivity of dry lower crustal rocks: Geophysics, v. 48, no. 1, p. 52–61.

Kaufmann, A. A., and Keller, G. V., 1981, The magnetotelluric sounding method: Amsterdam, Elsevier, 595 p.

—— , 1983, Frequency and time domain electromagnetic sounding: Amsterdam, Elsevier, 643 p.

Keller, G. V., 1963, Electrical properties in the deep crust: Institute of Electrical and Electronic Engineers on Ant. and Prop., April 11, v. 3, p. 344–357.

—— , 1964, Evaluation of geological effects on magnetotelluric signals: U.S. Geological Survey Open-File Report, 33 p.

—— , 1966, Dipole method for deep resistivity studies: Geophysics, v. 31, no. 6, p. 1088–1104.

—— , 1967, Electrical prospecting for oil: Quarterly, Colorado School of Mines, v. 63, no. 2, p. 1–268.

—— , 1968, Statistical study of electric fields from earth-return tests in the western states compared with natural electric fields: Institute of Electrical and Electronic Engineers Transactions PAS., PAS-87, no. 4, p. 1050–1057.

—— , 1971a, Natural-field and controlled-source methods in electromagnetic exploration: Geoexploration, v. 9, p. 99.

—— , 1971b, Electrical studies of the crust and upper mantle, *in* Heacock, J. G., ed., The structure and physical properties of the earth's crust: American Geophysical Union Geophysical Monograph 14, p. 107.

—— , 1975, DC resistivity methods for determining resistivity in the Earth's crust: Physics of the Earth and Planetary Interiors, v. 10, p. 201–208.

—— , 1982, Electrical properties of rocks and minerals, *in* Carmichael, R. S., ed., Handbook of physical properties of rocks, v. 1: Boca Raton, Florida, CRC Press, p. 217–272.

Keller, G. V., and Frischkecht, F. C., 1966, Electrical methods in geophysical prospecting: Oxford, Pergamon Press, 526 p.

Keller, G. V., and Furgerson, R. B., 1977, Determining the resistivity of a resistant layer in the crust, *in* Heacock, J. G., ed., The earth's crust: American Geophysical Union Geophysical Monograph 20, p. 440–469.

Keller, G. V., and Jacobson, J. J., 1983a, Deep electromagnetic soundings northeast of the Geysers steam field: Geothermal Resources Council Transactions, v. 7, p. 497–502.

—— , 1983b, Megasource electromagnetic survey in the Bruneau-Grandview area, Idaho: Geothermal Resources Council Transactions, v. 7, p. 505–510.

Keller, G. V., Anderson, L. A., and Pritchard, J. I., 1966, Geological survey investigation of the electrical properties of the crust and upper mantle: Geophysics, v. 31, no. 6, p. 1078–1087.

Kertz, W., 1954, Modelle fur erdmagnetics chen induzierte elektrische Strom in Undergrund: Nachr. Adkd. Wiss. Gottingen Math-Phys. Klasses, v. 11A, p. 101–110.

Khitarov, N. I., Slutsky, A. B., and Pugin, V. A., 1970, Electrical conductivity of basalts at high T-P and phase transitions under mantle conditions: Physics of Earth and Planetary Interiors, v. 3, p. 334–531.

Kolomijtseva, G. I., 1972, On electrical conductivity distribution in the earth's mantle by geomagnetic field secular variation data: Geomagnetism and Aeronomy, v. 12, p. 1082–1086.

Lahiri, B. N., and Price, A. T., 1939, Electromagnetic induction in non-uniform conductors, and the determination of the conductivity in the earth from terrestrial magnetic variations: Philosophical Transactions of the Royal Society, series A, v. 237, p. 509–531.

Lebedev, E. B., and Khitarov, N. I., 1964, Beginning of melting in granite and the electrical conductivity of its melts in relation to pressure of pore water: Geokhimiya, v. 3, p. 195–201.

Lilley, F.E.M., 1975, Magnetometer array studies; A review of the interpretation of observed fields: Physics of the Earth and Planetary Interiors, v. 10, no. 3, p. 231–240.

Losecke, W., and Muller, W., 1975, Two-dimensional magnetotelluric calculations for overhanging, high-resistivity structures: Journal of Geophysics, v. 41, p. 311–319.

Lubimova, E. A., and Feldman, I. S., 1970, Heat flow, temperature, and electrical conductivity of the crust and upper mantle in the USSR: Tectonophysics, v. 10, p. 245–281.

Madden, T. R., 1971, Resolving power of geoelectric measurements for locating resistive zones within the crust, *in* Heacock, J. G., eds., The structure and physical properties of the earth's crust: American Geophysical Union Geophysical Monograph 14, p. 95–104.

Madden, T. R., and Swift, C. M., Jr., 1969, Magnetotelluric studies of the electrical structure of the crust and upper mantle, *in* Part, P. J., ed., The earth's crust and upper mantle: American Geophysical Union Geophysical Monograph 13, p. 469–479.

McDonald, K. L., 1957, Penetration of the geomagnetic secular field through a mantle with variable conductivity: Journal of Geophysical Research, v. 63, p. 117–141.

Murase, T., and McBirney, A. R., 1973, Properties of some common igneous rocks and their melts at high temperatures: Geological Society of America Bulletin, v. 84, p. 3563–3592.

Noritomi, K., 1961, The electrical conductivity of rock and the determination of the electrical conductivity of the earth's interior: J. Min. Coll., Akita Univ., series A, v. 1, no. 1, p. 27–59.

Olhoeft, G. R., 1976, Electrical properties of rocks, *in* Strens, R.G.J., ed., The physics and chemistry of rocks and minerals: London, Wiley, p. 261–278.

—— , 1981, Electrical properties of granite with implications for the lower crust: Journal of Geophysical Research, v. 86, p. 931–936.

Oristaglio, M. L., and Hohmann, G. W., 1984, Diffusion of electromagnetic fields into a two-dimensional earth; A finite difference approach: Geophysics, v. 49, p. 870–894.

Parkhomenko, E. I., 1967, Electrical properties or rocks: New York, Plenum Press, 268 p.

—— , 1982, Electrical resistivity of minerals and rocks at high temperature and pressure: Reviews of Geophysics and Space Physics, v. 20, no. 2, p. 193–218.

Presnall, D. C., Simmons, C. L., and Porath, H., 1972, Changes in the electrical conductivity of a synthetic basalt during melting: Journal of Geophysical Research, v. 77, p. 5665–5672.

Price, A. T., 1949, The induction of electric currents in non-uniform thin sheets and shells: Quarterly Journal of Mechanics and Applied Mathematic, v. 2, p. 283–310.

—— , 1967a, Electromagnetic induction within the earth, *in* Matsushita, S., and Campbell, W., Physics of geomagnetic phenomena: New York, Academic Press, p. 235–298.

—— , 1967b, Magnetic variations and telluric currents, *in* Gaskell, T. F., ed., The earth's mantle: London, Academic Press, p. 125–170.

—— , 1970, The electrical conductivity of the earth: Quarterly Journal of the Royal Astronomical Society, v. 11, p. 23–42.

Reddy, I. K., Rankin, D., and Phillips, R. J., 1977, Three-dimensional modeling in magnetotelluric and magnetic variational sounding: Geophysical Journal of the Royal Astronomical Society, v. 51, p. 313–325.

Rikitake, T., 1966, Electromagnetism and the earth's interior: Amsterdam, Elsevier, 308 p.

Rostoker, G., 1966, Mid-latitude transition bays and their relation to the spatial movement of overhead current systems: Journal of Geophysical Research, v. 71, p. 79–95.

—— , 1972, Polar magnetic substorms: Reviews in Geophysics and Space Physics, v. 100, p. 157–211.

Runkorn, S. K., 1955, The electrical conductivity of the earth's mantle: EOS Transactions of the American Geophysical Union, v. 36, p. 191–198.

Scheube, H. G., 1958, Die Losungen der Dirichlestschen und Neumannschen Randwertaufgabe als Hilfsnmittle zue Rehandlung der Problemen des Endmagnetismus: Abhandlunger der Akad. Wiss. Gottingen, Math-Phys. Kl., Beitr. Int. Geophys. Johr., v. 4, p. 1-32.

Schmucker, U., 1960, Deep anomalies in electrical conductivity: Scripps Institution of Oceanography Report 61-83, 22 p.

—— , 1964, Anomalies of geomagnetic variations in the south-western United States: Journal of Geomagnetism and Geoelectricity, v. 15, p. 193–221.

Shankland, T. J., 1975, Electrical conduction in rocks and minerals; Parameters for interpretation: Physics of the Earth and Planetary Interiors, v. 10, p. 209–219.

Shankland, T. J., and Ander, M. E., 1983, Electrical conductivity, heat flow, and fluids in the lower crust: Journal of Geophysical Research, v. 88, p. 9457–9484.

Shankland, T. J., and Waff, H. S., 1977, Partial melting and electrical conductivity anomalies in the upper mantle: Journal of Geophysical Research, v. 82, p. 5409–5417.

Shankland, T. J., O'Connell, R. J., and Waff, H. S., 1981, Geophysical constraints on partial melt in the upper mantle: Reviews of Geophysics and Space Physics, v. 19, no. 3, p. 394–406.

Siebert, M., 1958, Die zeregung eines lokalen erdmagnetischen felds in ausseren und inneren anteil mit helfe des zweidimensionalen Fourier-theorem: Abhandlungen der Akad. Wiss. Gottingen, Math-Phys. Kl., Beitr. Int. Geophys. Jahr, v. 4, p. 33-38.

—— , 1962, Die Zerlegung eines zweidimensionalen Magnetfeldes in ausseren und inneren anteil mit helft der Cauchyschen Integralformel: Z. Geophys., v. 5, p. 231–236.

Siebert, M., and Kertz, W., 1957, Zur Zerlegung eines lokolen erdmagnetischen feldes in ausseren und inneren Antiel: Nachr. Akad. Wiss. Goltingen, Math Phys. Kl., v. 2a, p. 82–112.

Sochelnikov, V. V., 1968, On the conductivity determination of deep layers of the earth: Izvestiya AN SSSR, Fizika Zemli, v. 7, p. 65–71.

Stodt, J. A., Hohmann, G. W., and Ting, S. C., 1981, The telluric-magnetotelluric method in two-dimensional and three-dimensional environments: Geophysics, v. 46, no. 8, p. 1137.

Swift, C. M., 1971, Theoretical magnetotelluric and Turam response from two-dimensional inhomogeneities: Geophysics, v. 36, p. 38–52.

Tikhonov, A. N., 1950, On the determination of electrical characteristics of deep layers of the Earth's crust: Dokl. Akad. Nauk SSSR, v. 73, p. 295–297.

—— , 1965, Mathematical basis of the electromagnetic sounding: Zh. Vischisl. Mat. Fiz., v. 5, p. 207–211.

Tozer, D. C., 1959, The electrical properties of the earth's interior: Phys. Chm. Earth, v. 3, p. 414–436.

—— , 1969, Electrical conductivity of the mantle, *in* Hart, P. J., ed., The earth's crust and upper mantle: American Geophysical Union Geophysical Monograph 13, p. 618–621.

—— , 1970, Temperature, conductivity, compositions, and heat flow: Journal of Geomagnetism and Geoelectricity, v. 22, p. 34–43.

—— , 1981, The mechanical and electrical properties of earth's asthenosphere: Physics of Earth and Planetary Interiors, v. 25, p. 280–296.

Tyburczy, J. A., and Waff, H. S., 1983, Electrical conductivity of molten basalt and andesite to 25 kilobars pressure; Geophysical significance and implications for charge transport and melt structure: Journal of Geophysical Research, v. 88, no. B, p. 2413–2430.

Vanyan, L. L., 1981, Deep geoelectrical models; Geological and electromagnetic principles: Physics of Earth and Planetary Interiors, v. 25, p. 273–279.

Vanyan, L. L., and Berdichevskiy, M. N., 1977, The study of the asthenosphere of the East European Platform by electromagnetic sounding: Physics of Earth and Planetary Interiors, v. 12, p. 237–251.

Van Zijl, J.S.V., 1969, A deep Schlumberger sounding to investigate the electrical structure of the crust and upper mantle in South Africa: Geophysics, v. 34, no. 3, p. 450.

—— , 1975, A crustal geoelectrical model for South African Precambrian terrains based on deep Schlumberger sounding: Geophysics, v. 40, no. 4, p. 657.

Van Zijl, J.S.V., Hugo, P.L.V., and De Belloacq, J. H., 1970, Ultra deep Schlumberger sounding and crustal conductivity structure in South Africa: Geophysical Prospecting, v. 18, no. 4, p. 615.

Volarovich, M. P., and Parkhomenko, E. I., 1976, Electrical properties of rocks at high temperatures and pressures, *in* Adam, A., ed., Geoelectrical and geothermal studies: Budapest, Akademiai Kiado, p. 321–372.

Vozoff, K., 1972, The magnetotelluric method in the exploration of sedimentary basins: Geophysics, v. 37, no. 1, p. 98–141.

Waff, H. S., and Weill, D. F., 1975, Electrical conductivity of magmatic liquids; Effects of temperature, oxygen fugacity, and composition: Earth and Planetary Science Letters, v. 28, p. 254–261.

Wannamaker, P. E., Hohmann, G. W., and Ward, S. H., 1984, Magnetotelluric responses of three-dimensional bodies in layered earths: Geophysics, v. 49, no. 9, p. 1517.

Weidelt, R., 1975, Entwicklung und Erprobung eines Verfahrens zur dnversion zeveidimensionaler leftahig keitsstructures in E-polarization [Thesis]: Goltingen Universtat.

Word, D. R., Smith, H. W., and Bostick, F. X., Jr., 1971, Crustal investigations by the magnetotelluric tensor impedance method, *in* Heacock, J. G., ed., The structure and physical properties of the earth's crust: American Geophysical Union Geophysical Monograph 14, p. 145–167.

Yanovskiy, B. M., 1978, The earth's magnetism: Leningrad State University, 591 p.

Yukutake, T., 1959, Attenuation of geomagnetic secular variation through the conducting mantle of the earth: Tokyo University Bulletin of the Earthquake Research Institute, v. 37, p. 13–32.

Manuscript Accepted by the Society October 31, 1988

Printed in U.S.A.

Geological Society of America
Memoir 172
1989

Chapter 7

Laboratory techniques for determining seismic velocities and attenuations, with applications to the continental lithosphere

Nikolas I. Christensen and William W. Wepfer*
Department of Earth and Atmospheric Sciences, Purdue University, West Lafayette, Indiana 47907

ABSTRACT

The interpretation of seismic data is dependent on laboratory investigations of the elastic and anelastic properties of rocks. Important constraints on the composition of the continental lithosphere have been provided by comparing laboratory and field determined seismic velocities. Much less is known about the nature of attenuation in the crust and upper mantle. Furthermore, laboratory attenuations have not been studied as extensively as velocities. Nevertheless, the utilization of laboratory attenuation measurements to tie seismic data to the anelastic properties of rocks is promising. Laboratory velocity measurements are usually obtained with a pulse transmission technique, whereas measurements of seismic attenuation in the laboratory are determined by resonance techniques, ultrasonic pulse propagation, stress-strain hysteresis loop analysis, and torsion-pendulum oscillations. Velocities in rocks increase with increasing pressure, whereas attenuations decrease. The greatest changes, which occur over the first 100 MPa, are attributed to the closure of microcracks. Velocities decrease with increasing temperature. At temperatures below the boiling point of a rock's volatiles, attenuation appears to be temperature-independent, and above this, attenuation decreases. Increasing pore pressure, which lowers compressional and shear-wave velocities in sedimentary and crystalline rocks, produces a marked increase in Poisson's ratio. The influence of pore pressure on attenuation in crystalline rocks is at present poorly understood. At high pressures, velocities are primarily a function of mineralogy, whereas mineralogy may only be of secondary importance for attenuation. Anisotropy, which is common in velocities of many crustal and upper-mantle rocks, may also be an important property of rock attenuation. Attenuation may vary significantly with frequency for saturated rocks, but appears to be frequency-independent for dry rocks. The effect of frequency on velocity is minimal.

INTRODUCTION

Although many disciplines have contributed significantly to our knowledge of the nature of the continental crust and upper mantle, none has provided resolution comparable to seismologic studies. Since 1910, when A. Mohorovičić first presented evidence for a major seismic discontinuity in the Balkans at a depth of approximately 30 km, crustal thickness has been largely determined by seismic refraction investigations. These studies have also provided geophysicists with worldwide information on compressional wave velocities at various crustal depths, as well as upper-mantle velocities. Recently, significant data have become available on crustal velocity gradients, velocity reversals, shear-wave velocities, and anisotropy in the form of azimuthal variations and shear-wave splitting. Studies of the attenuation of seismic waves give additional insight into the nature of the continental lithosphere. In addition, large-scale reflection studies, which currently represent major geophysical efforts in many

*Present address: Amoco Production Company, Houston, Texas 77253.

Christensen. N. I., and Wepfer, W. W., 1989, Laboratory techniques for determining seismic velocities and attenuations, with applications to the continental lithosphere, *in* Pakiser, L. C., and Mooney, W. D., Geophysical framework of the continental United States: Boulder, Colorado, Geological Society of America Memoir 172.

countries, continue to provide new and exciting information on crustal structure and the nature of the Mohorovičić discontinuity.

Crustal and upper-mantle velocities and their attenuations are used to infer mineralogy, porosity, the nature of fluids occupying pore spaces, temperature at depth, and present or paleo-lithospheric stress reflected by mineral and crack orientation. Reflections within the continental crust and upper mantle originate from contrasts of acoustic impedances, defined as products of velocity and density. The interpretation of this seismic data is dependent on detailed knowledge of the elastic and anelastic properties of rocks provided by laboratory investigations. This chapter briefly reviews some of the techniques and major findings of laboratory seismology that apply to the exploration of the continental lithosphere.

VELOCITY MEASUREMENTS

In 1960, Francis Birch published the first comprehensive study of laboratory seismic velocities in crystalline rocks. Subsequent work by Birch and his research group (e.g., Birch, 1961, 1972; Simmons, 1964; Christensen, 1965, 1966) provided a foundation for future studies of rock elasticity pertinent to lithospheric velocities. At present, several laboratories throughout the world are actively engaged in studies of the seismic properties of rocks.

Although a more elaborate pulse matching scheme described by Mattaboni and Schreiber (1967) has been used, the technique for velocity measurements commonly employed in these laboratories is similar to or slightly modified from the pulse transmission method described by Birch (1960). This consists of determining compressional or shear-wave traveltimes through cylindrical rock specimens. Transducers are placed on the ends of the rock core. The sending transducer converts the input, an electrical pulse of 50 to 500V and 0.1 to 10 μsec width, to a mechanical signal, which is transmitted through the rock. The receiving transducer changes the wave to an electrical pulse, which is amplified and displayed on an oscilloscope screen (Fig. 1). Once the system is calibrated for time delays, the traveltime through the specimen is determined directly on the oscilloscope or with the use of a mercury delay line (Fig. 2). The major advantage of the delay line is that it increases the precision, especially for signals with slow rise times, because the gradual onset of the first arrival from the sample is approximated by the delay line.

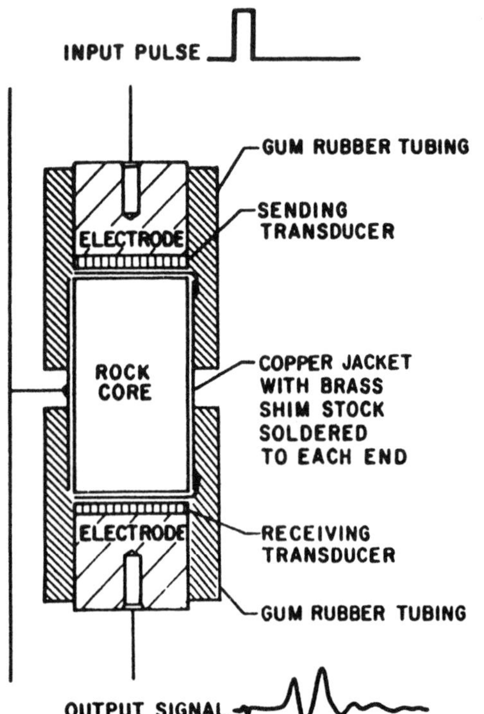

Figure 1. Rock sample assembly for high-pressure velocity measurements (Christensen, 1985).

The mercury delay line (see Christensen, 1985, Fig. 4) consists of a fixed and a moveable transducer mounted in a column of mercury. The signal from the delay line is displayed as a second trace on the oscilloscope. The distance between the two transducers is adjusted so that the delay line signal is superimposed on the signal from the sample. With proper calibration, the time of flight of the signal through the sample is then equivalent to that through the delay line; the rock velocity is simply calculated from the length of the mercury separating the transducers, the length of the sample, and the velocity of mercury.

Generally the rock specimens are cores 2.54 cm in diameter and 4 to 6 cm long. Bulk densities are calculated from dimensions and weights. For measurements at elevated pressures, the cores are jacketed with copper foil and rubber tubing to prevent high-pressure oil from entering microcracks and pores (Fig. 1). The core ends are either coated with silver conducting paint, or a strip of brass is spot-soldered to the copper jacket at each end to provide an electrical ground for the transducers. For measurements at high temperatures and pressures where gas is the pressure medium, the samples are usually encased in stainless steel. Velocity measurements as functions of confining and pore pressure require an additional pressure generating system and a more elaborate sample jacketing procedure, since the pore fluid must be isolated from the confining pressure fluid.

ATTENUATION MEASUREMENTS

Seismic wave attenuation has great potential as a tool to yield a better understanding of the anelastic properties, and hence the physical state, of rocks in the continental lithosphere. Because of this potential, an expanding body of laboratory work has concentrated on bringing to fruition the diagnostic capabilities of attenuation measurements.

Unlike velocity determinations, there is no common method used for measuring seismic attenuation in the laboratory. There is not even a standard definition of attenuation (O'Connell and Budiansky, 1978). The three parameters most often reported as the attenuation are the seismic quality factor Q, also referred to as the specific attenuation Q^{-1}, the attenuation coefficient α, and the

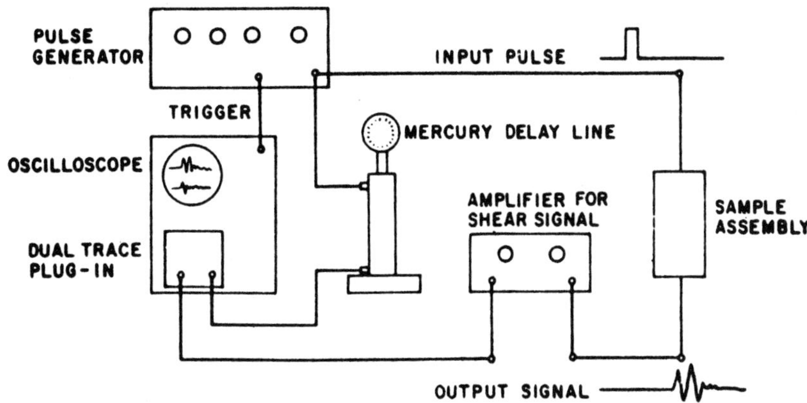

Figure 2. Electronics for velocity measurements (Christensen, 1985).

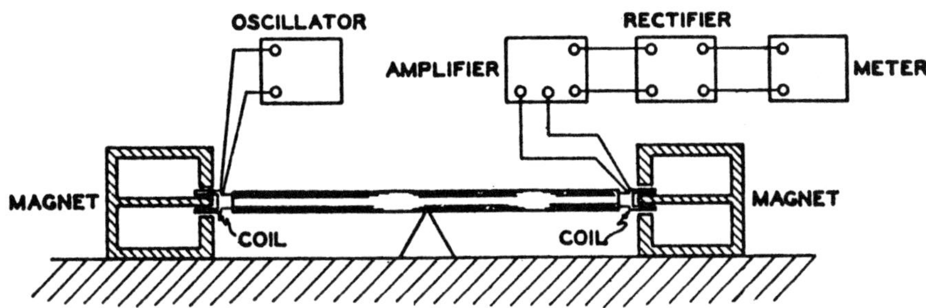

Figure 3. An example of a resonant bar apparatus (Born, 1941).

logarithmic decrement δ. These are related for low-loss materials ($Q > 10$) by

$$\frac{1}{Q} = \frac{\alpha V}{\pi f} = \frac{\delta}{\pi}, \tag{1}$$

where V is the phase velocity and f is the frequency (Johnston and Toksoz, 1981). In both the field and laboratory, difficulties arise in separating the intrinsic dissipation of the rock, i.e., processes by which seismic energy is converted into heat, from geometric spreading, transmission losses, scattering, and other factors. Nevertheless, the utilization of laboratory attenuation measurements to tie seismic data to the anelastic properties of rocks is promising, and the refinement of laboratory techniques and the theory concerning the mechanisms involved has yielded and will continue to supply valuable insights into the structure and composition of the continental crust and upper mantle.

The methods used in determining the attenuation can be separated into three categories: resonance techniques, ultrasonic pulse propagation, and low-frequency methods. Each method is evaluated in terms of the ease with which accurate results are obtained as functions of pressure (confining and pore), temperature, strain amplitude, and frequency.

The resonant bar method has had the widest application. Its use in measuring velocity and attenuation extends back to the earliest days of laboratory seismology. A diagram of the equipment used by Born (1941) is shown in Figure 3. The technique consists of fashioning a long, thin rod out of the material, clamping or suspending it at a vibrational node (typically the midpoint), and vibrating it at one of its resonant modes f_n. Extensional, torsional, and flexural motion may be employed, and the velocity for the first two is given by

$$V = \frac{2f_nL}{n} \quad n = 1, 2, \ldots, \tag{2}$$

where L is the length of the sample and n is the mode. Analogous to the Q of an electronic resonant circuit, the seismic Q is defined as

$$Q = \frac{f_n}{\Delta f}, \tag{3}$$

where Δf is the frequency spread between the 3 dB points below the resonance peak f_n. The attenuation may also be calculated from the decay of resonance after the driving force has been removed, which is characterized by the time constant of resonant decay τ (similar to that of an R-C circuit), and Q is then determined from

$$Q = \tau\pi f_n. \tag{4}$$

Practical considerations abound in the application of the resonant bar technique. The ratio of length to radius for the bar should be at least 10 to 1 (Tittmann, 1977); thus a 2.5-cm sample radius necessitates a length of at least 0.25 m, and a bar closer to 1 m in length is preferable. This also means that the sample must be well consolidated, have grain sizes smaller than the radius, and be relatively free of fractures. Such constraints impose a bias on the measurements. Application of confining and possibly pore pressure requires jacketing the sample, and this will change the resonant frequency of the rod. Damping of the resonance by the confining fluid lowers Q in the extensional and flexural cases (not significantly for torsional [Birch and Bancroft, 1938]); for this reason, a gas (such as helium because its long sonic wavelengths avoid standing waves in the pressure vessel [Gardner and others, 1964]) is often used as the confining pressure medium. An abundance of transducer configurations for exciting and measuring the vibrations have been used, with piezoelectric, electromagnetic, and electrostatic transducers being applied in many novel ways. Losses in the transducers may be significant. Mounting transducers on the ends of the sample affect the resonance peak, as will the end caps required for measurements at elevated pressures. This shift is purposely employed by Tittmann (1977) to end load the rod in the flexural mode with copper blocks, thereby lowering the resonant frequency. Most if not all of the above problems can either be correct or made negligible in determining the intrinsic Q of the sample.

The vibrational spectra of spherical rock specimens have also been employed to determine Q. Used by Birch (1975) and Mason and others (1978) on rocks, this technique has the distinct advantage of providing a wide spectrum of vibrational modes, as opposed to the limited number of modes available in any given rod configuration. The method consists of placing a sphere between a sending and receiving transducer and sweeping a variable-frequency oscillator very slowly through the frequency range. In his measurements, for example, Birch (1975) covered frequencies between about 25 and 250 kHz. Peaks corresponding to various modes are observed, and Q is calculated from equation (3). Tone bursts may also be applied with the subsequent decay of resonance giving Q (equation (4)). The spherical resonance method is tedious to implement because: (1) spheres of rock are difficult to fashion and are usually not isotropic, (2) support of the sphere at a vibrational node is attainable only for torsional modes, and (3) it is troublesome to apply pressure and temperature beyond room conditions. Birch also found that this method gives erratic results due to noise and asymmetrical peaks.

Resonance methods can routinely accommodate variable frequency attenuation measurements. They have been used in the 0.1- to 100-kHz frequency range, with most of the data occurring between 1 and 20 kHz. Variable strain has been analyzed directly (Winkler and Nur, 1982; Murphy, 1982) and indirectly (Johnston and Toksoz, 1980b). Temperature dependence has been measured under room conditions (Kissell, 1972; Attewell and Brentnall, 1964; Gordon and Davis, 1968) and at elevated pressures (pore and confining) (Jones and Nur, 1983). Measuring attenuation with resonance methods under pressure is fairly involved; it has been carried out by Gardner and others (1964) and Katahara and others (1982), for example.

Ultrasonic pulse propagation has been used only recently for attenuation measurements in rocks, although its use in nondestructive evaluation attenuation measurements in materials research extends back to the early 1950s (Roderick and Truell, 1952). Through-transmission uses separate source and receiver piezoelectric transducers on opposite ends of a cylindrical rock sample, similar to that described above for velocity measurements. Pulse-echo uses a single transducer for transmitting and detecting the signal, with the one or more echos observed on the oscilloscope having been reflected from the rock core face opposite the transducer. Unlike the resonance methods, no peak is observed from which to determine Q; rather, it is assumed that the amplitude reduction of the pulse with distance A(x) can be expressed in terms of a damped exponential

$$A(x) = A_0 e^{-\alpha x} \tag{5}$$

where A_0 is the initial amplitude and α is the attenuation coefficient; α is therefore determined in wave-propagation measurements, and equation (1) relates this to a Q value.

An advantage of wave-propagation methods is the comparative ease with which high-pressure measurements may be performed (Johnston and Toksoz, 1980a; Winkler, 1983, 1985; Ramana and Rao, 1974). Few variable strain and temperature attenuation measurements have been performed using pulse propagation. Measuring attenuation as a function of frequency is difficult for most ultrasonic pulse methods; Q values are typically reported at a single frequency, and when they are not, they are strictly band-limited. Furthermore, the frequencies used are all above 100 kHz and are normally about 1 MHz, placing them well outside the range of seismic field measurements. Only the technique of Winkler and Plona (1982) is designed to deliver Q as a function of frequency. Some or all of the following assumptions are made when ultrasonic wave propagation is used: Q is either frequency independent or slowly varying over the frequency range of interest, diffraction, beam spreading, coupling and transducer losses, and scattering can be corrected for or made negligible, and the ends of the sample are flat and parallel.

Through-transmission has been used in conjunction with spectral ratios (Toksoz and others, 1979; Sears and Bonner, 1981) and pulse broadening (Gladwin and Stacey, 1974; Ramana and Rao, 1974) to determine the attenuation of rocks. The spectral ratio technique is similar to the attenuation estimation procedure used in field seismology. When the output signal amplitude from sample, A_{sample}, is compared to a reference standard, A_{ref}, with an extremely high Q (such as aluminum, $Q > 10^5$), a linear equation is derived; plotting ln (A_{ref}/A_{sample}) vs. frequency yields a straight line whose slope equals $\pi L/(QV)$, where L is the sample length and V is the phase velocity. In addition to the above assumptions, this procedure assumes that the reference and sample assemblies are identical and that $Q_{ref} >>> Q_{sample}$.

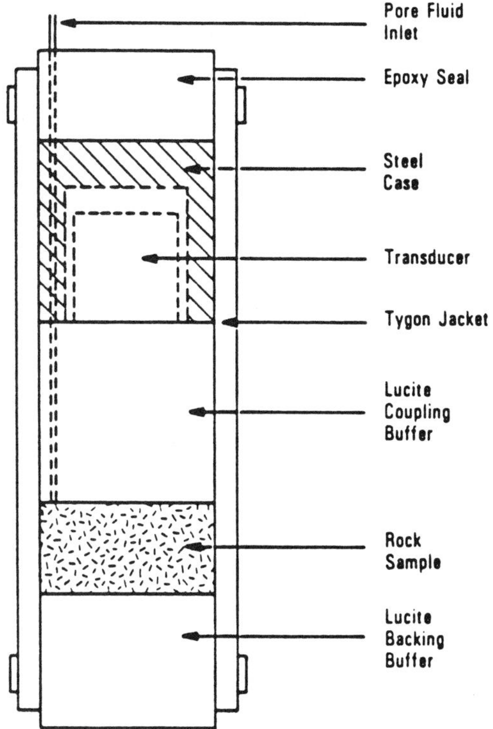

Figure 4. Pulse-echo sample assembly (Winkler and Plona, 1982).

Based on the constant Q assumption, Q may be derived from pulse broadening with increasing traveltime. Only the first arrival and the subsequent peak are needed for the Q calculation, thus making it easy to use on almost any wavetrain. A precise pick of the wave arrival is one of the primary difficulties in applying this method, and it has not seen extensive use.

Pulse-echo experimental procedures have implemented spectral analysis, as well as the exponential decay of multiple reverberations. Figure 4 shows the sample assembly employed by Winkler and Plona (1982) for spectral ratio attenuation determination. Two wavelets, the first reflected from the sample's front face and the second from the back, are separately windowed, and the phase and amplitude spectra are calculated for each. The phase velocity $V(\omega)$ is given by:

$$V(\omega) = \frac{2\omega L}{\Delta\phi}, \tag{6}$$

where L is the sample length and $\Delta\phi$ is the phase difference between the two pulses at frequency ω. The attenuation coefficient α, in decibels per centimeter, is calculated using

$$\alpha(\omega) = \frac{8.686}{L} \ln \left[\frac{A_1(\omega)}{A_2(\omega)} (1 - R_{12}^2(\omega)) \right] \tag{7}$$

where A is the amplitude, R is the reflection coefficient, and the subscripts on A indicate the first or second wavelet. The primary advantage of this method is that transducer problems are effectively suppressed, since the same one is both the source and the receiver. In addition, Q may be measured as a function of frequency, and the coupling buffer provides the reference (high Q) material. Problems include the need to find the reflection coefficient as a function of frequency $R(\omega)$, choosing an adequate buffer material, and the application of diffraction corrections. No use of this technique on igneous or metamorphic rocks has yet been reported. The amplitude reduction of echos entails fitting a damped exponential $A_0e^{-\alpha x}$ to the peaks of the decaying multiples, thereby determining α (Peselnick and Zeitz, 1959). An example of the echos obtained by Mason and Kuo (1971) is shown in Figure 5. This technique requires high Q and/or short samples since a number of multiples are needed for a good estimate of the attenuation.

Low-frequency methods seek to place attenuation measurements within the frequency range of field seismology. Two general approaches have been taken to achieve this: stress-strain hysteresis loop analysis, and the torsion pendulum. Stress-strain curve analysis uses cyclic loading and unloading of a rock sample to produce a stress vs. strain hysteresis loop, and Q is determined from the enery lost per cycle (Usher, 1962; Gordon and Davis, 1968; McKavanagh and Stacey, 1974; Brennan, 1981). The torsion pendulum requires fashioning a long, thin, cylindrical rod of rock, attaching a mass with a high moment of inertia to the lower end of the rod, and suspending it vertically from the upper end. Torsional oscillations are induced, and after the driving force is shut off, the amplitude decay of free vibrations gives the attenuation (Peselnick and Outerbridge, 1961; Woirgard and others, 1977; Murphy, 1982, 1984). A variant of this is the flexion pendulum of Woirgard and Guegen (1978) in which a vertical rod of rock is clamped at the bottom and weighted at the top, with flexural vibrations being used to determine *Q*.

These techniques yield attenuation data at frequencies comparable to those used in the field (0.001 to 1 Hz for stress-strain;

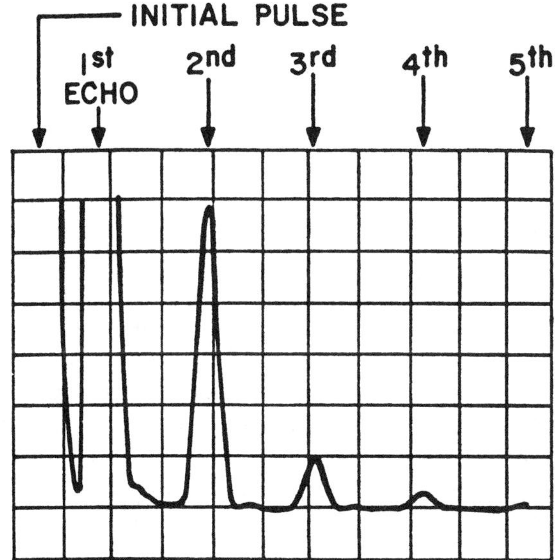

Figure 5. Pulse-echo decaying multiples. Time scale is 5 μsec per division (Mason and Kuo, 1971).

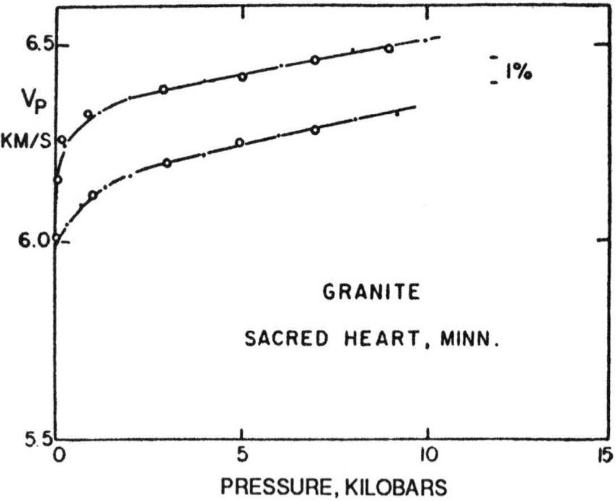

Figure 6. Compressional velocity vs. pressure for two specimens of granite (Birch, 1960). Velocity error bar of 1 percent is shown.

1 to 500 Hz for the torsion pendulum). Variable strain is easily incorporated, and use in a pressure vessel is discussed by Gordon and Davis (1968). Temperatures to 1,100°C were applied by Woirgard and Guegen (1978) to the flexion pendulum. In general, though, pressure-temperature (P-T) conditions in the earth are difficult to simulate with the low-frequency methods. Brennan and Stacey (1977) and others have shown that linearity for applied stresses (principle of superposition) holds only when the strain amplitude is less than 10^{-6}, and thus the application and detection of ultra-low stress and strain, respectively, is another source of problems.

DATA PERTINENT TO CRUSTAL AND UPPER-MANTLE COMPOSITION

Over the past three decades a large number of laboratory measurements have been reported (for a summary of velocity data, see Christensen, 1982). These have provided a basic understanding of many factors that influence velocity and attenuation of rocks believed to be abundant constituents of the continental lithosphere. The effects of some of these parameters are briefly reviewed in this section.

Pressure and temperature

The effect of confining pressure on velocities has been reported in a number of investigations. An example of data for a typical crystalline rock is shown in Figure 6. The characteristic shape of the curve of velocity vs. pressure is attributed to the closure of microcracks. As can be seen from Figure 6, much of the closure takes place over the first 100 MPa. Velocities measured in crystalline rocks at pressures to 3,000 MPa demonstrate that changes in velocity with pressure do not approach those of single crystals until the confining pressure is above 1,000 MPa. Even at these high pressures, solid contact between the mineral components is probably only approximate because some porosity has originated from anistropic thermal contraction of the minerals. The pressure derivatives of velocity for several rocks and minerals are given in Table 1. These values are probably reasonable for porosity-free rocks at lower crustal and upper mantle depths.

TABLE 1. COMPRESSIONAL VELOCITY PRESSURE AND TEMPERATURE DERIVATIVES*

Rock	$dV_p/\partial T$ $\times 10^{-3}$ km s^{-1}°C^{-1} at 200 MPa	$dV_p/\partial P$ km s^{-1} GPa^{-1} at 500-800 MPa
Serpentinite, Mid-Atlantic Ridge	-0.68	0.45
Granite, Cape Ann, Massachusetts	-0.39	0.25
Quartzite, Baraboo, Wisconsin	-0.45	0.30
Granulite, New Jersey Highlands	-0.49	0.20
Granulite, Saranac Lake, New York	-0.51	0.21
Basalt, East Pacific Rise	-0.39	0.16
Gabbro, Mid-Atlantic Ridge	-0.57	0.20
Granulite, Adirondacks, New York	-0.60	0.28
Amphibolite, Indian Ocean	-0.55	0.28
Anorthosite, Lake St. John, Quebec	-0.41	0.20
Granulite, Valle d'Ossola, Italy	-0.52	0.25
Dunite, Twin Sisters, Washington	-0.56	0.23
Eclogite, Nove' Dvory, Czechslovakia	-0.53	0.19

*Christensen, 1979.

Fewer data are available on the influence of temperature on rock velocities. It has been well known since the early work of Ide (1937) that the application of temperature to a rock at atmospheric pressure results in the creation of cracks that often permanently damage the rock (Fig. 7). Thus, reliable measurements of the temperature derivatives of velocity must be obtained at confining pressures high enough to prevent crack formation. In general, pressures of 200 MPa are sufficient for temperature measurements to 300°C.

A wide variety of techniques have been employed to measure the influence of temperature on rock velocities (for example, see Birch, 1943; Kern, 1978; Christensen, 1979). An example of data showing the influence of temperature on velocities is shown in Figure 8. Temperature derivatives of velocities for some common rocks and minerals are tabulated in Table 1.

Increasing temperature decreases velocities, whereas increasing pressure increases velocities. Thus, in a homogeneous crustal region, velocity gradients depend primarily on the geothermal gradient. The change of velocity with depth is given by:

$$\frac{dV}{dz} = \left(\frac{\partial V}{\partial P}\right)_T \frac{dP}{dz} + \left(\frac{\partial V}{\partial T}\right)_P \frac{dT}{dz} \tag{8}$$

where z is depth, T is temperature, and P is pressure. For regions with normal geothermal gradients (25° to 40°C/km), the change

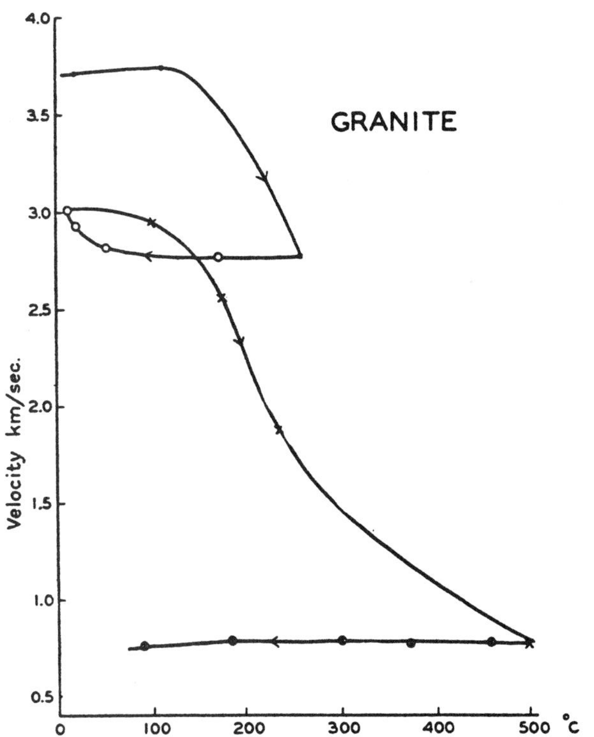

Figure 7. Compressional velocity vs. temperature for granite at atmospheric pressure (Ide, 1937).

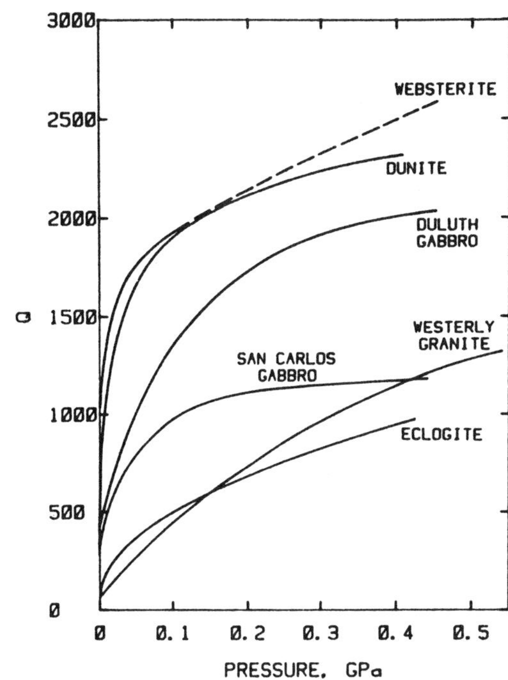

Figure 9. Shear-wave Q vs. pressure for several rocks (Katahara and others, 1982).

in compressional velocity with depth dV_p/dz is close to zero (Christensen, 1979). However, in high heat-flow regions, crustal velocity reversals are expected if compositional changes with depth are minimal.

All investigations have found that Q increases with increasing confining pressure. Beginning with Birch and Bancroft (1938), every experimental technique has observed a sharp increase in Q at low pressures, which then levels off at high pressures, a response similar to that observed for velocities. The form of the Q vs. P curve is generally attributed, therefore, to the closure of microcracks. This is illustrated in Figure 9 for several crustal and upper-mantle rocks, and pressure derivatives of Q for these rocks are given in Table 2. As with velocity measurements, few researchers have studied attenuation as a function of temperature for rocks of the lithosphere. At temperatures below the boiling point of the rock's volatiles, Q appears to be temperature independent, and above this, Q increases, indicating outgassing of pore fluids and/or thermal cracking (Johnston and Toksoz, 1980b; Tittmann and others, 1974). At the onset of partial melting, Q decreases (Spetzler and Anderson, 1968).

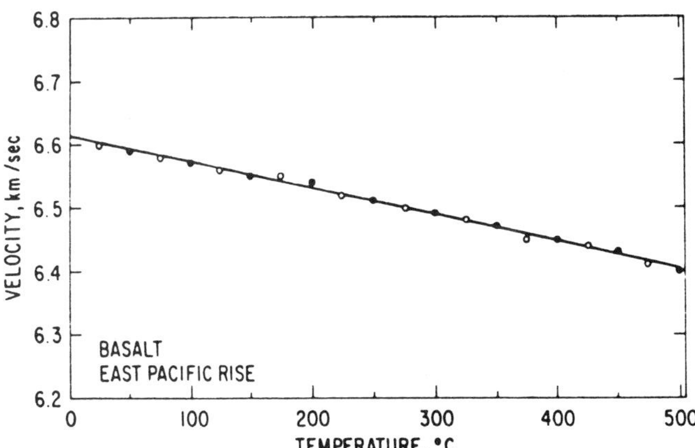

Figure 8. Compressional velocity vs. temperature at 200 MPa (Christensen, 1979).

Pore pressure

The influence of pore pressure on velocity and attenuation has been widely studied for sedimentary rocks; however, only a limited amount of data are available for crystalline rocks. Raising the pore pressure has approximately the same effect on velocities as lowering the confining pressure by the same amount (Todd and Simmons, 1972; Christensen, 1984, 1986). This is illustrated in Figure 10 for a gabbro, where velocities are shown as functions of confining pressure at atmospheric pore pressure and constant differential pressure (confining pressure minus pore pressure). The dramatic lowering of velocities with increasing pore pressure appears to be a common feature for crystalline continental rocks and may be one of many possible explanations for crustal low-velocity zones. Of significance, increases of pore pressure in

TABLE 2. ATTENUATION PRESSURE DERIVATIVES*

Rock	$\partial Q_s/\partial P$ (MPa^{-1} at 300-500 MPa)
Shikoku Eclogite	1.08
Westerly Granite	1.50
Duluth Gabbro	0.65
San Carlos Gabbro	0.17
Twin Sisters Dunite	0.57

*Katahara and others, 1982.

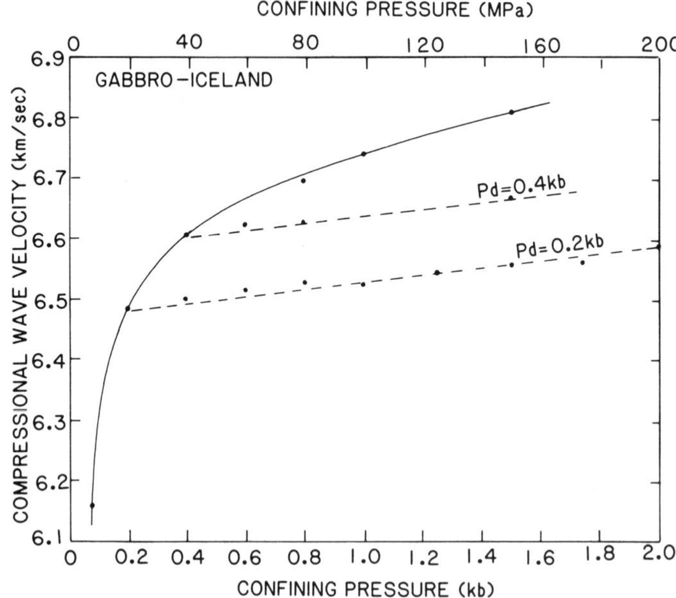

Figure 10. Velocities as functions of confining pressure and differential pressure *Pd* (Christensen, 1986).

crustal regions will be accompanied by marked increases in Poisson's ratios (Christensen, 1984). Although pore pressure attenuation data is sparse for crystalline rocks, the investigations on sedimentary rocks have found that fluid flow in microcracks is the mechanism responsible for the observed drop in Q with increasing saturation (Winkler and Nur, 1979; Clark and others, 1980). This effect should be seen in crystalline rocks as well.

Mineralogy

When pores and cracks are closed, velocities are primarily a function of mineralogy. For many suites of continental rocks, systematic changes in velocities with mineralogy have been observed (Birch, 1961; Christensen, 1965, 1966; Christensen and Fountain, 1975). In general, rocks rich in alkali feldspar, quartz, and mica have compressional wave velocities below 6.5 km/sec, whereas the mafic rocks with Ca-rich feldspar, amphibole, and pyroxene have compressional wave velocities close to 7.0 km/sec. Unaltered ultramafic rocks and eclogites usually have velocities in excess of 8.0 km/sec. A more detailed discussion of estimates of crustal and upper-mantle mineralogy from seismic velocities is presented in another chapter (Fountain and Christensen, this volume).

Vp/Vs ratios appear to be diagnostic of several common rock types (Christensen, 1972; Christensen and Fountain, 1975; Christensen and Salisbury, 1975). For example, Vp/Vs ratios are generally low in quartz-rich rocks and high in anorthosites and serpentinites.

Attenuation is dependent on the density and shape of microcracks and their state of saturation. Mineralogy may be only of secondary importance, although higher velocity rocks usually have higher Q values (Johnston, 1981). Additional studies are needed to further explore the relationship between attenuation and mineralogy.

Seismic anisotrophy

The interpretation of seismic data is greatly simplified if it is assumed that rocks beneath the Earth's surface behave as isotropic solids. Laboratory studies, however, clearly show that most rocks are anisotropic. Early laboratory measurements of compressional wave velocities from cores cut in three mutually perpendicular directions demonstrated that common continental crustal metamorphic rocks often have anisotropies greater than 10 percent (Birch, 1960; Christensen, 1965). In addition, it was observed that shear waves split into components with different polarizations and different velocities (Christensen, 1966). This shear-wave splitting, also referred to as acoustic double refraction and shear-wave birefringence, is observed in the laboratory by rotating the polarization directions of the shear-wave transducers relative to the rock fabric (Fig. 11).

Anisotropy originates from a number of mechanisms, the two most important being preferred orientation of minerals and alignment of cracks. Common continental crustal rocks such as schists, gneisses, and amphibolites are anisotropic due to orientation of micas and amphiboles (Christensen, 1965). In the continental upper mantle, olivine and pyroxene alignments are likely to produce anisotropy. At confining pressures less than approximately 20 MPa, laboratory anisotropy measurements are influenced by crack orientation as well as mineral orientation. In schistose rocks, crack orientation usually enhances mineral anisotropy at low pressures (Fig. 12). In addition, anisotropy is likely to originate from aligned stress-induced cracks in dilatancy zones in seismic regions (Crampin, 1978; Crampin and others, 1980; Crampin and McGonigle, 1981). Experimental studies of the effects of this phenomenon on seismic anisotropy are limited (Tocher, 1957; Nur, 1971; Gupta, 1973).

It is generally assumed, at least implicitly, that attenuation is independent of direction, and a dearth of experimental data on Q

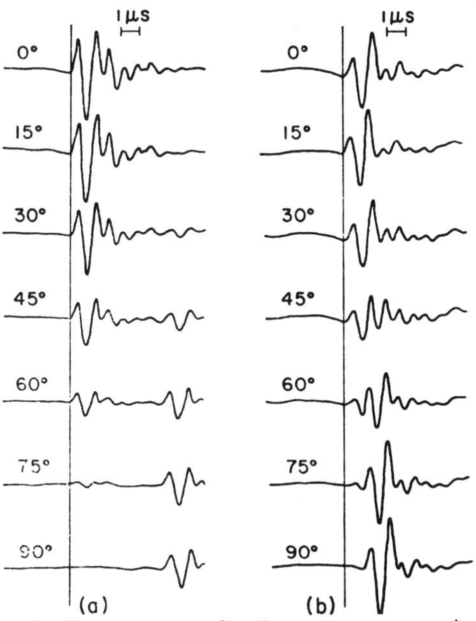

Figure 11. Oscilloscope traces for shear wave propagation through (a) slate and (b) dunite. The transducers are oriented at 0° to receive the higher velocity shear wave. Rotation of the transducers through 90° emphasizes the first arrival of the slower of the two shear waves (Christensen, 1971).

anisotropy neither confirms nor refutes this assumption. What little work that has been done (a taconite by Singh [1976], a granite by Lockner and others [1977], an oil shale by Johnston and Toksoz [1980a], and a deep-sea carbonate by Kim and others [1983]), however, points to a directional dependence of Q for some rocks. More measurements on igneous and metamorphic specimens are required to determine if Q anisotropy is an important factor in the Earth's crust.

Velocity-density relations

The dependence of velocity on density has been of interest in crustal gravity investigations where seismic data are also available. Several velocity-density relations for common rocks are listed in Table 3. These are usually expressed as a linear solution in the form $V = a + b\rho$, although for some rock suites a nonlinear solution may be more appropriate.

It is now understood that a single value dependence of velocity on density is valid only for limited compositions. Birch (1961) found that, to a first approximation, the compressional wave velocity can be expressed in terms of density and mean atomic weight. Mean atomic weights (m) were calculated from standard rock chemical analyses by

$$m = (\Sigma\, x_i m_i)^{-1} \quad (9)$$

where x_i is the proportion by weight of the *i*th oxide in the rock, and m_i is the mean atomic weight of the oxide (the formula

TABLE 3. VELOCITY-DENSITY LEAST-SQUARES SOLUTIONS OF THE FORM V = a + bρ*

Data Set	Pressure, MPa	a km/sec	b km/sec / g/cm³	r^2, %	Reference†
V_p (all rocks), m ≅ 21	10^3	-2.01	3.16	94	1
V_s (all rocks), m ≅ 21	10^3	-0.88	1.63	92	2
V_p, basalt	10^2	-4.44	3.64	95	3
V_s, balsat	10^2	-2.79	2.08	94	3

*V_p in km/sec, ρ in g/cm³, r = correlation coefficient.
†1. Birch, 1961; 2, Christensen, 1968; 3. Christensen and Salisbury, 1975.

weight divided by the number of particles in the formula). Most common rocks have mean atomic weights between 20 and 22, with the highest values for relatively iron-rich rocks. A velocity-density solution for m = 21 given by Birch (1960) and a similar calculation for shear velocities (Christensen, 1968) are presented in Table 3.

Frequency

Field-measured seismic velocities and attenuations are generally lower than those measured from sonic logs; these, in turn, are lower than ultrasonic velocities. The effect of frequency on velocity is slight, however, and is usually on the order of the uncertainty in the laboratory measurements (Birch, 1961). The bulk of the attenuation data indicates that Q is frequency-independent for dry rocks; for saturated and partially saturated samples, however, Q may vary significantly with frequency (Born, 1941; Wyllie and others, 1962; Tittmann and others, 1981).

CONCLUSIONS

Beginning with the pioneering work of Birch (1960), much has been learned about elastic wave propagation in rocks. Velocities are now available for most common rock types at pressures that exist in the continental crust and upper mante. Additional important contributions include laboratory measurements of shear-wave velocities, Vp/Vs ratios, and the influence of temperature and pore pressure on velocities.

For continental lithospheric studies, it is desirable in the future to obtain additional velocity data at high temperatures. This is especially true for shear-wave measurements. As more detailed crustal velocity data become available, it may become apparent that seismic anisotropy is an important crustal feature. Thus, systematic laboratory studies of compressional wave anisotropy and shear-wave splitting could be critical in understanding crustal composition, just as they have been in investigations of the

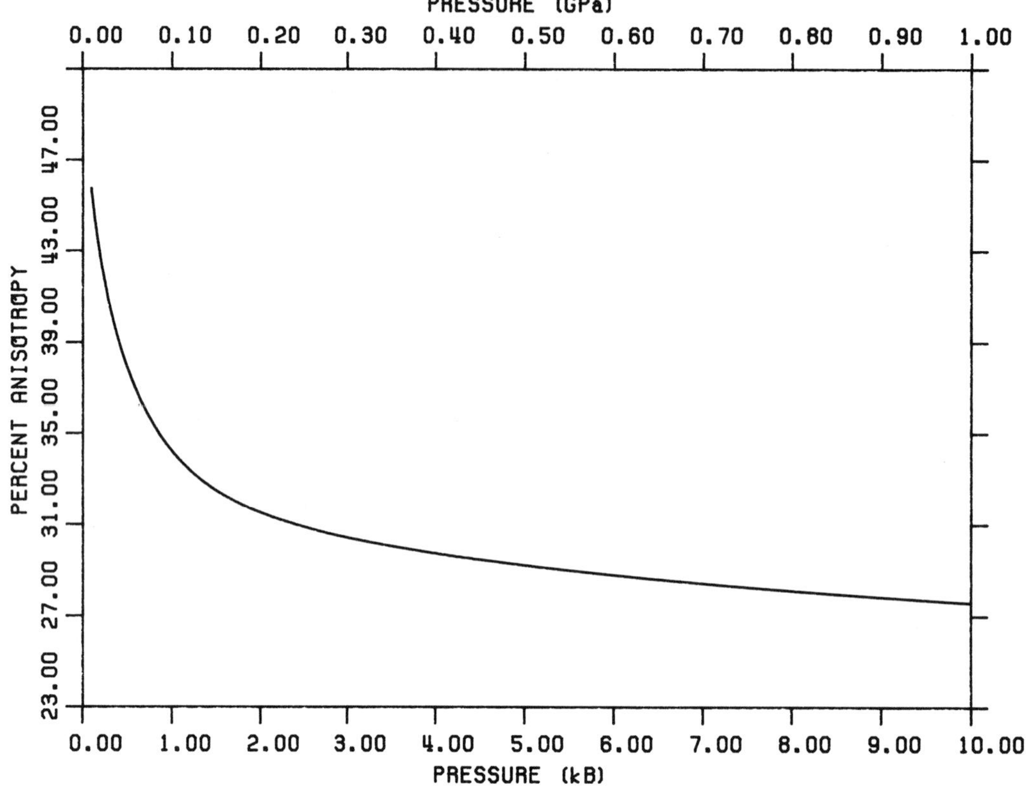

Figure 12. Seismic anisotropy $[(V_{max}-V_{min}]/V_{avg})$ as a function of pressure for a quartz-mica schist.

upper oceanic mantle. In lower crustal regions, anisotropy resulting from mineral orientation is likely, whereas in the shallow crust, crack anisotropy is thought to be important.

Many researchers have obtained high-quality attenuation data, and a number of loss-mechanism theories have proven quite successful. Although a variety of methods have been employed to measure Q, certain trends are evident. The magnitude may vary from one study to the next, but the relative effects of pressure, temperature, and saturation state are fairly consistent, and efforts to relate these conditions to the micro- and macrostructure of rocks should continue. Also, it would seem that the behavior of Q parallels some already well-established velocity characteristics such as dependence on confining and pore pressure and frequency. This raises some interesting questions: Is Q characterized by mineralogy at high pressures rather than cracks and their saturation condition? Is there a Q-density relation similar to the velocity-density relationship mentioned above? Is attenuation anisotropic in crustal and upper-mantle rocks? Are Qp/Qs ratios diagnostic, as suggested by Winkler and Nur (1979)? Is an effective pressure law applicable for Q? Finally, far too little laboratory attenuation work has been done on crystalline rocks. This must become a priority if significant use is to be made of seismic attenuation data of the continental crust and upper mantle.

ACKNOWLEDGMENTS

Preparation and publication of this work was supported by the Office of Naval Research (Contract N00014-89-J-1209) and by a Purdue University David Ross Fellowship.

REFERENCES CITED

Attewell, P. B., and Brentnall, D., 1964, Attenuation measurements on rocks in the frequency range 12 kc/s to 51 kc/s and in the temperature range of 100°K to 1150°K, *in* Spokes, E. M., and Christiansen, C. R., eds., Proceedings of the Sixth Symposium on Rock Mechanics: Rolla, University of Missouri, p. 330–357.

Birch, F., 1943, Elasticity of igneous rocks at high temperatures and pressures: Geological Society of America Bulletin, v. 54, p. 263–286.

—— , 1960, The velocity of compressional waves in rocks to 10 kilobars, Part 1: Journal of Geophysical Research, v. 65, p. 1083–1102.

—— , 1961, The velocity of compressional waves in rocks to 10 kilobars, Part 2: Journal of Geophysical Research, v. 66, p. 2199–2224.

—— , 1972, Numerical experiments on the velocities in aggregates of olivine: Journal of Geophysical Research, v. 77, p. 6385–6391.

—— , 1975, Velocity and attenuation from resonant vibrations of spheres of rock, glass, and steel: Journal of Geophysical Research, v. 80, p. 756–764.

Birch, F., and Bancroft, D., 1938, Elasticity and internal friction in a long column of granite: Seismological Society of America Bulletin,v. 28, p. 243–254.

Born, W. T., 1941, Attenuation constant of earth materials: Geophysics, v. 6, p. 132–148.

Brennan, B. J., 1981, Linear viscoelastic behavior in rocks, *in* Stacey, F. D., Paterson, M. S., and Nicholas, A., eds., Anelasticity in the earth: American Geophysical Union Geophysical Monograph 4, p. 13–22.

Brennan, B. J., and Stacey, F. D., 1977, Frequency dependence of elasticity of rock; Test of seismic velocity dispersion: Nature, v. 268, p. 220–222.

Christensen, N. I., 1965, Compressional wave velocities in metamorphic rocks at pressures to 10 kilobars: Journal of Geophysical Research, v. 70, p. 6147–6164.

——, 1966, Elasticity of ultrabasic rocks: Journal of Geophysical Research, v. 71, p. 5921–5931.

——, 1968, Chemical changes associated with upper mantle structure: Tectonophysics, v. 6, p. 331–342.

——, 1971, Shear wave propagation in rocks: Nature, v. 229, p. 549–550.

——, 1972, The abundance of serpentinites in the oceanic crust: Journal of Geology, v. 80, p. 709–719.

——, 1979, Compressional wave velocities in rocks at high temperatures and pressures, critical thermal gradients, and crustal low-velocity zones: Journal of Geophysical Research, v. 84, p. 6849–6857.

——, 1982, Seismic velocities, *in* Carmichael, R. S., ed., Handbook of physical properties of rocks, v. 2: Boca Raton, Florida, CRC Press, p. 1–228.

——, 1984, Pore pressure and oceanic crustal seismic structure: Geophysical Journal of the Royal Astronomical Society, v. 79, p. 411–424.

——, 1985, Measurements of dynamic properties of rock at elevated pressures and temperatures, *in* Pincus, H. J., and Hoskins, E. R., eds., Measurement of rock properties at elevated pressures and temperatures: Philadelphia, Pennsylvania, American Society for Testing and Materials ASTM STP869, p. 93–107.

——, 1986, Influence of pore pressure on oceanic crustal seismic velocities: Journal of Geodynamics, v. 5, p. 45–49.

Christensen, N. I., and Fountain, D. M., 1975, Constitution of the lower continental crust based on experimental studies of seismic velocities in granulite: Geological Society of America Bulletin, v. 86, p. 227–236.

Christensen, N. I., and Salisbury, M. H., 1975, Structure and constitution of the lower oceanic crust: Reviews of Geophysics and Space Physics, v. 13, p. 57–86.

Clark, V. A., Tittmann, B. R., and Spencer, T. W., 1980, Effect of volatiles on attenuation (Q^{-1}) and velocity in sedimentary rocks: Journal of Geophysical Research, v. 85, p. 5190–5198.

Crampin, S., 1978, Seismic-wave propagation through a cracked solid; Polarization as a possible dilatancy diagnostic: Geophysical Journal of the Royal Astronomical Society, v. 53, p. 467–496.

Crampin, S., and McGonigle, R., 1981, The variation of delays in stress-induced anisotropic polarization anomalies: Geophysical Journal of the Royal Astronomical Society, v. 64, p. 115–131.

Crampin, S., Evans, J. R., Ucer, S. B., Doyle, M., Davis, J. P., Yegorkine, G. V., and Miller, A., 1980, Observations of dilatancy-induced polarization anomalies; The potential for earthquake prediction: Nature, v. 286, p. 874–877.

Gardner, G.H.F., Wyllie, M.P.J., and Droschak, D. M., 1964, Effects of pressure and fluid saturation on the attenuation of elastic waves in sand: Journal of Petroleum Technology, v. 16, p. 189–198.

Gladwin, M. T., and Stacey, F. D., 1974, Anelastic degradation of acoustic pulses in rock: Physics of the Earth and Planetary Interiors, v. 8, p. 332–336.

Gordon, R., and Davis, L., 1968, Velocity and attenuation of seismic waves in imperfectly elastic rocks: Journal of Geophysical Research, v. 73, p. 3917–3935.

Guptam I. N., 1973, Premonitory variations in S-wave velocity anisotropy before earthquakes in Nevada: Science, v. 182, p. 1129–1132.

Ide, J. M., 1937, The velocity of sound in rocks and glasses as a function of temperature: Journal of Geology, v. 45, p. 689–716.

Johnston, D. H., 1981, Attenuation; A state-of-the-art summary, *in* Johnston, D. H., and Toksoz, M. N., eds., Seismic wave attention: Tulsa, Oklahoma, Society of Exploration Geophysicists, Geophysics Reprint Series No. 2, p. 123–135.

Johnston, D. H., and Toksoz, M. N., 1980a, Ultrasonic P and S wave attenuation in dry and saturated rocks under pressure: Journal of Geophysical Research, v. 85, p. 925–936.

——, 1980b, Thermal cracking and amplitude-dependent attenuation: Journal of Geophysical Research, v. 85, p. 937–942.

——, 1981, Definitions and terminology, *in* Johnston, D. H., and Toksoz, M. N., eds., Seismic wave attenuation: Tulsa, Oklahoma, Society of Exploration Geophysicists, Geophysics Reprint Series No. 2, p. 1–5.

Jones, T., and Nur, A., 1983, Velocity and attenuation in sandstone at elevated temperatures and pressures: Geophysical Research Letters, v. 10, p. 140–143.

Katahara, K. W., Manghnani, M. H., Devnani, N., and Tittmann, B. R., 1982, Pressure dependence of Q in selected rocks, *in* Akimoto, S., and Manghanani, M. H., eds., High pressure research in geophysics: Advances in Earth and Planetary Sciences, v. 12, p. 147–158.

Kern, H., 1978, The effect of high temperature and high confining pressure on compressional wave velocities in quartz-bearing and quartz-free igneous and metamorphic rocks: Tectonophysics, v. 44, p. 185–203.

Kim, D.-C., Katahara, K. W., Manghanani, M. H., and Schlanger, S. O., 1983, Velocity and attenuation anisotropy in deep-sea carbonate sediments: Journal of Geophysical Research, v. 88, p. 2337–2343.

Kissell, F. N., 1972, Effect of temperature on internal friction in rocks: Journal of Geophysical Research, v. 77, p. 1420–1423.

Lockner, D. A., Walsh, J. B., and Byerlee, J. D., 1977, Changes in seismic velocity and attenuation during deformation of granite: Journal of Geophysical Research, v. 82, p. 5374–5378.

Mason, W. P., and Kuo, J. T., 1971, Internal friction of Pennsylvania slate: Journal of Geophysical Research, v. 76, p. 2084–2089.

Mason, W. P., Marfurt, K. J., Beshers, D. N., and Kuo, J. T., 1978, Internal friction in rocks: Journal of the Acoustical Society of America, v. 63, p. 1596–1603.

Mattaboni, P., and Schreiber, E., 1967, Method of pulse transmission measurements for determining sound velocities: Journal of Geophysical Research, v. 72, p. 5160–5163.

McKavanagh, B., and Stacey, F. D., 1974, Mechanical hysteresis in rocks at low strain amplitudes and seismic frequencies: Physics of the Earth and Planetary Interiors, v. 8, p. 246–250.

Murphy, W. F., III, 1982, Effects of partial water saturation on attenuation in Massilon sandstone and Vycor porous glass: Journal of the Acoustical Society of America, v. 71, p. 1458–1468.

——, 1984, Seismic to ultrasonic velocity drift; Intrinsic absorption and dispersion in crystalline rocks: Geophysical Research Letters, v. 12, p. 85–88.

Nur, A., 1971, Effects of stress on velocity anistropy in rocks with cracks: Journal of Geophysical Research, v. 76, p. 2022–2034.

O'Connell, R. J., and Budiansky, B., 1978, Measure of dissipation in viscoelastic media: Geophysical Research Letters, v. 5, p. 5–8.

Peselnick, L., and Outerbridge, W. F., 1961, Internal friction in shear and shear modulus of Solenhofen limestone over a frequency range of 10^7 cycles per second: Journal of Geophysical Research, v. 66, p. 581–588.

Peselnick, L., and Zietz, X., 1959, Internal friction of fine-grained limestones at ultrasonic frequencies: Geophysics, v. 24, p. 285–296.

Ramana, Y. V., and Rao, M.V.M.S., 1974, Q by pulse broadening in rocks under pressure: Physics of the Earth and Planetary Interiors, v. 8, p. 337–341.

Roderick, R. L., and Truell, R., 1952, The measurement of ultrasonic attenuation in solids by the pulse technique and some results in steel: Journal of Applied Physics, v. 23, p. 267–269.

Sears, F. M., and Bonner, B. P., 1981, Ultrasonic attenuation measurements by spectral ratios utilizing signal processing techniques: Institute of Electrical and Electronic Engineers Transactions on Geoscience and Remote Sensing GE-19, p. 95–99.

Simmons, G., 1964, Velocity of shear waves in rocks to 10 kilobars, 1: Journal of

Geophysical Research, v. 69, p. 1123–1130.
Singh, V. P., 1976, Investigations of attenuation and internal friction of rocks by ultrasonics: International Journal of Rock Mechanics and Mining Sciences and Geomechanics Abstracts, v. 13, p. 69–74.
Spetzler, H., and Anderson, D. L., 1968, The effect of temperature and partial melting on velocity and attenuation in a simple binary system: Journal of Geophysical Research, v. 73, p. 6051–6060.
Tittmann, B. R., 1977, Internal friction measurements and their implications in seismic Q structure models of the crust *in* Heacock, J. G., ed., The earth's crust: American Geophysical Union Geophysical Monograph 20, p. 197–213.
Tittmann, B. R., Housley, R. M., Alers, G. A., and Cirlin, E. H., 1974, Internal friction in rocks and its relationship to volatiles on the moon, *in* Proceedings of the 5th Lunar Science Conference: Geochimica et Cosmochimica Acta, supplement 5, v. 3, p. 2913–2918.
Tittmann, B. R., Nadler, H., Clark, V. A., Ahlberg, L. A., and Spencer, T. W., 1981, Frequency dependence of seismic dissipation in saturated rocks: Geophysical Research Letters, v. 8, p. 36–38.
Tocher, D., 1957, Anisotropy in rocks under simple compression: EOS Transactions of the American Geophysical Union, v. 38, p. 89–94.
Todd, T., and Simmons, G., 1972, Effect of pore pressure on the velocity of compressional waves in low-porosity rocks: Journal of Geophysical Research, v. 77, p. 3731–3743.
Toksoz, M. N., Johnston, D. H., and Timur, A., 1979, Attenuation of seismic waves in dry and saturated rocks; 1, Laboratory measurements: Geophysics, v. 44, p. 681–690.
Usher, M. J., 1962, Elastic behavior of rock at low frequencies: Geophysical Prospecting, v. 10, p. 119–127.
Winkler, K. W., 1983, Frequency dependent ultrasonic properties of high-pressure sandstones: Journal of Geophysical Research, v. 88, p. 9493–9499.
—— , 1985, Dispersion analysis of velocity and attenuation in Berea sandstone: Journal of Geophysical Research, v. 90, p. 6793–6800.
Winkler, K. W., and Nur, A., 1979, Pore fluids and seismic attenuation in rocks: Geophysical Research Letters, v. 6, p. 1–4.
—— , 1982, Seismic attenuation; Effects of pore fluids and frictional sliding: Geophysics, v. 47, p. 1–15.
Winkler, K. W., and Plona, T. J., 1982, Technique for measuring ultrasonic velocity and attenuation spectra in rocks under pressure: Journal of Geophysical Research, v. 87, p. 10776–10780.
Woirgard, J., and Gueguen, Y., 1978, Elastic modulus and internal friction in enstatite, forsterite, and peridotite at seismic frequencies and high temperatures: Physics of the Earth and Planetary Interiors, v. 17, p. 140–146.
Woirgard, J., Sarrazin, Y., and Chaumet, H., 1977, An apparatus for the measurement of internal friction as a function of frequency between 10^{-6} Hz and 10 Hz: Review of Scientific Instruments, v. 48, p. 1322–1325.
Wyllie, M.R.J., Gardner, G.H.F., and Gregory, A. R., 1962, Studies of elastic wave attenuation in porous media: Geophysics, v. 27, p. 569–589.

MANUSCRIPT ACCEPTED BY THE SOCIETY OCTOBER 31, 1988

Printed in U.S.A.

Geological Society of America
Memoir 172
1989

Chapter 8

The crustal structure of the western continental margin of North America

Richard W. Couch
College of Oceanography, Oregon State University, Corvallis, Oregon 97331
Robin P. Riddihough
Geological Survey of Canada, 601 Booth Street, Ottawa, Ontario K1A 0E8, Canada

ABSTRACT

The continental margin of western North America largely coincides with recently active plate boundaries. The plate interactions along the margin vary in tectonic style from oblique convergence through transform faulting and translation to subduction. The margin also includes at least two major migrating triple plate junctions. Plate-tectonic and geologic reconstructions suggest that a similar range of tectonic processes has been active along this margin for many tens of millions of years. The crustal thickness of the continental margin as far as 200 km inland from the continental slope is almost everywhere considerably less than "normal" continental, and commonly about 20 km. It seems evident that these tectonic processes have either thinned preexisting continental crust or preserved and uplifted former oceanic crust. They have not produced the 30- to 40-km-thick crust normally associated with continental material.

INTRODUCTION

The continental margin of western North America from Baja California to the Queen Charlotte Islands of British Columbia is characterized by two principal plate-tectonic styles: strike-slip faulting and convergence (Fig. 1). Contemporary seismicity shows that most of the margin is active.

Along the west coast of the Queen Charlotte Islands, analyses of contemporary plate motions (Minster and Jordan, 1978) suggest a predominantly right-lateral strike-slip zone representing a north-northwest Pacific-American plate transform motion of 55 to 60 mm/yr. The same motion also occurs between Cape Mendocino and San Francisco. Between Cape Mendocino and northern Vancouver Island (the sites of complex, active triple plate junctions), the margin is subject to contemporary convergence between the small oceanic Juan de Fuca plate system and the North American plate (Riddihough, 1984). This northeast-southwest convergence varies between 10 and 45 mm/yr and is marked inland by the active Cascade volcanic range that stretches from Mt. Meager in British Columbia to Lassen Peak in northern California. From the San Francisco area southward, the principal contemporary boundary between the Pacific and American plates is located inland along the San Andreas Fault system. However, the continental margin is also seismically active and is characterized as a series of northwest-oriented subparallel "slivers" of Pacific plate. South of Point Conception, the continental margin widens into the California borderland, an extremely complex region of the Pacific plate containing faulting and block interaction, which may be converging northward with a subplate of the Pacific plate along the Transverse Ranges near Los Angeles and Santa Barbara.

Although there are variations in detail and timing, the reconstructed history of the margin over at least the last 50 m.y. (e.g., Atwater, 1970; Engebretson, 1982) contains the same tectonic elements. In fact, it seems likely that almost all parts of the present margin have been affected by both tectonic environments during this period. The changes in plate motion and boundaries during this period also suggest that most of the continental margin has been affected by the passage of one or more triple junctions and their extremely complex local tectonics. The present north Vancouver Island triple junction was probably stationary from 10 to 12 Ma, but has migrated northeastward since 1.5 Ma (Riddi-

Couch, R. W., and Riddihough, R. P., 1989, The crustal structure of the western continental margin of North America, *in* Pakiser, L. C., and Mooney, W. D., Geophysical framework of the continental United States: Boulder, Colorado, Geological Society of America Memoir 172.

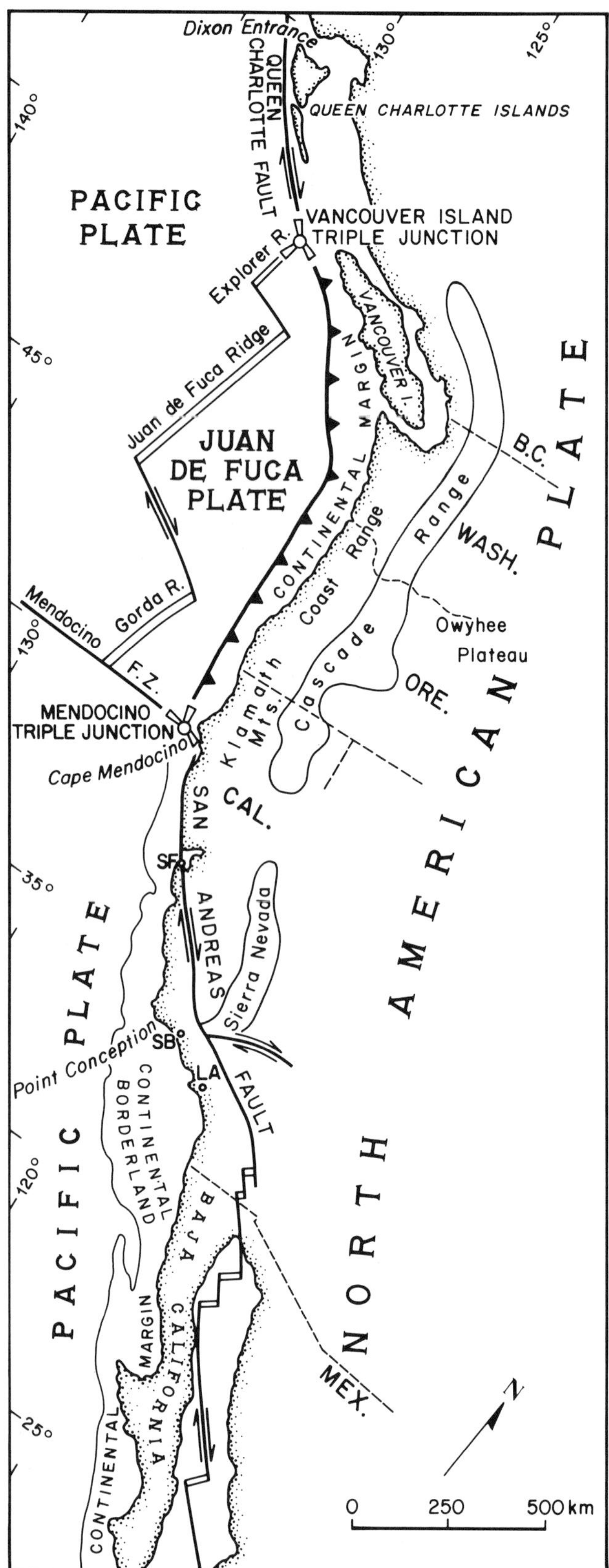

Figure 1. General tectonic and structural setting of the continental margin of western North America from Baja California to the Queen Charlotte Islands of British Columbia.

hough, 1977). The Cape Mendocino triple junction is migrating northward relative to North America at approximately 55 mm/yr (Atwater, 1970).

These plate-edge tectonics have not only formed and modified the structures of the margin, but have also affected a zone that may be more than 200 km wide. In this chapter we briefly describe the broad physical and structural parameters of the continental margin between northern Baja and the Queen Charlotte Islands, particularly the major crustal variations in this ocean-continent transition zone. The data base is variable in quality and reliability. However, we believe that the existing accumulated set of published crustal sections that have been based on a reconciliation of gravity and seismic data provide a satisfactory overall picture. This picture is not only consistent and remarkably similar along this margin, but significantly different from that seen along rifted continental margins.

Rather than just a traverse from south to north, the margins are considered under the headings in this chapter of Translational Margins, Triple Junctions, and Convergent Margins.

TRANSLATIONAL MARGINS

California Borderland: Crustal thickness

The borderland physiographic province extends from Punta Baja north of Vizcaino Bay in north-central Baja to Point Conception in southern California and is delimited on the east by the coast and on the west by the Patton Escarpment (Fig. 2). It is characterized by narrow, sinuous, generally northwest-southeast–trending ridges, separated by broad, flat-floored basins (Shepard and Emery, 1941; Emery, 1954). The gravity field consists of many high-amplitude, short-wavelength gravity anomalies, both positive and negative, that are generally elongate in the northwest-southeast direction (Oliver and others, 1980; Couch and others, 1989; Fig. 2). These short-wavelength anomalies generally coincide with bathymetric ridges and troughs; however, they also indicate that most of the ridges, in addition to being structural highs, include basement highs, and that the structural lows are also basement lows. Plawman (1978) has observed that, although high-density cores exist under most ridges, some show magnetic highs while others do not, suggesting marked changes in the composition of the basement blocks.

Emery (1954) plotted the depths of the ridge tops and the basin floors off southern California as a function of latitude; he showed that, on average, water depths within the borderland increase toward the south. Doyle and Gorsline (1977) extended these observations to include the borderland off northern Baja, and noted that the water continues to deepen southward and then shoals rapidly near the south end of the borderland. Figure 3 (center) shows the elevations of the ridge tops and basin floors within the borderland as a function of latitude. Examination of the gravity anomalies over the same area shows that, as a function of latitude (Fig. 3, top), on average they are more positive in the center of the borderland than at the north or south ends. Based on

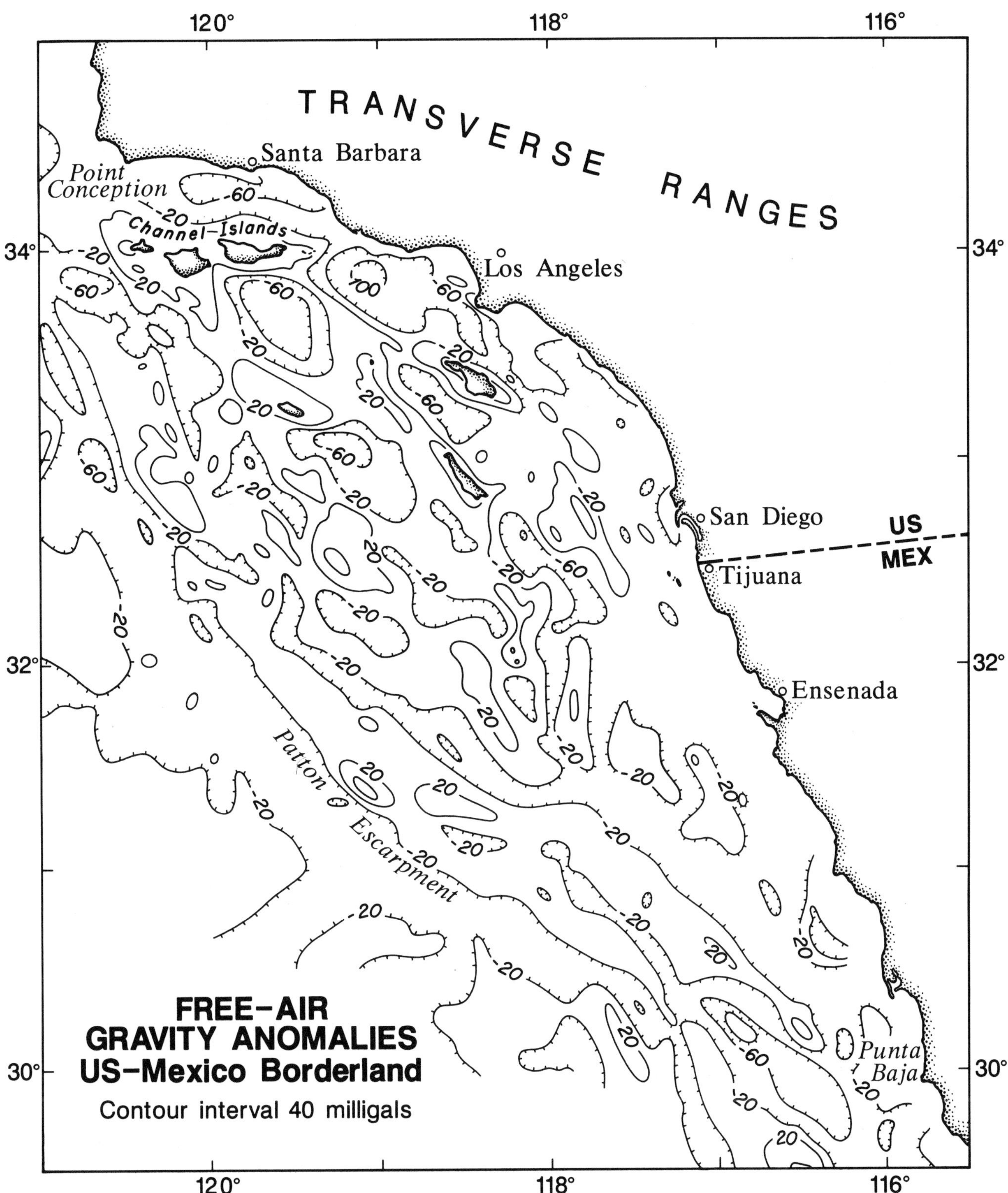

Figure 2. Free-air gravity anomalies of the Continental Borderland off southern California and northern Baja (after Couch and others, 1989). The negative anomaly associated with the Patton Escarpment marks the western boundary of the borderland.

averaged gravity anomalies and water depths, upper crustal densities of 2.4 g/cm^3, lower crustal and mantle densities of 3.0 and 3.3 g/cm^3, respectively, and horizons that are consistent with the refraction-determined depths reported by Shor and Raitt (1958) near 33°N and by Crandell and others (1983) near 34°24′N, computations show (Fig. 3, bottom) that the Moho, on average, shoals toward the south from 22 km near Point Conception to less than 14 km near 31°N. The Moho depth then increases southward to the southern extent of the borderland off Punta Baja. In general, as the average water depth increases within the borderland, the mantle shoals.

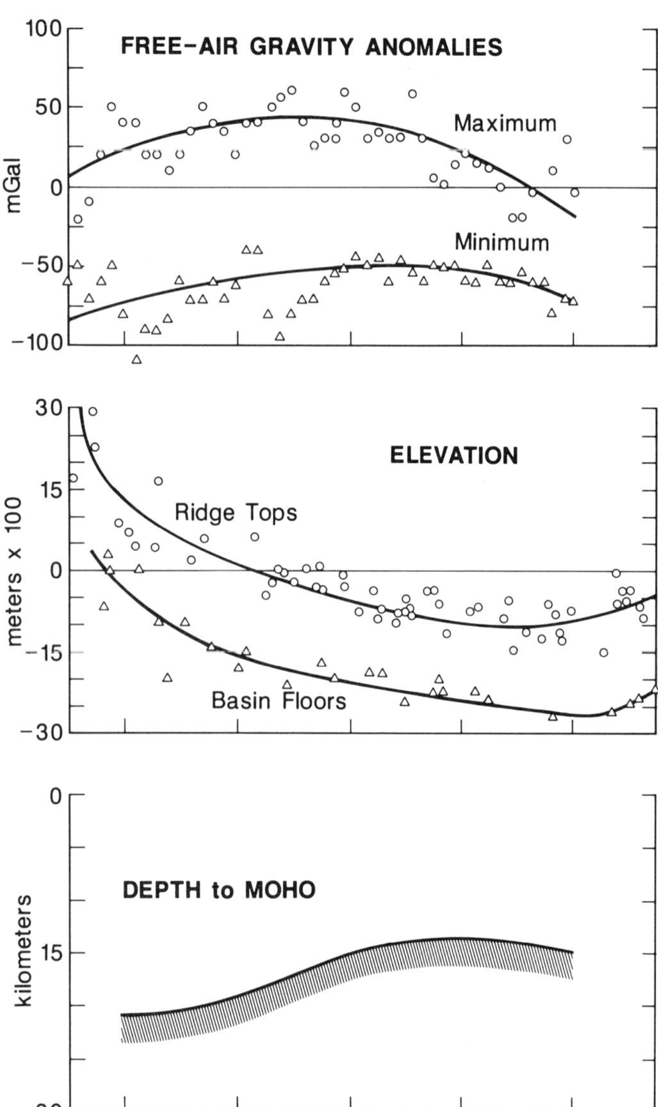

Figure 3. Free-air gravity anomalies, bathymetric, and topographic elevations, and average depth below sea level of the Moho as a function of latitude within the Continental Borderland of southern California and northern Baja. Moho depths derive from computations based on averaged (heavy lines) gravity anomalies, averaged water depths, an 8.36-km-thick upper crustal layer (that includes the sediments) of density 2.4 g/cm^3, a lower crustal layer of density 3.0 g/cm^3, and a mantle of density 3.3 g/cm^3. The refraction-determined depths of the Moho reported by Shor and Raitt (1958) near 33°N latitude and by Crandell and others (1983) near 34°24′N latitude also constrain the computed depths.

The expected translational direction of the Pacific plate relative to the North American plate in the vicinity of the borderland (Minster and Jordan, 1978) is approximately N30°W. This motion causes the faulted terranes of the borderland to move along trajectories that intersect the westwardly displaced Transverse Ranges of southern California (Fig. 2). As a consequence, slow underthrusting, or incipient subduction, occurs along the southwestern flank of the Transverse Ranges, and the process downwarping and apparent thickening of the crust occurs in the vicinity of the Santa Barbara Channel. Similarly, Humphreys and others (1984) and Sheffels and McNutt (1986) have proposed that intracontinental subduction occurs along the Transverse Ranges of southern California east of the borderland. Areal dilation associated with block rotation (Kammerling and Luyendyk, 1985) and basinal subsidence and filling also may have contributed to the increase in thickness of the crust south of the Channel Islands relative to the central borderland.

The difference between the average maxima and the minima of the gravity anomalies within the borderland (Fig. 3) apparently increases from south to north, whereas the average of the differences between the maxima and minima in bathymetric relief remains fairly constant. The short wavelengths of the gravity anomalies within the borderland indicate that most derive from upper crustal sources. Hence, the northward increase in anomaly amplitudes of the anomalies indicates that if there is no systematic change in crustal composition, deformation involving vertical displacement of the upper crustal rocks may also increase from south to north.

Regional gravity anomalies often indicate an increase in mass seaward of active trenches. This has been attributed to lithospheric flexure prior to subduction (e.g., Watts and others, 1976), but may also include a reduction in crustal thickness beneath the gravity high (e.g., Hayes, 1966; Couch and Woodcock, 1981). The average free-air gravity anomalies of the borderland (Fig. 3) suggest a similar effect south of the convergence zone of the Transverse Range. However, because the water depths increase southward, it is more probable that it is here associated with crustal thinning. The data do not indicate whether such thinning is occurring during translation or whether it occurred prior to translation during an earlier rifting phase.

Composition of the borderland crust

The stippled areas in Figure 4 outline the larger basins within the Continental Borderland. The basins are irregular in shape, but are generally elongate in a northwest-southeast direction. The nearshore basins generally parallel the coastline. The gravity minima over the basins indicate that sediment thicknesses are about 2 to 5 km, with the deeper Santa Monica, Ventura, Santa Cruz, and San Clemente basins having 6 km or more of fill. These thicknesses imply that, in places, the sediments make up

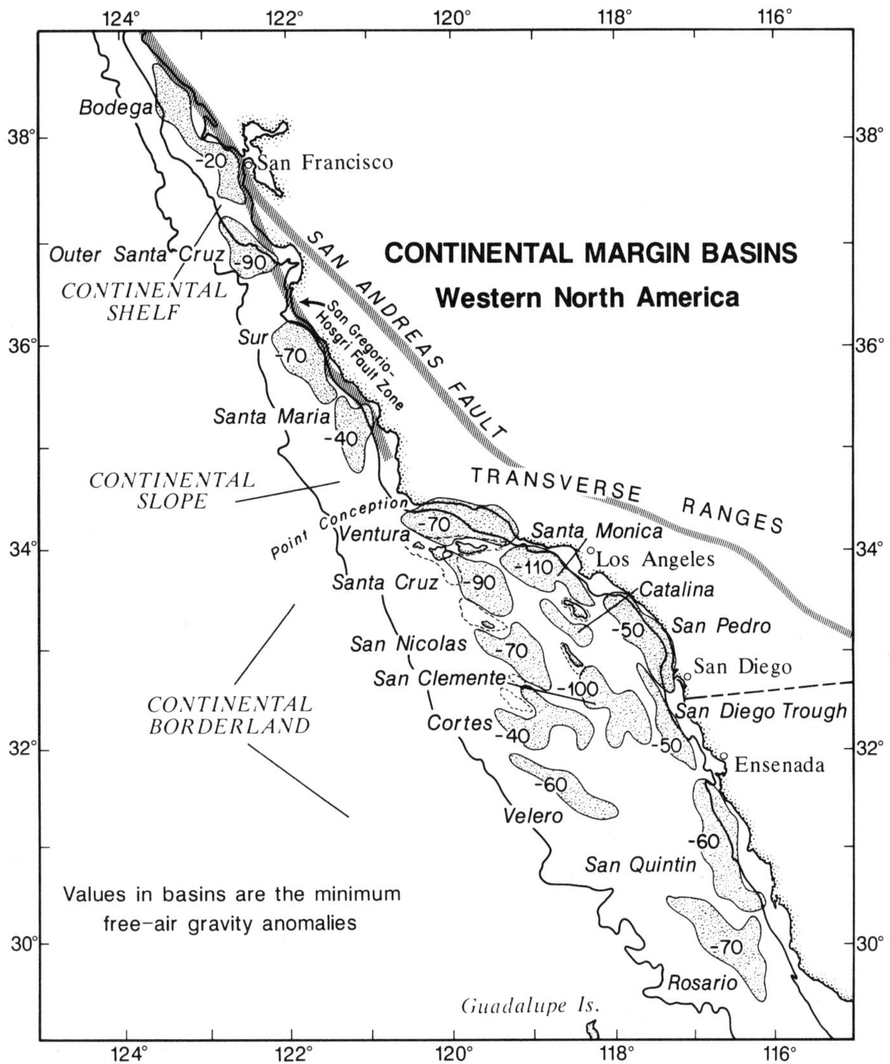

Figure 4. Crustal cross section of the Continental Borderland southwest of San Diego, California (after Shor and Raitt, 1958). These early seismic refraction measurements suggest that the crustal thickness within the borderland is about 20 to 25 km.

one-quarter to one-third of the thickness of the crust. Investigations by many researchers, including Vedder and others (1969, 1974, 1981) (Howell and others (1976, 1981), Vedder and Howell (1976), Hill (1976), Crouch (1979, 1981), Green and others (1987), and Teng and Gorsline (1987), of the highly folded and faulted sediments of the ridges and relatively flat-lying sediments of the basins have revealed a complex history of deposition and structural development within the borderland. The upper unconsolidated sediments (compressional wave velocities of less than 4 km/sec) decrease in thickness away from the coastline, whereas the consolidated sediments (velocities between 4 and 6 km/sec) appear to increase in thickness away from the continent (Fig. 5) (Shor and Raitt, 1958; Crandell and others, 1983).

The sediments that form the ridges and troughs of the outer borderland overlie materials with compressional wave velocities of 6.1 to 6.4 km/sec. These in turn overlie materials with velocities of 6.6 to 7.0 km/sec (Fig. 5) (Shor and Raitt, 1958; Crandell and others, 1983). Crustal models consistent with the seismic velocities and depths and the observed gravity and magnetic anomalies require densities of the rocks beneath the sediments of about 2.65 to 2.75 g/cm^3 and 2.9 to 3.0 g/cm^3 for the upper- and lower-crustal layers, respectively, and magnetizations of 0.003 to 0.007 emu/cm^3 for portions of both the upper- and lower-crustal layers (Plawman, 1978; Couch and others, 1989). Both the seismic refraction measurements and the crustal models suggest that, in some areas—particularly in the central and eastern sectors of the borderland—the materials with densities near 2.7 g/cm^3 and velocities between 6.1 and 6.4 km/sec are thin or absent, and the sediments overlie rock of higher crustal densities and velocities (Fig. 5). These high-density, high-velocity rocks comprise approximately one-half to two-thirds of the thickness of the crust of the borderland.

In addition to the seismic velocities, as indicated by Crandall and others (1983), the densities and magnetizations also imply that mafic and ultramafic rocks form the basement and subbasement beneath the sediments. Rock outcrops on islands within the borderland, and core and dredge samples (Vedder and Howell, 1976; Hill, 1976; Howell and Vedder, 1981; Vedder and others, 1981), indicate that Franciscan Complex rocks that include mélange, ophiolite, and other rock units of oceanic affinity form parts of the basement of the borderland. Plawman (1978) and Crouch (1981), using these data and other geophysical data, contend that large portions of the borderlands are floored with oceanic crustal rocks. Similarly, Couch and others, 1989) based on outcrop and geophysical data, we have proposed that Franciscan Complex rocks form a large part of the continental margin of Baja and extend from off the south tip of Baja into the borderland off southern California.

The high-velocity, high-density rocks of the lower crust of the borderland are probably gabbroic rocks of the lower oceanic crust; however, this lower crust is considerably thicker than the normal lower oceanic crust west of the borderland (Shor and Raitt, 1958) (Fig. 5). Imbrication of Mesozoic oceanic crust during the original accretion of these rocks somewhere south of their present location may be a cause of this thickening. Geophysical data also suggest (Plawman, 1978; Couch and others, 1989) that gabbroic rocks form, either as intrusions or as thrust slices, the cores of some of the ridges within the borderland.

Central and northern California

The continental margin of California between Point Conception and Cape Mendocino (Fig. 1) is a relatively narrow margin consisting of a shelf about 30 km wide and a slope about 30 km wide. The San Andreas fault system separating the North American and Pacific plates is oriented slightly oblique to the continental margin and passes offshore north of San Francisco (Figs. 1, 4). It then continues northwestward along the shelf near the coast, to the Mendocino triple junction (Currey and Nason, 1967). Earthquakes up to magnitude 7.0 have occurred on the offshore portion of the San Andreas in historic time (Richter, 1958, p. 472). Although located west of the main transform fault in southern California, kinematic considerations indicate that the San Gregorio–Hosgri fault zone (Fig. 4), oriented subparallel to the San Andreas but located offshore on the continental shelf between Point Conception and San Francisco, has also been active in Late Quaternary time (Weldon and Humphreys, 1986).

The Bodega, Outer Santa Cruz, Sur, and Santa Maria basins

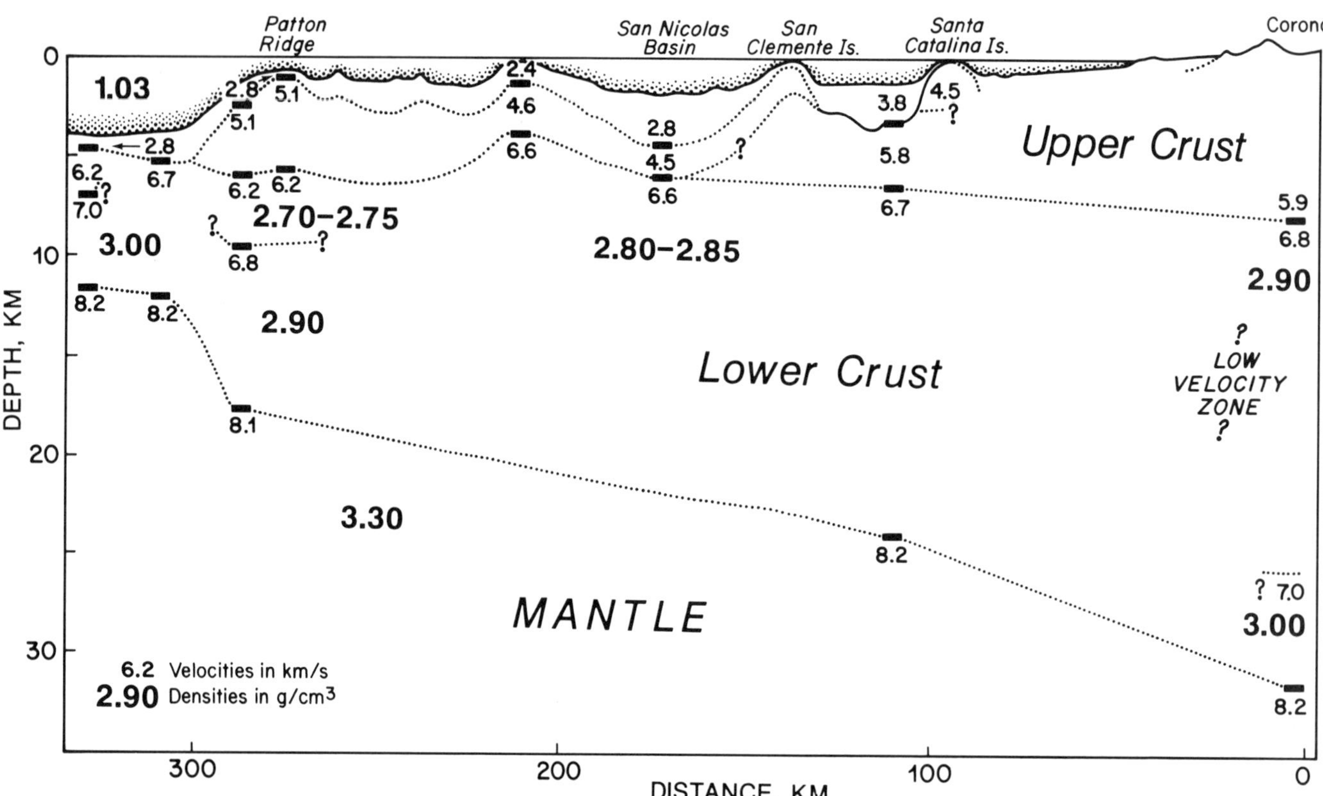

Figure 5. Sedimentary basins of the continental margin and Continental Borderland of southern California and northern Baja outlined by gravity anomalies. The numbers indicate the approximate free-air gravity minimum in milligals for each basin.

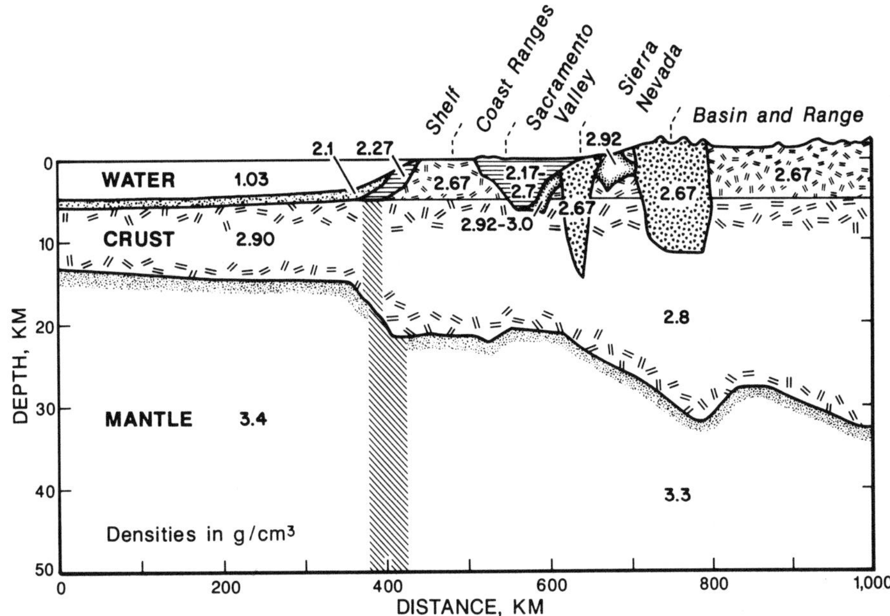

Figure 6. Model crustal cross section of the ocean-continent transition zone in central California (after Thompson and Talwani, 1964). This early cross section, based primarily on gravity measurements, suggests that the crust is about 22 km thick beneath the continental margin and coast ranges. (Later seismic determinations—e.g., Prodehl, 1979—have shown the crust to be 40 to 55 km thick beneath the Sierra Nevada.) The vertical band of hatched bars indicates lateral density changes in the lower crust and upper mantle.

(Fig. 4) occupy the continental margin between Point Conception and Cape Mendocino. Gravity anomaly minima indicate sediment thicknesses in these four large elongate basins are somewhat less than in similar-sized basins of the California Continental Borderland south of Point Conception. Estimated sediment thicknesses are 2, 9, 8, and 4 km, respectively.

South of San Francisco, near 37°N, crustal thickness as shown by Saleeby (1986) ranges from 25 to 28 km beneath the Coast Ranges and the western Great Valley 130 km east of the coastline. Crustal thickness then increases rapidly eastward from the central Great Valley to deeper than 50 km beneath the Sierra Nevada (Oliver and others, 1980). West of the San Andreas fault in this region, the Salinia block of the Pacific plate, with a total thickness of 25 km, has a lower layer with velocities of 6.4 to 6.6 km/sec, which Saleeby (1986) considered "enigmatic" and possibly representing the more mafic levels of the upper crystalline complex. The San Andreas fault is marked by a step in the Moho, and the accretionary marine sequences of Coast Ranges to the east are considered to be underlain by 15 km of oceanic crust.

The crustal cross section of the continental margin north of San Francisco computed by Thompson and Talwani (1964) shows (Fig. 6) a crust about 20 km thick in the vicinity of the continental shelf and coast ranges of north-central California. This is in good agreement with the seismic refraction results of Warren (1981), which showed a 20-km-thick crust north of San Francisco near Santa Rosa and Ukiah, California. However, Oppenheimer and Eaton (1984) could not reconcile earthquake traveltime data with this depth and preferred a Moho depth of 26 km. No observable long-wavelength changes in water depth, or in gravity, occur along the margin or west flank of the coastal mountains, hence the crustal thickness along the continental margin and near the coast between Point Conception and the southern Klamath Mountains is near 20 to 25 km.

Queen Charlotte Islands

The Queen Charlotte Islands continental margin is occupied by the present Pacific-American transform boundary. The continental margin (Fig. 7) is narrow, and the present zone of active faulting corresponds with a steep bathymetric scarp that drops from the narrow continental shelf to a 30-km-wide midslope terrace at depths of around 1,500 m. The outer edge of the terrace is subparallel and drops steeply to depths of 2,500 to 2,800 m. Seismicity along the margin is considerable. An earthquake (M = 8.0) occurred here in 1949. The level of seismicity has been estimated to correspond with the strain release expected from the predicted relative plate motions of 55 to 60 mm/yr (Hyndman and Weichert, 1983).

This broadly simple tectonic picture is, however, disturbed by evidence for an element of convergence across the margin. There is a 20° difference between the orientation of the fault trace and Pacific-American motion calculated from global plate motions (Minster and Jordan, 1978). Seismic profiling across the terrace also shows that it has many of the characteristics of an

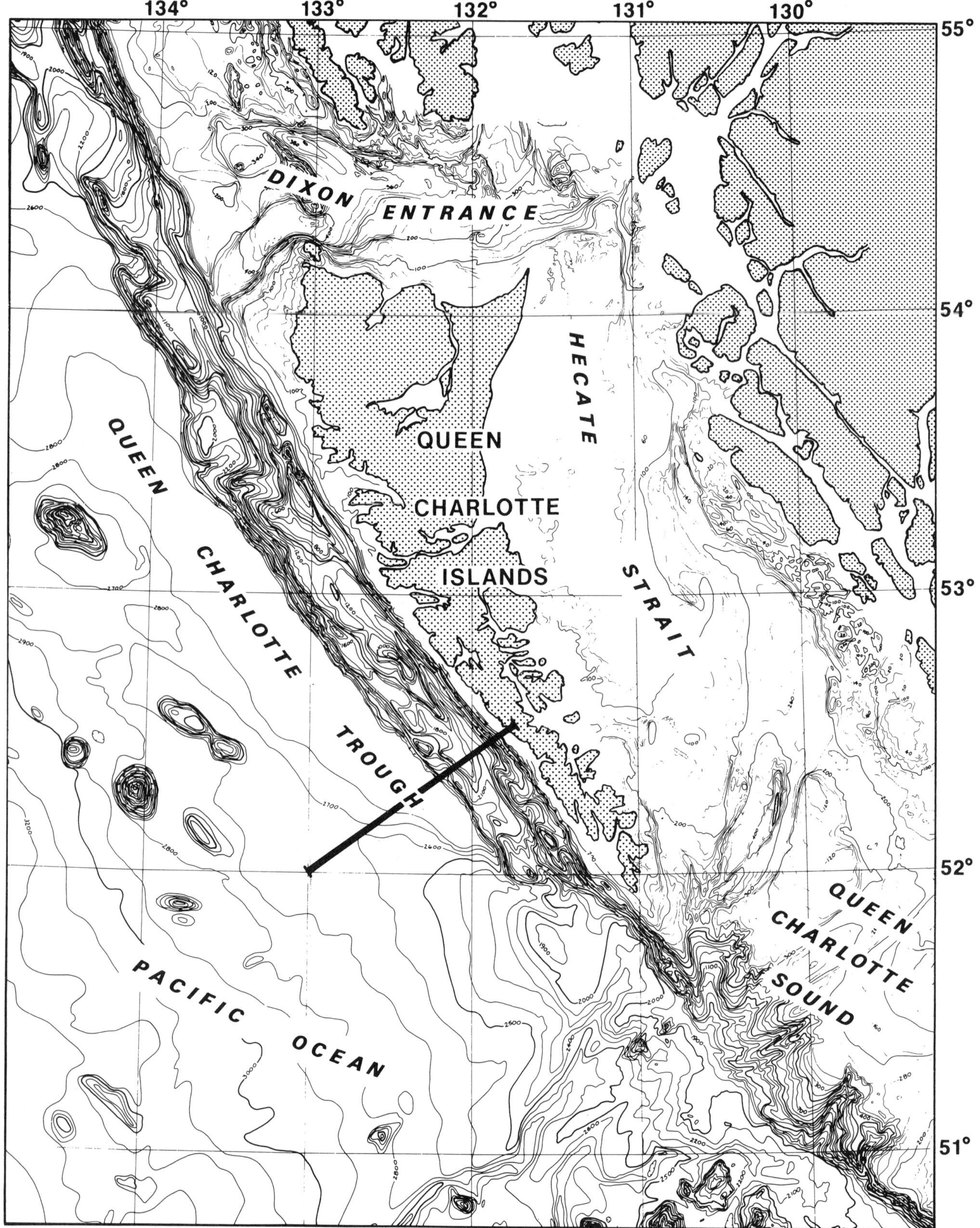

Figure 7. Bathymetric map of continental margin about the Queen Charlotte Islands, British Columbia. Heavy line indicates the location of the cross section shown in Figure 8.

accretionary prism of sediments, including compressive structures and folding (Riddihough and Hyndman, 1988). Other evidence, such as the recent uplift history of the Queen Charlotte Islands and the subsidence of Hecate Strait to the east, have led to a model of highly oblique underthrusting (Yorath and Hyndman, 1983). Plate reconstructions and dating of the observed uplifts, however, suggest that this may have been initiated only within the last 5 m.y.

The margin is characterized by a strong gravity gradient and a -80 mGal anomaly low over the terrace (Fig. 8). Interpretation of this gravity low and seismic refraction data (Horn and others, 1984; Yorath and others, 1985) show that the terrace is underlain by a strip of crust of oceanic thickness, which is faulted along both inner and outer margins and downdropped relative to the Pacific Ocean floor to the southwest. It is overlain by as much as 5 km of indurated sediments. Although this crust is apparently oceanic, the Pacific Ocean floor magnetic anomalies terminate against it along the outer edge of the terrace, and it has no magnetic expression.

Beneath the Queen Charlotte Islands, gravity and refraction data suggest a crustal thickness of around 20 km (Shor, 1962; Dehlinger and others, 1970; Riddihough and Hyndman, 1988). The estimates of as much as 5 m.y. of oblique convergence at a rate of near 1 mm/yr suggest that oceanic crust may underlie the Queen Charlotte Islands for approximately 50 km inland. However, there is little direct evidence for this. A mechanism has been suggested (Yorath and Hyndman, 1983) in which progressive underthrusting of the oceanic crust involves periodic seaward jumping of the principal transform dislocation so that it remains west of the Islands. If, as shown in Figure 8, a 10-km thickness of recent oceanic crust does underlie the Queen Charlotte Islands, the "lid" of Triassic and younger rocks (Wrangellia) may be less than 10 km thick. Considerable subcrustal erosion, caused by abrasion of the overriding plate by the subducting plate, may therefore have occurred.

An alternative explanation that may solve the "space" problem presented by oblique convergence is multiple "slicing" of the incoming oceanic crust into successive terrace "strips" and the effective removal of excess material by translation to the northwest (Riddihough and Hyndman, 1988). Under these circumstances, the 20 km of crust beneath the Queen Charlotte Islands would represent older accreted Wrangellia material.

TRIPLE JUNCTIONS

The western continental margin of North America contains two active triple plate junctions at the northern and southern ends of the Juan de Fuca plate system. Although such junctions are estimated to be transient in geological terms and have moved rapidly (Atwater, 1970; Engebretson and others, 1985), their complex local tectonics may be representative of the style of tectonic environments that has affected much of present continental margin in the past.

Northern Vancouver Island

Between Vancouver Island and the Queen Charlotte Islands, the Juan de Fuca Ridge system meets the continental margin at a ridge-trench-fault triple junction (Riddihough and others, 1980, 1984; Davis and Riddihough, 1982; Keen and Hyndman, 1979) (Fig. 9). The tectonics of this area are extremely complex. The Juan de Fuca Ridge system does not intersect the margin in a simple manner but breaks into a series of small, interconnected spreading centers and associated plate fragments. A recent history of the area, proposed by Riddihough and others (1980) and Davis and Riddihough (1982), hypothesized a major tectonic change at around 1 Ma. A ridge that had intersected the margin near Broks Peninsula (Fig. 9) on Vancouver Island stopped spreading and was replaced by new spreading centers to the northwest at the Tuzo Wilson Knolls and Dellwood Knolls. This isolated a former piece of Pacific plate beneath the Winona Basin (Figs. 9, 10b). As rates of spreading on the Dellwood and Tuzo Wilson Knolls are poorly controlled, the motion of this piece of plate with respect to the continental American plate is unknown but assumed to be small. Its isolation and landward tilting, coupled with high sedimentation rates from recent glaciation on land, have produced large thicknesses of sediments (Fig. 10b). The Winona Basin is estimated to contain as much as 6 km of sediments, all of which accumulated in the last 1 to 1.5 m.y., but which now have high seismic velocities and densities due to rapid lithification (Couch, 1969; Davis and Riddihough, 1982; Davis and Clowes, 1986).

Crustal thicknesses beneath northern Vancouver Island (primarily from gravity modeling) are of the order of 25 km (Fig. 10). However, the period of stability of the ridge intersection against the margin in this area (Riddihough, 1977) has apparently resulted in a series of small plutonic intrusions near Brooks Peninsula, the Alet Bay volcanic belt, dating from 4 to 9 Ma (Bevier and others, 1979). Although data are scant, it seems reasonable to expect a difference in the upper mantle beneath the margin on either side of this position that reflects the distinction between subduction (to the southeast) and nonsubduction (to the northwest).

The continental slope between Brooks Peninsula and the Tuzo Wilson Knolls is extremely steep, and from submersible and dredging operations (Tiffin and others, 1972; Yorath and others, 1977) is tectonically characterized by transcurrent faulting. However, during the period when the triple junction was located at Brooks Peninsula, the margin may have been subject to oblique underthrusting in a fashion similar to the present Queen Charlotte margin (Yorath and Hyndman, 1983). Crustal modeling across the margin (Yorath and Hyndman, 1983) (Fig. 10a) shows that the Moho beneath the continental shelf of Queen Charlotte Sound dips into the continent from a depth of less than 15 km to 25 km. The base of the crust could be continuous with the base of the offshore oceanic crust; however, there is no direct evidence that represents underthrusting. In fact, the recent history of Queen

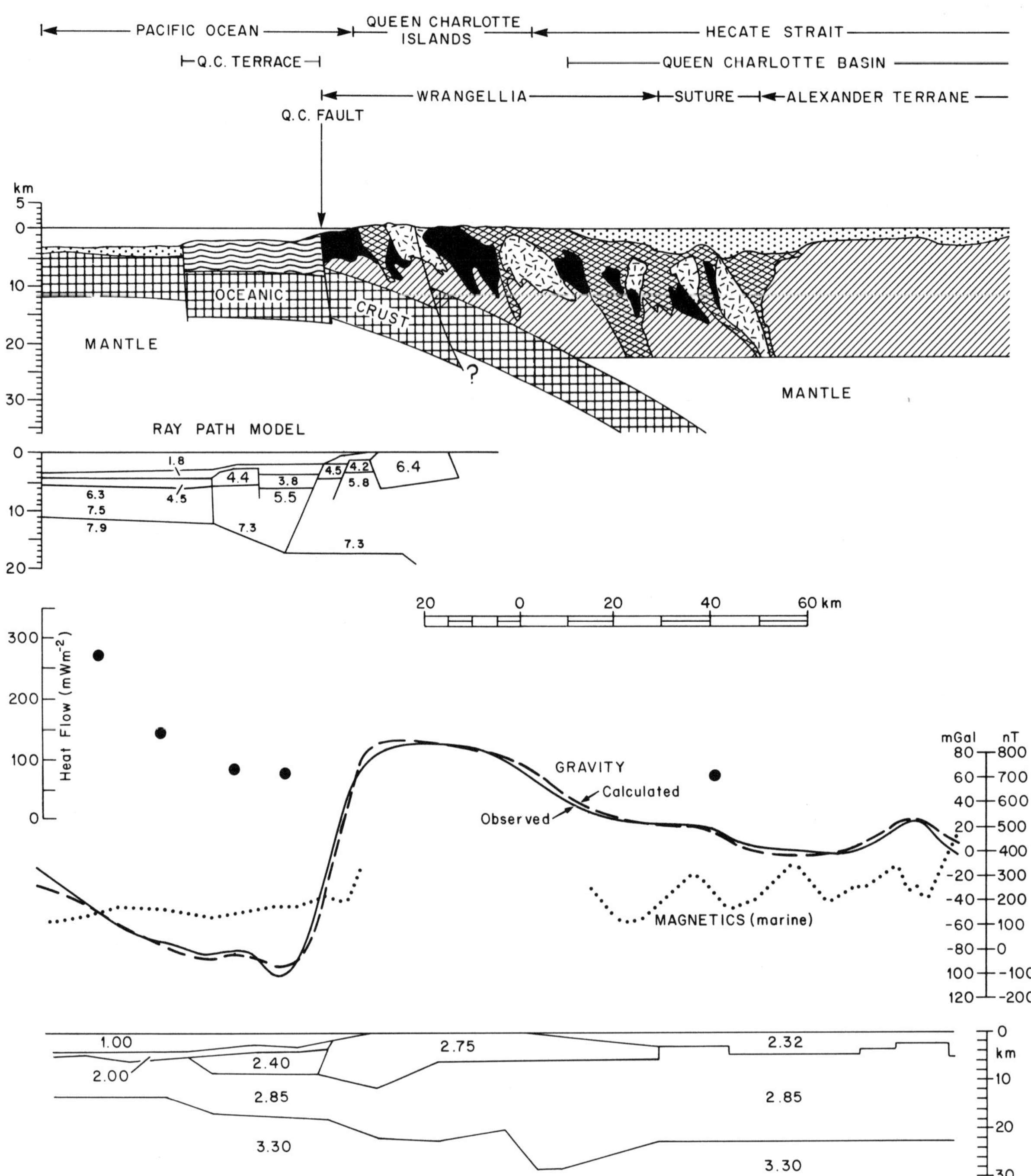

Figure 8. Crustal cross section of the continental margin across Moresby Island, British Columbia (after Yorath and others, 1985; continent-ocean transect B1). Upper section is geological; sediments shown dotted or wavy-lined where folded, basic rocks are block-intrusive, diagonal cross-hatched-extrusive, granites-conventional symbol, diagonal shading-crystalline crust. Ray path model numbers are velocities in kilometers per second. Lowest section is gravity model with densities in grams per cubic centimeter.

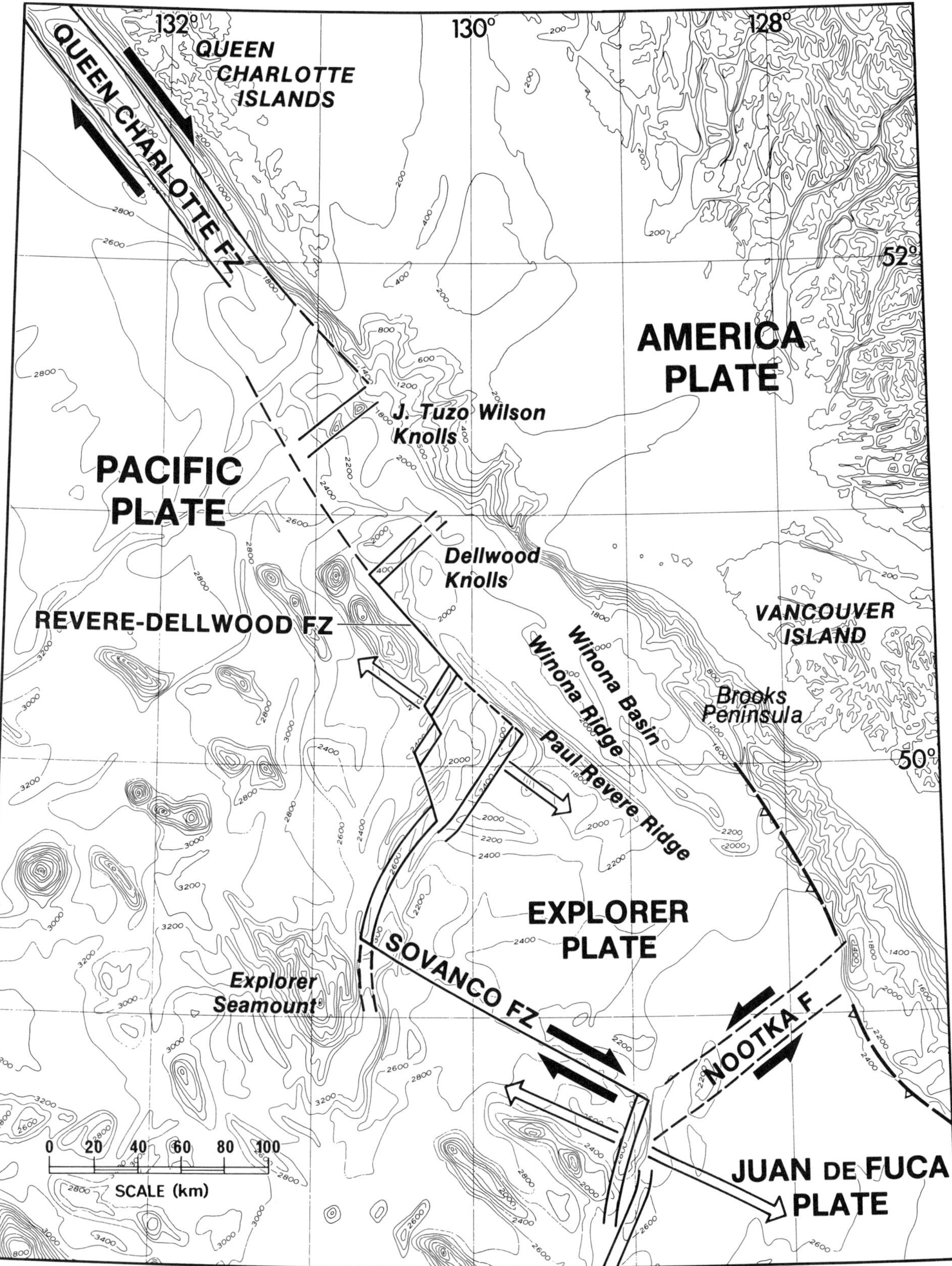

Figure 9. Bathymetric map of the continental margin and abyssal sea floor in the vicinity of the Queen Charlotte Islands and northern Vancouver Island. The map outlines the tectonic structures and plate motions of and about the Vancouver Island triple junction.

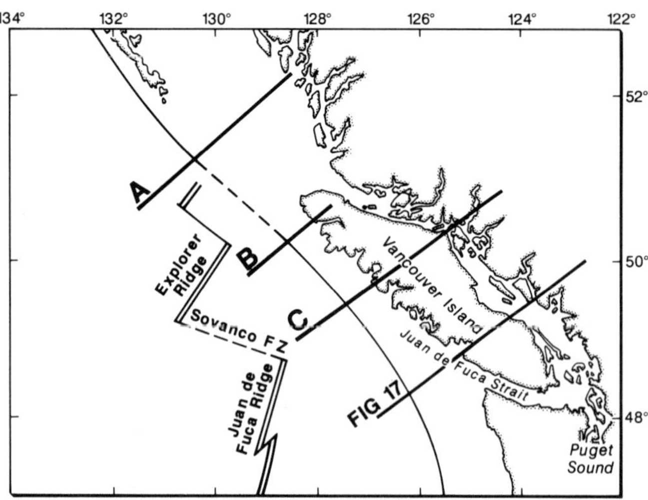

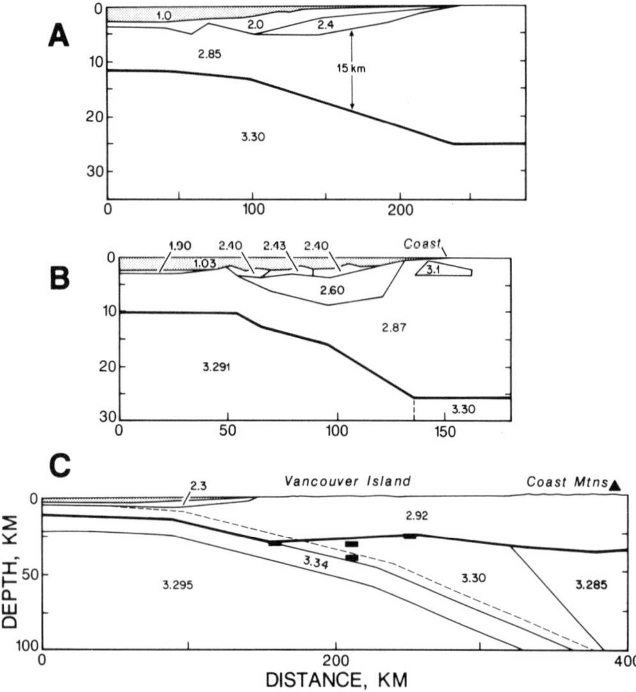

Figure 10. Crustal cross sections of the continental margin of and north of Vancouver Island, based on gravity and seismic refraction measurements (Riddihough, 1985 in Yorath and others, 1985). Cross section A traverses Queen Charlotte Sound between Vancouver Island and the Queen Charlotte Islands, section B (Davis and Riddihough, 1982) crosses the margin near the northern tip of Vancouver Island, and section C (Riddihough, 1979) crosses Vancouver Island just north of the Strait of Juan de Fuca. Numbers are densities given in grams per cubic centimeter; solid bars are seismic control points. The dashed horizon marks a sediment wedge.

Charlotte Sound is predominantly one of subsidence (Yorath and Hyndman, 1983). Some form of tensional or thermal thinning has been suggested by geologic reconstruction (Yorath and Chase, 1981).

Cape Mendocino

Cape Mendocino marks the junction of the Pacific, American, and Gorda plates (Figs. 1, 11). The junction is of the fault (San Andreas)–fault (Mendocino Fracture Zone)–trench (continental margin) type. Plate-tectonic geometry and reconstructions agree that the triple junction has moved northward along the margin at approximately 50 km/m.y. since its initiation at 20 to 25 Ma (Atwater, 1970). The margin south of Cape Mendocino, occupied by the Coast Ranges of California, was the site of the imbrication and subduction that produced the Franciscan Complex during late Mesozoic and early Tertiary time. The process whereby the triple junction migrated and the effects of this process on the margin have been discussed at length (e.g., Dickinson and Snyder, 1979a), but the detailed local tectonics and dynamics of the triple junction itself remain elusive.

One of the most significant processes that the northward progress of the triple junction is thought to produce is the enlarging of a "slab-window" beneath the continental margin. As proposed by Dickinson and Snyder (1979b), this is a triangular "hole" in the descending lithosphere, manifested as an area of continental margin beneath which the formerly descending lithosphere is replaced by asthenosphere as subduction ceases. Simply, as in northern Vancouver Island, the migrating triple junction marks the boundary between continental margin overlying subducting lithosphere and the continental margin that overlies only asthenosphere. Dickinson and Snyder (1979a) tracked the northward cut-off of Cascade volcanism as the subducting lithosphere disappeared, and Jachens and Griscom (1983) used gravity data to locate the present subcrustal southern margin of the Gorda plate, striking southeastward from Cape Mendocino.

The recent history and geometry of the triple junction is complicated by the extremely complex recent history of the Gorda plate. From its unique curved magnetic anomaly stripes (Raff and Mason, 1961) and internal seismicity (Couch, 1980), it is widely accepted (e.g., Silver, 1971a, b; Carlson and Stoddard, 1981; Wilson, 1986) that it is being internally deformed. Present Gorda-Pacific spreading parallel to the Blanco fracture zone would entail convergence along the Mendocino fracture zone. There is evidence to suggest that this has happened in the past (e.g., Couch, 1980), however, present very slow Gorda Ridge spreading in Escanaba Trough appears to be parallel to Mendocino Ridge. Thus the consensus is that some form of internal Gorda Plate adjustment is necessary. This may range from break-up into a limited number of rigid blocks (Riddihough, 1980) to pervasive shearing (Silver, 1971b) to plastic deformation (Wilson, 1986).

Resolution of these problems is essential to estimating the convergence history of the Gorda Plate with the North American

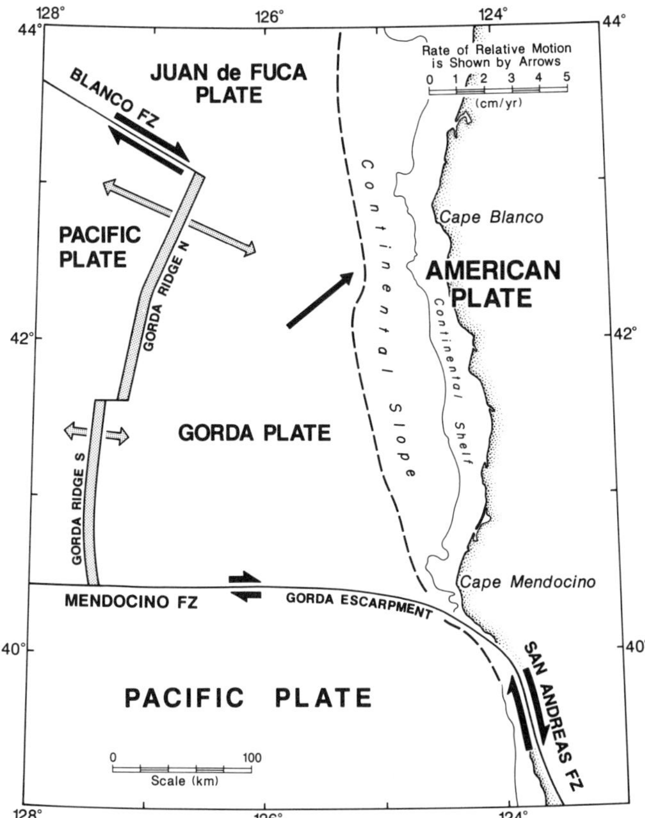

Figure 11. Tectonic map of sea floor and continental margin in the vicinity of Cape Mendocino. The map outlines the tectonic structures and plate motions of and about the Mendocino triple junction.

Plate. The existence of an accretionary wedge north of Cape Mendocino (Silver, 1971a) and the location of hypocenters defining the Gorda Plate beneath the margin (Smith and Knapp, 1980; Cockerham, 1984) testify to past convergence, but give little hint as to the direction and rate of present convergence. Riddihough (1980, 1984) has suggested that the Gorda Plate convergence ceased within the last 2 m.y.; Wilson (1986) wrote that it is still continuing.

The process of triple junction migration is poorly understood; however, recent and active NNW–SSE strike-slip faulting has been identified within the margin, which appears to represent northward extension of the San Andreas Fault (Herd, 1978; Fox 1976; Similia, 1980). Splitting of the margin into a series of slivers or subplates may thus be part of the migration process. This may explain why the transform boundary between the Pacific and American plates to the south in California is a broad (50 to 100 km) zone rather than a single fault.

Jachens and Griscom (1983) have interpreted negative isostatic gravity anomalies in the Coast Range mountains immediately south of Cape Mendocino to indicate the location and attitude of the subducted Gorda Plate. Their interpretation includes an asthenospheric window, as postulated by Dickinson and Snyder (1979b), beneath the continental margin and Coast Ranges south of Cape Mendocino, an overlying continental crust 20 km thick, and an abrupt change in crustal thickness at the junction of the Coast Ranges and the Klamath Mountains.

CONVERGENT MARGINS

From a historical point of view, the northeast Pacific may be considered the type section for plate tectonics. Studies of this area include the early observations of sea-floor magnetic "stripes" by Mason and Raff (1961); the interpretation of the marine magnetic anomalies as indicative of sea-floor spreading by Vine (1966), the concept of the transform fault by Wilson (1965), and the concept of plate motion by McKenzie and Parker (1967), Morgan (1968), and Le Pichon (1968), and more recently by Minster and Jordan (1978) and Engebretson and others (1984). Data from the area also contributed to the plate motion and interaction studies of Atwater (1970) and Riddihough (1984) and to the crustal imbrication models of Seely and others (1974) and Snavely (1987). Currently, studies in the area are yielding new concepts in sea-floor hydrothermal processes and metallogenesis (Lister, 1977; Hammond and others, 1984), the evolution of displaced terranes (Davis and others, 1978; Coney and others, 1980; Beck, 1980, 1984; Bruns, 1983), and geochemical processes, including gas formation in accretionary wedges along convergent plate boundaries (Kulm and others, 1986). The current convergent margin of the Juan de Fuca Plate (Fig. 1) stretches from northern Vancouver Island to Cape Mendocino. Although it lacks a bathymetric trench and ongoing thrust seismicity, it is being intensely studied as an example of a recent and active subduction complex.

Oregon

The Oregon margin (Fig. 1) is broadly north-south in orientation so that contemporary northeast-directed convergence and subduction is approximately 45° oblique. The Juan de Fuca Plate is presumed to descend beneath the margin, although few deep hypocenters have been recorded that can provide control upon its geometry.

Figure 12 shows the location of four geophysical cross sections of the south-central Oregon continental margin based on seismic, gravity, magnetic, and borehole information (Couch and Braman, 1979; Couch, 1980; Couch and Pitts, 1980; R. W. Couch and R. Foote, unpublished manuscript, 1984). The sections (Figs. 13a, b, 14, 15) show the oceanic crust subducting beneath the continental slope, a total crustal thickness of about 20 km in the vicinity of the shelf, and the sedimentary basins on the continental slope and shelf.

The shaded areas in the two sections of Figure 13 mark magnetic and crystalline basement rocks of early Eocene age that extend seaward from surface exposures on the continent. Younger sediments apparently have been carried beneath these basement rocks by the subducting oceanic plate. The basal layers of the basement rocks, as identified in outcrop, are basalts of the Siletz River Volcanics that were deposited on the sea floor (Snavely and Baldwin, 1948; Snavely and others, 1968, 1987; Baldwin, 1981). Sediments and turbidites of the Tyee and Flournoy Formations and younger marine sediments overlie the basalts. These rocks have been uplifted more than 3 km and accreted to the continent. The constitution of the layers that form the crust beneath the Siletz River Volcanics is apparently oceanic, and it has velocities (Dehlinger and others, 1965; Berg and others, 1966; Langston, 1977) and densities consistent with rocks of the lower oceanic crust.

Figure 13c depicts a hypothetical cross section of a convergent plate junction (Ernst, 1975) and shows the position and extent of the expected imbricated accretionary wedge abutted to, and thrust beneath, the leading edge of the continent. In the central Oregon sections (Figs. 13a, b), a similar structure occurs at the base of the slope where Miocene oceanic crust has carried Miocene and younger sediments beneath the Eocene basalts that now form the crystalline component of the continent's leading edge. The continent extends seaward to the toe of the slope. The minimal accreted sediments at the toe of the slope, however, suggest that either part of the post-Eocene subduction mélange

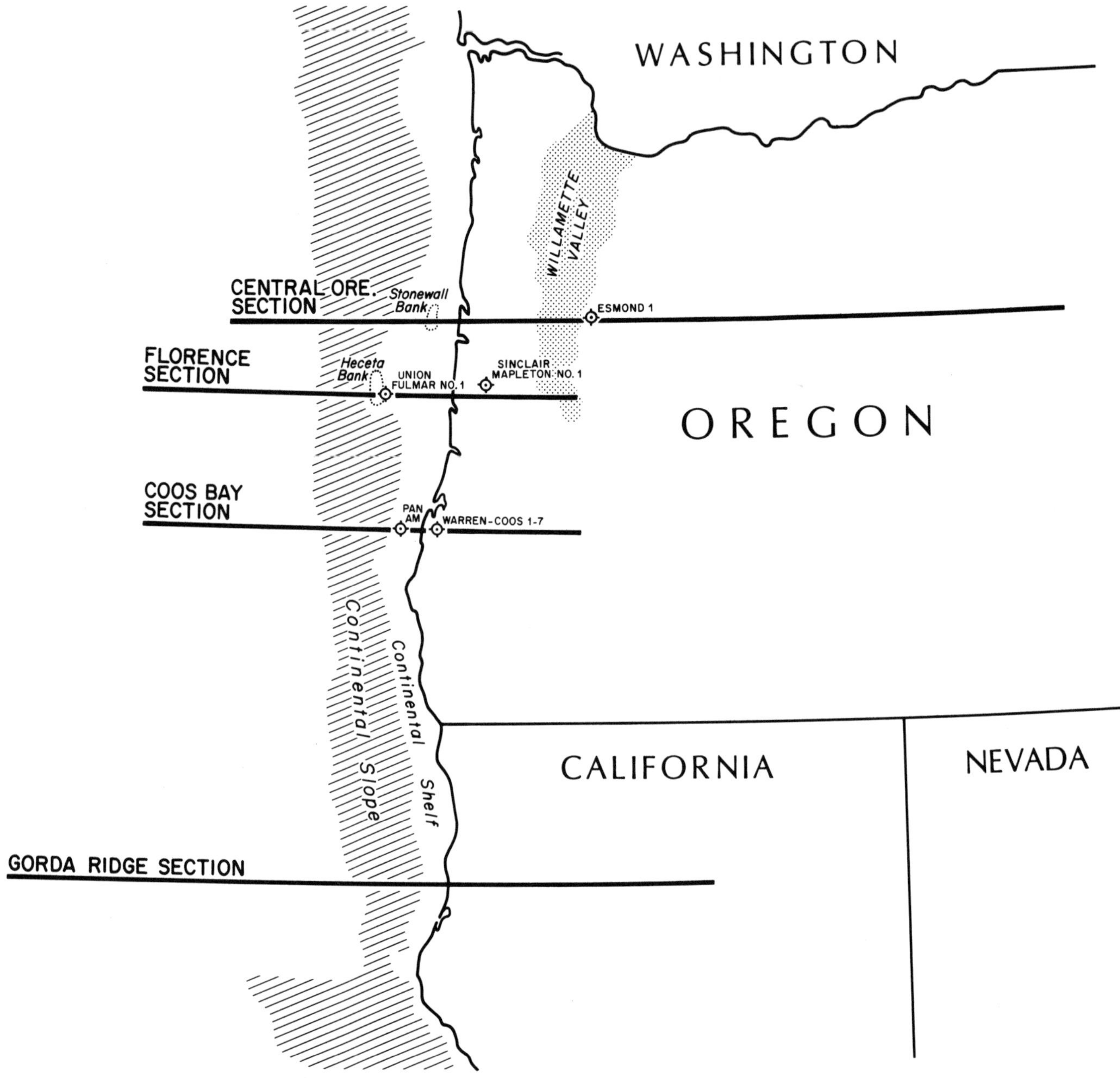

Figure 12. The Oregon continental margin. Map shows the locations of the cross sections shown in Figures 13, 14, and 15 as well as locations of wells included in the model cross sections.

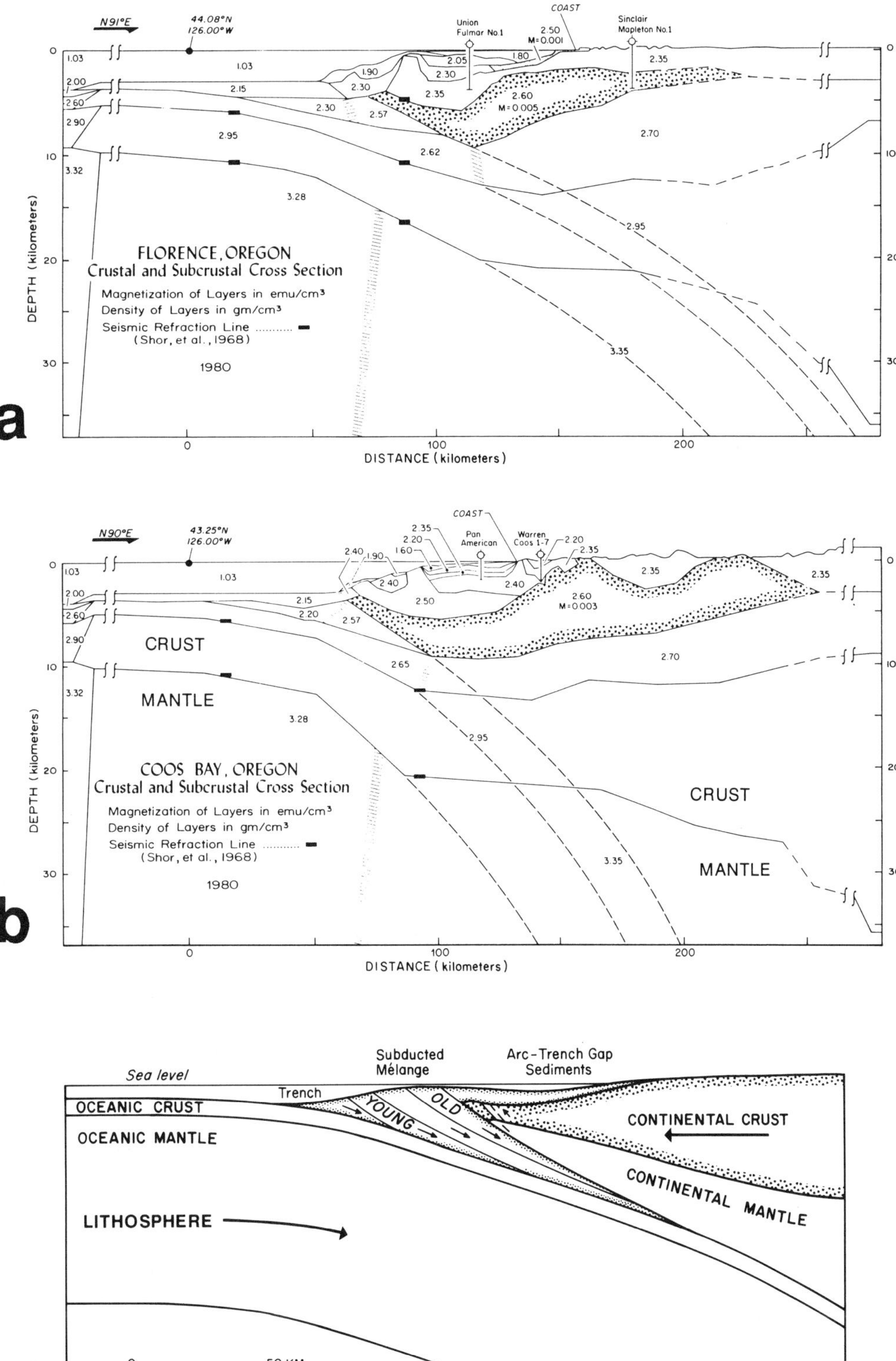

Figure 13. Crustal cross sections of the continental margin of Oregon near Florence (top) (Couch and Braman, 1979) and Coos Bay (middle) (Couch and Pitts, 1980). Stippled blocks outline magnetic crystalline basement. Hatched bars in the cross sections mark sites of lateral density changes in the mantle. The cross section at the bottom shows a hypothetical model of a convergent continental margin (after Ernst, 1975).

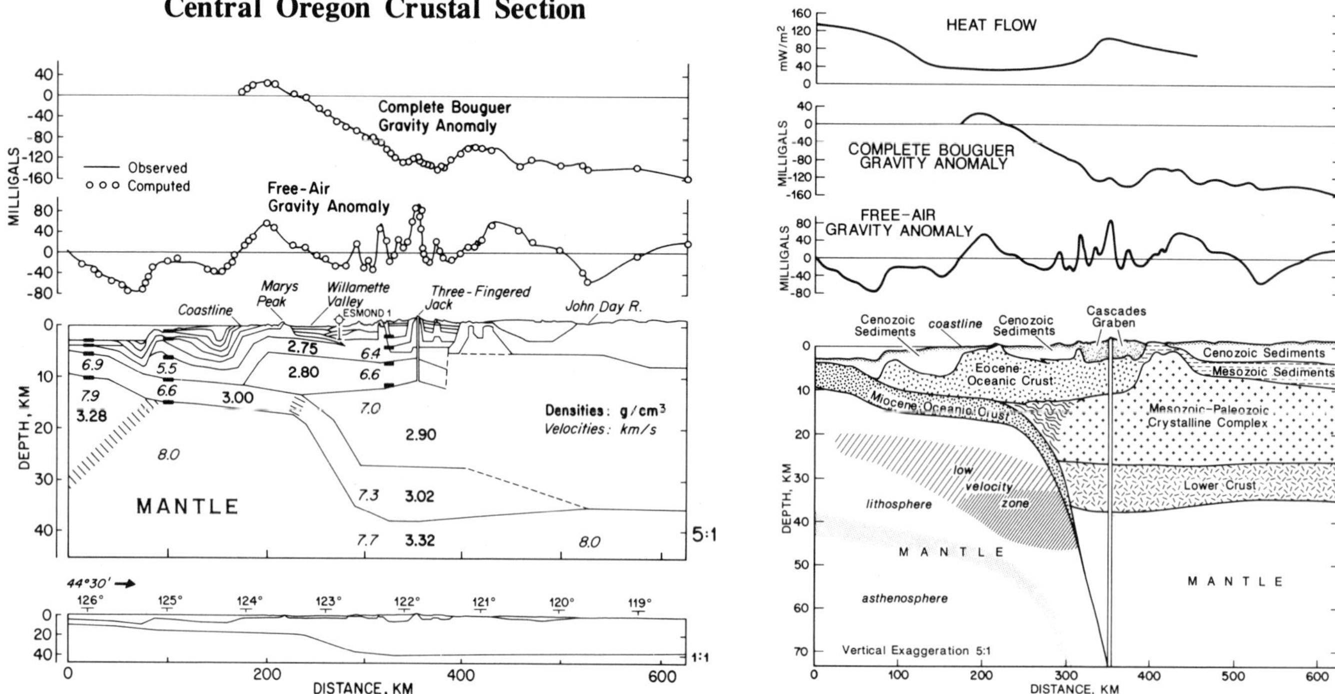

Figure 14. Crustal cross section of the ocean/continent transition zone in central Oregon. A geophysical model is on the left and a geologic interpretation of the model is on the right (from R. W. Couch and R. Foote, unpublished manuscript, 1984). In the geophysical model, heavy bars locate seismic refraction control, and hatched lines in the lower crust and mantle mark lateral density changes.

unit is missing or that the basement rock beneath the slope is an imbricate slice of oceanic crust younger than the Siletz River Volcanics, but older than the subducting Miocene crust.

Figure 14 shows a model cross section of the transition from ocean to continent in central Oregon (R. W. Couch and R. Foote, unpublished manuscript, 1984), constrained by seismic refraction and gravity data, surface geology, and borehole parameters. The central Oregon cross section indicates sediment thicknesses of approximately 5 km in the "trench" or trough at the base of the continental slope, approximately 7 km in the Newport basin on the continental shelf, and about 4 km in the Cascade graben along the axis of the volcanic arc. The section also shows a crustal thickness of about 10 km in the Cascadia abyssal plain, 15 to 20 km beneath the continental margin and the Coast Range, and about 40 km beneath the volcanic arc and high plateau.

The 180- to 200-km-wide crustal section between the continental slope and the Cascade volcanic arc is thus less than half the thickness of normal continental crust. The near-zero free-air gravity anomalies nevertheless indicate that the larger structures are in isostatic equilibrium. Bouguer gravity (Fig. 14) indicates that the major change in crustal thickness occurs beneath the Willamette Valley and the west flank of the Cascade Volcanic Arc, approximately 180 km east of the continental slope. Changes in the densities of the crustal rocks cannot accommodate the observed gravity anomalies; that is, the gravity observations alone require material of mantle density at relatively shallow depths beneath western Oregon.

Figure 14 includes a geological interpretation of the computed crustal cross section (R. W. Couch and R. Foote, unpublished manuscript. 1954). The 40-km-thick subducting lithosphere includes a low-velocity zone beneath the Coast Mountains and Willamette Valley (Dehlinger and others, 1965; Langston, 1977, 1981). It is not known whether the low-velocity zone extends west to the Juan de Fuca Ridge, is limited to the zone of interaction between ocean and continent, or is localized beneath western Oregon. Oceanic crust extends eastward beneath the older, accreted rocks of the continental margin and the mountains of the Coast Range and then descends beneath the volcanic arc and the continental rocks of the Owyhee Plateau. Although the total crustal thickness is 15 to 20 km thick west of the volcanic arc, the accreted "continental crust" is only 10 to 15 km thick; in the Newport Basin, it includes 7 km of post–early Eocene sediments and volcanics.

An alternative model that would fit the existing geophysical constraints would have the oceanic crust and uppermost mantle descend beneath the margin and Coast Range in a smooth arc to the same depth beneath the Cascade Range. Because of the necessity of dense rocks at shallow depths required by the gravity anomalies, this structural configuration would include material of mantle density, perhaps slices of late Eocene to early Miocene upper mantle, between the subducting crust and the Eocene

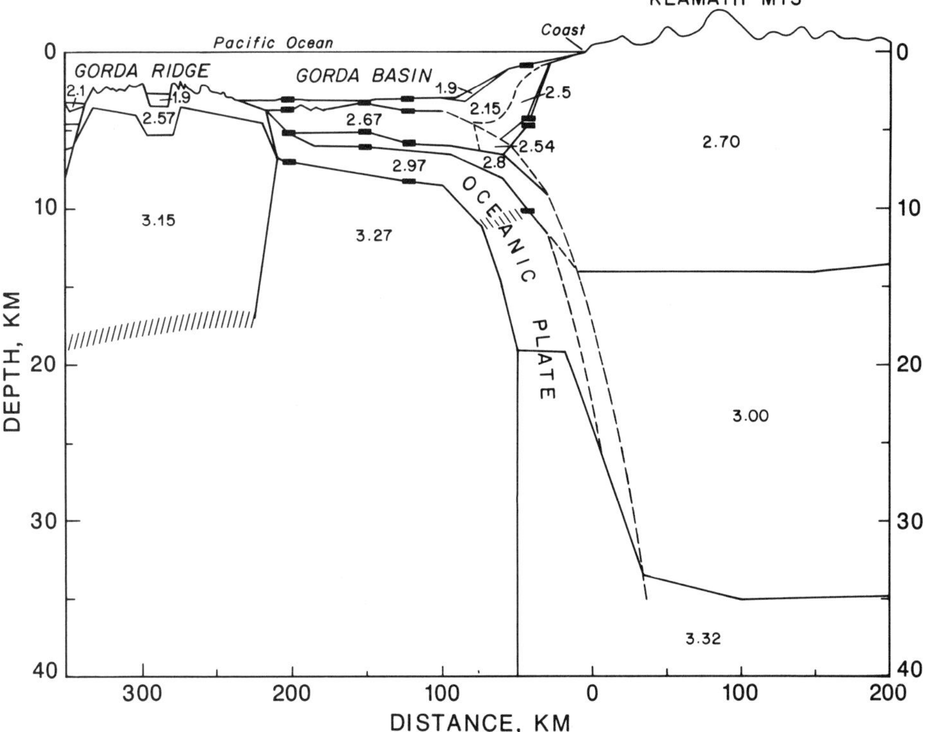

Figure 15. Crustal cross section extending from the Gorda Ridge to the Klamath Mountains. The section is located at 41°N latitude (after Couch, 1980). Numbers are densities in grams per cubic centimeters, solid bars are seismic control points (Shor and others, 1968).

crustal rocks of the Coast Range. This model would be similar to the interpretation of the continental margin structures of Vancouver Island described below.

In southern Oregon, the north-south Coast Range Eocene basalt province terminates against the Klamath Mountains. These mountains are regarded as formed of an older (Paleozoic to Mesozoic in age, with Late Jurassic intrusives) accreted series of tectonic slices or fragments of oceanic crust and ultramafic rocks (Irwin, 1981; Blake and others, 1985). North of Cape Mendocino, they are estimated to have a crustal thickness of the order of 30 to 35 km (Couch, 1980; Griscom, 1980). The ocean floor is occupied by the Gorda Plate, a subplate of the Juan de Fuca Plate, which has an extremely complex and poorly understood convergent history with the margin (see section on Mendocino Triple Junction, above).

Figure 15 shows part of a model cross section of the continental margin adjacent to the Klamath Mountains near 41°N (Couch, 1980). The model suggests that the transition zone between ocean and continent, composed primarily of the continental margin, is narrow, and that the Gorda Plate, with a relatively thin crust, dips beneath the margin and then turns down relatively sharply to accommodate the thicker crust beneath the Klamath Mountains complex. Earthquake hypocenters in the Gorda Plate beneath the coasta zone near Cape Mendocino (Smith and Knapp, 1980; Cockerham, 1984) suggest the dip of the descending plate is near 11° and the thickness of the overriding plate is 10 to 15 km near the coast.

Washington

The crustal structure of the Washington margin is similar to that in Oregon in many respects. In geological terms, the overlying American Plate at the margin is dominated by the Washington Coast Range and Olympic Mountains (Tabor and Cady, 1978), a terrane of oceanic basalts and sedimentary rocks of lower to middle Eocene age that has been identified as a former oceanic and seamount province (Duncan, 1982); it may have been accreted to the margin in late Eocene time. Its crust is primarily oceanic in thickness. Seismic results have been variously interpreted as indicating depths of 16 to 20 km to Moho (Tatel and Tuve, 1955; Taber and Smith, 1985), and the three gravity-model sections constrained by seismic refraction (Taber and Lewis, 1986), shown in Figure 16, confirm thicknesses of less than 20 km. The Olympic Peninsula is the site of a major gravity "low" that coincides with the location of considerable thicknesses of Eocene marine sediments underthrust beneath the Eocene Crescent basalts. These sediments are modeled in Figure 16a as a low-density wedge.

The geometry of the descending plate beneath the region is confirmed by seismologic observations that distinguish between

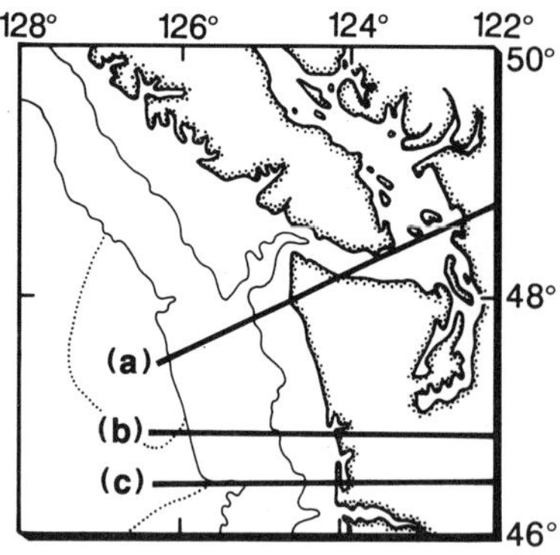

Figure 16. Model cross sections of the Washington continental margin. a, Section (Riddihough, 1979) passes just north of the Olympic Mountains. b, Section (McClain, 1981) passes through Grays Harbor south of the Olympics. c, Section (Riddihough, 1979) passes through Willapa Bay just north of the Columbia River. Numbers on the sections are model densities in grams per cubic centimeters. Dotted lines in sections a and c indicate a sediment wedge on the subducting plate.

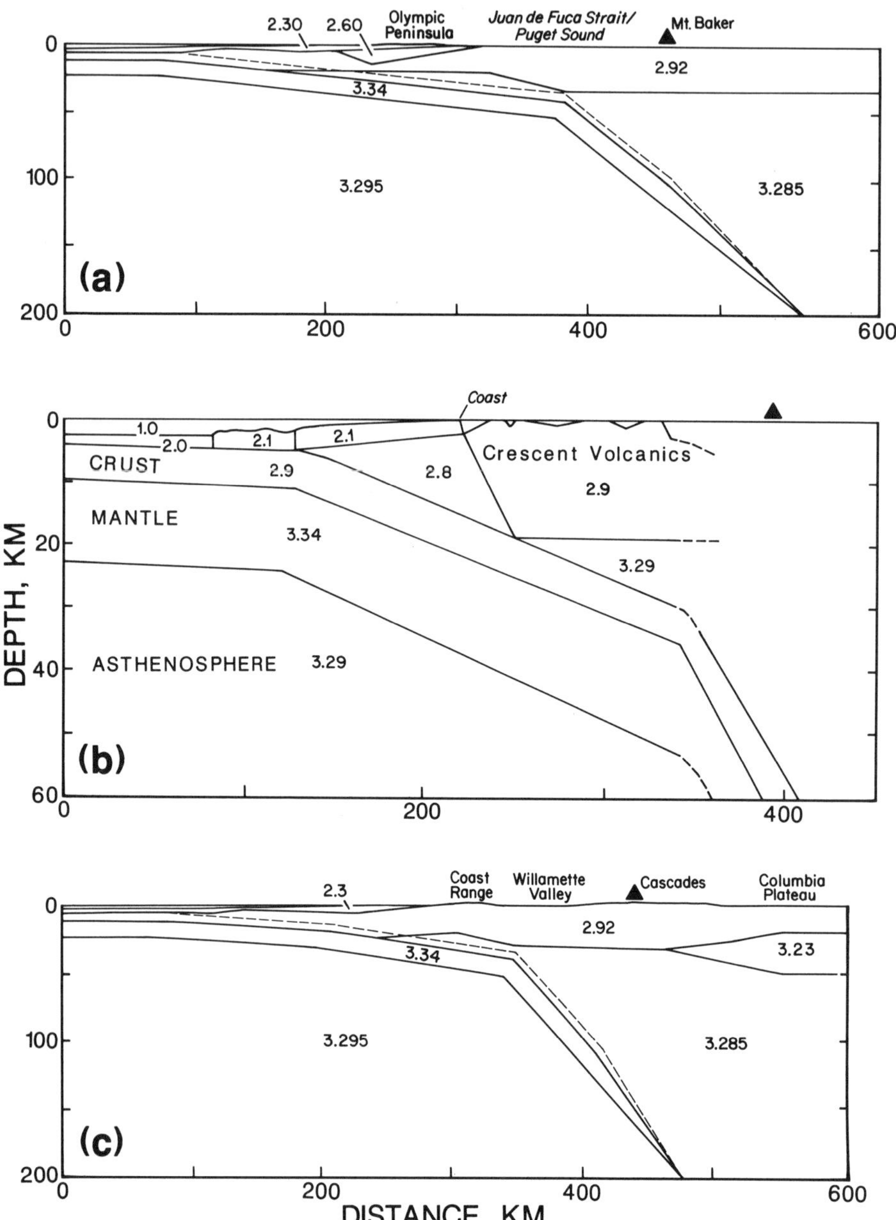

an upper-crustal suite of epicenters and a lower suite, presumed to occur in the downgoing slab (Crosson, 1976; Taber and Smith, 1985). The geometry of the slab at depths below about 80 km in Figures 16b and c is controlled by these hypocenters. A broad continental shelf and slope composed of considerable thicknesses of sediments shows good evidence of late Eocene to recent and contemporary compression, uplift, thrusting, and folding. Of particular interest is the seaward-dipping thrust faults at the foot of the continental slope identified by Silver (1972) and Seeley (1977) and more recently investigated by Lewis (1984) and Snavely (1987). Locally, recent leveling data show uplift decreasing inland eastward from the coast, a pattern that is typical of co-seismic uplift in other subduction zones and that has been interpreted here (Ando and Balazs, 1979; Riddihough, 1982) as suggesting aseismic subduction.

Southern Vancouver Island

The southern continental margin of Vancouver Island is almost exactly perpendicular to the northeasterly convergence of the Juan de Fuca Plate (Riddihough, 1984; Riddihough and others, 1984). The present rate of convergence is approximately 45 mm/yr, but reconstruction suggests that this represents a decrease over the last few million years.

The exposed geology of Vancouver Island is predominantly pre-Tertiary and identified as part of the Wrangellia terrane (Jones and others, 1977). In this respect it contrasts with the Eocene Coast Range of Washington and Oregon. Early unrevised refraction data (White and Savage, 1965) indicated that the crustal thickness of the island was of the order of 40 to 50 km. While this agreed with the then-current concepts of the probable thickness of older continental crust, it was in conflict with gravity interpretations (Couch, 1969; Stacey, 1973), which show that mantle-density materials must be located at depths of between 20 and 30 km beneath much of the island. The conflict was further addressed by Riddihough (1979), who suggested that material of high density and low P_n velocity might occur above a downgoing subducting Juan de Fuca Plate.

Refraction profiling, both along the axis and perpendicular to the continental margin, followed by deep seismic reflection profiling (Clowes and others, 1987; Ellis and others, 1983), has shown that the island is constructed from a series of thin, stacked thrust sheets and terranes of which the surface Wrangellia is one of the older parts (Fig. 17). The thickness of this lid is of the order of 10 to 15 km, and beneath it, Eocene and younger material is thrust as part of a massive underplating/accretionary process. This underthrust sequence is interpreted (Clowes and others, 1987) to consist of turbidites, oceanic basalts, and mafic rocks that have been offscraped and sutured to the overlying crust. The presently descending Juan de Fuca Plate is at a depth of 30 to 40 km beneath the island. The more mafic overlying sections of the accreted material seem to satisfactorily solve the gravity requirements of high-density material at depths of 20 to 30 km and can now be fitted into a much more detailed seismic refraction data set.

The contemporary accretionary complex exposed on the continental margin is interpreted to be continuous with the underplated region (Clowes and others, 1987), although alternative explanations involving discontinuous subduction episodes in which the major plate décollement "jumped" seaward and isolated a former plate are still feasible. Electrical conductivity models from magnetotelluric surveys across Vancouver Island (Kurtz and others, 1986) correlate closely with the descending plate geometry proposed and suggest that sediments moving down on the presently subducting plate have a high water content.

The detailed sections of Vancouver Island as shown in Figure 17 raise the question of the original thickness of older Wrangellia terranes and of whether major subcrustal erosion occurs. The presence of granites at the present surface, which have been formed at depth in the crust, suggests that both upper and lower layers may be removed either by subaerial and subcrustal erosion or shallow thrusting, and that much of the western Canadian margin may be constructed of stacked thrust sheets or "flakes" rather than vertical fault-bound crustal blocks.

CONTINENTAL MARGIN STRUCTURE: SIMILARITIES AND COMPARISONS

We have reviewed the broad crustal architecture of the translational tectonic regimes of the Borderland Province of northern Baja and southern California, the Queen Charlotte Islands, and in the broad sense, central and northern California, the triple junction tectonic regimes of Cape Mendocino and northern Vancouver Island, and the convergent regimes of Vancouver Island, Washington, and Oregon. As noted in the Introduction, during Cenozoic time, as triple plate junctions migrated, almost all of the continental margin experienced episodes of translational and convergent tectonics (Atwater, 1970); some areas may have experienced components of dilational or rift tectonics (e.g., Hall, 1981; Aydin and Page, 1984). In spite of this marked temporal and spatial tectonic variability along the margin, it is remarkable that the larger structural features are quite similar. Four aspects of the crustal structure and processes seem of particular importance to us.

Sedimentary basins

The 22 major sedimentary basins on the continental shelf, outlined in Figures 5 and 18, are about 70 to 200 km in length, with an average near 130 km; 10 to 50 km wide, with an average near 30 km; and are 2 to 10 km deep, with an average near 6 km. Smaller basins located along the continental slope and accretionary wedges at the toe of the slope (e.g., Kulm and Fowler, 1974) contain additional accumulations of Cenozoic sediments. The sediments compose about one-third, and the crystalline rocks about two-thirds, of the crust of the continental margin.

Sediments form about 6 km of the crustal thickness in the centers of the basins and less than 3 km between the basins. The dimensions and spacing of the basins, particularly those on the

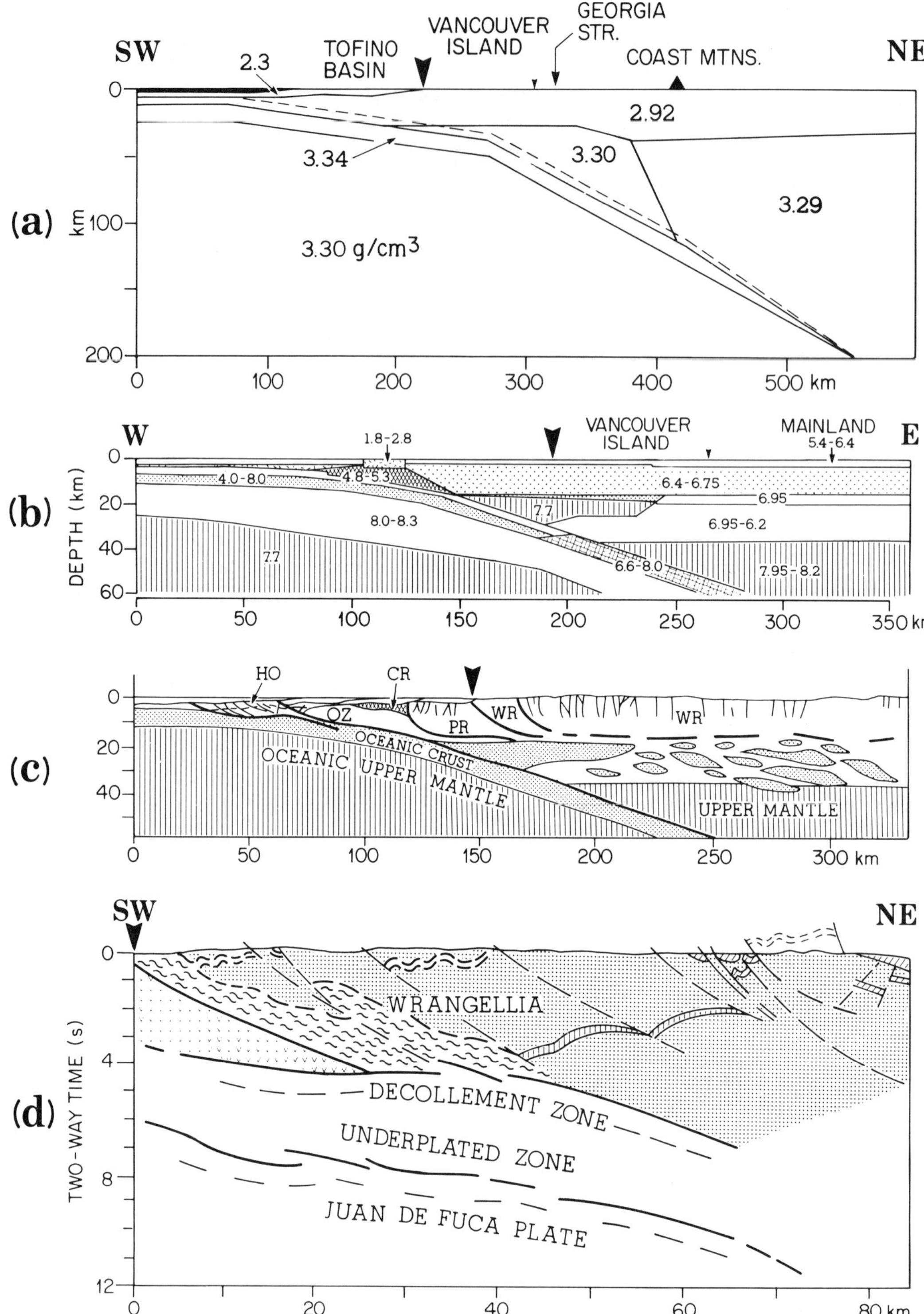

Figure 17. Model sections of southern Vancouver Island. Location of *the section* is shown in Figure 10. a, Section is a gravity model with densities in grams per cubic centimeters derived by Riddihough, 1979; sections b and c are modified from Monger and others, 1986. b, Seismic refraction model with velocities in kilometers per second. c, Geologic section with terranes identified as CR = Crescent, HO = Hoh, PLR = Pacific Rim, OZ = Ozette, WR = Wrangellia. d, Section is an interpretation of LITHOPROBE deep reflection seismic data from Yorath and others (1985). The large arrow marks the coast.

continental shelf north of Cape Mendocino, suggest the subsedimentary crustal rocks form an undulate structure parallel to the margin with an amplitude of a few kilometers and wavelengths of about 150 km. This seems to imply some laterally recurring regularity in the processes that have formed the continental margin and the post-Eocene basins of the margin. Such processes might include segmentation of the margin associated with subduction of the converging oceanic plate (Carr and others, 1974; Couch, 1975; Snavely, 1987), flexure of a thin crust under sedimentary loading, or shear fracturing oblique to the margin caused by strike-slip motion along the margin or by oblique subduction. The configurations of the basins of the Continental Borderland and northern California are more elongate and irregular than the basins north of Cape Mendocino. This may be because the basins of Baja and California have experienced a greater amount of shear than the basins of the Pacific Northwest.

Crustal thickness and erosion

The thickness of the crust of the western North American continental margin is uniformly near 15 to 20 km and generally increases in thickness from the continental slope to the continental interior. This thickness is intermediate between the juxtaposed 10-km-thick oceanic crust and 30- to 35-km-thick continental crust that occurs 200 to 300 km inland from the continental slope.

Figure 19 shows three cross sections of divergent or rifted margins: northern Alaska; the eastern United States off Woods Hole, Massachusetts (Worzel, 1965); and northwest Yucatan (Alvarado-Omana, 1987). These three passive-margin cross sections show a relatively wide continental shelf, a crust approximately 30 to 35 km thick at the seaward edge of the continental shelf, and a "transitional" crust that extends from the continental slope seaward about 200 km before it thins to oceanic crust. By comparison, the crustal cross sections of the borderlands (Fig. 5), central California (Fig. 6), central Oregon (summarized in Fig. 19), and off the Queen Charlotte Islands (Fig. 8) show a crust that abruptly increases in thickness from oceanic to about 15 to 20 km beneath the continental slope, and then increases in thickness to 30 to 40 km about 200 km landward of the slope. These two continental margin types are clearly distinct. The western continental margin of North America has formed largely by accretion and translation throughout Cenozoic time, and the configuration of its crust, in cross section, results from and reflects this tectonic history.

The analysis of Clowes and others (1987) indicates that the pre-Cenozoic Wrangellia terrane on Vancouver Island is part of a crust that is less than 20 km thick (Fig. 17c). Similarly, the pre-Cenozoic rocks of northern British Columbia and the Queen Charlotte Islands (Shor, 1962; Yorath and others, 1985) may form from 10 to 20 km of a 20- to 25-km-thick crust (Fig. 8). The expected crustal thickness in the vicinity of these older rocks would be about 35 km. The anomalously thin crust suggests that some process has thinned a normally thick continental crust.

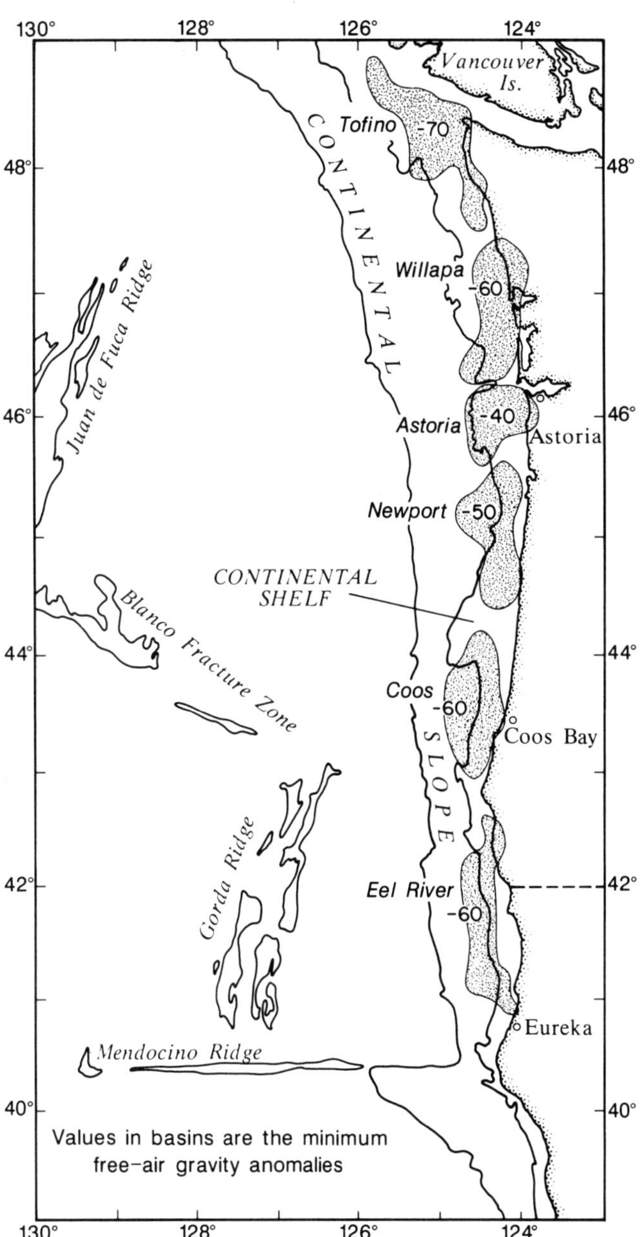

Figure 18. Sedimentary basins on the continental shelves of Oregon, Washington, and Vancouver Island, outlined by gravity anomalies. Numbers indicate approximate gravity anomaly minima.

Crustal thinning could occur by subcrustal erosion during an episode of plate convergence or by crustal extension and associated contemporary thermal and/or structural processes in the lower crust and uppermost mantle prior to and during continental rifting. In the latter situation, it also may be that the accreted portion of the margin with anomalously thin crust includes weaknesses that allow relatively easy separation and enhance the mobility of the newly accreted terranes of the continental margin.

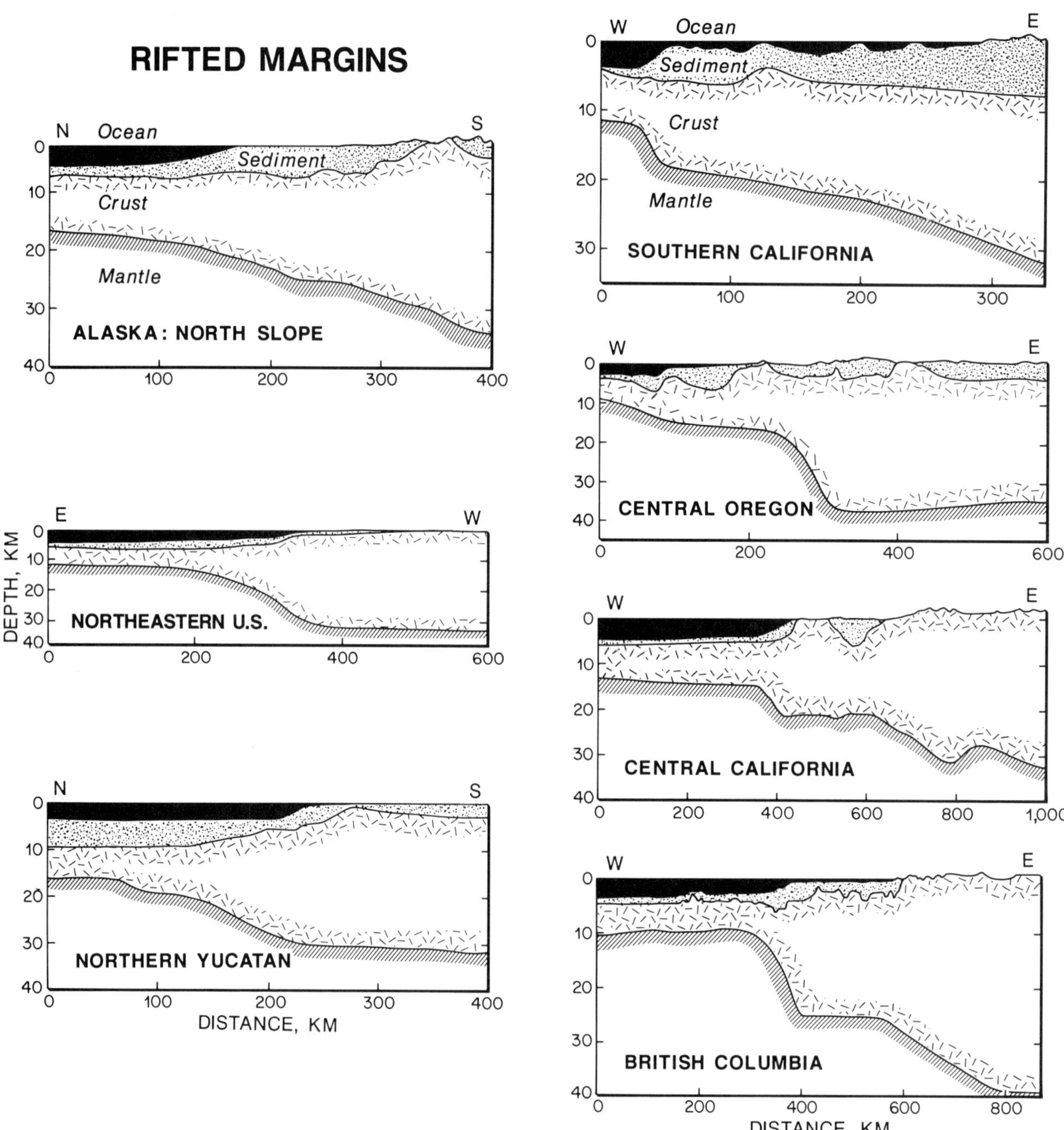

Figure 19. Cross sections of rifted and accretionary/translational continental margins. The Alaska North Slope section passes through the Arctic Wildlife Refuge; the northeastern U.S. section (Worzel, 1965) is near Woods Hole, Massachusetts; and the Yucatan (Alvarado-Omana, 1987) is located offshore north of Progreso, Mexico. The text describes the southern California (Shor and Raitt, 1958), central Oregon (R. W. Couch and R. Foote, unpublished manuscript, 1984), and central California (Thompson and Talwani, 1964) sections. The British Columbia section passes through the Queen Charlotte Islands (Dehlinger and others, 1970). In comparing, note differences in scales.

Shearing and lateral transport

Although the total thickness of the crystalline crust along the western continental margin of North America varies by only a few kilometers, the convergent margins of Oregon, Washington, and Vancouver Island contain relatively large accumulations of sediment on the continental slopes and shelves. By contrast, the translational margins of central California and northern British Columbia present a truncated appearance and contain relatively small amounts of sediments.

Crystalline rocks that floor the shelf basins in some places extend nearly to the base of the continental slope off central California, central Oregon (Figs. 4, 13), and British Columbia (Dehlinger and others, 1970). Pre-Cenozoic rocks very likely extend to the continental slope west of the Klamath Mountains (Fig. 15) and off southeasternmost Alaska. Although accreted rocks are predominant on these margins, the absence of the expected amount of slope sediments and the truncated appearance of the continental margin suggest that they have been sheared or abraded after formation. The long history of alternating accretionary and translational tectonics along these margins suggests that the sediments and possibly parts of the pre-Cenozoic crustal complex have been translated elsewhere. The earthquake and structural studies described above indicate that the current "terranes-in-motion" along the northeast Pacific margin include Baja proper, slices of the western continental margin of Baja, the California borderland, the many blocks that form southern California, slices of central and northern California west of and along the San Andreas Fault system (Weldon and Humphreys, 1986), crustal slices west of the Queen Charlotte Islands (Riddihough and Hyndman, 1986), and the continental margin terranes of southeastern Alaska (Bruns, 1983). The limited data available indicate that these terranes are moving or have moved recently northwestward along the continental margin. The process is thus clearly demonstrable from contemporary tectonics and deducible from older structures.

Accretion

Geological and geophysical studies suggest three major modes of continental accretion along the continental margins of eastern North America: (1) off-scraping of sediments from the oceanic plate by the continental plate during subduction (Byrne and others, 1966; Seely and others, 1974; Kulm and Fowler, 1974; Barnard, 1978); (2) imbrication of slices of oceanic crust, including the lower crust and occasional pieces of the upper mantle (Ernst, 1984a, b; Ellis and others, 1983; Spence and others, 1985; Clowes and others, 1987); and (3) the accretion of terranes by translation and/or convergence, such as the Washington/Oregon Coast Range and the California borderland (Blake, 1981; Duncan, 1982; Howell and others, 1986; Teng and Gorsline, 1987; Snavely, 1987).

Although the allochthonous terranes volumetrically constitute most of the material accreted to the continent, the terranes themselves are, in large part, formed of older accreted oceanic crust that has been expanded by the later addition of both marine and terrestrial sediments, and in some locales, metamorphosed. Accretion along this margin is thus an extremely complex process. Episodes of accretion are superimposed on earlier episodes, with intervening translation and probably related over or underthrusting. The migrations of the triple junctions that cause the changes provide additional complexity in the resultant structure.

CONCLUDING REMARKS

Considering the complexity of tectonic processes active along the western continental margin of North America, the uniformity of crustal thickness of about 20 km or less is remarkable. Although translational movements are extremely important along the margin and have played a critical role in the accumulation of the terrane that characterizes the western Cordillera, this thickness raises the question of how continental crust becomes thickened to its more normally accepted 30 to 40 km. If this margin is typical of Cordilleran orogeny, how is the crust finally thickened? What process or processes that we are *not* seeing along the margin are key to producing "standard" continental crust? What about continental collision, large-scale overthrusting, and mantle fractionation and underplating?

ACKNOWLEDGMENTS

We thank W. D. Mooney, E. Silver, and P. D. Snavely, Jr., for their many helpful comments in their review of this manuscript. S. Troseth and A Yeaple prepared the illustrations, and G. Walker prepared the manuscript. This research was supported in part by the Office of Naval Research under Contract N00014-84-C-0673 and in part by the National Aeronautics and Space Administration Grant NAG 5-788.

REFERENCES CITED

Alvarado-Omana, M., 1987, Gravity and structure of the south-central Gulf of Mexico, The Yucatan Peninsula, and adjacent areas from 17°N to 26°N and from 84°W to 93°W [M.S. thesis]: Corvallis, Oregon State University, 76 p.

Ando, M., and Balazs, E. I., 1979, Geodetic evidence for aseismic subduction of the Juan de Fuca plate: Journal of Geophysical Research, v. 84, p. 3023–3028.

Atwater, T., 1970, Implications of plate tectonics for the Cenozoic evolution of western North America: Geological Society of America Bulletin, v. 81, p. 3513–3536.

Aydin, A., and Page, B. M., 1984, Diverse Pliocene Quaternary tectonics in a transform environment; San Francisco Bay region, California: Geological Society of America Bulletin, v. 95, p. 1303–1317.

Baldwin, E. M., 1981, Geology of Oregon: Dubuque, Iowa, Kendall/Hunt Publishing Co, 170 p.

Barnard, W. D., 1978, The Washington continental slope: Quaternary tectonics and sedimentation: Marine Geology, v. 27, p. 79–114.

Beck, M. E., 1980, Paleomagnetic record of plate margin tectonic processes along the western edge of North America: Journal of Geophysical Research, v. 85, p. 7115–7131.

Beck, M. E., Jr., 1984, Has the Washington-Oregon Coast Range moved northward?: Geology, v. 12, p. 737–740.

Berg, J. W., Jr., and 10 others, 1966, Crustal refraction profile, Oregon Coast Range: Bulletin of the Seismological Society of America, v. 56, p. 1357–1362.

Bevier, M. L., Armstrong, R. L., and Souther, J. G., 1979, Miocene peralkaline volcanism in west-central British Columbia; Its temporal and plate tectonics setting: Geology, v. 7, p. 389–392.

Blake, M. C., 1981, The Franciscan Assemblage and related rocks in northern California; A reinterpretation, *in* Ernst, W. G., ed., The geotectonic development of California; Rubey volume 1: Englewood Cliffs, New Jersey, Prentice-Hall, p. 307–328.

Blake, M. C., Engebretson, D. C., Jayko, A. S., and Jones, D. L., 1985, Tectonostratigraphic terranes in southwest Oregon, *in* Howell, D. G., ed., Tectonostratigraphic terranes of the Circum-Pacific region: Houston Circum-Pacific Council for Energy and Mineral Resources Earth Science Series 1, p. 147–158.

Bruns, T. R., 1983, Model for the origin of the Yakutat Block; An accreting terrane in the northern Gulf of Alaska: Geology, v. 11, p. 718–721.

Byrne, J. V., Fowler, G. A., and Maloney, N. M., 1966, Uplift of the continental margin and possible continental accretion off Oregon: Science, v. 154, no. 3757, p. 1654–1656.

Carlson, R. L., and Stoddard, R. P., 1981, Deformation of the Gorda plate; A kinematic view [abs.]: EOS Transactions of the American Geophysical Union, v. 62, p. 1035.

Carr, M. J., Stoiber, R. E., and Drake, C. L., 1974, The segmented nature of some continental margins, *in* Burk, C. A., and Drake, C. L., eds., The geology of continental margins: New York, Springer-Verlag, p. 105–114.

Clowes, R. M., and 6 others, 1987, LITHOPROBE; Southern Vancouver Island; Cenozoic subduction complex imaged by deep seismic reflections: Canadian Journal of Earth Sciences, v. 24, p. 31–51.

Cockerham, R. S., 1984, Evidence for a 180-km-long subducted slab beneath northern California: Bulletin of the Seismological Society of America, v. 74, p. 569–576.

Coney, P. J., Jones, D. L., and Monger, J.W.H., 1980, Cordilleran suspect terranes: Nature, v. 288, p. 329–333.

Couch, R. W., 1969, Gravity and structures of the crust and subcrust in the northeast Pacific Ocean west of Washington and British Columbia [Ph.D. thesis]: Corvallis, Oregon State University, 179 p.

——, 1975, Segmentation of the continental margin of Central America: EOS Transactions of the American Geophysical Union, v. 55, no. 12, p. 1065.

——, 1980, Seismicity and crustal structure near the northern end of the San Andreas Fault system, *in* Steitz, R., and Sherborne, R., eds., Studies of the San Andreas Fault Zone in northern California: California Division of Mines and Geology Special Report 140, p 139–151.

Couch, R. W., and Braman, D. E., 1979, Geology of the continental margin near Florence, Oregon: Oregon Geology, v. 71, p. 171–179.

Couch, R. W., and Pitts, G. S., 1980, Structure of the continental margin near Coos Bay, Oregon, *in* Newton, V. C., ed., Prospects for oil and gas in the Coos Basin, Western Coos, Douglas, and Lane counties, Oregon: Portland, Oregon Department of Geology and Mineral Industries Oil and Gas Investigations 6, p. 23–26.

Couch, R., and Woodcock, S., 1981, Gravity and structure of the continental margins of southwestern Mexico and northwestern Guatemala: Journal of Geophysical Research, v. 86, no. B3, p. 1829–1940.

Crandall, G. J., Luyendyk, B. P., Reichle, M. S., and Prothero, W. A., 1983, A marine seismic refraction study of the Santa Barbara Channel, California: Marine Geophysical Researches, v. 6, p. 15–37.

Crosson, R. S., 1976, Crustal structure modelling of earthquake data; 2, Velocity structure of the Puget Sound region, Washington: Journal of Geophysical Research, v. 81, p. 3047–3054.

Crouch, J. K., 1979, Neogene tectonic evolution of the California borderland and western Transverse Ranges: Geological Society of America Bulletin, v. 90, p. 339–345.

——, 1981, Northwest margin of the California continental borderland; Marine geology and tectonic evolution: American Association of Petroleum Geologists Bulletin, v. 65, p. 191–218.

Curray, J. R., and Nason, R. D., 1967, San Andreas fault north of Point Arena, California: Geological Society of America Bulletin, v. 78, p. 413–418.

Davis, E. E., and Clowes, R. M., 1986, High velocities and seismic anisotropy in Pleistocene turbidites off western Canada: Geophysical Journal of the Royal Astronomical Society, v. 84, p. 381–400.

Davis, E. E., and Riddihough, R. P., 1982, The Winona Basin; Structure and tectonics: Canadian Journal of Earth Sciences, v. 19, p. 767–788.

Davis, G. A., Monger, J.W.H., and Burchfiel, B. C., 1978, Mesozoic constriction of the Cordilleran "collage," central British Columbia to central California, *in* Howell, D. G., and McDougall, K. A., eds., Mesozoic paleogeography of the western United States: Pacific Coast Paleography Symposium 2, p. 1–32.

Dehlinger, P., Chiburis, E. F., and Colvier, M. M., 1965, Local travel-time curves and their geologic implications for the Pacific Northwest states: Bulletin of the Seismological Society of America, v. 55, p. 587–607.

Dehlinger, P., Couch, R. W., McManus, D. A., and Gemperle, M., 1970, Northeast Pacific structure, *in* Maxwell, A., ed., The Sea, vol. 4, pt. 2: New York, Wiley-Interscience, p. 133–189.

Dickinson, W. R., and Snyder, W. S., 1979a, Geometry of triple junctions related to San Andreas transform: Journal of Geophysical Research, v. 84, p. 561–572.

——, 1979b, Geometry of subducted slabs related to San Andreas transform: Journal of Geology, v. 87, p. 609–627.

Doyle, L. J., and Gorsline, D. S., 1977, Marine geology of Baja California continental borderland, Mexico: American Association of Petroleum Geologists Bulletin, v. 61, p. 903–917.

Duncan, R. A., 1982, A captured island chain in the Coast Range of Oregon and Washington: Journal of Geophysical Research, v. 87, p. 10827–10837.

Ellis, R. M., and 15 others, 1983, The Vancouver Island Seismic Project: A CO-CRUST onshore-offshore study of a convergent margin: Canadian Journal of Earth Sciences, v. 20, p. 719–741.

Emery, K. O., 1954, General geology of the offshore area, southern California: California Division of Mines and Geology Bulletin, v. 170, p. 107–111.

Engebretson, D., 1982, Relative motions between oceanic and continental plates in the Pacific Basin [Ph.D. thesis]: Stanford, California, Stanford University, 211 p.

Engebretson, D. C., Cox, A., and Gordon, R. G., 1985, Relative motions between oceanic and continental plates in the Pacific Basin: Geological Society of America Special Paper 206, 64 p.

Ernst, W. G., 1975, Introduction, *in* Ernst, W. G., ed., Subduction zone metamorphism: Stroudsburg, Pennsylvania, Dowden, Hutchison and Ross, p. 1–14.

——, 1984a, California blueschists, subduction, and the significance of tectonostratigraphic terranes: Geology, v. 12, p. 436–440.

——, 1984b, Phanerozoic continental accretion and the metamorphic evolution of northern and central California: Tectonophysics, v. 100, p. 287–320.

Fox, K. F., 1976, Mélanges in the Franciscan Complex; A product of triple junction tectonics: Geology, v. 4, p. 737–740.

Green, H. G., Kennedy, M., Suarez, F., and Wong, V., 1987, Faulting–seismic reflection data from the California borderland, *in* Dauphin, J. P., ed., The Peninsular Province of California: American Association of Petroleum Geologists Memoir (in press).

Griscom, A., 1980, The Klamath Mountains province, *in* Oliver, H. W., ed., Interpretation of the gravity map of California and its continental margin: California Division of Mines and Geology Bulletin 205, p. 34–36.

Hall, C. A., 1981, San Luis Obispo Transform Fault and middle Miocene rotation of the western Transverse Ranges, California: Journal of Geophysical Research, v. 86, no. B2, p. 1015–1031.

Hammond, J. R., and 5 others, 1984, Discovery of high temperature hydrothermal venting on the Endeavor segment of the Juan de Fuca Ridge: EOS Transactions of the American Geophysical Union, v. 65, p. 1111.

Hayes, D. E., 1966, A geophysical investigation of the Peru-Chile Trench: Marine Geology, v. 4, p. 309–351.

Herd, D. G., 1978, Intracontinental plate boundary east of Cape Mendocino, California: Geology, v. 6, p. 721–725.

Hill, O. J., 1976, Geology of the Jurassic basement rocks, Santa Cruz Island, California, and correlation with other Mesozoic basement terranes in California, *in* Howell, D. G., ed., Aspects of the geologic history of the California continental borderland: American Association of Petroleum Geologists Pacific Section Miscellaneous Publication 24, p. 16–46.

Horn, J. R., Clowes, R. M., Ellis, R. M., and Bird, D. N., 1984, The seismic structure across an active oceanic/continental transform fault zone: Journal of Geophysical Research, v. 89, p. 3107–3120.

Howell, D. G., and Vedder, J. G., 1981, Structural implications of stratigraphic discontinuities across the southern California borderland, *in* Ernst, W. G., ed., The geotectonic development of California; Rubey volume 1: Englewood Cliffs, New Jersey, Prentice-Hall, p. 535–558.

Howell, D. G., McLean, H., and Vedder, J. G., 1976, Cenozoic tectonism on Santa Cruz Island, *in* Howell, D. G., ed., Aspect of the geologic history of the California continental borderland: American Association of Petroleum Geologists Pacific Section Miscellaneous Publication 24, p. 392–416.

Howell, D. G., Crouch, J. K., and Lee-Wong, F., 1981, Comparative study of rocks from Deep Sea Drilling Project Holes 467, 468, and 469, and the southern California borderland, *in* Yeats, R. S., and others, eds., Initial reports of the Deep Sea Drilling Project: Washington, D.C., U.S. Government Printing Office, v. 63, p. 907–918.

Howell, D. G., Champion, D. E., and Vedder, J. G., 1986, Terrane accretion, crustal kinematics, and basin evolution, southern California, *in* Ingersol, R. V., ed., Cenozoic basin development of coastal California; Rubey volume 6: Englewood Cliffs, New Jersey, Prentice-Hall, p. 242–258.

Humphreys, E., Clayton, R. W., and Hager, B. H., 1984, A tomographic image of mantle structure beneath southern California: Geophysical Research Letters, v. 11, p. 625–627.

Hyndman, R. P., and Weichert, D. H., 1983, Seismicity and rates of relative motion on the plate boundaries of western North America: Geophysical Journal of the Royal Astronomical Society, v. 72, p. 59–82.

Irwin, W. P., 1981, Tectonic accretion of the Klamath Mountains, *in* Ernst, W. G., ed., The geotectonic development of California; Rubey volume 1: Englewood Cliffs, New Jersey, Prentice-Hall, p. 29–49.

Jachens, R. C., and Griscom, A., 1983, Three-dimensional geometry of the Gorda Plate beneath northern California: Journal of Geophysical Research, v. 88, p. 9375–9392.

Jones, D. L., Silberling, N. J., and Hillhouse, J., 1977, Wrangellia; A displaced terrane in northwestern North America: Canadian Journal of Earth Sciences, v. 15, p. 2565–2577.

Kammerling, M. J., and Luyendyk, B. P., 1985, Paleomagnetism and Neogene tectonics of the northern Channel Islands, California: Journal of Geophysical Research, v. 90, no. B14, p. 12485–12502.

Keen, C. E., and Hyndman, R. D., 1979, Geophysical review of the continental margins of eastern and western Canada: Canadian Journal of Earth Sciences, v. 16, p. 712–747.

Kulm, L. D., and Fowler, G. A., 1974, Oregon continental margin structure and stratigraphy; A test of the imbricate thrust model, *in* Burk, C. A., and Drake, C. L., eds., The geology of continental margins: New York, Springer-Verlag, p. 261–283.

Kulm, L. D., and 13 others, 1986, Oregon margin subduction zone; Geologic framework, fluid venting, biological communities, and carbonate lithification observed by deep submersible: Science, v. 231, p. 561–566.

Kurtz, R. D., DeLaurier, J., and Gupta, J. C., 1986, Magneto-telluric sounding over Vancouver Island detects the subducting Juan de Fuca Plate: Nature, v. 321, p. 596–599.

Langston, C. A., 1977, Corvallis, Oregon, crustal and upper mantle receiver structure from teleseismic P and S waves: Bulletin of the Seismological Society of America, v. 67, no. 3, p. 713–724.

—— , 1981, Evidence for the subducting lithosphere under southern Vancouver Island and western Oregon from teleseismic P-wave conversions: Journal of Geophysical Research, v. 86, p. 3857–3866.

Le Pichon, X., 1968, Seafloor spreading and continental drift: Journal of Geophysical Research, v. 73, no. 12, p. 3661–3697.

Lewis, B.T.R., 1984, Deep-tow seismic reflection experiment, Washington margin, *in* Kulm, L. D., and others, eds., Western North American continental margin and adjacent ocean floor off Oregon and Washington: Woods Hole, Massachusetts, Marine Science Institute Ocean Margin Drilling Program; Regional Atlas Series, p. 27.

Lister, C.R.B., 1977, Qualitative models of spreading center processes including hydrothermal penetration: Tectonophysics, v. 37, p. 203–218.

Mason, R. G., and Raff, A. D., 1961, Magnetic survey off the west coast of North America, 32°N latitude to 42°N latitude: Geological Society of America Bulletin, v. 72, p. 1259–1266.

McKenzie, D. T., and Parker, R. L., 1967, The North Pacific; An example of tectonics on a sphere: Nature, v. 216, p. 1276–1280.

McLain, K. J., 1981, A geophysical study of accretionary processes on the Washington continental margin [Ph.D. thesis]: Seattle, University of Washington, 141 p.

Minster, J., and Jordan, T. H., 1978, Present-day plate motions: Journal of Geophysical Research, v. 83, p. 5331–5354.

Monger, J.W.H., Clowes, R. M., Price, R. A., Simony, P. S., Riddihough, R. P., and Woodsworth, G. J., 1986, Juan de Fuca Plate to Alberta Plains: Boulder, Colorado, Geological Society of America Continent/Ocean Transect B-2.

Morgan, W. J., 1968, Rises, trenches, great faults, and crustal blocks: Journal of Geophysical Research, v. 73, p. 1959–1982.

Oliver, H. W., and 7 others, 1980, Gravity map of California and its continental margin: California Division of Mines and Geology Geologic Data Map 3, Gravity Map of California.

Oppenheimer, D. H., and Eaton, J. P., 1984, Moho orientation beneath central California from regional earthquake travel times: Journal of Geophysical Research, v. 89, p. 10267–10282.

Plawman, T. L., 1978, Crustal structure of the continental borderland and adjacent portion of Baja California between latitudes 30°N and 33°N [M.S. thesis]: Corvallis, Oregon State University, 72 p.

Prodehl, C., 1979, Crustal structure of the western United States: U.S. Geological Survey Professional Paper 1088, 74 p.

Raff, A. D., and Mason, R. G., 1961, Magnetic survey of the west coast of North America, 40°N latitude of 50°N latitude: Geological Society of America Bulletin, v. 72, p. 1267–1270.

Richter, C. F., 1958, Elementary seismology: San Francisco, W. H. Freeman Co., 768 p.

Riddihough, R. P., 1977, A model for recent plate interactions off Canada's west coast: Canadian Journal of Earth Sciences, v. 14, p. 384–396.

—— , 1979, Gravity and structure of an active margin; British Columbia and Washington: Canadian Journal of Earth Sciences, v. 16, p. 350–363.

—— , 1980, Gorda Plate motions from magnetic anomaly analysis: Earth and Planetary Science Letters, v. 51, p. 163–170.

—— , 1982, Contemporary movements and tectonics on Canada's west coast: A discussion: Tectonophysics, v. 86, p. 319–341.

—— , 1984, Recent movements of the Juan de Fuca Plate system: Journal of Geophysical Research, v. 89, p. 6980–6994.

Riddihough, R. P., and Hyndman, R. D., 1988, The Queen Charlotte Islands margin, *in* Winterer, E. L., Hussong, D. M., and Decker, R. W., eds., The eastern Pacific Ocean and Hawaii: Boulder, Colorado, Geological Society of America, The Geology of North America, v. N (in press).

Riddihough, R. P., Currie, R. G., and Hyndman, R. D., 1980, The Dellwood Knolls and their role in the triple junction tectonics off northern Vancouver Island: Canadian Journal of Earth Sciences, v. 17, p. 577–593.

Riddihough, R. P., and 6 others, 1984, Geodynamics of the Juan de Fuca Plate, *in* Cabre, R., ed., Geodynamics of the eastern Pacific Region, Caribbean and Scotia Arcs: American Geophysical Union Geodynamics Series, v. 9, p. 5–21.

Saleeby, J. B., 1986, Central California offshore to Colorado Plateau: Boulder, Colorado, Geological Society of America Continent/Ocean Transect C-2.

Seeley, D. R., 1977, The significance of landward vergence and oblique structured trends on trench inner slopes, *in* Talwani, M., and Pitman, W. C., eds., Island arcs, deep sea trenches, and back-arc basins: American Geophysical Union, Maurice Ewing Series 1, p. 187–198.

Seeley, D. R., Vail, P. R., and Walton, G. G., 1974, Trench slope model, *in* Burk, C. A., and Drake, C. L., eds., The geology of continental margins: New York, Springer-Verlag, p. 249–260.

Sheffels, B., and McNutt, M., 1986, Role of subsurface loads and regional compensation in the isostatic balance of the Transverse Ranges, California; Evidence for intracontinental subduction: Journal of Geophysical Research, v. 90, no. B6, p. 6419–6431.

Shepard, F. P., and Emery, K. O., 1941, Submarine topography off the California coast; Canyons and tectonic interpretation: Geological Society of America Special Paper 31, 171 p.

Shor, G. G., Jr., 1962, Seismic refraction studies off the coast of Alaska; 1956–1957: Bulletin of the Seismological Society of America, v. 52, p. 37–57.

Shor, G. G., Jr., and Raitt, R. W., 1958, Seismic studies in the southern California borderland, *in* Geofisica Aplicada, tomo Z: 20th International Geologic Congress, Mexico, D.F., 1956, Trabajos, sect. 9, p. 243–259.

Shor, G. G., Dehlinger, P., Kirk, H. K., and French, W. S., 1968, Seismic refraction studies off Oregon and northern California: Journal of Geophysical Research, v. 73, p. 2175–2194.

Silver, E. A., 1971a, Transitional tectonics and late Cenozoic structure of the continental margin off northernmost California: Geological Society of America Bulletin, v. 82, p. 1–22.

——, 1971b, Tectonics of the Mendocino triple junction: Geological Society of America Bulletin, v. 82, p. 2965–2978.

——, 1972, Pleistocene tectonic accretion of the continental slope off Washington: Marine Geology, v. 13, p. 239–249.

Simila, G. W., 1980, Seismological evidence on the tectonics of the northern section of the San Andreas Fault zone in northern California: California Division of Mines and Geology Special Report 140, p. 131–137.

Smith, S. W., and Knapp, J. S., 1980, The northern termination of the San Andreas Fault, *in* Streitz, R., and Sherborne, R., eds., Studies of the San Andreas Fault zone in northern California: California Division of Mines and Geology Special Report 140, p. 153–164.

Snavely, P. D., 1987, Tertiary geologic framework, neotectonics, and petroleum potential of the Oregon-Washington continental margin, *in* Scholl, D. W., Grantz, A., and Vedder, J. G., eds., Geology and resource potential of the continental margin of western North America and adjacent ocean basins; Beaufort Sea to Baja California: American Association of Petroleum Geologists Memoir (in press).

Snavely, P. D., Jr., and Baldwin, E. M., 1948, Siletz River Volcanic series, northwestern Oregon: American Association of Petroleum Geologists Bulletin, v. 32, p. 806–812.

Snavely, P. D., Jr., MacLeod, N. S., and Wagner, H. C., 1968, Theoleiitic and alkalic basalts of the Eocene Siletz River; Volcanics, Coast Range: American Journal of Science, v. 266, p. 454–481.

Spence, G. D., Clowes, R. M., and Ellis, R. M., 1985, Seismic structures across the active subduction zone of western Canada: Journal of Geophysical Research, v. 90, p. 6754–6772.

Stacey, R. A., 1973, Gravity anomalies, crustal structure, and plate tectonics in the Canadian Cordillera: Canadian Journal of Earth Sciences, v. 10, p. 614–628.

Taber, J. J., and Lewis, B.T.R., 1986, Crustal structure of the Washington continental margin from refraction data: Bulletin of the Seismological Society of America, v. 76, no. 4, p. 1011–1024.

Taber, J. J., and Smith, S. W., 1985, Seismicity and focal mechanisms associated with the subduction of the Juan de Fuca Plate beneath the Olympic peninsula, Washington: Bulletin of the Seismological Society of America, v. 75, p. 237–249.

Tabor, R. W., and Cady, W. M., 1978, The structure of the Olympic Mountains, Washington; Analysis of a subduction zone: U.S. Geological Survey Professional Paper 1033, 38 p.

Tatel, H. E., and Tuve, M. A., 1955, Seismic exploration of a continental crust, *in* Poldervaart, A., ed., Crust of the earth: Geological Society of America Special Paper 62, p. 35–50.

Teng, S. L., and Gorsline, D. S., 1987, Stratigraphy of the continental borderland basins, southern California, *in* Dauphin, J. P., ed., The Peninsular Province of the Californias: American Association of Petroleum Geologists Memoir (in press).

Thompson, G. A., and Talwani, M., 1964, Geology of the crust and mantle, western United States: Science, v. 146, p. 1539–1549.

Tiffin, D. L., Cameron, B.E.B., and Murray, J. W., 1972, Tectonics and depositional history of the continental margin off Vancouver Island: Canadian Journal of Earth Sciences, v. 8, p. 1265–1281.

Vedder, J. G., and Howell, D. G., 1976, Neogene strata of the southern group of Channel Islands, California, *in* Howell, D. G., ed., Aspects of the geologic history of the California continental borderland: American Association of Petroleum Geologists Pacific Section Miscellaneous Publication 24, p. 47–52.

Vedder, J. G., Wagner, H. C., and Schoellhamer, J. E., 1969, Geologic framework of the Santa Barbara Channel region, *in* Geology, petroleum development, and seismicity of the Santa Barbara Channel region, California: U.S. Geological Survey Professional Paper 679-A, p. 1–11.

Vedder, J. G., and 6 others, 1974, Preliminary report on the geology of the continental borderland of southern California: U.S. Geological Survey Miscellaneous Field Studies Map MF-624, scale 1:1,000,000.

Vedder, J.G. , Crouch, J. K., and Lee-Wong, F., 1981, Comparative study of rocks from Deep Sea Drilling Project Holes 467, 468, and 469, *in* Yeats, R. S., and others, eds., Initial reports of the Deep Sea Drilling Project: Washington, D.C., U.S. Government Printing Office, v. 63, p. 907–918.

Vine, F. J., 1966, Spreading of the ocean floor; New evidence; Science, v. 154, p. 1405–1415.

Warren, D. H., 1981, Seismic-refraction measurements of crustal structure near Santa Rosa and Ukiah, California: U.S. Geological Survey Professional Paper 1141, p. 167–181.

Watts, A. B., Talwani, M., and Cochran, J. R., 1976, Gravity field of the northwest Pacific Ocean Basin and its margin, *in* Sutton, G. H., Manghnani, M. H., Moberly, R., and McAffee, E. U., eds., The geophysics of the Pacific Ocean Basin and its margin: American Geophysical Union Geophysical Monograph 19, p. 17–34.

Weldon, R., and Humphreys, E., 1986, A kinematic model of southern California: Tectonics, v. 4, no. 1, p. 33–48.

White, W.R.H., and Savage, J. C., 1965, A seismic refraction and gravity study of the earth's crust in British Columbia: Bulletin of the Seismological Society of America,v. 55, p. 463–486.

Wilson, D. S., 1986, A kinematic model for the Gorda deformation zone as a diffuse southern boundary of the Juan de Fuca Plate: Journal of Geophysical Research, v. 91, p. 10259–10269.

Wilson, J. T., 1965, Transform faults, ocean ridges, and magnetic anomalies southwest of Vancouver Island: Science, v. 150, p. 482–485.

Worzel, J. L., 1965, Pendulum gravity measurements at sea: 1936–1959: New York, John Wiley & Sons, 422 p.

Yorath, C. J., and Chase, R. L., 1981, Tectonic history of the Queen Charlotte Islands and adjacent areas; A model: Canadian Journal of Earth Sciences, v. 18, p. 1717–1739.

Yorath, C. J., and Hyndman, R. P., 1983, Subsidence and thermal history of Queen Charlotte Basin: Canadian Journal of Earth Sciences, v. 20, p. 135–159.

Yorath, C. J., and 7 others, 1985, LITHOPROBE, southern Vancouver Island; Seismic reflection sees through Wrangellia to the Juan de Fuca Plate: Geology, v. 13, p. 759–762.

Yorath, C. J., Tiffin, D. L., and Cameron, B.E.B., 1977, Submersible operation on the Pacific continental margin: Geological Survey of Canada Report of Activities Paper 77-1A, part A, p. 301–310.

Yorath, C. J., and 7 others, 1985, Intermontane Belt (Skeena Mountains) to Insular Belt (Queen Charlotte Islands): Boulder, Colorado, Geological Society of America Continent/Ocean Transect B-1.

Manuscript Accepted by the Society October 31, 1988

Printed in U.S.A.

Geological Society of America
Memoir 172
1989

Chapter 9

Regional crustal structure and tectonics of the Pacific Coastal States; California, Oregon, and Washington

Walter D. Mooney
U.S. Geological Survey, MS 977, 345 Middlefield Road, Menlo Park, California 94025
Craig S. Weaver
Geophysics Program, AK-50, U.S. Geological Survey, University of Washington, Seattle, Washington 98195

ABSTRACT

The Pacific Coastal States form a complex geologic environment in which the crust and lithosphere have been continuously reworked. We divide the region tectonically into the southern transform regime of the San Andreas fault and the northern subduction regime, and summarize the geophysical framework with contour maps of crustal thickness, lithospheric and seismicity cross sections, and results from site-specific geophysical studies. The uniformity of crustal thickness (30 ± 2 km) in southern California is remarkable, and appears to be primarily the result of crustal extension in the Mojave Desert and ductile shear of the lower crust along the plate transform boundary. Southern California seismicity defines a broad zone of deformation that extends from the Borderland to the Mojave Desert (about 300 km). The geophysical framework of central and northern California records magmatism and accretion associated with the Mesozoic and Cenozoic subduction, late Cenozoic transform faulting, and in the Basin and Range to the east, extension. The crust thickens from about 20 km at the coast to as much as 55 km in the Sierra Nevada, and thins to about 30 km in the Basin and Range. Cross sections of the crust show that seismic velocities and densities vary significantly over short distances perpendicular to the coast, reflecting processes that include the accretion of oceanic sediments and igneous crust, and significant lateral motion of crustal blocks. Maximum hypocentral depths in central California become deeper as the crust thickens to the west, but seismicity is low beneath the Great Valley and Sierra Nevada, which together appear to form a relatively undeforming block. The lower crust of the Pacific Coastal States has a high average seismic velocity (6.7 km/sec or greater), which probably is the product of tectonic underplating of oceanic crust and/or magmatic underplating by a basaltic melt.

The geophysical framework of the subduction regime is dominated by the subduction of the Gorda and Juan de Fuca plates, arc magmatism in the Cascade Range, and plateau volcanism and rifting in the back arc. As defined by earthquake hypocenters, the Juan de Fuca plate dips at a shallow angle (3°) within 50 km of the trench, increases to 10° beneath the continental shelf and coastal province, and plunges more steeply (25° dip) a short distance west of the Cascade Range. Whereas a true continental Moho exists from the Cascade Range to the east, the Moho is that of the subducting oceanic lithosphere west of the range. Crustal thickness increases from about 18 km at the coast to about 42 km beneath the Cascades Range, a distance of about 200 km. The crustal velocity structure and crustal thickness of the Cascades Range is relatively uniform along its axis. The velocity structure shows high velocities (greater than 6.5 km/sec) at

Mooney, W. D., and Weaver, C. S., 1989, Regional crustal structure and tectonics of the Pacific Coastal States; California, Oregon, and Washington, *in* Pakiser, L. C., and Mooney, W. D., Geophysical framework of the continental United States: Boulder, Colorado, Geological Society of America Memoir 172.

all depths greater than 10 km, indicating rocks of an intermediate-to-mafic composition, and a relatively low upper-mantle velocity of 7.7 ± 0.1 km/sec, indicating high temperatures. Seismological studies at the volcanic centers of the Cascades indicate that the dimensions of subsurface magmatic systems are small, on the order of a few kilometers. Some 1 to 6 km of Miocene and younger basaltic extrusives cover much of the back arc, thereby obscuring most of the pre-Miocene geology. However, geophysical data demonstrate the importance of Mesozoic compression and Cenozoic (particularly Eocene) extension, accompanied by magmatic underplating of the crust.

INTRODUCTION

The Pacific Coastal States (California, Oregon, and Washington; Fig. 1) are highly varied in their physiographic expression, geology, and tectonic history. The tectonic setting of the Pacific Coastal States may be divided at Cape Mendocino, California, into the southern *transform plate boundary* of the San Andreas fault system, and the northern *subduction boundary* where the Gorda and Juan de Fuca plates are consumed beneath the North American plate. Active-margin tectonics has dominated the evolution of the Pacific Coastal States since the Mesozoic era, and in this chapter we summarize the geophysical framework, based primarily on the interpretation of seismic refraction and reflection profiles, seismic network data, and gravity data. We emphasize the large-scale features of the crustal structure and rely mainly on work published since 1978, the time of the last comprehensive review (Hill, 1978).

For our discussion, the transform regime may be conveniently subdivided at the Garlock fault into a southern area of near-uniform crustal thickness in southern California, and a northern area of eastward-thickening crust in central and northern California. The subduction regime from Cape Mendocino to British Columbia is subdivided into fore arc, volcanic arc, and back arc. These divisions reflect the dominant roles that lateral translation, plate convergence, arc volcanism, and crustal extension have played in the geologic evolution of the Pacific Coastal States.

TRANSFORM FAULTING REGIME

Southern California: Gulf of California to the Garlock fault

Southern California is divided into four geologic provinces (Fig. 2). West of the San Andreas fault, the California Borderland and coastal province form a composite province consisting of the continental margin, Peninsular Range batholith, Los Angeles and Ventura basins, and Western Transverse Ranges (Fig. 2). The Mojave Desert east of the San Andreas fault is the interior province. Separating the coastal and interior provinces are the Central and Eastern Transverse Ranges, which obliquely cross the generally northwest-southeast–trending tectonic features of southern California (Fig. 2). The fourth province is the Salton Trough, a modern crustal pull-apart associated with the opening of the Gulf of California.

Geology. The California Borderland consists of tectonostratigraphic terranes that have been accreted to the continental margin by imbricate underthrusting during subduction, obduction, and translation associated with lateral plate motions (Coney and others, 1980; Howell and Vedder, 1981). Due to these plate processes, the Borderland contains numerous elevated blocks and ridges, as well as deep basins (Vedder, 1987). Mesozoic and younger Franciscan-type rocks (marine metasedimentary rocks, primarily graywacke and dark shale) are the dominant basement rock type. Tertiary marine sediments are locally abundant. Seismicity and the deformation of young sediments provide evidence for right-lateral strike-slip motion within the Borderland (Howell and Vedder, 1981). The geophysical setting of the entire Pacific coastal margin, including the California Borderland, is discussed in Couch and Riddihough (this volume).

The Peninsular Range batholith extends from the latitude of Los Angeles to southernmost Baja California, Mexico, a distance of 1,200 km. These Jurassic and Cretaceous plutonic rocks are intruded into predominately early Mesozoic rocks. Paleomagnetic data indicate that the batholith formed at a latitude 11° south of its present position (Hagstrum and others, 1985) and was translated northward in the Cretaceous or early Tertiary. Only a small portion of its motion is due to the Neogene opening of the Gulf of California.

The Los Angeles and Ventura basins may represent either pull-apart basins that developed between series of strike-slip faults, or synclines formed by large-scale compression. The Ventura basin contains a tremendous thickness (18 km) of marine sedimentary rocks ranging in age from Cretaceous to Pleistocene. It continues offshore as the Santa Barbara Channel basin, which has a thickness in excess of 10 km (Keller and Prothero, 1987); similar sedimentary thicknesses (10 to 12 km) occur beneath the Los Angeles basin.

The Transverse Ranges are subdivided into three ranges on the basis of geology and geography. The Western Transverse Range consists of marine sedimentary rocks of the Franciscan assemblage and Great Valley sequence, Mesozoic ophiolitic rocks, and Neogene sedimentary rocks (Hall, 1981). The Central Transverse Range consists of a core of Precambrian rocks with overlying Mesozoic marine metasedimentary rocks that were intruded by Mesozoic granitic rocks. The basement rocks of the central Transverse Range have been translated about 240 km to the northwest, principally by right-lateral slip on the San Andreas fault. The Eastern Transverse Range includes Precambrian plutonic rocks that intrude older gneisses (Crowell, 1981). The ongo-

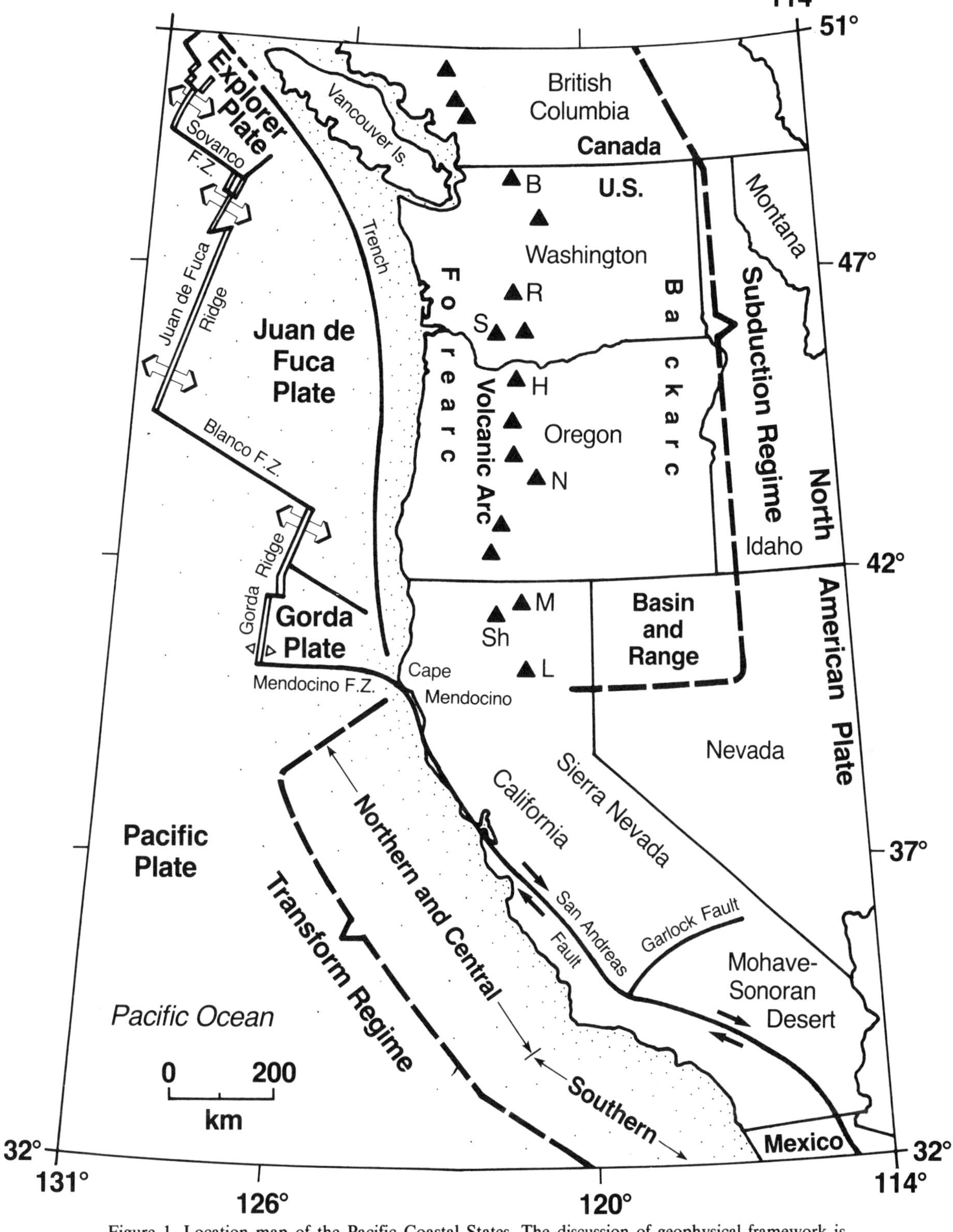

Figure 1. Location map of the Pacific Coastal States. The discussion of geophysical framework is divided into the transform regime south of Cape Mendocino, and the subduction regime to the north. Solid triangles mark active or dormant volcanoes of the Cascade Range: L = Lassen Peak; Sh = Mount Shasta; M = Medicine Lake Volcano; N = Newberry Volcano; H = Mount Hood; S = Mount St. Helens; R = Mount Rainer; B = Mount Baker.

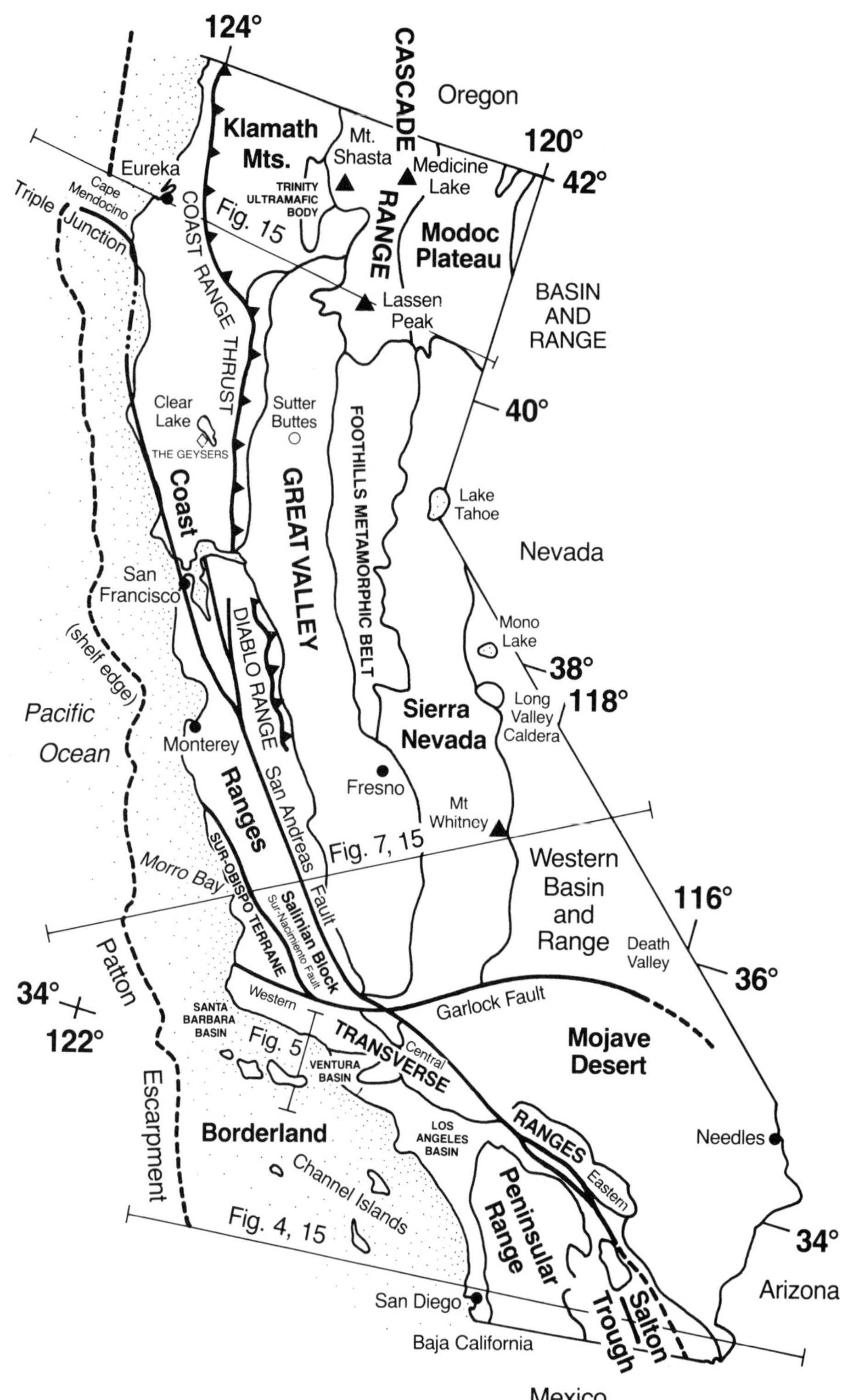

Figure 2. Location map of California geologic provinces, place names, and lithospheric/crustal cross sections shown in Figures 4, 5, 7, and 15. Geophysical framework of the transform regime (Salton Trough to Mendocino triple junction at 40.5°N) is divided by Garlock fault into southern California and central/northern California. Figure 3 shows a contour map of crustal thickness for this area.

ing uplift of the Transverse Ranges began within the past few million years and is commonly attributed to the convergence between the North American and Pacific plates across the "big bend" in the San Andreas fault (Fig. 2), although this concept is in dispute (Weldon and Humphreys, 1986), as discussed below.

The Mojave Desert of southern California encompasses the region east of the San Andreas fault and south of the Garlock fault, except for the eastern Transverse Range (Fig. 2). Extensive Quaternary deposits separate scattered exposures of Precambrian and Paleozoic metasedimentary rocks, and Mesozoic plutonic, volcanic, and sedimentary rocks (Burchfiel and Davis, 1972, 1981).

The Salton Trough lies on the plate boundary separating the Pacific and North American plates, and is the southern terminus of the San Andreas fault (Fig. 2). The trough is the northern onland extension of the Gulf of California spreading system and is a pull-apart basin which began opening at 4.5 Ma. (Larson and others, 1968; Moore and Buffington, 1968; Elders and others, 1972). More than 5 km of late Cenozoic sedimentary rocks are recognized in the Salton Trough from geologic and borehole information (Fuis and others, 1984).

Crustal structure of southern California. A contour map of the crustal thickness of California (Fig. 3) has been compiled from the references listed in Table 1 and combines results from seismic refraction and reflection profiles, analysis of seismic network data, and modeling of gravity data. Where multiple interpretations are in conflict, we chose that based on more reliable data. Contours between surveys were drawn by hand. Mooney (this volume) discusses data sources, analysis methods, and uncertainties in crustal seismology.

The primary feature of the crustal thickness map of southern California south of the Garlock fault is a nearly uniform thickness of 30 km ± 2 km. The thickest crust (32 km) occurs below the western and eastern Transverse Ranges, and the thinnest onland crust (28 km) occurs below the eastern Mojave Desert, beneath the Salton Trough, and along the coastline. Offshore, the crust thins gradually to 20 to 26 km across the California Borderland (Fig. 3). The average inland crustal thickness of 30 km is 6 km less than the average for the continental United States, but is approximately the same as the average for the Basin and Range province of Nevada, Utah, and Arizona (Braile and others, this volume; Pakiser, this volume). The relative uniformity of crustal thickness in southern California is remarkable in the light of the variable geologic histories of the surface rocks.

The Moho contour map, five published velocity-depth functions, and published gravity models have been used to make a lithospheric profile across southern California and the adjacent borderlands at the latitude of the Salton Trough (Fig. 4A); a similar cross section emphasizing details of the surficial and inferred deep geology is presented by Howell and others (1986). The present cross section illustrates the broader features of lithospheric structure from the Pacific basin to the eastern Mojave Desert, and incorporates the lateral variations in crustal structure evident in the five seismic velocity-depth functions (Fig. 4B,F). The most noteworthy features of the lithospheric cross section (Fig. 4A) are the shallow asthenosphere beneath the Salton Trough and the uniformly thin crust. The zone of asthenospheric upwelling, as defined by heat-flow and gravity data, is about 100 km wide at the base of the crust, and the lithosphere thickens to 60 to 70 km to either side (Iyer and Hitchcock, this volume).

Borderland and Coastal Province. The crust beneath the Pacific Ocean thickens abruptly from normal oceanic thickness (12 km) to a thickness of about 20 km at the Patton Escarpment (Fig. 2), and gradually reaches 29 km at the coastline (Fig. 3). Limited seismic refraction data (Shor and Raitt, 1958) and gravity modeling (Plawman, 1978) indicate that the Borderland is underlain by a dense middle and lower crust with a high compressional wave (V_p) seismic velocity (V_p = 6.7 to 7.2 km/sec) that almost certainly indicates an oceanic origin (Fig. 4C). East of the Borderland, the crust of the Peninsular Range is approximately 30 km thick, with seismic velocities in the range of 6.8 km/sec at a depth of only 10 km (Fig. 4D).

TABLE 1. LITHOSPHERIC AND CRUSTAL STUDIES IN CALIFORNIA*

Southern California

Shor and Raitt, 1958*; Hamilton, 1970*; Thatcher and Brune, 1973; Raikes, 1976; Hadley and Kanamori, 1977; Prodehl, 1979*; Raikes and Hadley, 1979; Raikes, 1980; Vetter and Minster, 1981; Ergas and Jackson, 1981; Fuis, 1981*; Nava and Brune, 1982; Walck and Minster, 1982; Fuis and others, 1982*, 1984; Hearn, 1984*; Humphreys and others, 1984; Cheadle and others, 1986*; Hearn and Clayton, 1986a, b*; Keller and Prothero, 1987*; McCarthy and others, 1987*.

Central California

Byerly, 1939; Healy, 1963*; Eaton, 1963, 1966; Hamilton and others, 1964*; Stewart, 1968; Carder and others, 1970; Carder, 1973; Healy and Peake, 1975; Hill, 1976; Pakiser, 1976; Steeples and Iyer, 1976; Prodehl, 1979*; Oliver and others, 1980; Pakiser and Brune, 1980; Zandt, 1981; Bolt and Gutdeutsch, 1982; Taylor and Scheimer, 1982; Walter and Mooney, 1982*; Feng and McEvilly, 1983; Blümling and Prodehl, 1983*; Mavko and Thompson, 1983; Crouch and others, 1984; Oppenheimer and Eaton, 1984*; Sanders, 1984; Walter, 1984; Wentworth and others, 1983; Wentworth and others, 1984; Blümling and others, 1985; Hill and others, 1985b; Luetgert and Mooney, 1985; Mooney and Colburn, 1985; Colburn and Mooney, 1986*; de Voogd and others, 1986; Hwang and Mooney, 1986*; Mooney and Ginzburg, 1986; Nelson and others, 1986; Trehu and Wheeler, 1986*; Zoback and Wentworth, 1986; Holbrook and Mooney, 1987*; Meltzer and others, 1987; Macgregor-Scott and Walter, 1988*.

Northern California

La Fehr, 1965; Shor and others, 1968*; Young and Ward, 1980; Oppenheimer and Herkenhoff, 1981; Spieth and others, 1981*; Warren, 1981; Shearer and Openheimer, 1982; Jachens and Griscom, 1983*; Oppenheimer and Eaton, 1984*; Stauber and Berge, 1985; Fuis and others, 1986; Eberhart-Phillips, 1986; Zucca and others, 1986*; Berge and Stauber, 1987.

*References used in constructing the crustal thickness contour map of Figure 3.

A cross section from the northern Borderland to the western Transverse Range and southern Coast Ranges shows a complex crustal structure with strong lateral variations (Keller and Prothero, 1987; Fig. 5). The crust thickens from about 23 km beneath the central Borderland to about 31 km beneath the western Transverse Range and southern Coast Ranges. The dual processes of convergence and translation are reflected in this cross section. The lower crust is interpreted to consist of tectonically underplated Mesozoic oceanic crust, which is in turn underlain by Neogene oceanic crust. Numerous faults, basins, and uplifted blocks provide evidence of the translational (wrench) tectonic setting that has controlled the late Cenozoic evolution of the Borderland and adjacent continental area.

Mojave Desert. The Mojave Desert forms a vast area of nearly uniform crustal thickness, and is the most featureless region on the Moho contour map of California (Fig. 3). Beneath

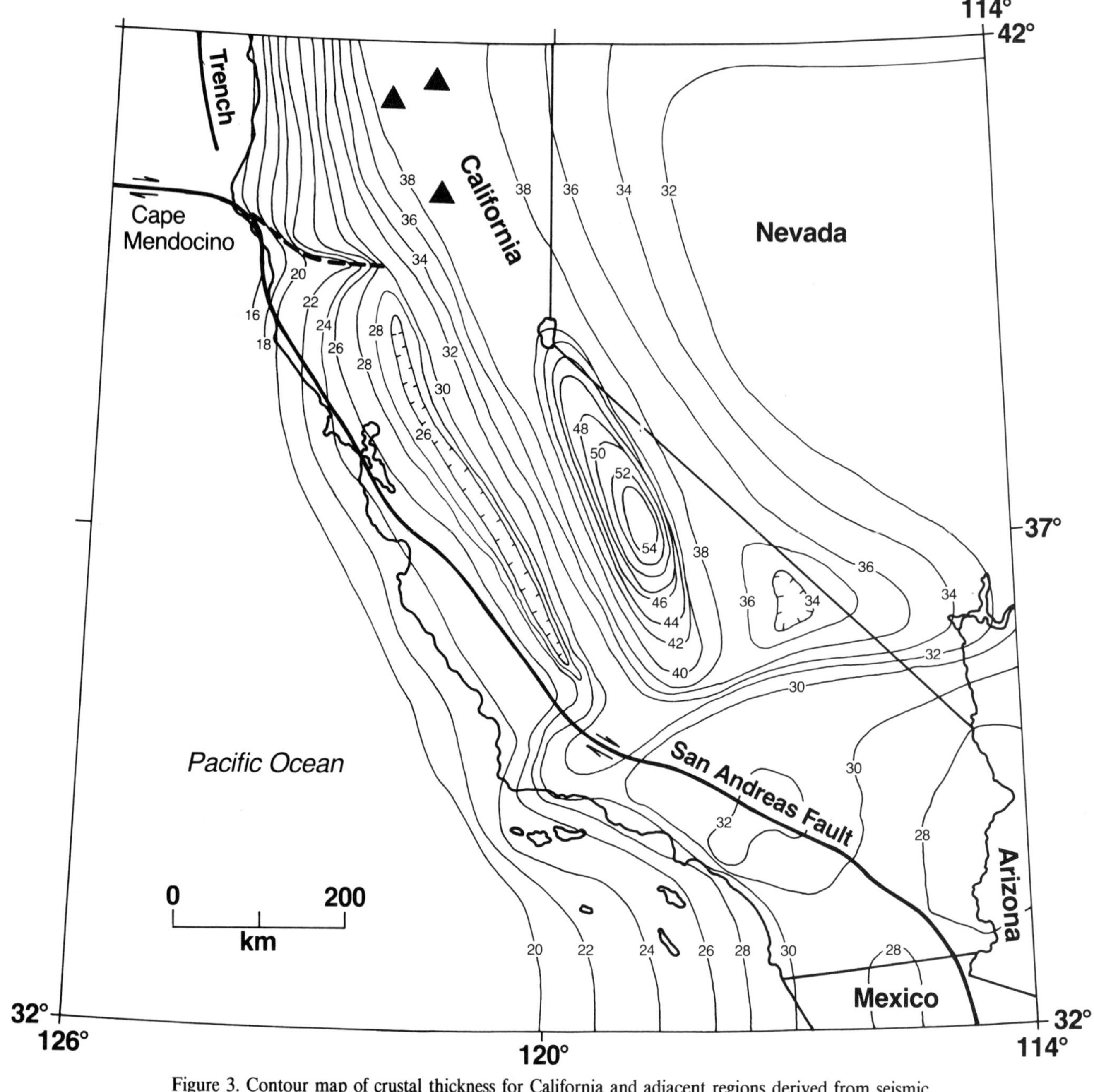

Figure 3. Contour map of crustal thickness for California and adjacent regions derived from seismic refraction, seismic reflection, seismic network, and gravity data (Table 1). Estimated error is 10 percent, or one to one and one-half contour intervals. Solid triangles mark volcanoes as identified in Figure 1. Southern California (south of the Garlock fault) has a crustal thickness of 30 ± 2 km, whereas central and northern California show a pronounced west-to-east crustal thickening.

the Mojave Desert the crust is 28 to 30 km thick, and the crustal seismic velocities are relatively low (e.g., V_p less than 6.5 km/sec above 20 km; Fig. 4F) for nearly the entire crust (Prodehl, 1979; Fuis, 1981; McCarthy and others, 1987). This velocity structure indicates that most of the crust consists of rocks of felsic to intermediate composition. A thin (2 to 5 km) basal crustal high-velocity layer ($V_p = 7.0 \pm 0.2$ km/sec) may be present in some places. The upper-mantle velocity in the Mojave Desert is about 7.9 km/sec, although variations of ±0.2 km/sec or more are reported from seismic network data (Vetter and Minster, 1981; Hearn, 1984).

Deep reflection data from the western and northern Mojave Desert show numerous low-angle, southwest-dipping, linear bands of reflections within the crust, which can be traced over several tens of kilometers (Fig. 6). With the exception of a reflection traceable to the surface at the Rand Thrust, the interpretation of these reflections is ambiguous. Prominent reflections also occur at midcrustal depths (about 15 km), at lower crustal depths (20 to 27 km), and at the crust-mantle transition (28 to 30 km; Fig. 6). Cheadle and others (1986) have proposed that many of these reflections represent deep crustal detachments related to either Mesozoic/early Cenozoic thrusts or low-angle normal faults caused by early Miocene northeast-southwest–directed crustal extension.

Deep seismic reflection data in the southeastern Mojave Desert near Needles, California (Fig. 2) reveal the geometry of

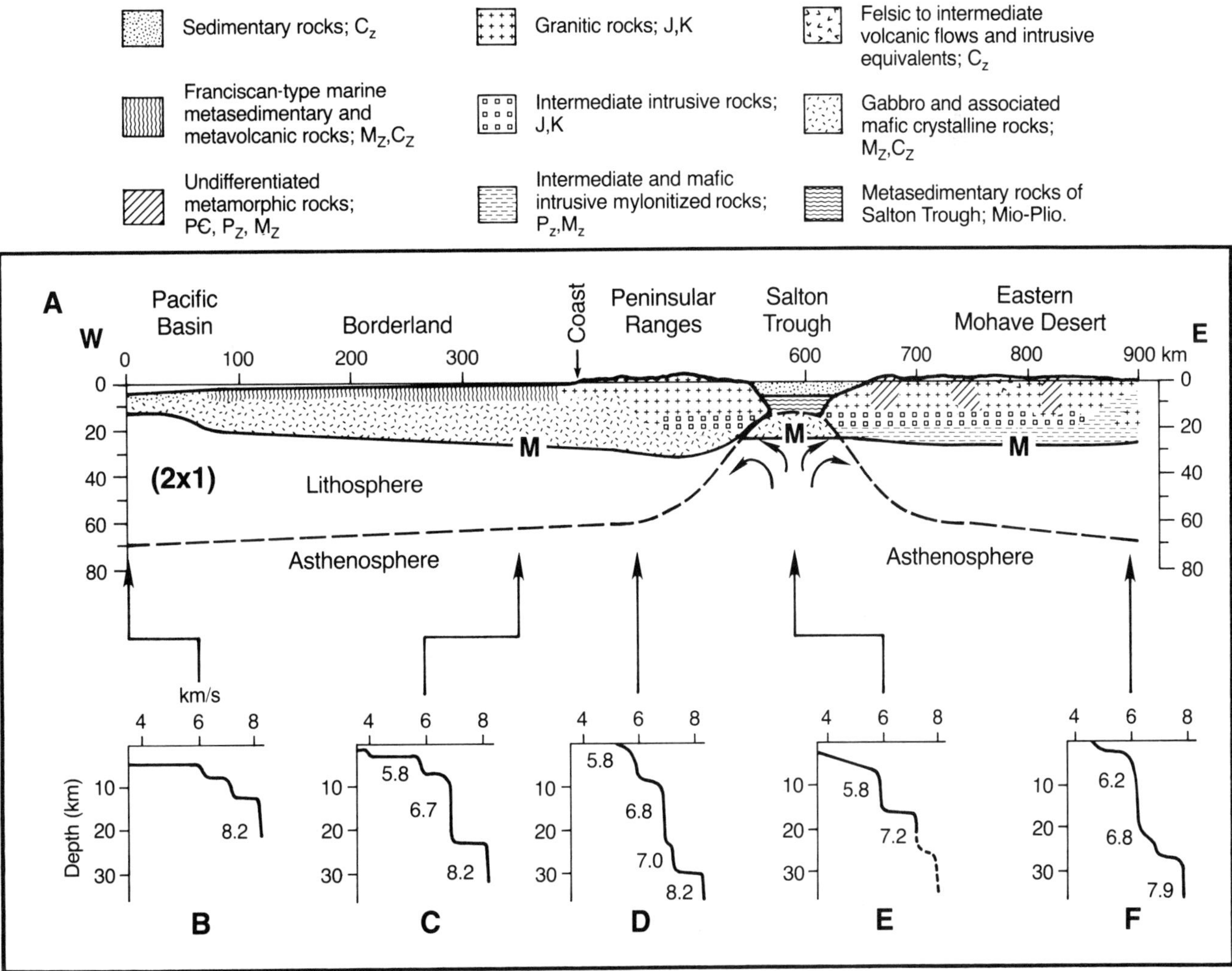

Figure 4. A, Lithospheric cross section with inferred crustal composition for southern California. Location in Figure 2. The continental crust has been thinned to 30 ± 2 km, and is 20 to 26 km thick beneath the California Borderland. B-F, Velocity-depth functions determined along this cross section. Lower crustal velocities of 6.7 km/sec and higher are found along the entire cross section and are interpreted to represent tectonically underplated oceanic crust and/or magmatic intrusions and underplated mafic rocks. The crust has completely rifted apart at the Salton Trough, and a new crustal section has been formed consisting of sedimentary rocks, metasedimentary rocks, and gabbro. References: B and C, Shor and Raitt, 1958; D, Thatcher and Brune, 1973, and Hearn, 1984; E, Fuis and others, 1984; F, Hearn, 1984, and McCarthy and others, 1987.

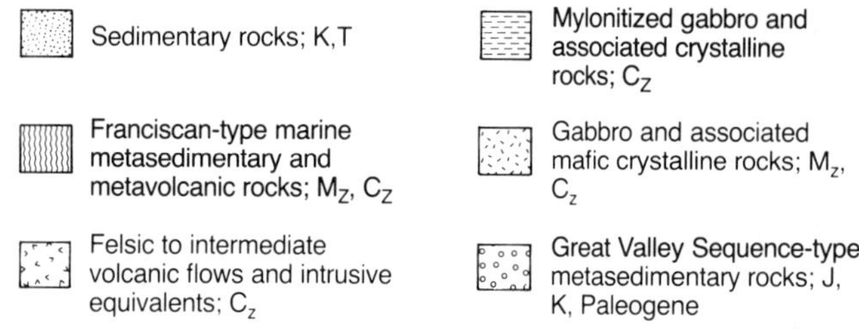

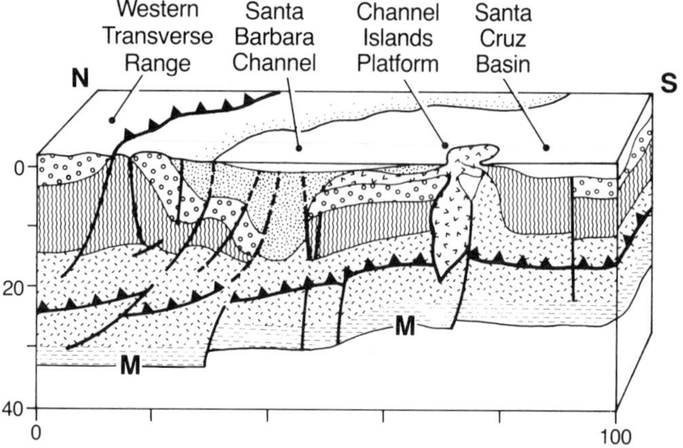

Figure 5. North-south crustal cross section with inferred composition from Borderlands to the Western Transverse Range (Keller and Prothero, 1987). Location in Figure 2. Franciscan and Great Valley Sequence rocks comprise much of the upper crust; Cretaceous and younger sediments are as thick as 15 km beneath the Santa Barbara Channel. This sedimentary trough continues onshore as the Ventura Basin (Fig. 2), where 18 km of sedimentary rocks occur; it was formed as a pull-apart basin. The lower crust is interpreted to consist of multiple sheets of tectonically underplated oceanic crust, younger under older.

the low-angle normal faults (detachment faults) that accommodate brittle upper crustal movements in the highly extended crust of the Mojave Desert (Okaya and others, 1985; Frost and Okaya, 1985; Morris and Frost, 1985). Reflections from these faults are detected to depths as great as 10 to 15 km. Opposite dips of adjacent packets indicate a complex crustal structure. Strong reflections continue to 9- to 10-sec two-way time (TWT), below which few reflections occur. The drop in reflection density at 9 to 10 sec suggests that the base of the crust is at a depth of 27 to 30 km, consistent with seismic refraction data (Fig. 3).

Transverse Ranges. The crustal structure of the Transverse Ranges is known principally from seismic network and gravity data. Although most mountain ranges with topography as prominent as the Transverse Ranges have crustal roots, the contour map of crustal thickness (Fig. 3) shows that the Eastern Transverse Range has only a small root (2 to 3 km), whereas the Central Transverse Range has no root. Thus, lateral translation occurs along the San Andreas fault in southern California without significant crustal thickening, whereas the uplift of the Transverse Ranges at the fault's "big bend" is generally considered to be the product of compression and crustal thickening associated with the convergence of the Pacific and North American plates. In order to reconcile this inconsistency, Weldon and Humphreys (1986) have proposed a revised kinematic model for southern California that calls for motion *parallel* to the San Andreas fault along the Transverse Ranges, and *not* parallel to the motion of the Pacific plate, as is commonly assumed (Atwater, 1970; Hill, 1982; Bird and Rosenstock, 1984). In this model, westward motion of the Sierra Nevada along the Garlock fault and strike-slip faulting offshore balance the slip vectors to produce slip parallel to the San Andreas fault north of the Garlock fault. Modern seismicity (Dewey and others, this volume) and deformation of young sediments observed in the California Borderland (Fig. 5) support the interpretation of active faulting and deformation off the coast of California. Sheffels and McNutt (1986) have proposed an isostatic model for the compensation of the (rootless) Transverse Ranges that involves flexure of a 50-km-thick elastic plate; they related their model to a high-velocity upper-mantle feature identified in the area using seismic tomography (Humphreys and others, 1984).

Salton Trough. A seismic refraction survey of the southern portion of the Salton Trough (the Imperial Valley) is interpreted as showing that along the axis of the valley, unmetamorphosed sedimentary rocks occupy the upper 5 km of the crust and are underlain to a depth of about 13 km by greenschist-facies metasedimentary rocks (Fuis and others, 1982, 1984). Thus, along the axis of the Imperial Valley, the entire 13 km of upper crust consists of sedimentary rocks that have been deposited within a pull-apart basin (Fig. 4). On the flanks of the valley, higher velocities (V_P greater than 6.0 km/sec) are interpreted to be the granitic basement that has rifted to form the crustal pull-apart.

Beneath the thick sedimentary upper crust of the Salton Trough is a lower crustal layer with a velocity of 7.2 km/sec. This layer apparently is not restricted to the axis of the valley; Hamilton (1970) reported a 7.1-km/sec layer at a depth of 14 km at a location 25 km west of the Salton Sea. Fuis and others (1982, 1984) have interpreted this layer as gabbro that has underplated the Salton Trough during rifting.

Despite the thick accumulation of sediments in the Salton Trough, the heat flow is high (100 mW/m^2 and greater), and the Bouguer gravity field is 10 to 50 mgal higher than over the flanking areas. Lachenbruch and others (1985) explained these observations with a lithospheric model that shows a shallow (hot) asthenosphere and a dense (3.1 g/cc) lower crust to compensate gravitationally for the overlying sedimentary and metasedimentary rocks. The key elements of their model have been incorporated in the cross section (Fig. 4A), which shows an upper crustal pull-apart (50 km wide) and the gabbroic lower crust (100 km wide).

The lithospheric structure of Figure 4A may be linked to a kinematic and evolutionary model of the Salton Trough. Lachenbruch and others (1985) have proposed that, as the Salton Trough has evolved over the past 4 to 5 m.y., the crustal thickness has approached a stationary value by a combination of sedimen-

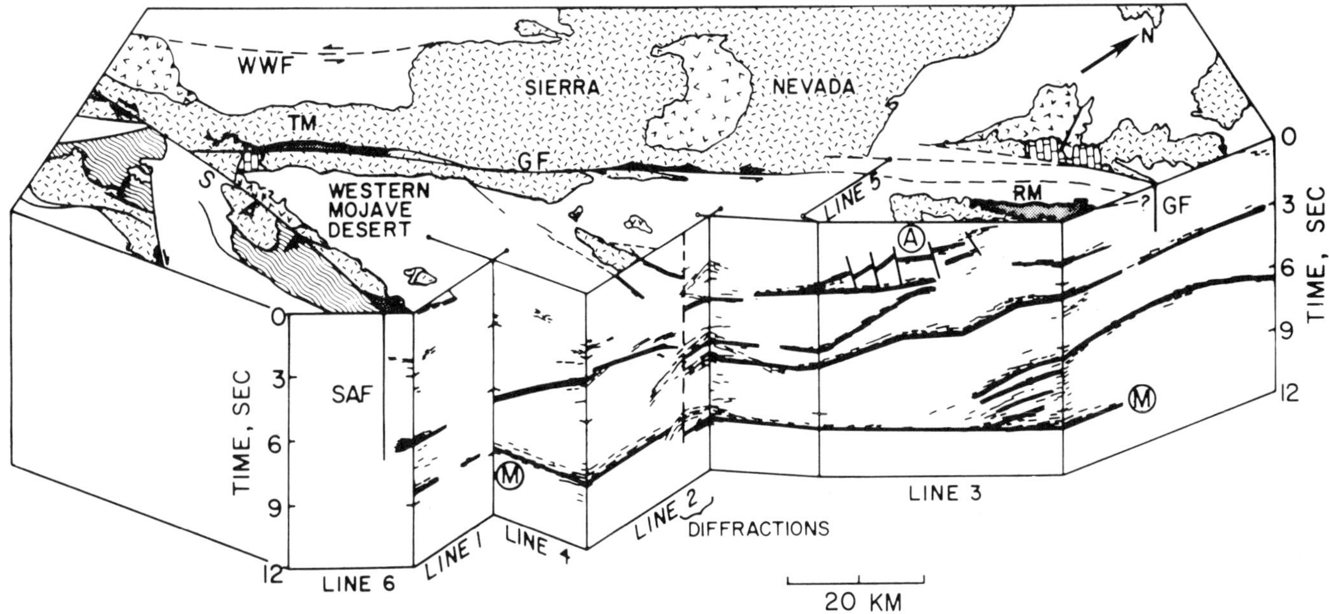

Figure 6. Composite line drawing of Mohave Desert seismic reflection data in a perspective plot (Cheadle and others, 1986). Location in Figure 2. Reflection times can be converted to approximate depth by multiplying by three. Reflection labeled A can be traced to the Rand Mountains (RM) thrust, and other reflections are evident in the middle and lower crust. These low-angle reflections may be Mesozoic thrusts reactivated during Cenozoic extension. Reflection labeled M is at the appropriate depth (30 km) for the Moho discontinuity (cf. Fig. 3). Other symbols: WWF = White Wolf fault; TM = Tehachapi Mountains; GF = Garlock fault; SAF = San Andreas fault.

tation and intrusion (or underplating) by gabbroic magma. They calculate an average extension rate during the formation of the trough of 20 to 50 percent/m.y., noting that slower rates are inadequate to account for the present composition of the crust, and faster rates would probably cause massive crustal melting. To achieve the differential plate velocities between the Pacific and North American plate with this strain rate, extension must have, on average, been distributed over a spreading region about 150 km wide (i.e., nearly the entire length of the Salton Trough), which is some 10 times wider than the zone of presently active seismicity. This implies that the present-day localized seismicity of the Salton Trough (Johnson and Hill, 1982) is an ephemeral phenomenon on the time scale of the trough's formation.

Central and northern California to Cape Mendocino

Geology. South of Cape Mendocino, central and northern California comprises four basic geologic provinces: the Coast Ranges, Great Valley, Sierra Nevada, and westernmost Basin and Range (Fig. 2). The Coast Ranges consist of broad belts of marine metasedimentary rocks (the Franciscan assemblage and Great Valley sequence), a large plutonic terrain of dominantly granitic composition (Salinian block), and Tertiary sedimentary and volcanic rocks.

South of the San Francisco Bay, the Coast Ranges are subdivided into three major geologic elements: the Diablo Range, the Salinian block, and the Sur-Obispo terrane located west of the Sur-Nacimiento fault zone (Fig. 2). The Diablo Range is an antiform with an exposed core of Mesozoic Franciscan assemblage virtually encircled by the coeval Great Valley sequence. The Coast Range fault separates the Great Valley sequence and ultramafic rocks (the Coast Range ophiolite) from the Franciscan assemblage. Despite their similarities in age and bulk composition, the Franciscan assemblage and the Great Valley sequence differ greatly in depositional histories. Whereas the Great Valley sequence shows regular bedding in normal stratigrahic sequence, the Franciscan assemblage is largely a mélange of metasedimentary and volcanic rocks, suggesting that it was deposited as an accretionary prism associated with a subduction zone, fore-arc basin, and volcanic arc (Dickinson, 1970). The coastal Sur-Obispo terrane also consists of Franciscan assemblage and Great Valley sequence rocks. The offshore geology is discussed by McCulloch (1987).

The Salinian block is bounded on the east by the San Andreas fault and on the west by the Sur-Nacimiento fault zones (Fig. 2). The basement of the Salinian block consists of felsic plutonic rocks that intrude amphibolite-grade metasedimentary rocks (Ross and McCulloch, 1979). Because of the similarity in composition and age to parts of the Sierra Nevada batholith, several investigators (e.g., Page, 1981) have suggested that the Salinian block was originally positioned between the Sierra Nevada and the Peninsular Range and has been translated northward 540 km along the San Andreas fault. More recently, paleomagnetic studies (Champion and others, 1984) have suggested that the Salinian block originated at a southern latitude some 2,500 km away from its current position.

The Coast Ranges north of the San Francisco Bay are composed of the Franciscan assemblage, Great Valley sequence, ultramafic rocks, and Tertiary sedimentary rocks. Granitic basement rocks are absent, but Pliocene and Pleistocene volcanic rocks occur locally (Fox and others, 1985). The offshore geology is discussed by Clarke (1987) and McCulloch (1987).

The Great Valley of California is a 700-km-long by 100-km-wide sedimentary basin situated between the granitic and metamorphic terrane of the Sierra Nevada and the Coast Ranges (Fig. 2). The sedimentary rocks of the Great Valley, which include the late Jurassic to late Cretaceous Great Valley sequence and overlying Cenozoic units, form an asymmetric syncline with a steeply dipping west limb and a gently dipping east limb (Dickinson, 1981). Models of the geologic history and deep crustal structure of the Great Valley have been strongly influenced by two concepts: (1) The standard evolutionary model for central California interprets the three geologic elements of the Franciscan assemblage–Great Valley–Sierra Nevada in terms of the common tectonic ensemble of an accretionary prism/fore-arc basin/magmatic arc. This model casts the Great Valley as the fore-arc basin, and suggests that it should be underlain by oceanic crust (Dickinson, 1970). (2) The pronounced aeromagnetic and gravity anomalies that run along the axis of the Great Valley (Kane and Godson, this volume) have been interpreted to indicate the existence of a suture between continental (i.e., Sierra Nevadan) crust to the east and oceanic affinity crust to the west (J. W. Cady, 1975; Blake and others, 1977).

The Sierra Nevada is a 600-km-long by 150-km-wide composite batholith that was emplaced over a period of nearly 100 m.y., from approximately 180 to 80 Ma. (Bateman and Eaton, 1967). The northwestern Foothills Metamorphic Belt consists of deformed and folded Paleozoic and Mesozoic metamorphic and ultramafic rocks (Saleeby, 1981), which may continue beneath the Great Valley. The uplift of the range to its present elevation occurred in late Cenozoic time. Geologic data (Christensen, 1966) indicate that about 10 m.y. ago the Sierra Nevada was 1,200 to 1,800 m high, with a gentle surface of only moderate relief. Uplift began sometime between 10 and 3.5 Ma, most likely around 10 Ma.

The western Basin and Range province of southeastern California occupies the triangular region east of the Sierra Nevada escarpment and north of the Mojave Desert block, as marked by the easterly extension of the Garlock fault (Fig. 2). The region is characterized by elongate ranges and intervening alluvial valleys, both with a general northward trend (Burchfiel and Davis, 1981; Nelson, 1981). Basin and Range faulting has been accompanied by a significant amount of extension, estimated to be between 50 and 100 km at this latitude (Hamilton and Myers, 1966; Wernicke and others, 1982). Davis and Burchfiel (1973) have suggested that the Basin and Range north of the Garlock fault has extended 60 km more than the Mojave block south of the fault.

The Long Valley Caldera–Mono Lake region of eastern California is an area of active seismicity and volcanism lying in the westernmost Basin and Range and at the eastern edge of the westward tilted block of the Sierra Nevada (Fig. 2). Long Valley Caldera is the southernmost and oldest member of a succession of three progressively younger volcanic complexes, each in a different stage of evolution (Bailey, 1982). The Long Valley depression is the product of a caldera collapse 0.7 m.y. ago following the eruption of the voluminous Bishop tuff, and intermittant volcanism has continued in the area to as recently as 700 yr ago (Bailey and others, 1976).

Crustal structure of central and northern California. The crustal thickness contour map is highly variable across central and northern California (Fig. 3). In central California, the crust generally thickens eastward, from about 24 km at the coast and 30 km in the Central Valley, to as much as 55 km beneath the Sierra Nevada before thinning to about 30 km in the Basin and Range (Pakiser, this volume). In the northern Coast Ranges, three-dimensional gravity models indicate that the contours swing sharply inland as the Mendocino triple junction is approached (Jachens and Griscom, 1983). Elsewhere in northern California, the contours of crustal thickness generally continue along geologic strike from central California, with the exception of the pronounced crustal thinning (from 54 to 40 km) in the Sierra Nevada near Lake Tahoe.

The region separating central and southern California (Garlock fault and southern Coast Ranges) is marked by roughly east-west–trending gravity contours. Clearly, distinct processes of crustal evolution have dominated in the two areas: south of the Garlock fault, uniform crustal thinning associated with Basin and Range extension affected the entire crust as far west as the coast, whereas north of the Garlock fault, part of the Mesozoic mag matic arc (the Sierra Nevada) remains intact.

The Moho contour map and six published velocity-depth functions have been used to make a crustal profile across central California and western Nevada (Fig. 7A). This cross section illustrates the broader features of the crustal structure, and the marked differences in seismic velocity-depth functions (Fig. 7B–G) demonstrate the strong lateral variations in crustal velocities. Complementary geologic cross sections for central and northern California, which use constraints from geophysical data, are presented by Bateman (1979), Suppe (1979), Etter and others (1981), Speed and Moores (1981), Wright and others (1981), and Saleeby and others (1986).

Coast Ranges. Crustal thickness of the California Coast Ranges varies from 18 to 30 km. The thinnest crust occurs along the coast (particularly at Cape Mendocino), and the thickest is along the Coast Range–Great Valley boundary and in the southern Coast Ranges (Fig. 3). West of the San Andreas fault, the crust appears to be everywhere thinner than or equal to 25 km, and is everywhere thicker than 25 km east of the fault. It is uncertain whether a step in crustal thickness occurs across the fault; none is indicated in Figure 3. Seismic velocity-depth functions determined for the Coast Ranges are shown in Figure 7B, C, and D.

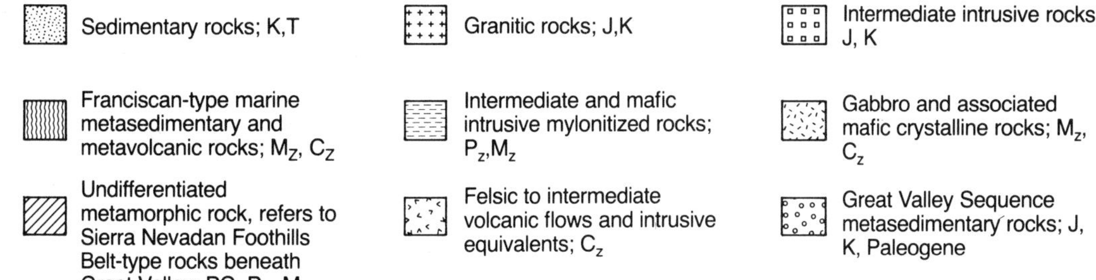

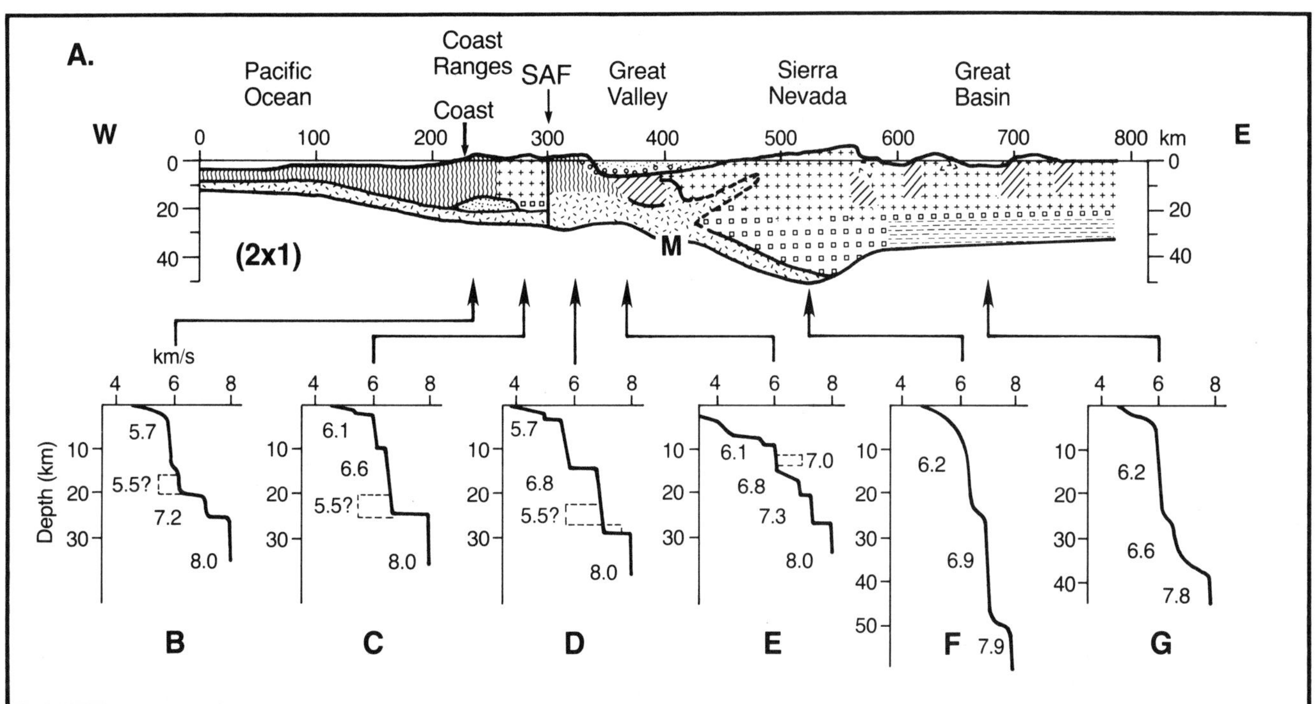

Figure 7. A, Crustal cross section with inferred composition for central California and western Nevada. Location in Figure 2. The crust thickens from 12 km in the Pacific basin to 24 to 32 km in the Coast Ranges and Great Valley, and attains a maximum of about 55 km in the southern Sierra Nevada. The crust is about 36 km thick in the westernmost Basin and Range (and about 30 km thick in the central Basin and Range; Pakiser, this volume). B-G, Velocity-depth functions determined along the cross section. Lower crustal low-velocity zones are locally present beneath the Coast Ranges and probably represent packages of subducted sediments. The lower crust east of the San Andreas fault and beneath the Great Valley consists of about 14 km of high-velocity rocks (Vp more than 6.7 km/sec), which are interpreted to consist of tectonically underplated oceanic crust and/or mafic intrusives and underplated rocks associated with the migration of the Mendocino triple junction in the late Cenozoic. References: B, Walter and Sharpless, 1987; C, Walter and Mooney, 1982, and Trehu and Wheeler, 1986; D, Walter and Mooney, 1982 and Blümling and Prodehl, 1983; E, Colburn and Mooney, 1986, and Holbrook and Mooney, 1987; F, Eaton, 1966, and Prodehl, 1979; G, Prodehl, 1979.

The crustal structure of the Sur-Obispo terrane of coastal central California is interpreted as consisting of 14 to 16 km of Franciscan assemblage (V_p = 5.6 to 5.9 km/sec) overlying about 6 to 8 km of underplated oceanic igneous crust (V_p = 7.2 km/sec; Fig. 7B) (Walter and Sharpless, 1987). Trehu and Wheeler (1986) have presented evidence for a pronounced lower crustal low-velocity zone (LVZ) beneath the Salinian and Sur-Nacimiento terranes. Their crustal model is based on a combined analysis of seismic reflection and refraction data, the former of which shows pronounced east-dipping reflectors between 4 and 10 sec TWT beneath the Salinian block. A wedge-shaped packet of reflectors correlates with the lower crustal seismic LVZ that Trehu and Wheller (1986) identified on the basis of wide-angle reflections in the refraction data. The modeled seismic velocity within the LVZ is 5.0 km/sec, occurring at a similar depth (19 ± 5 km) as an LVZ interpreted elsewhere in the Coast Ranges east of the San Andreas fault (Blümling and Prodehl, 1983; Blümling and others, 1985). Trehu and Wheeler (1986) have suggested that the east-dipping reflectors and associated LVZ consist of a package of deeply buried sediments that were subducted in the presently inactive offshore trench (Page and others, 1979). Lower crustal LVZs are not evident on all seismic refraction profiles in the Coast Range; rather, they appear to be local features that reflect the episodic subduction and crustal underplating of large packages of sedimentary rocks.

The crustal structure on opposite sides of the San Andreas fault of central California has been determined from a pair of parallel seismic refraction profiles. Not surprisingly, a contrast in structure has been identified across the fault as a result of the translational motion of major terranes. To the west, beneath the Salinian block, the average crustal thickness is 25 ± 2 km. The seismic velocity reaches 6.0 to 6.1 km/sec in the upper 2 to 3 km of the granitic upper crust and is 6.5 to 6.8 km/sec just above the Moho (Fig. 7C). To the east of the San Andreas fault in the Diablo Range, the crust is 28 ± 2 km thick, with the Franciscan assemblage forming a 14-km-thick, relatively low-velocity (V_p = 5.3 to 5.9 km/sec) upper crust (Fig. 7D). The seismic velocity of the 14-km-thick lower crust is 6.7 to 7.0 km/sec and may correspond to tectonically underplated oceanic crust (Walter and Mooney, 1982). The lower crust, however, is twice the thickness of normal oceanic crust and may consist of multiply underplated slabs.

A pronounced vertical upper crustal low-velocity zone (i.e., thick gouge) within the San Andreas fault zone has been reported at several sites in central California (e.g., Healy and Peake, 1975; Feng and McEvilly, 1983; Mooney and Colburn, 1985; Trehu and Wheeler, 1986). These observations were reviewed by Mooney and Ginzburg (1986), who argued that vertical LVZs are generally associated with creeping (weak) portions of the fault, whereas they appear to be absent from locked (strong) portions. However, the evidence is still inconclusive on this point.

The crustal structure of the northern Coast Ranges has been investigated with local seismic networks and limited seismic refraction data. There is some evidence that the crust thins from about 28 km to about 22 km just north of the San Francisco Bay area. Warren (1981) interpreted a crustal thickness of about 20 km in the Coast Ranges about 80 km north of the Bay based on wide-angle reflected arrivals; his data did not include P_n refractions to confirm this estimate. Zandt (1981) also reported thinner crust to the north of the Bay in his three-dimensional velocity inversion of teleseismic residuals. Oppenheimer and Eaton (1984) have reviewed the evidence for and against an anomalously thin crust in the northern Coast Ranges; based on their analysis of seismic network data, they favored a model with an average thickness of about 28 km (24 km at the coast and thickening to the east) and local variations in the lower crustal velocity structure. In constructing our contour map of crustal thickness (Fig. 3), we have relied most heavily in the northern Coast Ranges on the results of Oppenheimer and Eaton (1984) and the three-dimensional gravity modeling of Jachens and Griscom (1983) for the Mendocino area.

The Geysers–Clear Lake area. The Geysers–Clear Lake geothermal area of the northern Coast Ranges lies within the central belt of the Franciscan assemblage (Fig. 2). The area has been exploited for the production of electricity and is one of the world's largest commercial geothermal developments. The high heat flow (80 mW/m^2 and higher) and the occurrence of lavas ranging in age from 2.1 Ma to 10 ka (Donnelly-Nolan and others, 1981) suggest that magma is present within the midcrust beneath The Geysers–Clear Lake area. Magmatism in the area may be related to east-southeast crustal extension accompanying northward propagation of the San Andreas transform system between the Clear Lake region and Cape Mendocino over the past 3 m.y. The petrology of volcanic rocks in the area suggests mixing between mantle-derived melts and various crustal rocks (Hearn and others, 1981).

Geophysical investigations of The Geysers are summarized by Isherwood (1981). Gravity studies (Chapman, 1966, 1975) provided the first geophysical evidence for a possible magma chamber beneath the Clear Lake volcanics. Magnetic data (Isherwood, 1976), teleseismic P-wave delays (Iyer and others, 1979), seismic attenuation (Young and Ward, 1980), and heat-flow data (Lachenbruch and Sass, 1980) are interpreted to substantiate this model for the magma reservoir. Oppenheimer and Herkenhoff (1981) concluded from velocity-density modeling that a density decrease exceeding 15 percent is required within the crust just west of Clear Lake; they suggested that this decrease is caused by partially molten material somewhat more silicic than the surrounding medium. Eberhart-Phillips (1986) has determined the upper crustal velocity structure using an inversion method applied to earthquake and seismic refraction shot data. While she found some evidence for localized low velocities associated with fault zones, she found no other low velocities in the upper 7 km of the crust. This implies that the low-velocity, low-density zone identified by Oppenheimer and Herkenhoff (1981) is at a depth greater than 7 km, consistent with the earlier gravity model of Isherwood (1976), which showed a spherical (7-km radius) magma reservoir centered at a depth of 14 km in the same area.

The Great Valley. The deep structure of the Great Valley, as revealed by seismic reflection and refraction profiles, shows a complexity of crustal structure comparable to the surface geology of the Foothills Metamorphic Belt of the northwestern Sierra Nevada. A velocity-depth profile for the west flank of the Great Valley near the latitude of San Francisco (Fig. 2) shows 6 km of sedimentary rocks (V_p = 1.7 to 4.5 km/sec), a 2-km-thick basement layer (5.5 km/sec), and three crustal layers, each about 6 km thick (V_p = 6.1, 6.8, and 7.3 km/sec; Fig. 7E). Total crustal thickness is 26 ± 2 km, about 2 km less than in the Coast Range to the west. The geologic interpretation of the 6.1-km/sec layer is problematic as it is higher in velocity than typical Franciscan assemblage (V_p = 5.5 to 5.8 km/sec), but significantly lower in velocity than oceanic crustal rocks hypothesized to underlie the Great Valley (J. W. Cady, 1975). Colburn and Mooney (1986) have suggested that it correlates with the igneous and metamorphic rocks of the Sierra Nevada Foothills Metamorphic Belt, which have upper crustal velocities of 6.2 km/sec (Spieth and others, 1981).

A second axial Great Valley seismic refraction profile was recorded 10 to 15 km east of the west-flank profile, close to the geographic center of the valley, and just west of the steepest gradient of the Great Valley gravity and aeromagnetic anomaly (J. W. Cady, 1975; Blake and others, 1977). The velocity structure along this profile has important differences from the west-flank Great Valley profile, or from the Diablo Range farther to the west. A 7-km-thick upper crustal section with seismic velocities as high as 7.0 km/sec occurs beneath the valley sediments (Fig. 7E) and was interpreted by Holbrook and Mooney (1987) as an ophiolitic fragment. This body presumably continues to the east and is at least partially responsible for the Great Valley aeromagnetic anomaly.

Since the 7.0-km/sec body is not present beneath the west-flank of the Great Valley line, Holbrook and Mooney (1987) argued that the ophiolite is not connected with the Coast Range ophiolite; further, they suggested that the contrast in crustal velocity structure indicates that a fault or suture zone separates the basement of the Great Valley from that of the Coast Ranges (Fig. 7A).

In summary, the crustal structure of the Great Valley is complex and probably consists of Paleozoic metamorphic, magmatic arc, and crystalline oceanic rocks such as are exposed in the Foothills Metamorphic Belt of the Sierra Nevada (Fig. 2). The two lowermost crustal layers of the Great Valley axial profiles have seismic velocities of about 6.8 and 7.2 km/sec (Fig. 7E). These crustal layers are probably gabbroic and may be the result of tectonic doubling (imbricated tectonic underplating) of oceanic crust and/or magmatic underplating.

Coast Range–Great Valley transition. For 600 km along the east side of the Coast Ranges, the upper Jurassic and Cretaceous Franciscan assemblage and coeval Great Valley sequence are in contact along the Coast Range thrust (Fig. 2), which has been inferred to dip eastward through the crust (Bailey and others, 1964) and has been viewed as a subduction suture zone (Hamilton, 1969). Four seismic reflection profiles crossing the Coast Range–Great Valley transition have been interpreted; all but one of these has coincident seismic refraction data to provide complementary velocity control (Wentworth and others, 1983, 1984; Walter, 1984; Whitman and others, 1985). On the basis of these data, Wentworth and others (1984) have suggested that the east margin of the Franciscan assemblage forms a tectonic wedge that overlies Great Valley basement. In this interpretation the Franciscan tectonic wedge has thrust northeastward on the Great Valley basement and has concurrently uplifted the overlying Great Valley sequence. An important aspect of the wedge hypothesis is that the Coast Range thrust does not extend eastward to mantle depth, as initially proposed by Bailey and others (1964) and Irwin (1964). Instead, it is the roof thrust of the Franciscan wedge and it meets the sole thrust of the wedge at a blind tip beneath the Great Valley.

Sierra Nevada. A variety of geophysical observations made in the Sierra Nevada over the past 50 yr have been interpreted as showing that it has a thick crust (as thick as 55 km) beneath the region of highest elevation in the south, and a thinner crust (38 to 40 km) in the northern, less elevated part of the range (Fig. 3). The main features of the crustal velocity structure are relatively low velocities (V_p = 6.0 to 6.4 km/sec) in the upper 25 km and a thick, higher velocity layer (V_p = 6.9 km/sec) in the lower crust (Fig. 7F). The upper-mantle velocity is 7.8 to 8.0 km/sec (Eaton, 1963, 1966; Prodehl, 1979).

Oliver (1977) has reviewed gravity and magnetic investigations of the Sierra Nevada, and Pakiser and Brune (1980) have reviewed the evidence for the seismic structure. These investigators concluded that the gravity and seismic evidence strongly support the existence of a root that extends to a depth of about 55 km in the southern Sierra Nevada. Carder and others (1970) and Carder (1973) used unreversed seismic data to suggest that the range lacks a crustal root, but Pakiser and Brune (1980) reinterpreted this data as indicating a sheet or wedge of high-velocity rocks that rises to within a few kilometers of the surface on the west side of the range. Oliver (1977) previously suggested a gravity model across the Sierra Nevada that shows lower crustal rocks from the center of the range projecting steeply upward to shallow depth on the west, possibly connecting to the ophiolite belt of the western Foothills Metamorphic Belt.

Several lines of evidence support the concept of a heterogeneous crustal structure of the Sierra Nevada. Wide-angle reflection data from the central Sierra Nevada show a series of reflections coming from the lower crust, indicative of layering there (Pakiser and Brune, 1980). A seismic reflection section at 39.5°N shows a complex pattern of reflections with both eastward and westward dips (35° to 47°) that extend to 8-sec two-way time (TWT), or a depth of approximately 25 km (Nelson and others, 1986). Correlation with the surface geology suggests that the eastward-dipping reflections are from faults mapped at the surface in the Foothills Metamorphic Belt (Fig. 2; Clark, 1960; Speed and Moores, 1981). However, since these reflectors cannot be traced clearly through the 2- or 3-sec TWT of the seismic section to surface outcrops, their interpretation is equivocal. A seismic reflection survey on the west side of the Sierra Nevada at 37°N shows a pair of parallel planar reflectors with a 30° westward dip (Zoback and Wentworth, 1986). At the Sierra Nevada/eastern Great Valley boundary these reflections occur at 4.5- and 7.5-sec TWT, or a depth of approximately 14 and 23 km, respectively. The dip of these two prominent reflectors is just opposite that of the slab model of Pakiser and Brune (1980). However, a westward dip is consistent with the suggestion of Moores and Day (1984) and Day and others (1985) that the ophiolitic rocks of the western Sierra Nevada foothills are the product of obduction along west-dipping thrusts in the Nevadan orogeny. Spieth and others (1981) reported a thin high-velocity layer (7.0 km/sec) at a depth of 5 km beneath the Sierra Nevadan foothills at a location 300 km north of the reflection line of Zoback and Wentworth (1986). This seismic velocity is compatible with the presence of mafic and/or altered ultramafic (ophiolitic) rocks at depth, and may be compared with a layer with the same seismic velocity found beneath the axis of the Great Valley by Holbrook and Mooney (1987).

In summary, these studies suggest a complex Sierra Nevadan crustal structure. This complexity is the product of the evolution of the Sierra Nevada from a Paleozoic continental margin that was modified by Mesozoic and early Cenozoic subduction (including accretion, obduction, and arc plutonism), and late Cenozoic tectoincs involving Basin and Range extension (Saleeby, 1981).

The rapid uplift of the Sierra Nevada in the past 10 m.y. has been the subject of several geophysical models. Crough and Thompson (1977) suggested that the uplift is in response to the thermal thinning of the lithosphere caused by the northward migration of the Mendocino triple junction. Evidence for lithospheric thinning is presented by Mavko and Thompson (1983), who found—based on teleseismic data—50 km of thinning, from 110 km at the northern end of the Sierra Nevada to 60 km near Mono Lakes (Fig. 3). Chase and Wallace (1986) proposed an alternative mechanism for uplift, which involves late Cenozoic lithospheric faulting and subsequent isostatic adjustment for the Mesozoic Sierran crustal root.

California Basin and Range. The crust thins from 40 to 55 km in the Sierra Nevada to 30 to 36 km in the westernmost Basin and Range (Fig. 3). Below a depth of 3 to 5 km, the gross crustal structure of the western Basin and Range can be described with only two or three layers (V_p = 6.2, 6.6, and possibly 6.8 to 7.4 km/sec; Fig. 7G). Deep seismic reflection data in the Death Valley region of the southwestern Basin and Range provide evidence for the faulting that accommodates upper crustal extension. Sedimentary and volcanic rocks have collapsed along young high-angle normal faults that appear to cut older moderately dipping faults that have been rotated from high angle by extension. None of these faults appears to continue below the upper crust (6-sec TWT; 15-km depth); thus there appears to be a brittle upper crust and a ductile lower crust (de Voogd and others, 1986). The reflection data also show a high-amplitude, relatively broad band of reflections at 6 s (15 km) that are interpreted as a seismic "bright spot" associated with a midcrustal magma body (de Voogd and others, 1986).

The Long Valley Caldera region (Fig. 2) has been investigated as a natural "laboratory" of crustal deformation, seismicity, magmatism, and hydrothermal regimes (Hermance, 1983; Hill and others, 1985a). The shallow crustal structure of the region has been defined by both gravity (Kane and others, 1976) and seismic refraction data (Pakiser, 1976; Hill, 1976; Hill and others, 1985b). The refraction studies indicate that the upper 7 to 10 km of the crust consists of Mesozoic granitic rocks. These data also show clear secondary arrivals on profiles crossing the western part of the Long Valley caldera, which are interpreted as possible reflections from the top of a magma body at a depth of 7 to 8 km. No similar reflections have been identified beneath Mono Craters just south of Mono Lake (Fig. 2). Steeples and Iyer (1976) reported telesiesmic evidence for a large, low-velocity body at midcrustal depths beneath the western section of Long Valley caldera. Achauer and others (1986) reported similar evidence for a small, low-velocity volume, 200 to 600 km^3, beneath the Mono Craters area located 25 km to the north. Kissling and others (1984) used a geotomographic method to interpret 30,000 P-wave arrival times to determine the velocity structure beneath the caldera. Their results show a zone of low P-wave velocity beneath the southern part of the resurgent dome and the south moat of the caldera between 5 and 9 km below the surface. Evidence for the physical properties of the upper crust was provided by the detailed mapping of ray paths with anomalously high attenuation of S-waves (Ryall and Ryall, 1981; Sanders, 1984). These latter observations are generally interpreted to indicate a plexus of magma bodies. Constraints on the depth to the top (7 km) and bottom (17 km) of these bodies are provided by wide-angle reflections (Hill, 1976; Hill and others, 1985b; Luetgert and Mooney, 1985). These interpretations are summarized in Figure 8, which shows the composite crustal model derived from all data sources.

Mendocino triple junction. The transform plate boundary of the San Andreas fault system terminates at the Mendocino triple junction, where the subduction regime of the Pacific Northwest begins (Fig. 2). The triple junction and the incipient San Andreas fault were created when the East Pacific Rise reached the trench opposite the North American plate about 30 Ma. Since then, the triple junction has been migrating northwest along the coast of California from a location at a latitude near Los Angeles (Atwater, 1970; Atwater and Molnar, 1973; Dickinson and Snyder, 1971a, b). Since no subduction occurs along the transform fault but continues north of the migrating triple junction, an ever-enlarging triangular hole or asthenospheric window has developed wherein subducted lithosphere is missing. The northern edge of the window is at the Mendocino triple junction, and the slowly opening window, where subduction has ceased, is beneath the Coast Ranges.

Crustal and lithospheric structure in the area of the Mendocino triple junction has been investigated using several types of data: seismicity (Smith and Knapp, 1980; Cockerham, 1984; Walter, 1986), teleseismicity (Zandt and others, 1985), gravity (Jachens and Griscom, 1983), and heat flow (Lachenbruch and Sass, 1980; Zandt and Furlong, 1982; Furlong, 1984). The geometry of the subducting Gorda plate mapped by the earthquake hypocenters and the existence of an asthenospheric window to the south are consistent with the three-dimensional isostatic residual gravity modeling of Jachens and Griscom (1983). The asthenospheric window model is also consistent with the observation of Lachenbruch and Sass (1980) that the heat flow in the Coast Ranges south of Cape Mendocino has a relatively high value of about 80 mW/m^2 and decreases to about 40 mW/m^2 along the coast to the north. The transition to the high

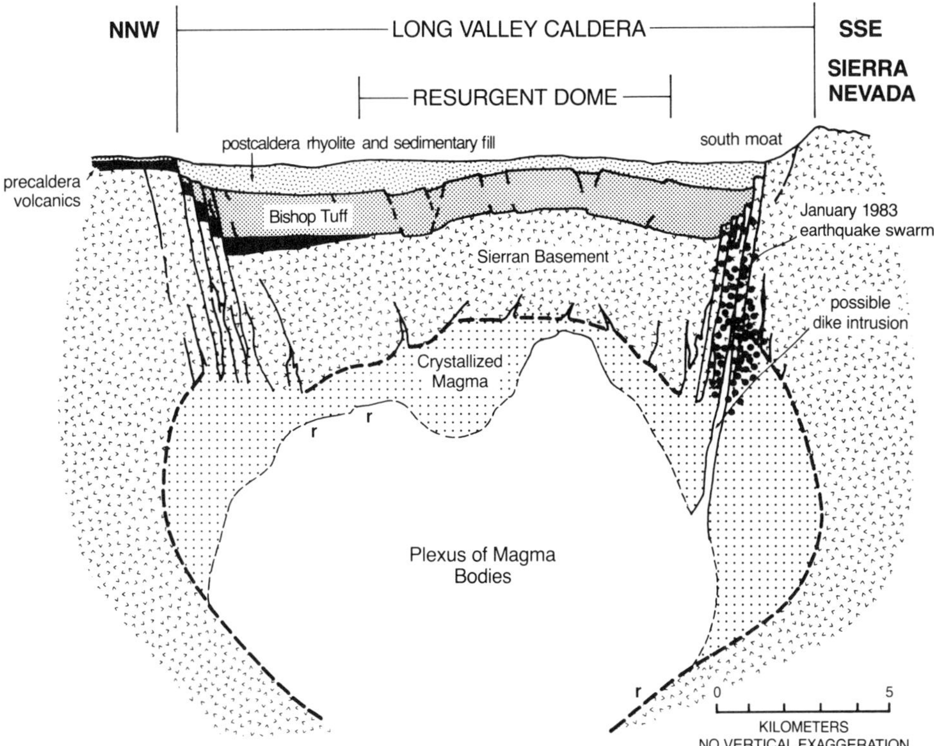

Figure 8. Schematic cross section through Long Valley Caldera in eastern California. Location in Figure 2. Data on which section is based are cited in the text. Area labeled "plexus of magma bodies" consists of magma bodies with estimated 2- to 3-km dimensions. r = Reflection points of wide-angle seismic reflections (see text).

heat-flow region occurs some 200 km south of Cape Mendocino; the triple junction has been migrating northward at a steady rate of 50 km/m.y., and thermal conductivity considerations imply that the source of the Coast Range heat-flow anomaly is at a depth of only about 20 km

Lachenbruch and Sass (1980) examined two hypotheses for the anomaly pattern, the first being that the high heat flow south of the Mendocino triple junction is due to conduction of heat from asthenospheric material at the base of the crust. In this case, as the triple junction migrates north, the upwelling asthenospheric material just south of the inland projection of the Mendocino transform fault would act as a buried one-sided spreading center that heats and thickens the North American crust by magmatic underplating. The second hypothesis is that the Coast Range heat-flow high is the result of shear heating at the base of the seismogenic zone. However, Lachenbruch and Sass (1980) show that in order to produce an appreciable heat-flow contribution without producing an (unobserved) heat-flow high along the San Andreas fault, it is necessary that the basal traction on the seismogenic layer be in the sense to resist (not drive) the fault motion.

The preferred model of Lachenbruch and Sass (1980)—heating by asthenospheric material at the base of the crust—is consistent with the summary of crustal structure presented in Figures 3 and 7: as the actively subducting slab migrates northward with the triple junction, the crust above the slab is left in place and is underplated by material derived from the asthenosphere. This process preserves the eastward-thickening crust of California produced by the Mesozoic and early Cenozoic subduction process. The primary crustal modification is the widespread addition of a lower crustal layer with a seismic velocity of about 7.0 km/sec beneath western California (Fig. 7; Zandt and Furlong, 1982; Furlong and Fountain, 1986). Scattered young volcanic rocks in the Coast Ranges (Fox and others, 1985) and the existence of The Geysers geothermal field provide additional evidence for recent basal crustal heating.

SUBDUCTION REGIME

North of Cape Mendocino, the tectonic framework of the Pacific Coastal States changes from the strike-slip regime of the San Andreas fault system to subduction of the Juan de Fuca plate beneath North America. Because the geologic setting of the Pacific Northwest has been dominated by active convergent-margin tectonics throughout Cenozoic time (Atwater, 1970; Riddihough, 1977), it is convenient to group the geologic provinces into three general areas: fore arc, volcanic arc, and back arc. The fore arc consists of the Klamath Mountains, the Coast Ranges of Oregon and Washington, the Olympic Mountains, and the Willamette Lowlands–Puget Sound Basin (Fig. 9). Our discussion of the volcanic arc includes the North Cascades of Washington, where little late Cenozoic volcanism has occurred, and the long, linear portion of the Cascade Range to the south, where late Cenozoic

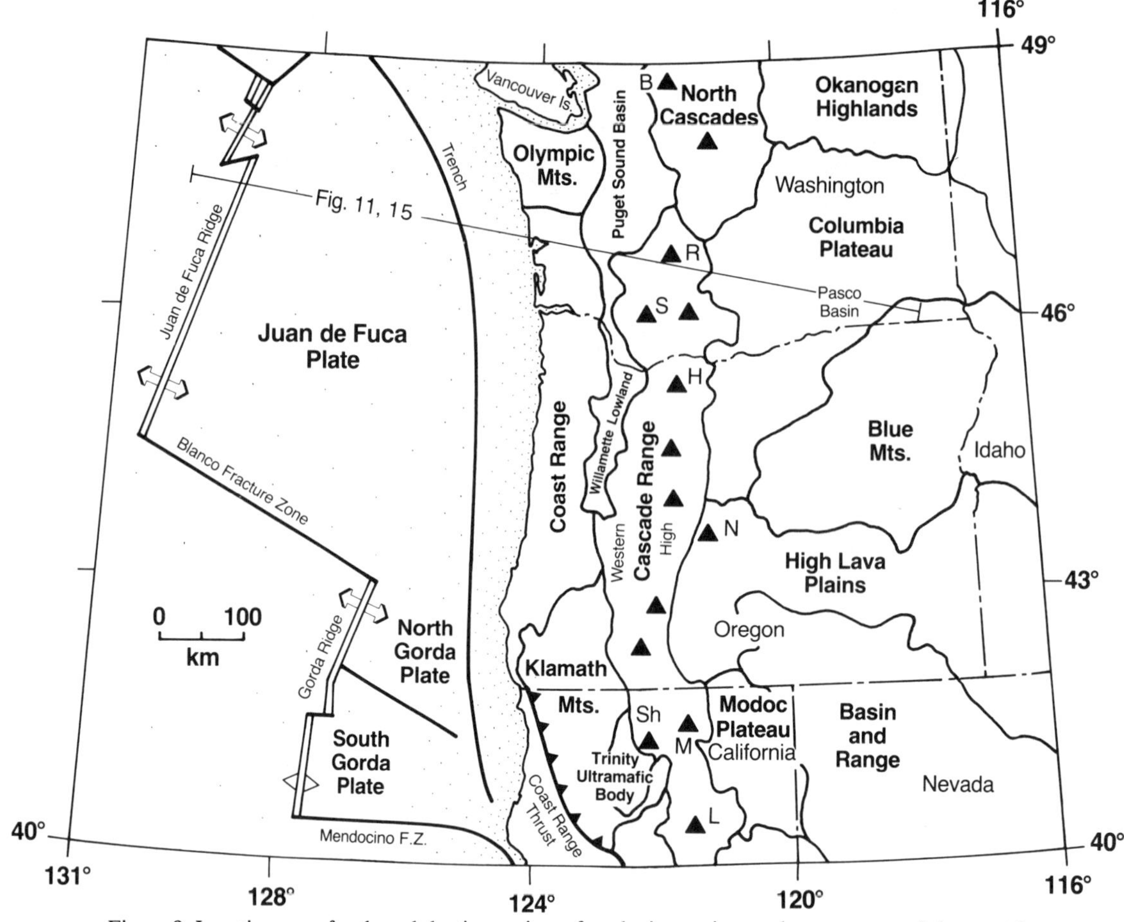

Figure 9. Location map for the subduction regime of geologic provinces, place names, and the crustal cross section shown in Figure 11. Solid triangles mark active or dormant volcanoes of the Cascade Range magmatic arc and are identified in Figure 1. The discussion of the geophysical framework is divided into fore arc, volcanic arc, and back arc. Contour map of crustal thickness appears in Figure 10.

volcanism is more extensive. Finally, we consider the back arc, a region consisting of diverse geologic terrains: to the south, the Modoc Plateau and the High Lava Plains, and to the north the Blue Mountains, Columbia Plateau, and the Okanogan Highlands.

Geology

The history of convergence between the North American and the Juan de Fuca plates has resulted in a complex fore-arc terrane made up of many accreted blocks. The offshore geology is described by Snavely (1987). Onshore, the Olympic Mountains and Coast Range of Washington and Oregon consist mainly of Miocene oceanic sediments and basalts (W. M. Cady, 1975). These sediments occur in a series of fault-bounded packages that were accreted to the continent during subduction (Tabor, 1972). The Olympic Mountains are interpreted to be oceanic basement obducted onto the continental margin. Paleomagnetic data from the Coast Range of Oregon indicate a significant rotation, perhaps as much as 70°, compared to inland provinces or the Klamath Mountains (Simpson and Cox, 1977; Magill and others, 1981; Beck and Engebretson, 1982). To the south, the Klamath Mountains contain by far the oldest rocks in the fore arc. The early Paleozoic Trinity ultramafic sheet (Irwin, 1981) overlies Paleozoic and Triassic metamorphic rocks in association with sedimentary and volcanic rocks of fore arc or arc affinity. The entire region is locally intruded with Jurassic and Cretaceous quartz diorite plutons. The Klamath Mountains are the product of the accretion of fragments of oceanic crust and island arcs (tectonostratigraphic terranes) along imbricate east-dipping thrust faults (Irwin, 1981).

East of the Olympic Mountains and the Coast Ranges lies the Puget-Willamette Lowland, a large depression covered by thick sedimentary sequences eroded from the Olympic Mountains during Miocene uplift and deposited during Pleistocene glaciation (Tabor, 1972; W. M. Cady, 1975). The north-south–trending Puget Sound portion of the lowland is cut by east-west–striking folds and faults estimated to be of mid-Cenozoic age, which are in most cases buried by the sedimentary sequences. Significant Bouguer gravity (as high as 100 mgal) and aeromagnetic anomalies are associated with these transverse structures (Gower and others, 1985).

The modern Cascade Range is the product of the most recent of repeated episodes of volcanism that have occurred along the margin of North America since late Paleozoic time (McBirney, 1978). Much of the earlier geologic history is poorly known because of the extensive late Cenozoic volcanic cover. The North Cascades of Washington are unique in that they expose high-grade Mesozoic metamorphic rocks (Misch, 1966). Late Cenozoic volcanism in the North Cascades is limited largely to stratovolcanoes at Mount Baker and Glacier Peak.

South of Mount Rainier, the Cascade Range has typically been considered to consist of two subprovinces. The Western Cascades are predominantly mid-Miocene in age, with a pulse of activity in late Miocene time (McBirney, 1978). In places, block faulting separates the Western Cascades of central Oregon from the High Cascades, the subprovince that includes all of the major Quaternary stratovolcanoes. The central High Cascades were interpreted by McBirney (1978) as being within a shallow crustal graben, with the downdropped area and the eastern side of the graben buried under Quaternary volcanic rocks; Smith and Taylor (1983), however, argued that the graben exists only north of central Oregon.

The back arc extends from the Modoc Plateau, California, the Okanogan Highlands, Washington (Fig. 9). The Modoc Plateau consists principally of Miocene to Recent basaltic flows and is part of the larger lava plains that cover much of eastern Oregon and Washington. The plateau grades into the Cascade Range to the west and the Sierra Nevada to the south (Fig. 9). To the north, the High Lava Plains of eastern Oregon have been extensively faulted on both large and small scales. Donath (1962) noted that there are 12 north-south–striking grabens, some as long as about 100 km, that dominate the topography of the area. Striking northwest across eastern Oregon are four right-lateral strike-slip fault zones that apparently serve to decouple extension of the heavily faulted Basin and Range from the Columbia Plateau, where there are relatively few young faults (Lawrence, 1976).

The Blue Mountains of northeastern Oregon (Fig. 9) consist largely of terranes of metamorphic rocks accreted to North America during Late Triassic to Early Cretaceous time (Brooks, 1979). Volcanism and sedimentation dominate the Cenozoic history of the range. The late Cenozoic history has been dominated by folding, faulting, and uplift.

The Columbia Plateau is a volcanic plateau that is covered with thick layers of Miocene flood basalt and thick sedimentary basins of Pliocene and Quaternary age. The basalts were erupted as tens to hundreds of individual flows, deposited between 17.5 and 6.0 Ma (Baksi and Watkins, 1973; Hooper, 1982). The source vents for the most recent flows are near the southeastern Washington-Idaho border (Swanson and others, 1975). Stratigraphic relations of the near-surface basalt flows indicate that these flows were extruded into a subsiding basin to the west of the source vents. Near the center of the Columbia Plateau, most individual basalt flows thicken, suggesting that the deepest part of the subsiding basin was near the present center of the plateau. The deeper geologic structure of the Columbia Plateau is unknown, as the thick basalt flows have buried all of the basement.

The Okanagon Highlands border the Columbia Plateau to the north. The Highlands consists of a series of north-south–striking ranges and river valleys and is characterized by three large metamorphic core complexes that contain high-grade gneisses whose protoliths are of Precambrian age (Price, 1981). Flanking the rocks of the core complexes are thick sequences of Eocene volcanic and sedimentary rocks. The entire region has been heavily folded and faulted throughout Mesozoic and early Cenozoic time (Misch, 1966; Fox and others, 1977). Several prominent grabens strike south from the Okanagon Highlands and appear to be buried by the Columbia River basalts at the northern margin of the Columbia Plateau.

Crustal structure in the subduction regime

The contour map of estimated crustal thickness beneath Washington, Oregon, and northern California (Fig. 10) relies almost entirely on studies unavailable for the Hill (1978) summary (Table 2). There are two major features in the crustal thickness map. The first is the pronounced eastward increase in

TABLE 2. CRUSTAL STUDIES IN THE SUBDUCTION REGIME

Line or Study No., Fig. 10	Reference	Method
1	Spence and others, 1985	Refraction
	Clowes and others, 1987	Reflection and refraction
2	Taber, 1983	Refraction
3	Taber and Lewis, 1986	Refraction
	Taber, 1983	Refraction
4	Shor and others, 1968	Refraction
5	Potter and others, 1986	Reflection
	Sanford and others, 1988	Reflection
6	Catchings and Mooney, 1988a	Refraction
	Glover, 1985	Refraction
7	Leaver and others, 1984	Refraction
8	Catchings and Mooney, 1988b	Refraction and gravity
9	Zucca and others, 1986	Refraction and gravity
	Fuis and others, 1986	Refraction and gravity
	Catchings, 1987	Refraction and gravity
10	Catchings and others, 1988	Refraction
11	Couch and Riddihough, this vol.	Gravity
12	Couch and Riddihough, this vol.	Gravity
13	EMSLAB Group, 1988	Electromagnetic
14	Crosson, 1976	Seismic network
	Zervas and Crosson, 1986	Seismic network
15	Rohay, 1982	Seismic network

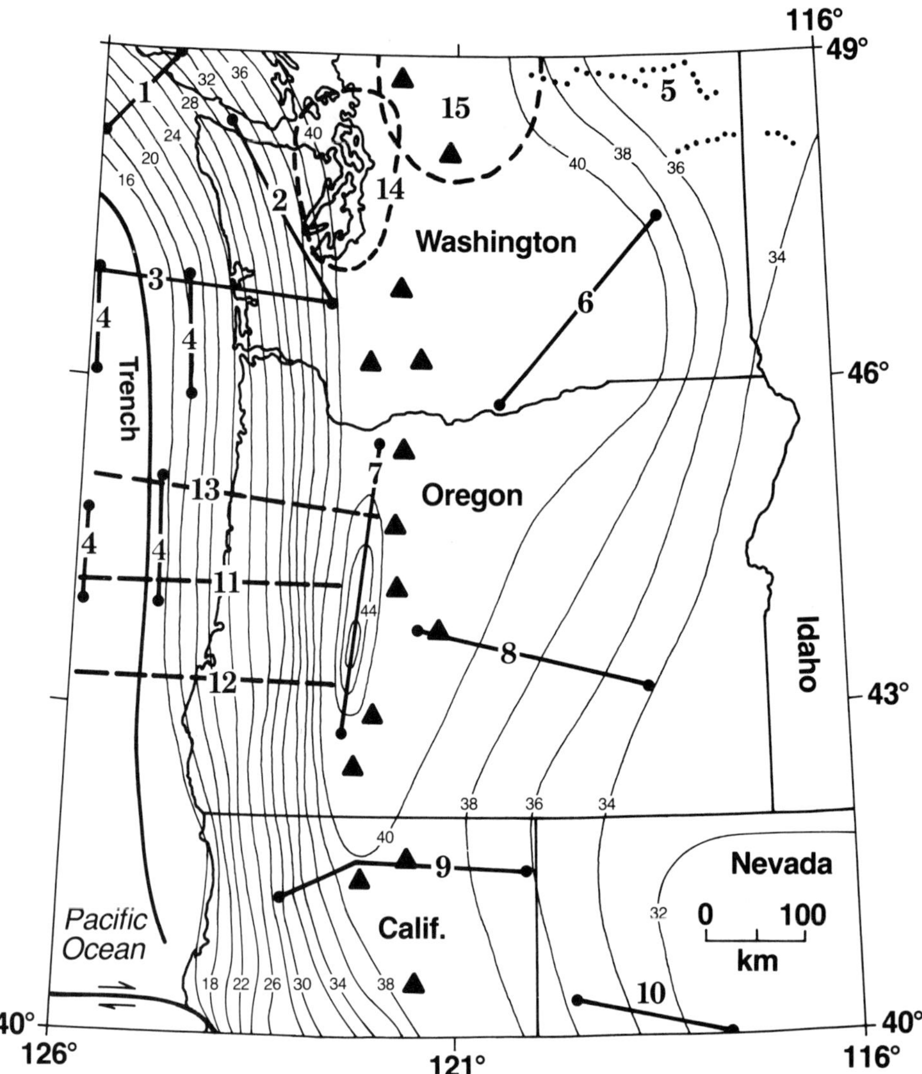

Figure 10. Contour map of crustal thickness for the subduction regime of the Pacific Coastal States. The crust thickens from 16 to 24 km at the coast to 40 km at the west flank of the Cascade Range. Central Oregon and Washington have a nearly uniform crustal thickness of 40 km; the crust thins in the back arc to about 34 km. Data sources 1 through 15 are listed in Table 2. Estimated errors are 10 percent, or one to two contour intervals.

crustal thickness from 16 km at the continental margin to about 40 km beneath the western flank of the Cascades Range. This crustal thickening is well constrained by refraction lines across Vancouver Island (line 1, Fig. 10) and across the Washington margin (line 3, Fig. 10). Gravity modeling along two profiles in Oregon (lines 11 and 12, Fig. 10) and preliminary interpretations of electrical and magnetic data collected by the EMSLAB experiment (EMSLAB Group, 1988) along a profile perpendicular to the northern Oregon coast (line 9, Fig. 10) are consistent with crustal thickening. The contours beneath the Klamath Mountains in southwestern Oregon and northwestern California lack seismic control. Nevertheless, the known thin oceanic crust and thick Cascades Range crust support the general trends represented by the contours.

The second major feature of the crustal thickness map is the presence of thick crust beneath the Cascade Range, the Puget Sound basin, and the Columbia Plateau (Fig. 10). Crustal thickness is estimated to be at least 38 km over this entire region, and locally reaches 46 km in the southern Oregon Cascade Range. A gradual eastward thinning of the crust occurs beneath the Basin and Range of southeastern Oregon and northeastern California (Fig. 10). The Moho shallows beneath the Okanagon Highlands in northeastern Washington, where a reflection profile (line 5, Fig. 10) has been interpreted as indicating a flat Moho at about 36 km depth beneath most of this province (Potter and others, 1986).

A cross section (Fig. 11) from the Juan de Fuca Ridge to the Columbia Plateau shows that both continental and oceanic Moho

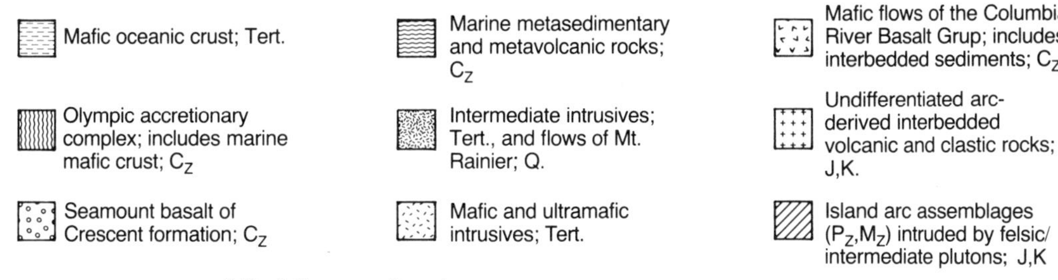

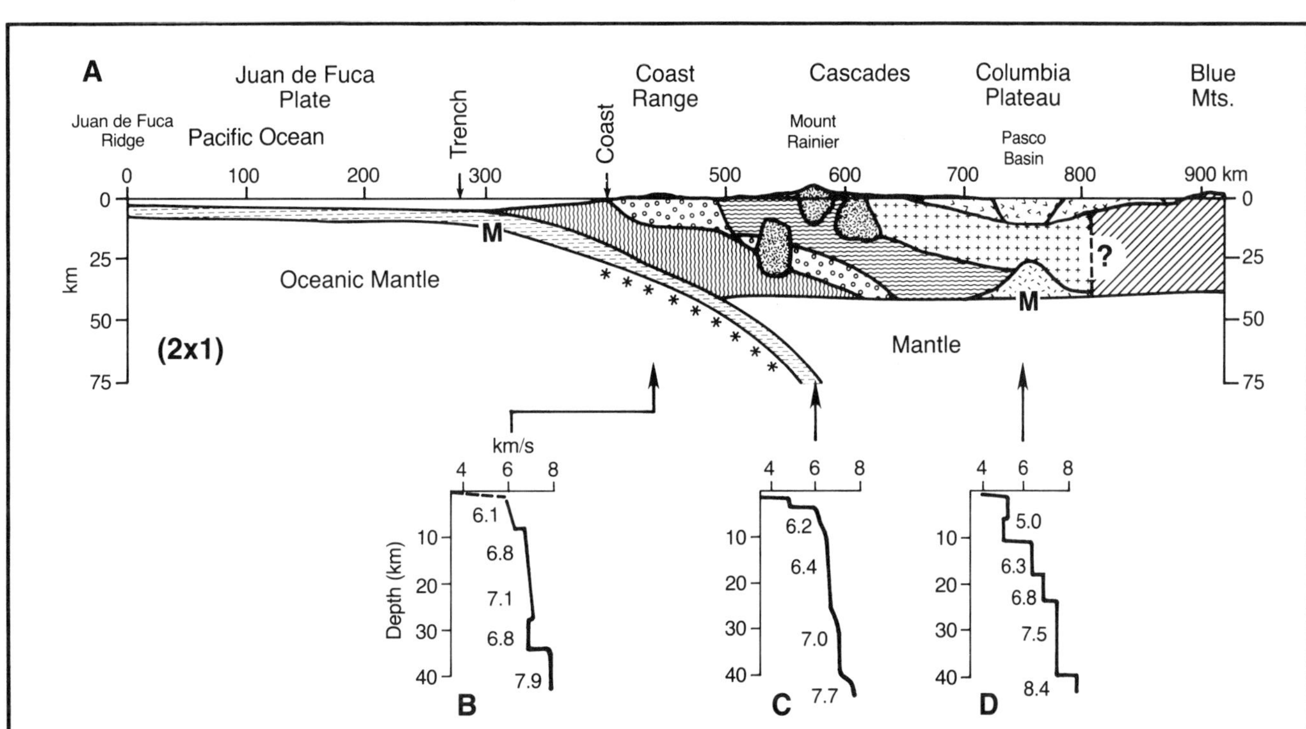

Figure 11. Crustal cross section from the Juan de Fuca plate to the Blue Mountains, modified from Cowan and Potter (1986). Location in Figure 9. The crust thickens rapidly as the Juan de Fuca plate subducts beneath the fore arc. The inferred crustal composition includes packages of accreted oceanic rocks, volcanic and intrusive rocks of the magmatic arc, an Eocene rift beneath the Columbia Plateau, and the allochthonous terrane making up the Blue Mountains. B-D, Velocity-depth functions determined along the cross section. The lower crust everywhere has a seismic velocity greater than 6.7 km/sec and is interpreted to consist of tectonically underplated oceanic crust and mafic intrusions and underplating. References: B, Taber, 1983, and Taber and Lewis, 1986; C, Leaver and others, 1984; D, Catchings and Mooney, 1988a.

depths are contoured in Figure 10 and that the subducting plate geometry along this profile is relatively simple. The Juan de Fuca plate dips at a shallow angle (3°) beneath the North American plate near the base of the continental slope, increases to about 11° 50 km east of the trench, and increases to about 25° in front of the Cascade Range (Fig. 11). At a depth of about 40 km, the Juan de Fuca plate intersects the continental Moho, and from this point eastward in southwestern Washington, the crustal thickness map (Fig. 10) reflects a continental Moho.

Fore arc. The most detailed profiles in the fore arc are the refraction/reflection studies across Vancouver Island (line 1, Fig. 10) and the refraction lines across the Olympic Mountains (lines 2, 3; Fig. 10). These investigations involve lines that are both approximately perpendicular and parallel to the coastal margin. Both experiments find that the break in the dip of the Juan de Fuca plate as it begins to subduct beneath North America occurs *landward* of the foot of the continental slope. A high-velocity (V_p = 7.7 km/sec) layer is found above the subducting plate beneath Vancouver Island in the depth range of 20 to 25 km (Spence and others, 1985; Clowes and others, 1987). This layer is interpreted to have one of two possible origins: either as a detached slab of oceanic lithosphere accreted in a discrete tectonic event, or as an imbricated package of mafic rocks derived by continuous accretion from the top of the subducting oceanic crust (Clowes and others, 1987). No such high-velocity zone has been identified beneath the Olympic Mountains.

East of the Olympic Mountains within thc Pugct Sound basin, Crosson (1976) estimated the crustal velocity structure from the inversion of local earthquake traveltime data. The crustal thickness is about 41 km (Fig. 10) with a P_n velocity of 7.7 km/sec. Crosson (1976) found evidence for a low-velocity zone at a depth between 35 and 41 km, where the velocity decreases from 7.2 km/sec to 6.9 km/sec. Berg and others (1966) estimated a crustal thickness of 16 km in western Oregon and Washington from sparse, unreversed refraction data; this estimate is incorporated into the teleseismic studies of Langston (1977, 1981). More recent teleseismic waveform modeling (Owens and others, 1988) indicates a 30-km-thick crust about 50 km from the coast in Washington, and is consistent with gravity modeling, seismicity data, and the crustal thickness contour map presented here (Fig. 10).

In the southernmost province of the fore arc, Zucca and others (1986) presented a pair of two-dimensional velocity models for the Klamath Mountains (western sector of line 9, Fig. 10). In agreement with geologic mapping (Irwin, 1964), the models are finely layered from the surface to at least 14 km depth and consist of a series of layers (V_p = 6.1 to 6.7 km/sec), ranging in thickness from 1 to 4 km, with alternating positive- and negative-velocity gradients. The base of the models at 14 km consists of a 7.0-km/sec layer that extends to unknown depth; crustal thickness has not been reliably determined beneath the Klamath Mountains. Fuis and others (1986) interpreted the velocity structure of the Klamath Mountains in terms of an imbricated stack of the thrust sheets consisting of oceanic rock layers of various compositions and ages (Fig. 12). By correlating the seismic model with an earlier crustal model derived from aeromagnetic data (Griscom, 1977), Fuis and others (1986) concluded that the Trinity ultramafic sheet extends to a depth of about 7 km below sea level and is underlain by Paleozoic metamorphic rocks (V_p = 6.5 to 6.7 km/sec), and sedimentary and volcanic rocks (V_p = 6.4 km/sec) that form a small seismic low-velocity zone. The basal 7.0-km/sec layer may consist of the ophiolitic basement of the North Fork terrain of Ando and others (1983).

Volcanic arc. The most detailed crustal investigations in the volcanic arc have been completed along the axis of the Oregon Cascade Range and in the vicinity of Mount Shasta and Medicine Lake Volcanos in California (Fig. 9). Although the Washington Cascades are less well studied, the available data do constrain both the crustal thickness and the upper-mantle seismic velocity, thereby allowing a discussion of the regional features of the arc.

Seismic refraction studies near Mount Shasta were part of a larger effort to characterize the crustal structure from the Klamath Mountains to the Modoc Plateau; the southern Cascade Range separates these two provinces (Fig. 9). Zucca and others (1986) presented an interpretation of the crustal structure from the eastern Klamath Mountains to the Modoc Plateau (line 9, Fig. 10), and a geologic cross section based on this structure is presented by Fuis and others (1986). The southern Cascades coincide with a significant transition in crustal structure separating the Klamath Mountains and the Modoc Plateau (Fig. 12). The evolution of the Klamath Mountains from fragments of oceanic crust and island arcs has previously been noted; the evolution of the Modoc Plateau is discussed below. The location of the Cascade Range just at the suture between these two areas suggests a relationship between the stable establishment of a magmatic arc and the local crustal structure; the Klamath Mountains/Modoc Plateau suture appears to have provided a convenient locus for the ascent of magma.

The total crustal thickness beneath the southern Cascades has not been reliably determined, but Zucca and others (1986) show that the Bouguer gravity can be fit with a flat crust/mantle boundary, rather than a boundary that deepens to the east as suggested by La Fehr (1965). A crustal thickness of 38 to 40 km is estimated, based on tentatively identified mantle arrivals and a comparison with the crustal thickness in the Oregon Cascades (Leaver and others, 1984).

An axial refraction profile along the Oregon Cascade Range shows that the crust beneath the range is thick and that there is relatively little lateral variation in the crustal velocity structure (line 7, Fig. 10; Leaver and others, 1984). The crust is about 44 km thick beneath the southern Cascade Range and thins to about 40 km to the north. An upper-mantle (P_n) velocity of 7.7 km/sec was interpreted using moderate-magnitude crustal earthquakes as sources (Fig. 11C). Beneath low-velocity surficial rocks, the upper and middle crust (3- to 29-km depth) show seismic velocities of 6.1 to 6.5 km/sec, and the lower crust (from 29 km to Moho depth) has a velocity of about 7.0 km/sec (Fig. 11C).

The crust of the Washington Cascades has been studied with earthquake and explosion traveltime data. A P_n velocity of 7.7 km/sec in the southern Washington Cascades is reported based on earthquake traveltime data, and the crustal thickness is estimated to be about 40 km, the same as northern Oregon (Leaver and others, 1984). In the North Cascades, Rohay (1982) modeled regional traveltime data to determine a 42-km-thick crust with a P_n velocity of 7.8 km/sec. Throughout the Washington Cascade Range, seismic velocities of 6.0 to 6.1 km/sec are observed within about 3 km of the surface (Leaver and Weaver, 1980; Rohay, 1982).

In summary, the available refraction data from the entire Cascade Range indicate a remarkably uniform average crustal structure. Relatively high crustal velocities, about 6.4 km/sec, are observed at shallow depths (10 km) from the Mount Shasta area to southern Washington, and this velocity may occur beneath the North Cascades as well. The crustal thickness is everywhere 42 $\pm$ 4 km, and the upper-mantle velocity is about 7.7 to 7.8 km/sec.

The small variation in crustal thickness (and velocity structure) in the Cascade Range contrasts with the strong change in the Moho depth in the fore arc (Fig. 10), and is perhaps unexpected in view of known variations in other geologic and geophysical parameters in the Cascades. Luedke and Smith (1982)

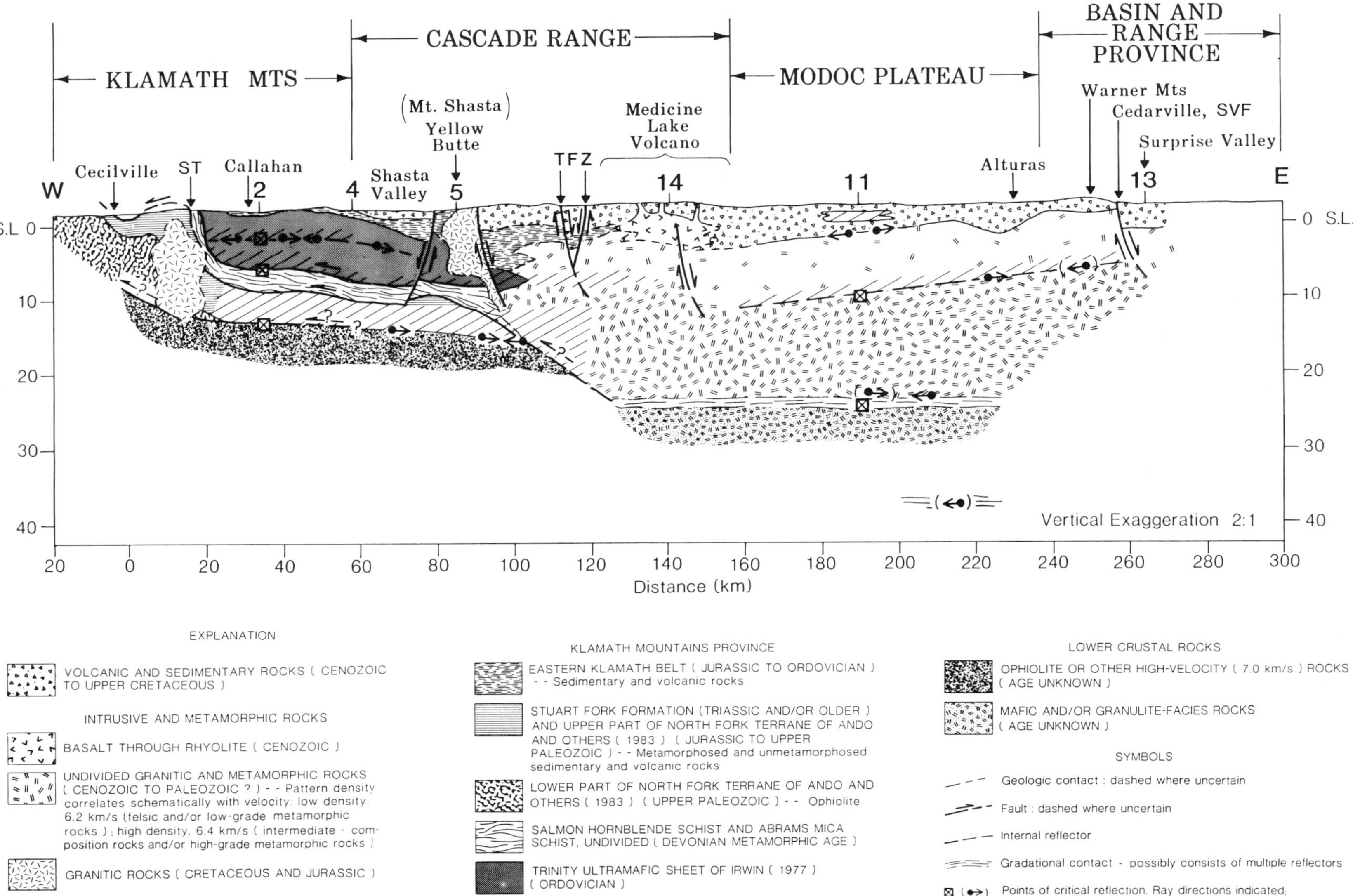

Figure 12. Crustal structure from the Klamath Mountains to the Modoc plateau (from Fuis and others, 1986). Profile 9 in Figure 10. The Klamath Mountains are underlain by a stack of oceanic layers. The Modoc Plateau is inferred to be underlain, beneath volcanic and sedimentary rocks, by crystalline igneous and metamorphic rocks. The Cascade Range, in between, is a complex suture region currently being intruded by magmas.

documented the great variations in its volcanic history over the last 15 m.y. Regional gravity studies show a broad Bouguer gravity low over the North Cascades, and a linear gravity low along the strike of the Oregon Cascades. Between these lows, at the approximate latitudes of Mount Rainer and Mount Hood, is a relative gravity high that essentially forms a saddle in the middle of the range (Williams and others, 1982). Heat flow increases dramatically south of Mount Hood, with values measured in excess of 100 mW/m^2 (Blackwell and Steele, 1983). Conductivity measurements along west-east profiles near Mount Shasta and near Mount Hood have been interpreted as indicating relatively flat-lying crustal structure (Stanley, 1984); however, transects near Mount Rainier and Mount St. Helens in Washington are interpreted as indicating a complex structure associated with accreted sediments that are juxtaposed against plutonic rocks (Stanley, 1984; Stanley and others, 1987). Finally, crustal seismicity is nonuniform within the Cascade Range; the only section with significant crustal earthquake activity is that between Mount Hood and Mount Rainier (Weaver and Malone, 1987).

We interpret the small variation in crustal velocity structure and crustal thickness in the Cascades to be due to the dominant influence of magmatic processes. These processes are in turn controlled by the pressure and temperature profiles in the crust and upper mantle. Excess crust (deeper than about 46 km) would be below the solidus for basalt and would therefore melt; thin crust (less than about 36 km) is thickened by intrusion and/or underplating. Just as extensional processes produce a crust of a nearly uniform thickness of 30 km (southern California: Fig. 3; the Basin and Range: Pakiser, this volume; Thompson and others, this volume), magmatic arcs produce crust with a characteristic velocity structure (Fig. 11C) and a thickness of about 40 km.

Crustal structure of volcanic centers. The crustal structure of the Quaternary andesitic and dacitic stratovolcanoes that lie along the crest of the Cascade Range (Mount Baker, Mount St. Helens, Mount Hood, Mount Shasta, and Lassen Peak) and the basaltic centers that lie to the east (Newberry and Medicine Lake Volcano) have been studied geophysically with varying detail. The crustal structure beneath the stratovolcanoes tends to be indistinguishable from the regional velocity structure (e.g., Rohay, 1982; Kohler and others, 1982; Berge and Stauber, 1987; Fig. 13). Leaver and others (1984) noted that the shallow crustal velocities determined along the Oregon Cascades axial seismic refraction profile (line 7, Fig. 10) are essentially the same as the local Cascades velocities identified in more detailed studies of the stratovolcanoes (e.g., Kohler and others, 1982). Thus, the available crustal seismic studies of the stratovolcanoes provide no compelling evidence for significant volumes of shallow intrusions beyond that which is found elsewhere along the Cascade Range. Furthermore, no simple expression of the Holocene magmatic system has yet been observed in the crustal structure beneath Mount Baker, Mount St. Helens, Mount Hood, Mount Shasta, and Lassen Peak.

The regional velocity structure around the basaltic centers of Newberry and Medicine Lake Volcanoes differs from that along the axis of the Cascades where the stratovolcanoes occur. Whereas a seismic velocity of 6.1 km/sec is encountered at a depth no greater than 2 to 4 km along the Cascades, at Newberry Volcano the 6.1-km/sec refractor is not observed until a depth of 8 km (Catchings and Mooney, 1988b). At Newberry Volcano, intrusions with seismic velocities of 5.6 to 5.8 km/sec occur within 1 km of the surface beneath the volcano (Fig. 13E). The regional crustal velocities around the intrusions are in the range of 4.0 to 5.3 km/sec, and are considerably lower than those beneath the Cascade Range stratovolcanoes. In part because of the lower velocity regional upper crustal structure, the basaltic centers stand

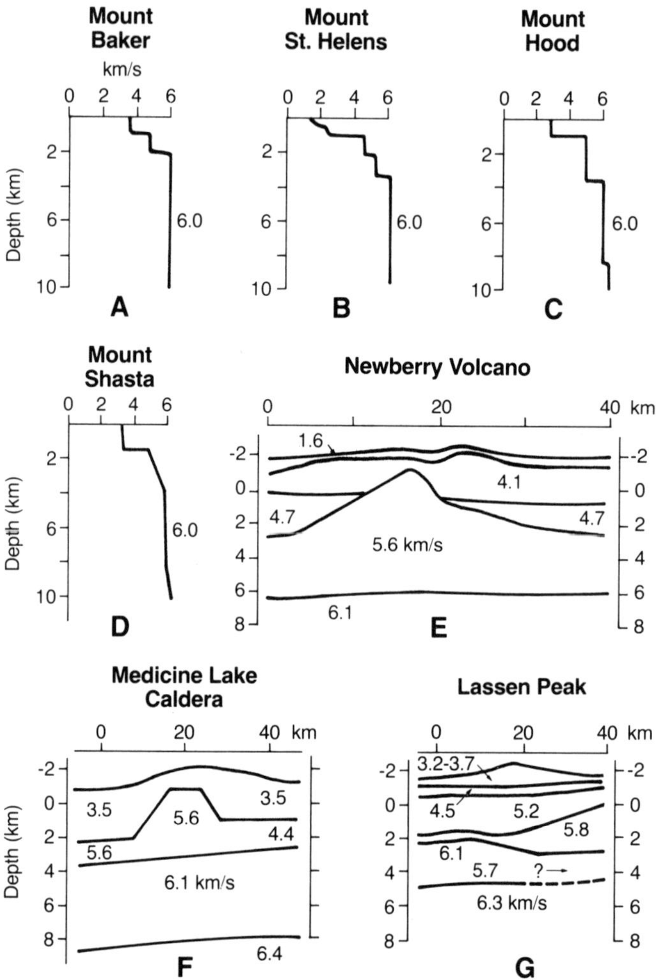

Figure 13. Upper crustal velocity structure of the stratovolcanoes and off-axis basaltic centers (Newberry and Medicine Lake Volcanoes). Seismic velocities of about 6.0 km/sec are found at shallow depth beneath the stratovolcanoes (A, B, C, D, G), but are not attained until a depth of 8 km and 5 km for Newberry Volcano and Medicine Lake Caldera, respectively. Because of these low upper crustal velocities, the crustal structure beneath the stratovolcanoes does not differ significantly from that found elsewhere along the axis of the High Cascades; there is no seismologic evidence for large (dimensions greater than a few kilometers) magmatic systems. References: A, Rohay, 1982; B, S. D. Malone, written communication, 1987; C, Kohler and others, 1982; D and F, Zucca and others, 1986; E, Catchings and Mooney, 1988b; G, Berge and Stauber, 1987.

out as relatively high-velocity intrusions (Fig. 13E,F). Beneath Medicine Lake Volcano, rocks with a velocity of 5.6-km/sec rise to within 1 km of the surface (Fig. 13F). At its base, the intrusion beneath Medicine Lake Volcano is approximately 20 km wide. The geometry of the intrusion identified from the seismic refraction study of Zucca and others (1986) agrees with the gravity modeling of Williams and Finn (1985) who modeled the Bouguer gravity over the volcano.

Returning to the Quaternary stratovolcanoes, the lack of resolvable variations in the shallow crustal structure indicates that the dimensions of the subsurface magmatic systems are small, on the order of a few kilometers at most. Geophysical studies at Mount St. Helens support this interpretation. Scandone and Malone (1985) and Shemeta and Weaver (1986) used seismicity to define an aseismic volume between active faults beneath the volcano. Between the crater floor and a depth of about 8 km, seismicity and petrologic considerations argue for a thin conduit, approximately 100 m in diameter. Below 8 km, the aseismic volume of possible magma storage is estimated to be less than 2 km in diameter and less than 4 to 5 km in depth extent. Based on petrologic considerations, magmatic processes involving intrusions by small batches of magma (approximately 1 km^3) are considered most probable at Mount St. Helens by Smith and Leeman (1987). Small magma volumes are consistent with the lack of a distinct signature in either gravity (Williams and others, 1987) or aeromagnetic data (Finn and Williams, 1987) over Mount St. Helens.

Contradicting the evidence for small magma volumes at the stratovolcanoes is the interpretation of heat-flow data for the Cascades. From about Mount Jefferson to Crater Lake, heat-flow measurements are almost double the continental average, typically in excess of 100 mW/m^2 (Blackwell and Steele, 1983). One model of this heat-flow data indicates that large volumes of partially molten rock should exist within about 8 km of the surface (Blackwell and others, 1982). The depth to the Curie point isotherm (570°C) in the central Oregon Cascade Range has been estimated at 9 km by Connard and others (1983), who noted that this depth was consistent with the heat-flow modeling for the depth to partial melt.

It may be possible to reconcile the differences between the seismic velocity/potential fields model and the heat-flow/Curie depth model with a thermal model wherein the Curie temperature is reached near a depth of 9 km, but not the melting point for dry intermediate-to-mafic rocks. In view of the nonuniqueness in the interpretation of seismic, gravity, heat-flow, and aeromagnetic data, the modeling of the thermal state and of magmatic processes in the Cascade Range remains uncertain.

Back arc. Four important new seismologic investigations of crustal and upper mantle structure of the back-arc province have been completed since the review by Hill (1978). Seismic refraction lines now exist across the Modoc Plateau, the High Lava Plains of east-central Oregon, and the Columbia Plateau, and a seismic reflection line was recorded across the Okanagon Highlands (lines 9, 8, 6, and 5; Fig. 10).

The crustal structure of the Modoc Plateau is known from seismic refraction profiles (Zucca and others, 1986; Catchings, 1987). The surficial flood basalts have seismic velocities of 4.5 to 5.2 km/sec (where deeper than about 2 km) and have an average thickness of 4.5 km (Fig. 12). Below the basalts the upper crust (4.5- to 25-km depth) shows seismic velocities of 6.2 to 6.4 km/sec, and the velocity increases to 7.0 km/sec at a depth of 25 km. Fuis and others (1986) compared the subvolcanic crustal structure of the Modoc Plateau with that of the Sierra Nevada (Eaton, 1966) and the Gabilan Range of west-central California (Walter and Mooney, 1982); they concluded that the plateau is underlain by granitic and metamorphic rocks generated in a magmatic arc environment. However, Catchings (1987) reinterpreted the same seismic refraction data as Zucca and others (1986) and concluded that the subvolcanic crustal structure is more similar to the extended crust of the northwestern Basin and Range (Thompson and others, this volume) than the Sierra Nevada or Gabilan Range. The interpretation of Catchings (1987) is consistent with the presence of Basin-and-Range–style extension on the plateau, as evidenced by the extensive block faulting (Macdonald, 1966), and with the chemistry of the surficial basalts, which show little crustal contamination by sialic material (McKee and others, 1983). Both interpretations give the crustal thickness as about 38 km (Fuis and others, 1986; Catchings, 1987).

The most reliable information regarding the crustal structure of eastern Oregon is a 180-km-long refraction line from the eastern High Cascades, across Newberry Volcano and the High Lava Plains. Catchings and Mooney (1988b) determined a crustal thickness of about 37 km and a P_n velocity of about 8.1 km/sec (Fig. 14A). As mentioned previously, seismic velocities in the upper crust in the vicinity of Newberry Volcano are much lower than those observed to the west along the Cascade Range refraction line, with an 8-km-thick section of velocity 4.1 to 5.6 km/sec. The midcrust (8- to 28-km depth) is characterized by velocities ranging from 6.1 to 6.5 km/sec, and an unusually high velocity (V_p = 7.4 km/s) is encountered in the lower crust. This layer has an average thickness of 7 km and may be the product of magmatic underplating and/or intrusion during extension (Catchings and Mooney, 1988b).

Catchings and Mooney (1988a) reported results from a 260-km-long refraction line across the Columbia Plateau, and Glover (1985) interpreted several lower resolution refraction lines. The crustal structure beneath the central Columbia Plateau is complex and shows features characteristic of continental rifts (Catchings and Mooney, 1988a). Above basement (V_p = 6.1 to 6.3 km/sec) are 5 to 12 km of Columbia River basalts and underlying sediments (Fig. 14B). Velocities above basement are relatively low (V_p = 5.1 to 5.6 km/sec), whereas velocities in the lower crust and upper mantle are atypically high (V_p = 7.5 to 8.4 km/sec). The basalts are 3 to 6 km thick along the profile, and are thickest near the center of the Columbia Plateau. The sedimentary layer beneath the basalts varies laterally in thickness, reaching its greatest thickness beneath the Pasco basin in an ap-

A

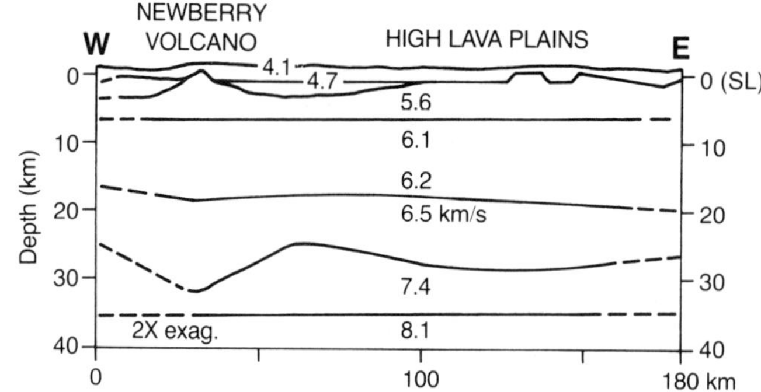

B

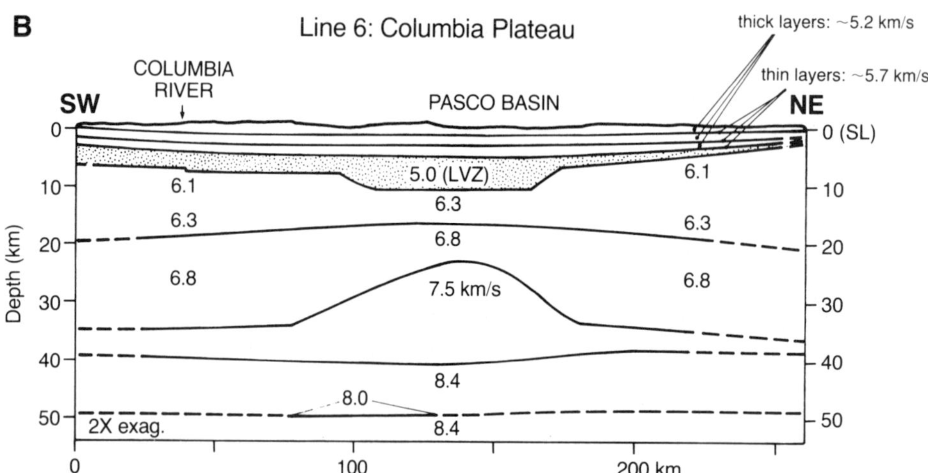

Figure 14. Crustal structure of east-central Oregon (Catchings and Mooney, 1988a) and the Columbia Plateau (Catchings and Mooney, 1988b) from seismic refraction measurements. Profile 8 in Figure 10. A, Beneath eastern Oregon, upper crustal seismic velocities are lower than found beneath the Cascades, and the basal crustal layer is interpreted to consist of underplated gabbro (in amphibolite or granulite metamorphic grade). B, A complex crustal structure is found beneath the Columbia Plateau. The Columbia River Basalt Group and underlying sediments are 5 to 12 km thick, and a subbasalt graben underlies the Pasco basin. The lower crust shows seismic velocities of 6.8 and 7.5 km/sec, with a thickened basal crustal layer (7.5 km/sec) that shows a geometry characteristic of continental rifts. The upper mantle velocity is anomalously high (8.4 km/sec) and may be indicative of seismic anisotropy.

parent deep graben (Fig. 14B). Between the basement and the upper mantle, two layers with velocities of 6.8 and 7.5 km/sec were identified; the 6.8-km/sec layer is thinned beneath the basement graben (Fig. 14). Where this thinning occurs, the 6.8-km/sec layer is replaced by the underlying 7.5-km/sec layer, creating the characteristic geometry and velocity structure recognized in other continental rifts (Mooney and others, 1983; Catchings and Mooney, 1988a).

Both the total crustal thickness and the upper-mantle velocity below the Columbia Plateau are noteworthy. The crust is approximately 40 km thick (Fig. 14B); previous estimates have ranged from 20 to 30 km (Hill, 1972). The reported P_n velocity along the refraction line was 8.4 km/sec, an atypically high upper-mantle velocity and considerably different from the mantle velocity of 7.9 to 8.1 estimated previously (Hill, 1972). Catchings and Mooney (1988a) have suggested that these previous studies may have misidentified updip apparent velocities from the lower crustal 7.5-km/sec layer as the Moho. The 8.4-km/s P_n velocity may represent the fast direction in an anisotropic upper mantle.

Seismic reflection lines across the Okanagon Highlands (line 5, Fig. 10) identified a highly reflective, complexly deformed crust, with a relatively flat Moho at depths of 33 to 35 km. The reflection data are interpreted as indicating a crust that has been reworked by both extensional and compressional processes (Potter and others, 1986; Sanford and others, 1988). The crust was thickened during Jurassic to Paleocene thrust faulting and then extended and thinned in Eocene time. Cross-cutting east- and west-dipping reflections are interpreted by Potter and others (1986) as zones of ductile deformation that accommodated the Eocene extension. Potter and others (1986) suggested that the nearly flat Moho developed during Eocene extension, a process that apparently also occurred in mid- to late Cenozoic time in the Basin and Range and the Modoc Plateau.

Seismicity considerations

Seismic refraction and reflection results have principally been used to describe the geophysical framework of the Pacific Coastal States, but earthquake hypocenters provide information that is not available from other methods. In the subduction regime, hypocentral cross sections indicate the geometry of the subduction plate, and in the transform region reveal relations between crustal structure and active tectonics. Accordingly, we have selected a subset of well-located earthquakes along 100-km-wide strips for four cross sections (Fig. 15). Three of these cross sections are coincident with our interpreted crustal sections (Figs. 4, 7, 11); a fourth cross section (Fig. 15B) crosses the Gorda plate and northern California just north of Cape Mendocino. Although hypocentral locations with statistical errors less than 5 km have been used in the cross sections, some location artifacts remain, and the cross sections must be interpreted with caution.

The two northern sections show the deep earthquakes characteristic of Benioff zones (Fig. 15A, B); it is likely that the more tightly defined Benioff zone beneath Washington reflects the higher density of seismic stations there as compared to the Cape Mendocino area. Beneath Washington, a comparison of the hypocentral depths with seismic refraction models of the continental margin indicate that these events occur below the Moho of the subducting plate (Taber and Smith, 1985). An analysis of all available deep earthquakes indicates that the plate dips at about 10° beneath the coast; that increases to about 25° inland (Weaver and Baker, 1988). Typically, the largest earthquakes (magnitude, 6.5 to 7.1) occur near this change in plate dip (Weaver and Baker, 1988). Beneath northern California, most of the recent earthquake activity is offshore, on the flat-lying section of the Gorda plate (Fig. 15B). (Note that in this cross section the earthquake distribution 50 km west of the trench has been artificially truncated because hypocenters with location errors greater than ±5 km have been rejected.) Inland, beneath northern California, the distribution of earthquakes is more complex than beneath western Washington. A clear separation between crustal events and those within the subducting Gorda plate does not occur until a depth of about 30 km beneath the Klamath Mountains (Fig. 15B). The deeper events indicate a dip of the subducting plate of about 25°, similar to western Washington. Both cross sections indicate that the plate is at a depth of about 80 to 100 km beneath the Cascade Range volcanoes.

The crustal earthquake distributions are considerably more complicated than the Benioff zone events; the reader is referred to Dewey and others (this volume) for a detailed discussion. Many of the earthquakes in Figure 15C, D are associated with the San Andreas fault and related strike-slip or dip-slip faults (Fig. 2). Control on earthquake hypocenters is relatively good across the entire central California section (Fig. 15C); in general, as crustal thickness increases from the coast to the western edge of the Great Valley, maximum hypocentral depths also increase. There are relatively few earthquakes beneath the Great Valley and the Sierra Nevada; the absence of deep earthquakes beneath the Sierra Nevada batholith at this latitude is particularly evident. This low level of seismicity suggests that the Great Valley and Sierra Nevada form a relatively undeforming block within the broader Pacific–North American plate boundary. In the western Basin and Range, maximum hypocentral depths are less than half the crustal thickness (Fig. 15C), reflecting a shallow depth to the brittle-ductile transition.

The southern California earthquake distribution (Fig. 15D) shows sharp lateral boundaries (vertical alignments of hypocenters) similar to central California (Fig. 15C). These alignments occur from the Borderland to the Salton Trough (a distance of about 300 km) and document the importance of translational tectonics in southern California. Hypocentral depths are most reliable between the coast and the San Andreas fault, and earthquakes there occur to a depth somewhat greater than half the crustal thickness.

Crustal earthquake distributions in the two northern cross sections show few vertical alignments of hypocenters (Fig. 15A, B). In Washington, the deepest crustal events are found near the point where the dip of the Juan de Fuca plate begins to steepen; considerably more deep crustal events occur at depths of 20 to 35 km beneath Puget Sound just to the north of the section shown in Fig. 15A (Crosson, 1976). Much of the crustal activity in Washington is concentrated within the volcanic arc, near Mount St. Helens. Hypocenters are shallower east of Mount St. Helens, possibly because of geologic structure associated with the southern Washington Cascades Conductor (Stanley and others, 1987). This conductive body is interpreted to be a region of compressed marine sedimentary rocks that localizes crustal slip to zones along its boundaries (Weaver and Malone, 1987). East of the Cascades the maximum depth of seismicity is about 20 km everywhere, and like the western Cascades, this depth is approximately half the crustal thickness.

DISCUSSION

We have divided this discussion of the geophysical framework of the Pacific Coastal States into the transform regime of the San Andreas fault and the subduction regime of the Gorda and Juan de Fuca plates. Our primary results are presented in the form of contour maps of crustal thickness (Figs. 3, 10), lithospheric (or crustal) cross sections (Figs. 4, 5, 7, 11), and seismicity cross sections (Fig. 15) that illustrate the geophysical evidence for the effects of subduction, transform faulting, arc magmatism, and back-arc extension on the structure, composition, and evolution of the lithosphere. Mesozoic and Cenozoic subduction at this active plate margin are expressed in the deep geology by: (1) the accretion of thick (as much as 18 km) packages of marine sediments (California Borderland, Coast Ranges, and Olympic Mountains), (2) arc magmatism (Sierra Nevada and Cascade

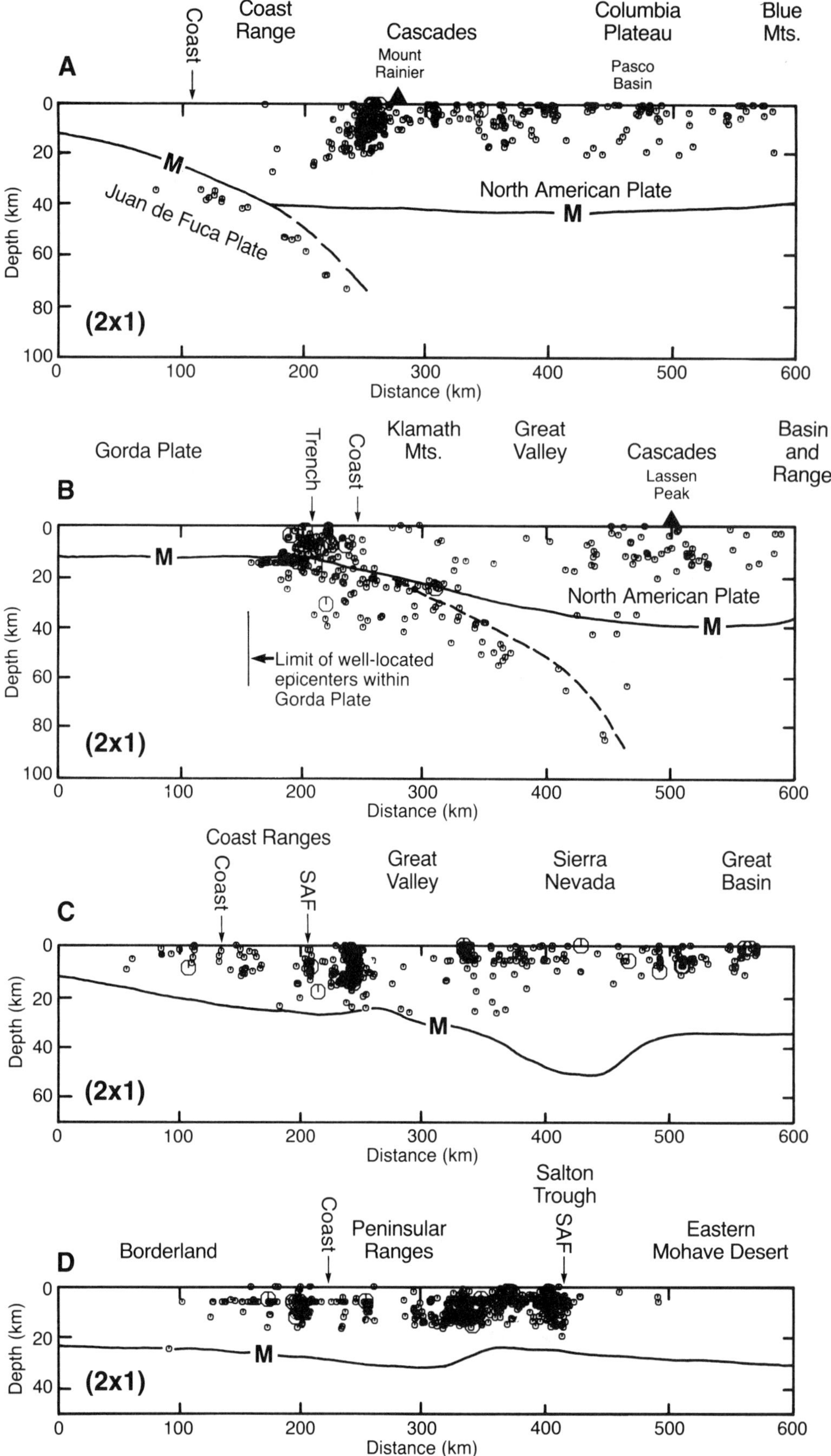
A
Coast
Coast Range
Cascades
Mount Rainier
Columbia Plateau
Pasco Basin
Blue Mts.
M
Juan de Fuca Plate
North American Plate
M
(2x1)
Depth (km)
Distance (km)
B
Gorda Plate
Trench
Coast
Klamath Mts.
Great Valley
Cascades
Lassen Peak
Basin and Range
M
North American Plate
M
Limit of well-located epicenters within Gorda Plate
(2x1)
Depth (km)
Distance (km)
Coast Ranges
C
Coast
SAF
Great Valley
Sierra Nevada
Great Basin
M
(2x1)
Depth (km)
Distance (km)
D
Borderland
Coast
Peninsular Ranges
Salton Trough
SAF
Eastern Mohave Desert
M
(2x1)
Depth (km)
Distance (km)

Figure 15. Cross sections of seismicity for the period 1980 to 1987. Hypocenters within 50 km have been projected onto the plane of the section. Vertical exaggeration is two to one. Earthquakes were selected with origin time RMS values less than 0.5 sec and hypocentral residuals less than 5 km. Above 35 km, all events are greater than magnitude 2.5, except for the Washington cross section, where the cut-off was magnitude 2.0; events below 35 km are greater than magnitude 1.5; two symbol sizes are used—larger size symbols indicate magnitudes of 4.0 or greater. The dashed lines represent the crustal thickness taken from Figures 3 and 10. A, Southern Washington section coincident with Figure 11; location in Figure 9. B, Northern California section north of Cape Mendocino (Figure 2). C, Central California section coincident with Figure 7; location in Figure 2. D, Southern California section coincident with Figure 4; location in Figure 2. Cross sections A and B both show a clear Benioff zone, but the distributions of crustal earthquakes in the two sections are quite different. Cross sections C and D primarily show earthquake clusters and vertically aligned hypocenters corresponding to major strike-slip faults. All four sections are discussed in the text.

Range), (3) the creation of fore-arc basins (Great Valley and Puget Sound/Willamette lowland), and (4) back-arc extension (northern California and eastern Oregon and Washington). In addition, there is abundant evidence for the tectonic underplating of the continental margin by marine sedimentary and igneous rocks. Tectonically underplated igneous rocks offset the more felsic composition of the accreted sedimentary rocks, and thus the subduction process presents at the smallest scale (vertical dimension, 16 to 22 km) the development of a "typical" continental crust, i.e., one that grades with depth in composition from felsic to increasingly mafic.

The contour map of crustal thickness in California (Fig. 3) reflects the influences of several tectonic processes. The eastward thickening of the crust in central and northern California is a product of Mesozoic and early Cenozoic subduction, much like that which is active in Oregon and Washington. Southern California has developed a nearly uniform crustal thickness (30 ± 2 km) principally as a result of ductile stretching of the lower crust during crustal extension (Mojave Desert and Salton Trough) and transform faulting (San Andreas and associated faults). An abrupt crustal thickening north of the Garlock fault is clearly evidenced in the contour map, as is a region of thin (16 to 24 km) crust associated with the Mendocino triple junction beneath the northern California Coast Ranges.

Recent and Pleistocene volcanism is abundant in the Pacific Coastal States, but details of the subsurface magmatic systems are elusive. The seismic structure beneath the stratovolcanoes of the Cascade Range is indistinguishable from the regional Cascade Range structure, indicating that the dimensions of the magmatic systems are small, on the order of a few kilometers. Whereas earlier studies hypothesized a single large magma chamber beneath some magmatic systems (e.g., Long Valley, Newberry Volcano, and Medicine Lake Volcano), the current evidence favors single or multiple small magma chambers.

The importance of magmatic underplating of the crust is evident from geophysical data reviewed here. In the Pacific Coastal States this underplating appears to occur in association with Basin-and-Range–style extension (Mojave Desert, eastern Oregon), localized rifting (Columbia Plateau and Okanogan Highlands), and onshore spreading centers (Salton Trough). In addition, the northward-migrating Mendocino triple junction appears to have underplated much of central and northern California. This magmatic underplating explains the presence of a thick, high-velocity layer in the lower crust of the California Coast Ranges and Great Valley, and is expressed at the surface by young extrusive rocks.

The geometry of the subducting plate in the Pacific Northwest is difficult to resolve because of the limited number of Benioff zone earthquakes. The plate's location has been best defined beneath Washington and northern Oregon, where it dips at about 10° at the coast and increases to about 25° in front of the Cascade Range. Plate geometry beneath central and southern Oregon is poorly determined. Seismicity in the overriding plate is concentrated at the volcanic centers of the Cascade Range; the deepest crustal earthquakes are found above the point where the Juan de Fuca plate steepens in dip. East of the Cascades, the maximum focal depths (20 km) are about half the crustal thickness.

Covering a land area amounting to only one-tenth of the conterminous United States, the Pacific Coastal states offer the opportunity to investigate all of the major processes of crustal and lithospheric evolution—subduction, rifting, volcanism, extension, accretion, compression, transform faulting, and tectonic and magmatic underplating. Continued studies will refine the general outline of the geophysical framework that has been presented here.

ACKNOWLEDGMENTS

We have benefited from informal discussions and critical reviews by a number of people. For review comments and/or for freely sharing their ideas and insights, we sincerely thank E. L. Ambos, H. Benz, T. M. Brocher, R. D. Catchings, J. P. Eaton, E. R. Flueh, G. S. Fuis, P. B. Gans, D. P. Hill, W. S. Holbrook, G.C.P. King, J. McCarthy, L. C. Pakiser, G. C. Rogers, R. C. Speed, G. A. Thompson, A. W. Walter, P. L. Ward, and J. E. Zollweg. R. Lester provided the earthquake hypocentral data for California, and the University of Washington Geophysics Program provided the data for Washington. The figures were patiently drafted by M. Tune and L. Neu of the University of Washington Health Sciences Division.

REFERENCES CITED

Achauer, U., Greene, L., Evans, J. R., and Iyer, H. M., 1986, Nature of the magma chamber underlying the Mono Craters area, eastern California, as determined from teleseismic travel time residuals: Journal of Geophysical Research, v. 91, p. 13873–13891.

Ando, C. J., Irwin, W. P., Jones, D. L., and Saleeby, J. B., 1983, The ophiolitic North Fork terrane in the Salmon River region, central Klamath Mountains, California: Geological Society of America Bulletin, v. 94, p. 236–252.

Atwater, T., 1970, Implications of plate tectonics for the Cenozoic tectonic evolution of western North America: Geological Society of America Bulletin, v. 81, p. 3513–3536.

Atwater, T., and Molnar, 1973, Relative motion of the Pacific and North American plates deduced from sea-floor spreading the Atlantic, Indian, and South Pacific Oceans; Proceedings of the Conference on Tectonic Problems of the San Andreas Fault System: Stanford, California, Stanford University Publications in Geological Science, v. 13, p. 136–148.

Bailey, E. H., Irwin, W. P., and Jones, D. L., 1964, Franciscan and related rocks: California Division of Mines and Geology Bulletin 183, p. 154–165.

Bailey, R. A., 1982, Other potential eruption centers in California; Long Valley, Mono Lake, Coso, and Clear Lake volcanic fields: California Division of Mines and Geology Special Publication 63, p. 17–28.

Bailey, R. A., Dalrymple, G. B., and Lanphere, M. A., 1976, Volcanism, structure, and geochronology of Long Valley caldera, Mono County, California: Journal of Geophysical Research, v. 81, p. 725–744.

Baksi, A. K., and Watkins, N. D., 1973, Volcanic production rates; Comparison of oceanic ridges, islands, and the Columbia Plateau basalts: Science, v. 180, p. 493–496.

Bateman, P. C., 1979, Map and cross section of the Sierra Nevada from Madera to the White Mountains, central California: Geological Society of America Map and Chart MC 28E, 2 sheets and 4 p. text.

Bateman, P. C., and Eaton, J. P., 1967, Sierra Nevada Batholith: Science, v. 158, p. 1407–1417.

Beck, M. E., Jr., and Engebretson, D., 1982, Paleomagnetism of small basalt exposures in the west Puget Sound area and speculation on the accretionary origin of the Olympic Mountains: Journal of Geophysical Research, v. 87, p. 3755–3760.

Berge, P. A., and Stauber, D. A., 1987, Seismic-refraction study of upper-crustal structure in the Lassen Peak area, northern California: Journal of Geophysical Research, v. 92, p. 10571–10579.

Bird, P., and Rosenstock, R. W., 1984, Kinematics of present crust and mantle flow in southern California: Geological Society of America Bulletin, v. 95, p. 946–957.

Blackwell, D. D., and Steele, J. L., 1983, A summary of heat flow studies in the Cascade Range: Transactions of the Geothermal Resources Council, v. 7, p. 233–236.

Blackwell, D. D., Bowen, R. G., Hull, D. A., Riccio, J., and Steele, J. A., 1982, Heat flow, arc volcanism, and subduction in northern Oregon: Journal of Geophysical Research, v. 87, p. 8735–8754.

Blake, M. C., Jr., Zietz, I., and Daniels, D. L., 1977, Aeromagnetic and generalized geologic map of parts of central California: U.S. Geological Survey Geophysical Investigations Map GP-918, scale 1:1,000,000.

Blümling, P., and Prodehl, C., 1983, Crustal structure beneath the eastern part of the Coast Ranges of central California from explosion-seismic and near-earthquake data: Physics of Earth Planetary Interiors, v. 31, p. 313–326.

Blümling, P., Mooney, W. D., and Lee, W.H.K., 1985, Crustal structure of southern Calaveras Fault Zone, central California, from seismic refraction investigations: Bulletin of the Seismological Society of America, v. 75, p. 193–209.

Bolt, B. A., and Gutdeutsch, R., 1982, Reinterpretation by ray tracing of the transverse refraction seismic profile through the California Sierra Nevada, Part 1: Bulletin of the Seismological Society of America, v. 72, p. 889–900.

Brooks, H. C., 1979, Plate tectonics and the geologic history of the Blue Mountains: Oregon Geology, v. 41, p. 71–80.

Burchfiel, B. C., and Davis, G. A., 1972, Structural framework and evolution of the southern part of the Cordilleran orogen, western United States: American Journal of Science, v. 272, p. 97–118.

——, 1981, Mojave Desert and environs, *in* Ernst, W. G., ed., The geotectonic development of California: Englewood Cliffs, New Jersey, Prentice-Hall, p. 215–252.

Byerly, P., 1939, Near earthquakes in Central California: Bulletin of the Seismological Society of America, v. 29, p. 427–462.

Cady, J. W., 1975, Magnetic and gravity anomalies in the Great Valley and western Sierra Nevada metamorphic belt, California: Geological Society of America Special Paper 168, 56 p.

Cady, W. M., 1975, Tectonic setting of the Tertiary volcanic rocks of the Olympic Peninsula, Washington: U.S. Geological Survey Journal of Research, v. 3, p. 573–582.

Carder, D. S., 1973, Trans-California seismic profile, Death Valley to Monterey Bay: Bulletin of the Seismological Society of America, v. 63, p. 571–586.

Carder, D. S., Qamar, A., and McEvilly, T. V., 1970, Trans-California seismic profile; Pahute Mesa to San Francisco Bay: Bulletin of the Seismological Society of America, v. 60, p. 1829–1846.

Catchings, R. D., 1987, Crustal structure of the northwestern United States [Ph.D. thesis]: Stanford, California, Stanford University, 183 p.

Catchings, R. D., and Mooney, W. D., 1988a, Crustal structure of the Columbia Plateau; Evidence for continental rifting: Journal of Geophysical Research, v. 93, p. 459–474.

——, 1988b, Crustal structure of east-central Oregon; Relationship between Newberry volcano and regional crustal structure: Journal of Geophysical Research, v. 93, p. 10081–10094.

Catchings, R. D., and 14 others, 1988, The 1986 PASSCAL Basin and Range Lithospheric Seismic Experiment: EOS Transactions of the American Geophysical Union, v. 69, p. 593–598.

Champion, D. E., Howell, D. G., and Gromme, C. S., 1984, Paleomagnetic and geologic data indicating 2,500 km of northward displacement for the Salinian and relate terranes, California: Journal of Geophysical Research, v. 89, p. 7736–7752.

Chapman, R. H., 1966, Gravity map of The Geysers area, California: California Division of Mines and Geology Mineral Information Service, v. 19, p. 148–149.

——, 1975, Geophysical study of the Clear Lake region, California: California Division of Mines and Geology Special Report, 116 p.

Chase, C. G., and Wallace, T. C., 1986, Uplift of the Sierra Nevada of California: Geology, v. 14, p. 730–733.

Cheadle, M. J., Czuchra, B. L., Byrne, T., Ando, C. S., Oliver, J. E., Brown, L. D., Kaufman, S., Malin, P., and Phinney, R. A., 1986, The deep crustal structure of the Mojave Desert, California, from COCORP seismic reflection data: Tectonics, v. 5, p. 293–320.

Christensen, M. N., 1966, Cenozoic crustal movements in the Sierra Nevada: Geological Society of America Bulletin, v. 77, p. 163–182.

Clark, L. D., 1960, Foothills fault system, western Sierra Nevada, California: Geological Society of America Bulletin, v. 71, p. 483–496.

Clarke, S. H., Jr., 1987, Geology of the California continental margin north of Cape Mendocino, *in* Scholl, D. S., Grantz, A., and Vedder, J. G., ed., Geology and resource potential of the continental margin of western North America and adjacent ocean basins; Beaufort Sea to Baja California: Circum-Pacific Council for Energy and Mineral Resources Earth Science Series, v. 6, p. 337–351.

Clowes, R. M., Brandon, M. T., Green, A. G., Yorath, C. J., Brown, A. S., Kanasewich, E. R., and Spencer, C., 1987, LITHOPROBE, southern Vancouver Island; Cenozoic subduction complex imaged by deep seismic reflections: Canadian Journal of Earth Sciences, v. 24, p. 31–51.

Cockerham, R. S., 1984, Evidence for a 180-km-long subducted plate beneath northern California: Bulletin of the Seismological Society of America, v. 74, p. 569–576.

Colburn, R. H., and Mooney, W. D., 1986, Two-dimensional velocity structure along the synclinal axis of the Great Valley, California: Bulletin of the Seismological Society of America, v. 76, p. 1305–1322.

Coney, P. J., Jones, D. L., and Monger, J.W.H., 1980, Cordilleran suspect terranes: Nature, v. 288, p. 329–333.

Connard, G., Couch, R., and Gemperle, M., 1983, Analysis of aeromagnetic measurements from the Cascade Range in central Oregon: Geophysics, v. 48, p. 376–390.

Cowan, D. S., and Potter, C. J., 1986, Juan de Fuca spreading ridge to Montana thrust belt: Boulder, Colorado, Geological Society of America Continent/Ocean Transect B-3, scale 1:500,000.

Crosson, R. S., 1976, Crustal modeling of earthquake data; 2, Velocity structure of the Puget Sound Region, Washington: Journal of Geophysical Research, v. 81, p. 3047–3054.

Crouch, J. K., Bachman, S. B., and Shay, J. T., 1984, Post-Miocene compressional tectonics along the central California margin, *in* Crouch, J. K., and others, eds., Tectonics and sedimentation along the California margin: Society of Economic Petrologists and Mineralogists Pacific Section, v. 38, p. 35–54.

Crough, S. T., and Thompson, G. A., 1977, Upper mantle origin of Sierra Nevada uplift: Geology, v. 5, p. 396–399.

Crowell, J. C., 1981, An outline of the tectonic history of southeastern California, *in* Ernst, W. G., ed., The geotectonic development of California: Englewood Cliffs, New Jersey, Prentice-Hall, p. 583–600.

Davis, G. A., and Burchfiel, B. C., 1973, Garlock fault; An intracontinental transform structure, southern California: Geological Society of America Bulletin, v. 84, p. 1407–1422.

Day, H. W., Moores, E. M., and Tuminas, A. C., 1985, Structure and tectonics of the northern Sierra Nevada: Geological Society America Bulletin, v. 96, p. 436–450.

de Voogd, B., Serpa, L., Brown, L., Hauser, E., Kaufman, S., Oliver, J., Troxel, B., Willemin, J., Wright, L. A., 1986, Death Valley bright spot; A midcrustal magma body in the southern Great Basin, California?: Geology, v. 14, p. 64–67.

Dickinson, W. R., 1970, Relations of andesite, granites, and derived sandstones to arc-trench tectonics: Reviews in Geophysics and Space Physics, v. 8, p. 813–860.

——, 1981, Plate tectonics and the continental margin of California, *in* Ernst, W. G., ed., The geotectonic development of California: Englewood Cliffs, New Jersey, Prentice-Hall, p. 1–28.

Dickinson, W. R., and Snyder, W. S., 1979a, Geometry of the triple junctions related to San Andreas transform: Journal of Geophysical Research, v. 84, p. 561–572.

——, 1979b, Geometry of subducted slabs related to San Andreas transform: Journal of Geology, v. 87, p. 609–627.

Donath, F. A., 1962, Analysis of Basin-Range structure, south-central Oregon: Geological Society of America Bulletin, v. 73, p. 1–16.

Donnelly-Nolan, J. M., Hearn, B. C., Jr., Curtis, G. H., and Drake, R. E., 1981, Geochronology and evolution of the Clear Lake volcanics, *in* McLaughlin, R. J., and Donnelly-Nolan, J. M., eds., Research in The Geysers–Clear Lake geothermal area, northern California: U.S. Geological Survey Professional Paper 1141, p. 47–60.

Eaton, J. P., 1963, Crustal structure from San Francisco, California, to Eureka, Nevada, from seismic-refraction measurements: Journal of Geophysical Research, v. 68, p. 5789–5806.

——, 1966, Crustal structure in northern and central California from seismic evidence; Geology of Northern California: California Division of Mines and Geology Bulletin 190, p. 419–426.

Eberhart-Phillips, D., 1986, Three-dimensional velocity structure in northern California Coast Ranges from inversion of local earthquake arrival times: Bulletin of the Seismological Society of America, v. 76, p. 1025–1052.

Elders, W. A., Rex, R. W., Meiday, T., Robinson, P. T., and Biehler, S., 1972, Crustal spreading in southern California: Science, v. 178, p. 15–24.

EMSLAB Group, 1988, The EMSLAB electromagnetic sounding experiment: EOS Transactions of the American Geophysical Union, v. 69, p. 89–99.

Ergas, R. A., and Jackson, D. D., 1981, Spatial variation of the crustal seismic velocities in southern California: Bulletin of the Seismological Society of America, v. 71, p. 671–689.

Etter, S. D., Fritz, D. M., Gucwa, P. R., Jordan, M. A., Kleist, J. R., Lehman, D. H., Raney, J. A., Worral, D. M., and Maxwell, J. C., 1981, Geologic cross sections, northern California coast ranges to northern Sierra Nevada, and Lake Pillsbury area to southern Klamath Mountains: Geological Society of America Map and Chart Series MC-28N, scale 1:250,000, 1 sheet and 8 p. text.

Feng, R., and McEvilly, T. V., 1983, Interpretation of seismic reflection profiling data for the structure of the San Andreas fault zone: Bulletin of the Seismological Society of America, v. 73, p. 1701–1720.

Finn, C., and Williams, D. L., 1987, An aeromagnetic study of Mount St. Helens: Journal of Geophysical Research, v. 92, p. 10194–10206.

Fox, K. F., Rinehart, C. D., and Engels, J. C., 1977, Plutonism and orogeny in north-central Washington; Timing and regional context: U.S. Geological Survey Professional paper 989, 27 p.

Fox, K. F., Fleck, R. J., Curtis, G. H., and Meyer, C. E., 1985, Implications of northwestwardly younger age volcanic rocks of west-central California: Geological Society of America Bulletin, v. 96, p. 647–654.

Frost, E. G., and Okaya, D. A., 1985, Geometry of detachment faults in the Old Woman–Turtle–Sacramento–Chemehuevi Mountains region of SE California [abs.]: EOS Transactions of the American Geophysical Union, v. 66, p. 978.

Fuis, G. S., 1981, Crustal structure of the Mojave Desert, California, *in* Howard, K. A., Carr, M. D., and Miller, D. M., eds., Tectonic framework of the Mojave and Sonoran Deserts, California and Arizona: U.S. Geological Survey Open-File Report 81-503, p. 36–38.

Fuis, G. S., Mooney, W. D., Healy, J. H., McMechan, G. A., and Lutter, W. J., 1982, Crustal structure of the Imperial Valley region, *in* The Imperial Valley, California, earthquake of October, 15, 1979: U.S. Geological Survey Professional Paper 1254, p. 25–49.

——, 1984, A seismic refraction survey of the Imperial Valley region, California: Journal of Geophysical Research, v. 89, p. 1168–1189.

Fuis, G. S., Zucca, J. J., Mooney, W. D., and Milkereit, B., 1986, A geological interpretation of seismic-refraction results in northeastern California: Geological Society of America Bulletin, v. 98, p. 53–65.

Furlong, K. P., 1984, Lithospheric behavior with triple junction migration; An example based on the Mendocino triple junction: Physics of Earth and Planetary Interiors, v. 36, p. 213–223.

Furlong, K. P., and Fountain, D. M., 1986, Continental crustal underplating; Thermal considerations and seismic-petrologic consequences: Journal of Geophysical Research, v. 91, p. 8285–8294.

Glover, D. W., 1985, Crustal structure of the Columbia Basin, Washington, from borehole and refraction data [M.S. thesis]: Seattle, University of Washington, 71 p.

Gower, H. D., Yount, J. C., and Crosson, R. S., 1985, Seismotectonic map of the Puget Sound region, Washington: U.S. Geological Survey Map I-1613, scale 1:250,000.

Griscom, A., 1977, Aeromagnetic and gravity interpretation of the Trinity ophiolite complex, northern California: Geological Society of America Abstracts with Programs, v. 9, p. 426–427.

Hadley, D., and Kanamori, H., 1977, Seismic structure of the Transverse Ranges, California: Geological Society of America Bulletin, v. 88, p. 1469–1478.

Hagstrum, J. T., McWilliams, M., Howell, D. G., and Gromme, S., 1985, Mesozoic paleomagnetism and northward translation of the Baja California Peninsula: Geological Society of America Bulletin, v. 96, p. 1077–1090.

Hall, C. A., 1981, Evolution of the western Transverse Ranges microplate; late Cenozoic faulting and basinal development *in* Ernst, W. G., ed., The geotectonic development of California: Englewood Cliffs, New Jersey, Prentice-Hall, p. 557–582.

Hamilton, R. M., 1970, Time-term analysis of explosion data from the vicinity of the Borrego Mts., California, earthquake of 9 April 1968: Bulletin of the

Seismological Society of America, v. 60, p. 367–381.

Hamilton, R. M., Ryall, A., and Berg, E., 1964, Crustal structure southwest of the San Andreas fault from quarry blasts: Bulletin of the Seismological Society of America, v. 54, p. 67–77.

Hamilton, W., 1969, Mesozoic California and the underflow of Pacific mantle: Geological Society of America Bulletin, v. 80, p. 2409–2430.

Hamilton, W., and Myers, W. B., 1966, Cenozoic tectonics of the western United States: Review of Geophysics, v. 4, p. 509–550.

Healy, J. H., 1963, Crustal structure along the coast of California from seismic-refraction measurements: Journal of Geophysical Research, v. 68, p. 5777–5787.

Healy, J. H., and Peake, L. G., 1975, Seismic velocity structure along a section of the San Andreas fault near Bear Valley, California: Bulletin of the Seismological Society of America, v. 65, p. 1177–1197.

Hearn, B. C., Donnelly-Nolan, J. M., and Goff, F. E., 1981, The Clear Lake Volcanics; Tectonic setting and magma sources, *in* McLaughlin, R. J., and Donnelly-Nolan, J. M., eds., Research in The Geysers–Clear Lake geothermal area, northern California: U.S. Geological Survey Professional Paper 1141, p. 24–25.

Hearn, T. M., 1984, P_n travel times in southern California: Journal of Geophysical Research, v. 89, p. 1843–1855.

Hearn, T. M., and Clayton, R. W., 1986a, Lateral velocity variations in southern California; 1, Results for the upper crust from Pg-waves: Bulletin of the Seismological Society of America, v. 76, p. 495–509.

—— , 1986b, Lateral velocity variations in southern California; 2, Results for the lower crust from P_n-waves: Bulletin of the Seismological Society of America, v. 76, p. 511–520.

Hermance, J. F., 1983, The Long Valley/Mono Basin volcanic complex in eastern California; Status of present knowledge and future research needs: Reviews of Geophysics and Space Physics, v. 21, p. 1545–1565.

Hill, D. P., 1972, Crustal and upper mantle structure of the Columbia Plateau from long range seismic-refraction measurements: Geological Society of America Bulletin, v. 83, p. 1639–1648.

—— , 1976, Structure of Long Valley caldera from seismic refraction experiments: Journal of Geophysical Research, v. 81, p. 745–753.

—— , 1978, Seismic evidence for the structure and Cenozoic tectonics of the Pacific Coast states, *in* Smith, R. B., and Eaton, G. P., eds., Cenozoic tectonics and regional geophysics of the western Cordillera: Geological Society of America Memoir 152, p. 145–174.

—— , 1982, Contemporary block tectonics; California and Nevada: Journal of Geophysical Research, v. 87, p. 5433–5440.

Hill, D. P., Bailey, R. A., and Ryall, A. S., 1985a, Active tectonic and magmatic processes beneath Long Valley Caldera, eastern California; An overview: Journal of Geophysical Research, v. 90, p. 11111–11120.

Hill, D. P., Kissling, E., Luetgert, J. H., and Kradolfer, U., 1985b, Constraints on the upper crustal structure of the Long Valley–Nono Craters volcanic complex, eastern California, from seismic refraction measurements: Journal of Geophysical Research, v. 90, p. 11135–11150.

Holbrook, W. S., and Mooney, W. D., 1987, The crustal structure of the axis of the Great Valley, California, from seismic refraction measurements: Tectonophysics, v. 140, p. 49–63.

Hooper, P. R., 1982, The Columbia River basalts: Science, v. 215, p. 1463–1468.

Howell, D. G., and Vedder, J. G., 1981, Structural implications of stratigraphic discontinuities across the southern California Borderland, *in* Ernst, W. G., ed., The geotectonic development of California: Englewood Cliffs, New Jersey, Prentice-Hall, p. 535–558.

Howell, D. G., and 7 others, 1986, Southern California to Socorro, New Mexico: Boulder, Colorado, Geological Society of America Continent/Ocean Transect C-3, scale 1:500,000.

Humphreys, E., Clayton, R. W., and Hager, B. H., 1984, A tomographic image of mantle structure beneath southern California: Geophysics Research Letters, v. 11, p. 625–627.

Hwang, L. J., and Mooney, W. D., 1986, Velocity and Q structure of the Great Valley, California, based on synthetic seismogram modeling of seismic refraction data: Bulletin of the Seismological Society of America, v. 76, p. 1053–1067.

Irwin, W. P., 1964, Late Mesozoic orogenies in the ultramafic belts of northwestern California and southwestern Oregon, *in* Geological Survey Research, 1964: U.S. Geological Survey Professional Paper 501-C, p. C1–C9.

—— , 1981, Tectonic accretion of the Klamath Mountains, *in* Ernst, W. G., ed., The geotectonic development of California: Englewood Cliffs, New Jersey, Prentice-Hall, p. 29–49.

Isherwood, W. F., 1976, Gravity and magnetic studies of The Geysers–Clear Lake geothermal region: Proceedings, United Nations Symposium on the Development of Geothermal Resources, v. 2, p. 1065–1073.

—— , 1981, Geophysical overview of The Geysers, *in* McLaughlin, R. J., and Donnelly-Nolan, J. M., ed., Research in The Geysers–Clear Lake geothermal area, northerm California: U.S. Geological Survey Professional Paper 1141, p. 83–96.

Iyer, H. M., Oppenheimer, D. H., and Hitchcock, T., 1979, Abnormal P-wave delays in The Geysers–Clear Lake geothermal area, California: Science, v. 204, p. 495–497.

Jachens, R. C., and Griscom, A., 1983, Three-dimension geometry of the Gorda plate beneath northern California: Journal of Geophysical Research, v. 88, p. 9375–9392.

Johnson, C. E., and Hill, D. P., 1982, Seismicity of the Imperial Valley, *in* The Imperial Valley, California, earthquake of October 15, 1979: U.S. Geological Survey Professional Paper 1254, p. 15–24.

Kane, M. F., Mabey, D. R., and Brace, R., 1976, A gravity and magnetic investigation of the Long Valley caldera, Mono County, California: Journal of Geophysical Research, v. 81, p. 754–762.

Keller, B., and Prothero, W., 1987, Western Transverse Ranges crustal structure: Journal of Geophysical Research, v. 92, p. 7890–7906.

Kissling, E., Cockerham, R. C., and Ellsworth, W. L., 1984, Three-dimensional structure of the Long Valley caldera, California, region by geotomography, *in* Hill, D. P., ed., Proceedings of Workshop 191, Active tectonic and magmatic processes beneath Long Valley Caldera, eastern California: U.S. Geological Survey Open-File Report 84-939, p. 188–220.

Kohler, W. M., Healy, J. H., and Wegener, S. S., 1982, Upper crustal structure of the Mount Hood, Oregon, region revealed by time-term analysis: Journal of Geophysical Research, v. 87, p. 339–355.

Lachenbruch, A. H., and Sass, J. H., 1980, Heat flow and energetics of the San Andreas fault zone: Journal of Geophysical Research, v. 85, p. 6185–6223.

Lachenbruch, A. H., Sass, J. H., and Galanis, S. P., Jr., 1985, Heat flow in southernmost California and the origin of the Salton Trough: Journal of Geophysical Research, v. 90, p. 6709–6736.

La Fehr, T. R., 1965, Gravity, isostasy, and crustal structure in the southern Cascade Range: Journal of Geophysical Research, v. 70, p. 5581–5597.

Langston, C. A., 1977, Corvaillis, Oregon, crustal and upper mantle receiver structure from teleseismic P and S waves: Bulletin of the Seismological Society of America, v. 67, p. 713–725.

—— , 1981, Evidence for the subducting lithosphere under southern Vancouver Island and western Oregon from teleseismic P wave conversions: Journal of Geophysical Research, v. 86, p. 3857–3866.

Larson, R. L., Menard, H. W., and Smith, S. M., 1968, Gulf of California; A result of ocean-floor spreading and transform faulting: Science, v. 161, p. 781–784.

Lawrence, R. D., 1976, Strike-slip faulting terminates the Basin and Range province in Oregon: Geological Society of America Bulletin, v. 87, p. 486–850.

Leaver, D. S., and Weaver, C. S., 1980, Refraction studies of the Mt. St. Helens region [abs.]: EOS Transactions of the American Geophysical Union, v. 62, p. 62.

Leaver, D. S., Mooney, W. D., and Kohler, W. M., 1984, A seismic refraction survey of the Oregon Cascades: Journal of Geophysical Research, v. 89, p. 3121–3134.

Luedke, R. G., and Smith, R. L., 1982, Map showing distribution, composition, and age of late Cenozoic volcanic centers in Oregon and Washington: U.S. Geological Survey Geological Investigations Map I–1091–D, scale 1:100,000.

Luetgert, J. H., and Mooney, W. D., 1985, Earthquake profiles from the Mammoth Lakes, California, earthquake swarm, January 1983; Implications for deep crustal structure: Bulletin of the Seismological Society of America, v. 75, p. 211–221.

Macdonald, G. A., 1966, Geology of the Cascade Range and Modoc Plateau, *in* Bailey, E. H., ed., Geology of northern California: California Division of Mines and Geology Bulletin 190, p. 65–96.

Macgregor-Scott, N., and Walter, A., 1988, Crustal velocities near Coalinga, California, modeled from a combined earthquake/explosion refraction profile: Bulletin of the Seismological Society America, v. 78, p. 1475–1490.

Magill, J. R., Cox, A. V., and Duncan, R., 1981, Tillamook volcanic series; Further evidence for tectonic rotation of the Oregon Coast Range: Journal of Geophysical Research, v. 86, p. 2953–2970.

Mavko, B. B., and Thompson, G. A., 1983, Crustal and upper mantle structure of the northern and central Sierra Nevada: Journal of Geophysical Research, v. 88, p. 5874–5892.

McBirney, A. R., 1978, Volcanic evolution of the Cascade Range: Earth and Planetary Sciences Annual Review, v. 6, p. 437–456.

McCarthy, J., Fuis, G., and Wilson, J., 1987, Crustal structure of the Wipple Mountains, southeastern California; A refraction study across a region of large continental extension: Geophysical Journal of the Royal Astronomical Society, v. 89, p. 119–124.

McCulloch, D. S., 1987, Regional geology and hydrocarbon potential of the offshore central California: *in* Scholl, D. S., Grantz, A., and Vedder, J. G., ed., Geology and resource potential of the continental margin of western North America and adjacent ocean basins; Beaufort Sea to Baja California: Circum-Pacific Council for Energy and Mineral Resources Earth Science Series, v. 6, p. 353–401.

McKee, E. H., Duffield, W. A., and Stern, R. J., 1983, Late Miocene and early Pliocene basaltic rocks and and their implications for crustal structure, northeastern California and south-central Oregon: Geological Society of America Bulletin, v. 94, p. 292–304.

Meltzer, A. S., Levander, A. R., and Mooney, W. D., 1987, Upper crustal structure, Livermore Valley and vicinity, California Coast Ranges: Bulletin of the Seismological Society of America, v. 77, p. 1655–1673.

Misch, P., 1966, Tectonic evolution of the northern Cascades of Washington State; A west-Cordilleran case history, *in* Gunning, H. C., ed., Special report: Canadian Institute of Mining and Metallurgy, v. 8, p. 101–148.

Mooney, W. D., and Colburn, R., 1985, A seismic refraction profile across the San Andreas, Sargent, and Calaveras faults, west-central California: Bulletin of the Seismological Society of America, v. 75, p. 175–191.

Mooney, W. D., and Ginzburg, A., 1986, Seismic measurements of the internal properties of fault zones, *in* Wang, C. Y., ed., The internal properties of fault zones: Pageoph, v. 124, p. 141–157.

Mooney, W. D., Andrews, M. C., Ginzburg, A., Peters, D. A., and Hamilton, R. M., 1983, Crustal structure of the northern Mississippi Embayment and a comparison with other continental rift zones: Tectonophysics, v. 94, p. 327–348.

Moore, D. G., and Buffington, E. C., 1968, Transform faulting and growth of the Gulf of California since the Late Pliocene: Science, v. 161, p. 1238–1241.

Moores, E. M., and Day, H. W., 1984, Overthrust model for the Sierra Nevada: Geology, v. 12, p. 416–419.

Morris, R. S., and Frost, E. G., 1985, Geometry of Mid-Tertiary detachment faulting overprinted on the Chocolate Mountains thrust system from seismic-reflection profiles in the Milpitas Wash region of southeastern California [abs.]: EOS Transactions of the American Geophysical Union, v. 66, p. 978.

Nava, F. A., and Brune, J. N., 1982, An earthquake-explosion reversed refraction line in the Peninsular Ranges of southern California and Baja California Norte: Bulletin of the Seismological Society of America, v. 72, p. 1195–1206.

Nelson, C. A., 1981, Basin and Range Province: *in* Ernst, W. G., ed., The geotectonic development of California: Englewood Cliffs, New Jersey, Prentice-Hall, p. 202–216.

Nelson, K. D., Zhu, T. F., Gibbs, A., Harris, R., Oliver, J. E., Kaufman, S., Brown, L., and Schweikert, R. A., 1986, COCORP deep seismic reflection profiling in the northern Sierra Nevada, California: Tectonics, v. 5, p. 321–333.

Okaya, D. A., McEvilly, T. V., and Frost, E. G., 1985, CALCRUST reflection profiling in the Mojava-Sonoran Desert [abs.]: EOS Transactions of the American Geophysical Union, v. 66, p. 978.

Oliver, H. W., 1977, Gravity and magnetic investigations of the Sierra Nevada batholith, California: Geological Society of America Bulletin, v. 88, p. 445–461.

Oliver, H. W., Chapman, R. H., Biehler, S., Robbins, S. L., Hanna, W. F., Griscom, A., Beyer, L. A., and Silver, E. A., 1980, Gravity map of California and its continental margin: California Division of Mines and Geology, scale 1:750,000.

Oppenheimer, D. H., and Eaton, J. P., 1984, Moho orientation beneath Central California from regional earthquakes travel times: Journal of Geophysical Research, v. 89, p. 10267–10282.

Oppenheimer, D. H., and Herkenoff, K. E., 1981, Velocity-density properties of the lithosphere from three-dimensional modeling at The Geysers–Clear Lake region, California: Journal of Geophysical Research, v. 86, p. 6057–6065.

Owens, T. J., Crosson, R. S., and Hendrickson, M. A., 1988, Constraints on the subduction geometry beneath western Washington from broadband teleseismic waveform modeling: Bulletin of the Seismological Society America, v. 78, p. 1319–1334.

Page, B. M., 1981, The southern Coast Ranges, *in* Ernst, W. G., ed., The geotectonic development of California: Englewood Cliffs, New Jersey, Prentice-Hall, p. 329–417.

Page, B. M., Wagner, H. C., McMulloch, D. S., Silver, E. A., Spotts, J. H., 1979, Geologic cross-section of the continental margin off San Luis Obispo, the southern Coast Ranges, and the San Joaquin Valley, California: Geological Society of America Map and Chart Series MC 28G, 1 sheet, with text, scale 1:250,000.

Pakiser, L. C., 1976, Seismic exploration of Mono Basin, California: Journal of Geophysical Research, v. 81, p. 3607–3618.

Pakiser, L. C., and Brune, J. N., 1980, Seismic models of the root of the Sierra Nevada: Science, v. 210, p. 1088–1094.

Plawman, T. L., 1978, Crustal structure of the continental borderland and adjacent portion of Baja California between latitudes 30°N and 30°N [M.S. thesis]: Corvallis, Oregon State University, 72 p.

Potter, C. J., Sanford, W. E., Yoos, T. R., Prussen, E. L., Keach, W., Oliver, J. E., Kaufman, S., and Brown, L. D., 1986, COCORP deep seismic reflection traverse of the interior of the North Americaan cordillera, Washington and Idaho; Implications for orogenic evolution: Tectonics, v. 5, p. 1007–1025.

Price, R. A., 1981, The Cordilleran foreland thrust and fold belt in the southern Canadian Rocky Mountains, *in* McClay, K. R., and Price, R. A., eds., Thrust and nappe tectonics: Geological Society of London Special Publication, p. 427–488.

Prodehl, C., 1979, Crustal structure of the western United States: U.S. Geological Survey Professional Paper 1034, 74 p.

Raikes, S. A., 1976, The azimuthal variation of teleseismic P-wave residuals for stations in southern California: Earth and Planetary Sciences Letters, v. 29, p. 367–372.

——, 1980, Regional variations in upper mantle structure beneath southern California: Geophysical Journal of the Royal Astronomical Society, v. 63, p. 187–216.

Raikes, S. A., and Hadley, D., 1979, The azimuthal variation of teleseismic P-residuals in southern California; Implications for upper-mantle structure: Tectonophysics, v. 56, p. 89–96.

Riddihough, R. P., 1977, A model for recent plate interactions off Canada's west coast: Canadian Journal Earth Sciences, v. 14, p. 384–396.

Rohay, A., 1982, Crust and mantle structure of the North Cascades Range, Washington [Ph.D. thesis]: Seattle, University of Washington, 163 p.

Ross, D. C., and McCulloch, D. S., 1979, Cross section of the southern Coast

Ranges and San Joaquin Valley from off-shore of Point Sur to Madera, California: Geological Society American Map and Chart Series MC-28H, scale 1:250,000.

Ryall, A., and Ryall, F., 1981, Attenuation of P and S waves in a magma chamber in Long Valley caldera, California: Geophysical Research Letters, v. 8, p. 557–560.

Saleeby, J. C., 1981, Ocean floor accretion and volcano-plutonic arc evolution of the Mesozoic Sierra Nevada, *in* Ernst, W. G., ed., The geotectonic development of California: Englewood Cliffs, New Jersey, Prentice-Hall, p. 130–181.

Saleeby, J. B., and 13 others, 1986, Central California offshore to Colorado Plateau: Boulder, Colorado, Geological Society of America Continent/Ocean Transect C-2, scale 1:500,000.

Sanders, C. O., 1984, Location and configuration of magma bodies beneath Long Valley, California, determined for anomalous earthquake signals: Journal of Geophysical Research, v. 89, p. 8287–8302.

Sanford, W. E., Potter, C. J., and Oliver, J. E., 1988, Detailed three-dimensional structure of the deep crust based on COCORP data in the Cordillera interior, north-central Washington: Geological Society America Bulletin, v. 100, p. 60–71.

Scandone, R., and Malone, S. D., 1985, Magma supply, magma discharge, and readjustment of the feeding system of Mount St. Helens during 1980: Journal of Volcanology and Geothermal Research, v. 22, p. 239–262.

Shearer, P. M., and Oppenheimer, D. H., 1982, A dipping moho and crustal low-velocity zone from P_n arrivals at The Geysers–Clear Lake, California: Bulletin of the Seismological Society of America, v. 72, p. 1551–1566.

Sheffels, B., and McNutt, M., 1986, Role of subsurface loads and regional compensation in the isostatic balance of the Transverse Ranges, California; Evidence for intracontinental subduction: Journal of Geophysical Research, v. 91, p. 6419–6431.

Shemeta, J. E., and Weaver, C. S., 1986, Seismicity accompanying the May 18, 1980, eruption of Mount St. Helens, Washington, *in* Keller, S.A.C., ed., Mount St. Helens; Five Years Later: Cheney, Eastern Washington University Press, p. 44–58.

Shor, G. G., Jr., and Raitt, R. W., 1958, Seismic studies in the southern California Borderland, *in* Geofisica Aplicada, Tomo Z: 20th International Geologic Congress, Mexico, D.F., 1956 (Trabajos), Section 9, p. 243–259.

Shor, G. G., Dehlinger, P., Kirk, H. K., and French, W. S., 1968, Seismic refraction studies off Oregon and northern California: Journal of Geophysical Research, v. 73, p. 2175–2194.

Simpson, R. W., and Cox, A., 1977, Paleomagnetic evidence for tectonic rotations of the Oregon Coast Range: Geology, v. 5, p. 585–589.

Smith, D. R., and Leeman, W. P., 1987, Petrogenesis of Mount St. Helens dacite magmas: Journal of Geophysical Research, v. 92, p. 10313–10334.

Smith, G. A., and Taylor, E. M., 1983, The central Oregon High Cascade graben; What?, where?, and when?: Transactions of the Geothermal Resource Council, v. 7, p. 275–279.

Smith, S. W., and Knapp, J. S., 1980, The northern termination of the San Andreas fault, *in* Streitz, R., and Sherburne, R., eds., Studies of the San Andreas Fault Zone in northern California: California Division of Mines and Geology Report 140, p. 153–164.

Snavely, P. D., Jr., 1987, Tertiary geologic framework, neotectonics, and petroleum potential of the Oregon–Washington continental margin, *in* Scholl, D. S., Grantz, A., and Vedder, J. G., eds., Geology and resource potential of the continental margin of western North America and adjacent ocean basins; Beaufort Sea to Baja California: Circum-Pacific Council for Energy and Mineral Resources Earth Science Series, v. 6, p. 305–335.

Speed, R. C., and Moores, E. M., 1981, Geologic cross section of the Sierra Nevada and the Great Basin along 40°N lat., northeastern California and northern Nevada: Geological Society of America Map and Chart Series MC-28L, 2 sheets and 12 p. text, scale 1:250,000.

Spence, G. D., Clowes, R. M., and Ellis, R. M., 1985, Seismic structure across the active subduction zone of western Canada: Journal of Geophysical Research, v. 90, p. 6754–6772.

Spieth, M. A., Hill, D. P., and Geller, R. J., 1981, Crustal structure in the northwestern foothills of the Sierra Nevada from seismic refraction experiments: Bulletin of the Seismological Society of America, v. 71, p. 1075–1087.

Stanley, W. D., 1984, Tectonic study of the Cascade Range and Columbia Plateau in Washington State based upon magnetotelluric soundings: Journal of Geophysical Research, v. 89, p. 4447–4460.

Stanley, W. D., Finn, C., and Plesha, J. L., 1987, Tectonics and conductivity structures in the southern Washington cascades: Journal of Geophysical Research, v. 92, p. 10179–10193.

Stauber, D. A., and Berge, P. A., 1985, Comparison the P-velocity structures of Mt. Shasta, California, and Newberry Volcano, Oregon [abs.]: EOS Transactions of the American Geophysical Union, v. 66, p. 25.

Steeples, D. W., and Iyer, H. M., 1976, Low-velocity zone under Long Valley as determined from teleseismic events: Journal of Geophysical Research, v. 81, p. 849–860.

Stewart, S. W., 1968, Preliminary comparison of seismic travel tome and inferred crustal structure adjacent to the San Andreas fault in the Diablo and Gabilan Ranges of central California, *in* Dickinson, W. R., and Grantz, A., eds., Geological problems of San Andreas Fault System Conference Proceedings: Stanford University Publications in Geological Sciences, v. 11, p. 218–230.

Suppe, J., 1979, Cross section of southern part of northern Coast Ranges and Sacramento Valley, California: Geological Society of America Map and Chart Series MC-28B, 1 sheet and 8-p. text, scale 1:250,000.

Swanson, D. A., Wright, T. L., and Helz, R. T., 1975, Linear vent systems and estimated rates of magma production and eruption for the Yakima basalt on the Columbia Plateau: American Journal of Science, v. 275, p. 877–905.

Taber, J. J., Jr., 1983, Crustal structure and seismicity of the Washington continental margin [Ph.D. thesis]: Seattle, University of Washington, 159 p.

Taber, J. J., and Lewis, B.T.R., 1986, Crustal structure of the Washington continental margin from refraction data: Bulletin of the Seismological Society of America, v. 76, p. 1011–1024.

Taber, J. J., Jr., and Smith, S. W., 1985, Seismicity and focal mechanisms associated with the subduction of the Juan de Fuca plate beneath the Olympic Peninsula, Washington: Bulletin of the Seismological Society of America, v. 75, p. 237–249.

Tabor, R. W., 1972, Age of Olympic metamorphism, Washington; K-Ar dating of low-grade metamorphic rocks: Geological Society of America Bulletin, v. 83, p. 1805–1816.

Taylor, S. R., and Scheimer, J. F., 1982, P-velocity models and earthquake locations in the Livermore Valley region, California: Bulletin of the Seismological Society of America, v. 72, p. 1255–1275.

Thatcher, W., and Brune, J. N., 1973, Surface waves and crustal structure in the Gulf of California region: Bulletin of the Seismological Society of America, v. 63, p. 1689–1698.

Trehu, A. M., and Wheeler, W. H., 1986, Possible evidence for subducted sedimentary materials beneath central California: Geology, v. 15, p. 254–258.

Vedder, J. G., 1987, Regional geology and petroleum potential of offshore southern California Borderland, *in* Scholl, D. S., Grantz, A., and Vedder, J. G., eds., Geology and resource potential of the continental margin of western North America and adjacent ocean basins; Beaufort Sea to Baja California: Circum-Pacific Council for Energy and Mineral Resources Earth Science Series, v. 6, p. 403–447.

Vetter, U., and Minster, J., 1981, P_n velocity anisotropy in Southern California: Bulletin of the Seismological Society of America, v. 71, p. 1511–1530.

Walck, M. C., and Minster, J. B., 1982, Relative array analysis of upper mantle lateral velocity variations in Southern California: Journal of Geophysical Research, v. 87, p. 1757–1772.

Walter, A. W., 1984, Velocity structure near Coalinga, California, *in* Rymer, M. J., and Ellsworth, W. L., eds., Mechanics of the May 2, 1983, Coalinga, California, earthquake: U.S. Geological Survey Open-File Report 85-44, p. 10–18.

Walter, A. W., and Mooney, W. D., 1982, Crustal structure of the Diablo and Gabilan Ranges, central California; A reinterpretation of existing data: Bulletin of the Seismological Society of America, v. 72, p. 1567–1590.

Walter, A. W., and Sharpless, S., 1987, Crustal velocity structure of the Sur-Obispo (Franciscan) Terrane between San Simeon and Santa Maria, California [abs.]: EOS Transactions of the American Geophysical Union, v. 68, p. 1366.

Walter, S. R., 1986, Intermediate-focus earthquakes associated with Gorda Plate subduction in Northern California: Bulletin of the Seismological Society of America, v. 76, p. 583–588.

Warren, D. H., 1981, Seismic-refraction measurements of crustal structure near Santa Rosa and Ukiah, California: U.S. Geological Survey Professional Paper 1141, p. 167–181.

Weaver, C. S., and Baker, G. E., 1988, Geometry of the Juan de Fuca plate beneath Washington and northern Oregon from seismicity: Bulletin of the Seismological Society of America, v. 78, p. 264–275.

Weaver, C. S., and Malone, S. D., 1987, Overview of the tectonic setting and recent studies of eruptions of Mount St. Helens, Washington: Journal of Geophysical Research, v. 92, p. 10149–10155.

Weldon, R., and Humphreys, E., 1986, A kinematic model of southern California: Tectonics, v. 5, p. 33–48.

Wentworth, C. M., Walter, A. M., Barton, J. A., and Zoback, M. D., 1983, Evidence on the tectonic setting of the 1983 Coalinga earthquakes from deep reflection and refraction profiles across the southeastern end of Kettleman Hills, *in* Bennett, J. H., and Sherburne, R. W., eds., The 1983 Coalinga, California, earthquake: California Division of Mines and Geology Special Publication 66, p. 113–126.

Wentworth, C. M., Blake, M. C., Jones, D. C., Walter, A. W., Zoback, M. D., 1984, Tectonic wedging associated with emplacement of the Fransciscan assemblage, California Coast Ranges, *in* Blake, M. C., Jr., ed., Franciscan geology of northern California: Society of Economic Paleontologists and Mineralogists, Pacific Section, v. 43, p. 163–173.

Wernicke, B., Spencer, J. E., Burchfiel, B. C., and Guth, P. L., 1982, Magnitude of extension in the southern Great Basin: Geology, v. 10, p. 489–502.

Whitman, D., Walter, A. W., Mooney, W. D., 1985, Crustal structure of the Great Valley, California; Cross profile [abs.]: EOS Transactions of the American Geophysical Union, v. 66, p. 973.

Williams, D. L., and Finn, C. A., 1985, Analysis of gravity data in volcanic terrain and gravity anomalies and subvolcanic intrusions in the Cascade Range and at other selected volcanoes, *in* Hinze, W. J., ed., The utility of regional gravity and magnetic anomaly maps: Tulsa, Oklahoma, Society of Exploration Geophysicists, p. 361–374.

Williams, D. L., Hull, D. A., Ackermann, H. D., and Beeson, M. H., 1982, The Mt. Hood region; Volcanic history, structure, and geothermal energy potential: Journal of Geophysical Research, v. 87, p. 2767–2781.

Williams, D. L., Abrams, G., Finn, C., Dzurisin, D., Johnson, D. J., and Denlinger, R., 1987, Evidence from gravity data for an intrusive complex beneath Mount St. Helens; Journal of Geophysical Research, v. 92, p. 10207–10222.

Wright, L. A., Troxel, B. W., Burchfiel, B. C., Chapman, R. H., and Labotka, T. C., 1981, Geologic cross section from the Sierra Nevada to the Las Vegas Valley, eastern California to southern Nevada: Geological Society of America Map and Chart Series MC-28M, 1 sheet and 15-p. text, scale 1:250,000.

Young, C., and Ward, R. W., 1980, Mapping seismic attenuation within geothermal systems using teleseisims with application to The Geysers–Clear Lake region: Journal of Geophysical Research, v. 85, p. 5227–5236.

Zandt, G., 1981, Seismic images of the deep structure of the San Andreas Fault system, central Coast Ranges, California: Journal of Geophysical Research, v. 86, p. 5039–5052.

Zandt, G., and Furlong, K. P., 1982, Evolution and thickness of the lithosphere beneath coastal California: Geology, v. 10, p. 376–381.

Zandt, G., Park, S. K., Ammon, C., and Oppenheimer, D., 1985, P-delays in northern California; Evidence on the southern edge of the subducting Gorda Plate [abs.]: EOS Transactions of the American Geophysical Union, v. 66, p. 958.

Zervas, C. E., and Crosson, R. S., 1986, P_n observation and interpretation in Washington: Bulletin of the Seismological Society of America, v. 76, p. 521–546.

Zoback, M. D., and Wentworth, C. M., 1986, Crustal studies in central California using an 800-channel seismic reflection recording system, *in* Baragangi, M., and Brown, L., eds., Reflection seismology; A global perspective: American Geophysical Union Geodynamic Series, v. 13, p. 183–196.

Zucca, J. J., Fuis, G. S., Milkereit, B., Mooney, W. D., and Catchings, R., 1986, Crustal structure of northern California from seismic-refraction data: Journal of Geophysical Research, v. 91, p. 7359–7382.

MANUSCRIPT ACCEPTED BY THE SOCIETY OCTOBER 31, 1988

Printed in U.S.A.

Geological Society of America
Memoir 172
1989

Chapter 10

Tectonic evolution of the Intermontane System; Basin and Range, Colorado Plateau, and High Lava Plains

R. Ernest Anderson
U.S. Geological Survey, MS 966, Box 25046, Denver Federal Center, Denver, Colorado 80225

ABSTRACT

Strong contrasts in the tectonic evolution of the three principal parts of the Intermontane System (Basin and Range, Colorado Plateaus, and High Lava Plains) are responsible for strong contrasts in crustal structure outlined elsewhere in this volume. Extremes in Cordilleran history are represented by the interior of the Colorado Plateaus and the Basin and Range province. The plateaus have enigmatically escaped strong deformation and magmatism since Precambrian time, whereas most parts of the Basin and Range have experienced repeated orogenesis, the youngest of which is continental magmatism and extensional deformation with dimensions and magnitudes that may not be exceeded anywhere in the world. Between these extremes lie the High Lava Plains, parts of which appear to be dominated by the passage of a single major pulse of magmatism and rifting that produced crust no older than late Cenozoic age. Numerous crustal-scale tectonic and magmatic events beginning in Archean time and extending into Cenozoic time are chronicled as events having a potential for molding crustal structure. They include protracted Archean and early Proterozoic south- or southwest-directed continental growth or cratonization by sparsely recorded processes of deformation, sedimentation, magmatism, and metamorphism, followed in middle Proterozoic time by epicratonal basin sedimentation and in the late Proterozoic by reshaping of the western continental margin by passive-margin rifting and foundering. Late Paleozoic and early Mesozoic continent-margin collisional events led to the obduction of the Roberts Mountains and Golconda allochthons and accretion of related terranes. A second event of continent-margin truncation and reshaping occurred in early Mesozoic time. It was followed during Mesozoic time by major east-directed subduction of oceanic lithosphere and associated accretions of volcanotectonic terranes to the west margin of the continent. Genetically related magmatism, metamorphism, and compressional tectonism reached far inboard of the continental margin; these were guided to some extent by preexisting tectonic features. During Cenozoic time, igneous activity of colossal magnitude swept through the Cordillera along regular paths without apparent regard for preexisting tectonic features. Coeval and subsequent extensional tectonism teamed with the magmatism to reshape the geophysical framework of much of the Basin and Range and High Lava Plains. Accordingly, it is a major challenge to recognize the fingerprint of ancient major crustal-scale events within the current geophysical framework.

Anderson, R. E., 1989, Tectonic evolution of the Intermontane System; Basin and Range, Colorado Plateau, and High Lava Plains, *in* Pakiser, L. C., and Mooney, W. D., Geophysical framework of the continental United States: Boulder, Colorado, Geological Society of America Memoir 172.

INTRODUCTION

The Intermontane System, as used here and outlined in Figure 1, constitutes the central part of the Cordilleran orogen from the United States–Mexico border northward to southeastern Oregon and southern Idaho. It consists of three provinces: the Colorado Plateaus, the Basin and Range, and the High Lava Plains. The Basin and Range is a structural entity consisting of at least four main physiographic sections (Fenneman, 1931). The Colorado Plateaus and High Lava Plains are physiographic entities each consisting of at least two main structural sections—the Colorado Plateaus of interior and transitional sections, and the High Lava Plains of the northeast-trending Snake River Plain and Yellowstone Plateau section and the west-northwest–trending lava plains section of southeastern Oregon (Christiansen and McKee, 1978). The United States part of the Basin and Range consists of northern and southern Basin and Range subsections with contrasting Cenozoic structural history and corresponding contrasts in physiography (Eaton, 1979).

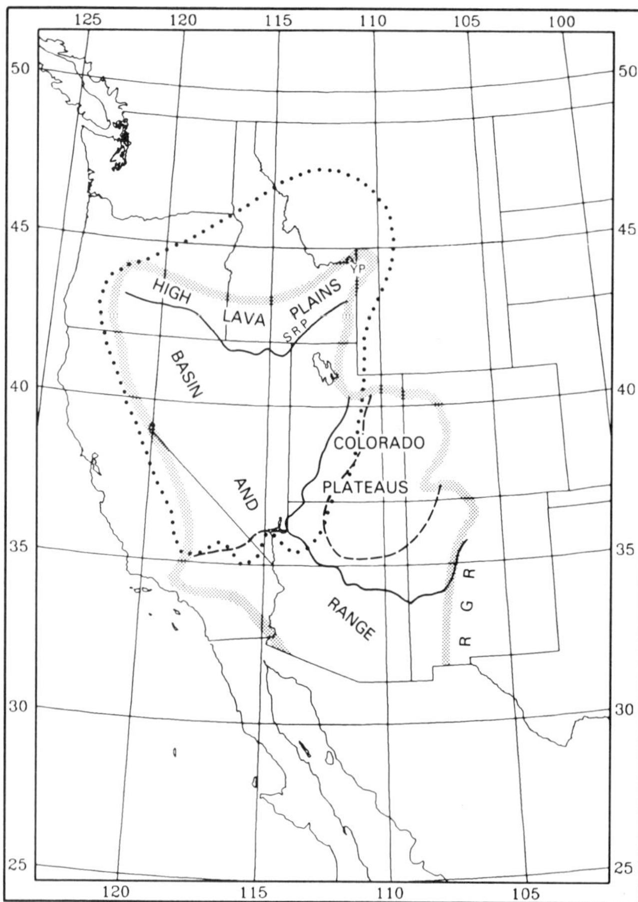

Figure 1. Outline of Intermontane System (patterned) and High Lava Plains, Basin and Range, and Colorado Plateaus provinces within that system. Dashed line in Colorado Plateaus is approximate boundary between the interior and transitional structural subsections of that province. Dashed line in Basin and Range separates northern and southern subprovinces. Dotted line is the boundary of the Cordilleran thermotectonic anomaly region from Eaton (1979).

Eaton (1979) emphasized that some of the fundamental geologic and geophysical properties of the northern Basin and Range (young faulting, high heat flow, gravitational intensity, and distribution of thermal springs) extend far beyond the physiographic boundaries of the province and thus define an enormous tectonophysical entity greater than 1×10^6 km^2 (Fig. 1) in area. Although not considered herein as part of the Intermontane System, the Rio Grande rift not only shares these properties with the northern Basin and Range, it has a similar Neogene tectonic history. In many important aspects the southern Basin and Range contrasts more with the northern Basin and Range than does the Rio Grande rift. These similarities and differences complicate characterization by blurring the boundaries of and the distinctions among the provinces. Despite this blurring, important distinctions in the intensity and form of Cenozoic tectonism and magmatism have left each of the three provinces included herein in the Intermontane System different from its neighbors, and many of their geophysical properties are of Cenozoic vintage.

The Intermontane System extends across fundamental cratonal boundaries of Archean and late Proterozoic age and embraces the effects of several episodes of pre-Cenozoic orogenesis, any or all of which could contribute to the current geophysical framework. This chapter summarizes the tectonic evolution of the Intermontane System and identifies major features, beginning with those of Archean age, that could contribute to the present-day geophysical framework.

PRE-CENOZOIC TECTONIC EVOLUTION

Archean rocks in the western United States are found mostly beyond the limits of the Intermontane System in the cores of major uplifted blocks that form the Wyoming age province (Peterman, 1979). Though details of the direction and rate of cratonization are lacking for this province, geologic and geochronologic studies point to the formation, by about 2.5 Ga, of a thick, stabilized sialic crust with a complex history of metamorphism, plutonism, and orogenesis possibly spanning 1 b.y. (Condie, 1976; Peterman, 1979). Stromatolite-bearing dolomitic marbles cut by 2.58-Ga granite in the Hartville uplift (Hofmann and Snyder, 1985) suggest that a craton-margin shelf environment may have predated Archean tectonism, metamorphism, and plutonism in part of the Wyoming age province. Highly metamorphosed Archean igneous and sedimentary rocks probably as old as 3 Ga are found in the Wasatch Mountains bordering the eastern Basin and Range and possibly at Antelope Island within the Basin and Range (Hedge and others, 1983). Known or inferred Archean rocks are widely exposed in the Basin and Range of northwestern Utah and south-central Idaho in the Raft River–Albian Mountains area (Armstrong, 1976; Compton and others, 1977). How far west from these exposures Archean rocks extend is not known, but it is possible that they formed the craton throughout much of northeastern Nevada, as indicated in Figure 2. It is therefore important to consider the possible influence of these ancient rocks on younger structures of the northern Basin and Range.

According to Karlstrom and others (1983) and Bruce Bryant (written communication, 1986), the crust in southern Wyoming and north-central Utah at the close of Archean time was deeply eroded and served as a source terrain for thick (as much as 13 km) early Proterozoic siliciclastic sediments that may have accumulated in fault-bounded troughs south of and parallel to the margin of the Archean craton. Bryant (1985) placed that margin along the present site of the Uinta Mountains (Fig. 2). South of this fundamental crustal boundary, an early Proterozoic terrane of sedimentary, volcanic, and subvolcanic intrusive rocks that had been metamorphosed to amphibolite grade about 1.8 to 1.7 Ga has been regarded as arc-related. These rocks presumably collided with the edge of the Archean craton (Hills and Houston, 1979). The early Proterozoic crust may have developed in southward-progressing accretionary steps or stages, as suggested by Stacey and Hedlund (1983) and Condie (1982) on the basis of lead-isotopic compositions. This pattern of crustal growth is in general agreement with Nd isotopic compositions of Mesozoic and Tertiary igneous rocks that suggest at least three generally southeastward-younging Precambrian age terrains (Farmer and DePaolo, 1984). Where the early Proterozoic terrain has been well studied, such as in Arizona (Anderson and Silver, 1976), two cycles of sedimentation and volcanism, each followed by deformation and metamorphism, are recognized (Burchfiel, 1979). The early Proterozoic crust may have grown by complex accretionary processes.

Early Proterozoic sedimentation at the south edge of the Archean craton was apparently followed by north-directed thrusting of the thick sedimentary prisms onto the Archean craton of the Wyoming age province (Karlstrom and others, 1983; Sears and others, 1982) and by intense metamorphism and anatectic melting (Karlstrom and others, 1983; Hedge and others, 1983). This episode of apparent control exerted by the early craton edge on sedimentation and deformation was followed by others. According to Bryant (1985), tectonism occurred along this trend in early, middle, and late Proterozoic and early Paleozoic time, as well as in latest Cretaceous and Tertiary time. The apparent tectonic endurance of such features not only makes them prime candidates for shaping the local geophysical framework but also candidates for controlling very young structures such as the segmentation of the Wasatch fault system, which is critical to earthquake-hazard assessments. They could also have exerted strong control on structural events along their westward projection into the Basin and Range (Fig. 2).

Middle Proterozoic rocks ranging in age from 1,600 to 850 m.y. old in the western United States mostly form assemblages of relatively unmetamorphosed epicratonal sedimentary and volcanic strata, such as the 20-km-thick Belt Supergroup in Idaho, Montana, and Washington. Within the area covered by this summary are the 7-km-thick Uinta Mountain Group of terrigenous sandstone, argillite, and conglomerate in northeastern Utah, the 3.7-km-thick Unkar and Chuar Groups of the Grand Canyon area, the 0.8-km-thick Troy Quartzite and Apache Group of eastern Arizona, and the 1.0- to 1.5-km-thick Crystal Spring and Beck Spring Formations of eastern California (Fig. 2). Environments of deposition range from possible crustal rifts for the thickest assemblages, through deep to shallow troughs for those of moderate thickness, to platform or shelf for the thinnest assemblages (Stewart, 1976). The Apache basin, for example, is a northwest-trending trough formed on 1.7-b.y.-old crystalline basement and filled with shallow-water marine and transitional sediments (Stevenson and Beus, 1982). Its formation is believed to have been halted by a widespread structural disturbance 820 to 770 m.y. ago that included block faulting with displacements of as much as 3.2 km (Elston and McKee, 1982). The sporadic distribution, varied orientations, association with mafic intrusive rocks, and known or inferred faulted margins of these depocenters led Stewart (1976) to postulate that they record widespread extensional tectonism unassociated with plate-margin rifting. He conceded that some of the troughs could have opened westward

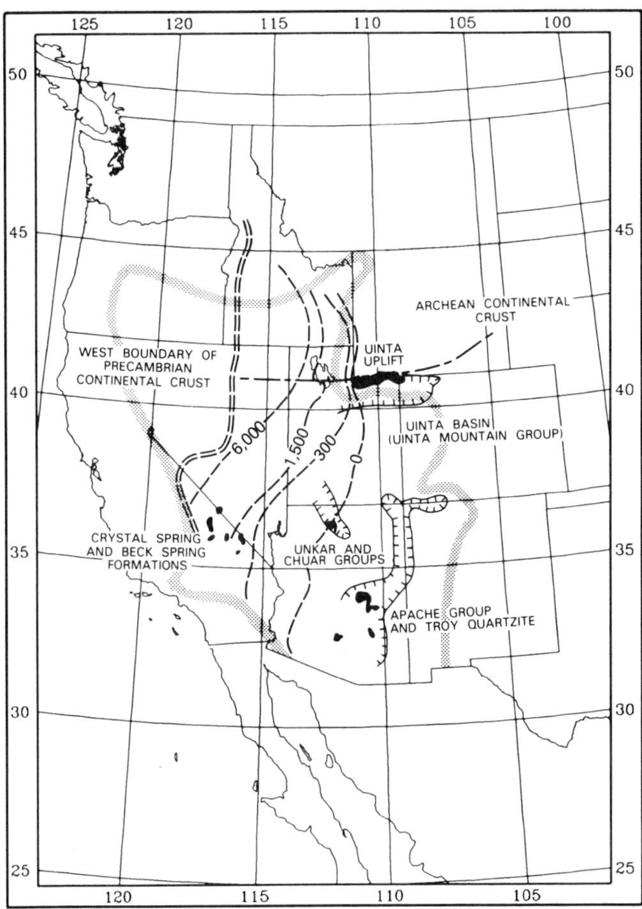

Figure 2. Map showing approximate west boundary of Precambrian continental crust (double-dashed line, from Farmer and DePaolo, 1984) and south boundary of Archean crust (short- and long-dashed line, from Bryant and Nichols, 1988) in the Intermontane System; distribution of middle Proterozoic epicratonic stratal assemblages (hachured lines, from Stewart, 1972), and isopachs of late Precambrian and Cambrian miogeoclinal rocks (dashed lines). Zero isopach is east limit of lower Cambrian strata from Stewart and Suczek (1977), other isopachs are thickness of upper Proterozoic and Cambrian strata from Stewart and Suczek (1977), Stewart (1972), and Bryant (1985).

into oceans, but such relationships remain undocumented. The great thickness of some of the assemblages suggests wholesale rifting of preexisting crust. Stewart (1972, 1976) summarized evidence indicating that some of these mid-Proterozoic assemblages were deformed prior to deposition of late Proterozoic rocks. Also, as emphasized by Burchfiel (1979), during the time of continental rifting, voluminous plutons were emplaced into Precambrian terranes extending from southeastern Canada to southeastern California. These magmas were generated by fusion of preexisting crust, a process possibly catalyzed by mantle diapirism (Anderson, 1983). They were emplaced anorogenically 1.34 to 1.49 b.y. ago, possibly during episodes of extensional tectonism associated with failed continental rifting (Anderson, 1983).

Stewart (1972) attached plate-tectonic significance to a major contrast in sedimentation pattern across an unconformity stretching for 4,000 km from Alaska and northern Canada to Mexico (Fig. 2). Although the unconformity and sedimentation patterns above it show some deflection where they cross the westward projection of the structural trough containing middle Proterozoic rocks of the Uinta Mountain Group (Bryant, 1985; Bryant and Nichols, 1988), in general the unconformity cuts across the structural grain of earlier Precambrian rocks, including structural troughs formed from the thick accumulations of middle Proterozoic strata mentioned in the previous paragraphs (M. W. Reynolds, written communication, 1986). For about 500 m.y. a westward-thickening wedge of Late Precambrian and early Paleozoic supracrustal detrital and carbonate strata accumulated above the unconformity on a slightly sinuous continental shelf in a miogeoclinal environment (Stewart, 1970). The lower part of the sedimentary sequence consists of diamictite, mudstone, argillite, sandstone, and conglomerate and intercalated volcanic rocks that may have accumulated in rift-valley basins (Fig. 3). These strata are overlain by carbonate and transitional assemblages. They were deposited under shallow-marine, intertidal, and supratidal shelf conditions and thicken westward from a few hundred meters on the craton east of the hingeline to almost 10 km about 400 km to the west (Stewart and Poole, 1974). Farther west, an assemblage of siliceous pelagic and hemipelagic strata consisting mostly of shale, sandstone, arkose, and conglomerate with minor chert, greenstone, and carbonate rock was presumably deposited on the continental slope and rise and oceanic crust (Fig. 3). This change in depositional patterns in Late Precambrian time is interpreted to have resulted from the rifting and drifting away in a passive-margin tectonic setting of a part of the original craton of unknown size (Stewart, 1972; Stewart and Poole, 1974; Burchfiel and Davis, 1975). As shown in Figure 3B, the continental borderland could have evolved by middle Paleozoic time into a subduction zone separated from the continental edge by an inner-arc basin.

That the broad area of shelf sediments accumulated in a tectonically stable environment is attested to by their remarkable lateral extent and lithologic uniformity (Stewart and Poole, 1974), as well as by the similar pattern of sedimentation and lateral retreat of the shelf-slope break in areas as distant as central Nevada and east-central California (Johnson and Murphy, 1984; Stevens, 1986). Other geologists have proposed that the main rifting event was much earlier—coinciding with formation of the mid-Proterozoic troughs. Whatever its age, the broad zone of preexisting sialic crust inboard of the rifted continental edge may have been attenuated beneath the thick prism of sediments if the dynamics of deformation were analogous to some modern rifted continental margins. Little is known of the type or distribution of basement structures produced during the hypothesized attenuation. Features such as the "ancient Ephram fault" (Moulton, 1975) and the Fillmore arch (Picha and Gibson, 1985) along the Cordilleran hingeline and uppermost Precambrian and Cambrian basalts west of the hingeline (Crittenden and others, 1971) are consistent with an extensional environment. Tectonic subsidence curves calculated by Bond and Kominz (1984) for early Paleozoic strata in the southern Canadian Rocky Mountains are consistent with the interpretation that the post-rift continental crust was thinned by thermal contraction during early development of the miogeocline. The curves resemble those of modern passive margins. The continental rifting and attenuation probably exerted strong controls on the location, trend, and form of subsequent structures such as the fold and thrust belt of southern Canada, the east limit of Mesozoic thrusting in Utah (Burchfiel and Hickcox, 1972; Picha and Gibson, 1985), and the sharp westward bend in the southern part of the Roberts Mountains allochthon in Nevada (Oldow, 1984).

Although the rifted edge of the continent is buried beneath supracrustal sediments or thrust faults, its position is known from isotopic and geologic data. Kistler and Peterman (1978) interpreted variations in initial $^{87}Sr/^{86}Sr$ ratios of Mesozoic granitic rocks in the western United States as reflecting differences in the source materials from which the granitic rocks were derived. Plutons with initial $^{87}Sr/^{86}Sr > 0.706$ are considered to have been derived largely from mafic lower continental crust of Precambrian age, whereas those with $^{87}Sr/^{86}Sr < 0.706$ are considered to have been derived from mafic crust of oceanic or transitional character that is of Paleozoic or younger age (Kistler and Peterman, 1978; Kistler, 1983). Zartman (1974) noted that variations in lead-isotopic compositions of Mesozoic and Cenozoic igneous rocks and hydrothermal ore deposits also reflect the geology of the source material from which the lead was derived. The same can be said for neodymium isotopes of plutonic rocks (Kistler, 1983). These interpretations have led to the establishment of a line of initial $^{87}Sr/^{86}Sr = 0.706$ (Fig. 2) and a lead-isotopic province boundary, both of which are in good agreement with the westernmost known distribution of Late Precambrian and lower Paleozoic passive-margin shelf and transitional strata (Ketner, 1977a, b). Oceanic crust is inferred to have been present west of the rifted continental edge, and an assemblage of siliceous strata was presumably deposited on it (Fig. 3). As noted above, it is also possible that the continental margin was not passive and that a subduction zone and associated volcanic arc existed outboard of the miogeocline during Paleozoic time. Regardless of the plate-tectonic model preferred, this well-located boundary of the

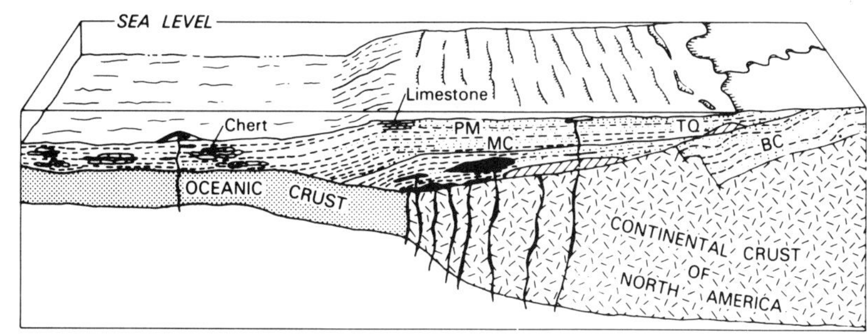

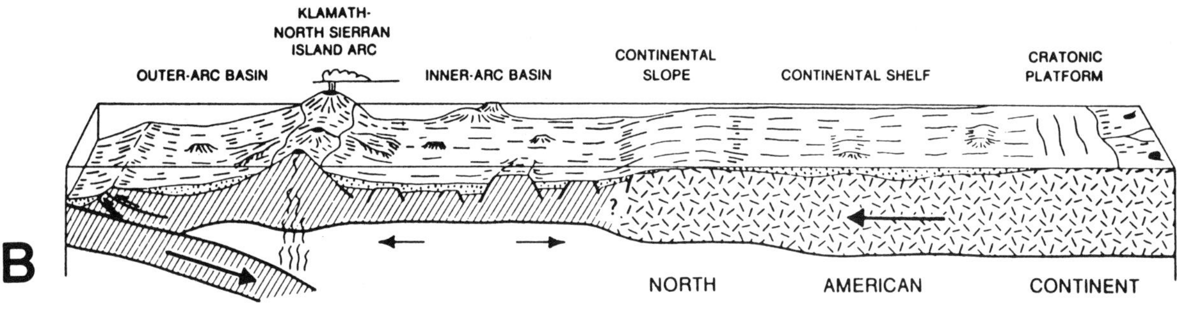

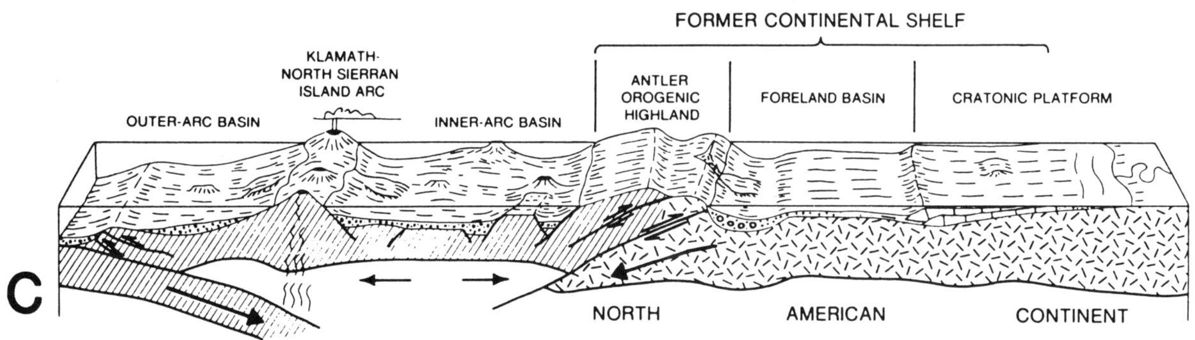

Figure 3. Cross sections. A, Diagrammatic representation of unconformity between the upper Proterozoic and Cambrian miogeoclinal sediments and volcanics (represented by the diamictite, crossruled; basalt, black; MC = McCoy Creek Group, TQ = Tintic Quartzite, PM = Prospect Mountain Quartzite) and underlying fault-bounded epicratonic sediments of middle Proterozoic age (represented by the Big Cottonwood Formation, BC) and the foundered and attenuated margin of the Precambrian craton. Note that horizontal distance is less and vertical exaggeration is greater than in B and C. From Stewart, 1972. B, Hypothetical model of relations between North American continent and island-arc and inner-arc basin systems during Silurian and Devonian time. From Poole and others, 1977. C, Hypothetical model of relations between the same tectonic elements as in B during Late Devonian and Mississippian time. From Poole and Sandberg, 1977.

crystalline continental crust appears to have had a strong influence on subsequent tectonic events as stated above and outlined in the following paragraphs.

During Late Devonian and Mississippian time, depositional patterns in the western part of the continental margin changed, from accumulation in isolated shallow basins resulting from vertical tectonics, to thick accumulations in the foreland trough of the Antler orogenic belt. As the Antler cordillera rose west of the shelf and continental rise, as much as 4.5 km of well-bedded flysch-like mudstone, siltstone, sandstone, conglomerate, and subordinate impure limestone accumulated on the preexisting shelf in a subsiding elongate foreland basin that migrated eastward as it grew (Poole, 1974; Poole and Sandberg, 1977). Sediment sources accordingly shifted from the craton on the east to the orogenic belt on the west. Evidence for the tectonic and depositional events is best preserved in western Nevada, where

Speed (1983) and Miller and others (1984) have outlined two collisional events that led to the obduction onto the western sediment-mantled continental edge of two allochthons composed of oceanic sedimentary and volcanic rocks. The obducted allochthons are the Roberts Mountains, emplaced during the Antler orogeny during early Mississippian time, and the Golconda, emplaced during the Sonoma orogeny during latest Permian and early Triassic time. The Roberts Mountains allochthon consists of a tectonic assemblage of pelagic, hemipelagic, turbiditic, and volcanic rocks of early Paleozoic age. The Golconda allochthon consists of similar rock suites that are of Mississippian to Permian age. Terrigenous clastics in these predominantly sedimentary assemblages were derived from both the volcanic arc and orogenic belt (Miller and others, 1984), and the actual volcanic arcs were not part of the obducted allochthons. Subsequent to each obduction, the arcs may have been cooled and contracted and were either subducted or now lie buried beneath thick Triassic flysch and arc volcanics (Speed, 1983). This largely buried tectonostratigraphic terrane accreted to the continent is known as Sonomia.

Emplacement of the allochthons was not accompanied by major mountain building, magmatism, or metamorphism in the adjacent continental crust or its supracrustal sediment cover. Also, the allochthons show only local shear strain and thrust duplexing and are generally devoid of oceanic lithosphere. The absence or paucity of these features led Speed (1983) to conclude that the obductions took place at a passive margin. To explain the absence in the obducted allochthon of oceanic lithosphere, Burchfiel and Davis (1975) proposed a combination of obduction of the sedimentary and volcanic assemblage and subduction of the underlying oceanic lithosphere. Also, Miller and others (1984) challenged the passive-margin characterization on the basis of stratigraphic and structural data. They suggested that short episodes of crustal shortening were separated by longer episodes of extension and/or transcurrent faulting analogous to active plate-margin deformation in the southwest Pacific.

Unlike the period following the Antler orogeny, when an oceanic paleogeography was restored west of the obducted terrain, the island arc west of the Golconda allochthon was welded to the continent as Sonomia, causing a major westward shift of the continental margin. Burchfiel and Davis (1972) inferred the newly accreted Sonomia to be thinner than normal continental crust. Juxtaposing an oceanic arc or arc-marginal crustal slab against a rifted continental edge should produce a crustal-scale geophysical signature unless subsequent cratonization or other dynamothermal events erased the contrasts in physical properties.

Late Paleozoic and early Mesozoic depositional patterns in the eastern part of the Cordilleran miogeocline and on the adjacent craton to the east were not obviously affected by the tectonic events in the Antler-Sonoman orogenic belt. Upper Paleozoic depositional patterns there were dominated by several basins and intervening northwest-trending highlands (Bissell, 1974; Stokes, 1979) that are commonly considered to be part of the ancestral Rocky Mountains. Three to 10 km of upper Paleozoic carbonate and clastic sediments accumulated in these basins, the deepest of

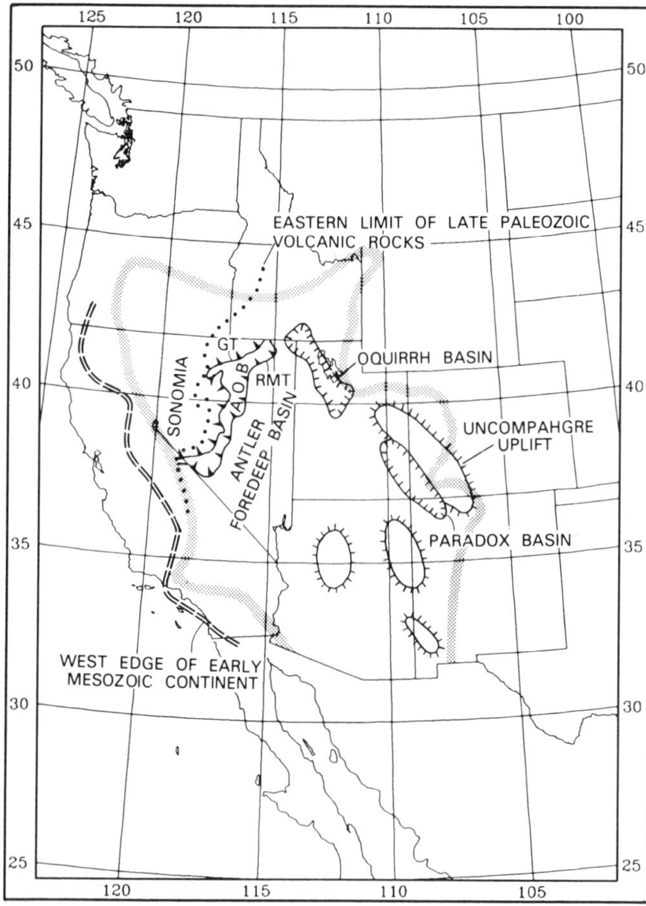

Figure 4. Map showing selected late Paleozoic and early Mesozoic tectonic features. Outlines of Pennsylvanian and early Permian Oquirrh basin from Jordan and Douglass (1980). Outline of Uncompahgre uplift, Paradox basin and other uplifts to the south from Kluth and Coney (1981). Trace of Roberts Mountain thrust (RMT), and Golconda Thrust (GT) from Schweickert and Lahren, 1987. Lines showing eastern limit of upper Paleozoic volcanic rocks and western edge of early Mesozoic continent from Burchfiel and Davis (1975).

which is the northwest-trending Oquirrh basin in northwestern Utah (Fig. 4). The extent to which the Oquirrh basin may be fault-controlled is unknown. On the basis of stratigraphic relations in the northwest part of the basin, Jordan and Douglass (1980) showed that as the basin matured, it deepened during Late Pennsylvanian and Early Permian time. Early basin deposits were cannibalized to form coarse conglomerates that were deposited in a trough inferred to have been bounded by northwest-trending high-angle faults. In the southeast part of the basin, Permian rocks in the Oquirrh Mountains contain structures related to slumping into the rapidly subsiding Oquirrh basin (Schurer, 1979). By contrast, the Paradox basin resembles a half-graben that is fault-bounded on the northeast against the Uncompahgre uplift (Baars and Stevenson, 1981). Faulting may have occurred along a reactivated basement fault. Based on the thickness of Pennsylvanian strata that accumulated in the half-graben and on coeval erosional stripping of the adjacent Uncompahgre uplift, throw on the fault exceeds 4 km and could be as much as 6 km. Kluth and Coney (1981) interpreted the Oquirrh and Paradox basins, together with the ancestral Rocky Mountains and other coeval northwest-trending intracratonic basement uplifts, as features resulting from the collision of North America and South America during the Ouachita-Marathon orogeny at the southeast margin of the continent. Whether such plate-tectonic interpretations are correct, crustal-scale features like the Oquirrh and Paradox basins should be kept in sharp focus when evaluating geophysical anomaly patterns and crustal structure.

Hamilton and Meyers (1966) and Hamilton (1969) emphasized that the major northeast-trending structural and paleogeographic trends associated with the late Precambrian to early Mesozoic miogeocline of the western United States are truncated on the west along a north-northwest–trending Mesozoic orogenic system in California. They suggested that the truncation, possibly of earliest Mesozoic age, resulted from the rifting and drifting away of a part of the North American continent of unknown size. Alternatively, very similar miogeoclinal rocks in Sonora, Mexico, suggest either continuous sedimentation in a sharply southeast-curving miogeocline and no truncation or that a major truncation resulted from large-displacement strike-slip faulting (Anderson and Silver, 1979; Stewart and others, 1984). Whatever type of structures were associated with it, the new orogenesis produced the second major reconfiguration of the continental margin.

Subsequent to the inferred truncation of the continent, a major change in Cordilleran paleogeography and tectonics began in Middle Triassic time and continued through the Mesozoic. During this time, the continent grew westward as assemblages of volcanic and sedimentary rocks of oceanic and island-arc affinities, as well as ophiolitic rocks, were accreted to it on the west. The accretionary rocks record thousands of kilometers of east-directed subduction of oceanic lithosphere beneath the continent in an Andean-type plate-convergence system (Hamilton, 1969; Dickinson, 1970; Burchfiel and Davis, 1975). Here we are not concerned with the newly accreted terranes of California but with the results of related magmatic and tectonic processes acting inboard from those terranes. A belt of arc-related plutonic and volcanic activity formed generally parallel to the old (late Paleozoic and earliest Mesozoic) accretionary trends in northern Nevada. To the south, at about the latitude of Lake Tahoe, the arc swings southeastward and cuts across those trends, as well as across the northeasterly trends of the miogeocline. From there it extends into the part of the craton where Precambrian crystalline rocks were thinly mantled by platform-facies sediments (Fig. 5; Burchfiel and Davis, 1972). Plutonism and metamorphism related to a relatively steeply dipping subduction system probably completed the process of cratonization of the old accreted terrains of northwestern Nevada.

East of the early Mesozoic arc, east-directed thrust faults developed in complex spatial patterns but with an overall eastward migration with time. As a result of the earlier transcurrent truncation of the continent, the newly generated thrusts involve oceanic and arc-related rocks in northwestern Nevada and adjacent California, miogeoclinal rocks to the southeast, and cratonal assemblages and basement rocks in southeastern California and adjacent Arizona (Burchfiel and Davis, 1975). In general, traces of east-directed thrust faults follow shifts in the locus of magmatic activity, suggesting a genetic relationship (Figs. 4, 5). The thrusting and magmatism had little effect on depositional patterns in the eastern part of the Cordilleran miogeocline in earliest Mesozoic time (Carr and Paull, 1983), but as they swept eastward they progressively obliterated the miogeocline. The craton of south-

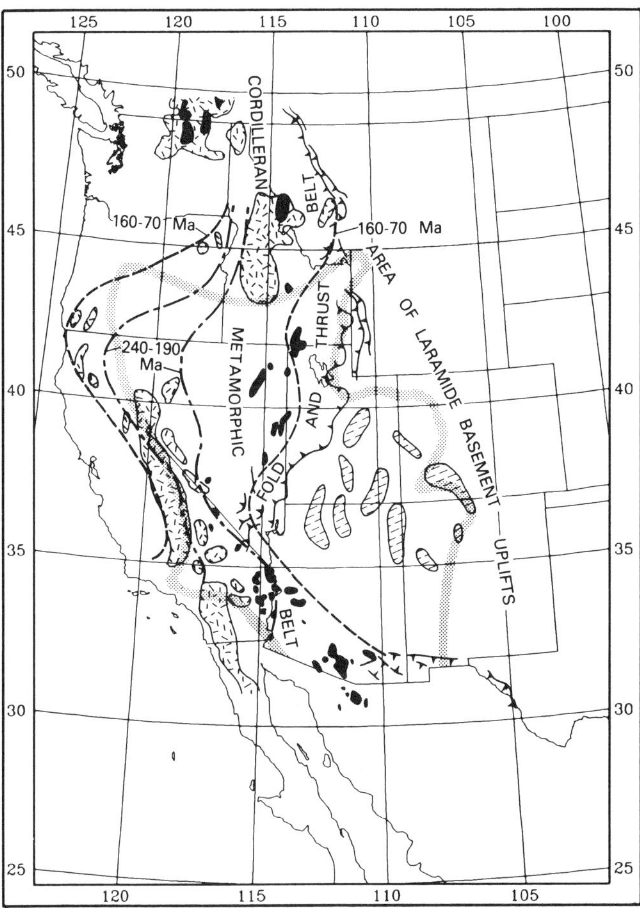

Figure 5. Map showing selected Mesozoic and earliest Cenozoic magmatic and tectonic features. The band of magmatic arc activity for the age range of 240 to 190 Ma (long-dash/short-dash lines) correlates roughly with the Sonomian orogeny and emplacement of the Golconda allochthon shown in Figure 4. The area occupied by magmatic arc activity from 190 to 160 Ma (not shown in figure) is contained within the 240- to 190-Ma area in the south but forms an envelope around it in the north (Burchfiel, 1979). The area occupied by magmatic arc activity from 160 to 70 Ma (dashed lines) generally envelopes both areas of earlier activity (from Burchfiel, 1979). Distribution of exposed batholithic rocks (hachure pattern) and areas of rocks overprinted by middle Mesozoic to early Tertiary regional metamorphism, ductile deformation and/or intrusion of syntectonic plutons (especially two-mica granites) (solid black) from Haxel and others (1984). Areas of uplifts and swells in and adjacent to the Colorado Plateaus (from Stewart, 1978).

western and southern Arizona was altered by arc-related volcanism and deformation (Kluth, 1983), but cratonal areas from northern Arizona northward were not seriously affected by the orogenic events. In the latter region, remnants of the Ancestral Rockies highlands, together with a major highland that rose in central Arizona and possibly renewed uplift along the Uinta axis, supplied detritus for Triassic and Jurassic shallow marine, alluvial plain, lacustrine, and eolian strata that blanketed much of the area that is now the Colorado Plateau (Blakey and Gubitosa, 1983; Blakey and others, 1983).

By late Mesozoic time the area of scattered plutonism in Nevada had swelled to enormous proportions and reached eastward into Utah (Fig. 5), probably as a result of accelerated convergence between the American and eastern Pacific plates and the attendant shallowing of the angle of subduction. Amphibolite-grade metamorphism of miogeoclinal and cratonal rocks occurred adjacent to or near some areas of plutonism. Late Jurassic to Late Cretaceous thrusting spread eastward beyond the east edge of the miogeocline in Nevada, Utah, Wyoming, and Idaho where miogeoclinal rocks, and in southern areas, basement rocks, were emplaced atop cratonal assemblages. The compressional deformation, which extended into Tertiary time in northern areas, produced highlands that shed enormous volumes of clastic debris eastward into the foredeep part of a giant inland seaway. In some western areas, foredeep deposits were involved in folding and thrusting. The source area was the Sevier orogenic belt (Armstrong, 1968), and the basin to the east was the Rocky Mountain geosyncline. Preexisting structural features such as the Oquirrh basin and Uinta axis produced strong local control of the timing, position, and eastward extent of individual thrusts (Burchfiel and Hickcox, 1972; B. Bryant, written communication, 1986).

The extent to which the upper crust was episodically modified by the combined effects of coeval thrusting and heating from mantle sources is displayed very well in latest Cretaceous and early Tertiary regional metamorphism and emplacement of anatectic granites of south-central Arizona (Haxel and others, 1984). These effects of metamorphism, ductile deformation, and plutonism are found in a discontinuous belt that stretches the full length of the United States Cordillera and is situated west of a fold-and-thrust belt that is devoid of similar effects. The timing and distribution of Cretaceous and early Tertiary orogenesis in the United States Cordillera differs from the patterns in Canada and Mexico (Armstrong, 1978). The differences reflect contrasts in the thermal weakening of the lithosphere far inland from the inferred arc-trench system, and may, in turn, reflect differences in the rate and geometry of subduction (Keith, 1978).

The style and distribution of latest Mesozoic deformation in the hinterland of the Sevier belt is largely obscured by younger disturbances. Although the hinterland has been suggested as an area of extension genetically related to the Sevier compressional structures (Hose and Danes, 1973), other investigators have interpreted it to be an area of coeval compression (Burchfiel and Hickcox, 1972; Christie-Blick, 1983; Haxel and others, 1984). Clastic strata in southeastern Arizona and southwestern New Mexico are interpreted to represent deposition in an extensional back-arc setting wholly different from the extensional environment proposed for the hinterland (Bilodeau and Lindberg, 1983).

CENOZOIC TECTONIC EVOLUTION

During early Tertiary time, systems of ancient basement faults were reactivated in a northeast-southwest compressional regime, resulting in the localization of numerous uplifts and monoclines that are clearly defined in cover rocks of the Colorado Plateau (Davis, 1978). This deformation coincided with the Laramide orogeny that produced numerous large basement-cored uplifts and adjacent basins in the adjoining Rocky Mountains (Fig. 5). Laramide deformation reached inland 1,200 to 1,500 km from a northwest-trending plate margin and was accompanied by scattered plutonism that represented the first important igneous activity in parts of Colorado and Arizona since Precambrian time. The eastward spread of deformation and plutonism may reflect a period of rapid convergence and low-angle subduction between the American and Farallon plates between 80 and 40 Ma (Dickinson and Snyder, 1979). Laramide uplifts probably developed in areas west of the Colorado Plateau in Arizona (Young, 1979; Kluth, 1983), Utah (Stanley and Collinson, 1979), and east-central Nevada (Fouch, 1979). Uplifts controlled sediment dispersal patterns in surrounding large lakes, such as Flagstaff and Uinta in Utah, and onto the area of the Colorado Plateau in Arizona. Unfortunately, the deformational style at basin margins is not known, and in some areas, such as east-central Nevada, the tectonic environment could have been extensional. Zoback and others (1981) noted the general difficulty in dating precisely the last episode of compressional deformation and the onset of extension over large areas of the western United States.

Coney (1978) emphasized that late Eocene time (50 to 40 Ma) marked an abrupt termination of compressional-style Laramide deformation throughout the entire North American Cordillera. He related the termination to a fundamental reorganization of plate motions at that time. Engebretson and others (1984) correlated the termination with an abrupt slowing of the plate-convergence rate and a flare-up of arc-related magmatism. This correlation is questionable because the magmatic flare-up is transgressive in time and space over 30 m.y. and more than 1,000 km (Zoback and others, 1981; Wernicke and others, 1987). The abrupt termination of compression is not reflected in the distribution patterns of volcanic and sedimentary rocks of possible extensional affinities. Near the beginning of Eocene time (55 Ma) volcanic-plutonic activity spread south across the Canadian border and by 50 to 40 Ma was active over a large part of the northwestern United States (Armstrong, 1978; Fox and others (1977). Patterns of metamorphism, mylonitization, disturbed isotopic systems, faulting, and stratal tilting associated with or directly following these igneous events are those that characterize a main phase of extensional tectonism (Wernicke and others, 1987; Hamilton, 1988). In eastern Washington, Eocene volcanics are concentrated in troughs, the axes of which are perpendicular to the direction of coeval extension (Fox and others, 1977; K. Fox, oral communication, 1986). Unfortunately, it is not known to what extent this widespread Eocene magmatism and extensional tectonism spread southward into the area now covered by the younger (Miocene) Columbia River basalt or volcanics of the Snake River Plain.

From middle Eocene through Miocene time, volcanism continued its sweep southward through Utah, Nevada, and eastern California (Fig. 6), terminating at a broad concave-northward arc between 37° and 38°N. (Stewart and others, 1977). Several major aeromagnetic anomalies in Nevada and Utah probably record positions of this south-moving arc. Calc-alkalic volcanic rocks accumulated to great thicknesses in depressions that commonly have a half-graben form and asymmetric growth-fault cross-profile, suggesting control by coeval extensional faulting. In central Nevada, nested calderas tend to form northwest- to west-northwest–trending troughs or belts (Burke and McKee, 1979;

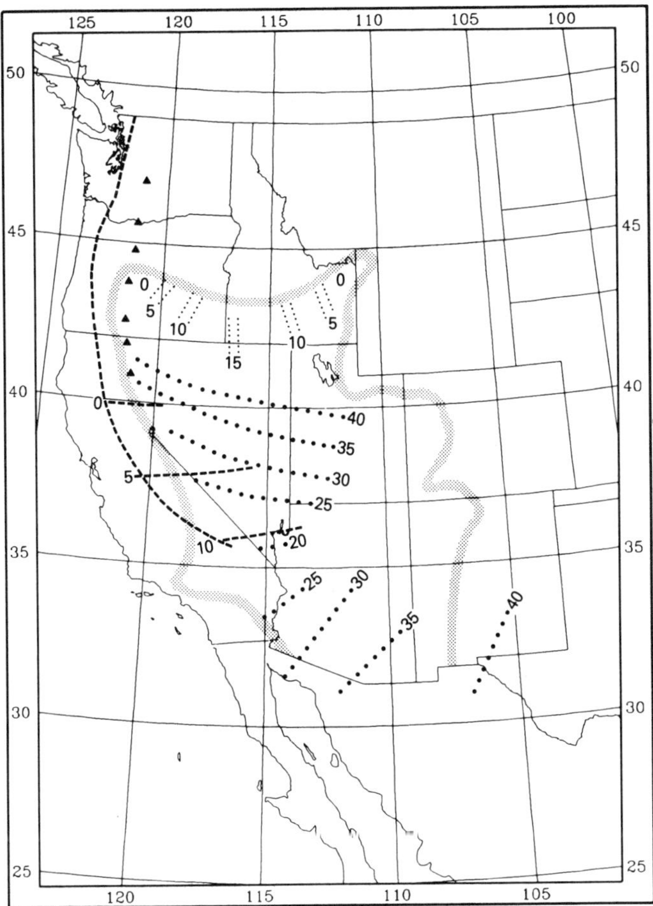

Figure 6. Patterns of migration of magmatism. Numbers are millions of years. Single dotted lines mark approximate times of inception of calc-alkaline volcanism between 400 and 20 Ma. Dashed lines mark southern boundary of northward-migrating cessation of arc volcanism; triangles show alignment of current volcanic arc (from Snyder and others, 1976). Double-dotted lines indicate approximate timing of basalt-rhyolite volcanism in the High Lava Plains.

Boden, 1986) that have no known preexisting structural control. These large features, as long as 100 km, may reflect far-field extensional paleostress conditions, with the least compressive stress oriented normal to their axes. Coeval with the southerly sweep of volcanism through Nevada and Utah from late Eocene to middle Miocene time, volcanism swept west out of southwest New Mexico across southern and northwest Arizona (Snyder and others, 1976; Fig. 6). The large-volume nested calderas and subjacent plutons of the Mogollon-Datil field of New Mexico formed during this magmatism (Elston, 1984). In many parts of the linear belts of isochronous rocks that mark the migratory trends, magmatism was cyclic, beginning with andesitic lavas and evolving through quartz latitic to rhyolitic ash-flow tuffs to high-silica rhyolitic ash-flow tuffs and lavas (McKee and Silberman, 1970; McKee, 1971; Noble, 1972). Tens of thousands of cubic kilometers of volcanic rock were erupted from scattered centers and nested calderas over periods of only a few million years, and cumulative volumes over the western Cordillera exceed 10^6 km^3 (Elston and Bornhorst, 1979). Extension beneath cauldron complexes may have been in the form of passively emplaced plutons and ductile thinning of the lithosphere.

Metamorphic tectonites that display regional kinematic coherence and evidence of simple shear associated with severe structural attenuation on low-angle fault systems (Wernicke, 1981) or at metamorphic core complexes (Crittenden and others, 1980) are found in or adjacent to some of these igneous terranes. Isotopic ages, together with isotopic systems reset to Cenozoic characteristics, show that the zones of extreme extension and midcrustal excision tend to follow igneous episodes closely in time (Coney, 1980). Elston (1984), Zoback and others (1981), and Eaton (1982) interpreted the igneous and tectonic events as manifestations of widespread extensional orogenesis, both the inception and main phases of which swept through the province along regular paths without regard to known preexisting fundamental tectonic features, such as the presence or absence of Precambrian crust, Cordilleran miogeocline, or Mesozoic volcanic-plutonic arc.

The early and middle Cenozoic igneous and tectonic activity occurred during a time of convergence and subduction of the Farallon plate to which most investigators relate them genetically (Coney, 1978; Lipman, 1980; Engebretson and others, 1984). Lipman (1980) summarized the numerous variables that can control the distribution and composition of subduction-related continental magmatism. Variables include preexisting structural flaws, irregularities or weaknesses above and/or below the Benioff zone, dip of the Benioff zone, rate of convergence, angle between plate boundaries and convergence direction, temperature and thickness of the subducted slab, and depth to magma generation. The varied patterns of migration and complex distribution of early and middle Cenozoic calc-alkaline volcanism in the Intermontane region can obviously be explained by choosing an interplay in time and space of these variables consistent with a smaller set of constraining parameters. As noted by Coney and Harmes (1984) and Wernicke and others (1987), there is a good correlation between the loci of extension and preexisting crustal thickening resulting from compressional deformation. Although the geographic correlation implies a genetic relationship between thickening and extension, the timing of extension is probably closely controlled by the thermal state of the lithosphere at the time of thickening (Sonder and others, 1987; Wernicke and others, 1987). Most likely the crust was thinned to less than its initial dimension during the magmatism despite possible large-volume additions from the lithospheric mantle. These relationships suggest an extensional environment that fostered magma production by lowering solidi as lithostatic pressures were lowered through extensional processes. The process could have been initiated by gravitational instability and buoyancy forces associated with a crust that was previously overthickened by compressional tectonism (Molnar and Chen, 1984; Sonder and others, 1987). The most important implication of these relationships is that main-phase Cenozoic extensional tectonism and magmatism in the Basin and Range are mainly a response to forces inherent in the Cordillera rather than to either plate-boundary stress relaxation or inland-distributed transform shear.

The late igneous events that are represented by the loci where the lateral migration was arrested (those of mid-Miocene age located between 37° and 38° in Nevada and Utah and between 34° and 36° in Arizona and California; Fig. 6) formed 10 m.y. or so after establishment of a transform boundary along the San Andreas system. Because they appear to be part of a continuous pattern of lateral migration, these late igneous and tectonic events are probably unrelated to the subduction process. Data on preferentially oriented dike swarms and fault-slip vectors indicate a uniformly oriented northeast-southwest extension direction over a large part of the Basin and Range during the latter stages of the magmatism. This direction is perpendicular to the inferred trend of the subduction-related trench, and as with the magmatism, persisted long after that trench began to be replaced by the lengthening San Andreas transform.

Many investigators have not associated the early and middle Cenozoic phase of predominantly calcalkalic igneous activity with extensional tectonism (Atwater, 1970; Scholz and others, 1971; McKee, 1971; Noble, 1972; Lipman and others, 1972; Stewart, 1978). Regardless of the far-field stress state, wherever these events occurred, they must have resulted in major reshaping of the geophysical framework of the crust and upper mantle. Volcanic calderas and cauldron complexes and subvolcanic batholiths that were emplaced at shallow structural levels during the passage of the magmatic pulses produce distinct geophysical anomaly patterns related to their internal physical properties (Lipman, 1980). Although these igneous structures resulted mostly from the reworking of preexisting crust (Farmer and Depaolo, 1983), they represent a colossal upward transfer of material from deep-source regions where magmas are generated and from which they are extracted. Crustal structure should reflect the passage of these thermomechanical magmatic pulses because there is a predictable density increase in deep-source regions, and a decrease at shallow levels of intrusion and extrusion.

Seismic-reflection profiles in the Basin and Range provide valuable insight into fault geometry, structural dynamics, and structural inheritance associated with the extensional processes (Allmendinger and others, 1983; Smith and Bruhn, 1984) and place constraints on deep-crustal structure (Klemperer and others, 1986). Some low-angle reflections in the middle to upper crust represent gently dipping zones of simple shear, along which the crust is severely attenuated, and some deep, laterally extensive, prominent reflections are presumably from the Moho. Coverage is inadequate to compare and contrast the reflective character and presumed structure of the middle and lower crust in areas of middle and late Cenozoic extreme extensional faulting with those of major coeval magma production. Such comparisons in concert with other geophysical evidence are necessary in order to evaluate the relative importance of magmatism and extensional faulting in molding the geophysical framework.

Strike-slip faults have long been recognized as important elements of Neogene extensional deformation—especially in the Great Basin region where examples such as those in the Death Valley, Las Vegas Valley, and Pahranagat Lakes areas have large displacements. As the number of detailed structural studies increases, the importance of strike-slip faults of intermediate and small displacements is becoming more fully appreciated. At least six different kinds of strike-slip faults, each with its own tectonic significance, are said to exist in the Basin and Range. The geophysical framework of areas cut by these faults may or may not be impacted by them, depending on which of the several kinds are present. Careful study and accurate categorization of strike-slip faults is an oft-neglected, though potentially important, part of understanding the geophysical framework of the Basin and Range.

In many parts of the Basin and Range it is possible to distinguish between an early episode of extension related to volcanism, plutonism, detachment faulting, and core-complex formation and a later episode responsible for the existing physiography produced by displacements on widely spaced, mainly steep normal faults and systems of strike-slip faults (Zoback and others, 1981; Eaton, 1982). This later episode is commonly referred to as Basin and Range extension and resulted in development of widespread sediment-filled basins separated by ranges. Although the episode generally began 15 to 10 m.y. ago, it is not the same age everywhere (Zoback and others, 1981). It began and ceased earlier in the southern Basin and Range of Arizona and California (Eberly and Stanley, 1978) than in the northern Basin and Range of Nevada and Utah, resulting in distinctly different physiography in those two parts of the province (Eaton, 1979). In other parts of the province, evidence for deeply penetrating, steep, young normal faults is absent. In the Sevier Desert and Raft River basins, for example, the depth penetration of most young block faults is limited by shallow detachment faults.

Thus, Basin and Range physiography is actually controlled by a hierarchy of faults that range from steep to shallow (Anderson and others, 1983). Also, in regions of intense young extension, such as the Death Valley region, the deformational mode controlling the gross physiography may be that of an evolving metamorphic core complex (Wernicke, 1981; Wernicke and others, 1986). Other contrasts in faulting geometry and mode further complicate Basin and Range extension. In the northern Basin and Range, for example, Wright (1976) recognized a large domain of predominantly normal faults and relatively low strain rate flanked on the west and south by a domain of mixed normal and strike-slip faults and assumed higher strain rate. On a somewhat smaller scale, Stewart (1980) recognized domains of contrasting tilt directions. Despite this complexity and variability, an episode of Basin and Range extension is widely recognized, and many investigators (the author excluded) associate it with the lengthening of the San Andreas transform system and the development of significant coupling between the Pacific and North American plates to form a broad zone of oblique extension.

The average composition of volcanic rocks changed as the style of extensional tectonism changed. Mafic eruptives were generally more mafic than those that had appeared earlier and have compositions that extend into the alkalic side of the boundary between alkalic and subalkalic rocks (Christiansen and Lipman, 1972; Elston, 1984). Also, later siliceous rocks are generally more siliceous than early ones, resulting in a strong tendency toward compositional bimodality. In the northern Basin and Range, the transition to bimodal volcanism began in the southeastern part and moved northwest in conjunction with (but lagging behind) the lengthening San Andreas transform fault (Snyder and others, 1976). The transition shows no clear pattern of migration in New Mexico and adjacent Arizona. In the Basin and Range, the magma production decreased significantly from that of early and mid-Cenozoic magmatism. In the High Lava Plains of Idaho and Oregon, by contrast, coeval bimodal volcanism represents a return of large-volume magmatism to that region. The two arms of the High Lava Plains owe their origins to coeval volcanic systems of bimodal basalt-rhyolite composition that propagated linearly from a starting area near the Idaho-Oregon-Nevada boundary 17 to 14 m.y. ago (Christiansen and McKee, 1978). The eastern arm propagated northeastward along a Precambrian zone of structural weakness to form the 500-km-long volcanic foundation for the Snake River Plain and Yellowstone Plateau. The western arm propagated west-northwest about 300 km across southeastern Oregon to form the Oregon High Lava Plains (Fig. 6). Christiansen and McKee (1978) interpreted the late Cenozoic magma production associated with the northeast and northwest propagations to have resulted directly from heating events linked to the intersection of outwardly (east-west) migrating zones of Basin and Range extension with the northern-boundary transverse structures that compensated for that extension. They suggested that magma production was augmented by concentrated faulting and stress relief at these intersections. On the basis of gravity, seismic, and geological data, Hamilton and Meyers (1966) interpreted the High Lava Plains to be a lava-filled tension rift in which the preexisting crust was thinned and ruptured by Cenozoic extension. They suggested that the original crust was completely rifted in the western part and thinned in the eastern part.

A clockwise change in least principal stress orientation from west-southwest–east-northeast to west-northwest–east-southeast seems to have occurred at about 10 Ma over a large part of the Basin and Range (Zoback and others, 1981). This change is mechanically consistent with the superposition of Pacific-margin dextral transform shear on the generally east-west intracontinental back-arc extension proposed by Atwater (1970). The change migrated as the western part of the North American plate began to "feel" the effect of significant coupling to the Pacific plate along the lengthening San Andreas transform system.

In summary, the Intermontane System embraces the extremes of Cordilleran history. It includes the interior section of the Colorado Plateaus that has enigmatically been bypassed by strong deformation and magmatism since Precambrian time. The system also embraces the High Lava Plains, parts of which may be underlain by crust no older than late Cenozoic. At the other extreme, large parts of the Basin and Range have experienced repeated orogenesis, the youngest of which is continental extension with dimensions and deformational magnitudes that may not be exceeded anywhere in the world. The combined effects of early and mid-Cenozoic extension and younger Basin-Range extension in many parts of the province exceeds 100 percent and locally exceeds 200 percent. As with large lateral motions now known to be integral parts of the processes of cratonization through plate-tectonic interactions, this newly appreciated large-magnitude extension requires large lateral motions and the involvement of major low-angle structures. The geometry and depth penetration of such structures and their relationship to rheologic layering in the crust is a subject of intense current interest. All aspects of the geophysical framework of the Intermontane System should be scrutinized closely in terms of plate-tectonic and extensional models that allow for lateral motions, the magnitudes of which have never before been appreciated.

ACKNOWLEDGMENTS

This report was improved by the thoughtful reviews of Bruce Bryant, Mitch Reynolds, John Stewart, Brian Wernicke, and Erick Frost. Russ Wheeler provided a helpful draft copy of a report he was preparing that deals with the tectonic framework of the Wasatch Front. Ted Barnhard helped prepare the illustrations.

REFERENCES CITED

Allmendinger, R. W., Sharp, J. W., von Tish, D., Serpa, L., Kaufman, S., Oliver, J., and Smith, R. B., 1983, Cenozoic and Mesozoic structure of the eastern Basin and Range from COCORP seismic-reflection data: Geology, v. 11, p. 532–536.

Anderson, J. L., 1983, Proterozoic anorogenic granite plutonism of North America: Geological Society of America Memoir 161, p. 133–154.

Anderson, T. H., and Silver, L. T., 1979, The role of the Mojave-Sonora megashear in the tectonic evolution of northern Sonora, *in* Anderson, T. H., and

Roldan-Quintana, Jaime, eds., Geology of northern Sonora: Geological Society of America Annual Meeting, 1979, Guidebook, p. 59–68.

Anderson, R. E., Zoback, M. L., and Thompson, G. A., 1983, Implications of selected subsurface data on the structural form and evolution of some basins in the northern Basin and Range province, Nevada and Utah: Geological Society of America Bulletin, v. 94, p. 1055–1072.

Armstrong, R. L., 1968, The Sevier orogenic belt in Nevada and Utah: Geological Society of America Bulletin, v. 79, p. 429–458.

——, 1976, The geochronometry of Idaho: Isochron/West, no. 15, p. 1–33.

——, 1978, Cenozoic igneous history of the U.S. Cordillera from lat 42° to 49°N: Geological Society of America Memoir 152, p. 265–282.

Atwater, T., 1970, Implications of plate tectonics for the Cenozoic tectonic evolution of western North America: Geological Society of America Bulletin, v. 81, p. 3513–3536.

Baars, D. L., and Stevenson, G. M., 1981, Tectonic evolution of the Paradox Basin, Utah and Colorado, *in* Geology of the Paradox Basin, 1981; 1981 Field Conference: Rocky Mountain Association of Geologists Guidebook, p. 23–31.

Bilodeau, W. L., and Lindberg, F. A., 1983, Early Cretaceous tectonics and sedimentation in southern Arizona, southwestern New Mexico, and northern Sonora, Mexico, *in* Reynolds, M. W., and Dolly, E. D., eds., Mesozoic paleogeography of the west-central United States: Rocky Mountain Section, Society of Economic Paleontologists and Mineralogists, p. 173–188.

Bissell, H. J., 1974, Tectonic control of late Paleozoic and early Mesozoic sedimentation near the hinge line of the Cordilleran miogeosynclinal belt, *in* Dickinson, W. R., ed., Tectonics and sedimentation: Society of Economic Paleontologists and Mineralogists Special Publication 22, p. 83–97.

Blakey, R. C., and Gubitosa, R., 1983, Late Triassic paleogeography and depositional history of the Chinle Formation, southern Utah and northern Arizona, *in* Reynolds, M. W., and Dolly, E. D., eds., Mesozoic paleogeography of the west-central United States: Rocky Mountain Section, Society of Economic Paleontologists and Mineralogists, p. 57–76.

Blakey, R. C., Peterson, F., Caputo, M. V., Geesaman, R. C., and Vorhees, B. J., 1983, Paleogeography of middle Jurassic continental, shoreline, and shallow marine sedimentation, southern Utah, *in* Reynolds, M. W., and Dolly, E. D., eds., Mesozoic paleogeography of the west-central United States: Rocky Mountain Section, Society of Economic Paleontologists and Mineralogists, p. 77–100.

Boden, D. R., 1986, Eruptive history and structural development of the Toquima caldera complex, central Nevada: Geological Society of America Bulletin, v. 97, p. 61–74.

Bond, G. C., and Kominz, M. A., 1984, Construction of tectonic subsidence curves for the early Paleozoic miogeocline, southern Canadian Rocky Mountains; Implications for subsidence mechanisms, age of breakup, and crustal thinning: Geological Society of America Bulletin, v. 95, p. 155–173.

Bryant, B., 1985, Structural ancestry of the Uinta Mountains, *in* Picard, M. D., ed., Geology and energy resources, Uinta Basin, Utah: Utah Geological Association Publication 12, p. 115–120.

Bryant, B., and Nichols, D. J., 1988, Late Mesozoic and early Tertiary reactivation of an ancient zone of crustal weakness along the Uinta Arch and its interaction with the Sevier orogenic belt, *in* Schmidt, C. J., and Perry, W. J., Jr., eds., Interaction of the Rocky Mountain foreland and Cordilleran thrust belt: Geological Society of America Memoir 171, p. 411–430.

Burchfiel, B. D., 1979, Geologic history of the central western United States: Nevada Bureau of Mines and Geology Report 33, p. 1–11.

Burchfiel, B. C., and Davis, G. S., 1972, Structural framework and evolution of the southern part of the Cordilleran Orogen, western United States: American Journal of Science, v. 272, p. 97–118.

——, 1975, Nature and controls of Cordilleran orogenesis, western United States; Extension of an earlier synthesis: American Journal of Science, v. 275-A, p. 363–396.

Burchfiel, B. C., and Hickcox, C. W., 1972, Structural development of central Utah, *in* Baer, J. L., and Callaghan, E., eds., Plateau–Basin and Range transition zone, central Utah, 1972: Salt Lake City, Utah Geological Association Publication 2, p. 55–66.

Burke, D. B., and McKee, E. H., 1979, Mid-Cenozoic volcano-tectonic troughs in central Nevada: Geological Society of America Bulletin, v. 90, p. 181–184.

Carr, R. T., and Paull, R. K., 1983, Early Triassic stratigraphy and paleogeography of the Cordilleran miogeocline, *in* Reynolds, M. W., and Dolly, E. D., eds., Mesozoic paleogeography of the west-central United States: Rocky Mountain Section, Society of Economic Paleontologists and Mineralogists, p. 39–55.

Christiansen, R. L., and Lipman, P. W., 1972, Cenozoic volcanism and plate-tectonic evolution of the western United States; Pt. 2, Late Cenozoic: Proceedings of the Royal Society of London, v. 271, p. 249–284.

Christiansen, R. L., and McKee, E. H., 1978, Late Cenozoic volcanic and tectonic evolution of the Great Basin and Columbia intermontane region, *in* Smith, R. B., and Eaton, G. P., eds., Cenozoic tectonics and regional geophysics of the western Cordillera: Geological Society of America Memoir 152, p. 283–315.

Christie-Blick, N., 1983, Structural geology of the southern Sheeprock Mountains, Utah; Regional significance: Geological Society of America Memoir 157, p. 101–124.

Compton, R. R., Todd, V. R., Zartman, R. E., and Naeser, C. W., 1977, Oligocene and Miocene metamorphism, folding, and low-angle faulting in northwestern Utah: Geological Society of America Bulletin, v. 88, p. 1237–1250.

Condie, K. C., 1976, The Wyoming Archean province in the western United States, *in* Windley, B. F., ed., The early history of the earth: New York, John Wiley and Sons, p. 499–510.

——, 1982, Plate tectonics model for Proterozoic continental accretion in the southwestern United States: Geology, v. 10, p. 37–42.

Coney, P. J., 1978, Mesozoic-Cenozoic Cordilleran plate tectonics, *in* Smith, R. B., and Eaton, G. P., eds., Cenozoic tectonics and regional geophysics of the western Cordillera: Geological Society of America Memoir 152, p. 33–50.

——, 1980, Cordilleran metamorphic core complexes; An overview, *in* Crittenden, M. D., Coney, P. J., and Davis, G. H., eds., Cordilleran metamorphic core complexes: Geological Society of America Memoir 153, p. 7–34.

Coney, P. J., and Harms, T. A., 1984, Cordilleran metamorphic core complexes; Cenozoic extensional relics of Mesozoic compression: Geology, v. 12, p. 550–554.

Crittenden, M. D., Jr., Schaeffer, F. E., Trimble, D. E., and Woodward, L. A., 1971, Nomenclature and correlation of some upper Precambrian and basal Cambrian sequences in western Utah and southeastern Idaho: Geological Society of America Bulletin, v. 82, p. 581–602.

Crittenden, M. D., Jr., Coney, P. J., and Davis, G. H., eds., 1980, Cordilleran metamorphic core complexes: Geological Society of America Memoir 153, 490 p.

Davis, G. H., 1978, Monocline fold pattern of the Colorado Plateau, *in* Matthews, V., ed., Laramide folding associated with basement block faulting in the western United States: Geological Society of America Memoir 151, p. 215–233.

Dickinson, W. R., 1970, Relations of andesitic volcanic chains, granitic batholith belts, and derivative graywacke-arkose facies to the tectonic framework of arc-trench systems: Reviews in Geophysics and Space Physics, v. 8, p. 813–860.

Dickinson, W. R., and Snyder, W. S., 1979, Geometry of subducted slabs related to San Andreas transform: Journal of Geology, v. 87, p. 609–628.

Eaton, G. P., 1979, Regional geophysics, Cenozoic tectonics, and geologic resources of the Basin and Range Province and adjoining regions, *in* Newman, G. W., and Goode, H. D., eds., 1979, Basin and Range Symposium: Denver, Colo., Rocky Mountain Association of Geologists and Utah Geological Association, p. 11–39.

——, 1982, The Basin and Range province; Origin and tectonic significance: Annual Review of the Earth and Planetary Sciences, v. 10, p. 409–440.

Eberly, L. D., and Stanley, T. B., Jr., 1978, Cenozoic stratigraphy and geologic history of southwestern Arizona: Geological Society of America Bulletin, v. 89, p. 921–940.

Elston, D. P., and McKee, E. H., 1982, Age and correlation of the late Proterozoic Grand Canyon disturbance, northern Arizona: Geological Society of America Bulletin, v. 93, p. 681–699.

Elston, W. E., 1984, Subduction of young oceanic lithosphere and extensional orogeny in southwestern North America during mid-Tertiary time: Tectonics, v. 3, no. 2, p. 229–250.

Elston, W. E., and Bornhorst, 1979, The Rio Grande rift in context of regional post-40 m.y. volcanic and tectonic event, *in* Riecker, R. E., ed., Rio Grande rift; Tectonics and magmatism: American Geophysical Union, p. 416–438.

Engebretson, D. C., Cox, A., and Thompson, G. A., 1984, Correlation of plate motions with continental tectonics—Laramide to Basin and Range: Tectonics, v. 3, no. 2, p. 115–119.

Farmer, G. L., and DePaolo, D. J., 1983, Origin of Mesozoic and Tertiary granite in the western United States and implications for pre-Mesozoic crustal structure; 1, Nd and Sr isotope studies in the geocline of the northern Great Basin: Journal of Geophysical Research, v. 88, p. 3379–3401.

——, 1984, Origin of Mesozoic and Tertiary granite in the western United States and implications of pre-Mesozoic crustal structure; 2, Nd and Sr isotopic studies of unmineralized and Cu and Mo-mineralized granite in the Precambrian craton: Journal of Geophysical Research, v. 89, no. B12, p. 10141–10160.

Fenneman, N. M., 1931, Physiography of western United States: New York, McGraw-Hill, 534 p.

Fouch, T. D., 1979, Character and paleogeographic distribution of upper Cretaceous(?) and Paleocene nonmarine sedimentary rocks in east-central Nevada, *in* Armentrout, J. M., Cole, M. R., and TerBest, H., Jr., eds., Cenozoic paleogeography of the western United States: Society of Economic Paleontologists and Mineralogists, 3rd Pacific Coast Paleogeography Symposium, Los Angeles, California, p. 97–111.

Fox, K. R., Jr., Rinehart, C. D., and Engels, J. C., 1977, Plutonism and orogeny in north-central Washington; Timing and regional context: U.S. Geological Survey Professional Paper 989, 27 p.

Hamilton, W., 1969, Mesozoic California and the underflow of Pacific Mantle: Geological Society of America Bulletin, v. 80, p. 2409–2430.

Hamilton, W., 1988, Tectonic setting and variations with depth of some Cretaceous and Cenozoic structural and magmatic systems of the western United States, *in* Ernst, W. G., ed., Metamorphism and crustal evolution of the western United States: Englewood Cliffs, New Jersey, Prentice-Hall, p. 1–40.

Hamilton, W., and Myers, B. W., 1966, Cenozoic tectonics of the western United States: Reviews of Geophysics, v. 4, no. 4, p. 509–552.

Haxel, G. B., Tosdal, R. M., May, D. J., and Wright, J. E., 1984, Latest Cretaceous and early Tertiary orogenesis in south-central Arizona; Thrust faulting, regional metamorphism, and granitic plutonism: Geological Society of America Bulletin, v. 95, p. 631–653.

Hedge, C. E., Stacey, J. S., and Bryant, B., 1983, Geochronology of the Farmington Canyon complex, Wasatch Mountains, Utah: Geological Society of America Memoir 157, p. 37–44.

Hills, F. A., and Houston, R. S., 1979, Early Proterozoic tectonics of the central Rocky Mountains, North America: University of Wyoming Contributions to Geology, v. 17, no. 2, p. 89–110.

Hofmann, H. J., and Snyder, G. L., 1985, Archean stromatolites from the Hartville uplift, eastern Wyoming: Geological Society of America Bulletin, v. 96, p. 842–849.

Hose, R. K., and Danes, Z. F., 1973, Development of late Mesozoic to early Cenozoic structures in the eastern Great Basin, *in* De Jong, K. A., and Scholten, R., eds., Gravity and tectonics: New York, John Wiley and Sons, p. 429–442.

Johnson, J. G., and Murphy, M. A., 1984, Time-rock model for Siluro-Devonian continental shelf, western United States: Geological Society of America Bulletin, v. 95, p. 1349–1359.

Jordan, T. E., and Douglass, R. C., 1980, Paleogeography and structural development of the Late Pennsylvanian to early Permian Oquirrh Basin, northwestern Utah, *in* Fouch, T. C., and Magathan, E. R., eds., Paleozoic paleogeography of the west-central United States; West-Central United States Paleogeography Symposium 1: Rocky Mountain Section Society of Economic Paleontologists and Mineralogists, p. 217–238.

Karlstrom, K. E., Flurkey, A. J., and Houston, R. S., 1983, Stratigraphy and depositional setting of the Proterozoic Snowy Pass Supergroup, southeastern Wyoming; Record of an early Proterozoic Atlantic-type cratonic margin: Geological Society of America Bulletin, v. 94, p. 1257–1274.

Keith, S. B., 1978, Paleosubduction geometries inferred from Cretaceous and Tertiary magmatic patterns in southwestern North America: Geology, v. 6, p. 515–521.

Ketner, K. B., 1977a, Late Paleozoic orogeny and sedimentation, southern California, Nevada, Idaho, and Montana, *in* Stewart, J. H., Stevens, C. H., and Fritche, A. E., eds., Paleozoic paleogeography of the western United States: Society of Economic Paleontologists and Mineralogists, Pacific Section, Pacific Coast Paleogeography Symposium 1, p. 363–369.

——, 1977b, Deposition and deformation of lower Paleozoic western facies rocks, northern Nevada, *in* Stewart, J. H., Stevens, C. H., and Fritche, A. E., eds., Paleozoic paleogeography of the western United States: Society of Economic Paleontologists and Mineralogists, Pacific Section, Pacific Coast Paleogeography Symposium 1, p. 251–258.

Kistler, R. W., 1983, Isotope geochemistry of plutons in the northern Great Basin: Geothermal Resources Council, Special Report No. 13, p. 3–8.

Kistler, R. W., and Peterman, Z. E., 1978, Reconstruction of crustal blocks of California on the basis of initial strontium isotopic compositions of Mesozoic granitic rocks: U.S. Geological Survey Professional Paper 1071, 17 p.

Klemperer, S. L., Hauge, T. A., Hauser, E. C., Oliver, J. E., and Potter, C. J., 1986, The Moho in the northern Basin and Range province, Nevada, along the COCORP 40°N seismic reflection transect: Geological Society of America Bulletin, v. 97, p. 603–618.

Kluth, C. F., 1983, Geology of the northern Canelo Hills and implications for the Mesozoic tectonics of southeastern Arizona, *in* Reynolds, M. W., and Dolly, E. D., eds., Mesozoic paleogeography of the west-central United States: Rocky Mountain Section, Society of Economic Paleontologists and Mineralogists, p. 159–171.

Kluth, C. F., and Coney, P. J., 1981, Plate tectonics of the Ancestral Rocky Mountains: Geology, v. 9, p. 10–15.

Lipman, P. W., 1980, Cenozoic volcanism in the western United States; Implications for continental tectonics, *in* Continental tectonics; Studies in geophysics: Washington, D.C., National Academy of Sciences, v. 161–174.

Lipman, P. W., Prostka, H. J., and Christiansen, R. L., 1972, Cenozoic volcanism and plate-tectonic evolution of the western United States; I, Early and middle Cenozoic: Royal Society of London Philosophical Transactions, v. 271, p. 217–248.

McKee, E. H., 1971, Tertiary igneous chronology of the Great Basin of western United States; Implications for tectonic models: Geological Society of America Bulletin, v. 82, p. 3497–3502.

McKee, E. H., and Silberman, M. L., 1970, Geochronology of Tertiary igneous rocks in central Nevada: Geological Society of America Bulletin, v. 81, p. 2317–2327.

Miller, E. L., Holdsworth, B. K., Whiteford, W. B., and Rodgers, D., 1984, Stratigraphy and structure of the Schoonover sequence, northeastern Nevada; Implications for Paleozoic plate-margin tectonics: Geological Society of America Bulletin, v. 95, p. 1063–1076.

Molnar, P., and Chen, W. P., 1983, Focal depths and fault plane solutions of earthquakes under the Tibetan plateau: Journal of Geophysical Research, v. 88, p. 1180–1196.

Moulton, F. C., 1975, Lower Mesozoic and upper Paleozoic petroleum potential of the hingeline area, central Utah, *in* Bolyard, D. W., ed., Symposium on deep drilling frontiers in the central Rocky Mountains: Rocky Mountain Association of Geologists, p. 87–97. Reprint, 1976, *in* Hill, J. G., ed., Symposium on geology of the Cordilleran hingeline: Rocky Mountain Asso-

ciation of Geologists, p. 219–229.
Noble, D. C., 1972, Some observations on the Cenozoic volcano-tectonic evolution of the Great Basin, western United States: Earth and Planetary Science Letters, v. 17, p. 142–150.
Oldow, J. S., 1984, Spatial variability in the structure of the Roberts Mountains allochthon, western Nevada: Geological Society of America Bulletin, v. 95, p. 174–185.
Peterman, Z. E., 1979, Geochronology and the Archean of the United States: Society of Economic Geologists Economic Geology, v. 74, p. 1544–1562.
Picha, F., and Gibson, R. I., 1985, Cordilleran hingeline—Late Precambrian rifted margin of the North American craton and its impact on the depositional and structural history, Utah and Nevada: Geology, v. 13, p. 465–468.
Poole, F. G., 1974, Flysch deposits of the Antler foreland basin, western United States, *in* Dickenson, W. R., ed., Tectonics and sedimentation: Society of Economic Paleontologists and Mineralogists Special Publication 22, p. 58–82.
Poole, F. G., and Sandberg, C. A., 1977, Mississippian paleogeography and tectonics of the western United States, *in* Stewart, J. H., Stevens, C. H., and Fritche, A. E., eds., Paleozoic paleogeography of the western United States: Society of Economic Paleontologists and Mineralogists, Pacific Section, Pacific Coast Paleogeography Symposium 1, p. 67–85.
Poole, F. G., Sandberg, C. A., and Boucot, A. J., 1977, Silurian and Devonian paleogeography of the western United States, *in* Stewart, J. H., Stevens, C. H., and Fritche, A. E., eds., Paleozoic paleogeography of the western United States: Society of Economic Paleontologists and Mineralogists, Pacific Section, Pacific Coast Paleogeography Symposium 1, p. 39–65.
Scholz, C. H., Barazangi, M., and Sbar, M. L., 1971, Late Cenozoic evolution of the Great Basin, western United States, as an ensialic interarc basin: Geological Society of America Bulletin, v. 82, p. 2979–2990.
Schurer, V. C., 1979, A Basin and Range chaos in the Oquirrh Mountains; Sedimentary or tectonic?: Rocky Mountain Association of Geologists–UGA, 1979 Basin and Range Symposium, p. 267–271.
Schweikert, R. A., and Lahren, M. M., 1987, Continuation of Antler and Sonoman orogenic belts to the eastern Sierra Nevada, California, and Late Triassic thrusting in a compressional arc: Geology, v. 15, p. 270–273.
Sears, J. W., Graff, P. J., and Holden, G. S., 1982, Tectonic evolution of lower Proterozoic rocks, Uinta Mountains, Utah and Colorado: Geological Society of America Bulletin, v. 93, p. 990–997.
Smith, R. B., and Bruhn, R. L., 1984, Intraplate extensional tectonics of the eastern Basin-Range; Inferences on structural style from seismic reflection data, regional tectonics, and thermal-mechanical models of brittle-ductile deformation: Journal of Geophysical Research, v. 89, p. 5733–5762.
Snyder, W. S., Dickinson, W. R., and Silberman, M. L., 1976, Tectonic implications of space-time patterns of Cenozoic magmatism in the western United States: Earth and Planetary Science Letters, v. 32, p. 91–106.
Sonder, L. J., England, P. C., Wernicke, B. P., and Christiansen, R. L., 1987, A physical model for Cenozoic extension of western North America, *in* Coward, M. P., and others, eds., Continental extensional tectonics: Geological Society Special Publication 18, p. 187–201.
Speed, R. C., 1983, Pre-Cenozoic tectonic evolution of northeastern Nevada, *in* The role of heat in the development of energy and mineral resources in the northern Basin and Range province: Geothermal Resources Council Special Report 13, p. 11–24.
Stacey, J. S., and Hedlund, D. C., 1983, Lead-isotopic compositions of diverse igneous rocks and ore deposits from southwestern New Mexico and their implications for early Proterozoic crustal evolution in the western United States: Geological Society of America Bulletin, v. 94, p. 43–57.
Stanley, K. O., and Collinson, J. W., 1979, Depositional history of Paleocene–lower Eocene Flagstaff Limestone and coeval rocks, central Utah: American Association of Petroleum Geologists Bulletin, v. 63, no. 3, p. 311–323.
Stevens, C. H., 1986, Evolution of the Ordovician through Middle Pennsylvanian carbonate shelf in east-central California: Geological Society of America Bulletin, v. 97, p. 11–25.
Stevenson, G. M., and Beus, S. S., 1983, Stratigraphy and depositional setting of the upper Precambrian Dox Formation in Grand Canyon: Geological Society of America Bulletin, v. 93, p. 163–173.
Stewart, J. H., 1970, Upper Precambrian and lower Cambrian strata in the southern Great Basin, California and Nevada: U.S. Geological Survey Professional Paper 620, 206 p.
——, 1972, Initial deposits in the Cordilleran geosyncline; Evidence of a late Precambrian (<850 m.y.) continental separation: Geological Society of America Bulletin, v. 83, p. 1345–1360.
——, 1976, Late Precambrian evolution of North America; Plate tectonics implication: Geology, v. 4, no. 1, p. 11–15.
——, 1978, Basin-range structure in western North America; A review, *in* Smith, R. B., and Eaton, G. P., eds., Cenozoic tectonics and regional geophysics of the western Cordillera: Geological Society of America Memoir 152, p. 1–13.
——, 1980, Regional tilt patterns of late Cenozoic basin-range fault blocks, western United States: Geological Society of America Bulletin, v. 91, p. 460–464.
Stewart, J. H., and Poole, F. G., 1974, Lower Paleozoic and uppermost Precambrian Cordilleran miogeocline, Great Basin, western United States, *in* Dickinson, W. R., ed., Tectonics and sedimentation: Society of Economic Paleontologists and Mineralogists Special Publication 22, p. 28–57.
Stewart, J. H., and Suczek, C. A., 1977, Cambrian and latest Precambrian paleogeography and tectonics in the western United States, *in* Stewart, J. H., Stevens, C. H., and Fritche, A. E., eds., Paleozoic paleogeography of the western United States: Society of Economic Paleontologists and Mineralogists, Pacific Section, Pacific Coast Paleogeography Symposium 1, p. 1–17.
Stewart, J. H., Moore, W. J., and Zietz, I., 1977, East-west patterns of Cenozoic igneous rocks, aeromagnetic anomalies, and mineral deposits, Nevada and Utah: Geological Society of America Bulletin, v. 88, p. 67–77.
Stewart, J. H., McMenamin, M.A.S., and Morales-Ramirez, J. M., 1984, Upper Proterozoic and Cambrian rocks in the Caborca region, Sonora, Mexico—Physical stratigraphy, biostratigraphy, paleocurrent studies, and regional relations: U.S. Geological Survey Professional Paper 1309, 36 p.
Stokes, W. L., 1979, Stratigraphy of the Great Basin region, *in* Newman, G. W., and Goode, H. D., eds., 1979 Basin and Range symposium: Rocky Mountain Association of Geologists and Utah Geological Association, p. 195–219.
Wernicke, B., 1981, Low-angle normal faults in the Basin and Range Province; Nappe tectonics in an extending orogen: Nature, v. 291, p. 645–648.
Wernicke, B. P., Hodges, K. V., and Walker, J. D., 1986, Geologic evolution of Tucki Mountain and vicinity, central Panamint Range, *in* Dunne, G. C., ed., Mesozoic and Cenozoic structural evolution of selected areas, east-central California: Geological Society of America Cordilleran Section Guidebook, p. 67–80.
Wernicke, B. P., Christiansen, R. L., England, P. C., and Sonder, L. J., 1987, Tectonomagmatic evolution of Cenozoic extension in the North American cordillera, *in* Coward, M. P., and others, eds., Continental extensional tectonics: Geological Society Special Publication 28, p. 203–221.
Wright, L., 1976, Late Cenozoic fault patterns and stress fields in the Great Basin and westward displacement of the Sierra Nevada block: Geology, v. 4, p. 489–494.
Young, R. A., 1979, Laramide deformation, erosion, and plutonism along the southwestern margin of the Colorado Plateau: Tectonophysics, v. 61, p. 15–47.
Zartman, R. E., 1974, Lead isotopic provinces of the western United States and their geologic significance: Economic Geology, v. 69, p. 792–805.
Zoback, M. L., Anderson, R. E., and Thompson, G. A., 1981, Cenozoic evolution of the state of stress and style of tectonism of the Basin and Range province of the western United States: Philosophical Transactions of the Royal Society of London, A-300, p. 407–434.

Manuscript Accepted by the Society October 31, 1988

Printed in U.S.A.

Geological Society of America
Memoir 172
1989

Chapter 11

Geophysics of the western Basin and Range province

George A. Thompson, Rufus Catchings, Erik Goodwin, Steve Holbrook, Craig Jarchow, Carol Mann, Jill McCarthy, and David Okaya
Geophysics Department, Stanford University, Stanford, California 94305

ABSTRACT

Continental extension in the Basin and Range province of the western United States has long been a subject of extensive research and debate. Many aspects of the evolution of the Basin and Range remain a mystery, including its detailed crustal structure, its fault geometries, and the nature of its highly reflective lower crust. In this chapter we address these questions with a reanalysis of existing refraction, reflection, and gravity data, supplemented with new seismic reflection data from representative regions within the province.

Our reanalysis of a seismic refraction profile, recorded between Fallon and Eureka, Nevada, in the early 1960s, leads to a preferred model in which the western Basin and Range is underlain by a lens of material transitional between crust and mantle with its base at an approximately constant depth of 34.5 km below our datum (1 km above sea level). In this model, the lens is centered beneath the Carson Sink in western Nevada, where it attains a maximum thickness of about 9 km and is characterized by an estimated refraction velocity of 7.5 km/sec, intermediate between usual lower crustal and upper mantle velocities. Gravity modeling along this Fallon to Eureka transect is improved by introducing moderate thinning (about 12 km) of the mantle lithosphere. On seismic reflection profiles, the base of the crust-mantle transition zone correlates roughly with an abrupt change from a strongly laminated lower crust into a seismically transparent upper mantle. We suggest that this transition zone represents an interlayering of mafic-ultramafic rocks formed during repeated episodes of partial melting, underplating of the crust by gabbroic magma, crystal settling, and subhorizontal shearing near the crust-mantle boundary. An alternate model, without the lens of material transitional between crust and mantle, satisfies the gravity data with a major (about 34 km) thinning of the mantle lithosphere.

We also address the geometry of faulting in the upper crust within the province with a review of representative seismic reflection profiles, and geodetic and earthquake studies. Although geodetic and earthquake data indicate that the Basin and Range is characterized by high-angle planar faults that penetrate to a depth of about 15 km, seismic reflection data suggest that listric faults and low-angle detachment surfaces are also important in the extension of the brittle upper crust. In contrast, the deeper (>15 km) crust is aseismic and is thought to deform ductilely. The transition from an upper crust that is comparatively transparent in seismic reflection records, to a middle crust (about 12 to 25 km) that in many places is highly laminated and reflective may indicate a rheologic change from brittle to ductile deformation with depth. As much as 10 km or more of upper crust has been removed tectonically and erosionally in places (the metamorphic core complexes), and therefore what was once midcrust is now at or near the surface. We address the prominent reflectivity in the midcrust by analyzing reflection

Thompson, G. A., Catchings, R., Goodwin, E., Holbrook, S., Jarchow, C., Mann, C., McCarthy, J., and Okaya, D., 1989, *in* Pakiser, L. C., and Mooney, W. D., Geophysical framework of the continental United States: Boulder, Colorado, Geological Society of America Memoir 172.

and borehole data in conjunction with synthetic seismogram modeling. Discontinuous, high-amplitude reflections in the midcrust are attributed to water-filled fracture zones and compositional layering on a scale of tens of meters. Geologically, primary lithologic variations (e.g., quartzofeldspathic to micaceous rocks) are enhanced by extension and subhorizontal shearing. Alignment of anisotropic minerals, especially micas, within zones of concentrated deformation may enhance velocity contrasts, producing the high-amplitude reflections observed.

Partial melting and inflation of the crust by mantle-derived basalts were probably important in generating the strongly reflective Moho at the base of the crust. Because this reflective boundary is subhorizontal across the province, we believe that it is a relatively young feature, indicative of the repeated reworking of the crust and upper mantle of this region of high heat flow and Cenozoic extension.

INTRODUCTION

The Basin and Range province, a unique region of broad continental rifting, is characterized by extensional fault-block mountains and deep, sediment-filled basins. In Nevada and Utah the province lies between the uplifted, comparatively unbroken blocks of the Sierra Nevada on the west and the Colorado Plateau on the east. To the south, in Arizona and New Mexico, the province sweeps around the Colorado Plateau to join the Rio Grande rift. The high elevations of the bounding regions (the Sierras and Colorado Plateau) are characteristic of rift margins worldwide.

During Cenozoic time, the northern Basin and Range province has tectonically widened about 250 km by the westward motion of the Sierra Nevada relative to the Colorado Plateau (for recent analyses, see Frei and others, 1984; Frei, 1986), and the present rate of extension is about 1 cm/yr (Thompson and Burke, 1973; Minster and Jordan, 1987). The total extension is comparable to that of such major (but narrower) rifts as the Red Sea (about 350 km total at the southeast end: Gettings and others, 1986) and the Gulf of California (about 400 km), and is far greater than that of the East African Rifts (Kenya Rift, 15 km: Baker and others, 1978).

Features characteristic of this extension include widespread seismicity (see Dewey and others, this volume), young Cenozoic fault scarps (e.g., Wallace, 1984), and abundant Cenozoic intrusive and extrusive igneous activity (Stewart and Carlson, 1976). In addition, the crust is thin (only 35 km or less throughout the Basin and Range) and the upper mantle P-wave velocities reported from refraction data are low (7.3 to 7.9 km/sec; see Pakiser, 1985, for a complete review). The heat flow in the province averages 1.5 to 2 times the value for stable continental areas (Lachenbruch and Sass, 1978), and is reflected in the volcanism and the reported low mantle velocities. Consistent with the heat flow, teleseismic delays of P-waves and proportionately larger delays of S-waves have been interpreted to indicate partial melting in the upper mantle (e.g., Romanowicz, 1979; see also Iyer and Hitchcock, this volume).

Despite the thin crust and the large amount of extension, the mean elevation in the province is high, 1.5 to 2 km in northern Nevada and Utah. The corresponding Bouguer anomaly (−150 to −200 mGal) indicates regional isostatic compensation but not compensation of individual ranges and basins, which have widths of only 10 to 30 km (Thompson and Burke, 1974). The source of this regional isostatic compensation dwells in the lower crust and within the upper mantle in some as-yet-unknown configuration, perhaps largely at the lithosphere-asthenosphere boundary. A model that incorporates a shallow, highly developed asthenosphere is consistent with the heat flow, teleseismic delays, and other evidence (e.g., Crough, 1983; Thompson and Zoback, 1979) and will be addressed below.

Many of these geophysical characteristics, including the seismicity, volcanism, thin crust, and high heat flow, are shared with other continental rifts and also with oceanic rifts. For these reasons, and because of the accessibility of the Basin and Range and its superb record of tectonic activity, the province deserves intensive study as a key to continental rifting processes. In this chapter we have chosen to highlight a few of the exciting and controversial avenues of geophysical research currently under study in the Basin and Range province. (For more comprehensive reviews of the geophysical and geologic characteristics of the province, readers are referred to Eaton, 1982, Stewart, 1978, Smith, 1978; Zoback and others, 1981; Thompson and Burke, 1974; Pakiser, 1985; Anderson, this volume). We begin by critically analyzing and interpreting representative data sets from the western part of the province, selected to illuminate the overall structure of the crust and upper mantle in this region. We also review the reflection, earthquake, and geodetic evidence that constrain the fundamental processes controlling extension in both the brittle upper crust and the ductile lower crust. Although much of our discussion pertains only to a few isolated studies located within the much broader Basin and Range province, we view the results from these studies as representative of the primary features of the crust and upper mantle throughout this broad zone of continental extension.

CRUSTAL STRUCTURE OF NORTHWESTERN NEVADA

The main features of the western Basin and Range province, including a thin crust and low upper-mantle velocity, have been interpreted from seismic refraction studies (e.g., Eaton, 1963; Hill

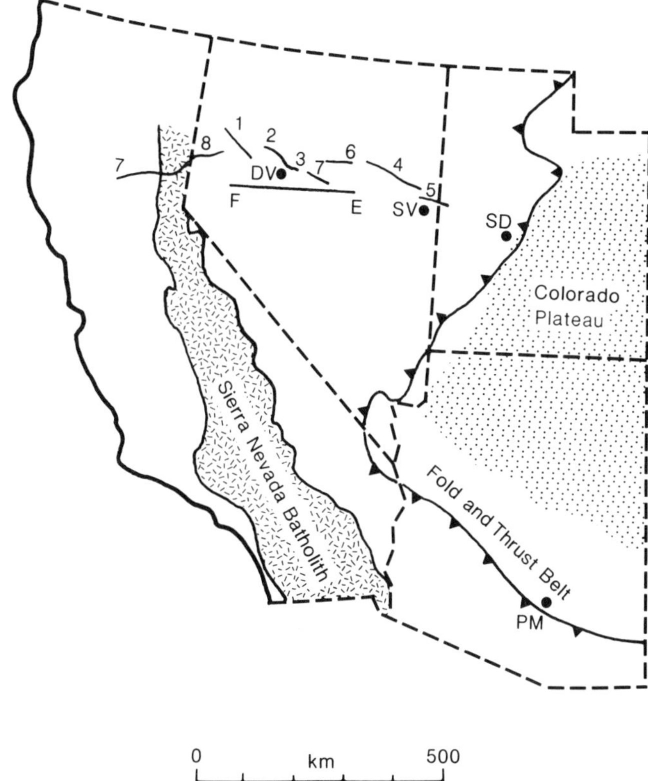

Figure 1. Location map of seismic data and areas discussed in this chapter. Lines from the COCORP 40°N Transect from west to east are: 8, 1, 2, 3, 7, 6, 4, and 5. F = Fallon; E = Eureka; DV = Dixie Valley; SV = Spring Valley; SD = Sevier Desert; PM = Picacho Mountains. The eastern extent of the Cordilleran fold-and-thrust-belt is delineated by thrust-fault markings.

and Pakiser, 1966, 1967; Prodehl, 1970, 1979; Stauber and Boore, 1978; Priestley and others, 1982; Stauber, 1980, 1983) and surface-wave dispersion studies (Priestley and Brune, 1978; Stauber, 1980). Perhaps the most important of these seismic studies for the northwestern Basin and Range is the Fallon-Eureka refraction profile recorded by the U.S. Geological Survey in the early 1960s (Eaton, 1963; Prodehl, 1979) (Fig. 1). Although the data collected along this profile represent sparse sampling of the seismic wavefield (there are gaps in spatial coverage of more than 25 km, some of which occur in areas critical for identifying crustal reflections and refractions), they are sufficient to provide a good, overall estimate of crustal structure across the northwestern Basin and Range province. Recently, deep seismic reflection profiles from industry and from the Consortium for Continental Reflection Profiling (COCORP) have complemented the Fallon-Eureka data with detailed information on the reflectivity and structure of the middle and deep crust (Allmendinger and others, 1983, 1987; Klemperer and others, 1986; Hauge and others, 1987; Potter and others, 1987).

Neither refraction nor reflection data alone are sufficient to reveal the correct crustal structure along the 40°N transect—the refraction data are inadequate because of sparse spatial sampling, resulting in limited resolution, whereas the reflection data lack velocity information for the lower crust and upper mantle, a necessity in defining boundary depths and crustal compositions. Thus, these data need to be integrated into one comprehensive model. The following discussion briefly reviews these existing refraction and reflection data sets before trying to reconcile them into such a model. We use gravity as a constraint on both the previous interpretations and the new interpretation presented here; we also use gravity as a check on a second, alternative model, which is equally consistent with the seismic observations.

In most previous work in the Basin and Range, the Moho was defined as the first major increase in seismic velocity to values of about 7.3 km/sec or higher. Although such velocities may be regarded as anomalously low P_n or upper-mantle velocities, we find evidence in the refraction data for a deeper transition, or discontinuity, to a normal mantle velocity of about 8.0 km/sec. This deeper transition correlates approximately with the base of layering in reflection sections, which has been called the "reflection Moho" (Klemperer and others, 1986). Between the traditional refraction Moho and the reflection Moho is a transitional layer that we call transitional crust-mantle. In the future, as new data become available, it may make better sense to identify the deeper discontinuity as both the reflection and the refraction Moho.

Basin and Range data sets and derived models

Fallon-Eureka refraction profile. Both Eaton (1963) and Prodehl (1979) interpreted the Fallon-Eureka refraction profile, using different methods of analysis and arriving at somewhat different results. Eaton proposed two possible crustal models (Fig. 2, top) the simpler of which has a uniform crust of 6.0 km/sec, about 23 km thick at Fallon and about 33 km at Eureka (all depths and thicknesses are given relative to a datum 1 km above sea level). Because this model predicts a 240-mGal change in the Bouguer gravity across the profile (assuming a density of 2.67 and 3.24 g/cm^3 for the crust and upper mantle, respectively), which greatly exceeds the 60-mGal value observed, we do not consider this model further. Eaton's second model includes a 6.6-km/sec lower-crustal layer, which increases the required crustal thickness by 2 to 3 km and reduces, but does not solve, the gravity problem. For both models, Eaton calculated an upper-mantle velocity of 7.8 km/sec. Prodehl (1979) reinterpreted the Fallon-Eureka data and estimated a thicker crust beneath both Eureka (35.5 km) and Fallon (30 km) and a slightly higher upper-mantle velocity (7.9 km/sec). His model also differs from Eaton's in the use of gradational boundaries and in the number and shape of crustal layers: he included, for example, a low-velocity zone beneath Eureka (Fig. 2, bottom).

The contrasting complexity of these two crustal models is primarily a result of the interpretation methods used to analyze the data. Eaton used a simple velocity-depth calculation such as that of Mota (1954) to determine the crustal structure. This

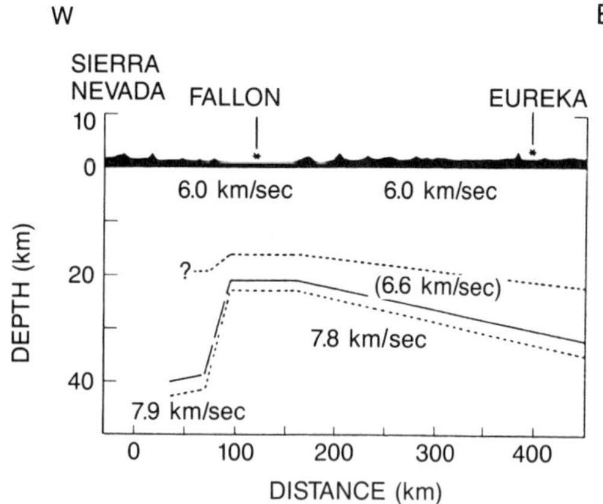

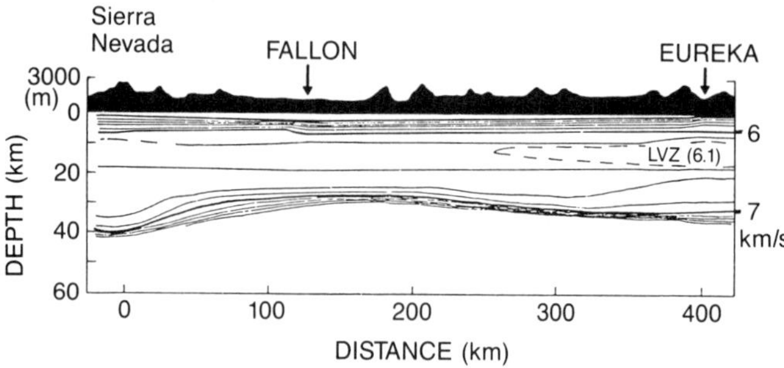

Figure 2. Two crustal models for the Fallon-Eureka profile proposed by Eaton (1963). Dashed lines define a two-layer crust with an upper crustal velocity of 6.0 km/sec and a lower crustal velocity of 6.6 km/sec, underlain by a 7.8-km/sec mantle velocity. An alternate model, represented by the solid line, defines a uniform crustal velocity of 6.0 km/sec overlying the 7.8-km/sec mantle. Bottom, Seismic model for the same profile determined by Prodehl (1979). Lines are velocity contours representing a gradational velocity structure. Velocity contours for 6.0 and 7.0 km/sec are labeled on the right side of the figure; the contour interval is 0.2 km/sec. Note the crustal low-velocity zone beneath Eureka and the significantly thicker crust beneath Fallon in this model.

method is reliable only in the case of planar layers with discrete velocity discontinuities. Prodehl used a method of interpretation developed by Giese (1966) that allows for vertical velocity gradients but still requires lateral homogeneity. Despite the differences in these two analysis techniques, both provide similar estimates of the gross crustal structure. This picture of crustal structure is further complemented by the COCORP reflection data, which show first-order similarities to the refraction models, as well as some tantalizing differences in detail.

COCORP Reflection Profile. In the early 1980s, COCORP conducted an east-west seismic reflection transect in nine segments at 40°N latitude across Nevada and California, about 100 km north of the Fallon-Eureka refraction profile (Fig. 1). The profiles reveal an upper crust characterized by fault-block mountains and asymmetric sedimentary basins, and a crudely laminated middle and lower crust characterized by subhorizontal, discontinuous reflections (Fig. 3). Reflectors in the lower crust are particularly strong and are prominent features on 80 percent of the stacked data (Hauge and others, 1987). The lower crustal laminated zone commonly spans more than a second and typically has a sharp planar base, referred to as the "reflection Moho" by Klemperer and others (1986). In many places the reflections from the lower crust coalesce into two distinct groups of reflections, an intracrustal event called "X" and a deeper Moho reflection, "M"; in other regions only the M event is present. Surprisingly, the reflection Moho appears to be continuous and gently undulating across the entire Great Basin and does not reflect the complex evolution of the region through Mesozoic and Paleozoic times (Klemperer and others, 1986). When corrected for the effects of sedimentary basins, and assuming a laterally constant crustal velocity, the reflection Moho gradually deepens eastward by approximately 7 km. Below the lower crustal reflective zone, the upper mantle appears seismically transparent.

Although individual segments of the COCORP survey lie generally 50 to 100 km north of the Fallon-Eureka refraction profile, the two experiments are close enough to be compared. In making this comparison, Klemperer and others (1986) correlated the increased reflectivity at the base of the crust with the strong velocity gradient modeled by Prodehl (1979). In addition, they noted that the Moho in Prodehl's model, when converted from depth to two-way traveltime, came closest to matching the reflection times of the reflection Moho recorded by COCORP. For these reasons they noted that the Prodehl refraction model was most nearly compatible with the reflection data, and they suggested that the reflection and refraction Mohos were approximately coincident. As a check on this correlation and as a test of the two grossly similar refraction models derived from the same data base, which in detail are actually different, we have chosen to reanalyze the Prodehl model using a more precise refraction algorithm written by J. H. Luetgert (written communicatin, 1985) and based on theory by Červený and others (1977).

Reanalysis of Prodehl model

Although Prodehl's (1979) model contains lateral variations, it was developed with a one-dimensional technique, whereby 1-D velocity-depth functions from different shotpoints were contoured into a two-dimensional model. This method can successfully model gross lateral velocity changes, but it is best suited as a starting point for more accurate, two-dimensional analysis. We have therefore digitized Prodehl's (1979) model (Fig. 4) and reanalyzed it with the 2-D ray-tracing technique of Červený and others (1977). This method allows for lateral velocity changes and is capable of modeling more realistic earth structures than the method used by Prodehl. Our reanalysis serves as a test of Prodehl's model as well as a motivation for the new model we present below.

Our results show that, although Prodehl's model accounts for most major refracted and reflected phases, some predicted

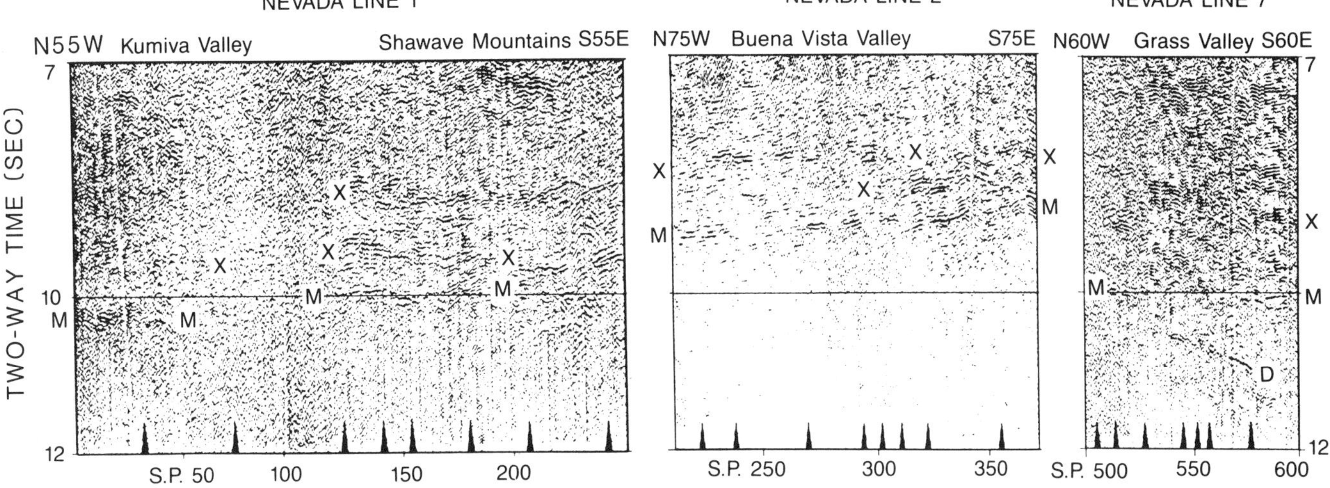

Figure 3. Data from COCORP lines 1, 2, and 7 showing the laminated character of the lower crust. The letters X and M, taken from Klemperer and others, (1986), enclose a packet of deep-crustal reflectors that are roughly consistent in depth and thickness with our 7.5-km/sec layer (after Klemperer and others, 1986).

phases are discontinuous, inconsistent with observed traveltimes, or not present in the data. From the Fallon shotpoint (Fig. 4A, B), some refracted arrivals in the 0- to 40-km range are more than 1 sec later than the calculated traveltimes, showing that Prodehl's model does not account for the low-velocity sediments of the Carson Sink. In addition, the model does not account for the first arrivals observed between 75 and 130 km or for the reflected phases in the two-second (reduced) time range (labeled b in Fig. 4A). These reflections provide critical constraints on the depth and geometry of intermediate and deep layers. Similarly, wide-angle reflections recorded between 200 and 275 km (1- to 4-sec reduced traveltime, Fig. 4A) are not accounted for in Prodehl's model.

Several discrepancies can also be observed in the interpretation of the phases from the Eureka shotpoint (Fig. 4C, D); (1) Calculated upper crustal arrivals are slightly later than the observed arrivals (0 to 70 km at about 1-sec reduced traveltime, Fig. 4C). (2) The high-amplitude reflections following these refracted phases in the 20- to 50-km distance range are not accounted for. (3) Prodehl's low-velocity zone located between 12 and 20 km depth near Eureka would prohibit refracted first arrivals from being observed between 70 and 130 km. These first-arrival refractions are, however, clearly observed on the seismic record. (4) Prodehl's model predicts that neither reflected nor refracted secondary arrivals should be observed in the range of 70 to 110 km because the gradient model does not allow energy in that range; these arrivals, however, are clearly present (labeled b, Fig. 4C). (5) Prodehl's model does not account for the wide-angle reflected energy observed beyond about 200 km (0- to 1-sec reduced time, Fig. 4c).

We also used 2-D gravity modeling to test Prodehl's (1979) crustal model. We constructed the gravity model (Fig. 5) by digitizing Prodehl's seismic model and assigning each layer a density based on its measured compressional wave velocity (Birch, 1961). The model shows that Prodehl's seismic refraction interpretation is reasonably consistent with the longest wavelength of the gravity data. The model predicts a 50-mGal drop in the anomaly between Fallon and Eureka, approximately what is observed. Prodehl's model provides a much better fit to the observed gravity data than Eaton's simple model, due largely to a thickening of lower crustal layers beneath Eureka. Prodehl's model is thus consistent with the gravity data; however, it incompletely accounts for several phases in the refraction data, as discussed above.

New model

Here we present a new model for the crust beneath northwestern Nevada. The new model (Fig. 6), which attempts to satisfy both the seismic refraction and reflection data, as well as the Bouguer gravity data, consists of six crustal and upper-mantle layers. Each layer contains velocity gradients, but each layer boundary is represented by a discrete velocity discontinuity. Although the new model is similar in general appearance to the models of both Eaton (1963) and Prodehl (1979), it differs in three specific ways.

First, we have accounted for the sedimentary basins (in particular, the Carson Sink) by including laterally variable, low-velocity layers in the uppermost crust. These layers are constrained only near the shotpoints, but we have extrapolated them across the model with the aid of geologic maps, short-wavelength gravity fluctuations, and the relative delays of refracted arrivals. The basins, particularly from 0 to 40 km east of Fallon, are important because they directly influence estimates of crustal thickness in the Fallon region. When basin delays are taken into account, the base of the crust is correspondingly shallower by as much as 2 to 3 km.

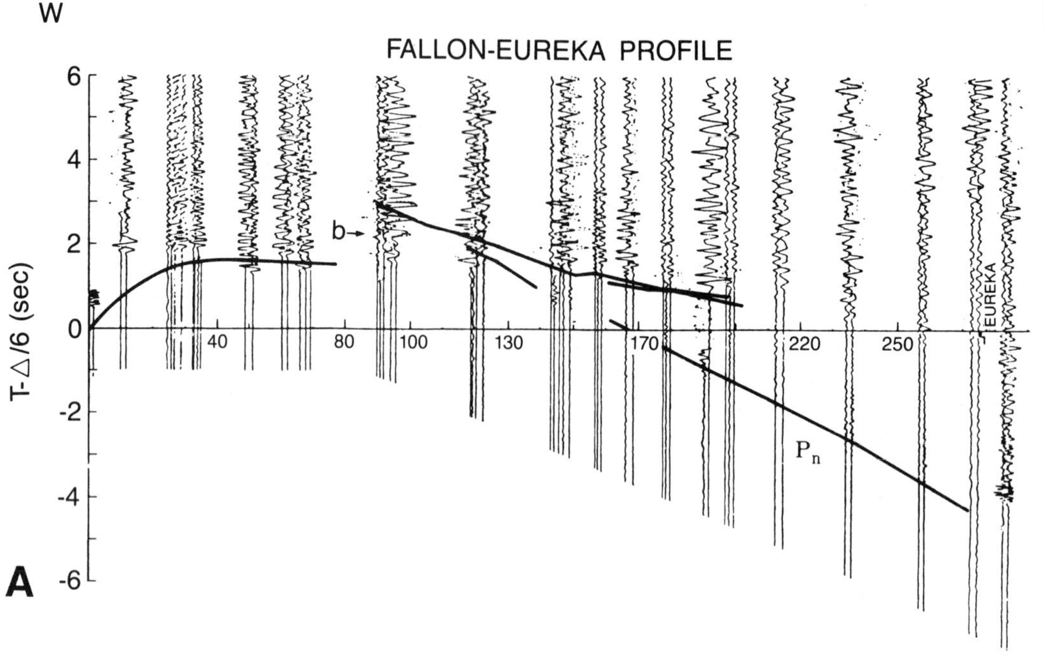

FALLON

EUREKA

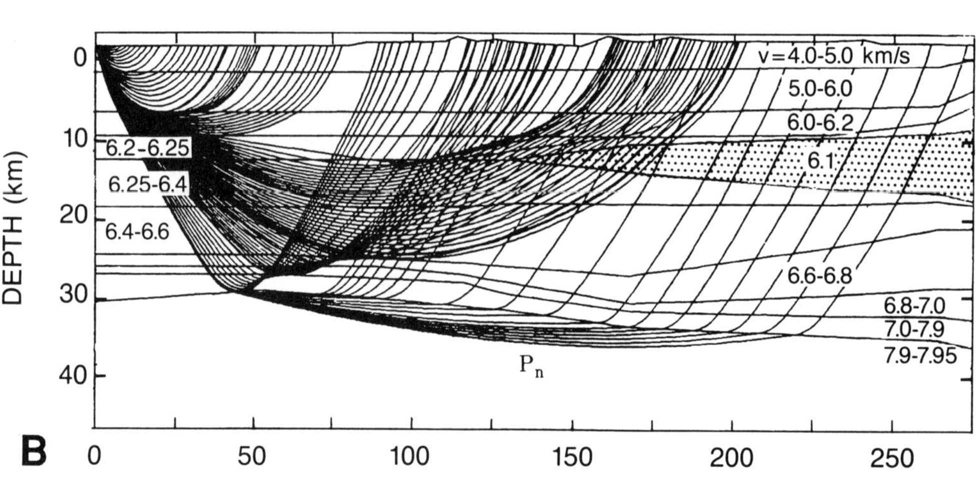

Figure 4. A, Fallon-Eureka seismic data plotted using a reducing velocity of 6.0 km/sec. The source was located at the Fallon shotpoint. Black lines show arrival times of phases predicted by Prodehl's (1979) seismic model. The events labeled b represent unmodeled midcrustal phases. B, Prodehl's (1979) gradational seismic model showing raypaths of important phases. Upper crustal low-velocity zone shown in stippled pattern. Depths are relative to a datum at sea level; layer velocities are in units of kilometers per second. C and D, Same as A and B, respectively, except the raypaths originate from the Eureka shotpoint.

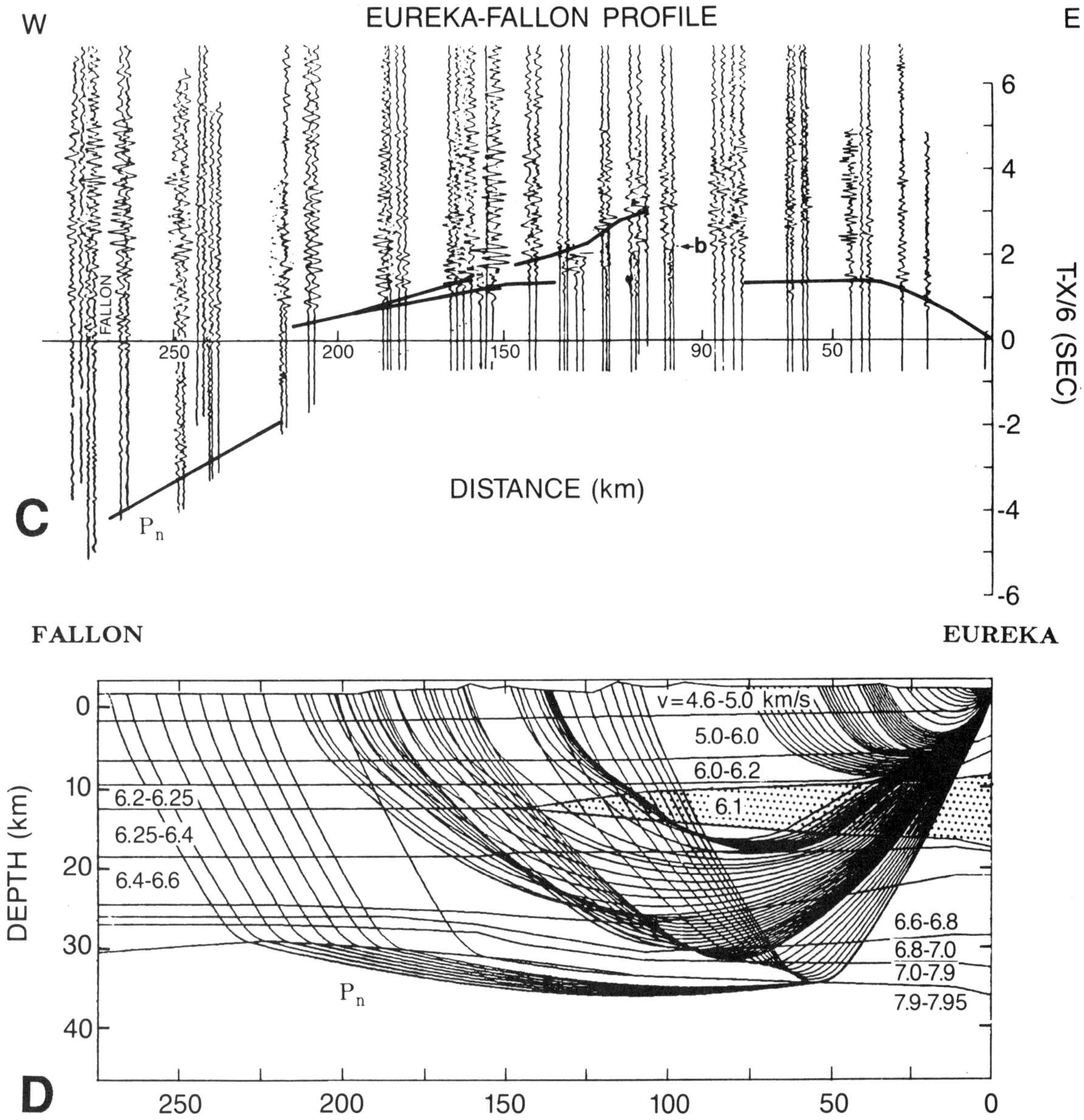

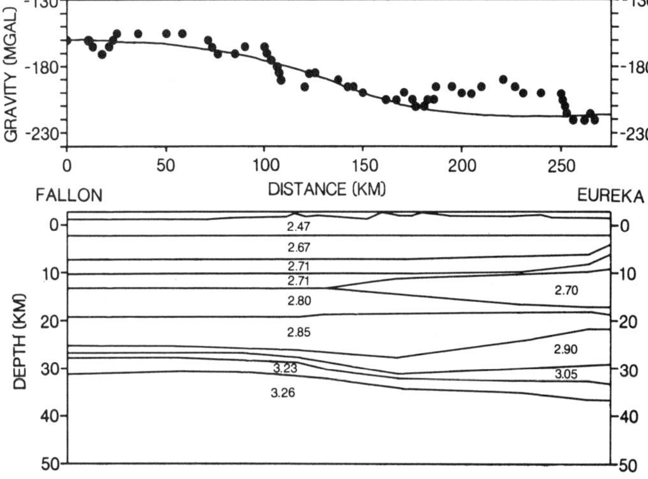

Figure 5. Gravity model derived from Prodehl's (1979) seismic model using the velocity-density relations of Birch (1961). Depths are relative to a datum at sea level. Velocities are in units of kilometers per second and densities are in units of grams per cubic centimeter. The small circles correspond to observed gravity obtained from the Society of Exploration Geophysicists' Gravity Map of the United States (1982). The solid line overlying the circles is the gravity anomaly predicted by the seismic model. Note that the model correctly predicts a 50-mGal drop in the gravity anomaly between Fallon and Eureka.

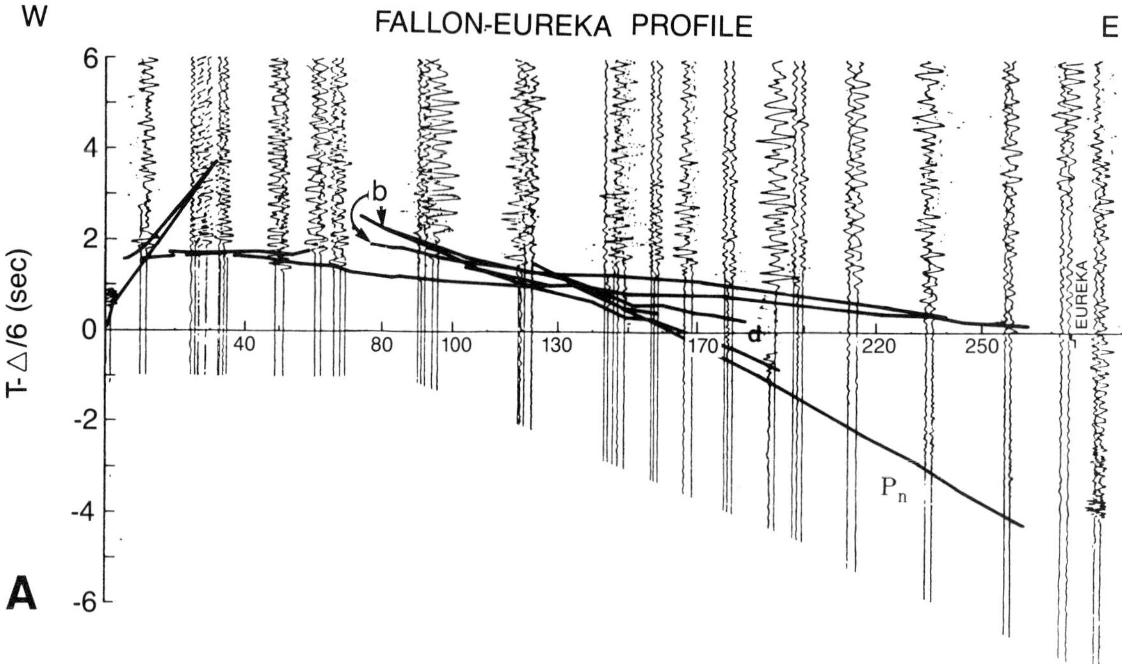

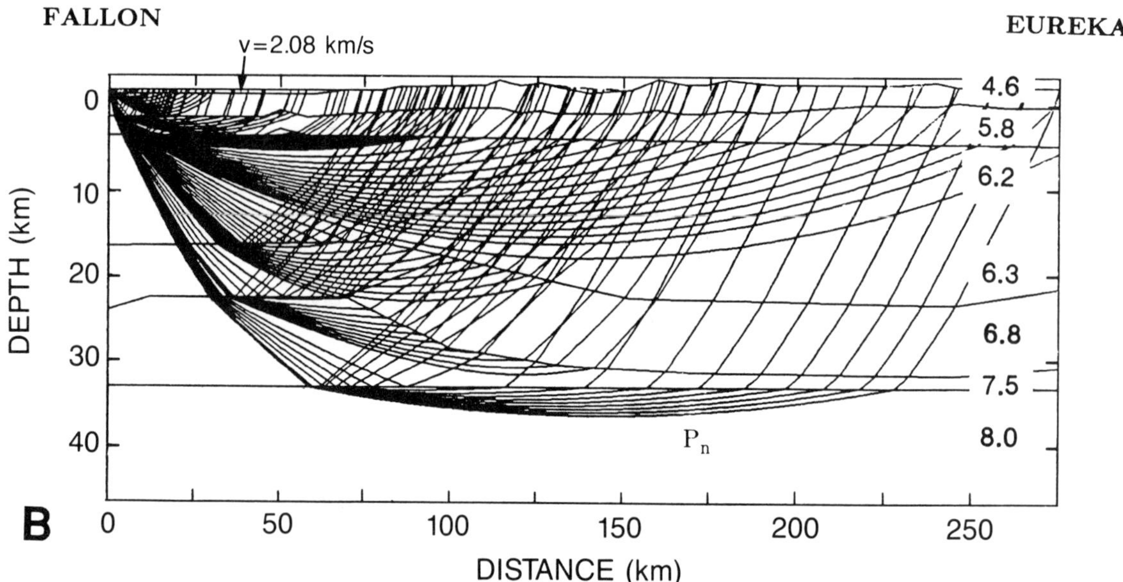

Figure 6. A, Fallon-Eureka refraction data plotted using a reducing velocity of 6.0 km/sec. The source was located at the Fallon shotpoint. Black lines show arrival times predicted by the new seismic model. Letters refer to phases discussed in the text. B, The new seismic model showing raypaths of important phases. Depths are relative to a datum at sea level. Layer velocities are in units of kilometers per second. Note the addition of more detail in the shallow velocity structure and the inclusion of the 7.5-km/sec transitional layer. C and D, Same as A and B, respectively, except data and predicted phases correspond to the Eureka shot.

Second, our interpretation of the velocity distribution in the middle and lower crust differs from that inferred by Eaton and Prodehl. We do not support the interpretation of a low-velocity zone in the midcrust (as reported by Prodehl) because refracted arrivals, although weak, are nevertheless observed continuously in the 75- to 130-km distance range. Additionally, we have included a lower crustal 6.8-km/sec layer in our model that corresponds approximately to the 6.6-km/sec layer modeled by Eaton, and to the "b" reflected and refracted phases noted by Prodehl (Figs. 2 and 4, respectively). We position this lower crustal layer at a depth ranging from 16.5 km beneath Fallon to about 21.5 km beneath Eureka; these depths are somewhat greater than those inferred by Eaton.

Finally, our revised model includes a deep, 7.5-km/sec

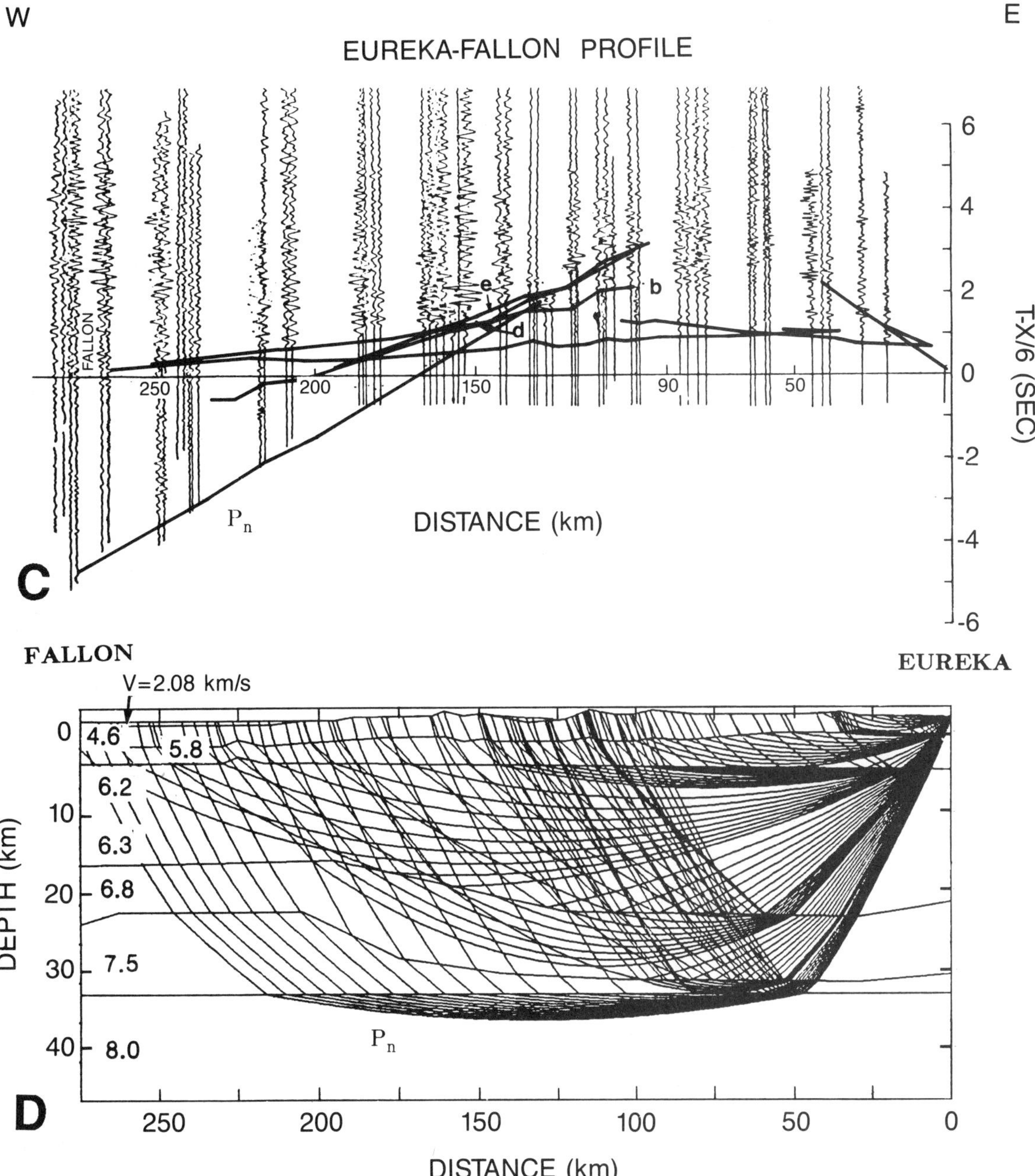

layer, transitional between the lower crust and normal upper mantle (see the refraction and reflection events labeled d, Fig. 6A, C). Because this intermediate layer, which we call transitional crust-mantle, is poorly sampled spatially, we are unable to determine its precise velocity. We have used 7.5-km/sec because that velocity appears to fit the observed refraction data best, although velocities of 7.3 to 7.6 km/sec also provide adequate fits. The depth to the 7.5 km/sec layer roughly corresponds to the Moho of previous workers. Beneath Fallon, the top of this transition is at 24.0 km, but it increases to 30 km beneath Eureka. Because of the relief on this layer, reflected arrivals from Fallon are absent beyond 155 km, and refracted arrivals are absent beyond 170 km. Beneath Eureka the layer is greatly thinned, making reflections and refractions from the bottom and top of the layer virtually indistinguishable. Where these events merge, we have labeled them collectively as e (Fig. 6C).

We have modeled the boundary between the base of this transition zone and the underlying mantle to be approximately flat at a depth of 33.5 km below sea level. As in the layer above, the normal upper-mantle velocity is difficult to determine precisely. The complex geometry of the lower crust, the uncertainty in shallow basin depths, and the poor spatial sampling permit a range of velocities between 7.8 and 8.1 km/sec. Nevertheless, we prefer a velocity of 7.9 to 8.0 km/sec, based on the fit to the observed refraction arrivals (Fig. 6).

Additional support for a crust-mantle transition zone comes

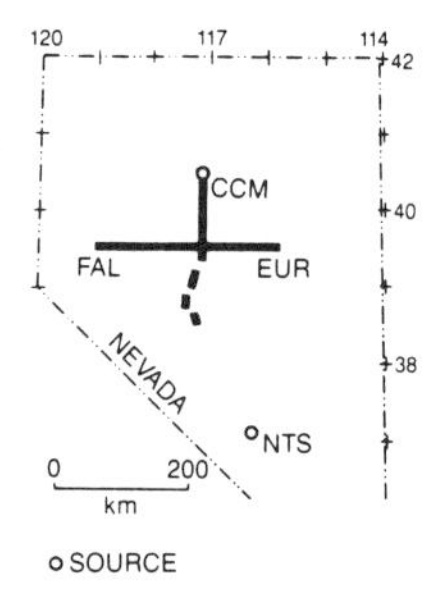

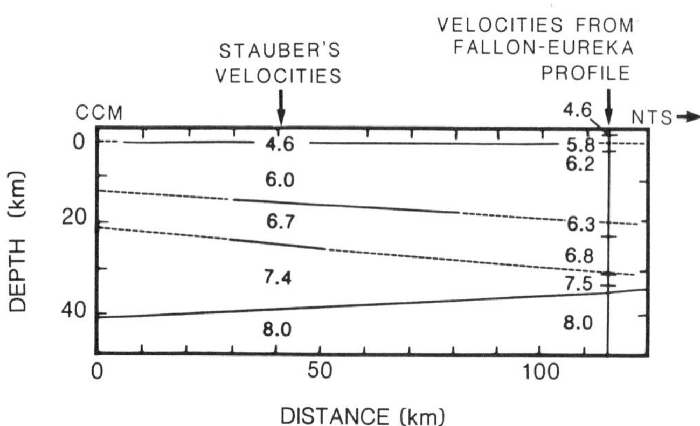

Figure 7. Top, Location map of Stauber's (1980) refraction line between Copper Canyon Mine (CCM) and the Nevada Test Site (NTS) relative to the Fallon-Eureka profile. Bottom, Stauber's alternate model for the profile between Copper Canyon and NTS. Stauber's velocities are shown on the left. Note approximate correspondence of velocities and depths between the new seismic model and Stauber's model. Velocities are in kilometers per second (after Stauber, 1980).

from work by Stauber (1980) in northwestern Nevada. Based on reversed and unreversed refraction profiles derived from quarry blasts, nuclear tests, and a down-hole shot, Stauber arrived at two possible crustal models. In his preferred model, the crust thins north of the Fallon-Eureka profile by as much as 6 km. In his alternative model, a northward-thickening layer with a velocity of 7.4 km/sec overlies normal mantle. Where our model of the Fallon-Eureka profile and Stauber's alternative model cross, they agree approximately in both depth and velocity. In particular, the top and bottom of his 7.4-km/sec layer correspond approximately to the upper and lower boundaries of our transitional crust-mantle layer (Fig. 7).

This boundary between the base of the transition zone and the underlying mantle can be correlated only roughly with the reflection Moho of Klemperer and others (1986) because the reflection and refraction lines are far from coincident. Nevertheless, the two-way traveltimes observed in the COCORP data agree to first order with those predicted by our model. However, according to Klemperer and others, the decreasing reflection time to the west (2.2 sec) implies a 7-km decrease in depth (assuming laterally constant crustal velocity), while a computer-generated 2-D reflection synthetic of our refraction model yields a nearly constant reflection time. This indicates either a northward thinning of the crust or a lateral change in crustal velocities. These possibilities can only be investigated with new data.

The new model offers a possible explanation for the zone of strong reflectivity in the lower crust (Fig. 3). This packet of reflections, including the Moho and X reflections of Klemperer and others, is several kilometers thick and is included within the welt of material we define as transitional crust-mantle. Hence, it is possible that these events represent abundant layering in the lower half of this "pillow," where mafic and ultramafic material may be interlayered.

It is apparent in this comparison that the reflection and refraction data are complementary. The packet of reflections in the lower crust is not evident in the refraction model, whereas the top of the transitional crust-mantle pillow can best be identified from refraction data. This difference is largely an effect of the contrasting resolution in the two techniques and serves to emphasize the importance of combining conventional reflection profiling with refraction/wide-angle reflection studies wherever possible.

Bouguer gravity of Nevada. To further examine this new crustal model, we tested it and an alternative against Bouguer gravity. The alternative model eliminates the transitional crust-mantle pillow in favor of thinning of the mantle lithosphere. Digitized Bouguer gravity profiles were constructed from the Gravity Map of the United States (Society of Exploration Geophysicists, 1982) along the Fallon-Eureka and COCORP transects. Since the anomaly varies little between the two transects, only the regional profile along 40°N is discussed here.

A one-dimensional Fourier transform reveals that there are four main components of the anomaly, with spatial wavelengths of about 590, 160, 46, and 32 km (Fig. 8). Comparison of the gravity with the geologic map of Nevada shows that the smaller wavelengths in the gravity field can be correlated with surface geologic features. For example, the two shortest wavelengths can be attributed to individual basins and ranges. The longest wavelength, 590 km, correlates with the regional elevation changes from the Lahontan Basin, centered near the Carson Sink, to the Bonneville Basin in the Sevier Desert (see also Eaton, 1980). Thus, the 590-km component must be associated with large lateral density variations in the lithosphere, such as changes in crustal thickness, in lithospheric thickness, or in upper-mantle density. We present two alternative models, one based on our new refraction seismic model with the addition of a slightly thinned upper-mantle lithosphere and the other based on earlier seismic models with no crust-mantle transition layer and with a greatly thinned upper-mantle lithosphere.

A gravity model derived directly from the new seismic model agrees reasonably well with the longest wavelength (590 km) of the observed gravity (Fig. 9), although the amplitude of the computed anomaly (approximately 65 mGal) is slightly large. The main contribution to the long-wavelength Bouguer gravity anomaly comes from two seismically constrained boundaries, the midcrustal (6.3 to 6.8 km/sec) and the transitional crust-mantle (6.8 to 7.5 km/sec) boundaries; this gravity model does not in-

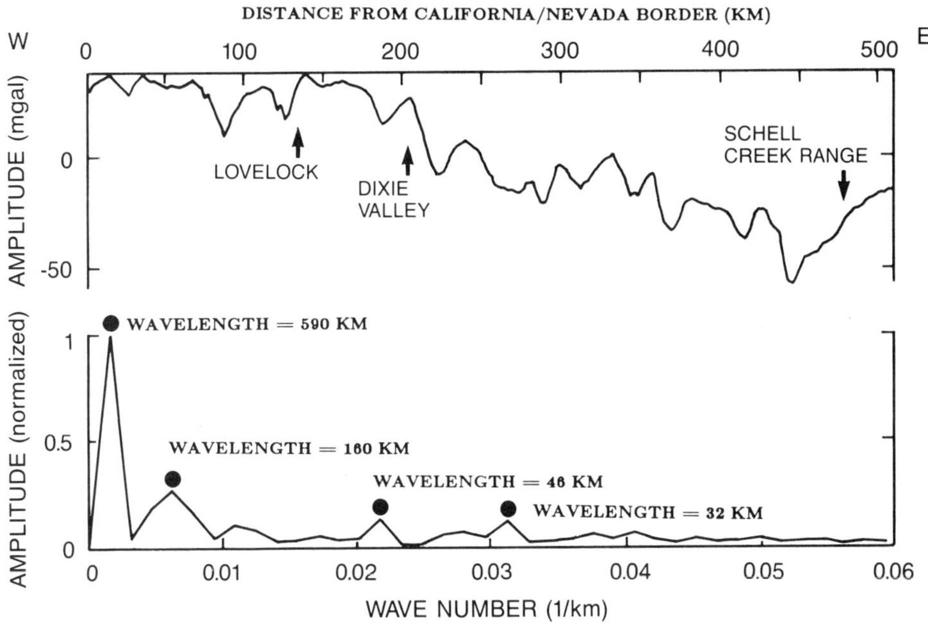

Figure 8. Top, Bouguer gravity profile with a constant value subtracted along COCORP 40°N transect. Data were obtained from the Society of Exploration Geophysicists' Gravity Map of the United States (1982). Bottom, Fourier transform of 40°N Bouguer gravity profile. Dots correspond to amplitude spectrum peaks discussed in the text.

clude the possible effects of variation in lithospheric thickness. By including a moderate amount (about 11 km) of lithospheric thinning in the western Nevada region, however, a slightly better fit between the observed and calculated gravity anomalies is obtained (Fig. 9). Seismic shear-wave and surface-wave investigations (Priestly and others, 1982; Taylor and Patton, 1986) suggest that the lithosphere/asthenosphere boundary is located at about 65 to 70 km beneath western Nevada, which is consistent with our preferred gravity model.

We have also developed an alternative model with no crust-mantle transition layer (Fig. 9). This model incorporates 34 km of relief on the lithosphere/asthenosphere boundary, a density of 3.30 g/cm^3 for the upper mantle lithosphere and 3.22 g/cm^3 for the asthenosphere. Although the resulting 40-km-thick lithosphere is 25 to 30 km thinner than that inferred by Priestley and others (1982) and Taylor and Patton (1986), shifting the boundary downward in the model while increasing the relief on the boundary will still permit a fit to the gravity. However, large relief on the boundary conflicts with the findings of Taylor and Patton.

Various models between our preferred model and the alternatives are also possible. Small modifications in densities, resulting in proportionally large modifications in density contrasts, would also influence the geometry of the models. Further tests of these two models will depend on better definition of the crust by new refraction experiments and perhaps by better resolution of the upper mantle by surface-wave and teleseismic body-wave observations.

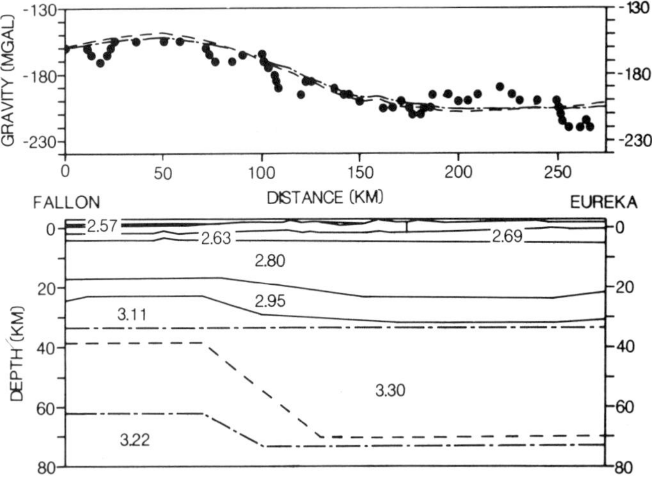

Figure 9. Gravity models derived from new seismic model (solid lines and dash-dot lines) and the alternate model (solid lines and short-dash lines) using the velocity-density relations of Birch (1961). Depths are relative to a datum 3 km above sea level. Layer velocities are in units of kilometers per second and densities are in units of grams per cubic centimeter. The circles correspond to gravity values obtained from the Society of Exploration Geophysicists' Gravity Map of the United States (1982). The lines overlying the circles are the gravity anomalies predicted by the two models.

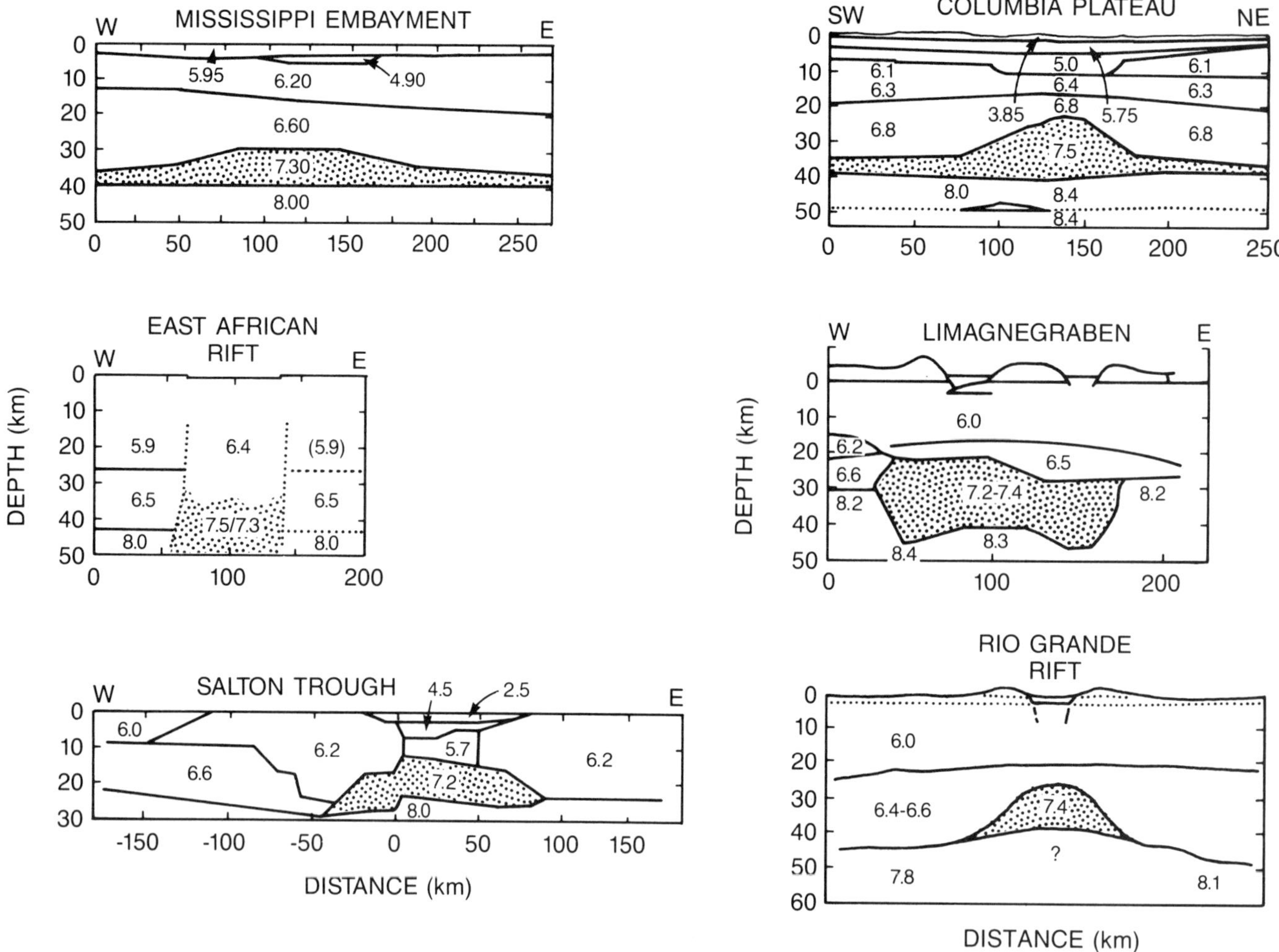

Figure 10. A comparison of seismic models for several rift zones that are interpreted to have anomalous (>7 km/sec) lower crustal layers (stippled). These layers may be similar to the transitional crust-mantle layer evident in our new seismic refraction model (after Mooney and others, 1983).

Transitional crust-mantle layer

Our reevaluation of refraction, reflection, and gravity data reveals a possible crustal structure for the Fallon-Eureka region similar to that reported in other continental rift zones. Several continental rifts are underlain by a deep "rift pillow" with a velocity of 7.3 to 7.6 km/sec overlying normal mantle with a velocity of about 8.0 km/sec (Mooney and others, 1983; Catchings 1987; Fig. 10). The Fallon-Eureka crustal structure is broadly consistent with these models. Superimposed on the block-fault basins and ranges that extend across the entire province (Stewart, 1971) are two long-wavelength topographic lows between the Sierra Nevada and the Colorado Plateau at 40°N. Both are marked by prominent regional gravity highs on Bouguer anomaly maps. The western of these two topographic lows, the Lahontan Basin, centered near Fallon in the Carson Sink, coincides with the region of our transitional crust-mantle, as well as with the region of unusually thin crust and low P_n velocity reported in previous refraction studies. Thus, the broad depression of the Lahontan Basin, together with the crust-mantle transition layer, defines a crustal structure similar to that reported in other continental rifts. Perhaps one indication that this area is tectonically extending more rapidly than other areas of the western Basin and Range is that historical earthquakes and Quaternary faulting are more pronounced here.

Although better constrained with the addition of the COCORP reflection results and with more sophisticated modeling techniques, our preferred interpretation has some similarities with Eaton's (1963) original model. We have modeled with more precision the geometry of the upper crustal basement surface and the midcrustal boundaries, and we have added a transitional crust-mantle layer. This layer, although still poorly constrained, may prove to be a critical element in the ultimate understanding of extensional processes. The confirmation of such a crust-mantle transitional layer (versus the alternative model of an upwelling in the asthenosphere), however, must await the analysis of the recently acquired PASSCAL reflection and refraction data collected approximately along the COCORP transect (Thompson and others, 1986; PASSCAL, 1988).

For the remainder of this chapter we focus in on smaller scale structures characteristic of the crust in the Basin and Range province. Specifically, we review the present-day understanding of fault geometries in the brittle upper crust, and present new evidence for extension-related ductile deformation in the lower crust.

THE UPPER CRUST: PRESENT-DAY FAULT GEOMETRIES

On a province-wide scale, Basin and Range normal faulting is remarkably uniform in both strike (approximately north-northeast) and fault spacing (15 ± 2.5 km) (Eaton, 1982). While the relationship between state of stress, surface fault trends, and slip directions is fairly clear, the relationship between the state of stress and subsurface fault geometry in the Basin and Range is more complex, and as yet it not understood. Individual faults are surprisingly variable in attitude, ranging from steep (60°-dipping) planar faults to subhorizontal (<12°) low-angle normal faults (e.g., Proffett, 1977; Wernicke, 1981, 1985; Anderson and others, 1983; Cape and others, 1983; Smith and Bruhn, 1984; Okaya and Thompson, 1985). Because an improved understanding of the subsurface fault geometries in the Basin and Range province is extremely important to our ultimate understanding of crustal rheology, the state of stress, and the way in which the entire crust extends, we here review the evidence documenting three fundamentally different styles of faulting.

The most powerful geophysical tool for delineating subsurface fault geometries is the seismic reflection technique. Anderson and others (1983) and Zoback and Anderson (1983) interpreted seismic reflection data from Dixie, Diamond, Railroad, Grass, Goshute, Mary's River, and Raft River Valleys, Nevada; the Sevier Desert Basin, Utah; the Fallon Basin, Nevada; and the Carson Desert, Nevada; they concluded that three basin styles exist in the Basin and Range province: (1) relatively simple sags associated with one or more major steep, planar normal faults; (2) prisms above tilted bedrock ramps that are displaced by moderately to deeply penetrating listric normal faults; and (3) assemblages of complexly deformed subbasins associated with sharply curving shallow listric faults and planar faults that sole in a detachment surface. To illustrate these three basin styles, we present seismic reflection lines from three basins: Dixie Valley, Nevada; Spring Valley, Nevada; and the Sevier Desert, Utah. In addition, we supplement this review with complementary geologic, geodetic, and seismicity data from these and other areas in the Basin and Range.

Steep, planar normal faults. Unmigrated and migrated versions of a seismic line from Dixie Valley (Okaya and Thompson, 1985) are shown in Figure 11. This seismic line illustrates the reflection character of a relatively simple basinal sag associated with one or more steeply dipping planar normal faults. Although

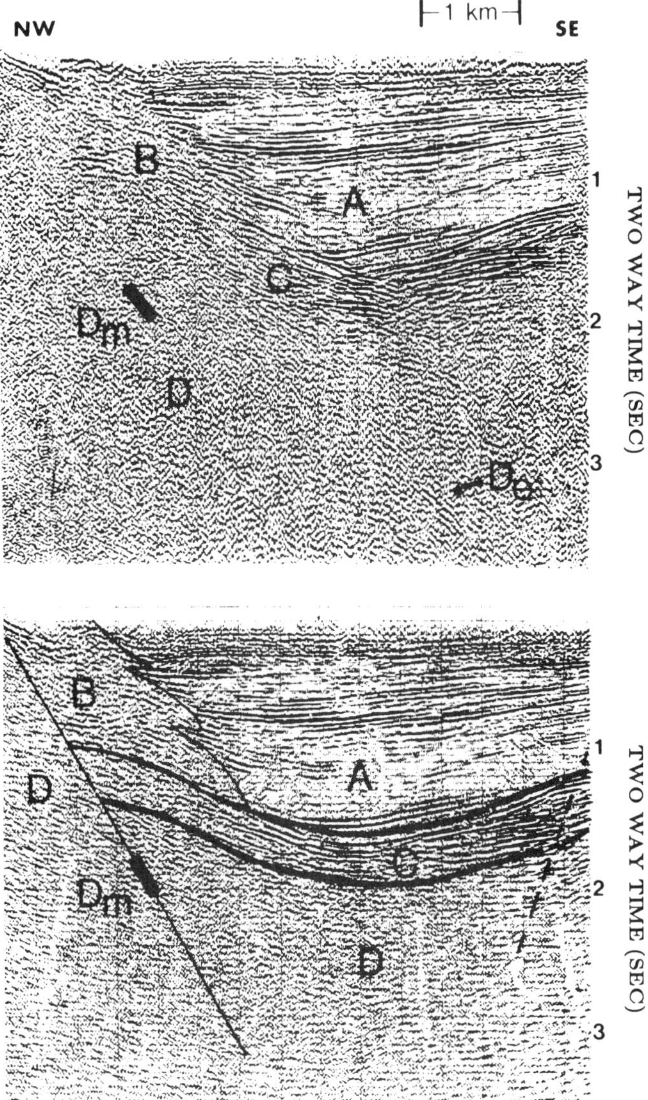

Figure 11. Unmigrated (top) and migrated (bottom) versions of a seismic line from Dixie Valley, Nevada. A = Lacustrine and playa deposits; B = alluvial fan; C = Tertiary volcaniclastic sequence; D = Mesozoic basement; D_e = fault plane reflection; D_m = hand migration of event D_e (from Okaya and Thompson, 1985).

the basin-bounding fault does not appear as a prominent reflection on these profiles, the basin strata are highly reflective and reveal a northwest-migrating basinal axis attributable to tilting or sagging of the basin during deposition (Okaya and Thompson, 1985). In the Dixie Valley study, the attitude of the basin-bounding fault plane was determined by hand-migrating a fault-plane reflection. The planar geometry suggested by the location of the migrated event is corroborated by other pieces of evidence: (1) gravity modeling, which suggests the same fault geometry; (2) surface observations of 50° dip, which connect the fault with the migrated event (event D_m in Fig. 11); (3) lack of stratal tilts,

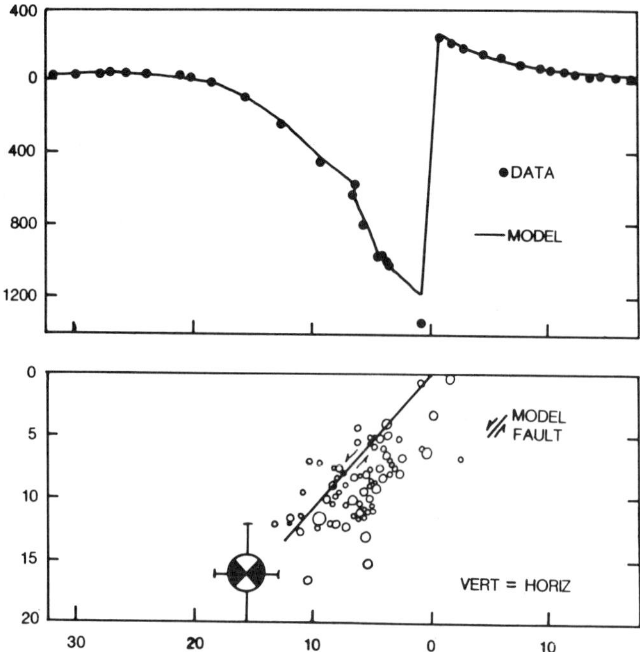

Figure 12. Plot of observed elevation changes (dots) and the predicted elevation changes (line) of the best-fit dislocation model for the Lost River Fault that ruptured during the Borah Peak, Idaho, earthquake. Bottom, plot of the dislocation model (line), together with the main shock and aftershocks of the Borah Peak earthquake (from Stein and Barrientos, 1985).

rendering listric geometries unlikely; and (4) earthquake data that indicate that the fault is planar to a depth of 15 km (Okaya and Thompson, 1985). Thus, Okaya and Thompson concluded that Dixie Valley is an asymmetric, tilted basin bounded to the northwest and internally cut by steeply dipping, planar normal faults.

Geodetic leveling studies provide independent constraints for the existence of deep, planar normal faults. In this method, surface deformation caused by dislocations in an elastic half-space is compared to differences in geodetic data measured before and after an earthquake. Stein and Barrientos (1985) have used this technique to constrain, with remarkable precision, the subsurface geometry of the Lost River fault, which ruptured during the earthquake at Borah Peak Idaho in 1983 (magnitude, 7.0). They found that the geodetic data in the vicinity of the fault is fit best by a planar dislocation with uniform dip-slip of 2.05 ± 0.10 m dipping 47° ± 2° southwest and extending to a depth of 13.3 ± 1.2 km (Fig. 12). Within the expected errors, this model fits the elevation changes at 94 percnt of the benchmarks. Stein and Barrientos (1985) also attempted to fit the data with a listric fault geometry, but the residuals for such a model were 2.0 times larger than those of the planar model.

Seismicity studies also provide evidence for steep, planar faults. Although earthquake studies generally have been only moderately successful in delineating subsurface Basin and Range fault geometries (simply because most Basin and Range earthquakes do not correlate with Quaternary faulting; Thompson and Burke, 1973; Smith, 1978; Zoback and others, 1981), a few notable exceptions do exist. In particular, studies conducted in the Fairview Peak–Dixie Valley, Nevada, region in 1980 and 1981 by the University of Nevada Seismological Laboratory (Vetter and others, 1983) documented that the seismogenic upper crust in this region is about 15 km thick, with a peak in seismic activity in the depth range of 10 to 12 km. Their results showed that the basin-bounding faults are approximately planar and maintain a dip of 50° to 60° to the base of the seismogenic crust. Similarly, observations by Smith and others (1985) of 43 of the largest aftershocks of the 1983 Borah Peak earthquake, and teleseismic waveform modeling by Okaya and Thompson (1985) of the 1954 Dixie Valley earthquake, suggest planar faulting that extends to depths of 15 km. Lastly, Jackson (1987), utilizing a worldwide review of focal depths and fault-plane solutions for large normal-fault earthquakes on continents, has shown that the overwhelming majority of earthquakes with magnitudes greater than 6.0 nucleate in the depth range of 6 to 15 km on faults dipping between 30° and 60°. Thus, steep, planar normal faulting in the top 15 km of the crust may be the most important factor controlling extension in the brittle, upper crust of the Basin and Range.

Listric normal faults. Planar, steeply dipping normal faults are the most abundant, but certainly are not the only type of fault observed in the Basin and Range province. A migrated seismic section from Spring Valley, Nevada (Fig. 1), shows a sedimentary prism that is structurally controlled by a listric normal fault (Gans and others, 1985; Fig. 13).

On the basis of seismic character, Gans and his co-workers divided the Spring Valley Basin fill into three zones (Fig. 14). All three zones thicken substantially toward the western end of the basin, suggesting rotational movement on the Schell Creek fault. This fault can be traced from surface outcrop to 1.5 sec beneath shotpoint 2750 along an east-dipping band of fault-plane reflections. Depth conversion indicates that the fault is planar to at least 2.5 km and has a dip of 35° to 50°E. Gans and his co-workers speculate that the Schell Creek fault may flatten at depth into the horizontal zone of reflections labeled L in Figure 14. If this interpretation is correct, then Spring Valley has developed above a listric normal fault and is a type B basin in the classification scheme of Anderson and others (1983) and Zoback and Anderson (1983).

In addition to Spring Valley, there are several other regions where listric fault geometries have been imaged on seismic reflection profiles in the Basin and Range. For example, Anderson and others (1983) noted a listric fault that controls extension in Goshute Valley, Nevada, while Smith and Bruhn (1984) documented abundant listric normal faulting in the eastern Great Basin near the Wasatch fault. Undoubtedly the most common occurrence of listric normal faults is in association with underlying zones of detachment. Cape and others (1983) have shown, for example, that numerous listric faults are common in the upper 5

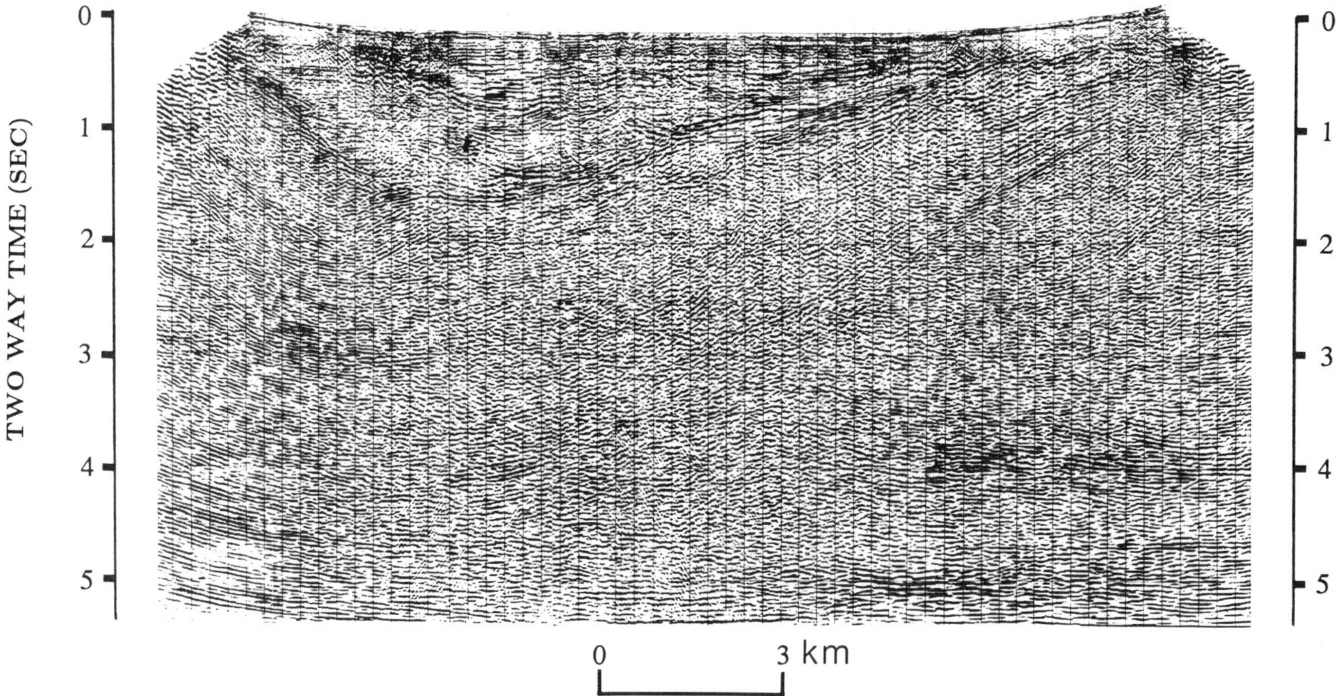

Figure 13. Migrated SOHIO Spring Valley seismic line (from Gans and others, 1985).

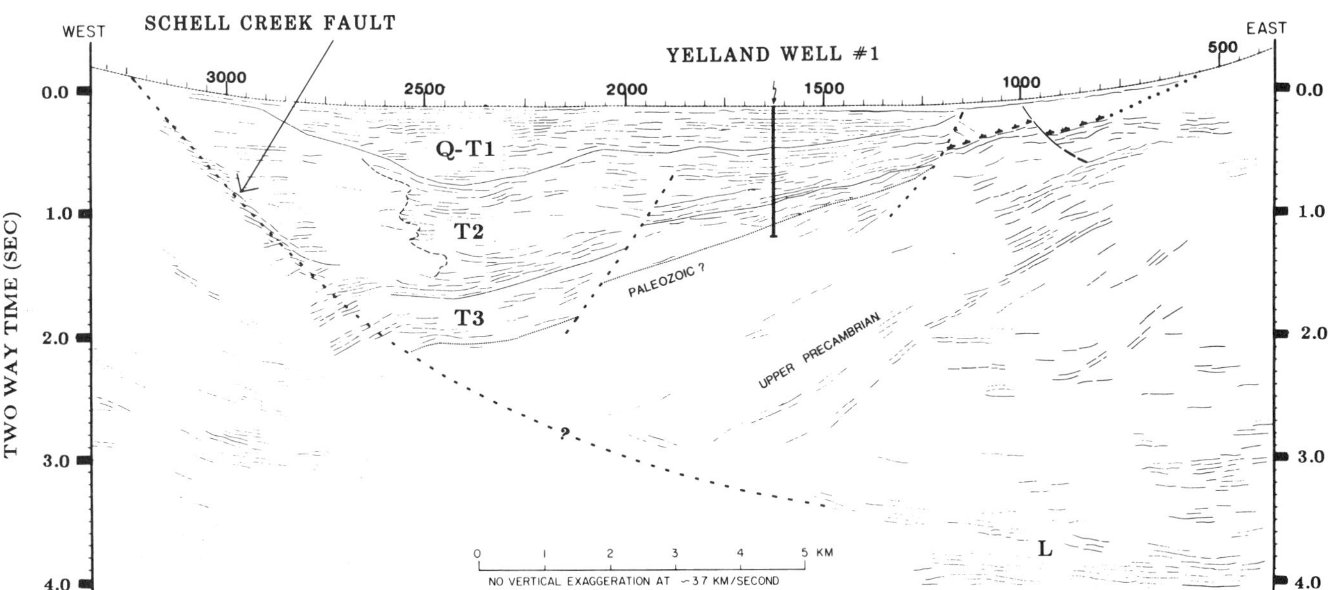

Figure 14. Interpretive line drawing of SOHIO Spring Valley seismic line. Q-T1 = Zone of subhorizontal reflection character composed of conglomerates, lacustrine limestone, and sandstone; T2 = same as zone Q-T1, except near the Schell Creek Fault, where the zone is nearly transparent and is composed of alluvial fan material; T3 = zone of high amplitude, somewhat continuous reflections that are correlated with volcanic rocks; L = zone of subhorizontal reflectors through which the basin-bounding Schell Creek Fault may flatten (from Gans and others, 1985).

km of the crust in the Rio Grande rift and can be observed to flatten abruptly into a zone of subhorizontal detachment at depth. Similarly, listric normal faults are also common in the upper plate of the Sevier Desert Detachment, where they are observed to sole onto the detachment surface (McDonald, 1976; Anderson and others, 1983; Van Tish and others, 1985).

Low-angle normal faults. Low-angle normal faults, commonly referred to as detachment faults, constitute the third and final example of major fault styles in the Basin and Range province. The unmigrated seismic reflection profile shown in Figure 15 from west-central Utah (Fig. 1) illustrates a well-bedded, wedge-shaped basin perched above a prominent event at 4.0 sec (CDP, 1700 to 1900). This event, interpreted as the Sevier Desert Detachment, is clearly evident throughout the entire western half and far eastern end of the line. The structural significance of the detachment has been analyzed thoroughly in the literature. McDonald (1976), the first to publish seismic reflection data revealing the detachment, interpreted it as a thrust with subsequent normal movement. As part of their 40°N transect, COCORP shot a 170-km east-west seismic line that traversed the Sevier Desert. On the COCORP lines, the detachment extends more than 120 km perpendicular to strike and can be traced from near the surface to a depth of 12 to 15 km (Allmendinger and others, 1983; Von Tish and others, 1985). Palinspastic restorations constructed by Sharp (1984) indicate that the Sevier Desert Detachment is mostly or entirely a new Cenozoic normal fault along which tens of kilometers of extension have taken place.

In addition to the Sevier Desert Detachment, several other examples of low-angle normal faults have recently been identified. These detachment faults, most commonly associated with metamorphic core complexes, have now been imaged in eastern Nevada (Hurich and others, 1985; McCarthy, 1986), southeastern California (Frost and Okaya, 1986a), and southern Arizona (Frost and Okaya, 1986b). Despite the now-common occurrence of low-angle normal faults identified either from surface exposures or from shallow reflection profiles, none have been shown to be active seismically. If our historical sampling window is long enough, this result requires that these features must move aseismically (Jackson, 1987). This conclusion is also supported by geologic studies that document the importance of high fluid pressures (Reynolds and Lister, 1987) and elevated temperatures (Miller and others, 1983; Lucchitta, 1985) during extension. Such factors could indeed influence the effective stress built up along the fault and its tendency to move aseismically.

Extensional models

While the geometry of the Sevier Desert detachment and the timing of its motion appear to be well known, the mechanism for forming such a low-angle normal fault remains enigmatic (Zoback and others, 1985). Indeed, the mechanism of extension for the *entire* Basin and Range province remains unclear. Several kinematic models for extension have been proposed. Wernicke (1981) suggested that extension occurs by "uniform-sense normal simple shear" along lithospheric-scale low-angle detachments (Fig. 16, top). According to Wernicke, the Sevier Desert detachment is one such low-angle fault. Another explanation for the low-angle detachments is the pure shear model (Fig. 16, middle), in which detachments are thought to mark the brittle-ductile transition in the crust (Smith, 1978; Rehrig and Reynolds, 1980; Miller and others, 1983). In this model, high-angle brittle normal faulting occurs above the transition, and ductile stretching along with intrusive dilation occurs below it. The "megaboudin" model (Fig. 16, bottom) incorporates aspects of both the pure shear and simple shear models (Davis and Coney, 1979; Hamilton, 1982; Kligfield and others, 1984), with deformation below the brittle-ductile transition occurring along anastomosing ductile shear zones separated by lenses, or "boudins," of brittle material.

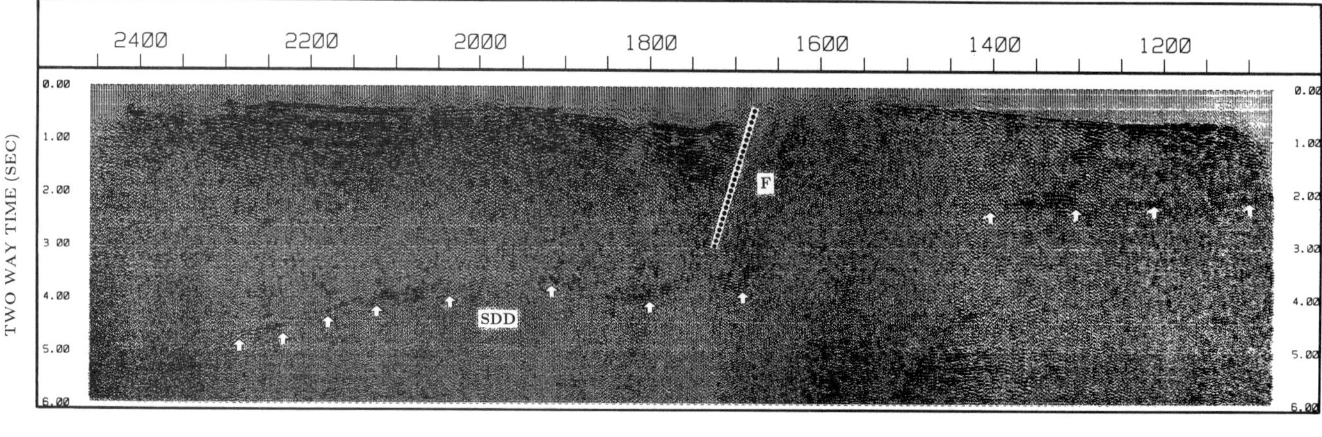

Figure 15. Unmigrated industry seismic section from Sevier Desert Basin. SSD = Sevier Desert Detachment; F = approximate location of the basin-bounding fault. The profile is oriented approximately west to east (left to right). Vertical exaggeration is 1.4 for rock velocities of 6.0 km/sec.

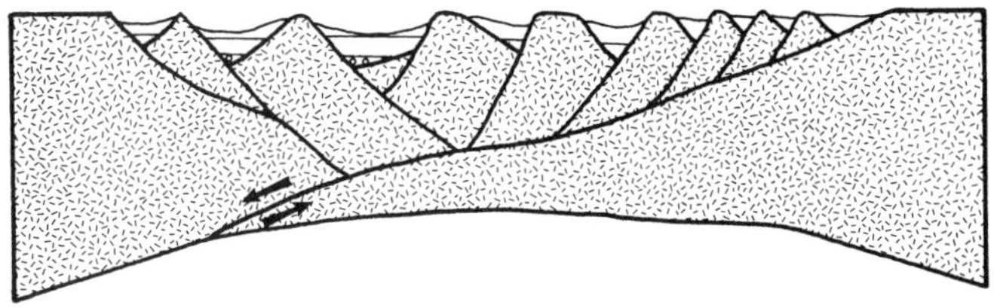

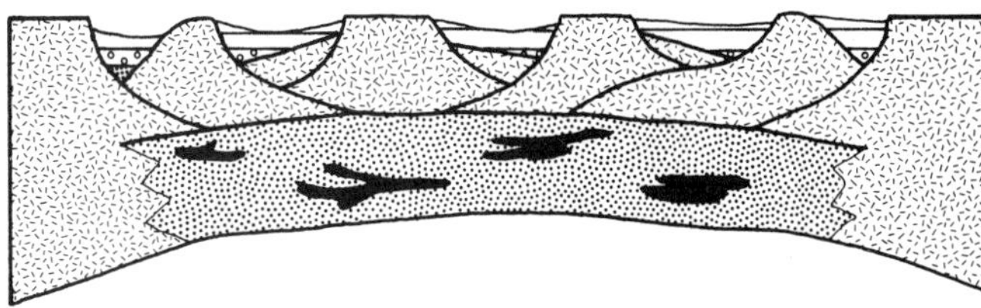

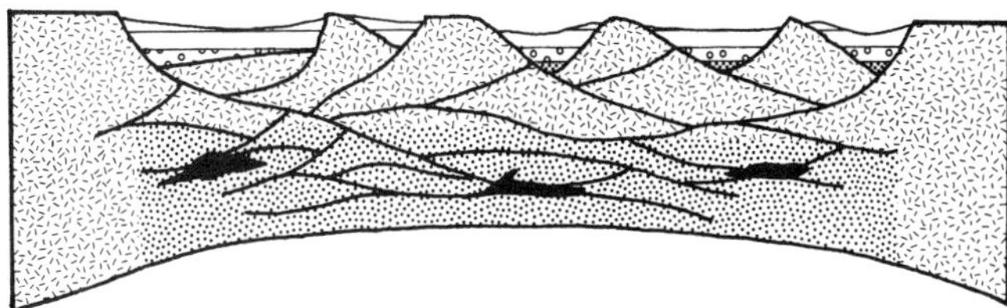

Figure 16. Generalized models for crustal extension in the Basin and Range province. Top, Simple shear model. The crust is thinned by crustal-penetrating shear zones that may offset the lithosphere-asthenosphere boundary. Middle, Pure shear model; extension in the upper crust on closely spaced faults is accommodated in the lower crust by stretching and intrusion. Bottom, Anastomosing shear zone model, a hybrid of the upper two models where the transition from brittle upper crust to ductile lower crust takes place over a region of anastomosing shear zones (from Allmendinger and others, 1987).

Our present knowledge of Basin and Range geology and geophysics does not allow us to state unequivocally which, if any, of these models is applicable to the province. Indeed, we may find, after much more geological and geophysical research has been conducted, that all three models are valid to varying degrees in different areas.

Although the rheologic change from a brittle seismogenic upper crust into a more ductile aseismic lower crust is constrained largely from seismicity studies, other geophysical techniques, such as seismic reflection profiling, may also be used to map this rheologic boundary (Klemperer and others, 1987; McCarthy and Thompson, 1988). Seismic reflection profiling in the Basin and Range province typically reveals a faulted, comparatively transparent upper crust underlain by a transition at a depth of 10 to 15 km or more into laminated subhorizontal reflections (Figs. 17, 18) that extend down to the Moho. This transition, located approximately at the base of the seismogenic zone, may mark the present-day brittle-ductile transition, although admittedly, the correlation between the two is imperfect. Nevertheless, the nature of the reflections below this transition and the rock lithologies responsible for their laminated yet discontinuous character can be best addressed in a few isolated regions of high extension where this reflective fabric rises in the crust and is either exposed at the surface or can be sampled in deep drillholes.

THE MIDDLE CRUST: REFLECTIVITY BENEATH HIGHLY EXTENDED TERRANES

Although the fairly even spacing of basins and ranges across eastern California, Nevada, Arizona, and western Utah suggests that the most recent period of Basin and Range extension has been broadly distributed across the province, geologic field studies (Proffett, 1977; Miller and others, 1983; Armstrong, 1982; Snoke, 1980) suggest that earlier periods of extension in the Oligocene and Miocene were more concentrated in isolated zones of high extension. These "highly extended terranes" are characterized by domino-style normal faults that dissect the brittle, upper crust and rotate strata to steep, often near-vertical, orientations (Proffett, 1977). Where uplift and tectonic denudation have been of sufficient magnitude (resulting in exposures of rocks originally at a depth of 5 to 15 km), these zones of extension serve as windows into the middle crust, exposing a deeper zone of décollement often underlain by subhorizontal, ductilely deformed rocks (Miller and others, 1983; McCarthy, 1986).

The Picacho Mountains of southern Arizona (Fig. 1) are one such example of a highly extended terrane. Located 70 km northwest of Tucson, this range is part of a west-northwest–trending corridor of intense mid-Tertiary extension that has affected several mountain ranges in southern Arizona and California. Extension within this corridor, measured in hundreds of percent, has unroofed rocks from original depths of at least 10 km and has exposed early to mid-Tertiary two-mica granites, indicative of midcrustal depths (Hamilton, 1982). Seismic reflection profiling in the Picacho region, together with a 5.5-km-deep drillhole located on the northeast flank of the range (Reif and Robinson, 1981), makes this highly extended terrane an ideal place to study the lithologies and deformational fabrics responsible for the unusual reflective character of midcrustal rocks in the Basin and Range (Goodwin and Thompson, 1988).

Three seismic reflection records are available from the Picacho Mountain region. The lines are parallel to one another, but together they provide a three-dimensional view of the reflectivity seen in highly extended areas. On all three of these lines, the upper 1.5 sec is seismically transparent (e.g., Fig. 17), while the rest of the record is dominated by high-amplitude, subhorizontal, and discontinuous reflections that extend down to the reflection Moho at 10 sec two-way traveltime (Fig. 18). In addition, the top of the zone of subhorizontal reflections can be seen to rise from a depth of nearly 11 km on the southwest end of one line to 3.7 km at the site of State A-1 well (Reif and Robinson, 1981). The reflective fabric also becomes deeper northeast of the well, defining a large, dome-shaped mass of highly reflective crust.

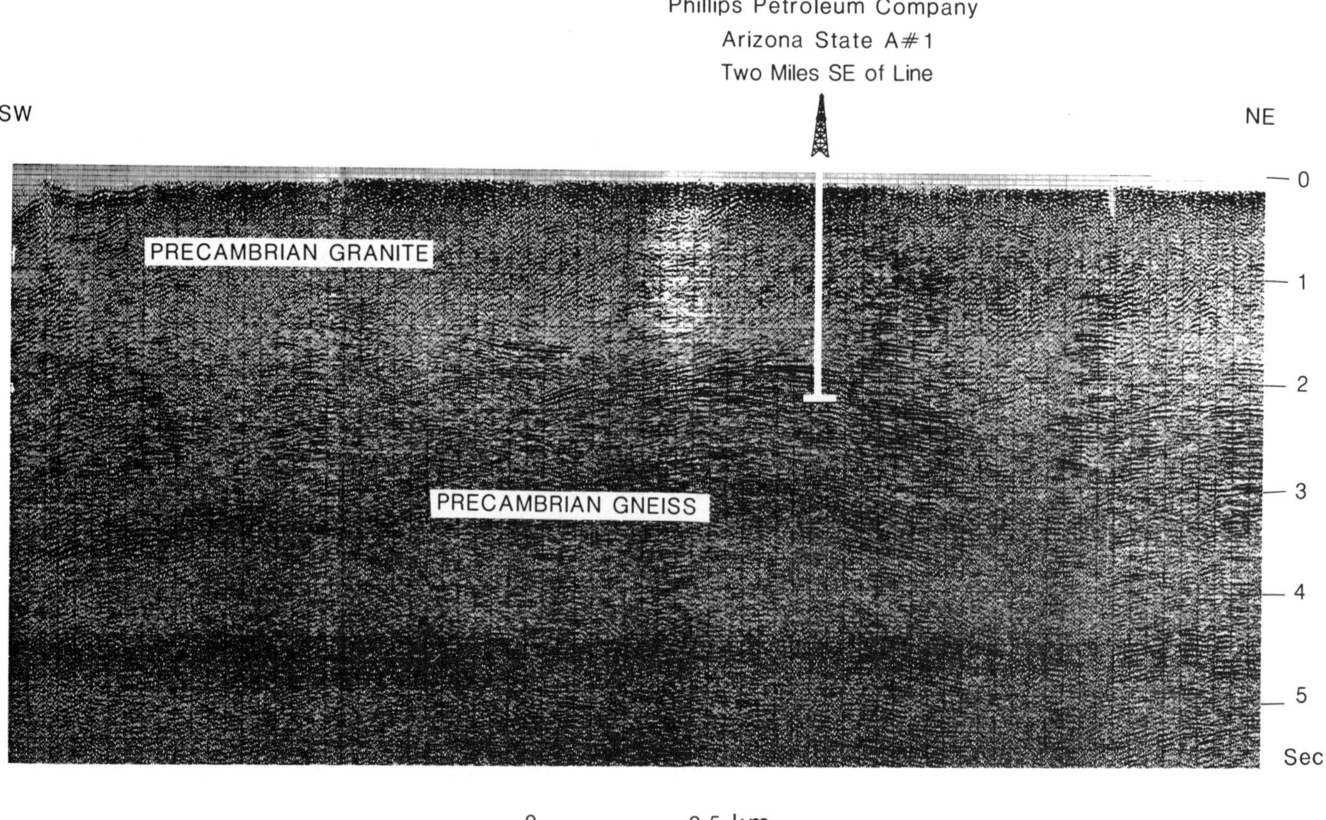

Figure 17. Forty-eight-fold reflection profile recorded by Phillips Petroleum northwest of the Picacho Mountain (PM) metamorphic core complex (Fig. 1). Precambrian granite exposed in the shallow section is replaced by Precambrian compositionally layered and ductilely deformed granodioritic gneiss at depth. The transition between these two rock types is marked by an intervening Tertiary intrusion.

Several simple, yet important, observations can be made from these seismic reflection data. First, the reflection amplitudes remain relatively constant throughout the section. This suggests that the impedance contrasts between the layers are of the same order. Acoustic impedance depends on both velocity and density, but if density varies directly with velocity, the velocity differences seen in the shallow crust could be the same as those responsible for reflections from the middle crust. Second, the reflections have the same discontinuous form from 1.5 to 10 sec. One might expect, because of Fresnel zone effects,* the reflections to become systematically more continuous with depth, but they do not. Apparently the observed seismic reflections are not coming from discrete reflecting surfaces but are the result of complex interference from many thin reflectors. Finally, several lines of evidence, most notably the observation that the reflection Moho is generally flat or smoothly undulating beneath the actively extending Basin and Range, suggest that the reflections in the lower crust and at the Moho are Cenozoic in age (Klemperer and others, 1986). If this interpretation is correct, then the reflectivity seen beneath these highly extended areas is largely a by-product of Cenozoic extension.

To quantify the reflectivity described above, 1-D and 2-D synthetic seismic modeling was carried out for comparison with the reflection data. We chose this approach to study the reflectivity because in the vicinity of the Picacho Mountains a deep drillhole provides unparalleled velocity information for the shallow crust. The Phillips Arizona State A-1, drilled to a total depth of 5.5 km, encountered three distinct crystalline units (Fig. 19): a detached and brittlely deformed upper plate, a brecciated and hydrothermally altered zone of detachment, and a ductilely deformed lower plate (Reif and Robinson, 1981). The synthetic seismograms reveal that the strong-amplitude reflections seen below 1.5 sec originate in the mylonitic gneiss of the lower plate—a surprising result first reported by Phillips Petroleum in 1981 (Reif and Robinson, 1981). One-dimensional modeling further illustrates that the reflection character from thinly layered models is sensitive to individual layers (Fig. 20). For the same

*The Fresnal zone is defined as the portion of a reflector from which reflected energy can arrive at a detector within the first one-half cycle of a reflection. It generally increases with depth because the frequency content of the data decreases while the velocity of the rock increases (see Mooney, this volume).

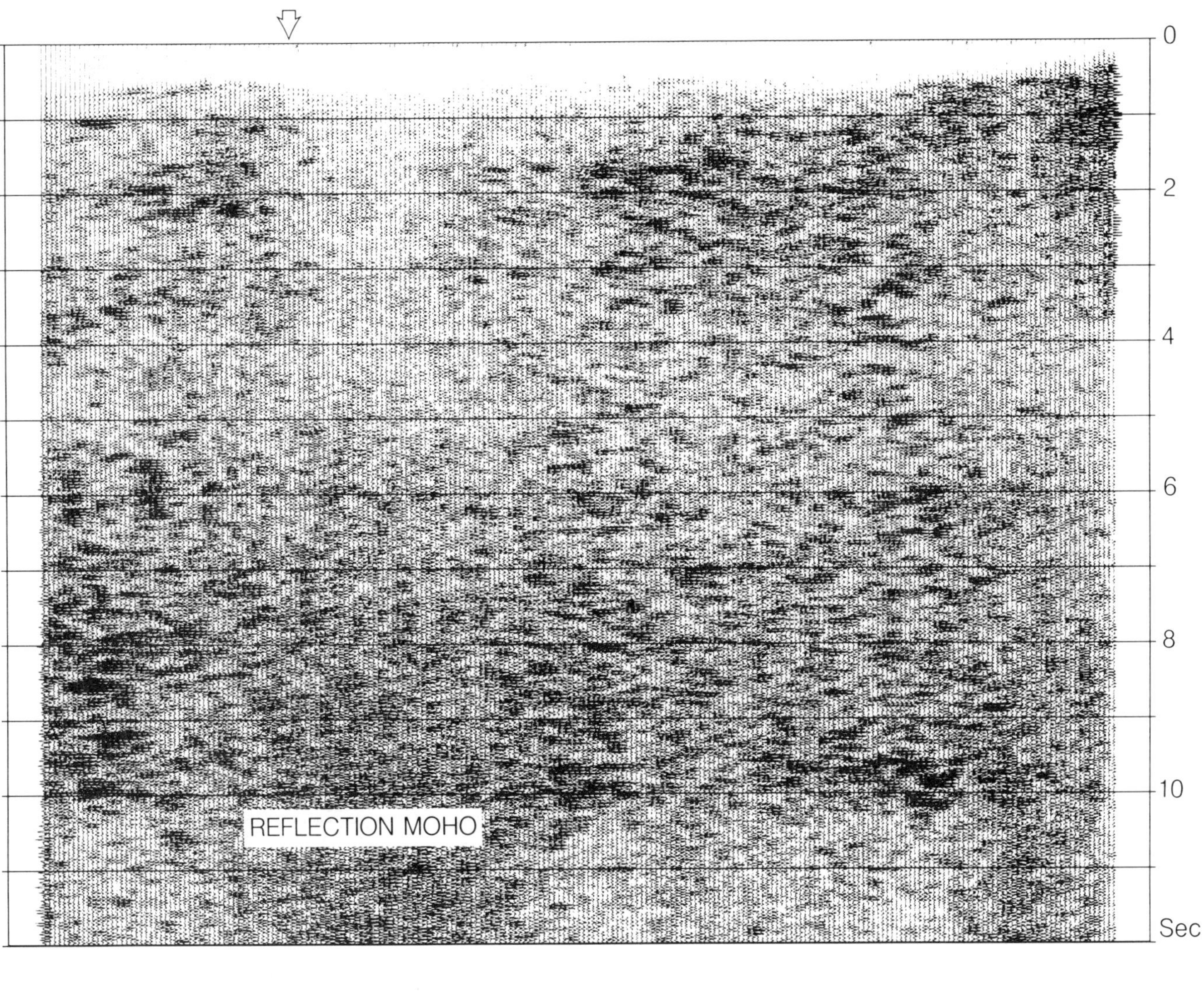

Figure 18. Phillips Line 40 recorrelated to 12 sec and conventionally processed. Beneath 1.5 sec the crust is highly reflective, especially in the vicinity of the reflection Moho at 10-sec two-way traveltime. After Goodwin and Thompson, 1988.

velocity perturbation, but at different depths, the effect on the synthetic seismogram can be quite different; changing the velocity of the layer at 3.9 km (Fig. 20D), for example, has less of an effect on the seismogram than changing the velocity of the layer at 5.2 km (Fig. 20C). Thus, it is not always apparent which layers will affect the synthetic seismic section most until it is forward modeled. In addition, the amplitudes of the synthetic seismic section are strongly dependent on the frequency of the source wavelet. A low-frequency synthetic seismogram is incapable of imaging thin velocity layering in the lower plate because it averages out the differences. This observation suggests that the reflections in the lower plate of the Picacho Mountains are a series of thin, laminated, high- and low-velocity layers that cause a frequency-dependent tuning.

Two-dimensional velocity models also were constructed based on the velocities measured in the Phillips well. The resulting wave-equation synthetic seismic sections are strikingly similar to the observed reflection records (Fig. 21). A necessary component of the 2-D model, however, is the presence of sharp lateral velocity changes to mimic the discontinuous reflections seen on Phillips Line 40. A velocity change greater than 5 percent along a layer is required to give the reflections their discontinuous form.

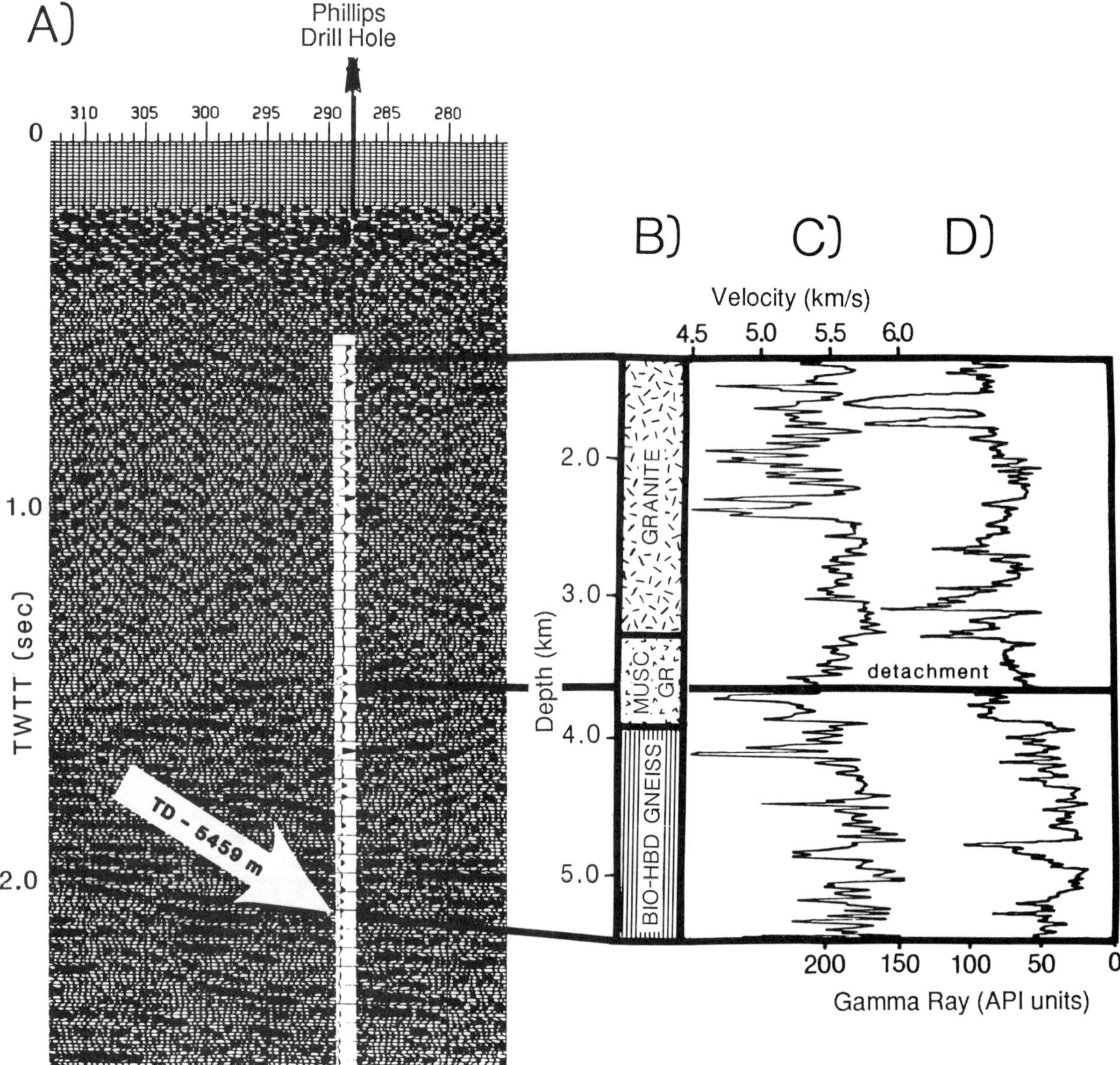

Figure 19. Comparison of seismic reflection data, synthetic seismogram, lithology log, and well logs from the Phillips Arizona State A-1 well. A, Portion of a seismic reflection profile recorded across the well with inset synthetic seismogram calculated from the velocity log. B, Lithologies encountered in the well as described from drill cuttings and cores. C, Velocity log measured in the well. D, Gamma-ray log recorded in the well. The State A-1 well, drilled to a depth of 5.5 km, penetrated an upper plate of brittlely deformed rocks, a hydrothermally altered zone of detachment, and a ductilely deformed lower plate. The synthetic seismogram shows that the reflections are coming from the lower-plate mylonitic gneisses. After Goodwin and Thompson, 1988.

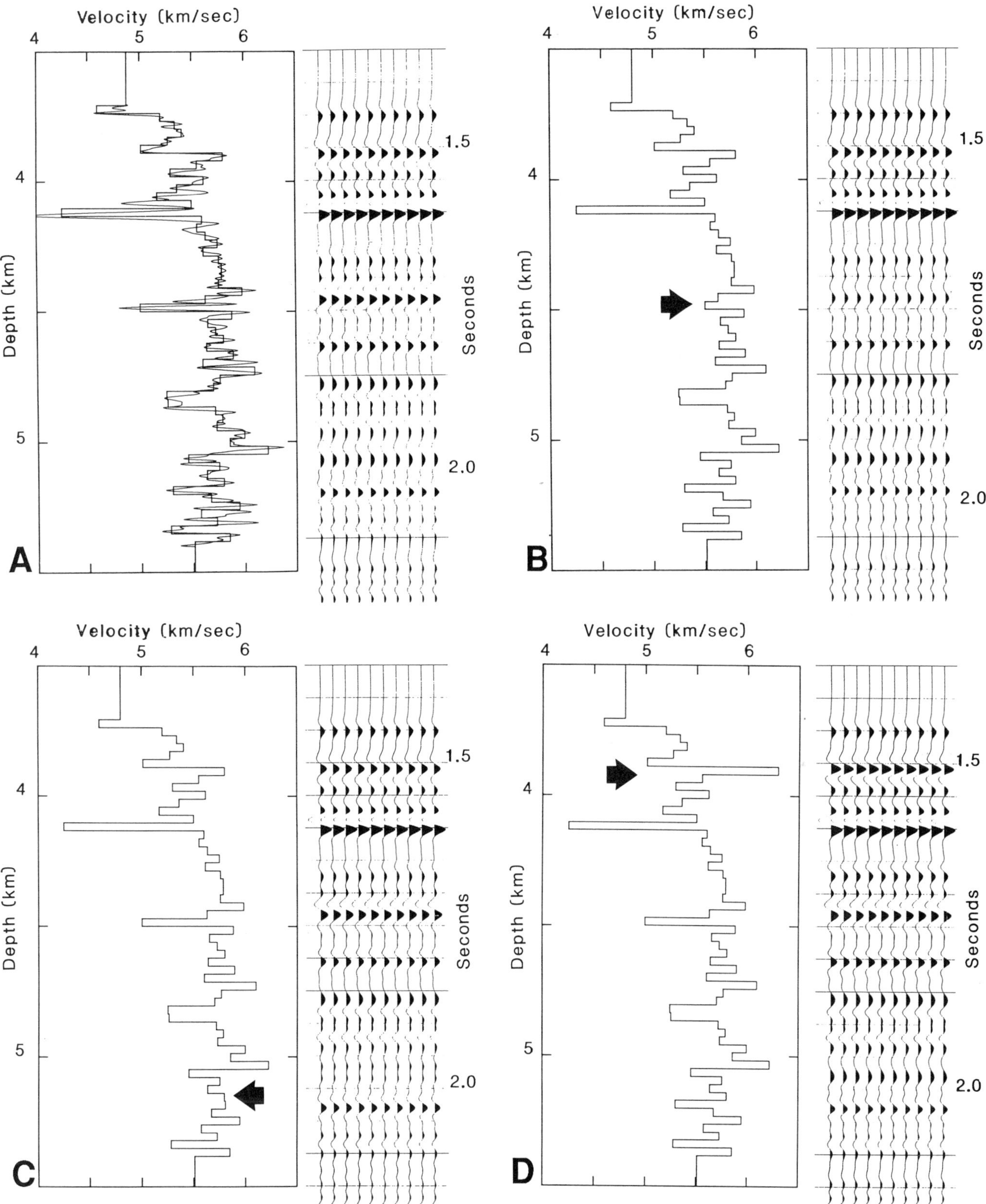

Figure 20. A, Simplified velocity log recorded from the lower plate and used in forward modeling. In B, C, and D, the layers at 4.4, 5.2, and 3.9 km, respectively, have been altered and forward modeled for comparison with A. Models B and C are examples of minor velocity changes causing a significant change in the reflectivity.

This form is well illustrated in models in which the velocity is allowed to vary by 5 and 10 percent (Fig. 22) laterally within a layer. When the velocity changes by 5 percent the reflections are continuous, and only subtle amplitude differences hint of a lateral velocity change. Once noise is added to the record it is unlikely that these amplitude variations would be recognizable. With a 10 percent lateral change in velocity the reflections acquire a discontinuous form; this discontinuous form becomes even more exaggerated when the velocity difference is further increased. If the velocity is changed too abruptly, however, diffractions will be prominent on the record.

Lateral variability is an important component of 2-D modeling; it is also readily evident in the geology of the cores of highly extended terranes. The lateral variability of lithologies is quite striking in the forerange of the Santa Catalina Mountains, for example, where alternate bands of dark- and light-colored gneiss (Keith and others, 1980) interfinger on a scale ranging from centimeters to tens of meters. Laterally the layers show abrupt terminations and are analogous to the layers incorporated in the 2-D models. This compositional layering is also evident in the topography of the forerange, where the cliffy areas tend to be pegmatitic, the highly eroded slopes gneissic, and the steep slopes a thinly laminted mixture of the two.

We have measured velocities on samples from the Santa Catalina forerange at confining pressures of as much as 1.2 kbar (about a 4-km depth). The velocities of each sample were measured in three perpendicular orientations to test for the presence of velocity anisotropy, a property common to foliated rocks (Jones and Nur, 1982). The results from one Precambrian gneiss suggest that velocity anisotropy could be an important contributor to the seismic record. Velocities of this rock were observed to vary by as much as 10 percent, which is equivalent to a reflection coefficient of 0.05. The reflection coefficients in the Phillips well are generally less than 0.02 (Fig. 23), meaning that, on average, only a small velocity difference of 5 to 10 percent between adjacent layers is necessary to generate the bulk of the reflections observed on the seismic profiles for the Picacho Mountains. Thus, stacked, thinly laminated, high- and low-velocity layers, whose origin could be either compositional or related to velocity anisotropy,

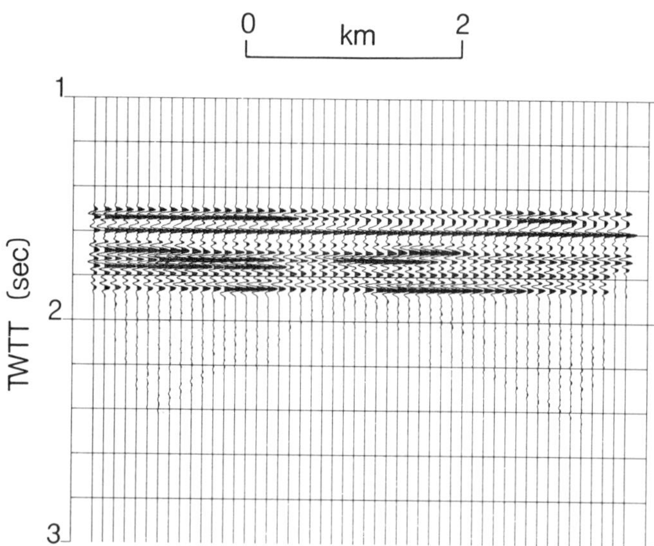

Figure 21. A 2-D wave-equation synthetic seismogram calculated from a velocity model based on the velocity log at the Phillips drillhole. The model requires lateral velocity changes between 5 and 10 percent that start and stop over distances of 100 m. The diffraction tails in the lower part of the record are quite small. After Goodwin and Thompson, 1988).

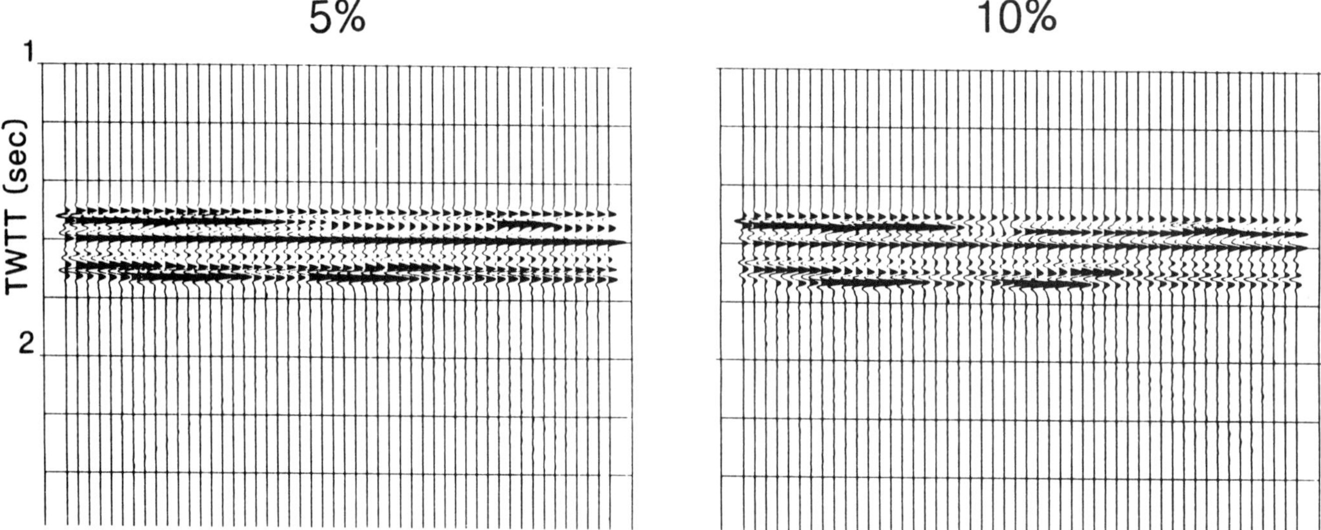

Figure 22. Two examples of synthetic seismograms showing the influence of lateral changes in the velocity model. A lateral velocity change of about 10 percent is required to make the reflections discontinuous. In A, the velocity within a layer changes by 5 percent and the reflections are nearly continuous. In B, it changes by 10 percent so that the discontinuous character becomes pronounced.

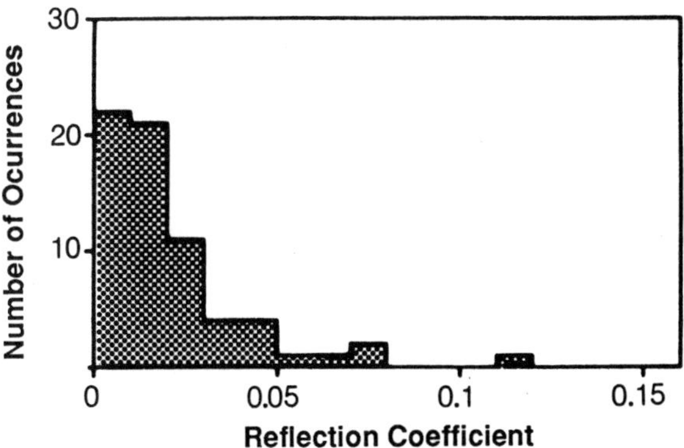

Figure 23. The reflection coefficient distribution in the lower plate of the Phillips well (where $\rho_1 = \rho_2$) shows the bulk of the coefficients are less than 0.05. A reflection coefficient of 0.05 corresponds to a velocity change of about 10 percent.

are sufficient to account for much of the reflectivity shown in Figure 18.

The larger reflections on the Phillips Line 40 (e.g., at 1.6 sec) require much higher velocity contrasts (0.4 km/sec or more) than those observed in most of the variations of composition and anisotropy; they therefore require an alternative interpretation. One explanation for these reflections is the combined influence of fractures (Reif and Robinson, 1981) and locally concentrated zones of subhorizontally oriented phyllosilicates. For example, because micaceous minerals are highly anisotropic, a micaceous-rich mylonitic rock with a subhorizontal foliation would be characterized by a 5 to 10 percent decrease in seismic velocity (measured in the vertical direction, perpendicular to foliation) with respect to the surrounding rock (Hurich and others, 1985; Fountain and others, 1984); similarly, the effect of fractures to reduce seismic velocity is a well-known property (e.g., Moos and Zoback, 1983). This interpretation seems to be supported by the Phillips well, which shows that not only are the major reflections typically associated with low-velocity zones, but that they also correspond to an increase in the gamma-ray response. Because the gamma-ray measurements are sensitive to the potassium content of the rock, a measured increase in the gamma-ray content is compatible with an increase in the amount of potassium-rich muscovite and biotite within these zones.

Based on these results, we infer that the crust in the area of the Phillips well, and by inference much of the middle crust in the Basin and Range, consists of a system of anastomosing shear zones, each with limited lateral extent and displacement, which collectively accommodate the extension mapped at the surface. In the vicinity of the highly extended terranes, where extension may be several hundred percent, the result is a deformed crust that is laminated on all scales. Part of the velocity layering can be attributed to alternating quartzofeldspathic and micaceous-rich rocks (like those seen in the Santa Catalina forerange), while in other areas, fractures and zones of aligned phyllosilicates are responsible for the seismic reflectivity. We emphasize that the highly extended areas offer a view of extension in the midcrust only, and that at some depth the mechanisms outlined above might be replaced by another. For example, perhaps the lower part of the seismic record is the result of compositional differences caused by intrusion into the lower crust (Meissner, 1973; Meissner and Wever, 1986). Alternatively, the high electrical conductivity and the suggestion that the depth to the reflectivity in the lower crust is related to heat flow are cited as evidence that pore fluids play a major role in lower crustal reflectivity (Klemperer and others, 1987). In any event, the presence of reflections from 1.5 sec to the Moho beneath the Picacho Mountains is an exciting result, indicating that these highly extended areas have been extensively reworked (Okaya and Thompson, 1986; Thompson and McCarthy, 1986) during Cenozoic extension.

DISCUSSION AND SUMMARY

On a lithospheric scale, the outstanding characteristic of the Basin and Range province is the thinness of the crust (generally less than 35 km) combined with the regionally high elevation (1 to 3 km across northern Nevada and Utah). Coupled with this characteristic and capable of explaining the high elevation, is a hot, thin lithosphere. The crust that overlies this anomalous lithosphere is also an active, changing element of the Basin and Range tectonic engine. The upper 15 km of the crust (the seismogenic zone) is broken up by a complex system of normal faults. Below 15 km, the deeper crust, as revealed in uplifted windows and in seismic reflection sections, has been extended along interconnected, subhorizontal, ductile shear zones, the ramping interconnections of which allow zones of high extension to pass laterally into zones of low extension.

Magmatic underplating and intrusion probably play a large part in the extension process (Thompson and McCarthy, 1986; Furlong and Fountain, 1986; Catchings, 1987). This magmatic underplating may give rise to the anomalous transitional crust-mantle layer that we have postulated in our seismic refraction, seismic reflection, and gravity-based crustal model. Such a transitional zone would most likely be composed of interlayered mafic and ultramafic rocks; it would appear laminated in reflection sections and would transmit refracted waves with velocities intermediate between "typical" crustal and mantle rocks. The observed velocities, which have commonly been reported as Pn, fall in the range 7.3 to 7.8 km/sec and have been variously termed anomalous upper mantle, altered lower crust, rift pillow, crust-mantle mix, etc. We prefer to interpret these results, pending new experiments with higher resolution, as representing a transition between crust and mantle, comparable to, but on a much larger scale than, that observed in ophiolites (Casey and others, 1981; Karson and others, 1984). Although granulite conditions usually prevail at the base of continental crust, as opposed to oceanic crust, the interlayering may be similar in composition and origin. Viewed in this way, the uppermost mantle—that sampled by seismic refraction surveys—is composed of mantle peridotite (petrologic mantle) *and* overlying cumulate dunites.

Above this, the transitional crust-mantle is composed of interlayered gabbro (or mafic granulite) and ultramafic cumulates. "Reflection Moho" corresponds to the base of the transitional zone.

Evidence for the geometry of young faults—those most closely associated with the fault-block mountains and basins—comes mainly from reflection and earthquake seismology. Some faults are planar, dip 50° or more, and penetrate the entire thickness of the seismogenic upper crust (about 15 km). A second fault type (listric) decreases in dip with depth and commonly bounds depositional wedges that thicken toward the fault. The reason for the contrasting behavior of the deep-going planar faults and the listric faults also is far from clear. Listric fault blocks may be comparable to giant slides moving toward deeper seated zones of extension in the crust (Cape and others, 1983) or toward magma chambers as in spreading ocean ridges (Harper, 1984; McCarthy and Mutter, 1986; McCarthy and others, 1988). A third type of fault is a regionally extensive subhorizontal detachment into which planar and listric normal faults sole at moderate depths. The Sevier Desert Detachment is perhaps the best example of this third fault type, dipping 12° and accommodating tens of kilometers of displacement. The mechanical basis for such movements is not understood.

In many parts of the province it is possible to distinguish older systems of intense faulting and extension that are not directly related to the present Basin and Range morphology (Anderson, this volume). These highly extended terranes, where uplifted and denuded, expose ductilely extended rocks of the midcrust. In one of these windows, the Picacho Mountains northwest of Tucson, Arizona, a 5.5-km drillhole and seismic reflection sections reveal the lithologies and deformational fabrics responsible for the reflective character of deep crustal rocks in the Basin and Range province. Rock property measurements and synthetic seismic modeling indicate that the reflectors are a series of thin (tens of meters), laminated, high- and low-velocity layers that demonstrate frequency-dependent tuning. Field and borehole data indicate that shear zones and mylonitic, micaceous gneisses alternating with mica-poor granitoid rocks have the requisite dimensions and velocity contrasts to produce the observed reflections.

Thus, we infer that the reflectivity in the middle and lower crust, particularly in the region associated with the metamorphic core complexes, is due in large part to textural and compositional variations associated with mineralization and mylonitization along deep ductile shear zones. The short, discontinuous nature of these seismic reflections requires that the ductile zones be strongly variable laterally, in contrast to what one would predict for a major, through-going low-angle normal fault. Similarly, because this reflectivity is common throughout much of the crust (often 4 to 10 sec two-way traveltime), we infer that extension in the ductile regime is distributed throughout the crust in a fashion similar to that proposed by Hamilton (1982) or Gans (1987). Nevertheless, the reflection evidence for a complex system of anastomosing ductile shear zones, coupled with the absence of seismically identified crustal-penetrating low-angle normal faults (Smithson and others, 1986; McCarthy, 1986; Jackson, 1987), is not readily reconcilable with the geologic evidence for large-scale extension along single, laterally extensive low-angle detachment surfaces (Howard and John, 1987; Davis and others, 1986; Wernicke, 1981; Reynolds and Spencer, 1985). Although it is possible that the low-angle detachments are restricted to the upper and middle crust only and do not project down into the ductile lower crust, or alternatively, that the ductile equivalents of these low-angle normal faults cannot be resolved seismically, neither of these solutions is entirely satisfying. We must therefore await further geological and geophysical studies of these highly extended regions in the Basin and Range before this dilemma can be resolved.

REFERENCES CITED

Allmendinger, R. W., and 7 others, 1983, Cenozoic and Mesozoic structure of the eastern Basin and Range from COCORP seismic reflection data: Geology, v. 11, p. 532–536.

Allmendinger, R. W., and 7 others, 1987, Overview of the COCORP 40°N transect, western U.S.A.; The fabric of an orogenic belt: Geological Society of America Bulletin, v. 98, p. 308–319.

Anderson, R. E., Zoback, M. L., and Thompson, G. A., 1983, Implications of selected subsurface data on the structural form of some basins in the northern Basin and Range province, Nevada and Utah: Geological Society of America Bulletin, v. 94, p. 1055–1072.

Armstrong, R. L., 1982, Cordilleran metamorphic core complexes; From Arizona to southern Canada: Annual Review of Earth and Planetary Science, v. 10, p. 129–154.

Baker, B. H., Crossley, R., and Goles, G. G., 1978, Tectonic and magmatic evolution of the southern part of the Kenya rift valley, *in* Neumann, E. R., and Ramberg, I. B., eds., Petrology and geochemistry of continental rifts: Dordrecht, Holland, D. Reidel Publishing Co., p. 29–50.

Birch, F., 1961, The velocity of compressional waves in rocks to 10 kilobars, Part 2: Journal of Geophysical Research, v. 66, p. 2199–2224.

Cape, C. D., McGeary, S., and Thompson, G. A., 1983, Cenozoic normal faulting and the shallow structure of the Rio Grande rift near Socorro, New Mexico: Geological Society of America Bulletin, v. 94, p. 3–14.

Casey, J. F., Dewey, J. F., Fox, P. J., Karson, J. A., and Rosencrantz, E., 1981, Heterogeneous nature of oceanic crust and upper mantle; A perspective from the Bay of Islands Ophiolite Complex, *in* Emiliani, C., ed., The sea, Volume 7: New York, John Wiley, p. 305–338.

Catchings, R. D., 1987, Crustal structure of the northwestern United States [Ph.D. thesis]: Stanford, California, Stanford University, 160 p.

Červený, V., Molotkov, I. A., and Psencik, I., 1977, Ray method in seismology: Prague, Karlova University, 214 p.

Crough, S. T., 1983, Rifts and swells; Geophysical constraints on causality: Tectonophysics, v. 94, p. 23–38.

Davis, G. H., and Coney, P. J., 1979, Geologic development of the Cordilleran metamorphic core complexes: Geology, v. 7, p. 120–124.

Davis, G. A., Lister, G. S., and Reynolds, S. J., 1986, Structural evolution of the Whipple and South Mountains shear zones, southwestern United States: Geology, v. 14, p. 7–10.

Eaton, G. P., 1980, Geophysical and geological characteristics of the crust of the

Basin and Range Province, *in* Burchfiel, C., Silver, E., and Oliver, J., eds., Continental tectonics, National Research Council Studies in Geophysics: Washington, D.C., National Academy of Sciences, p. 96–113.

—— , 1982, The Basin and Range Province; Origin and tectonic significance: Annual Review of Earth and Planetary Science, v. 10, p. 409–440A.

Eaton, J. P., 1963, Crustal structure from San Francisco, California, to Eureka, Nevada, from seismic refraction measurements: Journal of Geophysical Research, v. 68, p. 5789–5806.

Fountain, D. M., Hurich, C. A., and Smithson, S. B., 1984, Seismic reflectivity of mylonitic zones in the crust: Geology, v. 12, p. 195–198.

Frei, L. S., 1986, Additional paleomagnetic results from the Sierra Nevada; Further constraints on Basin and Range extension and northward displacement in the western United States: Geological Society of America Bulletin, v. 97, p. 840–849.

Frei, L. S., Magill, J. R., and Cox, A., 1984, Paleomagnetic results from the central Sierra Nevada; Constraints on reconstructions of the western United States: Tectonics, v. 3, p. 157–177.

Frost, E. G., and Okaya, D. A., 1986a, Geologic and seismologic setting of the CALCRUST Whipple line; Geometries of detachment faulting in the field and from reprocessed seismic-reflection profiles donated by CGG: Geological Society of America Abstracts with Programs, v. 18, p. 107.

—— , 1986b, Seismic-reflection view of the South Mountains, Arizona, detachment fault and its deeper crustal structure: Geological Society of America Abstracts with Programs, v. 18, p. 107.

Furlong, K. P., and Fountain, D. M., 1986, Continental crustal underplating; Thermal considerations and seismic-petrologic consequences: Journal of Geophysical Research, v. 91, p. 8285–8294.

Gans, P. G., 1987, An open-system, two-layer crustal stretching model for the eastern Great Basin: Tectonics, v. 6, p. 1–12.

Gans, P. G., Miller, E. L., McCarthy, J., and Ouldcott, M. L., 1985, Tertiary extensional faulting and evolving ductile-brittle transition zones in the northern Snake Range and vicinity; New insights from seismic data: Geology, v. 13, p. 189–193.

Gettings, M. E., Blank, H. R., Jr., Mooney, W. D., and Healey, J. H., 1986, Crustal structure of southwestern Saudi Arabia: Journal of Geophysical Research, v. 91, p. 6491–6512.

Giese, P., 1966, Towards a classification of the Earth's crust in the northern Alpine foreland, in the eastern Alps, and in parts of the western Alps, using characteristic refraction-traveltime curves, as well as a geological interpretation: Habilitationsschrift Freie Universität Berlin, 143 p.

Goodwin, E. B., and Thompson, G. A., 1988, The seismically reflective crust beneath highly extended terranes; Evidence for its origin in extension: Geological Society of America Bulletin, v. 100, p. 1616–1626.

Hamilton, W. B., 1982, Structural evolution of the Big Maria Mountains, northeastern Riverside County, southeastern California, *in* Frost, E. G., and Martin, D. L., eds., Mesozoic–Cenozoic tectonic evolution of the Colorado River region, California, Arizona, and Nevada: San Diego, Cordilleran Publishers, p. 1–28.

Harper, G. D., 1984, Tectonics of slow spreading mid-ocean ridges and consequences of a variable depth to the brittle/ductile transition: Tectonics, v. 4, p. 395–410.

Hauge, T., and 10 others, 1987, Crustal structure of western Nevada from COCORP deep seismic reflection data: Geological Society of American Bulletin, v. 98, p. 320–329.

Hill, D. P., and Pakiser, L. C., 1966, Crustal structure beneath Nevada Test Site and Boise, Idaho, from seismic refraction measurements, *in* The earth beneath the continents: American Geophysical Union Geophysical Monograph, v. 10, p. 391–419.

—— , 1967, Seismic-refraction study of crustal structure between the Nevada Test Site and Boise, Idaho: Geological Society of America Bulletin, v. 78, p. 685–704.

Howard, K. A., and John, B. E., 1987, Crustal extension along a rooted system of imbricate low-angle faults; Colorado River extensional corridor, California and Arizona, *in* Coward, M. P., Dewey, J. F., and Hancock, P. L., eds., Continental extensional tectonics: Geological Society of London Special Publication 28, p. 299–311.

Hurich, C. A., Smithson, S. B., Fountain, D. M., and Humphreys, M. C., 1985, Seismic evidence of mylonite reflectivity and deep structure in the Kettle Dome metamorphic core complex, Washington: Geology, v. 13, p. 577–580.

Jackson, J. A., 1987, Active normal faulting and crustal extension, *in* Coward, M. P., Dewey, J. F., and Hancock, P. L., eds., Contiental extensional tectonics: Geological Society of London Special Publication 28, p. 3–18.

Jones, T., and Nur, A., 1982, Seismic velocity and anisotropy in mylonites and the reflectivity of deep crustal fault zones: Geology, v. 10, p. 160–163.

Karson, J. A., Collins, J. A., and Casey, J. F., 1984, Geologic and seismic velocity structure of the crust/mantle transition in the Bay of Islands Ophiolite Complex: Journal of Geophysical Research, v. 89, p. 6126–6138.

Keith, S. B., Reynolds, S. J., Damon, P. E., Shafiqullah, M., Livingston, D. E., and Pushkar, P. D., 1980, Evidence for multiple intrusion and deformation within the Santa Catalina–Tortolita crystalline complex, southeastern Arizona, *in* Crittenden, M. D., Jr. Coney, P. J., and Davis, G. H., eds., Cordilleran metamorphic core complexes: Boulder, Colorado, Geological Society of America Memoir 153, p. 217–267.

Klemperer, S. L., Hauge, T. A., Hauser, E. C., Oliver, J. E., and Potter, C. J., 1986, The Moho in the northern Basin and Range Province, Nevada, along the COCORP 40°N seismic reflection transect: Geological Society of America Bulletin, v. 97, p. 603–618.

Klemperer, S. L., and the BIRPS group, 1987, Reflectivity of the crystalline crust; Hypotheses and tests: Geophysical Journal of the Royal Astronomical Society, v. 89, p. 217–222.

Kligfield, R., Crespi, J., Naruk, S., and Davis, G. H., 1984, Displacement and strain patterns of extensional orogens: Tectonics, v. 3, p. 557–609.

Lachenbruch, A. H., and Sass, J. H., 1978, Models of an extending lithosphere and heat flow in the Basin and Range Proavince, *in* Smith, R. B., and Eaton, G. P., eds., Cenozoic tectonics and regional geophysics of the Western Cordillera: Boulder, Colorado, Geological Society of America Memoir 152, p. 209–250.

Lucchitta, I., 1985, Heat and detachment in core-complex extension [abs.], *in* Lucchitta, I., Morgan, P., and Soderblom, L., eds., Heat and detachment in crustal extension on continents and planets: Houston, Lunar and Planetary Institute.

McCarthy, J., 1986, Reflection profiles from the Snake Range metamorphic core complex; A window into the mid-crust, *in* Barazangi, M., and Brown, L., eds., Reflection seismology; The continental crust: American Geophysical Union Geodynamics Series, v. 14, p. 281–292.

McCarthy, J., and Mutter, J. C., 1986, Relict magma chambers preserved within the Mesozoic North Atlantic crust [abs.]: EOS Transactions of the American Geophysical Union, v. 67, p. 1083.

McCarthy, J., and Thompson, G. A., 1988, Seismic imaging of extended crust with emphasis on the western United States: Geological Society of America Bulletin, v. 100, p. 1361–1374.

McCarthy, J., Mutter, J. G., Morton, J. L., Sleep, H. H., and Thompson, G. A., 1988, Relic magma chamber structures preserved within the Mesozoic North Atlantic crust?: Geological Society of America Bulletin, v. 100, p. 1423–1436.

McDonald, R. E., 1976, Tertiary tectonics and sedimentary rocks along the transition; Basin and Range province to plateau and thrust belt, Utah: Rocky Mountain Association of Geologists Symposium, p. 281–317.

Meissner, R., 1973, The "Moho" as a transition zone: Geophysical Surveys, v. 1, p. 195–216.

Meissner, R., and Wever, T., 1986, The nature and development of the crust according to deep reflection data from the German Variscides, *in* Barazangi, M., and Brown, L., eds., Reflection seismology; A global perspective: American Geophysical Union Geodynamics Series, v. 13, p. 31–42.

Miller, E. L., Gans, P. B., and Garing, J. D., 1983, The Snake Range Décollement; An exhumed mid-Tertiary ductile-brittle transition: Tectonics, v. 2, p. 239–263.

Minster, J. B., and Jordan, T. H., 1987, Vector constraints on western U.S. deformation from spacy geodesy, neotectonics, and plate motions: Journal of Geophysical Research, v. 92, p. 4788–4804.

Mooney, W. D., Andrews, M. C., Ginzburg, A., and Hamilton, R. M., 1983, Crustal structure of the northern Mississippi Embayment and a comparison with other continental rift zones: Tectonophysics, v. 94, p. 327–348.

Moos, D., and Zoback, M. D., 1983, In situ studies of velocity of fractured crystalline rocks: Journal of Geophysical Research, v. 88, p. 2345–2358.

Mota, L., 1954, Determination of dips and depths of geological layers by the seismic refraction method: Geophysics, v. 19, p. 242–254.

Okaya, D. A., and Thompson, G. A., 1985, Geometry of Cenozoic extensional faulting; Dixie Valley, Nevada: Tectonics, v. 4, p. 107–126.

—— , 1986, Involvement of deep crust in extension of Basin and Range Province, *in* Mayer, L., eds., Extensional tectonics of the southwestern United States; A perspective on processes and kinematics: Boulder, Colorado, Geological Society of America Special Paper 208, p. 15–22.

Pakiser, L. C., 1985, Seismic exploration of the crust and upper mantle of the Basin and Range province, *in* Drake, E. T., and Jordan, W. M., eds., Geologists and ideas; A history of North American geology: Boulder, Colorado, Geological Society of America, Centennian Special Volume 1, p. 453–469.

1986 PASSCAL Basin and Range Lithospheric Seismic Experimental Working Group, 1988, The 1956 PASSCAL Basin and Range Lithospheric Seismic Experiment: EOS American Geophysical Union Transactions, v. 69, p. 593, 596–598.

Potter, C. J., and 9 others, 1987, Crustal structure of north-central Nevada; Results from COCORP deep seismic profiling: Geological Society of America Bulletin, v. 98, p. 330–337.

Preistley, K. F., and Brune, J. N., 1978, Surface waves and the structure of the Great Basin of Nevada and western Utah: Journal of Geophysical Research, v. 83, p. 2265–2272.

Priestley, K. F., Ryall, A. S., and Fezie, G. S., 1982, Crust and upper mantle structure in the northwest Basin and Range province: Bulletin of the Seismological Society of America, v. 72, p. 911–923.

Prodehl, C., 1970, Seismic refraction study of crustal structure in the western United States: Geological Society of America Bulletin, v. 31, p. 2629–2645.

—— , 1979, Crustal structure of the western United States: U.S. Geological Survey Professional Paper 1034, 74 p.

Proffett, J. M., Jr., 1977, Cenozoic geology of the Yerington district, Nevada, and the implications for the nature and origin of Basin and Range faulting: Geological Society of America Bulletin, v. 88, p. 247–266.

Rehrig, W. A., and Reynolds, S. J., 1980, Geologic and geochronologic reconnaissance of a northwest-trending zone of metamorphic core complexes in southern and western Arizona, *in* Crittenden, M. D., Jr., Coney, P. J., and Davis, G. H., ed., Cordilleran metamorphic core complexes: Boulder, Colorado, Geological Society of America Memoir 153, p. 131–158.

Reif, D. M., and Robinson, J. P., 1981, Geophysical, geochemical, and petrographic data and regional correlation from the Arizona state A-1 well, Pinal County, Arizona: Arizona Geological Society Digest, v. 13, p. 99–109.

Reynolds, S. J., and Lister, G. S., 1987, Structural aspects of fluid-rock interactions in detachment zones: Geology, v. 15, p. 362–366.

Reynolds, S. J., and Spencer, J. E., 1985, Evidence for large-scale transport on the Bullard detachment fault, west-central Arizona: Geology, v. 13, p. 353–356.

Romanowicz, B. A., 1979, Seismic structure of the upper mantle beneath the United States by three-dimensional inversion of body wave arrival times: Geophysical Journal of the Royal Astronomical Society, v. 57, p. 479–506.

Sharp, J., 1984, West-central Utah; Palinspastically restored sections constrained by COCORP seismic reflection data [M.S. thesis]: Ithaca, New York, Cornell University, 60 p.

Smith, R. B., 1978, Seismicity, crustal structure, and intraplate tectonics of the interior of the western cordillera, *in* Smith, R. B., and Eaton, G. P., eds., Cenozoic tectonics and regional geophysics of the Western Cordillera: Boulder, Colorado, Geological Society of America Memoir 152, p. 111–144.

Smith, R. B., and Bruhn, R. B., 1984, Intraplate extensional tectonics of the eastern Basin and Range; Inferences on structural style from seismic reflection data, regional tectonics, and thermal-mechanical models of brittle-ductile deformation: Journal of Geophysical Research, v. 89, p. 5733–5762.

Smith, R. B., Richins, W. D., and Doser, D. I., 1985, The 1983 Borah Peak, Idaho, Earthquake; Regional seismicity, kinematics of faulting, and tectonic mechanism, *in* Stein, R. S., and Bucknam, R. C., eds., Proceedings of Workshop 28 on the Borah Peak, Idaho, Earthquake, Volume A: U.S. Geological Survey Open-File Report 85-290, p. 236–263.

Smithson, S. B., Johnson, R. A., and Hurich, C. A., 1986, Crustal reflections and crustal structure, *in* Barazangi, M., and Brown, L., eds., Reflection seismology; The continental crust: American Geophysical Union Geodynamics Series, v. 14, p. 21–32.

Snoke, A. W., 1980, Transition from infrastructure to suprastructure in the northern Ruby Mountains, Nevada, *in* Crittenden, M. D., Jr., Coney, P. J., and Davis, G. H., eds., Cordilleran metamorphic core complexes: Boulder, Colorado, Geological Society of America Memoir 153, p. 287–234.

Society of Exploration Geophysicists, 1982, Bouguer gravity map of the United States.

Stauber, D. A., 1980, Crustal structure in the Battle Mountain heat flow high in northern Nevada from seismic refraction and Rayleigh wave phase velocities [Ph.D. thesis]: Stanford, California, Stanford University, 240 p.

—— , 1983, Crustal structure in northern Nevada from seismic refraction data; The role of heat in the development of energy and mineral resources in the northern Basin and Range province: Geothermal Resources Council Special Report 13, p. 319–332.

Stauber, D. A., and Boore, D. M., 1978, Crustal thickness in northern Nevada from seismic refraction profiles: Bulletin of the Seismological Society of America, v. 68, p. 1049–1058.

Stein, R. S., and Barrientos, S. E., 1985, Planar high-angle faulting in the Basin and Range; Geodetic analysis of the 1983 Borah Peak, Idaho, earthquake: Journal of Geophysical Research, v. 90, p. 11355–11366.

Stewart, J. H., 1971, Basin and Range structure; A system of horst and grabens produced by deep-seated extension: Geological Society of America Bulletin, v. 82, p. 1019–1044.

—— , 1978, Basin-range structure in western North America; A review, *in* Smith, R. B., and Eaton, G. P., eds., Cenozoic tectonics and regional geophysics of the Western Cordillera: Geological Society of America Memoir 152, p. 1–31.

Stewart, J. H., and Carlson, J. E., 1976, Cenozoic rocks of Nevada: Nevada Bureau of Mines and Geology Map 52, scale 1: 1,000,000.

Taylor, S. R., and Patton, H. J., 1986, Shear-velocity structure from regionalized surface wave dispersion in the Basin and Range: Geophysical Research Letters, v. 13, p. 30–33.

Thompson, G. A., and Burke, D. B., 1973, Rate and direction of spreading in Dixie Valley, Basin and Range province, Nevada: Geological Society of America Bulletin, v. 84, p. 627–632.

—— , 1974, Regional geophysics of the Basin and Range province: Annual Review of Earth and Planetary Sciences, v. 2, p. 213–238.

Thompson, G. A., and McCarthy, J., 1986, A gravity constraint on the origin of highly extended terranes: Geological Society of America Abstracts with Programs, v. 18, p. 418.

Thompson, G. A., and Zoback, M. L., 1979, Regional geophysics of the Colorado Plateau: tectonophysics, v. 61, p. 149–181.

Thompson, G. A., Mooney, W. D., Priestley, K. F., and Smith, R. B., 1986, Nevada Basin-Range lithospheric seismic imaging experiment; PASSCAL, 1986: EOS American Geophysical Union Transactions, v. 67, p. 1096.

Vetter, V. R., Ryall, A. S., and Sanders, C. O., 1983, Seismological investigations of volcanic and tectonic processes in the western Great Basin, Nevada and eastern California: Geothermal Resources Council Special Report 13, p. 333–343.

Von Tish, D., Allmendinger, R. W., and Sharp, J., 1985, History of Cenozoic extension in the central Sevier Desert, west-central Utah, from COCORP seismic reflection data: American Association of Petroleum Geologists Bulletin, v. 69, p. 1077–1087.

Wallace, R. E., 1984, Patterns and timing of late Quaternary faulting in the Great Basin Province and relation to some regional tectonic features: Journal of Geophysical Research, v. 89, p. 5763–5769.

Wernicke, B., 1981, Low-angle normal faults in the Basin and Range province; Nappe tectonics in an extending orogen: Nature, v. 291, p. 645–648.

—— , 1985, Uniform-sense normal simple shear of the continental lithosphere: Canadian Journal of Earth Sciences, v. 22, p. 108–125.

Zoback, M. L., and Anderson, R. E., 1983, Style of Basin-Range faulting as inferred from seismic reflection data in the Great Basin, Nevada and Utah: Geothermal Resources Council Special Report 13, p. 363–381.

Zoback, M. L., Anderson, R. E., and Thompson, G. A., 1981, Cenozoic evolution of the state of stress and style of tectonism of the Basin and Range province of the western United States: Philosophical Transactions of the Royal Society of London, v. A 300, p. 407–434.

Zoback, M. L., Zoback, M. D., Thompson, G. A., McCarthy, J., and Smith, R. B., 1985, Scientific drilling in the Sevier Desert Basin, Utah; In situ study of the interaction between high- and low-angle normal faults: EOS American Geophysical Union Transactions, v. 66, p. 1096–1097.

MANUSCRIPT ACCEPTED BY THE SOCIETY OCTOBER 31, 1988

ACKNOWLEDGMENTS

This study is one product of a continuing workshop seminar in which we studied rifted regions, visiting areas as distant as the Aegean Sea and East Africa. In this account of the western Basin and Range province, we have tried to review the geophysics selectively and critically, especially in the light of new evidence from our own studies. We are grateful for discussions with other seminar participants including Heidi McGrew, Sam Joffe, Vicki Langenheim, and Thomas Reed. Walter Mooney, Jerry Eaton, John Proffett, Lou Pakiser, and Mary Lou Zoback were at times essential teachers and critics. Claus Prodehl provided seismic records from the Fallon-Eureka seismic refraction transect. Jim Luetgert provided software for seismic analysis. The U.S. Geological Survey acquired the Sevier Desert reflection profile from CGG and kindly provided processing facilities for its analysis. In addition, the reflectivity study was done in part on the computer facilities at the U.S. Geological Survey. Phillips Petroleum donated seismic and drillhole data from the Picacho Mountains, Arizona, while Sohio Petroleum collected and processed the seismic reflection data from Spring Valley, Nevada. Amos Nur provided access to rock velocity apparatus, and Petrophysics Inc. generously measured velocities for several rock samples. The work was partly supported by National Science Foundation Grants EAR-8306406 and EAR-8417309 and by the U.S. Geological Survey.

Printed in U.S.A.

Geological Society of America
Memoir 172
1989

Chapter 12

Geophysical and tectonic framework of the eastern Basin and Range–Colorado Plateau–Rocky Mountain transition

Robert B. Smith, Walter C. Nagy, Kelsey A. (Smith) Julander,* John J. Viveiros,* Craig A. Barker,* and Donald G. Gants*
Department of Geology and Geophysics, University of Utah, Salt Lake City, Utah 84112

ABSTRACT

Crustal structure of the eastern Basin-Range province and its transition to the Colorado Plateau–Rocky Mountain provinces has been influenced by several tectonic events: Precambrian margin rifting, Paleozoic sedimentation, late Mesozoic and early Cenozoic crustal compression and related thrust faulting, and middle to late Cenozoic extension and normal faulting. As a result of this complex history, the crust and upper mantle of this region is laterally heterogeneous across its 500-km east-west breadth.

The crust is ~30 km thick beneath the central Basin-Range and increases eastward across the transition to more than 40 km beneath the Colorado Plateau and middle Rocky Mountains. Despite the lateral variation in crustal thickness, the upper mantle has a generally uniform P-wave velocity of 7.9 km/sec. On the basis of surface-wave analyses, the lithosphere is estimated to be 65 km thick beneath the central Basin-Range and increases eastward to 80 km beneath the Colorado Plateau. An anomalously low, apparent velocity layer (~7.5 km/sec), at a depth of ~25 km was identified in the Basin and Range–Colorado Plateau transition from unreversed refraction profiles. This layer has been interpreted as a high-velocity zone at the top of a mantle bulge; it may, however, represent an ~10-km mantle upwarp of 7.9-km/sec upper-mantle material with the low apparent velocities resulting from down-dip ray propagation.

Seismic reflection profiling in the Basin and Range–Colorado Plateau–Rocky Mountain transition has been used principally to assess the geometry and structural style of Cenozoic basins and upper crustal faults. These seismic data reveal several asymmetric, eastward-tilted basins that are bounded by low- to high-angle (30° to 60°), planar to listric normal faults. An unusually widespread, 10° to 15° west-dipping reflection, identified as the Sevier Desert detachment, has been detected across a 190-km east-west width from near the surface in central Utah, westward to a depth of ~15 km near the Utah-Nevada border. This major structure extends over an area of 20,000+ km^2 and may have accommodated as much as 60 km of late Cenozoic crustal extension.

The Basin and Range–Colorado Plateau–Rocky Mountain transition is also coincident with the southern Intermountain seismic belt where diffusely distributed epicenters occur along a 100- to 200-km-wide, north-south–trending zone. Where precise hypocenters have been mapped with detailed seismograph networks, background seismicity does not in general correlate with mapped faults. Source studies of three historic M7$^+$ earthquakes in the Basin-Range have provided a hypothetical working model for large normal-faulting earthquakes in the region. These large events occurred on 40° to 60°

*Present addresses: Julander: Chevron U.S.A., San Ramon, California; Viveiros: Chevron U.S.A., Denver, Colorado; Barker: Tenneco Oil Col, Lafayette, Louisiana; Gants: Teledyne-Rockwell, Los Angeles, California.

Smith, R. B., Nagy, W. C., Julander, K. A., Viveiros, J. J., Barker, C. A., and Gants, D. G., 1989, Geophysical and tectonic framework of the eastern Basin and Range–Colorado Plateau–Rocky Mountain transition, *in* Pakiser, L. C., and Mooney, W. D., Geophysical framework of the continental United States: Boulder, Colorado, Geological Society of America Memoir 172.

planar normal faults and nucleated at midcrustal depths of $\sim$15 km, near the maximum depth of background seismicity. However, there is an intriguing paradox between the geometries of the seismogenic faults, associated with large M7$^+$ events, and the attitudes of some of the shallow normal faults identified from the seismic reflection data. Several faults identified on the reflection profiles reveal Quaternary planar, low-angle to listric normal faults that extend to depths of $<$6 km, unlike the deeper penetrating and steeper planar faults inferred from the large, M7+, normal-faulting earthquakes. Rheologic models for this region of active extension show that the M7+ earthquakes occurred near the transition between the brittle upper crust and a quasi-plastic middle crust, suggesting that the larger events require large stress drops to relieve strain in a more ductile medium.

Crustal extension rates have been derived from seismic moment tensors of historic earthquakes along the transition that vary from $<$1 to 5 mm/yr. An integrated east-west extension rate for the entire Basin-Range of $\sim$10 mm/yr is similar to Quaternary extension rates determined from geological and other geophysical data. The deformation rates of the transition area, however, are significantly smaller than those along the San Andreas fault system, where contemporary deformation may exceed 50 mm/yr.

INTRODUCTION

The eastern Basin-Range province and its transition to the Colorado Plateau and middle Rocky Mountain provinces (hereafter referred to as the BR-CP-RM transition; Fig. 1) is best understood within the framework of its regional crustal and lithospheric structure and tectonic history. This major intraplate region within the U.S. Cordillera has been dominated by Paleozoic sedimentation, late Mesozoic and early Cenozoic crustal compression, Cenozoic volcanism, and middle to late Cenozoic extension. The long history of sedimentation, deformation, and magmatism have profoundly influenced the composition and rheologic properties of the crust. Hence its regional geophysical signature is the integrated product of a complex tectonic evolution, and the region remains tectonically active today.

Initially, the western margin of North America was controlled by Precambrian rifting along a north-trending zone bounded on the east by the present-day BR-CP-RM transition, that marked a notable crustal boundary coincident with the present Wasatch Front (Bond and others, 1985). During the Paleozoic era, the transition marked the eastern edge of the western North American plate, accumulating as much as 12 km of sediments to the west and about 3 km of shelf sediments to the east. Beginning in the Early Cretaceous period ($\sim$140 Ma) and ending by the Late Cretaceous ($\sim$70 Ma), east-west compression associated with the Sevier and Laramide orogenies produced the major folds and thrusts of the Overthrust belt, structures that are also generally coincident with the present day BR-CP-RM transition. These compressional orogenic events likely increased the crustal thickness by underthrusting and/or overthrusting. As the eroding thrust belt progressed, magmatism was induced by the accumulation of high concentrations of heat-producing radioactive isotopes in the lower plate during the orogenies, thereby weakening the crust (Glazner and Bartley, 1985).

As early as mid-Miocene time ($\sim$45 Ma), intraplate extension associated with the Basin-Range epeirogeny was initiated (Dallmeyer and others, 1986). Crustal thinning likely accompanied widespread normal faulting during the development of the Basin-Range province, while the Colorado Plateau and the middle Rocky Mountains remained relatively stable. Inception of modern Basin-Range topography and concomitant high-angle normal faulting, however, did not generally develop before 10 Ma (Zoback and others, 1981). Lateral variations in crustal thicknesses as much as $\sim$15 km, active normal faulting, a major transition in regional heat flow, and an active earthquake regime now define the regional tectonic signature of this active extensional belt—the BR-CP-RM transition.

Area. This discussion focuses on the eastern Basin-Range area of western Utah, eastern Nevada, eastern Idaho, the northwestern Colorado Plateau, and parts of the middle and northern Rocky Mountain provinces of Utah and Wyoming (Fig. 1). Tectonically, it includes the extensional regimes of the central, northern, and eastern Basin-Range; the relatively stable, northern and western Colorado Plateau; and the middle Rocky Mountain foreland provinces (see Fig. 1).

Scope and Breadth. Seismic reflection profiling and the occurrences of the 1954 M7.1, Dixie Valley, Nevada; the 1959 M_L 7.5, Hebgen Lake, Montana; and the 1983 M_s 7.3, Borah Peak, Idaho, earthquakes have focused attention on the extensional tectonics and kinematics of the intraplate western U.S. Cordillera. This chapter principally addresses the crustal structure, seismicity, rheology, and contemporary tectonics of the BR-CP-RM transition. An evaluation of the kinematics of contemporary crustal extension based on the interpretation of seismic reflection profiles, earthquake data, and deformation rates determined from seismic moments of historic earthquakes will also be given.

Data from several unpublished theses, newly published papers, and data acquired in the late 1970s and early 1980s on regional crustal and lithospheric structure are incorporated in our

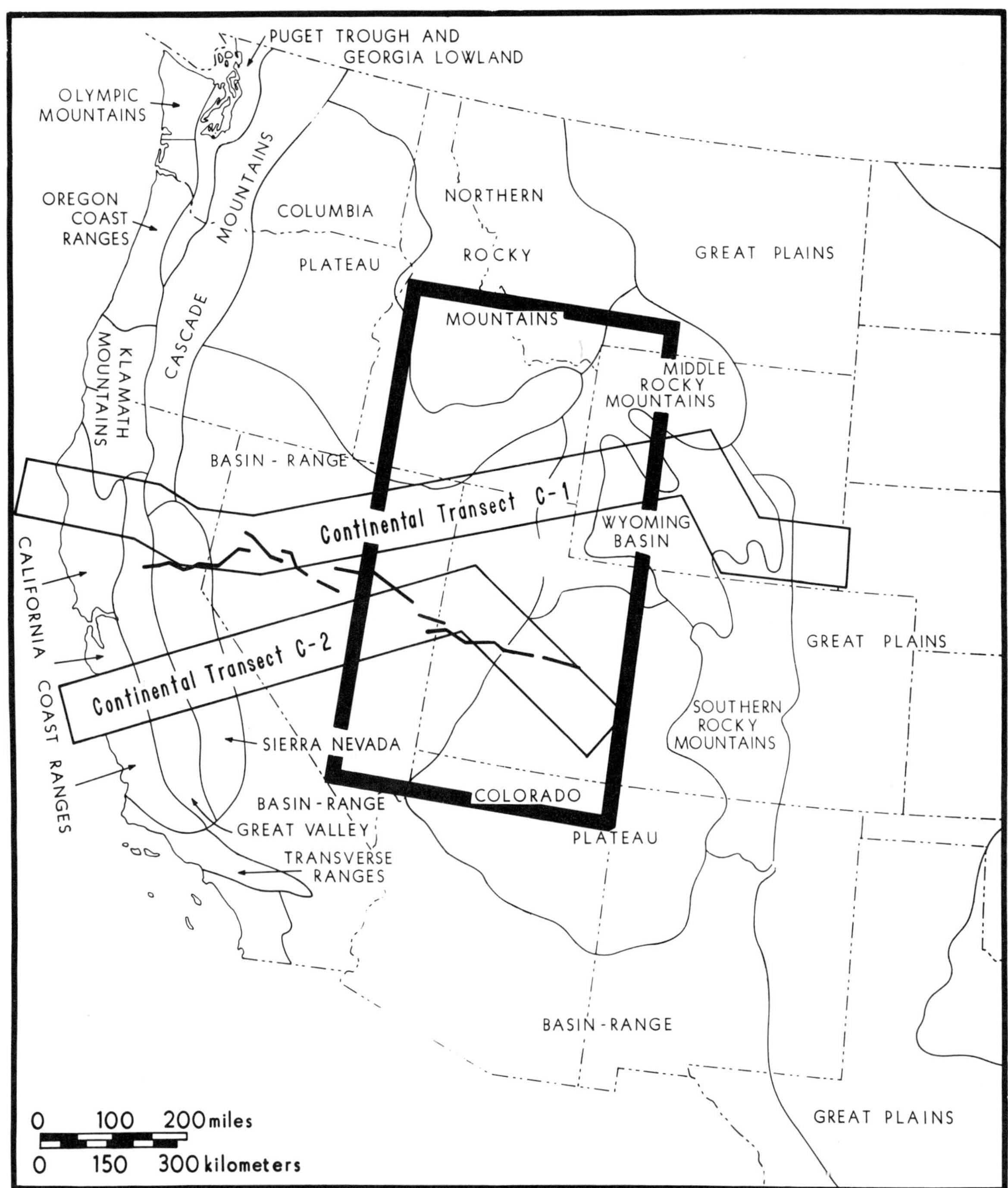

Figure 1. Area of discussion: Basin and Range–Colorado Plateau–Rocky Mountain (BR-CP-RM) transition. Locations of northern and southern lithospheric cross sections, C-1 (Blake and others, 1989) and C-2 (Saleeby and others, 1984), are shown in Figures 8 and 9, respectively. Physiographic provinces based on Fenneman (1946).

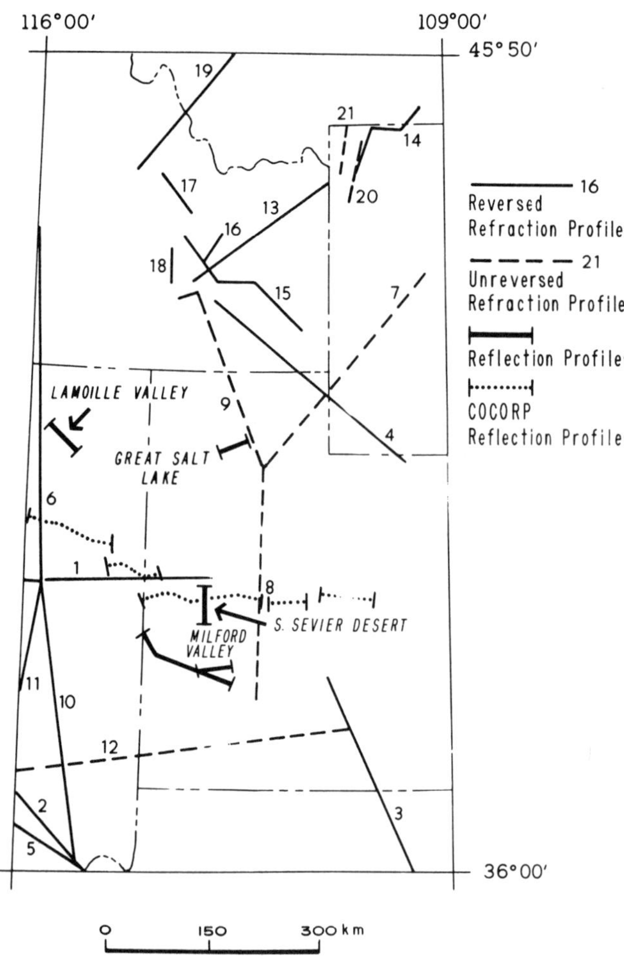

Figure 2. Locations of seismic refraction and/or wide-angle reflection profiles. Numbers correspond to refraction profiles indexed by author and source in Table 1. Note unreversed refraction profiles originating from large quarry blasts at the Bingham Canyon copper mine, Utah, at the vertex of lines 7, 8, and 9 (see discussion).

discussion. Twenty refraction profiles (Smith, 1989), several reflection profiles (Fig. 2; Table 1), and models from traveltime studies of local and regional earthquakes were used to produce generalized cross sections of crustal–upper-mantle structure.

The regional geophysical data discussed here pose interesting questions regarding crustal evolution. What data support the theory of large normal-faulting earthquake nucleation on steep- to low-angle and/or listric faults? What is the evidence for crustal low-velocity layers? Does a vertically layered brittle to quasi-plastic rheologic model of the crust reflect variations in seismic reflectivity and strength anisotropy? Can a quasi-plastic rheology limit the maximum depths of earthquake foci? Does crustal extension associated with historic seismicity account for the measured contemporary intraplate deformation, and how does it compare to earlier Quaternary deformation?

Accompanying discussions in this volume by Prodehl and Lipmann, Thompson and others, Pakiser and Anderson provide

TABLE 1. INDEX FOR REFRACTION DATA PRESENTED IN FIGURE 2

Reference Sources	Profile(s)
Braile and others, 1974, 1982	7, 13
Brokaw, 1985	14, 20, 21
Eaton, 1963	1
Elbring, 1984	16–18
Fauria, 1981	6
Hill and Pakiser, 1967	6
Johnson, 1965	5
Keller and others, 1975	8
Lehman and others, 1982	14
Martin, 1978	9
Mueller and Landisman, 1971	1
Prodehl, 1970, 1979	1–6, 10–12
Roller, 1964, 1965	2, 3
Sheriff and Stickney, 1984	19
Smith and others, 1982	13, 14
Sparlin and others, 1982	15
Willden, 1965	4

tectonic and geophysical information in the surrounding regions. The reader is referred to these chapters for additional insights into the crustal structure of the entire U.S. Cordillera and for comparison with the data presented here.

LITHOSPHERIC STRUCTURE

The lithosphere, as defined in this discussion, includes the crust and upper mantle. The continental crust, on the basis of seismic velocity structure and petrologic models, has generally been subdivided into four units (Mueller, 1977; Meissner, 1986): (1) a surface sedimentary layer, zero to several kilometers thick; (2) a well-defined upper crustal layer, ~6 km/sec, sometimes referred to as the "granitic" layer but which is likely more granidioritic in composition; (3) an intermediate-depth, but poorly defined ~6.5-km/sec layer thought to reflect a mixed migmatite layer; and (4) a lower crustal layer, ~6.7 km/sec, with an intermediate composition at a granulite metamorphic grade. Three of these units are resolved in the BR-CP-RM transition, but the intermediate-depth layer has not been unequivocally identified. To avoid confusion, we define the upper mantle as the zone below the crust with P-wave velocities greater than 7.6 km/sec and S-wave velocities greater than 4.4 km/sec. Based on these velocities, the upper mantle is thought to be predominantly composed of ultramafic rocks, such as peridotite, dunite, and garnet-bearing ecologite (Meissner, 1986).

Upper-Crustal, P_g, Velocity Distribution. The upper crust (exclusive of the sedimentary layer) is characterized by a pervasive layer thought to represent the crystalline basement composed primarily of supracrustal granitic material. The layer produces the well-defined P_g headwave branch. Figure 3 shows the regional distribution of P_g-velocities. The upper-crustal layer,

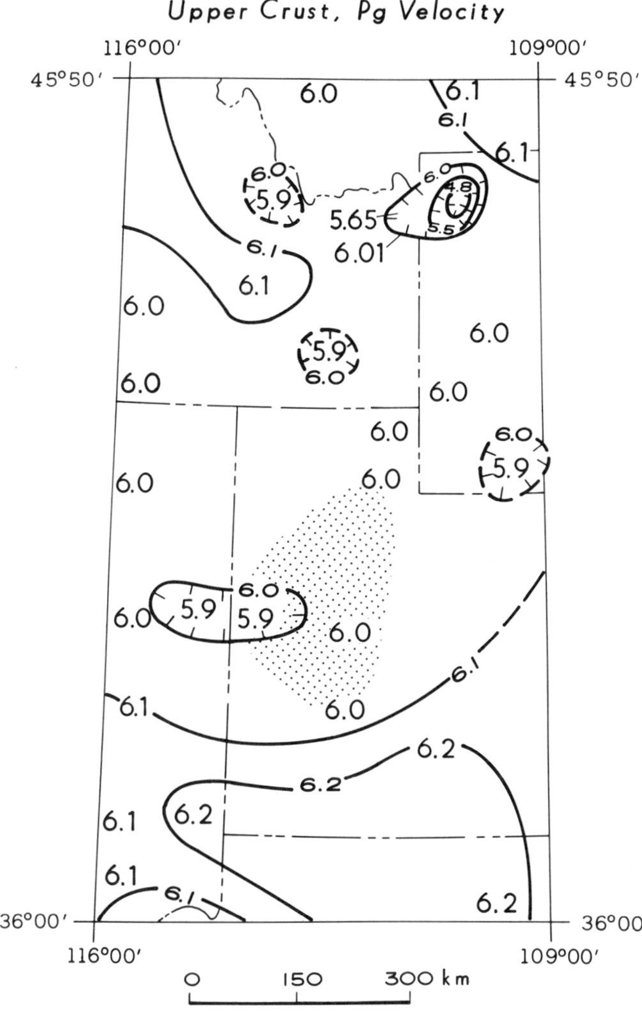

Figure 3. Upper crustal, P_g velocity, contour map of BR-CP-RM transition study area. Note the low velocities associated with the Quaternary Yellowstone caldera. Dots correspond to area of upper crustal low-velocity layer (see Fig. 9).

because of its close relationship with surface geology, fracture permeability, and surface heat flow is an indicator of the source and depth extent of tectonic features.

Regional P_g velocities, beneath the tectonically active eastern Basin and Range, are characteristically 5.9 to 6.0 km/sec, and correlate with regional heat flow of ~90 mWm^{-2}. The P_g velocities increase to 6.2 km/sec beneath the stable Colorado Plateau where the heat flow decreases to ~60 mWm^{-2}. The general association of moderate heat flow, <90 mWm^{-2}, pre-Cenozoic tectonism, and P_g velocities of 6.1 km/sec or higher, with high heat flow, >90 mWm^{-2}, and Cenozoic tectonism, with representative velocities of 6.0 km/sec or less, suggests a general correlation between seismic velocity and the degree of active tectonic extension. While higher temperatures are considered an attractive mechanism for velocity reduction in silicic rocks, laboratory studies (Spencer and Nur, 1976) show that composition and pore pressure are the dominant parameters controlling velocity. Thus, low P_g velocities may be influenced to a greater extent by excess pore fluids than by temperature. However, anomalous pore fluids may, in themselves, be produced by high-temperature metamorphism. For example, the Yellowstone caldera (Smith and others, 1982) and its associated hydrothermal systems with P_g velocities of 5.4 to 5.7 km/sec and very high heat flow, >1,500 mWm^{-2}, argues for this mechanism. The velocity of this layer for the entire BR-CP-RM averages 6.0 km/sec, near the continent-wide value.

Anomalous 7.5-km/sec layer. Cook (1962) and Pakiser (1963) were first to note an anomalous ~7.5-km/sec lower-crust–upper-mantle layer in the eastern Basin-Range. Cook (1962) interpreted the upper boundary of the 7.5-km/sec layer as the Moho based on unreversed refraction profiles and interpreted its composition as a mantle-crust mix. Unreversed refraction profiles in central and northern Utah (Fig. 2) recorded from quarry blast sources (Mueller and Landisman, 1971; Braile and others, 1974; Keller and others, 1975; Smith, 1978) and earthquake traveltime data recorded across a regional seismograph network in Utah (Pechmann and others, 1984; Loeb and Pechmann, 1986) also delineated the anomalous 7.5-km/sec velocity layer.

L. C. Pakiser (1987, personal communication) used a subset of the same earthquake data of Loeb (1986) in a two-station analysis of apparent velocities across the transition. Pakiser (1987, personal communication) has asserted that the 7.5-km/sec values are anomalously low because they are downdip ray paths from a mantle upwarp (Smith, 1978) centered beneath north-central Utah near the Bingham Canyon copper mine, the source location used in the refraction profiles of Braile and others (1974) and Keller and others (1975). The lack of reversed refraction profiles used to interpret the seismic refraction data leaves the interpretation of these low 7.4 to 7.6-km/sec apparent velocities unresolved.

Early refraction profiles used quarry blasts that could only be recorded to ~250 km, an insufficient distance to acquire the critically refracted head waves from layers deeper than 25 to 30 km. To investigate the anomalous crustal structure from longer shot offsets, Pechmann and others (1984) used travel times recorded on a regional seismic network in Utah from local earthquakes and nuclear blasts from Nevada and recognized both an arrival corresponding to a 7.5-km/sec layer, at distances between 120 to 250 km, and a branch corresponding to a 7.9-km/sec layer, at distances greater than 250 km. Based on forward traveltime modeling and a generalized linear inversion of earthquake traveltime data, Pechmann and others (1984) described the 7.5-km/sec and the 7.9-km/sec layers at depths of 25 and 40 km, as a "double-Moho" structure. They indicated that the breadth of the province-wide 7.9-km/sec layer would suggest a more universal definition of it as a simple Moho, the definition that we prefer.

Pechmann and others (1984) noted that relative P_n residuals increase from west to east across central Utah by more than 1.5 sec. Since only about 0.5 sec of this increase can be attributed to lateral variations in the overlying crust, Pechmann and others (1984) argued that the top of the 7.9-km/sec layer has significant eastward dip. Keller and others (1975) reported a P_n apparent velocity of 7.4 km/sec for refraction profile 8 (Fig. 2) shot southward from Bingham Canyon mine, Magna, Utah. Because this profile tends to lie parallel to the contours of P_n residuals for some sources, this profile may lie along a direction of small change in crustal structure, and the 7.4-km/sec apparent velocity may be close to the true velocity. Nonetheless, because of the sparse network data available to constrain the P_n residual contours, and the lack of reversed ray paths to determine true velocities and lateral changes in crustal structure, no firm conclusions are possible regarding the existence of a 7.5-km/sec layer beneath central Utah.

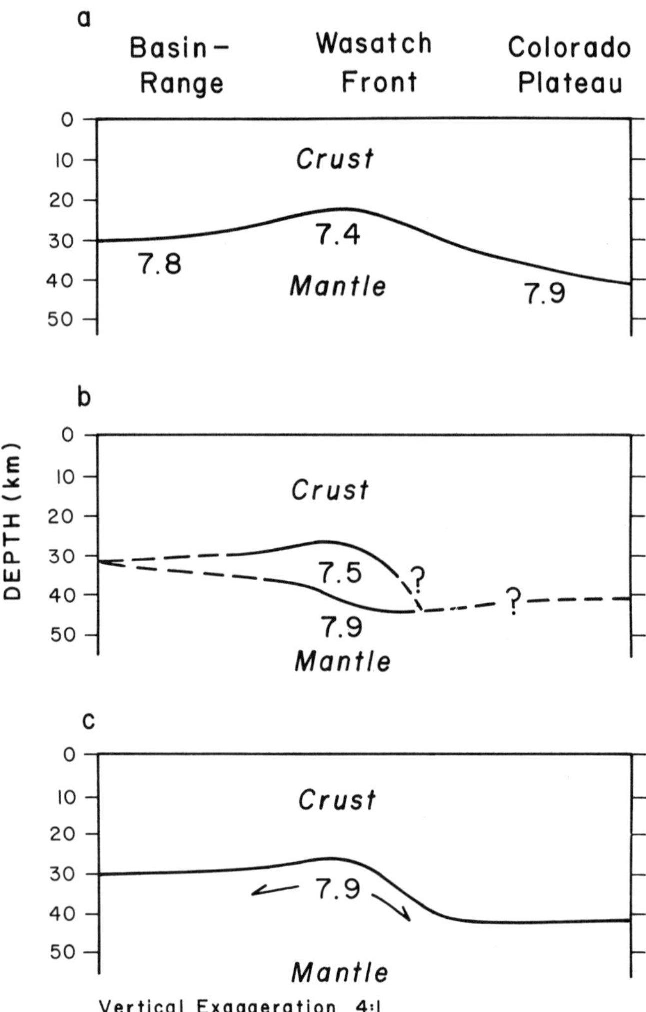

Figure 4. Models of generalized crust-mantle velocity structure for the BR-CP-RM transition. a, Interpretation of Mueller and Landisman (1971) and Keller and others (1975), defining 7.4-km/s layer. b, Interpretation of Loeb and Pechmann (1986), with an anomalous 7.5-km/sec lower crust and 7.9-km/sec upper mantle. c, Interpretation of Pakiser (1985), in which a mantle upwarp is inferred with a uniform upper-mantle velocity of 7.8 to 7.9 km/sec.

Three possible models thus exist for the eastern Basin-Range crust and upper mantle that could account for the reported anomalous velocities (Fig. 4). The model from unreversed profiles (Fig. 4a) that extended only to lengths of ~250 km (Mueller and Landisman, 1971; Keller and others, 1975) defines a shallow Moho below the Wasatch Front with a 7.4 to 7.5-km/sec velocity that increases in depth laterally to the east and west. This interpretation is inconsistent with traveltime observations at distances greater than 250 km. Figure 4b displays the interpretation by Loeb and Pechmann (1986) using the earthquake data with the anomalous 7.5-km/sec lower crustal layer and a 7.9-km/sec upper-mantle layer. Figure 4c shows the model of Pakiser (1985; 1987, personal communication) with a uniform 7.9-km/sec upper-mantle velocity and a ~10-km mantle upwarp beneath the Wasatch Front. Without detailed reversed refraction profiles it is not possible to properly choose between model b or c. Thus, whether the Moho, as defined by the top of the 7.9-km/sec layer, is at a depth of 25 to 30 km or at 35 to 45 km cannot yet be resolved.

Upper-mantle, P_n, velocity distribution. The upper-mantle, P_n branch, marks the pervasive Mohorovičić (Moho) discontinuity between the crust and upper mantle (Fig. 5). Its presence is based on the recognition of the refracted head wave, P_n, the wide-angle reflected wave, P_mP, and to a lesser extent on discontinuous, vertically incident reflections seen on some of the Consortium for Continental Reflection Profiling (COCORP) profiles. Seismic models of the crust–upper mantle boundary suggests that it ranges from a first-order velocity discontinuity, corresponding to an abrupt change in composition, to a second-order discontinuity, gradient, or laminated zone 1 to 3 km thick (Meissner, 1986). A laminated Moho structure suggests that it may be a mixed metamorphic zone with variable composition and state.

Central Nevada is characterized by 7.8 to 7.9-km/sec P_n values that continue eastward beneath the middle Rocky Mountains and Colorado Plateau (Fig. 5). This generally smooth distribution suggests a homogeneous upper mantle throughout the BR-CP-RM transition. If, however, the anomalous 7.5-km/sec layer is included as part of the lower crust (as discussed earlier), then it may represent an anomalously high-crustal velocity layer.

Crustal Thickness. A map of total crustal thickness (or equivalently the depth to the top of the mantle relative to sea level) is shown in Figure 6. On the basis of U.S. Geological Survey refraction profiles, the central Basin-Range has a thin, ~30-km thick crust. Using the unreversed profile interpretations from quarry blasts (Braile and others, 1974; Keller and others, 1975; Smith, 1978; Pakiser, 1985), the crust thins easterly to ~25 km at the BR-CP-RM transition (Fig. 6) and forms a mantle bulge. If the alternate solution of Loeb (1986) and Loeb and Pechmann (1986) for a uniform 7.9-km/sec upper mantle is chosen, the crust gradually thickens to the east, attaining a 45-km thickness beneath the BR-CP-RM transition zone.

In the northern BR-CP-RM transition, the Yellowstone–Snake River Plain volcanic province has a ~40-km-thick crust, similar to the thermally undeformed crust of the Rocky Mountains. However, in central Idaho the crust may thin to 30 to 35 km (Sheriff and Stickney, 1984), similar to the thin Basin-Range crust of Utah and Nevada. The crust then thickens eastward to ~40 km beneath the Colorado Plateau and to ~50 km below the middle Rocky Mountains.

Average Crustal Velocities. The total crustal column is the integrated product of thermal, magmatic, mechanical, and metamorphic evolution. To compare crustal evolution to tectonics, mean crustal velocities were determined (Fig. 7) by averaging the upper, intermediate, and lower crustal-layer velocities (comparable to interval velocities), weighted by their respective layer thicknesses. The sediment layer was found to be so laterally variable and the data on its thickness so sparse that it was omitted from this calculation.

Thick, silicic low-velocity upper crustal layers, velocity inversions, excess felsic lower crustal constituents, and a thin lower crust tend to decrease the average crustal velocity. Therefore, crust effected by youthful tectonism and large components of silicic intrusives, generally corresponds to low average crustal velocities. Greater quantities of mafic constituents, a thin silicic upper crust, basic intrusives, and thin to absent lower crustal layers (akin to an oceanic crust) correspond to higher average velocities.

Figure 7 displays the averaged crustal velocity in the BR-CP-RM transition. Earlier refraction interpretations, excluding the 7.5-km/sec layer (Fig. 7a), revealed an average crustal velocity of 6.1 to 6.2 km/sec in south-central Utah. However, there is a pronounced higher average crustal velocity of 6.4 to 6.5 km/sec (Fig. 7b), assuming the 7.5-km/sec layer is part of the crust. This

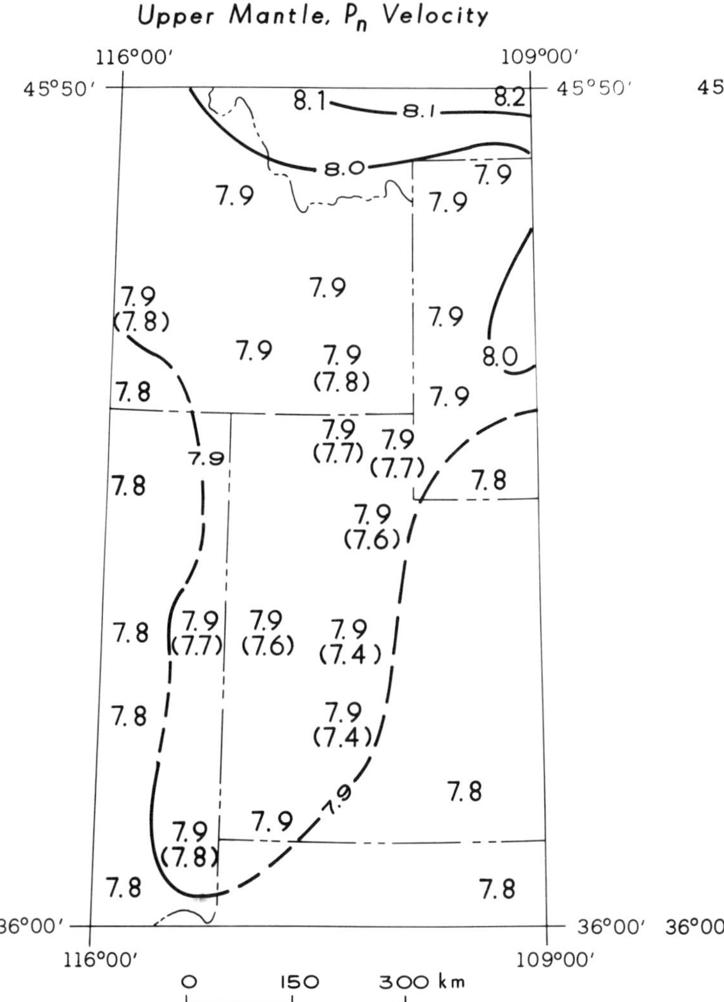

Figure 5. Upper-mantle, P_n velocity, contour map of BR-CP-RM transition study area. Velocities in parentheses were obtained form unreversed refraction profiles shot from Bingham copper mine, Utah (Braile and others, 1974; Keller and others, 1975). The contours roughly define the anomalous 7.5-km/sec lower crustal material of Loeb and Pechmann (1986), determined using regional earthquake traveltime data.

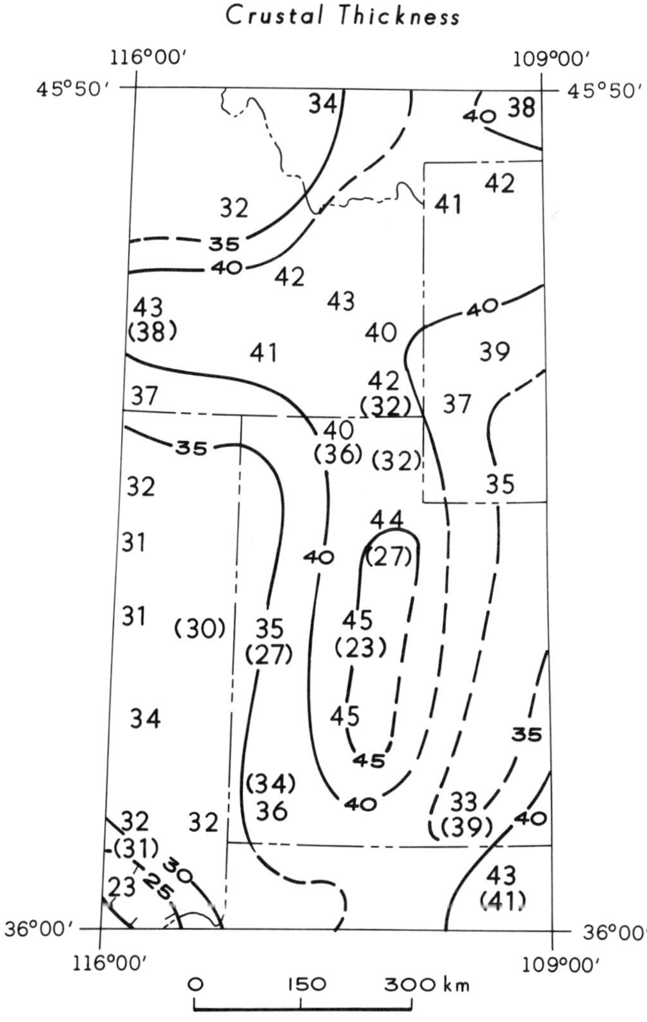

Figure 6. Crustal thickness map of BR-CP-RM transition study area (all depths relative to sea level). Thickness values in parentheses include the anomalous 7.5-km/sec lower crustal layer of Loeb and Pechmann (1986) and the approximate boundary of the mantle upwarp interpreted by Pakiser (1985) with a uniform 7.8- to 7.9-km/sec velocity.

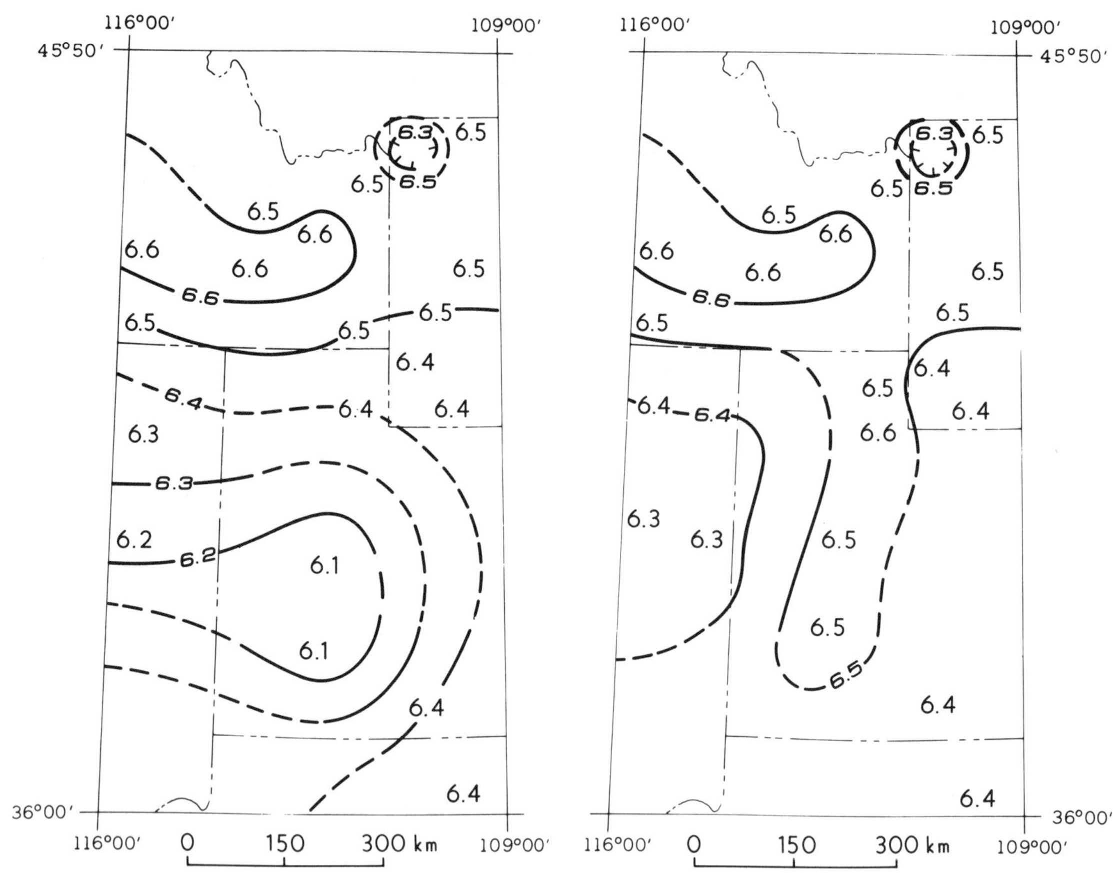

Figure 7. Average crustal velocity contour map (excluding sedimentary layers) of BR-CP-RM transition study area. (a) Left, Thickness-averaged velocity structure from unreversed refraction profiles. (b) Right, Averaged velocity structure adopting Loeb and Pechmann (1986) model (see also Fig. 4c).

increase is primarily from the contribution of the 7.5-km/sec lower crust, but also includes the presence of upper crustal low-velocity layers.

The southern Basin-Range values of 6.3 km/sec increase northeasterly toward the Rocky Mountains to values of 6.4 to 6.5 km/sec, an increase in the average crustal velocity that correlates with the general direction of crustal thickening and suggests that the ancient Precambrian crust may have imprinted the Paleozoic and Mesozoic evolution of the stable interior.

Regional geophysical profiles

Two east-west cross sections of the crust and upper mantle across the BR-CP-RM transition display the regional variations between upper crustal geology, velocity structure, topography, Bouguer gravity, and heat flow (Figs. 8, 9). These cross sections were generalized from profiles and maps containing geological and geophysical information gathered from unpublished seismic reflection profiles (Barker, 1986; Smith, 1984; Gants, 1985; Viveiros, 1986), the Continental Transects program (see Speed, 1989), and from the analysis of traveltime data from regional earthquakes recorded at the University of Utah Seismograph Stations network (Loeb, 1986). The northern profile, Figure 8, corresponds to Continental Transect C-1 (Blake and others, in preparation), and the southern profile, Figure 9, corresponds to Continental Transect C-2 (Saleeby and others, 1984).

The northern profile (Fig. 8) begins near the central Basin-Range in northern Nevada where the Moho depth has been identified at about 30 km with a velocity of 7.8 km/sec. If the trend of the 7.8 to 7.9-km/sec layer, identified along the Wasatch Front by Pechmann and others (1984) and Loeb and Pechmann (1986), marks the Moho, the crust then thickens to the east along the profile. Adopting the model of Loeb and Pechmann (1986), the Moho attains a depth of about 30 km beneath the central Basin-Range, then deepens to more than 40 km beneath the Sevier Desert in western Utah (southern profile, C-2, Fig. 9).

The upper crust of western Utah has a P-wave velocity reversal from 0.1 to 0.2 km/sec at depths between 7 and 10 km, near the depth of the Sevier Desert detachment, and marks the only well-identified crustal low-velocity layer. The regional Bouguer gravity high, decreasing in magnitude easterly beneath the western Colorado Plateau (Fig. 9), may reflect a high-density mantle wedge associated with active extension along the BR-CP-RM transition (Zoback and Lachenbruch, 1984) or a hypothetical eastward-dipping fault that cuts the entire crust and upper mantle (Wernicke, 1981). The general geophysical signature of the Basin-Range—the high regional heat flow of ~90 mWm^{-2}, a regional gravity high, and the anomalous 7.5- to 7.9-km/sec layer—are parameters consistent with an actively extending lithosphere, where heat transfer from the mantle coincides with magmatic intrusion and underplating that may in turn produce concomitant uplift and extension.

Crustal structure from seismic reflection data

In the past decade, seismic reflection data have been used to elucidate details about the crustal structure of the Basin-Range, Colorado Plateau, and Rocky Mountains provinces. In this section we review the evidence for the structural style and geometry of upper crustal faults imaged by the reflection method (see Fig. 2 for reflection profile locations).

Upper crustal structure. Seismic reflection data provided by the oil industry and by university seismic experiments have primarily been used to map the structural style and geometry of faults and basins in the upper crust of the BR-CP-RM transition. At shallow crustal depths, sedimentary layering and faults are resolvable at scales similar to structures mapped at the surface. The principal limitations of this method are the lack of velocity information and the presence of complicated two- and three-dimensional velocity structures that are difficult to migrate into their correct spatial coordinates. Ideally, reflection and refraction data are complementary and should be interpreted together. The refraction and wide-angle reflection data are useful for determining gross structure and providing velocity control. This velocity information can then be applied to the processing, particularly the migration of the higher vertical resolution reflection data.

Several examples of reflection data across Tertiary basins and normal faults in the Basin-Range reveal structural styles that differ from those inferred from surface geology. Zoback and Anderson (1983) and Anderson and others (1983) interpreted available industry reflection profiles primarily in Nevada and suggested three modes of extensional deformation for the central Basin-Range: (1) simple asymmetric sags bounded by ~60° dipping normal faults, (2) tilted ramps associated with moderately to deeply penetrating listric normal faults, and (3) assemblages of complexly deformed subbasins associated with both planar and listric normal faults that sole into low-angle detachments. Smith and Bruhn (1984) showed several examples of reflections for the transition region that revealed a range of structural styles from shallow to moderately dipping planar and listric normal faults bounding several asymmetric basins. Allmendinger and others (1983), Smith and Bruhn (1984), and Planke (1987) also interpreted reflection data for the Sevier Desert detachment of western Utah and concluded that there was little evidence for extensive penetration of the detachment reflector by shallow and younger normal faults.

Upper crustal structure and basin geometry

Northeastern Nevada. A 20-km-long, east-west, 12-fold *Vibroseis* reflection profile (Fig. 10) was recorded across Lamoille Valley (also called Mary's River Valley), Nevada (Fig. 2). Smith (1984) modeled true-amplitude reflection data and traveltimes for the Lamoille Valley using a two-dimensional ray-tracing algorithm. The model (Fig. 11) suggests a shallow, planar to gently dipping normal fault, dipping about 15° to 20° west near the surface, but decreasing to less than 10° at a depth of 4 km below the western Lamoille Valley. These data could not resolve whether the fault rotated continuously to its relatively flat geometry or whether it intersected a detachment at an oblique angle.

The fault geometry for the Lamoille Valley data is intriguing since its surface manifestation is a steep, west-dipping Quaternary scarp in unconsolidated alluvium on the west side of the Ruby Range. Several other reflection profiles for eastern Nevada (Effimoff and Pinezich, 1981; Anderson and others, 1983) display a variety of planar to listric normal faults, dipping from 40° to 60°, generally bounding asymmetric sedimentary basins.

Great Salt Lake, Utah. An example of a listric normal fault is seen on a reflection profile from the southern Great Salt Lake (Fig. 2) that was acquired in a 4- to 6-fold, high-resolution air-gun reflection survey. Viveiros (1986) processed and analyzed these data employing a detailed velocity analysis and a frequency-domain pre-stack migration technique to spatially resolve the fault geometry (Fig. 12).

A two-dimensional, asymptotic ray theory synthetic seismogram of these data (Fig. 12, lower) revealed a west-dipping reflection, interpreted as a normal fault zone, on the east side of the Great Salt Lake adjacent to the Precambrian rock exposed on the surface at Antelope Island (Viveiros, 1986). Asymmetric, eastward-tilted Tertiary sediments are truncated at the reflection and suggest a fault-zone discontinuity (Fig. 13) against the Precambrian basement. The interpreted fault dips ~60° near the surface and decreases to less than 10° at a depth of 5 km.

Southwestern Utah. A combined seismic refraction/reflection interpretation (Fig. 14) from the Milford Valley and the Mineral Range, southwestern Utah (Fig. 2), was made from a multiple-shot, reversed-refraction profile recorded at a 300-m-station spacing (Gertson, 1979) and a standard *Vibroseis* reflection survey interpreted by Barker (1986). Two-dimensional asymptotic ray theory modeling was applied to both the refraction data and the reflection data. Because of refraction ray-path multiplicity, Gertson (1979) applied an iterative time-term analysis for a two-dimensional velocity interpretation that was used as a starting model for the detailed ray-trace modeling. These models were perturbed by Barker (1986) in a "boot-strap" approach, first varying the geometry and velocities to fit the models of the reflection data, and then repeating the procedure for the refraction data until a suitable model was obtained for both data sets with the additional constraints from well logs.

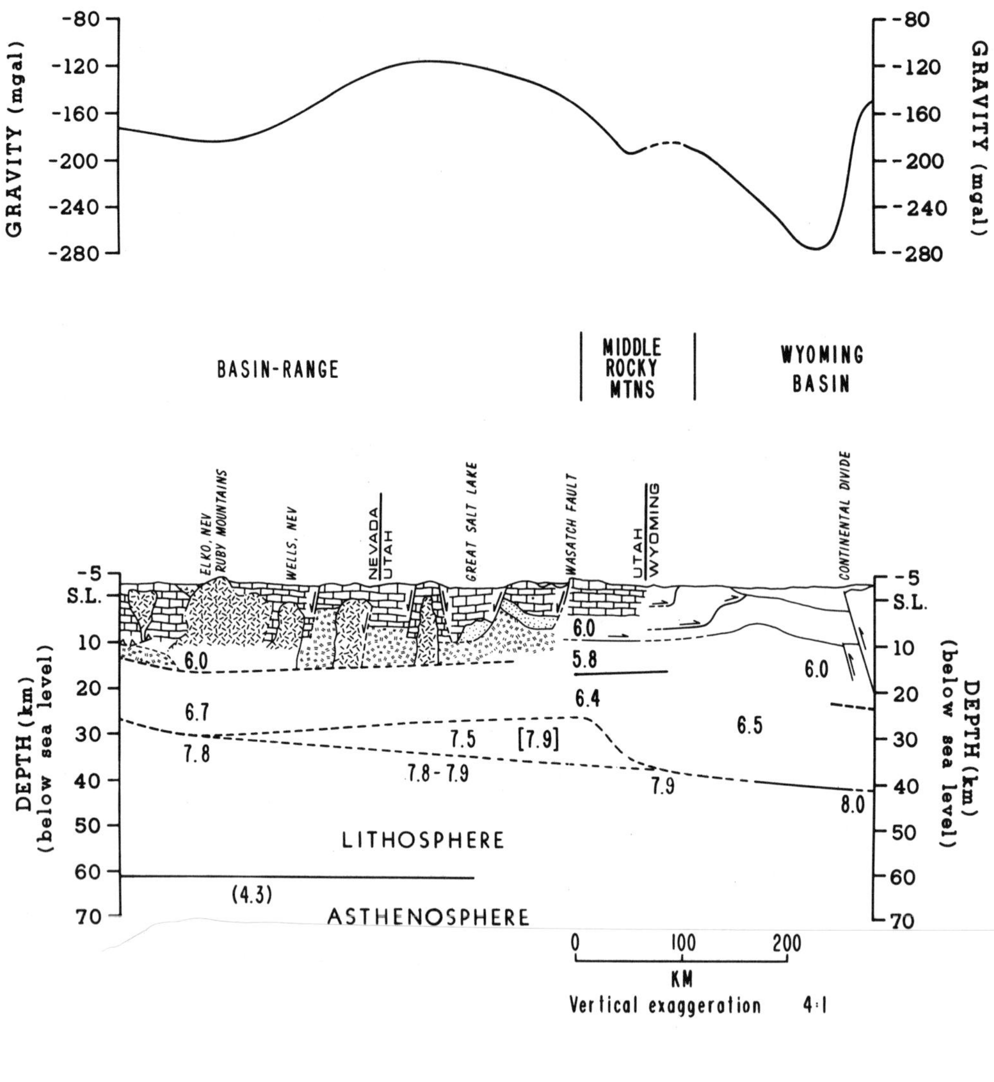

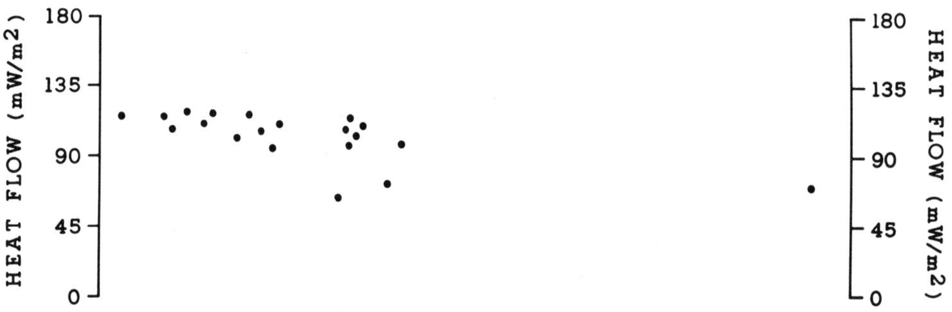

Figure 8. Cross section of continental lithosphere across the northern study area, C-1 (see Fig. 1 for location). Upper, Bouguer gravity anomaly along profile. Middle, Two-dimensional velocity cross section. Surface geology partly generalized from Continental Transects profile C-1 (Blake and others, 1989). P-wave velocity interpretation of Pakiser (1985) shown in brackets. P-wave velocities (kilometers per second) are in bold, S-wave velocities (kilometers per second) are in parentheses, densities (grams per cubic centimeter) are in italics. Lower: Heat-flow data (milliwatts per sq. meter). The sparse heat flow data for the middle Rocky Mountains are due to the lack of measurements for this area.

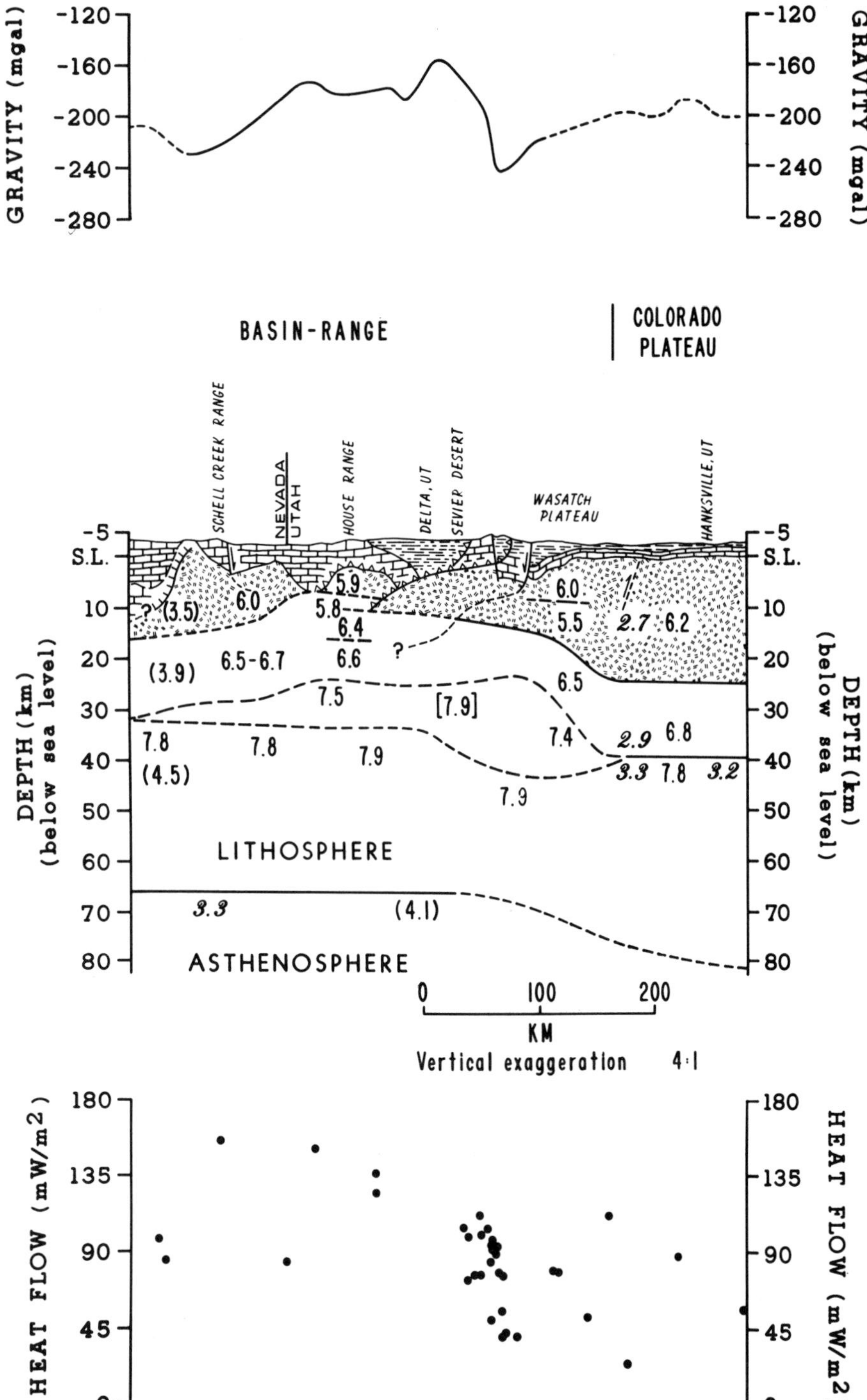

Figure 9. Cross section of continental lithosphere across the southern study area, C-2 (see Fig. 1 for cross-section location). Upper, Bouguer gravity anomaly along profile. Middle, Two-dimensional velocity cross section. Surface geology partly generalized from Continental Transects profile C-2 (Saleeby and others, 1984). P-wave velocity interpretation of Pakiser (1985) shown in brackets. P-wave velocities (kilometers per second) are in bold, S-wave velocities (kilometers per second) are in parentheses, densities (grams per cubic centimeter) are in italics. Lower: Heat-flow data (milliwatts per sq. meter).

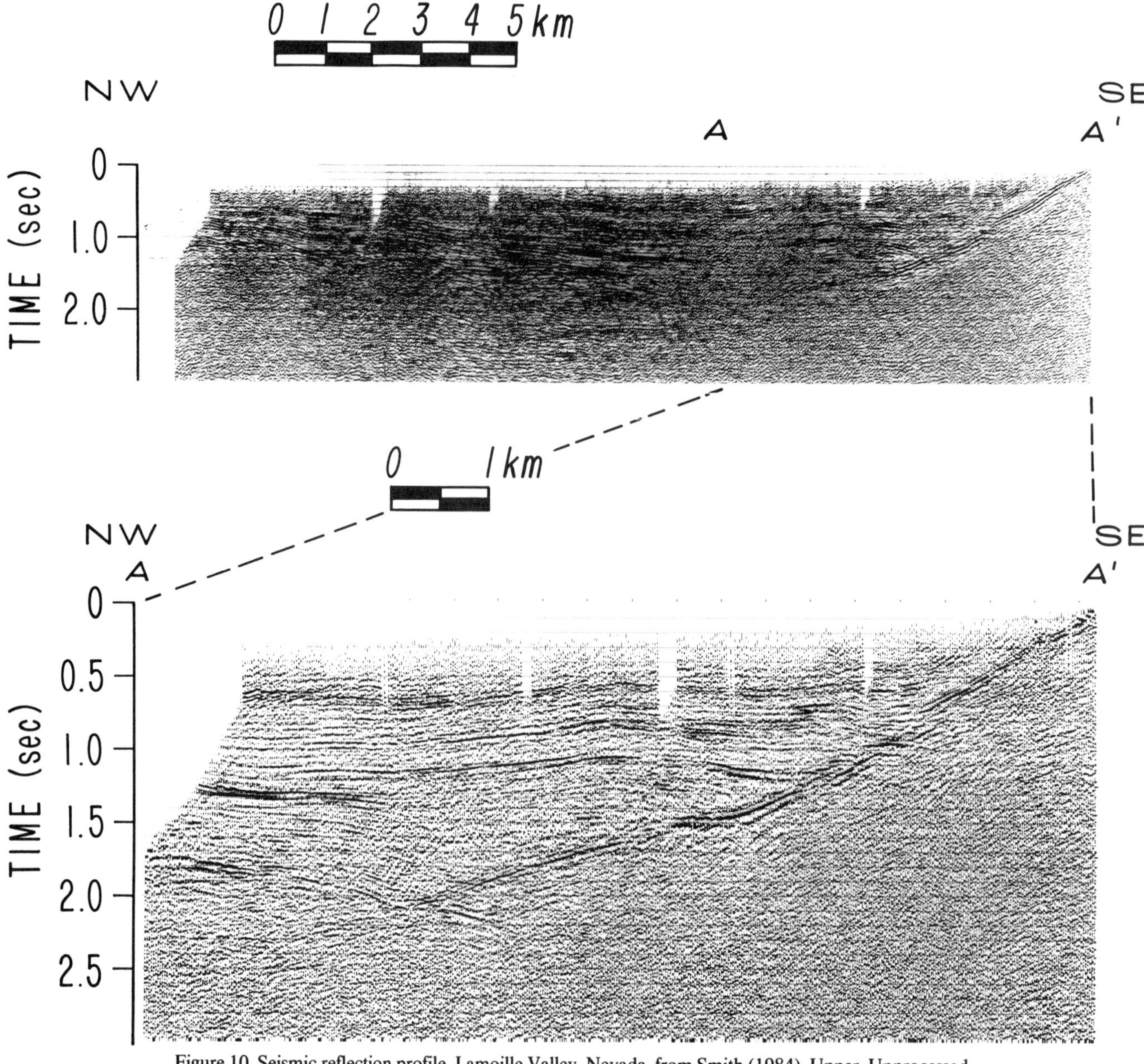

Figure 10. Seismic reflection profile, Lamoille Valley, Nevada, from Smith (1984). Upper, Unprocessed reflection profile. Lower, True-amplitude reprocessed profile for southeastern portion of line. See location in Figure 2.

The interpretations show eastward-dipping Tertiary sediments beneath the Milford Valley that terminate against a shallow, ~10°, west-dipping, planar normal fault on the west side of the Mineral Range (Figs. 14, 15). This fault geometry matched, within the errors of interpretation, the formation boundaries identified in nearby drillholes.

Using a regional reflection profile (Fig. 2) that extended from the south Milford Valley to the Nevada border (discussed in detail in the next section), a deeper reflector is interpreted to be a west-dipping low-angle fault (Fig. 15). This reflection appears to extend from beneath the Mineral Range at a <2-km depth to ~6 km beneath the western Milford Valley. This structure has a form similar to the overlying shallow low-angle fault and may be an adjacent en echelon normal fault that extends to the surface east of the Mineral Range.

Regional crustal structure from reflection data

Upper crustal low-velocity layer. The upper-crustal velocity structure of western Utah has been investigated using the joint interpretation of both seismic refraction and reflection data (Gants, 1985). Wide-angle reflections recorded on the U.S. Geo-

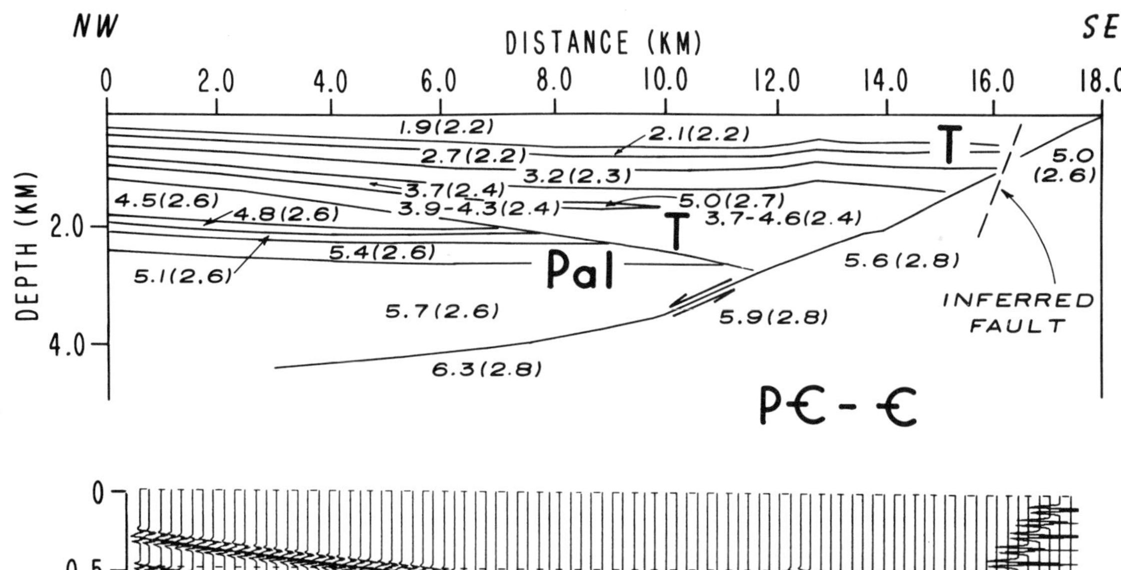

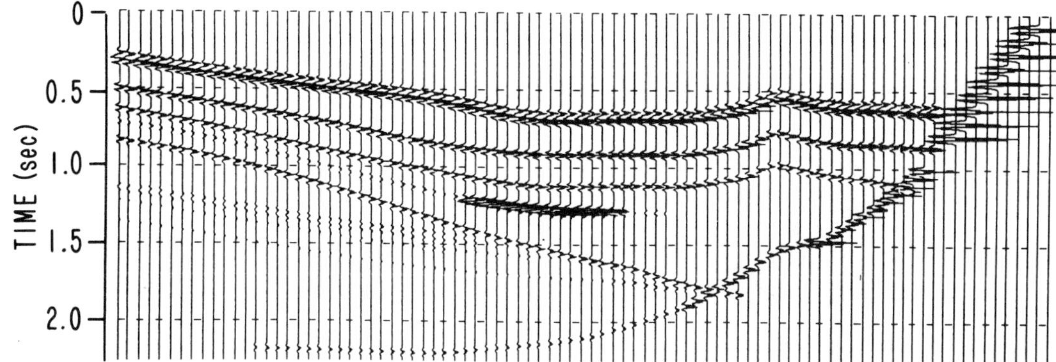

Figure 11. Interpreted two-dimensional velocity model (upper) and synthetic seismogram (lower) for Lamoille Valley, Nevada, by Smith (1984). Densities, in grams per cubic centimeters, are in parentheses. T = Tertiary, Pal = Paleozoic, PC-C = Precambrian-Cambrian.

logical Survey refraction line, DELTA-West (Fig. 2, profile 1), were first interpreted by Mueller and Landisman (1971) and revealed a 10-percent velocity reversal at depths between 9.5 and 11.5 km. Modeling of amplitudes using the reflectivity method by Müller and Mueller (1979) further supported the velocity reversal at these depths.

To investigate this anomalous layer, Gants (1985) reanalyzed the DELTA-West refraction profile in digital, true-amplitude form, and interpreted a nearby COCORP wide-offset line (Fig. 2). Gants incorporated two-dimensional asymptotic-ray theory traveltime fitting with true amplitudes for amplitude scaling (Fig. 16). His results showed that wide-angle reflections from COCORP, Utah line-5, originated from an interface at a 5.0-km depth, near the bottom of a layer with a 10-percent velocity decrease. Two-dimensional ray-trace models (Fig. 16) for the Delta-West profile, located about 50 km west of the cross-line, included the shallower structure and sedimentary units interpreted from the COCORP Utah east-west main line by Allmendinger and others (1983) and suggested a low-velocity layer that begins at a depth of 8 km and extends to 12 km with about a 10-percent velocity reduction.

Sevier Desert detachment. The Sevier Desert detachment has been interpreted from seismic reflection data in western Utah by several authors (McDonald, 1976; Allmendinger and others, 1983; Smith and Bruhn, 1984). Two-dimensional modeling of the Utah COCORP line (Fig. 2) and a COCORP cross-line, Utah line 5, (Fig. 17) by Gants (1985), suggested a 15° to 25° low-angle, west-dipping reflector that begins near the surface on the west flank of the Canyon Range and extends to a depth of about 15 km near the Nevada-Utah border over 150 km to the west. Whether the detachment is a reactivated thrust or an active normal fault has not yet been resolved, but the stratigraphic displacement across the fault, in a down-dip sense, may be as great as 60 km (Allmendinger and others, 1983; Smith and Bruhn, 1984). The total area of the Sevier Desert detachment may exceed 20,000 km^2 on the basis of its interpretation in adjacent seismic reflection profiles (Planke, 1987; Barker, 1986), suggesting that it underlies much of western Utah.

Upper crustal reflections suggestive of low-angle structures are also evident from a 150-km-long Vibroseis reflection profile in southwestern Utah (Fig. 2) that extended from southern Utah to the Nevada border (Fig. 18). Smith and Bruhn (1984) initially interpreted these data and suggested that a relatively flat reflector at depths of 8 to 10 km could be a detachment or thrust. Barker (1986) modeled the same reflection line as part of his investigation of the Milford Valley, Utah (Fig. 18), and interpreted low-

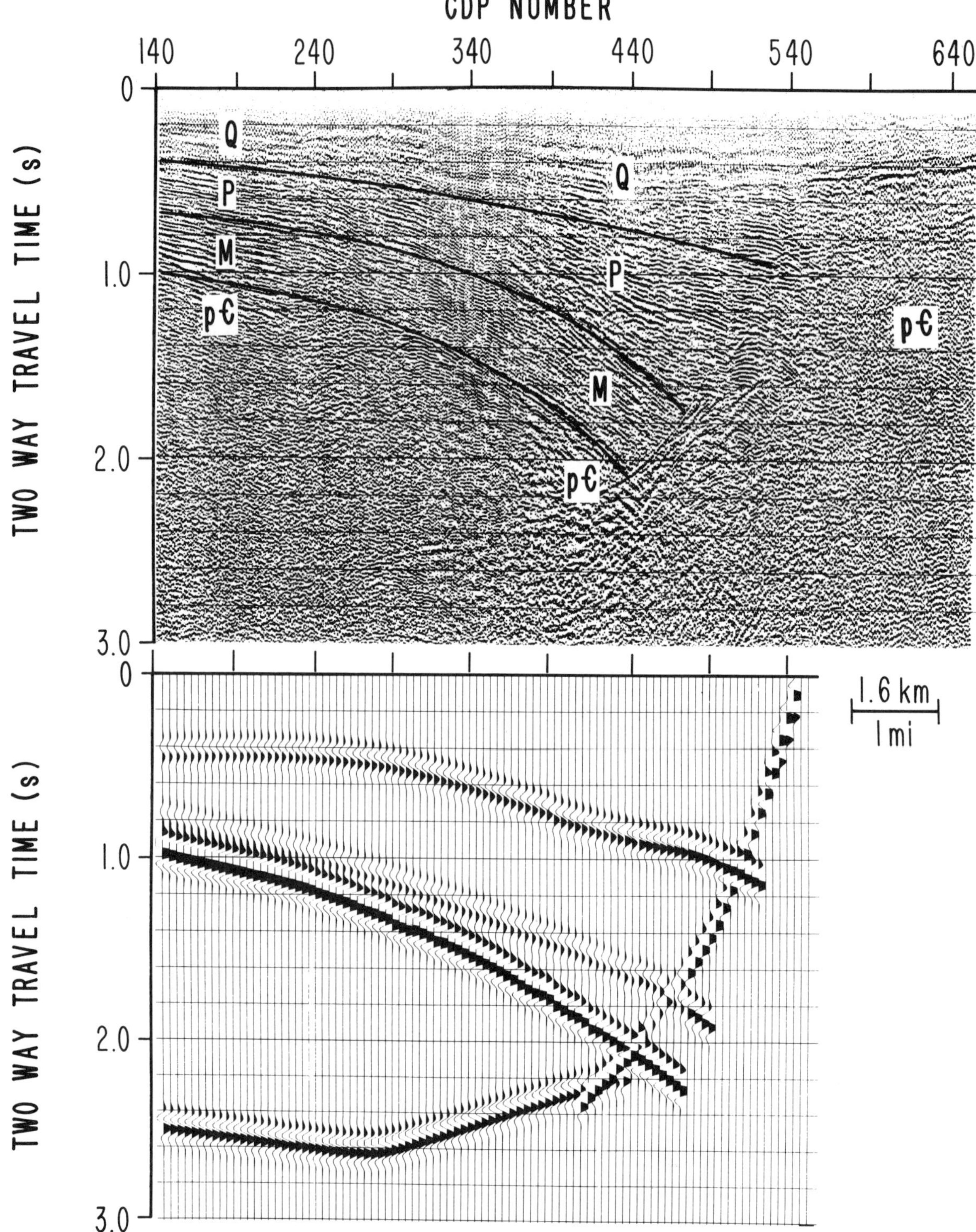

Figure 12. Southern Great Salt Lake (east-west) seismic reflection profile from Viveiros (1986). Upper, Interpreted seismic reflection profile. Lower, Synthetic seismogram from interpreted model. Q = Quaternary, P = Pliocene, M = Miocene, p€ = Precambrian. See profile location in Figure 2.

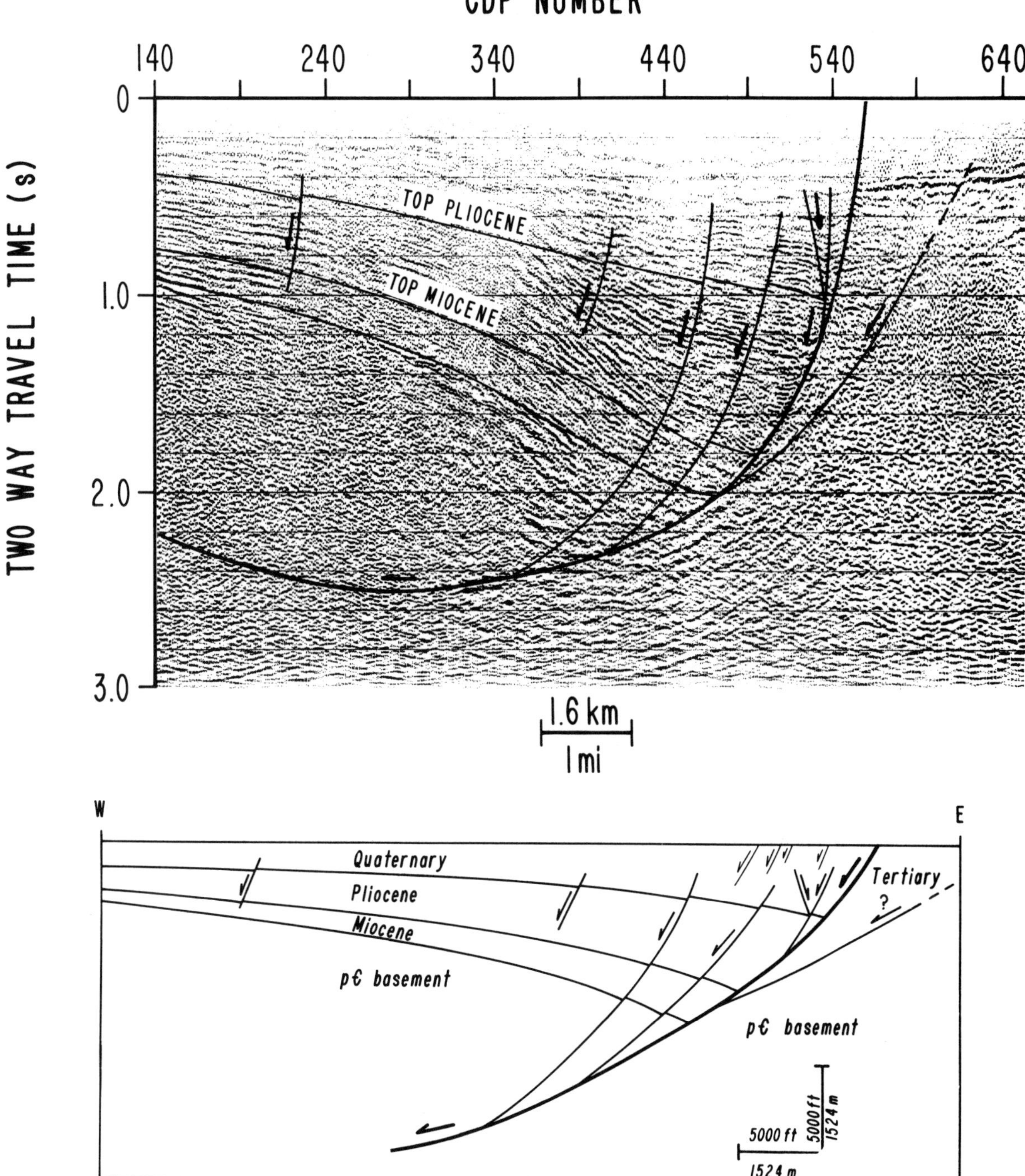

Figure 13. Geologic cross section interpreted from Great Salt reflection profile by Viveiros (1986). Upper, Reprocessed pre-stack migration reflection profile. Lower, Interpreted geologic cross section with stratigraphic ages and direction of faulting indicated.

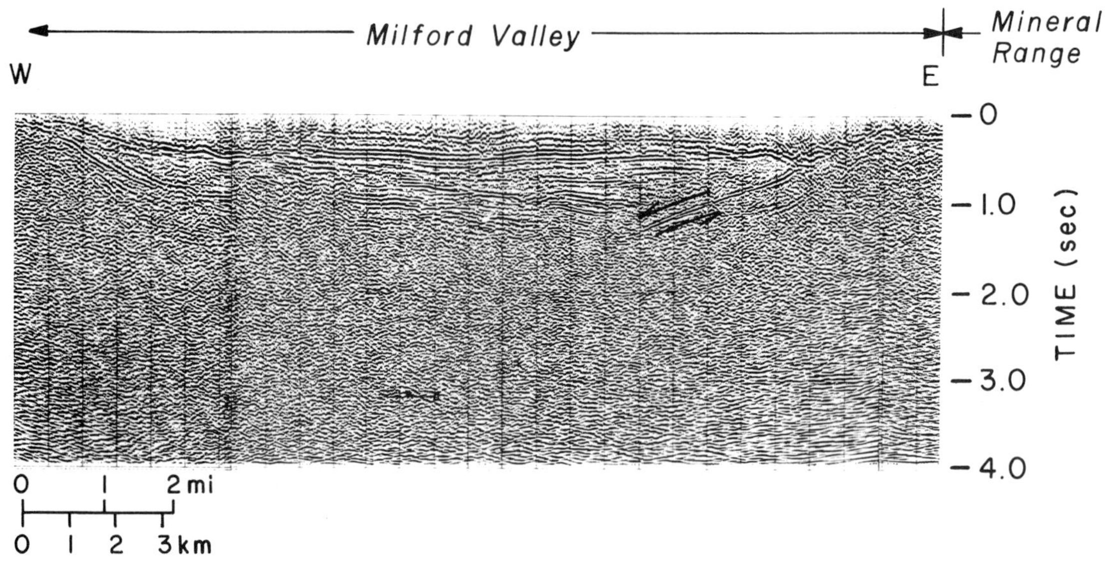

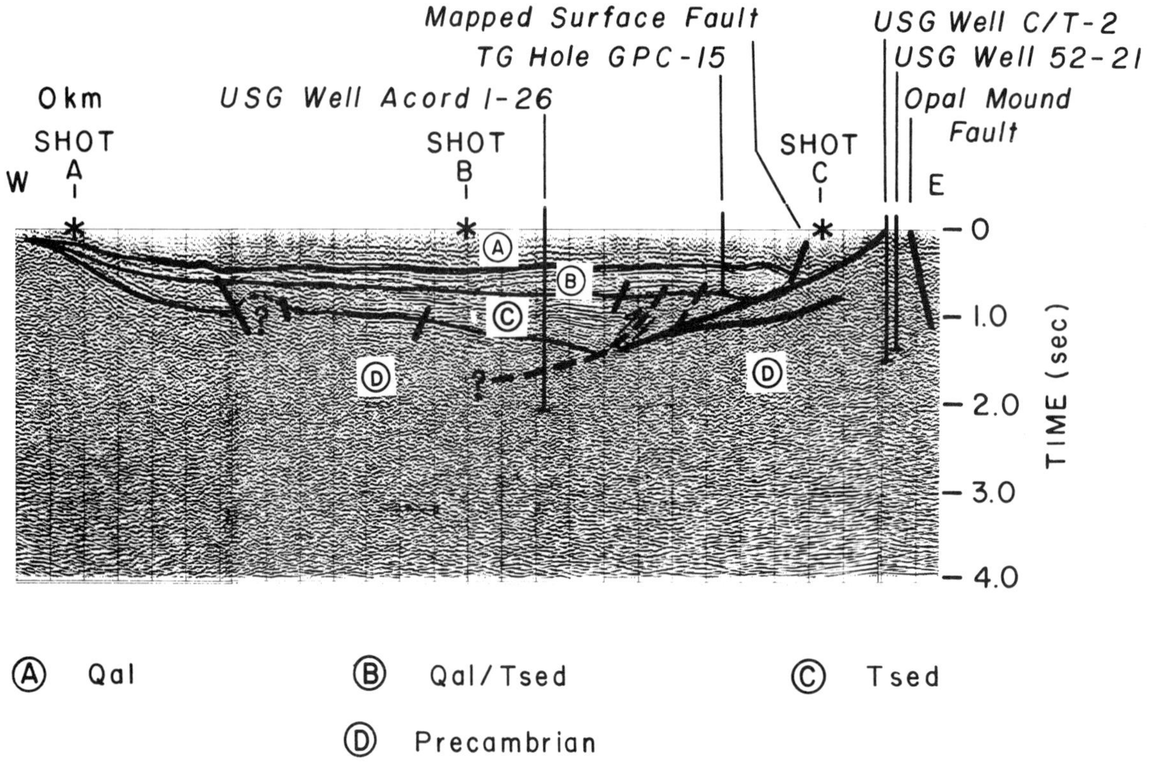

Figure 14. Reflection profile from Milford Valley, Utah, from Barker (1986). Upper: Observed reflection profile. Lower: Interpreted profile with drillhole locations indicating sources of sonic log information. See Figure 2 for location.

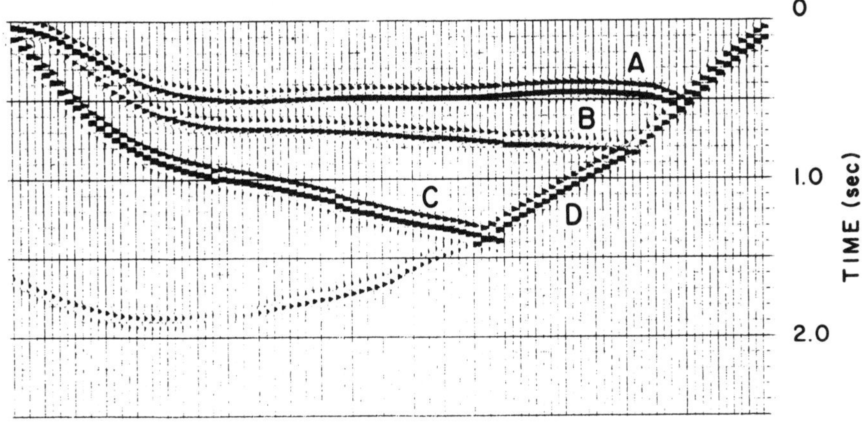

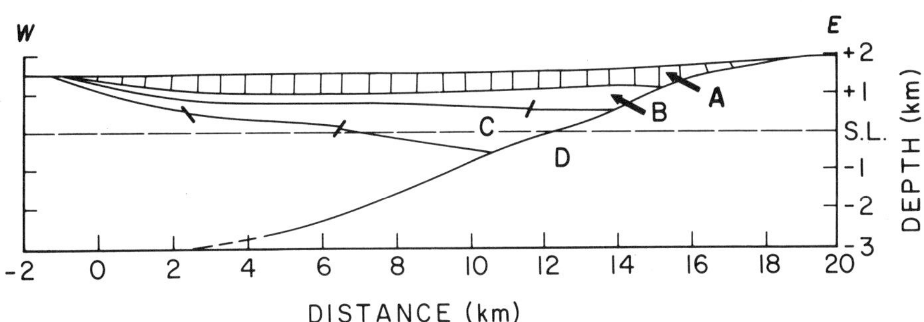

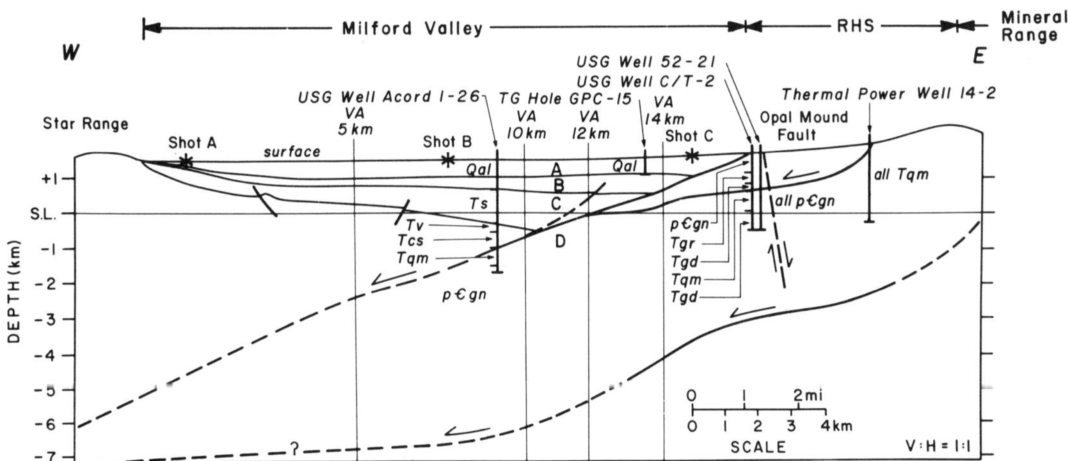

Figure 15. Milford Valley, Utah, upper crustal interpretation from Barker (1986). Upper, Synthetic seismogram. Middle, Model obtained from above synthetic seismogram. Lower, Interpreted geologic cross section from integrated reflection/refraction profiles. A = Q_{al}, Quaternary alluvium; B = Q_{al}, Quaternary semi-consolidated fill; C = T_{sed}, Tertiary sediments; D = Precambrian; RHS = Roosevelt Hot Springs geothermal area. See Figure 2 for location.

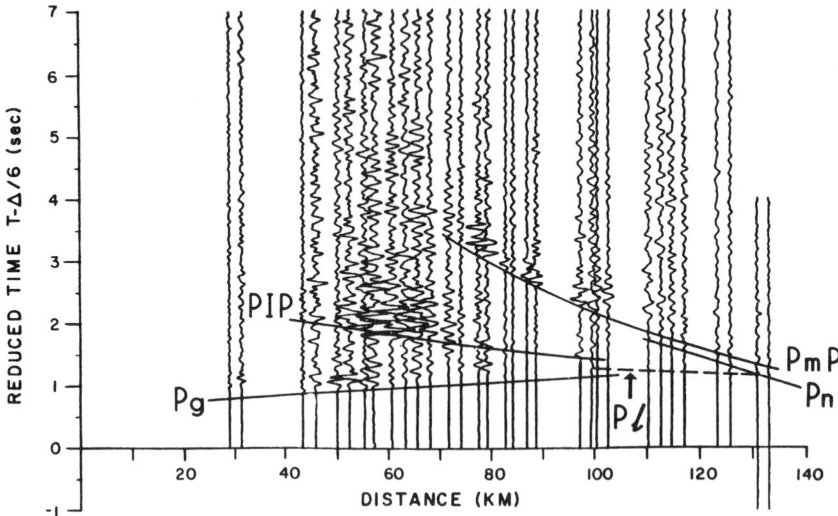

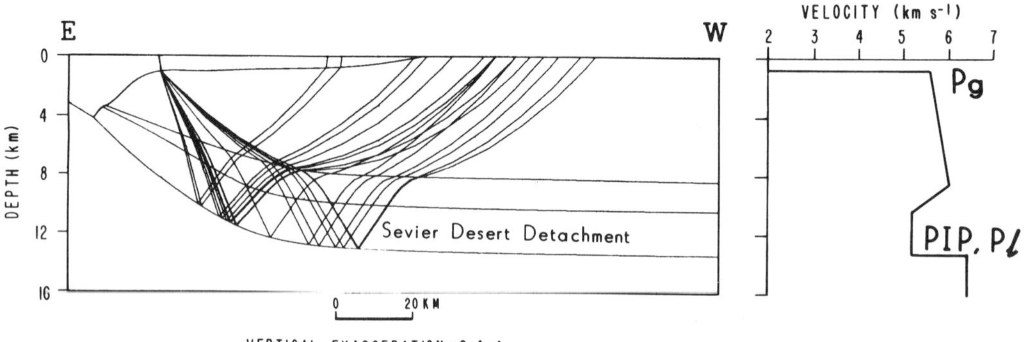

Figure 16. DELTA-West seismic refraction profile from Gants (1985). Upper: Observed (true-amplitude) data with interpreted branches and upper crustal velocity model. Lower, Upper crustal structure from reflection profile interpretation delineating Sevier Desert detachment. P_IP = wide-angle reflection, P_l = head wave from base of low-velocity layer, Pg = headwave from upper crust, P_n = headwave from crust-mantle boundary, and P_mP = wide-angle Moho reflection.

angle planar to listric faults along the west side of the San Francisco Mountains and the Mineral Range. Barker (1986) also suggested that the reflections may represent a midcrustal detachment that extends 90 km westward from the Beaver Valley, nearly to the Nevada border. This interpretation suggests that low-angle faults, extending from the surface on the east sides of asymmetric basins, may sole into deeper detachments. Note that the reflector, at a depth of ~8 to 9 km, is deeper than the Sevier Desert detachment identified 100 km north. If this reflection is a large-scale detachment, it may underlie a large area of southwestern Utah.

The upper crustal structure and its relationship to the hypothetical velocity structure and shear strength for the western Utah region are shown in Figure 19 (modified from Gants, 1985). This regional cross section includes the Sevier Desert detachment and other regional structures identified from both a Company General de Geophysics, line 302 (Smith and Bruhn, 1984), and the Utah COCORP line (Allmendinger and others, 1983). Note that the Sevier Desert detachment lies above the upper crustal low-velocity layer discussed earlier and is near the depth of the hypothetical transition from brittle to quasi-plastic flow.

Summary. The reflection data for the mid- to upper crust in the central and eastern Basin-Range reveal a persistence of 40° to 60° dipping, normal faults with both planar and listric geometries bounding asymmetric basins. These faults may truncate or sole into detachments. Some of these faults, however, may be underlain by deeper subhorizontal detachments, but the data are insufficient to resolve their location and geometry.

The COCORP and industry reflection profiles (Allmendinger and others, 1983; Smith and Bruhn, 1984; Klemperer and others, 1986) reveal the general features of the crustal structure of the BR-CP-RM transition. These include the presence of intermediate-crustal detachments, low-angle thrusts, and pervasive subhorizontal lower-crustal layering. In a review of the CO-

CORP 40°N transect across the western Colorado Plateau and the northern Basin-Range by Allmendinger and others (1987), prominent crustal reflections (Fig. 20) were interpreted as indicating: (1) complex dipping reflections and diffractions to depths as great as 45 km beneath the western Colorado Plateau: (2) intermediate- to lower-level seismic reflections within the Basin-Range crust, with west-dipping reflections in the eastern part of the province and subhorizontal reflections in the west; (3) a semi-continuous Moho beneath the Basin-Range; and (4) a general lack of mantle reflections beneath the Moho. The lower crust with its prominent band of flat reflectors at a depth of 15 to 30 km (Klemperer and others, 1986) may be an anastomosing shear zone, or it may indicate magmatic underplating that has added additional crustal material during the extensional process.

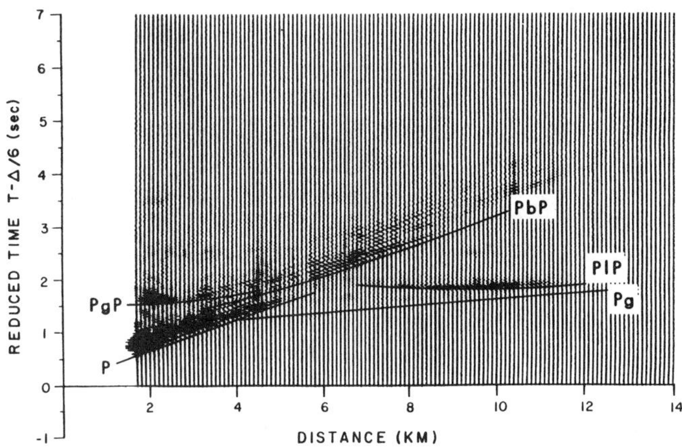

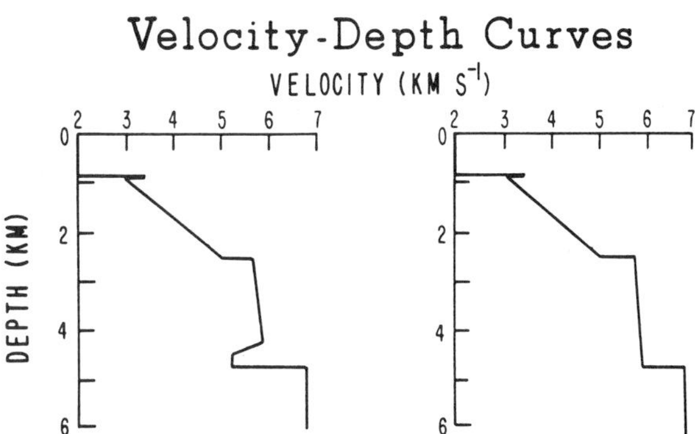

Figure 17. Refraction, wide-angle reflection profile (north-south), COCORP, UTAH line 5, across the Sevier Desert, western Utah, from Gants (1985). Upper, Wide-offset refraction profile synthetic seismogram. Lower, Modeled velocity-depth curves, with and without crustal low-velocity layer. See Figures 1 and 2 for profile location.

SEISMICITY

To show the historic seismicity of the BR-CP-RM transition, a new epicenter map of the study area was prepared (Fig. 21). Regional seismographic network data (described by Eddington and others, 1987) for Nevada, Idaho, Utah, and Wyoming were supplemented by catalog data from the University of Utah Seismograph Stations and the U.S. Geological Survey (E. R. Engdahl, personal communication, 1987).

The epicenter map (Fig. 21) includes data from historic earthquakes with a minimum magnitude threshold of approximately M = 4 prior to 1962 and M = 2 from 1962 through 1985. Regional seismicity is considered most complete along the southern Intermountain seismic belt and in Yellowstone, following the installation of regional seismograph networks in the mid-1970s.

The regional seismicity of the BR-CP-RM transition can be summarized as follows:

1. The regional earthquake patterns correspond to a broad north-south zone of intraplate extension with diffusely distributed epicenters across the BR-CP-RM transition and is part of the southern Intermountain seismic belt identified by Smith and Sbar (1974). Epicenters are scattered and, when accurately located with closely spaced portable seismographs, do not generally coincide with mapped Quaternary faults (Smith and Bruhn, 1984; Arabasz and Julander, 1986). Seismic slip for small- to moderate-magnitude earthquakes as revealed by focal mechanism studies in the northern region is dominated by normal faults of 30° to 60° dip (Smith and Lindh, 1978; Bjarnason, 1987), while in south-central Utah, slip predominates on fault planes with 30° to 60° dips and significant components of strike-slip displacement (Arabasz and Julander, 1986).

2. Nodal planes derived from focal mechanisms and complemented by source modeling of large Basin-Range earthquakes (Figs. 21, 22) show planar faults that dip from 40° to 60° for the 1959, M_L 7.5, Hebgen Lake, Montana, earthquake (Doser, 1985), ~50° for the 1983, M_S 7.3, Borah Peak, Idaho, earthquake (Doser and Smith, 1985); and ~60° for the 1954, M 7.1, Dixie Valley, Nevada, earthquake (Okaya and Thompson, 1985; Doser, 1986). Intermediate-magnitude events (Smith and Lindh, 1978) define nodal planes that vary in dip from 45° to 80°.

3. Three large (M7+) Basin-Range earthquakes provide a hypothetical working model (Fig. 22) for large normal-faulting earthquakes. This type of earthquake would nucleate at mid-crustal depths of 10 to 15 km, on 40° to 60° dipping, planar faults (Smith and Richins, 1984), and produce, with repeated large events, the subsidence and asymmetric stratal tilt of adjacent Tertiary basins. The large events also appear to nucleate near the maximum depth of aftershock activity and near the inferred depth of the brittle to quasi-plastic transition (Smith and Bruhn, 1984).

4. Return periods for large eastern Basin-Range earthquakes are several hundreds to thousands of years for M7+ earthquakes (Arabasz and Smith, 1981) versus tens to hundreds of years for large earthquakes on the San Andreas fault system (Schwartz and Coppersmith, 1984). Furthermore, the maximum-

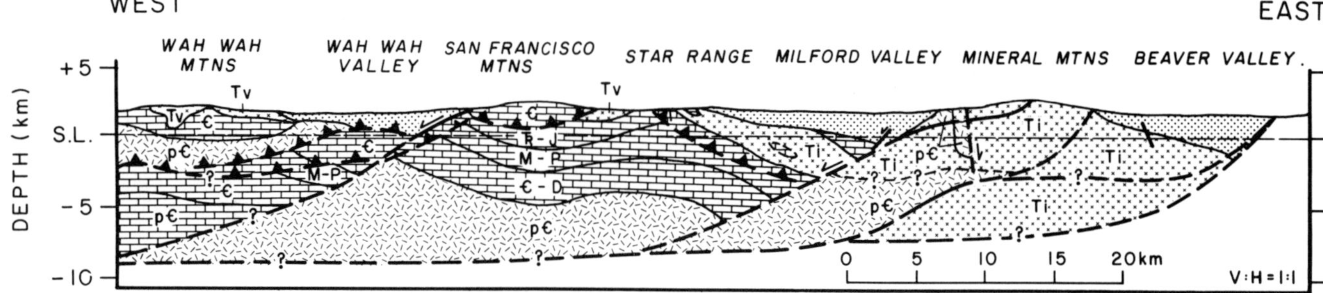

Figure 18. Interpreted, upper crustal cross section (east-west), southwestern Utah, from Barker (1986). Interpreted by Smith and Bruhn (1984) and Barker (1986). See Figure 2 for reflection profile location.

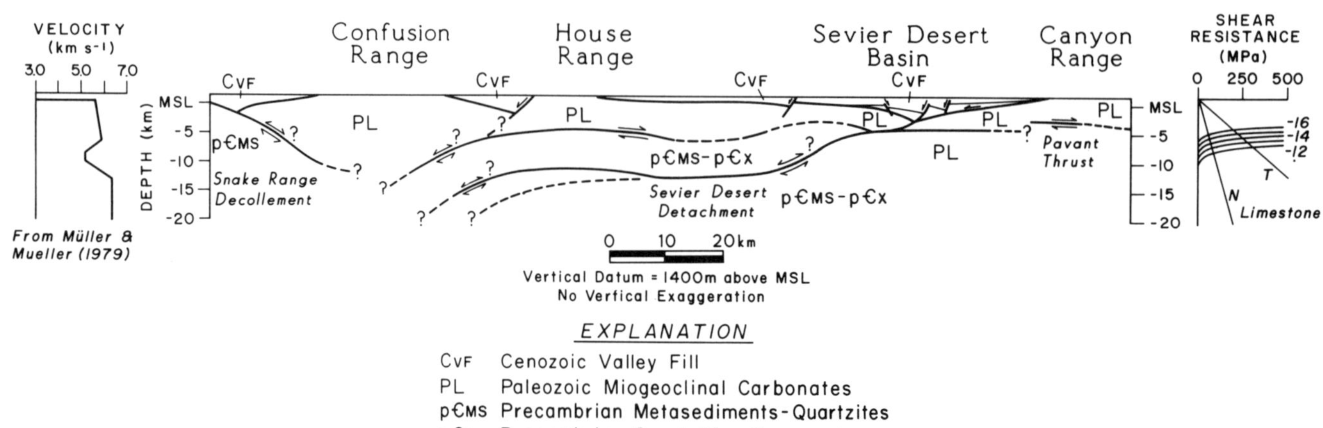

Figure 19. Interpreted upper crustal structure of western Utah with hypothetical shear-stress curves and velocity-depth profile interpreted from refraction, wide-angle reflection data of Gants (1985), Müller and Mueller (1979), and Smith and Bruhn (1984).

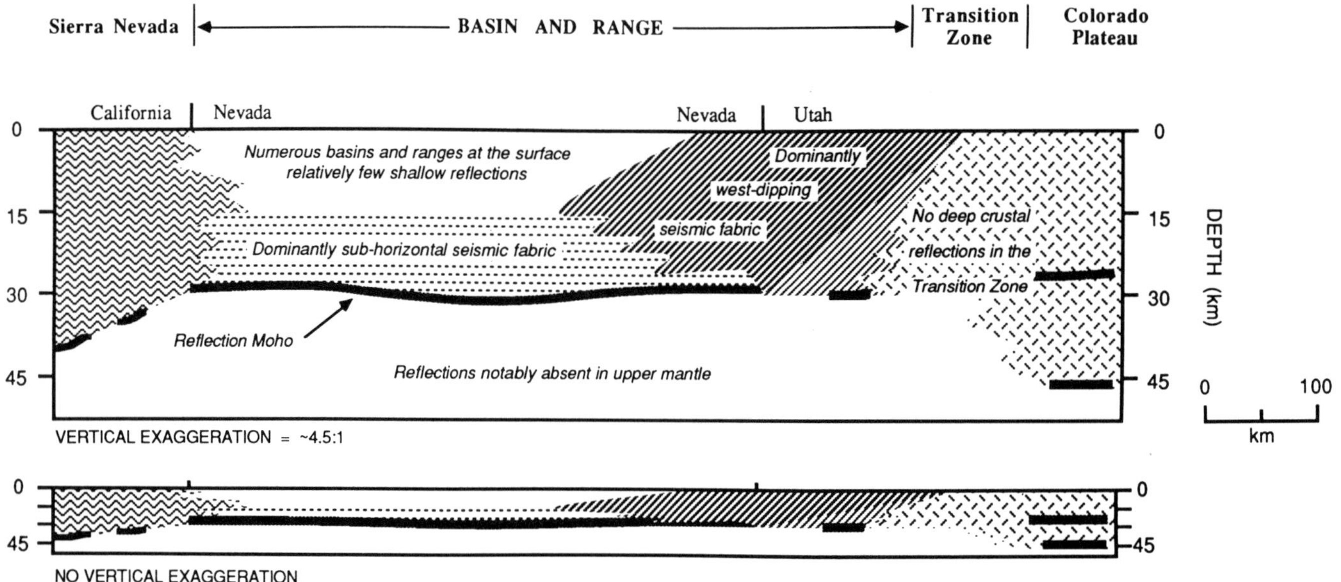

Figure 20. Interpreted east-west cross section of the Basin-Range crust based on the COCORP 40° N. Transect from Allmendinger and others (1987).

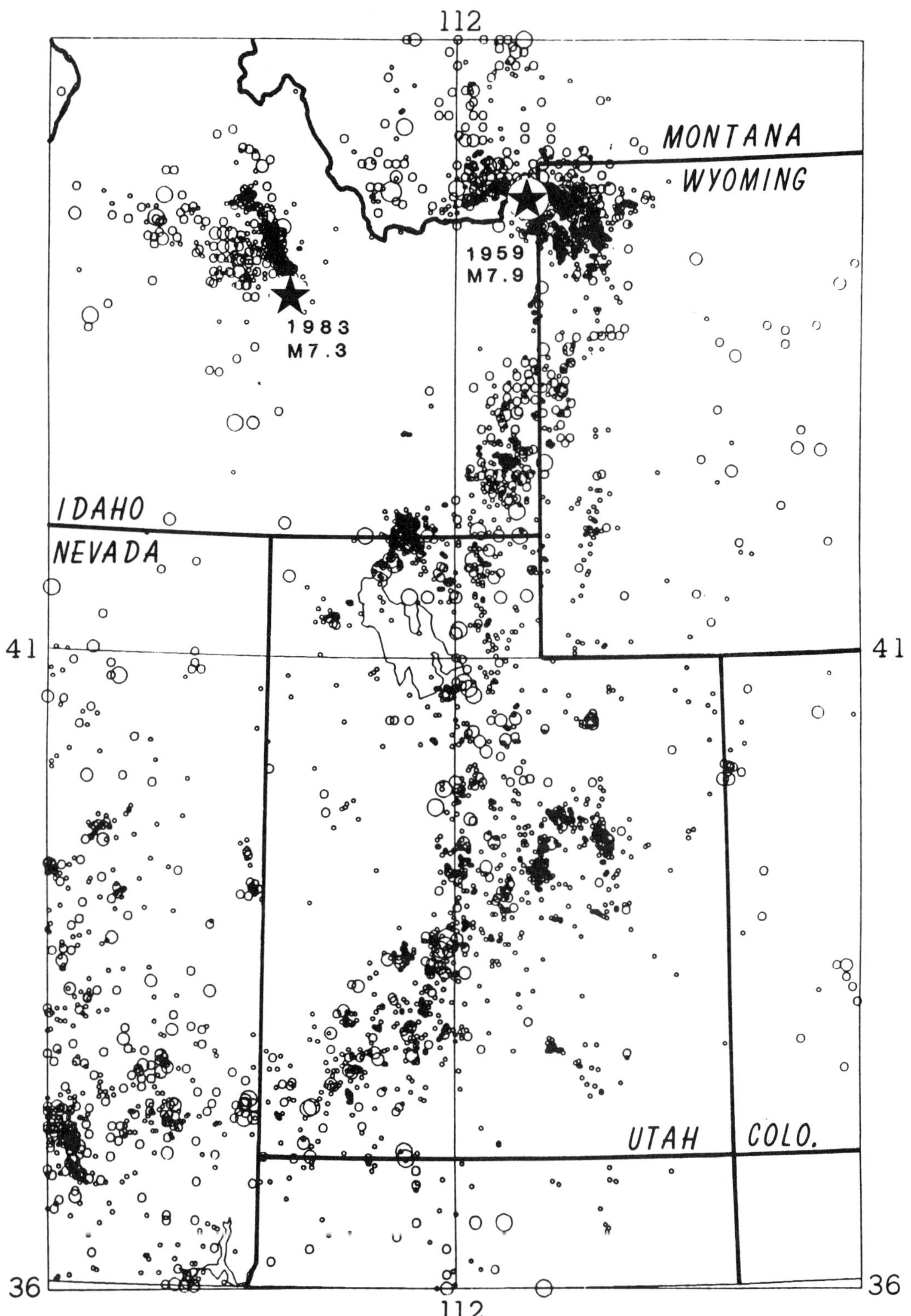

Figure 21. Earthquake epicenter map: ~1850–1985, for the southern and central Intermountain seismic belt. Minimum magnitude threshold: 1850–1961, M4.0; 1962–1985, M2.0. Data from the University of Utah Seismograph Stations and a DNAG compilation by E. R. Engdahl, U.S. Geological Survey, Denver, Colorado. Large, M7^{+}, earthquake epicenters are noted by stars.

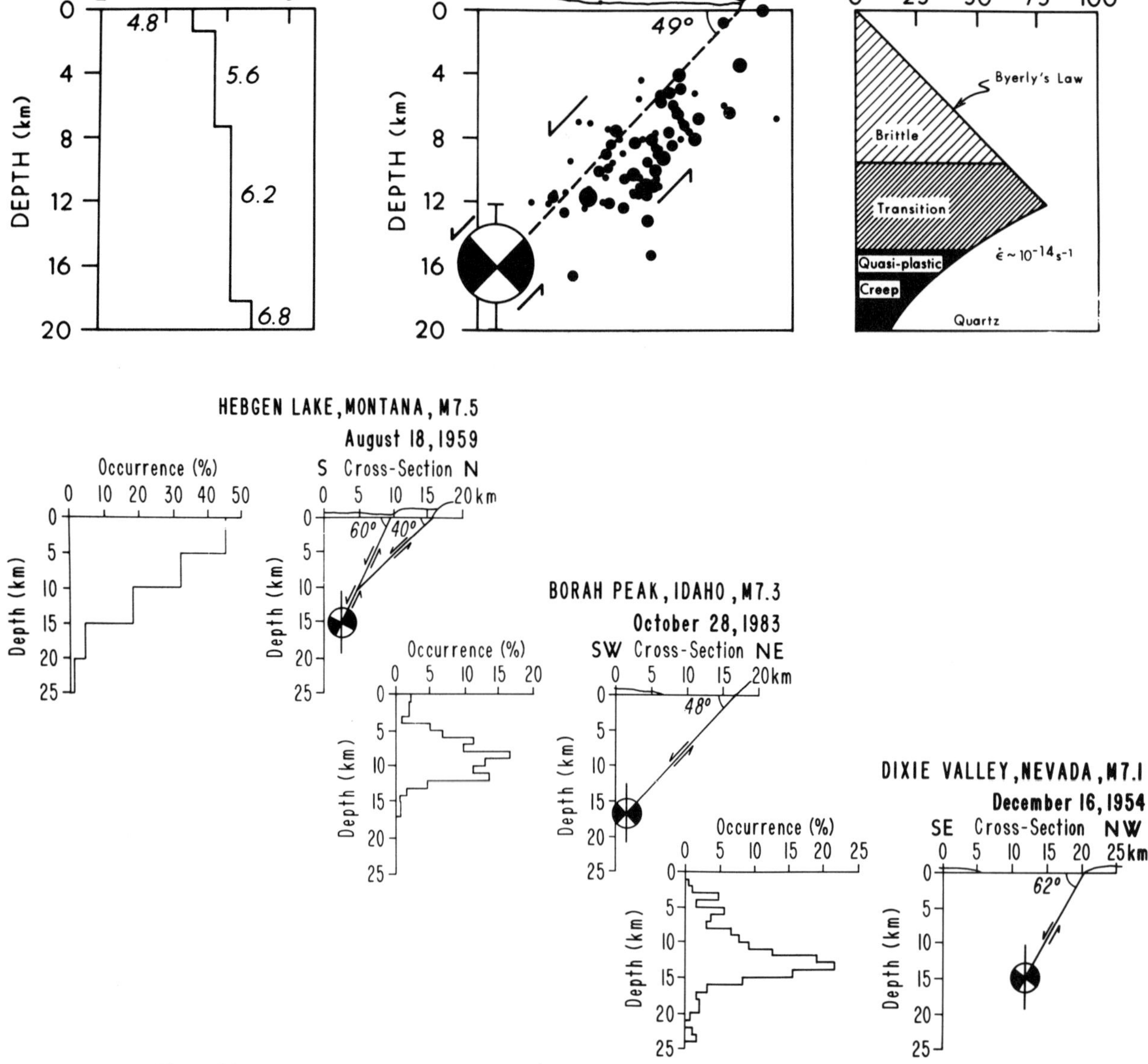

Figure 22. Hypothetical model for large, M7$^+$, Basin and Range earthquakes. Upper, Fault geometry and idealized stress-rheology model for the 1983, M_S7.3, Borah Peak, Idaho, earthquake (from Smith and others, 1985). Lower, Cross section of normal faults and focal depth frequencies associated with three large intraplate Basin-Range normal-faulting earthquakes.

magnitude earthquakes along the BR-CP-RM transition are approximately M 7.5, compared to M 8.5 along the San Andreas fault system.

In the past decade, accurate hypocenter data have been acquired, primarily by regional and portable seismic networks, that permit reliable focal-depth cross sections to be constructed for seismotectonic considerations. Smith and Bruhn (1984) hypothesized rheologic models for the Cordillera (Fig. 23) where the maximum focal depths, especially those in extensional regimes with vertical principal stress axes, correlate with the 80th percentile of the number of events (also see Sibson, 1982). This qualitative observation suggests that the depth of the transition from brittle to quasi-plastic flow is near the depth of maximum shear stress (Fig. 24). High heat flow in the eastern Basin-Range prevents focal depths from exceeding 10 to 15 km, while in the cooler Colorado Plateau, maximum focal depths exceed 40 km (Fig. 24).

The characteristics of the three well-studied M7$^+$ historic Basin-Range normal-faulting earthquakes thus define a working model (Fig. 22) for which the M7+ normal-faulting earthquakes nucleate at midcrustal depths on planar faults and where strain rates of ~10^4 sec^{-1} are necessary to achieve brittle fracture within

BASIN RANGE $q_o = 90\ mWm^{-2}$

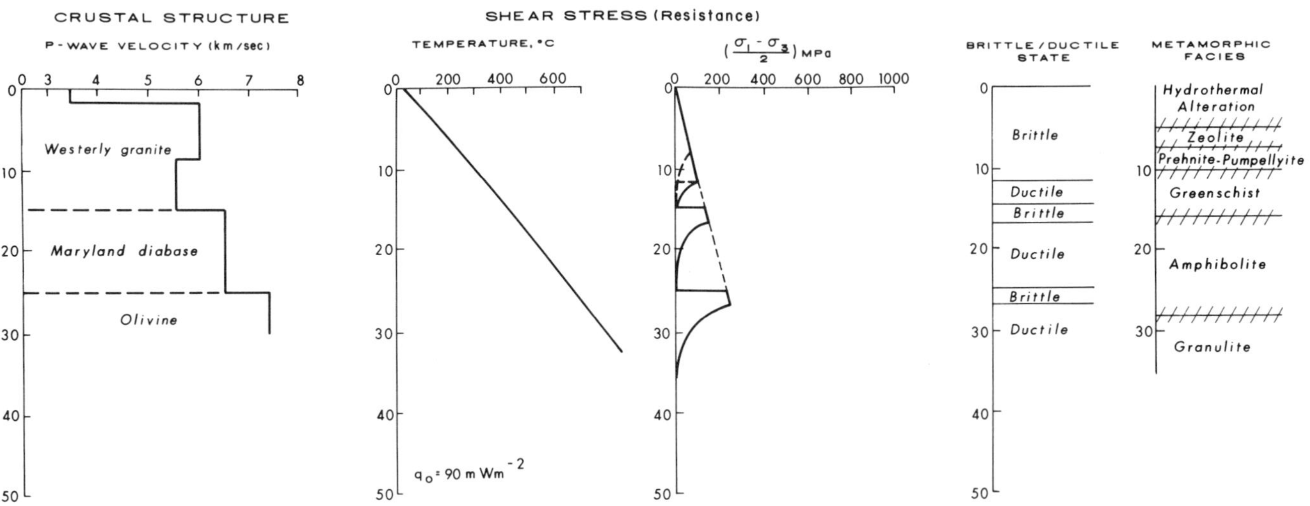

COLORADO PLATEAU $q_o = 60\ mWm^{-2}$

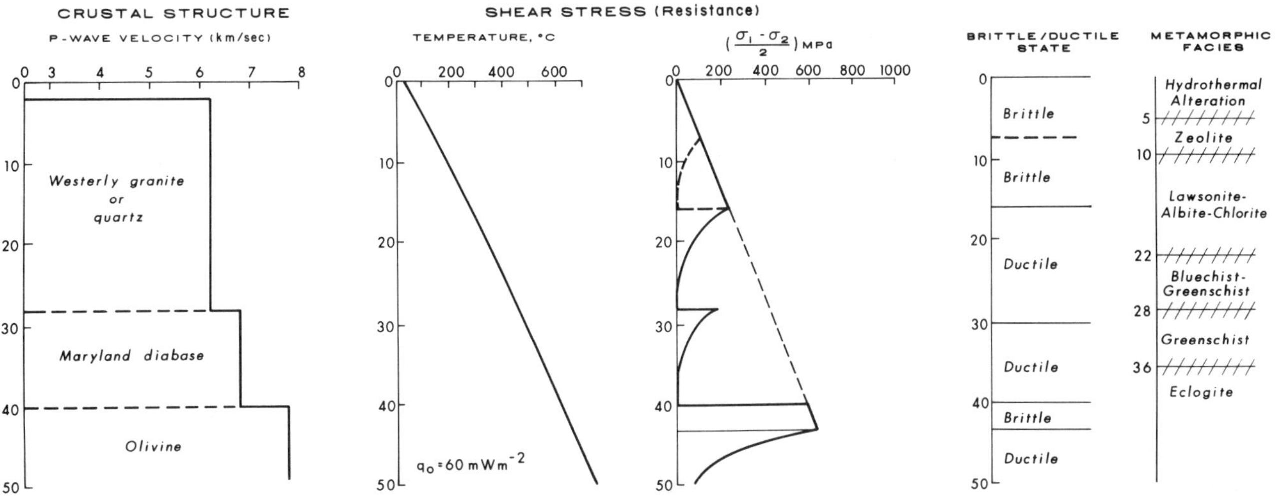

Figure 23. Basin-Range and Colorado Plateau rheologic models. Upper, Interpreted rheologic model of Basin-Range province. Lower, Interpreted rheologic model of Colorado Plateau province. See Smith and Bruhn (1984) for details of the rheologic modeling.

the more ductile intermediate-depth crustal material. This model has important ramifications for earthquake-hazard assessments in areas of active crustal extension where the nucleation process of the M7+ earthquakes may be associated with large faults that propagate upward from midcrustal depths, penetrate the entire seismogenic brittle layer, and break the surface.

CONTEMPORARY DEFORMATION

Two mechanisms are thought to accommodate lithospheric deformation: brittle failure associated with elastic strain release, and ductile flow. Focal mechanisms and fault-slip data on the principal stress directions provide information on brittle deformation (Smith, 1978, Zoback and Zoback, 1980). Terrestrial and satellite geodetic observations provide estimates of total strain associated with both brittle and ductile contributions of deformation.

Deformation from earthquakes. Brittle deformation associated with earthquakes can be estimated by summing earthquake moments to calculate the strain release. This method was outlined by Anderson (1979) and Doser and Smith (1982) for specific applications in the western United States. In recent work

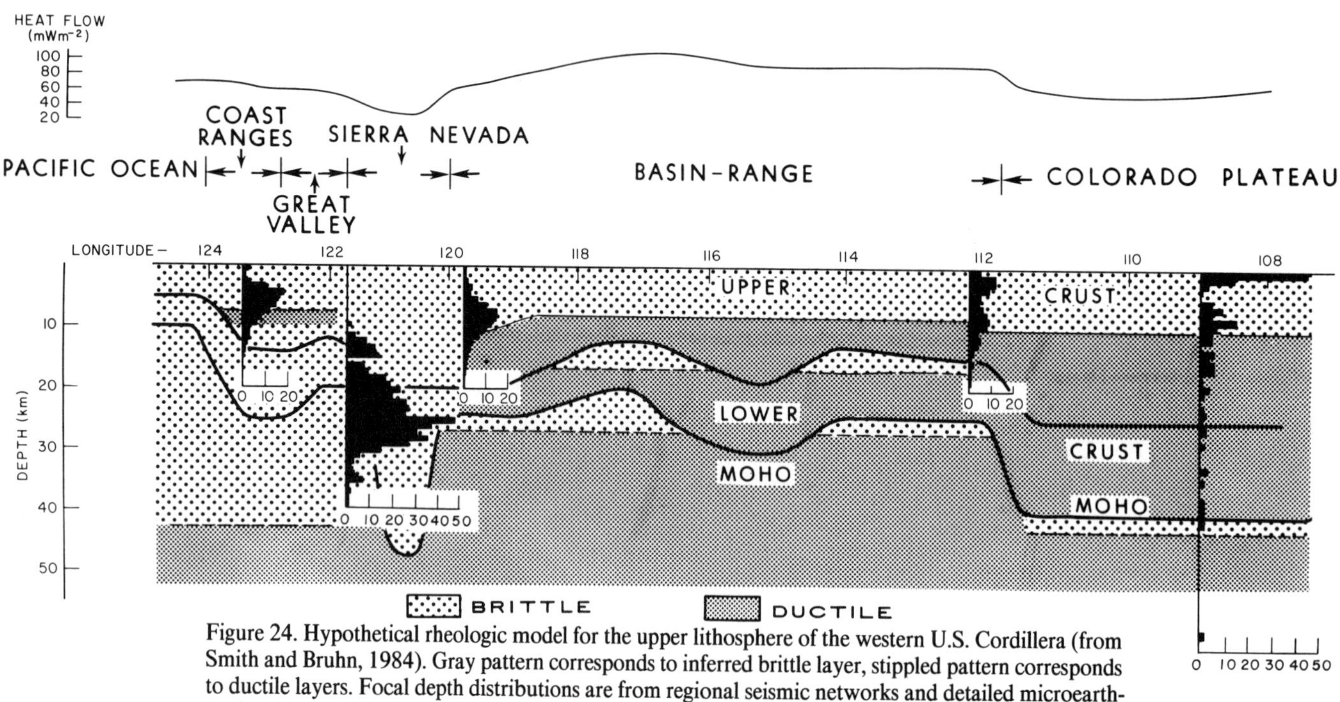

Figure 24. Hypothetical rheologic model for the upper lithosphere of the western U.S. Cordillera (from Smith and Bruhn, 1984). Gray pattern corresponds to inferred brittle layer, stippled pattern corresponds to ductile layers. Focal depth distributions are from regional seismic networks and detailed microearthquake surveys.

by Eddington and others (1987), horizontal components of regional strain rates for the western U.S. Cordillera were determined from historic seismicity and fault-plane solutions. Figure 25 is a map of the strain rate, displacement rates, and the directions of the principal strain tensors for the intraplate Cordillera, including the BR-CP-RM transition, derived from these data. The displacement rates ranged from less than 0.1 to more than 7.5 mm/yr for the Basin-Range, and as much as ~5 mm/yr for the transition.

For comparison, estimates of deformation on selected faults for the southern San Andreas fault system (Anderson, 1979) vary from 55 mm/yr, north of the Transverse Range, to 42 mm/yr to the south. Summing over larger areas of the Mojave Desert and parts of the southern San Andreas fault system yields up to 60 mm/yr of slip associated with north-south compression (Fig. 25). These large rates are primarily due to the contributions from two large historic earthquakes: the 1872, M 8.3, Owens Valley, California; and the 1857, M 8.3, Fort Tejon, California events.

Eastern Basin-Range deformation is characterized by north-south extension in central Idaho and in the Hebgen Lake–Yellowstone National Park region with rates as much as 4.7 mm/yr. Northwest- to east-west–directed extension, at rates from 0.1 to 1.5 mm/yr, extend southward along the Wasatch Front in Utah. A possible change from east-west extension to easterly compression occurs in southern Utah, in an area known to have significant Quaternary strike-slip faulting (Arabasz and Julander, 1986). The deformation then changes to a general northwest direction, with rates of deformation increasing to 7.5 mm/yr across central and western Nevada.

Basin-Range deformation rates, as deduced by historic earthquakes, are thus on the order of millimeters per year compared with tens of millimeters per year for the interplate San Andreas fault system. This difference also compares with the maximum-expected-magnitude earthquake of M ≅ 8.5 for the San Andreas fault system, compared with a maximum magnitude of M ≅ 7.5 for the intraplate Cordillera.

Province-wide Basin-Range deformation also reveals the contemporary kinematics of regional extension attributed to displacement associated with historic earthquakes. The deformation and strain rates (Eddington and others, 1987) are generalized along various azimuths in Figure 26 (profiles B-B′, B-B″, and C-C′). The deformation components along each profile were summed to obtain the integrated extension rates. Profile B-B′ across northern California, Nevada, and northern Utah exhibited a rate of 10 mm/yr, whereas profile B-B″ is a more general east-west line with a rate of 8.4 mm/yr. The southern line, C-C′, extends across southeastern California, southern Nevada, and southern Utah where the deformation rate diminishes to 3.5 mm/yr. However, if the 1872, M 8.3, Owens Valley, California, earthquake is included on profile C-C′, the deformation rate increases to ~30 mm/yr. This observation indicates that these rates must be interpreted with caution, because of the significant contribution of single large events.

The earthquake-related deformation rate in the northern Basin-Range thus appears to be more than twice that in the southern part of the province. This contemporary pattern suggests that the greatest opening of the Basin-Range to the north is similar to the late Cenozoic deformation that was deduced from geology by Wernicke and others (1982).

The earthquake-induced deformation rate of 8 to 10 mm/yr determined along the northern profiles (Fig. 26) also compares well with the deformation rates determined from other geophysi-

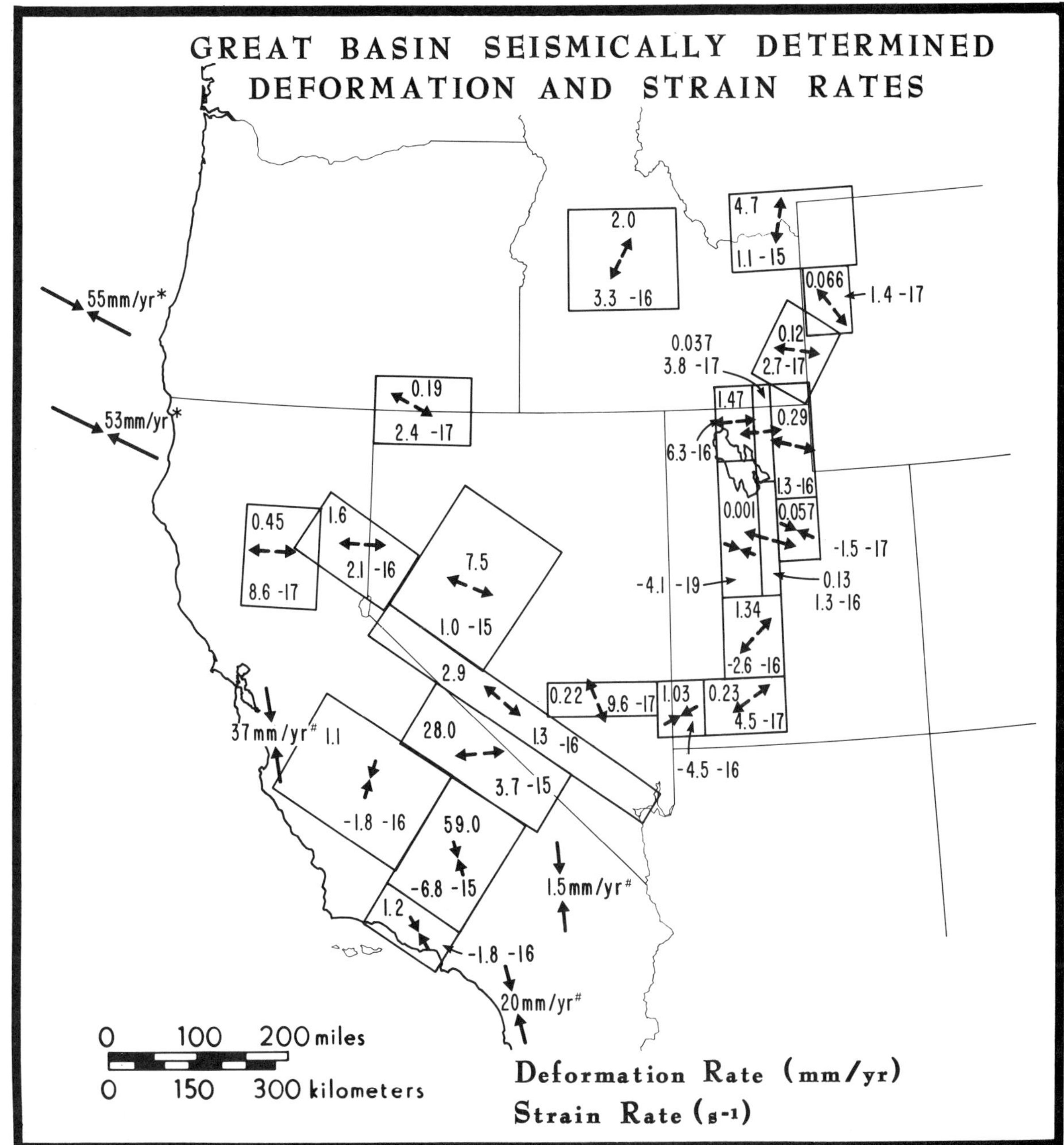

Figure 25. Basin-Range strain and deformation rates based on summation of seismic moment tensors from historic earthquakes ~1850 to 1983 (from Eddington and others, 1987). In each area, top value is the deformation rate in millimeters per year; arrow corresponds to direction of horizontal principal strain axis; bottom value is horizontal strain rate in sec^{-1} with second number to the power of 10; and asterisk indicates deformation rates from Hyndman and Wiechert (1983) and Anderson (1979).

cal and geological studies (Table 2). Prehistoric deformation was determined from interplate and intraplate models, oceanic geomagnetic data, heat flow, and geologic strain data. Geologically determined paleodeformation rates (Table 2) ranged from 1 to 20 mm/yr, except for Proffett's (1977) deformation rate of ~200 mm/yr for a restricted area of western Nevada. For example, Lachenbruch and Sass (1978) and Lachenbruch (1979) determined extension rates of 5 to 10 mm/yr for the Basin-Range using heat-flow constraints and thermal deformation models. A range of 1 to 20 mm/yr is consistent with the deformation produced by contemporary seismicity. Late Cenozoic rates of 1 to 20 mm/yr are of the same order as the contemporary earthquake-related deformation, and suggest that assessments of future seismicity (locations and general timing of likely large events) can be made using the late Cenozoic geologic information, such as fault locations, slip rates, and the magnitude of individual slip events.

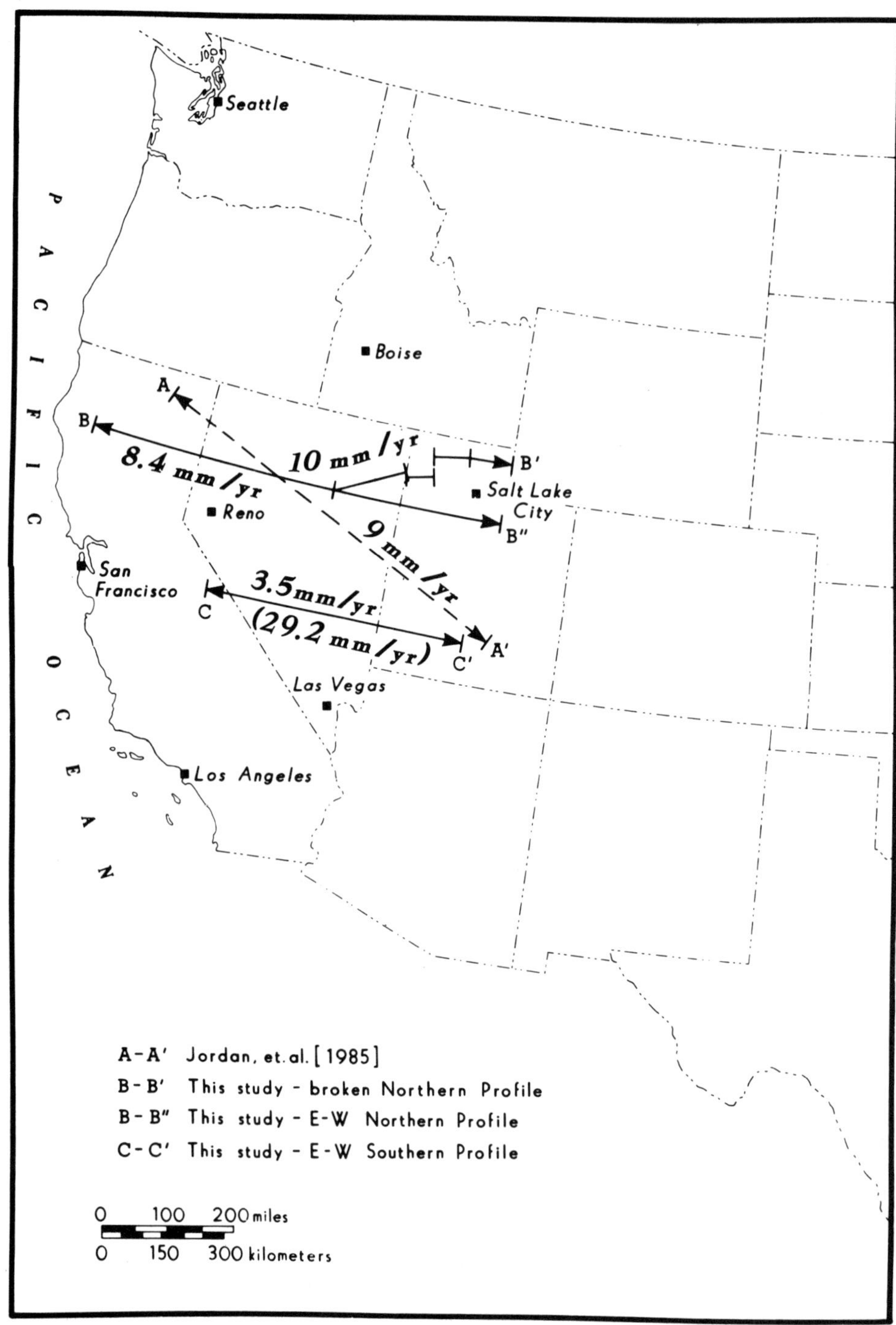

Figure 26. Basin-Range extension rates. Profiles B-B′, B-B″, and C-C′ are from Eddington and others (1987). Value in parentheses below C-C′ indicates deformation when the effect of the 1872, M ~8.3, Owens Valley, California, earthquake is included. A-A′ is the Basin and Range opening rate from Beroza and others (1985) and Minster and Jordan (1984), as determined from intraplate kinematic models of motion between the North American and Pacific plates.

This is a particularly useful idea in areas where historic seismicity is low or Quaternary faulting is currently quiescent.

Beroza and others (1985) have estimated deformation rates across the Basin-Range at 7.4 mm/yr (profile A-A′; Fig. 26) from North American and Pacific inter- and intraplate tectonic models constrained by satellite geodetic observations in California. The seismically determined deformation rate along profile B-B″ of 8.4 mm/yr compares remarkably well for these two different methods. This comparison also leads to the speculation that much of the contemporary Basin-Range extension is expressed as earthquake-generated brittle fracture in the upper 10 to 15 km of the lithosphere. Deeper extension may be expressed by ductile flow, but this mode of strain release does not appear to be dominant during the period of historic earthquake observations.

TABLE 2. BASIN–RANGE EXTENTION RATES*

Time	Opening Rate	Method
Late Cenozoic	3 - 20	Geological strain
Late Cenozoic	3 - 12	Heat flow
Holocene paleoseismicity	1 - 12	Fault-slip data
Historic seismicity	3.5 - 10	Historical earthquakes
Prehistoric (intraplate model)	<9	Intraplate models

*From Minster and Jordan, 1984; Beroza and others, 1985; and Eddington and others, 1987.

SUMMARY

The data and models presented here suggest general properties of the crustal structure, seismicity, and contemporary deformation of the BR-CP-RM transition. The Moho has an east-west variation in thickness of 10 km or more across the transition, but has a rather uniform upper-mantle P-wave velocity of 7.8 to 7.9 km/sec. The anomalously low-velocity, ~7.5-km/sec layer, reported for the upper mantle of the Wasatch Front on the basis of unreversed refraction profiles, may represent a low-velocity upper-mantle upwarp, a distinct high-velocity lower crustal layer, or it may be a true 7.9-km/sec layer with the low velocities produced from downdip ray propagation from a mantle bulge. Without properly reversed profiles, this important problem cannot yet be resolved.

A P-wave low-velocity layer at crustal depths of 8 to 12 km has been observed in central Utah that spatially correlates with the gently westward-dipping Sevier Desert detachment. Whether the low-velocity layer is directly related to low shear resistance or the hypothesized brittle to quasi-plastic zone cannot be ascertained, but a spatial correlation of these anomalous physical properties suggests that they may relate to the mechanics of extension.

Seismic reflection data from throughout the BR-CP-RM transition have revealed a range of structural styles dominated by asymmetric basins bounded by low- to high-angle dipping planar and listric normal faults. The listric and low-angle faults do not seem to penetrate beyond a depth of 6 km. The geometry and depths of the low-angle faults, however, do not fit the models of large, M7+, normal-faulting earthquakes in the region that show midcrustal nucleation depths of 15 km on 15° to 60° dipping planar faults—an intriguing paradox from the lower angle faults imaged by the reflection method.

Brittle deformation for the BR-CP-RM transition is characterized primarily by normal faulting on 30° to 60° dipping planes of moderate to large earthquakes. Extension rates deduced from the summation of seismic moment rates of historic earthquakes vary from 5 mm/yr in the Hebgen Lake–Yellowstone region, to a rate of 1.5 mm/yr along the Wasatch Front, and up to 7.5 mm/yr in central Nevada. Integrating these values across the northern Basin-Range gives an opening rate of ~10 mm/yr. These crustal deformation rates distinguish an important difference between the intraplate extension of the Basin-Range of as much as 10 mm/yr, and the interplate San Andreas fault system with deformation rates of 50 to 60 mm/yr, and suggest that the extensional mechanisms of the BR-CP-RM transition are nearly an order of magnitude slower than active interplate deformation.

ACKNOWLEDGMENTS

The information presented in this chapter represents a compilation of earthquake, lithospheric structure, and crustal deformation data gathered at the University of Utah for the past decade. Computer tabulations of lithospheric structure were prepared with the assistance of S. Willett, M. S. Bauer, S. Jackson, and R. D. Luzitano. J. B. Saleeby and M. C. Blake provided the early versions of Continental Transects C-1 and C-2. E. R. Engdahl, U.S. Geological Survey, Denver, Colorado, assisted with the compilation of the earthquake data. P. K. Eddington provided assistance with the seismic moment tensor data and contemporary deformation calculations. Discussions with L. W. Braile, J. C. Pechmann, M. L. Zoback, L. C. Pakiser and W. D. Mooney were appreciated. P. K. Eddington, R. W. Allmendinger, D. T. Loeb, and R. C. Speed allowed the use of their published and unpublished data. L. C. Pakiser and A. Ryall critically reviewed initial drafts of the manuscript. Drafting by S. Bromley and P. Onstott was invaluable. The manuscript was typed by J. Barlow and L. Burnett.

The research presented here was supported by National Astronautics and Space Administration Grant NAG-5-164, with computer time provided by the U.S. Geological Survey, Grant 14-08-0001-21983.

REFERENCES CITED

Allmendinger, R. W., and 7 others, 1983, Cenozoic and Mesozoic structure of the eastern Basin and Range province, Utah, from COCORP seismic-reflection data: Geology, v. 11, p. 532–536.

Allmendinger, R. W., and 6 others, 1987, Overview of the COCORP 40°N transect western U.S.A.: The fabric of an orogenic belt: Geological Society of America Bulletin, v. 98, p. 308–319.

Anderson, J. G., 1979, Estimating the seismicity from geological structure for seismic-risk studies: Bulletin of the Seismological Society of America, v. 69, p. 135–158.

Anderson, R. E., Zoback, M. L., and Thompson, G. A., 1983, Implications of selected subsurface data on the structural form and evolution of some basins in the northern Basin and Range province, Nevada and Utah: Geological Society of America Bulletin, v. 94, p. 1055–1072.

Arabasz, W. J., and Julander, D. R., 1986, Geometry of seismically active faults and crustal deformation within the Basin and Range–Colorado Plateau transition: Geological Society of America Special Paper 208, p. 43–74.

Arabasz, W. J., and Smith, R. B., 1981, Earthquake prediction in the Intermountain seismic belt—An intraplate extensional regime: *in* Earthquake Prediction; An International Review: American Geophysical Union, p. 248–258.

Barker, C. A., 1986, Upper-crustal structure of the Milford Valley and Roosevelt Hot Springs, Utah region, by seismic modeling [M.S. thesis]: Salt Lake City, University of Utah, 101 p.

Beroza, G. C., Jordan, T. H., Minster, J. B., Clark, T. A., and Ryan, J. W., 1985, VLBI vector position data; Application to western U.S. deformation: EOS Transactions of the American Geophysical Union, v. 66, p. 848.

Bjarnason, I. T., 1987, Contemporary tectonics of the Wasatch Front region, Utah, from earthquake focal mechanisms [M.S. thesis]: Salt Lake City, University of Utah, 79 p.

Bond, G. C., Christie-Blick, N., Kominz, M. A., and Devlin, W. J., 1985, An early Cambrian rift to post-rift transition in the Cordillera of western North America: Nature, v. 315, p. 742–746.

Braile, L. W., Smith, R. B., Keller, G. R., and Welch, R. M., 1974, Crustal structure across the Wasatch Front from detailed seismic studies: Journal of Geophysical Research, v. 79, p. 2669–2677.

Braile, L. W., and 9 others, 1982, The Yellowstone–Snake River Plain seismic profiling experiment; Crustal structure of the eastern Snake River Plain: Journal of Geophysical Research, v. 87, p. 2597–2609.

Brokaw, M. A., 1985, Upper crustal interpretation of Yellowstone determined from raytrace modeling of seismic refraction data [M.S. thesis]: Salt Lake City, University of Utah, 179 p.

Cook, K. L., 1962, The problem of the mantle-crust mix; Lateral inhomogeneity in the uppermost part of the earth's mantle: Advances in Geophysics, v. 9, p. 295–360.

Dallmeyer, R. D., Snoke, A. W., and McKee, E. H., 1986, The Mesozoic–Cenozoic tectonothermal evolution of the Ruby Mountains, East Humboldt Range, Nevada; A Cordilleran core complex: Tectonics, v. 5, p. 931–954.

Doser, D. I., 1985, Source parameters and faulting processes of the 1959 Hebgen Lake, Montana, earthquake sequence: Journal of Geophysical Research, v. 90, p. 4537–4555.

—— , 1986, Earthquake processes in the Rainbow Mountain–Fairview Park–Dixie Valley, Nevada, region 1954–1959: Journal of Geophysical Research, v. 91, p. 12572–12586.

Doser, D. I., and Smith, R. B., 1982, Seismic moment rates in the Utah region: Bulletin of the Seismological Society of America, v. 72, p. 525–551.

—— , 1985, Source parameters of the October 18, 1983, Borah Peak, Idaho, earthquake from body wave analysis: Bulletin of the Seismological Society of America, v. 75, p. 1041–1051.

Eaton, J. P., 1963, Crustal structure from San Francisco, California, to Eureka, Nevada, from seismic refraction measurements: Journal of Geophysical Research, v. 68, p. 5789–5806.

Eddington, P. K., Smith, R. B., and Renggli, C., 1987, Kinematics of Basin–Range intraplate extension, *in* Coward, M. P., Dewey, J. F., and Hancock, P. L., eds., Continental extensional tectonics: Geological Society of London Special Publication 28, p. 371–392.

Effimoff, I., and Pinezich, A. R., 1981, Tertiary structural development of selected valleys based on seismic data; Basin and Range province, northeastern Nevada, *in* Vine, F. J., and Smith, A. G., eds., Extensional tectonics associated with convergent plate boundaries: Philosophical Transactions of the Royal Society of London, series A, v. 300, p. 435–442.

Elbring, G. J., 1984, A method of inversion of two-dimensional seismic refraction data with applications to the Snake River Plain region of Idaho [Ph.D. thesis]: West Lafayette, Indiana, Purdue University, 123 p.

Fauria, T. J., 1981, Crustal structure of the northern Basin and Range and western Snake River Plain; A ray-trace travel-time interpretation of the Eureka, Nevada, to Boise, Idaho, seismic-refraction profile [M.S. thesis]: West Lafayette, Indiana, Purdue University, 91 p.

Fenneman, N. M., 1946, Physical divisions of the United States: U.S. Geological Survey Map, scale 1:7,000,000.

Gants, D. G., 1985, Geologic and mechanical properties of the Sevier Desert detachment as inferred by seismic and rheologic modeling [M.S. thesis]: Salt Lake City, University of Utah, 129 p.

Gertson, R. C., 1979, Interpretation of a seismic refraction profile across the Roosevelt Hot Springs, Utah, and vicinity [M.S. thesis]: Salt Lake City, University of Utah, 109 p.

Glazner, A. F., and Bartley, J. M., 1985, Evolution of lithospheric strength after thrusting: Geology, v. 13, p. 42–45.

Hill, D. P., and Pakiser, L. C., 1967, Seismic-refraction study of crustal structure between the Nevada test site and Boise, Idaho: Geological Society of America Bulletin, v. 78, p. 685–704.

Hyndman, R. D., and Wiechert, D. H., 1983, Seismicity and rates of relative motion on the plate boundaries of western North America: Geophysical Journal of the Royal Astronomical Society, v. 72, p. 59–82.

Johnson, L. R., 1965, Crustal structure between Lake Mead, Nevada, and Mono Lake, California: Journal of Geophysical Research, v. 70, p. 2863–2872.

Keller, G. R., Smith, R. B., and Braile, L. W., 1975, Crustal structure along the Great Basin–Colorado Plateau transition from seismic refraction studies: Journal of Geophysical Research, v. 80, p. 1093–1098.

Klemperer, S. L., Hauge, T. A., Hauser, E. C., Oliver, J. E., and Potter, C. J., 1986, The Moho in the northern Basin and Range province, Nevada, along the COCORP 40°N seismic reflection transect: Geological Society of America Bulletin, v. 97, p. 603–618.

Lachenbruch, A. H., 1979, Heat flow in the Basin and Range province and thermal effects of tectonic extension: Pure and Applied Geophysics, v. 117, p. 34–50.

Lachenbruch, A. H., and Sass, J. H., 1978, Models of an extending lithosphere and heat flow in the Basin and Range province: Geological Society of America Memoir 152, p. 209–250.

Lehman, J. L., Smith, R. B., Schilly, M. M., and Braile, L. W., 1982, Upper crustal structure of the Yellowstone caldera from seismic delay time analysis: Journal of Geophysical Research, v. 87, p. 2713–2730.

Loeb, D. T., 1986, The P-wave velocity structure of the crust-mantle boundary beneath Utah [M.S. thesis]: Salt Lake City, University of Utah, 126 p.

Loeb, D. T., and Pechmann, J. C., 1986, The P-wave velocity structure of the crust-mantle boundary beneath Utah from network travel-time measurements; Abstracts, 81st Annual Meeting of the Seismological Society of America: Earthquake Notes, v. 57, p. 10–11.

Martin, W. R., 1978, A seismic refraction study of the northeastern Basin and Range and its transition with the eastern Snake River Plain [M.S. thesis]: El Paso, University of Texas at El Paso, 40 p.

McDonald, R. E., 1976, Tertiary tectonics and sedimentary rocks along the transition; Basin–Range province to Plateau and Thrust belt provinces, Utah, *in* Hill, J. G., ed., Rocky Mountain Association of Geologists Symposium on the Geology of the Cordilleran hingeline: Rocky Mountain Association of Geologists, p. 281–317.

Meissner, R., 1986, The continental crust; A geophysical approach, *in* Donn, W. L., ed., International geodynamics series, v. 34: New York, Academic Press, 426 p.

Minster, J. B., and Jordan, T. H., 1984, Vector constraints on Quaternary deformation of the western United States east and west of the San Andreas fault, *in* Crouch, J. K., and Barbman, S. B., eds., Tectonics and sedimentation along the California margin: Pacific Section Society of Economic Paleontologists and Mineralogists, v. 38, p. 1–16.

Mueller, S., 1977, A new model of the continental crust, *in* Heacock, J., ed., The earth's crust: American Geophysical Union Monograph 20, p. 289–318.

Mueller, S., and Landisman, M., 1971, An example of the unified method of interpretation for crustal seismic data: Geophysical Journal of the Royal Astronomical Society, v. 23, p. 365–371.

Müller, G., and Mueller, S., 1979, Travel-time and amplitude interpretation of crustal phases on the refraction profile Delta-W, Utah: Bulletin of the Seismological Society of America, v. 69, p. 1121–1132.

Okaya, D. A., and Thompson, G. A., 1985, Geometry of Cenozoic extensional faulting; Dixie Valley, Nevada: Tectonics, v. 4, p. 107–126.

Pakiser, L. C., 1963, Structure of the crust and upper mantle in the western United States: Journal of Geophysical Research, v. 68, p. 5747–5756.

——, 1985, Seismic exploration of the crust and upper mantle of the Basin and Range province: Boulder, Colorado, Geological Society of america, The geology of North America Centennial Special Volume 1, p. 453–468.

Pechmann, J. C., Richins, W. D., and Smith, R. B., 1984, Evidence for a double Moho beneath the Wasatch front, Utah: EOS Transactions of the American Geophysical Union, v. 65, p. 988.

Planke, S., 1987, Cenozoic structures and evolution of the Sevier Desert Basin, west-central Utah, from seismic reflection data [M.S. thesis]: Salt Lake City, University of Utah, 163 p.

Prodehl, C., 1970, Seismic refraction study of crustal structure of the western U.S.: Geological Society of America Bulletin, v. 81, p. 2629–2646.

——, 1979, Crustal structure of the western United States: U.S. Geological Survey Professional Paper 1034, 74 p.

Proffet, J. M., Jr., 1977, Cenozoic geology of the Yerrington district, Nevada, and implications for the nature and origin of Basin and Range faulting: Geological Society of America Bulletin, v. 88, p. 247–266.

Roller, J. C., 1964, Crustal structure in the vicinity of Las Vegas, Nevada, from seismic and gravity observations, *in* Geological Survey Research 1964: U.S. Geological Survey Professional Paper 475-D, p. 108–111.

——, 1965, Crustal structure in eastern Colorado Plateau province from seismic-refraction measurements: Bulletin of the Seismological Society of America, v. 55, p. 107–119.

Saleeby, J. B., and 13 others, 1984, Continent-ocean transect, Corridor C2; Central California offshore to Colorado Plateau: Geological Society of America Centennial Continent/Ocean Transect C-2, scale 1:500,000.

Schwartz, D. P., and Coppersmith, R. J., 1984, Fault behavior and characteristic earthquakes; Examples from the Wasatch and San Andreas fault zone: Journal of Geophysical Research, v. 89, p. 5681–5698.

Sheriff, S. D., and Stickney, M. C., 1984, Crustal structure of southwestern Montana and east-central Idaho; Results of a reversed seismic-refraction line: Geophysical Research Letters, v. 11, p. 299–302.

Sibson, R. H., 1982, Fault zone models, heat flow, and the depth distribution of earthquakes in the continental crust of the United States: Bulletin of the Seismological Society of America, v. 72, p. 151–163.

Smith, K. A., 1984, Normal faulting in an extensional domain; Constraints from seismic reflection interpretation and modeling [M.S. thesis]: Salt Lake City, University of Utah, 165 p.

Smith, R. B., 1978, Seismicity, crustal structure, and intraplate tectonics of the interior of the western Cordillera, *in* Smith, R. B., and Eaton, G. P., eds., Cenozoic tectonics and regional geophysics of the western Cordillera: Geological Society of America Memoir 152, p. 111–144.

——, 1989, Lithospheric structure, seismicity, and contemporary deformation of the United States Cordillera, *in* Burchfiel, C., Lipman, P., and Zoback, M. L., eds., The Cordilleran Orogen; Conterminous U.S., Boulder, Colorado, Geological Society of America, The Geology of North America, v. G-3 (in press).

Smith, R. B., and Bruhn, R. L., 1984, Intraplate extensional tectonics of the eastern Basin–Range; Inferences on structural style from seismic reflection data, regional tectonics, and thermal-mechanical models of brittle/ductile deformation: Journal of Geophysical Research, v. 89, p. 5733–5762.

Smith, R. B., and Lindh, A., 1978, A compilation of fault plane solutions of the western United States, *in* Smith, R. B., and Eaton, G. P., Cenozoic tectonics and regional geophysics of the western Cordillera: Geological Society of America Memoir 152, p. 107–110.

Smith, R. B., and Richins, W. D., 1984, Seismicity and earthquake hazards of Utah and the Wasatch front; Paradigm and paradox, *in* Proceedings of Conference 26, A workshop on evaluation of regional and urban earthquake hazards and risk in Utah: U.S. Geological Survey Open-File Report 84-763, p. 73–112.

Smith, R. B., and Sbar, M. L., 1974, Contemporary tectonics and seismicity of the western United States with emphasis on the Intermountain seismic belt: Geological Society of America Bulletin, v. 85, p. 1205–1218.

Smith, R. B., and 9 others, 1982, The Yellowstone–eastern Snake River Plain seismic profiling experiment: Journal of Geophysical Research, v. 87, p. 1583–1596.

Smith, R. B., Richins, W. D., and Dosev, D. I., 1985, The Borah Peak earthquake; Seismicity, faulting kinematics, and tectonic mechanism, Workshop 27 on The Borah Peak earthquake: U.S. Geological Survey Open-File Report 85-290, 236–263.

Sparlin, M. A., Braile, L. W., and Smith, R. B., 1982, Crustal structure of the eastern Snake River Plain determined from ray-trace modeling of seismic refraction data: Journal of Geophysical Research, v. 87, p. 2619–2633.

Speed, R., 1989, Summary of continental transects, *in* Speed, R. C., ed., Phanerozoic evolution of North American continent-ocean transitions, Boulder, Colorado, Geological Society of America, The Geology of North America, v. CTV-1 (in press).

Spencer, J. E., and Nur, A., 1976, The effects of pressure, temperature, and pore water on velocities in Westerly Granite: Journal of Geophysical Research, v. 81, p. 899–920.

Viveiros, J. J., 1986, Cenozoic tectonics of the Great Salt Lake from seismic reflection data from the Basin-Range transition [M.S. thesis]: Salt Lake City, University of Utah, 98 p.

Wernicke, B. P., 1981, Low-angle normal faults in the Basin and Range province; Nappe tectonics in an extending orogen: Nature, v. 291, p. 645–648.

Wernicke, B. P., Spencer, J. C., Burchfiel, B. C., and Guth, P. L., 1982, Magnitude of extension in the southern Great Basin: Geology, v. 10, p. 499–502.

Willden, R., 1965, Seismic-refraction measurements of crustal structure between American Falls Reservoir, Idaho, and Flaming Gorge Reservoir, Utah, *in* Geological Survey Research 1965: U.S. Geological Survey Professional Paper 525-C, p. 44–50.

Zoback, M. D., and Lachenbruch, A. H., 1984, Upper mantle structure beneath the western U.S.: Geological Society of America Abstracts with Programs, v. 16, p. 705.

Zoback, M. L., and Anderson, R. E., 1983, Style of Basin–Range faulting as inferred from seismic reflection data in the Great Basin, Nevada and Utah: Geothermal Resources Council Special Report 13, p. 363–381.

Zoback, M. L., and Zoback, M. D., 1980, State of stress in the conterminous United States: Journal of Geophysical Research, v. 85, p. 6113–6156.

Zoback, M. L., Anderson, R. E., and Thompson, G. A., 1981, Cenozoic evolution of the state of stress and style of tectonism of the Basin and Range province of the western United States: Philosophical Transactions of the Royal Society of London, series A, v. 300, p. 407–434.

MANUSCRIPT ACCEPTED BY THE SOCIETY OCTOBER 31, 1988

Printed in U.S.A.

Geological Society of America
Memoir 172
1989

Chapter 13

Geophysics of the Intermontane system

L. C. Pakiser
U.S. Geological Survey, MS 966, Box 25046, Denver Federal Center, Denver, Colorado 80225

ABSTRACT

Crustal thicknesses in the Intermontane system—Colorado Plateau, Basin and Range province, and High Lava Plains of the Cordilleran region—have been reliably determined by an extensive network of seismic-refraction profiles as about 30 km in the Basin and Range and 42 km in both the Snake River Plain and the Colorado Plateau. The upper crust (velocity, 5.2 to 6.3 km/sec) averages about 20 km in thickness in the Basin and Range, 10 km in the Snake River Plain, and 28 km in the Colorado Plateau. The velocity in the lower crust ranges from 6.4 to possibly 7.5 km/sec. Upper-mantle velocities of 7.4 to 8.0 km/sec have been reported, but the lower P_n velocities are probably apparent downdip velocities. The true upper-mantle velocity throughout the Intermontane system is probably 7.8 to 7.9 km/sec.

Moho depths from seismic-reflection studies are about the same as those from refraction studies in the Basin and Range, but they are 5 to 10 km deeper in the Colorado Plateau. The patterns of crustal reflections are distinctly different in the western and eastern parts of the Basin and Range.

Geophysical studies indicate that the lithosphere is 50 to 65 km thick in the Basin and Range, 50 km in the Snake River Plain, and 90 to 100 km in the Colorado Plateau.

Bouguer gravity values range from −50 mGal in southwestern Arizona to −250 mGal in the Uinta basin of northeastern Utah. Gravity correlates inversely with topography in the Basin and Range, indicating isostatic equilibrium, but the Snake River Plain is slightly undercompensated. Long-wavelength gravity anomalies indicate that much of the isostatic compensation results from a mass deficiency in the mantle, probably related to variations in the thickness of the lithosphere. Prominent magnetic-high anomalies in the Snake River Plain and north-central Nevada result from mafic rock formed in Miocene rifts.

Heat flow is high throughout the Intermontane system, ranging from 1.5 heat-flow units (HFU) in central Nevada and the Colorado Plateau to more than 2.5 HFU in northwestern Nevada and the Snake River Plain. Electrical conductivity is anomalously high in the lower crust and upper mantle of the system.

Seismicity is distributed in broad bands in the Nevada and intermountain seismic zones, and epicenters commonly do not coincide with mapped faults; the larger earthquakes (magnitude greater than 6.0), however, occur only on planar, normal faults with dips of 50 to 60°. The stress field is compatible with lateral extension in the Basin and Range.

Examination of evidence for a "double Moho" in Utah suggests that the apparent velocities associated with the Moho propagation paths on which this model is based vary widely, indicating severe relief on the refractor surface; the evidence favors a single Moho, below which the velocity is 7.8 to 7.9 km/sec. Ray tracing and amplitude studies

Pakiser, L. C., 1989, Geophysics of the Intermontane system, *in* Pakiser, L. C., and Mooney, W. D., Geophysical framework of the continental United States: Boulder, Colorado, Geological Society of America Memoir 172.

leave unresolved the question of whether a transitional layer of velocity 7.5 km/sec exists in the lower crust of the Basin and Range.

Many geologic and geophysical characteristics are similar throughout the Intermontane system, but lithospheric models are distinctly different in the separate provinces, owing to the different geologic processes that have prevailed. The crust throughout the system prior to Cenozoic extension and volcanism probably consisted of an upper layer of silicic gneisses and schists overlying a lower crust of silicic to intermediate composition in the granulite facies. During extension, the crust of the Basin and Range was thinned and invaded by gabbroic magma from the mantle, resulting in a lower crust consisting of as much as two-thirds mafic material. During crustal rifting or extension in the Snake River Plain, the crust was extensively invaded by mafic magma from the mantle, and a silicic fraction was fused and erupted as rhyolite. The Snake River Plain crust is now predominantly mafic, and the 10-km-thick upper crust consists primarily of sedimentary and volcanic rocks.

Resolution of remaining unsolved problems must await thorough reanalysis of available data and new field experiments based on coincident application of different geophysical methods, combined with geologic studies and deep drilling.

INTRODUCTION

The Intermontane system consists of the Colorado Plateau, Basin and Range province, and High Lava Plains of the central part of the Cordilleran region of the western United States, as described geologically by Anderson (this volume). This chapter provides an overview of the geophysical framework of the Intermontane system and discusses some unsolved problems of the region, emphasizing uncertainties concerning the nature of the lower-crust to upper-mantle transition zone and crustal structure in the vicinity of the Wasatch Front.

This overview is not, and is not intended to be, a comprehensive review of the lithospheric structure and tectonics of the Intermontane system. Significant portions of such a review are provided in the foregoing chapters by Anderson, Thompson and others, and Smith and others. Rather, this overview is intended as a commentary on the geophysics of the Intermontane system and provides information not contained in the previous chapters. I have borrowed—with attribution—and greatly abbreviated portions of other contributions to this Memoir to make this overview more or less complete.

GEOPHYSICAL FRAMEWORK

Our understanding of the lithospheric structure of the Intermontane system comes mainly from seismic investigations, but gravity, magnetic, electrical, and heat-flow studies have also contributed significantly to our understanding of the geophysical framework of the region.

Seismic structure of the lithosphere

Crustal-thickness and velocity models of the different provinces of the Intermontane system have been reliably determined by an extensive network of seismic-refraction profiles, as discussed by Smith (1978, 1982), Smith and others (1982), Braile and others (1982), Pakiser (1985), and Braile and others (this volume). A few of these profiles have been selected (Fig. 1) to illustrate the crustal structure of the region.

The crust ranges in thickness from possibly as little as 25 km in the Basin and Range of northwestern Nevada, west-central Utah, and southern Arizona, to more than 40 km in the Snake River Plain of southern Idaho and the Colorado Plateau of Utah and Arizona (Fig. 2). The average crustal thickness of the Basin and Range is about 30 km, and the 35-km contour nearly coincides with the boundaries of the province, thus separating the Basin and Range distinctly from adjacent provinces (Fig. 2). The crust thickens to more than 50 km in the Sierra Nevada to the west. The crust thickens over horizontal distances of about 50 to 100 km from the thin crust of the Basin and Range to the thicker crusts of adjacent provinces.

The configuration of the Moho in Utah (Fig. 2) differs from the model in Smith and others (this volume, Figs. 5-9), which is based on the model of Pechmann and others (1984), Loeb and Pechmann (1986), and Loeb (1986). They interpret traveltimes of seismic waves recorded on the University of Utah seismic network from earthquake sources and nuclear explosions at the Nevada Test Site (NTS) to suggest the existence of a "double-Moho" beneath the Wasatch Front: an upper Moho at a depth of 25 to 30 km, below which the velocity is 7.5 km/sec, and a lower Moho at a depth of about 40 km, below which the velocity is 7.9 km/sec. Smith and others (this volume) regard only the deeper 7.9-km/sec refractor as the Moho. I reinterpreted the same traveltime data to suggest a single Moho, below which the velocity is about 7.9 km/sec, and to determine the crustal-thickness contours in Utah (Fig. 2). The resulting map is very similar to that of Smith (1978), which reveals a crust with nearly symmetrical variations in crustal thickness about a north-trending axis in eastern Nevada. This axis of symmetry is virtually identical to the axis of symmetry of the boundaries of the Great Basin.

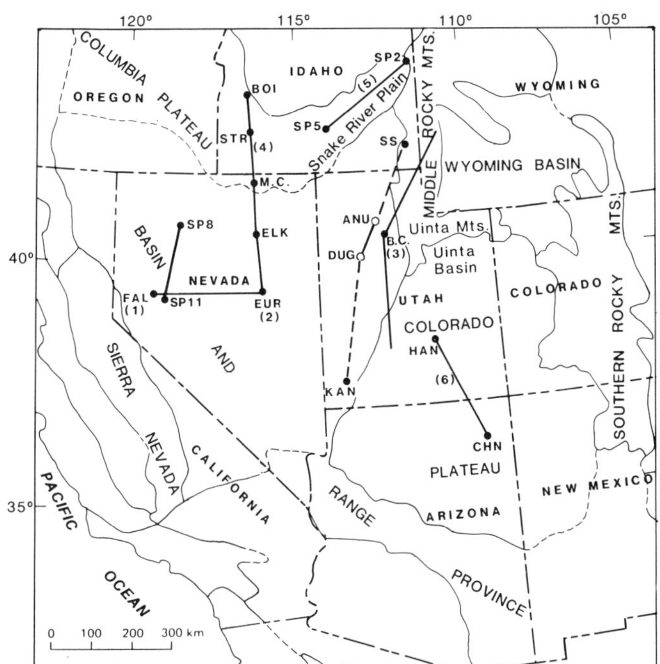

Figure 1. Map of the western United States showing locations of key seismic profiles in the Basin and Range province, Snake River Plain, and Colorado Plateau. Solid dots are seismic sources, open circles are stations DUG and ANU in University of Utah seismic network. Solid lines are seismic profiles discussed in text, dashed lines connect earthquake sources with DUG and ANU. FAL = Fallon shotpoint; EUR = Eureka shotpoint; B.C. = Bingham Canyon copper mine source; ELK = Elko shotpoint; M.C. = Mountain City shotpoint; STR = C. J. Strike Reservoir shotpoint; BOI = Boise shotpoint; SP2 and SP5 = shotpoints in eastern Snake River Plain; SP8 and SP11 = shotpoints from PASSCAL experiment in the western Basin and Range; HAN = Hanksville shotpoint; CHN = Chinle shotpoint. KAN = location of April 5, 1981, earthquake (M_L, 4.6) near Kanaraville; SS = location of October 14, 1982, earthquake (M_L, 3.9) near Soda Springs. Numbers in parentheses under lettered shotpoints are locations of seismic models in Figure 3.

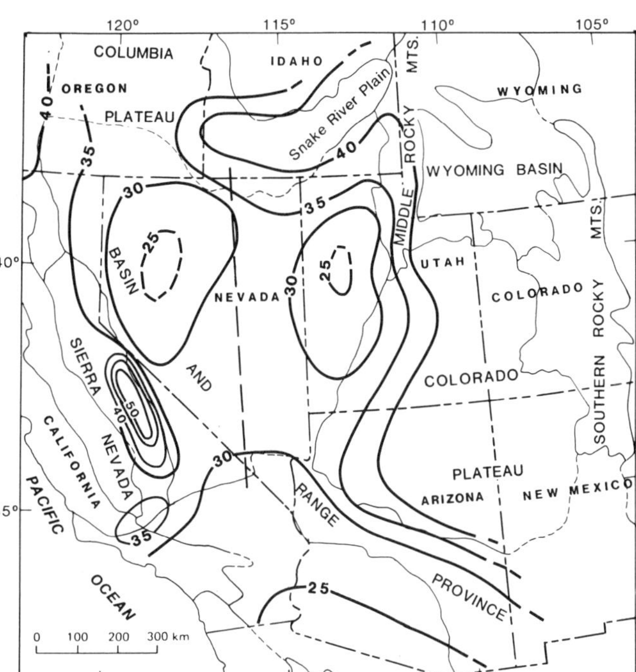

Figure 2. Map showing contours of crustal thickness; contour interval = 5 km. Axis of symmetry in eastern Nevada is shown as long dashes. Contours are dashed where depths are uncertain. Contours are modified from Figure 6-2 in Smith (1978) using data from Mooney and Weaver (this volume) in the Pacific coastal states, Thompson and others (this volume) in northwestern Nevada, Braile and others (1982) in the eastern Snake River Plain, and my reinterpretation of data provided by J. C. Pechmann (written communication, 1986) in Utah.

Crustal velocity models

I have compiled (Pakiser, 1985) essentially all published crustal velocity models in the Basin and Range province and from them traced the historical evolution of seismic models in the Basin and Range to the Colorado Plateau transition zone and at the NTS from the earliest ones of Tatel and Tuve (1955) and Carder and Bailey (1958) to the recent ones of Müller and Mueller (1979) and Taylor (1983). For this review, I have selected the velocity models at Fallon and Eureka, Nevada, of Eaton (1963) and Thompson and others (this volume) and the models at the Bingham Canyon, Utah, copper mine of Braile and others (1974) and Keller and others (1975) to illustrate velocity structure in the Basin and Range; the models of Hill and Pakiser (1966) and Braile and others (1982) for the Snake River Plain; and the model of Roller (1965) for the Colorado Plateau. These models reveal distinctly different crustal characteristics for each province (Fig. 3). The velocity of the upper crust is characteristically 5.2 to 6.3 km/sec. The velocity of the lower crust is characteristically 6.4 to 6.8 km/sec, but may range upward to values as high as 7.5 km/sec.

Basin and Range. Comparison of the preferred velocity model of Thompson and others (this volume; Figs. 1, 3, model 1a) with the model of Eaton (1963; Fig. 3, model 1b) reveals that they are virtually identical above the 24-km-deep Moho of Eaton, but the preferred model of Thompson and others contains a 10-km-thick transitional layer of velocity 7.5 km/sec above the 34-km-deep Moho. The alternative crustal model of Thompson and others without the transitional layer in the lower crust and with the Moho (not shown) at 24 km is compatible with the seismic and gravity data, but to account for the gravity anomaly it requires lithospheric thinning of 20 km or more between Eureka and Fallon (see Fig. 9 in Thompson and others, this volume). Stauber (1980, 1983) has suggested similar thinning of the lithosphere in the area of the Battle Mountain heat flow high in northwestern Nevada.

The velocity models at Eureka (Fig. 1) are very similar, except that the preferred model of Thompson and others (Fig. 3, model 2a) contains a 4-km-thick transitional layer in the lower crust, whereas the 1963 model of Eaton (Fig. 3, model 2b) does not. The depth to Moho in both models is 34 km. However, the

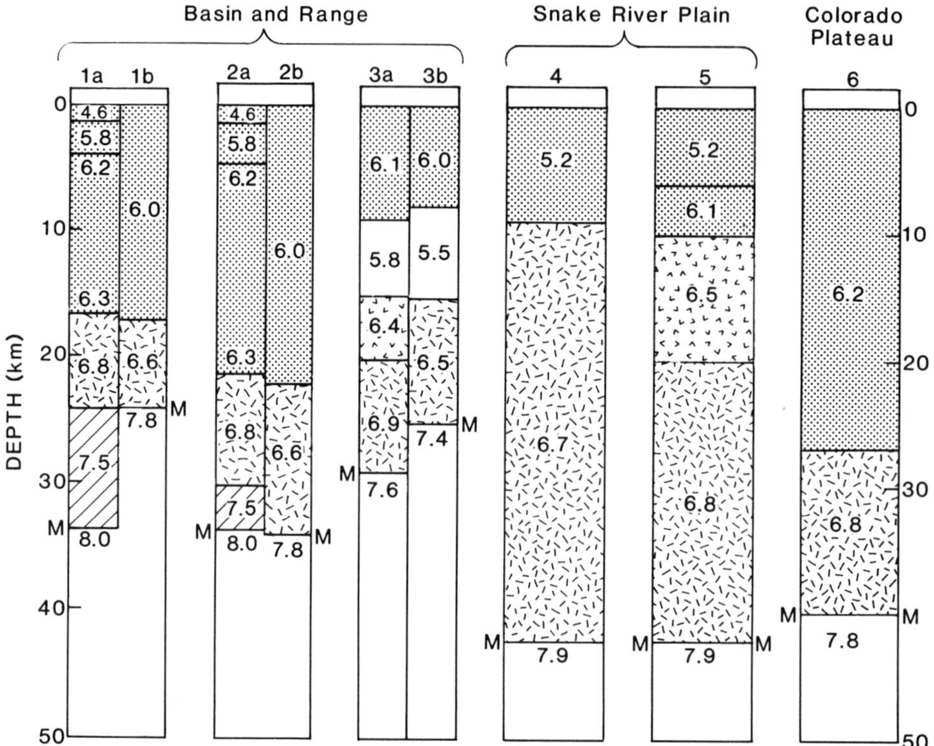

Figure 3. Crustal models representative of Basin and Range, Snake River Plain, and Colorado Plateau. Models 1a and 2a are from Thompson and others (this volume); 1b and 2b, from Eaton (1963); 3a, from Braile and others (1974); 3b, from Keller and others (1975); 4, from Hill and Pakiser, 1966); 5, from Braile and others (1982); 6, from Roller (1965). M = Moho. Velocities are in kilometers per second. Locations of models are shown in Figure 1. Scale above sea level (0 depth) is same as scale below sea level.

depth to Moho at Eureka is reduced to only 30 km in the Thompson and others model if the transitional layer is eliminated, and this alternative model can be reconciled with the seismic and gravity data. The models of Thompson and others (this volume) and Eaton (1963) are nearly the same as the Hill and Pakiser (1966) model at Eureka (Fig. 4).

The velocity models of Braile and others (1974) and Keller and others (1975) at the Bingham Canyon copper mine (Figs. 1, 3) are similar, except that the Moho in the Braile and others model (Fig. 3, model 3a) is 4 km deeper than the Moho in the Keller and others model (Fig. 3, 3b), partly because of the higher velocity in the lower 9 km of the crust in the former. Both models contain a low-velocity layer in the upper crust at depths ranging from 8 to 15 km. The two models are not strictly comparable, however, because the profile interpreted by Braile and others extends northeast from Bingham Canyon across the middle Rocky Mountains into Wyoming, whereas the profile of Keller and others extends south into the Colorado Plateau (Fig. 1).

Velocities in the uppermost mantle in the Basin and Range models of Figure 3 range from 7.4 km/sec in the Keller and others model at Bingham Canyon to 8.0 km/sec in the preferred Thompson and others models at Fallon and Eureka. The low P_n velocities reported for profiles radiating outward from Bingham Canyon may be apparent velocities resulting from downdip propagation from the Moho high in west-central Utah (Fig. 2). The differences in the P_n velocities between Fallon and Eureka of 7.8 vs. 8.0 km/sec reported by Eaton (1963) and Thompson and others (this volume) are not large and can be related to variations in the thicknesses of crustal layers, most probably in the thickness of the 7.5-km/sec transitional layer in the lower crust.

From examination of essentially all the evidence for P_n velocities in the Basin and Range province (Pakiser, 1985), I have concluded that the characteristic velocity in the uppermost mantle of the province is 7.8 to 7.9 km/sec.

Snake River Plain. The western Snake River Plain P-wave velocity model (Figs. 1; 3, model 4) of Hill and Pakiser (1966) is nearly identical to the eastern Snake River Plain model (Fig. 3, model 5) of Braile and others (1982), except for the somewhat more detailed crustal velocity layering revealed in the latter. The depth to Moho is 43 km, and the P_n velocity is 7.9 km/sec in both models.

The profile interpreted by Hill and Pakiser is the only one available in the western Snake River Plain. Many more crustal details were revealed in the major eastern Snake River Plain–Yellowstone seismic profiling experiment of 1978 (Smith, 1982; Smith and others, 1982). For example, Sparlin and others (1982) derived a velocity model for a southeast-trending profile that is nearly at right angles to the profile along the axis of the eastern

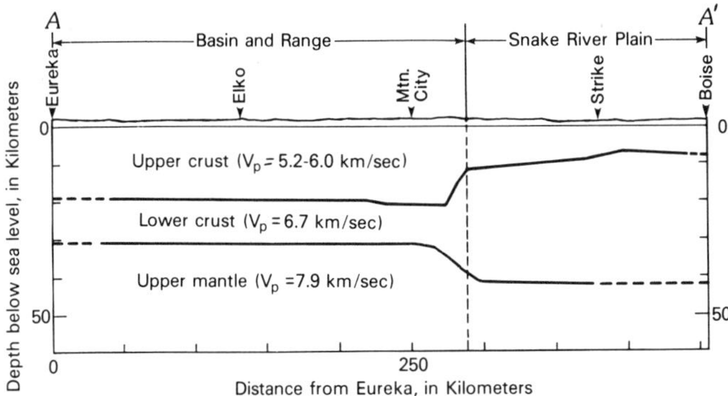

Figure 4. Crustal structure in the northern Basin and Range and western Snake River Plain between Eureka, Nevada, and Boise, Idaho. V_p = compressional wave velocity. Scale above sea level is the same as scale below sea level. Locations of shotpoints are shown in Figure 1. Adapted from Figure 15 in Hill and Pakiser (1966). The physiographic boundary of Basin and Range–Snake River Plain transition zone (Fig. 1) is about 60 km south of structural boundary in this figure.

Snake River Plain interpreted by Braile and others (1982). The resulting model is consistent with that of Braile and others, but it indicates that the upper crust is laterally heterogeneous. In particular, the model is bounded by a fault on the northwest margin of the plain, with a down-to-the-southeast offset of more than 4 km.

Greensfelder and Kovach (1982) have interpreted Rayleigh-wave–dispersion curves and teleseismic S-wave delays to derive an S-wave crustal model of the eastern Snake River Plain. They compared the resulting model with results from refraction studies, heat flow, volcanic history, and geology. Their model contains a crustal layer of high velocity at depths of 8 to 20 km in which the P- and S-wave velocities are 6.5 and 3.9 km/sec, respectively. This layer overlies a 21-km-thick lower crust in which the P-wave velocity is higher (6.8 km/sec) but the S-wave velocity is lower (3.6 km/sec). The upper-mantle S-wave velocity is 4.1 to 4.2 km/sec, much less than the 4.5 to 4.6 km/sec that would be expected from the P_n velocity of 7.9 km/sec. The low (3.6 km/sec) S-wave velocity in the lower crust, together with the high (6.8 km/sec) P-wave velocity, could be the result of partial melting, but Greensfelder and Kovach preferred to explain the low S-wave velocity as resulting from depletion of quartz caused by fusion of rhyolitic magmas generated by intrusion of basaltic magma from the mantle that was erupted 10 to 6 m.y. ago.

Quartz has a very low Poisson's ratio, so a decrease in quartz content with depth could cause an increase in Poisson's ratio, thus producing an inversion in S-wave but not P-wave velocity with depth. Conversely, Priestley and Orcutt (1982) found that distances of the Moho-reflection caustic are at least 20 km closer to explosion sources in the eastern than in the western Snake River Plain. This requires a significant P-wave low-velocity layer in the lower crust, consistent with the S-wave low-velocity layer of Greensfelder and Kovach. The low-velocity layer disappears to the southwest and is missing in the western Snake River Plain. The low P- and S-wave velocities of Priestley and Orcutt suggest elevated temperatures in the lower crust of the eastern Snake River Plain, and are compatible with the progression of volcanism from southwest to northeast of about 3.5 cm/yr.

Colorado Plateau. The crustal structure of the Colorado Plateau is also discussed by Prodehl and Lipman (this volume). The crust of Roller's (1965) P-wave velocity model at Hanksville in the Colorado Plateau (Figs. 1; 3, model 6) is 41 km thick, which is about the same as the 43-km-thick crust of Hill and Pakiser (1966) and Braile and others (1982) in the Snake River Plain (Fig. 3, models 4 and 5). However, the velocity distribution within the crust is distinctly different. The upper crust (P-wave velocity, 5.2 km/sec) is only about 10 km thick in the Snake River Plain, as compared with the 27-km-thick upper crust (6.2 km/sec) in the Colorado Plateau, suggesting distinctly different processes of crustal evolution in the two provinces. Depth to Moho in Roller's crustal model at Chinle, Arizona (Fig. 1), is 43 km.

Warren (1969) interpreted a U.S. Geological Survey seismic refraction profile, centered near the Mogollon rim on the Colorado Plateau southern boundary in the vicinity of Payson, Arizona, to indicate crustal thickness and velocity layering similar to that of Roller (1965). Perhaps the velocity structure of the crust is uniform between Hanksville and Payson and is representative of the Colorado Plateau as a geologic unit.

Seismic-reflection studies

Seismic-reflection studies of crustal structure in the western United States are discussed elsewhere in this volume (Smith and others, Smithson and Johnson, Thompson and others) and are not reviewed in detail here. Reflection surveys by the Consortium for Continental Reflection Profiling (COCORP) of the Basin and Range crust establish two general features (Allmendinger and others, 1983, 1987): an asymmetric pattern of seismic-reflection events with west-dipping reflections in the eastern part of the province and subhorizontal reflections in the west, and a pronounced reflection Moho at depths of 30 ± 2 km (locally, 34 km) with no resolvable offsets and no clear sub-Moho reflections (Smith and others, Fig. 20, this volume).

In the eastern Basin and Range, many normal faults terminate downward at widely ranging depths at regional low-angle detachment faults with as much as tens of kilometers of Cenozoic displacement. Low-angle reflections, normal to the structural trend, extend over horizontal distances of as much as 120 km and penetrate to depths of 15 to 20 km (Allmendinger and others, 1983). The Sevier Desert detachment zone in central Utah, for example, extends from a surface zone of normal faulting to a depth of 12 to 15 km at an average west dip of about 12°. Allmendinger and others (1983) have suggested 30 to 60 km of horizontal extension along the Sevier Desert detachment.

In the western Basin and Range, the brittle upper crust tends to be transparent, and the generally aseismic middle to lower crust is highly reflective. Allmendinger and others (1987) concluded that the layered lower crust is probably produced by processes accompanied by thermal anomalies: magmatism, enhanced ductility, and extension.

In northwestern Nevada, the reflection Moho (Klemperer and others, 1986) is clearly not a first-order discontinuity, which is the usual simplifying assumption of refraction modeling. The Moho reflection group is resolved in many places into two distinct reflections as much as 1.2 sec apart, and elsewhere, the separate reflections merge. Klemperer and others (1986) concluded that the Moho is a young feature in the Basin and Range province, possibly related to velocity contrasts caused by Cenozoic magmatism and extension. Thompson and others (this volume) associate these processes with underplating by gabbroic magma.

The "reflection Moho" in the Colorado Plateau (Allmendinger and others, 1987; Hauser, 1988) appears to be at a depth of 45 to 50 km, and therefore as much as 10 km deeper than the refraction Moho. Perhaps the reflection Moho comes from a discontinuity within the upper mantle, or results from late-arriving intercrustal multiples.

Upper-mantle structure

Velocity-depth models for the upper mantle of the Basin and Range have been determined by Niazi and Anderson (1965), Johnson (1967), Archambeau and others (1969), Burdick and Helmberger (1978), Priestley and Brune (1978, 1982), Priestley and Orcutt (1982), Priestley and others (1980), Stauber (1980, 1983), and Mavko and Thompson (1983); see Iyer and Hitchcock, this volume, for a comprehensive review of upper-mantle structure. Most models place the base of the lithosphere at a depth of 60 to 65 km, although Archambeau and others (1969) have suggested that the subcrustal lithosphere is very thin or missing. The lithosphere might be 20 km or more thinner in the area of the Battle Mountain heat-flow high in northwestern Nevada (Stauber, 1980, 1983) than in central Nevada (Priestley and Brune, 1978). In western Nevada a similar thinning of the lithosphere is required to reconcile the alternative seismic model of Thompson and others (this volume) with the gravity data.

According to a model proposed by Zoback and Lachenbruch (1984), the lithosphere thickens from 50 to 65 km beneath the Basin and Range to 90 to 100 km beneath the Colorado Plateau. This model is consistent with gravity, seismic, surface-elevation, and heat-flow data. The upper-mantle lid beneath the 43-km-thick crust of the Snake River Plain is only 10 km thick or less, based on surface-wave dispersion studies by Priestley and Brune (1982) and Greensfelder and Kovach (1982).

Gravity and magnetic investigations

Gravity and magnetic investigations of the lithosphere of the Intermontane system and adjacent provinces are discussed elsewhere in this Memoir by Jachens and others, by Kane and Godson, and in the separate chapters on regional investigations of the crust and upper mantle of the continental United States.

A highly generalized Bouguer gravity map of the Intermontane system (Fig. 5) crudely reflects variations in crustal thickness (Fig. 2). It displays the same bilateral symmetry about a north-trending axis in eastern Nevada (Eaton and others, 1978); the areas of thin crust in northwestern Nevada, west-central Utah, and southern Arizona are expressed by relatively high gravity (Figs. 2, 5). The thicker crust of the Snake River Plain is expressed by a Bouguer gravity anomaly that is more than 100 mGal higher in the western Snake River Plain than in central Nevada. The gravity field in central Nevada, where the crust is about 30 km thick, is about the same (less than –200 mGal) as it is in the western Colorado Plateau of Utah, where the crust is about 40 km thick. The gravity declines to less than –250 mGal in the Uinta Basin of northeastern Utah, and rises to more than –50 mGal in southwestern Arizona (Fig. 5). The Uinta Mountains, north of the basin, are expressed by a gravity high.

The gravity field of the Basin and Range correlates inversely with topography (Fig. 6), indicating that the province is isostatically compensated (Mabey, 1960). Topography is also symmetrical about a north-trending axis in eastern Nevada. Mabey (1966) showed that the regional gravity in the Basin and Range can be predicted within 5 mGal from the Bouguer correction, demonstrating nearly complete regional isostatic compensation. The combination of high topography, low gravity, and thin crust in the Basin and Range led me (Pakiser, 1963) to conclude that

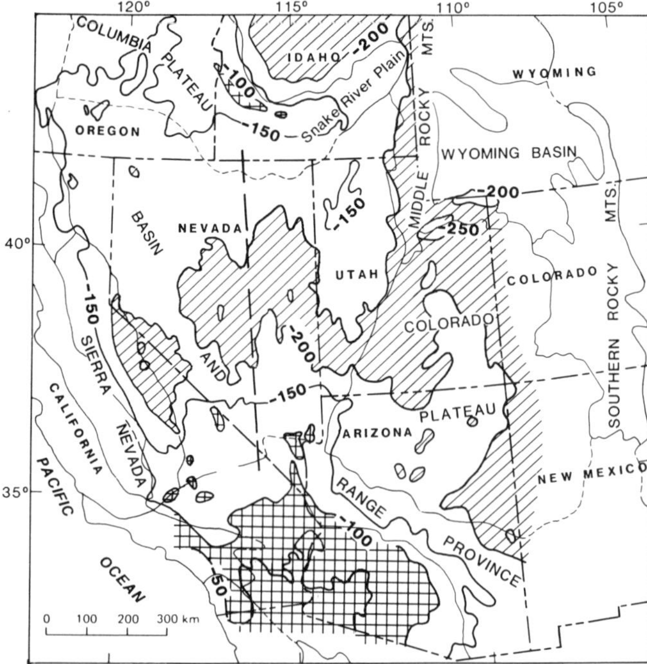

Figure 5. Map showing variations in Bouguer gravity. Contour interval = 50 mGal. Areas of relatively high gravity (above –100 mGal) and low gravity (below –200 mGal) are shown by different patterns. Axis of symmetry in eastern Nevada is shown by long dashes. Generalized from gravity map of Society of Exploration Geophysicists (1982), scale 1:2,500,000.

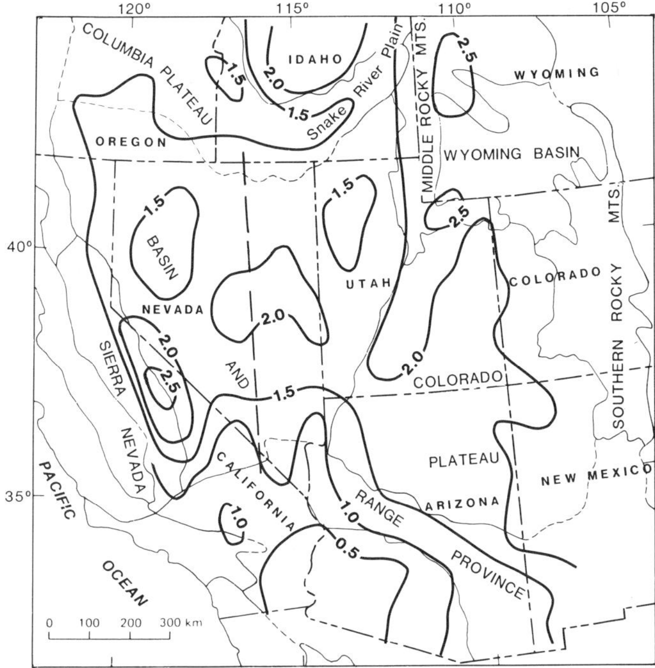

Figure 6. Map showing average altitude of surface above sea level based on 1° × 1° elements; contour interval = 0.5 km. Axis of symmetry in eastern Nevada is shown by long dashes. Data are from Strange and Woollard (1964). Adapted from Figure 8-5 in Blackwell (1978).

much of the isostatic compensation is caused by low-density rocks in the upper mantle. These rocks are now recognized as being primarily in the asthenosphere.

In the Snake River Plain, Bouguer gravity-anomaly values are systematically about 35 mGal higher than those predicted from the Bouguer correction (Mabey, 1966). Hill (1963) interpreted three pronounced en echelon gravity highs in the western Snake River Plain (merged into a single high in the generalized gravity map; Fig. 5) as representing mafic bodies possibly about 0.3 g/cm^3 more dense than the surrounding material and having a combined volume of more than 30,000 km^3. He concluded that the Snake River Plain is undercompensated isostatically by an amount proportional to the mass of these bodies. Mabey (1975) studied gravity and magnetic anomalies over the western Snake River Plain and found that the plain is underlain by a dense magnetic crustal layer buried under a thin upper crust, which is consistent with the seismic model of Hill and Pakiser (1966). Mabey suggested that the dense magnetic material is mafic rock formed by rifting in Miocene time. A north-northwesterly trending magnetic high in north-central Nevada, similar in form to the gravity highs in the western Snake River Plain, has been interpreted by Mabey and others (1978) and Zoback and Thompson (1978) as the expression of a Miocene rift.

Jachens and others (this volume) interpret the northeast-trending gravity-anomaly fabric in the Colorado Plateau to express primarily lithologic variations in the Precambrian basement, whereas the west-northwest- to north–trending anomalies of the Basin and Range express the grain of the linear mountain ranges related to the extensional history of the province. They note that short-wavelength anomalies can be related to geologic features mapped at the surface, whereas intermediate-wavelength anomalies result from variations in average crustal density.

Kane and Godson (this volume) conclude that short-wavelength gravity anomalies are primarily caused by density variations in the crust and that long-wavelength anomalies are primarily caused by density variations in the upper mantle. The gravity field of the Basin and Range is characterized by a dominant long-wavelength gravity low that results from the thick, shallow asthenosphere there; a more positive long-wavelength gravity field to the east results from the eastward thickening of the lithosphere near the axis of the Rocky Mountains (Kane and Godson, this volume).

Heat-flow and electrical studies

Heat flow in the Intermontane system has been discussed by Blackwell (1978) and Lachenbruch and Sass (1978), and is reviewed by Morgan and Gosnold in this volume. A generalized heat-flow map (Fig. 7) reveals that the characteristic heat flow in the region is about 2 HFU. Areas of exceptionally high heat flow (greater than 2.5 HFU) are found in the Snake River Plain and in the Battle Mountain area of northwestern Nevada. Lower heat flow (less than 1.5 HFU) is found in the Eureka area of east-central Nevada and in the Colorado Plateau. Low heat flow in the Eureka area is attributed to the cooling effect of downward circulation of ground water (Lachenbruch and Sass, 1978; Morgan and Gosnold, this volume). Blackwell (1978) found that areas of high heat flow correlate well with areas of Cenozoic volcanism, plutonism, and thermal spring activity within a Cordilleran thermal anomaly zone.

The same symmetry about a north-trending axis in eastern Nevada, previously noted for the crustal-thickness, gravity, and topographic contours, is evident for the heat-flow contours (Fig. 7). The Great Basin is partially outlined by the 1.5-HFU contour in California to the west and in central Utah and southwestern Wyoming to the east.

Lachenbruch and Sass (1978) have suggested that moderate lithospheric extension rates in the Basin and Range province can account for the high heat flow without calling on anomalous flux from the asthenosphere. They concluded that the anomalous heat flow increases roughly 1 HFU for every 1- to 2-percent increase in extension rates per million years. This is consistent with extension rates from other sources of evidence (Smith and others, this volume).

The electrical properties of the crust and upper mantle of the Intermontane system are discussed by Keller (this volume, Part 2). Research during the last two decades has disclosed that the electrical conductivity of the lower crust and upper mantle is anomalously high in the intermountain area of the west, and that this anomalous conductivity may be caused primarily by high

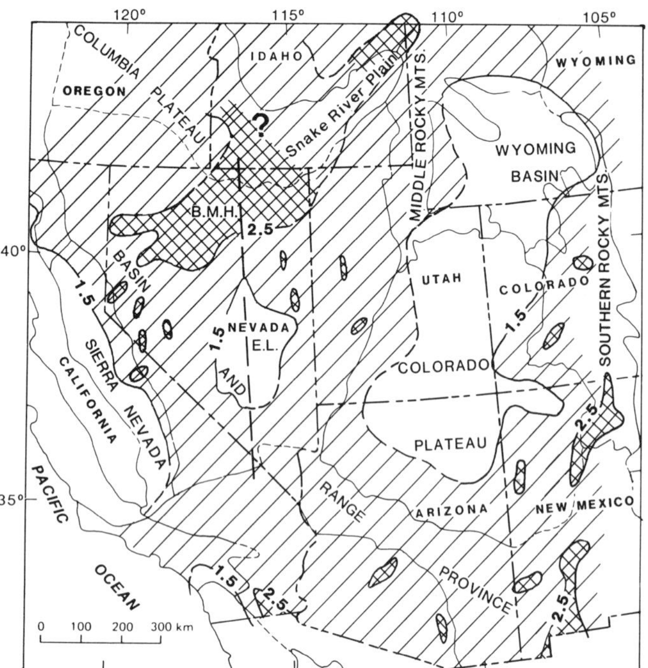

Figure 7. Map showing variations in surface heat flow. Contours are for 1.5 and 2.5 heat-flow units (HFU) (cal/cm^2 sec). Areas of heat flow above 1.5 and 2.5 HFU are shown by different patterns. Areas below 1.5 HFU are unpatterned. Axis of symmetry in eastern Nevada is shown by long dashes. B.M.H. = Battle Mountain heat-flow high; E.L. = Eureka heat-flow low. Adapted from Figure 9-1 in Lachenbruch and Sass (1978).

temperatures resulting from high heat flow. The absence of strong magnetic anomalies in much of the Intermontane system suggests shallow Curie temperatures limiting the depth extent of magnetized bodies (Blakely and Connard, this volume).

Seismicity and stress

The seismicity of the Intermontane system is discussed by Dewey and others (this volume), Smith (1978), Smith and others (this volume), and Thompson and others (this volume). The seismicity of the region is characterized by earthquakes that occur in broad zones, as much as 150 km wide, in the Nevada and intermountain seismic zones. Epicenters are scattered; in many places they do not coincide with mapped faults. Earthquakes occur within the brittle crust, and their depths rarely exceed 20 km.

Thompson and others (this volume) use seismic-reflection data to define three geometries for deeply penetrating faults in the Basin and Range: (1) steep- to moderate-dipping, planar, normal faults; (2) listric normal faults; and (3) low-angle normal faults. Earthquakes of magnitude 6.0 or larger seem to occur only on planar, normal faults with dips of 50 to 60°, and to nucleate at depths of 6 to 15 km (Barrientos and others, 1987; Jackson, 1987; Thompson and others, this volume).

The direction of the least horizontal stress in the Basin and Range province is oriented approximately east-west, parallel to the direction of extension, whereas in the relatively stable Colorado Plateau it is oriented north-south (Zoback and Zoback, this volume).

TRANSITION ZONES

The lithosphere of the Intermontane system has been explored as thoroughly as that of any region of the continental United States, as the extensive networks of seismic-refraction (Braile and others, this volume) and seismic-reflection profiles (Smithson and Johnson, this volume) suggest. Nevertheless, some major problems remain unsolved, awaiting new and better planned field experiments and more thorough analysis of existing and new data, which will lead to more reliable lithospheric models. Two of these unsolved problems are discussed herein; others are discussed by Anderson and others, Thompson and others, and Smith and others in earlier chapters.

Wasatch Front region

The Basin and Range to Middle Rocky Mountains–Colorado Plateau transition zone (Fig. 1) has been studied intensively by seismic-refraction profiling for more than a third of a century. Beginning with the pioneering work of Tuve and his co-workers in 1954 (Tatel and Tuve, 1955), many refraction profiles have been recorded in the region, but unfortunately none is reversed. Lacking evidence to the contrary for individual profiles, all are interpreted on the assumption of horizontal velocity discontinuities within the crust and at the Moho. I disagree, however, (Pakiser, 1985), as published profiles in the vicinity of the Wasatch Front seem to radiate outward from sources at or near the top of a Moho structural high (Fig. 2). Thus the propagation paths of P_n waves are probably downdip, and the resulting apparent velocities are probably less (7.4 to 7.6 km/sec) than the true refractor velocity.

To obtain the "double-Moho" model in the vicinity of the Wasatch Front, Pechmann and others (1984), Loeb and Pechmann (1986), and Loeb (1986) plotted traveltimes from earthquake and nuclear sources to stations in the University of Utah seismic network for each source separately and for all sources combined. They found that least-squares fits of the traveltime data in the distance range of 120 to 250 km for waves propagating along the upper Moho define an average velocity of 7.5 km/sec. Beyond 250 km to about 700 km, the average least-squares velocity is 7.9 km/sec (lower Moho). Assuming that these velocities represent true refractor velocities, they performed a modified time-term inversion to derive the model described by Smith and others (this volume). Smith and others, however, regarded only the deeper 7.9-km/sec refractor as the Moho.

The evidence for the "double Moho" is unconvincing because the traveltime variations for local sources result mainly from variations in depths to the upper Moho refractor, whether the velocity is 7.5 km/sec or some larger number more appro-

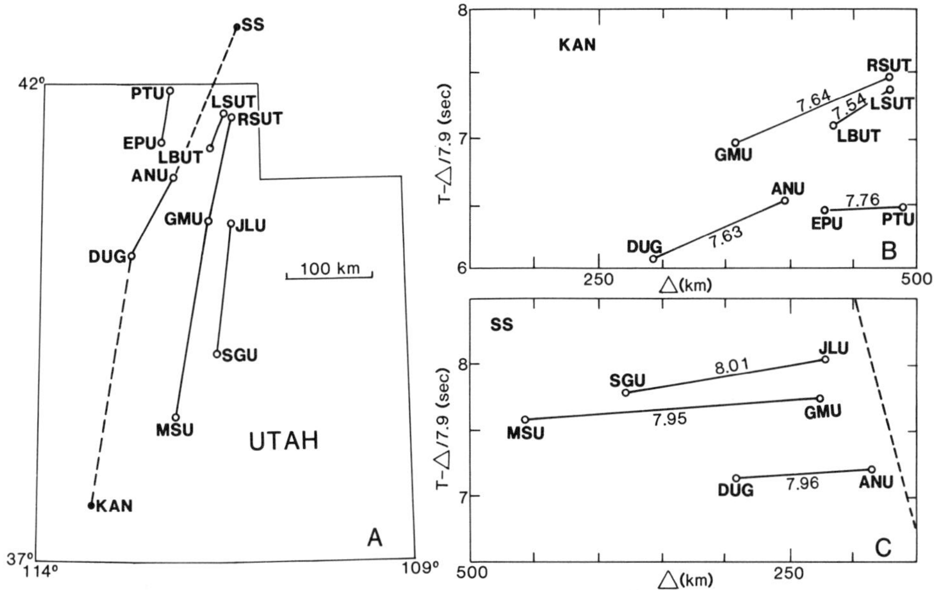

Figure 8. Map and travel-time graphs showing a portion of data used to obtain Moho contours in the Wasatch Front region (Fig. 2). A, locations of KAN, DUG, ANU, and SS (defined in Fig. 1), and station pairs EPU-PTU, LBUT-LSUT, MSU-GMU, GMU-RSUT, and SGU-JLU from University of Utah seismic network. B, Travel times of station pairs from KAN. C, Traveltimes of station pairs from SS. Apparent velocities for station pairs are in kilometers per second. Data are from J. C. Pechmann (written communication, 1986).

priate for the upper mantle; therefore, least-squares fitting of the traveltime data in the 120- to 250-km distance range is inappropriate. Propagation paths from more distant sources (for example, the earthquakes labeled KAN and SS; Figs. 1, 8) are roughly parallel to the Wasatch Front and to the seismic network stretching along it, as well as to the Moho contours (Fig. 2), so least-squares fits of these traveltimes, perhaps fortuitously, yield a fairly reliable P_n velocity, even though the scatter of the traveltime data beyond 250 km is very large.

I plotted traveltimes between pairs of stations along the same or nearly the same azimuths from the local sources, and have concluded that apparent velocities determined from them—which range from about 7 to 10 km/sec—are primarily indicators of severe relief on the refractor surface.

P_n traveltimes between the station pair DUG-ANU (Figs. 1, 8) are partially reversed by earthquake sources near Kanaraville, Utah (KAN), and near Soda Spring, Idaho (SS). The distance between DUG and ANU is about 100 km, but the true subsurface reversal of the Moho paths is only about 30 km because of opposing displacements of the points of emergence from the Moho of waves from KAN and SS. Nevertheless, this station pair is the nearest approach to a "reversed profile" in the Wasatch Front region of any of the many data sets that have been recorded and interpreted.

The apparent velocities between DUG and ANU from KAN and SS are 7.63 (downdip) and 7.96 km/sec (updip), respectively (Fig. 8), indicating an average north dip of about 2°. If, for example, the velocity of the lower crust is 6.4 km/sec, the true upper-mantle velocity determined from the apparent velocities is 7.8 km/sec, comparable to the 7.9-km/sec velocity determined by Pechmann and Loeb. Apparent velocities from KAN and SS for four unreversed station pairs nearly parallel to and east of DUG-ANU (LBUT-LSUT, MSU-GMU, GMU-RSUT, and SGU-JLU) are nearly the same as those determined for DUG-ANU, thus supporting the gentle north dip and a value of 7.8 to 7.9 km/sec as the true upper-mantle velocity (Fig. 8). The apparent velocity for the station pair EPU-PTU north of DUG-ANU is higher than other velocities from KAN, suggesting nearly horizontal propagation along the Moho.

The depth to Moho increases from west to east by about 10 km between the station pairs shown (Fig. 8) at dips normal to the contours of about 7°. Propagation of P_n waves along the station pairs examined (Fig. 8) is nearly parallel to the Moho contours (Fig. 2). To arrive at the Moho contours in Utah (Fig. 2), I arbitrarily assigned a depth-to-Moho of 25 km beneath DUG and calculated depths beneath 24 other network stations from their delay-time differences from DUG, assuming an upper-mantle velocity of 7.9 km/sec.

This analysis does not preclude a transitional layer a few kilometers thick in the lower crust similar to that proposed by Thompson and others (this volume), but detection of such a layer is beyond the resolving power of the available data.

Lower-crust to upper-mantle transition zone

In their reinterpretation of the Fallon to Eureka seismic-refraction profile initially described by Eaton (1963), Thompson

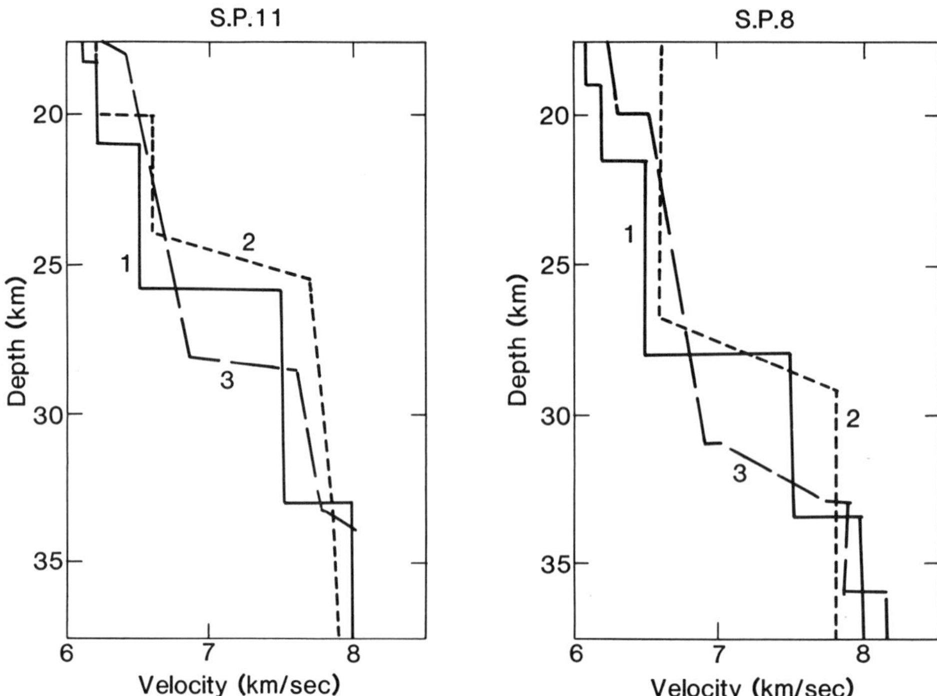

Figure 9. Preliminary velocity-depth models of lower-crust-to-upper-mantle transition zone for SP8 and SP11 from PASSCAL 1986 western Basin and Range seismic experiment. Depths are below surface (approximately 1.5 km above sea level). Locations of shot points are shown in Figure 1. Model 1 for both shot points is from R. D. Catchings (written communication, 1988); model 2 is from W. S. Holbrook (written communication, 1988); model 3 is from H. M. Benz (written communication, 1988).

and others (this volume) were able to fit to the seismic record sections two rather different models: one, with 4- to 10-km-thick transitional layer of velocity 7.5 km in the lower crust, and the other without the transitional layer (Fig. 3). The model without the transitional layer is almost the same as the much earlier model of Eaton. This suggests that, with high-quality but sparse data like that of the Fallon-Eureka profile, the modern ray-tracing techniques used by Thompson and others are unable to resolve the details of the lower-crust to upper-mantle transition zone, and that other evidence, such as reflection or gravity data, is needed to discriminate among models. Even with evidence from other methods, uncertainties about the validity of the models remain.

More recently, in 1986, two perpendicular refraction/wide-angle reflection profiles were recorded in the northern Basin and Range under the auspices of the Program for Array Seismic Studies of the Continental Lithosphere (PASSCAL), along with several densely spaced reflection spreads (Whitman and Catchings, 1988). The northeast- and northwest-trending profiles intersect east of Lovelock, Nevada. The northwest-trending profile coincides in part with the COCORP seismic reflection line in the area (Allmendinger and others, 1987). The station spacing of the refraction profiles is close (1 to 2 km), and the quality of the data is high. Preliminary results from the northeast-trending profile only (Fig. 1) are considered here.

Three different investigators have been able to create three different, yet similar, preliminary crustal models for this profile (Fig. 9). R. D. Catchings (written communication, 1988) used ray tracing to derive a preliminary model that includes a 4- to 8-km-thick transitional layer in the lower crust (velocity, 7.5 km/sec). The upper-mantle velocity below the 33-km-deep Moho is 8.0 km/sec (Fig. 9); (see also Whitman and Catchings, 1988). Catchings developed a similar model for the northwest-trending profile (not shown in Figs. 1, 9). The discontinuities between layers are shown as first-order discontinuities (Fig. 9), but they actually are zones of steep velocity gradients (R. D. Catchings, written communication, 1988).

W. S. Holbrook (written communication, 1988) has also used ray tracing to derive a model for the northeast-trending profile. His model does not contain a separate transitional layer, but it does contain a 2- to 3-km-thick gradient zone between the 6.6-km/sec lower crust and the 7.8-km/sec upper mantle. The crustal thickness in the Holbrook model, which is very similar to that of Eaton (1963), ranges from about 30 km in the northeast to about 26 km in the southwest (Fig. 9). His model also contains a 3- to 5-km-thick low-velocity zone (not shown in Fig. 9) at a depth of about 40 km in the upper mantle in the northeast portion of the profile (see also Holbrook and others, 1987).

Using a different technique—amplitude modeling that employs high-frequency, reflectivity synthetic seismograms—H. M. Benz (written communication, 1988) has derived a model for the northeast-trending profile that is somewhat similar to both Catchings' and Holbrook's models. Benz's model contains a narrow gradient zone at a depth of about 28 to 32 km in which the velocity jumps from 6.8 to 7.6 km/sec, and a gradient zone or discontinuity at a depth of about 33 to 36 km in which the velocity jumps from 7.8 to 8.0 to 8.2 km/sec (Fig. 9). The deeper jump could be considered as being within the upper mantle (see also Benz and Smith, 1987).

Although the described models of Catchings, Holbrook, and Benz for the PASSCAL profiles are preliminary and research on them is continuing, they are each compatible with the refraction/wide-angle reflection data and the COCORP reflection results (Allmendinger and others, 1987) for models intermediate between those of SP8 and SP11. Although similar to each other overall, there seems to be no compelling reason to prefer one over the other. The issue of whether there is or is not a transitional layer in the lower crust remains unresolved.

DISCUSSION

Eaton and others (1978) and Eaton (1979) have described a large central province in the Cordilleran region that includes the Intermontane system and extends northeast to include a portion of the northern Rocky Mountains. This broad area shares some fundamental geologic and geophysical properties with the Basin and Range: young faulting, high heat flow, low gravitational intensity, low P_n velocity, high electrical conductivity in the lower crust and upper mantle, and numerous thermal springs. Eaton has termed this region the Cordilleran thermotectonic anomaly region. Whereas Eaton emphasized the similarities in the different provinces of the region, it is important also to call attention to their differences.

The Basin and Range province is characterized by a thin crust, averaging about 30 km thick, overlying a Moho that is nearly horizontal. In the western Basin and Range, the brittle upper crust is seismically transparent, except for shallow, near-surface basin structures, and the middle to lower crust that produces conspicuous discontinuous subhorizontal reflections. In the eastern Basin and Range, both the upper and lower crust produce conspicuous west-dipping reflections that are related to structural elements such as the Sevier Desert detachment fault. Few deep crustal reflections have been recorded in the Basin and Range to Colorado Plateau transition zone. The lithosphere of the Basin and Range is thin, averaging about 60 km, and the heat flow exceeds 2 HFU except in the Eureka area of central Nevada. Extensional processes of Cenozoic age have had a strong influence on the geophysical framework of the province.

The crusts of the Colorado Plateau and Snake River Plain have about the same thickness, 40 to 45 km, but the proportions of upper to lower crust in the Colorado Plateau are about the same as those of the Basin and Range, whereas the upper crust is thin in the Snake River Plain. The lithosphere is about 90 to 100 km thick beneath the Colorado Plateau, but much thinner (about 50 km) beneath the Snake River Plain.

These differences in lithospheric structure reflect differences in the tectonic development of the separate provinces of the Intermontane system and imply different crustal compositions.

The crust of the Colorado Plateau may be representative of the crust of the entire Intermontane system prior to the Cenozoic extension of the Basin and Range and the volcanic activity in the Snake River Plain. The 27- to 28-km-thick crystalline upper crust of the Colorado Plateau (P-wave velocity, 6.1 to 6.2 km/sec) probably consists of predominantly silicic gneisses and schists similar in composition to the core complexes of the Basin and Range. However, few exposures of such rocks crop out in the region. The lower crust of the Colorado Plateau (velocity, 6.8 km/sec) probably consists of rocks of silicic to intermediate composition in the granulite facies.

The Basin and Range crust was thinned and invaded with mafic magma from the upper mantle during Cenozoic extension. Valasek and others (1987) have proposed a compositional model for the Basin and Range crust based on temperature-corrected velocities from wide-angle and normal-incidence seismic reflections in the Ruby Mountains core complex of Nevada. The 9-km-thick upper layer (velocity, 6.2 km/sec) consists of quartzofeldspathic rocks such as metasedimentary rocks, migmatites, and deformed granites. The upper layer is underlain by a 7-km-thick midcrustal layer (velocity, 6.4 km/sec) that consists of quartzofeldspathic rocks interlayered with mafic intrusives, possibly as amphibolite, that may make up as much as one-third of the midcrust. The 9-km-thick lower crust (velocity 6.7 to 6.8 km/sec) consists predominantly of mafic rocks from the mantle, interlayered with material of quartzofeldspathic composition, possibly constituting one-third of the total. The rocks of the lower crust are probably in the granulite facies. The 3-km-thick transition zone between the lower crust and upper mantle (velocity, 7.4 to 7.8 km/sec) may consist of a mafic residuum or layered cumulates from the overlying gabbroic layer, perhaps resulting from underplating. As much as one-third of the Basin and Range crust may have been formed by intrusion or underplating by gabbroic magma from the upper mantle (Valasek and others, 1987).

The crust of the Snake River Plain consists almost entirely of rocks of velocity 6.5 to 6.8 km/sec in which mafic material predominates. Gabbroic magma from the mantle was injected into the crust as a result of extension between the northern Basin and Range and the Idaho batholith, thus causing fusion of granitic material of the original crust and resulting in silicic volcanism (Hamilton, 1963; Hill and Pakiser, 1966; Hamilton and Myers, 1966; Braile and others, 1982). The uppermost 10 km of the crust of the Snake River Plain probably consists of interlayered sedimentary and volcanic rocks, ranging in composition from rhyolite to basalt. The lower crust of the eastern Snake River Plain may still contain a partially molten fraction of silicic composition corresponding to the low-velocity zone of Greensfelder and Kovach (1982) and Priestley and Orcutt (1982).

CONCLUSIONS

Our knowledge of the geophysical framework of the Intermontane system has been steadily expanding during the past three decades. The pace of geophysical research has been strongly accelerating during the past 10 years, especially as a result of COCORP and other seismic-reflection studies. Geophysical studies have led to important new insights into the nature of the geologic structure and the tectonic and magmatic processes occurring in the crust and upper mantle of the region. Many of these advances are outlined in this overview and in the preceding chapters.

Our knowledge of the P-wave velocity structure of the lithosphere of the Intermontane system can be considered as possibly more complete than that of any other region of the continental United States, but the S-wave velocity structure is still but poorly known. Variations in the thickness of the lithosphere within the region are known only in a very general way.

Gravity, magnetic, heat-flow, and electrical investigations have significantly expanded our knowledge of the structure and dynamic processes of the lithosphere of the Intermontane system, but much remains to be learned about the density and electrical-conductivity structure and temperature distribution in the deep crust and upper mantle.

Our knowledge of the seismicity and stress field of the Intermontane system rivals that of any other region of the United States, excepting the San Andreas fault of California, but the confident assessment of earthquake hazards and possible prediction of earthquakes are largely matters for future research.

We can take considerable pride in what we have learned about the geophysics of the Intermontane system, but it is a matter of concern that such wide-ranging and conflicting values have been reported for P_n velocities, for example, owing largely to poor design of some seismic field experiments. As a result, disparate velocity models are proposed for transition zones such as the Wasatch Front region. More nearly definitive answers to problems and questions—such as the nature of transition zones, the origin of reflections in the deep crust, the S-wave velocity structure of the lithosphere, density distribution in the crust and upper mantle, composition of the lower crust, and the nature of faulting—must await the results of more thorough analysis of available data and of new field experiments based on coincident application of many geophysical methods, combined with geologic studies and deep drilling (Mooney and Pakiser, this volume).

ACKNOWLEDGMENTS

I thank R. E. Anderson, H. M. Benz, R. D. Catchings, G. R. Keller, W. D. Mooney, J. C. Pechmann, R. B. Smith, A. Walter, and M. L. Zoback for reviews of the manuscript and helpful discussions, and E. M. Omdahl for preparing the manuscript and R. J. Ricotta for drafting the figures. I am grateful to H. M. Benz, R. D. Catchings, and W. S. Holbrook for permission to publish their preliminary velocity models of the lower-crust-to-upper-mantle transition zone in northwestern Nevada. Finally, I am especially grateful to J. C. Pechmann for permission to use unpublished data from the University of Utah seismic network in my development of an alternate velocity model of the crust and upper mantle of Utah.

REFERENCES CITED

Allmendinger, R. W., and 7 others, 1983, Cenozoic and Mesozoic structure of the eastern Basin and Range province, Utah, from COCORP seismic-reflection data: Geology, v. 11, p. 532–536.

Allmendinger, R. W., and 7 others, 1987, Overview of the COCORP 40°N Transect, western United States; The fabric of an orogenic belt: Geological Society of America Bulletin, v. 98, p. 308–319.

Archambeau, C. B., Flinn, E. A., and Lambert, D. G., 1969, Fine structure of the upper mantle: Journal of Geophysical Research, v. 74, p. 5825–5865.

Barrientos, S. E., Stein, R. S., and Ward, S. N., 1987, Comparison of the 1959 Hebgen Lake, Montana, and the 1983 Borah Peak, Idaho, earthquakes from geodetic observations: Seismological Society of America Bulletin, v. 77, p. 784–808.

Benz, H. M., and Smith, R. B., 1987, Amplitude and ray trace modeling of vertical-incidence to wide-angle refraction/reflection data from the 1986 Basin-Range PASSCAL seismic experiment [abs.]: EOS Transactions of the American Geophysical Union, v. 68, p. 1360.

Blackwell, D. D., 1978, Heat flow and energy loss in the western United States, *in* Smith, R. B., and Eaton, G. P., eds., Cenozoic tectonics and regional geophysics of the western Cordillera: Geological Society of America Memoir 152, p. 175–208.

Braile, L. W., Smith, R. B., Keller, G. R., and Welch, R. M., 1974, Crustal structure across the Wasatch front from detailed seismic refraction studies: Journal of Geophysical Research, v. 79, p. 2669–2677.

Braile, L. W., and 9 others, 1982, The Yellowstone–Snake River Plain seismic profiling experiment; Crustal structure of the eastern Snake River Plain: Journal of Geophysical Research, v. 87, p. 2597–2609.

Burdick, L. J., and Helmberger, D. V., 1978, The upper mantle P velocity structure of the western United States: Journal of Geophysical Research, v. 83, p. 1699–1712.

Carder. D. S., and Bailey, L. F., 1958, Seismic wave traveltimes from nuclear explosions: Seismological Society of America Bulletin, v. 48, p. 377–398.

Eaton, G. P., 1979, Regional geophysics, Cenozoic tectonics, and geologic resources of the Basin and Range province and adjoining regions, *in* Newman, G. W., and Goode, H. D., eds., 1979, Basin and Range Symposium, Denver, Colorado: Rocky Mountain Association of Geologists and Utah Geological Association, p. 11–39.

Eaton, G. P., Wahl, R. R., Prostka, H. J., Mabey, D. R., and Kleinkopf, M. D., 1978, Regional gravity and tectonic patterns; Their relation to late Cenozoic epeirogeny and lateral spreading in the western Cordillera, *in* Smith, R. B., and Eaton, G. P., eds., Cenozoic tectonics and regional geophysics of the western Cordillera: Geological Society of America Memoir 152, p. 61–91.

Eaton, J. P., 1963, Crustal structure from San Francisco, California, to Eureka, Nevada, from seismic-refraction measurements: Journal of Geophysical Research, v. 68, p. 5789–5806.

Greensfelder, R. W., and Kovach, R. L., 1982, Shear wave velocities and crustal structure of the eastern Snake River Plain, Idaho: Journal of Geophysical Research, v. 87, p. 2643–2653.

Hamilton, W., 1963, Overlapping of late Mesozoic orogens in western Idaho: Geological Society of America Bulletin, v. 74, p. 779–788.

Hamilton, W., and Myers, W. B., 1966, Cenozoic tectonics of the western United States: Reviews of Geophysics, v. 4, p. 509–549.

Hauser, E. C., 1988, Reflection Moho at 50 km (16 s) beneath the Colorado Plateau on COCORP deep reflection data [abs.]: EOS Transactions of the American Geophysical Union, v. 69, p. 496.

Hill, D. P., 1963, Gravity and crustal structure in the western Snake River Plain, Idaho: Journal of Geophysical Research, v. 68, p. 5807–5819.

Hill, D. P., and Pakiser, L. C., 1966, Crustal structure between the Nevada Test Site and Boise, Idaho, from seismic-refraction measurements, *in* Steinhart, J. S., and Smith, T. J., eds., The earth beneath the continents: American Geophysical Union Monograph 10, p. 391–419.

Holbrook, W. S., Catchings, R. D., Mooney, W. D., Jarchow, C. M., and Thompson, G. A., 1987, A two-dimensional velocity model of the upper lithosphere in the northern Basin and Range province of Nevada [abs.]: EOS Transactions of the American Geophysical Union, v. 68, p. 1361.

Jackson, J. A., 1987, Active normal faulting and crustal extension, *in* Coward, M. P., Dewey, J. F., and Hancock, P. L., eds., Continental extensional tectonics; Geological Society Special Publication 28: Blackwell Scientific Publications, p. 3–18.

Johnson, L. R., 1967, Array measurements of P velocities in the upper mantle: Journal of Geophysical Research, v. 72, p. 6309–6325.

Keller, G. R., Smith, R. B., and Braile, L. W., 1975, Crustal structure along the Great Basin–Colorado Plateau transition from seismic refraction studies: Journal of Geophysical Research, v. 80, p. 1093–1098.

Klemperer, S. L., Hauge, T. K., Hauser, E. C., Oliver, J. E., and Potter, C. J., 1986, The Moho in the northern Basin and Range province, Nevada, along the COCORP 40°N seismic-reflection transect: Geological Society of America Bulletin, v. 97, p. 603–618.

Lachenbruch, A. H., and Sass, J. H., 1978, Models of an extending lithosphere and heat flow in the Basin and Range province, *in* Smith, R. B., and Eaton, G. P., eds., Cenozoic tectonics and regional geophysics of the western Cordillera: Geological Society of America Memoir 152, p. 209–250.

Loeb, D. T., 1986, The P-wave velocity structure of the crust-mantle boundary beneath Utah [M.S. thesis]: Salt Lake City, University of Utah, 126 p.

Loeb, D. T., and Pechmann, J. C., 1986, The P-wave velocity structure of the crust-mantle boundary beneath Utah from network traveltime measurements [abs.]: Earthquake Notes, v. 57, p. 10–11.

Mabey, D. R., 1960, Regional gravity survey of part of the Basin and Range province: U.S. Geological Survey Professional Paper 400-B, p. B283–B-285.

——, 1966, Relation between Bouguer gravity anomalies and regional topography in Nevada and the eastern Snake River Plain, Idaho: U.S. Geological Survey Professional Paper 550-B, p. B-108–B-110.

——, 1975, Interpretation of a gravity profile across the western Snake River Plain, Idaho: Geology, v. 3, p. 53–55.

Mabey, D. R., Zietz, I., Eaton, G. P., and Kleinkopf, M. D., 1978, Regional magnetic patterns in part of the Cordillera in the western United States, *in* Smith, R. B., and Eaton, G. P., eds., Cenozoic tectonics and regional geophysics of the western Cordillera: Geological Society of America Memoir 152, p. 93–106.

Mavko, B. B., and Thompson, G. A., 1983, Crustal and upper mantle structure of the northern and central Sierra Nevada: Journal of Geophysical Research, v. 88, p. 5874–5892.

Muller, G., and Mueller, S., 1979, Traveltime and amplitude interpretation of crustal phases on the refraction profile Delta-W, Utah: Seismological Society of America Bulletin, v. 69, p. 1121–1132.

Niazi, M., and Anderson, D. L., 1965, Upper mantle structure of western North America from apparent velocities of P waves: Journal of Geophysical Research, v. 70, p. 4633–4640.

Pakiser, L. C., 1963, Structure of the crust and upper mantle in the western United States: Journal of Geophysical Research, v. 68, p. 5747–5756.

——, 1985, Seismic exploration of the crust and upper mantle of the Basin and Range province, *in* Drake, E. T., and Jordan, W. M., eds., Geologists and ideas; A history of North American geology: Geological Society of America Centennial Special Volume 1, p. 453–469.

Pechmann, J. C., Richins, W. D., and Smith, R. B., 1984, Evidence for a "Double Moho" beneath the Wasatch front, Utah [abs.]: EOS Transactions of the American Geophysical Union, v. 65, p. 988.

Priestley, K., and Brune, J., 1978, Surface waves and the structure of the Great Basin of Nevada and western Utah: Journal of Geophysical Research, v. 83, p. 2265–2272.

——, 1982, Shear wave structure of the southern volcanic plateau of Oregon and Idaho and the northern Great Basin of Nevada from surface wave dispersion: Journal of Geophysical Research, v. 87, p. 2671–2675.

Priestley, K. F., and Orcutt, J. A., 1982, Extremal traveltime inversion of explosion seismology data from the eastern Snake River Plain, Idaho: Journal of Geophysical Research, v. 87, p. 2634–2642.

Priestley, K. F., Orcutt, J. A., and Brune, J. N., 1980, Higher-mode surface waves and structure of the Great Basin of Nevada and western Utah: Journal of Geophysical Research, v. 85, p. 7166–7174.

Roller, J. C., 1965, Crustal structure in the eastern Colorado Plateau province from seismic-refraction measurements: Seismological Society of America Bulletin, v. 55, p. 107–119.

Smith, R. B., 1978, Seismicity, crustal structure, and intraplate tectonics of the interior of the Western Cordillera, *in* Smith, R. B., and Eaton, G. P., eds., Cenozoic tectonics and regional geophysics of the Western Cordillera: Geological Society of America Memoir 152, p. 111–144.

——, 1982, Preface to Yellowstone–Snake River Plain Symposium papers: Journal of Geophysical Research, v. 87, p. 2581.

Smith, R. B., and 9 others, 1982, The 1978 Yellowstone–Eastern Snake River Plain seismic profiling experiment; Crustal structure of the Yellowstone region and experimental design: Journal of Geophysical Research, v. 87, p. 2583–2596.

Society of Exploration Geophysicists, 1982, Gravity anomaly map of the United States.

Sparlin, M. A., Braile, L. W., and Smith, R. B., 1982, Crustal structure of the eastern Snake River Plain determined from ray trace modeling of seismic refraction data: Journal of Geophysical Research, v. 87, p. 2619–2633.

Stauber, D. A., 1980, Crustal structure in the Battle Mountain heat-flow high in northern Nevada from seismic refraction profiles and Rayleigh wave phase velocities [Ph.D. thesis]: Stanford, California, Stanford University, 315 p.

——, 1983, Crustal structure of northern Nevada from seismic refraction data, *in* The role of heat in the development of energy and mineral resources in the northern Basin and Range province: Geothermal Resources Council Special Report 13, p. 319–332.

Strange, W. E., and Woollard, G. P., 1964, The use of geologic and geophysical parameters in the evaluation, interpretation, and production of gravity data: St. Louis, Missouri, U.S. Air Force Aeronautical Chart and Information Center, Contract AF 23(601)–3879 (Phase 1).

Tatel, H. E., and Tuve, M. A., 1955, Seismic exploration of a continental crust, *in* Poldervaart, A., ed., Crust of the Earth: Geological Society of America Special Paper 62, p. 35–50.

Taylor, S. R., 1983, Three-dimensional crust and upper mantle structure at the Nevada Test Site: Journal of Geophysical Research, v. 88, p. 2220–2232.

Valasek, P. A., Hawman, R. B., Johnson, R. A., and Smithson, S. B., 1987, Nature of the lower crust and Moho in eastern Nevada from "wide-angle" reflection measurements: Geophysical Research Letters, v. 14, p. 1111–1114.

Warren, D. H., 1969, A seismic-refraction survey of crustal structure in central Arizona: Geological Society of America Bulletin, v. 80, p. 257–282.

Whitman, D., and Catchings, R. D., 1988, PASSCAL Basin and Range lithospheric structure [abs.]: EOS Transactions of the American Geophysical Union, v. 69, p. 259.

Zoback, M. L., and Lachenbruch, A. H., 1984, Upper mantle structure beneath the western U.S.: Geological Society of America Abstracts with Programs, v. 16, p. 705.

Zoback, M. L., and Thompson, G. A., 1978, Basin and Range rifting in northern Nevada; Clues from a mid-Miocene rift and its subsequent offsets: Geology, v. 6, p. 111–116.

MANUSCRIPT ACCEPTED BY THE SOCIETY OCTOBER 31, 1988

Printed in U.S.A.

Geological Society of America
Memoir 172
1989

Chapter 14

Crustal structure of the Rocky Mountain region

Claus Prodehl
University, Geophysikalisches Institut, Hertzstrasse 16, 7500 Karlsruhe 21, West Germany
Peter W. Lipman
U.S. Geological Survey, Box 25046, Denver Federal Center, Denver, Colorado 80225

ABSTRACT

The Rocky Mountain system constitutes the eastern side of the broad Cordilleran orogen of central North America. This chapter summarizes seismic and other geophysical data that constrain crustal thickness and structure in the region. The crustal thickness ranges from less than 40 to more than 50 km. It is thickest in southern Montana and in Colorado, and is 10 to 15 km thinner in the intervening middle Rocky Mountains. Crustal thickness along the Rio Grande rift, essentially a prong of basin-and-range structure that projects northward into the southern Rocky Mountains, is less than 35 km, and transitions from adjacent thicker crust may be abrupt. Velocities in the upper crust are generally 6 km/sec or less to depths of at least 20 km; lower crustal velocities are 6.4 to 7.4 km/sec. Minor inversions in velocity structure occur only in the upper crust. The lower crust is separated from the mantle by a transition zone several kilometers thick, except possibly in the Yellowstone Plateau and the Rio Grande rift. Upper-mantle velocities beneath the Moho are less than 8 km/sec through most of the region.

With the exception of the Basin and Range province to the west, the surrounding areas have similar crustal thickness; thus, no distinct crustal root is concentrated below the Rocky Mountains. The major crustal contrast with the Great Plains and Colorado Plateau provinces is a higher mean velocity of 6.2 km/sec of the upper crust and an upper-mantle velocity of greater than 8 km/sec east of the Rocky Mountain front. The longitudinal variation in total crustal thickness may be partly a relict from cratonization in Archean time, as suggested by the lack of change in crustal thickness across the Rocky Mountain front. Late Mesozoic to early Tertiary Laramide compressional uplifts and magmatism, widespread in the Rocky Mountain region, can be correlated with decreased crustal velocities and densities in the upper crust. Decreased rigidity of the lithosphere and permanently modified Moho structure, evident in the Rio Grande rift, possibly encroached gradually northward into the southern Rocky Mountains. The geologically diverse provinces of the Rocky Mountain region offer especially favorable opportunities to evaluate the evolution of Cordilleran crustal structure through time: some features appear to represent little-modified Precambrian craton, whereas others record overprinting and crustal modification by successive Phanerozoic tectonic and magmatic events.

Prodehl, C., and Lipman, P. W., 1989, Crustal structure of the Rocky Mountain region, *in* Pakiser, L. C., and Mooney, W. D., Geophysical framework of the continental United States: Boulder, Colorado, Geological Society of America Memoir 172.

INTRODUCTION

The Rocky Mountains face the Great Plains to the east, the Columbia Plateau and the Great Basin of the Basin and Range province to the west, and the Colorado Plateau and southeastern Basin and Range province to the southwest. This region is divided into: (1) the northern Rocky Mountains, extending from Idaho and western Montana into Canada; (2) the middle Rocky Mountains in Wyoming, southern Idaho, and Utah; and (3) the southern Rocky Mountains in Wyoming, Colorado, and New Mexico (Fig. 1). The Yellowstone Plateau and adjacent Snake River Plain separate the northern Rocky Mountains from the middle Rocky Mountains, and the Wyoming basins lie between the middle Rocky Mountains to the west and the southern Rocky Mountains to the east in Wyoming and adjacent parts of northern Colorado. The Rio Grande Rift is essentially a prong of basin-range structure that projects northward into the southern Rocky Mountains.

The number of deeply penetrating seismic crustal studies that have been made in the Rocky Mountain region (Fig. 1) is modest compared to some other areas described in this volume, e.g., the Basin and Range province. The northern Rocky Mountains were studied by seismic methods mainly in the late 1950s. A detailed investigation of the Yellowstone Plateau area in 1978 and 1980 is essentially limited to upper crustal structure. Reconnaissance seismic studies of the middle and southern Rocky Mountains in the 1960s provide the best regional study currently available for the Rocky Mountain area. COCORP (Consortium for Continental Reflection Profiling) surveys are available for parts of the Wind River thrust and the Laramie Range. In the Rio Grande rift a series of mostly nonreversed profiles reveals the generalized crustal structure of the southern rift. Only a limited area where a magma body has been detected near Socorro, New Mexico, has been investigated by COCORP surveys.

This chapter reviews all available seismic data, makes comparisons with constraints from other geophysical methods and from surface geology, and interprets the crustal structure of the region. In spite of the rather heterogeneous data sets, we do have a generalized picture of crustal structure of the Rocky Mountain region, which may serve as a basis for a preliminary interpretation of the historic evolution of the crust. In contrast to the more active intermontane and Pacific regions of the Cordillera to the west, where present crustal structure can be expected to be dominated by geologically relatively recent events, the geologically diverse provinces of the Rocky Mountain region offer exceptional opportunities to evaluate the evolution of Cordilleran crustal structure through time.

GEOLOGIC FRAMEWORK

Introduction

The Rocky Mountain system constitutes the eastern side of the broad Cordilleran orogen, which first began to develop its present width in late Mesozoic time. In this region, Cordilleran features overprint and truncate older sedimentation and tectonic regimes related to formation of the North American craton and subsequent events. The older structural events have markedly influenced the orientation and style of Cordilleran tectonics, resulting in a complexly overlapping mosaic of upper crustal structural provinces within the Rocky Mountains. Over the past 100 m.y., the Rocky Mountain region was first blocked out by compressional deformation involving folds, thrusts, and block uplifts in late Mesozoic and early Tertiary time. These events were followed by a complex sequence of extensional tectonic events that began in the Eocene in northern regions and spread to the southern Rocky Mountains in middle and late Tertiary time. These structural provinces within the Rocky Mountains are now commonly ascribed to interactions between the western margin of the American plate and various Pacific plates. Such interactions have been complex over the past 100 m.y., resulting in the exceptional width (1,000 to 1,500 km) of the Cordilleran zone in comparison with many plate-margin orogenic belts. The eastern margin of the Rocky Mountain system, the Rocky Mountain Front, is thus the major active tectonic boundary within the American plate.

Figure 1. Index map of seismic refraction surveys and deep seismic reflection lines in the Rocky Mountains and adjacent areas. Physical divisions from Fenneman (1946). ●—● = seismic refraction lines and shotpoints; ●●●● = deep seismic reflection lines (COCORP surveys). Stipple segments between A and A′ indicate approximate location of cross section in Figure 23. **Shotpoints: 11, 16, 27-43:** U.S. Geological Survey lines (Jackson and others, 1963; Prodehl, 1970, 1979; Prodehl and Pakiser, 1980; Stewart and Pakiser, 1962; Warren, 1975 and unpublished): 11 = Boise, 16 = Delta, 27 = American Falls Reservoir, 28 = Bear Lake, 29 = Flaming Gorge Reservoir, 30 = Hanksville, 31 = Chinle, 32 = Lumberton, 33 = Cochetopa, 34 = Wolcott, 35 = Sinclair, 36 = Climax, 37 = Gnome, 38 = Lamar, 39 = Guernsey, 40 = Agate, 41 = Hale, 42 = Brewster, 43 = Selden. **BC1:** Greenbush Lake, British Columbia. Earth Physics Branch of the Department of Mines, Energy and Resources, Canada, 1969 (Hales and Nation, 1973). **M1-M5:** Montana survey of University of Wisconsin, Princeton University, and Carnegie Institution of Washington, 1959 (Meyer and others, 1961a): M1 = Cliff Lake, M2 = Sailor Lake, M3 = Missouri River, M4 = Fort Peck Reservoir, M5 = Acme Pond. **M6-M8:** USGS LASA calibration 1966 (Borcherdt and Roller, 1967): M6 = Grey Cliff, M7 = Billings, M8 = Angela. **LASA:** USGS crustal calibration survey 1968 (Warren and others, 1972): area indicated by large circle. **Y1-Y13:** Yellowstone Park–eastern Snake River Plain seismic profiling experiment. University of Utah, Purdue University, and others: Y1-Y11 = shotpoints 1-11 of Smith and others (1982, Fig. 1), Y12 = Gay Mine, Y13 = Condo Mine. **U1:** Bingham (Braile and others, 1974; Keller and others, 1975). **R1–R5:** Rio Grande rift crustal studies: R1 = GASBUGGY (Toppozada and Sanford, 1976), R2 = Dicethrow III (Olsen and others, 1979), R3 = Santa Rita (Cook and others, 1979), R4 = Millrace, R5 = Distant Runner (Sinno and others, 1986). **B1-B2:** Southern Basin and Range province profiles, University of Arizona (Gish and others, 1981, Harden, 1982): B1 = Tyrone, B2 = Jackpile. **W1-6 and N1-4 COCORP surveys:** W1-2 = Wyoming lines 1, 1A, and 2 (Smithson and others, 1979); W3-6 = Wyoming lines 3-6 (Brewer and others, 1982); N1-4 = New Mexico lines 1, 1A, 2, 2A, 3, 4 (Brown and others, 1979). **VELA UNIFORM stations mentioned in text:** V = Vernal, Utah; D = Durango, Colorado; R = Raton, New Mexico.

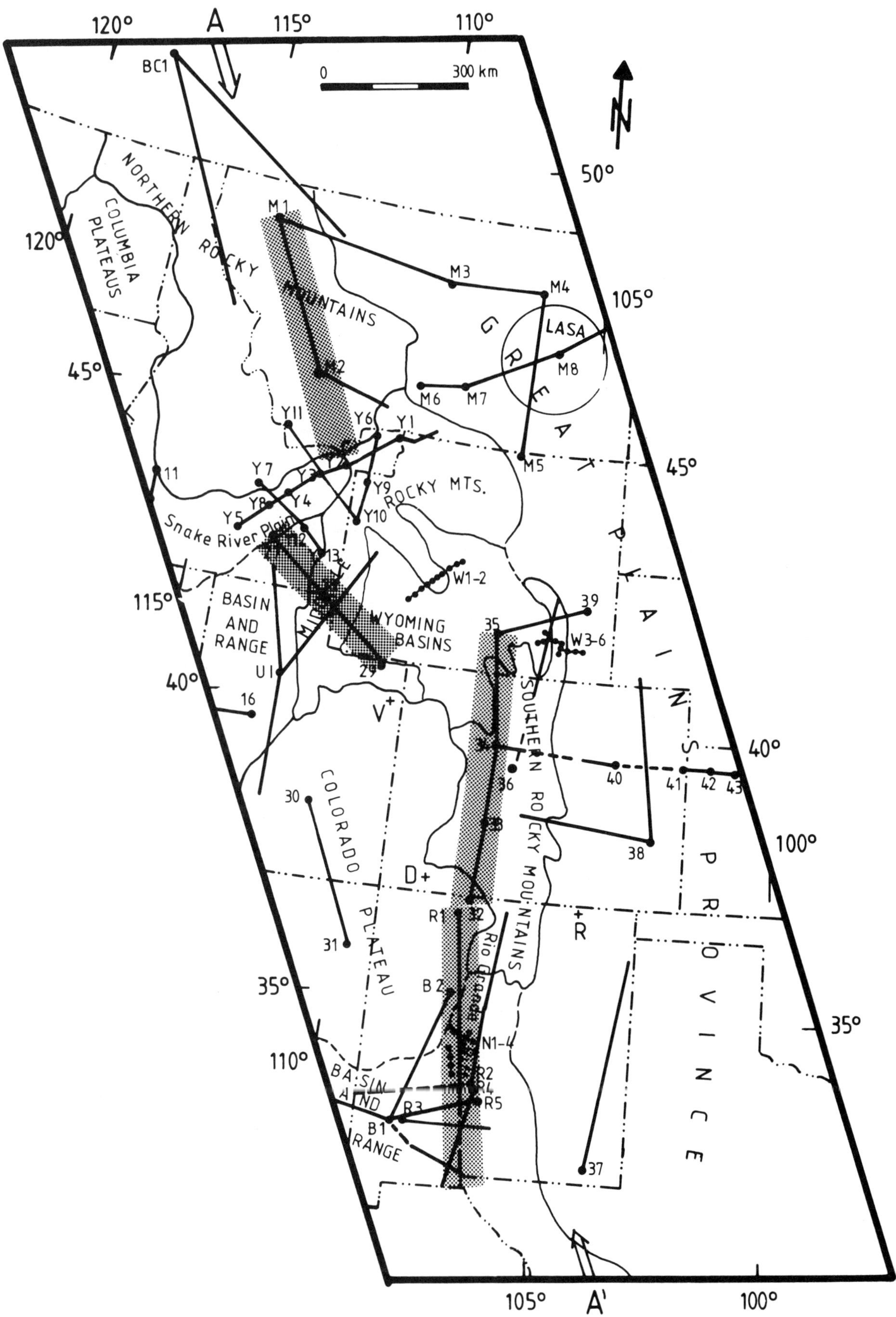

A
A'
0
300 km
N
120°
115°
110°
105°
100°
50°
45°
40°
35°
BC1
NORTHERN ROCKY MOUNTAINS
COLUMBIA PLATEAUS
M1
M3
M4
M6
M7
M8
M5
LASA
GREAT PLAINS PROVINCE
Y1
Y6
Y7
Y8
Y4
Y5
Y9
Y10
Y11
11
ROCKY MTS.
Snake River Plain
W1-2
BASIN AND RANGE
WYOMING BASINS
U1
16
29
35
39
W3-6
V
SOUTHERN ROCKY MOUNTAINS
36
40
41
42
43
30
COLORADO PLATEAU
38
D
R1
32
R
31
Rio Grande
B2
N1-4
R2
R5
R3
B1
37
BASIN AND RANGE

As physiographically defined (Fig. 1), the Rocky Mountains include diverse sedimentary, igneous, and tectonic elements varying widely in age, geographic distribution, and geological significance. Included are the margin of the Precambrian crystalline craton; an extensive cratonic sedimentary sequence developed along a passive margin in latest Precambrian and early Paleozoic time; the igneous and tectonic record of successive plate-convergence events, probably beginning in middle Paleozoic time and culminating in a fully developed Andean margin late in Mesozoic time; eastward Laramide migration of igneous and tectonic features related to plate convergence in early Cenozoic time; and gradual termination of subduction-related activity and inception of extensional and transform tectonic regimes later in Cenozoic time.

Major present-day geomorphic elements of the Rocky Mountain system include: (1) the northern Rocky Mountains, (2) the Yellowstone Plateau and associated parts of the eastern Snake River Plain, (3) the middle Rocky Mountains, (4) Wyoming basins, (5) the southern Rocky Mountains, and (6) the Rio Grande rift and associated southeastern basin-range region (Fig. 1). In brief, the northern Rocky Mountains constitute the mountainous parts of Idaho and Montana that were blocked out largely by foreland folding, thrusting, volcanism, and plutonism in Late Cretaceous to early Tertiary time. The Yellowstone Plateau–eastern Snake River Plain region constitutes a late Cenozoic locus of bimodal rhyolitic and basaltic volcanism that trends northeast, nearly perpendicular to the earlier compressional structures of the northern and middle Rocky Mountains. The middle Rocky Mountains and Wyoming basins are dominated by large basement-cored compressional uplifts and intervening broad basins that were blocked out early in the Cenozoic and became loci for thick sedimentary accumulations in middle to late Cenozoic time. In comparison with the middle Rocky Mountains, the southern Rocky Mountains in Colorado and northern New Mexico constitute a narrower and more intensely compressed zone of early Tertiary uplifts between the Colorado Plateau and Great Plains provinces. The Rio Grande rift is a locus of intracontinental extension along the axis of the southern Rocky Mountains; the rift has accommodated northwesterly movement of the Colorado Plateau away from the interior of the craton in the late Cenozoic, and it merges with the broader extensional region of the Basin and Range province in southern New Mexico.

Cordell (1978) and Eaton (1986) consider the southern Rocky Mountains and Rio Grande rift as a tectonic unit. They showed that the southern Rocky Mountains and the Rio Grande rift constitute the crestal range of a mammoth, continental arch-like feature that Eaton has named the Alvarado Ridge. Its axis trends south from Casper, Wyoming, to El Paso, Texas.

Some geophysical signatures within the Rocky Mountain system, such as heat flow, reflect largely the younger geologic elements; others, such as aeromagnetic anomalies, are strongly influenced by the Precambrian structural grain established at the time of cratonization. The seismic crustal structure that is the focus of this chapter is sensitive to both early established structural elements, such as the gross geometry of the Precambrian upper crust, and to more recently active structural elements, such as shallow crustal structure, Moho depth, and physical properties of the lithospheric mantle, which owe their present character largely to interrelated thermal and extensional events in late Cenozoic time. The accompanying brief summary of the geologic evolution of the Rocky Mountain region, taken largely from Burchfiel and others (1982) and Lipman (1982), provides a framework within which to interpret more detailed aspects of the geophysical expression.

Precursor framework

Precambrian crystalline rocks exposed in the Rocky Mountains are the westward extension of the Archean and early Proterozoic terranes of the North American craton. Near the Colorado-Wyoming border, Archean gneiss, schist, and granite having ages greater than 2.5 Ga are in abrupt contact with several sequences of metamorphosed sedimentary, volcanic, and plutonic rocks in the southern Rocky Mountains that yield ages of 1.8 to 1.6 Ga. The southern terrane lacks evidence of older Archean basement and thus apparently is a complex of accretional elements that added new material to the North American craton. Younger anorogenic plutonic sequences were intruded at about 1.4 to 1.5 and 1.0 Ga. Structural trends are generally northeast and are paralleled by locally prominent shear zones and loci of Phanerozoic igneous activity such as those of the Colorado mineral belt, the Jemez zone, and the eastern Snake River Plain–Yellowstone trend.

Unconformably overlying the Precambrian crystalline rocks are thick local sequences of largely terrigenous clastic sedimentary rocks of late Precambrian age, including the Belt Supergroup in Montana and Idaho, and the Uinta Mountain Group and the Uncompahgre Formation in western Colorado and northeastern Utah. These sedimentary rocks may record initial rifting and development of a passive continental margin along the western side of the Rocky Mountain region for which the record becomes more clear in Paleozoic time. The early Paleozoic miogeoclinal sequence in the Rocky Mountain region widely includes a thin incomplete succession of basal terrigenous clastic rocks, overlain by carbonate rocks, that thicken westward into a more complete continental shelf sequence.

During late Paleozoic time, sedimentation in the cratonal and eastern shelf region was disrupted by development of the Ancestral Rocky Mountains. The uplifts and basins caused by this weak compression are distributed northwestward across the southern Rocky Mountain region as far west as eastern Utah, resulting in erosional stripping of older Paleozoic sedimentary rocks and accumulation of thick local terrigenous deposits of late Paleozoic age. These events may reflect foreland deformation related to collisional tectonics along the Ouachita orogenic belt to the southeast.

Initial development of an Andean-type margin along the western side of the American continent in early Mesozoic time represents a significant change in the tectonic development of the Cordilleran region, but the major initial effect across most of the Rocky Mountain region was merely to terminate marine sedimentation. Lower Mesozoic cratonic sedimentary rocks are thin and discontinuous red beds that largely lack fossil control, making reconstruction of the depositional history difficult. By Early Cretaceous time, however, encroachment of a wide seaway into the continental interior produced a widespread marine sequence that grades westward into thick foredeep clastics. These sediments were shed from and partly involved in folds and thrusts that were developing along the west margin of the Rocky Mountain region.

Eastward migration of tectonic and igneous activity

During late Mesozoic and early Cenozoic time, the foreland fold and thrust belt terminated in the early Cordillera, Laramide uplifts and basins developed within the Rocky Mountains, and the patterns of igneous activity migrated eastward. These significant changes may have reflected increased rates of convergence between the American and Farallon plates between 80 and 40 Ma.

From Late Cretaceous through early Tertiary time, folds and thrusts encroached increasingly eastward across the northern Rocky Mountains. In the middle and southern Rockies, growth of large compressional uplifts was accompanied by subsidence of broad basins that were filled with subaerial and lacustrine sediments. These Laramide structures partly rejuvenated those of the Ancestral Rocky Mountains and extended as far east as the Black Hills in more subdued style. Northeast-directed compression and crustal shortening in the middle Rocky Mountains implies that the Colorado Plateau block was shifted northward along oblique-slip structures now obscured by features of the Rio Grande rift. Along the eastern side of the Rocky Mountain region, marine sedimentation continued on the craton through the end of Cretaceous time, cut marine sandstones and shales grade westward into thick orogenic sediments along the eastern margin of the growing orogenic belt.

At about the same time as Laramide deformation was initiated, the Andean magmatic belt, which had been continuous in the western Cordillera until about 80 Ma, began to sweep eastward, became discontinuous, and finally became inactive in mid-Tertiary time. Eastward sweep and extinction of magmatic activity was diachronous across the region, culminating in the northern Rocky Mountains about 70 Ma, as recorded by the Boulder batholith and associated volcanic rocks, but continuous in the middle and southern Rockies until about 60 Ma.

Transition to extensional tectonics

From Eocene to Holocene time, igneous activity, tectonics, and Holocene sedimentation in the Rocky Mountains record a gradual transition from the plate-convergence regime that dominated Cordilleran tectonics for several hundred million years to a gradually enlarging extensional regime that reflects destruction of the Farallon plate and beginning of transform motion between Pacific and American plates.

Much of the Rocky Mountain region was tectonically relatively quiescent in middle Tertiary time, but initial extensional deformation began as early as the Eocene in the northern Rockies, and somewhat later along the Rio Grande rift in the southern Rocky Mountains. Associated early grabens were broad, diffuse, and oriented parallel to plate boundaries farther to the west; in places, stratal rotations were severe, presumably reflecting listric faults that flattened with depth. This deformation was associated with continued arc volcanism to the west and is interpreted as an intracontinental type of back-arc extension. The orientation and style of extension changed at about 20 Ma, as documented especially well in the southern Rocky Mountains and along the Rio Grande rift. This change involved a gradual transition from back-arc extension to extension of basin-range type, involving high-angle faulting associated with fundamentally basaltic or bimodal volcanism. Major grabens, as along the Rio Grande rift, formed at the crests of broad uplifts that project well beyond the rift structures. In both episodes of Tertiary extension, the deformation geometry was broadly similar across the entire width of the Cordillera, and the major stress boundary within the American plate is along the Rocky Mountain front.

Middle Tertiary igneous activity emplaced mainly calc-alkaline volcanic rocks that show complex space-time compositional patterns that have been interpreted as retrograde activity of the Andean arc as the descending slab slowed, steepened, and deformed internally. Major volcanic regions of this type in the Rocky Mountain region include the Lowland Creek, Challis-Absoraka, San Juan, and Mogollon Datil fields. As this seemingly arc-related volcanism shifted south and west, contrasting assemblages of dominantly basaltic and alkalic activity began in the late Oligocene in areas of initial extensional deformation, such as in the Rio Grande rift and in the northern Rocky Mountains. In addition to their significance for plate-tectonic interpretations, these rocks are associated with many important base- and precious-metal ore deposits. Later Cenozoic rift-related volcanism is dominantly basaltic and bimodal basalt and rhyolite, as exemplified by volcanism of the Snake River–Yellowstone trend and along the Rio Grande rift.

Eocene basin-fill sediments within the Rocky Mountains reflect a continuation of post-Laramide orogenic sedimentation. To the east on the Great Plains, remnants of the thin, widespread postorogenic sediments also indicate continued erosion of the Laramide uplifts. Younger grabens, associated with the initiation of extensional deformation, contain thick accumulations of continental sediments that are lithologically variable, reflecting proximity to rugged local topography and varying tectonic histories of the different basins.

Increased uplift rates in the entire Rocky Mountain region and adjacent parts of the High Plains and the Colorado Plateau during the past 5 m.y. appear to have been associated in space and time with increased volcanism. Maximum regional uplift has occurred along the southern Rocky Mountains sector of the Rio

Grande rift and the Yellowstone area, both margins of especially intense volcanism. Significant thermal perturbation of the lower crust and upper mantle is a likely consequence.

Active seismicity

Present-day seismic activity in the interior Cordilleran United States is relatively low, compared to that of California and western Nevada, but several loci are evident (Smith, 1978, Plate 6-1).

The intermountain seismic belt extends north through Utah, Idaho, and Wyoming, turns northwest at Yellowstone National Park, and continues through western Montana (Smith and Sbar, 1974). Loci of greatest activity (Smith, 1978, Fig. 6-6) are: (1) the western part of the middle Rocky Mountains, near the border between southern Idaho and central Wyoming; (2) western Yellowstone Park and Hebgen Lake; and (3) near Flathead Lake, Montana.

Another distinct zone of seismicity surrounds the Colorado Plateau. According to Smith (1978), the interior of the plateau is a relatively rigid block that accommodates uplift or rotation mainly near its margins. Some epicenters in western and central Colorado are earthquakes induced by fluid injection in oil wells of the Rangely field and in waste disposal wells at the Rocky Mountain Arsenal near Denver.

Another weak seismic zone trends along the Rio Grande rift north into the southern Rocky Mountains as far as Wyoming. Detailed sesimic studies have documented areas of concentrated microearthquake activity within the rift, especially from Belen to Socorro, and west of Espanola, New Mexico. Both these areas have been interpreted as associated with "modern magma bodies" at middle to upper crustal levels. The small number of shocks within the rift, their distribution, suggests that the rift may not be spreading at present, in agreement with geodetic measurements (Sanford and others, 1979).

NORTHERN ROCKY MOUNTAINS

The northern Rocky Mountains are a geologically complex region in which Proterozoic to Mesozoic miogeoclinal sedimentary rocks overlie Archean basement. The sediments were intricately thrust faulted and folded in late Mesozoic time, widely blanketed by volcanics and intruded by coeval plutons in early Tertiary time, and disrupted by northeast-trending extensional structures in middle to late Tertiary time. Modern detailed crustal studies have not been made in the northern Rocky Mountains. Two deep seismic refraction surveys provide only reconnaissance data: (1) a reversed line recorded in 1959 in western Montana (Asada and Aldrich, 1966; Meyer and others, 1961a); and (2) two unreversed long-range profiles made in 1969, which extend south-southeast from a shotpoint in Canada (Hales and Nation, 1973). For the 1959 line (M1-M2 in Fig. 1), only traveltimes are available, but a record section was published for one of the 1969 profiles (BC1 to the south-southwest). The station separation is rather large for both lines: 15 to 20 stations distributed over a distance of 400 km. The best existing interpretation of crustal structure for the region is a fence diagram (Fig. 2), published by McCamy and Meyer (1964, Fig. 1). Only the shotpoints at Cliff Lake and Sailor Lake (M1 and M2 in Fig. 1) are within the Rocky Mountains proper.

In the fence diagram, the crustal section between Cliff Lake (M1) and Sailor Lake (M2) is divided into an upper crust 20 to 25 km thick with a 6.0-km/sec velocity, and a lower crust with a 7.4-km/sec velocity that thickens from 15 to 20 km in the north to 25 to 30 km in the south. Near the base of the lower crust, an additional high-velocity layer (7.6 km/sec) is indicated to decrease in thickness southward from about 5 km near Cliff Lake to 1 to 2 km near Sailor Lake. Total crustal thickness increases southward from 44 km near Cliff Lake to 50 km near Sailor Lake. Meyer and others (1961a) did not see the additional thin lower crustal layer at the base of the crust.

The two 1969 seismic profiles, recorded from a shotpoint at Greenbush Lake, British Columbia (BC1, in Fig. 1), were called the Rocky Mountain profile and the Great Plains profile by Hales and Nation (1973). The Rocky Mountain profile extends south-southeast, approximately following the continental divide near the Idaho-Montana border, and thus is nearly parallel to the profile from Cliff Lake to Sailor Lake. Crustal information is obtained only at 40 to 100 km from the shotpoint, which is about 100 km north of the Canadian border. Hales and Nation reported a crustal thickness of 37 km, in agreement with the northwestward shallowing of the Moho in western Montana found by McCamy and Meyer (1964).

The two studies differ considerably, especially in the interpreted crustal velocities. From amplitude studies, Hales and Nation inferred a velocity decrease from 6.5 to 5.8 km/sec at depths of 5 to 20 km in the upper crust. For the lower crust, they postulated a velocity of 6.4 km/sec, interpreting secondary large-amplitude phases as refracted waves from an intermediate crustal boundary. McCamy and Meyer (1964) generally found higher velocities in the lower crust (6.9 to 7.4 km/sec) in the northern Rocky Mountain profile. Their average velocity for upper crust is 5.8 to 6.1 km/sec. An additional intermediate layer (6.6 km/sec) is present only on the profile extending east from Cliff Lake.

The reliability of the early results is difficult to evaluate. Then, it was common to pick arrival times, plot them on a graph, and correlate a maximum number of arrivals by a series of straight lines. Intercept times and apparent velocities were used to calculate depth and dip for planar layers with discrete velocity discontinuities (e.g., Mota, 1954). In another study, in Mexico, Meyer and others (1961b) demonstrated the problem of correlating secondary arrivals. Depending on how many traveltime branches an author finds to fit the later arrivals, a more or less complicated internal structure results. In the most extreme models, crustal thickness differs by as much as 7 km, varying from a 37-km one-layer model to a 44-km three-layer crystalline crust. More recently, by the use and publication of record sections instead of simple arrival-time plots, the reliability and possible

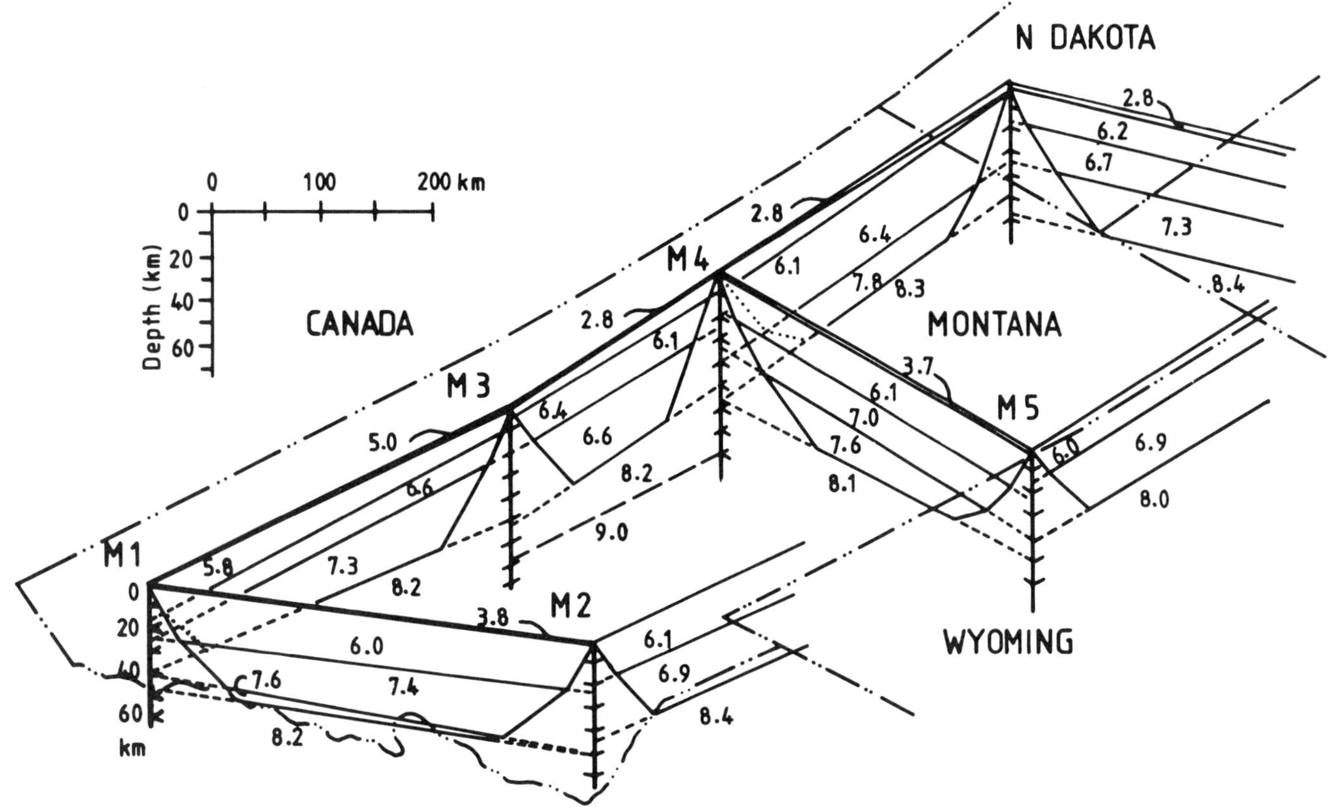

Figure 2. Fence diagram showing crustal model for northern Rocky Mountains and adjacent Great Plains in Montana (McCamy and Meyer, 1964, Fig. 1). M1, . . shotpoints; 6.0, . . . velocities in kilometers per second.

ambiguities of interpretation can be checked more easily by the reader. Today, secondary large-amplitude phases are being interpreted almost exclusively as wide-angle reflections.

The uncertainties of this early interpretive procedure are also illustrated in Figure 2, where velocities, numbers of layers, and even depths of layers do not correspond where they meet at a single shotpoint. The only reliable correlation is that of the first-arrival P phases with the resulting depth determination for the crystalline basement, and the general trend of dip and relative depth of the Moho. The absolute depth of the Moho, however, depends on the average crustal velocity used, which again depends on the internal crustal structure based on secondary arrivals, and may thus be uncertain by ± 5 km. Because all the models for the northern Rocky Mountains and adjacent Great Plains were based on the same interpretive procedures, however, the relative differences in crustal thickness may be fairly reliable.

The crustal thickness under the Great Plains of Montana is generally greater than under the adjacent northern Rocky Mountains, and the crust thins from southeast to northwest along the axis of the northern Rocky Mountains. The velocity structures of McCamy and Meyer (1964) have obvious similarities to those found by more recent crustal studies in the southern Rocky Mountains and adjacent Great Plains of Colorado. Accordingly, their crustal model, in spite of its uncertainties, is likely not too far from reality. The results of Hales and Nation (1973) are difficult to combine with the earlier crustal model for Montana: (1) the 1969 profiles were unreversed; (2) the interpretive method used by Meyer and others (1961a, b) did not permit low-velocity zones; and (3) as noted earlier, the Hales and Nation model applies to an area 100 km north of the Canadian border.

The general deepening of the crust-mantle boundary parallel to the continental divide from northwest to southeast coincides broadly with the increasingly negative Bouguer gravity (Lyons and O'Hara, 1982). A magnetic anomaly maximum in the area of Glacier National Park (Zietz, 1982) does not correlate in any obvious way with the available seismic data and corresponding crustal structure.

Seismic refraction investigations, aiming to calibrate the Large Aperture Seismic Array (LASA) located in the Great Plains of Montana, indicate a decreasing thickness of the upper crust (6.1 km/sec), from near the mountain front toward the east (Borcherdt and Roller, 1976). This geometry, which agrees with the earlier results of McCamy and Meyer (1964), may correlate with the higher heat flow to the west, under the Northern Rocky Mountains (Blackwell, 1978). All earthquakes reported for the Northern Rocky Mountains (Freidline and others, 1976; Sbar and others, 1972) have focal depths less than 18 km, and thus are well within the brittle upper crust.

In summary, the crust of the northern Rocky Mountains has been interpreted to be about 50 km thick, consisting of 20 to 25 km of upper crust and 25 to 30 km of lower crust. A transitional layer may be present between lower crust and upper mantle. No conventional crustal root is present. The differences in crustal structure, evidently, are mainly in the upper crust. Comparatively low velocities, high heat flow, and low Bouguer gravity coincide with high average elevation; seismicity is concentrated at depths of less than 20 km. These geophysical properties of the upper crust may be related to a mainly silicic composition and a brittle character through most of the upper crust.

YELLOWSTONE PLATEAU

The Yellowstone Plateau, consisting largely of silicic lavas and ash-flow tuffs of Pleistocene age, lies at the northeast end of the eastern Snake River Plain, where basaltic volcanism presently dominates. Together these features constitute a northeast-migrating locus of bimodal basaltic and rhyolitic volcanism that crosses the trend of the northern and middle Rocky Mountains. This locus has been interpreted by many as the trace of a mantle hot spot, which at present lies beneath the Yellowstone Park area (Smith and Christiansen, 1980). In contrast to the northern Rocky Mountains, the deep structure of the Yellowstone Plateau and the associated eastern Snake River Plain region has been intensively studied recently (Smith, 1982; Smith and others, 1982; Braile and others, 1982; Daniel and Boore, 1982; Pelton and Smith, 1982; Pitt and Hutchinson, 1982; Sparlin and others, 1982).

In 1978, 15 shots were recorded by nearly 200 stations, of which 8 shots and 150 stations were used to generate 5 profiles in the Yellowstone Park region (Smith and others, 1982). These observations include one reversed line (shotpoints Y1 and Y2, in Fig. 1), for which good records were obtained in the northeast direction to a distance of 180 km and in the southwest direction to a distance of 290 km. Both lines were recorded well beyond Yellowstone Park, in the Beartooth Mountains to the northeast, and in the eastern Snake River Plain to the southwest (Smith and others, 1982).

In a detailed analysis of the upper crustal structure, the average thickness of the crust was determined to be about 43 km, with a mean crustal velocity of 6.3 km/sec (Schilly and others, 1982). Depths to the top of the upper crustal crystalline basement vary from 5 km near Island Park to 1 km at the northeast side of the Yellowstone Plateau; this is interpreted as progressive thinning of the Quaternary volcanic rocks to the northeast. The upper crustal layer averages 10 km thick, with a velocity of 6.0 km/sec. In addition, the seismic data indicate a large lateral inhomogeneity in which P-wave velocities decrease at least 10 percent, interpreted as a low-velocity body beneath the northeastern Yellowstone Plateau.

Delay-time analysis for all crystalline basement P_g arrivals of the Yellowstone array (Lehman and others, 1982) yielded an interpretation of the three-dimensional velocity distribution and the upper crustal layering beneath the Yellowstone caldera (Fig. 3). Outside the caldera the crystalline upper crust has a velocity of 6.05 km/sec, but inside it the velocity decreases to 5.7 km/sec. This low-velocity area extends 15 km beyond the caldera to the northeast, in a distribution broadly similar to that shown by the Bouguer gravity low (Smith and Christiansen, 1980). Smaller areas of low P-wave velocity, 4 km/sec, occur beneath the northwestern caldera rim and in the southwestern caldera near Upper and Midway Geyser basins (Fig. 3). The 6.05-km/sec velocity is interpreted as characteristic of the Precambrian granitic gneisses exposed north and south of the Yellowstone caldera. The 5.7-km/sec velocity is interpreted as a 10-km-thick hot body of granitic composition. Alternative interpretations were offered for the 4.0-km/sec low-velocity bodies. The northeastern anomaly, which is associated with a –20-mgal gravity anomaly, may reflect a large steam-dominated system or a silicic body containing 10 to 50 percent partial melt. The southwestern low-velocity area, which lacks a corresponding gravity signature, may reflect a zone of high seismic attenuation associated with a water-saturated portion of the caldera ring-fracture zone.

Detailed crustal velocity-depth models (Fig. 3), derived from the two long southwest-northeast seismic profiles (Smith and others, 1982) indicate that beneath an approximately 2-km-thick near-surface layer having highly variable velocities (3.0 to 4.8 km/s), an upper crustal layer 3 to 4 km thick has velocities of 4.9 to 5.5 km/sec. An intermediate crustal layer 8 to 10 km thick (6.5-km/sec velocity) is underlain by relatively homogeneous lower crust (6.7 to 6.8 km/sec) about 25 km thick. Total crustal thickness is about 43 km, and the upper-mantle velocity below the Moho is 7.9 km/sec.

Smith and others (1982) emphasized that the P^* branches (later P arrivals from which the lower crustal properties were determined) are similar in apparent velocity, intercept times, and wave forms for all the profiles. A map of the migrated P^* refracting paths provides evidence that the P^* velocity of 6.8 km/sec beneath the caldera is similar to that outside the Yellowstone area, including the northern Rocky Mountains and easternmost Snake River Plain.

The upper-mantle structure beneath Yellowstone Park was investigated by observing teleseismic events, using an array of 26 telemetered seismic stations and three groups of portable stations operated within a 150-km circle around the Yellowstone caldera (Iyer, 1975, 1979). Maximum teleseismic delays were about 1.5 sec inside the caldera, and the delays were large over a 100-km-wide area around the caldera. These delays were interpreted by Iyer as due to the presence of low-velocity material with horizontal dimensions similar to the caldera (40 × 80 km) near the surface and extending to 200 to 250 km beneath the caldera. Iyer has estimated that the P-wave velocity within the anomalous body is lower than in the surrounding rocks by about 15 percent in the upper crust and by 5 percent in the lower crust and upper mantle.

A high-velocity middle crust (6.5 km/sec) extends from a depth of 10 to 20 km beneath the Yellowstone Plateau. In comparison, in the northern and middle Rocky Mountains (Figs. 2, 6), upper crustal velocities of about 6.0 km/sec reach depths of about 20 km. Similar relatively high velocities were observed in the Snake River Plain below a depth of 10 km (Braile and others, 1982; Fauria, 1981; Hill and Pakiser, 1967; Prodehl, 1979). Braile and others (1982) interpreted this intermediate-depth high-velocity layer as follows: Intense intrusion of mantle-derived basaltic magma into the upper crust generated explosive silicic volcanism, associated caldera collapse, and regional uplift. As the "hot spot" moved to the northeast, subsequent cooling of the intruded upper crust produced the 6.5-km/sec intermediate-depth layer. Notably, earthquake hypocenters along a profile across the Yellowstone caldera are shallower than 12 km above the high-velocity intermediate crustal layer (Smith and others, 1974). From this observation we would expect the brittle layer to be thinner under the Yellowstone area than under the Snake River Plain to the southwest.

A large negative residual gravity anomaly of –60 mgal, relative to the regional Bouguer anomaly field of –200 mgal, coincides with the caldera (Eaton and others, 1975). This anomaly can be interpreted entirely by low densities in the upper 16 km of crust: thin layers of rhyolite flows, caldera fill, and combinations of caldera fill, altered basement rock and rhyolite magma. Similarly, Lehman and others (1982) recalculated gravity models that agree with the detailed seismic velocity structure derived from the delay-time analysis, where all gravity anomalies are explained by velocity and density contrasts within the uppermost 10 km of the crust.

According to Smith and others (1982), the large lateral upper crustal inhomogeneities suggest that the thermal-mechanical processes have not yet equilibrated and that the degree of melting and magma composition could vary beneath the Yellowstone Plateau. In contrast, available data show little evidence for strong lateral variations in the seismic structures of the lower crust across the Yellowstone region and Snake River Plain (Braile and others, 1982; Smith and others, 1982), suggesting that the lower crust may have been little affected by the thermal processes in the upper crust or that thermal effects are counterbalanced by emplacement of dense mafic magma.

Iyer (1979) postulated that the low-velocity body of the upper mantle down to 250 km depth consists partly of molten rock, with a high proportion of partial melt at shallow depths, which is responsible for the observed Yellowstone volcanism. It cannot be distinguished whether the Yellowstone melting anomaly is associated with a deep source such as a plume, chemical plume, or gravitational anchor.

MIDDLE ROCKY MOUNTAINS

The middle Rocky Mountains consist largely of the Idaho-Wyoming overthrust belt: ranges of miogeoclinal Paleozoic and Mesozoic sedimentary rocks, which have been deformed into

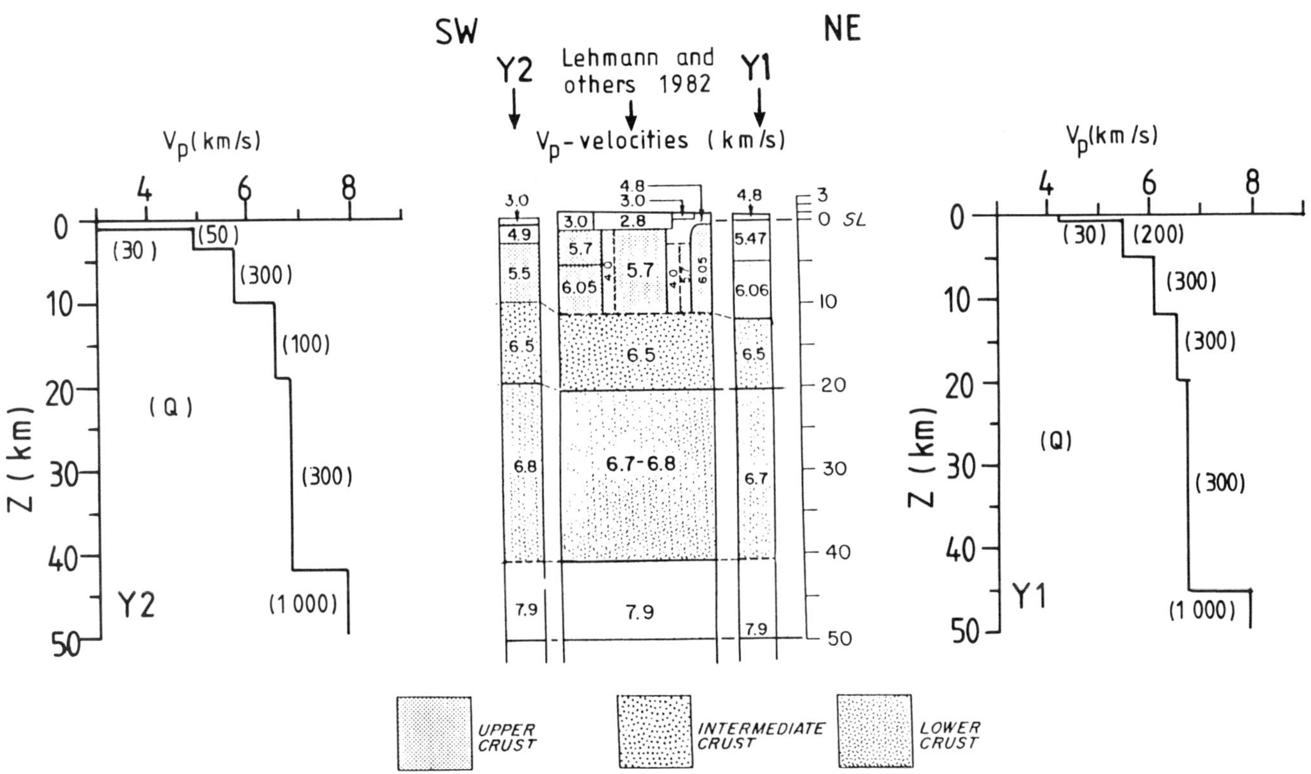

Figure 3. Generalized P-wave velocity model for the Yellowstone region and velocity-depth models for profiles 2 NE and 1 SW traversing the Yellowstone Plateau. (300), . . . apparent Q (attenuation); 5.7, . . . velocities in kilometers per second (Smith and others, 1982, parts of Figs. 6-8).

closely spaced folds and thrust slices without exposing any Precambrian basement (King, 1977). The crustal structure of this region has been studied by two seismic experiments (Fig. 1). In 1963 the U.S. Geological Survey obtained a reversed refraction line between Flaming Gorge Reservoir, Utah, and American Falls Reservoir, Idaho, with an intermediate shotpoint at Bear Lake (Willden, 1965). This line is crossed, close to its intermediate shotpoint, by a nearly perpendicular nonreversed 1971 refraction profile extending northeast from Bingham, Utah (Braile and others, 1974). Both experiments used modern high-precision instruments and obtained high-quality data, from which crustal structural information was derived. Both lines are 340 km long and extend into the Basin and Range province at their western ends.

A limitation of the Bingham line, having only a single shotpoint (U1, in Fig. 1), was partly overcome by using data from discrete distance intervals to derive separate structural models (Braile and others, 1974); thickness and velocity variations could thus be inferred, despite the flat-layered earth approximation required by the interpretive technique. For the Flaming Gorge–American Falls line, the data for each profile (Fig. 4) were first interpreted separately by a flat-layer earth approximation, then compiled into a two-dimensional cross section (Fig. 5; Prodehl, 1970, 1979). The traveltimes of the curves a, a-b, b, c, and d in Figure 4 were calculated from the model in Figure 5b, and are plotted on the record sections for comparison. Where they compare favorably, the model is supported. Where they are just weakly suggested by the data, the model is doubtful. In addition to the generalized model of Figure 5a, Figure 5b indicates which depth and distance ranges are actually controlled by the observed phases a, a-b, b, c, and d of Figure 4.

The model of Figure 5b might be refined by applying a ray-tracing technique, but the data do not justify more sophisticated modelling. At workshops held by the Commission of Controlled Source Seismology to check and compare interpretive techniques (Ansorge and others, 1982; Finlayson and Ansorge, 1984; Mooney and Prodehl, 1984), the problem of phase correlation (i.e., the quality of data) was shown to be the primary constraint on the reliability of a model, independent of interpretive method. It was also shown that, as soon as phase correlation is assured, an optimum fit of theoretical (i.e., calculated from the model) and observed position of phases to be interpreted can be obtained, if depth, velocity, and velocity gradient of a reflecting boundary zone as well as the velocity of the overlying layers are chosen carefully. The accuracy of velocity and depth determination is then on the order of ±0.1 km/sec and ±1 km, respectively (Jacob and others, 1985).

Crustal thickness of the middle Rocky Mountains is 40 km, averaging 10 km thinner than in the northern and southern Rocky Mountains. Typical basement velocities are found at depths of 4 km and more. A 5-km-thick low-velocity zone below a thin high-velocity layer at a depth of 15 to 20 km, where velocity decreases from 6.5 to 5.8 km, is typical for the 20-km depth range on the American Falls–Flaming Gorge line. This feature is not seen elsewhere in the Rocky Mountain region. An average velocity of 7.0 km/sec characterizes the lower crust between 20 to 25 and 40 km.

Toward Bingham, the crust appears to thin abruptly by about 10 km near the transition from the middle Rocky Mountains into the Basin and Range province (Braile and others, 1974). No analogous change in crustal thickness was detected between Bear Lake and American Falls (Fig. 5). This area in southeastern Idaho, which according to Fenneman (1946) is part of the Basin and Range province, is underlain by a crust 40 to 45 km thick, as is found also under the western and eastern Snake River Plain (Prodehl, 1970, 1979; Braile and others, 1982). Consequently, basin and range crust may not exist between Bear Lake and American Falls Reservoir, but terminates near the Idaho/Utah state border (Fig. 1).

An east-west cross section showing seismicity from Cache Valley to the Bear River Range, northern Utah, shows hypocenters clustered along a narrow zone beneath the western margin of the middle Rocky Mountains and confined to the upper crust at depths of 4 to 16 km (Smith and Sbar, 1974, Fig. 7). From these and other microearthquake investigations, Smith (1977) concluded that the northern intermountain seismic belt is a major zone of combined extension and shear strain, oriented subparallel to the San Andreas fault.

Braile and others (1974) reported that crustal thickening, from less than 30 km in the eastern Basin and Range province to more than 40 km in the middle Rocky Mountains, occurs 40 km east of the Wasatch Front rather than at it. According to Braile and others (1974, Fig. 12), the zone of crustal thickening coincides approximately with the eastern limit of Basin and Range faulting (Eardley, 1962), the center of the intermountain seismic belt (Smith and Sbar, 1974), the boundary between high and normal heat flow (Roy and others, 1972), the deepening of the Curie magnetic boundary (Shuey and others, 1973), and the center of a shallow high-conductivity layer in the upper mantle (Porath, 1971). Gough (1984) correlated the increased upper-mantle conductivity with heat flow and inferred that partly melted low-Q upper mantle underlies most of the central-western United States, from the Basin and Range province to the southern Rocky Mountains. Mantle upwelling was inferred beneath the Wasatch Front, as well as in the southern Rocky Mountains and Rio Grande rift. However, a more refined seismic investigation will be necessary to reveal details of crustal structure in the transition zone from the eastern Basin and Range province to the middle Rocky Mountains. Recent COCORP data, for example (see Fig. 6 in Allmendinger and others, 1987), suggest that the thin Basin and Range province crust does not extend east of the Wasatch Front, but that the crust thickens gradually from the Basin and Range province to the Colorado Plateau in the transition zone south of the Bingham-northeast line interpreted by Braile and others (1974). Pakiser (1985) reexamined the evidence for very low P_n velocities and for the eastward extension of the thin Basin and Range province crust in the vicinity of the Wasatch Front. He concluded that reported velocities as low as 7.4 to

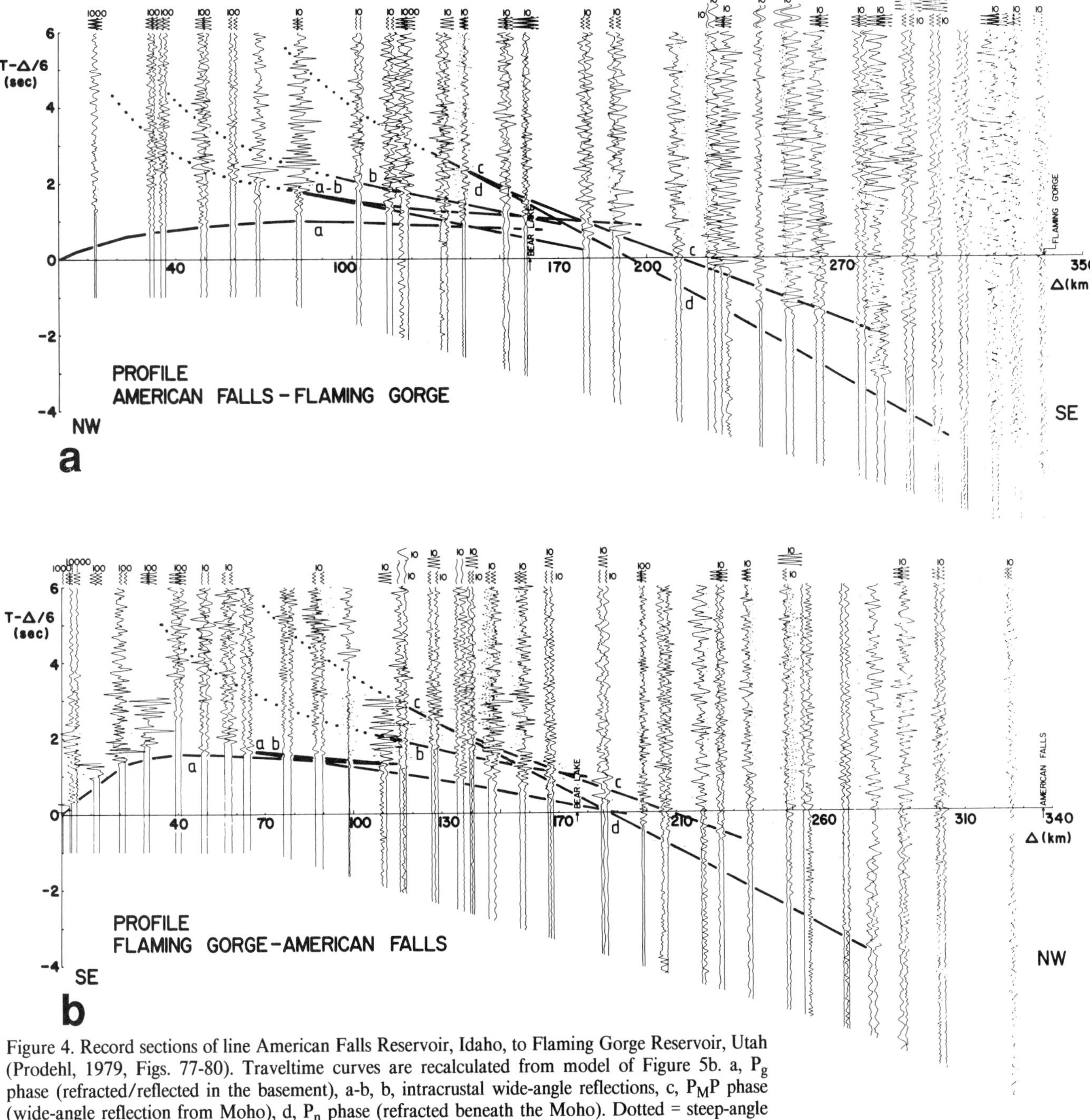

Figure 4. Record sections of line American Falls Reservoir, Idaho, to Flaming Gorge Reservoir, Utah (Prodehl, 1979, Figs. 77-80). Traveltime curves are recalculated from model of Figure 5b. a, P_g phase (refracted/reflected in the basement), a-b, b, intracrustal wide-angle reflections, c, P_MP phase (wide-angle reflection from Moho), d, P_n phase (refracted beneath the Moho). Dotted = steep-angle reflections. a, American Falls Reservoir to Flaming Gorge Reservoir. b, Flaming Gorge Reservoir to American Falls Reservoir.

7.6 km/sec near the Wasatch Front were really apparent velocities resulting from downdip propagation from a structural high centered near Bingham Canyon, and he expressed doubt that the thin crust extends east of the Wasatch Front.

The narrow low-velocity zone at a depth of 20 km, where velocity drops from 6.5 to 5.8 km/sec and which is not found elsewhere in the Rocky Mountain region, could be interpreted tentatively as the base of the Idaho-Wyoming overthrust belt. These thrusts may involve eastward gravity-caused movements, following vertical uplifts of silicic crust in Mesozoic to Cenozoic time, as Eardley (1968) proposed, but with the gliding plane at a much deeper crustal level. The lower crust has been left undisturbed, similar to the uniform lower crust of the adjacent Yellowstone Plateau and eastern Snake River Plain.

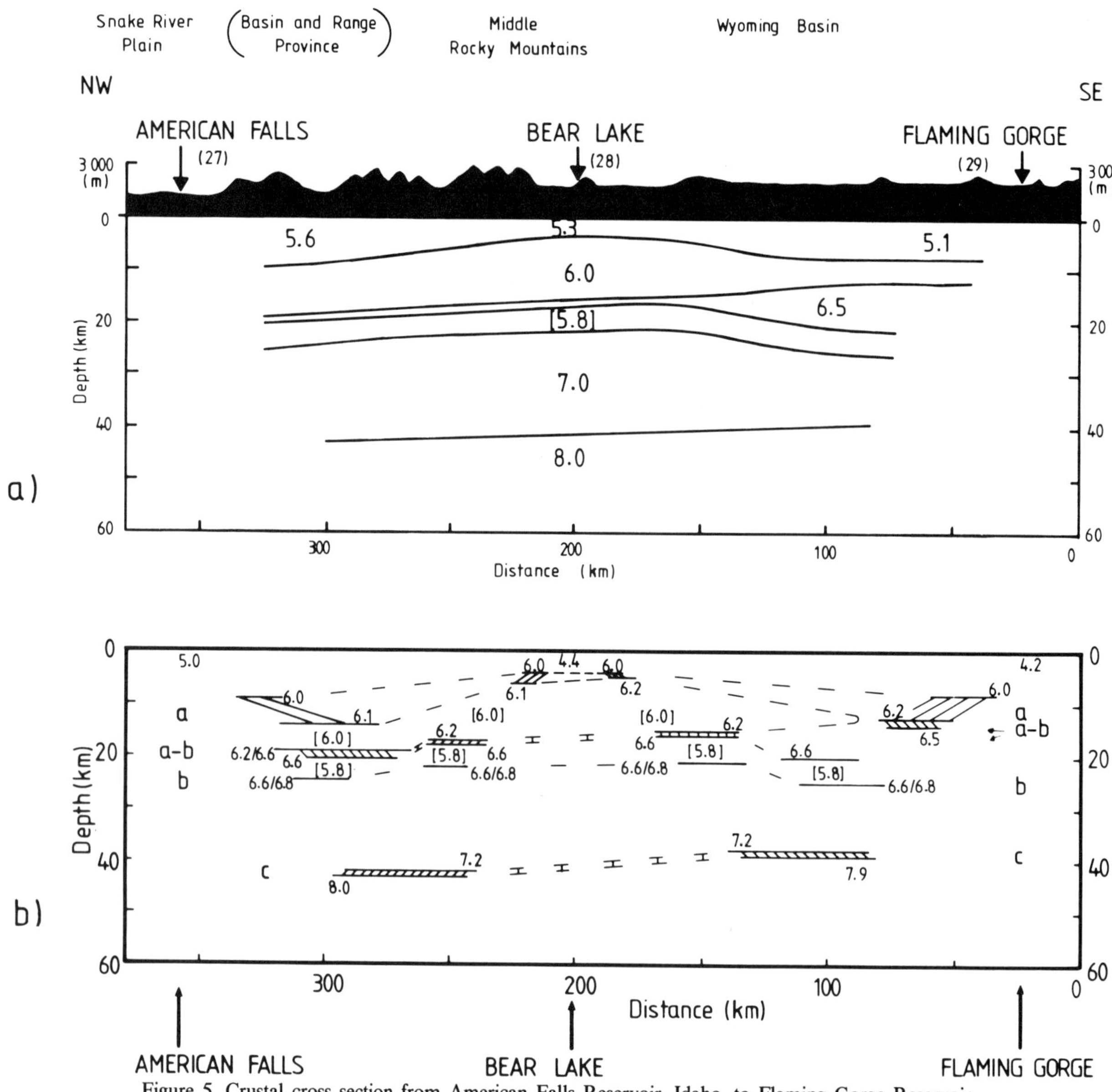

Figure 5. Crustal cross section from American Falls Reservoir, Idaho, to Flaming Gorge Reservoir, Utah. a, Generalized model. b, Main features of crustal structure. Here, depth ranges are marked by a, a-b, etc., from which the corresponding phases respectively reflections a, a-b, b, c, and d (correlated in Fig. 4a-b) are actually returned to the surface. Velocities in kilometers per second.

WYOMING BASINS

Vertical deformation during the Late Cretaceous–early Tertiary Laramide orogeny was intense in the Wyoming and Colorado foreland east of the Cordilleran geosyncline of the middle Rocky Mountains (Brewer and others, 1980). Here, Laramide basement was uplifted as much as 6 km, and adjacent basins subsided as much as 7.5 km (Gries, 1983) along deeply penetrating faults, whose character has been studied by seismic reflection methods by COCORP and by the petroleum industry. Many of the reflection surveys penetrate only slightly into crystalline basement and mainly reveal details of the sedimentary sequence, but COCORP surveys in 1976 and 1977 along the South Pass area at the southeastern end of the Wind River Mountains recorded reflections through much of the crust.

Seismic surveys also identified large thrusts, in which Phanerozoic sedimentary rocks underlie Precambrian basement along flanks of the uplifted blocks (Gries, 1983). An exceptionally documented example is the east-trending Uinta Mountains in Utah: Precambrian rocks of the mountain block are flanked by the Green River Basin to the north containing 7.5 km of Phanerozoic sedimentary fill, and by the Uinta basin to the south containing more than 9 km of sedimentary fill (Fig. 6). Similarly, the Granite Mountains in Wyoming are flanked by the Wind River basin to

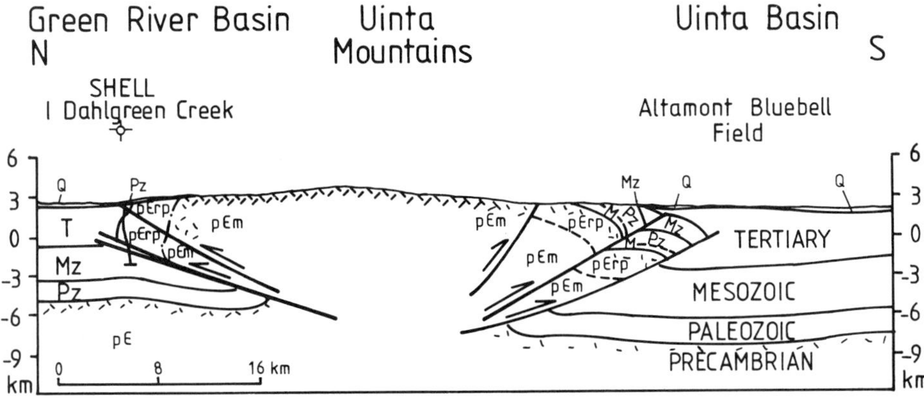

Figure 6. Simplified north-south cross section across the Uinta Mountains from the Green River Basin to the Uinta Basin, showing major movement on the thrust faults flanking the Uinta Mountains and sedimentary thickness of adjacent basins (Gries, 1983, Fig. 17). p€rp = Red Pine shale, p€m = Mutual Formation.

the north and the Hanna basin to the south, which contain more than 14.5 and 3 km of Phanerozoic sedimentary fill, respectively (Gries, 1983, Fig. 16). Other structures with similar thrust overlaps include the east-trending Owl Creek Range, the north-trending Laramie Range, the northwest-trending Wind River Mountains, and the Casper Arch thrust (Gries, 1983; Skeen and Ray, 1983).

Seismic profiles across the Casper Arch thrust and the Pacific Creek anticline, southwest of the Wind River Range, provide nearly continuous sections through Laramide basement uplifts, where anticlines or monoclines in the sedimentary section overlie reflective thrust faults in the Precambrian crystalline basement (Thompson and Hill, 1986).

None of the long refraction lines, which explore crustal structure to Moho depths, extend completely across the Wyoming basins and uplifts (Fig. 1). Some upper crustal properties, such as total sedimentary thickness and average velocity, can be determined, but almost no information on the lower crustal structure is available. The two seismic lines traversing the middle Rocky Mountains terminate to the east in the Green River basin. Basement rocks (6-km/sec velocity) extend as deep as 9 km (Prodehl, 1979), in agreement with more detailed seismic reflection studies (Hansen, 1965; Gries, 1983). An intracrustal discontinuity is present at a depth of 22 km; the total crustal thickness is about 41 km northwest of Flaming Gorge Reservoir (Fig. 5; Prodehl, 1979).

One refraction line traversing the southern Rocky Mountains terminates at shotpoint Sinclair in the Hanna Basin (35, in Fig. 1). The few data recorded in the basin indicate a depth to basement rocks of as much as 9 km north of the Sinclair shotpoint. The cross sections of Gries (1983, Figs. 16, 23) for two lines traversing the Granite Mountains show sedimentary thicknesses of about 3 and 6 km at the northern margins of the Hanna and Great Divide Basins, about 60 to 90 km north and northwest of shotpoint Sinclair.

Information on structures deeper within the crust is available only from two COCORP surveys (Fig. 1). One (W1-2) traverses the southeastern end of the Wind River Mountains, between the Green River and Wind River basins in southwestern Wyoming (Smithson and others, 1979; Brewer and others, 1980; Oliver, 1982; Lynn and others, 1983; Sharry and others, 1986). The other (W3-6) extends across the northern end of the Laramie Range and the flanking Laramie and Denver basins in southeastern Wyoming; it is discussed in the section on the southern Rocky Mountains.

The Wind River survey includes 158 km of deep seismic reflection profiling that recorded as much as 20 sec of two-way traveltimes (Fig. 7); the thrust fault flanking the Wind River uplift can be traced to at least a depth of 24 km, and possibly 36 km, with a fairly uniform dip of 30° to 35°. Additional reflectors present in the seismic data that are subparallel to the main Wind River thrust could represent other faults. At least 21 km of crustal shortening is interpreted along the thrust, but no large-scale folding is evident in the basement (Brewer and others, 1980; Smithson and others, 1979). The base of the sedimentary rocks beneath the thrust, though poorly defined in the data, is estimated at a depth of 9 km (Smithson and others, 1979).

These results were generally supported by Sharry and others (1986), who reprocessed the COCORP seismic line. They concluded that the dip of the Wind River thrust flattens at 22 km depth, and they interpreted two additional detachment zones at depths of 10 to 13 km and 33 to 34 km. They interpreted the crust below the Wind River thrust to be faulted into lens-shaped packages characterized by a duplex structure at a depth of 17 to 26 km. Smithson and others (1979) also interpreted a deep reflection structure at 6 to 8 sec, corresponding to a depth of about 20 km, as "a probably multiply deformed structure of a complexity typical of structures found in exposed basement rocks."

A detailed Bouguer gravity survey along the seismic line shows an increase from –250 mgal at the flank of the Green River basin to –160 mgal in the core of the Wind River Range; a crustal model combining both seismic and gravity data suggests that most of the Bouguer gravity anomaly is due to effects of Precambrian crust against low-density sedimentary rocks of the basins

(Smithson and others, 1979). In this model, dense material must also have been emplaced at depth beneath the uplift. The Moho does not appear as a strong or continuous feature in the reflection data, and the gravity model (Smithson and others, 1979, Fig. 18) extends only to a displaced lower crustal layer of high density (2.89 g/cm^3) at a depth of about 28 km, underlying upper crust (2.73 g/cm^3).

Wide-angle seismic data, recorded on a fixed array during the 1977 COCORP Wind River survey, have yielded a two-dimensional model (Fig. 8) with a depth range of 25 km, which extends from the center of the Wind River Mountains into the Wind River basin (Wen and McMechan, 1985). Besides mapping the top of the Precambrian basement underlying the Wind River basin and cropping out in the Wind River Range, these authors identified an intrabasement reflector at a depth of 5 km, a low-velocity region from 6 to 15 km, a high-reflectivity laminar zone (possibly an intrusive feature) at a depth of 15 km, a second low-velocity region from 17 to 23 km, and a laminar transition zone at a depth between 23 and 25 km that separates the middle crust (mean velocity, about 6 km/sec) from the lower crust (mean velocity, about 7 km/sec). This result seems to correspond to the conclusion (Sharry and others, 1986) that all visible deformation occurred above 33 km, mostly at a depth of 17 to 26 km.

In a 1969 seismic reflection survey, near-vertical reflections were observed with 3-km spreads at five sites along a 100-km line: one site in the Green River basin, another at South Pass, and three sites in the Wind River basin (Perkins and Phinney, 1971). Reflections were seen at times as great as 14 sec. The best reflectors were at about 11 km depth, but deeper reflectors include a possible arrival from about 35 km, interpreted as possibly the Moho.

From short-period Rayleigh waves, Keller and others (1976) showed an S-wave model for the upper crust of the Uinta basin. S-wave velocities are generally low to a depth of 8 km; they generally represent sedimentary fill, underlain by 9 km of upper crustal basement in which S-wave velocities are equivalent to 6.1- to 6.4-km/sec P-wave velocities.

In conclusion, complex crustal structures can be resolved in much detail, and thrust zones seen at the surface may be traced to great depths. This has been shown for the Wind River thrust, which was traced to the base of the upper crust but evidently does not continue into the lower crust. The result may be viewed as evidence that tectonic movements or thrusting may originate as deep as at the base of the upper crust. Here, eventually, a velocity inversion may be produced, as was discussed in the previous section, and was also indicated by Wen and McMechan (1985), who showed velocities as low as 5.6 to 6.0 km/sec below a depth of 20 km (Fig. 8) in the Wind River Uplift area. Wen and McMechan discussed as reasonable causes horizontal compression and thrusting, including possible local intrusion of high-velocity material (7.0 km/sec) into upper crustal levels near depths of 15 km.

SOUTHERN ROCKY MOUNTAINS

The crustal structure of the southern Rocky Mountains seems to be similar to that of the northern Rocky Mountains. An

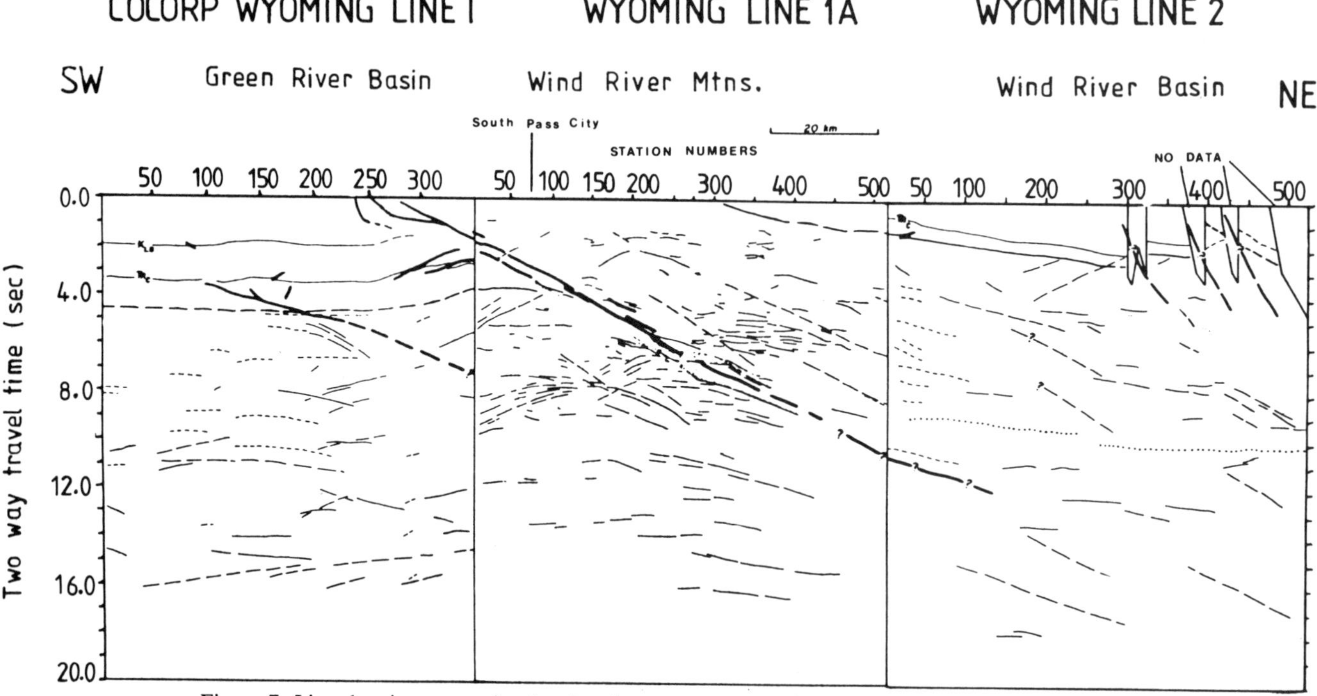

Figure 7. Line drawing composite showing the structure and reflections of the Wind River thrust depicted from the COCORP Wyoming lines 1, 1A, and 2 (Smithson and others, 1979, Fig. 12).

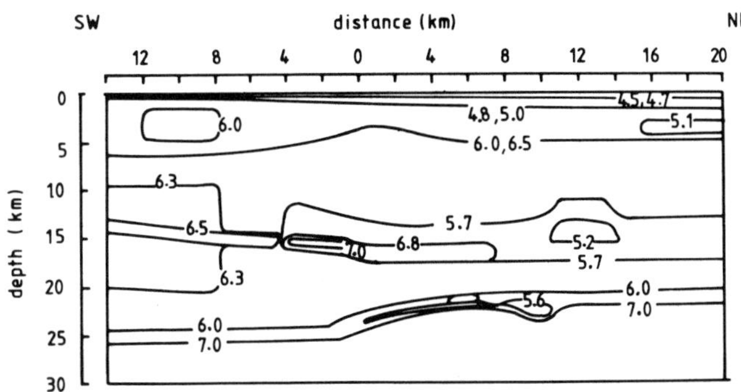

Figure 8. Wind River velocity model of wide-angle seismic data (Wen and McMechan, 1985, Fig. 6). Velocities in kilometers per second.

upper crust, about 25 km thick, with an average velocity near 6.0 km/sec, overlies an equally thick lower crust. Because of more recent data, details of the velocity structure of the upper and lower crust can be determined more accurately than that in the northern Rocky Mountains. Although the data are not perfect, the mean crustal velocity here is lower than in the middle and northern Rocky Mountains. The Moho is found at depths ranging from 48 to 52 km in the south to 40 km north of the Colorado-Wyoming state border.

The southern Rocky Mountains are a structurally complex region of north-trending Laramide basement uplifts separated by shallow basins locally referred to as "parks." The basement of the southern Rocky Mountains is mostly early Proterozoic (1.6 to 1.8 Ma). A major Precambrian age boundary near the Colorado-Wyoming line, named the Cheyenne belt (Karlstrom and Houston, 1984), separates Proterozoic from Archean basement. Much of the southern Rocky Mountains were blanketed by andesitic to rhyolitic volcanic rocks in middle Tertiary time. The present elevated topography of the region is largely due to late Cenozoic regional uplift associated with extensional deformation and development of the Rio Grande rift axially within the southern Rocky Mountains.

The area has been studied seismically more thoroughly than the remainder of the Rocky Mountain region. A 1979 COCORP reflection survey crosses the Rocky Mountain front at the Laramie Range in southern Wyoming (Allmendinger and others, 1982; Brewer and others, 1982; Johnson and Smithson, 1986). Two refraction profiles run from the southern Rocky Mountains into the Great Plains: a reversed profile across the Laramie Range from Sinclair to Guernsey, Wyoming, and an incomplete profile across the Front Range, between Wolcott and Agate, Colorado. A nonreversed line from Lamar, Colorado, reaches into the southern Rocky Mountains, but crustal information from it is limited to the Great Plains province (Haggag, 1974; Prodehl, 1977). The subparallel north-trending ranges west of North, Middle, and South Parks are covered by a longitudinal multiply reversed and overlapping refraction line from the Hanna basin in southern Wyoming across the San Juan Mountains near the Colorado–New Mexico border (Prodehl and Pakiser, 1980). Parallel to that line, a nonreversed profile extends from Climax, Colorado, through the Front and Laramie Ranges (Jackson and Pakiser, 1965; Prodehl and Pakiser, 1980).

The COCORP reflection profiles across the Laramide Mountains in southeastern Wyoming (W3-6, Fig. 1) show the presence of several en echelon reflectors dipping west 20° to 50° and traceable to depths of 10 to 12 km (Fig. 9); some may be traced into surface faults flanking the Laramie Mountains (Brewer and others, 1982). These reflectors were interpreted as thrust faults; their distributi on and orientation suggest that the mountains were uplifted by horizontal crustal shortening during the Laramide orogeny. None of the reflectors is as striking as the Wind River thrust, perhaps because of smaller relative movements on the Laramide frontal faults, on the order of 3 km or less. The geometry of faulting below 10 to 12 km is obscure from the COCORP data. According to Brewer and others (1982), the moderate dip of the thrusts in the upper 10 to 12 km and their inferred distribution under the mountains and basins indicate that horizontal compression was the dominant mode of deformation in the upper crust.

A major crustal feature traced in the COCORP data is the Mullen Creek–Nash Fork shear zone, part of the boundary between Archean and Proterozoic crust (Cheyenne belt), which seemingly controls a north-south lateral variation in Laramide tectonic style (Allmendinger and others, 1982). The reflection data from the Laramie Mountains and the Laramide basin suggest that this shear zone dips about 55° to the southeast (SZ, in Fig. 9). Northwest of the shear zone, the seismic basement is also characterized by southeast-dipping features. Complex reflections extend to 15 km or deeper under the Laramie basin; these are interpreted as structural or erosional truncations in metasedimentary rocks overlying the Archean basement (Allmendinger and others, 1982). Short, discontinuous reflections 35 to 40 km deep north of the shear zone, and laterally continuous flat reflections south of the shear zone at about 48 km were interpreted as the crust-mantle transition.

An unpublished refraction line, recorded by the U.S. Geological Survey in 1965, crosses the Laramie Range 40 to 50 km north of the COCORP survey; shotpoints are Sinclair in the Hanna basin (see previous section) and Guernsey in the Denver basin on the Great Plains to the east (35 and 39, in Fig. 1). A varying thickness of sedimentary rocks is indicated by the first arrivals of the P_g wave (Fig. 10), although these have not been evaluated in detail. The traveltime curves shown on the record sections in Figure 10, which are calculated from the model in Figure 11b, suggest that some of the larger-amplitude phases may be reflections from a midcrustal boundary (b) and from the Moho (c). Corresponding arrivals are only clear on some of the records, and therefore the correlation of these phases is not certain. P_n phases are only seen to distances of 180 to 190 km. Evidently, the upper crust is thinner west of the Laramie Range than to the east (Fig. 11). The crust is 38 km thick under the Laramie Range and surrounding basins, which may correspond to weak deep events

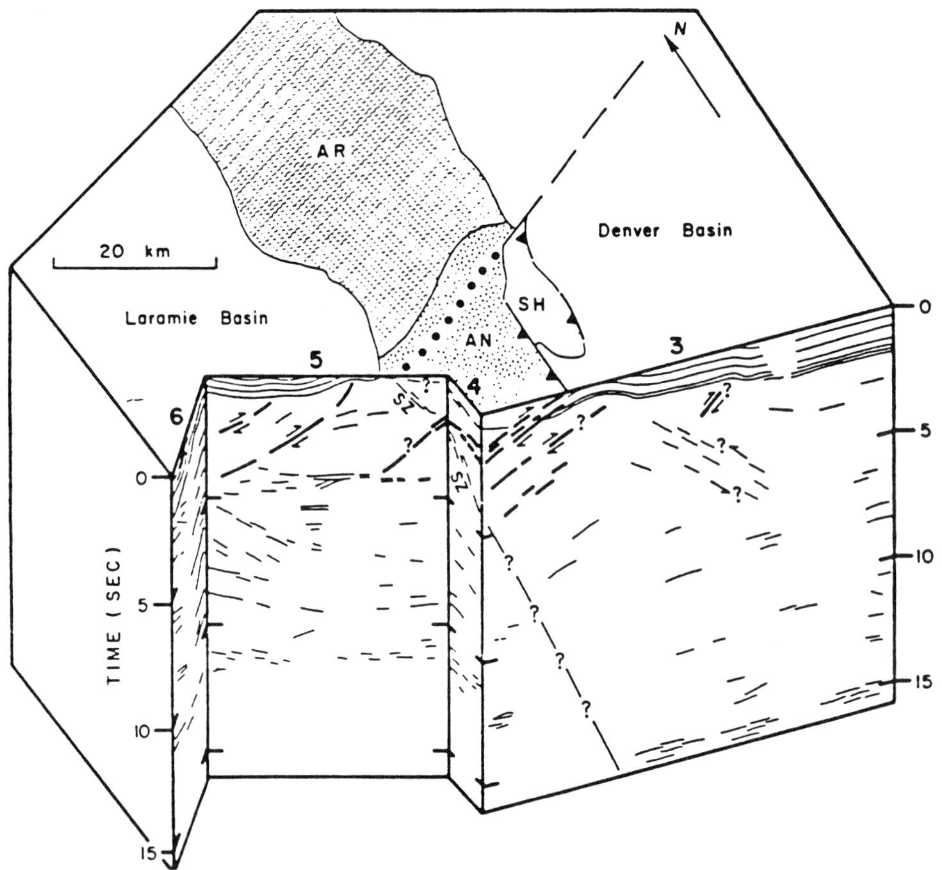

Figure 9. Block diagram showing major interpreted features of the Laramie Mountains, from the COCORP Wyoming lines 3-6 (Brewer and others, 1982, Fig. 12). SZ = inferred position of Mullen Creek–Nash Fork shear zone (dashed and dotted at surface). AN = anorthosite; SH = Sherman Granite; AR = Archean basement.

at 11 to 12.5 sec on COCORP lines W4-6 (Allmendinger and others, 1982). The Guernsey record section contains no evidence for deeper events corresponding to the continuous band of events at 15 to 16 sec on the eastern part of COCORP line W3, which is about 50 km south of the refraction line.

In the Front Range region of Colorado, a very shallow reflection survey indicates that steep reverse faults flank the mountains north of Golden (Davis and Young, 1977). Data were recorded only up to a few seconds, and the short lines barely cross into exposed Precambrian rocks.

A previously unpublished and incomplete seismic refraction profile, recorded by the U.S. Geological Survey in 1965, extends east across the Front Range between shotpoints at Wolcott in the Rocky Mountains and Agate in the Great Plains (Fig. 1). This profile crosses the Climax line only 10 km north of Climax. Stations were located at 130 to 280 km from shotpoints to record deep crustal reflections (Fig. 12); because observations are lacking at P_g distances, upper crustal structures can only be estimated. It is assumed that the upper crust beneath the Front Range should have a velocity structure similar to that of the neighboring Sawatch Range (see below, Fig. 15), and that the average velocity beneath the Front Range is also near 6 km/sec (Fig. 14). The average velocity under the Great Plains is taken from the Lamar lines, 150 km to the south (Haggag, 1974; Prodehl, 1977). The base of the upper crust may be near 20 km beneath the Great Plains and near 30 km beneath the Front and Sawatch Ranges. Reflections from a midcrustal boundary (phase b, in Fig. 12), are clear on both record sections. Scarce crust-mantle reflections (phase c) suggest Moho depths of 48 km under the Front Range and 50 km beneath the Denver basin near the city of Denver (Fig. 13). A weak P_n phase can be seen in both directions. The crustal structure obtained on the line from Lumberton to Sinclair near Wolcott (see below, Fig. 15) is similar to that from Wolcott to the east. It is therefore suggested that the crust near Climax is similar (dashed lines in Fig. 13a), but decreases in thickness farther north (Fig. 14).

A nonreversed refraction line through the northernmost Front Range and the Laramie Range (north from 36, in Fig. 1), based on blasting at the Climax molybdenum mine, was observed at shotpoint distances of 120 to 360 km (Jackson and Pakiser, 1965; Prodehl and Pakiser, 1980, Fig. 7). The data show exceptionally good reflections b and c from an intracrustal boundary

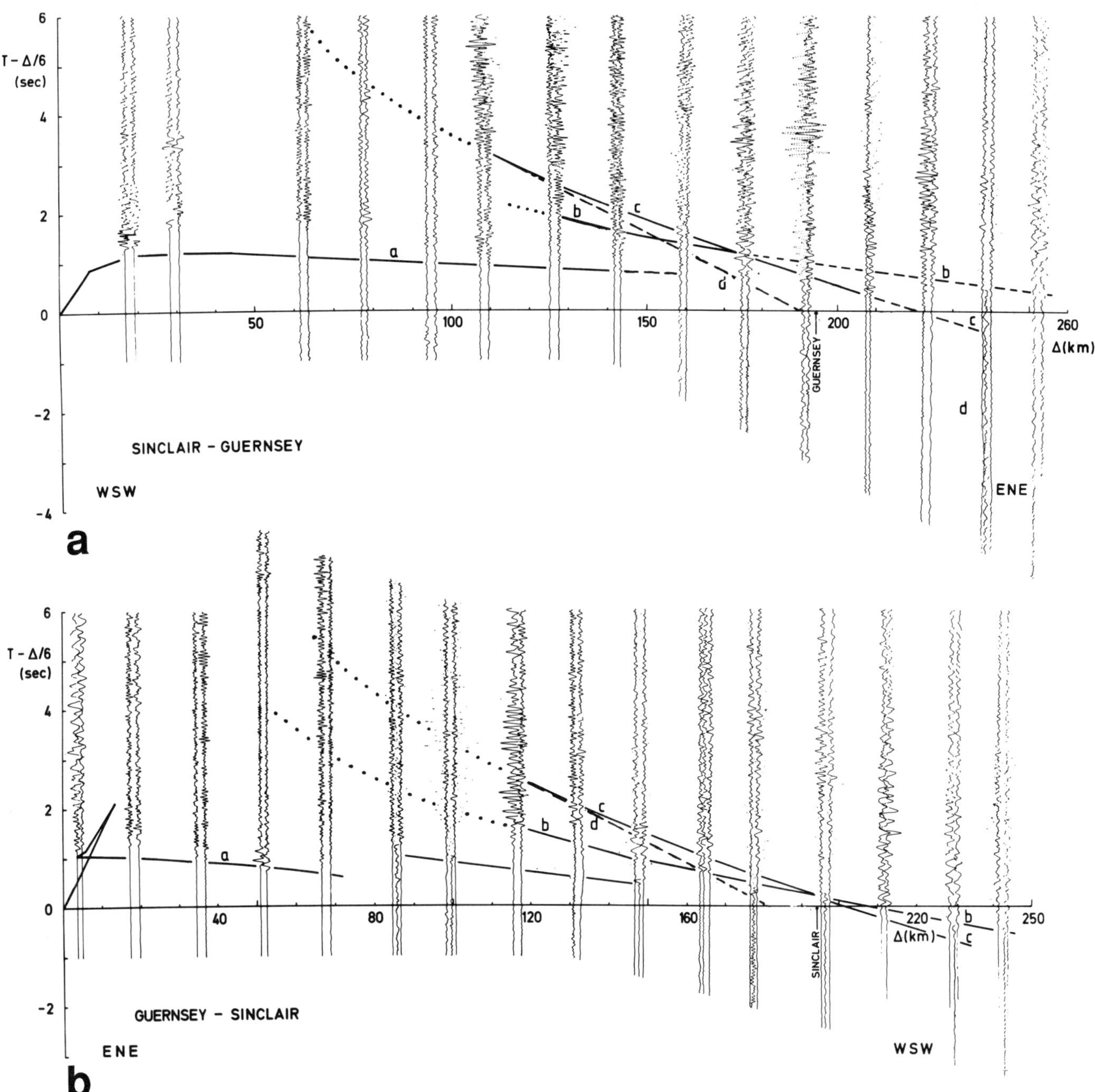

Figure 10. Record sections of the line from Sinclair to Guernsey, Wyoming. Traveltime curves are recalculated from model of Figure 11b. Explanation of correlated phases in Figure 4. a, Sinclair to Guernsey. b, Guernsey to Sinclair.

near a depth of 34 km and the Moho at 52 km, assuming an average velocity of 6 km/sec for the upper crust to a depth of about 30 km (Jackson and Pakiser, 1965). The lower crust has a mean velocity of 6.7 km/sec. The crust-mantle transition is 46 to 52 km deep in northern Colorado, but early arrivals beyond a distance of 190 km (Jackson and Pakiser, 1965) suggest that the crust thins to about 40 km beneath the Laramie Range, which is in agreement with the COCORP results (Brewer and others, 1982) and the Sinclair-Guernsey refraction line.

A generalized longitudinal crustal section along the Laramie and Front Ranges, based on all available seismic data, shows an abrupt decrease in crustal thickness near the northern end of the high mountains of the Front Range (Fig. 14). This change is in agreement with the approximate beginning of early P_n arrivals in the Climax-N record section (Jackson and Pakiser, 1965). Based on this observation and in slight disagreement with Allmendinger and others, who placed the change of crustal thickness north of Laramie where the Cheyenne belt crosses the Laramie Range

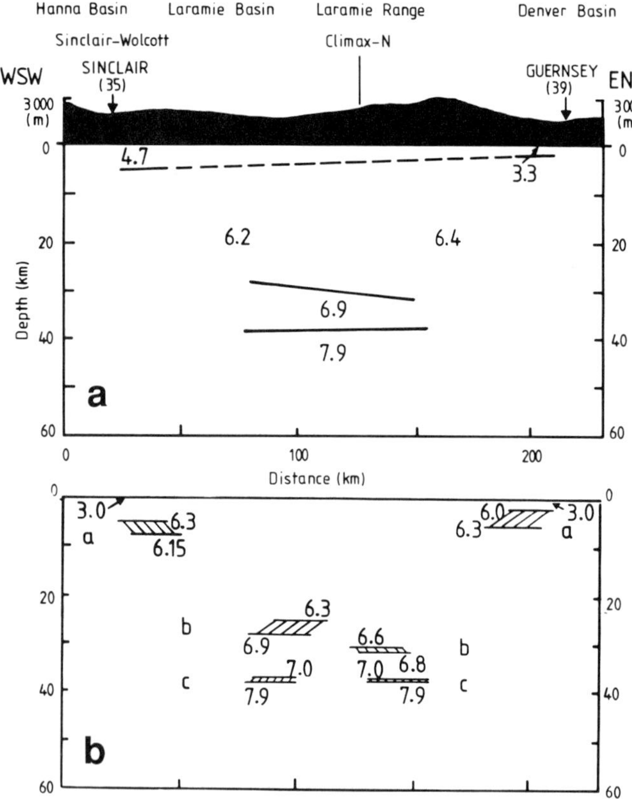

Figure 11. Crustal structure from Sinclair to Guernsey, Wyoming. Velocities in kilometers per second. a, Generalized model. b, Cross section of Figure 11a showing at which distance ranges reflectors resp. boundaries with large positive velocity gradient (dashed ranges) are actually controlled by the phases b and c, as correlated in the corresponding record sections of Figure 10.

(Johnson and others, 1984, Fig. 1; Allmendinger and others, 1982, Fig. 11), we interpret the boundary between thin (39 km) and thick (48 km) crust to be about 50 km farther south, crossing the Laramie Range near the Colorado-Wyoming border. Such a relation also agrees better with the 40-km-thick crust beneath the Park Range (Fig. 15), discussed below.

The most complete refraction survey is the line from Sinclair, Wyoming, to Lumberton, New Mexico (Fig. 1), traversing the Park Range, Sawatch Mountains, La Garita Mountains, and San Juan Mountains of western Colorado (Prodehl and Pakiser, 1980). The data, recorded in 1965, are based on intermediate drillhole shots at Wolcott and Cochetopa, as well as the endpoints. The quality of the records is variable, especially at large distances. The data are especially poor on the 200-km-long Sinclair-Wolcott section and the crustal section (Fig. 15) lacks resolution north of Wolcott. On the Wolcott-Lumberton section, several stations were closely spaced near the wide-angle distance range for reflections from the Moho, permitting detailed observations for this distance range, and allowing a phase-to-phase correlation of the most important wide-angle reflections (Prodehl and Pakiser, 1980, Figs. 2-5).

The resulting crustal section (Fig. 15) shows four units:

1. The uppermost crust is characterized by a gradually increasing velocity, from 4 to 5 km/sec near surface to 6.1 to 6.2 km/sec at depths of 4 to 9 km.

2. The main upper crustal unit extends to depths of about 28 km. Velocities are generally low, about 6 km/sec, but a thin, higher velocity layer (6.2 km/sec) between 12 and 15 km depth is suggested from Lumberton to north of Wolcott.

3. The lower crust is not separated from upper crust by a distinct reflection-producing boundary, in contrast with that in the Climax profile across the Front Range. Instead, the lower crust is marked by a zone of increasing velocity, from about 6 km/sec at the base of the upper crust at a depth of about 28 km, to 7.1 to 7.3 km/sec just above the Moho.

4. The Moho rises from 48 km between Wolcott and Lumberton to only 40 km north of Sinclair, but the site of crustal thinning could not be located closely from the available data.

The interpretation of relatively thin crust beneath Wyoming and thicker crust beneath Colorado also is consistent with scattered traveltime information from individual stations that recorded the Climax explosions: the P_n arrivals for the VELA-Uniform stations at Vernal, Utah, and Raton, New Mexico, are late compared to arrivals recorded along the Laramie Range. In contrast, the arrival at Durango, southwestern Colorado, is early, suggesting either a decreased crustal thickness or an increased average velocity west of the Front Range (Jackson and Pakiser, 1965; Prodehl and Pakiser, 1980, Fig. 7).

Despite the limitation that some lines were recorded only at large distances, an attempt has been made to present the crustal structure of the southern Rocky Mountains in a three-dimensional fence diagram (Fig. 16). It is viewed at the same angle as that published for the Cordilleran United States west of the Rocky Mountains (Prodehl, 1970, 1979). The low velocity of about 6 km/sec, which is suggested for the upper crust below the high mountains of central Colorado, may not extend north of the Front Range or beneath the Laramie Range. The mean velocity of the upper crust increases toward the Great Plains. The mean velocity of the lower crust is also lower under the high mountains than outside. Crustal thickness is greatest beneath central Colorado and decreases northward; it remains constant across the Rocky Mountain front into the Great Plains.

As has been discussed above for the middle Rocky Mountains (Fig. 5), the models shown in Figure 16 might be refined by applying a ray-tracing technique, but the data do not justify more sophisticated modeling. The problem of phase correlation (i.e., the quality of data) is the primary constraint on the reliability of a model, independent of interpretive method.

As the models of Johnson and others (1984) show, rise of the Moho by about 10 km along the line from Lumberton to Sinclair is compatible with a 50-mgal decrease in the Bouguer gravity anomaly, which averages about –290 mgal between Lumberton (32, in Fig. 1) and Wolcott (34), in contrast to about –240 mgal between Sinclair (35) and Wolcott (Fig. 26; Lyons and O'Hara, 1982). Similarly, thinning of the crust from 52 to 54

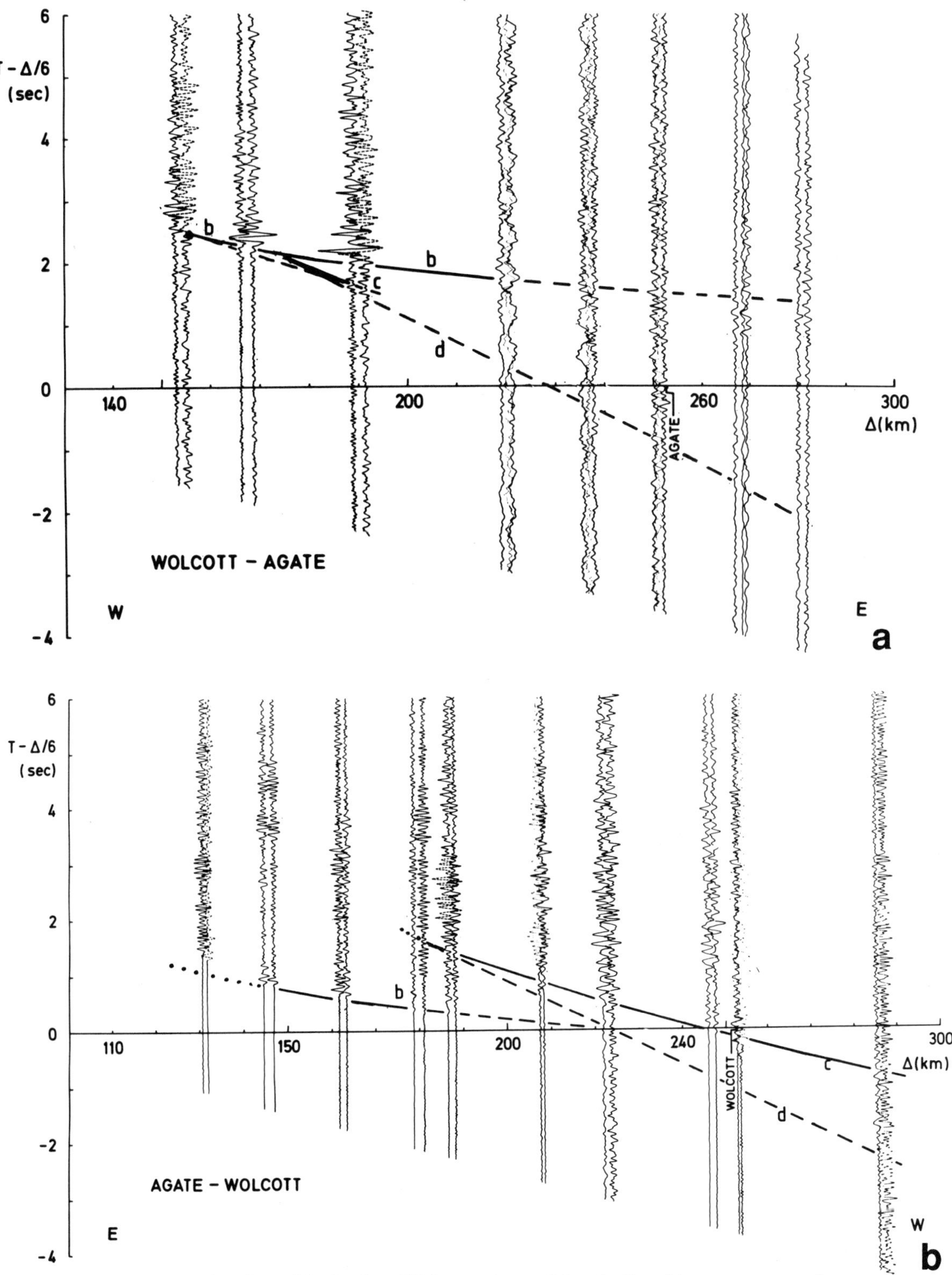

Figure 12. Record sections of the line from Wolcott to Agate, Colorado. Traveltime curves are recalculated from model of Figure 13b. Explanation of correlated phases on Figure 4. a, Wolcott to Agate. b, Agate to Wolcott.

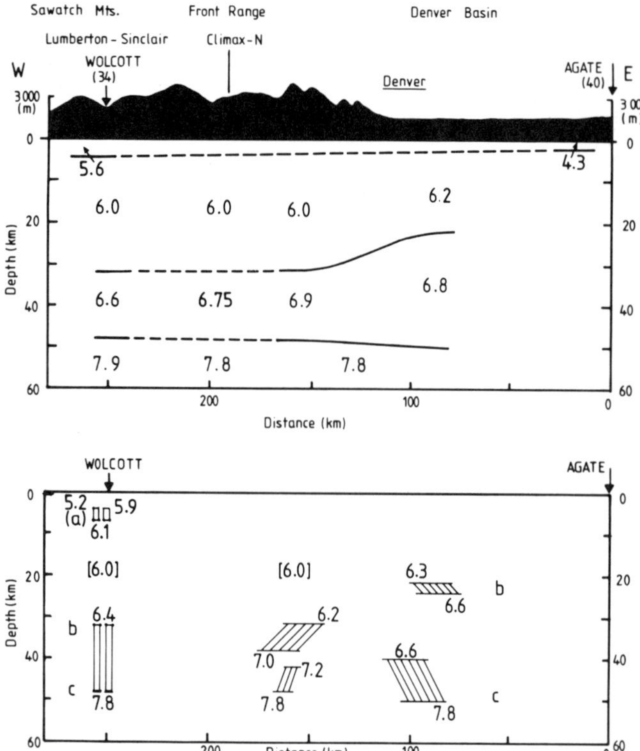

Figure 13. Crustal structure from Wolcott to Agate, Colorado. Velocities in kilometers per second. a, Generalized model. b, Cross section of Figure 13a showing at which distance ranges reflectors resp. boundaries with large positive velocity gradient (dashed ranges) are actually controlled by the phases b and c, as correlated in the corresponding record sections of Figure 12.

km in the Front Range to about 37 km in the Laramie Range is compatible with a decrease of about 100 mgal in the Bouguer anomaly: about –290 mgal in the Front Range north of Climax (36, in Fig. 1), and about –190 mgal in the Laramie Range. The lower average surface altitudes north of Wolcott and in the Laramie Range (Fig. 27; Diment and Urban, 1981) also suggest that the region is isostatically compensated (Prodehl and Pakiser, 1980). This change of geophysical parameters is obviously connected to the presence of the Precambrian Cheyenne belt (Karlstrom and Houston, 1984), but the available seismic data indicate that the geologically determined location of the Cheyenne belt coincides with crustal thinning in the area of the Sierra Madre, yet disagrees by about 50 km in the area of the Laramie Range.

The gravity profiles of Johnson and others (1984) across three Precambrian-cored Laramide uplifts within the Cheyenne belt in Wyoming show a northward increase in gravity values of 50 to 100 mgal. The change in crustal thickness implied by this increase is thought to represent a Proterozoic structure, rather than a Laramide feature, because the gravity gradient is spatially related to the Cheyenne belt, which has been immobile since about 1,650 Ma. The northward gradient is also perpendicular to gravity gradients associated with other local Laramide uplifts and subperpendicular to regional, longer wavelength Laramide gradients (Johnson and others, 1984).

While the broad variations in Bouguer gravity are generally compatible with the gross crustal structure in the southern Rocky Mountains, such as the low mean crustal and upper-mantle velocities and the thick crust in central and southern Colorado (Figs. 24, 25), some variations may be related to more local geologic features (Prodehl and Pakiser, 1980). Thus, the pronounced gravity low in the San Juan Mountains cannot be explained in terms of isostatic compensation of a crustal root. Rather, a low-density igneous complex within the upper crust, consisting of numerous coalescing calderas underlain by a concealed batholith, may be the principle cause of the gravity low (Plouff and Pakiser, 1972). Similarly, a gravity low associated with the Colorado Mineral belt, which extends northeastward from Climax to Denver, is probably the expression of a silicic batholith of Tertiary age (Case, 1967; Tweto, 1975).

Heat-flow measurements indicate that the upper crust in the southern Rocky Mountains is a zone of high temperatures, supporting the seismic and gravity evidence for low mean velocity and density of the upper crust (Blackwell, 1978). The low velocity and high temperature probably are related to widespread Cenozoic magmatism, which may have been initiated in Late Cretaceous time by the migration of a subduction zone into the region (Lipman and others, 1971).

As noted earlier, the region of high electrical conductivity and heat flow, attributed to a partly melted, low-Q upper mantle beneath much of the western United States (Gough, 1984), terminates eastward along the eastern front of the southern Rocky Mountains and Rio Grande rift. In this region the high-conductivity layer rises to about 45 km (near the base of the crust), similar to that observed along the Wasatch Front (Porath, 1971). This upwelling should cause a decrease in P_n velocity beneath the Moho and may explain why P_n phases are weak or missing in the southern Rocky Mountains.

In summary, the crust of the southern Rocky Mountains is similar to that of the northern Rocky Mountains: upper and lower crust are each about 25 km thick. The Moho is found at a depth of 50 km throughout Colorado, but rises to 40 km in Wyoming.

In contrast to seismic crustal studies in the Hercynian and Alpine mountain systems of central Europe (e.g., Gajewski and Prodehl, 1985, 1987; Hirn and Perrier, 1974; Prodehl, 1977, 1981, 1984), no substantial velocity inversion is present beneath the southern Rocky Mountains. Thus, the observed increased heat flow cannot be attributed to any widespread partial melting in the upper crust, but may rather be due to increased heat production caused by a low-density upper crust of silicic composition. The extreme negative Bouguer gravity strongly supports the suggestion that rocks of silicic composition extend to depths of 25 km. The lack of major natural seismicity, disregarding the areas with induced seismicity near Rangely and near Denver, Colorado, may suggest that, because of its high temperature, the upper crust is less brittle than normal.

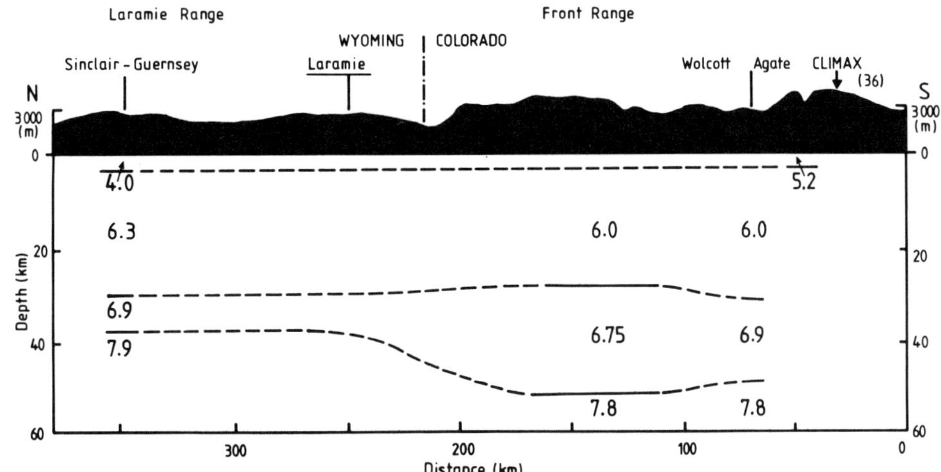

Figure 14. Generalized crustal structure from Climax, Colorado, toward Douglas, Wyoming, through the Front and Laramie Ranges. Dashed lines = estimated interpolation.

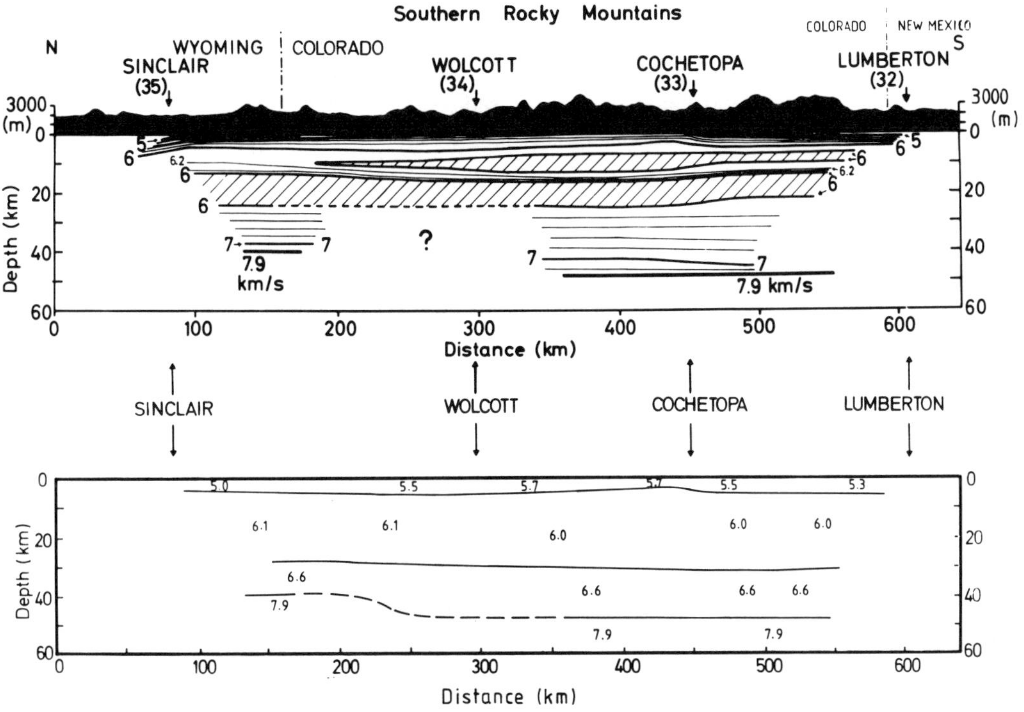

Figure 15. Crustal cross section from Sinclair, Wyoming to Lumberton, New Mexico. Velocities in kilometers per second. a, Cross section showing lines of equal velocity (Prodehl and Pakiser, 1980, Fig. 6). b, Generalized crustal structure.

The Cheyenne belt marks a change in crustal structure. Its surface expression agrees well with a change in crustal structure at the Sierra Madre, but is 50 km farther north in the Laramie Range than the suggested crustal thinning. The apparent contradiction could be solved if one assumed that the suggested continuation of the Mullen Creek–Nash Fork zone to the west changes its angle of incidence from less than 45° in the Laramide Range to near 90° in the Sierra Madre.

RIO GRANDE RIFT

The main characteristic of the Rio Grande rift is its thin crust: 40 km in the north and less than 30 km in the south. Details of the change in crustal structure beneath southern Rocky Mountains and Rio Grande rift are not known. The velocity in the 20-km-thick upper crust averages around 6.0 km/sec; velocity inversions evidently do not exist. A possible exception is within a proposed magma chamber near a depth of 20 km in the middle of the rift near Socorro, New Mexico, but velocity values there are not known. The change in crustal thickness results from a decreased thickness of the lower crust; its average velocity is 6.4 to 6.5 km/sec in the north and center of the rift and 6.6 to 6.7 km/sec in the south. Upper-mantle velocity is 7.7 to 7.8 km/sec. In the central part of the rift the crust-mantle boundary beneath the axis of the rift is 10 km shallower than for the neighboring areas of the Great Plains to the east and the Colorado Plateau to the west. Farther south, the depth variations become asymmetric: crustal thickness increases 20 km toward the Great Plains to the east, but less than 5 km toward the southern Basin and Range province to the west.

The predominantly Neogene Rio Grande rift projects from New Mexico northward into Colorado through the core of the southern Rocky Mountains, constituting an extensional overprint on Laramide compressional tectonic features (Tweto, 1975, 1979; Eaton, 1986). To the south, the rift merges with the eastern Basin and Range province. En echelon rift grabens accumulated as much as 6 km of sedimentary fill, interleaved with variable volumes of dominantly basaltic volcanics. Rifting began in late Oligocene time as a broad diffuse zone of distributed extension within the area of continental-arc volcanism that blanketed the southern Rocky Mountains; this episode of within-arc extension is interpreted as an intracontinental analog of extensional processes associated with circum-Pacific oceanic arcs (Elston and Bornhorst, 1979; Lipman, 1980). After middle Miocene time, the rifting became restricted to a narrower zone, and the strain geometry changed, apparently reflecting termination of convergence at the western margin of the American plate. Because of the close tectonic and geographic interplay between the northern half of the rift and the southern Rocky Mountains, we review the crustal structure of the entire Rio Grande rift, even though some of the highest quality structural data were obtained south of the Rocky Mountains proper.

The structure of the Rio Grande rift has been investigated by several refraction profiles. In central New Mexico, the nonreversed line DICETHROW III, recorded in 1976, runs north-south for 300 km (Olsen and others, 1979). Subparallel to and west of this line, a 550-km profile was recorded with a total of 16 stations from the GASBUGGY nuclear test in 1967 (Toppozada and Sanford, 1976). The upper crust of the Jemez Mountains volcanic field, astride the western flank of the Rio Grande rift in north-central New Mexico, was studied by means of a series of short seismic refraction profiles with maximum recording distances of 100 km (Olsen and others, 1986; Ankeny and

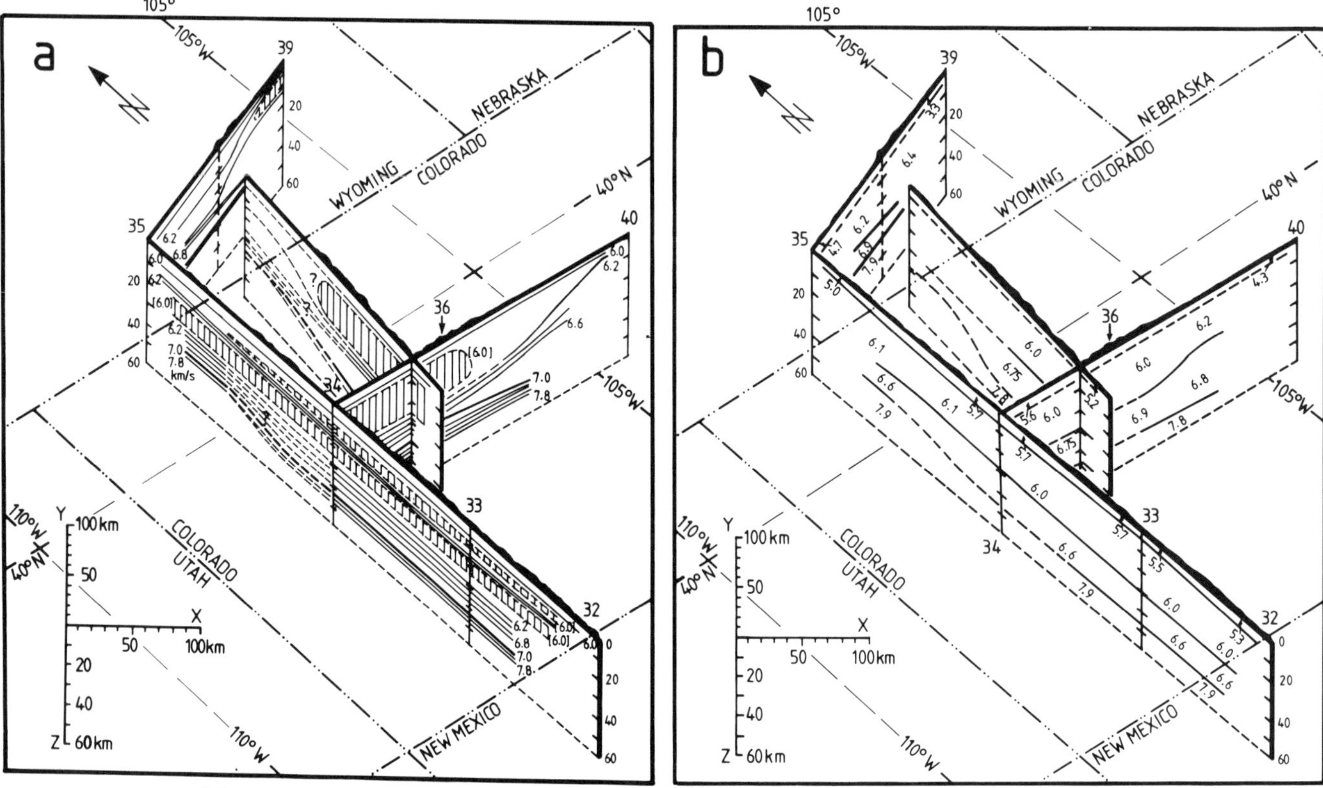

Figure 16. Fence diagram showing crustal structure under Colorado and southern Wyoming. The diagram is viewed from southwest to northeast, approximately parallel to the line from Phoenix, Arizona to Denver, Colorado. The depth z is exaggerated 2:1 versus the horizontal direction y (southwest to northeast). The cross sections are drawn along the lines shown in Figures 11, 13, 14, and 15. The shotpoints are numbered according to Figure 1. Dashed lines show extrapolated and uncertain results. a, Model showing lines of equal velocity. Contour interval = 0.2 kilometers per second. b, Generalized model showing average layer velocities.

others, 1986). In southern New Mexico, a nonreversed refraction profile 220 km long crosses the rift from west to east (Cook and others, 1979). Recently, three regional profiles were collected by Sinno and others (1986) in the southern rift: a reversed east-west profile across the rift, a nonreversed north-south axial profile, and a nonreversed northwest-southeast profile. A detailed COCORP reflection study near Socorro, New Mexico (Brown and others, 1979), focused on mapping the top of an inferred magma body, initially identified from near-earthquake shear waves (Sanford and others, 1973). The seismic body-wave studies were supplemented by surface-wave studies (Keller and others, 1979b; Sinno and Keller, 1986) and gravity interpretations (Daggett and others, 1986). Finally, a teleseismic deep-seismic sounding survey revealed an anomalous asthenospheric structure beneath the rift (Spence and others, 1979; Davis and others, 1984; Parker and others, 1984).

Crustal structure under the northernmost part of the Rio Grande rift proper, the San Luis Valley, is not known. The one seismic reflection study here (Davis and Stoughton, 1979) covered only shallow structures and therefore cannot evaluate whether the deeper crust is influenced by rifting processes.

Under the San Juan Mountains immediately to the west of the San Luis Valley, a thick crust of 48 km is found (Fig. 15). This is different from other rifts, such as the Rhinegraben (Edel and others, 1975; Fuchs and others, 1987) or the Limagnegraben (Hirn and Perrier, 1974) with similar topographically and structurally high margins. There, distinct thinning of lower crust and mantle upwarping extend well beneath the flanks, far beyond the limits of the graben areas.

The mostly northerly profile in the Rio Grande rift area showing a structural change is the GASBUGGY line (R1, in Fig. 1). As only 16 stations were available to record the explosion over a distance of 550 km, the resulting crustal model is crude. Toppozada and Sanford (1976) interpreted a 40-km-thick crust, including 19 km of upper crust and 21 km of lower crust. The lower crustal velocity is 6.5 km/sec throughout, but the upper crustal velocity is 6.15 km/sec west of Albuquerque and 5.8 km/sec farther south within the rift. These results indicate that crustal thickness decreases by about 8 km near the Colorado–New Mexico border, related to southward termination of the Rocky Mountains.

A detailed investigation of upper crustal structure beneath the Jemez Mountains volcanic area (Ankeny and others, 1986; Olsen and others, 1986) defined a low-velocity zone (5.6 to 5.8 km/sec at a depth of 2 to 12 km within surrounding Precambrian crystalline rocks (5.9 km/sec). The low-velocity zone is centered beneath the Valles Caldera. It is interpreted as a partially molten silicic pluton overlying an inferred basaltic intrusion, most likely at the base of the upper crust (Ankeny and others, 1986, Fig. 9).

Farther south, the more detailed DICETHROW III profile, defined by 40 recording stations, follows the eastern flank of the rift from the shotpoint in the Jorrecado del Muerto basin (R2, in Fig. 1) northward almost to the Colorado border (Olsen and others, 1979). Prominent reflections were recorded from a mid-crustal boundary at a depth of 21 km, and from the crust-mantle boundary at 34 km. P_n arrivals were also clear between a distance of 180 and 320 km, indicating an apparent velocity of 7.6 to 7.8 km/sec. Velocities in the upper and lower crust are 6.0 and 6.4 km/sec, respectively. The sedimentary cover, with a mean velocity of 4.3 km/sec, is about 3 km thick. This crustal model applies approximately to the area of Abo Pass, 30 km east of the inferred Socorro magma body (Sanford and others, 1977; Rinehart and others, 1979).

The Socorro "magma body" was originally detected by analysis of phases from microearthquakes interpreted as shear-wave reflections (S_xS) and phases converted from shear to longitudinal waves during reflection at interface x (S_xP) (Sanford and Long, 1965; Sanford and others, 1973, 1977). The upper surface of the 2-km-thick laccolith extends from about 10 km south of Socorro 60 km northward toward Belen. It is approximately 8 km wide in the east-west direction, and its top is 18 to 20 km deep (Sanford and others, 1977), close, but possibly not identical to, the intracrustal discontinuity in the region (Olsen and others, 1979). Magnetotelluric data indicate the presence of a conductive layer (4 to 8 Ohm-m) beneath Santa Fe at a depth of 10 to 17 km and beneath El Paso at 21 to 28 km (Hermance, 1982, Figs. 17, 18). The latter one is situated on Schmucker's anomaly (1970; see also Weidelt, 1975), which seemed to indicate a central rift conductor. Such a central rift conductor continuing to the north along the Rio Grnade rift has, however, not been confirmed by the array study of Porath and Gough (1971). An extensive magnetotelluric array study in the area of the supposed Socorro magma body (Jiracek and others, 1983, 1987) actually could not detect a high-conductivity layer within the crust, possibly because the amount of magma was too low to be measured.

COCORP surveys in 1975–1976 collected 155 km of 24-fold reflection data in the area of the inferred magma body near Socorro (N1-4, in Fig. 1) (Brown and others, 1979, 1980). In the resulting schematic cross section (Fig. 17), the top of the basement within the Albuquerque basin has substantial relief, and Tertiary antithetic normal faults are evident on the east side of the rift. The underlying upper crust is characterized by marked lateral and vertical variation in reflection character, including relatively long correlative features, and distinct bands of short, discontinuous reflections and seismically transparent zones. Most prominent coherent reflections were recorded from a depth of approximately 20 km (Brown and others, 1980), corresponding with the top of the postulated magma body (Sanford and others, 1977). Finally, multiple reflectors at 10 to 12 sec were recorded from depths of 30 to 36 km, appropriate for the Moho. This laminated band of reflectors, varying in number and amplitude, indicates a complex transition zone from crust to mantle. Brown and others deduced a scale of a few hundred meters or less for the seismic discontinuities defined by the 10- to 15-Hz reflections. This is a fine structure similar to that proposed for the lower crust from a combined interpretation of near-vertical and wide-angle reflection data in southern Germany (Fuchs and others, 1987; Gajewski and Prodehl, 1987).

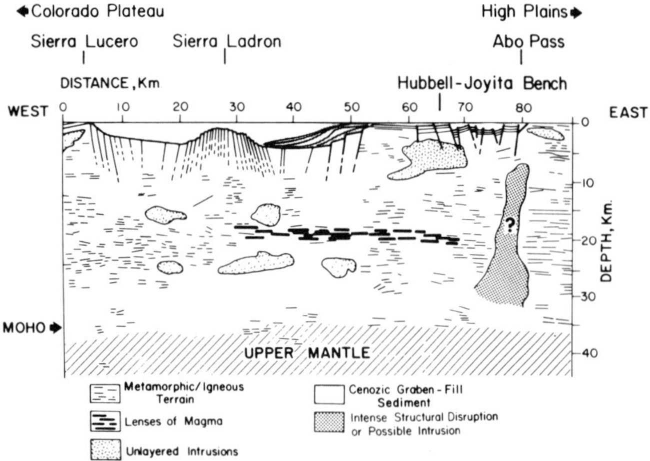

Figure 17. Schematic cross section of COCORP across the Rio Grande rift, approximately along New Mexico lines 1 and 1A (Brown and others, 1980, Fig. 14).

Based on these refraction and reflection studies, as well as on surface-wave investigations, Keller and others (1979b) inferred a schematic cross section of the crust across the Rio Grande rift in the Albuquerque–Santa Fe area (Fig. 18). The lower crust is thinned beneath the rift axis, and a mantle upwarp is inferred.

Three recently observed intersecting profiles (Sinno and others, 1986) supply good information on the crustal structure of the southern rift. The reversed east-west line shows a 7.7-km/s-P_n velocity and a subhorizontal 32-km-deep Moho. Results from the north-south axial line and the northwest-southeast line indicate an apparent P_n velocity of 7.9 to 8.0 km/sec and crustal thinning to the south and southeast. Sinno and others (1986) interpreted these lines together as indicating 4 to 6 km of crustal thinning from the northern rift and a regional P_n velocity of approximately 7.7 km/sec (Fig. 19). They also suggested that the Rio Grande rift is a crustal feature that is separate and distinct from the Basin and Range province (Fig. 24).

For other rifts, a high-velocity layer with velocities of 7.0 to 7.5 km/sec is found between the lower crust and uppermost mantle. This high-velocity layer is thought to represent a rift cushion: either an anomalously low-velocity upper mantle (Mueller and others, 1973; Hirn and Perrier, 1974; Baker and Wohlenberg, 1971; Maguire and Long, 1976; Berckhemer and others, 1975), or high-velocity lowermost crust, forming a crust-mantle transition zone (Edel and others, 1975; Prodehl, 1981; Mooney and others, 1983; KRISP Working Group, 1987; Fuchs and others, 1987). In the Rio Grande rift, a single nonreversed refraction profile 220 km long (Cook and others, 1979), observed from mine blasts near Santa Rita (R3, in Fig. 1), shows strong arrivals at distances of 40 to 130 km from the shotpoint that were attributed to such a 7.4-km/sec layer below 26 km depth. As other available data do not require the existence of this layer, it seems unlikely that the interpretation by Cook and others (1979) is the only possible one.

The mantle unwarp beneath the southern Rio Grande rift is similar to one found farther north (Olsen and others, 1979). A P_n velocity of 7.7 km/sec suggests elevated upper-mantle temperatures (Sinno and others, 1986). The Rio Grande rift has thinner crust and lower P_n velocity than the adjacent provinces (Figs. 18, 19), and the crust thins southward along the rift (see Fig. 23). The two-dimensional models of Sinno and others (1986) are consistent with available seismic and gravity data and do not require anisotropy in the upper mantle beneath the southern rift (Daggett and others, 1986). The observed crustal thinning may record permanent modification of the Moho structure, a relation perhaps valid beneath rifts in general.

Recent surface-wave studies (Sinno and Keller, 1986) along propagation paths centered within the rift are consistent with more limited earlier data (Keller and others, 1979b). The resulting averaged earth models are consistent with available seismic refraction data and show increases in Poisson's ratio from 0.24 in the upper crust to 0.27 in the lower crust and to 0.29 in the upper mantle; these results imply that the rigidity of the lithosphere may decrease with depth in the southern Rio Grande rift. The interpretation that the upper mantle below the rift has a low rigidity and presumably high temperatures agrees with heat-flow results (Sinno and Keller, 1986). Similar Poisson's ratios have been observed recently in southwest Germany (Holbrook and others, 1987, 1988). These results also are consistent with the partly melted low-Q upper mantle inferred by Gough (1984).

Deep-penetrating seismic studies of upper-mantle structure were based on observations of teleseismic delays. The first survey used a 24-element seismometer array across the Rio Grande rift recording 22 teleseismic events (Spence and others, 1979, 1982). In 1982–1983, about 40 teleseismic events were recorded by 20 mobile stations along a 1,000-km linear array from Moab, Utah,

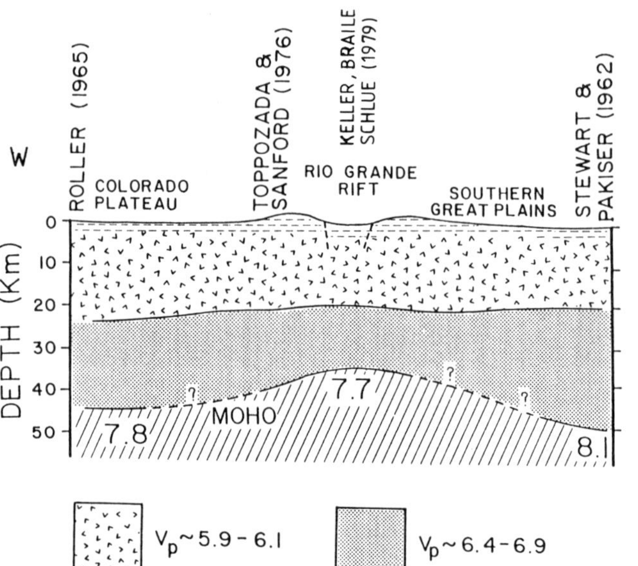

Figure 18. Schematic cross section across the Rio Grande rift in the Albuquerque–Santa Fe area, based on refraction and surface-wave results (Keller and others, 1979b, Fig. 6).

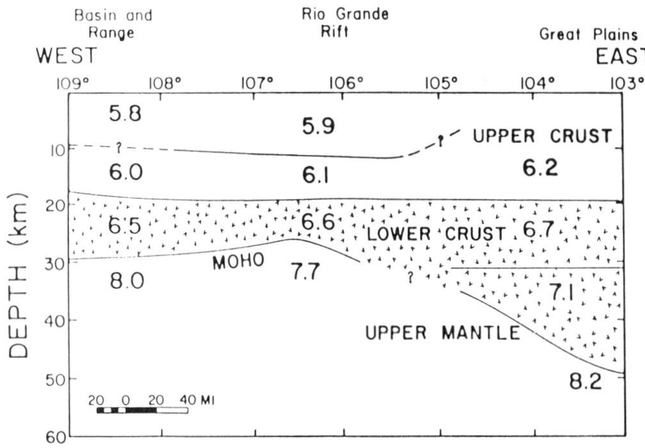

Figure 19. Schematic cross section through the southern Rio Grande rift (Sinno and others, 1986, Fig. 18).

southeastward to Odessa, Texas, which crosses the Rio Grande rift at about a 45° angle at Valles Caldera near Los Alamos, New Mexico (Davis and others, 1984; Parker and others, 1984). P-wave arrivals show maximum delays of as much as 1.5 sec, centered on the rift, with measurable delays extending several hundred kilometers to either side of the rift. Variations in crustal velocity are considered too small to explain the delay pattern fully (Davis and others, 1984). Inversion of the data suggests a low-velocity low-density body, located well below the Moho. A 70-km upper and 100-km lower boundary is proposed for this body, characterized by a 6 to 8 percent reduction in velocity, and centered beneath the Rio Grande rift near Los Alamos.

Although the data density is sparse, first-arrival times are early for the stations most distant from the 1967 GASBUGGY explosion in several directions at a distance as great as 500 km, except to the east (Warren and Jackson, 1968). These results may be consistent with early observations of an 8.4 to 8.5-km/sec layer at a depth of about 100 km, on a profile from Lake Superior to Arizona (Roller and Jackson, 1966). Similarly, apparent first-arrival velocities are 8.4 km/sec at distances of 390 to 1,100 km east from the Nevada Test Site (Ryall and Stuart, 1963). These results correspond to the lower boundary of the low-velocity upper-mantle area of Davis and others (1984).

Baldridge and others (1984) concluded that deformation of the large region associated with the Rio Grande rift must have perturbed the entire lithosphere (Fig. 20). The P-wave velocity in the lower crust beneath the rift (6.5 km/sec) is slower than lower crustal velocities (6.7 to 6.8 km/sec) to the east and west. Strong wide-angle reflections imply the presence of a thin, tabular low-rigidity layer at midcrustal depths of about 20 km in the vicinity of the Socorro "magma body;" this anomalous layer is thought to be more extensive than the magma body itself (Olsen and others, 1982). This feature may constitute local expression of the Conrad discontinuity (Baldridge and others, 1984). The model also suggests that the crust has thinned to about 33 km under the rift axis, from about 45 km beneath the Arizona–New Mexico border and 50 km under the Great Plains to the east. The sub-Moho velocity along the rift is 7.6 to 7.7 km/sec, lower than under the Great Plains but similar to that widely present in the Basin and Range province. A broad, long-wavelength gravity low, extending 500 km across the state of New Mexico at the latitude of Albuquerque, is correlated with the shape of the low-velocity zone in the upper mantle (Baldridge and others, 1984). In possible disagreement with other authors, as discussed below in more detail, the axis of this asthenospheric anomaly is modeled about 200 km west of the rift axis, similar to the asymmetrical high heat flow associated with the rift zone (Reiter and others, 1979, 1986). Not taken into account in this model is the anomalous low-velocity mantle below 70 km depth, inferred by Spence and others (1979, 1982) and Davis and others (1984).

WESTERN BORDER OF THE RIO GRANDE RIFT

The western border of the Rio Grande rift is defined by the San Juan Mountains and the Sawatch Range of the southern Rocky Mountains, the eastern Colorado Plateau, and the southeastern Basin and Range province. The crustal structure of the southern Rocky Mountains has been discussed already. For the eastern Colorado Plateau, the only available seismic refraction profile is from Hanksville, Utah, to Chinle, Arizona (30, 31 in Fig. 1; Roller, 1965; Prodehl, 1970, 1979). For the southeasternmost Basin and Range province and adjacent parts of the Colorado Plateau, control is provided by lines from Tyrone to Jackpile, New Mexico (B1, B2 in Fig. 1), and from Tyrone west (Harden, 1982; Gish and others, 1981).

The geophysical character of the transition from the Basin and Range province to the Colorado Plateau is discussed elsewhere (Smith and others, this volume). Only the refraction data bearing on the western margin of the Rio Grande rift are summarized here. The interpretive model for the Hanksville-Chinle line, slightly modified from Prodehl (1979), shows the crust-mantle boundary at a depth of 40 km, rising slightly from north-northwest to south-southeast (Figs. 21, 22), in agreement with Roller (1965). The average velocity of the upper crust is about 6.1 km/sec, and that of the lower crust is about 6.8 km/sec, with the midcrustal boundary at a depth of about 25 km. Because of limited data, it is unknown whether the crust thins abruptly or more gradually at the border of the Rio Grande rift. The P_n velocity under the Colorado Plateau is only 7.6 km/sec (Prodehl, 1979), almost the same as under the adjacent rift. Surface-wave dispersion studies suggest a slightly thicker crust (45 km) (Keller and others, 1979a).

The crust is 35 km thick north of Tyrone, New Mexico, under the southeastern Colorado Plateau and adjacent Basin and Range province (Harden, 1982; Jaksha, 1982), but the thickness decreases to the west (Gish and others, 1981). Crustal thickness correlates well with Bouguer gravity for this area (Harden, 1982, Figs. 4, 9). Harden's data suggest that an isolated area of thick

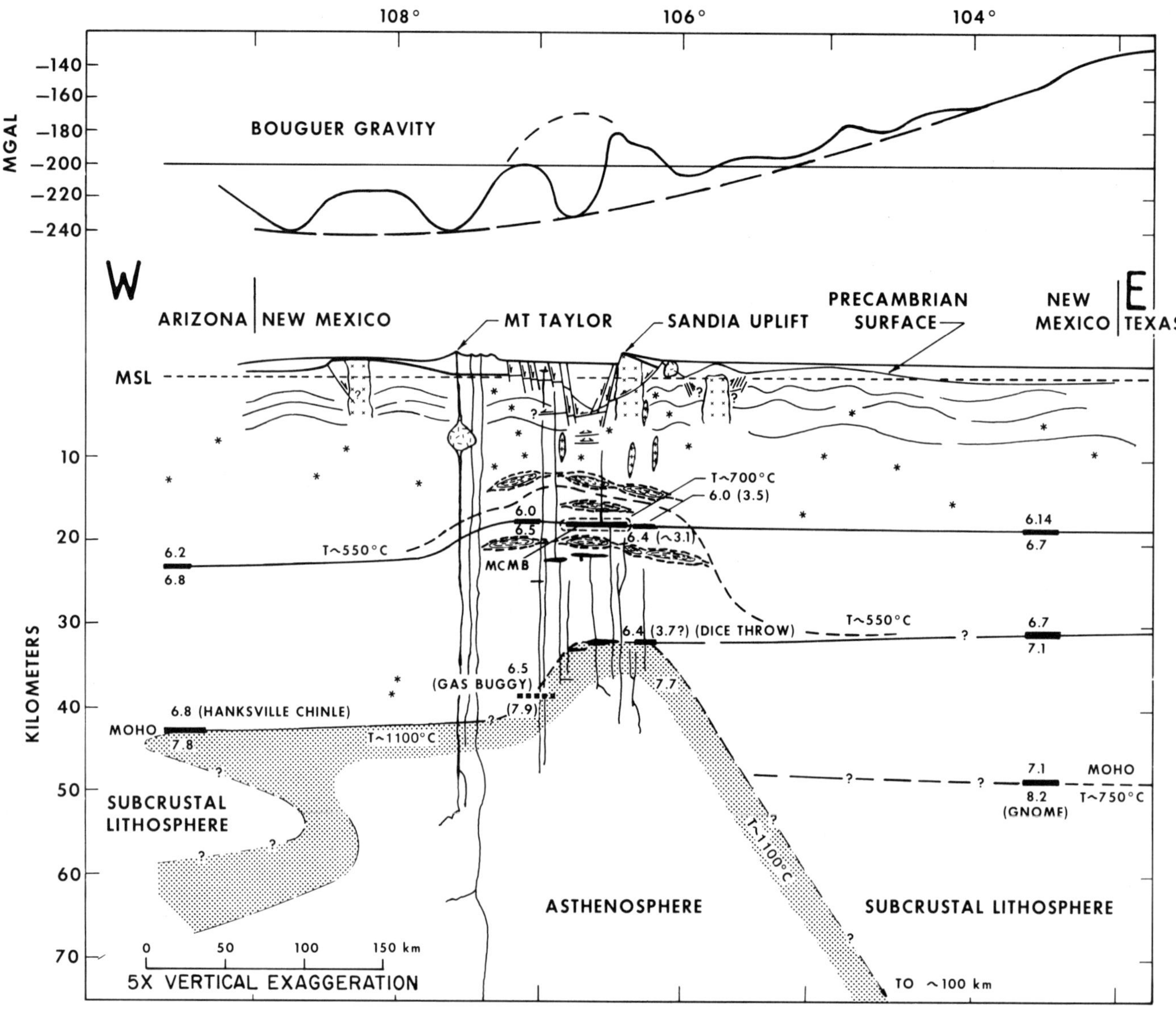

Figure 20. Lithospheric structure of the Rio Grande rift, as shown by a cross section through the northern Albuquerque-Belen basin (Baldridge and others, 1984, Fig. 2).

crust or a batholithic mass near Silver City, New Mexico, may separate the Rio Grande rift from the Basin and Range province. The P_n velocity west of the rift is poorly known.

The exceptionally low P_n velocity between Hanksville and Chinle suggests that the asthenosphere might be close to or even reach the crust-mantle boundary, despite the generally low heat flow (Blackwell, 1978). As discussed above, Baldridge and others (1984) concluded that the asthenosphere reaches Moho depths underneath the Rio Grande rift and the adjacent Colorado Plateau to the west, on the basis of a 500-km-broad long-wavelength gravity low (Fig. 20).

In contrast, surface-wave dispersion studies of Keller and others (1979a) suggest a lithosphere thickness of 60 km for the Colorado Plateau, 30 km under the rift and 100 km under the Great Plains to the east, in agreement with the electrical conductivity data. A more likely interpretation (Zoback and Lachenbruch, 1984) is that the lithosphere thickens from 50 to 65 km beneath the Basin and Range province to 90 to 100 km beneath the Colorado Plateau, which is consistent with gravity, seismic, surface elevation, and heat-flow data (see also Pakiser [1985] for a discussion of the Basin and Range to Colorado Plateau transition zone).

Hermance (1982) considered that the mantle from the base of the crust to depths greater than 100 km is grossly similar beneath the Colorado Plateau and the Rio Grande rift, as indicated by available seismic and magnetotelluric data. Long-period magnetotelluric data from the Colorado Plateau, near Farmington, New Mexico, define a resistive crystalline crust (2,000 Ohm-m) down to 28 km, underlain by a less resistive zone (13 Ohm-m) down to at least 130 km (Hermance, 1982, Fig. 20). In

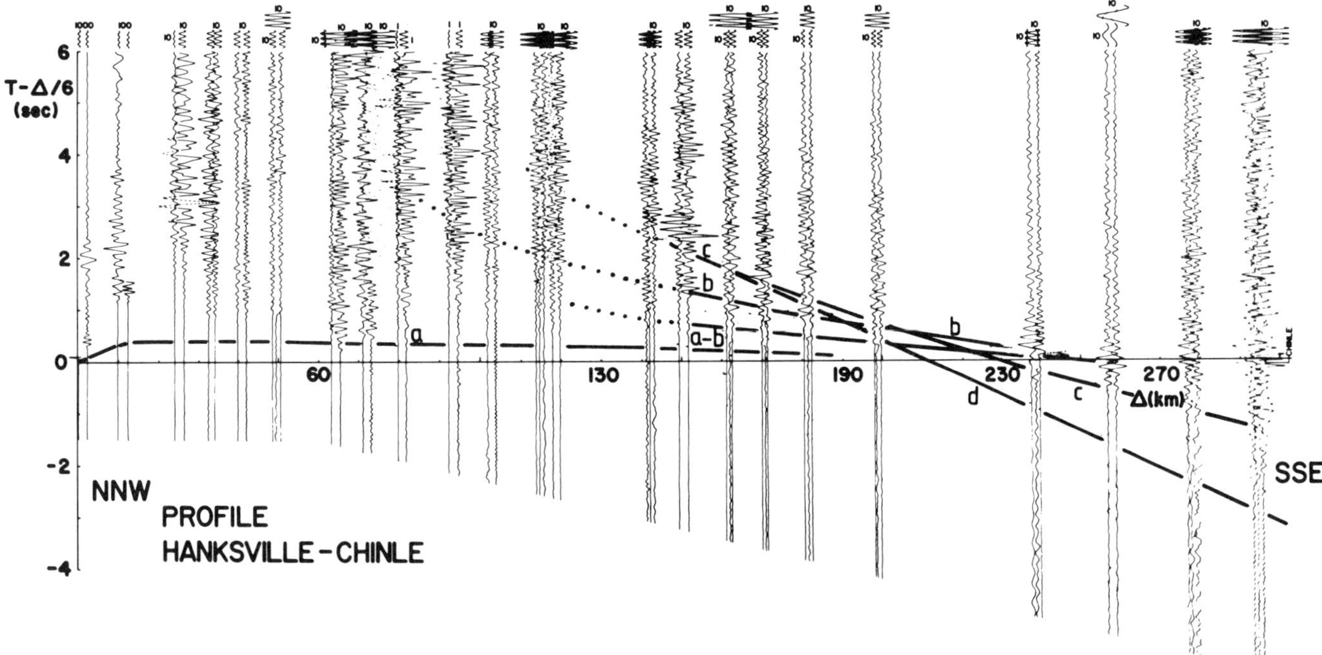

Figure 21. Record section from Hanksville, Utah, to Chinle, Arizona. Traveltime curves as recalculated from model of Figure 22b. Explanation of correlated phases in Figure 4.

contrast, the seismically determined crustal thickness of the Colorado Plateau is about 40 km (Roller, 1965; Prodehl, 1979), requiring that the transition from resistive crust to conductive mantle occurs within a high-velocity lower crustal layer about 12 km thick.

The velocity structure of the Colorado Plateau, as judged from the line from Hanksville to Chinle, is similar to that of the Great Plains east of the Rocky Mountain front. Other geophysical parameters such as similar Bouguer gravity, low heat flow, and moderate to low seismicity confirm this view. These results are in general agreement with the geological conclusion that the Colorado Plateau is a stable block within the tectonically active Cordillera. The plateau has remained essentially unchanged since it became separated from the stable North American continent by the formation of the southern Rocky Mountains and Rio Grande rift. The exceptionally low P_n velocity is puzzling. However, as Figure 22 demonstrates, the 7.6-km/sec velocity beneath the Colorado Plateau is not based on true subsurface-path reversal. If the Moho dips down slightly from the center of the Hanksville to Chinle profile, the true P_n velocity could be somewhat greater.

STRUCTURAL SUMMARY

A generalized crustal section (for location, see Fig. 1) has been compiled for the Rocky Mountains and the Rio Grande rift (Fig. 23). This section approximately follows the crest of the northern Rocky Mountains, but is offset by about 200 km to the southwest in the Yellowstone–Snake River Plain region into the middle Rocky Mountains. After being offset again by about 250

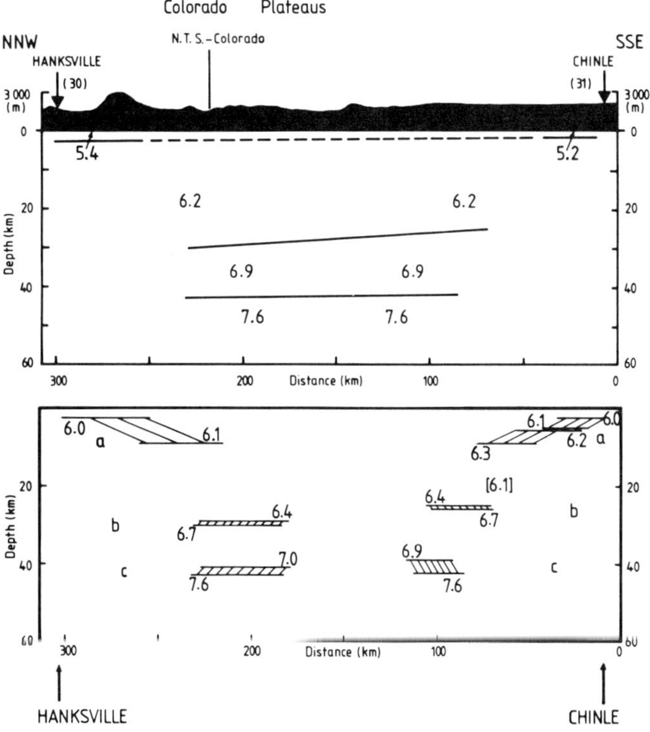

Figure 22. Crustal structure of the central Colorado Plateau. Velocities in kilometers per second. a, Generalized model. b, Cross section of Figure 22a showing at which distance ranges reflectors, respectively boundaries with large positive velocity gradient (dashed ranges) are actually controlled by the phases b and c, as correlated in the corresponding record section of Figure 21.

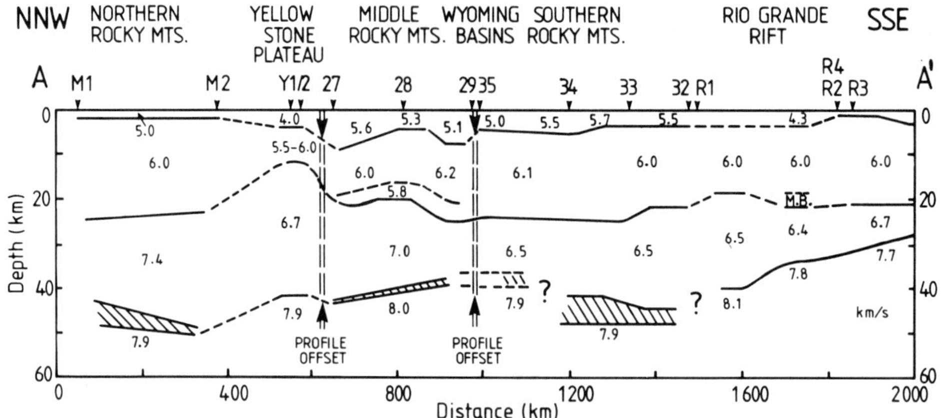

Figure 23. Simplified crustal cross section through Rocky Mountains and Rio Grande rift. For location AA′, see Figure 1. Between the Yellowstone Plateau and Middle Rocky Mountains, the line is offset by about 200 km to the southwest; between the Middle and Southern Rocky Mountains, the line is offset by about 250 km to the northeast. 6.0, . . . velocities in kilometers per second; M1, . . . shotpoints (see Fig. 1). M.B. = "magma body" of Socorro; dashed area = crust-mantle transition zone.

km to the east in the Wyoming basins and uplift area, the section follows the axis of the southern Rocky Mountains and the Rio Grande rift, i.e., the axis of Eaton's (1986) Alvarado Ridge.

The crust has a maximum thickness of 45 to 50 km under the northern and central southern Rocky Mountains; it is 38 to 41 km thick under the Yellowstone Plateau, the middle Rockies, and the northernmost southern Rocky Mountains. A zone of transitional velocity is evident between lower crust and upper mantle everywhere except under the tectonically active areas of the Yellowstone Plateau and the Rio Grande rift. The velocity of the lower crust ranges from 6.4 to 7.4 km/sec, but the higher velocities are based on early interpretations. Velocity inversions are minor in the upper crust; either the depth range through which an inversion is postulated is narrow, as in the middle Rocky Mountains, or the velocity inversion is small, as in the southern Rocky Mountains. The interpretive methods applied to data for the northern Rocky Mountains were not able to consider inversion zones, but the low mean velocity of the upper crust (6 km/sec) may indicate a fine structure similar to that in the southern Rocky Mountains. Nowhere has a velocity inversion been postulated within the lower crust of the Rocky Mountains or the Rio Grande rift.

Crustal thickness in the Rocky Mountain region ranges from slightly less than 40 km to slightly more than 50 km (Fig. 24). The crust is thickest in the northern Rocky Mountains (southern Montana) and in the southern Rocky Mountains (central Colorado). The crustal thickness is about 10 to 15 km less in the middle Rocky Mountains and the Wyoming basins than to the north or south. Similar trends apply to the Great Plains province to the east: thick mountain crust is accompanied by thick plains crust. The crust even thickens slightly eastward across the mountain front in eastern Montana and in eastern Colorado. The thick crust of eastern New Mexico (50 km), as indicated by the GNOME data (Stewart and Pakiser, 1962), evidently has no counterpart to the west, where the crust of the Rio Grande rift is less than 35 km thick. In the Rio Grande rift the crustal thickness changes gradually from more than 40 km in the north to less than 30 km in the south near the Mexico–U.S. border.

West of the Rocky Mountains, crustal thickness changes gradually. The eastern Snake River Plain and the adjacent prong of the northeasternmost Basin and Range province, as well as the Colorado Plateau, are, on the average, 40 km thick. In central Utah, west of the Wasatch Mountains, however, the crust thins abruptly from about 40 km to less than 30 km. The Rio Grande rift is bordered to the northwest by the slightly thicker (40 km) Colorado Plateau crust, but to the southwest beneath the southeastern Basin and Range province, the crust thins to less than 30 km (Gish and others, 1981; Keller and others, 1979a; Warren, 1969).

Trends of average crustal P-wave velocity (Fig. 25) generally correlate well with crustal thickness (Fig. 24). Thin crust is characterized by a low crustal velocity and vice versa. The highest velocities are beneath the northern Rocky Mountains and adjacent Great Plains of Montana and beneath the southern Rocky Mountains and Great Plains of Colorado. The Snake River Plain also shows a high mean crustal velocity, decreasing only slightly toward Yellowstone Park.

Upper-mantle velocities immediately beneath the Moho are low for the Basin and Range province, Colorado Plateau, and Rio Grande rift (7.4 to 7.8 km/sec), suggesting that the asthenosphere is shallow beneath the entire region. Contrary to the interpretation of Gough (1984), the upper-mantle velocity is higher beneath the southern Rockies. Upper-mantle velocities are near 7.9 km/sec below most of the Rocky Mountains, including the Yellowstone–eastern Snake River Plain area. Velocities are as high as 8.0 km/sec locally under areas such as the middle Rockies and the transition from the southern Rockies to the Rio Grande rift. They exceed 8 km/sec under the entire Great Plains.

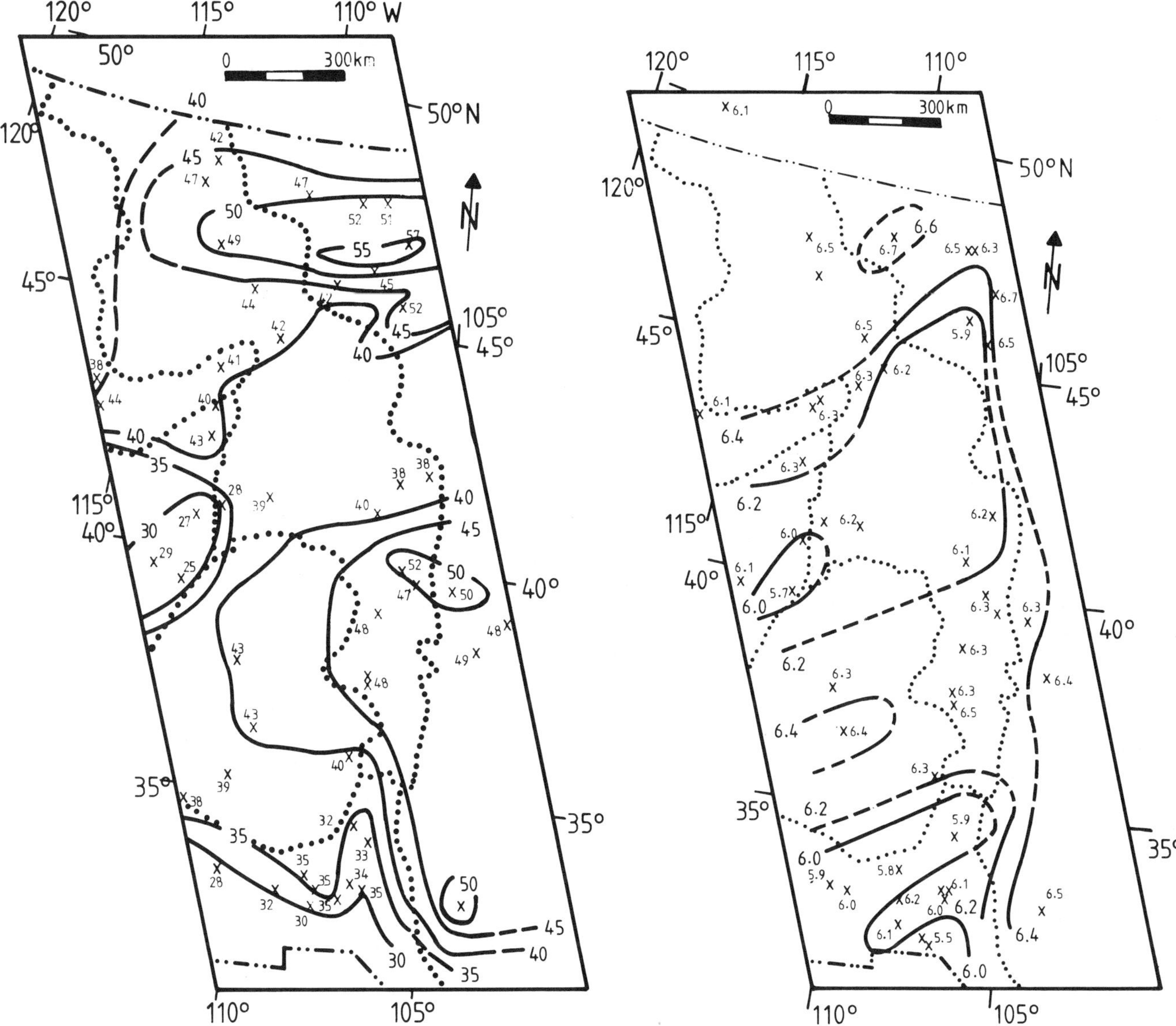

Figure 24. Contour map of total crustal thickness underneath the Rocky Mountains, Rio Grande rift, and adjacent areas. Contour interval = 5 km below surface. Observed values are plotted at half their observation distance from the corresponding shotpoint. Dotted lines = physical divisions from Fenneman (1946); see Figure 1.

Figure 25. Averaged crustal P-wave velocities of the Rocky Mountains and the Rio Grande rift and adjacent areas. Contour interval, 0.2 kilometers per second. Further explanations on Figure 24.

Because these velocities are usually determined from first-arrival data, and the profiles are usually 300 km long, the results even for Montana should be reliable.

The Bouguer gravity and average elevation maps (Figs. 26, 27) show similar trends. The highest elevations (more than 3,000 m) coincide broadly with the lowest gravity (less than –300 mgal), centered in the southern Rocky Mountains. The largest deviation from this correlation occurs at the transition to the eastern Basin and Range province, where Bouguer gravity remains constant at about –200 mgal, whereas elevation decreases to less than 1,500 m. Crustal structure also correlates with the trends of Bouguer gravity and average elevation within the Cordillera, except for the Basin and Range province as discussed elsewhere (Prodehl, 1979; Smith and others, this volume; Thompson and others, this volume). The Great Plains to the east are characterized by an abrupt increase in gravity, accompanied by decreased average elevation (Figs. 26, 27), even though the crustal thickness equals that of the Rocky Mountains (Fig. 24). Though the crustal thickness of the northern and southern Rockies and adjacent Great Plains is about the same (50 km), the

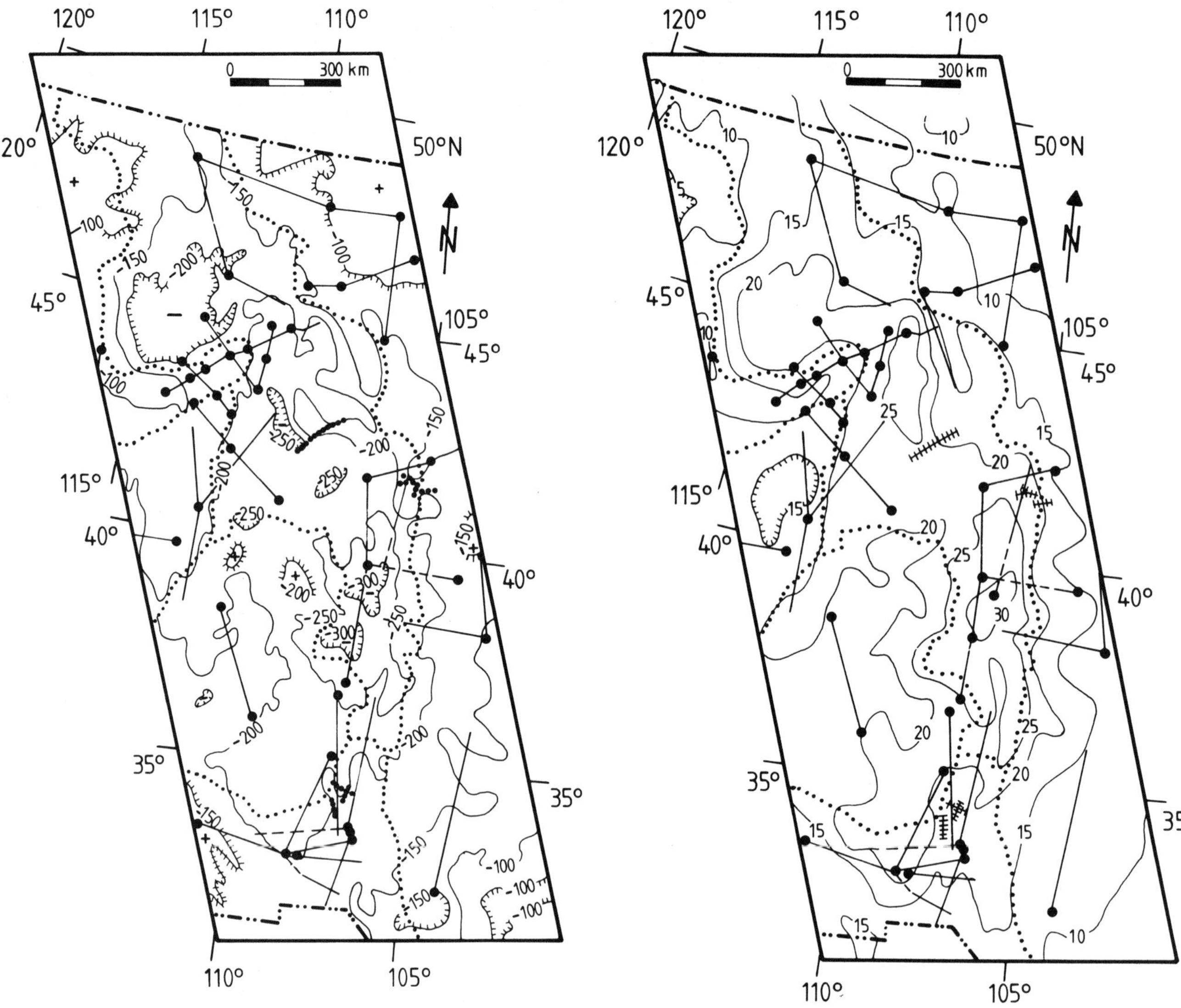

Figure 26. Bouguer gravity anomaly map of the Rocky Mountains, the Rio Grande rift, and adjacent areas. Contour interval = 50 mgal. Simplified after Lyons and O'Hara (1982). Physical divisions (dotted lines) and seismic profiles from Figure 1.

Figure 27. Average elevation map of the Rocky Mountains, the Rio Grande rift, and adjacent areas. Average elevation contours on the surface of the solid Earth. Contour interval = 500 m. Contours are labeled in hundreds of meters. Simplified after Diment and Urban (1982). Physical divisions (dotted lines) and seismic profiles from Figure 1.

Bouguer gravity does not correlate well with average elevation in the northern Rockies. A relatively thin upper crust characterized by a low mean velocity and a thicker lower crust characterized by high velocities are able to compensate the relatively high gravity with the moderate average elevation.

While there is little correlation between magnetic intensity (Zietz, 1982) and seismic crustal structure, heat-flow data (Blackwell, 1978, Fig. 8-2) show a fair correlation with seismic data, particularly for the upper crust. High heat flow corresponds to thin low-velocity upper crust, especially in the Yellowstone–Snake River Plain area. Relatively high heat flow in the Rocky Mountains and Rio Grande rift also corresponds to a moderately low upper crustal velocity (about 6.0 km/sec), similar to that of the broader Cordilleran thermal anomaly. In contrast, the low heat flow of the Colorado Plateau also is associated with low upper crustal velocity.

The results compiled in this chapter agree well with the trends in crustal thickness compiled by Allenby and Schnetzler (1983), although some details differ. Their map shows that the thick crust under the Great Plains of eastern Montana and Colorado extends as far as eastern Nebraska and Kansas, in accordance with geologic results that the eastern extension of the basement terrane exposed in mountain uplifts can be traced into eastern Nebraska and Kansas as well (P. K. Sims, written communication, 1986). The thin crust of the Wyoming basins is evidently local.

The main features of crustal structure have been known for at least 25 yr (Pakiser, 1963; Pakiser and Steinhart, 1964). No distinct crustal root is concentrated below the Rocky Mountains. The major crustal contrast with the Great Plains province is the seismic velocity in the upper crust, as weakly shown in the older data for Montana and clearly shown in the better data for Colorado. The upper-crustal velocity is about 6 km/sec beneath the Rocky Mountains to depths of about 20 km or more; it is substantially higher, about 6.2 km/sec, under the Great Plains and the Colorado Plateau. A second contrast is the uppermost mantle velocity: less than 8 km/sec west of the Rocky Mountain front, and 8 km/sec or more to the east. The Rio Grande rift is expressed by shallowing of the crust-mantle boundary to a depth of less than 30 km, but more data are needed to outline the rift structure in detail.

The available refraction data delineate the gross crustal structures beneath the Rocky Mountains, but only the Yellowstone–Snake River Plain experiment and a few profiles in the Rio Grande rift have been done since 1970. Many features of the crustal structure that could be determined by modern methods can only be inferred. The existing modern reflection surveys in Wyoming and New Mexico would benefit from support by high-resolution long-range refraction experiments. The power of combined use of refraction and reflection surveys has recently been demonstrated in southern Germany (Fuchs and others, 1987; Gajewski and others, 1985, 1987). Especially needed are reversed long-range refraction lines, which run subparallel to major features such as the continental divide in Montana, the middle Rocky Mountains from Yellowstone to Salt Lake City, along the Rocky Mountain front from Laramie south through the Sangre de Cristo Mountains, and detailed investigation of the Rio Grande rift by several subparallel lines within the rift and along the flanks. An additional high-priority need is an east-west traverse through the Wyoming basins.

CHANGING CRUSTAL STRUCTURE THROUGH TIME

Because analytical and modeling problems are complex, seismologists have concentrated on interpreting the geometry of present-day velocity and density structure of the crust in the western United States, rather than the geologic origins of the observed features. An ultimate goal, though, should be to correlate the development of observed structural features of the crust with regional geologic evolution, at least on a broad scale commensurate with the geologic summary presented earlier in this work. The geologically diverse provinces of the Rocky Mountain region offer especially favorable opportunities to evaluate the evolution of Cordilleran crustal structure through time: some features appear to represent little modified Precambrian craton, whereas others record overprinting and crustal modification by successive Phanerozoic tectonic and magmatic events. In contrast, for the more active Intermontane and Pacific regions of the Cordillera to the west, the intensity of successive accretionary, deformation, and magmatic events has largely obscured the geophysical crustal expression of earlier features, and effects of geologically recent events can be expected to dominate the present crustal structure. The following preliminary interpretations for the Rocky Mountain region are accordingly offered in an attempt to stimulate more thinking on the historic evolution of the crust of the eastern Cordillera.

The thick and relatively dense crust on the High Plains just east of the northern Rocky Mountains (Figs. 24, 26) appears to constitute Archean craton that has been little modified since about 2.6 Ga. The crust of this age province thins from about 50 to 40 km to the south, toward the boundary (Cheyenne belt) with Proterozoic crust near the Colorado-Wyoming line. This thinning may be inherited from an original oceanward wedging of the continental margin, analogous to that characterizing the present Pacific margin (Couch and others, this volume). The increased crustal thickness of Proterozoic crust on the High Plains farther south may similarly be a little-modified feature of original accretion to the craton. The dominant structural grain of both the Archean and Proterozoic basement underlying the High Plains remains northeasterly, paralleling the accretion boundaries, as indicated by trends of the major regional aeromagnetic anomalies (Zietz, 1982; Kane and Godson, this volume).

Late Precambrian rift truncation of the craton along the western margin of the northern Rocky Mountains, and resulting formation of a passive continental margin, may be expressed by the decreased crustal thickness westward across the northern Rockies and into eastern Washington and Oregon (Fig. 24), although the structure of this region has been somewhat complicated by extensional deformation since middle Tertiary time. Early Paleozoic extensional features, which trend roughly east across the Rocky Mountains (Larson and others, 1985), are not clearly delineated in available crustal structural data.

Late Paleozoic uplifts of the Ancestral Rocky Mountains in Colorado and New Mexico seemingly mark the first major development of a northwest-trending structural grain in the region, a trend that dominates much of the later Phanerozoic events. Within the Rocky Mountains, crustal structures related to formation of the Ancestral Rockies have been so overprinted by later features that no seismic expression seems distinguishable. The geophysical expression of correlative structures beneath Cenozoic cover on the southern High Plains should have been little modified subsequent to formation, but seismic data on crustal structure are limited for this region. The increase in crustal thickness to more than 45 km in this sector of the High Plains may be partly due to the late Paleozoic compressional deformation.

Late Mesozoic to early Tertiary Laramide compressional uplifts and magmatism were widespread in the Rocky Mountain region, which owes its initial structural identity to these events. Though crustal effects of Laramide events have been widely overprinted by later Cenozoic events, Laramide compressional features appear relatively little modified in the eastern parts of the central Rockies and Wyoming basins. There, despite the compressive nature of the deformation, the crust is similar in thickness to the Archean craton to the east, but lower seismic velocities and densities in the upper crust indicate that the major effect of Laramide compression has been to tectonically thicken the lower density upper crust without appreciably affecting the lower crust. Intense magmatism in the northern Rocky Mountains also appears to have reduced the upper crustal seismic velocities and densities there, as recorded by the Cliff Lake–Sailor Lake profile (Fig. 2), which extends across the Boulder batholith and Elkhorn Mountain volcanic fields—the most prominent Laramide magmatic features of the region. This change seems likely to reflect emplacement of batholithic bodies of silicic magma in the upper crust, as also reflected by gravity lows (Biehler and Bonini, 1969; Kane and Godson, this volume). Although total crustal thickness is greater than in the nonmagmatic middle Rocky Mountains, this variation may be a relict from cratonization in Archean time, as suggested by the lack of change in crustal thickness across the Rocky Mountain front (Fig. 24).

The voluminous middle Tertiary magmatism, which is widely superimposed on Laramide features in the southern Rocky Mountains, appears to have had much the same effect on crustal structure as the Laramide magmatism: decreased crustal velocities, especially in the upper crust, without significant change in total crustal thickness (Fig. 16). These changes are also strongly expressed by large Bouguer gravity lows.

Late Tertiary extension and associated thermal effects are especially pronounced in the southern Rocky Mountains, where the Rio Grande rift essentially constitutes a northward-projecting prong of basin-range structure. Crustal thickness is reduced by 5 to 10 km (Fig. 20), and velocities of the lower crust are relatively low. These changes in crustal structure are being interpreted as related to attenuation of the cratonic crust, intrusion of mantle-derived basaltic magma into the lower crust, and presence of remaining partial melt in the upper mantle. Such crustal extension and magmatism is accompanied by high heat flow and current seismicity. Analogous extension structures are developed less strongly in adjacent parts of the southern Rocky Mountains and across much of the northern Rocky Mountains, somewhat complicating interpretation of crustal structure over much of the region.

A final important structural problem is the broad 1- to 2-km uplift of the entire eastern Cordillera and adjacent High Plains in late Cenozoic time, as indicated by rugged present topography in the Rocky Mountains, erosion of middle Tertiary sedimentary deposits on the High Plains, and deposition farther to the east, as well as by paleobotanic data (Scott, 1975; Taylor, 1975; Axelrod and Bailey, 1976). We recognize no compositional or structural features within the crust to account for this regional event, and accordingly, we infer that it is related to changes deeper within the mantle. Attractive possibilities include thermal expansion resulting from delayed heating of the initially cool oceanic subducted slab that in middle Tertiary time is believed to have been subducted far eastward along a low-angle subduction system (Lipman and others, 1971; Coney and Reynolds, 1977). A related alternative may be heating of the upper mantle and lower crust by asthenospheric mantle that first could come in contact with the eastern Cordilleran region through an enlarging "slab window" (Dickinson and Snyder, 1979) that has developed along the plate margin over the past 30 m.y. as subduction was progressively terminated between the Pacific and American plates along the enlarging San Andreas transform.

Thermally driven uplift of the western United States over the past 10 m.y., in close association with basaltic magmatism and extensional tectonism, has been on a scale (several kilometers) comparable to the thermal subsidence related to cooling of oceanic crust with time and distance from the active spreading center. Analogies with the topography and geophysical expression of slow-spreading oceanic ridges seem especially apt for the Rio Grande rift and adjacent broad shoulder regions from the Colorado Plateau as far east as the Great Plains (Cordell, 1978; Eaton, 1986).

ACKNOWLEDGMENTS

The chapter would not have been completed without the encouragement, assistance, and insistence of L. C. Pakiser. A two-day meeting in March 1986, organized by the editors of this volume, improved the general layout of our contribution. We have benefited from preprints of papers and maps provided by V. Haak, G. R. Keller, R. W. Simpson, and R. B. Smith. Unpublished data recorded in 1965 by the U.S. Geological Survey are from the files of the USGS Office of Earthquake Studies in Menlo Park, California. Finally, we thank G. R. Keller, L. C. Pakiser, P. K. Sims, S. W. Smithson, and an anonymous reviewer for review and constructive criticism of our manuscript.

REFERENCES CITED

Allenby, J. R., and Schnetzler, C. C., 1983, United States crustal thickness: Tectonophysics, v. 93, p. 13–31.

Allmendinger, R. W., Brewer, J. A., Brown, L. D., Kaufman, S., Oliver, J. E., and Houston, R. S., 1982, COCORP profiling across the Rocky Mountain front in southern Wyoming: Part 2, Precambrian basement structure and its influence on Laramide deformation: Geological Society of America Bulletin, v. 93, p. 1253–1263.

Allmendinger, R. W., and 7 others, 1987, Overview of the COCORP 40°N transect, western United States; The fabric of an orogenic belt: Geological Society of America Bulletin, v. 98, p. 308–314.

Ankeny, L. A., Braile, L. W., and Olsen, K. H., 1986, Upper crustal structure beneath the Jemez Mountains volcanic field, New Mexico, determined by three-dimensional simultaneous inversion of seismic refraction and earthquake data: Journal of Geophysical Research, v. 91, p. 6188–6198.

Ansorge, J., Prodehl, C., and Bamford, D., 1982, Comparative interpretation of explosion seismic data: Journal of Geophysics, v. 51, p. 69–84.

Asada, T., and Aldrich, L. T., 1966, Seismic observations and explosions in Montana, *in* Steinhart, J. S., and Smith, T. S., eds., The earth beneath the continent: American Geophysical Union Geophysical Monograph 10, p. 382–390.

Axelrod, D. I., and Bailey, H. P., 1976, Tertiary vegetation, climate, and altitude of the Rio Grande depression, New Mexico—Colorado: Evolution, v. 22, p. 595–611.

Baker, B. H., and Wohlenberg, J., 1971, Structure and evolution of the Kenya rift valley: Nature, v. 229, p. 538–542.

Baldridge, W. S., Olsen, H. K., and Callendar, J. F., 1984, Rio Grande rift; Problems and perspectives, *in* Baldridge, W. S., Dickerson, P. W., Riecker, R. E., and Zidek, J., eds., Rio Grande rift; Northern New Mexico: Socorro, New Mexico Geological Society, p. 1–12.

Berckhemer, H., and 9 others, 1975, Deep seismic soundings in the Afar region and on the highland of Ethiopia, *in* Pilger, A., and Roesler, A., eds., Afar depression of Ethiopia: Stuttgart, Schweizerbart, p. 89–107.

Biehler, S., and Bonini, W. B., 1969, A regional gravity study of the Boulder Batholith, Montana: Geological Society of America Memoir 115, p. 401–422.

Blackwell, D. D., 1978, Heat flow and energy loss in the western United States, *in* Smith, R. B., and Eaton, G. P., eds., Cenozoic tectonics and regional geophysics of the western Cordillera: Geological Society of America Memoir 152, p. 175–208.

Borcherdt, C. A., and Roller, J. C., 1967, Preliminary interpretation of a seismic-refraction profile across the Large Aperture Seismic Array, Montana: U.S. Geological Survey Technical Letter NCER 2, 53 p.

Braile, L. W., Smith, R. B., Keller, G. R., Welch, R. M., and Meyer, R. P., 1974, Crustal structure across the Wasatch Front from detailed seismic refraction studies: Journal of Geophysical Research, v. 79, p. 2669–2677.

Braile, L. W., and 9 others, 1982, The Yellowstone-Snake River Plain seismic profiling experiment: Crustal structure of the eastern Snake River Plain: Journal of Geophysical Research, v. 87, p. 2597–2609.

Brewer, J. A., Smithson, S. B., Oliver, J. E., Kaufman, S., and Brown, L. D., 1980, The Laramide orogeny; Evidence from COCORP deep crustal seismic profiles in the Wind River Mountains, Wyoming: Tectonophysics, v. 62, p. 165–189.

Brewer, J. A., Allmendinger, R. W., Brown, L. D., Oliver, J. E., and Kaufman, S., 1982, COCORP profiling across the Rocky Mountains front in southern Wyoming; Part 1, Laramide structure: Geological Society of America Bulletin, v. 93, p. 1242–1252.

Brown, L. D., and 7 others, 1979, COCORP seismic reflection studies of the Rio Grande rift, *in* Rieker, R. E., ed., Rio Grande rift; Tectonics and magmatism: Washington, D.C., American Geophysical Union, p. 169–184.

Brown, L. D., Chapin, C. E., Sanford, A. R., Kaufman, S., and Oliver, J. E., 1980, Deep structure of the Rio Grande rift from seismic reflection profiling: Journal of Geophysical Research, v. 85, p. 4773–4800.

Burchfiel, B. C., Lipman, P. W., Eaton, G. P., and Smith, R. B., 1982, The Cordillera orogen; Conterminous United States section, *in* Palmer, A. R., ed., Perspectives in regional geologic synthesis: Geological Society of America DNAG Special Publication 1, p. 91–98.

Case, J. E., 1967, Geophysical ore guides along the Colorado mineral belt: U.S. Geological Survey Open-File Report, 16 p.

Coney, P. J., and Reynolds, S. J., 1977, Cordilleran Benioff zones: Nature, v. 270, p. 403–406.

Cook, F. A., McCullar, D. B., Decker, E. R., and Smithson, S. B., 1979, Crustal structure and evolution of the southern Rio Grande rift, *in* Rieker, R. E., ed., Rio Grande rift; Tectonics and magmatism: Washington, D.C., American Geophysical Union, p. 195–208.

Cordell, L., 1978, Regional geophysical setting of the Rio Grande rift: Geological Society of America Bulletin, v. 89, p. 1073–1090.

Daggett, P. H., Keller, G. R., Morgan, P., and Wen, C.-L., 1986, Structure of the southern Rio Grande rift from gravity interpretation: Journal of Geophysical Research, v. 91, p. 6157–6167.

Daniel, R. G., and Boore, D. M., 1982, Anomalous shear wave delays and surface wave velocities at Yellowstone caldera, Wyoming: Journal of Geophysical Research, v. 87, p. 2731–2744.

Davis, P. M., Parker, E. C., Evans, J. R., Iyer, H. M., and Olsen, K. H., 1984, Teleseismic deep sounding of the velocity structure beneath the Rio Grande rift, *in* Baldridge, W. S., Dickerson, P. W., Riecker, R. E., and Zidek, J., eds., Rio Grande rift; Northern New Mexico: Socorro, New Mexico Geological Society, p. 29–38.

Davis, T. L., and Stoughton, D., 1979, Interpretation of seismic-reflection data from the northern San Luis Valley, south-central Colorado, *in* Riecker, R. E., ed., Rio Grande rift; Tectonics and magmatism: Washington, D.C., American Geophysical Union, p. 185–194.

Davis, T. L., and Young, K.L.Y., 1977, Seismic investigation of the Colorado Front Range zone of Golden, Colorado: Rocky Mountain Association of Geology Symposium, p. 77–88.

Dickinson, W. D., and Snyder, W. S., 1979, Geometry of triple junctions and subducted slabs related to San Andreas transform: Journal of Geophysical Research, v. 84, p. 561–572.

Diment, W. H., and Urban, T. C., 1981, Average elevation map of the conterminous United States; Gilluly average method: U.S. Geological Survey Geophysical Investigations Map GP-933, scale 1:2,500,000.

Eardley, A. J., 1962, Structural geology of North America, 2nd ed.: New York, Harper and Row, p. 327–350.

—— , 1968, Major structures of the Rocky Mountains of Colorado and Utah: University of Missouri at Rolla Journal, v. 1, p. 79–99.

Eaton, G. P., 1986, A tectonic redefinition of the southern Rocky Mountains: Tectonophysics, v. 132, p. 163–193.

Eaton, G. P., and 7 others, 1975, Magma beneath Yellowstone National Park: Science, v. 188, p. 787–795.

Edel, J. B., Fuchs, K., Gelbke, C., and Prodehl, C., 1975, Deep structure of the southern Rhinegraben area from seismic-refraction investigations: Journal of Geophysics, v. 41, p. 333–356.

Elston, W. E., and Bornhorst, T. J., 1979, The Rio Grande rift in context of regional post-40 m.y. volcanic and tectonic events, *in* Riecker, R. E., ed., Rio Grande rift; Tectonics and magmatism: Washington, D.C., American Geophysical Union, p. 416–438.

Fauria, T. J., 1981, Crustal structure of the northern Basin and Range and western Snake River Plain; A ray-trace travel-time interpretation of the Eureka, Nevada, to Boise, Idaho, seismic-refraction profile [M.S. thesis]: West Lafayette, Indiana, Purdue University, 91 p.

Fenneman, N. M., 1946, Physicl divisions of the United States: U.S. Geological Survey, scale 1:7,000,000.

Finlayson, D. M., and Ansorge, J., eds., 1984, Proceedings of a workshop on Interpretation of Seismic Wave Propagation in Laterally Heterogeneous Structures: Canberra, Australia, Bureau of Mineral Resources Report 258, 207 p.

Freidline, R. A., Smith, R. B., and Blackwell, D. D., 1976, Seismicity and contemporary tectonics of the Helena, Montana, area: Seismological Society of America Bulletin, v. 66, p. 81–95.

Fuchs, K., and 7 others, 1987, Crustal evolution of the Rhinegraben area; 1, Exploring the lower crust in the Rhinegraben by unified geophysical experiments: Tectonophysics, v. 141, p. 261–275.

Gajewski, D., and Prodehl, C., 1985, Crustal structure beneath the Swabian Jura, southwest Germany, from seismic refraction investigations: Journal of Geophysics, v. 56, p. 69–80.

—— , 1987, Crustal evolution of the Rhinegraben area; 2, Crustal structure of the Black Forest, southwest Germany, from seismic-refraction investigations: Tectonophysics, v. 142, p. 27–48.

Giese, P., and Prodehl, C., 1976, Main features of crustal structure in the Alps, *in* Giese, P., Prodehl, C., and Stein, A., eds., Explosion seismology in central Europe; Data and results: Berlin-Heidelberg, Springer, p. 347–375.

Gish, D. M., Keller, G. R., and Sbar, M. L., 1981, A refraction study of deep crustal structure in the Basin and Range; Colorado Plateau of eastern Arizona: Journal of Geophysical Research, v. 86, p. 6029–6038.

Gough, D. I., 1984, Mantle upflow under North America: Nature, v. 311, p. 428–432.

Gries, R., 1983, North-south compression of Rocky Mountain foreland structures, *in* Lowell, J. D., ed., Rocky Mountains foreland basins and uplifts: Rocky Mountain Association of Geologists Symposium, p. 9–32.

Haggag, I., 1974, Die Geschwindigkeitsverteilung von Kompressionswellen in zwei Gebieten der westlichen USA, abgeleitet aus refraktionsseismischen Messungen [thesis]: Karlsruhe, West Germany, University, Geophysics Institute, 98 p.

Hales, A. L., and Nation, J. B., 1973, A seismic refraction survey in the northern Rocky Mountains; More evidence for an intermediate crustal layer: Geophysical Journal of the Royal Astronomical Society, v. 35, p. 381–399.

Hansen, W. R., 1965, Geology of the Flaming Gorge area, Utah-Colorado-Wyoming: U.S. Geological Survey Professional Paper 490, 196 p.

Harden, S. H., 1982, A seismic refraction study of west-central New Mexico [M.S. thesis]: El Paso, University of Texas, 58 p.

Hermance, J. F., 1982, Magnetotelluric and geomagnetic deep-sounding studies in rifts and adjacent areas; Constraints on physical processes in the crust and upper mantle, *in* Palmason, G., ed., Continental and oceanic rifts: American Geophysical Union Geodynamics Series, v. 8, p. 169–192.

Hill, D. P., and Pakiser, L. C., 1967, Seismic-refraction study of crustal structure between the Nevada Test Side and Boise, Idaho: Geological Society of America Bulletin, v. 78, p. 685–704.

Hirn, A., and Perrier, G., 1974, Deep seismic sounding in the Limagnegraben, *in* Illies, J. H., and Fuchs, K., eds., Approaches to taphrogenesis: Stuttgart, Schweizerbart, p. 329–340.

Holbrook, W. S., Gajewski, D., and Prodehl, C., 1987, Shear-wave velocity and Poisson's ratio structure of the upper lithosphere in southwest Germany: Geophysical Research Letters, v. 14, p. 231–234.

Holbrook, W. S., Gajewski, D., Krammer, A., and Prodehl, C., 1988, An interpretation of wide-angle compressional and shear ware data in southwest Germany; Poisson's ratio and petrological implications: Journal of Geophysical Research, v. 93, p. 12081–12106.

Iyer, H. M., 1975, Anomalous delays of teleseismic P waves in Yellowstone National Park: Nature, v. 253, p. 425–427.

——, 1979, Deep structure under Yellowstone National Park, U.S.A.: A continental "hot spot": Tectonophysics, v. 56, p. 165–197.

Jackson, W. H., and Pakiser, L. C., 1965, Seismic study of crustal structure in the southern Rocky Mountains: U.S. Geological Survey Professional Paper 525-D, p. 85–92.

Jackson, W. H., Stewart, S. W., and Pakiser, L. C., 1963, Crustal structure in eastern Colorado from seismic-refraction measurements: Journal of Geophysical Research, v. 68, p. 5767–5776.

Jacob, A.W.B., Kaminski, W., Murphy, T., Phillips, W.E.A., and Prodehl, C., 1985, A crustal model for a northeast-southwest profile through Ireland: Tectonophysics, v. 113, p. 75–113.

Jaksha, L. H., 1982, Reconnaissance seismic refraction-reflection surveys in southwestern New Mexico: Geological Society of America Bulletin, v. 93, p. 1030–1037.

Jiracek, G. R., Gustafson, E. P., and Mitchell, P. S., 1983, Magnetotelluric results opposing magma origin of crustal conductors in the Rio Grande rift: Tectonophysics, v. 94, p. 299–326.

Jiracek, G. R., Rodi, W. L., and Vanyan, L. L., 1987, Implications of magnetotelluric modeling for the deep crust environment of the Rio Grande rift: Physics of Earth and Planetary Interiors, v. 45, p. 179–192.

Johnson, R. A., and Smithson, S. B., 1986, Interpretive processing of crustal seismic reflection data; Examples from Laramie Range COCORP data, *in* Barazangi, M., and Brown, L., eds., Reflection seismology; A global perspective: American Geophysical Union Geodynamic Series, v. 13, p. 197–208.

Johnson, R. A., Karlstrom, K. E., Smithson, S. B., and Houston, R. S., 1984, Gravity profiles across the Cheyenne belt; A Precambrian crustal suture in southeastern Wyoming: Journal of Geodynamics, v. 1, p. 445–472.

Karlstrom, K. E., and Houston, R. S., 1984, The Cheyenne belt; Analysis of a Proterozoic nature in southern Wyoming: Precambrian Research, v. 25, p. 415–446.

Keller, G. R., Smith, R. B., and Braile, L. W., 1975, Crustal structure along the Great Basin—Colorado Plateau transition from seismic refraction studies: Journal of Geophysical Research, v. 80, p. 1093–1098.

Keller, G. R., Smith, R. B., Braile, L. W., Heaney, R., and Shurbet, D. H., 1976, Upper crustal structure of the eastern Basin and Range, northern Colorado Plateau, and middle Rocky Mountains from Rayleigh-wave dispersion: Bulletin of the Seismological Society of America, v. 66, p. 869–876.

Keller, G. R., Braile, L. W., and Morgan, P., 1979a, Crustal structure, geophysical models, and contemporary tectonism of the Colorado Plateau: Tectonophysics, v. 61, p. 131–147.

Keller, G. R., Braile, L. W., and Schlue, J. W., 1979b, Regional crustal structure of the Rio Grande rift from surface wave dispersion measurements, *in* Riecker, R. E., ed., Rio Grande rift; Tectonics and magmatism: Washington, D.C., American Geophysical Union, p. 115–126.

King, P. B., 1977, The evolution of North America, rev. ed.: Princeton, New Jersey, Princeton University Press, 197 p.

KRISP Working Group, 1987, Structure of the Kenya rift from seismic refraction: Nature, v. 325, p. 239–242.

Larson, E. E., Patterson, P. E., Curtis, G., Drake, R., and Mutschler, F. E., 1985, Petrologic, paleomagnetic, and structural evidence of a Paleozoic rift system in Oklahoma, New Mexico, Colorado, and Utah: Geological Society of America Bulletin, v. 96, p. 1364–1372.

Lehman, J. A., Smith, R. B., and Schilly, M. M., 1982, Upper crustal structure of the Yellowstone caldera from seismic delay time analyses and gravity correlations: Journal of Geophysical Research, v. 87, p. 2713–2730.

Lipman, P. W., 1980, Cenozoic volcanism in the western United States; Implications for continental tectonics, *in* Studies in geophysics; Continental tectonics: Washington, D.C., National Academy of Sciences, p. 161–174.

——, 1982, Tectonic setting of the mid- to late Tertiary in the Rocky Mountain region; A review, *in* Proceedings, The genesis of Rocky Mountain ore deposits; Changes with time and tectonics: Society of Exploration Geophysicists Denver Region, p. 125–132.

Lipman, P. W., Christiansen, R. L., and Prostka, H. J., 1971, Evolving subduction zones in the western United States, as interpreted from igneous rocks: Science, v. 148, p. 821–825.

Lynn, H. B., Quam, S., and Thompson, G. A., 1983, Depth migration and interpretation of the COCORP Wind River, Wyoming, seismic reflection data: Geology, v. 11, p. 462–469.

Lyons, P. L., and O'Hara, N. W., compilers, 1982, Gravity anomaly map of the United States, exclusive of Alaska and Hawaii: Society of Exploration Geophysicists, scale 1:2,500,000.

Maguire, P.K.H., and Long, R. E., 1976, The structure on the western flank of the Gregory rift, Kenya; Part 1, The crust: Geophysical Journal of the Royal Astronomical Society, v. 44, p. 661–675.

Martin, W. R., 1978, A seismic refraction study of the northeastern Basin and Range and its transition with the eastern Snake River Plain [M.S. thesis]: El Paso, University of Texas.

McCamy, K., and Meyer, R. P., 1964, A correlation method of apparent velocity measurement: Journal of Geophysical Research, v. 69, p. 691–699.

Meyer, R. P., Steinhart, J. S., and Bonini, W. E., 1961a, Montana 1959, *in* Steinhart, J. S., and Meyer, R. P., eds., Explosion studies of continental structure: Washington, D.C., Carnegie Institution of Washington Publication 622, p. 305–343.

Meyer, R. P., Steinhart, J. S., and Woollard, G. P., 1961b, Central Plateau, Mexico, 1957, *in* Steinhart, J. S., and Meyer, R. P., eds., Explosion studies of continental structure: Washington, D.C., Carnegie Institution of Washington Publication 622, p. 199–225.

Mooney, W. D., and Prodehl, C., eds., 1984, Proceedings of the 1980 workshop of the International Association of Seismology and Physics of the Earth's Interior on the seismic modeling of laterally varying structures; Contributions based on data from the 1978 Saudi Arabian refraction profile: U.S. Geological Survey Circular 937, 158 p.

Mooney, W. D., Andrews, M. C., Ginzburg, A., Peters, D. A., and Hamilton, R. M., 1983, Crustal structure of the northern Mississippi embayment and a comparison with other continental rift zones: Tectonophysics, v. 94, p. 327–348.

Mota, L., 1954, Determination of dips and depths of geological layers by the seismic refraction method: Geophysics, v. 19, p. 242–254.

Mueller, S., Peterschmitt, E., Fuchs, K., Emter, D., and Ansorge, J., 1973, Crustal structure of the Rhinegraben area: Tectonophysics, v. 20, p. 381–391.

Oliver, J., 1982, Tracing surface features to great depths; A powerful means for exploring the deep crust: Tectonophysics, v. 81, p. 257–272.

Olsen, K. H., Keller, G. R., and Stewart, J. N., 1979, Crustal structure along the Rio Grande rift from seismic refraction profiles, *in* Riecker, R. E., ed., Rio Grande rift; Tectonics and magmatism: Washington, D.C., American Geophysical Union, p. 127–143.

Olsen, K. H., Cash, D. J., and Stewart, J. N., 1982, Mapping the northern and eastern extent of the Socorro midcrustal magma body by wide-angle seismic reflections: New Mexico Geological Society Guidebook 33, p. 179–185.

Olsen, K. H., and 6 others, 1986, Jemez Mountains volcanic field, New Mexico; Time term interpretation of the CARDEX seismic experiment and comparison with Bouguer gravity: Journal of Geophysical Research, v. 91, p. 6175–6187.

Pakiser, L. C., 1963, Structure of the crust and upper mantle in the western United States: Journal of Geophysical Research, v. 68, p. 5747–4756.

——, 1985, Seismic exploration of the crust and upper mantle of the Basin and Range province: Geological Society of America DNAG Centennial Special Volume 1, p. 453–469.

Pakiser, L. C., and Steinhart, J. S., 1964, Explosion seismology in the western hemisphere, *in* Odishaw, H., ed., Research in geophysics; Volume 2, Solid earth and interface phenomena: Cambridge, Massachusetts Institute of Technology Press, p. 123–147.

Parker, E. C., Davis, P. M., Evans, J. R., Iyer, H. M., and Olsen, K. H., 1984, Upwarp of anomalous asthenosphere beneath the Rio Grande rift: Nature, v. 312, p. 354–356.

Pelton, J. R., and Smith, R. B., 1982, Contemporary vertical surface displacement in Yellowstone National Park: Journal of Geophysical Research, v. 87, p. 2745–2761.

Perkins, W. E., and Phinney, R. A., 1971, A reflection study of the Wind River uplift, Wyoming, *in* Heacock, J. G., ed., The structure and physical properties of the Earth's crust: American Geophysical Union Geophysical Monograph 14, p. 41–50.

Pitt, A. M., and Hutchison, R. A., 1982, Seismicity and heat flux at Mud Volcano, Yellowstone National Park, Wyoming: Journal of Geophysical Research, v. 87, p. 2762–2766.

Plouff, D., and Pakiser, L. C., 1972, Gravity study of the San Juan Mountains, Colorado: U.S. Geological Professional Paper 800-B, p. 183–190.

Porath, H., 1971, Magnetic variation anomalies and seismic low-velocity zone in the western United States: Journal of Geophysical Research, v. 76, p. 2643–2648.

Porath, H., and Gough, D. I., 1971, Mantle conductive structures in the western United States from magnetometer array studies: Geophysical Journal of the Royal Astronomical Society, v. 22, p. 261–275.

Prodehl, C., 1970, Seismic refraction study of crustal structure of the western United States: Geological Society of America Bulletin, v. 81, p. 2629–2646.

——, 1977, The structure of the crust-mantle boundary beneath North America and Europe as derived from explosion seismology, *in* Heacock, J. G., ed., The earth's crust: American Geophysical Union Geophysical Monograph 20, p. 349–369.

——, 1979, Crustal structure of the western United States: U.S. Geological Survey Professional Paper 1034, 74 p.

——, 1981, Structure of the crust and upper mantle beneath the central European rift system: Tectonophysics, v. 80, p. 255–269.

——, 1984, Structure of the Earth's crust and upper mantle, *in* Fuchs, K., and Soffel, H., eds., Landolt-Börnstein; Numerical data and functional relationships in science and technology; Series 5, v. 2a, Geophysics of the solid Earth, the moon, and the planets: Berlin-Heidelberg, Springer, p. 97–206.

Prodehl, C., and Pakiser, L. C., 1980, Crustal structure of the southern Rocky Mountains from seismic measurements: Geological Society of America Bulletin, v. 91, p. 147–155.

Reiter, M., Mansure, A. J., and Shearer, C., 1979, Geothermal characteristics of the Rio Grande rift within the southern Rocky Mountain complex, *in* Riecker, R. E., ed., Rio Grande rift; Tectonics and magmatism: Washington, D.C., American Geophysical Union, p. 253–267.

Reiter, M., Eggleston, R. E., Broadwell, B. R., and Minier, J., 1986, Estimates of terrestrial heat flow from deep petroleum tests along the Rio Grande rift in central and southern New Mexico: Journal of Geophysical Research, v. 91, p. 6225–6245.

Rinehart, E. J., Sanford, A. R., and Ward, R. M., 1979, Geographic extent and shape of an extensive magma body at mid-crustal depths in the Rio Grande rift near Socorro, New Mexico, *in* Riecker, R. E., ed., Rio Grande rift; Tectonics and magmatism: Washington, D.C., American Geophysical Union, p. 237–251.

Roller, J. C., 1965, Crustal structure in the eastern Colorado Plateau province from seismic-refraction measurements: Bulletin of the Seismological Society of America, v. 55, p. 107–119.

Roller, J. C., and Jackson, W. H., 1966, Seismic wave propagation in the upper mantle; Lake Superior, Wisconsin, to central Arizona: Journal of Geophysical Research, v. 71, p. 5933–5941.

Roy, R. F., Blackwell, D. D., and Decker, E. R., 1972, Continental heat flow, *in* Robertson, E., ed., The nature of the solid earth: New York, McGraw-Hill, p. 506–543.

Ryall, A., and Stuart, D. J., 1963, Travel times and amplitudes from nuclear explosions, Nevada Test Side to Ordway, Colorado: Journal of Geophysical Research, v. 68, p. 5821–5835.

Sanford, A. R., and Long, L. T., 1965, Microearthquake crustal reflections, Socorro, New Mexico: Bulletin of the Seismological Society of America, v. 55, p. 579–586.

Sanford, A. R., Alptekin, O. S., and Toppozada, T. R., 1973, Use of reflection phases on microearthquake seismograms to map an unusual discontinuity beneath the Rio Grande rift: Bulletin of the Seismological Society of America, v. 63, p. 2021–2034.

Sanford, A. R., and 6 others, 1977, Geophysical evidence for a magma body in the crust in the vicinity of Socorro, New Mexico, *in* Heacock, J. G., ed., The earth's crust: American Geophysical Union Geophysical Monograph 20, p. 385–403.

Sanford, A. R., Olsen, K. H., and Jaksha, L. H., 1979, Seismicity of the Rio Grande rift, *in* Riecker, H. E., ed., Rio Grande rift; Tectonics and magmatism: Washington, D.C., American Geophysical Union, p. 145–168.

Sbar, M. L. Barazangi, M., Dorman, J., Scholz, C. H., and Smith, R. B., 1972, Tectonics of the intermountain seismic belt, western United States; Microearthquake seismicity and composite fault plane solutions: Geological Society of America Bulletin, v. 83, p. 13–28.

Schilly, M. M., Smith, R. B., Braile, L. W., and Ansorge, J., 1982, The 1978 Yellowstone—eastern Snake River Plane seismic profiling experiment; Data and upper crustal structure of the Yellowstone region: Journal of Geophysical Research, v. 87, p. 2692–2704.

Schmucker, U., 1970, Anomalies of geomagnetic variation in the southwestern United States: Bulletin of the Scripps Institution of Oceanography, v. 13, p. 33–53.

Scott, G. R., 1975, Cenozoic surfaces and deposits in the southern Rocky Mountains: Geological Society of America Memoir 144, p. 227–248.

Sharry, J., Lungan, R. T., Janovich, D. B., Jones, G. M., Hill, N. R., and Guidish, T. M., 1986, Enhanced imaging of the COCORP seismic line, Wind River Mountains, *in* Barazangi, M., and Brown, L., eds., Reflection seismology; A global perspective: American Geophysical Union Geodynamics Series 13, p. 223–236.

Shuey, R. T., Schellinger, D. K., Johnson, E. H., and Alley, L. B., 1973, Aeromagnetics and the transition between the Colorado Plateau and Basin and Range provinces: Geology, v. 1, p. 107–110.

Sinno, Y. A., and Keller, G. R., 1986, A Rayleigh wave dispersion study between El Paso, Texas, and Albuquerque, New Mexico: Journal of Geophysical Research, v. 91, p. 6168–6174.

Sinno, Y. A., Daggett, P. H., Keller, G. R., Morgan, P., and Harder, S. H., 1986, Crustal structure of the southern Rio Grande rift determined from seismic refraction profiling: Journal of Geophysical Research, v. 91, p. 6143–6156.

Skeen, R. C., and Ray, R. R., 1983, Seismic models and interpretation of the Caspar Arch thrust; Application to Rocky Mountain foreland structure, *in* Lowell, J. D., ed., Rocky Mountain foreland basins and uplifts: Rocky Mountain Association of Geologists Symposium, p. 99–124.

Smith, R. B., 1977, Intraplate tectonics of the western North American plate: Tectonophysics, v. 37, p. 323–336.

—— , 1978, Seismicity, crustal structure, and intraplate tectonics of the interior of the western Cordillera, *in* Smith, R. B., and Eaton, G. P., eds., Cenozoic tectonics and regional geophysics of the western Cordillera: Geological Society of America Memoir 152, p. 111–144.

—— , 1982, Preface to Yellowstone—Snake River Plain symposium papers: Journal of Geophysical Research, v. 87, p. 2581.

Smith, R. B., and Christiansen, R. L., 1980, Yellowstone Park as a window on the Earth's interior: Scientific American, v. 242, p. 104–117.

Smith, R. B., and Sbar, M. L., 1974, Contemporary tectonics and seismicity of the western United States with emphasis on the intermountain seismic belt: Geological Society of America Bulletin, v. 85, p. 1205–1218.

Smith, R. B., Shuey, R. T., Freidline, R. O., Otis, R. M., and Alley, L. B., 1974, Yellowstone Hot Spot; New magnetic and seismic evidence: Geology, v. 2, p. 451–455.

Smith, R. B., and 9 others, 1982, The 1978 Yellowstone–eastern Snake River Plain seismic profiling experiment; Crustal structure of the Yellowstone region and experiment design: Journal of Geophysical Research, v. 87, p. 2583–2596.

Smithson, S. B., Brewer, J. A., Kaufman, S., Oliver, J. E., and Hurich, C. A., 1979, Structure of the Laramide Wind River uplift, Wyoming, from COCORP deep reflection data and from gravity data: Journal of Geophysical Research, v. 84, p. 5955–5972.

Sparlin, M. A., Braile, L. W., and Smith, R. B., 1982, Crustal structure of the eastern Snake River Plain determined from ray-trace modeling of seismic refraction data: Journal of Geophysical Research, v. 87, p. 2619–2633.

Spence, W., Gross, R. S., and Jaksha, L. H., 1979, P-wave velocity structure beneath the central Rio Grande rift: EOS Transactions of the American Geophysical Union, v. 60, p. 953–954.

—— , 1982, P-wave velocity structure beneath the Jemez lineament, New Mexico: EOS Transactions of the American Geophysical Union, v. 63, p. 1117–1118.

Stewart, S. W., and Pakiser, L. C., 1962, Crustal structure in eastern New Mexico interpreted from the GNOME explosion: Bulletin of the Seismological Society of America, v. 52, p. 1017–1030.

Taylor, R. B., 1975, Neogene tectonism in south-central Colorado: Geological Society of America Memoir 144, p. 211–226.

Thompson, G. A., and Hill, J. L., 1986, The deep crust in convergent and divergent terranes; Laramide uplifts and Basin–Range rifts, *in* Barazangi, M., and Brown, L., eds., Reflection seismology; The continental crust: American Geophysical Union Geodynamics Series 14, p. 243–256.

Thompson, G. A., and Zoback, M. L., 1979, Regional geophysics of the Colorado Plateau: Tectonophysics, v. 61, p. 149–181.

Toppozada, T. R., and Sanford, A. R., 1976, Crustal structure in central New Mexico interpreted from the Gasbuggy explosion: Bulletin of the Seismological Society of America, v. 66, p. 877–886.

Tweto, O., 1975, Laramide (late Cretaceous–early Tertiary) orogeny in the southern Rocky Mountains: Geological Society of America Memoir 144, p. 1–44.

—— , 1979, The Rio Grande rift system in Colorado, *in* Riecker, R. E., ed., Rio Grande rift; Tectonics and magmatism: Washington, D.C., American Geophysical Union, p. 33–56.

Warren, D. H., 1968, Transcontinental geophysical survey (35°–39°N) seismic refraction profiles of the crust and upper mantle from 100° to 112° longitude: U.S. Geological Survey Miscellaneous Geological Investigations Map I-533-D, scale 1:1,000,000.

—— , 1969, A seismic-refraction survey of crustal structure in central Arizona: Geological Society of America Bulletin, v. 80, p. 257–282.

—— , 1975, Record sections for the seismic refraction profile Agate–Concordia, eastern Colorado and western Kansas: U.S. Geological Survey Open-File Report 75-380, 86 p.

Warren, D. H., and Jackson, W. H., 1968, Surface seismic measurements of the project GASBUGGY explosion at intermediate distance ranges: U.S. Geological Survey Open-File Report, 45 p.

Warren, D. H., Healy, J. H., Bohn, J., and Marshall, P. A., 1972, Crustal calibration of the Large-Aperture Seismic Array (LASA), Montana: U.S. Geological Survey Open-File Report, 163 p.

Weidelt, P., 1975, Inversion of two-dimensional conductivity structures: Physics of Earth and Planetary Interiors, v. 10, p. 282–291.

Wen, J., and McMechan, G. A., 1985, Application of two-dimensional synthetic seismogram modeling to interpretation of wide-angle seismic data from Wind River, Wyoming: Journal of Geophysical Research, v. 90, p. 3617–3625.

Willden, R., 1965, Seismic refraction measurements of crustal structure between American Falls Reservoir, Idaho, and Flaming Gorge Reservoir, Utah: U.S. Geological Survey Professional Paper 525-C, p. 44–50.

Zietz, I., 1982, Composite magnetic anomaly map of the United States; Part A, Conterminous United States: U.S. Geological Survey Geophysical Investigations Map 954-A, scale 1:2,500,000.

Zoback, M. L., and Lachenbruch, A. H., 1984, Upper mantle structure beneath the western U.S.: Geological Society of America Abstracts with Programs, v. 16, p. 705.

MANUSCRIPT ACCEPTED BY THE SOCIETY OCTOBER 31, 1988

Printed in U.S.A.

Geological Society of America
Memoir 172
1989

Chapter 15

Crustal structure of the continental interior

L. W. Braile
Department of Earth and Atmospheric Sciences, Purdue University, West Lafayette, Indiana 47907

ABSTRACT

Seismic refraction profiles and deep seismic reflection profiling data have been compiled for the continental interior of the United States and adjacent southern Canada. A total of 58 refraction profiles and 17 deep reflection profiling studies are available in the midcontinental area. Statistics derived from the refraction data indicate that the crust of the continental interior is characterized by a crustal thickness of about 43 km, an average crustal compressional seismic velocity of 6.5 km/sec, and a nearly uniform uppermost mantle compressional velocity (P_n) of 8.1 km/sec. The crust displays a generally layered character with compressional velocities near 6.1 km/sec for the upper 10 to 20 km and about 6.8 km/sec for the lower crust. In some areas, generally near the southern and western margins of the relatively undeformed craton, a high-velocity (approximately 7.0 to 7.4 km/sec) lower crustal layer 10 to 20 km thick is present just above the Moho. These areas also display higher than average mean crustal velocity and crustal thickness. Contour diagrams of crustal seismic data also identify a crustal structure anomaly consisting of higher average crustal velocity and crustal thickness associated with the Midcontinent Rift System. Seismic reflection profiling data, which are primarily available in areas of the midcontinent that are known to be anomalous, provide improved resolution of structural features of the continental crust, particularly in areas characterized by the presence of thick (5 to 20 km or more) basins filled with Proterozoic sedimentary or volcanic rocks. For other parts of the crystalline crust, seismic reflections are, in general, laterally discontinuous and commonly dip moderately, relative to shallow reflections and reflections from the underlying crust-mantle boundary.

INTRODUCTION

The crystalline crust of the continental interior of North America is exposed in only a relatively small number of locations in the United States. However, broad exposures of Superior, Churchill, and Grenville Province rocks crop out in the Canadian Shield just north of the U.S. boundary. The crystalline rocks of the cratonic basement of North America represent a complex geologic history. In addition to the exposures in Canada, in the United States where Paleozoic sedimentary rocks cover the craton, geophysical studies, subsurface petrologic and isotopic age data on rocks, and rare outcrops of the basement have been used to identify Precambrian tectonic features such as the Midcontinent Rift System and the Reelfoot Rift in the Mississippi Embayment, basement geologic provinces, and geochronologic boundaries. Recent reviews of geological, geochemical, and geophysical data on the midcontinent region have been published by Lidiak (1982), Van Schmus and Bickford (1981), Denison and others (1984), Hinze and Braile (1988) and Hamilton (this volume). Because of the limited exposures of the crystalline crust in the continental interior, geophysical studies, particularly seismic profiling, are the primary source of information on the structure and physical properties of the continental crust.

In this chapter, I review and analyze the regional crustal seismic data for the midcontinent area of the United States and southern Canada. The area of interest is bounded approximately by the Appalachian Mountains on the east, the Front Range of the Rocky Mountains on the west, and the Gulf Coastal Plain on the south. This definition of the midcontinent province of central North American coincides approximately with the edge of the

Braile, L. W., 1989, Crustal structure of the continental interior, *in* Pakiser, L. C., and Mooney, W. D., Geophysical framework of the continental United States: Boulder, Colorado, Geological Society of America Memoir 172.

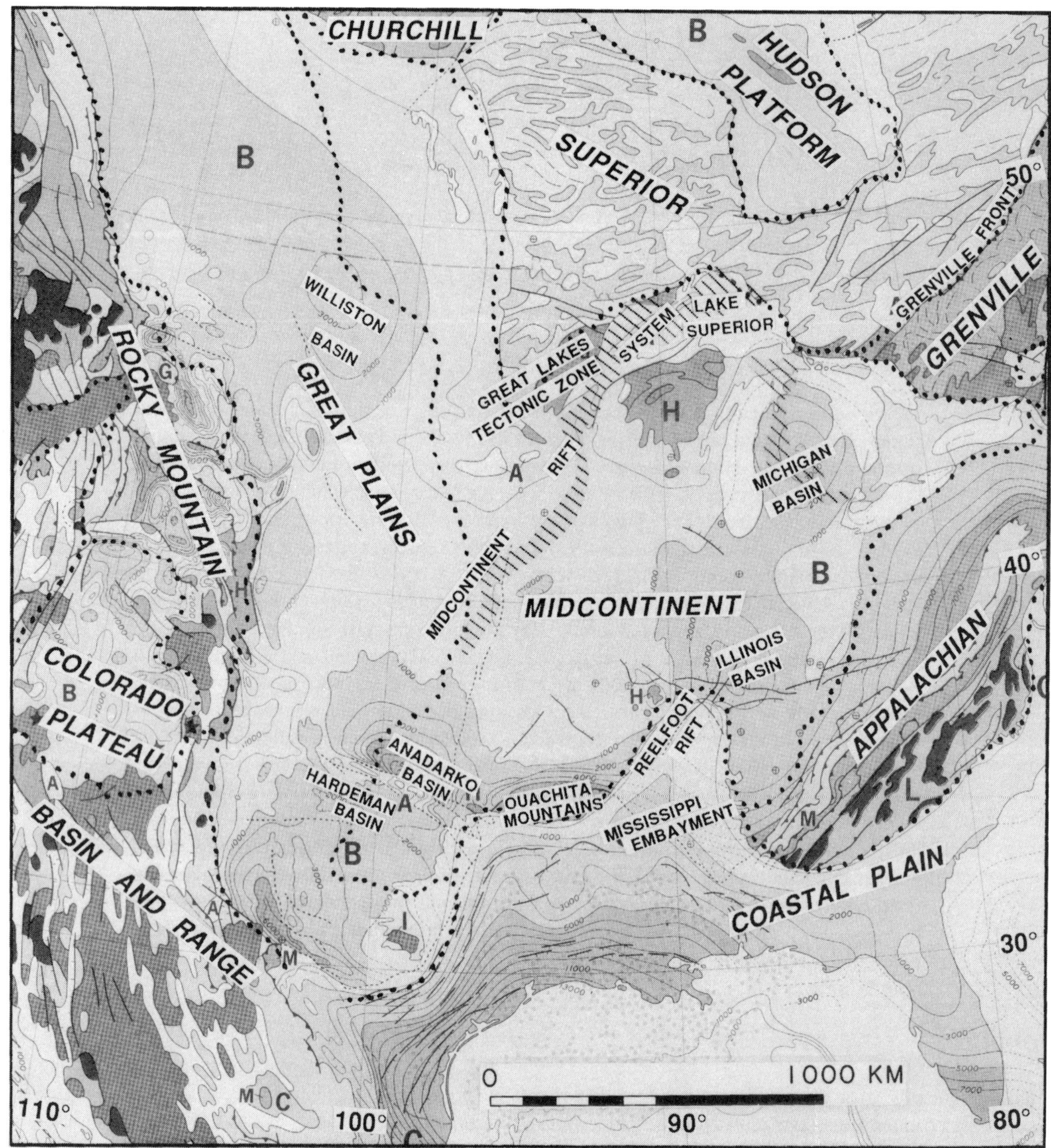

Figure 1. Index map of the midcontinent region of the conterminous United States and southern Canada. Principal tectonic units, geologic provinces, and locations of major basins are shown. Contours are basement depth (m). Tectonic map is from King and Edmonston, (1972). Province boundaries for the United States are from Fenneman (1946) and for Canada from Douglas and Price (1972).

stable craton, which has been relatively undeformed since Precambrian time. Figure 1 is an index map of the midcontinent region showing major tectonic and geographic features.

Because the crustal seismic data for the midcontinent area are relatively small in volume and poorly distributed, our analysis of the nature of the Earth's crust in the midcontinent will be limited largely to a regional characterization. A few areas of the midcontinent have sufficient data coverage to allow a more detailed view. One objective of this study is a comprehensive and up-to-date review of crustal seismic data for the midcontinent region. For this purpose, a number of previously published crustal seismic studies are described and illustrated. This selective review serves not only to describe the main characteristics of the continental crust in the midcontinent region, but also to illustrate the type and quality of observed seismic data and analyses that have been used to interpret the nature of the continental crust. An additional objective of this paper is to provide a summary and synthesis of the current state of knowledge of the midcontinental crust. For this purpose, individual crustal seismic studies have been correlated along selected west to east profiles to provide interpretive cross-sectional diagrams of the crust and uppermost mantle seismic velocity structure. Additionally, contour maps of important crustal structure parameters have also been prepared to illustrate the regional variations of crustal structure throughout the midcontinent province.

In this study, I have used primarily the results of crustal seismic refraction profiles that have been conducted in the midcontinent region in the last 40 years. Although other seismic data are available for interpretation of the crust of the midcontinent region, they are generally so widely spaced that they provide limited information for a regional summary and comparison. Velocity structure interpreted from analysis of converted phases from teleseismic arrivals has been presented by Jordan and Frazer (1975) for the Canadian Shield, Kurita (1973) for the central midcontinent region, and Owens and others (1984) and Zandt and Owens (1986) for the Cumberland Plateau. With the exception of the Mississippi Embayment region, little information on velocity structure is available from regional earthquake data because of the relatively small number of events and widely spaced networks. Apparent velocities from regional earthquake arrivals provide useful information on the velocity of crustal layers or of the upper mantle in some parts of the midcontinent. However, the sparse distribution of earthquakes and the ambiguity of interpretation due to large station spacings and the poorly known hypocentral locations limit the use of such data for crustal seismic studies. Recently, deep seismic reflection data have become available for a number of regions in the midcontinent. These data provide improved resolution on specific features of crustal structure, particularly for shallow to intermediate depths (0-20 km) within the crust where seismic wave-energy penetration is greatest and signal-to-noise ratios are highest. Analysis of the reflection data for the midcontinent region and comparison with refraction results have been included here for several areas. Surface-wave dispersion analysis provides a possible additional source of crustal seismic data for the midcontinent region. However, these data are again generally widely spaced and are primarily useful for province-wide analyses. The available surface-wave dispersion data for the entire United States and southernmost Canada continental region have been compiled and interpreted in a regional summary by Braile and others (this volume).

SEISMIC DATA

A compilation, organized according to provinces, of crustal seismic profiles for the midcontinent region is contained in Table 1, and their locations are illustrated in Figure 2. The numbers assigned to the profiles in Table 1 are keyed to the provinces and are consistent with the numbers used in the continental compilation presented by Braile and others (this volume). Only profiles of sufficient length and number of seismograph stations to provide velocity structure information on a significant portion of the continental crust have been used. The midcontinent area includes approximately half of the total surface area of the continental crust of the United States (including the continental margins). However, the midcontinental crustal seismic data represent only about one-fourth to one-third of the data that are available for the entire continental United States. In addition to the relative sparsity of data for the midcontinental crust, the available seismic profiles are very irregularly distributed. A substantial number of profiles are located in three specific areas: the Williston basin, the Lake Superior region, and the upper Mississippi Embayment. Large areas of the midcontinent contain almost no crustal seismic data. These areas include the central Great Plains (South Dakota, Nebraska, Kansas), the northern Great Plains (Alberta and Saskatchewan), the midcontinent area just south of the Great Lakes (Iowa, Illinois, Indiana, Ohio, Michigan, and western Pennsylvania), and the Canadian Shield just north of the Great Lakes (Fig. 2). A large number of the crustal seismic profiles for the midcontinental region are in areas of known anomalous crust, such as the Williston basin, Midcontinent Rift System of Lake Superior, the northern Mississippi Embayment, and along the Grenville Front. A relatively small number of profiles appear to be located on the "normal" craton. Similarly, most deep seismic reflection profiles (Fig. 2 and Table 2) have been located over known crustal structure anomalies.

Much of the available seismic refraction/wide-angle reflection data for the midcontinent region were recorded over 20 years ago and thus predate modern methods of data collection, processing, and interpretation. Some of the older data and interpretations are excellent. Others, however, are poor because of the use of analog recording methods, of large station spacings, and of exclusively first-arrival interpretation. Recent crustal seismic data in the midcontinent region have been collected in the Canadian CO-CRUST project in the Williston basin and Grenville regions and by the U.S. Geological Survey in the upper Mississippi Embayment. The midcontinental crustal seismic data have been interpreted by a variety of methods, sometimes resulting in quite different interpretations even for profiles within almost the same geographic area or for adjacent surveys. The time-term method was used for interpretation of many of the older crustal seismic data sets. In this method (Mooney, this volume) a two-dimensional array of seismic stations is analyzed for first-arrival times to obtain the velocity of prominent refractors and approximate depths to the refractor. Because of the wide spacing of stations and the two-dimensional array, the method is not generally conducive to display in the form of a cross-sectional profile. Although the refractor velocities are often well determined by this technique, the wide distribution of stations, restrictive assumptions of the method, and ambiguity in the shallow velocity structure can limit the reliability of interpretation of depths to the refractor. An example of the use of the time-term method for analysis of the crustal velocity structure in the midcontinent region is the study by Berry and West (1966) for the Lake Superior region.

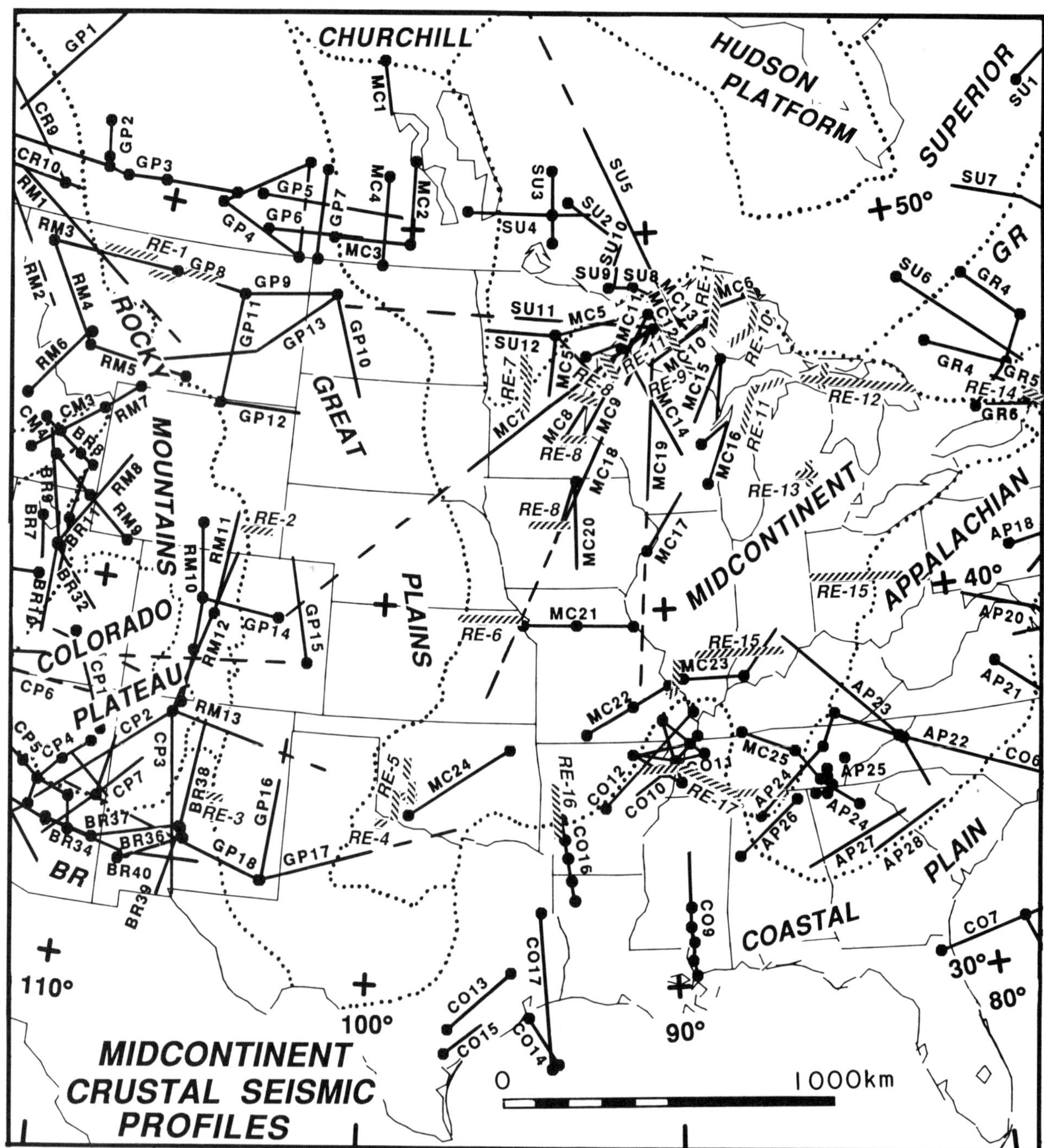

Figure 2. Index map of the midcontinent and adjacent regions showing locations of crustal seismic data. Solid lines are refraction profiles. Profiles are dashed for large source-receiver distances where only limited information on crustal velocity structure is available. Dots are shotpoint locations. Some shotpoint locations are not shown for closely spaced shotpoints. Some data sets recorded in arrays instead of long profiles are illustrated schematically. Numbers refer to index of profile names contained in Table 1. Locations of deep seismic reflection profiles are shown by ruled pattern. Numbers adjacent to the reflection profiles refer to index of profile names contained in Table 2. Dotted lines are province boundaries (Table 1) from Fenneman (1946) for the United States and Douglas and Price (1972) for Canada. The Grenville province is labeled GR, and the Basin and Range province is labeled BR.

TABLE 1. SEISMIC REFRACTION PROFILES FOR THE MIDCONTINENT UNITED STATES AND ADJACENT AREAS

No.*	Profile	Reference
AP18	Milroy - E	Katz, 1954
AP20	N. Middle Atlantic States - ECOOE	James and others, 1968; Merkel and Alexander, 1969
AP21	S. Middle Atlantic States - ECOOE	James and others, 1968
AP22	S. Profile ECOOE	Hales and others, 1968
AP23	E. Tennessee	Steinhart and Meyer, 1961; Tatel and Tuve, 1955
AP24	E. Tennessee	Warren, 1968; Prodehl and others, 1984
AP25	S. Appalachian Time Terms	Long and Liow, 1986
AP26	S. Appalachian	Long and Liow, 1986
AP27	N. Georgia	Dorman, 1972
AP28	Georgia–South Carolina	Kean and Long, 1980
BR7	E. Basin and Range	Berg and others, 1960
BR8	American Falls–Flaming Gorge	Willden, 1965
BR9	Bingham - N	Martin, 1978
BR10	Bingham - S	Keller and others, 1975
BR11	Bingham - NE	Braile and others, 1974
BR32	N. Utah	Tatel and Tuve, 1955; Steinhart and Meyer, 1961
BR34	Globe–Tyrone	Gish and others, 1981
BR36	Tyrone–Distant Runner	Sinno and others, 1986
BR37	Dice Throw - W	Jaksha, 1982
BR38	Dice Throw	Olsen and others 1979
BR39	Mill Race–U.S.–Mexican Border	Sinno and others, 1986
BR40	Southern Rio Grande Rift	Cook and others, 1979
CM3	Eastern Snake River Plain	Braile and others, 1982b
CM4	Conda - SP7, Y-SRP	Sparlin and others, 1982
CO6	S. Profile ECOOE	Hales and others, 1968
CO7	Jacksonville - E	Hersey and others, 1959
CO9	S. Mississippi	Warren and others, 1966
CO10	Mississippi Embayment Axial	Mooney and others, 1983
CO11	Mississippi Embayment	Ginzburg and others, 1983
CO12	Cape Girardeau–Little Rock	McCamy and Meyer, 1966
CO13	Cleveland–Victoria, Texas	Cram, 1961
CO14	Galveston	Ewing and others, 1955
CO15	Victoria, Texas	Dorman and others, 1972
CO16	Ouachita–PASSCAL	Keller and others, 1989; Jardine, 1988
CO17	SE Texas	Hales and others, 1970
CP1	Hanksville–Chinle	Roller, 1965
CP2	Gasbuggy - SW	Warren and Jackson, 1968
CP3	Gasbuggy - S	Toppozada and Sanford, 1976
CP4	Gila Bend–Sunrise	Warren, 1969; Healy and Warren, 1969
CP5	Blue Mountain–Bylas	Warren, 1969; Healy and Warren, 1969
CP6	Bilby - SE	Archambeau and others, 1969
CP7	Arizona–New Mexico	Tatel and Tuve, 1955; Steinhart and Meyer, 1961
CR9	Rocky Mountain Trench	Bennett and others, 1975; Spence and others, 1977
CR10	Kaiser–Highland Valley	Cumming and others, 1979
GP1	Edzoe–Fort McMurray	Mereu and others, 1976
GP2	Ripple Rock	Richards and Walker 1959
GP3	Greenbush Lake - E	Chandra and Cumming, 1972
GP4	S. Saskatchewan	Kanasewich and Chiu, 1985; Morel-a-L'Huissier and others, 1987
GP5	Saskatchewan - EDZOE	Bates and Hall, 1975
GP6	Assiniboia–Carlyle, Saskatchewan	Hajnal and others, 1984
GP7	SE. Saskatchewan	Hajnal and others, 1984; Morel-a-L'Huissier and others, 1987
GP8	Big Sandy River–Fort Peck	McCamy and Meyer, 1964
GP9	Fort Peck–Garrison	McCamy and Meyer, 1964
GP10	Garrison - S	McCamy and Meyer, 1964

TABLE 1. SEISMIC REFRACTION PROFILES FOR THE MIDCONTINENT UNITED STATES AND ADJACENT AREAS (continued)

No.*	Profile	Reference
GP11	Fort Peck–Acme Pond	McCamy and Meyer, 1964
GP12	Acme Pond - E	McCamy and Meyer, 1964
GP13	Greycliff–Charleson	Warren and others, 1972
GP14	Wolcott–Agate	Prodehl and Lipman, this volume
GP15	E. Colorado	Jackson and others, 1963
GP16	Gnome - N	Mitchell and Landisman, 1971
GP17	Gnome - E	Romney and others, 1962
GP18	Gnome - W	Romney and others, 1962
GR4	Ontario–Quebec	Mereu and others 1986 a, b; Huang and others, 1986
GR5	Grenville–Appalachians	Taylor and others, 1980
GR6	Tahawus - W	Katz, 1954
MC1	W. Manitoba	Hall and Hajnal, 1973
MC2	Superior–Churchill - NS	Green and others, 1980; Delandro and Moon, 1982 Morel-a-L'Huissier and others, 1987
MC3	Superior–Churchill - EW	Green and others, 1980; Delandro and Moon, 1982 Morel-a-L'Huissier and others, 1987; Hajnal, 1986
MC4	W. Manitoba	Hajnal and others, 1984; Morel-a-L'Huissier and others, 1987
MC5	N. Minnesota	Tuve, 1953; Steinhart and Meyer, 1961; Tatel and others, 1953; Tuve and others, 1954
MC6	Lake Superior	Berry and West, 1966; Smith and others, 1966a, b
MC7	Lake Superior–Colorado	Roller and Jackson, 1966
MC8	Gambler High	Cohen and Meyer, 1966
MC9	Gambler Low	Cohen and Meyer, 1966
MC10	Keweenaw	Steinhart and Meyer, 1961
MC11	Western Lake Superior Basin	Luetgert and Meyer, 1982
MC12	Western Lake Superior Basin	Luetgert and Meyer, 1982
MC13	Western Lake Superior Basin	Luetgert and Meyer, 1982
MC14	Central Wisconsin	Steinhart and Meyer, 1961
MC15	Central Wisconsin	Slichter, 1951; Steinhart and Meyer, 1961
MC16	Central Wisconsin	Slichter, 1951; Steinhart and Meyer, 1961
MC17	Wisconsin	Slichter, 1951; Steinhart and Meyer, 1961
MC18	Lake Superior–Wichita	Green and Hales, 1968
MC19	Lake Superior–Little Rock	Green and Hales, 1968
MC20	Iowa	Cohen, 1966
MC21	St. Joseph–Hannibal	Stewart, 1968
MC22	Hercules–St. Genevieve	Stewart, 1968
MC23	S. Indiana–S. Illinois	Baldwin, 1980
MC24	Manitou–Chelsea	Mitchell and Landisman, 1970; Tryggvason and Qualls, 1967
MC25	E. Tennessee	Warren, 1968; Prodehl and others, 1984
RM1	Great Plains	Hales and Nation, 1973
RM2	Rocky Mountains	Hales and Nation, 1973
RM3	Cliff Lake–Big Sandy River	McCamy and Meyer, 1964
RM4	Cliff Lake–Sailor Lake	McCamy and Meyer, 1964
RM5	Sailor Lake - E	McCamy and Meyer, 1964
RM6	SW Montana	Sheriff and Stickney, 1984
RM7	Yellowstone	Smith and others, 1982
RM8	Bingham - NE	Braile and others, 1974
RM9	American Falls-Flaming Gorge	Willden, 1965
RM10	S. Rocky Mountains	Jackson and Pakiser, 1965; Prodehl and Pakiser, 1980
RM11	Front Range	Jackson and Pakiser, 1965; Prodehl and Pakiser, 1980
RM12	Gasbuggy - N	Warren and Jackson, 1968
RM13	Gasbuggy - E	Warren and Jackson, 1968

TABLE 1. SEISMIC REFRACTION PROFILES FOR THE MIDCONTINENT UNITED STATES AND ADJACENT AREAS (continued)

No.*	Profile	Reference
SU1	Superior	Berry and Fuchs, 1973
SU2	English River–Wabigoon	Young and others, 1986
SU3	Northwest Ontario	Hall and Hajnal, 1969
SU4	Manibota–Ontario	Hall and Hajnal, 1973
SU5	Superior–Churchill	Mereu and Hunter, 1969
SU6	Kirkland Lake	Hodgson, 1953
SU7	Grenville	Mereu and Jobidon, 1971
SU8	Quetico–Shebandowan	Young and others, 1986
SU9	Highway 599-R	Young and others, 1986
SU10	Highway 599-C/S	Young and others, 1986
SU11	W. of Lake Superior	Lewis and Meyer, 1968
SU12	N. Minnesota	Tuve, 1953; Steinhart and Meyer, 1961; Tatel and others, 1953; Tuve and others, 1954

*Abbreviations for Midcontinent United States and Adjacent provinces: Appalachian = AP, Basin and Range = BR, Columbia Plateau = CM, Coastal Plain = CO, Colorado Plateau = CP, Cordillera = CR, Great Plains = GP, Grenville = GR, Midcontinent = MC, Rocky Mountain = RM, Superior = SU.

Traditional seismic refraction data recorded along profiles have largely been interpreted using travel-time calculations with plane-layered velocity models. More recently, ray-trace travel-time methods (Mooney, this volume) for laterally inhomogeneous velocity structures and comparison of observed seismic record sections with synthetic seismograms (Braile and Smith, 1975) have been utilized for interpretation of midcontinent crustal seismic data.

In this chapter, I have generally attempted to utilize the crustal velocity models as presented in the published literature (Table 1). Some interpretation and interpolation of the data have been necessary for the velocity structure correlation between seismic profiles and in regions where closely spaced profiles have been interpreted as different or even contradictory crustal models. Additional discussion on the characteristics of the seismic data, methods of data collection, processing and interpretation, and procedures for compilation and analysis of crustal refraction profiles are contained in the papers by Braile and others (this volume) and Mooney (this volume).

Deep seismic reflection profiling data for the continental interior region of the United States are listed in Table 2 and their locations are shown in Figure 2. Most of these data are from the COCORP (Consortium for Continental Reflection Profiling) program and were recorded during the last 15 years. Most of the deep reflection profiles in the continental interior have been designed to study specific, known geologic features, and the profiles are thus located in anomalous crustal regions. Additional discussion of deep reflection profiling results for the eastern United States is provided by Phinney and Roy-Chowdhury (this volume). Reflection profiling data for the midcontinent region of Canada are compiled and analyzed by Green and Clowes (1983).

TABLE 2. DEEP SEISMIC REFLECTION PROFILES IN U.S. CONTINENTAL INTERIOR (GREAT PLAINS AND MIDCONTINENT PROVINCES)

No.	Profile	Reference
RE1	COCORP-Montana	Latham and others, 1988
RE2	COCORP-Laramie	Brewer and others, 1982; Allmendinger and others, 1982
RE3	COCORP-Abo Pass	Brown and others, 1980
RE4	COCORP-Hardeman	Oliver and others, 1976
RE5	COCORP-Oklahoma	Brewer and others, 1981, 1983
RE6	COCORP-Kansas	Brown and others, 1983; Serpa and others, 1984
RE7	COCORP-Minnesota	Gibbs and others, 1984; Gibbs, 1986
RE8	Midcontinent Rift-Minnesota-Wisconsin Iowa	Chandler and others, 1989
RE9	Midcontinent Rift-Keweenaw Peninsula	Hinze and others, 1989
RE10	GLIMPCE-Lake Superior, Lines B and F	Behrendt and others, 1988; Fox, 1988
RE11	GLIMPCE-Lake Superior, Lake Michigan	Cannon and others, 1989; Fox, 1988
RE12	GLIMPCE-Lake Huron	Green and others, 1988; Smith, 1988
RE13	COCORP-Michigan Basin	Brown and others, 1982; Zhu and Brown, 1986
RE14	COCORP-Adirondack	Klemperer and others, 1985
RE15	COCORP-Ohio, Illinois Basin	Pratt and others, 1989
RE16	COCORP-Arkansas	Nelson and others, 1982; Lillie and others, 1983
RE17	COCORP-Reelfoot-Tennessee-Alabama	COCORP Atlas, 1988

CHARACTERISTIC MIDCONTINENT CRUSTAL MODELS

Crustal structure statistics for the midcontinent region have been computed for the crustal models from interpreted seismic profiles listed in Table 1. The model statistics are given in Table 3. For the average compressional wave velocity of the crystalline crust, the sedimentary layer is not included in the calculation. For models containing sharp gradient zones such as the transition between two homogeneous regions of velocity, the center of the gradient zone is interpreted to correspond to the boundary between the two layers. In computing the compressional wave velocity of the lower crustal layer, the normal lower crustal layer (with a velocity of approximately 6.5 to 6.8 km/sec) was included, as well as the high-velocity lower crust (velocities of approximately 7.0 to 7.4 km/sec, where present. Examination of the crustal velocity statistics shows that the average crustal thickness of the midcontinent is approximately 42.5 km, the average seismic velocity of the crust is approximately 6.5 km/sec, and the average upper-mantle velocity is approximately 8.1 km/sec. All of these numbers are slightly higher than the corresponding values for the entire continent as computed by Braile and others (this volume). Standard deviations for the midcontinent statistics are also smaller than the corresponding standard deviations for the entire continent, illustrating as expected, a more uniform average crustal thickness and velocity in the craton as compared to the surrounding portions of the continent that have been tectonically active in Phanerozoic time.

Although the average compressional wave velocity is given as approximately 6.5 km/sec in Table 3, the interpreted velocity models of the crust are not uniform. In fact, the crust is predominantly layered with an upper-crustal layer of approximately 6.14 km/sec and a lower-crustal layer that averages approximately 6.82 km/sec. The upper-crustal velocity in the midcontinent region is somewhat higher than the normal value of about 6.0 km/sec, which is representative of the western United States. This may be a result of composition or temperature characteristics of the two regions or could simply be the result of the fact that the crystalline basement in the midcontinent region is representative of a more deeply eroded crystalline crust and, therefore, of rocks that were formed at greater depth or subjected to greater pressure and temperature conditions. Although it is convenient to talk about the continental crust of the midcontinent area as being layered and containing an upper-crustal and a lower-crustal layer as well as a high-velocity basal lower-crustal layer in some areas, this concept of layering does not necessarily imply perfectly planar layers of homogeneous velocity nor abrupt interfaces between the layers. Some variation in velocity is often present within the layers, and some interpreted models contain prominent velocity gradients within the layers. Boundaries between the layers (or zones of relatively homogeneous velocity) can be sharp or can be velocity transition zones several hundred meters to a few kilometers thick. However, considering the resolution of the seismic refraction and wide-angle reflection data and the characteristic wavelengths of seismic waves for these data, the transition zones can generally be modeled as simple interfaces. A further discussion of the concept and interpretation of layering is given below.

TABLE 3. CRUST AND UPPER-MANTLE MODEL STATISTICS; CONTINENTAL INTERIOR, NORTH AMERICA

Variable* (x)	Sample Size (n)	Mean ($\bar{x}$)	Standard Deviation (s_x)	Units
H_c	97	42.5	6.12	km
$\overline{V}_p$	79	6.51	0.17	km/sec
P_n	97	8.10	0.11	km/sec
V_{uc}	66	6.14	0.14	km/sec
V_{lc}	67	6.82	0.23	km/sec
Z_{lc}	66	15.94	6.30	km

*H_c = Crustal thickness (depth to Moho); $\overline{V}_p$ = average compressional wave velocity of the crystalline crust; P_n = upper-mantle compressional wave velocity (Moho velocity); V_{uc} = compressional wave velocity of the "upper crustal layer"; V_{lc} = compressional wave velocity of the "lower crustal layer"; Z_{lc} = depth to the top of the lower crust.

Representative midcontinental crustal models are illustrated in Figure 3. Models for both average or "normal" cratonic crust and anomalous crust are shown. The average midcontinent model is derived from the mean value statistics presented in Table 3. The average shield model is obtained from eleven Superior Province crustal profiles. The northern Missouri model from Stewart (1968) is one of the few profiles recorded away from the anomalous regions of the Williston basin, the Lake Superior region of the Midcontinent Rift System, and the upper Mississippi Embayment; it appears to approximate a standard crustal model with seismic velocity structure similar to the average midcontinental model and the average shield model. The Great Plains model (Jackson and others, 1963) has a significantly thicker crust than the average midcontinental model. Most models for the Great Plains region, however, also contain a high-velocity lower-crustal layer just above the upper mantle. The Mississippi Embayment model (McCamy and Meyer, 1966) displays a thinned upper-crustal layer, a thick, low-velocity sedimentary layer associated with Mississippi Embayment sediments, and a high-velocity lower-crustal layer just above the Moho. The representative midcontinent crustal models displayed in Figure 3 provide a useful means of comparison with other crustal seismic models.

CRUSTAL STRUCTURE OF MIDCONTINENT SUBPROVINCES

In this section, I describe data and crustal model interpretations for various subprovinces within the midcontinent region of the United States and southern Canada. These examples serve to

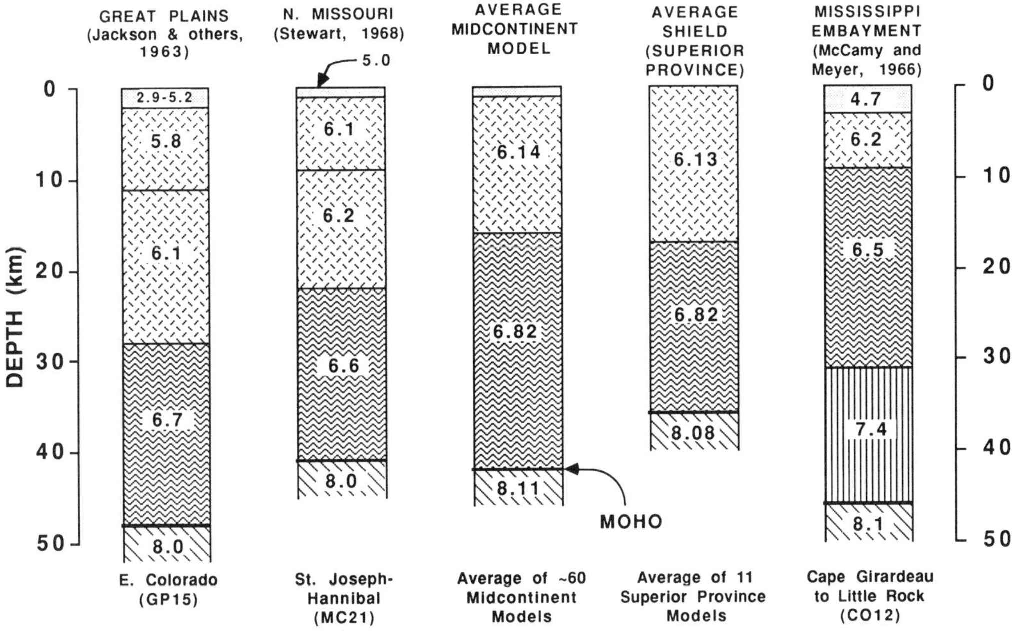

Figure 3. Comparison of midcontinental crustal velocity models. Numbers on models are compressional wave velocities in km/sec. Province codes (Table 1) and numbers below the models refer to profile locations in Figure 2 and references in Table 1.

illustrate the variety of data quantity and quality and the different interpretation methods that are representative of the crustal seismic studies that provide information on the crust of the midcontinent craton. The crustal models presented in these sections also illustrate the variation of crustal structure throughout the midcontinent region and distinctive features of the velocity models that characterize the crust within each of the subprovinces.

Canadian Shield

Examples of seismic refraction data and an interpreted crustal model for the Canadian Shield area are shown in Figure 4. The data consist of a reversed refraction profile recorded during the 1977 CO-CRUST experiment on the edge of the Williston basin in southern Canada. The record sections are vertical-component, reduced traveltime seismic data along a north-south profile, with an average station spacing of approximately 8 km. Signal-to-noise ratio is generally good to excellent, and prominent first-arrivals indicate refracted traveltime branches from upper crustal, lower crustal, and upper mantle depths. The interpreted plane-layered crustal model (Green and others, 1980) is illustrated beneath the seismic data. Computed traveltimes for the velocity model are shown as straight lines on the record sections. Except for the relatively large dip on the top of the lower-crustal layer, the velocity model shown in Figure 4 is fairly typical of Superior Province models. An upper crustal velocity of slightly more than 6 km/sec, a lower-crustal velocity of about 6.5 km/sec and an upper-mantle velocity slightly greater than 8 km/sec are characteristic of normal crust within the craton, as illustrated in Figure 3.

Data and an interpreted model from a reversed refraction profile recorded in central Wisconsin (Steinhart and Meyer, 1961) are shown in Figure 5. Observed and theoretical traveltimes are illustrated in the traveltime plot, and the interpreted model from which the theoretical traveltimes are computed is shown beneath the traveltime curves. The refraction data shown here are similar to much of the data recorded in the early 1960's. Average spacing of seismic recording for this experiment was approximately 15 km, and the plane-layered velocity-structure interpretation relies primarily on first-arrival traveltimes. A fairly typical cratonic crust of about 40 km thick is interpreted in central Wisconsin.

Reversed refraction profile record sections for the Superior Province in eastern Canada near the Grenville Front and interpreted velocity structures for the two shotpoints are illustrated in Figure 6 (Berry and Fuchs, 1973). These data were recorded with a fairly large seismograph station spacing (approximately 25 km), and the recorded seismograms are plotted in record section form with true amplitudes (amplitudes of seismograms multiplied by distance for convenient scaling). The seismic data have been interpreted using the travel-times (computed from the illustrated models and plotted on the record section) and comparison of the observed amplitudes and waveforms with synthetic seismogram calculations using the reflectivity method (Fuchs and Müller, 1971). The synthetic seismogram comparison is particularly useful in matching the observed secondary arrivals, which are prom-

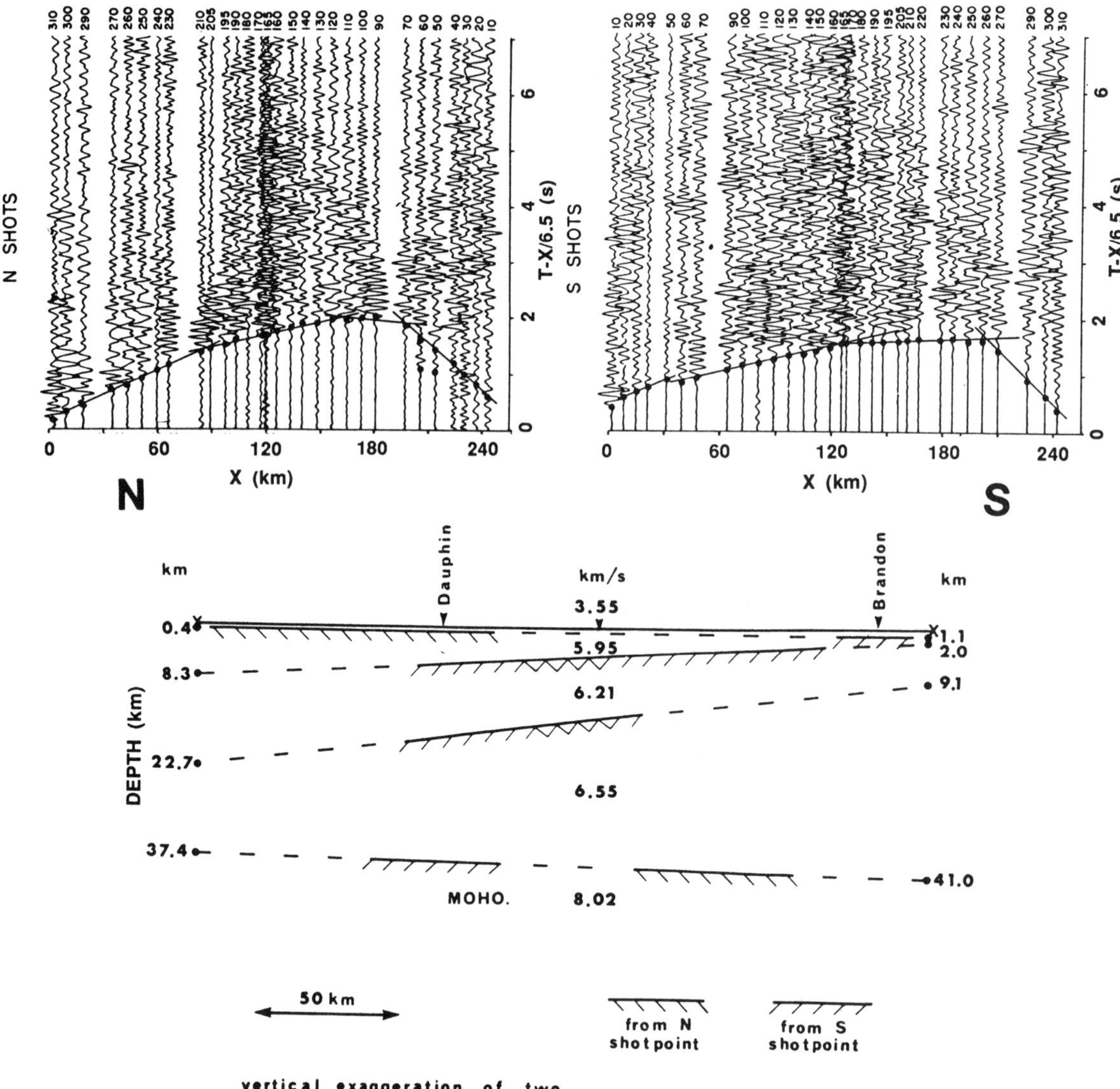

Figure 4. Example of reversed refraction profile seismic data (north-south profile MC2, Fig. 2) and resulting crustal velocity model. Numbers in the model are compressional wave velocities in km/sec and depths of dipping layer interfaces at the north and south end of the model. From Green and others, 1980.

inent on the record section, and in constraining the velocity structure of the lower crust. The wide-angle reflections from the sharp increase in velocity at the Moho (PmP) are particularly useful in defining velocities in the deeper part of the crust. Again, a crust approximately 44 km thick above an upper mantle with a compressional wave velocity of more than 8 km/sec is interpreted from these data for the Superior Province of eastern Canada.

Midcontinent Rift System

A number of seismic experiments have been performed in the area of the Midcontinent Rift System (Van Schmus and Hinze, 1985) in the last 25 years. Most of the crustal seismic data on the Midcontinent Rift System (MRS) are in the Lake Superior portion of the rift and were obtained during the 1963 and 1964 cooperative Lake Superior experiments. These experiments took advantage of the efficiency and effectiveness of detonating explosives in Lake Superior to record long-line refraction profiles in a variety of directions from the shotpoints. Additional seismic data primarily emphasizing upper-mantle propagation paths were recorded in the 1966 EARLY RISE experiment from shots in Lake Superior (Warren and others, 1967). A number of interpretations based on the 1963 and 1964 Lake Superior experiment data are contained in the AGU Monograph (Volume 10) edited by Steinhart and Smith (1966).

Much of the seismic data for the Lake Superior area were interpreted using the time-term method. Velocity determinations

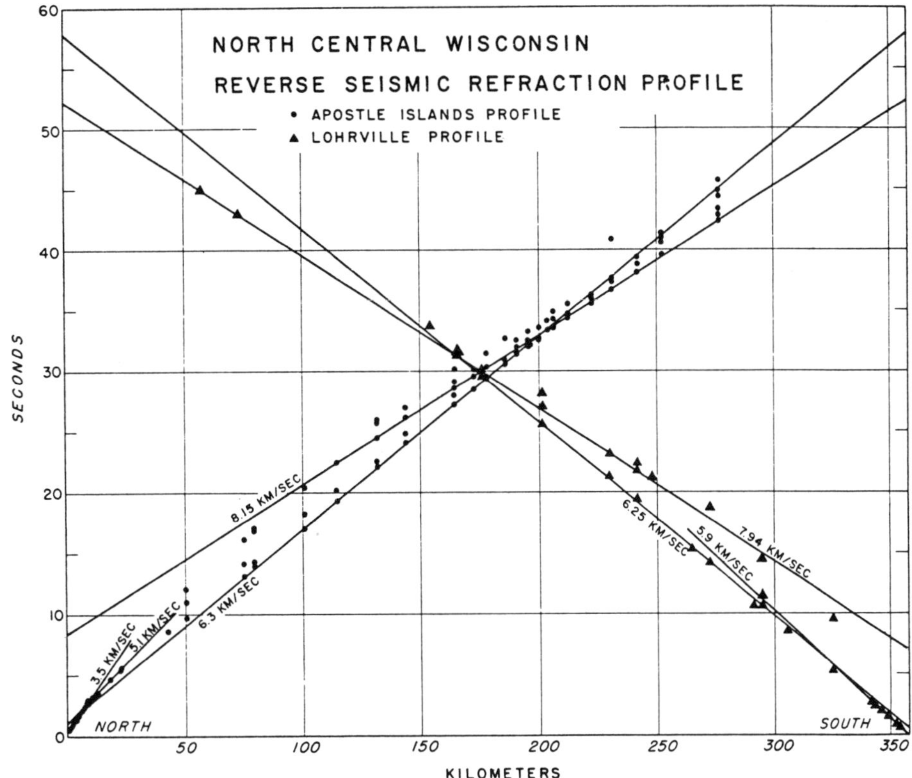

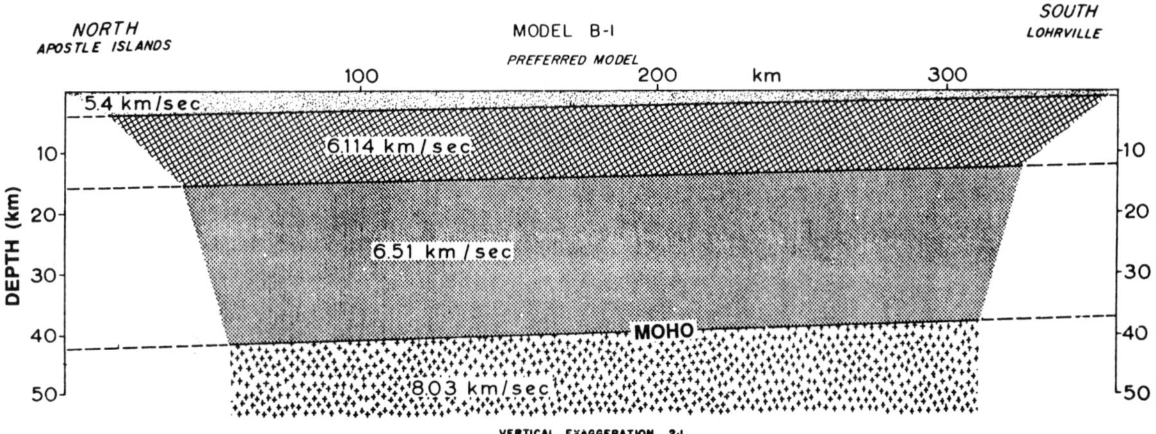

Figure 5. Traveltime data for central Wisconsin reversed refraction profile (profile MC14, Fig. 2) and interpreted velocity model. Numbers on the traveltime plot give apparent velocities in km/sec. Numbers on the velocity model are compressional wave velocities in km/sec. From Steinhart and Meyer, 1961.

from the time-term method for the "upper refractor" (depth of about 5 km) were approximately 6.6 km/sec and for the Moho were approximately 8.1 km/sec. An interpretive crustal model for an east-west profile beneath the western Lake Superior Basin is shown in Figure 7. The striking features of the velocity structure interpretation in the Lake Superior area are that beneath a thick section of sedimentary rocks in the Lake Superior Basin, the crust appears to be anomalously thick and of high velocity (approximately 6.8 km/s) throughout the entire depth range of the crust. Thus, a typical upper-crustal layer (with a velocity of about 6.1 km/sec and average thickness of about 12 to 15 km) is missing beneath the MRS in the Lake Superior area. More recently, Halls (1982) has compiled the traveltime data in the Lake Superior area, including the 1963 and 1964 Lake Superior experiment data, the EARLY RISE data, and adjacent refraction profile interpretations. Halls (1982) used a time-term approach to determine the apparent crustal thickness in the Lake Superior area (Figure 8). Although, as recognized by Halls (1982), one

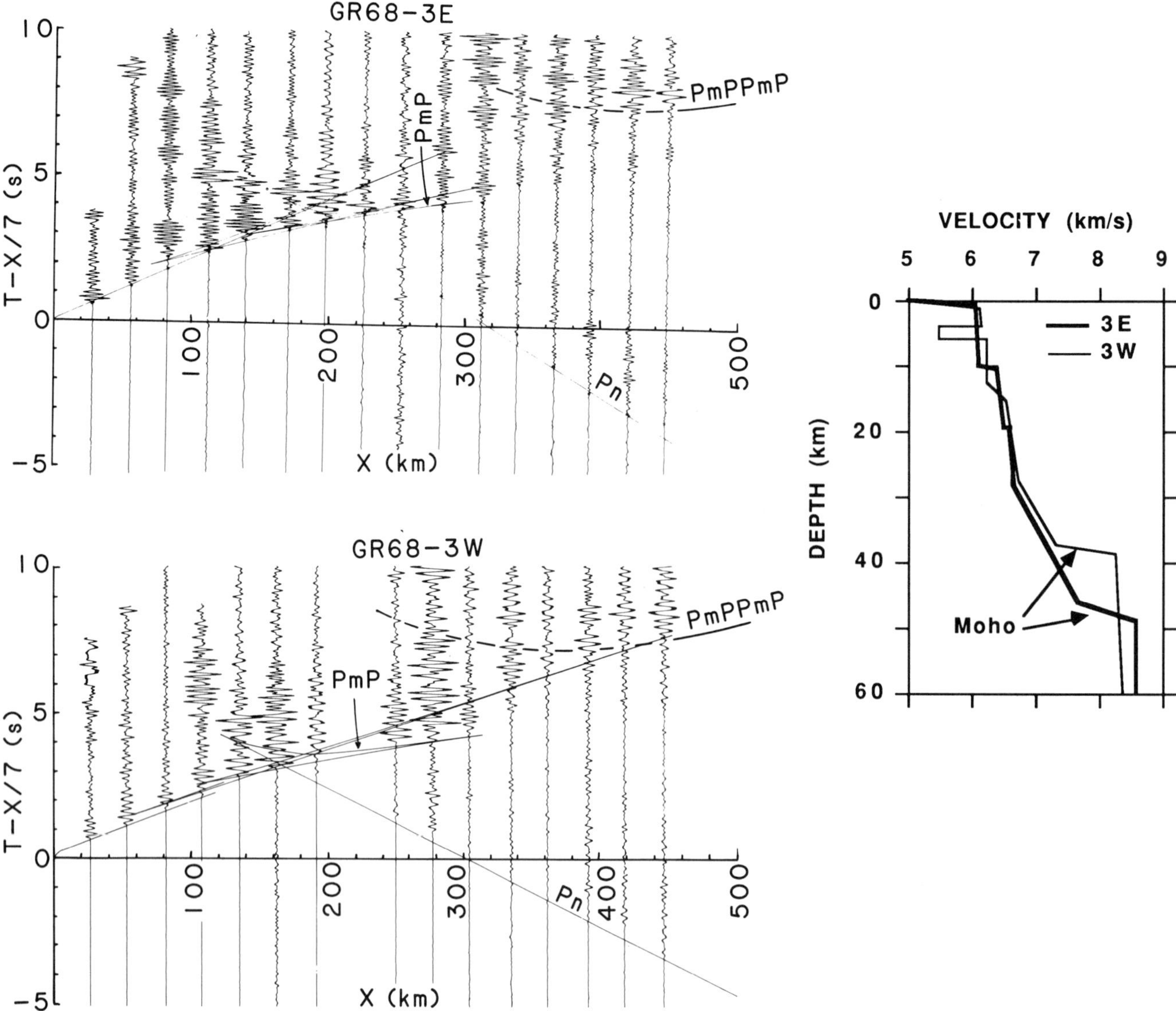

Figure 6. Example of seismic refraction record sections (profile SU1, Table 1) and interpreted crustal and upper-mantle velocity model for the Superior Province region of eastern Canada. Traveltime curves superimposed on the seismic record sections are for the velocity structure shown at the right. Phase PmP is the wide-angle reflection from the Moho. Phase PmPPmP is a multiple of the Moho reflection. Phase Pn is the head wave from the uppermost mantle. Modified from Berry and Fuchs, 1973.

must interpret the apparent crustal thickness map cautiously because of relatively sparse spacing of stations and the ambituities of the interpretation caused by assumptions of the time-term method, it is clear that the Lake Superior Basin is underlain by an anomalously thick crust more than 50 km thick and that the normal crustal thickness adjacent to the Lake Superior region is about 35 to 40 km in the surrounding shield areas.

South of Lake Superior along the Midcontinent Rift System, two unreversed refraction profiles were recorded by the University of Wisconsin Geophysics Group in 1963 and 1964 using shotpoints in Lake Superior. The locations of the profiles with respect to the MRS, which is delineated by the pronounced gravity high, are shown in Figure 9 (Cohen and Meyer, 1966). The velocity structures interpreted from first-arrival times for the two refraction profiles are also shown in Figure 9. These models agree with other interpretations of crustal thickening beneath the MRS and of the thinning or absence of a typical 6.1-km/sec basement beneath the sedimentary layer.

Because of the prominent gravity high associated with the Midcontinent Rift System, gravity modeling has been an important contributor to an understanding of the crustal structure of the MRS. Density models that are computed to match the observed gravity anomaly over the MRS and that utilize constraints provided by the available crustal seismic data and surface geology have demonstrated a pervasive structural anomaly beneath the MRS. The areas that have been most thoroughly studied are the

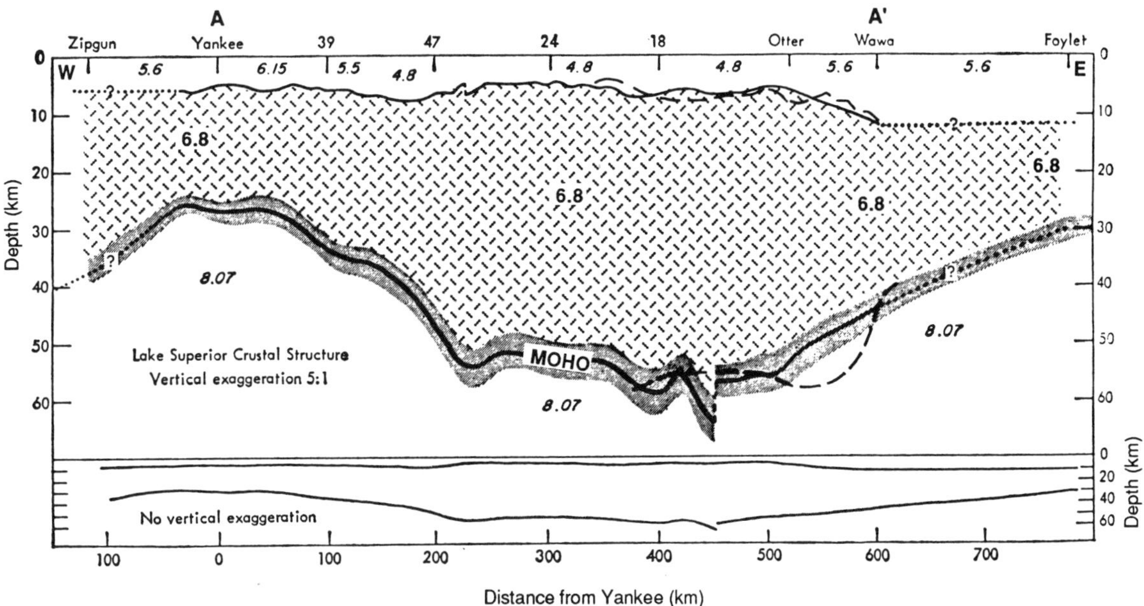

Figure 7. Interpreted cross section (profile MC6, Table 1) from west-southwest to east-northeast along Lake Superior (Smith and others, 1966). Location of the A-A′ profile is shown in Figure 8. Numbers on model are seismic compressional wave velocities in km/sec.

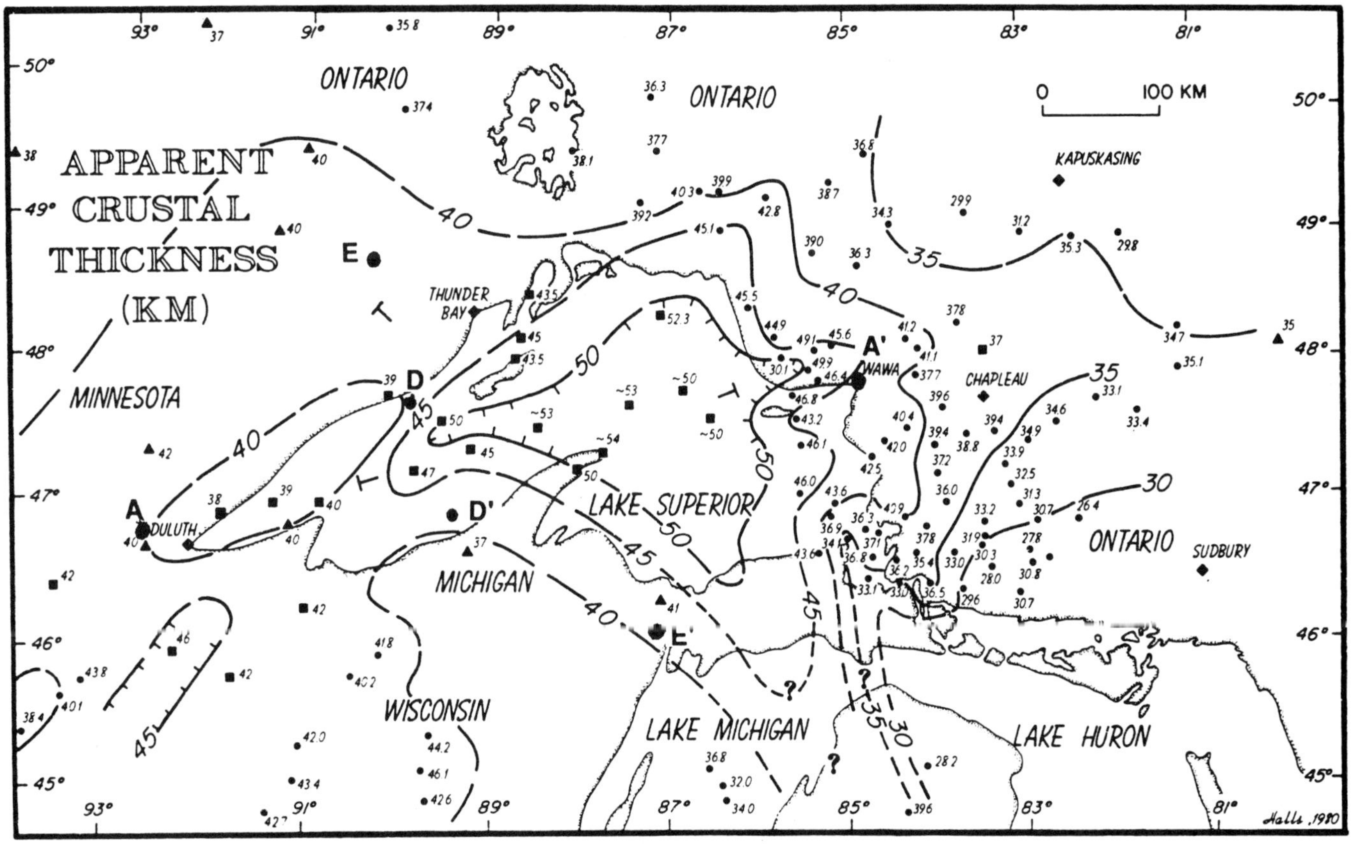

Figure 8. Contour map of apparent crustal thickness interpreted from a time-term analysis of Moho refraction arrivals in the Lake Superior area. Contours are depth to Moho in km. From Halls, 1982.

western Lake Superior region and the Midcontinent Rift System just to the south of Lake Superior in Minnesota and Wisconsin. Examples of crustal velocity and density models and computed gravity anomalies are shown in Figures 10 and 11. These models demonstrate the principal features of the anomalous crust of the Midcontinent Rift System. Beneath the MRS, high-density material extends from lower crustal depths into the upper crust. The excess mass of the crust caused subsidence and crustal thickening, creating flanking sedimentary basins that contribute to relative negative gravity anomalies along the edges of the Midcontinent Gravity High. An important conclusion of the gravity modeling is that although the models represent non-unique solutions to the observed anomaly, it is possible to match the observed data utilizing a density model in which the majority of density contrast is at upper-crustal depths (Figs. 10 and 11). Thus, it is conceivable that the Keweenawan rifting event involved sufficient splitting and extension of the upper crust to produce a subsiding basin, which was filled with extrusive volcanic rocks to depths as great as 20 km. Of course, the Keweenawan magmatic event probably resulted in profound changes at lower crustal depths as well by intrusion of mafic magmas. However, significant velocity and density contrasts between models of the lower crust beneath the Midcontinent Rift System and surrounding areas are not required by the available data.

Recently, deep-crustal seismic reflection profiling data have become available for several profiles recorded across the Midcontinent Rift System from the Lake Superior region to as far south as Kansas (Fig. 2). Interpretation of these reflection record sections for the Minnesota, Wisconsin, and Lake Superior region by Behrendt and others (1988), Fox (1988), Chandler and others (1989), and Hinze and others (1989), and for the Michigan Basin area by Brown and others (1982) and Zhu and Brown (1986) provides a detailed structural interpretation of the crust beneath the MRS. The southern portion of the MRS in Kansas has been

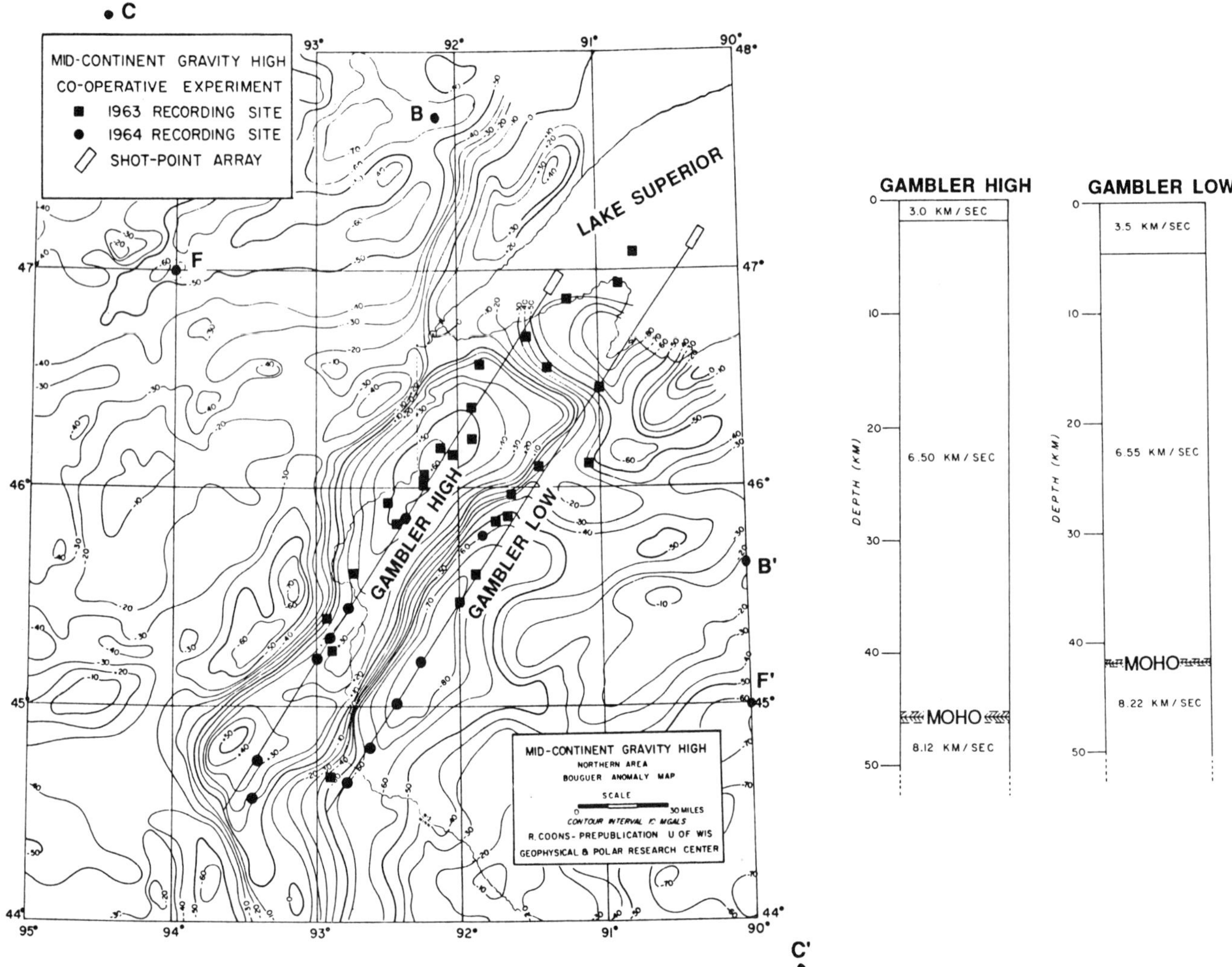

Figure 9. Index map showing the locations of shotpoints and recording sites for midcontinent gravity high seismic refraction experiment in Wisconsin and Minnesota. Crustal models are interpretations of the Gambler High (MC8, Fig. 2) and Gambler low (MC9, Fig. 2) profiles. Contours are gravity anomalies associated with the midcontinent gravity high. Modified from Cohen and Meyer, 1966.

studied by Brown and others (1983) and Serpa and others (1984) using reflection profiling data. Examples of MRS reflection data and interpretations of the data are shown in Figures 12, 13, 14, and 15. A regional view of the details of the structure of the Midcontinent Rift System has recently been provided by seismic reflection and refraction data from the GLIMPCE (Great Lakes International Multidisciplinary Program on Crustal Evolution) experiment. An example of an interpretation of some of these data is shown in Figure 14. Principal features of the crustal model which are revealed by the reflection profiling data are the presence of a deep (approximately 15 to 30 km) basalt-filled basin overlain by a layer of sedimentary rocks (Figure 14). The basalt basin clearly shows thickening of the sequence of layers toward the center and at least two major stages of extrusive deposition (Hinze and others, 1989). Sedimentary wedges in half-basins that flank the Midcontinent Rift System contain up to about 4 km of sedimentary rock of contemporaneous or younger ages. The basalt-filled basins were subsequently subjected to compression, and thrust faults bound the edges of the basin, with basalts overthrust above some of the younger sedimentary rocks in the flanking basin.

An interpreted crustal structure cross section of the Midcontinental Rift System in the Minnesota to Wisconsin area is shown in Figure 16. This interpretation utilizes the crustal refraction data in Minnesota and central Wisconsin adjacent to the MRS, as well as the Gambler High and Gambler Low refraction data (Cohen and Meyer, 1966). Additionally, the reflection interpretations (Behrendt and others, 1988; Hinze and others, 1989) provide detail on the shallow structure beneath the Midcontinent Gravity Anomaly. It appears that the gravity anomaly is due primarily to the basalt-filled basin produced during rifting. The rifting process apparently broke the entire upper-crustal layer because the basalts extend to depths greater than 20 km. This process implies substantial uplift and extension of the crust at the time of rifting, followed by subsidence during extrusive activity that filled a broad rift basin, and crustal thickening as a result of the subsidence and infilling of the basalt basin. Compression after deposition of the basalts caused thrust faulting at the edges of the basin. The detailed structure of the lower crust beneath the basalt basin is unknown because the reflection data either do not extend to lower-crustal depths or display discontinuous and weak reflections that cannot be interpreted with certainty. The refraction data show no indication of large velocity contrasts although it is expected that the lower crust would be intruded (as schematically illustrated in Figure 16) during the rifting and extrusive event. These lower-crustal effects may not be observed because the velocities and densities of intruded mafic rocks would be approximately the same as the lower crust.

Northern Mississippi Embayment

A reversed refraction profile in the northern Mississippi Embayment region was recorded by McCamy and Meyer (1966) along the western flank of the Embayment (Profile CO12 on Figure 2 and Table 1). The interpreted model is shown in Figure 17 (McCamy and Meyer, 1966). Seismograph spacing for these refraction profiles averaged approximately 10 km. The velocity structure was derived by analysis of first- and later-arrival refracted traveltimes. The Mississippi Embayment model (Fig. 17) contains a sedimentary wedge thickening to the south associated with Cretaceous and younger Mississippi Embayment sediments. A typical upper-crustal layer (with a velocity of about 6.2 km/sec) is present, but is anomalously thin. Beneath this upper

A

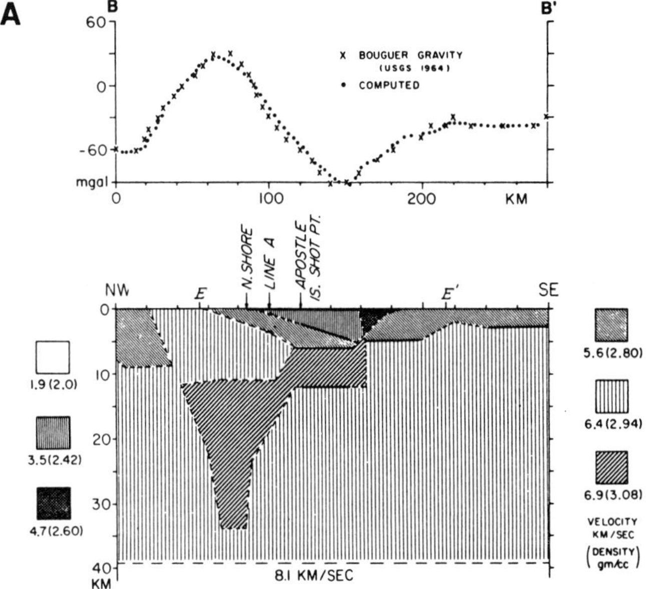

B

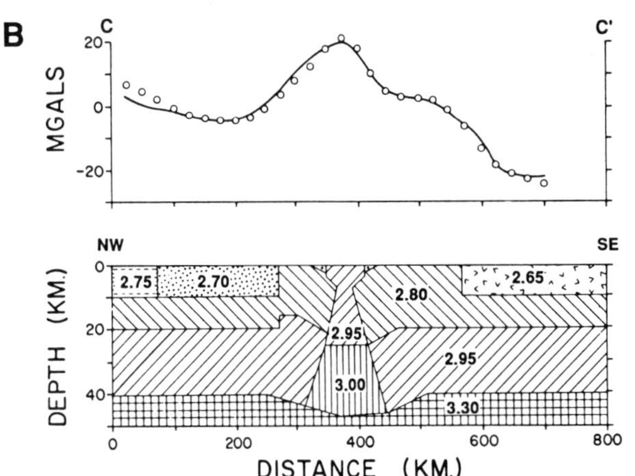

Figure 10. Examples of density models used to fit the gravity data over the Midcontinent Rift System in the Minnesota to Wisconsin area. The density models are constrained by seismic refraction observations where data are available. A, Northwest-southeast cross section over Lake Superior (from Ocola and Meyer, 1973). Profile B-B′ location is shown in Figure 9. B, Northwest-southeast cross-section from Minnesota to Wisconsin (from Chandler and others, 1982). Numbers in the model are densities in g/cm^3. Profile C-C′ location is shown in Figure 9.

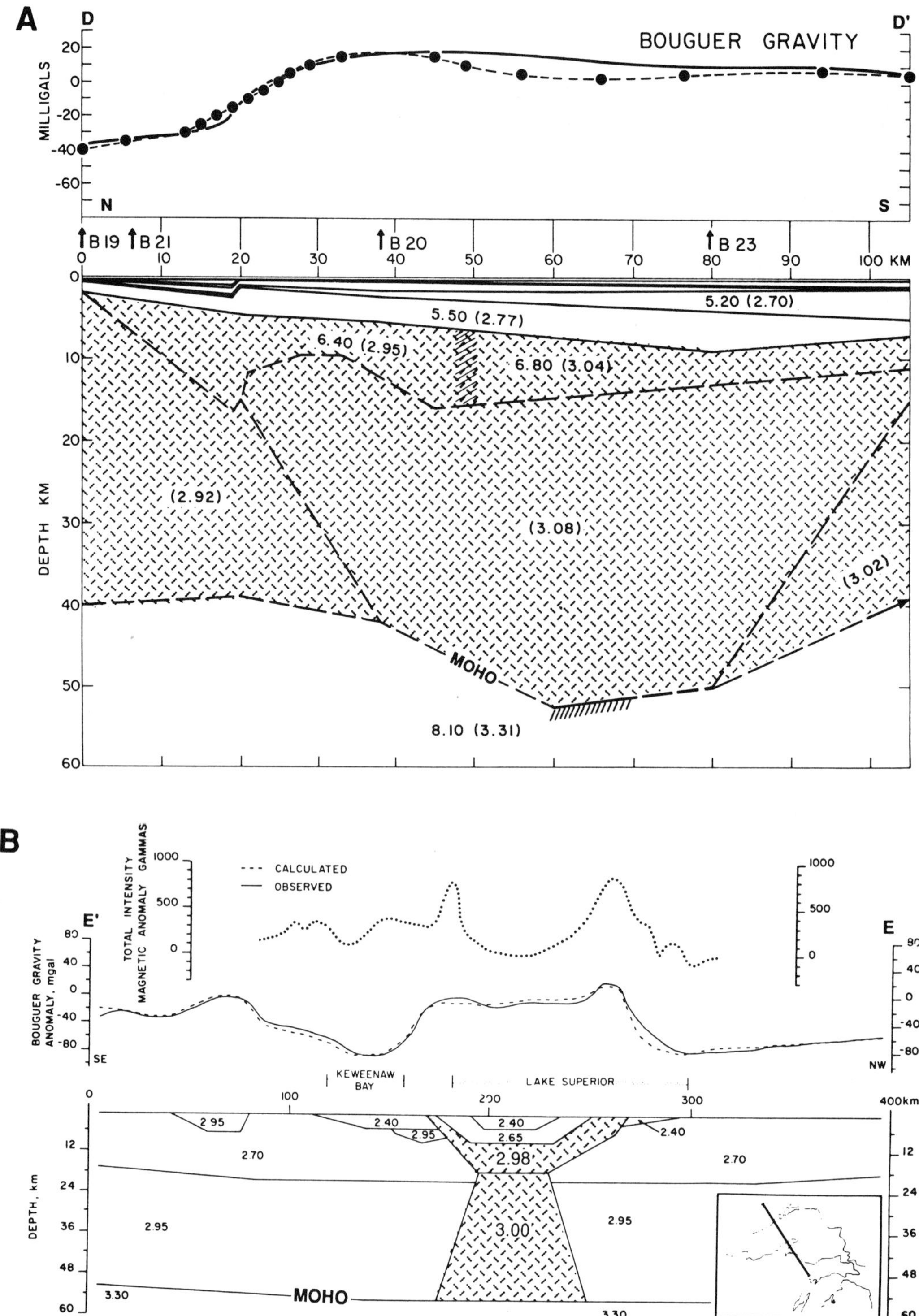

Figure 11. Example of gravity models constrained by seismic refraction data where available for western Lake Superior. A, North-south cross section and observed (dots) and computed (solid line) Bouguer gravity data over a section of western Lake Superior. Numbers in the interpreted geologic cross section are compressional wave velocities in km/sec and densities (in parentheses) in g/cm^3 (from Luetgert and Meyer, 1982). Profile D-D′ location is shown in Figure 8. B, Southeast-northwest cross section over western Lake Superior showing interpreted crustal density model and observed (solid line) and computed (dashed line) Bouguer gravity anomaly values. Numbers in the model are densities in g/cm^3 (from Hinze and others, 1982). Profile E-E′ location is shown in Figure 8.

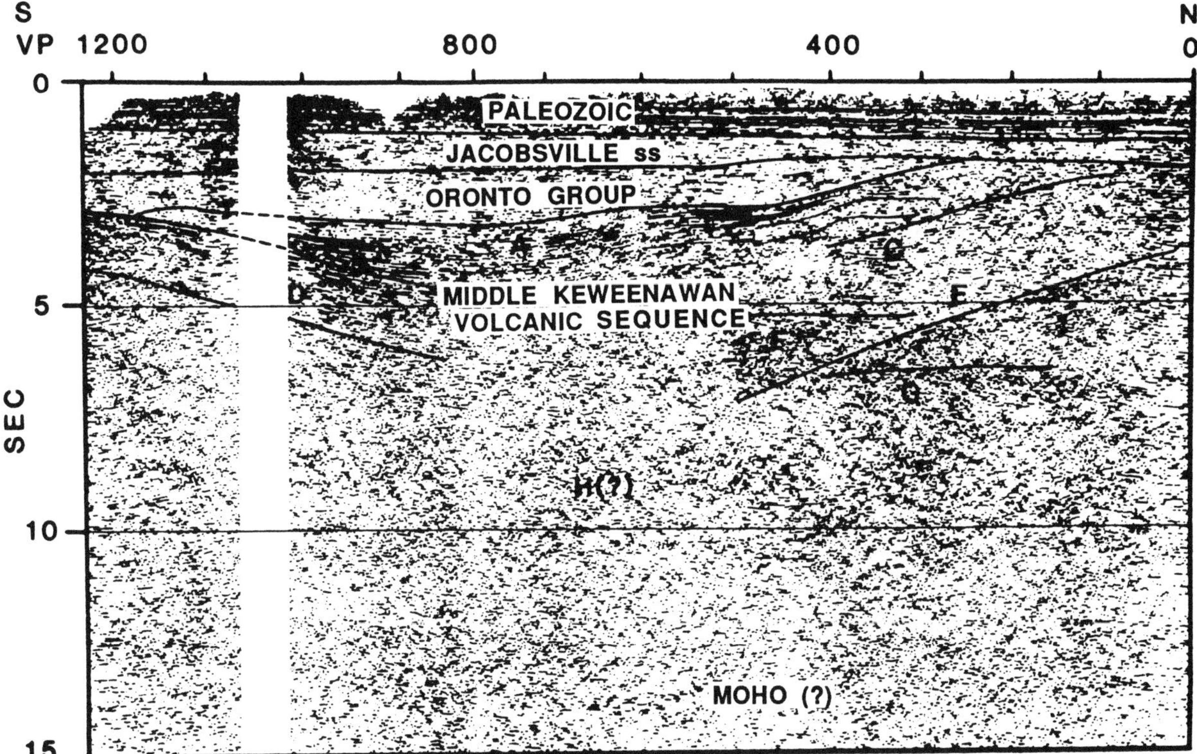

Figure 12. Reprocessed reflection data for the Michigan Basin (RE13, Figure 2) showing prominent reflectors from 3 to 6 sec two-way time (about 8 to 18 km depth), interpreted as Keweenawan volcanics and sediments beneath the Michigan Basin. From Zhu and Brown, 1986.

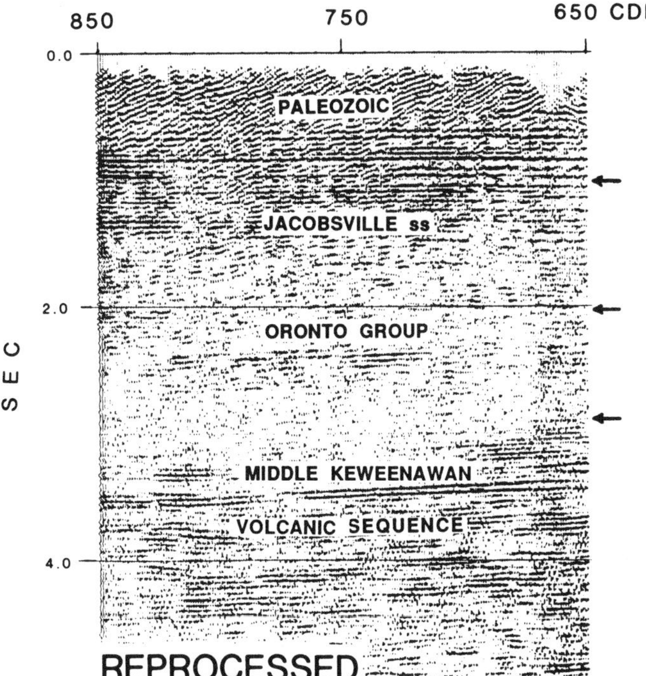

Figure 13. Enlarged reflection profiling record section from the COCORP Michigan Basin data (Figure 12) showing reflections from sedimentary rocks (above about 3 sec) and prominent reflections from middle Keweenawan volcanic rocks. Arrows indicate approximate boundaries between units. From Brown, 1986.

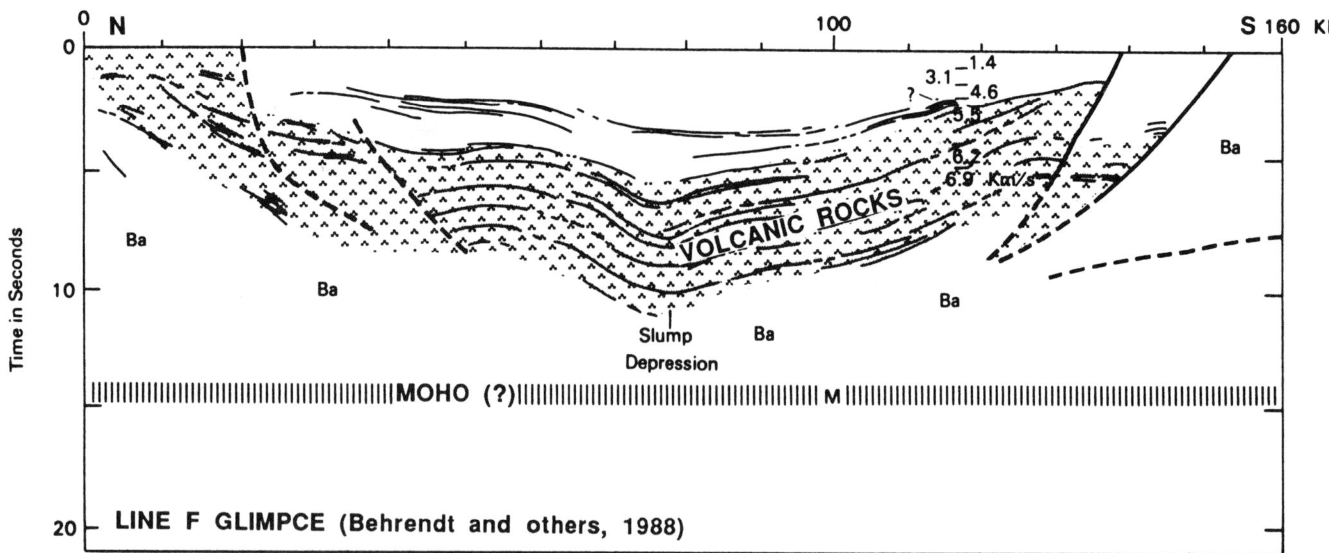

Figure 14. Interpreted line drawing of reflections from GLIMPCE profile F (RE10, Fig. 2). Shaded area interpreted as Keweenawan volcanic rocks of the Midcontinent Rift System beneath Lake Superior. Numbers are velocities from refraction data.

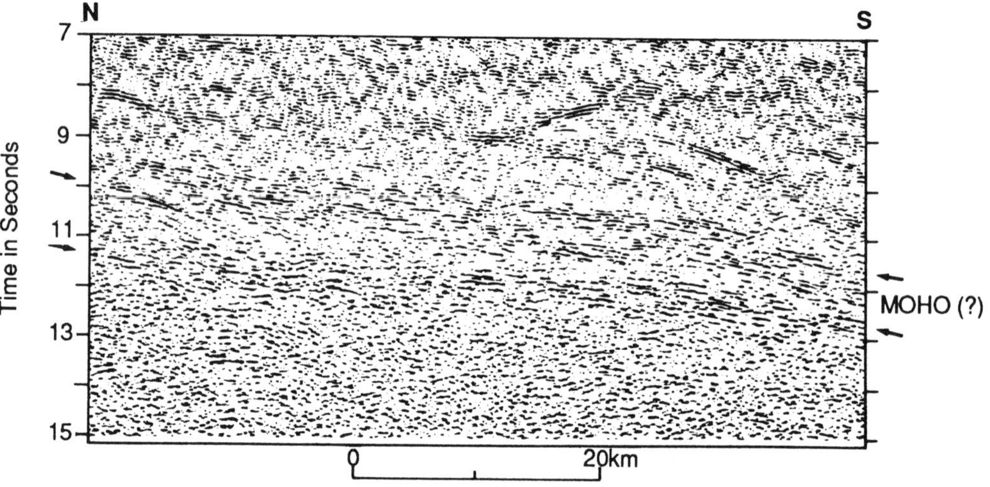

Figure 15. Example of deep reflection data (portion of Line B (RE10, Figure 2) of GLIMPCE experiment) showing reflections that may represent the Moho dipping southward toward the axis of the Lake Superior Basin (Fig. 14). From Behrendt and others, 1988.

crustal layer, a thick layer of typical lower-crustal velocity is present underlain by a high-velocity (7.4 km/sec) lower-crustal layer just above the upper mantle. The presence of the high-velocity lower-crustal layer has been considered to be a characteristic of the Mississippi Embayment and an anomalous feature. However, as discussed below, this high-velocity lower crust appears to be more widespread than previously thought. A thinned upper crust and uplifted lower-crustal rocks (or intrusion of high-velocity rocks into the upper crust) beneath the Mississippi Embayment appears to be the principal anomalous structure of the Embayment. This uplift of lower-crustal rocks is probably the cause of the long-wavelength positive regional gravity anomaly that trends northeastward along the Embayment (Cordell, 1977; Hildenbrand, 1985).

Ervin and McGinnis (1975) suggested that the Mississippi Embayment is underlain by a late Precambrian rift and that the rifting event produced the high-velocity lower-crustal layer. Recent geological and geophysical data in the northern Mississippi Embayment area have supported the rift interpretation (Hildenbrand and others, 1977; Zoback and others, 1980; Zoback and Zoback, 1981; Kane and others, 1981; Hamilton and Zoback, 1982; Braile and others, 1982a, 1986; Mooney and others, 1983). In 1980, the U.S. Geological Survey performed an extensive crustal seismic study (Mooney and others, 1983; Ginzburg and

Figure 16. Interpreted crustal velocity structure across the Midcontinent Rift System in north-central Minnesota to central Wisconsin. The location of the F-F′ profile is shown in Figure 9. Heavy lines indicate interpretations from crustal refraction data. The schematic geologic structure in the 0 to 20-km depth range beneath the midcontinent gravity anomaly follows the interpretation of Behrendt and others (1988) and Hinze and others (1989) based on deep reflection profiling data. The reflection interpretation is projected from the Lake Superior region. Numbers in the crustal model are compressional wave velocities in km/sec. Question marks and schematic representation of crustal structure beneath the volcanic rocks of the Midcontinent Rift System indicate that velocities and interpretation beneath the volcanic rocks are speculative.

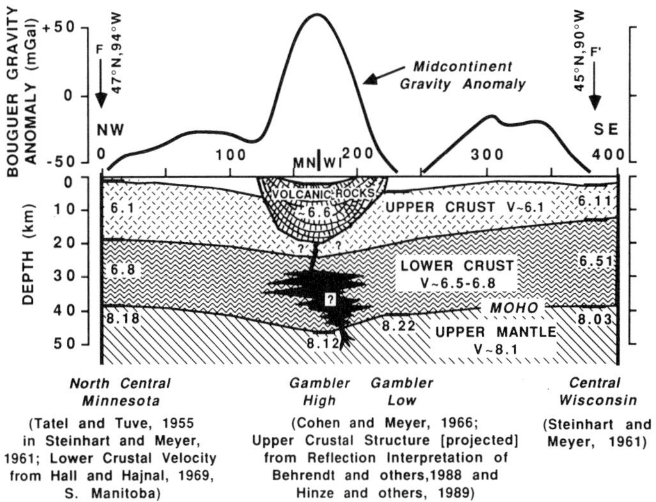

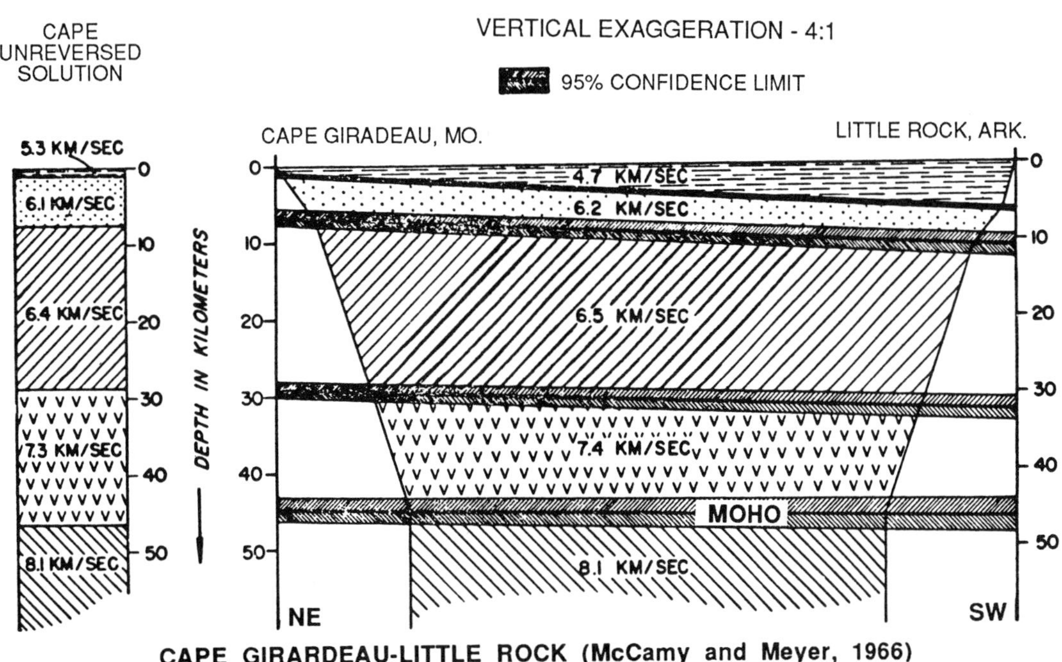

Figure 17. Interpreted crustal model for Cape Girardeau, Missouri, to Little Rock, Arkansas. From McCamy and Meyer, 1966.

others, 1983) in the Reelfoot Rift area of the northern Mississippi Embayment (Fig. 18). Their seismic experiment involved a number of shotpoints and seismograph recording profiles parallel to the inferred rift, along the axis of the rift and across the rift. The data were recorded on a 120-instrument matched array of seismographs arranged in profiles with approximately 2 km station spacing. Interpretation of the data involved traveltime, amplitude, and synthetic seismogram modeling using ray-trace calculations in two-dimensional velocity structures. Examples of the seismic data, sample ray-trace calculations, theoretical traveltime curves, and the interpreted models are shown in Figures 19 and 20 (Ginzburg and others, 1983). These examples also illustrate modern methods of crustal seismic data collection and interpretation that can provide improved resolution of crustal-velocity models. The dense coverage of stations and number of shotpoints and the two-dimensional interpretation procedure utilizing ray-trace cal-

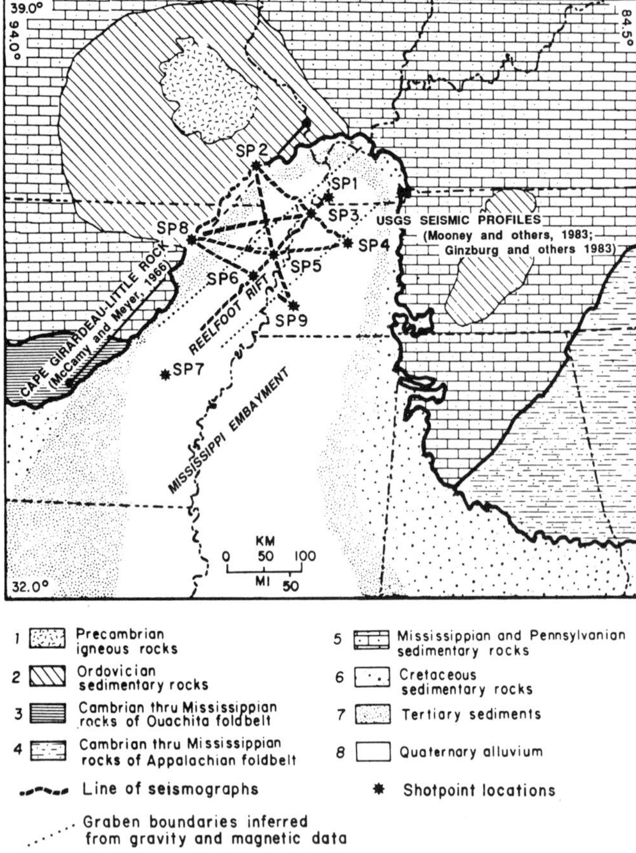

Figure 18. Index map showing locations of shotpoints and recording profiles for the Upper Mississippi Embayment seismic refraction experiment (Mooney and others, 1983; Ginzburg and others, 1983) and the Cape Girardeau to Little Rock profile of McCamy and Meyer (1966).

culations and synthetic seismograms have provided a detailed view of the crustal structure in the northern Mississippi Embayment region. Because of the limited distance range of recording in the 1980 experiment, the upper-mantle seismic velocity and depth to the Moho are only determined for a few recorded profiles. However, the crustal models presented by Ginzburg and others (1983) and Mooney and others (1983) provide confirmation of the presence of a thinned upper-crustal layer beneath the Reelfoot Rift in the northern Mississippi Embayment (Figure 18) and the presence of a high-velocity (approximately 7.3 km/sec) lower-crustal layer just above the upper mantle.

As discussed below, the high-velocity lower-crustal layer is not limited to the Mississippi Embayment area, but occurs extensively along the western and southern edges of the U.S. craton. The principal anomalous feature of the Mississippi Embayment crust is the uplift of the deep-crustal layers and thinning of the upper crust, which was followed by subsidence and resultant crustal thickening in Phanerozoic time. The tectonic evolution and crustal structure of the northern Mississippi Embayment area are discussed by Zoback and others (1980), Kane and others (1981), Mooney and others (1983), and Braile and others (1982a, 1986).

INTERPRETIVE CROSS SECTIONS OF THE MIDCONTINENT CRUST

West to east cross-sections of the midcontinental crust are shown in Figures 21 and 22. The cross-sections were prepared by compiling the crustal refraction models that cross or are adjacent to 37°N latitude (Figure 21) and 49°N latitude (Figure 22) across the midcontinent region. The 49°N cross section extends from the western edge of the relatively undeformed craton eastward to the central portions of the craton represented by the Superior and Grenville provinces of south-central and southeastern Canada. The 37°N cross section is adjacent to the southern margin of the relatively undeformed craton and crosses the anomalous regions of the Southern Oklahoma aulacogen associated with the Anadarko basin (Figure 1), and the Mississippi Embayment. The crustal seismic models and appropriate reference information for the interpreted cross-sections are listed in Tables 4 and 5. For each crustal model shown in the cross-sections, compressional wave velocity is listed, along with a heavy line to indicate the depths of major interfaces or transition zones indicated on the velocity model. Correlations between crustal layers are shown by dashed lines where they are interpolated between published velocity models. Dotted lines show areas of less certain correlation.

Although there are a few fairly wide gaps in coverage along the 37°N and 49°N profiles, correlations of the velocity models along the cross sections indicate a relatively consistent crustal velocity layering in the midcontinent region. Only a few inconsistencies in crustal velocities appear on the cross-sectional diagrams. For example, no intermediate velocity associated with lower crustal rocks (approximately 6.6 km/sec) is reported for crustal model D on Figure 21. The absence of this layer confuses the correlation between adjacent velocity models. On the 49°N profile (Fig. 22), crustal model B contains a lower-crustal velocity of 6.6 km/sec. In this area of the Williston Basin (Fig. 1), both in eastern Montana and western North Dakota, and in southern Canada, all of the other interpreted velocity models (in addition to the models referred to in Fig. 22) contain a high-velocity lower-crustal layer with a velocity of about 7.2 km/sec. Because data of model B were interpreted primarily from first-arrival traveltimes, it is possible that the high-velocity lower-crustal layer is actually present in the area of model B, but is a layer hidden in the velocity structure and never is represented by a first-arrival traveltime branch. Similarly, in the easternmost Williston basin, Hajnal and others (1984) showed a pinching out of the high-velocity lower-crustal layer between models D and E on Figure 22. However, this high-velocity lower-crustal layer is present farther to the east (model F of Figure 22) and therefore could be continuous across this region.

Upper-crustal velocities of 6.0 to about 6.3 km/sec are characteristic of the entire midcontinent except for the Lake Superior region (Fig. 22) where a deep, basalt-filled basin has ap-

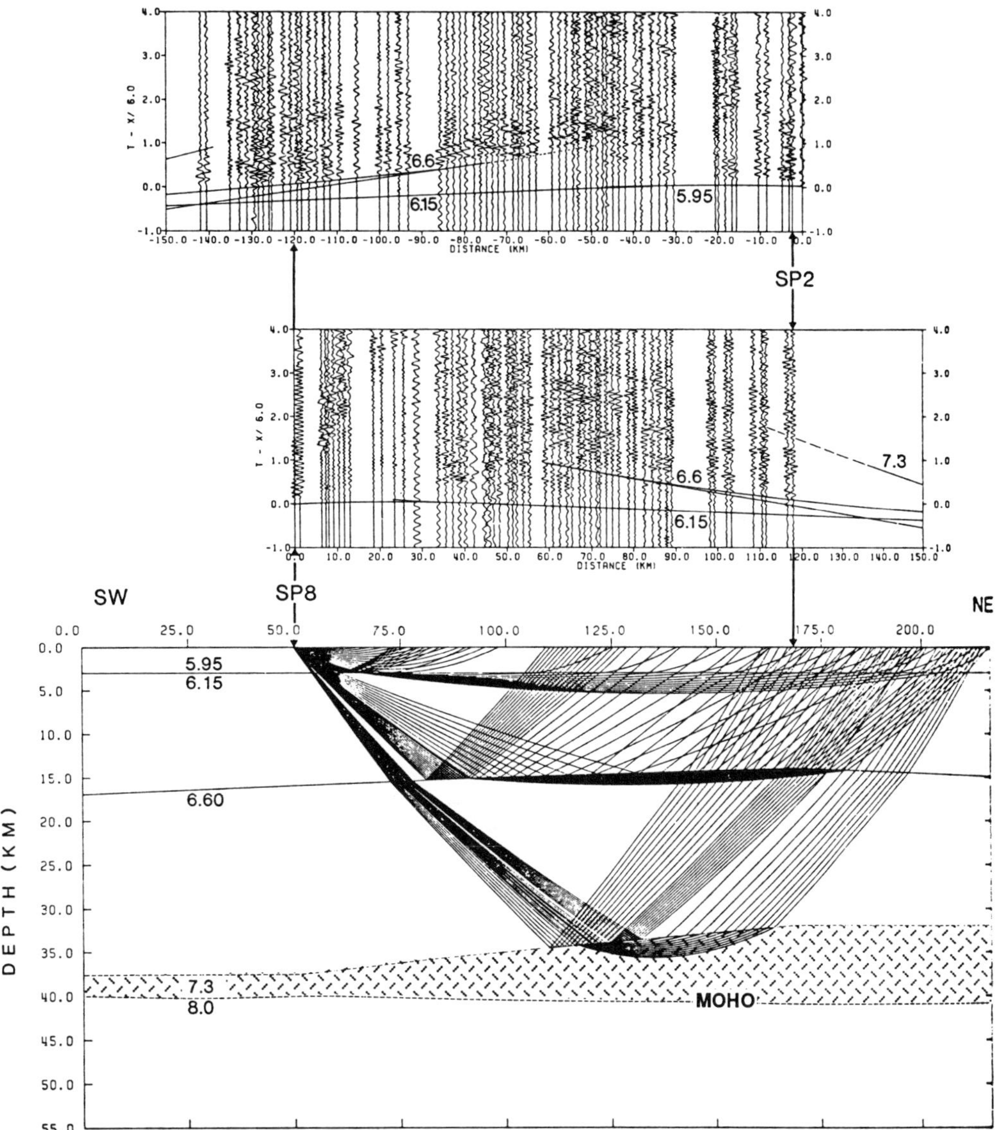

Figure 19. Observed seismic data for the reversed refraction profile (shotpoint 2 to shotpoint 8) and the reverse shot, and crustal velocity model showing rays traced from shotpoint 8 to the northeast for the western flank profile (Figure 18), northern Mississippi Embayment refraction experiment. Numbers in crustal model are compressional wave velocities in km/sec. From Ginzburg and others, 1983.

parently replaced all or part of the upper crust. The upper-crustal layer is generally about 15 km thick, although it appears to be thinned beneath the eastern part of the Williston basin (Fig. 22) and beneath the Mississippi Embayment (Fig. 21). A lower-crustal layer with velocities of approximately 6.4 to 6.8 km/sec is present nearly everywhere in the midcontinent region. A high-velocity lower-crustal layer (with velocities of approximately 7.1 to 7.4 km/sec) is present beneath nearly all of the 37°N-latitude cross section and the western portion of the 49°N-latitude cross section beneath the Williston basin. As we have seen, the central part of the midcontinent craton is best represented in the crustal seismic data by models for the Superior Province (Figs. 3 and 22), which do not contain this high-velocity lower-crustal layer. As seen on the midcontinent average crustal models (Fig. 3) and on the interpreted cross-sections (Fig. 22), the central part of the craton consists of a two-layered crust with a total thickness of about 35 to 40 km. Near the margins of the craton within the midcontinent, the crust is significantly thicker (about 45 to 50 km) and is generally characterized by the presence of the high-velocity lower-crustal layer just above the Moho.

CONTOUR MAPS OF CRUSTAL STRUCTURE

Contour maps of crustal thickness, upper-mantle seismic velocity, and average crustal velocity have been prepared for the midcontinent region and are illustrated in Figure 23. The crustal

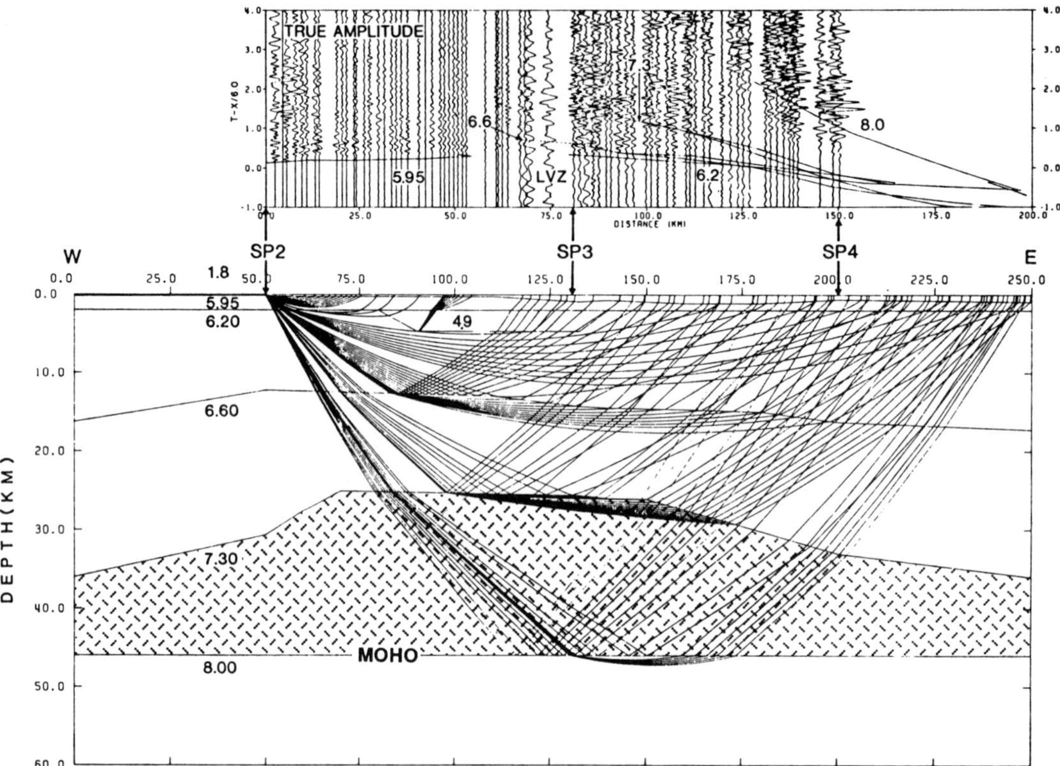

Figure 20. Seismic refraction data from shotpoint 2 and crustal model with rays traced from shotpoint 2 to the east along profile across the Reelfoot Rift (Fig. 18). Numbers in crustal model are compressional wave velocities in km/sec. A low-velocity (4.9 km/sec) layer interpreted as rift sedimentary rocks and an upwarp of the high-velocity lower crust are centered beneath the Reelfoot Rift (from Ginzburg and others, 1983.

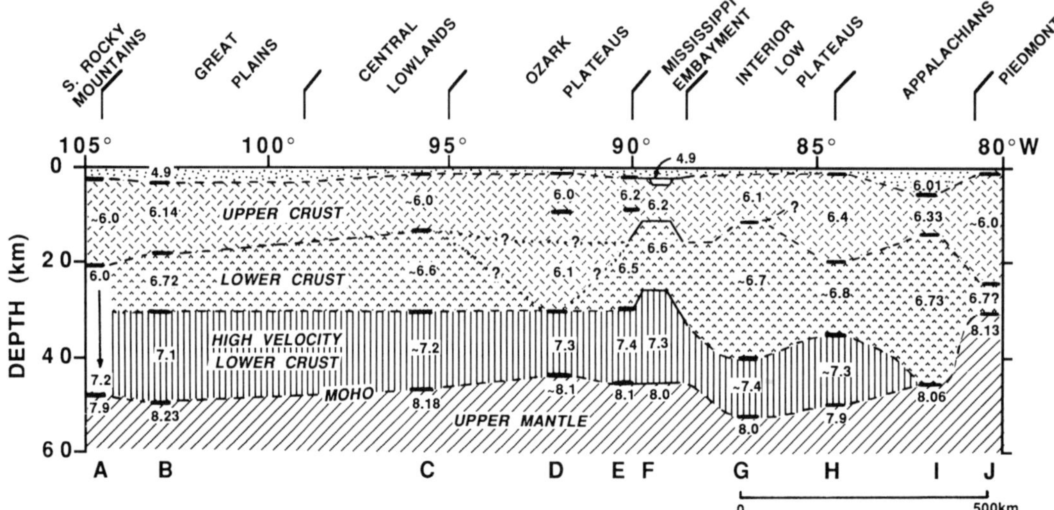

Figure 21. Interpreted crustal structure and correlations from refraction data in the midcontinent region along 37°N latitude. Numbers in the crustal model are compressional wave velocities in km/sec. Solid lines indicate interpretations from refraction data. Short, heavy lines indicate interpreted depths from seismic profiles that intersect the cross section. Dashed lines show interpreted correlations. Dotted lines indicate less certain correlations. Letters below the cross-sectional diagram indicate the interpreted refraction profiles listed in Table 4. Provinces shown above the model are from Fenneman (1946).

thickness contour map (Figure 23B) shows that a broad region of the midcontinent is characterized by a crust approximately 40 km thick. Anomalously thick crust, which may be associated with the high-velocity lower-crustal layer, exists in the western part of the midcontinent in the Great Plains region and along the southern edge of the midcontinent. The crust thickens locally beneath the Midcontinent Rift System and along the Grenville Front.

An upper-mantle seismic velocity of greater than 8 km/sec is present beneath almost the entire midcontinent region (Fig. 23C). Higher-than-normal upper-mantle velocities (greater than 8.2 km/sec) are present beneath the Williston Basin.

The average compressional wave velocity of the crystalline crust in the midcontinent area (Fig. 23D) is normally about 6.5 km/sec. Higher-than-normal average crustal velocities are present along the southern margin of the midcontinent and beneath the Williston basin. These areas are also characterized by the presence of the high-velocity lower-crustal layer, as illustrated in Figures 21 and 22. The Midcontinent Rift System also displays a slightly higher average crustal velocity.

Because of the relatively small number of crustal seismic velocity models for the entire midcontinent and the uneven distribution of these models, it is likely that these contour maps (Fig. 23) will change as future data are added.

LAYERING AND HOMOGENEITY OF THE MIDCONTINENTAL CRUST

The general uniformity and continuity of seismic velocity structure, as deduced primarily from refraction profiles, leads to an interpretation of an approximately layered crust. Most published models, and the interpretations presented here, represent the crust as nearly horizontal layers of nearly homogeneous velocity separated by interfaces. On a smaller scale, geologic exposures and reflection profiling data demonstrate that these crustal zones or layers are structurally and compositionally complex. However, over scales of seismic wavelengths (a few hundred meters to a few kilometers for crustal refraction and wide-angle reflection data) the layers appear to be nearly homogeneous in velocity, although this does not necessarily imply homogeneity in composition or petrology. Vertically, the continental crust in the midcontinent region is definitely separated into relatively uniform zones of approximately constant velocity, which are separated by nearly horizontal boundaries. Prominent layering and moderately dipping interfaces are well defined on many reflection profiling record sections. This layering does not imply that the boundaries between lower and higher velocity crustal rocks are interfaces. Although these are often shown as discontinuities in velocity, the

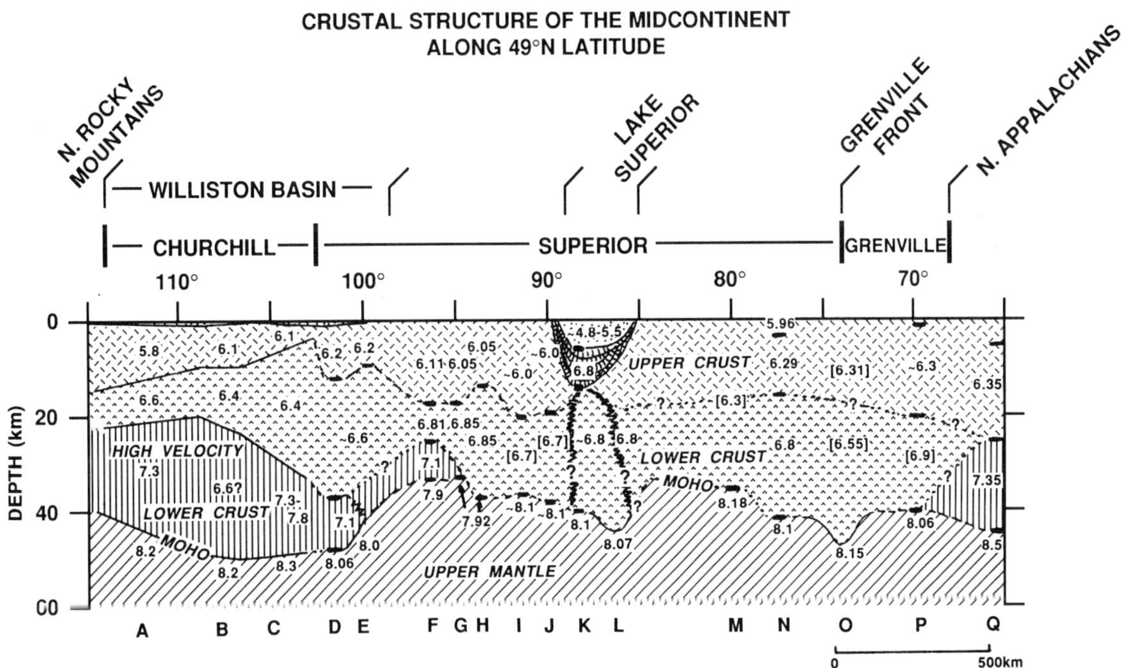

Figure 22. Interpreted crustal structure and correlations from refraction data in the midcontinent region along 49°N latitude. Numbers in the crustal model are compressional wave velocities in km/sec. Solid lines indicate interpretations from refraction data. Short, heavy lines indicate interpreted depths from seismic profiles that intersect the cross section. Dashed lines show interpreted correlations. Dotted lines indicate less certain correlations. Values enclosed in square brackets indicate averages where complicated velocity structure or gradients have been interpreted. Letters below the cross-sectional diagram indicate the interpreted refraction profiles listed in Table 5.

TABLE 4. PROFILES FOR 37° CROSS SECTION

Profile	No.	Location/Comment	Reference
A	RM10	Southern Rocky Mountains Lumberton, Colorado	Prodehl and Pakiser, 1980
B	GP16	Great Plains, Gnome-North Also, a similar interpretation by Mitchell and Landisman (1971)	Stewart and Pakiser, 1962
C	MC24	Chelsea, Oklahoma Also, a similar interpretation by Tryggvason and Qualls (1967)	Mitchell and Landisman, 1970
D	MC22	Hercules, Missouri Southern Missouri Profile	Stewart, 1968
E	CO12	W. Flank Mississippi Embayment Cape Girardeau, Missouri	McCamy and Meyer, 1966
F	CO10	Mississippi Embayment Reelfoot Rift	Mooney and others, 1983; Ginzburg and others, 1983
G	MC25	Cumberland Plateau Profiles Campbell, Tennessee	Prodehl and others, 1984
H	AP24	Cumberland Plateau Profiles Burnside, Kentucky	Prodehl and others, 1984
I	AP23	E. Tennessee	Steinhart and Meyer, 1961
J	AP22 AP28	ECOOE, Southern Line Southern Piedmont Refraction Line (projected)	Hales and others, 1968; Kean and Long, 1980

transition region may be a few hundred meters to a few kilometers thick and imply a gradual change in composition or petrology, resulting in a change in velocity with depth. Within regions of relatively homogeneous composition, the velocity as a function of depth may change as a result of pressure and temperature changes with depth. These effects may cause small positive or negative velocity gradients even in layers of homogeneous composition.

Relatively little crustal refraction seismic data are available to investigate in detail the question of homogeneity and layering of the midcontinental crust. The refraction profiles demonstrate a homogeneity and layering on a horizontal scale of tens to hundreds of kilometers. These interpretations are from data of relatively long seismic wavelength and for propagation paths that are nearly horizontal (refracted head waves and wide angle reflections). Thus, considerable horizontal "averaging" of the crustal velocity structure is inherent in the interpretation of most crustal seismic refraction data. Analysis of travel times of crustal seismic data is sensitive to large-scale (tens of kilometers horizontally and a few kilometers vertically) velocity variations. Amplitude data and related synthetic seismogram modeling methods are more useful in defining detailed velocity structure within nearly homogeneous layers or across velocity transition zones between layers (Banda and others, 1982; Braile and Chiang, 1986). However, few detailed results utilizing these methods for studies of the crust in the midcontinent region are available.

RESULTS FROM REFLECTION PROFILING DATA

Crustal structure studies for the continental interior region from deep seismic reflection profiling data are limited to a relatively small number of areas (Fig. 2, Table 2) and are largely aimed at specific, anomalous geologic features. However, these studies have recently provided important information on crustal structure because of the higher resolution of the reflection method. Recent reviews of continental crustal characteristics as

TABLE 5. PROFILES FOR 49° CROSS SECTION

Profile	No.	Location/Comment	Reference
A	RM3	Central Montana Cliff Lake–Missouri River	Steinhart and Meyer, 1961; McCamy and Meyer, 1964
B	GP8	Eastern Montana Missouri River–Fort Peck Reservoir, 6.6-km/sec velocity above Moho is questioned because all other profiles in area display high-velocity lower crust	Steinhart and Meyer, 1961; McCamy and Meyer, 1964
C	GP9	Eastern Montana Fort Peck Reservoir–Garrison	Steinhart and Meyer, 1961; McCamy and Meyer, 1964
D	GP6	Williston Basin 1977 CO-CRUST 1979 CO-CRUST	 Green and others, 1980 Hajnal and others, 1986
E	MC3	Williston Basin 1979 CO-CRUST	Hajnal and others, 1984; Hajnal, 1986
F	SU4	Eastern Manitoba	Hall and Hajnal, 1973
G	SU4	Western Ontario	Hall and Hajnal, 1973
H	SU3	Western Ontario	Hall and Hajnal, 1973
I	SU9	Ontario Line 599-R, lower-crust velocity is average of gradient zone	 Young and others, 1986
J	SU10	Ontario Line CS, lower-crust velocity is average of gradient zone	Young and others, 1986
K	MC12	Western Lake Superior	Luetgert and Meyer, 1982
L	MC6	Eastern Lake Superior	Berry and West, 1966; Smith and others, 1966b
M	SU6	Superior Province, Ontario	Hodgson, 1953
N	GR4	Grenville Front Area 1982 CO-CRUST	Huang and others, 1986; Mereu and others, 1986
O	GR3	Grenville Front	Mereu and Jobidon, 1971
P	GR2	Grenville Province Line 1	Berry and Fuchs, 1973
Q	AP4	Tracadie–East Point	Ewing and others, 1966

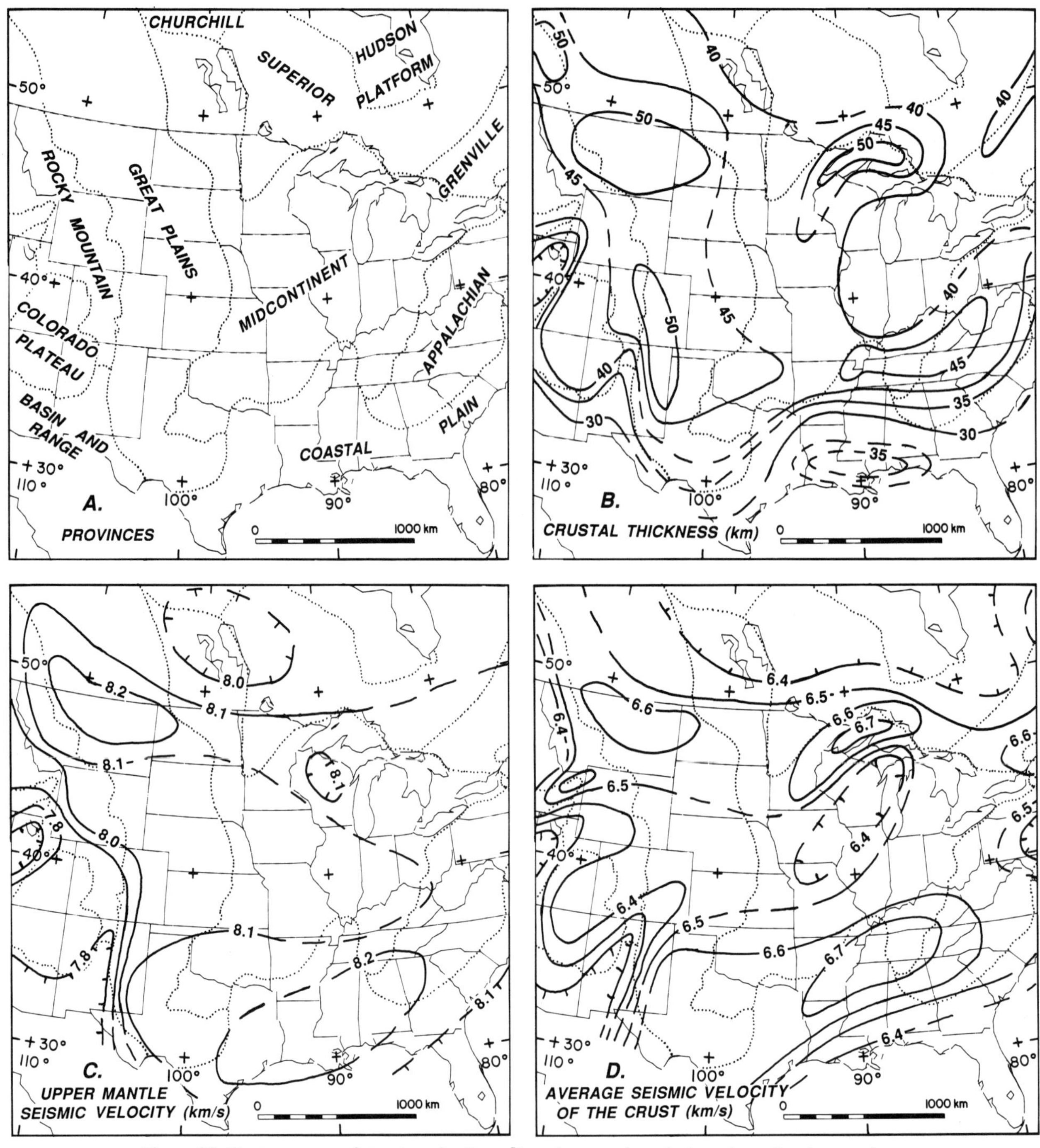

Figure 23. Contour maps of structure for the midcontinent region. A, Location of major province boundaries for the continental interior and adjacent regions. B, Contour map of crustal thickness. C, Contour map of upper-mantle compressional seismic velocity (P_n). D, Contour map of average compressional seismic velocity of the crust.

defined by deep reflection data have been published by Brown (1987), Allmendinger and others (1987), and Hauser and Oliver (1987). Within the midcontinent region, reflection data have been used to recognize subhorizontal layered reflectors in deep sedimentary basins such as the Hardeman basin in Oklahoma (Oliver and others, 1976; Brewer and others, 1981) and the Illinois basin (Pratt and others, 1989). Structures associated with thrust faulting involving both the sedimentary rocks and the crystalline crust have been delineated in a number of regions utilizing deep reflection data. Large-scale overthrusting of a thick section of sedimentary rocks is evident in the buried Ouachita system of southern Arkansas (Nelson and others, 1982; Lillie and others,

1984). Along the Grenville Front in Lake Huron (Green and others, 1988; Smith, 1988) and in Ohio (Pratt and others, 1989) and across the Great Lakes tectonic zone in Minnesota (Gibbs and others, 1984; Gibbs, 1986), deep reflection data show prominent dipping reflections that can be traced from near the surface to lower-crustal depths. These reflective dipping zones are interpreted as faults associated with crustal overthrusting along continental sutures. An example of an interpretation of the dipping crustal reflections is shown in Figure 24.

Reflection data over the Midcontinent Rift System were discussed above. In these anomalous regions, the deep reflection data provide excellent resolution for structures and sedimentary and volcanic rock layering at shallow to intermediate depths within the crust. Volcanic-filled basins extend along the Midcontinent Rift System underlying the Michigan basin (Brown and others, 1982; Zhu and Brown, 1986) and the Lake Superior basin (Behrendt and others, 1988; Cannon and others, 1989; Hinze and others, 1989) and along the southern "arm" of the MRS in Minnesota, Wisconsin, Iowa (Chandler and others, 1989), and Kansas (Brown and others, 1983; Serpa and others, 1984).

The highly reflective lower crust that has been commonly observed in reflection data for the western United States does not seem to be present in the continental interior (Allmendinger and others, 1987). Reflections at middle to lower crustal depths in the midcontinent tend to be dipping and are in many cases interpreted as diffractions. Laterally continuous events are rare (Allmendinger and others, 1987). The crust-mantle boundary in the continental interior also appears to be less reflective than in the western United States (Brown, 1987). Reflections from the Moho in the midcontinent are commonly poorly defined and discontinuous. Because the lower crust is not particularly reflective, the Moho in the midcontinent region cannot always be identified by an abrupt decrease in reflections with depth.

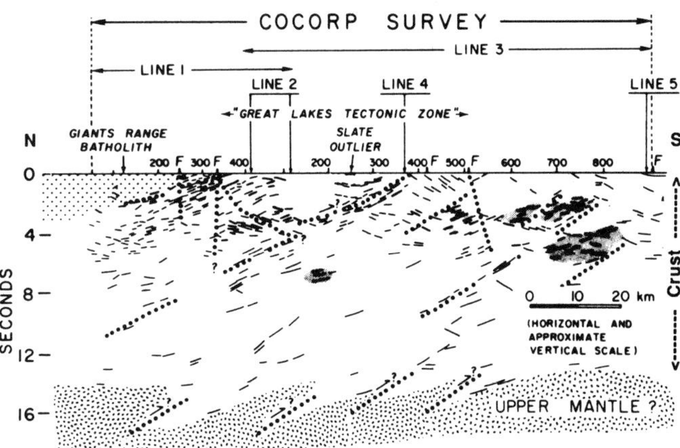

Figure 24. Line drawing showing interpreted reflections from COCORP deep reflection profiling data (profile RE7 in Fig. 2) across the Great Lakes Tectonic Zone (Fig. 1). From Gibbs and others, 1984.

ACKNOWLEDGMENTS

I am grateful to Bill Hinze, Walter Mooney, Lou Pakiser, and Randy Keller for helpful discussions and reviews of the manuscript. This research was partially supported by grants from the Earth Sciences Division of the National Science Foundation.

REFERENCES CITED

Allmendinger, R. W., Brewer, J. A., Brown, L. D., Kaufman, S., Oliver, J. E., and Houston, R. S., 1982, COCORP profiling across the Rocky Mountain Front in southern Wyoming; Part 2, Precambrian basement structure and its influence on Laramide deformation: Geological Society of America Bulletin, v. 93, p. 1253–1263.

Allmendinger, R. W., Nelson, K. D., Potter, C. J., Barazangi, M., Brown, L. D., and Oliver, J. E., 1987, Deep seismic reflection characteristics of the continental crust: Geology, v. 15, p. 304–310.

Archambeau, C. B., Flinn, E. A., and Lambert, D. G., 1969, Fine structure of the upper mantle: Journal of Geophysical Research, v. 74, p. 5825–5865.

Baldwin, J. L., 1980, A crustal seismic refraction study in southwestern Indiana and southern Illinois [M.S. thesis]: West Lafayette, Indiana, Purdue University, 99 p.

Banda, E., Deichmann, N., Braile, L. W., and Ansorge, J., 1982, Amplitude study of the Pg phase: Journal of Geophysics, v. 51, p. 153–164.

Bates, A., and Hall, D. H., 1975, Upper mantle structure in southern Saskatchewan and western Manitoba from project Edzoe: Canadian Journal of Earth Sciences, v. 12, p. 2134–2144.

Behrendt, J. C., and 7 others, 1988, Crustal structure of the Midcontinent rift system; Results from GLIMPCE deep seismic reflection profiles: Geology, v. 16, p. 81–85.

Bennett, G. T., Clowes, R. M., and Ellis, R. M., 1975, A seismic refraction survey along the southern Rocky Mountain trench, Canada: Seismological Society of America Bulletin, v. 65, p. 37–54.

Berg, J. W., Jr., Cook, K. L., Narans, H. D., Jr., and Dolan, W. M., 1960, Seismic investigation of crustal structure in the eastern part of the Basin and Range province: Seismological Society of America Bulletin, v. 50, p. 511–535.

Berry, M. J., and Fuchs, K., 1973, Crustal structure of the Superior and Grenville provinces of the northeastern Canadian Shield: Seismological Society of America Bulletin, v. 63, p. 1393–1432.

Berry, M. J., and West, G. F., 1966, A time-term interpretation of the first-arrival data of the 1963 Lake Superior experiment, *in* Steinhart, J. S., and Smith, T. J., eds., The earth beneath the continents: American Geophysical Union Geophysical Monograph 10, p. 166–180.

Braile, L. W., and Chiang, C. S., 1986, The continental Mohorovicic discontinuity; Results from near-vertical and wide-angle seismic reflection studies, *in* Barazangi, M., and Brown, L., eds., Reflection seismology; A global perspective: American Geophysical Union Geodynamics Series Volume 13, p. 257–272.

Braile, L. W., and Smith, R. B., 1975, Guide to the interpretation of crustal refraction profiles: Geophysical Journal of the Royal Astronomical Society, v. 40, p. 145–176.

Braile, L. W., Smith, R. B., Keller, G. R., Welch, R. M., and Meyer, R. P., 1974, Crustal structure across the Wasatch Front from detailed seismic refraction studies: Journal of Geophysical Research, v. 79, p. 2669–2766.

Braile, L. W., Keller, G. R., Hinze, W. J., and Lidiak, E. G., 1982a, An ancient rift complex and its relation to contemporary seismicity in the New Madrid seismic zone: Tectonics, v. 1, p. 225–237.

Braile, L. W., and 9 others, 1982b, The Yellowstone–Snake River Plain seismic profiling experiment; Crustal structure of the eastern Snake River Plain: Journal of Geophysical Research, v. 87, p. 2597–2609.

Braile, L. W., Hinze, W. J., Keller, G. R., Lidiak, E. G., and Sexton, J. L., 1986, Tectonic development of the New Madrid Rift Complex, Mississippi Embayment, North America: Tectonophysics, v. 131, p. 1–21.

Brewer, J. A., Brown, L. D., Steiner, D., Oliver, J. E., Kaufman, S., and Denison, R. E., 1981, Proterozoic basin in the southern Midcontinent of the United States revealed by COCORP deep seismic reflection profiling: Geology, v. 9, p. 569–575.

Brewer, J. A., Allmendinger, R. W., Brown, L. D., Oliver, J. E., and Kaufman, S., 1982, COCORP profiling across the Rocky Mountain Front in southern Wyoming; Part 1, Laramide structure: Geological Society of America Bulletin, v. 93, p. 1242–1252.

Brewer, J. A., Good, R., Oliver, J. E., Brown, L. D., and Kaufman, S., 1983, COCORP profiling across the southern Oklahoma aulacogen; Overthrusting of the Wichita Mountains and compression within the Anadarko Basin: Geology, v. 11, p. 109–114.

Brown, L. D., 1986, Aspects of COCORP deep seismic profiling, *in* Barazangi, M., and Brown, L., eds., Reflection seismology; A global perspective: American Geophysical Union Geodynamics Series Volume 13, p. 209–222.

Brown, L. D., 1987, Proterozoic tectonic elements in the U.S. mapped by COCORP deep seismic profiling, *in* Kroner, A., ed., Proterozoic lithospheric evolution: American Geophysical Union Geodynamics Series, v. 17, p. 69–83.

Brown, L. D., Chapin, C. E., Sanford, A. R., Kaufman, S., and Oliver, J., 1980, Deep structure of the Rio Grande Rift from seismic reflection profiling: Journal of Geophysical Research, v. 85, p. 4773–4800.

Brown, L., Jensen, L., Oliver, J., Kaufman, S., and Steiner, D., 1982, Rift structure beneath the Michigan Basin from COCORP profiling: Geology, v. 10, p. 645–649.

Brown, L. D., and 7 others, 1983, Intracrustal complexity in the United States midcontinent; Preliminary results from COCORP surveys in northeastern Kansas: Geology, v. 11, p. 25–30.

Cannon, W. F., and 11 others, 1989, The North American Midcontinent Rift beneath Lake Superior from GLIMPCE seismic reflection profiling: Tectonics, v. 8, p. 305–332.

Chandler, V. W., Bowman, P. L., Hinze, W. J., and O'Hara, N. W., 1982, Long-wavelength gravity and magnetic anomalies of the Lake Superior region, *in* Wold, R. J., and Hinze, W. J., eds., Geology and tectonics of the Lake Superior Basin: Geological Society of America Memoir 156, p. 223–237.

Chandler, V. W., McSwiggen, P. L., Morey, G. B., Hinze, W. J., and Anderson, R. R., 1989, Interpretation of seismic reflection, gravity, and magnetic data across the middle Proterozoic Midcontinent rift system in western Wisconsin, eastern Minnesota, and central Iowa: American Association of Petroleum Geologists (in press).

Chandra, N. N., and Cumming, G. L., 1972, Seismic refraction studies in western Canada: Canadian Journal of Earth Sciences, v. 9, p. 1099–1109.

Cohen, T. J., 1966, Explosion seismic studies of the midcontinent gravity high [Ph.D. thesis]: Madison, University of Wisconsin, 329 p.

Cohen, T. J., and Meyer, R. P., 1966, The midcontinent gravity high; Gross crustal structure, *in* Steinhart, J. S., and Smith, T. J., eds., The earth beneath the continents: American Geophysical Union Geophysical Monograph 10, p. 141–165.

Cook, F. A., McCullar, D. B., Decker, E. R., and Smithson, S. B., 1979, Crustal structure and evolution of the southern Rio Grande Rift, *in* Riecker, R. E., ed., Rio Grande Rift; Tectonics and magmatism: American Geophysical Union, p. 195–208.

Cordell, L., 1977, Regional positive gravity anomaly over the Mississippi Embayment: Geophysical Research Letters, v. 4, p. 285–287.

Cram, I. H., Jr., 1961, Crustal structure refraction survey in south Texas: Geophysics, v. 26, p. 560–573.

Cumming, W. B., Clowes, R. M., and Ellis, R. M., 1979, Crustal structure from a seismic refraction profile across southern British Columbia: Canadian Journal of Earth Sciences, v. 16, p. 1024–1040.

Delandro, W., and Moon, W., 1982, Seismic structure of Superior-Churchill Precambrian boundary zone: Journal of Geophysical Research, v. 87, p. 6884–6888.

Denison, R. E., Lidiak, E. G., Bickford, M. E., and Kisvarsanyi, E. B., 1984, Geology and geochronology of Precambrian rocks in the central interior region of the United States, *in* Harrison, J. E., and Peterman, Z. E., eds., Correlation of Precambrian rocks of the United States and Mexico: U.S. Geological Survey Professional Paper 1241-C, p. C-1–C-20.

Dorman, J., Worzel, J. L., Leyden, R., Cook, T. N., and Hatziemmanuel, M., 1972, Crustal section from seismic refraction measurements near Victoria, Texas: Geophysics, v. 37, p. 225–236.

Dorman, L. M., 1972, Seismic crustal anisotropy in northern Georgia: Seismological Society of America Bulletin, v. 62, p. 39–45.

Douglas, R.J.W., and Price, R. A., 1972, Nature and significance of variations in tectonic styles in Canada, *in* Price, R. A., and Douglas, R.J.W., eds., Variations in tectonic styles in Canada: Geological Society of Canada Special Paper 11, p. 625–688.

Ervin, C. P., and McGinnis, L. D., 1975, Reelfoot rift; Reactivated precursor to the Mississippi Embayment: Geological Society of America Bulletin, v. 86, p. 1287–1295.

Ewing, M., Worzel, J. L., Ericson, D. B., and Heezen, B. C., 1955, Geophysical and geological investigations in the Gulf of Mexico, Part 1: Geophysics, v. 20, p. 1–18.

Ewing, M., Dainty, A. M., Blanchard, J. E., and Keen, J. J., 1966, Seismic studies on the eastern seaboard of Canada; The Appalachian system I: Canadian Journal of Earth Sciences, v. 3, p. 89–109.

Fenneman, N. M., 1946, Physical divisions of the United States: U.S. Geological Survey Map, scale 1:7,000,000.

Fox, A. J., 1988, An integrated geophysical study of the southeastern extension of the Midcontinent Rift System [M.S. thesis]: West Lafayette, Indiana, Purdue University, 112 p.

Fuchs, K., and Müller, G., 1971, Computation of synthetic seismograms with the reflectivity method and comparison with observations: Geophysical Journal of the Royal Astronomical Society, v. 23, p. 417–433.

Gibbs, A. K., 1986, Seismic reflection profiles of Precambrian crust; A qualitative assessment, *in* Barazangi, M., and Brown, L., eds., Reflection seismology; The continental crust: American Geophysical Union Geodynamics Series 14, p. 95–106.

Gibbs, A. K., Payne, B., Setzer, T., Brown, L. D., Oliver, J. E., and Kaufman, S., 1984, Seismic reflection study of the Precambrian crust of central Minnesota: Geological Society of America Bulletin, v. 95, p. 280–294.

Ginzburg, A., Mooney, W. D., Walter, A. W., Lutter, W. J., and Healy, J. H., 1983, Deep structure of northern Mississippi Embayment: American Association of Petroleum Geologists Bulletin, v. 67, p. 2031–2046.

Gish, D. M., Keller, G. R., and Sbar, M. L., 1981, A refraction study of deep crustal structure in the Basin and Range; Colorado Plateau of eastern Arizona: Journal of Geophysical Research, v. 86, p. 6029–6038.

Green, A. G., and Clowes, R. M., 1983, Deep geology from seismic reflection studies in Canada: First Break, v. 1, p. 24–33.

Green, A. G., and 7 others, 1980, Cooperative seismic surveys across the Superior-Churchill boundary zone in southern Canada: Canadian Journal of Earth Sciences, v. 17, p. 617–532.

Green, A. G., and 9 others, Milkereit, B., Davidson, A., Spencer, C., Hutchinson, D. R., Cannon, W. F., Lee, M. W., Agena, F., Behrendt, J. C., and Hinze, W. J., 1988, Crustal structure of the Grenville front and adjacent terranes: Geology, v. 16, p. 788–792.

Green, R.W.E., and Hales, A. L., 1968, The travel times of P waves to 30° in the central United States and upper mantle structure: Seismological Society of America Bulletin, v. 58, p. 267–289.

Hajnal, Z., 1986, Crustal reflection and refraction velocities; A comparison, *in* Barazangi, M., and Brown, L., eds., Reflection seismology; The continental crust: American Geophysical Union Geodynamics Series 13, p. 247–256.

Hajnal, Z., and 6 others, 1984, An initial analysis of the earth's crust under the Williston Basin; 1979 COCRUST experiment: Journal of Geophysical Research, v. 89, p. 9381–9400.

Hales, A. L., and Nation, J. B., 1973, A seismic refraction survey in the northern Rocky Mountains; More evidence for an intermediate crustal layer: Geophysical Journal of the Royal Astronomical Society, v. 35, p. 381–399.

Hales, A. L., Helsey, C. E., Dowling, J. J., and Nation, J. B., 1968, The east coast onshore-offshore experiment; 1, The first arrival phases: Seismological Society of America Bulletin, v. 58, p. 757–819.

Hales, A. L., Helsey, C. E., and Nation, J. B., 1970, Crustal structure study of Gulf Coast of Texas: American Association of Petroleum Geologists Bulletin, v. 54, p. 2040–2057.

Hall, D. H., and Hajnal, Z., 1969, Crustal structure of northwestern Ontario; Refraction seismology: Canadian Journal of Earth Sciences, v. 6, p. 81–99.

——, 1973, Deep seismic studies in Manitoba: Seismological Society of America Bulletin, v. 63, p. 883–910.

Halls, H. C., 1982, Crustal thickness in the Lake Superior region, *in* Wold, R. J., and Hinze, W. J., eds., Geology and tectonics of the Lake Superior Basin: Geological Society of America Memoir 156, p. 239–243.

Hamilton, R. M., and Zoback, M. D., 1982, Tectonic features of the New Madrid seismic zone from seismic-reflection profiles, *in* McKeown, F. A., and Pakiser, L. C., eds., Investigations of the New Madrid, Missouri, earthquake region: U.S. Geological Survey Professional Paper 1236, p. 55–82.

Hauser, E. C., and Oliver, J. E., 1987, A new era in understanding the continental basement; The impact of seismic reflection profiling, *in* Fuchs, K. and Froidevaux, C., eds., Composition, Structure and Dynamics of the Lithosphere—Asthenosphere System: American Geophysical Union Geodynamics Series, v. 16, p. 1–32.

Healy, J. H., and Warren, D. H., 1969, Explosion seismic studies in North America, *in* Hart, P. J., ed., The earth's crust and upper mantle: American Geophysical Union Geophysical Monograph 13, p. 208–220.

Hersey, J. B., Bunce, E. T., Wyrick, R. F., and Dietz, T. F., 1959, Geophysical investigation of the continental margin between Cape Henry, Virginia, and Jacksonville, Florida: Geological Society of America Bulletin, v. 70, p. 437–465.

Hildenbrand, T. G., 1985, Rift structure of the northern Mississippi embayment from the analysis of gravity and magnetic data: Journal of Geophysical Research, v. 90, p. 12607–12622.

Hildenbrand, T. G., Kane, M. F., and Stauder, W., 1977, Magnetic and gravity anomalies in the northern Mississippi Embayment and their spatial relation to seismicity: U.S. Geological Survey Map 914, scale 1:1,000,000.

Hinze, W. J., and Braile, L. W., 1988, Geophysical aspects of the craton; U.S., *in* Sloss, L. L., ed., Sedimentary cover—North American craton, U.S.: Boulder, Colorado, Geological Society of America, Geology of North America, v. D-2, p. 5–24.

Hinze, W. J., Braile, L. W., and Chandler, V. W., 1989, A geophysical profile of the southern margin of the Midcontinent Rift System in western Lake Superior: Tectonics (in press).

Hinze, W. J., Wold, R. J., and O'Hara, N. W., 1982, Gravity and magnetic anomaly studies of Lake Superior, *in* Wold, R. J., and Hinze, W. J., eds., Geology and tectonics of the Lake Superior basin: Geological Society of America Memoir 156, p. 203–221.

Hodgson, J. H., 1953, A seismic survey of the Canadian Shield; 1, Refraction studies based on rock-bursts at Kirkland Lake, Ontario: Ottawa, Ontario, Dominion Observatory Publication 16, p. 111–163.

Huang, H., Spencer, C., and Green, A., 1986, A method for the inversion of refraction and reflection travel times for laterally varying velocity structures: Seismological Society of America Bulletin, v. 76, p. 837–846.

Jackson, W. H., and Pakiser, L. C., 1965, Seismic study of crustal structure in the southern Rocky Mountains: U.S. Geological Survey Professional Paper 525-D, p. D-85-D-92.

Jackson, W. H., Stewart, S. W., and Pakiser, L. C., 1963, Crustal structure in eastern Colorado from seismic-refraction measurements: Journal of Geophysical Research, v. 68, p. 5767–5776.

Jaksha, L. H., 1982, Reconnaissance seismic refraction-reflection surveys in southwestern New Mexico: Geological Society of America Bulletin, v. 93, p. 1030–1037.

James, D. E., Smith, T. J., and Steinhart, J. S., 1968, Crustal structure of the Middle Atlantic states: Journal of Geophysical Research, v. 73, p. 1983–2007.

Jardine, W. G., 1988, Seismic study in the Ouachita system of southern Arkansas and the adjacent Gulf Coastal Plain [M.S. thesis]: West Lafayette, Indiana, Purdue University, 112 p.

Jordan, T. H., and Frazer, L. N., 1975, Crustal and upper mantle structure from Sp phases: Journal of Geophysical Research, v. 80, p. 1504–1518.

Kanasewich, E. R., and Chiu, S.K.L., 1985, Least-squares inversion of spatial seismic refraction data: Seismological Society of America Bulletin, v. 75, p. 865–880.

Kane, M. F., Hildenbrand, T. G., and Hendricks, J. D., 1981, A model for the tectonic evolution of the Mississippi Embayment and its contemporary seismicity: Geology, v. 9, p. 563–567.

Katz, S., 1954, Seismic study of crustal structure in Pennsylvania and New York: Seismological Society of America Bulletin, v. 44, p. 303–325.

Kean, E. A., and Long, L. T., 1980, A seismic refraction line along the axis of the southern Piedmont and crustal thickness in the southeastern United States: Earthquake Notes, v. 51, p. 3–13.

Keller, G. R., Smith, R. B., and Braile, L. W., 1975, Crustal structure along the Great Basin–Colorado Plateau transition from seismic refraction studies: Journal of Geophysical Research, v. 80, p. 1093–1097.

Keller, G. R., Lidiak, E. G., Hinze, W. J., and Braile, L. W., 1983, The role of rifting in the tectonic development of the Midcontinent, U.S.A.: Tectonophysics, v. 94, p. 391–412.

Keller, G. R., and 6 others, 1989, The Paleozoic continent-ocean transition in the Ouachita Mountains imaged from PASSCAL wide-angle reflection-refraction data: Geology, v. 17, p. 119–122.

King, P. B., and Edmonston, G. J., 1972, Generalized tectonic map of North America: U.S. Geological Survey Miscellaneous Geologic Investigation Map I-688, scale 1:15,000,000.

Klemperer, S. L., Brown, L. D., Oliver, J. E., Ando, C. J., Czuchra, B. L., and Kaufman, S., 1985, Some results of COCORP seismic reflection profiling in the Grenville-age Adirondack Mountains, New York State: Canadian Journal of Earth Sciences, v. 22, p. 141–153.

Kurita, T., 1973, Regional variations in the structure of the crust in the central United States from P-wave spectra: Seismological Society of America Bulletin, v. 63, p. 1663–1687.

Latham, T. S., and 5 others, 1988, COCORP profiles from the Montana plains: the Archean cratonic crust and a lower crustal anomaly beneath the Williston basin: Geology, v. 16, p. 1073–1076.

Lewis, B.T.R., and Meyer, R. P., 1968, A seismic investigation of the upper mantle to the west of Lake Superior: Seismological Society of America Bulletin, v. 58, p. 565–596.

Lidiak, E. G., 1982, Basement rocks of the main interior basins of the midcontinent: University of Missouri, Rolla, Journal, v. 3, p. 5–24.

Lillie, R. J., and 7 others, 1983, Crustal structure of Ouachita Mountains, Arkansas; A model based on integration of COCORP reflection profiles and regional geophysical data: American Association of Petroleum Geologists Bulletin, v. 67, p. 907–931.

Long, L. T., and Liow, J., 1986, Crustal thickness, velocity structure, and the isostatic response function in the southern Appalachians, *in* Barazangi, M., and Brown, L., eds., Reflection seismology; The continental crust: American Geophysical Union Geodynamics Series 14, p. 215–222.

Luetgert, J. H., and Meyer, R. P., 1982, Structure of the western base of Lake Superior from cross structure refraction profiles, *in* Wold, R. J., and Hinze, W. J., eds., Geology and tectonics of the Lake Superior Basin: Geological Society of America Memoir 156, p. 245–255.

Martin, W. R., 1978, A seismic refraction study of the northeastern Basin and Range and its transition with the eastern Snake River Plain [M.S. thesis]: El Paso, University of Texas, 41 p.

McCamy, K., and Meyer, R. P., 1964, A correlation method of apparent velocity measurement: Journal of Geophysical Research, v. 69, p. 691–699.

——, 1966, Crustal results of fixed multiple shots in the Mississippi Embayment, *in* Steinhart, J. S., and Smith, T. J., eds., The earth beneath the continents; American Geophysical Union Geophysical Monograph 10, p. 166–180.

Mereu, R. F., and Hunter, J. A., 1969, Crustal and upper mantle under the Canadian Shield from project Early Rise data: Seismological Society of America Bulletin, v. 59, p. 147–165.

Mereu, R. F., and Jobidon, G., 1971, A seismic investigation of the crust and Moho on a line perpendicular to the Grenville Front: Canadian Journal of Earth Sciences, v. 8, p. 1553–1583.

Mereu, R. F., Majumdar, S. C., and White, R. E., 1976, The structure of the crust and upper mantle under the highest ranges of the Canadian Rockies from a seismic refraction survey: Canadian Journal of Earth Sciences, v. 14, p. 196–208.

Mereu, R., and 11 others, 1986a, The 1982 COCRUST seismic experiment across the Ottawa-Bonnechere graben and Grenville Front in Ottawa and Quebec: Geophysical Journal of the Royal Astronomical Society, v. 84, p. 491–514.

Mereu, R., Wang, D., and Kuhn, O., 1986b, Evidence for an inactive rift in the Precambrian from a wide-angle reflection survey across the Ottawa-Bonnechere graben, *in* Barazangi, M., and Brown, L., eds., Reflection seismology; The continental crust: American Geophysical Union Geodynamics Series 14, p. 127–134.

Merkel, R. H., and Alexander, S. S., 1969, Use of correlation analysis to interpret continental margin ECOOE refraction data: Journal of Geophysical Research, v. 74, p. 2683–2697.

Mitchell, B. J., and Landisman, M., 1970, Interpretation of crustal section across Oklahoma: Geological Society of America Bulletin, v. 81, p. 2647–2656.

——, 1971, Geophysical measurements in the southern great plains, *in* Heacock, J. G., ed., The structure and physical properties of the Earth's crust: American Geophysical Union Geophysical Monograph 14, p. 77–93.

Mooney, W. D., Andrews, M. C., Ginzburg, A., Peters, D. A., and Hamilton, R. M., 1983, Crustal structure of the northern Mississippi Embayment and a comparison with other continental rift zones: Tectonophysics, v. 94, p. 327–348.

Morel-a-L'Huissier, P., Green, A. G., and Pike, C. J., 1987, Crustal refraction surveys across the Trans-Hudson Orogen/Williston Basin of south-central Canada: Journal of Geophysical Research, v. 92, p. 6403–6420.

Nelson, K. D., and 7 others, 1982, COCORP seismic reflection profiling in the Ouachita Mountains of western Arkansas: Geometry and geologic interpretation: Tectonics, v. 1, p. 413–430.

Ocola, L. D., and Meyer, R. P., 1973, Central North American rift system; 1, Structure of the axial zone from seismic and gravimetric data: Journal of Geophysical Research, v. 78, p. 5173–5194.

Oliver, J., Dobrin, M., Kaufman, S., Meyer, R., and Phinney, R., 1976, Continuous seismic reflection profiling of the deep basement, Hardeman County, Texas: Geological Society of America Bulletin, v. 87, p. 1537–1546.

Olsen, K. H., Keller, G. R., and Stewart, J. N., 1979, Crustal structure along the Rio Grande Rift from seismic refraction profiles, *in* Riecker, R. E., ed., Rio Grande Rift; Tectonics and magmatism: American Geophysical Union, p. 127–143.

Owens, T. J., Zandt, G., and Taylor, S. R., 1984, Seismic evidence for an ancient rift beneath the Cumberland Plateau, Tennessee; A detailed analysis of broadband teleseismic P waveforms: Journal of Geophysical Research, v. 89, p. 7783–7795.

Pratt, T., and 7 others, 1989, Major Proterozoic basement features of the eastern midcontinent of North America revealed by recent COCORP profiling: Geology, v. 17, p. 505–509.

Prodehl, C., and Pakiser, L. C., 1980, Crustal structure of the southern Rocky Mountains from seismic measurements: Geological Society of America Bulletin, v. 91, p. 147–155.

Prodehl, C., Schlittenhardt, J., and Stewart, S. W., 1984, Crustal structure of the Appalachian Highlands in Tennessee, Tectonophysics, v. 109, p. 61–76.

Richards, T. C., and Walker, D. J., 1959, Measurement of the thickness of the earth's crust in the Albertan plains of western Canada: Geophysics, v. 24, p. 262–284.

Roller, J. C., 1965, Crustal structure in the eastern Colorado Plateau province from seismic-refraction measurements: Seismological Society of America Bulletin, v. 55, p. 107–119.

Roller, J. C., and Jackson, W. H., 1966, Seismic-wave propagation in the upper mantle; Lake Superior, Wisconsin, to Denver, Colorado, *in* Steinhart, J. S., and Smith, T. J., eds., The earth beneath the continents: American Geophysical Union Geophysical Monograph 10, p. 270–275.

Romney, C., Brooks, B. G., Mansfield, R. H., Carder, D. S., Jordan, J. N., and Gordon, D. W., 1962, Travel times and amplitudes of principal body phases recorded from GNOME: Seismological Society of America Bulletin, v. 52, p. 1057–1074.

Serpa, L., and 6 others, 1984, Structure of the southern Keweenawan Rift from COCORP surveys across the Midcontinent Geophysical Anomaly in northeastern Kansas: Tectonics, v. 3, p. 367–384.

Sheriff, S. D., and Stickney, M. C., 1984, Crustal structure of southwestern Montana and east-central Idaho; Results of a reversed seismic refraction line: Geophysical Research Letters, v. 11, p. 299–302.

Sinno, Y. A., Daggett, P. H., Keller, G. R., Morgan, P., and Harder, S. H., 1986, Crustal structure of the southern Rio Grande rift determined from seismic refraction profiling: Journal of Geophysical Research, v. 91, p. 6143–6156.

Slichter, L. B., 1951, Crustal structure in the Wisconsin area: Arlington, Virginia, Office of Naval Research Report N9, NR86200.

Smith, J. G., 1988, An integrated geophysical study of the Grenville Front in Lake Huron [M.S. thesis]: West Lafayette, Indiana, Purdue University, 122 p.

Smith, R. B., and 9 others, 1982, The 1978 Yellowstone-eastern Snake River Plain seismic profiling experiment; Crustal structure of the Yellowstone region and experiment design: Journal of Geophysical Research, v. 87, p. 2583–2596.

Smith, T. J., Steinhart, J. S., and Aldrich, L. T., 1966a, Lake superior crustal structure: Journal of Geophysical Research, v. 71, p. 1141 1172.

——, 1966b, Crustal structure under Lake Superior, *in* Steinhart, J. S., and Smith, T. J., eds., The earth beneath the continents: American Geophysical Union Geophysical Monograph 10, p. 181–197.

Sparlin, M. A., Braile, L. W., and Smith, R. B., 1982, Crustal structure of the eastern Snake River Plain determined from ray-trace modeling of seismic refraction data: Journal of Geophysical Research, v. 87, p. 2619–1633.

Spence, G. D., Clowes, R. M., and Ellis, R. M., 1977, Depth limits on the M discontinuity in the southern Rocky Mountain Trench, Canada: Seismological Society of America Bulletin, v. 67, p. 543–546.

Steinhart, J. S., and Meyer, R. P., 1961, Explosion studies of continental structure: Carnegie Institute of Washington Publication 622, 409 p.

Steinhart, J. S., and Smith, T. J., eds., 1966, The earth beneath the continents: American Geophysical Union Geophysical Monograph 10, 663 p.

Stewart, S. W., 1968, Crustal structure in Missouri by seismic-refraction methods: Seismological Society of America Bulletin, v. 58, p. 291–323.

Stewart, S., and Pakiser, L., 1962, Crustal structure in eastern New Mexico interpreted from the Gnome explosion: Seismological Society of America Bulletin, v. 52, p. 1017–1030.

Tatel, H. E., and Tuve, M. A., 1955, Seismic exploration of a continental crust: Geological Society of America Special Paper 62, p. 35–50.

Tatel, H. E., Adams, L. H., and Tuve, M. A., 1953, Studies of the earth's crust using waves from explosions: American Philosophy Society Proceedings, v. 97, p. 658–669.

Taylor, S. R., Toksoz, M. N., and Chaplin, M. P., 1980, Crustal structure of the northeastern United States; Contrasts between Grenville and Appalachian Provinces: Science, v. 208, p. 595–597.

Toppozada, T. R., and Sanford, A. R., 1976, Crustal structure in central New

Mexico interpreted from the GASBUGGY explosion: Seismological Society of America Bulletin, v. 66, p. 877–886.

Tryggvason, E., and Qualls, B. R., 1967, Seismic refraction measurements of crustal structure in Oklahoma, Journal of Geophysical Research, v. 72, p. 3788–3740.

Tuve, M. A., 1953, The earth's crust: Yearbook of the Carnegie Institution of Washington, v. 52, p. 103–108.

Tuve, M. A., Tatel, H. E., and Hart, P. J., 1954, Crustal structure from seismic exploration: Journal of Geophysical Research, v. 59, p. 415–422.

Van Schmus, W. R., and Bickford, M. E., 1981, Proterozoic chronology and evolution of the midcontinent region, North America, *in* Kroner, A., ed., Precambrian plate tectonics: Amsterdam, Elsevier, p. 261–296.

Van Schmus, W. R., and Hinze, W. J., 1985, The Midcontinent Rift System: Annual Reviews of Earth and Planetary Sciences, v. 13, p. 345–383.

Warren, D. H., 1968, Transcontinental geophysical survey (35°–39°N) seismic refraction profiles of the crust and upper mantle: U.S. Geological Survey Maps I-532-D, I-533-D, I-534-D, and I-535-D, scale 1:1,000,000.

—— , 1969, Seismic-refraction survey of crustal structure in central Arizona: Geological Society of America Bulletin, v. 80, p. 257–282.

Warren, D. H., and Jackson, W. H., 1968, Surface seismic measurements of the project GASBUGGY explosion at intermediate distance ranges: U.S. Geological Survey Open-File Report 1023, 45 p.

Warren, D. H., Healy, J. H., and Jackson, W. H., 1966, Crustal seismic measurements in southern Mississippi: Journal of Geophysical Research, v. 71, p. 3437–3458.

Warren, D. H., Healy, J. H., Hoffmann, J. C., Kempe, R., Rauula, S., and Stuart, D. J., 1967, Project Early Rise; Traveltimes and amplitudes: U.S. Geological Survey Technical Letter 6, 150 p.

Warren, D. H., Healy, J. H., Bohn, J., and Marshall, P. A., 1972, Crustal calibration of the large aperture seismic array (LASA), Montana: U.S. Geological Survey Open-File Report 1671, 163 p.

Willden, R., 1965, Seismic-refraction measurements of crustal structure beneath American Falls Reservoir, Idaho, and Flaming Gorge Reservoir, Utah: U.S. Geological Survey Professional Paper 525-C, p. C-44-C-50.

Young, R. A., Wright, J., and West, G. F., 1986, Seismic crustal structure northwest of Thunder Bay, Ontario, *in* Barazangi, M., and Brown, L., eds., Reflection seismology; The continental crust: American Geophysical Union Geodynamics Series 14, p. 143–155.

Zandt, G., and Owens, T. J., 1986, Comparison of crustal velocity profiles determined by seismic refraction and teleseismic methods: Tectonophysics, v. 128, p. 155–161.

Zhu, T., and Brown, L. D., 1986, Consortium for continental reflection profiling Michigan surveys; Reprocessing and results: Journal of Geophysical Research, v. 91, p. 11477–11495.

Zoback, M. D., and Zoback, M. L., 1981, State of stress and intraplate earthquakes in the United States: Science, v. 213, p. 96–104.

Zoback, M. D., and 5 others, 1980, Recurrent intraplate tectonism in the New Madrid seismic zone: Science, v. 209, p. 971–976.

Manuscript Accepted by the Society October 31, 1988

Printed in U.S.A.

Geological Society of America
Memoir 172
1989

Chapter 16

Geophysical framework of the Appalachians and adjacent Grenville Province

Steven R. Taylor
Earth Sciences Department, Lawrence Livermore National Laboratory, University of California, Livermore, California 94550

ABSTRACT

A number of geophysical studies, including nonseismic (gravity, magnetics, resistivity, and heat flow) and seismic (refraction, teleseismic P-wave residuals, P_n time terms, reflection, inversion of teleseismic receiver functions, and surface-wave dispersion) results, are combined with geological observations to obtain a picture of the subsurface structure in the Appalachians. Based on differences in geologic history and geophysical structure, the Appalachians can be divided across strike into a northern and southern segment separated by a line marking the approximate northward extension of Alleghanian deformation. This boundary is located approximately in the vicinity of southern New England. Along strike, the mountain chain can be divided into belts of either continental or oceanic affinity. This along-strike boundary exists in the vicinity of the Inner Piedmont/Charlotte belt (or possibly into the slate belt) in the southern Appalachians, and in the Connecticut Valley synclinorium in the northern Appalachians.

In the northern Appalachians, the Grenville Province (exposed in the Adirondacks) is characterized by a relatively uniform crust on the basis of refraction models. However, detailed analysis of reflection profiles and structure derived from inversion of crustal receiver functions illustrates a more complex picture. A zone of prominent reflectors is seen to exist at depths of approximately 18 to 26 km, which correlates well with a high-velocity anomaly inferred from the receiver functions. Separate studies of crustal receiver functions also suggest the existence of relatively low shear velocities and possibly a high Poisson's ratio in the lower crust. These structures correlate with a highly conductive lower crust inferred to exist from electromagnetic sounding. The lack of Moho reflections and P_s converted phases suggests that the crust-mantle boundary in the region is gradational.

Farther east, the crust appears to be thicker beneath the Taconic thrusts where the thrust (at ~2-km depth) and underlying shelf sequence have an aggregate thickness of about 4.5 km. The reflector at a depth of 4.5 km continues beneath the Green Mountains, suggesting that they are allochthonous. Gravity, refraction, and reflection data suggest a fundamental change in crustal character east of the Precambrian outliers and serpentinite belt in the Connecticut Valley synclinorium. A regional gravity high, found to the east of the Precambrian outliers, indicates that there is a deep-seated increase in crustal density. This is further supported by the existence of high velocities in the lower crust in the central orogenic belt, observed from refraction data. In this transitional zone, there is a thick series of east-dipping reflectors observed on reflection profiles extending through much of the crust and flattening beneath the Bronson Hill anticlinorium and Merrimac synclinorium. Moho reflections are observed on the Consortium for Continental Reflection Profiling (COCORP) profiles, and magnetotelluric data suggests that the lower crust is fairly resistive. Data from P_n time terms and teleseismic P-wave

Taylor, S. R., 1989, Geophysical framework of the Appalachians and adjacent Grenville Province, *in* Pakiser, L. C., and Mooney, W. D., Geophysical framework of the continental United States: Boulder, Colorado, Geological Society of America Memoir 172.

residuals suggest that the crust is slightly thicker in this central belt. Farther east, refraction models and traveltime residuals indicate that the crust thins beneath the Avalon block, and the high-velocity lower crust is absent.

In the southern Appalachians, the Grenville crust shows many similarities to that in the north, including the gradational crust-mantle boundary and absence of high velocities in the lower crust. However, the crust appears to be much thicker on the average and approaches thicknesses of 50 km in some localities beneath the Valley and Ridge and western Blue Ridge Province.

COCORP reflection profiling illustrates that the Valley and Ridge, Blue Ridge, and Inner Piedmont are allochthonous and have been thrust over an early Paleozoic shelf sequence. The magnitude of thrusting may be at least 175 km and points out the importance of thin-skinned tectonics in continental collision episodes. The root-zone for the southern Appalachian décollement has not yet been identified and is a topic of considerable controversy. Gravity, magnetic, reflection, and refraction data all suggest that a fundamental change in crustal character occurs in the vicinity of the Inner Piedmont–Charlotte belt boundary. However, it remains unclear whether this region represents a suture zone separating North America from accreted island-arc sequences or a décollement extending farther east beneath the slate belt and into the Coastal Plain. Similar to the western part of the Connecticut Valley synclinorium in the northern Appalachians, a zone of east-dipping reflectors is observed near the Inner Piedmont–Charlotte belt boundary. However, in the northeast Georgia profile the reflectors do not appear to penetrate the crust, and a shallow-dipping reflector can be traced farther east beneath the Charlotte belt/slate belt. The southeast-dipping reflectors penetrate much of the crust on the southernmost COCORP line in southwestern Georgia where North American rocks are juxtaposed with rocks that were probably once part of the African continent. A well-defined Moho reflection is observed on reflection profiles to the east and southeast of these dipping reflectors.

The gravity and magnetic data all suggest that the crust to the east of the Inner Piedmont–Charlotte belt boundary is of oceanic affinity. In contrast to the northern Appalachians, the southern refraction data indicates that the crust thins dramatically beneath the Charlotte belt–Carolina slate belt, and no evidence of high velocities is observed in the lower crust. Interestingly, the Charlotte belt and slate belt rocks are thought to be of Avalon affinity, and the crustal velocity structure shows many similarities to the Avalon block in the northern Appalachians.

INTRODUCTION

The Appalachian Mountains consist of an eroded core of a Paleozoic mountain chain that extends at least 3,000 km from the southeastern United States to Newfoundland. The structure of the orogenic belt contains the records of a number of major tectonic events including rifting events, arc-continent, and continent-continent collisional episodes.

The tectonic histories of ancient mountain belts such as the Appalachians are often interpreted in terms of the reigning evolutionary models for neotectonic belts such as the Himalayas, the Alps, or the Zagros, where geologic and geophysical controls are either lacking or poorly known at best. In contrast, as detailed below, the geology and geophysics of older mountain belts such as the Appalachians are often known in such great detail that it is difficult to formulate even a simple first-order evolutionary model because of the numerous complications and seemingly conflicting observations encountered (i.e., there are too many constraints). Because ancient, deeply eroded orogenic belts are probably representative of the deeply eroded cores of more recent mountain chains, they provide a window into the subsurface processes involved with mountain building. Therefore, the geological and geophysical information obtained from older belts such as the Appalachians should be used for interpreting the processes involved with the formation of the younger belts.

In this chapter, the geology and tectonic history of the Appalachians are first reviewed. The following sections contain a review of nonseismic geophysical observations (gravity, magnetics, resistivity, and heat flow), and seismic results (refraction, teleseismic P-wave residuals, P_n time terms, reflection, inversion of teleseismic receiver functions, and surface-wave dispersion). The conclusion section attempts to synthesize much of the available geophysical information in order to derive a picture that is consistent with the geologic constraints.

GEOLOGIC SETTING OF THE APPALACHIAN MOUNTAIN BELT

In this section, we briefly review the geologic setting of the northern and southern Appalachians, the Adirondacks, and the Grenville Province. More detailed summaries can be found in McLelland and Isachsen (1980) and Dewey and Burke (1973) for the Grenville Province; Rodgers (1970), Bird and Dewey (1970), and Osberg (1978) for the northern Appalachians; and Rodgers (1970), Hatcher (1978), and Secor and others (1986) for the southern Appalachians. A generalized geologic map of eastern North America is shown in Figure 1, and the major provinces are shown in Figure 2. Table 1 lists the major Appalachian orogenic episodes, and the nature and extent of their deformation.

The Precambrian (~1.1 Ga) Grenville Province, exposed in eastern Canada and the Adirondacks, extends southward in the subsurface west of the Appalachian mountain belt. The rocks of the Grenville Province are largely remobilized older basement of Superior or Hudsonian age and consist mainly of thick sequences of high-grade metasediments extensively intruded by granitic and anorthositic rocks. The Adirondack Mountains consist of a core of Precambrian Grenville rocks (~1.1 Ga) surrounded by gently dipping Paleozoic rocks (Fig. 1). Four major intersecting fold sets have been mapped in the Adirondacks, giving the region a complex fold-interference pattern (McLelland and Isachsen, 1980). It has been postulated that the Adirondacks and associated Grenville basement represent deeper levels of an ancient "Tibetan Plateau" characterized by crustal thickening and shortening behind a continental collision zone (Dewey and Burke, 1973; McLelland and Isachsen, 1980).

The southern Appalachians extend from central Alabama to about 41°N latitude in southernmost New York state and can be divided across strike into three main structural provinces (Rodgers, 1970). The westernmost section includes the marginal fold and thrust belt of the Valley and Ridge Province, which involves Alleghanian (late Carboniferous–Permian; 270 to 220 Ma) deformation of Paleozoic miogeoclinal rocks. The Blue Ridge Province is located to the east of the Valley and Ridge Province and consists of an allochthonous belt involving Precambrian (1.0 to 1.1 Ga) basement and younger rocks. The late Precambrian and Paleozoic metasedimentary and metavolcanic rocks become more intensely metamorphosed from west to east across the Blue Ridge Province, which separates platform rocks of the Valley and Ridge Province from metavolcanic and metasedimentary rocks and intrusives of the Piedmont Province. The Blue Ridge can be divided across strike into a western and eastern block that are separated by a major (Taconic?) fault system (Hayesville-Fries fault; Wehr and Glover, 1985). The western block consists mainly of Grenville basement nonconformably overlain by late Precambrian–early Paleozoic continental to shallow marine metasediments. The eastern block consists of coeval metamorphosed turbidite sequences intercalated with mafic and ultramafic igneous rocks that are the same as those of the Inner Piedmont (Hatcher, 1978). The southern portion of the Blue Ridge is broken by numerous westward-directed thrust faults. Possible Grenville-age terranes are also found farther east, extending into the Piedmont as far east as the eastern Carolina slate belt (Pine Mountain belt, Sauratown Mountains, Baltimore gneiss domes, and Rayleigh belt; Farrar, 1985).

It has been suggested that Grenville rocks of the Blue Ridge, Pine Mountain belt, and Sauratown Mountains (located within or just east of the Inner Piedmont) form the late Precambrian edge of the rifted continental margin and that the Goochland terrane was rifted and subsequently reattached to North America during the Taconic orogeny (Rankin, 1976). Alternatively, the Goochland terrane may have never rifted from North America, and the Carolina slate belt may have been thrust westward from the east over the Goochland terrane during the Taconic orogeny (Farrar, 1985). It is unclear from geological observations whether Grenville basement underlies the (allochthonous?) Carolina slate belt/Charlotte belt. More likely, the Charlotte belt/Carolina slate belt is a part of Avalon accreted terrane (Williams and Hatcher, 1982).

The allochthonous nature of the Blue Ridge zone is further indicated by the existence of fensters in the Blue Ridge thrust sheet such as the Grandfather Mountain window in North Carolina, where low-grade metamorphosed platform sedimentary rocks and carbonates are found (Hatcher, 1978). Tectonic slivers of relatively unmetamorphosed carbonates have been observed in the Brevard fault zone, which marks the southeast boundary of the Blue Ridge and gives additional evidence that lower Paleozoic continental margin deposits extend beneath the Blue Ridge (Hatcher, 1971, 1978).

The Piedmont Province is located to the east of the Brevard fault zone and can be subdivided into a number of different zones based on differences in metamorphic grade and dominant lithology (Inner Piedmont, Charlotte belt, Carolina slate belt). In general, the Piedmont consists of metamorphosed eugeoclinal rocks of lower and middle Paleozoic age which are associated with Paleozoic plutons of variable composition. The Inner Piedmont is bounded by the Brevard zone to the west and the Kings Mountain belt to the east and is thought to be part of the large suspect Piedmont terrane of Williams and Hatcher (1982). Rocks of the Inner Piedmont consist of late Precambrian to early Paleozoic highly deformed pelitic to quartzofeldspathic sedimentary and mafic volcanic sequences regionally metamorphosed from upper greenschist to upper amphibolite facies.

The Charlotte and Carolina slate belts, located to the east of the Piedmont, are grouped as part of the late Precambrian–early Paleozoic Avalon terrane by Williams and Hatcher (1982). Both provinces contain a thick sequence of volcanic rocks and associated sedimentary rocks metamorphosed to greenschist grade in the slate belt and upper amphibolite grade in the Charlotte belt. The volcanic rocks are mainly mafic and are thought to represent oceanic island-arc assemblages (cf. Farrar, 1985). The boundary between the Charlotte belt and Carolina slate belt is thought to be a metamorphic gradient rather than a structural contact, and ramping of underlying thrust surfaces may have moved high-

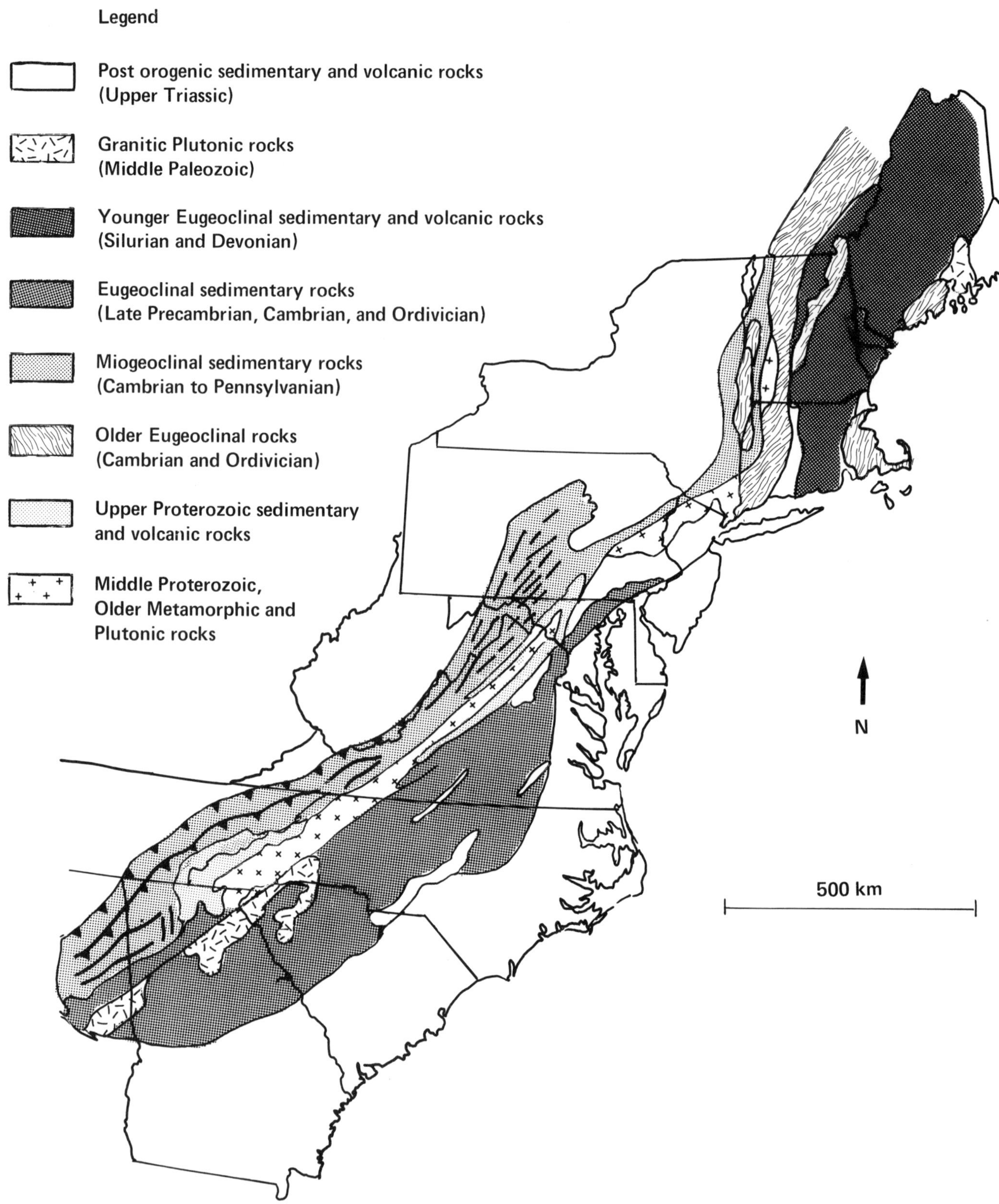

Legend

Post orogenic sedimentary and volcanic rocks (Upper Triassic)

Granitic Plutonic rocks (Middle Paleozoic)

Younger Eugeoclinal sedimentary and volcanic rocks (Silurian and Devonian)

Eugeoclinal sedimentary rocks (Late Precambrian, Cambrian, and Ordivician)

Miogeoclinal sedimentary rocks (Cambrian to Pennsylvanian)

Older Eugeoclinal rocks (Cambrian and Ordivician)

Upper Proterozoic sedimentary and volcanic rocks

Middle Proterozoic, Older Metamorphic and Plutonic rocks

N

500 km

Figure 1. Generalized geologic map of part of the Appalachian orogen; generalized from King, 1969.

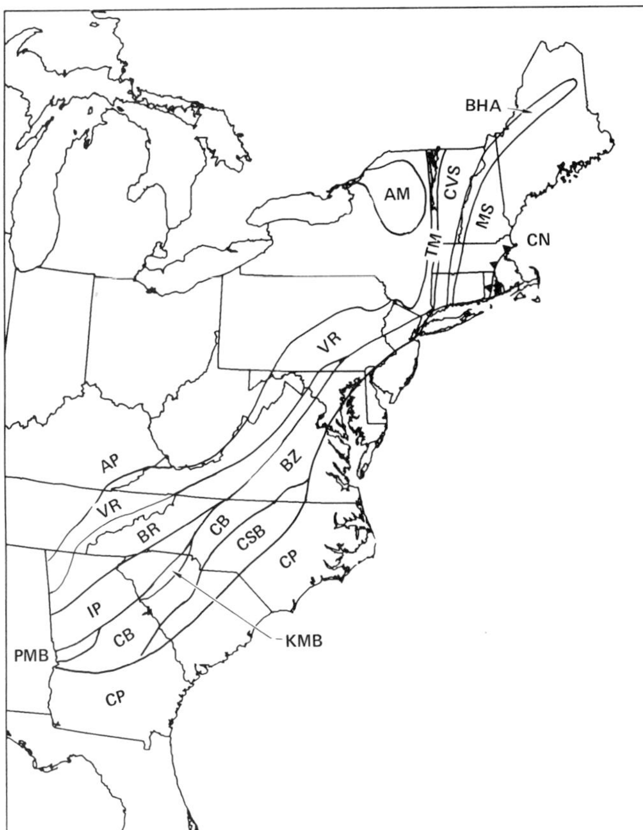

Figure 2. Major geologic provinces in the Appalachian mountain belt. AM = Adirondack Mountains; TM = Taconic Mountains; CVS = Connecticut Valley synclinorium; BHA = Bronson Hill anticlinorium; MS = Merrimac synclinorium; CN = Clinton–Newbury–Bloody Bluff fault zone; AP = Appalachian Plateau; VR = Valley and Ridge; BR = Blue Ridge; BZ = Brevard fault zone; IP = Inner Piedmont; KMB = Kings Mountain belt; PMB = Pine Mountain belt; CB = Charlotte belt; CSB = Carolina slate belt; CP = Coastal Plain.

grade rocks of the Charlotte belt to higher structural levels (Dallmeyer and others, 1986). The southern Appalachians contain two zones of ultramafic bodies: a well-defined belt associated with the Blue Ridge, and an eastern belt of irregularly distributed bodies in the Piedmont (Misra and Keller, 1978).

The northern Appalachians can be divided into three major tectonic units (Bird and Dewey, 1970; Naylor, 1975): a western belt and an eastern belt, possibly representing the margins of two once-convergent continental masses, surrounding a central orogenic belt composed mainly of eugeoclinal lithologies (Fig. 1). The western unit is mainly underlain by rocks of the Precambrian Grenville Province, which are exposed in the Adirondacks and outlying massifs such as the Green Mountains, Berkshires, and the Hudson Highlands. In contrast to the southern Appalachians, no Grenville basement has been found east of the central Connecticut Valley synclinorium (Zartman and Naylor, 1984). Unconformably overlying the Grenville basement is an Eocambrian to Cambrian platform sequence that grades upward into a Lower Ordovician clastic sequence. Found above are the Taconic klippes which consist primarily of deep-water shales, sandstones, and graywackes (cf. Stanley and Ratcliffe, 1985). Paleontologic evidence suggests that they were deposited contemporaneously with the shelf sediments and represent continental slope and rise deposits (Zen, 1972) that may have been incorporated within an accretionary wedge (Stanley and Ratcliffe, 1985). The Taconic allochthons consist of a series of imbricate thrusts with a west-to-east, young-to-old stacking order that formed as the result of deformation under horizontal compression rather than gravity sliding.

The central orogenic belt consists of a number of broad structural warps. The Connecticut Valley synclinorium is found to the east of the previously discussed Precambrian massifs and can be traced from Connecticut through Quebec to the Gulf of St. Lawrence. The Connecticut Valley synclinorium contains a thick, highly metamorphosed eugeoclinal sequence divided into two members separated by a major Middle Ordovician unconformity. The lower two-thirds of the Connecticut Valley synclinorium consists mainly of mica schist and mica quartzite with lenses of greenstone schist and amphibolite and are probably representative of continental rise deposits. The upper third of the sequence contains more pelitic and carbonate rocks and fewer volcanic rocks (except in the easternmost section). A linear serpentinite and ultramafic belt of Ordovician age follows the western flanks of the Connecticut Valley synclinorium (Chidester, 1968). Most of the ultramafic rocks are interpreted to be altered fault slivers of oceanic crust and upper mantle material imbricated with accretionary wedge materials (Stanley and Ratcliffe, 1985).

East of Connecticut Valley synclinorium lies the Bronson Hill anticlinorium, which consists of a chain of elliptical gneissic domes (Oliverian Plutonic series in New Hampshire) associated with a sequence of metavolcanic and metasedimentary rocks. The structure can be traced from Connecticut through northern New Hampshire and is probably continuous with the Boundary Mountains anticlinorium in Maine. Mantling the domes is a series of mafic metavolcanic rocks associated with felsic metavolcanics and metasediments (Ammonoosuc Volcanics in New Hampshire) of Middle Ordovician age or older (440 ± 30 m.y.; Naylor, 1975). Unconformably overlying the Ammonoosuc is a series of Silurian to Lower Devonian highly metamorphosed clastics with some carbonates and volcanic rocks. Leo (1985) concluded that the Ammonoosuc Volcanics were formed as part of a continental-margin arc from partial melting of a basaltic source.

Eastward of the Bronson Hill anticlinorium lies the Merrimack synclinorium, a major northeast-trending tectonic feature extending from eastern Connecticut through Maine and into New Brunswick. It is the site of thick accumulations of Ordovician to Lower Devonian metasedimentary rocks typically metamorphosed to sillimanite grade as far north as south-central Maine. These metasedimentary rocks (Devonian Littleton Formation) can be correlated with Devonian strata at the top of the Bronson Hill anticlinorium and the Connecticut Valley synclinorium. The Merrimac synclinorium also contains large volumes of intrusive

TABLE 1. OROGENIC MOVEMENTS IN THE APPALACHIAN REGION*

Orogenic Episode and Approximate Date	Known Area of Influence	Maximum Manifestation
Appalachian Movements		
Palisades		
Late Triassic (Carnian–Norian) 190–200 m.y.	Belt along central axis of already completed mountain	Fault throughs, broad warping, basaltic lava, dike swarm
Alleghany		
Pennsylvanian and/or Permian (WSestphalian and later) 230–260 m.y.	West side of central and southern Appalachians, southeast side of northern Appalachians, perhaps also in Carolina Piedmont	Strong folding, also middle-grade metamorphism and granite intrusion, at least in southern New England
Early Ouachita		
Mid-Mississippian through early Pennsylvanian (Viséan to early Westphalian)	Only in southernmost Appalachians in central Alabama	Clastic wedge, also possible broad east-west structures that invluenced later deformation
Acadian		
Devonian, mainly Middle, but episodic into Mississippian (Emsian–Eifelian) 360–400 m.y.	Whole of northern Appalachians, except along northwest edge; as far southwest of Pennsylvania	Medium- to high-grade metamorphism, granite intrusion
Salinic		
Late Silurian (Ludlow)	Local on northwest side northern Appalachians	Mily angular unconformity, minor clastic wedge
Taconic		
Middle (and Late) Ordovician (Caradoc, locally probably older) 450–500 m.y.	Generally on northwest side of northern Appalachians, locally elsewhere; an early phase in Carolinas and Virginia, perhaps general in Piedmont province	Strong angular unconformity, gravity slides(?), at least low-grade metamorphism, granodioritic and ultramafic intrusion
Penobscot		
Early Ordovician or older (arenig or older)	Local on northwest side of northern Appalachians	Strong angular unconformity, slaty cleavage, possible some intrusion
Avalonian		
Latest Precambrian	Southeastern Newfoundland, Cape Breton Island, southern New Brunswick; probably also central and southern Appalachians (Florida?)	Probably some deformation, uplift of sources of coarse arkosic debris, gravity slides(?)
Late Precambrian about 580 m.y.	Southeastern Newfoundland, Cape Breton Island, southern New Brunswick; perhaps eastern Massachusetts	Mostly low-grade metamorphism, granitic intrusion
Grenville (pre-Appalachian) Movements		
Late Precambrian 800–1,100 m.y.	Eastern North America including western part of Appalachian region	High-grade metamorphism, granitic and other intrusions

*Modified from Rodgers (1970)

rocks belonging to the Lower and Middle Devonian New Hampshire Plutonic Series and the Mesozoic White Mountain Magma series (Chapman, 1976). Granites of the New Hampshire Plutonic series, which were emplaced during the middle Devonian Acadian orogeny, are exposed throughout much of the central orogenic belt of New England and give age dates of about 360 Ma (Naylor, 1975).

On the eastern flank of the Merrimac synclinorium is a major northeast-trending thrust belt (Clinton-Newbury, Bloody Bluff, and Lake Char faults) extending from southern Connecticut through eastern Massachusetts (Skehan, 1968). Magnetic anomalies associated with the rocks in the thrust belt suggest that the faults continue offshore in an east-northeast direction into the Gulf of Maine (Alvord and others, 1976) and possibly into New Brunswick (Nelson, 1976). In eastern Massachusetts, the northwest-dipping thrusts greatly offset metamorphic isograds, and no stratigraphic units can be traced across them. As discussed by Skehan and Murray (1980), the Clinton-Newbury and Bloody Bluff fault zones may mark the Acadian suture between the North American plate and Avalonia, although some interplate motion may have occurred during the Alleghanian orogeny. Deformation and metamorphism in the Merrimac synclinorium probably delineate a zone of maximum intensity of the Middle Devonian Acadian orogeny (Rodgers, 1970).

The eastern basement is exposed to the east of the above-described thrust belt. These units are probably correlative with rocks of the Avalon Zone in Newfoundland and southeastern New Brunswick (Bird and Dewey, 1970; Nelson, 1976). The region in eastern Massachusetts is characterized by plutonic, metasedimentary, and metavolcanic rocks, metamorphosed mainly to chlorite grade and ranging in age from late Precambrian to Carboniferous. Unfossiliferous strata and scattered age dating have made geological interpretations enigmatic (Naylor, 1975). However, according to Naylor (1975), no rocks have been assigned an age older than 650 Ma, which is significantly younger than the Grenville-age rocks in the western belt.

Although the northern and southern Appalachians exhibit a number of similarities, there are a number of geologic differences as well. The northern equivalent of the Valley and Ridge Province is older and less developed; it involved the westward thrusting of eugeoclinal rocks (the Taconic klippe in New York State) over miogeoclinal rocks during the Ordovician Taconic orogeny. Uplifts exposing Precambrian rocks, such as the Green Mountains in Vermont and the Berkshires in western Massachusetts, involve Grenville basement and are located in the vicinity of the transition of the miogeocline to eugeocline. However, in the northern Appalachians, the uplifts are not as extensive as those in the Blue Ridge, and Precambrian rocks are not exposed along the Sutton Mountain anticlinorium in Quebec. To the east of the Precambrian uplifts in the northern Appalachians is a north-to-northeast–trending belt of serpentinites located on strike with the Newfoundland ophiolites. In contrast to the southern Appalachians, few ultramafic rocks are found to the east of the serpentinite belt in New England except along coastal Maine (Osberg, 1978), and correlatives of Grenville basement are not found east of the central Connecticut Valley synclinorium. Except for a few exposures in coastal Maine, Paleozoic carbonate rocks typical of stable continental margins are rarely found to the east of the Precambrian inliers in the northern Appalachians (Williams, 1979; Osberg, 1978).

PALEOZOIC EVOLUTION OF THE APPALACHIANS

A possible scenario for the evolution of the Appalachians is briefly outlined in this section. Figure 3 shows cross sections illustrating the possible tectonic evolution for both the northern and southern Appalachians. These figures are based mainly on summaries of the tectonic evolution for various segments of the Appalachians found in Rodgers (1970), Osberg (1978), Stanley and Ratcliffe (1985), and Robinson and Hall (1980), among others, for the northern Appalachians, and in Odom and Fullagar (1973), Hatcher (1978), and Secor and others (1986) for the southern Appalachians. The models are extremely simplified to match the large-scale geophysical and geologic constraints and do not account for possible lateral deformation and smaller accreted blocks. Because of uncertainties in the timing of collisional episodes, arrival of accreted terranes, subduction polarities, and complications caused by lateral deformation, Figure 3 should be considered speculative at best and is included simply to give a rough cartoon of the major Paleozoic events in the Appalahcians. Table 1 lists major orogenic episodes in the Appalachians and the maximum manifestation in the area of influence (from Rodgers, 1970).

Currently, it is believed that a continental rifting stage initiated approximately 820 Ma, leading to the formation of the Iapetus Ocean (proto-Atlantic; Rankin, 1976). The late Precambrian and Cambrian geology of the western belt is characterized by the establishment of an Atlantic-type, stable continental margin.

Late Precambrian lithologies in the Avalon belt also indicate a rifting stage with the development of an active continental margin (Kennedy, 1976). Geochemical, paleomagnetic, and paleontologic evidence suggests that the western and eastern (Avalonia) belts were located on opposite sides of the Iapetus Ocean (Strong and others, 1974; McKerrow and Cocks, 1977; Kent and Opdyke, 1978).

Early or Mid-Ordovician through Permian times are characterized by the episodic closing of the Iapetus Ocean. The Bronson Hill anticlinorium was a site of major volcanic activity in this time period, as shown by the presence of thick volcanic sequences. The curvature of the Bronson Hill anticlinorium (convex to the northwest) and the asymmetrical distribution of volcanics (Osberg, 1978) in the central mobile belt suggest that an eastward-dipping Benioff zone existed at this time. The Charlotte belt and Carolina slate belt are thought to have been part of an island-arc sequence of Cambrian age that originated away from North America and was accreted in early or middle Paleozoic time (Dallmeyer and others, 1986).

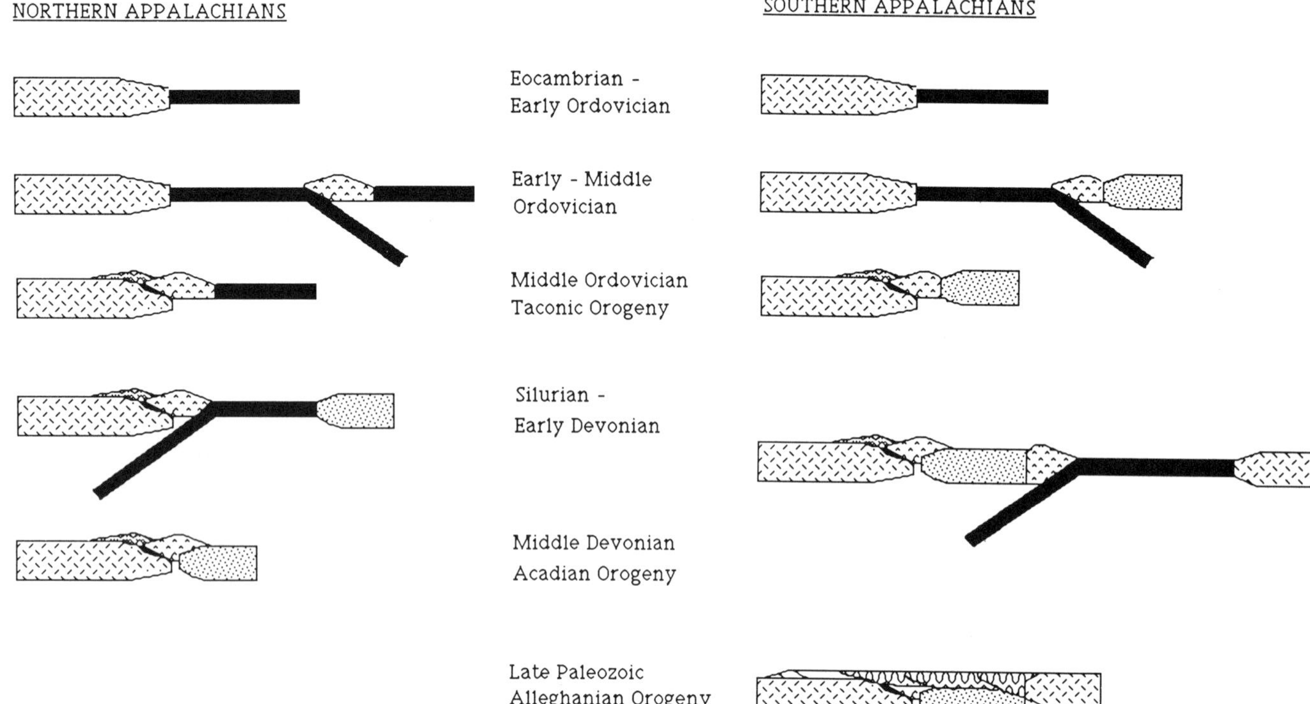

Figure 3. Schematic illustration of plate-tectonic model for the evolution of the northern and southern Appalachians. Profiles taken in vicinity of COCORP lines 1 and 2 (see Fig. 6a).

Major deformation occurring in Middle and Late Ordovician time marked the climax of the Taconic orogeny, which affected rocks within and west of the Bronson Hill anticlinorium. Evidence based on styles of deformation and metamorphism indicates that the Taconic orogeny probably was an episode of arc-continent collision. Numerous island-arc segments and inner-arc basins now located within the central mobile belt were probably involved in the deformation.

Rocks of the Inner Piedmont are thought to be continental slope and rise sediments deposited off the coast of North America, which were deformed and metamorphosed in early and/or middle Paleozoic time (Secor and others, 1986). The Inner Piedmont rocks may have been caught between the Craton and accreted island arcs of the Charlotte and slate belts (Secor and others, 1986), although a middle Paleozoic collision cannot be ruled out.

Silurian to Early Devonian time was a period of relative quiescence and erosion of the highlands (Boucot, 1968). In the Early Devonian period there is evidence of increased tectonic activity, as indicated by the deposition of vast thicknesses of turbidite sequences across the major synclinoria (Littleton Formation of New Hampshire) and renewed volcanism along arcs in the central mobile belt. This episode of major deformation climaxed in mid-Devonian time and is called the Acadian orogeny. It spanned a period of approximately 30 m.y. (Naylor, 1971). The styles of deformation associated with the Acadian orogeny show many parallels with those documented in other continental convergence zones (Dewey, 1977). Thus, the Acadian orogeny was probably a period of continental collision between North America and the Avalon block now exposed in southeastern New England and the eastern Piedmont. However, as discussed by O'Hara and Gromet (1985), the Avalon terrane may have been accreted in two stages during the Acadian and Alleghanian orogenies.

The Acadian orogeny in the southern Appalachians does not appear to have been as widely felt as in the northern Appalachians (Rodgers, 1970). Geochemical analysis suggests that early Alleghanian (315 Ma) granites east of the Carolina slate belt appear to have been derived from partial melting of sialic Precambrian basement (Dallmeyer and others, 1986). Thus, the Avalon terrane in the southern Appalachians appears to be composed of rocks of both oceanic and continental affinity that converged on North America in early and middle Paleozoic time (rather than middle and late Paleozoic time as observed in the northern Appalachians).

Alleghanian deformation in Pennsylvanian to Permian times strongly affected rocks in southeastern New England and the southern Appalachians. The Alleghanian orogeny probably represents the final closure of the Iapetus Ocean by the collision of Africa with North America. This orogenic episode resulted in much of the thin-skinned thrusting evident on COCORP profiles of the southern Appalachians. As discussed by Secor and others (1986), the Alleghanian orogeny was a period of continental collision involving a number of separate deformation episodes. It has been suggested (Dewey and Kidd, 1974) that the Alleghanian deformational belt truncates the Appalachian-Caledonian belt in southern Connecticut and Rhode Island and is correlative with the Hercynian belt in central Europe. Strike-slip faulting in the northern Appalachians may have been important in late Paleozoic time (Rodgers, 1970; Arthaud and Matte, 1977; Ballard and Uchupi, 1975), although the sense of motion is not well defined.

GRAVITY

Gravity and magnetic anomalies generally show very good correlations with geologic features in the eastern United States. Their prominent features are described in this section, and their fits into the tectonic framework of the eastern United States are discussed below. Detailed gravity and aeromagnetic maps are available over much of the eastern U.S. and are shown in companion chapters in this volume (Kane and Godson, this volume; Simpson and others, this volume).

The prominent features observed in the regional gravity field in the northeastern United States are a low over the Taconic klippe in eastern New York, a low over the White Mountains in northern New Hampshire, and a northeast-trending low over the Connecticut Valley synclinorium in Vermont and western Maine. A north-northeast–trending gravity high is found in western Connecticut, western Massachusetts, western Vermont, and extending into Quebec; it is associated with the Precambrian uplifts and the serpentinite belt. There is also a gravity high found in the northern Adirondacks and along the Atlantic coast (Simmons, 1964). As noted by Kane and others (1972), more detailed maps show a discontinuity in regional trends where broad northeast-striking gravity elements of eastern and northern New England are separated from the more north-trending anomalies of western New England along a line extending from Rhode Island to north-central Vermont.

In the Adirondack Mountain region, Simmons (1964) observed that prominent gravity lows occur over the anorthosite bodies, presumably due to their lower density (2.72 g/cm^3) relative to surrounding Precambrian rocks (2.82 g/cm^3). The steep gradients at the edge of the anorthosite bodies suggest that the intrusions are tabular in shape, with thicknesses of approximately 3 to 4.5 km and with local roots 10 to 15 km deep. The gravity high in the northern Adirondacks (the Plattsburg anomaly) is covered by Paleozoic sedimentary rocks and is thought to be due to a buried intrusive mass at a depth of approximately 3 km. Regional gravity suggests small variations (~3 km of relief) in the depth to the Moho and a 2° westward dip.

The regional gravity high found to the east of the Taconic klippe that follows the Green–Sutton–Notre Dame Mountains uplift is thought to be due to a deep-seated change in the configuration and/or density of the lower crust and may be due to the buried eastern edge of Grenville continental crust (Diment and others, 1972). Smaller scale gravity highs in this region are associated with belts of late Precambrian–Early Cambrian metavolcanic rocks (Tibbit Hill volcanics) and mafic and ultramafic rocks associated with the ophiolite and serpentinite belt [which are actually found approximately 10 to 25 km southeast of the axis of the regional gravity high (Kumarapeli and others, 1981)].

The gravity lows over the central part of New England are thought to be due to an increase in the abundance of low-density felsic intrusive rocks of the Acadian-age New Hampshire Plutonic series. Detailed modeling of these plutons suggests that they are thin, tabular bodies less than 2.5 km thick (Nielson and others, 1976).

Bouguer gravity has been extensively studied in the southern Appalachians, especially in conjunction with recent COCORP results. The main feature of the Bouguer gravity in the southern Appalachians is the Piedmont gravity gradient, which is a paired negative-positive anomaly extending much of the length of the mountain belt. To the west, extremely low gravity values (–100 to –50 mgal) occur over the Appalachian Plateau and Blue Ridge, increasing to +40 mgal in the eastern Piedmont. The Piedmont gravity gradient has been interpreted as a transition from thick to thin crust (from the northwest to the southeast; Cook, 1984), a southeast-dipping suture zone extending through the crust and separating thick granitic crust to the west from mafic crust to the east (Hatcher and Zietz, 1980; Hutchinson and others, 1983), a ramp structure representing the Precambrian Grenville passive margin overthrust by the Blue Ridge and Inner Piedmont (Cook, 1984), and a change in crustal density associated with buried low-density sialic Grenville crust and mafic crust of accreted microplates to the southeast of the Charlotte belt and Carolina slate belt (Dainty and Frazier, 1984).

It has been suggested that the north-trending gravity high associated with the Green–Sutton–Notre Dame Mountains correlates with similar but smaller features in the Blue Ridge of the southern Appalachians, rather than the high located to the east of the Piedmont gravity gradient. The central New England gravity low is similar to the Piedmont gravity low, and the Gulf of Maine gravity high correlates with the Carolina slate-belt high (Diment and others, 1972). This implies that a major structural offset between the northern and southern Appalachians occurs in the vicinity of southern New York (line EF in Fig. 12, Diment and others, 1972).

MAGNETICS

Excellent aeromagnetic data are available over much of the Appalachians with line spacing of less than 1.6 km (Zietz and others, 1972; Zietz and Gilbert, 1980). Because a dipolar field falls as r^{-3} as opposed to r^{-2} for the gravity data, the magnetic anomalies show a very impressive correlation with surface geologic features (see Fig. 2 of Kane and Godson, this volume). Aeromagnetic data have been crucial for mapping important structural and stratigraphic contacts where they are often obscured by glacial sediments or water and are therefore very important for local interpretations.

In general, felsic intrusive rocks are characterized by low magnetic expression, while metasedimentary/metavolcanic sequences may produce alternating high and low anomalies associated with interstratified felsic and mafic sequences. Prominent features observed on the aeromagnetic maps are the Clinton-Newbury and Bloody Bluff fault zone in eastern Massachusetts and eastern Connecticut (Alvord and others, 1976; Taylor and others, 1980a), rocks of the White Mountain Plutonic Series, diabase dikes associated with Triassic rifting, and the Oliverian domes and gneiss domes of the Bronson Hill anticlinorium (Griscom and Bromery, 1968). Magnetic highs are also associated with certain metasedimentary/metavolcanic sequences and the serpentinite/ophiolite belt in western New England (Kumarapeli and others, 1981). In general, Cambrian-Ordovician rocks have a stronger magnetic signature than Silurian-Devonian rocks.

In the southern Appalachians, two major features are observed. King and Zietz (1978) have identified a major magnetic lineament (the New York–Alabama lineament) extending 1,600 km from the Mississippi embayment to northern Pennsylvania. The lineament is located to the west of side of the regional gravity low situated just west of the Blue Ridge. Magnetic anomalies to the northwest of the lineament generally trend to the north, and anomalies to the southeast are northeast-trending. King and Zietz (1978) have interpreted the New York–Alabama lineament as an ancient strike-slip fault associated with large-scale lateral motions experienced during the continental collision episode of the Alleghanian orogeny.

A second major magnetic feature in the southern Appalachians is the change from long-wavelength regionally high anomalies of the Blue Ridge and Inner Piedmont to short-wavelength variations of the Charlotte/slate belts (Hatcher and Zietz, 1980). Many of the high-amplitude, short-wavelength anomalies of the Charlotte/slate belts correlate with mafic intrusions and mafic volcanic rocks. The magnetic anomalies show regionally low values in South Carolina and northeast Georgia that increase to the north in North Carolina and Virginia, indicating a northward increase in magnetic material in the upper crust. Hatcher and Zietz (1980) interpreted this change in magnetic character as due to a fundamental change in crustal character from granitic crust beneath the Inner Piedmont to mafic crust beneath the Charlotte belt/Carolina slate belt. They suggested that this boundary is a late Taconic or early Acadian cryptic suture separating an accreted terrane from North America.

As found in the northern Appalachians, there is typically a good correlation of high-amplitude, short-wavelength magnetic anomalies with gravity highs generally associated with near-surface mafic rocks. The location of the change in magnetic character in the Inner Piedmont is generally slightly east of the Piedmont gravity gradient. Horton and others (1984) observed that magnetic anomalies associated with the Piedmont and Valley and Ridge (including the New York–Alabama lineament) are truncated by part of the Brunswick magnetic low in westernmost Georgia and Alabama. Based on this observation and other geological considerations (such as drillhole information), they suggested that the lineament is a transcurrent fault truncating the northeast-trending Appalachian structures and may mark the late Paleozoic suture zone between North America and Africa and/or intervening accreted terranes.

RESISTIVITY

A number of resistivity studies have been carried out in the Appalachians utilizing a variety of techniques, including magnetotellurics (Kasameyer, 1974), controlled-source electromagnetic sounding (Connerney and others, 1980; Thompson and others, 1983), geomagnetic sounding (Bailey and others, 1978), and dipole sounding (Keller, 1966). Resistivity measurements in the northeastern United States suggest the presence of a highly conductive lower crust in the Adirondack Mountains in New York state (Connerney and others, 1980), while a resistive lower crust underlies a slightly conductive, approximately 15-km-thick upper crust in New England (Kasameyer, 1974). Using geomagnetic sounding, Bailey and others (1978) found evidence for high telluric current flow in central New England, which they attributed to a 200° thermal anomaly and low resistivity at the base of the crust. However, the anomaly they observed may actually be an effect caused by highly conductive formations found in the Merrimack synclinorium, such as the Brimfield Schist, which contains abundant graphite and iron sulfides, and channels currents to the north (Madden, 1971).

In the Adirondack Mountains, Connerney and others (1980) and Greenhouse and Bailey (1981) favored the hypothesis that the increased conductivity in the lower crust is due to the existence of hydrated minerals such as serpentine or amphibole. However, Shankland and Ander (1982) argued that free water can exist in the lower crust, resulting in high conductivities.

In the Georgia Piedmont, Thompson and others (1983) inferred the existence of a high-conductivity layer extending to a depth of approximately 15 km. They noted that this high-conductivity zone lies within a zone of layered reflections observed on COCORP profiles (Cook and others, 1979), which are interpreted to be overthrust shelf sediments. Connate water trapped within these sedimentary layers could increase the observed conductivity at these depths.

HEAT FLOW

Heat-flow measurements in the eastern U.S. are few and are subject to many effects that are difficult to correct for, such as climatic variations, lateral heat transfer, thermal conductivity, and radiogenic heat production. Using measurements collected in the northeastern United States, Birch and others (1968) discovered the linear relationship between heat flow and heat production (radioactivity), suggesting an enrichment of radioactive elements in the upper crust. This linear relation is a useful tool for isolating mantle heat flow (reduced heat flow) from the near-surface radiogenic component.

Diment and others (1972) summarized early heat-flow measurements and made corrections for Pleistocene climatic variations (Fig. 1 of Morgan and Gosnold, this volume). The low heat-flow values in the Adirondack Mountains are mainly confined to the Precambrian anorthosites, which are very low in radioactive elements (Birch and others, 1968). More recent measurements, which include measurements of radioactivity, were made by Jaupart and others (1982) in New Hampshire; they found a relatively uniform heat-flow field of approximately 1.1 HFU (1 HFU = 42 mW/m^2) upon which local highs of ~1.5 HFU are superimposed. Heat-flow maps of eastern North America indicate that central New England is characterized by a north-northeast–trending zone of high heat flow. However, Jaupart and others (1982) suggested that this pattern may be fortuitous because measurements were taken from the highly radioactive White Mountain Plutonic Series. Jaupart and others (1982) found that metasedimentary rocks in the same region as the radioactive plutons have heat-flow values closer to the mean heat flow for old continental terranes. They suggested that surface heat-flow measurements in New England reflect variations in near-surface radiogenic heat production and that no deep-seated thermal perturbations from the most recent orogenic events (Alleghanian orogeny, ~275 Ma) or magmatic episodes (intrusion of the White Mountain Plutonic Series <200 Ma) are evident in the heat-flow data. Further complications in the interpretation of the New England heat-flow data are due to effects of horizontal heat conduction caused by lateral variations in heat production (Jaupart, 1983).

Igneous rocks containing relatively high concentrations of radioactive heat-producing elements also occur in the southern Appalachians (Costain and others, 1986). In general, late Paleozoic (254 to 330 Ma) synmetamorphic to postmetamorphic granitic plutons show high heat-flow values (50 to 80 mW/m^2) relative to premetamorphic plutons, which have values similar to the country rock. This may be due to the dispersal of radioactive materials during metamorphic episodes.

Heat-flow versus heat-production relations indicated that the Piedmont and Coastal Plain may be two different heat-flow provinces. In the Piedmont province, Costain and others (1986) suggested that the heat-producing granites are truncated at a depth of approximately 8 km by the southern Appalachian décollement observed on nearby reflection profiles. Farther east beneath the Coastal Plain, a higher reduced heat flow was observed, suggesting either a higher heat flux from the lower crust and upper mantle or that the heat-producing granites are not truncated at depth by the décollement, which is rooted to the west. The existence of these intrusive bodies beneath the insulating sedimentary cover of the Atlantic Coastal Plain may provide an important low-temperature geothermal resource (Costain and others, 1980).

SUMMARY OF REFRACTION MODELS IN EASTERN NORTH AMERICA

Velocity structures based on regional traveltimes suggest that significant differences in crustal structure exist between the Grenville and Appalachian Provinces and along the strike of the Appalachians. The locations of the refraction studies described below are shown in Figure 4a. Notable differences in lower-crustal velocities between the northern and southern Appalachians provide important clues regarding the tectonic history of the Appalachian orogenic belt. In this section, refraction results for the northern Appalachians, the Adirondacks, and the southern Appalachians are reviewed.

Early refraction models in the northern Appalachians generally consisted of scattered lines using timed explosions (Katz, 1955; Steinhart and others, 1962; Nakamura and Howell, 1964; Taylor, 1980) or earthquake traveltimes (Leet, 1941; Taylor and others, 1980) recorded on both portable and permanent stations. Refraction models for the northern and southern Appalachians and the Grenville Province are shown in Figures 4b and c.

The earliest refraction studies in the northern Appalachians were summarized by Leet (1941), who studied traveltimes from timed quarry blasts and earthquakes recorded at early stations in Weston and Harvard, Massachusetts. Using both primary and secondary arrivals for both P- and S-waves, V_p and V_s velocity models were calculated from distances ranging from 19 to 950 km. The average crustal thickness was 36 km, and V_p increased from 6.1 km/sec at the surface to 7.2 km/sec at the base of the crust. A fairly high P_n velocity of 8.43 km/sec was estimated, and Poisson's ratio was about 0.25 throughout the crust. It appears that some of the S-wave arrival-time picks were taken from different arrivals within the L_g wavetrain, which is a complicated higher mode crustal shear phase, and assigning the apparent velocities to various depth intervals is tenuous.

A refraction model in the central part of the New England Appalachians was derived from timed quarry blasts using a simple layered-earth model (Taylor and Toksoz, 1979; Taylor, 1980). The model consisted of a 39-km-thick crust with P_n velocity of 8.1 km/sec and a high-velocity lower crustal layer with V_p = 7.3 km/sec. The model was expanded by inversion of regional P- and S-wave traveltimes for 170 regional earthquakes recorded by stations of the Northeastern United States Seismic Network (Taylor and others, 1980). Again, the model was limited by the crude parameterization, but as summarized by Taylor and Toksoz (1982), the northern Appalachians appear to be composed of a relatively thick (40 km) crust. Upper crustal velocities are typically 5.8 to 6.1 km/sec, and the lower crust is characterized by relatively high velocities of 7 km/sec. As discussed below, the velocity model derived for the northern Appalachians showed some interesting contrasts with those computed in the same fashion for the Grenville Province.

Two models presented by Chiburis and Ahner (1979) in southeastern New England and Nakamura and Howell (1964) in eastern Maine suggest that the crust thins, and that a high-velocity lower crustal layer is missing along the Atlantic coast. As discussed below, the region where the Chiburis and Ahner (1979) model was compiled is located in a region characterized by a zone of early P_n arrival times and apparently crustal thinning (Fig. 5). Interestingly, these surveys may have sampled rocks of the eastern block (or Avalon zone) discussed above. Nakamura and Howell (1964) also found evidence for a fairly abrupt crust-mantle transition zone in easternmost Maine.

In 1984, detailed refraction lines were collected in Maine

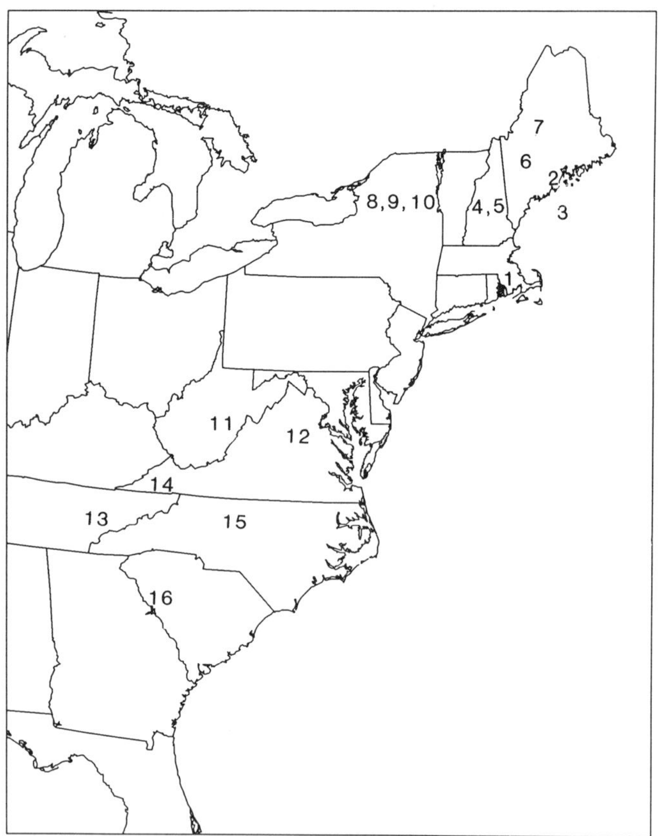

Figure 4a. Location of refraction studies in the Appalachians. Numbers refer to references: 1 = Chiburus and Ahner (1979); 2 = Klemperer and Leutgert (1986); 3 = Nakamura and Howell (1964); 4 = Leet (1941); 5 = Taylor and others (1980b); 6 = Steinhart and Meyer (1961); 7 = Leutgert (1985); 8 = Katz (1955); 9 = Aggarwal, *in* Schnerk and others (1976); 10 = Taylor and others (1980b); 11, 12 = Bollinger and others (1980); 13 = Borcherdt and Roller (1966) and Prodehl and others (1984); 14, 15 = Warren (1968) and Steinhart and Meyer (1961); 16 = Kean and Long (1980) and Lee and Dainty (1982).

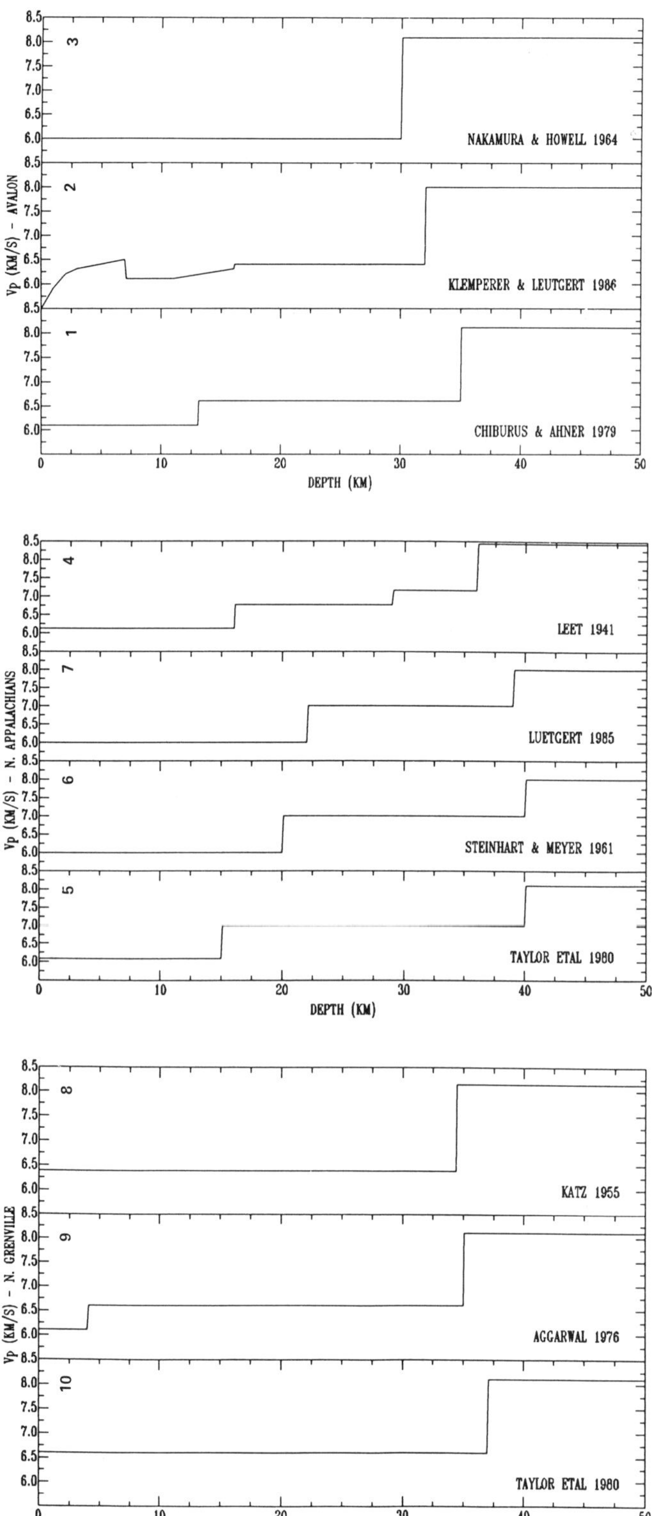

Figure 4b. P-velocity models for refraction profiles in the northern Appalachians divided into three groups: Grenville (left); central New England (middle); and easternmost New England (Avalon; right).

and southeastern Quebec by the U.S. Geological Survey (Luetgert, 1985; Klemperer and Luetgert, 1986). Because upper crustal velocities are relatively high in this area, with small velocity gradients, refracted rays do not penetrate deeply into the crust. To obtain better resolution of the structure in the mid- to lower crust, Klemperer and Luetgert (1986) interpreted the data using both ray tracing and near-vertical reflection techniques. The upper crust shows localized low-velocity zones with velocities ranging from 5.7 to 6.3 km/sec. Relatively high velocities of 6.8 to 7.2 km/sec are observed in the lower crust (Luetgert, 1985; Klemperer and Luetgert, 1986). Luetgert and others (1987) observed a prominent PmP phase reflected from a sharp Moho, which allowed them to construct detailed crustal thickness profiles in the region. The PmP phase was observed to arrive progressively later from the southeast (10-sec two-way traveltime, TWTT) to the northwest (12-sec TWTT). Analysis of the reflected arrivals using synthetic seismograms indicated that the crust increased in thickness from 32 km near the coast to 42 km in northwest Maine.

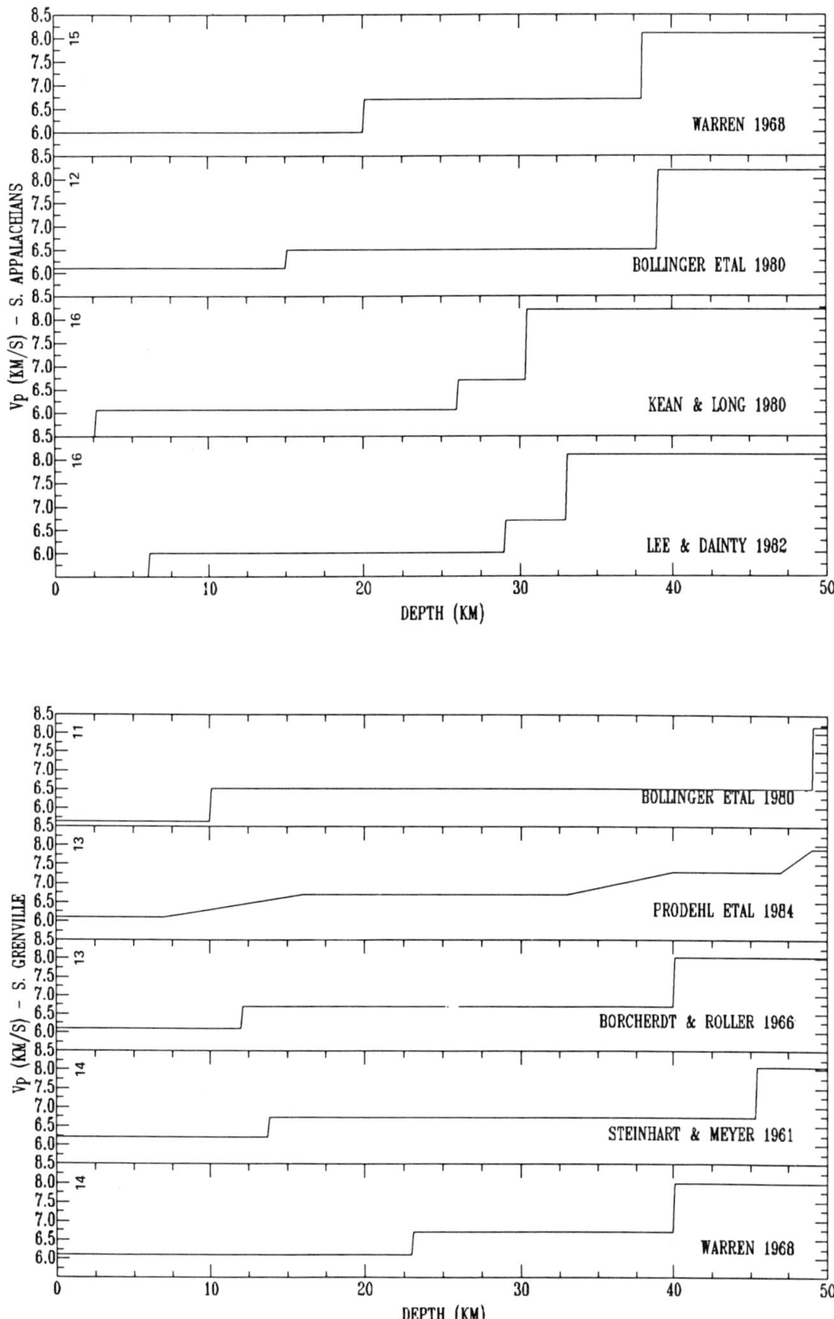

Figure 4c. P-velocity models for refraction profiles in the southern Appalachians divided into two groups: Grenville (left); and Piedmont (right). Numbers for each velocity model correspond to those in map: Figure 4a.

In eastern Maine, rocks of possible Avalon affinity found east of the Norumbega fault zone show high velocities of >6.4 km/sec at depths of 4 to 7 km, with slightly lower velocities of ~6.1 to 6.4 km/sec between 7 and 14 km depth (Klemperer and Luetgert, 1986). Normal-movement velocity analysis successfully enhanced possible Moho reflections at 10 to 11-sec two-wave traveltime in this eastern belt. There was also evidence of as much as 10 percent velocity anisotropy in the uppermost crust with the high velocities parallel to the structural trends in the region.

Refraction models from the Grenville Province appear to be fairly similar along its length from eastern Canada to the southeastern United States. In eastern Canada the high-velocity lower crustal layer is absent or weakly developed (Dainty and others, 1966; Berry and Fuchs, 1973). It also appears that the Grenville crust thickens from about 36 km at its eastern edge to about 45 km at its western edge in the vicinity of the Grenville front. Detailed seismic studies in the La Malbaie Region, Quebec, indicate a 43-km-thick crust with an average upper-crustal velocity of

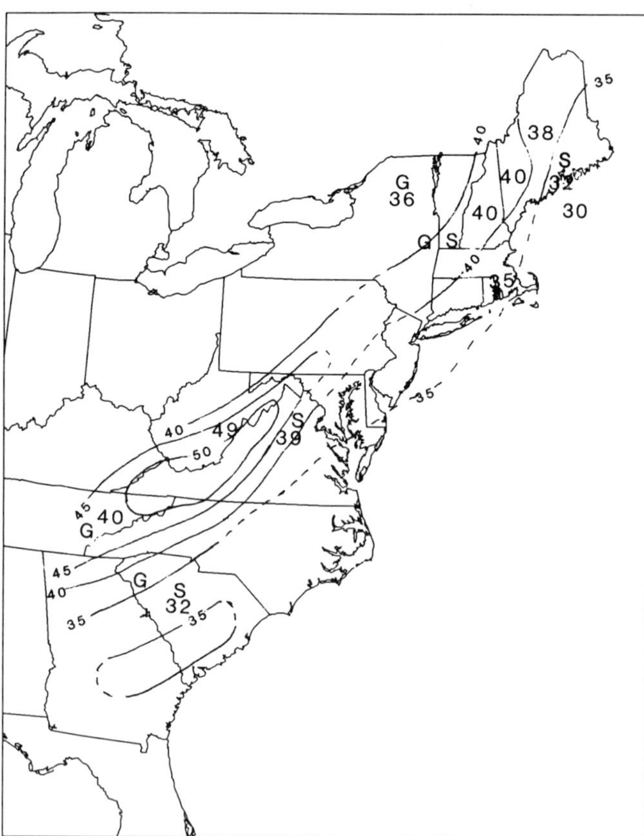

Figure 5. Estimates of crustal thickness (in kilometers) in the Appalachians based on constraints from P_n time terms (Taylor and Toksoz, 1982a; James and others, 1968) and refraction models shown in Figure 4. Large numbers correspond to approximate crustal thicknesses from refraction studies; G and S correspond to regions where a gradational or a sharp crust-mantle boundary has been reported, respectively.

about 6.5 km/sec and lower-crustal velocities of about 6.8 km/sec, with a thin 7.1-km/sec layer at the base of the crust and P_n velocities of 8.1 km/sec (Lyons and others, 1980).

In New York state, the Grenville crust appears to be uniform, about 36 km thick, with velocities ranging from 6.4 to 6.6 km/sec (Katz, 1955; Aggarwal *in* Schnerk and others, 1976; Taylor and others, 1980). The velocity models of Katz (1955) were determined from refraction lines extending from quarries in the central Adirondacks. The traveltime profiles were very straight, indicating a uniform, unlayered crust, having average velocities of about 6.4 km/sec. The velocity of the Marcy anorthosite body was 6.6 km/sec, and the average crustal thickness was determined to be 35 km, with upper-mantle velocities of 8.1 km/sec. Taylor and others (1980) also noted the apparent homogeneity of the Adirondack crust, especially in comparison with regional traveltimes in the New England Appalachians. They estimated the crust to be approximately 37 km thick, with an average velocity of 6.6 km/sec, overlying an upper mantle with $V_p = 8.1$ km/sec.

Farther south, Katz (1955) found a fairly uniform crust for a refraction line extending across east-central Pennsylvania with average P velocity of 6.0 km/sec and a thickness of 33 km. It should be noted that the lines constructed by Katz (1955) were coarsely sampled and had few P_n arrivals past the expected crossover distance. West of the Blue Ridge, refraction lines with rather coarse station spacing were modeled using two layers. The Grenville crust appears to have P velocities of 6.2 km/sec in the upper crust; the lower crust has a P velocity of about 6.7 km/sec, which is low relative to the northern Appalachians (Steinhart and Meyer, 1961).

In the southern Appalachians west of the Blue Ridge, a number of detailed refraction studies are available that sample mainly Grenville basement. Early studies by Steinhart and Meyer (1961), Borcherdt and Roller (1966), and Warren (1968) indicated a two-layer crust with upper-crustal velocities of 6.1 to 6.2 km/sec (to depths of 12 to 23 km) and lower-crustal velocities of 6.7 km/sec. Crustal thickness in this region were reported to be 40 to 45 km, with P_n velocities of 8.0 to 8.1 km/sec.

The refraction study of Borcherdt and Roller (1966) involved the collection of two crossed, reversed refraction profiles, each about 400 km in length, in the vicinity of the Cumberland Plateau Observatory (CPO), Tennessee. The velocity models were derived using homogeneous dipping layers to match traveltimes from first arrivals. The absence of clear secondary arrivals was noted (especially in contrast to the western United States), suggesting gradational velocity discontinuities. To the northeast, approximately 20 km from CPO, a well-defined increase of velocities from depths of about 10 to 20 km was inferred. This correlates reasonably well with a mid-crustal high-velocity anomaly observed from inversion of crustal receiver functions (Owens and others, 1984).

Prodehl and others (1984) reinterpreted refraction lines of Borcherdt and Roller (1966) beneath the Cumberland Plateau, using stacking techniques to enhance the record sections and ray-tracing to match both primary and secondary arrivals. Upper-crustal velocities of 6.1 to 6.2 km/sec were found overlying a 6.7- to 6.8-km/sec lower crust. A gradational boundary was postulated to separate the two layers between 7 and 10 km depth. From the lack of a clear PmP phase, a thick, gradational Moho was inferred to exist between 34 and 47 km depth where velocities increased from 6.7 to 8.0 km/sec. To the northeast of CPO, velocities in the upper 20 km were slightly greater than those beneath CPO, but a high-velocity anomaly of the sort inferred by Owens and others (1984) was not observed. Interestingly, Prodehl and others (1984) noted that the crustal structure perpendicular to the strike of the Appalachian grain was more uniform than that parallel to it. No evidence of a crustal "root" was observed beneath the Blue Ridge, as had been inferred from previous studies and gravity information. In fact, the crust appears to thicken slightly to the west beneath the stable continental platform. As discussed below, the velocity models of Prodehl and others (1984) generally match extremely well with those derived from inversion of crustal receiver functions (Owens and others, 1984).

More recent results involving traveltimes from earthquakes and quarry blasts elucidate the variations in crustal structure across the southern Appalachians. Analysis of travel times from regional earthquakes suggests that a fairly thick crust (~40 to 45 km) occurs beneath the Cumberland Plateau and thickens to 45 to 50 km beneath the Valley and Ridge and Blue Ridge (Kean and Long, 1980; Bollinger and others, 1980; Carts and Bollinger, 1981). To the east of the Blue Ridge, in the Piedmont, slate belts, and coastal plain, the crust thins dramatically.

Bolinger and others (1980) measured P- and S-wave traveltimes from regional earthquakes and timed quarry blasts in the Valley and Ridge, Blue Ridge, and Piedmont Provinces of Virginia. Crustal thicknesses of about 49 km were found beneath the Appalachian Plateau and Valley and Ridge with a 10-km surface layer, with V_p = 5.6 km/sec overlying a 39-km-thick lower crust with V_p = 6.5 km/sec. P_n and S_n velocities were found to be 8.2 and 4.8 km/sec, respectively, and 10 percent velocity anisotropy was observed in the upper crust of the Valley and Ridge with the fast direction parallel to the structural trend. The crust appeared to thin slightly to 39 km beneath the Blue Ridge and 31 km beneath the central Piedmont. A 15-km-thick upper crustal layer with V_p and V_s of 6.1 and 3.5 km/sec was observed to overlie a lower crustal layer with V_p and V_s of 6.5 and 3.8 km/sec, respectively. On the basis of traveltimes and apparent velocities, secondary arrivals thought to be Moho reflections from beneath the Blue Ridge were observed.

Carts and Bollinger (1981) studied regional traveltimes in the southeastern United States from a well-recorded earthquake and found that the travel times could be fit using a single-layer crust with V_p = 6.3 km/sec and V_s = 3.7 km/sec overlying an upper mantle with P_n and S_n velocities of 8.1 and 4.6 km/sec, respectively. Crustal thicknesses were estimated to be 40 km beneath the Valley and Ridge and central U.S., 33 km beneath the Piedmont, and 31 km beneath the Coastal Plain.

Refraction models of Kean and Long (1980) and synthetic seismograms of Lee and Dainty (1982) indicate that the crust beneath the Piedmont and Charlotte and Carolina Slate belts is only 33 km thick. Long (1979) and Kean and Long (1980) inferred that the crust again thickens to the southeast in the Coastal Plain to 35 to 40 km and thins again to 29 km toward the coast. In contrast to the northern Appalachians, no evidence of high compressional velocities ($\geqslant$7.0 km/sec) is seen in the lower crust. Within and west of the Blue Ridge, lower crustal velocities are approximately 6.7 km/sec, similar to those typical of Grenville crust found to the north. P_n velocities of 8.2 km/sec were observed in the region.

East of the Blue Ridge, beneath the Piedmont and slate belts, the velocities in the lower crust appear to be very low (6.0 to 6.5 km/sec), and only a thin 6.7-km/sec layer is inferred to exist at the base of the thin crust (cf. Kean and Long, 1980). Kean and Long (1980) also saw evidence of a fairly high-amplitude Moho reflection beneath the southern Piedmont, suggesting an abrupt crust-mantle boundary.

In summary, refraction models collected along the Appalachians show a number of variations both parallel and perpendicular to the strike of the mountain belt. In both the northern and southern Appalachians, the crust appears to be thin near the coastline (30 to 35 km), with fairly low velocities occurring in the lower crust (6.0 to 6.6 km/sec). In the central highland part of the belt, the crust thickens to 40 to 50 km. Relatively high velocities ($\geqslant$7.0 km/sec) are observed in the lower crust beneath the northern Appalachians, which are conspicuously absent in corresponding regions in the southern Appalachians. This observation has important implications regarding differences in the tectonic evolution between the northern and southern Appalachians. To the west of the Appalachians, in the Grenville Province, the crust appears to thin to 35 to 40 km, and velocities of 6.6 to 6.7 km/sec are observed in the lower crust.

TELESEISMIC P-WAVE RESIDUALS AND P_n TIME TERMS

Analysis of teleseismic P-wave residuals and P_n time terms are useful tools for determining lateral variations in crust and upper-mantle structure. Both techniques basically calculate traveltime delays (residuals) with respect to a wavefront propagating across a seismic network. In a region characterized by low seismic velocities, traveltime delays are large relative to regions of high velocity. Teleseismic P-wave residuals are affected by both crust and upper-mantle velocity structure; the basic technique is described in Taylor and Toksoz (1979). P_n time terms are sensitive mainly to variations in crustal thickness and/or velocity and variations in P_n velocity (cf. Willmore and Bancroft, 1960). Although the two techniques are useful for studying lateral variations in velocity structure, they provide only limited information about vertical variations unless a good cross-fire of rays can be obtained from a number of azimuths. In this case, it is possible to perform an inversion of the traveltime residuals for a three-dimensional image (cf. Aki and others, 1977). By combining P_n time terms and teleseismic P-wave residuals, it is possible to isolate crustal effects from those of the upper mantle.

James and others (1968) studied P_n time terms from more than 100 shots at 150 portable stations located in the Middle Atlantic states. They calculated average P_n velocities of 8.15 km/sec in the northern part of their study area and 8.25 km/sec in the southern part. The crust beneath the Piedmont and Coastal Plain is relatively uniform, with average thicknesses of 35 km. The crust thickens to 50 to 55 km beneath the Valley and Ridge in the northern part of the area and beneath the Blue Ridge in the southern part. Bouguer gravity values based on estimated crustal thicknesses, an assumed linear velocity-depth function, and different velocity-density relations were calculated that showed a reasonably good agreement with observed gravity for values of $d\rho/dV_p$ = 0.20 to 0.23. By comparison with velocity-density relations for different rock types, it was concluded that a mafic lower crust composed of amphibolite overlies an upper mantle composed of peridotite. However, the presence of intermediate-mafic granulites in the lower crust cannot be ruled out on the basis of these data.

Absolute teleseismic P-wave residuals were measured from nine nuclear explosions (three from the Aleutians, six from Novaya Zemlya) across a number of stations in the eastern United States (Fletcher and others, 1978). Stations in the Valley and Ridge and Blue Ridge generally had large positive residuals that correlated well with the large positive P_n time terms of James and others (1968). Interestingly, P-wave residuals at the Cumberland Plateau station CPO (where RSCP is now located) are –0.1 sec to the northwest and –1.0 sec to the northeast, which agrees well with the high velocities observed in the midcrust to the northeast of RSCP by Owens and others (1984). Positive residuals in the Coastal Plain were attributed to thick sequences of low-velocity sediments.

Absolute P-wave residuals in the northeastern United States were more negative than those of the southeastern United States, indicating a slightly thinner crust or higher crustal velocities in the northeastern United States. Large, negative residuals were observed in Quebec and central Vermont, which were interpreted to be due to the existence of ultramafic rocks in the crust.

Teleseismic P-wave residuals were studied in the Piedmont Province of Georgia and South Carolina (Volz, 1979). From a three-dimensional inversion following the technique of Aki and others (1977), a low-velocity zone dipping approximately 30° to the southeast was observed and interpreted to represent an ancient subduction zone. The intersection of the upper part of the low-velocity anomaly correlates reasonably well with the Piedmont gravity gradient and the zone of east-dipping reflectors observed on COCORP reflection profiles discussed below (Long and Liow, 1986).

In the northern Appalachians, Taylor and Toksoz (1979) measured teleseismic P-wave residuals from 68 events recorded at 50 stations. To isolate crustal and upper-mantle effects, P_n time terms from 170 regional earthquakes were calculated across the same network (Taylor, 1980; Taylor and Toksoz, 1982). Although the P_n time terms were not as well constrained as the average teleseismic P-wave residuals for each station, the two sets of measurements showed the same basic patterns.

Large positive residuals (late arrivals) occur throughout central and northern New Hampshire, southern Maine, and eastern Vermont. As observed by Fletcher and others (1978), these contrast sharply with large negative residuals in western Vermont, northern New York, southeastern New York, and southwestern Connecticut. Shallow, localized structural differences should result in rapidly varying trends between adjacent stations. However, the slowly varying distribution of residuals suggests that the observed variations are probably due to deep, regional structures such as differences in crustal thickness and upper-mantle velocity.

The average relative P-wave residuals show a good correlation with Bouguer gravity values in the northeastern United States (Taylor and Toksoz, 1979). However, the stations located to the west of the Precambrian outliers showed more negative gravity values for a given residual than those in New England above Paleozoic basement, which suggests that the two regions show differences in both crustal velocities and crustal thickness.

Estimates of crustal thickness were obtained based on average relative teleseismic residuals and P_n time terms, which showed the same basic patterns (Fig. 5). The variations in thickness for the teleseismic residuals were less than those estimated from P_n time terms because variations in average crustal velocities were accounted for in the former case. The contours in Figure 5 are very generalized and are mainly constrained by P_n time terms and values from refraction studies.

Overall, the crust beneath the northern Appalachian Province appears to be 3 km thicker than that of the Grenville Province exposed in the Adirondacks. Average crustal velocities estimated from the P-wave residuals differ very little but may be slightly lower in many parts of the Appalachians. However, as discussed above, refraction results indicate that the vertical distribution of velocities differs markedly between the two provinces, with the Adirondacks showing fairly uniform velocities and the northern Appalachians showing velocities that are low in the upper crust and high in the lower crust. There is a region of greater crustal thickness beneath eastern Vermont, New Hampshire, and southern Maine. In western Vermont and western Connecticut the crust thins rapidly to the west. These features are located just to the east of the Taconic thrust belt and in western Vermont, and the gradients show an impressive correlation with the serpentinite belt. Slight crustal thinning is observed in eastern Massachusetts; it is not clear if this represents a trend that may continue eastward toward the outer edge of the continental margin. Alternatively, it may represent the contact between the central orogenic belt and the eastern belt described earlier. The average crustal thickness of 36 km in northern New York agrees well with refraction models of Katz (1955) and those of Aggarwal in Schnerk and others (1976).

The traveltime residuals measured by Taylor and Toksoz (1982) were inverted for three-dimensional crust and upper-mantle structure following the technique of Aki and others (1977). It was found that structures down to at least 200 km can be correlated with large-scale geologic and tectonic features. The Grenville upper mantle is characterized by velocities that are approximately 2 percent higher than those beneath the northern Appalachians; these velocities are highest beneath the Adirondack dome. A relatively low-velocity anomaly dips to the northwest beneath the central mobile belt of the Appalachians and shows a spatial correlation with the Bronson Hill–Boundary Mountains anticlinorium in New Hampshire and Maine.

REFLECTION PROFILING

An exciting new phase of seismic exploration in the Appalachians has been undertaken with the acquisition of near-vertical reflection profiles. The profiles include those collected by the Consortium for Continental Reflection Profiling (COCORP) from the Adirondacks, New England, and the southern Appa-

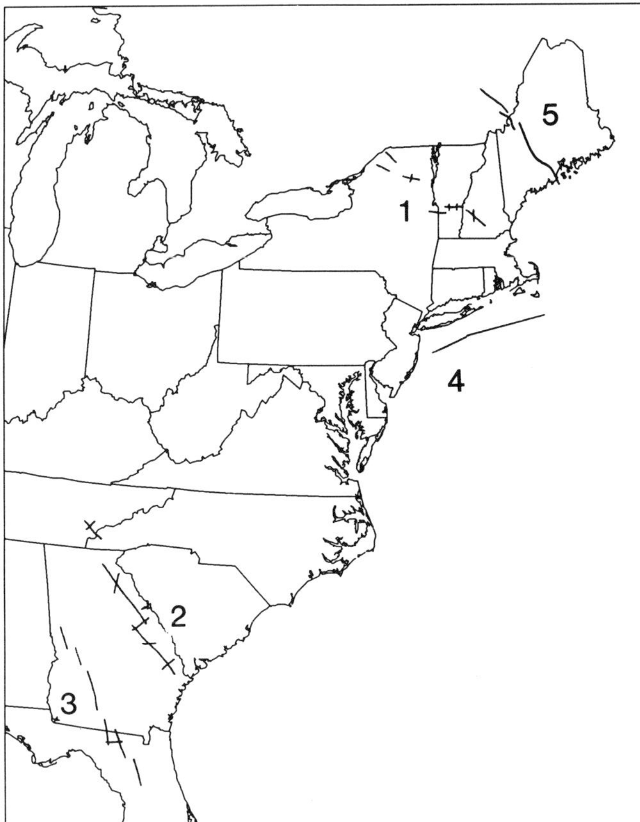

Figure 6. Location of reflection profiles in the Appalachians that are shown in Figure 7. 1 = Northern Appalachian traverse (cf. Ando and others, 1983); 2 = southern Appalachian northeast Georgia traverse (cf., Cook and others, 1979); 3 = southern Appalachian southwest Georgia traverse (cf. Nelson and others, 1985a); 4 = Long Island Platform (cf. Phinney, 1986; Hutchinson and others, 1986); 5 = Quebec-Maine traverse (cf. Stewart and others, 1986).

lachians; the Ministere de l'Energie et des Ressources (MERQ) from Quebec; the U.S. Geological Survey in Maine and the Long Island Platform; and Virginia Polytechnic Institute in the central Appalachians (Fig. 6). This section briefly reviews the results of these informative studies; the details regarding experimental design and interpretation are covered in a separate chapter of this memoir.

The earliest study of Clark and others (1978) was carried out cross the Brevard fault zone in North Carolina. The Brevard fault zone was interpreted to be a shallow southeast-dipping zone of ductile deformation. More importantly, a prominent set of reflectors from a depth of approximately 6 km was observed and interpreted to be an overthrust Cambrian sedimentary shelf sequence correlative with miogeoclinal rocks outcropping in the Valley and Ridge Province located more than 75 km to the northwest.

The first extensive reflection observations were collected by COCORP along a 348-km profile across the Valley and Ridge, Blue Ridge, Inner Piedmont, and western Charlotte belt in Tennessee, North Carolina, and Georgia (Cook and others, 1979). The major findings included a prominent subhorizontal band of reflections with a gentle eastward dip at a depth of approximately 5 to 7 km beneath the Valley and Ridge and 9 to 12 km beneath the westernmost Charlotte belt of the eastern Piedmont (Fig. 7a). Because the reflecting horizon beneath the Valley and Ridge could be correlated with Cambrian-Ordovician shelf sequences outcropping to the west, it was suggested that the autochthonous sedimentary rocks were overthrust from the east by igneous and metamorphic crystalline rocks mainly during the late Paleozoic Alleghanian orogeny. It was also confirmed that the Brevard fault is an east-dipping splay of the main décollement. Few reflections are observed below a depth of 8 km beneath and west of the Inner Piedmont.

To the east of the Inner Piedmont, the character of the crustal reflections changes significantly beneath the Kings Mountain belt and westernmost Charlotte belt of the eastern Piedmont. A complex pattern of reflections is observed, and a broad zone of east-dipping reflections exists; as discussed below, these dipping reflections could represent the root zone for the décollement (cf. Iverson and Smithson, 1982) or an overthrust continental margin (Cook and others, 1979). Farther east, a subhorizontal reflective layer is again observed at a depth of about 15 km. Cook and others (1979) preferred the hypothesis that this reflecting horizon is a continuation of the overthrust sedimentary sequence that extends farther east beneath the slate belt, suggesting that most of the southern Appalachians is allochthonous. A discontinuous but horizontal Moho reflection is observed east of the Inner Piedmont at about a 10- to 11-sec two-way traveltime (~30 to 33 km depth), indicating an abrupt crust-mantle velocity transition. The lack of Moho reflections in the northwestern part of the line is consistent with the gradational crust-mantle boundary inferred from crustal receiver functions (Owens and others, 1984) and refraction profiles (Prodehl and others, 1984).

The southern Appalachians COCORP traverse was extended across Georgia into the Coastal Plain and confirmed earlier observations discussed above (Cook and others, 1981). The added COCORP line was offset approximately 30 km to the southwest of the original profile. The southeast-dipping reflections beneath the Charlotte belt were again observed between depths of about 10 and 17 km. These were interpreted to be either depositional or tectonic (imbricate thrusts) features in late Precambrian and/or early Paleozoic shelf-edge strata (Cook and others, 1981) or from root-zone thrusts dipping steeply into the crust (Hatcher and Zietz, 1980).

The subhorizontal reflectors extending southeast into the Coastal Plain were again observed, and Cook and others (1981) noted low stacking velocities associated with these reflections (especially along the eastward projection of the Augusta fault). Based on these low velocities and the apparent continuity of the midcrustal reflectors with the interpreted overthrust sedimentary sequences to the northwest, Cook and others (1981) preferred the hypothesis of a continuous décollement extending into the Coastal Plain.

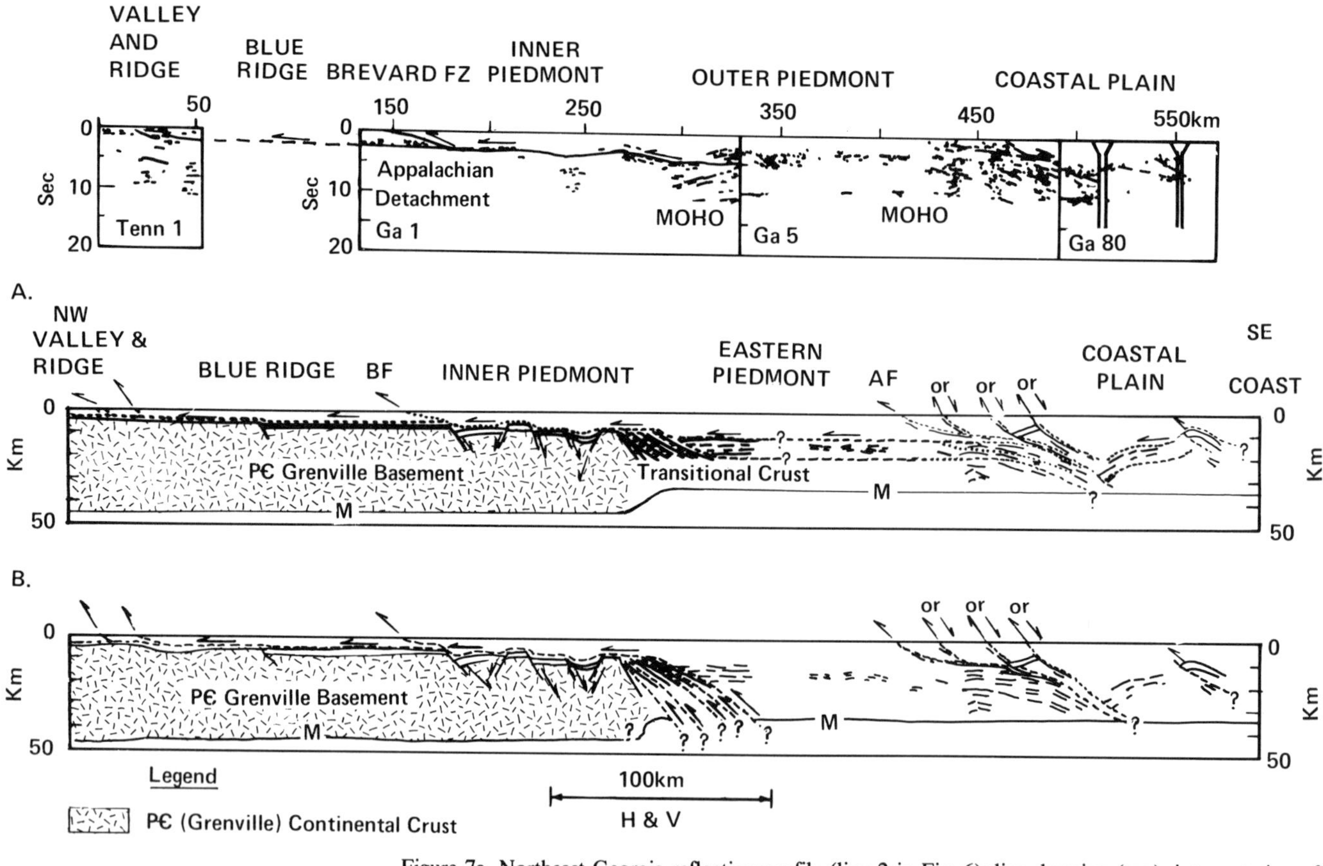

Figure 7a. Northeast Georgia reflection profile (line 2 in Fig. 6): line drawing (top), interpretation of subhorizontal detachment extending to the Coastal Plain (middle; Cook and others, 1981), interpretation of near-vertical suture and decollement root zone at the eastern edge of the Inner Piedmont (cf. Hatcher and Zietz, 1980; Iverson and Smithson, 1982), modified from Cook and others (1981).

Based on a comparison of reflection profiles available at the time, Harris and Bayer (1979) suggested that southern Appalachian décollement extended as far east as the present-day continental shelf, and extrapolated the results to the New England Appalachians. The extrapolation to the northern Appalachians was questioned on geological grounds by Williams (1980) and on geophysical grounds by Taylor and Toksoz (1982). Williams (1980) argued that the geologic and structural evidence for an allochthonous orogenic belt in the northern Appalachians is poor; notably the lack of an Alleghanian fold-and-thrust belt in the north and the presence of essentially autochthonous Silurian and younger cover rocks across the orogen in the north but not in the south. Based on a compilation of refraction data along the Appalachians, Taylor and Toksas (1982) noted a difference in the velocities of the lower crust beneath the crystalline basement in the northern and southern Appalachians. The New England COCORP traverse illustrated that the magnitude of thin-skinned thrusting was indeed greater in the southern than in the northern Appalachians (Ando and others, 1983).

The extension of the southern Appalachian décollement into the Coastal Plain came into question in a series of papers by Iverson and Smithson (1982, 1983a, b) initiating a "root-zone" controversy that will probably not be resolved without drilling some well-placed, deep exploration boreholes. Iverson and Smithson (1982, 1983a, b) proposed that the east-dipping reflections beneath the Charlotte belt represent the root-zone from which the southern Appalachian décollement originates. In this model, the subhorizontal reflector located to the southeast of the root-zone is related to the structure beneath an accreted island arc, rather than the Blue Ridge and Piedmont décollement.

Iverson and Smithson (1983a) based this interpretation on five main lines of evidence. They documented a dramatic change in character of the reflections across the east-dipping reflective zone beneath the Kings Mountain belt. True-amplitude seismic data illustrated the contrast between the strong, continuous reflections from the base of the allochthon and the complex, relatively weak reflections obtained from beneath the slate belt. Migration of the dipping events further illustrates that the southeast-dipping reflections actually extend beneath the subhorizontal reflections to the southeast and are therefore probably not from a continuous reflector at the base of a décollement. The change in character was interpreted by Iverson and Smithson (1983a) to represent a

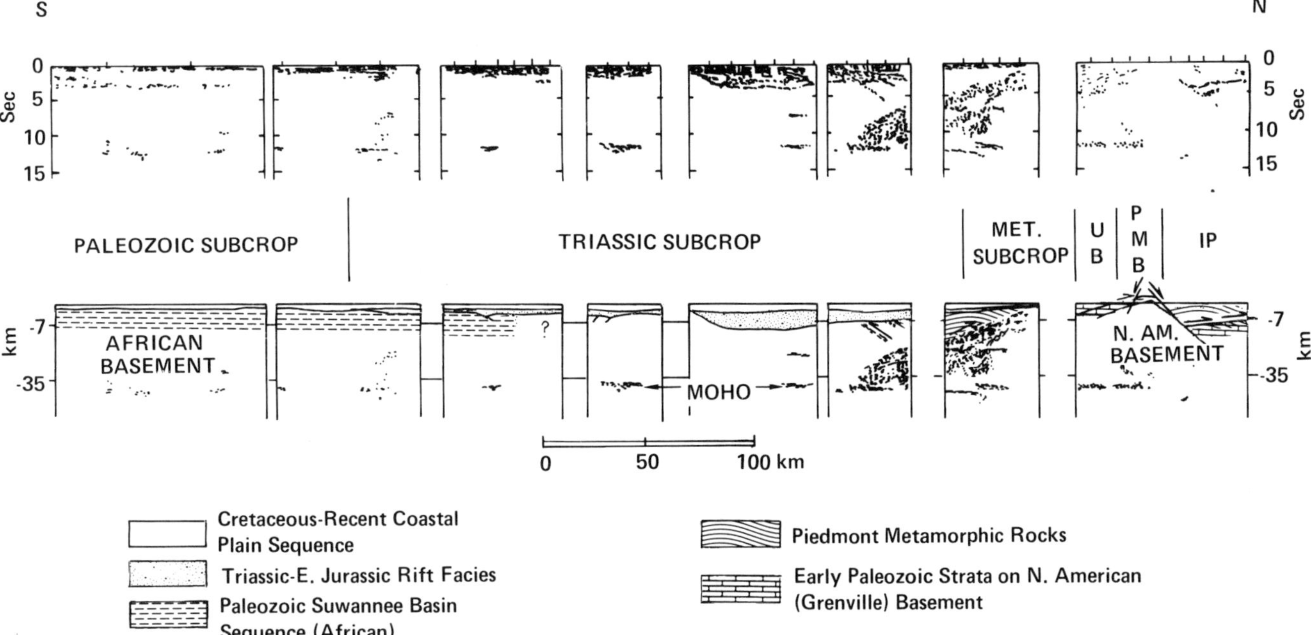

Figure 7b. Southwest Georgia reflection profile (line 3 in Fig. 6): line drawing (top) and interpretation bottom), from Nelson and others (1985a).

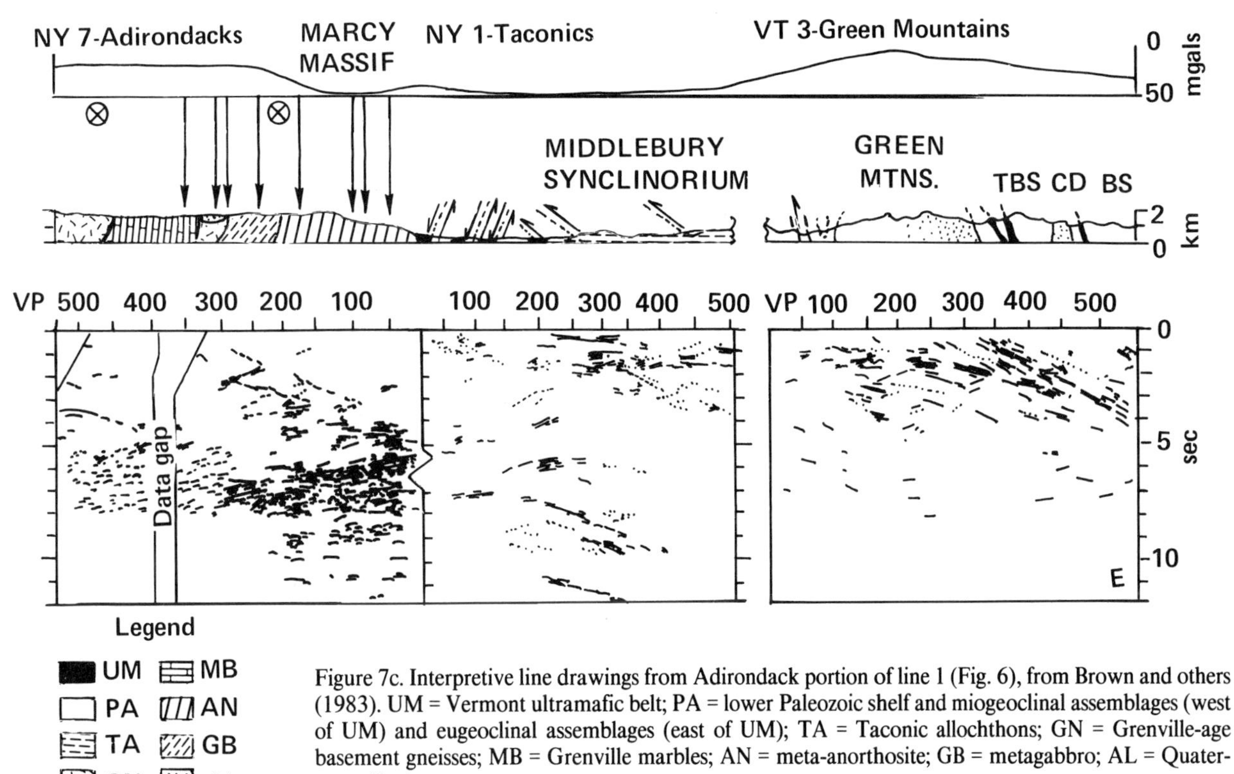

Figure 7c. Interpretive line drawings from Adirondack portion of line 1 (Fig. 6), from Brown and others (1983). UM = Vermont ultramafic belt; PA = lower Paleozoic shelf and miogeoclinal assemblages (west of UM) and eugeoclinal assemblages (east of UM); TA = Taconic allochthons; GN = Grenville-age basement gneisses; MB = Grenville marbles; AN = meta-anorthosite; GB = metagabbro; AL = Quaternary alluvium.

transition of strong reflections from the overthrust shelf rocks to complex, semi-continuous reflections from highly deformed crystalline rocks and mylonite zones occurring beneath an accreted island arc. Iverson and Smithson (1982) also noted that the east-dipping reflections were similar to those observed by Ando and others (1983) east of the Green Mountains in New England, who suggested the presence of a large thrust ramp—which is structurally different than the two models proposed by Iverson and Smithson (1982) and Cook and others (1979).

Iverson and Smithson (1983a) also identified the Piedmont gravity gradient as representing a fundamental change in crustal geometry, thus supporting their arguments for a décollement root-zone (although they recognized the nonuniqueness problems involved with these data; cf. Hutchinson and others, 1983). Problems involving palinspastic reconstruction and mechanical problems with large overthrust sheets were also pointed out by Iverson and Smithson (1982). The suggestion that the Kings Mountain belt represents a major structural boundary between the Inner Piedmont and the Charlotte belt is further supported by isotopic evidence (LeHuray, 1986).

In subsequent papers by Cook (1983, 1984a, b), the original models preferred by Cook and others (1979, 1981) involving extension of the southern Appalachian décollement to the Coastal Plain were modified and closely match the buried continental margin transition zone of Ando and others (1983). In these models, Grenville continental crust extends approximately to the east side of the Inner Piedmont; farther east the accreted crust is of oceanic or transitional (island arc) affinity. The principal difference with the "root-zone" model of Iverson and Smithson (1982) is related to the attitude of the Inner Piedmont–Charlotte belt transition; in other words, whether the transition is a steep, lithosphere-penetrating suture or an east-dipping ramp structure. By comparison with reflection data from present-day rifts, passive margins, and convergent plate boundaries, Lillie (1984) concluded that the east-dipping reflectors represent the edge of the rifted continental margin that is preserved intact beneath the décollement. In both cases, a fundamental difference in the structure of the lower crust exists on both sides of the boundary, but available geophysical and geologic evidence is insufficient to distinguish between the two hypotheses.

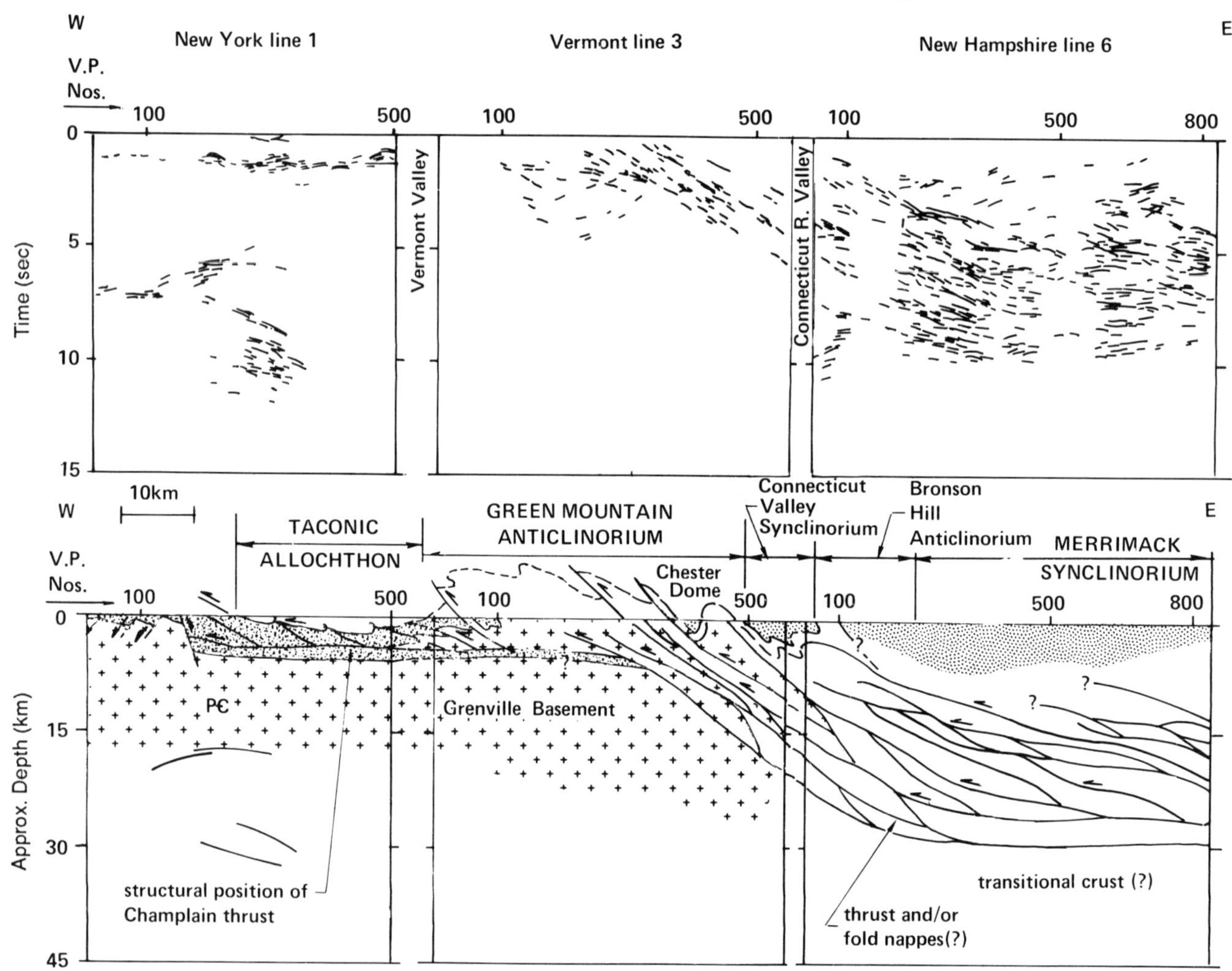

Figure 7d. Line drawing and interpretation of New England portion of line 1 (Fig. 6), from Ando and others (1984).

The most recent COCORP results in the southern Appalachians are from a line collected across the southeastern edge of the Inner Piedmont to the Coastal Plain in southwestern Georgia and northern Florida (Nelson and others, 1985a, b). The southeast-dipping reflections are again observed to the southeast of the Inner Piedmont and Pine Mountain belt (Fig. 7b). However, in this profile, the reflections extend through the crust along an approximately 50-km-wide zone. The discontinuous, laminated, subhorizontal Moho reflections are again seen to occur to the southwest at depths of about 33 to 36 km (11- to 12-sec two-way traveltime) and truncate the dipping reflectors. Drillhole information suggests that the south-dipping reflections represent a crustal penetrating Alleghanian suture separating known African basement from North American (Grenville) basement. Nelson and others (1985b) also noted that the postulated suture coincides with the Brunswick magnetic anomaly, which diverges from the Piedmont gravity gradient in western Georgia and continues eastward offshore where it meets the northeast-trending East Coast Magnetic anomaly. It is suggested that the Piedmont gravity gradient marks the southeast limit of Grenville basement and that the two anomalies are separated by Paleozoic accreted terranes (Charlotte belt, slate belt, etc.) caught between North America and Africa. Lateral motions during the Alleghanian orogeny appear to have truncated these structures to the southwest (Horton and others, 1984).

COCORP lines traversing the Adirondacks and New England Appalachians yielded a number of interesting similarities and contrasts with results from the southern Appalachians. The principal finding in the Adirondacks was a prominent zone of layered reflections occurring between depths of 18 and 26 km (termed the Tahawus complex; Brown and others, 1983; Klemperer and others, 1985; Fig. 7c). The wedge-shaped zone occurs beneath the Marcy anorthosite massif and consists of northwest-dipping reflections (~10° dip) overlying a subhorizontal set of reflections.

There is little evidence of coherent Moho reflections, which is consistent with results from inversion of crustal receiver functions at RSNY (Owens and others, 1987). The upper crust is relatively transparent and shows some cross-cutting, discontinuous reflectors that are weak relative to those from the Tahawus complex.

A number of different hypotheses have been formulated regarding the origin of the Tahawus complex, including igneous layering (cumulate layers beneath anorthosites or sills from Precambrian rifting), gneissic layering, layered metasedimentary rocks and associated metavolcanic rocks, or magma at depth (Brown and others, 1983; Klemperer and others, 1985). It has also been noted that the high-conductivity zone of Connerney and others (1980) correlates well with the Tahawus complex. Negative P-wave traveltime residuals (Taylor and Toksoz, 1979) and high velocities at these depths (Owens and others, 1987) rule out the possibility of magma bodies. The preferred models involve a buried metasedimentary sequence or mafic or ultramafic cumulates (associated with the anorthosites), although the high velocities at these depths may tend to favor the latter hypothesis.

The extension of the COCORP transverse into New England has been described by Ando and others (1983, 1984). Beneath the Taconic belt, a prominent reflector is observed at a depth of about 4.5 km, which is thought to represent the contact between the autochthonous shelf carbonate rocks and overlying flysch sequence with the underlying Grenville basement (Fig. 7d). A weak reflection at about 2 km depth may be from the base of the Taconic allochthon. The 4.5-km reflector can be traced beneath the Green Mountains, suggesting that the Precambrian crystalline rocks have been thrust over the shelf rocks.

To the east of the Green Mountains, beneath the Connecticut Valley synclinorium, the reflection signature changes dramatically, showing many similarities to the east-dipping reflections observed to the south beneath the easternmost Inner Piedmont. The east-dipping reflectors have a complicated geometry and dip

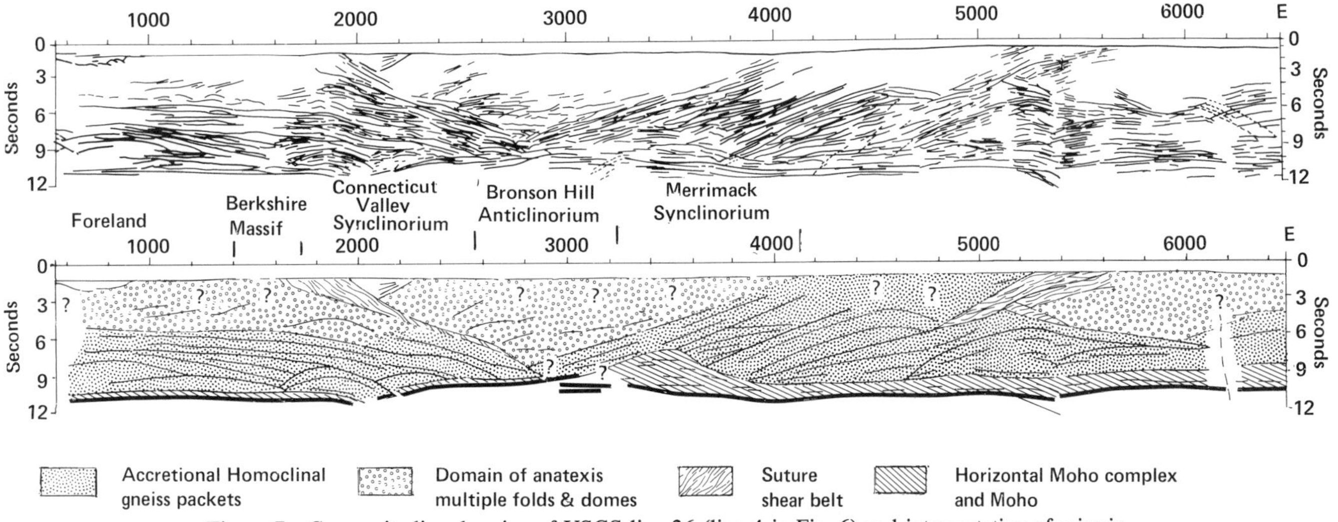

Figure 7e. Composite line drawing of USGS line 36 (line 4 in Fig. 6) and interpretation of seismic section, modified from Phinney (1986).

of about 40°. They extend to depths of about 14 km and project back to the eastern part of the Green Mountains. The reflectors flatten somewhat to the east but extend throughout much of the crust to depths of 30 km beneath the Bronson Hill anticlinorium and Merrimac synclinorium.

Ando and others (1983, 1984) interpreted the observations to be consistent with preserved continental margin transition-zone eugeosynclinal rocks overthrust from the east. In this model, the east-dipping reflections (east of the Green Mountains) are from ramping of thrust sheets within the Taconic root zone. Additional geophysical measurements, such as gravity, P-wave residuals, and regional traveltimes (cf. Taylor and Toksoz, 1982), suggest that a fundamental change in crustal structure occurs in this region. However, it is unclear whether the east-dipping events represent a ramp structure or a lithosphere-penetrating suture, as suggested by Hatcher and Zietz (1980) and Iverson and Smithson (1982) for the southern Appalachians, and Taylor and Toksoz (1980, 1982) for the northern Appalachians.

The earliest reflection results in the northern Appalachians were collected in Quebec (St. Julien and Hubert, 1975; St. Julien and others, 1983). These profiles suggested extensive northwestward thrusting with progressively deeper levels of detachment to the southeast. These lines have been continued into Maine; preliminary results suggest that the décollement extends at increasingly deeper levels farther east beneath the Connecticut Valley Synclinorium and Chains Lake Massif (from 3.5 sec two-way traveltime in the northwest to 7.8 sec in the southeast; Stewart and others, 1986). Thus, as was observed in southern New England, it appears that Grenville basement extends beneath the Connecticut Valley Synclinorium and is covered by about 24 km of Taconian thrust sheets.

Multichannel reflection lines collected by the USGS across the Long Island Platform have added important insight into the structure across the narrow, intensely deformed part of the southern New England Appalachians, which includes data from the craton and Avalon zone, and the accreted terranes caught in between (Phinney, 1986; Hutchinson and others, 1986; Fig. 6). The accreted terranes form a wedge-shaped "keystone" structure separating the craton from Avalonia (Fig. 7e). Crustal penetrating sutures dipping at about 25° to 42° to the east and west, respectively, separate the craton from the accreted terranes and Avalonia. The Moho is generally observed as a sharp boundary at about 9.5- to 12-sec TWTT, with some laminations in the eastern part, and appears to be shallower in the central portion. It is not clear if this feature is real, or caused by velocity pullup due to high-velocity material in the lower crust within the accreted terranes.

SHEAR-VELOCITY STRUCTURE FROM CRUSTAL RECEIVER FUNCTIONS

The detailed crustal and upper-mantle shear-velocity structure has been investigated in a few localities in the Appalachians by modeling teleseismic waveforms. The technique consists of isolating a "receiver function" beneath a seismic station that basically represents the impulse response of the crust and upper mantle in the station vicinity (cf. Langston, 1979; Owens and others, 1983). The receiver function can then be modeled or inverted for a layered shear-velocity structure (Taylor and Owens, 1984). By utilizing broadland seismic data, resolution of velocity discontinuities can be made to within 2.5 km, and lateral variations in the vicinity of a particular station can be inferred by modeling receiver functions from different azimuths. The technique complements refraction and reflection studies because of the detailed vertical velocity information it provides in a small region.

In the southern Appalachians, Langston and Isaacs (1981) used long-period seismic P-waveforms to obtain a crustal thickness of approximately 41 km beneath SCP in the Valley and Ridge Province. The thickness of 41 km obtained from the P-waveforms is significantly greater than the 33 km from refraction results (Katz, 1955). Estimates of crustal thickness obtained from teleseismic receiver functions are often greater than those from refraction results that use only simple two- or three-layer velocity models, especially in regions with gradational crust-mantle boundaries. In these cases, the thickness derived from refraction results often correlates with the top of the transition between crust and mantle obtained from the receiver functions. Because the study was restricted to long-period data, no detailed crustal structure was obtainable.

A detailed investigation of a large number of broadband teleseismic P-waveforms at the Regional Seismic Test Network (RSTN) station RSCP located in the Cumberland Plateau region in eastern Tennessee yielded a number of interesting results (Owens and others, 1984). The region is characterized by a gradational, possibly laminated crust-mantle boundary lying between depths of 40 and 55 km. As discussed by Zandt and Owens (1986), this result is remarkably consistent with the refraction models obtained by Prodehl and others (1984) in the same region; the velocity models are compared in Figure 8.

Velocity models obtained from three back azimuths delimited a midcrustal high-velocity body located to the northeast of RSCP between approximately 15 and 22 km depth. This high-velocity structure is coincident with the southwestern terminus of the East Continent Gravity High, which is thought to mark the boundary of an extensive late Precambrian rift system (Keller and others, 1982). The uppermost crust at RSCP is composed of an approximately 1.7-km-thick layer of low-velocity Pennsylvanian sediments, which complicate the early part of the receiver function.

Two investigations of crustal structure have been made in the Grenville Province to the north of the Adirondacks. Jordan and Frazer (1975) studied in the S_p (S- to P-conversions) from two deep-focus South American earthquakes recorded at a number of long-period Canadian stations. Using forward modeling, the crustal thickness was estimated to be 35 km with surprisingly low shear velocities (V_s = 3.4 km/sec) in the lower crust. By comparing their estimated V_s with P-velocities from nearby refraction

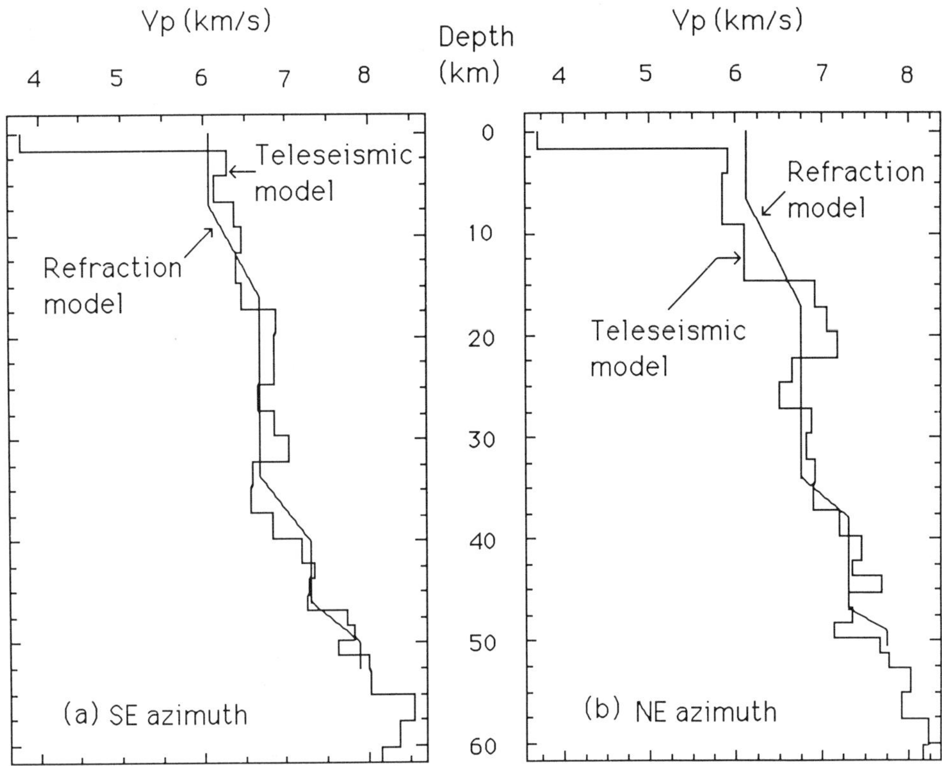

Figure 8. Comparison of P-velocity models in the Cumberland Plateau, Tennessee, from modeling of refraction data (Prodehl and others, 1984) and inversion of teleseismic receiver functions and station RSCP (Owens and others, 1984).

models (Berry and Fuchs, 1973), a very high Poisson's ratio, 0.33, was calculated. Comparison with expected values of Poisson's ratio and V_S for rocks at 10 kbar suggested that hydrated ultramafic rocks (serpentinized peridotite) may exist in the lower crust of the Grenville Province.

A detailed investigation of the crustal structure at the RSTN station RSNY located in the Adirondacks has provided some interesting comparisons with nearby reflection and refraction results (Owens, 1984; Owens and others, 1987). In general, receiver functions estimated at RSNY are simple (particularly to the northeast) relative to those from tectonically active regions or from regions with a near-surface sedimentary layer, such as at RSCP (Owens and others, 1983). This suggests that the crust in the Adirondacks is relatively uniform. In general, average crustal velocities derived from inversion of the RSNY receiver functions from four back azimuths were similar to average velocities obtained from regional traveltime studies (cf. Taylor and Toksoz, 1980).

However, to the southeast (and possibly southwest) of RSNY, high shear-wave velocities (V_S ~3.9 to 4.0) were observed between depths of approximately 18 and 26 km beneath the Marcy anorthosite massif. This zone of relatively high velocities at midcrustal depths correlates well with a zone of prominent reflectors (Tahawus complex of Klemperer and others, 1985) obtained from COCORP lines located about 60 km to the southeast of RSNY (Fig. 9; Brown and others, 1983). Below these depths, low shear velocities ranging from 3.6 to 3.7 km/sec were observed between depths of 26 and 40 km in reasonably good agreement with Jordan and Frazer (1975).

At RSNY, a gradational crust-mantle boundary was observed between about 42 and 50 km, particularly to the southeast and southwest. At RSNY and other RSTN statins, it is commonly observed that the crustal thickness obtained from refraction techniques correlates well with the top of the crust-mantle transition zone obtained from inversion of receiver functions. This is an improvement over previous traveltime studies in the Adirondacks using simple layered earth models incapable of resolving gradational crust-mantle transitions.

SHEAR-VELOCITY STRUCTURE FROM SURFACE WAVES

A few studies of surface-wave dispersion have been made in the eastern United States that help place constraints on crustal thickness and upper-mantle velocity. However, because of the long wavelengths involved with these surface-wave analyses (~25 to 200 km), resolution of detailed structures is very limited. In fact, crustal thicknesses probably can be determined only to within ±5 km for typical inversion of phase velocity data.

The earliest studies involved tripartite measurement of Rayleigh-wave phase velocities in the New York–Pennsylvania area over the Valley and Ridge and Adirondacks region (Oliver and others, 1961; Dorman and Ewing, 1962). The crustal thickness was estimated to be 37 to 38 km, with an average crustal shear velocity of 3.6 km/sec and 4.7 km/sec in the uppermost mantle. An upper-mantle low-velocity zone with V_s = 4.3 km/sec was inferred to lie between a depth of 84 and 100 km.

Taylor and Toksoz (1982) measured interstation phase and group velocities in the northeastern United States and Canada in the period range of 15 to 50 sec. Although the differences in the dispersion between paths in the Appalachian and Grenville Provinces were not significant under consideration of measurement errors, separate inversions yielded velocity models consistent with previous traveltime studies. The Appalachians are characterized by a slightly thicker crust with a higher velocity lower layer relative to the Grenville Province. The Grenville Province was characterized by a 35-km-thick crust with velocities increasing from 3.5 km/sec in the upper 15 km to 3.8 km/sec in the lower crust. The Appalachian dispersion curves were fit well with V_s = 3.5 km/sec in the upper 20 km and V_s = 4.0 km/sec in the lower 20 km, with a total crustal thickness of 40 km. The upper-mantle velocity was about 4.5 to 4.6 km/sec in both regions. Frequency-wavenumber power spectra across a short-period seismic network in central New England also provided evidence of high shear velocities in the mid- to lower crust. Little evidence was seen for a low-velocity zone in the upper mantle.

Mitchell and Herrmann (1979) measured single-station phase and group velocities for Rayleigh and Love waves in the eastern United States in the period range of 4 to 70 sec. An isotropic earth model was found to fit the data that basically consisted of a two-layer crust. Beneath a low-velocity sedimentary surface layer, the shear velocity appears to gradually increase from 3.4 to 3.7 km/sec to a depth of about 25 km. The lower crust is characterized by velocities of 3.8 to 4.0 km/sec, with a crustal thickness of about 40 km. The upper-mantle shear velocity is 4.8 km/sec and overlies a poorly resolved upper-mantle low-velocity zone between depths of about 70 and 110 km. There was also some suggestion of weak polarization anisotropy in the lower crust, with $V_{SH} > V_{SV}$.

In the southern Appalachians, Long and Mathur (1972) measured Rayleigh wave phase velocities in the period range of 15 to 50 sec across the central United States, Valley and Ridge, Blue Ridge, and western Piedmont. A 40-km-thick crust with V_s = 3.5 km/sec in the upper 20 km and V_s = 3.8 km/sec in the lower crust was estimated to overlie an upper mantle with V_s = 4.6 km/sec.

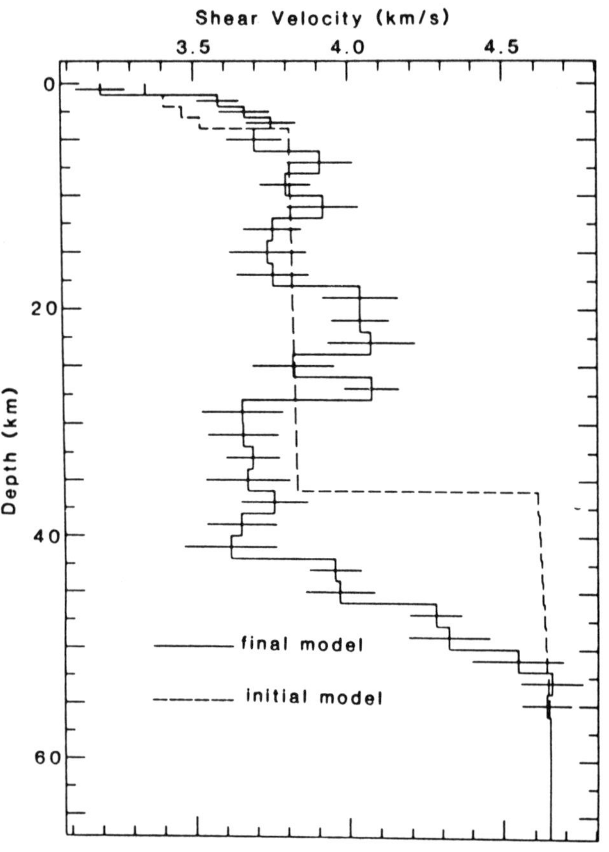

DISCUSSION

One of the most significant results of geophysical studies in the Appalachians is the pronounced difference in crustal structure between the Precambrian Grenville Province (North American craton) and the accreted terranes of the Paleozoic Appalachian Province. As elucidated by the combination of various geophysical observations (highlighted by the reflection results), the boundary between the early Paleozoic margin of North America and the accreted terranes is represented by a moderately dipping ramp structure rather than a near-vertical suture penetrating the crust and upper mantle. This ramp structure is often interpreted to flatten into a complicated series of imbricate thrusts in the mid- to lower crust beneath the accreted belts and acts as a root-zone for

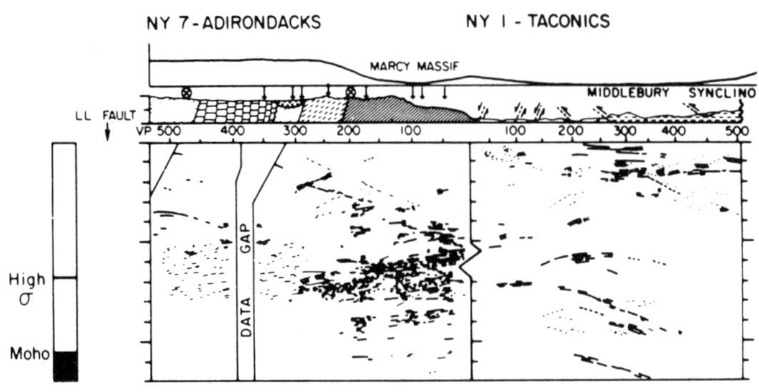

Figure 9. Velocity structure from inversion of teleseismic receiver functions at station RSNY in the Adirondacks (left; Owens and others, 1986) and COCORP line drawing from profile located approximately 60 km to the south (right; Brown and others, 1983).

a thin-skinned continent-directed décollement (cf. Cook, 1986). The nature of the reflective character, velocity structure, and gravity and magnetic signature changes considerably between the craton margin and accreted terranes. The observed contrasts in crustal structure between the two orogenic belts are probably the result of variations in petrology, water content, and (less likely) temperature, due to differences in tectonic evolution.

Electrical measurements in the northeastern United States suggest the presence of a highly conductive lower crust in the Adirondack Mountains in New York state (Connerney and others, 1980), and a resistive lower crust underlying a slightly conductive 15-km-thick upper crust in New England (Kasameyer, 1974). The differences in lower crustal conductivity between the two regions may be due to the presence of either free water or hydrated minerals. The slightly conductive 15-km-thick upper layer in New England correlates well with the 15-km-thick upper layer observed from refraction studies and probably corresponds to metamorphosed eugeoclinal rocks of the major synclinoria. This implies that the observed differences in velocity and conductivity of the lower crust between the two belts may be the result of a hydrated lower crust beneath the Grenville Province. Although the rocks of the lower crust may be compositionally similar, a hornblende-granulite facies assemblage beneath the Grenville would yield lower velocities than a pyroxene-granulite assemblage beneath the Appalachians (Christensen and Fountain, 1975). However, conductivity measurements on amphiboles at lower crustal conditions are lacking (Connerney and others, 1980), and it is not clear if the existence of hornblende granulites (which generally contain less than 1 percent water; Manghnani and others, 1974) in the lower crust will explain both the conductivity measurements and the observed seismic velocities.

The existence of low shear-velocities in the lower crust beneath the Grenville Province is also supported by the study of the S_p phase across eastern Canada by Jordan and Frazer (1975), who found evidence for a lower crustal layer with a very high Poisson's ratio of 0.33. As suggested by Jordan and Frazer (1975), the existence of serpentinite in the lower crust of the Grenville Province may explain their observations of a low-velocity layer with a high Poisson's ratio and would be consistent with the observed high deep-crustal conductivities. Owens and others (1987) also found evidence for low shear velocities in the lower crust beneath the Adirondacks, although they had little information on P-velocities.

The low velocities and high conductivity in the lower crust of the Adirondacks could also be due to the presence of aqueous fluids. Based on comparisons of laboratory and field conductivity data, Shankland and Ander (1983) argued that free water in amounts of only 0.01 to 0.1 percent in fracture porosity is sufficient to explain high conductivities in the lower crust of stable environments. Using the theoretical calculations of O'Connell and Budiansky (1977) for fluid-saturated media with interconnected cracks and reasonable choices of crack density, distribution of aspect ratios, and fluid viscosity, it is possible to reduce the velocities by the observed amounts with fluid contents substantially less than 1 percent. As discussed by Fyfe (1986), a number of mechanisms exist that could cause fluid release of these amounts into the lower crust, even in regions that appear to be anhydrous, such as the Adirondacks.

Temperature differences may affect the velocities observed in the lower crust. However, at temperatures and pressures representative of the lower crust in older geologic belts, the effect of temperature on seismic velocities is small relative to pressure (Christensen, 1979). As discussed by Taylor and Toksoz (1979), temperature differences may be important in the upper mantle and may account for the observed teleseismic P-wave delays.

As an alternative explanation to those based on physical properties such as water content or temperature, the rocks of the Grenville and Appalachian Provinces may show contrasts in their chemistry and petrology that are caused by differences in their tectonic evolution. As was noted by Taylor and Toksoz (1982), rocks of the lower crust in the Grenville Province may be very similar to the quartzo-feldspathic gneisses found on the surface. The homogeneous character of the crust in this portion of the Grenville Province is consistent with the hypothesis that the crust underwent substantial reactivation, thickened, and became vertically uniform during the Grenville orogeny (Dewey and Burke, 1973). Subsequent to the thickening, the crust was eroded to relatively deep levels, as evidenced by the surface exposure of granulite terrains (Putmam and Sullivan, 1979).

In contrast, the lower crust in the Appalachians exhibits velocities more representative of mafic rocks such as pyroxene-garnet granulites. Geochemical models suggest that some members of the White Mountain Plutonic series in central New Hampshire were formed by reaction of fractionated mantle-derived alkali basalt with metamorphosed tholeiitic (oceanic) basalt at the base of the crust (Loiselle, 1978). At temperatures and pressures representative of the lower crust, a tholeiitic basalt will alter to a garnet or pyroxene granulite (Green and Ringwood, 1972) and would be characterized by relatively high velocities ($\geq$7.0 km/sec; Christensen and Fountain, 1975).

As discussed by Taylor and Toksoz (1980) the upper 15-km layer in the northern Appalachians probably corresponds to rocks that have been subjected to a high degree of compression and crustal shortening during the Taconic and Acadian orogenies. The metasedimentary rocks exposed in the central orogenic belt of the northern Appalachians were probably subduction mélanges and back-arc and fore-arc basin deposits from an accreted island arc. As discussed by Dickerson and Seely (1979), oceanic crust may often be trapped beneath a fore-arc basin; it is expected that these and back-arc basin mafic rocks will be caught up in the crust during collision episodes. Additionally, oceanic arcs appear to be composed of mafic cumulates from upper crustal magmas and preexisting oceanic crust upon which the arc was formed (Kay and Kay, 1985). This is further supported by geochemical studies of volcanic rocks exposed in the Bronson Hill Anticlinorium, which appear to have been produced by partial melting or fractional crystallization of basaltic source rocks (Leo, 1985). Thus, accreted island arcs are expected to have a mafic lower crust, and involvement of ocean floor within the Appalachians could account for the higher velocities found in the lower crust relative to the predominantly ensialic crust of the Grenville Province.

Based on teleseismic P-wave residuals, P_n residuals, and gravity, it appears that the transition zone between Grenville and Appalachian crust in the northeastern U.S. occurs in the vicinity of the Precambrian uplifts and the serpentinite belt. This north-northeast–trending belt may mark the suture zone between the craton and the Appalachian orogen. At many locations along this postulated suture, particularly in northwest Vermont, geologic and geophysical features, such as high crustal velocities, a linear gravity high, a serpentinite belt, Precambrian uplifts, and the Taconic thrusts, show many similarities to the Ivera zone in northern Italy (Giese and Prodehl, 1976; Fountain, 1976). However, the northern Appalachian reflection profiles suggest that this suture is actually more of a ramp structure that dips farther into the Connecticut Valley synclinorium and may continue to the western edge of the Bronson Hill anticlinorium (Ando and others, 1983; Stewart and others, 1986).

A second suture appears to be located in southeastern New England, separating rocks of the accreted terranes from Avalon rocks. This feature is marked by the Clinton–Newbury–Bloody Bluff fault zone (CN-BB F.Z.) in eastern Massachusetts (see Fig. 2) and other structures such as the Norumbega fault zone in eastern Maine (Loiselle and Ayuso, 1979). In eastern Massachusetts, the CN-BB F.Z. separates rocks that are probably correlative with Avalon rocks from those of the central mobile belt. The northwest-dipping fault zone is also well-marked by such features as the offset of metamorphic isograds, a strong magnetic signature, and a cataclastic zone containing mylonite that is up to 1.5 km thick. A pronounced gravity anomaly is also associated with the fault zone (Taylor and others, 1980b), and the Bouguer gravity is relatively high over the Avalon basement. The fault zone continues offshore, and on the basis of gravity and magnetic data, trends either east-northeast or curves toward the east, where it transforms into a right-lateral strike-slip fault (Simpson and others, 1980). It may have become a transform fault in late Paleozoic time (Ballard and Uchupi, 1975; Arthaud and Matte, 1977), which may have obliterated any evidence of ophiolites that would have resulted from Devonian collisional episodes.

Recently acquired geologic evidence suggests that the Avalon terrane in southeastern New England can be divided into two distinct zones that were accreted to North America during either Ordovician or Devonian time, and during the Alleghanian orogeny (O'Hara and Gromet, 1985). The deep crustal structure of the Avalon block appears to differ from that of the central belt. Refraction models, P_n, and teleseismic P-wave residuals indicate that the crust of the eastern block is probably thinner than that of the central belt and may be missing a high-velocity lower-crustal layer. This is consistent with geochemical models of Avalon plutonic and volcanic rocks, suggesting that the Avalon arc system was originally formed on rifted continental crust (Kennedy, 1976).

Three-dimensional inversion of traveltime data illustrates that structures down to perhaps 200 km and greater can be correlated with large-scale geologic features (Taylor and Toksoz, 1979). This has the important implication that major orogenic belts have effects, extending well into the lithosphere, which are stable for extended periods of time, perhaps as long as a billion years. The lateral variations in seismic properties of the crust and upper mantle beneath the northeastern United States are very small relative to those observed over similar distances in active tectonic regions such as central Asia or the western United States. Poupinet (1979) showed that absolute P-residuals are a linear function of the square root of age inside stable continental plates. This indicates that the crust and upper mantle become increasingly more uniform with age during evolution toward a state of equilibrium.

Recent COCORP seismic reflection profiling in the southern Appalachians indicates that the platform rocks overlying the Grenville basement can be traced beneath the Blue Ridge and continue at least 150 km to the east (Cook and others, 1979; Fig. 7a). The extent of this thin-skinned thrusting is problematic, and arguments can be made for its extension to the Inner Piedmont/Charlotte belt boundary (cf. Iverson and Smithson, 1982; Hatcher and Zietz, 1980) or farther east into the Carolina slate belt (cf. Cook, 1984b). This appears to be one of the major unresolved questions at this time; it appears that the answer may be somewhere between the postulated steeply dipping crustal-penetrating suture and the extension of the décollement into the accreted terranes, i.e., a moderately dipping ramp structure (Ando and others, 1983; Cook, 1986).

Another important question is the difference in crustal structure between the central accreted belts in the northern and southern Appalachians. The central belt in the northern Appalachians is characterized by a thicker crust with velocities in the lower portions that are higher than those in the Charlotte belt and Carolina slate belt in the southern Appalachians. The reflective character is also slightly different in the two belts: imbricate, discontinuous reflectors appear throughout much of the crust along the northern profile, while subhorizontal reflectors are confined to a narrow zone in the midcrust in the south (Fig. 7a, e).

For the Inner Piedmont, the low velocities in the lower crust are easily explained because of the presence of Grenville basement beneath the décollement. However, the thin crust with low average velocities throughout the apparently mafic crust of the Charlotte belt/Carolina slate belt raises a number of questions. The low velocities may support the hypothesis of Cook (1984b) that the décollement continues into the Charlotte belt/Carolina slate belt and covers transitional (rift stage) basement. Alternatively, because the Charlotte belt/Carolina slate belt is thought to be part of the Avalon accreted terrane (Williams and Hatcher, 1982), their structure should be similar to the Avalon crust to the north, which is indeed the case.

Geochemical studies of plutonic and volcanic rocks of the Avalon block in both the northern and southern Appalachians suggest that the terrane formed as a continental arc or an island

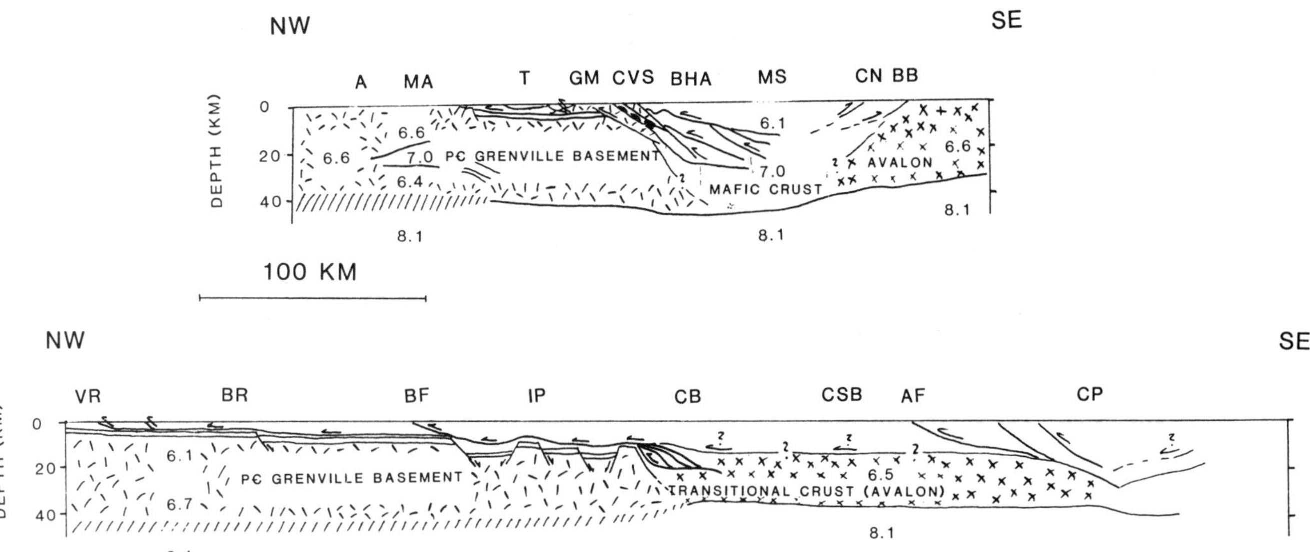

Figure 10. Summary figure showing interpretation of geophysical data in the northern Appalachians (top) and southern Appalachians (bottom). Profiles taken parallel to COCORP lines 1 and 2 in Figure 6; numbers represent seismic velocities in each region based on refraction studies. Gradational crust–mantle boundary shown by short diagonal lines at base of crust, abrupt Moho by solid line.

arc built on rifted continental crust (cf. Rast, 1980; Bland and Blackburn, 1980). Further, Loiselle and Ayuso (1980) noted that plutonic rocks of Avalon affinity to the southeast of the Norumbega fault zone in southeastern Maine are of more continental composition than the calc-alkaline plutonic rocks of the central orogenic belt of the Merrimac synclinorium.

Because the northern and southern Appalachians show many contrasts in structural style and tectonic history, it is not surprising that the reflection results in the northern Appalachians contrast with those in the southern. Although the profiles show some similarities in the vicinity of the ramp zone separating the craton from accreted belts, the magnitude of the thin-skinned thrusting is much more extensive in the south than the north (Fig. 7a, e).

The tectonic evolution of the northern and southern Appalachians shows a number of similarities up through the mid-Devonian Acadian orogeny. However, the southern Appalachians have an additional late Paleozoic orogenic episode (the Alleghanian orogeny), which resulted in westward thrusting of the Blue Ridge, Inner Piedmont, and possibly Charlotte belt/Carolina slate belt (Cook and others, 1979). In the northern Appalachians, the Alleghanian orogeny was only felt in southeastern New England and by strike-slip faulting farther north. It has been suggested that the Alleghanian deformational belt truncates the Appalachian-Caledonian belt in southern Connecticut and Rhode Island and is correlative with the Hercynian belt in central Europe (Dewey and Kidd, 1974). The extensive crustal shortening of the late Paleozoic Alleghanian marginal fold-and-thrust belt of the Valley and Ridge Province is confined mainly to the southern Appalachians. Slices of relatively unmetamorphosed carbonates are found to the east of the Blue Ridge in the Brevard fault zone and the Grandfather Mountain window (Hatcher, 1978). The ultramafic rocks found in the southern Appalachians are diffuse and irregularly distributed throughout the Piedmont zone, indicating their involvement in the numerous thrusts of the allochthonous crystalline belt (Misra and Keller, 1978). In contrast, the ultramafics of the northern Appalachians form a narrow north-northeast–trending belt, and lower Paleozoic platform rocks are rarely found east of the Precambrian outliers in the northern Appalachians.

As pointed out by Williams (1980), Alleghanian deformation involves Carboniferous rocks in the Piedmont of the southern Appalachians. In contrast, autochthonous Silurian and younger cover rocks unconformably overlie rocks of the Ordovician Taconic thrust belt and can be traced eastward into the central orogenic belt of the northern Appalachians. This implies that any major décollement in the northern Appalachians would have to be a pre-Silurian structure.

CONCLUSIONS

Geophysical studies have radically improved our understanding of the tectonic processes involved with the formation of the Appalachian Mountain belt and Adirondacks. As illustrated by the above discussion, a number of geophysical techniques can be effectively combined to produce a relatively coherent picture of the subsurface structure in a region while placing fairly tight constraints on geological interpretations of its tectonic history. Figure 10 illustrates the major features inferred from geophysical observations and some geological interpretation for the northern and southern Appalachians. The profiles are taken along lines coincident with COCORP lines 1 and 2 (Fig. 6). A number of differences and similarities between the northern and southern Appalachians are observed on the profiles that correspond to differences and similarities in their tectonic evolution. The Precambrian Grenville province is similar in both regions and is characterized by a fairly uniform crust with velocities of 6.6 to 6.7 km/sec in the lower part. The crust-mantle boundary appears to be gradational in both regions.

The magnitude of overthrusting appears to be much greater in the southern than the northern Appalachians, due to the influence of the Alleghanian orogeny in the southern but not the northern Appalachians. The Precambrian outliers of the Green Mountains and Berkshires are allochthonous and were probably transported during the Taconic orogeny. In contrast, the Precambrian outliers of the Blue Ridge were transported much greater distances during the Alleghanian and perhaps Taconic orogeny.

Geologically, the crystalline accreted terranes in both the northern and southern Appalachians show some similarities; however, structural differences in the subsurface are observed. In the north, the central terranes show a number of east-dipping reflectors that penetrate into the lower crust where high velocities are observed (~7.0 km/sec). The lower crust of the central belts in the southern Appalachians is not as reflective, and the crust appears to be very thin and of low velocity (~6.5 km/sec). In both the north and the south, the central accreted terranes are separated from the westward-lying Grenville basement by east-dipping reflectors and a change in gravity and magnetic character. Also, both belts have an abrupt crust-mantle boundary.

These central belts appear to be a series of oceanic and continental arc systems that were accreted to North America during the closure of the Iapetus Ocean in Paleozoic time. The observed differences in lower-crustal structure may be related to whether the arcs were built on oceanic or continental crust. In the north, the central belts of the Connecticut Valley synclinorium (east of the Chester domes), Bronson Hill anticlinorium, and Merrimac synclinorium may have been built on oceanic basement, which may account for the higher velocities in the lower crust. In contrast, the rocks of the Charlotte belt and slate belt appear to be of Avalon affinity and were continental arcs showing many similarities in velocity structure to Avalon terrane in the northern Appalachians.

It is interesting to note the relative importance of the different geophysical techniques for piecing together different portions of the same puzzle. Clearly, the recent COCORP results have done the most to reshape our thinking about the crustal structure in the Appalachians. However, without constraints from other geophysical techniques, a number of important questions would remain unanswered. Inversion of station receiver functions can provide detailed information regarding velocities associated with identified reflectors. Refraction models have been fundamental in determining changes in crustal structure in a number of specific regions and improving our understanding of vertical variations in crustal structure. Lateral variations in structure are effectively studied using teleseismic P-wave residuals, P_n time terms, gravity, and to some extent, magnetic observations. Conductivity measurements and detailed velocity models obtained from receiver functions have been shown to provide some important constraints on the interpretation of various reflections and crustal velocities determined from refraction experiments. Although models obtained from surface-wave dispersion measurements are consistent with refraction models, the method appears to be most useful in regions where other seismic measurements are unobtainable. This is the case in many tectonically active regions where surface-wave dispersion measurements can be used to define low-velocity zones, particularly in the upper mantle.

ACKNOWLEDGMENTS

Detailed reviews of the manuscript by Robert Hatcher, Douglas Rankin, Jerry Sweeney, and Walter Mooney are greatly appreciated. I thank Bob Schock and Norm Burkhard at Lawrence Livermore National Laboratory for allocating time for me to work on this project. This work was performed under the auspices of the U.S. Department of Energy by the Lawrence Livermore National Laboratory under Contract W-7405-ENG-48.

REFERENCES CITED

Aki, K., Christoffersson, A., and Huseby, E. S., 1977, Determination of the three-dimensional seismic structure of the lithosphere: Journal of Geophysical Research, v. 82, p. 277–296.

Alvord, D. C., Bell, K. G., Pease, M. H., and Barosh, P. J., 1976, The expression of bedrock geology between the Clinton-Newbury and Blocky-Bluff fault zones, northeastern Massachusetts: U.S. Geological Survey Journal of Research, v. 4, p. 601–604.

Ando, C. J., Cook, F. A., Oliver, J. E., Brown, L. D., and Kaufman, S., 1983, Crustal geometry of the Appalachian orogen from seismic reflection studies, *in* Hatcher, R. D., Jr., Williams, H., and Zeitz, I., Contributions to the tectonics and geophysics of mountain chains: Geological Society of America Memoir 158, p. 83–101.

Ando, C. J., and 11 others, 1984, Crustal profile of mountain belt; COCORP deep seismic reflection profiling in New England Appalachians and implications for architecture of convergent mountain chains: American Association of Petroleum Geologists Bulletin, v. 68, no. 7, p. 819–837.

Arthaud, F., and Matte, P., 1977, Late Paleozoic strike-slip faulting in southern Europe and northern Africa; Result of a right-lateral shear zone between the Appalachians and the Urals: Geological Society of America Bulletin, v. 88, p. 1305–1320.

Bailey, R. C., Edwards, R. N., Garland, G. D., and Greenhouse, J. P., 1978, Geomagnetic sounding of eastern North American and the White Mountain heat flow anomaly: Geophysical Journal of the Royal Astronomical Society, v. 55, p. 499–502.

Ballard, R. D., and Uchupi, E., 1975, Triassic rift structure in Gulf of Maine: American Association of Petroleum Geologists Bulletin, v. 59, p. 1041–1072.

Berry, M. J., and Fuchs, K., 1973, Crustal structure of the Superior and Grenville

Provinces of the northeastern Canadian shield: Seismological Society of America Bulletin, v. 63, p. 1393–1432.

Birch, F., Roy, R. F., and Decker, E. R., 1968, Heat flow and thermal history in New England and New York, *in* Zen, E-an, White, W. S., Handley, J. B., and Thompson, J. B., Studies of Appalachian geology; Northern and Maritime: New York, Interscience, p. 437–451.

Bird, J. M., and Dewey, J. F., 1970, Lithosphere plate–continental margin tectonics and the evolution of the Appalachian orogen: Geological Society of America Bulletin, v. 81, p. 1031–1060.

Bland, A. E., and Blackburn, W. H., 1980, Geochemical studies on the greenstones of the Atlantic seaboard volcanic province, south-central Appalachians, *in* Wones, D. R., ed., Proceedings, The Caledonides in the U.S.A.; IGCP Project 27: Blacksburg, Virginia Polytechnic Institute and State University Memoir 2, p. 263–278.

Bollinger, G. A., Chapman, M. C., and Moore, T. P., 1980, Central Virginia regional seismic network; Crustal velocity structure in central and southwestern Virginia: U.S. Nuclear Regulatory Agency NUREG CR-1217, 187 p.

Borcherdt, R. D., and Roller, J. C., 1966, A preliminary summary of a seismic-refraction survey in the vicinity of the Cumberland Plateau Observatory, Tennessee: U.S. Geological Survey Technical Letter 43, 31 p.

Boucot, A. J., 1968, Silurian and Devonian of the northern Appalachians, *in* Zen, E-An, ed., Studies of Appalachian geology; Northern and Maritime: New York, Interscience, p. 83–94.

Brown, L., and 7 others, 1983, Adirondack-Appalachian crustal structure; The COCORP northeast traverse: Geological Society of America Bulletin, v. 94, p. 1173–1184.

Carts, D. A., and Bollinger, G. A., 1981, A regional crustal velocity model for the southeastern United States: Seismological Society of America Bulletin, v. 71, p. 1829–1847.

Chapman, C. A., 1976, Structural evolution of the White Mountain magma series: Geological Society of America Memoir 146, p. 281–300.

Chiburus, E. F., and Ahner, R. O., 1979, Northeastern U.S. Seismic Network Bulletin 15, 7 p.

Chidester, A. H., 1968, Evolution of the ultramafic complexes of northwest New England, *in* Zen, E-An, ed., Studies of Appalachian geology; Northern and Maritime: New York, Interscience, p. 343–354.

Christensen, N. I., 1979, Compressional wave velocities in rocks at high temperatures and pressures, critical thermal gradients, and crustal low-velocity zones: Journal of Geophysical Research, v. 84, p. 6849–6865.

Christensen, N. I., and Fountain, D. M., 1975, Constitution of the lower continental crust based on experimental studies of seismic velocities in granite: Geological Society of America Bulletin, v. 86, p. 227–236.

Clark, H. B., Costain, J. K., and Glover, L., 1978, Structural and seismic reflection studies of the Brevard ductile deformation zone near Rosman, North Carolina: American Journal of Science, v. 278, p. 419–441.

Connerney, J.E.P., Nekut, E., and Kuckes, A. F., 1980, Deep crustal electrical conductivity in the Adirondacks: Journal of Geophysical Research, v. 85, p. 2603–2614.

Cook, F. A., 1983, Some consequences of palinspastic reconstruction in the southern Appalachians: Geology, v. 11, p. 86–89.

——, 1984a, Geophysical anomalies along strike of the southern Appalachian Piedmont: Tectonics, v. 3, p. 45–61.

——, 1984b, Towards an understanding of the southern Appalachian Piedmont crustal transition; A multidisciplinary approach: Tectonophysics, v. 109, p. 77–92.

——, 1986, Continental evolution by lithospheric shingling, *in* Barazangi, M., and Brown, L., eds., Reflection seismology; The continental crust: American Geophysical Union Geodynamics Series, v. 14, p. 13–21.

Cook, F. A., Albough, D. S., Brown, L. D., Kaufman, S., Oliver, J. E., and Hatcher, R. D., Jr., 1979, Thin-skinned tectonics in the crystalline southern Appalachians; COCORP seismic-reflection profiling of the Blue Ridge and Piedmont: Geology, v. 7, p. 563–567.

Cook, F. A., Brown, L. D., Kaufman, S., Oliver, J. E., and Peterson, T. A., 1981, COCORP seismic profiling of the Appalachian orogen beneath the Coastal Plain of Georgia: Geological Society of America Bulletin, part 1, v. 92, p. 738–748.

Costain, J. K., Glover, L., and Sinha, A. K., 1980, Low-temperature geothermal resources in the eastern United States: EOS Transactions of the American Geophysical Union, v. 61, p. 1–3.

Costain, J. K., Speer, J. A., Glover, L., Perry, L., Dashevsky, S., and McKinney, M., 1986, Heat flow in the Piedmont and Atlantic Coastal Plain of the southeastern United States: Journal of Geophysical Research, v. 91, p. 2123–2135.

Dainty, A. M., and Frazier, J. E., 1984, Bouguer gravity in northeastern Georgia; A buried suture, a surface suture, and granites: Geological Society of America Bulletin, v. 95, p. 1168–1175.

Dainty, A. M., Keen, C. D., Keen, M. J., and Blancard, J. E., 1966, Review of geophysical evidence on crust and upper-mantle structure on the eastern seaboard of Canada: American Geophysical Union Monograph 10, p. 349–369.

Dallmeyer, R. D., Wright, J. E., Secor, D. T., and Snoke, A. W., 1986, Character of the Alleghanian orogeny of the southern Appalachians; Part 2, Geochronological constraints on the tectonothermal evolution of the eastern Piedmont in South Carolina: Geological Society of America Bulletin, v. 97, p. 1329–1344.

Dewey, J. F., 1977, Suture zone complexities; A review: Tectonophysics, v. 40, p. 53–67.

Dewey, J. F., and Burke, K.C.A., 1973, Tibetan, Viriscan, and Precambrian basement reactivation; Products of continental collision: Journal of Geology, v. 81, p. 683–692.

Dewey, J. F., and Kidd, W. S., 1974, Continental collisions in the Appalachian Caledonian orogenic belt; Variations related to complete and incomplete suturing: Geology, v. 2, p. 543–546.

Dickerson, W. R., and Seely, D. R., 1979, Structure and stratigraphy of forearc regions: American Association of Petroleum Geologists Bulletin, v. 63, p. 2–31.

Diment, W. H., Urban, T. C., and Revetta, F. A., 1972, Some geophysical anomalies in the eastern United States, *in* Robertson, E., ed., The nature of the solid earth: New York, McGraw-Hill, p. 544–574.

Dorman, J., and Ewing, M., 1962, Numerical inversion of seismic surface wave dispersion data and crust-mantle structure in the New York-Pennsylvania area: Journal of Geophysical Research, v. 67, no. 13, p. 5227–5241.

Farrar, S. S., 1985, Tectonic evolution of the easternmost Piedmont, North Carolina: Geological Society of America Bulletin, v. 96, p. 362–380.

Fletcher, J. B., Sbar, M. L., and Sykes, L. R., 1978, Seismic trends and travel-time residuals in eastern North America and their tectonic implications: Geological Society of America Bulletin, v. 89, p. 1656–1676.

Fountain, D. M., 1976, The Ivrea-Verbano and Stronta-Ceneri zones, northern Italy; A cross section of the continental crust; New evidence from seismic velocities of rock samples: Tectonophysics, v. 33, p. 145–165.

Fyfe, W. S., 1986, Fluids in deep continental crust, *in* Reflection seismology; The continental crust: American Geophysical Union Geodynamics Series, v. 14, p. 33–39.

Giese, P., and Prodehl, C., 1976, Main features of crustal structure in the Alps, *in* Giese, P., Prodehl, C., and Stein, A., eds., Explosion seismology in central Europe; Data and results: New York, Springer, p. 347–375.

Green, D. H., and Ringwood, A. E., 1972, A comparison of recent experimental data on the gabbro-garnet granulite-eclogite transition: Journal of Geology, v. 80, p. 277–288.

Greenhouse, J. P., and Biley, R. C., 1981, A review of geomagnetic variation measurements in the eastern United States; Implications for continental tectonics: Canadian Journal of Earth Sciences, v. 18, p. 1268–1289.

Griscom, A., and Bromery, R. W., 1968, Geologic interpretation of aeromagnetic data for New England, *in* Zen, E-An, White, W. S., Hadley, J. B., and Thompson, J. B., Studies of Appalachian geology; Northern and Maritime: New York, Interscience, p. 425–436.

Harris, L. D., and Bayer, K. C., 1979, Sequential development of the Appalachian orogen above a master décollement; A hypothesis: Geology, v. 7, p. 568–572.

Hatcher, R. D., Jr., 1971, Structural, petrologic, and stratigraphic evidence favoring a thrust solution to the Brevard problem: American Journal of Science, v. 270, p. 177–202.

—— , 1978, Tectonics of the western Piedmont and Blue Ridge, southern Appalachians; Review and speculation: American Journal of Science, v. 78, ... p. 276–304.

Hatcher, R. D., Jr., and Zietz, I., 1980, Tectonic implications of regional aeromagnetic and gravity data from the southern Appalachians, *in* Wones, D. R., ed., Proceedings, The Caledonides in the U.S.A.; IGCP Project 27: Blacksburg, Virginia Polytechnic Institute and State University Memoir 2, p. 235–244.

Horton, J. W., Zietz, I., and Neathery, T. L., 1984, Truncation of the Appalachian Piedmont beneath the Coastal Plain of Alabama; Evidence from new magnetic data: Geology, v. 12, p. 51–55.

Hutchinson, D. R., Grow, J. A., and Klitgord, K. D., 1983, Crustal structure beneath the southern Appalachians; Nonuniqueness of gravity modeling: Geology, v. 11, p. 611–615.

Hutchinson, D. R., Grow, J. A., Klitgord, K. D., and Detrick, R. S., 1986, Moho reflections from the Long Island Platform, eastern United States, *in* Barazangi, M., and Brown, L., eds., Reflection seismology; The continental crust: American Geophysical Union Geodynamics Series, v. 14, p. 173–187.

Iverson, W. P., and Smithson, S. B., 1982, Master décollement root zone beneath the southern Appalachians and crustal balance: Geology, v. 10, p. 241–245.

—— , 1983a, Reprocessed COCORP southern Appalachian reflection data; Root zone to Coastal Plain: Geology, v. 11, p. 422–425.

—— , 1983b, Reprocessing and reinterpretation of COCORP southern Appalachian profiles: Earth and Planetary Science Letters, v. 62, p. 75–90.

James, D. E., Smith, T. J., and Steinhart, J. S., 1968, Crustal structure of the Middle Atlantic states: Journal of Geophysical Research, v. 73, p. 1983–2007.

Jaupart, C., 1983, Horizontal heat transfer due to radioactivity contrasts; Causes and consequences of the linear heat flow relation: Geophysical Journal of the Royal Astronomical Society, v. 75, p. 411–435.

Jaupart, C., Mann, J. R., and Simmons, G., 1982, A detailed study of the distribution of heat flow and radioactivity in New Hampshire, U.S.A.: Earth and Planetary Science Letters, v. 59, p. 267–287.

Jordan, T. H., and Frazer, L. N., 1975, Crustal and upper-mantle structure from SP phases: Journal of Geophysical Research, v. 80, p. 1504–1518.

Kane, M. F., Yellin, M. J., Bell, K. G., and Zietz, I., 1972, Gravity and magnetic evidence of lithology and structure in the Gulf of Maine region: U.S. Geological Investigations Map 726-B.

Kasameyer, P. W., 1974, Low-frequency magnetotelluric survey of New England [Ph.D. thesis]: Cambridge, Massachusetts Institute of Technology, 201 p.

Katz, S., 1955, Seismic study of crustal structure in Pennsylvania and New York: Seismological Society of America Bulletin, v. 45, p. 303–325.

Kay, S. M., and Kay, R. W., 1985, Role of crystal cumulates and the oceanic crust in the formation of the lower crust of the Aleutian arc: Geology, v. 13, p. 461–464.

Kean, A. E., and Long, L. T., 1980, A seismic refraction line along the axis of the southern Piedmont and crustal thicknesses in the southeastern United States: Earthquake Notes, v. 51, p. 3–13.

Keller, G. R., Bland, A. E., and Greenberg, J. K., 1982, Evidence in the eastern midcontinent region, United States: Tectonics, v. 1, p. 213–223.

Keller, G. V., 1966, Dipole method for deep resistivity studies: Geophysics, v. 31, p. 1088–1104.

Kennedy, M. J., 1976, Southeastern margin of the northeastern Appalachians; Late Precambrian orogeny on the continental margin: Geological Society of America Bulletin, v. 87, p. 1317–1325.

Kent, D. V., and Opdyke, N. D., 1978, Paleomagnetism of the Devonian Catskill red beds; Evidence for motion of the coastal New England–Canadian Maritime region relative to cratonic North America: Journal of Geophysical Research, v. 83, p. 1441–1450.

King, A. F., 1980, The birth of the Caledonides; Late Precambrian rocks of the Avalon peninsula, Newfoundland, and their correlatives in the Appalachian orogen, *in* Wones, D. R., ed., Proceedings, The Caledonides in the U.S.A.; IGCP Project 27: Blacksburg, Virginia Polytechnic and State University Memoir 2, p. 3–8.

King, E. R., and Zietz, I., 1978, The New York–Alabama lineament; Geophysical evidence for a major crustal break in the basement beneath the Appalachian basin: Geology, v. 6, p. 312–318.

King, P. B., 1969, Tectonic map of North America: U.S. Geological Survey Map G-67154, scale 1:5,000,000.

Klemperer, S. L., and Luetgert, J. H., 1987, Seismic reflection NMO and CDP processing applied to conventional refraction data from coastal Maine: Seismological Society of America Bulletin, v. 77, p. 614–630.

Klemperer, S. L., Brown, L. D., Oliver, J. E., Ando, C. J., Czuchra, B. L., and Kaufman, S., 1985, Some results of COCORP seismic reflection profiling in the Grenville-age Adirondack mountains, New York State: Canadian Journal of Earth Sciences, v. 22, p. 141–153.

Kumarpeli, P. S., Goodacre, A. K., and Thomas, M. D., 1981, Gravity and magnetic anomalies of the Sutton Mountains region, Quebec and Vermont; Expressions of rift volcanics related to the opening of Iapetus: Canadian Journal of Earth Sciences, v. 18, p. 680–692.

Langston, C. A., 1979, Structure under Mount Rainier, Washington, inferred from teleseismic body waves: Journal of Geophysical Research, v. 84, p. 4749–4762.

Langston, C. A., and Isaacs, C. M., 1981, A crustal thickness constraint for central Pennsylvania: Earthquake Notes, v. 52, p. 13–22.

Lee, C. K., and Dainty, A. M., 1982, Seismic structure of the Charlotte and Carolina Slate Belts of Georgia and South Carolina: Earthquake Notes, v. 53, p. 23–38.

Leet, D., 1941, Trial travel times for northeastern America: Seismological Society of America Bulletin, v. 31, p. 325–334.

LeHuray, A. P., 1986, Isotopic evidence for a tectonic boundary between the Kings Mountain belt and Inner Piedmont belts, southern Appalachians: Geology, v. 14, p. 784–787.

Leo, G. W., 1985, Trondhjemite and metamorphosed quartz keratophyre tuff of the Ammonoosuc volcanics (Ordovician), western New Hampshire and adjacent Vermont and Massachusetts: Geological Society of America Bulletin, v. 96, p. 1493–1507.

Lillie, R. J., 1984, Tectonic implications of subthrust structures revealed by seismic profiling of Appalachian-Ouachita orogenic belt: Tectonics, v. 3, no. 6, p. 619–646.

Loiselle, M., 1978, Geochemistry and petrogenesis of the Belknap Mountains complex and Plinz Range, White Mountain series, New Hampshire [Ph.D. thesis]: Cambridge, Massachusetts Institute of Technology, 320 p.

Loiselle, M. C., and Ayuso, R. A., 1979, Geochemical characteristics of granitoids across the Merrimack synclinorium, eastern and central Maine, *in* Wones, D. R., ed., Proceedings, The Caledonides in the U.S.A.; IGCP Project 27: Blacksburg, Virginia Polytechnic Institute and State University Memoir 2, p. 117–121.

Long, L. T., 1979, The Carolina Slate Belt; Evidence of a continental rift zone: Geology, v. 7, p. 180–184.

Long, L. T., and Liow, J. S., 1986, Crustal thickness, velocity structure, and the isostatic response function in the southern Appalachians, *in* Barazangi, M., and Brown, L., Reflection seismology; The continental crust: American Geophysical Union Geodynamics Series, v. 14, p. 215–222.

Long, L. T., and Mathur, U. P., 1972, Southern Appalachian crustal structure from the dispersion of Rayleigh waves and refraction data: Earthquake Notes, v. 43, p. 31–39.

Luetgert, J. H., 1985, Depth to Moho and characterization of the crust in the northern Appalachians from 1984 Maine-Quebec seismic refraction data: EOS Transactions of the American Geophysical Union, v. 66, p. 1074.

Luetgert, J. H., Mann, C. E., and Klemperer, S. L., 1987, Wide-angle deep crustal reflections in the northern Appalachians: Geophysical Journal of the Royal Astronomical Society, v. 89, p. 183–188.

Lyons, J. A., Forsyth, D. A., and Mair, J. A., 1980, Crustal studies in the La Malbaie region, Quebec: Canadian Journal of Earth Sciences, v. 17, p. 478–490.

Madden, T. R., 1971, The resolving power of geoelectic measurements for delineating resistive zones within the crust, *in* The structure and physical properties of the earth's crust: American Geophysical Union Monograph 14, p. 95–105.

Manghnani, M. H., Ramananantoandro, R., and Clark, S. P., 1974, Compressional and shear wave velocities in granulite facies rocks and eclogites to 10 kbar: Journal of Geophysical Research, v. 79, p. 5427–5446.

McKerrow, W. S., and Cocks, L.R.M., 1977, The location of the Iapetus Ocean suture in Newfoundland: Canadian Journal of Earth Sciences, v. 14, p. 488–495.

McLelland, J., and Isachsen, Y., 1980, Structural synthesis of the southern and central Adirondacks; A model for the Adirondacks as a whole and plate-tectonics interpretations: Geological Society of America Bulletin, Part 1, v. 91, p. 68–72; Part 2, v. 91, p. 208–292.

Misra, K. C., and Keller, F. B., 1978, Ultramafic bodies in the southern Appalachians; A review: American Journal of Science, v. 278, p. 389–418.

Mitchell, B. J., and Herrmann, R. B., 1979, Shear velocity structure in the eastern United States from the inversion of surface-wave group and phase velocities: Seismological Society of America Bulletin, v. 69, no. 4, p. 1133–1148.

Nakamura, Y., and Howell, B. F., 1964, Maine seismic-experiment frequency spectra of refraction arrivals and the nature of the Mohorovičić discontinuity: Seismological Society of America Bulletin, v. 54, p. 9–18.

Naylor, R. S., 1971, Acadian orogeny; An abrupt and brief event: Science, v. 172, p. 558–560.

——, 1975, Age provinces in the northern Appalachians: Earth and Planetary Science Letters Annual Review, v. 3, p. 387–400.

Nelson, A. E., 1976, Structural elements and deformational history of rocks in eastern Massachusetts: Geological Society of America Bulletin, v. 87, p. 1377–1383.

Nelson, K. D., and 8 others, 1985a, New COCORP profiling in the southeastern United States; Part 1, Late Paleozoic suture and Mesozoic rift basin: Geology, v. 13, p. 714–718.

Nelson, K. D., McBride, J. H., Arnow, J. A., Oliver, J. E., Brown, L. D., and Kaufman, S., 1985b, New COCORP profiling in the southeastern United States; Part 2, Brunswick and East Coast magnetic anomalies, opening of the north-central Atlantic Ocean: Geology, v. 13, p. 718–721.

Nielson, D. L., Clark, R. G., Lyons, J. B., England, E. J., and Burns, D. J., 1976, Gravity models and mode of emplacement of the New Hampshire plutonic series, *in* Lyons, P. C., and Brownlow, A. H., eds., Studies in New England geology: Geological Society of America Memoir 146, p. 301–318.

O'Connell, R. J., and Budiansky, B., 1977, Viscoelastic properties of fluid-saturated cracked solids: Journal of Geophysical Research, v. 82, p. 5719–5735.

Odom, A. L., and Fullagar, P. D., 1973, Geochronologic and tectonic relationships between the inner Piedmont, Brevard zone, and Blue Ridge belts, North Carolina: American Journal of Science, v. 273-A, p. 133–149.

O'Hara, K., and Gromet, L. P., 1985, Two distinct late Precambrian (Avalonian) terranes in southeastern New England and their late Paleozoic juxtaposition: American Journal of Science, v. 285, p. 673–709.

Oliver, J., Kovach, R., and Dorman, J., 1961, Crustal structure of the New York-Pennsylvania area: Journal of Geophysical Research, v. 66, p. 215–225.

Osberg, P. H., 1978, Synthesis of the geology of the northeastern Appalachians, U.S.A.; IGCP Project 27: U.S. Geological Survey Professional Paper 78-13, p. 137–147.

Owens, T. J., 1984, Crustal structure of the Adirondack Mountains determined from broad-band teleseismic waveform modeling: Earthquake Notes, v. 55, p. 78.

Owens, T. J., Taylor, S. R., and Zandt, G., 1983, Isolation and enhancement of the response of local seismic structure from teleseismic P-waveforms: Lawrence Livermore National Laboratory Report UCID-19809, 33 p.

——, 1984, Crustal structure from teleseismic P-waveforms; Preliminary results at RSCP, RSSD, and RSNY: Lawrence Livermore National Laboratory Report UCID-19859, 28 p.

Owens, T. J., Zandt, G., and Taylor, S. R., 1984, Seismic evidence for an ancient rift beneath the Cumberland Plateau, Tennessee; A detailed analysis of broad-band teleseismic P-waveforms: Journal of Geophysical Research, v. 89, p. 7783–7795.

Owens, T. J., Taylor, S. R., and Zandt, G., 1987, Summary of crustal structure at RSTN sites determined from broadband teleseismic P-waveforms: Seismological Society of America Bulletin, v. 77, p. 631–662.

Phinney, R. A., 1986, A seismic cross section of the New England Appalachians; The orogen exposed, *in* Barazangi, M., and Brown, L., Reflection seismology; The continental crust: American Geophysical Union Geodynamics Series, v. 14, p. 157–172.

Poupinet, G., 1979, On the relation between P-wave travel-time residuals and the age of continental plates: Earth and Planetary Science Letters, v. 43, p. 149–161.

Prodehl, C., Schlittenhardt, J., and Stewart, S. W., 1984, Crustal structure of the Appalachian highlands in Tennessee: Tectonophysics, v. 109, p. 61–76.

Putman, G. W., and Sullivan, J. W., 1979, Granitic pegmatites as estimators of crustal pressures; A test in the eastern Adirondacks, New York: Geology, v. 7, p. 549–553.

Rankin, D. W., 1976, Appalachian salients and recesses; Late Precambrian continental breakup and the opening of the Iapetus Ocean: Journal of Geophysical Research, v. 81, p. 5605–5619.

Rast, N., 1980, The Avalonian plate in the northern Appalachians and Caledonides, *in* Wones, D. R., ed., Proceedings, The Caledonides in the U.S.A.; IGCP Project 27: Blacksburg, Virginia Polytechnic Institute and State University Memoir 2, p. 63–66.

Robinson, P., and Hall, L. M., 1980, Tectonic synthesis of southern New England, *in* Wones, D. R., ed., Proceedings, The Caledonides in the U.S.A.; IGCP Project 27: Blacksburg, Virginia Polytechnic Institute and State University Memoir 2, p. 73–82.

Rodgers, J., 1970, The tectonics of the Appalachians: New York, Wiley-Interscience, 271 p.

Schnerk, R., Aggarwal, Y. P., Golisono, M., and England, F., 1976, Regional Seismological Bulletin of the Lamont-Doherty Network: Palisades, New York, Lamont-Doherty Geological Observatory, 15 p.

Secor, D. T., Snoke, A. W., and Dallmeyer, R. D., 1986, Character of the Alleghanian orogeny in the southern Appalachians; Part 3, Regional tectonic relations: Geological Society of America Bulletin, v. 97, p. 1345–1353.

Shankland, T. J., and Ander, M. E., 1983, Electrical conductivity, temperatures, and fluids in the lower crust: Journal of Geophysical Research, v. 88, p. 9475–9484.

Simmons, G., 1964, Gravity survey and geological interpretation, northern New York: Geological Society of America Bulletin, v. 75, p. 81–98.

Simpson, R. W., Shride, A. F., and Bothner, W. A., 1980, Offshore extension of the Clinton-Newbury and Bloody Bluff fault systems of northeastern Massachusetts, *in* Wones, D. R., ed., Proceedings, The Caledonides in the U.S.A.; IGCP Project 27: Blacksburg, Virginia Polytechnic Institute and State University Memoir 2, p. 229–233.

Skehan, J. W., 1968, Fracture tectonics of southeastern New England as illustrated by the Wachusett-Marlborough Tunnel, *in* Zen, E-An, White, W. S., Hanley, J. B., and Thompson, J. B., eds., Studies in Appalachian geology; Northern and Maritime: New York, Wiley-Interscience, p. 281–290.

Skehan, J. W., and Murray, D. P., 1980, Geologic profile across southeastern New England: Tectonophysics, v. 69, p. 285–319.

Stanley, R. S., and Ratcliffe, N. M., 1985, Tectonic synthesis of the Taconian orogeny in western New England: Geological Society of America Bulletin, v. 96, p. 1227–1250.

Steinhart, J. S., and Meyer, R. P., 1961, Explosion studies of continental structure: Carnegie Instition of Washington Publication 622, p. 409.

Steinhart, J. S., Green, R., Asada, T., Rodriguez, B. A., Aldrich, L. T., and Tauve, M. A., 1962, Seismic studies: Carnegie Institution of Washington Yearbook

61, p. 221–234.

Stewart, D. B., and 8 others, 1986, The Quebec–western Maine seismic reflection profile; Setting and first-year results, *in* Barazangi, M., and Brown, L., eds., Reflection seismology; The continental crust: American Geophysical Union Geodynamics Series, v. 14, p. 189–200.

St. Julien, P., and Hubert, C., 1975, Evolution of the Taconian orogen in the Quebec Appalachians: American Journal of Science, v. 275A, p. 337–362.

St. Julien, P., Slivitsky, A., and Feininger, T., 1983, A deep structural profile across the Appalachians in southern Quebec, *in* Hatcher, R. D., Jr., Williams, H., and Zeitz, I., Contributions to the tectonics and geophysics of mountain chains: Geological Society of America Memoir 158, p. 103–113.

Strong, D. F., Dickson, W. L., O'Driscoll, L. F., Kean, B. F., and Stevens, R. K., 1974, Geochemical evidence for east-dipping Appalachian subduction zone in Newfoundland: Nature, v. 248, p. 37–39.

Taylor, S. R., 1980, Crust and upper mantle structure of the northeastern United States [Ph.D. thesis]: Cambridge, Massachusetts Institute of Technology, 288 p.

Taylor, S. R., and Owens, T. J., 1984, Frequency-domain inversion of receiver functions for crustal structure: Earthquake Notes, v. 55, p. 5–12.

Taylor, S. R., and Toksoz, M. N., 1979, Three-dimensional crust and upper-mantle structure of the northeastern United States: Journal of Geophysical Research, v. 84, p. 7627–7644.

——, 1982a, Crust and upper-mantle structure in the Appalachian orogenic belt; Implications for tectonic evolution: Geological Society of America Bulletin, v. 93, p. 315–329.

——, 1982b, Structure in the northeastern United States from inversion of Rayleigh wave phase and group velocities: Earthquake Notes, v. 53, no. 4, p. 5–24.

Taylor, S. R., Simmons, G., and Barosh, P., 1980a, A gravity survey of the Clinton-Newbury and Bloody Bluff fault zones in eastern Massachusetts: Geological Society of America Abstracts with Programs, v. 12, p. 86.

Taylor, S. R., Toksoz, M. N., and Chaplin, M. P., 1980b, Crustal structure of the northeastern United States; Contracts between the Grenville and Appalachian provinces: Science, v. 208, p. 595–597.

Thompson, B. G., Nekut, A., and Kuckes, A. F., 1983, A deep crustal electromagnetic sounding in the Georgia Piedmont: Journal of Geophysical Research, v. 88, p. 9461–9473.

Volz, W., 1979, Travel time perturbations in the crust and upper mantle in the southwest Piedmont [M.S. thesis]: Atlanta, Georgia Institute of Technology, 198 p.

Warren, D. J., 1968, Transcontinental geophysical survey (35°–39°N) seismic refraction profiles of the crust and upper mantle from 74° to 87°W longitude: U.S. Geological Survey Miscellaneous Investigations Map I-535-D.

Wehr, F., and Glover, L., 1985, Stratigraphy and tectonics of the Virginia–North Carolina Blue Ridge; Evolution of a late Proterozoic–early Paleozoic hinge zone: Geological Society of America Bulletin, v. 96, p. 285–295.

Williams, H., 1979, Appalachian orogen in Canada: Canadian Journal of Earth Sciences, v. 16, p. 792–807.

——, 1980, Comments on "Thin-skinned tectonics in the crystalline southern Appalachians; COCORP seismic-reflection profiling of the Blue Ridge and Piedmont" by Cook and others: Geology, v. 8, p. 211–212.

Williams, H., and Hatcher, R. D., Jr., 1982, Suspect terranes and accretionary history of the Appalachian orogen: Geology, v. 10, p. 530–536.

Willmore, P. L., and Bancroft, A. M., 1960, The time term approach to refraction seismology: Geophysical Journal, v. 3, p. 419–432.

Zandt, G., and Owens, T. J., 1986, Comparison of crustal velocity profiles determined by seismic refraction and teleseismic methods: Tectonophysics, v. 128, p. 155–161.

Zartman, R. E., and Naylor, R. S., 1984, Structural implications of some radiometric ages of igneous rocks in southeastern New England: Geological Society of America Bulletin, v. 95, p. 522–539.

Zen, E., 1972, The Taconide zone and the Taconic orogeny in the western part of the northern Appalachian orogen: Geological Society of America Special Paper 135, 72 p.

Zietz, I., and Gilbert, F., 1980, Aeromagnetic map of part of the southeastern United States: U.S. Geological Survey Geophysical Investigations Map GP-936, scale 1:2,000,000.

Zietz, I., Gilbert, F., and Kirby, J. R., 1972, Northeastern United States regional aeromagnetic maps: U.S. Geological Survey Open-File map, 13 sheets, scale 1:250,000.

Manuscript Accepted by the Society October 31, 1988

Printed in U.S.A.

Geological Society of America
Memoir 172
1989

Chapter 17

Atlantic and Gulf of Mexico continental margins

A. M. Trehu* and K. D. Klitgord
U.S. Geological Survey, Woods Hole, Massachusetts 02543
D. S. Sawyer and R. T. Buffler**
Institute for Geophysics, University of Texas at Austin, Austin, Texas 78751

ABSTRACT

The U.S. Atlantic and Gulf of Mexico continental margins are thickly sedimented passive margins that formed when Pangea split apart during Middle Jurassic time to create the Atlantic Ocean and Gulf of Mexico. Keys to understanding the process of continental breakup and its relation to preexisting structure are found in the structure of the crust beneath the sediment-filled basins and adjacent platforms and embayments that outline the margins. Because of the great thickness of post-rift sedimentary rock in the basins and the presence of massive reef carbonates and salt layers and diapirs, the crustal structure beneath the basins in the region of the transition between oceanic and continental crust is poorly known at present. Recent advances in seismic reflection and refraction data collection techniques (both sources and receivers), however, are just beginning to yield new data to look at the crustal structure in this important region. One of the most important results on the deep structure of continental margins obtained in recent years is recognition of a thick, high-velocity (7.2 to 7.5 km/sec) layer at the base of the crust beneath the U.S. Atlantic continental margin as well as beneath several other margins worldwide. This layer is observed beneath both extended continental crust and early oceanic crust and has been interpreted to indicate that extensive intrusive magmatism was associated with the late stage of rifting and early sea-floor spreading. Only a weak suggestion of such a layer has been observed beneath the Gulf of Mexico margin, although this may be in part due to the difficulty of observing lower crustal arrivals because of the extensive presence of salt in the shallow section.

INTRODUCTION

During the Paleozoic construction of Pangea, a broad band of accreted terranes accumulated between the continental plates of North America and South America/Africa (Williams and Hatcher, 1983). Rift basins formed within this broad zone during early Mesozoic time as Pangea began to break apart. Crustal modification by extension continued along a narrower zone that became the site of the present deep marginal basins, and eventually oceanic crust was formed by sea-floor spreading just seaward of these marginal basins. Keys to understanding the process of continental breakup and its relationship to preexisting structure are found in the structure of the crust under the sediment-filled basins and adjoining basement platforms and embayments that presently outline the margin.

The present U.S. Atlantic and Gulf of Mexico continental margins encompass a series of sedimentary basins and crustal platforms and embayments (Fig. 1). Because the basement structure has little correlation with the bathymetry, these features have been recognized primarily on the basis of drillholes and numerous geophysical studies. The stratigraphy, structure, and geologic history of the basins have recently been treated in detail in Poag

*Present address: College of Oceanography, Oregon State University, Corvallis, Oregon 97331.
**Present address: Department of Geology and Geophysics, Rice University, Houston, Texas 77251.

Trehu, A. M., Klitgord, K. D., Sawyer, D. S., and Buffler, R. T., 1989, Atlantic and Gulf of Mexico continental margins, *in* Pakiser, L. C., and Mooney, W. D., Geophysical framework of the continental United States: Boulder, Colorado, Geological Society of America Memoir 172.

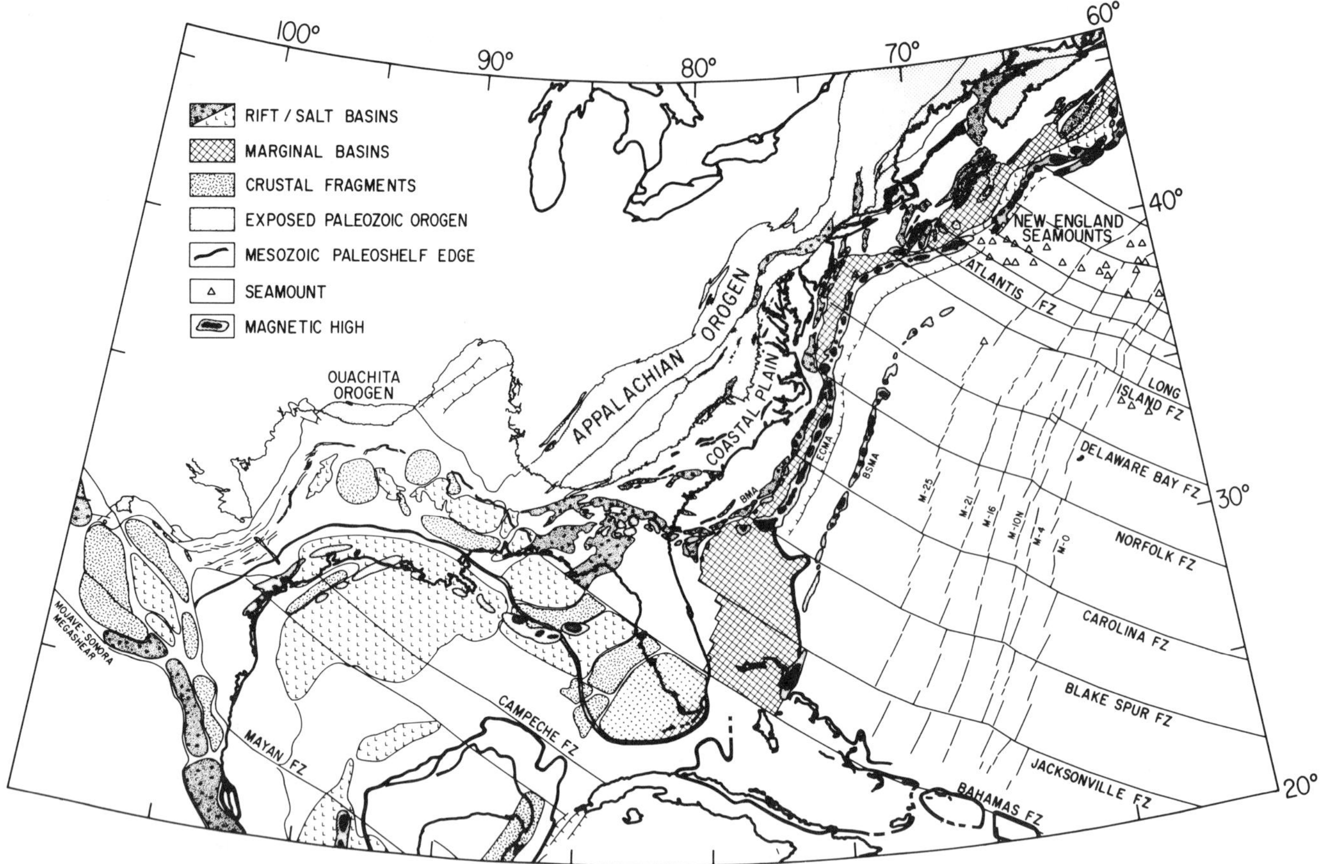

Figure 1. Map of the U.S. Atlantic and Gulf of Mexico continental margins showing the location of structural features and geophysical lineaments. Modified from Pilger (1981), Klitgord and others (1984, Fig. 1; 1988; Plate 2c), and Buffler and Sawyer (1985, Fig. 2).

(1985) and in volumes in the Geological Society of America *Geology of North America* series on the western North Atlantic Ocean (Vogt and Tucholke,1986), the U.S. Atlantic continental margin (Sheridan and Grow, 1988), and the Gulf of Mexico continental margin (A. Salvador, unpublished data).

In this chapter, the current state of knowledge of the crustal structure of the U.S. Atlantic and Gulf of Mexico continental margins is summarized. The discussion is necessarily weighted toward the Atlantic margin because more information on the crustal structure is available. A summary of the basement structural framework is presented in the second section based on geophysical techniques integrated with drillhole data and geologic mapping. In the third section, we evaluate what can be said about the structure of the crystalline crust from seismic reflection and refraction, magnetic, gravity, and subsidence data, and discuss the limitations of the data and modeling techniques. Broad regions of salt and carbonate bank deposits severely limit our ability to resolve the deeper crustal structure beneath both the Atlantic and Gulf of Mexico margins, but recent improvements in data collection and processing technology are slowly enabling us to penetrate these boundaries. We take a historical approach in order to emphasize the evolution of knowledge of margin structure and the interplay between technological and geological advances.

BASEMENT STRUCTURAL FRAMEWORK

Atlantic continental margin

The major crustal units of the Atlantic margin (continental crustal along the landward edge of the margin, deep marginal basins overlying transitional crust, and oceanic crust seaward of the marginal sedimentary basins; Fig. 2a) have been identified on the basis of characteristic signatures seen in magnetic anomaly patterns (Klitgord and Behrendt, 1979), structural features seen on seismic reflection lines (e.g., Klitgord and others, 1988), gravity models (e.g., Grow and others, 1979; Sheridan and others, 1988), and geographically limited large-offset seismic data (e.g., Sheridan and others, 1979; LASE study group, 1986; Trehu and others, 1986, 1989). A detailed summary of the structural framework of the U.S. Atlantic margin is given in Klitgord and others (1988).

A major basement feature of the margin, found along its entire length, is the "hinge zone" (Fig. 2b, c), within which the basement deepens steeply to the east by a series of down-dropped fault blocks. This hinge zone marks the western boundary of a series of sedimentary basins filled with thick wedges of Mesozoic and younger sedimentary rock that lap onto the block-faulted edge of continental crust to the west. The character of block faulting at the hinge zone varies along the margin and may reflect the influence of older crustal structure on the Mesozoic rifting (Bally, 1981; Klitgord and others, 1988). Half-graben structures with seaward-dipping border faults are found at the Georges Bank hinge zone, whereas faulted blocks with landward-dipping faults form the hinge zone in the Baltimore Canyon trough.

Landward of the hinge zone, shallow basement platforms and embayments contain early Mesozoic rift basins filled with Upper Triassic to Lower Jurassic clastic sedimentary rocks intruded by Jurassic igneous rock (Manspeizer, 1981, 1985; Robinson and Froelich, 1985; Klitgord and Hutchinson, 1985; Hutchinson and others, 1986b; Manspeizer and Cousminer, 1988). Rifting in this region ceased by Middle Jurassic time, and any further crustal modification was a secondary effect of continued extension in the marginal basins to the east. The amount of crustal thinning associated with this early rifting remains uncertain but was probably small. We refer to the crust in this region just landward of the hinge zone as "rifted continental crust."

Seaward of the hinge zone, the thick sedimentary wedge of the marginal basins obscures the basement structure. These marginal basins are known as the Scotian basin, Georges Bank basin, Baltimore Canyon trough, Carolina trough, Blake Plateau basin, and Bahamas basin (Fig. 1). Offsets between the basins are aligned with major oceanic fracture zones and represent zones of transform motion during rifting. North of the Blake Plateau, the seaward margin of these basins is marked by the East Coast magnetic anomaly (ECMA) (Klitgord and Behrendt, 1979). Although this prominent linear magnetic anomaly is generally thought to be near the landward edge of oceanic crust (Keen, 1969; Emery and others, 1970; Klitgord and Behrendt, 1979; Sheridan and others, 1979; Hutchinson and others, 1983; Alsop and Talwani, 1984), the details of its origin remain controversial (Klitgord and others, 1988). Why a similar anomaly is not present along the seaward edge of the Blake Plateau or along the southern edge of the U.S. Gulf of Mexico margin also remains a question. We refer to the crust between the hinge zone and the ECMA as "transitional" crust; it is also often referred to as "rift-stage" crust. This crust is derived from continental crust that was extensively deformed by brittle fracturing of the upper crust and by ductile stretching, underplating, and/or igneous intrusion of the lower crust during the final stage of rifting before sea-floor spreading began around 175 Ma (Klitgord and Schouten, 1986).

Although the existence of Atlantic marginal basins was suspected on the basis of seismic reflection and refraction profiles shot during the 1930s to 1960s (Drake and others, 1959; Emery and others, 1970), little was known of their internal structure and stratigraphy until the 1970s, when a series of multichannel seismic reflection lines was acquired across the margin by the U.S. Geological Survey (USGS) and others (Schlee and others, 1976; Folger and others, 1979). These studies revealed a number of structural similarities and differences among the basins. For example, basin widths range from about 50 km for the Carolina trough to more than 350 km for the Blake Plateau basin. Postrift sedimentary thicknesses vary from about 7 km in the Georges Bank basin to more than 13 km in the Baltimore Canyon trough and Blake Plateau basin (Klitgord and others, 1988). A zone of salt diapirs is found along the seaward margin of the Carolina trough (Dillon and others, 1983), Georges Bank basin (Schlee and Klitgord, 1988), and Scotian basin (Austin and others, 1980;

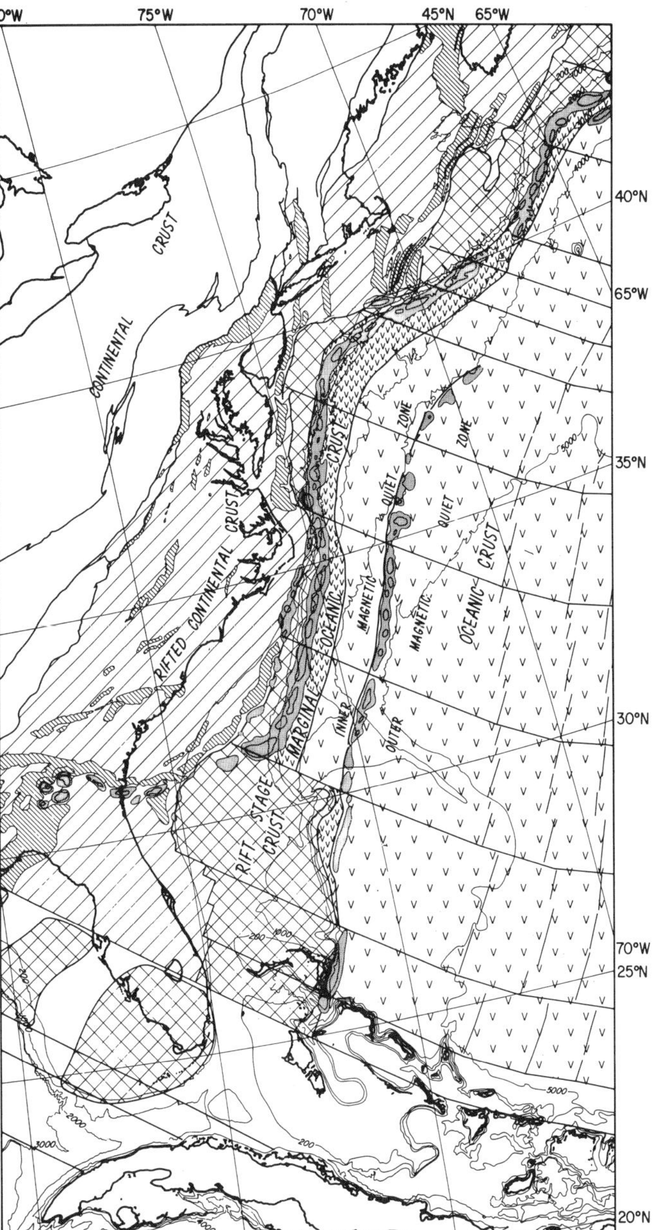

Figure 2A. Map of the U.S. Atlantic continental margin showing the boundaries of crustal provinces discussed in the text (from Klitgord and others, 1988; Figure 15).

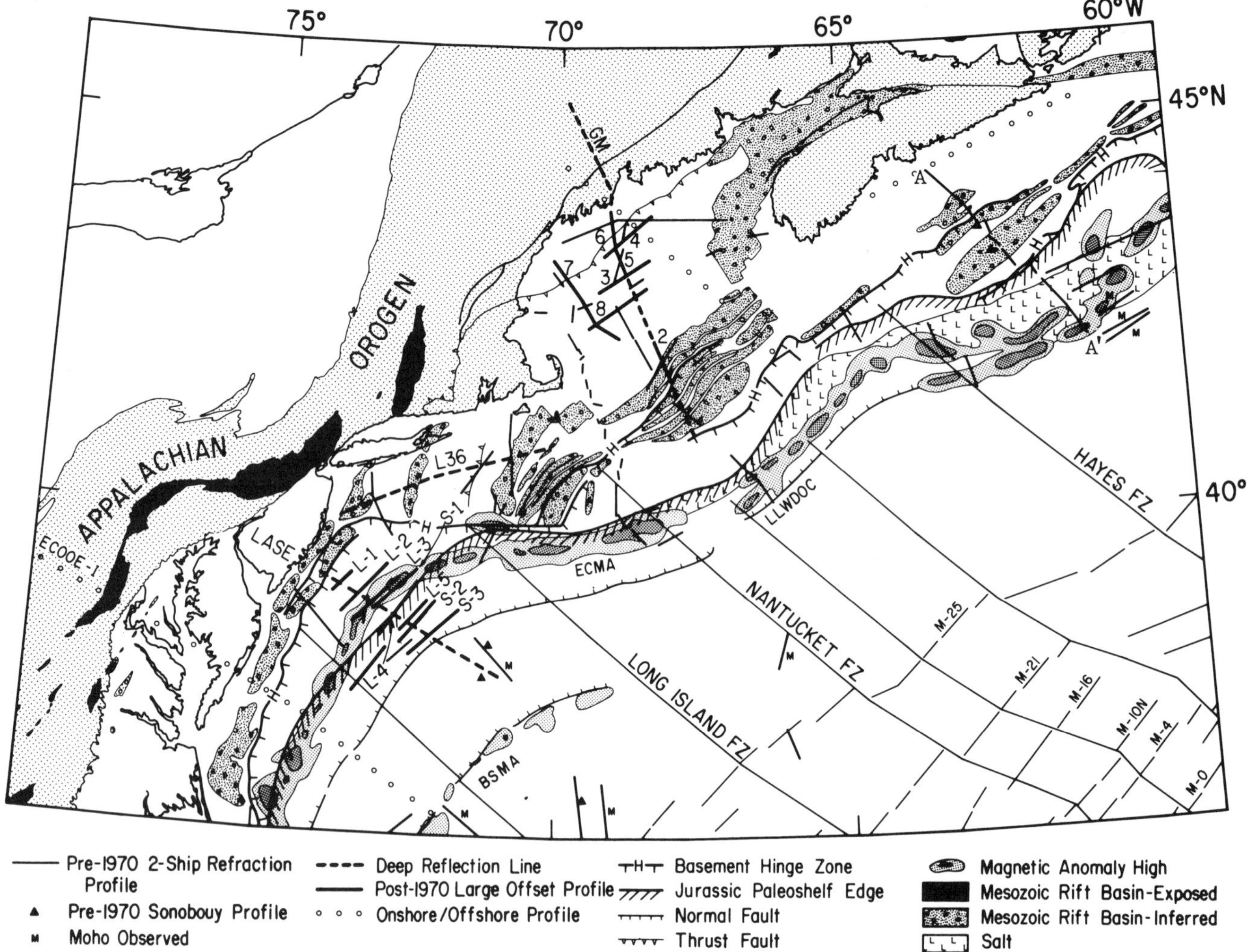

Figure 2B. Map of the U.S. Atlantic continental margin north of 36°N showing the location of seismic refraction studies and major tectonic features. The letter M next to a pre-1970 profile indicates that Moho arrivals were observed; for post-1970 data, results are discussed in the text. A–A′ and B–B′ refer to the composite cross sections shown in Figure 6. ECMA = East Coast magnetic anomaly; BSMA = Blake Spur magnetic anomaly; BMA = Brunswick magnetic anomaly; BE = Blake Escarpment; LLWDOC = landward limit of well-defined oceanic crust; ECOOE = East Coast onshore-offshore experiment. Lines from recent experiments are numbered to facilitate referencing specific results in the text. LASE = Large Aperture Seismic Experment; GM = Gulf of Maine experiment; CT = Carolina trough experiment; NAT = North Atlantic transect.

Jansa and Wade, 1975), whereas seismic evidence of salt is absent in the Blake Plateau basin and only three salt diapirs have been mapped in the Baltimore Canyon trough (Grow, 1980). These differences are probably due to differences in preexisting crustal structure and variations in sediment source regimes.

Because of the thick sediment-fill in the basins, the structure of the underlying crystalline crust (and, in places, even the depth to the pre-rift basement) is poorly known. The problem of imaging the basement using seismic techniques is increased by the presence of high-velocity material, interpreted to represent a prograding carbonate bank-reef complex marking the paleoshelf edge, on the seaward side of the basins (Schlee and others, 1976; Grow and others, 1979; Sheridan and others, 1979; Poag, 1985). This carbonate structure effectively masks the deeper structure in the region of the transitional to oceanic crustal boundary. Until recently, most of our information on crustal structure in this region was from magnetic depth-to-basement estimates and poorly constrained gravity studies. This lack of knowledge of the thickness and composition of the transitional crust beneath the basins is a major uncertainty that limits further progress on modeling the process of rifting and subsequent subsidence history (Celerier, 1986). Recent large-aperture seismic experiments, however, are beginning to yield additional constraints (LASE study group, 1986; Trehu and others, 1986, 1989).

The region between the ECMA and the Blake Spur magnetic anomaly (BSMA) is referred to as the inner magnetic quiet

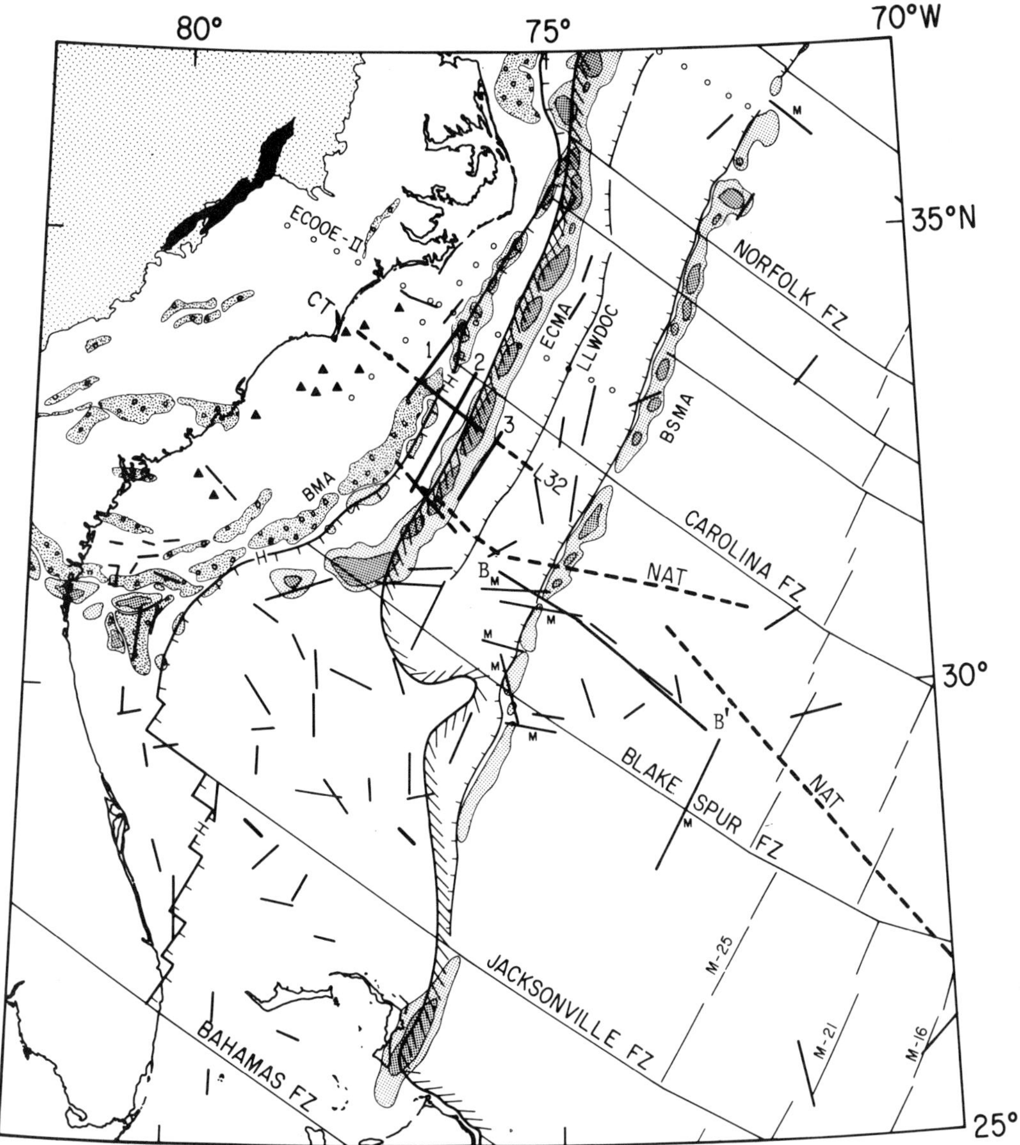

Figure 2C. Map of the U.S. Atlantic continental margin south of 37°N showing the location of seismic-refraction studies and major tectonic features. See Figure 2B caption for explanation of symbols.

zone because of the low amplitude of the sea-floor spreading magnetic anomalies and has generally been treated as a single crustal unit (Vogt, 1973; Sheridan and others, 1979). In this chapter, we divide this region into two parts. The landward limit of well-defined oceanic crust is defined on the basis of the hyperbolic reflections characteristic of oceanic basement and crustal velocity measurements and is found tens of kilometers seaward of the ECMA (Klitgord and Grow, 1980; Sheridan and others, 1979; Klitgord and others, 1988). At this boundary (Fig. 2a, b, c), the basement steps down to the northwest, and the acoustic reflector J3 laps onto the basement from the west (Klitgord and Grow, 1980). This basement step is called the J3 scarp by Klitgord and others (1988). Offshore Georges Bank, seaward-dipping reflectors characterize the upper part of the crust in the region between the ECMA and the J3 scarp. South of Georges Bank, the very reflective sedimentary horizon, J3, masks the seismic character of the underlying basement between the ECMA and the J3 scarp. On other passive margins with less sediment cover than the U.S. Atlantic margin, a zone of thick crust with oceanic velocities and characterized by seaward-dipping reflectors has been observed between the marginal basin zone and the ocean crust and has been interpreted to represent lava flows interbedded with clastic and volcanic sedimentary material that were formed during the early stages of sea-floor spreading (Hinz, 1981; Mutter and others, 1984; Mutter, 1985). This interpretation is consistent with results from the Ocean Drilling Program on the Norwegian margin (Leg 104 Scientific Party, 1986). This region between the ECMA and the J3 scarp is referred to herein as "marginal oceanic crust."

The inner magnetic quiet zone is bounded on the east by the

BSMA. This anomaly coincides with a basement ridge in the oceanic crust (Klitgord and Grow, 1980; Tucholke and others, 1982) and has been interpreted to reflect an eastward jump of the spreading center approximately 170 Ma (Vogt, 1973; Klitgord and Grow, 1980; Klitgord and Schouten, 1986). The BSMA converges with the ECMA seaward of the Georges Bank basin (Klitgord and Schouten, 1986). The amplitude of the BSMA is greatest along the seaward edge of the Blake Plateau, where the ECMA is absent. In this region, the BSMA may represent the boundary between transitional crust and oceanic crust (Sheridan and others, 1981).

Gulf of Mexico

The Gulf of Mexico region (Fig. 3) contains a broad Mesozoic sedimentary basin, called the Gulf basin, that includes the present Gulf of Mexico basin and the adjacent rift basins. Regionalization of crustal provinces in much of the Gulf basin is speculative because the basement structure is masked over broad regions by carbonate bank deposits or zones of salt mobilization.

The Bahamas fracture zone, which trends northwest across southern Florida and the West Florida shelf, forms the northeast margin of the Gulf basin and is a Jurassic transform boundary that joined the Gulf of Mexico spreading center to the Atlantic spreading center (Klitgord and others, 1984). Onshore, it joins into the Gulf Rim fault zone, which is a composite fault and graben system that trends northwest across southern Alabama, central Mississippi, northern Louisiana, and central Texas; it forms the northern and northwestern boundary of the Gulf basin (Anderson, 1979; Martin, 1980; Pilger, 1981). This fault zone has been defined to include the Luling, Mexia, Talco, South Arkansas, Pickens, and Gilbertown fault and graben systems. The fault zone across the Florida Panhandle connecting the Gilbertown fault zone with the Bahamas fracture zone has been referred to as the Jay fault by Smith (1983). Near the Texas-Oklahoma border, the Gulf Rim fault zone swings around to the southwest and forms the western edge of the Gulf basin. On the basis of plate-tectonic reconstructions, Klitgord and Schouten (1986) predicted that the northeast part of the Gulf Rim fault zone was created primarily by shear motion, whereas the northwest part was created primarily by extension. In an alternative model, Pindell (1985) predicted initial extension along the northern and northeastern segments of the Gulf Rim and shearing along the northwestern segment.

During early Mesozoic time, a 200-km-wide zone of uplifted Paleozoic blocks and intervening salt basins developed just south of the Gulf Rim fault zone (Woods and Addington, 1973; Pilger, 1978, 1981; Klitgord and others, 1984; Pindell, 1985; Klitgord and Schouten, 1986). On the West Florida shelf, this region includes three basins (the South Florida, Tampa, and Appalachicola basins) and two Paleozoic arches (Sarasota and Middle Ground) overlain by a massive late Mesozoic and Tertiary carbonate platform (Ball and others, 1988; Klitgord and others, 1984). Onshore from Mississippi to Texas, there are at least five basins (Conecuh, Manila, Mississippi Salt, North Louisiana, and East Texas Salt basins) and five Paleozoic uplifted blocks (Wiggins Arch, Jackson Dome, Monroe Uplift, LaSalle Arch, and Sabine Uplift) (Anderson, 1979; Pilger, 1981). Buffler and Sawyer (1985) concluded that this region is underlain by thick transitional crust; it corresponds to the zone of rifted continental crust on the Atlantic margin.

No well-defined hinge zone separating rifted continental crust from rift-stage crust has been identified along the Gulf of Mexico margin, although Corso (1987) has suggested that one exists beneath the Lower Cretaceous shelf edge complex that developed along the southern edge of the zone of Paleozoic blocks and salt basins. This paleo-shelf edge is located along the West Florida Escarpment and extends onshore across southern Louisiana and southeastern Texas (Anderson, 1979).

The Gulf Coast Salt Dome Province is the region bounded to the north by the Lower Cretaceous paleo-shelf edge and to the south by the Sigsbee Escarpment (Buffler, 1983a, b). A thick wedge of Tertiary clastic deposits migrated seaward across this salt diapir province, mobilizing the salt and forming a complex pattern of basins, channels, and diapirs that characterizes the Texas-Louisiana slope region (Martin, 1980; Bouma, 1983). The eastern end of the northern Gulf Coast Salt Dome Province is overlain by the broad Plio-Pleistocene Mississippi Fan. The West Florida Salt basin is the southeastern continuation of this salt dome province along the base of the West Florida Escarpment.

Because of the thick sediments and extensive mobilization of salt, the underlying basement and crustal structure in this region is unknown. A broad zone of transitional crust between the Lower Cretaceous shelf edge and the Sigsbee Escarpment has been inferred from the extent and thickness of the sedimentary basins (Buffler and others, 1980, 1981, 1984; Buffler and Sawyer, 1985) and from magnetic anomaly patterns (Klitgord and others, 1984, 1988; Klitgord and Schouten, 1986). Two seismic refraction lines (Cram, 1961; Dorman and others, 1972) and Rayleigh wave–dispersion and S_n data (Keller and Shurbet, 1975; Shurbet, 1976) recorded at stations on the Gulf coastal plain also suggest the presence of thinned crust in this region. The estimated crustal thickness of 8 to 20 km (Buffler and Sawyer, 1985), however, is poorly constrained because the total sedimentary thickness is poorly known.

The deep central Gulf of Mexico basin is located seaward of the Sigsbee Escarpment, West Florida Escarpment, Campeche Escarpment, and Mexican Ridges (Shaub and others, 1984). Seismic refraction data indicate that most of this region is underlain by oceanic crust, although some transitional crust may be present in the southeastern part of the deep Gulf of Mexico (Buffler and others, 1981, 1984; Ibrahim and others, 1981). Paleogeographic plate reconstructions indicate that this crust was probably formed during Late Jurassic time (Buffler and others, 1980; Klitgord and Schouten, 1986), but there are no basement samples and the overlying Jurassic(?) sediments have not been drilled.

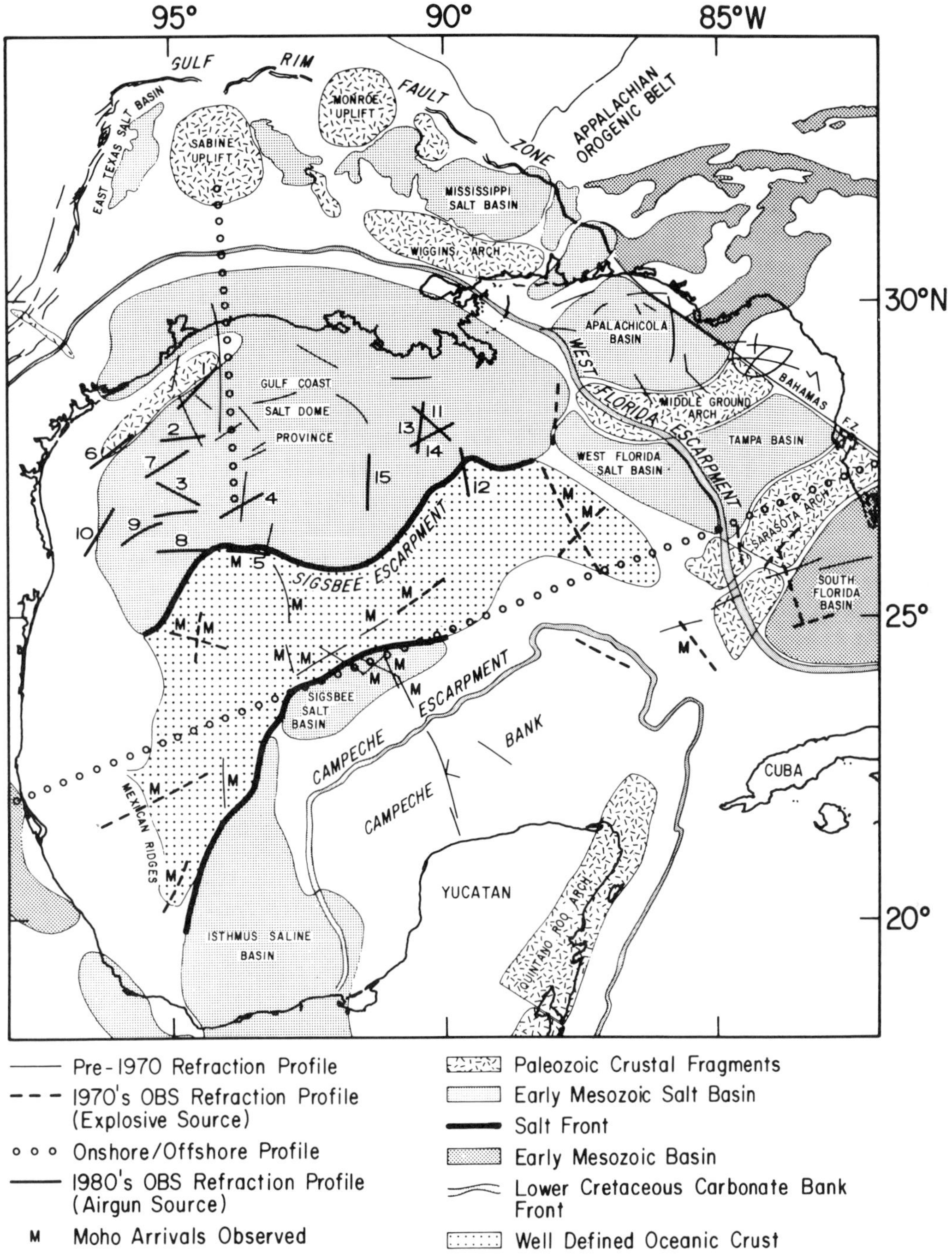

Figure 3. Map of the Gulf of Mexico showing the location of crustal provinces and major tectonic features. Outlines of tectonic boundaries adapted from Anderson (1979), Martin (1980), Klitgord and others (1984, 1988), Buffler and Sawyer (1985), and Klitgord and Schouten (1986). Locations of seismic refraction profiles discussed in the text are also shown. The letter M next to pre-1980 profiles indicates that Moho arrivals were observed; post-1980 profiles are numbered and discussed in the text.

CRUSTAL STRUCTURE FROM GEOPHYSICAL MEASUREMENTS AND MODELS

Seismic refraction data: pre-1970

Beginning in 1935 and continuing through the 1950s, the Geological Society of America (GSA) and the Office of Naval Research (ONR) sponsored a series of seismic refraction experiments on the Atlantic continental margin north of Cape Hatteras under the direction of Maurice Ewing. The results of these studies were published in a series of papers in the GSA *Bulletin* (Bentley and Worzel, 1956; Carlson and Brown, 1955; Drake and others, 1954; Ewing and others, 1937, 1939, 1940, 1950; Katz and others, 1953; Katz and Ewing, 1956; Officer and Ewing, 1954; Oliver and Drake, 1951; Press and Beckmann, 1954) and synthesized by Drake and others (1959). The Atlantic margin south of Cape Hatteras was studied by Hersey and others (1959), Antoine and Henry (1969), Sheridan and others (1966), and Dowling (1968). Additional data from the oceanic crustal region were presented by Houtz and Ewing (1963) and Ewing and Ewing (1959). Crustal structure beneath the Gulf of Mexico was investigated by Antoine and Ewing (1963), Antoine and Harding (1965), Ewing and others (1960, 1962), Hales and others (1970), Swolfs (1967), and Worzel and Watkins (1973). The locations of the profiles are shown in Figures 2 and 3. Only profiles penetrating several kilometers below basement have been included. Profiles showing arrivals from the Monorovičić discontinuity (Moho) are accompanied by the letter M. Comprehensive compilations of these data, including tables of geographic coordinates and solutions, have recently been published as part of the Ocean Margin Drilling (OMD) series of atlases (Locker and Chatterjee, 1984; Houtz, 1984; Houtz and Bryan, 1984; Houtz and Rabinowitz, 1984).

Most of these early experiments used explosive sources and two ships. One ship steamed along the line shooting, and the other ship held station to receive the signals from the shots on a hydrophone (Fig. 4a). A few experiments used a single shooting ship and sonobuoy receivers (Fig. 4b). Shot spacings were typically 3 to 8 km, and maximum offsets were generally 50 to 75 km, although a few profiles were as long as 100 km and many of the shallow-water shelf profiles were only about 25 km long. Signals generally were recorded on analog magnetic tape and camera records having a limited dynamic range.

The general method of interpreting these data was described in Ewing and others (1939). The data were presented in the form of P-wave first-arrival traveltimes as a function of direct water-wave traveltime or range. Based on the assumption that the observed seismic signals represent head waves that had traveled in an earth composed of a stack of homogeneous layers, the velocities, thicknesses, and dips (for reversed data) of the layers were determined by fitting lines through the data points. The problems of this method of interpretation have been discussed by Drake and others (1959) and Purdy and Ewing (1986). In the presence of lateral velocity variations, errors in inferred velocity structure can be induced by forcing the solution to fit the model of a reversed profile for a stack of homogeneous, dipping layers because the portion of a layer represented by first arrivals on the two profiles may not coincide. Additionally, sediment velocities are generally best represented by vertical gradients (Houtz and Ewing, 1963), and the velocities attributed to the series of layers that approximates this vertical gradient are usually arbitrary. Moreover, the method does not allow a velocity decrease with depth. In most cases, however, the spatial density of the data was not adequate to permit resolution of lateral velocity variations, gradients, or low-velocity zones.

The nonuniqueness inherent in this type of interpretation is illustrated in Fig. 5 (from Purdy and Ewing, 1986). Traveltimes were calculated for a model in which a layer of sediment overlays a rough basement. Isovelocity contours in the crust followed the basement surface. Synthetic traveltime data were generated by tracing rays through this model. Four different model data sets were constructed simulating shot spacings of 1 and 3 km with receivers at two different locations. The synthetic traveltime data were then inverted using standard methods to determine the velocity structure. The data sets that had 3-km shot spacing were interpreted as a stack of homogeneous layers with velocity discontinuities between layers, whereas the 1-km-spaced data were interpreted as a stack of layers having gradients within the layers and no velocity discontinuities between layers. Four very different solutions were obtained for these model studies (Fig. 5a). Note that this example includes only the error due to model parameterization and incomplete sampling of the traveltime curve; it does not include any errors in timing or navigation, which may have been quite large during some of these experiments.

In spite of these limitations, most of what was known about the structure of the Atlantic and Gulf of Mexico margins prior to the early 1970s was derived from this type of data (e.g., Fig. 6). Figure 6a, across the Scotian Basin, shows two sedimentary basins separated by a basement ridge (Officer and Ewing, 1954). Subsequently collected seismic reflection data indicated that the basement surface, defined by velocities greater than 5 km/sec, does indeed represent crystalline basement landward of the hinge zone. Seaward of the hinge zone, however, the 5- to 6-km/sec velocities reflect carbonate sedimentary units, and the supposed basement ridge is a carbonate bank. Similar high seismic velocities (5 to 6 km/s) at a shallow depth (2 to 3 km) on the Campeche Shelf may also be caused by carbonate bank deposits, as suggested by the presence of such material in nearby wells.

Lower-crustal and upper-mantle information was only obtained in deep water seaward of the continental slope (Fig. 6b, Hersey and others, 1959), and little can be said about the continent-ocean crustal boundary from these data. Note, however, the 7.2-km/sec lower oceanic crustal layer adjacent to the margin. The presence of such a layer has been substantiated by recent seismic data (see section on recent large-offset seismic data), but its significance for crustal evolution is not yet resolved.

Results of these experiments are not discussed further in this

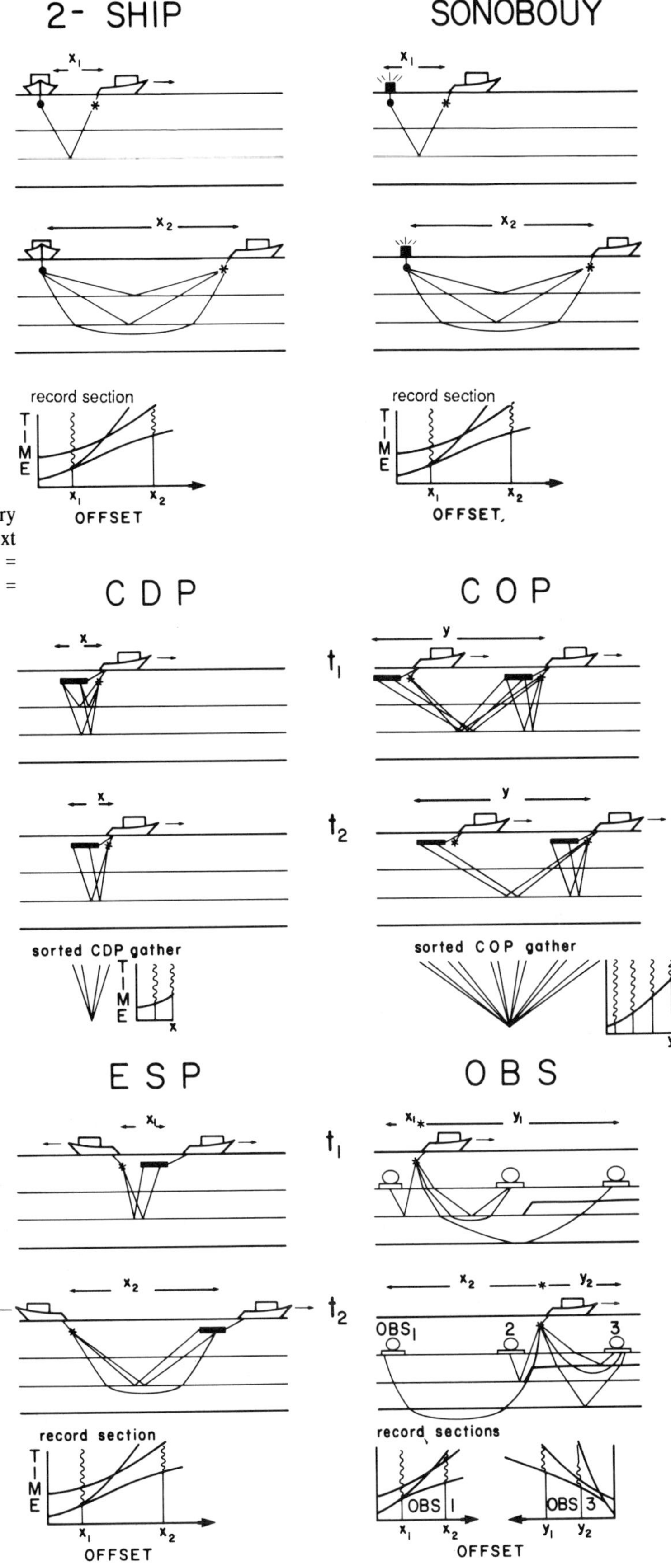

Figure 4. Schematic illustration of the source-receiver geometry for various methods of collecting offshore seismic data. See text for further explanation. CDP = common depth point; COP = constant offset profile; ESP = expanding spread profile; OBS = ocean-bottom seismometer.

chapter, although they were instrumental in guiding subsequent studies. The structural models of the marginal basins that were derived from these studies have been superceded by models derived from the higher resolution techniques discussed in the following sections of this chapter. For the oceanic crustal domain, however, these data remain a primary source of information and have recently been included in the review by Purdy and Ewing (1986).

Onshore-offshore experiments: 1960s

In the 1960s, several onshore-offshore experiments (Figs. 2b, c, 3) were conducted, during which shots as large as 6 tons were exploded and recorded by stations on land at ranges of as much as 800 km on the Atlantic margin (Steinhart and others, 1962; Hales and others, 1968; Barrett and others, 1964) and 1,500 km in the Gulf of Mexico (Hales and others, 1970a, b). The northeast-southwest–trending Gulf of Mexico profile (Fig. 3) is the most thoroughly studied of these profiles (Hales and others, 1970a). Because the crustal structure along much of the profile could be assumed on the basis of previous refraction experiments, this profile provided good control of upper-mantle structure. Moho velocities of 7.9 and 8.1 km/sec were determined beneath the continental and oceanic crust, respectively, and a step in velocity to about 8.6 km/sec at a depth of 57 km was reported. This is similar to results obtained more recently elsewhere in the deep ocean basins (e.g., Orcutt and Dorman, 1977; LADLE Study Group, 1983). A velocity decrease was inferred below 75 km on the basis of amplitude patterns. It should be noted that Hales and others (1970) was one of the first papers in which complete record sections showing the seismograms were presented, and in which the amplitude information was used to discriminate between several models that fit the traveltimes equally well.

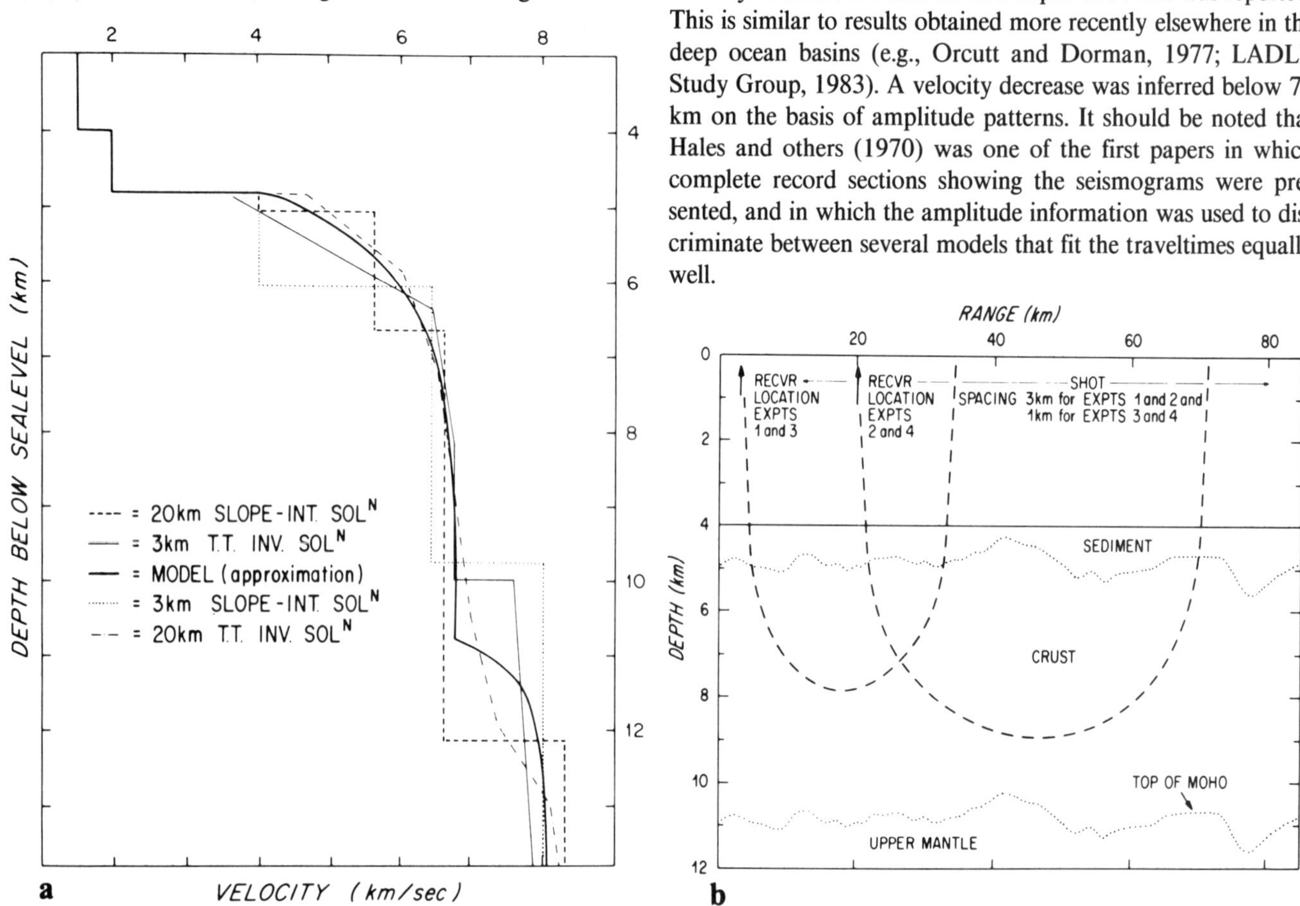

Figure 5. From Purdy and Ewing, 1986. A, The heavy solid line is the model velocity vs. depth function defined in B; the four other curves represent solutions for the velocity-depth function obtained from synthetic data sets calculated for various receiver positions and shot spacings. 20 km SLOPE-INT SOLN = solution obtained using the slope-intercept method for synthetic data calculated for a surface receiver located 20 km from the left boundary of the model and shots spaced 3 km apart. 3 km TT INV SOLN = solution obtained using a traveltime inversion method for synthetic data calculated for a surface receiver located 3 km from the left boundary of the model and shots spaced 1 km apart. 3 km SLOPE-INT SOLN = slope-intercept solution for receiver at 3 km and 3-km shot spacing; 20 km TT INV SOLN = traveltime inversion solution for receiver at 20 km and 1-km shot spacing. B, Velocity model for calculating synthetic traveltime data. Seafloor is flat at 4-km depth. Basement topography was determined from a spline fit to a typical Atlantic Ocean basin basement topographic profile. The Moho is conformal to the basement topography, defining a 6-km-thick crust. The velocity within the crust was defined by $V = V_0 + G_0 \int_0^z \exp(-\gamma z)dz$ where z is measured from the basement surface. For $0 < z < 6$ km, $V_0 = 4$ km/sec, $G_0 = 2.8$ sec^{-1}, and $\gamma = 1.0$ km^{-1}. For $z > 6$ km, $G_0 = 2.5$ sec^{-1} and $\gamma = 2.0$ km^{-1}. V_0 and G_0 are the velocity and velocity gradient at the basement, and γ is an arbitrary constant defining the rate of decrease of the gradient with depth.

Because the Atlantic margin lines were shorter and the crustal structure along the lines was less well known than for the Gulf of Mexico profile, the data were very scattered, and original interpretations were tenuous. For the ECOEE-II line, a single 30-km-thick crustal layer with a velocity of 6.03 overlying a Moho with a velocity of 8.13 km/sec was determined; for the ECOOE-I line, a two-layer crust with velocities of 5.8 and 6.3 km/sec and a thickness of 25 km was obtained (Hales and others, 1968). Assuming that the crustal and upper-mantle velocities were known, James and others (1968) derived a seaward thinning of the crust of about 10 km (from 40 to 30 km) beneath the coastal plain and continental shelf from a time-term analysis. In Maine and the Gulf of Maine, a two-layer crust with velocities of 6.05 and 6.8 km/sec and a thickness of 34 km overlying an 8.1-km/sec Moho was obtained (Steinhart and others, 1962). The traveltime data, however, are compatible with a wide range of models, ranging from a single 30-km-thick layer with a velocity of 6.0 km/sec overlying an 8.0-km/sec Moho (Nakamura and Howell, 1964) to a crust within which the velocity gradually increases from 6.0 to 8.1 km/sec over a thickness of about 45 km (Steinhart and others, 1962). To discriminate among these different models requires densely spaced, large-dynamic-range data, so that amplitudes of both refracted arrivals and wide-angle reflections can be matched. Data meeting these specifications have recently been collected along several sections of the Atlantic and Gulf of Mexico margins and are discussed in the section on recent large-aperture seismic data.

Lewis and Meyer (1977) reexamined the ECOOE-II data, adding constraints imposed by offshore seismic reflection data in order to study the upper-mantle structure, and determined that the velocity increased from 8.0 to 8.5 km/sec at a depth of 70 km beneath both continental and oceanic crust. More of the data from the east coast onshore-offshore experiments deserve reconsideration in light of recent results on sediment and crustal velocity structure in these regions. This is particularly important because changes in the procedures for obtaining permits to shoot explosives offshore now make it a very time-consuming, and in many cases futile, procedure. It is unlikely that additional offshore data will be acquired in the near future from shots large enough to provide very long-range data that will contain information on upper-mantle structure.

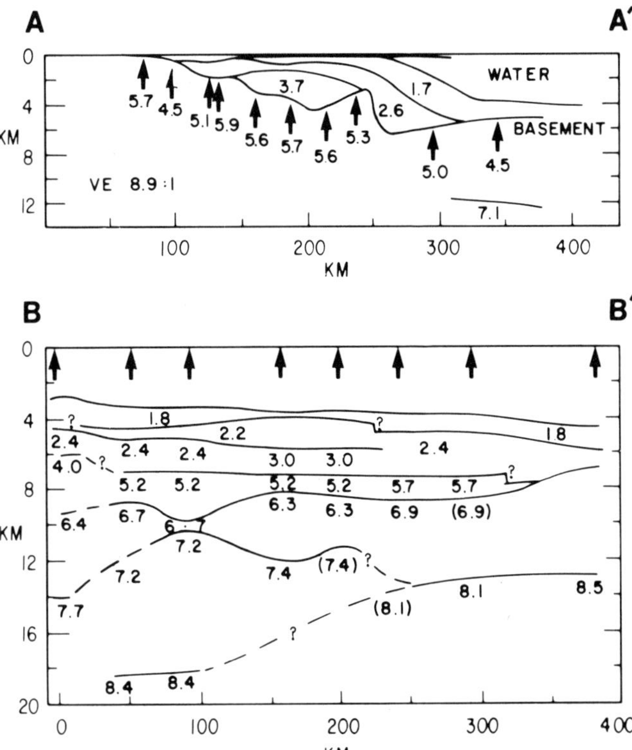

Figure 6. Representative composite cross sections based on early seismic refraction data. A, The Scotian basin, adapted from Drake and others, 1954; see A-A′ in Figure 2B for location. B, The oceanic crust seaward of the Carolina trough; from Hersey and others, 1959, see B-B′ in Figure 2C. Each section was constructed from a series of refraction profiles shot across the margin. Seismic velocities are indicated in units of kilometers per second. Arrows indicate the midpoint of the individual profiles.

Multichannel seismic reflection data

During the 1970s, the U.S. Geological Survey contracted the acquisition of a grid of multichannel common-depth-point (CDP) seismic reflection lines along the U.S. Atlantic continental margin (Fig. 7). Additional multichannel data have been collected along the Atlantic margin by Lamont-Doherty Geological Observatory (LDGO), Woods Hole Oceanographic Institution (WHOI), University of Texas Marine Science Institute (UTMSI), and others. Most of the publicly available seismic reflection data in the Gulf of Mexico were collected by the University of Texas Institute for Geophysics (UTIG) (Fig. 8). These studies went far toward elucidating the structure of the sedimentary basins that had been suggested on the basis of the early refraction experiments. The results are discussed in detail in the Geological Society of America *Geology of North America* volumes on the Atlantic continental margin (Sheridan and Grow, 1988) and Gulf of Mexico basin (A. Salvador, unpublished data).

Atlantic continental margin. Results from the CDP seismic studies on the Atlantic margin are summarized in a series of cross sections shown in Fig. 9. These cross sections were constructed from regional seismic stratigraphic studies tied to local seismic stratigraphic studies and Continental Offshore Stratigraphic Test (COST) well results (e.g., Poag, 1982, 1985; Klitgord and others, 1988). Velocities for converting time sections to depth sections were determined primarily from downhole velocity measurements and from root-mean-squared (rms) velocities obtained from coherence and normal-moveout analyses of the multichannel reflection data. In the deeper parts of the basins, velocities are poorly known.

The structure across the Georges Bank basin (Klitgord and others, 1982; Schlee and Klitgord, 1988) is best illustrated on seismic line 19 (Fig. 9a). The basement hinge zone is a series of

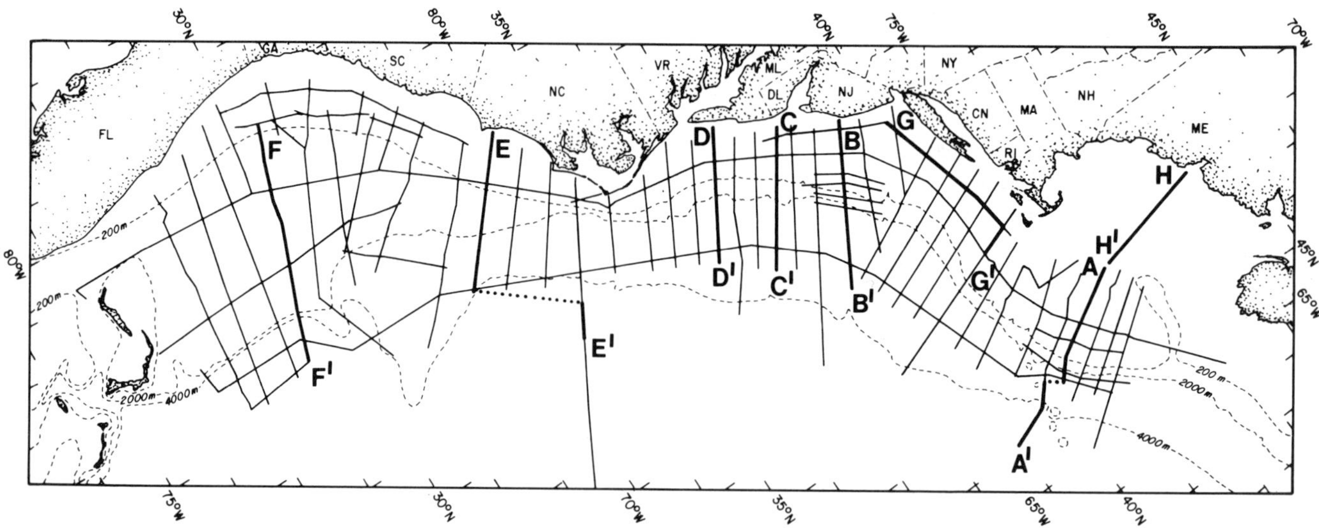

Figure 7. Location of multichannel seismic reflection lines collected by the USGS during the 1970s on the Atlantic continental margin. Lines forming the basis for the cross sections shown in Figures 9 and 10 are indicated by heavy lines and labeled.

block-faulted half-graben structures bounded by seaward-dipping faults. A salt layer masks structures deeper than about 8 km in the main basin. The post-rift unconformity coincides with the top of the salt in the main basin and shallows both landward and seaward. It laps westward onto the base of the basement hinge zone and undulates upward over each fault block structure. Near the shelf edge, it rises upward to the east over a series of Cretaceous intrusive bodies, fault blocks, and possible salt diapir structures. A narrow zone of salt diapirs and a broad zone of seaward-dipping reflectors separate the paleoshelf edge and ECMA from the landward limit of well-defined oceanic crust.

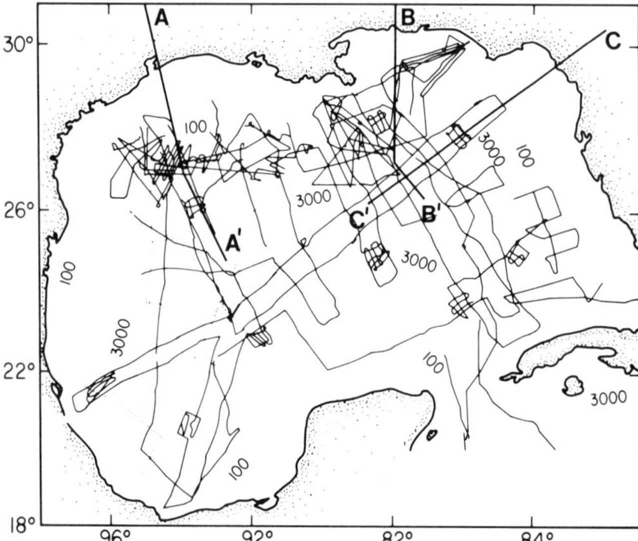

Figure 8. Location of multichannel seismic-reflection lines collected by UTIG in the Gulf of Mexico. Locations of cross sections shown in Figure 11 are shown by heavy lines and labeled. Solid lines are from industry-sponsored cruises, dotted lines are from NSF-sponsored cruises, and dashed lines were collected for site surveys related to DSDP Leg 77.

The Baltimore Canyon trough is a deep sedimentary basin that fills an indentation in the margin between the Long Island and Carolina platforms (Fig. 1). The basin narrows to the south, and its structure varies considerably along its length (Grow, 1980; Grow and others, 1988; Klitgord and others, 1988). Cross sections along seismic lines 25 (Fig. 9b) and 10 (Fig. 9c) are representative of the northern part of the basin. A steep hinge zone and landward-dipping block faults are characteristic of the western edge of the northern basin (Klitgord and others, 1988). The post-rift unconformity is at a depth of over 13 km, and the seaward end of this surface terminates at a fault known as the East Coast boundary fault that is coincident with the ECMA (Alsop and Talwani, 1984; Klitgord and others, 1988). The carbonate bank–paleoshelf edge complex migrated seaward of this boundary fault during Jurassic time and masks the deeper structure over the zone corresponding to where salt diapirs and seaward-dipping reflectors are observed off Georges Bank. The crustal structure along this line was recently investigated during the Large Aperture Seismic Experiment and is discussed in the section on recent large-aperture seismic data.

In the southern part of the Baltimore Canyon Trough (Fig. 9d), a large rift basin, located just landward of the hinge zone, has internal sedimentary structures indicative of an east-dipping border fault on its western edge (Klitgord and Hutchinson, 1985). Basement structures between the hinge zone and the ECMA are masked by a thick volcaniclastic wedge that underlies the post-rift unconformity and by the carbonate units of the paleoshelf edge. Between the ECMA and the landward limit of well-defined oceanic crust, a very reflective sedimentary surface (J3 of Klitgord and Grow, 1980) forms the acoustic basement and laps onto oceanic basement at the J3 scarp.

The Carolina trough is the simplest of the basins of the U.S. Atlantic margin. It appears to have formed by simple pulling

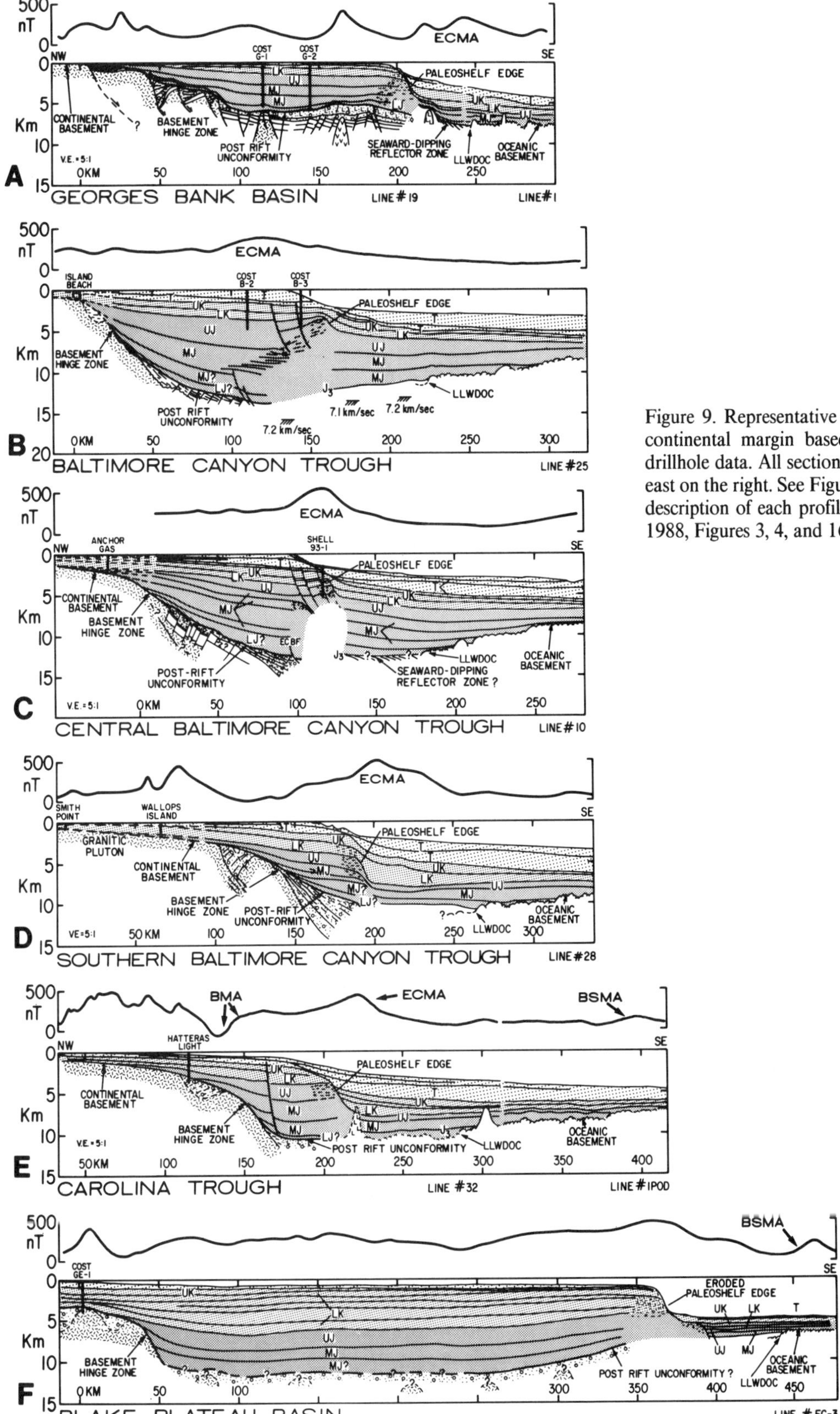

Figure 9. Representative cross sections across the U.S. Atlantic continental margin based primarily on seismic reflection and drillhole data. All sections are oriented with west on the left and east on the right. See Figure 7 for section locations. See text for a description of each profile. (Modified from Klitgord and others, 1988, Figures 3, 4, and 16a.)

apart of North America and Africa parallel to the trend of preexisting Paleozoic structures, without complications induced by oblique or transform motion (Klitgord and Behrendt, 1979; Hutchinson and others, 1983; Dillon and Popenoe, 1988). The structure of the basin along its entire 400-km length is summarized in Figure 9e. Landward of the hinge zone, a rift graben is shown beneath the magnetic low of the Brunswick magnetic anomaly (BMA) (Hutchinson and others, 1983). The BMA is coincident with the basement hinge zone along the entire length of the Carolina trough. The basin, however, shows a clear image on only one seismic line (line 32). Whether the seismic evidence for this graben is real or an artifact due to diffractions from a rough basement has not been resolved. Its acoustic character is similar to that of other rift basins landward of the hinge zone north of Cape Hatteras (Klitgord and Hutchinson, 1985; Hutchinson and others, 1986b). The BMA swings onshore in Georgia where it has been associated with the late Paleozoic suture zone between North America and Africa (Popenoe and Zietz, 1977; Klitgord and others, 1983; Daniels and others, 1983; Nelson and others, 1985a, b). Onshore, rift basins are clearly imaged beneath the magnetic low of the BMA (Nelson and others, 1987). Nelson and others (1985b) proposed that the offshore segment of the BMA also represents the suture between the North American and African plates before the opening of the Atlantic Ocean.

At the hinge zone, basement deepens sharply and is lost beneath a strong reflector (the post-rift unconformity) that flattens out at a depth of about 12 km. A fault in the sedimentary units near the hinge zone has been interpreted to be a growth fault, formed as a result of the seaward migration of salt (Dillon and others, 1983). Beneath the seaward edge of the basin near the ECMA, basement is obscured by the paleoshelf edge and a zone of salt diapirs (Dillon and others, 1983). Recent data from this region suggest the presence of seaward-dipping reflectors just seaward of the salt diapir zone (J. C. Mutter, personal communication, 1986). Results of a recent large-aperture experiment designed to determine the crustal structure beneath the deepest part of the basin, as well as beneath the adjacent oceanic and continental platform regions, are discussed in the section on recent large-aperture seismic data.

The Blake Plateau basin contrasts sharply with the other basins of the margin. It is a 350-km-wide sedimentary plateau underlain by as much as 12 km of nearly flat-lying Jurassic and Cretaceous carbonate and clastic sedimentary fill (Shipley and others, 1978; Dillon and others, 1983, 1985; Dillon and Popenoe, 1988). The landward edge of the basin is defined by a well-developed hinge zone and a rift basin located just landward of the hinge zone (Klitgord and others, 1983). The seaward edge is marked by the Blake Escarpment, a steep erosional scarp (Paull and Dillon, 1980) that has truncated the paleoshelf edge (Fig. 2). Interlayered high- and low-velocity layers, corresponding to massive limestone and less consolidated clastic layers, result in many peg-leg multiples and P-wave to S-wave conversions (Trehu, 1984). As a result of this complicated velocity structure, the seismic-reflection data are very noisy, and basement has not been recognized beneath most of the basin. A north-south–trending basement ridge that has been recognized on several lines beneath the southeastern part of the plateau (Sheridan and others, 1981; A. M. Trehu, unpublished data, 1986), and a slight westward dip of the reflectors beneath the middle and outer part of the plateau suggests that the deepest part of the basin is in the northwest region (Dillon and others, 1983, 1985).

Deep crustal information (i.e., from the crystalline crust beneath the sediments) in the USGS CDP seismic lines is limited both by acquisition source size and by data-processing techniques that were designed to enhance the overlying sedimentary units. None of the CDP seismic lines that cross the marginal basins contain any direct information on the deep crustal and Moho structure beneath the basins, although reprocessing of some of these lines may yield useful crustal information. A few of the CDP lines over the basement platforms landward of the hinge zone, however, did penetrate through the crust to Moho (Behrendt and others, 1983; Hutchinson and others, 1985, 1986a; Phinney, 1986). An interpretation of USGS line 36 on the Long Island platform (Fig. 10a) shows the structure of continental crust landward of the hinge zone and has been discussed by Hutchinson and others (1985, 1986a, b) and Phinney (1986). A seismic reflector interpreted to be the Moho is observed at 9.5 to 11 sec along the line. An upwarp (in time) of this reflection is associated with a west-dipping fault, the Block Island fault, that extends through the crust and may represent a major detachment along which crustal extension occurred during rifting. The velocity information needed to determine the crustal thickness and to resolve whether the apparent crustal thinning associated with the Block Island fault is real or an artifact of lateral velocity heterogeneity is not presently available. Four shallow rift basins are also observed along this line, and their internal stratigraphy and relation to the post-rift unconformity provide constraints on the amount and distribution of uplift and erosion associated with the late stages of rifting and the initiation of sea-floor spreading.

In 1984, a seismic reflection line was contracted by the USGS across the Gulf of Maine especially to study deep-crustal structure (Hutchinson and others, 1987, 1988). This line is a continuation of a seismic reflection and refraction transect that begins in the Precambrian craton in Canada and extends to the oceanic crust. The landward portion of this line has been discussed by Stewart and others (1986). The offshore reflection line extends from the coast to the hinge zone, where it joins USGS CDP line 19 that crosses the Georges Bank basin. The Moho is observed at a depth of about 10.5 sec along the line (Fig. 10b), suggesting that the crust landward of the hinge zone does not thin significantly as the hinge zone is approached. More than 150 km west of the hinge zone, eastward-dipping bands of reflections are observed to cut through the entire crust and are interpreted to represent Paleozoic thrust faults. Less than 150 km from the hinge zone, eastward-dipping reflections in the upper crust are truncated in the lower crust, which is characterized by subhorizontal reflections. This can be interpreted to be the signature of either Paleozoic crustal melting or Mesozoic extension that modified the

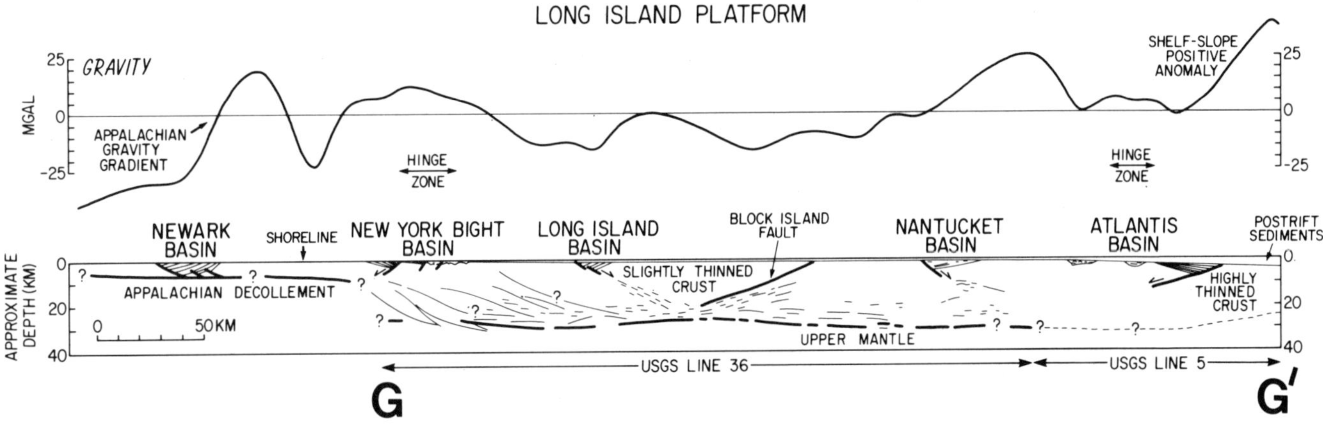

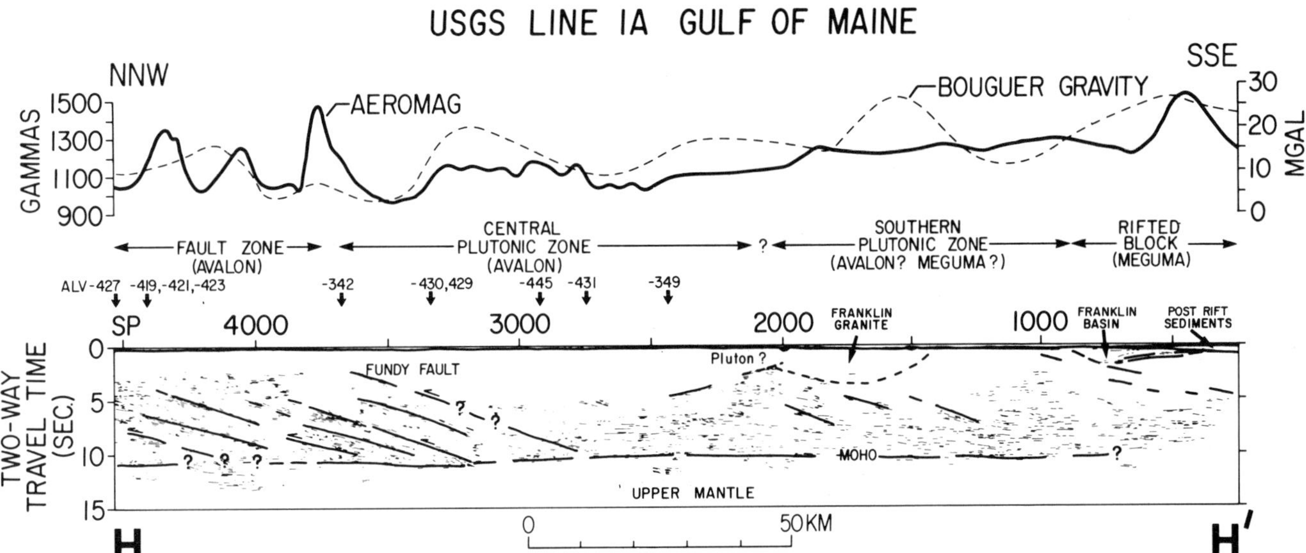

Figure 10. Top: schematic interpreted line drawing of USGS line 36 on the Long Island Platform. Location of section is shown in Figure 7. From Hutchinson and others, 1986a, Figure 4. Bottom: schematic interpreted line drawing of USGS Deep Crustal line across the Gulf of Maine. Location of section is shown in Figure 7. From Hutchinson and others, 1988, Figure 11.

texture of the lower crust (Hutchinson and others, 1987, 1988). The contract reflection line was supplemented in 1985 by a grid of intermediate-resolution CDP reflection profiles and several large-aperture seismic lines that will be discussed in the section on recent large-aperture seismic data.

Gulf of Mexico continental margin. The Gulf of Mexico margin cannot be separated into a series of discrete marginal basins like the Atlantic margin. This is at least partly due to our poor knowledge of the basement structure beneath the Gulf Coast Salt Dome province. For example, the possible boundary between the early Mesozoic salt basins found around the rim of the Gulf basin and the large sedimentary basins of the northern Gulf of Mexico (analogous to the Atlantic margin hinge zone) is obscured by salt structures and by the Lower Cretaceous carbonate bank shelf-edge complex. The main features of the U.S. margin of the Gulf of Mexico are illustrated in three cross sections shown in Fig. 11 (adapted from Martin and Buffler, 1984; Winkler and Buffler, 1984). Velocities for converting time sections to depth were derived from velocities obtained from seismic refraction experiments as well as from root-mean-squared velocities determined from CDP seismic reflection data. As with the Atlantic margin sections, velocities are, in general, poorly known for the deeper parts of the sections.

Figure 11a crosses the margin near Galveston, Texas, and is oriented approximately parallel to the Early to Middle Jurassic rifting direction predicted by plate tectonic reconstructions (Klitgord and others, 1988). Figures 11b and c cross the margin in Alabama and Florida and represent transects across what is thought to have been a broad Mesozoic shear zone (Klitgord and others, 1984; Klitgord and Schouten, 1986). An alternative model by Buffler and Sawyer (1985), Pindell (1985), and Dunbar and Sawyer (1987) for the rifting and early sea-floor-

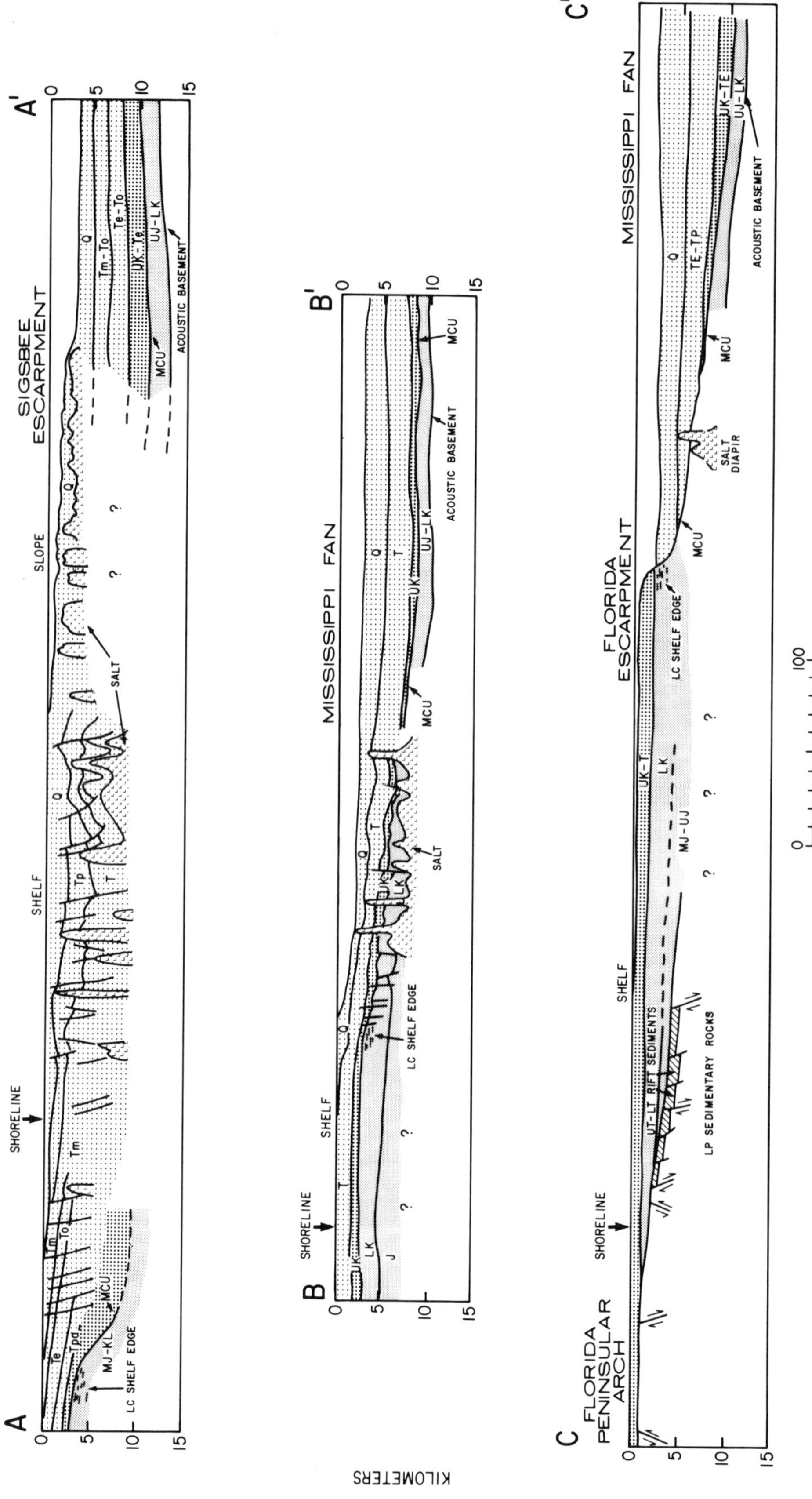

Figure 11. Regional cross sections of the Gulf of Mexico based on seismic reflection and drillhole data. Adapted from Martin and Buffler, 1984; Winkler and Buffler, 1984. Locations of sections are shown in Figure 8.

spreading history of the Gulf region includes north-south rifting and predicts more extension along the northern Gulf Rim (Fig. 11b, c) and more shearing along the western rim (Fig. 11a). The existing seismic structural information does not enable us to eliminate either of these possible models. A discussion of the plate kinematic constraints on these two models is given in Klitgord and others (1988).

In all three cross sections, the basement structure between the Lower Cretaceous shelf edge and well-defined oceanic crust is poorly determined because of thick sedimentary sections, the extensive deformation associated with the mobilization of Jurassic salt, and the zone of Cretaceous carbonate bank–shelf edge deposits, all of which limit the penetration of seismic energy. Transitional crust and its boundary with oceanic crust is expected in this region. Landward of the Lower Cretaceous shelf edge, a series of rift basins can be seen on the Florida shelf near the shoreline until basement is lost beneath the highly reflective carbonates of the shelf (e.g., Ball and others, 1983; Klitgord and others, 1984). The deep-water part of the Gulf of Mexico that lies seaward of the Sigsbee, Campeche, and West Florida Shelf Escarpments, and Sigsbee and West Florida Salt Basins is underlain by a thick section of Late Jurassic to Recent sedimentary rocks (Shaub and others, 1984). This is the region of well-defined oceanic crust based on the interpretation of seismic reflection and refraction data (Buffler and Sawyer, 1985).

Magnetic data

Magnetic anomaly data are important for crustal studies because they can be used to: (1) map the regional distribution of shallow crustal rock types sampled in outcrop or in wells, (2) infer upper-crustal rock types where basement samples are not available, (3) map the regional extent of structural features recognized by field mapping or with seismic data, (4) estimate depth-to-magnetic-basement, (5) estimate paleomagnetic parameters of rock bodies, and (6) correlate offshore structures with structures mapped onshore. Aeromagnetic surveys of the U.S. Atlantic coastal plain and continental margin provide sufficiently detailed areal coverage for undertaking most of the types of studies mentioned above. Far less extensive coverage is publicly available for the Gulf of Mexico coastal plain and continental margin. There are scattered small aeromagnetic surveys onshore and a sparse set of marine surveys offshore.

Atlantic continental margin. Regional mapping of geologic units (e.g., granitic plutons) and tectonic features (e.g., fault zones) of the Atlantic continental margin and coastal plain on the basis of magnetic data, generally in conjunction with gravity data, has been attempted in a number of studies. These include investigations in the New England area by Kane and others (1972), Harwood and Zietz (1976), Alvord and others (1976), Simpson and others (1980), Klitgord and others (1982), and Hutchinson and others (1985, 1986b, 1987, 1988); in the middle Atlantic region by Higgins and others (1974), Pavlides and others (1974), Sumner (1979), Klitgord and Hutchinson (1985), Phillips (1985), and Klitgord and others (1988); in the southeastern U.S. by Popenoe and Zietz (1977), Hatcher and others (1977), Daniels and Zietz (1978), Hatcher and Zietz (1979), Klitgord and others (1983, 1984), Hutchinson and others (1983), Daniels and others (1983), and Higgins and Zietz (1983); for the entire coastal plain by Zietz and Gibert (1980), Zietz and others (1980), and Zietz (1982); and for the entire Atlantic margin by Taylor and others (1968) and Klitgord and Behrendt (1977, 1979). The upper surface of crystalline rock along the entire U.S. Atlantic margin has been mapped by Klitgord and Behrendt (1979) using depth-to-magnetic-basement estimates to interpolate basement depths between locations where basement depth was identified from drillhole and seismic reflection data.

As discussed in the first section of this chapter, three large-amplitude magnetic lineaments that are closely associated with major crustal boundaries are observed along the U.S. Atlantic continental margin (Fig. 2; see also Kane and Godsen, this volume). These lineaments are the East Coast, Brunswick, and Blake Spur magnetic anomalies. The nature of the source of each is still being debated, but as geophysical markers they provide ideal boundaries for aggregating other data sets and focusing correlation studies. A detailed discussion of the sources of these anomalies is given in Klitgord and others (1988).

The ECMA is close to the boundary between transitional and oceanic crust. It has variously been interpreted to result from dikes (Emery and others, 1970; Alsop and Talwani, 1984), to be an edge effect due to the juxtaposition of continental crust and sedimentary units (Keen, 1969; Alsop and Talwani, 1984), or to be an edge effect due to the juxtaposition of oceanic crust and sedimentary units (Taylor and others, 1968; Klitgord and Behrendt, 1979; Hutchinson and others, 1983). The ability of different models to reproduce the shape of the anomaly reflects the nonuniqueness of the magnetic data. In fact, the anomaly may be due to all of these effects to varying degrees along the margin. Figure 12a shows the model of Alsop and Talwani (1984) across the Baltimore Canyon trough. The primary source for the anomaly in this model is a fault bounding a graben into which a block of highly magnetized material with an inclination of 60° has been emplaced. A sidelobe of the anomaly, which is not characteristic of the ECMA along most of its length, is modeled by a second intrusive block. In comparison, Figure 12b shows the magnetic model of Hutchinson and others (1983) for the Carolina trough. In this model, the primary source of the anomaly is the edge effect of juxtaposing reversely magnetized oceanic crust with an inclination of 20 against a weakly magnetized continental crustal unit. Both models predict the observed magnetic anomalies.

The BMA is located over the basement hinge zone along the landward edge of the Carolina trough (Klitgord and Behrendt, 1979). In this region, it has been tentatively interpreted to be associated with what is interpreted as a rift graben on USGS seismic reflection profile 32 (Hutchinson and others, 1983). This anomaly swings onshore across southern Georgia, where it is also called the Altamah magnetic anomaly and has been interpreted to be the magnetic expression of the late Paleozoic suture zone

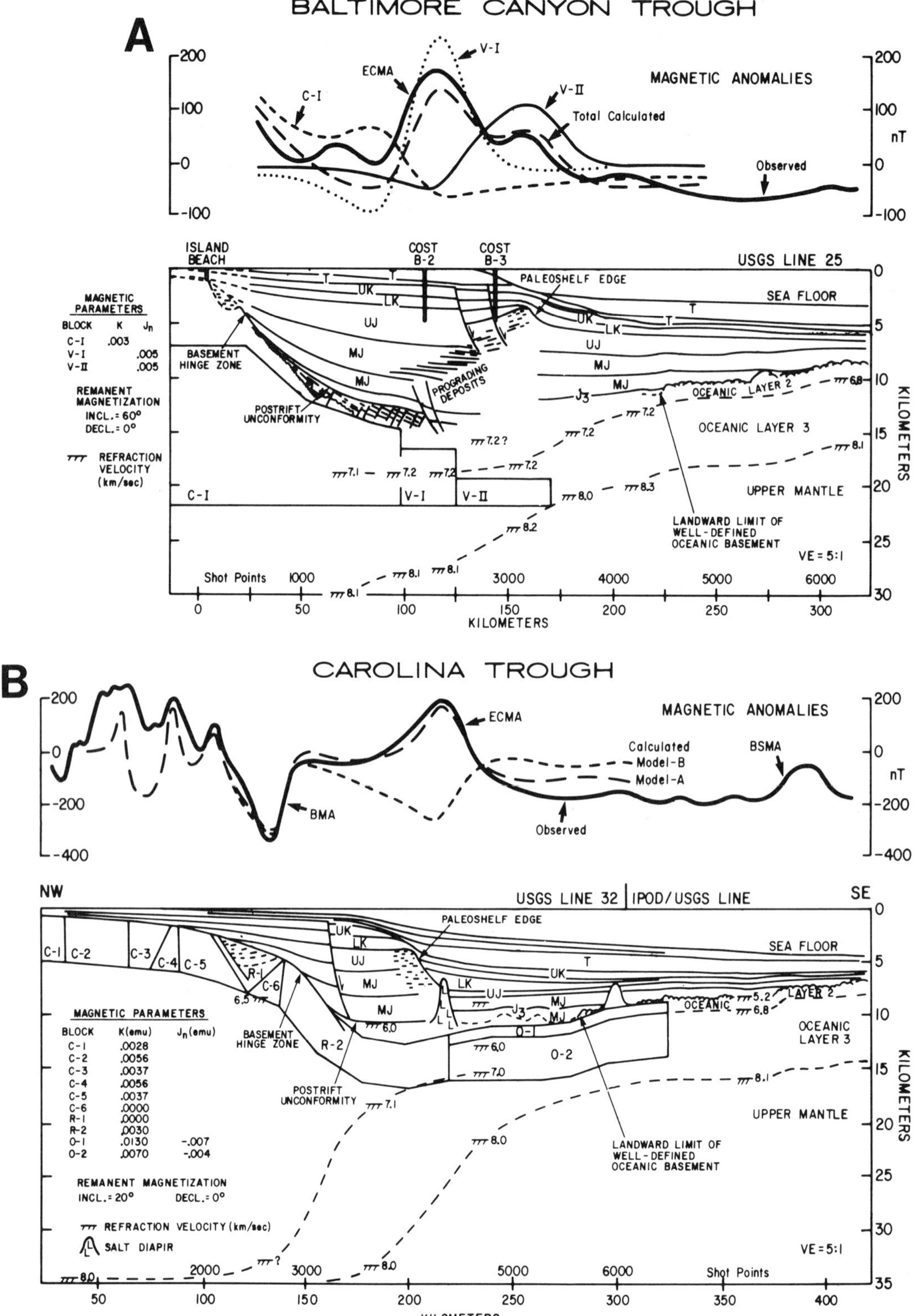

Figure 12. A, Model of the ECMA along USGS line 25 and the LASE COP line calculated by Alsop and Talwani (1984). Contributions from individual components of the model are shown. B, Model of the ECMA along USGS line 32 calculated by Hutchinson and others (1983). From Klitgord and others (1988, Fig. 16). ECMA, BMA, and BSMA as defined in the text.

between a fragment of African crust (the Suwannee terrane) and North America (Popenoe and Zietz, 1977; Daniels and others, 1983; Klitgord and others, 1983, 1984; Higgins and Zietz, 1983; Nelson and others, 1985a, b).

The BSMA is located within the oceanic crust (Sheridan and others, 1979; Klitgord and Grow, 1980) and has been associated with a major spreading-center shift in Middle Jurassic time (Vogt, 1973; Klitgord and Grow, 1980). This anomaly separates the inner and outer magnetic quiet zones along the U.S. Atlantic margin (Rabinowitz, 1974), and coincides with the most seaward of a series of basement steps that parallel the margin (Klitgord and Grow, 1980; Klitgord and others, 1988). Low-amplitude, linear magnetic anomalies, most of which can be attributed to basement relief within the inner magnetic quiet zone, characterize the zone between the BSMA and the ECMA (Klitgord and Grow, 1980).

Gulf of Mexico continental margin. Few similar magnetic studies exist for the Gulf of Mexico margin because of the sparse data sets and paucity of basement samples and deep seismic reflection data to help constrain the interpretations of magnetic data. A few studies have been conducted of the eastern Gulf of Mexico (Gough, 1967; Klitgord and others, 1983, 1984; Daniels and others, 1983), and coastal plain (Zietz, 1982), and several authors have treated the entire Gulf region (Martin, 1980; Hall and others, 1984a; Pilger and others, 1984). In the region between the Gulf Rim fault zone and the Cretaceous paleo-shelf, magnetic anomalies have been useful for mapping the extent of uplifted basement blocks. A prominent northwest-trending magnetic high that extends from the Tampa basin to the Mississippi delta, seaward of the paleoshelf edge, may mark the boundary between oceanic and transitional crust; it may thus be the Gulf basin analogue of the ECMA (Klitgord and others, 1984). More detailed studies require additional magnetic surveys and better seismic control on crystalline basement.

Gravity data

In the absence of seismic data measuring the crustal thickness beneath the thick sediments of the marginal basins, two-dimensional modeling of the gravity anomalies along profiles across the margin has been useful for obtaining an estimate of crustal thickness in the zone of the ocean-continent transition (Sheridan and others, 1988).

Atlantic continental margin. More than 40,000 km of marine gravity data have been collected along the Atlantic margin (Grow and others, 1979). The dominant feature controlling the distribution of the free-air gravity anomalies is the topography of the continental slope and rise. When this effect is removed, a strong correlation with the basin structure of the margin can be seen in the isostatic gravity anomalies (Karner and Watts, 1982; Simpson and others, this volume). However, a strong correlation is also observed between the isostatic residual gravity map and sedimentary features such as the paleoshelf edge, indicating that the lateral density variations within these sedimentary features must be better known before reliable inferences can be made about the deeper crustal structure from the gravity data.

The first attempt to model the variation in crustal thickness across the margin using gravity data was by Worzel and Shurbet (1955), who showed that the crust thinned from a continental thickness of 35 km to an oceanic thickness of about 8 km along a series of transects between Georges Bank and Cape Hatteras. The width of the transition zone varies from 50 to 200 km along that portion of the margin.

Incorporating the data from the CDP seismic reflection lines collected during the 1970s, Grow and others (1979), Kent (1979), Hutchinson and others (1983), and Swift and others (1987) modeled the ocean-continent transition zone in more detail. Step-like changes in crustal thickness were modeled beneath the hinge zone and the ECMA in both the Baltimore Canyon and Carolina troughs, whereas crustal thinning was more gradual beneath Georges Bank. The Blake Plateau was modeled as a broad zone of intermediate thickness crust. The results of these models are summarized in Figure 13 and compared to crustal thickness estimates obtained from "total tectonic subsidence" calculations (discussed in the section on thermo-mechanical modeling).

A troubling aspect of these models that casts doubt on the significance of the inferred abrupt steps in crustal thickness is that the steps in general underlie regions in which the depth to basement is poorly known or in which near-surface structures, in particular the paleoshelf edge, may result in lateral density variations that have not been represented correctly in the model. An example of the possible magnitude of this effect is shown in Figure 14 in which two models of the Baltimore Canyon trough along CDP seismic reflection line 25 are compared. Figure 14a is from Grow (1981); Figure 14b is from C. Keen and I. Reid (unpublished manuscript, 1986). The deeper crustal structure in Figure 14b is based on refraction data collected during LASE (Large Aperture Seismic Experiment; see section on recent large-aperture seismic data). The free-air gravity anomaly can be matched by modeling the paleoshelf edge as a relatively high-density body without requiring an abrupt step in crustal thickness at the ECMA. It should be noted that probably neither of the models in Figure 14 is correct because the LASE data indicate that the paleoshelf edge is underlain by lower-velocity (and consequently lower density) material. This figure does, however, illustrate the need for a systematic study to evaluate which components of the gravity signature may be attributed to sedimentary structures and which must be attributed to deeper structure. For additional discussion of errors in the gravity modeling due to uncertainties in sediment densities, see Barton (1986).

Gulf of Mexico continental margin. A gravity map of the Gulf of Mexico has been published by Pilger and Angelich (1984) and Hall and others (1984b). Few gravity-modeling studies, however, have been done across the Gulf of Mexico margin because of the poor control on sediment thickness. Keller and Shurbet (1975) modeled the gravity anomaly along a transect across the coastal plain near Galveston, Texas, and obtained a

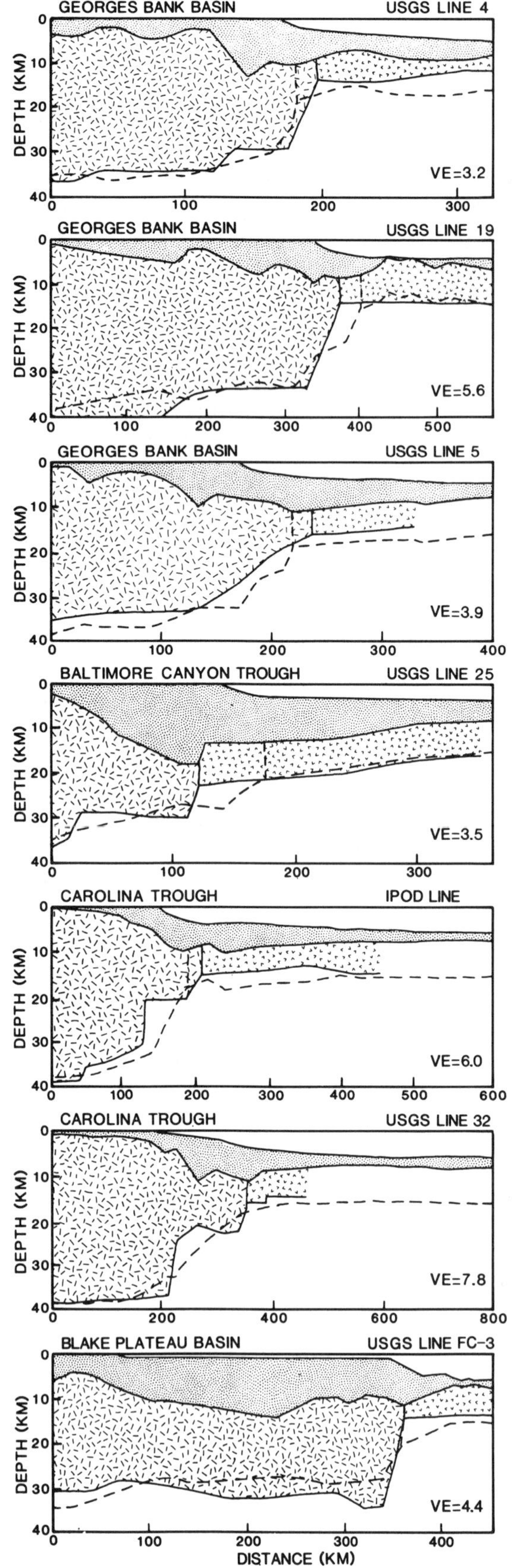

seaward thinning of the crust from about 40 km landward of the Gulf Rim fault zone to less than 10 km near the coast. They interpreted this result to mean that the continent-ocean boundary was located landward of the coastline and did not consider the probable existence of an extensive zone of transitional crust. In the eastern Gulf, Klitgord and others (1984) obtained a Moho depth of about 20 km beneath the South Florida and Tampa basins and a Moho depth of 30 km beneath the intervening arches, based on the observation of gravity lows over the arches and highs over the basins. Hall and others (1983) also used gravity anomalies to define the extent of crustal blocks in a reconstruction of the plate-tectonic history of the Gulf of Mexico.

Subsidence data and thermo-mechanical modeling

The CDP seismic reflection data and associated gravity and magnetic studies of the 1970s stimulated a great deal of effort toward modeling the thermo-mechanical evolution of the Atlantic margin and comparing the predicted subsidence of the margin to that determined from drillholes and seismic stratigraphic studies (e.g., Sleep, 1971; McKenzie, 1978; Steckler and Watts, 1978; Royden and Keen, 1980; Keen and Barrett, 1981; Watts, 1981; Sawyer and others, 1983; Watts and Thorne, 1984; Keen, 1985; Sawyer, 1988). A detailed discussion of these models is beyond the scope of this chapter. Because the thermo-mechanical models provide a framework for interpreting the deep-crustal seismic observations and designing new experiments, however, a brief summary is provided.

Because the sedimentary basins along the Atlantic and Gulf margins are too thick to be accounted for by simply loading the crust (Watts, 1981), it is generally assumed that some form of tectonic subsidence must be an important component in their development. Sleep (1971) originally proposed that the tectonic subsidence was due to crustal thinning following thermal uplift and erosion. Although this mechanism may explain certain intracratonic basins, it cannot produce the amount of subsidence observed along the Atlantic continental margin. Falvey (1974) discussed the added component of subsidence due to metamorphic effects in the lower crust.

Most models being considered today represent variations of the pure-shear uniform-extension model presented by McKenzie (1978; Fig. 15a). The initial condition is a lithosphere, including a crustal layer, that is in thermal equilibrium. The crust and lithosphere are then stretched, and hot asthenosphere wells up from below. This stretching phase is accompanied by initial subsidence to maintain isostatic equilibrium. Because of the difficulty of estimating the age of nonmarine rift sedimentary units and the poor definition of synrift sedimentary thicknesses, the initial tec-

Figure 13. Crustal thickness calculated from gravity models (shaded region) compared to that calculated from "total tectonic subsidence" (dashed line is Moho; adapted from Sawyer, 1985). See text for further discussion.

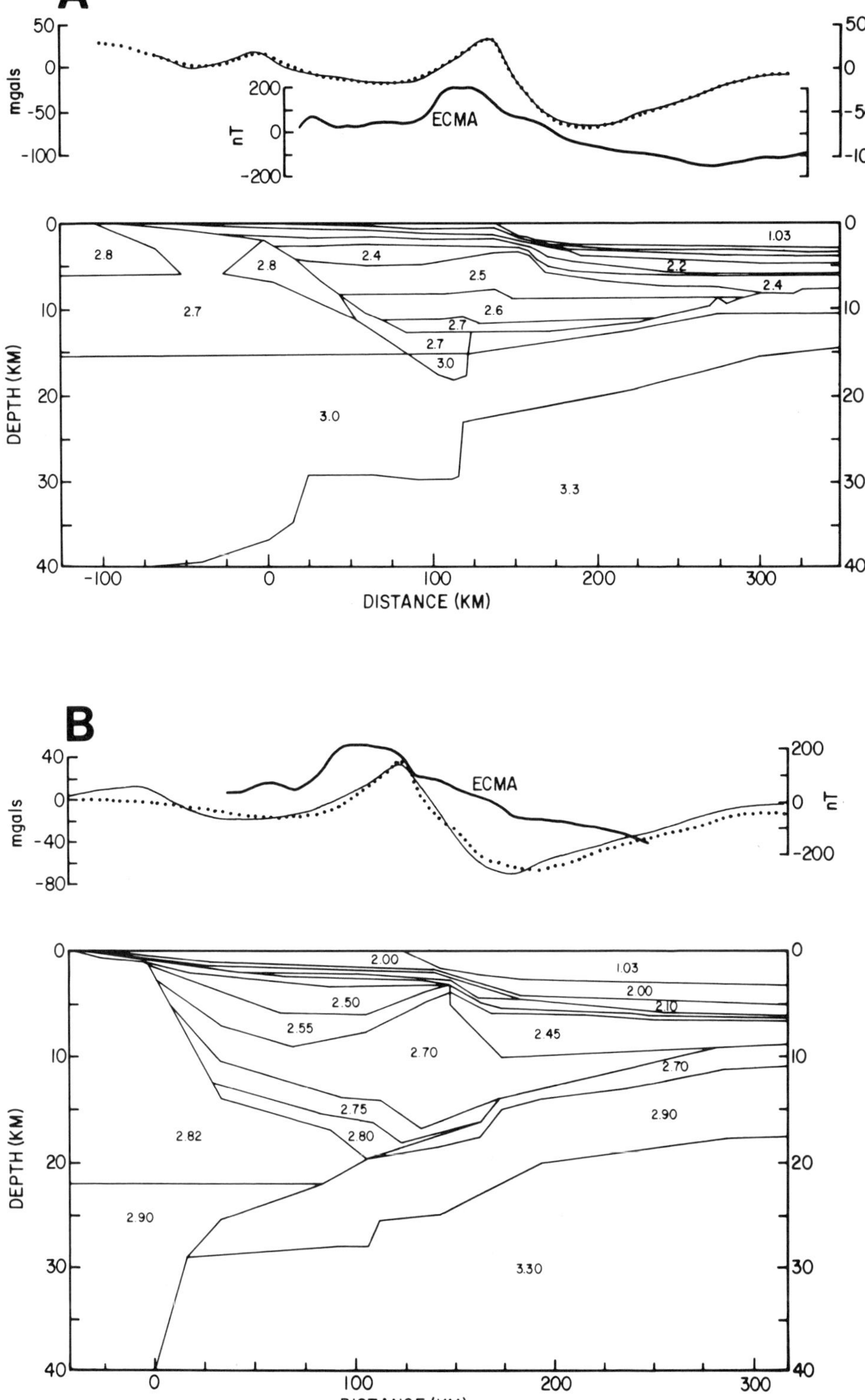

Figure 14. Comparison of two gravity models across the Baltimore Canyon trough along USGS line 25 and the LASE COP line. A, From Grow (1981). B, From C. Keen and I. Reid, unpublished manuscript, 1985. See text for further explanation.

tonic subsidence is usually the most difficult parameter to measure. Thermal subsidence occurs as the asthenosphere cools and thermal equilibrium is reestablished. This model predicts a subsidence history for any point of the margin that is a function of the amount of crustal thinning (Fig. 15b).

A number of variations on this model have been proposed to address discrepancies between data and the model or to make the model more geologically reasonable. Jarvis and McKenzie (1980), examined the effect of a finite time for the initial extension. Royden and Keen (1980) and Beaumont and others (1982) looked at the effect of depth-dependent extension, in which the upper crust extends by a different amount than the lower crust and upper mantle. Royden and others (1980) included the thermal effect of dike injection into the crust. Beaumont and others (1982) modeled the effects of melt segregation and accumulation below the crust. Keen (1985) and Buck (1986) included the effect of convective flow in the asthenosphere. These variations in the model lead to predicted differences in the thermo-mechanical history of the margin that have only a subtle effect on the final basin configuration, but which can result in large differences in the early thermal history and petroleum potential of the basin.

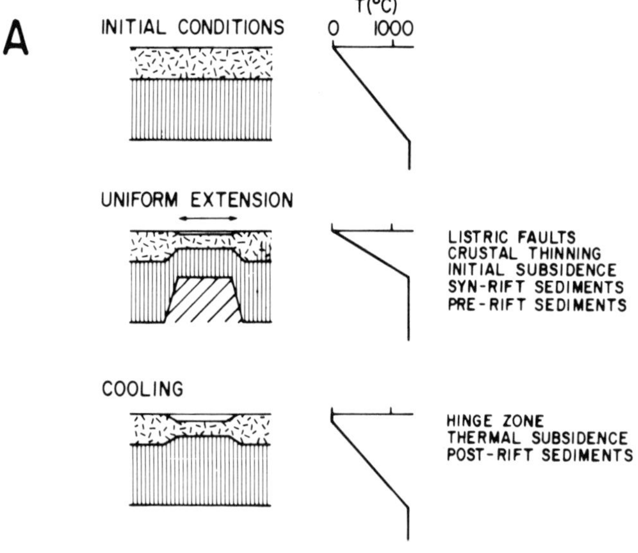

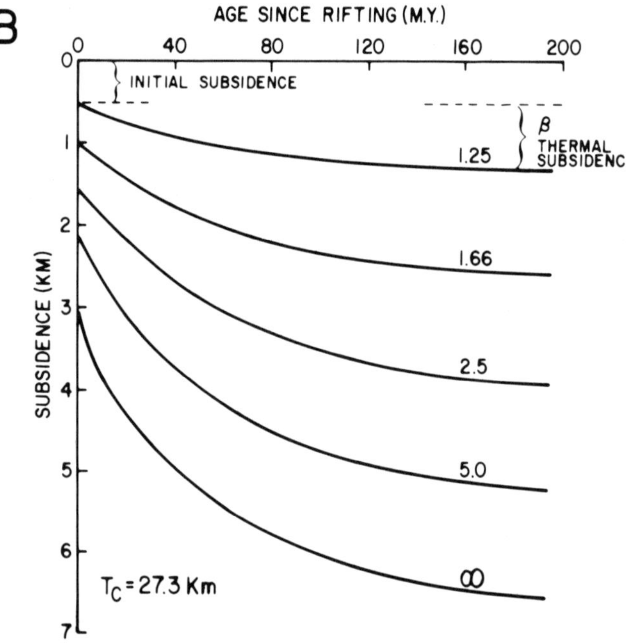

Figure 15. A, Schematic diagram showing the principal features of the stretching model of McKenzie (1978). From Watts, 1981. B, Subsidence as a function of time since rifting for McKenzie's stretching model. From Watts, 1981.

Studies of continental extension regimes, such as the Basin and Range province, have led to the recent development of a different type of model for extension known as the simple-shear model (Bally, 1981; Wernicke, 1985; Lister and others, 1986; LePichon and Barbier, 1987). Crustal failure in this model is not symmetric but takes place along low-angle detachment faults. This model predicts the development of asymmetric crustal structures and subsidence histories for conjugate margins that developed as upper- and lower-plate margins (Lister and others, 1986). The lower-plate margin, comprising rocks that were originally below the detachment surface and remnants of the upper plate, has the more complex structure. The upper-plate margin has a relatively simple block-faulted structure and is composed of rocks originally above the detachment surface, plus underplated material. A discussion of how this model explains certain structural variations found along the Atlantic passive continental margin is found in Klitgord and others (1988). The implications of this model for the subsidence history of the margin are only beginning to be explored quantitatively (Buck and others, 1988).

Atlantic continental margin. Subsidence analyses of wells, using the backstripping technique developed by Watts and Ryan (1976), Steckler and Watts (1978), and Watts and Steckler (1981), have been used to test the subsidence models. In the backstripping analysis, the sedimentary layers are progressively removed, correcting for the effects of compaction and water depth (including sea-level changes), to remove the part of the subsidence that is due to loading. The remaining subsidence as a function of time is attributed to initial and thermal subsidence and compared to model predictions. Backstripping results from several wells on the U.S. Atlantic margin are shown in Figure 16. The errors are quite large due to uncertainties in the loading response of the lithosphere. Smaller errors are due to uncertainties in compaction parameters, sediment ages, and paleowater depths. Although the results from the U.S. Atlantic margin are roughly consistent with the pure-shear uniform-extension model, we cannot readily distinguish among the model variations on the basis of these data. Watts and Thorne (1984) predicted the stratigraphy across and landward of the hinge zone as a function of various model parameters and suggested using the stratigraphy in this region to discriminate among the models.

A drawback of the backstripping method is that the coverage provided is very limited due to the small number of deep wells. An alternative approach has been suggested by Sawyer

(1985), who estimated the crustal stretching factor β from the "total tectonic subsidence," defined as the amount of subsidence calculated from unloading the entire margin given the sediment thickness and an average sediment density (assumed where not known). Unfortunately, the thickness of the synrift sediments is not known in any of the U.S. marginal basins. From the β values obtained, he then predicted the crustal thickness along the entire U.S. Atlantic margin. These results were compared to those obtained from the gravity data in Figure 13. The agreement is quite good, although the Moho steps are generally less sharp than in the gravity models. This agreement is not surprising because both methods of estimating the crustal thickness are affected by the same uncertainties, primarily the depth to basement and the lateral density variations in the sediments. The sharpness of changes in the Moho depth predicted by the total tectonic subsidence model, moreover, is dependent on the assumed flexural parameters.

In the Carolina trough, Celerier (1986) has combined the backstripping approach, which provides the history of the basin at a single point, with the total tectonic subsidence approach, which provides broader areal coverage, by essentially backstripping the entire basin using detailed seismic stratigraphic data in

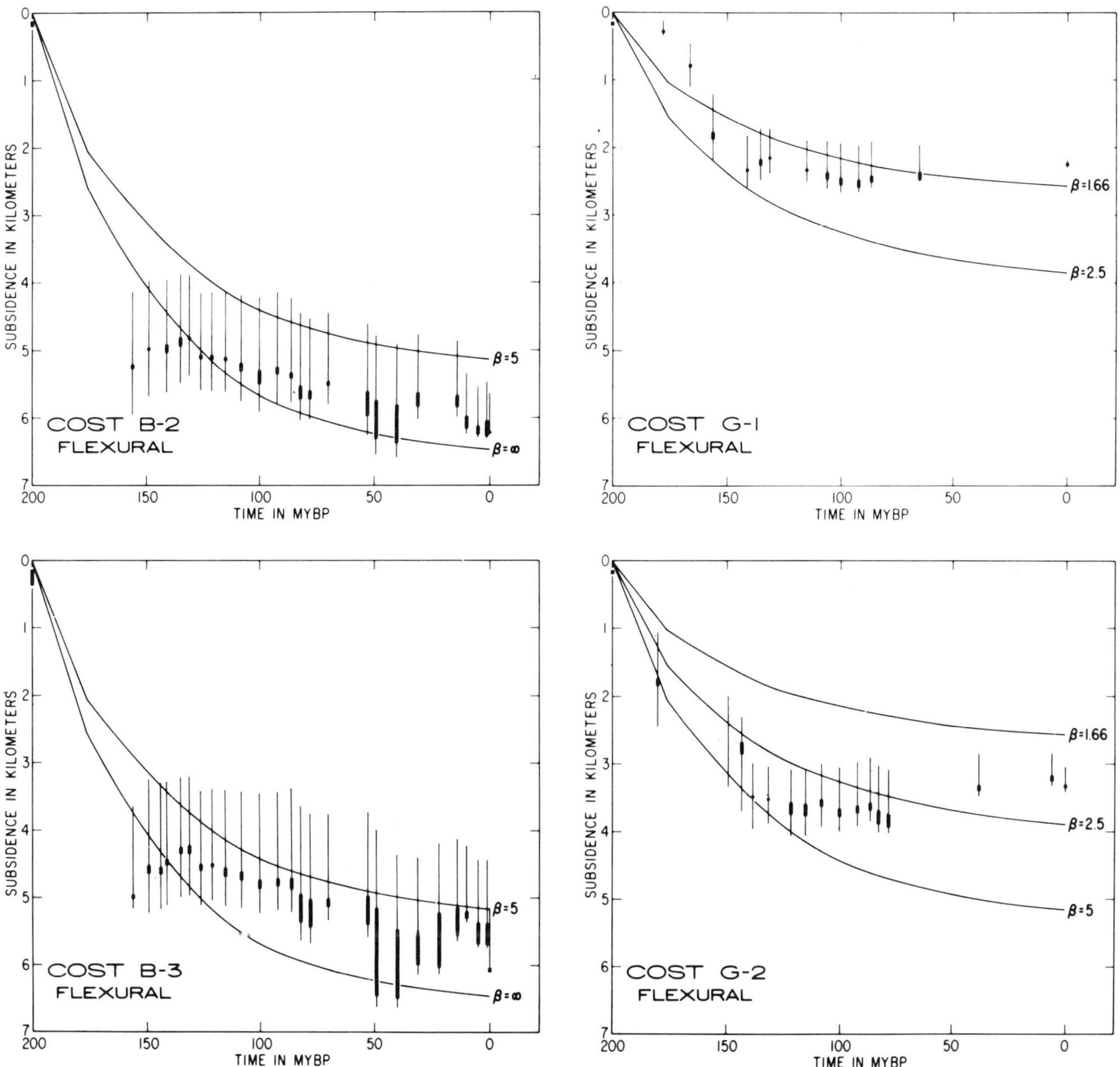

Figure 16. Subsidence history for four deep drillholes on the Atlantic continental margin. The vertical bars represent observed values and associated errors. Predicted subsidence curves for appropriate values of β ar superimposed. From Sawyer and others, 1983. β is defined as the initial crustal thickness divided by the final crustal thickness.

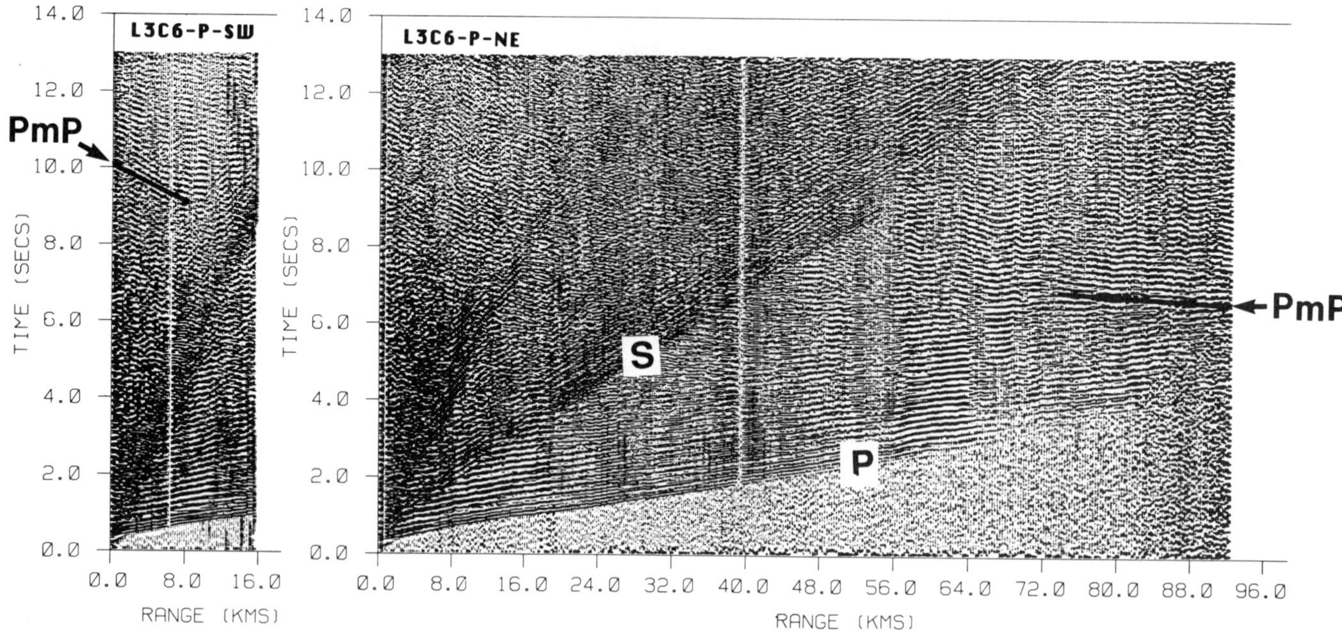

Figure 17. Example of a record section of recent large-aperture data collected by the USGS in the Gulf of Maine in 1985 (hydrophone component from an OBS along line 3 in Fig. 2A). A reduction velocity of 8 km/sec has been applied. The source was two 33-liter (2,000 in^3) airguns fired simultaneously every 2 min, yielding approximately 200-m shot spacing. The data were bandpass filtered (5 to 15 Hz) to remove 20-Hz noise generated by whales that completely masked the data at long ranges. Direct P-, S-wave refracted arrivals and near-vertical and wide-angle Moho reflections (PmP) are labeled. From Trehu, 1987.

lieu of wells. He has demonstrated the relationship between the many different sources of uncertainty in the data and the number of free parameters in the model. Celeirier concluded that the flexural response of the crust and nonuniform extension must be included in the model to correctly simulate the early stratigraphic history, but that the effects of lateral heat flow and thermal blanketing of sediments, though significant, cannot be resolved when data uncertainties are considered.

Although the attempts to model the thermo-mechanical history of the Atlantic continental margin have yielded many insights about the process of rifting, additional data are needed to discriminate among the models. Of particular importance are: (1) reinterpretation and reprocessing of seismic reflection data to increase the resolution of the pre- and post-rift stratigraphy across the hinge zone in order to better estimate the amount and timing of uplift and subsidence in this region; (2) additional seismic reflection data collection and reprocessing of existing data to define the thickness of the synrift deposits and the depth to basement where it is presently unknown; (3) large-aperture seismic data to measure the crustal thickness and velocity beneath the margin and test the gravity and subsidence models; (4) and further knowledge of sediment compaction properties.

Gulf of Mexico continental margin. In the Gulf of Mexico, applicability of the backstripping and total tectonic subsidence methods is limited because of the sparse information on total sediment thickness. Yet total tectonic subsidence analysis for the Gulf basin has been used to estimate the distribution of extended continental crust (Buffler and Sawyer, 1985; Dunbar and Sawyer, 1987). These authors concluded that the zone of extended continental crust is highly asymmetrical: broad in the north and narrow in the south. This conclusion, however, may be partly an artifact of assuming a thin sediment cover over the Campeche shelf region based on high velocities (5 to 6 km/sec) observed in refraction data. Drilling data indicate that these high velocities are due to massive carbonates rather than to crystalline rocks.

Recent large-aperture seismic data

The importance of having large dynamic range and dense spatial sampling (several traces/wavelength) for large-offset as well as for near-vertical incidence seismic data has become increasingly clear. This combination of data permits the use of data-processing techniques such as: (1) frequency-wavenumber (f/k) filtering to remove coherent noise having a different phase velocity from the signals of interest; (2) stacking to increase the signal-to-noise ratio, and (3) tau-p transformation to unravel triplication points in the traveltime curve (e.g., Diebold and Stoffa, 1981). With closely spaced shots, very low-amplitude arrivals can be recognized from background noise on the basis of correlation of phases even though the amplitude of the signal may be very small. The information contained in amplitude-range variations and in secondary arrivals, particularly large-amplitude wide-angle reflections, can be exploited more reliably (Fig. 17).

This technique also permits the use of relatively small sources (e.g., large airguns or small explosive charges) to obtain data at large offsets, eliminating the permitting problems that contributed to the decline of large-offset seismic studies along the U.S. continental margins during the late 1960s and 1970s.

Coordination of CDP reflection data acquisition with refraction/wide-angle reflection data acquisition is also an important feature of recent studies. The reflection data provide information on lateral variations in structure with a resolution that cannot be matched by refraction data, whereas the refraction/wide-angle reflection data provide information on the velocity structure that is needed to properly process the reflection data and to help determine the geologic meaning of observed reflections. Several recent experiments, both onshore and offshore, have demonstrated the value of combining the two types of data (see review by Mooney and Brocher, 1987).

An alphabet soup of techniques has arisen to meet the need for densely spaced large-aperture data (Fig. 4). The constant offset profile (COP) is a modification of the common-depth-point (CDP) technique in which normal streamers on two ships are used to simulate a much longer streamer (Stoffa and Buhl, 1979). The longer effective offset should lead to better velocity determinations than can be obtained from conventional CDP data. The offset can also be adjusted on the basis of previous data to exploit large-amplitude, wide-angle reflections to target particular features.

Large-aperture refraction/wide-angle reflection data can be obtained using either: (1) a single ship to provide the sound source, and microprocessor driven, digital ocean-bottom seismometers (OBS) to record; or (2) two ships in an expanding spread profile (ESP) configuration where the two ships move apart from each other as one ship shoots and the other receives on a multichannel hydrophone streamer. Each method has advantages and disadvantages. An important advantage of ESP over OBS is the unlimited data storage capacity that permits very dense data coverage and the resulting signal enhancement techniques (limited only by the speed at which the data acquisition computer can process the data and the computer operators can change tapes). OBSs are limited in this respect by the capacity of their internal recorders and battery supplies; recent advances in recording technology, however, are leading to large increases in internal storage capacity. The common-depth-point geometry of an ESP averages lateral heterogeneity in a way that permits rigorous one-dimensional interpretation techniques to be used to model the structure directly under the CDP (Diebold and Stoffa, 1981). This characteristic, however, can be a disadvantage as well as an advantage because information about lateral heterogeneity is lost, and in the presence of lateral heterogeneity, the solution may be biased.

The advantages of OBSs are the better control on lateral velocity variations obtained from having overlapping measurements along the line (Fig. 4) and the added flexibility in terms of experiment layout (e.g., two-dimensional recording arrays) and location (e.g., lakes and other enclosed regions where large seismic vessels with long hydrophone streamers cannot maneuver). Moreover, OBSs can record three orthogonal components of ground motion, as well as pressure, whereas only pressure is recorded in an ESP. Although an OBS experiment nominally requires 1.5 times more ship time than an ESP (i.e., three passes over the line for deployment, shooting, and recovery versus one pass each for two ships), in practice this disadvantage is often balanced by the logistical difficulty of running two ships simultaneously.

The locations of recent large-aperture experiments are shown in Figures 2 and 3. One of the first of these experiments was the 1981 Large Aperture Seismic Experiment (LASE), during which ESPs and COPs were shot across the Baltimore Canyon trough by scientists from Lamont-Doherty Geological Observatory, Woods Hole Oceanographic Institution, the Bedford Institute of Oceanography, and the University of Texas Institute for Geophysics (LASE Study Group, 1986). A similar program of COPs and ESPs was shot by LDGO, UTIG, and BGR (Bundesanstalt für Geowissenschaften und Rohstoffe, West Germany) seaward of the Carolina trough to study the structure of oceanic crust (Mutter and the NAT Study Group, 1985). The data pertaining to the continental margin collected during this latter experiment have not yet been processed. An ESP program was conducted by the USGS and the UTIG in the Gulf of Mexico (Phillips and others, 1984).

OBS lines have been shot by the USGS, the Naval Oceanographic Research and Development Agency (NORDA) and the Scripps Institution of Oceanography (Scripps) in the Carolina trough (Trehu and others, 1986, 1989), by the USGS in the Gulf of Maine (Hutchinson and others, 1987, 1988), by BIO off Nova Scotia and Newfoundland (Keen and Barrett, 1981; Keen and Cordsen, 1981; Reid and Keen, 1987), and by UTIG in the Gulf of Mexico (Ebeniro and others, 1985, 1988; Nakamura and others, 1985a, b).

Atlantic continental margin. During LASE, a COP seismic line was shot across the Baltimore Canyon trough along USGS CDP seismic line 25; five ESPs were shot perpendicular to the COP (Fig. 2). The results are summarized in Figure 18a (adapted from LASE Study Group, 1986). Results from three sonobuoy refraction lines shot by the USGS in 1976 (Sheridan and others, 1979) are also shown. Although the data collection and analysis techniques used by Sheridan and others (1979) were similar to those described for pre-1970 experiments, the experiment is discussed in this section because it was planned specifically to address questions posed in response to CDP data and because the results complement those of LASE.

The most striking result of LASE was the apparent continuity of a layer having a velocity of 7.2 to 7.5 km/sec and extending beneath both the marginal basin (presumed to be floored by transitional crust) and oceanic basement. Such a layer is not characteristic of typical old Atlantic crust as defined by Purdy (1983). As mentioned above, however, a 7.2-km/sec layer was observed in many of the old refraction profiles located near the margin, and is frequently observed beneath continental rift

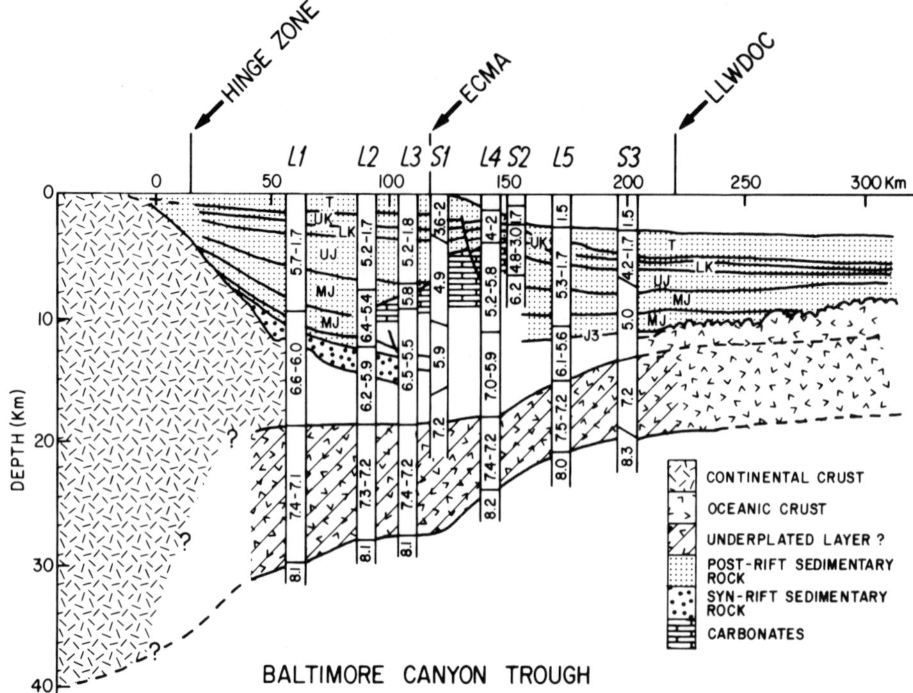

Figure 18A. Interpretive cross section constructed from the results of LASE. The velocity profiles obtained from the ESPs (L1-L5) are superimposed (adapted from LASE Study Group, 1986; seismic stratigraphy from Klitgord and others, 1988). Results from Sheridan and others (1979) have also been included (S1-S3).

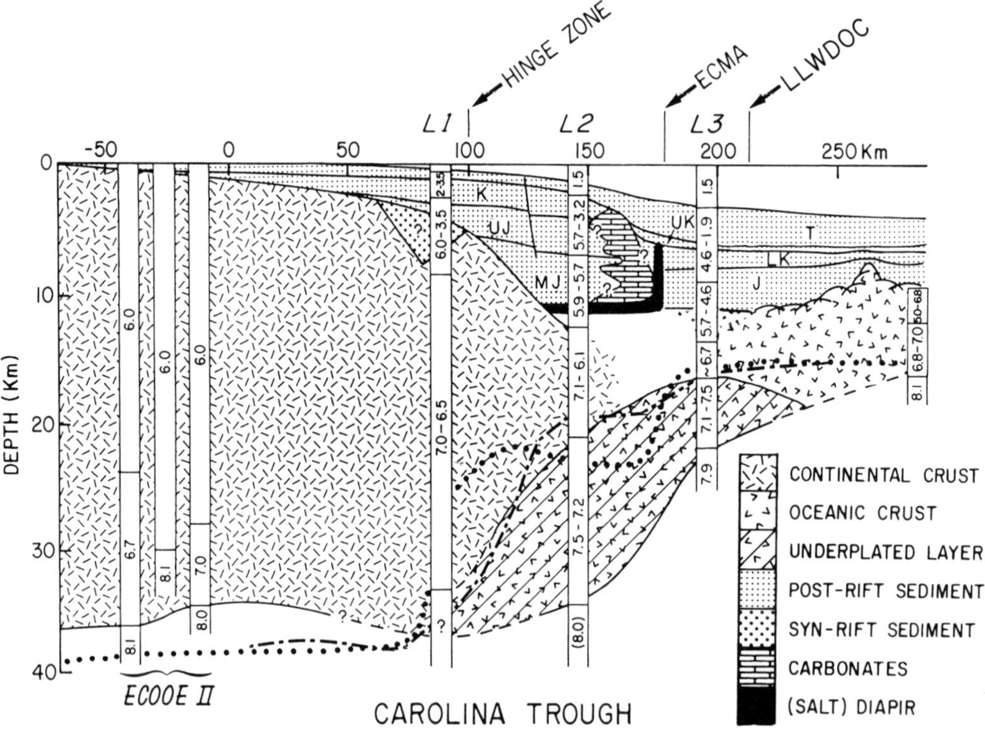

Figure 18B. Interpretive cross section of the Carolina trough based on data from the 1985 USGS large-aperture experiment (L1-L3) (from Tréhu and others, 1989). A "typical" old Atlantic ocean basin velocity structure (Purdy and Ewing, 1986) and three different interpretations of the East Coast onshore-offshore experiment (Hales and others, 1968; James and others, 1968; Lewis and Meyer, 1977) are also included. Also shown are the Moho from gravity (dotted line; from Hutchinson and others, 1983) and thermo-mechanical modeling (dashed line; from Celerier, 1986).

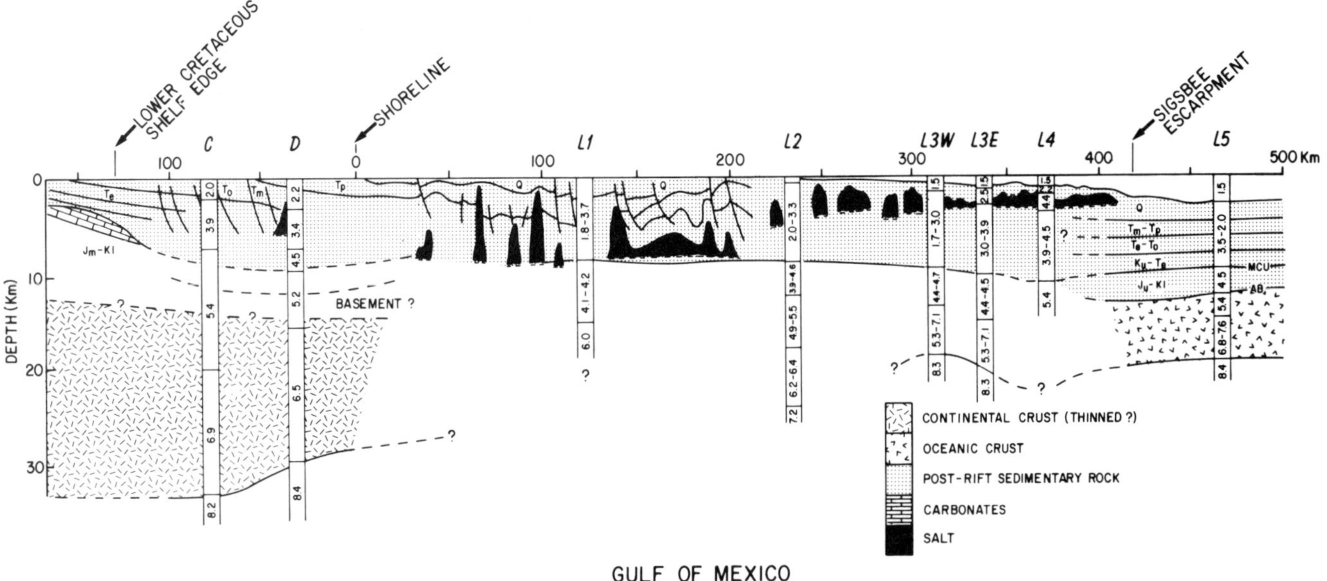

Figure 18C. Cross section of the northern Gulf of Mexico margin constructed from the results of a recent airgun-OBS large-aperture experiment (L1-L5) (from Ebeniro and others, 1985). Because L3 exhibited marked lateral variation along the line, velocity-depth functions representing the eastern and western ends of the line are shown. Onshore velocity-depth functions (C and D) are from Cram (1961) and Dorman and others (1972), respectively.

zones (Mooney and Brocher, 1987). It has been suggested that the 7.2-km/sec layer represents a zone of underplated material that accumulated beneath highly stretched continental crust and early oceanic crust (LASE Study Group, 1986), resulting in the crustal provinces that we refer to as transitional and marginal oceanic crust. An alternative explanation is that the apparent continuity of this layer beneath the ECMA is fortuitous, and that the 7.2-km/sec layer represents the pre-rift continental lower crust landward of the ECMA and represents oceanic lower crust seaward of the ECMA. High-resolution large-offset data landward of the hinge zone are needed to resolve this question. It should be noted that the model presented by Hales and others (1968) for the ECOOE-I line did not contain a high-velocity lower crustal layer, although such a layer might be present but not resolvable in those data. A high-velocity lower crustal layer has been observed farther inland along the axis of the Appalachian mountains (Taylor, this volume).

In 1985, the USGS, NORDA, and Scripps shot a series of four OBS lines in the region of the Carolina trough using 100-lb explosive shots spaced 1 km apart to a maximum offset of 160 km (Trehu and others, 1986, 1989). Lines 1, 2, and 3 (Fig. 2c) were shot parallel to structure, with one line in each of the offshore crustal provinces defined in the introduction: line 1 landward of the hinge zone; line 2 along the deepest part of the Carolina trough between the hinge zone and the ECMA; and line 3 between the ECMA and the landward limit of well-defined oceanic basement. Line 4 was shot across the Carolina trough along USGS CDP line 32. Results based on modeling the traveltimes and amplitudes observed along lines 1, 2 and 3 are summarized in Figure 18b (from Trehu and others, 1989). A lens of 7.2- to 7.5-km/sec material that reaches a maximum thickness of about 15 km is observed beneath lines 2 and 3. This layer is either very deep or absent landward of the hinge zone (line 1). It is interpreted to represent magmatic material underplated to the base of the crust during the latest stages of rifting and early sea-floor spreading.

In 1985, the USGS conducted a series of OBS lines in the Gulf of Maine, landward of the hinge zone (Trehu, 1987). These data complement a deep crustal (15-sec record length) CDP line contracted by the USGS in 1984 and a grid of upper- and mid-crustal (5-sec record length) CDP lines shot in parallel with the OBS experiment. Signals from two 33-liter (2,000 in^3) airguns were recorded to 110-km range with a shot spacing of 100 to 300 m. That strong near-vertical and wide-angle reflections (PmP on Fig. 17) were observed from this relatively weak source in a noisy, shallow-water environment was somewhat surprising and bodes well for future attempts to determine deep crustal structure along continental margins using airgun sources. Large variations in the amplitude patterns among the lines indicate laterally heterogeneous layering in the mid- and lower crust, but the similarity of the traveltime and range at which the Moho reflections are observed indicates that the average crustal structure is fairly homogeneous, with an average velocity of about 6.4 km/sec and a thickness of 32 km.

Gulf of Mexico continental margin. Between 1974 and 1982, UTIG conducted an OBS program in the Gulf of Mexico using explosive shots (1 to 120 lb each) at 2.5-km intervals along 110-km-long lines (Ibrahim and others, 1981). Interpretive techniques were similar to those discussed in the section on pre-1970 seismic refraction data. Most of these lines were located over oceanic crust in the center of the Gulf of Mexico and showed first arrivals from Moho (Fig. 3). Shot spacing was inadequate to resolve fine details of the crustal structure or interpret low-amplitude second arrivals.

Since 1983, UTIG has conducted a series of experiments along the Gulf of Mexico margin over the region of transitional crust using large airgun sources and dense shot spacing (Nakamura and others, 1985a, b; Ebeniro and others, 1985, 1988). These OBS lines are numbered in Figure 3. Lines 2, 3, 4, and 5 were also shot as ESPs by UTIG and USGS (Phillips and others, 1984). A section summarizing the results of lines 1 through 5 is shown in Figure 18c. Data from several older onshore refraction profiles are also shown (Cram, 1961; Dorman and others, 1971). This section shows that the oceanic crust underlies the deep Gulf of Mexico basin and may extend landward of the Sigsbee escarpment and that thinned continental crust is found near the coast. The position and structure of the continent-ocean crustal transition, however, remains elusive. The widely present, shallow, mobilized salt complicates all types of seismic work on the northern margin and ongoing research at UTIG is focusing on using large-aperture and near-vertical seismic data to better resolve the distribution of the salt and to image the structures beneath it.

SUMMARY

Over the past 50 yr, the crustal structure of the U.S. Atlantic and Gulf of Mexico continental margins has been investigated using a variety of geophysical techniques. Integrating these results has helped to focus the limited direct information from seismic measurements and the more extensive indirect information based on other geophysical techniques, so that some broad inferences can be drawn.

The Atlantic continental margin can be divided into several crustal zones on the basis of characteristic magnetic anomaly patterns, structural features seen on seismic reflection lines, gravity models, and limited large-aperture seismic data (Fig. 2a). Seaward of the landward limit of well-defined oceanic basement, the crustal structure has been relatively well determined on the basis of many seismic-refraction and reflection experiments. The results are summarized in Purdy and Ewing (1986). The upper crustal surface is characterized by hyperbolic reflections due to the rough basement topography. The crustal column comprises two layers: a 2- to 3-km-thick upper layer in which the velocity is both laterally and vertically variable, generally increasing from about 4 to 6.4 km/sec; and a 4.5- to 5.5-km-thick lower layer in which the velocity is nearly homogeneous (6.7 to 7.2 km/sec). Unusually thin crust with anomalously low velocity has been associated with fracture zones.

West of the landward limit of well-defined oceanic basement, the characteristic reflection signature of oceanic basement is lost beneath a highly reflective sedimentary layer (J3) along most of the margin (Klitgord and Grow, 1980). This boundary is therefore referred to as the J3 scarp (Klitgord and others, 1988). On Georges Bank, seaward-dipping reflectors are observed in the upper crust in the region between the J3 scarp and the ECMA (Klitgord and others, 1988). Seaward-dipping reflectors have also been tentatively identified on the NAT line crossing the Carolina trough (J. Mutter, 1986, personal communication) and on several other USGS seismic lines that cross the zone between the ECMA and the J3 scarp seaward of the Baltimore Canyon and Carolina troughs (Klitgord and others, 1988). Similar seaward-dipping reflectors on other margins have been interpreted to represent basalt flows interbedded with clastic and volcanic sedimentary material that were formed during the early stages of sea-floor spreading (Hinz, 1981; Mutter and others, 1984; Mutter, 1985; White and others, 1987). Data from recent large-aperture seismic experiments indicate that the crustal velocity and thickness in this region are suggestive of oceanic-type crust, except that the lower crust is unusually thick and its velocity is anomalously high (7.2 to 7.4 km/sec; LASE Study Group, 1986; Trehu and others, 1986, 1989; White and others, 1987). The seaward extent of this high-velocity lower crust has not been determined. This region between the ECMA and the J3 scarp is referred to as the zone of marginal oceanic crust (Fig. 2a).

The structure of the crust between the ECMA and the hinge zone (rift-stage crust in Fig. 2a) is poorly known because of the thick sedimentary fill; even the depth to basement is unknown throughout most of this zone. This represents a serious gap in our knowledge of continental margin structure because the signature of the last stages of continental rifting should be found in this region. The width of the zone between the ECMA and the hinge zone varies from about 50 km across the Carolina trough to 350 km across the Blake Plateau. In the absence of direct seismic measurements, thermo-mechanical (total tectonic subsidence) and gravity models have been used to predict crustal structure. Gravity studies suggest a 10- to 20-km-thick crust in this region with major steps in crustal thickness occurring at the ECMA and the hinge zone. The total tectonic subsidence study suggests a similar but more gradual change in crustal thickness. The uncertainties inherent in the gravity and total tectonic subsidence models are similar because both methods depend heavily on poorly defined sediment thickness information. Large differences between the crustal structure predicted by these models and that obtained from recent seismic studies confirm these uncertainties.

Data from recent large-aperture experiments suggest that the 7.2- to 7.5-km/sec lower crustal layer extends beneath both the Baltimore Canyon and Carolina troughs and increases considerably in thickness in this region (LASE Study Group, 1986; Trehu and others, 1986, 1989). A lower crustal layer with nearly the same velocity and lateral extent as that beneath the Carolina trough has also been observed beneath the Hatton Bank in the northeast Atlantic (White and others, 1987), suggesting that it may be a common characteristic of passive margins. It has been suggested that this layer represents material underplated to the crust during the final stages of the rifting process (LASE Study Group, 1986; Buck and Mutter, 1986; White and others, 1987; Mutter and others, 1988; Trehu and others, 1989). This result suggests that the ECMA coincides with an upper crustal boundary but not with a major throughgoing crustal boundary. It may represent a lateral change from older, very extended continental rocks to younger volcanic rocks in the upper crust overlying a continuous lower crust formed by magmatic processes during the late rifting and early seafloor-spreading phases of margin evolution. The implications for thermo-mechanical models of the rifting process of a thick underplated layer are only beginning to be explored. Buck and Mutter (1986) and Mutter and others (1988) have recently suggested that enough melt to generate both a thick underplated layer and voluminous outpourings of basalt should be generated as a result of convection driven by steep lateral thermal gradients resulting from rapid rifting. White and others (1987) attributed the presence of large volumes of melt to proximity to a hot spot during rifting.

Landward of the hinge zone (rifted continental crust in Fig. 2a), where the sediment cover is thin, crustal information is available from several deep seismic reflection and refraction lines. Measurements of total crustal thickness, however, are sparse. The crustal thickness appears to be about 30 to 35 km wherever it can be estimated. Beneath the Carolina platform, the high-velocity lower crustal layer observed to the east is either thin or absent, on the basis of a large-aperture seismic profile. Beneath the Long Island platform, the crust appears to thin beneath a basin formed during the early stage of rifting, but this apparent crustal thinning is based only on an upwarp of the Moho observed on a seismic reflection profile. Confirmation of this effect awaits large-offset data from the region to permit a better determination of possible lateral velocity variations. In the Gulf of Maine, the average crustal velocity is about 6.4 km/sec, and the thickness is about 32 km. Amplitude differences among the lines indicate that the mid- to lower crust consists of laterally variable high- and low-velocity layering. A thick, high-velocity lower crust, however, is not present. Further analysis of recently collected large-offset data and acquisition of additional data from this region are needed to better resolve possible variations in crustal thickness and to determine the relation between preexisting structure and the final configuration of continental breakup. More detailed interpretations of the stratigraphy associated with pre- and synrift grabens and with the post-rift unconformity are also needed in order to define the history of uplift and subsidence in this region and thus help constrain thermo-mechanical models of rifting.

Because of the thickness of the regional sedimentary section, the extensive salt distribution, and broad distribution of massive carbonate platform deposits, the Gulf of Mexico continental margin cannot be precisely regionalized into zones of rifted continental crust, rift-stage crust, and marginal oceanic crust. A region of oceanic crust in the central Gulf of Mexico has been outlined on the basis of refraction experiments (Ibrahim and others, 1978; Buffler and Sawyer, 1985). The boundary between normal oceanic crust and marginal oceanic crust may be located beneath the salt diapir provinces, where basement structure is poorly defined on reflection profiles. A counterpart of the ECMA is found only in the region of the West Florida shelf (Klitgord and Schouten, 1986). The broad region of rifted continental crust south and west of the Gulf Rim fault zone and Bahamas fracture zone suggests that the distinction between rifted continental crust and rift-stage crust as defined for the Atlantic margin may not exist along this margin. Alternatively, this boundary may be hidden beneath the Lower Cretaceous carbonate bank–shelf edge complex. Recent large-aperture seismic experiments (Phillips and others, 1984; Ebeniro and others, 1986, 1988) have confirmed that the crust landward of the Sigsbee Escarpment is thin (8 to 12 km), and Rayleigh wave, S_n wave, onshore refraction, and gravity data suggest that thin crust extends beneath the continental shelf and coastal plain (Keller and Shurbet, 1975; Shurbet, 1976).

Although many questions remain about the deep crustal structure beneath the U.S. Atlantic and Gulf of Mexico continental margins, recent advances in instrumentation and interpretation techniques have provided the means to collect the necessary information, and answers should be forthcoming within the next several years.

REFERENCES CITED

Alsop, L. E., and Talwani, M., 1984, The east coast magnetic anomaly: Science, v. 226, p. 1189–1191.

Alvord, D. C., Bell, K. G., Pease, M. H., Jr., and Barosh, P. J., 1976, The aeromagnetic expression of bedrock geology between the Clinton-Newbury and Bloody Bluff fault zones, northeastern Massachusetts: U.S. Geological Survey Journal of Research, v. 4, no. 5, p. 601–604.

Anderson, E. G., 1979, Basic Mesozoic study in Louisiana, the northern coastal region, and the Gulf basin province: Louisiana Geological Survey Folio series no. 3.

Antoine, J. W., and Ewing, J., 1963, Seismic refraction measurements on the margins of the Gulf of Mexico: Journal of Geophysical Research, v. 68, p. 1975–1996.

Antoine, J. W., and Harding, J. L., 1965, Structure beneath continental shelf, northeastern Gulf of Mexico: American Association of Petroleum Geologists Bulletin, v. 49, p. 157–171.

Antoine, J. W., and Henry, V. J., 1969, Seismic refraction study of shallow part of continental shelf off Georgia coast: American Association of Petroleum Geologists Bulletin, v. 49, p. 601–609.

Austin, J. A., Uchupi, E., Shaughnessy, D. R., and Ballard, R. D., 1980, Geology of the New England passive margin: American Association of Petroleum Geologists Bulletin, v. 64, p. 501–526.

Ball, M. N., Martin, R. G., Fook, R. Q., and Applegate, A. V., 1988, Structure and stratigraphy of the western Florida Shelf: U.S. Geological Survey Open File Report 88–439, 61 p.

Bally, A. W., 1981, Atlantic-type margins, *in* Geology of passive continental margins; History, structure, and sedimentological record, with special emphasis on the Atlantic margin: American Association of Petroleum Geologists Education Course Note Series 19, p. 1-1–1-48.

Barrett, D. L., Berry, M., Blanchard, J. E., Keen, M. J., and McAllister, R. E., 1964, Seismic studies on the eastern seaboard of Canada; The Atlantic coast of Nova Scotia: Canadian Journal of Earth Sciences, v. 1, p. 10–22.

Barton, P. J., 1986, The relationship between seismic velocity and density in the continental crust; A useful constraint?: Geophysical Journal of the Royal Astronomical Society, v. 87, p. 195–200.

Beaumont, C., Keen, C. E., and Boutilier, R., 1982, On the evolution of rifted continental margins; Comparison of models and observations for the Nova Scotia margin: Geophysical Journal of the Royal Astronomical Society, v. 70, p. 667–716.

Behrendt, J. C., and Klitgord, K. D., 1980, High sensitivity aeromagnetic survey of the U.S. Atlantic continental margin: Geophysics, v. 45, p. 1813–1846.

Behrendt, J. C., Hamilton, R. M., Ackermann, H. D., Henry, V. J., and Bayer, K. C., 1983, Marine multichannel seismic-reflection evidence for Cenozoic faulting and deep crustal structure near Charleston, South Carolina, *in* Gohn, G. S., ed., Studies related to the Charleston, South Carolina, earthquake of 1886; Tectonics and seismicity: U.S. Geological Survey Professional Paper 1313, p. J1–J29.

Bentley, L. R., and Worzel, J. L., 1956, Geophysical investigations in the emerged and submerged Atlantic Coastal Plain; Part 10, Continental slope and rise south of the Grand Banks: Geological Society of America Bulletin, v. 67, p. 1–18.

Bouma, A. H., 1983, Intraslope basins in northern Gulf of Mexico; A key to ancient submarine canyons and fans, *in* Watkins, J. S., and Drake, C. L., eds., Studies in continental margin geology: American Association of Petroleum Geologists Memoir 34, p. 567–581.

Buck, W. R., 1986, Small-scale convection induced by passive rifting: The cause for uplift of rift shoulders: Earth and Planetary Science Letters, v. 77, p. 362–372.

Buck, W. R., and Mutter, J. C., 1986, Convestive partial melting; 2, Controls on basalt volume at rifted margins: EOS Transactions of the American Geophysical Union, v. 67, p. 1192.

Buck, W. R., Martinez, F., Steckler, M. S., and Cochran, J. R., 1988, Thermal consequences of lithospheric extension; pure and simple: Tectonics, v. 7, p. 213–234.

Buffler, R. T., 1983a, Structure of the Sigsbee Scarp, Gulf of Mexico, *in* Bally, A. W., ed., Seismic expression of structural styles; A picture and work atlas: American Association of Petroleum Geologists Studies in Geology 15, v. 2, p. 2.3.2-50.

——, 1983b, Structure and stratigraphy of the Sigsbee Salt Dome area, deep south-central Gulf of Mexico, *in* Bally, A. W., ed., Seismic expression of structural styles; A picture and work atlas: American Association of Petroleum Geologists Studies in Geolcgy 15, v. 2, p. 2.3.2-56.

Buffler, R. T., and Sawyer, D. S., 1985, Distribution of crust and early history, Gulf of Mexico basin: Gulf Coast Association of Geological Socictics Transactions, v. 35, p. 333–344.

Buffler, R. T., Watkins, J. S., Worzel, J. L., and Shaub, F. J., 1980, Structure and early geologic history of the deep central Gulf of Mexico, *in* Pilger, R., ed., Proceedings of a Symposium on the Origin of the Gulf of Mexico and the Early Opening of the Central North Atlantic: Baton Rouge, Louisiana State University, p. 3–16.

Buffler, R. T., Shaub, F. J., Huerta, R., and Ibrahim, A. K., 1981, A model for the early evolution of the Gulf of Mexico basin; Proceedings, 26th International Geological Congress, Geology of Continental Margins Symposium, Paris, July 1980: Oceanologica Acta, p. 129–136.

Buffler, R. T., and others, 1984, Initial reports of the Deep Sea Drilling Project: Washington, D.C., U.S. Government Printing Office, v. 77, 747 p.

Carlson, R. O., and Brown, M. N., 1955, Seismic-refraction profiles in the submerged Atlantic coastal plain near Ambrose lightship: Geological Society of America Bulletin, v. 66, p. 969–976.

Celerier, B., 1986, Models for the evolution of the Carolina trough and their limitations [Ph.D. thesis]: Cambridge, Massachusetts Institute of Technology, 206 p.

Corso, W., 1987, Development of the Early Cretaceous northwest Florida carbonate platform [Ph.D. thesis]: Austin, University of Texas at Austin, 180 p.

Cram, I. H., 1961, A crustal structure refraction survey in south Texas: Geophysics, v. 26, p. 560–573.

Daniels, D. L., and Zietz, I., 1978, Geological interpretation of geomagnetic maps of the Coastal Plain region of South Carolina and parts of North Carolina and Georgia: U.S. Geological Survey Open-File Report 78-261, 47 p.

Daniels, D. L., Zietz, I., and Popenoe, P., 1983, Distribution of subsurface lower Mesozoic rocks in the southeastern United States, as interpreted from regional aeromagnetic and gravity maps, *in* Gohn, G. S., ed., Studies related to the Charleston, South Carolina, earthquake of 1886; Tectonics and seismicity: U.S. Geological Survey Professional Paper 1313, p. K1–K23.

Diebold, J. B., and Stoffa, P. L., 1981, The travel-time equation, tau-p mapping and inversion of common midpoint data: Geophysics, v. 46, p. 238–254.

Dillon, W. P., and Popenoe, P., 1988, The Blake Plateau basin and Carolina trough, *in* Sheridan, R. E., and Grow, J. A., eds., Atlantic continental margin, U.S.: Boulder, Colorado, Geological Society of America, The Geology of North America, v. I-2, p. 291–328.

Dillon, W. P., and 6 others, 1983, Growth faulting and salt diapirism; Their relationship and control in the Carolina Trough, east North America, *in* Watkins, J. S., and Drake, C. L., eds., Studies in continental margin geology: American Association of Petroleum Geologists Memoir 34, p. 21–48.

Dillon, W. P., Paull, C. K., and Gilbert, L. E., 1985, History of the Atlantic continental margin off Florida; The Blake Plateau basin, *in* Poag, C. W., ed., Geologic evolution of the United States Atlantic Margin: New York, Van Nostrand Reinhold Company, p. 189–216.

Dorman, J., Worzel, J. L., Leyden, R., Crook, T. N., and Hartziemmanuel, M., 1972, Crustal section from seismic refraction measurements near Victoria, Texas: Geophysics, v. 37, p. 325–336.

Dowling, J. J., 1968, The East Coast onshore-offshore experiment; 2, Seismic refraction measurements on the continental shelf between Cape Hatteras and Cape Fear: Seismological Society of America Bulletin, v. 58, p. 821–834.

Drake, C. L., Wokrzel, J. L., and Beckmann, W. C., 1954, Geophysical investigations in the emerged and submerged Atlantic coastal plain; Part 9, Gulf of Maine: Geological Society of America Bulletin, v. 65, p. 957–970.

Drake, C. L., Ewing, M., and Sutton, G. H., 1959, Continental margins and geosynclines of the East Coast of North America north of Cape Hatteras: Physics and Chemistry of the Earth, v. 3, p. 110–198.

Dunbar, J. A., and Sawyer, D. S., 1987, Implications of continental crust extension for plate reconstruction; An example from the Gulf of Mexico: Tectonics, v. 6, p. 739–755.

Ebeniro, J. O., Sawyer, D. S., Nakamura, Y., Shaub, F. J., and O'Brien, W. P., Jr., 1985, Deep structure of the shelf and slope of the northern Gulf of Mexico: Part B, An ocean-bottom seismograph–air gun experiment: University of Texas Institute for Geophysics Technical Report 41, 37 p.

Ebeniro, J. O., Nakamura, Y., Sawyer, D. S., and O'Brien, W. P., Jr., 1988, Sedimentary and crustal structure of the northwestern Gulf of Mexico: Journal of Geophysical Research, v. 93, p. 9075–9092.

Emery, K. O., Upchupi, E., Phillips, J. D., Bowin, C. O., Bunce, E. T., and Knott, S. T., 1970, Continental rise off eastern North America: American Association of Petroleum Geologists Bulletin, v. 54, p. 44–108.

Ewing, J. I., and Ewing, M., 1959, Seismic refraction measurements in the Atlantic Ocean basins, in the Mediterranean Sea, on the Mid-Atlantic Ridge, and in the Norwegian Sea: Geological Society of America Bulletin, v. 70, p. 291–318.

Ewing, M., Carry, A. P., and Rutherford, H. M., 1937, Geophysical investigations in the emerged and submerged Atlantic Coastal Plain; Part 1, Methods and results: Geological Society of America Bulletin, v. 48, p. 753–802.

Ewing, M., Woolard, G. P., and Vine, A. C., 1939, Geophysical investigations in the emerged and submerged Atlantic Coastal Plain; Part 3, Barnegat Bay, New Jersey: Geological Society of America Bulletin, v. 50, p. 257–296.

——, 1940, Geophysical investigations in the emerged and submerged Atlantic Coastal Plain; Part 4, Cape May, New Jersey: Geological Society of America Bulletin, v. 51, p. 1821–1840.

Ewing, M., Worzel, J. L., Steenland, N. C., and Press, F., 1950, Geophysical investigations in the emerged and submerged Atlantic Coastal Plain; Part 5, Woods Hole, New York, and Cape May sections: Geological Society of America Bulletin, v. 61, p. 877–892.

Ewing, J. I., Antoine, J., and Ewing, M., 1960, Geophysical measurements in the western Caribbean Sea and in the Gulf of Mexico: Journal of Geophysical Research, v. 65, p. 4087–4126.

Ewing, J. I., Worzel, J. L., and Ewing, M., 1962, Sediments and oceanic structural history of the Gulf of Mexico: Journal of Geophysical Research, v. 67, p. 2509–2527.

Folger, D. W., Dillon, W. P., Grow, J. S., Klitgord, K. D., and Schlee, J. S., 1979,

Evolution of the Atlantic continental margin of the United States, *in* Talwani, M., Hay, W., and Ryan, W.B.F., eds., Deep drilling results in the Atlantic Ocean; Continental margins and paleoenvironment: American Geophysical Union Maurice Ewing series 3, p. 87–108.

Gough, D. I., 1967, Magnetic anomalies and crustal structure in eastern Gulf of Mexico: American Association of Petroleum Geologists Bulletin, v. 51, p. 200–211.

Grow, J. A., 1980, Deep structure and evolution of the Baltimore Canyon through in the vicinity of the COST No. B-3 well, *in* Scholle, P. A., ed., Geological studies of the COST B-3 well, U.S. Mid-Atlantic continental slope area: U.S. Geological Survey Circular 833, p. 117–124.

—— , 1981, The Atlantic margin of the United States, *in* Bally, A. W., ed., Geology of passive continental margins; History, structure, and sedimentologic record, with special emphasis on the Atlantic margin: American Association of Petroleum Geologists Education Course Note Series 19, p. 3-1–3.44.

Grow, J. A., Bowin, C. O., and Hutchinson, D. R., 1979, The gravity field of the U.S. Atlantic continental margin: Tectonophysics, v. 59, p. 27–52.

Grow, J. A., Klitgord, K. D., and Schlee, J. S., 1988, Structure and evolution of Baltimore Canyon Trough, *in* Sheridan, R. E., and Grow, J. A., eds., Atlantic continental margin, U.S.: Boulder, Colorado, Geological Society of America, The Geology of North America, v. I-2, p. 269–290.

Hales, A. L., Helsley, C. E., Dowling, J. J., and Nation, J. B., 1968, The East Coast onshore-offshore experiment; 1. The first-arrival phases: Seismological Society of America Bulletin, v. 58, p. 757–783.

Hales, A. L., Helsley, C. E., and Nation, J. B., 1970, Traveltimes for an oceanic path: Journal of Geophysical Research, v. 75, p. 7362–7381.

Hall, D. J., Cabanaugh, T. D., Watkins, J. S., and McMillen, K. J., 1983, The rotational origin of the Gulf of Mexico based on regional gravity data, *in* Watkins, J. S., and Drake, C. L., eds., Studies in continental margin geology: American Association of Petroleum Geologists Memoir 34, p. 155–166.

Hall, S., Shepherd, A., and Titus, M., 1984a, Magnetic anomalies of the eastern Gulf of Mexico, *in* Buffler, R. T., and others, eds., Gulf of Mexico, Atlas 6, Ocean Margin Drilling Program, Regional Atlas Series: Woods Hole, Massachusetts, Marine Science International, sheet 3, scale 1 in. = 37.5 mi.

Hall, S., Shepherd, A., Titus, M., Snow, R., Pilger, R., and Angelich, M. T., 1984b, Gravity anomalies, *in* Buffler, R. T., and others, eds., Gulf of Mexico, Atlas 6, Ocean Margin Drilling Program, Regional Atlas Series: Woods Hole, Massachusetts, Marine Science International, sheet 2, scale 1 in. = 37.5 mi.

Harwood, D. S., and Zietz, I., 1976, Geologic interpretation of an aeromagnetic map of southern New England: U.S. Geological Survey Geophysical Investigation Map GP-906, scale 1:250,000.

Hatcher, R. D., Jr., and Zietz, I., 1979, Tectonic implications of regional aeromagnetic data from the southern Appalachians, *in* Wones, D. R., ed., The Caledonides in the U.S.A.; Proceedings of the Second Symposium of the International Geological Correlation Program, Caledonide Orogen Project: Blacksburg, Virginia Polytechnic Institute and State University Memoir 2, p. 235–244.

Hatcher, R. D., Jr., Howell, D. E., and Talwani, P., 1977, Eastern Piedmont fault system; Speculations on its extent: Geology, v. 5, p. 636–640.

Hersey, J. B., Bunce, E. T., Wyrick, R. F., and Dietz, F. T., 1959, Geophysical investigation of the continental margin between Cape Henry, Virginia, and Jacksonville, Florida: Geological Society of America Bulletin, v. 70, p. 437–466.

Higgins, M. W., and Zietz, I., 1983, Geologic interpretation of geophysical maps of the pre-Cretaceous "basement" beneath the Coastal Plain of the southeastern United States, *in* Hatcher, R. D., Jr., Williams, H., and Zietz, S., eds., Contributions to the tectonics and geophysics of mountain chains: Geological Society of America Memoir 158, p. 125–130.

Higgins, M. W., Zietz, I., and Fisher, G. W., 1974, Interpretation of aeromagnetic anomalies bearing on the origin of Upper Chesapeake Bay and river course changes in the central Atlantic seaboard region; Speculations: Geology, v. 2, p. 73–76.

Hinz, K., 1981, A hypothesis on terrestrial catastrophes; Wedges of very thick oceanward dipping layers beneath passive continental margins; Their origin and paleoenvironmental significance: Geologisches Jahrbuch Reihe E, Geophysik, v. 22, p. 3–28.

Houtz, R. E., 1984, Seismic velocity structure and tables; *in* Uchupi, E., and Shor, A. N., eds., Eastern North American continental margin and adjacent ocean floor; Ocean Margin Drilling Program Regional Data Synthesis Series, Atlas 3: Woods Hole, Massachusetts, Marine Science International, sheets 6 and 7, scale 1 in. = 22 mi.

Houtz, R. E., and Bryan, G. M., 1984, Seismic velocity structure and tables, *in* Bryan, G. M., and Heirtzler, J. R., eds., Eastern North American continental margin and adjacent ocean floor; Ocean Margin Drilling Program, Regional Data Synthesis Series, Atlas 5: Woods Hole, Massachusetts, Marine Science International, sheets 7 and 8, scale 1 in. = 25 mi.

Houtz, R. E., and Ewing, J. I., 1963, Detailed sedimentary velocities from seismic refraction profiles in the western North Atlantic: Journal of Geophysical Research, v. 68, p. 5233–5258.

—— , 1964, Sedimentary velocities of the western North Atlantic margin: Seismological Society of America Bulletin, v. 54, p. 867–895.

Houtz, R. E., and Rabinowitz, P. D., 1984, Seismic velocity structure and tables, *in* Ewing, J. I., and Rabinowitz, P. D., eds., Eastern North American continental margin and adjacent ocean floor; Ocean Margin Drilling Program, Regional Data Synthesis Series, Atlas 4: Woods Hole, Massachusetts, Marine Science International, sheets 7 and 8, scale 1 in. = 23 mi.

Hutchinson, D. R., Grow, J. A., Klitgord, K. D., and Swift, B. A., 1983, Deep structure and evolution of the Carolina Trough, *in* Watkins, J. S., and Drake, C. L., eds., Studies in continental margin geology: American Association of Petroleum Geologists Memoir 34, p. 129–152.

Hutchinson, D. R., Klitgord, K. D., and Detrick, R. S., 1985, Block Island fault; A Paleozoic crustal boundary on the Long Island Platform: Geology, v. 13, p. 875–879.

Hutchinson, D. R., Grow, J. A., and Klitgord, K. D., and Detrick, R. S., 1986a, Moho reflections from the Long Island Platform, eastern United States, *in* Barazangi, M., and Brown, L., eds., Reflection seismology; The continental crust: American Geophysical Union Geodynamics Series, v. 14, p. 173–188.

Hutchinson, D. R., Klitgord, K. D., and Detrick, R. S., 1986b, Rift basins of the Long Island platform: Geological Society of America Bulletin, v. 97, p. 688–702.

Hutchinson, D. R., Klitgord, K. D., and Trehu, A. M., 1987, Structure of the lower crust beneath the Gulf of Maine: Geophysical Journal of the Royal Astronomical Society, v. 89, p. 189–194.

Hutchinson, D. R., Klitgord, K. D., Lee, M. W., and Trehu, A. M., 1988, U.S.G.S. deep seismic reflection profile across the Gulf of Maine: Geological Society of America Bulletin, v. 100, p. 172–184.

Ibrahim, A. K., Carye, J., Latham, G., and Buffler, R. T., 1981, Crustal structure in the Gulf of Mexico from OBS refraction and multichannel reflection data: American Association of Petroleum Geologists Bulletin, v. 65, p. 1207–1229.

James, D. E., Smith, T. J., and Steinhart, J. S., 1968, Crustal structure of the Middle Atlantic states: Journal of Geophysical Research, v. 73, p. 1983–2007.

Jansa, L. F., and Wade, J. A., 1975, Geology of the continental margin off Nova Scotia and Newfoundland, *in* Van der Linden, W.J.M., and Wade, J. A., eds., Offshore geology of eastern Canada: Geological Survey of Canada Paper 74-30, p. 51–105.

Jarvis, J. G., and McKenzie, D. P., 1980, Sedimentary basin formation with finite extension rates: Earth and Planetary Science Letters, v. 48, p. 42–52.

Kane, M. F., Yellin, M. J., Bell, K. G., and Zietz, I., 1972, Gravity and magnetic evidence of lithology and structure in the Gulf of Main region: U.S. Geological Survey Professional Paper 726, p. B1–B22.

Karner, G. D., and Watts, A. B., 1982, On isostasy at Atlantic-type continental margins: Journal of Geophysical Research, v. 87, p. 2923–2948.

Katz, S., and Ewing, M., 1956, Seismic-refraction measurements in the Atlantic Ocean basin, west of Bermuda: Geological Society of America Bulletin,

v. 67, p. 475–510.

Katz, S., Edwards, R. S., and Press, F., 1953, Seismic refraction profile across the Gulf of Maine: Geological Society of America Bulletin, v. 64, p. 249–252.

Keen, C. E., 1985, The dynamics of rifting; Deformation of the lithosphere by active and passive driving forces: Geophysical Journal of the Royal Astronomical Society, v. 80, p. 95–120.

Keen, C. E., and Barrett, D. L., 1981, Thinned and subsided continental crust on the rifted margin of eastern Canada; Crustal structure, thermal evolution, and subsidence history: Geophysical Journal of the Royal Astronomical Society, v. 65, p. 443–466.

Keen, C. E., and Cordsen, A., 1981, Crustal structure, seismic stratigraphy, and rift processes of the continental margin off eastern Canada; Ocean bottom seismic refraction results off Nova Scotia: Canadian Journal of Earth Sciences, v. 18, p. 1523–1538.

Keen, M. J., 1969, Possible edge effect to explain magnetic anomalies off the eastern seaboard of the U.S.: Nature, v. 222, p. 72–74.

Keller, G. R., and Shurbet, D. H., 1975, Crustal structure of the Texas Gulf coastal plain: Geological Society of America Bulletin, v. 86, p. 807–810.

Kent, K. M., 1979, Two-dimensional gravity model of the southeast Georgia Embayment-Blake Plateau [M.S. thesis]: Newark, University of Delaware, 89 p.

Klitgord, K. D., and Behrendt, J. C., 1979, Basin structure of the U.S. Atlantic margin, *in* Watkins, J. S., Montadert, L., and Dickerson, P. W., eds., Geological and geophysical investigations of continental margins: American Association of Petroleum Geologists Memoir 29, p. 85–112.

Klitgord, K. D., and Grow, J. A., 1980, Jurassic seismic stratigraphy and basement structure of western Atlantic magnetic zone: American Association of Petroleum Geologists Bulletin, v. 64, p. 1658–1680.

Klitgord, K. D., and Hutchinson, D. R., 1985, Distribution and geophysical signatures of early Mesozoic rift basins beneath the U.S. Atlantic continental margin, *in* Robinson, G. R., and Froelich, A. J., eds., Proceedings of the 2nd U.S. Geological Survey Workshop on the Early Mesozoic Basins of the Eastern North America: U.S. Geological Survey Circular 946, p. 45–61.

Klitgord, K. D., and Schouten, H., 1986, Plate kinematics of the central Atlantic, *in* Vogt, P. R., and Tucholke, B. E., eds., The western Atlantic region: Boulder, Colorado, Geological Society of America, The Geology of North America, v. M., p. 351–378.

Klitgord, K. D., Schlee, J. S., and Hinz, K., 1982, Basement structure, sedimentation, and tectonic history of the Georges Bank basin, *in* Scholle, P. A., and Wenkan, C. R., eds., Geological studies of the COST Nos. G-1 and G-2 Wells, United States North Atlantic Outer Continental Shelf: U.S. Geological Survey Circular 861, p. 160–186.

Klitgord, K. D., Dillon, W. P., and Popenoe, P., 1983, Mesozoic tectonics of the southeastern United States Coastal Plain and continental margin, *in* Gohn, G. S., eds., Studies related to the Charleston, South Carolina, earthquake of 1886; Tectonics and seismicity: U.S. Geological Survey Professional Paper 1313, p. P1–P15.

Klitgord, K. D., Popenoe, P., and Schouten, H., 1984, Florida; A Jurassic transform plate boundary: Journal of Geophysical Research, v. 89, p. 7753–7772.

Klitgord, K. D., Hutchinson, D. R., and Schouten, H., 1988, U.S. Atlantic continental margin; Structural and tectonic framework, *in* Sheridan, R. E., and Grow, J. A., eds., Atlantic continental margin, U.S.: Boulder, Colorado, Geological Society of America, The Geology of North America, v. I-2, p. 19–53.

LADLE Study Group, 1983, A lithosphere seismic refraction profile in the western North Atlantic Ocean: Geophysical Journal of the Royal Astronomical Society, v. 75, p. 23–69.

LASE Study Group, 1986, Deep structure of the U.S. East Coast passive margin from large aperture seismic experiments (LASE): Marine and Petroleum Geology, v. 3, p. 234–242.

Leg 104 Scientific Party, 1986, Reflector identified, glacial onset seen: Geotimes, March, p. 12–15.

LePichon, X., and Barbier, F., 1987, Passive margin formation by low-angle faulting within the upper crust; The northern Bay of Biscay margin: Tectonics, v. 6, p. 133–150.

Lewis, B.T.R., and Meyer, R. P., 1977, Upper mantle velocities under the East Coast margin of the U.S.: Geophysical Research Letters, v. 4, p. 341–344.

Lister, G. S., Etheridge, M. A., and Symonds, P. A., 1986, Detachment faulting and the evolution of passive continental margins: Geology, v. 14, p. 246–250.

Locker, S. D., and Chatterjee, S. K., 1984, Seismic velocity structure, *in* Buffler and others, eds., Gulf of Mexico, Ocean Margin Drilling Program, Regional Atlas Series: Woods Hole, Massachusetts, Marine Science International, Atlas 6, sheet 4.

Manspeizer, W., 1981, Early Mesozoic basins of the central Atlantic passive margins, *in* Bally, A. W., ed., Geology of passive continental margins; History, structure, and sedimentologic record, with special emphasis on the Atlantic margin): American Association of Petroleum Geologists Education Course Note Series 19, p. 4-1–4-60.

——, 1985, Early Mesozoic history of the Atlantic passive margin, *in* Poag, C. W., ed., Geologic evolution of the United States Atlantic margin: New York, Van Nostrand Reinhold Co., p. 1–23.

Manspeizer, W., and Cousminer, H. L., 1988, Late Triassic-early Jurassic synrift basins of the U.S. Atlantic margin, *in* Sheridan, R. E., and Grow, J. A., Atlantic continental margin, U.S.: Boulder, Colorado, Geological Society of America, The Geology of North America, v. I-2, p. 197–216.

Martin, R. G., 1980, Distribution of salt structures in the Gulf of Mexico; Map and descriptive text: U.S. Geological Survey Miscellaneous Field Studies Map MF-1213.

Martin, R. G., and Buffler, R. T., 1984, Alabama to Cuba, *in* Buffler, R. T., and others, Gulf of Mexico, Ocean Margin Drilling Program, Regional Atlas Series: Woods Hole, Massachusetts, Marine Science International, Atlas 6, sheet 25.

McKenzie, D. P., 1978, Some remarks on the development of sedimentary basins: Earth and Planetary Science Letters, v. 40, p. 25–32.

Mooney, W. D., and Brocher, T. M., 1987, Coincident seismic reflection/refraction studies of the continental lithosphere; A global review: Reviews of Geophysics, v. 25, p. 7237–7242.

Mutter, J. C., 1985, Seaward dipping reflectors and the continent-ocean boundary at passive continental margins; Proceedings of the 1983 IUGG Conference: Tectonophysics, v. 114, p. 117–131.

Mutter, J. C., Talwani, M., and Stoffa, P. L., 1984, Evidence for a thick oceanic crust adjacent to the Norwegian margin: Journal of Geophysical Research, v. 89, p. 483–502.

Mutter, J. C., and NAT Study Group, 1985, Multichannel seismic images of the oceanic crust's internal structure; Evidence for a magma chamber beneath the Mesozoic Mid-Atlantic Ridge: Geology, v. 9, p. 629–632.

Mutter, J. C., Buck, W. R., and Zehnder, C. M., 1988, Convective melting: A model for the formation of thick basaltic sequences during the initiation of spreading: Journal of Geophysical Research, v. 93, p. 1031–1048.

Nakamura, Y., and Howell, B. F., Jr., 1964, Maine seismic experiment; Frequency spectra of refraction arrivals and the nature of the Mohorovicic discontinuity: Seismological Society of America Bulletin, v. 54, p. 9–18.

Nakamura, Y., Sawyer, D. S., Ebeniro, J. O., O'Obrien, W. P., Jr., Shaub, F. J., and Oberst, J., 1985a, Crustal structure of the Green Canyon area, northern Gulf of Mexico; An ocean-bottom seismograph-airgun experiment: University of Texas Institute for Geophysics Technical Report 38, 26 p.

Nakamura, Y., Shaub, F. J., Mackenzie, K., and Oberst, J., 1985b, Crustal structure of the northwestern Gulf of Mexico, offshore Texas: University of Texas Institute for Geophysics Technical Report 39, 33 p.

Nelson, K. D., and 8 others, 1985a, New COCORP profiling in the southeastern United States; Part 1, Late Paleozoic suture and Mesozoic rift basin: Geology, v. 13, p. 714–717.

Nelson, K. D., McBride, J. H., Arnow, J. A., Oliver, J. E., Brown, L. D., and Kaufman, S., 1985b, New COCORP profiling in the southeastern United States; Part 2, Brunswick and East Coast magnetic anomalies, opening of the north-central Atlantic Ocean: Geology, v. 13, p. 718–721.

Nelson, K. D., McBride, J. H., Arnow, J. A., Willie, L. D., Oliver, J. E., and Kaufman, S., 1987, Results of recent COCORP profiling in the southeastern United States: Geophysical Journal of the Royal Astronomical Society, v. 89, p. 141–146.

Officer, C. B., and Ewing, M., 1954, Geophysical investigations in the emerged and submerged Atlantic coastal plain; Part 7, Continental shelf, continental slope, and continental rise south of Nova Scotia: Geological Society of America Bulletin, v. 65, p. 653–670.

Oliver, J. E., and Drake, C. L., 1951, Geophysical investigations in the emerged and submerged Atlantic coastal plain; Part 6, The Long Island area: Geological Society of America Bulletin, v. 62, p. 1287–1296.

Orcutt, J. A., and Dorman, L. M., 1977, An oceanic long-range explositions experiment: Journal of Geophysics, v. 43, p. 257–263.

Paull, C. K., and Dillon, W. P., 1980, Erosional origin of the Blake escarpment; An alternative hypothesis: Geology, v. 8, p. 538–542.

Pavlides, L., Sylvester, K. A., Daniels, D. L., and Bates, R. G., 1974, Correlation between geophysical data and rock types in the Piedmont and coastal plain of northeast Virginia and related areas: U.S. Geological Survey Journal of Research, v. 2, no. 5, p. 569–580.

Phillips, J. D., 1985, Aeromagnetic character and anomalies inthe Gettysburg basin and vicinity, Pennsylvania; A preliminary appraisal: U.S. Geological Survey Circular 946, p. 133–135.

Phillips, J. D., and 13 others, 1984, Deep structure of the shelf and slope along the Gulf of Mexico transect; Two-ship expanding spread profiles [abs.]: EOS Transactions of the American Geophysical Union, v. 65, p. 1007.

Phinney, R. A., 1986, A seismic cross section of the New England Appalachians; The orogen exposed, *in* Barazangi, M., and Brown, L., eds., Reflection seismology; The continental crust: American Geophysical Union Geodynamics Series, v. 14, p. 157–172.

Pilger, R. H., Jr., 1978, A closed Gulf of Mexico, pre-Atlantic Ocean plate reconstruction and the early rift history of the Gulf and North Atlantic: Gulf Coast Association of Geological Societies Transactions, v. 28, p. 385–393.

—— , 1981, The opening of the Gulf of Mexico; Implications for the northern Gulf Coast: Gulf Coast Association of Geologic Societies Transactions, v. 3, p. 377–381.

Pilger, R. H., and Angelich, M. T., 1984, Gravity anomalies, *in* Buffler, R. T., and others, eds., Gulf of Mexico, Atlas 6, Ocean Margin Drilling Program, Regional Atlas Series: Woods Hole, Massachusetts, Marine Science International, sheet 2.

Pilger, R. H., Rubin, D. S., and Kauth, L. M., 1984, Magnetic anomalies of the eastern Gulf of Mexico, *in* Buffler, R. T., and others, eds., Gulf of Mexico, Atlas 6, Ocean Margin Drilling Program, Regional Atlas Series: Woods Hole, Massachusetts, Marine Science International, sheet 3.

Pindell, J. L., 1985, Alleghenian reconstruction and subsequent evolution of the Gulf of Mexico, Bahamas, and proto Caribbean: Tectonics, v. 4, p. 1–39.

Poag, C. W., 1982, Foraminiferal and seismic stratigraphy, paleoenvironments, and depositional cycles in the Georges Bank basin, *in* Scholle, P. A., and Wenkam, C. R., eds., Geological studies of the COST Nos. G-1 and G-2 Wells, U.S. North Atlantic outer continental margin: U.S. Geological Survey Circular 861, p. 43–91.

—— , ed., 1985, Geologic evolution of the United States Atlantic margin: New York, Van Nostrand Reinhold Co., 383 p.

Popenoe, P., and Zietz, I., 1977, The nature of the geophysical basement beneath the Coastal Plain of South Carolina and northeastern Georgia, *in* Rankin, D. W., ed., Studies related to the Charleston, South Carolina, earthquake of 1886; A preliminary report: U.S. Geological Survey Professional Paper 1028, p. 119–137.

Press, F., and Beckmann, W. C., 1954, Geophysical investigations in the emerged and submerged Atlantic coastal plain; Part 6, The Long Island area: Geological Society of America Bulletin, v. 62, p. 1287–1296.

Purdy, G. M., 1983, The seismic structure of 140 m.y. old crust in the western central Atlantic Ocean: Geophysical Journal of the Royal Astronomical Society, v. 7, p. 115–138.

Purdy, G. M., and Ewing, J. I., 1986, Seismic structure of the ocean crust, *in* Tucholke, B. E., and Vogt, P. R., eds., Western Atlantic region: Boulder, Colorado, Geological Society of America, The Geology of North America, v. M., p. 313–330.

Rabinowitz, P. D., 1974, The boundary between oceanic and continental crust in the western North Atlantic, *in* Burk, C., and Drake, C., eds., The geology of continental margins: New York, Springer-Verlag, p. 67–84.

Reid, I., and Keen, C., 1987, Deep crustal structure beneath extensional basins of the Grand Banks region [abs.]: EOS Transactions of the American Geophysical Union, v. 678, p. 1356.

Robinson, G. R., and Froelich, A. J., eds., 1985, Proceedings of the Second U.S. Geological Survey Workshop on the Early Mesozoic Basins of Eastern North America: U.S. Geological Survey Circular 946, 147 p.

Royden, L., and Keen, C. E., 1980, Rifting process and thermal evolution of the continental margin of eastern Canada determined from subsidence curves: Earth and Planetary Science Letters, v. 51, p. 343–361.

Sawyer, D. S., 1985, Total tectonic subsidence; A parameter for distinguishing crust type at the U.S. continental margin: Journal of Geophysical Research, v. 90, p. 7751–7770.

—— , 1988, Thermal evolution, *in* Sheridan, R. E., and Grow, J. A., eds., Atlantic continental margin, U.S.: Boulder, Colorado, Geological Society of America, The Geology of North America, v. I-2, p. 417–428.

Sawyer, D. S., Toksoz, M. N., Sclater, J. G., and Swift, B. A., 1983, Thermal evolution of the Baltimore Canyon trough and Georges Bank basin, *in* Watkins, J. S., and Drake, C. L., eds., Studies in continental margin geology: American Association of Petroleum Geologists Memoir 34, p. 743–764.

Schlee, J. S., and Klitgord, K. D., 1988, Georges Bank basin; A regional synthesis, *in* Sheridan, R. E., and Grow, J. A., eds., Atlantic continental margin: Boulder, Colorado, Geological Society of America, The Geology of North America, v. I-2, p. 234–268.

Schlee, J., and 6 others, 1976, Regional geologic framework off northeastern United States: American Association of Petroleum Geologists Bulletin, v. 60, p. 926–951.

Shaub, F. J., Buffler, R. T., and Parsons, J. G., 1984, Seismic stratigraphic framework of the deep central Gulf of Mexico basin: American Association of Petroleum Geologists Bulletin, v. 68, p. 1790–1802.

Sheridan, R. E., and Grow, J. A., eds., 1988, Atlantic continental margin, U.S.: Boulder, Colorado, Geological Society of America, The Geology of North America, v. I-2, 610 p.

Sheridan, R. E., Drake, C. L., Nafe, J. E., and Hennion, J., 1966, Seismic refraction study of the continental margin east of Florida: American Association of Petroleum Geologists Bulletin, v. 60, p. 1972–1991.

Sheridan, R. E., Grow, J. A., Behrendt, J. C., and Bayer, K. C., 1979, Seismic refraction study of the continental edge of the eastern United States: Tectonophysics, v. 59, p. 1–26.

Sheridan, R. E., Crosby, J. T., Bryan, G. M., and Stoffa, P. L., 1981, Stratigraphy and structure of the southern Blake Plateau, northern Florida Straits, and northern Bahama Platform from multichannel seismic reflection data: American Association of Petroleum Geologists Bulletin, v. 65, p. 2571–2593.

Sheridan, R. E., Grow, J. A., and Klitgord, K. D., 1988, Geophysical data, *in* Sheridan, R. E., and Grow, J. A., eds., Atlantic continental margin: Boulder, Colorado, Geological Society of America, The Geology of North America, v. I-2, p. 177–196.

Shurbet, D. H., 1976, Conversion of S_n at a continental margin: Seismological Society of America Bulletin, v. 66, p. 327–330.

Simpson, R. W., Shride, A. F., and Bothner, W. A., 1980, Offshore extension of the Clinton-Newbury and Bloody Bluff fault systems of northeastern Massachusetts *in* Wones, D., ed., The Caledonides in the U.S.A.: Blacksburg, Virginia Polytechnic Institute Memoir 2, p. 229–233.

Sleep, N. H., 1971, Thermal effects of the formation of Atlantic continental margins by continental breakup: Geophysical Journal of the Royal Astronomical Society, v. 24, p. 325–350.

Smith, D. L., 1983, Basement model for the panhandles of Florida: Gulf Coast Association of Geological Societies Transactions, v. 33, p. 203–208.

Steckler, M. S., and Watts, A. B., 1978, Subsidence of the Atlantic-type continental margin off New York: Earth and Planetary Science Letters, v. 41, p. 1–13.

Steinhart, J. S., Green, R., Asada, T., Rodriguez, B., Aldrich, L. T., and Tuve, M. A., 1962, The Maine seismic experiment: Carnegie Institution of Washington Yearbook, v. 63, p. 221–231.

Stewart, D. B., and 8 others, 1986, The Quebec-western Maine seismic reflection profile; Setting and first year results, *in* Barazangi, M., and Brown, L., eds., Reflection seismology; The continental crust: American Geophysical Union Geodynamics Series, v. 14, p. 189–200.

Swift, B. A., Sawyer, D. S., Grow, J. A., and Klitgord, K. D., 1987, Subsidence, crustal structure, and thermal evolution of Georges Bank basin: American Association of Petroleum Geologists Bulletin, v. 71, p. 702–718.

Swolfs, H. S., 1967, Seismic refraction studies in the southwestern Gulf of Mexico [M.A. thesis]: College Station, Texas A&M University, 42 p.

Taylor, P. T., Zietz, I., and Dennis, L. S., 1968, Geologic implications of aeromagnetic data for the eastern continental margin of the United States: Geophysics, v. 33, no. 5, p. 755–780.

Trehu, A. M., 1984, Effects of bottom sediments and near-surface sediment structure on ocean bottom seismometer data [abs.]: EOS Transactions of the American Geophysical Union, v. 65, p. 1014.

——, 1987, Data report for large offset OBS data collected during cruise GYRE-85-11 in the Gulf of Maine: U.S. Geological Survey Open-File report 87-644, 56 p.

Trehu, A. M., and Wheeler, W. H., 1987, Possible evidence for subducted sedimentary materials beneath central California: Geology, v. 15, p. 254–258.

Trehu, A. M., Miller, G., Gettrust, J., Ballard, A., Dorman, L., Schreiner, A., 1986, Deep crustal structure beneath the Carolina trough [abs.]: EOS Transactions of the American Geophysical Union, v. 67, p. 1103.

Trehu, A. M., Ballard, A., Dorman, L. N., Gettrust, J. F., Klitgord, K. D., and Schreiner, A., 1989, Structure of the lower crust beneath the Carolina trough, U.S. Atlantic continental margin: Journal of Geophysical Research (in press).

Tucholke, B. E., Houtz, R. E., and Ludwig, W. J., 1982, Sediment thickness and depth to basement in western North Atlantic Ocean basin: American Association of Petroleum Geologists Bulletin, v. 66, p. 1384–1935.

Vogt, P. R., 1973, Early events in the opening of the North Atlantic, *in* Tarling, D. H., and Runcorn, S. K., eds., Implications of continental drift in the earth sciences, v. 2: Orlando, Florida, Academic Press, p. 693–712.

Vogt, P. R., and Tucholke, B. E., eds., 1986, The western North Atlantic region: Boulder, Colorado, Geological Society of America, The Geology of North America, v. M, 696 p.

Watts, A. B., 1981, The U.S. Atlantic continental margin; Subsidence history, crustal structure, and thermal evolution, *in* Bally, A. W., ed., Geology of passive continental margins; History, structure, and sedimentologic record, with special emphasis on the Atlantic margin: American Association of Petroleum Geologists Education Course Note Series 19, p. 2-1–2.75.

Watts, A. B., and Ryan, W.B.F., 1976, Flexure of the lithosphere and continental margin basins: Gectonophysics, v. 36, p. 24–44.

Watts, A. B., and Steckler, M. S., 1981, Subsidence and tectonics of Atlantic-type continental margins: Oceanologica Acta, no. SP, p. 143–153.

Watts, A. B., and Thorne, J., 1984, Tectonics, global changes in sea level, and their relationship to stratigraphical sequences at the U.S. Atlantic continental margin: Marine and Petroleum Geology, v. 1, p. 319–339.

Wernicke, B., 1985, Uniform-sense normal simple shear of the continental lithosphere: Canadian Journal of Earth Sciences, v. 22, p. 108–125.

White, R. S., Spence, G. D., Fowler, S. F., McKenzie, D. P., Wesbrook, G. K., and Bowen, A. N., 1987, Magmatism at rifted continental margins: Nature, v. 330, p. 439–444.

Winkler, C. D., and Buffler, R. T., 1984, Mexico to Florida, *in* Buffler, R. T., and others, eds., Gulf of Mexico, Atlas 6, Ocean Margin Drilling Program, Regional Atlas Series: Woods Hole, Massachusetts, Marine Science International, p. 25.

Woodruff, C. M., Jr., 1980, Regional tectonic features of the inner Gulf Coast: Oil and Gas Journal, v. 78, p. 264–275.

Woods, R. D., and Addington, J. W., 1973, Pre-Jurassic geologic framework of northern Gulf Basin: Gulf Coast Association of Geological Societies Transactions, v. 23, p. 92–108.

Worzel, J. L., and Shurbet, G. L., 1955, Gravity anomalies at continental margins: Proceedings of the National Academy of Sciences, U.S.A., v. 41, p. 458–469.

Worzel, J. L., and Watkins, J. S., 1973, Evolution of the northern Gulf Coast deduced from geophysical data: Gulf Coast Association of Geological Societies Transactions, v. 23, p. 84–91.

Zietz, I., 1982, Composite magnetic anomaly map of the United States: U.S. Geological Survey Geophysical Investigations Map GP-954-A, scale 1:2,500,000.

Zietz, I., and Gilbert, F. P., 1980, Aeromagnetic map of part of the southeastern United States: U.S. Geological Survey Geophysical Investigations Map GP-936, scale 1:2,000,000.

Zietz, I., Gilbert, F. P., and Kirby, J. R., Jr., 1980, Aeromagnetic map of Delaware, Maryland, Pennsylvania, West Virginia, and part of New Jersey and New York: U.S. Geological Survey Geophysical Investigations Map GP-927, scale 1:2,000,000.

Manuscript Accepted by the Society October 31, 1988

ACKNOWLEDGMENTS

We thank C. Keen, J. Grow, and J. Behrendt for thorough reviews. Discussions with M. Purdy, J. Ewing, J. Mutter, R. White, and D. Hutchinson are gratefully acknowledged. Patty Forrestel and Jeff Zwinakis drafted the figures that are new to this manuscript. D.S.S. appreciates the support of National Science Foundation Grant OCE-8401621.

Printed in U.S.A.

Geological Society of America
Memoir 172
1989

Chapter 18

A crust/mantle structural framework of the conterminous United States based on gravity and magnetic trends

Martin F. Kane and Richard H. Godson
U.S. Geological Survey, Box 25046, Denver Federal Center, Denver, Colorado 80225

ABSTRACT

We use a short-wavelength (<250 km) gravity map and a magnetic map to define anomaly-trend zones in the conterminous United States. The zones appear to define major regions of coherent structure within the crust. The main orientations of the zone boundaries are northeasterly and northwesterly, except along the western Cordillera, where they trend parallel and normal to the western coastline and to the strike of the underlying subduction zone. Correlations between geophysical and geologic features along the eastern and western margins of the United States suggest that a major cause of the zone and boundary trends is plate interaction at the continental margins. Linear gravity highs, commonly linked into extensive chains, are interpreted as the expressions of horsts that occupy the central parts of rift systems. Major crustal deformation appears to have occurred along the boundaries of the horsts over long periods of geologic time. The oldest basement of the central cratonic region may be overprinted in many places by younger tectonic events so that primary features of the oldest basement are obscured. We also observe that the structural pattern of the continent thus defined exhibits a crude symmetry about the midcontinent rift system. The pattern appears to have evolved from 1.5 to 1.0 Ga. We point out several specific geologic feature–geophysical anomaly correlations, although a comprehensive treatment of these is not attempted.

We identify two types of regional magnetic anomalies, which are associated with regions underlain by thick sedimentary sections, and which may be linked to a shallow mantle.

A comparison of a gravity map of long wavelengths (>1,000 km) with similarly filtered maps of crustal thickness and P_n-velocity shows a significant correlation between major gravity features and the lateral distribution of mantle-surface velocity (gravity highs/high velocity, and gravity lows/low velocity). There does not appear to be a correlation with crustal thickness except at the continental boundaries and possibly in the northeastern United States. We further show that the great gravity low of the western United States corresponds laterally with an S-wave velocity low in the uppermost mantle. Using the vertical extent of the low-velocity zone as the vertical dimension of the gravity anomaly source, we estimate a decrease in the density of the source of less than 1 percent. The corresponding decrease in the shear modulus is about 17 percent. The positive correlation between gravity features and mantle velocities implies that the isostatic compensation of the first-order topographic features of the continent is in the uppermost part of the mantle. The large decrease in the shear modulus suggests that the likely cause of the isostatic compensation in the western United States is an abnormally

Kane, M. F., and Godson, R. H., 1989, A crust/mantle structural framework of the conterminous United States based on gravity and magnetic trends, *in* Pakiser, L. C., and Mooney, W. D., Geophysical framework of the continental United States: Boulder, Colorado, Geological Society of America Memoir 172.

high uppermost-mantle temperature that produces thermal buoyancy of the elevated topographic mass.

We use a combination of geophysical crustal zones, mantle properties, and regional geology to propose that the conterminous United States be divided into four tectonic regions: (1) an Appalachian province that includes the eastern Gulf Coastal Plain and Florida, (2) a western Gulf Coastal Plain province, (3) a central cratonic province that extends from the Appalachians to the Rocky Mountain front, and (4) a western province extending west from the Rocky Mountain front.

INTRODUCTION

In 1982, 1:2,500,000-scale gravity (Lyons and O'Hara, 1982) and magnetic (Zietz, 1982) maps of the conterminous United States were published using the data bases then available. These were the first maps of the gravity and magnetic fields of the United States that showed sufficient detail to distinguish anomalies caused by moderate-size, shallow geologic bodies from those caused by broader and/or deeper sources of less certain origin. Hinze and Zietz (1985) and Kane and Godson (1985) contributed the first synoptic interpretations of the maps as a whole. Kane and Godson (1985) used a gravity map composed of wavelengths <250 km to divide the gravity field of the conterminous United States into a series of zones. The zones, which were defined on the basis of dominant anomaly trends and distinctive anomaly patterns, showed a reasonable correlation with basement geology.

In this chapter we apply the same technique to produce a magnetic zone map of the United States. We combine it with a slightly modified version of the gravity zone map to arrive at a provisional structural framework of the crust. We then explore some of the geological implications of the framework. Finally, we examine the correspondence among very long-wavelength (>1,000 km) gravity anomalies, crustal thicknesses, P_n-velocity, and mantle S-wave velocity. We group the zones into a broader framework, derived from mantle properties, certain aspects of the structural framework, and information derived from geological and other geophysical studies.

We recognize that the process of defining the gravity and magnetic zones is subjective, particularly with respect to the zone boundaries. We believe, however, that the provisional division of the combined gravity and magnetic fields into lateral zones by a relatively simple technique is an important first step in the synoptic interpretation of the geophysical maps of a continental region.

DATA BASES

The gravity data used in this report are the Bouguer gravity values described by Godson (1985). The data were interpolated to a 4-km grid from a total base of about 1 million stations. The data reduction used a density of 2.67 g/cm^3 for the Bouguer correction; terrain corrections were made for topography 0.9 km or more from the stations.

The magnetic maps are based on data digitized from the contours of an updated version of the eastern half of the magnetic map of the United States (Zietz, 1982) and from the contours of magnetic maps of the western United States (Bond and Zietz, 1987). The data were organized into a uniform 2-km grid after application of various editing procedures to minimize base-level changes among the constituent data sets.

PROPERTIES AND CAUSES OF GRAVITY AND MAGNETIC ANOMALIES

Properties of anomalies

Although gravity and magnetic features can arise from the same source, their appearances are usually different because of the different way in which the anomalies are generated. In some cases a difference will exist between co-located gravity and magnetic anomalies because they do not arise from the same part of the source. In the following section we explore some of the differences between gravity and magnetic anomalies by examining the anomaly characteristics of a wide source or plate; that is, a source whose width is at least a few times greater than its thickness and its depth.

The gravity attraction of a horizontally inclined plate is directly proportional to the solid angle subtended by the edges of a plane that passes horizontally through the center of the plate (Nettleton, 1942). According to this relation, the peak of the anomaly is located over the center of the plate, and the amplitude of the anomaly is proportional to the density contrast and thickness of the plate. For a plate that is also elongate, the axis of the anomaly and that of the elongation of the plate are colinear, and the centers of the gradients that form the anomaly sides indicate the approximate locations of the edges of the plate. An approximate depth to the center of the plate is given by a relation between the amplitude and maximum gradient of the anomaly (Bott and Smith, 1958; Bancroft, 1960; Kane and Bromery, 1968).

In contrast to gravity, the magnetic attraction of a horizontally inclined plate (Nettleton, 1942) is proportional to the difference between the solid angle subtended by the top of the plate and that subtended by the bottom. For induced magnetization of an elongate plate, the dipole nature of the magnetizing field causes a latitude-dependent displacement of the anomaly axes away from the axis of the plate (a low to the north and a high to the south for normal polarity), although the anomaly and source axes still remain parallel (Vacquier and others, 1951). At middle latitudes the combined effect of the solid-angle difference and the dipole field is a reduction in the magnetic intensity over the center of the plate and a concentration of contour closures over plate edges that are oriented at a substantial angle to the direction of the magnetizing field. The subtraction of the solid angle subtended by the top of the plate from that subtended by the bottom acts somewhat like a high-pass filter so that the short-wavelength content of the data is enhanced. The straightforward horizontal spatial relations that exist between gravity anomalies and their sources do not apply, and there is no simple relation between anomaly amplitude and the thickness and magnetization of the source. Two offsetting advantages of the magnetic method are a more direct indication of the shape of the upper parts of magnetic igneous and crystalline rock units and a more precise estimate of the depth of the anomaly source.

For gravity and magnetic anomalies, the decrease in the amplitude of component wavelengths with increase in depth to source is inversely proportional to the magnitude of the wavelength; that is, the short-wavelength components of an anomaly damp out with depth at a faster rate than the long-wavelength components (Kane and Godson, 1985, Fig. 3). This property underlies the frequently used qualitative observation of gravity and magnetic interpretation that short-wavelength features are shallow in origin. It also leads to the use of filtering to separate the effects of shallow sources (short-wavelength features) from those of deep or laterally extensive sources (long-wavelength features).

Causes of anomalies

Density and magnetization, rock properties that give rise to gravity and magnetic anomalies, differ significantly in their relation to rock composition. Density is a bulk mineral property that undergoes relatively small changes during geologic cycling or reworking processes. The nonporous crystalline rocks, which represent the bulk of the Earth's crust, have a typical average density value of about 2.8 g/cm^3 except where a large percentage of the rock mass is of felsic composition (Smithson, 1971). The density contrast of the crystalline rock units that form significant volumes in the crust has an effective range of about 0.1 to 0.3 g/cm^3 and an average value of about 0.2 g/cm^3 (Kane and Bromery, 1968).

The magnetization of a rock resides in its accessory minerals, of which the most important is magnetite. Susceptibility, which is the predominant cause of rock magnetism, causes the magnetization axis of the mineral to be aligned in the direction of the Earth's present main field. Magnetization is commonly associated with igneous rocks, but it is controlled more generally by the thermal history of a rock. The remanent magnetization of a mineral is fixed in the direction of the Earth's field when the mineral is cooled below the Curie temperature; the remanent magnetization is removed when heated above this temperature. Metamorphism can also change the magnetization of a rock, but the magnitude of the effect is usually small. In contrast to density, rock magnetization can range through several orders of magnitude. In addition, remanent magnetization may be aligned in a direction substantially different from that of the Earth's present main field.

For the most part, broadly distributed (100 km or more) near-uniform layers of sedimentary strata do not contribute significantly to the anomalies shown in Figures 1 and 2. On the gravity map, the effects of such strata have been largely removed by the filtering process. The presence of the strata still has an indirect effect in that the sources of anomalies beneath the strata are generally at a greater depth below the measurement surface than those in areas of exposed basement. The principal effects of the greater depth are a reduction in the sharpness of the anomaly gradients and a decrease in the anomaly amplitudes, particularly for those anomalies that are relatively narrow (Kane and Godson, 1985). Sedimentary units rarely contain sufficient volumes of magnetic minerals to cause observable anomalies, except where nonmagnetic iron-bearing minerals are altered into magnetic ones by the thermal effect of intrusions. Because magnetic anomalies are generally narrower than gravity anomalies to begin with, the effect of an increase in source depth is more pronounced for magnetic anomalies than for gravity. For this reason, magnetic anomalies in areas underlain by thick accumulations of sedimentary rock are usually subdued and commonly absent. Most of the anomalies of the gravity map (Fig. 1) can be attributed to sources in the crystalline basement, whereas those of the magnetic map (Fig. 2) arise from sources in the crystalline basement, volcanic units within a kilometer or so of the surface, or igneous intrusions.

Felsic plutons, which have typical average densities of about 2.7 g/cm^3, commonly cause gravity lows (Bott and Smithson, 1967; Kane and Bromery, 1968). Thick piles of felsic volcanic rocks can cause gravity lows because of retained porosity and/or because the densities of the minerals are similar to those of felsic plutons. Sedimentary rock units have low densities primarily because of porosity, but those that fill broad basins (100 km or more in width) show little or no apparent gravity anomaly (Fig. 1) either because of the filtering or because their deficient mass is offset isostatically. Basin deposits with narrow widths (typically less than 100 km), however, commonly cause pronounced lows. Presumably, the relative mass deficiency of the basin strata is not offset immediately below the basin by a compensating high-density mass. Densities of masses of mafic and ultramafic plutonic and metamorphic rock range from about 2.9 to about 3.3 g/cm^3 (Judd and Shakoor, 1981); such masses cause pronounced gravity highs where present in sufficient volume. Densities of mafic volcanic piles are difficult to evaluate because they commonly contain porous volcanic units and interlayered sedimentary rocks. Where porosity has been removed by induration, however, mafic volcanic piles produce moderate to strong positive gravity anomalies because of the abundance of high-density minerals.

Magnetic anomalies that are prominent on small-scale maps of the continent are typically associated with large masses of igneous rock and with crystalline basement. The most pronounced anomaly patterns are usually caused by exposed volcanic rock units, but anomalies that arise from this source damp out sharply with increasing depth. Mafic and ultramafic plutons are commonly strongly magnetic; some of the most intense magnetic anomalies are observed over ultramafic masses. Some felsic plutons are magnetic, either because magnetite is a primary constituent or because it develops from other, nonmagnetic iron-bearing minerals during metamorphism. Many felsic plutons are not magnetic in themselves but, as noted above, they can cause magnetic minerals to be formed in the surrounding host rock by the thermal effects of their emplacement.

In a previous paper (Kane and Godson, 1985), we used a residual gravity map in which the long-wavelength component (>250 km) had been filtered out; we showed that a majority of the prominent anomalies displayed by the map arise from sources within the Earth's crust. In the magnetic field, the high dynamic range of magnetic properties, the variability of accessory mineral concentration, and the natural high-pass filtering lead to an enhancement of short-wavelength features of shallow origin. In general, sources of spatial magnetic anomalies are believed to reside in the Earth's crust because the estimated temperature below the crust is too high for magnetism to be retained.

Gravity maps from which isostatic or polynomial regional fields have been removed generally retain a more precise representation of the residual field anomalies than does a short-wavelength residual map based on filtering. Derivative-type gravity maps, such as those of the horizontal and vertical gradients, bring out the gradients but do not show the anomaly amplitudes. Filtered maps, such as that shown in Figure 1, retain the sense of the amplitude and enhance both the axes and the gradients of elongate anomalies. For these reasons, filtered maps are best suited for analysis based on anomaly trend.

GEOPHYSICAL ZONES

Zones

The primary criterion that we use to define the gravity and magnetic zones is anomaly trend or, in a few places, the lack of dominant trend. As noted previously, trend is best defined by anomaly axes, which for gravity features are aligned along the source axes, and for magnetic features, are parallel to the source axes but displaced toward the source edges. Expression of trends by gravity gradients is less consistent and usually less intense than that expressed by gravity anomaly axes. For wide sources, the gravity gradients are associated with the source edges. The primary cause of the orientation of the axes and edges of geologic bodies is structure. We conclude, therefore, that the definition of zones using gravity and magnetic trends can be regarded as a first-order method of defining structural zones.

Boundaries

The most easily defined type of zone boundary is where parallel trends in both the gravity and magnetic fields terminate abruptly against trends of a different orientation. Another type of boundary is defined by a linear geologic feature, such as a rift, which causes a prominent linear anomaly or linear system of anomalies. In some places one field may indicate a sharply delineated boundary where the other does not. In others, the locations indicated by the two fields may not closely agree, which indicates either that the anomalies are arising from different parts of a common source or that the anomaly sources are different.

Despite the complexities, we believe that the general positions of the boundaries and the areal extent of the zones represent significant divisions of the gravity and magnetic fields and the geologic sources that give rise to them. Independent analyses would undoubtedly combine some zones and divide others, but the general partitioning based on anomaly trends would probably be similar.

ZONE MAPS

Gravity zone map

A modified version of the gravity zone map of our previous paper is shown superposed on the gravity map (Fig. 1; color version in back cover pocket). The modifications are as follows. First, we have added a small subzone in the northeastern region (southwest New England) where anomalies trend north in sharp contrast to adjacent anomalies (Kane and others, 1972). Second, we have modified the boundaries of the zone of north-trending anomalies that extends from the Great Lakes to northern Alabama. Reexamination has indicated that the modified boundaries better define the limits of the north-trending anomalies. We have also extended the western of these two boundaries southward to the Gulf coast, dividing the Gulf coastal zone into eastern and western parts. Third, we have altered the southern boundaries of the gravity pattern of the MGA (midcontinent gravity anomaly) to show narrowing of the rift anomaly in two discrete steps. Fourth, we have simplified the zones and boundaries of the westernmost United States. Previously, we had controlled the partitioning of the zones by the northerly coastal gravity trends, but we now conclude that continuity with the gravity patterns to the east is more consistent with the procedure used elsewhere.

We also qualify the character of a boundary by a diamond-shaped symbol where anomalies indicate that a major structure is associated with the boundary or by using a dashed line where the location is less certain. Dashed lines are also used to subdivide some of the zones. As before, we show the MGA as a separate zone.

The principal conclusion of our earlier report (Kane and Godson, 1985) was that linear gravity highs, which are commonly linked into extensive chains, were the dominant feature of the short-wavelength (<250 km) gravity map. We interpreted many of these gravity highs to be expressions of the central parts of rift systems that were uplifted along boundary faults, which are characterized in many places by reverse movement. Several linear highs in the continental interior were shown to be located within

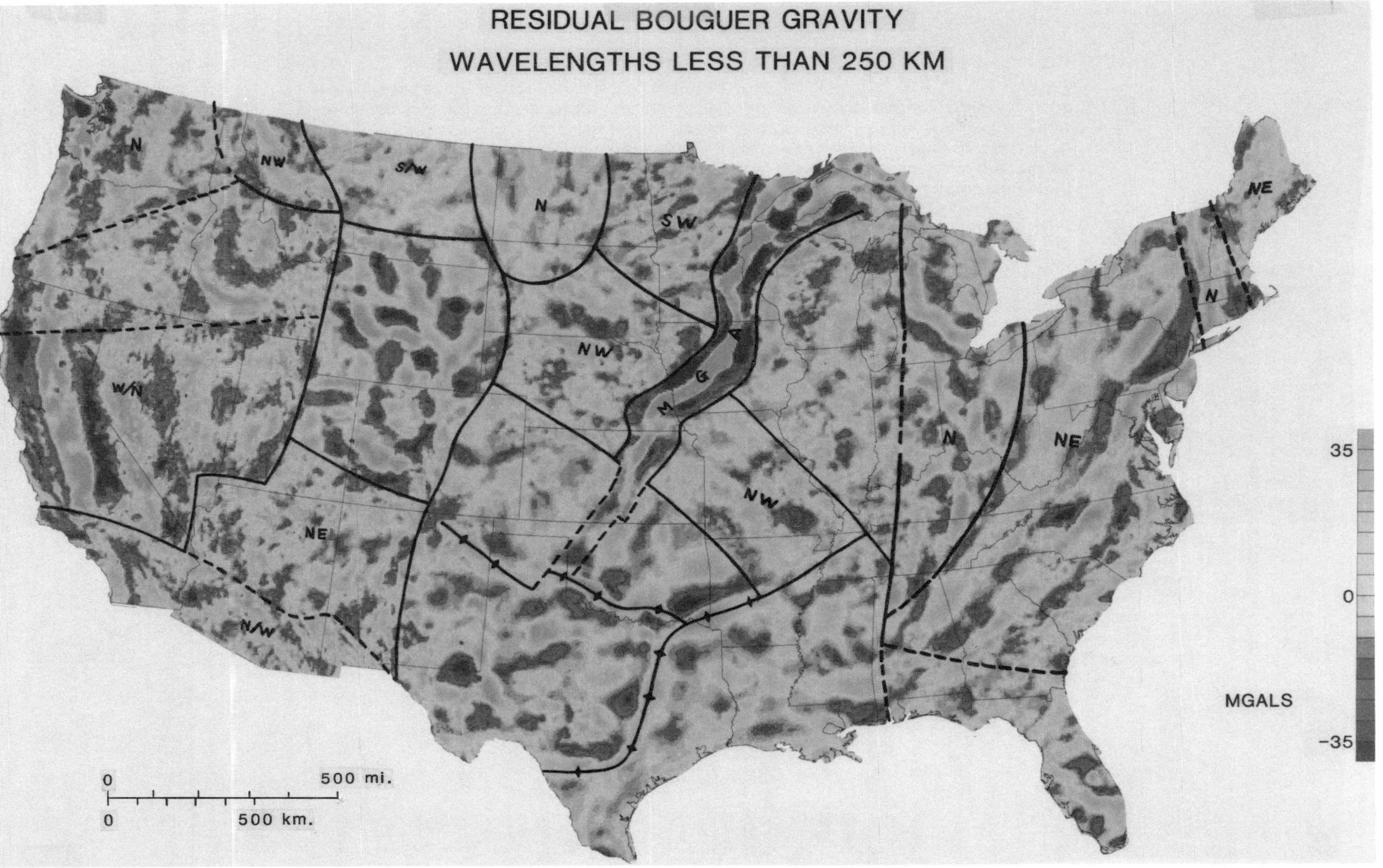

Figure 1. Gravity zone map of the conterminous United States derived from short-wavelength (<250 km) Bouguer gravity map. Dashed lines indicate subdivisions within zones or where zone boundary locations are less clear. Diamond marks along zone boundary indicate that boundary is underlain by a major structure. Trends within zones are designated by initials of cardinal directions (N = north; NW = northwest; W/N = west of north, etc.). Modified after Kane and Godson (1985).

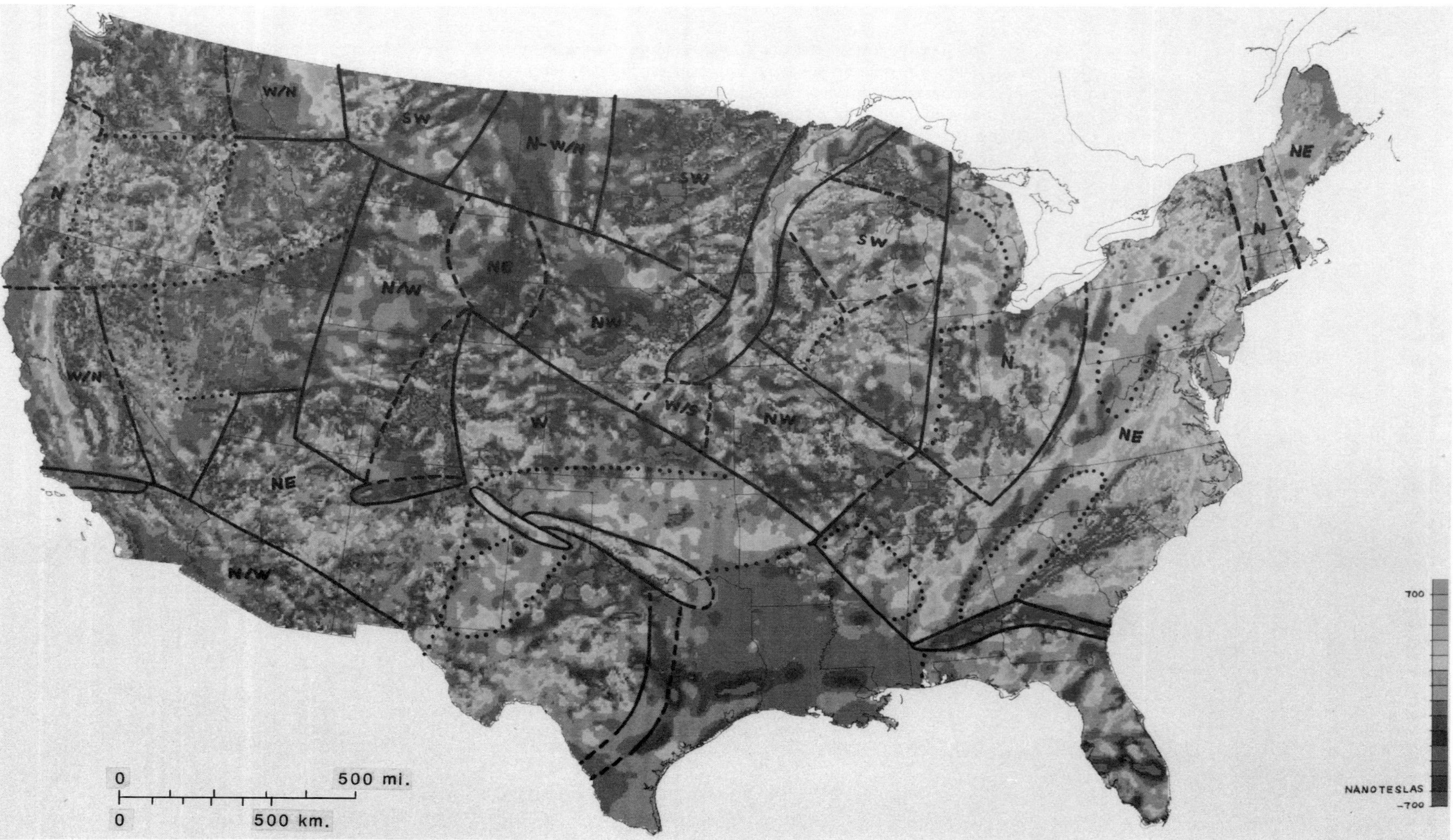

Figure 2. Total-field magnetic zone map of the conterminous United States. Solid lines enclosing broad areas indicate major zones; solid lines enclosing elongate areas indicate single major underlying geologic structures of similar areal extent; dashed lines indicate where zone boundaries are less clear or subdivisions within the zones based on trends; dotted lines outline areas typified by low-amplitude magnetic highs of broad lateral extent or subzone boundaries based on anomaly characteristics other than trend; anomaly trend directions are as indicated in Figure 1.

the boundaries of sedimentary basins. This association was explained by the crustal stretching model for basins (McKenzie, 1978). Extensive chains of highs in the eastern and south-central regions were attributed to continental shelf rift zones that failed during continental collisions along the eastern and southern margins of the continent. We noted that strong west-of-north and northerly trends in the western regions parallel the continental margin. We concluded that plate collisions along continental margins cause extensive linear geologic structures and associated linear gravity anomalies that are aligned parallel to the collision zones. We also noted a number of correlations between specific geologic features and gravity anomalies.

Magnetic Zone Map

Figure 2 (color version in back cover pocket) shows a magnetic zone outline superposed on the magnetic map. The map is similar to that compiled by Zietz (1982), except that this map includes coverage of several small areas not available earlier. The smaller scale of Figure 2 also effectively excludes many of the smaller anomalies of Zietz's map.

Like the gravity zone map, we used dashed lines to delineate trend subzones and to denote boundaries where the supporting magnetic evidence is less definitive. In addition, we used dotted lines to outline broad elongate magnetic anomalies whose sources should have similar broad lateral dimensions. We also use a dotted line to outline areas whose general magnetic expression contrasts with that of surrounding regions in ways other than trend. Of particular importance in this latter group are areas that enclose low-amplitude highs of broad extent.

Hinze and Zietz (1985) pointed out that magnetic anomalies in the eastern regions are generally more numerous and intense than those in the western regions. They concluded that the principal cause of the subdued anomalies in the western regions is an abnormally shallow Curie isotherm and related low crustal magnetization. Magnetic expression in the Cordilleran and Appalachian systems was generally associated with the major structural patterns of the orogens, except for anomalies related to subal-lochthonous crystalline units in the Appalachians and those caused by westerly striking faults and plutons in the Cordillera. In addition, they noted that the magnetic field defines the buried eastern and southern edges of the Precambrian craton in the eastern region, delineates numerous provinces in the intracratonic region, and shows that these provinces are transected by conspicuous linear structures, notably the midcontinent rift and the Wichita uplift.

Composite gravity–magnetic zone map

A comparison of Figures 1 and 2 shows that the gravity and magnetic zone patterns are similar. Except for the westernmost areas, the trends of the boundaries are typically east-of-north and north-of-west; in a few places they are northerly. In the west, the boundaries are more typically northerly and south-of-west. In most zones the gravity and magnetic data indicate the same trend direction. There is one zone where the trend is specified only by the gravity field and four subzones where it is indicated only by the magnetic field.

Figure 3 is a composite of the gravity and magnetic zones superposed on a geologic feature map. The zones are designated by letter for ease of reference. In outlining the zones we have given heaviest weight to the gravity evidence by including all gravity-indicated boundaries. In a few places we have based the boundary location on the magnetic field contrast or on the outline of a magnetic anomaly. We include trend directions for all zones indicated by both the gravity and magnetic expression, as well as one subzone (O3) that is based only on gravity. Subzones based on magnetic data are excluded. Where the magnetic and gravity data agree on the boundary location, we use a solid line; where the locations and azimuths of the boundaries are reasonably proximate (usually within 20° and 100 km), we use a dashed line; and where either has been used alone to define the boundary, we use a dashed line with the letter G or M to indicate a gravity or magnetic source, respectively. Where subzones are included, their boundaries are shown by dashed lines.

DISCUSSION OF ZONES

We describe each of the zones and many of the corresponding geologic features, most of which are taken from the tectonic map of North America (King, 1969). We begin with zones MGA, A, and O because results of their interpretation apply to the interpretation of the remaining zones.

Zone MGA

A system of gravity anomalies (Fig. 1) indicates that the MGA extends 1,700 km southwestward from the United States–Canada border to the northern edge of the Wichita uplift. The MGA lies near the center of the cratonic core of the United States, and its source is covered by sedimentary strata except at its northern end. Throughout most of its length, the gravity anomaly system is made up of a chain of gravity highs flanked by lows. The average width of the chain is about 100 km, and the highs are somewhat wider than the flanking lows. The peak Bouguer gravity value of one of the highs is more than 70 mGal, the values of the lows are less than –110 mGal, and the average Bouguer gravity field in the surrounding region is about –30 mGal (Lyons and O'Hara, 1982). These values yield relative maximum amplitudes for the highs and lows of about +100 and –80 mGal, respectively. Assuming crustal density contrasts of ± 0.3 g/cm^3, the high and low anomalies indicate minimum source thicknesses (Kane and Bromery, 1968) of about 8 and 6 km, respectively. Thus the sources form substantial parts of the thickness of the Earth's crust.

The lateral extent of the magnetic anomaly system is similar to that of the gravity anomaly system, but it is made up of a more complex group of highs and lows whose peak values are only moderately higher than those of the surrounding magnetic field.

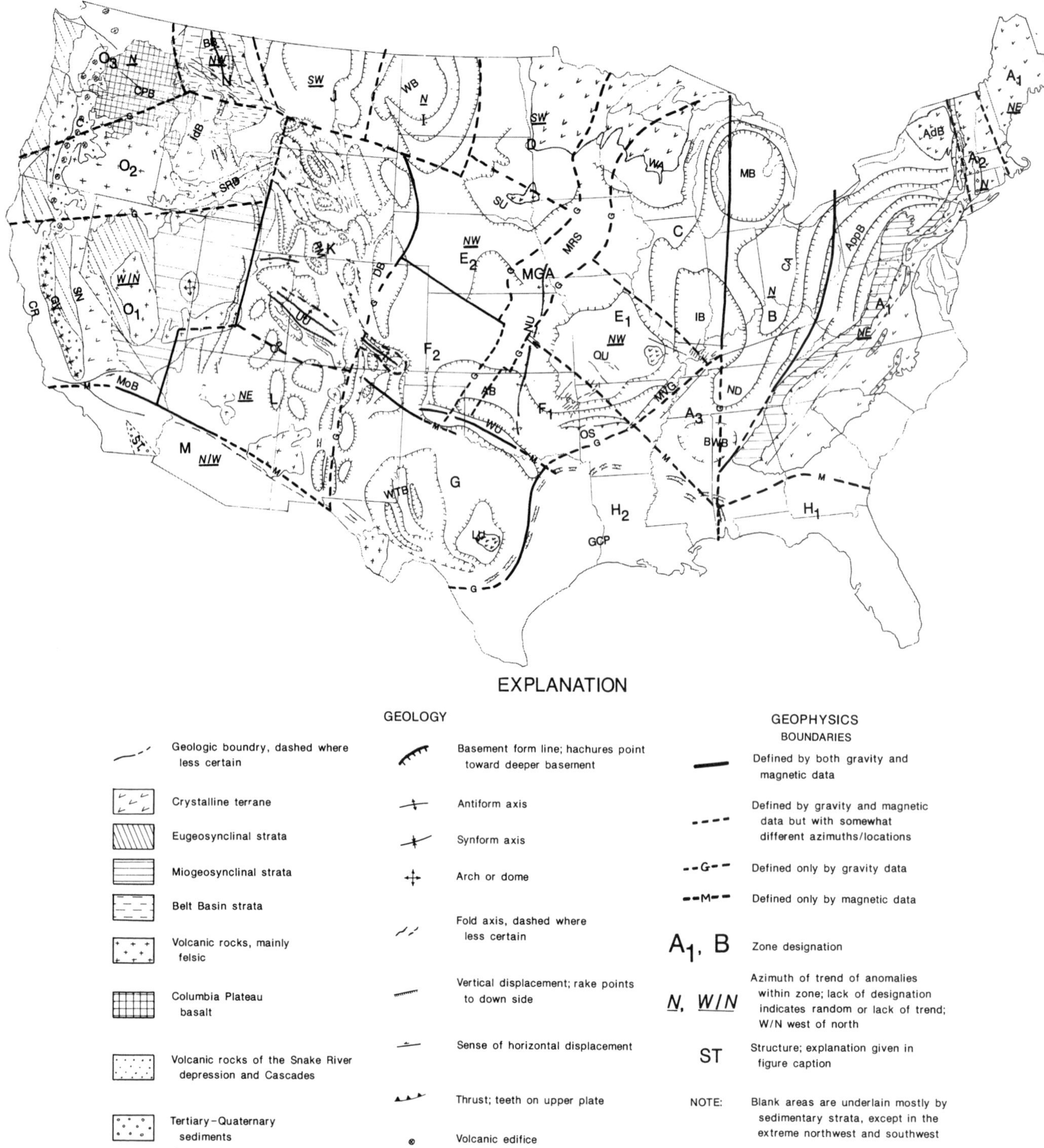

Figure 3. Composite of gravity and magnetic zones superposed on map of geologic features. Geology features from King (1969), except as noted. AB = Anadarko basin; AdB = Adirondacks block; AppB = Appalachian basin; BB = Belt basin; BWB = Black Warrior basin; C = Cascades; CA = Cincinnati arch; CR = Coast Ranges; CP = Colorado Plateau; CPB = Columbia Plateau basalt; DB = Denver basin; GCP = Gulf Coastal Plain; GV = Great Valley; IB = Illinois basin; IdB = Idaho batholith; LU = Llano uplift; MB = Michigan basin; MRS = Midcontinent Rift System (King and Zietz, 1971); MVG = Mississippi Valley graben (Kane and others, 1981); MoB = Mojave block; ND = Nashville dome; NU = Nemaha uplift; OS = Ouachita structure; OU = Ozark uplift; RM = Rocky Mountains; ST = Salton trough; SN = Sierra Nevada; SRD = Snake River depression; SU = Sioux uplift; UU = Uncompahgre uplift (Larsen and others, 1985); WA = Wisconsin arch; WTB = West Texas basin; WU = Wichita uplift.

The magnetic highs have a distinctive curvilinear shape that generally corresponds to the outlines of the gravity highs. The trend of the magnetic anomaly system as a whole is discordant with the trends of magnetic features in the surrounding region.

The basement rocks that are exposed in the northern part of the MGA show that the anomaly systems coincide with a rift of Precambrian age (Leith and others, 1935; White, 1966; Halls, 1966; Thiel, 1956; King and Zietz, 1971; Wold and Hinze, 1982). The gravity high is associated with a central structural block of interlayered mafic volcanic and sedimentary strata that is uplifted along faults that dip, in many places, inward beneath the block. Layered mafic volcanic rock may be the source of the gravity high, although Chase and Gilmer (1973) have shown that subjacent intrusive masses of mantle-derived rocks could be an alternate or supplementary source. The apparent causes of the flanking gravity lows are basins filled with clastic sedimentary rocks (Thiel, 1956). The magnetic highs also correlate with the volcanic units; their outlines indicate the lateral extent of the top of the volcanic flows (King and Zietz, 1971). The typical anomaly over the sedimentary basins is a magnetic low that is generally devoid of smaller anomalies.

The surface of the basement deepens gradually southwestward from the outcrop area for a distance of about 1,400 km and then slopes sharply downward beneath the Anadarko basin. The gravity anomaly pattern has a pronounced riftlike form for much of its length, but it narrows abruptly where it crosses the northern edge of the Anadarko basin. Over the deepest part of the basin, the central gravity high is no longer evident. The magnetic highs generally decrease in amplitude southwestward. They lose their characteristic shape in the same area where the gravity pattern begins to narrow. Magnetic expression of the rift is not apparent over the Anadarko basin.

The geophysical and geologic evidence indicates that the cratonic core of the continent was split about 1.1 Ga (Silver and Green, 1972; Van Schmus and others, 1982) by a rift that attained a length of at least 1,700 km. As summarized by Wold and Hinze (1982), the events included: (1) an outpouring of large volumes of volcanic rocks along the central part of the rift zone; (2) accumulation of clastic sediments into central basins and interlayering of these sediments with the centrally placed volcanic units; (3) uplift of the central part of the rift as horsts along steep, generally reverse, boundary faults; and (4) at least minor movement along the faults in post-Precambrian time. If it is assumed that the sedimentary strata in the uppermost parts of the horsts were originally deposited near the floor of the basin, then the amount of uplift may have been several kilometers or more (Chase and Gilmore, 1973, Fig. 3). The crustal thinning that is implied by the rifting and the nearly contemporaneous clastic sedimentation is consistent with the sedimentary basin model of McKenzie (1978). Central horsts, which were raised along reverse boundary faults, are a typical feature of major sedimentary basins (Milanovsky, 1981). We noted (Kane and Godson, 1985) that central linear gravity highs, which are associated in many places with basement uplifts, are a common feature of the major basins of the conterminous United States.

Zone A

The geophysical anomalies of zone A predominantly trend to the northeast, which is an orientation that has long been recognized as a fundamental geologic characteristic of the Appalachians. The zone is divided into three subzones: A1, which includes most of the Appalachians; A2, which is a small area typified by northerly trends and abutted by A1; and A3, which lies outside the Appalachians but has northeast magnetic trends.

The gravity pattern in the region southwest of subzone A2 is dominated by two parallel northeast-trending chains of elongate highs. Over most of their length the chains are separated by a series of gravity lows of moderate to high intensity and flanked outwardly by lows that are less intense as well as less continuous. The gravity highs of the chain to the northwest are symmetrical and relatively simple in shape; the amplitudes of these highs are moderate and the gradients are low. In contrast, the gravity highs of the chain to the southeast are complex in shape and asymmetric toward the northwest; that is, the gradients on the northwest sides of the highs are steeper than those on the southeast; the amplitudes of these highs are higher, and the gradients are steeper, than those of the gravity highs of the chain to the northwest. The contrasting characteristics of the two sets of anomalies indicate that the sources of the southeast chain are shallower or more massive than those of the northwest and they form interfaces with the surrounding rock units that dip at steeper angles. In the eastern parts of New York and Pennsylvania, the two chains are separated by an intense elongate high, which is flanked to the southeast by a large, arcuate, intense gravity low. The gravity pattern east of A2 also trends to the northeast, but the anomalies are less intense than those to the southwest and have a much more diffuse and subdued appearance.

Subzone A2 is dominated by an intense north-trending gravity high that appears to be a continuation of the gravity-high chain in southeasternmost New York and southern New Jersey. The lesser anomalies, mainly lows, also trend north, and the gravity pattern of the subzone as a whole is sharply outlined against the northeast trends of the features to the east and southwest (Kane and others, 1972).

The trends of the magnetic anomalies in subzones A1 and A2 are generally consistent with the orientations of the gravity anomalies, but direct correlation between the individual magnetic and gravity anomalies is lacking. Many of the short-wavelength magnetic features must reflect the detailed structure of the near-surface basement rock units. There are two broad areas within subzone A1 that are typified by broad, low-amplitude regional highs and relatively few local anomalies. The location of the southern boundary of zone A is based on a prominent cross-cutting magnetic feature, the Brunswick anomaly (Pickering and others, 1977).

Subzone A3 lies outside the Appalachian region. The trends of the magnetic anomalies are northeast like those of zones A, but the trends of the gravity anomalies are random. Several low-amplitude regional magnetic highs are present in the southwest part of the subzone.

The northwest chain of gravity highs and parallel flanking lows that extends from western New York to Mississippi correlates with the Appalachian basin on the northeast and crosses a mixed terrane of basin sedimentary strata and allochthonous sedimentary and crystalline rock units (Cook and others, 1979; Harris and Bayer, 1979) on the southwest. Although some geologic features of the allochthon correlate with some secondary elements of the gravity pattern, the trends of the main components of the allochthon are somewhat more westerly than those of the gravity field. In addition, the thickness and density contrasts within the allochthon, especially the sedimentary part, are not sufficient to account for the amplitudes of the major gravity anomalies. These relationships indicate that the pattern of prominent highs and lows must be associated with the basin, and in particular, with the basement terrane that underlies it. The gravity pattern over the Appalachian basin is similar to that of the midcontinent rift, especially if allowance is made for the much greater depth of the basement terrane beneath the basin. The sedimentary strata of the basin were deposited as a continental shelf sequence of lower Paleozoic age (Rodgers, 1968). The similarity of the basin/shelf gravity pattern to that of the midcontinent rift suggests that the former continental shelf is underlain by a central rift system (Kane, 1983). The correspondence of the shelf with the rift also implies that the rift gravity pattern indicates the extent of the shelf (and the basin) beneath the allochthon. In the area where the basin surface is exposed, the basin strata are largely undeformed, which indicates that much of the continental shelf has been preserved.

Immediately southwest of subzone A2, the southeast edge of the former continental shelf (Rodgers, 1968) is near the basin-crystalline terrane boundary, whereas in the far southwest it is beneath the allochthon (Cook and others, 1979; Harris and Bayer, 1979). In either location, the continental shelf edge lies northwest of much of the crystalline terrane so that the terrane must be part of a continental margin or an assemblage of magmatic arcs that was left accreted to North America after formation of the Atlantic Ocean.

The chain of gravity highs that extends from northern Vermont to central Alabama lies over crystalline terrane. It resembles the gravity-high chains of the Appalachian basin and the MGA, if allowance is made for the structural deformation that characterizes the terrane. Deformation of the sources of the gravity highs is implied by the steep gradients and discontinuous sharp outlines of the anomalies. Taken together, the geophysical and geologic evidence suggests that the source of this chain of gravity highs may be an exhumed rift system (Long, 1979; Kane, 1983) that underlay a continental shelf segment left accreted to the North America plate after the continental breakup of Mesozoic time (Kane, 1983). The gravity highs of the chain are asymmetric, with steep linear gradients on their northwest sides and irregular, more gentle ones on their southeast sides. Steep linear gradients of this kind are caused by steeply dipping planar density interfaces and are typically attributed to faults. The properties of the gravity highs indicate that their sources are dense rock masses raised up to the northwest along the entire length of the chain on steeply dipping faults. The shapes, densities, and structural setting of these sources resemble those of the central horsts described for the midcontinent rift system.

The gravity-high chain lies just to the southwest of a regional gravity gradient that has been interpreted as the boundary between the two continental plates (Cook and Oliver, 1981; Kane, 1983) that collided in lower to middle Paleozoic time (Hatcher, 1978). The southeast edge of the early Paleozoic continental shelf (Rodgers, 1968) is also located in the vicinity of the gradient.

The intense elongate high in southeastern New York and northeastern Pennsylvania correlates with thick clastic sedimentary strata of Devonian age and may be related to a local tensional regime associated with the continental collision of the same age (Kane, 1983).

We interpret the gravity field of zone A as the expression of the crustal structure of two separate continental margins, which developed in the beginning of the Paleozoic and came together as the result of a continent-continent collision in early to middle Paleozoic time. The geometry of the continental shelf that now underlies the Appalachian basin, and of the underlying rift indicated by the gravity anomalies, was controlled by the shape of the craton margin at the time of shelf formation. The original pattern of the rift may be deformed to some degree by the later tectonics, principally the plate collision, but the lack of major deformation of the sedimentary strata and the continuity of the gravity pattern suggest that the rift is well preserved. The complex pattern formed by the combination of subzone A2 and the remainder of A1 to the east can be attributed to a variety of causes, including an assemblage of exotic terranes and island arcs (Osberg, 1978) and the effects of a prominent outlier of the craton (the Adirondacks block) on the shape of the collision zone.

The deformation of the crystalline terrane southwest of subzone A2 appears to have been caused by subduction of the outer part of the North American margin beneath the continental plate that collided from the east and/or southeast. The southeast-directed subduction apparently led to uplift of the underthrust margin, shedding of sedimentary strata and thin sheets of the crystalline terrane from the uplifted landmass (Elliot, 1976) over part of the North American shelf, intrusion of the crystalline terrane by magmatic arcs that rose from the subducted plate (Hamilton, 1981), and probable moderate to severe lateral deformation of the elevated crust that accommodated the direction of collision and the shapes of the colliding margins. The faulted northwest margins of the interpreted rift-system horsts may have coalesced to form the primary zone of crustal failure during the continental collision, that is, the zone along which part of the accreted continental shelf was uplifted. The geometry of the structure of zone A would have been controlled by the orientation of the collision zone, an orientation dependent in part on the prior trends of the margins. Much of the former North American continental shelf, which now forms the Appalachian basin and which appears to be intact, must have been protected by a fortuitous combination of the shapes of the continental margin, the direction of approach between the two continental plates, and the polarity of the subduction zone.

The two areas of broad low-amplitude magnetic highs, Al and A3, correlate with regions that are underlain by thick accumulations of sedimentary strata. The whole of subzone A3 correlates closely with the outline of the Mississippi embayment.

Zone O

The western region is divided into three subzones according to contrast in anomaly patterns and minor changes in trend.

The gravity expression of subzone O1 is made up of an array of alternating highs and lows that strike west of north. The low-gravity field of the coastal region correlates with the Coast Ranges and the westernmost edge of the Great Valley, the adjacent high correlates with the Great Valley and a tract of eugeosynclinal strata that lies along the west border of the Sierra Nevada, and the next adjacent low mainly correlates with the felsic plutonic rocks of the Sierra Nevada. The Great Valley high is attributed to a segment of mafic crust (Cady, 1975) that underlies the valley floor and the eugeosynclinal rocks to the east. The sharply defined west border of the gravity high lies near the valley axis, which is offset toward the western side of the valley. A prominent linear magnetic high also follows the valley axis. The Sierra Nevada gravity low is caused, at least in part, by the felsic plutonic rocks that make up most of the mass of the batholith. The irregularly shaped gravity low that lies east of the Sierra Nevada low correlates with widespread exposures of Tertiary felsic volcanic rocks on the north, including abundant calderas, and with valley-fill deposits and Tertiary-age felsic plutonic rocks on the south (Stewart, 1980). Linear magnetic anomalies, which also trend west of north, are present in the western part of the zone, but elsewhere the magnetic anomaly shapes and trends are variable.

The coastal gravity low and the Great Valley high of O1 continue into O2, but the anomalies in this area are less intense and more irregular in outline and do not correlate closely with the exposed rock units. The gravity anomalies of the central part of subzone O2, which is underlain almost entirely by felsic and mafic volcanic rocks of Tertiary age, are of low amplitude and very irregular. In the eastern region, which encompasses the southern two-thirds of Idaho, the gravity highs correlate on the northwest with part of the Columbia Plateau basalts, on the south with basalts of the Snake River downwarp, and on the northeast with the Precambrian rock units in the Idaho batholith region. The western two-thirds of the prominent arcuate gravity low of this region coincides with felsic plutonic rocks of the Idaho batholith. The magnetic anomalies in O2 are generally of small amplitude and extent and are very irregular. In the eastern part of the subzone, however, there is a correlation between a low-amplitude anomaly pattern and the Columbia Plateau and Snake River downwarp basalts.

The gravity anomalies of O3 have an overall northern trend that is more sharply defined on the west than on the east. The irregularly shaped linear high over the coastal region corresponds with a tract of sedimentary rocks of Tertiary age that contain pillow basalts. The irregular low to the east correlates in large part with the Cascade volcanic belt. There are no clear-cut gravity anomaly–geology correlations in the eastern region of the subzone, which is underlain mostly by the Columbia Plateau basalts. The gravity highs of the region may reflect variations in the thickness of the basalts. The magnetic field shows many minor anomalies that are probably caused by the numerous local volcanic features.

The gravity trends of zone O are parallel with the orientation of the coastline, with the parallelism diminishing from the coast inward toward the continental interior. Because the shape of the coastline was formed by the subduction process, the parallelism of the coastline and the anomalies implies that the trends of the anomalies and their sources are also caused by the orientation of the subduction zone. The western gravity low of subzone O1 corresponds to a mélange of sedimentary rocks that developed at the juncture of the underthrust and overriding plates (Hamilton, 1979). The mafic rock indicated by the adjacent Great Valley gravity high may be a strip of oceanic crust left attached to the continental crust because of the difference in thickness between oceanic and continental crust (W. Hamilton, oral communication, 1981). The extensions of the Great Valley high into subzones O2 and O3 may also be caused by similar strips of oceanic crust. The Sierra Nevada batholith, of Jurassic and Cretaceous age, can be attributed to a magmatic arc that arose from the underthrust plate (Hamilton, 1981). Broad tracts of felsic volcanic rocks, like those that correlate with the northern part of the gravity low east of the Sierra Nevada low, typically indicate the presence of underlying felsic plutons. These rocks may also constitute a magmatic arc, but one that is younger and less thick than the Sierra Nevada batholith. Similarly, the extensions of the Sierra Nevada low into subzones O2 and O3 can be interpreted as expressions of magmatic arcs.

We conclude that the sources of the geophysical anomalies of the western part of zone O were generated by the subduction of the oceanic plate beneath western North America and that the trends of the sources and their associated gravity anomalies reflect the geometry of the subduction zone.

Zone B

Zone B is characterized by north-trending geophysical anomalies that are more pronounced in the gravity field than in the magnetic field. Gravity highs, flanked in most places by lows, trend north and are arranged in an en echelon series that has an overall north trend. The series is more irregular, and the anomalies narrower and less pronounced, than in the gravity-high chains of zone A. Two of the gravity highs in the northern third of the zone trend northwest, but the axis of the northern one turns back to the north to form a "dogleg" shape. The typical magnetic expression of zone B does not extend as far south as the gravity expression, and the magnetic trend directions are not as consistent. The most clearly defined magnetic trends are along the west margin of the zone. Some of the magnetic anomalies in the interior of the zone are aligned slightly east of north.

Zone B is underlain by the Cincinnati arch–Nashville dome in its southern two-thirds and by much of the Michigan basin in its northern third. The series of gravity highs does not correlate closely with the axis of the Cincinnati arch–Nashville dome, but the "dogleg"-shaped gravity high on the north, which is the central geophysical feature of the Michigan basin, indicates the presence of a central rift. The low-amplitude regional magnetic highs correlate in large part with the Michigan basin. Keller and others (1982) investigated the prominent linear gravity high near the south part of the boundary that separates zones A and B and concluded that it is caused by a rift.

Zones C and D

Zones C and D are separated by the MGA. The gravity anomalies of zone C are of moderate amplitude and random orientation. The most pronounced gravity anomaly is a high that forms most of the northwest margin of the zone. The gravity pattern of zone D is made up of narrow linear anomalies of moderate amplitude and regular southwest trend and contrasts with the random pattern of zone C. The magnetic pattern of zone D also shows a southwest trend throughout. The magnetic field of zone C is divided into four subzones: a northern subzone with low-amplitude highs and lows of random trend, a central subzone with southwest trends like those of zone D, a southwestern subzone with small, randomly oriented anomalies, and a southeastern subzone that exhibits broad, low-amplitude regional highs.

Crystalline basement rocks, which are exposed adjacent to the northern part of the midcontinent rift, extend east and west to underlie substantial parts of the northwest area of C and the northeast area of D (King, 1969). The trend of the structures of the exposed basement in zone D agrees closely with that of the gravity and magnetic patterns. The Bouguer gravity field (Lyons and O'Hara, 1982) in the central magnetic subzone of zone C (Fig. 2) shows the presence of a major gravity low, a feature that has been largely removed from the SWL gravity map (Fig. 1) by the filtering process. The northern part of the gravity low is underlain by exposures of felsic plutonic rock, which indicate that the source of the low is a batholith. The northern magnetic subzone of zone C is underlain by igneous rock units that form part of the east-trending arm of the midcontinent rift and by older crystalline rocks along the south shore of Lake Superior. The regional magnetic highs of the southeastern magnetic subzone correlate with part of the Illinois basin. Much of the southwestern subzone is underlain by part of the margin of the midcontinent rift.

The tectonic events reflected by the batholith, the basin, and the parts of the rift that underlie zone C must have greatly altered the preexisting basement. The southwest-trending magnetic trends in the central subzone of C are similar to trends of the pre-rift basement of zones C and D and may be inherited from the structural pattern that existed prior to the later tectonic events. Like the geology, the conformity of the magnetic trends of the central subzone of C with those of zone D suggest that the rift may have split a basement terrane that was once continuous.

Zone E

E1 and E2, although separated by the MGA, are designated subzones because they are typified by similar broad, alternating linear gravity highs and lows of northwest trend. The trends of the magnetic field are more variable than those of the gravity field, but northwest-trending magnetic anomalies are numerous. The southwest boundaries of the subzones appear to connect across the midcontinent rift. The gravity expression of the rift narrows abruptly where it is intersected by the boundaries.

Much of zone E1 is underlain by the Ozark uplift, which exposes broad areas of felsic volcanic and plutonic rock in its eastern corner. The exposures are located over the northern part of the major gravity low (Guinness and others, 1982) in the southeast part of E1. The gradient-amplitude relation (Kane and Bromery, 1968) shows that the maximum depth to the center of the source is about 10 km. The correspondence of the felsic igneous rocks with the gravity low and the upper crustal maximum depth to the center of the source strongly imply that the source is a felsic pluton. The gravity effect of the pluton may be augmented by the felsic volcanic rock units that have a similar bulk mineral density. Another broad, elongate gravity low, located in the northwest part of E2 along the southwest edge of the Sioux uplift, also indicates the presence of a felsic pluton (Klasner and King, 1986). The gravity highs on either side of the low are exaggerated by the filtering process (Kane and Godson, 1985).

As noted by Bickford and others (1986), the northwest-trending geophyscial anomalies of zone E are a prominent but puzzling feature of the midcontinent. If it is assumed that the gravity lows of the zone are caused by felsic igneous rocks, then the geology/gravity pattern resembles that of subzone O1 in western Nevada where the lows were attributed to a magmatic arc. The resemblance suggests that the northwest-trending anomalies may be the expression of magmatic arcs generated by a subduction zone located as much as 500 to 1,000 km or more beyond the southwest boundary of the zone. The well-defined location of the southwest boundary and the marked change in the geophysical expression of the midcontinent rift at its intersection with the boundary suggests that the boundary marks a major crustal division in the central cratonic region.

Zone E corresponds in part with the 1.63 to 1.80 Ga Central Plains orogen of Sims and Peterman (1986) and with the central part of the Tennessee-Montana trend of Hatcher and others (1987). The zone extends beyond both of these terranes on the northeast but does not reach as far southwest or west as the Central Plains orogen or as far northwest or southeast as the Tennessee-Montana trend. The predominant age of the felsic volcanic and plutonic rocks in the vicinity of the gravity low in the southeast part of the zone is 1.42 to 1.50 Ga (Bickford and others, 1986). Igneous rocks of this age are also present in the basement farther northwest along the zone as well as granitic-rhyolitic rocks in the age groups 1.34 to 1.40 Ga (Bickford and others, 1986) and 1.63 to 1.70 Ga (Sims and Peterman, 1986). Taken together, the evidence favors the presence of a magmatic arc of 1.42 to 1.50 Ga, but magmatic arcs at other times cannot be ruled out.

Zones F and G

Zones F and G are divided by the Wichita uplift. They are discussed together because of a moderate similarity in their gravity patterns and a lack of strong contrasts in their magnetic patterns. There is no preferred orientation in the gravity fields of either of the zones. The gravity anomalies of F1 and G are fairly broad and of moderate amplitude; those of F2 are more variable in lateral extent and generally of less amplitude than those of F1 and G. The similarities suggest that the midcontinent rift and the Wichita uplift may divide a basement terrane that was once continuous.

The magnetic patterns of zones F and G are made up of broad low-amplitude features except in the western part of F2 and the northeast part of G where there are anomalies of smaller lateral extent and high amplitudes. The magnetic expression of Zone F is divided along an east-trending line into a field on the north that is characterized by a majority of westerly oriented anomalies and a field on the south that is made up primarily of broad, low-amplitude regional highs. Similar regional highs are present in the northeast part of zone G. As in the other zones, the broad magnetic highs correlate with regions that are underlain by thick accumulations of sedimentary strata.

Subzone F1 is defined on the southeast by the southeast limit of the broad, low-amplitude magnetic highs. The location of the boundary is also based on a gravity high located just south of the Ouachita thrust belt. The gravity high continues southwest to the juncture of the Ouachita and Wichita structures; from there it trends south to form the east boundary of zone G. A broad, but faintly expressed, magnetic anomaly also traces out the east boundary of zone G. The northwest and west sides of the gravity high are marked by steep gradients.

F1 and F2 are separated from zone G by a belt of pronounced linear gravity and magnetic anomalies that trend north-of-west to the Rocky Mountain front. The east half of the belt correlates with the Wichita uplift, but there is no geologic evidence of a similar structure underlying the west half. The west half of the belt is offset south of and overlaps the west end of the east half. The overlaps are interrupted by an apparent southward extension of the west boundary of the midcontinent rift into zone G. The Wichita uplift is bound along both sides by reverse faults that dip inward beneath the uplift (Brewer and others, 1983). To the north the uplift is thrust over the sedimentary strata of the Anadarko basin. Seismic reflection profiles (Brewer and others, 1981) also show the presence of very thick stratified rocks south of the uplift.

West of the center of zone G, a linear, northwest-trending gravity high forms the east side of a "doughnut"-shaped gravity high that encloses an intense, circular gravity low. The linear high correlates with the uplifted central platform of the west Texas basin.

The extensive gravity highs along the southeast and east boundaries of zones F and G separate the continental interior from the Coastal Plain. The highs display steep gradients on the cratonward side in much the same way as do the anomalies of the gravity-high chain over the crystalline terrane of the Appalachians. We suggest, as we did for the Appalachian region, that highs along the boundaries of zones F and G are caused by rift-related horsts with steeply dipping faults along their cratonward sides. The rifts presumably were part of the infrastructure of the continental shelves of the precollision continental margin, and the rift-horst faults coalesced to form the principal zone of crustal failure during the collision.

Zone H

Subdivision of this zone is based on a contrast in the breadth of the gravity anomalies. The contrast in breadth of the anomalies indicates a similar contrast in the breadth of their geologic sources, which implies that the geologic units making up the basement of subzones H1 and H2 are substantially different. The lower amplitudes of the gravity anomalies and the almost complete absence of magnetic anomalies of H2 can be explained by the much greater thickness of sedimentary strata of the western Gulf coast region.

Subzone H1 is separated from zone A by a prominent magnetic low, the Brunswick anomaly. The low has been attributed to a suture zone (Nelson and others, 1985) that accreted an exotic terrane to North America in late Paleozoic time. The gravity pattern of H2 resembles those of the zones to the north and west, but, as noted above, it contrasts with that of H1, a probable accreted terrane. These relationships suggest that the contact between the terrane accreted in late Paleozoic time and the autochthonous basement of the continent to the west lies near the boundary between H1 and H2. The H1/H2 boundary also marks the approximate location of a change from moderately thick to very thick sedimentary strata and of the east limb of a chain of narrow grabens that is concentric to the very thick sedimentary deposits of the western Gulf Coastal Plain. The combined evidence suggests that the division between H1 and H2 marks a major structural boundary in the underlying crust.

Zones I and J

The gravity and magnetic anomalies of zone I trend north in the eastern part of the zone and northwest in the central and western part. The dominant gravity feature is a northwest-trending high located in the southwest part of the zone. In contrast to the gravity, the major magnetic anomaly is a north-trending composite linear high in the eastern part of the zone. The gravity and magnetic anomalies of zone J trend southwest and correlate more closely than those of zone I.

The deepest part of the Williston basin lies within zone I. The basin is underlain in the western part of zone I by a basin platform that correlates closely with the major gravity high. An antiform structure in strata of Cretaceous age, bound on its southwest side by a fault, is exposed directly over the basement platform. The surface structure indicates that the fault-bounded basement platform, which has substantial vertical relief, remained active from its inception through the Mesozoic era. The anomaly trends of zone I extend northward into Canada where they correlate with elements of the Trans-Hudson orogen (Green and others, 1985; Klasner and King, 1986; Thomas and others, 1987).

The floor of the Williston basin is topographically complex and is shallower under the eastern third of zone J than under zone I (King, 1969). The shape of the basement surface, which presumably reflects some aspect of basement structure, shows little correlation with the trends of the gravity and magnetic fields. There are a number of intense magnetic highs in the western part of zone J, which correlate with igneous plutonic and volcanic features of Tertiary age.

The magnetic field of the Williston basin is notable for the absence of the broad magnetic highs that are apparent over thick sedimentary strata elsewhere in the continental United States.

Zones K and L

Zones K and L include the eastern part of the high altitude region of the western United States. The gravity anomalies of zone K form a distinctive pattern of alternating highs and lows with intense amplitudes and sharp boundaries. The gravity highs are commonly linked together into curved and disjointed chains. The gravity anomalies of zone L are narrower and much less pronounced than those of K, and show a crude northeast orientation. In addition, the zone is overprinted by abundant small anomalies that give the gravity pattern a diffuse appearance.

The magnetic anomalies of zone K are divided into three subzones: a northeast subzone with northeast trends; a western subzone with north-of-west trends; and a southeast subzone that has narrow, intersecting linear anomalies. The magnetic anomalies of zone L are generally oriented northeast, but the trend pattern is not as clearly defined as that of the gravity.

The gravity highs of zone K correlate with the various uplifted basement blocks that form the ranges of the Rocky Mountains and are caused mainly by the density contrast between the indurated crystalline rocks of the mountain blocks and the porous basin fill. The gravity lows correlate with intermontane basins except for two lows in the southeast corner, which are underlain by broad exposures of felsic plutonic and volcanic rocks. These lows indicate the presence of underlying felsic plutons, one partly exposed and the other hidden by volcanic cover. The magnetic anomalies of zone K are associated mainly with igneous rocks and with rock units within the crystalline basement.

Zone L is underlain primarily by sedimentary rock units but also by a few areas of scattered felsic intrusions on the north and several broad areas of volcanic rock units in the south. Although there are several well-defined basins within the zone, they show little gravity expression, probably because they are developed in sedimentary strata that have small density contrasts with the basin fill. The magnetic anomalies generally correlate with the igneous rock units.

Zone K includes the Rocky Mountain province and the northern third of the Colorado Plateau; zone L includes the remainder of the Colorado Plateau. The eastern boundaries of K and L are located near the Rocky Mountain front, except that the south part of the eastern boundary of K passes southwestward through the middle of the Denver basin, and the eastern boundary of L lies just to the west of the front. The proximity of the western boundaries of K and L to the Wasatch front is closer than that of the eastern boundaries to the Rocky Mountain front.

The geophysical data show that the deep structure of the northern third of the Colorado Plateau differs significantly from the two-thirds to the south. The geophysical patterns indicate that either a separate basement structural province or a very broad transition zone underlies the northern part of the plateau. Larsen and others (1985) concluded that this region is the site of a broad southeast-trending rift system, which continues southeastward to connect with the Wichita uplift. The geophysical expression over the structure proposed by Larsen and others (1985) is compatible with that of a rift, if allowance is made for distortion of the rift pattern by subsequent tectonic events.

Zone M

Zone M lies between zone L and subzone O1 on the north and the United States–Mexico border on the south. Both gravity and magnetic trends are northwest, in distinct contrast to the northeasterly trends of L and the more northerly trends of O1. Although the gravity anomalies are broader and of greater amplitude, the trends of the magnetic field are more uniform. Most of the eastern half of the northern boundary is based on the contrast with the magnetic patterns of zones O and L (Sumner, 1985), which is more distinct than the contrast of the gravity patterns. The west end of the boundary is based on a broad magnetic feature that interrupts the northerly trends of the geophysical features of the coastal region.

The geology that underlies the eastern two-thirds of zone M is made up of a large number of exposures of granitic and crystalline rocks separated by broader areas of alluvium. Continuity of the gravity anomalies that correlate with rock units in the exposed bedrock is interrupted by the alluvium whose variable thickness has an unpredictable effect on the gravity expression. The alluvium has relatively little effect on the continuity of the magnetic anomalies so that they give a more complete picture of the trends than do the gravity anomalies. Faults are numerous along the northwest boundary of the zone.

Farther to the west, a triangular gravity high, which correlates with the Salton trough, is probably caused by the presence of oceanic crust below the trough sediments. The gravity and magnetic anomalies of the coastal region do not correlate closely with the underlying rock units. The disparity is probably caused by the structural complexity that characterizes the region.

The northern boundary of zone M is discordant with the structure of the Mojave block and passes through its center rather than along its north or south edge where well-defined west-trending structures are located. Farther to the west, however, the geophysically indicated boundary corresponds with the trend and location of the underlying structure.

Zone N

The broad anomalies that characterize zone N predominantly trend to the northwest. The gravity pattern is made up of a central northwest-trending high that is flanked by two lows. There is an overprint of small anomalies, particularly in the areas of the lows, and the central high has a "dogleg" shape that resembles the shape of the central high of the Michigan basin. The magnetic field is composed mainly of a broad high-amplitude high in the east and broad low-amplitude lows in the west.

Zone N corresponds closely with the main northern tract of Belt basin sedimentary strata of Precambrian age (Harrison and others, 1974). The gravity high in the center of the zone is located over the Purcell antiform, which is similar in shape and location to the central platforms that are present in other interior basins. A narrow asymmetrical graben filled with sediments of Cenozoic age follows the northeast edge of the Purcell antiform and indicates recent deformation. Numerous faults are present along the south boundary of zone N; some of these may extend eastward along the south boundary of zone J. Any extension to the west, however, is hidden by the Columbia Plateau basalts.

CRUSTAL STRUCTURE

Boundary framework

The boundaries of the composite geophysical zones in the central craton form a crude orthogonal framework centered about the midcontinent rift. The orthogonality is best displayed by the intersection of the MGA, which trends N30°E, with the southwest border of zone E, which trends N60°W. The northwest trend does not appear to extend southeast of zones C, E, and F, but the northeast trend is present in the southeastern boundary of zone B and is the dominant trend within zone A. Both trends are present in the boundaries of the zones immediately west of the craton, but farther west, comparative trends of the framework are northerly and south-of-west.

The northwest-oriented boundaries are more consistent as a group than the northeast to west-of-north ones, but their trends become more westerly beyond the Rocky Mountain front. With one exception, they extend from the east border of zone B to the Wasatch front and beyond. The exception is the southwest boundary of zone E, which on the basis of the geophysical evidence, does not appear to continue beyond the Rocky Mountain front. A colinear belt of geologic structures, however, begins just west of the front and continues to the western border of zone K. The western ends of the group as a whole fall near or within the western continental margin of late Proterozoic–early Paleozoic time (Armin and Mayer, 1983).

In many places, the elements of the boundary framework correspond to major structural features like the midcontinent rift and the continent–continental shelf transition zone on the west side of the Appalachian basin. In some places, the geophysical trends within a zone conform with one or more of its boundaries, which suggests that the source of the zone and boundary trends is the same. In others, the zone and boundary trends are discordant, which indicates that their sources are different.

We conclude that the geophysical boundary framework is the expression of a primary structural framework of the continental crust. The orthogonality and extent of the framework suggests that it was impressed on the continent during a relatively short period by a major global tectonic event. In some places, the tectonic event controlled the dominant structural trends of both the zones and the boundaries, but in other places it controlled only the boundaries, which cut across the preexisting structures of the zones.

Age of the boundary/structural framework

The age of the northeast-trending midcontinent rift, the central element of the proposed structural framework, is about 1.1 Ga. The age of the structures of zone E, whose northeast and southwest boundaries appear to extend to the Wasatch front, is based on that of the interpreted magmatic arc terrane; that is, 1.3 to 1.7 Ga, with the evidence favoring an age of about 1.4 to 1.5 Ga. The southwest boundary of K and F, which includes the Wichita uplift, extends from the northwest corner of the Gulf Coastal Plain to the Wasatch front. The part of the boundary west of the Rocky Mountain front lies southwest of and parallel to the broad Uncompahgre uplift (Larsen and others, 1985). They concluded that the uplift is part of a rift zone that extends southeastward to include the Wichita uplift. They gave evidence of igneous activity and basin development in its western part at about 1.5 Ga. Brewer and others (1981) concluded that a basin developed on the south side of the Wichita uplift between 1.2 and 1.6 Ga.

The west border of zone B is the boundary of a region that appears to contain a system of north-oriented rifts with an estimated age of more than 1.0 Ga (Keller and others, 1982). We infer the age of the boundary to be the same as that of the rifts because of the similarity in trend. The eastern boundary of zone B was formed during the initial development about 0.6 Ga of the continental shelf that underlies the Appalachian basin. The west and northwest borders of subzone H2, the western Gulf Coastal Plain, are characterized by linear gravity highs. We have suggested that these highs are the expressions of rifts that underlie the central parts of continental shelves whose age is probably similar to that of the shelf that underlies the Appalachian basin.

The eastern two-thirds of the northern border of zone M is underlain by faults with relatively recent movement. Similarly, the south-of-west–trending subzone boundaries within the western region appear to be associated with the recent subduction along the west coast of the continent. The east and west borders of zones K and L correspond with the boundary fault zones of the Rocky Mountains and Colorado Plateau. The most recent activity on the fault zones is very young and effectively masks evidence of any boundary structures of an earlier age. Because of their trends, however, we suspect that the position of the present fault zones are located near structural boundaries whose ages are similar to those of the central craton, that is, 1.0 to 1.5 Ga.

Two types of anomalous magnetic features

In the descriptions of the zones, we noted the presence of broad, low-amplitude magnetic highs over many areas that are underlain by thick accumulations of sedimentary strata. Areally small, moderate- to high-amplitude anomalies are generally sparse over many of the areas of the broad highs. The correlation with the thick sedimentary strata suggests that the anomalies may be related to thin crust/mantle upwarp like that implied by the sedimentary basin model of McKenzie (1978). The broad magnetic highs could be the magnetic expression of mantle upwarp or of broad swarms of dikes emplaced in the lower crust as a result of the upwarp. The depth is such that the cumulative effect of the individual anomalies of a sufficient number of dikes would be much the same as the anomaly caused by a broad mantle upwarp.

A broad region of the magnetic field of the western United States, which includes the western part of subzone O1 and much of the northwestern part of zone K, contrasts with the magnetic field of adjacent regions because of a lack of small magnetic anomalies (Fig. 2). The region correlates laterally with that of the thinnest crust and lowest mantle surface velocity (Braile and others, this volume). The correlation suggests that mantle conditions may be affecting the sources of the small magnetic anomalies either by thermal alteration of their magnetization or by control of their emplacement.

VERY LONG-WAVELENGTH (VLWL) GEOPHYSICAL FEATURES

Most of the range of the amplitude of the Bouguer gravity field of the United States is in the long-wavelength part, especially in wavelengths that exceed 1,000 km (Fig. 4B; color version inside back cover pocket). A comparison with a similarly filtered terrain map (Fig. 4A) shows a marked inverse correlation between the major features of the two maps. The correlation is best demonstrated in the western United States where the outline of the high-altitude region corresponds closely with that of the major VLWL gravity low. It is unlikely that the source of the gravity low is in the crust because its magnitude (250 mGal) is so large that it would require a density reduction of 0.2 g/cm^3 (Kane and Bromery, 1968) for the entire crust. Moreover, the crust in the region of the gravity low is abnormally thin (Braile and others, this volume) so that high-density mantle is elevated above its typical depth in such a way as to offset, at least partially, the effect of any reduction in crustal density. It seems clear that the source of the gravity low must be in the mantle.

Figure 5 shows a simplified VLWL gravity map (panel B) and filtered versions of crustal thickness (panel A) and P_n velocity (panel C) maps compiled by Braile and others (this volume). The upper cutoff of the filter was chosen as 1,000 km, like that for the VLWL gravity map. Comparison of panels A and B shows that there is a correlation between thinning continental crust and increasing regional gravity along the continental-oceanic boundaries and a weak correlation between low gravity and thick crust in the northeast part of the United States. Elsewhere, there is little or no correlation or, as in part of the western United States, an opposite correlation of thin crust and low gravity field. In contrast, there is a significant direct correlation between the VLWL gravity field and the map of the P_n velocity over much of the western United States and over the central Gulf coast region. In the east-central United States, there is a moderate correlation between an east-trending P_n velocity low and a gravity low composed of two minima connected by a saddle. The comparisons indicate that the major source of the VLWL gravity field within the conterminous region of the United States is in the mantle (for a review of previous similar conclusions, see Pakiser, 1985) and that it begins at the mantle surface.

A large S-wave velocity decrease (Fig. 6) is present in the upper mantle in the active tectonic region of the western United States compared to stable tectonic regions of the North American plate (Grand and Helmberger, 1984; Helmberger and others, 1985). The active tectonic region corresponds generally with the region of low P_n velocity and low gravity. Figure 6 shows a decrease in S-wave velocity of 0.4 to 0.5 km/sec from the top of the mantle to a depth of about 160 km, a decrease of about 0.25 km/sec from 160 to 200 km, and a disappearance of the decrease at about 400 km. The average decrease is 0.27 km/sec with about 75 percent of the cumulative decrease taking place between 40 and 200 km. The breadth of the gravity and seismic anomalies indicate that the width of the anomalous mantle region is between 1,000 and 1,500 km. If we assume that the vertical and horizontal extents of the low-velocity layer and the source of the gravity low are the same, we can use the dimensions of the low-velocity zone and the amplitude of the VLWL gravity low to calculate the density contrast (Kane and Bromery, 1968).

We use the central region of the United States as a reference area and compare its geophysical and depth quantities with those over the center of the VLWL gravity low to the west. The average gravity value of the reference region is about –40 mGal, compared to maximum amplitude of the gravity low of about –230 mGal, a difference of –190 mGal. We neglect a 2-km difference

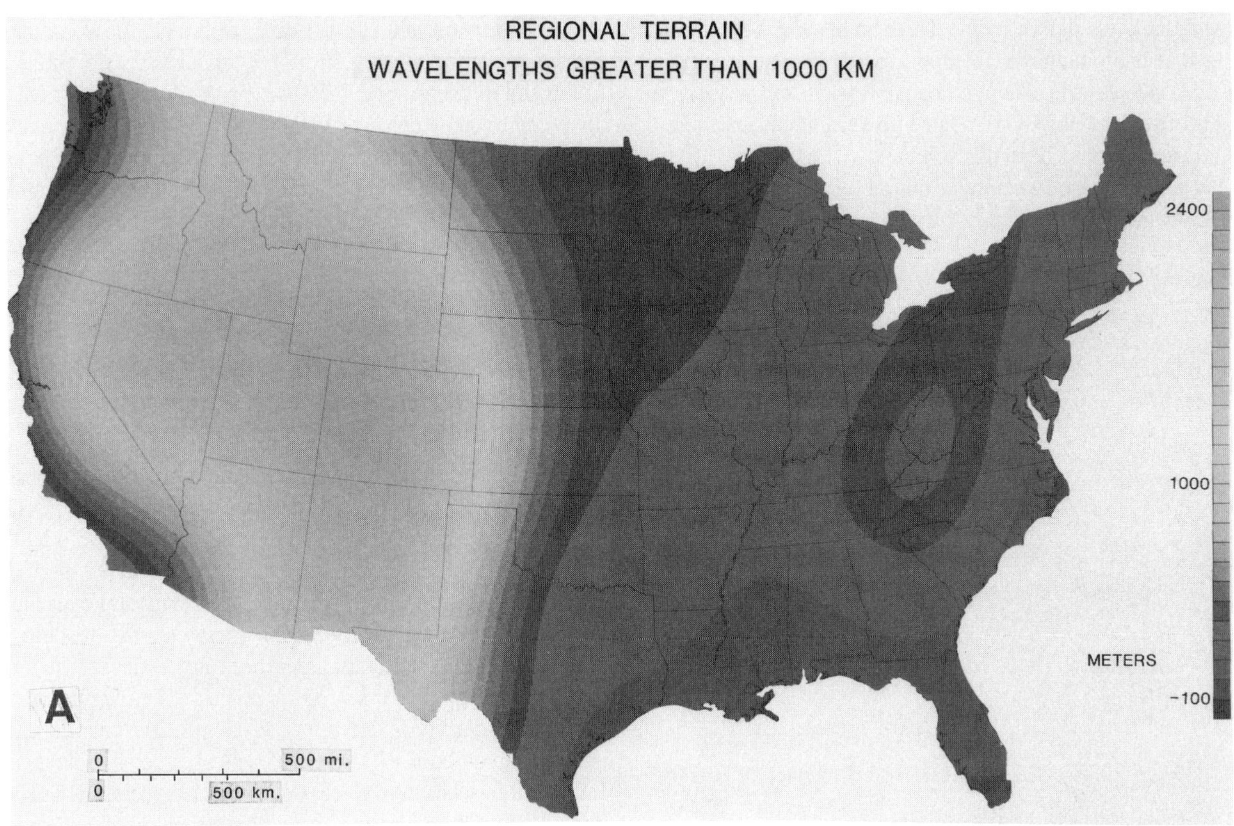

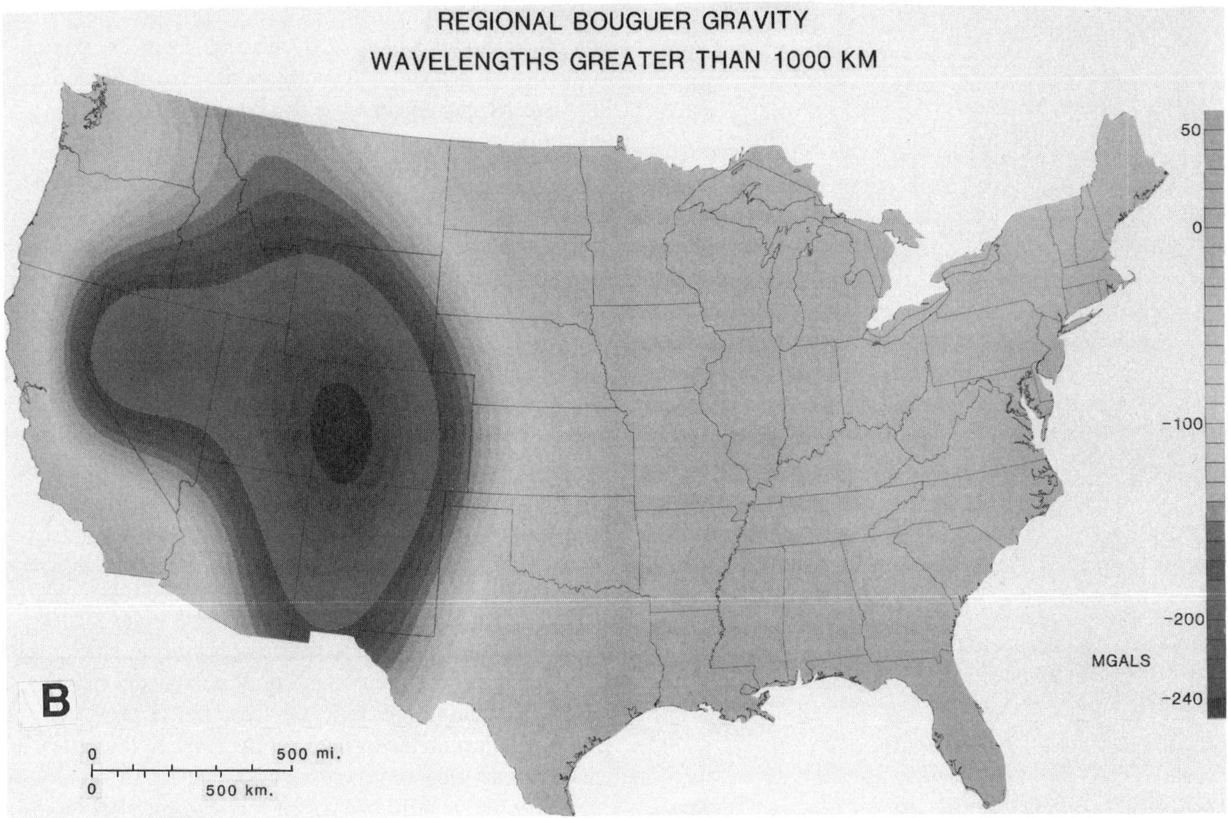

Figure 4. Comparison of terrain and gravity maps composed of wavelengths 1,000 km and longer. A, Regional terrain. B, Regional Bouguer gravity.

in crustal thickness shown by the filtered map (Fig. 5B). We also assume that a probable decrease in average crustal density implied by a decrease in average crustal P-wave velocity for the region is offset by a closely correlated upwarping of dense mantle material (Braile and others, this volume).

Because of the quasi-circular outline of the gravity and velocity lows and the large width-to-thickness ratio, we use the gravity attraction of a circular disk (Nettleton, 1942) to estimate the density of the source. With –190 mGal as the amplitude, 360 km as the thickness, and 1,250 km as the diameter, we calculate an average source density of –0.020 g/cm^3. A similar calculation can be made by assuming that the positive mass of the topography of the high-altitude region is offset by the negative mass of the low-density region in the mantle and that the lateral limits of the two regions are the same. Using an average elevation of 2.3 km (Fig. 4A) and a density for the topography of 2.67 g/cm^3, and dividing by 360 km, we calculate an average source density of –0.017 g/cm^3.

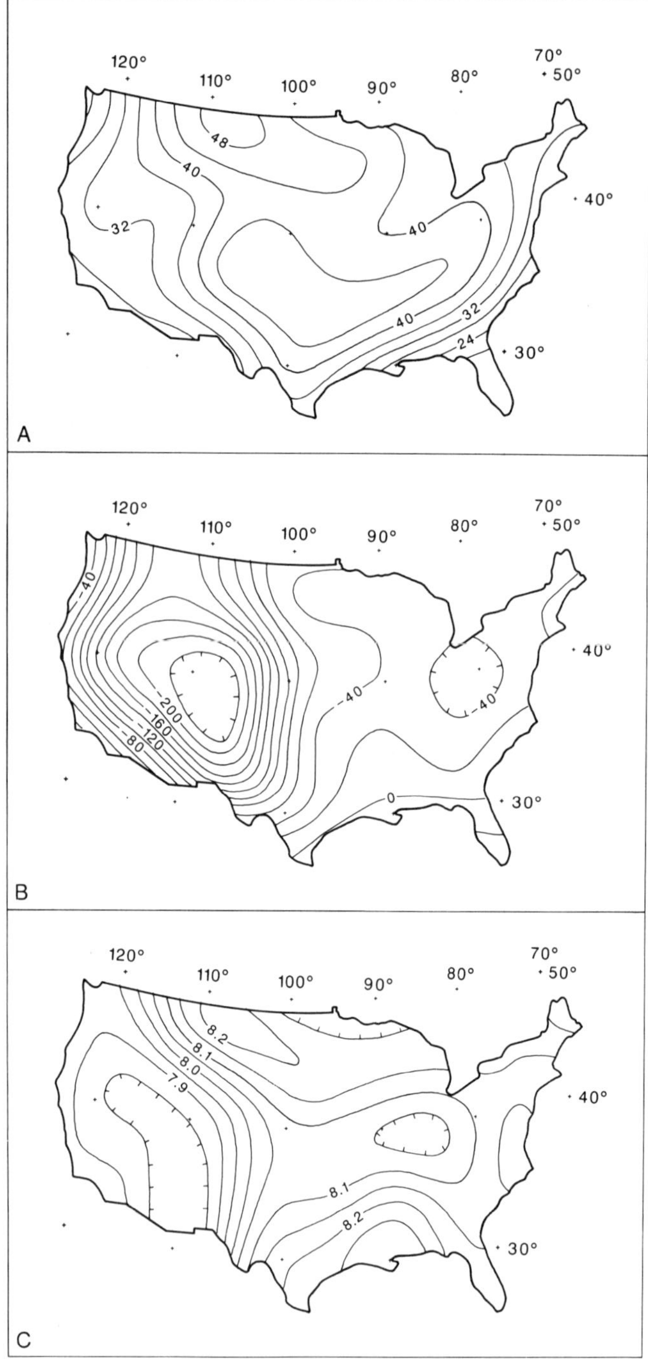

Figure 5. Comparison of long-wavelength (>1,000 km) features of the conterminous United States. A, Crustal thickness (in kilometers). B, Bouguer gravity (in milligals). C, P_n velocity (in kilometers per second).

If we assume a linear relation between the S-wave velocity decrease and the density decrease, we can calculate a vertical density distribution that gives a more meaningful figure for the uppermost part of the mantle. Distribution of the average value of –0.020 g/cm^3, according to the velocity model (Fig. 6), leads to a density decrease of about –0.033 g/cm^3 in the upper 120 km of the mantle, a decrease of about –0.018 g/cm^3 in the next 40-km interval, and gradual disappearance of the decrease over the lowest 200-km interval. According to these calculations, about 75 percent of the isostatic compensation of the surface mass takes place in the upper 160 km of the mantle. The magnitude of the decrease in density in the upper 120 km of the mantle is small, amounting to about 1.0 percent of typical upper-mantle densities. It is small in relation to the S-wave velocity decrease in the same interval, which is about 8 percent. Use of these values in the velocity equation for S-waves gives a corresponding reduction in the shear modulus of about 17 percent. The reduction of P_n is a little over 4 percent, and substitution of the above values in the P-wave velocity equation yields a decrease in the bulk modulus of about 3 percent.

The small decrease in density and the large change in the shear modulus suggest that the cause is a change in physical state rather than in composition. A plausible cause is an anomalously high upper-mantle temperature, which is probably the result of interaction of the underthrust oceanic plate with the mantle, and which possibly caused to partial melting. A possible cause of the gravity high and the high mantle-surface velocity observed in the central Gulf coastal region is a cooling of upwelling mantle associated with formation of the crust under the Gulf of Mexico.

DISCUSSION

A principal conclusion of our report is that the predominant gravity and magnetic trends of the conterminous United States, and their associated structures, have been controlled by global plate collisions and, to a lesser extent, by plate separations along the margins of the present and prior continental margins. These global tectonic events appear to have had a profound and far-reaching effect on the structure of the crust. In the eastern United States, plate separation and collision appear to be the primary factors in the control of the structural/geophysical trends of the entire Appalachians. In the western United States, the effect of the present subduction zone reaches to the Rocky Mountain

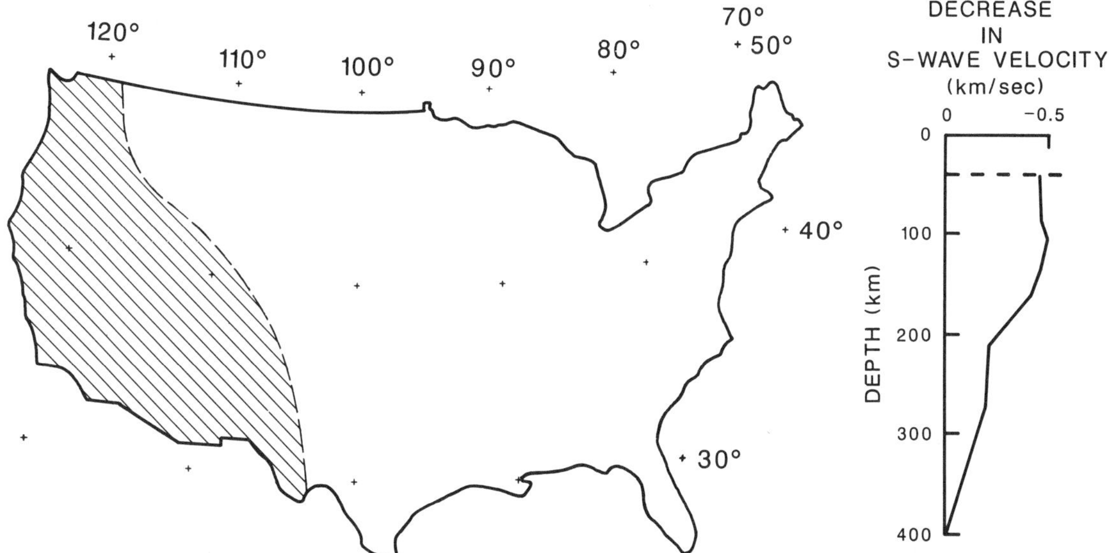

Figure 6. S-wave low-velocity zone in the western United States. Diagram on right shows velocity decrease with depth beginning at surface of mantle (40 km) relative to remainder of United States. Adapted from Grand and Helmberger (1984) and Helmberger and others (1985).

front, more than a third of the continental width. We detect evidence of structural boundaries of an earlier age in the east half of the uplifted western region, but in the region of the previous continental shelf farther to the west (Armin and Mayer, 1983), we found little evidence of geophysical patterns that we could attribute to presubduction structures. On the south, we see evidence of a subduction zone whose effects reach into the middle of the craton. It is only in the northern cratonic region that we found geophysical evidence of structures that can be clearly attributed to ancient tectonic events, that is, early Proterozoic or older.

Another principal conclusion is that the structural framework of the central craton, as expressed by the dominant gravity and magnetic trends, was probably established at 1.1 to 1.5 Ga, that of the areas to the east and south about 0.6 Ga, and that of the western region at 0.2 Ga or later. Evidence of post-Mesozoic movements along some of the elements of the primary structural framework suggests that the framework has accommodated crustal deformation at least since the mid-Proterozoic. There are, however, many transverse structures, such as the Nemaha ridge (Fig. 3), the 38th Parallel fault system (Heyl, 1972), and the New York–Alabama lineament (King and Zietz, 1978), that appear to link elements of the primary framework and that probably also yield when the stress field is oriented in the proper direction. Many of the transverse structures may be inherited from an older framework that was altered or destroyed by the tectonic events of 1.1 to 1.5 Ga. If so, the use of gravity and magnetics in the central craton to delineate terranes older than 1.5 Ga will be more difficult than previously recognized. Knowledge of the younger overprint may point to ways to overcome some of the difficulties. The magnetic pattern of the central subzone of zone C suggests that structural patterns of older terranes may be preserved in very short-wavelength features. It is possible that these patterns can be distinguished by use of filtering techniques or by more effective application of the magnetic method, which is generally more sensitive to critical short-wavelength features than the gravity method.

As concluded in previous reports (Kane and others, 1981; Kane and Godson, 1985), the central parts of rift systems appear to be regions of major crustal deformation, particularly along continental margins.

We propose that the gravity-magnetic zones of Figure 3 can be grouped into four principal tectonic regions. The division is based partly on the geophysical and geologic features of the regions and partly on the character of the mantle. The first region is the eastern region, which is made up almost entirely of the Appalachian province. The primary tectonic characteristic is the dominant northeast-striking structural grain displayed by the structures of the exposed basement and the trends of the geophysical fields. The grain was inherited in part from the continental shelf that developed about 0.6 Ga and was reinforced in the continent-continent collision that followed in early to middle Paleozoic time. The present tectonic grain of the region is the result of that collision and the continental separation that followed in Mesozoic time. Included in the region is the eastern Gulf coast and Florida, which appear to have been accreted to the Appalachian region in late Paleozoic time (Nelson and others, 1985).

The second region is the central and western Gulf coast region. The nature of the crustal structure in this region is not clearly indicated in the gravity and magnetic fields because the basement is deeply buried by the young sediments. As noted above, thermal effects in the mantle probably played the major role in the generation of its present structure, which began to form in early Mesozoic time.

The third region is the central craton, which extends from the western boundaries of the first two regions to the Rocky Mountain front. The trends of the boundaries and of the southern two-thirds of the zones were probably imprinted beginning at 1.5 Ga and ending at 1.0 Ga. Its western margin, however, has undoubtedly been altered by the Laramide orogeny of Cretaceous time.

The fourth region is the remainder of the continent west of the Rocky Mountain front. The pre-Mesozoic structure of this region has been severely overprinted by orogenic events related to the underthrusting of the oceanic plate along its western margin (Bird, 1984). The geometry of the geologic and geophysical patterns of the westernmost region suggests that the subduction zone between the Pacific and North America plates is the primary source of the tectonic patterns. Farther east, the relation of structure to the subduction zone is less direct. The region between the Rocky Mountain and Wasatch fronts probably forms a transition zone that incorporates tectonic patterns of the craton with those that arose from the underthrust plate. The age of the structural patterns in the western half are Mesozoic or younger, whereas those of the eastern half probably inherit many of their trends from structures formed at the same time as those of the central craton. A major factor in the tectonics of this region is the low-density, S-wave low-velocity zone that we attribute to a relatively high-temperature region in the uppermost mantle and associate with the underthrust plate. Accordingly, the elevated western terrane is supported by the thermal buoyancy of a thick heated zone in the mantle and should recede when the heat source dies out and the residual heat is dissipated.

ACKNOWLEDGMENTS

We thank our colleagues T. G. Hildenbrand and R. W. Simpson, Jr., who provided major assistance in the compilation and processing of the national gravity and magnetic data sets and maps, and whose ongoing discussions of concepts have been valuable. L. W. Braile kindly provided early versions of his seismic maps. Special thanks are due to P. K. Sims, who gave valuable insights on the geology of the midcontinent region, and to L. C. Pakiser, who contributed vital information for calculation of the density model of the mantle and, as always, considerable other guidance.

REFERENCES CITED

Armin, R. A., and Mayer, L., 1983, Subsidence analysis of the Cordilleran miogeocline; Implications for the timing of late Proterozoic rifting and amount of tension: Geology, v. 11, p. 702–705.

Bancroft, A. M., 1960, Gravity anomalies over a buried step: Journal of Geophysical Research, v. 65, p. 1630–1631.

Bickford, M. E., Van Schmus, W. R., and Zietz, I., 1986, Proterozoic history of the midcontinent region of North America: Geology, v. 14, p. 492–496.

Bird, P., 1984, Laramide crustal thickening event in the Rocky Mountain foreland and Great Plains: Tectonics, v. 3, p. 741–758.

Bond, K. R., and Zietz, I., 1987, Composite magnetic anomaly map of the western United States west of 96° longitude: U.S. Geological Survey Map GP-977, scale 1:2,500,000.

Bott, M.H.P., and Smith, R. A., 1958, The estimation of the limiting depth of gravitating bodies: Geophysical Prospecting, v. 6, p. 1–10.

Bott, M.P.H., and Smithson, S. B., 1967, Gravity investigations of subsurface shape and mass distributions of granite batholiths: Geological Society of America Bulletin, v. 78, p. 859–878.

Brewer, J. A., Brown, L. D., Steiner, D., Oliver, J. E., Kaufman, S., and Denison, R. E., 1981, Proterozoic basin in the southern midcontinent of the United States revealed by COCORP deep seismic reflection profiling: Geology, v. 9, p. 569–575.

Brewer, J. A., Good, R., Oliver, J. E., Brown, L. D., and Kaufman, S., 1983, COCORP profiling across the southern Oklahoma aulacogen; Overthrusting of the Wichita Mountains and compression within the Anadarko basin: Geology, v. 11, p. 109–114.

Cady, J. W., 1975, Magnetic and gravity anomalies in the Great Valley and western Sierra Nevada metamorphic belt, California: Geologic Society of America Special Paper 168, 56 p.

Chase, C. G., and Gilmer, T. H., 1973, Precambrian plate tectonics; The midcontinent gravity high: Earth and Planetary Science Letters, v. 21, p. 70–78.

Cook, F. A., and Oliver, J. E., 1981, The late Precambrian–early Paleozoic continental edge in the Appalachian orogen: American Journal of Science, v. 281, p. 993–1008.

Cook, F. A., Albaugh, D. S., Brown, L. D., Kaufman, S., Oliver, J. E., and Hatcher, R. D., Jr., 1979, Thinned-skinned tectonics in the crystalline Appalachians; COCORP seismic-reflection profiling of the Blue Ridge and the Piedmont: Geology, v. 7, p. 563–567.

Elliot, D., 1976, The motion of thrust sheets: Journal of Geophysical Research, v. 81, p. 949–963.

Godson, R. H., 1985, Preparation of a digital grid of gravity anomaly values of the conterminous United States, *in* Hinze, W. J., and others, eds., The utility of regional gravity and magnetic anomaly maps: Tulsa, Oklahoma, Society of Exploration Geophysicists, p. 38–45.

Grand, S. P., and Helmberger, D. V., 1984, Upper mantle shear structure of North America: Geophysical Journal of the Royal Astronomical Society, v. 76, p. 399–438.

Green, A. G., Hajnal, Z., and Weber, W., 1985, An evolutionary model of the western Churchill province and western margin of the Superior province in Canada and the north-central United States: Tectonophysics, v. 116, p. 281–322.

Guinness, E. A., Arvidson, R. E., Strebeck, J. W., Schulz, K. J., Davies, G. F., and Leff, C. E., 1982, Identification of a Precambrian rift through Missouri by digital imaging processing of geophysical and geological data: Journal of Geophysical Research, v. 87, p. 8529–8545.

Halls, H. C., 1966, A review of the Keweenawan geology of the Lake Superior region, *in* Steinhart, S. J., and Smith, T. J., eds., The Earth beneath the continents; A volume of geophysical studies in honor of Merle A. Tuve: American Geophysical Union Geophysical Monograph 10 (National Academy of Science–National Research Council Publication 1467), p. 3–27.

Hamilton, W., 1979, Tectonics of the Indonesian region: U.S. Geological Survey Professional Paper 1078, 345 p.

——, 1981, Crustal evolution by arc magmatism: Philosophical Transactions of the Royal Society of London, v. 301, p. 279–291.

Harris, L. D., and Bayer, K. C., 1979, Sequential development of the Appalachian orogen above a master décollement; A hypothesis: Geology, v. 7, p. 568–572.

Harrison, J. E., Griggs, A. B., and Wells, J. D., 1974, Tectonic features of the Precambrian Belt basin and their influence on post-Belt structures: U.S. Geological Survey Professional Paper 866, 15 p.

Hatcher, R. D., Jr., 1978, Synthesis of the southern and central Appalachians,

U.S.A., *in* Tozer, E. T., and Schenk, P. E., eds., Caledonian-Appalachian Orogen of the North Atlantic region: Geological Survey of Canada Paper 78-13, p. 149–157.

Hatcher, R. D., Jr., Zietz, I., and Litehiser, J. J., 1987, Crustal subdivisions of the eastern and central United States and a seismic boundary hypothesis for eastern seismicity: Geology, v. 15, p. 528–532.

Helmberger, D. V., Engen, G., and Grand, S. P., 1985, Upper-mantle cross section from California to Greenland: Journal of Geophysics, v. 58, p. 92–100.

Heyl, A. V., 1972, The 38th Parallel lineament and its relationship to ore deposits: Economic Geology, v. 67, p. 879–894.

Hinze, W. J., and Zietz, I., 1985, The composite magnetic-anomaly map of the conterminous United States, *in* Hinze, W. J., and others, eds., The utility of regional gravity and magnetic anomaly maps: Tulsa, Oklahoma, Society of Exploration Geophysicists, p. 1–24.

Judd, W. R., and Shakoor, A., 1981, Density, *in* Touloukian, Y. S., and Ho, C. Y., eds., Physical properties of rocks and minerals: New York, McGraw-Hill, 548 p.

Kane, M. F., 1983, Gravity evidence of crustal structure in the United States Appalachians, *in* Schenk, P. E., ed., Regional trends in the geology of the Appalachian-Caledonian-Hercynean-Mauritanide Orogen: Dordrecht, Reidel, NATO Advanced Study Institute Series C; Mathematics and Physical Sciences, v. 116, p. 45–54.

Kane, M. F., and Bromery, R. W., 1968, Gravity anomalies in Maine, *in* Zen, E-an, and others, eds., Studies of Appalachian geology; Northern and Maritime: New York, John Wiley & Sons, p. 415–424.

Kane, M. F., and Godson, R. H., 1985, Features of a pair of long-wavelength (>250 km) and short-wavelength (<250 km) Bouguer gravity maps of the United States, *in* Hinze, W. J., and others, eds., The utility of regional gravity and magnetic anomaly maps: Tulsa, Oklahoma, Society of Exploration Geophysicists, p. 46–61.

Kane, M. F., Simmons, G., Diment, W. H., Fitzpatrick, M. M., Joyner, W. B., and Bromery, R. W., 1972, Bouguer gravity and generalized geologic map of New England and adjoining areas: U.S. Geological Survey Geophysical Investigations Map GP-839, scale 1:1,000,000.

Kane, M. F., Hildenbrand, T. G., and Hendricks, J. D., 1981, Model for the tectonic evolution of the Mississippi embayment and its contemporary seismicity: Geology, v. 9, p. 563–568.

Keller, G. R., Bland, A. E., and Greenberg, J. K., 1982, Evidence for a major late Precambrian tectonic event (rifting?) in the eastern midcontinent region, United States: Tectonics, v. 1, p. 213–223.

King, E. R., and Zietz, I., 1971, Aeromagnetic study of the midcontinent gravity high of the central United States: Geological Society of America Bulletin, v. 82, p. 2187–2208.

——, 1978, The New York–Alabama lineament; Geophysical evidence for a major crustal break in the basement beneath the Appalachian basin: Geology, v. 6, p. 213–318.

King, P. B., compiler, 1969, Tectonic map of North America: U.S. Geological Survey, scale 1:5,000,000.

Klasner, J. S., and King, E. R., 1986, Precambrian basement geology of North Dakota and South Dakota: Canadian Journal of Earth Sciences, v. 23, p. 1083–1102.

Larsen, E. E., Patterson, P. E., Curtis, G., Drake, R., and Mutschler, F. E., 1985, Petrologic, paleomagnetic, and structural evidence of a Paleozoic rift system in Oklahoma, New Mexico, Colorado, and Utah: Geologic Society of America Bulletin, v. 96, p. 1364–1372.

Leith, C. K., Lund, R. J., and Leith, A., 1935, Precambrian rocks of the Lake Superior region: U.S. Geological Survey Professional Paper 184, 34 p.

Long, L. T., 1979, The Carolina slate belt; Evidence of a continental rift zone: Geology, v. 7, p. 180–184.

Lyons, P. L., and O'Hara, N. W., 1982, Gravity anomaly map of the United States, exclusive of Alaska and Hawaii: Society of Exploration Geophysicists, scale 1:2,500,000.

McKenzie, D., 1978, Some remarks on the development of sedimentary basins: Earth and Planetary Science Letters, v. 40, p. 25–32.

Milanovsky, E. E., 1981, Aulacogens of ancient platforms; Problems of their origin and tectonic development: Tectonophysics, v. 73, p. 213–248.

Nelson, K. D., and 8 others, 1985, New COCORP profiling in the southeastern United States; Part 1, Late Paleozoic suture and Mesozoic rift basin: Geology, v. 13, p. 714–718.

Nettleton, L. L., 1942, Gravity and magnetic calculations: Society of Exploration Geophysicists, v. 7, p. 293–309.

Osberg, P. H., 1978, Synthesis of the geology of the northeastern Appalachians, U.S.A., *in* Tozer, E. T., and Schenk, P. E., eds., Caledonian–Appalachian Orogen of the North Atlantic region: Geological Survey of Canada Paper 78-13, p. 137–147.

Pakiser, L. C., 1985, Seismic exploration of the crust and upper mantle of the Basin and Range province: Geological Society of America Centennial Special Volume 1, p. 453–469.

Pickering, S., Higgins, M., and Zietz, I., 1977, Relation between the southeast Georgia embayment and the onshore extent of the Brunswick anomaly [abs.]: EOS Transactions of the American Geophysical Union, v. 58, p. 432.

Rodgers, J., 1968, The eastern edge of the North American continent during Cambrian and early Ordovician, *in* Zen, E-an, and others, eds., Studies of Appalachian geology; Northern and Maritime: New York, John Wiley & Sons, p. 141–149.

Silver, L. T., and Green, J. C., 1972, Time constants for Keweenawan igneous activity: Geological Society of America Abstracts with Programs, v. 4, p. 665.

Sims, P. K., and Peterman, Z. E., 1986, Early Proterozoic Central Plains orogen; A major buried structure in the north-central United States: Geology, v. 14, p. 488–491.

Smithson, S. B., 1971, Densities of metamorphic rocks: Geophysics, v. 36, p. 690–694.

Stewart, J. H., 1980, Geology of Nevada: Reno, University of Nevada, Nevada Bureau of Mines and Geology Special Publication 4, 136 p.

Sumner, J. S., 1985, Crustal geology of Arizona as interpreted from magnetic, gravity, and geologic data, *in* Hinze, W. J., and others, eds., The utility of regional gravity and magnetic maps: Tulsa, Oklahoma, Society of Exploration Geophysicists, p. 164–180.

Thiel, E., 1956, Correlation of gravity anomalies with the Keweenawan geology of Wisconsin and Minnesota: Geological Society of America Bulletin, v. 67, p. 1079–1100.

Thomas, M. D., Sharpton, V. L., and Grieve, R.A.F., 1987, Gravity patterns and Precambrian structure in the North American Central Plains: Geology, v. 15, p. 489–492.

Vacquier, V., Steenland, N. C., Henderson, R. G., and Zietz, I., 1951, Interpretation of aeromagnetic maps: Geological Society of America Memoir 47, 151 p.

Van Schmus, W. R., Green, J. C., and Halls, H. C., 1982, Geochronology of Keweenawan rocks of the Lake Superior region; A summary, *in* Wold, R. J., and Hinze, W. J., eds., Geology and tectonics of the Lake Superior Basin: Geological Society of America Memoir 156, p. 165–171.

White, W. S., 1966, Tectonics of the Keweenawan basin, western Lake Superior region: U.S. Geological Survey Professional Paper 524-E, 23 p.

Wold, R. J., and Hinze, W. J., eds., 1982, Geology and tectonics of the Lake Superior basin: Geological Society of America Memoir 156, 280 p.

Zietz, Isidore, compiler, 1982, Composite magnetic anomaly map of the United States; Part A, Conterminous United States: U.S. Geological Survey, scale 1:2,500,000.

MANUSCRIPT ACCEPTED BY THE SOCIETY OCTOBER 31, 1988

Printed in U.S.A.

Geological Society of America
Memoir 172
1989

Chapter 19

Isostatic residual gravity and crustal geology of the United States

R. C. Jachens, R. W. Simpson, and R. J. Blakely
U.S. Geological Survey, 345 Middlefield Road, Menlo Park, California 94025
R. W. Saltus
U.S. Geological Survey, Box 25046, Denver Federal Center, Denver, Colorado 80225

ABSTRACT

A new isostatic residual gravity map of the conterminous United States presents continent-wide gravity data in a form that can be readily used, with geologic information and other geophysical data, in studies of the composition and structure of the continental crust. This map was produced from the gridded gravity data on which the recently released *Gravity Anomaly Map of the United States* is based. About 1 million onland and 0.8 million offshore gravity observations interpolated to a 4- by 4-km grid serve as the basis for both maps. The Airy-Heiskanen model of isostatic compensation of topography applied to topographic and bathymetric data averaged over 5- by 5-min compartments was used to remove, to first order, the large, long-wavelength Bouguer gravity anomalies caused by deep density distributions that support topographic loads. The parameters used in the Airy-Heiskanen model were topographic density, 2.67 g/cm^3; sea-level crustal thickness, 30 km; and density contrast across the base of the model crust, 0.35 g/cm^3.

Many of the conspicuous short-wavelength anomalies (widths less than several hundred kilometers) on the isostatic residual gravity map correlate with mapped or near-surface geologic features, and primarily reflect shallow-density distributions rather than any departures from isostatic equilibrium. In general, gravity highs occur over (1) mafic igneous bodies emplaced in rift or magmatic arc settings or as isolated intrusions controlled by structures; (2) accreted slices of mafic oceanic, island-arc, or transitional crust; and (3) uplifted crystalline basement. Gravity lows are found over (1) thick bodies of felsic intrusive or extrusive rocks; (2) sedimentary deposits in extensional, convergent, or transform settings; and (3) depressed crystalline basement. Anomalies with widths as much as 1,000 km or more also appear to reflect crustal properties in many cases—several broad gravity highs are associated with crust having a high average seismic wave velocity, and comparable broad gravity lows occur over areas of low average seismic velocity.

Alternative ways of viewing the isostatic residual gravity data provide additional information about density distributions in the crust. The first-vertical derivative map accentuates gravity anomalies over shallow, abrupt density changes at the expense of those resulting from deeper or more gradual density transitions. The maximum horizontal gradient map contains information about the locations of pronounced density boundaries. Two-dimensional spectral analysis of the gravity data provides a quantitative means for identifying dominant fabrics in the gravity field and for distinguishing various terranes from each other.

Neither Bouguer nor isostatic residual gravity anomalies are particularly well suited for practical modeling of deep structure in conjunction with deep seismic information.

Jachens, R. C., Simpson, R. W., Blakely, R. J., and Saltus, R. W., 1989, Isostatic residual gravity and crustal geology of the United States, *in* Pakiser, L. C., and Mooney, W. D., Geophysical framework of the continental United States: Boulder, Colorado, Geological Society of America Memoir 172.

However, a scheme in which the entire Earth outside the area of interest is approximated by laterally homogeneous layers and isostatically compensated topography, and in which the area of interest is modeled using the seismic constraints applied in a two-and-one-half-dimensional geometry, holds promise for exploiting useful features of both the Bouguer and isostatic residual gravity anomalies.

INTRODUCTION

The first isostatic residual gravity map of the conterminous United States (Hayford and Bowie, 1912) was published in color as a companion to the first Bouguer gravity anomaly map. Both maps were based on 89 pendulum gravity measurements that, remarkably, sampled nearly every major gravity anomaly across the country. The latest isostatic residual gravity map, presented in Figure 1 (in pocket inside back cover), is based on about 1 million onland and 0.8 million offshore gravity observations compiled for the *Gravity Anomaly Map of the United States* (Society of Exploration Geophysicists, 1982; Godson, 1985). For onland areas, approximately 95 percent of all 5- by 5-min cells contain at least one gravity observation, and many parts of the country are covered by more densely distributed observations.

The new gravity data base provides a unique and powerful tool for studying the geophysical framework of the conterminous United States and its offshore regions. Only the aeromagnetic data base rivals gravity in its regional extent and in the density and uniformity of coverage. Both gravity and aeromagnetic observations serve to place the more detailed, yet more scattered, information from seismic experiments into a regional context. New techniques for manipulating, enhancing, and filtering continent-wide gravity data sets, and new modes of presentation made possible by computers (including gray-scale, color, and shaded relief maps, or combinations of these) provide unprecedented opportunities to examine the gravity data in different ways and to increase the usefulness of the gravity data for studies of continental structure and tectonic evolution.

The first isostatic residual gravity maps were prepared by geodesists who were interested primarily in verifying the existence of isostasy and in inferring the details of the isostatic compensation mechanism. Studies along these lines—to explore the nature of the isostatic mechanism—have continued to the present day. A second line of investigation made use of the anomalies on isostatic residual gravity maps to study the geology of the crust. Gilbert (1913) recognized early the utility of isostatic residual maps for this purpose, and Woollard (1936, 1966, 1968) devoted much of his career to exploring the relation of isostatic anomalies to geologic features and to seismic parameters.

Our own purposes have been along these latter lines: to use isostatic residual gravity anomalies to illuminate the configurations and relationships of geologic bodies in the crust. In the discussion that follows, we refer to the gravity field caused by the masses that isostatically compensate topography as the *isostatic regional gravity* field, and to the gravity anomalies that remain after the isostatic regional field has been subtracted from the Bouguer gravity field—that is, after the isostatic correction has been applied—as the *isostatic residual gravity* field. In the past, isostatic residual gravity maps have been called simply *isostatic gravity maps* or *isostatic anomaly maps.* We insert the word "residual" partly to be more explicit and partly in the hope that its unfamiliar ring will remind the reader to avoid some of the pitfalls that commonly arise in the interpretation of such maps. One major pitfall is the tendency to interpret all isostatic residual gravity anomalies in terms of "undercompensation" or "overcompensation," a practice that is generally unproductive for reasons discussed below.

Purpose in making the isostatic residual gravity map

Our goal in producing a new isostatic residual gravity map was to present the gravity data in a form that could be used closely with geologic maps and other available geophysical data to study the geology and structure of the mid- to upper crust. Bouguer gravity anomaly maps are not well suited to this purpose for continent-wide areas because the gravity anomalies caused by near-surface bodies of geologic interest often are distorted or even concealed by large-amplitude, long-wavelength anomalies related to isostasy. (Wavelength, as used here, is roughly twice the breadth of a high or low anomaly.) This is true particularly in mountainous areas such as the western United States and in the Appalachians, but also in adjacent flatter areas such as the midcontinent (Jachens and Griscom, 1985; Simpson and others, 1986).

Most of the long-wavelength part of the Bouguer gravity field is caused by the deep crustal and upper-mantle density distributions that support the topography in a manner compatible with the principle of isostasy (Simpson and others, 1986). Areas of high topography are underlain by mass deficiencies that give rise to regional Bouguer gravity lows, whereas areas of low topography, such as ocean basins, are associated with high Bouguer gravity values. This inverse correlation between elevation and Bouguer gravity is readily seen in Figure 2a. By approximating the compensating masses with a simple model, computing the gravitational effect, and removing it from the Bouguer gravity data, anomalies caused by shallow-density distributions of geologic interest are remarkably enhanced. This process of applying an isostatic correction attempts to remove the observed correlation of gravity with topography, and results in a much narrower range of gravity anomaly values than before the correction was applied (Fig. 2b), suggesting that the isostatic model successfully predicts, at least to first order, an important density distribution. Not all long-wavelength anomalies are suppressed by the isostatic

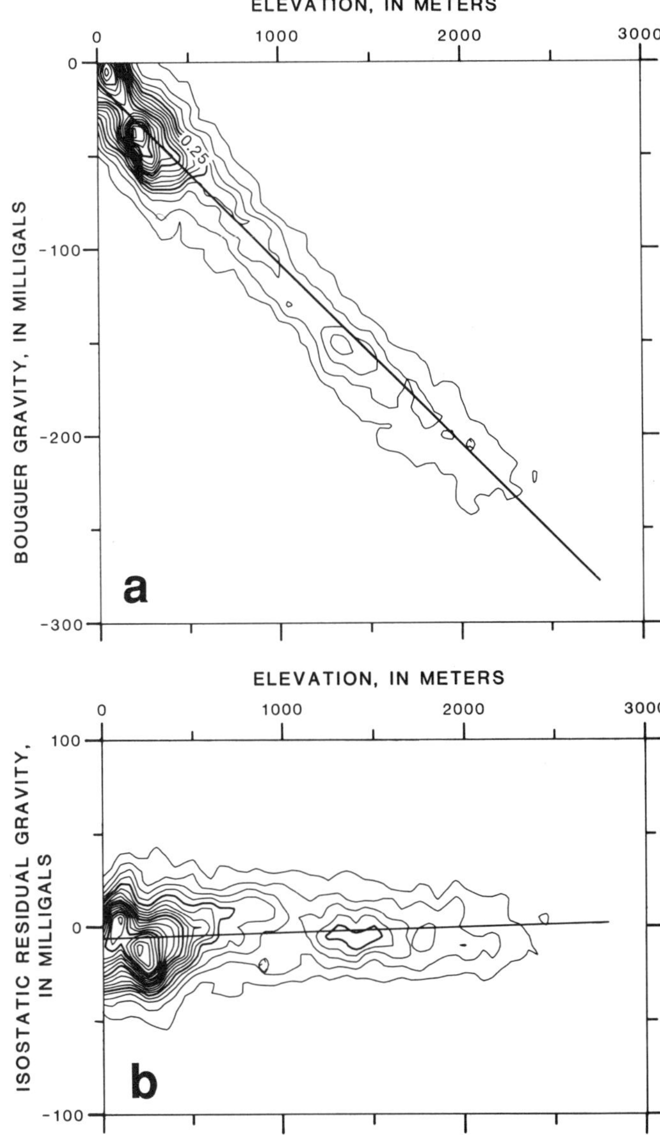

Figure 2. a, Inverse correlation of onland Bouguer gravity anomaly values with elevation for the conterminous United States. Bouguer gravity and topography grids described in the text were sampled at 20-km intervals to obtain the values for this two-dimensional histogram. The numbers of grid points falling into cells 5 mGal tall by 50 m wide were counted; these numbers were normalized by dividing by the largest number, and the results were contoured at 0.05 (5 percent) intervals after a small amount of smoothing. Closed high at about 250 m, which falls slightly below the main trend, is caused by broad areas of low Bouguer value in the midcontinent, which can also be seen on the isostatic residual map (Fig. 1). A regression line $y' = a + bx$ is shown, for which $a = -11.5$ mGal and $b = -0.0942$ mGal/m; standard error of estimate, 24.0; correlation coefficient, −0.942. b, Isostatic residual gravity values versus elevations for the conterminous United States; constructed as in Figure 2a. A regression line $y' = a + bx$ is shown, for which $a = -6.7$ mGal and $b = 0.0039$ mGal/m; standard error of estimate, 18.5; correlation coefficient, 0.148 (after Simpson and others, 1986).

correction, only those that are correctly related to the surface topography by the chosen isostatic model.

Experience with isostatic residual gravity maps indicates that they serve well as a vehicle for both qualitative and quantitative interpretation of gravity anomalies from mid- to upper-crustal sources. As such, they are an effective tool for extrapolating surface geology into the subsurface and for placing detailed but sparsely distributed geophysical information into a regional context. Although isostatic residual gravity data work best for delineating sources in roughly the top 10 to 15 km of the crust, we show in a later section that, with slight modification, these data can also be used to model deeper sources by incorporating independent constraints from seismic or other deep-sounding geophysical techniques.

In the sections that follow, we first describe the data sets and procedures used to generate the isostatic residual gravity map and then discuss the correlation between the gravity anomalies and geology by means of some specific examples familiar to us. We then present three alternative ways of viewing the gravity data: as a first-vertical derivative map; as a maximum horizontal gradient map; and in the spectral domain. We conclude by suggesting a method for quantitative modeling of the isostatic residual gravity data with constraints based on deep seismic information.

The reader will find many similarities between this chapter and an earlier paper on the same topic (Simpson and others, 1986). Despite this duplication, we and the editors decided that a book on the geophysical framework of the United States would benefit from a contribution dealing with the isostatic residual gravity field. In order to permit this chapter to stand alone, we have included some sections and figures from the earlier paper, generally in abbreviated form. In the rest of the chapter, we have attempted to focus more on the combined use of the gravity data with geological and other geophysical information, particularly seismic data.

MAP PREPARATION

The construction of an isostatic residual gravity map requires Bouguer gravity values to define the gravity field and topographic elevations to define the geometry of the compensating masses in the chosen model.

Gravity data

The gridded gravity-data set described by Godson and Scheibe (1982) and Godson (1985) that was assembled to prepare the *Gravity Anomaly Map of the United States* (Society of Exploration Geophysicists, 1982) was used in the preparation of this isostatic residual gravity map. Onland, the grid was constructed from Bouguer gravity values, calculated using a reduction density of 2.67 g/cm^3 and terrain corrected by computer for topography between 0.895 and 166.7 km from a given station. Offshore, the grid was constructed from free-air gravity values. Before gridding, the observations were projected with an Albers

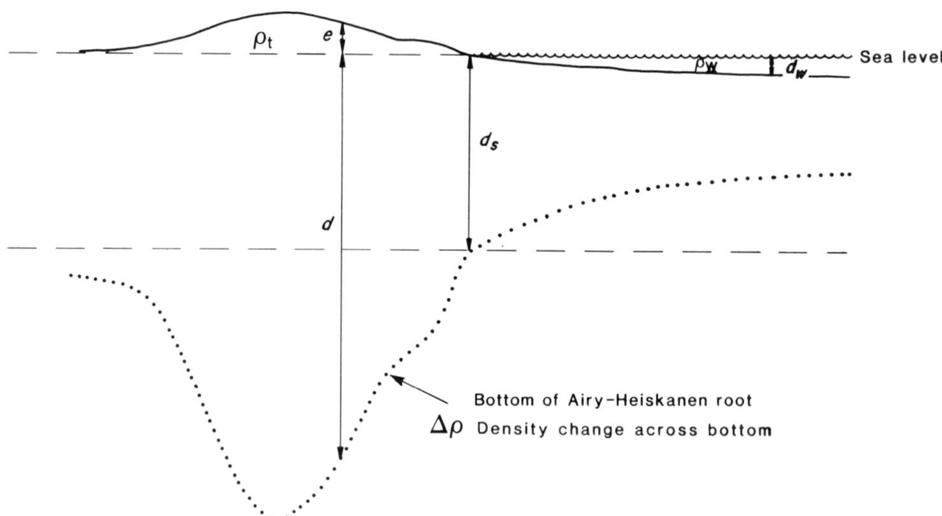

Figure 3. Geometry of compensating masses in Airy-Heiskanen local-compensation model. Key: e = elevation of topographic surface above sea level; d = depth to bottom of root; d_s = depth to bottom of root for sea-level elevation; ρ_t = density of topography; d_w = depth of water in ocean areas; ρ_w = density of water; $\Delta\rho$ = density contrast at depth across bottom of root (after Simpson and others, 1986).

equal-area conic projection (Snyder, 1982), with a central meridian at longitude 96°W and standard parallels at latitudes 29.5° and 45.5°N. In the gridding process, data within a search radius of 40 km of a given gridpoint were used to aid interpolation in areas of sparse data coverage (Godson and Scheibe, 1982). The resulting grid has a 4-km interval between gridpoints in the projected coordinate system.

The distribution of the 1 million onland gravity observations and 0.8 million offshore observations is shown as an inset map on the *Gravity Anomaly Map of the United States* (Society of Exploration Geophysicists, 1982) and in Godson (1985). This data-distributiom map should be consulted to ascertain the quality of data coverage in particular areas of interest. The 4-km interval between gravity gridpoints means that gravity features with widths less than about 10 km are not accurately portrayed on this map, even in areas where the original gravity observations were more densely distributed. In other areas, where the original data were sparsely distributed, even larger features may be missing or distorted.

Topographic data

Bathymetric and topographic data sets were obtained from the National Geophysical Data Center (NGDC, NOAA, Code E/GC12, 325 Broadway, Boulder, Colorado 80303). Offshore, the 5- by 5-min synthetic bathymetric profiling system (SYNBAPS) data set was used (NOAA Announcement 81-MGG-14); onland, the 5- by 5-min North American data set was used (NOAA Announcement 1980 SE-V). Details on the process of combining these two data sets were presented by Simpson and others (1983b).

The 5- by 5-min topographic-bathymetric data were projected using an Albers projection to match the coordinate system of the gravity-data grid. In the projection process, the 5- by 5-min unprojected data grid was converted by linear interpolation to an 8- by 8-km rectilinear projected grid. Less resolution is needed in the grid defining the compensating masses because their gravity effect is smoothed by distance. Tests described in Simpson and others (1983a) suggest that 5-min topographic data are adequate to calculate the attraction of the compensating masses for regional maps, although 3-min data would be more suitable for more detailed maps.

Method of calculation

We used an Airy-Heiskanen model of local compensation for our isostatic model. As explained below, all isostatic models give similar results to first order, and for our purposes the application of *any* isostatic correction was more important than the exact details of the model chosen. The calculation of an isostatic regional field in the local Airy-Heiskanen model of compensation (Heiskanen and Moritz, 1967) requires values for three parameters: the density ρ_t of the topographic load, the model crustal thickness d_s for sites at sea level, and the density contrast $\Delta\rho$ across the base of the model crust (Fig. 3). These parameters determine the depth d to the base of the model crust under land areas by the relation

$$d = d_s + e\,(\rho_t/\Delta\rho), \qquad (1)$$

where e is the elevation of the topographic surface. For the Great Lakes, the weight of the lake water was added to the topographic load in the calculation.

In the Airy-Heiskanen formalism, oceanic crustal columns with water depth d_w (taken to be positive here) have a negative load (weight deficiency) at the top caused by the presence of water of density ρ_w rather than rock (Heiskanen and Moritz, 1967). Negative compensation is provided by an upward protrusion of denser material that has its top at a depth:

$$d = d_s - d_w \left(\frac{\rho_t - \rho_w}{\Delta\rho} \right). \quad (2)$$

In many published figures depicting the geometry of the Airy-Heiskanen compensation model, it appears that the entire crustal column above the bottom of the root must have the same density as the topography. No such restriction, however, is required by the equations; the density of the model crust may change as a function of depth, provided only that the superposed changes with depth are horizontally uniform and that the density contrast across the bottom of the root is constant regardless of depth. The density ρ_t serves only to define the weight of the surface load; the model does not require that this density extend beneath the topography.

To prepare the isostatic residual gravity map, we used parameters that give a root geometry that over large parts of the conterminous United States is close to the geometry of the M-discontinuity as determined by seismic refraction experiments ([Eq. 1] in Woolard, 1968) a topographic load density of 2.67 g/cm^3, a depth to the bottom of the root of 30.0 km for sites at sea level, and a density contrast across the base of the model crust of 0.35 g/cm^3. The M-discontinuity is an appropriate depth for the bottom of an Airy root (for want of better information), because it is a major velocity discontinuity and probably also an important density discontinuity. However, especially in the western United States, major discrepancies exist between our calculated depth to the bottom of the Airy-Heiskanen root and the depth to the seismically determined M-discontinuity (Braile and others, this volume). No simple isostatic model is likely to provide an accurate description of the geometries of the compensating masses for an entire continental area.

Calculation of attraction of the compensating masses at a point on the Earth's surface was performed in two steps. The attraction on a flat Earth out to a distance of 166.7 km was calculated by using a program (Simpson and others, 1983a) based on the fast Fourier transform algorithm developed by Parker (1972). This result was combined with a published regional field for the attraction of both topography and compensation beyond 166.7 km to a distance of 180° on a spherical Earth (Kärki and others, 1961). Although a mismatch exists between the model parameters for the published result of Kärki (d_s = 30 km; $\Delta\rho$ = 0.6 g/cm^3) and those used for calculation of the root attraction inside 166.7 km (d_s = 30 km; $\Delta\rho$ = 0.35 g/cm^3), simple calculations suggest that this mismatch causes long-wavelength errors of no more than 5 mGal for most of the onland conterminous United States. These long-wavelength errors will not affect the relative shapes of typical anomalies in Figure 1. Offshore, the maximum mismatch for the very deepest parts of the map area may reach 10 mGal for our choice of parameters.

The isostatic regional gravity field obtained by this procedure is the gravitational attraction of the compensating mass system at sea level. Because most gravity stations lie above sea level, the proper isostatic correction for a given observation site is obtained by upward continuation of the sea-level result. This upward-continuation adjustment is generally small, given the depth to the base of the model crust and the consequent smoothness of the isostatic regional field. Even in areas of extreme topographic relief, such as the Sierra Nevada of California, the adjustment rarely exceeds 5 mGal. Such an upward continuation has been applied to the isostatic regional field, and details of the correction method were discussed by Simpson and others (1983a).

The final step in the production of the isostatic residual gravity map was subtraction of the isostatic regional field from the Bouguer gravity map. The grid of Godson and Schiebe (1982) contains free-air gravity values offshore; these values could have been retained with no Bouguer or isostatic corrections applied because the free-air anomaly can be viewed as an isostatic residual gravity anomaly with a depth of compensation at sea level. We chose, instead, to continue the Airy-Heiskanen model offshore, and a Bouguer correction was applied to the offshore data by using the 5-min bathymetric data set to determine water depths. Because of the averaging inherent in the bathymetric data set and because of the 40-km search radius used in preparation of the Godson and Scheibe (1982) gravity grid, the spatial mismatch between the original free-air gravity observation and the Bouguer correction applied is potentially quite large. We estimate that in areas of extreme relief on the sea bottom, this mismatch could result in errors as large as 40 mGal, although for most oceanic areas the error is probably less than 10 mGal, or one contour interval. These mismatch errors appear as small, circular high-amplitude anomalies, about 40 km or less wide, in areas where bottom depths change rapidly. The reality of offshore anomalies can be tested by comparing the data on this map with the original free-air offshore data shown on the *Gravity Anomaly Map of the United States* (Society of Exploration Geophysicists, 1982). We think that the patterns in the offshore data on the isostatic residual gravity map are of sufficient interest to warrant keeping these data on this map in spite of the problems. The user should be aware, however, of possible inconsistencies, especially in areas where the water depths change rapidly.

For land areas, isostatic residual gravity values are subject to errors in the Bouguer gravity values caused by uncertainty in the station elevation, terrain correction, and gravity measurement that probably combine to yield an uncertainty of less than 2 mGal for most observations. Additional errors are introduced by ignoring the correction for terrain within 0.895 km of the observation site (probably less than 1 mGal for most stations but tens of milligals for a few stations). Errors are also introduced in the calculation of the isostatic regional field (generally less than 1 mGal but as large as 5 mGal locally). We expect that most of the isostatic residual gravity values calculated onland are accurate to within 5 mGal, or half a color contour interval. Areas of extreme topographic relief are most likely to contain larger errors.

Choice of isostatic model

The Airy-Heiskanen isostatic model clearly is only a simple approximation to a very complex phenomenon. In this model, isostatic support of topography is provided by the variations in thickness of a model crust overlying a denser substratum. The geometry of the base of this model crust is an amplified mirror image of the surface topography. This model approximates reasonably well the actual differences in crustal thickness between the continents and ocean basins and predicts the crustal roots that are present beneath some mountain ranges such as the Sierra Nevada (Bateman and Eaton, 1967). However, in some other areas of the United States, such as the Basin and Range province, the crustal thickness as shown by Braile and others (this volume) is a rather poor reflection of the topography, suggesting that other isostatic models would be more appropriate. Although it would be highly desirable to incorporate additional geophysical data into the construction of a more realistic isostatic model, it appears to us that neither the quality nor the quantity of such data is yet adequate to warrant such an application on a continent-wide scale.

We have also confined our investigations to models using *local* compensation mechanisms. There are situations in which the compensating support is likely to be *distributed* over a broad area by the flexural strength of the crust rather than lying directly under the topographic load (Banks and others, 1977). Although studies of particular tectonic settings in which flexural support is likely to be important require the incorporation of this mechanism in some form (e.g., Karner and Watts, 1983), continent-wide studies over varied tectonic provinces are probably well advised to use the simplest isostatic models possible. McNutt (1980) examined the response functions for the western and eastern halves of the country separately, and concluded that the compensation in the west was essentially local, whereas compensation in the east did require a nonlocal mechanism.

Fortunately, for the purpose of generating an isostatic residual gravity map to enhance the anomalies caused by shallow geologic sources, the lack of a completely realistic isostatic model does not pose a serious problem. Most isostatic models produce residual gravity maps of very similar appearance, because the total compensating mass for all models is the same—equal in magnitude (but opposite in sign) to the mass of the topographic load—and because the compensating masses generally are sufficiently deep that their gravitational effects observed at the Earth's surface are smoothed by distance. Tests reported by Saltus (1984), Jachens and Griscom (1985), and Simpson and others (1986) indicate that gravitational attractions predicted by various models based on a wide variety of parameters differ by only a small percentage of the total computed effect, usually less than 10 to 15 percent even in the most extreme cases. The implication is that any isostatic model consistently applied will serve the first-order goal of illuminating shallow density contrasts of geologic importance. Because even these slight differences may become important when attempting to quantitatively interpret gravity anomalies where additional constraints on deep structure are available, we outline in a later section a method for modifying the isostatic residual gravity data to permit the incorporation of additional constraints at the modeling stage.

SOURCES OF THE ISOSTATIC RESIDUAL GRAVITY ANOMALIES

Caution in the interpretation of isostatic residual gravity anomalies

We have already presented the following warning in other publications that discuss the isostatic residual gravity map of the United States (Jachens and others, 1985; Simpson and others, 1986) but feel it is worth repeating here because of widespread misconceptions about the physical significance of anomalies on an isostatic residual gravity map. A common tendency is to attribute anomalies on an isostatic residual gravity map to "isostatic imbalances," that is, to loads that are not completely compensated. In this view, anomalous isostatic gravity highs might imply "undercompensation," whereas isostatic gravity lows might indicate "overcompensation." Worldwide gravity data confirm that, in general, isostasy obtains over most of the Earth's surface (Bowin, 1985), but establishing whether isostatic equilibrium exists in the vicinity of any given anomaly is extremely difficult from gravity data alone. The reason is that even density inhomogeneities within the crust that are in complete local isostatic equilibrium will produce anomalies on an isostatic residual gravity map because the inhomogeneities are closer to the observation surface than their compensating masses (Gilbert, 1913; Woollard, 1966).

The point is best made by reference to a simple example. In Figure 4a, a uniform mountain resting on a uniform crust is supported by a local Airy-Heiskanen root. On an isostatic residual gravity map, no anomaly would appear, because the effect of the mountain is removed by the Bouguer correction, and the effect of its root is removed by the isostatic correction. In Figure 4b, a dense mass in the upper crust is also supported by a local root. Because there is no topography in this case, neither the effect of the mass nor of its compensation has been removed. The compensating mass distribution, because it lies deeper than the mass which it supports, will produce a gravity anomaly that is broader and of lower amplitude than that produced by the mass. For the example in Figure 4b, the net result is a substantial anomaly even though the mass is completely compensated. In principle, the presence of the compensating masses could be inferred from characteristics of the total isostatic anomaly—a central anomaly of one sign flanked by broader anomalies of the opposite sign, with the integral of the total anomaly over the surface of the Earth equal to zero. In practice, however, the flanking anomalies are low in amplitude and commonly masked by neighboring anomalies, so that they cannot be clearly distinguished. Thus, isostatic residual gravity anomalies alone do not reveal much about the presence or absence or style of isostatic equilibrium for individual source bodies.

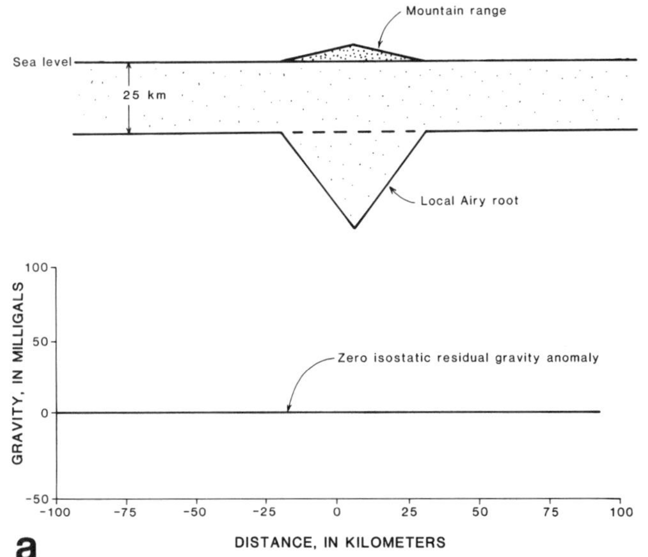

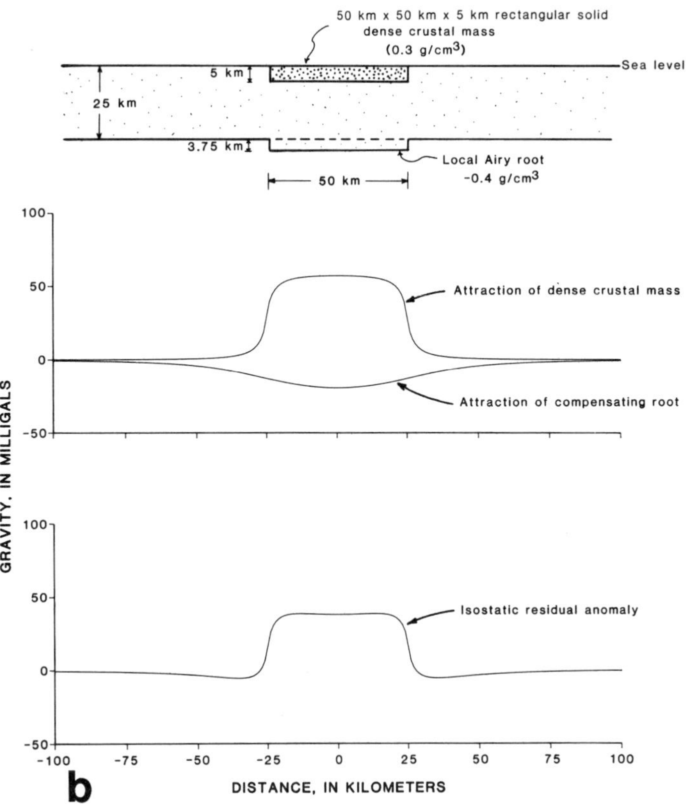

Figure 4. a, A zero-mGal isostatic residual anomaly occurs over a constant density mountain that is completely compensated in the local Airy-Heiskanen model. Bouguer correction removes attraction of the mountain range above sea level, and isostatic correction removes attraction of the root, the existence of which is inferred from the topography. b, A large, nonzero isostatic residual anomaly occurs over the center of a dense shallow rectangular mass in the crust, even though it is completely compensated by a local Airy-Heiskanen root. In this case, there is no Bouguer correction because there is no topography above sea level. Also, there is no isostatic correction for a root, even though one may exist, because the usual isostatic correction only considers the compensating masses under topographic loads. The amount of excess mass in the dense, shallow body is compensated by an equal deficiency of mass in the root; however, the greater depth of the root greatly smooths the gravity low that accompanies the mass deficiency. Thus, when the gravity low is subtracted from the high produced by the dense body, the high remains quite evident in the residual, even though by Gauss's theorem both the low and the high must have equal but opposite volumes under their mathematical surfaces (after Simpson and others, 1986).

Correlation between geology and isostatic residual gravity

Our aim in constructing the isostatic residual gravity map shown in Figure 1 was to present the gravity data in a form that emphasized anomalies caused by upper crustal sources. To illustrate that our goal has been achieved to some degree, Figure 5 shows the isostatic residual gravity data as contours on a simplified geologic map of Wyoming and vicinity. The geologic structure of the crust beneath this area is dominated by a series of sediment-filled basins and basement uplifts that formed during the Laramide orogeny in Late Cretaceous to early Tertiary time (Hamilton, 1981). All the uplifts are faithfully reflected as isostatic gravity highs, whereas most of the basins are characterized by low gravity values. The gravity relief from the uplifts to the adjacent basins, on the order of 50 to 80 mGal, is due in large part to the density contrast between the Precambrian and Paleozoic rocks that form the cores of the uplifts and the younger sedimentary rocks that fill the basins (Malahoff and Moberly, 1968; Hurich and Smithson, 1982).

As earlier studies have shown, the gravity anomalies associated with Laramide uplifts and basins are also clearly evident on the Bouguer gravity map of this region (Malahoff and Moberly, 1968; Society of Exploration Geophysicists, 1982). However, on the Bouguer gravity map they are superposed on a broad regional gravity field that increases by more than 100 mGal from southwest to northeast across the area shown in Figure 5. To compare the anomalies over the various uplifts with each other or to interpret them quantitatively, this regional field must be removed. Insofar as this regional field reflects compensation to topography, the application of an isostatic correction (which accounts for about 120 mGal of gravity change across the area) is more appropriate than more arbitrary regional-residual separation procedures such as polynomial fitting or wavelength filtering. Such techniques may still be needed after an isostatic correction to separate anomalies caused by different bodies in the subsurface.

Another place where a strong correlation between mapped geology and isostatic residual gravity can be readily demonstrated is in the Coast Ranges and Klamath Mountains of northern California (Fig. 6). This area is composed of a series of imbricate thrust slices generally separated by subhorizontal- to gently east-dipping regional thrust faults (Irwin, 1981). Extensive density sampling (Jachens and others, 1986a) indicates that the major

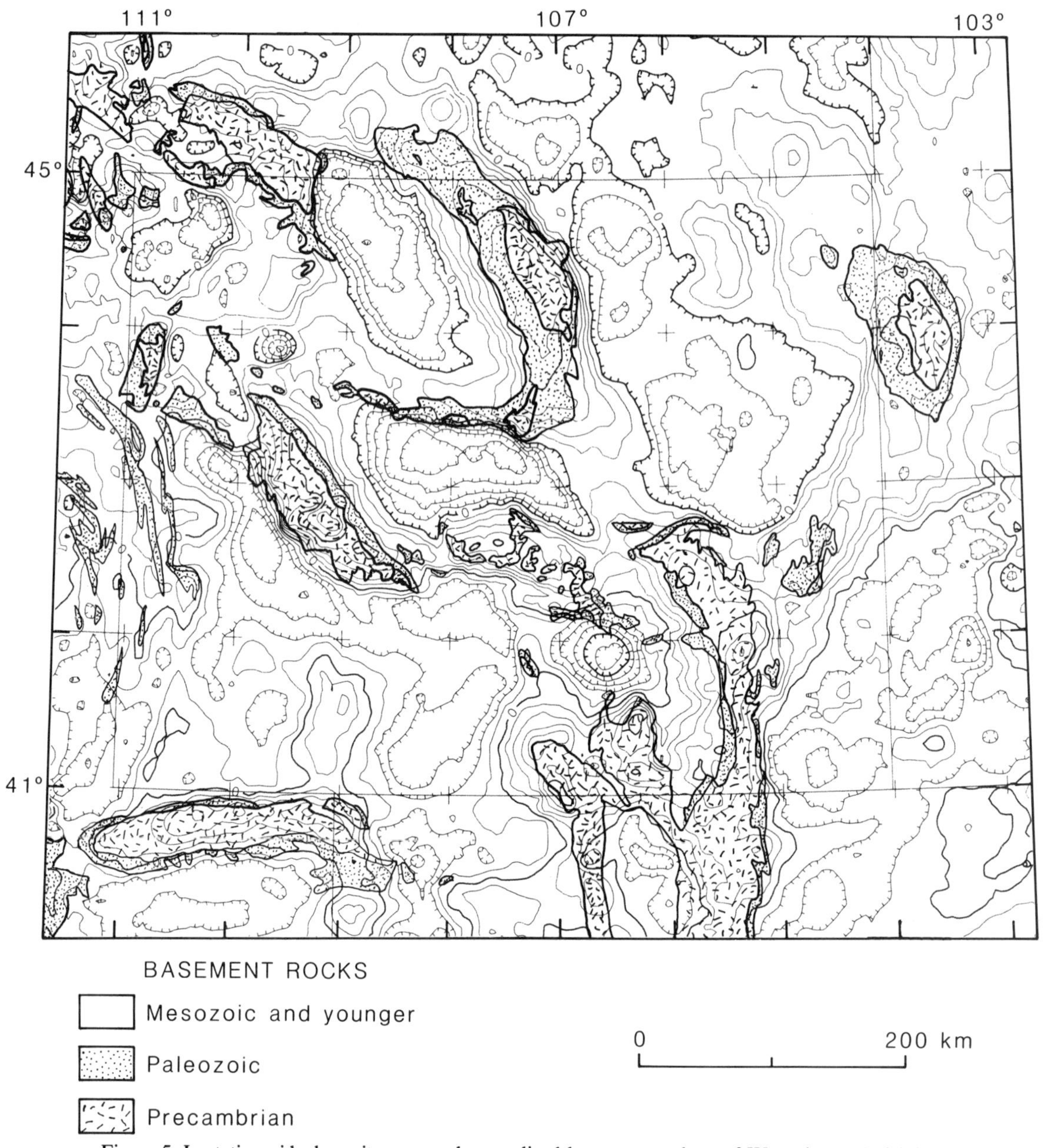

Figure 5. Isostatic residual gravity map and generalized basement geology of Wyoming and vicinity. Hachures indicate direction of decreasing gravity. Contour interval = 10 mGal.

geologic units in this area can be divided by density into two groups. One group includes chiefly sedimentary and metasedimentary rocks that yield densities in the range of 2.60 to 2.65 g/cm^3. The second group, which includes ophiolitic rocks, mafic plutonic and volcanic rocks, metavolcanic rocks, and mélanges containing mafic and ultramafic rocks, is characterized by densities generally greater than 2.75 g/cm^3.

The isostatic residual gravity over the area is closely related to the densities of the rocks exposed at the surface (Fig. 6). Generally, gravity values are less than –25 mGal where low-density rocks are exposed but are considerably higher over exposures of higher density rocks. In addition, most fault contacts between low-density and high-density units lie near the bottoms of the gravity gradients that separate the areas of low and high gravity. This relationship is consistent with the geological inferences that the high-density units are tabular bodies overlying the low-density units and separated from them by low-angle faults.

On a more regional scale, many conspicuous anomalies displayed in Figure 1 are closely correlated with surface or near-surface geology. A number of these anomalies are listed in Table 1 and keyed to Figure 7. This table is a slightly updated version of the one given in Simpson and others (1986) and represents an attempt to categorize the anomalies according to probable litho-

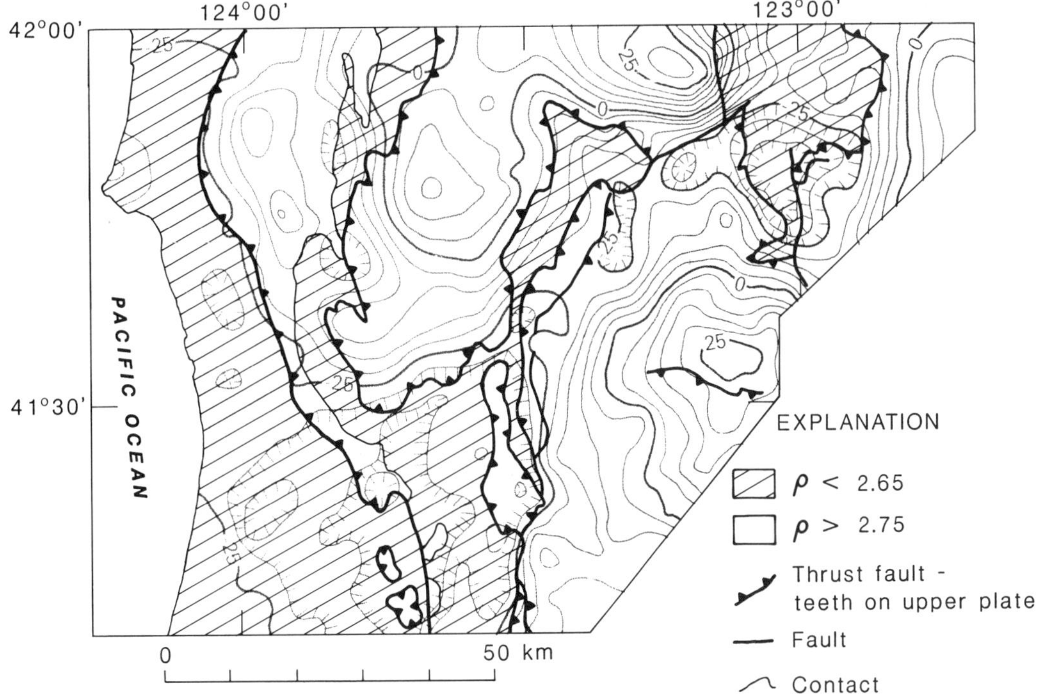

Figure 6. Isostatic residual gravity map and generalized density map of northwestern California. Hachures indicate direction of decreasing gravity. Contour interval = 5 mGal.

logic source and geologic-tectonic setting. Our main purposes in presenting this table here are to illustrate the close correlation between the residual gravity anomalies and known geology, and to provide a tool that will help in the interpretation of other anomalies in Figure 1 whose sources are concealed. We have used this table as the basis for speculations about the sources of some of the concealed source anomalies on Figure 1 (Simpson and others, 1986).

One outgrowth of our efforts to construct Table 1 was a list of papers that present quantitative and qualitative interpretation of gravity anomalies in the conterminous United States. Although we realize that this bibliography is far from complete, we nevertheless believe that even in its incomplete state, it can serve as a useful starting point for anyone attempting to incorporate gravity into their interpretations of crustal geology and geophysics. Thus we include the bibliography in Appendix A of Simpson and Jachens (this volume), together with index maps showing the locations of profiles and areas discussed in the papers.

Another result of our work detailed in Table 1 was the realization that, in order to understand many of the gravity anomalies, it was necessary to be able to compare the gravity data directly with other kinds of geological and geophysical information. To facilitate such comparisons by others, we have published the isostatic residual gravity map in contoured form on a clear-film base at a scale of 1:2,500,000 (Jachens and others, 1985). Other maps for which this can serve as an overlay include: (1) *Geologic Map of the United States* (King and Beikman, 1974); (2) *Basement Rock Map of the United States* (Bayley and Muehlberger, 1968); (3) *Tectonic Map of the United States* (U.S. Geological Survey and American Association of Petroleum Geologists, 1961); (4) *Gravity Anomaly Map of the United States* (Society of Exploration Geophysicists, 1982); (5) *Magnetic Anomaly Map of the United States* (Zietz, 1982); (6) *Generalized Structural, Lithologic, and Physiographic Provinces in the Fold and Thrust Belts of the United States* (Bayer, 1983).

Causes of longer wavelength isostatic residual gravity anomalies

The correlations between gravity and geology discussed in the preceding paragraphs and presented in Table 1 generally apply to anomalies and crustal features with breadths up to several hundred kilometers. Anomalies with these characteristic wavelengths are commonly an indication of an excess or deficit of dense material in the crust, rather than any departure from isostatic equilibrium (see Fig. 4a). For intermediate-wavelength (IWL) isostatic residual anomalies with breadths of 500 to 1,000 km or more, the questions of source depth and of isostatic equilibrium must be more carefully explored. Although the sources of these anomalies could be shallow-seated and completely compensated, such anomalies could also indicate deep-seated sources (possibly below the lithosphere) or areas that are out of isostatic balance (Woollard, 1966). Examples of such anomalies are the broad area of low isostatic residual values centered over northern Missouri and adjacent states and extending, perhaps, from Texas to Wisconsin (Fig. 1) and the broad areas of high values over the

TABLE 1. SOURCES OF GRAVITY HIGHS AND LOWS LESS THAN SEVERAL HUNDRED KILOMETERS WIDE

Explanation of columns:

O. source of density contrast
  x geologic-tectonic setting
    (x1) selected examples and references

SOURCES OF HIGHS

1. Mafic Igenous bodies–mostly autochthonous gabbro, basalt, and diorite

  a. Rift setting
    (a1) Midcontinent gravity high–Craddock and others (1963), King and Zietz (1971), Ocola and Meyer (1973), Chase and Gilmer (1973), Green (1983).
    (a2) Wichita Mountains–Mitchell and Landisman (1970), Pruatt (1976), Powell and Phelps (1977), Brewer and others (1981; see also (f3)
    (a3) Michigan high–Hinze and others (1975), Brown and others (1982)
    (a4) East-continent gravity high–Keller and others (1982), Lidiak and others (1985)
    (a5) East Coast gravity high (partly edge effect also)–Rabinowitz (1974), Rabinowitz and LeBrecque (1977), Karner and Watts (1982)
    (a6) Snake River Plain–µabey (1976), Sparlin and others (1982)
    (a7) Appalachian high (? in part)–Cook and Oliver (1981)

  b. Magmatic arc
    (b1) Parts of Sierra Nevada–Oliver and Mabey (1963), Griscom and Oliver (1980)
    (b2) Peninsular Ranges–Kovach and others (1962), Griscom and Oliver (1980)

  c. Isolated intrusions–location probably controlled by structure
    (c1) South GEorgia triplet
    (c2) Plutons flanking Mississippi embayment graben–Hildenbrand and others (1982)

2. Mafic Crust–mostly allochthonous accreted oceanic or island-arc basement or transitional basement formed during rifting

  d. Large pieces
    (d1) Oregon and Washington Coast Ranges–Snavely and others (1980)
    (d2) Parts of the Appalachians–Hutchinson and others (1983), Thomas (1983)
    (d3) Southeastern United States suture zone–Nelson and others (1985a, b)
    (d4) Ouachita orogenic belt–Thomas (1985)

  e. Small pieces and tectonic slivers
    (e1) Mafic and ultramafic thrust slices in Klamath Mountains and Sierra Nevada–LaFehr (1966), Jachens and others (1986)
    (e2) Great Valley high–Cady (1975)

3. Uplifted Crystalline Basement–(dense by virtue of original depth?)

  f. Uplifted by collision or compression on well-defined structures
    (f1) Laramide ranges (in part)–Smithson and others (1979), Hamilton (1981), Hurich and Smithson (1982), Malahoff and Moberly (1968)
    (f2) Parts of Appalachians–Gristom (1963), Thomas (1983)
    (f3) Wichita and Arbuckle Mountains uplift (?)–see references for (a2)

  g. Cause of uplift uncertain (arching or doming?)
    (g1) Adirondack Mountains–Simmons (1964), King (1977)
    (g2) Llano Uplift–Barnes and others (1954), King (1977)

SOURCES OF LOWS

4. Felsic Igneous Rocks (including volcaniclastic sediment)

  h. Batholiths and plutons
    (h1) Idaho batholith
    (h2) Central Wisconsin–Hinze (1959), Aiken and others (1983), Klasner and others (1985)
    (h3) Eastern Sierra Nevada–Oliver and Mabey (1963), Griscom and Oliver (1980)
    (h4) Mojave Desert–Jachens and others (1986b)

  i. Volcanic piles–commonly in volcanic depressions
    (i1) San Juan volcanic field–Plouff and Pakiser (1972)
    (i2) Cascade Range–LaFehr (1965), Blakely and others (1985)

  j. Calderas
    (j1) Yellowstone–Eaton and others (1975)
    (j2) Long Valley–Kane and others (1976)
    (j3) Timber Mountain–Kane and others (1981)

5. Sedimentary rocks

  k. Extensional settings–grabens and half-grabens
    (k1) Rio Grande rift (also volcanic rocks)–Decker and Smithson (1975), Ramberg and others (1978), Cordell (1978, 1982)
    (k2) Basins in Basin and Range province–Thompson (1959), Anderson and others (1983)
    (k3) Flanking lows on midcontinent gravity high–see (a1)
    (k4) Triassic basins along East Coast–Sumner (1977)–not shown in Figure 7

  m. Convergent settings–subduction-related accretionary prisms and basins
    (m1) Offshore Washington and Oregon–Dehlinger and others (1968), Snavely and others (1980)
    (m2) Great Valley of California–Byerly (1966), Suppe (1979)
    (m3) Puget Sound–Stuart (1961)

TABLE 1. SOURCES OF GRAVITY HIGHS AND LOWS LESS THAN SEVERAL HUNDRED KILOMETERS WIDE (continued)

SOURCES OF LOWS (continued)

5. Sedimentary Rocks (continued)
   - n. Convergent settings–collision and compression-related basins
     - (n1) Laramide basins–Strange and Woollard (1964), Case and Keefer (1966), Hurich and Smithson (1982)
     - (n2) Ouachita low–Lyons (1961), Nicholas and Rozendal (1975), Nelson and others (1982), Lillie and others (1983)
     - (n3) Appalachian basin–King (1977), Karner and Watts (1983)
     - (n4) Martinsburg basin, Sevier basin–Shanmugam and Lash (1982)
   - o. Transform related basins
     - (o1) Local basins along San Andreas fault–Chapman and Griscom (1980)
     - (o2) California offshore basins–Harrison and others (1966), Blake and others (1978), Beyer (1980)

6. Depressed Crystalline Basement–(less dense than adjacent basements by virtue of crustal layering?)
   - p. Basement downdropped in extensional settings
     - (p1) Basement under sedimentary basins of type k
   - q. Basement depressed by subduction
     - (q1) Basement under sedimentary basins of type m
     - (q2) Downgoing Gorda plate–Jachens and Griscom (1983)
   - r. Basement depressed by collision and compression
     - (r1) Basement under sedimentary basins of type n
   - s. Basement depressed in transform settings
     - (s1) Basement under sedimentary basins of type o

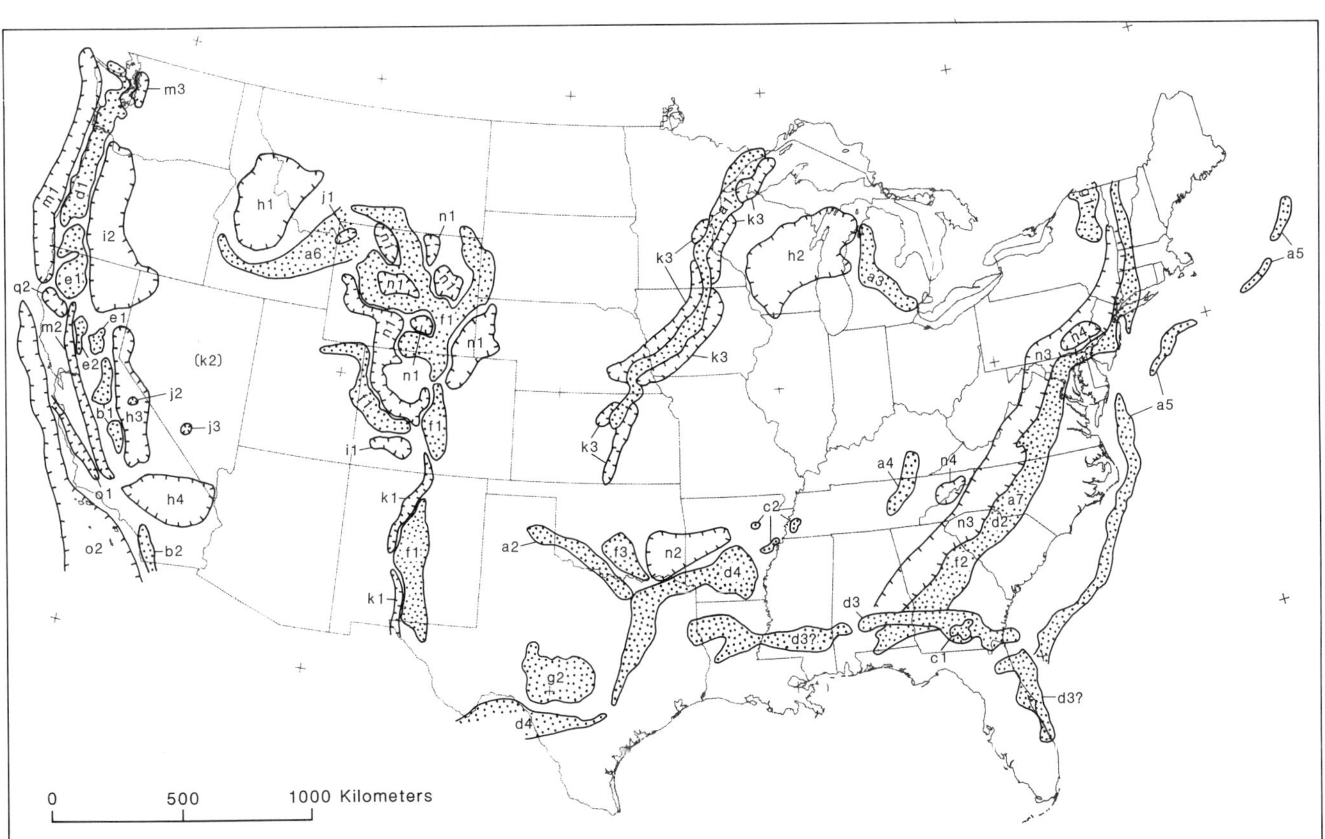

Figure 7. Index map showing locations of isostatic residual gravity anomalies less than approximately 250 km wide that are categorized in Table 1. Highs are stippled, lows are hachured.

eastern Dakotas and Minnesota and over eastern Arkansas and nearby states in the Mississippi Embayment. If these types of anomalies can be shown to be caused by compensated sources within the lithosphere, then the isostatic residual gravity map may constitute a powerful tool for distinguishing large provinces with different crustal and upper-mantle chemistries and, hence, with different modes of origin.

George Woollard, in a series of papers spanning many years (1936, 1962, 1966, 1968, 1972, 1976) pointed out the correlation of isostatic anomalies with geologic bodies and suggested that even long-wavelength isostatic anomalies correlated with crustal geology and seismic parameters. He explored in considerable detail the correspondences between gravity anomalies, seismic velocities, and depths to velocity interfaces. The results of his studies strongly suggest that many IWL residual anomalies are at least partly caused by shallow crustal sources. This suggestion is supported by comparisons between residual gravity and geology in the midcontinent, which show that the edges of some of the IWL anomalies coincide with known Precambrian tectonic features (Simpson and others, 1986). The step in regional base level of gravity that occurs across the southern end of the midcontinent gravity high (al in Fig. 7) probably marks the transition between concealed extensions of older Precambrian provinces to the northwest that are exposed in surrounding areas (Dutch, 1983; Green and others, 1985; Klasner and King, 1986) and one or more felsic Precambrian terranes that probably were sutured to the northwestern terranes sometime after 1,700 Ma (Bickford and others, 1981b; Dutch, 1983; Thomas and others, 1984). Similarly, the northeast-trending gravity gradient between the regional gravity low centered over northern Missouri and the high over the Mississippi Embayment approximately follows the boundary drawn by Bickford and others (1981a) between a terrane to the northwest consisting of abundant granite and metavolcanic and metasedimentary rocks, and a terrane to the southeast consisting of rhyolitic flows, ash-flow tuff, and epizonal granite plutons. These close correlations between the IWL residual anomalies and Precambrian terrane boundaries lends support to the idea that some of the IWL anomalies partly reflect crustal density distributions.

A new compilation and presentation of the seismic refraction data (Braile and others, this volume) provides additional support for the belief that the IWL residual gravity anomalies reflect density distributions that are in part shallow. Most of the first-order features shown in the contour map of average seismic velocity of the crust (Braile and others, this volume, Fig. 5) are spatially correlated with characteristic isostatic residual gravity base levels (Fig. 1). Areas of high average crustal velocity centered over Montana and over the Mississippi Embayment coincide with broad regions of high gravity values (roughly +15 to +25 mGal), whereas the intervening region extending from Wisconsin through Missouri and eastern Kansas that is characterized by lower average crustal velocity corresponds to a broad gravity low that averages –15 mGal or less. Low crustal velocities in the southwestern United States and in the extreme northwestern corner of the country also are areas of broad gravity lows. Finally, the region of extremely low average crustal velocity centered over Nevada and Utah roughly coincides with a region of low gravity values in Nevada, but the velocity feature appears to extend farther to the east than the gravity low.

The crustal density variations implied by the average crustal velocity variations are crudely comparable to the density variations required to produce the regional gravity anomalies seen in Figure 1. Based on an empirical relationship between P-wave velocity and density (Hill, 1978), in the velocity range of 6 to 7 km/sec, each increase of 0.1 km/sec implies a density increase of 0.044 g/cm^3. Thus the average crustal velocity data given by Braile and others (this volume) imply typical regional differences in average crustal density of about 0.1 g/cm^3 and extreme differences as great as 0.2 g/cm^3. These values of average crustal density differences are the same magnitude as the differences that Simpson and others (1986) found were necessary to account for the IWL residual gravity anomalies in the United States, based on tests of an extremely simplified model for isostatic compensation of intracrustal density variations.

Therefore, both geologic and seismic evidence suggest that not only do the short-wavelength isostatic residual gravity anomalies reflect compensated density distributions within the crust but many of the longer wavelength anomalies may also.

ALTERNATIVE REPRESENTATIONS OF THE ISOSTATIC RESIDUAL GRAVITY DATA

The first vertical derivative map

Vertical derivatives of potential field maps commonly are used to accentuate short-wavelength anomalies at the expense of the longer wavelengths of the field. Although such procedures tend to render the maps less useful for quantitative interpretations, suppression of long-wavelength features and sharpening of short-wavelength anomalies facilitate comparing and distinguishing trends and anomaly fabrics in various domains. Vertical derivative maps also help to focus attention on anomalies resulting from abrupt lateral changes in near-surface densities and so are useful for comparing gravity with mapped geology. Figure 8 (in pocket inside back cover) shows the first vertical derivative of the isostatic residual gravity field (upward continued 10 km to suppress some of the short-wavelength noise generated by the derivative).

Many distinct changes in fabric and trends of anomalies across major structural or tectonic boundaries are prominent in Figure 8. The Appalachian suture zone, marked by a nearly continuous 3,000-km-long gravity high (a7, d2, f2 in Fig. 7), represents a collisional boundary (Griscom, 1963; Thomas, 1983) between plates displaying markedly contrasting patterns on the vertical derivative map. From southeast to northwest across this suture, a pattern of relatively short-wavelength, somewhat randomly oriented anomalies changes abruptly to a pattern of linear, longer wavelength anomalies trending northeast, subparallel to the trend of the Appalachian Orogen. Some of this change probably reflects the fact that crystalline basement rocks are deeply buried beneath sedimentary rocks of the Appalachian geosyncline just northwest of the Appalachian gravity high, but this can be only part of the explanation because the longer wavelength pattern persists to the northwest where the sedimentary rocks are thin. A more subtle change in anomaly character marks the transition from the Colorado Plateau to the southern Basin and Range province in Arizona and New Mexico. The Colorado Plateau in the northeastern part of Arizona and adjacent areas of Utah, Colorado, and New Mexico is characterized by a northeast-trending anomaly fabric (Fig. 8) that reflcts lithologic variations in the concealed Precambrian basement (Case and Joesting, 1972; Sumner, 1985). In contrast, a strong west-northwest–trending anomaly grain is evident in southern and southeastern Arizona and southwestern New Mexico. This grain is caused by linear mountain ranges and intermontane basins that resulted from west-southwestward extension in the southern Basin and Range province (Stewart, 1978).

Concealed boundaries between large domains also may be inferred from Figure 8. The residual gravity field over northern Minnesota and eastern North Dakota is characterized by alternating linear east-northeast–trending high- and low-gravity anomalies, about 50 km wide. In Minnesota where Archean basement is exposed, most gravity lows occur over granitic rocks, whereas the gravity highs coincide with volcanic or metasedimentary rocks (King and Beikman, 1974). On the basis of the continuation of individual anomalies, and more generally, on the similarity of the gravity fields over northern Minnesota and eastern North Dakota, we infer that the Archean basement of northern Minnesota extends westward beneath the Cretaceous and younger sedimentary rocks of North Dakota to about longitude 100°W. A major structural or lithologic boundary in the Precambrian basement beneath central North Dakota is suggested by the change from narrow east-northeast–trending gravity anomalies in eastern North Dakota to much broader north- to northwest-trending anomalies in the western part of the state. This boundary has been identified by Green and others (1985) as the probable western margin of the Superior craton, with allochthonous terranes of oceanic affinities belonging to the Churchill structural province lying to the west.

Applying the first vertical derivative is equivalent to a pseudomagnetic transform: the first vertical derivative of a gravity anomaly is proportional to the magnetic anomaly that would be observed if dense material were replaced by magnetic material in exact proportion, and if the magnetization direction and the local geomagnetic-field direction were vertical. Because of this property, the map can be usefully compared with maps of aeromagnetic anomalies, despite the fact that, even in ideal situations, some differences exist between the two types of maps because the direction of the local geomagnetic field generally is not vertical. Several important rock groups may be distinguished by the coincidence (or noncoincidence) of aeromagnetic and gravity anomalies, although such identifications must be made with caution because density and magnetization do not uniquely define the various rock types (see compilations of rock property measurements in Telford and others [1976]; Carmichael [1982]; Johnson and Olheoft [1984]). Mafic igneous rocks generally are both dense and magnetic, and large mafic bodies sometimes display associated gravity highs and large magnetic anomalies. Such combinations of anomalies can be seen by comparing Figures 1 and 8 with the aeromagnetic map of the United States (Zietz, 1982; Kane and Godson, this volume) over (1) the Duluth Complex in Minnesota (north end of al in Fig. 7), (2) the mafic plutonic rocks of the Peninsular Ranges batholith in southern California (b2 in Fig. 7), (3) the stratiform gabbro complex in the Wichita Mountains of Oklahoma and Texas (a2 in Fig. 7), and (4) the mafic volcanic rocks of the Coast Ranges of Oregon and Washington (dl in Fig. 7). Similar combinations of anomalies in other areas where mafic rocks are not exposed probably indicate the presence of mafic rocks at depth.

Horizontal gradients as an aid to source-body definition

One enhancement technique that we have found particularly useful for extracting geologic information from gravity data is a method based on horizontal gradients. For near-surface bodies with near-vertical contacts, the maximum horizontal gradient of gravity as measured along a profile will occur nearly over the contact. If a contour map is made of the amplitude of the horizontal gradient (i.e., of the magnitude of the slope of the gravity field at a point without regard to the direction of the slope), then lines drawn along "ridge tops" will approximately outline the density boundaries of the source bodies. This procedure has been automated (Blakely and Simpson, 1986) and can be applied to any gridded data set to produce outlines such as those shown in Figure 9. The lines so obtained will not lie directly over contacts if the original assumptions are invalid, that is, if the contacts do not dip steeply or if the source body itself is too deep.

The coincidence of the lines of maximum horizontal gradient with mapped contacts is quite striking at larger scales. The extensions of the lines beyond exposed contacts, and the places where such lines deviate from mapped geology often give useful information about the extrapolations of surface geology into the subsurface.

Spectral analysis to quantify regional anomaly patterns and crustal fabrics

Much can be learned about the sources of individual anomalies by careful modeling studies that use all available geologic and geophysical constraints. For regional gravity data where a great many anomalies combine to form complicated patterns, the problem of interpretation is less straightforward. The geometric shape and distribution of density within a causative mass determine the shape of its gravity anomaly, but shapes of individual anomalies are sometimes difficult to determine in large regional collections of data. The far-field components of each anomaly, for

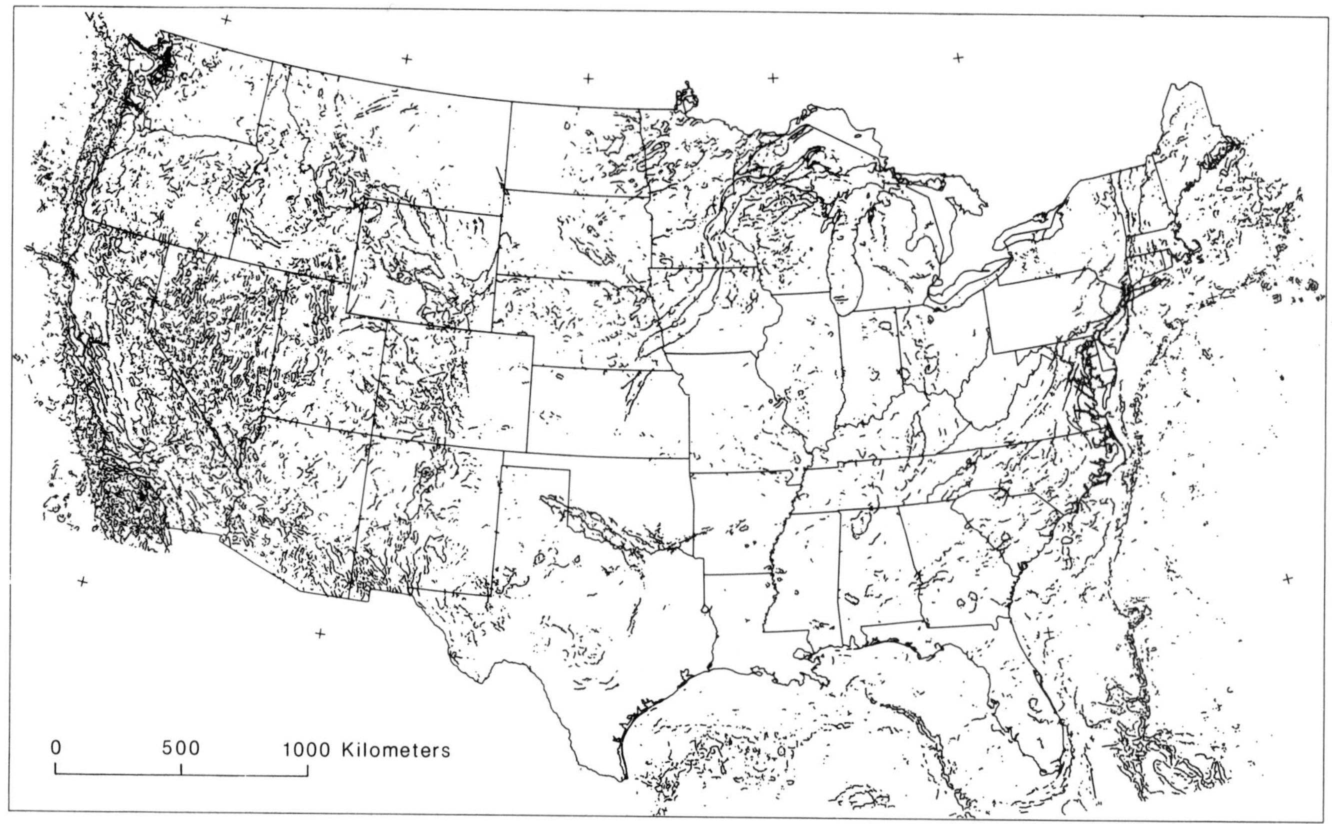

Figure 9. Locations of maximum horizontal gradient in the isostatic residual gravity field as determined by program MAXSPOT (Blakely and Simpson, 1986). Sinuous sequences of dots indicate important lateral density boundaries.

example, blend and interfere with those of all neighboring anomalies. A classic technique to sort out spectral properties (i.e., significance of certain wavelengths) of measured data is with Fourier analysis where data measured in *x, y* coordinates are remapped into wavenumber space with coordinates k_x, k_y. Wavenumbers k_x and k_y are inversely proportional to wavelengths in the *x* and *y* directions, respectively; i.e.,

$$k_x = 2\pi/\lambda_x$$
$$k_y = 2\pi/\lambda_y$$

We symbolize this remapping by:

$$g(x, y) \leftrightarrow \tilde{g}(k_x, k_y),$$

where $g(x, y)$ in the present example is measured isostatic residual data and $\tilde{g}$ (k_x, k_y) is its Fourier transform. Specific wavelength components of the *entire collection* of anomalies are mapped to unique locations in the two-dimensional Fourier domain: long wavelengths near the origin, shorter wavelengths farther from the origin.

Figure 10 shows an example from the Basin and Range province. The pattern of isostatic residual anomalies (Fig. 10a) is mapped into a "bulls-eye" pattern in the Fourier domain (Fig. 10b). (The contour interval in Fig. 10b is 1.0, but contours represent ln $|\tilde{g}(k_x, k_y)|^2$; hence, each contour represents an amplitude 0.69 *times* less than the next highest contour.) Amplitudes of $\tilde{g}$ (k_x, k_y) are perfectly symmetrical through the origin, as expected for the Fourier transform of real data. The amplitudes also have a high degree of radial symmetry with a marked and continuous decrease as k_x and k_y increase; i.e., each long wavelength (small k_x and k_y) is more important to the shape of $g(x, y)$ than all shorter wavelengths. The rate of attenuation of $|\tilde{g}(k_x, k_y)|$ away from the origin is approximately exponential and depends in large part on an important fact of potential theory: those parts of a gravity anomaly (or any potential field) caused by the short-wavelenth attributes of a density distribution are smoothed more than their long-wavelength counterparts; the shorter the wavelength, the greater the smoothing. This smoothing translates to the Fourier domain as an exponential decay of $|\tilde{g}(k_x, k_y)|$.

If lateral variations in crustal density were perfectly isotropic in horizontal directions, $|\tilde{g}(k_x, k_y)|$ would have perfect radial symmetry, and the contours in Figure 10b would be circular. Deviations from radial symmetry in $|\tilde{g}(k_x, k_y)|$ (Fig. 10b) are related to systematic patterns in $g(x, y)$ and consequently represent systematic variations in crustal density (Simpson and others, 1986). It is difficult, however, to see these perturbations in Figure 10b above the strong influence of exponential decay. To emphasize them, we normalize $|\tilde{g}(k_x, k_y)|$ by dividing by its radially symmetric part; e.g.,

$$A_o(k_x, k_y) = |\tilde{g}(k_x, k_y)|/R(k)$$

where

$$R(k) = \int_0^{2\pi} |\tilde{g}(k_x, k_y)| d\theta$$

$$k = \sqrt{k_x^2 + k_y^2}$$

and

$$\theta = \arctan(k_y/k^x).$$

Figure 10c shows ln $|A_o(k_x, k_y)|^2$ with contour interval of 0.1. (Each contour represents an amplitude 0.95 *times* less than the next highest level.) Maxima in this map indicate specific trends and wavelengths in the density of the Basin and Range province that dominate all others. Bracewell (1965) showed that maxima in $|A_o(k_x, k_y)|$ correspond to "corrugations" in $g(x, y)$. A maximum located a distance δ from the origin and at an angle ϕ measured counterclockwise from the k_y axis represents a corrugation in $g(x, y)$ with wavelength $\lambda = 2\pi/\delta$ and trend ϕ measured counterclockwise from the x axis.

For example, Figure 10c shows a significant maximum located between $70° < \phi < 90°$ and centered roughly at $\delta = 0.25$ km^{-1}. This maximum corresponds to anomalies in $g(x, y)$ (Fig. 10a) with wavelengths of approximately 25 km and trends between 12°W and 8°E. These north-trending anomalies are fundamentally related to the extensional tectonic history of the Basin and Range province. Quite a different pattern exists in other parts of the isostatic residual gravity map and are manifested in $A_o(k_x, k_y)$ by a different pattern of maxima (Simpson and others, 1986).

MODELING ISOSTATIC RESIDUAL GRAVITY DATA WITH SEISMIC CONSTRAINTS

The close correlation between the gravity anomalies in Figure 1 and near-surface geology suggests that isostatic residual gravity can serve as an effective tool for extending geologic information into the subsurface and for linking it with results based on other geophysical methods. Integration of gravity and seismic information is especially powerful because of the relationship between densities and seismic velocities. The availability of good gravity coverage on a continent-wide scale adds a further dimension to the utility of gravity data in integrated interpretations. With such coverage, the gravity can be used to interpolate between widely spaced seismic profiles, to help identify those features of seismic interpretations that are of regional importance as opposed to those that are of only local interest, and to place the sparsely distributed detailed seismic information into a regional context. Perhaps more important, the regional coverage gives the

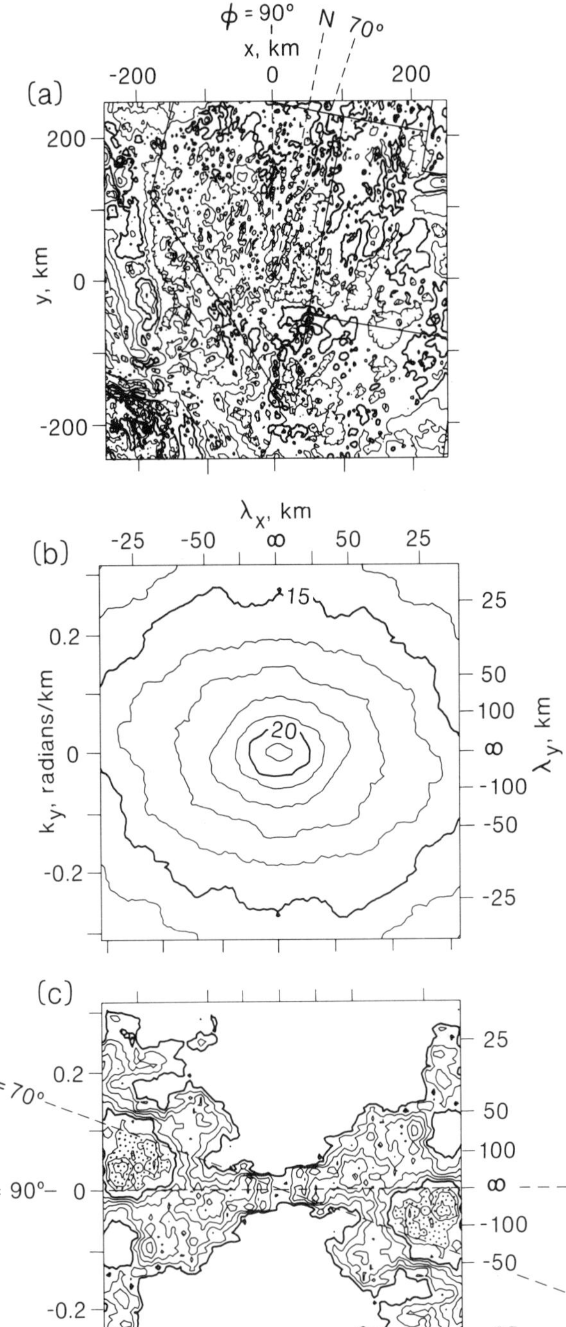

Figure 10. Example of spectral analysis of isostatic residual gravity data. a, Isostatic residual gravity anomalies in the Basin and Range province. Contour interval = 20 mGal. b, Smoothed, two-dimensional Fourier transforms of the isostatic residual data in a. Contours represent logarithms of squared amplitudes, with a contour interval of 1.0. c, Normalized and smoothed amplitude spectra, calculated by dividing the amplitude spectra by its average radial spectra. Contours represent logarithms of squared amplitudes, with a contour interval of 0.1. Levels less than 0 not shown. Stippled areas correspond to high-amplitude region with $70° < \phi < 90°$ and $\delta \approx 0.25$ km^{-1}.

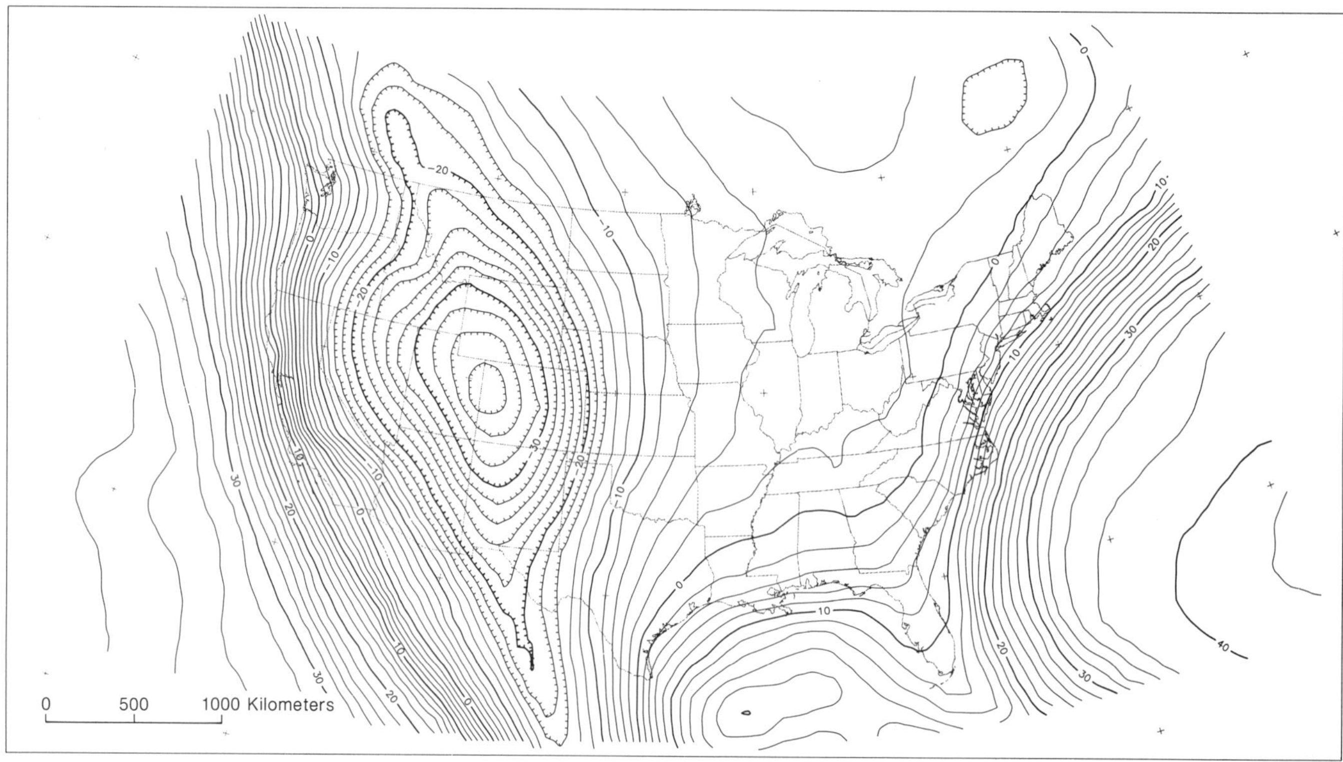

Figure 11. Map of the far field part of the isostatic correction. Contour level at any point represents the attraction of all compensating masses and topographic masses lying at distances greater than 166.7 km (as measured along the surface of the Earth) from the point. Data from Kärki and others (1961). Contour interval = 2 mGal.

interpreter the freedom to seek answers to specific questions about an interpretation in those places where the data best address the questions, instead of being restricted to answering all questions based on data along a few profiles that usually are chosen before the nature of the questions is well understood. Thus, the gravity data should play a major role in any crustal investigations, not only at the interpretational stage, but also at the planning stage.

Quantitative interpretation of the isostatic residual gravity data with seismic constraints depends somewhat on the depth to which reliable seismic information is available. If the seismic interpretation is confined to roughly the upper 10 km of the crust, then it has been our experience that in many areas the isostatic gravity anomalies can be modeled directly, without the need for removing any additional regional gradients (e.g., Jachens and Griscom [1985], Blakely and McKee [1985], Jachens and others [1986a]), although slight base-level adjustments may be necessary to offset the effects of the long-wavelength anomalies that remain in the data. For consistency, these models might include, in addition to bodies of anomalous density in the upper crust, deeper density distributions that would isostatically compensate them (e.g., see Fig. 4b). The precise shape of these deep-density distributions is not crucial, but the gravity effect of them may provide valuable insights during the interpretation. Griscom and Grafft (1982, as shown in Jachens and Griscom, 1985) presented an example of such an interpretation over the Ventura basin in southern California.

If deep crustal seismic information is available, then gravity modeling with seismic constraints in areas with pronounced topographic relief should be approached in a somewhat different manner. The isostatic residual gravity data shown in Figure 1 are not particularly appropriate for modeling deep crustal structure because the gravity effects of a crude approximation to the deep-density structure have already been removed from the data by performing the isostatic reduction. No such approximation has been removed from Bouguer gravity data. Modeling based on these data may not be appropriate either, however, because in many cases the gravitational effects of isostatic compensation of topography far removed from the area of interest can be substantial. Some sense of the possible magnitude of such effects can be seen in the map of the United States (Fig. 11), which shows the combined effects of worldwide topography and its compensation from all areas from 166.7 km out to 180° from an observation point (after Kärki and others, 1961). Including such distant sources in any local gravity model would be a formidable task.

Even on a more local scale, modeling along profiles could be severely influenced by isostatic effects arising from nearby sources that do not lie in the plane of the profile. For example, any gravity model along a profile confined to the low-lying areas of the Great Valley of California must account for the large gravitational influence of the deep crustal root beneath the adjacent Sierra Nevada Mountains. For profiles that are not perpendicular to the trend of the Sierra Nevada, three-dimensional modeling techniques would be required.

We propose the following scheme to model gravity anomalies due to deep structure while incorporating seismic constraints. In order to retain useful features of both Bouguer and isostatic residual anomalies, we propose that gravity data along seismic profiles are subjected to a "partial isostatic reduction" that accounts for worldwide topography and topographic compensation except within a slot centered about the profile. Thus the entire Earth outside the slot is approximated by laterally homogeneous layers and isostatically compensated topography. For the region within the slot, standard two-and-one-half–dimensional models (Cady, 1980) centered on the profile, terminated at the edges of the slot, and constrained by seismic data, can be used to fit the "partially" reduced isostatic residual gravity data.

The scheme described above is similar in philosophy to that used by Glennie (1932, 1936) in generating the "normal warp" gravity anomaly. This anomaly, which results from Bouguer gravity values corrected for topographic and isostatic effects from areas between 166.7 km and the antipode, has been used to study crustal structural over areas so large that the topographic and isostatic effects from distant regions are not constant.

The "partially" reduced gravity data can be generated relatively easily from isostatic residual gravity data by computing, at each observation point, the isostatic correction for topography within the slot (using the same isostatic model that had been used to perform the original reduction) and subtracting it from the residual gravity data. For the Airy-Heiskanen model, a computer program that uses elevations averaged over 3- by 3-min compartments is available for the computation (Jachens and Roberts, 1981). Figure 12 shows an example of the differences between the Bouguer anomaly, the isostatic anomaly, and the partial isostatic anomaly for a slot 60 km wide. It is quite clear from the figure that the choice of reduction procedure has a substantial effect on the modeling process. Although we have not yet applied this technique to a practical modeling problem, we present it here in order to stimulate discussion of better ways to combine and integrate gravity data with increasingly high-quality seismic constraints.

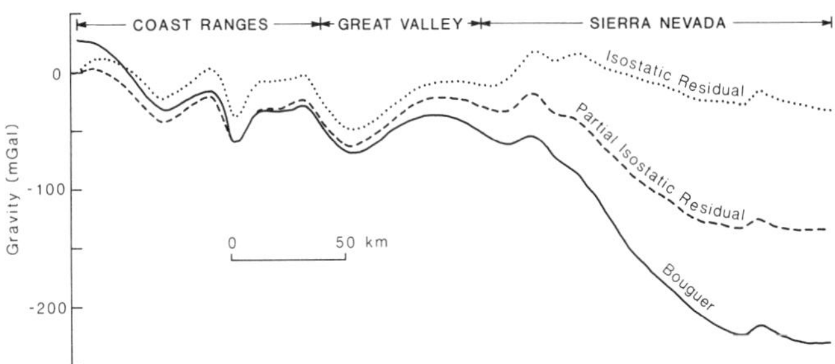

Figure 12. Example for a northeast-trending profile from the coast of California near Monterey to the Sierra Nevada showing the Bouguer gravity anomaly, the isostatic residual gravity anomaly, and the partial isostatic residual gravity anomaly. Slot = 60 km wide.

CONCLUSIONS

For the purpose of displaying anomalies, trends, and patterns caused by crustal features of geological interest, the choice of isostatic model is less important than the application of an isostatic correction of some sort. The differences between models are of less interest than the overall patterns if the goal is to highlight shallow-density contrasts of geologic importance.

It may not be possible to determine from gravity data alone whether an individual isostatic residual anomaly has a source that is in local isostatic balance or imbalance. In the absence of information to the contrary, the most productive approach for interpreting such anomalies is probably to assume that they are completely compensated and to use relevant geological and geophysical data to construct a density model of the source bodies. If the bodies are not in isostatic balance, then their imbalance should become apparent in the modeling process if enough additional geological and geophysical constraints are available.

Many short-wavelength isostatic residual anomalies on the map of the conterminous United States can be attributed to geologic features mapped at the surface or in the shallow subsurface. We have tentatively classified the sources of such anomalies less than several hundred kilometers wide into a small number of geologic-tectonic categories as an aid in the interpretation of highs and lows in regions where the source bodies are not exposed at the surface. Most positive anomalies can be attributed to mafic rocks in rift settings or to accreted or upthrusted dense, commonly mafic, material. Most lows can be explained in terms of sedimentary deposits, generally in convergent tectonic settings, to felsic igneous rocks, both extrusive as in calderas and intrusive as in batholiths, and to downwarped crustal sections. Many anomalies are caused by structures of Precambrian age; some may be caused by reactivation of ancient structures by later tectonic events.

Some of the intermediate-wavelength anomalies appear to correlate both with geologic features exposed at the surface and with variations in average crustal velocity. These anomalies probably delineate terranes of different origin with the sign of the anomaly over them implying mafic or felsic chemistries.

The greatest challenge for future applications of gravity investigations to understanding the tectonic framework of the United States lies in developing new and better ways of integrating the gravity data with geologic, seismic, and other geophysical information.

ACKNOWLEDGMENTS

We thank James Case, Andrew Griscom, Carl Wentworth, and Mark Zoback for their long-standing encouragement and interest in the preparation of these maps. Case and Griscom also gave us much help in the construction of Table 1, and many others freely shared their ideas and insights into the meanings of isostatic anomalies. We also thank William Hinze for drawing our attention to the work of Glennie.

REFERENCES CITED

Aiken, O. W., Keller, G. R., and Hinze, W. J., 1983, Geologic significance of surface gravity measurements in the vicinity of the illinois deep drill holes: Journal of Geophysical Research, v. 88, p. 7307–7314.

Anderson, E. R., Zoback, M. L., and Thompson, G. A., 1983, Implications of selected subsurface data on the structural form and evolution of some basins in the northern Basin and Range province, Nevada and Utah: Geological Society of America Bulletin, v. 94, p. 1055–1072.

Banks, R. J., Parker, R. L., and Huestis, S. P., 1977, Isostatic compensation on a continental scale; Local versus regional mechanism: Geophysical Journal of the Royal Astronomical Society of London, v. 51, p. 431–452.

Barnes, V. E., Romberg, F., and Anderson, W. A., 1954, Geology and geophysics of Blanco and Gillespie Counties, Texas, *in* Barnes, V. E., and Bell, W. C., eds., Cambrian Field Trip; Llano Area, March 19–20, 1954: San Angelo Geological Society, p. 78–90.

Bateman, P. C., and Eaton, J. P., 1967, Sierra Nevada batholith: Science, v. 158, p. 1407–1417.

Bayer, K. C., 1983, Generalized structural, lithologic, and physiographic provinces in the fold and thrust belts of the United States: U.S. Geological Survey Map, scale 1:2,500,000.

Bayley, R. W., and Mueuhlberger, W. R., 1968, Basement rock map of the United States: U.S. Geological Survey Map, scale 1:2,500,000.

Beyer, L. A., 1980, Offshore southern California, *in* Oliver, H. W., ed., Interpretation of the gravity map of California and its continental margin: California Division of Mines and Geology Bulletin 205, p. 8–15.

Bickford, M. E., Harrower, K. L., Hoppe, W. J., Nelson, B. K., Nusbaum, R. L., and Thomas, J. J., 1981a, Rb-Sr and U-Pb geochronology and distribution of rock types in the Precambrian basement of Missouri and Kansas: Geological Society of America Bulletin, v. 92, p. 323–341.

Bickford, M. E., Sides, J. R., and Cullers, R. L., 1981b, Chemical evolution of magmas in the Proterozoic terrane of the St. Francois Mountains, southeastern Missouri; 1, Field, petrographic, and major element data: Journal of Geophysical Research, v. 86, p. 10365–10386.

Blake, M. C., Jr., and 7 others, 1978, Neogene basin formation in relation to plate-tectonic evolution of the San Andreas fault system, California: American Association of Petroleum Geologists Bulletin, v. 62, p. 344–372.

Blakely, R. J., and McKee, E. H., 1985, Subsurface structural features of the Saline Range and adjacent regions of eastern California as interpreted from isostatic residual gravity anomalies: Geology, v. 13, p. 781–785.

Blakely, R. J., and Simpson, R. W., 1986, Locating edges of source bodies from magnetic or gravity anomalies: Geophysics, v. 51, p. 1494–1498.

Blakely, R. J., Jachens, R. C., Simpson, R. W., and Couch, R. W., 1985, Tectonic setting of the southern Cascade Range as interpreted from its magnetic and gravity fields: Geological Society of America Bulletin, v. 96, p. 43–48.

Bowin, C., 1985, Global gravity maps and the structure of the earth, *in* Hinze, W. J., ed., The utility of regional gravity and magnetic anomaly maps: Society of Exploration Geophysicists, p. 88–101.

Bracewell, R., 1965, The Fourier transform and its applications: New York, McGraw-Hill, 381 p.

Brewer, J. A., Brown, L. D., Steiner, D., Oliver, J. E., Kaufman, S., and Denison, R. E., 1981, Proterozoic basin in the southern Midcontinent of the United States revealed by COCORP deep seismic reflection profiling: Geology, v. 9, p. 569–575.

Brown, L., Jenson, L., Oliver, J., Kaufman, S., and Steiner, D., 1982, Rift structure beneath the Michigan basin from COCORP profiling: Geology, v. 10, p. 645–649.

Byerly, P. E., 1966, Interpretations of gravity data from the central Coast Ranges and San Joaquin Valley, California: Geological Society of America Bulletin, v. 77, p. 83–94.

Cady, J. W., 1975, Magnetic and gravity anomalies in the Great Valley and western Sierra Nevada metamorphic belt, California: Geological Society of America Special Paper 168, 56 p.

——, 1980, Calculation of gravity and magnetic anomalies of finite-length right polygonal prisms: Geophysics, v. 45, p. 1507–1512.

Carmichael, R. S., 1982, Magnetic properties of minerals and rocks, *in* Carmichael, R. S., ed., Handbook of physical properties of rocks: Boca Raton, Florida, CRC Press, v. II, p. 229–287.

Case, J. E., and Joesting, H. R., 1972, Regional geophysical investigations in the central Colorado Plateau: U.S. Geological Survey Professional Paper 736, 31 p.

Case, J. E., and Keefer, W. R., 1966, Regional gravity survey of the Wind River basin, Wyoming: U.S. Geological Survey Professional Paper 550-C, p. C120–C128.

Chapman, R. H., and Griscom, A., 1980, Coast Ranges, *in* Oliver, H. W., ed., Interpretation of the gravity map of California and its continental margin: California Division of Mines and Geology Bulletin 205, p. 24–27.

Chase, C. G., and Gilmer, T. H., 1973, Precambrian plate tectonics; The midcontinent gravity high: Earth and Planetary Science Letters, v. 21, p. 70–78.

Cook, F. A., and Oliver, J. E., 1981, The late Precambrian–early Paleozoic continental edge in the Appalachian orogen: American Journal of Science, v. 281, p. 993–1008.

Cordell, L., 1978, Regional geophysical setting of the Rio Grande rift: Geological Society of America Bulletin, v. 89, p. 1073–1090.

——, 1982, Extension in the Rio Grande rift: Journal of Geophysical Research, v. 87, p. 8561–8569.

Craddock, C., Thiel, E. C., and Gross, B., 1963, A gravity investigation of the Precambrian of southeastern Minnesota and western Wisconsin: Journal of Geophysical Research, v. 68, p. 6015–6032.

Decker, E. R., and Smithson, S. B., 1975, Heat flow and gravity interpretation across the Rio Grande rift in southern New Mexico and west Texas: Journal of Geophysical Research, v. 80, p. 2542–2552.

Dehlinger, P., Couch, R. W., and Gemperle, M., 1968, Continental and oceanic structure from the Oregon Coast westward across the Juan de Fuca ridge: Canadian Journal of Earth Sciences, v. 5, p. 1079–1090.

Dutch, S. I., 1983, Proterozoic structural provinces in the north-central United States: Geology, v. 11, p. 478–481.

Eaton, G. P., and 7 others, 1975, Magma beneath Yellowstone National Park: Science, v. 188, p. 787–796.

Gilbert, G. K., 1913, Interpretation of anomalies of gravity: U.S. Geological Survey Professional Paper 85-C, p. 27–37.

Glennie, E. A., 1932, Gravity anomalies and the structure of the Earth's crust: Survey of India Professional Paper 27.

——, 1936, Gravity anomalies in the United States: Journal of Geology, v. 44, p. 765–782.

Godson, R. H., 1985, Preparation of a digital grid of gravity-anomaly values of the conterminous United States, *in* Hinze, W. J., ed., The utility of regional gravity and magnetic anomaly maps: Society of Exploration Geophysicists, p. 38–46.

Godson, R. H., and Scheibe, D. M., 1982, Description of magnetic tape containing conterminous U.S. gravity data in gridded format: U.S. Department of Commerce, National Technical Information Service PB82-254789 (magnetic tape with description), 5 p.

Green, A. G., Hajnal, Z., and Weber, W., 1985, An evolutionary model of the western Churchill province and western margin of the Superior province in Canada and north-central United States: Tectonophysics, v. 116,

p. 281–322.

Green, J. C., 1983, Geologic and geochemical evidence for the nature and development of the Middle Proterozoic (Keweenawan) midcontinent rift of North America: Tectonophysics, v. 94, p. 413–437.

Griscom, A., 1963, Tectonic significance of the Bouguer gravity field of the Appalachian System: Geological Society of America Special Paper 73, p. 163–164.

Griscom, A., and Grafft, K. S., 1982, A model of the Ventura basin, California, from gravity data: Geological Society of America Abstracts with Programs, v. 14, p. 168.

Griscom, A., and Oliver, H. W., 1980, Isostatic gravity highs along the west side of the Sierra Nevada and the Peninsular Ranges batholiths, California [abs.]: EOS Transactions of the American Geophysical Union, v. 61, p. 1126.

Hamilton, W., 1981, Plate-tectonic mechanism of Laramide deformation: University of Wyoming Contributions to Geology, v. 19, p. 87–92.

Harrison, J. C., von Huene, R. E., and Corbato, C. E., 1966, Bouguer gravity anomalies and magnetic anomalies off the coast of southern California: Journal of Geophysical Research, v. 71, p. 4921–4941.

Hayford, J. F., and Bowie, W., 1912, The effect of topography and isostatic compensation upon the intensity of gravity: U.S. Coast and Geodetic Survey Special Publication 10, 132 p.

Heiskanen, W. A., and Mortiz, H., 1967, Physical geodesy: San Francisco, W. H. Freeman, 364 p.

Hildenbrand, T. G., Kane, M. F., and Hendricks, J. D., 1982, Magnetic basement in the upper Mississippi Embayment region; A preliminary report, *in* McKeown, F. A., and Pakiser, L. C., eds., Investigations of the New Madrid, Mississippi, earthquake region: U.S. Geological Survey Professional Paper 1236-E, p. 39–53.

Hill, D. P., 1978, Seismic evidence for the structure and Cenozoic tectonics of the Pacific coast states, *in* Smith, R. B., and Eaton, G. P., eds., Cenozoic Tectonics and regional geophysics of the Western Cordillera: Geological Society of America Memoir 152, p. 145–174.

Hinze, W. J., 1959, A gravity investigation of the Baraboo syncline region: Journal of Geology, v. 67, p. 417–446.

Hinze, W. J., Kellogg, R. L., and O'Hara, N. W., 1975, Geophysical studies of basement geology of southern peninsula of Michigan: American Association of Petroleum Geologists Bulletin, v. 59, p. 1562–1584.

Hurich, C. A., and Smithson, S. B., 1982, Gravity interpretation of the southern Wind River Mountains, Wyoming: Geophysics, v. 47, p. 1550–1561.

Hutchinson, D. R., Grow, J. A., and Klitgord, K. D., 1983, Crustal structure beneath the southern Appalachians; Nonuniqueness of gravity modeling: Geology, v. 11, p. 611–615.

Irwin, W. P., 1981, Tectonic accretion of the Klamath Mountains, *in* Ernst, W. G., ed., The geotectonic development of California: Englewood Cliffs, New Jersey, Prentice-Hall, p. 29–49.

Jachens, R. C., and Griscom, A., 1983, Three-dimensional geometry of the Gorda plate beneath northern California: Journal of Geophysical Research, v. 88, p. 9375–9392.

—— , 1985, An isostatic residual map of California; A residual map for interpretation of anomalies from intracrustal sources, *in* Hinze, W. J., ed., The utility of regional gravity and magnetic anomaly maps: Society of Exploration Geophysicists, p. 347–360.

Jachens, R. C., and Roberts, C. W., 1981, Documentation of a FORTRAN program, "isocomp," for computing isostatic residual gravity: U.S. Geological Survey Open-File Report 81-574, 26 p.

Jachens, R. C., Simpson, R. W., Saltus, R. W., and Blakely, R. J., 1985, Isostatic residual gravity anomaly map of the United States (exclusive of Alaska and Hawaii): National Oceanic and Atmospheric Administration, Geophysical Data Center Map on clear film, scale 1:2,500,000.

Jachens, R. C., Barnes, C. G., and Donato, M. M., 1986a, Subsurface configuration of the Orleans fault; Implications for deformation in the western Klamath Mountains, California: Geological Society of America Bulletin, v. 97, p. 388–395.

Jachens, R. C., Simpson, R. W., Griscom, A., and Mariano, J., 1986b, Plutonic belts in southern California defined by gravity and magnetic anomalies: Geological Society of America Abstracts with Programs, v. 18, p. 120.

Johnson, G. R., and Olhoeft, G. R., 1984, Density of rocks and minerals, *in* Carmichael, R. S., ed., Handbook of physical properties of rocks: Boca Raton, Florida, CRC Press, v. III, p. 1–38.

Kane, M. F., Mabey, D. R., and Brace, R. L., 1976, A gravity and magnetic investigation of the Long Valley caldera, Mono County, California: Journal of Geophysical Research, v. 81, p. 754–762.

Kane, M. F., Webring, M. W., and Bhattacharyya, B. K., 1981, A preliminary analysis of gravity and aeromagnetic surveys of the Timber Mountain area, southern Nevada: U.S. Geological Survey Open-File Report 81-189, 40 p.

Kärki, P., Kivioja, L., and Heiskanen, W. A., 1961, Topographic-isostatic reduction maps for the world for the Hayford zones 18-1, Airy-Heiskanen system, T = 30 km; Helsinki, Publications of the Isostatic Institute of the International Association of Geodesy, no. 35, 23 p.

Karner, G. D., and Watts, A. B., 1982, On isostasy at Atlantic-type continental margins: Journal of Geophysical Research, v. 87, p. 2923–2948.

—— , 1983, Gravity anomalies and flexure of the lithosphere at mountain ranges: Journal of Geophysical Research, v. 88, p. 10449–10477.

Keller, G. R., Bland, A. E., and Greenberg, J. K., 1982, Evidence for a major late Precambrian tectonic event (rifting?) in the eastern midcontinent region, United States: Tectonics, v. 1, p. 213–223.

King, E. R., and Zietz, I., 1971, Aeromagnetic study of the midcontinent gravity high of the central United States: Geological Society of America Bulletin, v. 82, p. 2187–2208.

King, P. B., 1977, The evolution of North America: Princeton, New Jersey, Princeton University Press, 197 p.

King, P. B., and Beikman, H. M., 1974, Geologic map of the United States: U.S. Geological Survey Map, scale 1:2,500,000.

Klasner, J. S., and King, E.R., 1986, Precambrian basement geology of North and South Dakota, Canadian Journal of Earth Sciences, v. 23, p. 1083–1102.

Klasner, J. S., King, E. R., and Jones, W. J., 1985, Geologic interpretation of gravity and magnetic data for northern Michigan and Wisconsin, *in* Hinze, W. J., ed., The utility of regional gravity and magnetic anomaly maps: Society of Exploration Geophysicists, p. 267–286.

Kovach, R. L., Allen, C. R., and Press, F., 1962, Geophysical investigations in the Colorado delta region: Journal of Geophysical Research, v. 67, p. 2845–2871.

LaFehr, T. R., 1965, Gravity, isostasy, and crustal structure in the southern Cascade Range: Journal of Geophysical Research, v. 70, p. 5581–5597.

—— , 1966, Gravity in the eastern Klamath Mountains, California: Geological Society of America Bulletin, v. 77, p. 1177–1190.

Lidiak, E. G., Hinze, W. J., Keller, G. R., Reed, J. E., Braile, L. W., and Johnson, R. W., 1985, Geologic significance of regional gravity and magnetic anomalies in the east-central Midcontinent, *in* Hinze, W. J., ed., The utility of regional gravity and magnetic anomaly maps: Society of Exploration Geophysicists, p. 287–307.

Lillie, R. J., and 7 others, 1983, Crustal structure of Ouachita Mountains, Arkansas; A model based on integration of COCORP reflection profiles and regional geophysical data: American Association of Petroleum Geologists Bulletin, v. 67, p. 907–931.

Lyons, P. L., 1961, Geophysical background of Arkoma basin tectonics: Tulsa Geological Society Digest, v. 29, p. 94–104.

Mabey, D. R., 1976, Interpretation of a gravity profile across the western Snake River plain, Idaho: Geology, v. 4, p. 53–55.

Malahoff, A., and Moberly, R., Jr., 1968, Effects of structure on the gravity field of Wyoming: Geophysics, v. 33, p. 781–804.

McNutt, M., 1980, Implications of regional gravity for state of stress in the Earth's crust and upper mantle: Journal of Geophysical Research, v. 85, p. 6377–6396.

Mitchell, B. J., and Landisman, M., 1970, Interpretation of a crustal section across Oklahoma: Geological Society of America Bulletin, v. 81, p. 2647–2656.

Nelson, K. D., and 7 others, 1982, COCORP seismic reflection profiling in the Ouachita Mountains of western Arkansas; Geometry and geologic interpretation: Tectonics, v. 1, p. 413–430.

Nelson, K. D., and 8 others, 1985a, New COCORP profiling in the southeastern United States; Part 1, Late Paleozoic suture and Mesozoic rift basin: Geology, v. 13, p. 714–717.

Nelson, K. D., McBride, J. H., Arnow, J. A., Oliver, J. E., Brown, L. D., and Kaufman, S., 1985b, New COCORP profiling in the southeastern United States; Part 2, Brunswick and east coast magnetic anomalies, opening of the north-central Atlantic Ocean: Geology, v. 13, p. 718–721.

Nicholas, R. L., and Rozendal, R. A., 1975, Subsurface positive elements within Ouachita foldbelt in Texas and their relation to Paleozoic cratonic margin: American Association of Petroleum Geologists Bulletin, v. 59, p. 193–216.

Ocola, L. C., and Meyer, R. P., 1973, Central North American rift system; 1, Structure of the axial zone from seismic and gravimetric data: Journal of Geophysical Research, v. 78, p. 5173–5194.

Oliver, H. W., and Mabey, D. R., 1963, Anomalous gravity field in east-central California: Geological Society of America Bulletin, v. 74, p. 1293–1298.

Parker, R. L., 1972, The rapid calculation of potential anomalies: Geophysical Journal of the Royal Astronomical Society of London, v. 31, p. 447–455.

Plouff, D., and Pakiser, L. C., 1972, Gravity study of the San Juan Mountains, Colorado: U.S. Geological Survey Professional Paper 800-B, p. B183–B190.

Powell, B. N., and Phelps, D. W., 1977, Igneous cumulates of the Wichita province and their tectonic implications: Geology, v. 5, p. 52–56.

Pruatt, M. A., 1976, Geophysical interpretations, *in* Powell, B. N., and Fischer, J. F., eds., Plutonic igneous geology of the Wichita magmatic province, Oklahoma: Geological Society of America South Central Section, 10th Annual Meeting Field Trip Guidebook 2, p. 4–7.

Rabinowitz, P. D., 1974, The boundary between oceanic and continental crust in the western North Atlantic, *in* Burk, C. A., and Drake, C. L., eds., The geology of continental margins: New York, Springer-Verlag, p. 67–84.

Rabinowitz, P. D., and LeBrecque, J. L., 1977, The isostatic gravity anomaly; Key to the evolution of the ocean-continent boundary at passive continental margins: Earth and Planetary Science Letters, v. 35, p. 145–150.

Ramberg, I. B., Cook, F. A., and Smithson, S. B., 1978, Structure of the Rio Grande rift in southern New Mexico and west Texas based on gravity interpretation: Geological Society of America Bulletin, v. 89, p. 107–123.

Saltus, R. W., 1984, A description of colored gravity and terrain maps of the southwestern cordillera: U.S. Geological Survey Open-File Report 84-95, 8 p.

Shanmugam, G., and Lash, G. G., 1982, Analogous tectonic evolution of the Ordovician foredeeps, southern and central Appalachians: Geology, v. 10, p. 562–566.

Simmons, G., 1964, Gravity survey and geological interpretation, northern New York: Geological Society of America Bulletin, v. 75, p. 81–98.

Simpson, R. W., Jachens, R. C., and Blakely, R. J., 1983a, AIRYROOT; A Fortran program for calculating the gravitational attraction of an Airy isostatic root out to 166.7 km: U.S. Geological Survey Open-File Report 83-883, 66 p.

Simpson, R. W., Saltus, R. W., Jachens, R. C., and Godson, R. H., 1983b, A description of colored isostatic gravity maps and a topographic map of the conterminous United States avilable as 35mm slides: U.S. Geological Survey Open-File Report 83-884, 16 p.

Simpson, R. W., Jachens, R. C., Blakely, R. J., and Saltus, R. W., 1986, A new isostatic residual gravity map of the conterminous United States with a discussion on the significance of isostatic residual anomalies: Journal of Geophysical Research, v. 91, p. 8348–8372.

Smithson, S. B., Brewer, J. A., Kaufman, S., Oliver, J. E., and Hurich, C. A., 1979, Structure of the Laramide Wind River uplift, Wyoming, from COCORP deep-reflection data and from gravity data: Journal of Geophysical Research, v. 84, p. 5955–5972.

Snavely, P. D., Jr., Wagner, H. C., and Lander, D. L., 1980, Interpretation of the Cenozoic geologic history, central Oregon continental margin; Cross-section summary: Geological Society of America Bulletin, v. 91, p. 143–146.

Snyder, J. P., 1982, Map projections used by the U.S. Geological Survey: U.S. Geological Survey Bulletin 1532, 313 p.

Society of Exploration Geophysicists, 1982, Gravity anomaly map of the United States (exclusive of Alaska and Hawaii): Society of Exploration Geophysicists, scale 1:2,500,000.

Sparlin, M. A., Braile, L. W., and Smith, R. B., 1982, Crustal structure of the eastern Snake River plain determined from ray trace modeling of seismic refraction data: Journal of Geophysical Research, v. 87, p. 2619–2633.

Stewart, J. H., 1978, Basin-range structure in western North America; A review: Geological Society of America Memoir 152, p. 1–31.

Strange, W. E., and Woollard, G. P., 1964, The use of geologic and geophysical parameters in the evaluation, interpolation, and prediction of gravity: Hawaii Institute of Geophysics Report 64-17, 4-1–4-42.

Stuart, D. J., 1961, Gravity study of crustal structure in western Washington: U.S. Geological Survey Professional Paper 424-C, p. C273–C276.

Sumner, J. R., 1977, Geophysical investigation of the structural framework of the Newark-Gettysburg Triassic basin, Pennsylvania: Geological Society of America Bulletin, v. 88, p. 935–942.

——, 1985, Crustal geology of Arizona as interpreted from magnetic, gravity, and geologic data, *in* Hinze, W. J., ed., The utility of regional gravity and magnetic anomaly maps: Society of Exploration Geophysicists, p. 164–180.

Suppe, J., 1979, Structural interpretation of the southern part of the northern Coast Ranges and Sacramento Valley, California; Summary: Geological Society of America Bulletin, v. 90, p. 327–330.

Telford, W. M., Geldart, L. P., Sheriff, R. E., and Keys, D. A., 1976, Applied geophysics: Cambridge, Cambridge University Press, 860 p.

Thomas, J. J., Shuster, R. D., and Bickford, M. E., 1984, A terrane of 1,350- to 1,400-m.y.-old silicic volcanic and plutonic rocks in the buried Proterozoic of the mid-continent and in the Wet Mountains, Colorado: Geological Society of America Bulletin, v. 95, p. 1150–1157.

Thomas, M. D., 1983, Tectonic significance of paired gravity anomalies in the southern and central Appalachians: Geological Society of America Memoir 158, p. 113–124.

Thomas, W. A., 1985, The Appalachian-Ouachita connection; Paleozoic orogenic belt at the southern margin of North America: Annual Reviews of Earth and Planetary Science, v. 13, p. 175–199.

Thompson, G. A., 1959, Gravity measurements between Hazen and Austin, Nevada; A study of Basin-Range structure: Journal of Geophysical Research, v. 64, p. 217–229.

U.S. Geological Survey and American Association of Petroleum Geologists, 1961, Tectonic map of the United States exclusive of Alaska and Hawaii: U.S. Geological Survey Map, scale 1:2,500,000.

Woollard, G. P., 1936, An interpretation of gravity-anomalies in terms of local and regional geologic structures: EOS Transactions of the American Geophysical Union, Part 1, p. 63–74.

——, 1962, The relation of gravity anomalies to surface elevation, crustal structure, and geology: University of Wisconsin, Geophysical and Polar Research Center Research Report 62–9, 356 p.

——, 1966, Regional isostatic relations in the United States, *in* Steinhart, J. S., and Smith, T. J., eds., The earth beneath the continents: American Geophysical Union Geophysical Monograph 10, p. 557–594.

——, 1968, The interrelationship of the crust, the upper mantle, and isostatic gravity anomalies in the United States, *in* Knopoff, L., Drake, C. L., and Hart, P. J., eds., The crust and upper mantle of the Pacific area: American Geophysical Union Geophysical Monograph 12, p. 312–341.

——, 1972, Regional variations in gravity, *in* Robertson, E. C., ed., The nature of the solid earth: New York, McGraw-Hill, p. 463–505.

——, 1976, Regional changes in gravity and their relation to crustal parameters, *in* Hitoshi, A., and Iizuka, S., eds., Volcanoes and tectonosphere: Tokyo, Tokai University Press, 370 p.

Zietz, I., 1982, Composite magnetic anomaly map of the United States; Part A, Conterminous United States: U.S. Geological Survey Geophysical Investigations Map GP-954A, scale 1:2,500,000.

Manuscript Accepted by the Society October 31, 1988

Printed in U.S.A.

Geological Society of America
Memoir 172
1989

Chapter 20

Electrical structure of the crust and upper mantle beneath the United States; Part 2, Survey of data and interpretation

George V. Keller
Department of Geophysics, Colorado School of Mines, Golden, Colorado 80401

ABSTRACT

Great volumes of data are available about the properties of the near surface of the Earth from borehole surveys and from direct-current and controlled-source electromagnetic soundings. A significant result derived from these data is a surface conductance map for the continental United States, showing that the conductance in the near-surface section ranges from <1 S/m in exposed shield areas to >5,000 S/m in the deep sedimentary basins. The existence of a surface layer with high conductance significantly interferes with our capability to probe into the mantle with electrical geophysical methods.

Information about the more conductive regions in the lower part of the crust and in the mantle is obtained from magnetotelluric and geomagnetic deep soundings. Perhaps 10,000 magnetotelluric soundings have been carried out in the last decade, but only about 1,200 are available in the public domain. Results are available from six networks of temporary magnetic observatories, using from several tens to 50 recording instruments. These various data sets have been obtained primarily from the western part of the United States, with only very sparse data available from the East, and virtually no data from the Southeast.

A striking result is the observation that, in the Basin and Range province of the western United States, anomalously high values of conductivity are observed at depths as shallow as 10 km in the crust. It has been hypothesized that these conductive zones are caused either by partial melting of crustal rocks in regions of high heat flow, or by the presence of significant amounts of water in fractured rock.

There appears to be a lack of correspondence between some of the main features of the geoelectric section and the general features of the interior of the Earth, which have been developed from seismic and gravity data. The base of the sedimentary and weathered section at the surface of the crust is the only important boundary in the conductivity profile that coincides with a corresponding seismic and density boundary. The other important geoelectric features of the crust and outer mantle seem discordant with the Earth's structure as we know it from seismic and gravity data.

INTRODUCTION

The data base that exists for compiling information on the electrical structure of the crust and upper mantle beneath the conterminous 48 United States is extensive, but not uniform in coverage. It consists of a very large volume of information obtained from well logs, along with a variety of crustal-scale electrical probes carried out with direct-current (DC) methods, controlled-source electromagnetic methods, the magnetotelluric (MT) method, and the geomagnetic deep-sounding method. Results obtained with these various methods tend to overlap only slightly; each method used seems to provide results specifically

Keller, G. V., 1989, Electrical structure of the crust and upper mantle beneath the United States; Part 2, Survey of data and interpretation, *in* Pakiser, L. C., and Mooney, W. D., Geophysical framework of the continental United States: Boulder, Colorado, Geological Society of America Memoir 172.

Figure 1. Areas of increased thickness of conductive sedimentary section. Sixty-six recognized basins are indicated.

about one or another characteristics of the overall conductivity profile described earlier in Part 1.

The conductivity profile can be divided into a sequence of zones from the surface of the Earth inward, each zone characterized by some conductivity feature. The surface zone, zone 1, consists of water-bearing rocks and the ocean waters, which are relatively conductive. Zone 2 is the crystalline basement, where the absence of water causes the conductivity to be low. Zone 3 is a zone where rock is rendered conductive by elevated temperatures, and which can be recognized at depths ranging from tens to hundreds of kilometers, depending on the thermal regime. Rock conductivity continues to increase with depth, with deeper zones being identified, but only in the upper three zones can we also recognize lateral changes that are in some way related to geologic features.

INFORMATION FROM GEOPHYSICAL WELL LOGS

Zone 1 has been well explored by drilling, particularly in areas where it consists of potentially hydrocarbon-bearing sedimentary rocks. Many areas of increased thickness of the conductive sedimentary section have been recognized, as shown in Figure 1, where some 66 basins are indicated. The basins occupy about half the surface area, but much of the rest of the country is covered by a relatively thinner sedimentary veneer. Only 5 to 10 percent of the area of the United States has outcropping crystalline rocks belonging to zone 2 of the conductivity profile.

Well log compilations used here come from two sources. One source is an earlier study of conductivity logs from the Denver-Julesburg, Gulf Coast, and Appalachian basins, as well as the Colorado Plateaus, for which integral conductance values

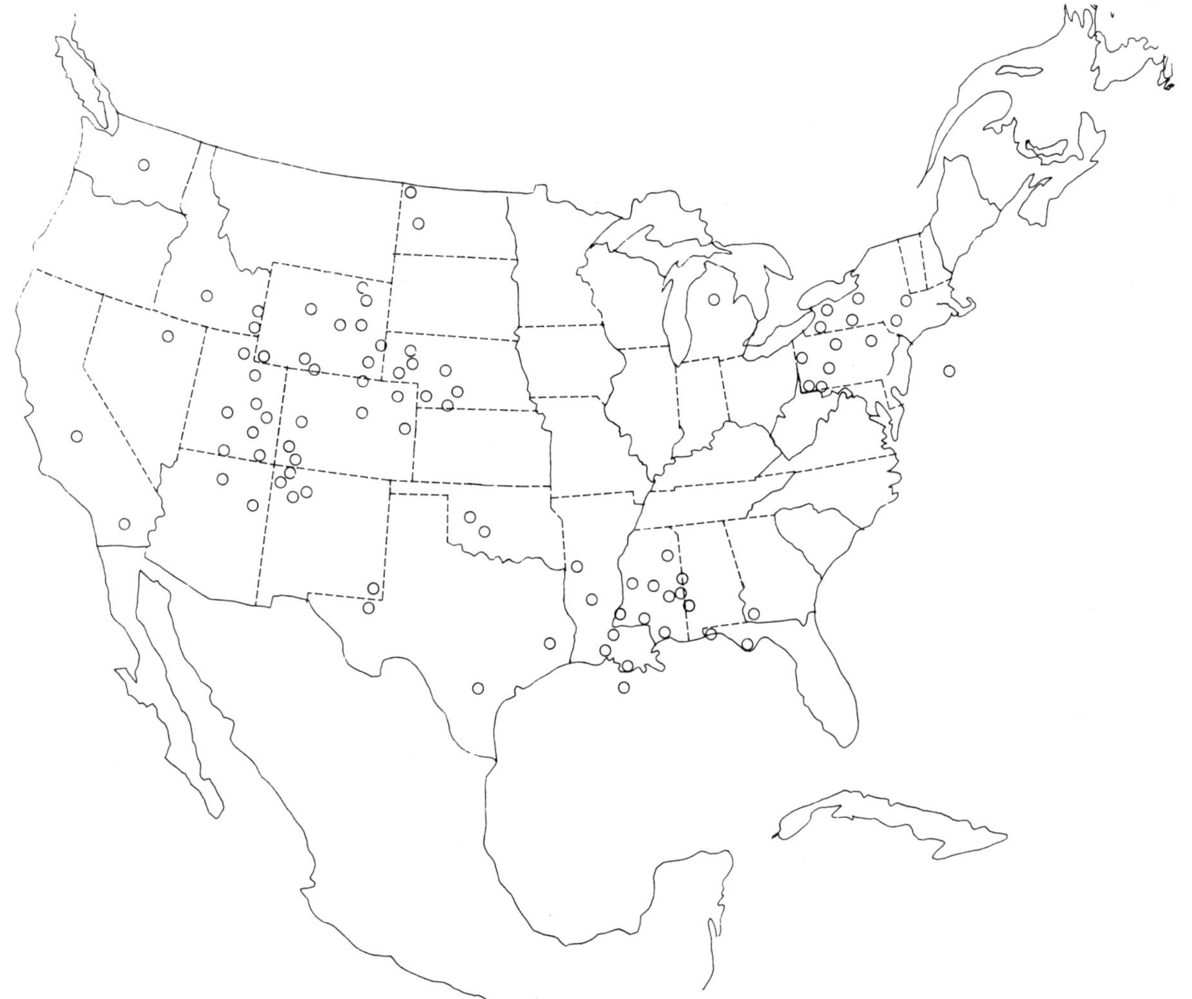

Figure 2. Locations of wells from which electric logs were used for the study of conductance in the surface layer.

have been compiled for about 200 wells (Keller, 1967). The second source is a recent study (Keller, 1987) in which an additional 100 wells, widely distributed among the various sedimentary basins, were included. The locations of wells for which integral conductance compilations have been done are shown in Figure 2. A contoured conductance map based on these data is shown in Figure 3. Because the number of data contained is small, the contours have been forced to follow the outlines of the known basins.

The high conductance present in the sedimentary cover in the Gulf Coast area, where it reaches 6,000 Siemans, and in the intermountain basins of the intermountain west, where values of 1,000 to 2,000 Siemans are common, should be noted. These conductances pose a significant problem to the use of electrical probing methods as they currently exist in studying the electrical characteristics of zones 2 and 3. Conversely, the conductance in the basins in the central and eastern United States is moderate, and does not vary rapidly from location to location. This part of the country is particularly favorable for the quantitative use of deep electrical sounding methods.

CONTROLLED-SOURCE SURVEYS

As noted in Part 1, controlled-source surveys include direct-current soundings, which have the capability of detecting the transverse resistance in zone 2 if arrays with dimensions approaching 100 km are used. They also include various types of electromagnetic induction soundings, which have the capability of detecting the top of zone 3 in areas where it is unusually shallow or conductive.

Early crustal-scale direct-current resistivity surveys were carried out by the U.S. Geological Survey at the locations indicated in Figure 4 (Keller and others, 1966; Anderson and Keller, 1966; Jackson, 1967). Extensive sets of DC sounding data were obtained in the western United States as a result of grounding tests of the Northwest-Southwest Intertie, a high-voltage direct-current transmission line connecting the Pacific Northwest with

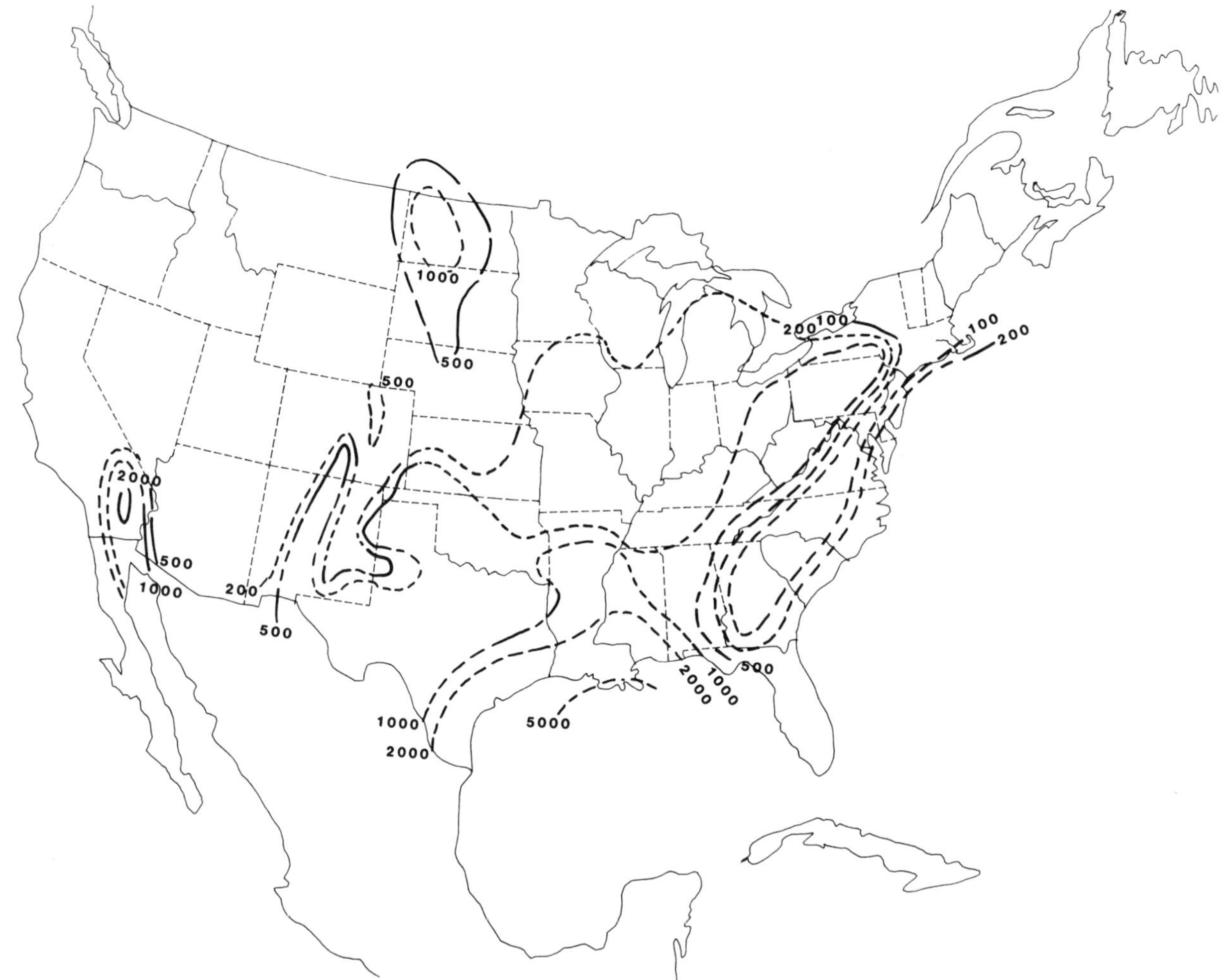

Figure 3. Conductance map of the United States based on electric log compilations. Contour values are given in Siemans.

the southern California and Nevada area (Keller, 1968; Cantwell and others, 1965). During these tests, currents as much as several hundred amperes were passed over the line and returned through the ground at terminals near Portland, Oregon; Tracey, California; and Boulder City, Nevada (see Fig. 4). Many hundreds of measurements of electrical field strength were made around each terminal at distances ranging from a few tens of meters to several hundred kilometers. The data were used to construct Schlumberger-type sounding curves for interpretation.

During 1973 and 1974, extensive electrical surveys, including DC soundings, were carried out in north-central Wisconsin to map the electrical properties of the crust (Keller and Furgerson, 1977; Sternberg and Clay, 1977; Sternberg, 1979; see map in Fig. 4). The surface in the area consists of a cover of glacial till with a thickness of no more than 100 m, which contributes a conductance for zone 1 of less than 1 Siemans. The suboutcrop in the area is mostly Precambrian granite, with minor amounts of metamorphic rocks composing the top part of zone 2. The area is among the most ideal in the continental United States for exploring zone 2 without interference from zone 1.

In 1975, extensive Schlumberger and dipole mapping surveys were performed in western Virginia, as a part of a regional exploration program for geothermal energy. In the area of the survey, the outcropping rocks are sedimentary units ranging in age from Devonian and older. The conductance of zone 1 amounts to only a few Siemans in this area, making it favorable for the use of direct-current methods in studies of zone 2.

A typical set of electric field measurements for a crustal scale resistivity sounding is shown in Figure 5. (These are data from measurement of the effect of High-voltage Direct Current (HVDC) grounding current in the vicinity of Boulder City, Nevada.) The scatter is reasonably typical of that observed in dipole mapping surveys of crustal scale. Average apparent resistivity curves from sets of dipole mapping data in the various areas surveyed are shown in Figure 6. These curves can be used to determine an effective conductance for zone 1 and an effective resistance for zone 2 from the position of the maximum. These values are listed in Table 1.

In addition to the direct-current dipole mapping surveys listed above, a number of attempts have been made to use

Figure 4. Locations of deep DC resistivity surveys.

1. Pacific Northwest (Keller, 1968)
2. Central Valley of California (Keller, 1968)
3. Southern Nevada (Keller, 1968)
4. Colorado Plateau (Jackson, 1966)
5. Nebraska-Iowa area (Anderson and Keller, 1966)
6. Wisconsin (Keller and Furgerson, 1977).
7. Virginia
8. Central Pennsylvania (Anderson and Keller, 1966)
9. Adirondacks of New York (Anderson and Keller, 1966)
10. Maine (Anderson and Keller, 1966).

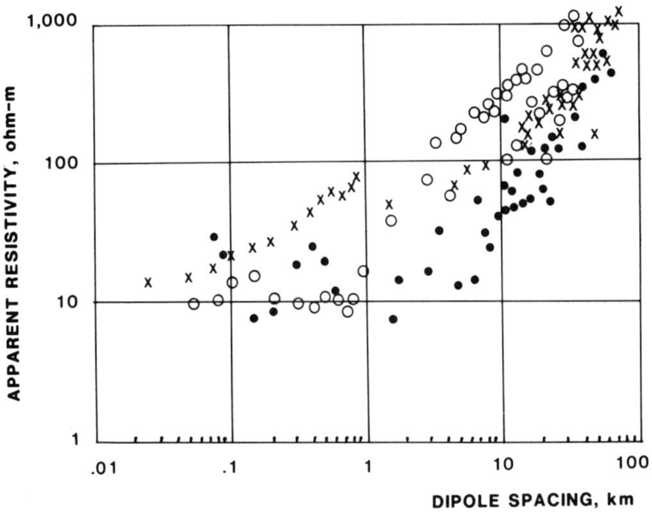

Figure 5. Three DC data sets obtained with the dipole method in the central United States (from Anderson and Keller, 1966).

TABLE 1. CONDUCTANCES AND RESISTANCES OF THE CRUST FROM DIRECT-CURRENT SOUNDINGS

Location	Conductance of Zone 1 (seimans	Resistance of Zone 2 (x10**6 ohms)
Adirondack Mountains, New York	0.1	260
Maine and New Hampshire	0.2	180
Nebraska-Iowa	15	150
Colorado Plateaus	120	150
Southern California	45	5
Central California	200	4
Wisconsin	1	190
Southern Nevada	60	200
Columbia Plateau, Washington, and Oregon	100	160

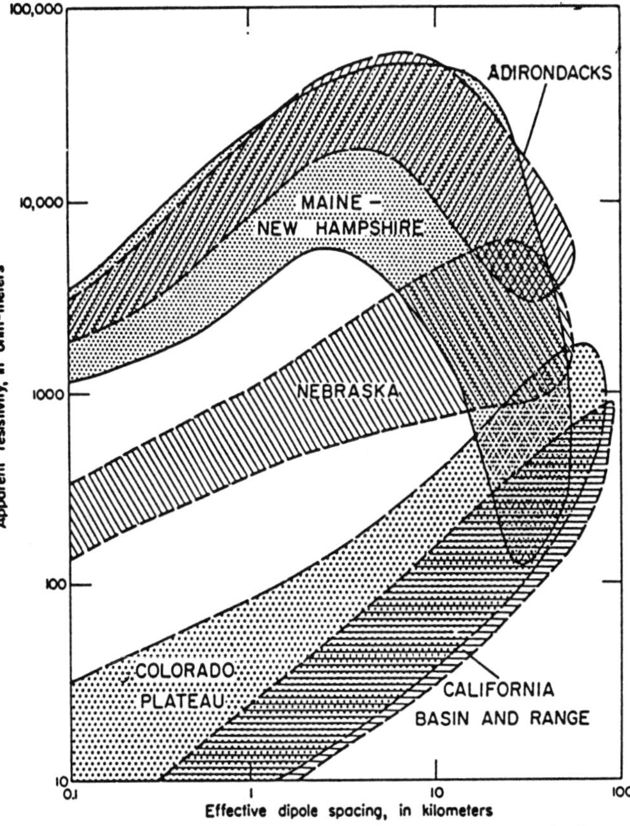

Figure 6. Summary of some dipole resistivity measurements (Anderson and Keller, 1966).

controlled-source electromagnetic methods to map the boundary between zones 2 and 3. Because it has been difficult until recently to provide a sufficiently strong field to penetrate to depths of many tens of kilometers in the Earth, these preliminary controlled-source electromagnetic surveys have been carried out in areas particularly favorable to such investigations, such as areas of anomalously shallow occurrence of the top of zone 3 or areas where zone 1 is poorly developed.

One of the more interesting of such efforts was based on the use of the Pacific Northwest-Southwest High-voltage Direct-Current (HDVC) Intertie as a source (Skohan, 1980; Lienert and Bennett, 1977; Lienert, 1979; Towle, 1980). In February 1971, the Sylmar terminal of the HVDC line was damaged by an earthquake and put out of service for nearly a year. In the summer of 1971, during testing of a new installation, the Bonneville Power Administration made the HVDC line available to the Colorado School of Mines (CSM) and the University of Texas at Dallas (UTD) for experiments related to electromagnetic induction in the Earth. For these experiments, the HVDC line (see Fig. 7) was energized with a square-wave current having an amplitude of approximately 300 amps and a period of 600 sec. The line, about 1,000 km in length, provided a uniquely strong source of low-frequency electromagnetic energy for investigating the subsurface along the route of the HVDC line.

The path of the HVDC line, as shown in Figure 7, parallels the supposed boundary between the Basin and Range province for most of its length, with the Sierra Nevada batholith lying to the west. This experiment offered an opportunity to study the effect of relatively high heat flow in the Basin and Range province on electrical properties at depths up to a few tens of kilometers. The experimental procedure used differed between the northern half of the line (where measurements were made by the CSM team), and the southern half of the line (where measurements were made by the UTD team). For measurements made in the north, the transient and magnetic fields due to the steps in current transmitted along the HVDC line were recorded over a bandpass from DC to 20 Hz in the time domain. At each location, it was assumed that a sounding was being made independent of the measurements at other sites. On the other hand, over the southern part of the line, measurements were made with recording magnetometers that had a bandpass of 0 to 1 Hertz. In interpretation, the combined responses along a traverse of station were used to determine the subsurface structure.

The conductivity structures recognized over the northern part of the line can be summarized as follows. Quite high values of conductivity were observed over surface expressions of the intermontane basins; the surface layer is moderately conductive, even in the areas where Tertiary volcanic rocks are exposed. Beneath the conductive surface sequence, the basement consists of highly resistive basement rocks, which in this area are metasedimentary or metavolcanic rocks of Mesozoic or older age. At locations far enough removed from the source line that the electromagnetic field penetrated to great depths, a conductive third zone was identified. However, in areas mapped as Known

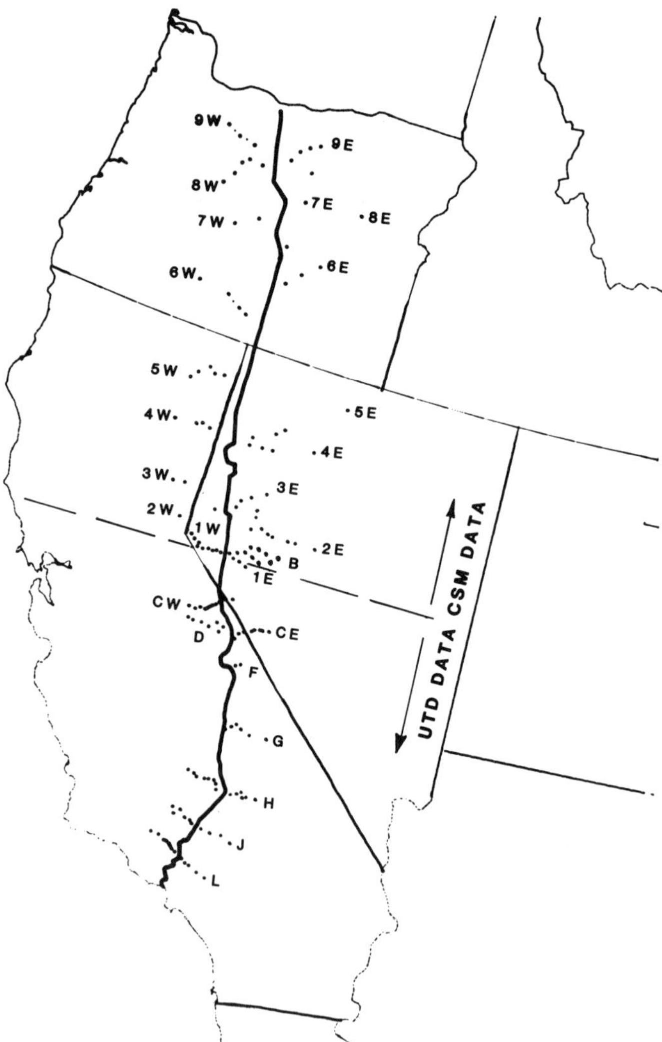

Figure 7. Location of the Pacific Northwest-Southwest High Voltage Direct Current Intertie, and locations where various components of the electromagnetic field were measured (Lienert and Bennett, 1977; Skokan, 1980). Recording locations formed a series of traverses across the line, as indicated by the letter and number designators (9E, L).

Geothermal Resource Areas (KGRAs), the conductivity increased monotonically with depth.

Typical stitched cross sections of conductivity values determined with one-dimensional inversions of the individual recordings are shown in Figure 8. Consider profile 8 in Figure 8, a profile that lies outside the KGRAs. The section here shows a conductive first layer resting on a resistive second layer; only at the far western end of the profile can the presence of a deep conductive zone be recognized. In contrast, consider traverse 4, which crosses the Black Rock Desert of northwestern Nevada, an area of extensive surface manifestations of geothermal activity. In the center of this traverse, a conductive zone is present that extends from the surface to great depth. Lower conductivities that

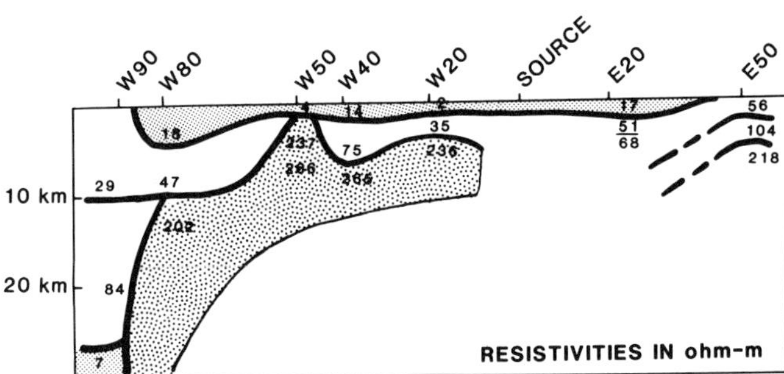

Figure 8. Pseudosection compiled from one-dimensional inversions of signals recorded along profile 8E-8W (see Fig. 7). Individual stations are indicated by a code showing direction (E or W from line) and the distance from the line (90 km, 80 km, etc.). Numbers within the cross section are resistivities in ohm-meters obtained by inversion (Skokan, 1980).

are more characteristic of nonthermal areas are present both to the east and the west.

As noted above, data acquisition and interpretation procedures for the southern half of the line were different (Lienert, 1979). These traverses are indicated by letter designators (A, B, etc.) in Figure 7. The results along the various lines were as follows:

Line AB. The northernmost of the lines on which magnetometers were used for recording lies entirely within the Basin and Range province, a short distance south of Reno. The data provide strong evidence for a zone of high conductivity with its upper boundary at a depth of 18 to 25 km.

Line CW-CE. This line is situated along the northern shore of Mono Lake. The data indicate a crust of low average conductivity consistent with a conductivity of 0.03 S/m starting at a depth of 17 km.

Line D-E. Along this traverse, the observations give clear evidence of a more conductive crust to the east than to the west.

Line G. This line lies at the southeast end of the Sierra Nevada batholith, near Coso. Observations show a complex electrical structure, with a pronounced conductivity low near the center of the traverse, and high conductivities at the two ends of the line.

Line H-I. Data from this profile, which lies at the southern end of the Sierra Nevada batholith, clearly show lower conductivities and greater depths to zone 3. An interpretation yields a depth of 38 km with a conductivity of 0.04 S/m.

Line J-K. Data from this line also show a relatively great depth to the conductive rocks in zone 3, with an interpretation of 74 km.

Connerney and others (1980) and Connerney and Kuckes (1980) have described the use of a bipolar controlled-source electromagnetic method in the Adirondack Mountains of New York (see Fig. 4). The source consisted of a loop of wire 4.5 km in diameter; the field was detected at distances ranging from 20 to 62 km using fluxgate magnetometers. Interpretation of coupling measurements made over the frequency band from 0.05 to 400 Hz indicate a conductivity profile in which the conductivity increases progressively, with depth reaching probable values as great as 0.04 to 0.08 S/m at depths beyond 23 km (see Fig. 9). This zone appears to have an integral conductance of the order of 400 S, with a probable thickness of 6 to 12 km, below which the conductivity returns to more normal crustal values.

Another bipolar electromagnetic sounding has been reported for the Georgia piedmont (Thompson and others, 1983; see Fig. 4 for location), although in this case the source loop was placed with its axis horizontal. The results indicate that the surface rocks have conductivities characteristic of zone 2 of the conductivity profile, with values in the range 0.00002 to 0.00025 S/m. Beginning at a depth of 15 km, the conductivity appears to increase to more than 0.001 S/m. A possible explanation of this conductive zone is that it represents sedimentary rock trapped beneath an overthrust plate during the Appalachian orogeny.

MAGNETOTELLURIC DATA

Far more information has been obtained about the Earth's conductivity profile using the natural-source methods than the controlled-source methods. Extensive magnetotelluric surveys have been carried out in the United States in the past decade (1977 to 1986), largely in support of resource exploration programs for oil and gas or for geothermal energy. It is likely that 10,000 magnetotelluric surveys have been carried out in this period, although not all are useful for crustal studies and not all are available in the public domain. Results suggest that magnetotelluric soundings need to be carried to a period of several thousand seconds in order to provide information on the upper part of zone 3; some results indicate that it may well be necessary to carry magnetotelluric soundings to periods of several tens of thousands of seconds. Because the effort and time required to

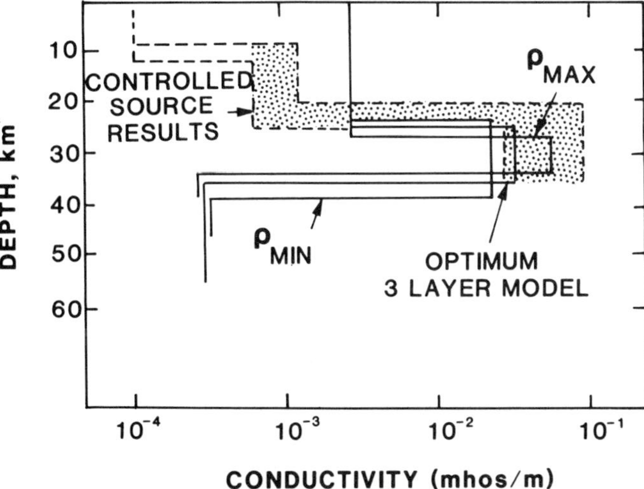

Figure 9. Comparison of conductivity profiles obtained with a controlled-source electromagnetic sounding method and a natural field method in the Adirondack area of New York (Connerney and Kuckes, 1980).

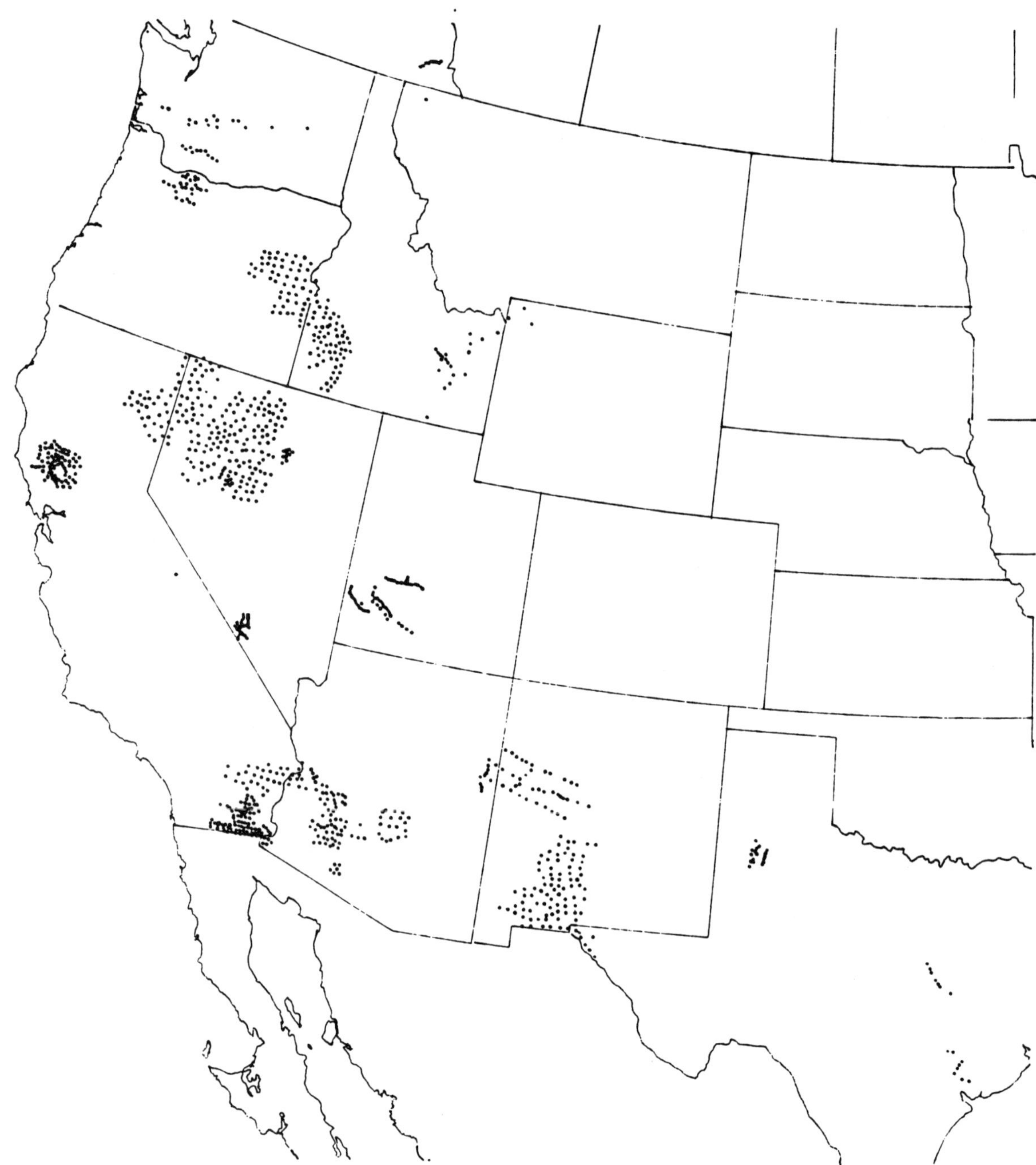

Figure 10. MT stations in the western United States.

carry out a single magnetotelluric sounding increase markedly with the depth or period desired, one normally sees commercial surveys terminated at periods of hundreds of seconds, after penetration past drillable depths has been achieved. In this study, fewer than 2,000 soundings were identified that were carried to periods of a few thousand seconds; only four were identified that extended to periods of a few tens of thousands of seconds. Almost all of the available magnetotelluric soundings included in this summary were carried out in the western half of the United States. This biasing of coverage results from the fact that commercial concentrations of oil and gas or geothermal energy are most likely to occur in the West, compounded further by the fact that the complicated structure of the western United States has appeared to attract the interests even of those investigators with purely academic motivation.

The locations of MT stations used in this compilation are shown in Figure 10. It should be noted that in some areas, MT stations were occupied at closely adjacent locations; in these cases, not all the individual locations could be shown on Figure 10.

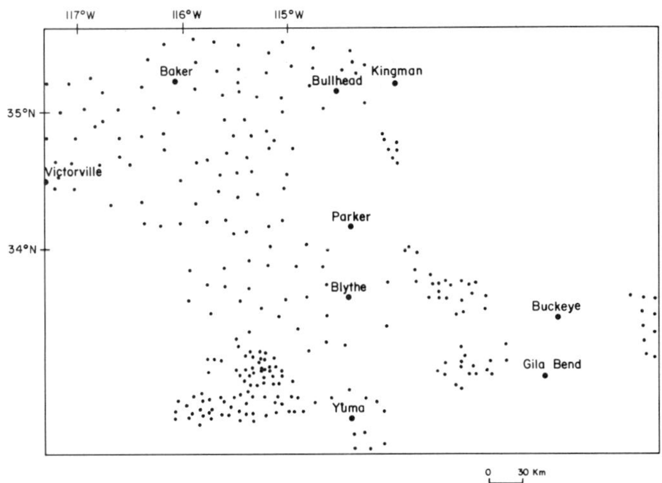

Figure 11. Southwestern United States MT location map (Group Seven, 1982).

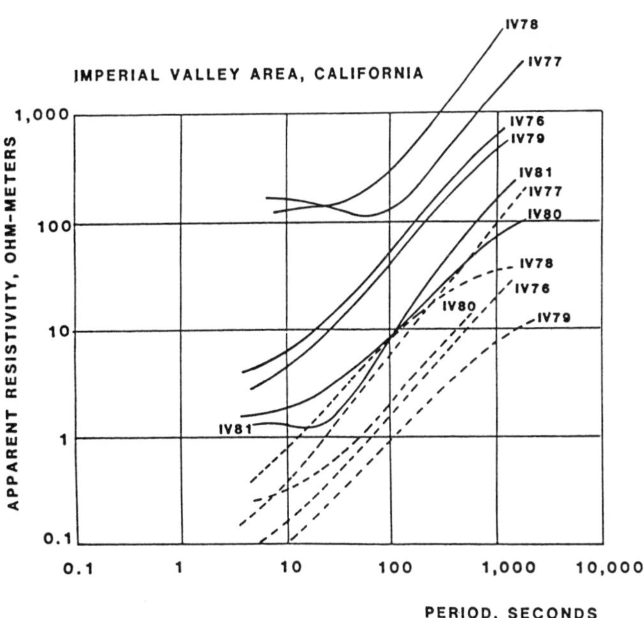

Figure 12a. Typical MT soundings from the southwestern United States, showing the presence only of a shallow conductive zone at the surface of the crust. Maximum rotated resistivity values are shown as solid curves, minimum values as dashed curves.

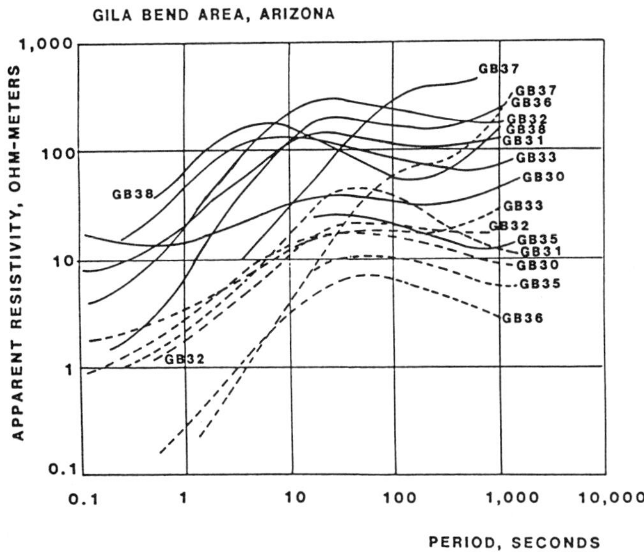

Figure 12b. Typical MT soundings from the Gila Bend area, Arizona. In contrast to the curves in Fig. 12a, these curves suggest the presence of a deep conductive zone in the crust.

Southern California and Arizona

An extensive survey has been carried out in southern California and southwestern Arizona (Group Seven, 1980, 1982; Williston-McNeil and Associates, 1979), as indicated by the location map in Figure 11. The area covered in this survey lies on the boundary between two major tectonic provinces: the Basin and Range province to the east, and the Imperial Valley—which is thought to reflect the extension of a crustal rift under the North American continent—to the west. In the southern California part of the area, the MT curves are dominated by the conductive sedimentary section in the Imperial Valley, which according to well survey data, has a conductance of 1,000 to 2,000 Siemans. Even at periods of 2,000 sec, these curves do not roll over to indicate the existence of more conductive rocks in zone 3. The rising branch of the MT curves (see Fig. 12a) is appropriate for S values of 500 to 1,500 Siemans, while the curves do not roll over even for apparent resistivity/period combinations where a rollover would be observed for a depth of 50 km to the conductive part of zone 3.

MT curves from the Basin and Range part of the survey area have a markedly different character than those from the Imperial Valley; a summary of the curves is shown in Figure 12b. Their shape is more complex, indicating the existence of a significant conductive zone in the upper part of the crust. A number of them indicated penetration to a final conductive zone at the longest periods observed. The upper boundary of this conductive zone occurs at depths between 10 and 20 km. The conductance of the zone is 500 to 1,000 Siemans, where the MT sounding penetrates to more resistive rocks beneath.

The electrical structure in the area covered by this survey is complicated, and so one does not expect one-dimensional interpretations to provide more than a qualitative indication of the subsurface. Stations are located too far apart to permit a detailed interpretation. As an intermediate approach to evaluating these data, the following procedure was used to merge the collection of MT curves into a form that would reduce the problems imposed by static shifts and lateral effects from rapid changes in the conductivity of the near-surface rocks. A grid was superimposed on the survey area, and all MT curves that would lie within a given

PSEUDO DEPTH = 8km

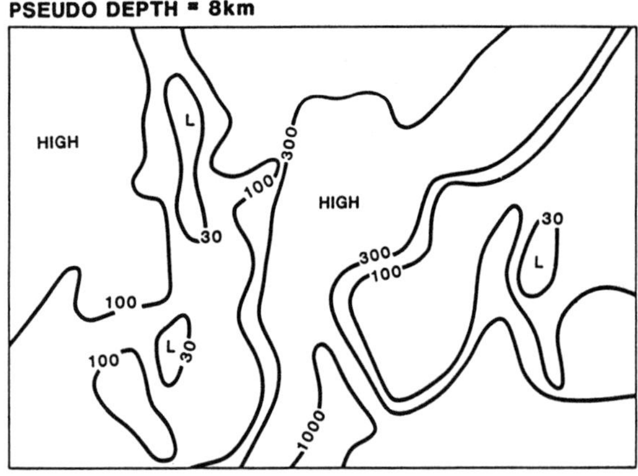

PSEUDO DEPTH = 16km

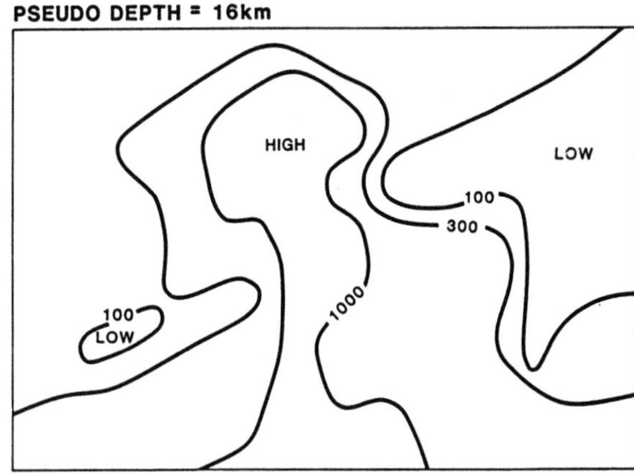

PSEUDO DEPTH = 30 km

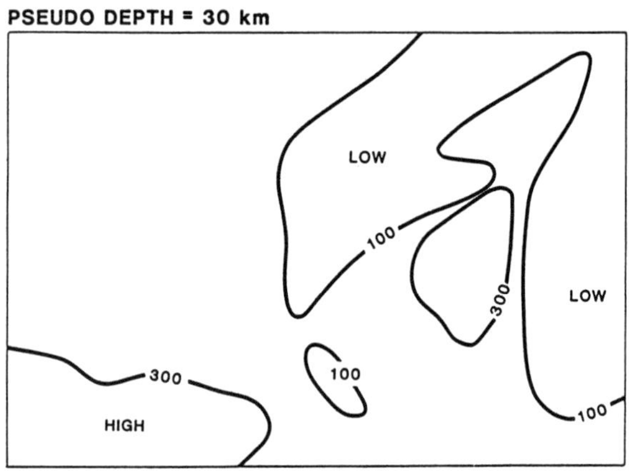

Figure 13. Apparent resistivity maps of southern California and western New Mexico at skin depths of 8, 16, and 30 km. The map area and geographic reference points are the same as in Fig. 11. Contoured values are expressed in ohm-meters.

search radius of each node point on this grid were averaged together. The search radius was selected to be 30 km, inasmuch as the effects we wish to observe lie at such depths or deeper, so lateral averaging should not unduly smooth out deep features, but should smooth surface effects. After average MT curves had been computed for each grid point, a new value of incremental resistivity was computed for each skin depth, using the method proposed by Schmucker (1984). This process permits the compilation of maps that show a reasonable resistivity to expect at a given depth in the Earth, which is equal to the skin depth for which the calculation was carried out. Such skin-depth slices are shown in Figure 13. Additional MT soundings have been reported for an area in Mexico close to the California-Mexico boundary (Gamble and others, 1980, 1981).

Northwestern Nevada

Northwestern Nevada became the site of intensive geothermal exploration and development when the U.S. Department of Energy funded an industry-coupled exploration program. It is an area in which numerous hot springs exist, and many heat-flow studies indicate regionally high heat flow. The area has come to be known as the Battle Mountain heat-flow high, although a better name might be the northwest Nevada thermal belt. A location map showing station sites of a regional survey is shown in Figure 14.

Using the merging method described for the southern California–Arizona data set, plan maps of effective resistivity at skin depths of 12, 20, and 30 km were prepared (see Fig. 15). In the southeastern half of the area covered, the resistivity remains very low at all depths in the crust; it is in the range of 0.1 to 0.3 S/m in the Carson Sink–Dixie Valley area. To the northwest, the effective resistivity values are considerably greater and more typical of what one would expect in a normal crust. Figure 16 shows a contour map of values for the integral conductance for layers in the upper part of the crust, derived from the MT curves. In the Dixie Valley and the Carson Sink, phenomenally large conductances exist within the crust, with values exceeding 10,000 Siemans.

Southern Rio Grande Valley

The Rio Grande Valley is now widely recognized as being an important continental rift, with major changes in the nature of the crust and mantle being associated with it. A regional survey (Group Seven, 1981a-c) was carried out at the sites indicated on the map in Figure 17. The area of this survey extends southward from Socorro in central New Mexico into the western tip of Texas in the vicinity of El Paso. Most of the curves in this survey exhibited a rollover at longer periods, indicating penetration to a final conductive layer. The depths to this zone were least in the southern part of the valley, around El Paso—10 to 15 km—and deeper to the north—20 to 25 km. Contour maps of apparent resistivity at depth are shown in Figure 18. A contour map of the integral conductance determined for the crust is shown in Figure 19.

Extensive magnetotelluric soundings have been reported from the northern part of the Rio Grande Rift, and parts of northern New Mexico and Arizona to the west of the rift. The area covers the Valley Caldera, and volcanic regions in the north-

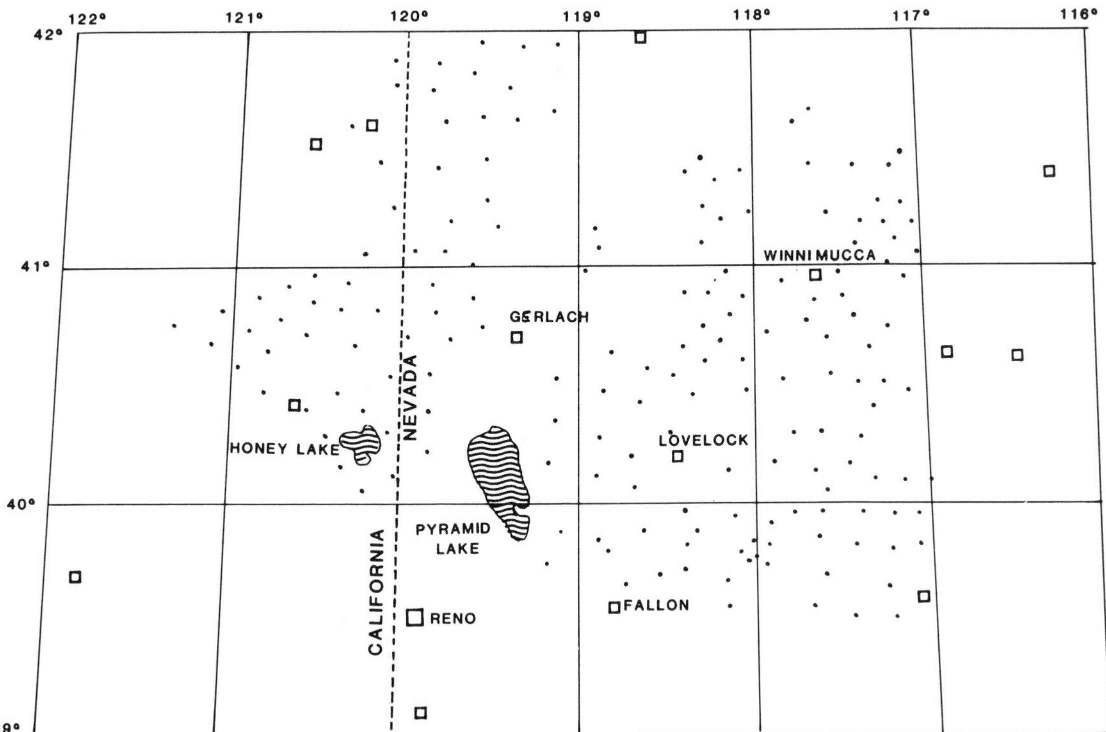

Figure 14. Locations of MT stations in northwestern Nevada and nearby areas in California.

ern New Mexico and Arizona (Jiracek and others, 1979; Ander and others, 1980; Hermance and Pederson, 1980; Aiken and Ander, 1981; Ander, 1981, 1984; Pedersen and Hermance, 1976, 1978; Keshet and Hermance, 1986). In some cases, closely spaced stations permitted detailed interpretation using two-dimensional modeling. For example, Ander (1984) produced the interpreted electrical cross section shown in Figure 20, which again indicates the presence of rock with conductivities in the range from 0.014 to 0.2 S/m at depths ranging from 15 to 25 km. The deep conductive zone becomes more conductive to the southeast, in the direction of the Jemez lineament.

Hermance and Pedersen (1980) reported on the basis of several stations a zone of anomalously high conductivity (S = 2,200 Siemans) at a depth of 10 to 17 km in the vicinity of Santa Fe.

Jiracek and others (1986) reported on interpretation of MT soundings from the central part of the Rio Grande Rift, between Belen and Soccoro. The electrical cross section resulting from this effort is shown in Figure 21. It is striking that a conductive layer is recognized, beginning at a depth of 10 km beneath the rift and to the west, and that this conductive zone is bottomed by more resistant rock beyond a depth of 25 km. East of the rift, the section is quite different. A thin conductive zone may be present between a depth of 22 and 25 km, but beyond 25 km the conductivity is low: 0.005 S/m.

Southwestern Idaho and eastern Oregon

Temperatures observed in some deep wildcat oil wells drilled in the Snake River Plain, along with some surface thermal manifestations, attracted geothermal exploration interest to the area in the late 1970s. Stations included in a regional survey are shown in Figure 22 (Group Seven, 1981a-c). The survey area covered the lower Snake River Plain, extending from Mountain Home in central Idaho into central Oregon.

Typical MT curves for this area are shown in Figure 23. As observed in other areas in the intermontane west, a striking feature of these curves is the indication of an important conductive section in the upper crust. Near Mountain Home, the depth to the top of this zone is between 3 and 12 km with the conductance of the zone 500 to 2,000 Siemans. In the western part of Oregon, the depth to the conductive layer is 8 to 20 km, while the conductance is 1,000 to 3,000 Siemans. Figure 24 shows a contour map of the conductance in the crust.

Stanley (1982) and Stanley and others (1977) have described the results of MT soundings carried out in the eastern Snake River Plain, at locations ranging from Yellowstone Caldera on the northeast to the Raft River Geothermal Area, on the Utah-Idaho border to the southwest. These curves show clearly expressed maximums at a period several tens of seconds and with maximum apparent resistivities of 50 to 100 ohm-meters; one

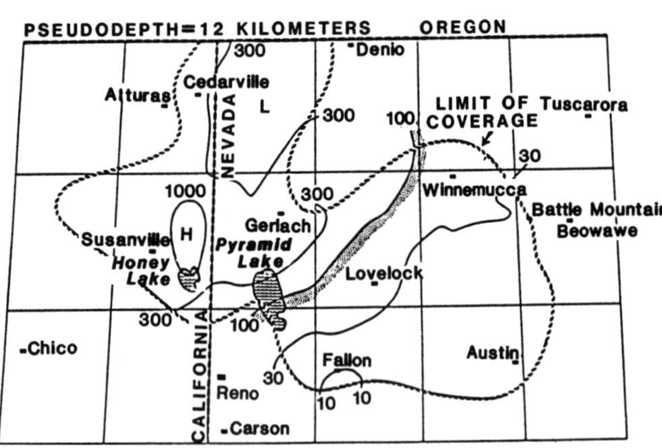

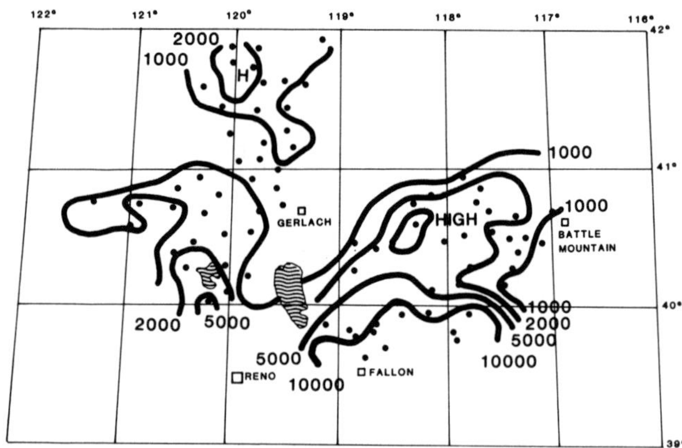

Figure 16. Total conductance in the crust based on MT soundings in northwestern Nevada and nearby areas of California. Contour levels are expressed in Siemans.

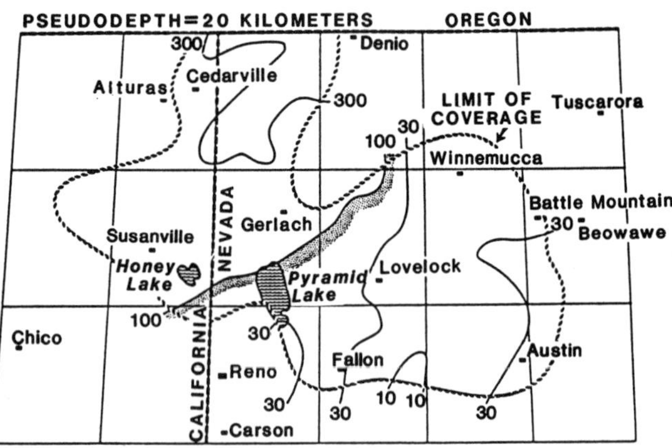

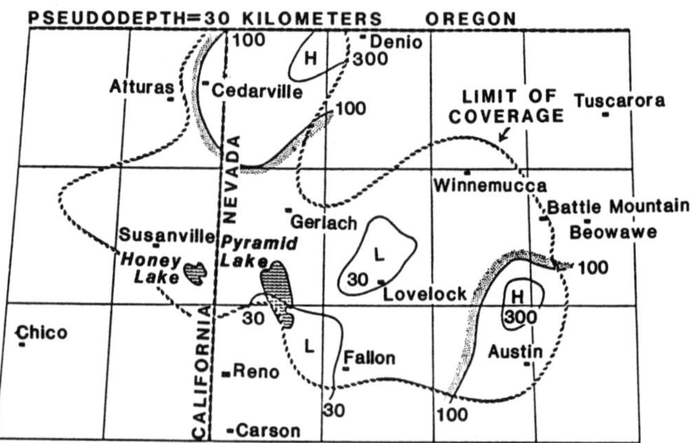

Figure 15. Apparent resistivity contour maps at skin depths of 12, 20 and 30 km, using MT soundings from the Nevada-California area defined in Fig. 14. Solid contours are resistivities in ohm-meters. The dashed lines indicate the areas of coverage.

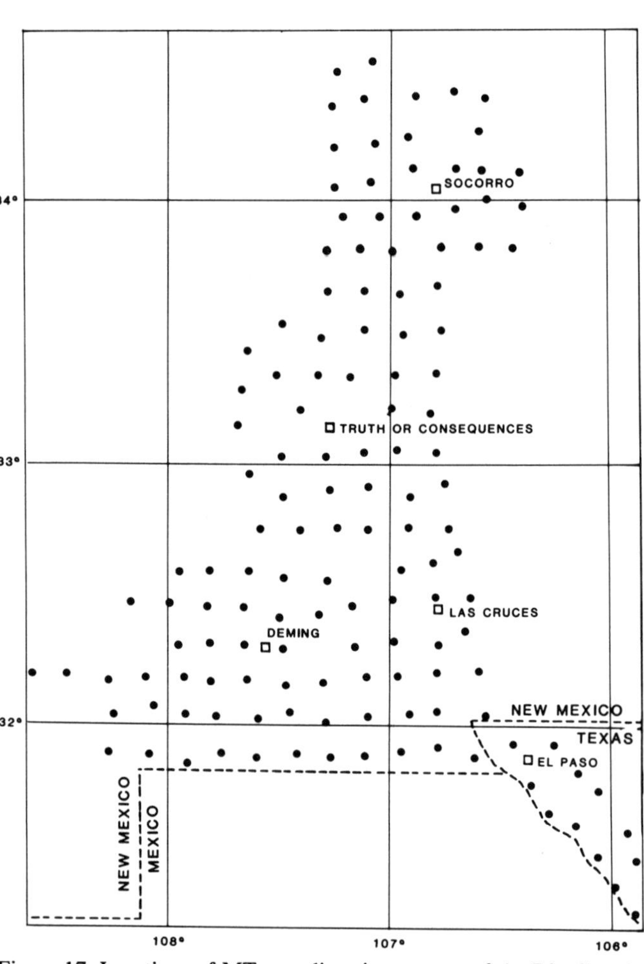

Figure 17. Locations of MT soundings in a survey of the Rio Grande Valley in southern New Mexico (Group Seven, 1982).

Figure 18. Apparent resistivity contour map at a skin depth of 8 kilometers for the southern part of the Rio grande Valley in New Mexico.

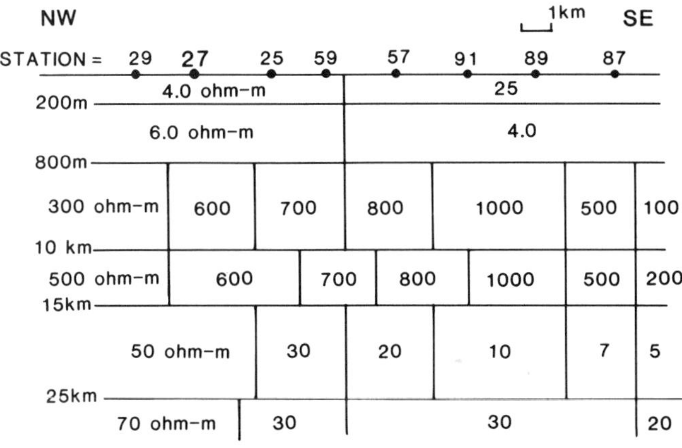

Figure 20. Finite-difference two-dimensional interpretation of a short profile of MT soundings in northwestern New Mexico (Ander, 1981). Resistivities within cells are expressed in ohm-meters.

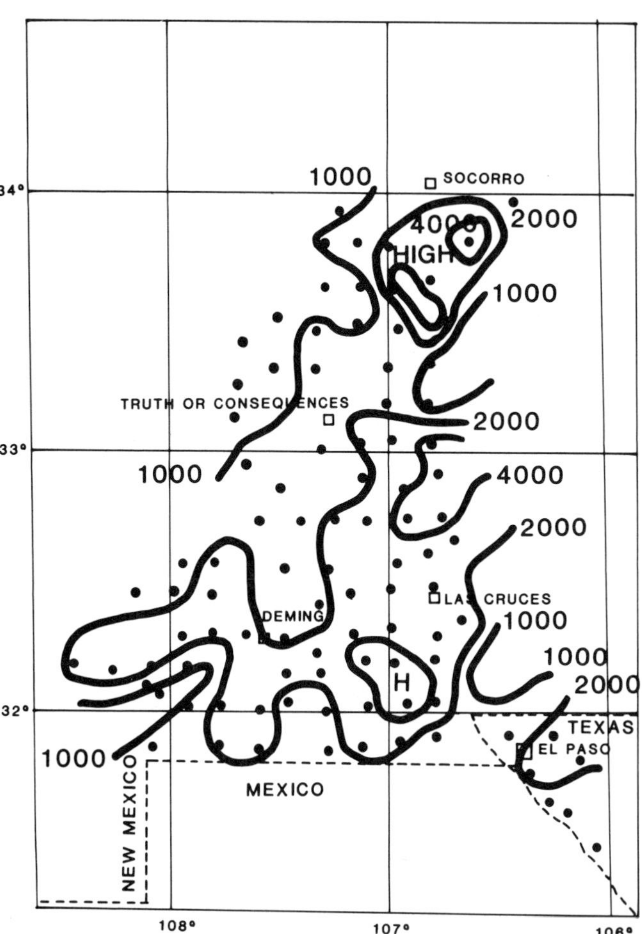

Figure 19. Total conductance in the upper part of the crust in the Rio grande Valley area of southern New Mexico. Contour levels are expressed in Siemans.

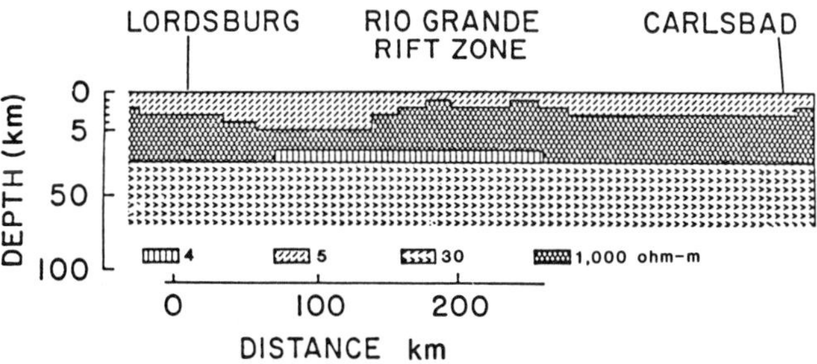

Figure 21. Modelled two-dimensional cross section that satisfies MT soundings crossing the Rio grande Rift (Jiracek and others, 1979). Note the break in the vertical scale.

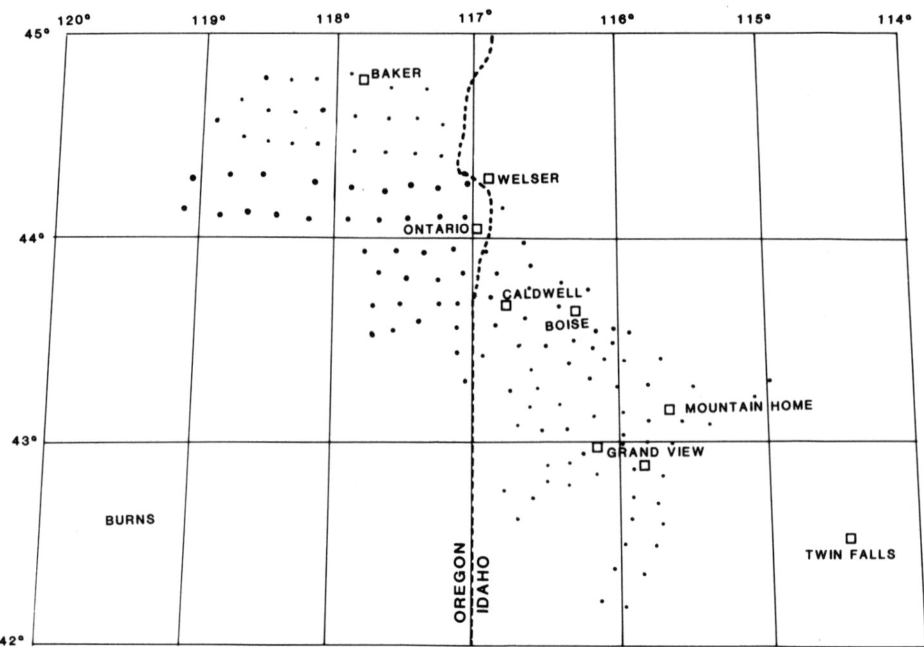

Figure 22. Location of MT stations in the Snake River Plains area of Idaho and Oregon (Group Seven, 1981).

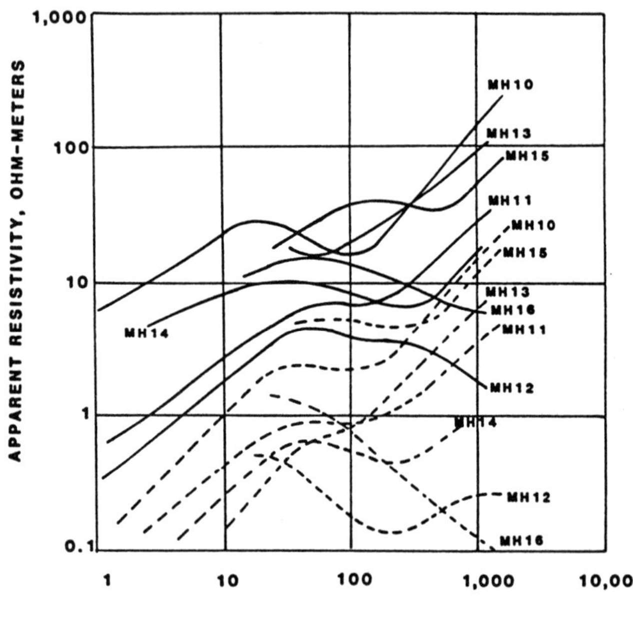

Figure 23. Typical MT curves from the Snake River Plains. Solid curves are values of maximum apparent resistivity. Dashed curves are values of minimum apparent resistivity.

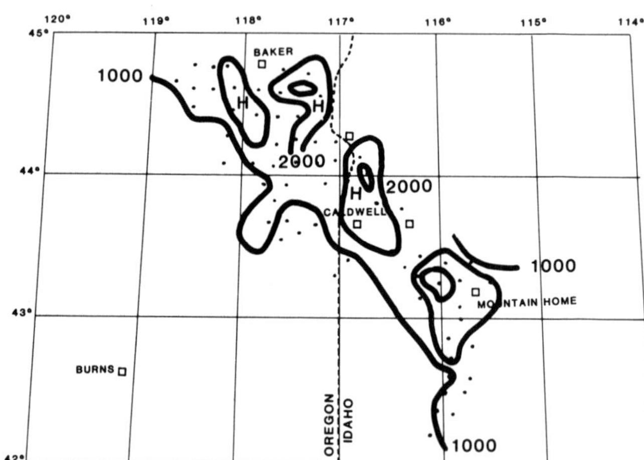

Figure 24. Total conductance in the upper part of the crust, based on MT soundings in the Snake River Plain. Contour levels are expressed in Siemans.

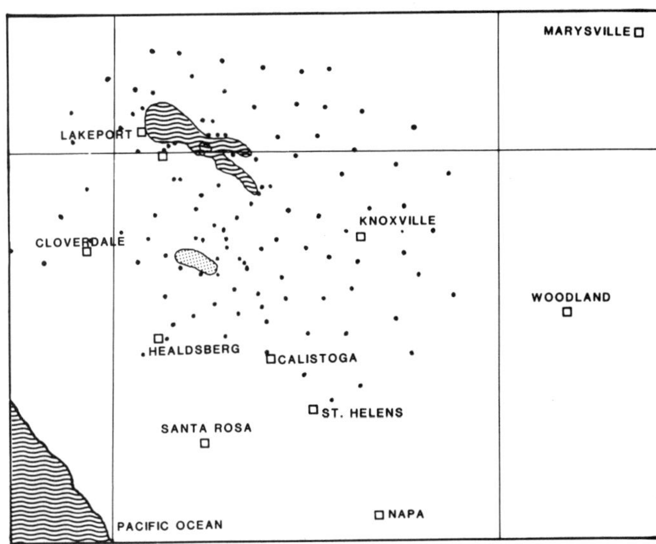

Figure 25. Locations of MT soundings in northern California, in the vicinity of The Geysers Steam Field.

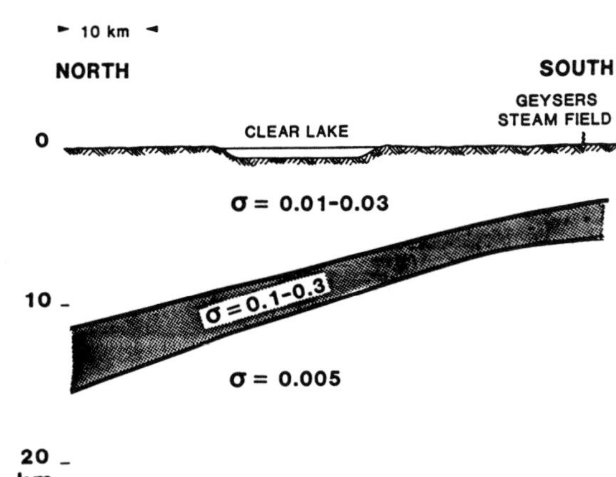

Figure 26. Electrical conductivity cross section through the Geysers area, California, based on one-dimensional inversions of MT soundings. The Steam Field is the dotted area south of Clear Lake.

dimensional inversions yield depths of 15 to 20 km to a crustal conductive zone with a conductivity of 0.1 to 1 S/m.

Central California

Many magnetotelluric soundings have been carried out in the vicinity of The Geysers, an area of geothermal steam production in central California. A gravity anomaly mapped in the area is thought to reflect a hot batholithic rock mass at relatively shallow depths in the crust. A map showing locations of MT soundings in the greater Geysers area is shown in Figure 25 (Group Seven, 1981a-c). Magnetotelluric soundings at The Geysers indicate the presence of quite conductive rocks in the near surface, with maximum conductivity being reached at depths of 2 to 4 km. The conductance of these rocks amounts to 1,000 to 2,000 Siemans. Few MT curves indicate a rollover from their rising segment, indicating a resistive lower unit at least to a depth of 50 km. A typical electrical cross section of The Geysers is shown in Figure 26.

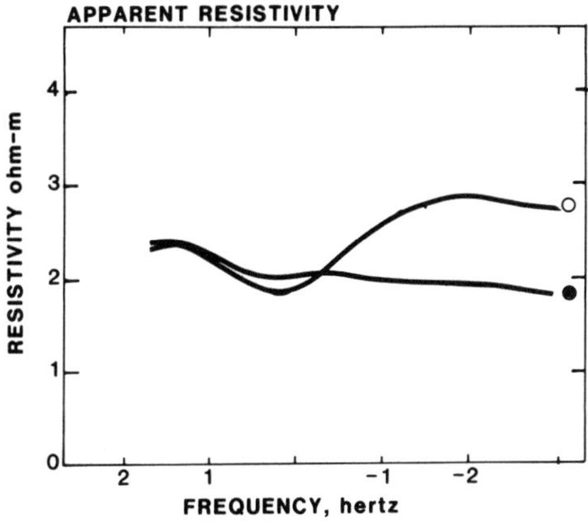

Figure 27. Typical MT sounding curve set for a location on the Columbia Plateau (Washington state), from Prieto and others (1985). The horizontal scale is expressed as the base-10 logarithm of frequency.

Pacific Northwest-Cascades area

During the period 1983 to 1985, a major effect was undertaken to determine the electrical structure of the West Coast of the United States, both offshore and onshore Washington State, and in adjacent parts of British Columbia and Oregon. Unfortunately, the results of this ambitious undertaking were not available at the time of writing of this report. However, earlier studies had been completed by Stanley (1979, 1984), Prieto and others (1985), and Goldstein and others (1978, 1982).

Stanley (1984) described two magnetotelluric profiles that were recorded in Washington to study the electrical structure of the Cascade Range. One profile, some 400 km in length, crossed the Cascade Range midway between Mount Rainier and Mount Adams. A second shorter profile extended from southwest of Mount St. Helens to southeast of Mount Adams.

Prieto and others (1985) described the use of MT soundings performed in eastern Washington for determining the structures in zone 1 that are of interest in hydrocarbon exploration. However, an example of a typical MT sounding curve (see Fig. 27) yields a maximum apparent resistivity of about 800 ohm-meters at a period of 100 sec, indicating the presence of relatively more conductive rocks at a depth of 30 to 40 km.

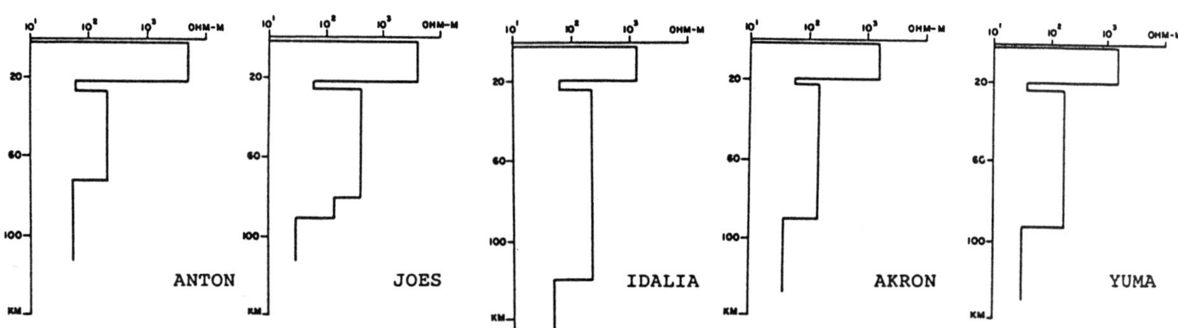

Figure 28. Comparison of one-dimensional inversions of MT soundings in eastern Colorado (Chaipayungan and Landisman, 1977).

Colorado Plateau area

The Colorado Plateaus represent a relatively undisturbed block of crust largely surrounded by geologic provinces in which the crust has undergone considerable extension (the Basin and Range and the Rio Grande Rift). Early MT soundings by Plouff (1966), using a primitive scalar MT system, yielded results that indicated the presence of a conductive zone beginning at a depth of 30 to 35 km, even though the quality of data was not up to modern standards. A more recent effort (Pedersen and Hermance, 1981) shows a conductive region to begin at a depth of 28 km near Farmington, New Mexico. The interpreted resistivity of 0.065 S/m in this conductive zone is said to persist at least to a depth of 130 km.

High Plains and Central Plains

Almost all of the MT data available for this compilation were obtained west of the Front Range of the Rocky Mountains. However, some studies have been carried out to the east, notably those by Mitchell and Landisman (1971) and Chaipayungpun and Landisman (1977).

Six MT soundings on the High Plains in eastern Colorado were reported by Chaipayungpun and Landisman (1977). These soundings were interpreted by an automated inversion process to yield the piecewise constant conductivity profiles shown in Figure 28. Although these inversions show a thin conductive zone at a depth of 20 km, an examination of the field data suggests that such a zone is undetectable, and that the significant transition to more conductive rocks occurs at a depth of 60 to 70 km, with the conductivity at greater depths being of the order of 0.03 S/m.

In MT soundings made in the northern panhandle of Texas, Mitchell and Landisman (1971) found depths of 50 to 70 km to conductive rocks beneath the resistant crust.

Canadian Shield

The only MT soundings reported for the eastern half of the United States are a set of five stations observed in Wisconsin (Dowling, 1970). These soundings yielded a generally lower conductivity profile than observed elsewhere, as might be expected. Even so, there is an order of magnitude increase in conductivity at a depth of 12 km, while possible further increases in conductivity could be seen on data from two of the soundings at a depth of 70 km.

GEOMAGNETIC DEEP-SOUNDING DATA

The geomagnetic deep-sounding method provides information on variations in the characteristics of zone 3. Wide coverage of the United States, except for the southeastern quarter, has been reported. Most of these surveys were accomplished using large arrays of simultaneously recording magnetometers; a few data are available from small arrays. Coverage is shown on the map in Figure 29.

Early magnetic variation studies in the southwestern United States have been reported by Schmucker (1964) and Swift (1967). Early possible interpretations of magnetic variations with periods of 3,600 to 7,200 sec are shown in Figure 30, from Schmucker (1964) and Swift (1967). In both studies, there is a marked transition in the electrical properties of the upper mantle between southern and west Texas.

Studies in the western United States

During the summer of 1968, the University of Texas at Dallas and the University of Alberta operated a large array of magnetometers in the western United States. The study area extended from central Wyoming in the north to the Mexican border on the south and from central Nevada on the west to western Nebraska on the east (the area is indicated on the map in Fig. 29; Porath and Dziewonski, 1971a, b; Gough, 1974; Caner, 1971; Bingham and others, 1985; Alabi and others, 1975; Porath, 1971a, b. In analyzing the magnetic variations, both the numerical approach of separating the internal field from the total field (Porath and others, 1970), and the empirical approach of removing a regional from the observed data (Reitzel and others, 1970; Porath and Gough, 1971) were used. A major conductive lineament in the mantle was recognized beneath the Wasatch front and the southern Rockies, as well as a regional area of high

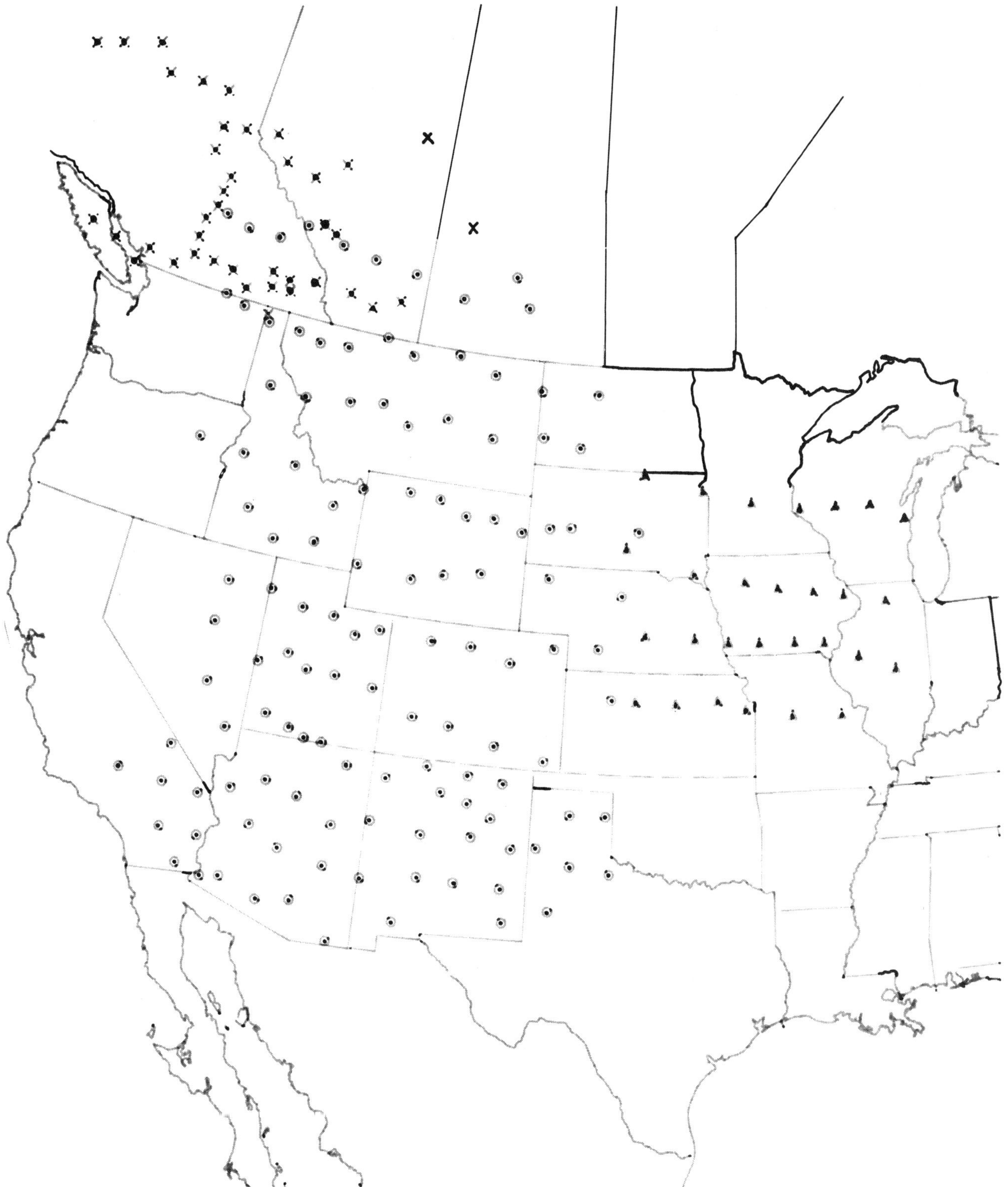

Figure 29. Locations of magnetometer arrays in geomagnetic deep soundings in the western United States.

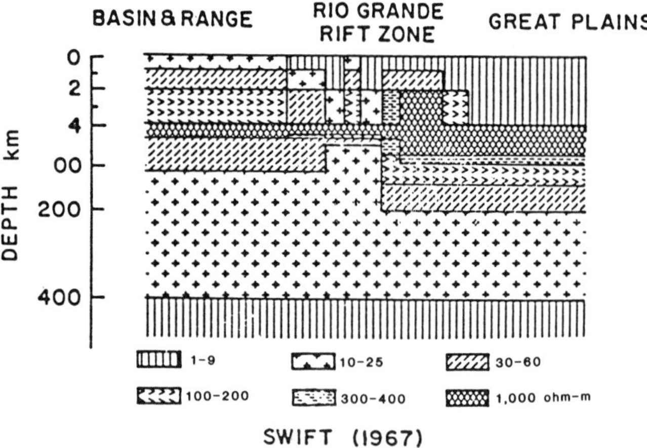

Figure 30. Resistivity cross section which models magnetometer array data across the southern Rocky Mountains. Note the break in the vertical scale.

conductivity under the Basin and Range province. The electrical structures appear to be grossly two-dimensional, extending from the northern to the southern limits of the area surveyed.

A cross section along the 38th Parallel is shown in Figure 31. This cross section is neither unique nor a best estimate in any sense; it is based on the calculation of the surface of a perfectly conductive region at depth on which flowing currents could explain the observed behavior of the magnetic field of internal origin. Use of the model does not imply the existence of a boundary at the depths shown in Figure 31. Rather, when conductivity increases with depth, the highest induced current density will be observed near a skin depth. At progressively longer periods, the apparent surface on which the currents appear to accumulate will move downward slowly.

Studies in the northwest United States

Magnetometer studies were carried out in an area extending from the Dakotas to Oregon and Washington in the United States, and from Saskatchewan to British Columbia in Canada (Camfield and Gough, 1975; Law and Greenhouse, 1981; Porath and others, 1971, Camfield and others, 1970; Law and others, 1980; Lajoie and Caner, 1970; Pham, 1980; Gough and others,

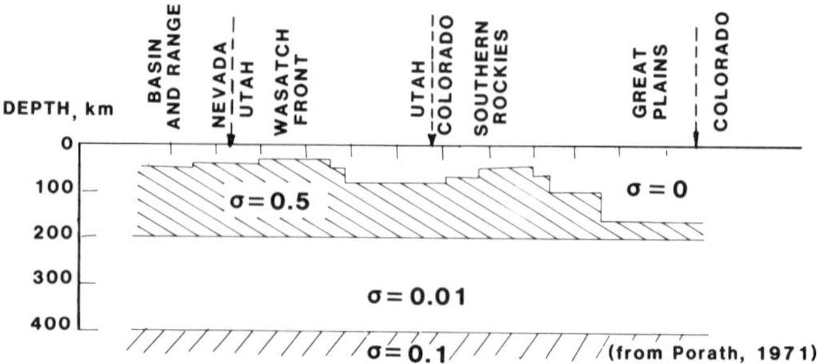

Figure 31. Interpreted conductivity cross section along the 38th Parallel, based on geomagnetic deep sounding data.

1982; Gough and Camfield, 1972). A prominent anomaly in magnetic induction has been given the name of "North American Central Plains Anomaly." The anomaly runs from the Black Hills along the boundary between Montana and the Dakotas; it is strong but narrow. Because of its narrow width, the anomaly is most certainly not caused by conductivity structures in the mantle but rather from conductivity structures in the crust. There is no obvious correlation with gravity or static, magnetic field anomalies; the suggestion was made by Camfield and others (1970) that graphite in metamorphic rocks of the basement might be the origin.

In addition to the Central Plains anomaly, increased conductivities in the mantle are evident west of the Front Range of the northern Rockies, and in two local areas of small anomalous field behavior at the Front Range and immediately west of the Rocky Mountain Trench. A cross section along the Canada-U.S. boundary is shown in Figure 32.

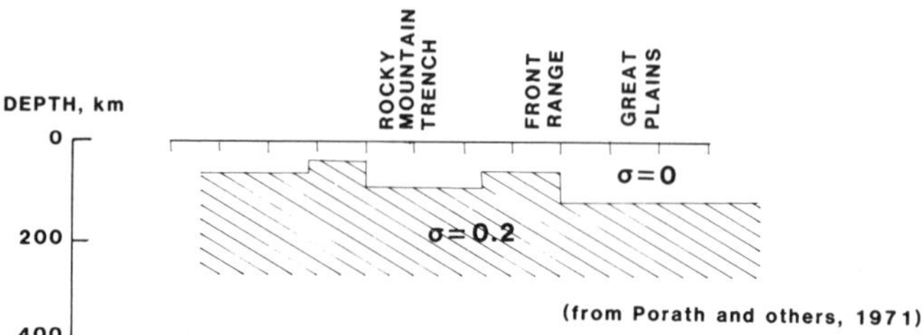

Figure 32. Interpreted conductivity cross section along the U.S.-Canadian border based on geomagnetic deep sounding data.

Caner and others (1967), reporting on a magnetic variation survey in western Canada, noted a transition from a western area of high conductivity in the mantle to an eastern area of lower conductivity. The transition zone is near the Rocky Mountain Trench.

Law and others (1980) have reported on magnetic variation studies carried out in central and western Washington. Stations near the Cascade volcanic zone showed anomalous behavior. It is thought that the anomaly is caused by a long but narrow conductive zone running north-south in close proximity to the volcanic trend. The conductor appears to terminate before reaching the Canadian border.

Northeastern United States and southeastern Canada

Greenhouse and Bailey (1981), Bailey and others (1974, 1978), and Cochran and Hyndman (1974) have described a magnetometer array study of an area extending from the St. Lawrence Seaway and Newfoundland on the north to Virginia–West Virginia on the south. Connerney and Kuckes (1980) reported on measurements made in the Adirondacks. In the north, the induction field is dominated by a strong coastal effect, suggesting that the rocks in the continental shelf are unusually conductive; recent drilling for oil and gas in this area have revealed thick sequences of conductive sediment offshore. Numeri-

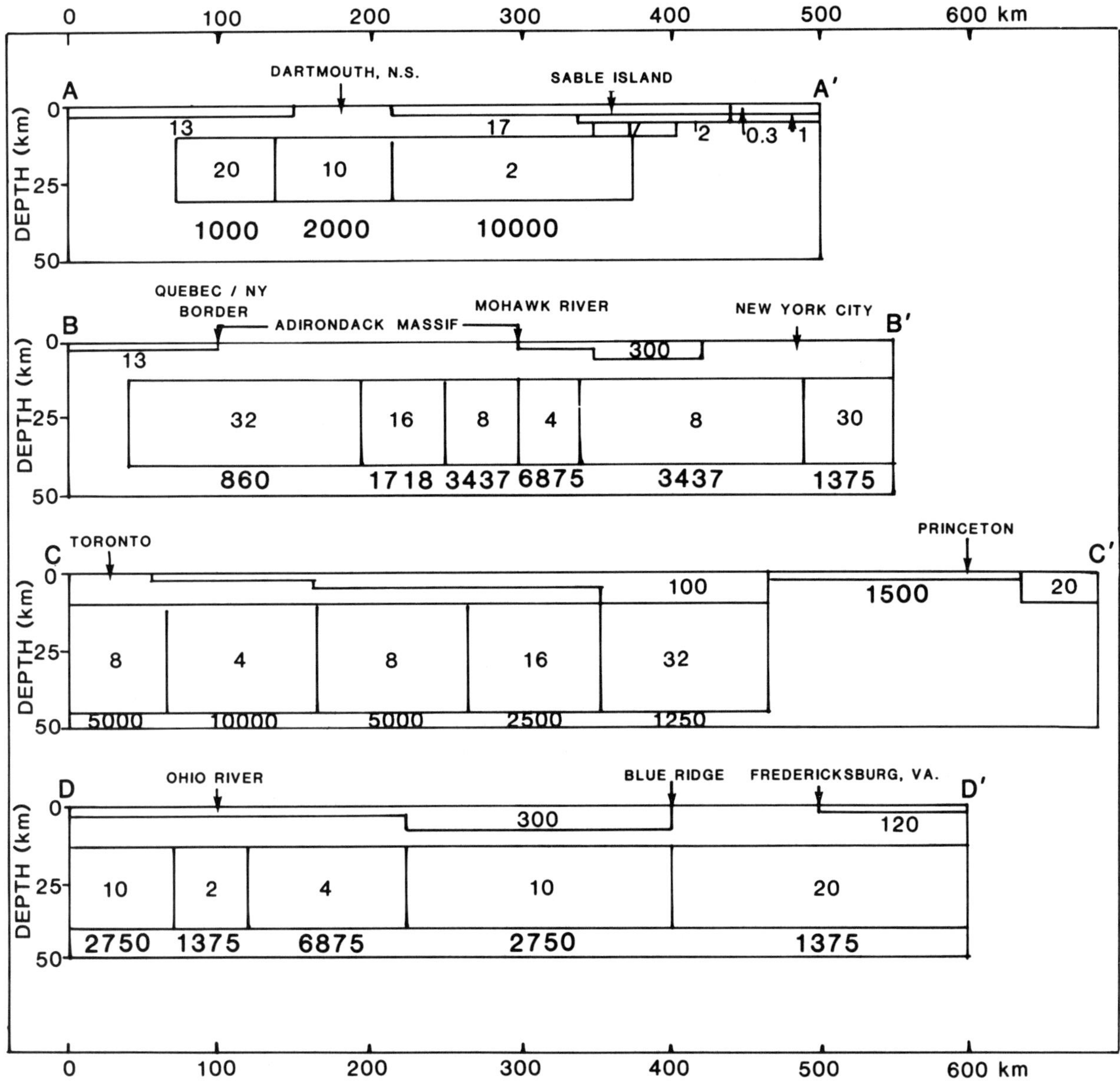

Figure 33. Finite difference cross sections that fit magneto-variational data in the northeastern United States (Greenhouse and Bailey, 1981). Numbers on cross sections are the resistivity values assigned to cells in the model.

cal models have been computed for sections as indicated in Figure 29 (profiles D-D′ through G-G′). These are shown in Figure 33.

DISCUSSION

Substantial information has been obtained on the electrical structure of the crust and mantle under the United States over the past several decades. Widely diverse methods have been used, and interpretations of data in terms of subsurface electrical structures have not been standardized. As a result, it is not surprising that it is difficult to recognize generalizations about the electrical structure.

First, let us examine the various conductivity profiles that have been observed with the magnetotelluric method. A comparison of profiles from many areas of the United States is given in Figure 34. From these data, it is clear that the extensional areas in the western United States, including the Basin and Range province and the Rio Grande Rift, are characterized by surprisingly high conductivity values at surprisingly shallow depths. To the west of the Basin and Range province, and to the east of the Front Range of the Rockies, as well as in the Colorado Plateau province, depths to conductive zones beneath the crust are greater. Not much information is available from the eastern regions of the U.S., but what few data there are show a considerably lower conductivity at shallow mantle depths than in any of the western areas.

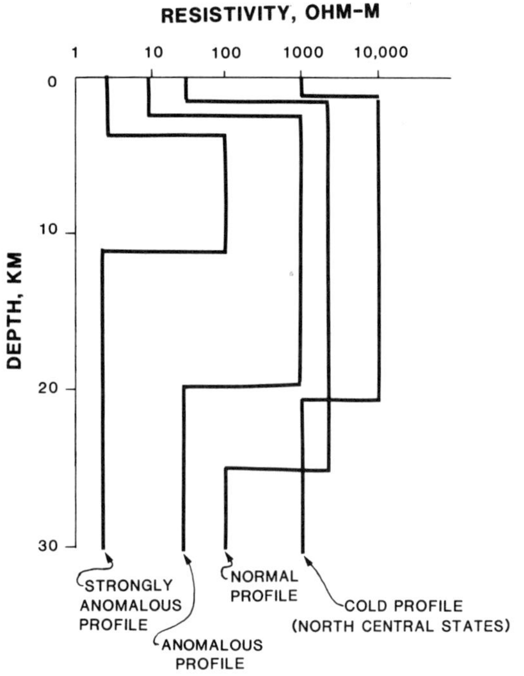

Figure 34. Idealized conductivity-depth profiles for various regions in the conterminous 48 United States.

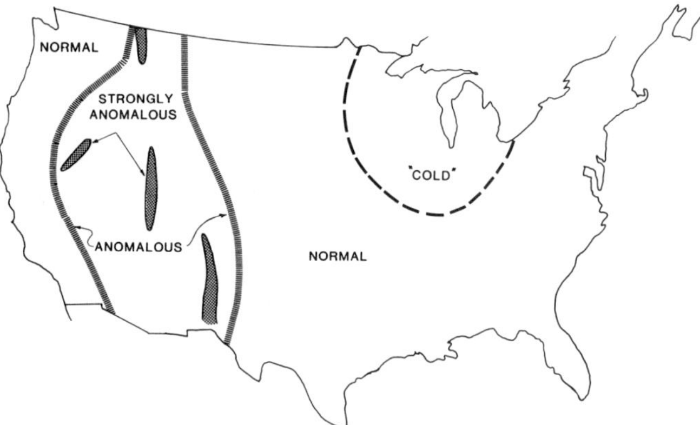

Figure 35. Probable geographic distribution of the idealized conductivity-depth profiles shown in Figure 34.

The MT and GDS methods are incapable of providing information on the resistive part of the crust, and the information available from DC surveys is sparse. Yet, what information there is indicates no similar marked difference in the resistant crust between the areas considered.

Two explanations have been advanced for the existence of high conductivity in the lower crust: high heat flow causing near or partial melting, and fracture formation with significant water content. Rarely, graphitic zones have been suggested as a cause. As noted earlier, both molten rock-forming minerals and highly saline water have conductivities in the same range, 1 to 100 S/m, so either can cause a rock conductivity of 0.1 S/m, when present in concentrations of 2 to 3 percent (see Fig. 3 in Part 1 of this study). A significant difference is that the two conductive materials exist over different temperature ranges; molten silicate minerals probably would not be present below temperatures of 800° to 1,200°C, while saline water probably would not be present above temperatures of 400° to 600°C. Therefore, the choice is between extensive regions of high heat flow which would cause the mantle or lower crust to reach temperatures of 800° to 1,200° at depths of only 15 to 25 km, or zones of very extensive microfracturing in the lower crust must exist at more or less normal temperatures (400° to 600°C). Both explanations seem to require a set of conditions in the lower crust that cannot be easily accepted.

Extensive fracturing at depth but not observed in the shallow crust requires some explanation. One normally expects crystalline rocks to form in equilibrium with their pressure and temperature, with no leftover void space, and then to be subject to progressive microfracturing as the pressure and temperature are reduced through weathering; this leads to differential expansion and contraction. However, G. R. Jiracek (personal communication, 1986) has suggested that fracturing can occur in a zone of plasticity in the lower crust, thus explaining the water content. Another possible explanation is through phase changes at the boundary between the mantle and the crust, which will generate porosity and yield liquid water to fill it. Therefore, in areas of crust formation, particularly if characterized by a regional rise, as is the case in the Basin and Range, magmatically underplated new crust (from mantle) will form a layer at the bottom of the old crust, which is porous and wet.

The mantle is believed to become progressively more conductive until the transition at 600 to 800 km is reached. However, the results of magnetic variation surveys indicate that profound lateral changes in the electrical properties of the upper mantle are present at depths as much as 350 km. The changes are characterized in GDS results by major conductive features, as shown in the summary map in Figure 35. The roots of the Rocky Mountains and the Wasatch front in particular appear to be reflected by important electrical structures extending to more than a depth of 300 km. This seems difficult to explain, when one expects the movement of the lithosphere to decouple the surface tectonic features from their original roots in the deeper mantle.

A further factor to consider is the lack of correspondence between many of the main features of the geoelectric section and the general features of the interior of the Earth, which have been developed from seismic and gravity data. The base of the sedimentary and weathered section at the surface of the crust is the only important boundary in the conductivity profile, which coincides with a corresponding seismic and density boundary. The other important geoelectric features of the crust and upper mantle seem discordant with the Earth's structure as we know it from seismic and gravity data. It is to be hoped that resolution of this problem will lead to a more complete understanding of the Earth.

REFERENCES CITED

Aiken, C.L.V., and Ander, M. E., 1981, A regional strategy for geothermal exploration with emphasis on gravity and magnetotellurics: Journal of Volcanology and Geothermal Research, v. 9, p. 1–27.

Alabi, A. O., Camfield, P. A., and Gough, D. I., 1975, The North American Central Plains conductivity anomaly: Geophysical Journal of the Royal Astronomical Society, v. 43, p. 815–833.

Ander, M. E., 1981, Magnetotelluric exploration in Arizona and New Mexico for hot dry rock geothermal energy using SQUID magnetometers, *in* Weinstock, H., and Overton, W. C., Jr., SQUID applications to geophysics: Society of Exploration Geophysicists, p. 61–65.

—— , 1984, A detailed magnetotelluric audiomagnetotelluric study of the Jemez volcanic zone, New Mexico: Journal of Geophysical Research, v. 89, no. B5, p. 3335–3353.

Ander, M. E., Goss, R., Strangway, D., Hillebrand, C., Laughlin, A. W., and Hudson, C., 1980, Magnetotelluric audiomagnetotelluric study of the Zuni hot dry rock prospect, New Mexico: Geothermal Resources Council Transactions, v. 4, p. 8.

Anderson, L. A., and Keller, G. V., 1966, Experimental deep resistivity probes in the central and eastern United States: Geophysics, v. 31, no. 6, p. 1105.

Bailey, R. C., Edwards, R. N., Garland, G. D., Kurtz, R., and Pitcher, D., 1974, Electrical conductivity studies over a tectonically active area in eastern Canada: Journal of Geomagnetism and Geoelectricity, v. 26, p. 125–146.

Bailey, R. C., Edwards, R. N., Garland, G. D., and Greenhouse, J. P., 1978, Geomagnetic sounding of eastern North America and the White Mountains heat flow anomaly: Geophysical Journal of the Royal Astronomical Society, v. 55, p. 499–502.

Bingham, D. K., Gough, D. I., and Ingham, M. R., 1985, Conductive structures under the Canadian Rocky Mountains: Canadian Journal of Earth Sciences, v. 22, p. 385–398.

Camfield, P. A., and Gough, D. I., 1975, Anomalies in daily variation fields and structure under northwestern United States and southwestern Canada: Geophysical Journal of the Royal Astronomical Society, v. 41, p. 193–218.

Camfield, P. A., Gough, D. I., and Porath, H., 1970, Magnetometer array studies in the northwestern United States and southwestern Canada: Geophysical Journal of the Royal Astronomical Society, v. 22, p. 201–221.

Caner, B., 1971, Quantitative interpretation of geomagnetic depth-sounding data in western Canada: Journal of Geophysical Research, v. 76, p. 7202–7216.

Caner, B., Cannon, W. H., and Livingston, C. E., 1967, Geomagnetic depth sounding and upper mantle structure in the Cordillera region of western North America: Journal of Geophysical Research, v. 72, p. 6335–6351.

Cantwell, T., Nelson, P., Webb, J., and Orange, A. S., 1965, Deep resistivity measurements in the Pacific Northwest: Journal of Geophysical Research, v. 70, no. 8, p. 1931–1937.

Chaipayungpun, W., and Landisman, M., 1977, Crust and upper mantle near the western edge of the Great Plains, *in* Heacock, J. G., ed., The earth's crust: American Geophysical Union Geophysics Monograph 20, p. 553–575.

Cochran, N. A., and Hyndman, R. D., 1974, Magnetotelluric and magnetovariational studies in Atlantic Canada: Geophysical Journal of the Royal Astronomical Society, v. 20, p. 385–406.

Connerney, J.E.P., and Kuckes, A. F., 1980, Gradient analysis of geomagnetic fluctuations in the Adirondacks: Journal of Geophysical Research, v. 85, p. 2615–2624.

Connerney, J.E.P., Nekut, A., and Kuckes, A. F., 1980, Deep crustal electrical conductivity in the Adirondacks: Journal of Geophysical Research, v. 85, p. 2603–2614.

Dowling, F. L., 1970, Magnetotelluric measurements across the Wisconsin Arch: Journal of Geophysical Research, v. 75, no. 14, p. 2683–2698.

Gamble, T. D., Goubau, W. M., Goldstein, N. E., and Clarke, J., 1980, Reference magnetotellurics at Cerro Prieto: Geothermics, v. 9, p. 49–63.

Gamble, T. D., Goubau, W. M., Goldstein, N. E., Miracky, R., Stark, M., and Clarke, J., 1981, Magnetotelluric studies at Cerro Prieto: Geothermics, v. 10, nos. 3-4, p. 169–182.

Goldstein, N. E., Mozley, E., Gamble, T. D., and Morrison, H. F., 1978, Magnetotelluric investigations at Mt. Hood, Oregon: Geothermal Resources Council Transactions, v. 2, p. 219–221.

Goldstein, N. E., Mozley, E., and Wilt, M., 1982, Interpretation of shallow electrical features from electromagnetic and magnetotelluric surveys at Mount Hood, Oregon: Journal of Geophysical Research, v. 87, no. B4, p. 2815–2828.

Gough, D. I., 1974, Electrical conductivity under western North America in relation to heat flow, seismology, and structure: Journal of Geomagnetism and Geoelectricity, v. 26, p. 105–123.

Gough, D. I., and Camfield, P. A., 1972, Convergent geophysical evidence of a metamorphic belt through the Black Hills of South Dakota: Journal of Geophysical Research, v. 77, p. 3168–3170.

Gough, D. I., Bingham, D. K., Ingham, M. R., and Alabi, A. O., 1982, A regional magnetometer array study: Canadian Journal of Earth Sciences, v. 19, p. 1680–1690.

Greenhouse, J. P., and Bailey, R. C., 1981, A review of geomagnetic variation measurements in the eastern United States; Implications for continental tectonics: Canadian Journal of Earth Sciences, v. 18, p. 1268–1289.

Group Seven, Inc., 1977, A survey of the geothermal potential of the Escalante desert, Utah: Group Seven, Inc., unpublished internal report, 36 p.

—— , 1980, Imperial Valley, California, magnetotelluric survey: Group Seven, Inc., unpublished internal report, 11 p., 2 plates.

—— , 1981a, Western Nevada magnetotelluric survey: Group Seven, Inc., unpublished internal report, 24 p., 6 plates.

—— , 1981b, Clear Lake, California, magnetotelluric survey: Group Seven, Inc., unpublished internal report 22 p., 6 plates.

—— , 1981c, Snake River Plain/southeast Oregon magnetotelluric survey: Group Seven, Inc., unpublished internal report, 24 p., 6 plates.

—— , 1982, Southwestern United States magnetotelluric survey: Group Seven, Inc., unpublished internal report, 24 p., 6 plates.

Hermance, J. F., and Pedersen, J., 1980, Deep structure of the Rio Grande Rift: Journal of Geophysical Research, v. 85, p. 3899–3912.

Jackson, D. B., 1966, Deep resistivity probes in the southwestern United States: Geophysics, v. 31, no. 6, p. 1123.

Jiracek, G. R., Ander, M. E., and Holcombe, H. T., 1979, Magnetotelluric soundings of crustal conductive zones in major continental rifts, *in* Riecker, R. E., ed., Rio Grande Rift; Tectonics and magmatism: American Geophysical Union, p. 209–222.

Keller, G. V., 1967, Electrical prospecting for oil: Quarterly Journal of the Colorado School of Mines, v. 63, no. 2, p. 1–268.

—— , 1968, Statistical study of electric fields from earth-return tests in the western states compared with natural electric fields: Institute of Electrical and Electronics Engineers Transactions PAS, v. PAS-87, no. 4, p. 1050–1057.

—— , 1987, Conductance map of the United States; Progress report: Physics of the Earth and Planetary Interiors, v. 45, p. 216–225.

Keller, G. V., and Furgerson, R. B., 1977, Determining the resistivity of a resistant layer in the crust, *in* Heacock, J. G., ed., The earth's crust: Geophysical: American Geophysical Union Monograph 20, p. 440–469.

Keller, G. V., Anderson, L. A., and Pritchard, J. I., 1966, Geological survey investigations of the electrical properties of the crust and upper mantle: Geophysics, v. 31, no. 6, p. 1078–1087.

Keshet, Y., and Hermance, J. F., 1986, A new regional electrical model for the southern section of the Rio Grande Rift, and the adjacent Basin and Range and Great Plains: Journal of Geophysical Research, v. 91, B6, p. 6359–6366.

Lajoie, J. J., and Caner, B., 1970, Geomagnetic induction anomaly near Kootenay Lake; A strip-like feature in the lower crust?: Canadian Journal of Earth Sciences, v. 7, p. 1568–1579.

Law, L. K., and Greenhouse, J. P., 1981, Geomagnetic variation sounding of the asthenosphere beneath the Juan de Fuca Ridge: Journal of Geophysical Research, v. 86, p. 967–978.

Law, L. K., Auld, D. R., and Booker, J. R., 1980, A geomagnetic variation anomaly coincident with the Cascade volcanic belt: Journal of Geophysical Research, v. 85, no. B10, p. 5297–5302.

Lienert, B. R., 1979, Crustal electrical conductivities along the eastern flank of the Sierra Nevadas: Geophysics, v. 44, no. 11, p. 1830–1845.

Lienert, B. R., and Bennett, D. J., 1977, High electrical conductivities in the lower crust of the northwestern Basin and Range; An application of inverse theory to controlled-source deep-magnetic-sounding experiment, *in* Heacock, J. G., ed., The earth's crust; Geophysics: American Geophysical Union Monograph 20, p. 531–552.

Mitchell, B. J., and Landisman, M., 1971, Electrical and seismic properties of the earth's crust in the southwestern Great Plains: Geophysics, v. 36, p. 368–381.

Pedersen, J., and Hermance, J. F., 1976, Towards resolving the absence or presence of an active magma chamber under the southern Rio Grande Rift: EOS Transactions of the American Geophysical Union, v. 57, p. 1014.

——, 1978, Evidence for molten material at shallow to intermediate crustal levels beneath the Rio Grande Rift at Santa Fe [abs.]: EOS Transactions of the American Geophysical Union, v. 59, p. 390.

——, 1981, Deep electrical structure of the Colorado Plateau as determined from magnetotelluric measurements: Journal of Geophysical Research, v. 86, no. B3, p. 1849–1857.

Pham, V. N., 1980, Magnetotelluric surveys of the Mount Meager region and the Squamish Valley, British Columbia: Ottawa, Ontario, Earth Physics Branch, Energy and Resources Open-File Report, 46 p.

Plouff, D., 1966, Magnetotelluric soundings in the southwestern United States: Geophysics, v. 31, no. 6, p. 1145–1152.

Porath, H., 1971a, Magnetic variation anomalies and seismic low-velocity zone in the western United States: Journal of Geophysical Research, v. 76, p. 2643–2648.

——, 1971b, A review of the evidence on low resistivity layers in the earth's crust, *in* Heacock, J. G., ed., The earth's crust; Geophysics: American Geophysical Union Geophysical Monograph 14, p. 127–144.

Porath, H., and Dziewonski, A., 1971a, Crustal resistivity anomalies from geomagnetic deep sounding studies: Reviews in Geophysics and Space Physics, v. 9, p. 891–915.

——, 1971b, Mantle conductive structures in the western United States from magnetometer array studies: Geophysical Journal of the Royal Astronomical Society, v. 19, p. 237–260.

Porath, H., and Gough, D. I., 1971, Mantle conductive structures in the western United States from magnetometer array studies: Geophysical Journal of the Royal Astronomical Society, v. 22, p. 261–275.

Porath, H., Oldenburg, D. W., and Gough, D. I., 1970, Separation of magnetic variation field and conductive structures in the western United States: Geophysical Journal of the Royal Astronomical Society, v. 19, p. 237–260.

Porath, H., Gough, D. I., and Camfield, P. A., 1971, Conductive structures in the northwestern United States and southwest Canada: Geophysical Journal of the Royal Astronomical Society, v. 23, p. 387–398.

Prieto, C., Perkins, C., and Berkman, E., 1985, Columbia River basalt plateau; An integrated approach to interpretation of basalt covered areas: Geophysics, v. 50, no. 12, p. 2709–2619.

Reitzel, J. S., Gough, D. I., Porath, H., and Anderson, C. W., 1970, Geomagnetic deep sounding and upper mantle structure in the western United States: Geophysical Journal of the Royal Astronomical Society, v. 19, p. 213–235.

Schmucker, U., 1964, Anomalies of geomagnetic variations in the southwestern United States: Journal of Geomagnetism and Geoelectricity, v. 15, p. 193–221.

——, 1984, Deep electromagnetic sounding in geothermal exploration, *in* Rapolla, A., and Keller, G. V., eds., Geophysics of geothermal areas; State of the art and future development: Golden, Colorado School of Mines Press, p. 73–126.

Skokan, C. K., 1980, Interpretation of long-line time-domain electromagnetic data from northwestern United States: Golden, Colorado School of Mines Report, 25 p.

Stanley, W.E.D., 1979, A regional magnetotelluric survey of the Cascade Mountains region: U.S. Geological Survey Open-File Report 7914, 122 p.

——, 1982, Magnetotelluric soundings on the Idaho National Engineering Laboratory facility, Idaho: Journal of Geophysical Research, v. 87, no. B4, p. 2683–2691.

——, 1984, Tectonic study of Cascade Range and Columbia Plateau in Washington state based upon magnetotelluric soundings: Journal of Geophysical Research, v. 89, no. B6, p. 4447–4460.

Stanley, W. D., Boehl, J. E., Bostick, F. X., and Smith, H. W., 1977, Geothermal significance of magnetotelluric sounding in the eastern Snake River Plain-Yellowstone region: Journal of Geophysical Research, v. 82, p. 2501–2514.

Sternberg, B. K., 1979, Electrical resistivity structure of the crust in the southern extension of the Canadian Shield; Layered earth models: Journal of Geophysical Research, v. 84, no. B1, p. 212–228.

Sternberg, B. K., and Clay, C. S., 1977, Flambeau anomaly; A high conductivity anomaly in the southern extension of the Canadian Shield, *in* Heacock, J. G., ed., The earth's crust; Geophysics: American Geophysical Union Geophysical Monograph 20, p. 501–530.

Swift, C. M., 1967, Magnetotelluric investigation of an electrical conductivity anomaly in the southwestern United States [M.S. thesis]: Cambridge, Massachusetts Institute of Technology, 211 p.

Thompson, B. G., Nekut, A., and Kuckes, A. F., 1983, A deep crustal electromagnetic sounding in the Georgia Piedmont: Journal of Geophysical Research, v. 88, no. B11, p. 9461–9473.

Towle, J. N., 1980, Observations of a direct current concentration on the eastern Sierran front; Evidence for shallow crustal conductors on the eastern Sierran front and beneath the Coso Range: Journal of Geophysical Research, v. 85, no. B5, p. 2484–2490.

Williston, McNeil, and Associates, 1979, Nevada National Test Site magnetotelluric survey: Williston, McNeil, and Associates, unpublished internal report, 88 p.

MANUSCRIPT ACCEPTED BY THE SOCIETY OCTOBER 31, 1988

Printed in U.S.A.

Geological Society of America
Memoir 172
1989

Chapter 21

Paleomagnetism of North America; The craton, its margins, and the Appalachian Belt

Rob Van der Voo
Department of Geological Sciences, University of Michigan, Ann Arbor, Michigan 48109-1063

ABSTRACT

This chapter examines, with a new set of reliability criteria, the data base of paleomagnetic poles available for the North American craton. If it is deemed desirable that all paleopoles used in the construction of an apparent polar wander path meet these criteria, then these criteria can be considered as very stringent; indeed, only four Phanerozoic results pass all seven criteria. On the other hand, more than 115 pass three or more of the criteria. Thus, the usefulness of these criteria lies in the flexibility with which they can be applied, the ease with which researchers can apply their own weighting of importance to the data base, and the ease with which a given polar wander path can be constructed (and constrained) for different purposes with different reliability filters.

The polar wander paths, so constructed, are only as interesting as the use that is being made of them. In this review, paleopoles obtained from the thrust belt margins of the craton are discussed in terms of rotations of thrust sheets and blocks such as in the Wyoming-Montana overthrust belt and the Colorado Plateau, whereas the Appalachian belt is discussed in terms of displaced and exotic terranes. New insights in the cratonic data base, particularly as they pertain to the now-recognized ubiquitous remagnetizations, have clarified some previously controversial issues, such as the hypothesis of a displaced Acadia terrane that was postulated to have moved during Carboniferous time. With the presently available reference poles, displacements can now be documented only for pre–Middle Devonian time. The Avalon basement terranes, in particular, have yielded a reliable data set that reveals their allochthonous nature for early Paleozoic time. It is paleomagnetically plausible to postulate that Avalon and Hercynian Europe constituted an Armorica plate that collided with the North American craton by Early Devonian time. Ordovician and older paleopoles from Avalon show high southerly paleolatitudes and have strong affinities with those for Hercynian Europe, as well as Gondwanaland, suggesting that Armorica was located adjacent to Gondwanaland before it broke off in Ordovician time and traveled northward.

INTRODUCTION

The applications of paleomagnetic techniques have evolved considerably in the last 20 yr. Whereas megascale substantiation of the drift of the major continents was one of the major goals in the 1950s and 1960s, paleomagnetic workers are now much more involved in mesoscale problems: the documentation and delineation of microplates and displaced terranes, the use of paleomagnetic vectors as structural markers, or the dating of regional diagenetic events through the paleomagnetic overprints that these events have left in the rock record.

Thus, and with great enthusiasm, has the paleomagnetic community begun to sample the rocks in the mobile belts, because it is precisely there that the effects of displaced terranes, deformation, and diagenesis are to be expected. Unfortunately, this has gradually resulted in an imbalance in the quality of the

Van der Voo, R., 1989, Paleomagnetism of North America; The craton, its margins, and the Appalachian Belt, *in* Pakiser, L. C., and Mooney, W. D., Geophysical framework of the continental United States: Boulder, Colorado, Geological Society of America Memoir 172.

data: for every pole or paleomagnetic direction to be used for dating diagenesis or to document displacement or tilt, a set of high-quality reference poles or directions is needed. In most cases these reference directions should be those of the craton, and in essence it is no overstatement to claim that a detailed and reliable cratonic apparent polar wander path is the first prerequisite for sound paleomagnetic interpretations of any sort. This seemingly innocuous statement actually hides a large problem: the craton, with its inherent lithospheric stability and typical lack of paleomagnetically suitable rocks, is the worst place for a paleomagnetic study. Not only are the preferred rock types (red beds, igneous intrusive, and extrusive rocks) generally lacking, there is also a dearth of possibilities to test the stability of the magnetization and to constrain its age through fold and conglomerate tests or reversal stratigraphy because of the generally incomplete nature of the sections. Thus, the first issue discussed in this review is the quality and availability of cratonic paleopoles.

Outside the stable cratonic interior—but where presumptions of coherence with the craton still exist (such as in the forelands and marginal fold belts of the Rocky Mountains, the Appalachians, and the Ouachita belt)—structural control becomes the overriding issue in any paleomagnetic study. For stratified rocks, a magnetization of prefolding age yields an inclination that can be restored with respect to the paleohorizontal regardless of any known or undetected rotations. Since paleolatitude is directly determined from the inclination, it, too, is independent of any rotation. Thus, irrespective of thrust sheet rotations or other structural complexities (such as introduced by plunging folds), the paleolatitudes obtained from rocks in thrust belt margins are likely to be representative of the craton. Issues related to the cratonic margins are discussed in the third section of this review.

Displaced terranes in the western Cordillera form the subject of a separate contribution by Beck (this volume), whereas those suspected for the Appalachians are discussed in this contribution. There has been much discussion recently about the sense, age, and location of any large lateral offsets of Late Paleozoic age in the northern Appalachians, but as shown in the fourth section, it appears that this controversial issue is being gradually resolved because of improved insights into the paleomagnetic data base.

REFERENCE POLES FROM THE CRATON

Reliability criteria

The most important prerequisite for modern paleomagnetic work is a reliable and reasonably detailed apparent polar wander path for the craton. For the Phanerozoic of North America, a cratonic pole path has existed from the very beginning of paleomagnetic studies (Runcorn, 1956; Irving, 1956; Collinson and Runcorn, 1960; Cox and Doell, 1960). In gross outline (i.e., with a precision of perhaps 20° of arc for the Mesozoic and 40° for the Paleozoic), the earlier paths are still valid. In detail, however, the cratonic path has been modified considerably over the years by the addition of new studies and, even more importantly, by the elimination of older studies that no longer pass modern reliability criteria. Unfortunately, these reliability criteria are not universally accepted; rather, they form a subjective high- or low-pass filter, with the severity of the criteria dependent on the goals to be achieved. Because of their subjective nature it must be explicitly stated that the criteria given below are strictly my own; moreover, very few poles pass all criteria. Thus, their application results in a relative quality index: given equal weight for each criterion, the "reliability" of a paleopole increases with increasing number of criteria satisfied.

The following seven criteria constitute the basis for my quality index:

1. *Well-determined age.* The rocks for which the paleopole is obtained should be dated with great precision, i.e., more than ± 20 Ma or ± 8 percent on a numerical (radiometric) scale for Paleozoic times. For stratigraphically dated rocks, ages should be defined within a series (e.g., Lower, Middle, or Upper Devonian). If ages are not known to within a period, results should not be used at all.

2. *Sufficient quantity.* The number of samples and sites studies should be sufficient to average out secular variation of the geomagnetic field and errors due to random sampling, orientation, and bedding measurement, and must provide a satisfactory basis for statistical treatment of the measured directions. Cones of confidence (α_{95}) or precision parameters (k), as a measure of such "errors," can be constrained to be within specified limits (e.g., McElhinny and Embleton, 1976). My own choice is for α_{95} to be less than 16° and for k to be more than 10.0, whereas the number of samples should be 25 or more.

3. *Demagnetization.* Laboratory demagnetizations and directional analysis must have been performed and be published in sufficient detail to convince others that magnetic overprints have been removed, and that uncontaminated, characteristic directions of magnetization have been isolated. If no demagnetization has been performed, results should not be used.

4. *Field tests.* Stability tests, particularly those that constrain the age of magnetization (fold-, contact-, conglomerate test), are essential for poles to be anchor points on the pole path. Such tests must be statistically significant for results to pass this criterion.

5. *Tectonic coherence.* Structural control for tilt and tectonic rotation, including a presumption of coherence with the craton or block involved, should be complete.

6. *Reversals.* The presence of reversals, although by no means proof of a primary magnetization, is a guarantee that enough time has lapsed during the acquisition process of that magnetization to: (1) preclude instantaneous ancient remagnetization; (2) average out short-lived nondipole components of the geomagnetic field (secular variation); and (3) average out a systematic bias, such as caused by a small, unrecognizable secondary overprint superimposed on antipodal characteristic directions.

7. *No suspicion of remagnetization.* This criterion is fulfilled when there is no similarity of a result to paleopoles for a younger age than the rocks under study. If this criterion is not fulfilled, then an ancient but younger remagnetization can hardly ever be

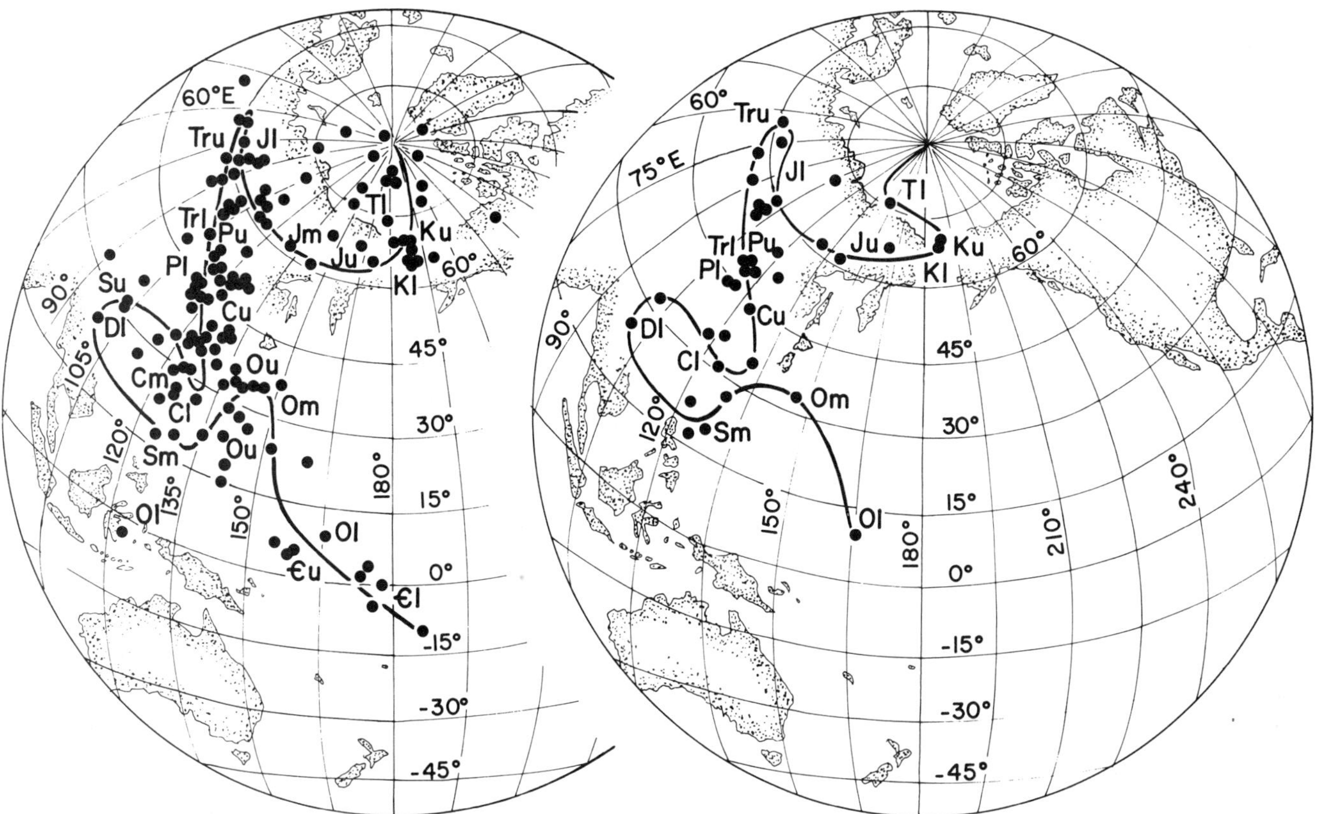

Figure 1. Apparent polar wander paths constructed for the North American craton on the basis of paleopoles that meet a minimum of three reliability criteria (left) or six reliability criteria (right), as explained in the text. The poles and number of criteria they satisfy are given in Table 1.

precluded. If it is suspected that a pole is based on a remagnetization, it should be included in a pole list only if the age constraints on the (re)magnetization are established by independent means.

In a list of Phanerozoic paleopoles for cratonic North America (Table 1), it can be seen at a glance that almost no paleopoles satisfy all seven criteria. For Paleozoic time, for example, only two (Lower Carboniferous and Silurian) poles have a quality index of 7.

With a quality index (Q) based on how many of these seven criteria are met, it becomes possible to construct apparent polar wander paths according to a specified minimum Q. Examples of pole paths with different Qs are given in Figure 1. It is clear that for very high Q—i.e., on the most stringent basis—there are very few poles available, which precludes pole-path construction techniques that rely on an abundant data base, such as the moving-window averaging of Irving and Irving (1982). It is a matter of debate whether it is better to use a large data base (but with a low-Q pass filter) under the assumption that errors are random and will average out, or use a sparse but highly reliable data base. I prefer to construct paths based on the more reliable poles only, because I believe that errors will not be random and that the bias thus introduced is typically greater than the purely statistical confidence limits associated with the pole path. However, a high Q is no guarantee that a paleopole is "perfect," a problem exemplified by the Late Devonian paleopoles discussed later.

The Remagnetization Problem

The presence of remagnetizations in North American rocks was emphasized early by Creer (1968), but awareness of this possibility has increased enormously in the last five years, and consequently, has caused a further reexamination of the existing data base. Several critical formations, especially those of Paleozoic age, have been resampled. As a result, the available data base for pole path construction has decreased: some poles were found to be based on multivectorial remanent magnetizations and were not previously recognized as such despite demagnetization analysis (e.g., Kent and Opdyke, 1985); others have now been shown to be due to complete remagnetization in times much younger than the rocks themselves (e.g., McCabe and others, 1984). Intuitively, it seems clear that the older the rock formation studied, the higher the chance that such remagnetizations have occurred. Nevertheless, Mesozoic and Cenozoic formations with no obvious signs of even low-grade metamorphism are now also known to be partially or completely remagnetized (e.g., Schwartz and Van der Voo, 1984; Reynolds and others, 1985; Hornafius, 1985), and it appears that tests that preclude remagnetization or constrain the age of magnetization will assume an ever-increasing importance.

The Paleozoic carbonate rocks of the Appalachians and cratonic interior have been particularly disappointing. Magnetiza-

TABLE 1. PHANEROZOIC POLE POSITIONS FOR THE NORTH AMERICAN CRATON

Rock Unit, Location	Age	Pole Position Lat, Long	k	α_{95}	Reliability 1	2	3	4	5	6	7	Q	Reference
Cenozoic (Paleocene, Eocene) Results													
Beaverhead Flows, Montana	Tl	66N,239E	15	6		x	x				x	3	Hanna (1967)
Flagstaff Intrusives, Colorado	Tl	68N,189E	8	14			x	x			x	3	McMahon and Strangway (1968)
Spanish Peaks Dikes, Colorado, New Mexico	25	81N,211E	6	13			x		x			*	Larson and Strnagway (1969)
Mistastin Lake Volcanics, Labrador	39	86N,118E		2	x	x	x		x			4	Currie and Larochelle (1969)
Monterey Intrusives, Virginia	47	88N,46E	52	9	x	x	x		x	x		5	Lovlie and Opdyke (1974)
Monterey Intrusives, Virginia, West Virginia	47	86N,244E	21	10	x	x	x		x			4	Ressetar and Martin (1980)
Absaroka Flows, Wyoming	48	84N,177E	15	8	x	x	x		x	x		5	Shive and Pruss (1977)
Rattlesnake Hills Volcanics, Wyoming	48	79N,146E	14	10	x	x	x		x			4	Sheriff and Shive (1980)
Combined Eocene Intrusives, Montana	51	82N,170E	19	4	x	x	x		x	x		5	Diehl and others (1980)
Cape Dyer Basalts, Baffin Island	58	83N,305E	155	6	x		x			x		3	Deutsch and others (1971)
Golden Basalt Flow 2, Colorado	59	81N,64E	9	11	x		x		x			3	Larson and others (1969)
Nacimiento Formation, New Mexico	59	76N,148E	22	3	x	x	x		x	x	x	6	Butler and Taylor (1978)
Gringo Gulch Volcanics, Arizona	62	77N,201E	678	1	x	x	x				x	4	Barnes and Butler (1980)
Combined Paleocene Intrusives, Montana	64	82N,181E	20	5	x	x	x		x	x		5	Diehl and others (1980)
Mesozoic Results													
Roskruge Volcanics, Arizona	72	74N,176E		6	x	x	x			x	x	5	Vugteveen and others (1981)
Elkhorn Volcanics, Montana	78	69N,189E	60	5	x	x	x	x		x	x	6	Hanna (1967, 1973)
Mesaverde Group, Wyoming, Utah	83	65N,198E	4	13	x		x		x		x	4	Kilbourne (1969)
Niobrara Formation, Wyoming, Colorado, Kansas	87	64N,188E	170	4	x	x	x		x		x	5	Shive and Frerichs (1974)
Magnet Cove Intrusions, Arkansas	98	65N,187E		9	x	x	x		x		x	5	Sharon and Hsü (1969)
Isachsen Diabase, Northwest Territories	109	69N,180E	20	8	x		x		x		x	4	Larochelle and others (1965)
Monteregian Hills Intrusion, Quebec	126	71N,190E		2	x	x	x		x	x	x	6	Larochelle (1969)
Mt. Ascutney Gabbro, Vermont	130	64N,187E	335	14	x		x		x		x	4	Opdyke and Wensink (1966)
Upper Morrison Formation, Colorado	149	68N,162E	663	4	x	x	x		x	x	x	6	Steiner and Helsley (1975)
Stump Formation, Wyoming	150	64N,170E	18	15	x	x	x	x				4	Schwartz and Van der Voo (1984)
Canelo Hills Volcanics, Arizona	151	63N,132E	33	6	x	x	x	x		x	x	6	Kluth and others (1982)
Lower Morrison Formation, Colorado	154	61N,142E	549	5	x	x	x		x	x	x	6	Steiner and Helsley (1975)
White Mountain Intrusion, Vermont, New Hampshire	162	79N,167E			x	x	x		x	x		*	Van Alstine (1979)
Summerville Sandstone, Utah	165	67N,110E	51	4	x		x		x	x	x	5	Steiner (1978)
Corral Canyon Rocks, Arizona	172	62N,116E	50	6	x	x	x			x	x	5	May and others (1986)
Twin Creek Limestone, Wyoming	172	68N,145E	32	9	x	x	x	x				4	McCabe and others (1982)
Diabase Dikes, Anticosti Island	178	76N,85E		1	x		x		x		x	4	Larochelle (1971)
Newark Volcanics II, Connecticut-Maryland	179	65N,103E	92	1	x	x	x		x		x	5	Smith and Noltimier (1979)
New Jersey Volcanics, New Jersey	190	63N,108E	49	4	x	x	x		x		x	5	Opdyke (1961)
Connecticut Valley Volcanics, Connecticut, Massachusetts	193	65N,87E	31	11	x	x	x		x		x	5	De Boer (1968)
Piedmont Dikes, Georgia, South Carolina, North Carolina	195	66N,86E	49	3	x	x	x				x	4	Dooley and Smith (1982)

TABLE 1. PHANEROZOIC POLE POSITIONS FOR THE NORTH AMERICAN CRATON (continued)

Rock Unit, Location	Age	Pole Position Lat, Long	k	α_{95}	Reliability 1	2	3	4	5	6	7	Q	Reference
Mesozoic Results (continued)													
Newark Volcanics I, Connecticut, Maryland	195	63N,83E	56	2	x	x	x		x		x	5	Smith and Noltimier (1979)
Kayenta Formation, Utah	197	61N,83E	51	10	x	x	x		x		x	5	Johnson and Nairn (1972)
Kayenta Formation, Utah	197	62N,75E	81	7	x	x	x		x	x	x	6	Steiner and Helsley (1974)
Nova Scotia Basic Rocks, combined	200	72N,105E	149	7	x	x	x	x	x		x	6	Irving and others (1976): 8.165
Wingate Formation, Utah	203	59N,63E		8	x	x	x		x		x	5	Reeve (1975)
Diabases, Pennsylvania	204	62N,105E	118	2	x	x	x		x		x	5	Beck (1972)
Holyoke, Granby Flows, Massachusetts	204	55N,88E	41	11	x		x		x		x	4	Irving and Banks (1961)
Newark Basin, West Limb, New Jersey	206	51N,49E	58	13	x	x	x	x			x	5	Van Fossen and others (1986)
Manicouagan Structure, Quebec	215	60N,89E	58	5	x	x	x		x		x	5	Irving and others (1976): 8.135
Newark Basin, East Limb, New Jersey	215	62N,115E	41	11	x	x	c		c			4	Van Fossen and others (1986)
Chinle Formation, Utah	222	61N,64E	892	4	x	x	x	x	x	x	x	7	Reeve (1975)
Chinle Formation, New Mexico	222	58N,79E	62	4	x	x	x		x	x	x	6	Reeve and Helsley (1972)
Upper Maroon Formation, Colorado	Trl	60N,102E	26	9		x	x	x	x	x	x	6	McMahon and Strangway (1968)
Moenkopi Formation, Utah	Trl	57N,89E	85	5	x	x	x		x	x	x	6	Helsley (1969)
Upper Moenkopi Formation, Colorado	Trl	55N,103E	102	3	x	x	x		x	x	x	6	Helsley and Steiner (1974)
Upper Moenkopi drillcore, Colorado	Trl	57N,100E	368	6	x	x	x		x	x	x	6	Baag and Helsley (1974)
Moenkopi Formation, Arizona	Trl	58N,101E	10	2	x	x	x	x	x	x	x	7	Purucker and others (1980)
Chugwater Formation, Wyoming	Trl	49N,112E	78	6	x	x	x		x	x	x	6	Van der Voo and Grubbs (1977)
Chugwater Formation, Wyoming	Trl	47N,114E	999	2	x	x	x		x	x	x	6	Shive and others (1984)
Chugwater Formation, Wyoming	Trl	45N,115E	47	4	x	x	x		x	x	x	6	Herrero-Bervera and Helsley (1983)
Ankareh Formation, Wyoming	Trl	51N,105E	143	4		x	x		x		x	4	Grubbs and Van der Voo (1976)
Paleozoic Results													
Hoskinnini Tongue, Arizona	P	50N,121E	23	7			x		x		x	3	Farrell and May (1969)
Basic Sill, Prince Edward Island	Pu	51N,111E			x		x		x		x	4	Larochelle (1967)
Ochoan Redbeds, Southwest U.S.	Pu	55N,119E	39	15	x	x	x		x	x	x	6	Peterson and Nairn (1971)
Guadalupian Redbeds, Southwestern U.S.	Pu	51N,125E	214	5	x	x	x		x	x	x	6	Peterson and Nairn (1971)
Pegmatite dikes, Connecticut	255	35N,126E	63	10	x	x			x		x	*	De Boer and Brookins (1972)
Toroweap Formation, Arizona	Pu	47N,103E	36	7			x		x		x	3	Farrell and May (1969)
Elephant Canyon Formation, Utah	Pl	44N,120E	27	5	x	x	x		x		x	5	Gose and Helsley (1972)
Ingelside Formation, Colorado	Pl	46N,122E	147	2	x	x	x		x		x	5	Diehl and Shive (1979)
Cutler Formation, Colorado	Pl	44N,120E		6	x	x	x		x		x	5	Helsley (1971)
Cutler Formation, Utah	Pl	44N,116E	21	2	x	x	x		x		x	5	Gose and Helsley (1972)
Fountain and Lykins Formations, Colorado	Pl	48N,119E	6	13	x		x	x	x		x	5	McMahon and Strangway (1968)
Minturn and Maroon Formations, Colorado	Pl	43N,115E	19	3	x	x	x		x		x	5	Miller and Opdyke (1985)
Leonardian Redbeds, Southwest U.S.	Pl	45N,115E	20	15	x	x	x		x	x	x	6	Peterson and Nairn (1971)
Casper Formation, Wyoming	Pl	51N,123E	129	2	x	x	x		x		x	5	Diehl and Shive (1981)
Wolfcampian Redbeds, Southwest U.S.	Pl	41N,118E	46	11	x	x	x		x		x	5	Peterson and Nairn (1971)
Helderberg Synfolding, New York–Virginia	Pl	49N,115E	64	–	x	x	x	x	x		x	6	Scotese (1985)
Mauch Chunk Synfolding, Pennsylvania	Pl	51N,118E	35	5	x	x	x	x			x	5	Kent and Opdyke (1985)
Dunkard Formation, Virginia	Cu	44N,123E	162	5	x	x	x		x	x	x	6	Helsley (1965)
Casper Formation, Wyoming	Cu	46N,129E	40	2	x	x	x		x		x	5	Diehl and Shive (1981)
Mason Creek, Illinois	Cu	37N,131E	45	8	x	x	x		x		x	5	Scotese (1985)
Buffalo Siltstone, Pennsylvania	Cu	27N,123E	13	6	x		x		x		x	*	Payne and others (1981)

TABLE 1. PHANEROZOIC POLE POSITIONS FOR THE NORTH AMERICAN CRATON (continued)

Rock Unit, Location	Age	Pole Position Lat, Long	k	α_{95}	Reliability 1	2	3	4	5	6	7	Q	Reference
Paleozoic Results(continued)													
Bush Creek Limestone, Pennsylvania	Cu	36N,124E	13	4	x		x		x		x	4	Payne and others (1981)
Prince Edward Island Redbeds	Cu	42N,133E	36	6	x	x	x				x	4	Roy (1966)
Prince Edward Island Redbeds	Cu	41N,126E	433	6	x	x	x				x	4	Black (1964)
Tormentine Formation, Prince Edward Island	Cu	41N,132E	107	4	x		x				x	3	Roy (1966)
Hurley Creek Formation, New Bunswick	Cu	39N,125E	58	10	x		x				x	3	Roy and others (1968)
Bonaventure Formation, New Brunswick	Cu	38N,133E	20	10	x		x				x	3	Roy (1966)
Cumberland Group, Nova Scotia	Cu	36N,125E	85	5	x	x	x	x			x	5	Roy (1969)
Morien Group, Nova Scotia	Cu	40N,131E	32	6	x	x	x	x			x	5	Scotese and others (1984)
Riversdale Group, Nova Scotia	Cu	36N,122E	91	6	x	x	x			x	x	5	Roy (1977)
Pomquet and Lismore Formations, Nova Scotia	Cm	29N,122E			x		x			x	x	4	Scotese and others (1984)
Barachois Group, West Newfoundland	Cm	34N,143E	107	7	x	x	x	x	x		x	6	Murthy (1985)
New Brunswick Volcanics	Cm	21N,135E	49	10	x	x	x			x	x	5	Seguin and others (1985)
Minudie Point, Nova Scotia	Cm	36N,122E	91	6	x	x	x			x	x	5	Roy (1977)
Hopewell Group, New Brunswick	Cm	34N,118E	32	7	x	x	x	x		x	x	6	Roy and Park (1969)
Cheverie Formation, Nova Scotia	Cl	24N,152E	59	6	x	x	x	x			x	*	Spariosu and others (1984)
Maringouin and Shepody, Nova Scotia	Cl	36N,122E	89	3	x	x	x	x		x	x	6	Roy and Park (1974)
Mauch Chunk Formation, Pennsylvania	Cl	31N,125E	16	8	x	x	x	x		x	x	6	Kent and Opdyke (1985)
Deer Lake Formation, West Newfoundland	Cl	22N,122E	40	9	x	x	x	x	x	x	x	7	Irving and Strong (1984)
Spout Falls Formation, West Newfoundland	Cl	29N,140E	61	7	x	x	x		x		x	5	Murthy (1985)
Windsor Group, Nova Scotia	Cl	36N,137E	118	6	x	x	x	x			x	5	Scotese and others (1984)
Horton Group, Nova Scotia	Cl	32N,136E	87	8	x	x	x					3	Scotese and others (1984)
Jeffreys Village Member, West Newfoundland	Cl	27N,131E	54	8	x	x	x		x	x	x	6	Murthy (1985)
Catskill Redbeds, New York	Du	47N,117E	116	5	x	x	x		x			*	Kent and Opdyke (1978)
Catskill Redbeds, Pennsylvania, Maryland	Du	44N,124E	109	5	x	x	x	x		x		*	Van der Voo and others (1979)
Catskill Redbeds, south, Pennsylvania	Du	26N,124E	16	16	x		x			x	x	4	Miller and Kent (1986a)
Catskill Redbeds, north, Pennsylvania	Du	33N,90E	165	7	x	x	x	x			x	5	Miller and Kent (1986b)
Peel Sound Formation, Northwest Territories	D1	25N,99E	66	9	x	x	x		x	x	x	6	Dankers (1982)
Bloomsburg Formation, Pennsylvania	Su	32N,102E	35	9	x	x	x	x		x	x	6	Roy and others (1967)
Rose Hill Formation, Pennsylvania-Virginia	Sm	19N,129E	18	6	x	x	x	x		x	x	6	French and Van der Voo (1979)
Wabash Reef Limestone, Indiana	Sm	17N,125E	74	5	x	x	x	x	x	x	x	7	McCabe and others (1985)
Beemerville Complex, New Jersey	435	35N,126E	23	20	x							1	Proko and Hargraves (1973)
Ringgold Gap sediments, Georgia	Sl/Ou	28N,142E	62	7	x	x	x			x	x	5	Morrison and Ellwood (1986)
Cordova Secondary, Ontario	446	31N,102E	18	11	x	x	x		x	x		5	Dunlop and Stirling (1985)
Juniata Formation, Pennsylvania-Virginia	Ou	32N,114E	53	5	x	x	x	x		x		5	Van der Voo and French (1977)
Steel Mountain Secondary, West Newfoundland	451	23N,139E	22	13		x	x		x			3	Murthy and Rao (1976)
Chapman Ridge Formation, Tennessee	Om	27N,112E	38	15	x		x	x				3	Watts and Van der Voo (1979)
Martinsburg Shale, Pennsylvania	Om	36N,160E	5	14	x		x			x	x	*	Hower (1979)

TABLE 1. PHANEROZOIC POLE POSITIONS FOR THE NORTH AMERICAN CRATON (continued)

Rock Unit, Location	Age	Pole Position Lat, Long	k	α_{95}	Reliability 1	2	3	4	5	6	7	Q	Reference
Paleozoic Results(continued)													
Moccasin-Bays Formations, Tennessee	Om	33N,147E	135	6	x	x	x	x		x	x	6	Watts and Van der Voo (1979)
St. George Formation, West Newfoundland	Ol	30N,119E	910	3	x				x			2	Deutsch and Rao (1977a)
St. George Formation, West Newfoundland	Ol	26N,126E	202	7	x	x	x		x			4	Beales and others (1974)
Oneota Dolomite, Iowa, Minnesota, Wisconsin	O1	10N,166E	18	12	x	x	x		x	x	x	6	Jackson and Van der Voo (1985)
Buckingham Dike, Quebec	497	3S,123E		7		x	x	x	x		x	5	Dankers and Lapointe (1981)
Black, Unaweap Canyon dikes, Colorado	497	37N,102E		9	x	x				x		3	Larson and others (1985)
Lamotte Formation (recalculated), Missouri	Єui	1N,168E			x				x		x	*	Van der Voo and others (1976)
Point Peak Member (Wilberns Formation) Texas	Єu	6N,159E	171	5	x	x	x		x		x	5	Van der Voo and others (1976)
Morgan Creek–Welge (Wilberns) Texas	Єu	24N,151E	20	4	x	x	x		x		x	5	Watts and others (1980a)
Taum Sauk Limestone (Bonneterre), Missouri	Єu	4S,176E	12	7	x	x	x		x		x	5	Dunn and Elmore (1985)
Lion Mountain Member (Riley Formation) Texas	Єu	27N,146E	13	10	x		x		x			3	Watts and others (1980a)
Cap Mountain Member (Riley Formation) Texas	Єu	33N,140E	8	8	x		x		x	x		4	Watts and others (1980a)
Hickory Member (Riley Formation) Texas	Cm	34N,145E	12	4	x	x	x		x			4	Watts and others (1980a)
Wichita Granites, Oklahoma	525	4N,164E	3	18	x						x	*	Spall (1968, 1970)
McClure Mountain Complex, Colorado	535	18N,142E			x	x					x	3	Lynnes and Van der Voo (1984)
Colorado Intrusives I	Є	15N,142E	22	10		z	z			z	z	4	French and others (1977)
Colorado Intrusives II	Є	5N,174E	39	8			x			x	x	3	French and others (1977)
Colorado Intrusives III	Є	48N,107E	16	15			x			x		2	French and others (1977)
Rome-Waynesboro Formations, Tennessee-Maryland	Єm	36N,150E	21	10	x	x	x	x				4	Watts and others (1980b)
Tapeats Sandstone, Arizona	Єll	5N,158E	15	3	x	x	x		x	x		5	Elston and Bressler (1977)
Bradore Sandstone, West Newfoundland	Єl	23N,160E	110	5	x	x	x					3	Deutsch and Rao (1977b)
Unicoi Basalts, Virginia, Tennessee	Єl	0N,178E	24	14	x		x			x		3	Brown and Van der Voo (1982)
Buckingham flows I, Quebec	572	6N,154E		14	x		x					2	Dankers and Lapointe (1981)
Buckingham flows II, Quebec	573	1N,173E		6	x	x	x					3	Dankers and Lapointe (1981)
Buckingham flows III, Quebec	574	10S,186E		13	x		x				x	3	Dankers and Lapointe (1981)

Note:

The entries in this table are for the North American craton, including its thrust belt margins insofar as no suspicions exist about relative displacements; obviously, rotations can occur there and this mandates that reliability criterion 5 cannot be checked as being met. The craton outline used is that of Van der Voo (1981), while for Carboniferous and younger time New England and the Canadian Maritime provinces have also been included.

The reliability criteria (1-7) and Q are explained in the text. An asterisk is placed in the Q column when the result is not to be used for apparent polar wander path construction, as explained in the text; an "x" mark under any of the seven reliability columns means that this particular criterion is being met.

Abbreviations: T = Tertiary, Tr = Triassic, P = Permian, C = Carboniferous, D = Devonian, S = Silurian, O = Ordovician, Є = Cambrian, u = upper, m = middle, l = lower; Lat = latitude, Long = longitude; k and α_{95} are the precision parameter and radius of the cone of 95 percent confidence, respectively (Fisher, 1953).

tions as much as hundreds of millions of years younger than the rocks themselves have frequently been obtained. The carriers of the remagnetizations are magnetites, which is a common carrier for unremagnetized rocks as well, and the rocks cannot be distinguished macroscopically. Therefore, it appears to be impossible thus far to predict whether a given formation is remagnetized until a complete paleomagnetic laboratory study has been performed in a time-consuming and somewhat unrewarding effort. Table 2 presents a list of all cratonic Paleozoic carbonate formations that are inferred to have been remagnetized. Whether the remagnetizations occurred through viscous remanence acquisition during a period of ancient uplift and cooling from only moderate temperatures (Kent, 1985), with the carriers of the magnetization being completely reset, or whether the carriers grew authigenically and thus acquired a chemical remanent magnetization (McCabe and others, 1983) at a time much later than deposition cannot be decided at this time. When the paleopoles for the remagnetized carbonate rocks are plotted (Fig. 2) and compared with the pole path of Figure 1, it becomes clear that the remagnetizations are all of late Paleozoic age. The locations of the remagnetized rocks (Fig. 3) are widespread and often far-removed from the cratonic margins. This raises interesting speculations about the cause of the remagnetizations; if the magnetic carriers can be shown to be chemically grown at the time the magnetizations were acquired, fluids migrating over large distances could be held responsible (e.g., Oliver, 1986). The basis for the following discussion (Table 1) of the data base for Mesozoic, Paleozoic, and Precambrian times is a compilation of published paleopoles, based on demagnetization studies with a Q of 3 or more, or previously included in selective pole lists.

Mesozoic and early Tertiary reference poles

Compilations of selected paleopoles for all or part of the Mesozoic and Cenozoic have been published at regular intervals (McElhinny, 1973; Van der Voo and French, 1974; Steiner, 1978; Vugteveen and others, 1981; Van der Voo, 1981; Irving and Irving, 1982; Harrison and Lindh, 1982; Kluth and others, 1982; Diehl and others, 1983; Gordon and others, 1984; Shive and others, 1984; Van der Voo and others, 1984; May and Butler, 1986; see also Gordon, 1984). The paleopoles selected for this chapter (Table 1) are either very recently published or have been obtained from such previous lists.

For the periods of Mesozoic and Cenozoic time, the available paleopoles have always been fairly abundant, with the exception of the Jurassic. However, through recent work by paleomagnetists at the University of Arizona (Kluth and others, 1982; May and Butler, 1986), the Jurassic polar wander path

TABLE 2. PALEOMAGNETIC DATA FROM COMPLETELY REMAGNETIZED PALEOZOIC CARBONATE ROCKS OF NORTH AMERICA*

Rock Unit, Location	Rock Age	Paleopole	k	α_{95}	Reference
St. Joe Limestone, Arkansas	Cl	39N,132E	–	–	Scott (1979)
Barnett Formation, Texas	Cl	49N,119E	905	3	Kent and Opdyke (1979)
Leadville Formation, Colorado	Cl	46N,123E	86	9	Horton and others (1984)
Salem Limestone, Indiana	Cl	45N,125E	–	–	C. McCabe (personal communication, 1986)
Greenbrier Group, Pennsylvania	Cl	43N,131E	62	8	Chen and Schmidt (1984)
Columbus and Delaware Limestone, Ohio	Du	46N,120E	41	2	Martin (1975)
Martin Formation, Arizona	Du	56N,109E	66	2	Elston and Bressler (1977)
Temple Butte Formation, Arizona	Du	53N,115E	83	2	Elston and Bressler (1977)
Tully Limestone, New York	Dm	49N,116E	86	4	Kent (1985)
Onondaga Limestone, New York	Dl	40N,121E	39	8	Kent (1979)
Helderberg sequence, New York–Virginia	Su–Dl	49N,115E	64	–	Scotese (1985)
Mayville Dolomite, Wisconsin	Sl	37N,122E	39	8	Kean and Voltz (1989)
Trenton Limestone, New York-Quebec	Ou	53N,127E	98	3	McCabe and others (1984)
Ouachita Carbonates, Oklahoma	Ol	46N,129E	66	3	Steiner (1973)
Kindblade Formation, Oklahoma	Ol	43N,128E	306	4	Elmore and others (1985)
Allentown Dolomite, Pennsylvania	€u	47N,114E	81	8	Stead and Kodama (1984)
Bonneterre Dolomite, Missouri	€u	43N,126E	120	5	Wisniowiecki and others (1983)
Nolichucky Formation, Tennessee	€u	40N,120E	-	–	Gillett (1982)
Muav Limestone, Arizona	€m	55N,110E	41	3	Elston and Bressler (1977)
Leithsville Formation, Pennsylvania	€m	53N,113E	95	8	Stead and Kodama (1984)
Abrigo Formation, Arizona	€m	59N,89E	302	3	Elston and Bressler (1977)

*For explanation of abbreviations, see Table 1.

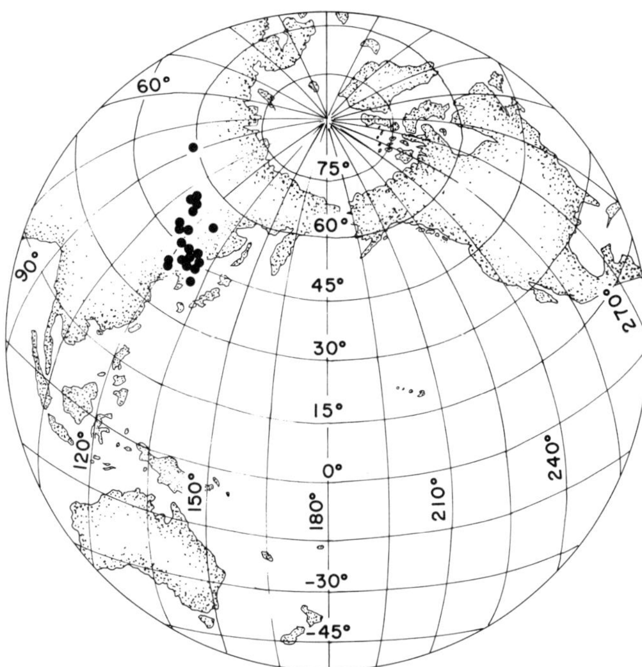

Figure 2. Paleopoles (Table 2) from remagnetized Paleozoic carbonate rocks of the North American cratonic interior.

segment is now well defined in spatial coordinates and reasonably well calibrated in time. Jurassic paleopoles published in the 1960s, such as those for the White Mountain intrusives (Opdyke and Wensink, 1966; see also Van Alstine, 1979), are now regarded as unreliable. While these particular poles still satisfy many reliability criteria, they are commonly regarded as contaminated by an unrecognized present-day overprint; such poles have been marked by an asterisk in Table 1, indicating that they should not be used for pole path construction. May and Butler (1986) have noted that the newly added paleopoles for the Jurassic define paleolatitudes for cratonic North America that are more southerly than previously thought (e.g., Irving and Irving, 1982); this, in turn, has serious implications for displaced-terrane hypotheses and models dealing with interactions between the North American craton and Pacific oceanic plates (see Beck, this volume).

For the Triassic, a large number of sedimentary (mostly red bed) sequences have yielded paleomagnetic results of good quality. However, the acquisition age of the magnetization has been a matter of debate. Because many studies were carried out for magnetostratigraphic purposes, a difference of as much as 10 m.y. between deposition and the age of magnetization, as suggested by Larson and colleagues (1982; see also Larson and Walker, 1982) may assume an extraordinary importance. Conversely, for purposes of polar wander path construction, a delay in magnetization acquisition of some 10 m.y. is within the typical temporal resolution of Triassic and earlier times. There are several Triassic paleopoles for well-dated igneous rocks, moreover, so that the Triassic path segment can be considered with a fair amount of confidence. It appears that the typical paleomagnetic red-bed problem related to slow, postdepositional growth of hematite, and hence an uncertainty about the age of magnetization, is not so critical for pole path construction purposes insofar as the Triassic is concerned. For other continents this is often not the case (Ballard and others, 1986; Van der Voo, 1988), and the remagnetization problems of early Mesozoic and older rocks are compounded in intercontinental comparisons by uncertainties in the time scale used (see Van der Voo and others, 1984, for a discussion related to temporal uncertainties and their implications for Pangea reconstructions).

Polar wander paths have most recently been constructed for the Mesozoic and Tertiary of the North American craton by Harrison and Lindh (1982), Irving and Irving (1982), and notably by Gordon and others (1984). The technique of Gordon and others is based on the assumption, not required for conventional path construction, that path segments represent small-circle arcs around a paleomagnetic Euler pivot. In turn, this presumes that the motion of a continent with respect to the dipole (rotation) axis is constant for the duration of this path segment. If true, this implies that the polar wander path arc is the inverted representation of the motion of the continent, not unlike the arc of a hot-spot trace that results from the mantle volcanism that pierces

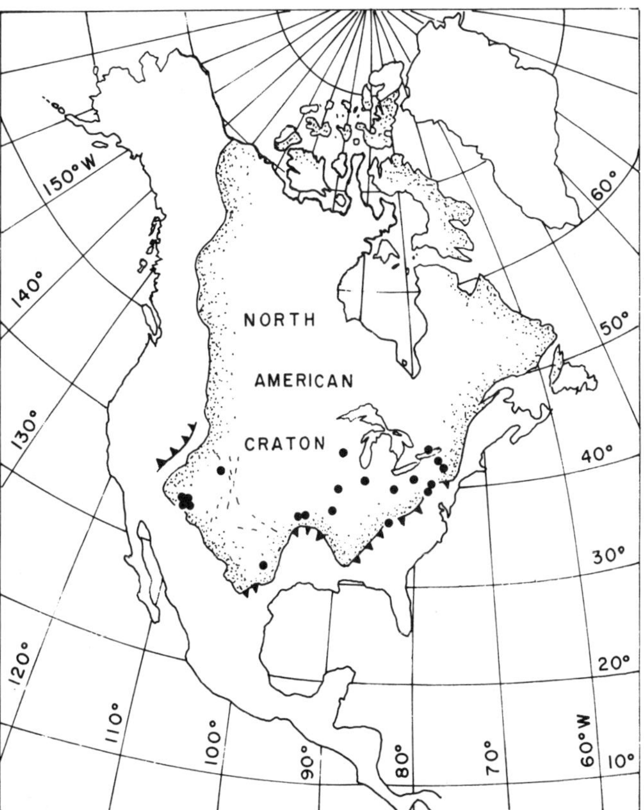

Figure 3. Locations of the carbonate rocks from the North American cratonic interior that have yielded the Late Paleozoic remagnetizations of Table 2 and Figure 2.

progressively through the lithospheric plate as it moves. With this assumption, path construction would then be constrained to the fitting of small-circle segments.

A comparison of a conventionally constructed (albeit smoothed) polar wander path and the small-circle path constructed by Gordon and others (1984) is shown in Figure 4 for North America. While to first approximation the two paths are very similar, there are important differences in detail (especially at 200 Ma). One reason for the differences lies in the fact that the conventional path was constructed with moving-window averaging (Irving and Irving, 1982), with the result that isolated poles carry much less weight. That there are three paleomagnetic poles (for the Wingate and Chinle Formations and the Newark Basin sediments [western limb]; Table 1) located just where the arc segments of Figure 4 intersect in a cusp at about 200 to 220 Ma constitutes strong support for the small-circle (Euler pivot) method. As more high-quality paleomagnetic data become available, a test of the underlying assumption (constant motion), on which these small-circle path segments are based, becomes possible. If the assumption appears to be viable, it is not unreasonable to anticipate that eventually we will be able to invert the Euler pivot data set simultaneously for most continents to arrive at a global set of kinematic parameters.

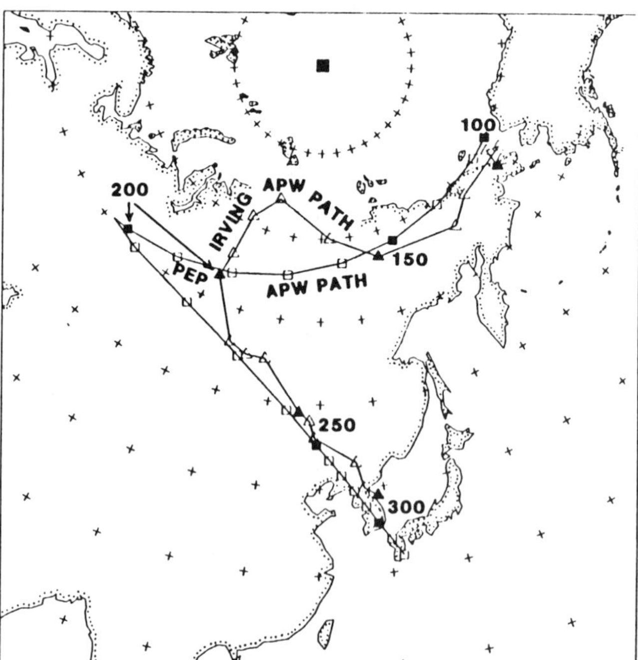

Figure 4. North American apparent polar wander paths at 10-m.y. intervals for the period between 100 Ma and 300 Ma in stereographic projection. Triangles indicate the path of Irving and Irving (1982); squares indicate the best-fit path from Gordon and others (1984), showing a cusp at about 200 Ma. From Gordon and others (1984); reprinted with permission of the authors.

Paleozoic reference poles

Well-determined paleopoles have been available for Late Carboniferous and Early Permian time ever since the first paleomagnetic studies were carried out on North American rocks (Graham, 1954; Doell, 1955; Runcorn, 1956). More sophisticated laboratory techniques have improved the resolution of multicomponent magnetizations, and more recently determined paleopoles are located about 5° closer to the North American craton than the earlier ones. This minor shift results from a more complete removal of secondary, often northerly and downward, overprints superimposed on the typical (reversed) southerly and shallow late Paleozoic direction (e.g., Van der Voo and McCabe, 1985). At present the Late Carboniferous/Early Permian average paleomagnetic pole appears to be the best determined ancient pole location for the North American craton. Exceptions are formed, however, by two poles (marked by an asterisk in Table 1), indicating their lack of suitability for path construction: the Permian Pegmatite dike and the Upper Carboniferous Buffalo Siltstone poles. They appear to be anomalous, possibly due to an unrecognized normal polarity overprint.

For pre-Pennsylvanian times, however, the state of our knowledge reveals a quantum jump down from the quality levels for younger times. This is mostly due to the scarcity of suitable rock types, as well as to the often more complex history in terms of diagenesis, deformation, or even metamorphism. One very important additional problem, namely that of ancient yet secondary magnetizations, has already been noted. It seems inevitable that at some point the North American path must loop back on itself, with the intersection point being representative of the ancient geomagnetic field at two rather different times. Yet there is every reason to be suspicious of any such loops, especially those previously proposed for some periods in Paleozoic time. Recent work (Miller and Kent, 1986a, b), for instance, on the Late Devonian Catskill red beds of Pennsylvania, suggests that the loop between the Late Devonian and Late Permian of the path of Van der Voo (1981) displays Permian remagnetizations and should be discounted, as already argued by Irving and Strong (1984, 1985). Thus these Late Devonian paleopoles have been marked with an asterisk in Table 1, indicating that they should not be used for polar wander path construction. Similarly, there is still no firm evidence for primary magnetizations in Middle Cambrian rocks, so that the Cambrian loop of Watts and others (1980a) may also be a composite effect of Cambrian remanence plus late Paleozoic remagnetizations (e.g., Gillett, 1982; Lynnes and Van der Voo, 1984; Larson and others, 1985).

Despite all these drawbacks of Paleozoic paleomagnetism, there are also some pairs or groups of internally consistent and apparently well-determined anchor poles, notably for Late Cambrian, Late Ordovician, Silurian, and Mississippian time. These poles satisfy five or six of the reliability criteria defined in the previous chapter, as can be seen in Table 1, and most of them do not resemble poles for younger periods so that strong suspicions about remagnetizations do not exist. Figure 1 includes the Paleozoic path segment that can be constructed on the basis of the available paleopoles, as given in Table 1. The Paleozoic portion of this table constitutes an update of previous compilations (Van der Voo and others, 1980, 1984; Van der Voo, 1981; Irving and Irving, 1982). Besides the Late Devonian ones, a few other poles are noted by asterisks and have not been used for polar wander path construction. These are the Early Carboniferous Cheverie and Middle Ordovician Martinsburg poles, which are suspected to be anomalous due to a regional rotation, and the Late Cambrian Lamotte and Wichita Granites poles, which appear to be much inferior to the other Late Cambrian poles listed.

The Precambrian

For Precambrian time there is a decrease in paleomagnetic resolution by another order of magnitude beyond that noted for Paleozoic time. Some of the causes are the same: scarcity of suitable rocks, complex thermal or deformational histories, and remagnetization (documented early on by the work of Reid [1972] for 1.8-Ga rocks). Three additional difficulties appear for the Precambrian: (1) the concept of "craton" appears to have uncertain validity for early Proterozoic or Archean times, (2) the sparser data base and the inevitable polarity ambiguity (north vs. south pole) may lead to multiple and equally valid polar wander path proposals, and (3) the typical precision in radiometric dating shows increasing error limits with increasing age.

As a result of these difficulties, there are no possibilities for immediate gratification in Precambrian paleomagnetic research; instead, such work seems to be a lifetime labor, preferably by teams of geochronologists, petrologists, structural geologists, and paleomagnetists working together to arrive at well-defined paleopoles. It is not surprising, therefore, that currently most American paleomagnetic studies are concerned with younger rocks in such areas as Alaska, the Cordillera, or the Rocky Mountains, to delineate displaced terranes and determine their displacement histories.

Precambrian paleomagnetic research, insofar as it is oriented toward tectonics, typically addresses the following types of questions: Was plate tectonics operative in middle Proterozoic and earlier times? If so, were the plate velocities, the nature of plate boundaries, and the products of plate interactions similar or quite different from those of Phanerozoic times? Are intracontinental orogenic belts always the sutures resulting from continent-continent collisions, or were there ensialic orogenies? Are apparent velocities of continental blocks with respect to the pole representative of real plate velocities, or was there significant true polar wander? Are climatic indicators (such as glacial relicts) diagnostic of latitudinal position, or were there dramatically different conditions in terms of atmosphere, climate, and obliquity? In addition, there are the geomagnetically oriented questions that address issues related to dynamo theory and reversals, and questions related to crustal evolution (e.g., those involving Precambrian dike swarms, as discussed by Halls, 1982).

In terms of results, the best known Precambrian pole paths for North America (or segments thereof) have been published mostly by Canadian workers (e.g., Robertson and Fahrig, 1971; Spall, 1971; Irving and Park, 1972; Larson and others, 1973; Irving and Naldrett, 1977; McGlynn and Irving, 1978; Irving, 1979; Roy, 1983). Roy's compilation of some of the published paths is reproduced here in Figure 5, together with the preferred path by Van der Voo (1982), based on the compilation of Irving (1979). This figure shows the complexities of Precambrian apparent polar wander, as well as the inevitable differences of opinion between various researchers.

The North American path is by far the best determined path for any of the continents, and is generally constructed on the assumption that the poles used are all valid for one and the same continental plate, i.e., that the area represented retained rigidity and coherence for the time since the magnetizations were acquired. For middle and late Proterozoic time, there is substantial evidence to corroborate this assumption for North America (Elston and Scott, 1973; Irving and Lapointe, 1975; Pullaiah and Irving, 1975; Irving and McGlynn, 1976, 1981); the same has been argued for other continents (e.g., McElhinny and McWilliams, 1977). This assumption is not necessarily warranted for early Proterozoic or Archean time, but there are at present not enough paleopoles to argue definitively against it (cf. Roy and others, 1978, versus Seyfert and Cavanaugh, 1978).

The assumption stated above does not imply that all Proterozoic provinces of North America must have belonged to the North American craton throughout their Precambrian history; what it does imply is that the magnetizations found in these provinces were generally acquired after the collision, if any, of the province with the cratonic nucleus. An illustrative example of this aspect of Precambrian paleomagnetism can be found in the paleomagnetic publications of the 1970s that deal with the Grenville Province. Irving and others (1972) initially described paleopoles from Grenville rocks that were in apparent disagreement with those from contemporaneous rocks in the rest of the Canadian Shield. They proposed a reconstruction in which the Grenville and the Superior provinces were separated by about 10,000 km at about 1,100 Ma. Further results (e.g., Buchan and Dunlop, 1973) could not yet refute this idea, until it was realized that the Grenville paleopoles were all based on magnetizations acquired during uplift and cooling *after* the Grenville orogeny (McWilliams and Dunlop, 1975), and that these magnetizations could not, therefore, provide any information about preorogenic convergence or synorogenic continent-continent collisions. With further age dating, especially with $^{40}Ar/^{39}Ar$ techniques, and with the realization that magnetic and isotopic unblocking spectra reveal many similarities (York, 1978), it became firmly established that the Grenville paleopoles were younger than initially thought and that they were indeed representative of the North American craton as a whole (Buchan and others, 1977; McWilliams and Dunlop, 1978; Berger and others, 1979; Watts, 1981; Dunlop and Stirling, 1985, and other references therein).

Returning to global tectonics, the path for North America by itself can hardly provide any answers to the questions posed above because most of the North American paleopoles appear to

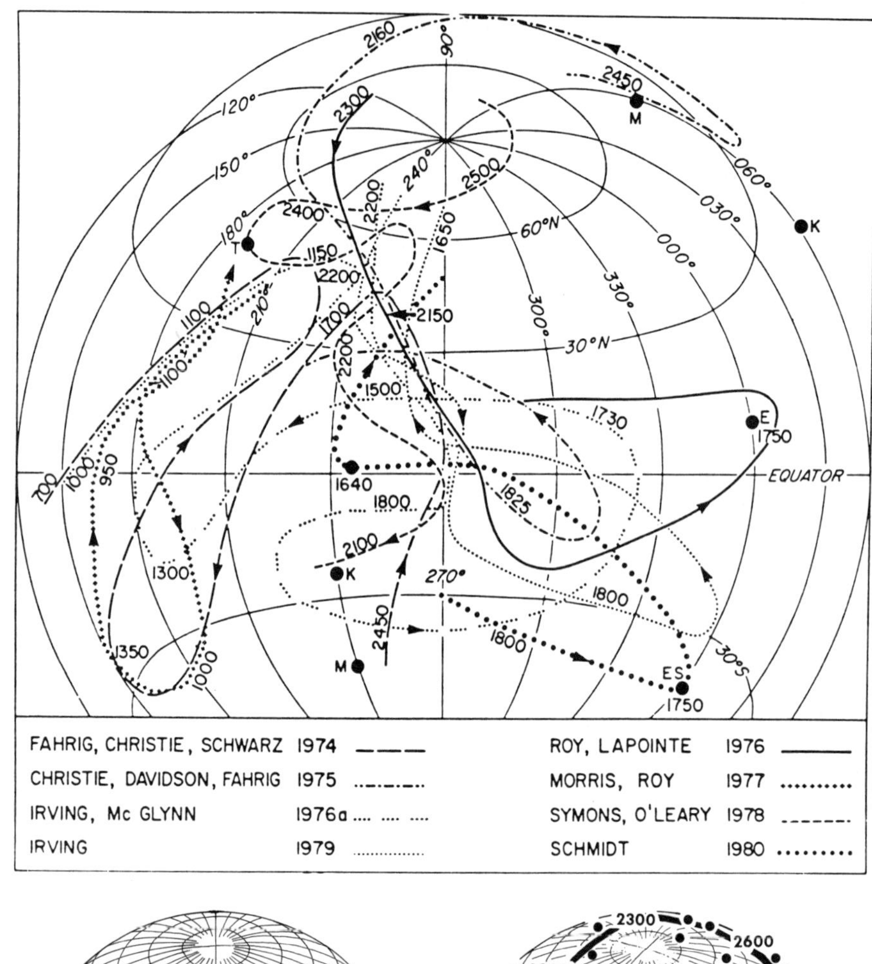

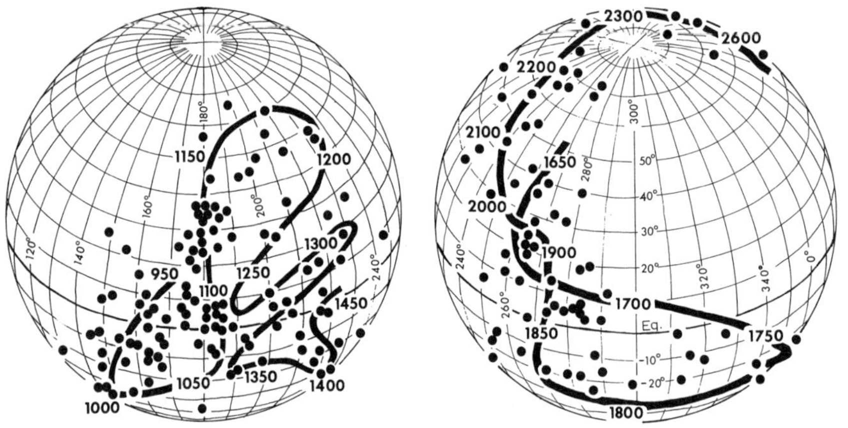

Figure 5. Superposition of eight polar wander paths for the Precambrian of North America, proposed by the different authors as indicated, in Lambert equal-area projection. From Roy (1983); reprinted with permission of the author and Elsevier Publishing Co. For comparison, the Precambrian apparent polar wander path of Van der Voo (1982), based on a compilation by Irving (1979), is shown below in two parts, with the individual pole positions shown as dots.

be based on magnetizations acquired after the consolidation of the province with the craton. Instead, we must turn to comparisons between paths for North America and those for other continental blocks. Earlier versions of the continental reconstructions that resulted from such comparisons have been published (see, for example, those by Ueno and others [1975], Morel and Irving [1978], and Irving and McGlynn [1979]). Yet because the polar wander paths from the other continents are poorly known, it would be premature at this time to expect thoroughly documented plate tectonic interpretations. About the best that can be expected are case histories in which paleopoles with similar ages have been obtained for two separate but well-studied cratonic blocks (e.g., Ueno and others, 1975; Onstott and Hargraves, 1981; Pesonen and Neuvonen, 1981; Onstott and others, 1984).

It is of interest to note that, in these more recent papers, which describe apparently much better dated and more reliable paleomagnetic results than was the case earlier, conclusions have been reached in support of relative plate motions between the blocks studied, in contrast with the cautious or controversial interpretations of earlier workers (e.g., the supercontinent model of Piper, 1976). In the coming decade I predict renewed interest in such Precambrian questions, coupled with sophisticated geochronological techniques which will lead to major advances in our understanding of Precambrian tectonics.

THRUST BELT MARGINS OF THE CRATON

The thrust belt margins of the craton, such as those found in the Canadian Rocky Mountains, the Valley and Ridge Province of the Appalachians or the Ouachita fold and thrust belt, show geologic coherence with the craton yet have obviously been deformed and therefore did not retain rigidity with respect to the craton. Where stratified rocks are involved, at least the tilt, i.e., rotation about horizontal axes, is rather readily recognized and can be corrected.

Rotations about vertical axes, including those resulting from plunging folds (MacDonald, 1980), are not easily quantifiable a priori. Such rotations would affect paleomagnetic declinations, and hence, paleomagnetic reference poles. Thus, it is important to recognize that paleomagnetic directions from thrust-fold belts may have only limited value for cratonic reference pole paths. This problem may be capitalized on by using paleomagnetic techniques to quantify such rotations about a vertical axis when appropriate reference directions are available. It should be noted that inclinations, and hence, paleomagnetic latitudes, are not affected by any rotations about a vertical axis; paleolatitude results from the cratonic margins can be used to confirm and amplify results from the cratonic interiors. Because of the much greater variety and suitability of the rock types typically found in the thrust-fold belts, and because paleomagnetic constraints such as provided by fold or conglomerate tests are more readily available there, this paleolatitudinal confirmation is extremely important. When this data base becomes large, it may eventually be possible to use a scalar set of contemporaneous inclinations for contouring purposes in order to arrive at a best-fit paleolatitude pattern, and, hence, a best-fit paleopole.

Thrust sheet rotations

Rotations about a vertical axis are paleomagnetically recognized through the declination (D) deviation. Statistical techniques are well developed for testing the significance of such declination deviations (Beck, 1980; Demarest, 1983). They are based on a comparison between mean directions with their cones of 95 percent confidence (alpha) from the rotated block (with subscript x) and those predicted for the area of interest from the appropriate reference block (with subscript o). Significantly discordant directions are present when the rotation r exceeds the error limit Δr (Beck, 1980; McWilliams, 1984):

$$r = D_o - D_x$$
$$\Delta r = (\Delta D_o^2 + \Delta D_x^2)^{1/2}$$
$$\Delta D_o = \sin^{-1} (\sin \alpha_o / \cos I_o)$$
$$\Delta D_x = \sin^{-1} (\sin \alpha_x / \cos I_x),$$

where I represents the inclination. A similar expression can, of course, be used to test the significance of a paleomagnetically determined rotation of a microplate or displaced terrane.

For the eastern overthrust belts of the United States, few significant thrust sheet rotations had been documented in published form until recently. In many segments of the Appalachian Valley and Ridge Province the declination did not appear to deviate significantly from those expected, i.e., those determined by extrapolation from the craton (Schwartz and Van der Voo, 1983). However, some of these Appalachian studies were based on magnetizations that are now recognized as secondary. Thus, Kent and Opdyke (1985), Kent (1986), and Miller and Kent (1986a), in restudying some of these formations, have found indications of thrust sheet rotations in Pennsylvania.

Substantial rotations, moreover, have been shown in the Wyoming and Helena, Montana salients where declination deviations of up to 90° occur (Grubbs and Van der Voo, 1976; Eldredge and Van der Voo, 1988). The rotation patterns thus determined are rather complex and most pronounced near the thrust front of the overthrust belts (Fig. 6). This led these investigators to postulate that the rotations are caused by foreland buttressing effects and that they are the greatest when the angle between the direction of thrusting and the buttress edge is the most oblique.

In the Mediterranean-Alpine and Hercynian belts of Europe, similar phenomena have been documented (Carey, 1958; Channell and others, 1980; Eldredge and others, 1985; Van der Linden, 1985; Bachtadse and Van der Voo, 1986; Perroud, 1986); most patterns appear to support causal theories involving buttressing effects.

Colorado Plateau: Is it rotated?

Suggested originally by Hamilton (1981), a clockwise rotation of the Colorado Plateau has recently been tested through the use of paleomagnetic data. The handicap in this investigation, however, is that the amount of rotation is small, suggested to be 3.9° by some (Bryan and Gordon, 1986) and between 9° and 14° by others (e.g., Steiner, 1986) on the basis of paleomagnetic analyses. Whatever the choice between these figures, the amount is comparable to the typical analytical error limits on individual paleomagnetic directions; the analysis thus becomes one of trying to reduce the error limits by testing the cumulative data set, rather than pole by pole. Moreover, for much of Phanerozoic time, a clockwise rotation of the Colorado Plateau simply displaces the apparent paleopole along the polar wander path in a younging direction, and a minor age difference between the plateau and cratonic pole pairs to be compared may have the same effect as a rotation. The offset for the smaller amount of 3.9° between the Colorado Plateau poles and the poles from the rest of the craton appears to be systematic, and the movements are inferred to be of post–Middle Cretaceous age (Bryan and Gordon, 1986).

APPALACHIAN BELT

After the paleomagnetic success stories about displaced terranes and microplates in the Mediterranean (Van der Voo and Zijderveld, 1969) and the Canadian and U.S. Cordillera (Hillhouse, 1977; Jones and others, 1977; see also Beck, this volume), it was only a matter of time before paleomagnetic techniques would be applied to a search for displaced terranes in the older mountain belts of North America. The Appalachian belt is an obvious target for such a search, not in the least because exotic terranes had been identified there already on the basis of (mostly) paleontological evidence (e.g., Wilson, 1966).

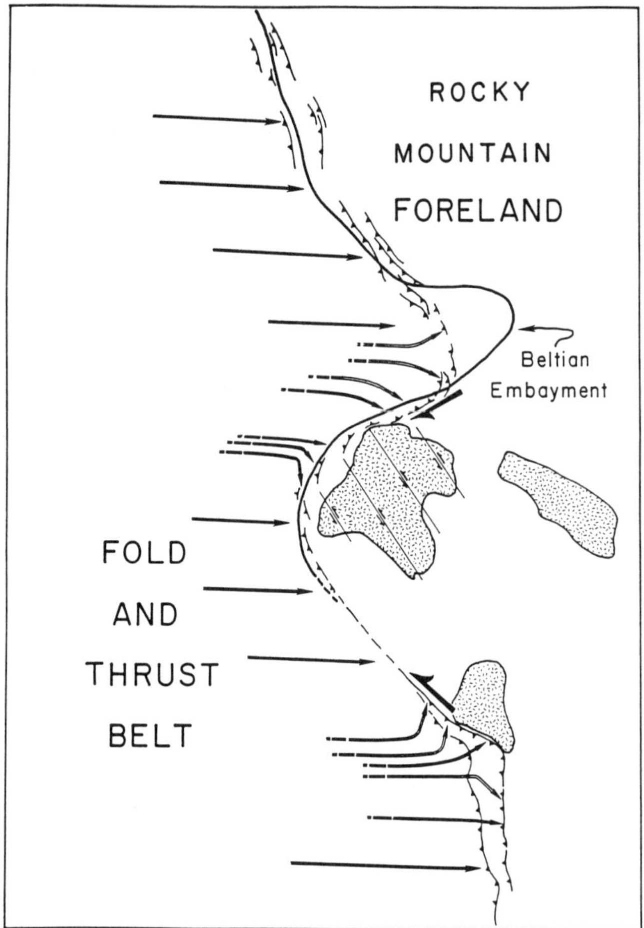

Figure 6. Schematic sketch of the Wyoming and Montana Rocky Mountains, showing the Helena (north) and Wyoming (south) salients. The stippled regions are foreland uplifts; the heavy line denotes the Precambrian Beltian embayment. The two heavy half-arrows show the sense of shear present in the intervening promontory; the long straight arrows show the thrust sheet displacement direction from west to east. The shorter, and locally bent, arrows represent rotated declinations measured in the Cretaceous Kootenai Formation in Montana and the Triassic Ankareh Formation in Wyoming. From Eldredge and Van der Voo (1988).

One could hypothetically imagine the following sequence of events for the final arrival of an exotic terrane: (1) convergence between the terrane and the craton, orthogonal or oblique, with the intervening ocean disappearing through subduction; (2) docking of the terrane through continent-continent collision, with the formation of a suture; (3) postdocking strike-slip displacement of the terrane, along faults that do not necessarily coincide with the suture (a step, that is not an essential one); and (4) final welding of the terrane, with no further displacements. Preceding these four steps, individual subterranes may have amalgamated well before convergence or docking (steps 1 and 2); they may have previously formed part of larger continental plates that disintegrated; or they may have constituted island arcs, seamounts, or oceanic plateaus with sialic crust (Ben Avraham and others, 1981; Nur, 1983). In the identification of such terranes, paleomagnetic techniques can be used to identify or constrain the allochthonous nature, provided that terrane paleolatitudes were significantly different from those for the craton to which they were welded. Furthermore, paleomagnetism can constrain the rates of convergence and the timing of first docking, as well as later (postdocking) displacements (step 3) and final welding (step 4).

Because of the inevitable incompleteness of the Paleozoic paleomagnetic record for either the craton or the suspect terranes themselves, a model based on paleomagnetism alone cannot be fully successful without taking into account several lines of geologic evidence. Foremost among such geologic constraints are the known orogenic phases, such as the Taconic (Mid- to Late Ordovician), Acadian (Early to Mid-Devonian) and Alleghenian (Mid- to Late Carboniferous) orogenies, on the assumption that continent-continent collisions (step 2) are recognizable in major deformation phases. Secondarily, faunal, paleogeographic, or facies information can be important building stones in any model (e.g., Neuman, 1984).

The evidence for or against postdocking displacements and for allochthony, respectively, are discussed below insofar as paleomagnetic data are available. On geological grounds, evidence for displaced terranes in the Appalachians has been presented previously; readers are referred to these reviews (e.g., in Hatcher and others, 1983).

Late Paleozoic displacements: Are they to be discounted?

In the Appalachians, the Avalon basement of eastern Newfoundland has lent its name to a set of terranes with significant geologic similarities. These Avalonian terranes are now recognized along the Canadian Maritime coast in Nova Scotia and New Brunswick, in the Boston Basin, and in the Piedmont Province of the southern Appalachians (Williams, 1979; Williams and Hatcher, 1983; Samson and others, 1982). For the northern Appalachians, it has long been known that deep-water marine sedimentation ceased by Devonian time; therefore, in all likelihood, subduction and docking of the Avalon terrane must have been during or preceding the Acadian orogeny. There was no a priori geological reason to assume any late Paleozoic displacements of the Avalonian terranes, so it came as a considerable surprise that Late Devonian and Early Carboniferous paleomagnetic data (Kent and Opdyke, 1978, 1979) appeared to indicate a paleolatitude difference of about 20° between the craton and the Avalonian terranes. This paleolatitude difference seemed to exist not only for Avalonian regions but also between the North American craton and Europe (Van der Voo and others, 1979). Such a difference had already been noted for earlier times between North America and Great Britain south of the Scottish Lewisian basement (Morris, 1976; Piper, 1979). Thus, Kent and Opdyke (1978) and Van der Voo and others (1979) quite logically assumed that the inferred major displacement (of about 2,000 km) occurred in Carboniferous time, rather than in Early Devonian time as postulated earlier (e.g., Piper, 1979).

Resistance to this idea arose primarily because of two issues. First, some objected to the idea that this displacement was not only so late in the Paleozoic but also that it was sinistral, whereas most of the geologic evidence for strike-slip movements indicated a dextral sense of motion (e.g., Webb, 1969; Bradley, 1982) during Carboniferous time. Second, major difficulties arose about the location of the fault(s) responsible for the displacement. Specific delineations attempted on paleomagnetic grounds, separating provinces with "North American" from those with "European/Acadian" paleolatitudes (e.g., Van der Voo and Scotese, 1981; Kent, 1982; Spariosu and Kent, 1983), met with increasing disbelief, supported by geological and, increasingly, paleomagnetic arguments (Roy and Morris, 1983; Smith and Watson, 1983; Briden and others, 1984; Irving and Strong, 1984, 1985).

A key element in the geographic array of possible terranes is western Newfoundland (the Long Range). This area has a Grenville-type Precambrian basement: although displacement with respect to the adjacent craton in Labrador cannot be excluded entirely, it has always seemed rather unlikely. However, when Opdyke (1981) first reported Early Carboniferous results from the Barachois Formation in western Newfoundland, the sampling area appeared to be displaced just as much as the more easterly terranes. Puzzled by this result, Kent and Opdyke embarked in 1983 on a resampling and renewed laboratory study of the formation that had provided the main reference pole for the craton, the Early Carboniferous Mauch Chunk Formation of Pennsylvania's Valley and Ridge Province (D. V. Kent, personal communication, 1986). Independently, Roy and Morris (1983) and Irving and Strong (1984, 1985) had begun to formulate their hypotheses that all, even the most reliable, of the Late Devonian and Early Carboniferous reference poles from the craton were actually based on Permian remagnetizations. For Early Carboniferous time, this has now been shown most convincingly (Kent and Opdyke, 1985). For the Late Devonian the results are not very conclusive, although it appears certain that the previously publshed reference pole from the folded Catskill red beds is based on secondary, albeit ancient and pre-folding, magnetizations (Miller and Kent, 1986a, b).

In retrospect, it is no exaggeration to state that many paleomagnetists, including some of the later visionary proponents of remagnetization (e.g., Irving, 1979), were perhaps too easily fooled and that these remagnetizations should have been recognized as such early on, as indeed they were in other formations (Roy and others, 1967; French and Van der Voo, 1977; Kent, 1979; Scotese and others, 1982; Hodych and others, 1984; Perroud and Van der Voo, 1984). On the other hand, and in defense of the original ideas, the reference results (Catskill, Mauch Chunk) seemed most reliable, because they fulfilled many of the typical criteria: normal and reversed polarities, complete thermal as well as chemical and alternating-field demagnetization, good structural control, and even positive fold tests.

At this time there is no longer much paleomagnetic support of major displacements of "Acadia" or other terranes in New England and the Canadian Maritime provinces, insofar as Late Devonian and younger time is concerned. This notwithstanding, a few puzzling Early Carboniferous paleolatitude anomalies remain: for the New Brunswick volcanics (Seguin and others, 1985: about 10° farther south than expected) and for the east Newfoundland St. Lawrence Granite, alaskites, and associated dikes (Irving and Strong, 1985: 14° farther south). The bulk of the data, particularly those for Europe and North America, however, does now show agreement within the limits of statistical uncertainty. For earlier times, the main conclusions of Morris (1976), Piper (1979), and Van der Voo (1983) about a major displacement between Europe and North America and between the Avalon terranes and the craton as well, are still valid, but this displacement could perhaps better be seen as a docking, rather than a postdocking, displacement.

Early Paleozoic Appalachian displaced terranes

Since there are no longer any paleomagnetic indications for post–Middle Devonian (postdocking) displacements of large magnitude in the Appalachians, the paleomagnetic evidence for allochthony and the constraints on the time of docking itself should be examined.

The Avalon basement areas of Newfoundland, Cape Breton Island in Nova Scotia, and the Boston basin have all provided latest Precambrian or early Paleozoic paleomagnetic data for structurally well-documented and well-dated stratified rocks that reveal paleolatitudes significantly higher than those predicted by extrapolation from the craton (Irving and Strong, 1985; Van der Voo and Johnson, 1985; Johnson and Van der Voo, 1985, 1986; Fang Wu and others, 1986). In addition, there are several studies for granitic or mafic plutons in the Avalonian and adjacent (Dunnage, Gander) zones of New England and the Canadian Atlantic provinces with high paleolatitudes, although neither paleomagnetic laboratory observations nor structural control lends

TABLE 3. LATEST PRECAMBRIAN–EARLY PALEOZOIC PALEOMAGNETIC DATA FROM AVALON*

Formation	Symbol	Age	Paleopole	P-Lat	dp	Reference
Nova Scotia						
Cape Breton Granites	CB	650–490	37N,176E	8	2	Rao and others (1981)
Cape Breton Granites	CB1	410 ± 50	16N,319E	55	6	Rao and others (1981)
Cape Breton Granites	CB2	460 ± 60	32N,263E	59	7	Rao and others (1981)
Fourchu Group	F1	610–540	56N,225E	45	6	Johnson and Van der Voo (1986)
Bourinot Volcanics	BV	~540	21N,340E	49	11	Johnson and Van der Voo (1985)
Bourinot Sediments	BS	~540	33N,354E	47	11	Johnson and Van der Voo (1985)
Dunn Point Volcanics	DP	437+15	2N,316E	42	5	Van der Voo and Johnson (1985)
Newfoundland						
Marystown, Calmer	MG	625–600	32N,231E	36	6	Irving and Strong (1985)
Marytstown, Famine Back Cove	MG	625–600	1N,342E	34	5	Irving and Strong (1985)
Marystown, Garnish	MG	625–600	5N,329E	34	7	Irving and Strong (1985)
Boston Basin						
Roxbury Formation	RF	600–570	13N,267E	55	8	Fang Wu and others (1986)

*Abbreviations: P-Lat = paleolatitude in degrees, dp = error associated with the paleolatitude in degrees (McElhinny, 1973).

great reliability to these results (Lapointe, 1979; Roy and others, 1979; Rao and others, 1981; Weisse and others, 1985). The results of Upper Precambrian, Cambrian, and Ordovician rocks published thus far from the Avalonian terranes are listed in Table 3. While it seems to be clear that the docking of Avalon was of pre–Late Devonian age (Johnson and Van der Voo, 1983), the available Silurian results at present are not conclusive enough to give specific details.

The intermediate to high paleolatitudes have been uniformly interpreted as Southern Hemisphere ones, in order to minimize subsequent Paleozoic drift of the Avalon terranes with respect to the craton or the pole. It is of interest to note that paleomagnetic observations on early Paleozoic rocks from southern England, Wales, and most of Hercynian Europe indicate intermediate to high southern paleolatitudes as well (e.g., Perroud and others, 1984); it has long been known that the assembled Gondwana continent was generally located at the South Pole (e.g., McElhinny and Embleton, 1976). Van der Voo (1979, 1982) and co-workers (e.g., Perroud and others, 1984; Van der Voo and others, 1985a, b) have argued that all these areas, including Gondwana, formed a coherent land mass during early Paleozoic time, which disintegrated, possibly in Ordovician time. These authors have postulated the existence of an Armorica plate, loosely composed of Hercynian Europe, the Avalonian terranes, and southern Great Britain, which drifted northward during Silurian time to collide with the combined North America and Baltica continents. This collision resulted in the formation of the Old Red Continent in Devonian time (Van der Voo, 1983; Perroud and Bonhommet, 1984).

Whether Armorica was one coherent land mass during this northward drift or comprised a mosaic of island elements (e.g., Neuman, 1984) is beyond the resolution of the currently available paleomagnetic data, given that rapid drift of the elements resulted in rapid apparent polar wander. This in turn causes small differences in age to result in relatively large paleolatitude differences with an inherent loss of resolution.

The evidence for the coherence between Avalon, the Armorican Massif, and Gondwana can be found in their characteristic paleolatitudes. Because a comparison between components of mobile belts (such as between Cape Breton Island in Nova Scotia and the Armorican Massif in France) may be hampered by relative rotations, it is more rigorous to start an analysis of relative positions in terms of paleolatitudes than with paleopoles. For the North American craton, the Avalonian terranes, the Armorican Massif, and Gondwana (for an offshore Moroccan location), a comparison of paleolatitudes vs. time is shown in Figure 7 (from Johnson and Van der Voo, 1986). Error bars are shown for the age uncertainties as well as the statistically determined paleolatitude uncertainties. It can be seen at a glance that North America's early Paleozoic equatorial paleolatitudes do not agree with those for the other blocks, whereas the patterns for Avalon, the Armorican Massif, and Gondwana are very similar. When the comparison is broadened to include the full paleomagnetic information, i.e., on the basis of pole path segments, the Armorican Massif and Gondwana are very similar (Hagstrum and others, 1980; Perigo and others, 1983), whereas the paleopoles from Cape Breton Island reveal a similar rate of apparent polar wander and can be brought into agreement with the paths of the other blocks through a simple rotation. The higher the paleolatitudes shown by these blocks, the less there is ambiguity in terms of relative paleolongitude: as argued by Perroud and others (1984) and by Van der Voo and others (1985b), the only likely location of the Armorican Massif is off the coast of northwestern Africa.

All the available Devonian paleolatitudes for Avalon or

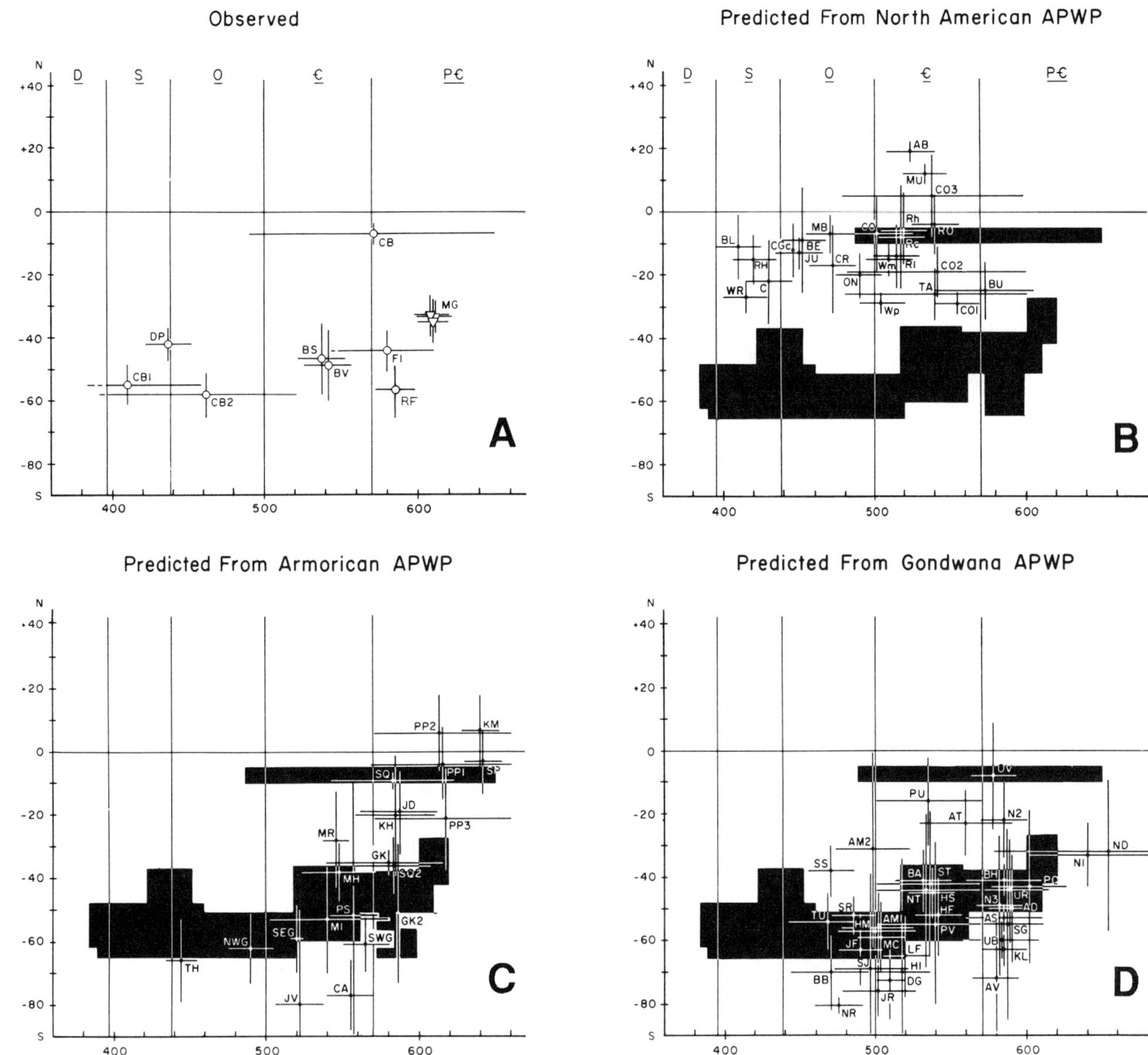

Figure 7. Latest Precambrian and Early Paleozoic paleolatitudes plotted vs. age for the Nova Scotia Avalon basement terrane. The observed paleolatitudes for Avalon (Table 3) are represented by open symbols in A and their age and paleolatitude uncertainties are shown by the error bars. The field of the error bars of A is repeated in black in the other three diagrams for ease of comparison. The letter abbreviations next to each symbol represent individual pole positions listed in Johnson and Van der Voo (1986). Predicted paleolatitudes, represented by error crosses in the remaining diagrams, were calculated from paleomagnetic data as available for the continents shown, assuming the fits of Bullard and others (1965) and Smith and Hallam (1970) between North America (including Avalon), the European and Gondwanaland continents. B, Paleolatitudes extrapolated from the North American cratonic data. C, Paleolatitudes predicted from data from the Armorican Massif, France. D, Paleolatitudes predicted from Gondwanaland paleopoles. Modified from Johnson and Van der Voo (1986), with the addition of the paleolatitude of the Roxbury Formation (RF) from the Boston basin area.

Armorica are near-equatorial and in good agreement with those for North America, whereas those for Gondwana are still relatively high (Kent, 1982; Perroud and Bonhommet, 1984; Perroud and others, 1984; Hurley and Van der Voo, 1987). This has led to the idea that Armorica (including Avalon) rifted apart from Gondwana and collided with North America and Baltica during or before Devonian time. Since in the latest Ordovician, Armorica's paleolatitudes are still very high, the demand for enough time to let Armorica drift northward toward equatorial North America mandates that Armorica and North America could have collided at the earliest in Late Silurian time. Given the two constraints on this collision just mentioned and the distribution of Siluro-Devonian orogenic belts, it appears that the Acadian orogeny must be ascribed to this collision. The convergence process (step 1) of the Avalonian terranes is therefore in process during pre-Devonian time, which may explain the significant paleolatitude differences between Avalon and the North American craton mentioned earlier for those times.

Finally, paleomagnetic studies farther south, in the central and southern Appalachians, have generally not succeeded in discovering pre-Carboniferous paleomagnetic directions. Exceptions are found for the Ordovician of the Delaware Piedmont area (Rao and Van der Voo, 1980; Brown and Van der Voo, 1983). These results, however, were obtained from metamorphic rocks and do not compare in reliability with those from the well-dated and stratified rocks of the Avalon basement in Canada. Nevertheless, these results reveal very high paleolatitudes; this knowledge could also be used as support for an Armorican or peri-Gondwana affinity of these areas. In the absence of critical Devonian data for this region, it cannot be decided whether the Piedmont Province docked at the same time as the Avalon terranes of Canada (see Hatcher and Odom, 1980; Williams and Hatcher, 1983), or whether its collision with North America occurred later, during the Alleghenian orogeny. In subsurface Lower Paleozoic rocks from Florida, similar steeply inclined "peri-Gondwana"-type directions have been reported (Jones and others, 1983), confirming the allochthony of this area suggested previously by faunal affinities (Cramer, 1970, 1971).

Remaining problems

If the collision between the Avalonian terranes and the craton in the northern Appalachians is Acadian, what does the Taconic orogeny represent? Late Ordovician or Silurian paleomagnetic results from the zones between Avalon and the craton could shed light on this problem, but too few results exist, and at this time, only nonpaleomagnetic evidence is available. It seems likely that the Taconic orogeny was caused by the collision of the craton with an island arc (Williams, 1979), but this awaits paleomagnetic confirmation.

To the south, well-dated pre-Carboniferous paleomagnetic data from the Piedmont between Alabama and Virginia would be highly desirable in order to resolve the paleoposition of the Avalon-type areas in the Piedmont. The timing of the movement on the Piedmont overthrusts is not at all clear; in addition to a Late Carboniferous component of movement, earlier components must be considered (Hatcher and others, 1983). Here, too, the cause of the Taconic orogenic phase remains unclear, although Taconic deformation is not as evident as in the northern Appalachians. A curious dilemma exists because of the aforementioned steep directions for the Ordovician from Delaware: why is the main deformation in Delaware and adjacent Pennsylvania of Late Ordovician (Taconic) age (Wagner and Crawford, 1975), while the paleomagnetic directions indicate a large paleolatitude difference between the Delaware rocks and the craton, indicating that a Taconic collision is most unlikely? Could it be that Taconic deformation occurred on two far-apart margins of the Ordovician Iapetus Ocean?

CONCLUSIONS

The reference poles from the craton, supported by paleolatitude information from the adjacent foreland and overthrust belts, provide the basic prerequisite paleomagnetic data for global paleoreconstructions and for microplate, displaced-terrane and structural analyses. Our knowledge, while probably much more accurate than a few years ago, is expanding at a slow rate, especially because of the increasing realization that remagnetizations are more common than previously thought.

In the last decade, the abundant and paleomagnetically suitable rocks of the forelands and overthrust belts have provided many paleomagnetic results thought to be reliable because of constraints provided by fold or conglomerate tests, but with a declination uncertainty caused by the possibility of local rotations. Such rotations, however, do not affect paleolatitude determinations for stratified rocks and can be used to advantage when a known reference direction is available. Thus, interesting determinations of thrust sheet and block rotations have become possible, especially in the northern Rocky Mountains and the Colorado Plateau.

In the Appalachian belt, displacements of outboard terranes, such as the Avalon basement terranes of the Canadian Atlantic provinces and New England, have been documented for Early Paleozoic times; however, earlier postulates for Late Paleozoic strike-slip offsets appear to have been based on erroneous (remagnetized) reference poles. The Avalon terranes show the same intermediate to high early Paleozoic paleolatitude patterns as the Armorican Massif in France and the Moroccan and Sahara areas of Africa; the postulate of an early Paleozoic juxtaposition of these elements is supported by the available paleomagnetic data. It is suggested that Avalon and most of Hercynian Europe, including the Armorican Massif, formed part of the same plate—Armorica—in latest Ordovician through Silurian time. Armorica is thought to have rifted away from Gondwana in middle Paleozoic time, with a northward drift during Silurian time, ending in continent-continent collision in Devonian time, thus producing the Acadian orogeny.

ACKNOWLEDGMENTS

I thank Ben van der Pluijm, Chris R. Scotese, Henry N. Pollack, Dennis V. Kent, Valerian Bachtadse, and Niels Abrahamsen for their critical reading of the manuscript. Bill Muehlberger and an anonymous reviewer provided valuable comments. This work has been supported by National Science Foundation Grant EAR-84-07007.

REFERENCES CITED

Baag, C.-G., and Helsley, C. E., 1974, Remanent magnetization of a 50 m core from the Moenkopi Formation: Journal of Geophysical Research, v. 79, p. 3308–3320.

Bachtadse, V., and Van der Voo, R., 1986, Paleomagnetic evidence for crustal and thin-skinned rotations in the European Hercynides: Geophysical Research Letters, v. 13, p. 161–164.

Ballard, M. M., Van der Voo, R., and Halbich, I. W., 1986, Remagnetizations in Late Permian and Early Triassic rocks from southern Africa and their implications for Pangea reconstructions: Earth and Planetary Science Letters, v. 79, p. 412–418.

Barnes, A. E., and Butler, R. F., 1980, A Paleocene paleomagnetic pole from the Gringo Gulch Volcanics: Geophysical Research Letters, v. 7, p. 545–548.

Beales, F. W., Carracedo, J. C., and Strangway, D. W., 1974, Paleomagnetism and the origin of Mississippi Valley–type ore deposits: Canadian Journal of Earth Sciences, v. 11, p. 211–223.

Beck, M. E., 1972, Paleomagnetism of Upper Triassic diabase from southeastern Pennsylvania; Further results: Journal of Geophysical Research, v. 77, p. 5673–5687.

——, 1980, Paleomagnetic record of plate-margin tectonic processes along the western edge of North America: Journal of Geophysical Research, v. 85, p. 7115–7131.

Ben Avraham, Z., Nur, A., Jones, D. L., and Cox, A. V., 1981, Continental accretion and orogeny; From oceanic plateau to allochthonous terranes: Science, v. 213, p. 47–54.

Berger, G. W., York, D., and Dunlop, D. J., 1979, Calibration of Grenvillian palaeopoles by $^{40}Ar/^{39}Ar$ dating: Nature, v. 277, p. 46–48.

Black, R. F., 1964, Paleomagnetic support of the theory of rotation of the island of Newfoundland: Nature, v. 202, p. 945–948.

Bradley, D. C., 1982, Subsidence in Late Paleozoic basins in the northern Appalachians: Tectonics, v. 1, p. 107–123.

Briden, J. C., Turnell, H. B., and Watts, D. R., 1984, British paleomagnetism, Iapetus Ocean, and the Great Glen Fault: Geology, v. 12, p. 428–431.

Brown, P. M., and Van der Voo, R., 1982, Paleomagnetism of the latest Precambrian/Cambrian Unicoi basalts from the Blue Ridge, northeast Tennessee and southwest Virginia; Evidence for Taconic deformation: Earth and Planetary Science Letters, v. 60, p. 407–414.

Brown, P. M., and Van der Voo, R., 1983, A paleomagnetic study of Piedmont metamorphic rocks in northern Delaware: Geological Society of America Bulletin, v. 94, p. 815–822.

Bryan, P., and Gordon, R. G., 1986, Rotation of the Colorado Plateau; An analysis of paleomagnetic data: Tectonics, v. 5, p. 661–667.

Buchan, K. L., and Dunlop, D. J., 1973, Magnetisation episodes and tectonics of the Grenville Province: Nature Physical Science, v. 246, p. 28–30.

Buchan, K. L., Berger, G. W., McWilliams, M. O., York, D., and Dunlop, D. J., 1977, Thermal overprinting of natural remanent magnetization and K/Ar ages in metamorphic rocks: Journal of Geomagnetism and Geoelectricity, v. 29, p. 401–410.

Bullard, E. C., Everett, J. E., and Smith, A. G., 1965. A symposium on continental drift; 4, The fit of the continents around the Atlantic: Philosophical Transactions of the Royal Society, v. 258, p. 41–51.

Butler, R. F., and Taylor, L. H., 1978, A middle Paleocene paleomagnetic pole from the Nacimiento Formation, San Juan basin, New Mexico: Geology, v. 6, p. 495–498.

Carey, S. W., 1958, A tectonic approach to continental drift; A symposium on continental drift: Hobart, Tasmania, p. 177–355.

Channell, J.E.T., Catalano, R., and D'Argenio, B., 1980, Paleomagnetism and deformation of the Mesozoic continental margin in Sicily: Tectonophysics, v. 61, p. 391–407.

Chen, D. L., and Schmidt, V. A., 1984, Paleomagnetism of the Middle Mississippian Greenbrier Group in West Virginia, U.S.A., *in* Van der Voo, R., Scotese, C. R., and Bonhommet, N., eds., Plate reconstruction from Paleozoic paleomagnetism: American Geophysical Union Geodynamics Series, v. 12, p. 48–62.

Collinson, D. W., and Runcorn, S. K., 1960, Polar wandering and continental drift: Evidence from paleomagnetic observations in the United States: Geological Society of America Bulletin, v. 71, p. 915–958.

Cox, A. V., and Doell, R. R., 1960, Review of paleomagnetism: Geological Society of America Bulletin, v. 71, p. 645–768.

Cramer, F. H., 1970, Middle Silurian continental movement estimated from phytoplankton-facies transgression: Earth and Planetary Science Letters, v. 10, p. 87–93.

——, 1971, Position of the north Florida Paleozoic block in Silurian time; Phytoplankton evidence: Journal of Geophysical Research, v. 76, p. 4754–4757.

Creer, K. M., 1968, Palaeozoic palaeomagnetism: Nature, v. 219, p. 246–250.

Currie, K. L., and Larochelle, A., 1969, A paleomagnetic study of volcanic rocks from Mistastin Lake, Labrador, Canada: Earth and Planetary Science Letters, v. 6, p. 309–315.

Dankers, P., 1982, Implications of Early Devonian poles from the Canadian Arctic archipelago for the North American apparent polar wander path: Canadian Journal of Earth Sciences, v. 19, p. 1802–1809.

Dankers, P., and Lapointe, P., 1981, Paleomagnetism of Lower Cambrian volcanics and a cross-cutting Cambro-Ordovician diabase dyke from Buckingham (Quebec): Canadian Journal of Earth Sciences, v. 18, p. 1174–1186.

DeBoer, J., 1968, Paleomagnetic differentiation and correlation of the Late Triassic volcanic rocks in the Central Appalachians (with special reference to the Connecticut Valley): Geological Society of America Bulletin, v. 79, p. 609–626.

DeBoer, J., and Brookins, D. G., 1972, Paleomagnetic and radiometric age determination of (Permian) pegmatites in the Middletown district (Connecticut): Earth and Planetary Science Letters, v. 15, p. 140–144.

Demarest, H. H., 1983, Error analysis for the determination of tectonic rotation from paleomagnetic data: Journal of Geophysical Research, v. 88, p. 4321–4328.

Deutsch, E. R., and Rao, K. V., 1977a, New paleomagnetic evidence fails to support rotation of western Newfoundland: Nature, v. 266, p. 314–318.

——, 1977b, Paleomagnetism of the Lower Bradore sandstones, and the rotation of Newfoundland: Tectonophysics, v. 33, p. 337–357.

Deutsch, E. R., Kritjansson, and May, B. T., 1971, Remanent magnetism of Lower Tertiary lavas on Baffin Island: Canadian Journal of Earth Sciences, v. 8, p. 1542–1552.

Diehl, J. F., and Shive, P. N., 1979, Paleomagnetic studies of the Early Permian Ingelside Formation of northern Colorado: Geophysical Journal of the Royal Astronomical Society, v. 56, p. 271–282.

——, 1981, Paleomagnetic results from the Late Carboniferous/Early Permian Casper Formation; Implications for northern Appalachian tectonics: Earth and Planetary Science Letters, v. 54, p. 281–292.

Diehl, J. F., Beske-Diehl, S., Beck, M. E., and Hearn, B. C., 1980, Paleomagnetic results from Early Eocene intrusions, north-central Montana; Implications for North American apparent polar wandering: Geophysical Research Letters, v. 7, p. 541–544.

Diehl, J. F., Beck, M. E., Beske-Diehl, S., Jacobson, D., and Hearn, B. C., 1983, Paleomagnetism of the Late Cretaceous-early Tertiary north-central Montana alkalic province: Journal of Geophysical Research, v. 88, p. 10593–10609.

Doell, R. R., 1955, Paleomagnetic study of rocks from the Grand Canyon of the Colorado River: Nature, v. 176, p. 1167.

Dooley, R. E., and Smith, W. A., 1982, Paleomagnetism of early Mesozoic diabase dikes in the South Carolina Piedmont: Tectonophysics, v. 90, p. 283–307.

Dunlop, D. J., and Stirling, J. M., 1985, Post-tectonic magnetizations from the Cordova gabbro, Ontario, and Palaeozoic reactivation in the Grenville province: Geophysical Journal of the Royal Astronomical Society, v. 91, p. 521–550.

Dunn, W. J., and Elmore, R. D., 1985, Paleomagnetic and petrographic investigation of the Taum Sauk Limestone, southeast Missouri: Journal of Geophysical Research, v. 90, p. 11469–11483.

Eldredge, S., and Van der Voo, R., 1988, Paleomagnetic study of thrust sheet rotations in the Helena and Wyoming salients of the northern Rocky Mountains, *in* Perry, W. J., and Schmidt, C., eds., Interaction of the Rocky Mountain Foreland and Cordilleran Thrust belt: Geological Society of America Memoir 171, p. 319–332.

Eldredge, S., Bachtadse, V., and Van der Voo, R., 1985, Paleomagnetism and the orocline hypothesis: Tectonophysics, v. 119, p. 153–179.

Elmore, R. D., Dunn, W., and Peck, C., 1985, Absolute dating of a diagenetic event using paleomagnetic analysis: Geology, v. 13, p. 558–561.

Elston, D. P., and Bressler, S. L., 1977, Paleomagnetic poles and polarity zonation from Cambrian and Devonian strata of Arizona: Earth and Planetary Science Letters, v. 36, p. 423–433.

Elston, D. P., and Scott, G. R., 1973, Paleomagnetism of some Precambrian basaltic flows and red beds, eastern Grand Canyon, Arizona: Earth and Planetary Science Letters, v. 18, p. 253–265.

Fang Wu, Van der Voo, R., and Johnson, R. J., 1986, Eocambrian paleomagnetism of the Boston Basin; Evidence for displaced terrane: Geophysical Research Letters, v. 13, p. 1450–1453.

Farrell, W. E., and May, B. T., 1969, Paleomagnetism of Permian red beds from the Colorado Plateau: Journal of Geophysical Research, v. 74, p. 1495–1504.

Fisher, R. A., 1953, Dispersion on a sphere: Proceedings of the Royal Society, v. A217, p. 295–305.

French, A. N., and Van der Voo, R., 1979, The magnetization of the Rose Hill Formation at the classical site of Graham's fold test: Journal of Geophysical Research, v. 84, p. 7688–7696.

French, R. B., and Van der Voo, R., 1977, Remagnetization problems with the paleomagnetism of the middle Silurian Rose Hill Formation of the central Appalachians: Journal of Geophysical Research, v. 82, p. 5803–5806.

French, R. B., Alexander, D. H., and Van der Voo, R., 1977, Paleomagnetism of upper Precambrian to lower Paleozoic intrusive rocks from Colorado: Geological Society of America Bulletin, v. 88, p. 1785–1792.

Gillett, S. L., 1982, Paleomagnetism of the Late Cambrian *Crepicephalus-Aphelaspis* trilobite zone boundary in North America; Divergent poles from isochronous strata: Earth and Planetary Science Letters, v. 58, p. 383–394.

Gordon, R. G., 1984, A Paleocene North American paleomagnetic pole incorporating declination-only data: Geophysical Research Letters, v. 11, p. 477–480.

Gordon, R. G., Cox, A. V., and O'Hare, S., 1984, Paleomagnetic Euler poles and the apparent polar wander and absolute motion of North America since the Carboniferous: Tectonics, v. 3, p. 499–537.

Gose, W. A., and Helsley, C. E., 1972, Paleomagnetic and rock magnetic studies of the Permian Cutler and Elephant Canyon Formations in Utah: Journal of Geophysical Research, v. 77, p. 1534–1548.

Graham, J. W., 1954, Rock magnetism and the Earth's magnetic field during Paleozoic time: Journal of Geophysical Research, v. 59, p. 215–222.

Grubbs, K. L., and Van der Voo, R., 1976, Structural deformation of the Idaho-Wyoming overthrust belt (U.S.A.), as determined by Triassic paleomagnetism: Tectonophysics, v. 33, p. 321–336.

Hagstrum, J. T., Van der Voo, R., Auvray, B., and Bonhommet, N., 1980, Eocambrian-Cambrian paleomagnetism of the Armorican Massif, France: Geophysical Journal of the Royal Astronomical Society, v. 61, p. 489–517.

Halls, H. C., 1982, The importance and potential of mafic dyke swarms in studies of geodynamic processes: Geoscience Canada, v. 9, p. 145–154.

Hamilton, W., 1981, Plate tectonic mechanism of Laramide deformation: Contributions to Geology of the University of Wyoming, v. 19, p. 87–92.

Hanna, W. F., 1967, Paleomagnetism of Upper Cretaceous volcanic rocks of southwestern Montana: Journal of Geophysical Research, v. 72, p. 595–610.

—— , 1973, Paleomagnetism of the Late Cretaceous Boulder Batholith, Montana: American Journal of Science, v. 273, p. 778–802.

Harrison, C.G.A., and Lindh, T., 1982, A polar wander curve for North America during the Mesozoic and Cenozoic: Journal of Geophysical Research, v. 87, p. 1903–1920.

Hatcher, R. D., Jr., and Odom, A. L., 1980, Timing of thrusting in the southern Appalachians, U.S.A.; Model for orogeny?: Journal of the Geological Society of London, v. 137, p. 321–327.

Hatcher, R. B., Jr., Williams, H., and Zietz, I., eds., 1983, Contributions to the tectonics and geophysics of mountain chains: Geological Society of America Memoir 158, 223 p.

Helsley, C. E., 1965, Paleomagnetic results from the Lower Permian Dunkard Series of West Virginia: Journal of Geophysical Research, v. 70, p. 413–424.

—— , 1969, Magnetic reversal stratigraphy of the Lower Triassic Moenkopi Formation of western Colorado: Geological Society of America Bulletin, v. 80, p. 2431–2450.

—— , 1971, Remanent magnetization of the Permian Cutler Formation of western Colorado: Journal of Geophysical Research, v. 76, p. 4842–4848.

Helsley, C. E., and Steiner, M. B., 1974, Paleomagnetism of the Lower Triassic Moenkopi Formation: Geological Society of America Bulletin, v. 85, p. 457–464.

Herrero-Bervera, E., and Helsley, C. E., 1983, Paleomagnetism of a polarity transition in the Lower(?) Triassic Chugwater Formation, Wyoming: Journal of Geophysical Research, v. 88, p. 3506–3522.

Hillhouse, J. W., 1977, Paleomagnetism of the Triassic Nikolai Greenstone, McCarthy Quadrangle, Alaska: Canadian Journal of Earth Sciences, v. 14, p. 2578–2592.

Hodych, J. F., Buchan, K. L., and Patzold, J. P., 1984, Paleomagnetic dating of the transformation of oolitic geothite to hematite in iron ore: Canadian Journal of Earth Sciences, v. 21, p. 127–130.

Hornafius, J. S., 1985, Neogene tectonic rotation of the Santa Ynez Range, western Transverse Ranges, California, suggested by paleomagnetic investigation of the Monterey Formation: Journal of Geophysical Rsearch, v. 90, p. 12503–12522.

Horton, R. A., Geissman, J. W., and Tschauder, R. J., 1984, Paleomagnetism and rock magnetism of the Mississippian Leadville (carbonate) Formation and implications for the age of sub-regional dolomitization: Geophysical Research Letters, v. 11, p. 649–652.

Hower, J. C., 1979, A Lower Ordovician geomagnetic pole from Pennsylvania: Earth and Planetary Science Letters, v. 44, p. 65–72.

Hurley, N. F., and Van der Voo, R., 1987, Paleomagnetism of Upper Devonian reefal limestones, Canning basin, western Australia: Geological Society of America Bulletin, v. 98, p. 138–146.

Irving, E., 1956, Palaeomagnetic and palaeoclimatological aspects of polar wandering: Geofisica Pura e Applicata, v. 33, p. 23–41.

—— , 1979, Paleopoles and paleolatitudes of North America and speculations about displaced terrains: Canadian Journal of Earth Sciences, v. 16, p. 669–694.

Irving, E., and Banks, M. R., 1961, Paleomagnetic results from the Upper Triassic lavas of Massachusetts: Journal of Geophysical Research, v. 66, p. 1935–1939.

Irving, E., and Irving, G. A., 1982, Apparent polar wander paths, Carboniferous through Cenozoic and the assembly of Gondwana: Geophysical Surveys, v. 5, p. 141–188.

Irving, E., and Lapointe, P. L., 1975, Paleomagnetism of Precambrian rocks of Laurentia: Geoscience Canada, v. 2, p. 90–98.

Irving, E., and McGlynn, J. C., 1976, Proterozoic magnetostratigraphy and the tectonic evolution of Laurentia: Philosophical Transactions of the Royal

Society of London, Series A, v. 280, p. 433–468.

——, 1979, Palaeomagnetism in the Coronation Geosyncline and arrangement of continents in the middle Proterozoic: Geophysical Journal of the Royal Astronomical Society, v. 58, p. 309–336.

——, 1981, On the coherence, rotation, and palaeolatitude of Laurentia in the Proterozoic, *in* Kroener, A., ed., Precambrian plate tectonics: Amsterdam, Elsevier, p. 561–598.

Irving, E., and Naldrett, A. J., 1977, Paleomagnetism in Abitibi greenstone belt, and Abitibi and Matachewan diabase dikes; Evidence of the Archean geomagnetic field: Journal of Geology, v. 85, p. 157–176.

Irving, E., and Park, J. K., 1972, Hairpins and superintervals: Canadian Journal of Earth Sciences, v. 9, p. 1318–1324.

Irving, E., and Strong, D. F., 1984, Palaeomagnetism of the early Carboniferous Deer Lake Group, western Newfoundland; No evidence for mid-Carboniferous displacement of "Acadia": Earth and Planetary Science Letters, v. 69, p. 379–390.

——, 1985, Paleomagnetism of rocks from Burin Peninsula, Newfoundland; Hypothesis of Late Paleozoic displacement of Acadia criticized: Journal of Geophysical Research, v. 90, p. 1949–1962.

Irving, E., Park, J. K., and Roy, J. L., 1972, Palaeomagnetism and the origin of the Grenville Front: Nature, v. 236, p. 344–346.

Irving, E., Tanczyk, J., and Hastie, J., 1976, Catalogue of paleomagnetic directions and poles, fourth issue; Mesozoic results, 1954–1975: Ottawa, Geomagnetic Service of Canada Geomagnetic Series, v. 6, 70 p.

Jackson, M., and Van der Voo, R., 1985, A Lower Ordovician paleomagnetic pole from the Oneota Dolomite, Upper Mississippi Valley: Journal of Geophysical Research, v. 90, p. 10449–10461.

Johnson, A. H., and Nairn, A.E.M., 1972, Jurassic paleomagnetism: Nature, v. 240, p. 551–552.

Johnson, R.J.E., and Van der Voo, R., 1983, Paleomagnetism of the Mid-Devonian Fisset Brook Formation, Cape Breton Island, Nova Scotia [abs.]: EOS Transactions of the American Geophysical Union, v. 64, p. 216.

——, 1985, Middle Cambrian paleomagnetism of the Avalon terrane in Cape Breton Island, Nova Scotia: Tectonics, v. 4, p. 629–651.

——, 1986, Paleomagnetism of the Late Precambrian Fourchu Group, Cape Breton Island, Nova Scotia: Canadian Journal of Earth Sciences, v. 23, p. 1673–1685.

Jones, D. L., Silberling, N. J., and Hillhouse, J. W., 1977, Wrangellia; A displaced terrane in northwestern North America: Canadian Journal of Earth Sciences, v. 14, p. 2565–2577.

Jones, D. S., MacFadden, B. J., Opdyke, N. D., and Smith, D. L., 1983, Paleomagnetism of Lower Paleozoic rocks of the Florida basement [abs.]: EOS Transactions of the American Geophysical Union, v. 64, p. 690.

Kean, W. F., and Voltz, C. E., 1989, Paleomagnetic studies of rocks at the Ordovician-Silurian boundary in Wisconsin: Neda Volume, Geoscience Wisconsin, in press.

Kent, D. V., 1979, Paleomagnetism of the Devonian Onondaga limestone revisited: Journal of Geophysical Research, v. 84, p. 3576–3588.

——, 1982, Paleomagnetic evidence for post-Devonian displacement of the Avalon Platform (Newfoundland): Journal of Geophysical Research, v. 87, p. 8709–8716.

——, 1985, Thermoviscous remagnetization in some Appalachian limestones: Geophysical Research Letters, v. 12, p. 805–808.

——, 1986, Separation of pre-folding and secondary magnetizations from the Bloomsburg Formation from the southern limb of the Pennsylvania reentrant [abs.]: EOS Transactions of the American Geophysical Union, v. 67, p. 270.

Kent, D. V., and Opdyke, N. D., 1978, Paleomagnetism of the Devonian Catskill red beds; Evidence for motion of the coastal New England–Canadian Maritime region relative to cratonic North America: Journal of Geophysical Research, v. 83, p. 4441–4450.

——, 1979, The Early Carboniferous paleomagnetic field for North America and its bearing on the tectonics of the northern Appalachians: Earth and Planetary Science Letters, v. 44, p. 365–372.

——, 1985, Multicomponent magnetizations from the Mississippian Mauch Chunk Formation of the central Appalachians and their tectonic implications: Journal of Geophysical Research, v. 90, p. 5371–5383.

Kilbourne, D. E., 1969, Paleomagnetism of some rocks from the Mesaverde Group, southwestern Wyoming and northeastern Utah: Geological Society of America Bulletin, v. 80, p. 2069–2074.

Kluth, C. F., Butler, R. F., Harding, L. E., Shafiqullah, M., and Damon, P. E., 1982, Paleomagnetism of Late Jurassic rocks in the northern Canelo Hills, southeastern Arizona: Journal of Geophysical Research, v. 87, p. 7079–7086.

Lapointe, P. L.,1979, Paleomagnetism and orogenic history of the Botwood group and Mount Peyton Batholith, central Mobile Belt, Newfoundland: Canadian Journal of Earth Sciences, v. 16, p. 866–876.

Larochelle, A., 1967, Paleomagnetic directions of a basic sill in Prince Edward Island: Geological Survey of Canada Paper 67–39, 6 p.

——, 1969, Paleomagnetism of the Monteregian Hills; Further new results: Journal of Geophysical Research, v. 74, p. 2571–2575.

——, 1971, Note on the paleomagnetism of two diabase dikes, Anticosti Island, Quebec: Geological Association of Canada Proceedings, v. 23, p. 73–76.

Larochelle, A., Black, R. F., and Wanless, R. K., 1965, Paleomagnetism of the Isachsen diabase rocks: Nature, v. 208, p. 179–182.

Larson, E. E., and Strangway, D. W., 1969, Magnetization of the Spanish Dike swarm, Colorado, and Shiprock Dike, New Mexico: Journal of Geophysical Research, v. 74, p. 1505–1514.

Larson, E. E., and Walker, T. R., 1982, A rock magnetic study of the lower massive sandstone, Moenkopi Formation (Triassic), Gray Mountain area, Arizona: Journal of Geophysical Research, v. 87, p. 4819–4836.

Larson, E. E., Mutschler, F. E., and Brinkworth, G., 1969, Paleocene virtual geomagnetic poles determined from volcanic rocks near Golden, Colorado: Earth and Planetary Science Letters, v. 7, p. 29–32.

Larson, E. E., Reynolds, R., and Hoblitt, R., 1973, New virtual and paleomagnetic pole positions from isotopically dated Precambrian rocks in Wyoming, Montana, and Arizona; Their significance in establishing a North American apparent polar wander path: Geological Society of America Bulletin, v. 84, p. 3231–3248.

Larson, E. E., Walker, T. R., Patterson, P. E., Hoblitt, R. P., and Rosenbaum, J. G., 1982, Paleomagnetism of the Moenkopi Formation, Colorado Plateau; Basis for long-term model of acquisition of chemical remanent magnetization in red beds: Journal of Geophysical Research, v. 87, p. 1081–1106.

Larson, E. E., Patterson, P. E., Curtis, G., Drake, R., and Mutschler, F. E., 1985, Petrologic, paleomagnetic, and structural evidence of a Paleozoic rift system in Oklahoma, New Mexico, Colorado, and Utah: Geological Society of America Bulletin, v. 96, p. 1364–1372.

Løvlie, R., and Opdyke, N. D., 1974, Rock magnetism and paleomagnetism of some intrusions from Virginia: Journal of Geophysical Research, v. 79, p. 343–349.

Lynnes, C. S., and Van der Voo, R., 1984, Paleomagnetism of the Cambro-Ordovician McClure Mountain alkalic complex, Colorado: Earth and Planetary Science Letters, v. 71, p. 163–172.

MacDonald, W. D., 1980, Net tectonic rotation, apparent tectonic rotation, and the structural tilt correction in paleomagnetic studies: Journal of Geophysical Research, v. 85, p. 3659–3669.

Martin, D. L., 1975, A paleomagnetic polarity transition in the Devonian Columbus limestone of Ohio; A possible stratigraphic tool: Tectonophysics, v. 28, p. 125–134.

May, S. R., and Butler, R. F., 1986, North America Jurassic apparent polar wander; Implications for plate motion, paleogeography, and Cordilleran tectonics: Journal of Geophysical Research, v. 91, p. 11519–11544.

May, S. R., Butler, R. F., Shafiqullah, M., and Damon, P. E., 1986, Paleomagnetism of Jurassic volcanic rocks in the Patagonia mountains, southeastern Arizona; Implications for the North American 170 Ma reference pole: Journal of Geophysical Research, v. 91, p. 11545–11555.

McCabe, C., Van der Voo, R., and Wilkinson, B. H., 1982, Paleomagnetic and rock magnetic results from the Twin Creek Formation (Middle Jurassic), Wyoming: Earth and Planetary Science Letters, v. 60, p. 140–146.

McCabe, C., Van der Voo, R., Peacor, D. R., Scotese, C. R., and Freeman, R., 1983, Diagenetic magnetite carries ancient yet secondary magnetizations in some Paleozoic sedimentary carbonates: Geology, v. 11, p. 221–223.

McCabe, C., Van der Voo, R., and Ballard, M. M., 1984, Late Paleozoic remagnetization of the Trenton Limestone: Geophysical Research Letters, v. 11, p. 979–982.

McCabe, C., Van der Voo, R., Wilkinson, B. H., and Devaney, K., 1985, A Middle/Late Silurian paleomagnetic pole from limestone reefs of the Wabash Formation (Indiana, USA): Journal of Geophysical Research, v. 90, p. 2959–2965.

McElhinny, M. W., 1973, Paleomagnetism and plate tectonics: Cambridge, Cambridge University Press, 358 p.

McElhinny, M. W., and Embleton, B.J.J., 1976, Precambrian and early Palaeozoic palaeomagnetism in Australia: Philosophical Transactions of the Royal Society of London, v. A 280, p. 417–431.

McElhinny, M. W., and McWilliams, M. O., 1977, Precambrian geodynamics; A paleomagnetic view: Tectonophysics, v. 40, 137–159.

McGlynn, J. C., and Irving, E., 1978, Multicomponent magnetization of the Pearson Formation (Great Slave Supergroup, N.W.T.) and the Coronation Loop: Canadian Journal of Earth Sciences, v. 15, p. 642–654.

McMahon, B. E., and Strangway, D. W., 1968, Stratigraphic implications of paleomagnetic data from Upper Paleozoic–Lower Triassic red beds of Colorado: Geological Society of America Bulletin, v. 79, p. 417–428.

McWilliams, M. O., 1984, Confidence limits on net tectonic rotation: Geophysical Research Letters, v. 11, p. 825–827.

McWilliams, M. O., and Dunlop, D. J., 1975, Precambrian paleomagnetism; Magnetizations reset by the Grenville orogeny: Science, v. 190, p. 269–272.

——, 1978, Grenville paleomagnetism and tectonics: Canadian Journal of Earth Sciences, v. 15, p. 687–695.

Miller, J. D., and Kent, D. V., 1986a, Paleomagnetism of the Upper Devonian Catskill Formation from the southern limb of the Pennsylvania salient; Possible evidence of oroclinal rotation: Geophysical Research Letters, v. 13, p. 1173–1176.

——, 1986b, Synfolding and prefolding magnetizations in the Upper Devonian Catskill Formation of eastern Pennsylvania; Implications for the tectonic history of Acadia: Journal of Geophysical Research, v. 91, p. 12791–12803.

Miller, J. D., and Opdyke, N. D., 1985, Magnetostratigraphy of the Red Sandstone Creek section, Vail, Colorado: Geophysical Research Letters, v. 12, p. 133–136.

Morel, P., and Irving, E., 1978, Tentative paleocontinental maps for the early Phanerozoic and Proterozoic: Journal of Geology, v. 86, p. 535–561.

Morris, W. A., 1976, Transcurrent motion determined paleomagnetically in the northern Appalachians and Caledonides and the Acadian orogeny: Canadian Journal of Earth Sciences, v. 13, p. 1236–1243.

Morrison, J., and Ellwood, B. B., 1986, Paleomagnetism of Silurian-Ordovician sediments from the Valley and Ridge province, northwest Georgia: Geophysical Research Letters, v. 13, p. 189–192.

Murthy, G. S., 1985, Paleomagnetism of certain constituents of the Bay St. George sub-basin, western Newfoundland: Physics of the Earth and Planetary Interiors, v. 39, p. 89–107.

Murthy, G. S., and Rao, K. V., 1976, Paleomagnetism of Steel Mountain and Indian Head anorthosites from western Newfoundland: Canadian Journal of Earth Sciences, v. 13, p. 75–83.

Neuman, R. B., 1984, Geology and paleobiology of islands in the Ordovician Iapetus Ocean; Review and implications: Geological Society of America Bulletin, v. 95, p. 1188–1201.

Nur, A., 1983, Accreted terranes: Reviews of Geophysics and Space Physics, v. 21, p. 1779–1785.

Oliver, J., 1986, Fluids expelled tectonically from orogenic belts; Their role in hydrocarbon migration and other geologic phenomena: Geology, v. 14, p. 99–102.

Onstott, T. C., and Hargraves, R. B., 1981, Proterozoic transcurrent tectonics: Nature, v. 289, p. 131–136.

Onstott, T. C., Hargraves, R. B., York, D., and Hall, C., 1984, Constraints on the motions of South American and African Shields during the Proterozoic; 1, $^{40}Ar/^{39}Ar$ and paleomagnetic correlations between Venezuela and Liberia: Geological Society of America Bulletin, v. 95, p. 1045–1054.

Opdyke, N. D., 1961, The paleomagnetism of the New Jersey Triassic; A field study of the inclination error in red sediments: Journal of Geophysical Research, v. 66, p. 1941–1949.

——, 1981, Paleomagnetism of the Barachois Formation and the extent of the Acadian displaced terrane [abs.]: EOS Transactions of the American Geophysical Union, v. 62, p. 264.

Opdyke, N. D., and Wensink, H., 1966, Paleomagnetism of rocks from the White Mountain plutonic-volcanic series in New Hampshire and Vermont: Journal of Geophysical Research, v. 71, p. 3045–3051.

Payne, M. A., Shulik, S. J., Donahue, J., Rollins, H. B., and Schmidt, V. A., 1981, Paleomagnetic poles for the Carboniferous Brush Creek limestone and Buffalo siltstone from southwestern Pennsylvania: Physics of the Earth and Planetary Interiors, v. 25, p. 113–118.

Perigo, R., Van der Voo, R., Auvray, B., and Bonhommet, N., 1983, Paleomagnetism of Late Precambrian–Cambrian volcanics and intrusives of the Armorican Massif, France: Geophysical Journal of the Royal Astronomical Society, v. 75, p. 235–260.

Perroud, H., 1986, Paleomagnetic evidence for tectonic rotations in the Variscan Mountain Belt: Tectonics, v. 5, p. 205–214.

Perroud, H., and Bonhommet, N., 1984, A Devonian pole for Armorica: Geophysical Journal of the Royal Astronomical Society, v. 77, p. 839–845.

Perroud, H., and Van der Voo, R., 1984, Secondary magnetizations from the Clinton-type iron ores of the Silurian Red Mountain Formation, Alabama: Earth and Planetary Science Letters, v. 67, p. 391–399.

Perroud, H., Van der Voo, R., and Bonhommet, N., 1984, Paleozoic evolution of the Armorica plate on the basis of paleomagnetic data: Geology, v. 12, p. 579–582.

Pesonen, L. J., and Neuvonen, K. J., 1981, Palaeomagnetism of the Baltic Shield; Implications for Precambrian tectonics, *in* Kroner, A., ed., Precambrian plate tectonics: Amsterdam, Elsevier, p. 623–648.

Peterson, D. N., and Nairn, A.E.M., 1971, Palaeomagnetism of Permian red beds from the southwestern United States: Geophysical Journal of the Royal Astronomical Society, v. 23, p. 191–205.

Piper, J.D.A., 1976, Palaeomagnetic evidence for a Proterozoic supercontinent: Philosophical Transactions of the Royal Society of London, v. A 280, p. 469–490.

——, 1979, Aspects of Caledonian palaeomagnetism and their tectonic implications: Earth and Planetary Science Letters, v. 44, p. 176–192.

Proko, M. S., and Hargraves, R. B., 1973, Paleomagnetism of the Beemerville (New Jersey) alkaline complex: Geology, v. 1, p. 185–186.

Pullaiah, G., and Irving E., 1975, Paleomagnetism of the contact aureole and late dikes of the Otto stock, Ontario, and its application to early Proterozoic apparent polar wandering: Canadian Journal of Earth Sciences, v. 12, p. 1609–1618.

Purucker, M. E., Elston, D. P., and Shoemaker, E. M., 1980, Early acquisition of characteristic magnetization in red beds of the Moenkopi Formation (Triassic) Gray Mountain, Arizona: Journal of Geophysical Research, v. 85, p. 997–1012.

Rao, K. V., and Van der Voo, R., 1980, Paleomagnetism of a Paleozoic anorthosite from the Appalachian Piedmont, northern Delaware; Possible tectonic implications: Earth and Planetary Science Letters, v. 47, p. 113–120.

Rao, K. V., Seguin, M. K., and Deutsch, E. R., 1981, Paleomagnetism of Siluro-Devonian and Cambrian granititic rocks from the Avalon zone in Cape Breton Island, Nova Scotia: Canadian Journal of Earth Sciences, v. 18, p. 1187–1210.

Reeve, S. C., 1975, Paleomagnetic studies of sedimentary rocks of Cambrian and Triassic age [Ph.D. thesis]: Dallas, University of Texas, 426 p.

Reeve, S. C., and Helsley, C. E., 1972, Magnetic reversal sequence in the upper portion of the Chinle Formation, Montoya, New Mexico: Geological Society of America Bulletin, v. 83, p. 3795–3812.

Reid, A. B., 1972, A paleomagnetic study at 1800 million years in Canada [Ph.D. thesis]: Calgary, University of Alberta.

Ressetar, R., and Martin, D. L., 1980, Paleomagnetism of Eocene igneous intrusives in the Valley and Ridge Province, Virginia and West Virginia: Canadian Journal of Earth Sciences, v. 17, p. 1583–1587.

Reynolds, R. L., Fishman, N. S., Hudson, M. R., Karachewski, J. A., and Goldhaber, M. B., 1985, Magnetic minerals and hydrocarbon migration; Evidence from Cement (OK), North Slope (AK), and the Wyoming-Idaho-Utah Thrust Belt [abs.]: EOS Transactions of the American Geophysical Union, v. 66, p. 867.

Robertson, W. A., and Fahrig, W. F., 1971, The great Logan paleomagnetic loop; The polar wandering path from Canadian Shield rocks during the Neohelikian Era: Canadian Journal of Earth Sciences, v. 8, p. 1355–1372.

Roy, J. L., 1966, Désaimantation thermique et analyse statistique de sédiments Carbonifères et Permiens de l'est du Canada: Canadian Journal of Earth Sciences, v. 3, p. 139–161.

——, 1969, Paleomagnetism of the Cumberland Group and other Paleozoic formations: Canadian Journal of Earth Sciences, v. 6, p. 663–669.

——, 1977, La position stratigraphique determinée paléomagnétiquement de sédiments carbonifères de Minudie Point, Nouvelle Ecosse: à propos de l'horizon répère magnétique du Carbonifère: Canadian Journal of Earth Sciences, v. 14, p. 1116–1127.

——, 1983, Paleomagnetism of the North American Precambrian; A critical look at the data base: Precambrian Research, v. 19, p. 319–348.

Roy, J. L., and Morris, W. A., 1983, A review of paleomagnetic results from the Carboniferous of North America; The concept of Carboniferous geomagnetic field horizon markers: Earth and Planetary Science Letters, v. 65, p. 167–181.

Roy, J. L., and Park, J. K., 1969, Paleomagnetism of the Hopewell Group, New Brunswick: Journal of Geophysical Research, v. 74, p. 594–604.

——, 1974, The magnetization process of certain red beds; Vector analysis of chemical and thermal results: Canadian Journal of Earth Sciences, v. 11, p. 437–471.

Roy, J. L. Opdyke, N. D., and Irving, E., 1967, Further paleomagnetic results from the Bloomsburg Formation: Journal of Geophysical Research, v. 72, p. 5075–5086.

Roy, J. L., Robertson, W. A., and Park, J. K., 1968, Stability of the magnetization of the Hurley Creek Formation: Journal of Geophysical Research, v. 73, p. 697–702.

Roy, J. L., Morris, W. A., Lapointe, P. L., Irving, E., Park, J. K., and Schmidt, P. W., 1978, Apparent polar wander paths and the joining of the Superior and Slave provinces during early Proterozoic time; Comment and reply: Geology, v. 6, p. 132–133.

Roy, J. L., Anderson, P., and Lapointe, P. L., 1979, Paleomagnetic results from three rock units of New Brunswick and their bearing on the Lower Paleozoic tectonics of North America: Canadian Journal of Earth Sciences, v. 16, p. 1210–1227.

Runcorn, S. K., 1956, Paleomagnetic comparisons between Europe and North America: Geological Association of Canada Proceedings, v. 9, p. 77–85.

Samson, S. L., Secor, D. T., Jr., Snoke, A. W., and Palmer, A. R., 1982, Geological implications of recently discovered Middle Cambrian trilobites in the Carolina slate belt: Geological Society of America Abstracts with Programs, v. 14, p. 607.

Schwartz, S. Y., and Van der Voo, R., 1983, Paleomagnetic evaluation of the orocline hypothesis in the central and southern Appalachians: Geophysical Research Letters, v. 10, p. 505–508.

Schwartz, S. Y., and Van der Voo, R., 1984, Paleomagnetic study of thrust sheet rotation during foreland impingement in the Wyoming-Idaho overthrust belt: Journal of Geophysical Research, v. 89, p. 10077–10086.

Scotese, C. R., 1985, The assembly of Pangea; Middle and Late Paleozoic paleomagnetic results from North America [Ph.D. thesis]: Chicago, University of Chicago, 339 p.

Scotese, C. R., Van der Voo, R., and McCabe, C., 1982, Paleomagnetism of the Upper Silurian and Lower Devonian carbonates of New York State; Evidence for secondary magnetizations residing in magnetite: Physics of the Earth and Planetary Interiors, v. 30, p. 385–395.

Scotese, C. R., Van der Voo, R., Johnson, R. E., and Giles, P. S., 1984, Paleomagnetic results from the Carboniferous of Nova Scotia, *in* Van der Voo, R., Scotese, C. R., and Bonhommet, N., eds., Plate reconstructions from Paleozoic paleomagnetism: American Geophysical Union Geodynamics Series, v. 12, p. 63–81.

Scott, G. R., 1979, Paleomagnetic studies of the Early Carboniferous St. Joe limestone, Arkansas: Journal of Geophysical Research, v. 84, p. 6277–6285.

Seguin, M.-K., Singh, A., and Fyffe, L., 1985, New paleomagnetic data from Carboniferous volcanics and red beds from central New Brunswick: Geophysical Research Letters, v. 12, p. 81–84.

Seyfert, C. K., and Cavanaugh, M. D., 1978, Apparent polar wander paths and the joining of the Superior and Slave provinces during early Proterozoic time; Comment and reply: Geology, v. 6, p. 133–135.

Sharon, L., and Hsu, I.-C., 1969, Paleomagnetic investigations of some Arkansas alkalic igneous rocks: Journal of Geophysical Research, v. 74, p. 2774–2779.

Sheriff, S. D., and Shive, P. N., 1980, The Rattlesnake Hills of central Wyoming revisited; Further paleomagnetic results: Geophysical Research Letters, v. 7, p. 589–592.

Shive, P. N., and Frerichs, W. E., 1974, Paleomagnetism of the Niobrara Formation in Wyoming, Colorado, and Kansas: Journal of Geophysical Research, v. 79, p. 3001–3009.

Shive, P. N., and Pruss, E. F., 1977, A paleomagnetic study of basalt flows from the Absaroka Mountains, Wyoming: Journal of Geophysical Research, v. 82, p. 3039–3048.

Shive, P. N., Steiner, M. B., and Huycke, D. T., 1984, Magnetostratigraphy, paleomagnetism, and remanence acquisition in the Triassic Chugwater Formation of Wyoming: Journal of Geophysical Research, v. 89, p. 1801–1815.

Smith, A. G., and Hallam, A., 1970, The fit of the southern continents: Nature, v. 225, p. 139–144.

Smith, D. I., and Watson, J., 1983, Scale and timing of movements on the Great Glen Fault, Scotland: Geology, v. 11, p. 523–526.

Smith, T. E., and Noltimier, H. C., 1979, Paleomagnetism of the Newark-trend igneous rocks of the north-central Appalachians and the opening of the central Atlantic Ocean: American Journal of Science, v. 279, p. 778–807.

Spall, H., 1968, Paleomagnetism of basement granites of southern Oklahoma and its implications; Progress report: Oklahoma Geological Survey Notes, v. 28, p. 65–80.

——, 1970, Paleomagnetism of basement granites in southern Oklahoma; Final report: Oklahoma Geological Survey Notes, v. 30, p. 136–150.

——, 1971, Precambrian apparent polar wandering; Evidence from North America: Earth and Planetary Science Letters, v. 10, p. 273–280.

Spariosu, D. J., and Kent, D. V., 1983, Paleomagnetism of the Lower Devonian Traveler Felsite and the Acadian orogeny in the New England Appalachians: Geological Society of America Bulletin, v. 94, p. 1319–1328.

Spariosu, D. J., Kent, D. V., and Keppie, J. D., 1984, Late Paleozoic motions of the Meguma terrane, Nova Scotia; New paleomagnetic evidence, *in* Van der Voo, R., Scotese, C. R., and Bonhommet, N., eds., Plate reconstructions from Paleozoic paleomagnetism: American Geophysical Union Geodynamics Series, v. 12, p. 82–98.

Stead, R. J., and Kodama, K. P., 1984, Paleomagnetism of the Cambrian rocks of the Great Valley of east-central Pennsylvania; Fold test constraints on the age of magnetization, *in* Van der Voo, R., Scotese, C. R., and Bonhommet, N., eds., Plate reconstructions from Paleozoic paleomagnetism: American Geophysical Union Geodynamics Series, v. 12, p. 120–130.

Steiner, M. B., 1973, Late Paleozoic partial remagnetization of Ordovician rocks from southern Oklahoma: Geological Society of America Bulletin, v. 84, p. 341–346.

——, 1978, Magnetic polarity during the Middle Jurassic as recorded in the

Summerville and Curtis Formations: Earth and Planetary Science Letters, v. 38, p. 331–345.
—— , 1986, Rotation of the Colorado Plateau: Tectonics, v. 5, p. 649–660.
Steiner, M. B., and Helsley, C. E., 1974, Magnetic polarity sequence of the Upper Triassic Kayenta Formation: Geology, v. 2, p. 191–194.
—— , 1975, Reversal patterns and apparent polar wander for the Late Jurassic: Geological Society of America Bulletin, v. 86, p. 1537–1543.
Ueno, H., Irving, E., and McNutt, R. H., 1975, Paleomagnetism of the Whitestone anorthosite and diorite, the Grenville polar track, and relative motions of the Laurentian and Baltic Shield: Canadian Journal of Earth Sciences, v. 12, p. 209–226.
Van Alstine, D. R., 1979, Apparent polar wander with respect to North America since the late Precambrian [Ph.D. thesis]: Pasadena, California Institute of Technology, 358 p.
Van der Linden, W.J.M., 1985, Looping the loop; Geotectonics of the Alpine-Meditteranean region: Geologie en Mijnbouw, v. 64, p. 281–295.
Van der Voo, R., 1979, Paleozoic assembly of Pangea; A new plate tectonic model for the Taconic, Caledonian, and Hercynian orogenies [abs.]: EOS Transactions of the American Geophysical Union, v. 60, p. 241.
—— , 1981, Paleomagnetism of North America; A brief review, *in* McElhinny, M. W., and Valencio, D. A., eds., Paleoreconstruction of the continents: American Geophysical Union and Geological Society of America Geodynamics Series, v. 2, p. 159–176.
—— , 1982, Pre-Mesozoic paleomagnetism and plate tectonics: Annual Reviews of Earth and Planetary Sciences, v. 10, p. 191–220.
—— , 1983, Paleomagnetic constraints on the assembly of the Old Red Continent: Tectonophysics, v. 91, p. 271–283.
—— , 1988, Triassic-Jurassic plate migrations and paleogeographic reconstructions in the Atlantic domain, *in* Manspeizer, W., ed., Triassic-Jurassic rifting: Amsterdam, Elsevier, Part A, p. 29–40.
Van der Voo, R., and French, R. B., 1974, Apparent polar wander for the Atlantic-bordering continents; Late Carboniferous to Eocene: Earth Science Reviews, v. 10, p. 99–119.
—— , 1977, Paleomagnetism of the Late Ordovician Juniata Formation and the remagnetization hypothesis: Journal of Geophysical Research, v. 82, p. 5796–5802.
Van der Voo, R., and Grubbs, K. L.,1977, Paleomagnetization of the Triassic Chugwater red beds revisited (Wyoming, USA): Tectonophysics, v. 41, p. T27–T33.
Van der Voo, R., and Johnson, R.J.E., 1985, Paleomagnetism of the Dunn Point Formation (Nova Scotia); High paleolatitudes for the Avalon terrane in the Late Ordovician: Geophysical Research Letters, v. 12, p. 337–340.
Van der Voo, R., and McCabe, C., 1985, Implications of remagnetized limestone paleomagnetic poles for Late Paleozoic APW of cratonic North America [abs.]: EOS Transactions of the American Geophysical Union, v. 66, p. 875.
Van der Voo, R., and Scotese, C. R., 1981, Paleomagnetic evidence for a large (c. 2000 km) sinistral offset along the Great Glen Fault during Carboniferous time: Geology, v. 9, p. 583–589.
Van der Voo, R., and Zijderveld, J.D.A., 1969, Paleomagnetic research in the western Mediterranean: Transactions of the Royal Geological and Mining Society of the Netherlands, v. 26, p. 121–138.
Van der Voo, R., French, R. B., and Williams, D. W., 1976, Paleomagnetism of the Wilberns Formation (Texas) and the Late Cambrian paleomagnetic field for North America: Journal of Geophysical Research, v. 81, p. 5633–5638.
Van der Voo, R., French, A. N., and French, R. B., 1979, A paleomagnetic pole position from the folded Upper Devonian Catskill red beds, and its tectonic implications: Geology, v. 7, p. 345–348.
Van der Voo, R., Jones, M., Gromme, C. S., Eberlein, G. D., and Churkin, M., Jr., 1980, Paleozoic paleomagnetism and northward drift of the Alexander terrane, southeast Alaska: Journal of Geophysical Research, v. 85, p. 5281–5296.
Van der Voo, R., Peinado, J., and Scotese, C. R., 1984, A paleomagnetic reevaluation of Pangea reconstructions, *in* Van der Voo, R., Scotese, C. R., and Bonhommet, N., eds., Plate reconstructions from Paleozoic paleomagnetism: American Geophysical Union Geodynamics Series, v. 12, p. 11–26.
Van der Voo, R., Perroud, H., and Bonhommet, N., 1985a, Reply *to* Comment on "Paleozoic evolution of the Armorica plate on the basis of paleomagnetic data": Geology, v. 13, p. 380–381.
—— , 1985b, Reply *to* Comment on "Paleozoic evolution of the Armorica plate on the basis of paleomagnetic data": Geology, v. 13, p. 589–590.
Van Fossen, M. C., Flynn, J. J., and Forsythe, R. D., 1986, Paleomagnetism of Early Jurassic rocks, Watchung Mountains, Newark Basin; Evidence for complex rotations along the border fault: Geophysical Research Letters, v. 13, p. 185–188.
Vugteveen, R. W., Barnes, A. E., and Butler, R. F., 1981, Paleomagnetism of the Roskruge and Gringo Gulch Volcanics, southeast Arizona: Journal of Geophysical Research, v. 86, p. 4021–4028.
Wagner, M. E., and Crawford, M. L., 1975, Polymetamorphism of the Precambrian Baltimore Gneiss in southeastern Pennsylvania: American Journal of Science, v. 275, p. 653–682.
Watts, D. R., 1981, Paleomagnetism of the Fond du Lac Formation and the Eileen and Middle River sections with implications for Keweenawan tectonics and the Grenville problem: Canadian Journal of Earth Sciences, v. 18, p. 829–841.
Watts, D. R., and Van der Voo, R., 1979, Paleomagnetic results from the Ordovician Moccasin, Bays, and Chapman Ridge Formations of the Valley and Ridge Province, eastern Tennessee: Journal of Geophysical Research, v. 84, p. 645–655.
Watts, D. R., Van der Voo, R., and Reeve, S. C., 1980a, Cambrian paleomagnetism of the Llano uplift, Texas: Journal of Geophysical Research, v. 85, p. 5316–5330.
Watts, D. R., Van der Voo, R., and French, R. B., 1980b, Paleomagnetic investigations of the Cambrian Waynesboro and Rome formations in the Valley and Ridge Province of the Appalachian mountains: Journal of Geophysical Research, v. 85, p. 5331–5343.
Webb, G. W., 1969, Paleozoic wrench faulting in Canadian Appalachians: American Association of Petroleum Geologists Memoir 12, p. 754–786.
Weisse, P. A., Haggerty, S. E., and Brown, L. L., 1985, Paleomagnetism and magnetic mineralogy of the Nahant Gabbro and tonalite, eastern Massachusetts: Canadian Journal of Earth Sciences, v. 22, p. 1425–1435.
Williams, H., 1979, Appalachian orogen in Canada: Canadian Journal of Earth Sciences, v. 16, p. 792–807.
Williams, H., and Hatcher, R. B., Jr., 1983, Appalachian suspect terranes, *in* Hatcher, R. D., Jr., Williams, H., and Zietz, I., eds., Contributions to the tectonics and geophysics of mountain chains: Geological Society of America Memoir 158, p. 33–53.
Wilson, J. T., 1966, Did the Atlantic close and then re-open?: Nature, v. 211, p. 676–681.
Wisniowiecki, M., Van der Voo, R., McCabe, C., and Kelly, W. C., 1983, A Pennsylvanian paleomagnetic pole from the mineralized Late Cambrian Bonneterre Formation, southeast Missouri: Journal of Geophysical Research, v. 88, p. 6540–6548.
York, D., 1978, A formula describing both magnetic and isotopic blocking temperatures: Earth and Planetary Science Letters, v. 39, p. 89–93.

Manuscript Revised November 24, 1986
Manuscript Accepted by the Society October 31, 1988

Printed in U.S.A.

Geological Society of America
Memoir 172
1989

Chapter 22

Paleomagnetism of continental North America; Implications for displacement of crustal blocks within the Western Cordillera, Baja California to British Columbia

Myrl E. Beck, Jr.
Department of Geology, Western Washington University, Bellingham, Washington 98225

ABSTRACT

A compilation of 89 paleomagnetic poles for westernmost North America—southern British Columbia to northern Mexico—demonstrates that the Cordillera is composed to a very important extent of displaced crustal blocks. The distribution of paleopoles also shows (with decreasing degree of certainty) that (1) plate interactions along the leading edge of North America are responsible for terrane displacement; (2) since about 125 Ma, displacement relative to stable North America was northward, with clockwise rotation; (3) prior to 125 Ma (and beginning at perhaps 155 Ma), terranes moved generally southward relative to North America, and rotated counterclockwise; and (4) displaced terranes tend to fragment, the farther they travel. Complications in the interpretation of paleomagnetic data arise from several sources, principally uncertainty as to hemisphere of origin for displaced terranes older than middle Cretaceous, uncertainty as to the correct shape of the North American reference curve for rocks older than middle Cretaceous, and questions as to the reliability of the data themselves. No effective, immediate way to rid paleomagnetism of these questions and uncertainties exists. For that reason, interpretation of paleomagnetic data in terms of regional tectonics requires large data sets and tends to generate highly useful (but broad-brush) general models, not the detailed and specific reconstructions geologists might prefer.

INTRODUCTION

This chapter summarizes paleomagnetic evidence from the Western Cordillera of North America that has shown that most crustal blocks on the western edge of the range did not form where they are now. Even those that did form near their present location seem to have suffered small displacements, including ubiquitous block rotations. As there seems to be no reason to regard the Cordillera as unique in either origin or setting, these observations may apply to most mountain belts. The resulting unexpected increase in degrees of freedom seems certain to require a thorough restructuring of orogenic theory.

A dramatic shift toward general acceptance of high levels of lateral crustal mobility hit the earth sciences shortly after World War II. It began in the 1950s with the paleomagnetic verification of continental drift and progressed in the early 1960s through oceanographic and geophysical evidence for renewal, lateral spreading, and destruction of the ocean floor. By the late 1960s, as plate tectonics, it had become perhaps the most widely embraced explanatory tool that geology has ever enjoyed. Among the crustal features explicable by this new tool were orogenic belts.

An outgrowth or secondary consequence of plate tectonics surfaced in the early 1970s, arising mainly from regional paleomagnetic and stratigraphic studies along the western edge of North America. The set of ideas referred to has been variously called "microplate tectonics," the "terrane concept," or (less correctly) "accretionary tectonics." These ideas were formulated

Beck, M. E., Jr., 1989, Paleomagnetism of continental North America; Implications for displacement of crustal blocks within the Western Cordillera, Baja California to British Columbia, *in* Pakiser, L. C., and Mooney, W. D., Geophysical framework of the continental United States: Boulder, Colorado, Geological Society of America Memoir 172.

with the Western Cordillera of North America in mind, although some of the concepts clearly were foreshadowed by work in the Mediterranean (Zijdervelt and Van der Voo, 1973, provide a summary). The ideas themselves are simple, sweeping, and even elegant; the evidence on which they are based (some summarized in this chapter) is often dramatic. It shows that essentially the entire western edge of North America inland a variable distance ranging up to several hundred kilometers is in some sense allochthonous ("displaced" or even "rearranged" may be better). Separate displaced blocks ("terranes") may be as small as individual thrust sheets, or they may comprise blocks of crust with lateral dimensions of several hundred kilometers. "Rearrangement" may consist of attachment of exotic crustal elements to North America (accretion), of movement of crustal blocks along the edge of the continent (coastwise translation), or of large-scale in situ block rotation. Accretion, coastwise translation, and rotation are important new variables that must be considered in any useful analysis of orogeny.

Microplate tectonics in the North American Cordillera had simultaneous, parallel, and cooperative development in stratigraphy and paleomagnetism. Of the two, stratigraphy has been much the more useful in defining terranes (considered here as synonymous with the individual displaced crustal blocks, although this does not quite do justice to the "official" definition contained in Beck and others, 1980). An early terrane map is reproduced in Figure 1. This map is dramatically out of date; a newer version (Silberling and Jones, 1984) describes more than 100 separate terranes for the continental United States (Canada to Mexico) alone.

Stratigraphic and structural studies may suggest the nature of displacements a terrane has experienced, but it is paleomagnetism that has demonstrated that some terranes have moved extreme distances (as much as several thousand kilometers), and that systematic, regionally pervasive block rotations have taken place. Because this chapter focuses on paleomagnetism, and because of space considerations, the geological aspects of microplate tectonics are not discussed here. Two early publications that detail these aspects are Jones and others (1972) and Monger and Ross (1971). More recent summaries include Coney and others (1980), Jones and others (1982), and Howell (1985). Relevant early paleomagnetic publications include Irving and Yole (1972), Beck and Noson (1972), and Packer and Stone (1972). The first published attempt at a microplate synthesis from a paleomagnetic point of view was Beck (1976). Owing to a striking miscarriage of the referee system, an earlier paper by D. B. Stone and D. R. Packer never saw the light of day, although "preprints" circulating throughout the geological community were influential. The present paper brings the paleomagnetic account of Cordilleran microplate tectonics up to date.

Elsewhere I have compiled and interpreted paleomagnetic data for the western edge of North America from Baja California to southern British Columbia (Beck, 1989). In this chapter I use that compilation (summarized in Table 1), but concentrate on the nature of the paleomagnetic record and on tectonic processes, rather than geologic history. That subject is treated more thoroughly in Beck (1989). This chapter, then, is less concerned with what happened than with how it happened and how we know. I also describe how paleomagnetic techniques are used to study tectonics, and point out a few problems and limitations. A discussion of reference curves is also given, supplementing the treatment of Van der Voo (this volume).

PALEOMAGNETIC DATA

From a paleomagnetic standpoint, the western edge of North America is the best-studied plate margin in the world. Table 1 summarizes the relevant data, selected using criteria summarized elsewhere (Beck, 1980, 1989). Paleomagnetic studies are beset by many sources of potential error, and nowhere is error more likely to materialize than among the complicated rocks and geology of orogenic belts. Thus data selection for tectonic interpretation in orogenic belts is not only vital but also difficult. Van der Voo (this volume) and the Appendix in this chapter deal with matters of data selection and reliability. In consulting Table 1, the reader is warned that not all entries are of equal reliability; the original publications should be studied in all cases. Because so many things can go wrong with any paleomagnetic study, tectonic conclusions based on only a single pole, however well determined, have diminished credibility. The interpretations that follow are based on groups of independently determined poles, all of which point to the same conclusions.

Figure 2 shows the sampling locations for the data of Table 1, and Figures 3 through 6 are plots of Cordilleran paleomagnetic poles and the appropriate North America reference poles for four time slices. Taken together they indicate the following (my interpretations):

1. The western edge of the continent has been extensively rearranged by large-scale block displacement. In situ rotations about vertical axes, north-south translations, or both, have been active to such an extent that essentially the entire edge of the continent can be regarded as allochthonous. Because paleomagnetism cannot detect purely east-west motions, it has little to say about whether particular blocks are exotic (originally not part of North America) or simply displaced fragments of the North America shoreline. To make the latter distinction requires geological input.

2. Most crustal fragments have been displaced as if by dextral shear. Relative to the craton, Cretaceous and younger blocks have moved northward and rotated clockwise to an impressive extent (Figs. 3 through 5). Jurassic and older rocks (Fig. 6) may have moved relatively southward prior to about mid-Cretaceous time (ca. 125 Ma), but this is still uncertain (see below).

3. Because displacement of crustal blocks is common at the edge of the continent and far less so inland, it is reasonable to suppose that displacement is driven and controlled by relative motions between major plates.

4. According to Engebretson and others (1985), relative velocities along the western edge of North America have been

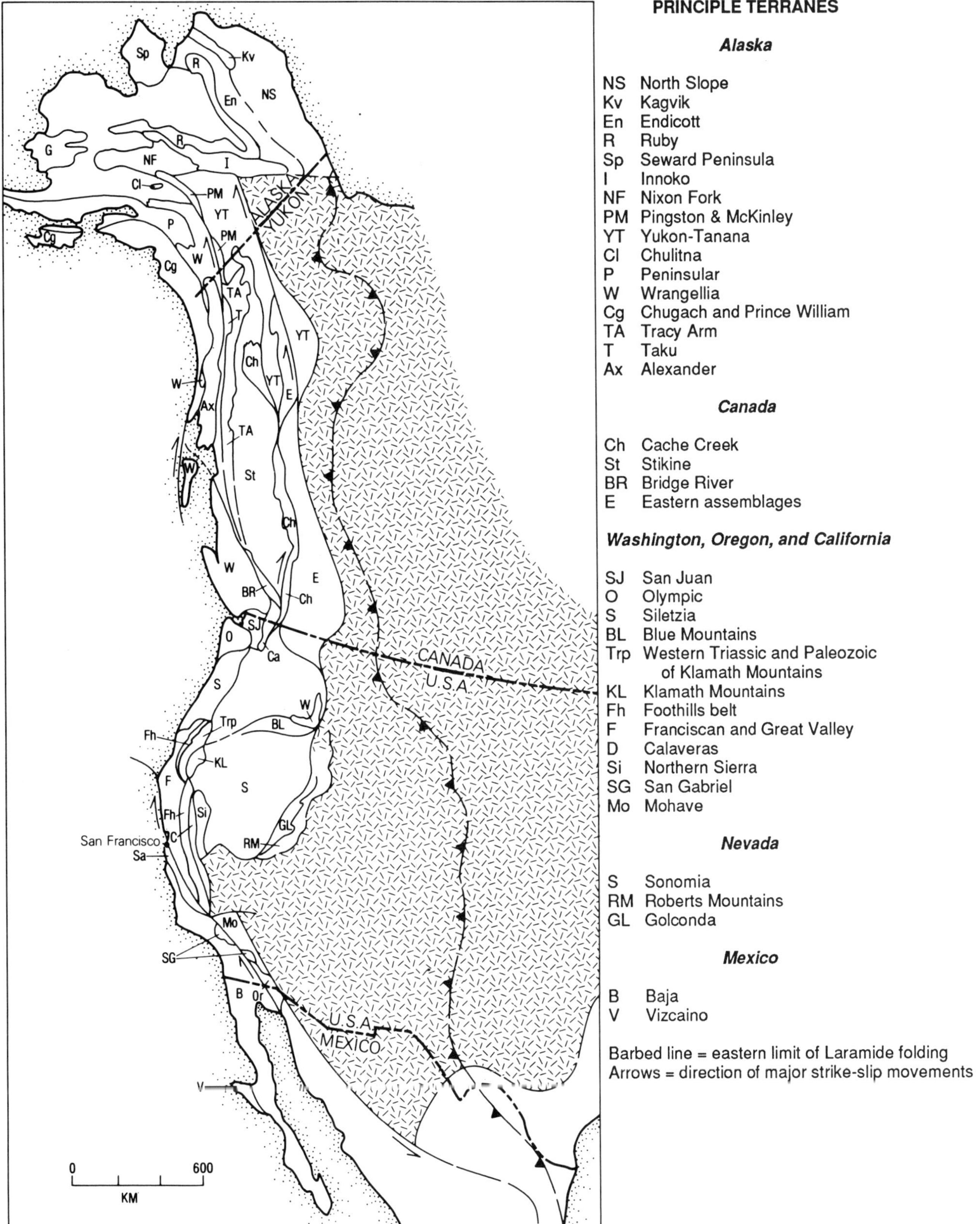

Figure 1. Terranes map for North America. From Beck and others (1980). Many of these terranes have been minutely subdivided; see Silberling and Jones (1984) and Howell (1985).

TABLE 1. SELECTED PALEOMAGNETIC POLES, WESTERN CORDILLERA

Unit*	Age (Ma)	Paleomagnetic Pole N/E	A_{95}	Reference
Neogene				
CR	10	84.2/218.2	3.0	Symons (1969a, b)
SA	10	85.6/144.2	12.0	Choiniere and Swanson (1979)
GR	15	85.8/207/4	8.2	Choiniere and Swanson (1979)
IG	15	84.5/210.9	5.4	Hooper and others (1979)
SO	5	79.4/83.8	8.9	Maniken (1972)
KK	20	78.3/223.2	8.6	Kanter and McWilliams (1982)
PB	15	66.3/145.9	4.3	Calderone and Butler (1984)
VC	25	62.4/323.6	5.3	Terres and Luyendyk (1985)
VW	25	29.0/303.9	6.5	Terres and Luyendyk (1985)
MNO	15	19.0/308.3	2.6	Hornafius (1985)
MNY	10	60.6/32.4	2.8	Hornafius (1985)
WT	20	25.1/322.3	3.4	Kamerling and Luyendyk (1985)
SL	20	75.8/123.8	3.2	Calderone and Butler (1984)
Paleogene				
MF	60	72/263	10.8	Hicken and Irving (1977)
HT	45	79.8/226.0	12.4	Symons (1977a)
SBB	30?	86.5/308.8	7.2	Irving and others (1985)
CIC	30	85.1/206.4	4.7	Beck and others (1982)
SP	50	79.4/304.2	3.0	Fox and Beck (1985)
SG	40	69.5/150.9	5.7	Symons (1973a)
PT	50	74.5/187.7	18.6	Beck and Engebretson (1982)
BR	50	86.6/324.6	8.8	Beck and Engebretson (1982)
BH	55	77.3/309.7	6.1	Globerman and others (1982)
WH	50	73.3/326.4	5.3	Wells and Coe (1985)
GV2	40	77.2/5.7	14.1	Wells and Coe (1985)
GV1	40	75.5/345.5	5.4	Beck and Burr (1979)
OV	35	70.0/317.9	3.2	Bates and others (1981)
TV	55	65.5/312.7	6.9	Magill and others (1981)
EI	40	58.4/313.2	10.4	Beck and Plumley (1980)
SV	55	45/307	8.0	Simpson and Cox (1977)
CF	40	83.9/277.6	7.4	Gromme and others (1987)
OI	30	66.8/324.8	8.3	Beck and Plumley (1986)
WC3	30	70.7/284.8	7.5	Magill and Cox (1980)
YB	35	58/308	20.0	Simpson and Cox (1977)
TF	45	49/308	11.0	Simpson and Cox (1977)
WC2	30	74.1/302.2	13.0	Magill and Cox (1980)
WC1	30	86.9/347.1	6.9	Beck and others (1986)
NA	30	80.5/120.5	12.1	Irving and others (1976), pole 11-442
PD	60	-----	---	Champion and others (1984)
MR	25	46.4/329.3	9.2	Greenhaus and Cox (1979)
GG	60	77.0/201.0	1.4	Barnes and Butler (1980)
Cretaceous				
AX	100?	76/327	5.9	Monger and Irving (1980)
HK	70	73.5/323.2	10.7	Symons (1977a)
CS	110	70/3	7.0	Symons (1977b)
HS	90	83.4/129.2	11.7	Symons (1973b)
SBA	100	65.0/345.1	6.2	Irving and others (1985)
TS	75?	27.5/297.5	7.1	Beck (1975)
MS	90	67.9/34.6	3.4	Beck and others (1981)
JM	80?	74.6/190.8	3.9	Russell and others (1982)
KE	135?	56.9/332.0	7.4	Acache and others (1982)
GRN	70?	81.4/177.6	5.7	Mankinen and Irwin (1982)
BB	140	57.6/194.8	7.7	Gromme and others (1967)
GRS	70?	68.0/156.8	12.4	Mankinen and Irwin (1982)
MD	90	76.9/234.0	5.7	Geissman and others (1984)
SNC	90	71.2/180.9	3.6	Gromme and Merrill (1965); Frei and others (1984)
PP	70	-----	---	Champion and others (1984)
SS	85	71.9/321.8	3.5	Kanter and McWilliams (1982)
FM	85	54.4/18.4	7.8	McWilliams and Howell (1982)
LW	80	57.9/149.6	4.8	Fry and others (1985)
PRC	85	84.7/346.2	3.2	Hagstrum and others (1985)
RO	70	73.6	176.0	Vugteveen and others (1981)
AL	1157	81.3/336.1	8.5	Hagstrum and others (1985)
LB	90?	79.5/316.6	6.3	Hagstrum and others (1985)
VFC	100?	82.8/13.2	3.6	Hagstrum and others (1985)
Jurassic and Triassic				
TKA	225	38/133	8.3	Monger and Irving (1980)
TKS	225	24/146	8.7	Monger and Irving (1980)
HZ3	205	40/144	29.5	Monger and Irving (1980)
HZ2	205	17/185	21.6	Monger and Irving (1980)
HZ1	190	70/57	13.1	Monger and Irving (1980)
TI	145	58.4/121.8	9.2	Symons (1983a)
GB	200	52.3/13.3	6.7	Symons (1983b)
KFX	230	-23/232 23/52	3.8	Yole and Irving (1980)

TABLE 1. SELECTED PALEOMAGNETIC POLES, WESTERN CORDILLERA (continued)

Unit*	Age (Ma)	Paleomagnetic Pole N/E	A_{95}	Reference
		Jurassic and Triassic (continued)		
NVA	200?	52.7/348.7	5.1	Symons (1985a)
IIB	190? 130?	71.0/35.0	7.6	Symons (1985b)
TU	175?	49.2/119/3	18.3	Symons (1974)
CM	200	57.3/11.8	3.8	Symons and Litalien (1984)
SD3	145?	60.2/295.3	5.8	Hillhouse and others (1982)
SD2	230	46.6/130.0 -46.6/310.0	13.9	Hillhouse and others (1982)
SD1	230	7.6/170.3 -7.6/350.3	17.0	Hillhouse and others (1982)
OJ	145?	68.7/322.0	6.9	Wilson and Cox (1980)
RTC	145?	71.1/159.9	15.7	Hannah and Verosub (1980)
HL	200? 155?	35.1/135.2	5.9	Hudson and Geissman (1984)
SJ	155	64.5/165.2	8.1	Bogen and others (1985)
FF	150?	26/310	9.8	Gromme and Gluskoter (1965)
ST	160	27.1/13.9	5.4	McWilliams and Howell (1982)
CH	150	62.2/130/3	4.2	Kluth and others (1982)
SH	215	29.7/22.0	14.9	Hagstrum and others (1985)

*Column identifies unit sampled, as shown in Figure 1.

strongly north-oblique and convergent, or dextral-transform, since about the beginning of the Cretaceous. This agrees well with the pattern of displacement of crustal blocks seen in the paleomagnetic literature (Beck, 1976). By analogy, the evidence summarized below and in Beck (1989) implies that motions relative to North America (RTNA) had a southerly (left-oblique or sinistral) component prior to about middle Cretaceous time. This interpretation also agrees with, but is not required by, the plate reconstructions of Engebretson and others (1985).

5. Most Cretaceous batholiths located near the continental margin seem to have formed south of about present-day latitude 40°N, even though currently they are distributed as far north as Alaska. In fact, Irving and others (1985) have shown that, on reasonable assumptions, the point of origin of most (perhaps all) Cretaceous batholithic rocks studied so far was within a few degrees of present northern Baja California. Beck (1986) and Beck and others (1981a) speculated on plate configurations that might be responsible for this distribution.

6. North-south transport RTNA was of two sorts: transport of exotic terranes as passengers on oceanic plates, and coastwise transport of detached slivers of the North American continental margin. Exotic terranes accreted to North America also tended to fragment and move along the continental margin (e.g., dispersal of the Wrangellia terrane during or after accretion; Irving and others [1985]). The same process also affected arc and fore-arc units that were constructed on the edge of the continent (Champion and others, 1984; Beck, 1986).

Evidence for displacement of crustal blocks RTNA can be seen in Figures 3 through 6. These represent reliable paleomagnetic poles for units Triassic through Miocene in age from the western edge of North America (Fig. 2 shows sampling areas; Table 1 gives data and references). Not included in this compilation are several interesting studies on Cretaceous limestone blocks from the Franciscan Formation of California; these (e.g., Tarduno and others, 1986) clearly are far-traveled non-North American rock fragments that may yield information on terrane trajectories but are difficult to interpret otherwise. Several studies (PD, PP, EU) listed in Table 1 are not plotted in Figure 3 through 6; these represent units that have been so disrupted internally that a meaningful mean declination cannot be calculated. Nevertheless, such units yield useful inclination and paleolatitude data.

Reference poles for stable North America are shown by solid symbols in Figures 3 through 6. Autochthonous rock units properly dated and studied should have paleomagnetic poles that fall close to the North American apparent polar wander (APW) path defined by these poles. It is clear from these illustrations that many (most) poles from the edge of the continent do not fall near the North American APW path. This is the paleomagnetic basis for the contention that the edge of the continent is composed largely of allochthonous blocks. Also obvious from Figures 3 through 5 (but not Fig. 6) is that Cretaceous and younger rock units tend to have been transported northward and rotated clockwise, RTNA. This follows because the poles in Figures 3 through 5 are displaced systematically to the right of the reference path (viewed from the sampling area), and also tend to be farther away from the sampling area than are time-equivalent North American poles. Irving (1979) has given an excellent graphical explanation of this kind of interpretation.

Figure 6 shows paleomagnetic poles for pre-Cretaceous rocks distributed on both sides of the reference path. Uncertainties about polarity make this plot less reliable than Figures 3 through 5 (see below). However, if Figure 6 is correct, it indicates that some early Mesozoic units have been rotated counterclockwise RTNA. Furthermore, many Triassic and Jurassic poles are slightly closer to the sampling area than are equivalent APW poles, arguing for southward transport. This plot is part of the support for interpretation 4, above.

Analyzing the sense of displacement of allochthonous crustal blocks for Jurassic and earlier time is a bit more complicated than for Cretaceous and Tertiary time. The most intractable prob-

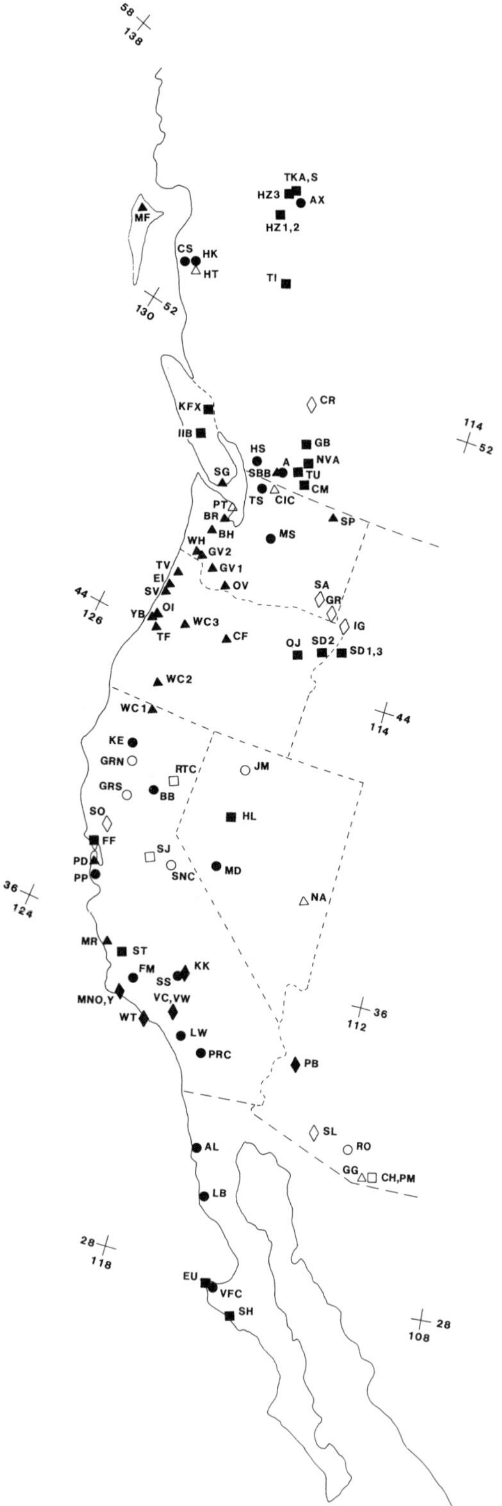

Figure 2. Sampling localities for paleomagnetic studies, western Cordillera, keyed to Table 1. Diamonds indicate Neogene units; triangles, Paleogene; circles, Cretaceous; squares, pre-Cretaceous. Open symbols indicate concordant result; all others are discordant in declination, inclination, or both.

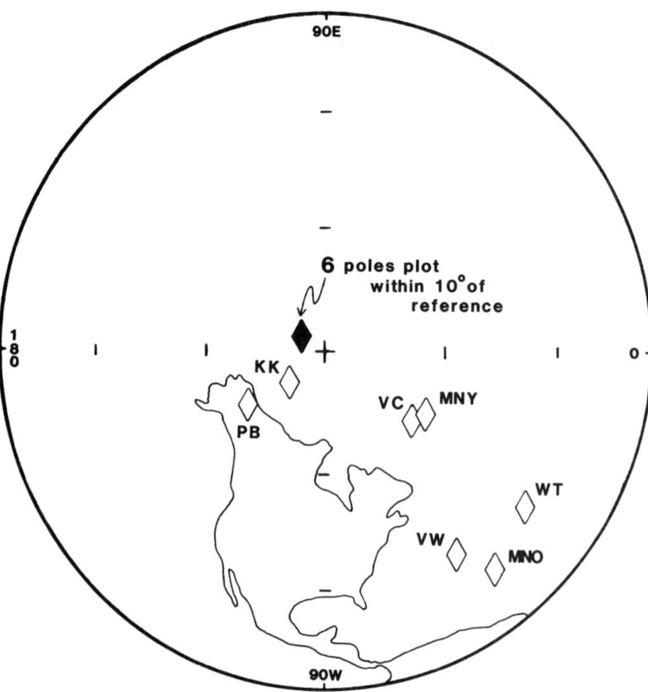

Figure 3. Equal-area projection of paleomagnetic poles for Neogene units, Table 1. Solid symbol is North American reference pole; from Diehl and others (1983).

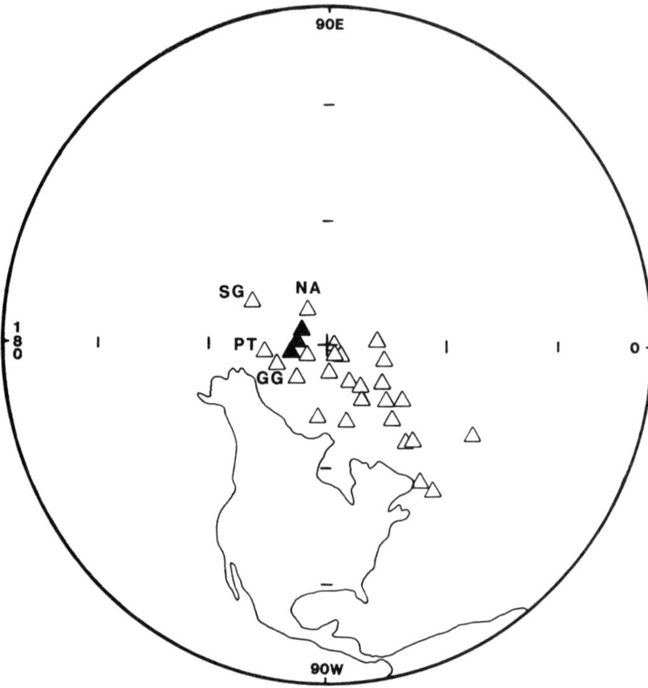

Figure 4. Paleomagnetic poles for Paleogene units, Table 1. Solid symbols are reference poles for three periods (Paleocene, early to middle Eocene, late Eocene); from Diehl and others (1983) and Beck (1984b).

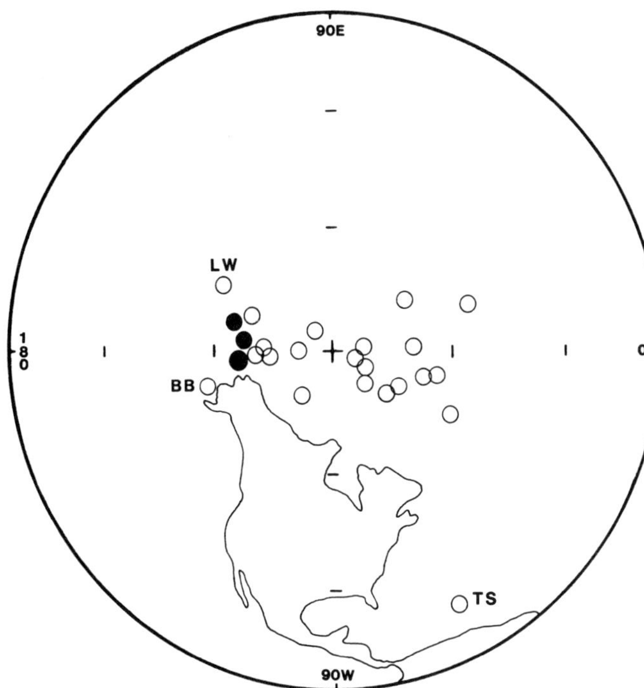

Figure 5. Paleomagnetic poles for Cretaceous units, Table 1. Solid symbols are reference poles, for 140, 130, and (large symbol) 120 to 70 Ma. From Irving and Irving (1982).

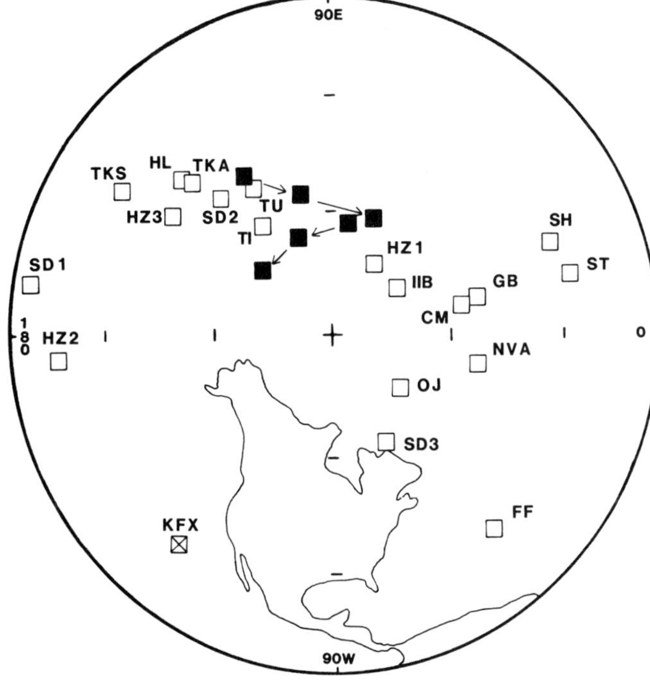

Figure 6. Paleomagnetic poles for Triassic and Jurassic units, Table 1. Solid symbols are reference poles ato 20-m.y. intervals 150 to 250 Ma. From Gordon and others (1984).

lem concerns geomagnetic polarity; prior to about 125 Ma, polarity transitions were common, and therefore the hemisphere of origin of allochthonous rocks of pre-Barremian age cannot be determined from the paleomagnetic directions alone. Rocks of mid- to Late Cretaceous age, by contrast, are almost certain to have acquired a normal-polarity magnetization, and thus the hemisphere of origin of such rocks follows directly from the inclination of the remanent magnetization. Rocks younger than about 70 Ma also can have either polarity, but it is unlikely (although not impossible in some cases) that rocks this young have moved to their present location alongside cratonic North America from south of the equator.

Another difficulty encountered in using paleomagnetic directions to analyze block displacements for pre-Early Cretaceous rocks concerns the APW path for stable North America. As described in the Appendix, determining the displacement of a crustal block with respect to North America from paleomagnetic directions involves comparison of directions of remanent magnetization or paleomagnetic poles from the allochthonous block with "expected" directions or reference poles from the craton. For this reason, tectonic interpretations based on paleomagnetism can be no better than the quality of the reference curve. For North America the reference APW path seems to be reasonably reliable for mid-Cretaceous and later time. For earlier Mesozoic time, however, differences of as much as 12° exist between leading compilations; these differences have important implications for Cordilleran tectonics. The problem of Triassic/Jurassic APW for stable North America and the sense of displacement of allochthonous crustal blocks RTNA in early Mesozoic time is discussed in the next section. See Van der Voo (this volume) for a general discussion of the North American APW path.

ALTERNATIVE NORTH AMERICAN APW PATHS: TECTONIC CONSEQUENCES

Figure 7 shows two recent APW paths for the North American craton, from about the beginning of Mesozoic time to the mid-Cretaceous. Error limits (of the order of 5 to 10°) are omitted for simplicity. The path shown by solid lines and triangles is the 30-Ma running-average path of Irving and Irving (1982), which has been used by most recent investigators. The second path (dashed lines and circles) is from Gordon and others (1984). Both APW paths fundamentally are based on the set of paleomagnetic results for stable North America, but they reflect different criteria for data selection, different time scales for converting biostratigraphic to absolute ages, and radically different computation methods. Nevertheless, they agree reasonably well except for a period around 200 Ma, where they differ by nearly 12°, mostly in longitude, and another at 170 Ma, where they are about 9° apart, mostly in latitude. After about 120 Ma the two paths are identical for most purposes. Differences for Jurassic time, however, are very important.

In order to assess the validity of the two curves it is necessary to understand their underlying assumptions. Both are at-

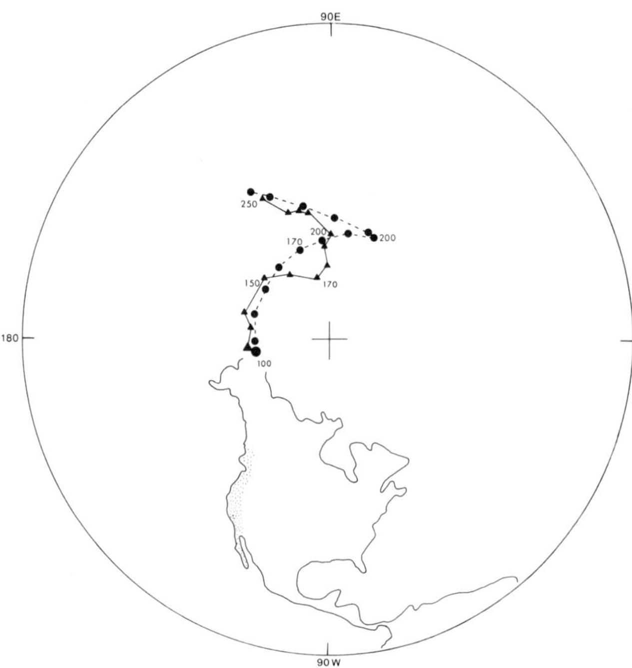

Figure 7. Reference poles after Irving and Irving (1982), shown by triangles, and after Gordon and others (1984, circles).

tempts to chart the motion of a large, internally undeformed block of crust (in this case, the North American craton) relative to the spin-axis of the earth. Both rely on the geocentric axial dipole hypothesis of paleomagnetism (Irving, 1964, provides an explanation), which allows magnetic directions found in rocks to be used to locate the sampling site in paleolatitude and orient it relative to the paleomeridian (e.g., Irving, 1964; Tarling, 1983). However, this is a complicated task, and inaccuracies can occur (Van der Voo, this volume). For instance, too little sampling may result in random error caused by failure to average out geomagnetic secular variation and dipole wobble. Failure to allow properly for the effects of postmagnetization structural displacements (folding, faulting, etc.) also can result in error. Rocks that are poor magnetic recorders, or whose age is not well known, may add to the scatter. Many of the differences between the Irving and Irving (1982) and Gordon and others (1984) APW paths arise from differences in the way this inevitable scatter is dealt with.

In the method used by Irving (Irving and Irving, 1982; Irving, 1977, 1979), the available data are first "filtered," using several uncomplicated and objective tests for paleomagnetic and age reliability. Results that pass the filter are next arranged chronologically; biostratigraphic ages are converted to absolute ages using some selected time scale (for instance, van Eysinga, 1975, was used in Irving, 1979). Finally, random error in both pole position and age of magnetization is minimized by averaging over some time window, using the method of Fisher (1953). No weighting is attempted; in effect, all studies not eliminated by filtering, and that fall within the designated time window (30 Ma in Irving and Irving, 1982) are treated as independent and equally valid estimates of the true position of the paleopole. The Irving method has many attractive attibutes, including minimizing bias in the selection of data, not relying on a specific model for the cause of APW, and providing a conventional and easily applied estimate of the confidence interval about the mean pole. However, if APW is a spasmodic (as opposed to smooth) process, as many studies have suggested (e.g., Briden, 1967; Irving and Park, 1972), then the Irving method unavoidably will obscure the APW path near points of rapid directional change. Also, because there is an inevitable general tendency for secondary magnetizations to bias collections of poles toward the present spin-axis, to the extent that the Irving method (or any method) admits insufficiently "filtered" poles to the data set, the APW path calculated will be deflected systematically toward higher latitudes.

The APW path of Gordon and others (1984) was also constructed using "filtered" data for the North American craton. In this case, filtering includes much recalculation and weighting and was significantly more demanding than in the Irving method; for example, for the period 200 to 100 Ma, the compilation of Gordon and others (1984) has 12 entries, whereas that of Irving and Irving (1982) has 19. The most important difference in the two paths, however, lies in the method of "averaging." The APW path of Gordon and others (1984) is explicitly model-dependent, in that it assumes that APW is caused by plate motion and that the absolute motion of large plates can be modeled reasonably well for long periods of time by steady-state rotation about a stationary Euler pole. Their "averaging" process consists first of visually identifying smooth segments of the APW path (tracks) that might reasonably represent small circles about a stationary "paleomagnetic Euler pole" (PEP), and then determining the pole. Next, the velocity of the pole along the tracks is determined by actual measurement, plotted against time, and smoothed. Tracks are separated by points of abrupt directional change (cusps) that, according to the model, are caused by first-order changes in the boundary conditions that drive plate motion. For the purposes of this chapter, the important cusps recognized by Gordon and others (1984) are those that occur at approximately 205 and 125 Ma. Although it is clear that there were abrupt changes in the polar velocity along the 205-to-125-Ma track, Gordon and others (1984) concluded that they cannot be resolved reliably, and so constructed their APW path by assuming a constant polar velocity along the track.

The chief advantage that the PEP method has over the moving-average method lies in its ability to retain good definition at cusps. This is, in fact, a supremely important property from the point of view of tectonics if, as seems likely, first-order reorganizations in plate motions are reflected in important orogenic events (e.g., Beck, 1984a). The chief disadvantage of the PEP method, however, is clearly its dependence on a specific, and very simple, plate-tectonic model. Errors will result from the Gordon and others (1984) approach to defining APW to the extent that the underlying assumptions of the method are violated. Absolutely steady-state rotation about an absolutely stationary Euler pole requires that the driving forces for plate motion remain totally invariant. It seems unreasonable to expect totally invariant geologic conditions of any sort for periods of more than about 10^6 to 10^7 years. Thus PEP tracks that extend nearly 10^8 yr must be regarded as first-order approximations, as was clearly recognized by Gordon and others (1984).

May and Butler (1986) also used the PEP approach to analyze APW, and have produced an alternate APW path for the Jurassic (Fig. 8). They used an even more critically filtered data set, although, unlike Gordon and others (1984), they did not weight the data. They also used several new poles that were unavailable to Gordon and others (1984). Perhaps because of these differences, May and Butler recognized an additional cusp in the APW path at about 150 Ma. The difference between the two PEP poles at this cusp (150 Ma) is 7°, whereas the difference between the poles of Irving and Irving (1982) and those of Gordon and others (1984) at 150 Ma is much less (2.6°), because their analyses each have "smoothed out" the 150-Ma cusp.

The choice of time scale for converting biostratigraphic ages to absolute ages makes a surprisingly large difference in the resulting APW path. The time scale used in Irving and Irving (1982) is not specified, but it may have been the van Eysinga (1975) scale used in Irving (1979). Gordon and others (1984) and May and Butler (1986) used the scale of Harland and others (1982). Another compilation might make use of even newer time scales (e.g., the DNAG time scale; Palmer, 1983). The Harland and DNAG scales differ by only about 5 m.y. or less through the beginning of Mesozoic time; however, differences of 15 m.y. between the van Eysinga scale and the two newer scales are common from about Toarcian time backward through the Mesozoic. This can make a very important difference to the placement of a biostratigraphically dated pole along the APW track. For instance, the rate of angular progression determined by May and Butler (1986) for their J1-J2 track (ca. 205 to 150 Ma) is 0.7°/m.y. Thus, a 15-m.y. error in age results in an error in placement of the pole of more than 10°, which could translate into erroneous "paleopositioning" of allochthonous crustal fragments by as much as 1,200 km. Much of the difference between the various APW paths can be attributed to differences in time scales.

Some "worst case" examples illustrate the limits of uncertainty that result from use of different APW paths and time scales.

1. Pole HZ2 (Monger and Irving, 1980) is from the Hazelton Formation of British Columbia, the age of which is cited as Sinemurian, or about 205 Ma, using the Harland and others (1982) time scale. Assuming that the terrane ("Stikinia") including the Hazelton rocks originated in the Northern Hemisphere, and using the APW path of Gordon and others (1984), the HZ2 rocks have been rotated about 108° counterclockwise and translated southward (in the ancient coordinate system) by nearly 10°. There are, of course, large error limits on these calculations. Rotation (R) and poleward translation (P) statistics (Beck and others, 1986) for this calculation are:

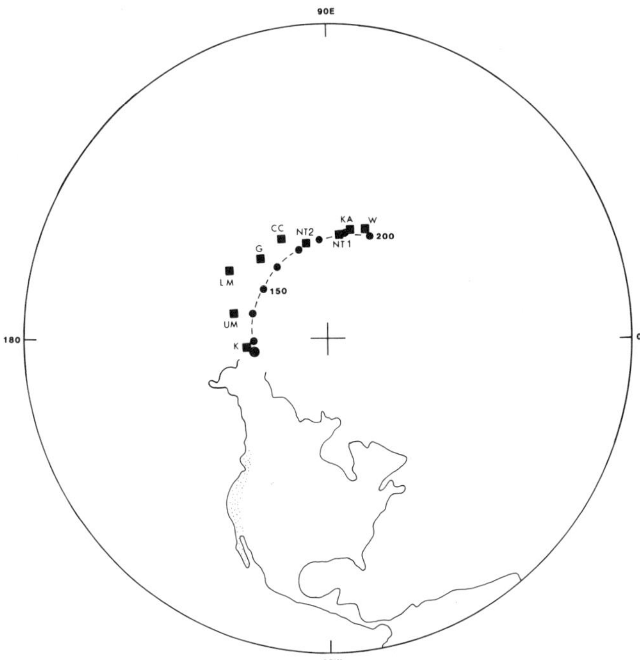

Figure 8. Contrast between reference polar path of Gordon and others (1984) for Jurassic time (circles) and selected reliable poles from May and Butler (1987), shown by squares. Symbols identified in Table 2.

$$R = -108.5 \pm 23.9°$$
$$P = -9.9 \pm 19.4°$$

where the confidence limits are at the 95 percent level. Lacking convincing evidence to the contrary, one might conclude from this calculation that the HZ2 rocks are allochthonous, that they have rotated through a large angle relative to the stable interior of the continent, and that they may have moved southward (away from the early Jurassic pole) by perhaps 1,000 km. The latter conclusion would be tentative owing to the large confidence limits on *P;* however, if other determinations in the same terrane gave similar values for *P*, it probably would be accepted.

Using the van Eysinga (1975) time scale and the Irving and Irving (1982) APW path, the result is quite different. According to the van Eysinga scale the Sinemurian HZ2 rocks are about 190 m.y. old. A calculation based on the Irving 190-Ma pole gives the following values:

$$R = -98.2 \pm 20.8$$
$$P = -1.4 \pm 22.2.$$

From these statistics, one would still conclude that a large rotation had occurred. However, the sampling area now appears to be essentially concordant in paleolatitude; only if other paleomagnetic studies in the same terrane found large values of *P* would one be tempted to conclude that any relative north-south translation had taken place. In fact, the conclusions of Monger and Irving (1980), based on paleomagnetic directions from five localities, was that the Stikine terrane had moved relatively northward by more than 1,000 km since Hazelton time.

2. Pole SJ (Table 1; Bogen and others, 1985) is from the western foothills of the Sierra Nevada and is based on remagnetized marine volcanics and related rocks. Remagnetization is dated indirectly as about Kimmeridgian or Oxfordian, or roughly 155 Ma on all three time scales. For 155 Ma, there is no significant difference between the Irving and Irving and Gordon and others APW paths, but the May and Butler (1986) pole is distinct because those authors recognized a second Jurassic cusp (J2) at about 150 Ma. From the May and Butler data for 155 Ma, I compute a reference pole at 62.6N/127.6E; this is the vector sum of poles for the 151-Ma Glance Conglomerate (Kluth and others, 1982) and the 171-Ma Corral Canyon volcanics (May and others, 1986), the former being given a relative weight of 3. Rotation and poleward-translation statistics for pole SJ are as follows:

Using the 155-Ma reference pole of Irving and Irving (1982):

$$R = 9.7 \pm 18.6$$
$$P = -11.8 \pm 15.8.$$

Using the May and Butler (1986) pole:

$$R = 5.0 \pm 10.2$$
$$P = -16.1 \pm 8.2.$$

The main difference here is the interpretation placed on P. Both values of P are negative, suggesting possible southward transport. However, using the Irving and Irving APW path, the value of P is not significant at 95 percent probability, but with the May and Butler curve it is. If the May and Butler curve is correct—and the results obtained by Bogen and others (1985) are reliable—there is strong evidence that allochthonous rocks now composing part of the western foothills belt of the Sierra Nevada came from the north.

DIRECTION OF RELATIVE MOTION OF ALLOCHTHONOUS TERRANES DURING JURASSIC TIME

In both examples discussed in the last section, the paleomagnetic data could be interpreted to indicate that allochthonous terranes now part of the Western Cordillera of North America have moved relatively southward. Conversely, Cretaceous paleomagnetic data (illustrated in Fig. 5) strongly support northward movement, and most tectonic interpretation based on paleomagnetic and/or plate-motion data emphasize this sense of displacement (e.g., Beck, 1986; Champion and others, 1984; Hagstrum and others, 1985; Irving and others, 1985). Nevertheless, it seems to me that the balance of evidence now favors southward transport of allochthonous terranes RTNA prior to about 125 Ma. As discussed more fully elsewhere (Beck, 1989), results of plate-motion studies (e.g., Engebretson and others, 1985) and geological studies (e.g., Oldow and others, 1984) suggest, or at least permit, relative southward transport in pre–mid-Cretaceous time. This section will review the paleomagnetic evidence, relying on Figures 9 and 10.

Figure 9 can be interpreted to suggest that there was a general period of southward motion of allochthonous terranes RTNA during (Late?) Jurassic and Early Cretaceous time, followed by a period of northward movement during late Cretaceous and early Tertiary time. Figure 9 was constructed from the paleomagnetic data of Table 1 and the reference poles summarized in Table 2. Western Cordilleran paleomagnetic studies were first assigned to four general regions. Interior British Columbia and Washington (roughly "Superterrane I" of Monger and others, 1982, and neighboring terranes that may have shared most of its history) are shown in Figure 9a. Data from "Wrangellia" (Jones and others, 1977) and related rocks are summarized in Figure 9b. Figures 9c and d show results for coastal California and Baja California and the Sierra Nevada block, respectively. Not all the studies from Table 1 are included in Figure 9. In particular, "inboard" regions (e.g., parts of Nevada and Arizona) are not represented, and poles for rocks younger than about 30 Ma are excluded. Some poles on Figure 9 may be misassigned (placed in the wrong region), but it seems to me that the general pattern shown by the illustration is clear regardless of how the data are grouped.

Figure 9 looks like a conventional plot of paleolatitude vs. time, but instead it shows the statistic P (poleward transport) for each study, together with its 95-percent confidence limits, plotted at the estimated age of the sampled rocks. Because negative P indicates relative southward transport, the data are plotted with P increasing downward. This allows the diagrams to be read as a road log (with the east-west dimension missing); a downward trend indicates southward relative displacement, and vice versa. If a symbol's error bar crosses the $P = 0$ line, no statistically significant north-south relative movement (at the 95 percent confidence limit) is indicated, although this does not prove that no such movement ever occurred. For example, in Figure 9a, symbols for the time period 150 to 75 Ma all plot below the $P = 0$ line, whereas symbols for 50 Ma and later plot above the line, but have error bars that cross it. This can be read as evidence of general northward relative transport affecting the region during the period 75 to 50 Ma. It also suggests that there has been little, if any, relative north-south movement since 50 Ma, and that, if some did occur, it was most likely to the south.

Clearly there is no reason to regard the four regions of Figure 9 as rigid or even quasi-rigid blocks that must show the same history of tectonic transport throughout. What is implied in Figure 9 is that major parts of each region shared roughly the same tectonic setting since perhaps mid-Jurassic time, and therefore probably experienced roughly the same general history of tectonic displacements. As used here, a "region" can be severely

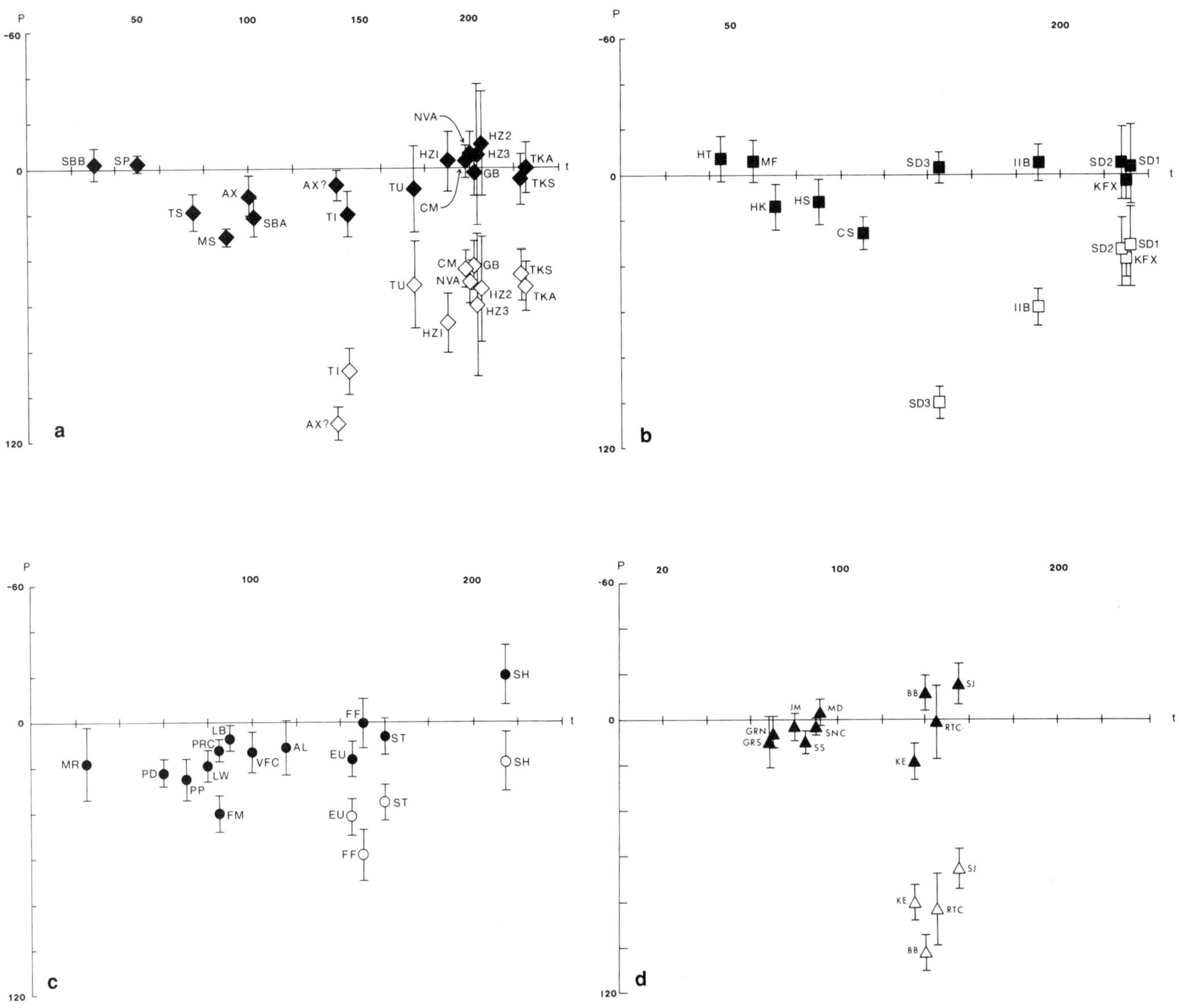

Figure 9. Poleward transport (*P*) vs. age (*t*), calculated for the paleomagnetic data in Table 1, using reference poles summarized in Table 2. Positive (negative) *P* indicates northward (southward) transport, relative to the North American craton; see Appendix. a, "Superterrane I" (interior British Columbia and adjacent areas). b, "Wrangellia" (Seven Devils terrain of Oregon and coastal British Columbia). c, Coastal California and Baja California. d, Sierra Nevada block.

deformed internally, but it should not contain parts that moved in opposite directions. Obviously Figure 9 is not intended for use in making detailed palinspastic reconstructions!

Earth scientists accustomed to examining paleomagnetic data for tectonic purposes will be familiar with the "hemisphere problem," expressed in Figure 9 by the fact that two symbols are plotted for all Jurassic and Early Cretaceous poles. This reflects the fact that the magnetic polarity, and hence the hemisphere of origin, of rocks older than about 125 Ma cannot be determined from the magnetic directions alone (see Beck, 1989, for a more complete discussion). In Figure 9 the open symbols represent the Southern Hemisphere alternative for each pole older than 125 Ma. For some poles (e.g., BB in Figure 9d) the southern option is clearly absurd. However, in view of the fact that relative plate velocities well above 100 km/m.y. are not uncommon (e.g., Engebretson and others, 1985), some of the "southern option" symbols in Figure 9 certainly could be correct.

The pattern shown by Figures 9a, b, and c are all similar in that they have a cluster of northern-option Jurassic (and later Triassic) poles slightly above the $P = 0$ line, and another cluster of mid- to Late Cretaceous poles below the line. For the most part the negative *P*-values of the Jurassic poles are not significant (at 95 percent confidence); most of the Cretaceous values, however, are significant. Figure 9d differs in that its Cretaceous entries are

TABLE 2. JURASSIC AND CRETACEOUS REFERENCE POLES FOR STABLE NORTH AMERICA*

Pole	Age	Lat.	Long.	A_{95}	Calculation Method	Reference
K	80-130	68.0N	186.0E	2.2		Mankien (1978)
UM*	145	64.6	164.2	3.98		Steiner and Helsley (1975)
LM*	149	58.6	146.2	4.2		Steiner and Helsley (1975)
G	151	62.7	131.5	6.3		Kluth and others (1982)
CC	171	61.8	116.0	6.2		May and others (1986)
NT2	179	65.3	103.2	1.4		Smith and Noltimier (1979)
NT1	195	63.0	83.2	2.3		Smith and Noltimier (1979)
KA*	197	61.9	78.1	6.3		Helsley and Steiner (1974)
W*	203	59.6	70.4	8.0		Reeve (1975)
M	215	58.8	89.9	5.8		Robertson (1967) Larochelle and Currie (1967)
C	225	57.7	79.1	7.0		Reeve and Helsley (1972)
	80-130	68.0	186.0	2.2	K	
	135	67.2	178.1	2.8	2K + 1UM	
	140	66.1	170.8	3.3	1K + 2UM	
	145	64.6	164.2	3.9	UM	
	150	60.9	139.3	5.3	LM + G	
	155	62.7	128.3	4.6	4G + CC	
	160	62.5	124.4	6.3	11G + 9CC	
	165	62.2	120.5	6.2	3G + 7CC	
	170	61.8	116.0	6.2	CC	
	175	63.7	110.0	3.8	CC + NT2	
	180	65.3	103.2	1.4	NT2	
	185	64.8	95.3	1.7	5NT2 + 8NT1	
	190	64.0	89.1	2.0	11NT1 + 5NT2	
	195	63.0	83.2	2.3	NLT1	
	200	60.8	74.1	7.2	K + W	
	205	59.6	70.4	8.0	W	
	210	59.5	81.9	6.6	5W + 7M	
	215	58.8	89.9	5.8	M	
	220	58.4	84.4	6.4	M + C	
	225	57.7	79.1	7.0	C	

*Poles listed in top section are those selected as most reliable by May and Butler (1986); these are used in lower section to calculate reference poles at 5-m.y. intervals by linear interpolation, with the weightings shown. (For instance, the 140-Ma pole is the vector sum of the K pole [130 Ma] and twice the Upper Morrison [UM] pole [145 Ma]). Poles marked with asterisk have been corrected for 3.8° of clockwise rotation of the Colorado Plateau (May and Butler, 1986, Table3).

not significantly displaced at 95 percent confidence, although they also lie below the $P = 0$ line. What seems fairly convincing from the four diagrams taken together is that terranes were transported southward RTNA prior to about 125 Ma, and northward thereafter. One might construct alternative scenarios by judicious selection of southern option poles, or by ignoring some of the data, but it is difficult to use Figure 9 to sustain a model of continuous northward transport.

Figure 10 provides additional evidence of southward transport RTNA prior to about 125 Ma. It shows the paleolatitude of a point on the coastline of North America at the Oregon-California border (specifically, 42°N, 124°W), calculated from the reference poles of Table 2. The value of Figure 10 is that it shows the north-south component of absolute velocity of a point on the western edge of the continent, measured using the spin-axis of the Earth as a reference. The outstanding feature shown by Figure 10 is an extremely rapid increase in latitude between about 160 and 130 Ma. During that time the western edge of North America had a strong northward component of motion, probably at least 10 cm/yr (100 km/m.y.). If displacement of allochthonous terranes is a result of relative motion between neighboring plates, then a priori, relative southward transport during this period of time would be expected. To carry terranes north RTNA during the interval 160 to 130 Ma, an oceanic plate would have had to move northward even faster than North America, that is, at an absolute velocity greater than 10 cm/yr. Certainly speeds this high are possible, and several instances have actually been observed (Engebretson and others, 1985). However, given the usual range of plate velocities, it seems much more likely that North America outran its oceanic neighbors during this

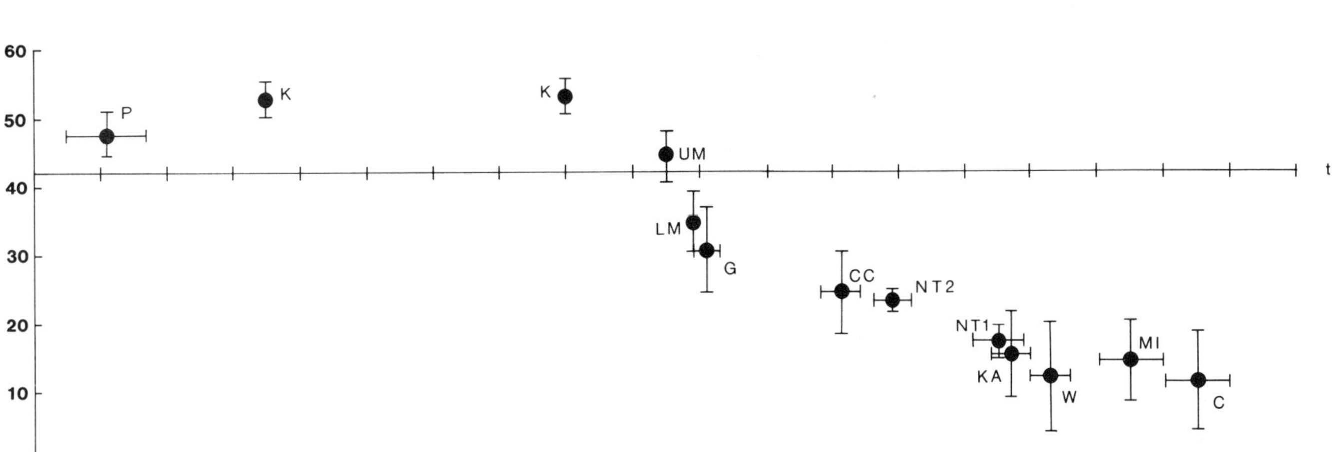

Figure 10. Paleolatitude of a point currently located at 42N, 124W, calculated from the reference poles of Table 2 by standard method. Note abrupt northward trend from about 155 to about 135 Ma. Cretaceous reference pole taken as invariant during the interval 130 to 80 Ma. Paleocene reference pole (*P*) from Diehl and others (1983).

period, causing relative tectonic transport to the south. There is some confirmation of this supposition in the plate motion studies of Engebretson and others (1985; see discussion in Beck, 1989).

There are two ways around the implications of Figure 10. One is that the reference poles used in the calculations of paleolatitude are wrong. As explained in Table 2, these are based on the set of Jurassic and Early Cretaceous poles selected by May and Butler (1987) as most reliable. These poles (10 in all, plus the Cretaceous reference pole of Mankinen, 1978) are unlikely to be discarded by future compilers and thus will form an important component of any future APW reference path. Additional data plus new methods of combining them eventually may produce a somewhat different path, but it seems unlikely that the rapid northward component of motion of western North America will disappear in the process. Beck (1989) described a set of similar calculations using the reference poles of Irving and Irving (1982) and Gordon and others (1984). The Late Jurassic to Early Cretaceous northward dash of North America also is apparent using these reference paths, although it is perhaps a little less dramatic.

D. C. Engebretson has pointed out (oral communication, 1986) that the "northward dash" of the last paragraph could represent true polar wander rather than plate motion. That is, it could reflect a relatively sudden shift of the entire Earth, or perhaps the entire lithosphere, relative to the spin-axis (e.g., Goldreich and Toomre, 1969). If this were the case, then the oceanic plates fronting North America also would have had a rapid base velocity to the north, and so northward and southward relative motion between them and North America would be equally likely. However, if polar wandering is to blame, there ought to be a similar record of rapid motion in APW paths for other plates, although the record would be modified by the effect of individual plate motions. As described below, there is no evidence of such a process.

A recent examination of paleomagnetic poles for South America (Beck, 1987b) shows no episode of rapid polar shift; very little APW occurred during the interval. Eurasia's polar shift matches North America's rather well, but this is not surprising in view of the fact that the two continents were attached until Late Cretaceous time (Smith and Briden, 1977). For the same reason, South America's small APW during Late Jurassic–Early Cretaceous time is duplicated in Africa. Finally, the compilation of Irving and Irving (1982) shows a 50-m.y. gap in the paleomagnetic record for Australia for just the interval in question, so that continent (and India as well) makes no contribution to solution of the problem. Thus the paleomagnetic data base for deciding whether North America's rapid poleward movement during the interval 160 to 130 Ma was a worldwide phenomenon or the result of particularly rapid movement on the part of North America alone is reduced to two independent measurements, obviously too few to judge. What is needed are APW paths for oceanic plates (particularly those of the Pacific basin), but these are not available. Hot-spot tracks can be used to chart "absolute" motion of oceanic plates, but comparing paleolatitude determinations based on paleomagnetic data with those based on hot spots can be misleading (see, for instance, a comparison of estimates of true polar wander since the early Tertiary using both methods; Jurdy, 1981). Thus the hypothesis that North America's rapid northward movement from 160 to 130 Ma represents true polar wander cannot be tested with the information in hand. If true polar wander is a relatively rare occurrence, then the most probable scenario for North America's Late Jurassic–Early Cretaceous poleward movement involves simple plate motion, not a shift of the entire lithosphere relative to the pole. In that case, oceanic plates off the western edge of North America most likely moved relatively southward during that period, carrying the bulk of allochthonous terranes with them.

TECTONIC PROCESSES

It seems to me that there can be little doubt that interaction of western North America with plates of the Pacific basin is responsible for nearly all of the tectonic displacements shown by the paleomagnetic record. This follows from two observations: abundant, large-scale block displacements appear to be confined largely to the continental margin; and patterns of displacement of allochthonous crustal blocks in the western Cordillera match directions of relative motion deduced from magnetic anomalies in the Pacific basin. My confidence in the ability of plate interactions to move paleomagnetic sampling areas has been unwavering since the early 1970s (although during the same period my conception of how this might be accomplished has "wavered" considerably). As my ideas on the subject are abundantly published (Beck, 1976, 1980, 1986), this section is limited to discussion of a few new ideas. Most of the comments that follow pertain to allochthonous crustal blocks moving alongside North America (coastwise transport); these can consist of exotic terranes brought to North America by oceanic plates and added to its western margin during subduction, or they can be fragments of North America itself. Because North America's western margin has trended roughly north-south since early Mesozoic time, paleomagnetists are denied the ability to determine whether allochthonous terranes are exotic, since east-west displacements leave no paleomagnetic record. To separate components of poleward displacement and rotation acquired by exotic terranes into portions acquired before and after accretion requires input from strutural geology, stratigraphy, and plate reconstruction studies.

Effect of curvature of the continental margin

It seems fairly certain that oblique subduction can detach slivers from the leading edge of a continent and transport them along the continental margin (e.g., Fitch, 1972). It also seems likely that this mechanism has been responsible for much of the north-south movement of allocthonous terranes in western North America (Beck, 1983, 1986). According to analysis given in the last-cited papers, the angle of obliquity (the angle between the vector of relative motion and the normal to the coastline) is an important parameter in determining whether coastwise transport of detached slivers occurs. Equally important is the buttress effect: a detached sliver can move for long distances only if there is no immobile buttress of undetached continental lithosphere to prevent it. Thus, the concave-outward outline of the southwestern margin of South America probably has prevented extensive coastwise transport of detached crustal blocks (Beck, 1987). The North American continent tends to be convex toward the west, which favors tectonic transport, as shown below.

Figure 11 is a cartoon that is meant to show how the shape of the continent might have affected tectonic transport. The continental outline is based on today's shape, except that a little of the curvature introduced by Basin and Range expansion has been removed. Admittedly the western edge of North America may have had quite a different outline in the Mesozoic. The arrows shown depict relative plate velocities; oceanic plates are converging with North America and subducting under it in the direction shown. For the sake of simplicity, the convergence arrows are drawn parallel to one another; this of course can only be approximately true, and then only if the Euler pole of relative motion is very far from the continental margin. Figure 11a is modeled on the plate-tectonic setting of North America in Late Cretaceous time; Figure 11b shows something like the situation that might have prevailed during Late Jurassic and Early Cretaceous time. Given this configuration is is likely that:

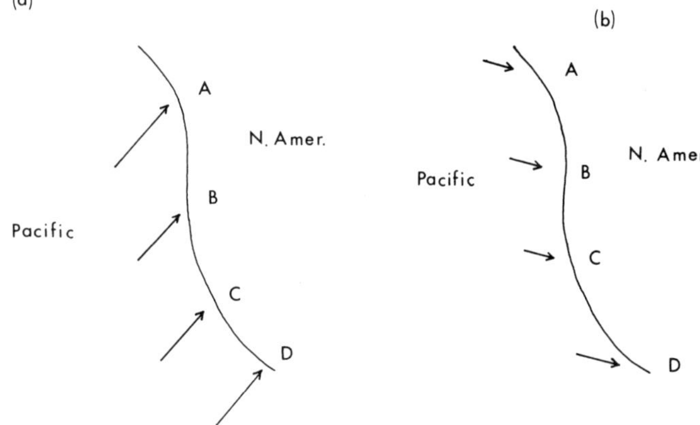

Figure 11. Effect of the shape of the continental margin on tectonics. Letters A through D represent areas with contrasting tectonic settings; see text. a, Cretaceous; b, Jurassic.

1. Batholiths would be generated in profusion in regions C and D during middle or Late Cretaceous time (because of rapid, near-normal subduction). However, much of this material might also have been detached and transported northward, to and past region B, but not much past region A.

2. Fore-arc deposits created in the same regions (C and D) and at the same time would share the same relative transport history as the batholithic rocks.

3. Not as much batholithic and fore-arc material would be created during Late Jurassic time (because of much slower relative motion). Region B would be the best place to look for such rocks, although some of them may have been swept relatively southward shortly after birth (and then perhaps northward again during Late Cretaceous and early Tertiary time).

4. Region D should have experienced important left-lateral faulting during the earlier period, which would cease when the convergence vectors changed direction.

Paleomagnetic and geologic evidence summarized elsewhere (Beck, 1983, 1986, 1989) indicates that many of these "probabilities" are real; that is, that they describe, in a general way and with some exceptions, conditions actually encountered in the Western Cordillera.

Effect of location of the Euler pole on tectonic transport

If the Euler pole describing relative motion between adjoining oceanic and continental plates is located in the ocean basin, then net tectonic transport can be opposite to the direction of subduction-related vergence and subduction-induced coastwise transport. This is illustrated in Figure 12, which shows an exotic terrane being swept relatively northward on an oceanic plate, but because of curvature of the small circles describing relative motion between the two plates, being accreted with vergence to the southeast. If the rocks at location A (Fig. 12) are studied paleomagnetically, they will show net northward transport ($P > 0$), but structures within and adjacent to them will indicate transport to the south. If the Euler pole is on the continental side of the margin, such an apparent contradiction in transport direction cannot arise. Obviously this mechanism applies mainly to allochthonous terranes that are truly exotic; terranes that originate as part of the continent might also behave in this way (block B, Fig. 12), but the circumstances required are improbable.

The balance of evidence suggests that western North America probably has not been subject to this sort of complication since mid-Jurassic time. Euler poles describing relative motion between the Farallon and North American plates lay in the (present day) central-southern Pacific prior to about 135 Ma (Engebretson and others, 1985); potentially this could yield the situation shown in Figure 12 (northward net displacement but with southward vergence). However, the geologic and paleomagnetic record calls for the opposite scenario (for some early Jurassic and late Triassic rocks): a small amount of net southward displacement, but with northward movement along the edge of the continent after mid-Cretaceous time. According to the diagrams of Engebretson and others (1985), the only other plate that interacted with western North America until late Tertiary time was the Kula plate, and Kula–North America poles all lie well east of the edge of the continent. Finally, it is always possible that western North American tectonics has been influenced by interaction with small, ephemeral plates the evidence for which all has been destroyed by subduction (e.g., Tarduno and others, 1986). Such entities can be conjured up and caused to work any amount of mischief on the geologic record, but they do not seem to me to be entirely necessary. It appears to me that most of what we know of Western Cordilleran tectonics can be accounted for reasonably well by known or inferred behavior of the Farallon, Kula, and Pacific plates.

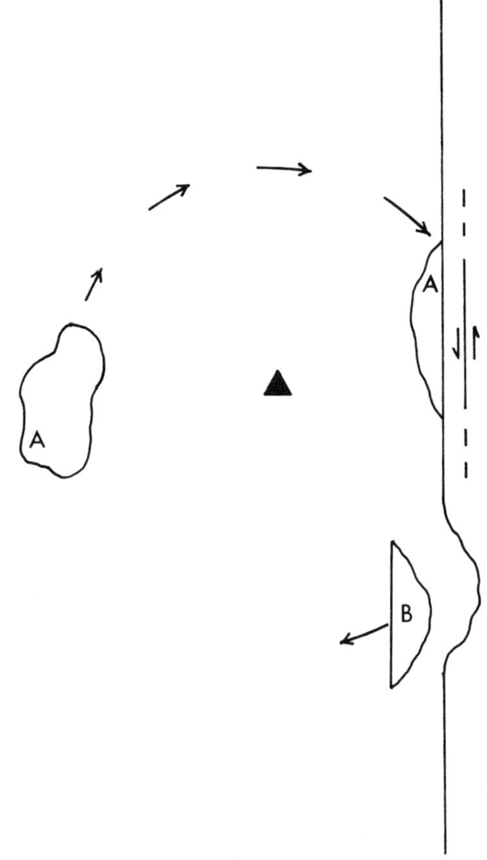

Figure 12. Map showing effect of location of Euler pole. Configuration shown illustrates net northward relative transport of an exotic terrane (A) that nevertheless accretes with vergence to the south. Block B shows the same scenario involving a detached piece of the continent. See text.

Rotations and the scale of displaced crustal blocks

Most of the studies cited thus far as evidence of relative north-south displacements also require rotations, because their paleomagnetic declinations diverge significant from expected declinations calculated from reference poles for stable North America (statistical tests are described in the Appendix). A plot (not shown) of rotation values illustrates that Cretaceous and Paleogene units have rotated predominantly clockwise, whereas Jurassic and Triassic units show about equal amounts of clockwise and counterclockwise rotation (assuming a Northern Hemisphere origin). This pre-Cretaceous pattern is what might be predicted for rocks that have been translated to the south in a zone of left-lateral shear, then later northward by dextral shear. Some of the rotations (e.g., KFX) are so large that they could be either clockwise or counterclockwise.

The amount of rotation experienced by the various units has a bearing on the questions of the mechanism of tectonic displacement and the size of the individual displaced crustal blocks. As mentioned earlier, there are two ways to transport allochthonous crustal blocks; as passengers on oceanic plates, and as independent crustal blocks in a shear zone at the edge of the continent. There seems to be no obvious predictive relationship between rotation (R) and poleward transport (P) for truly exotic terranes; too many variables are involved, such as the location of the Euler pole and the points of origin and accretion. However, in the case of crustal blocks that are constrained to travel along the edge of the continent, the path at least is known, and for these there is some value in comparing R and P.

Figure 13 is a plot of rotation (R) vs. poleward transport (P) for selected units of Cretaceous age from the westernmost Cordillera, based on the pole positions tabulated in Table 1. I have omitted inboard units from Nevada (JM, MD), as well as the result for the Twin Sisters Dunite (TS) which obviously has been affected by local tectonics. Most of the rocks on which the remaining entries are based probably originated at the edge of North America and then were displaced northward along its margin. Some of the sedimentary units and smaller plutons conceivably could be accreted, but not many.

The gently curved line (the TR line, for "translation rotation") shown in Figure 13 represents an estimate of the amount of rotation that a crustal fragment would acquire simply by moving northwestward along the continental margin, without any additional element of local rotation. It was designed for a starting point at Guaymas, Republic of Mexico; curves for other starting points would be slightly (but not significantly) different. The steps in deriving the TR line are as follows:

1. The Cretaceous reference pole was taken as 68°N, 186°E.
2. Both Guaymas (28°N, 249°E) and the refrence pole were rotated clockwise around a pole at 49°N, 286°E, in 5° increments. This yielded fictitious "locations" and "observed poles" for the displaced "Guaymas terrane."
3. R and P values were obtained for each "location," using equations given in the Appendix. These determine TR (Fig. 13).

The choice of 49°N, 249°E is arbitrary, but it is the best choice available. 49°N, 249°E is the Pacific–North America instantaneous relative motion pole of Minster and Jordan (1978). Much of the western edge of North America today is composed of (or lies substantially parallel to) transform faults that are small circles about this pole. If North America in Cretaceous and early Tertiary time was shaped like it is today, rotation about this pole simulates "coastwise transport" of allochthonous terranes. To the extent that North America had a different shape, the curve in Figure 13 will be wrong (but not, I think, the point it illustrates). Note that by using this technique I am not implying that the allochthonous terranes moved with the Pacific plate.

Most points on Figure 13 plot well above TR, implying that they experienced local (clockwise) rotation, in addition to the rotation acquired during tectonic transport. For instance, Cretaceous plutons from near Prince Rupert, British Columbia (HK), should have acquired about 18° of rotation as a consequence of moving northward along the continental margin; its remaining 32° of clockwise rotation must represent additional clockwise rotation about a nearby axis. By contrast, because it plots essentially on the curve, the displaced Cretaceous pole for the Mt. Stuart batholith (MS) can be accounted for without any need to invoke local rotations.

There are some significant groupings in Figure 13. For instance, units from coastal Baja and southern California plot together about 15° to 20° above TR. This implies that these units belong to a crustal block that has undergone displacement without much internal disruption, but it also suggests that the crustal

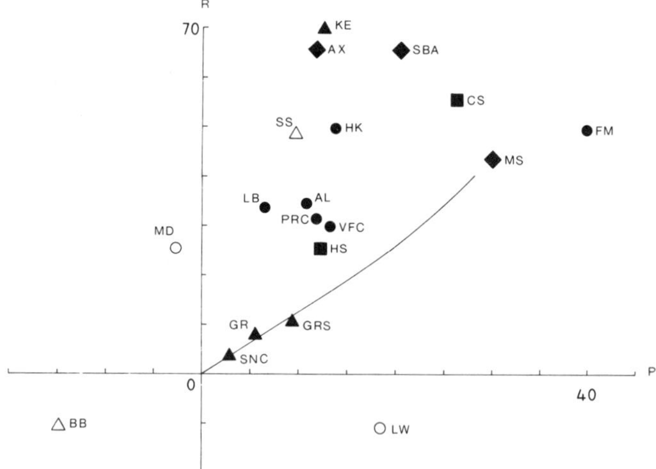

Figure 13. Translation-rotation. Rotation (R) plotted against poleward transport (P) for selected units of Cretaceous age. Line shown is the amount of rotation acquired as a result of pure "coastwise transport"—translation parallel to the North American margin. See text.

block rotated through a substantial angle in addition to the rotation it experienced in moving north along the continental margin. Thus it makes sense to "reconstruct" southern California/Baja California in Cretaceous time by sliding it along the continental margin to its correct relative paleolatitude. However, the additional 15° to 20° of local clockwise rotation also would have to be accommodated somehow, perhaps by reorienting the coastline. The aberrant location of FM and LW from the same region argues that these units were not part of the block. Likewise, SNC, GRN, and GRS form a group that lies on the TR line, which I take to indicate that a small amount of rotation was acquired by a quasi-rigid Sierran block in the process of moving north a few hundred kilometers. (Note that Sierran pole BB plots near the southward extension of the TR line.) By contrast, Cretaceous poles from the Northwest (Superterrane I and Wrangellia) have P and R values that are highly scattered and that plot far above the TR line (local rotations range up to 50°). This seems to indicate that Cretaceous sampling areas from British Columbia have rotated independently, although the size of the individual coherent rotated blocks cannot be resolved with current data. From Figure 13 there seems to be a vague tendency for large P values to be associated with scattered R, which may simply mean that crustal blocks undergoing coastwise displacement tend to fragment if they move too far. This tendency also is seen in the scattered mean declinations shown by some Jurassic rock units, for instance EU and separate sampling localities for the Hazelton Formation (HZ1–3).

CONCLUSIONS

Paleomagnetic evidence indicates that nearly the entire western edge of North America consists of displaced crustal blocks. From geologic evidence, some of these are recognized as accreted and exotic, but most appear to be displaced fragments of North America itself. The bulk of available evidence argues for relative northward motion since mid-Cretaceous time, perhaps with relative southward motion during the Jurassic and Early Cretaceous. The paleomagnetic evidence also supports a model in which large (ca. 1,000 km) blocks move coherently through short distances, but fragment when transported for thousands of kilometers. This tendency for coherent crustal blocks to become smaller the farther they travel may help to account for the recent proliferation of terranes recognized in the Western Cordillera.

ACKNOWLEDGMENTS

Thanks to R. Burmester for keeping the laboratory going and to D. Engebretson for teaching some of my courses. Both Burmester and Engebretson, as well as R. Butler, R. Speed, and E. Irving, have contributed ideas through many pleasant conversations. J. Hillhouse, D. Jones, and D. Howell reviewed the manuscript, and W. Mooney provided useful editorial advice. Both Howell and Mooney were extremely helpful. D. Jones compiled the map reproduced as Figure 1.

APPENDIX: CONCORDANCE/DISCORDANCE STATISTICS AND PALEOMAGNETIC RECONSTRUCTIONS

Given a reference curve and a set of reliable, accurately dated paleomagnetic poles, it is an easy matter to test for most kinds of "allochthoneity"; that is, to determine which poles belong to rock units that probably have moved, relative to the craton, since the rocks became magnetized, and which probably do not. This section summarizes the relevant mathematical procedures, following Beck (1980) and Beck and others (1986). It should be reemphasized that these tests (paleomagnetic concordance/discordance tests) are only as reliable as the input data—the reference pole and the paleomagnetic pole being tested. Reference poles were discussed earlier, and it seems clear that North American reference poles for mid-Cretaceous and later time are extremely reliable, but poles for earlier times are not. It is obviously well beyond the scope of this chapter to review paleomagnetic procedures that lead to reliable (and therefore usefully testable) pole positions (see Hillhouse, this volume, and references therein). However, for what they are worth I briefly summarize my personal set of guidelines, largely empirical, a set with which few if any other practicing paleomagnetists would agree absolutely. In stating these "guidelines" I intend to err in the direction of caution, and the reader should be so warned; certainly not every study that fails these tests should be ignored.

A user's guide to useful paleomagnetic data

Under most circumstances a paleomagnetic study needs the following to be useful for tectonic analysis (other purposes have other requirements);

1. It must be a study of rocks that have correctly and accurately recorded the direction of some ancient geomagnetic field.
2. The time at which the rocks acquired their magnetization must be known.
3. The paleomagnetic study must extend to rocks magnetized over enough time to average nondipole elements of the ancient field to (near) zero.
4. The postmagnetization structural history of the rocks must be known or reliably inferred. Regional stratigraphic relations of the rocks at the sampling site also are important.

In order to be satisfied that these requirements have a high probability of being met, I usually ask myself the following questions:

1. What kind of rock was studied? Igneous rocks seem to have the fewest complications; volcanic rocks especially so. Sedimentary rocks also can be useful, but they are subject to more complications. For instance, some red beds (but not all) appear to have been magnetized long after the detrital part of the rock accumulated, and over a long period of time. Some detrital sedimentary rocks may acquire inclinations that are too flat (the inclination error problem), although most seemingly do not. Metamorphic rocks suffer from all sorts of potential complications, including anisotropy caused by foliation and lineation, doubt about the age of the magnetization, and questions about postmagnetization structural history. For anything but young igneous rocks I expect to see a forthright discussion of the potential problems, and if I do not see it I am suspicious.

2. What stability tests were performed? Because all rocks potentially can acquire secondary magnetizations that obscure or distort their "signal," the burden of proof is on the investigator to show that this has not happened, or that it has been dealt with adequately. Thus, magnetic cleaning of some type is required for every study, and complicated procedures involving both thermal and alternating field demagnetization are reassuring in the case of most sedimentary rocks and are absolutely essential for metamorphic rocks, rocks with complicated tectonic histories, and very old rocks. If the rocks are old, at all complicated, and folded, I want to see a fold test; if none is proffered (many situations prevent a fold test), I am suspicious. It tends to be reassuring for several reasons if the data show several polarity transitions; this is especially true if the mean of normal and reverse groups are close to 180° apart.

3. How much scatter is there? Too little scatter sometimes means that the nondipole part of the geomagnetic field has not been averaged to zero. On the other hand, too much scatter may imply various kinds of error, but if the error is random, the mean direction may not be affected significantly. Dealing with rapidly cooled igneous rocks, I like to see values of the statistic k (best estimate of the precision parameter) of roughly 15 to 75 (corresponding to angular standard deviations in the 9° to 21° range). For the case of slowly magnetized rocks (e.g., plutonic igneous rocks, perhaps some red beds, metamorphic rocks) the situation is more complicated because much of the normal variability of the geomagnetic field may be averaged out within-site. For such cases, between-site k values much higher than 75 might be expected, but unless the author makes a straightforward argument for slow magnetization, I am suspicious. Reversals also help convince me that enough time has been "sampled" to average out the nondipole field. Great stratigraphic thicknesses also help, especially if accompanied by proof that the magnetization is primary.

4. What does the scatter look like? Here one looks for a familiar paleomagnetic blemish, streaking. Streaking is an elongate distribution of site-mean directions; it can signify a great number of things, almost all unfortunate. For instance, it can indicate that magnetic cleaning was not done properly. It can signal problems with the structural correction. It can even mean that the author "sampled too much time," and has caught apparent polar wander at work. We do not know everything we should about causes and consequences of the geomagnetic secular variation, but mostly it seems to produce a near-circular pattern of either directions or poles. Highly elongate distributions may be useful, but they provoke suspicion.

5. When was the magnetization acquired? This is essential, if only

to determine which reference pole to use in the concordance/discordance calculations. In young volcanic and hypabyssal rocks there is seldom much reason to doubt that the magnetization has the same age as the rock, as given by appropriate radiometric measurements. In older rocks, rocks that have been deeply buried, rocks that have been strongly tectonized, and rocks adjacent to plutons, there is every reason to suspect partial or total remagnetization. Fold tests, reversals, and the proper kind of scatter help test for a primary magnetization. Large thicknesses of rock with the same polarity throughout and showing little between-site scatter are suspect. Strongly discordant radiometric ages also may point to remagnetization. If a rock's age of magnetization is much younger than its geologic age, many problems can ensue. For instance, it may be difficult or impossible to deduce just when the rock did become magnetized. Similarly, making corrections for structural tilt may be difficult. Remagnetized rocks are not automatically useless, but they are hard to interpret.

6. What does the author know about the postmagnetization structural history of his rocks? More paleomagnetic studies come to grief in this area than in all the others combined. Obviously, if the rock unit studied has been tilted after magnetization, its direction needs to be corrected for the effect of tilt. To do this properly requires precise knowledge of strike and dip at each site. The possibility that the fold plunges must be considered and dealt with, as also the possibility that the observed dip is primary (original) and not tectonic. A fold test (if one is possible) usually can distinguish whether the magnetization is pre- or postfolding. If the latter is true, then all evidence of the paleohorizontal may be lost. This is certainly true of structureless plutonic rocks and most metamorphic rocks. Tectonic interpretations for such rocks are much more difficult to advance convincingly.

Assuming that useful paleomagnetic data have been obtained (from a rock unit that is suspected of being allochthonous), the next step is to test it against a reference pole or direction. In the material that follows the subscript *o* refers to observed poles or directions (the study being tested), and the subscript *r* indicates reference values. Comparisons can be made using either directions or poles, and there seems to be little to choose between the two methods. Nearly all of the mathematical relationships that follow depend on the dipole formula ($\tan I = 2 \cot p$), and the laws of sine and cosine from spherical trigonometry. The following quantities are involved: D = declination; I = inclination; p = colatitude (angular distance from sampling site to pole); A_{95} = radius of the circle of 95 percent confidence about a mean pole, calculated using virtual geomagnetic poles; α_{95} = radius of the circle of 95 percent confidence about a mean direction, calculated using directions; λ, ϕ = latitude and longitude of a sampling site; λ', ϕ' = latitude and longitude of a pole.

In the equations below, the 95 percent confidence level is used, but circles at any confidence level may be substituted.

The expected direction

If a site has always been rigidly attached to a landmass for which an APW path is available, then it is useful to inquire what magnetic direction one should expect to find for rocks at that site. If directions actually found do not match the reference directions, some relative motion may be implied. Referring to Figure 14 (top),

$$N = \phi' - \phi$$
$$\cos p_r = \sin \lambda \sin \lambda' + \cos \lambda \cos \lambda' \cos N$$
$$\tan I_r = 2 \cot p_r$$
$$\sin D_r = \sin N \cos \lambda' / \sin p_r.$$

The 95 percent error limit on the expected declination, ΔD_r, is related to the confidence circle about the mean pole (A_{95} and the reference colatitude (p_r), as shown in Figure 14 (bottom)

$$\sin \Delta D_r = \sin A_{95} / \sin p_r.$$

The error limit on the inclination is determined by the same parameters:

$$\Delta I_r = 2 A_{95} / (1 + 3 \cos^2 p_r).$$

This is derived by differentiating the dipole formula with respect to p, and is approximate only for large A_{95}.

Calculating rotation (R) *and flattening* (F), *using directions*

Once a reference direction is obtained, declinations can be compared directly to obtain the rotation of the site (relative to the stable land mass that provides the reference pole):

$$R = D_o - D_r.$$

As defined, positive R implies clockwise rotation. The quantity F, the "flattening" of inclination, is calculated similarly:

$$F = I_r - I_o.$$

In this case, positive F implies relative northward transport.

Obviously, there are error limits on R and F. These are given by

$$\Delta R = 0.8(\Delta D_r^2 + \Delta D_o^2)^{1/2}$$
$$\Delta F = 0.8\,(\Delta I_r^2 + \Delta I_o^2)^{1/2}.$$

The factor 0.8 in these equations is approximate, and is introduced to compensate for the fact that ΔD and ΔI as defined herein delineate a rectangular confidence region that is larger than the desired circular region (Demarest, 1982). Figure 15 shows how ΔD_o and ΔI_o are calculated:

$$\sin \Delta D_o = \sin \alpha_{95} / \cos I_o.$$
$$\Delta I_o = \alpha_{95}.$$

Calculating rotation and poleward translation (P), *using poles*

In many cases it is convenient to work exclusively with poles. Many paleomagnetic studies, particularly early ones, report only paleomagnetic poles, and for these this method is much quicker. Studies that report mean paleomagnetic pole positions are of two types: some (particularly precomputer-age examples) calculate a mean direction and confidence circle, then map to pole space. The result of transforming a circle about a mean direction into pole space is to convert it into an oval. An alternative method is to convert each site-mean direction into a VGP, and then average the VGP. This gives a circular confidence region about the mean pole. R and P are calculated in the same way for both cases, but the error limits (ΔR and ΔP) are different.

Figure 16 (top) shows how R and P are calculated.

$$\cos R = (\cos s - \cos p_o \cos p_r) / \sin p_o \sin p_r$$
$$P = p_o - p_r.$$

In these equations, p_o is calculated directly from I_o using the dipole formula. The variables s and p_r can be calculated using vector methods, spherical trigonometry, or by simply measuring off the stereonet. The

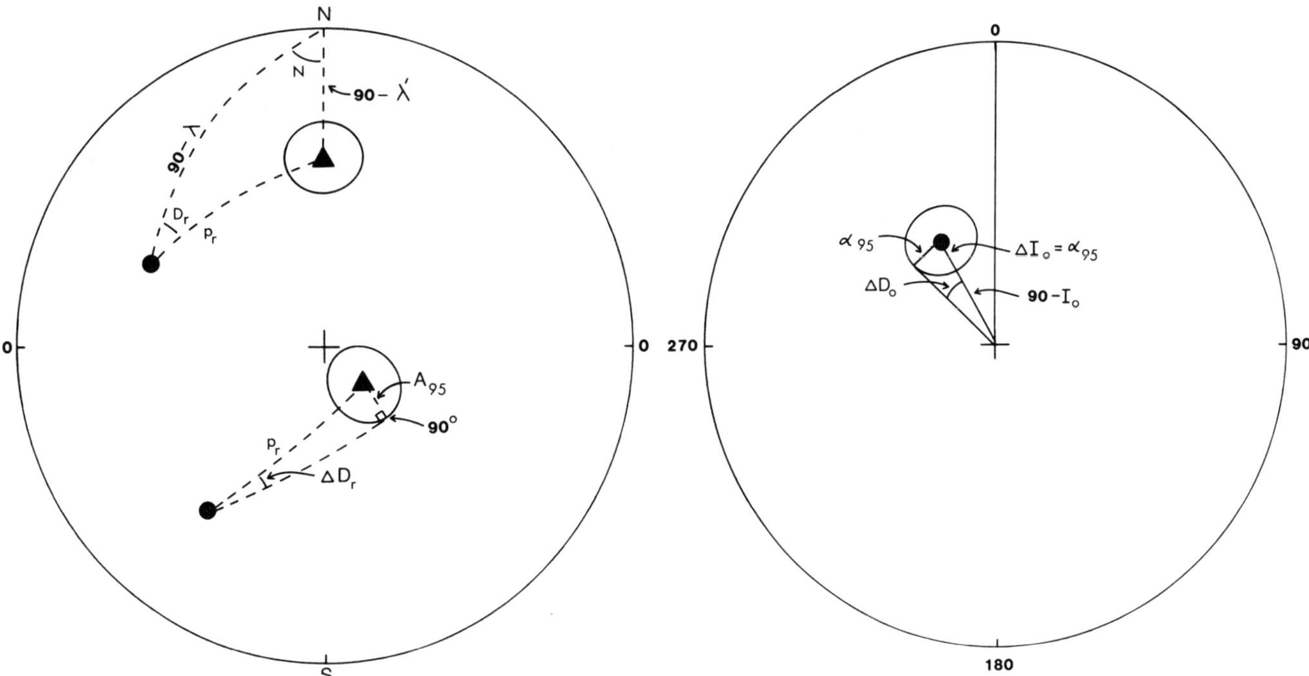

Figure 14. Calculation of expected declination (D_r) and its confidence limits (ΔD_r).

Figure 15. Stereographic projection illustrating calculation of limit on observed declination (ΔD_o).

parameter P is the poleward displacement; it gives the angular distance that the allochthonous crustal block appears to have moved toward (positive) or away (negative) from the ancient pole, relative to the stable landmass for which the reference pole was calculated. For North America this is nearly, but not quite, the same as the north-south displacement.

Confidence limits on R and P are calculated differently, depending on which method was used to obtain the mean pole. Case (1), in which the mean pole was calculated directly from a mean direction, is illustrated in the lower half of Figure 16. The confidence area about the reference pole is a circle, but the equivalent area about the observed pole is an oval, with semi-axes: $\sin \Delta m = \sin \alpha_{95} \sin p_o / \cos I_o$, in the direction of the declination (normal to the paleomeridian), and $\Delta p = 2\alpha_{95}/(1 + 3\cos^2 I_o)$ in the direction of the inclination (along the paleomeridian). Thus

$$\Delta R = 0.8\,(\Delta D_o^2 + \Delta D_r^2)^{1/2}$$
$$\sin \Delta D_o = \sin \Delta m / \sin p_o$$
$$\sin \Delta D_r = \sin A_{95} / \sin p_r, \text{ and}$$
$$\Delta P = 0.8\,(\Delta P_o^2 + \Delta P_r^2)^{1/2}$$
$$\Delta P_o = \Delta p$$
$$\Delta P_r = A_{95}.$$

In the case where the observed pole is the mean of a set of VGP, the confidence interval about it is a circle. Thus,

$$\sin \Delta D_o = \sin A'_{95} / \sin p_o,$$
$$\sin \Delta D_r = \sin A_{95} / \sin p_r,$$
$$\Delta P_o = A'_{95},$$
$$\Delta P_r = A_{95},$$

where A'_{95} is the radius of the circle of confidence about the observed pole.

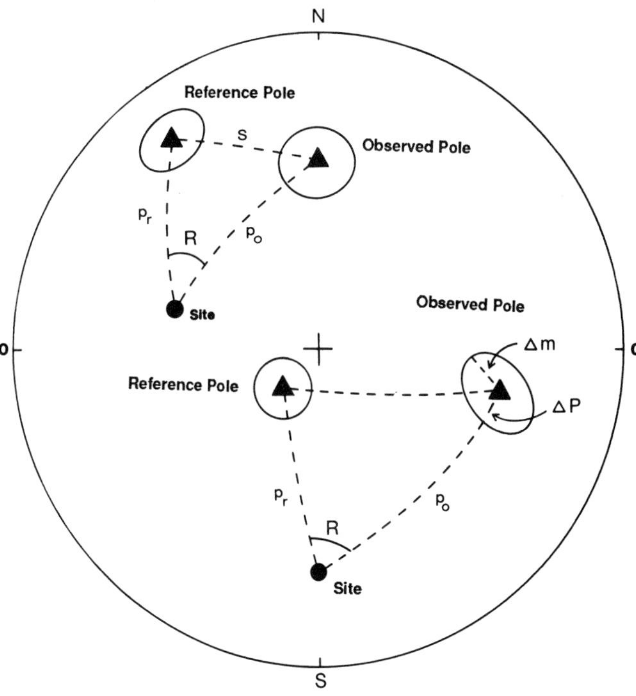

Figure 16. Stereographic projection illustrating calculation of poleward transport (P) and rotation (R) and their respective confidence limits, using poles. See text.

REFERENCES CITED

Achache, J., Cox, A., and O'Hare, S., 1982, Paleomagnetism of the Devonian Kennett limestone and the rotation of the eastern Klamath Mountains, California: Earth and Planetary Science Letters, v. 61, p. 365–380.

Barnes, A. E., and Butler, R. F., 1980, A Paleocene paleomagnetic pole for the Gringo Gulch volcanics: Geophysical Research Letters, v. 7, p. 545–548.

Bates, R. G., Beck, M. E., Jr., and Burmester, R. F., 1981, Tectonic rotations in the Cascade Range of southern Washington: Geology, v. 9, p. 184–189.

Beck, M. E., Jr., 1975, Remanent magnetism of the Twin Sisters dunite intrusion and implications for the tectonics of the western Cordillera: Earth and Planetary Science Letters, v. 26, p. 263–268.

—— , 1976, Discordant paleomagnetic pole positions as evidence of regional shear in the western Cordillera of North America: American Journal of Science, v. 276, p. 694–712.

—— , 1980, Paleomagnetic record of plate-margin tectonic processes along the western edge of North America: Journal of Geophysical Research, v. 85, p. 7115–7131.

—— , 1983, On the mechanism of tectonic transport in zones of oblique subduction: Tectonophysics, v. 93, p. 1–11.

—— , 1984a, Introduction to the special issue on correlations between plate motions and Cordilleran tectonics: Tectonics, v. 3, p. 103–105.

—— , 1984b, Has the Washington-Oregon Coast Range moved northward?: Geology, v. 12, p. 737–740.

—— , 1986, Model for late Mesozoic early Tertiary tectonics of coastal California and western Mexico and speculations on the origin of the San Andreas fault: Tectonics, v. 5, p. 49–64.

—— , 1988, Paleomagnetic data for active plate margins of South America: Journal of South American Earth Science, v. 1, p. 39–52.

—— , 1989, Tectonic significance of paleomagnetic results for the western conterminous United States, *in* Burchfiel, B. C., Lipman, P. W., and Zoback, M. L., eds., The Cordilleran orogen; Conterminous U.S.: Boulder, Colorado, Geological Society of America, The Geology of North America, v. G-3 (in press).

Beck, M. E., Jr., and Burr, C. E., 1979, Paleomagnetism and tectonic significance of the Goble volcanic series, southwestern Washington: Geology, v. 7, p. 175–179.

Beck, M. E., Jr., and Engebretson, D. C., 1982, Paleomagnetism of small basalt exposures in the west Puget Sound area, Washington, and speculations on the accretionary origin of the Olympic Mountains: Journal of Geophysical Research, v. 87, p. 3755–3760.

Beck, M. E., Jr., and Noson, L., 1972, Anomalous paleolatitudes in Cretaceous granitic rocks: Nature Physical Sciencese, v. 235, p. 11–13.

Beck, M. E., Jr., and Plumley, P. W., 1980, Paleomagnetism of intrusive rocks in the Coast Range of Oregon; Microplate rotations in middle Tertiary time: Geology, v. 8, p. 573–577.

Beck, M. E., Jr., Cox, A., and Jones, D. L., 1980, Penrose Conference Report: Mesozoic and Cenozoic microplate tectonics of western North America: Geology, v. 8, p. 454–456.

Beck, M. E., Jr., Burmester, R. F., and Schoonover, R., 1981, Paleomagnetism and tectonics of the Cretaceous Mt. Stuart batholith of Washington; Translation or tilt: Earth and Planetary Science Letters, v. 56, p. 336–342.

—— , 1982, Tertiary paleomagnetism of the north Cascade Range, Washington: Geophysical Research Letters, v. 9, p. 515–518.

Beck, M. E., Burmester, R. F., Craig, D. E., Gromme, C. S., and Wells, R. E., 1986, Paleomagnetism of middle Tertiary volcanic rocks from the western Cascade Series northern California: Journal of Geophysical Research, v. 91, p. 8219–8230.

Bogen, N. L., Kent, D. V., and Schweickert, R. A., 1985, Paleomagnetism of Jurassic rocks in the western Sierra Nevada metamorphic belt and its bearing on the structural evolution of the Sierra Nevada block: Journal of Geophysical Research, v. 90, p. 4627–4638.

Briden, J. C., 1967, Recurrent continental drift of Gondwanaland: Nature, v. 215, p. 1334–1339.

Calderone, G., and Butler, R. F., 1984, Paleomagnetism of Miocene volcanic rocks from southwestern Arizona; Tectonic implications: Geology, v. 12, p. 627–630.

Champion, D. E., Howell, D. G., and Gromme, C. S., 1984, Paleomagnetic and geological data indicating 2500 km of northward displacement for the Salinean and related terranes, California: Journal of Geophysical Research, v. 89, p. 7736–7752.

Choiniere, S. R., and Swanson, D. A., 1979, Magnetostratigraphy and correlation of Miocene basalts of the northern Oregon Coast and Columbia Plateau, southeastern Washington: American Journal of Science, v. 279, p. 755–777.

Coney, P. J., Jones, D. L., and Monger, J.W.H., 1980, Cordilleran suspect terranes: Nature, v. 288, p. 329–333.

Demarest, H. H., Jr., 1983, Error analysis for the determination of tectonic rotation from paleomagnetic data: Journal of Geophysical Research, v. 88, p. 4321–4328.

Diehl, J. F., Beck, M. E., Jr., Beske-Diehl, S., Jacobson, D., and Hearn, B. C., Jr., 1983, Paleomagnetism of the Late Cretaceous and early Tertiary north-central Montana alkalic province: Journal of Geophysical Research, v. 88, p. 10593–10609.

Engebretson, D. C., Cox, A., and Gordon, R. G., 1985, Relative motions between oceanic and continental plates in the Pacific Northwest: Geological Society of America Special Paper 206, 59 p.

Fisher, R. A., 1953, Dispersion on a sphere: Proceedings of the Royal Society of London, v. A217, p. 295–305.

Fitch, T. J., 1972, Plate convergence, transcurrent faults, and internal deformation adjacent to southeast Asia and the western Pacific: Journal of Geophysical Research, v. 77, p. 4432–4460.

Fox, K. R., Jr., and Beck, M. E., Jr., 1985, Paleomagnetic results for Eocene volcanic rocks from northeastern Washington and the Tertiary tectonics of the Pacific Northwest: Tectonics, v. 4, p. 323–341.

Frei, L. S., Magill, J. R., and Cox, A., 1984, Paleomagnetic results from the central Sierra Nevada; Constraints on reconstruction of the western United States: Tectonics, v. 3, p. 157–178.

Fry, J. G., Bottjer, D. J., and Lund, S. P., 1985, Magnetostratigraphy of displaced Upper Cretaceous strata in southern California: Geology, v. 13, p. 648–651.

Geissman, J. W., Callian, J. T., Oldow, J. S., and Humphries, S. E., 1984, Paleomagnetic assessment of oroflexural deformation in west-central Nevada and significance for emplacement of allochthonous assemblages: Tectonics, v. 3, p. 179–200.

Globerman, B. H., Beck, M. E., Jr., and Duncan, R. A., 1982, Paleomagnetism and tectonic significance of Eocene basalt from the Black Hills, Washington Coast Range: Geological Society of America Bulletin, v. 93, p. 1151–1159.

Goldreich, P., and Toomre, A., 1969, Some remarks on polar wandering: Journal of Geophysical Research, v. 74, p. 2555–2567.

Gordon, R. G., Cox, A., and O'Hare, S., 1984, Paleomagnetic Euler poles and the apparent polar wander and absolute motion of North America since the Carboniferous: Tectonics, v. 3, p. 499–537.

Greenhaus, M. R., and Cox, A., 1979, Paleomagnetism of the Morro Rock–Islay Hill complex as evidence for crustal block rotations in central coastal California: Journal of Geophysical Research, v. 84, p. 2393–2400.

Gromme, C. S., and Gluskoter, H. J., 1965, Remanent magnetization of spilite and diabase in the Franciscan Formation, western Marin County, California: Journal of Geology, v. 73, p. 74–94.

Gromme, C. S., and Merrill, R. T., 1965, Paleomagnetism of Late Cretaceous granitic plutons in the Sierra Nevada, California; Further results: Journal of Geophysics Research, v. 70, p. 3407–3420.

Gromme, C. S., Merrill, R. T., and Verhoogen, J., 1967, Paleomagnetism of Jurassic and Cretaceous plutonic rocks in the Sierra Nevada and its significance for polar wandering and continental drift: Journal of Geophysical Research, v. 72, p. 5661–5684.

Gromme, C. S., Beck, M. E., Jr., Wells, R. E., and Engebretson, D. C., 1986, Paleomagnetism of the Tertiary Clarno Formation of central Oregon and its significance for the tectonic history of the Pacific Northwest: Journal of Geophysical Research, v. 91, p. 14089–14103.

Hagstrum, J. T., McWilliams, M., Howell, D. G., and Gromme, C. S., 1985, Mesozoic paleomagnetism and northward translation of the Baja California peninsula: Geological Society of America Bulletin, v. 96, p. 1077–1090.

Hannah, J. L., and Verosub, K. L., 1980, Tectonic implications of remagnetized upper Paleozoic strata of the northern Sierra Nevada: Geology, v. 8, p. 520–524.

Harland, W. B., Cox, A., Llewellyn, P. G., Pickton, C.A.G., Smith, A. G., and Walters, R., 1982, A geologic time scale: Cambridge, England, Cambridge University Press, 131 p.

Helsley, C. E., and Steiner, M. B., 1973, Paleomagnetics of the Lower Triassic Moenkopi Formation: Geological Society of America Bulletin, v. 85, p. 475–464.

Hicken, A., and Irving, E., 1977, Tectonic rotation in western Canada: Nature, v. 268, p. 219–220.

Hillhouse, J. W., Gromme, C. S., and Vallier, T. L., 1982, Paleomagnetism and Mesozoic tectonics of the Seven Devils volcanic arc in northeastern Oregon: Journal of Geophysical Research, v. 87, p. 3777–3794.

Hooper, P. R., Knowles, C.L.R., and Watkins, N. D., 1979, Magnetostratigraphy of the Imnaha and Grande Ronde basalts in the southeast part of the Columbia Plateau: American Journal of Science, v. 279, p. 737–754.

Hornafius, J. S., 1985, Neogene tectonic rotation of the Santa Ynez Range, western Transverse Ranges, California, suggested by paleomagnetic investigations of the Monterey Formation: Journal of Geophysical Research, v. 90, p. 12503–12522.

Howell, D. G., ed., 1985, Tectonostratigraphic terranes of the Circum-Pacific Region: Houston, Circum-Pacific Council for Energy and Mineral Resources, 851 p.

Hudson, M. R., and Geissman, J. W., 1984, Preliminary paleomagnetic data from the Jurassic Humboldt lopolith, west-central Nevada; Evidence for thrust belt rotation in the Fencemaker allochthon: Geophysical Research Letters, v. 11, p. 828–831.

Irving, E., 1964, Paleomagnetism and its application to geological and geophysical problems: New York, John Wiley & Sons, 399 p.

——, 1977, Drift of the major continental blocks since the Devonian: Nature, v. 270, p. 304–309.

——, 1979, Paleopoles and paleolatitudes of North America and speculations about displaced terrains: Canadian Journal of Earth Sciences, v. 16, p. 669–694.

Irving, E., and Irving, G. A., 1982, Apparent polar wander paths Carboniferous through Cenozoic and the assembly of Gondwana: Geophysical Surveys, v. 5, p. 141–188.

Irving, E., and Park, J. K., 1972, Hairpins and superintervals: Canadian Journal of Earth Sciences, v. 9, p. 1318–1324.

Irving, E., and Yole, R. W., 1972, Paleomagnetism and the kinematic history of mafic and ultramafic rocks in fold mountain belts: Ottawa, Canada, Earth Physics Branch Publications, v. 42, p. 87–95.

Irving, E., Tanczyk, E., and Hastie, J., 1976, Catalogue of paleomagnetic directions and pole positions; 5, Cenozoic results: Ottawa, Canada, Earth Physics Branch, Department of Energy, Mines and Resources, v. 6, 70 p.

Irving, E., Woodsworth, G. J., Wynne, P. J., and Morrison, A., 1985, Paleomagnetic evidence for displacement from the south of the Coast Plutonic Complex, British Columbia: Canadian Journal of Earth Sciences, v. 22, p. 584–598.

Jones, D. L., Irwin, W. P., and Ovenshine, A. T., 1972, Southeast Alaska; A displaced continental fragment?: U.S. Geological Survey Professional Paper 800-B, p. 211–217.

Jones, D. L., Silberling, N. J., and Hillhouse, J., 1977, Wrangellia; A displaced terrane in northwestern North America: Canadian Journal of Earth Sciences, v. 14, p. 2565–2577.

Jones, D. W., Cox, A., Coney, P. J., and Beck, M. E., Jr., 1982, The growth of western North America: Scientific American, v. 247, no. 5, p. 70–84.

Jurdy, D. M., 1981, True polar wander: Tectonophysics, v. 74, p. 1–16.

Kamerling, M. J., and Luyendyk, B. P., 1985, Paleomagnetism and Neogene tectonics of the northern Channel Islands, California: Journal of Geophysical Research, v. 90, p. 12485–12502.

Kanter, L. R., and McWilliams, M., 1982, Rotation of the southernmost Sierra Nevada, California: Journal of Geophysical Research, v. 87, p. 3819–3830.

Kluth, C. F., Butler, R. F., Harding, L. E., Shafiqullan, M., and Damon, P. E., 1982, Paleomagnetism of Late Jurassic rocks in the northern Canello Hills, southeastern Arizona: Journal of Geophysical Research, v. 87, p. 7079–7086.

Larochelle, A., and Currie, K. L., 1967, Paleomagnetic study of igneous rocks from the Manicouagan structure, Quebec: Journal of Geophysical Research, v. 72, p. 4163–4169.

Magill, J., and Cox, A., 1980, Tectonic rotation of the Oregon western Cascades: Salem, Oregon, Department of Geology and Mineral Industries Special Paper 10, 67 p.

Magill, J., Cox, A., and Duncan, R., 1981, Tillamuck Volcanic Series; Further evidence for tectonic rotation of the Oregon Coast Range: Journal of Geophysical Research, v. 86, p. 2953–2970.

Mankinen, E. A., 1972, Paleomagnetism and potassium-argon ages of the Sonoma Volcanics, California: Geological Society of America Bulletin, v. 83, p. 2063–2072.

——, 1978, Paleomagnetic evidence for a late Cretaceous deformation of the Great Valley sequence, Sacramento Valley, California: U.S. Geological Survey Journal of Research, v. 6, p. 383–390.

Mankinen, E. A., and Irwin, W. P., 1982, Paleomagnetic study of some Cretaceous and Tertiary sedimentary rocks of the Klamath Mountains province, California: Geology, v. 10, p. 82–87.

May, S. R., and Butler, R. F., 1986, North American Jurassic apparent polar wander; Implications for plate motion, paleogeography, and Cordilleran tectonics: Journal of Geophysical Research, v. 91, p. 11519–11544.

May, S. R., Butler, R. F., Shafiqullah, M., and Damon, P. E., 1986, Paleomagnetism of Jurassic volcanic rocks in the Patagonia Mountains, southeastern Arizona; Implications of the North American 170 Ma reference pole: Journal of Geophysical Research, v. 91, p. 11545–11557.

Minster, J. B., and Jordan, T. H., 1978, Present-day plate motions: Journal of Geophysical Research, v. 83, p. 5331–5354.

Monger, J.W.H., and Irving, E., 1980, Northward displacement of northcentral British Columbia: Nature, v. 285, p. 289–293.

Monger, J.W.H., and Ross, C. A., 1971, Distribution of Fusulinaceans in the western Canadian Cordillera: Canadian Journal of Earth Science, v. 8, p. 259–278.

Monger, J.W.H., Price, R. A., and Tempelman-Kluit, D. J., 1982, Tectonic accretion and the origin of the two major metamorphic and plutonic welts in the Canadian Cordillera: Geology, v. 10, p. 70–75.

McWilliams, M., and Howell, D. G., 1982, An exotic origin for terranes of western California: Nature, v. 297, p. 215–217.

Oldow, J. S., Ave Lallemant, H. G., and Schmidt, W. J., 1984, Kinematics of plate convergence deduced from Mesozoic structures in the western Cordillera: Tectonics, v. 3, p. 201–228.

Packer, D. R., and Stone, D. B., 1972, An Alaskan Jurassic paleomagnetic pole and the Alaskan orocline: Nature Physical Science, v. 237, p. 25–26.

Palmer, A. R., 1983, The decade of North American geology time scale: Geology, v. 11, p. 503–504.

Reeve, S. C., 1975, Paleomagnetic studies of sedimentary rocks of Cambrian and Triassic age [Ph.D. thesis]: Dallas, University of Texas, 426 p.

Reeve, S. C., and Helsley, C. E., 1972, Magnetic reversal sequence in the upper portion of the Chinle Formation, Montoya, New Mexico: Geological Society of America Bulletin, v. 83, p. 3975–3812.

Robertson, W. A., 1967, Manicouagan, Quebec, paleomagnetic results: Canadian Journal of Earth Sciences, v. 4, p. 641–649.

Russell, R. J., Beck, M. E., Jr., Burmester, R. F., and Speed, R. C., 1982, Cretaceous magnetizations in northwestern Nevada and tectonic implications: Geology, v. 10, p. 423–428.

Silberling, N. J., and Jones, D. L., 1984, Lithotectonic terrane maps of the North

American Cordillera: U.S. Geological Survey Open-File Report 84-523, scale 1:2,500,000.

Simpson, R. W., and Cox, A., 1977, Paleomagnetic evidence for tectonic rotation of the Oregon Coast Range: Geology, v. 5, p. 585–589.

Smith, A. G., and Briden, J. C., 1977, Mesozoic and Cenozoic paleocontinental maps: Cambridge, Cambridge University Press, 63 p.

Smith, T. E., and Noltimier, H. C., 1979, Paleomagnetism of the Newark trend igneous rock of the north-central Appalachians and the opening of the central Atlantic Ocean: American Journal of Science, v. 279, p. 778–807.

Steiner, M. B., and Helsley, C. E., 1975, Reversal pattern and apparent polar wander for the Late Jurassic: Geological Society of America Bulletin, v. 86, p. 1537–1543.

Symons, D.T.A., 1969a, Paleomagnetism of four late Miocene gabbroic plugs in south-central British Columbia: Canadian Journal of Earth Sciences, v. 6, p. 653–662.

——, 1969b, Paleomagnetism of the Late Miocene plateau basalts in the Cariboo region of British Columbia: Geological Survey of Canada Paper 69-43, p. 16.

——, 1973a, Paleomagnetic zones in the Oligocene East Sooke Gabbro, Vancouver Island, British Columbia: Journal of Geophysical Research, v. 78, p. 5100–5109.

——, 1973b, Concordant Cretaceous paleolatitudes from felsic plutons in the Canadian Cordillera: Nature Physical Science, v. 241, p. 59–61.

——, 1974, Paleomagnetism of the Lower Jurassic Tulameen ultra-mafic-gabbro complex, British Columbia: Geological Survey of Canada Paper 74-1, Part B, p. 177–183.

——, 1977a, Geotectonics of Cretaceous and Eocene plutons in British Columbia; A paleomagnetic fold test: Canadian Journal of Earth Sciences, v. 14, p. 1246–1262.

——, 1977b, Paleomagnetism of Mesozoic plutons in the westernmost Coast Plutonic Complex of British Columbia: Canadian Journal of Earth Sciences, v. 14, p. 2127–2139.

——, 1983a, Further paleomagnetic results from the Jurassic Topley Intrusives in the Stikinia subterrane of British Columbia: Geophysical Research Letters, v. 10, p. 1065–1068.

——, 1983b, New paleomagnetic data for the Triassic Guichon batholith of south-central British Columbia and their bearing on Terrane I tectonics: Canadian Journal of Earth Sciences, v. 20, p. 1340–1344.

——, 1985a, Paleomagnetism of the Triassic Nicola Volcanics and geotectonics of the Quesnellia subterrane of Terrane I, British Columbia: Journal of Geodynamics, v. 2, p. 229–244.

——, 1985b, Paleomagnetism of the West Coast Complex and the geotectonics of the Vancouver Island segment of the Wrangellian subterrane: Journal of Geodynamics, v. 2, p. 211–228.

Symons, D.T.A., and Litalien, C. R., 1984, Paleomagnetism of the Lower Jurassic Copper Mountain intrusions and the geotectonics of Terrane I, British Columbia: Geophysical Research Letters, v. 11, p. 685–688.

Tarduno, J. A., McWilliams, M., Sliter, W. V., Cook, H. E., Blake, M. C., Jr., and Premoli-Silva, I., 1986, Southern hemisphere origin of the Laytonville Limestone of California: Science, v. 231, p. 1425–1428.

Tarling, D. H., 1983, Paleomagnetism: London, Chapman and Hall, 379 p.

Terres, R. R., and Luyendyk, B. P., 1985, Neogene tectonic rotation of the San Gabriel Region, California, suggested by paleomagnetic vectors: Journal of Geophysical Research, v. 90, p. 12467–12484.

Van Eysinga, F.W.B., compiler, 1975, Geological timetable, 3rd ed.: Amsterdam, Elsevier Publishing Company.

Vugteveen, R. W., Barnes, A. E., and Butler, R. F., 1981, Paleomagnetism of the Roskruge and Gringo Gulch Volcanics, southeastern Arizona: Journal of Geophysyical Research, v. 86, p. 4021–4028.

Wells, R. E., and Coe, R. S., 1985, Paleomagnetism and geology of Eocene volcanic rocks of southwest Washington; Implications for mechanisms of tectonic rotation: Journal of Geophysical Research, v. 90, p. 1925–1947.

Wilson, D., and Cox, A., 1980, Paleomagnetic evidence for tectonic rotation of Jurassic plutons in Blue Mountains, eastern Oregon: Journal of Geophysical Research, v. 85, p. 3681–3689.

Yole, R. W., and Irving, E., 1980, Displacement of Vancouver Island; Paleomagnetic evidence from the Karmutsen Formation: Canadian Journal of Earth Sciences, v. 17, p. 1210–1228.

Zijderveld, J.D.A., and Van der Voo, R., 1973, Paleomagnetism in the Mediterranean area, *in* Implications of continental drift for the earth sciences, v. 1: New York, Academic Press, p. 133–161.

Manuscript Submitted October 1985
Manuscript Accepted by the Society October 31, 1988

Printed in U.S.A.

Geological Society of America
Memoir 172
1989

Chapter 23

Heat flow and thermal regimes in the continental United States

Paul Morgan
Geology Department, Box 6030, Northern Arizona University, Flagstaff, Arizona 86011
William D. Gosnold
Department of Geology and North Dakota Mining and Mineral Resources Institute, University of North Dakota, Grand Forks, North Dakota 58202

ABSTRACT

Heat flow and associated lithospheric thermal regimes in the continental United States are intricately related to the tectonic evolution and physical properties of the continental lithosphere. More than 2,000 published heat-flow values are now available for the United States, although the distribution of these data is very uneven, with most data sites in the western United States. The data define a single basic thermal regime in all provinces east of the Rocky Mountains and variable thermal regimes in provinces in the western United States. "Normal" heat flow east of the Rockies is interpreted to reflect a stable thermal regime with regional variations associated with redistribution of heat by ground-water convection and heterogeneity in upper crustal heat production. With the exception of Lake Superior in the Canadian Shield, there is a general trend of increasing surface heat flow with increasing crustal thickness in this stable region. Heat flow in the western United States is generally high, and the surface heat-flow pattern is not everywhere controlled by province boundaries. Calculated geotherms for much of the region suggest temperatures near the solidus close to the Moho for much of the region, with major crustal melting predicted by steady-state geotherm calculations for hot thermal subprovinces. In the Basin and Range province, extension may have been an important factor in producing or maintaining the high heat flow, but elsewhere—and perhaps in the Basin and Range also—convection of heat into the crust by mantle-derived melts not simply related to extension appears to be the primary heat source. These melts are probably related to subduction, but the exact mechanism of their origin is unclear. Low heat flow in the eastern Snake River Plain and the "Eureka Low" appears to result from downward convection by ground-water recharge. Low heat flow in the Sierra Nevada and Northwest Pacific coastal provinces appears to be related to subduction. Relatively normal heat flow in the Colorado Plateau is inconsistent with the elevation of the plateau, and an increase in heat flow with depth is predicted beneath this province. High heat flow in the California Coast Ranges is related to growth of the San Andreas fault zone and local segments of extension along its length. These thermal regimes are directly related to variations in lithospheric thickness across the United States, and regional elevation differences are associated with both crustal thickness and geotherm variations. Increased understanding of these thermal regimes and their evolution is leading to a better understanding of the tectonic evolution, physical properties, and lithospheric structure of the continental United States.

Morgan, P., and Gosnold, W. D., 1989, Heat flow and thermal regimes in the continental United States, *in* Pakiser, L. C., and Mooney, W. D., Geophysical framework of the continental United States: Boulder, Colorado, Geological Society of America Memoir 172.

INTRODUCTION

In no geophysical parameter is the range in recent tectonic activity in the continental United States more dramatically manifested than heat flow, the outward flow of heat across the Earth's surface. Variations in surface heat flow clearly delineate contrasting thermal regimes in the different tectonic environments. However, although regions with recent tectonic activity commonly have higher heat flow than more stable regions, no direct correlation between surface heat flow, the geotherm, and "tectonic age" (the age of the last major tectonic and thermal activity) can be made comparable to the oceanic age–heat flow–elevation relationship. Part of the problem lies in the chemical heterogeneity of continental crust. Typically from a few percent to more than 65 percent of continental surface heat flow is derived from radiogenic decay of uranium, thorium, and potassium isotopes in the upper crust, and this chemical heterogeneity can give a wide range of surface heat flow even if the "tectonic" heat-flow component is constant (Rao and others, 1982; Morgan, 1985). The balance of the problem lies in the complex and often protracted deformational processes of continents. Creation of new oceanic crust can be treated as an instantaneous and predictable event at the ridge crest (Parsons and Sclater, 1977). Continental deformation is often distributed in both space and time and results in a variety of thermal regimes, the details of which are unpredictable. Thermal models must therefore be tailored to the chemical and deformational heterogeneity of the continents, and surface heat flow cannot be analyzed in isolation from other parameters.

Another important factor in understanding heat-flow data is the recognition that only conducted heat flow is generally represented in a surface heat-flow measurement. A significant portion of heat transfer may be convective, especially in areas of young tectonics. For example, relatively simple models have been used successfully to describe the gross thermal evolution of oceanic lithosphere from its creation at mid-ocean ridges to its thermal maturity in ocean basins (e.g., Parsons and Sclater, 1977). These models only match the surface heat-flow data, however, in areas where sediments are thought to prevent convection of sea water through the crust, generally remote from the ridges and where the thermal variations are subdued. The most convincing proof of the thermal models is their success in predicting the depth evolution of oceanic lithosphere in terms of the isostatic response of the lithosphere to an increase in density (and probably thickness) as a function of cooling. Near mid-ocean ridges, heat is redistributed in the upper crust; much of it is lost by the thermally induced convection of sea water through the upper crust to which the gross lithospheric thermal structure is insensitive, but to which heat-flow measurements are very sensitive.

In continental environments, heat is redistributed by convection in the near-surface, driven both by thermally induced buoyancy where the thermal gradients and rock permeabilities are high, and by any ground-water movements driven by a hydraulic gradient. This redistribution of heat must be included in any analysis of continental heat flow. Several studies suggest that continental elevation is related to thermal structure in a manner similar to sea-floor elevations, but a more complex model that considers the variable thickness (and possibly density) of continental crust, crustal radiogenic heat production, and complexities in the continental geotherm evolution is required (Crough and Thompson, 1976; McKenzie, 1978; Brott and others, 1981; Morgan, 1983; Lachenbruch and others, 1985; Morgan and Burke, 1985). In only rare instances is it possible to trace the details of the elevation history of a continental region, and this parameter has limited use in elucidating continental thermal histories. Thermal, tectonic, and elevation histories of the lithosphere of the United States are intricately linked, and we are perhaps just beginning to understand their interrelation. It is also probable that the lithospheric thermal regime and deformational style are interrelated; thus, structural, as well as elevation, histories, may be useful in elucidating thermal histories. In the context of this volume, heat flow and thermal regimes are relevant to many of the geophysical parameters discussed and are commonly cited in geological and geophysical investigations in the U.S., such as, for example, magnetotellurics (Chaipayungpun and Landisman, 1977), magnetics (Mayhew, 1982a, b), integrated studies of crustal stress (Zoback and Zoback, 1980), general geophysical and geological models (Eaton, 1980), geochemistry and petrology (Kay and Kay, 1980), gravity (Keller and others, 1979), and seismology (Smith, 1978; Black and Braile, 1982). In addition, heat-flow data constrain models of young tectonic processes; for example, formation of the Basin and Range province (Lachenbruch and Sass, 1978; Blackwell, 1978), large-scale ground-water flow systems, such as the Snake River Plain (Brott and others, 1978, 1981), and possibly even crustal chemistry in stable regions (Morgan, 1985).

Three goals are attempted in this review: (1) presentation of the current heat-flow data set for the United States; (2) discussion of these data in terms of regional thermal regimes and their interrelations with other geophysical parameters; and (3) discussion of the thermal regimes in terms of regional tectonics and magmatism. These topics have been discussed recently on a global scale for continental lithosphere in reviews by Vitorello and Pollack (1980), Sclater and others (1980, 1981), Pollack (1982), Morgan (1984), and Morgan and Sass (1984). In general, this review can be regarded as an update of the excellent recent reviews of U.S. heat flow, which include Lachenbruch and Sass (1977, 1978), Blackwell (1978), and Sass and others (1981). The U.S. heat-flow data set has grown too large to be discussed in detail in the space available in this chapter, and the reader is strongly advised to seek original references where local aspects of heat flow are important. U.S. geothermal gradient data have recently been compiled by Guffanti and Nathenson (1980) and Nathenson and Guffanti (1988). We attempt here to follow the spirit of the title of the Memoir and present the thermal framework of the continental United States, providing regional information for the general reader, the interrelation among heat flow and other geophysical parameters, and discussion of the geological implications of the thermal regimes.

HEAT FLOW IN THE CONTINENTAL UNITED STATES

Terrestrial heat flow is defined as the product of the thermal gradient and thermal conductivity given by:

$$q_s = K\delta T/\delta z \tag{1}$$

where q_s is surface heat flow, $\delta T/\delta z$ is the vertical thermal gradient defined to be positive downward, and K is the thermal conductivity of the rock units in which the gradient is measured. On land, thermal gradient measurements are most commonly made in boreholes, and conductivities are measured in the laboratory on samples recovered from the boreholes. Techniques for these measurements have been described by Birch (1950), Beck (1965), Sass and others (1971), and Kappelmeyer and Haenel (1974). Measurements also can be made in the sediments of some deep lakes using modifications of the marine techniques described by Langseth (1965). The distribution of heat-flow data thus depends, to a large extent, on the distribution of boreholes and lakes suitable for heat-flow determinations. There is a sampling bias toward the areas of Cenozoic tectonic activity.

More than 2,000 heat-flow determinations have now been published for the continental United States; these data are summarized in Figure 1. The sampling bias is evident in these data, with approximately 80 percent of the data being in the western third of the country, and approximately 35 percent of the data from the eastern two-thirds coming from Lake Superior.

Heat-flow values represented by the data plotted in Figure 1 are near-surface conductive heat-flow values, representing the vertical component of conductive heat flow over the depth interval of the geothermal gradient measurement. As discussed by Lachenbruch and Sass (1977) and below, near-surface heat-flow values may not be representative of heat flow at depth due to transient thermal conditions (i.e., heat-flow changes with depth and time), nonconductive heat transfer (thermal convection by ground-water movements, magmatism, or tectonic deformation), and shallow crustal heat production. The most significant source of shallow crustal heat production is radiogenic heat generation by the spontaneous decay of ^{232}Th, ^{235}U, ^{238}U, and ^{40}K (Birch and others, 1968; Roy and others, 1968; Lachenbruch, 1968). The discovery by Birch and others (1968) of a linear relationship between surface heat flow and surface heat generation in plutonic rocks in the same tectonic setting allows the removal of the upper crustal heat generation component of surface heat flow. This linear relationship is generally written in the form:

$$q_s = q^* + A_s b, \tag{2}$$

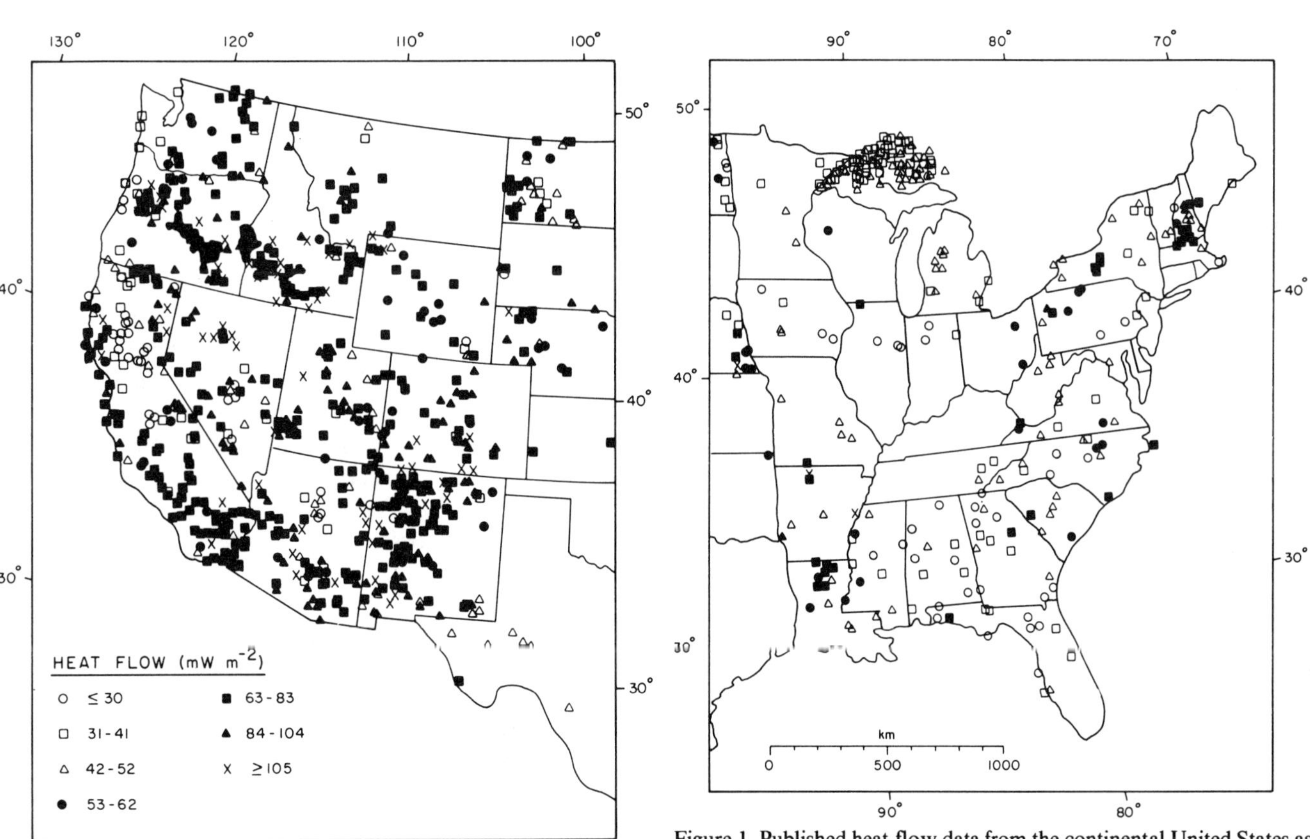

Figure 1. Published heat-flow data from the continental United States as of 1986. Data source, J. H. Sass (personal communication, 1986) with minor updates from Gosnold (1988) and D. L. Smith (personal communication, 1986).

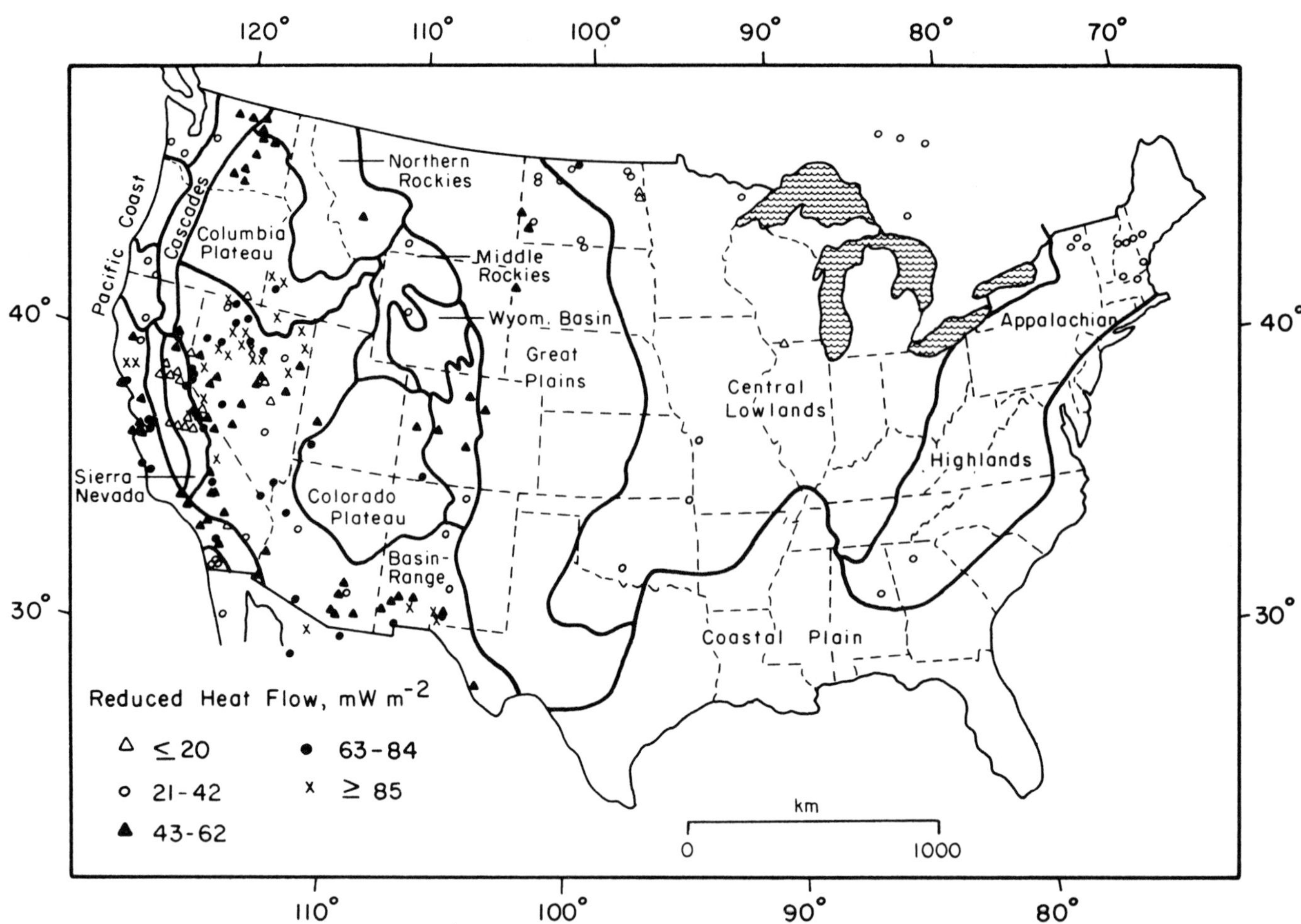

Figure 2. Published reduced heat-flow values [q^* from equation (2)] and major physiographic provinces in the continental United States. Modified from Sass and others (1981).

where q_s is the measured surface heat flow, A_s is the heat production of the plutonic rocks in which the measurement was made, and q^* and b are constants defining the relationship. The intercept on the heat-flow axis, q^* or the "reduced heat flow," is the heat-flow equivalent to zero upper crustal heat production, and is interpreted to represent the heat flow from below the upper crustal heat-producing layer. Heat flow represented by q^* includes heat generated from radiogenic heat sources in the lower crust and upper mantle not represented by the surface heat production, heat convected within the lithosphere during tectonism and magmatism, and heat from below the lithosphere. The slope of the relationship, b, has the dimensions of length and is interpreted to be a depth scaling factor describing the vertical extent of the upper crustal heat-producing layer. Both the reduced heat flow and the depth scaling factor vary between tectonic provinces, but b is usually within the range of 10 ± 6 km (e.g., Vitorello and Pollack, 1980; Morgan, 1984). Figure 2 shows the subset of heat-flow data for the United States for which both heat flow and surface heat production values are available; reduced heat-flow values have been calculated for this figure using equation (2) and assuming $b = 10$ km. As before, there is a very uneven data distribution, with most of the data coming from the western third of the country.

The U.S. heat-flow data set is now discussed briefly province by province to generalize the thermal characteristics of each province to constrain the different thermal regimes. For a more complete discussion of an earlier subset of these data, the reader is referred to Sass and others (1981).

Provinces east of the Rocky Mountains

There are five physiographic provinces east of the Rocky Mountains (the Rockies), but as shown by the data summary given in Table 1, these provinces are thermally indistinguishable statistically. The combined data set for these five provinces is plotted as a histogram in Figure 3a, and has a mean and standard deviation of 51 ± 20 mW m^{-2} (number of data, $n = 530$). The subset of these data for which both heat-flow and heat-production data are available is plotted as a graph in Figure 4. If the Great Plains data are excluded, the remaining data define a rough linear trend between heat flow and heat production. This trend has been defined by taking the complete data set, applying a linear least-squares regression, rejecting data falling off the line by more than three standard errors in the intercept of the line, and repeating the regression. The resulting line is shown in Figure 4 after the rejection of 13 data points and suggests a reduced heat flow of 31 ± 1 mW m^{-2} (n = 20) for the United States east of the Rockies with a heat-production depth parameter of 8.1 ± 0.4 km. These values are very close to published analyses of these data (e.g., Roy and others, 1968). The mean heat flow east of the Rockies is close to the mode of the data set, and there is no significant difference in the mean of the heat flow from the heat-flow vs. heat-production subset of the data (53 ± 18 mW m^{-2}, n = 33). This result suggests that this data set is reasonably characterized by its mean, and this value will be used to characterize the thermal regime of the U.S. east of the Rockies. Scatter in the Great Plains heat-flow–heat-production data is considered in the discussion of the thermal regime of this region.

Rocky Mountains provinces

Heat-flow data for the combined Rocky Mountains provinces have a mean value of 149 ± 181 mW m^{-2} (n = 110) and are shown in a histogram in Figure 3b. This mean does not characterize the data that have a mode around 75 mW m^{-2} but includes a large component of high values in excess of 150 mW m^{-2}. Heat-flow–heat-production data from these provinces are plotted in Figure 5 and clearly do not suggest a linear relation between these parameters. However, the mean heat flow of this subset of the data, 76 ± 15 mW m^{-2} (n = 11), is close to the mode of the complete data set, suggesting that a background thermal regime may exist characterized by a heat flow of about 75 mW m^{-2} upon which subregional anomalies are superimposed.

Colorado Plateau

In contrast to the Rocky Mountains provinces, the Colorado Plateau heat-flow data set has a low mean of 68 ± 26 mW m^{-2} (n = 117), very close to the mode of the data set, as shown in Figure 3c. This value is used to characterize the thermal regime of the Plateau. Only three heat-flow–heat-production pairs have been published for the Colorado Plateau; these data are plotted on Figure 5. No linear relation is suggested by these few data.

Basin and Range province

Heat-flow data from the Basin and Range province have a mean of 113 ± 120 mW m^{-2} (n = 232), significantly different from the mode of the data set, which is about 85 mW m^{-2} (Fig. 3d). Heat-flow–heat-production data from this province do not define a linear relation, as shown in Figure 6 (however, see also below), but the mean of this subset of the data, 86 ± 36 mW m^{-2} (n = 66), is close to the mode of the data set. Like the Rocky Mountains provinces heat flow, this suggests that a heat flow of about 85 mW m^{-2} may characterize a "background" for this province upon which subregional anomalies are superposed.

TABLE 1. MEAN HEAT FLOW BY PHYSIOGRAPHIC PROVINCE*

Province	Mean Heat Flow ± Standard Deviation (mW m^{-2}(n))
Central Lowlands	59 ± 22 (119)
Canadian Shield	41 ± 9 (174)
Great Plains	66 ± 26 (87)
Coastal Plain	42 ± 17 (71)
Appalachian Highlands	53 ± 14 (79)
All Provinces East of Rocky Mountains	51 ± 20 (530)
Rocky Mountain Provinces	149 ±181 (110)
	[76 ± 15 (11)][a]
Colorado Plateau	68 ± 26 (117)
Basin and Range	113 ±120 (232)
	[86 ± 36 (66)][b]
Columbia Plateau	104 ± 58 (378)
Cascade Range	101 ±108 (57)
Sierra Nevada	54 ± 31 (23)
	[36 ± 13 (9)][c]
Salton Trough	141 ± 69 (104)
(Central Valley)	(112 ± 44 (27)][d]
Pacific Coastal	46 ± 44 (202)
	[76 ± 29 (35)][e]
	[63 ± 19 (136)][f]

*First mean heat-flow values given for each province are based on all data from province (n = number of data). Means in brackets are based on subsets of the data as follows: [a, b, d, e] = means of data from heat-flow–heat-production sites only; [c] = mean of data from heat-flow–heat-production sites that closely follow linear regression only; [f] = mean of data within range 10 to 110 mW m^{-2} only. Data source J. H. Sass (personal communication, 1986).

Columbia Plateau

Several subprovinces are included in the Columbia Plateau province; these data are shown in a histogram in Figure 3e. This data set is multimodal, with the primary mode at about 85 mW m^{-2}, a secondary mode around 125 mW m^{-2}, and a significant number of values above 150 mW m^{-2}. The mean heat flow for this province, 104 ± 58 mW m^{-2} (n = 378), is significantly displaced from the primary mode and probably has little regional significance. There is no significant correlation between heat flow and surface heat production in this province, but very few data are available for this plot (Fig. 7). The data set somewhat resembles the Basin and Range data set (Fig. 3d) and has a similar primary mode. The multimodal character of this data set reflects in part the different tectonic subprovinces in the Columbia Plateau.

Cascade Range

In common with the Columbia Plateau, the Cascade Range data set is distinctly multimodal (Fig. 3f), and little significance

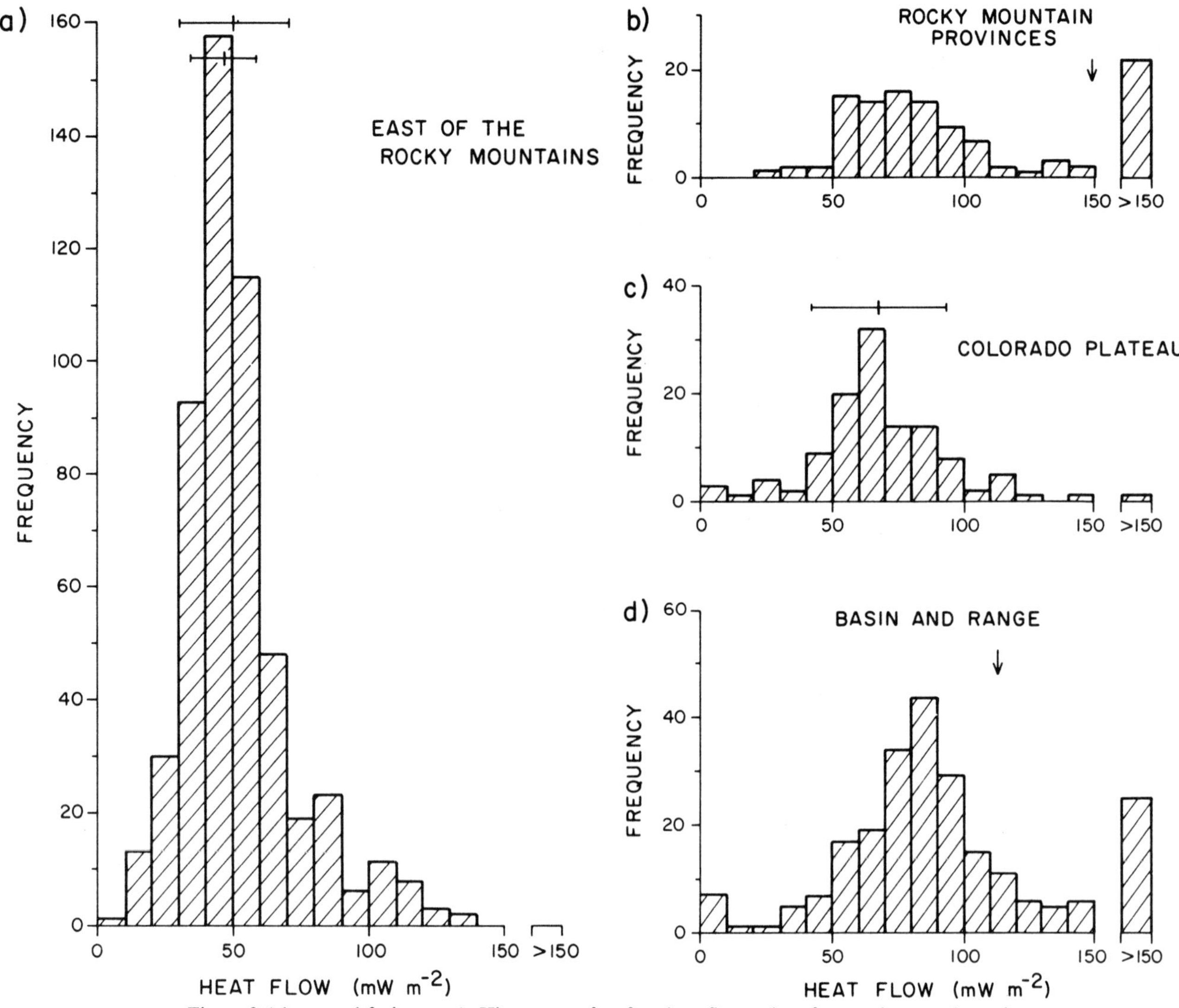

Figure 3 (above and facing page). Histograms of surface heat-flow values for provinces and province groups listed in Table 1 and discussed in the text. Data means and standard deviations are indicated by bars and means with large standard deviations are indicated by arrows (see Table 1). Data are divided as follows: a, Provinces east of the Rocky Mountains. b, Rocky Mountains provinces. c, Colorado Plateau. d, Basin and Range province. e, Columbia Plateau. f, Cascade Range. g, Sierra Nevada. h, Coastal provinces except Salton trough. i, Salton Trough.

should be ascribed to the mean of 101 ± 108 mW m^{-2} (n = 57). There is no correlation among the few heat-flow–heat-production data pairs available for this province (Fig. 7). As before, the multimodal character of this data set reflects different tectonic subprovinces, and the Cascades cannot be characterized by a single "background" heat-flow mean. A primary division between high and low to normal heat flow exists between the High Cascade and Western Cascade Ranges.

Sierra Nevada

In contrast to the adjacent Basin and Range province, heat flow in the Sierra Nevada is typically low, with a mean of 54 ± 31 mW m^{-2} (n = 23). The data set does not display a well-defined mode, however, as shown in Figure 3g, and the significance of this mean is not clear. Part of the heat-flow–heat-production data set for the Sierra Nevada displays a good linear correlation with most of the divergent points coming from the Basin and Range–Sierra Nevada transition zone (Fig. 8). The regression line plotted in Figure 8 has a slope of 9.9 ± 0.3 km and a heat-flow intercept of 16.7 ± 0.6 mW m^{-2} (n = 9). This regression probably characterizes the thermal regime of the main part of the Sierra Nevada, and this subset of the data has a mean of 36 ± 13 mW m^{-2} (n = 9). Most of the higher heat-flow values come from the transition zone and represent an intermediate thermal regime between the Sierra Nevada and Basin and Range provinces.

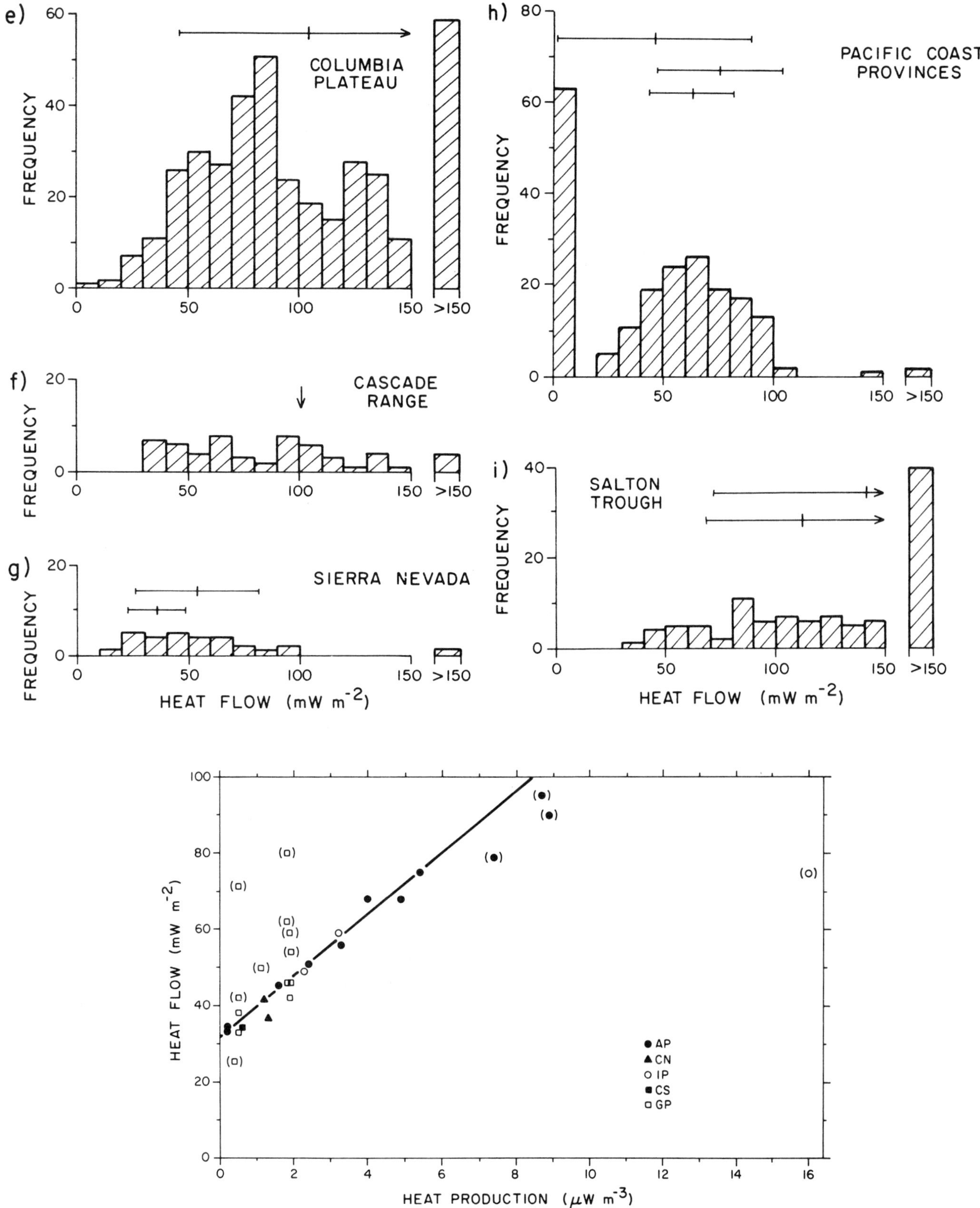

Figure 4. Heat-flow–heat-production data pairs for provinces east of the Rocky Mountains. Key to data points: AP = Appalachian Highlands; CN = Coastal Plains; IP = Interior Platform; CS = Canadian Shield; GP = Great Plains.

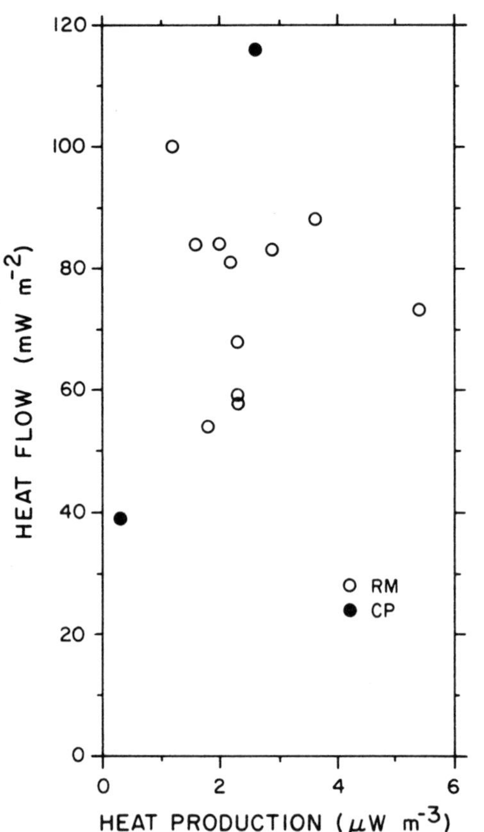

Figure 5. Heat-flow–heat-production data pairs for Rocky Mountains provinces (RM) and Colorado Plateau (CP).

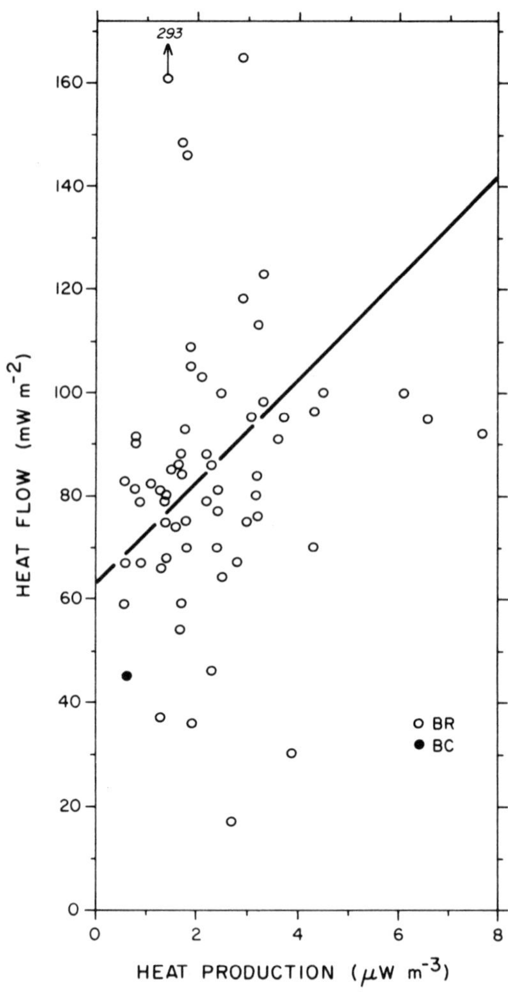

Figure 6. Heat-flow–heat-production data pairs for Basin and Range province (BR) and Arizona Transition Zone (BC).

Pacific Coastal provinces

A variety of tectonic provinces are included in this province group, which excluding the Salton trough, has a heat-flow mean of 46 ± 44 mW m^{-2} (n = 202). A large number of the values in this province are very low (Fig. 3h), and reflect only the shallow thermal regime and local ground-water movement. If these data are ignored, the data set has a well-defined mode around 65 mW m^{-2} (Fig. 3h). Examination of the spatial distribution of these values indicates a bimodal distribution with low to normal heat flow in the Klamath Mountains and Coastal Ranges of the Pacific Northwest and high heat flow in the California Coastal Ranges. Heat-flow–heat-production data pairs from this province group and the individual provinces show no correlation (Fig 9); the mean of this subset of the data is 76 ± 29 mW m^{-2} (n = 35).

Heat flow from the Salton trough in southern California is generally high, with a significant number of values in excess of 150 mW m^{-2}. The mean of these data is 141 ± 69 mW m^{-2} (n = 104), but this mean probably has no regional significance and there is no clearly defined mode in the data (Fig. 3i). Heat-flow–heat-production data pairs show no correlation for this region, as shown in Figure 10, and the mean heat flow for this subset of these data is 112 ± 44 mW m^{-2} (n = 27). It is clear that the Salton trough is characterized by a hot thermal regime, but there are significant local variations.

Limitations of available data

Although heat flow is an important geophysical and tectonic parameter, making heat-flow determinations is not routine, and many local factors can affect a measurement. These factors include topography and variations in ground-surface temperature (Birch, 1950; Roy and others, 1972; Blackwell and others, 1980), lateral variations in thermal conductivity and conductivity anisotropy (Simmons, 1961; McBirney, 1963; Jaeger, 1965; Lee and Henyey, 1974; Lee, 1975; Jensen, 1983), ground-water flow in the measurement interval (Drury and Lewis, 1983; Drury, 1984; Majorowicz and others, 1984; Drury and others, 1984; Beck and Shen, 1985), local heat production (e.g., Lovering, 1948; Hamza and Beck, 1975), past climate (Birch, 1948; Allis, 1978; Beck, 1977, 1982; Shen and Beck, 1983), and the effects of uplift, erosion, and sedimentation (Benfield, 1949; Birch, 1950; England and Richardson, 1980). Combinations of these factors, which probably contribute to spurious estimates of regional heat flow in the northern Basin and Range province, have been eloquently discussed by Blackwell (1983). Criteria can be placed on the

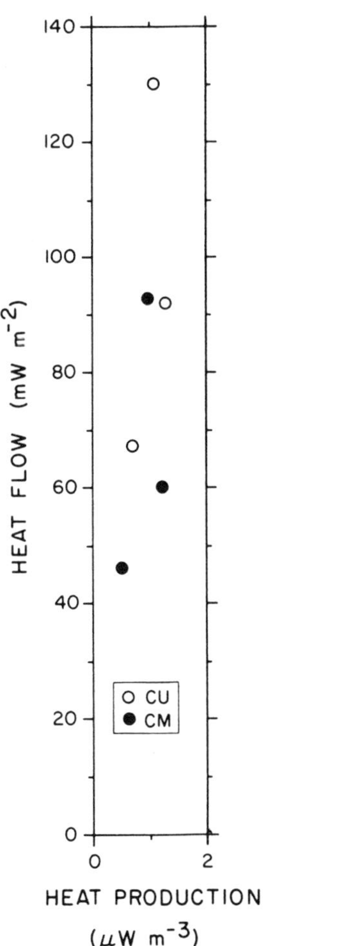

Figure 7. Heat-flow–heat-production data pairs for Columbia Plateau (CU) and Cascade Mountains (CM).

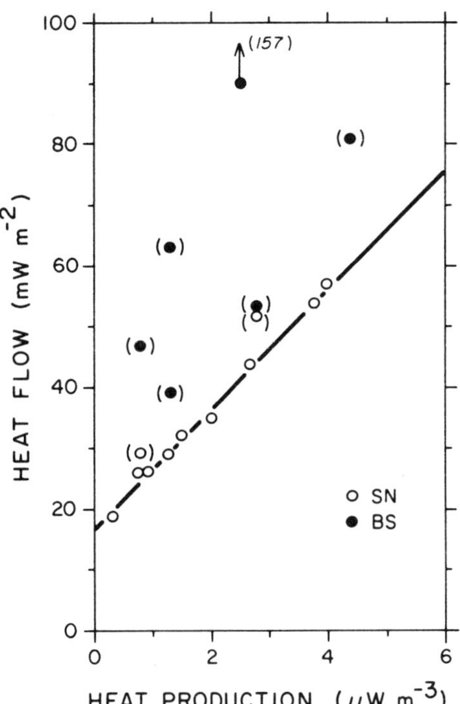

Figure 8. Heat-flow–heat-production data pairs for Sierra Nevada (SN) and Sierra Nevada–Basin and Range transition (BS).

requirements for a heat-flow determination (Jessop, 1983; Chapman and others, 1984), and statistical methods can be used to analyze data from multiple sites (Lewis and Beck, 1977; Reiter and others, 1985). Often in regional analysis the choice is between imperfect data or no data, in which case the former is preferable, although their limitations must always be recognized. Even where apparently high-quality data are available from a single site, they may not always be regionally representative (Swanberg and Morgan, 1980).

Implicit in the nature of heat-flow data is lateral variation in surface heat flow due to variations in crustal heat generation, transport of heat by ground-water flow, and lateral heterogeneity in tectonic and magmatic processes. Even in areas of the western U.S. where the data density is relatively high, there are usually not sufficient data to identify and characterize the processes responsible for these lateral heat variations. The summary of the data given above has attempted to elucidate regional characteristics of the data to constrain regional thermal regimes, and extreme caution should be exercised in using the means in Table 1.

FACTORS CONTROLLING THE THERMAL REGIME IN THE LITHOSPHERE

One of the primary uses of heat-flow data is to constrain the thermal regime of the lithosphere. For this use it is important to appreciate the factors that control heat transfer in the lithosphere; for convenience, these factors have been divided into three groups: (1) shallow factors, which are important in the uppermost crust where ground-water flow and lateral thermal conductivity variations are important; (2) crustal factors, which include crustal thickness and radiogenic heat production; and (3) tectonothermal factors, which include thermal convection by magmas and lithospheric deformation, regional erosion, and sedimentation, and thinning and thickening of the lithosphere.

Shallow factors

On a regional basis the three basic factors controlling the shallow thermal regime are the regional heat flow, thermal conductivity, and heat transfer by ground-water movements. For this review we assume (perhaps incorrectly) that we have sufficient data to constrain the regional heat flow, and thus we consider only the effects of conductivity and ground-water convection.

Thermal conductivities of rocks control the vertical thermal gradient for a given heat flow and depend on a number of parameters including temperature, mineralogy, and porosity. At the surface they typically range from less than 1 W m^{-1} K^{-1} for unconsolidated sediments to greater than 6 W m^{-1} K^{-1} for halite, dolomite, and quartzite (Roy and others, 1981). Most of this variation comes from the effects of low-conductivity fluids filling pore spaces and high-conductivity monomineralic aggregates, both typically found at shallow depths. Crystalline rocks likely to form a significant volume of the crust and upper mantle have a smaller range in mean thermal conductivity at temperatures as high as 1,000°C, 2.0 to 3.5 W m^{-1} K^{-1}, and thus, shallow-conductivity variations have the greatest potential effect on the lithospheric thermal regime. Deeper conductivity variations will be discussed with crustal factors below.

Crystalline rocks typically have a higher thermal conductivity than porous sediments; thus, for the same heat flow, lower thermal gradients occur on basement outcrops than in sedimentary basins (Blackwell and Chapman, 1977). It is not uncommon for the basement/sediment contrast to be about a factor of 2, especially for younger basins, which will approximately result in a factor of 2 higher temperature at the base of the sediments relative to the equivalent depth in the basement. Beneath a deep basin this can have a significant effect on upper crustal temperatures (temperature differences of the order of 100°C may be expected beneath a 5-km-deep basin with a background heat flow of 60 mW m^{-2}). Thermal refraction in the Basin and Range province is discussed in detail by Blackwell (1983). Refraction temperature differences are attenuated with depth and are probably insignificant in the lower crust and upper mantle. A clear example of the effect of near-surface thermal conductivity variations on the thermal regime is given by the new U.S. thermal gradient map by Nathenson and Guffanti (1988) shown in Figure 11. Areas of high thermal gradient on this map in the central and eastern U.S. in general do not have correspondingly high heat flow (Fig. 1), and are primarily the result of low-sediment thermal conductivities.

Thermal convection by ground-water flow is a major problem in the interpretation of heat-flow data, and major thermal

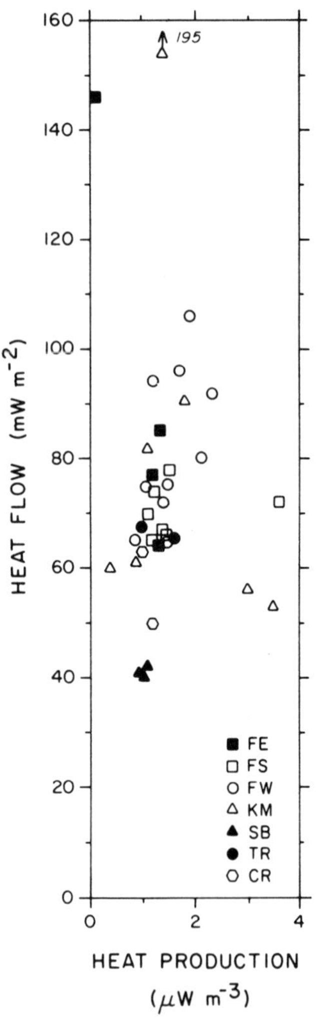

Figure 9. Heat-flow–heat-production data pairs for Pacific Coastal provinces excluding the Salton trough. Key to provinces: FE, FS, FW = San Andreas fault east, south, and west, respectively. KM = Klamath Mountains; SB = Central Valley–Coast Ranges; CR = Pacific Northwest Coastal Ranges; TR = Transverse Ranges.

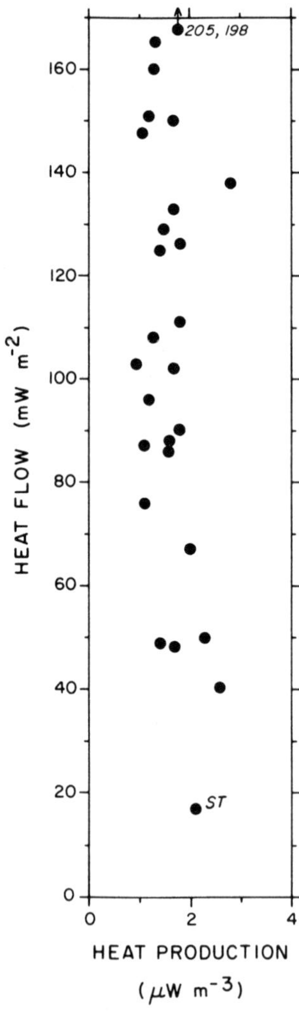

Figure 10. Heat-flow–heat-production data pairs for Salton trough (ST).

provinces, such as the "Eureka Low" in Nevada, may be upper crustal in origin and reflect heat transfer by ground-water flow (Lachenbruch and Sass, 1977). As a first approximation, as ground water merely redistributes the heat, assuming that the actual measurements are not affected by the flow (i.e., that convection does not occur over the gradient measurement interval), areas of low heat flow associated with recharge and areas of high heat flow associated with discharge will balance each other in a regional heat-flow analysis. However, as most areas are undersampled by heat-flow data, and discharge of hot water to the surface is not measured by conventional heat-flow measurements, a simple average of heat-flow values may not represent the regional heat flow (e.g., Brott and others, 1981). Significant transient disturbances can occur during the development and waning of convection systems, and errors in regional heat-flow estimates can result from ground-water convection (see Lachenbruch and Sass, 1977, for more complete discussion).

In terms of the crustal thermal regime, ground-water convection can significantly modify temperatures in convecting zones, and as with shallow-conductivity contrasts, these modifications can be conducted with attenuation into the underlying crust. These effects can be very important locally, especially in areas where heat transport is dominated by convection, such as over upper crustal intrusions (e.g., the Yellowstone caldera: Morgan and others, 1977) or large areas of ground-water recharge (e.g., the "Eureka Low": Lachenbruch and Sass, 1977), but are generally expected to be attenuated in the lower crust and upper mantle.

Crustal factors

Shallow factors generally cause a redistribution of the regional heat flow or result in shallow modification of the thermal regime, but a significant portion of continental heat flow is generated within the crust, resulting in a vertical variation in heat flow. Conductivity contrasts between the crust and mantle also control the lithospheric thermal regime. Unfortunately, neither the crustal heat production nor deep thermal conductivities are well defined.

The linear relation between surface heat flow and heat generation [equation (2)] allows an estimate of the amount of crustal heat generation but does not constrain its distribution with depth (Lachenbruch, 1968). Lateral variations in the crustal inventory of heat production result in large part in scatter of heat flow in stable areas, and must also result in variations in the geotherm and deep thermal regimes within "thermal" provinces (Rao and others, 1982; Morgan, 1984; Morgan and Sass, 1984).

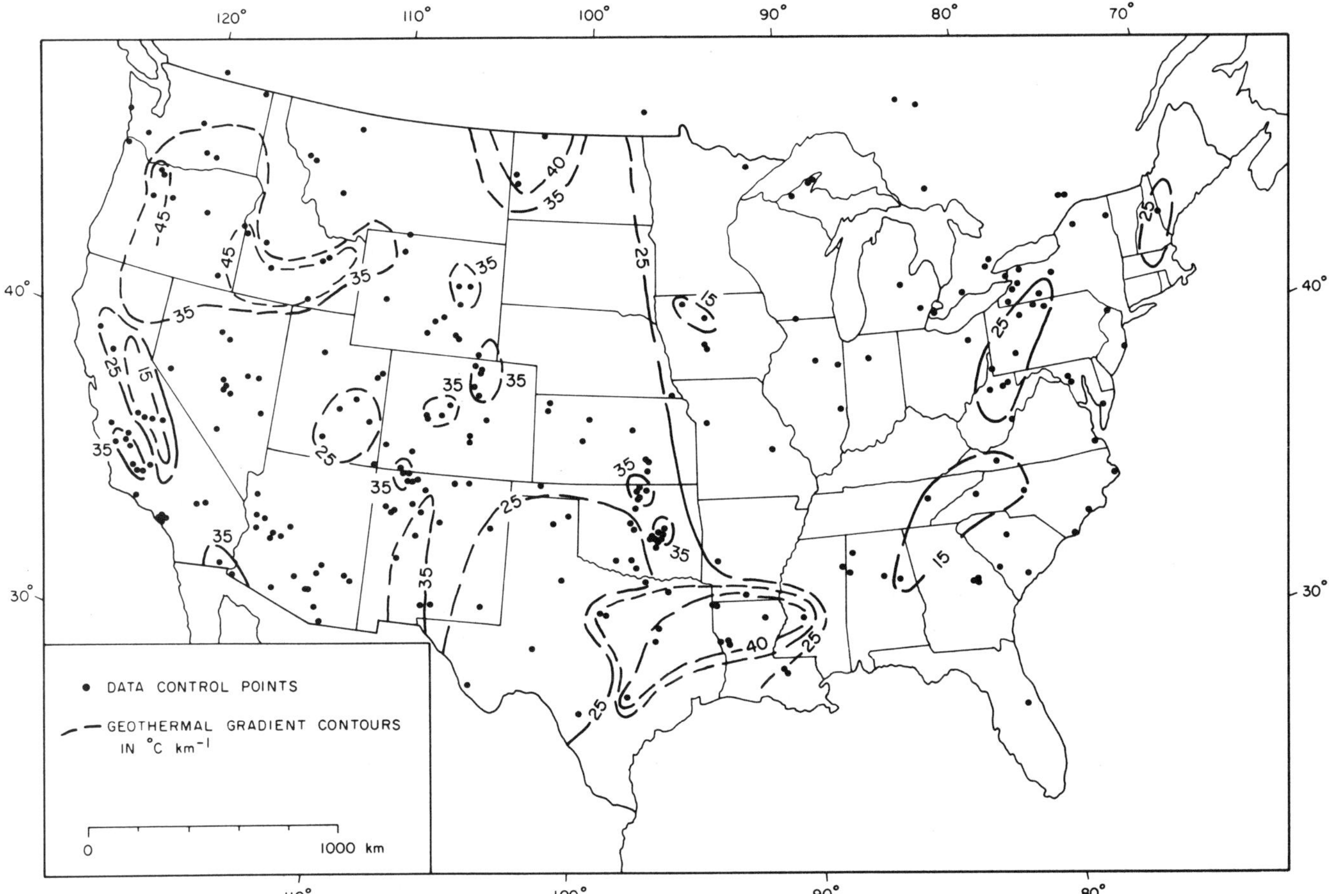

Figure 11. Regional geothermal gradient contours in the continental United States. Contour interval 10°C km^{-1} unless otherwise noted. Redrawn from Nathenson and Guffanti (1988).

However, uncertainty as to the vertical distribution of this heat generation results in uncertainty in the geotherm of as much as a factor of 2 or more in the effect of crustal heat production on the geotherm (Lachenbruch, 1970; Blackwell, 1971; Costain, 1978; Gosnold and Swanberg, 1980; Jaupart and others, 1981; Morgan and Sass, 1984). In calculation of geotherms for this review, we have used two model distributions of crustal heat production: the step model in which upper crustal heat generation is uniform from the surface (A_S) to depth b where it drops to zero, and the exponential model in which heat production decreases exponentially from a value of A_S at the surface with an exponential depth scaling factor b (Lachenbruch, 1970). There is a factor of 2 difference in the temperature increase associated with crustal heat generation predicted by these two models (exponential model gives a hotter geotherm), and while the range of geotherms between the two geotherms predicted by these models does not cover geotherms predicted by all possible distributions, most likely distributions predict geotherms within this range. Predictions using these models should be useful for relative comparisons of different thermal regimes but may be in error in absolute predictions of the geotherm.

Thermal conductivities in the lower crust and upper mantle are somewhat poorly constrained because of uncertainties in composition and anisotropy, as well as uncertainties in the effects of temperature and, to a lesser extent, pressure on conductivities. Compositionally, the lower crust is likely to have a lower thermal conductivity than the upper mantle (Roy and others, 1972, 1981; Smithson and Decker, 1974), and thus thick crust will result in a higher mantle geotherm than thin crust for the same background heat flux.

Thermal conductivities probably decrease with increasing depth and temperature in the crust below the upper crustal compositional heterogeneity as lattice, or phonon, conductivity has a negative temperature coefficient for most reasonable mineralogies. Conductivities in the mantle probably increase with increasing depth and temperature, since above about 500°C, radiative, or photon, conductivity that has a positive temperature coefficient becomes the dominant parameter controlling the conductive gradient (Birch and Clark, 1940; Clark, 1957; Fujisawa and others, 1968; Kanamori and others, 1968; Schatz and Simmons, 1972; MacPherson and Schloessin, 1982). For relatively cool geotherms, typical of stable regions, it is possible to choose mean conductivities for the crust and mantle that predict geotherms not significantly different from those predicted using temperature-dependent conductivity models (Morgan, 1984). For hotter regimes, the differences between geotherms predicted by uncorrected-mean and temperature-dependent conductivity models are larger (Buntebarth, 1984, p. 10–12), but as conduction is commonly not the dominant mechanism of heat transfer in these regimes, the conductivity uncertainties are of second-order importance. For primarily comparative purposes in this review, we have used mean conductivity models.

Tectonothermal factors

Crustal factors include the most important parameters controlling the geotherm in stable areas in which a condition of steady-state conduction may be assumed. Stable areas may be defined as regions that have not been affected by significant tectonic or magmatic activity for the past 100 to 250 m.y. or so (the thermal relaxation time constant of the lithosphere; Lachenbruch and Sass, 1977; Morgan, 1984; Morgan and Sass, 1984). In areas of more recent tectonism and magmatism, however, the assumption of steady-state conduction through the lithosphere is usually invalid, and more complex models must be considered. Magmatism always results in a hotter geotherm as heat is convected upward with magma ascent. Tectonic activity can raise or lower the geotherm.

Deformation of the lithosphere commonly occurs at rates faster than heat can be transferred by conduction; thus, any net vertical movement of the lithosphere will result in transient thermal convection of heat. Using the generally valid assumption that the lithosphere is incompressible, extensional tectonics result in a thinning of the preextension lithospheric column, which if maintained in isostatic balance, implies a net upward movement of material in the lithosphere, and upward convection of heat (McKenzie, 1978; Lachenbruch and Sass, 1978). During compressional tectonics the lithosphere tends to be thickened, the net movement of material is downward, and surface heat flow is reduced (Oxburgh and Turcotte, 1974; Bickle and others, 1975; Toksöz and Bird, 1977). There are responses to these changes in lithospheric thickness, however, which reduce the effect of deformational convection: thinning of the buoyant crust during extension commonly causes downwarp, and the deposition of cold sediments in the resulting basin reduces surface heat flow; upwarp and erosion following compressional thickening increase surface heat flow. Changes in crustal thickness associated with extension or compression may result in changes in the thickness parameter associated with upper crustal radiogenic heat production and/or the vertical distribution of the heat-producing elements. Finally, and perhaps most significantly, magmatism is commonly associated with tectonic deformation, resulting in a strong upward convective flux of heat. The thermal heterogeneity of stable continental crust (the starting point for thermal models of deformation) and the thermal complexity of the thermal processes associated with continental tectonics make it difficult quantitatively to generalize tectonothermal factors (Morgan, 1984). Models discussed below with application to U.S. thermal provinces are thought to be conceptually valid, but detailed numerical predictions must be regarded with caution.

HEAT-FLOW PROVINCES AND THERMAL REGIMES

The factors controlling the thermal regime discussed above affect the geotherm and the validity of extrapolation of the near-

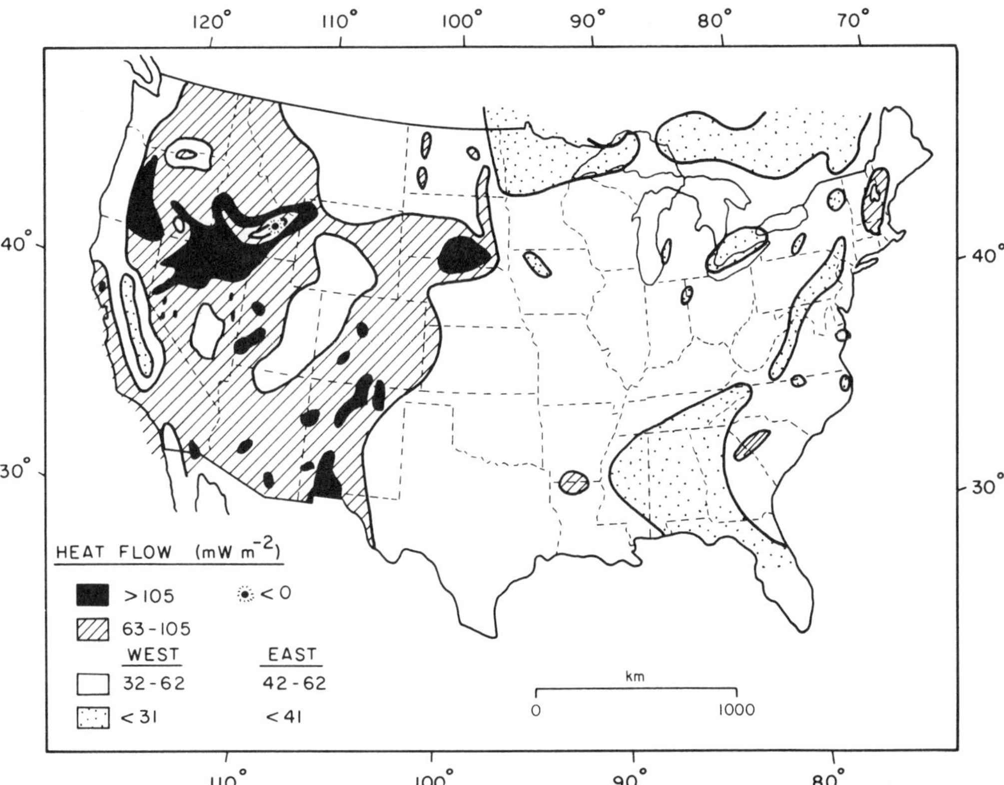

Figure 12. Generalized heat flow in the United States. Modified from Sass and others (1981).

surface data to depth in all areas of the United States. Different factors have the most significance in the interpretation of the heat-flow data presented above for different provinces. A generalized heat-flow map of the United States to accompany this discussion, modified from Sass and others (1981), is given in Figure 12.

Provinces east of the Rocky Mountains

Heat flow east of the Rocky Mountains is generally low; we group several provinces in this category, including the Central Lowlands, the Canadian Shield, the Great Plains, the Coastal Plain, and the Appalachian Highlands. Previous compilations of heat-flow data, which have included the Great Plains and Midcontinent, have suggested that at least part of the Great Plains has a thermal signature similar to the adjacent Rocky Mountains Provinces, indicating perhaps a significant tectonothermal component of heat flow in this region (Sass and others, 1981). New data and a reevaluation of existing data for the Great Plains by Gosnold (1989), however, clearly demonstrate that many of the high heat-flow values from the Great Plains are either overestimates due to poorly constrained thermal conductivities, or high values associated with water flow. With this revised interpretation, it is clear that the Great Plains data do not require a significant tectonothermal component to the surface heat flow. Arguments presented below, based on elevation data, preclude a similar thermal structure beneath the Rocky Mountains and Great Plains. The Great Plains and provinces east are best considered as a single thermal province with regional variations in surface heat flow associated with near-surface and crustal factors.

Heat flow is not statistically different in the provinces east of the Rocky Mountains, but differences in mean heat-flow values correlate with gross crustal structure variations. Mean heat flow generally increases with crustal thickness, as shown in Figure 13, with the possible exception of the Canadian Shield (Lake Superior). Rao and others (1982) and Morgan (1984) have shown that in stable provinces (last tectonothermal event Paleozoic or older), reduced heat flow is relatively uniform, with a global mean around 27 mW m^{-2}, and that most variation in surface heat flow in these provinces is associated with variations in crustal heat production. The correlation between mean heat flow and crustal thickness indicated in Figure 13 suggests that total crustal heat production increases with crustal thickness, perhaps through increasing of the effective thickness of the heat-production layer during compressional thickening of the crust. Alternatively, the thicker crust could be associated with an upper crust richer in heat production than thinner crust, but there is no evidence in bulk compositional data for the crust to support this hypothesis (see, however, discussion of Lake Superior data, below).

The increase in crustal heat production with crustal thickness could be represented by a thickening of the upper crustal heat-producing layer [increase in b in equation (2)], which would

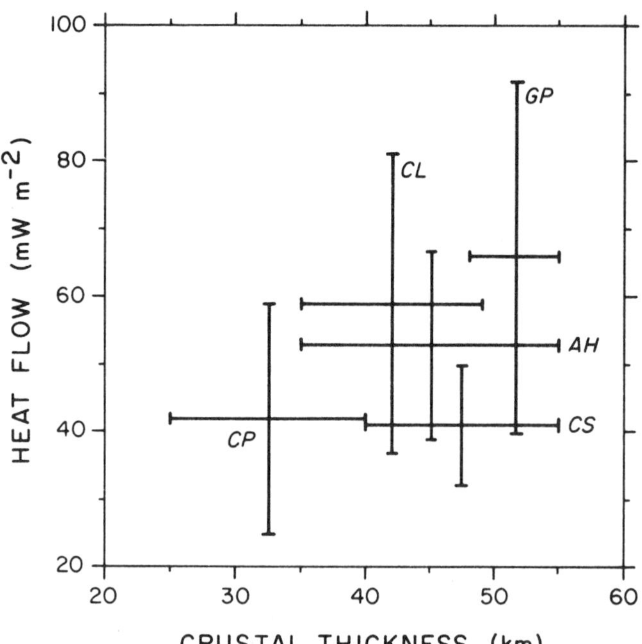

Figure 13. Mean heat flow and standard deviation (from Table 1) vs. typical crustal thickness and range (from Allenby and Schnetzler, 1983) for provinces east of the Rocky Mountains. Key to provinces: AH = Appalachian Highlands; CP = Coastal Plains; CL = Central Lowlands; CS = Canadian Shield; GP = Great Plains.

result in a change in slope in the heat-flow–heat-production relation, repetition in the crustal section in thrust slices, and/or thickening of the lower crust with heat production not represented by surface heat-production data that would change the intercept (q^*) in the heat-flow–heat-production relationship. Currently available data are insufficient to distinguish among these different mechanisms, but it appears likely that the general increase in mean heat flow with crustal thickness is associated with a larger component of crustal heat generation in the thicker crust.

Heat-flow data from the Lake Superior portion of the Canadian Shield clearly do not follow the heat-flow–crustal thickness trend and have the lowest mean heat flow for some of the thickest crust. It is possible that data from this region are unreliable as they were made by oceanographic methods, and there may be unappreciated perturbations that skew the whole data set. However, the data may be consistent with this unusual segment of crust. The crust of the Lake Superior region is also anomalous with respect to its density and seismic structure (Smith and others, 1966; O'Brien, 1967) and elevation, and is interpreted to be more mafic in bulk composition than typical continental crust. As mafic rocks are generally depleted in uranium, thorium, and potassium with respect to more silicic rocks, it seems reasonable to suggest that the Lake Superior crust is depleted in heat-producing isotopes relative to typical continental crust and that low heat flow from this thick crustal section is the result of its unusually mafic bulk composition.

Numerous variations in heat flow can also be identified within individual provinces and account for much of the scatter represented by the standard deviations of the means in Table 1. Some of these variations are again thought to result from heterogeneity in crustal heat generation. Smith and others (1981) and Costain and others (1986) identify heat-flow variations in the Appalachian Mountains and Coastal Plains that roughly correlate with lithology and are most easily explained by corresponding variations in crustal radiogenic heat production. Where heat-flow and heat-production data pairs are available, they commonly do not suggest a simple single linear relation, but deviations are thought to result at least in part from sublocal variations in the effective depth parameter for the vertical scaling of upper crustal heat production. Thus, as suggested by Roy and others (1972), the reduced heat flow [q^* in equation (2)] may be more important in characterizing a heat-flow province than the slope of the heat-flow–heat-production relation. In some areas of metamorphic lithology, it is apparent that surface heat-production values may not be representative of upper crustal heat production, and postplutonic tectonism may have changed the effective thickness of the upper crustal heat generation in other regions (Costain and others, 1986).

Local thermal anomalies have also been reported for northern Louisiana and the Michigan basin associated with lithologically controlled crustal heat-production variations (Smith and Dees, 1982a, b; Speece and others, 1985). Very low-surface heat-flow values, less than 20 mW m^{-2}, from the Charlotte belt of crystalline rocks and in the Paleozoic sedimentary terrains of northeastern Alabama and northwestern Georgia in the southern Appalachian Mountains cannot be explained simply in terms of low crustal heat generation (Smith and others, 1981). Either the reduced heat flow is anomalously low in these areas or there is regional downward convection of heat through these terrains by ground-water recharge. Near-surface water flow is indicated in many of the temperature profiles from these sites, and we consider these anomalies to be of shallow crustal origin.

Ground-water flow is clearly responsible for much of the lateral variation of heat flow in the Great Plains (Gosnold, 1989). The locations of the thermal anomalies are determined by the aquifer flow patterns, often related to structure in the crystalline basement, and are generally related to the west-to-east slope of the Great Plains. The most widespread and largest amplitude anomaly occurs in southern South Dakota and northern Nebraska. Another large-amplitude anomaly occurs on the eastern flank of the Denver basin, and small anomalies occur on structures such as the Billings and Nesson anticlines in the Williston basin. A generalized cross section of the aquifers and plot of heat flow for four sites in north-central North Dakota, showing the relationship between surface heat flow and ground-water flow, is shown in Figure 14. Ground-water flow in the Great Plains is driven primarily by the piezometric gradient, and similar regional flow and heat-flow variations have been reported to the north in Canada (Majorowicz and others, 1985). In contrast, thermal anomalies in the Gulf of Mexico coastal plain appear to be caused

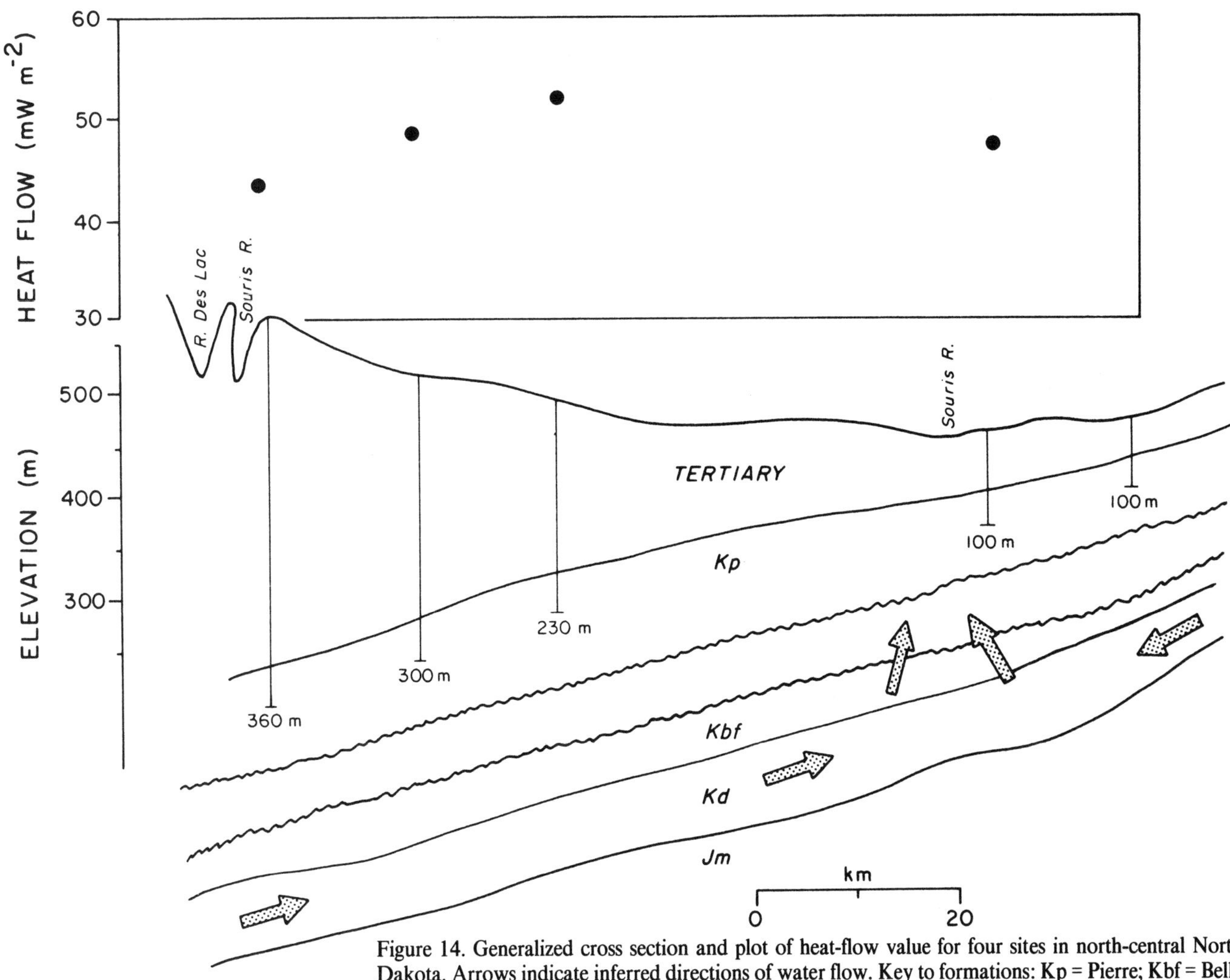

Figure 14. Generalized cross section and plot of heat-flow value for four sites in north-central North Dakota. Arrows indicate inferred directions of water flow. Key to formations: Kp = Pierre; Kbf = Belle Fourche; Kd = Dakota; Jm = Morrison. Redrawn from Gosnold (1989).

by ground-water convection driven by sediment compaction (Wallace and others, 1979; Bethke, 1986). In all cases, however, the thermal anomalies are upper crustal in origin, and while they may represent a transient imbalance in heat loss and heat supply to the crust, no deep thermal disturbance is implied.

In calculation of representative geotherms for the stable portion of the continental United States east of the Rocky Mountains, it has been assumed that the primary regional variables affecting surface heat flow relevant to the deep geotherm are crustal radiogenic heat production and crustal thickness. Geotherms have been calculated for the models described in Figure 15, assuming steady-state one-dimensional heat flow. Many assumptions are necessary in the calculation of a geotherm from surface heat-flow data; the effects of some of these assumptions on the geotherm are shown in curves 1 to 4 in Figure 15, all calculated for the same surface heat flow of 51 mW m^{-2}, the mean heat flow for provinces east of the Rocky Mountains. All curves were calculated using a reduced heat flow of 31 mW m^{-2}, and mean surface heat production [A_s in equation (2)] was calculated from the difference in mean surface heat flow and reduced heat flow, and the effective depth scaling parameter [$A_s = [q_s - q^*]/b$ from equation (2)]. Curves 1 and 2 show the geotherm assuming an exponential decrease in radiogenic heat production from the surface with effective depth parameters of 10 km (global average) and 8.1 km (calculated from regression of data in Fig. 4), respectively. Curve 3 shows the geotherm for the same parameters as curve 2, but assuming a step distribution of heat production from the surface. In curve 4 we assume that not all crustal heat generation is represented in the heat-flow–heat-generation regression; i.e., part of the reduced heat flow is derived from crustal heat production (e.g., Roy and others, 1972; Allis, 1979). For this calculation a lower crustal component of 7.5 mW m^{-2} was assumed in the reduced heat flux, and the geotherm was calculated by assuming the same surface heat generation as in curve 2, and exponential decrease in heat generation with depth, but an increased effective depth scaling parameter of 11.43 km.

Geotherms for individual provinces, 5 to 7 in Figure 15, were calculated using the same basic assumptions as used for the

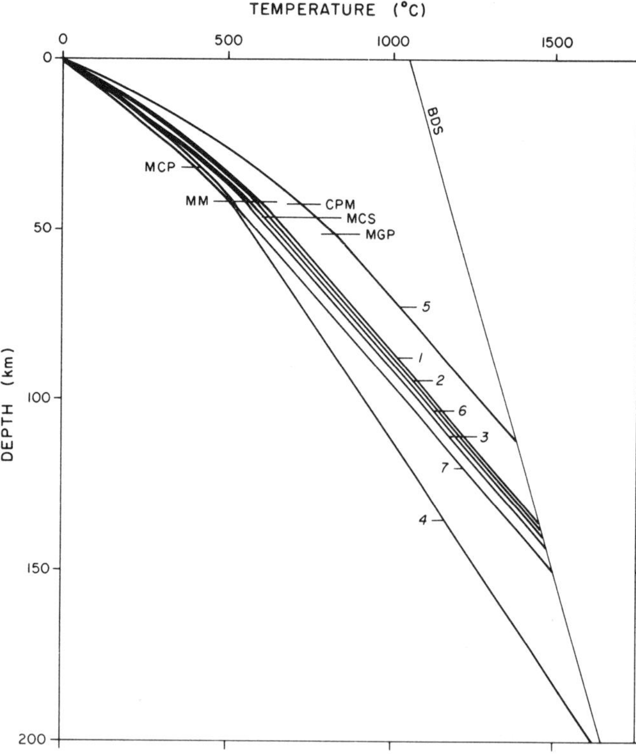

Figure 15. Geotherms calculated for the stable central and eastern U.S. and the Colorado Plateau based on the assumption of steady-state conduction. Curves were calculated with the Smithson and Decker (1974) conductivity model for the lithosphere and the following heat-flow and heat-production parameters:

Curve	1	2	3	4	5	6	7
q_s, mW m^{-2}	51	51	51	51	66	41	42
A_s, μW^{-3}	2.0	2.47	2.47	2.47	2.47	1.23	2.47
b, km	10.0	8.1	8.1	11.43	14.17	8.1	4.5
Distribution	e	e	s	e	e	e	e

where *e* represents an exponential heat production distribution and *s* is a step distribution. Curves 1 through 4 represent possible mean geotherms for all the provinces. Curves 5, 6, and 7 represent geotherms for the Great Plains, Canadian Shield (Lake Superior), and Atlantic Coastal Plain, respectively. Mean crustal thicknesses taken from the crustal thickness map of Allenby and Schnetzler (1983) are indicated by MCP, MM, CPM, MCS, and MGP for the Atlantic Coastal Plain, the mean of all provinces, the Colorado Plateau, the Canadian Shield (Lake Superior), and the Great Plains, respectively. All curves assume a reduced heat flow q^* of 31 mW m^{-2} (Roy and others, 1972). For Curves 1 through 3 and 5 through 7, q^* is assumed to be conducted through the mantle lithosphere. For Curve 4, a lower crustal component of 7.5 mW m^{-2} is assumed to contribute to q^* (Roy and others, 1972; Allis, 1979). Maximum lithospheric thicknesses for these assumptions are indicated by the intersections of the geotherms with the basalt dry solidus (BDS), taken from Lachenbruch and Sass (1978). See text for additional discussion.

calculation of the geotherm in curve 2 with Moho depths appropriate to each province, and different surface heat generation and/or effective depth scaling parameters to match the mean province heat-flow values. Curves were calculated using assumptions that maximized the geotherm differences among provinces; the curves shown in Figure 15 represent the maximum likely regional variations among provinces. Curve 5 was calculated for the mean Great Plains heat flow of 66 mW m^{-2}, assuming that higher heat flow in this province results from an increase in the effective depth scaling parameter to 14.17 km. Curve 6 was calculated for a surface heat flow of 41 mW m^{-2}, a "normal" effective depth scaling parameter of 8.1 km, but a low surface heat production of 1.23 μW m^{-3} to represent the dominantly mafic crust beneath Lake Superior in the Canadian Shield. Curve 7 was calculated assuming a thinned effective depth scaling parameter of 4.45 km to represent the thinned crust and low mean heat flow in the Coastal Plain.

It is clear from the different geotherms calculated for the same surface heat flow (Curves 1 through 4, Fig. 15) that uncertainty in the absolute position of the geotherm increases with depth. Additional uncertainties can be demonstrated by using models with different depth distributions of heat generation, including a component of mantle heat generation and more complex conductivity models. However, we consider the geotherms shown in Figure 15 a useful indicator of the magnitude of the uncertainty in the geotherm. Of the same order of magnitude as this uncertainty is the difference between mean geotherms in different "stable" provinces. For the curves shown, a range in mantle lithosphere temperatures at similar depths between provinces of about 250°C is predicted. Even larger differences are predicted for Moho temperatures, which are strongly depth dependent, the means ranging from about 420°C for the Coastal Plains to about 830°C for the Great Plains. A range in lithospheric thickness of about 40 km is also predicted based on the depth of intersection of the geotherms with the basalt dry solidus. Despite considerable uncertainty in the geotherm at depth, significant differences in the mean geotherms, Moho temperatures, and lithospheric thicknesses are predicted among different provinces east of the Rocky Mountains, although they all belong to essentially the same thermal regime. Thermal heterogeneity within provinces may result in an even greater range in geotherms and related properties than predicted for the province "means" discussed here.

Rocky Mountains provinces

Heat flow in the Rocky Mountains provinces is generally high. Using the modal heat flow for these provinces of 75 mW m^{-2} and assuming "average" crustal heat-production parameters (A_s = 2.0 μW m^{-3}, b = 10 km), the geotherm shown in Curve 1 of Figure 16 was calculated. This geotherm implies a reduced heat flow of 55 mW m^{-2}, close to the reduced heat flow of 58 mW m^{-2} originally calculated for the Basin and Range province by Roy and others (1968, 1972). Temperatures near the basalt dry solidus are predicted just below the Moho, and temperatures exceed the hydrous solidi for muscovite granite, tonalite, and amphibolite in the lower crust. Thus, if these temperatures are realistic, anhydrous conditions must exist in the lowermost crust. Higher surface heat-production values and/or a larger effective heat-production depth parameter may be applicable in some areas of the Rockies, but for all reasonable steady-state models of the geotherm for this region, temperatures near the basalt dry solidus are predicted near the Moho (Decker and others, 1984). Lower temperatures are predicted if a convective component of heat transfer through magmatic intrusion or crustal underplating is assumed in modeling the geotherm (see Basin and Range, below). However, whatever reasonable assumptions are made, temperatures at or close to the solidus are predicted at or just below the Moho.

There is considerable heat-flow variation within the Rocky Mountains provinces, as described in Sass and others (1981; see Fig. 2 for province boundaries). Heat flow is high in the northern Rocky Mountains, with a mean similar to the mode, and a geotherm probably similar to that shown in Curve 1 of Figure 16. Heat flow in the middle Rocky Mountains is generally similarly high, although scattered lower values occur such as in the Beartooth Plateau. Heat flow in the Wyoming basin between the middle and southern Rocky Mountains is normal (54 to 67 mW m^{-2}; Decker and others, 1980), and is perhaps best characterized by a geotherm similar to the Great Plains geotherm. Heat flow is generally high, with some very high values in the southern Rocky Mountains in Colorado and New Mexico; a large number of values are in excess of 150 mW m^{-2}. Such high heat-flow values can only be explained by steady-state conduction with unrealistically large melt fractions in the crust, as illustrated by curve 2 of Figure 16. The only reasonable conclusion to this "melting paradox," which has been noted also in parts of the Basin and Range province by Lachenbruch and Sass (1978), is that the geotherm cannot be represented by steady-state conductive models.

Some high heat-flow values in the Rocky Mountains provinces are undoubtedly caused by shallow redistribution of heat by ground-water flow. This effect is perhaps best documented in the Rio Grande rift basins of the southern Rocky Mountains (Morgan and others, 1981, 1986a), and it is probably the cause of many local anomalies. The high elevation of the Rockies suggests that they are probably primarily a recharge area, and regional heat flow is expected to be reduced, not increased, on average, by ground-water flow. Tectonothermal effects associated with Laramide compression in the Rocky Mountains are insufficient to explain even the modal high heat flow for these provinces, because, assuming Moho temperatures close to the solidus at the end of Laramide tectonism (about 40 Ma), most of the transient thermal effects of this tectonism would be dissipated now. Blackwell (1978) and Morgan and others (1986b) have suggested that subduction-related volcanism, plutonism, and related extension that climaxed in mid-Tertiary time in the western U.S. may have created a thermal event responsible for the background high heat flow in the Basin and Range province. If similar magmatic activity occurred 15 to 20 m.y. ago in the Rocky Mountains provinces, it could explain their high modal heat flow by cooling of a very hot lithosphere. Eocene andesitic volcanism and related plutonism was common in the northern and middle Rocky Mountains, and Oligocene and Early Miocene andesitic and rhyolitic volcanism and plutonism in the southern Rocky Mountains (Stewart and Carlson, 1978) would be expected to have some residual thermal anomaly. However, unless magmatic activity continued after the dates indicated by the surface evidence of

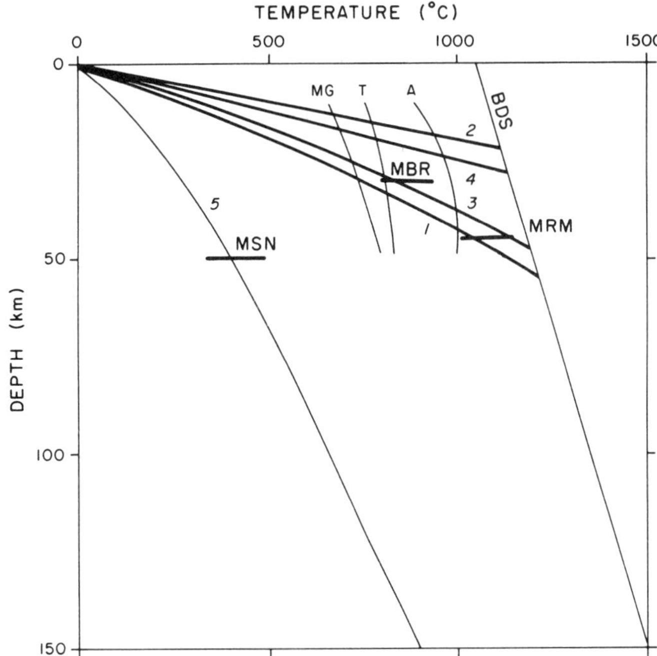

Figure 16. Geotherms calculated for the Rocky Mountains, Basin and Range, and Sierra Nevada provinces using similar assumptions to those described in the caption to Figure 15. Heat-flow and heat-production parameters:

Curve	1	2	3	4	5
q_s, mW m^{-2}	75	150	85	113	36
A_s, μW m^{-3}	2.0	2.0	2.0	2.0	1.9
b, km	10	10	10	10	9.9

All curves were calculated for an exponential distribution of crustal heat production with all reduced heat flow assumed to be conducted through the mantle lithosphere. Curves 1 and 2 represent typical (modal) and high Rocky Mountains heat flow, respectively. Curves 3 and 4 represent typical (modal) and high Basin and Range heat flow, respectively. Curve 5 represents the Sierra Nevada geotherm. Approximate crustal thicknesses are indicated by MBR, MRM, and MSN for the Basin and Range, Rocky Mountains, and Sierra Nevada provinces, respectively (from Allenby and Schnetzler, 1983). Possible "wet" crustal solidi are indicated by MG, T, and A, for muscovite granite, tonalite, and amphibolite, respectively (from Fyfe, 1978). BDS = basalt dry solidus (Lachenbruch and Sass, 1978). See text for additional discussion.

these mid-Tertiary events, an additional heat source is required to explain the high modal heat flow and locally very high heat flow.

In the southern Rocky Mountains, reduced heat-flow values in the range of 80 to 90 mW m^{-2} have been determined in the area of the northern physiographic terminus of the Rio Grande rift coincident with a negative Bouguer gravity anomaly (Leadville area: Decker and others, 1984). These data are consistent with the emplacement of low-density, young intrusions (<5 Ma) in the upper crust in this region. Most young volcanism in this region is basaltic in composition, and Decker and others (1984) suggest that the low-density intrusions may be the product of crustal melts generated in response to basaltic crustal plutonism that has not reached the surface. This basaltic activity appears to be focused on, but not limited to, the Rio Grande rift, and may be related to, but is not simply a response to, crustal extension. Late Miocene to Holocene basaltic magmatic activity throughout this region is thought to have thickened and heated the crust and may be the main source of the high regional heat flow. A sharp (less than 50 km) transition from high to normal heat flow near the Colorado-Wyoming border is consistent with a crustal heat source for the high heat flow (Decker and others, 1984), and sharp heat-flow transitions are common in the western U.S. (Blackwell, 1978).

A very young example of this crustal magmatic activity is the Socorro magma chamber, probably representing a 1- to 2-km Holocene basaltic thickening of the crust over an area of 1,200 km^2 in central New Mexico (Sanford and others, 1977). Heat flow is very high where young magmatism is most intense in the Valles caldera, New Mexico (Baldridge and others, 1984). Although Neogene basaltic volcanism is not common in the northern and middle Rocky Mountains, active seismicity and young faulting (Smith, 1978) indicate tectonic rejuvenation of these provinces. This tectonic rejuvenation could be associated with widespread basaltic plutonic activity that was prevented from reaching the surface by density and stress factors, but which has residual high heat flow from mid-Tertiary magmatism. Thus, the primary source of high heat flow in the Rocky Mountains provinces is likely to be Cenozoic crustal magmatism.

Colorado Plateau

The mean heat flow for the Colorado Plateau of 68 ± 26 mW m^{-2} is very similar to the Great Plains, and so, to a first approximation, this province may be characterized by the "mean" Great Plains geotherm (Curve 5, Fig. 15). Slightly thinner crust beneath the Colorado Plateau would result in slightly lower Moho temperatures than for the Great Plains: in the range of 650° to 750°C. However, Cenozoic volcanism and major Neogene uplift of the plateau suggest that the plateau is not a stable province similar to provinces east of the Rockies (Hunt, 1956; Lucchitta, 1979; Hamblin and others, 1981; Morgan and Swanberg, 1985). The Colorado Plateau sits at a higher mean elevation than the Great Plains despite its slightly thinner crust, which from isostatic arguments outlined below is taken to indicate a thinner and hotter lithosphere beneath the plateau than beneath the Great Plains. This thinner lithosphere can only be reconciled with the similar mean surface heat flow of the two provinces by a relative increase in heat flow with depth beneath the plateau.

Low surface heat flow in some sectors of the plateau is almost certainly the result of shallow ground-water recharge (e.g., southwestern margin), but silica heat-flow estimates in this region indicate normal heat flow, and shallow ground-water convection is not thought to reduce significantly the mean plateau surface heat flow (Swanberg and Morgan, 1985). Heat-production data from the plateau are very limited due to its thick sedimentary cover, but there is no compositional evidence to suggest that the plateau crust is significantly depleted in upper crustal radiogenic heat production, resulting in normal surface heat flow for above-normal reduced heat flow. Erosion following uplift of the plateau is only locally significant and would cause a transient increase, rather than a decrease, in surface heat flow. Therefore we must conclude that, to reconcile the surface heat-flow mean with the plateau elevation and recent uplift, a transient heating of the lower lithosphere beneath the plateau must have occurred, the effects of which have not yet fully reached the surface.

Models of Cenozoic heating and thinning of the plateau lithosphere have been presented by Bird (1979), Bodell and Chapman (1982), and Morgan and Swanberg (1985). They show the effects of thinning and heating of the plateau lithosphere, causing relatively rapid uplift, but a significant delay in conduction of the deep thermal anomaly to the surface. The predictions of one of these models for lithospheric thinning, uplift, and surface heat flow as a function of time is shown in Figure 17. As can be seen from this example, the rapid rise in surface heat flow lags rapid uplift by the order of 20 m.y. (actual lag depends on lithospheric thickness; see Lachenbruch and Sass, 1977; Bodell and Chapman, 1982; Morgan, 1983), and a heating and thinning event starting 10 to 60 m.y. ago beneath the plateau would be compatible with the Neogene uplift but would result in only a minor surface heat-flow anomaly. Lack of control on the pre-uplift surface heat flow and lack of constraint on the rate and mechanism of lithospheric heating and thinning limits rigorous testing of these models, but they all suggest that rapid recent uplift of the plateau with a "normal" surface heat flow implies a relative increase in heat flow with depth and a higher geotherm at depth than predicted for the Great Plains with similar surface heat flow (Curve 5, Fig. 15).

Basin and Range province

Despite the much thinner crust in the Basin and Range province than the Rocky Mountains provinces and more obvious importance of extensional tectonics, the magnitude and scatter of surface heat flow in these two provinces are very similar. The modal heat flow for the Basin and Range of about 85 mW m^{-2} suggests a Basin and Range "modal" steady-state geotherm slightly higher than the Rocky Mountains provinces "modal" geotherm, and Moho temperatures close to or exceeding the probable lower crustal solidus are suggested (Curve 3, Fig. 16). Lachenbruch and Sass (1977, 1978) and Sass and others (1981) have identified thermal subprovinces within the Basin and Range province, namely the "Eureka Low" in the Great Basin of Nevada, a region of submodal heat flow (<63 mW m^{-2}), and the "Battle Mountain High" in northern Nevada and extending into the Snake River Plain of Idaho, a region of heat flow well in excess of the mode (>104 mW m^{-2}). Several smaller areas of high heat flow have been identified within the Basin and Range province, as shown in Figure 12, the most easterly of which are associated with the Rio Grande rift in New Mexico and continue across the Basin and Range–southern Rocky Mountains physiographic boundary along the Rio Grande rift. Low heat flow in the "Eureka Low" is thought to be the result of downward convection of heat by ground-water recharge (Lachenbruch and Sass, 1977). High heat flow in the hotter subprovinces implies reduced heat flow 300 percent higher than in stable provinces and super-

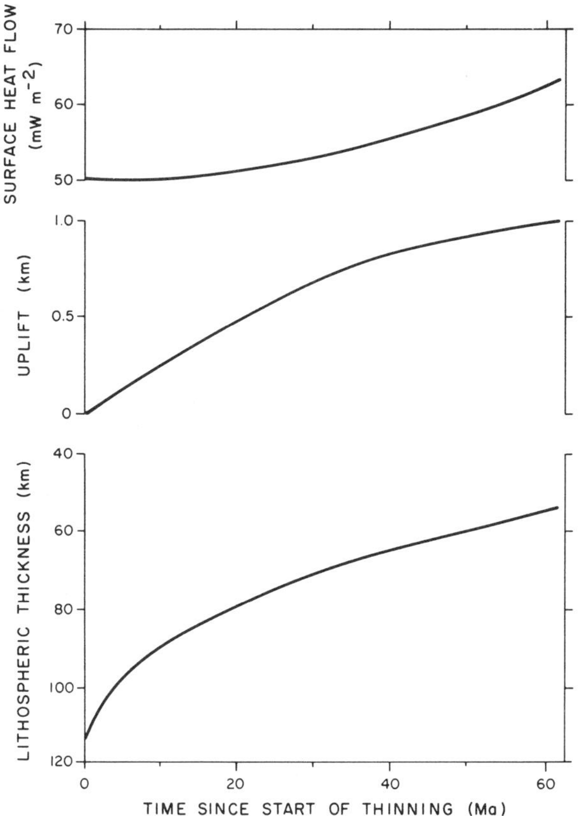

Figure 17. Example of uplift and surface heat flow predicted by moving plane source model of lithospheric thinning. Model parameters: initial surface heat flow, 50 mW m^{-2}; surface radiogenic heat production, 2 μW m^{-3}; heat-production depth parameter 10 km; heat-production depth distribution, average of exponential decrease with depth and constant heat production to 10 km; crustal thickness, 40 km; crustal thermal conductivity, 2.5 W m^{-1} K^{-1}; volume coefficient of expansion, 3×10^{-5} K^{-1}; lithosphere-asthenosphere boundary temperature, 1,200°C; initial lithospheric thickness, 115 km; increase in heat flux at base of lithosphere, 6 mW m^{-2}; mantle diffusivity, 1 mm^2 s^{-1}. From Morgan and Swanberg (1985).

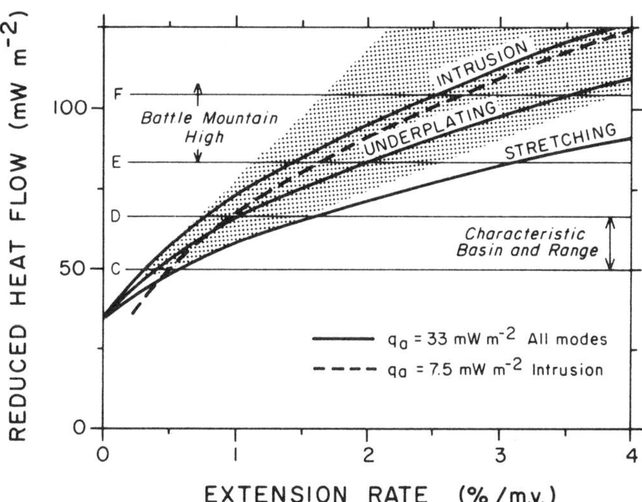

Figure 18. Variations of reduced heat flow with extension rate for extension models in dynamic equilibrium in which the crustal thickness is maintained constant. Crustal thickness is maintained by solid-state crystal accretion to the base of the crust (stretching), crustal underplating (underplating), and intrusion of basalt into the crust (intrusion). The assumed heat flux from the asthenosphere is given by q_a for the curves. From Lachenbruch and Sass (1978).

solidus temperatures in the lower crust for all compositions for geotherms calculated using steady-state assumptions (Curve 4, Fig. 16; Lachenbruch and Sass, 1978).

As with the Rocky Mountains provinces, it is impossible to find a satisfactory shallow or steady-state crustal origin for the high heat flow in the Basin and Range province, and the primary source of anomalously high heat flow must be tectonothermal in origin. Lachenbruch and Sass (1978) have developed a series of analytical models that demonstrate the effects of extension on the geotherm with application to the Basin and Range. Conceptually, these models are based on the upward convection of heat by the net upward movement of material in the lithosphere during the stretching of a constant-thickness crust. This crust is maintained constant in thickness by solid-state crystal accretion to the base of the crust, or by upward migration of magma and subsequent release of latent heat of solidification by extension-related underplating or intrusion of the crust. The results of these model calculations for a state of dynamic thermal equilibrium during continuous lithospheric extension (extension is assumed to have been continuous for sufficient time for temperatures to be steady, i.e., 15 to 20 m.y.) are shown in Figure 18. Similar calculations for instantaneous lithospheric stretching and subsequent lithospheric cooling have been presented by McKenzie (1978).

Detailed analyses of extension in the Basin and Range (Eaton, 1982; Blackwell, 1983; Morgan and others, 1986b) suggest that caution must be used in applying the Lachenbruch and Sass (1978) thermal models of extension, because extension has been neither continuous nor at a constant rate throughout the Neogene. In many areas there is evidence for at least two phases of extension, an early, high-strain phase possibly associated with subduction-related magmatism, and a later, lower strain phase associated with smaller Quaternary extension in many areas. Extension must have played an important role in producing or maintaining the modern high heat flow in the Basin and Range province. However, as high heat flow also occurs in areas with little evidence of the amount of extension predicted to cause the observed heat-flow anomalies (e.g., in the southern Rocky Mountains and Columbia Plateau), magmatism, in addition to that included in the constant-crustal thickness extensional models of Lachenbruch and Sass (1978), seems likely. There is evidence of preextension subduction-related magmatism, which would have pre-heated the crust prior to extension, lessening the time required for the extension-related thermal anomaly to be established, and post- or nonextension-related magmatism, which may have prolonged and perhaps enhanced high heat flow after major extension. We continue consideration of this problem in discussion of the thermal regime of the Columbia Plateau.

Columbia Plateau

Subprovinces within the Columbia Plateau include the Columbia Basin, the Blue Mountains, the High Lava Plains, and the Owyhee Upland. We also include the Eastern and Western Snake River Plains in our discussion of this province. The individual thermal characteristics of this area have been discussed by Sass and others (1981). In general, this province is a region of high heat flow and is thermally continuous with the Basin and Range to the south, the northern and middle Rocky Mountains to the east, and the Cascade Range to the west (Fig. 12). The primary mode in the heat-flow data from this province is 85 mW m^{-2}, and is identical to that for the Basin and Range. A secondary higher mode around 125 mW m^{-2} corresponds to the hotter subprovinces in the Basin and Range, the largest of which, the "Battle Mountain High," is continuous across the boundary between these provinces. Steady-state geotherms predicted for the Basin and Range Province and shown in Figure 16 are thus equally applicable (or inapplicable!) to the Columbia Plateau. Blackwell (1969, 1978) named the continuous zone of high heat flow in the western U.S. which extends from the trans-Mexico volcanic belt to central-western Canada and includes the Columbia Plateau, the Cordilleran thermal anomaly zone. The similarity in heat-flow values along the Cordilleran thermal anomaly zone suggests that similar thermal processes may be operating along its length.

Basaltic and rhyolitic volcanism younger than 17 Ma is common throughout the Columbia Plateau (Stewart and Carlson, 1978), and the province has moderate to low seismicity (Smith, 1978). Heat flow tends to be highest in the areas of youngest volcanism and tectonism, suggesting a genetic relationship. Blackwell (1978) documented a relationship between high heat flow and young volcanism, and suggested that the modal heat flow of 85 mW m^{-2} is characteristic of areas in the Cordilleran thermal anomaly zone with volcanism older than 17 Ma. In addition, heat-flow–heat-production data from these regions define a reasonable linear relationship with a reduced heat flow of about 58 mW m^{-2}, which can be explained by cooling from a major subduction-related thermal event starting at 15 to 20 Ma (Blackwell, 1978; Morgan and others, 1986b).

Highest heat flow is in the Yellowstone caldera, where the Columbia Plateau intersects the middle Rocky Mountains and the site of the largest post-Miocene explosive volcanism in the U.S. Late and Middle Cenozoic extension in the Columbia Plateau is not alone sufficient to explain the generally high heat flow and cannot explain lateral variations of heat flow within the province. Some of this variation is due to complex aquifer systems in the volcanic and plutonic near-surface rocks, the most prominent example of which is the 200-km-long by 50-km-wide zone of low heat flow along the eastern Snake River Plain (Fig. 12). Generally, high heat flow and some of the lateral variability in heat flow are thought to be due to magmatism associated with subduction beneath the west coast. Subregional variations may be attributed to variations in timing and magnitude of crustal magmatism.

The Yellowstone–Snake River Plain volcanism and heat flow are of particular significance within this province. A progression of the onset of silicic volcanism east-northeast along the eastern Snake River Plain toward Yellowstone at about 35 mm/yr for the past 14 to 18 m.y. indicates a progressive thermal history along the plain (Armstrong and others, 1975), which W. J. Morgan (1972) has suggested marks the trace of a mantle "hot spot" or plume. Brott and others (1978, 1981) presented heat-flow data from the western and eastern Snake River Plain and an analysis of gravity and elevation data. They showed that low, shallow heat flow in the eastern plain is caused by extensive lateral water flow in the Snake River aquifer, and that regional heat flow below the aquifer is expected to be between 100 and 150 mW m^{-2}, a result supported by limited deep well data. The elevation profile along the plain is consistent with lithospheric cooling after a progressive major heating event, with heating currently occurring beneath the Yellowstone Plateau (Fig. 19; Brott and others, 1981). Heat-flow data from Yellowstone Lake within the caldera (Morgan and others, 1977) are consistent with the regional heat-flow estimate for the caldera from geochemical analyses of about 2,000 mW m^{-2} (Fournier and others, 1976), which combined with the 2,500 km^2 area of the caldera, gives a caldera heat loss of approximately 5 percent of the total heat flow from the Pacific Northwest (Sass and others, 1981). The primary source of this high heat flow is thought to be basaltic intrusions from the mantle, which eventually raise the geotherm sufficiently to cause melting of the silicic component of the middle and upper crust. Other young calderas in the western U.S., with a smaller but similar and significant component of heat derived from mantle magmatism, include Long Valley (Lachenbruch and others, 1985; Sorey, 1985; Blackwell, 1985) and the Valles caldera (Tomczyk and Morgan, 1987); these systems demonstrate the effectiveness of this subcrustal heat source. In between these local foci of mantle magmatism, distributed mantle-derived magma-

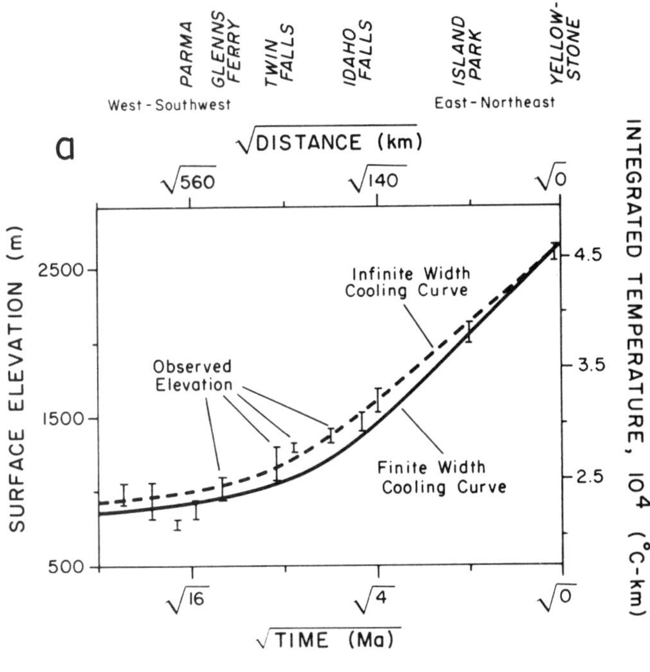

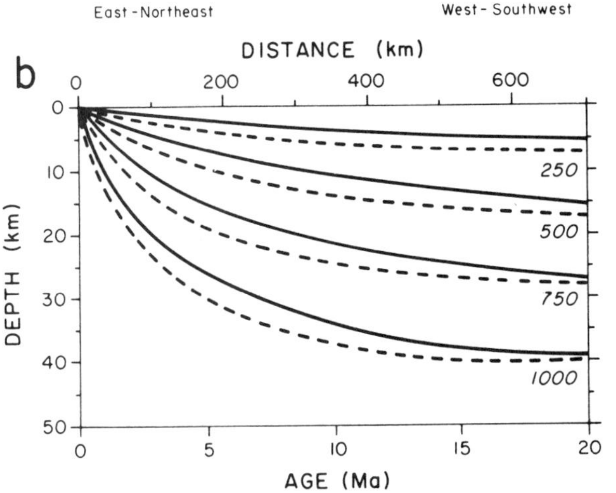

Figure 19. a, Integrated temperature-elevation curves for the infinite-width constant-temperature moving-source plane and the finite-width constant-temperature moving-source plane (dashed and solid lines, respectively). Average observed elevation along the axis of the Snake River Plain is shown in the figure. The elevations and integrated temperatures are scaled for comparison. The top scale is square root of distance; the bottom scale is square root of time. Various towns and other locations along the Snake River Plain are given below the graph. b, Predicted subsurface isotherms (250°, 500°, 750°, and 1,000°C) below the axis (solid lines) and margins (dashed lines) of the Snake River Plain. From Brott and others (1981).

tism could make a significant contribution to regional high heat flow in the Cordilleran thermal anomaly zone.

Cascade Range

The Cascade Range forms the western boundary of the Cordilleran thermal anomaly zone in the Pacific Northwest, and very high heat flow follows the High Cascade Range in Oregon (Blackwell and others, 1982). This high heat flow is clearly related to active arc volcanism. From gravity data and the width of the heat-flow transition zone in this region, Blackwell and others (1982) postulate the presence of a hot, low-density region about 60 km wide in the crust below 7 to 10 km, probably a zone of temporary residence for magmas derived from the subduction zone. Thus, again, mantle-derived magmas appear to be the source of the high heat flow. Much of the local variability in heat flow in the Cascade Range appears to be the result of shallow redistribution of heat by ground-water convection, dramatic effects of which have been measured by Williams and Von Herzen (1983) in Crater Lake, Oregon.

Active arc volcanism in the Oregon Cascades migrated 40 to 60 km westward during the last 6 to 10 m.y., associated with a shallowing of the angle of subduction responsible for the volcanism. Heat flow in the Western Cascades was presumably high during arc volcanism, similar to the modern High Cascades, but is now low, and the thermal transition from high to low heat flow occurs between the active and extinct magmatic arcs.

Sierra Nevada

The western margin of the high heat-flow zone in the southwestern United States is marked by an abrupt transition to low heat flow between the Basin and Range and Sierra Nevada provinces (Fig. 12). Heat-flow–heat-production data pairs from the Sierra Nevada yield the lowest regional reduced heat flow for the United States and a correspondingly cool geotherm, as shown by curve 5 in Figure 16. There is no evidence to suggest that this low heat flow is the result of the removal of heat by ground-water recharge as the surface heat-flow values are internally consistent with uniform reduced heat flow over a large area (Lachenbruch, 1968), and the granitic rocks of the Sierra Nevada batholith are not deficient in radiogenic heat production. There is evidence for Neogene uplift and erosion of the Sierra, which would be expected to increase surface heat flow, so the only viable mechanism for the low heat flow in this province is a deep tectonothermal mechanism.

The most widely accepted explanation for low heat flow in the Sierra Nevada is the lingering effect of a deep heat sink caused by downward convection of heat associated with subduction beneath this region, ending since 20 Ma (Blackwell, 1971; Roy and others, 1972). Very low heat flow is commonly measured in the arc-trench gap in active subduction systems, and as cooling associated with subduction is expected to extend to a depth of at least 50 to 100 km in this region, it is not unreasonable to expect low

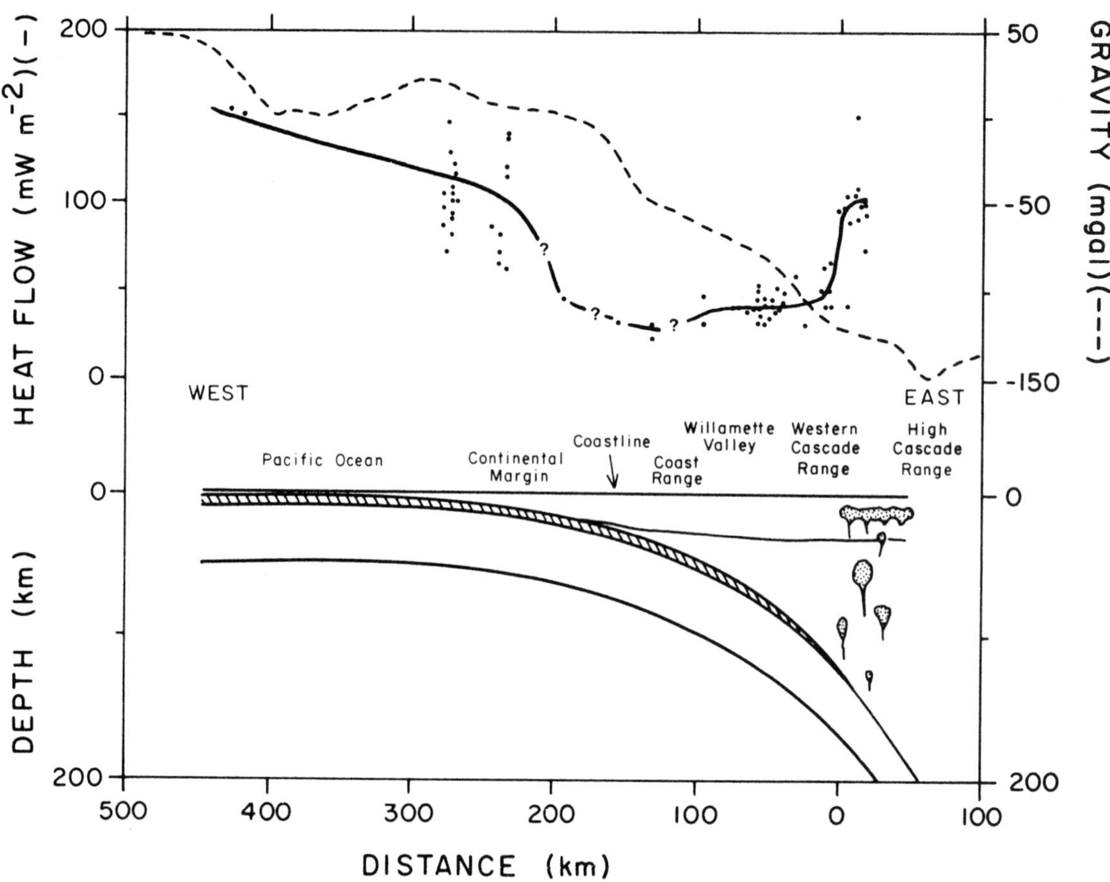

Figure 20a. Cross section of the subduction zone, gravity, and heat-flow profiles in the Pacific Northwest.

surface heat flow for 20 m.y. or more after subduction has ceased. Low heat flow associated with this subduction probably extended east of the Sierra Nevada during active subduction, but the sharp modern Sierra–Basin and Range thermal transition now marks the western boundary of mantle-derived magmatism and extension. Chase and Wallace (1986) have suggested that Neogene uplift of the Sierra has resulted from the release of an overcompensated erosional remnant by lithospheric faulting associated with Basin and Range extension. Alternatively, or perhaps in addition, Neogene uplift of the Sierra suggests that the overthickened subduction-cooled lithosphere of the Sierra is now heating and thinning, but perhaps like the Colorado Plateau, the effects of this deep heating have yet to reach the surface. It is expected that at depth the true geotherm is hotter than that predicted by steady-state calculations.

Pacific Coastal provinces

Contrasting thermal regimes are found in the Pacific Coastal provinces, as shown in Figure 12, with high heat flow along much of the California Coast Ranges and low heat flow along the Pacific margin in the Klamath Mountains and Coast Ranges of northernmost California, Oregon, and Washington. Heat flow in this latter region is typically around 40 mW m^{-2} and suggests a geotherm similar to the geotherm predicted for Lake Superior (Curve 6, Fig. 15). Low heat flow in this region is not thought to be related to low crustal heat production, however, but to absorption of heat by subduction beneath the Pacific coast. The low heat flow continues inland to a sharp transition to high heat flow just west of the active volcanic arc (e.g., Blackwell and others, 1982). This pattern of low heat flow west of high heat flow continues into Canada (Hyndman, 1976; Lewis and others, 1985) and reappears to the south in southern Mexico (Ziagos and others, 1985). A cross section across the northern Oregon Cascades is shown in Figure 20 (from Blackwell and others, 1982), showing the typical relation among surface heat flow, Bouguer gravity, crustal temperatures, and subduction in this zone.

High heat flow in the California Coast Ranges occurs along a section of the Pacific margin in which active subduction is absent. East-west convergence between the Pacific and North American plates is accommodated in this region by strike-slip motion along the San Andreas transform fault zone, which has been growing since about 20 Ma by northward migration of the Mendicino triple junction (Atwater, 1970). As the San Andreas fault zone has grown in length, the subducting slab has been cut off east of the fault, allowing the growth of a "slabless window"

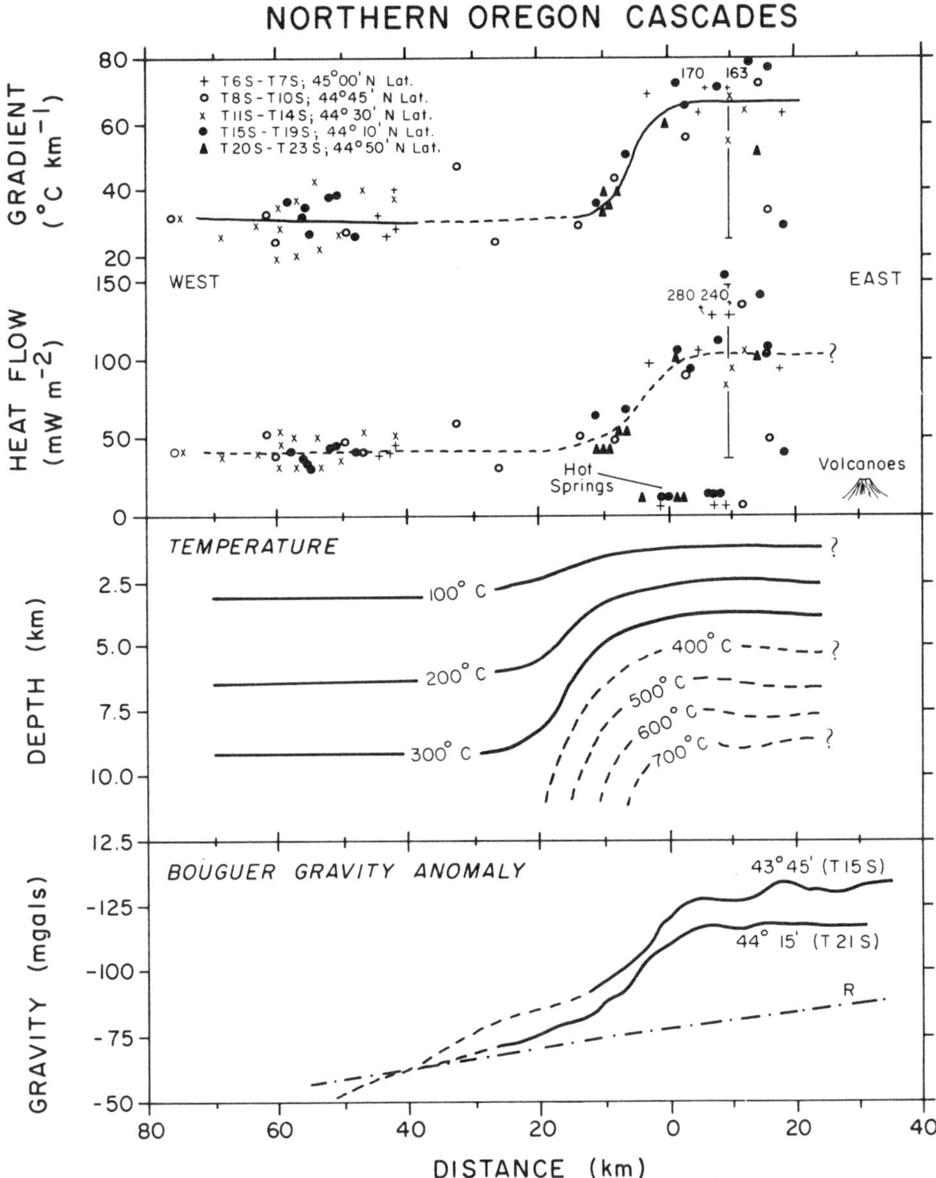

Figure 20b. Geothermal gradient, heat flow, interpreted crustal temperatures, and regional Bouguer gravity values for the western part of the northern Oregon Cascade Range. Heat-flow data between latitudes 43°15′N and 45°05′N are projected onto the profile. The zero distance reference is the 100 mW m^{-2} heat-flow contour line. From Blackwell and others (1982).

(Lachenbruch and Sass, 1980). High heat flow in this zone could be caused by either distributed frictional heating along the San Andreas fault zone or from the effect of the growth of the "slab-less window." Lachenbruch and Sass (1980) presented heat-flow data that suggest no evidence for local frictional heating of the main fault trace at any latitude over a 1,000-km distance from Cape Mendicino to San Bernardino, although the regional heat flow is high. The regional thermal anomaly is therefore thought to be related to distributed frictional heating at depth, or from the truncation of subduction.

Lachenbruch and Sass (1980) and Zandt and Furlong (1982) modeled the thermal anomaly as the migration of the Mendicino triple junction, the truncation of subduction causing a "hole" in the lithosphere beneath which hot asthenosphere rises, resulting in an increase in surface heat flow by a factor of 2 within 4 m.y. of the passage of the triple junction. After the initial upwelling of the asthenosphere into this "hole," the lithosphere slowly cools and thickens. This cooling results in a lowering of the geotherm and a decrease in surface heat flow and elevation away from the triple junction, somewhat analogous to cooling and sinking of the Snake River Plain after passage of the Yellowstone "hot spot." Model results from Lachenbruch and Sass (1980), shown in Figure 21, suggest that the plate thickness (the level to which the asthenosphere rises) is 20 km beneath the

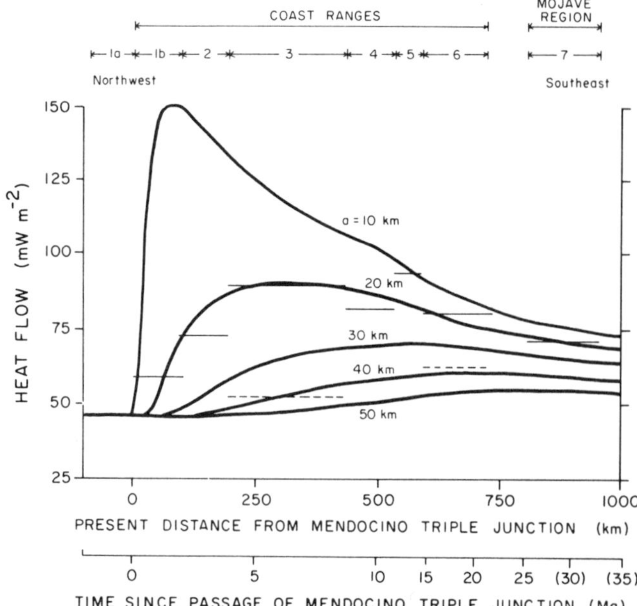

Figure 21. Heat flow as a function of distance from the Mendicino Triple Junction for selected initial thicknesses (a) of the North American plate. Horizontal solid lines are mean heat-flow values in the San Andreas fault zone along the path of the Mendicino Triple Junction. Horizontal dashed lines are mean (corrected) heat-flow values interior to the adjacent Great Valley. From Lachenbruch and Sass (1980).

Coast Ranges, thickening to about 40 km beneath the Great Valley to the east.

The region of most anomalous heat flow in the Pacific Coastal provinces is the region of southern California immediately north of the Gulf of California in which offsets in the transform zone occur. Right-stepping offsets in this dextral strike-slip fault system cause zones of extension in the region of the offset ("pull-apart" basins), which may be expected to behave thermally in the same manner as other zones of extension. One such zone, the Salton trough, has been studied in detail and modeled by Lachenbruch and others (1985) showing a complex geotherm heated at depth by upward convection in response to extension and depressed near the surface by the rapid deposition of sediments into the basin formed by extension. These relationships are illustrated in Figure 22 (from Lachenbruch and others, 1985). The models of the Salton trough differ from those of the Basin and Range (Fig. 18; Lachenbruch and Sass, 1978) in that isostatic balance is maintained in the crustal section for the Salton trough models as extension progresses, and the resulting crustal structure is constrained by gravity and seismic data. In addition, the relatively constant strain across the San Andreas fault zone during the 4- to 5-m.y. evolution of this basin suggests more direct application of the dynamic equilibrium assumptions of these models in this region than in the Basin and Range. Magmatic additions to the crust simply in response to extension may be sufficient to explain all thermal and evolutionary aspects of the Salton trough.

IMPLICATIONS OF THERMAL REGIMES

Many physical properties of the lithosphere are temperature-dependent, including its thickness, and we discuss briefly here the implications to these properties of the different U.S. thermal regimes discussed above. A common definition of the lithosphere is the outer layer of the Earth in which heat is dominated by conduction (Morgan, 1984), and although this definition may not be strictly applicable to areas of active tectonism or magmatic activity, it can be applied as a first approximation to the U.S. geotherms. These geotherms are expected to remain conductive only below appropriate crustal or mantle solidi temperatures, and thus a wide range in thicknesses is predicted for the U.S. lithosphere for the different thermal regimes (Figs. 15, 16). Thick lithosphere is predicted for the stable provinces east of the Rocky Mountains, but significant differences in lithospheric thickness (±20 km, at least) may be expected to be associated with differences in crustal heat generation and thickness. Lateral variations in lithospheric thickness may be expected over horizontal distances of the same order as the lithospheric thicknesses. In the regions of more recent tectonic activity in the western U.S., much thinner lithosphere may be expected, probably not much greater than the crustal thickness in many places (Fig. 16). Thicker lithosphere is predicted for the Colorado Plateau and Sierra Nevada, but not as thick as predicted by the geotherms in Figures 15 and 16 because transient deep heating is expected in both of these provinces.

Surface elevation is also related to the geotherm, among other parameters, under isostatic conditions, as lithospheric density is related to temperature through thermal expansion. Heating or cooling of the lithosphere results in decreases or increases in density and uplift or downwarp, respectively (Crough and Thompson, 1976). Surface elevation has been derived as a function of lithospheric thickness for different crustal thicknesses by Morgan and Burke (1985), as shown in Figure 23, using a number of simplifying assumptions including a constant crust-mantle density contrast. These curves work reasonably well for most of the continental U.S. (±500 m) with the thermal regimes discussed above. For example, the elevation of the Great Plains is consistent with their thick crust only if a relatively thick (stable) lithosphere is assumed. Relatively high elevations in the Basin and Range for the thin crust in this province are consistent with hot, thin lithosphere. Local departures from good agreement can be found, however, such as the low elevation of thick crust in the Lake Superior region, which must be explained in terms of an above-"normal" density crust in addition to cold, thick lithosphere, and high elevation in the Colorado Plateau, which is consistent with thinner lithosphere than predicted by the steady-state conductive geotherm.

It is particularly instructive to consider the evolution of regional elevation in the western U.S. in terms of evolution of the geotherm. For example, the Colorado Plateau was at sea level prior to Laramide compression, suggesting a very cold geotherm and thick lithosphere for this province at this time and/or a thinner crust than the modern crust. There is increasing evidence

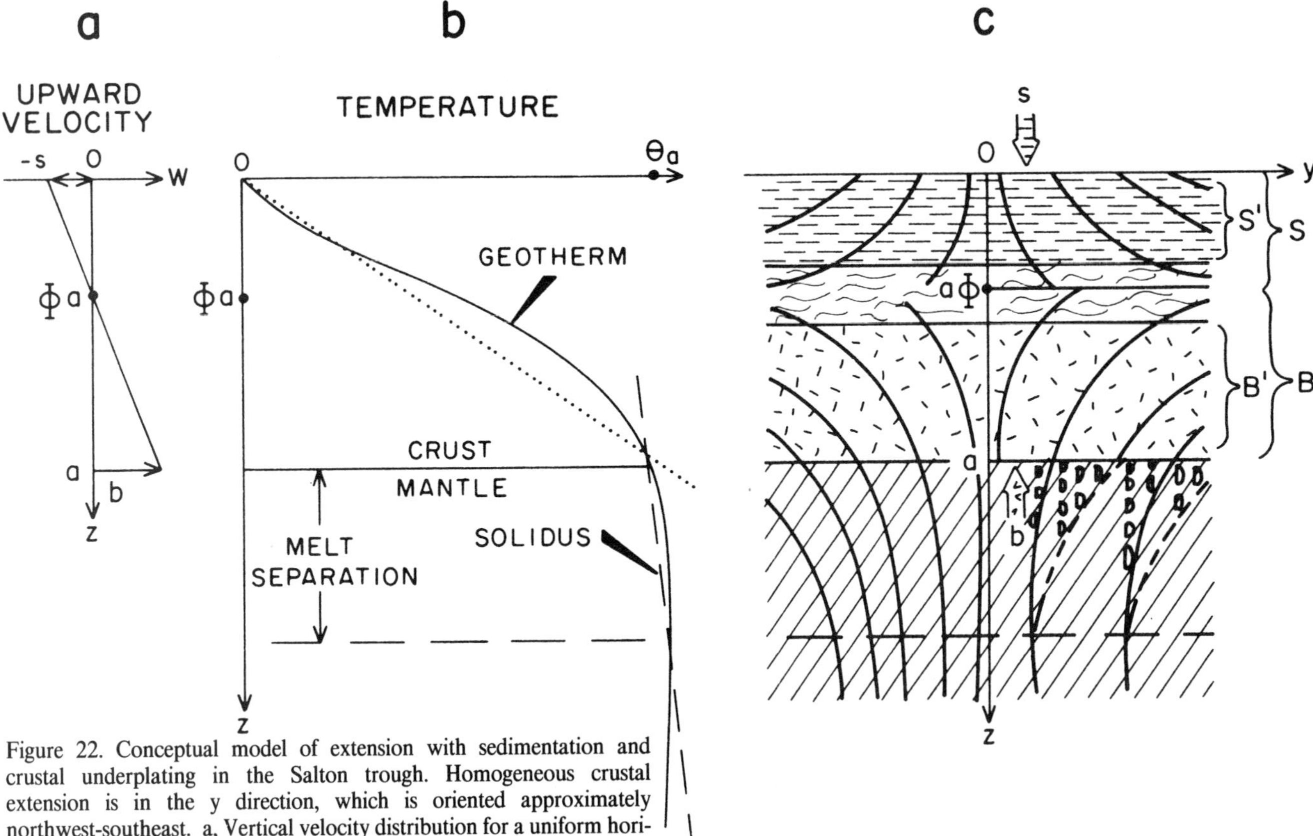

Figure 22. Conceptual model of extension with sedimentation and crustal underplating in the Salton trough. Homogeneous crustal extension is in the y direction, which is oriented approximately northwest-southeast. a, Vertical velocity distribution for a uniform horizontal extensional string rate y. b, Steady-state geothermal model. c, Kinematic model. Solid arrows represent average particle trajectories, dashed arrows represent refractory depleted mantle, and tear drops represent separated basaltic melt that rises and freezes to crust at rate b. Latent heat is liberated at base of crust $z = a$. Subsidence is compensated by distributed sedimentation at rate *s*. S′ and B′ are transient thicknesses of sediment and mantle-derived crystalline rock, respectively; their steady state thicknesses are S and B. Illustrated for a crust whose total thickness is equal to stationary value. ϕ = dimensionless sedimentation rate. From Lachenbruch and others (1985).

for uplift of much of the western U.S. during the past 5 to 10 m.y. (Morgan and Swanberg, 1985), even though extensional tectonics have predominated and thinning of the buoyant crust and subsidence may have been expected. Late Miocene to Recent magmatic thickening of the crust is consistent with this uplift and high heat flow in many regions (e.g., Morgan and others, 1986b), suggesting that magmatism and extension may not be simply related.

Another important fundamental parameter related to the geotherm is lithospheric strength, and the lithosphere is predicted to be significantly weaker where the geotherm is hot and the lithosphere thin. Integrated lithospheric strength as a function of strain rate for different geotherms predicted by different reduced heat-flow values is shown in Figure 24 (from Lynch and Morgan, 1987), and more than an order of magnitude decrease in lithospheric strength is predicted as the reduced heat flow increases from "stable" values of 25 to 33 mW m^{-2} to values characteristic of tectonically active regions (reduced heat flow >60 mW m^{-2}). Localization of Cenozoic deformation in the western U.S. may have been caused by preheating and weakening of the lithosphere by pre-deformational magmatism. Perhaps the Colorado Plateau was deformed less than adjacent provinces because of its cool geotherm and strong lithosphere (Morgan and Swanberg, 1985).

The depth of the brittle-ductile transition(s) is also related to the geotherm: the hotter the geotherm the shallower the transition(s). Morgan and others (1986b) have suggested that contrasting Neogene extensional styles in the Rio Grande rift and possibly other areas of the Basin and Range may be related to evolution of the geotherm. Eaton (1982) and others have related the maximum depth of earthquakes, which presumably can occur only in the brittle portion(s) of the lithosphere, to the depth of the brittle-ductile transition and regional heat flow. Chase and Wallace (1986) have recently explained young uplift and tilting of the Sierra Nevada in terms of the rupture by Basin and Range faulting of a thick, cool lithospheric plate that held the Sierras down by flexural strength.

Seismic velocities and electrical conductivity both show a strong temperature dependence, and thus regional variations in lower crustal and upper-mantle seismic velocities and conductivities are expected between different thermal regimes. There is, in fact, excellent correlation between P_n velocities and predicted

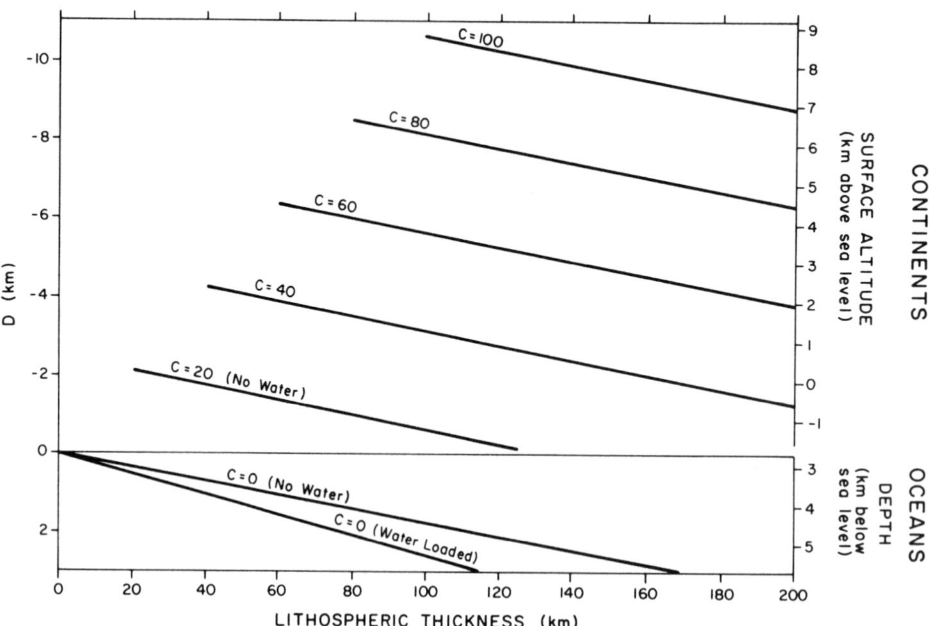

Figure 23. Variations in surface altitude as a function of lithospheric and crustal thickness variations. Parameter D is the difference in surface altitudes of the different lithosphere columns and a zero-thickness lithosphere column and is positive downward. C = crustal thickness in kilometers. Continent surface altitudes were calculated assuming the zero-thickness lithosphere altitude is represented by the midocean ridge altitude corrected for the effect of water loading. Ocean depths were calculated using the same reference column without the water-loading correction. The following parameter values were used: asthenosphere density = 3.2 Mg m^{-3}; Moho density contrast = 0.4 Mg m^{-3}; sea-water density = 1.0 Mg m^{-3}; volume coefficient of thermal expansion for lithosphere = 3×10^{-5} $°C^{-1}$; asthenosphere temperature = 1,200°C. Modified from Morgan and Burke (1985).

Moho temperatures (Black and Braile, 1982) and between the depth to a zone of high electrical conductivity and surface heat flow (Warren and others, 1969; Chaipayungpun and Landisman, 1977), although the relation between electrical conductivity and temperature is complex (Shankland and Waff, 1977; Shankland and Ander, 1983). As density is also related to temperature, regional Bouguer gravity anomalies may be expected to be associated with different thermal regimes, but as most long-wavelength anomalies are strongly dependent on crustal thickness and compositional density variations, the heterogeneity of the crust must be removed before the thermal anomalies would be apparent. Gravity anomalies should result from lateral thermal contrasts in the lithosphere and associated changes in depth to the top of the asthenosphere. Depth to the Curie isotherm is also dependent on the geotherm. Mayhew (1982a, b) has correlated thickness of the magnetic crust with regional thermal regimes.

Finally, there is obviously a close relationship between magmatism and the geotherm (Blackwell, 1978). High geotherms in the western U.S. are undoubtedly primarily the result of upward convective heat transfer into the crust by mantle-derived magmas. Tectonic convection (e.g., extension) may trigger magmatism if the geotherm is close to the solidus, and conversely, raising of the geotherm by magmatic activity may localize deformation.

CONCLUDING REMARKS

Thermally, the United States can be divided into two parts by the boundary between the Great Plains and Rocky Mountains provinces, as indicated by earlier geophysical studies. To the east a stable thermal regime has been established, with heat-flow variations caused primarily by heterogeneity in crustal heat generation and redistribution of heat by ground-water convection. Hot thermal regimes dominate the United States west of the boundary and are related to Cenozoic tectonic and magmatic activity. The continuity of thermal regimes across physiographic and tectonic boundaries suggests that mantle-derived magmatism may be the most important mechanism responsible for the high heat flow, although locally (e.g., the Salton trough), tectonically induced magmatism may dominate. Some subregional areas of low heat flow are related to ground-water flow (e.g., the "Eureka Low" and eastern Snake River Plain), but low heat flow in the Sierra Nevada and coastal Pacific Northwest is thought to be related to subduction.

Heat-flow data represent only one component of energy loss from the Earth, and other components may be significant, especially in the western U.S. Energy is also dissipated through regional uplift, volcanism, thermal springs, and seismic energy release, and may be stored in the lithosphere during transient

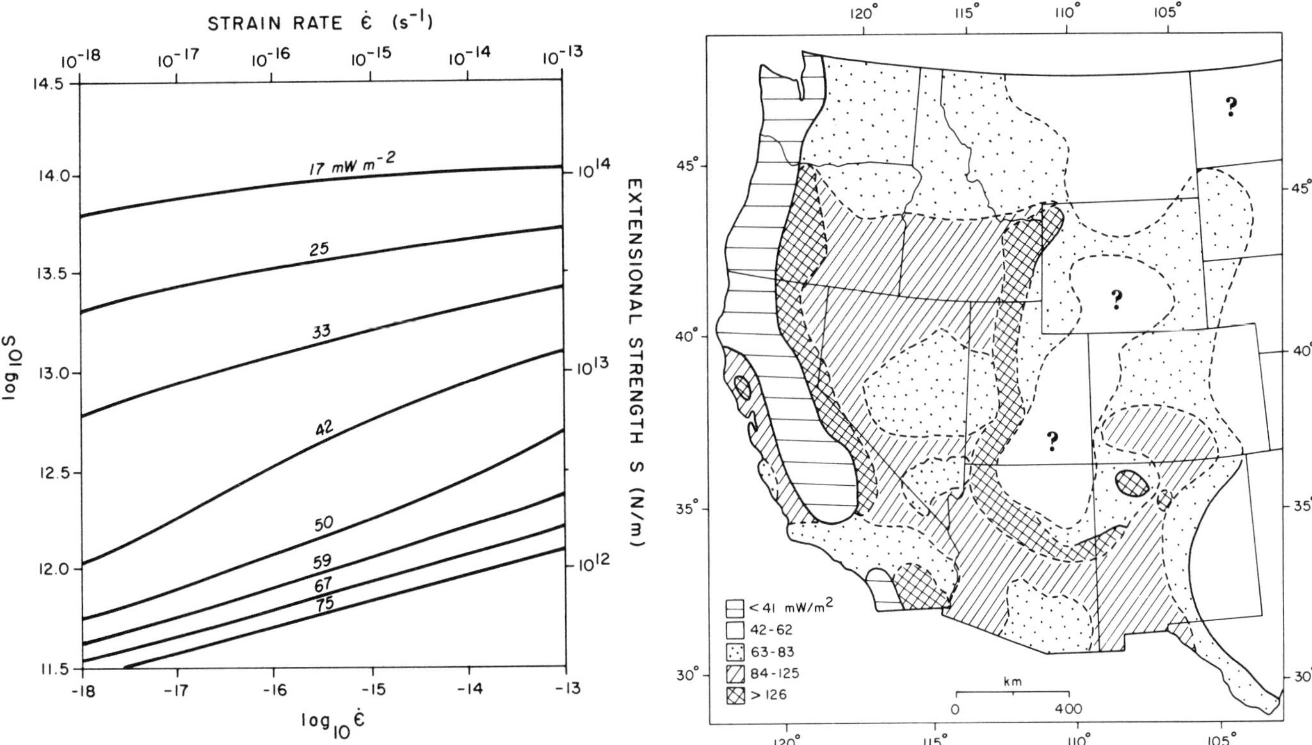

Figure 24. Integrated extensional lithospheric strength as a function of strain rate for lithospheres with a 20-km-thick silicic upper crust, a 20-km-thick mafic lower crust, and different thermal regimes. Reduced heat-flow values for the lithospheres represented by each curve are indicated on the curves. From Lynch and Morgan (1987).

Figure 25. Energy-flux map for the western United States. From Blackwell (1978).

thermal events. Blackwell (1978) has attempted to include these components in an energy-flux map of the western U.S., as shown in Figure 25. The heat-flow data base for the U.S. has grown dramatically during the past 10 years, and in some places is approaching the density and quality necessary to address some of the fundamental problems of the tectonic evolution of the U.S. (Blackwell, 1983). The major energy source for high heat flow in the western U.S. is still a mystery, although it appears to be related to subduction. Only through expansion and careful analysis of the heat-flow data base and careful consideration of all energy-loss components and thermal implications of the geotherm will the complex Cenozoic history of the western U.S. and subtle variations in lithospheric structure of the stable central and eastern U.S. be fully understood and appreciated.

ACKNOWLEDGMENTS

A review paper is the agglomeration of the ideas of many people under the pen of the reviewers. We gratefully acknowledge the receipt of reprints from our heat-flow colleagues for use in this review, many of which space and style have not allowed us to reference, although their contents have contributed to our conclusions. J. H. Sass kindly provided a copy of the U.S. Geological Survey heat-flow data file for our analysis. We thank W. H. Diment, D. D. Blackwell, and L. C. Pakiser for constructive reviews of a draft of this manuscript, but we remain responsible for any errors or misconceptions remaining herein. M. K. Morgan, S. Sabala Foreman, and E. Mead assisted with preparation of the manuscript.

REFERENCES CITED

Allenby, R. J., and Schnetzler, C. C., 1983, United States crustal thickness: Tectonophysics, v. 93, p. 13–31.

Allis, R. G., 1978, The effect of Pleistocene climatic variations on the geothermal regime in Ontario; A reassessment: Canadian Journal of Earth Sciences, v. 15, p. 1875–1879.

——, 1979, A heat production model for stable continental crust: Tectonophysics, v. 57, p. 151–165.

Armstrong, R. L., Leeman, W. P., and Malde, H. E., 1975, K-Ar dating, Quaternary and Neogene volcanic rocks of the Snake River Plain, Idaho: American Journal of Science, v. 275, p. 225–251.

Atwater, T., 1970, Implications of plate tectonics for the Cenozoic tectonic evolution of western North America: Geological Society of America Bulletin, v. 81, p. 3513–3536.

Baldridge, W. S., Olsen, K. H., and Callender, J. F., 1984, Rio Grande rift;

Problems and perspectives: New Mexico Geological Society Guidebook 35, p. 1–12.

Beck, A. E., 1965, Techniques for measuring heat flow on land, *in* Lee, W.H.K., ed., Terrestrial heat flow, American Geophysical Union Geophysical Monograph 8, p. 24–57.

——, 1977, Climatically perturbed temperature gradients and their effect on regional and continental heat flow means: Tectonophysics, v. 41, p. 17–39.

Beck, A. E., and Shen, P. Y., 1985, Temperature distribution in flowing liquid wells: Geophysics, v. 50, p. 1113–1118.

Benfield, A. E., 1949, The effect of uplift and denudation on underground temperatures: Journal of Applied Physics, v. 20, p. 66–70.

Bethke, C. M., 1986, Inverse hydrologic analysis of the distribution and origin of Gulf Coast-type geopressured zones: Journal of Geophysical Reearch, v. 91, p. 6535–6545.

Bickle, M. J., Hawkesworth, C. J., England, P. C., and Athey, D. R., 1975, A preliminary thermal model for regional metamorphism in the eastern Alps: Earth and Planetary Sciences Letters, v. 26, p. 13–28.

Birch, F., 1948, The effects of Pleistocene climatic variations upon geothermal gradients: American Journal of Science, v. 246, p. 729–760.

——, 1950, Flow of heat in the Front Range, Colorado: Geological Society of America Bulletin, v. 61, p. 567–630.

Birch, F., and Clark, H., 1940, The thermal conductivity of rocks and its dependence upon temperature and composition: American Journal of Science, v. 238, p. 613–635.

Birch, F., Roy, R. F., and Decker, E. R., 1968, Heat flow and thermal history of New York and New England, *in* Zen, E., White, W. S., Hadley, J. B., and Thompson, J. B., Jr., eds., Studies of Appalachian geology; Northern and Maritime: New York, Interscience, p. 437–451.

Bird, P., 1979, Continental delamination and the Colorado Plateau: Journal of Geophysical Research, v. 84, p. 7561–7571.

Black, P. R., and Braile, L. W., 1982, P_n velocity and cooling of the continental lithosphere: Journal of Geophysical Research, v. 87, p. 10557–10568.

Blackwell, D. D., 1969, Heat flow in the northwestern United States: Journal of Geophysical Research, v. 74, p. 992–1007.

——, 1971, The thermal structure of the continental crust, *in* Heacock, J. G., ed., The structure and physical properties of the Earth's crust: American Geophysical Union Geophysical Monograph 14, p. 169–184.

——, 1978, Heat flow and energy loss in the western United States, *in* Smith, R. B., and Eaton, G. P., eds., Cenozoic tectonics and regional geophysics of the western Cordillera: Geological Society of America Memoir 152, p. 175–208.

——, 1983, Heat flow in the northern Basin and Range province: Geothermal Resources Council Special Report 13, p. 81–92.

——, 1985, A transient model of the geothermal system of the Long Valley caldera, California: Journal of Geophysical Research, v. 90, p. 11229–11241.

Blackwell, D. D., and Chapman, D. S., 1977, Interpretation of geothermal gradient and heat flow data for Basin and Range geothermal systems: Transactions of the Geothermal Resources Council, v. 1, p. 19–20.

Blackwell, D. D., Steele, J. L., and Brott, C. A., 1980, The terrain effect on terrestrial heat flow: Journal of Geophysical Research, v. 85, p. 4757–4772.

Blackwell, D. D., Bowen, R. G., Hull, D. A., Riccio, J., and Steele, J. L., 1982, Heat flow, arc volcanism, and subduction in northern Oregon: Journal of Geophysical Reseach, v. 87, p. 8735–8754.

Bodell, J. M., and Chapman, D. S., 1982, Heat flow in the north-central Colorado Plateau: Journal of Geophysical Research, v. 87, p. 2869–2884.

Brott, C. A., Blackwell, D. D., and Mitchell, J. C., 1978, Tectonic implications of the heat flow of the western Snake River Plain, Idaho: Geological Society of America Bulletin, v. 89, p. 1697–1707.

Brott, C. A., Blackwell, D. D., and Ziagos, J. P., 1981, Thermal and tectonic implications of heat flow in the eastern Snake River Plain, Idaho: Journal of Geophysical Research, v. 86, p. 11709–11734.

Buntebarth, G., 1984, Geothermics: Berlin, Springer-Verlag, 144 p.

Chaipayungpun, W., and Landisman, M., 1977, Crust and upper mantle near the western edge of the Great Plains, *in* Heacock, J. G., ed., The Earth's crust: American Geophysical Union Geophysical Monograph 20, p. 553–575.

Chapman, D. S., Keho, T. H., Bauer, M. S., and Picard, M. D., 1984, Heat flow in the Uinta basin determined from bottom hole temperature (BHT) data: Geophysics, v. 49, p. 453–466.

Chase, C. G., and Wallace, T. C., 1986, Uplift of the Sierra Nevada of California: Geology, v. 14, p. 730–733.

Clark, S. P., 1957, Absorption spectra of some silicates in the visible and near infrared: American Mineralogist, v. 42, p. 732–742.

Costain, J. K., 1978, A new model for the linear relationship between heat flow and heat generation: EOS Transactions of the American Geophysical Union, v. 59, p. 392.

Costain, J. K., Speer, J. A., Glover, L., III, Perry, L., Dashevsky, S., and McKinney, M., 1986, Heat flow in the Piedmont and Atlantic Coastal Plain of the southeastern United States: Journal of Geophysical Research, v. 91, p. 2123–2135.

Crough, S. T., and Thompson, G. A., 1976, Thermal model of continental lithosphere: Journal of Geophysical Research, v. 81, p. 4857–4862.

Decker, E. R., and Smithson, S. B., 1975, Heat flow and gravity interpretation across the Rio Grande rift in southern New Mexico and West Texas: Journal of Geophysical Research, v. 80, p. 2542–2552.

Decker, E. R., Baker, K. R., Bucher, G. J., and Heasler, H. P., 1980, Preliminary heat flow and radioactivity studies in Wyoming: Journal of Geophysical Research, v. 85, p. 311–321.

Decker, E. R., Bucher, G. J., Buelow, K. L., and Heasler, H. P., 1984, Preliminary interpretation of heat flow and radioactivity in the Rio Grande rift zone in central and northern Colorado: New Mexico Geological Society Guidebook 35, p. 45–50.

Drury, M. J., 1984, Heat flow and heat generation in the Churchill Province of the Canadian Shield, and their paleotectonic significance: Tectonophysics, v. 115, p. 25–44.

Drury, M. J., and Lewis, T. J., 1983, Water movement within Lac du Bonnet batholith as revealed by detailed thermal studies of three closely spaced boreholes: Tectonophysics, v. 95, p. 337–351.

Drury, M. J., Jessop, A. M., and Lewis, T. J., 1984, The detection of groundwater flow by precise temperature measurements in boreholes: Geothermics, v. 13, p. 163–174.

Eaton, G. P., 1980, Geophysical and geological characteristics of the crust of the Basin and Range province, *in* Studies in geophysics, continental tectonics: Washington, D.C., National Academy of Sciences, p. 96–113.

——, 1982, The Basin and Range province; Origin and tectonic significance: Annual Reviews of Earth and Planetary Sciences, v. 10, p. 409–440, 1982.

England, P. C., and Richardson, S. W., 1980, Erosion and the age dependence of continental heat flow: Geophysical Journal of the Royal Astronomical Society, v. 62, p. 421–437.

Fournier, R. O., White, D. E., and Truesdell, A. H., 1976, Convective heat flow in Yellowstone National Park; Proceedings of the Second United Nations Symposium on the Development and Use of Geothermal Resources: Washington, D.C., U.S. Government Printing Office, v. 1, p. 731–739.

Fujisawa, H., Fujii, N., Mizutani, H., Kanamori, H. and Akimoto, S., 1968, Thermal diffusivity of Mg_2SiO_4, Fe_2SiO_4, and NaCl at high pressures and temperatures, Journal of Geophysical Research, v. 73, p. 4727–4733.

Fyfe, W. S., 1978, The evolution of the earth's crust; Modern plate tectonics to ancient hot spots: Chemical Geology, v. 23, p. 89–114.

Gosnold, W. D., 1989, Heat flow in the Great Plains of the United States, Journal of Geophysical Research (in press).

Gosnold, W. D., and Swanberg, C. A., 1980, A new model for the distribution of crustal heat sources: EOS Transactions of the American Geophysical Union, v. 61, p. 387.

Guffanti, M., and Nathenson, M., 1980, Preliminary map of temperature gradients in the conterminous United States: Transactions of the Geothermal Resources Council, v. 4, p. 53–56.

Hamblin, W. K., Damon, P. E., and Bull, W. B., 1981, Estimates of vertical crustal strain rates along the western margins of the Colorado Plateau: Geology, v. 9, p. 293–298.

Hamza, V. M., and Beck, A. E., 1975, Analysis of heat flow data; Vertical variations of heat flow and heat producing elements in sediments: Canadian Journal of Earth Sciences, v. 12, p. 996–1005.

Hunt, C. B., 1956, Cenozoic geology of the Colorado Plateau: U.S. Geological Survey Professional Paper 279, 99 p.

Hyndman, R. D., 1976, Heat flow in the inlets of southwestern British Columbia: Journal of Geophysical Research, v. 81, p. 337–349.

Jaeger, J. C., 1965, Application of the theory of heat conduction to geothermal measurements, *in* Lee, W.H.K., ed., Terrestrial heat flow: American Geophysical Union Geophysical Monograph 8, p. 7–23.

Jaupart, C., Sclater, J. G., and Simmons, G., 1981, Heat flow studies; Constraints on the distribution of uranium, thorium, and potassium in the continental crust: Earth and Planetary Science Letters, v. 52, p. 328–344.

Jensen, J. C., 1983, Calculation on the thermal conditions around a salt diapir: Geophysical Prospecting, v. 31, p. 481–489.

Jessop, A. M., 1983, The essential ingredients of a continental heat flow determination: Zentralblatt für Geologie und Paläontologie, v. 1, p. 70–79.

Kanamori, H., Fujii, N., and Mizutani, H., 1968, Thermal diffusivity measurements of rockforming minerals from 300° to 1100°K: Journal of Geophysical Research, v. 73, p. 595–605.

Kappelmeyer, O., and Haenel, R., 1974, Geothermics with special reference to application: Berlin, Gebrüder Borntraeger, 238 p.

Kay, R. W., and Kay, S. M., 1980, Chemistry of the lower crust; Inferences from magmas and xenoliths, *in* Studies in geophysics, continental tectonics: Washington, D.C., National Academy of Sciences, p. 139–150.

Keller, G.R ., Braile, L. W., and Morgan, P., 1979, Crustal structure, geophysical models, and contemporary tectonism of the Colorado Plateau: Tectonophysics, v. 61, p. 131–147.

Lachenbruch, A. H., 1968, Preliminary thermal model of the Sierra Nevada: Journal of Geophysical Research, v. 73, p. 6877–6989.

——, 1970, Crustal temperature and heat production; Implications of the linear heat-flow relation: Journal of Geophysical Research, v. 75, p. 3291–3300.

Lachenbruch, A. H., and Sass, J. H., 1977, Heat flow and the thermal regime of the crust, *in* Heacock, J. G., ed., The Earth's crust; Its nature and physical properties: American Geophysical Union Geophysical Monograph 20, p. 626–675.

——, 1978, Models of an extending lithosphere and heat flow in the Basin and Range province, *in* Smith, R. B., and Eaton, G. P., eds., Cenozoic tectonics and regional geophysics of the western Cordillera: Geological Society of America, Memoir 152, p. 209–250.

——, 1980, Heat flow and energies of the San Andreas fault zone: Journal of Geophysical Research, v. 85, p. 6185–6222.

Lachenbruch, A. H., Sass, J. H., and Galanis, S. P., Jr., 1985, Heat flow in southernmost California and the origin of the Salton trough: Journal of Geophysical Research, v. 90, p. 6709–6736.

Langseth, M. G., 1965, Techniques of measuring heat flow through the ocean floor, *in* Lee, W.H.K., ed., Terrestrial heat flow: American Geophysical Union Geophysical Monograph 8, p. 58–77.

Lee, T.-C., 1975, Focusing and defocusing of heat flow by a buried sphere: Geophysical Journal of the Royal Astronomical Society, v. 43, p. 635–641.

Lee, T.-C., and Henyey, T. L., 1974, Heat-flow refraction across dissimilar media: Geophysical Journal of the Royal Astronomical Society, v. 39, p. 319–333.

Lewis, T. J., and Beck, A. E., 1977, Analysis of heat-flow data; Detailed observations in many holes in a small area: Tectonophysics, v. 41, p. 41–59.

Lewis, T. J., Jessop, A. M., and Judge, A. S., 1985, Heat flux measurements in southwestern British Columbia; The thermal consequences of plate tectonics: Canadian Journal of Earth Sciences, v. 22, p. 1262–1273.

Lovering, T. S., 1948, Geothermal gradients, recent climatic change, and rate of sulphide oxidation in the San Manuel District, Arizona: Economic Geology, v. 43, p. 1–20.

Lucchitta, I., 1979, Late Cenozoic uplift of the southwestern Colorado Plateau and adjacent lower Colorado River region: Tectonophysics, v. 61, p. 63–95.

Lynch, H. D., and Morgan, P., 1987, The tensile strength of the lithosphere and the localization of extension, *in* Coward, M. P., Dewey, J. F., and Hancock, P. L., eds., Continental extensional tectonics: Geological Society of London Special Publication 28, p. 53–65.

MacPherson, W. R., and Schloessin, H. H., 1982, Lattice and radiative thermal conductivity through high p, T polymorphic structure transition melting points: Physics of Earth and Planetary Interiors, v. 29, p. 58–68.

Majorowicz, J. A., Jones, F. W., Lam, H. L., and Jessop, A. M., 1984, The variability of heat flow both regional and with depth in southern Alberta, Canada; Effect of ground-water flow: Tectonophysics, v. 106, p. 1–24.

——, 1985, Terrestrial heat flow and geothermal gradients in relation to hydrodynamics in the Alberta Basin, Canada: Journal of Geodynamics, v. 4, p. 265–283.

Mayhew, M. A., 1982a, An equivalent layer magnetization model for the United States derived from satellite altitude magnetic anomalies: Journal of Geophysical Research, v. 87, p. 4837–4845.

——, 1982b, Application of satellite magnetic anomaly data to Curie isotherm mapping: Journal of Geophysical Research, v. 87, p. 4846–4854.

McBirney, A. R., 1963, Conductivity variations and terrestrial heat flow distribution: Journal of Geophysical Research, v. 68, p. 6323–6329.

McKenzie, D. P., 1978, Some remarks on the development of sedimentary basins: Earth and Planetary Science Letter, v. 40, p. 25–32.

Morgan, P., 1983, Constraints on rift thermal processes from heat flow and uplift: Tectonophysics, v. 94, p. 277–298.

——, 1984, The thermal structure and thermal evolution of the continental lithosphere: Physics and Chemistry of the Earth, v. 15, p. 107–193.

——, 1985, Crustal radiogenic heat production and the selective survival of ancient continental crust: Journal of Geophysical Research, v. 90, supplement, p. C561–C570.

Morgan, P., and Burke, K., 1985, Collisional plateaus: Tectonophysics, v. 119, p. 137–151.

Morgan, P., and Sass, J. H., 1984, Thermal regime of the continental lithosphere: Journal of Geodynamics, v. 1, p. 143–166.

Morgan, P., and Swanberg, C. A., 1985, On the Cenozoic uplift and stability of the Colorado Plateau: Journal of Geodynamics, v. 3, p. 65–85.

Morgan, P., Blackwell, D. D., Spafford, R. E., and Smith, R. B., 1977, Heat flow measurements in Yellowstone Lake and the thermal regime of the Yellowstone caldera: Journal of Geophysical Research, v. 82, p. 3719–3732.

Morgan, P., Harder, V., Swanberg, C. A., and Daggett, P. H., 1981, A groundwater convection model for Rio Grande rift geothermal resources: Transactions of the Geothermal Resources Council, v. 5, p. 193–196.

Morgan, P., Harder, V., and Giordano, T. H., 1986a, Heat and fluid flow in the Rio Grande rift; A possible modern thermal analogue of a Mississippi Valley-type ore-forming system, *in* Geology in the real world; The Kingsley Dunham volume: London, Institute of Mining and Metallurgy, p. 295–305.

Morgan, P., Seager, W. R., and Golombek, M. P., 1986b, Cenozoic thermal, mechanical, and tectonic evolution of the Rio Grande rift: Journal of Geophysical Research, v. 91, p. 6263–6276.

Morgan, W. J., 1972, Convection plumes and plate motions: American Association of Petroleum Geologists Bulletin, v. 56, p. 203–213.

Nathenson, M., and Guffanti, M., 1988, Geothermal gradients in the conterminous United States: Journal of Geophysical Research, v. 93, p. 6437–6450.

O'Brien, P.N.S., 1967, Quantitative discussion on seismic amplitudes produced by explosions in Lake Superior: Journal of Geophysical Research, v. 72, p. 2569–2575.

Oxburgh, E. R., and Turcotte, D. L., 1974, Thermal gradients and regional metamorphism in overthrust terrains with special reference to the Eastern Alps: Schweizerische Mineralogische und Petrographische Mitteilungen, v. 54, p. 641–661.

Parsons, B., and Sclater, J. G., 1977, An analysis of the variation of ocean floor bathymetry and heat flow with age: Journal of Geophysical Research, v. 82, p. 803–827.

Pollack, H. N., 1982, The heat flow from the continents: Annual Reviews of Earth and Planetary Sciences, v. 10, p. 459–481.

Rao, R.U.M., Rao, G. V., and Reddy, G. K., 1982, Age dependence of continental heat flow; Fantasy and fact: Earth and Planetary Science Letters, v. 59, p. 288–302.

Reiter, M., Minear, J., and Gutjhar, A., 1985, Variance analysis of estimates and measurements of terrestrial heat flow: Geothermics, v. 14, p. 499–509.

Roy, R. F., Decker, E. R., Blackwell, D. D., and Birch, F., 1968, Heat generation of plutonic rocks and continental heat flow provinces: Earth and Planetary Science Letters, v. 5, p. 1–12.

Roy, R. F., Blackwell, D. D., and Decker, E. R., 1972, Continental heat flow, *in* Robertson, E. C., ed., The nature of the solid earth: New York, McGraw-Hill, p. 506–543.

Roy, R. F., Beck, A. E., and Touloukian, Y. S., 1981, Thermophysical properties of rocks, *in* Touloukian, Y. S., Judd, W. R., and Roy, R. F., eds., Physical properties of rocks and minerals: New York, McGraw-Hill, p. 409–502.

Sanford, A. R., Mott, R. P., Shuleski, P. J., Rinehart, E. J., Caravella, F. J., Ward, R. M., and Wallace, T. C., 1977, Geophysical evidence for a magma body in the crust in the vicinity of Socorro, New Mexico, *in* Heacock, J. G., ed., The Earth's crust; Its nature and physical properties: American Geophysical Union Geophysical Monograph 20, p. 385–403.

Sass, J. H., Lachenbruch, A. H., and Monroe, R. J., 1971, Thermal conductivity of rocks from measurements on fragments and its application to heat flow determinations: Journal of Geophysical Research, v. 76, p. 3391–3401.

Sass, J. H., Blackwell, D. D., Chapman, D. S., Costain, J. K., Decker, E. R., Lawver, L. A., and Swanberg, C. A., 1981, Heat flow from the crust of the United States, *in* Touloukian, Y. S., Judd, W. R., and Roy, R. F., eds., Physical Properties of rocks and minerals: New York, McGraw-Hill, p. 503–548.

Schatz, J., and Simmons, G., 1972, Thermal conductivity of earth materials at high temperatures: Journal of Geophysical Research, v. 77, p. 6966–6983.

Sclater, J. G., Jaupart, C., and Galson, D., 1980, The heat flow through oceanic and continental crust and the heat loss of the earth: Reviews of Geophysics and Space Physics, v. 18, p. 269–311.

Sclater, J. G., Parsons, B. and Jaupart, C., 1981, Oceans and continents; Similarities and differences in the mechanisms of heat loss: Journal of Geophysical Research, v. 86, p. 11535–11552.

Shankland, T. J., and Ander, M. E., 1983, Electrical conductivity, temperatures, and fluids in the lower crust: Journal of Geophysical Research, v. 88, p. 9475–9484.

Shankland, T. J., and Waff, H. S., 1977, Partial melting and electrical conductivity anomalies in the upper mantle: Journal of Geophysical Research, v. 82, p. 5409–5417.

Shen, P. Y., and Beck, A. E., 1983, Determination of surface temperature history from borehole temperature gradients: Journal of Geophysical Research, v. 88, p. 7385–7493.

Simmons, G., 1961, Anisotropic thermal conductivity: Journal of Geophysical Research, v. 66, p. 2269–2270.

Smith, D. L., and Dees, W. T., 1982a, Heat flow in the Gulf Coastal Plain: Journal of Geophysical Research, v. 87, p. 7687–7693.

——, 1982b, Indicators of low-temperature geothermal resources in northern Louisiana and central Mississippi: Journal of Volcanology and Geothermal Research, v. 14, p. 389–393.

Smith, D. L., Gregory, R. G., and Emhof, J. W., 1981, Geothermal measurements in the southern Appalachian Mountains and southeastern Coastal Plains: American Journal of Science, v. 281, p. 281–298.

Smith, R. B., 1978, Seismicity, crustal structure, and intraplate tectonics of the interior of the western Cordillera, *in* Smith, R. B., and Eaton, G. P., eds., Cenozoic tectonics and regional geophysics of the western Cordillera: Geological Society of America Memoir 152, p. 111–144.

Smith, T. J., Steinhart, J. S., and Aldrich, L. T., 1966, Lake Superior crustal structure: Journal of Geophysical Research, v. 71, p. 1141–1172.

Smithson, S. B., and Decker, E. R., 1974, A continental crustal model and its geothermal applications: Earth and Planetary Science Letters, v. 22, p. 215–225.

Sorey, M. L., 1985, Evolution and present state of the hydrothermal system in Long Valley caldera: Journal of Geophysical Research, v. 90, p. 11219–11228.

Speece, M. A., Bowen, T. D., Folcik, J. L., and Pollack, H. N., 1985, Analysis of temperatures in sedimentary basins; The Michigan basin: Geophysics, v. 50, p. 1318–1334.

Stewart, J. H., and Carlson, J. E., 1978, Generalized maps showing distribution, lithology, and age of Cenozoic igneous rocks in the western United States, *in* Smith, R. B., and Eaton, G. P., eds., Cenozoic tectonics and regional geophysics of the western Cordillera: Geological Society of America Memoir 152, p. 263–282.

Swanberg, C. A., and Morgan, P., 1980, The silica heat flow technique; Assumptions and applications: Journal of Geophysical Research, v. 85, p. 7206–7214.

Toksöz, M. N., and Bird, P., 1977, Modelling of temperatures in continental convergence zones: Tectonophysics, v. 41, p. 181–193.

Tomczyk, T., and Morgan, P., 1987, Evaluation of the thermal regime of the Valles caldera, New Mexico, U.S.A., by downward continuation of temperature data: Tectonophysics, v. 134, p. 339–345.

Vitorello, I., and Pollack, H. N., 1980, On the variation of continental heat flow with age and the thermal evolution of continents: Journal of Geophysical Research, v. 85, p. 983–995.

Wallace, R. H., Jr., Kraemer, T. F., Taylor, R. E., and Wesselman, J. B., 1979, Assessment of geopressured-geothermal resources in the northern Gulf of Mexico basin: U.S. Geological Survey Circular 790, p. 132–155.

Warren, R. E., Sclater, J. G., Vacquier, V., and Roy, R. F., 1969, A comparison of terrestrial heat flow and transient geomagnetic fluctuations in the southwestern United States: Geophysics, v. 34, p. 463–478.

Williams, D. L., and Von Herzen, R. P., 1983, On the terrestrial heat flow and physical limnology of Crater Lake, Oregon: Journal of Geophysical Research, v. 88, p. 1094–1104.

Zandt, G., and Furlong, K. P., 1982, Evolution and thickness of the lithosphere beneath coastal California: Geology, v. 10, p. 376–381.

Ziagos, J. P., Blackwell, D. D., and Mooser, F., 1985, Heat flow in southern Mexico and the thermal effects of subduction: Journal of Geophysical Research, v. 90, p. 5410–5420.

Zoback, M. L., and Zoback, M. D., 1980, State of stress in the conterminous United States: Journal of Geophysical Research, v. 85, p. 6113–6156.

MANUSCRIPT ACCEPTED BY THE SOCIETY OCTOBER 31, 1988

Printed in U.S.A.

Geological Society of America
Memoir 172
1989

Chapter 24

Tectonic stress field of the continental United States

Mary Lou Zoback
U.S. Geological Survey, 345 Middlefield Road, Menlo Park, California 94025
Mark D. Zoback
Department of Geophysics, Stanford University, Stanford, California 94305

ABSTRACT

The orientation and relative magnitudes of in situ tectonic stress in the continental United States have been inferred from a variety of indicators, including earthquake focal mechanisms, stress-induced elliptical borehole enlargement ("breakouts"), hydraulic fracturing stress measurements, and young fault slip and volcanic alignments. The data come from a wide range of depths (0.1 to 20 km) and have been assigned a quality ranking according to their reliability as tectonic stress indicators. The data show that regionally consistent orientations persist throughout the seismic "brittle" upper crust. Stress provinces are defined on the basis of uniform stress orientations and relative stress magnitudes (style of faulting).

The sources of stress for all the major stress provinces are believed to be linked either directly or indirectly to plate-tectonic processes. Most of the central and eastern United States is part of a broad "midplate" stress province (which includes most of Canada and possibly the western Atlantic basin) characterized by NE- to ENE-oriented maximum horizontal compression. This orientation range coincides with both absolute plate motion and ridge push directions for North America. With the exception of the San Andreas region and most of the Pacific Northwest area, the remainder of the western United States (generally, the thermally elevated region from the high Great Plains to the west and including the Basin and Range province) is characterized by extensional tectonism. Within areas of "classic" basin-range structure, both those currently active and those largely quiescent, the least horizontal stress is oriented approximately E-W (between WNW and ENE). In the Colorado Plateau interior and the southern Great Plains, the least horizontal stress is NNE, roughly perpendicular to that in surrounding areas. The state of stress in these two regions may be related to pronounced lateral variations in lithosphere thickness beneath them.

New focal-mechanism data in western Washington and northern Oregon define a region of NE compression apparently associated with subduction of the Juan de Fuca plate. In the coastal region of central California there appears to be a zone approximately 100 to 125 km wide on both sides of the San Andreas fault in which the maximum horizontal stress is oriented NE, nearly perpendicular to the fault itself. While compression orthogonal to the San Andreas fault explains late Tertiary–Quaternary folding and reverse faulting subparallel to the San Andreas, this NE compression seems incompatible with right-lateral slip on the N40°W–trending San Andreas fault. The observed stress field, however, is consistent with current estimates of the direction of relative plate motions, provided that the shear strength of the San Andreas is appreciably lower than the level of far-field shear stress in the crust.

Zoback, M. L., and Zoback, M. D., 1989, Tectonic stress field of the continental United States, *in* Pakiser, L. C., and Mooney, W. D., Geophysical framework of the continental United States: Boulder, Colorado, Geological Society of America Memoir 172.

INTRODUCTION

In this chapter we present an update of our earlier compilation (Zoback and Zoback, 1980) of stress-orientation data for the continental (conterminous) United States. In the earlier work we showed that the modern state of stress has regional uniformity and is primarily tectonic in origin. The data we consider come from a wide range of depths and indicate consistent stress orientations throughout the upper crust (approximately the upper 20 km).

The data presented here are a subset of a new compilation of stress data for all of North America we are preparing for the Decade of North American Geology (DNAG) Neotectonics volume. Several changes have been made to our earlier compilation. The data base has been considerably enlarged, from 226 to more than 400 points, and we have slightly revised our criteria for reliable tectonic stress field indicators. In addition, all of the data points have been assigned a quality ranking, and a few points have been deleted from the original data set based on the revised criteria and quality-ranking procedure. This chapter focuses on a new, quality-ranked stress orientation map of the United States. A detailed listing of the individual data points in the new data base is not included here; it will be part of the DNAG Neotectonics volume.

As we pointed out in our previous study, the stress data permit definition of stress provinces, regions characterized by generally uniform stress orientations and relative stress magnitudes, or style of faulting. Our description and discussion of these stress provinces parallels the discussion in the seismicity chapter (Dewey and others, this volume), which divides the United States into four major plate-tectonic provinces. Correlations between the stress field and plate-tectonic interactions and other regional geophysical parameters are also addressed.

Principal stress orientation indicators

Data on principal stress orientations come from four main types of indicators: earthquake focal mechanisms, elliptical wellbore enlargements, or "breakouts," in situ stress measurements, and geological data on young volcanic vent alignments and fault offsets. The assumptions, difficulties, and uncertainties associated with each type of indicator have been discussed earlier in some detail (Zoback and Zoback, 1980). One of the primary factors accounting for the marked increase in data in this compilation compared to our previous one is the large number of new orientations from wellbore breakout analyses (e.g., Gough and Bell, 1981, 1982; Dart, 1985; Plumb and Cox, 1987; Dart and Zoback, 1987). The physical process leading to the occurrence of these breakouts is now fairly well understood (Gough and Bell, 1981; Plumb and Hickman, 1985; Zoback and others, 1985a), and the widespread availability of data from wells drilled for petroleum exploration makes this technique a valuable data source in the new compilation.

We have revised the age criteria for the geologic (fault-slip) indicators. Previously, in the eastern United States we used fault offsets of Tertiary or Quaternary age as compiled by Prowell (1983). We have now restricted our consideration to faults in the eastern United States that show offsets of Miocene or younger strata. The age criteria for geological data from the western United States remains the same, < 5 Ma, although most data are younger than 3 Ma.

A new category of in situ stress measurements, petal centerline or core-induced fractures (IS-PC, Table 1), has been included. These planar, vertical or very steeply dipping fractures are observed in oriented cores and are believed to be extensional fractures formed in advance of a downcutting drill bit; their orientation is thus thought to record the maximum horizontal compressive stress direction (Kulander and others, 1977, 1979; Dean and Overbey, 1980; Ganga Rao and others, 1979). Evans (1979) has examined oriented cores from 13 gas wells in Pennsylvania, Ohio, West Virginia, Kentucky, and Virginia, and determined petal centerline fracture orientations for hundreds of meters of core for most well. The inferred maximum horizontal stress orientations are consistent within wells, between nearby wells, and with adjacent hydraulic fracturing results and focal mechanisms (Bollinger and Wheeler, 1982), and so these data have been included in our compilation.

In addition to enlarging the data set, an additional contribution of this study is a quality-ranking system that is intended to characterize how accurately a particular data point records the tectonic stress field. Four qualities have been assigned, A through D, with A the highest quality and D the lowest. Whenever possible, the quality ranking was based on a statistical analysis of the accuracy of the data. A brief discussion of the rationale for the ranking follows; a detailed description of the ranking for each type of indicator will be included in Zoback and Zoback (1989).

In general, the A determinations come from averages of closely spaced data or from multiple and/or very high-quality observations at a single locality. We believe these data record the horizontal principal stress orientations in the upper crust (to seismogenic depths) to an accuracy of 10° to 15°.

The B determinations are intermediate in quality. Typically they represent one or two determinations at a single site; they may be very high-quality determinations but lack the statistical strength of multiple measurements. Note in Table 1 that even a well-constrained single-event focal mechanism rates only a B quality, because of the uncertainty in inferring principal stress orientations from even well-determined P- and T-axes (cf. McKenzie, 1969). The estimated accuracy of these B category stress orientation determinations is between 15° and about 20° to 25°.

The C quality determinations are the lowest quality measurements considered to be reliable indicators of the orientation of the tectonic stress field (probably $\pm 25°$). Either these data have some inherent uncertainty or inconsistency in the measurements themselves, or there is not enough information available to constrain stress orientations accurately from the data. Still, these data are included in the compilation because we believe they provide some reliable information about the stress field.

As indicated in Table 1, two general classes of information are included in the D quality ranking: (1) data that may be more indicative of local rather than tectonic stresses (e.g., G-FS, IS-OC, IS-HF), and (2) single observations of questionable quality (FM, IS-BO, G-VA) or multiple observations at a single site with a broad scatter or bimodal distribution of orientations (IS-BO, IS-OC). For some parts of the world these data may be the only type of information available; however, given the overall quality and density of the United States data set we see no reason to plot these questionable points. Thus, the stress map included in this chapter includes only A, B, and C quality data. Although the D-quality determinations are not plotted, they have been retained in the data base.

STRESS MAP OF THE CONTINENTAL UNITED STATES

Using the quality-ranked data set described above, we have prepared a map of maximum horizontal principal stress orientations (Fig. 1). In regions of known or inferred extensional stress regimes, the stress axes plotted in Figure 1 are actually the intermediate principal stress, S_2. The length of the orientation arrow is scaled proportional to its quality, and the center symbol designates the type of stress indicator. To emphasize broad-scale orientation patterns we have consistently plotted the maximum horizontal compressive stress orientation regardless of the actual state of stress (compressional or extensional) suggested by the indicator. The type of faulting or stress regime is given for each point in the detailed data set included in Zoback and Zoback (1989). Figure 2 is an index map showing the names of physiographic provinces shown on the stress maps in Figures 1 and 3.

Figure 3 is a generalized stress map showing average principal stress orientations, stress regime, and stress province boundaries. The stress provinces are characterized by a relatively uniform stress field and stress regime. The boundaries between some provinces are sharp (<75 km; e.g., between the Rio Grande Rift and southern Great Plains); between others, they may be broad and transitional (e.g., about 300 km wide between the San Andreas and Cordilleran extensional stress provinces).

Nearly all of the central and eastern United States is characterized by a compressive stress regime (reverse and strike-slip faulting, with the vertical stress, S_v, less than one or both of the horizontal stresses); in these areas in Figure 3 the maximum horizontal stress is indicated by inward-pointing arrows. Large portions of the western United States, however, are currently undergoing extensional tectonism (S_v = maximum principal stress). For the known or inferred extensional areas on the generalized stress map (Fig. 3), the orientation of the minimum horizontal stress (S_3) is indicated by outward-pointing arrows. Regions dominated by strike-slip tectonism (S_v = intermediate stress), such as the San Andreas fault system, are obviously transitional in character and typically contain evidence of reverse and some normal faulting subsidiary to the main strike-slip deformation. There are a few examples of extensional (normal faulting) points in the areas of compressional tectonics; however, we are aware of no data documenting major reverse faulting (other than a M = 3.7 reverse-faulting earthquake in northwestern Arizona: Brumbaugh, 1980) within the areas of extensional tectonics.

Discussion and description of the stress provinces shown in Figure 3 are divided into four sections covering the "plate-tectonic" provinces (San Andreas transform, Rocky Mountain/ Intermountain intraplate, Cascade convergent, and midplate central and eastern United States) as defined and discussed in the chapter on seismicity elsewhere in this Memoir (Dewey and others, this volume). The reader is referred to Dewey and others (this volume) for a detailed summary of active deformation in these regions. The four major plate-tectonic provinces generally coincide with stress provinces; however, several of the plate-tectonic provinces include more than one stress province. It is not our attempt to describe each stress province exhaustively; the reader is referred to our previous paper (Zoback and Zoback, 1980) for further description and discussion of these provinces, which we still consider generally valid except where noted.

SAN ANDREAS TRANSFORM PROVINCE

This province accommodates most of the relative motion between the Pacific and North American plates. The dominant structural feature is the San Andreas fault system, with deformation primarily concentrated along NW-striking, right-lateral strike-slip faults. Surprisingly, the overall orientation of the maximum horizontal stress (S_{Hmax}) in central California is roughly northeasterly, particularly when stress indicators from right-lateral strike-slip focal mechanisms along the San Andreas fault system (with N-trending P-axes) are ignored (Zoback and others, 1987). The pattern of generally NE compression in central California has recently been confirmed by more than 100 stress-induced wellbore breakouts from the California Coast Ranges and eastern Great Valley area (Mount and Suppe, 1987; Mount, in press). These data are not shown in Figure 1. Focal mechanisms for moderate-size earthquakes in the central California Coast Ranges—i.e., events that occur on faults subsidiary to and west of the San Andreas system proper—show a combination of reverse and strike-slip faulting with NNE- to NE-trending P-axes (Eaton and Rymer, 1989). Inversion of all these "off-San Andreas" focal mechanisms in central California (including the 1983 M = 6.7 Coalinga earthquake that occurred east of the San Andreas) to find a mean best-fitting deviatoric stress tensor (using the technique described by Angelier, 1984) yielded a rather good fit with a maximum horizontal compressive stress direction of N34°E. Similarly, focal mechanisms for events along the Coast Ranges–Great Valley boundary in north-central California show reverse and strike-slip faulting with NE- to E-trending P-axes (Wong and others, 1988). In southern California, the direction of maximum principal stress is more northerly. Jones (1988) analyzed groups of focal mechanisms within 10 km of the fault and found P-axes trending generally northerly, 60° to 70° oblique to the WNW strike of the San Andreas in this region.

TABLE 1. QUALITY RANKING SYSTEM FOR STRESS ORIENTATIONS

	A	B	C	D
Focal Mechanism (FM)	Average P-axis or formal inversion of four or more single-event solutions in close geographic proximity (at least one event M≥4.0, other events M≥3.0)	Well-constrained single-event solution (M≥4.5) or average of two well-constrained single-event solutions (M≥3.5) determined from first motions and other methods (e.g., moment tensor wave-form modeling, or inversion)	Single-event solution (constrained by first motions only, often based on author's quality assignment) (M ≥ 2.5) Average of several well-constrained composites (M >2.0)	Single composite solution Poorly constrained single event solution Single event solution for M<2.5 event
Welbore Breakout (IS-BO)	Ten or more distinct breakout zones in a single well with S.D. ≤12° and/or combined length >300 m Average of breakouts in two or more wells in close geographic proximity with combined length >300 m and S.D. ≤12°	At least six distinct breakout zones in a single well with S.D.≤20° and/or combined length >100 m	At least four distinct breakouts with S.D.<25° and/or combined length >30 m	Less than four consistently oriented breakouts or <30 m combined length in a single well Breakouts in a single well with S.D.≥25°
Hydraulic Fracture (IS-HF)	Four or more hydrofrac orientations in single well with S.D.≤12°, depth >300 m Average of hydrofrac orientations for two or more wells in close geographic proximity, S.D.≤12°	Three or more hydrofrac orientations in a single well with S.D. < 20° Hydrofrac orientations in a single well with 20°<S.D.<25°	Hydrofrac orientations in a single well with 20°<S.D. <25°. Distinct hydrofrac orientation change with depth, deepest measurements assumed valid One or two hydrofrac orientations in a single well	Single hydrofrac measurement at < 100-m depth
Petal Centerline Fracture (IS-PO)			Mean orientation of fractures in a single well with S.D.<20°	
Overcore (IS-OC)	Average of consistent (S.D. ≤ 12°) measurements in two or more boreholes extending more than two excavation radii from the excavation wall, and far from any known local disturbances, depth >300 m	Multiple consistent (S.D.<20°) measurements in one or more boreholes extending more than two excavation radii from excavation well, depth >100 m	Average of multiple measurements made near surface (depth >5-10 m) at two or more localities in close proximity with S.D.≤25° Multiple measurements at depth >100 m with 20°<S.D.<25°	All near-surface measurements with S.D. >15°, depth <5 m All single measurements at depth Multiple measurements at depth with S.D.>25°
Fault Slip (G-FS)	Inversion of fault-slip data for best-fitting mean deviatoric stress tensor using Quaternary-age faults	Slip direction on fault plane, based on mean fault attitude and multiple observations of the slip vector. Inferred maximum stress at 30° to fault	Attitude of fault and primary sense of slip known, no actual slip vector	Offset coreholes Quarry pop-ups Postglacial surface fault offsets

TABLE 1. QUALITY RANKING SYSTEM FOR STRESS ORIENTATIONS (continued)

	A	B	C	D
Volcanic Vent Alignment* (G-VA)	Five or more Quaternary vent alignments or "parallel" dikes with S.D. <12°	Three or more Quaternary vent alignments or "parallel" dikes with S.D. <20°	Single well-exposed Quaternary dike Single alignment with at least five vents	Volcanic alignment inferred from less than five vents

S.D. = standard deviation.

*Volcanic alignments must be based, in general, on five or more vents or cinder cones. Dikes must not be intruding a regional joint set.

Zoback and others (1987) explained the observed stress field adjacent to the San Andreas as a consequence of several superimposed effects: current relative plate motions (resulting in a component of convergence across the San Andreas), extension in the Basin and Range province to the east, and a shear strength for the San Andreas fault that is appreciably lower than the level of far-field shear stresses in the crust. The inferred low shear strength for the San Andreas fault is consistent with the implication of relatively low heat flow all along the fault zone (Lachenbruch and Sass, 1973, 1980).

The western boundary of the San Andreas province lies offshore somewhere within the Pacific plate. Available stress information extends only as far west as the zone of high seismicity associated with the fault zone itself; however, seismic reflection data indicate abundant late Neogene folds and faults striking subparallel to the San Andreas on the continental shelf (McCulloch, 1987). The boundary with the Basin and Range province to the east lies somewhere within the seismically quiescent Great Valley because both young fault offsets and focal mechanism data indicate a combination of normal and strike-slip faulting within the Sierra Nevada–Great Basin boundary zone. Focal mechanisms of the 1975 M = 5.7 Oroville earthquake, as well as smaller events in the western Sierra Nevada farther to the south, consistently show normal faulting on north-striking planes; B-axes (assumed to be approximately equal to the maximum horizontal stress direction) vary from N15°W to N11°E (Langston and Butler, 1976; Eaton and Simirenko, 1980; Eaton and others, 1981; J. P. Eaton, written communication, 1986). Within the central Sierra Nevada block, a single hydrofracture measurement and geological observations of conjugate small-displacement strike-slip fault sets (Lockwood and Moore, 1979) suggest a strike-slip stress regime with a NNE to NE S_{Hmax} orientation. However, recent focal mechanisms from the Durwood Meadows area of the southern Sierra Nevada (35.92°N; 118.317°W) indicate nearly pure normal faulting on N-S–striking nodal planes (Jones and Dollar, 1986). Similarly, Lake Tahoe is situated in a late Cenozoic N-S–striking graben within the northern Sierra Nevada block.

Deformation east of the Sierran front (the western physiographic boundary of the northern Basin and Range province) includes both strike-slip and normal faulting with a N to NNE S_{Hmax} orientation and is discussed in more detail in the next section. Because of the contrast between compressional (strike-slip and reverse faulting) deformation in the California Coast Ranges and the focal mechanisms indicating extensional (normal) faulting within the Sierra Nevada and the paucity of data in the intervening Great Valley, we have approximated the eastern boundary of the San Andreas stress province along the axis of the Great Valley in central California. In southern California, the Garlock fault and the western Mojave block contain strike-slip faults subparallel to the San Andreas system and are also included in the San Andreas stress province.

ROCKY MOUNTAIN/INTERMONTANE INTRAPLATE PROVINCE

As the title implies, the deformation in the region included in this broad province occurs within the North American plate; however, it is strongly influenced by interplate (specifically North American–Pacific plate) interaction. Geological data indicate that interaction between these two plates has been a major factor in deformation of the region since middle Tertiary time (Atwater, 1970; Atwater and Molnar, 1973; Stewart, 1978; Eaton, 1979; Zoback and others, 1981). Regionally high elevations (typically 1 to 2+ km above sea level) and a relatively thin crust characterize the entire region, from the Rocky Mountains west to the Sierra Nevada and Cascade Ranges. Various lines of geophysical data indicate that the source of this elevation is an anomalous warm upper mantle (or equivalently thin lithosphere) (Thompson and Burke, 1974; Eaton, 1980).

This plate-tectonic province includes three distinct stress provinces: Cordillera extensional, Colorado Plateau interior, and the southern Great Plains. In our earlier work we attempted further subdivision in the northern half of the region based on scanty and generally poor-quality data. The current synthesis is substantially simplified. Note that the Pacific Northwest region

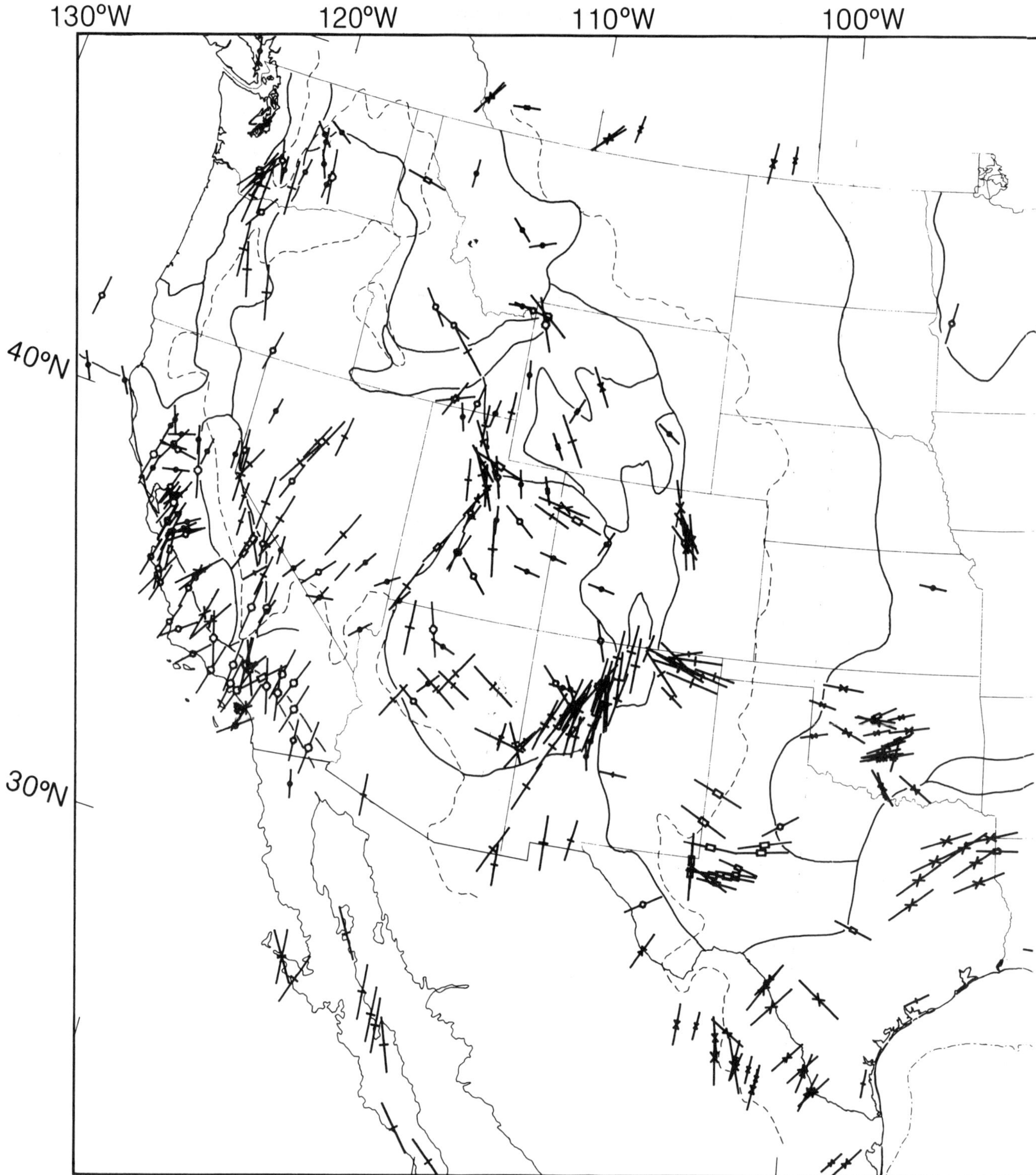

Figure 1. Map of maximum horizontal compressive stress orientations. Solid lines define physiographic province boundaries shown in Figure 2. Quality of data indicated by weight and length of line. As described in the text, three qualities (A, B, C) are plotted. Dashed line marks the 1,100-m elevation contour based on 1° average elevations. The dot-dash line offshore from the eastern United States is the 200-m bathymetric contour, approximating the shelf-slope break.

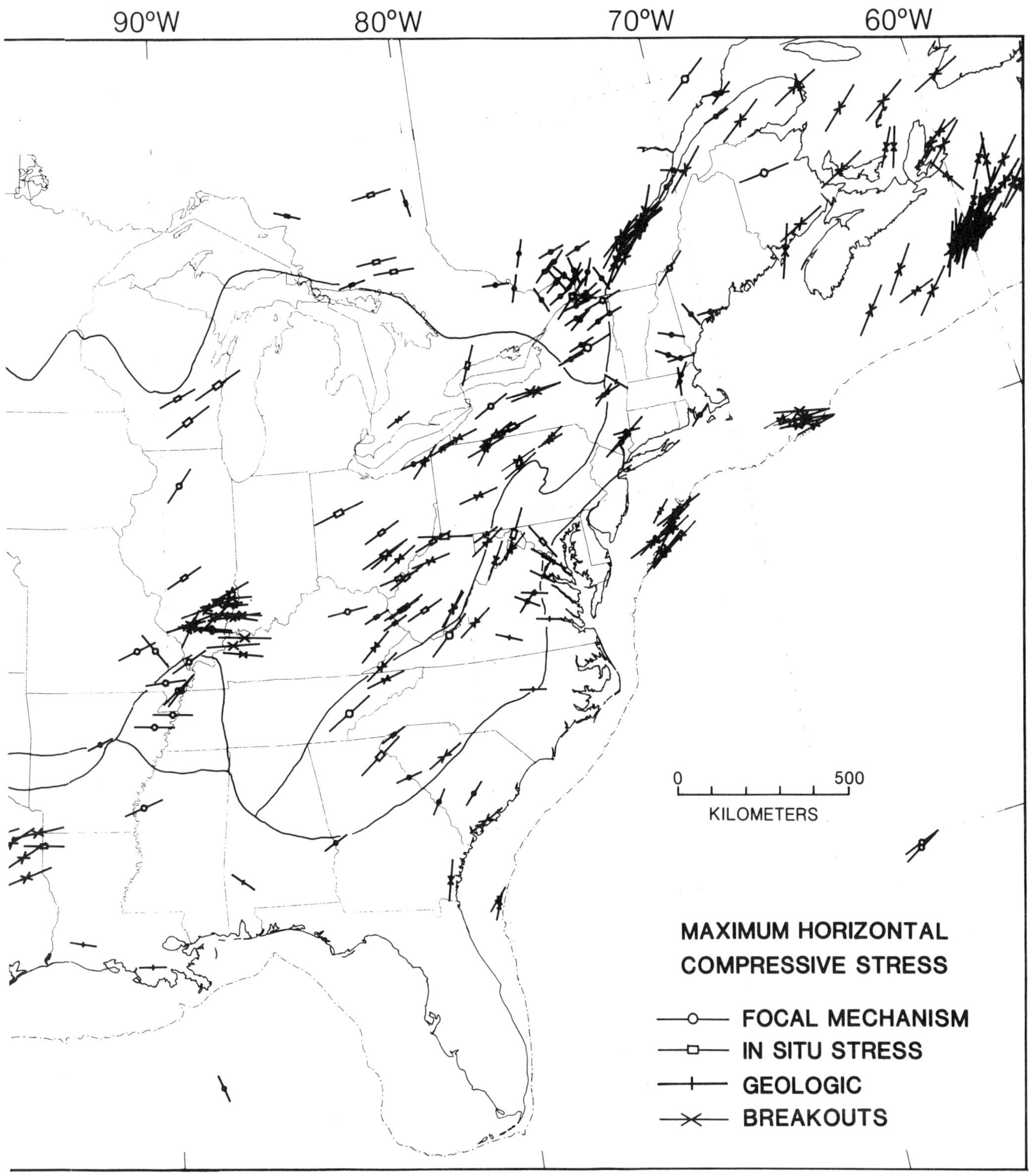
90°W
80°W
70°W
60°W
0
500
KILOMETERS
MAXIMUM HORIZONTAL COMPRESSIVE STRESS
FOCAL MECHANISM
IN SITU STRESS
GEOLOGIC
BREAKOUTS

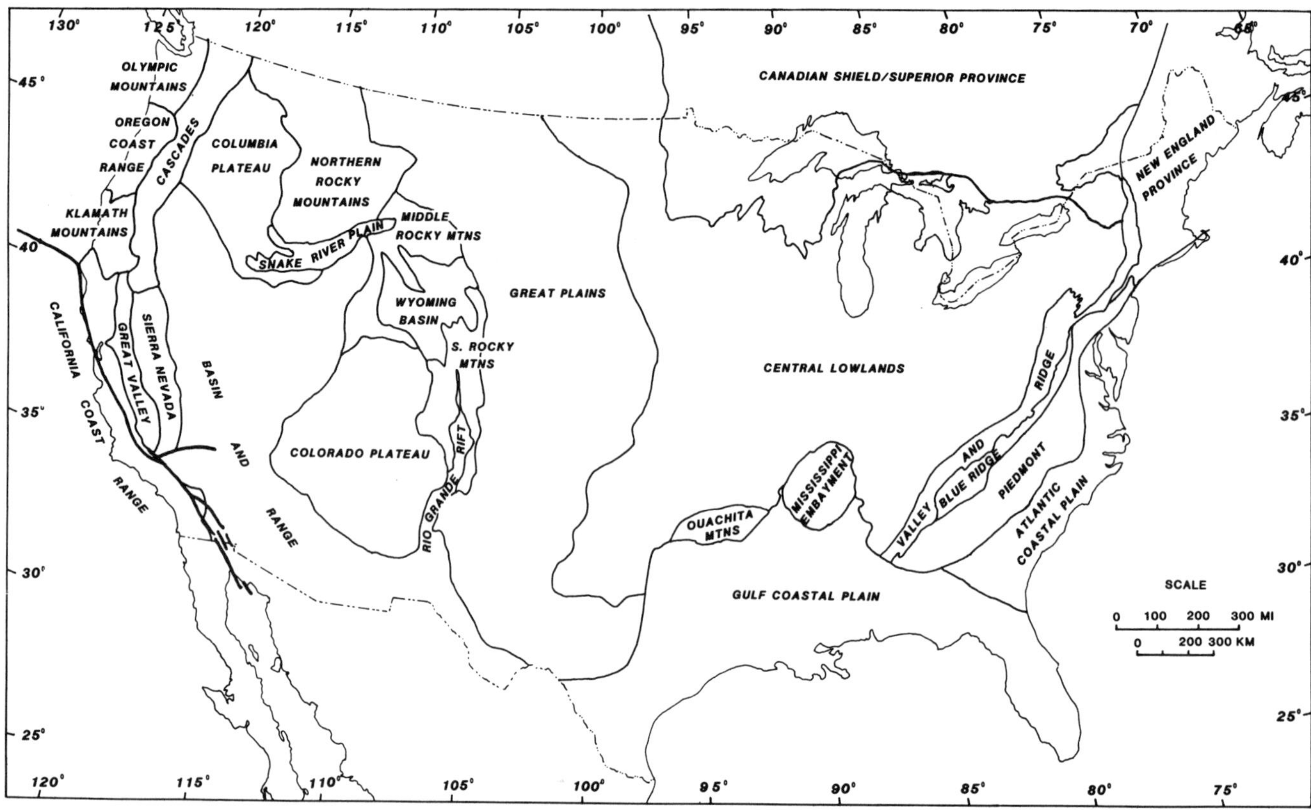

Figure 2. Index map showing physiographic provinces.

(including the Cascade convergent zone) is covered separately in the next section.

Cordilleran extensional province

On the basis of the current stress data set we have greatly increased the size of the "Basin and Range/Rio Grande Rift" stress province to the extent that the earlier name no longer seems appropriate. The new simplified "Cordilleran extensional" province characterized by WNW to ENE–trending S_{hmin} orientations (Fig. 3) includes areas of classic "basin and range" structure (generally north-trending normal faults) in Nevada and parts of Utah, Oregon, Arizona, New Mexico, Colorado, Idaho, and Montana, but also includes adjacent parts of southwestern Wyoming and the Denver basin that do not exhibit characteristic basin-range structure and are considered physiographically to be part of the Rocky Mountains. These latter two areas are tentatively included in this Cordilleran extensional province because the data indicate horizontal stress orientations consistent with nearby regions of the Basin and Range and because available focal mechanisms (although limited) suggest normal faulting. In addition, these two areas (as well as most of the rest of the Cordilleran extensional stress province) coincide with a broad zone of high regional elevation (see the 1,100-m elevation contour, Fig. 1) and heat flow (Eaton, 1979).

The typically north-striking normal faults found within and bounding the range blocks throughout much of this region indicate extensional tectonism that began locally as early as 38 Ma (Elston and Bornhorst, 1979; Reynolds, 1979; Zoback and others, 1981; Ruppel, 1982). Deformation responsible for the modern physiography and structure in both the northern Basin and Range ("Great Basin"—primarily Nevada and western Utah) and the Rio Grande Rift physiographic provinces is probably considerably younger, less than 10 Ma (Stewart, 1978; Zoback and others, 1981; Golombek and others, 1983). The southern Basin and Range province (Arizona and southwestern New Mexico) has been tectonically quiescent for about the past 10 m.y. (Eberly and Stanley, 1978), although moderate, low-level seismicity still persists in this region (Brumbaugh, 1987). Areas of basin and range structure north of the Snake River Plain in Idaho and Montana remain relatively active to the present time, as indicated by widespread evidence of latest Quaternary (last 15,000 yr) faulting (Scott and others, 1985) as well as the 1983 M = 7.3 Borah Peak normal fault earthquake.

As shown in Figure 1, the data indicate S_{Hmax} (intermediate principal stress) orientations ranging between about N and NE in the Rio Grande Rift and most of the Basin and Range province south of the Snake River Plain about 40°N. The data within and north of the Snake River Plain, southern Wyoming, and east-central Colorado indicate S_{Hmax} orientations between N and NW. The least horizontal stress (extension) directions shown in Figure 3 are orthogonal to the orientations plotted in Figure 1.

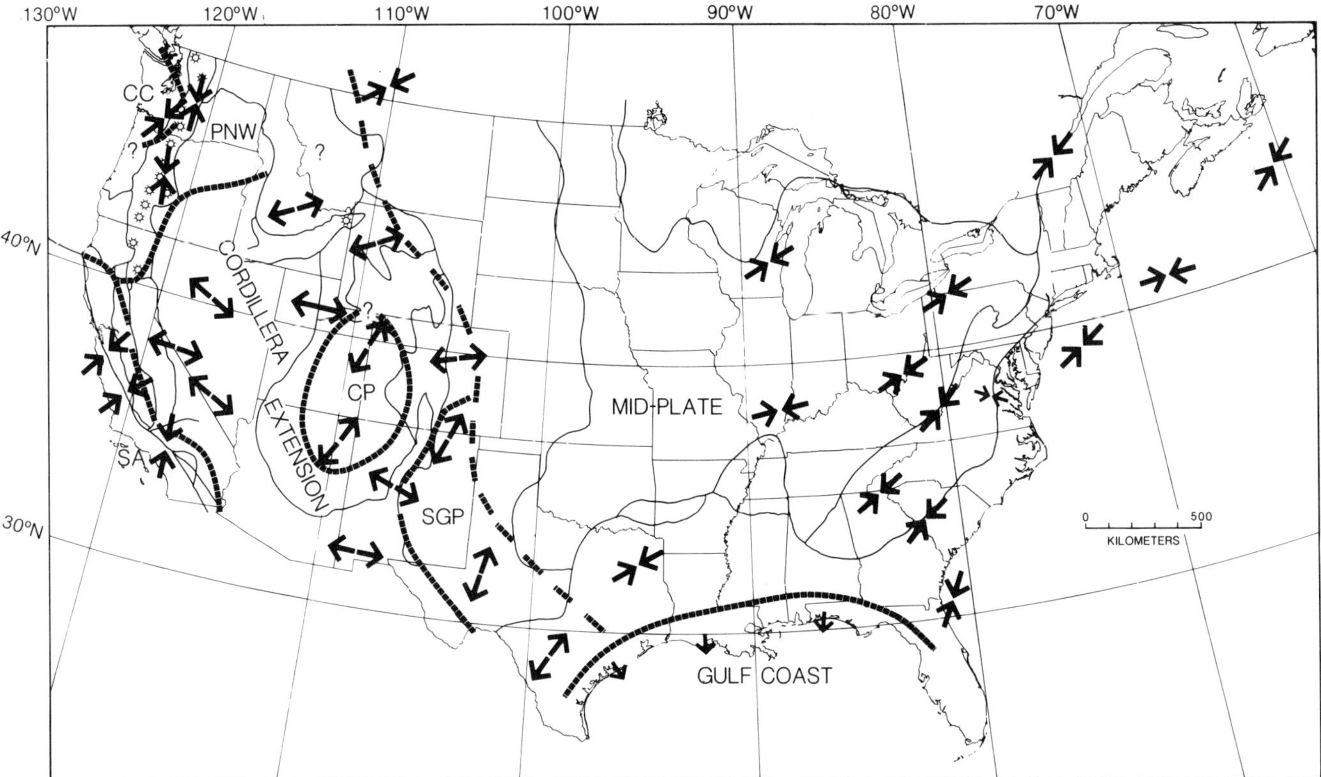

Figure 3. Generalized stress map for the continental United States. Outward-pointing arrows are given for areas characterized by extensional deformation. Inward-pointing areas are shown for regions dominated by compressional tectonism (thrust and strike-slip faulting). Stress provinces are delineated by the thick dashed lines: CC = Cascade convergent province; PNW = Pacific Northwest; SA = San Andreas province; CP = Colorado Plateau interior; SGP = Southern Great Plains.

There also appears to be a tendency for stress orientations near the physiographic boundary between the Basin and Range/Rio Grande Rift provinces and the Colorado Plateau and North Rocky Mountains province to be aligned parallel and perpendicular to the local trend of the boundary and often topography.

A small local anomaly in this regional stress pattern occurs in the area directly north of Yellowstone where earthquake focal-mechanism data suggest approximately N-S extension (E-W S_{Hmax} orientations). This area, as well as part of the adjacent Idaho batholith, was included in the Hegben Lake/Centennial Valley stress province of Zoback and Zoback (1980). Continuity of the regional S_{Hmax} orientations between N and NW around this local anomaly and a revision, based on surface-wave modeling (Patton, 1985), of the single focal mechanism in the Idaho batholith, resulted in elimination of the Hegben Lake/Centennial Valley stress province. Thus, in contrast to our previous interpretation, it now appears that a region of relatively uniform, approximately E-W extensional stresses surrounds the Colorado Plateau interior.

The regional variations in the pattern of stress in the northern Basin and Range province (Nevada and Utah) shown in Figure 1 are summarized in a recent paper (Zoback, 1989). The abundance of data in this well-studied area permit the establishment of constraints on the relative magnitude of the principal stresses, as well as on their orientations. As noted previously (Wright, 1976; Zoback and Zoback, 1980), the presence of both strike-slip and normal surface faulting and focal mechanisms along the western margin of the northern Basin and Range province suggest a stress field of the form:

$$S_v \approx S_{Hmax} >> S_{hmin} \text{ or}$$
$$S_1 \approx S_2 >> S_3.$$

In such a stress field it is possible for normal, strike-slip, or oblique fault slip to occur, depending primarily on the orientation of the fault plane (cf. Angelier, 1979, Fig. 4). Thus, in many areas a variety of strike-slip and normal-faulting focal mechanisms are observed that are simply associated with slip on preexisting faults of varying orientations reactivated in response to the same regional stress field. This appears to be the present situation at the Nevada Test Site area in southern Nevada (Hamilton and Healy, 1969; Stock and others, 1985).

Detailed observations of faulting in the region of combined normal and strike-slip faulting along the western side of the province (including the Sierra Nevada–northern Basin and Range boundary zone and the Walker Lane belt of western Nevada, see Stewart, 1988) can constrain both stress orientation and relative magnitudes. Limits on variations in the stress field are simply illustrated with field observations from a single site, the Owens Valley area of easternmost central California, the site of M ≈ 7.7 earthquake in 1872. Although a prominent NNW–trending east-facing vertical scarp (typically 1 m high) was formed during the earthquake, a recent careful field investigation of the fault zone indicates that for most of the fault trace the dominant sense of offset was right-lateral strike-slip, with an estimated ratio of lateral to vertical offset during the 1872 event ranging between 4:1 and 10:1 and averaging 6:1 (Beanland and Clark, 1989; Lubetkin and Clark, 1988). For much of its length the fault trends NNW through Owens Valley, and is located 5 to 20 km east of the subparallel Independence fault, which is a part of the Sierran frontal normal fault zone. Late Pleistocene and Holocene slip on the Independence fault has been dominantly dip-slip; no evidence of lateral slip has been detected (Gilespie, 1982).

An analysis of stress and slip directions (Zoback, 1989) indicates that the two observed contrasting styles of offset on the subparallel faults can be explained with large fluctuations in the relative magnitude of S_{Hmax}. For a consistent S_{hmin} orientation of N80°W (suggested from nearby data), the nearly pure normal dip-slip on the Independence fault requires that S_{Hmax} must be close in magnitude to S_{hmin} ($S_v >> S_{Hmax} \approx S_{hmin}$), whereas the large component of strike-slip offset on the Owens Valley fault zone in 1872 requires a regime in which $S_{Hmax} \approx S_v >> S_{hmin}$. Thus, it appears that the approximatly N-S–oriented principal stress (S_{Hmax}) has had large temporal fluctuations in relative magnitude, possibly accompanied by small horizontal rotations of the principal stresses. The time scale of these fluctuations may be tens of thousands of years.

Currently the Wasatch front region of north-central Utah appears to be one area within the Basin and Range province where the two horizontal stresses have approximately equal magnitudes ($S_{Hmax} \approx S_{hmin}$). Evidence for this comes from conflicting hydrofracture orientations from nearby wells (Haimson, 1981, 1984; Zoback, 1984, 1989), stress magnitudes measured by hydraulic fracturing (Haimson, 1984, 1985), poorly defined and inconsistently oriented wellbore breakouts (Zoback, 1984), and nearly pure normal dip-slip on faults of a wide variety of trends (both surface-slip vectors and focal mechanisms) (Zoback, 1983, 1984, 1989). Superposed striae on a fault surface near Salt Lake City indicate temporal variations, suggesting that the modern stress regime characterized by approximately equal horizontal stresses ($S_{Hmax} \approx S_{hmin}$) may not be the representative long-term pattern of stresses in this area (Zoback, 1989).

Colorado Plateau interior

The available data, though rather sparse, indicate that the interior portion of the Colorado Plateau is characterized by a distinct WNW S_{Hmax} orientation, orthogonal to that in the surrounding Cordilleran extensional province. New focal-mechanism from the Plateau interior, however, suggest that the stress regime is extensional (Wong and Humphrey, 1989) and not compressional as previously reported (Thompson and Zoback, 1979; Zoback and Zoback, 1980). These new, small-magnitude events show primarily normal faulting with some shallow strike-slip faulting events (Wong and Humphrey, 1989). True tectonic compression (reverse faulting) is observed only in an area of relatively shallow (generally, <3 km), apparently mining-induced seismicity in east-central Utah, and P-axes within this region show a 50° variation in strike (Wong, 1985).

The previously proposed compressive Plateau interior stress regime (Thompson and Zoback, 1979; Zoback and Zoback, 1980), which was based primarily on the mining-induced events in east-central Utah and relatively shallow strike-slip stress indicators in the Rangely/Piceance basin area of northwestern Colorado, now seems best reinterpreted as an extensional stress regime characterized by a combination of normal and strike-slip faulting, albeit with local variations in relative stress magnitudes. The difficulty in interpreting the tectonic stress field in this region and understanding the apparent variations in the stress field may be related to generally low differential stresses suggested by the absence of major faulting or seismicity within the Plateau interior (Zoback and Zoback, 1980). All the data do, however, convincingly demonstrate that the maximum horizontal stress in this region is WNW, approximately orthogonal to that in the surrounding regions.

This stress province is distinctly smaller than the Colorado Plateau physiographic province, as Basin and Range–style extensional tectonism has encroached upon the margins of the Plateau proper (Best and Hamblin, 1978; Thompson and Zoback, 1979; Zoback and Zoback, 1980). The transition between the Cordilleran extensional province and the Plateau interior may be quite broad (100 to 150 km), particularly in Arizona where the nearly orthogonal S_{Hmax} orientations of the two provinces are observed in closed proximity in several areas (see Fig. 1). Aldrich and Laughlin (1984) have studied the plateau interior boundary in westernmost Arizona and New Mexico and have suggested that it is coincident with a 50-km-wide Precambrian province boundary.

Southern Great Plains

The southern Great Plains stress province appears to form a boundary zone between the active extensional tectonism of the western Cordillera and the relatively stable midplate region of the central and eastern United States. Contrasts in stress orientations and a consistent indication of extensional tectonism (basaltic volcanism and normal fault focal mechanisms) distinguish this area from the relatively stable midcontinent region adjacent to the east.

This province generally coincides with the major topographic gradient (about 100 m/225 km) separating the thermally elevated western Cordillera from the midcontinent area. A lack of data preclude delineation of the northern extent of this province;

however, on the basis of stress orientations, the Denver basin area of central Colorado does not appear to be part of this province. Changes in crustal thickness are not sufficient to explain this major topographic gradient in the southern Great Plains area (Braile, this volume); thus much of the decay in topography is probably due to eastward lithospheric thickening beneath the central United States (Zoback and Lachenbruch, 1984; Iyer, this volume).

PACIFIC NORTHWEST AND CASCADE CONVERGENCE ZONE

The active andesitic volcanoes of the Cascade Range are a manifestation of subduction of the Juan de Fuca or "Cascadia" plate. In early Tertiary time the Juan de Fuca plate was being subducted beneath most of the western Cordillera of Canada and the United States (Atwater, 1970; Engebretson and others, 1984). The convergence rate between the Juan de Fuca and North American plates, however, was greater than the spreading rate along the Juan de Fuca–Pacific ridge (East Pacific Rise); hence the Juan de Fuca plate was gradually consumed, bringing the North American and Pacific plates into contact. The relative motion between the Pacific and North American plates resulted in development of the San Andreas and Queen Charlotte (Canada) transform systems beginning about 30 Ma. At present these transform systems dominate the 2,600-km Pacific–North American plate boundary between Alaska and the Gulf of California. The Cascade volcano chain extends roughly 1,000 km from Mt. Lassen in California through Oregon and Washington to Garibaldi Peak in southern British Columbia. Thus, the current zone of subduction represents less than 40 percent of the length of the western North America plate boundary between Mexico and Alaska.

Whereas the volcanic belt appears to be a clear indication of subduction, supporting geophysical data are more ambiguous. The trench offshore is poorly developed bathymetrically and seismically, although refraction data clearly indicate its presence (Taber, 1983; Taber and Smith, 1985). Deep earthquakes (>30 km) do occur beneath northwestern Washington and southern British Columbia; these are primarily normal-faulting events believed related to the bending in the subducted slab (Taber and Smith, 1985). The subduction zone may be poorly defined geophysically because of the limited size of the subducted plate and its extremely young age (oldest magnetic anomalies in oceanic crust at the current "trench" location are only 9 to 10 m.y. old).

Evidence on the state of stress in the crust of this region is sparse and restricted to relatively few upper-crustal earthquake focal mechanisms and volcanic vent alignments, most of which indicate a roughly N-S S_{Hmax} orientation in central Oregon and a NNE orientation in southern Washington (Fig. 1). However, Riddihough (1977) has estimated the Juan de Fuca–North American convergence direction to be N50°E; geodetic data from the Puget Sound region indicate a maximum compressive strain accumulation in a N71°E ± 6° direction (Savage and others, 1981). The existing data, together with the lack of focal mechanisms from the coastal regions of Washington and Oregon, led previous workers to include that the entire Pacific Northwest region is a single stress province characterized by N-S compression (Smith, 1977; Zoback and Zoback, 1980; Sbar, 1982) and apparently unrelated to the NE-SW convergence with the Juan de Fuca plate. Recent data, however, suggest reinterpretation of this region.

Cascade convergent zone. Weaver and Smith (1983), Yelin and Crosson (1982), and Weaver and others (1987) have presented focal mechanisms in western Washington and Oregon with distinct northeast-trending (average about N40°E) P-axes. These mechanisms are all strike-slip events; one is beneath Puget Sound (Yelin and Crosson, 1982) and several occur along the 90-km-long NNW–trending St. Helens seismic zone along the west side of the Cascades (Weaver and Smith, 1983; see Dewey and others, this volume). A common characteristic of all these southwestern Washington and northwestern Oregon strike-slip earthquakes is that one of the nodal planes strikes between N-S and N20°W. Thus, it is unlikely that these events could be consistent with the general pattern of N-S compression inferred from events farther to the east and from volcanic vent alignments to the south. Several focal mechanisms in southern Vancouver Island, Canada also indicate NE to ENE compression, suggesting that the Cascade convergent zone may extend into southernmost Canada (see Fig. 3) (Adams, 1987).

Weaver and Michaelson (1985) and Weaver and Smith (1983) have suggested that the subducting Juan de Fuca plate beneath the Pacific Northwest may be divided into two segments separated by a broad NE-striking boundary between Mt. Hood (Oregon) and Mt. Rainier (Washington). Differences in earthquake distributions and late Cenozoic and Quaternary volcanism led Weaver and Michaelson (1985) to suggest relatively low-angle subduction and strong coupling in the northern segment (resulting in the observed NE compression) in contrast to the southern segment where the subducting slab may have broken off and coupling between the two plates is weak. Thus, the observed NE compression in western Washington may define a distinct stress province characterized by strike-slip faulting and a NE S_{Hmax} orientation related to Juan de Fuca–North American convergence.

Pacific Northwest. The balance of the Pacific Northwest is included in a stress province characterized by N-S compression but whose boundaries are poorly defined. As discussed in Zoback and Zoback (1980), deformation in this region consists of both strike-slip and reverse focal mechanisms and late Tertiary to Quaternary folding along E-W axes in the Columbia basin to the east.

The N-S orientation of the maximum horizontal stress indicates that the state of stress in this broader region, encompassing most of the Pacific Northwest, may be more controlled by the larger scale transform motion between the Pacific and North American plates than by the Juan de Fuca–North American convergence. Spence (1989) has suggested that collision of the northwestward-moving Pacific plate with the Gorda–Juan de

Fuca–Explorer plate system causes northward compression of this offshore plate system that is transferred to the overriding North American plate, creating the N-S compression in the Pacific Northwest.

MIDPLATE CENTRAL AND EASTERN UNITED STATES

Consideration of available focal mechanisms, in situ stress measurements, and geological observations led Sbar and Sykes (1973) to suggest that an area extending from west of the Appalachian Mountains to the middle of the continent from southern Illinois to southern Ontario was characterized by a relatively uniform NE- to E-trending maximum compressive stress. Additional hydraulic fracturing stress measurements in the Great Lakes area by Haimson (1977) further documented the consistent NE to ENE orientation of the maximum horizontal stress in the midcontinent. Expanding on this, in our 1980 paper we concluded that a large portion of the central U.S. (roughly from the Great Plains east to the Appalachians) was characterized by a relatively uniform compressive stress field with a S_{Hmax} oriented NE to ENE. Thus, in addition to the uniformity of stress orientation, there seems to be uniformity of relative stress magnitudes. New data have expanded this province to include much of Canada (Adams, in press) as well as the eastern seaboard of the United States (Zoback and others, 1986).

The data available for the eastern seaboard region as of 1980 consisted primarily of young fault offsets and some poorly constrained earthquake focal mechanisms and ambiguous hydraulic fracturing data. The largest and most consistent body of data was the young reverse faults on the Coastal Plain (Prowell, 1983; Wentworth and Mergner-Keefer, 1983). These faults typically strike N to NE, dip steeply, and offset sedimentary rocks of Late Cretaceous to Miocene age. Because of the poor exposure of these faults, only the vertical component of offset is generally detected. However, true offsets, when observed, are typically dip-slip (D. Prowell, oral communication, 1984), although minor lateral offsets of both right-lateral and left-lateral sense have been reported.

The general N to NE strike of the faults with Miocene and younger offset led Zoback and Zoback (1980) and Wentworth and Mergner-Keefer (1983) to suggest a NW S_{Hmax} direction for the Atlantic Seaboard. Further support for a distinct "Atlantic Coast" stress province came from a compilation of earthquake focal mechanisms near the New York–New Jersey area that consistently indicated reverse faulting on NE-striking nodal planes (Yang and Aggarwal, 1981).

A large body of new data are incompatible with a separate Atlantic Coast stress province characterized by NW compression. The new data include a large number of wellbore breakouts both onland and on the continental shelf (Plumb and Cox, 1986; Dart and Zoback, 1987); in situ stress (hydraulic fracturing) studies at several localities (Zoback and others, 1985b; M. D. Zoback and others, 1986); and better constrained earthquake focal mechanisms, particularly for the New York–New Jersey area (Seboroski and others, 1982, Houlday and others, 1984; Quittmeyer and others, 1985; see also the discussion of the validity of some of the earlier focal mechanisms in M. L. Zoback and others, 1986, p. 307) and the Charleston area (Talwani, 1982). Analysis of rather variable focal mechanisms in the New England area indicates a mean best-fitting regional stress tensor with an S_{Hmax} orientation of about E-W (Gephart and Forsyth, 1985). Thus, the bulk of the data consistently indicate a S_{Hmax} orientation of NE to nearly E-W for the eastern seaboard and continental shelf (Fig. 1).

We have retained in the data set stress orientations inferred from the young reverse faults that offset Miocene and younger rocks. These data, a large number of which are in eastern Virginia (note the small area with NW compression indicated on Figs. 1, 3), appear incompatible with the new data on the regional stress field. Because most of these faults strike within 20° or less of the inferred S_{Hmax} direction (based on nearby wellbore breakouts in northwestern Virginia, and West Virginia), it is unlikely that this incompatibility can be explained by large amounts of undetected strike-slip movement on these faults.

Thus, stress orientation data from the Great Plains east to the Atlantic continental margin indicate a generally consistent S_{Hmax} orientation between NE and E and averaging about ENE; a similar pattern is noted throughout most of Canada (Gough and others, 1983; Hasegawa and others, 1985; M. L. Zoback and others, 1986). This compressive midplate stress province may also extend eastward throughout much of the western Atlantic basin, based on a few focal mechanisms from that area (M. L. Zoback and others, 1986), most notably the 1978 Bermuda rise event with a P-axis oriented N60°E (Stewart and Helmberger, 1981; Nishenko and Kafka, 1982).

SOURCES OF STRESS

We believe that the majority of the broad, regionally uniform in situ stress field, as mapped with the various types of indicators, is tectonic in origin. Discussion of the sources of stress can be neatly divided into two broad regions: a central and eastern United States midplate region and the western Cordillera region (western Great Plains and regions to the west).

Central and eastern United States

As mentioned above, the central and eastern United States appear part of a broad midplate compressive stress province that includes most of Canada and possibly also much of the western North Atlantic basin—to within about 250 km of the Mid-Atlantic ridge (M. L. Zoback and others, 1986). S_{Hmax} orientations within the province vary between NE and E and average about ENE. While localized stresses may be important in places, the overall uniformity in the midplate stress pattern suggests a far-field source. As the North American plate essentially lacks any attached subducting slab, the most likely plate-driving forces are ridge push and basal drag. It is unlikely that the stress field in the central and eastern United States is influenced by any resistive shear forces on the San Andreas transform; shear stresses generated by Pacific–North American interaction are probably not transmitted through the thermally weakened, actively deforming western Cordillera.

Reding (1984) investigated the state of stress in the North American plate with a single-plate elastic finite-element analysis, using both ridge-push models (in which ridge-push forces were distributed over the oceanic lithosphere) and driving basal-drag models. Net torque on the plate was balanced by resistance provided by either pinning the western plate margin or by allowing resistive basal-drag forces (in the ridge-push case). The predicted maximum horizontal stress orientations were ENE, in good agreement with the observations, and were nearly identical for both the ridge-push and driving basal-drag models. Thus, the stress orientation data base for the North American plate does not allow discrimination between these two models.

Nevertheless, there are two general observations that lead us to prefer ridge push over basal drag as the primary source of stress in the midplate region. In a lower-mantle reference frame, cratonic North America is moving southwesterly (Minster and Jordan, 1978). Minster and others (1974) noted that the speed of a plate varies inversely with its continental area and suggested that deep roots beneath continents increase the basal drag and hence slow the lateral movement. Drag-related compression associated with thick cratonic lithosphere moving southwesterly through relatively passive asthenosphere presumably would be most pronounced along the leading southwestern margin of the craton (southern Great Plains area). However, the observed N-S extensional stress regime in the southern Great Plains stress province demonstrates that drag-related compression at the leading edge of the continent is not occurring. Furthermore, seismic activity generally increases from west to east within the craton (from Great Plains region east) rather than decrease from this "leading edge," which would appear to argue against anomalously high compressive stress associated with the motion of the North American lithosphere with respect to the asthenosphere.

Western United States

The state of stress in the western United States is believed to be closely linked to plate interactions along the western margin of the continent, presently the NW-directed relative motion of the Pacific plate with respect to the North American plate and the NE convergence between the Juan de Fuca and North American plates. Prior to the late Cenozoic development of the San Andreas transform fault (post-30 Ma), subduction occurred beneath much of the western United States (Atwater, 1970; Atwater and Molnar, 1973; Engebretson and others, 1985). Extensional tectonism in the Basin and Range and Rio Grande Rift provinces and regions north of the Snake River Plain was initiated during this period of subduction, primarily in an "intra-arc" setting, as inferred from contemporaneous calc-alkaline magmatism (Eaton, 1978, 1979; Elston and Bornhorst, 1979; Reynolds, 1979; Zoback and others, 1981). This earlier history of subduction and extensive invasion of the western United States lithosphere by magmatism profoundly influenced the lithosphere and upper-mantle structure of the region as inferred from the regional topographic high, which begins in the central Great Plains and extends westward to include the Sierra Nevada range (see 1,100 m contour in Fig. 1). This elevation anomaly is probably best developed in the northern Basin and Range, where broad regions of crust with a thickness of 30 to 33 km stand at an average elevation of about 1,700 m above sea level. In areas within this elevation anomaly where thermal equilibrium has probably been obtained (notably the Basin and Range and Rio Grande Rift province, and excluding the Sierra Nevada and Colorado PLateau), heat-flow values are high (averaging 80 to 100 m w/m^2) (Lachenbruch and Sass, 1978).

It appears that the modern state of stress throughout the western United States is influenced not only by present-day plate motions, but also—in some complex manner—by lateral variations in upper-mantle structure that are probably related to earlier plate interactions. Two regions characterized by extensional tectonism oriented nearly orthogonal to surrounding regions, the Colorado Plateau interior and the southern Great Plains, are both areas of proposed relatively abrupt lithospheric thickening (Porath, 1971; Thompson and Zoback, 1978; Zoback and Lachenbruch, 1984; Iyer, this volume). The least horizontal stress orientation in both these regions is generally perpendicular to the extension direction in the surrounding actively extending areas.

Of course, the most obvious manifestation of the influence of modern plate interactions is the deformation along the San Andreas fault. The relative transform motion between the North American and Pacific plates results in a strike-slip stress regime. Based on global plate kinematics for the last 3 m.y., Minster and Jordan (1978) predicted a relative velocity in central California between the Pacific–North American plates of 56 ± 3 mm/yr, N35° ± 2°W (RM2); they estimated the actual slip vector from Holocene geological data in the same region to be 34 ± 3 mm/yr, N41° ± 2°W (Minster and Jordan, 1984). Including estimates of Basin and Range extension, Minster and Jordan (1984) concluded that deformation west of the San Andreas fault must include 4 to 13 mm/yr of crustal shortening orthogonal to the fault. As noted above, regional earthquake, geological, and well-bore breakout data in the Coast Ranges both east and west of the fault suggest a general NE–SW trend for S_{Hmax}, approximately perpendicular to the San Andreas. The compression across the San Andreas fault may be related to a change in relative motion between the North American and Pacific plates tied to recent changes in the absolute motion of the Pacific plate (Cox and Engebretsen, 1985; Pollitz, 1986). Cox and Engebretson (1985) indicated that there has been a 17° clockwise rotation of the Pacific–North American relative motion vector in the past 5 m.y., which would result in a component of convergence across the San Andreas in central California.

DISCUSSION

The stress data base has been significantly improved as a result of several factors. First, the establishment of wellbore breakouts as reliable indicators of horizontal stress orientation (Bell and Gough, 1979; Gough and Bell, 1981, 1982; Hickman and others, 1985; Zoback and others, 1985a) has greatly increased the size of the data base, particularly for the eastern United States. These data are interpreted from high-resolution dipmeter logs, a log frequently run in petroleum exploration wells, or from acoustic borehole televiewer logs, a less widely used but more sensitive instrument (Zoback and others, 1985a). In each well, the statistical weight of multiple determinations of stress orientations, often made over a significant range of depths, makes this technique one of the best indicators of horizontal stress orientation. Thus, this stress indicator offers the opportunity to obtain, relatively inexpensively, high-quality new data on stress orientations wherever there are petroleum exploration wells.

A second factor that has greatly improved the data base is the introduction of the quality-ranking system (Table 1). The significance of single points in areas of sparse data coverage is now much clearer. In addition, interpretation of sometimes conflicting data in regions of dense coverage is also simplified.

The use of higher quality earthquake focal mechanisms has also improved the data base. Better constrained mechanisms are frequently reported now as a result of the use of wave-form modeling (both body and surface waves), P/SV ratios, and systematic grid searches seeking all possible solutions. While all these tools help to improve the focal mechanisms, the inherent ambiguity in inferring stress orientation from P- and T-axes remains a problem.

The pattern of stress provinces inferred from the new data base are somewhat simplified from our earlier study, both in the western and eastern United States. One of the most significant changes in our interpretation of the stress data is the elimination of an Atlantic Coast stress province characterized by NW compression. The new data indicate a remarkable uniform S_{Hmax} orientation (between NE and E) for all of the U.S. east of the Great Plains, including the continental margin. This uniform midplate stress province includes most of Canada east of the westernmost Cordillera.

Tertiary vertical offsets on generally NE-striking steep reverse faults in the Atlantic Coastal Plain reported by Prowell (1983) are inconsistent with this uniform midplate NE to ENE S_{Hmax} orientation. As noted previously, their strike precludes their activation as primarily strike-slip faults even if large lateral offsets on these faults have gone undetected. At present we can offer no good explanation for these NW-striking reverse faults other than to note that nowhere have these young faults been identified as being seismically active. It has been suggested that they may represent deformation in an earlier stress regime; however, we consider this highly unlikely as the inferred source of the broad midplate stress field is believed to be plate tectonic in origin, and there have not been major changes in plate geometry in post-Miocene time. Most of these young fault offsets have been observed near the Fall Line, the boundary between the coastal-plain sediments and basement of the Piedmont, generally a major break in topography. Possibly these primarily vertical offsets on steep faults parallel to the Appalachian trend are a result of the same mechanism responsible for a hypothesized young uplift of parts of the Appalachian Mountains (Hack, 1979). In the central Virginia area where the faults are best documented, focal mechanisms of microearthquakes seem to indicate both NE- and NW-striking P-axes (Munsey and Bollinger, 1985), although there has been some dispute of the validity of the composite mechanisms with NW P-axes (Nelson and Talwani, 1985).

CONCLUSIONS

An almost two-fold increase in the number of data points substantiates earlier studies that indicate regionally uniform stress orientations throughout the upper crust. These data and adoption of a quality-ranking system have enabled us to modify and somewhat simplify previously defined stress provinces in the continental United States. The most important differences between this data compilation and interpreted stress provinces reported here and those of Zoback and Zoback (1980) are as follows:

1. General ENE compression associated with the midcontinent stress field now appears to extend all the way to the Atlantic continental margin and possibly well into the western Atlantic basin. A distinct Atlantic Coastal Plain stress province (characterized by northwest compression) is not supported or justified by the available data.

2. Data defining NE compression associated with subduction of the Juan de Fuca plate are observed in western Washington and Oregon, and these areas make up the Cascade convergence zone stress province.

3. The extensional stress field associated with the Basin and Range and Rio Grande Rift is now observed in central Colorado and western Wyoming. Thus, an approximately E-W–oriented extensional stress field is generally associated with the broad uplifted region of the western Cordillera and may completely surround the Colorado Plateau.

4. The previously reported WNW S_{Hmax} orientation within the Colorado Plateau interior has been further substantiated; however, the plateau interior now appears to be characterized by an extensional, not a compressive, stress regime (Wong and Humphrey, 1989).

Earthquake focal mechanisms, geological, and wellbore breakout data adjacent to the San Andreas fault suggest generally northeast (fault-normal) compression in central California, in contrast to a N-trending S_{Hmax} orientation that would be inferred from the right-lateral strike-slip faulting along the NW-striking fault itself. The fault-normal compression may be the result of very low shear strength on the San Andreas fault coupled with a clockwise change in Pacific–North American relative plate motion in the past 5 m.y., as well as extensional processes in the Basin and Range.

In many regions the substantially enlarged data set confirms the generally broad-scale, tectonic origins of the stress field in the various provinces defined. The expanded data set also defines numerous localized, relatively small-scale variations in the stress field within individual provinces. Improved resolution and future study of these variations will yield new insight into the forces that deform the crust.

ACKNOWLEDGMENTS

We express our gratitude to the many workers who have contributed data to this compilation. Since this work does not include a listing of individual data points, many of these contributors have gone unacknowledged. Special thanks to Ivan G. Wong for providing numerous new data still in press, and to Bob Herrmann, Lou Pakiser, and an anonymous reviewer for helpful comments and criticisms. We also thank Mara E. Schiltz for maintaining the stress data base and generating the stress map.

REFERENCES CITED

Adams, J., 1987, Canadian crustal stress data; A compilation to 1987: Geological Survey of Canada Open-File Report 1622, 130 p.

—— , 1989, Crustal stresses in eastern Canada, *in* Gregersen, S., and Basham, P. W., eds., Earthquakes at North Atlantic passive margins. Neotectonics and postglacial rebound, Proc. NATO Advanced Workshop, Vordingborg, Denmark: Kluwer Academic Publishers, in press.

Aldrich, M. J., Jr., and Laughlin, A. W., 1984, A model for the tectonic development of the southeastern Colorado Plateau boundary: Journal of Geophysical Research, v. 89, p. 10207–10218.

Angelier, J., 1984, Tectonic analysis of fault slip data sets: Journal of Geophysical Research, v. 89, p. 5835–5848.

Atwater, T., 1970, Implications of plate tectonics for the Cenozoic tectonic evolution of western North America: Geological Society of America Bulletin, v. 81, p. 3513–3536.

Atwater, T., and Molnar, P., 1973, Relative motion of the Pacific and North American plates deduced from seafloor spreading in the Atlantic, Indian, and south Pacific Oceans, *in* Kovach, R. L., and Nur, A., eds., Proceedings of the Conference on Tectonic Problems of the San Andreas Fault System, 13: Stanford, California, Stanford University Publications, p. 136–148.

Beanland, S., and Clark, M. M., 1989, The Owens Valley fault zone, eastern California, and surface rupture associated with the 1872 earthquake: U.S. Geological Survey Bulletin, in press.

Bell, J. S., and Gough, D. I., 1979, Northeast-southwest compressive stress in Alberta; Evidence from oil wells: Earth and Planetary Science Letters, v. 45, p. 475–482.

Best, M. G., and Hamblin, W. K., 1978, Origin of the northern Basin and Range province; Implications from the geology of its eastern boundary: Geological Society of America Memoir 152, p. 313–340.

Bollinger, G. A., and Wheeler, R. L., 1982, The Giles County, Virginia, seismogenic zone; Seismological results and geological interpretations: U.S. Geological Survey Open-File Report 82-585, 136 p.

Brumbaugh, D. S., 1980, Analysis of the Williams, Arizona, earthquake of November 4, 1971: Bulletin of the Seismological Society of America, v. 70, p. 885–891.

—— , 1987, A tectonic boundary for the southern Colorado Plateau: Tectonophysics, v. 136, p. 125–136.

Cox, A., and Engebretson, D. C., 1985, Change in motion of Pacific plate at 5 m.y. BP: Nature, no. 313, p. 472–474.

Dart, R., 1985, Horizontal-stress directions in the Denver and Illinois basins from the orientations of borehole breakouts: U.S. Geological Survey Open-File Report 85-733, 41 p.

Dart, R., and Zoback, M. L., 1987, Principal stress orientations on the Atlantic continental shelf inferred from the orientations of borehole elongations: U.S. Geological Survey Open-File Report 87-283, 43 p.

Dean, C. S., and Overbey, W. K., Jr., 1980, Possible interaction between thin-skinned and basement tectonics in the Appalachian basin and its bearing on exploration for fractured reservoirs in the Devonian shale, *in* Wheeler, R. L., and Dean, C. S., eds., Proceedings, Symposium on Western limits of detachment and related structures in the Appalachian foreland, Chattanooga, Tennessee, April 6, 1978: Morgantown, West Virginia, U.S. Department of Energy Report DOE/METC/SP-80/23, p. 3–29.

Eaton, G. P., 1979, Regional geophysics, Cenozoic tectonics, and geologic resources of the Basin and Range province and adjoining regions, *in* Newman, G. W., and Goode, H. D., eds., 1979 Basin and Range Symposium: Rocky Mountain Association of Geologists and Utah Geological Association, p. 11–39.

—— , 1980, Geophysical and geological characteristics of the crust of the Basin and Range province, *in* Continental tectonics: Washington, D.C., National Academy of Sciences, p. 96–113.

Eaton, J. P., 1982, The Basin and Range province; Origin and tectonic significance: Annual Review of Earth and Planetary Science, p. 409–440.

Eaton, J. P., and Rymer, M. J., 1989, Regional seismotectonic model for the southern Coast Ranges, and the May 2, 1983, Coalinga earthquake: U.S. Geological Survey Professional Paper 1487 (in press).

Eaton, J. P., and Simerenko, M., 1980, Report on microearthquake monitoring in the vicinity of Auburn Dam; July 1977-June 1978: U.S. Geological Survey Open-File Report 80-604, 34 p.

Eaton, J. P., McHugh, C. A., and Lester, F. W., 1981, Report on microearthquake monitoring in the vicinity of Auburn Dam, California; November 1976-March 1977: U.S. Geological Survey Open-File Report 81-244, 20 p.

Eberly, L. D., and Stanley, T. B., Jr., 1978, Cenozoic stratigraphy and geologic history of southwestern Arizona: Geological Society of America Bulletin, v. 89, p. 921–940.

Elston, W. E., and Bornhorst, T. J., 1979, The Rio Grande rift in context of regional post-40 m.y. volcanic and tectonic events, *in* Reicker, R. E., ed., Rio Grande Rift; Tectonics and magmatism: Washington, D.C., American Geophysical Union, p. 416–438.

Engebretson, D. C., Cox, A., and Thompson, G. A., 1984, Correlation of plate motions with continental tectonics; Laramide to Basin-Range: Tectonics, v. 3, p. 115–119.

Engebretson, D. C., Cox, A., and Gordon, R. A., 1985, Relative motions between oceanic and continental plates in the Pacific Basin: Geological Society of America Special Paper 206, 64 p.

Evans, M. A., 1979, Fractures in oriented Devonian shale cores from the Appalachian basin [M.S. thesis]: Morgantown, West Virginia University, 278 p.

Ganga Roa, H.V.S., Advani, S. H., Chang, P., Lee, S. C., and Dean, C. S., 1979, In-situ stress determination based on fracture responses associated with coring operations: Proceedings, 20th U.S. Symposium on Rock Mechanics, Austin, Texas, p. 683–690.

Gephart, J. W., and Forsyth, D. D., 1985, On the state of stress in New England as determined from earthquake focal mechanisms: Geology, v. 13, p. 70–76.

Gillespie, A. R., 1982, Quaternary glaciation and tectonism in the southeastern Sierra Nevada, Inyo County, California [Ph.D. thesis]: Pasadena, California Institute of Technology, 695 p.

Golombek, M. P., McGill, G. E., and Brown, L., 1983, Tectonic and geologic evolution of the Espanola basin, Rio Grande Rift; Structure, rate of extension, and relation to the state of stress in the western United States: Tectonophysics, v. 94, p. 483–507.

Gough, D. I., and Bell, J. S., 1981, Stress orientations from oil well fractures in Alberta and Texas: Canadian Journal of Earth Sciences, v. 18, p. 638–645.

—— , 1982, Stress orientation from borehole wall fractures with examples from Colorado, east Texas, and northern Canada: Canadian Journal of Earth Sciences, v. 19, p. 1958–1970.

Gough, D. I., Fordjor, C. K., and Bell, J. S., 1983, A stress province boundary and tractions on the North American plate: Nature, v. 305, p. 619–621.

Hack, J. T., 1979, Rock control and tectonism—Their importance in shaping the Appalachian Highlands: U.S. Geological Survey Professional Paper 1126-B, 17 p.

Haimson, B. C., 1977, Crustal stress in the continental United States as derived from hydrofracturing tests, *in* Heacock, J. C., ed., The Earth's crust: American Geophysical Union Geophysical Monograph Series, v. 20, p. 575–592.

——, 1981, Hydrofracturing studies in drillhole DH-101, Fifth water underground power plate site, Diamond Fork Power System–Bonneville Unit, Central Unit Projects, report: Provo, Utah, Bureau of Reclamation, 29 p.

——, 1984, Stress measurements in the Wasatch hinterland complement existing tectonic and seismicity data: EOS (Transactions of the American Geophysical Union, v. 65, p. 1118–1119.

——, 1985, Stress measurements at the Jordanelle Dam Site, central Utah, using wireline hydrofracturing: Final Technical Report submitted to U.S. Geological Survey, Contract No. 14-08-0001-2189, 34 p.

Hamilton, R. M., and Healy, J. H., 1969, Aftershocks of the Benham nuclear explosion: Bulletin of the Seismological Society of America, v. 59, p. 2271–2281.

Hasegawa, H. S., Adams, J., and Yamazaki, K., 1985, Upper crustal stresses and vertical stress migration in eastern Canada: Journal of Geophysical Research, v. 90, p. 3637–3648.

Hickman, S. H., Healy, J. H., and Zoback, M. D., 1985, In situ stress, natural fracture distribution, and borehole elongation in the Auburn geothermal well, Auburn, New York: Journal of Geophysical Research, v. 90, 5497–5512.

Houlday, M., Quittmeyer, R., and Mrotek, K., and Statlon, C. T., 1984, Recent seismicity in north- and east-central New York State: Earthquake Notes, v. 55, no. 2, p. 16–20.

Jones, L. M., 1988, Focal mechanisms and the state of stress along the San Andreas Fault in southern California: Journal of Geophysical Research, v. 93, p. 8869–8891.

Jones, L. M., and Dollar, R. S., 1986, Evidence of Basin and Range extensional tectonics in the Sierra Nevada; The Durrwood Meadows swarm, Tulare County, California, 1983–1984: Bulletin of the Seismological Society of America, v. 76, p. 439–461.

Kulander, B. R., Dean, S. L., and Barton, C. C., 1977, Fractographic logging for determination of pre-core and core-induced fractures—Nicholas Combs No. 7239 well, Hazard, Kentucky: U.S. Energy Research and Development Agency MERC/CR-77/3, 44 p.

Kulander, B. R., Barton, C. C., and Dean, S. L., 1979, The application of fractography to core and outcrop fracture investigations: U.S. Department of Energy Report METC/SP-79/3, 174 p.

Lachenbruch, A. H., and Sass, J. H., 1973, Thermo-mechanical aspects of the San Andreas fault system, *in* Kovach, R. L., and Nur, A., eds., Proceedings of the Conference on the Tectonic Problems of the San Andreas Fault System: Palo Alto, California, Stanford University Press, p. 192–205.

——, 1980, Heat flow and energetics of the San Andreas fault zone: Journal of Geophysical Research, v. 85, p. 6185–6222.

Langston, C. A., and Butler, R., 1976, Focal mechanism of the August 1, 1975, Oroville, earthquake: Bulletin of the Seismological Society of America, v. 66, p. 1111–1120.

Lockwood, J. P., and Moore, J. G., 1979, Regional extension of the Sierra Nevada, California, on conjugate microfault sets: Journal of Geophysical Research, v. 84, p. 6041–6049.

Lubetkin, L.K.C., and Clark, M. M., 1988, Late Quaternary activity along the Lone Pine fault, eastern California: Geological Society of America Bulletin, v. 100, p. 755–766.

McCulloch, D. S., 1987, Regional geology and hydrocarbon potential of offshore central California, *in* Scholl, D. W., Grantz, A., and Vedder, J. G., Geology and resource potential of the continental margin of North America and adjacent ocean basins; Beaufort Sea to Baja California: Houston, Texas, Circum-Pacific Council for Energy and Mineral Resources, Earth Science Series, v. 6, p. 353–401.

McKenzie, D. P., 1969, The relationship between fault plane solutions for earthquakes and the directions of the principal stresses: Seismological Society of America Bulletin, v. 59, p. 591–601.

Minster, J. B., and Jordan, T. H., 1978, Present-day plate motions: Journal of Geophysical Research, v. 83, p. 5331–5360.

Minster, J. B., and Jordan, T. H., 1984, Vector constraints on Quaternary deformation of the western United States east and west of the San Andreas fault, *in* Crouch, J. K., and Bachman, S. B., eds., 1984, Tectonics and sedimentation along the California margin: Pacific Section, Society of Economic Paleontologists and Mineralogists, v. 38, p. 1–16.

Minster, J. B., Jordan, T. H., Molnar, P., and Haines, E., 1974, Numerical modeling of instantaneous plate tectonics: Geophysical Journal of the Royal Astronomical Society, v. 36, p. 541–576.

Mount, V.F., in press, Present-day stress directions in California: Journal of Geophysical Research, in press.

Munsey, J. W., and Bollinger, G. A., 1985, Focal mechanism analyses for Virginia earthquakes: Bulletin of the Seismological Society of America, v. 75, p. 1613–1636.

Nelson, K., and Talwani, P., 1985, Reanalysis of focal mechanism data for the central Virginia seismogenic zone: Earthquake Notes, v. 56, no. 3, p. 76.

Nishenko, S. P., and Kafka, A. L., 1982, Earthquake focal mechanisms and the intraplate setting of the Bermuda Rise: Journal of Geophysical Research, v. 87, p. 3929–3941.

Patton, H. J., 1985, P-wave fault-plane solutions in the generation of surface waves by earthquakes in the western United States: Geophysical Research Letters, v. 12, p. 518–521.

Plumb, R. A., and Cox, J. W., 1987, Stress directions in eastern North America determined to 4.5 km from borehole elongation measurements: Journal of Geophysical Research, v. 92, p. 4805–4816.

Plumb, R. A., and Hickman, S. H., 1985, Stress-induced borehole elongation; A comparison between the four-arm dipmeter and borehole televiewer in the Auburn geothermal well: Journal of Geophysical Research, v. 90, p. 5513–5521.

Pollitz, F. F., 1986, Pliocene change in Pacific-plate motion: Nature, v. 320, p. 738–741.

Porath, H., 1971, Magnetic variation anomalies and seismic low-velocity zone in the western United States: Journal of Geophysical Research, v. 83, p. 2265–2272.

Prowell, D. G., 1983, Index of faults of Cretaceous and Cenozoic age in the eastern United States: U.S. Geological Survey Miscellaneous Field Studies Map MF–1269, scale 1:2,500,000.

Quittmeyer, R. C., Statton, C. T., Mrotek, K. A., and Houlday, M., 1985, Possible implications of recent microearthquakes in southeastern New York State: Earthquake Notes, v. 56, no. 2, p. 35–42.

Reding, L. M., 1984, North American plate stress modeling; A finite element analysis [M.S. thesis]: Tucson, University of Arizona, 111 p.

Reynolds, M. W., 1979, Character and extent of basin-range faulting, western Montana and east-central Idaho: Rocky Mountain Association of Geologists and Utah Geological Association 1979 Basin and Range Symposium, p. 185–193.

Riddihough, R. P., 1977, A model for recent plate interactions off Canada's west coast: Canadian Journal of Earth Sciences, v. 14, p. 384–396.

Ruppel, E. T., 1982, Cenozoic block uplifts in east-central Idaho and southwest Montana: U.S. Geological Survey Professional Paper 1224, 24 p.

Savage, J. C., Lisowski, M., and Prescott, W. H., 1981, Geodetic strain measurements in Washington: Journal of Geophysical Research, v. 86, p. 4929–4940.

Sbar, M. L., 1982, Delineation and interpretation of seismotectonic domains in western North America: Geophysical Research Letters, v. 10, p. 177–180.

Sbar, M. L., and Sykes, L. R., 1973, Contemporary compressive stress and seismicity in eastern North America; An example of intraplate tectonics: Geological Society of America Bulletin, v. 84, p. 1861–1882.

Scott, W. E., Pierce, K. L., and Hait, M. H., Jr., 1985, Quaternary tectonic setting of the 1983 Borah Peak Earthquake, central Idaho *in* Stein, R. S., and Bucknam, R. C., eds., Proceedings of Workshop 28 on the Borah Peak, Idaho earthquake: U.S. Geological Survey Open-File Report 85-290, p. 1–16.

Seborowski, D. P., Williams, G., Kelleher, J. A., and Statton, C. T., 1982, Tectonic implications of recent earthquakes near Annsville, New York: Bulletin of the Seismological Society of America, v. 72, p. 1606–1609.

Smith, R. B., 1977, Intraplate tectonics of the western North American plate: Tectonophysics, v. 37, p. 323–336.

Spence, W., Stress origins and earthquake potentials in Cascadia: Journal of Geophysical Research, v. 94, p. 3076–3088.

Stewart, G. S., and Helmburger, D. V., 1981, The Bermuda earthquake of March 24, 1978; A significant oceanic intraplate event: Journal of Geophysical Research, v. 86, p. 7027–7036.

Stewart, J. H., 1978, Basin and range structure in western North America; A review: Geological Society of America Memoir 152, p. 1–31.

—— , 1988, Tectonics of the Walker Lane belt, western Great Basin; Mesozoic and Cenozoic deformation in a zone of shear, *in* Ernst, W. G., ed., Metamorphism and crustal evolution, western conterminous; Rubey volume, 7: Englewood Cliffs, New Jersey, Prentice-Hall, p. 683–713.

Stock, J. M., Healy, J. H., Hickman, S. H., and Zoback, M. D., 1985, Hydraulic fracturing stress measurements at Yucca Mountains, Nevada, and relationship to the regional stress field: Journal of Geophysical Research, v. 90, p. 8691–8706.

Taber, J. J., 1983, Crustal structure and seismicity of the Washington continental margin [Ph.D. thesis]: Seattle, University of Washington, 175 p.

Taber, J. J. and Smith, S. W., 1985, Seismicity and focal mechanisms associated with the subduction of the Juan de Fuca plate beneath the Olympic Peninsula, Washington: Bulletin of the Seismological Society of America, v. 75, p. 237–249.

Talwani, P. D., 1982, Internally consistent pattern of seismicity near Charleston, South Carolina: Geology, v. 10, p. 654–658.

Thompson, G. A., and Burke, D. B., 1974, Regional geophysics of the Basin and Range province: Earth and Planetary Science Annual Reviews, v. 2, p. 213–238.

Thompson, G. A., and Zoback, M. L., 1979, Regional geophysics of the Colorado Plateau: Tectonophysics, v. 61, p. 149–181.

Weaver, C. S., and Michaelson, C. A., 1985, Seismicity and volcanism in the Pacific Northwest; Evidence for the segmentation of the Juan de Fuca plate: Geophysical Research Letters, v. 12, p. 215–218.

Weaver, C. S., and Smith, S. W., 1983, Regional tectonic and earthquake hazard implications of a crustal fault zone in southwestern Washington: Journal of Geophysical Research, v. 88, p. 10371–10383.

Weaver, C. S., Grant, W. C., and Shemata, J. E., 1987, Local crustal extension at Mount St. Helens, Washington: Journal of Geophysical Research, v. 92, p. 10170–10178.

Wentworth, C. M., and Mergner-Keefer, M., 1983, Regenerate faults of small cenozoic offset; Probable earthquake sources in the southeastern United States: U.S. Geological Survey Professional paper 1313, p. 51–520.

Wong, I. G., 1985, Mining-induced earthquakes in the Book Cliffs and eastern Wasatch Plateau, Utah, U.S.A.: International Journal of Rock Mechanics Mining Science and Geomechanical Abstracts, v. 22, no. 4, p. 263–270.

Wong, I. G., Ely, R. W., and Kollman, A. C., 1988, Contemporary seismicity and tectonics of the northern and central Coast Ranges–Sierran block boundary zone, California: Journal of Geophysical Research, v. 93, p. 7813–7833.

—— , 1989, Contemporary seismicity, faulting, and the state of stress in the Colorado Plateau: Geological Society of America Bulletin, v. 101, no. 9 (in press).

Wright, L., 1976, Late Cenozoic fault patterns and stress fields in the Great Basin and westward displacement of the Sierra Nevada block: Geology, v. 4, p. 489–494.

Yang, J. P., and Aggarwal, Y. P., 1981, Seismotectonics of northeastern United States and adjacent Canada: Journal of Geophysical Research, v. 86, p. 4981–4998.

Yelin, T. S., and Crosson, R. S., 1982, A note on the south Puget Sound basin magnitude 4.6 earthquake of 11 March 1978 and its aftershocks: Bulletin of the Seismological Society of America, v. 72, p. 1033–1038.

Zoback, M. D., and Zoback, M. L., 1989, State of stress in North America, *in* Slemmons, D. B., Engdahl, E. R., Blackwell, D., and Schwartz, D., eds., Neotectonics of North America: Boulder, Colorado, Geological Society of America, Decade Map Volume (in press).

Zoback, M. D., Tsukahara, H., and Hickman, S., 1980, Stress measurements in the vicinity of the San Andreas fault; Implications for the magnitude of shear stress at depth: Journal of Geophysical Research, v. 85, p. 6157–6173.

Zoback, M. D., Moos, D., Mastin, L., and Anderson, R. N., 1985a, Wellbore breakouts and in-situ stress: Journal of Geophysical Research, v. 90, p. 5523–5530.

Zoback, M. D., Prescott, W. H., and Kruger, S. W., 1985b, Evidence for lower crustal ductile strain localization in southern New York: Nature, v. 317, p. 705–707.

Zoback, M. D., Moss, D., Coyle, B. and Anderson, R. N., 1986, In-situ and physical property measurements in Appalachian site survey boreholes: American Association of Petroleum Geologists Bulletin, v. 70, p. 666.

Zoback, M. D., Zoback, M. L., Mount, V. S., Suppe, J., Eaton, J. P., Healy, J. H., Oppenheimer, D., Reasenberg, P., Jones, L., Raleigh, C. B., Wong, I. G., Scotti, O., Wentworth, C. M., 1987, State of stress of the San Andreas fault system: Science, v. 238, p. 1105–1111.

Zoback, M. L., 1983, Structure and Cenozoic tectonism along the Wasatch fault zones, Utah: Geological Society of America Memoir 157, p. 3–27.

—— , 1984, Constraints on the in-situ stress field along the Wasatch front: U.S. Geological Survey Open-File Report 84-763, p. 286–309.

Zoback, M. L., 1989, State of stress and modern deformation of the northern Basin and Range province: Journal of Geophysical Research,v. 94, p. 7105–7128.

Zoback, M. L., and Lachenbruch, A. H., 1984, Upper mantle structure beneath the western United States: Geological Society of America Abstracts with Programs, v. 16, p. 705.

Zoback, M. L., and Zoback, M. D., 1980, State of stress in the conterminous United States: Journal of Geophysical Research, v. 85, p. 6113–6156.

Zoback, M. L., Anderson, R. E., and Thompson, G. A., 1981, Cainozoic evolution of the state of stress and style of tectonism of the Basin and Range province of the western United States, *in* Vine, F. J., and Smith, A. G., organizers, Extensional tectonics associated with convergent plate boundaries: Royal Society of London, p. 189–216.

Zoback, M. L., Nishenko, S. R., Richardson, R. M., Hasegawa, H. S., and Zoback, M. D., 1986, Mid-plate stress, deformation, and seismicity, *in* Vogt, P. R., and Tucholke, B. E., eds., The western North Atlantic region: Boulder, Colorado, The Geology of North America, v. M, p. 297–312.

Manuscript Accepted by the Society October 31, 1988

Printed in U.S.A.

Geological Society of America
Memoir 172
1989

Chapter 25

Earthquakes, faults, and the seismotectonic framework of the contiguous United States

J. W. Dewey
U.S. Geological Survey, Box 25046, Denver Federal Center, Denver, Colorado 80225
David P. Hill and W. L. Ellsworth
U.S. Geological Survey, 345 Middlefield Road, Menlo Park, California 94025
E. R. Engdahl
U.S. Geological Survey, Box 25046, Denver Federal Center, Denver, Colorado 80225

ABSTRACT

Recent decades have seen dramatic improvement in the ability of earth scientists to resolve the geometries of earthquake sources and to relate these geometries to local and global tectonic processes. Well-studied seismogenic faults consist of individual segments that are offset from each other along strike or that have different strikes. The boundaries between segments help control or limit rupture propagation during large earthquakes. On many segments of the plate-bounding San Andreas fault system, fault orientation and structure at depth are well represented by the surface fault trace. In the Cordillera away from the San Andreas system, faults that have produced large earthquakes seem commonly to be expressed at the surface by traces that are distorted representations of the faults at depth. In the United States east of the Cordillera, it has not been possible to associate earthquakes with geologically mapped faults as directly as in the western United States. The following observations, however, suggest that many eastern earthquakes occur on reactivated pre-Cenozoic faults: (1) epicenters in some sources coincide with mapped pre-Cenozoic faults; (2) in other sources, planar zones of hypocenters and instrumentally inferred fault planes parallel regional structural trends; and (3) some pre-Cenozoic faults show Cenozoic displacement. Global tectonic models have been extended to account for the distribution and focal depths of earthquakes that occur in California away from the principal plate boundary and to account for the orientations of regional stress tensors east of the plate boundary. Plate-tectonic models have also led to the hypothesis that a large thrust-interface earthquake might someday occur in the Pacific Northwest subduction zone, notwithstanding the absence of such earthquakes in the historic record.

INTRODUCTION

Our review of the seismicity of the contiguous United States considers the hypocentral distributions and focal mechanisms of earthquakes and their relation to the tectonic framework of the United States. In particular, we summarize current interpretations of regional patterns of seismicity within the context of plate-tectonic models by focusing on selected individual seismic sources for which the association of seismicity with geologically mapped faults or other geologic structures seems particularly clear-cut. These aspects of the review reflect widely accepted advances made in recent decades in the understanding of the relationship of earthquakes to geology. In addition, we consider situations where the association of seismicity with geologic structure is still obscure. Some of these situations correspond to earthquake source zones that cannot yet be associated with well-defined geologic

Dewey, J. W., Hill, D. P., Ellsworth, W. L., and Engdahl, E. R., 1989, Earthquakes, faults, and the seismotectonic framework of the contiguous United States, *in* Pakiser, L. C., and Mooney, W. D., Geophysical framework of the continental United States: Boulder, Colorado, Geological Society of America Memoir 172.

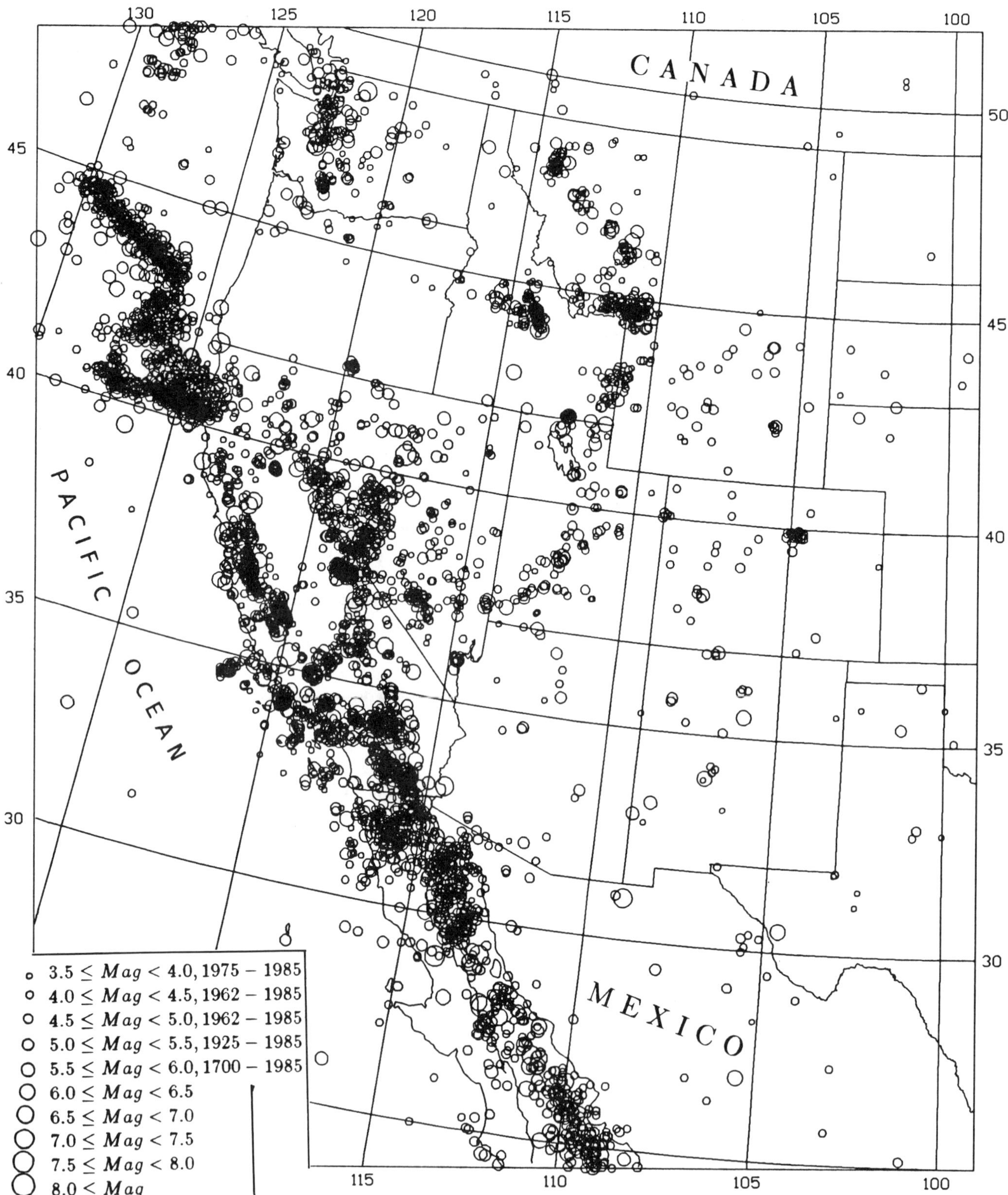

Figure 1. Seismicity of the contiguous 48 states. Note that different time periods are used for earthquakes in different magnitude ranges. Earthquakes of magnitude less than 5.5 are plotted only for time periods during which earthquakes of similar magnitude are likely to have been catalogued if they occurred beneath most of the land area of the United States. Earthquakes are plotted from the database used for

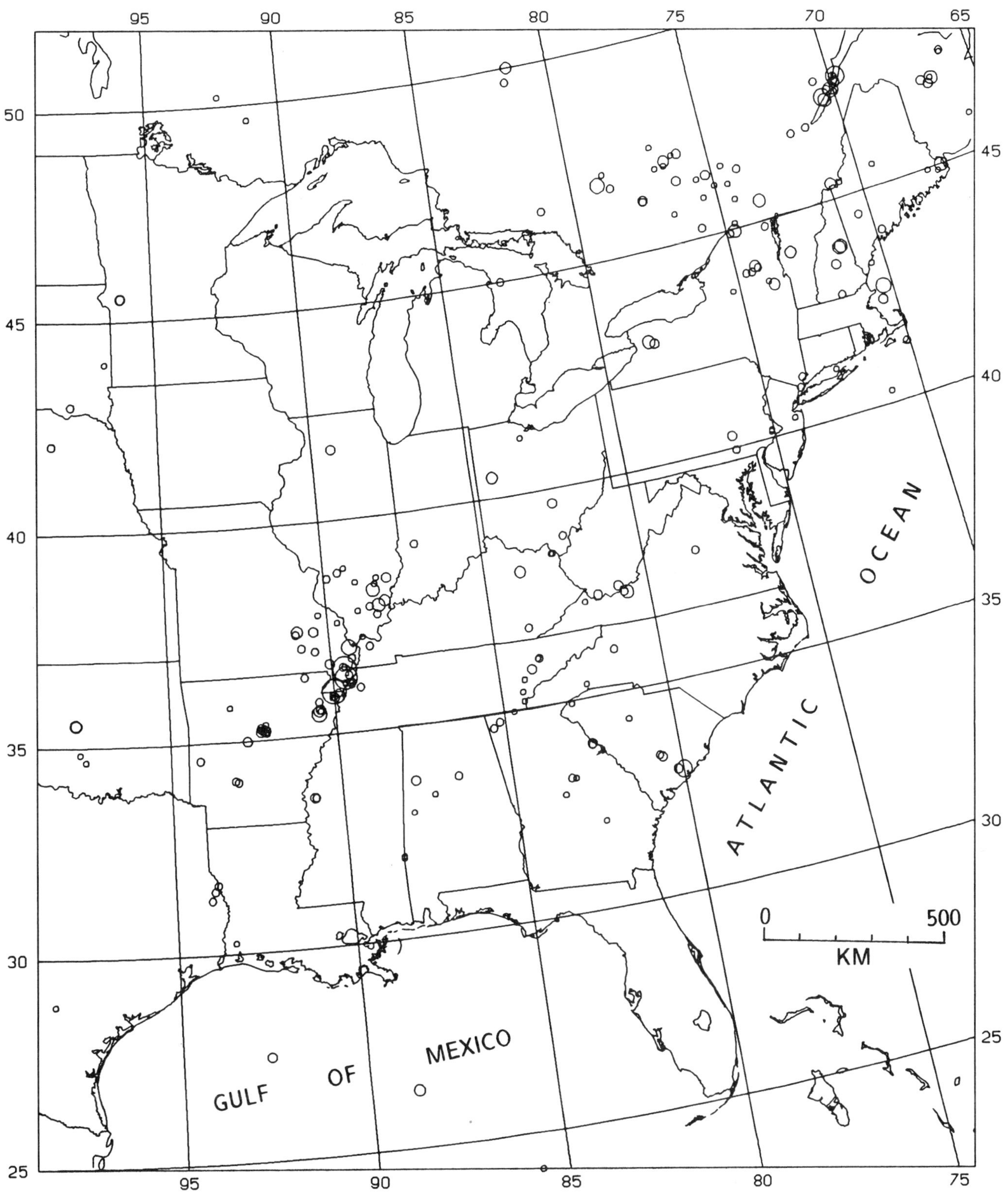

the Geological Society of America, Decade of North American Geology (DNAG), seismicity map of North America (Engdahl and Rinehart, 1988), using the same conventions to assign magnitudes to plotted events. Where more than one earthquake has been assigned to the same epicenter (to within 4 km), only a single symbol for the largest earthquake is plotted.

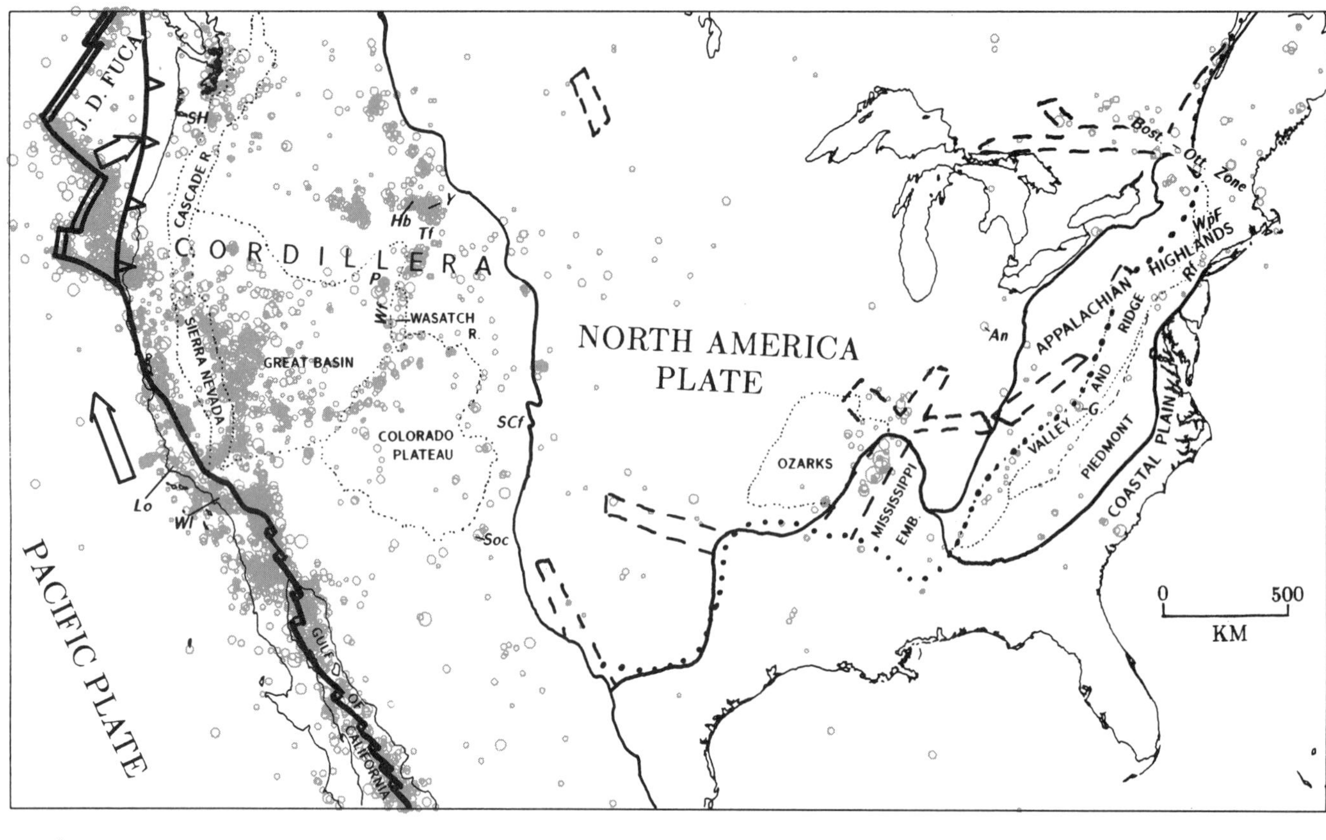

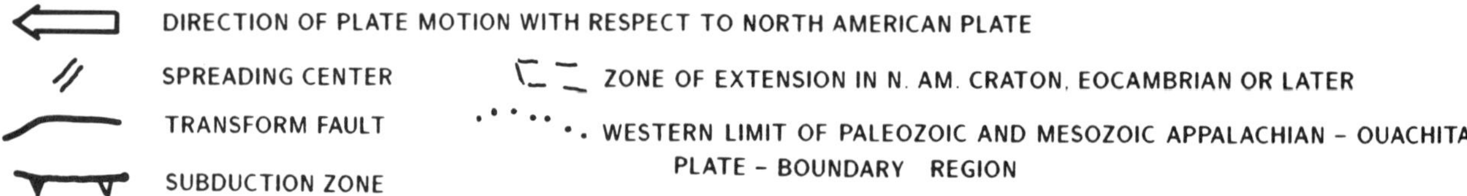

Figure 2. Plates, plate boundaries, geologic provinces, and physiographic entities that are used in the text to describe the distribution of earthquakes. Epicenters are those of Figure 1. Abbreviations of seismic sources and faults whose locations are not shown in other figures are as follows: An = Anna, Ohio; G = Giles County, Virginia; Hb = Hebgen Lake, Montana; Lo = Lompoc, California; P = Pocatello Valley, Idaho; Rf = Ramapo fault, New Jersey; SCf = Sangre de Cristo fault, Colorado; SH = Mount St. Helens, Washington; Soc = Socorro, New Mexico; Tf = Teton fault, Wyoming; Wf = Wasatch fault, Utah; Wl = Wilmington oil field, California; WpF = Wappingers Falls, New York; Y = Yellowstone National Park, Wyoming. Zones of Eocambrian or later extension in the North American craton are taken from Keller and others (1983).

structures; others correspond to well-mapped faults or plate-tectonic boundaries that apparently have the potential for producing major earthquakes but have not yet produced such earthquakes during the historical period.

DISTRIBUTION OF EARTHQUAKES ACROSS THE CONTINENT

The level of seismic activity is much higher overall in the western states than in the central and eastern states (Fig. 1). This first-order variation of seismicity is due to the proximity of the western states to late Cenozoic boundaries of major tectonic plates. Several of the individual plate-boundaries show high levels of seismicity (Fig. 2). Seismicity extending more than a thousand kilometers from the plate boundaries into the Cordillera (Fig. 2) appears also to be a consequence of conditions on the principal boundaries themselves (Atwater, 1970; Scholz and others, 1971). The central and eastern states, in contrast, have been well removed from major plate boundaries throughout Cenozoic time.

Some known or postulated seismic sources of the western United States were quiescent during the period for which small and moderate earthquakes are plotted in Figure 1. The low levels of seismicity on several sections of the Pacific–North American plate boundary in California (Fig. 2) correspond to source regions of historic great earthquakes. The low level of earthquake activity along the convergent boundary between the Juan de Fuca and

North American plates in the Pacific Northwest is a subject of controversy; it, too, may be a temporary quiescence, or it may reflect aseismic subduction.

The seismicity of the United States east of the Cordillera, though much lower on average than that of the Cordillera and Pacific Coast, is nonetheless important as an indicator of midplate tectonic processes and as a potential hazard to human beings. The largest historic earthquakes in the central and eastern United States had magnitudes greater than 6.5 and probably involved fault rupture through tens of kilometers of the Earth's crust. Their recurrence in future decades would cause extensive damage. As in the western United States, some potential sources of strong earthquakes in the eastern or central states are probably not characterized in Figure 1 by clusters of small and moderate shocks.

Nearly all of the earthquakes in Figure 1 occurred within the Earth's crust, with most occurring within the upper crust. Factors that influence variations in the maximum depth of earthquake occurrence (thickness of the seismogenic crust) include rock type, pore fluid pressure, mode of faulting, mechanical process responsible for seismogenic stresses, strain rate, and temperature (Sibson, 1982; Meissner and Strehlau, 1982; Zoback and Hickman, 1982). An inverse correlation of the thickness of the seismogenic crust with temperature is especially striking at regional scales (Sibson, 1982; Chen and Molnar, 1983). Continental U.S. earthquakes in regions of high heat flow tend to occur in the uppermost 10 km of the Earth's crust. Regions with crustal earthquakes at depths of 20 km or more are usually characterized by crustal convergence, which depresses geotherms, or by crust that has had hundreds of millions of years to cool since it last experienced significant tectonism. Influences of crustal properties other than temperature are implied by local variations of maximum focal depth within regions of approximately uniform heat flow (Sbar and Sykes, 1977; Bollinger and others, 1985; Doser and Kanamori, 1986).

Shocks with well-determined focal depths of 40 km or greater are limited to the subducted lithosphere of the Juan de Fuca plate system and to isolated patches of the uppermost mantle beneath the western foothills of the Sierra Nevada (Marks and Lindh, 1978), the Wasatch Range (Zandt and Richins, 1980), and the Colorado Plateau (Wong and others, 1984). The deepest earthquakes instrumentally recorded in the 48 contiguous states have depths between 80 and 90 km.

SAN ANDREAS FAULT SYSTEM

The seismically active transform-fault boundary between the Pacific and North American plates passes through coastal California between the Salton Trough and Cape Mendocino (Wilson, 1965; McKenzie and Parker, 1967; Atwater, 1970; Figs. 2, 3). In central and southern California, from 60 to 70 percent of the relative plate motion is accounted for by right-lateral shear on the San Andreas fault and its major inland branches, which include the Hayward and Calaveras faults in central California and the San Jacinto fault in southern California (Minster and Jordan, 1978; Prescott and others, 1981; Weldon and Humphreys, 1986). Some of the remaining relative plate motion can reasonably be attributed to right-lateral strike-slip motion along northwest-trending coastal faults such as the Hosgri and San Gregorio faults in central California (Fig. 3, Gawthrop, 1978; Ellsworth and others, 1981) and offshore faults of the continental borderlands in southern California (Weldon and Humphreys, 1986).

In northern California, 60 to 70 percent of the total relative plate motion is accommodated by slip on the San Andreas fault and shear in the Coast Ranges (Prescott and Yu, 1986). North of San Francisco, geodetic and geologic data do not reveal how long-term right-lateral shear in the Coast Ranges east of the San Andreas is partitioned between slip on major faults—such as the Rodgers Creek, Maacama, and Green Valley fault zones (Herd, 1978)—and deformation of the blocks between the faults (Prescott and Yu, 1986). In addition, there is no direct evidence in this region for offshore faults that would account for the 30 to 40 percent of relative plate motion that does not occur on the San Andreas or in the Coast Ranges (Prescott and Yu, 1986).

The northwest-trending lineations of epicenters associated with right-lateral faults are superimposed on a background of more widely scattered epicenters and clusters of epicenters (Fig. 3; Allen and others, 1965). Many of California's damaging earthquakes occur on faults that are not characterized by predominantly right-lateral displacements; these faults may be of second-order importance from a global tectonic standpoint, but may exert a first-order influence on local geologic structure and topography. In addition, a number of small background earthquakes probably occur on faults that even locally are geologically inconspicuous.

Beneath much of coastal California and the San Andreas fault system, the seismogenic crust is about 12 km thick (Fig. 4), or slightly less than half the thickness of the structural crust defined by the depth to the Mohorovičić discontinuity (Moho). Deformation of the lower crust as well as the upper mantle beneath most of the San Andreas system must involve stable frictional sliding (Tse and Rice, 1986) or ductile or quasi-plastic aseismic deformation (Meissner and Strehlau, 1982; Sibson, 1982). Some of the relatively deep hypocenters of the Cape Mendocino region (Fig. 4a) occur in lithosphere of the Juan de Fuca plate that has been subducted beneath the North American plate (see the section on Pacific Northwest Subduction Zone); others may be due to crustal convergence across the plate boundary extending west of Cape Mendocino (Seeber and others, 1970). The 20- to 25-km thickness of the seismogenic zone beneath the Transverse Ranges (Fig. 4b) is likely a consequence of depressed geotherms resulting from crustal convergence.

Major strike-slip faults

Over much of their lengths, major strike-slip faults of the San Andreas system have configurations at seismogenic depth that are well represented by the fault traces at the ground surface. Focal mechanisms of earthquakes on the faults usually have

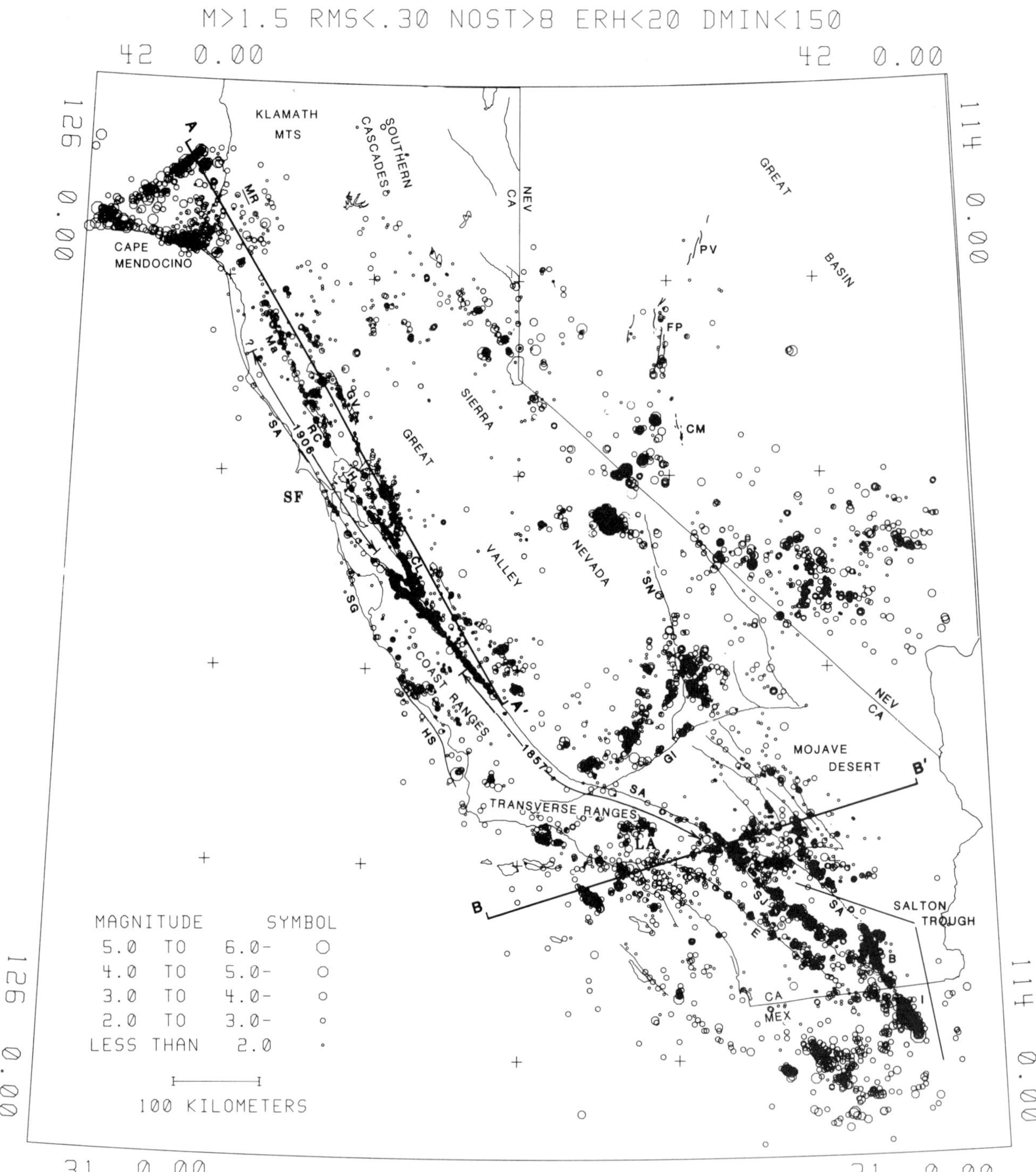

Figure 3. Map showing earthquakes that occurred in California and western Nevada during 1980–1981 (modified from Hill and others, 1983), together with selected faults. A-A′ and B-B′ define depth sections in Figure 4. Lines indicating major active faults include: B = Brawley; CL = Calaveras; E = Elsinore; Gl = Garlock; GV = Green Valley; H = Hayward; HS = Hosgri; I = Imperial; Ma = Maacama; MR = Mad River; RC = Rodgers Creek; SA = San Andreas; SG = San Gregorio; SJ = San Jacinto; SN = Sierra Nevada. Sections of the San Andreas fault labeled 1857 and 1906 correspond approximately to the source regions of the great earthquakes of those years; faulting in 1906 may have extended north to 41°N. Other abbreviations are: CM = Cedar Mountain; FP = Fairview Peak; LA = Los Angeles; PV = Pleasant Valley; SF = San Francisco.

nodal planes with strikes that are within 10° of the local strikes of the faults at the surface (Bolt and others, 1968; Eaton and others, 1970; Allen and Nordquist, 1972; Ellsworth, 1975; Reasenberg and Ellsworth, 1982). Where the faults have produced many small earthquakes, the hypocenters of these shocks define planar surfaces that project to the ground surface near the geologically mapped traces (Eaton and others, 1970; Ellsworth, 1975; Bakun and others, 1980; Reasenberg and Ellsworth, 1982). Fault segments defined by offsets or changes in strike of the surface trace commonly correspond to deep-seated segments that rupture completely in moderate or large earthquakes, and the boundaries of the surface segments correspond to points of rupture initiation or termination at depth (Bakun and others, 1980; Sibson, 1986). In contrast, the secondary faults that transfer slip between adjacent well-defined segments of a strike-slip fault may not be clearly revealed at the ground surface (Weaver and Hill, 1978/79; Reasenberg and Ellsworth, 1982; Sibson, 1986). The relation of mapped surface faults to deep-seated faults is also obscure at several locations where fault zones intersect, such as in the San Gorgonio Pass region of southern California (Allen, 1981; Nich-

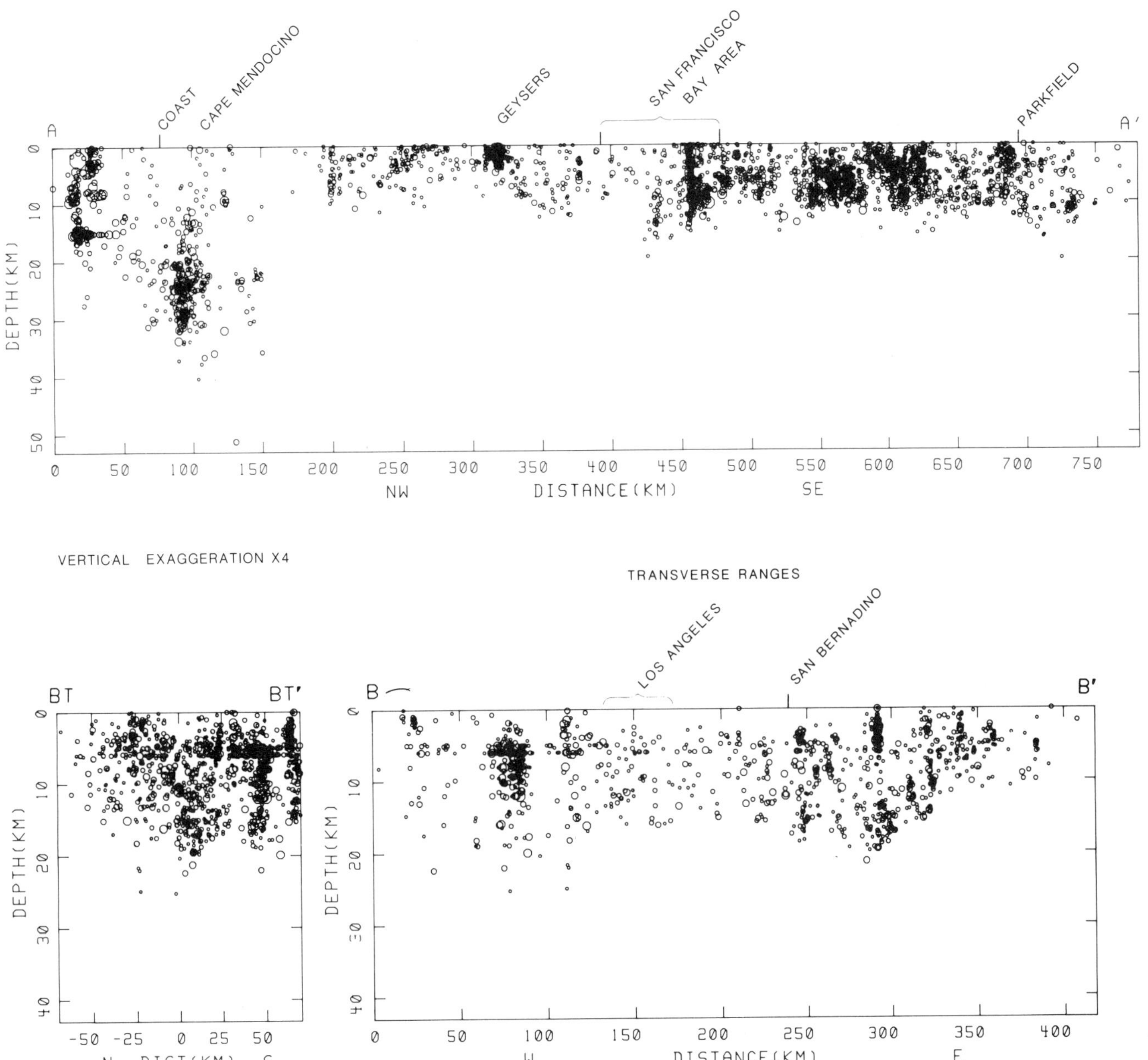

Figure 4. (A-A′) Cross section along the San Andreas fault in northern and central California and in the Cape Mendocino region (Fig. 3). Hypocenters are projected onto the plane of the profile for earthquakes within 50 km on either side of the profile. B-B′, Cross section along the Transverse Ranges in southern California (Fig. 3). Hypocenters are projected onto the plane of the profile for earthquakes within 75 km on either side of the profile. BT-BT′, Vertical section perpendicular to B-B′.

olson and others, 1986a) and the southeastern San Francisco Bay area (Ellsworth and others, 1982).

The frequencies of different-size earthquakes vary along the lengths of the major strike-slip faults. The mode of seismic release on a given segment of fault may be influenced by the orientation of the segment with respect to the direction of relative plate motion, by the composition of the wall rock (Allen, 1968), or by the way in which the segment is separated from neighboring segments (Segall and Pollard, 1980; King, 1983; Sibson, 1986).

At the largest scale, the San Andreas fault proper may be considered as having four sections that correspond to parts of the fault that respectively have and have not broken in great historical earthquakes (Byerly, 1937; Allen, 1968; Wallace, 1970). The sections that broke in the great earthquakes of 1857 and 1906 are, over most of their lengths, quiescent down to the smallest earthquakes that can be routinely detected by existing seismic networks. The map of earthquake epicenters in California for 1980 and 1981 (Fig. 3) illustrates this quiescence, which has persisted for at least the five decades during which the region has been monitored by seismographs of the University of California at Berkeley and the California Institute of Technology (Hileman and others, 1973; Bolt and Miller, 1975). These two sections appear to be locked. They show no evidence of aseismic slip (creep), and geodetic surveys show that they are accumulating shear strain energy at rates predicted by long-term geologic displacement rates (Savage, 1983). In contrast, the stretch of the San Andreas fault in central California between the 1906 and 1857 breaks is characterized by aseismic slip and frequent small to moderate earthquakes. Geodetic measurements indicate that little shear strain is accumulating in the crustal blocks on either side of the fault; most of the long-term 32-mm/yr slip along this section of the fault is accounted for by current creep rates (Savage and Burford, 1973; Thatcher, 1979; Lisowski and Prescott, 1981). Thus, although the central creeping section of the San Andreas fault occasionally produces earthquakes with magnitudes of about 5½, this section of the fault may be unable to store sufficient strain energy to generate much larger earthquakes.

The long sections of the San Andreas fault defined by the presence or absence of historic great earthquakes consist of smaller segments, which in turn, may be divided into still smaller segments. In detail, surface exposures of major branches of the San Andreas fault system consist of multiple fault strands within zones of fractured rock that vary in width from a few hundred meters to a few kilometers. Along much of the fault system, displacements within the fault zones over the last 1,000 to 10,000 yr have tended to occur repeatedly on the same narrow strands, which are generally less than a meter wide and commonly only a few centimeters wide (Wallace, 1968; Clark and others, 1972; Allen, 1981; Sieh and Jahns, 1984). The maximum continuous length of these currently active fault strands mappable at the surface is about 18 km (Wallace, 1973).

Sieh (1978a, 1981) has proposed that each individual fault segment slips by approximately the same amount in all earthquakes. In Sieh's model, which he terms the uniform-slip model,

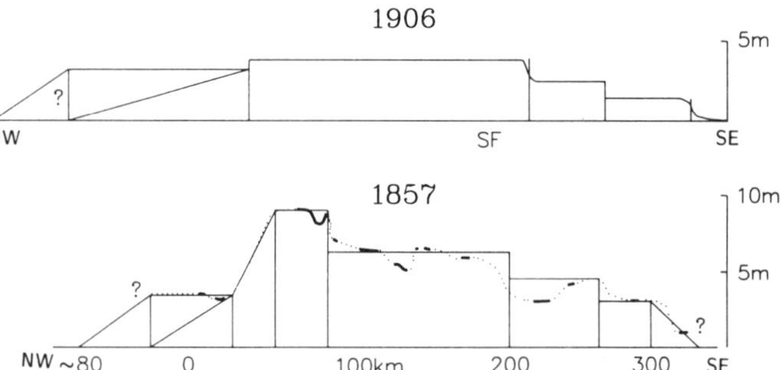

Figure 5. Simplified slip functions for the 1857 and 1906 earthquakes (after Fig. 18 of Sieh, 1978a). Scales are the same for both earthquakes. The origin of the horizontal distance scale for the 1857 shock is arbitrarily set at the presumed epicenter approximately 23 km south-southeast of Parkfield (Fig. 6). SF = San Francisco. Heavy lines for the 1857 shock indicate well-controlled slip; dotted lines indicate interpolated and extrapolated slip. Large rectangles and triangles represent geometrically simple approximations to the slip distribution.

the characteristic displacement on a given fault segment might, in one seismic cycle, occur in an earthquake that was confined to that fault segment, whereas in another seismic cycle the same displacement might occur in a great earthquake that involved many fault segments. Schwartz and Coppersmith (1984) hypothesized that characteristic displacements on fault segments tend to occur in earthquakes that are similar throughout their source lengths; such an earthquake is referred to as the characteristic earthquake for its fault segment.

Great 1857 and 1906 San Andreas earthquakes. The uniform-slip model was inspired by the variation of displacement along the approximately 360-km rupture zone of the great 1857 Fort Tejon earthquake (Fig. 5; Sieh, 1978a, 1981). Two moderate foreshocks occurred a few hours before the 1857 main shock in the northern part of the main-shock source region (near Parkfield, Fig. 6). Sieh (1978b) has suggested that the main-shock rupture also began at the north on a segment of the fault that normally experiences frequent small and moderate earthquakes, and that rupture propagated southeastward into sections of the fault that normally experience few small and moderate earthquakes. Detailed studies of geomorphology and offsets in Quaternary stratigraphy along the 1857 rupture imply that the central segment of the fault trace that experienced 8 m of slip in the 1857 earthquake had 8 m or more of slip in earlier, late Holocene, earthquakes, and had recurrence times between great earthquakes of 240 to 450 yr (Sieh and Jahns, 1984). The southern section of the 1857 break, which experienced 3- to 4-m offsets in 1857, has experienced similar offsets in prehistoric earthquakes, on average, every 145 yr.

Fault displacements in the 1906 San Francisco earthquake also varied along the several-hundred kilometer length of the rupture zone (Fig. 5; State Earthquake Investigation Commission, 1908; Thatcher, 1975). Relatively little, however, is known, about displacements on prior ruptures along the 1906 fault seg-

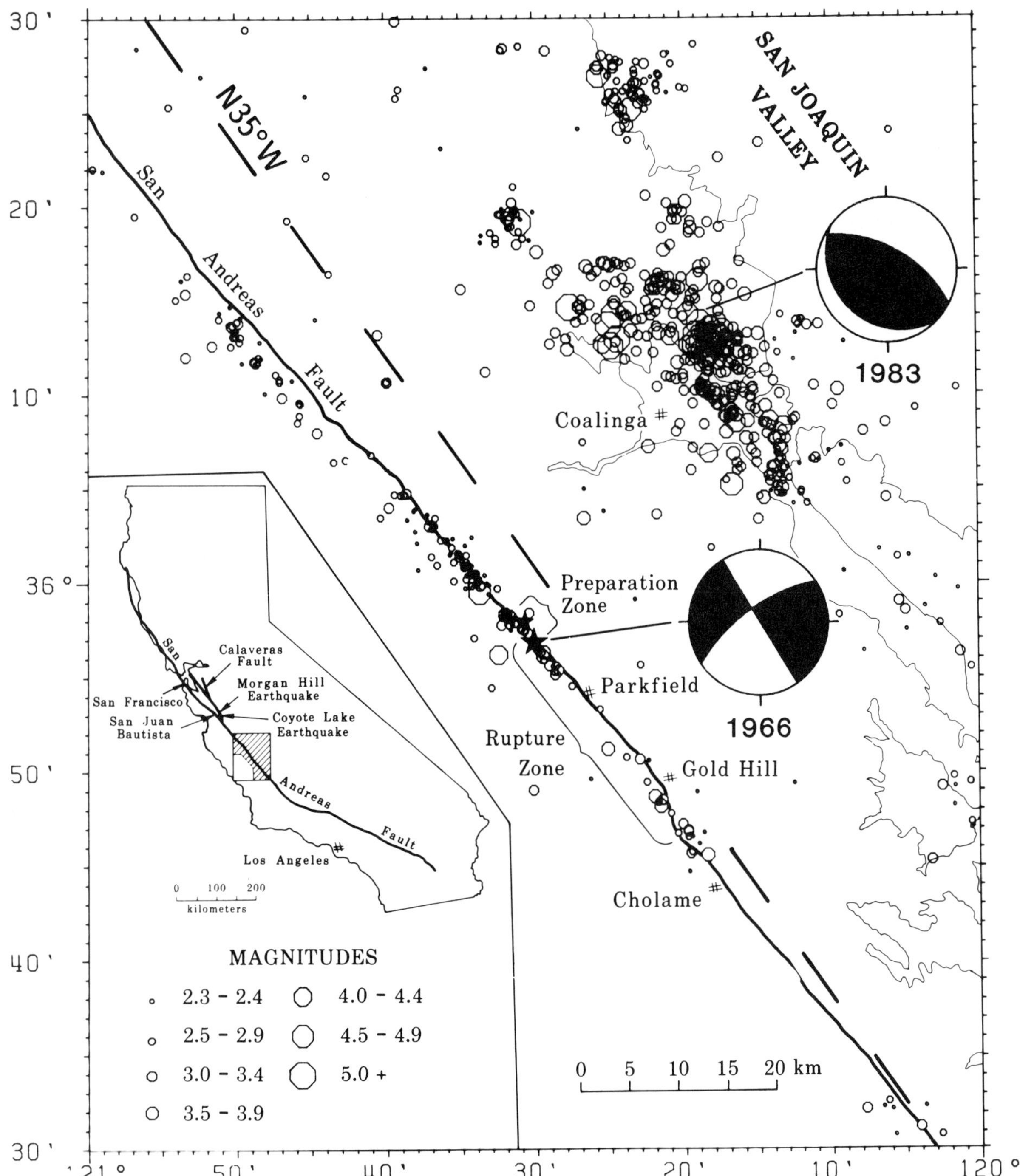

Figure 6. Map showing epicenters in the Parkfield and Coalinga regions for the period 1975–1984, modified from Bakun and Lindh (1985). Focal mechanisms of the Parkfield earthquake of June 28, 1966 (Bolt and others, 1968), and the Coalinga earthquake of May 2, 1983 (Choy, 1985), are represented in lower-hemisphere projections of the focal sphere, with quadrants of compressional P-wave first-motions in black. Brackets along the fault show the preparation zone and rupture zone of characteristic Parkfield earthquakes. Epicenters within 20 km north and east of Coalinga are predominantly aftershocks to the 1983 earthquake. Coalinga aftershocks are plotted only if magnitude 3 or larger. N35°W is the direction of pure transform motion between the Pacific and North American plates (Minster and Jordan, 1978).

ment, so the uniform-slip model cannot yet be tested on that segment.

Moderate earthquakes along the San Andreas system. Five magnitude 5 to 6 earthquakes that have occurred on the central segment of the San Andreas fault near Parkfield (Fig. 6) since the great 1857 earthquake illustrate the influence of fault segmentation in controlling the rupture of characteristic earthquakes. The Parkfield shocks are located in a transition zone between the central, creeping, section of the fault and the "locked" (quiescent) segment that broke in 1857. The shocks are notable for the regularity of their occurrence. Moderate Parkfield earthquakes have occurred in 1881, 1901, 1922, 1934, and 1966, implying an average recurrence interval since 1857 of 21 to 22 yr (Bakun and Lindh, 1985). Furthermore, seismograms written on instruments common to the 1922, 1934, and 1966 events are nearly identical; each of the instrumentally recorded shocks apparently initiated about 8 km northwest of Parkfield and propagated southward 20 to 25 km to near Cholame (Bakun and McEvilly, 1984). Bakun and McEvilly (1984) and Bakun and Lindh (1985) have proposed that the regularity of occurrence of the earthquakes is due to a uniform tectonic loading rate, to a uniform stress drop in each earthquake, and to the rupture length in each shock being constrained by a left-stepping bend in the fault to the north, where the shocks nucleate, and a right-stepping bend or offset to the south. Bakun and McEvilly (1984) and Bakun and Lindh (1985) forecasted that the next Parkfield earthquake would occur in a 10-yr time interval centered on 1988.

Studies of the August 6, 1979, Coyote Lake earthquake (M_L = 5.9, BRK*) and the April 24, 1984, Morgan Hill earthquake (M_L = 6.2, BRK) along the Calaveras fault in central California provide additional evidence for influence of segmentation on fault rupture in moderate earthquakes.

The Coyote Lake main shock ruptured from its epicenter, near Coyote Lake, southeastward to a small right-stepping offset in the subsurface fault beneath San Felipe Lake (Fig. 7). Aftershocks located immediately south of the San Felipe offset began 4 hr after the main shock and probably represent a delayed response of that segment of the fault to high stresses resulting from the main-shock rupture to the north (Reasenberg and Ellsworth, 1982). Focal mechanism solutions for aftershocks on the main-shock rupture surface have strikes that are rotated slightly counterclockwise to the strike of the overall aftershock zone, suggesting that the main-shock rupture surface itself comprised smaller segments that cannot be resolved in the distribution of hypocenters (Reasenberg and Ellsworth, 1982).

As with the Coyote Lake earthquake, the 1984 Morgan Hill earthquake propagated from north to south, terminating at the northern end of the rupture surface of the Coyote Lake earthquake. Termination of the Morgan Hill earthquake is probably related to an inferred left-stepping offset in the Calaveras Fault (Fig. 7) and to the stress barrier resulting from strain release of the Coyote Lake earthquake. In their analysis of the strong motion waveform data from the Morgan Hill earthquake, Hartzell and Heaton (1986) found evidence for uneven slip along the main shock fault surface. The greatest displacements (80 to 100 cm) were concentrated at depths of 6 to 8 km near the northern (hypocentral) and southern ends of the rupture surface, with areas of 20 cm of slip in between. Aftershocks of the earthquake surround the areas of greatest displacement, which produced no aftershocks (Cockerham and Eaton, 1984).

Bakun and others (1984) pointed out that the 1979 Coyote Lake and the 1984 Morgan Hill earthquakes are apparently similar to moderate shocks that occurred on the same fault segments in 1897 and 1911. Thus, for their respective fault segments, these earthquakes may be characteristic earthquakes with recurrence times of about 75 yr.

Imperial fault earthquakes. The Imperial fault (Fig. 8) is one of a series of right-stepping, strike-slip fault segments in the transition between the system of ridges and transform faults beneath the Gulf of California and the San Andreas fault system to the northwest (Fig. 2). The Imperial fault may be viewed as a short transform fault between zones of crustal extension that are sites of geothermal activity and repeated earthquake swarms (Lomnitz and others, 1970; Johnson and Hadley, 1976; Weaver and Hill, 1978/79).

The Imperial fault earthquakes of May 19, 1940 (M_L = 6.7, PAS), and October 1, 1979 (M_L = 6.6, PAS), illustrate the variability of rupture processes that may be associated with major shocks on the same fault segment. The seismic moment of the 1940 earthquake was nearly 10 times that of the 1979 event (Kanamori and Regan, 1982). The 1940 event began near the northern end of the Imperial fault (Doser and Kanamori, 1986) and apparently ruptured the entire length of the fault (Fig. 8). The 1979 event began near the center of the Imperial fault and ruptured predominantly to the north (Johnson and Hutton, 1982; Archuleta, 1984; Hartzell and Heaton, 1983; Silver and Masuda, 1985). The earthquakes produced similar, though not identical, surface displacements (maximum about 1 m) along the northern section of the fault, but the 1940 earthquake produced 5 to 6 m of displacement to the south, in a region where surface rupture was not observed in 1979 (Sharp, 1982). The difference between the 1940 and 1979 earthquakes might be taken either as a measure of the variation between characteristic earthquakes on a fault segment or as evidence for failure of the characteristic earthquake model for the part of the Imperial fault that ruptured in both shocks. On that section of the fault, the 1940 and 1979 data are consistent with the uniform slip model.

Reverse faulting in the Transverse and Coast Ranges

Much of the seismicity occurring off the San Andreas fault is related to a component of crustal convergence, normal to the strike of the fault system, which is highest in the latitudes of the Transverse Ranges, but which is present as well in the Coast Ranges of central and northern California. Crustal convergence in

---

*Abbreviations for most types of magnitude are as suggested by Nuttli and others (1978). M, unsubscripted, is used for magnitudes of Gutenberg and Richter (1954) or Richter (1958). Abbreviations for individual seismographic observatories providing magnitudes are given by Presgrave and others (1985).

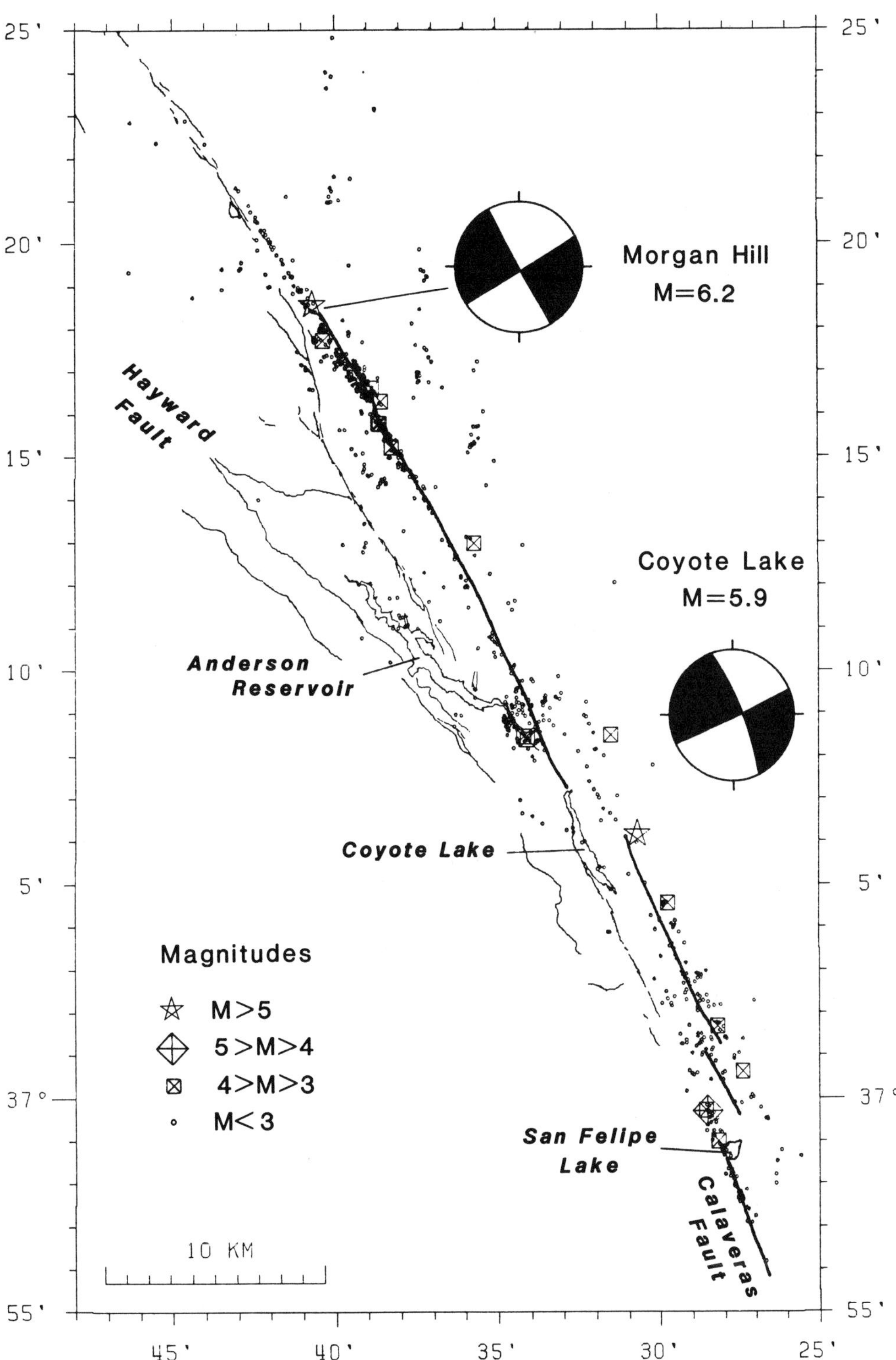

Figure 7. Map showing epicenters of aftershocks to the Coyote Lake earthquake of August 6, 1979 (Reasenberg and Ellsworth, 1982), and the Morgan Hill earthquake of April 24, 1984 (Bakun and others, 1984). Light lines are mapped faults; heavy lines indicate subsurface rupture zones inferred from aftershock distributions and focal mechanism solutions. Stars indicate main shock epicenters. Main-shock focal-mechanisms (Liu and Helmberger, 1983; Hartzell and Heaton, 1986) are represented as in Figure 6.

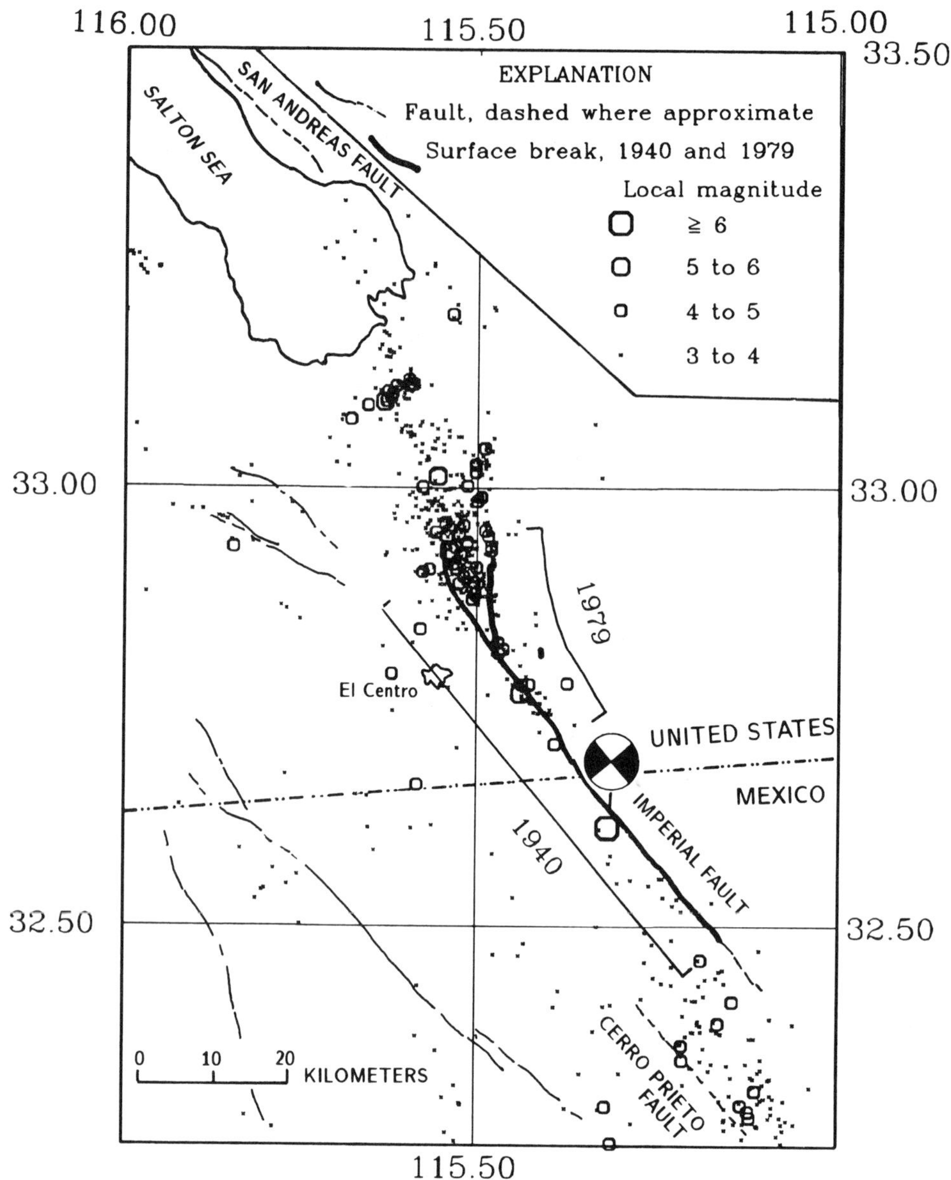

Figure 8. Seismicity of the Imperial fault region, 1974–1985, from the catalog of the California Institute of Technology. Brackets indicate extent of surface faulting in 1940 (Richter, 1958) and 1979 (Sharp and others, 1982). Focal mechanism is for the 1979 main shock (Hartzell and Heaton, 1983).

the Transverse Ranges is caused by the pronounced left-stepping bend in the San Andreas fault between the Salton Trough and the southern Coast Ranges, although details of the kinematics of crustal displacements remain controversial (Bird and Rosenstock, 1984; Weldon and Humphreys, 1986). Crustal convergence north of the Transverse ranges may be due in part to a 3° to 5° counterclockwise cant of the San Andreas fault in central and northern California with respect to the azimuth of pure transform motion between the North American and Pacific plates (Fig. 6) and in part to the westward movement of the Sierra Nevada due to crustal spreading in the Basin and Range province (Hill, 1982; Minster and Jordan, 1984; Weldon and Humphreys, 1986).

Focal mechanisms of earthquakes occurring within the Transverse Ranges show a mix of thrust and strike-slip mechanisms generally consistent with north-south shortening (Lee and others, 1979; Webb and Kanamori, 1985; Nicholson and others, 1986a). Some of the deeper earthquakes beneath the Transverse Ranges have focal mechanisms with subhorizontal fault planes, suggesting that the shallower, more steeply dipping, thrust faults may be rooted in a décollement surface at depths of 12 to 13 km (Corbett and Johnson, 1982; Webb and Kanamori, 1985; Nicholson and others, 1986a). The July 21, 1952, Kern County earthquake (M_L = 7.7, PAS; 35.0°N, 119.0°W; Richter, 1958) and the February 9, 1971, San Fernando earthquake (M_L = 6.6,

PAS; 34.4°N, 118.4°W; U.S. Geological Survey and National Oceanic and Atmospheric Administration, 1971) are examples of damaging earthquakes in the Transverse Ranges that involved reverse or oblique-reverse faulting.

The occurrence on May 2, 1983, of an M_L = 6.7 (BRK) earthquake near the town of Coalinga (Fig. 6) emphasizes the importance of crustal convergence and uplift in the Coast Ranges north of the Transverse Ranges (Rymer and Ellsworth, 1985). Fault-plane solutions for the Coalinga main shock imply reverse faulting on a northwest-striking plane or planes, with crustal material being transported nearly perpendicular to the direction of tectonic transport on the neighboring San Andreas fault (Fig. 6). Stein and King (1984), Choy (1985), and Fielding and others (1984) proposed that the main shock involved slip on a high-angle reverse fault dipping to the northeast, whereas Eaton (1985a, b) and Wentworth and others (1985) argued for slip on a low-angle reverse (thrust) fault that dips to the southwest. Eberhart-Phillips and Reasenberg (1985) suggested that the main shock may have involved both low-angle and high-angle reverse faults.

The Coalinga main shock has not been associated with a fault that is geologically mapped at the ground surface. The event occurred, however, beneath a young anticline that was further deformed in a self-similar manner (crustal uplift of 0.5 m) during the earthquake (Stein, 1985). Geological recognition of other buried reverse faults such as produced the Coalinga earthquake may depend on recognizing the geologically recent folding associated with the underlying faulting (Stein and King, 1984).

Conjugate strike-slip faults

East-northeast–trending, left-lateral, strike-slip faults splay out in a conjugate relation to the San Andreas fault in southern California. The largest of the left-lateral faults is the Garlock fault (Fig. 3). Kinematically, these faults accommodate the lateral displacement of crustal blocks away from the zones of convergence associated with the left-stepping bends of the plate boundary (Hill and Dibblee, 1953; Hill, 1982; Bird and Rosenstock, 1984). In addition, the eastern section of the Garlock fault appears to have functioned as a transform fault between the extending Great Basin crust to the north and the Mojave block to the south (Davis and Burchfiel, 1973; Weldon and Humphreys, 1986). On the basis of the geologically determined long-term slip rate across the Garlock fault (about 10 mm/yr) and the fact that some sections of the fault appear not to have ruptured in the past 500 yr, Astiz and Allen (1983) regarded the Garlock fault as a seismic gap capable of producing large earthquakes in the future.

Left-lateral strike-slip faults appear also to contribute to the background seismicity pattern in the vicinity of major right-lateral faults in California (Ellsworth, 1975; Warren, 1979; Nicholson and others, 1986a, b). Nicholson and others (1986a, b) suggested that short conjugate faults near the southern San Andreas define the northeast-trending boundaries to rectangular crustal blocks that are sandwiched between major right-lateral branches of the fault system (Fig. 9). These blocks would rotate in a clock-wise direction within the right-lateral deformation field of the subparallel right-lateral faults. An important implication of this model is that the local stress field in the vicinity of the block corners would be perturbed where the corners rotate into or away from a thoroughgoing right-lateral fault. A block corner that rotates into a strike-slip fault would produce a patch of higher-than-normal compressive stress across the fault plane, corresponding to a barrier or asperity that might control the rupture during a large earthquake on the strike-slip fault. Extensional stresses generated where block corners rotate away from the throughgoing strike-slip faults may account for the occurrence of small, normal-fault earthquakes in the generally compressive tectonic regime of the eastern Transverse Ranges.

PACIFIC NORTHWEST SUBDUCTION ZONE

North of Cape Mendocino, the Juan de Fuca plate is situated between the Pacific and North American plates (McKenzie and Parker, 1967; Atwater, 1970). The Juan de Fuca plate consists of youthful, oceanic lithosphere. Its boundaries with the Pacific plate are transform faults or spreading ridges (Fig. 2). The boundary of the Juan de Fuca plate with the North American plate is a zone of convergence across which the Juan de Fuca plate is moving northeast with respect to the North American plate at a velocity of 30 to 40 mm/yr (Riddihough, 1977; Nishimura and others, 1984). The plate convergence is accommodated by oblique subduction of the Juan de Fuca plate.

The Pacific Northwest has several types of seismic zones that are characteristic of subduction zones worldwide, but some elements of the northwest subduction zone have historically been inactive, while their counterparts worldwide tend to be highly active. Landward-dipping Wadati-Benioff zones beneath western Washington and northern California (Fig. 10; Nowroozi, 1973; Bolt, 1979; Smith and Knapp, 1980; Crosson, 1983; Cockerham, 1984; Taber and Smith, 1985; Walter, 1986) are similar to Wadati-Benioff zones worldwide that are found only in areas of current or recent subduction. Crustal seismicity extends hundreds of kilometers into the overriding plate in western Washington (Fig. 10; Crosson, 1972; Weaver and Smith, 1983) as it does in some other subduction zones worldwide (Yoshii, 1979; Reyners, 1980; Biswas and others, 1986). In contrast to most subduction zones worldwide, however, there has not yet been identified an earthquake of any size whose location and focal mechanism indicate that it occurred as the result of thrust faulting at shallow depth on the interface between the Juan de Fuca and North American plates. The historical quiescence of western Oregon (Fig. 1) is also unusual for a boundary region between converging plates.

Earthquakes occurring within the Juan de Fuca plate

Crustal earthquake activity within the Juan de Fuca plate is most intense in the southern segment of the plate, a segment also

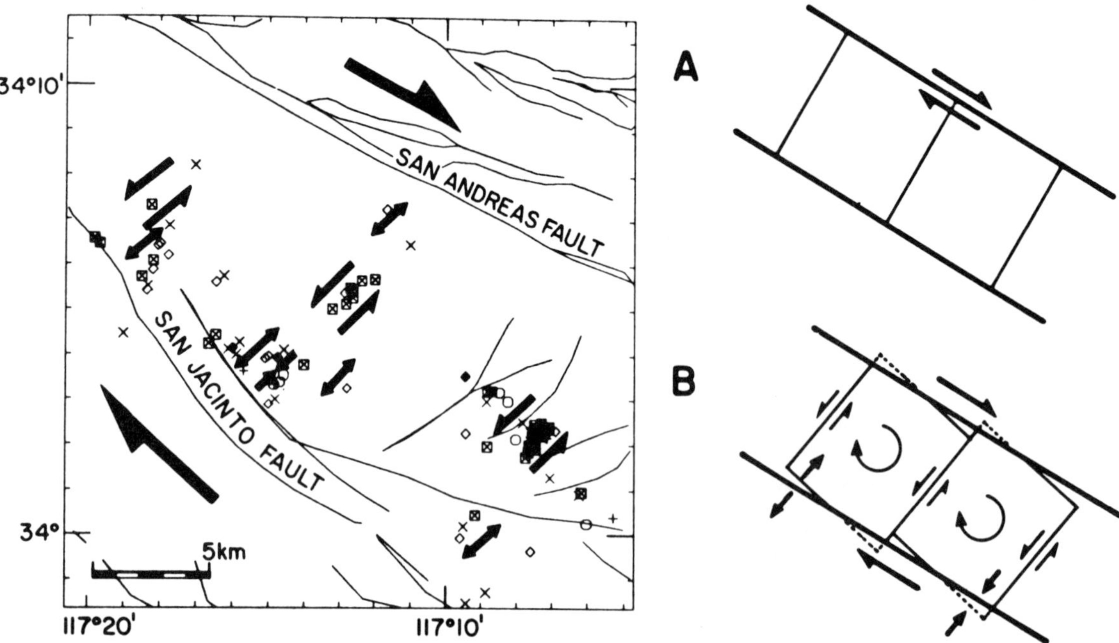

Figure 9. Diagram from Nicholson and others (1986a), illustrating their hypothesis for the origin of earthquakes that occur between the San Jacinto and San Andreas faults. A, Right-slip occurs on San Jacinto and San Andreas faults in large earthquakes. B, between the occurrence of large events, these faults are locked, causing rotation of the small blocks between them and earthquakes on the boundaries of the blocks. Epicentral symbols indicate types of focal mechanisms: circles indicate thrust; diamonds, normal; crosses, strike-slip; boxes with crosses, oblique strike-slip.

referred to as the Gorda plate. At shallow depths within the southern Juan de Fuca plate, earthquakes are predominantly strike-slip and characterized by the same north-south P-axes and east-west T-axes as characterize San Andreas fault earthquakes to the south (Bolt and others, 1968; Seeber and others, 1970; Simila and others, 1975). Detailed marine geophysical studies and studies of individual earthquake sequences, however, show that the predominant active faults within the Gorda plate are not parallel to faults of the San Andreas system but are northeast-trending, left-lateral faults (Silver, 1971; Smith and Knapp, 1980). Aftershocks of one such earthquake, that of November 8, 1980 (M_S = 7.2, USGS) define the 120-km-long lineament seen at 41.1°N, 124.3°W in Figure 3 (Eaton, 1981; Smith and others, 1981). Because these sequences occur both landward and seaward of the accretionary wedge, and in the case of the 1980 earthquake cross-cut it, they are evidence for the continuity beneath the overriding North American plate of a subducting plate in which the tension axis is oriented downdip. In contrast, the North American plate is broken by numerous northwest-trending thrust faults, such as those of the Mad River fault zone (Fig. 3; Carver and others, 1982), whose orientation and sense of slip are consistent with stresses originating from the northeastward motion of the Juan de Fuca plate against the North American plate.

Seismicity within the subducted Juan de Fuca plate extends to depths of between 80 and 90 km in western Washington and northern California (Fig. 10). The two largest instrumentally recorded earthquakes in western Washington, the shocks of April 13, 1949 (M = 7.0, Gutenberg and Richter, 1954) and April 29, 1965 (m_b = 6.5, U.S. Coast and Geodetic Survey), occurred in the subducted plate. Focal mechanisms of earthquakes in the Wadati-Benioff zone beneath western Washington indicate that these shocks result from intraplate stress, rather than from seismic slip at the thrust interface between the plates (Isacks and Molnar, 1971; Rogers, 1983; Taber and Smith, 1985).

Mount St. Helens seismicity

Earthquakes associated with the 1980 eruptions of Mount St. Helens (Fig. 2) illustrate the interplay between earthquakes and volcanic processes in the Cascadia volcanic arc. A seismic swarm beneath the summit of the mountain gave the first notice of the volcano's reawakening, beginning a week before the first eruption. Seismicity preceding the May 18 eruption was concentrated in the vicinity of the volcano (Endo and others, 1981). The catastrophic eruption of May 18 was accompanied by a moderate earthquake. Within two weeks following the eruption, seismicity had spread tens of kilometers away from the mountain (Weaver and others, 1981). Although most of the Mount St. Helens earthquakes had waveforms similar to those of tectonic earthquakes elsewhere in western Washington (Weaver and others, 1981), a subset had a characteristic "low-frequency" waveform at local seismographs of the sort that is commonly recorded only on

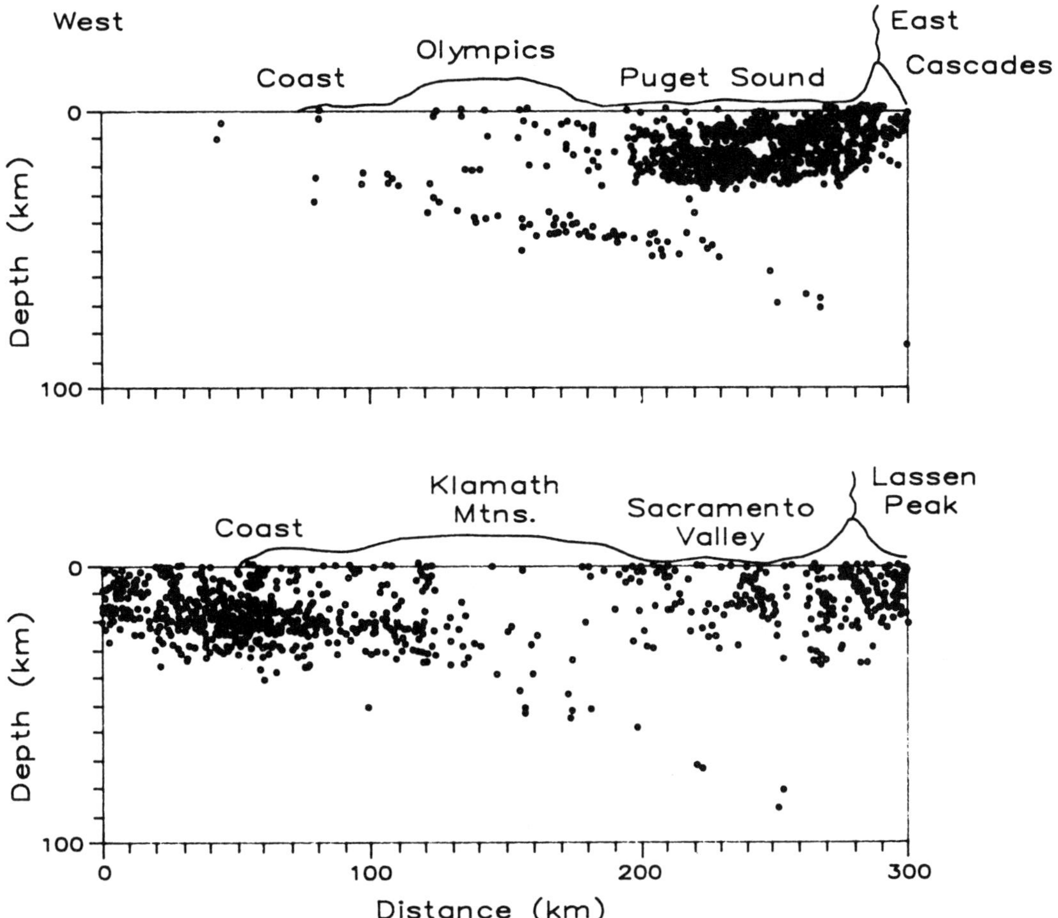

Figure 10. Cross sections of Pacific Northwest Wadati-Benioff zones. Top, Western Washington (Taber and Smith, 1985). Bottom, Northern California (Walter, 1986). Note the similarities in both the dips of the subduction zones and the maximum focal depths.

active volcanoes. These low-frequency earthquakes may reflect the direct influence of magma in the earthquake process (Endo and others, 1981; Malone and others, 1981; Qamar and others, 1983). Finally, the earthquake that accompanied the May 18 eruption appears to have been caused partly (Malone and others, 1981; Kanamori and Given, 1982) or wholly (Kanamori and others, 1984) by the rockslide and directed explosion that constituted the eruption itself.

Quiescence of the interplate-thrust zone

The lack of identified seismogenic thrust faulting on the interface between the Juan de Fuca plate and the North American plate poses a question of major importance in seismic hazard evaluations. Does the lack of such earthquakes reflect aseismic underthrusting (Crosson, 1972; Ando and Balazs, 1979; Adams, 1984), or does it reflect a temporary quiescence that will someday be followed by great thrust-fault earthquakes (Savage and others, 1981; Heaton and Kanamori, 1984)? Heaton and Hartzell (1986) summarized data from other subduction zones to argue that the observed historic quiescence of the thrust interface is not decisive evidence against the possibility of future great thrust-interface earthquakes. Furthermore, Atwater (1987) found strong evidence in Holocene tidal marsh sediments for great, prehistoric, thrust-fault earthquakes along the coast of Washington. The situation is complicated by an apparent segmentation of the subduction zone (Silver, 1978; Weaver and Smith, 1983; Weaver and Michaelson, 1985): conceivably, some segments are locked and storing strain energy, to be released in future thrust-fault earthquakes, whereas other segments are being subducted aseismically (Weaver and Smith, 1983).

CORDILLERA EAST OF THE SIERRA NEVADA

The Cordillera east of the Sierra Nevada (Fig. 2) is characterized by predominantly extensional tectonics, with most earthquakes occurring as the result of normal, oblique-normal, or strike-slip faulting. Directions of extension in most of the Cordillera are east-west or southeast-northwest (Zoback and Zoback, this volume). Several models would explain the extension as primarily a response to lessening compressive stress across the western boundary of the North American plate (Scholz and others, 1971; Stewart, 1978; Sbar, 1982), corresponding to a progressive decrease in the length of Pacific Coast subduction zones from mid-Cenozoic time to the present (Atwater, 1970). Other models postulate an effect in the Cordillera from distributed, northwest-trending, right-lateral shear between the Pacific plate and the interior of the North American plate, which would result in relative extension across north-trending faults (Atwater, 1970; Hill, 1982).

The seismogenic zone in most of the Cordillera east of the Sierra Nevada extends to a depth of about 15 km (Smith, 1978). The Yellowstone National Park (Fig. 2) region provides an example of the effect of elevated geotherms on thinning the seismogenic zone; the seismogenic zone in regions of highest heat flow within the park extends to a depth of only 5 km or so (Trimble and Smith, 1975; Smith and others, 1977; Pitt and others, 1979; Pitt and Hutchinson, 1982). Earthquakes with depths exceeding 40 km and extending down to nearly 90 km have occurred in the upper mantle beneath the Wasatch Mountains (Fig. 2) and the Colorado Plateau (Fig. 2) (Zandt and Richins, 1980; Wong and others, 1984). Upper-mantle earthquakes that cannot be associated with subducted lithosphere, such as these Cordilleran shocks, have been recognized in other parts of the world, but they are uncommon and poorly understood (Chen and Molnar, 1983).

Characteristics of faulting in large Cordilleran earthquakes

Like their strike-slip counterparts in the San Andreas fault system, well-studied normal and oblique-normal faults in the Cordillera are made up of smaller segments that are either offset from each other along strike or that have different strikes (Schwartz and Coppersmith, 1984). As in the San Andreas system, there is evidence that fault segmentation may partly determine the sizes of the largest, or characteristic, earthquakes on a normal or oblique normal fault (Schwartz and Coppersmith, 1984).

The zone of surface faulting associated with a major normal or oblique-normal earthquake may have a somewhat different orientation or dimensions than the orientation or dimensions of the fault plane at depth that is inferred from seismologic or geodetic observations. These discrepancies are in most cases probably due to the modification of the main shock dislocation by near-surface rock that has different mechanical properties than rock at hypocentral depths. On a broad scale, the seismologic, geodetic, and geologic data are commonly mutually consistent with the hypothesis that the zone of surface faulting is the outcrop of the buried fault that produced the main shock, rather than a secondary effect arising from sediment compaction or aseismic settling of crustal blocks.

The Borah Peak earthquake of October 28, 1983 (M_S = 7.3, USGS; Fig. 11), provided examples of the complexities that may accompany even comparatively simple normal faulting. The main shock involved primarily oblique-normal slip on a single plane whose strike and sense of displacement are similar to those of the surface fault scarps (Crone and others, 1987; Doser and Smith, 1985; Richins and others, 1987; Scott and others, 1985; Stein and Barrientos, 1985; Dewey, 1987). The causative fault is a conspicuous range-front fault that had been mapped and trenched prior to the earthquake (Hait and Scott, 1978). The most recent scarp-forming event prior to the Borah Peak earthquake had occurred in late Pleistocene or early Holocene time (Scott and others, 1985) and produced surface displacements similar to those of the 1983 earthquake (Schwartz and Crone, 1985). Complexities in the Borah Peak earthquake process include large variations in the amount of displacement from point to point on the fault plane (Ward and Barrientos, 1986), the occurrence of some surface faulting that does not extend to seismogenic depths (Ward and Barrientos, 1986), and late aftershock activity extending north of the main-shock rupture surface by an

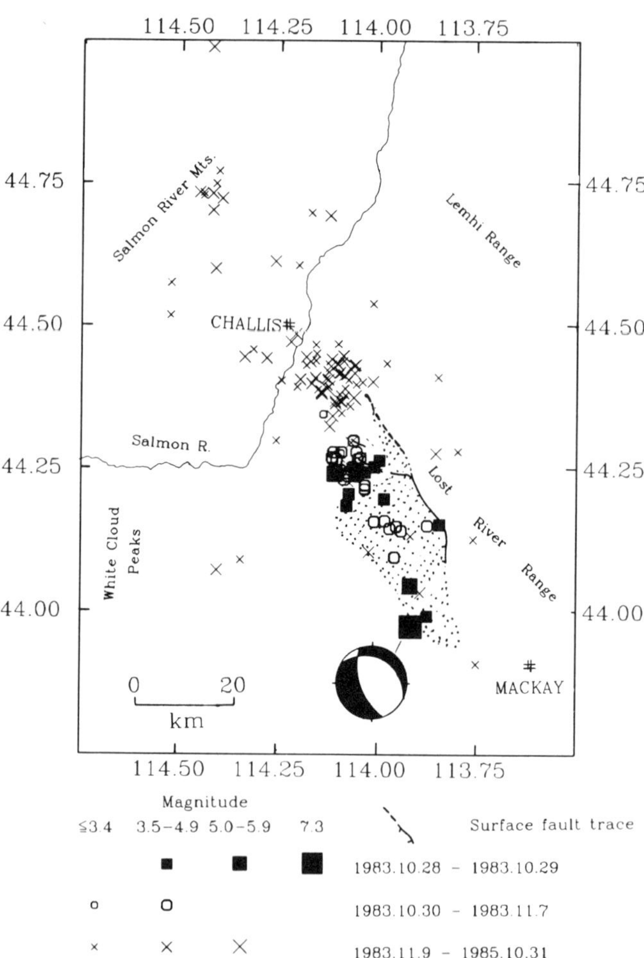

Figure 11. Epicentral region of the Borah Peak, Idaho, earthquake of October 28, 1983 (after Dewey, 1987). Main shock focal mechanism (U.S. Geological Survey, 1984) is represented as in Figure 6. Surface fault traces are from Crone and others (1987). Stippled region is surface projection of fault rupture at depth (variable slip planar fault model of Ward and Barrientos, 1986).

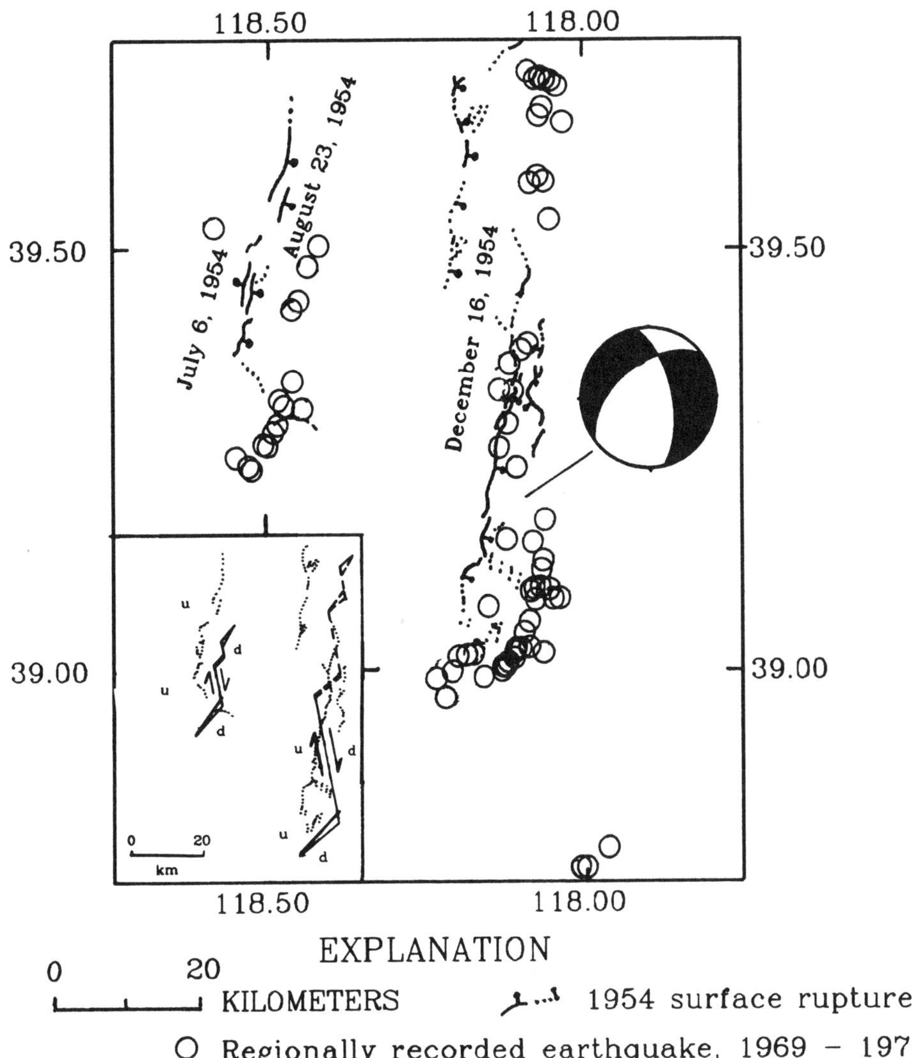

Figure 12. Epicentral region of the central Nevada earthquakes of 1954. Focal mechanism (portrayed as in Fig. 6) is that of the Fairview Peak earthquake (Romney, 1957). Regionally recorded earthquakes are those of Ryall and Malone (1971). Dates are those on which surface faulting occurred. Inset shows relation of deep-seated fault model of Ryall and Malone (solid and dashed lines) to surface faulting (dotted lines).

amount equal to the length of the rupture itself (Fig. 11; Zollweg and Richins, 1985; Dewey, 1987).

The Fairview Peak, Nevada, earthquake of December 16, 1954 (M = 7.1, Richter, 1958; Fig. 12), is an example of a major earthquake that apparently involved slip on several segments with different orientations, with the strikes of fault segments near the surface differing by tens of degrees from the strikes of fault segments at depth. The earthquake was the largest of a sequence of central Nevada shocks that began in July 1954; the sequence included four earthquakes of magnitude greater than 6.5, each of which was associated with surface fault rupture (Richter, 1958). The body-wave radiation pattern implies that the Fairview Peak earthquake nucleated at a depth of approximately 15 km and involved right-lateral, normal-oblique slip on a fault striking about N11°W (Romney, 1957; Okaya and Thompson, 1985; Doser, 1986). This strike corresponds to one of several intersecting faults identified by Stauder and Ryall (1967) and by Ryall and Malone (1971) in the distribution and focal mechanisms of aftershocks recorded more than a decade after the main shock (Fig. 12). The surface fault scarps associated with the Fairview Peak earthquake (Slemmons, 1957) trend north-northeast, about 20° clockwise of the strike of the main-shock fault suggested by seismologic data (Fig. 12). Snay and others (1985) inverted geodetic observations to infer a coseismic slip on a deep-seated fault (depth greater than 6 km) parallel to the seismologically determined main shock fault plane; they also infer slip on upper crustal faults (depths less than 6 km) parallel to, and downdip from, the surface scarps.

Other major Cordilleran earthquakes for which surface fault traces appear to have been distorted representations of deep-seated main-shock faults were the Pleasant Valley, Nevada, earthquake of October 3, 1915 (M = 7¾, Gutenberg and Richter, 1954; Wallace, 1984; Fig. 3), the Cedar Mountain, Nevada, earthquake of December 21, 1932 (M = 7.3, Richter, 1958; Fig. 3), and the Hebgen Lake earthquake of August 18, 1959 (M_L = 7.1, PAS; Myers and Hamilton, 1964; Fig. 2).

Major late Cenozoic faults that have not historically produced large earthquakes

One of the most important results from the expansion of regional seismographic networks in the 1970s is the demonstration that many conspicuous normal or oblique-normal, late Cenozoic faults in the Cordillera are quiescent even at the small and moderate magnitude levels. Examples of conspicuous Holocene or late Cenozoic normal faults that have been monitored by local networks (including temporary microearthquake networks) and found to be largely aseismic are as follows: the Wasatch fault, Utah (Fig. 2; Arabasz and Smith, 1981); the Teton fault in Wyoming (Fig. 2; Doser and Smith, 1983); the Sangre de Cristo fault in Colorado (Fig. 2; Keller and Adams, 1976; Kirkham and Rogers, 1981; McCalpin, 1982); the La Jencia fault, near Socorro, New Mexico (Fig. 2; Machette, 1986), and many faults in Nevada and adjacent eastern California (Van Wormer and Ryall, 1980).

A subset of major late Cenozoic faults that have high levels of small-earthquake seismicity on or near their fault planes are the faults that have produced major earthquakes in the past several decades. Such faults typically remain seismically active for many decades following the occurrence of a large earthquake (Ryall, 1977; Wallace, 1981).

These seismicity characteristics suggest that many major faults of the Cordillera have seismic cycles involving relatively short periods of high activity—a large earthquake and associated aftershocks—separated by much longer periods of quiescence (Ryall, 1977). In the Cordillera east of the Sierra Nevada, geologically inferred recurrence times of most major late Cenozoic faults are a few thousand years or longer (Wallace, 1981). Accordingly, our seismicity data apply to only small parts of the seismic cycles on individual faults. The inference of long-quiescence seismic cycles is reasonable, however, in light of similar behavior on much more active, plate-boundary faults (see the section on the San Andreas fault system). The inference is also consistent with historically long quiescences observed for major seismogenic faults in extensional terranes of the Middle East and China (Allen, 1975).

One family of major late Cenozoic faults that may not be seismogenic are the low-angle, listric, normal faults that are prominent in geologic and seismic sections of the shallow crust. Many authors (e.g., Jackson and McKenzie, 1983; Smith and Bruhn, 1984; Stein and Barrientos, 1985; Arabasz and Julander, 1986) have emphasized the paucity of evidence in hypocenter distributions and focal mechanism orientations for seismogenic faulting on low-angle normal faults at any depth in the Cordillera east of the Sierra Nevada or in continental extensional terranes worldwide.

Earthquakes that occur away from the major late Cenozoic Cordilleran faults

The level of small and moderate earthquake activity is commonly high within tens of kilometers of major Holocene normal faults, even if the faults themselves are quiescent. Focal mechanisms of most small and moderate off-fault earthquakes in the Cordillera are consistent with these shocks resulting from a regional extensional stress field in which the axis of least compressive stress has the same orientation as that controlling the large earthquakes and the conspicuous normal or oblique-normal late Cenozoic faults (e.g., Smith and Lindh, 1978; Stickney, 1978; Vetter and Ryall, 1983; Wong and others, 1984; Krueger-Knuepfer and others, 1985; Jones and Dollar, 1986; Arabasz and Julander, 1986). Strike-slip as well as normal-slip focal mechanisms commonly occur in the same geographic area, implying either variation in the orientation of maximum compressive stress (Vetter and Ryall, 1983) or variations in the orientation of preexisting planes of weakness in a tectonic stress field in which the axes of maximum compressive stress and of intermediate stress are approximately equal (Zoback, 1989).

The Pocatello Valley, Idaho, earthquake of March 28, 1975 (M_S = 6.0, USGS) is an example of a damaging earthquake that did not occur on any of the geologically prominent faults in its epicentral region. The main-shock focal mechanism (Bache and others, 1980), geodetically measured deformation in the epicentral region, and the general trend of locally recorded aftershocks (Arabasz and others, 1981) are best fit by a normal fault that strikes northeast, oblique to the northerly trend of surface topography and to the northerly strikes of previously mapped faults in the region (Fig. 13). The causative fault may extend from depth upward to the subsurface basement beneath Pocatello Valley (Arabasz and others, 1981). Arabasz and Julander (1986) hypothesized that other seismogenic midcrustal faults in Utah are truncated above by shallow low-angle detachments and are therefore not mappable in near-surface geology.

Seismicity associated with molten magma bodies in the Cordilleran extensional terrane

Several loci of high seismicity in the Cordillera east of the Sierra Nevada and Cascade ranges are associated with geologically recent volcanism. As is commonly observed in volcanic regions worldwide (Richter, 1958), earthquakes near Cordilleran magmatic centers tend to occur in swarms. Over a period of months or years, shocks tend to be distributed over many faults rather than concentrated on a single fault.

The Yellowstone caldera, northwestern Wyoming (Fig. 2), and the Long Valley caldera, eastern California (Fig. 14), are

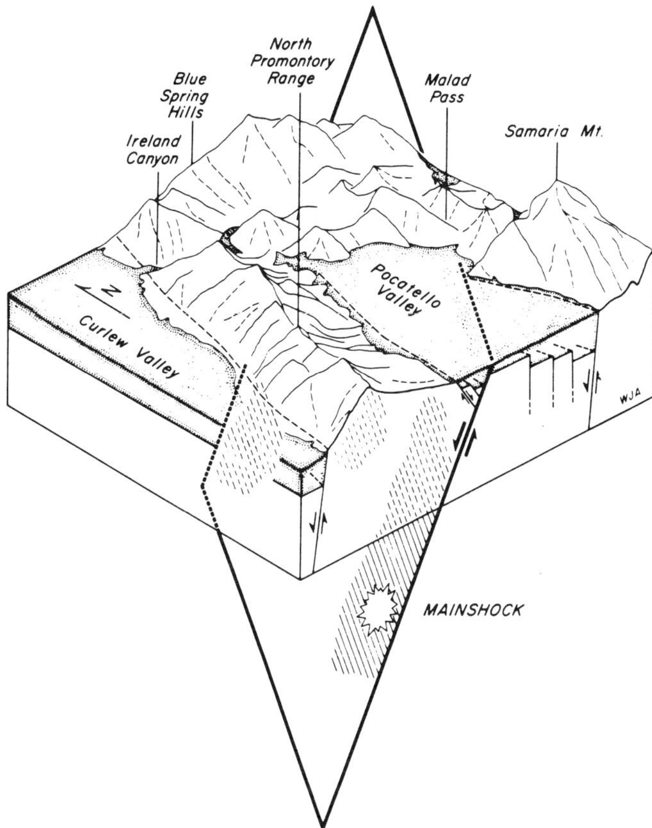

Figure 13. Schematic representation of causative fault of Pocatello Valley earthquake of March 28, 1975 (Arabasz and others, 1981). The dimensions of the crustal block are about 20 km north-south, 25 km east-west, and 7 km vertically. The location of Pocatello Valley is shown in Figure 2.

outstanding examples of seismically active magmatic centers. In addition to numerous small earthquakes, both of these large silicic calderas have in recent years or decades experienced uplift of their central resurgent domes by tens of centimeters and earthquakes of magnitude greater than 6 nearby (Smith and Christiansen, 1980; Hill and others, 1985). Other large crustal magma bodies with high levels of earthquake activity are near Socorro, New Mexico (Fig. 2; Sanford and others, 1979; Jaksha and Sanford, 1986), and at several locations in eastern California and western Nevada besides the Long Valley caldera (Van Wormer and Ryall, 1980; Walter and Weaver, 1980).

At Long Valley, California (Fig. 14), the recent earthquake activity has been directly linked with a new intrusion of magma into the upper crust (Savage and Cockerham, 1984). Focal mechanisms and lineations of epicenters do not have a clear relation to faults mapped at the surface. The west-northwest–trending band of intense seismicity within the Long Valley caldera (Fig. 14) is associated with right-lateral strike-slip displacements; Savage and Cockerham (1984) and Rundle and Whitcomb (1984) proposed different mechanisms by which injection of magma may have perturbed the local stress field so as to favor such faulting. The three north-northeast–striking lineations in the Sierra Nevada south of the caldera (Fig. 14) were activated during a sequence of four M_L (BRK) = 6.0 earthquakes in May 1980. Focal mechanisms from two of the magnitude 6^+ earthquakes and many of the smaller events indicate that these lineations correspond to nearly vertical fault planes with predominantly left-lateral slip (Vetter and Ryall, 1983; Lide and Ryall, 1985). One of the magnitude six earthquakes and an M_L = 5.8 (BRK) earthquake that occurred 20 km southeast of Crowley Lake in October, 1978, had focal mechanisms that could not be fit by a simple double-couple fault-plane solution. Julian (1983) proposed that these earthquakes had compensated linear-vector-dipole (CLVD) mechanisms and that they resulted from the rapid, forceful injection of magma into north-northeast–striking cracks (dikes). Controversy over this hypothesis of a non-shear-failure focal mechanism continues (Julian and Sipkin, 1985; Lide and Ryall, 1985; Wallace, 1985).

CENTRAL AND EASTERN UNITED STATES

The United States east of the Cordillera is a midplate tectonic environment. Historical and instrumental epicenters do not define continuous belts of deformation that connect with major plate boundaries. Rather, they are concentrated in zones with dimensions of hundreds of kilometers superimposed on a background of isolated epicenters and small clusters of epicenters (Fig. 1). Historically active seismic zones show a general spatial association with zones of Paleozoic extension in the interior of the North American craton and with regions that experienced high deformation during periods of Paleozoic and early Mesozoic time when the eastern United States was near a plate boundary (Fig. 2). This association may be due to the old zones of deformation containing numerous unhealed faults, some of which are favorably oriented to slip in response to the current stress field (e.g., Sbar and Sykes, 1973; Sykes, 1978). Alternatively, the old rifts and near-plate–boundary regions may concentrate seismicity by virtue of an anomalous bulk property within the crust or upper mantle, such as higher than average porosity (Costain and others, 1986) or greater ductility below the seismogenic layer (Zoback, 1983).

In most of the eastern and central United States, the stress field is characterized by a nearly horizontal, northeast- through east-striking axis of maximum compressive stress (Sbar and Sykes, 1973; Herrmann, 1979; Haimson and Doe, 1983; Zoback and Zoback, this volume). Most earthquakes have strike-slip, oblique-reverse, or reverse fault mechanisms. The uniformity of stress-tensor orientation over a broad area of the eastern and central United States suggests that the stress field arises from forces that drive or resist plate motions. Finite-element modeling of tectonic plates (Richardson and others, 1979) and qualitative reasoning (Zoback and Zoback, 1981) point to ridge-push forces and/or drag forces at the base of the North American plate as the most likely sources of the midplate stress field. Departures from the average stress field of the eastern and central United States

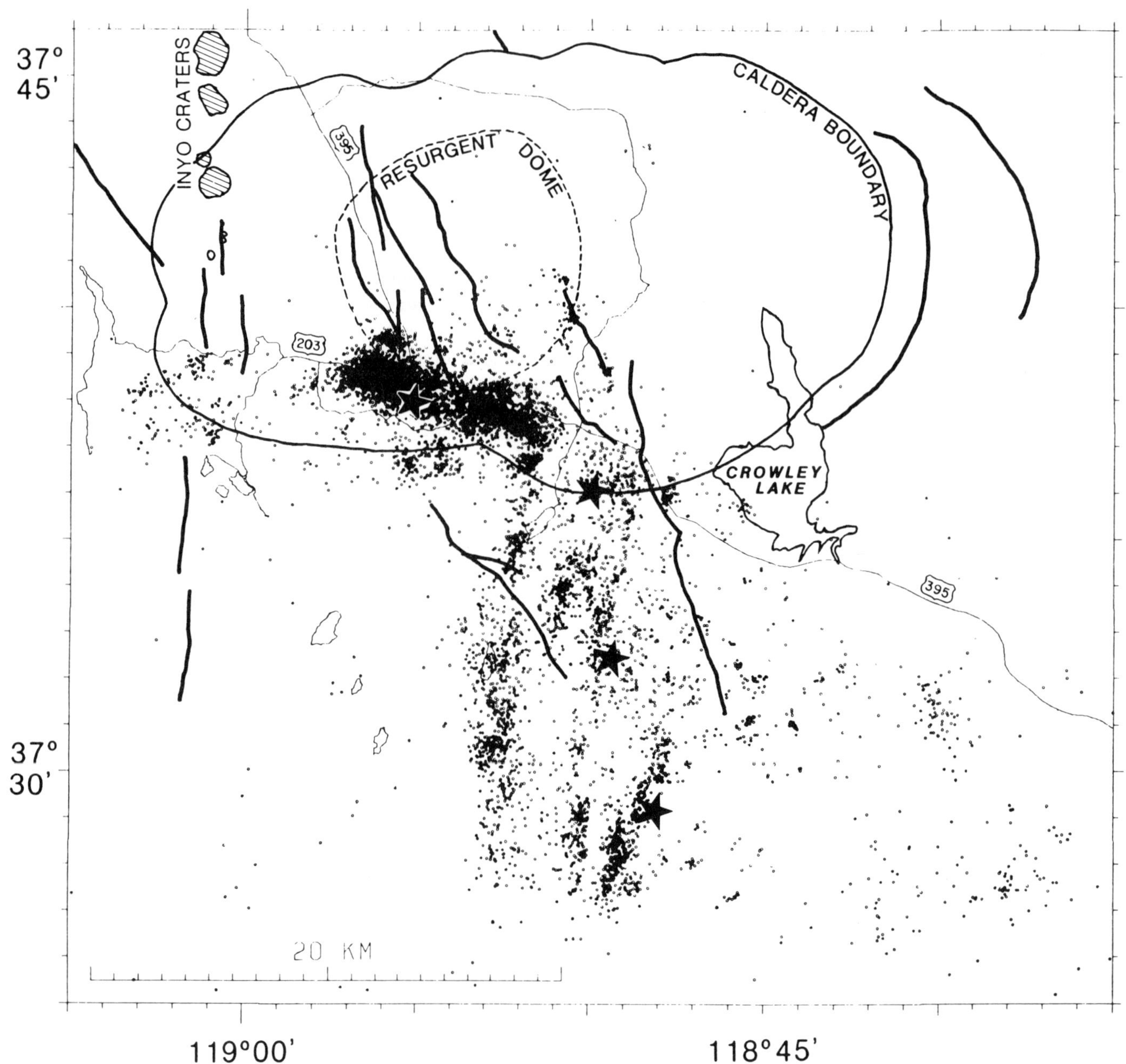

Figure 14. Map showing epicenters of earthquakes of M>1 in the vicinity of Long Valley caldera for the period June 1, 1982, through July 31, 1984. Stars indicate epicenters of the four M>6 earthquakes that occurred on May 25–27, 1980. Heavy lines are faults, light lines are roads. Rhyolite domes of the Inyo Craters were erupted 500 to 600 yr ago.

may be due to density contrasts within the lithosphere or to a variety of loading or unloading phenomena (Richardson and others, 1979; Nunn, 1985).

Inferences on the geologic characteristics of seismogenic faults in the eastern and central United States are typically based on indirect evidence. Surface faulting has not been observed to accompany any U.S. earthquake east of the Cordillera, and geologic conditions at most locations are not favorable to recognition of small amounts of youthful displacement such as might characterize a fault whose large earthquakes have very long interoccurrence times (Prowell, 1983). Large faults that were originally formed before Cenozoic time have long been prime suspects for seismogenic faults in the eastern and central United States, on the assumption that they would persist as zones of weakness. A number of such faults are now known to have slipped in Cenozoic time, while the region was in a midplate tectonic regime as it is today (York and Oliver, 1976; Howard and others, 1978; Prowell, 1983; Wentworth and Mergner-

Keefer, 1983). Although those faults with recognized Cenozoic displacements have not been clearly shown to be sources of present-day earthquakes, their presence in the eastern and central United States strongly suggests that some large, pre-Cenozoic faults have been reactivated and are potentially seismogenic. In addition, there are examples of several historically active eastern or central U.S. seismic zones for which subsurface geophysical data suggest reactivated pre-Cenozoic faults as earthquake sources. Most sources east of the Cordillera, however, have not been convincingly linked with a specific geologically mapped or geophysically imaged fault. Much work remains on identifying characteristics of pre-Cenozoic faults that would be indicative of a current capability to generate earthquakes. It is possible that some strong earthquakes have occurred on previously minor faults or fractures that would not have been conspicuous even if well exposed.

The maximum reliably determined focal depths of central and eastern U.S. earthquakes are 25 to 30 km (Chen and Molnar, 1983), but these depths are known from only relatively few sources. Special relocations of shocks in the most active part of the Mississippi Embayment (Nicholson and others, 1984; Andrews and others, 1985) show them to occur, with few exceptions, above 15 km. The seismogenic zone in active parts of the Valley and Ridge province extends to 25 or 30 km (Bollinger and others, 1985; Johnston and others, 1985), but beneath the South Carolina Coastal Plain it does not appear to extend below 15 km (Tarr and Rhea, 1983). Some persistent sources of small earthquakes in upstate New York and New England are confined to the uppermost few kilometers of the Earth's crust (Sbar and Sykes, 1977; Pulli and Guenette, 1981; Ebel and others, 1982). Bollinger and others (1985) hypothesized that systematic differences in focal depths between the Valley and Ridge province and the Piedmont/Coastal Plain region reflect differences in the compositions of the source rock or in strain rate. "In situ" stress measurements suggest that the shallow focal depths of some midplate regions are due to high shear stresses existing only near the free surface (Zoback and Hickman, 1982).

Amarillo-Wichita uplift and Anadarko basin

Earthquakes of southern Oklahoma and the Texas panhandle appear to be broadly associated with the Amarillo-Wichita uplift and the Anadarko basin (Fig. 15; Gordon, 1983). The Meers fault, an element of the Paleozoic fault system separating the Anadarko basin from the Amarillo-Wichita uplift, was reactivated at least once in the Holocene to produce a surface scarp approximately 26 km long with a maximum displacement of nearly 5 m. The most recent fault-slip probably caused an earthquake of magnitude greater than 6.5 (Gilbert, 1982; Donovan and others, 1983; Slemmons and others, 1985; Madole and Rubin, 1985). The region within 10 km of the Holocene scarp has been free of shocks larger than magnitude 3 for at least 25 yr (Fig. 15; Lawson, 1985). This section of the Meers fault thus appears similar to fault segments in the western United States that are

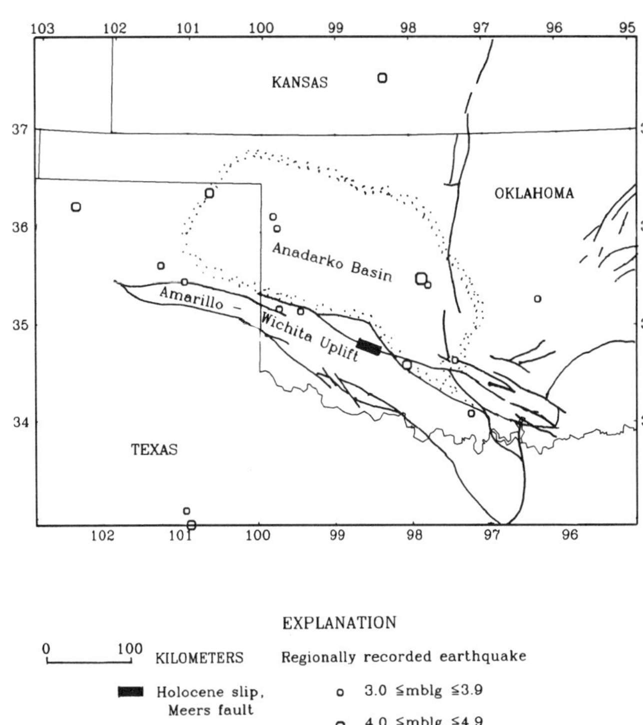

Figure 15. Seismicity of the Amarillo-Wichita Uplift and Anadarko basin. Epicenters are those of instrumentally recorded earthquakes (m_bL_g = 3.0 or greater) occurring through 1980 (Gordon, 1983). Stippled outline of Anadarko basin encloses 2,500 m, below sea level, contour on Precambrian basement.

currently quiescent but that have earlier produced large earthquakes (see the sections above on the San Andreas fault system and the Cordillera east of the Sierra Nevada). The largest historical earthquake in the region of Figure 15, the El Reno, Oklahoma, earthquake of April 9, 1952 (35.5°N, 97.8°W), had an m_b (L_g) magnitude of 5.0 (Gordon, 1983).

Mississippi Embayment

The Mississippi Embayment seismic zone (Fig. 16) was the site of the largest earthquakes to have affected the United States east of the Cordillera in historic time. Nuttli (1973) estimated from intensity observations that the three largest earthquakes in the New Madrid sequence of 1811 and 1812 had magnitudes greater than 7, and that 15 of the aftershocks of these events had magnitudes greater than 6.

Seismicity in the Mississippi Embayment is associated with a number of discrete faults and fault segments within a broader northeast-trending zone. Instrumentally recorded earthquakes define several linear trends with different strikes within the overall zone (Fig. 16; Stauder and others, 1976; Stauder, 1982; Gordon, 1983; Andrews and others, 1985). The focal mechanisms of these

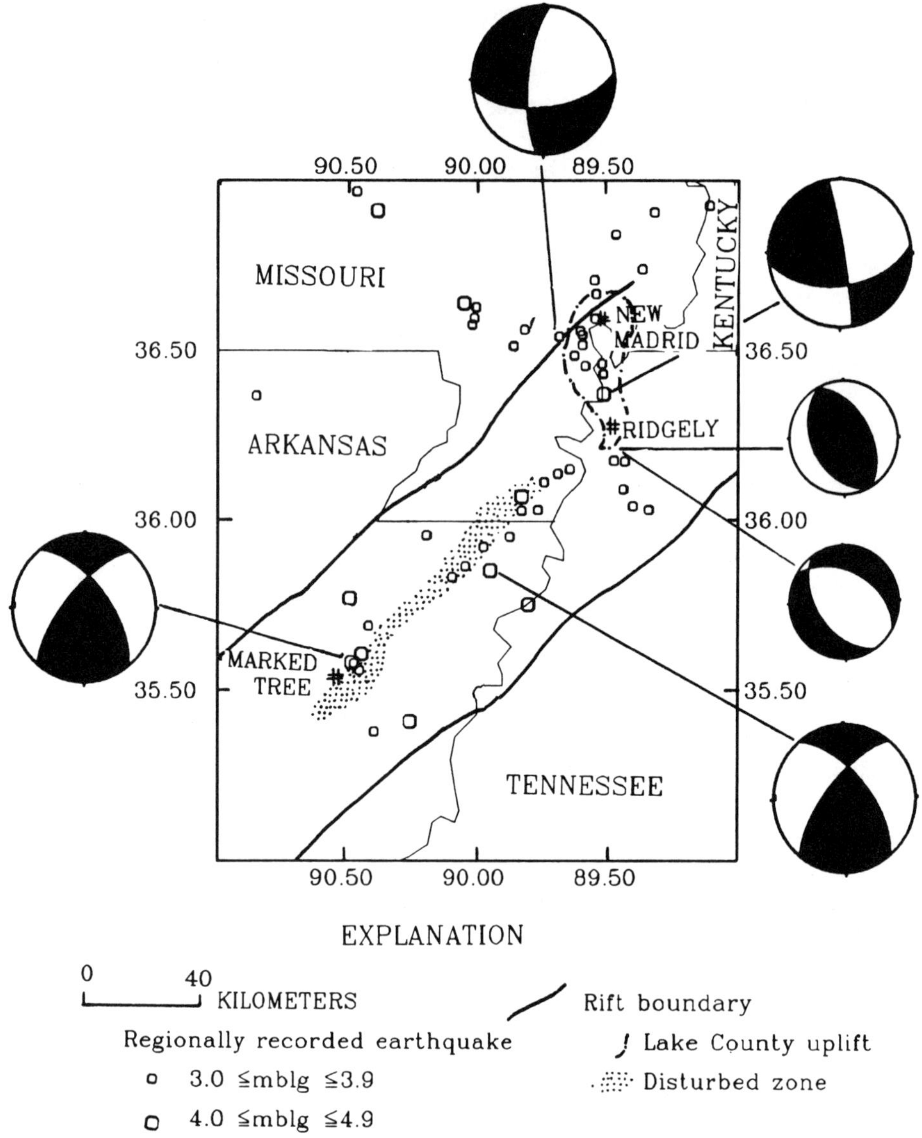

Figure 16. The Mississippi Embayment seismic zone. Epicenters of regionally recorded earthquakes are those of Gordon (1983). Focal mechanisms represented by larger symbols are single-event mechanisms of Herrmann and Canas (1978); those represented by smaller symbols are composite mechanisms of Nicholson and others (1984). Focal mechanism conventions of Figure 6 apply. Disturbed zone is that found in subsurface reflectors by Crone and others (1985). Lake County uplift is as mapped by Russ (1982). Rift boundary is as defined by Hildenbrand (1985).

shocks usually have one nodal plane parallel to the linear trends defined by the epicenters, as is expected if the linear trends correspond to individual fault zones (Fig. 16; Herrmann and Canas, 1978; Nuttli, 1979; Nicholson and others, 1984; Andrews and others, 1985). Intensity data imply that the largest shocks of the 1811–1812 sequence occurred at several distinct locations within the overall northeast-trending zone (Nuttli, 1973, 1979), and it is reasonable to suppose that individual faults and fault segments defined in the present-day seismicity may each correspond to the source of one of the large shocks of 1811–1812.

The seismogenic faults of the Mississippi Embayment probably existed prior to the 1811–1812 earthquake sequence; some may have originally formed in Precambrian time. The northeast-trending subzone of epicenters between Marked Tree, Arkansas, and Ridgely, Tennessee (Fig. 16), is on the axis of a late Precambrian–early Paleozoic rift (Ervin and McGinnis, 1975; Hildenbrand and others, 1982; Hildenbrand, 1985). Northeast-striking faults with post–Upper Cretaceous displacements of 50 to 80 m have been identified in seismic reflection data at the northern end of the Marked Tree–Ridgely lineament (Zoback and others, 1980). That lineament is also well correlated with a narrow band of disrupted reflection horizons in seismic data, situated beneath horizons corresponding to the base of the Paleozoic, which Crone and others (1985) have interpreted as being due to pre-Late Cretaceous intrusive activity and associated faulting (Fig. 16).

The north-northwest-trending subzone of seismicity between Ridgely, Tennessee, and New Madrid, Missouri, is coincident with the Lake County uplift, which has risen approximately 10 m in Holocene time (Russ, 1982); much of the uplift occurred between 2,000 yr BP and the occurrence of the 1811–1812 earthquakes (Russ, 1982). Focal mechanisms of shocks beneath the Lake County uplift imply a number of different fault orientations and styles of fault displacement (Herrmann and Canas, 1978; O'Connell and others, 1982; Nicholson and others, 1984; Andrews and others, 1985). Nicholson and others (1984) found that the distribution of hypocenters and the focal mechanisms at the southern end of the uplift are well explained in terms of primary reverse faulting on a northwest-striking plane, with normal faulting occurring as secondary deformation within the hanging wall of the primary reverse fault.

The recent persistence of small and moderate earthquakes on the faults that are thought to have slipped during the 1811–1812 earthquakes contrasts with the quiescence that tends to settle upon major Holocene faults of the western United States in the decades following large earthquakes. In addition, the 600-yr recurrence times of large Mississippi Embayment earthquakes implied by geologic data for the past 2,000 yr (Russ, 1982) are far shorter than the recurrence times suggested by the low rate of average deformation throughout Cenozoic time (McKeown, 1982; Crone and Brockman, 1982). It is possible that the whole episode of high deformation of the past several thousand years is best viewed as the active part of a very long seismic cycle whose quiescent period may last millions of years.

The intense seismicity on the presumed 1811–1812 faults is bordered by a diffuse zone of seismicity in eastern Missouri, southern Illinois, and southern Indiana that may be associated in general with rift structures branching northward from the Mississippi Embayment (Figs. 1, 2; Braile and others, 1982). Gordon (1983) noted that a swarm of shallow earthquakes in southernmost Illinois had a location and focal mechanism consistent with slip on a fault mapped by Heyl and McKeown (1978). Focal mechanisms in the Ozarks of Missouri and Arkansas have commonly involved a component of normal faulting (Mitchell, 1973; Haar and others, 1984), which is unusual for shocks of midplate North America.

Isolated sources northwest of the Appalachians

Isolated sources of moderate-magnitude (5.5 or less, Nuttli, 1979; Street and Turcotte, 1977) earthquakes near Attica, New York (Fig. 17), and Anna, Ohio (Fig. 2), may be occurring on faults that have experienced less than 100 m of vertical displacement in Paleozoic time or later. The Attica source lies within kilometers of the Clarendon-Linden fault zone; focal mechanism solutions for two regionally recorded events have nodal planes that strike parallel to the fault zone (Fig. 17). The fault zone has

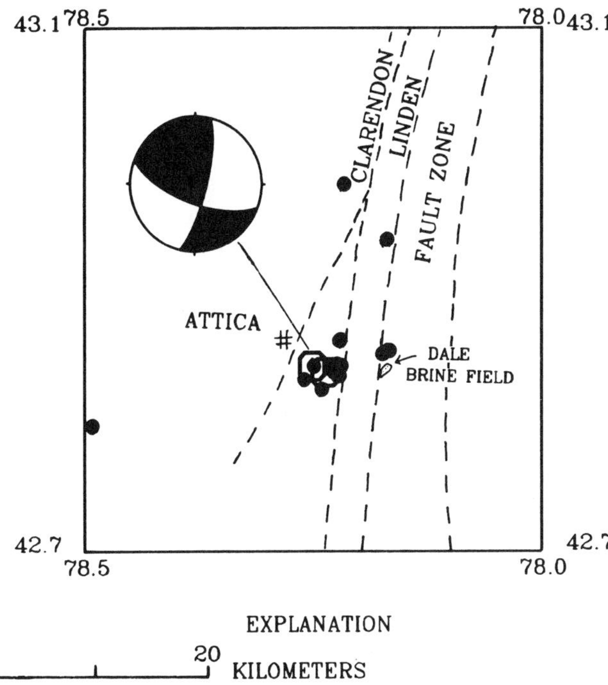

Figure 17. Instrumentally recorded earthquakes from the Attica, New York, region. Clarendon-Linden fault zone is from Van Tyne (1975), as shown in Fletcher and Sykes (1977). Epicenters of regionally recorded earthquakes are from Dewey and Gordon (1984); epicenters of microearthquakes are from Fletcher and Sykes (1977). Focal mechanism is from Herrmann (1978) and is for both regionally recorded shocks; it is represented as in Figure 6. Epicenters of shocks in the Dale Brine Field, a site of induced seismicity, are not plotted individually.

from 30 to 90 m of post-Devonian vertical displacement (Hutchinson and others, 1979). At Anna, intersecting faults with maximum post-Ordovician vertical displacements of tens of meters have been inferred from structural contours on Paleozoic strata and from proprietary seismic reflection data (Stone and Webster Engineering Corporation, 1976). Epicenters of regionally recorded (Dewey and Gordon, 1984) and locally recorded (Pollack and Christensen, 1984) earthquakes, and a composite focal mechanism for the locally recorded shocks (Pollack and Christensen, 1984), are consistent with these earthquakes occurring as slip on one or more of the inferred faults.

Southern Valley and Ridge

Instrumentally recorded earthquakes in the southern Appalachians are concentrated beneath the Valley and Ridge province, near the western edge of that part of the Appalachians in which thin-skin overthrusting has carried eugeosynclinal or miogeosynclinal rock over Precambrian basement or platform deposits on Precambrian basement. The earthquakes have focal depths concentrated between 8 and 16 km, below the 3- to 6-km depth of the Appalachian detachment in the active region (Bollinger and Wheeler, 1983; Johnston and others, 1985). The geologic characteristics of the source regions of the earthquakes are therefore obscured by the zone of late Paleozoic folding and thrust faulting above the detachment. Bollinger and Wheeler (1983) suggested that shocks in the Giles County, Virginia, region (Fig. 2) occur on reactivated faults that originally formed prior to, or coincident with, the emplacement of the overthrust sheet. The largest historical earthquake in the southern Valley and Ridge was the magnitude 5.8 (Nuttli and others, 1979) Giles County, Virginia, earthquake of May 31, 1897.

South Carolina Coastal Plain

The destructive Charleston, South Carolina, earthquake of August 31, 1886, probably had a magnitude of 6.5 to 7.0 (Nuttli and others, 1979). Most data on the 1886 earthquake and on the seismicity of the South Carolina Coastal Plain are consistent with the source of the earthquake having had dimensions of tens of kilometers and having been situated about 30 km west of Charleston, close to the region of maximum intensity (Dutton, 1889; Bollinger, 1983; Fig. 18). These data include the far-field intensity observations of the 1886 main shock (Nuttli and others, 1979) and the epicenters of microearthquakes recorded in the past decade from the South Carolina Coastal Plain (Talwani, 1982; Tarr and Rhea, 1983). The latter define a zone whose position, overall north-northeast trend, and length are remarkably similar to the orientation and length of the zone of maximum intensities in 1886 (Fig. 18). The observation that aftershocks in the years immediately following 1886 occurred over a large area of the South Carolina Coastal Plain, rather than exclusively in the region of maximum intensities (Seeber and Armbruster, 1987), does not, however, have a straightforward interpretation in terms of a source with dimensions of tens of kilometers. Seeber and Armbruster (1987) cited this observation as support for their hypothesis (Seeber and Armbruster, 1981) that the 1886 earthquake occurred as the result of slip over hundreds of kilometers of a nearly horizontal detachment surface.

Post–Late Cretaceous vertical deformation has been small in the 1886 meizoseismal region. Seismic reflection surveys (Hamilton and others, 1983; Behrendt and others, 1983) show evidence for several faults that have experienced only modest (tens of meters) vertical displacement in Cenozoic time (Fig. 18). However, Obermeier and others (1985) and Talwani and Cox (1985) have documented the occurrence of at least two prehistoric earthquakes in the past 3,000 yr that produced sandblows at a site that also produced sandblows in 1886. The frequency of large Holocene earthquakes that is implied by the sandblow observations appears to be substantially higher than the frequency of large earthquakes implied by post–Late Cretaceous deformation (e.g., Wentworth and Mergner-Keefer, 1983). In the last several thousand years, the South Carolina region may have been in a time of high seismicity, anomalous with respect to the Cenozoic as a whole, similar to that suggested for the Mississippi Embayment (see the section on the Mississippi Embayment).

Middle Atlantic states

Historic seismicity within the coastal regions of the Middle Atlantic states, downstate New York through central Virginia, has consisted of small and moderate shocks, with the largest magnitude being about 5 (Nuttli and others, 1979; Nottis, 1983). Several zones of small earthquakes are in the general vicinity of major faults that were active at least as late as Mesozoic time, when the region was last near a major plate boundary.

The Stafford and Brandywine faults of central Virginia (Fig. 19) have been viewed as possibly seismogenic (Wentworth and Mergner-Keefer, 1983) because they have experienced late Cenozoic displacements (Mixon and Newell, 1977). Characteristics of small central Virginia earthquakes are inconsistent with these earthquakes occurring on the faults, but leave open the possibility that the quiescence of the faults is only temporary, like that of some of the major faults of the western United States. The small earthquakes have occurred diffusely over an equidimensional

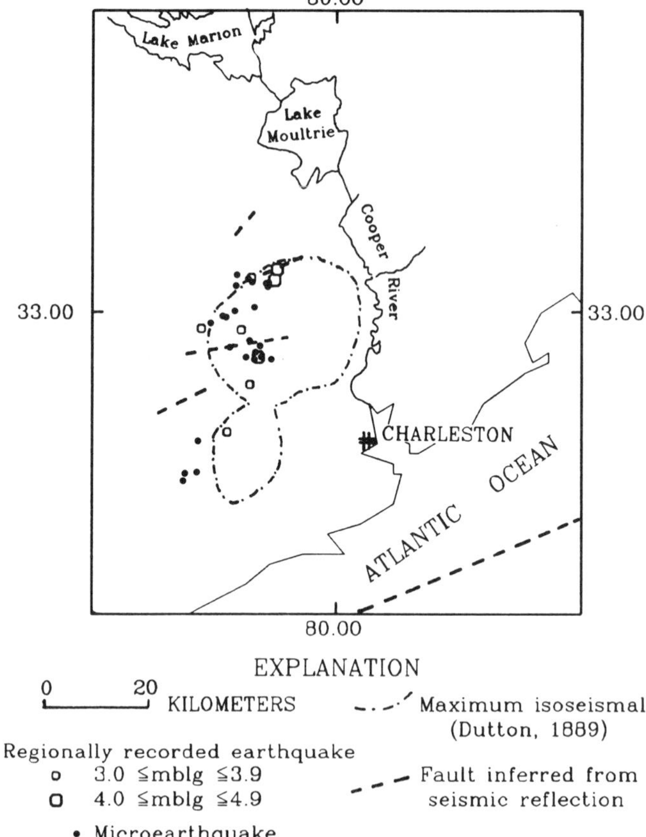

Figure 18. The region of strongest shaking in the 1886 South Carolina earthquake (Dutton, 1889), and epicenters of recent instrumentally recorded earthquakes (Dewey, 1983; Tarr and Rhea, 1983). Inferred faults are those of Hamilton and others (1983) and Behrendt and others (1983).

area, having dimensions of about 100 km (Fig. 19; Bollinger and Sibol, 1985), rather than being concentrated on lineations that are parallel to the strikes of the geologically mapped faults. Focal mechanisms imply many different fault orientations (Munsey and Bollinger, 1985), but most do not have north-northeast–striking nodal planes with reverse slip such as would be expected from earthquakes on the Stafford and Brandywine faults (Mixon and Newell, 1977). However, about half have nearly horizontal P-axes that trend west-northwest through west. A west-northwest through west-trending axis of maximum compressive tectonic stress would generally be favorable to reverse motion on north-northeast–striking dip-slip faults such as the Stafford and Brandywine.

Several recent studies have contested the judgement of Aggarwal and Sykes (1978) that the Ramapo fault in northern New Jersey (Fig. 2) is active. The fault was a basin-bounding normal fault in Mesozoic time; Aggarwal and Sykes suggested that it had been reactivated as a reverse fault in the current tectonic regime. A number of shocks have hypocenters near the fault surface (Kafka and others, 1985), and P-wave radiation patterns are consistent with many of the small earthquakes occurring as slip on the fault (Yang and Aggarwal, 1981). The first-motion data are also consistent, however, with reverse faulting on planes striking at a large angle to the Ramapo fault, and for some earthquakes the data require such mechanisms (Seborowski and others, 1982; Quittmeyer and others, 1985). In situ stress measurements obtained near the fault show that, at least at shallow depths (1 km and less), stresses are so oriented and of sufficient magnitude to cause faulting on northwest-striking planes, but not on planes parallel to the Ramapo fault (Zoback and others, 1985). Finally, studies of the fabric of the fault gouge suggest that the fault has not been reactivated since Mesozoic time (Ratcliffe, 1984).

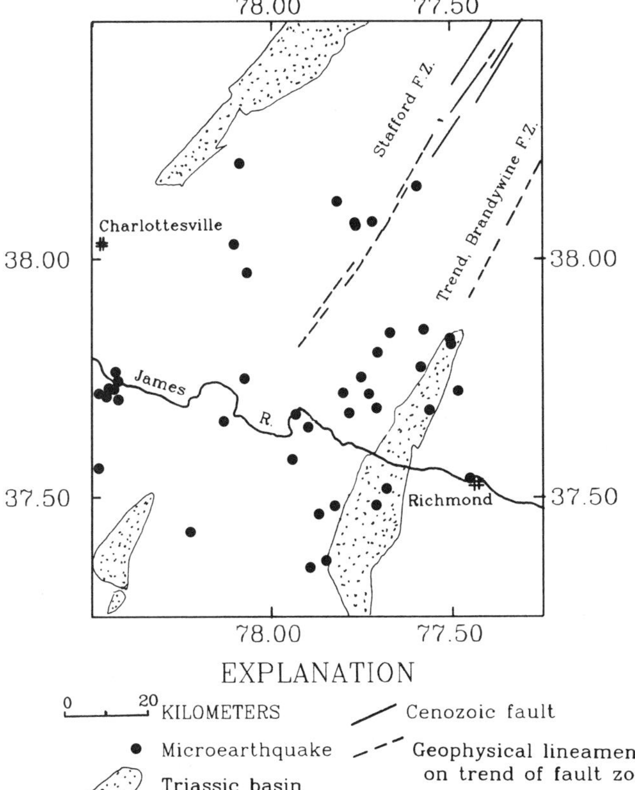

Figure 19. Central Virginia seismic zone. Microearthquake epicenters are from Bollinger and Sibol (1985). Faults and geophysical lineaments are as compiled by Mixon and Newell (1977).

New England and northeastern New York

Focal mechanisms of shocks in New England have diverse orientations (Pulli and Toksoz, 1981). If the shocks occur in response to a uniform tectonic stress field, the wide range of focal mechanism orientations suggests that they occur on preexisting planes of weakness, rather than as slip in previously intact rock (Gephart and Forsyth, 1985). In addition, some of these shocks occur near mapped or inferred faults (Ebel and others, 1982; Ebel, 1983). However, there is not yet a case in New England in which both distributions of earthquake hypocenters and focal mechanisms define planes that coincide with a previously recognized, geologically mapped fault.

Earthquake epicenters in central New England, upstate New York, and adjacent Canada are sometimes viewed as defining a single, northwest-trending belt of seismicity, usually called the Boston-Ottawa zone (Figs. 1, 2; Diment and others, 1972; Sbar and Sykes, 1973). The zone is transverse to the strike of the Appalachian Highlands, parallel to the direction of the early opening of the Atlantic, and on the continental projection of a major Mesozoic transform fault (Fletcher and others, 1978). The Boston-Ottawa zone does not correspond to a few, sharply defined geologic features, but appears to be associated with a broad region in which magmatism postdated the local opening of the Atlantic. Sykes (1978) noted that the zone's location and orientation with respect to the continental margin and its association with post-rifting magmatism make it similar to several other important midplate seismic sources worldwide. He suggested that the Boston-Ottawa zone, and similar zones situated elsewhere in midplate regions near the landward ends of transform faults, contain large numbers of preexisting fractures that determined the position of the transform faults in the first stages of sea-floor spreading. These fractures remained active into the period of sea-floor spreading, and they produce earthquakes sporadically in response to the present midplate stress regime. The largest historic earthquake in the U.S. part of the Boston-Ottawa zone was probably the Cape Ann, Massachusetts, earthquake of 1755, the intensity distribution of which suggests a magnitude of about 6 (Street and Lacroix, 1979).

EARTHQUAKES TRIGGERED BY HUMAN ACTIVITY

There is strong evidence that seismicity at several tens of localities in the United States has been induced by human activity. The variety of human disturbances to the Earth's surface or upper crust that have occasionally triggered earthquakes (Table 1) implies that there are a number of different mechanisms involved. In some cases, human intervention acts principally to locally increase shear stress in a volume of rock, whereas in other cases the human intervention changes the frictional properties of a fault surface (Simpson, 1986). For the latter, weakening of a previously locked fault (Healy and others, 1968) and local strengthening of a fault that previously was failing aseismically (Denlinger and Bufe, 1982; Pennington and others, 1986) have been suggested at different localities.

In many cases of induced seismicity, the human-caused increases in shear stresses or changes in fault strengths must have been small. Shocks associated with some quarrying or mining operations are thought to have been induced by changes in vertical load of less than 1 MPa (Pomeroy and others, 1976; Yerkes and others, 1983; Wong, 1984). Changes in strength of a preexisting fault due to the load and increased pore pressure from a large, newly impounded reservoir are about 1 MPa (Bell and Nur, 1978). These changes in stress or strength are orders of magnitude less than the strength of rock and imply that the rock was stressed to near the failure point prior to the human disturbance that triggered the earthquakes (Zoback and Hickman, 1982; Simpson, 1986).

CONCLUSIONS

Interpretation of regional seismicity by plate tectonic models. For two decades, it has been generally accepted that the historically active, northwest-striking, right-lateral faults of the San Andreas system constitute a transform boundary between interior North America and the Pacific. Recent work in coastal California accounts for the occurrence and focal mechanisms of earthquakes off the major right-lateral faults, and for variations in the thickness of the seismogenic zone, as consequences of differences in strike between the major right-lateral faults and the direction of pure transform motion between the North American and Pacific plates. Important uncertainties remain, however, in identifying the geologic structures that accommodate the fraction of relative plate motion not accounted for by strain measurements in the San Andreas system.

The hypothesis that the tectonics of the Pacific Northwest are dominated by the subduction of the Juan de Fuca plate system beneath the North American plate also dates from the beginning of the plate-tectonic era. The hypothesis has been buttressed in the past decade by identification of Wadati-Benioff zones extending to depths of between 80 and 90 km beneath northern California and western Washington. The outstanding plate-tectonics problems in the Pacific Northwest involve the historic lack of seismicity in coastal Oregon and on the thrust interface between the Juan de Fuca and North American plate; at issue is whether this quiescence is permanent or will be followed by large earthquakes at these locations.

Global tectonic models have been used to explain regional extension in the Cordillera east of the Sierra Nevada and the orientation of the regional stress-tensor in the central and eastern United States. It is likely that many source zones in the central and eastern United States are reactivated structures that originally formed near plate boundaries.

Characteristics of sources in which association of earthquakes with geologic structures seems clear. Well-observed seismogenic faults are physically segmented; individual segments are either offset from each other along strike or have different strikes. The segmentation occurs at many scales, so that a fault that would be considered continuous at one scale comprises smaller segments when viewed in more detail. The physical segmentation leads to along-strike variations in seismic behavior of individual faults. Evidence from recent well-studied earthquakes in California and Idaho supports the hypothesis that physical segmentation and along-strike variations in strength predispose shocks on each section of fault to have a characteristic rupture displacement and rupture length.

The mechanical processes associated with an individual earthquake are complex to the extent that interpretation of a single type of data on the basis of a single-plane fault model may be misleading. The expressions in surface geology of some well-studied faults are distorted representations of the faults at depth that are defined by seismologic or geodetic data. Aftershocks frequently occur away from main-shock rupture surfaces; the distribution of aftershocks are usually blurred representations of main-shock faults.

Some of the faults that have produced the largest earthquakes in California and the Cordillera have been characterized by long periods of quiescence between large earthquakes. The absence of observed small or moderate earthquakes on other geologically young faults therefore does not constitute evidence that the faults will remain inactive in the future.

Sources in which association of earthquakes with geologic structure is currently obscure. In the western United States, numerous small and moderate earthquakes occur away from major mapped late Cenozoic faults. The focal mechanisms of most of these earthquakes are similar to focal mechanisms of large earthquakes on the major faults, implying that the shocks are due to movement on small faults or fractures under the same stress fields that cause the large earthquakes. A minority of off-fault shocks have reliably determined mechanisms that imply local stress fields in which the principal stress axes have different orientations than in the stress fields producing movement on major nearby faults. Some of these anomalous focal mechanisms have been explained as the result of rotations of rigid crustal blocks. Antithetic or secondary faulting may also produce focal mechanisms that are substantially different than focal mechanisms on the primary faults.

Preexisting faults that were originally formed in Mesozoic or earlier time, and that have been active in Cenozoic time with very low long-term slip rates, are probably responsible for many of the moderate and large earthquakes of the eastern and central United States. In no U.S. source region east of the Cordillera, however, is

TABLE 1. PROBABLE HUMAN-INDUCED SEISMICITY, U.S.A.*

Location	Latitude (N)	Longitude (W)	Triggering Activity	Reference
The Geysers, California	38.8°	122.8°	Geothermal steam production	Majer and McEvilly (1979); Denlinger and Bufe (1982); Eberhart-Phillips and Oppenheimer (1984)
Oroville, California	39.4°	121.5°	Reservoir impoundment	Lahr and others (1976); Morrison and others (1976)
Lompoc, California†	34.6°	120.4°	Surface quarrying	Yerkes and others (1983)
Wilmington Oil Field, California†	33.8°	118.2°	Primary hydrocarbon production	Kovach (1974); Yerkes and Castle (1976)
Nevada Test Site	37.2°	116.3°	Nuclear explosions	Boucher and others (1969); Hamilton and Healy (1969); Hamilton and others (1972)
Wallace, Idaho	47.5°	115.9°	Underground silver mining	Blake (1972)
Lake Mead, Nevada–Arizona	36.0°	114.8°	Reservoir impoundment	Carder (1945); Rogers and Lee (1976)
East-central Utah	39.5° 39.5°	111.0° 110.5°	Underground coal mining	Smith and others (1974); Wong (1984)
Rangely, Colorado	40.1°	108.9°	Fluid injection for secondary oil recovery	Raleigh and others (1972)
Denver, Colorado	39.9°	104.9°	Fluid injection for waste disposal	Evans (1966); Healy and others (1968)
West Texas	32.0° 33.0°	103.0° 100.9°	Fluid injection for secondary oil recovery	Rogers and Malkiel (1979); Harding (1981); Gordon (1983)
South Texas	28.8°	98.3°	Primary hydrocarbon production	Pennington and others (1986)
Northwestern South Carolina and neighboring Georgia	33.0° to 35.0°	81.0° to 84.0°	Reservoir impoundment	Talwani (1977); Talwani and others (1979)
Dale, New York§	42.4°	78.1°	Fluid injection for salt recovery	Fletcher and Sykes (1977)
Wappingers Falls, New York†	41.6°	73.9°	Surface quarrying	Pomeroy and others (1976)

*Listed are sites whose epicenters are plotted in Figure 1 and sites representing the range of human activities that induce earthquakes.
†Earthquakes too small to be plotted in Figure 1. Plotted as letter in Figure 2.
§Plotted in Figure 17.

the relation between natural earthquakes and geologic structure established as clearly as in well-understood source regions of the western United States. In areas where faults are suggested from seismologic data, the suspect faults are either not exposed at the surface or have not been shown from geologic evidence to have slipped in Cenozoic time. Major exposed faults that have geologic evidence of late-Cenozoic displacement do not have associated planar zones of earthquakes with focal mechanisms implying slip on the faults. The absence of clear links between individual earthquakes and individual faults leaves open the possibility that the mechanical relationship between earthquakes and faults is different east of the Cordillera than in well-understood sources of the western United States.

ACKNOWLEDGMENTS

We thank Clarence Allen, Geoff King, Lou Pakiser, Jim Taggart, and Carl Wentworth for their helpful criticisms of the manuscript. Kerry Sieh, Bill Bakun, Craig Nicholson, and Walt Arabasz gave us permission to use figures from their papers.

REFERENCES CITED

Adams, J., 1984, Active deformation of the Pacific Northwest continental margin: Tectonics, v. 3, p. 449–472.

Aggarwal, Y. B., and Sykes, L. R., 1978, Earthquakes, faults, and nuclear power plants in southern New York and northern New Jersey: Science, v. 200, p. 425–429.

Allen, C. R., 1968, The tectonic environments of seismically active and inactive areas along the San Andreas fault system, *in* Dickinson, W. R., and Grantz, A., eds., Proceedings of the Conference on Geologic Problems of the San Andreas Fault System: Stanford, California, Stanford University Publications in the Geological Sciences, v. 11, p. 70–82.

——, 1975, Geological criteria for evaluating seismicity: Geological Society of American Bulletin, v. 86, p. 1041–1057.

——, 1981, The modern San Andreas fault, *in* Ernst, W. G., ed., The geotectonic development of California: Englewood Cliffs, New Jersey, Prentice-Hall, p. 511–534.

Allen, C. R., and Nordquist, J. M., 1972, Foreshock, main shock, and larger aftershocks of the Borrego Mountain earthquake: U.S. Geological Survey Professional Paper 787, p. 16–30.

Allen, C. R., St. Amand, P., Richter, C. F., and Nordquist, J. M., 1965, Relationship between seismicity and geologic structure in the southern California region: Bulletin of the Seismological Society of America, v. 55, p. 753–797.

Ando, M., and Balazs, E. I., 1979, Geodetic evidence for aseismic subduction of the San Juan de Fuca plate: Journal of Geophysical Research, v. 84, p. 3023–3028.

Andrews, M. C., Mooney, W. D., and Meyer, R. P., 1985, The relocation of microearthquakes in the northern Mississippi Embayment: Journal of Geophysical Research, v. 90, p. 10223–10236.

Arabasz, W. J., and Julander, D. R., 1986, Geometry of seismically active faults and crustal deformation within the Basin and Range–Colorado Plateau transition in Utah, *in* Mayer, L., ed., Extensional tectonics of the southwestern United States; A perspective on processes and kinematics: Geological Society of America Special Paper 208, p. 43–74.

Arabasz, W. J., and Smith, R. B., 1981, Earthquake prediction in the Intermountain Seismic Belt; An intraplate extensional regime, *in* Simpson, D. W., and Richards, P. G., eds., Earthquake prediction; An international review: American Geophysical Union Maurice Ewing Series 4, p. 248–258.

Arabasz, W. J., Richins, W. D., and Langer, C. J., 1981, The Pocatello Valley (Idaho-Utah border) earthquake sequence of March to April 1975: Bulletin of the Seismological Society of America, v. 71, p. 803–826.

Archuleta, R. J., 1984, A faulting model for the 1979 Imperial Valley earthquake: Journal of Geophysical Research, v. 89, p. 4559–4585.

Astiz, L., and Allen, C. R., 1983, Seismicity of the Garlock fault, California: Bulletin of the Seismological Society of America, v. 73, p. 1721–1734.

Atwater, B. F., 1987, Evidence for great Holocene earthquakes along the outer coast of Washington State: Science, v. 236, p. 942–944.

Atwater, T., 1970, Implications of plate tectonics for the Cenozoic evolution of western North America: Geological Society of America Bulletin, v. 81, p. 3513–3536.

Bache, T. C., Lambert, D. G., and Barker, T. G., 1980, A source model for the March 28, 1975, Pocatello Valley earthquake from time-domain modeling of teleseismic P waves: Bulletin of the Siesmological Society of America, v. 70, p. 405–418.

Bakun, W. H., and Lindh, A. G., 1985, The Parkfield, California, earthquake prediction experiment: Science, v. 229, p. 619–623.

Bakun, W. H., and McEvilly, T. V., 1984, Recurrence models and Parkfield, California, earthquakes: Journal of Geophysical Research, v. 89, p. 3051–3058.

Bakun, W. H., Stewart, R. M., Bufe, C. G., and Marks, S. M., 1980, Implication of seismicity for failure of a section of the San Andreas fault: Bulletin of the Seismological Society of America, v. 70, p. 185–201.

Bakun, W. H., Clark, M. M., Cockerham, R. S., Ellsworth, W. L., Lindh, A. G., Prescott, W. H., Shakal, A. F., and Spudich, P., 1984, The 1984 Morgan Hill, California, earthquake: Science, v. 225, p. 288–291.

Behrendt, J. C., Hamilton, R. M., Ackermann, H. D., Henry, V. J., and Bayer, K. C., 1983, Marine multichannel seismic-reflection evidence for Cenozoic faulting and deep crustal structure near Charleston, South Carolina: U.S. Geological Survey Professional Paper 1313, p. J1–J29.

Bell, M. L., and Nur, A., 1978, Strength changes due to reservoir-induced pore pressure and stresses and application to Lake Oroville: Journal of Geophysical Research, v. 83, p. 4469–4483.

Bird, P., and Rosenstock, R. W., 1984, Kinematics of present crust and mantle flow in southern California: Geological Society of America Bulletin, v. 95, p. 946–957.

Biswas, N. N., Aki, K., Pulpan, H., and Tytgat, G., 1986, Characteristics of regional stresses in Alaska and neighboring areas: Geophysical Research Letters, v. 13, p. 177–180.

Blake, W., 1972, Rock-burst mechanics: Quarterly of the Colorado School of Mines, v. 67, no. 1, 64 p.

Bollinger, G. A., 1983, Speculations on the nature of seismicity at Charleston, South Carolina: U.S. Geological Survey Professional Paper 1313, p. T1–T11.

Bollinger, G. A., and Sibol, M. S., 1985, Seismicity, seismic reflection studies, gravity, and geology of the central Virginia seismic zone; Part 1, Seismicity: Geological Society of America Bulletin, v. 96, p. 46–57.

Bollinger, G. A., and Wheeler, R. L., 1983, The Giles County, Virginia, seismic zone: Science, v. 219, p. 1063–1065.

Bollinger, G. A., Chapman, M. C., Sibol, M. S., and Costain, J. K., 1985, An analysis of earthquake focal depths in the southeastern U.S.: Geophysical Research Letters, v. 12, p. 785–788.

Bolt, B. A., 1979, Seismicity of the western United States, *in* Hatheway, A. W., and McClure, C. R., Jr., eds., Geology in the siting of nuclear power plants: Geological Society of America Reviews in Engineering Geology, v. 6, p. 95–107.

Bolt, B. A., and Miller, R. D., 1975, Catalogue of earthquakes in northern California and adjoining areas 1 January 1910-31 December 1972: Berkeley, University of California, Seismographic Stations, 567 p.

Bolt, B. A., Lomnitz, C., and McEvilly, T. V., 1968, Seismological evidence on the tectonics of central and northern California and the Mendocino escarpment: Bulletin of the Seismological Society of America, v. 58, p. 1724–1767.

Boucher, G., Ryall, A., and Jones, A. E., 1969, Earthquakes associated with underground nuclear explosions: Journal of Geophysical Research, v. 74, p. 3808–3820.

Braile, L. W., Keller, G. R., Hinze, W. J., and Lidiak, E. G., 1982, An ancient rift complex and its relation to contemporary seismicity in the New Madrid seismic zone: Tectonics, v. 1, p. 225–237.

Byerly, P., 1937, Earthquakes off the coast of northern California: Bulletin of the Seismological Society of America, v. 27, p. 73–96.

Carder, D. S., 1945, Seismic investigations in the Boulder Dam area, 1940–1944, and the influence of reservoir loading on earthquake activity: Bulletin of the Seismological Society of America, v. 35, p. 175–192.

Carver, G. A., Stephens, T. A., and Yound, J. C., 1982, Quaternary reverse and thrust faults, Mad River fault zone, *in* Late Cenozoic history and forest geomorphology of Humboldt County, California: Guidebook for Friends of the Pleistocene 1982 Pacific Cell Field Trip, p. 93–100.

Chen, W.-P., and Molnar, P., 1983, Focal depths of intracontinental and intraplate earthquakes and their implications for the thermal and mechanical properties of the lithosphere: Journal of Geophysical Research, v. 88, p. 4183–4214.

Choy, G. L., 1985, Source parameters of the Coalinga, California, earthquake of May 2, 1983, inferred from broadband body waves, *in* Rymer, M. J., and Ellsworth, W. L., eds., Mechanics of the May 2, 1983, Coalinga earthquake: U.S. Geological Survey Open-File Report 85-44, p. 83–131.

Clark, M. M., Grantz, A., and Rubin, M., 1972, Holocene activity of Coyote Creek fault as recorded in sediments of Lake Cahuilla: U.S. Geological Survey Professional Paper 787, p. 112–130.

Cockerman, R. S., 1984, Evidence for a 180-km-long subducted slab beneath northern California: Bulletin of the Seismological Society of America, v. 74, p. 569–576.

Cockerham, R. S., and Eaton, J. P., 1984, The April 24, 1984, Morgan Hill earthquake and its aftershocks; April 24 through September 30, 1984, *in* The 1984 Morgan Hill, California, earthquake: California Division Mines and Geology Special Publication 68, p. 215–236.

Corbett, E. J., and Johnson, C. E., 1982, The Santa Barbara, California, earthquake of 13 August, 1978: Bulletin of the Seismological Society of America, v. 72, p. 2201–2226.

Costain, J. K., Bollinger, G. A., and Speer, J. A., 1986, Hydroseismicity; A hypothesis for intraplate seismicity near passive rifted margins: Earthquake Notes, v. 57, no. 1, p. 13.

Crone, A. J., and Brockman, S. R., 1982, Configuration and deformation of the Paleozoic bedrock surface: U.S. Geological Survey Professional Paper 1236, p. 115–135.

Crone, A. J., McKeown, F. A., Harding, S. T., Hamilton, R. M., Russ, D. P., and Zoback, M. D., 1985, Structure of the New Madrid seismic source zone in southeastern Missouri and northeastern Arkansas: Geology, v. 13, p. 547–550.

Crone, A. J., Machette, M. N., Bonilla, M. G., Lienkaemper, J. J., Pierce, K. L., Scott, W. E., and Bucknam, R. C., 1987, Surface faulting accompanying the Borah Peak earthquake and segmentation of the Lost River fault, central Idaho: Bulletin of the Seismological Society of America, v. 77, p. 739–770.

Crosson, R. S., 1972, Small earthquakes, structure, and tectonics of the Puget Sound region: Bulletin of the Seismological Society of America, v. 62, p. 1133–1171.

——, 1983, Review of seismicity in the Puget Sound region from 1970 through 1978, *in* Yount, J. C., and Crosson, R. S., eds., Earthquake hazards of the Puget Sound region, Washington: U.S. Geological Survey Open-File Report 83-19, p. 6–18.

Davis, G. A., and Burchfiel, B. C., 1973, Garlock fault; An intracontinental transform structure, southern California: Geological Society of America Bulletin, v. 84, p. 1407–1422.

Denlinger, R. P., and Bufe, C. G., 1982, Reservoir conditions related to induced seismicity at the Geysers steam reservoir, northern California: Bulletin of the Seismological Society of America, v. 72, p. 1317–1327.

Dewey, J. W., 1983, Relocation of instrumentally recorded pre-1974 earthquakes from the South Carolina region: U.S. Geological Survey Professional Paper 1313, p. Q1–Q9.

——, 1987, Instrumental seismicity of central Idaho: Bulletin of the Seismological Society of America, v. 77, p. 819–836.

Dewey, J. W., and Gordon, D. W., 1984, Map showing recomputed hypocenters of earthquakes in the eastern and central United States and adjacent Canada, 1925–1980: U.S. Geological Survey Miscellaneous Field Studies Map MF-1699, 39 p.

Diment, W. H., Urban, T. C., and Revetta, F. A., 1972, Some geophysical anomalies in the eastern United States, *in* Robertson, E. C., ed., The nature of the solid Earth: New York, McGraw-Hill, p. 544–572.

Donovan, R. N., Gilbert, M. C., Luza, K. V., Marchini, D., and Sanderson, D., 1983, Possible Quaternary movement on the Meers fault southwestern Oklahoma: Oklahoma Geology Notes, v. 43, p. 124–133.

Doser, D. I., 1986, Earthquake processes in the Rainbow Mountain–Fairview Peak–Dixie Valley, Nevada, region 1954–1959: Journal of Geophysical Research, v. 91, p. 12572–12586.

Doser, D. I., and Kanamori, H., 1986, Depth of seismicity in the Imperial Valley region (1977–1983) and its relation to heat flow, crustal structure, and the October 15, 1979, earthquake: Journal of Geophysical Research, v. 91, p. 675–688.

Doser, D. I., and Smith, R. B., 1983, Seismicity of the Teton–southern Yellowstone region: Bulletin of the Seismological Society of America, v. 73, p. 1369–1394.

——, 1985, Source parameters of the 28 October 1983 Borah Peak, Idaho, earthquake from body wave analysis: Bulletin of the Seismological Society of America, v. 75, p. 1041–1051.

Dutton, C. E., 1889, The Charleston earthquake of August 31, 1886: U.S. Geological Survey, Ninth Annual Report, 1887–1888, p. 203–528.

Eaton, J. P., 1981, Distribution of aftershocks of the November 8, 1980, Eureka earthquake: Earthquake Notes, v. 52, no. 1, p. 44–45.

——, 1985a, Regional seismic background of the May 2, 1983, Coalinga earthquake, *in* Rymer, M. J., and Ellsworth, W. L., eds., Mechanics of the May 2, 1983, Coalinga earthquake: U.S. Geological Survey Open-File Report 85-44, p. 44–60.

——, 1985b, The May 2, 1983, Coalinga earthquake and its aftershocks; A detailed study of the hypocenter distribution and of the focal mechanisms of the larger aftershocks, *in* Rymer, M. J., and Ellsworth, W. L., eds., Mechanics of the May 2, 1983, Coalinga earthquake: U.S. Geological Survey Open-File Report 85-44, p. 132–201.

Eaton, J. P., O'Neill, M. E., and Murdock, J. N., 1970, Aftershocks of the 1966 Parkfield-Cholame, California, earthquake; A detailed study: Bulletin of the Seismological Society of America, v. 60, p. 1151–1197.

Ebel, J. E., 1983, A detailed study of the aftershocks of the 1979 earthquake near Bath, Maine: Earthquake Notes, v. 54, no. 4, p. 27–40.

Ebel, J. E., Vudler, V., and Celata, M., 1982, The 1981 microearthquake swarm near Moodus, Connecticut: Geophysical Research Letters, v. 9, p. 397–400.

Eberhart-Phillips, D., and Oppenheimer, D. H., 1984, Induced seismicity in the Geysers geothermal area, California: Journal of Geophysical Research, v. 89, p. 1191–1207.

Eberhart-Phillips, D., and Reasenberg, P., 1985, Hypocenter locations and constrained fault-plane solutions for Coalinga aftershocks, May 2-24, 1983; Evidence of a complex rupture geometry, *in* Rymer, M. J., and Ellsworth, W. L., eds., Mechanics of the May 2, 1983, Coalinga earthquake: U.S. Geological Survey Open-File Report 85-44, p. 202–224.

Ellsworth, W. L., 1975, Bear Valley, California, earthquake sequence of February-March 1972: Bulletin of the Seismological Society of America, v. 65, p. 483–506.

Ellsworth, W. L., Lindh, A. G., Prescott, W. H., and Herd, D. G., 1981, The 1906 San Francisco earthquake and the seismic cycle, *in* Simpson, D. W., and Richards, P. G., eds., Earthquake prediction; An international review: American Geophysical Union Maurice Ewing Series 4, p. 126–140.

Ellsworth, W. L., Olson, J. A., Shijo, L. N., and Marks, S. N., 1982, Seismicity and active faults in the eastern San Francisco Bay region: California Division of Mines and Geology Special Publication 62, Proceedings of a Conference on Earthquake Hazards in the Eastern San Francisco Bay Area, p. 83–91.

Endo, E. T., Malone, S. D., Noson, L. L., and Weaver, C. S., 1981, Locations, magnitudes, and statistics of the March 20-May 18 earthquake sequence: U.S. Geological Survey Professional Paper 1250, p. 93–107.

Engdahl, E. R. (compiler), and Rinehart, W. A. (preparer), 1988, Seismicity map of North America, Continent-Scale Map-004, Decade of North American Geology, Geological Society of America, scale 1:5,000,000, 4 sheets.

Ervin, G. P., and McGinnis, L. D., 1975, Reelfoot Rift; Reactivated precursor to the Mississippi Embayment: Geological Society of America Bulletin, v. 86, p. 1287–1295.

Evans, D. , 1966, The Denver area earthquakes and the Rocky Mountain disposal well: Mountain Geologist, v. 3, p. 23–36.

Fielding, E., Barazangi, M., Brown, L., Oliver, J., and Kaufman, S., 1984, COCORP seismic profiles near Coalinga, California; Subsurface structure of the western Great Valley: Geology, v. 12, p. 268–273.

Fletcher, J. B., and Sykes, L. R., 1977, Earthquakes related to hydraulic mining and natural seismic activity in western New York State: Journal of Geophysical Research, v. 82, p. 3767–3780.

Fletcher, J. W., Sbar, M. L., and Sykes, L. R., 1978, Seismic trends and traveltime residuals in eastern North America and their tectonic implications: Geological Society of America Bulletin, v. 89, p. 1656–1676.

Gawthrop, W. H., 1978, Seismicity and tectonics of the central California coastal region, *in* Silver, E. A., and Normark, W. R., eds., Special Report: California Division of Mines and Geology, v. 137, p. 45–56.

Gephart, J. W., and Forsyth, D. D., 1985, On the state of stress in New England as determined from earthquake focal mechanisms: Geology, v. 13, p. 70–72.

Gilbert, M. C., 1982, Geologic setting of the eastern Wichita Mountains with a brief discussion of unresolved problems: Oklahoma Geological Survey Guidebook, v. 21, p. 1–30.

Gordon, D. W., 1983, Revised hypocenters and correlation of seismicity and tectonics in the central United States [Ph.D. thesis]: St. Louis, Missouri, St. Louis University, 199 p.

Gutenberg, B., and Richter, C. F., 1954, Seismicity of the earth and associated phenomena, 2nd ed.: Princeton, New Jersey, Princeton University Press, 310 p.

Haar, L. C., Fletcher, J. B., and Mueller, C. S., 1984, The 1982 Enola, Arkansas, swarm and scaling of ground motion in the eastern United States: Bulletin of the Seismological Society of America, v. 74, p. 2463–2482.

Haimson, B. C., and Doe, T. W., 1983, State of stress, permeability, and fractures in the Precambrian granite of northern Illinois: Journal of Geophysical Research, v. 88, p. 7355–7371.

Hait, M. H., Jr., and Scott, W. E., 1978, Holocene faulting, Lost River Range, Idaho: Geological Society of America Abstracts with Programs, v. 10, p. 217.

Hamilton, R. M., and Healy, J. H., 1969, Aftershocks of the Benham nuclear explosion: Bulletin of the Seismological Society of America, v. 59, p. 2271–2281.

Hamilton, R. M., Smith, B. E., Fischer, F. G., and Papanek, P. J., 1972, Earthquakes caused by underground nuclear explosions on Pahute Mesa, Nevada Test Site: Bulletin of the Seismological Society of America, v. 62, p. 1319–1341.

Hamilton, R. M., Behrendt, J. C., and Ackermann, H. C., 1983, Local multichannel seismic-reflection evidence for tectonic features near Charleston, South Carolina: U.S. Geological Survey Professional Paper 1313, p. I1-I18.

Harding, S. T., 1981, Induced seismicity Cogdell Canyon Reef oil field; Summaries of technical reports, XI: U.S. Geological Survey Open-File Report 81-167, p. 452–455.

Hartzell, S. H., and Heaton, T. H., 1983, Inversion of strong ground motion and teleseismic data for the fault rupture history of the 1979 Imperial Valley, California, earthquake: Bulletin of the Seismological Society of America, v. 73, p. 1553–1583.

——, 1986, Rupture history of the 1984 Morgan Hill, California, earthquake from the inversion of strong motion records: Bulletin of the Seismological Society of America, v. 76, p. 649–674.

Healy, J. H., Rubey, W. W., Griggs, D. T., and Raleigh, C. B., 1968, The Denver earthquakes: Science, v. 161, p. 1301–1310.

Heaton, T. H., and Hartzell, S. H., 1986, Source characteristics of hypothetical subduction earthquakes in the northwestern United States: Bulletin of the Seismological Society of America, v. 76, p. 675–708.

Heaton, T. H., and Kanamori, H., 1984, Seismic potential associated with subduction in the northwestern United States: Bulletin of the Seismological Society of America, v. 74, p. 933–941.

Herd, D. G., 1978, Intracontinental plate boundary east of Cape Mendocino, California: Geology, v. 6, p. 721–725.

Herrmann, R. B., 1978, A seismological study of two Attica, New York, earthquakes: Bulletin of the Seismological Society of America, v. 68, p. 641–651.

——, 1979, Surface wave focal mechanisms for eastern North American earthquakes with tectonic implications: Journal of Geophysical Research, v. 84, p. 3543–3552.

Herrmann, R. B., and Canas, J.-A., 1978, Focal mechanism studies in the New Madrid seismic zone: Bulletin of the Seismological Society of America, v. 68, p. 1095–1102.

Heyl, A. V., and McKeown, F. A., 1978, Preliminary seismotectonic map of central Mississippi Valley and environs: U.S. Geological Survey Miscellaneous Field Studies Map MF-1011, scale 1:500,000.

Hildenbrand, T. H., 1985, Rift structure of the northern Mississippi Embayment from the analysis of gravity and magnetic data: Journal of Geophysical Research, v. 90, p. 12607-12622.

Hildenbrand, T. G., Kane, M. F., and Hendricks, J. D., 1982, Magnetic basement in the upper Mississippi Embayment region; A preliminary report: U.S. Geological Survey Professional Paper 1236, p. 39–53.

Hileman, J. A., Allen, C. R., and Nordquist, J. M., 1973, Seismicity of the southern California region 1 January 1932 to 31 December 1972: Pasadena, California Institute of Technology Earthquake Research Association Contribution 2384, 83 p.

Hill, D. P., 1982, Contemporary block tectonics; California and Nevada: Journal of Geophysical Research, v. 87, p. 5433–5450.

Hill, D. P., Eaton, J. P., Cockerham, R. S., Ellsworth, W. L., and Johnson, C. E., 1983, Seismicity along the Pacific–North American plate boundary in California and western Nevada, 1980: U.S. Geological Survey Open-File Report 83-174, 23 p.

Hill, D. P., Bailey, R. A., and Ryall, A. S., 1985, Active tectonic and magmatic processes beneath Long Valley caldera, eastern California; An overview: Journal of Geophysical Research, v. 90, p. 11111-11120.

Hill, M. L., and Dibblee, T. W., 1953, San Andreas, Garlock, and Big Pine faults; A study of the character, history, and significance of their displacements: Geological Society of America Bulletin, v. 77, p. 435–438.

Howard, K. A., and 12 others, 1978, Preliminary map of young faults in the United States as a guide to possible fault activity: U.S. Geological Survey Miscellaneous Field Studies Map MF-916, 2 sheets, scales 1:5,000,000 and 1:7,500,000.

Hutchinson, D. R., Pomeroy, P. W., Wold, R. J., and Halls, H. C., 1979, A geophysical investigation concerning the continuation of the Clarendon-Linden fault across Lake Ontario: Geology, v. 7, p. 206–210.

Isacks, B., and Molnar, P., 1971, Distribution of stresses in the descending lithosphere: Reviews of Geophysics and Space Physics, v. 9, p. 103–174.

Jackson, J., and McKenzie, D., 1983, The geometrical evolution of normal fault systems: Journal of Structural Geology, v. 5, p. 471–482.

Jaksha, L. H., and Sanford, A. R., 1986, Earthquakes near Albuquerque, New Mexico, 1976–1981: Journal of Geophysical Research, v. 91, p. 6293–6303.

Johnson, C. E., and Hadley, D. M., 1976, Tectonic implications of the Brawley earthquake swarm, Imperial Valley, California, January 1975: Bulletin of the Seismological Society of America, v. 66, p. 1133–1144.

Johnson, C. E., and Hutton, L. K., 1982, Aftershocks and preearthquake seismicity: U.S. Geological Survey Professional Paper 1254, p. 59–76.

Johnston, A. C., Reinbold, D. J., and Brewer, S. I., 1985, Seismotectonics of the southern Appalachians: Bulletin of the Seismological Society of America, v. 75, p. 291–312.

Jones, L. M., and Dollar, R. S., 1986, Evidence of Basin-and-Range extensional tectonics in the Sierra Nevada; The Durrwood Meadows swarm, Tulare County, California: Bulletin of the Seismological Society of America, v. 76, p. 439–461.

Julian, B. R., 1983, Evidence for dyke intrusion earthquake mechanisms near Long Valley, California: Nature, v. 303, p. 323–325.

Julian, B. R., and Sipkin, S. A., 1985, Earthquake processes in the Long Valley caldera area, California: Journal of Geophysical Research, v. 90, p. 11155-11169.

Kafka, A. L., Schlesinger-Miller, E. A., and Barstow, N. L., 1985, Earthquake activity in the greater New York City area; Magnitudes, seismicity, and geologic structures: Bulletin of the Seismological Society of America, v. 75, p. 1285–1300.

Kanamori, H., and Given, J. W., 1982, Analysis of long-period seismic waves excited by the May 18, 1980, eruption of Mount St. Helens; A terrestrial monopole: Journal of Geophysical Research, v. 87, p. 5422–5432.

Kanamori, H., and Regan, J., 1982, Long period surface waves: U.S. Geological Survey Professional Paper 1254, p. 55–58.

Kanamori, H., Given, J. W., and Lay, T., 1984, Analysis of seismic body waves excited by the Mt. St. Helens eruption of May 18, 1980: Journal of Geophysical Research, v. 89, p. 1856–1866.

Keller, G. R., and Adams, H. E., 1976, A reconnaissance microearthquake survey of the San Luis Valley, southern Colorado: Bulletin of the Seismological Society of America, v. 66, p. 345–347.

Keller, G. R., Lidiak, E. G., Hinze, W. J., and Braile, L. W., 1983, The role of rifting in the tectonic development of the Midcontinent, USA: Tectonophysics, v. 94, p. 391–412.

King, G., 1983, The accommodation of large strains in the upper lithosphere of the earth and other solids by self-similar fault systems; the geometrical origin of b-value: Pure and Applied Geophysics, v. 121, p. 761–815.

Kirkham, R. M., and Rogers, W. P., 1981, Earthquake potential in Colorado: Colorado Geological Survey Bulletin 43, 171 p.

Kovach, R. L., 1974, Source mechanisms for Wilmington oil field, California, subsidence earthquakes: Bulletin of the Seismological Society of America, v. 64, p. 699–711.

Kruger-Knuepfer, J. L., Sbar, M. L., and Richardson, R. M., 1985, Microseismicity of the Kaibab Plateau, northern Arizona, and its tectonic implications: Bulletin of the Seismological Society of America, v. 75, p. 491–505.

Lahr, K. M., Lahr, J. C., Bufe, C. G., and Lester, F. M., 1976, The August 1975 Oroville earthquakes: Bulletin of the Seismological Society of America, v. 66, p. 1085–1099.

Lawson, J. E., 1985, Seismicity at the Meers fault: Earthquake Notes, v. 55, no. 1, p. 2.

Lee, W.H.K., Yerkes, R. F., and Simirenko, M., 1979, Recent earthquake activity and focal mechanisms in the western Transverse Ranges, California: U.S. Geological Survey Circular 799-A, p. 1–26.

Lide, C. S., and Ryall, A. S., 1985, Aftershock distribution related to the controversy regarding mechanisms of the May 1980, Mammoth Lakes, California, earthquakes: Journal of Geophysical Research, v. 90, p. 11151-11154.

Lisowski, M., and Prescott, W. H., 1981, Short-range distance measurements along the San Andreas fault system in central California: Bulletin of the Seismological Society of America, v. 71, p. 1607–1624.

Liu, H.-L., and Helmberger, D. V., 1983, The near-source ground-motion of the 6 August 1979 Coyote Lake, California, earthquake: Bulletin of the Seismological Society of America, v. 73, p. 201–218.

Lomnitz, C., Mooser, F., Allen, C. R., Brune, J. N., and Thatcher, W., 1970, Seismicity and tectonics of northern Gulf of California region, Mexico; Preliminary results: Geofisica Internacional, v. 10, no. 2, p. 37–48.

Machette, M. N., 1986, History of Quaternary offset and paleoseismicity along the La Jencia fault, central Rio Grande Rift, New Mexico: Bulletin of the Seismological Society of America, v. 76, p. 259–272.

Madole, R. F., and Rubin, M., 1985, Holocene movement on the Meers Fault, southwest Oklahoma: Earthquake Notes, v. 55, no. 1, p. 1–2.

Majer, E. L., and McEvilly, T. V., 1979, Seismological investigations at The Geysers geothermal field: Geophysics, v. 44, p. 246–269.

Malone, S. D., Endo, E. T., Weaver, C. S., and Ramey, J. W., 1981, Seismic monitoring for eruption prediction: U.S. Geological Survey Professionaal Paper 1250, p. 93–107.

Marks, S. M., and Lindh, A. G., 1978, Regional seismicity of the Sierran foothills in the vicinity of Oroville, California: Bulletin of the Seismological Society of America, v. 68, p. 1103–1115.

McCalpin, J. P., 1982, Quaternary geology and neotectonics of the west flank of the northern Sangre de Cristo mountains, south-central Colorado: Colorado School of Mines Quarterly, v. 77, 97 p.

McKenzie, D., and Parker, R. L., 1967, The North Pacific; An example of tectonics on a sphere: Nature, v. 216, p. 1276.

McKeown, F. A., 1982, Overview and discussion: U.S. Geological Survey Professional Paper 1236, p. 1–14.

Meissner, R., and Strehlau, J., 1982, Limits of stresses in continental crusts and their relation to the depth-frequency distribution of shallow earthquakes: Tectonics, v. 1, p. 73–89.

Minster, J. B., and Jordan, T. H., 1978, Present-day plate motions: Journal of Geophysical Research, v. 83, p. 5331–5354.

——, 1984, Vector constraints on Quaternary deformation of the western United States east and west of the San Andreas fault, *in* Crouch, J. K., and Bachman, S. B., eds., Tectonics and sedimentation along the California margin: Pacific Section Society of Economic Paleontologists and Mineralogists, v. 38, p. 1–16.

Mitchell, B. J., 1973, Radiation and attenuation of Rayleigh waves from the southeastern Missouri earthquake of Oct. 21, 1965: Journal of Geophysical Research, v. 78, p. 886–899.

Mixon, R. B., and Newell, W. L., 1977, Stafford fault system; Structures documenting Cretaceous and Tertiary deformation along the Fall Line in northeastern Virginia: Geology, v. 6, p. 437–440.

Morrison, P. W., Stump, B. W., and Uhrhammer, R., 1976, The Oroville earthquake sequence of August 1975: Bulletin of the Seismological Society of America, v. 66, p. 1065–1084.

Munsey, J. W., and Bollinger, G. A., 1985, Focal mechanism analysis for Virginia earthquakes (1978–1984): Bulletin of the Seismological Society of America, v. 75, p. 1613–1636.

Myers, W. B., and Hamilton, W., 1964, Deformation accompanying the Hebgen Lake earthquake of August 17, 1959: U.S. Geological Survey Professional Paper 435, p. 55–97.

Nicholson, C., Simpson, D. W., Singh, S., and Zollweg, J. E., 1984, Crustal studies, velocity inversions, and fault tectonics; Results from a microearthquake survey in the New Madrid seismic zone: Journal of Geophysical Research, v. 89, p. 4545–4558.

Nicholson, C. L., Seeber, L., Williams, P., and Sykes, L. R., 1986a, Seismicity and fault kinematics through the eastern Transverse Ranges, California; Block rotation, strike-slip faulting, and shallow angle thrusts: Journal of Geophysical Research, v. 91, p. 4891–4908.

——, 1986b, Seismic evidence for conjugate slip and block rotation within the San Andreas fault system, southern California: Tectonics, p. 629–648.

Nishimura, C., Wilson, D. S., and Hey, R. N., 1984, Pole of rotation analysis of present-day Juan de Fuca plate motion: Journal of Geophysical Research, v. 89, p. 10,283–10,290.

Nottis, G. N., ed., 1983, Epicenters of northeastern United States and southeastern Canada, onshore and offshore; Time period 1534–1980: New York State Museum Map and Chart Series Number 38, 2 map sheets and 38 pages text.

Nowroozi, A. A., 1973, Seismicity of the Mendocino escarpment and the aftershock sequence of June 26, 1968; Ocean bottom seismic measurements: Bulletin of the Seismological Society of America, v. 63, p. 441–456.

Nunn, J. A., 1985, State of stress in the northern Gulf Coast: Geology, v. 13, p. 429–432.

Nuttli, O. W., 1973, The Mississippi Valley earthquakes of 1811 and 1812; Intensities, ground motion, and magnitudes: Bulletin of the Seismological Society of America, v. 63, p. 227–248.

——, 1979, The seismicity of the Central United States, *in* Geology in the siting of nuclear power plants: Geological Society of America Reviews in Engineering Geology, v. 4, p. 67–93.

Nuttli, O. W., Bolt, B. A., Flinn, E. A., and Savage, J. C., 1978, Nomenclature and terminology for seismology; Editors handbook 3, Association of Earth Sciences Editors: Society of Exploration Geophysicists, 10 p.

Nuttli, O. W., Bollinger, G. A., and Griffiths, D. W., 1979, On the relation between Modified Mercalli Intensity and body-wave magnitude: Bulletin of the Seismological Society of America, v. 69, p. 893–909.

Obermeier, S. F., Gohn, G. S., Weems, R. E., Gelinas, R. L., and Rubin, M., 1985, Geologic evidence for recurrent moderate to large earthquakes near Charleston, South Carolina: Science, v. 227, p. 408–411.

O'Connell, D. R., Bufe, C. G., and Zoback, M. D., 1982, Microearthquakes and faulting in the area of New Madrid, Missouri–Reelfoot Lake, Tennessee: U.S. Geological Survey Professional Paper 1236, p. 31–38.

Okaya, D. A., and Thompson, G. A., 1985, Geometry of Cenozoic extensional faulting; Dixie Valley, Nevada: Tectonics, v. 4, p. 107–125.

Pennington, W. D., Davis, S. D., Carlson, S. M., Dupree, J., and Ewing, T. E., 1986, The evolution of seismic barriers and asperities caused by the depressuring of fault planes in oil and gas fields of south Texas: Bulletin of the Seismological Society of America, v. 76, p. 939–948.

Pitt, A. M., and Hutchinson, R. A., 1982, Hydrothermal changes related to earthquake activity at Mud Volcano, Yellowstone National Park, Wyoming: Journal of Geophysical Research, v. 87, p. 2762–2766.

Pitt, A. M., Weaver, C. S., and Spence, W., 1979, The Yellowstone Park earthquake of June 30, 1975: Bulletin of the Seismological Society of America, v. 69, p. 187–205.

Pollack, H. N., and Christensen, D. H., 1984, Geophysical investigations of the western Ohio–Indiana region: Nuclear Regulatory Commission NUREG/CR-4339, 82 p.

Pomeroy, P. W., Simpson, D. W., and Sbar, M. L., 1976, Earthquakes triggered by surface quarying; The Wappingers Falls, New York sequence of June, 1974: Bulletin of the Seismological Society of America, v. 66, p. 685–700.

Prescott, W. H., and Yu, S.-Beih, 1986, Geodetic measurement of horizontal deformation in the northern San Francisco Bay region, California: Journal of Geophysical Research, v. 91, p. 7475–7484.

Prescott, W. H., Lisowski, M., and Savage, J. C., 1981, Geodetic measurement of crustal deformation on the San Andreas, Hayward, and Calaveras faults near San Francisco, California: Journal of Geophysical Research, v. 86, p. 10853–10869.

Presgrave, B. W., Needham, R. E., and Minsch, J. H., 1985, Seismograph station codes and coordinates, 1985 edition: U.S. Geological Survey Open-File Report 85-714, 385 p.

Prowell, D. C., 1983, Index of Cretaceous and Cenozoic faults in the eastern United States: U.S. Geological Survey Miscellaneous Field Studies Map MF-1269.

Pulli, J. J., and Guenette, M. J., 1981, A note on the Chelmsford-Lowell, Massachusetts earthquakes of 1980 and 1938: Earthquake Notes, v. 52, no. 2, p. 3–11.

Pulli, J. J., and Toksoz, M. N., 1981, Fault plane solutions for northeastern United States earthquakes: Bulletin of the Seismological Socioety of America, v. 71, p. 1875–1882.

Qamar, A., St. Lawrence, W., Moore, J. N., and Kendrick, G., 1983, Seismic signals preceding the explosive eruption of Mt. St. Helens, Washington on 18 May 1980: Bulletin of the Seismological Society of America, v. 73, p. 1797–1813.

Quittmeyer, R. C., Statton, C. T., Mrotek, K. A., and Houlday, M., 1985, Possible implications of recent microearthquakes in southern New York state: Earthquake Notes, v. 56, p. 35–42.

Raleigh, C. B., Healy, T. H., and Bredehoeft, J. D., 1972, Faulting and crustal stress at Rangely, Colorado: American Geophysical Union Monograph 16, p. 275–284.

Ratcliffe, N. M., 1984, Northeastern U.S. seismicity and tectonics, *in* Summaries of technical reports, v. 18, U.S. Geological Survey Open-File Report 84-268, p. 69–70.

Reasenberg, P., and Ellsworth, W. L., 1982, Aftershocks of the Coyote Lake, California, earthquake of August 6, 1979; A detailed study: Journal of Geophysical Research, v. 87, p. 10637–10655.

Reyners, M., 1980, A microearthquake study of the plate boundary, North Island, New Zealand: Geophysical Journal Royal Astronomical Society, v. 63, p. 1–22.

Richardson, R. M., Solomon, S. C., and Sleep, N. H., 1979, Tectonic stress in the plates: Reviews of Geophysics and Space Physics, v. 17, p. 981–1019.

Richins, W. D., and 6 others, 1987, The 1983 Borah Peak, Idaho earthquake and its aftershocks: Bulletin of the Seismological Society of America, v. 77, p. 694–723.

Richter, C. F., 1958, Elementary seismology: San Francisco, Freeman, 768 p.

Riddihough, R. P., 1977, A model for recent plate interactions off Canada's west coast: Canadian Journal of Earth Sciences, v. 14, p. 384–396.

Rogers, A. M., and Lee, W.H.K., 1976, Seismic study of earthquakes in the Lake Mead, Nevada-Arizona, region: Bulletin of the Seismological Society of America, v. 66, p. 1657–1681.

Rogers, A. M., and Malkiel, A., 1979, A study of earthquakes in the Permian basin of Texas–New Mexico: Bulletin of the Seismological Society of America, v. 69, p. 843–865.

Rogers, G. C., 1983, Some comments on the seismicity of the northern Puget Sound–southern Vancouver Island region, *in* Yount, J. C., and Crosson, R. S., eds., Earthquake hazards of the Puget Sound region, Washington: U.S. Geological Survey Open-File Report 83-19, p. 19–39.

Romney, C., 1957, Seismic waves from the Dixie Valley–Fairview Peak earthquakes: Bulletin of the Seismological Society of America, v. 47, p. 301–319.

Rundle, J. B., and Whitcomb, J. H., 1984, A model for deformation in Long Valley, California, 1980–1983: Journal of Geophysical Research, v. 89, p. 9371–9380.

Russ, D. P., 1982, Style and significance of surface deformation in the vicinity of New Madrid, Missouri: U.S. Geological Survey Professional Paper 1236, p. 95–114.

Ryall, A., 1977, Earthquake hazard in the Nevada region: Bulletin of the Seismological Society of America, v. 67, p. 517–532.

Ryall, A., and Malone, S. D., 1971, Earthquake distribution and mechanism of faulting in the Rainbow Mountain–Dixie Valley–Fairview Peak area, central Nevada: Journal of Geophysical Research, v. 76, p. 7241–7248.

Rymer, M. J., and Ellsworth, W. L., 1985, Mechanics of the May 2, 1983, Coalinga earthquake; An introduction, *in* Rymer, M. J., and Ellsworth, W. L., eds., Mechanics of the May 2, 1983, Coalinga earthquake: U.S. Geological Survey Open-File Report 85-44, p. 1–3.

Sanford, A. R., Olsen, K. H., and Jaksha, L. H., 1979, Seismicity of the Rio Grande Rift, *in* Riecker, R. E., ed., Rio Grande Rift; Tectonics and magmatism: Washington, D.C., American Geophysical Union, p. 145–168.

Savage, J. C., 1983, Strain accumulation in the western United States: Annual Reviews Earth Planetary Sciences, v. 11, p. 11–43.

Savage, J. C., and Burford, R. O., 1973, Geodetic determination of relative plate motion in central California: Journal of Geophysical Research, v. 78, p. 832–845.

Savage, J. C., and Cockerham, R. S., 1984, Earthquake swarm in Long Valley caldera, California, January 1983; Evidence for dike inflation: Journal of Geophysical Research, v. 89, p. 8315–8324.

Savage, J. C., Lisowski, M., and Prescott, W. H., 1981, Geodetic strain measurements in Washington: Journal of Geophysical Research, v. 86, p. 4929–4940.

Sbar, M. L., 1982, Delineation and interpretation of seismotectonic domains in western North America: Journal of Geophysical Research, v. 87, p. 3919–3928.

Sbar, M. L., and Sykes, L. R., 1973, Contemporary compressive stress and seismicity in eastern North America; An example of intra-plate tectonics: Geological Society of America Bulletin, v. 84, p. 1861–1882.

—— , 1977, Seismicity and lithospheric stress in New York and adjacent areas: Journal of Geophysical Research, v. 82, p. 5771–5786.

Scholz, C. H., Barazangi, M., and Sbar, M. L., 1971, Late Cenozoic evolution of the Great Basin, western United States, as an ensialic interarc basin: Geological Society of America Bulletin, v. 82, p. 2979–2990.

Schwartz, D. P., and Coppersmith, K. J., 1984, Fault behavior and characteristic earthquakes; Examples from the Wasatch and San Andreas fault zones: Journal of Geophysical Research, v. 89, p. 5681–5698.

Schwartz, D. P., and Crone, A. J., 1985, The 1983 Borah Peak earthquake; A calibration event for quantifying earthquake recurrence and fault behavior on Great Basin normal faults, *in* Stein, R. S., and Bucknam, R. C., eds., Workshop 28 on the Borah Peak, Idaho, earthquake: U.S. Geological Survey Open-File Report 85-290, p. 264–285.

Scott, W. E., Pierce, K. L., and Hait, M. H., 1985, Quaternary tectonic setting of the 1983 Borah Peak earthquake, Central Idaho: Bulletin of the Seismological Society of America, v. 75, p. 1053–1066.

Seborowski, K. D., Williams, G., Kelleher, J. A., and Statton, C. J., 1982, Tectonic implications of recent earthquakes near Annsville, New York: Bulletin of the Seismological Society of America, v. 72, p. 1601–1609.

Seeber, L., and Armbruster, J. G., 1981, The 1886 Charleston, South Carolina, earthquake and the Appalachian detachment: Journal of Geophysical Research, v. 86, p. 7874–7894.

—— , 1987, The 1886–89 aftershocks of the Charleston, South Carolina, earthquake; A widespread burst of seismicity: Journal of Geophysical Research, v. 92, p. 2663–2696.

Seeber, L., Barazangi, M., and Nowroozi, A., 1970, Microearthquake seismicity and tectonics of coastal northern California: Bulletin of the Seismological Society of America, v. 60, p. 1669–1699.

Segall, P., and Pollard, D. D., 1980, Mechanics of discontinuous faults: Journal of Geophysical Research, v. 85, p. 4337–4350.

Sharp, R. V., 1982, Comparison of 1979 surface faulting with earlier displacements in the Imperial Valley: U.S. Geological Survey Professional Paper 1254, p. 213–221.

Sharp, R. V., and 14 others, 1982, Surface faulting in the central Imperial Valley: U.S. Geological Survey Professional Paper 1254, p. 119–143.

Sibson, R. H., 1982, Fault zone models, heat flow, and the depth distribution of earthquakes in the continental crust of the United States: Bulletin of the Seismological Society of America, v. 72, p. 151–163.

—— , 1986, Rupture interaction with fault jogs, *in* Das, S., Boatwright, J., and Scholz, C. H., eds., Earthquake source mechanics: American Geophysical Union Geophysical Monograph 37, Maurice Ewing Series 6, p. 157–167.

Sieh, K. E., 1978a, Slip along the San Andreas fault associated with the great 1857 earthquake: Bulletin of the Seismological Society of America, v. 68, p. 1421–1448.

—— , 1978b, Central California foreshocks of the great 1857 earthquake: Bulletin of the Seismological Society of America, v. 68, p. 1731–1749.

—— , 1981, A review of geological evidence for recurrence times of large earthquakes, *in* Simpson, D. W., and Richards, P. G., eds., Earthquake prediction; An international review: American Geophysical Union, Maurice Ewing Series 4, p. 181–207.

Sieh, K. E., and Jahns, R. H., 1984, Holocene activity of the San Andreas fault at Wallace Creek, California: Geological Society of America Bulletin, v. 95, p. 883–896.

Silver, E. A., 1971, Tectonics of the Mendocino triple junction: Geological Society of America Bulletin, v. 82, p. 2965–2978.

—— , 1978, Geophysical studies and tectonic development of the continental margin of the western United States, lat 34 deg to 48 deg N, *in* Smith, R. B., and Eaton, G. P., eds., Cenozoic tectonics and regional geophysics of the western Cordillera: Geological Society of America Memoir 152, p. 251–262.

Silver, P., and Masuda, T., 1985, A source extent analysis of the Imperial Valley earthquake of October 15, 1979, and the Victoria earthquake of June 9, 1980: Journal of Geophysical Research, v. 90, p. 7639–7651.

Simila, G. W., Peppin, W. A., and McEvilly, T. V., 1975, Seismotectonics of the Cape Mendocino, California area: Geological Society of America Bulletin, v. 86, p. 1399–1406.

Simpson, D. W., 1986, Triggered earthquakes: Annual Review of Earth and Planetary Sciences 1986, v. 14, p. 21–42.

Slemmons, D. B., 1957, Geological effects of the Dixie Valley–Fairview Peak, Nevada, earthquakes of December 16, 1954: Bulletin of the Seismological Society of America, p. 353–375.

Slemmons, D. B., Ramelli, A. R., and Brocoum, S. J., 1985, Earthquake potential of the Meers fault, Oklahoma [abs.]: Earthquake Notes, v. 55, p. 1.

Smith, R. B., 1978, Seismicity, crustal structure, and intraplate tectonics of the interior of the western Cordillera, *in* Smith, R. B., and Eaton, G. P., eds., Cenozoic tectonics and regional geophysics of the western Cordillera: Geological Socioety of America Memoir 152, p. 111–144.

Smith, R. B., and Bruhn, R. L., 1984, Intraplate extensional tectonics of the eastern basin-range; Inferences on structural style from seismic reflection data, regional tectonics, and thermal-mechanical models of britle-ductile deformation: Journal of Geophysical Research, v. 89, p. 5733–5762.

Smith, R. B., and Christiansen, R. L., 1980, Yellowstone Park as a window on the Earth's interior: Scientific American, v. 242, no. 2, p. 104–117.

Smith, R. B., and Lindh, A. G., 1978, Fault-plane solutions of the western United States; A compilation, *in* Smith, R. B., and Eaton, G. P., eds., Cenozoic tectonics and regional geophysics of the western Cordillera: Geological Society of America Memoir 152, p. 107–109.

Smith, R. B., Winkler, P. L., Anderson, J. G., and Scholz, C. H., 1974, Source mechanisms of microearthquakes associated with underground mines in eastern Utah: Bulletin of the Seismological Society of America, v. 64, p. 1295–1317.

Smith, R. B., Shuey, R. T., Pelton, J. R., and Bailey, J. P., 1977, Yellowstone hot spot; Contemporary tectonics and crustal properties from earthquake and aeromagnetic data: Journal of Geophysical Research, v. 82, p. 3665–3676.

Smith, S. W., and Knapp, J. S., 1980, The northern termination of the San Andreas fault: California Division Mines Geology Special Report 140, p. 153–164.

Smith, S. W., McPherson, R. C., and Severy, N. I., 1981, The Eureka earthquake of 1980; Breakup of the Gorda plate: Earthquake Notes, v. 52, no. 1, p. 44.

Snay, R. A., Cline, M. W., and Timmerman, E. L., 1985, Dislocation models for the 1954 earthquake sequence in Nevada, *in* Stein, R. S., and Bucknam, R. C., eds., Workshop 28 on the Borah Peak, Idaho, earthquake: U.S. Geological Survey Open-File Report 85-290, p. 535–555.

State Earthquake Investigation Commission, Lawson, A. C., Chairman, 1908, Report on the California Earthquake of April 18, 1906: Carnegie Institution of Washington, vol. 1, 450 p.; vol. 2, 192 p.; atlas.

Stauder, W., 1982, Present-day seismicity and identification of active faults in the New Madrid Seismic Zone: U.S. Geological Survey Professional Paper 1236, p. 21–30.

Stauder, W., and Ryall, A., 1967, Spatial distribution and source mechanisms of microearthquakes in central Nevada: Bulletin of the Seismological Society of America, v. 57, p. 1317–1345.

Stauder, W., Kramer, M., Fischer, G., Schaefer, S., and Morrissey, S. T., 1976, Seismic characteristics of southeast Missouri as indicated by regional telemetered microearthquake array: Bulletin of the Seismological Society of America, v. 66, p. 1953–1964.

Stein, R. S., 1985, Evidence for surface folding and subsurface fault slip from geodetic changes associated with the 1983 Coalinga, California, earthquake, *in* Rymer, M. J., and Ellsworth, W. L., eds., Mechanics of the May 2, 1983, Coalinga earthquake: U.S. Geological Survey Open-File Report 85-44, p. 225–253.

Stein, R. S., and Barrientos, S. E., 1985, Planar high-angle faulting in the Basin and Range; Geodetic analysis of the 1983 Borah Peak, Idaho, earthquake: Journal of Geophysical Research, v. 90, p. 11355–11366.

Stein, R. S., and King, G.C.P., 1984, Seismic potential revealed by surface folding; 1983 Coalinga, California, earthquake: Science, v. 224, p. 869–871.

Stewart, J. H., 1978, Basin-range structure in western North America; A review, *in* Smith, R. B., and Eaton, G. P., eds., Cenozoic tectonics and regional geophysics of the western Cordillera: Geological Society of America Memoir 152, p. 1–31.

Stickney, M. C., 1978, Seismicity and faulting of central western Montana: Northwest Geology, v. 7, p. 1–9.

Stone and Webster Engineering Corporation, 1976, Faulting in the Anna, Ohio, region: Preliminary Safety Analysis Report submitted to the U.S. Nuclear Regulatory Commission, Wisconsin Utilities Project, Koshkonong Nuclear Plant, Units 1 and 2, Amendment 12, Appendix 2I, 11 p.

Street, R. L., and Lacroix, A., 1979, An empirical study of New England seismicity: Bulletin of the Seismological Society of America, v. 69, p. 159–175.

Street, R. L., and Turcotte, F. T., 1977, A study of northeastern North American spectral moments, magnitudes, and intensities: Bulletin of the Seismological Society of America, v. 67, p. 599–614.

Sykes, L. R., 1978, Intraplate seismicity, reactivation of preexisting zones of weakness, alkaline magmatism, and other tectonism postdating continental fragmentation: Reviews of Geophysics and Space Physics, v. 16, p. 621–688.

Taber, J. J., and Smith, S. W., 1985, Seismicity and focal mechanisms associated with the subduction of the Juan de Fuca plate beneath the Olympic Peninsula, Washington: Bulletin of the Seismological Society of America, v. 75, p. 237–249.

Talwani, P., 1977, Stress distribution near Lake Jocassee, South Carolina: Pure and Applied Geophysics, v. 115, p. 275–281.

—— , 1982, Internally consistent pattern of seismicity near Charleston, South Carolina: Geology, v. 10, p. 654–658.

Talwani, P., and Cox, J., 1985, Paleoseismic evidence for recurrence of earthquakes near Charleston, South Carolina: Science, v. 229, p. 379–381.

Talwani, P., Stevenson, D., Amick, D., and Chiang, J., 1979, An earthquake swarm at Lake Keowee, South Carolina: Bulletin of the Seismological Society of America, v. 69, p. 825–841.

Tarr, A. C., and Rhea, S., 1983, Seismicity near Charleston, South Carolina, 1973–1978: U.S. Geological Survey Professional Paper 1313, p. R1–R17.

Thatcher, W., 1975, Strain accumulation on the northern San Andreas fault since 1906: Journal of Geophysical Research, v. 80, p. 4873–4880.

—— , 1979, Systematic inversion of geodetic data in central California: Journal of Geophysical Research, v. 84, p. 2283–2295.

Trimble, A. B., and Smith, R. B., 1975, Seismicity and contemporary tectonics of the Hebgen Lake–Yellowstone region: Journal of Geophysical Research, v. 80, p. 733–741.

Tse, S. T., and Rice, J. R., 1986, Crustal earthquake instability in relation to the depth variation of frictional slip properties: Journal of Geophysical Research, v. 91, p. 9452–9472.

U.S. Geological Survey, 1984, Preliminary determination of epicenters, monthly listing, October 1983: U.S. Geological Survey, 20 p.

U.S. Geological Survey and National Oceanic and Atmospheric Administration, 1971, The San Fernando, California, earthquake of February 9, 1971: U.S. Geological Survey Professional Paper 733, 254 p.

Van Tyne, A., 1975, Subsurface investigation of the Clarendon-Linden structure, western New York: New York State Museum and Science Service Open-File Report.

Van Wormer, J. D., and Ryall, A. S., 1980, Sierra Nevada–Great Basin boundary zone; Earthquake hazard related to structure, active tectonic processes, and anomalous patterns of earthquake occurrence: Bulletin of the Seismological Society of America, v. 70, p. 1557–1572.

Vetter, U. R., and Ryall, A. S., 1983, Systematic change of focal mechanism with depth in the western Great Basin: Journal of Geophysical Research, v. 88, p. 8237–8250.

Wallace, R. E., 1968, Notes on stream channels offset by the San Andreas fault, southern Coast Ranges, California, *in* Dickinson, W. R., and Grantz, A., eds., Proceedings of the Conference on Geologic Problems of the San Andreas Fault System: Stanford, California, Stanford University Publications in the Geological Sciences, v. 11, p. 6–21.

—— , 1970, Earthquake recurrence intervals on the San Andreas fault: Geological Society of America Bulletin, v. 81, p. 2875–2890.

—— , 1973, Surface fracture patterns along the San Andreas fault: Stanford, California, Stanford University Publications Geological Sciences, v. 13, p. 248–250.

—— , 1981, Active faults, paleoseismology, and earthquake hazards in the western United States, *in* Richards, P. G., and Simpson, D. W., eds., Earthquake prediction; An international review: American Geophysical Union Maurice Ewing Series 4, p. 209–216.

—— , 1984, Faulting related to the 1915 earthquakes in Pleasant Valley, Nevada: U.S. Geological Survey Professional Paper 1274A, 33 p.

Wallace, T. C., 1985, A reexamination of the moment tensor solutions of the 1980 Mammoth Lakes earthquakes: Journal of Geophysical Research, v. 90, p. 11171–11176.

Walter, A. W., and Weaver, C. S., 1980, Seismicity of the Coso Range, California: Journal of Geophysical Research, v. 85, p. 2441–2458.

Walter, S. R., 1986, Intermediate-focus earthquakes associated with Gorda plate subduction in northern California: Bulletin of the Seismological Society of America, v. 76, p. 583–588.

Ward, S. N., and Barrientos, S. E., 1986, An inversion for slip distribution and fault shape from geodetic observations of the 1983, Borah Peak, Idaho, earthquakes: Journal of Geophysical Research, v. 91, p. 4909–4919.

Warren, D. H., 1979, Fault-plane solutions for microearthquakes preceding the Thanksgiving Day, 1974, earthquake at Hollister, California: Geophysical Research Letters, v. 6, p. 633–636.

Weaver, C. S., and Hill, D. P., 1978/79, Earthquake swarms and local crustal spreading along major strike-slip faults in California: Pure and Applied Geophysics, v. 117, p. 51–64.

Weaver, C. S., and Michaelson, C. A., 1985, Seismicity and volcanism in the Pacific northwest; Evidence for the segmentation of the Juan de Fuca plate: Geophysical Research Letters, v. 12, p. 215–218.

Weaver, C. S., and Smith, S. W., 1983, Regional tectonic and earthquake hazard implications of a crustal fault zone in southwestern Washington: Journal of Geophysical Research, v. 88, p. 10371–10383.

Weaver, C. S., Grant, W. C., Malone, S. D., and Endo, E. T., 1981, Post-May 18 seismicity; Volcanic and tectonic implications: U.S. Geological Survey Professional Paper 1250, p. 109–121.

Webb, T. H., and Kanamori, H., 1985, Earthquake focal mechanisms in the eastern Transverse Ranges and San Emigdio Mountains, southern California and evidence for a regional décollement: Bulletin of the Seismological Society of America, v. 71, p. 737–757.

Weldon, R., and Humphreys, E., 1986, A kinematic model of southern California: Tectonics, v. 5, p. 33–48.

Wentworth, C. M., and Mergner-Keefer, M., 1983, Regenerate faults of small Cenozoic offset; Probable earthquake sources in the southeastern United States: U.S. Geological Survey Professional Paper 1313, S1–S20.

Wentworth, C. M., Zoback, M. D., and Bartow, J. A., 1985, Tectonic setting of the 1983 Coalinga earthquakes from seismic reflection profiles; A progress report, *in* Rymer, M. J., and Ellsworth, W. L., eds., Mechanics of the May 2, 1983, Coalinga earthquake: U.S. Geological Survey Open-File Report 85-44, p. 19–30.

Wilson, J. T., 1965, A new class of faults and their bearing on continental drift: Nature, v. 4995, p. 343.

Wong, I. G., 1984, Mining-induced seismicity in the Colorado Plateau, western United States, and its implications for the siting of an underground high-level nuclear waste repository, *in* Gay, N. C., and Wainwright, E. H., eds., Proceedings of the 1st International Congress on Rockbursts and Seismicity in Mines, Johannesburg, 1982: South African Institute of Mining and Metallurgy Symposium Series No. 6, p. 147–152.

Wong, I. G., Cash, D. J., and Jaksha, L. H., 1984, The Crownpoint, New Mexico, earthquakes of 1976 and 1977: Bulletin of the Seismological Society of America, v. 74, p. 2435–2449.

Yang, J.-P., and Aggarwal, Y. P., 1981, Seismotectonics of northeastern United States and adjacent Canada: Journal of Geophysical Research, v. 86, p. 4981–4998.

Yerkes, R. F., and Castle, R. O., 1976, Seismicity and faulting attributable to fluid extraction: Engineering Geology, v. 10, p. 151–167.

Yerkes, R. F., Ellsworth, W. L., and Tinsley, J. C., 1983, Triggered reverse fault and earthquake due to crustal unloading, northeast Transverse Ranges, California: Geology, v. 11, p. 287–291.

York, J. E., and Oliver, J. E., 1976, Cretaceous and Cenozoic faulting in eastern North America: Geological Society of America Bulletin, v. 87, p. 1105–1114.

Yoshii, T., 1979, A detailed cross-section of the deep seismic zone beneath northeastern Honshu, Japan: Tectonophysics, v. 55, p. 349–360.

Zandt, G., and Richins, W. D., 1980, An upper mantle earthquake beneath the middle Rocky Mountains in northeast Utah [abs.]: Earthquake Notes, v. 50, p. 69–70.

Zoback, M. D., 1983, Intraplate earthquakes, crustal deformation, and in-situ stress, *in* Hays, W. W., and Gori, P. L., eds., The 1886 Charleston, South Carolina, earthquake and its implications for today: U.S. Geological Survey Open-File Report 83-843, p. 169–178.

Zoback, M. D., and Hickman, S., 1982, In-situ study of the physical mechanisms controlling induced seismicity at Monticello Reservoir, South Carolina: Journal of Geophysical Research, v. 87, p. 6959–6974.

Zoback, M. D., and Zoback, M. L., 1981, State of stress and intraplate earthquakes in the United States: Science, v. 213, p. 96–104.

Zoback, M. D., Hamilton, R. M., Crone, A. J., Russ, D. P., McKeown, F. A., Brockman, S. R., 1980, Recurrent intraplate tectonism in the New Madrid seismic zone: Science, v. 209, p. 971–976.

Zoback, M. D., Anderson, R. N., and Moos, D., 1985, In-situ stress measurements in a 1 km-deep well near the Ramapo fault zone [abs.]: EOS Transactions of the American Geophysical Union, v. 59, p. 110.

Zoback, M. L., 1989, State of stress and modern deformation on the northern Basin and Range Province: Journal of Geophysical Research, v. 94, p. 7105–7128.

Zollweg, J. E., and Richins, W. D., 1985, Later aftershocks of the 1983 Borah Peak, earthquake and related activity in central Idaho, *in* Stein, R. S., and Bucknam, R. C., eds., Workshop 28 on the Borah Peak, Idaho, Earthquake: U.S. Geological Survey Open-File Report 85-290, p. 345–356.

Manuscript Accepted by the Society October 31, 1988

Printed in U.S.A.

Geological Society of America
Memoir 172
1989

Chapter 26

Crustal structure of the western U.S. based on reflection seismology

Scott B. Smithson
Department of Geology and Geophysics, University of Wyoming, Laramie, Wyoming 82071
Roy A. Johnson
Department of Geosciences, Gould-Simpson Building, University of Arizona, Tucson, Arizona 85721

ABSTRACT

Interpretation of crustal reflection profiles shows contrasting crustal styles and Moho from the craton to the Cordilleran belt. Crustal deformation, however, may be determined from highly reflective ductile fault (mylonite) zones that occur in many geologic settings, and fractures and chemical alteration may cause shallow detachments to reflect, e.g., the Sevier Desert and the Picacho Mountains detachments. The oldest Archean crust in Minnesota consists of a stack of nappes 30 km thick, and the underlying Moho, which is nonreflective, may be gradational. Younger Archean is sutured to this along a complex, moderately dipping zone marked by a mylonite. Proterozoic crust in Kansas and the Colorado Plateau is characterized by arcuate crossing reflections that can be caused by a combination of folding and intrusion. The Mojo is generally nonreflective except in the extended terranes of the Basin and Range, Rio Grande rift, northwest Cordillera, and Mojave-Sonoran Desert. The best crustal reflections are found in the extended terrane of the Basin and Range, where the subhorizontal reflection geometry is probably caused by ductile flow (metamorphism) under simple shear, resulting in a strong compositional layering and lensing of the deep crust and Moho. This young Moho contrasts sharply with the nonreflective Moho under the craton that may represent a gabbro-eclogite phase change. Underplating by gabbroic magma may have taken place in such diverse tectonic settings as the Minnesota Archean, the Oklahoma Proterozoic, and the Basin and Range Cenozoic crust. No evidence for Moho offset has been found in any of the areas studied. A complex crustal reflection pattern and distinct Moho reflection in the northwest Cordillera may be related to moderate extension conditioned by less (than Basin and Range) thermal input to the crust. Compression caused folding and thrust faulting in the Wyoming foreland. Basaltic magma chambers may have been detected beneath the Rio Grande rift and Death Valley; apparently, basaltic magma is more likely to generate reflections than granitic magma. Questions exist about how much of the exposed Precambrian crust in southern California, where a midcrustal detachment has been proposed, is allochthonous. Reflection profiling reveals a crust that is both vertically and horizontally heterogeneous.

Smithson, S. B., and Johnson, R. A., 1989, Crustal structure of the western U.S. based on reflection seismology, *in* Pakiser, L. C., and Mooney, W. D., Geophysical framework of the continental United States: Boulder, Colorado, Geological Society of America Memoir 172.

Figure 1. Map showing location of crustal reflection data in the United States. This report covers the western United States and Minnesota, while Arkansas is included in the Appalachians (Phinney, this volume).

INTRODUCTION

The rapidly growing data base of deep-crustal reflection profiles easily represents the most complete source of information on the nature of the deep crust. No other method besides drilling has the potential for revealing structure and composition at so fine a scale, yet problems remain in the application of the reflection method and in the interpretation of results. The problem is twofold: the target of the crystalline crust is a much more complex target than most sedimentary basins, and interpretations are usually minimally constrained in the absence of other subsurface control. Thus, the question is how to interpret any given reflection or how to obtain an interpretable data set. We begin with some general considerations, followed by a review of results from the western United States (Fig. 1).

TECHNIQUE

Reflections in crystalline crust

The seismic-reflection method is applied routinely to map structures in both horizontal to subhorizontal sedimentary rocks and in more strongly deformed areas such as fold and thrust belts. The situation is, however, more complex in crystalline crust, mainly for the following three reasons: (1) reflection coefficients are much smaller in crystalline rocks (about 0.13 vs. 0.30 for maximum values in crystalline and sedimentary rocks, respectively); (2) interfaces in crystalline rocks are far more contorted and discontinuous due to the effects of deformation and intrusion because these rocks generally have been deformed much more ductilely than sedimentary rocks; and (3) lack of drilling or exposure of objectives in crystalline rocks means lack of "ground truth" to confirm, modify, or reject interpretations.

The geometry of crystalline rocks is generally complex (Figs. 2, 3; Smithson and others, 1977a, Fig. 1); the seismic response of these rocks may be complex or weak, or both. The strongest reflections are generated by constructive interference from layers involving velocity reversals (Fuchs, 1969; Smithson and others, 1977a; Fountain and others, 1984) so that layered metamorphic sequences and possibly layered igneous intrusions should generate good reflections. If reflections are generated from layering, then the reflection wavelet will be multicyclic instead of a single wavelet, and constructive interference may generate reflections two to four times stronger than reflections from a single interface (a simple single wavelet). These multicyclic reflections will tend to be more common in crustal reflection sections.

Complex geometry—such as folded, irregular, or corrugated contacts or lenses—tends to interrupt the continuity of reflections and produce arcuate reflections and diffractions (Figs. 2, 3; Smithson, 1979, 1986; Smithson and others, 1980; Wong and others, 1982). Such data must be migrated or moved updip on the seismic section in order to achieve a "proper" image of the subsurface geometry; a good image is not necessarily achieved for complex targets. Because mylonites may be thick, relatively planar, and have moderate dip, they are probably the best reflectors in the crust (Smithson and others, 1979; Fountain and others, 1984); recent reflection profiling over mylonites demonstrates that they may be remarkably reflective (Hurich and others, 1985). Thus, mylonites may be mapped in seismic sections and used to resolve crustal-scale deformation of the crust (Smithson and others, 1986). Although in petroleum exploration, "sideswipe" or seismic events out of the plane of the recording profile are commonly considered noise, in the complex geometries found in the Precambrian, out-of-the-plane events may be the rule rather than the exception and must be considered in the interpretation (Blundell and Raynard, 1986); such events are considered

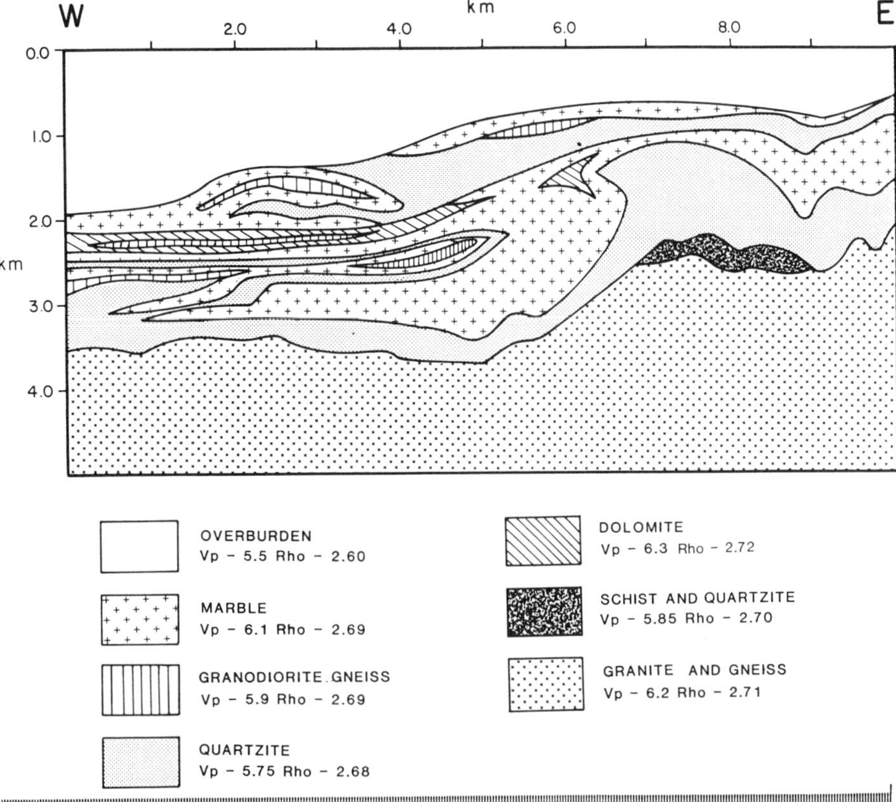

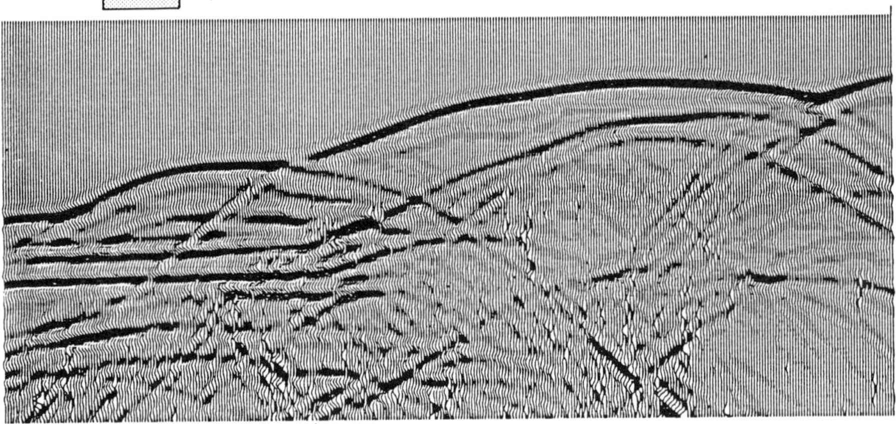

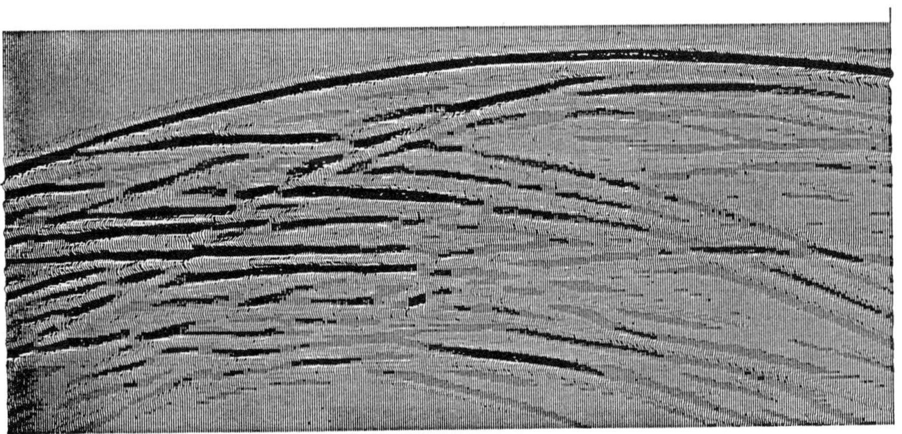

Figure 2. Seismic model of nappe structure observed in the infrastructure of the Ruby Mountains, Nevada. Synthetic seismogram was generated on a computer from model at a depth of 2 km (middle). The same model at a depth of 20 km shows how the reflection response changes with depth. Note the arcuate events with weak diffraction tails (after Hurich and Smithson, 1987).

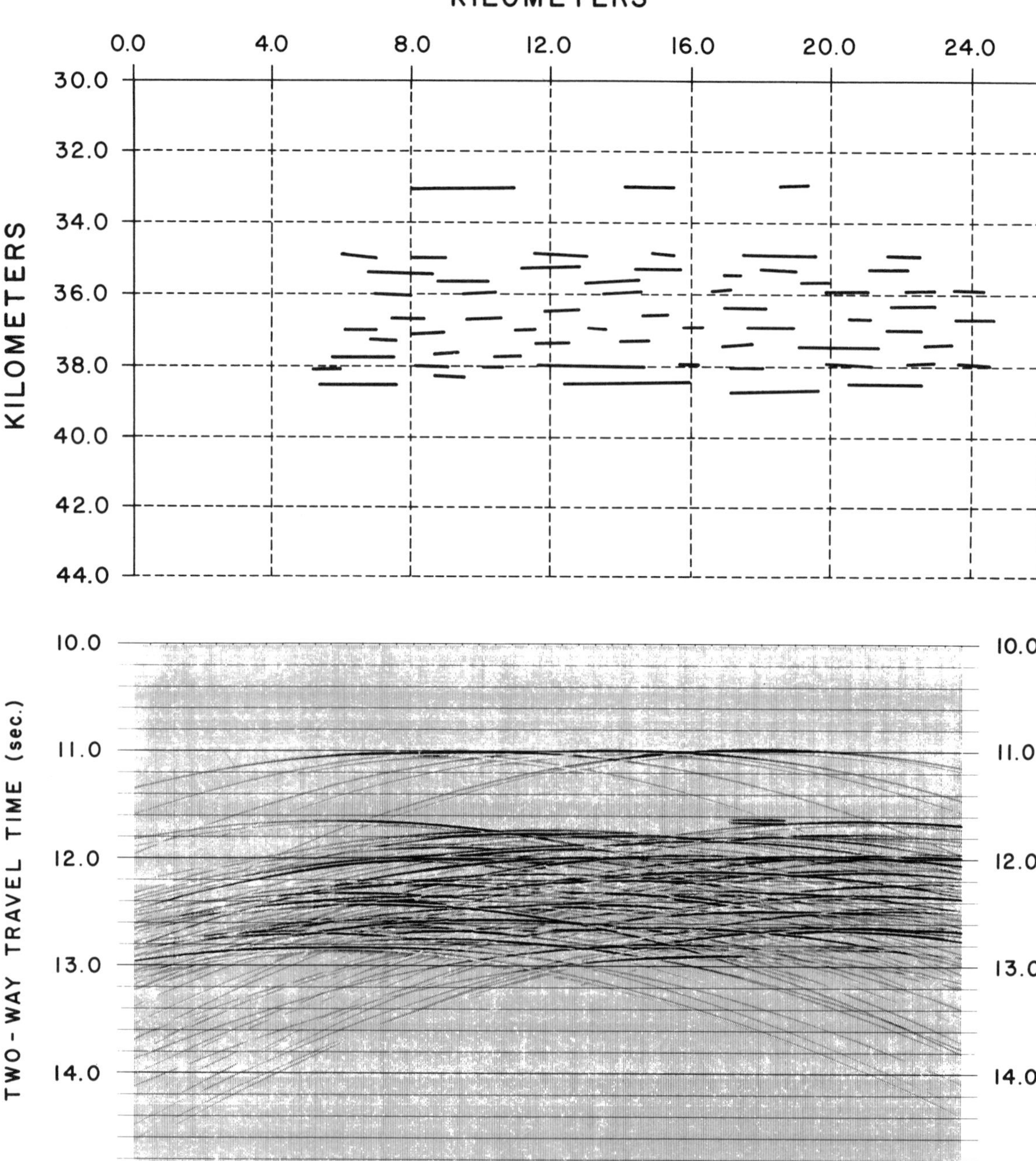

Figure 3. Model of a series of randomly spaced reflector segments at lower crustal depths (top). The reflectors, which are generally shorter than the Fresnel zone, represent short reflective segments in a complex terrane such as boudins in a gneiss terrane or in an extended terrane (bottom). The model response is dominated by short, discontinuous, and arcuate events similar to those observed in the lower crust in crustal reflection profiles (after Hurich and Smithson, 1987).

as signal unless they follow a near-surface path, but they will confuse interpretation.

Recording

The choice of recording parameters can make or break a data set because they determine what can be done in all the succeeding steps. Ideally, the recording experiment is based on the geologic problem in an area and specific processing considerations, if known a priori. Velocity determination is a critical step in interpretation that will be strongly affected by the recording geometry. Basically, an idealized recording system with an infinite number of channels and infinite dynamic range would solve most recording problems and allow "perfect" recording experiments. The lack of ideal recording means that almost any recording experiment becomes a *compromise* between competing demands, e.g., close station spacing for resolving high dip vs. long offset for resolving velocities. Vibrators give added flexibility for wide-angle recording to determine velocities (Kubichek and others, 1984; Liu and others, 1986; Valasek and others, 1987).

Major sources of uncertainty that enter into the interpretation of crustal reflection data are the effects of recording conditions and processing on the final interpretation. The effects of recording and processing are seen on all scales from individual shot gathers to entire lines hundreds of kilometers long. Vertical strips of no reflection, so common in reflection profiles recorded on land, must reflect the effects of recording or processing such as severe static effects (Allmendinger and others, 1983, Fig. 2); in general, such abrupt changes in reflection character can hardly represent actual geologic structure. Areas in which data is strongly affected by recording conditions are the Rio Grande Rift; Parkfield; Nevada; the Sierra Nevada; and Death Valley. Other aspects of seismic reflection profiling are discussed by Mooney (this volume).

Resolution

When we point out the importance of mylonites formed by ductile faulting for generating good crustal reflections, geologists continually ask, "How thick a mylonite zone can you detect—30 cm?" Unfortunately, the answer is no; seismic techniques lack this resolving power by several orders of magnitude and probably always will. The highest amplitude reflections are generated by tuning from layers with thicknesses on the order of one-quarter wavelength. As the thickness decreases below one-quarter wavelength, the amplitude of a single reflection is decreased by destructive interference, until at a thickness of about one-tenth wavelength the reflection is usually too weak to be seen (Widess, 1973). For lower signal-to-noise (S/N) ratios this minimum thickness will be larger (Widess, 1973). This means that for a typical wavelet of 20 Hz and a velocity of 6 km/sec, a maximum-amplitude reflection from a single layer imbedded in a different medium occurs when it is 75 m thick. Minimum thickness that might be seen is 30 m, but only in very good S/N data. Increased frequency content decreases the resolvable layer thickness.

Horizontal resolution, which is often neglected, is much poorer. It is generally defined as the radius of the first Fresnel zone (Sheriff and Geldart, 1982), i.e., a reflection is generated from an area rather than from a point. This gives the minimum horizontal spacing of features that may be resolved. Fresnel zone-radius values for the same parameters as used above are 1.7 km at 10 km depth and 3.0 km at 30 km depth (Sheriff and Geldart, 1982). This means that a reflector that contains a gap may appear continuous if the gap is smaller than a Fresnel zone. Clearly, horizontal resolution is a more important constraint in the deep crust for crustal scale imaging than vertical resolution.

Resolving power depends on frequency; thus, attaining high frequencies is important. However, high frequencies by themselves are not sufficient. Band width (in terms of octaves) is also critical. Recording data within a band from 100 to 200 Hz is not high resolution because this is only one octave. For example, a 100- to 200-Hz wavelet (one octave) is high frequency but will have many side lobes, thereby reducing resolution. About 2½ octaves (i.e., 10 to 50 Hz) is considered minimal. Therefore, the low frequencies as well as the highs are necessary for a desirable wavelet.

AREAS

Laramide uplifts in Wyoming

Wind River Mountains. The history of interpretation of the Consortium for Continental Reflection Profiling (COCORP) Wind River data set in Wyoming (Fig. 1) illustrates some of the problems in the acquisition and interpretation of crustal reflection interpretation. The Wind River seismic profile was the first strictly problem-oriented seismic data set recorded by COCORP and has been the crustal reflection data set most extensively studied by different groups. The experiment was designed to resolve the controversy over the origin of the broad, basement-cored asymmetrical uplifts of the Laramide foreland: were these formed by compression involving crustal shortening along gently dipping thrust faults or by vertical movements along high-angle reverse faults (Smithson and others, 1978, Fig. 1, p. 648)? In all of these uplifts, the oversteepened flank is faulted along a moderately dipping thrust fault that has been postulated by the proponents of vertical uplift to steepen into a vertical reverse fault at depth (Prucha and others, 1965). Of all the Laramide uplifts, the Wind River uplift is the largest, exhibiting a vertical separation of 10 to 12 km on the basement. The expectation was that, with so much offset of the crystalline basement, deep reflections might be displaced but correlatable across the fault, thereby revealing its geometry deep in the crust. Surprisingly, the fault zone itself appeared to give a continuous reflection from near the surface to deep (7 to 8 sec) within the crust (Fig. 4) and was readily correlated with the thrust fault where drilling showed that Precambrian rocks were in contact with overturned Cretaceous

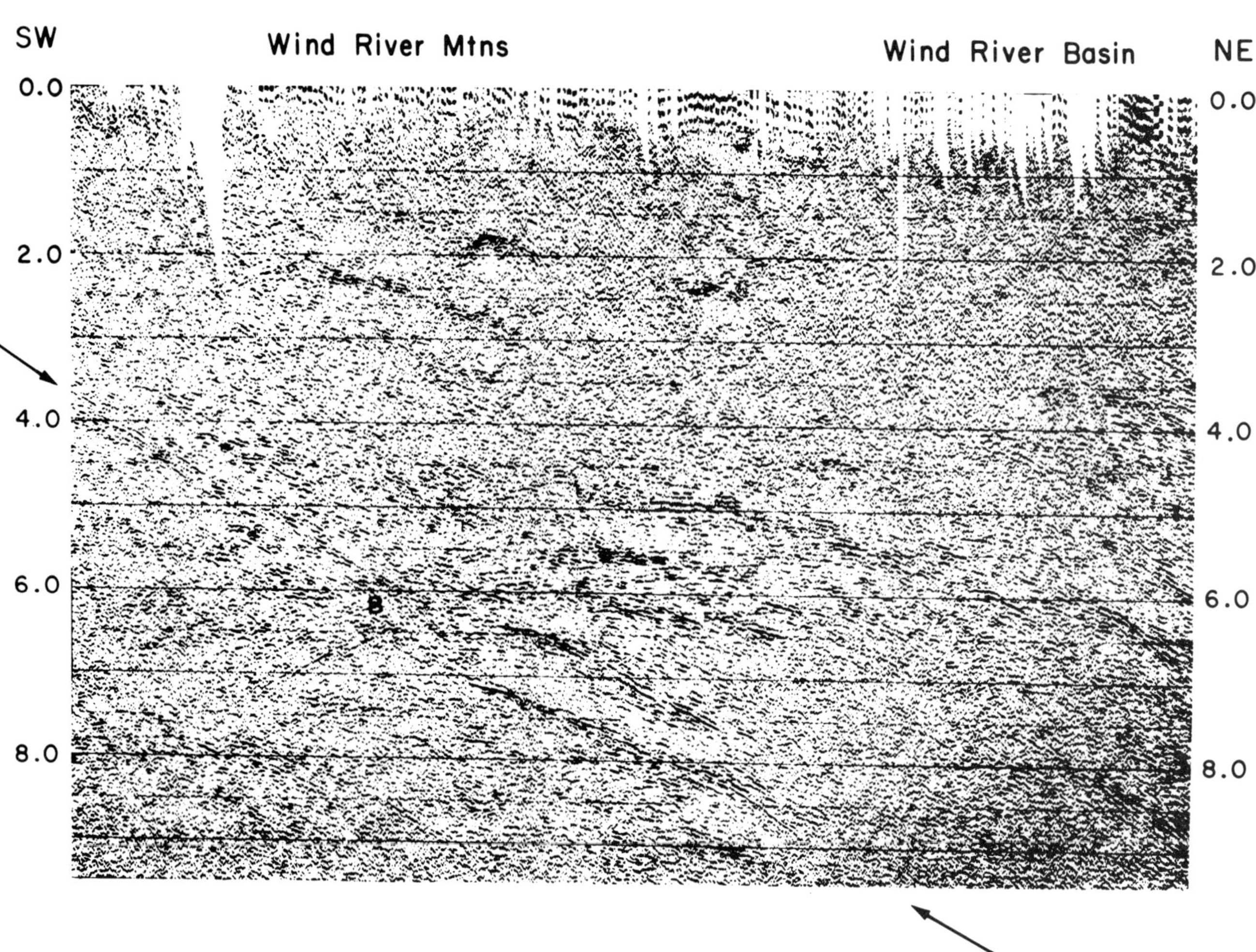

Figure 4. Original COCORP processed seismic section across the Wind River thrust. Thrust reflection (B) is marked by arrows. Sedimentary rocks are cut off at about 4.2 sec. Note continuation of thrust reflection in crystalline crust.

sedimentary rocks. This relationship established the practice of interpreting deep-crustal reflections by correlating them with surface geology, an approach that is still very successful today.

The Wind River fault thus has been interpreted on the basis of continuous reflections as a thrust fault dipping at about 42° and continuing to a depth of at least 24 km (Smithson and others, 1979). The postulated fault-zone reflection immediately led to two questions: could the reflection simply be an anomalous event, and why would the fault zone reflect? Geophysicists in industry, experienced in reflection seismology, pointed out that the reflection might be an anomalous event such as those often found in complicated structural environments. Seismic modeling by Sacrison (1978) had demonstrated that anomalous reflections could be encountered in complex structural environments such as Laramide folds. Until reasons for the high reflectivity of crustal fault zones could be demonstrated, the answer to this question was that the continuity of the reflection from the known fault zone with the deeper event had to be more than coincidental.

The crustal fault-zone reflection might be explained by one of two competing hypotheses (Smithson and others, 1979; Zawislak and Smithson, 1981): (1) juxtaposition of different crystalline rock types such as granite and gabbro, or (2) a ductile deformation zone (mylonite or shear zone) that was a gneissic zone several hundred meters wide. In the second case, the fault zone itself would cause the reflection. The seismic identification of the Wind River thrust led to two major results besides the identification of the thrust at depth: (1) the interpretation of many deep crustal reflections as fault zones (Allmendinger and others, 1983; Bortfeld and others, 1985; Cheadle and others, 1986a), and (2) attempts to explain crustal fault-zone reflectivity (Smithson and others, 1979; Wong and others, 1982; Jones and Nur, 1982, 1984; Fountain and others, 1984; Hurich and others, 1985). As it

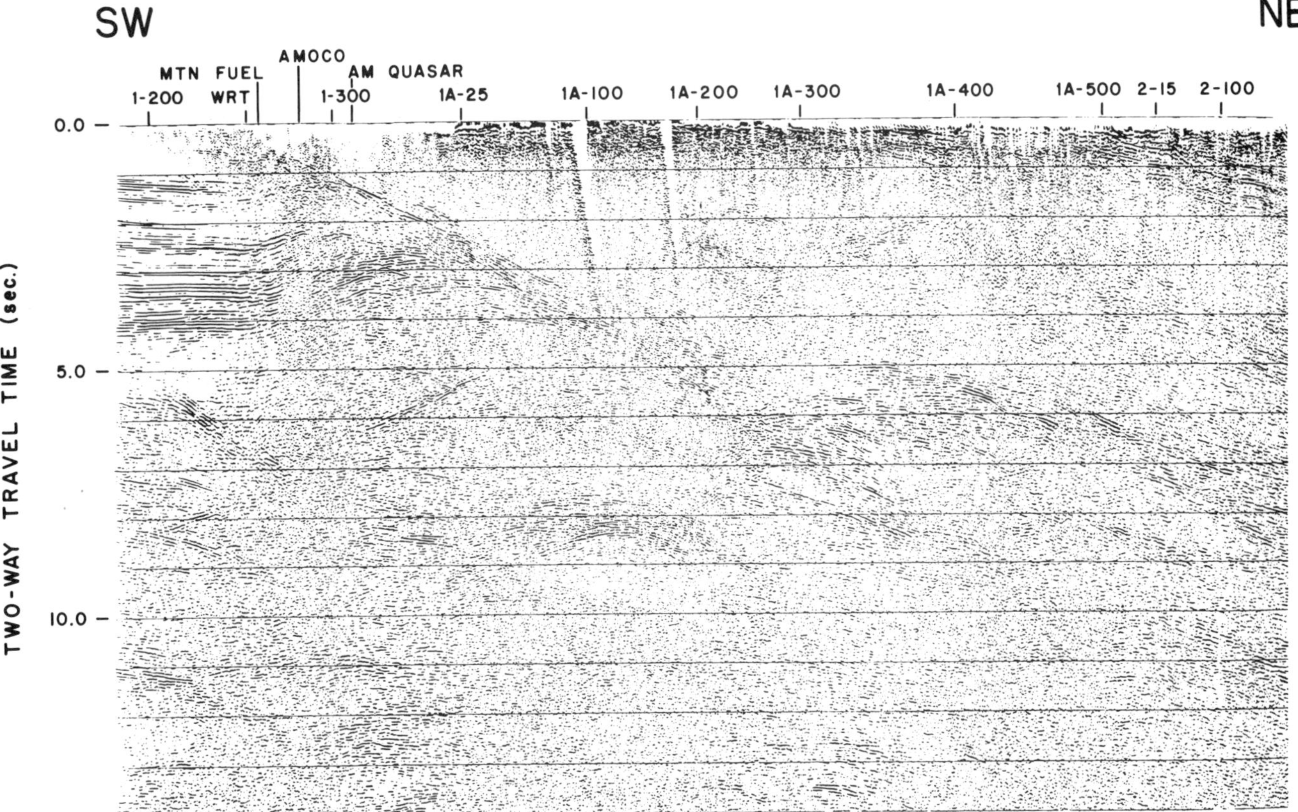

Figure 5. Seismic section reprocessed by Gulf Oil Co. (Sharry and others, 1986), showing increased resolution and less geometric distortion through processing. Sedimentary rocks of the Green River Basin are visible on the left under the Precambrian overhang.

turns out, from the beginning, the multicyclic character of the fault zone reflections was the tipoff that the fault-zone reflections were coming from a layered sequence, thereby eliminating the first hypothesis. Because mylonite zones are characterized by strongly layered fabrics caused by ductile deformation, the possibility that reflections were caused by velocity anisotropy had to be considered. The principal cause of anisotropy in common crystalline rocks is the preferred orientation of biotite or other phyllosilicates that have extreme velocity anisotropy of about 65 percent (Alexandrov and Rhyzova, 1961), which can result in velocity anisotropies in rocks of 10 to 20 percent (Birch, 1960; Kern and Fakhimi, 1975; Jones and Nur, 1982). Therefore, the first seismic model for mylonite reflectivity took into account the effect of fabric as well as compositional layering by using velocities measured for biotite-rich anisotropic gneisses (Wong and others, 1982, p. 102). This resulted in a synthetic seismogram qualitatively similar to the observed reflections; discontinuous layering was used to simulate the high-amplitude multicyclic reflections. Mylonite-zone reflectivity was finally demonstrated in the field by following a mylonite zone from the surface to depth in a reflection profile on the flank of a core complex. This showed that strong, continuous reflections could be generated from layering in a mylonite zone 1 km thick (Hurich and others, 1985, Fig. 26), demonstrating how the Wind River thrust could generate reflections from the middle crust, which presumably was within the ductile zone at a depth of about 15 km or more during Laramide deformation. The identification of reflections from the Wind River thrust demonstrated that Laramide uplifts were compressional features and showed that deep fault zones themselves were reflective. The fact that faults could be followed through much of the crust opened the possibility of tracing crustal-scale deformation with reflection profiles (Smithson and others, 1986).

Based on the geometry and character of the reflection itself, the thrust fault is interpreted to have an apparent dip of about 35° and consist of a layered zone of discontinuous mylonitic rocks, which could consist of a range of crustal rock types, including peridotite. A maximum true dip of about 42° was determined based on the strike of gravity contours (Smithson and others, 1979), but using the same gravity data and a reprocessed and interpolated seismic section (Fig. 5), Sharry and others (1986) suggested a true dip of 35°. Lynn and others (1983) proposed that the main fault had splays and parallel faults, presenting a more complicated geometry than the earlier interpretations. The fault reflection can be traced continuously to a depth of about 21 km (7 sec) and fairly safely to a depth of about 24 km (8 sec). From here, some workers (Lynn and others, 1983; Sharry and others, 1986) have continued the fault to a depth of about 32 km (10 sec), but this is probably based on confusing multiples with the fault reflection (Zawislak and Smithson, 1981). This point is emphasized by the fact that dipping events are so common between 6 and 12 sec that one could put the fault anywhere over an interval 18 km deep. Industry cross-lines, providing valuable 3-D control, indicate that the fault undulates (McLeod, 1981) so that strikes and dips of the fault may change significantly along the strike.

The fault shows some convex-downward curvature in several migrated interpretations (Smithson and others, 1979; Lynn and others, 1983; Sharry and others, 1986; Thompson and Hill, 1986), with the amount of curvature depending on the velocity function used for migration. Based on gravity interpretation (Hurich and Smithson, 1982), the fault probably displaces more dense rocks in the middle crust, but since no evidence has been found to suggest that it displaces the Moho, the deep trajectory of the thrust is uncertain. A strongly convex-downward event attributed to the Pacific Creek thrust (Lynn and others, 1983) probably results from migration of sideswipe, because this event has too many cycles to be generated by a mylonite zone associated with the small (~200 m) vertical separation on this thrust.

The attitude of the thrust-fault reflection clearly shows that Laramide foreland uplifts were formed by crustal shortening (compression) rather than simple vertical movements, a shortening probably caused by plate convergence (Brewer and others, 1980). The shortening caused folding of the basement, which eventually resulted in thrusting of the steep flank of the uplifted Phanerozoic rocks that, at the southeast end of the Wind River uplift, wrap around the Precambrian core and define a fold plunging 10°SE. The Phanerozoic rocks are cut by numerous high-angle faults and are displaced slightly across these faults. Southwestward-dipping sedimentary rocks are preserved above the Precambrian basement along the oversteepened southwest flank of the Wind River uplift. These observations indicate that the basement has indeed been folded on a large scale to explain the rotation observed in overlying sedimentary rocks while it has been displaced along high-angle faults on a small scale to accommodate this strain. This style of basement deformation is found in other Wyoming uplifts. Crustal thickening has been achieved by the approximately 20 km of shortening observed along the major thrust, and the Wind River Mountains have become a classic example of thick-skinned deformation of the continental crust. To interpret the deep structure of the crust, data from a 3-km drillhole (Smithson and Ebens, 1971) were incorporated with up-plunge exposures to propose complex folding (Smithson and others, 1980); Sharry and others (1986) interpreted the same features as a crustal duplex.

The area has also become a classic one for studying the effect of multiple reflections on deep-crustal reflection interpretation (Smithson and others, 1979; Wallace, 1980; Zawislak and Smithson, 1981; Jones, 1985). The reasons for the interpretation of events at long traveltime on lines 1 and 2 as multiples have been given in detail (Smithson and others, 1979; Zawislak and Smithson, 1981), the major ones being correspondence of deep events with near-surface structure in sedimentary rocks and the fact that Archean geology of the core of the uplift is far too complex to give the observed deep seismic response. Thus *in some areas,* the sedimentary section may generate multiples strong enough to mask primary reflections from the underlying crystalline crust.

Laramie Mountains. The Laramie Mountains are a Laramide foreland basement uplift similar to the Wind River Mountains except that the vergence and thrusting are eastward, and the displacement (vertical separation) of 1.5 km is much smaller. The COCORP crustal-reflection profile across the area was intended to study an uplift that forms part of the Rocky Mountain front and a Precambrian crustal boundary, the Cheyenne belt, which separates the Archean Wyoming Province to the north from Proterozoic crust to the south (Brewer and others, 1982; Allmendinger and others, 1982). The COCORP profile starts in the Denver-Julesburg basin on the east, crosses the Rocky Mountain front at the Laramie Mountains, which are cored primarily by the Precambrian Laramie anorthosite complex, and continues into the Laramie basin, crossing the Cheyenne belt. Major interpretations from this profile were the moderate 30° dip of the flanking thrust of the Laramie Mountains, the steep dip of the Cheyenne belt to the southeast, and the great (12 km) change in thickness of Moho depth across the Cheyenne belt, which may cause different Laramide structural styles on either side of the belt (Brewer and others, 1982; Allmendinger and others, 1982).

Examination of the COCORP-processed seismic section reveals a discrepancy between the geology around the Laramie Mountains thrust fault and the seismic section where an overthrust, plunging syncline does not appear in the seismic section underneath the thrust. In addition, events dipping 30° westward used to fix the dip of thrusting (Allmendinger and others, 1982) do not displace the Precambrian surface, and thus have been interpreted as Precambrian features (Johnson and Smithson, 1985). Reprocessing of this seismic data reveals that the original mute allowed refracted energy to slip through, creating a processing artifact that, if honored, demands a steeply dipping thrust fault (Johnson and Smithson, 1985). Reprocessing also reveals a reflection generated from a thrust-fault contact between anorthosite and underlying sedimentary rocks. This reflection shows the near-surface dip of the thrust to be 30° to 35°. Importantly, this thrust-fault reflection does not continue through the crystalline crust, in contrast with the Wind River thrust reflection. Thus the Laramie thrust does not seem to be reflective in a manner similar to the Wind River thrust. Even though the displacement is only several kilometers, this could form a mylonite zone 100 to 200 m wide if the depth range were appropriate for ductile deformation; in fact, however, the depth range was only slightly more than 5 km, and the surface expression of the fault is a crush zone several tens of meters wide. This fault zone, therefore, probably does not reflect because it is characterized by brittle deformation.

For this reason, the Laramie thrust represents the other end of the spectrum of fault-zone reflectivity (in contrast to the Wind River thrust) and appears to be nonreflective because of its shallow crustal level rather than its total displacement. Most Laramide thrusts that penetrate the basement on the flanks of folds seem to fit into this category (Sacrison, 1978; Bally, 1983; Thompson and Hill, 1986); however, deeper seismic data from the Laramie Mountains reveal sharp dipping events at 15 km that project into the position of the thrust fault near the surface (Iltis, 1983). These events could represent reflections from a fault zone that passed from brittle deformation at 5 km burial to ductile deformation at 13 km burial.

The Laramie Mountains thrust shows less movement than most other Laramide thrusts, partly because reverse faulting along the Rocky Mountain front is distributed along a number of faults from the eastern flank of the Laramie Mountains out into the Denver-Julesberg basin. In addition, the plunging anticline in front of the Laramie Mountains clearly shows that the basement has been folded and thrust-faulted in response to compression. Thus, folding as well as accompanying thrust faulting characterized the style of deformation for the basement in Wyoming during the Laramide.

Field records from this area show strong reverberations generated within sedimentary rocks of the basins. Therefore, Moho events picked at 16 sec (~48 km) (Allmendinger and others, 1982; Hale and Thompson, 1982) are probably multiples confused as deep reflections, especially since no discrete Moho event is found at 16 sec, but rather, a general ringing is observed. If this were the Moho, the 16-sec traveltime would place it about 10 km deeper than other data suggest (Jackson and Pakiser, 1965), and would cause a stronger southward gravity gradient than is observed (Johnson and others, 1984).

Casper Arch. Industry seismic lines across the Casper Arch have been interpreted by Thompson and Hill (1986) to show a number of convex-downward reflective basement thrusts that pass upward into folds. The faults dip from 29° to 63°, and dips between 34° and 42° are most common. They suggest that the faults flatten and disappear into the ductile middle crust.

Wyoming Archean-Proterozoic suture

A Proterozoic suture zone in southeastern Wyoming, the Cheyenne belt, marks the border between the Archean Wyoming province, which includes the 3.1- to 2.7-Ga basement rocks of the Wind River Mountains and the northern Medicine Bow Mountains, and 1.7-Ga crust to the south (Karlstrom and Houston, 1984). The suture zone itself consists of a steeply dipping mylonite zone from 1 to 7 km wide (Karlstrom and Houston, 1984). This area has been the site of extensive geophysical studies, including acquisition of COCORP seismic lines (Allmendinger and others, 1982) and seismic reflection and gravity data by the University of Wyoming (Smithson and others, 1977b; Johnson and others, 1984). Gravity interpretation indicated that charnockitic syenite associated with the Laramie anorthosite complex was underlain at several kilometers depth by mafic rock (Hodge and others, 1973), and later seismic work showed refractions coming from a high-velocity (mafic) zone. Both seismic-reflection interpretation (Smithson and others, 1977b; Allmendinger and others, 1982) and gravity interpretation (Hodge and others, 1973) suggest that the Laramie anorthosite itself is about 4 to 6 km thick. Reflections from depths of 4 to 18 km are found beneath the vast 1.4-Ga-old Sherman granite and indicate the presence of heterogeneities that could be xenoliths or the base of the intrusion. A zone of reflections resembling an unconformity on COCORP line 5 migrates into the shape of a subhorizontal saucer (Smithson and others, 1985, p. 27, Fig. 4) and could come from a layered mafic intrusion, as indicated by modeling (Wong and others, 1982; p. 96, Fig. 4).

The shear zone marking the Proterozoic suture has been identified on the COCORP sections by Allmendinger and others (1982), who postulated that it dips about 55° southeast. Although reflections from the Moho were also picked from the COCORP data, the events at these traveltimes are highly questionable, and the evidence for a change in crustal structure across the shear zone comes solely from gravity data (Johnson and others, 1984), which suggests a change in depth from about 38 km on the north (Archean) to 47 km on the south (Proterozoic). The gravity field decreases to the southeast at the suture zone, and the interpretation of this observation is that crust thickens to the south and/or crustal density decreases to the south (Johnson and others, 1984; Fig. 4, p. 454). Because of Laramide effects, some long-wavelength thickening of the crust occurs to the south, but an abrupt change in crustal thickness or change in crustal density or both is associated with the boundary and has apparently persisted since middle Proterozoic time (Johnson and others, 1984). The southern Proterozoic province is believed to consist of deeply eroded roots of a migrating chain of island arcs and a continental margin that evolved from an early Proterozoic Atlantic-type passive margin to a convergent margin in the middle Proterozoic with the accretion of island-arc terranes (Karlstrom and Houston, 1984).

Minnesota

Some of the oldest Archean crust in North America is found in the Precambrian of Minnesota, which has dates of 3.6 to 3.8 Ga (Goldich and Hedge, 1974). Here the ancient Minnesota Valley gneiss terrane is bounded on the north by the Great Lakes tectonic zone (Sims and others, 1980) and a series of late Archean greenstone belts. The Great Lakes tectonic zone was presumed to be a steeply dipping Archean border, but COCORP crustal-reflection profiles (Gibbs and others, 1984) show a gently dipping reflection beneath the zone (Fig. 6). The fault zone dips 25 to 30° northward, consists of parallel imbricate zones, and probably represents an Archean suture (Gibbs and others, 1984). The presence of high-angle and strike-slip faults is inferred in areas of no reflections. The Moho is interpreted to be offset by faults (Gibbs and others, 1984).

The largest continuous reflection can be traced on Minnesota

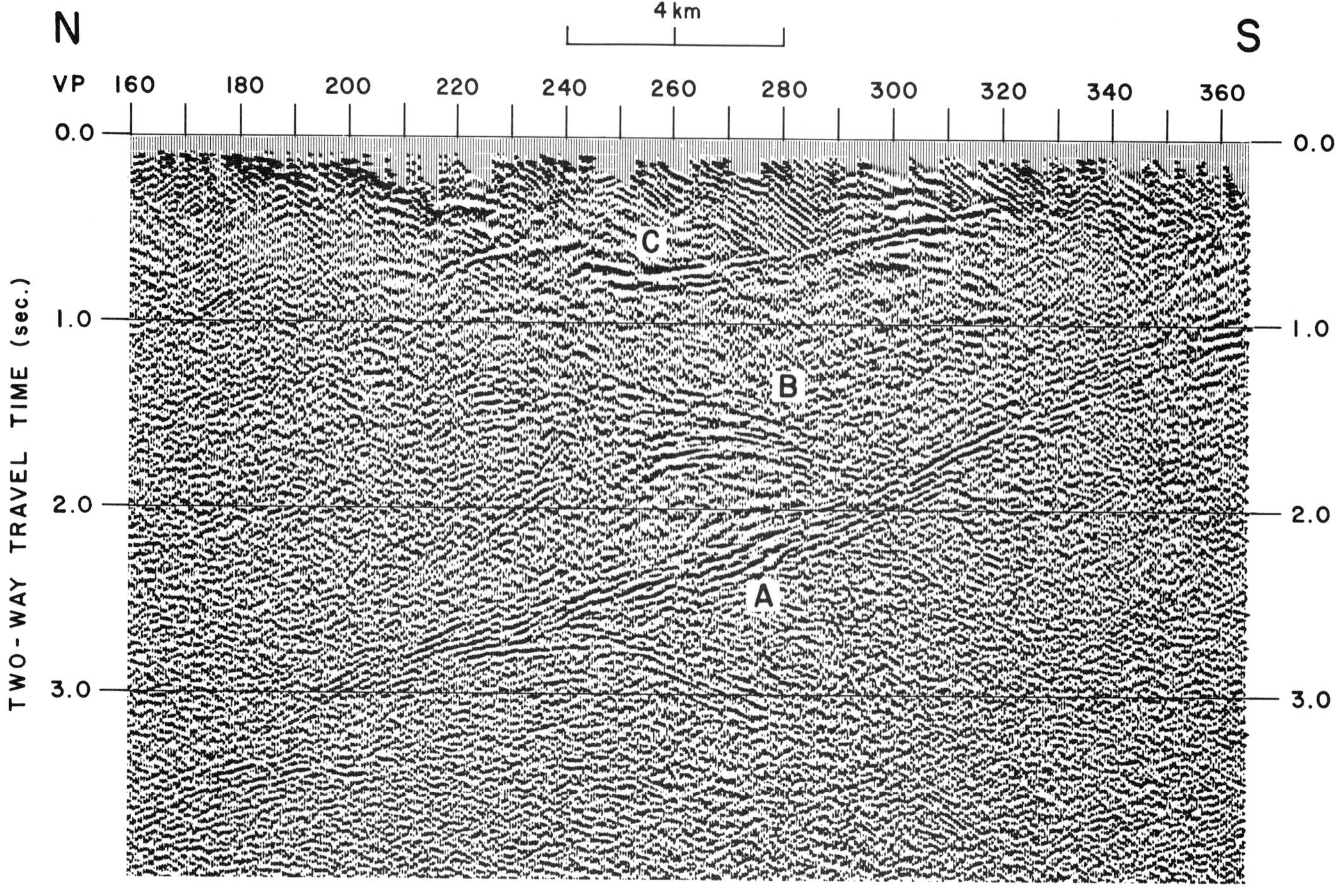

Figure 6. COCORP Line 3 from Minnesota across the Great Lakes tectonic zone. Reprocessed at University of Wyoming. A, Reflection from Archean suture that is probably a mylonite. B, Reflection from postulated recumbent fold in greenstone terrane. C, Reflection from base of Proterozoic Animikie basin showing thrust faulting of northern margin.

Line 1 and Line 3 for about 50 km (Fig. 6) and on a strike line, Line 2, with a true dip of approximately 30° to a depth of 25 km. The reflection intersects the surface at the proposed boundary of Sims and others (1980) between the greenstone-granite terrane and the gneiss terrane. This major feature of the COCORP survey is interpreted to be the contact between the two terranes. The contact itself is interpreted to be a 2-km-thick mylonite zone that might broaden into a zone of more uniform ductile deformation at midcrustal depths. The character of the reflections matches the seismic response computed for mylonite zones in modeling studies done by Wong and others (1982) and Fountain and others (1984), and is consistent with mylonite reflections observed in other seismic profiles (Hurich and others, 1985). Drilling into the basement recovered a core of gabbroic mylonite (D. Southwick, personal communication, 1986). An event (B in Fig. 6) just above the mylonite reflection migrates to form a recumbent fold and adds some support to the existence of a compressive regime during the formation of the zone. It could also be argued that the fold was formed earlier (but still during a horizontal regime), and that the fault zone is actually a listric normal fault placing younger units on older units (Wernicke, 1985), as found in the Basin and Range where mylonites occurring along such faults are known to be good reflectors.

On Line 3, a Proterozoic basin, indicated by a 3-km-deep reflection, extends for a lateral distance of about 13 km between CDP 350 and CDP 650 (Fig. 6, event C) and is offset in places. Drill cores near the surface (D. Southwick, personal communication, 1986) found Animikie slate with dips between 24° and 55°. Thrusting into the bottom and oversteepened flank of the Animikie basin accompanied folding and cleavage formation in the sedimentary and volcanic rocks of the basin. Movement within the Archean basement was taken up along a number of discrete slip surfaces (faults) so that folding (deformation) of Archean features such as the suture-zone reflection are not observed. The shape of the basin as revealed by the seismic section suggests that strong folding and cleavage formation within the basin were accommodated by thrusting along discrete breaks in the underlying basement.

The ancient gneiss terrane at the south end of Line 3 yields some of the most interesting reflection results because of the implications for early crustal genesis. In the ancient gneiss terrane, a distinctive reflection pattern is seen, mostly in the form of broad

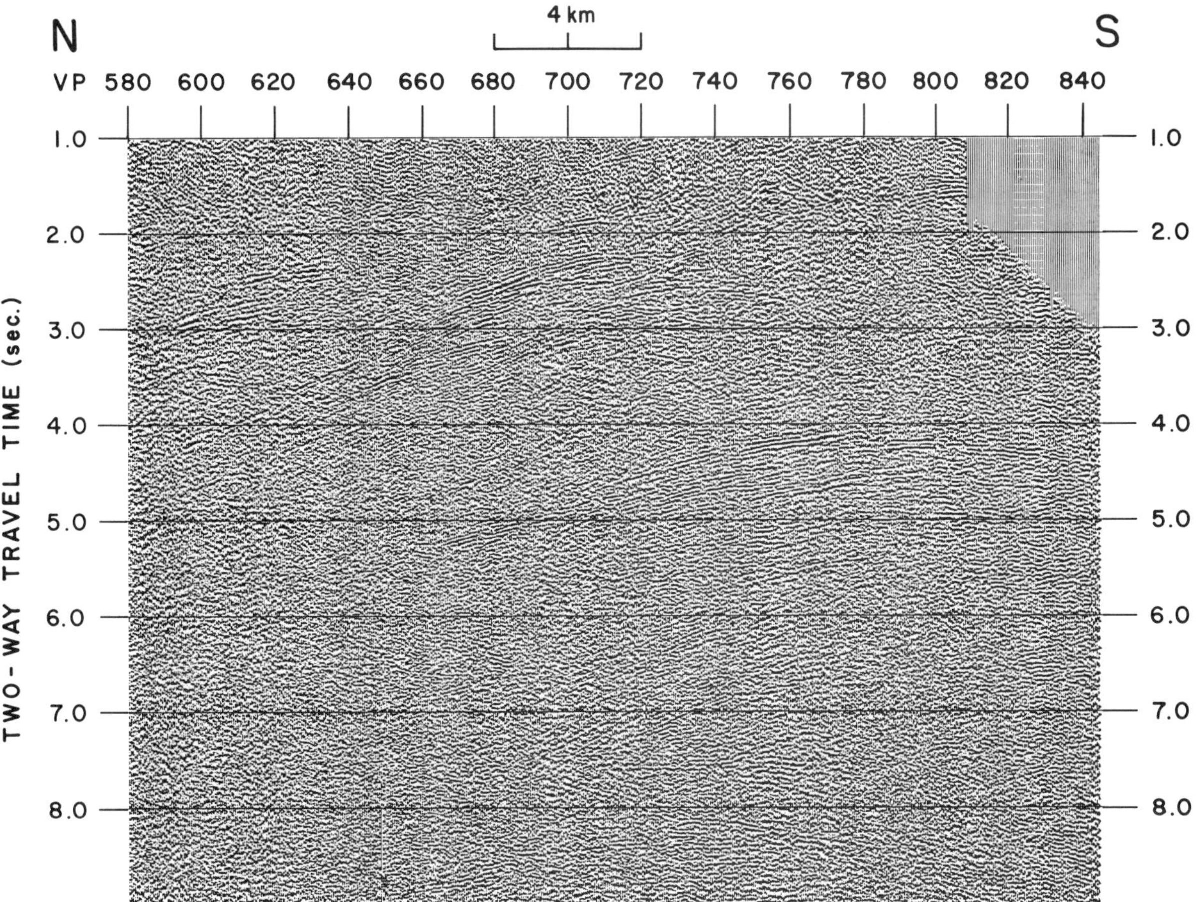

Figure 7. COCORP Line 3 from Minnesota showing reflections in ancient gneiss terrane. Reprocessed at University of Wyoming. Gently convex upward events from 1.5 to 9 sec are interpreted as a stack of nappes.

arcs (Fig. 7). The geometry of the events indicates complexly folded and faulted rocks. Seismic-reflection data from the southern end of COCORP line 3 show packets of reflections that extend almost continuously from a depth of about 2 to 30 km (0.8 to 9.5 sec) (Fig. 7). The reflections come from a series of layered zones (apparently discontinuous) that are as much as 15 km long and 3 to 4.5 km (1 to 1.7 sec) thick, separated by transparent zones 1 to 3 km (0.3 to 1.5 sec) thick. Short (1 to 2 km) discontinuous subhorizontal reflections are found from 30 to about 35 km (9 to 11 sec). No distinct Moho reflection can be seen.

Based on metamorphic facies, the present depth of exposure is about 20 km. The reflection pattern indicates extreme crustal heterogeneity down to a depth of 30 km. Comparison with surface geology and seismic modeling suggests that the rocks are recumbently folded and thrust into nappes, probably during a 3.55-Ga event (Bauer, 1980). The parallelism of reflections that approach possible fold hinges in places is an expression of early recumbent folding found in most old crystalline terranes. The broad arching evident in the reflections is manifested in exposures in the Minnesota River Valley and is ascribed to the 3.1-Ga event, as is the granulite facies metamorphism. The layering is almost certainly the result of transposition so that layering is not direct evidence for the presence of supracrustal rocks; the layering could represent any combination of igneous and sedimentary precursors in what is now a metamorphic sequence.

The reflections thus represent a stack of nappes of highly variable composition, probably originally an igneous and sedimentary mix that extends from near the present surface to a depth of 30 km (Fig. 8). The reflection pattern almost certainly indicates a horizonal (compressive) tectonic regime to this depth. The folded layered rock sequences most plausibly represent a large-scale migmatite, possibly a restite, from which the ubiquitous granites were "sweated out." Such reflections would best be generated from alternating felsic and mafic layers (rocks with maximum contrast in reflectivity), but the effect of differentiation through partial melting can result in some unusual compositions and physical properties, e.g., mineralogies rich in garnet and aluminum silicate (Winkler, 1972), so that high-density, high-velocity rocks occur within the mafic layers.

The short, subhorizontal reflections (Fig. 8) below 30 km (9 sec) may indicate the effects of crustal underplating; they could also indicate horizontal tectonic movements under increased duc-

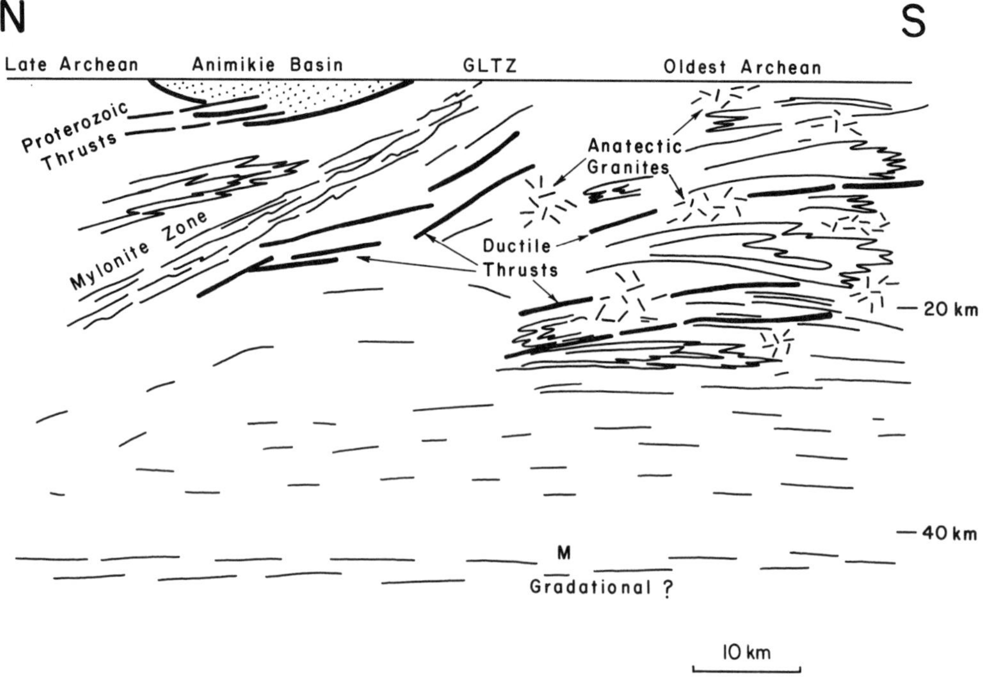

Figure 8. Interpretative cross section of Archean crustal structure in Minnesota showing suture, Animikie basin, and ancient gneiss terrane. Archean basement is remobilized along discrete thrusts to deform supracrustal rocks in Animikie basin. Complex, moderately dipping suture zone can be followed geophysically to about 20 km. Oldest Archean crust passes from a thick (about 30 km) stack of nappes interspersed with anatectic granites into a subhorizontal tectonic regime in lowermost crust above either a gradational Moho or a sharp Moho with a small change in composition.

tility compared with the rocks above. If so, the underplating took place in only the lowermost 5 to 10 km of the Archean crust, implying that the deformed early Archean crust was 45 to 50 km thick before underplating, although one can postulate earlier injections that became interfolded with the nappes to thicken the crust before the 3.55-Ga event.

Moho reflections are not identified. Generally, the deepest events are short discontinuous events between 9 and 12 sec, particularly underneath the ancient gneiss terrane (CDPs 1250–1650). Deep reflections simply die out after these traveltimes, so that no distinct Moho reflection is seen at vertical incidence. No compelling seismic evidence for the sawtooth Moho proposed by Gibbs and others (1984) is found, and such a Moho configuration is inconsistent with seismic models and observed gravity anomalies. The lack of a clear vertical-incidence Moho reflection in the presence of other deep events is peculiar and may be telling us something about the nature of the ancient Moho, which has been detected at about 42 km from refraction data (Greenhalgh, 1981); a strong wide-angle Moho reflection has recently been recorded by the University of Wyoming seismograph crew. A plausible explanation for this is that the Moho is formed by a phase change (O'Reilly and Griffin, 1985; Smithson, 1989) resulting in a gradation from crustal to mantle rocks (Fig. 8).

Magnetic data were used to constrain larger scale structural features because of the different magnetic signatures for younger vs. older Archean terranes. On average, the intensity of anomalies in southern Minnesota is higher than in northern Minnesota. Lidiak (1974) found that the more siliceous granulite-facies gneisses in southern Minnesota have higher magnetizations than more mafic rocks in Manitoba. The observed magnetic profile was fit by a single block having a thickness of 20 km and an edge that projects (dips) upward at 30°, representing the ancient gneiss terrane of southern Minnesota. Thus the magnetic interpretation strengthens the seismic interpretation and suggests a moderately dipping contact between two distinct terranes. The ancient gneiss could not have been incorporated and remobilized into the greenstone terrane unless its magnetic signature was also reworked by recrystallization of magnetic minerals under new oxidation conditions.

Kansas

COCORP crustal-reflection profiling has helped delineate the extent of Proterozoic rifts in Kansas that are correlated with exposed Keweenawan rifting in northern Michigan (Serpa and others, 1984). The COCORP seismic profiles in Kansas cross the Mid-Continent gravity high, a rift structure consisting of half grabens that is an extension of the Keweenawan, and these profiles contrast sharply with the Michigan seismic data. in Kansas the seismic profiles show two shallow, asymmetric basins associated with rifting (Serpa and others, 1984). Both basins appear to be bordered by gently dipping faults on their west flanks and to have tilted westward along these faults. Deposits in these basins range from 3 km thick on the east to 8 km thick on the west. By analogy with the Keweenawan Rift, deposits are inferred to consist of red beds and mafic flows, which cause some of the 80-mgal positive gravity anomaly. Because of the positive gravity anomaly and the high stacking velocities, mafic intrusions are postulated to form a significant part of the crust beneath the rift. As opposed to the seismically transparent deep crust underneath Michigan, the seismic sections in Kansas are covered with arcuate and criss-crossing events, which are in striking contrast with many other deep-crustal seismic sections. These seismic sections exhibit a complicated wavefield at traveltimes corresponding to lower crustal depths (Fig. 9); the wavefield consists of arcuate crossing events that cannot successfully be migrated to separate domains. This suggests that at least some of these arrive from out of the plane of the section, but whether these events originate from deep within the crust or from the near surface is uncertain. This is a fundamental problem in the interpretation of many such seismic sections. If these events are coming from depth, then this seismic section represents an important example of a complicated (not plane-layered) deep-crustal section.

Serpa and others (1984, p. 373) suggested that at least one of these events is arriving vertically from a reflector in the upper mantle because of its behavior on crossing seismic lines. The Moho is not represented by discrete reflections, but a general decrease in seismic energy at 14 to 15 sec (Fig. 9) has been postulated to mark the crustal boundary (Serpa and others, 1984). The arcuate and crossing events are, in fact, quite similar to complex patterns from seismic models generated for typical igneous and tectonic features (folded and intruded) in crystalline rocks (Smithson, 1979; Wong and others, 1982). Does this mean that, for the first time, we are seeing typical geologically complex crustal structure in the seismic data from Kansas?

Further information on the crust about 50 km from the seismic line is provided from the xenolith population in diatremes (Brookins and Meyer, 1974). The xenoliths are predominantly gabbro and mafic granulites with relatively small amounts of more felsic rocks. With the predominantly mafic xenolith population, it is somewhat perplexing that the crust appears so reflective unless the deep crust has been sampled selectively by the diatreme. Nevertheless, the seismic sections of Kansas represent the seismic response of typical heterogeneous crystalline crust consisting of nonhorizontal folded metamorphic rocks and igneous intrusives.

Oklahoma and Hardeman County, Texas

COCORP data from Hardeman County, Texas (Oliver and others, 1976) and adjacent areas in Oklahoma (Brewer and others, 1981, 1983) are unique in showing the strongest, most continuous crustal reflections yet found (Fig. 10). Brewer and others (1981) interpreted them as coming from a sedimentary sequence, possibly containing volcanics, that reveals a vast, previously unknown Proterozoic basin. The layered sequence of rocks has also been called a "granite batholith" associated with a layered igneous complex (Lynn and others, 1981), but no intrusive relations are observed, and reflections have all the characteristics of a supracrustal (depositional) sequence. The area has undergone an extensive history of faulting, with late Precambrian faults apparently reactivated into late Paleozoic time (Brewer and others, 1981, 1983). Precambrian normal faulting resulted in subsidence along a rift to emplace a mafic layered igneous complex in the rift followed by uplift and erosion of the subsided block along the same faults. During Pennsylvanian time, the Wichita block was overthrust onto the Anadarko basin.

Wavelet processing provides an estimate of reflection coefficients and indicates a sequence of quartzofeldspathic rocks interlayered with basalt or dolomite, i.e., a maximum contrast in reflectivity and little chance for porosity. Basalt seems to be the more likely choice based on the general environment. These strong reflections resemble those from the sedimentary rocks and mafic igneous rocks in the Karroo basin (Fatti, 1972) and come from a layered succession in the Hardeman basin believed to be a 10- to 15-km-thick sequence of interlayered volcanic and sedimentary rocks that may extend over a vast area. The deep crust generates many arcuate events (Schilt and others, 1981) of the type that would come from any complex structure (Smithson, 1979; Smithson and others, 1980). These have been interpreted as originating from point diffractors between depths of 10 and 22 km (Schilt and others, 1981), but reflections from synformal features such as mafic intrusions might be a more likely explanation because of the relatively high amplitude of these events. The deep crust slightly to the north, however, seems to be transparent. Interpretation of a crustal refraction line (Mitchell and Landisman, 1970) indicates unusually thick crust with a high velocity of 7.2 km/sec in the lower crust. We interpret these data, because of the high-velocity, thickened lower crust, to indicate a mafic overthickened lower crust; in fact, the unusually thick crust isostatically supports a mass surplus not in the topography but in the lower crust. This dense lower crust is postulated to have formed by underplating by gabbroic intrusions; such underplating may have formed contemporaneously with subsidence to form the thick Proterozoic depositional basin (Fig. 11) (Smithson and others, 1985).

Colorado Plateau

The Colorado Plateau and its transition to the Basin and Range Province have been profiled by COCORP (Allmendinger and others, 1986) as part of their 40°N transect across the western United States; the lines extend eastward from the Basin and Range lines in the Sevier Desert. The western flank of this area represents the hingeline of the ancient North American continent

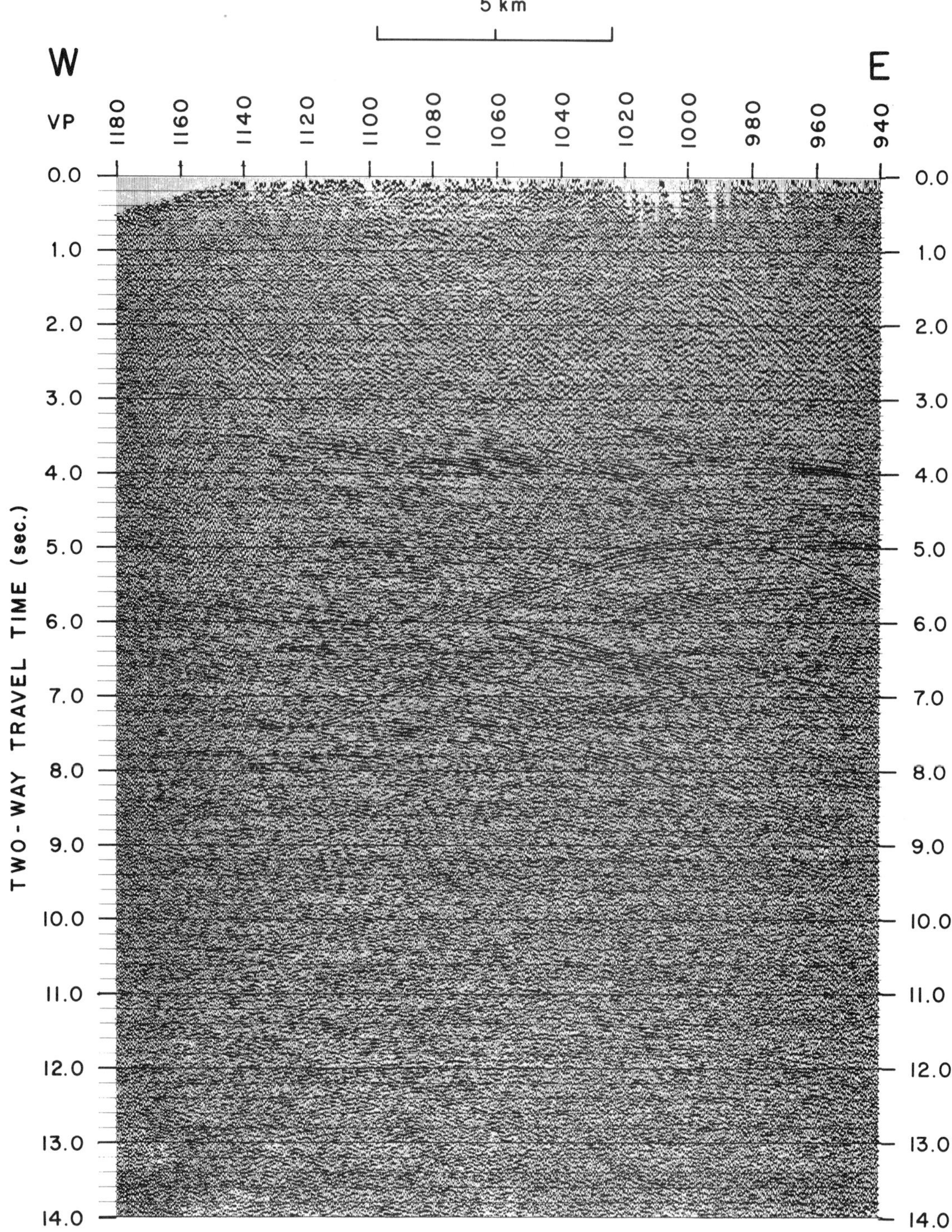

Figure 9. COCORP Line 1 from Kansas showing criss-crossing arcuate events above a transparent zone with no distinct Moho reflection.

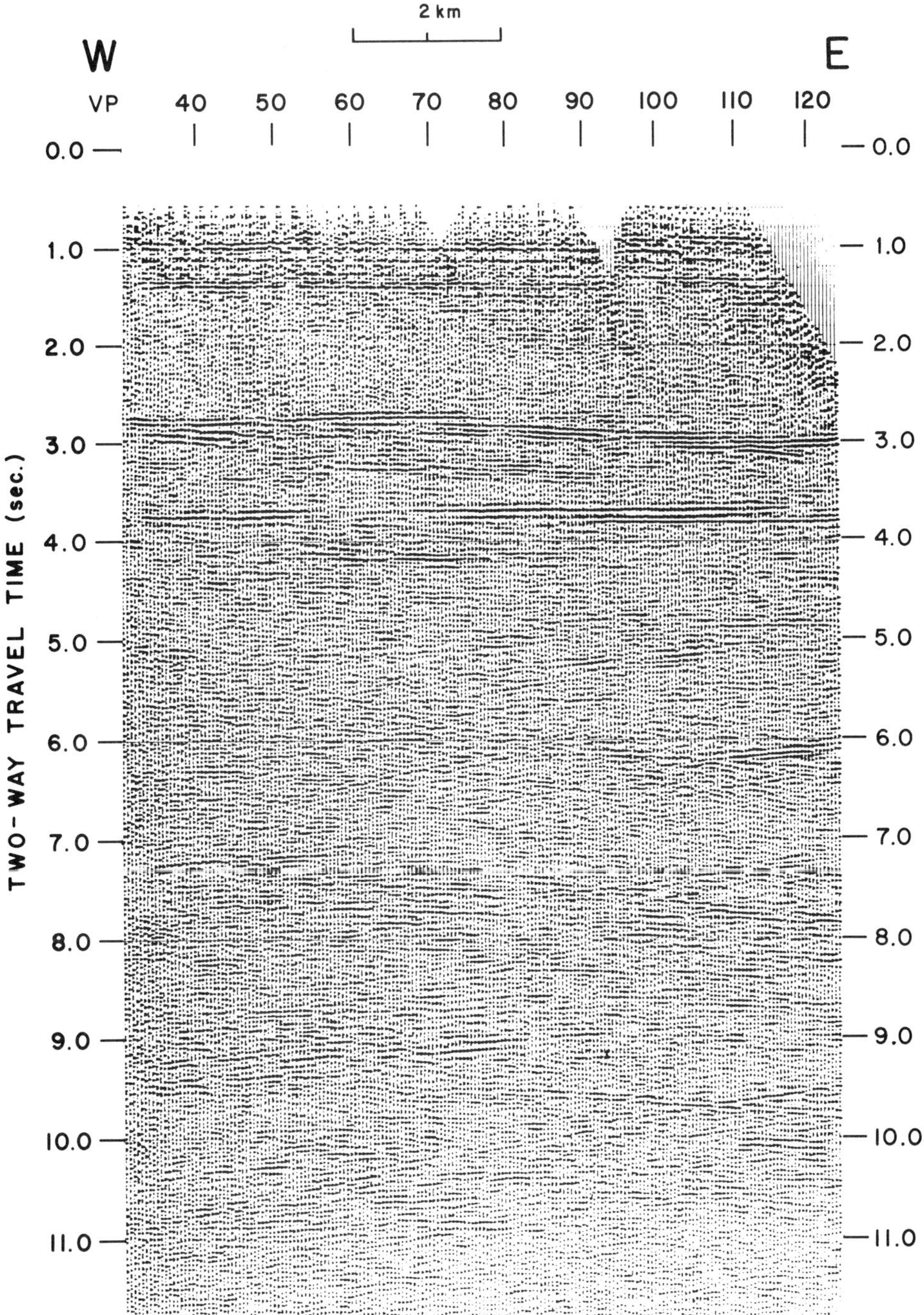

Figure 10. COCORP Line 1 from Hardeman County, Texas, the original COCORP survey, showing strong horizontal reflections from a Proterozoic basin above arcuate events in the middle and lower crust. No distinct Moho reflection.

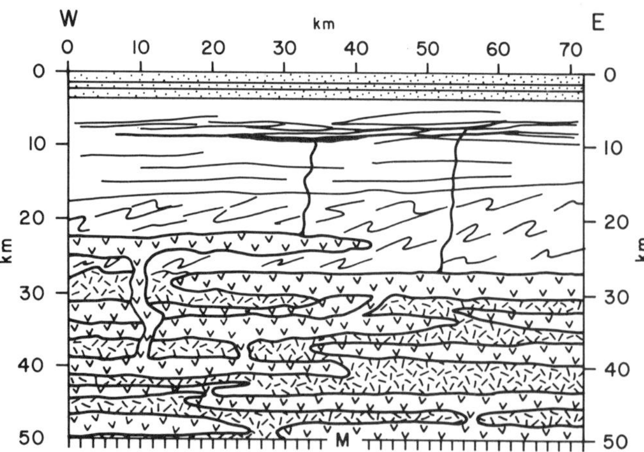

Figure 11. Interpretive cross section of crustal structure in southwest Oklahoma and northern Texas. Thick Proterozoic basin filled with interlayered sedimentary rocks and volcanics is underlain by normal crust underplated and thickened by coalescing gabbroic sills.

where late Precambrian and Paleozoic sedimentary rocks thicken greatly westward (Stokes, 1976) and were affected by extensive (in time and space) deformation. The Colorado Plateau represents an enigmatic cratonic block, surrounded by Laramide compressional zones and later Cenozoic extension. About 9 km of Phanerozoic sedimentary rocks are deposited above basement in the Plateau.

The seismic data consist of COCORP Utah Lines 3 and 4. Line 3 shows reflections primarily along the east side of the line, with some normal faulting in the top of the basement and possible thrusting in sedimentary rocks (Allmendinger and others, 1986, Fig. 3); however, the crustal part of the seismic section is essentially transparent. This may be a data acquisition problem. Line 4 exhibits a zone of essentially flat-lying reflections in the sedimentary section (down to 1.4 sec) underlain by a transparent zone from 1.4 to 4.0 sec. Below this is a series of arcuate, dipping, criss-crossing events from 4 to 14 sec (Allmendinger and others, 1986, Fig. 4). A subhorizontal, multicyclic reflection extends for 28 km from VP 470 to 750. A weak multicyclic reflection extends discontinuously for 20 km on the east end of the line.

Line 3 tells us little about the crust. Line 4 is dominated by the arcuate crossing events, which may be from out of the plane since no cross-line control is available. In any case, they strongly resemble the events found in Proterozoic crust in Kansas (Serpa and others, 1984). If not sideswipe, then they are the typical expression of complex, nonparallel structure such as we expect to find in the Precambrian, i.e., complex superposed deformation, intrusion, and faulting (Smithson, 1979; Smithson and others, 1980). Xenoliths rather far from this line in southeastern Utah indicate that the deep crust is mafic, high velocity, and heterogeneous, which may explain the arcuate events. The event at 14.9 sec may be the Moho at about 45 km unless it is an upper-mantle discontinuity beneath a gradational Moho. If this event is the Moho, then its arrival time is consistent with the depths and crustal velocities determined by refraction seismology (Prodehl, 1979). The event at 8.7 sec comes from a depth of about 28 km. It may be the top of the lower crust (a local "Conrad" discontinuity), a detachment zone, or just about anything. Age relations between this event and the arcuate events are not at all clear (Allmendinger and others, 1986). Thus the nature of the lithosphere in the Colorado Plateau remains an enigma. Clearly, the cratonic crust in the plateau is thicker and different in structure from the recently deformed crust of the Basin and Range (Allmendinger and others, 1986).

Rio Grande rift

One of the classic studies of crustal-reflection profiling by COCORP was conducted in the Rio Grande rift near Socorro, New Mexico; it illustrates the importance of using other geophysical data to constrain a reflection interpretation. The Rio Grande rift is a Cenozoic graben consisting of a series of individual basins trending north-south. Extension at about 26 Ma followed Laramide thrusting. The impetus for a deep-crustal reflection profile came from a series of studies by Sanford and his co-workers (Sanford and others, 1973, 1977), which suggested the presence of a magma chamber at midcrustal depth based on reflected phases from microearthquakes and other geophysical data such as heat flow. The COCORP survey (Brown and others, 1979, 1980) revealed a multicyclic high-amplitude reflection at 8 sec, which corresponds approximately with the same horizon found by S-S reflections from microearthquakes (Sanford and others, 1977). Based on this correspondence and the reflection amplitude, the event was called a reflection from an inferred magma chamber. Brocher (1981a) worked with true reflection amplitudes to demonstrate that these reflections were generated by alternating 30- to 40-m-thick layers and that the reflectivity, which was higher than for normal rocks, could be produced by a small amount of melt within the interstices of solid rock. Rinehart and others (1979) mapped this reflection over a wide area using microearthquakes. Olsen and others (1982) showed that this event was visible as a wide-angle reflection over great lateral extent in crustal refraction data; they recompiled the properties of the inferred magma chamber for both refraction and reflection. Jurdy and Brocher (1980) and Brocher (1981b) worked out the shallow structure of the rift, which is so critical for interpreting deeper reflections.

A fairly clear event appears at 12 sec under the rift. This may correspond to the Moho (Brown and others, 1979, p. 173, Fig. 2). If so, it represents another case where the Moho under an extensional area seems to give a stronger reflection.

These studies nicely illustrate an approach to interpretation of crustal reflection data. Without the additional geophysical data to constrain the interpretation and the careful analysis of a reflection amplitude anomaly by Brocher (1981a), high amplitudes by themselves could not have been used to suggest the presence of magma because unusually high-amplitude reflections are also found east of the rift under the High Plains and from some layered sequences in general.

This COCORP data set also provides both information on faulting and an example of a typical interpretation problem. Early interpretations of the rift structure (Brown and others, 1979) suggested that the rift was bordered by high-angle normal faults, and that the eastern margin, marked by a large transparent zone, was the site of either an intrusion or a fault. Later, Cape and others (1983) suggested the presence of listric normal faulting related to rifting. Recently, deVoogd and others (1986) have carried the listric-faulting hypothesis further and proposed that some of these faults may be reactivated Laramide thrusts. They also reprocessed and reanalyzed the data and suggested that the seismic transparent zone under the eastern shoulder of the rift was due to changed coupling of the source (recording conditions) rather than geology; i.e., intrusion or faulting. Another possibility for the transparent zone is a degraded stack of the data caused by the extreme lateral velocity variations. Here we have an example of the reliance of the geological interpretations on the recording conditions and the parameters used in processing; what originally was considered to be evidence for igneous intrusion now appears to be a degraded data quality caused by poor coupling and/or inappropriate stacking velocities.

Northern Basin and Range

The structure of the Basin and Range Province in Nevada and western Utah is characterized by Mesozoic compression that formed recumbent folds and thrusts and Cenozoic extension that generated listric and high-angle normal faults, and that was accompanied by volcanism and granitic intrusion. The dominating feature of the Basin and Range is the sediment-filled low-velocity basins separating ranges of high-velocity Paleozoic or basement rocks; this marked contrast in near-surface velocity structure causes severe static problems, making any long, continuous interfaces in this area difficult to detect.

Interpretations of the extensive COCORP lines in Nevada (Allmendinger and others, 1987; Hauge and others, 1987; Potter and others, 1987; Hauser and others, 1987) emphasize a planar, highly reflective Moho and lower crust, relatively nonreflective upper crust, and range-bounding normal faults that cannot be traced to the midcrust; thus they do not appear to cut the Moho. Hauge and others (1987) and Potter and others (1987) have suggested that either the Moho has not been cut by faulting accompanying extension, or that it has been reconstituted and is therefore a young feature. A different depth was found for the reflection Moho and refraction Moho in western Nevada (Allmendinger and others, 1987). The horizontal lower-crustal seismic fabric is apparently related to magmatism and ductile deformation during which intrusions were localized along high-strain zones. Allmendinger and others (1987, Fig. 7, p. 316) summarized four modes of extension: (1) crustal penetrating horst and graben, (2) subhorizontal decoupling zone, (3) anastomosing shear zones surrounding lenses, and (4) a crustal penetrating shear zone.

Nevada upper-crust extensional faults. The geometry of Cenozoic extensional faults in the northern Basin and Range is indicative of the mechanisms causing intracontinental rifting. Early theories (Stewart, 1971) to explain Basin and Range structure could be grouped into three major styles: (1) the range-valley structures were related to gravity sliding off regional highs, similar to landslides; (2) range-valley structures were related to strike-slip movements within the crust; and (3) range-valley structures were related to lower-crustal ductile extension with a brittle upper-crustal response. Recent theories have concentrated on this third style.

Thompson (1959) suggested graben formation associated with igneous intrusion at depth, and Cloos (1968), while analyzing Gulf Coast fracture patterns, simulated with clay models the formation of tilted, highly asymmetric grabens rooted in horizontal "detachment" surfaces. Stewart (1980) suggested this variety of upper-crustal fault geometries: (1) asymmetric or symmetric horst-graben structures, (2) tilt blocks with bounding planar faults to depth, and (3) tilt blocks with bounding listric faults rooted in either ductile lower crust or horizontal slip ("detachment") surfaces.

Effimoff and Pinezich (1981), using industry profiles from several valleys, interpreted shallow listric faults separating the ranges and valleys. Hastings (1979) suggested planar high-angle faulting in the upper few kilometers separating the Carson Sink and Stillwater Range. Using shallow industry profiles, gravity modeling, and microearthquake distributions, Okaya and Thompson (1985) found planar high-angle faulting to a depth of 15 km separating the Stillwater Range and Dixie Valley. Smith and Bruhn (1984) migrated industry data to reveal clear listric faults underneath Great Salt Lake and the Wasatch front. Zoback and Anderson (1983) reinterpreted several profiles used by Effimoff and Pinezich (1981), finding not just listric faulting but a range of fault and basin geometries. These geometries support Stewart's (1980) contention that the upper crust extended with not one style of basin-range structure, but several.

In COCORP data the effects of strong, near-surface velocity variations are manifested in lack of continuity and generally weak reflections, either from recording or processing problems, or both. This results in a number of vertical strips on seismic sections showing no reflections, clearly a result of the technique, not the geology. The basins are generally poorly imaged while the intervening ranges are not imaged at all. The fact that basins with reflective sedimentary layering are poorly imaged indicates that the data have extreme problems and that events that do appear must be very strong. The ranges in eastern Nevada typically consist of Paleozoic sedimentary rocks that should generate clear seismic reflections. The lack of reflections from these rocks beneath the Cenozoic sediments in the intervening basins is one of the mysteries of this area.

Field evidence suggests the following structural relations: (1) horsts and grabens, with the grabens displaying symmetry, asymmetry, or graben-within-graben internal relations; (2) tilt blocks where the high points are the ranges, the basins are pas-

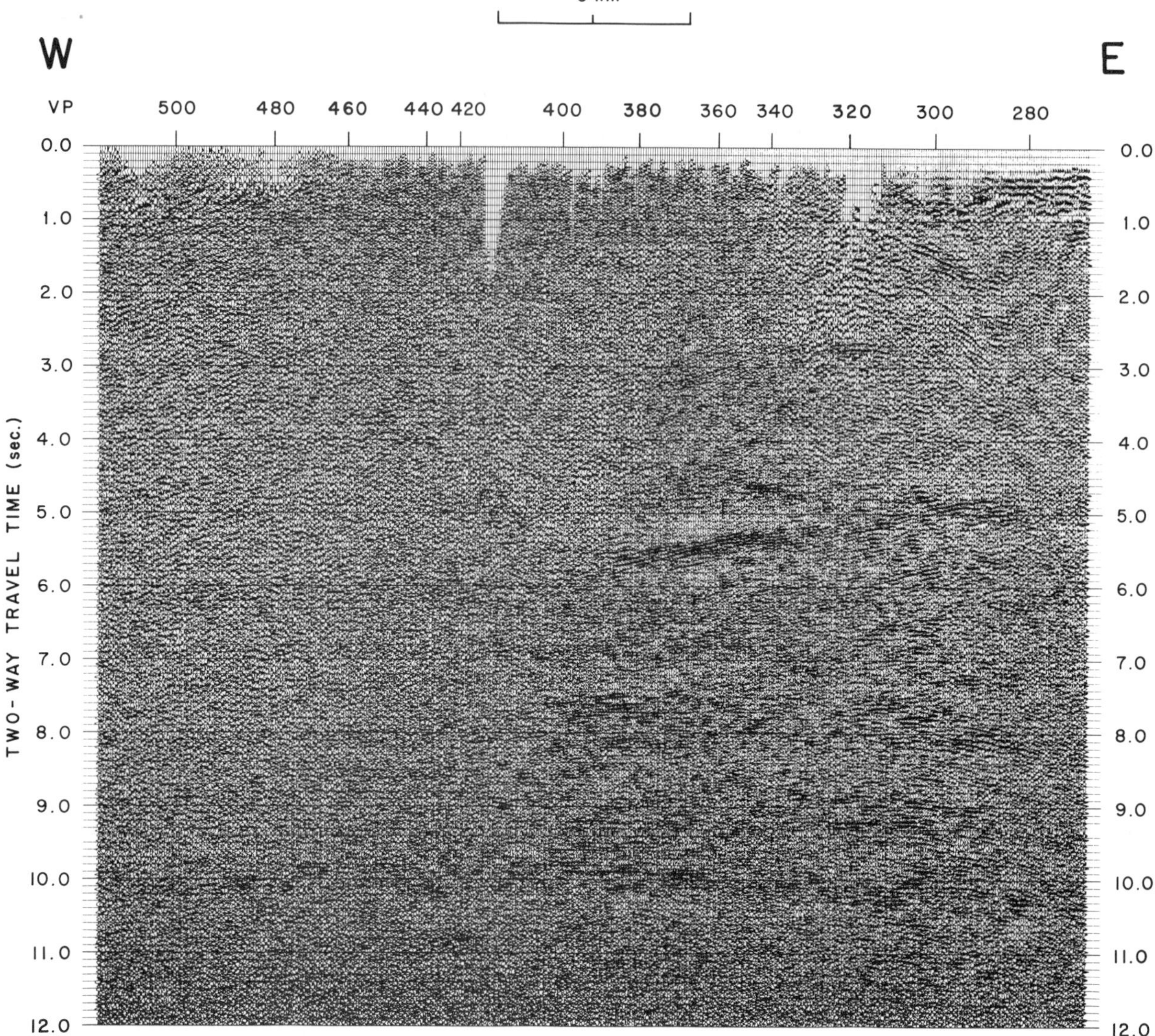

Figure 12. COCORP Line 4 from Nevada across the Cherry Creek Range. Reprocessed at the University of Wyoming. Note strong reflections at 5 and 8 to 10 sec passing laterally in an almost totally transparent section.

sively formed, and faults separating the blocks are planar to depth (domino-style); (3) tilt blocks separated by listric faults that root in a ductilely extending lower crust or in a horizontal ("detachment") surface. This suggests that several styles of extension are active in the upper crust in the Basin and Range, or that the seismic studies have not been able to image sufficiently the fault geometries. Neither available industry profiles nor COCORP transect profiles adequately image the bounding faults.

Nevada mid- to lower crust. The crust in Nevada contains some of the strongest subhorizontal reflections found. Most events are short (1 to 2 km) and discontinuous, and the most continuous events (5 to 10 km) are found in the midcrust and lower crust. Probably the fewest events occur between about 1 to 3 sec, although the relationship may be reversed in some places, with more events in the upper crust and few at depth. Some strong, west-dipping events are found at midcrustal levels (5 to 7 sec) on COCORP Line 4 (Fig. 12). The most consistent events, which cover about one-third of the sections, are a multicyclic event or series of events between 9 and 11 sec with a duration of about 0.5 sec (Fig. 14). This has been identified as the reflection from the Moho (Hauser and others, 1987; similarly seen by Okaya, 1986) or a pair of reflections. In any case, this event is consistent in its reappearance within a certain time range and with a multicyclic character. In some places, the event is distorted by near-surface velocity contrasts; in others it appears to dip. This event may be accompanied by another event about 1 sec higher that has been called a second Moho. This event may be weaker, but displays a similar character. The lowermost event around 10 sec is best

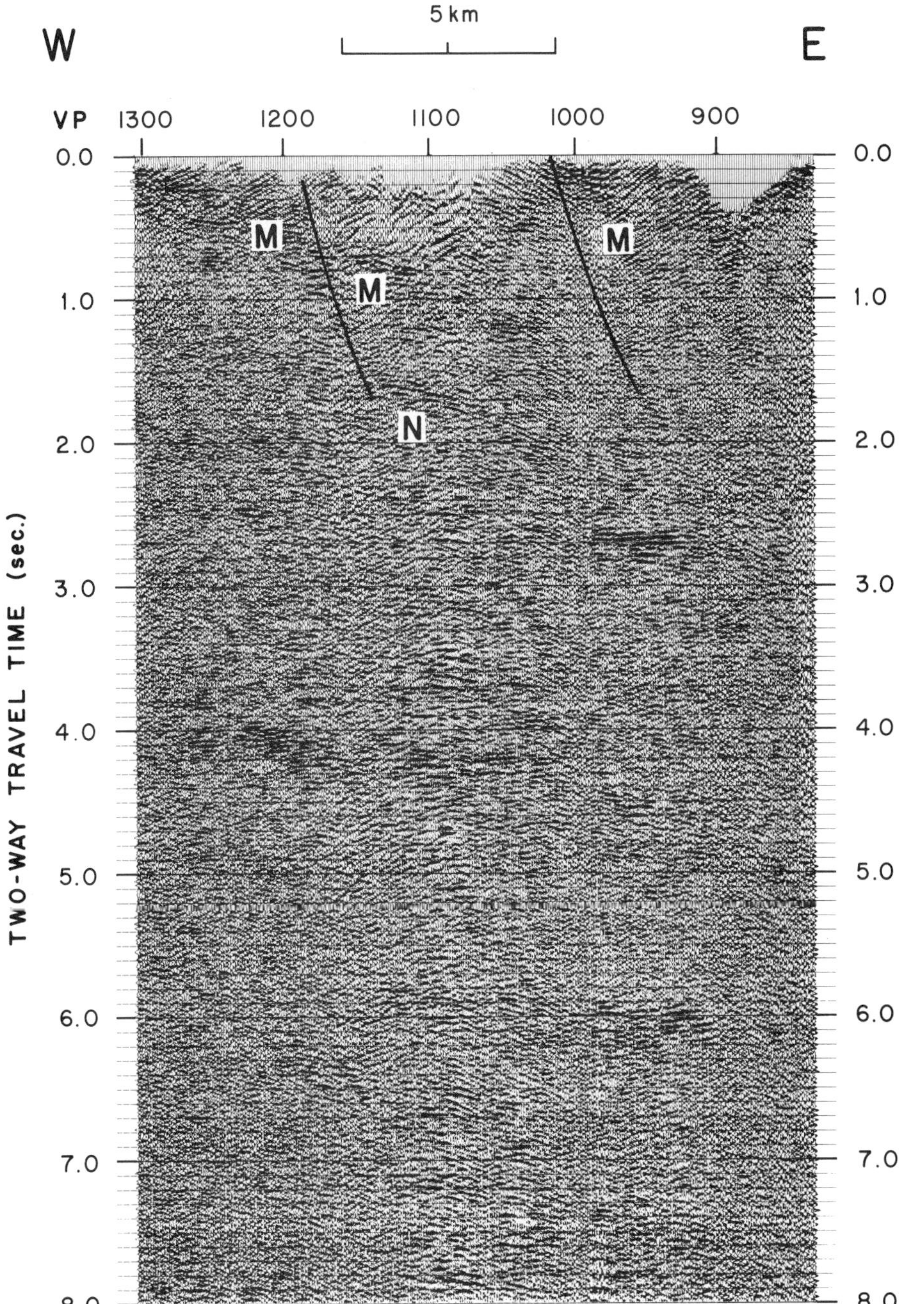

Figure 13. Forty-eight–fold CDP seismic profile recorded over the northern Ruby Mountains by the University of Wyoming seismograph crew. M = mylonite reflection offset by Basin-and-Range faulting, N = reflection interpreted from a nappe based on down-plunge projection. Numerous deeper reflections are probably related to ductile shear.

interpreted as a Moho reflection; the event about 1 sec above it is most likely a lower-crustal reflection (Valasek and others, 1987).

Metamorphic core complexes: Ruby Mounatins, Snake Range, Albion Range. Metamorphic core complexes are domes in crystalline rocks of a lower plate separated from a younger, lower grade upper plate by a mylonite zone and an extensional listric detachment fault. They are located along the Cordilleran hinterland (Crittenden and others, 1980). Metamorphic core complexes typify the eastern Basin and Range and have been the subject of widespread recent study (Reif and Robinson, 1981; Snoke and Lush, 1984; Gans and others, 1985; Hurich and others, 1985; Hauser and others, 1987; Valasek and others, 1989). Several seismic reflection lines, together with wide-angle data, have been recorded across the Ruby Mountains core complex by

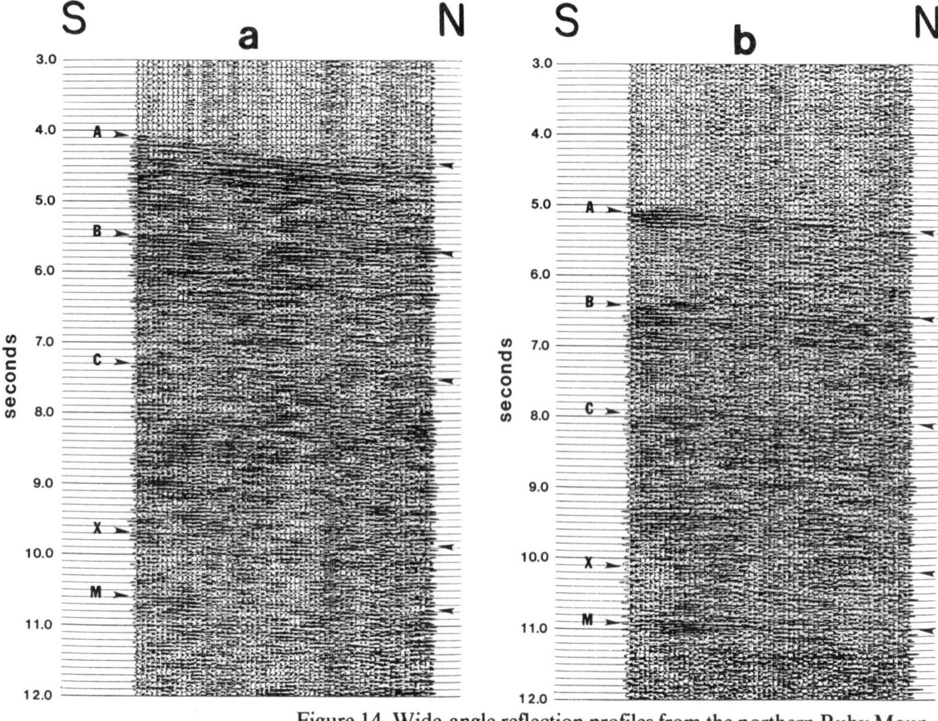

Figure 14. Wide-angle reflection profiles from the northern Ruby Mountains recorded by the University of Wyoming seismograph crew. Note numerous events down to the Moho reflection at 11 sec.

the University of Wyoming 70 km north of the COCORP profile. A CDP profile across the plunge of the core complex images the mylonite zone at shallow depth as well as reflections through the entire crust and particularly strong events at 2.8, 4, and 6 sec (Fig. 13). This profile reveals Basin and Range faults cutting the mylonite zone (Valasek and others, 1989); as in the Kettle dome (Hurich and others, 1985), it shows the mylonite zone to be a good reflector. Other strong upper- to middle-crustal reflections may also be mylonite zones formed by crustal-scale extension. The seismic section offers a strong contrast with the almost totally transparent COCORP section to the south, and shows the upper crust as well as the lower crust to be highly reflective. This illustrates the fundamental dilemma of whether we are looking at differences in geology between the northern and southern Ruby Mountains, or whether we are looking at recording/processing effects. We believe that we are primarily seeing the effects of recording and/or processing.

Wide-angle recording at offsets of 18 to 50 km provides moveouts large enough to estimate velocities to the base of the crust (Valasek and others, 1987) (Fig. 14). Several crossing wide-angle profiles were incorporated into a seismic model that matches the traveltime and moveout of the various reflections, although it does not attempt to match the detailed layering implied by the multicyclic character within the bands (Table 1).

The strong layering demonstrated by seismic sections suggests that most of the crust is anisotropic where the amount and type of anisotropy depend on mineralogy, and reflection travel paths would largely lie in the slowest direction. Because of the

TABLE 1. VELOCITY STRUCTURE IN RUBY MOUNTAINS CORE COMPLEX*

Discontinuity	Depth (km)	Velocity (km/sec)	T-corrected velocity (km/sec)
	0.0		
		3.80	
	1.5		
		5.50	
A	3.0		
		6.10	6.20
B	12.5		
		6.35	6.4–6.5
C	19.5		
		6.6	6.7–6.8
X	28.0		
		7.2–7.5	7.4–7.8
M	31.0		
		7.8	7.9–8.1

*Velocity model matching traveltimes and moveouts for five bands of reflections identified on 27.6-km offset shot gather (Fig. 14). Velocity for the surface layer is an average for a multilayered zone. Right-hand column gives temperature-corrected velocities.

strong multicyclic reflections (Fig. 14), most of the various interval velocities (Table 1) must be averages for interlayered high- and low-velocity rocks. Based on laboratory velocity determinations (Christensen, 1982) velocities of 6.1 to 6.2 km/sec correspond to quartzofeldspathic rocks; this agrees with observa-

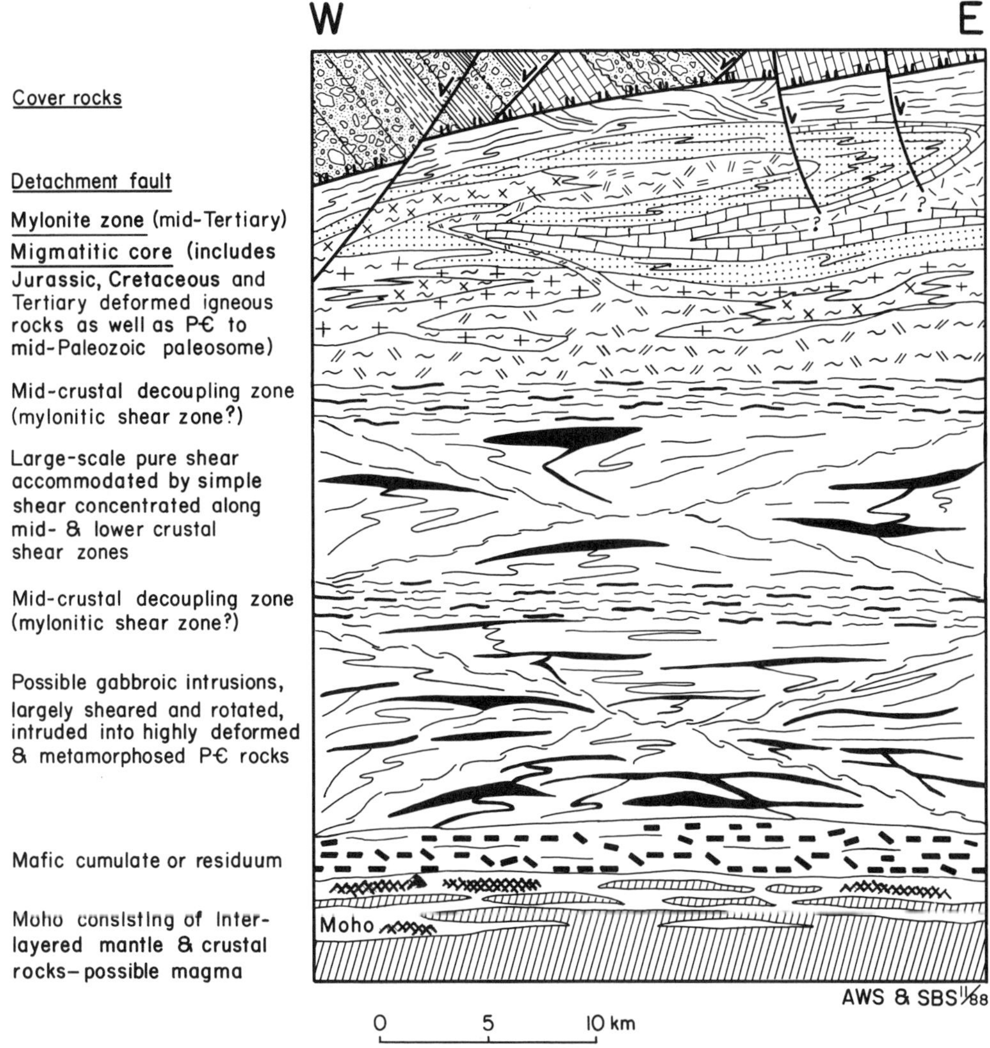

Figure 15. Interpretive cross section of crust in the northern Ruby Mountains of the Basin and Range. Upper plate overlies detachment fault and mylonite zone. Subhorizontal nappes of the migmatitic core (infrastructure) are underlain by deformed granitic intrusions and anastomosing mylonite zones (concentrated shear). These pass downward into sheared mafic rocks (possible intrusions) and sheared lower crust and Moho. Layered Moho is accentuated by shear accompanying extension. (From Valasek and others, 1989.)

tions of interlayered quartzites, schists, marbles, and layered to lenticular granites in the upper crust. Velocities of 6.4 to 6.5 km/sec correspond to slightly more mafic, but not intermediate (andesitic), compositions. These rocks probably are mostly quartzofeldspathic (deformed granites and metasediments) in which about one-third of the material is mafic, possibly intrusive. The next zone from a depth of 20 to 28 km has model velocities of 6.7 to 6.8 km/sec. If it is in granulite facies, which seems likely (Sandiford and Powell, 1986), then it is rather more felsic than gabbro, possibly about one-third granitic material. This is the zone that would have to be the primary locus for underplating accompanied by partial melting of the overlying more felsic rocks to produce the granitic intrusions in the upper crust. The lowermost 3 km in the crust (the zone between the "X" reflection and the Moho) has a relatively high velocity of 7.4 to 7.8 km/sec that lies between normal velocities for lower crust and upper mantle. This zone may consist of mafic residuum or layered cumulates from the overlying gabbroic layer and thus could be the result of underplating.

The high reflectivity of the Moho could be caused by interlayered peridotitic rocks and more felsic rocks, or even magma (Smithson, 1989). At most, one-third of the present crustal section could have been formed by underplating by gabbroic magma, but the amount is probably less than this because some gabbroic rocks would have been present in the preexisting lower crust.

An interpretive cross section (Fig. 15) is based on available seismic reflection data and surface geology projected down-

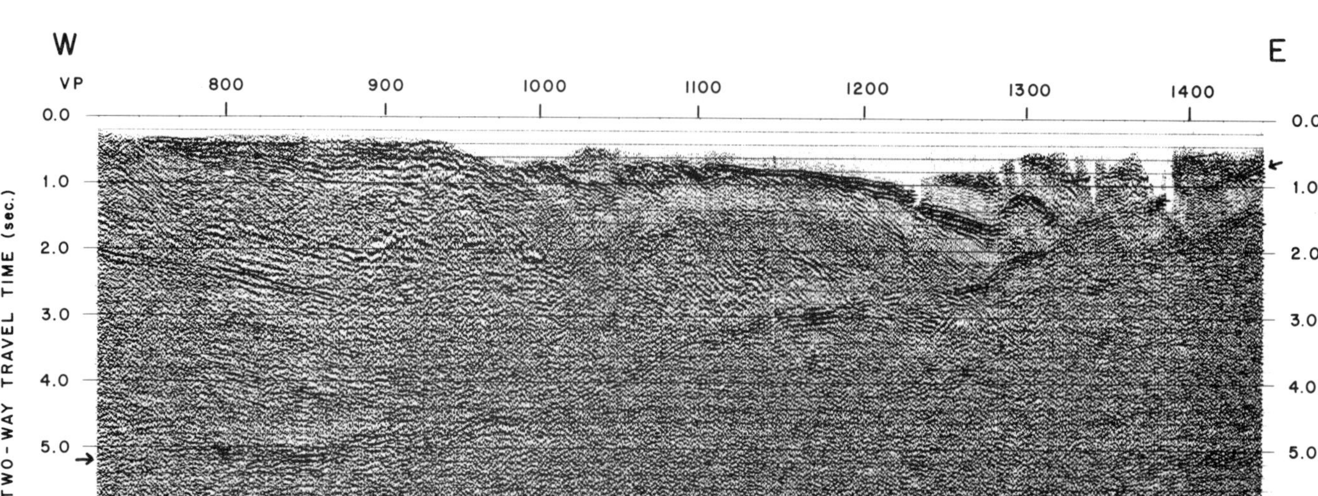

Figure 16. COCORP Utah Line 1 showing the spectacular continuous reflection from the Sevier Desert detachment (arrows). Note strength and continuity of the reflection that can be followed from the surface to depth.

plunge, and shows a near-surface mylonite zone representing Cenozoic extension above nappes of the infrastructure formed during Mesozoic compression and denuded during the later extension. This is underlain by a zone of concentrated shear (anastomosing mylonites) above a zone of intrusion and distributed shear. The generally reflective crust, the subhorizontal reflections, and the strong midcrustal, lower crustal, and Moho reflections are attributed to ductile flow accompanying extension that has generated sharp subhorizontal layering into which intrusions are drawn out. In other words, the seismic profiles show the effect of *extensional* regional metamorphism on the deep crust.

The Snake Range core complex in easternmost Nevada has been studied using industry profiles and a COCORP line (Gans and others, 1985; McCarthy, 1986). These studies show that the northern Snake Range detachment forms an antiformal surface originating as a local dislocation between a brittle upper plate and a ductile lower plate. Brittle normal faults of the upper plate sole into the detachment to the east and cut the detachment to the west of the Snake Range (Gans and others, 1985). The lower crust extends ductilely, accompanied by intrusion, to impart the subhorizontal layering indicated by reflection data (McCarthy, 1986).

Sevier Desert. In COCORP Utah Line 1 across the Sevier Desert, the reflection pattern has some similarities and some differences with the Nevada data. Near the eastern end of the line, a multicyclic Moho-depth reflection steps up to the east from 11.2 to 10.2 sec, but a shallow reflection interpreted as the Sevier Desert detachment (Allmendinger and others, 1983) dips to the west across most of the section (Fig. 16) and is underlain by a second parallel reflection in the western part of the section. In addition, a reflection dipping east from the Snake Range may mark the Snake Range detachment. A dense group of reflections occurs between 6 and 8 sec beneath VP 1400. The strong multicyclic reflection marking the Sevier Desert detachment is generated from a zone of silicification and brecciation in Paleozoic rocks in the near surface. Analysis of first breaks on field records indicates that a significant velocity contrast (lithology contrast) exits across the detachment where the Sevier Desert detachment reflection is very shallow. At depth, the Sevier Desert detachment may pass into a mylonite zone.

Sevier Desert detachment. The Sevier Desert detachment (Fig. 16), which has experienced low-angle slip due to extension (Allmendinger and others, 1983), was first interpreted from seismic reflection data as a possibly reactivated thrust fault by McDonald (1976). Low-angle normal faults in the Basin and Range have been argued to result from both Tertiary reactivation of Mesozoic thrust surfaces and as primary extensional features without any Mesozoic compressional precursor (Coney, 1980).

Most interpretations agree that no significant disruption of the Sevier Desert detachment by high-angle Basin and Range faulting has occurred. On COCORP Utah Line 1 (Fig. 16), apparent offset of the Sevier Desert detachment by high-angle faults at VPs 1290 and 1330 are attributed to velocity pull-up due to overlying Basin and Range structures (Allmendinger and others, 1983; Von Tish and others, 1985). In contrast, seismic modeling suggests that the Sevier Desert detachment could be offset in this area by at least two high-angle normal faults with individual throws of about 500 m. Models that avoid offset of the Sevier Desert detachment rely on abrupt lateral velocity variations associated with overlying structures to cause all of the apparent offset

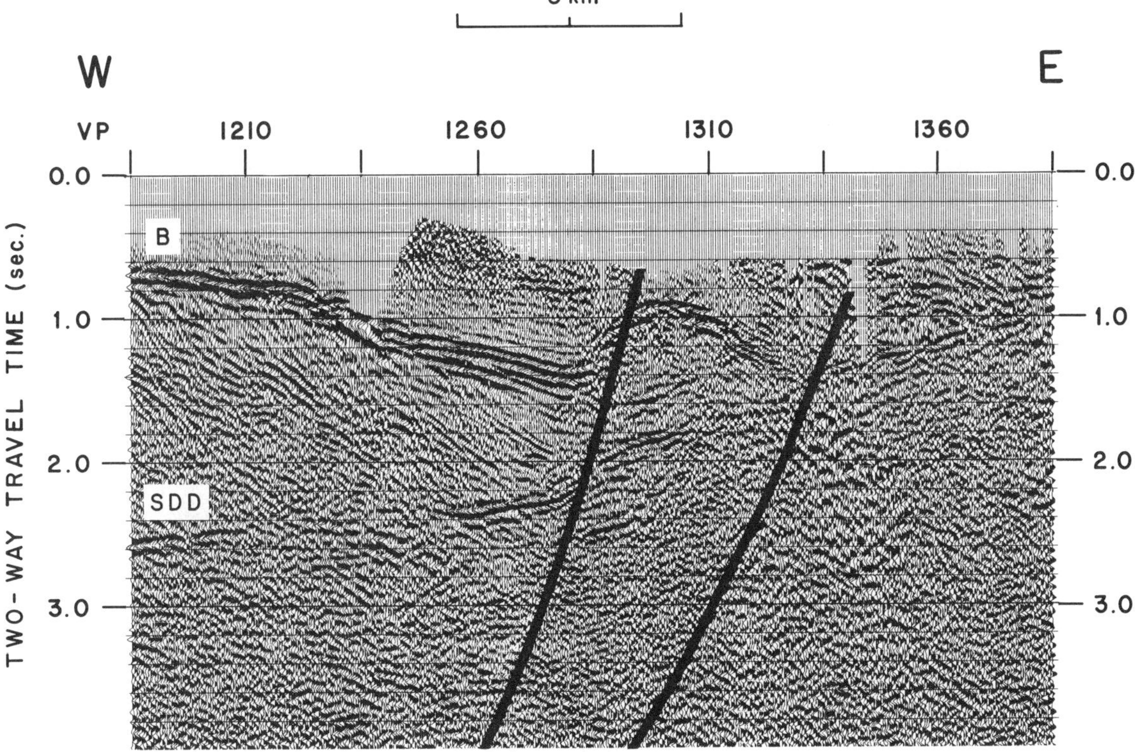

Figure 17. Reprocessed profile from COCORP Utah Line 1 interpreted with high-angle normal faults cutting a Pliocene basalt reflection (B) and the Sevier Desert detachment reflection (SDD). Width of section is about 20 km.

of the reflection. Such models can be constructed to give synthetic seismic responses that mimic the observed reflections, but obtaining a good fit is difficult for geologically reasonable models. A better fit is achieved by a ~500-m fault offset of the Sevier Desert detachment combined with reasonable velocity variations due to overlying Basin and Range structure (Fig. 17). A basalt layer, dated at 4.2 Ma, causes a prominent marker reflection in the seismic data that clearly has been faulted (event B, Fig. 17). Throw evident at this level at VP 1290 is greater than throw inferred from modeling of the deeper Sevier Desert detachment interface, suggesting that early displacement was accommodated along a listric fault that soled into the Sevier Desert detachment or was truncated by it. Later stages of high-angle displacement at this fault breached the detachment, so that normal slip along the Sevier Desert detachment ceased after 4.2 Ma.

Sevier Desert Moho. The nature of Moho structure beneath the Sevier Desert on COCORP Utah Line 1 is not immediately clear because of broken Moho reflections. Depth conversion places Moho-depth reflections, which step up in time to the east by about 1 sec (Fig. 18a), just below the range of the crust-mantle transition zone (25 to 30 km) as determined by Smith (1978) from crustal refraction studies. However, Moho topography sufficient to result in abrupt 1-sec time shifts, as seen in the data, are not suggested by refraction or gravity results (Smith, 1978; Eaton and others, 1978). No evidence of Moho faulting is apparent in these data, as perhaps it is in the case of the Outer Isles thrust and Flannan thrust on British MOIST data (Smythe and others, 1982). The deep reflections beneath the Sevier Desert are interpreted instead to be from a flat Moho (or upper-mantle reflectors), and appear to be offset by velocity pull-up due to lateral velocity variations in the upper crust (Allmendinger and others, 1983).

Significant improvement in Moho reflection-data quality and continuity has been obtained for these Moho events through the application of what we call "megastatics" (Smithson and others, 1987). Megastatics is a process in which large static shifts are removed from seismic traces in an attempt to remove velocity effects and simultaneously to correct for lateral velocity variations that degrade signal-to-noise enhancement expected of common-midpoint stacking. The velocity model used for reprocessing of the Sevier Desert Moho reflections accounts for lateral velocity variations due to Basin and Range structure and due to juxtaposition of younger, lower velocity rocks and older, higher velocity rocks across the Sevier Desert detachment. Corrections range from near 0.0 to more than 0.9 sec. Since reflection events are corrected for large lateral time shifts, time migration after stack (Fig. 18b) produces more useful results than can be obtained by time migration of the original stacked data.

Although there is evidence for gentle eastward dip of Moho reflections (Fig. 18b), dip is dependent on detailed knowledge of

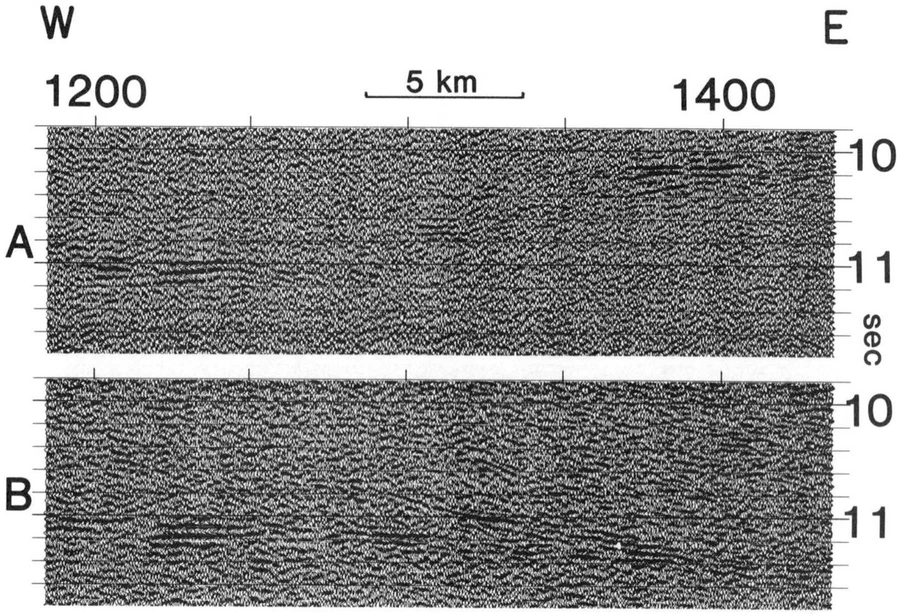

Figure 18. A, Moho (or sub-Moho) reflections from beneath the Sevier Desert, Utah, from reprocessed COCORP data of Line 1. Note that the events step up to the east (right) from 11.2 to 10.2 sec in three segments. Discontinuities in the Moho reflection are due to upper-crustal velocity variations, not true Moho structure. Width of section is about 30 km. B, Migration of section shown in Figure 22 after "megastatics" processing. Reflections exhibit significantly improved continuity. Based on these events, Moho is at a depth of about 32 km, and dips gently to the east.

the overlying velocity structure in spite of the improved image. Unrecognized lateral velocity variations would change the dip, and even gentle west dip of the Moho could be possible based on these data. True east dip is supported by older refraction data (Smith, 1978) that indicate general eastward crustal thickening toward the Colorado Plateau.

A Moho depth of 31 km was determined by Liu and others (1986) from an expanding-spread experiment conducted in conjunction with COCORP profiling in the Sevier Desert. Other estimates of crustal thickness in western Utah generally have been less, ranging from 25 to 30 km (Smith, 1978; see Pakiser, 1985, for a review). However, if the Moho reflection picked by Liu and others (1986) is correct, the expanding-spread data provide the best constraint on local crustal thickness in western Utah.

Southern Basin and Range

Mojave-Sonoran Desert. The central Mojave-Sonoran Desert has been subjected to mid-to-late Cenozoic regional extension that was usually accommodated by low-angle (detachment) faults with upper-plate brittle and lower-plate ductile deformation. These regional detachment faults are found within a belt of metamorphic core complexes (Crittenden and others, 1980) and in a larger surrounding region that also underwent high extension.

The Whipple Mountains metamorphic core complex is ringed by detachment faults of mid-Tertiary age (Davis and others, 1980). Within the lower plate of this mountain range are mylonitic rocks thought to have formed at midcrustal levels during the extension forming the detachment faults. Crustal reflection profiles, collected by CALCRUST west of the Whipple Mountains, reveal a highly reflective mid- to lower crust that may be related to mylonites (Henyey and others, 1987). Packets of refletions suggest a stacked, lensoid geometry created as the lower crust deformed ductilely under low-angle simple shear. The constant crustal thickness under the domed core complex implies that material was added at the base of the crust after doming. That this material is also seismically reflective may indicate lower-crustal igneous and tectonic processes not manifested at the surface (Henyey and others, 1987).

Sonoran Desert. A series of related core complexes similar to the Whipple Mountains is found in Arizona. A Phillips-Anschutz well and seismic lines provide contrasting information with subsurface control on one of the metamorphic core complexes in Arizona. The well, the Phillips Arizona State A-1, was drilled through almost 5,500 m of crystalline basement rocks to test the hypothesis that the basement rocks were overthrust onto deformed sedimentary rocks (Reif and Robinson, 1981). The drill hole never left crystalline rocks, and the intrabasement reflections were attributed to brecciated zones and gneissic foliation associated with a nearby core complex in the Picacho Mountains (Reif and Robinson, 1981) where the geology has been studied by Yeend (1976) and Rehrig and Reynolds (1980).

The seismic line through the well and the VSP have been reprocessed and analyzed at the University of Wyoming. The reprocessed seismic line has fewer short reflections because coherency filtering was omitted, resulting in a seismic section with

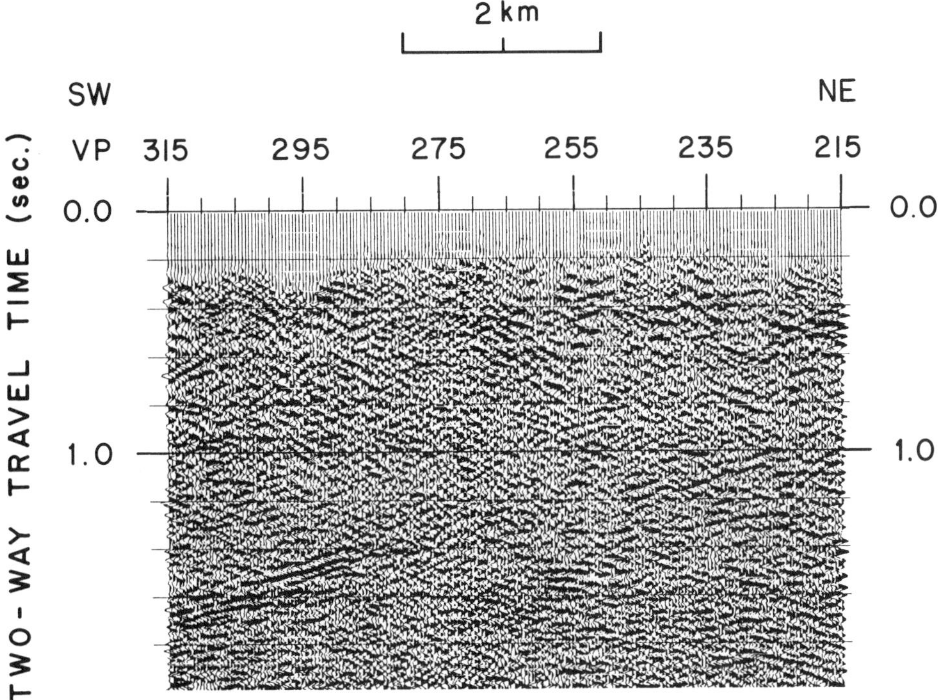

Figure 19. Seismic section that passes Phillips Arizona State 1-A well, processed at University of Wyoming. Fault zone represented by dipping event between 1 and 1.6 sec. Reprocessing without coherence filter has greatly changed the character of the reflection.

more character and less difference between the upper and lower part of the section (Fig. 19); this provides a good example of how processing can affect interpretation. A VSP shows reflections originating from contacts between mylonite and granite and from detachment zones where fracturing is common. The shear zones may be detachment faults associated with core complexes.

The dipping reflection in the reprocessed seismic section (Fig. 19) is interpreted as marking the detachment zone above the mylonite (Fig. 20). The mylonite itself appears to vary laterally in reflection character, probably because the predominantly granitic mylonites are less reflective. Reflectivity of the relatively uniform rock types seems to be controlled by fractures. Formation of phyllosilicates under retrograde conditions may additionally increase the reflectivity (Goodwin and Thompson, 1988). The structure is interpreted to be a rather transparent Precambrian Oracle granite, underlain by Tertiary granite, and fractured Tertiary mylonite on top of a less reflective infrastructure (Fig. 20). These data provide an example of reflections generated from the brittle zone in a fractured and altered detachment fault.

Mojave Desert

The Mojave area is a wedge-shaped block between the Garlock and San Andreas faults over which COCORP has conducted an extensive recording experiment (Cheadle and others, 1986a,b). The area is covered with a thin layer of Cenozoic deposits concealing a dominantly granitic basement that represents the remains of an Andean-type island arc (Sierran granite) overprinting earlier Precambrian crust (Hamilton, 1969). Mesozoic east-vergent thrusts are found in the eastern part of the block; these may be related to the evolution of the island arc (Burchfiel and Davis, 1972). The Rand schist is a paradoxical unit underlying 85-Ma granite and Precambrian gneiss and may itself be an allochthonous correlative of the Pelona-Orocopia Schist (Cheadle and others, 1986a). Detachment faulting, accompanied by extensive volcanism, took place in the early Miocene, and since mid-Miocene, strike-slip movements on the San Andreas Faults have created north-south compression within the Mojave block.

The COCORP seismic sections are highly variable in the amount and extent of reflections. Line 3, the longest profile, and Line 2, which intersects the southwest end of Line 3, contain numerous events. The other lines have few events, and Line 7 across the San Andreas fault has almost none. Line 3 shows a series of relatively continuous events between 2 and 7 sec. Of particular interest are two events at 3 and 6 sec that appear to be almost continuous beneath the Garlock fault (Cheadle and others, 1986b, p. 298, Fig. 3a) and two shallow (1 to 6 sec) events that dip southwest from VP 500. The seismic sections contain many linear dipping events that are probably reflected refractions and arcuate events that may be diffractions or side-arriving energy.

The COCORP interpretation (Cheadle and others, 1986b, p. 313, Fig. 11) is based on three alternative models involving Mesozoic thrusting, late Cretaceous to early Cenozoic thrusting, and Miocene extension. As with most deep-crustal reflection

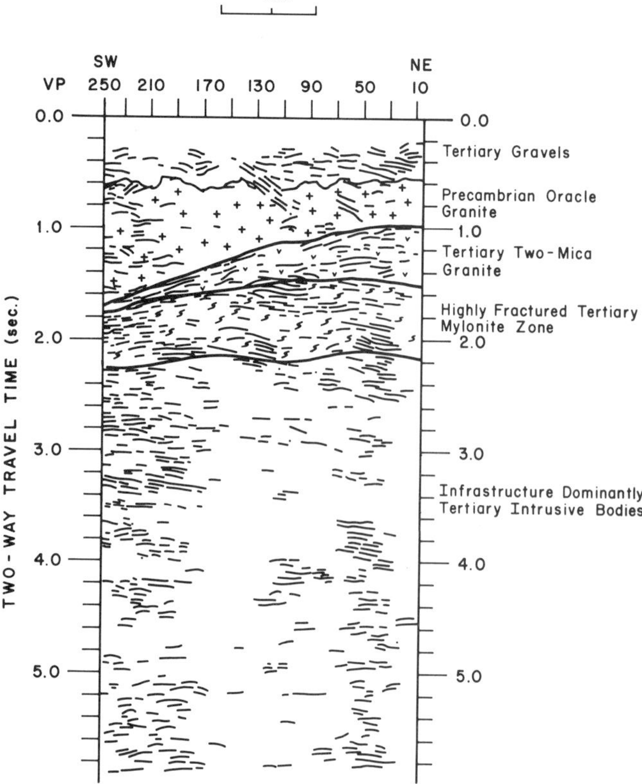

Figure 20. Interpretation of crustal section near Picacho Mountains, Arizona.

data, constraints are minimal, especially where most of the crystalline geology is covered by Cenozoic sediments. The multicyclic nature of these reflections, with a duration of 0.5 to 1 sec, suggests layered sequences ranging from 1.5 to 3 km thick; the problem is what these represent. The shallow, dipping reflections south of VP 500 were interpreted as the Rand thrust and a possible later detachment fault (Cheadle and others, 1986b, p. 302). Sporadic events between 9 and 10 sec are attributed to the Mojo, which appears to rise to the southwest (Cheadle and others, 1986a). The Moho is apparently a good reflector beneath the Mojave block, so that the discontinuous nature of Moho events may simply be caused by recording conditions. The reflection Moho seems to be somewhat shallower than the refraction Moho (Prodehl, 1979). Most of the reflections above the Moho are interpreted as moderately dipping faults, either Mesozoic to early Cenozoic thrusts or mid-Cenozoic extensional faults (see Mooney, this volume), depending on whether the compressional or extensional alternatives are favored. This interpretation shows the Garlock fault terminated at a depth of less than 9 km (above the 3-sec reflection). Thus the Garlock fault must either flatten with depth or continue downward into a subhorizontal detachment zone that decouples movement on the fault from the lower crust. This interpretation also points out that most geologic features represented in the reflection section dip southwest.

Other interpretations are possible; these generally fall between some combination of thrusting and extension postulated by Cheadle and others (1986a). There is one major exception that illustrates an important rule in evaluating crustal reflection data: "don't rely on line drawings." An examination of the original seismic sections reveals that reflections are generally short and choppy, lacking any distinctive character. This means that they are correlated with difficulty, and that such correlations are unusually subjective. The best examples of this are the 3- and 6-sec events that are correlated continuously beneath the Garlock fault. Both events are missing in the vicinity of the fault; the 3-sec event has a 10-km gap here, while the 6-sec event shows both a gap and an offset that renders correlation uncertain. Thus, no compelling reason exists to terminate the Garlock fault above 9 km depth. This remains, however, one possible interpretation.

As with much of the western United States, a critical problem is to distinguish between structures caused by extension and those caused by compression in this area where they are superposed. Here, as pointed out by Cheadle and others (1986a), most features in the seismic section dip southwest. We suggest that these events primarily reflect the effect of the compression that caused the eastward-verging structures. These events could represent the effect of southwest-dipping ductile deformation zones (mylonites) interleaved with subparallel metamorphic rocks along thrusts that dissect earlier metamorphic layering to give the discontinuous reflection pattern. This pattern is, however, undoubtedly affected by recording conditions and processing. These data provide an interesting contrast with the northern Sierra Nevada COCORP traverse because those data show almost no events across the same granite-batholith terrain; the data from the Mojave block show more. In the core-complex crustal environment, we have suggested that extension produces a more reflective crust and Moho (Hurich and others, 1985). This may or may not be true here. This crust is thinner than typical crust associated with Andean-type batholiths and may have been considerably altered by extension. Reflections beneath the granite terrain can either be from mylonites associated with strong tectonic transport or from metamorphosed supracrustal rocks beneath the floor of granitic diapirs.

Death Valley

Burchfiel and Stewart (1966) first recognized Death Valley as a pull-apart basin and related its formation to movement along the Death Valley and Furnace Creek strike-slip faults. The structure is complex and includes thrust faults, strike-slip faults, low-angle extensional faults, turtleback surfaces, and horsts and grabens (Stewart, 1983). COCORP crustal-reflection profiles cross the Death Valley area about 30 km north of the Mojave block.

The seismic reflection sections are highly variable in the number and nature of reflection events. They are characterized by short ($\leq$1 km) events that range from horizontal to moderately dipping. Shallow-dipping events to 2 s on Line 9 mark a half-

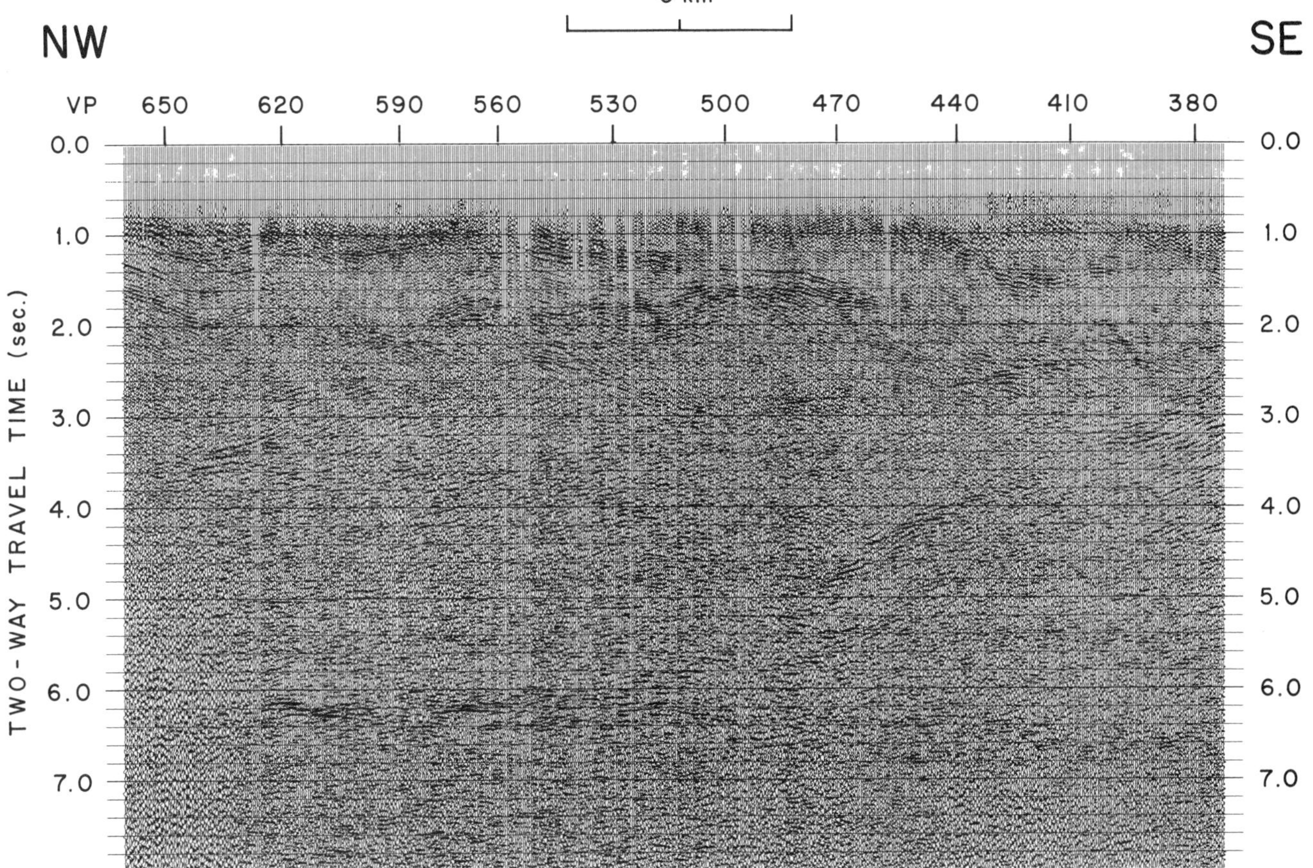

Figure 21. COCORP Line 11 from Death Valley showing a postulated magma reflection at 6 sec and a dipping event proposed to be a reflection from a feeder dike that goes from the magma chamber to a cinder cone on the surface. Note the en echelon appearance of the dike reflection.

graben bordered by a listric fault (Serpa and others, 1988), and Line 11 shows a relatively continuous packet of reflections interpreted as a thick (2 sec) succession of folded and/or faulted sedimentary rocks running the length of the valley (Fig. 21). The most striking feature is a midcrustal reflection at 6 sec, which is horizontal and in places high in amplitude. Because of the amplitude similarities with the Socorro midcrustal reflection and the presence of a cinder cone nearby on the surface, this event is postulated to be generated by a basaltic magma chamber (de Voogd and others, 1986). Furthermore, a dipping event that extends from above the high-amplitude reflection to near the surface by the cinder cone was suggested as a reflection from a feeder dike that is now consolidated. Thus this data set is proposed to image an active basaltic magma chamber beneath a pull-apart basin (de Voogd and others, 1986). Other results include half-graben geometries for the basins and a Moho reflection around 10 sec (Serpa and others, 1988). The area is compared to the Mojave block by de Voogd and others (1986), and a midcrustal reflection is correlated with a subhorizontal detachment fault extending from the Mojave underneath Death Valley. The Moho may be formed from intrusions during extension in a manner similar to that in the Basin and Range (Serpa and others, 1988).

The "magma reflection," which is characterized by high amplitude (Fig. 21), resembles the reflection from a postulated, generally accepted magma chamber in the Socorro area. The interpretation of magma is based on the high amplitude of the reflection and the similarity with the data from Socorro. Like this reflection, the one from Death Valley is multicyclic (Fig. 21). This means that it is generated by some form of layering, which leads to two conclusions. If a magma chamber is present, it is a series of closely spaced sills, and because of the layering, a stronger reflection may be generated by constructive interference. A basic question is whether the amplitude is so high that the reflection *must* be generated by magma and whether magma would have consolidated, thereby decreasing the acoustic-impedance contrast.

The "feeder reflection" is much more problematic. Close inspection (Fig. 21) shows that it is made up of a series of short en

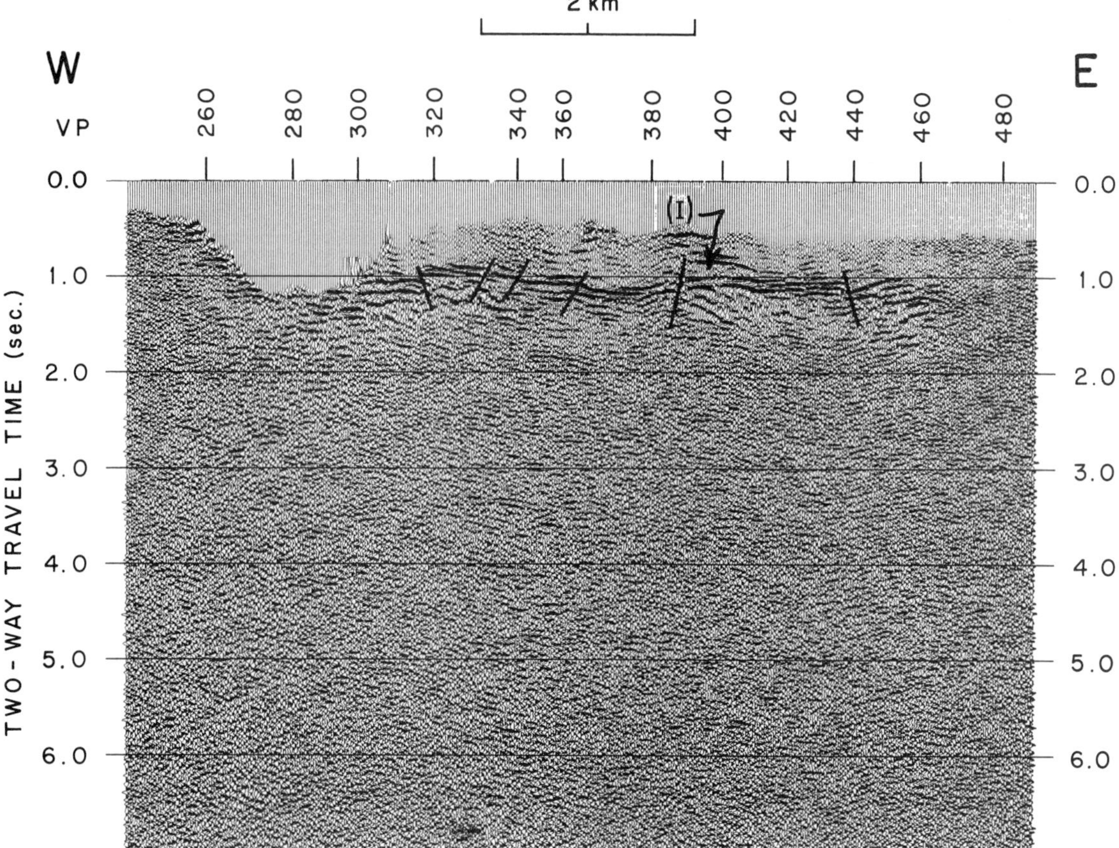

Figure 22. Seismic profile across the northern part of Long Valley caldera recorded by the University of Wyoming. Note postulated intrusion (I) and faults in Bishop tuff.

echelon reflections that dip more gently than the overall event. In our opinion, the event is questionable, and its position may be pure coincidence.

The Moho may be at 10 sec in this area; however, the events at 10 sec are sparse and weak, so not much information is provided on the Moho. Strong Moho events like those found in Nevada seem to be lacking.

Long Valley caldera

Reflection profiling in areas that produce poor data quality—such as volcanic terranes—represents a special challenge that puts exceptional demands on recording techniques, although these areas may be of particular tectonic interest. One of these is Long Valley, California, a young caldera where magma appears to be within 10 km of the surface (Rundle and others, 1985). Here, reflection profiling by the University of Wyoming, using vibrators, has shown that shallow faulted structure within the caldera may be mapped, that four vibrators have sufficient energy to observe refractions at ranges of 18 km even though this caldera environment is highly attenuating, and that weak reflections are observed from depths approximately the same as determined by other methods for a magma chamber (Rundle and others, 1985). CDP profiling in the caldera has allowed delineation of the fault system and a possible intrusion within the Bishop tuff (Fig. 22) that may influence the hydrology and therefore the geothermal anomalies within the caldera. Deeper reflections (about 3 to 4 sec) may be related to magma, but those events are weak and the correlation with magma is uncertain. Thus careful application of recording and processing techniques allows subsurface mapping by seismic reflections even in the most inhospitable environment of a young caldera.

San Andreas fault

A reflection profile across the San Andreas fault in Bear Valley, 180 km southeast of San Francisco, shows striking mid- and lower-crustal reflections in the granitic Gabilan block to the west and very few reflections in the Franciscan rocks to the east (Feng and McEvilly, 1983). Low velocities in the fault zone and lateral velocity variations across the fault zone were recognized and used to show that Moho depth across the fault is similar even though reflection times to the Moho were distinctly different. A COCORP reflection profile across the fault at Parkfield (McBride and Brown, 1986) has events so few and questionable as to defy interpretation.

Central California

Seismic lines, using an 800-channel recording system, were recorded from the Coast Range to the Sierra foothills (Zoback and Wentworth, 1986; Wentworth and others, 1987) and had the unusual advantage of detailed crustal refraction coverage. The good quality of the data illustrates the desirability of a large-channel recording system (Zoback and Wentworth, 1986, Fig. 7, p. 190) for both improvement of data quality and velocity control. Interpretation of seismic data suggests that a wedge of Franciscan rocks has been driven between the Great Valley Sequence and basement. This implies that the Coast Range thrust is cut off against basement at the eastern tip of the wedge. Dipping reflections on the west flank of the Sierra may be thrust faults along which the island-arc rocks of the foothills were obducted.

Farther to the south, a COCORP profile near Coalinga imaged a thick succession of the Great Valley and Tertiary sedimentary rocks but nothing in the crust beneath these sediments (Fielding and others, 1984).

Northern Cordillera

A COCORP traverse in the Pacific Northwest crosses the pre-Mesozoic continental margin, accreted in Jurassic through Paleocene time, and the hinterland of the Cordilleran orogeny (Potter and others, 1986). This area includes crystalline domes, core complexes, and large allochthonous masses resulting from a long history of Mesozoic and Paleocene compression and Eocene extension that affected Precambrian crust of the early North American continental margin. The area in which core complexes are traversed is the southern extension of the crystalline Shuswap terrane in Canada (Monger and others, 1982; Brown and Read, 1983) and includes two core complexes of the northern type.

Seismic-reflection data of surprisingly high quality given the complexity of the crystalline terrane was revealed by Hurich and others (1985), which led to the COCORP survey. The COCORP sections show alternating strips of abundant arcuate reflections and transparent zones, which are typical of recording/processing conditions rather than actual geology. The parts of the seismic sections that do contain reflections are almost overwhelming in their richness. Bundles of convex-upward reflections are interspersed with more linear, discrete events and are underlain by discontinuous and sporadically dipping reflections at around 11 sec (Fig. 23). The linear events are interpreted as faults, either thrust or extensional, and presumably ductile (Potter and others, 1986), based on cross-cutting relations of the reflections and east dip for extensional faults and west dip for thrusts; however, fault interpretation is difficult in most cases.

The short University of Wyoming seismic line (Fig. 24) shows that the base of a detachment fault and underlying mylonite zone are good reflectors along the eastern flank of the Kettle dome (Hurich and others, 1985). The reflection character of the mylonite zone along the flank of this core complex is related to the specific rock types involved (Fountain and others, 1987). Good reflections were found down to 6 sec, and the generally high reflectivity of the crust was noted. These reflections were attributed to the effects of superposed compression and extension, and while the difficulty in separating effects of extension from compression in the reflection wave field was pointed out, the prominent reflections were attributed to extension.

Numerous arcuate reflection packets in the COCORP data are caused by layered sequences of complex geometry, as much as about 6 km in thickness. They are probably too thick to be typical mylonite zones unless the entire crust has been subjected to homogeneous slip. They could represent deformed supracrustal rocks, deformed igneous intrusions, or possibly even layered igneous intrusions. Their significance is important because they characterize the lower crut in a polygenetic, reconstituted terrane.

The 11-sec event probably is a Moho reflection. The COCORP interpretation (Potter and others, 1986) emphasizes that this is a flat interface and is therefore probably a young feature. That the Moho is a young or reworked structure seems likely; however, the details of the Moho reflection show that it dips in places and is rather complex (Fig. 23) so that it need not be flat locally. It arrives consistently around 11 sec on a regional scale.

NATURE OF THE CRUST

Intrusions

Seismically transparent zones have typically been interpreted as massive igneous intrusions (plutons). Seismic sections may appear transparent for many other reasons (Smithson, 1979). Because they are not generally layered, plutons will not be as good reflectors as mylonites and layered metamorphic sequences, but results suggest that reflection data will give information about plutons, particularly their lower contacts. A number of seismic data sets indicate that reflections come from the bottom of, or beneath, large intrusions ranging from granites to anorthosites to layered intrusions. Reflections have been found from beneath granite batholiths (Smithson and others, 1977b; Lynn and others, 1981), from the base of Laramie anorthosite (Smithson and others, 1977b; Allmendinger and others, 1982), from beneath Adirondack anorthosite (Brown and others, 1983) and from within layered intrusions in the Wichita Mountains (Widess and Taylor, 1959). The strength of reflections from layered mafic intrusions is unknown; however, the interlayered granites and gabbros of the Wichita Mountains in Oklahoma generate good reflections and demonstrate the reflectivity of gabbroic sills within quartzofeldspathic rocks (Widess and Taylor, 1959). Reflections from the tops of batholiths have generally not been verified, although these features constitute important geologic targets. Roof zones of intrusions will probably require special recording and processing techniques before they can be imaged routinely. Intrusions have been suggested as a reason for the strong reflectivity of the Basin-and-Range crust (Klemperer and others, 1986).

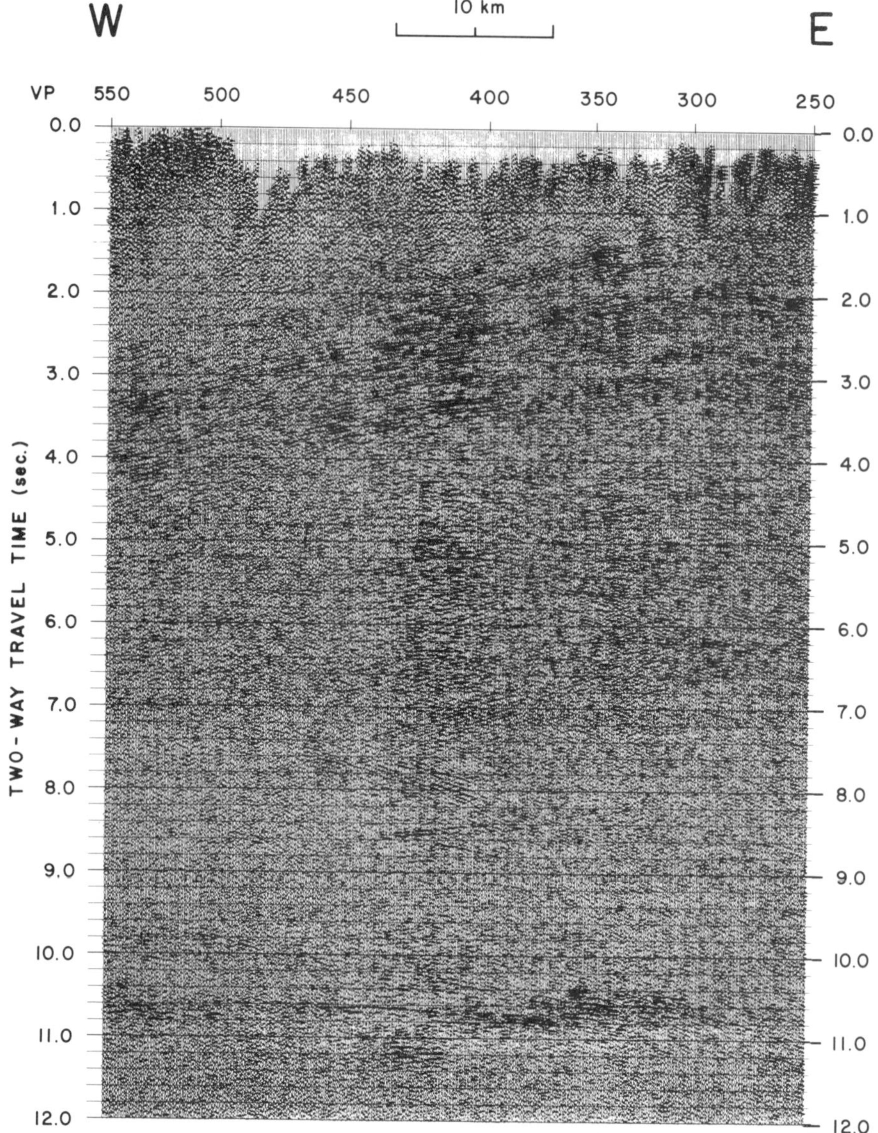

Figure 23. COCORP Line 5 from northeast Washington showing strong upper-crustal reflections, complex arcuate events from lower crust, and irregular Moho reflection.

The Moho

The Moho is generally not identifiable as a discrete reflection in seismic-reflection profiles across Precambrian terranes in the craton. For example, in the Kansas COCORP profiles, a change in seismic character without a discrete reflection may occur at approximate Moho depth (Serpa and others, 1984). One possibility is that the Precambrian Moho is a gradational—either compositionally or mineralogically (phase change)—change that is a poor reflector at vertical incidence (Smithson, 1989). Alternatively, the Moho could be represented by a small step in velocity (no layering) that does not produce a strong enough reflection to be visible (Smithson and others, 1977a, p. 120, Fig. 1). A Moho reflection has been picked at 16 sec (about 48 km) under the Proterozoic crust of the Laramie Range, Wyoming (Allmendinger and others, 1982; Hale and Thompson, 1982), but because of the ringy nature of the seismic section at this depth and because this places the Moho far deeper than other data suggest, there is little to recommend it as a true Moho reflection. In general, the Moho appears to be resolved more frequently in wide-angle reflection profiles (Meissner, 1973). The Moho beneath the Precambrian in the United States is nonreflective at vertical incidence and reflective at wide angles in Minnesota and Wyoming. This may be because of the gradational nature of the Precambrian Moho (Smithson, 1989), and special recording experiments may have to be designed to resolve the question.

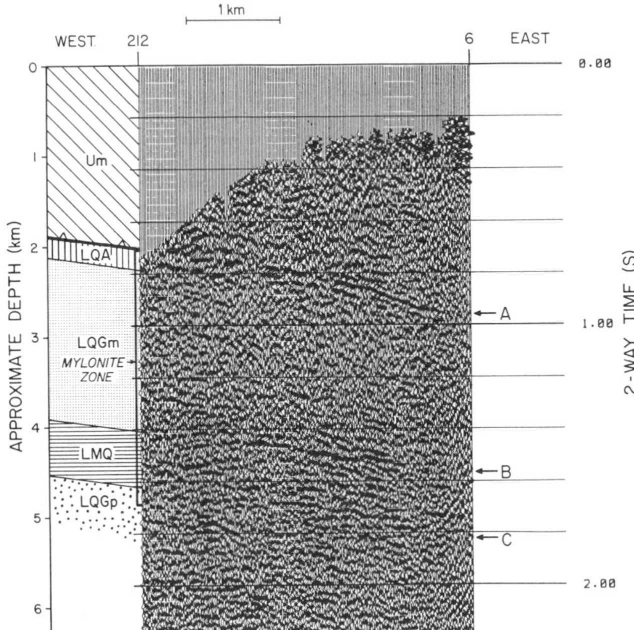

Figure 24. Unmigrated 48-fold seismic section recorded by the University of Wyoming showing correlation between down-dip projection of exposed mylonite zone and seismic data. Horizontal:vertical scale is 1:1. A, Reflections from Kettle River fault and upper part of mylonite zone. B, Reflections from interlayered marble, quartzite, paragneiss, and schist. C, Probably top of porphyroclastic gneiss body of unknown thickness; transition between mylonite zone and infrastructure of core complex occurs at about this level. Depths are determined assuming constant velocity of 6.0 km/sec.

The Moho beneath extensional terranes is quite a different story. Beneath the Basin and Range, the Moho appears as a sharp, high-amplitude, multicyclic reflection. Klemperer and others (1986) suggested that the Moho in this region is a young feature. Layering at the Moho may have been emphasized by ductile flow in the lowermost crust during the extreme Basin and Range extension. A young, sharp Moho is also found below the complex terrane in Washington (Potter and others, 1986). The Moho is a mechanical discontinuity in some areas, as shown by layering oblique to the Moho in the lower crust. Thus the strikingly different reflection response of the Moho in different tectonic environments must indicate a difference in the nature of the Moho zone (Smithson, 1989).

Lower crust

Lower crust may be classified on the basis of its reflection response as follows (Smithson, 1986): (1) nonreflective lower crust; (2) planar, near-horizontal reflections; (3) broadly arched reflections; and (4) arcuate, criss-crossing events. The poor reflectivity of the lower crust in some areas may be only apparent; lack of reflections may be caused by recording/processing conditions, steep dips, strong folding, compositional gradation, or compositional homogeneity (Smithson, 1986). If a volume of the lower crust is nonreflective because it is truly homogeneous, then it is almost certainly an igneous intrusion since sedimentary and metamorphic rocks are rarely homogeneous over large volumes. Planar near-horizontal reflections could be caused by intrusions of gabbroic sills in more felsic rocks, discrete mylonite zones, homogeneous ductile deformation of a heterogeneous lower crust, recumbently folded metamorphic rocks, or possibly a layered mafic intrusion. If underplating occurs by intrusion of gabbro as discrete sills into lower-velocity material, the reflection response would probably consist of subparallel, subhorizontal reflections; i.e., class 2. If, however, the sills coalesce, probably no visible reflections would be generated, and the lower (or lowermost) crust would appear transparent, but would give a higher (gabbroic) lower-crustal velocity (Meissner, 1967); i.e., class 1. Another possibility that is likely in a lower-crustal environment is that a sill differentiates into mafic layers like the Palisades sill and generates weak subhorizontal reflections; visibility would depend on the details of layering and signal/noise.

Supracrustal rocks would produce multicyclic and probably broadly arched reflections, i.e., class 3, or if late folding was stronger, would form crossing arcuate events, i.e., class 4, and a metasedimentary component is common (Smithson and Brown, 1977; Kay and Kay, 1981; Fountain and Salisbury, 1981).

Mylonites are excellent reflectors (Fountain and others, 1984; Hurich and others, 1985; Smithson and others, 1985), and this reflectivity is probably largely due to layering on all scales and to their relatively planar nature. Two of the mylonite zones discussed were exposed at what was formerly midcrustal levels (corresponding to sillimanite grade) and thus extended into the lower crust. In other areas, mylonites probably extend through the crust and into the upper mantle; examples include the Outer Isles thrust observed in the MOIST profiles (Brewer and Smythe, 1984). Both mylonites and supracrustal sequences have been deformed so much that layering would not have any primary significance, and large amounts of intrusive igneous component could be incorporated into what is now a strongly layered sequence. Rather similarly, if the lower crust were subjected to roughly homogeneous simple shear, mixed rocks that had highly variable and irregular geometries could be drawn into semicontinuous layers parallel to the slip direction. Thus, lower-crustal layering may have several origins, but in its most advanced development is likely related to strain or extreme strain. Mylonites formed during extension may combine with more homogeneous strain in the ductile deep crust to produce very reflective zones of high strain (mylonites) and intervening zones of moderate strain in which earlier features are rotated toward the horizontal as the crust extends. This same strain pattern may have created multilayered reflectors for the Moho in the Basin and Range–Rio Grande rift, the best lower-crustal and Moho reflections in the United States.

A major question is whether ancient Archean crust is truly different from Proterozoic crust; i.e., was it thin and then thickened later by crustal underplating, as suggested by Fyfe (1974)? Because of inadequate contrast in acoustic impedance, underplating by either thick (greater than 100 m) sills or coalescing thinner sills of gabbro may not produce good reflections from the lower crust unless they are differentiated into sharp layers, so areas with no reflections or weak subhorizontal reflections at lower-crustal depths could be candidates for crustal underplating. Lower crust may be affected and altered by a wide range of processes after initial formation, so that in many areas the lower crust may be polygenetic; i.e., orogenic crust undergoing later extension, early Archean crust thickening by underplating, or earlier crust differentiating by anatexis. Seismic-reflection sections show a wide range of lower crustal responses ranging from seemingly transparent (homogeneous?), to layered, to highly folded, lithologically complex structures indicative of complex genetic processes.

CONCLUSIONS

The reflection pattern of the Nevada Basin and Range forms a sharp contrast compared with the craton and compressional regimes in the western United States. The best lower-crustal and Moho reflections in the United States are related to ductile flow and metamorphism of the deep crust during extension (Hurich and others, 1985; McCarthy, 1986; Smithson, 1986; Allmendinger and others, 1987), producing strong subhorizontal layering and lensing (Smithson, 1986; Henyey and others, 1987; Valasek and others, 1987) (Figs. 3, 15). Possible igneous intrusions (Klemperer and others, 1986) are metamorphosed and drawn out into the layering, thereby accentuating it. This is caused by concentrated and distributed simple shear on a small scale, and large-scale pure shear in the lower crust (Valasek and others, 1989). The Moho in the Basin and Range is approximately flat and is probably a young feature (Klemperer and others, 1986; Allmendinger and others, 1987). The Moho does not seem to be offset by extensional faulting, suggesting that faults pass into distributed ductile deformation in the deep crust; seismic modeling, however, reveals that through-going listric faults that displace the Moho are difficult to recognize. Although intrusions and magmatism may have caused crustal underplating and have been proposed as sources of strong crustal reflections (Klemperer and others, 1986; Potter and others, 1987), the physical mechanism for generating these reflections remains unclear.

The crust in cratonic and compressive regimes is generally less reflective and may be characterized by crossing arcuate events (Serpa and others, 1984) caused by folded, faulted, and intruded layered rocks; i.e., a geometrically and compositionally complex sequence of metamorphic and igneous rocks. Seismic modeling (Figs. 2, 3) shows that these arcuate events may be caused by combinations of folding, lensing (boudinage), and in certain cases, intrusion. Metamorphic rocks probably predominate (Smithson and Brown, 1977; Fountain and others, 1987) and give rise to multicyclic reflections where their dip is not too steep. While a good Moho reflection is found in the reconstituted crust of the Pacific Northwest, vertical-incidence Moho reflections are generally lacking in the craton; this observation suggests that the Moho in the craton is very different and may be compositionally gradational or mineralogically gradational, such as the gabbro-eclogite phase change. The arcuate reflection pattern in the crust of northeast Washington indicates that extensional overprinting has not been as strong here as in the Basin and Range, yet the fact that a discrete Moho reflection is visible may be related to the later extensional history. Lower ductility in the deep crust of northeast Washington related to less thermal input may be the difference.

Many crustal reflections have been interpreted as fault zones. Although ductile fault zones (mylonites) are demonstrated to be excellent reflectors, not every strong crustal reflection is likely to be a fault. In addition, fracturing and alteration (phyllosilicates, carbonate, silica) may cause faults to reflect, as demonstrated by the Picacho Mountains and Sevier Desert detachments. Layered metamorphic sequences and possibly layered igneous intrusions will also generate good reflections, and many of the segments of multicyclic reflections are undoubtedly from metamorphic rocks such as supracrustal sequences, high-grade gneisses, and migmatites. The importance of fluids in generating deep-crustal reflections (Jones and Nur, 1982) remains unknown. Commonly these layered metamorphic sequences contain evidence for at least a small sedimentary component; however, layering by itself is not evidence for a metasedimentary origin. Rather, it is evidence of deformation.

The oldest crust in Minnesota consists of a stack of nappes almost 30 km thick underlain by a lowermost crust of different structural style that could involve underplating. The younger Archean has been thrust onto the oldest gneiss terrane and reactivated during Proterozoic orogenesis along discrete thrusts to deform Animikie rocks. The mid- to lower Proterozoic crust of Kansas and eastern Utah probably consists of geometrically complex metamorphic rocks and igneous intrusions. Unusually thick crust in southwest Oklahoma may have been underplated by mafic magma, possibly causing subsidence that allowed deposition of 12 km of volcanic and sedimentary rocks in a Proterozoic basin.

Standard COCORP reflection profiles across the San Andreas fault show little about crustal structure or the fault, probably because of the complexity and steep dip of the target. The Garlock fault may be terminated at a depth of 9 km, or it may continue much deeper; evidence for terminating it at 9 km is not compelling. Connected with the first possibility is a postulated midcrustal detachment horizon underlying much of southeast California. Based on the underlying extent of the Pelona-Orocopia-Rand schists, a major question still exists about whether the entire upper crust in southern California is allochthonous. A Conrad discontinuity is generally lacking in most crustal data sets; a midcrustal detachment (Serpa and others, 1988), if present, could be considered a local Conrad in eastern California. Gabbroic rocks, several kilometers thick and occurring locally at midcrustal depths, might be interpreted as a Conrad discontinuity (Smithson, 1978) but it would be of restricted extent. Sills filled with basaltic magma probably cause reflections under the Rio Grande rift and possibly under Death Valley; rhyolitic magma may not generate such good reflections. Thick, arcuate packets of reflections in the lower crust of reconstituted terrane in Washington represent some combination of deformed igneous and supracrustal rocks in the lower crust. As crustal reflection profiles accumulate, not surprisingly, they emphasize the horizontal and vertical heterogeneity of the crust (Smithson and Brown, 1977). Detailed velocity profiles are a necessity to develop a better understanding of crustal structure. Improved recording experiments and interpretational techniques will be used in the future to answer the problems posed in this review.

ACKNOWLEDGMENTS

Financial support was received from National Science Foundation Grants EAR-8400227, EAR-8512083, EAR-8511919, and EAR-8519153. Computing was carried out on the DISCO VAX 11/780 of the Program for Crustal Studies. We are particularly indebted to an anonymous reviewer who rewrote one section of the paper, and to several other reviewers for corrections and improvements.

REFERENCES CITED

Alexandrov, K. S., and Rhyzhova, T. V., 1961, Elastic properties of rock forming minerals; 2, Layered silicates: U.S.S.R., Bulletin of the Academy of Science Geophysics Series, v. 9, p. 1165–1168.

Allmendinger, R. W., and 5 others, 1982, COCORP profiling across the Rocky Mountain front in southern Wyoming; Part 1, Precambrian basement structure and its influence on Laramide deformation: Geological Society of America Bulletin, v. 93, p. 1253–1263.

Allmendinger, R. W., and 7 others, 1983, Cenozoic and Mesozoic structures of the eastern Basin and Range province, Utah, from COCORP seismic-reflection data: Geology, v. 11, p. 532–536.

Allmendinger, R. W., and 6 others, 1986, Phanerozoic tectonics of the Basin and Range-Colorado Plateau transition from COCORP data and geologic data, *in* Barazangi, M., and Brown, L., eds., Reflection seismology; The continental crust: American Geophysical Union Geodynamics Series, v. 14, p. 257–268.

Allmendinger, R. W., and 7 others, 1987, Overview of the COCORP 40° transect, western United States; The fabric of an orogenic belt: Geological Society of America Bulletin, v. 98, p. 308–319.

Bally, A. W., 1983, Seismic expression of structural styles: American Association of Petroleum Geologists Studies in Geology Series 15, v. 3, 375 p.

Bauer, R. L., 1980, Multiphase deformation in the Granite Falls–Montevideo area, Minnesota River Valley, *in* Morey, G. B., and Hanson, G. N., Selected studies of Archean gneisses and lower Proterozoic rocks, southern Canadian shield: Geological Society of America Special Paper 182, p. 1–17.

Birch, F., 1960, The velocity of compressional waves in rocks to 10 kilobars, 1: Journal of Geophysical Research, v. 65, p. 1083–1102.

Blundell, D. J., and Raynard, B., 1986, Modeling lower crust reflections observed on BIRPS profiles, *in* Barazangi, M., and Brown, L., Reflection seismology; A global perspective: American Geophysical Union Geodynamics Series, v. 13, p. 287–297.

Bortfeld, R. K., and others, 1985, First results and preliminary interpretation of deep-reflection seismic recordings along profile DEKORP 2-South: Journal of Geophysics, v. 57, p. 137–163.

Brewer, J. A., and Smythe, D. K., 1984, MOIST and the continuity of crustal reflector geometry along the Caledonian-Appalachian orogen: Journal of the Geological Society of London, v. 141, p. 105–120.

Brewer, J. A., Smithson, S. B., Oliver, J. E., Kaufman, S., and Brown, L. D., 1980, The Laramide orogeny; Evidence from COCORP deep crustal seismic reflection profiles in the Wind River Mountains, Wyoming: Tectonophysics, v. 62, p. 165–189.

Brewer, J. A., and 5 others, 1981, Proterozoic basin in the southern Midcontinent of the United States revealed by COCORP deep seismic reflection profiling: Geology, v. 9, p. 569–575.

Brewer, J. A., Allmendinger, R. W., Brown, L. D., Oliver, J. E., and Kaufman, S., 1982, COCORP profiling across the northern Rocky Mountain front in southern Wyoming; Part 1, Laramide structure: Geological Society of America Bulletin, v. 93, p. 1242–1252.

Brewer, J. A., Good, R., Oliver, J. E., Brown, L. D., and Kaufman, S., 1983, COCORP profiling across the southern Oklahoma aulacogen: Overthrusting of the Wichita Mountains and compression within the Anadarko Basin: Geology, v. 11, p. 109–114.

Brocher, T. M., 1981a, Geometry and physical properties of the Socorro, New Mexico, magma bodies: Journal of Geophysical Research, v. 86, p. 9420–9432.

——, 1981b, Shallow velocity structure of the Rio Grande Rift north of Socorro, New Mexico; A reinterpretation: Journal of Geophysical Research, v. 86, p. 4960–4970.

Brookins, D. G., and Meyer, H.O.A., 1974, Crustal and upper mantle stratigraphy beneath eastern Kansas: Geophysical Research Letters, v. 1, p. 269–273.

Brown, L. D., and 7 others, 1979, COCORP seismic reflection studies of the Rio Grande Rift, *in* Riecker, R. E., ed., Rio Grande Rift: Tectonics and Magmatism, p. 169–184.

Brown, L. D., Chapin, C. E., Sanford, A. R., Kaufman, S., and Oliver, J., 1980, Deep structure of the Rio Grande Rift from seismic reflection profiling: Journal of Geophysical Research, v. 85, p. 4773–4800.

Brown, L. D., and 7 others, 1983, Intracrustal complexity in the U.S. midcontinent: Preliminary results from COCORP surveys in NE Kansas: Geology, v. 11, p. 25.

Brown, R. L., and Read, P. B., 1983, Shuswap terrane of British Columbia; A Mesozoic "core complex": Geology, v. 11, p. 164–168.

Burchfiel, B. C., and Stewart, J. H., 1966, "Pull-apart" origin of the central segment of Death Valley, California: Geological Society of America Bulletin, v. 77, p. 439–440.

Burchfiel, B. C., and Davis, G. A., 1972, Structural framework and evolution of the southern part of the Cordilleran orogen, western United States: American Journal of Science, v. 272, p. 97–118.

Cape, C. D., McGeary, S., and Thompson, G. A., 1983, Cenozoic normal faulting and the shallow structure of the Rio Grande Rift near Socorro, New Mexico: Geological Society of America Bulletin, v. 94, p. 3–14.

Cheadle, M. J., and 6 others, 1986a, Geometries of deep crustal faults; Evidence from the COCORP Mojave survey, *in* Barazangi, M., and Brown, L., Reflection seismology; Continental crust: American Geophysical Union Geodynamics Series, v. 14, p. 305–312.

——, 1986b, The deep crustal structure of the Mojave Desert, California, from COCORP seismic reflection data: Tectonics, v. 5, p. 293–320.

Christensen, N. I., 1982, Seismic velocities, *in* Carmichael, R. S., ed., Handbook of physical properties of rocks, v. 2: Boca Raton, Florida, CRC Press, p. 2–228.

Cloos, E., 1968, Experimental analysis of Gulf Coast fracture patterns: American Association of Petroleum Geologists Bulletin, v. 52, p. 420–444.

Coney, P. J., 1980, Cordilleran metamorphic core complexes; An overview, *in* Crittenden, M. D., Jr., Coney, P. J., and Davis, G. H., eds., Cordilleran metamorphic core complexes: Geological Society of America Memoir 153, p. 7–34.

Crittenden, M. D., Jr., Coney, P. J., and Davis, G. H., eds., 1980, Cordilleran metamorphic core complexes: Geological Society of America Memoir 153, 490 p.

Davis, G. A., Anderson, J. L., Frost, E. G., and Shackelford, T. J., 1980, Mylonitization and detachment faulting in the Whipple-Bucksin-Rawhide Mountains terrane, southeastern California and western Arizona, *in* Crittenden, M. D., Jr., Coney, P. J., and Davis, G. H., eds., Cordilleran metamorphic core complexes: Geological Society of America Memoir 153, p. 79–131.

De Voogd, B., Brown, L. D., and Merey, C., 1986, Nature of the eastern boundary of the Rio Grande Rift from COCORP surveys in the Albuquerque Basin, New Mexico: Journal of Geophysical Research, v. 91, p. 6305–6320.

Eaton, G. P., Wahl, R. R., Porstka, H. J., Mabey, D. R., and Kleinkopf, M. D., 1978, Regional gravity and tectonic patterns; Their relation to late Cenozoic epeirogeny and lateral spreading in the western Cordillera, *in* Smith, R. B., and Eaton, G. P., eds., Cenozoic tectonics and regional geophysics of the western Cordillera: Geological Society of America Memoir 152, p. 51–91.

Effimoff, I., and Pinezich, A. R., 1981, Tertiary structural development of selected valleys based on seismic data, Basin and Range province, northeastern Nevada: Philosophical Transactions of the Royal Society of London, series A, v. A300, p. 435–442.

Fatti, J. L., 1972, The influence of dolerite sheets on reflection seismic profiling in the Karroo basin: Transactions of the Geological Society of South Africa, v. 75, part 2, p. 71–76.

Feng, R., and McEvilly, T. V., 1983, Interpretation of seismic reflection profiling data for the structure of the San Andreas fault zone: Bulletin of the Seismological Society of America, v. 73, p. 1701–1720.

Fielding, E., Barazangi, M., Brown, L., Oliver, J., and Kaufman, S., 1984, COCORP seismic profiles near Coalinga, California; Subsurface structure of the western Great Valley: Geology, v. 12, p. 268–274.

Fountain, D. M., and Salisbury, M. H., 1981, Exposed cross-sections through the continental crust; Implications for crustal structure, petrology, and evolution: Earth and Planetary Science Letters, v. 56, p. 263–277.

Fountain, D. M., Hurich, C. A., and Smithson, S. B., 1984, Seismic reflectivity of mylonite zones in the crust: Geology, v. 12, p. 195–198.

Fountain, D. M., McDonough, D. T., and Gorham, J. M., 1987, Seismic reflection models of continental crust based on metamorphic terrains: Geophysical Journal of the Royal Astronomical Society, v. 89, p. 61–67.

Fuchs, K., 1969, On the properties of deep crustal reflections: Journal of Geophysics, v. 35, p. 133–149.

Fyfe, W. S., 1974, Archean tectonics: Nature, v. 249, p. 338–340.

Gans, P. B., Miller, E. L., McCarthy, J., and Ouldcott, M. L., 1985, Tertiary extensional faulting and evolving ductile-brittle transition zones in the northern Snake Range and vicinity: Geology, v. 13, p. 189–194.

Gibbs, A. K., and 5 others, 1984, Seismic reflection study of the Precambrian crust of central Minnesota: Geological Society of America Bulletin, v. 95, p. 280–294.

Goldich, S. S., and Hedge, C. E., 1974, 3,800 myr granitic gneiss in southwestern Minnesota: Nature, v. 252, p. 467–468.

Goodwin, E. B., and Thompson, G. A., 1988, The seismically reflective crust beneath highly extended terranes; Evidence for its origin in extension: Geological Society of America Bulletin, v. 100, p. 1616–1626.

Greenhalgh, S. A., 1981, Seismic investigations of crustal structure in east-central Minnesota: Physics of Earth and Planetary Interiors, v. 25, p. 372–389.

Hale, L. D., and Thompson, G. A., 1982, The seismic reflection character of the Mohorovičić discontinuity: Journal of Geophysical Research, v. 87, p. 4625–4635.

Hamilton, W., 1969, Mesozoic California and the underflow of the Pacific mantle: Geological Society of America Bulletin, v. 80, p. 2409–2430.

Hastings, D. D., 1979, Results of exploratory drilling, northern Fallon basin, western Nevada, *in* Newman, G. W., and Goode, H. D., eds., Basin and Range symposium: Rocky Mountain Association of Geologists and Utah Geological Association, p. 515–522.

Hauge, T. A., and 10 others, 1987, Crustal structure of western Nevada from COCORP deep seismic-reflection data: Geological Society of America Bulletin, v. 98, p. 320–329.

Hauser, E., and 9 others, 1987, Crustal structure of eastern Nevada from COCORP deep seismic-reflection data: Geological Society of America Bulletin, v. 98, p. 833–844.

Henyey, T. E., Okaya, D. A., Frost, E. G., and McEvilly, T. V., 1987, CALCRUST (1985) seismic reflection survey, Whipple Mountains detachment terrane, California; An overview: Geophysical Journal of the Royal Astronomical Society, v. 89, p. 111–119.

Hodge, D. S., Owen, L. B., and Smithson, S. B., 1973, Gravity interpretation of Laramie anorthosite complex, Wyoming: Geological Society of America Bulletin, v. 84, p. 1451–1464.

Hurich, C. A., an Smithson, S. B., 1982, Gravity interpretation of the southern Wind River Mountains, Wyoming: Geophysics, v. 47, no. 11, p. 1550–1561.

——, 1987, Compositional variation and the origin of deep crustal reflections: Earth and Planetary Science Letters, v. 85, p. 416–426.

Hurich, C. A., Smithson, S. B., Fountain, D. M., and Humphreys, M. C., 1985, Seismic evidence of mylonite reflectivity and deep structure in the Kettle Dome metamorphic core complex, Washington: Geology, v. 13, p. 577–580.

Iltis, S. T., 1983, Processing and interpretation of seismic reflection data from the Precambrian of the central Laramie Range, Albany County, Wyoming [M.S. thesis]: Laramie, University of Wyoming, 94 p.

Jackson, W. H., and Pakiser, L. C., 1965, Seismic study of crustal structure in the southern Rocky Mountains: U.S. Geological Survey Professional Paper 525-D, p. 85–92.

Johnson, R. A., and Smithson, S. B., 1985, Thrust faulting in the Laramie Mountains, Wyoming, from reanalysis of COCORP data: Geology, v. 13, p. 534–537.

Johnson, R. A., Karlstrom, K. E., Smithson, S. B., and Houston, R. S., 1984, Gravity profiles across the Cheyenne belt, a Precambrian crustal suture in southeastern Wyoming: Journal of Geodynamics, v. 1, p. 445–471.

Jones, T. D., 1985, Nature of seismic reflections from the crystalline basement: COCORP Wind River line, Wyoming: Journal of Geophysical Research, v. 90, p. 6783–6791.

Jones, T. D., and Nur, A., 1982, Seismic velocity and anisotropy in mylonites and the reflectivity of deep crustal faults: Geology, v. 10, p. 260–263.

——, 1984, Nature of seismic reflections from deep crust-fault zones: Journal of Geophysical Research, v. 89, p. 3153–3171.

Jurdy, D. M., and Brocher, T. M., 1980, Shallow velocity structure of the Rio Grande Rift near Socorro, New Mexico: Geology, v. 8, p. 185–189.

Karlstrom, K. E., and Houston, R. S., 1984, The Cheyenne belt; Analysis of a Proterozoic suture in southern Wyoming: Precambrian Research, v. 25, p. 415–446.

Kay, R. W., and Kay, S. M., 1981, The nature of the lower continental crust; Inferences from geophysics, surface geology, and crustal xenoliths: Reviews of Geophysics and Space Physics, v. 19, p. 271–297.

Kern, H., and Fakhimi, M., 1975, Effect of fabric anisotropy on compressional wave propagation in various metamorphic rocks for the range 20-700°C at 2 kbars: Tectonophysics, v. 28, p. 227–244.

Klemperer, S. L., Hauge, T. A., Hauser, E. C., Oliver, J. E., and Potter, C. J., 1986, The Moho in the northern Basin and Range province, Nevada, along the COCORP 40°N seismic-reflection transect: Geological Society of America Bulletin, v. 97, p. 603–618.

Kubichek, R. F., Humphreys, M. C., Johnson, R. A., and Smithson, S. B., 1984, Long-range recording of VIBROSEIS data; Simulation and experiment: Geophysical Research Letters, v. 11, p. 809–812.

Lidiak, E. G., 1974, Magnetic characteristics of some Precambrian basement rocks: Zeitschrift fur Geophysik, v. 40, p. 549–564.

Liu, C., Zhu, T., Farmer, H., and Brown, L., 1986, An expanding spread experiment during CORCORP's field operation in Utah, *in* Barazangi, M., and Brown, L., eds., Reflection seismology; The continental crust: American Geophysical Union Geodynamics Series, v. 13, p. 234–246.

Lynn, H. B., Hale, L. D., and Thompson, G. A., 1981, Seismic reflections from the basal contacts of batholiths: Journal of Geophysical Research, v. 86, p. 10633–10638.

Lynn, H. B., Quam, S., and Thompson, G. A., 1983, Depth migration and interpretation of the COCORP Wind River, Wyoming, seismic reflection data: Geology, v. 11, p. 462–469.

MacLeod, M. K., 1981, The Pacific Creek anticline; Buckling above a basement thrust fault, *in* Boyd, D. W., ed., Rocky Mountain Foreland basement tectonics: Laramie, University of Wyoming Contributions to Geology, v. 19, p. 143–161.

McBride, J. H., and Brown, L. D., 1986, Reanalysis of the COCORP deep seismic reflection profile across the San Andreas Fault, Parkfield, California: Bulletin of the Seismological Society of America, v. 76, p. 1668–1686.

McCarthy, J., 1986, Reflection profiles from the Snake Range metamorphic core complex; A window into the mid-crust: American Geophysical Union Geodynamic Series, v. 14, p. 281–292.

McDonald, R. E., 1976, Tertiary tectonics and sedimentary rocks along the transition, Basin and Range province to plateau and thrust belt province, Utah: Rocky Mountain Association of Geologists Symposium, p. 281–317.

Meissner, R., 1967, Zum aufbau der Erdkruste, Ergebinisse der Weitwinkelmessungen im bayerischen Molassenbecken: Gerlands Beitrage zur Geophysik, Tiel 2, v. 76, p. 295–314.

——, 1973, The "Moho" as a transition zone: Geophysical Surveys, v. 1, p. 195–216.

Mitchell, B. J., and Landisman, M., 1970, Interpretation of a crustal section across Oklahoma: Geological Society of America Bulletin, v. 81, p. 2647–2656.

Monger, J.W.H., Price, R. A., and Templeman-Kluit, D. J., 1982, Tectonic accretion and the origin of two major metamorphic and plutonic welts in the Canadian Cordillera: Geology, v. 10, p. 70–75.

Okaya, D. A., 1986, Seismic profiling of the lower crust; Dixie Valley, Nevada, *in* Barazangi, M., and Brown, L., eds., Reflection seismology; The continental crust: American Geophysical Union Geodynamics Series, v. 14, p. 269–281.

Okaya, D. A., and Thompson, G. A., 1985, Geometry of Cenozoic extensional faulting; Dixie Valley, Nevada: Tectonics, v. 4, p. 107–125.

Oliver, J. E., Dobrin, M., Kaufman, S., Meyer, R., and Phinney, R., 1976, Continuous seismic reflection profiling of the deep basement, Hardeman County Texas: Geological Society of America Bulletin, v. 87, p. 1537–1546.

Olsen, K. H., Cash, D. J., and Stewart, J. H., 1982, Mapping the northern and eastern extent of the Socorro midcrustal magma body by wide-angle seismic reflections: 33rd Field Conference, Albuquerque County, New Mexico Geological Society Guidebook 2, p. 179–185.

O'Reilly, S. Y., and Griffin, W. L., 1985, A xenolith-derived geotherm for southeastern Australia and its geophysical implications: Tectonophysics, v. 111, p. 41–63.

Pakiser, L. C., 1985, Seismic exploration of the crust and upper mantle of the Basin and Range province, *in* Drake, E. T., and Jordan, W. M., eds., Geologists and ideas; A history of North American geology: Geological Society of America, Centennial Special Volume 1, p. 453–469.

Potter, C. J., and 7 others, 1986, COCORP deep seismic reflecton traverse of the interior of the North American Cordillera, Washington and Idaho; Implications for orogenic evolution: Tectonics, v. 5, p. 1007–1026.

Potter, C. J., and 9 others, 1987, Crustal structure of north-central Nevada; Results from COCORP deep seismic profiling: Geological Society of America Bulletin, v. 98, p. 330–337.

Prodehl, C., 1979, Crustal structure of the western United States: U.S. Geological Survey Professional Paper 1037, 74 p.

Prucha, J. J., Graham, J. A., and Nickelson, R. P., 1965, Basement controlled deformation in Wyoming province of Rocky Mountain foreland: American Association of Petroleum Geologists Bulletin, v. 49, p. 966–992.

Rehrig, W. A., and Reynolds, S. J., 1980, Geologic and geochronologic reconnaissance of a northwest-trending zone of metamorphic core complexes in southern and western Arizona, *in* Crittenden, M. D., Jr., Coney, P. J., and Davis, G. H., Cordilleran metamorphic core complexes: Geological Society of America Memoir 153, p. 159–175.

Reif, D. M., and Robinson, J. P., 1981, Geophysical, geochemical, and petrographic data and regional correlation from the Arizona state A-1 well, Pinal County, Arizona: Arizona Geological Society Digest, v. 13, p. 99–109.

Rinehart, E. J., Sanford, A. R., and Ward, R. M., 1979, Geographic extent and shape of an extensive magma body at midcrustal depths in the Rio Grande rift near Socorro, New Mexico, *in* Riecker, R. E., ed., Rio Grande rift; Tectonics and magmatism: American Geophysical Union, p. 237–252.

Rundle, J. B., and others, 1985, Seismic imaging in Long Valley, California, by surface and borehole techniques; An investigation of active tectonics: EOS Transactions of the American Geophysical Union, v. 66, p. 194–200.

Sacrison, W. R., 1978, Seismic interpretation of basement block faults and associated deformation, *in* Matthews, V., III, ed., Laramide folding associated with basement block faulting in the western United States: Geological Society of America Memoir 151, p. 39–49.

Sandiford, M., and Powell, R., 1986, Deep crustal metamorphism during continental extension; Modern and ancient examples: Earth and Planetary Science Letters, v. 79, p. 151–158.

Sanford, A. R., Alptekin, O. S., and Toppozada, T. R., 1973, Use of reflection phases on micro-earthquake seismograms to map an unusual discontinuity beneath the Rio Grande rift: Bulletin of the Seismological Society of America, v. 63, p. 2021–2034.

Sanford, A. R., and 6 others, 1977, Geophysical evidence for a magma body in the crust in the vicinity of Socorro, New Mexico, *in* Heacock, J. G., ed., The Earth's crust; Its nature and physical properties: American Geophysical Union Geophysical Monograph Series 20, p. 385–403.

Schilt, F. S., Kaufman, S., and Long, G. H., 1981, A three-dimensional study of seismic diffraction patterns from deep basement sources: Geophysics, v. 46, p. 1673–1683.

Serpa, L., and 6 others, 1984, Structure of the southern Keweenawan rift from COCORP surveys across the Midcontinent geophysical anomaly in northeastern Kansas: Tectonics, v. 3, p. 367–384.

Serpa, L., and 6 others, 1988, Structure of the central Death Valley pull-apart basin and vicinity from COCORP profiles in the southern Great Basin: Geological Society of America Bulletin, v. 100, p. 1437–1450.

Sharry, J., and 5 others, 1986, Enhanced imaging of the COCORP seismic line, Wind River Mountains, *in* Barazangi, M., and Brown, L., eds., Reflection seismology; A global perspective: American Geophysical Union Geodynamics Series, v. 13, p. 223–236.

Sheriff, R. E., and Geldart, L. P., 1982, Exploration seismology: Cambridge, Cambridge University Press, v. 1, 253 p.

Sims, P. K., Card, K. D., Morry, G. B., and Peterman, Z. E., 1980, The Great Lakes tectonic zone; A major crustal feature in North America: Geological Society of America Bulletin, v. 91, p. 690–698.

Smith, R. B., 1978, Seismicity, crustal structure, and interplate tectonics of the interior of the western Cordillera, *in* Smith, R. B., and Eaton, G. P., eds., Cenozoic tectonics and regional geophysics of the western Cordillera: Geological Society of America Memoir 152, p. 111–144.

Smith, R. B., and Bruhn, R. B., 1984, Intraplate extensional tectonics of the eastern Basin-Range; Inferences on structural style from seismic reflection data, regional tectonics, and thermal mechanical models of brittle-ductile deformation: Journal of Geophysical Research, v. 89, p. 5735–5762.

Smithson, S. B., 1978, Modeling continental crust; Structural and chemical constraints: Geophysical Research Letters, v. 5, p. 749–752.

——, 1979, Aspects of continental structure and growth; Targets for scientific deep drilling: University of Wyoming Contributions to Geology, v. 17, p. 65–75.

——, 1986, A physical model of the lower crust from North America based on seismic reflection data, *in* Dawson, J. B., Carswell, D. A., Hall, J., and Wedepohl, K. H., eds., The nature of the lower continental crust: Geological Society of London Special Publication 24, p. 23–34.

——, 1989, Contrasting types of lower crust, in Mereu, R. F., Mueller, S., and Fountain, D. M., eds., Processes of Earth's lower crust: IUGG Geophysics Monograph 51, v. 6, p. 53–63.

Smithson, S. B., and Brown, S. K., 1977, A model for lower continental crust: Earth and Planetary Science Letters, v. 35, p. 134–144.

Smithson, S. B., and Ebens, R. J., 1971, Interpretation of data from a 3.05-kilometer borehole in Precambrian crystalline rocks, Wind River Mountains, Wyoming: Journal of Geophysical Research, v. 76, p. 7079–7087.

Smithson, S. B., Shive, P. N., and Brown, S. K., 1977a, Seismic velocity, reflections, and structure of the crystalline crust, *in* Heacock, J. G., ed., The Earth's crust; Its nature and physical properties: American Geophysical Union Geophysical Monograph, v. 20, p. 254–270.

——, 1977b, Seismic reflections from Precambrian crust: Earth and Planetary Science Letters, v. 35, p. 134–144.

Smithson, S. B., Brewer, J. A., Kaufman, S., Oliver, J. E., and Hurich, C. A., 1978, Nature of the Wind River thrust, Wyoming, from COCORP deep reflection data and from gravity data: Geology, v. 6, p. 648–652.

——, 1979, Structure of the Laramide Wind River uplift, Wyoming, from COCORP deep reflection data and gravity data: Journal of Geophysical Research, v. 84, p. 5955–5972.

Smithson, S. B., Brewer, J. A., Kaufman, S., Oliver, J. E., and Zawislak, R. L., 1980, Complex Archean lower crustal structure revealed by COCORP crustal refelection profiling in the Wind River Range, Wyoming: Earth and Planetary Science Letters, v. 46, p. 295–305.

Smithson, S. B., Pierson, W. R., Wilson, S. L., and Johnson, R. A., 1985, Seismic reflection results from Precambrian crust, *in* Tobi, A. C., and Touret, J.L.R., eds., The deep Proterozoic crust in the North Atlantic provinces: NATO Advanced Study Institutes Series, Series C; Mathematical and Physical Sciences, v. 158, p. 21–37.

Smithson, S. B., Johnson, R. A., and Hurich, C. A., 1986, Crustal reflections and crustal structure, *in* Barazangi, M., and Brown, L., eds., Reflection seismology; The continental crust: American Geophysical Union Geodynamics Series, v. 14, p. 21–33.

Smithson, S. B., Johnson, R. A., Hurich, C. A., Valasek, P. A., and Branch, C., 1987, Deep crustal structure and genesis from contrasting reflection patterns; An integrated approach: Geophysical Journal of the Royal Astronomical Society, v. 89, p. 67–73.

Smythe, D. K., and 6 others, 1982, Deep structures of the Scottish Caledonides revealed by the MOIST reflection profile: Nature, v. 299, p. 338–340.

Snoke, A. W., and Lush, A. P., 1984, Polyphase Mesozoic-Cenozoic deformational history of the northern Ruby Mountains–East Humboldt Range, Nevada, *in* Lintz, Joseph, Jr., ed., Western geological excursions: Geological Society of America Annual Meeting, Mackay School of Mines, Reno, Nevada, v. 4, p. 232–260.

Stewart, J. H., 1971, Basin and Range structure; A system of horsts and grabens produced by deep-seated extension: Geological Society of America Bulletin, v. 57, p. 1317–1345.

——, 1980, Regional tilt patterns of late Cenozoic basin-range fault blocks, western United States: Geological Society of America Bulletin, v. 91, part 1, p. 460–464.

——, 1983, Extensional tectonics in the Death Valley area, California; Transport of the Panamint Range structural block 80 km northwestward: Geology, v. 11, p. 153–157.

Stokes, W., 1976, What is the Wasatch Line?: Rocky Mountain Geological Association Symposium, p. 11–25.

Thompson, G. A., 1959, Gravity measurements between Hazen and Austin, Nevada; A study of basin-range structure: Journal of Geophysical Research, v. 64, p. 217–229.

Thompson, G. A., and Hill, J. L., 1986, The deep crust in convergent and divergent terranes; Laramide uplifts and Basin-Range rifts, *in* Barazangi, M., and Brown, L., eds., Reflection seismology; The continental crust: American Geophysical Union Geodynamics Series, v. 14, p. 243–257.

Valasek, P. A., Hawman, R. B., Johnson, R. A., and Smithson, S. B., 1987, Nature of the lower crust and Moho in eastern Nevada from "wide-angle" reflection measurements: Geophysical Research Letters, v. 145, p. 1111–1114.

Valasek, P. A., Snoke, A. W., Hurich, C. A., and Smithson, S. B., 1989, Nature and origin of seismic reflection fabric, Ruby–East Humboldt metamorphic core complex, Nevada: Tectonics, v. 8, p. 391–415.

Von Tish, D. B., Allmendinger, R. W., and Sharp, J. W., 1985, History of Cenozoic extension in central Sevier Desert, west-central Utah, from COCORP seismic reflection data: American Association of Petroleum Geologists Bulletin, v. 69, p. 1077–1087.

Wallace, M., 1980, Deep basement reflections in Wind River Line 1: Geophysical Research Letters, v. 7, p. 729–732.

Wentworth, C. M., Zoback, M. D., Griscom, A., Jachens, R. C., and Mooney, W. D., 1987, A transect across the Mesozoic accretionary margin of central California: Geophysical Journal of the Royal Astronomical Society, v. 89, p. 105–111.

Wernicke, B., 1985, Uniform-sense normal simple shear of the continental lithosphere: Canadian Journal of the Earth Sciences, v. 22, p. 108–125.

Widess, M. B., 1973, How thin is a thin bed?: Geophysics, v. 38, p. 1176–1180.

Widess, M. B., and Taylor, G. L., 1959, Seismic reflections from layering within the Precambrian basement complex, Oklahoma: Geophysics, v. 24, p. 417–425.

Winkler, H. G., 1972, Petrogenesis of metamorphic rocks: New York, Springer-Verlag, 334 p.

Wong, Y. K., Smithson, S. B., and Zawislak, R. L., 1982, The role of seismic modeling in deep crustal reflection interpretation, Part 1: University of Wyoming Contributions to Geology, v. 20, p. 91–109.

Yeend, W., 1976, Reconnaissance geologic map of the Picacho Mountains, Arizona: U.S. Geological Survey Miscellaneous Field Studies Map MF-778, scale 1:62,500.

Zawislak, R. L., and Smithson, S. B., 1981, Problems and interpretation of COCORP deep seismic reflection data, Wind River Range, Wyoming: Geophysics, v. 46, p. 1684–1701.

Zoback, M. D., and Wentworth, C., 1986, Crustal studies in central California using an 800-channel seismic reflection recording system: American Geophysical Union Geodynamics Series, v. 14, p. 183–196.

Zoback, M. L., and Anderson, R. E., 1983, Style of Basin-Range faulting as inferred from seismic reflection data in the Great Basin, Nevada and Utah: Geothermal Research Council Special Report 13, p. 363–381.

Manuscript Accepted by the Society October 31, 1988

Printed in U.S.A.

Geological Society of America
Memoir 172
1989

Chapter 27

Reflection seismic studies of crustal structure in the eastern United States

Robert A. Phinney and Kabir Roy-Chowdhury
Department of Geological and Geophysical Sciences, Princeton University, Princeton, New Jersey 08540; ARCO Oil & Gas Company, Plano, Texas 75075

ABSTRACT

Since 1978, several seismic reflection transects have been run across the Appalachian and Ouachita orogens; several special topical reflection lines in the craton have shown the complexity and variability of the Precambrian basement. New methods of signal extraction and display make it possible to reprocess these stacked sections to provide standardized crustal images. The limitations to image fidelity induced by the presence of noise and unresolvable structure variations, as well as the limitations inherent in the physics of wave propagation, are discussed in a tutorial section to provide tools for the understanding of reflection profiles. Of particular importance is the need for the interpreter to work interactively with the final processing step in which processed digital data are transformed into a graphical image.

Reprocessed sections for a set of transects across the Appalachians provide a framework for a rediscussion of the structure and evolution of the orogen. Features that appear on several lines are (1) an east-dipping suture that separates the exotic terranes of the central and eastern regions from the post-Taconian North American continent; (2) west of the suture, thin-skinned, west-directed thrusting of elements of the post-Taconian margin onto the edge of the craton; (3) east of the suture, a compressed central package extending to near Moho; and (4) the east-dipping boundary produced by Alleghanian collision of an Avalonian terrane. Of particular interest is the indication of a partially rifted fragment of Grenville basement, which forms an antiformal buttress just west of the principal suture. On the lines best showing the eastern Avalonian terrane is an indication that the collision resulted in midcrustal delamination, leading to west-directed overthrusting and underplating of the crust by the colliding terrane. The relationship between Moho reflection strength and Mesozoic extensional faults leads to the inference that the extension extends regionally throughout the interior of the orogen and out to the present continental margin, and that the extension was responsible for the enhanced reflectivity of Moho and lower crust.

INTRODUCTION

Reflection seismology is now established as a key tool for the study of continental crustal structure. Starting in an organized way in 1974–1975 with the Consortium for Continental Reflection Profiling (COCORP) program, reflection profiling for scientific purposes has produced more than 12,000 km of line in the conterminous United States, including the shallow-water portions of the continental margins. Profiling in the eastern and central U.S. has been carried out largely by contractor crews under the direction of university investigators and government agencies (Table 1), and has met with a degree of success not at all anticipated a dozen years ago. Although the total length of section produced does not appear impressive when displayed at a small map scale (Fig. 1), it is quite appreciable when measured in terms of a typical 35-km crustal thickness.

Phinney, R. A., and Roy-Chowdhury, K., 1989, Reflection seismic studies of crustal structure in the eastern United States, *in* Pakiser, L. C., and Mooney, W. D., Geophysical framework of the continental United States: Boulder, Colorado, Geological Society of America Memoir 172.

TABLE 1. PRIMARY LITERATURE REFERENCES FOR CRUSTAL SEISMIC REFLECTION PROGRAMS; EASTERN AND CENTRAL U.S.*

1. Quebec–Maine–Gulf of Maine corridor
U.S. Geological Survey, Geological Survey of Canada
(a) Quebec SOQUIP line [QMT]
St. Julien and others, 1983
LaRoche, 1983†
(b) Maine reflection profiling: [QMT], 1984–85
Spencer and others, 1987
Stewart and others, 1986
(c) Maine refraction program: [MRP], 1985
Luetgert, 1985
Luetgert and others, 1987
Hennet, 1989
(d) Gulf of Maine marine reflection profiling: [GMM], 1985
Hutchinson and others, 1987

2. Adirondack–New England Transect
COCORP§: 1980–81
Brown and others, 1983a: [NEC]
Ando and others, 1984: New Hampshire [NEC]
Klemperer and others, 1985: Adirondack [ADK]
Klemperer and others, 1983†

3. Long Island Platform marine multichannel survey [LIP]
U.S. Geological Survey, 1979
Hutchinson and others, 1985, 1986a, b
Phinney, 1986
Grow and others, 1983†

4. Newark Basin rift tectonics program [NBT]
U.S. Geological Survey, 1984
Ratcliffe and others, 1986

5. Great Valley wide angle study [GVR]
Princeton University, 1982–84
Hawman and Phinney, 1985
Hawman, 1988

6. James River corridor, Virginia [JRC]
U.S. Geological Survey, Virginia Tech: 1980–81
Harris and others, 1982
Pratt and others, 1987

7. Tennessee–North Carolina (Grandfather Mt.) profile [GMP]
U.S. Geological Survey, 1979
Harris and Bayer, 1979
Harris and others, 1981

8. Southern Appalachian region: reflection profiling
(a) COCORP transect, 1978 [GAA]
Cook and others, 1979
Cook and others, 1983†
(b) COCORP transect, 1979 [GAB]
Cook and others, 1981
(c) U.S. Geological Survey profiles, 1981
Behrendt, 1986
(d) COCORP Charleston program, 1978 [CHP]
Schilt and others, 1983
(e) U.S. Geological Survey Charleston program [CHP]
Behrendt and others, 1981
(f) COCORP Georgia–Florida transect 1983-85
Nelson and others, 1985a, b [GFT]
(g) ADCOH** site study, crystalline overthrust belt, 1985 [ADC]
Coruh and others, 1985, 1987
Phinney and Roy-Chowdhury, 1985
Roy-Chowdhury and Phinney, 1987

9. Mississippi embayment [MIS]
U.S. Geological Survey, 1978–79
Zoback and others, 1980
Mooney and others, 1983

10. Ouachita transect
(a) COCORP reflection profiling, 1981 [OUC]
Nelson and others, 1982
Lillie and others, 1983a
Lillie and others, 1983b†
(b) PASSCAL‡ imaging experiment, 1986 [OUP]
Braile and others, 1986

11. Hardeman Basin–Wichita Mts.–Anadarko Basin
Pan American Petroleum Corporation
Widess and Taylor, 1959
(a) COCORP, 1975 [HRC]
Oliver and others, 1976
(b) COCORP, 1979–80 [WMT]
Brewer and others, 1981
Brewer and others, 1983
Good and others, 1983†

12. Kansas–midcontinent geophysical anomaly [KAN]
COCORP, 1981
Brown and others, 1983b
Setzer and others, 1983†

13. Minnesota Archean profiling [MIN]
COCORP, 1979
Gibbs and others, 1984

14. Michigan Basin [MBA]
COCORP, 1978
Brown and others, 1982
Zhu and Brown, 1986

*For each area, the information includes: Research Institution, with year(s) data acquired; principal references describing results; bracketed symbols [...] are key to Figure 1.
†Reflection data appearing in AAPG seismic atlas are recommended for detailed study.
§Consortium for Continental Reflection Profiling, operated under NSF support by Cornell University.
‡Program for Array Seismic Studies of the Continental Lithosphere, a program element of the Incorporated Research Institutions for Seismology, supported by NSF under the Continental Lithosphere Program.

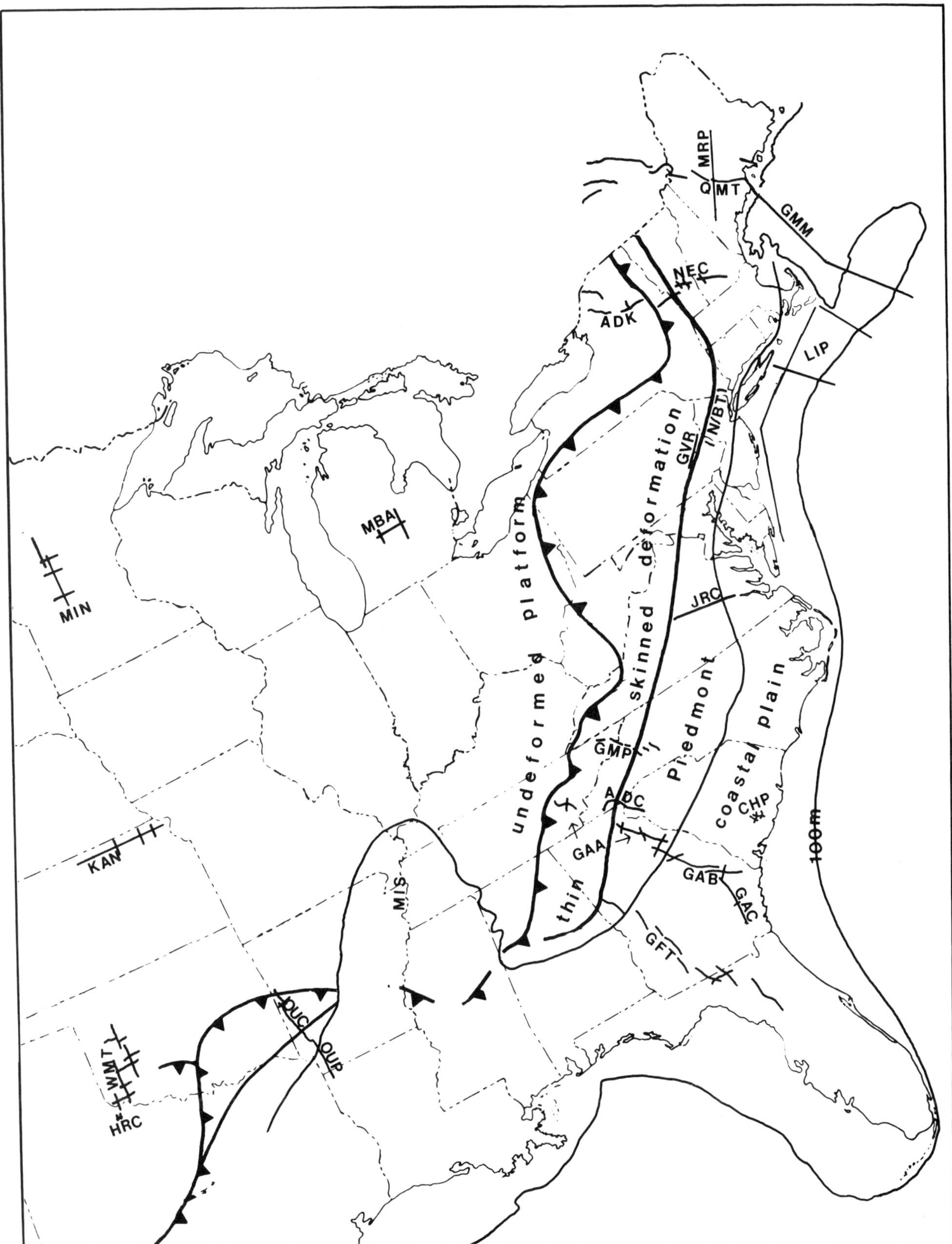

Figure 1. Eastern United States (based on fold and thrust belts of the U.S.; USGS, scale 1:2,500,000, 1984) with reflection survey lines cited in the text; key to the lines in Table 1.

A companion article in the present volume (Taylor, this volume) provides an integrated view of crustal structure in the eastern U.S., based on the full spectrum of geophysical data and the correlative geological literature. Crustal reflection profiling, however, is still sufficiently novel as a technique, and its interpretation still evolving, that it is appropriate to take a closer look at the reflection data separately. In this chapter, we discuss a number of issues relating to interpretation of the results, and do so with reference to a set of reflection lines that cross the Appalachian-Ouachita orogen. The first part of the chapter is in the form of an essay on the subject of the limitations and interpretation of crustal seismic reflection sections to enable a better appreciation of the discussion that follows. The aim is to stimulate the reader to work with high-quality reproductions of the data in order to become knowledgeable by practice. A second aim is to emphasize the importance of the final processing steps, in which a digital representation of the section is converted to a graphical display of the subsurface image.

BASIC REFLECTION SEISMIC PROCESSING

We consider the conventional geometry for reflection profiling: an array of geophone groups (generally cabled together) is used to record signals from a shotpoint or vibrator point within or slightly off-end the array. The seismic traces recorded from one such source-point constitute a source gather (e.g., Fig. 2), and in two-dimensional display show the different seismic waves arriving at the array. The arrivals in a source gather may, but need not, include reflections, refractions, ground-roll, shear waves, air waves, converted waves, multiples and the like. The collection of all source gathers forms a three-dimensional data set (source, receiver group, time), which is a sampled representation of all the seismic wavefields produced in the experiment.

The goal of seismic data processing is to convert this raw 3-D data manifold into a 2-D cross-sectional representation of subsurface structure. Figure 3 gives a schematic overview of, and Table 2 lists in more detail, the steps involved in this transformation, some mandatory and some optional. These are not explained further here, as they form the subject matter of several recent texts (Sheriff and Geldart, 1982, 1983; Waters, 1987; Claerbout, 1985a, b). Of all the steps listed in Table 2, migration and stacking are the most important. Migration converts the observed reflection waveforms (functions of distance and time) to an image of the subsurface (a function of distance and depth), i.e., places the reflectors in their spatially correct locations. Stacking improves the signal-to-noise ratio (SNR) by multiply illuminating the subsurface reflecting regions from different angles and summing the observations after correcting for the different path lengths. The conventional processing stream (Table 2) does the stacking before migration. Migration of the unstacked data, followed by stacking, is a more correct procedure, but involves two orders of magnitude more computation and data handling, and is thus used only in very special cases.

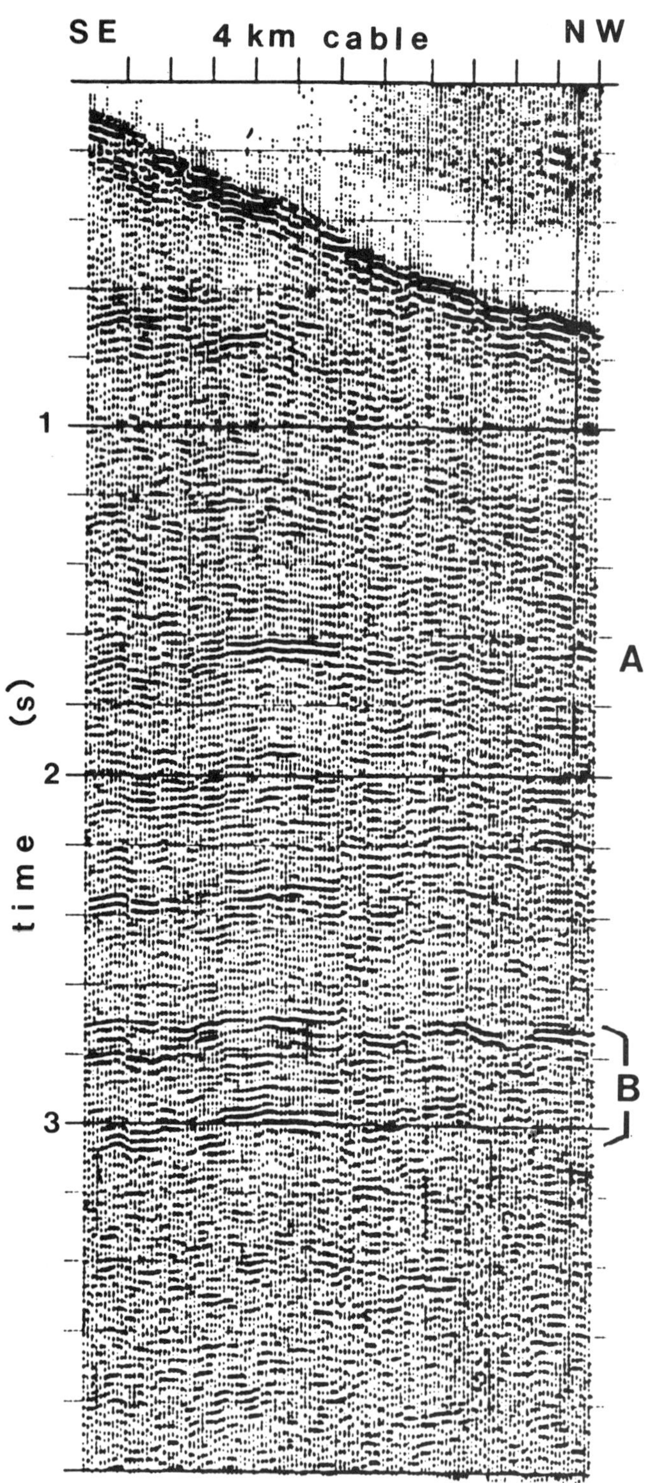

Figure 2. A source gather from the ADCOH high-resolution study (Phinney and Roy-Chowdhury, 1987) on the southern Appalachian allochthon 5 km southeast of the Brevard Fault Zone. Clear reflections are seen from the basal sequence of the parautochthonous platform deposits (B: ~3 sec) and from the Brevard Fault Zone (A: ~1.6 sec). Pronounced static shifts of more than a cycle produce the jittery looking arrivals of the otherwise coherent signals for all times.

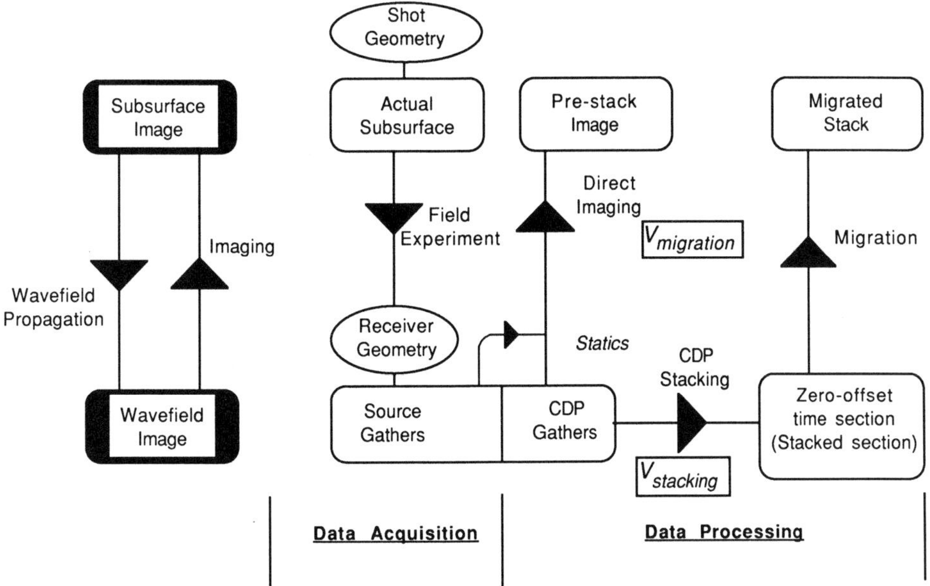

Figure 3. Diagram showing the relationships between different seismic processing methods and images of the subsurface they produce.

PROBLEMS PECULIAR TO CRUSTAL REFLECTION SEISMOLOGY

Reflection profiling represents the first appearance in continental geophysics of a technique that can produce an image of the subsurface structure (seismic section). As such, the latter has been widely accepted as a finished representation for use in geological interpretation. At the same time, the acquisition of field data and the processing of data to produce reflection images have become the concern of a rather specialized cadre of geophysicists. It is not uncommon for the interpreter to be unfamiliar with the various steps and the effect of choosing the parameters on the final result (section). Thus, while reflection profiles are widely available in the geological literature, it is not easy to deduce from the literature what constitutes signal, what constitutes noise, what the possible distortions may be, and in general, what information is actually inherent in the data. Unlike sedimentary basins, the crustal column, in general, may not possess strong layering or contrasts of velocity and density. Consequently, energy reflected from intracrustal interfaces is likely to be discontinuous and weak. The seismic cable used in crustal studies often has the same length as that used for shallower hydrocarbon prospecting. The subsurface illumination is, consequently, mostly near-vertical (especially for the lower crust). The small normal moveouts that result make it difficult to obtain reliable estimates of crustal velocity. In the following four subsections, we discuss some of these issues that arise while dealing with seismic sections of the Earth's crust.

Velocity

Unfortunately, the unexpected degree of success in imaging crustal structure by reflection methods has not been accompanied by equally good determinations of crustal velocities. Cook and others (1979) used stacking velocity variations along the line of section to conclude that the crust in one area is, in some sense, faster than in another. However, the yield of geologically useful velocity information from crustal reflection profiling has not been particularly rewarding. In complex crystalline basement rocks, many of the techniques that yield velocity in flat-lying sedimentary situations are hard to apply: measurement of variations in normal moveout, or integration from surface to depth of the reflectivity function obtained from deconvolved traces. Both methods are particularly hampered by noise. For deep reflections, using conventional recording geometry, significant variations in normal moveout time are often masked by noise. The reflectivity analysis has an intrinsic tendency to drift away from the correct solution as depth increases, which is exacerbated by the lower SNR at greater depths.

Velocity information, if available, may be put to several different uses: (1) to generate the correct normal moveouts for stacking, (2) to apply a time-depth scaling factor to a stacked section to place reflections at approximately correct depths, (3) to migrate a line (place reflections at correct depth and position), and (4) to infer geophysical or tectonophysical results about the composition and history of the area. Each of these instances is affected in a different way by the sources of error in velocity estimation.

Velocity (more precisely, the speed of propagation of seismic waves) has a meaning in our context only when considered over a volume or along a path, and is related to the measurement of a time delay associated with one or more paths. Newer techniques designed to obtain deep crustal velocity exploit certain paths. By making special arrangement to record signals in the "wide-angle" regime, at offsets well in excess of 10 km, adequate

TABLE 2. CONVENTIONAL SEISMIC PROCESSING STEPS*

1. Produce Shot Gathers	
a. Demultiplex†	Sort field data into sequential shot traces
b. Vibroseis Correlation†	Convert Vibroseis signals back to equivalent shot data
2. Initial Cleanup	
c. Frequency Filter	De-emphasis of cominant noise frequencies
d. Resampling	Fewer time points/second, faster processing
e. Trace Editing§	Visual inspection, removal, or degaining of bad traces
f. True Amplitude Recovery§	Remove (approximately) effects of wavefront spreading and decay
g. Datum statics§	Correct for shot/receiver elevations
h. F-K filtering	Remove shear waves, ground roll, air waves
3. Prepare CDP Gathers	
i. Sort†	Rearrange traces into CDP gathers
j. Trace Equalization	Balance average trace amplitudes
k. Deconvolution, Wavelet Processing	Multiple suppression, restoration of impulsive source
4. CDP Processing	
l. Mute§	Suppress refraction arrivals
m. Velocity Analysis§	Determine optimum moveout curve for stacking
n. Residual Statics§	Remove time shifts due to local surface conditions
o. Normal Moveout†	Correct time scale to zero offset
p. Stack†	Add traces in CDP gather. Produce one output trace for that midpoint.
5. Post-stack Processing	
q. Deconvolution	Whiten the spectrum, remove multiples.
r. Frequency Filter§	Emphasize frequency band with highest signal/noise
6. Imaging the Zero-offset Wavefield	
s. Automatic Gain Control	Amplitude equalization for plot
t. Coherency Filter	Empahasize coherent signals
u. Plot†	Convert digital data to graphic display
7. Imaging the Subsurface	
v. Migration§	Direct transformation of wavefield into subsurface representation
w. Inversion	More elaborate construction of subsurface image, often by iteration.
x. Plot†	Convert subsurface image to graphic display

*Steps represent the required minimum = †; steps are basic practice = §

reflection moveout is obtained to constrain deep velocities. This idea is not as easy as it might seem, due to the difficulty of uniquely identifying a single reflection across such a range of distances—a consequence of the great complexity of the crust at scales comparable to a wavelength. More generally, by analyzing the compatibility of pre-stack migrated images run with different subsurface velocity models, bounds may be placed on permissible models. Both methods involve the reconciliation of signal traveltimes for many different paths involving large source-receiver offsets and dipping reflector surfaces. Eventually, a basic tradeoff must be made between statistical reliability and spatial resolution. For reliability, the paths must be long, and, for resolution, short. The entire area of such multiple-path methods is a current research topic of much interest, in which supercomputers, computer graphics, and parallel processing are playing a major role.

Traditional refraction methods, using traveltimes and amplitudes of first arrivals, can be extremely useful where adequate positive velocity gradients exist. In many areas of continental crust, however, the velocity-depth function oscillates about a nearly constant average velocity, and no refraction arrivals propagate horizontally in the bulk of the crustal section. In recent years, use of near-critical reflections has added substantially to the velocity information obtainable from refraction surveys. In current work, the distinction between refraction and reflection surveys is becoming increasingly blurred. In an example of such a combined experiment, Braile and others (1986) used a 400-channel array of portable instruments, 100 km long, to record narrow- and wide-angle reflections, as well as refractions, from a series of explosive sources. Preliminary inversion of the refraction times and some of the important wide-angle reflections have enabled this group to develop a two-dimensional crustal model that reconciles all the times. The factor limiting the resolution in this study was the spacing of the source points (~10 km), which although a great improvement over traditional refraction studies, should probably be reduced to 5 km to adequately exploit the potential resolution of the method.

Data processing techniques adequate for extraction of velocity structure from such new kinds of field experiments represent an area of active research. When large numbers of sources and receivers are used, and when the sensor spacing is close enough that coherent signals can be unambiguously identified, the data may be reduced to a tabulation of a large number of path traveltimes. Determination of the velocity structure may then be based on the widely developed tomographic methods, used mainly in medical imaging, but now being brought to bear in seismology. When the source and/or sensor spacings are small enough that the recorded wavefields are not spatially aliased, then direct imaging of the wavefield data using migration before stack becomes possible. In either case, the processing demands significant computer capability.

Hawman and Phinney (1985) and Hawman (1988) have shown how limited segments of continuously sampled wavefields at large offsets can yield useful information about deep crustal velocity. Shots from quarry blasts in the Great Valley of eastern Pennsylvania and on the Scranton Gravity High were recorded on a 12-element array (100-m spacing). The array waveforms were scanned for coherent energy at all possible ray-parameters and the detected signals composited into a master p-τ diagram of the wide-angle reflections from deep and midcrust for each of the two regions (Fig. 4, top). The resultant velocity-depth inversions represent good composite one-dimensional solutions for the 100-km-long region of signal propagation (Fig. 4, bottom). The results in this case are quite interesting, for the entire mid-to-lower crust is found to have an anomalously high velocity >6.8 km/sec. The technique is probably the best available for localizing major crustal velocity layering, without resorting to very large-scale experiments of the sort discussed above.

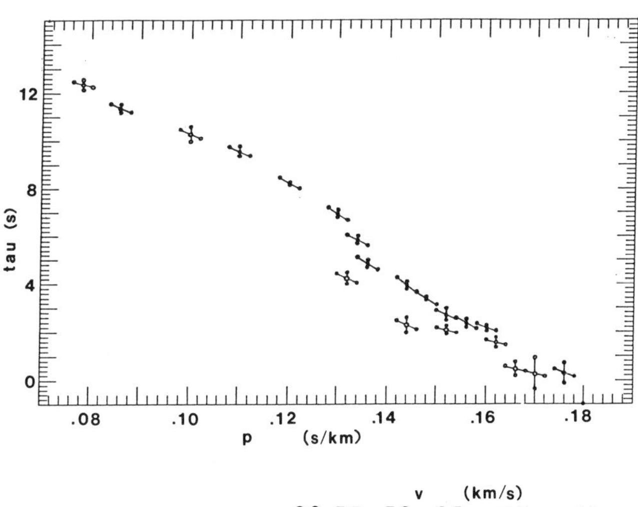

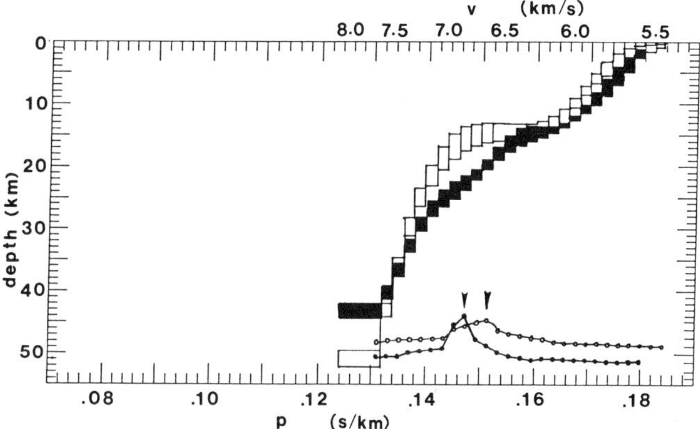

Figure 4. Top: Composite τ display from wide-angle array recordings of quarry blasts in the Great Valley, Pennsylvania (filled circles) and on the Scranton Gravity High, Pennsylvania (open circles) after Hawman and Phinney (1985). The vertical bars indicate standard errors in τ, and the sloping bars indicate the range of p averaged. Bottom, Velocity models obtained from the above datasets by Backus-Gilbert inversion, with 1-σ errors (filled boxes indicate Great Valley; open boxes indicate Scranton Gravity High). One example each of a typical resolving kernel for a target p (arrow) is also shown.

Migration before stack

In the stacking process, each output trace is obtained by summing n input traces. This ratio is known as the *fold* of the stack. By stacking, therefore, the volume of data to be subsequently handled and processed is reduced by n, which may be somewhere between 10 and 200, typically 24 for crustal reflection data. Conventional processing does the stack as early as possible, and makes feasible the routine processing of fairly large data sets on large minicomputers or workstations. The computationally intensive migration step is done after stack in the conventional protocol (Table 2). Unfortunately, some key assumptions that go into this conventional processing begin to fail when strong lateral velocity variations and structural variations occur in the subsurface. The most general, and broadly justifiable way to obtain a correct subsurface image turns out to require that migration be carried out on the unstacked data. Stacking then consists of a sum or more general composite function of ~n images. Migration before stack may be performed on source gathers or on common offset sections; in any event, the computational effort is ~n times greater than the conventional procedure. Application of migration before stack to crustal reflection data lies in the near future, as adequate computer power and storage capacities become available.

We associate a velocity with migration; it is the true velocity or velocity structure through which reflected seismic waves must pass on their two-way trip to and from the target. Since, in general, the true velocity structure is unknown, the before-stack migration approach described here must be applied iteratively in a sequence of steps aimed at optimizing the image. It is exactly in the philosophy of conventional velocity analysis, and drives the large computational load up even more. Most of the actual experience with before-stack migration comes from the oil companies, which routinely apply it to particularly difficult or complex situations, such as overthrust belts and salt domes.

Stacking

Common depth point (CDP) stacking is the basis of all conventional reflection processing. Each trace in the output stacked section is obtained from a compositing step involving all the field traces having a particular source-receiver midpoint on the ground. This collection of traces is called a CDP gather, or a common midpoint gather (CMP), and is usually displayed with the source-receiver offset as the x-axis. The traveltime curve for a given reflection on a CDP gather is always stationary at 0 offset, and almost always can be approximated by a hyperbola, for all but the most steeply dipping reflectors. The n (fold) traces are corrected for the hyperbolic moveout, using stacking velocity as a parameter, and summed, producing a single composite trace equivalent to the zero-offset trace for primary reflections. When a section is produced, generating a panel of such stacked traces, one for each midpoint, it is normally a good approximation to the section obtained from an "echo sounder" experiment, in which a co-located source and receiver are used to collect a reflection trace along a line of many closely spaced surface positions. To the extent that this process produces good signal quality and that the assumptions behind the CDP stack are met, the CDP section may be migrated into an image of the subsurface. Otherwise, the much more elaborate migration before stack, mentioned in the last section, must be employed.

CDP stacking is well established in industry as a fairly routine process, including the removal of static time shifts due to near-surface effects and the selection of a stacking velocity function. It is the process used for all crustal reflection sections or marine multichannel lines published to date. It would be a mistake, however, to think that these procedures can be optimally applied without the exercise of a great deal of expert discretion in the choice of algorithms and parameters. This is so, even more decisively, with deep crustal data, where procedures and parameters routinely applied by industry are not necessarily applicable, and where the SNR is at best mediocre by comparison with industry experience.

Issues that need to be addressed in stacking crustal data include the following:

1. Shot gathers must be edited to remove or suppress noisy traces. These are often caused by proximity to cultural noise or by poor electrical or mechanical connection at the sensor. Such noise frequently exceeds both ambient ground noise and signal, producing a distinctly non-Gaussian noise field. Stacking of gathers that contain even a few such large-amplitude noise traces is guaranteed to give a poor result. Often, noise reduction algorithms, such as diversity stack, trace balancing, or sliding window methods, can replace much of the manual labor of editing.

2. Gathers must be inspected to assess the characteristics of the different elements in the data: (a) P-wave reflections; (b) P- and S-wave refractions; (c) surface waves; (d) background noise; (e) scattered signals, such as surface waves generated by P-coupling; and (f) the presence of obvious multiple reflections. A protocol for reducing the interfering signals (b-f) is then designed, which may include bandpass filtering, f-k filtering, refraction muting, deconvolution, or coherency filtering.

3. The severity of the residual statics problem can be estimated by inspection of the refraction arrival (e.g., Fig. 2). Unless residual statics are sufficiently minimized, not only do the signals not sum coherently when stacked, but methods for eliminating undesirable wave types (see previous paragraph) may not function well. Residual statics algorithms lead to perhaps the most complex computer programs, and many workers may not have access to a good selection of methods, even with commercial processing packages.

4. Stacking is an iterative process, in which a sequence of trial stacking velocity models is used. Each time through, the residual statics and some of the noise suppression steps must be rerun.

Klemperer and Brown (1985) made an interesting study of the effectiveness of noise suppression processing before stack. CDP stacking is itself a filter specifically tuned to eliminate signals on the gather that do not have the characteristic normal moveout of a true reflection. These authors found that, in a surprising number of cases, CDP stacking without prestack noise suppression was nearly as effective as processing that included that step. This might be expected in cases where the noise amplitudes are approximately Gaussian and the data are not spatially aliased. On the other hand, Zhu and Brown (1986) reprocessed data from the Michigan Basin along the lines discussed here, and achieved a significant improvement in the section.

It must be reemphasized here that stacking velocity analysis results in velocities that stack the data well, i.e., produce a visually pleasing stack-section. For horizontal interfaces (layer-cake model), interval or layer velocities can be derived from the stacking velocities by using the so-called Dix equation (Waters, 1987). Estimates of crustal velocity obtained by this procedure are seldom reliable because first differences of the stacking velocity model amplify errors beyond acceptable limits.

Migration after stack

Migration of a CDP stacked section is the conventional method of producing a subsurface image (Fig. 3). Many variants of the basic wavefield transformation described by Claerbout (1985a, b) are now available. In a synthetic example, we show (Fig. 5) the generation of a migrated image of a buried structure

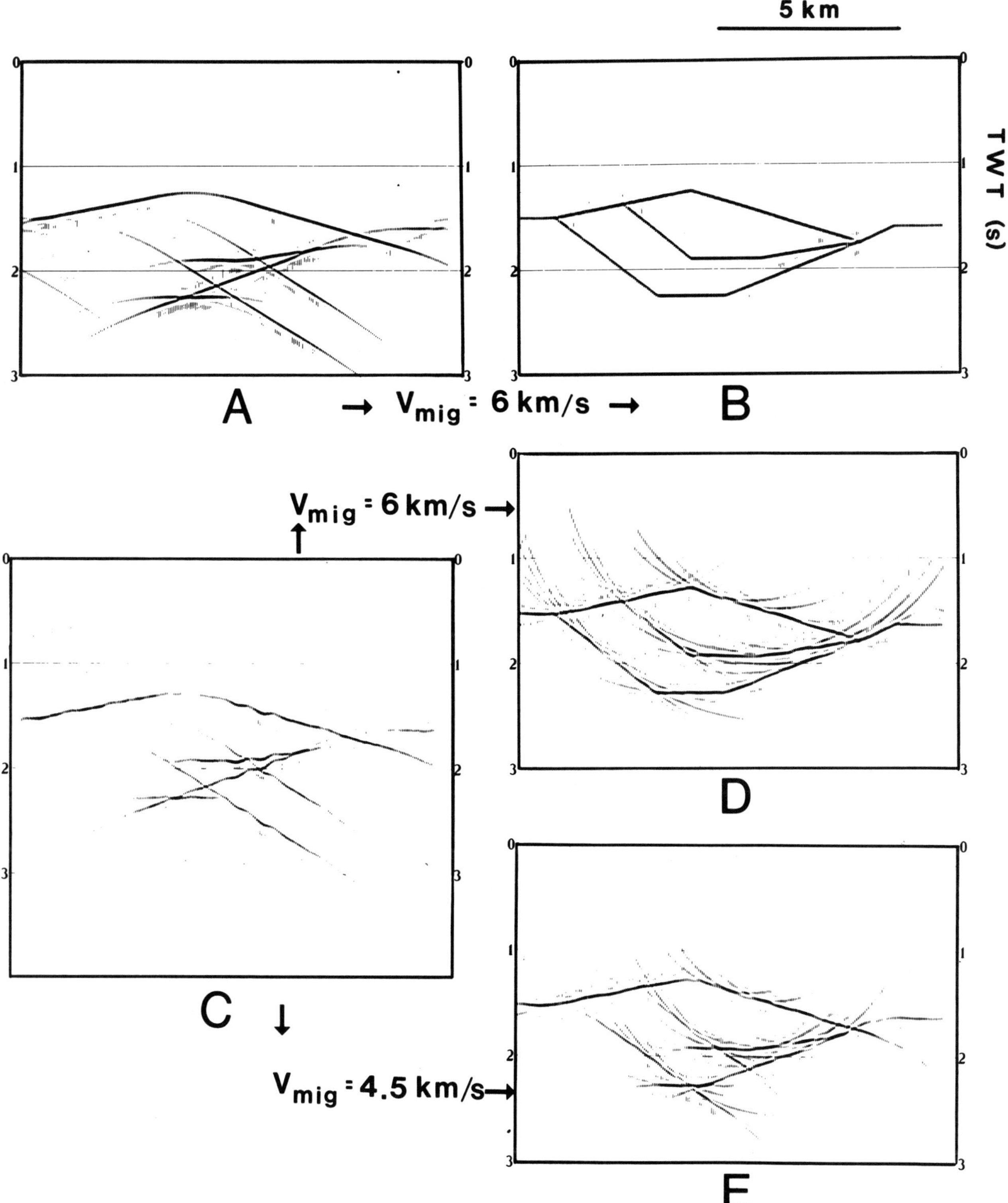

Figure 5. Effect of time and amplitude jitter on poststack migration. The synthetic model, with $\nu = 6$ km/sec everywhere, is reconstructed (B) by migration of the time section (A). The section is then contaminated (C) with random amplitude (≤50 percent) and time (≤50 msec) statics. The migrated image (D) is now substantially contaminated by "smiles." Migration at 4.5 km/sec gives a result (E), which, although technically undermigrated, is a viable candidate for display, since the smiles are minimized over much of the image.

(B) from the zero-offset time section (A), using the correct constant model velocity of 6.0 km/sec. In practice, with good signal quality, migration must always be undertaken provisionally with a number of different velocity models. The final choice is then based on the appearance of the resulting image, and on any a priori information available about the true subsurface velocities.

Undesirable contamination of the migrated image results whenever the input data contain components that are not legitimate images of a reflected wavefield. Migration of deep crustal data often produces disappointing results. This is produced by amplitude and time-static fluctuations in the time section (Warner, 1987). The degradation that can be produced is illustrated in Figure 5. Each trace in the synthetic time section (A) has been multiplied by a random amplitude distortion function ($\leqslant$50 percent) and a time jitter function ($\leqslant$50 msec) to produce the distorted section (C). Migration of (C) at 6 km/sec yields the image (D) that shows the desired image strongly overlain by "smiles." Only the rectilinearity of the desired image provides a basis for judging this result, a constraint not available in real applications. When migration is accomplished with a reduced velocity of 4.5 km/sec (undermigration!), the visual effect of the "smiles" is reduced (E), while some geometric distortion has crept into the deeper part of the model. A worse looking result could be produced by adding ordinary band-limited noise to the traces.

Migration of stacked data may also break down for geometrical reasons. The true 3-D subsurface model must be independent of some strike azimuth, and the line of section must be along the dip direction. The validity of much reconnaissance crustal profiling is presumed on the choice of a dip direction based on surface geology. When the true model is three-dimensional, any line of profile can have "sideswipes," reflections from off the plane of section. Often an experienced interpreter can identify or rule out sideswipes on the basis of the overall geometry of the various reflectors. And, as noted above, many steeply dipping structures with lateral velocity variations fall outside the range of applicability of CDP stacking, and imaging must be done on the unstacked data.

In this chapter we display (Figs. 12-16, 18, 19) deep crustal sections from the Appalachians, where the line of section can be chosen as a dip line. Regional data on average crustal velocities invariably give values above 6 km/sec for rocks below the sediment cover. We have standardized on a display that combines coherency processing (below) with a moderate undermigration at 4.5 km/sec.

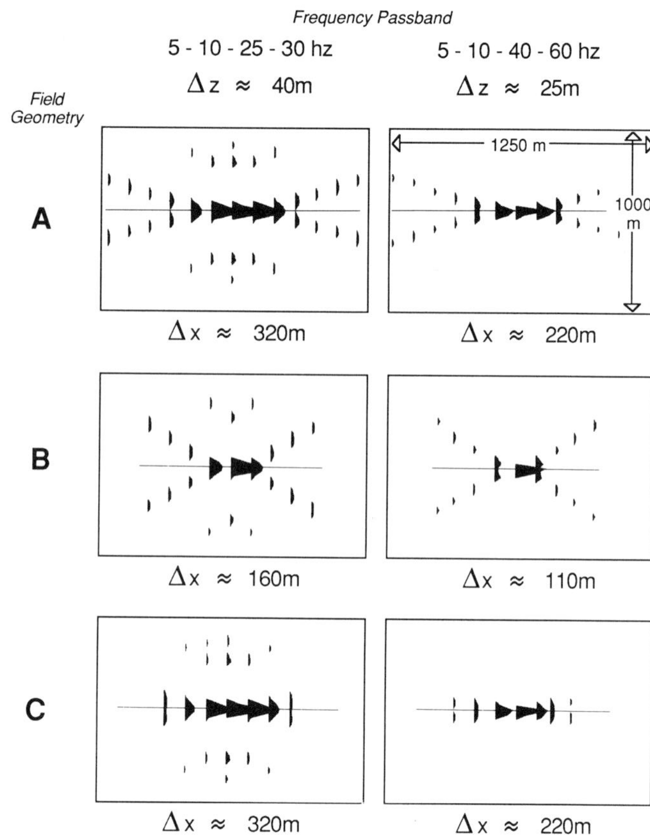

Figure 6. Limits to resolving power. The resolving power may be estimated by determining the half width of the imaging (migration) function for a particular experiment. In these examples, a point target at 15-km depth is imaged using different experimental parameters. The displays all show only a part of the imaged area, 1.25 km × 1.00 km centered on the target. Left, Zero phase source wavelet, trapezoidal passband 5-10-25-30 Hz. Right, passband 5-10-40-60 Hz. A, Single fold, zero offset field geometry, source points between –5 and +5 km, at 100-m intervals. B, Same as A, with source points between –10 and +10 km, at 200-m intervals. C, 50-channel split spread receiver array, 10 km end to end. Source point is moved from –5 to 5 km, at 100-m intervals. Half-width estimates, Δx and Δz are given for each case.

Resolution

The wave nature of the seismic signal limits in a fundamental way the ability to resolve fine-scale detail. Waters (1987) has provided a basic account of this topic. In the discussion that follows, we briefly touch on some aspects of this subject; a full treatment is beyond the scope of this review.

The limits on resolving power determined by the wave propagation process may be illustrated by computing the "imaging function" for some simple cases (Fig. 6). This is the image computed by a migration-before-stack applied to synthetic data for a point reflector at some depth. It is dependent on the geometry of the survey, particularly the lateral extent (aperture) of the source and receiver coverage above the target, and on the frequency content of the source signal. Resolving power may be defined as the ability to distinguish two identical targets that are slightly separated. For these examples, it is determined by the shape and width of the central peak (the remaining features are windowing effects of the finite frequency bandwidth and experiment aperture). Figure 6 shows estimates of the vertical (Δz) and horizontal (Δx) resolving length, defined as the appropriate half-width of the main peak.

On theoretical grounds, the vertical and horizontal resolving lengths are predicted to have the forms:

$$\Delta z \sim \left(\frac{v}{f}\right); \quad \Delta z \sim \left(\frac{L}{Z}\right)\left(\frac{v}{f}\right),$$

where v is the velocity, f is a frequency representing the high end of the main frequency band, Z is the depth of the target, and L is the horizontal extent of the surface coverage. From the model results in Figure 6, the constants may be estimated, giving:

$$\Delta z = \frac{1}{6}\left(\frac{v}{f}\right); \quad \Delta x = \left(\frac{L}{Z}\right)\left(\frac{v}{f}\right).$$

For typical deep crustal data, these rules give ~40 m for vertical resolution and ~320 m for horizontal resolution of a 15-km-deep target. When noise, velocity variability, and statics are incorporated, the results should be considered worse by at least a factor of two. It is emphasized that the horizontal resolution estimates here are valid only in the case that the data have been correctly migrated.

REPROCESSING

We have described special difficulties in interpreting deep reflection seismic sections. We now consider how reprocessing of existing data can substantially enhance the information in the section. While complete reprocessing of the field gathers is a valid option, the manipulation of stacked sections can be an equally important tool for the interpreter.

The standard distribution CDP stack is a baseline for interpretation and evaluation. In discussing any particular stack, we can distinguish several potential levels of reanalysis: (1) presentation of the stack, from the setting of display parameters or display mode, to filtering and noise removal by coherency processing; (2) poststack migration or other imaging procedures; (3) restacking, by revision of the stacking protocol and improvement of such steps as residual statics, velocity analysis, and coherency processing; and complete reprocessing in a migration before stack procedure. Levels (1) and (2) represent areas where great improvement in the quality of the result can be obtained, as we demonstrate in some of the examples in this chapter. These require computational power of the sort found in a microcomputer-based workstation. Restacking (3) is practical with similar hardware, at least for limited segments of section.

Current practice in reinterpretation seems to fall into one of two categories: workers with little or no processing capability are restricted to published or distributed copies of section stacks, or workers with processing capability put much effort into restacking—which involves much data handling and computation. From the examples in this chapter, we believe that all workers should take advantage of tools for poststack reprocessing and image enhancement. The scope of this Memoir does not allow us to develop a comparable analysis of interpretational issues involved in migration and stacking.

Converting the stack to an image

Considering the simplest level of processing, in which a stacked section is converted to a display, three types of operations are of interest: linear filtering for SNR improvement, coherency filtering (nonlinear) for signal detection, and selection of a display algorithm and parameters.

Linear filtering. At times, the stacked section has a low SNR, which can be significantly improved by ordinary frequency filtering to remove frequencies that have very little signal. COCORP Georgia Line 1 (Cook and others, 1979) was the first long line to image the southern Appalachian overthrust over most of its extent. The fragment shown in Figure 7 shows both the horizontal detachment and the splay that outcrops as the Brevard Fault Zone. The stacked section is corrupted by shear-wave noise generated at the low-frequency end of the 8- to 32-Hz vibrator sweep. After digital filtering to remove frequencies below 15 Hz, both the horizontal detachment and the BFZ are clearly visible.

Two-dimensional frequency-wavenumber filtering (F-K) is most commonly used on field gathers to remove interfering waves that are not reflections. On stacked data, such issues arise often in marine work, where side reflections in the water or nearby ships can produce strong linear artifacts crossing the section, and F-K filtering is needed. On marine profiles across continental margins, water-bottom multiples from intermediate water depths interfere severely with primary reflections from the subsurface, and fairly clever schemes for F-K filtering are required.

Coherency filtering. In the stacked section, a reflection arrival can be defined as energy that is spatially phase-coherent and has an apparent slowness[1] appropriate to compressional waves incident from below at a permitted range of angles. The process of preparing crustal stacked sections for interpretation often consists of a manual "squinting-and-picking" operation wherein wavelets that are aligned across a group of neighboring traces are picked. The result is an interpretive line drawing; while it incorporates the skill of the interpreter, there is no way to verify its consistency short of having several interpreters rework the data. The line drawing makes it possible to develop a migrated image using ray-theoretical mapping of the line segments (Warner, 1986); this procedure is extremely fast computationally, and is not contaminated by nonphysical "smiles."

Kong and others (1985) described a procedure for extracting coherent signals automatically, and producing a section that is largely cleared of uncorrelated background noise. This "coherency filter" is nonlinear and data-dependent; it is based on an assessment of the semblance function across n-trace panels of data, and the use of the semblance measure to weight the reconstruction of the section from its p-τ representation. In the second half of this chapter we show seismic sections that have been

[1]Strictly, *slowness* is the reciprocal of speed of propagation. *Apparent slowness,* also called *horizontal slowness* (often simply called *slowness*), is the reciprocal of the *apparent velocity* of the wave across a geophone array and is the same as the *ray parameter* for a horizontally stratified Earth.

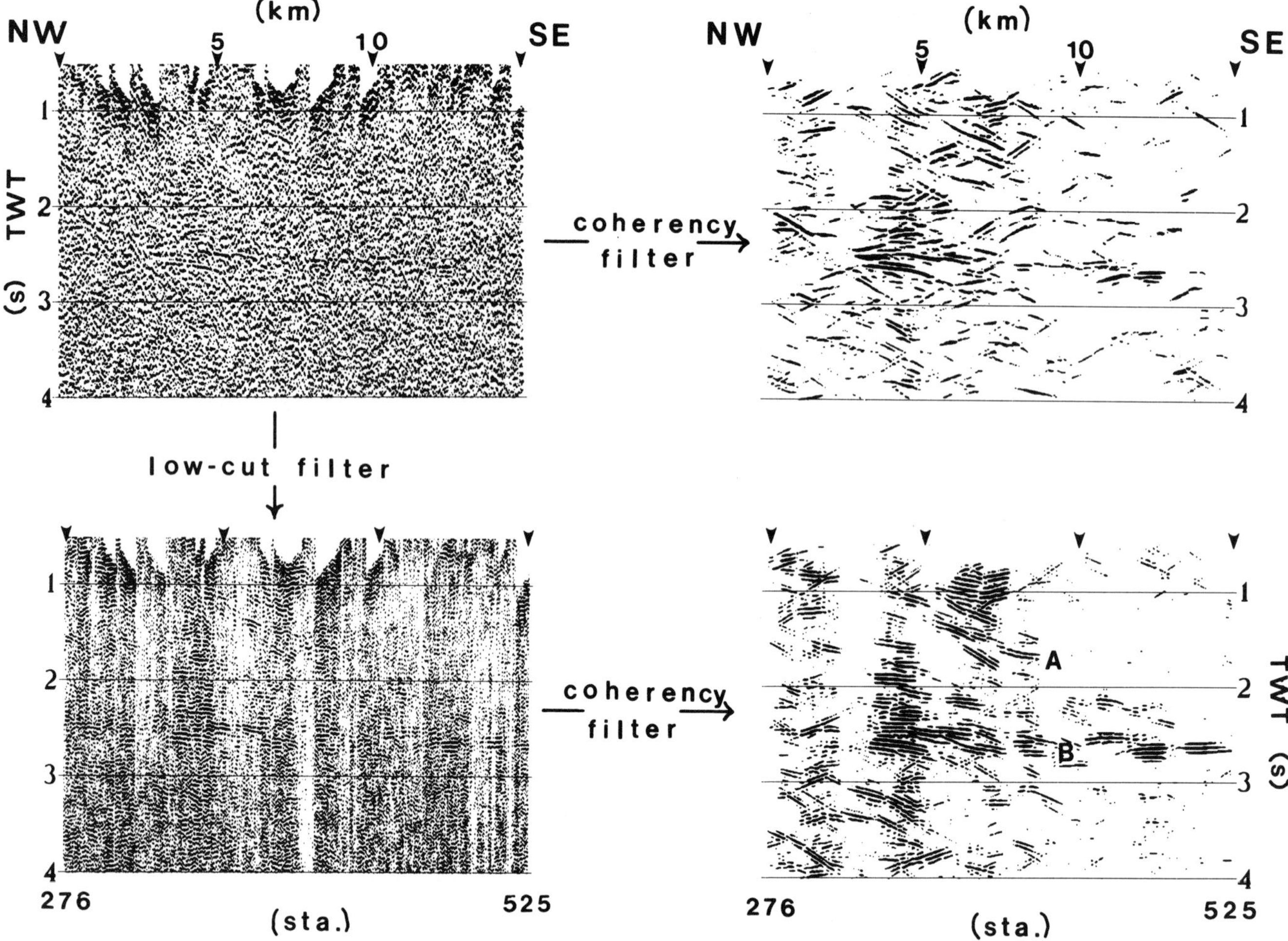

Figure 7. Fragment of stacked section from COCORP Georgia Line 1 (Cook and others, 1979), showing seismic image of Appalachian overthrust. A = Brevard Fault Zone. B = Basal reflector (probably middle Cambrian Rome Formation). Upper left is original stack, optimized for display of reflections; upper right is coherency filtered version. At lower left, the data have been low cut filtered to suppress shear wave noise ~8 to 14 Hz; lower right is coherency filtered version.

cleaned up by means of the coherency filter, and pose these as examples of the potential usefulness of such an adaptive nonlinear filter in many areas of reflection processing. Zheng and Brown (1986) have described a coherency filter based on similar ideas, which is now used by COCORP for producing standard distribution displays. Their filter incorporates a final step in which the filter output is mixed back with the original data, thus producing a section that has more of the traditional "feel" of a section than those that appear here.

The coherency filter is not a magic black box. It is characterized by its tunable parameters, which must be chosen by the interpreter to correspond to an a priori assessment of how much spatial coherence characterizes a real signal in a given case. The parameters are: the number of traces across which coherence is measured, the range of apparent slowness that correspond to real signal, the statistic used to measure coherence (e.g., semblance), and the mapping of this statistic into a weight function. An example of this processing (Fig. 8) shows a section of COCORP Georgia Line 5 (Cook and others, 1981). Readers are urged to study carefully the two halves of this figure and assess the performance of the filter on patches with different visual coherency.

Validation of a particular choice of parameters is based on two tests: (1) Does the finished image come reasonably close to showing all the phase lineups that the interpreter considers legitimate by a careful study of the conventional display, and have as few artifacts and false alarms as possible? (2) Does the processing pass a null test? In Figure 9, we have removed phase coherency from half of a test panel, and insist that the processed result show negligible energy in the randomized area. Coherency processing is a distinctly nonlinear process, but is so in precisely the same sense that conventional display is nonlinear. It is thus of some importance in which order coherency processing is combined with bandpass or F-K filtering. A significant difference is seen when coherency processing is used to extract signals from the example in Figure 7; it is preferable to improve the SNR with the low-cut filter before applying the coherency process. Many factors come

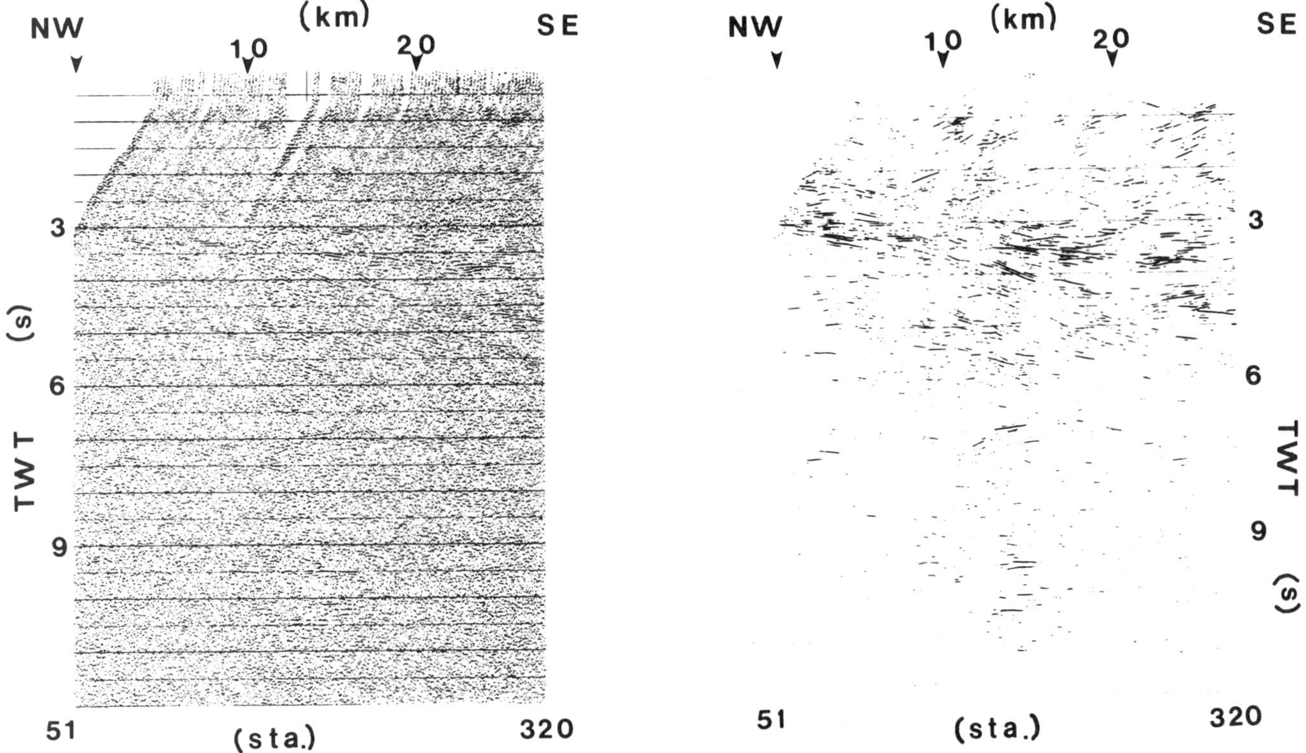

Figure 8. Left, Portion of released version of the COCORP Georgia Line 5 CDP stack (Cook and others, 1981). The coherency processed version (right) is displayed with a threshold that retains a visible noise background consisting of a scattering of weak, spatially incoherent events.

into play when we combine coherency processing with other algorithms. For example, to a stacked section, apply coherency filtering, then poststack migration. The effect (Kong and others, 1985) is to remove from the input to the migration much of the noise that gives rise to excessive smiles in the migrated image (see also Fig. 5 and the related discussion).

The performance of a coherency processor depends very much on the SNR of the input. When the input data are generated by a bandlimited random process with no trace-to-trace correlation, the processor is certain to produce, on statistical grounds alone, a nonzero output section. For sections with poor SNR, then, we have chosen to display the output with a visible noise background, to permit the interpreter to visually assess the signal and noise characteristics. The following figures in this chapter form a series with progressively poorer SNR: Figure 12a-d, Long Island Platform (USGS offshore Line 36); Figure 16, New Hampshire (COCORP New England Line 6); Figure 19, Georgia (COCORP southern Appalachian Line 5); Figure 15, Vermont (COCORP New England Lines 1 and 3); and Figure 13-14. Adirondack (COCORP New England Lines 10, 7, and 11).

In this chapter all sections are presented with coherency processing as described here. In addition, most have been migrated afterward at 4.5 km/sec. Naturally, coherent artifacts of the stacking process will be preserved as faithfully as genuine signals. Many of the events between 0.5 and 1.0 sec in Figures 13 and 14, for example, may well be artifacts introduced by shear-wave refractions and surface waves that were not removed by muting.

Presentation. A frequently overlooked transformation of the data is that which maps a digital representation of a section into a graphical image. It is normal for a section to present quite different visual appearances for different choices of display mode and parameters. Moreover, displays that are most useful in chart size for the interpreter seldom reproduce well on the scales required for publication. The format adopted by the American Association of Petroleum Geologists (AAPG) for its three-volume seismic atlas is the most successful attempt yet to put usable seismic sections in the open literature (Bally, 1983).

The seismic interpreter conventionally looks on a stacked section for phase peaks that line up when viewed at some angle. This approach is dependent on the "variable area" mode of display. Unfortunately, it leads to visual artifacts that can mislead the interpreter. The visual sense that a series of phases line up to form a signal is strongest when the filled-in positive peaks overlap to produce a continuous black stripe. This can only occur when: the apparent slowness of the signal is close enough to zero (i.e., the arrivals are subhorizontal), and the peak amplitude is equal or greater than the trace spacing. The visual impression that a signal is present is thus a highly nonlinear function of signal amplitude and apparent slowness, and can be "cooked" by the choice of trace spacing or signal amplitude.

These effects can be seen in the left and center panels of

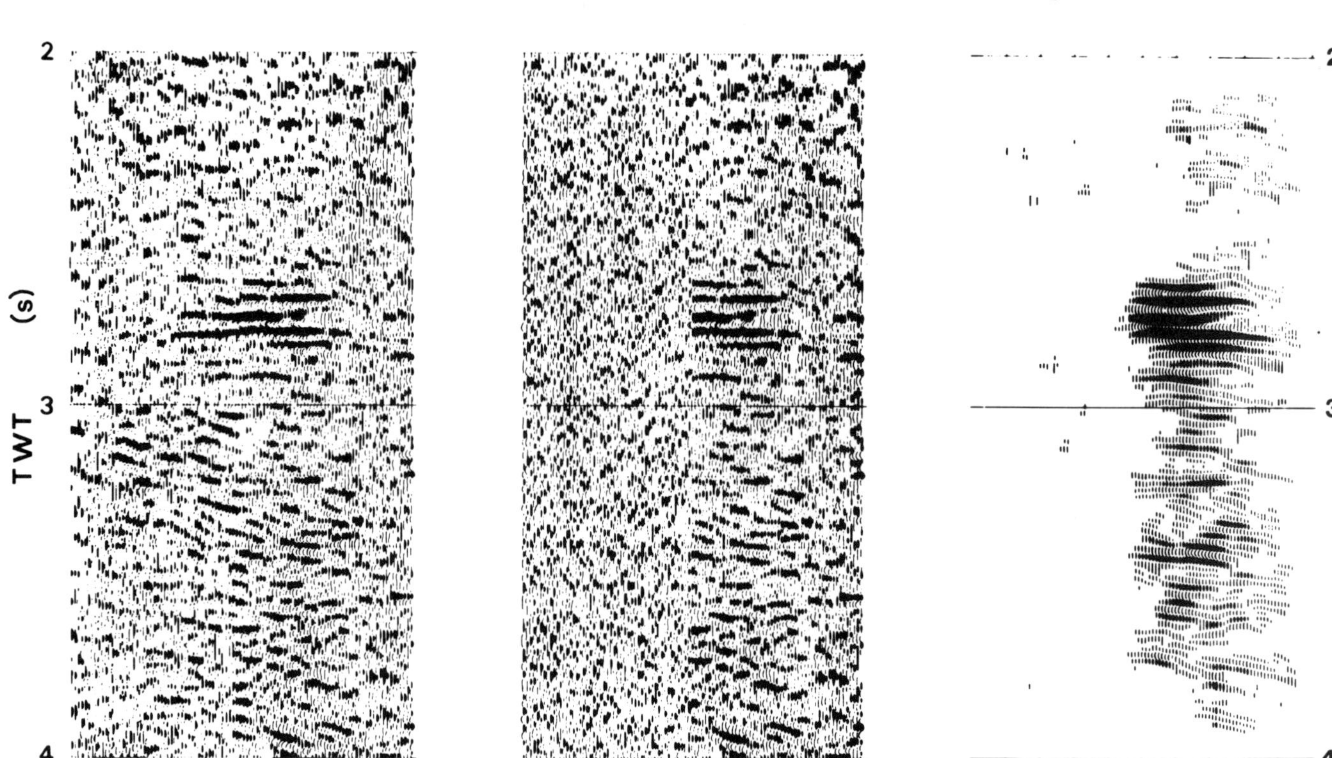

Figure 9. Validation of coherency filtering (null test). For a test fragment from a stacked section (left), half the traces have been replaced by noise traces with the same power spectrum (center) before applying the filter. The result shows no signal energy from the randomized half of the panel (right).

Figure 10, which differ in the choice of trace spacing and amplitude scaling. Other modifications of the conventional display also affect the appearance of a section. In the figure, we have suppressed data points that are zero or negative. It is also possible to suppress positive values less than some threshold or to nonlinearly rescale data values; for example:

$$\log |value| \text{ or } value^2\, sgn\,(value).$$

Again referring to the two examples, the effective dynamic range (ratio of trace spacing to 200 dpi raster spacing) is only of the order of 10 to 20 db. Thus, although a conventional stack display is a familiar and perhaps well-understood mode of communication, it is far from being a complete, unbiased, or objective presentation of a section. The selection and publication of a section with particular display parameters thus constitutes an inescapable interpretive step. In the image-processing literature, our practice of variable area wiggle trace display is considered quite inferior to a raster format with variable gray or color scaling.

The implication of these comments is that, given the capability now available in microcomputers and low-end workstations, every worker who seeks to reinterpret a seismic line should consider obtaining the stacked tapes and try several different choices of display parameters.

Future trends

At present, restacking of crustal reflection data remains a laborious task for most research groups outside of industry, due principally to the labor of iterating the stacking process. A major factor is the difficulty of handling ~50 reels of magnetic tape in a small facility. The availability of 600-mbyte streaming tapes and 3-gigabyte optical disks or videocassettes may relieve this problem in the very near future.

It has been fortunate that the continental crust commonly shows near-horizontal structures, for the CDP stack is not a particularly good tool for working with structures much steeper than about 40°. In industry, some successes are reported in imaging steep or overturned flanks of salt domes and steep faults by resorting to direct imaging (migration) of individual source gathers into partial images (Fig. 3) before compositing. These computationally intensive techniques have not yet been used in any degree for processing of crustal reflection data. It is expected that they will be more practicable on supercomputers now available to the research community. For the new generation of deep-

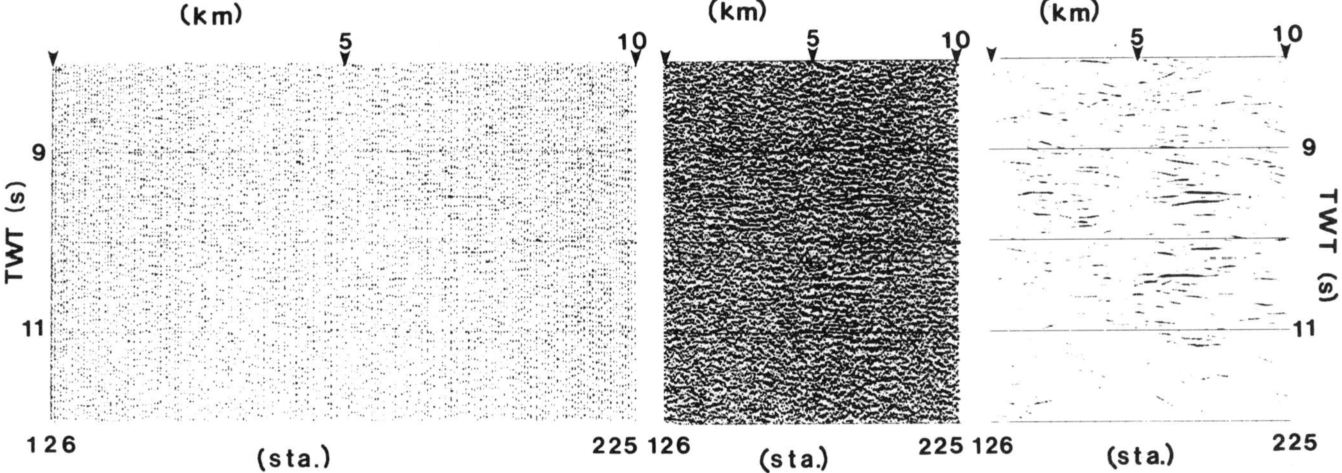

Figure 10. Presentation of stacked data. This example uses a 100-trace by 4-sec panel from the stacked section of COCORP Georgia Line 5 (Cook and others, 1981). Left, Variable area display with positive-fill, amplitudes scaled to the inter-trace separation. Middle, Horizontal scale is cut in half; amplitudes increased. Right, Same display parameters; data has been subjected to coherency processing (Kong and others, 1985) to emphasize arrivals with high semblance over several neighboring traces (12-trace window).

imaging experiments using explosives, migration before stack will involve the imaging of source gathers. For marine reflection studies, where the multifold CDP geometry is retained, we should see migration before stack applied to common offset sections or slant-stack sections.

PRESENTATION AND DISCUSSION OF SELECTED SECTIONS

Many of the reflection profiles listed in Table 1 and shown in Figure 1 cross some part of the Appalachian orogen. A companion chapter in this volume (Taylor, this volume) addresses these in the context of other geophysical information, such as refraction, gravity, heat flow, and teleseismic waveform analysis. In this section, we, too, focus on these lines because of the possibility of developing a regional story by comparison. We have reprocessed the stacked data for the Appalachian lines specifically to improve SNR and to enhance their visual clarity and usefulness to the nonspecialist for reference. These are annotated to facilitate tracking some of the most interesting results, and to give the reader some sense of the basis on which current interpretations rest.

Of the projects listed in Table 1, the principal transects of the Appalachian orogen are: [1], the Quebec-Maine transect; [2], the Adirondack-New England transect; [3], the Long Island Platform Line 36; [6], the James River corridor; [8a, b], the COCORP Georgia transect; [8f], COCORP Georgia-Florida transect; and [10], the Ouachita transect. Most of our presentation is based on transects [2], [3], and [8a, b], which have been digitally reprocessed to bring out the coherent information. Transect [10] is discussed only briefly, and transects [1], [6], and [8f] were not available for reprocessing at the time this work was done. All the sections [Figs. 12–16, 18, 19] are displayed on uniform horizontal and vertical scales designed to produce a 1:1 cross section at 6.0-km/sec velocity. Locations on the sections are given (in kilometers) from the left end (west northwest, or north) of the given line, e.g., *km:225.*

USGS lines on the Long Island Platform

The excellent sections obtained by the BIRPS program in the United Kingdom from marine profiles around the British Isles are well-known examples of studying continental crust by going to sea (Warner, 1986; Matthews and Cheadle, 1986; Brewer and others, 1983b; Brewer and Smythe, 1983). In the U.S., the marine lines acquired on the Atlantic margin in the 1970s by the U.S. Geological Survey (USGS) are of similar high quality, and provide a reference framework for looking at the noisier land data. (For a review of the entire USGS program, which was directed toward the sedimentary basins on the margin, see Grow and others, 1979). The 320-km transect by Line 36 cuts WSW–ENE across the extension of the Appalachian trends of southern New England (Fig. 11); the reprocessed, migrated section is shown in Figure 12a–d. It was acquired as an inshore baseline for lines recorded across the margin for study of the offshore basins. Extensive discussions of this line were given by Hutchinson and others (1985, 1986a, b), and Phinney (1986).

Description and interpretation. On Line 36, the uniform 1.0-sec thickness of Neogene sediment lying atop the Appalachian basement serves as an ideal interface for imaging the crust, without the attenuation loss and image distortion often produced by thicker, more complex sedimentary overburden, or by a heterogenous outcropping basement surface. The line cuts across the structural grain of southern New England at about 45° to the

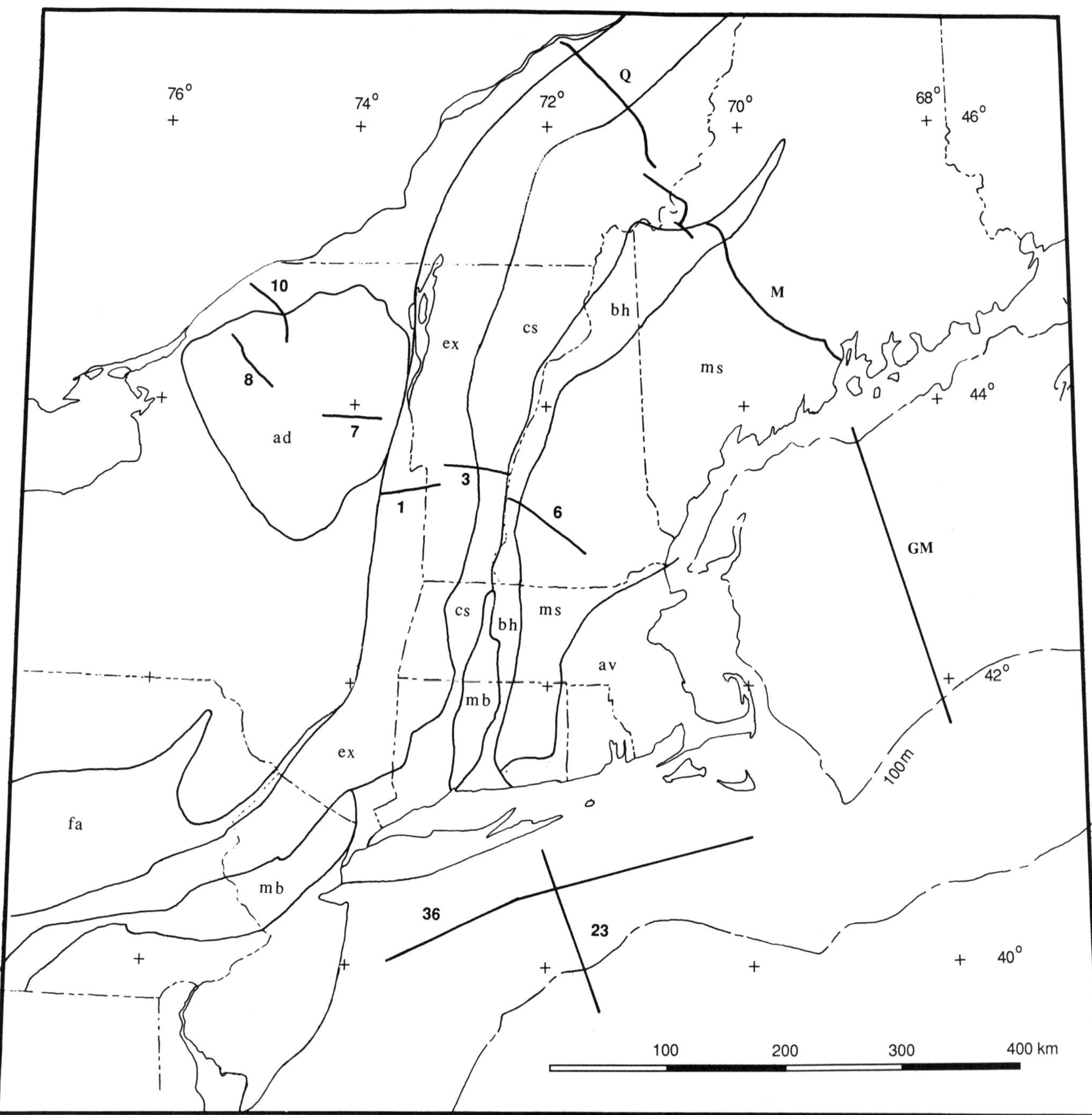

Figure 11. Location map for northern Appalachian reflection lines discussed in the text: Quebec–Maine–Gulf of Maine Transect (SOQUIP/USGS/GSC); Adirondack Lines 7, 8, and 10 (COCORP); New England Appalachian Lines 1, 3, and 6 (COCORP), and multichannel marine lines 23 and 36 on the Long Island Platform (USGS). ad = Adirondacks; fa = folded Appalachians; ex = external thrust belt including Grenville basement massifs; cs = Connecticut Valley synclinorium; mb = Mesozoic basins; bh = Bronson Hill anticlinorium; ms = Merrimack synclinorium; av = Avalonian terranes.

regional N15°E strike. Thus, an apparent dip of 30° on the ~1:1 section (Fig. 12a–d) would represent a true dip of about 50°. The generally continuous nature of many of the reflections on this line makes it possible to approximate the true cross section by post-stack migration. The following discussion generally pertains to the entire section (Fig. 12a–d), with reference to specific figures being made wherever necessary.

The reflections from the sediment-basement interface can be followed continuously across the tops of the rift basins; they are thus produced by the basal Neogene sediments, not by the basement surface proper. Below, weak, patchy horizontal events between 1 and 3 sec are due to multiple reflections in the water-sediment stack, and mask weak dipping primary reflections from within basement. The basement surface is broken by three Mesozoic rift basins (Hutchinson and others, 1986b). The complex New York Bight basin (Figure 12a, C) is seen on cross lines to widen and deepen toward the south, as the shelf edge is approached. The Long Island basin (Fig. 12b, F) and the Nantucket basin (Fig. 12d, R) both show half-graben structure, with the active fault on the west. These basins serve to distort the ray paths of deep reflections, with the result that the generally continuous deep structural fabric seen elsewhere on this line appears more fragmented due to focusing and defocusing beneath the basins. Examples are X in Figure 12d, and the Moho under basins C and F.

Moho can be followed as a nearly horizontal, nearly continuous reflection band around 9 to 11 sec, and serves as the base of the strongly laminated, variably dipping packages that make up mid- and lower crust. If we regard the Moho as a package of rocks, rather than a simple boundary, it can be regarded as a lower crustal horizontal boundary layer (Phinney, 1986), with noticeably variable thickness. From *km:0* eastward to *km:130,* it is quite thin, ~0.3 sec, whereas east of *km:130* it appears to be 1 to 2 sec thick, depending on the identification of the transition above to the more dipping reflectors of the mid-crust. This thick Moho can be most readily seen at *km:160-180, km:220,* and *km:280-300.* Patchy horizontal reflections below the interpreted Moho depth are probably due to multiples in the water sediment column, and are distinguished from the true Moho reflections by the much better continuity of the latter across the whole section.

Moho has nearly constant depth from *km:0* to *km:100,* then jumps upward at *km:115.* Farther east, it is extremely smooth, as it deepens under *km:180-240* (*Q-Q-Q*) and thins under *km:280-300* (*V*). It is necessary to assess whether variations in reflection time of Moho are due to velocity variations in the overlying crust or are genuine images of depth variations. Phinney (1986) argued that the thinning at points such as *L2* (Fig. 12b) and V (Fig. 12d) could not be artifacts of high average crustal velocity because, if an average linear relation between velocity and density is assumed, the required high crustal density would yield a large positive gravity anomaly, which is not observed.

In broadest outline, the crust along this line consists of a mid- to lower crust that is strongly layered, with generally low dips, and an upper crust that is much less reflective, but that shows weak indications of moderate to steep dips (Fig. 12b, J). This situation is found in many areas; it is a matter of current debate whether it is "normal" or not. Here, from 1 to 3 sec, multiples of the water sediment column mask any primary events; moreover, for nonhorizontal structures the validity of the approximations used in CDP stacking breaks down at shallower depths, and the blank region around 3 to 5 sec may be a processing problem. Despite such cautions, the weight of the evidence from this and other lines suggests that a real difference in reflectivity is generally involved. The underlying cause is not established, although several proposals have been advanced, including the role of water, of compositional layering, of the scale of layering in tectonically deformed rocks, and of igneous intrusions. For example, where the transition between nonreflective upper crust and reflective lower crust is well imaged in places as a dipping boundary, the contrast must be in the rocks (N-P between *km:140-180*; S-T; Fig. 12d). Brown (1986) concluded, on the basis of a large number of COCORP profiles, that the transparent upper-crust–reflective lower-crust paradigm is a simplification that describes at best half the available lines.

The most widely discussed analog is found in the metamorphic core complex belt of the Basin and Range, where often strongly deformed and faulted upper-plate rocks are in detachment contact with a mylonitic, layered, higher grade lower basement plate. The seismic data in the Basin and Range show the same characteristic transparent upper crust and layered lower crust, while the regional tectonic setting is dominated by extension. There, Klemperer and others (1986) and Hearn (1986) have argued for the importance of magmatic underplating in making the Moho strongly reflective. In the Appalachians, the crust we see now is the result of several episodes of Paleozoic compression (with inferred extension), terminated by thinning and rifting in the Mesozoic. Whether the last extensional phase, which is necessarily a small overprint on the gross collisional structure, is sufficient to have determined the seismic structure at depth remains an open question. Perhaps the most satisfactory version of this idea involves the injection of magma into the lower crust during early phases of extension. The intrusives are emplaced in existing fault zones or layered gneisses, and substantially enhance the seismic reflectivity. Neither the volume fraction of newly injected magma nor the crustal strain induced in the crust by extension need be large.

In interpreting crustal reflection profiles, it is tempting to look for transparent zones and consider these to be geologically significant; they could occur over a large, uniform batholith or in a place where reflectors are nearly vertical. However, the extremely variable quality of typical land data makes this quite risky, for many of the "no data" gaps are due to problems of signal strength (e.g., Figs. 13 through 16). On this line, however, we identify two places where a through-cutting zone of no signal is a plausible candidate for the locus of a through-cutting strike-slip dislocation. At E (Fig. 12a), is a rather abrupt change in the direction of the reflectors, which appears in the CDP stack as crossing diffractions. At W (Fig. 12d), a noticeable change in the

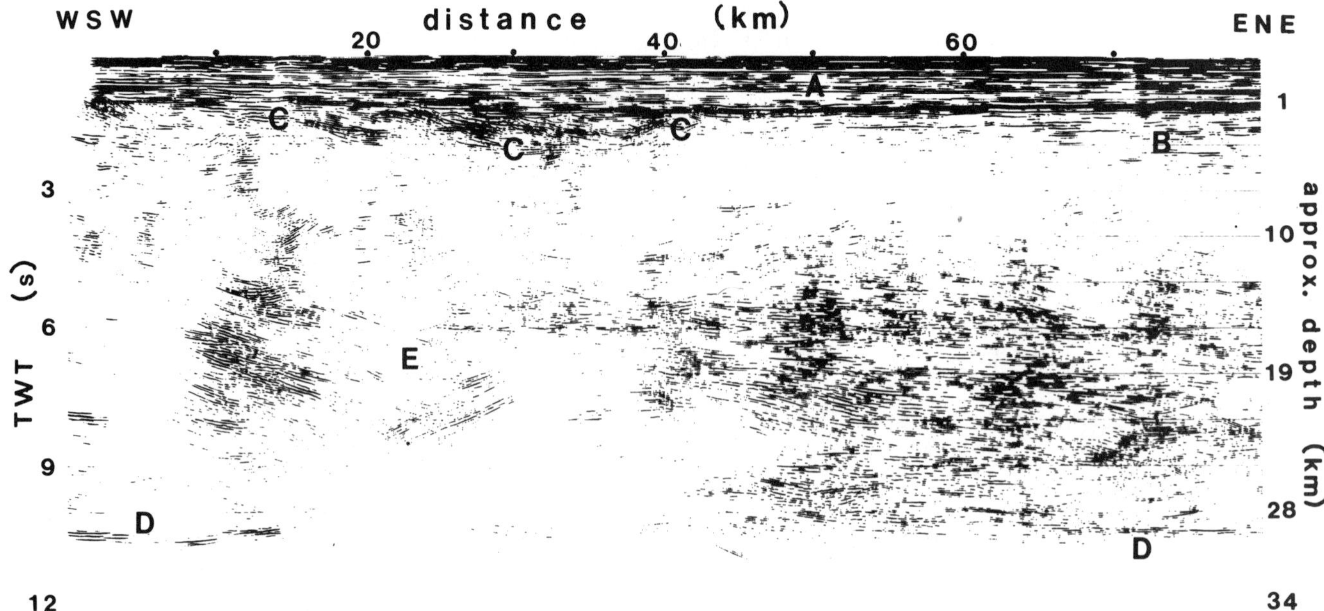

Figure 12a. USGS marine line 36, Long Island Platform, *km: 0-80.* The contractor stacked section (Phinney, 1986; Hutchinson and others, 1986) has been coherency filtered and migrated at V_m = 4.5 km/sec. Two-way reflection time has been converted to depth using 2 km/sec in the sediments and 6 km/sec in the crystalline basement. A = Neogene sedimentary section, 0 to 1 sec. B = Multiples of Neogene section. C = Complex (Triassic?) rift basin. Strength and continuity of deeper events are seriously degraded by this basin, in comparison with the region east of *km:40.* D = Nearly continuous, flat, thin (<0.3 s) Moho event. E = Pronounced discontinuity in fabric in middle crust shows appearance expected of a vertical, transcurrent fault cutting the crust below *km:21.*

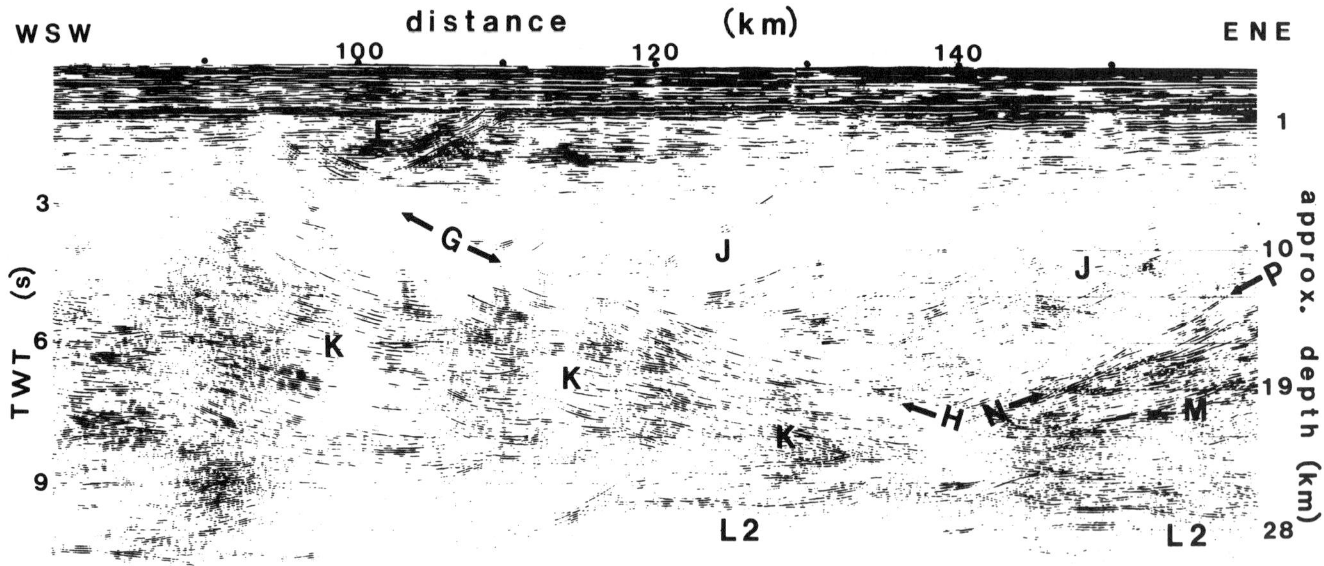

Figure 12b. USGS marine line 36, Long Island Platform, *km:80-160.* The contractor stacked section has been coherency filtered and migrated at V_m = 4.5 km/sec. F = Simple (Triassic?) half graben, with hinge on the east and active wall on the west. G-H: Major discontinuity between weak, west-dipping events (J) and the large wedge of strong, slightly east-dipping events (K) that extends to the base of the crust. L1 = Flat Moho at 10.3 sec with transition to L2, flat Moho at 9.2 sec. M = Deep package of strong reflectors, west-dipping above, and merging into flat Moho below. See Figure 12c also. N-P = Major discontinuity between regions M below and J above.

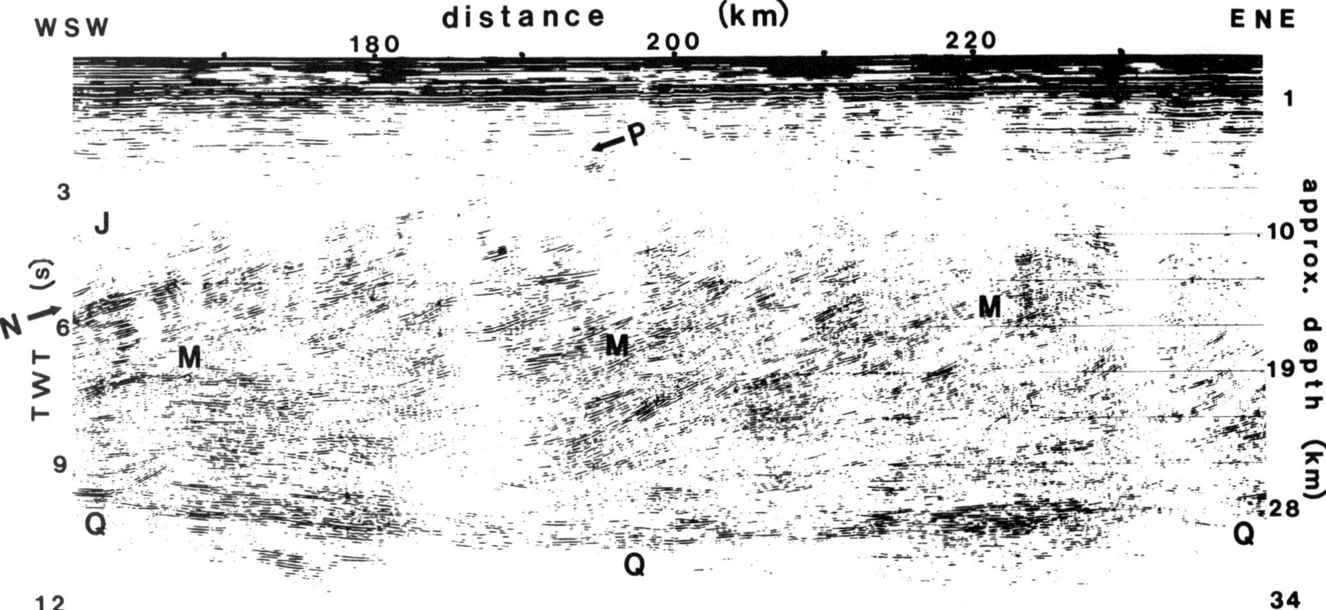

Figure 12c. USGS marine line 36, Long Island Platform, *km:160-240*. The contractor stacked section has been coherency filtered and migrated at V_m = 4.5 km/sec. Q = Moho sags from 9.5 sec (west) to 10.5 sec (center) and rises again to ~9.5 sec (east). M = Entire crust below N-P consists of strong reflectors in conformable geometry: horizontal, and conformable to Moho, in lower crust, and west-dipping above.

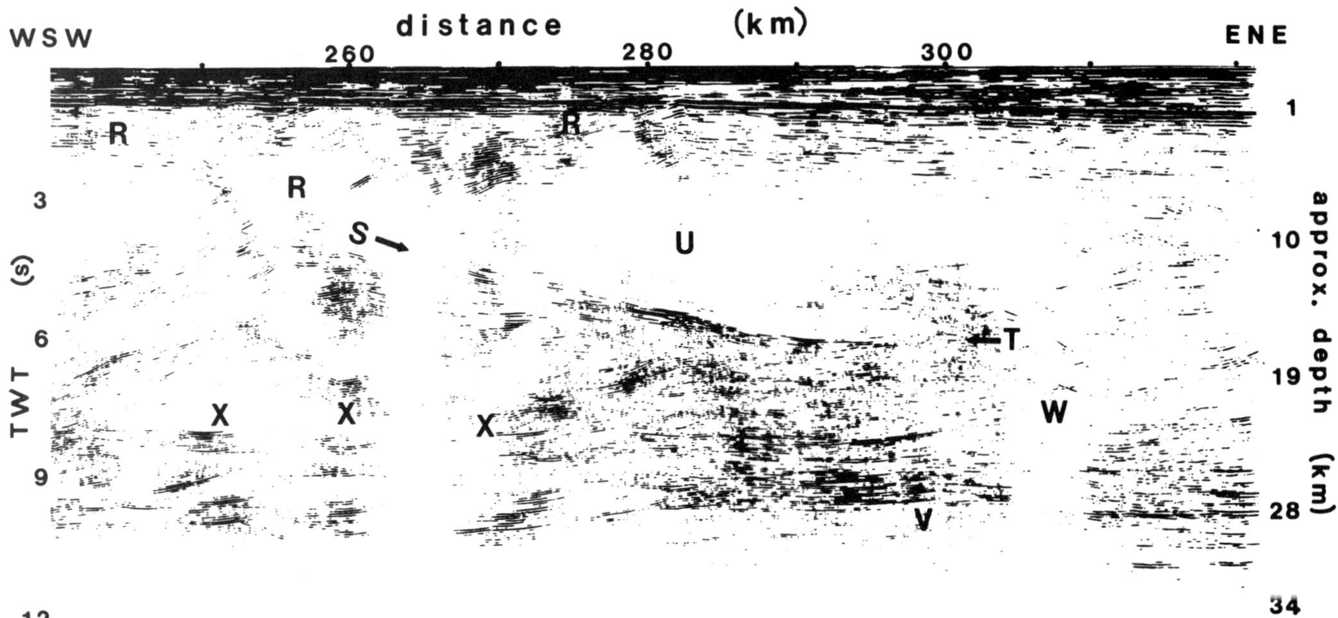

Figure 12d. USGS marine line 36, Long Island Platform, *km:240-320*. The contractor stacked section has been coherency filtered and migrated at V_m = 4.5 km/sec. R = (Triassic?) half graben with hinge on the east and active wall on the west. S-T = Major east-dipping reflective discontinuity between transparent zone above (U) and the predominantly subhorizontal package that forms the lower crust. V = Moho rises to 9 sec at *km:300*. W = Blank zone through crust, candidate for location of vertical, through-cutting transcurrent fault. X = Alternating vertical stripes of strong and weak signal occur directly beneath the graben, which produces the distortion by bending seismic waves on their way from and to the surface.

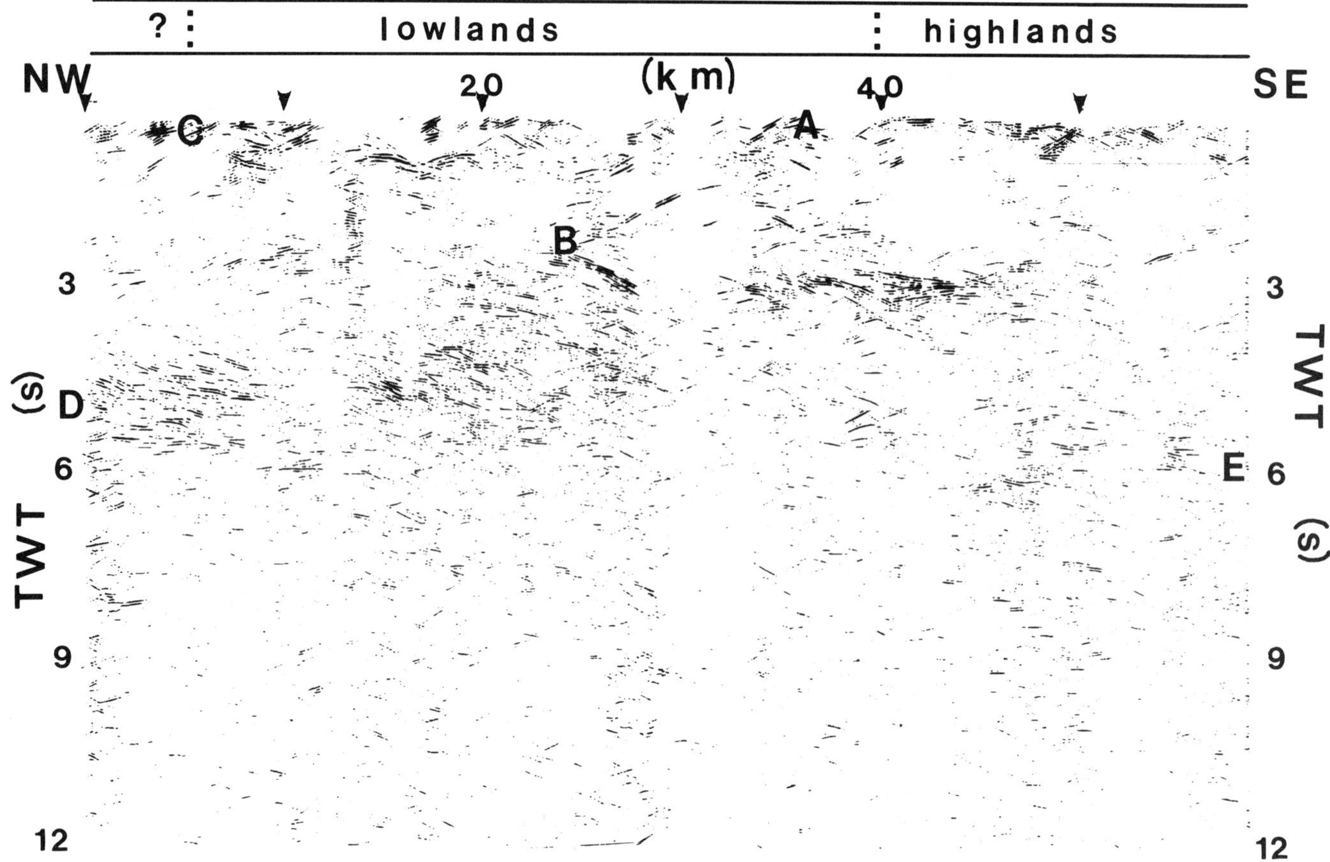

Figure 13. COCORP Adirondack Line 10 stack with coherency filtering. A-B-C = St. Lawrence Lowlands Grenvillian supracrustals appear as a keystone-shaped package 2.5 sec deep, between *km:5-40.* Coherent reflections between 3 and 5 sec lack continuity and context for interpretation. D-E = Possibly continuous band of reflectors between 5 and 6 sec. Below 6 sec is 100 percent noise! In Figures 13 through 16, the geologic annotations at top are obtained from state geologic maps, scale 1:250,000, and are limited in precision by the format we are using. In many places, the indicated surface geology is applicable to a very shallow zone only, due to the large amount of horizontal transport suggested by the seismic sections.

general appearance of the reflections occurs across the data gap. In both cases, it is an abrupt change in dip direction that validates the interpretation, not merely the data gap, which could be due, at E for example, to the overlying rift basin.

Geological correlation and discussion. Line 36 cuts across the southernmost extension of the New England Appalachians. After projecting the line to a northwest-southeast dip line, there are some expected ambiguities in establishing an offshore-onshore correlation. The case of the following correlation was argued by Phinney (1986). Observation of the V-shaped wedge J (Fig. 12b) on both Line 36 and Line 23 establishes a regional strike of N15°E for that feature, and an angle of 45° between Line 36 and the northwest-southeast structural dip line. This wedge is then correlated with the corresponding onshore belt in central Massachusetts and Connecticut (Rodgers, 1985), and is supported by recently published structural models for that belt. (Specific difficulties exist in detail, which we judge to be explainable by plausible variations along strike, and will not be examined in this chapter.)

1. The highly deformed Cambro-Ordovician metasediments, remobilized Grenville-age basement, and granitic intrusives of western Connecticut, with Taconian deformation and Acadian overprint, are correlated with *km:0–90.* In outcrop, the inferred upper story of westward thin-skinned thrusts is wholly lacking, having been removed by erosion. Similarly, Figure 12a shows no indication of a horizontal detachment at ~1 to 2 s, such as reflector B in Vermont (Fig. 14).

2. The Mesozoic Hartford half-graben in central Connecticut and the offshore basins at C, F, and R are part of the same regional extensional system; the Hartford basin, however, is faulted on its eastern boundary, whereas basins F and R have western boundary faults.

3. The Bronson Hill–Merrimack belt of eastern Massachusetts and Connecticut is correlated, on the basis of regional dips

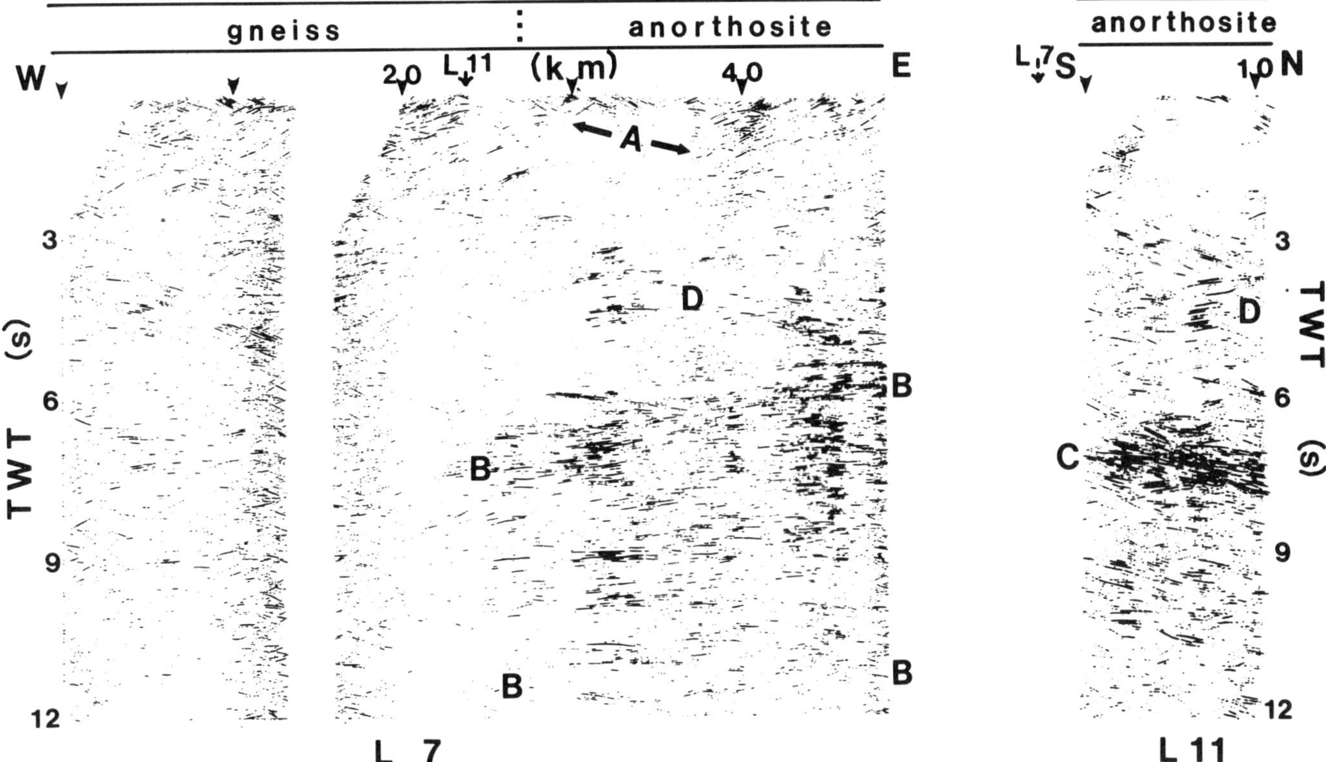

Figure 14. COCORP Adirondack Lines 7 (left) and 11 (right) stack with coherency filtering. Lines intersect along southern edge of the Marcy anorthosite (Klemperer and others, 1985). A = Approximate position of the base of the anorthosite is inferred from gravity studies; west-dipping reflector around *km:40-45* is only visible evidence on this section. B = Lower half of crust is imaged (*km:24-40*) as a layered, reflective, lens-shaped package, horizontal at the bottom. C = On line 11, the reflections from the top of this package are anomalously strong, the result of focusing of wavefronts with several different angles of incidence. D = Reflections above 6 sec suggest additional lens-shaped packages, but poor coverage and continuity do not permit specific interpretation.

and high metamorphic grade, with the V-shaped keystone J (*km:90-200*), which we have inferred to be a tectonically shortened and thickened interterrane.

4. The Avalonian Hope Valley and Esmond-Dedham terranes of southeastern New England (O'Hara and Gromet, 1985) are correlated with the rest of the section, east of *km:200* (excluding the Nantucket basin (R), which does not extend onshore). The Hope Valley Shear Zone, proposed by these authors as a suture of Alleghanian accretion, can be correlated with the east-dipping fault R-S-T.

The key to these correlations is the close resemblance between the keystone-shaped wedge J and the inferred structure of the Bronson Hill–Merrimack belt in Massachusetts, as proposed by Hatcher (1981) and Zen and others (1983). We regard the appearance of an eastward-dipping suture like G-H (Fig. 12b) in every Appalachian seismic line (Fig. 21) as a demonstration of the importance of this feature. In detail, the outcrop geology in southern Connecticut suggests that the Bronson Hill–Merrimack keystone is underlain and uplifted by an east-west regional arch of the eastern Hope Valley basement; we therefore propose that the Massachusetts and Long Island Platform sections are representative, and southern Connecticut is an uplifted, unroofed, and very deeply exposed part of the keystone. Additional complexities that lie offshore and are not now known could easily modify this picture.

The interior of the keystone has few reflectors, generally with steep west dips (>~30°), concordant with the eastern boundary N-P except in a narrow western boundary zone. Thus, J appears to be simply superposed on its eastern basement M. Its western boundary is a significant dislocation between two different dip domains. If it is a shortened and thickened supracrustal package squeezed between the eastern basement M and the western basement K, then the observed dips suggest that J originated on or adjacent to M, and was obducted against the suture G-H by delamination accompanying the west-directed subduction of M. The western boundary G-H extends to the surface at the position of the active boundary fault of the half graben F; this inferred suture was active both in compression and in subsequent extension. The geometry and proposed evolution of the keystone J are very similar to proposals by Needham and Knipe (1986) for the

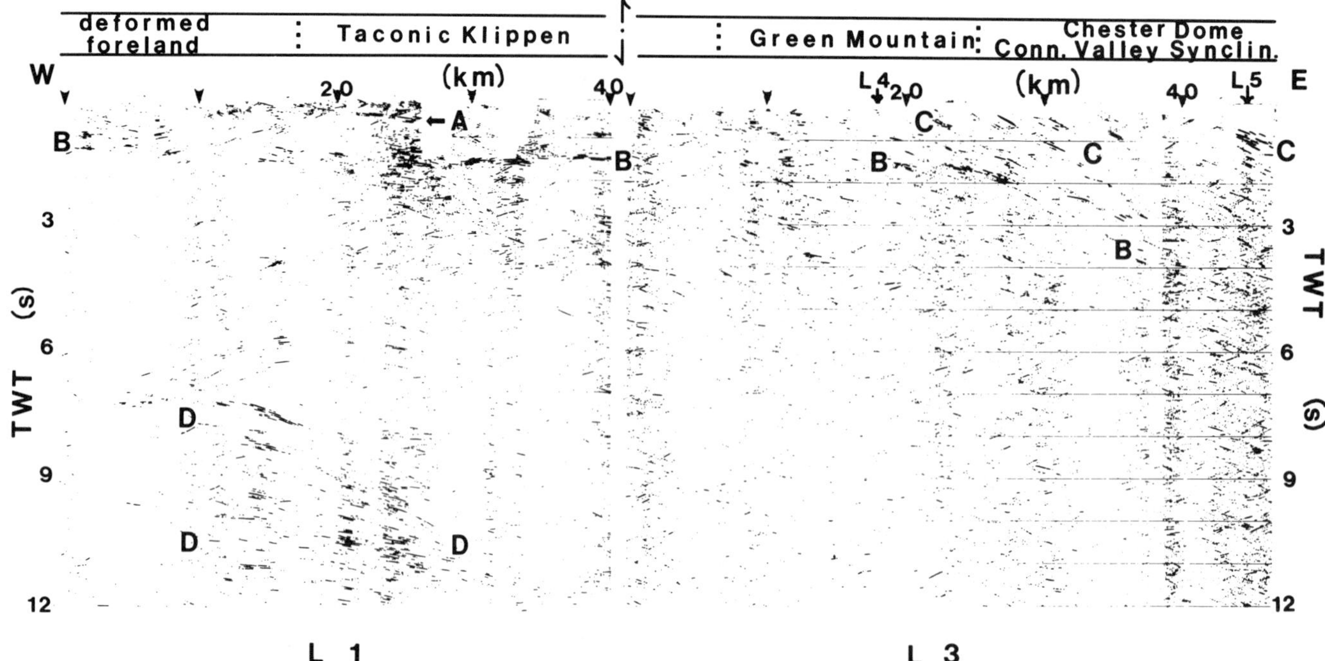

Figure 15. COCORP New England Lines 1 and 3 stack with coherency filtering. Although displayed together, Line 3 is offset 20 km north of Line 1. These lines form a transect from the Taconic foreland across the Green Mountains to the Connecticut River (Brown and others, 1983a; Brown and others, 1984). The section is extremely noisy; surface geology suggests interpretations: A = Nearly horizontal event from the base of the Taconic allochthons. B = Horizontal event at 1.4 sec from top of Grenville basement, becomes a weak, east-dipping event about *km:25,* Line 3. C = East-dipping events between *km:15-45* on Line 3 show that the Green Mountain basement massif, the Connecticut Valley synclinorium, and the Chester dome are rooted in structures that dip about 30°E. D = A reflective, lens-shaped package between 7 and 12 s on Line 1, which may correlate with B on Adirondack Line 7 (Fig. 14).

Southern Highlands of Scotland. We associate the significant thinning of the crust under the keystone J with the extension recorded by normal faulting and graben formation along the fault G-H at the time of Mesozoic rifting.

Farther east (*km:240-300*), the east-dipping reflector R-S-T is similar to G-H: it is the active western boundary of an extensional basin; the crust thins under T, and it separates regions of different reflection character. The interior domain U is even more transparent than J. Rather than rooting nearly at Moho, this feature terminates listrically against a horizontally laminated lower crust with noticeably high reflection strength; this association supports the idea that extension gives rise to enhanced lower crustal reflectivity.

We interpret the broad-arched massifs K and M as the cores of distinct terranes. This is confirmed by differences in the character of Moho already noted. West of the suture G-H, K consists of many nearly flat-lying mid- to lower crustal reflectors. In this area (*km:40-80*) the seismic section is so nearly horizontal that it lacks diagnostic features that might help in establishing an interpretation. The complex onshore geology could easily produce the transparent upper crust, but little else can be inferred from the seismic data. In this area, the strike of the line becomes more nearly parallel to regional strike than elsewhere, and we conjecture that many of these reflectors may, like those of the Taconian root zone in Vermont (Fig. 15), be dipping downward from northwest to southeast. The antiformal massif M thickens eastward to make up at least 80 percent of the crust at *km:230.* In its strongly layered character, its westward dip, and its basement relation to the keystone wedge J, it may be correlated with the strongly tectonized Hope Valley terrane onshore (O'Hara and Gromet, 1985). Given the limited outcrop area of the Avalonian, arguments for identifying the transparent domain U on the basis of onshore geology are at best suggestions. However, the eastern Avalonian basement (the Esmond-Dedham terrane) is dominated by little-metamorphosed Eocambrian granitic plutons (O'Hara and Gromet, 1985). If U is a comparable crustal fragment, then the fault R-S-T would correlate with the Hope Valley Shear Zone, and we would have an image of the late Paleozoic accretionary suture proposed by O'Hara and Gromet (1985).

The importance of such a second suture lies in the imprint that both Acadian and Alleghanian orogenic activity, along with Carboniferous plutonism and rift basins (Gromet, 1987), have left on rock isotopic signatures in central and eastern New England. It suggests that, although an initial Acadian collision may have occurred along the suture G-H, later Paleozoic orogenic activity may have occurred on both sutures and entrained two separate colliding terranes along with the Bronson Hill–Merrimack interterrane belt J. We propose that the antiformal basement massif M

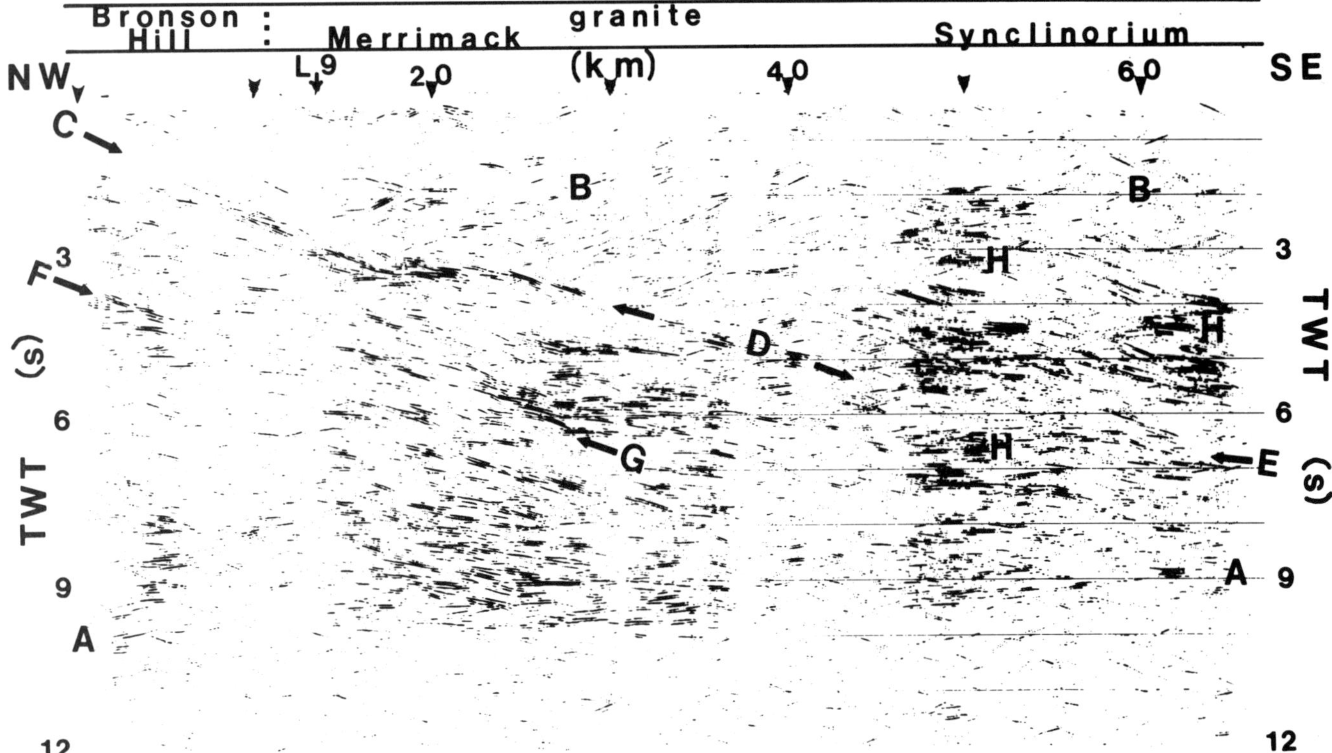

Figure 16. COCORP New England Line 6 coherency filtered stack migrated at 4.5 km/sec. Line 6 continues the transect of Figure 15 southeast across New Hampshire. It is interpreted in light of the very similar, but better quality, section on the Long Island Platform (Fig. 12b). A = Moho rises from 10 sec on the northwest to 9 sec on the southeast. B = Merrimack synclinorium, with folded metasediments and sialic plutons shows west-dipping structures in the center (*km:40*) at a depth of ~4 sec. C-D-E = Reflector interpreted as a fault. In the east, it can be followed to *km:60* ~7 sec. Bronson Hill anticlinorium is thrust westward and above C-D-E. F-G = Reflector interpreted as a fault, and as the probable eastern extension of the Grenville basement surface B (Fig. 15). H = Chaotic bright spots throughout the deepest part of the Merrimack synclinorium and into basement.

correlates with the Avalonian Hope Valley terrane. It appears structurally very much like the metamorphic core complexes of the Basin and Range. It is highly metamorphosed with strong subhorizontal tectonic layering and forms a less mobile basement beneath detached supracrustals. It represents a mature stage of tectonic accretion of crustal material in the mid- to lower crustal levels. Internally, the characteristic sigmoidal step-up forms over the layered Moho (*km:190-220*) suggest (R. D. Hatcher, personal communication, 1987) that this massif is in effect a crustal-scale duplex, tectonically assembled by internal thrust duplication in a lower plate being overridden by a higher level thrust sheet. If it is to have served as a relatively rigid accreting basement to the Bronson Hill–Merrimack keystone in Acadian time, then it probably experienced the indicated strong internal deformation earlier.

In the following sections we consider a series of COCORP VIBROSEIS transects. This pioneering program has been engaged in long reconnaissance traverses in the conterminous U.S. since 1978 (Brown and others, 1986). The data we show here have been coherency enhanced and redisplayed to bring out the signals as well as possible. However, SNR varies a great deal, and the section quality seen on the marine data is not achievable. On land, near-surface heterogeneity of a crystalline basement, crooked roads, and noisy or skipped records combine to seriously degrade signal quality and increase the cost of data acquisition and processing. In particular, the lack of clear reflectors on a portion of a section may as easily be due to poor SNR as to the lack of subsurface reflectors.

COCORP Adirondack and New England lines

The 1980-1981 transect from the St. Lawrence River to eastern New Hampshire (with gaps) crossed both the Adirondacks and the New England Appalachians (Figs. 11, 13 through 16). Except for Line 6 (Fig. 16) in New Hampshire, SNR is too low to permit migration. The noise threshold for the display has been set to admit some incoherent background to assist the interpreter.

The Adirondack lines show disappointingly little, and are representative of the kind of deep reflection section often obtained over Precambrian basement. Readers may use this discussion for their own interpretation of the data or may prefer agnosticism. Line 10 (Fig. 13) crosses the Colton-Carthage mylonite zone, which separates the central Adirondack highlands and the St. Lawrence lowlands (McLelland and Isachsen, 1986; Klemperer and others, 1985). Both terranes are a part of the Grenville Province; upper amphibolite facies metasediments and metavolcanics characterize the lowlands, and granulite facies metaplutonic and metasedimentary rocks characterize the highlands. Multiphase folding is dominated by regional-scale reclined to recumbent isoclinal folds. Near the mylonite zone, these folds in the lowlands verge southeast.

On the seismic section we seek some indication of reflections between 0 and 3 sec that might be correlated with these features. The earliest arrivals (ca. 0.5 sec) are patchy and too discontinuous to interpret; they probably consist mostly of reflections from complex near-surface structure. The geometry of a regional seismic survey (line length, 10 km) is not conducive to imaging the shallow structure. Knowledge of the location and importance of the Colton-Carthage zone (*km:40*) permits us to identify a weak west-dipping event A-B. The bright east-dipping event B would migrate up and westward, and is interpreted as part of an east-dipping boundary C-B on the west flank of the feature A-B-C. With this interpretation, the Adirondack lowlands would be this V-shaped package A-B-C with a depth of ~8 km and a breadth of ~35 km. A couple of hyperbolic arcs at *km:15-20, 1.0 sec* look like real data, but would need to be resurveyed with the right field parameters to make any sense. The remainder of the section shows some interesting complexity between 3 and 6 sec, but has insufficient continuity or surface control to be useful. A subhorizontal alignment D-E at ~5 to 6 sec, weakly indicated, could be the target of attempted restacking of the data. Nothing of any perceptible significance can be seen below 6 sec. With a refraction Moho at ~36 km (Taylor, this volume), the only statement we can make is that there is negligible coherent energy below 9 sec, which corresponds to ~28 km depth. This section, and, indeed, most of the COCORP profiles on Precambrian crust, do not fit the common generalization that the upper crust is transparent and the lower crust strongly layered and reflective.

In Figure 14 are displayed two crossing lines at the southern margin of the Marcy anorthosite in the high peaks region of the Adirondack highlands. Gravity data, combined with field mapping, suggest that the anorthosite, although regionally extensive, is a fairly thin sheet, perhaps 3 to 5 km thick (Simmons, 1964). The expected position of the base of the anorthosite on Line 7 is marked by A, a weak west-dipping reflection at ~1 sec, but it is not visible on the cross line 11. On Line 7, the strong package of reflectors B at depth, (the Tahawus complex of Klemperer and others, 1985), is very similar to the kind of characteristic lower crustal structures seen on the Long Island Platform (Fig. 12). Modest dips near the top (~7 sec) become horizontal toward the weakly imaged base (~12 sec), which is the only clue we have from these sections to the depth of Moho under the Adirondacks. On the cross line, this package appears as the complex bright spot C near 7 sec; it is visibly composed of arrivals with several different slopes and is bright due to focusing of these arrivals. If migrated, the bright spot would be spread out into image elements well off the ends of the line.

The west half of Line 7 shows virtually nothing we can work with. A data gap is seen, however, between *km:14* and *17*; at the edges of this data gap appear tantalizing suggestions of imageable reflections. This suggests that the complexity of the surface geology, and the failure of the moveout curve to be sufficiently hyperbolic, caused the stacking process to degrade the result for the western half of Line 7 by summing signals out of phase (overstacking!). Near the data gap, the stacking fold becomes very small, and the stacked signal is consequently better. This is a case where reprocessing of the field data might provide a better section.

COCORP New England traverse

This traverse begins about 50 km south of Adirondack Line 7, and extends east and southeast across Vermont and New Hampshire to a point near Nashua, N.H. (Figs. 15, 16). Brown and others (1983a) and Ando and others (1984) studied this cross section, giving an interpretation very similar to the present rediscussion of the enhanced data. Lines 1 and 3 extend across the Taconic foreland and klippen, the Green Mountain massif, and the Connecticut Valley synclinorium (Fig. 15). The SNR on this section is just adequate to produce some useful signals; along with the well-studied surface geology and the higher quality Line 6 in New Hampshire, enough context is available for an interpretation. A reflective, strongly layered package between 8 and 12 sec lies completely within Grenville basement D; it is similar to B on Line 7, and lines up unexpectedly well with that feature.

For interpretation of the shallow part of Figure 15, we make reference to the recent compilation and synthesis of Taconic structure by Stanley and Ratcliffe (1985). Reflector A, at 0.3 sec (~0.6 km) marks the Giddings Brook Thrust, above which lie the Taconic klippen; these consist of Cambro-Ordovician deep-water clastics that have been transported westward above the slightly deformed carbonate platform rocks of the early Ordovician passive margin of North America. On the seismic section, the reflectors B form a horizontal alignment at ~3.0 sec, and are interpreted as the base of the platform rocks lying on Grenville basement. This surface is traced eastward to ~*km:25* on Line 3, beneath the Green Mountain massif, where it rolls over to an eastward dip, reaching at least 4 sec at the end of the line (Connecticut River). Weak, east-dipping reflectors C extend beneath the Green Mountain massif and the Connecticut Valley synclinorium, defining the principal structural direction in the eastern part of the allochthon.

The continuation of the traverse eastward into New Hampshire on Line 6 (Fig. 16) extends across the Bronson Hill anticlinorium and into the Merrimack synclinorium[2]. The fault that

[2]Now divided into the central Maine–Kearsarge synclinorium and the Merrimack Trough, based on relations in southwestern Maine. At the scale at which we are dealing with the seismic section this distinction is not yet useful.

defines the contact between Grenville basement and allochthonous rocks of the orogen (Fig. 15, B) can be followed from 4 sec (F) to at least 6 sec (G). Parallel to F-G is the feature C-D-E?, which appears as an east-dipping western basement to the Bronson Hill–Merrimack belt, and is correlative with G-H of Figure 12b, which we interpreted as a major suture. Moho (A-A) is identifiable, and shallows from 10 to 9 sec across the section, just as on the Long Island Platform line (Fig. 12b). The interior of the Bronson Hill–Merrimack belt B shows weak, west-dipping events, which roll over into a narrow east-dipping boundary zone near the suture C-D-E (as seen in J in Fig. 12b). The Bronson Hill anticlinorium has been carefully mapped along much of its length, and consists in detail of multiple gneiss-cored domes with steep axial planes that have been refolded in places from original west vergence to east vergence (Robinson, 1979). In the seismic section, it is seen to be recumbent, lying above the suture C-D-E, and beneath the Merrimack stack. In the upper part of the Merrimack, the reflections are extremely weak, and we cannot establish any correlation with the plutons seen in surface exposure. It is difficult to draw any real inferences about the bright, patchy, horizontal reflections H to the east at the base of the Bronson Hill–Merrimack belt. Although quite easily attributable to complex structures, their position at the base of the terrane, with its ubiquitous intrusive granite sheets, suggests the possibility that these might be related mafic intrusives. We also do not see any indication at the east end of the line, of the strong, west-dipping lower crustal wedge, such as M on Line 36, which would have indicated Hope Valley basement. Indeed, the widening northward of the interior terranes (Fig. 11), suggests the V-shape that they assume in the south must give way to a broader structure toward the north; a west-dipping eastern contact with Avalonian basement is expected to lie another 40 km off the section.

Grenville basement begins to thin under the eastern edge of the Green Mountain massif (Fig. 15), and appears to vanish somewhere in the strongly deformed deep crust east of *km:30* on Line 6 (Fig. 16). Grenville rocks in the Green Mountains and in the Chester dome seem to have been tectonically plucked off the continental margin at different levels along the surface F-G. At surface outcrop, the supracrustals transported above fault F-G appear in eastern Vermont as the deformed metasediments of the Connecticut Valley synclinorium. Cook and others (1979) proposed that east-dipping reflectors near the edge of the Grenville basement in Georgia may be due to continental slope sediments, covered in situ by allochthonous crystalline thrust sheets. The example in New England indicates that their interpretation of this feature may be broadly correct, but that the reflections are produced by strongly deformed metamorphic rocks, not by original sedimentary layering.

In current thinking about northern Appalachian geology, the North American Precambrian basement and the exotic Avalonian terranes bracket the medial Gander terrane (Williams and Hatcher, 1983), called Bronson Hill–Merrimack belt in this chapter. From the prevalence of Acadian metamorphic ages in this terrane, it is generally considered to have initially accreted to North America at that time; on Line 6, C-D-E is the suture. Recent argon-dating work, however, has yielded many Alleghanian uplift ages for both basement and cover, along with Carboniferous plutons through much of this area, suggesting that collisional tectonics continued to operate over much of later Paleozoic time (Gromet, 1989). From the high grades of metamorphism on this transect, it is inferred that the crust was, at the time of accretion, at least 10 km thicker than today, even after allowance for possible heating by the plutons. The Bronson Hill anticlinorium, which is regionally consistent as a tectonic element, appears on the seismic section as a west-verging recumbent nappe. In its position at the base of the accreted terranes, it appears to have functioned as a boundary layer on the obduction suture between the Merrimack rocks and the pre-Acadian North American margin.

Farther north, the more recent refraction-reflection traverse from Quebec, across Maine and the Gulf of Maine (Table 1, 1) has produced a substantial quantity of good data (LaRoche, 1983; Stewart and others, 1986; Unger and others, 1987; Hutchinson and others, 1987). At this writing, many of the data and interpretations are not in final release and publication, and we restrict discussion to a later paragraph with consideration of innovations in approach and technique.

Southern Appalachian transect, 1978–1979

COCORP lines across the southern Appalachians in 1978–1979 from the Tennessee Valley and Ridge province to the coastal plain were the first to establish the importance of reflection profiling in studying the structure of an orogenic belt (Cook and others, 1979, 1981). In the vicinity of the Blue Ridge of Georgia, Tennessee, and South Carolina, the Appalachian orogen is at its widest, 600 km, with more than 350 km exposed. The idea that the Blue Ridge province, with its Grenville-age basement slices and late Precambrian sediments and volcanics, had been transported westward more than 100 km on a master décollement was proposed by Hatcher (1971, 1972) and subsequently supported by a short reflection line (Clark and others, 1978). The 1978 COCORP work (Lines 1-4) not only imaged the master detachment, but provided enough lateral continuity to demonstrate westward transport of the Blue Ridge by at least 150 km along this detachment. The 1979 work (Lines 5–8) crossed the interior belts of the orogen and their eastward extension under the coastal plain (Cook and others, 1981). These lines showed considerable deep-rooted structure from the Kings Mountain Belt (denoted by MLCZ on Fig. 17) eastward, but left some ambiguity about the location of the root zone for the thin-skinned thrusting and the eastward limit of Grenville basement. Moreover, the identification in the eastern Piedmont of rocks of Avalonian association, and the determination there of late Paleozoic metamorphic ages, has highlighted the question of where and in what form can be found the suture between the interior belts of the Piedmont and exotic Avalonian or African terranes. The 1983–1984 COCORP transect in western Georgia and Florida crossed this

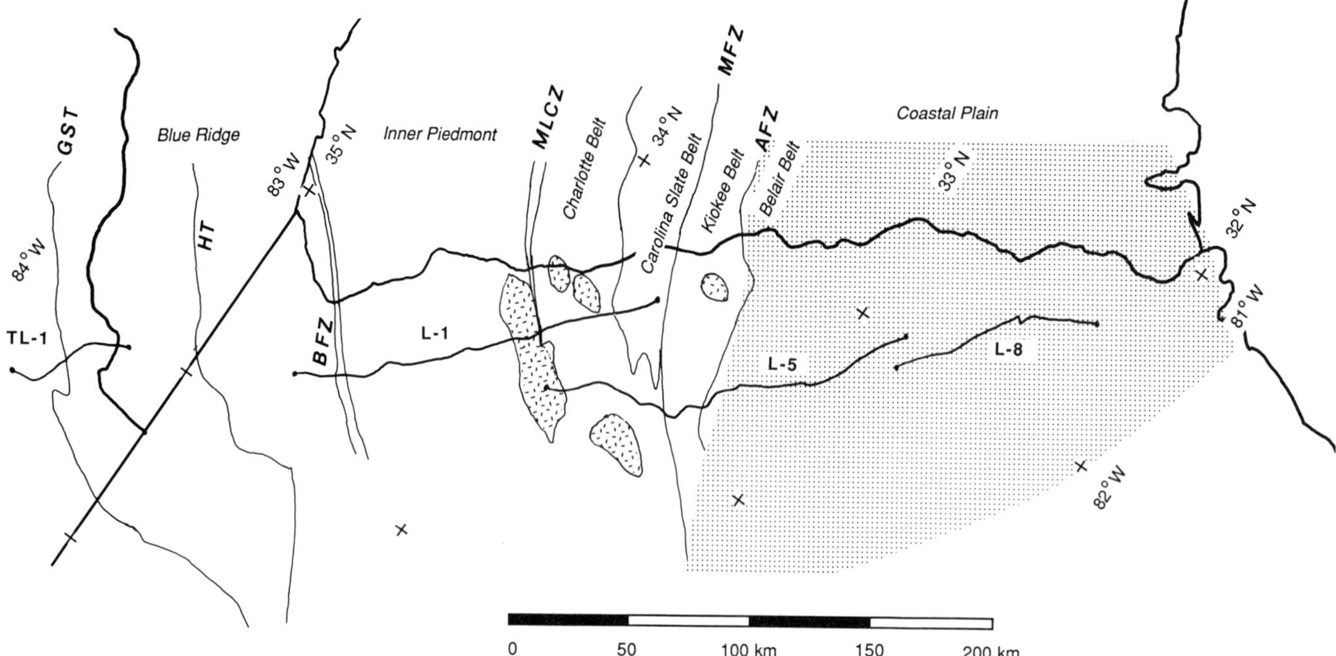

Figure 17. Map of principal 1978–1979 southern Appalachian COCORP lines: Tennessee Line 1 (TL-1) and Georgia Lines 1, 5, and 8. Major lithotectonic belts as indicated. Principal tectonic boundaries: GST = Great Smoky Thrust; HT = Hayesville Thrust; BFZ = Brevard Fault Zone; MLCZ = Middleton-Lowndesville Cataclastic Zone; MFZ = Modoc Fracture Zone; AFZ = Augusta Fault Zone. Hachured areas indicate plutons; stippled area indicates coastal plain.

suture at a key place, but left the problem open for the rest of the southern Appalachians northeast of the transect. All these issues have remained controversial; in the next paragraphs we suggest solutions based on post-stack reprocessing of the 1978–1979 transect and on important geological and geochronological work since 1981.

Description and interpretation of the seismic section. The stacked sections from COCORP Lines 1, 5, and 8 have been coherency filtered, migrated (at 4.5 km/sec), and redisplayed into a summary image of the subsurface information. The relationship of these lines to the regional geology is shown in Figure 17. The processed images are given in Figures 18 and 19a–d. Lines 5 and 8 have been joined into a single composite, since their end points line up quite well along regional strike. Since this profile extends across four frames at the standard scale we have adopted, we show an interpretive cross section in Figure 20 for clarity, and have marked the major boundaries of Figure 20 in Figures 18 and 19a–d in overlay. Display parameters are chosen so that the weakest events are at what we judge to be the noise level. In Figure 20, same small and capital letters denote corresponding lithotectonic units on Line 1 (top) and the composite Line 5-8 (bottom), respectively. When the discussion pertains to both parts of the figure, both letters are used (e.g., the zone e/E).

The first impression of this section is of numerous reflection events that are scattered, in places quite densely, over the entire section. Few of these events behave according to the familiar model for a reflection: as a clean, impulsive wavelet that can be followed for substantial distances across section and distinguished from other reflection energy on the basis of amplitude and phase. In contrast, most behave like weakly backscattered energy from a structure defined by the local dip of lamination. We thus seek to develop an interpretation by identifying domains having smoothly behaved dip directions and supplementing this with identification of the more distinct, spatially continuous reflections. Interpretation thus involves a further stage of image analysis at a higher level than the simple *n*-trace coherency filter. We now look for coherent behavior on the scale of part or all of the section, and possibly along curved lines. We have constructed the solid curves of Figure 20 in two ways: (1) certain simple, clear, fairly impulsive events may be combined into a single smooth curve that connects across gaps due to poor S/N; and (2) smooth curves may be constructed so as to be tangent to local dip directions, as integral curves of the dip field—we single out those that separate two domains with different signal character or dip behavior. The distinction is not sharply drawn, for some integral curves of the dip field will be seen to define fairly clear events. Until this kind of interpretation is automated, in the fashion of the coherency filter, the judgment of the interpreter will be a substantial factor in determining the result. The data are presented in a form that will permit the reader to try alternative interpretations based on the primary data, the field of signals and dip directions. The shorter, lighter lines on Figure 20 are meant to suggest event density and dip on the section.

As an example of the first case, the Moho appears on the

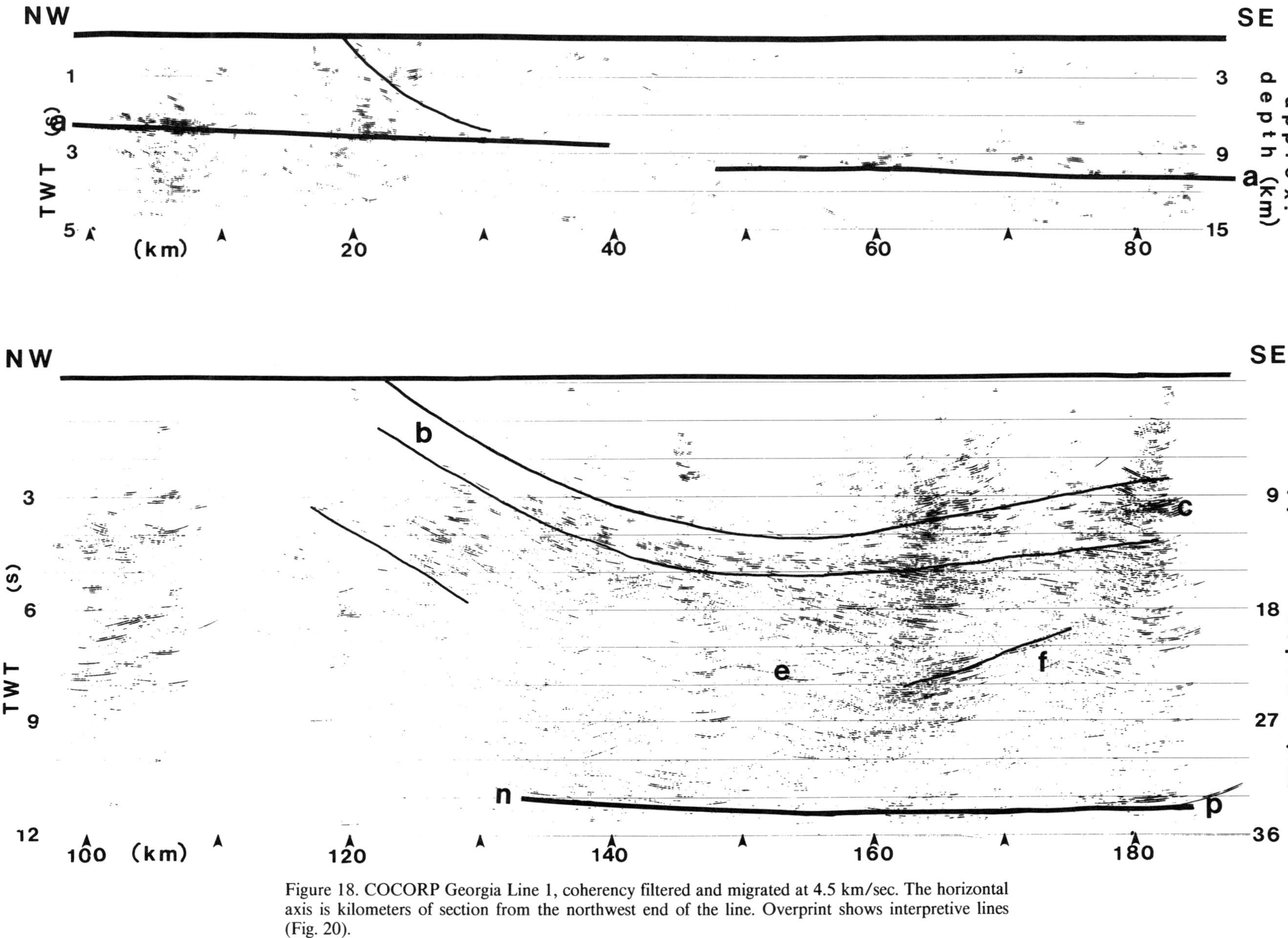

Figure 18. COCORP Georgia Line 1, coherency filtered and migrated at 4.5 km/sec. The horizontal axis is kilometers of section from the northwest end of the line. Overprint shows interpretive lines (Fig. 20).

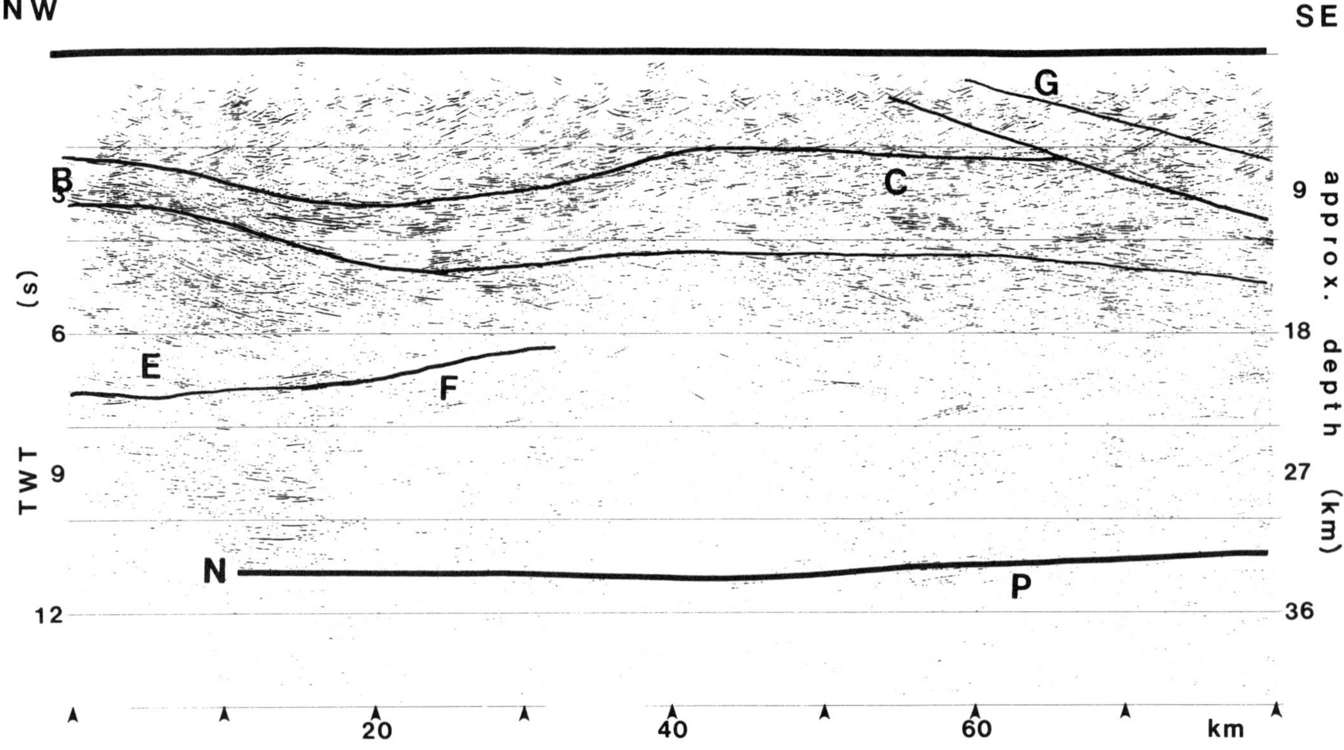

Figure 19a. COCORP Georgia Line 5, after coherency filtering and migration at 4.5 km/sec. *km:0-80* from the northwest end of line. Overprint shows interpretive lines (Fig. 20).

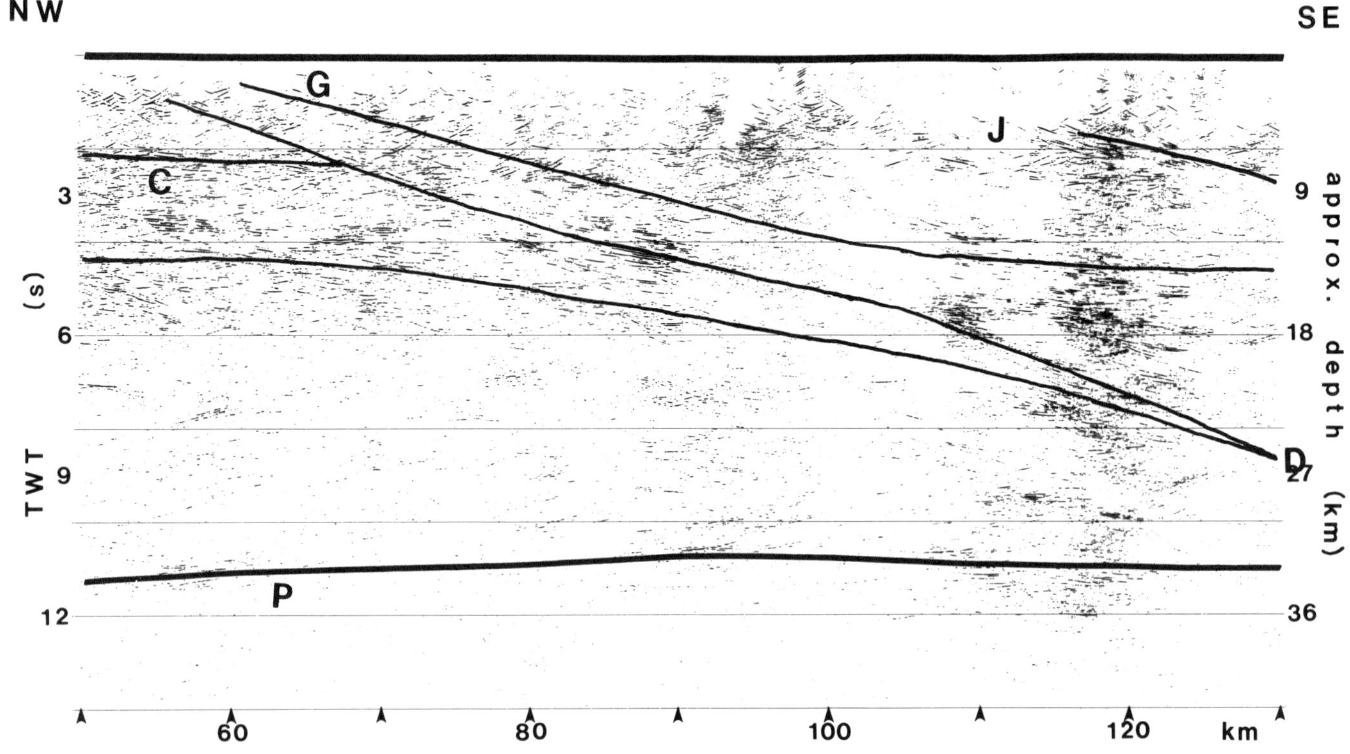

Figure 19b. COCORP Georgia Line 5, after coherency filtering and migration at 4.5 km/sec. *km:50-130* from the northwest end of line. Overprint shows interpretive lines (Fig. 20).

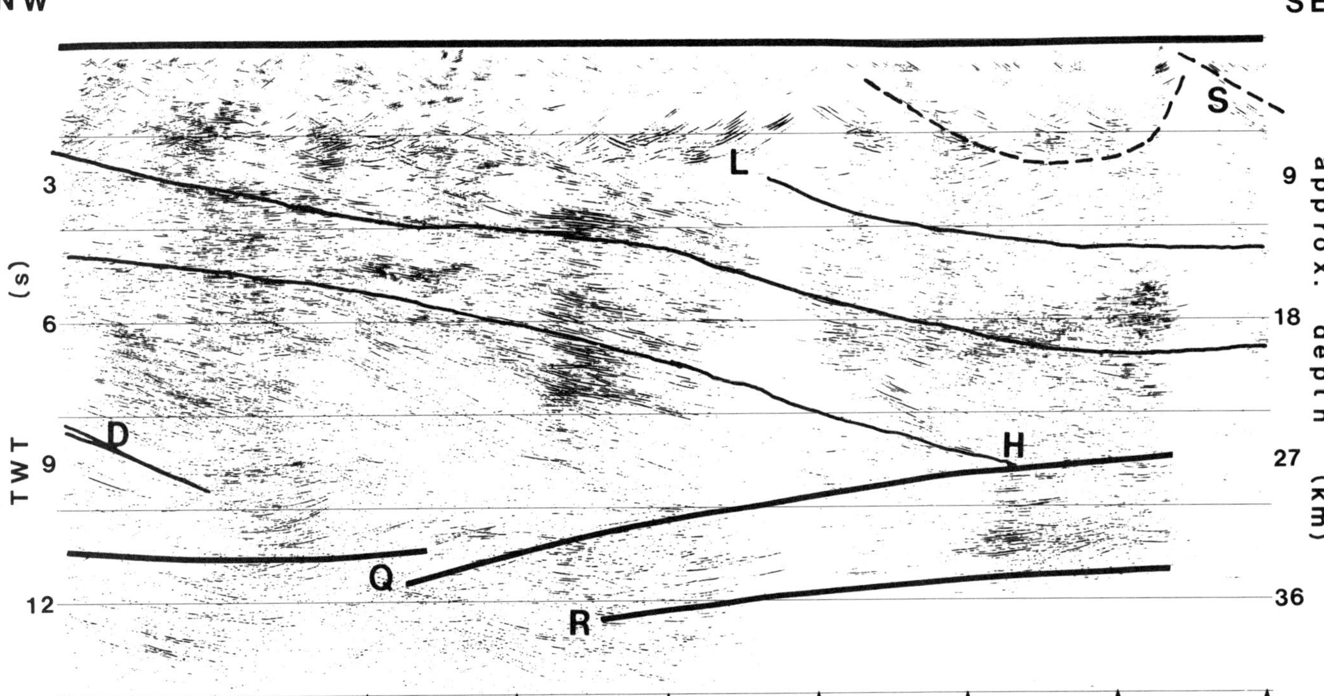

Figure 19c. COCORP Georgia Lines 5 and 8, concatenated after coherency filtering and migration at 4.5 km/sec. *km:130-210* from the northwest end of Line 5. Overprint shows interpretive lines (Fig. 20).

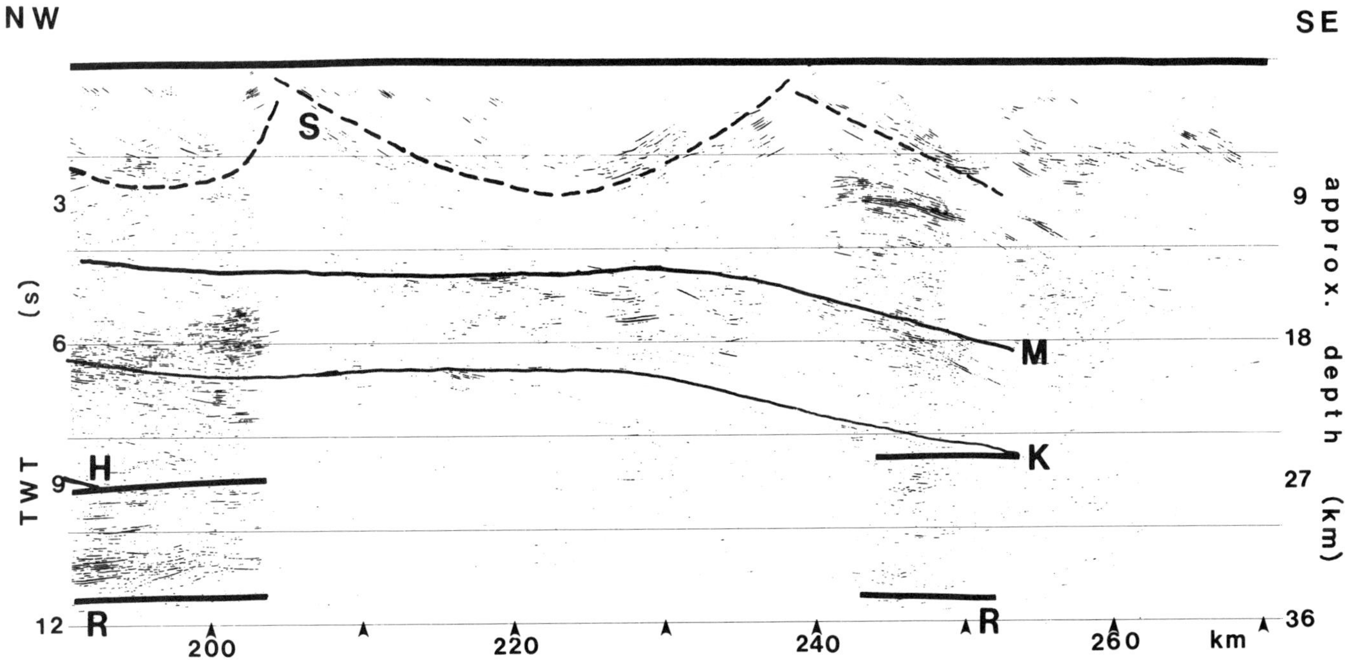

Figure 19d. COCORP Georgia Lines 5 and 8, concatenated after coherency filtering and migration at 4.5 km/sec. *km:190-270* from the northwest end of Line 5. Overprint shows interpretive lines (Fig. 20).

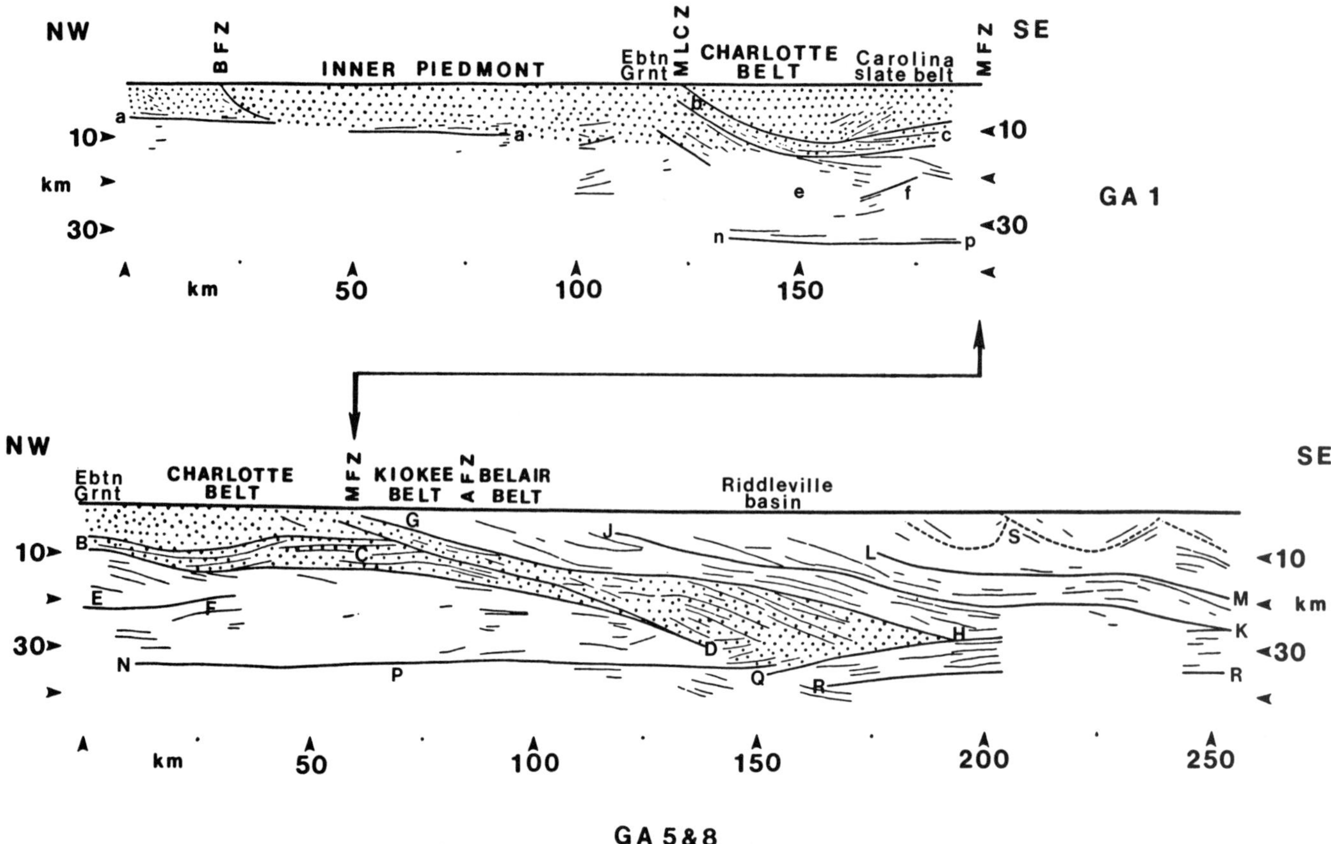

Figure 20. Composite interpretation of COCORP Georgia Lines 1, 5, and 8 (Figs. 18, 19a-d). See text for method used to create this interpretation. Letters are place markers referred to in the text. Dotted pattern = interior of Taconian orogen, extruded westward to form Alleghanian thrust sheets. Eastern packages, east of G-H-Q: Avalonian continental collider. Western basement, west of a-B-D: Grenville basement.

eastern portion of Line 1 as separated weak, but consistent, horizontal arrivals at ~11 sec near *km:135, 160, and 180* (Fig. 18). In Figure 19d, the lines Q-H and R-R define a slightly west-dipping basal package; Q-H is a boundary between two dip domains, while R-R is a boundary between a dip domain (above) and a transparent domain. The latter is a common way of defining the reflection Moho. In Figure 19a, B-C is an integral curve of the smoothly varying dip elements in the layer and separates that layer from an upper slab where fragmentary reflection events suggest steep dips and complex structure. The entire mid-crust east of *km:50* is full of east-dipping events, and many integral curves could be sketched in. G-H (Figs. 19b, c, 20) correlates at the surface with the Modoc Fracture Zone (MFZ), shows a few good events near the surface, defines the base of a complex surface zone, and then passes into an easily traced set of laminar reflections. With similar behavior, J-K (Figs. 19b, c, 20) correlates with the Augusta Fault, and L-M (Figs. 19c, d, and 20) passes into the western boundary of the Riddleville basin[3].

East of *km:180* (Figs. 19c, d, and 20), dip behavior in the top 10 to 15 km is less chaotic than elsewhere in the near-surface, and shows a sequence of alternations, which we interpret as a set of open folds. The dashed line is a representative integration of the dip directions, which shows the form of the folding, but is not proposed as a boundary. The unchaotic appearance of the dip directions in this upper region helps validate our interpretation that the complex shallow dip behavior under the Charlotte Belt (*km:0* to *km:50*) is not entirely due to stacking problems but reflects complex structure.

Discussion of geology and tectonics. Table 3 is a summary of the regional geology along the line of the transect, and is based principally on the Glover and others (1983) summary of recent geochronologic information. The region is divided into a series of basement belts, characterized by general amphibolite facies metamorphism and internally by ubiquitous polyphase ductile deformation. The boundaries of these belts are identified in the field as mylonitic shear/fault zones of regional extent. Granitic plutons are intrusive into these belts; along this particular section are the Acadian-age Elberton Granite and several small Alleghanian-age plutons. The Modoc Fracture Zone divides Taconian and Alleghanian metamorphic ages. The Blue Ridge and Inner Piedmont belts are allochthonous; this large composite

[3]Only a hint of the base of the Riddleville is seen in our processing at *km:154* and 5-km depth. See Peterson and others, 1984).

TABLE 3. GEOLOGICAL SUMMARY SOUTHERN APPALACHIAN TRANSECT COCORP LINES 1, 5, AND 8*

northwest to southeast

Principal Base,emt	Intrusive or Supracrustal	Fracture/ Fault Zones and location on line	Structure and Lithostratigraphy	Age of Regional Metamorphism
Eastern Blue Ridge				Taconian [and Acadian overprint]
		Brevard Fault Zone [L 1, *km:20*]		
Inner Piedmont	Chauga Belt Elberton Granite [Acadian]		epiclastics and metavolcanics: mica schists, quartzites, amphibolites quartzofeldspathis paragneisses	Taconian [and Acadian overprint]
		Middleton-Lowndesville Fracture Zone [L1, *km:170;* L5-8,--]		
Charlotte Belt	Carolina Slate Belt Small plutons [Alleghanian]			Taconian
		Modoc Fracture Zone [L1,*km:230*; L5-8,*km:55*]	[Alleghanian ductile deformation front]	
Kiokee Belt				Alleghanian
		Augusta Fault Zone [L5-8,*km:65*]		
	Belair Belt			
Coastal Plain Onlap				

L1; L5-8

*Based on Glover and others (1983 and Hatcher and others (1980).

crystalline thrust sheet thickens eastward to about 12 km near the Middleton-Lowndesville cataclastic zone. The location of the root zone for this thrusting has been controversial, with Hatcher and Zietz (1980) and Iverson and Smithson (1982) preferring the vicinity of the cataclastic zone and Cook and others (1981), suggesting that the thin-skinned regime extends eastward at least another 100 km. The Carolina Slate Belt, the most extensive of the supracrustal belts in the Carolinas, extends southward across Line 1, but ends before reaching Line 5, probably due to northward plunge of an open synformal package. The supracrustal Kings Mountain Belt, also significant in the Carolinas, narrows to a ~1-km-wide fracture zone at Line 1 and does not appear to extend as far south as Line 5; we adopt the terminology of Rozen (1981) and Griffin (1981) and call it the Middleton-Lowndesville Cataclastic Zone (MLCZ). In the discussion of the seismic data that follows, we refer to the composite interpretation (Fig. 20) and use key letters that appear on the sections (Figs. 18, 19a–d).

We consider the cross-sections in Figure 20 in light of three main questions: (1) the extent of the great southern Appalachian crystalline allochthon, in which rocks with Taconian metamorphic ages were thrust into their current geometry during Alleghanian orogeny; (2) the nature of the basement beneath the eastern part of the allochthon; and (3) the geometry of the eastern terrane, which was brought into its current position by the Alleghanian orogeny and drove the Taconian rocks westward over North American basement.

The allochthon. Since the discovery of the regional extent of the Blue Ridge–Inner Piedmont crystalline thrust sheet by Cook and others (1979), controversy has continued over the position, to the east, of the root zone for this thrusting. The set of east-dipping reflectors near b at *km:125–160* has a plausible geometry for this root zone, and extrapolates to the MLCZ at the surface. Harris and others (1981) and Lillie (1984) considered these reflections to be strata preserved within Precambrian half grabens. Hatcher and Zietz (1980) proposed the Kings Mountain Belt for the root zone on the basis of magnetic data. Iverson and Smithson (1982) reprocessed parts of the COCORP seismic data and supported this proposal. Cook and others (1979, 1981) preferred to locate the root zone further east, based on the presence of horizontal layered reflections east of the MLCZ at ~3 to 4 sec. Evidence from the reprocessed sections permits us to propose a resolution of this issue. Of particular interest is the 50-km zone of overlap when Lines 1 and 5–8 are projected onto regional dip.

In this zone a well-layered subhorizontal package, ~3 to 8 km thick, assumes the form of an open syncline directly under the Charlotte Belt. Beneath the syncline, a zone marked e/E is bounded on the east by a west-dipping reflection f/F and has the form of a deep pillow or wedge with bottom at a depth of ~20 km. The west limb of the syncline turns up to the surface on Line 1 where it coincides with the cataclastic zone, but it flattens out on Line 5 at a depth of 8 to 10 km beneath the conjectural extension of this zone (*km:20*). We infer that the layering continues westward without break into the Inner Piedmont area where there is general agreement about the existence of the main detachment at a depth of ~10 km. The main detachment can be followed eastward along the path B-C-D, where it roots near the Moho, ~130 km east of the cataclastic zone. This solution is similar to that preferred by Cook and others (1981). We conclude that the Middleton-Lowndesville Cataclastic Zone is a late splay off the main detachment, similar to the Brevard Fault Zone farther west (Line 1, *km:20–25*). Additional support for this interpretation is the contrast in dip behavior between the band B-C, where layering is nearly horizontal, and the superjacent package forming the Charlotte belt, where layering is variable in dip and discontinuous; this would characterize an overthrust geometry, with the surface package having been compressed and transposed into steep folds above a detachment. The Carolina Slate Belt (CSB) cannot be deeply rooted, given the continuous subhorizontal layering beneath. Indeed, it is questionable whether the group of west-dipping events at *km:150-165* just above the detachment on Line 1 is easily reconciled with the shallow synform inferred for the CSB on the basis of surface outcrop. We suggest that the CSB, which does not cross Lines 5–8, is probably rooted too shallow on Line 1 (<3 km) to have been properly imaged.

The change in appearance of the basal detachment from a single reflector (west of *km:60* on Line 1) to a thick layered package (B/b-C/c) may be detectable west of the cataclastic zone in the pair of reflectors at *km:100,* between 8 and 11 km deep. There is insufficient evidence to judge whether the master décollement lies at the top or bottom of this package. The evident detachment of strongly deformed Charlotte Belt rocks above the layered reflectors does not prevent the master décollement from lying deeper. Indeed, the ADCOH profiling of the main thrust mass in the Blue Ridge (just off the left end of Line 1) has shown the same compound geometry: a major low-angle thrust within the thrust sheet, with a transposed surface package above and a laminated package below (Coruh and others, 1987; Hatcher and others, 1987). In the event that B/b-C/c has not traveled very far, its role as a cover package to the basement beneath would bear investigation. It would be a possible root for the late Precambrian rift-related sequences found in the allochthonous western Blue Ridge.

In Figure 20, we have marked (dot pattern) the great thrust sheet, having Taconian metamorphic ages and an Alleghanian emplacement. This pattern is extended downward to the detachment and eastward into the subsurface on Lines 5–8 to a deep, V-shaped package. This is interpreted to be the pre-Alleghanian root of the Taconian orogenic welt (*km:120–180*). If these Taconian thrust sheets are retrodeformed back into a V-shaped keystone, centered at *km:150,* and with the notch at the depth of the present-day Moho, the resulting feature would be roughly ~150 km wide at the surface.

The basement. The picture just presented implies that the basement underlying the main thrust sheet be pre-Ordovician, possibly Grenvillian. The easternmost point of this basement would be under D, at *km:140* on Line 5–8. We argue that it is in fact Grenville basement and that the large antiformal arch (crest at *km:50,* Line 5–8) is a fragment of North American basement

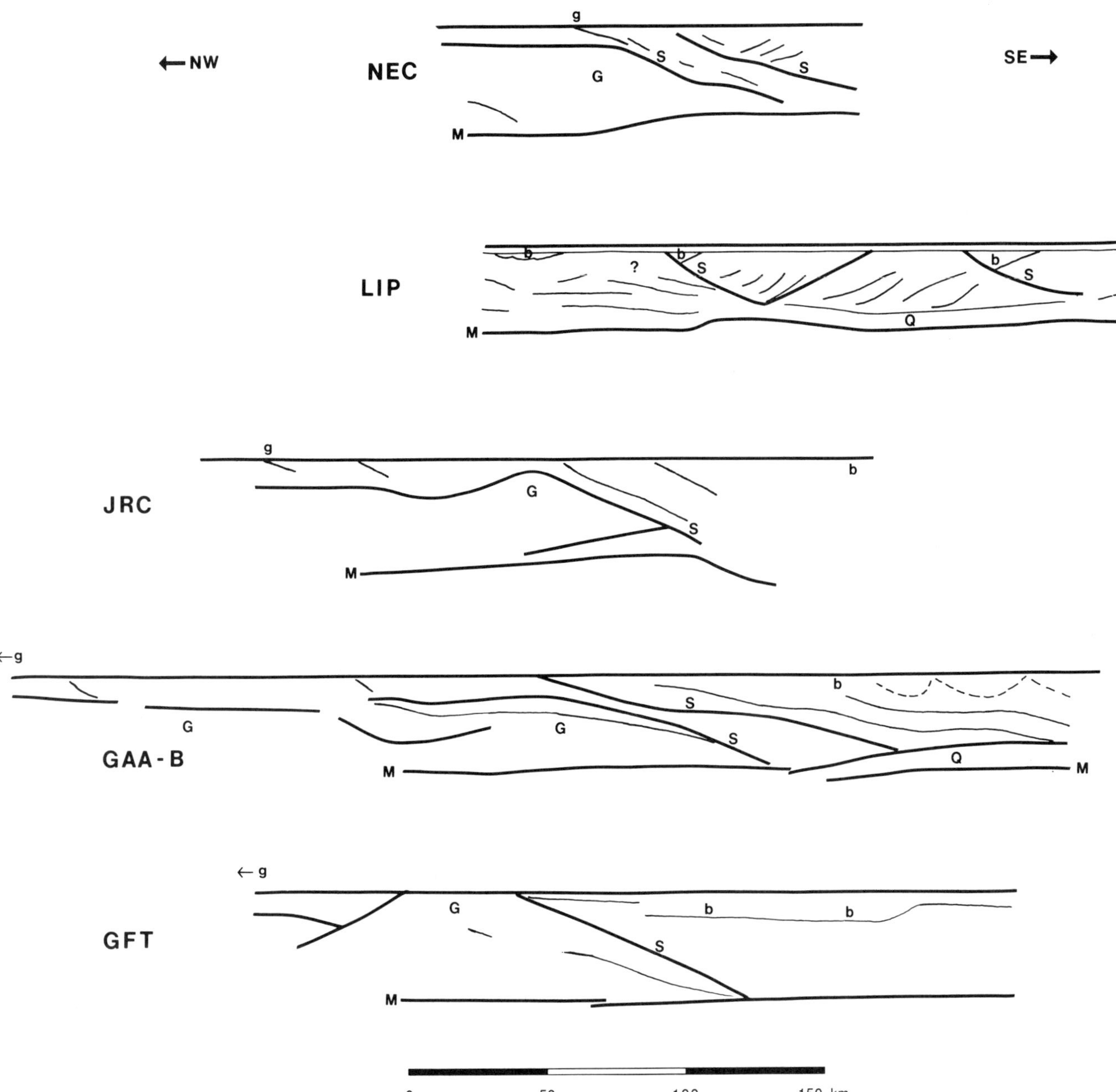

Figure 21. Comparative anatomy of five cross sections of the Appalachian orogen. NEC (New England COCORP transect), LIP (Long Island Platform), and GAA-B (Georgia COCORP transects) are rediscussed in this chapter; JRC (James River Corridor) is adapted from Pratt and others (1987), and GFT (Georgia-Florida COCORP transect, unmigrated) from Nelson and others (1985a) neither are reinterpreted here. g = westernmost surface appearance of thrust slices of Grenville basement. b = locations of Mesozoic rift basins. Features for comparison: G = Grenville basement; M = Moho; S = major crust-cutting sutures; Q = eastern accreted crust shows thick horizontal lower crustal boundary layer.

that was partially rifted away from the continent during late Precambrian rifting. The argument is made by reference to the interpretation by Pratt and others (1987) of the traverse in central Virginia and the latest COCORP lines in western Georgia (Nelson and others, 1985a, b, 1987b), which have been redrawn for comparison in Figure 21. Both lines show a major basement high just west of the principal crust-penetrating suture S. In Virginia, this is exposed as the Grenville-age Goochland terrane, and in western Georgia, as the Pine Mountain Belt (Clark, 1952; Nelson and others, 1987a), consisting of Grenville basement with metamorphosed cover rocks. In Virginia, the synform west of the Goochland terrane exposes rift-related volcanics and clastics. We suggest that these may be more or less in place and locate a failed rift. On Georgia Lines 1 and 5–8, we propose that the basement feature E/e, which has the form of a covered basin or pillow, is a similar late Precambrian rift that did not completely open. It may be noted that Long (1979), based on gravity data, argued that the Carolina Slate Belt represents a paleo-rift. The seismic section leads us to conclude that the slate belt is not rooted, and that the observed gravity field is due to the failed rift underlying the thrust sheet.

The eastern terrane and the Alleghanian suture. The change from Taconian to Alleghanian metamorphic ages at the Modoc Fracture Zone is a critical constraint in understanding the eastern part of Line 5–8. A major interpretive boundary, G-H, is traceable eastward from the fracture zone to a point 140 km away (*km:200*), where it splits off from a pronounced lower crustal layered package Q-H-K-R. This extension of the fracture zone to depth is proposed as the Alleghanian suture. The crustal section of the eastern terrane can be seen at the eastern end of the line, where the mid- to lower crust is composed of strongly layered, nearly horizontal packages, and the upper crust (nearly 20 km thick) has been partially detached and folded. These broad folds show some indication of becoming tighter and more overturned to the west (*~km:180–200*), but the proposed structures become smaller and more difficult to image. The entire section above the lowest layer appears to have delaminated at H; the upper section is thrust upward along the fracture zone, and the lowest layer is subducted. A group of east-dipping reflections (Fig. 19c: *km:130–170, 12–13s*; Fig. 20) is not as easily integrated into a large structure as the west-dipping package Q-H-K-R, but suggests that a reversal of this subduction may have delaminated a fragment of the North American crust.

The terrane that collided with North America during the Alleghanian collision and drove the Blue Ridge thrust sheet into place thus has the appearance of a stack of mega-layers. The whole stack above the Modoc Fracture Zone has been tilted eastward, exposing deeper levels in the terrane at the western end of the section. The interior boundaries (whose exact locations might be chosen differently by different interpreters) are major detachments themselves. J-K can be followed to the outcrop of the Augusta Fault Zone, a major crustal fault, and L-M bounds the folded package above from the laminar package below. Moreover, L-M is arguably continuous with the active western border fault of the Triassic Riddleville basin (Peterson and others, 1984).

The Avalonian affinity of the eastern terranes in the Piedmont is now quite well established. The geometry of collision of Avalonian crust in Alleghanian time can be compared in Figure 21, for the Long Island Platform (LIP), the Virginia profile (JRC), and the COCORP Georgia lines under discussion (GAA-B). In each case there is nearby correlative age dating to identify the eastern terrane as Avalonian (Zartman and Naylor, 1984; Gromet, 1987; Glover and others, 1983). In each case, the terrane is delaminated and thrust upward on an east-dipping suture, and a lower crustal layer (Q) appears with the geometry required for the subducting crust/lithosphere below. The failure of the west Georgia line (GFT) to show this geometry is not surprising since some major changes in subsurface geology occur along strike between GAA-B and GFT (Nelson and others, 1985a; Tauvers and Muehlberger, 1987), and that collider is considered to be a fragment of "Africa."

Comments on the Moho and rifting

It is a matter of general observation that Moho reflections in the continental interior are much harder to observe than those from either the Basin and Range or the Appalachians. The dominance of Neogene rifting in the structure of the Basin and Range has made it possible to identify the key tectonic features of continental crust undergoing extension: (1) the formation of detached, normal-faulted brittle upper crust over a metamorphic lower crust that undergoes ductile extension, with a region of mylonitization at the boundary; (2) the occurrence, in some areas, of extension-related mafic volcanism; and (3) the appearance of a strongly reflective Moho on seismic reflection profiles (Klemperer and others, 1986; Brown, 1986). Although the deeper geometry of many of the normal faults is controversial, many examples have been found of normal faults that move on reactivated thrusts of Laramide or earlier origin, as well as those which have clearly cross-cut older thrusts.

In the Appalachians, late, postorogenic extension does not dominate the modern crustal structure, but appears (with some exceptions) as a minor modifier of collisional structures built up during the Paleozoic orogenies. The importance of extension should not be underestimated, however, since many areas in the orogen stood well above sea level for $\sim 10^8$ yr in late Paleozoic time. Modern thick-crustal analogs (Andes, Tibet) show gravitationally driven extension in parts of the mountain belt as a necessary response to the geometries produced by collision (Tapponier and others, 1981; Armijo and others, 1986; Allmendinger and Eremchuk, 1987). Carboniferous basins in New England and the Maritime Provinces show intraorogenic extension going on during late Paleozoic time.

On the Appalachian seismic sections, the Mesozoic basins are signposts for extension. Some of these have opened along preexisting thrusts or sutures. On the Long Island Platform line, the overall extension is probably greatest of the sections discussed in this chapter, for the line is quite close to the edge of the continental shelf. This is manifest in the slightly thinner crust, with elevation below sea level, the consistently bright Moho and lower crustal reflections, and the number and size of half-grabens imaged. The exceptional brightness of the lower crust under the eastern suture/normal fault confirms the relation between rifting and lower crustal/Moho reflectivity and supports the suggestion that this is due to intrusion of mafic magmas beneath and into the lower crust.

On all the sections in Figure 21, the westernmost appearance of supportable Moho reflections occurs about 50 km west of the surface trace of the westernmost suture. This is approximately the locus of the Appalachian gravity gradient (Society of Exploration Geophysicists, 1982) separating low Bouguer (~–75 mgal) regions in the interior from high Bouguer (~0 mgal) gravity in the Piedmont and Coastal Plain. We suggest that the entire Appalachian orogen east of this line underwent general extension and thinning of the order of 5 to 10 percent in the course of the opening of the Atlantic. The consequence would be that nearly every major thrust, detachment, or suture may have been reactivated in normal faulting. This should be assumed, even if the subsequent stripping of the surface by erosion has eliminated the characteristic fault basins.

A final implication that we draw from these ideas concerns the nature of the Moho itself. We believe that the continental Moho is generally a zone of mechanical detachment between crust and mantle. This follows from its role as a major chemical boundary. It will have undergone shear and relative motion any time a tectonic episode induces substantial vertical or horizontal motion. Initially, continental Moho in many places may be brought into being by the tectonic juxtaposition of an accretionary wedge or complex above subducting oceanic lithosphere. It is then subject to modification at subsequent times, being a natural boundary for the localization of strain and the relative movement of crust and mantle. This tendency is enhanced by the natural contrast in mechanical properties, which is a consequence of the large compositional contrast. Such a Moho should be considered a mechanical boundary layer, and may be several kilometers thick, appearing as a laminated lower crustal layer rather than as a simple sharp boundary.

Recent work in the Southern Appalachians

A period of emphasis in the 1970s on the Charleston earthquake led to COCORP and USGS programs, including local profiles in the Charleston area and regional lines both onshore and offshore (references in Table 1). A regionally important Jurassic sill near the base of the coastal plain sediments has made it difficult to build a story about the tectonics of this area based on the local lines alone. The picture has become somewhat clearer, with the acquisition, by COCORP, of a regional north-south line in western Georgia/northern Florida and a line along the Georgia coastal plain (Nelson and others, 1985a,b, 1987b). A major Triassic rift basin and a through-cutting suture of the crust were found, oriented east-west, which correlates with the Brunswick Magnetic Anomaly (BMA). This feature swings east and then northeast, as it passes offshore into the East Coast Magnetic Anomaloy. Nelson and others (1985a,b) and McBride and Nelson (1987) have used this correlation to argue that the BMA is the location of the Alleghanian suture of Africa to North America. Tauvers and Muehlberger (1987), using deep well data of Chowns and Williams (1983) and other evidence, have pointed to complexities in the coastal plain basement subcrop, which cast doubt on features of this model. The west Georgia suture bears a strong resemblance to the one we have shown on Lines 5–8, and appears to us to play the same role as the locus of Alleghanian collisions against North America. However, the distinctly different appearances on the seismic sections of the respective colliders (south and southeast of the suture), as well as many differences in the regional geophysical and geologic context suggest that the Alleghanian colliders are quite different in west Georgia and along the South Carolina border.

COCORP Ouachita transect

The Ouachita Mountains in Arkansas are the result of a Carboniferous collision of a north-traveling terrane with the south-facing Paleozoic passive margin (Nelson and others, 1982; Lillie and others, 1983a,b). They are characterized by a deformed (northern) foreland, an overthrust zone, a medial arch (the Benton uplift) that exposes lower Paleozoic platform rocks, and a southern zone of Carboniferous deep-water orogenic sediments. What makes this section particularly interesting in comparison with the Appalachians is that the core of the Ouachita orogen is not eroded to the same midcrustal level (~15 km) as the Appalachians. The Benton uplift, for example, rather than appearing in outcrop as a basement-cored allochthonous massif (Blue Ridge, Green Mountains), is deeply mantled by platform sediments. The COCORP lines show hardly any reflections from within the crystalline crust (below 20 km), but provide an excellent shallower section showing the structure assumed by the sediment mantle in response to the compression of the orogen and the movement and uplift of the core of the Benton. Because of the thickness of the sediments, the tectonically buried edge of the Paleozoic continent is even less accessible than in the Appalachians, there being no relevant exposures of deeper rocks. The PASSCAL Ouachita experiment (see below) was designed to look for this deeply buried transition in a southward extension of the COCORP line.

COCORP Michigan Basin

An early COCORP survey in the Michigan basin sought to find the structure of the Proterozoic basement beneath the very broad Paleozoic basin (Brown and others, 1982). Gravity and magnetic data clearly indicate the extension of the Keweenawan rift system under this area. This line is most instructive because of the great improvement produced by a careful reprocessing (Zhu and Brown, 1986). While the original results imaged the Keweenawan basin, the signal quality was poor, little detail was visible, and no crustal reflections were seen at all from beneath 14 km. The reprocessing involved a careful look at the field gathers, with the design of F-K and deconvolutional filters to remove signal-generated noise, followed by detailed velocity analysis. The revised section shows the 24-km-deep basin clearly, permitting the lower volcanic rocks to be distinguished from the upper clastic sediments. Weak lower crustal reflections include a distinct Moho event at 14 sec (42 km).

NEW TECHNIQUES IN CRUSTAL REFLECTION

Increasingly, conventional crustal reflection profiling is being extended by application of new field geometries, sources, and processing methods. In addition, reflection studies are now undertaken in concert with other geophysical work, and, as important, with correlative geologic mapping effort.

Quebec–Maine–Gulf of Maine

The Quebec–Maine–Gulf of Maine transect (Fig. 11), organized jointly by the U.S. Geological Survey and the Geological Survey of Canada, is illustrative of a new approach to crustal profiling, in which reflection profiling is combined with wide-angle refraction studies, earthquake data, gravity data, and special field work to approach the problem in a multi-institution effort. The main profiles were run in three stages:

1. A conventional line from the St. Lawrence River southeast to the Maine border was acquired on behalf of the Province of Quebec for hydrocarbon assessment purposes (LaRoche, 1983; St. Julien and others, 1983). On this line, the original Cambro-Ordovician rifted margin underlying the platform rocks is manifested through a series of step-down normal faults, deepening toward the southeast. The most distal of the Taconian allochthons were imaged above the platform rocks.

2. An 800-channel sign-bit VIBROSEIS line was run by the USGS across Maine in 1984-1985. This line spans a full cross section of the interior regions of the northern Appalachian orogen. Starting in the Taconian allochthons, it crosses the more easterly allochthonous units, the thinning out of Grenville basement, and the Gander terrane, ending in the coastal volcanic belt with its Avalonian affinities (Stewart and others, 1986; Spencer and others, 1987; Unger and others, 1987). The data are characterized by an unusually high fold and by an unusual amount of fine-scale detail in the stacked section. Major west-dipping structures that cross-cut the crust under the Merrimack synclinorium are clearly visible in this high-quality data set (The digital data have not been released at the date of revision of this chapter). A particularly important preliminary result is the successful imaging of two of the plutons crossed by the line, which appear as tabular, transparent zones, with a well-layered substrate, and a thickness of ~6 to 10 km. With nearly an order of magnitude more field data than conventional surveys, reprocessing this kind of data may be particularly difficult.

3. A marine multichannel survey was run across the Gulf of Maine, in combination with sonobuoy refraction work, to connect up with earlier USGS lines on Georges Bank (Hutchinson and others, 1987). This line is a first look at the easterly elements of the northern Appalachian orogen, the offshore extension of the Avalon terrane and the Meguma terrane, and provides a connection between the geology of the Bay of Fundy–Nova Scotia area and the Long Island Platform–southeast Massachusetts area. Excellent deep crustal reflections were reported, although the presence of glacial sediments on an irregular bedrock surface made the imaging of the crust here somewhat more difficult than was the case on the Long Island Platform.

A refraction survey, both parallel to the reflection lines (northwest-southeast), and perpendicular, along the axis of the Merrimack synclinorium (southwest-northeast), was conducted by the USGS (Luetgert, 1985). Five hundred to four thousand pound dynamite shots were spaced at 40-km intervals and fired into a 120-element array of refraction instruments at 800-m spacing. In this area, the nearly constant crustal velocity function has made it difficult to learn much from the first arrivals. Wide-angle reflections from deep crust and Moho, however, were readily detected, and could be used to develop a regional picture of crustal structure (Luetgert and others, 1987). A shot recorded broadside to the Merrimack axial deployment gave excellent fan coverage of deep crustal structure.

In a piggyback deployment, a 48-channel reflection array with 100-m group spacing was used to record seven large dynamite shots along the Merrimack synclinorium axis at distances of 12 to 150 km (Hennet, 1989). The close station spacing permitted an unaliased recording of the wavefield across the array, and demonstrates that nearly all the arriving energy at these large offsets consists of deep reflections. Figure 22 (right panel) shows one of these source gathers; wide-angle reflections are obtained from all depths in the mid- to lower crust (i.e., ~18 to ~21 sec). A semblance analysis of normal moveout of the energy around 20 sec (left panel) shows an apparent crustal rms velocity ~6.8 km/sec (valid on the assumption of flat layering). Figure 23 shows two of the 48-channel recordings along with a portion of the refraction recordings in CDP coordinates: time is NMO corrected, and traces are displayed at source-receiver midpoints that are collected in equidistant bins. It corresponds to a single-fold CDP section. The unaliased signals from the reflection arrays B show clearly the continuity and apprent velocity of the reflected waves; the lack of continuity of wave phase across the refraction stations A forces us to fall back on the wave amplitudes, which are much less consistent from trace to trace.

Several conclusions can be drawn from this study:

1. Dynamite sources of 500 to 2,000 lb provide sufficent reflection strength for crustal and even upper-mantle reflections out to offsets of 150 km or more. With complete coverage by reflection stations spaced less than 250 m apart, excellent single-fold CDP sections can be obtained for depths of 40 km or more.

2. Unaliased wavefield recording makes it possible to identify and utilize reflections that show poor coherence on conventional refraction spreads.

3. Signal strength and moveout at long offsets is adequate to infer good rms crustal velocities.

PASSCAL Ouachita experiment

A similar study was a large-scale north-south profile 200 km long in the southern Ouachitas, aimed at looking for the deep structure of the tectonically buried Ouachita margin on the southern extension of the COCORP Ouachita line (Braile and others, 1986). The experiment was sponsored by PASSCAL (Program for Array Seismic Studies of the Continental Lithosphere) as a demonstration of how data of the sort described above could be gathered systematically; 400 portable seismic recorders spaced

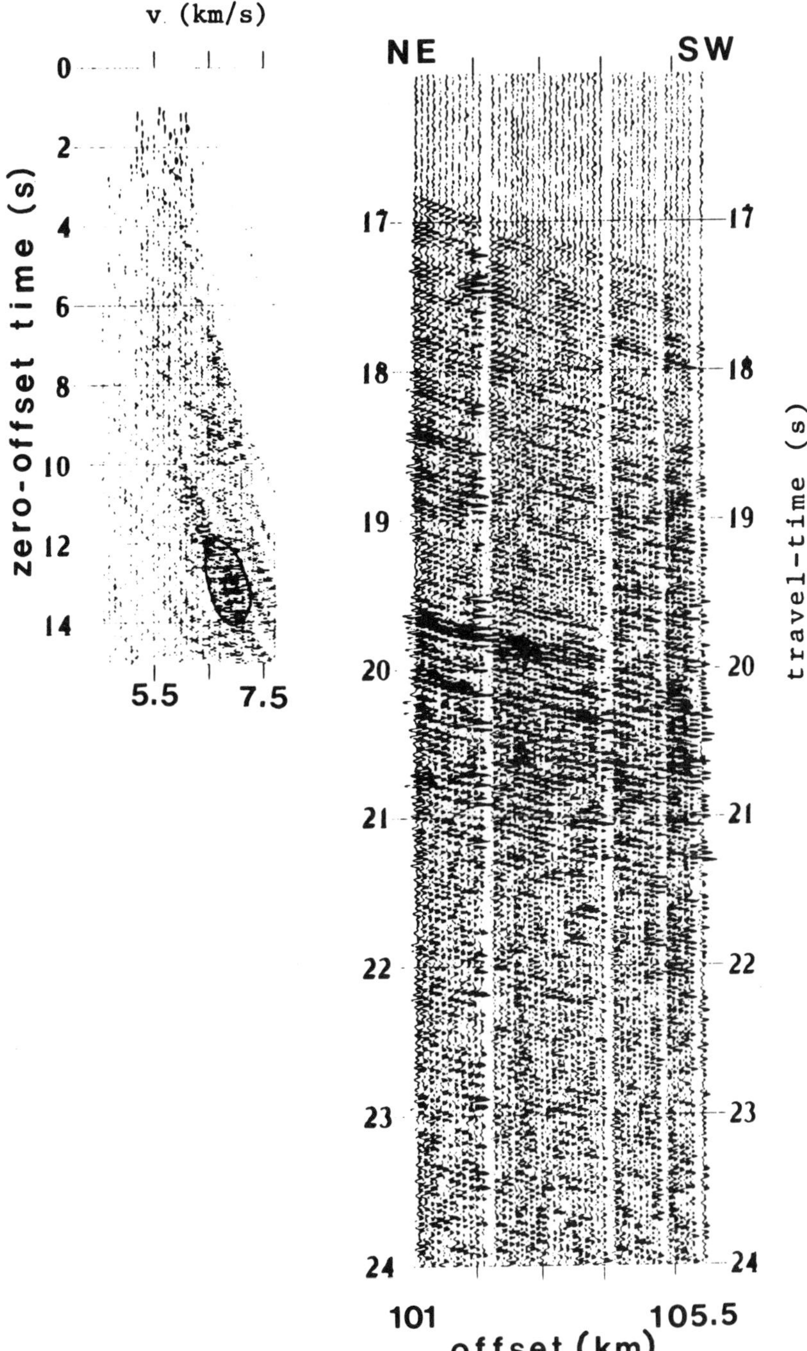

Figure 22. Wide-angle source gather from central Maine (Bottcher and Phinney, 1985), with a dynamite source fired into a 4.7-km, 48-channel spread at 101 km near offset. Strong reflection (at 19.7 sec on channel 1) is identified as PmP, a near-critical reflection from the vicinity of the Moho. Strong reflections in the 20- to 21-sec interval must be subcrustal. Semblance display in normal moveout velocity space (left) indicates an abnormally high rms crustal velocity around 6.8 km/s.

250 m apart gave an unaliased wavefield recording of 100-km aperture for each source; shots were spaced ~10 km. The data provide, for the first time, a test bed for combining the imaging and velocity information inherent in continuously recorded wide offset data. The goal is to develop a set of migration before stack procedures for long-offset data that will yield both a migrated subsurface image and a spatially variable velocity model.

FUTURE DIRECTIONS

We have emphasized the increasing importance of new data-acquisition techniques and processing ideas, particularly in the context of the increased availability of computing power and data-acquisition facilities adequate for the job. Similarly, new strategies are emerging for selecting areas for study. The original COCORP program of reconnaissance profiling will now be increasingly directed to the craton, where coverage has been so slight that very few generalizations are available to us after a decade. This will take the form of extending existing transects into the midcontinent. The value of specific problem-oriented reflection surveys optimized toward a particular target is now demonstrated (Coruh and others, 1987); the ADCOH program to site a proposed ultra-deep drillhole in the southern Appalachian overthrust has based its strategy on a detailed VIBROSEIS cross section. Programs of this sort are needed to fill in many gaps in our understanding of the deep structure of orogenic belts. Finally, reflection profiling can be optimized for even high-resolution studies, using shorter cables, single vibrators, shorter lines, higher frequencies and special processing (Roy-Chowdhury and Phinney, 1987). This technology is now available to individual investigators who wish to solve a specific field problem.

As this chapter illustrates, the availability now of numerous crustal reflection profiles from many different settings makes it possible to base inferences on the intercomparison of different lines, with critical similarities and differences. Moreover, few of these lines have been reimaged using state-of-the-art processing techniques. In consequence, major opportunities exist for reprocessing and reintepreting existing lines. For the student of Appalachian structure, for example, many profiles in the western cordillera, the Canadian Appalachians, the British Caledonides, and Europe are now available for careful comparison.

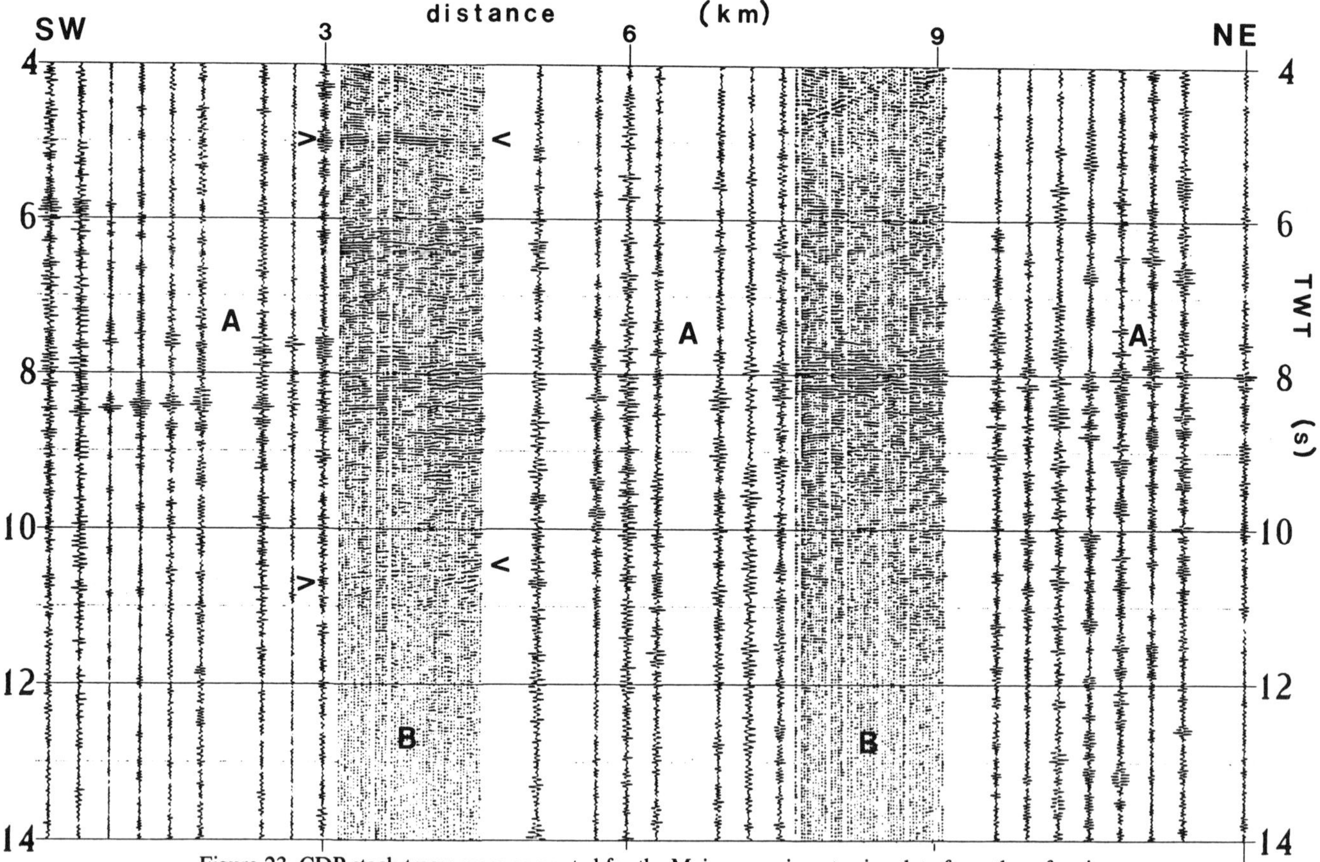

Figure 23. CDP stack traces were computed for the Maine experiment using data from the refraction stations spaced at 800 m (Stewart and others, 1986) and data from the reflection spread spaced at 100 m. Fold varies between 1 and 3. The display demonstrates the importance of close station spacing in establishing the phase coherence and slowness (and therefore the identification) of reflected signals. Moreover, it demonstrates that the high signal/noise ratio produced by explosive sources makes it possible to obtain reliable images of lower crust and Moho at very low fold.

ACKNOWLEDGMENTS

We thank Sharyn Magee for helping us with the reprocessing of the seismic data and the display and reproduction of the sections. This research was supported by the funds of the Department of Geological and Geophysical Sciences, Princeton University, and by National Science Foundation Grants EAR-8307597, EAR-8417894, and EAR-8520349. While we alone are responsible for the statements and interpretations contained in this essay, it has benefitted from comments by John Unger, Larry Brown, and three anonymous reviewers.

REFERENCES CITED

Allmendinger, R., and Eremchuk, J., 1987, Southward extension in the southernmost Altiplano-Puna plateau; Quaternary deformation of the Pasto Ventura region: EOS Transactions of the American Geophysical Union, v. 68, p. 415.

Ando, C. J., and 11 others, 1984, Crustal profile of Mountain Belt; COCORP deep seismic reflection profiling in New England Appalachians and implications for architecture of convergent mountain chains: American Association of Petroleum Geologists Bulletin, v. 68, p. 819–837.

Armijo, R., Tapponier, P., Mercier, J. L., and Tong-Lin, H., 1986, Quaternary extension in southern Tibet; Field observations and tectonic implications: Journal of Geophysical Research, v. 91, p. 13803–13872.

Bally, A. W., ed., 1983, Seismic expression of structural styles: American Association of Petroleum Geologists Studies in Geology Series 15, 3 vols.

Behrendt, J. C., 1986, Structural interpretation of multichannel seismic reflection profiles crossing the southeastern United States and adjacent continental margin; Decollements, faults, Triassic(?) basins, and Moho reflections, *in* Barazangi, M., and Brown, L., eds., Reflection seismology; The continental crust: American Geophysical Union Geodynamics Series, v. 14, p. 201–214.

Behrendt, J. C., Hamilton, R. M., Ackermann, H. D., and Henry, V. J., 1981, Cenozoic faulting in the vicinity of the Charleston, South Carolina, 1886 earthquake: Geology, v. 9, p. 117–122.

Braile, L. W., McMechan, G. A., and Keller, G. R., 1986, PASSCAL Ouachita lithospheric seismic study; Data and preliminary results: EOS Transactions of the American Geophysical Union, v. 67, p. 1101.

Brewer, J. A., and Smythe, D. K., 1983, The Maine and Outer Isle seismic traverse, *in* Bally, A. W., Seismic expression of structural styles: American Association of Petroleum Geologists Studies in Geology Series 15, v. 3, p. 3.2.1-23–3.2.1-28.

Brewer, J. A., and 5 others, 1981, Proterozoic basin in the southern Midcontinent of the United States revealed by COCORP deep seismic reflection profiling: Geology, v. 9, p. 569–575.

Brewer, J. A., Good, R., Oliver, J. E., Brown, L. D., and Kaufman, S., 1983a, COCORP profiling across the southern Oklahoma aulacogen; Overthrusting of the Wichita Mountains and compression within the Anadarko Basin: Geology, v. 11, p. 109–114.

Brewer, J. A., and 5 others, 1983b, BIRPS deep seismic reflection studies of the British Caledonides: Nature, v. 305, p. 206–210.

Brown, L. D., 1986, Lower continental crust; Variations mapped by COCORP deep seismic profiling: Annales Geophysicae, v. 5B, p. 325–330.

Brown, L. D., Jensen, L., Oliver, J., Kaufman, S., and Steiner, D., 1982, Rift structure beneath the Michigan Basin from COCORP profiling: Geology, v. 10, p. 645–649.

Brown, L. D., and 7 others, 1983a, Adirondack-Appalachian crustal structure; The COCORP northeast traverse: Geological Society of America Bulletin, v. 94, p. 1173–1184.

Brown, L. D., and 7 others, 1983b, Intracrustal complexity in the United States midcontinent; Preliminary results from COCORP surveys in northeastern Kansas: Geology, v. 11, p. 25–30.

Brown, L. D., Barazangi, M., Kaufman, S., and Oliver, J., 1986, The first decade of COCORP; 1974–1984, *in* Barazangi, M., ed., Reflection seismology; A global perspective: American Geophysical Union Geodynamics Series, v. 13, p. 107–120.

Chowns, T. M., and Williams, C. T., 1983, Pre-Cretaceous rocks beneath the Georgia coastal plain; Regional implications, *in* Gohn, G. S., ed., Studies related to the Charleston, South Carolina, earthquake of 1886; Tectonics and seismicity: U.S. Geological Survey Professional Paper 1313, p. L1–L42.

Claerbout, J. F., 1985a, Fundamentals of geophysical data processing: London, Blackwell Scientific Publications, 274 p.

——, 1985b, Imaging the earth's interior: London, Blackwell Scientific Publications, 398 p.

Clark, H. B., Costain, J. K., and Glover, L., 1978, Structural and seismic reflection studies of the Brevard ductile deformation zone near Rosman, North Carolina: American Journal of Science, v. 278, p. 419–441.

Clark, J., 1952, Geology and mineral resources of the Thomaston Quadrangle: Georgia Geological Survey Bulletin 59, 103 p.

Cook, F., and 5 others, 1979, Thin-skinned tectonics in the crystalline southern Appalachians; COCORP seismic reflection profiling of the Blue Ridge and Piedmont: Geology, v. 7, p. 563–567.

Cook, F. A., Brown, L. D., Kaufman, S., Oliver, J. E., and Peterson, T. A., 1981, COCORP seismic profiling of the Appalachian orogen beneath the coastal plain of Georgia: Geological Society of America Bulletin, v. 92, p. 738–748.

Cook, F. A., Brown, L. D., Kaufman, S., and Oliver, J. E., 1983, The COCORP southern Appalachian traverse, *in* Bally, A. W., ed., Seismic expression of structural styles: American Association of Petroleum Geologists Studies in Geology Series 15, v. 3, p. 3.2.1.-1–3.2.1.-6.

Coruh, C., and 9 others, 1985, Preliminary results from seismic reflection regional lines; Appalachian deep drill hole project site investigation: EOS Transactions of the American Geophysical Union, v. 66, p. 1075.

Coruh, C., and 5 others, 1987, Results from regional VIBROSEIS profiling; Appalachian ultra-deepcore hole site study: Geophysical Journal of the Royal Astronomical Society, v. 89, p. 147–156.

Gibbs, A. K., and 5 others, 1984, Seismic reflection study of the Precambrian crust of central Minnesota: Geological Society of America Bulletin, v. 95, p. 280–294.

Glover, L., III, Speer, J. A., Russell, G. S., and Farrar, S. S., 1983, Ages of regional metamorphism and ductile deformation in the central and southern Appalachians: Lithos, v. 16, p. 223–243.

Good, R., Brown, L. D., Oliver, J. E., and Kaufman, S., 1983, COCORP deep seismic reflection traverse across the southern Oklahoma aulacogen, *in* Bally, A. W., ed., Seismic expression of structural styles: American Association of Petroleum Geologists Studies in Geology Series 15, p. 3.2.2-33–3.2.2-37.

Griffin, V. S., Jr., 1981, The Lowndesville belt north of the South Carolina–Georgia border, *in* Horton, J. W., Butler, J. R., and Milton, D. J., eds., Geological investigations of the Kings Mountain Belt and adjacent areas in the Carolinas: Carolina Geological Survey Field Trip Guidebook, p. 166–173.

Gromet, L. P., 1987, Avalonian terranes of late Paleozoic tectonism in southeastern New England; Constraints and problems, *in* Dallmeyer, R. D., ed., Terranes in the Circum-Atlantic Paleozoic orogens: Geological Society of America Special Paper 230, p. 193–212.

Grow, J. A., Matlick, R. E., and Schlee, J. S., 1979, Multichannel seismic depth sections and interval velocities over the outer continental shelf and upper continental slope between Cape Hatteras and Cape Cod, *in* Watkins, J. S., Montedert, L., and Dickerson, P. W., eds., Geological investigations of continental margins: American Association of Petroleum Geologists Memoir 29, p. 65–83.

Grow, J. A., Hutchinson, D. R., Klitgord, K. D., Dillon, W. P., and Schlee, J. S., 1983, Representative multichannel seismic profiles over the U.S. Atlantic margin, *in* Bally, A. W., ed., Seismic expression of structural styles: American Association of Petroleum Geologists Studies in Geology Series 15, v. 2, p. 2.2.3-1–2.2.3-17.

Harris, L., and Bayer, K., 1979, Eastern projection of Valley and Ridge beneath metamorphic sequences of the Appalachian orogen: American Association of Petroleum Geologists Bulletin, v. 63, p. 1579.

Harris, L. D., Harris, A. G., deWitt, W., Jr., and Bayer, K. C., 1981, Evaluation of southern eastern overthrust belt beneath Blue Ridge–Piedmont thrust: American Association of Petroleum Geologists Bulletin, v. 65, p. 2497–2505.

Harris, L. D., deWitt, W., Jr., and Bayer, K. C., 1982, Intepretive seismic profile along Interstate I-64 from Valley and Ridge to the Coastal Plain in central Virginia: US.. Geological Survey Oil and Gas Investigation Chart OC-123.

Hatcher, R. D., Jr., 1971, Structural, petrologic, and stratigraphic evidence favoring a thrust solution to the Brevard problem: American Journal of Science, v. 270, p. 177–202.

——, 1972, Developmental model for the southern Appalachians: Geological Society of America Bulletin, v. 81, p. 933–940.

——, 1981, Thrusts and nappes in the North American Appalachian orogen, *in* McClay, C. R., and Price, N. J., eds., Thrust and nappe tectonics: Geological Society of London Special Publication 9, p. 491–499.

Hatcher, R. D., Jr., and Zietz, I., 1980, Tectonic implications of regional aeromagnetic and gravity data, *in* Wones, D. R., ed., Proceedings, The Caledonides in the U.S.A.: Blacksburg, Virginia Polytechnic Institute Department of Geological Science Memoir 2, p. 235–244.

Hatcher, R. D., Jr., Costain, J., Coruh, C., Phinney, R., and Williams, R., 1987, Tectonic implications of new Appalachian Ultradeep Core Hole (ADCOH) seismic reflection data from the crystalline southern Appalachians: Geophysical Journal of the Royal Astronomical Society, v. 89, p. 157–162.

Hawman, R. B., 1988, Wide-angle reflection studies of the crust and upper-mantle beneath eastern Pennsylvania [Ph.D. thesis]: Princeton, New Jersey, Princeton University.

Hawman, R. B., and Phinney, R. A., 1985, Crustal structure of southern Maine and eastern Pennsylvanian; Extremal inversion of wide-angle reflections: EOS Transactions of the American Geophysical Union, v. 66, p. 308.

Hearn, T., 1986, Reflection and refraction Mohos; A travel time discrepancy in western Nevada: EOS Transactions of the American Geophysical Union, v. 67, p. 1101.

Hennet, C. G., 1989, Coincident wide-angle reflection and refraction studies of the crustal structure in central Maine [Ph.D. thesis]: Princeton, New Jersey, Princeton University.

Hutchinson, D. R., Klitgord, K. D., and Detrick, R. S., 1985, The Block Island Fault; A Paleozoic crustal boundary on the Long Island platform: Geology, v. 13, p. 875–879.

Hutchinson, D. R., Grow, J. A., Klitgord, K. D., and Detrick, R. S., 1986a, Moho reflections from the Long Island platform, eastern United States, *in* Barazangi, M., and Brown, L., eds., Reflection seismology; The continental crust: American Geophysical Union Geodynamics Series, v. 14, p. 173–188.

Hutchinson, D. R., Klitgord, K. D., and Detrick, R. S., 1986b, Rift basins of the Long Island platform: Geological Society of America Bulletin, v. 97, p. 688–702.

Hutchinson, D. R., Trehu, A. M., and Klitgord, K. D., 1987, Structure of the lower crust beneath the Gulf of Maine: Geophysical Journal of the Royal Astronomical Society, v. 89, p. 189–194.

Iverson, W. P., and Smithson, S. B., 1982, Master decollement root zone beneath southern Appalachians and crustal balance: Geology, v. 10, p. 241–245.

Klemperer, S. L., and Brown, L. D., 1985, Simulations of noise rejection and mantissa-only recording; An experiment in high-amplitude noise reduction with COCORP data: Geophysics, v. 50, p. 709–714.

Klemperer, S. L., Brown, L. D., Oliver, J. E., Ando, C., and Kaufman, S., 1983, Crustal structure in the Adirondacks, *in* Bally, A. W., ed., Seismic expression of structural styles: American Association of Petroleum Geologists Studies in Geology, Series 15, v. 1, p. 1.5-12–1.5-16.

Klemperer, S. L., and 5 others, 1985, Some results of COCORP seismic reflection profiling in the Grenville-age Adirondack Mountains, New York state: Canadian Journal of Earth Sciences, v. 22, p. 141–153.

Klemperer, S. L., Hauge, E. C., Hauser, E. C., Oliver, J. E., and Potter, C. J., 1986, The Moho in the northern Basin and Range province, Nevada, along the COCORP 40°N seismic reflection transect: Geological Society of America Bulletin, v. 97, p. 603–618.

Kong, S. M., Phinney, R. A., and Roy-Chowdhury, K., 1985, A nonlinear signal detector for enhancement of noisy seismic record sections: Geophysics, v. 50, p. 539–550.

LaRoche, P. J., 1983, Appalachians of southern Quebec seen through seismic line 2001, *in* Bally, A. W., ed., Seismic expression of structural styles: American Association of Petroleum Geologists Studies in Geology Series 15, v. 3, p. 3.2.1-7–3.2.1-22.

Lillie, R. J., 1984, Tectonic implications of subthrust structures revealed by seismic profiling of Appalachian-Ouachita orogenic belt: Tectonics, v. 3, p. 619–646.

Lillie, R. J., and 7 others, 1983a, Crustal structure of Ouachita Mountains, Arkansas; A model based on integration of COCORP reflection profiles and regional geophysical data: American Association of Petroleum Geologists Bulletin, v. 67, p. 907–931.

Lillie, R. J., and 5 others, 1983b, Subsurface structure of the Ouachita Mountains, Arkansas, from COCORP deep seismic reflection profiles, *in* Bally, A. W., ed., Seismic expression of structural styles: American Association of Petroleum Geologists Studies in Geology Series 15, v. 3, p. 3.4.1-83–3.4.1-87.

Long, L. T., 1979, The Carolina slate belt; Evidence of a continental rift zone: Geology, v. 7, p. 180–184.

Luetgert, J. H., 1985, Depth to Moho and characterization of the crust in the northern Appalachians from 1984 Maine-Quebec seismic refraction data: EOS Transactions of the American Geophysical Union, v. 66, p. 1174.

Luetgert, J. H., Mann, C. E., and Klemperer, S. L., 1987, Wide-angle deep crustal reflections in the northern Appalachians: Geophysical Journal of the Royal Astronomical Society, v. 89, p. 183–188.

Matthews, D. H., and Cheadle, M. J., 1986, Deep reflections from the Caledonides and Variscides west of Britain and comparison with the Himalayas, *in* Barazangi, M., and Brown, L., eds., Reflection seismology; The continental crust: American Geophysical Union Geodynamics Series, v. 13, p. 5–19.

McBride, J. H., and Nelson, K. D., 1988, Integration of COCORP deep reflection and magnetic anomaly data in the southeastern United States; Implications for origin of the Brunswick and East Coast magnetic anomalies: Geological Society of America Bulletin, v. 100, p. 436–445.

McLelland, J. M., and Isachsen, Y. W., 1986, Synthesis of geology of the Adirondack Mountains, New York, and their tectonic setting within the southwestern Grenville province, *in* Moore, J. M., Davidson, A., and Baer, A. J., eds., The Grenville Province: Geological Association of Canada Special Paper 31, p. 75–94.

Mooney, W. D., Andrews, M. C., Ginzburg, A., Peters, D., and Hamilton, R. M., 1983, Crustal structure of the northern Mississippi embayment and a comparison with other continental rift zones: Tectonophysics, v. 94, p. 327–348.

Needham, D. T., and Knipe, R. J., 1986, Accretion- and collision-related deformations in the southern Uplands accretionary wedge, southwestern Scotland: Geology, v. 14, p. 303–306.

Nelson, K. D., and 7 others, 1982, COCORP seismic reflection profiling in the Ouachita Mountains of western Arkansas; Geometry and geologic interpretation: Tectonics, v. 1, p. 413–430.

Nelson, K. D., and 8 others, 1985a, New COCORP profiling in the southeastern United States; Part 1, Late Paleozoic suture and Mesozoic rift basin; Geology, v. 13, p. 714–717.

Nelson, K. D., and 5 others, 1985b, New COCORP profiling in the southeastern United States; Part 2, Brunswick and east coast magnetic anomalies; Opening of the north-central Atlantic Ocean: Geology, v. 13, p. 718–721.

Nelson, K. D., Arnow, J. A., Giguere, M., and Schamel, S., 1987a, Normal-fault boundary of an Appalachian basement massif?; Results of COCORP profiling across the Pine Mountain belt in western Georgia: Geology, v. 15, p. 832–836.

Nelson, K. D., and 6 others, 1987b, Results of recent COCORP profiling in the southeastern United States: Geophysical Journal of the Royal Astronomical Society, v. 89, p. 141–146.

O'Hara, K. D., and Gromet, L. P., 1985, Two distinct late Precambrian (Avalonian) terranes in southeastern New England and their late Paleozoic juxtaposition: American Journal of Science, v. 285, p. 673–709.

Oliver, J., Dobrin, M., Kaufman, S., Meyer, R., and Phinney, R., 1976, Continuous seismic reflection profiling of the deep basement, Hardeman County, Texas: Geological Society of America Bulletin, v. 87, p. 1537–1546.

Peterson, T., Brown, L., Cook, F., Kaufman, S., and Oliver, J., 1984, Structure of the Riddleville Basin from COCORP seismic data and implications for reactivation tectonics: Journal of Geology, v. 92, p. 261–271.

Phinney, R. A., 1986, A seismic cross section of the Appalachians; The orogen exposed, *in* Barazangi, M., and Brown, L., eds., Reflection seismology; The continental crust: American Geophysical Union Geodynamics Series, v. 14, p. 157–172.

Phinney, R. A., and Roy-Chowdhury, K., 1985, High-resolution seismics for evaluating the proposed site for the southern Appalachian deep drill hole: EOS Transactions of the American Geophysical Union, v. 66, p. 987.

Pratt, T. L., Coruh, C., and Costain, J. K., 1987, A geophysical study of the earth's crust in central Virginia and its implications for Appalachian crustal structure: Journal of Geophysical Research (in press).

Ratcliffe, N. M., Burton, W. D., D'Angelo, R. M., and Costain, J. K., 1986, Low-angle extensional faulting, reactivated mylonites, and seismic reflection geometry of the Newark basin margin in eastern Pennsylvania: Geology, v. 14, p. 766–770.

Robinson, P., 1979, Bronson Hill anticlinorium and Merrimack synclinorium in central Massachusetts, *in* Skehan, J. W., and Osberg, P. H., eds., The Caledonides in the U.S.A.; Geological excursions in the northeast Appalachians: Weston, Massachusetts, Boston College Department of Geology and Geophysics, p. 126–174.

Rodgers, J., compiler, 1985, Bedrock geological map of Connecticut: Connecticut Geological and Natural History Survey.

Roy-Chowdhury, K., and Phinney, R. A., 1987, Improved resolution of reflections from the crystalline upper crust: Geophysical Journal of the Royal Astronomical Society, v. 89, p. 35–40.

Rozen, R. W., 1981, The Middletown-Lowndesville cataclastic zone in the Elberton East Quadrangle, Georgia, *in* Horton, J. W., Butler, J. R., and Milton, D. J., eds., Geological investigations of the Kings Mountain Belt and adjacent areas in the Carolinas: Carolina Geological Survey Field Trip Guidebook, p. 174–180.

Schilt, F. S., Brown, L. D., Oliver, J. E., and Kaufman, S., 1983, Subsurface structure near Charleston, South Carolina; Results of COCORP profiling in the Atlantic Coastal Plain, *in* Gohn, G. S., ed., Studies related to the Charleston, South Carolina, earthquake of 1886; Tectonics and seismicity: U.S. Geological Survey Professional Paper 1313, p. H1–H19.

Setzer, T., and 5 others, 1983, A COCORP seismic reflection profile in northeastern Kansas, *in* Bally, A. W., ed., Seismic expression of structural styles: American Association of Petroleum Geologists Studies in Geology Series 15, v. 3, p. 2.2.1-7–2.2.1-12.

Sheriff, R. E., and Geldart, L. P., 1982, Exploration seismology; Volume 1, History, theory, and data acquisition: Cambridge, Cambridge University Press, 253 p.

——, 1983, Exploration seismology; Volume 2, Data-processing and interpretation: Cambridge, Cambridge University Press, 221 p.

Simmons, G., 1964, Gravity survey and geological interpretation, northern New York: Geological Society of America Bulletin, v. 75, p. 81–98.

Society of Exploration Geophysicists, 1982, Gravity anomaly map of the United States: Society of Exploration Geophysicists, U.S. Geological Survey, Defense Mapping Agency, and National Oceanic and Atmospheric Administration, scale 1:2,500,000.

Spencer, C., Green, A., and Luetgert, J., 1987, More seismic evidence on the location of Grenville basement beneath the Appalachians of Quebec-Maine: Geophysical Journal of the Royal Astronomical Society, v. 89, p. 177–182.

Stanley, R., and Ratcliffe, N., 1985, Tectonic synthesis of the Taconian orogeny in western New England: Geological Society of America Bulletin, v. 96, p. 1227–1250.

St.-Julien, P., Slivitsky, A., and Feininger, T., 1983, A deep structural profile across the Appalachians of southern Quebec, *in* Hatcher, R. D., Jr., Williams, H., and Zietz, I., eds., Contributions to the tectonics and geophysics of mountain chains: Geological Society of America Memoir 158, p. 103–111.

Stewart, D. B., and 8 others, 1986, The Quebec–western Maine seismic reflection profile; Setting and first-year results, *in* Barazangi, M., and Brown, L., eds., Reflection seismology; The continental crust: American Geophysical Union Geodynamics Series, v. 14, p. 189–200.

Tapponier, P., Mercier, J. L., Armijo, R., Tong-Lin, H., and Ji, Z., 1981, Field evidence for active normal faulting in Tibet: Nature, v. 294, p. 410–414.

Tauvers, P. R., and Muehlberger, W. R., 1987, Is the Brunswick magnetic anomaly really the Alleghanian Suture?: Tectonics, v. 6, p. 331–342.

Unger, J. D., Stewart, D. B., and Phillips, J. D., 1987, Interpretation of migrated seismic reflection profiles across the northern Appalachians in Maine: Geophysical Journal of the Royal Astronomical Society, v. 89, p. 171–176.

Warner, M. R., 1986, Deep seismic reflection profiling of the continental crust at sea, *in* Barazangi, M., and Brown, L., eds., Reflection seismology; The continental crust: American Geophysical Union Geodynamics Series, v. 13, p. 281–286.

——, 1987, Migration; Why doesn't it work for deep continental data?: Geophysical Journal of the Royal Astronomical Society, v. 89, p. 21–25.

Waters, K. H., 1987, Reflection seismology, 3rd ed.: New York, John Wiley & Sons, 538 p.

Widess, M. B., and Taylor, G. L., 1959, Seismic reflections from layering within the Precambrian basement complex, Oklahoma: Geophysics, v. 17, p. 417–425.

Williams, H., and Hatcher, R. D., Jr., 1983, Appalachian suspect terranes, *in* Hatcher, R. D., Jr., Williams, H., and Zietz, I., eds., Contributions to the tectonics and geophysics of mountain chains: Geological Society of America Memoir 158, p. 33–53.

Zartman, R. E., and Naylor, R. S., 1984, Structural implications of some radiometric ages of igneous rocks in southeastern New England: Geological Society of America Bulletin, v. 95, p. 522–539.

Zen, E., Goldsmith, R., Ratcliffe, N. M., Robinson, P., and Stanley, R. S., 1983, Bedrock geological map of Massachusetts: U.S. Geological Survey, scale 1:250,000.

Zheng, L., and Brown, L. D., 1986, Application of coherency enhancement to deep reflection data: EOS Transactions of the American Geophysical Union, v. 67, p. 303.

Zhu, T., and Brown, L. D., 1986, COCORP Michigan surveys; Reprocessing and results: Journal of Geophysical Research, v. 91, p. 11477–11496.

Zoback, M. D., and 5 others, 1980, Recurrent intraplate tectonism in the New Madrid seismic zone: Science, v. 209, p. 971–976.

MANUSCRIPT ACCEPTED BY THE SOCIETY OCTOBER 31, 1988

Printed in U.S.A.

Geological Society of America
Memoir 172
1989

Chapter 28

Seismic properties of the crust and uppermost mantle of the conterminous United States and adjacent Canada

L. W. Braile and W. J. Hinze
Department of Earth and Atmospheric Sciences, Purdue University, West Lafayette, Indiana 47907
R.R.B. von Frese
Department of Geology and Mineralogy, Ohio State University, Columbus, Ohio 43210
G. Randy Keller
Department of Geological Sciences, University of Texas at El Paso, El Paso, Texas 79968

ABSTRACT

Seismic refraction profiles for the conterminous United States and adjacent Canada have been compiled from published and unpublished sources. The crustal models derived from these profiles were used to compile data on upper-mantle seismic velocity (P_n), crustal thickness (H_c) and average seismic velocity of the crystalline crust ($\overline{V_p}$). These data indicate continent-wide averages of $P_n = 8.02$ km/sec, $H_c = 36.1$ km, and $\overline{V_p} = 6.44$ km/sec. Comparison of compressional wave parameters with shear-wave data derived from surface-wave dispersion models at 95 North American locations indicates an average value for Poisson's ratio of 0.258 for the crust and 0.270 for the uppermost mantle. Contour maps illustrating lateral variations in crustal thickness, upper-mantle velocity, and average seismic velocity of the crystalline crust show a number of features correlative with geologic and tectonic provinces. Comparison of the distribution of seismic parameters with a smoothed free-air anomaly map indicates that a complicated mechanism of isostatic compensation exists for the North American continent involving both lateral density changes in the crust and upper mantle and variations in the thickness of the crust. Several features on the seismic contour maps also are correlative with regional magnetic anomalies.

INTRODUCTION

Seismic data are a major component of investigations into the nature, composition, and configuration of the continental crust. Knowledge of the seismic properties of the crust has proven useful in studies of such diverse topics as basin development (e.g., Green, 1977), identification of tectonic provinces (e.g., Pakiser and Zietz, 1965), and characterization of geothermal anomalies (e.g., Bartelsen and others, 1982). A variety of seismic techniques have been employed in these studies, but refraction profiling and measurements of surface wave dispersion are among those methods most commonly used for regional analyses. In this study, we have compiled refraction and surface-wave results for the conterminous United States and adjacent Canada (latitude 25° to 60°N) for the purpose of analyzing regional variations in seismic properties of the crust and uppermost mantle and their relation to tectonic features and other geophysical data. Although several aspects of our approach are different, this study can in part be considered an extension and update (our compilation includes previously unpublished results and models published through 1987) of the overviews of Steinhart and Meyer (1961), Herrin and Taggart (1962), Pakiser and Steinhart (1964), James and Steinhart (1966), Kanasewich (1966), McConnell and others (1966), Herrin (1969), Healy and Warren (1969), Warren and Healy (1973), Berry (1973), and Allenby and Schnetzler (1983). The accumulated seismic data provide useful statistical data on the seismic properties of the continental crust and contour maps of these properties for the central portion of North America.

Braile, L. W., Hinze, W. J., von Frese, R.R.B., and Keller, G. R., 1989, Seismic properties of the crust and uppermost mantle of the conterminous United States and adjacent Canada, *in* Pakiser, L. C., and Mooney, W. D., Geophysical framework of the continental United States: Boulder, Colorado, Geological Society of America Memoir 172.

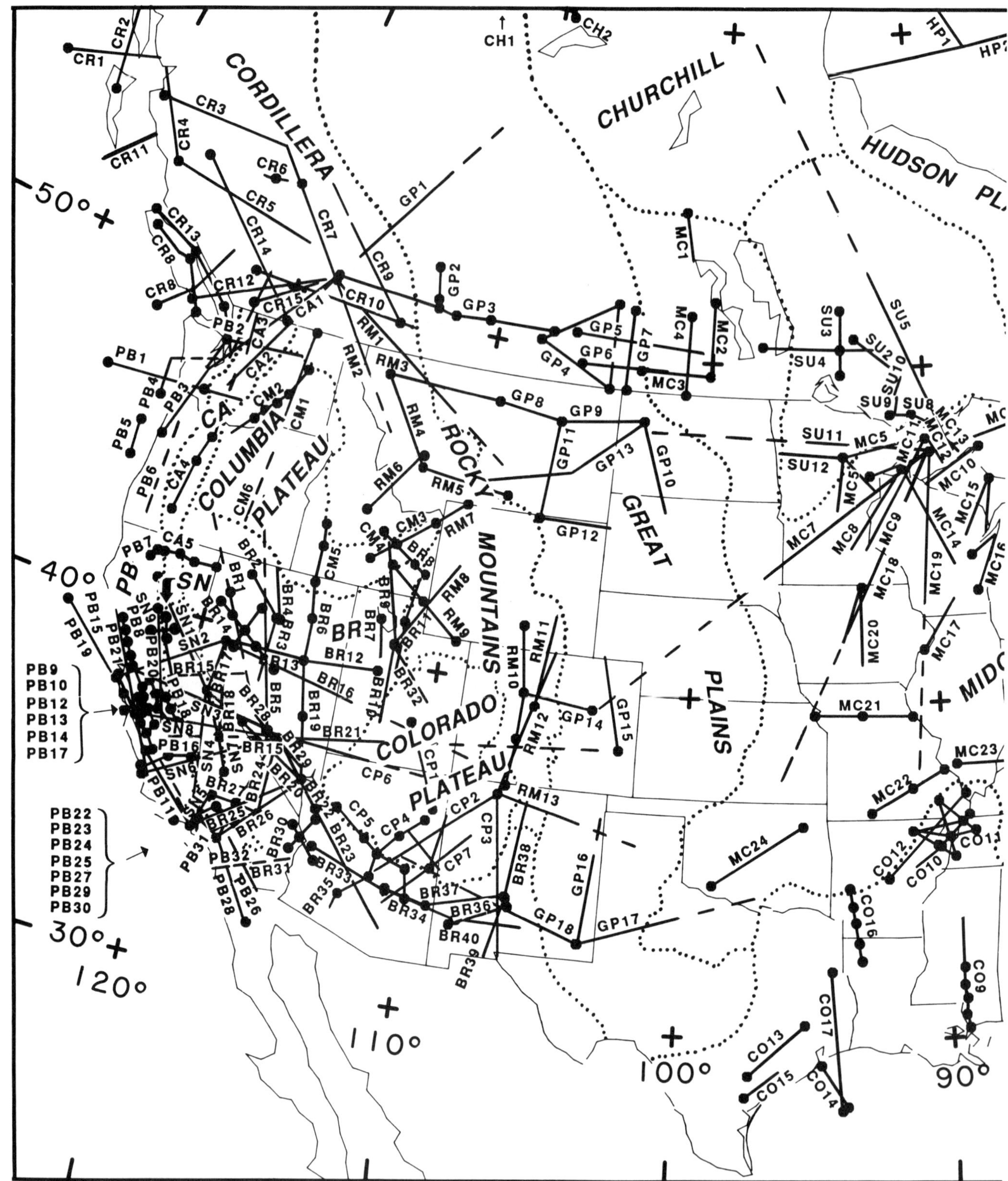

Figure 1. Index map showing locations of deep crustal seismic refraction profiles (solid lines). Shotpoint locations are shown by dots. Some shotpoint locations are schematic because of a large number of shots in a small area. Profiles are dashed where only a general location of an array is indicated and for profiles with large source-receiver distance where observations contain limited information on crustal structure. Province boundaries are from Fenneman (1946) for the United States and from Douglas and Price

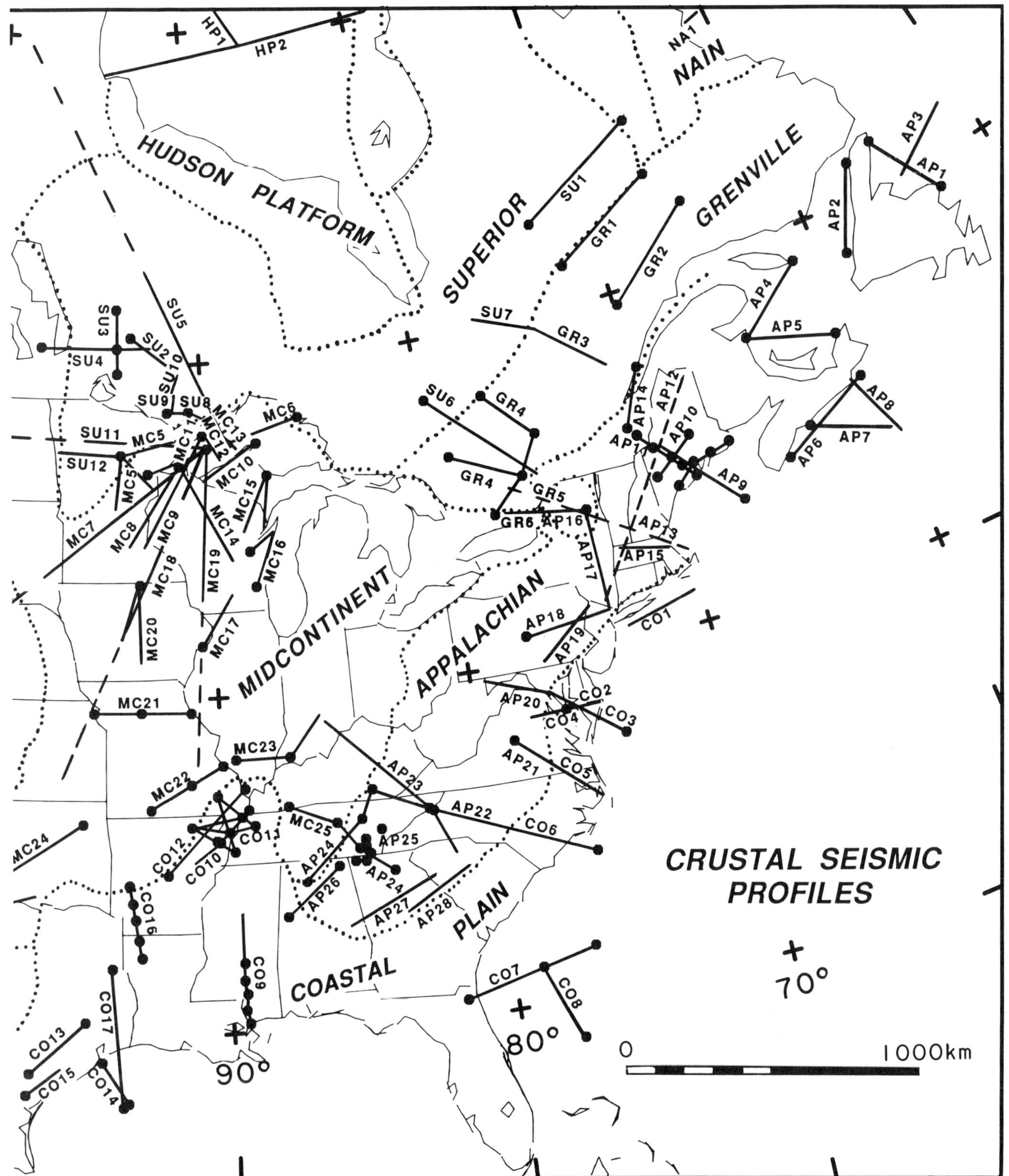

(1972) for Canada. PB, CA, SN, and BR indicate the Pacific Border, Cascades, Sierra Nevada and Basin and Range Provinces, respectively. The numbers adjacent to the seismic refraction profiles refer to the profile names and references listed in Table 2.

TABLE 1. ABBREVIATIONS FOR NORTH AMERICAN PROVINCES

Appalachian	AP
Basin and Range	BR
Cascade	CA
Churchill	CH
Coastal Plain	CO
Colorado Plateau	CP
Columbia Plateau	CM
Cordillera	CR
Great Plains	GP
Grenville	GR
Hudson Platform	HP
Midcontinent	MC
Nain	NA
Pacific Border	PB
Rocky Mountain	RM
Sierra Nevada	SN
Superior	SU

Data that have become available in the last decade have significantly improved the statistics of crustal velocity structure parameters for much of North America and provided better coverage in many regions. However, several provinces are characterized by relatively few crustal seismic measurements.

DATA COMPILATION AND PRESENTATION

In our compilation of refraction results, values for the thickness of the crust (H_c), the compressional wave velocity of the uppermost mantle (P_n), and the average compressional wave velocity of the crystalline crust ($\overline{V_p}$) were tabulated. Results for continental shelves were included, but no oceanic data were considered. H_c is defined as the total thickness of all layers extending from the surface to the Mohorovičić discontinuity (Moho). The P_n velocity is the velocity of the compressional head wave traveling in the uppermost mantle just beneath the Moho. Although P_n velocities are commonly reported and have been demonstrated to have tectonic significance (e.g., Pakiser and Zietz, 1965), this quantity, when measured by seismic refraction studies, represents limited penetration into the upper mantle. Thus, the P_n velocity values tabulated here may differ from values inferred from earthquake observations and seismic delay time studies (Herrin, 1969; Romanowicz, 1979; Dziewonski and Anderson, 1983) where the propagation paths sample a much larger region of the mantle.

The average velocity of the crust ($\overline{V_p}$) has been shown to be an indicator of mean crustal composition (Smithson and others, 1981). It was calculated using the formula:

$$\overline{V_p} = \frac{\sum_{i=1}^{n} h_i}{\sum_{i=1}^{n} (h_i/V_i)},$$

where h_i and V_i are the thickness and velocity, respectively, of the ith layer for an n-layered crust (exclusive of the sedimentary layer).

Crustal models for 247 refraction profiles were compiled. The locations of these profiles are shown in Figure 1 along with physiographic province boundaries (Table 1). Sources of the seismic data are provided in Table 2. In our compilation, we have utilized refraction profile data almost exclusively. Near-vertical reflection data often provide excellent structural information, but velocities are difficult to determine. Crustal velocity structure from local earthquake array data are available for a limited number of locations, and, in general, provide results that are similar to corresponding refraction models. Crustal models derived from teleseismic transfer function studies from single stations (Fernandez and Careaga, 1968; Kurita, 1973; Langston, 1977; Burdick and Langston, 1977; Owens and others, 1987) were not included. However, models derived from these studies show generally good agreement with refraction models where comparisons are possible (Zandt and Owens, 1986). Histograms of the seismic data are shown in Figure 2, and statistical summaries of the seismic data are given in Table 3. The 337 values of H_c have a mean value of 36.10 km; the 225 values of $\overline{V_p}$ have a mean value of 6.435 km/sec; the 320 values of P_n have a mean value of 8.018 km/sec.

Contour maps of these values were constructed for interpretation and comparison with other geophysical and geological data. However, a discussion of the limitations of the contouring process is appropriate first. Our comments are not intended as criticisms of particular studies, but as reminders that such factors as logistics, restricted funding, and limited numbers of instruments prevent the collection of an ideal data set.

The most obvious point is that the distribution of profiles is uneven (Fig. 1). Coverage in the western United States is sufficient to give an adequate regional picture in most areas, but coverage is very sparse in the cratonic areas to the east. Another consideration is that there are large variations in the quantity and quality of data along individual profiles. In some cases, the station spacing is approximately 3 km, while in others the station spacing is more than 30 km. Signal-to-noise ratios for individual records is also highly variable. Finally, the data represented in Figure 1 were gathered over a period of almost four decades when rapid changes in instrumentation and recording technology, as well as interpretation methods, were occurring.

There are several other cautions that generally apply when interpreting results of seismic profiling experiments. Anisotropy may be a factor in regard to P_n velocity determinations (Bamford, 1973; Bamford and others, 1979). However, the averaging inherent in the contouring process and the approximately random orientation of the profiles suggest that any effect of anisotropy on regional determinations would be small. Differing interpretive assumptions (i.e., sharp velocity discontinuities vs. gradients; planar interfaces vs. laterally complex interfaces) also yield different earth models. Another minor factor with regard to P_n velocities is that most studies assume for seismic modeling purposes that the Earth is flat. Consideration of a spherical Earth would generally reduce the reported P_n velocity values by 0.03 to 0.06 km/sec (Black and Braile, 1982). Interpretation techniques as-

TABLE 2. SEISMIC REFRACTION PROFILES FOR NORTH AMERICA

No.	Profile	Reference
AP1	St. Anthony–Cape Freels	Ewing and others, 1966
AP2	Mainland–Ferolle Point	Ewing and others, 1966
AP3	NE Newfoundland	Keen and Hyndman, 1979, Fenwick and others, 1968; Dainty and others, 1966
AP4	Tracadie–East Point	Ewing and others, 1966
AP5	Tracadie–Cheticamp	Ewing and others, 1966
AP6	Port Hebert–Cole Harbour	Barrett and others, 1964
AP7	Hubley Lake–SE	Barrett and others, 1964
AP8	ESE Nova Scotia	Keen and Hyndman, 1979; Dainty and others, 1966
AP9	Maine	Steinhart and others, 1964
AP10	Maine	Luetgert, 1985; Unger and others, 1987
AP11	Maine	Luetgert, 1985; Luetgert and other, 1987; Unger and others, 1987
AP12	NE United States	Taylor and Toksoz, 1979
AP13	Grenville–Appalachians	Taylor and others, 1980; Taylor and Toksoz, 1982
AP14	La Malbaie	Lyons and others, 1980
AP15	New England	Phinney, 1986
AP16	Tahawus–W	Katz, 1954
AP17	Tahawus–S	Katz, 1954
AP18	Milroy–E	Katz, 1954
AP19	E. Penn. and N.J.	Redmond and Kodama, 1984; Sienko, 1982; Langston and Isaacs, 1981
AP20	N. Middle Atlantic States–ECOOE	James and others, 1968; Merkel and Alexander, 1969
AP21	S. Middle Atlantic States–ECOOE	James and others, 1968
AP22	S. Profile ECOOE	Hales and others, 1968
AP23	E. Tennessee	Steinhart and Meyer, 1961; Tatel and Tuve, 1955
AP24	E. Tennessee	Warren, 1968; Prodehl and others, 1984
AP25	S. Appalachian Time Terms	Long and Liow, 1986
AP26	S. Appalachian	Long and Liow, 1986
AP27	N. Georgia	Dorman, 1972
AP28	Georgia–South Carolina	Kean and Long, 1980
BR1	Tonapha–Bend	Priestley and others, 1982
BR2	Copper Canyon Mine–Quinn River	Stauber, 1983
BR3	Copper Canyon Mine–S	Stauber, 1983
BR4	Copper Canyon Mine	Stauber and Boore, 1978
BR5	NTS–Winnemucca	Stauber and Boore, 1978
BR6	Eureka–Boise	Hill and Pakiser, 1966
BR7	E. Basin and Range	Berg and others, 1960
BR8	American Falls–Flaming Gorge	Willden, 1965
BR9	Bingham–N	Martin, 1978
BR10	Bingham–S	Keller and others, 1975
BR11	Bingham–NE	Braile and others, 1974
BR12	Delta–W	Keller and others, 1975; Mueller and Landisman, 1971; Müller and Mueller, 1979
BR13	Fallon–Eureka	Eaton, 1963
BR14	E. Basin and Range–PASSCAL	Catchings and others, 1988
BR15	NTS–Central California	Kind, 1972
BR16	SHOAL, Fallon–SE	Archambeau and others, 1969
BR17	Fallon–Mono Lake	Prodehl, 1979
BR18	Fallon–China Lake	Prodehl, 1979
BR18	Eureka–Lake Mead	Prodehl, 1979
BR20	Lake Mead–Mono Lake	Johnson, 1965
BR21	NTS–Ordway	Ryall and Stuart, 1963
BR22	NTS–Kingman	Diment and others, 1961
BR22	Kingman–NTS	Roller, 1964; Prodehl, 1979
BR23	NTS–Tucson	Langston and Helmberger, 1974
BR24	NTS–Ludlow	Gibbs and Roller, 1966
BR25	Lake Mead–Santa Monica Bay	Roller and Healy, 1963
BR26	California–Nevada	Press, 1960
BR27	Mojave–Ludlow	Prodehl, 1979
BR28	Beatty–Skull Mountain	Hoffman and Mooney, 1984
BR29	NTS	Taylor, 1983
BR30	Whipple Mountains	McCarthy and others, 1987
BR31	Salton Trough	Fuis and others, 1984
BR32	N. Utah	Tatel and Tuve, 1955; Steinhart and Meyer, 1961
BR33	Parker–Glove, Arizona	Sinno and others, 1981
BR34	Globe–Tyrone	Gish and others, 1981
BR35	Gila Bend–Sunrise	Warren, 1969; Healy and Warren, 1969
BR36	Tyrone–Distant Runner	Sinno and others, 1986
BR37	Dice Throw–W	Jaksha, 1982
BR38	Dice Throw	Olsen and others, 1979
BR39	Mill Race–U.S.-Mexican Border	Sinno and others, 1986
BR40	Southern Rio Grande Rift	Cook and others, 1979
CA1	Greenbush Lake–Tumwater	Johnson and Couch, 1970
CA2	Greenbush Lake–Longmire	Johnson and Couch, 1970
CA3	North Cascades	Rohay, 1982
CA4	Oregon Cascades	Leaver and others, 1984
CA5	Northeastern California	Zucca and others, 1986
CH1	Yellowknife NW–SE	Clee and others, 1974
CH2	Great Slave Lake	Barr, 1971
CM1	Columbia Plateau	Hill, 1972
CM2	Columbia Plateau	Catchings and Mooney, 1988
CM3	Eastern Snake River Plain	Braile and others, 1982

TABLE 2. SEISMIC REFRACTION PROFILES FOR NORTH AMERICA (continued)

No.	Profile	Reference
CM4	Conda–SP7, Y-SRP	Sparlin and others, 1982
CM5	Eureka–Boise	Hill and Pakiser, 1966
CM6	E. Oregon	Dehlinger and others, 1965
CO1	Maryland Coast	Phinney, 1986
CO2	Chesapeake Bay	Tuve, 1951; Steinhart and Meyer, 1961; Tatel and others, 1953; Tuve and others, 1954
CO3	N. Middle Atlantic States–ECOOE	James and others, 1968; Merkel and Alexander, 1969
CO4	Washington, D.C.–E	Katz and Ewing, 1956
CO5	S. Middle Atlantic States–ECOOE	James and others, 1968
CO6	S. Profile ECOOE	Hales and others, 1968
CO7	Jacksonville–E	Hersey and others, 1959
CO8	Blake Plateau	Hersey and others, 1959
CO9	S. Mississippi	Warren and others, 1966
CO10	Mississippi Embayment Axial	Mooney and others, 1983
CO11	Mississippi Embayment	Ginzburg and others, 1983
CO12	CapeGirardeau–Little Rock	McCamy and Meyer, 1966
CO13	Cleveland–Victoria, Texas	Cram, 1961
CO14	Galveston	Ewing and others, 1955
CO15	Victoria, Texas	Dorman and others, 1972
CO16	Ouachita–PASSCAL	Keller and others, 1989; Jardine, 1988
CO17	SE Texas	Hales and others, 1970
CP1	Hanksville–Chinle	Roller, 1965
CP2	Gasbuggy–SW	Warren and Jackson, 1968
CP3	Gasbuggy–S	Toppozada and Sanford, 1976
CP4	Gila Bend–Sunrise	Warren, 1969; Healy and Warren, 1969
CP5	Blue Mountain–Bylas	Warren, 1969; Healy and Warren, 1969
CP6	Bilby–SE	Archambeau and others, 1969
CP7	Arizona–New Mexico	Tatel and Tuve, 1955; Steinhart and Meyer, 1961
CR1	Dixon Entrance	Shor, 1962
CR2	Bird Lake	Johnson and others, 1972
CR3	Bird Lake–Prince George	Forsyth and others, 1974
CR4	Ripley Bay	Johnson and others, 1972
CR5	Ripley Bay–Greenbush Lake	Berry and Forsyth, 1975
CR6	Quesnel	Berry and Forsyth, 1975
CR7	Greenbush Lake–McLeod Lake	Berry and Forsyth, 1975
CR8	Vancouver Island	Ellis and others, 1983; McMechan and Spence, 1983; Spence and others, 1985; Clowes and others, 1986
CR9	Rocky Mountain Trench	Bennett and others, 1975; Spence and others, 1977
CR10	Kaiser–Highland Valley	Cumming and others, 1979
CR11	Queen Charlotte Islands	Horn and others, 1984
CR12	Nitinat–Greenbush Lake	Berry and Forsyth, 1975
CR13	Vancouver Island	White and Savage, 1965
CR14	Puntchesakut Lake–Osoyos	White and others, 1968
CR15	Greenbush–Hope	White and others, 1968
GP1	Edzoe–Fort McMurray	Mereu and others, 1976
GP2	Ripple Rock	Richards and Walker, 1959
GP3	Greenbush Lake–E	Chandra and Cumming, 1972
GP4	S. Saskatchewan	Kanasewich and Chiu, 1985; Morel-a-L'Huissier and others, 1987
GP5	Saskatchewan–EDZOE	Bates and Hall, 1975
GP6	Assiniboia–Carlyle, Saskatchewan	Hajnal and others, 1984
GP7	SE Saskatchewan	Hajnal and others, 1984; Morel-a-L'Huissier and others, 1987
GP8	Big Sandy River–Fort Peck	McCamy and Meyer, 1964
GP9	Fort Peck–Garrison	McCamy and Meyer, 1964
GP10	Garrison–S	McCamy and Meyer, 1964
GP11	Fort Peck–Acme Pond	McCamy and Meyer, 1964
GP12	Acme Pond–E	McCamy and Meyer, 1964
GP13	Greycliff–Charleson	Warren and others, 1972
GP14	Wolcott–Agate	Prodehl and Lipman, 1988
GP15	E. Colorado	Jackson and others, 1963
GP16	Gnome–N	Mitchell and Landisman, 1971
GP17	Gnome–E	Romney and others, 1962
GP18	Gnome–W	Romney and others, 1962
GR1	Front	Berry and Fuchs, 1973
GR2	Grenville	Berry and Fuchs, 1973
GR3	Grenville	Mereu and Jobidon, 1971
GR4	Ontario–Quebec	Mereu and others, 1986a, b; Huang and others, 1986
GR5	Grenville–Appalachians	Taylor and others, 1980
GR6	Tahawus–W	Katz, 1954
HP1	Hudson Bay NW–SE	Hobson and others, 1967
HP2	Hudson Bay E–W	Hobson and others, 1967
MC1	W. Manitoba	Hall and Hajnal, 1973
MC2	Superior–Churchill–NS	Green and others, 1980; Delandro and Moon, 1982; Morel-a-L'Huissier and others, 1987
MC3	Superior–Churchill–EW	Green and others, 1980; Delandro and Moon, 1982; Morel-a-L'Huissier and others, 1987
MC4	W. Manitoba	Hajnal and others, 1984; Morel-a-L'Huissier and others 1987

TABLE 2. SEISMIC REFRACTION PROFILES FOR NORTH AMERICA (continued)

No.	Profile	Reference
MC5	N. Minnesota	Tuve, 1953; Steinhart and Meyer, 1961; Tatel and others, 1953; Tuve and others, 1954
MC6	Lake Superior	Berry and West, 1966; Smith and others, 1966
MC7	Lake Superior–Colorado	Roller and Jackson, 1966
MC8	Gambler High	Cohen and Meyer, 1966
MC9	Gambler Low	Cohen and Meyer, 1966
MC10	Keweenaw	Steinhart and Meyer, 1961
MC11	Western Lake Superior Basin	Luetgert and Meyer, 1982
MC12	Western Lake Superior Basin	Luetgert and Meyer, 1982
MC13	Western Lake Superior Basin	Luetgert and Meyer, 1982
MC14	Central Wisconsin	Steinhart and Meyer, 1961
MC15	Central Wisconsin	Slichter, 1951; Steinhart and Meyer, 1961
MC16	Central Wisconsin	Slichter, 1951; Steinhart and Meyer, 1961
MC17	Wisconsin	Slichter, 1951; Steinhart and Meyer, 1961
MC18	Lake Superior–Wichita	Green and Hales, 1968
MC19	Lake Superior–Little Rock	Green and Hales, 1968
MC20	Iowa	Cohen, 1966
MC21	St. Joseph–Hannibal	Stewart, 1968a
MC22	Hercules–St. Genevieve	Stewart, 1968a
MC23	S. Indiana–S. Illinois	Baldwin, 1980
MC24	Manitou–Chelsea	Mitchell and Landisman, 1970; Tryggvason and Qualls, 1967
MC25	E. Tennessee	Warren, 1968; Prodehl and others, 1984
NA1	Labrador Coast	Keen and Hyndman, 1979; van der Linden and Srivastava, 1975
PB1	Washington Coast–Olympic Mountains	Taber, 1983; Taber and Smith, 1985; Taber and Lewis, 1986; Crosson, 1976
PB2	Puget Sound	Tuve, 1954; Steinhart and Meyer, 1961; Tatel and others, 1953; Tuve and others, 1954
PB3	Oregon Coast Range	Berg and others, 1966
PB4	Oregon Coast	Shor and others, 1968
PB5	Oregon Coast	Shor and others, 1968
PB6	W. Oregon	Dehlinger and others , 1965
PB7	Northeastern California	Zucca and others, 1986
PB8	Geysers–Clear Lake	Warren, 1981; Shearer and Oppenheimer, 1982
PB9	San Francisco–Paraiso	Hamilton and others, 1964
PB10	San Francisco–Camp Roberts	Healy, 1963
PB11	Camp Roberts–Santa Monica Bay	Healy, 1963
PB12	Diablo Range	Stewart, 1968b; Healy and Peake, 1975; Blumling and Prodehl, 1983; Walter and Mooney, 1982; Steppe and Crosson, 1978
PB13	Gabilan Range	Stewart, 1968b; Healy and Peake, 1975; Blumling and Prodehl, 1983; Walter and Mooney, 1982; Steppe and Crosson, 1978
PB14	Great Valley	Colburn and Mooney, 1986; Hwang and Mooney, 1986
PB15	Central California	Oppenheimer and Eaton, 1984
PB16	SJ-6 Profile	Trehu and Wheeler, 1987
PB17	Calaveras Fault	Blumling and others, 1985; Hartzell and Heaton, 1986
PB18	Central Great Valley	Holbrook and Mooney, 1987
PB19	Chase V–S. California	Lomnitz and Bolt, 1967
PB20	Chico–S. California	Lomnitz and Bolt, 1967
PB21	Coast Range	Kind, 1972
PB22	S. California	Hauksson, 1987
PB23	Los Angeles Basin	Hauksson, 1987
PB24	S. California	Kanamori and Hadley, 1975
PB25	Mojave–Transverse Ranges–Peninsular Ranges	Hadley and Kanamori, 1977
PB26	Borrego Mountain	Hamilton, 1970
PB27	Point Mugu	Stierman and Ellsworth, 1976
PB28	S. California–Baja California	Nava and Brune, 1982
PB29	S. California–Mojave	Vetter and Minster, 1981
PB30	S. California	Hearn, 1984
PB31	Corona, S. California	Gutenberg, 1952
PB32	Salton Trough	Fuis and others, 1984
RM1	Great Plains	Hales and Nation, 1973
RM2	Rocky Mountains	Hales and Nation, 1973
RM3	Cliff Lake–Big Sandy River	McCamy and Meyer, 1964
RM5	Sailor Lake–E	McCamy and Meyer, 1964
RM6	SW Montana	Sheriff and Stickney, 1984
RM7	Yellowstone	Smith and others, 1982
RM8	Bingham–NE	Braile and others, 1974
RM9	American Falls–Flaming Gorge	Willden, 1965
RM10	S. Rocky Mountains	Jackson and Pakiser, 1965; Prodehl and Pakiser, 1980
RM11	Front Range	Jackson and Pakiser, 1965; Prodehl and Pakiser, 1980
RM12	Gasbuggy–N	Warren and Jackson, 1968
RM13	Gasbuggy–E	Warren and Jackson, 1968

TABLE 2. SEISMIC REFRACTION PROFILES FOR NORTH AMERICA (continued)

No.	Profile	Reference
SN1	Shasta Reservoir-Mono Lake	Eaton, 1966
SN2	San Francisco–Fallon	Eaton, 1963
SN3	Sierra Nevada, Payute Mesa–S.F. Bay	Bateman and Eaton, 1967; Carder and others, 1970; Pakiser and Brune, 1980; Bolt and Gutdeutsch, 1982
SN4	Mono Lake–Santa Monica Bay	Prodehl, 1979
SN5	China Lake–Santa Monica Bay	Prodehl, 1979
SN6	NTS–San Luis Obispo	Prodehl, 1979
SN7	Mono Lake–China Lake	Eaton, 1966
SN8	Trans-California	Carder, 1973
SN9	NW Sierra Foothills	Spieth and others, 1981
SU1	Superior	Berry and Fuchs, 1973
SU2	English River–Wabigoon	Young and others, 1986
SU3	Northwest Ontario	Hall and Hajnal, 1969
SU4	Manitoba–Ontario	Hall and Hajnal, 1973
SU5	Superior–Churchill	Mereu and Hunter, 1969
SU6	Kirkland Lake	Hodgson, 1953
SU7	Grenville	Mereu and Jobidon, 1971
SU8	Quetico–Sherbandowan	Young and others, 1986
SU9	Highway 599-R	Young and others, 1986
SU10	Highway 599-C/S	Young and others, 1986
SU11	W. of Lake Superior	Lewis and Meyer, 1968
SU12	N. Minnesota	Tuve, 1953; Steinhart and Meyer, 1961; Tatel and others, 1953; Tuve and others, 1954

sume a one- or two-dimensional Earth; departures from this idealization can produce observed apparent velocities that vary considerably from the true velocities within the Earth. This is particularly true of unreversed refraction profiles. In refraction profiling experiments, velocity determinations involve averaging over considerable horizontal distances; thus the results cannot truly be plotted at a point for contouring. Also, vertical sampling in the Earth is seldom uniform because of differences in station spacing, shot-receiver distances along the profiles, and numbers and locations of sources. For example, profile CP3 (Fig. 1) mostly provides information about P_n velocity because the profile was 550 km long and had an average station spacing of more than 32 km. On the other hand, profile CM3 (Fig. 1) mostly provides information about crustal velocity structure because it was 250 km long and had an average station spacing of about 2.5 km. Finally, the problems of velocity inversions (low-velocity layers) are well known and may have an effect of H_c and $\overline{V}_p$ determinations.

With these considerations in mind, the values of H_c, P_n, and $\overline{V}_p$ were contoured and are presented in Figures 3, 4, and 5, respectively. Our procedure was to determine the values of H_c, P_n, and $\overline{V}_p$ from velocity models presented by various authors as listed in Table 2. Inferred sedimentary layers were deleted from the models in the calculation of $\overline{V}_p$ so that the value determined indicates the average velocity of the crystalline continental crust. For crustal velocity models in which the Moho is inferred to be a transition zone with velocity increasing over a small depth range, the depth to the Moho was chosen as the center point of this gradient zone, and the P_n velocity was selected as the velocity immediately beneath the Moho transition zone. For crustal models along profiles for which lateral thickness and/or velocity variations were inferred by the authors, multiple observations of H_c, $\overline{V}_p$, and P_n were initiall plotted on a map at locations along the profile inferred to be most representative of the crustal structure. For example, P_n velocities along a reversed seismic refraction profile of several hundred kilometers in length are most representative of the upper-mantle velocity near the center of the profile and in general do not necessarily indicate the upper-mantle velocity beneath the shotpoint locations. Several seismic refraction profiles have been reinterpreted after the initial crustal model was published. In these cases, we have tried to use the most recent interpretations. Finally, interpretations along the same profile, or on intersecting profiles, or closely spaced observations may be contradictory (Ansorge and others, 1982). In these cases, we have simply averaged the various values.

Any contouring operation involves subjectivity, and the production of the maps shown in Figures 3, 4, and 5 was certainly no exception. However, our goal was to depict regional variations for a large portion of the North American continent. For certain regions where data coverage is sufficient (for example, the Basin and Range province), a more rigorous contouring procedure could be employed. More detailed contour maps of crustal thickness and P_n velocity have been presented by Smith (1978) for the western United States, Hearn (1984) for southern California, and Taylor and Toksoz (1979) for the northeastern United States; and of crustal thickness by Halls (1982) for the Lake Superior regions, Pakiser (1985) for the Basin and Range, and Keller and others (1979a) for the Colorado Plateau area. Additionally, more detailed contour maps of crustal parameters for various regions are contained in this volume. The contours were initially drawn by hand in such a way that major variations based on only one observation were not honored exactly. Although no attempt was made to display the reliability of the data on the contour maps, a comparison of the contour data with the spatial distribution of observations (Fig. 1) provides an approximate indication of reliability.

No discussion of the seismic properties of the crust is really complete without considering shear waves, and several profiling experiments have deployed horizontal seismometers in an effort to record shear-wave arrivals (e.g., Braile and others, 1974; Keller

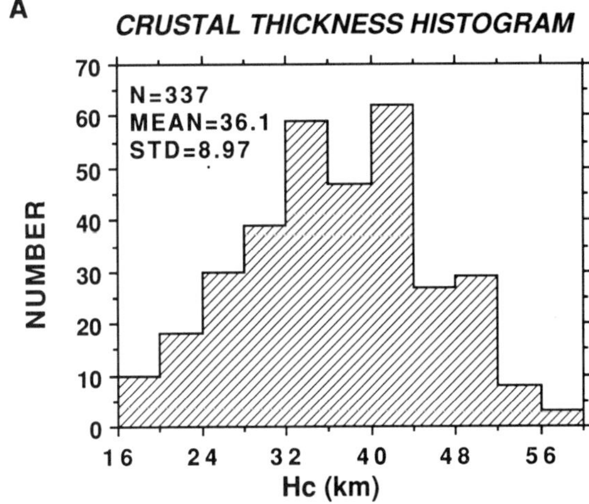

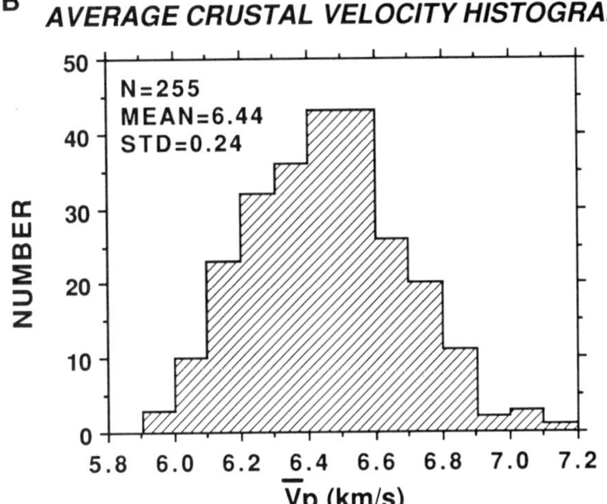

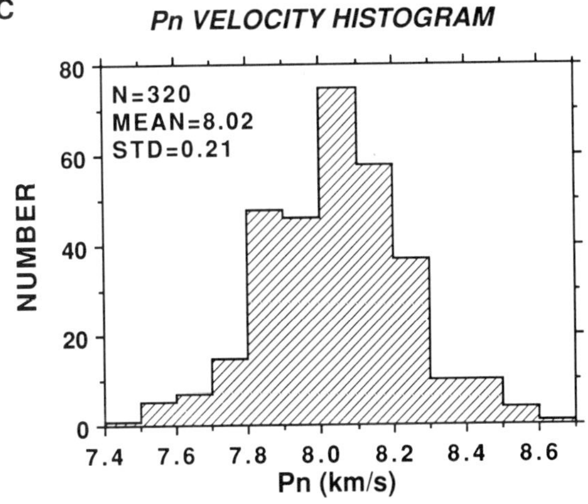

Figure 2. Histograms. (A), Crustal thickness (H_c); (B), average crustal velocity ($\overline{V}_p$); (C), uppermost mantle seismic velocity (P_n) determined from the seismic refraction profiles illustrated in Figure 1 and listed in Table 2. N refers to the number of observations. Std = the standard deviation.

and others, 1975). To date, these efforts have met with limited success, provide little coverage, and are not enough of a factor to consider in this study. However, a significant body of information on shear-wave velocity structure is available from studies of surface-wave dispersion. These studies primarily concern Rayleigh waves, which are most sensitive to shear-wave velocity although they are also slightly affected by variations in density and compressional wave velocity (Der and others, 1970).

A variety of techniques have been employed to determine dispersion across arrays of three or more stations, between two stations, or between the source and a single station (see Kovach, 1978, and Dziewonski and Hales, 1972, for reviews of surface-wave concepts and techniques). Earth models determined from such studies represent averages across the array, between two stations, or along the propagation path from the source to a single station. We have compiled the results from 95 dispersion studies at locations shown in Figure 6 and tabulated in Table 4. The point at which these results were plotted (Fig. 6) is the center of the array or the midpoint of the propagation path. We were able to determine 88 values of H_c' (thickness of the crust from shear-wave data), 67 values of S_n (upper-mantle shear-wave velocity), and 76 values of $\overline{V}_s$ (average shear-wave velocity of the crystalline crust). Note that the S_n velocity used here is derived from surface-wave dispersion experiments and thus represents an average shear-wave velocity for a considerable portion of the upper mantle. Regional S_n velocities derived from seismograph array recordings of earthquakes and explosions (Brune and Dorman, 1963; Huestis and others, 1973; Ibrahim and Nuttli, 1967) are generally consistent with the surface-wave dispersion results. In order to compare the compressional-wave and shear-wave results, the contours in Figure 3, 4, and 5 were interpolated to provide values of H_c, P_n, and $\overline{V}_p$ at the points on Figure 6 where values H_c', S_n, and $\overline{V}_s$ were available. There are several limitations to this procedure, but because the experiments involved do not spatially coincide, an interpolation was required.

The values of H_c and H_c' are compared in Figure 7. If all the corresponding values of H_c and H_c' agreed, they would fall on the diagonal line across Figure 7. There is considerable scatter, which is to be expected considering the interpolation applied and differences in interpretation techniques, but the mean values of H_c and H_c' are very close (36.1 ± 8.97 km and 36.7 ± 6.8 km, respectively), and few points are more than 5 km from the line which indicates $H_c = H_c'$.

A major goal of the shear-wave velocity compilation was to obtain information on Poisson's ratio values in the crust and upper mantle. Thus, Figure 8 was prepared by plotting corresponding pairs of $\overline{V}_p$ and $\overline{V}_s$ values and corresponding pairs of P_n and S_n values. These data are plotted in two groups and indicate a mean Poisson's ratio for the crust (σ_c) of 0.258 and a mean Poisson's ratio for the upper mantle (σ_m) of 0.270 for continental North America. These observations show that the practice of assuming Poisson's ratio has a value of $\sigma = 0.25$ is generally valid for crustal rocks, but that Poisson's ratio is slightly greater in the upper mantle than in the continental crust.

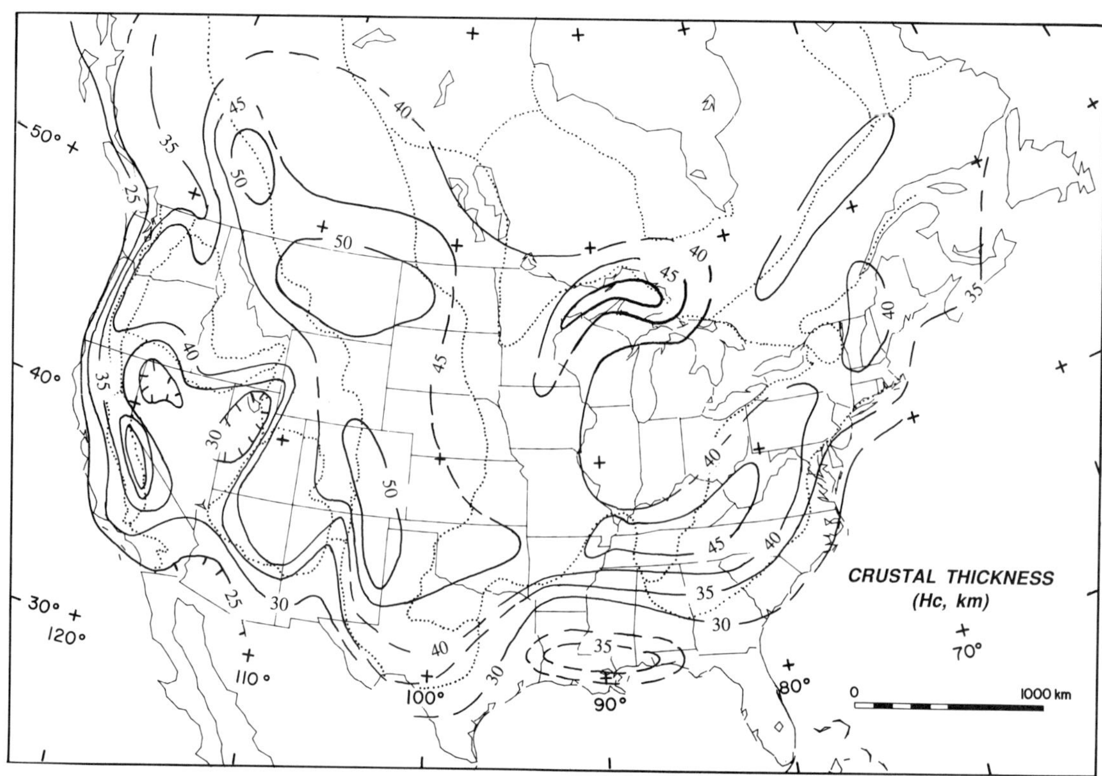

Figure 3. Contour map of crustal thickness for a portion of the North American continent. Numbers show crustal thickness in kilometers measured from the surface to the inferred Moho discontinuity. Contour interval = 5 km.

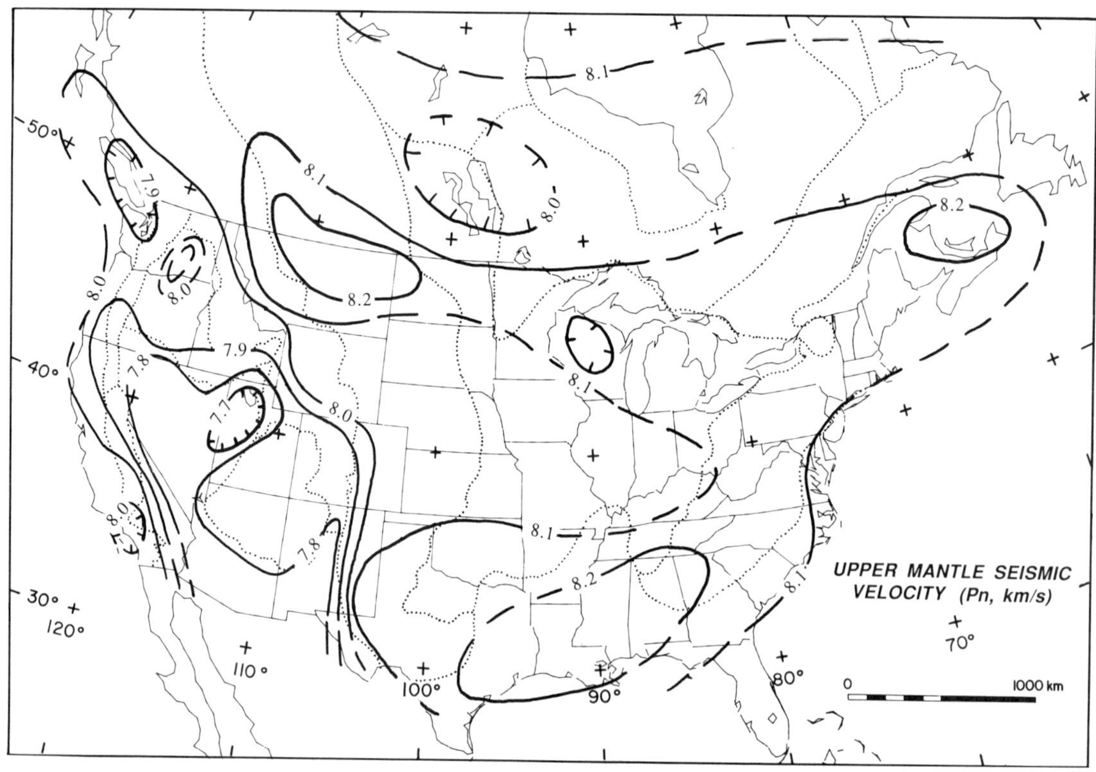

Figure 4. Contour map of upper-mantle seismic velocity (P_n). Contours give inferred P_n velocity in kilometers per second. Contour interval = 0.1 km/sec.

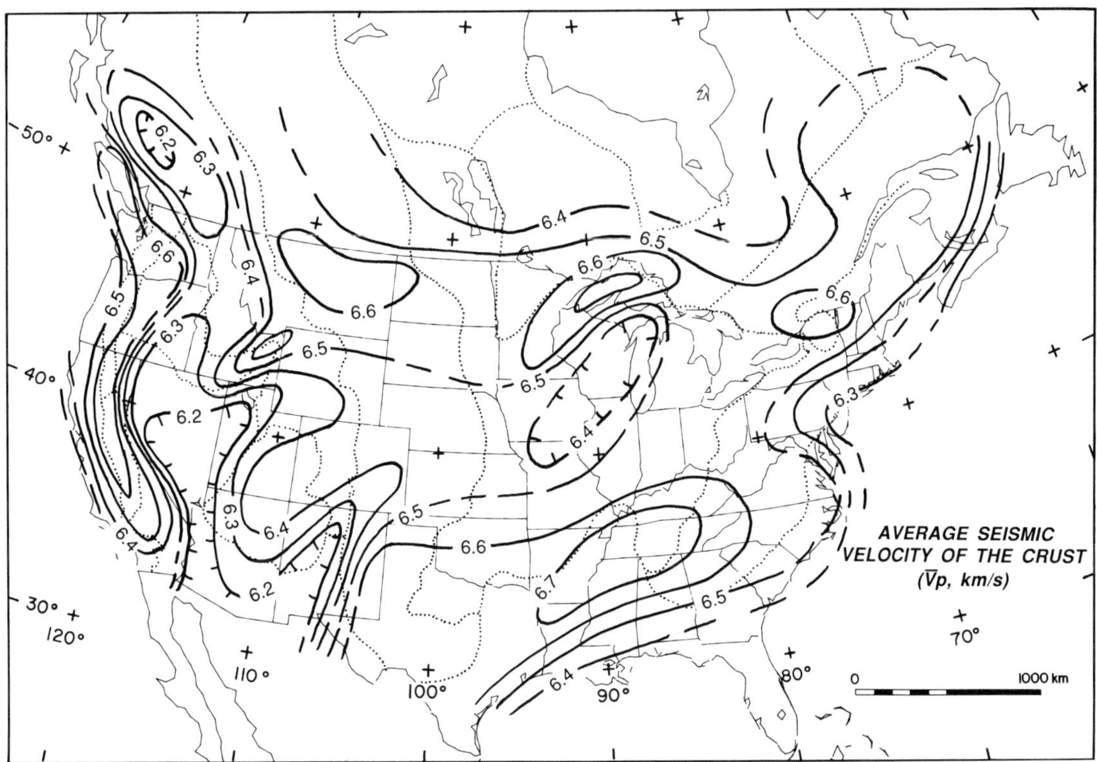

Figure 5. Contours of average seismic velocity of the crust ($\overline{V}_p$) for a portion of the North American continent. Contours give values of $\overline{V}_p$ in kilometers per second. Contour interval = 0.1 km/sec.

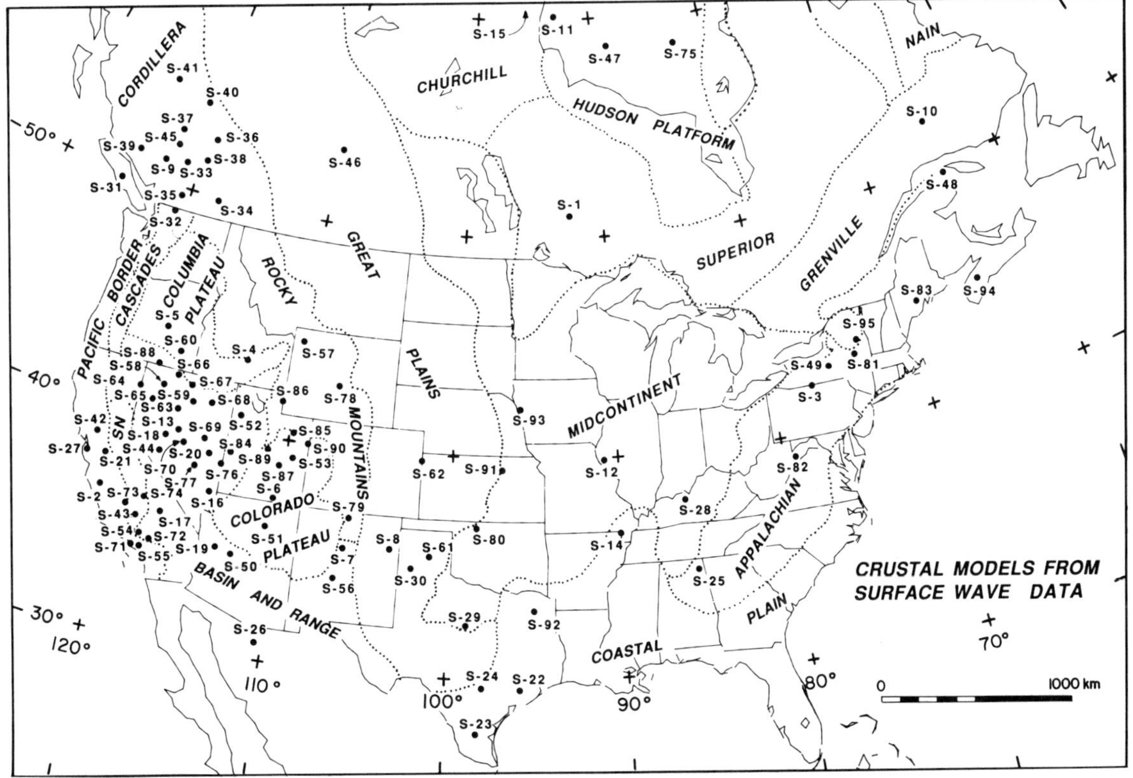

Figure 6. Location of approximate centers (dots) of surface-wave dispersion arrays or two-station path profiles used in the study of crustal and uppermost mantle structure of North America. Numbers refer to locations and references listed in Table 4. SN = Sierra Nevada Province.

REGIONAL VARIATIONS IN SEISMIC PROPERTIES OF THE CRUST AND UPPER MANTLE

A number of prominent regional variations in seismic properties are evident on the contour maps shown in Figures 3, 4, and 5, and the major relationships with tectonic features (Fig. 9) are generally consistent with those identified by previous workers (e.g., Pakiser and Steinhart, 1964; Allenby and Schnetzler, 1983). Examples of these relationships are:

1. Using the 104th meridian of longitude as a dividing line, there is a significant contrast in crustal properties between the eastern and western portions of the area. Relatively thick crust and high velocities characterize the eastern portion of the continent where the crust averages about 42 km in thickness and 6.5 km/sec in P-wave velocity, and the P_n velocity averages about 8.1 km/sec. The crustal thickness varies considerably in the western United States, but with the exception of the Montana area and the Pacific Border, Cascades, and Sierra Nevada provinces, both P_n and average crustal velocities are low.

2. The Basin and Range province in the western part of the United States is one of the most anomalous regions in continental North America in terms of crustal seismic properties. Both a very thin crust (≅ 30 km) and anomalously low upper-mantle velocities (≅ 7.8 km/sec) are evident.

3. The Great Plains is prominent in that it represents the area of North America whose average crustal thickness is greatest. Crustal thickness values in this area average about 48 km.

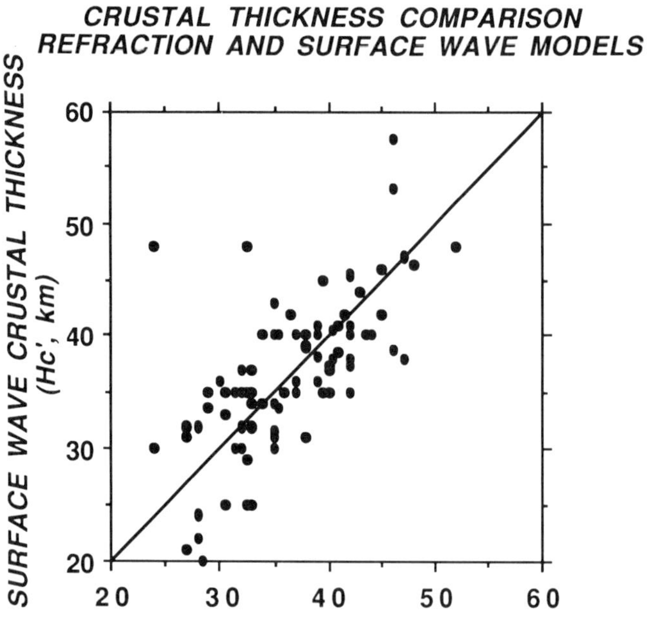

CRUSTAL THICKNESS COMPARISON
REFRACTION AND SURFACE WAVE MODELS

SURFACE WAVE CRUSTAL THICKNESS (Hc', km)

(Hc, km)
REFRACTION CRUSTAL THICKNESS

Figure 7. Scatter diagram indicating the relationship of crustal thickness as determined from refraction data (H_c) to crustal thickness as determined from surface-wave dispersion models (H_c') at the same locations. H_c observations were interpolated from the contour map at the location of the surface wave dispersion crustal thickness determinations (H_c').

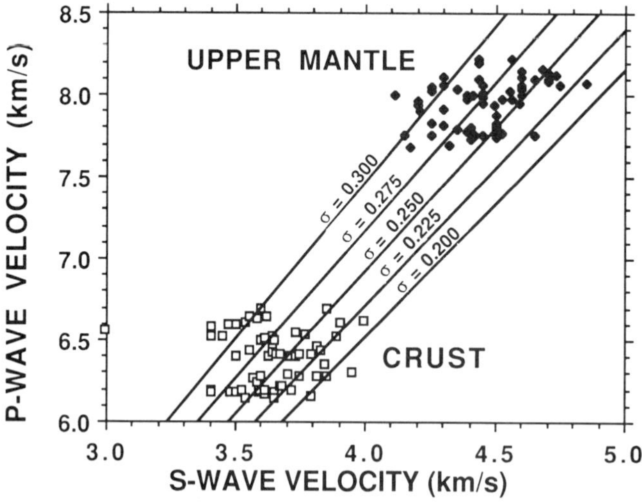

Figure 8. Diagram illustrating the relationship between compressional and shear-wave velocities for the crust and uppermost mantle. Compressional wave velocities are determined from the refraction data. Shear-wave velocities are from the surface-wave dispersion models. Velocities and Poisson's ratio values are shown by open squares for the crust and by solid diamonds for the upper mantle. σ = Poisson's ratio.

4. There is a general correlation between the distribution of P_n velocities and crustal thickness. Higher P_n velocities are usually associated with thicker crust. The Colorado Plateau may be the major exception to this relationship.

5. Crustal structure on the Pacific and Atlantic continental margins is distinctly different. For the Atlantic margin, the crust thins gradually toward the oceanic plate and is underlain by normal upper-mantle velocities. In contrast, the crust at the Pacific continental margin is complicated by the presence of adjacent mountain ranges and is largely underlain by lower velocity upper mantle.

6. There appears to be no simple relationship between crustal thickness and regional topography (Figs. 3, 10). Many areas have relatively thick crust without a corresponding regional elevation high. Similarly, areas such as the Basin and Range province, which has a regionally high elevation, are underlain by thin crust. Exceptions to these observations are the Sierra Nevada Mountains where prominent mountain "roots" are present and the Appalachian Mountains where a small "root" seems to be present.

There are several additional features evident on the contour maps presented in Figures 3, 4, and 5 which have not previously been described. Examples are:

1. The distribution of average crustal velocity ($\overline{V_p}$) as shown in Figure 5 has not been previously analyzed. Pakiser and Robinson (1966, 1967) and Smithson and others (1981) have discussed the importance of this average velocity as an indicator of crustal composition, and Smithson and others (1981) noted

TABLE 3. CRUST AND UPPER MANTLE MODEL STATISTICS: CONTINENTAL NORTH AMERICA

VARIABLE* (x)	SAMPLE SIZE (N)	MEAN ($\bar{X}$)	STANDARD DEVIATION (S_x)	UNITS
H_c	337	36.10	8.97	km
$\bar{V}_p$	255	6.435	0.235	km/sec
P_n	320	8.018	0.205	km/sec
H_c'	88	36.70	6.80	km
$\bar{V}_s$	67	3.639	0.163	km/sec
S_n	76	4.471	0.165	km/sec
σ_c	64	0.258	0.034	......
σ_m	70	0.270	0.023	......

*Refraction profile results:

H_c - Crustal thickness

$\bar{V}_p$ - Average compressional wave velocity of the crystalline crust

P_n - Upper mantle compressional wave velocity (Moho velocity)

Surface wave dispersion results:

H_c' - Crustal thickness

$\bar{V}_s$ - Average shear wave velocity of the crystalline crust

S_n - Upper mantle shear wave velocity (from dispersion models)

Poisson ratio results:

σ_c - Poisson ratio for the crystalline crust

σ_m - Poisson ratio for the upper mantle

that the average velocity of the crust is an important parameter for understanding the genesis of the continental crust. The average seismic velocity of the crust shows several prominent anomalies correlative with geologic features. The Basin and Range province and Rio Grande Rift, where low average crustal velocities are evident, are perhaps the most prominent features on Figure 5. The Snake River Plain in the western United States and the Midcontinent Rift System extending southwest from Lake Superior also show up as distinct, $\bar{V}_p$ anomalies. In addition, high average seismic velocities of the crust are present in the Montana area of the northern Great Plains and a region in the southeastern portion of the United States that contains several old rift zones (Reelfoot Rift, Southern Oklahoma Aulacogen, East Continent Gravity High). Distinct differences in average seismic velocity of the crust are also evident between the Pacific and Atlantic continental margins. In the area of the Atlantic continental margin, the average crustal velocity decreases gradually toward the ocean. Adjacent to the Pacific Ocean, the continental margin is characterized by an area of high average velocity of the crust. This area coincides with regions of Cenozoic volcanism and accreted terranes (Howell, 1985). The relative highs over the Grenville region in eastern Canada and the Colorado Plateau correlate with the tectonic provinces.

2. An interesting correlation of seismic properties is noted in the Montana area of the upper Great Plains in which thick crust, high P_n velocity, and high average crustal seismic velocities all correlate. Basins associated with the Southern Oklahoma Aulacogen and Mississippi Embayment also display a similar, but less pronounced, correlation of seismic properties.

3. A weakly defined east-west–trending low in P_n velocity and average crustal velocity is associated with the midwestern United States.

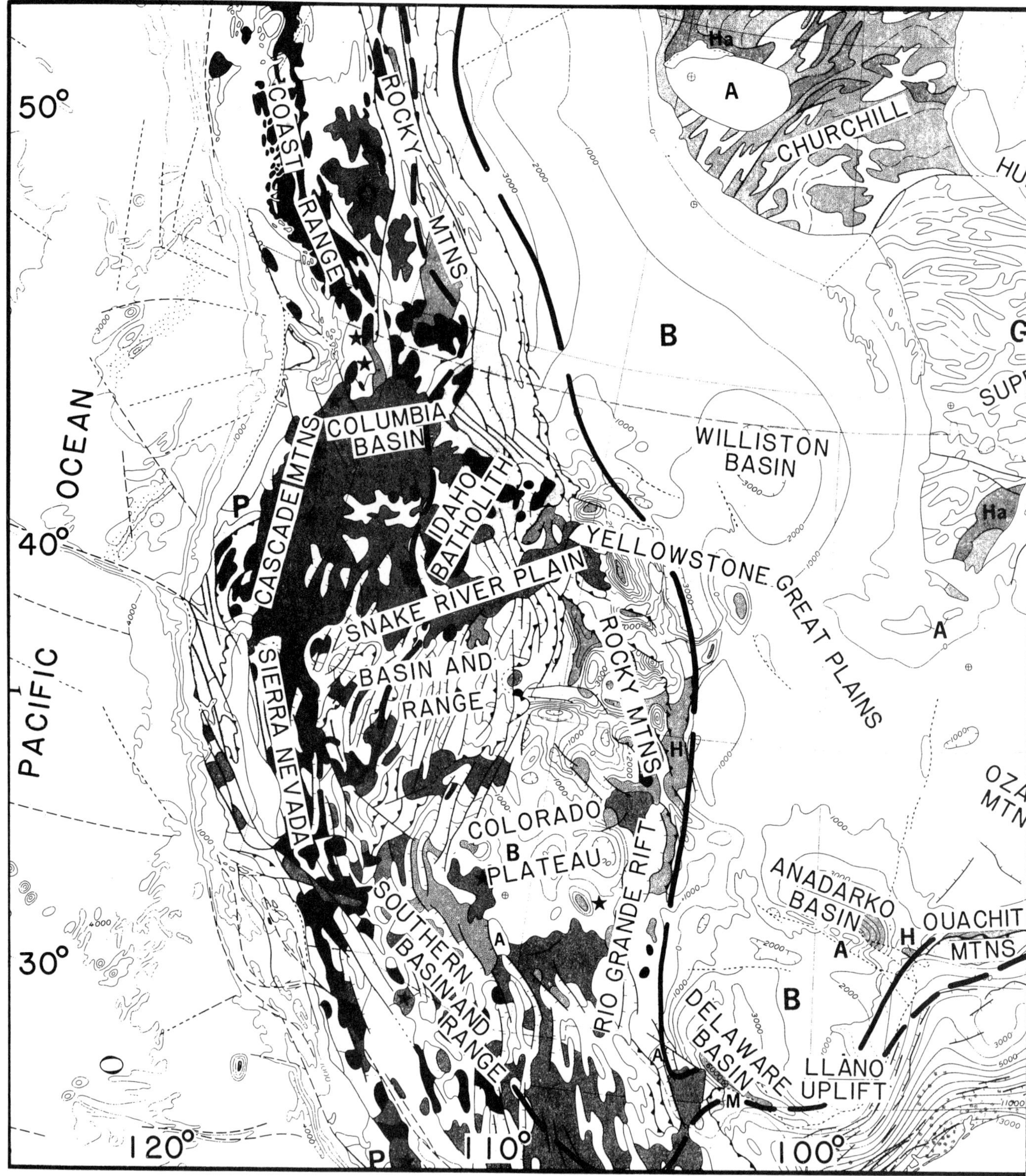

Figure 9. Index map of a portion of the North American continent showing principal tectonic units, geologic provinces, and locations of major basins and uplifts. The long dashed line shows the approximate limit of the North American continental craton which has remained relatively undeformed since Precambrian time. Short dashed line gives the approximate limit of the continental craton whose edges have suffered deformation since Precambrian time. Tectonic map from King and Edmonston (1972).

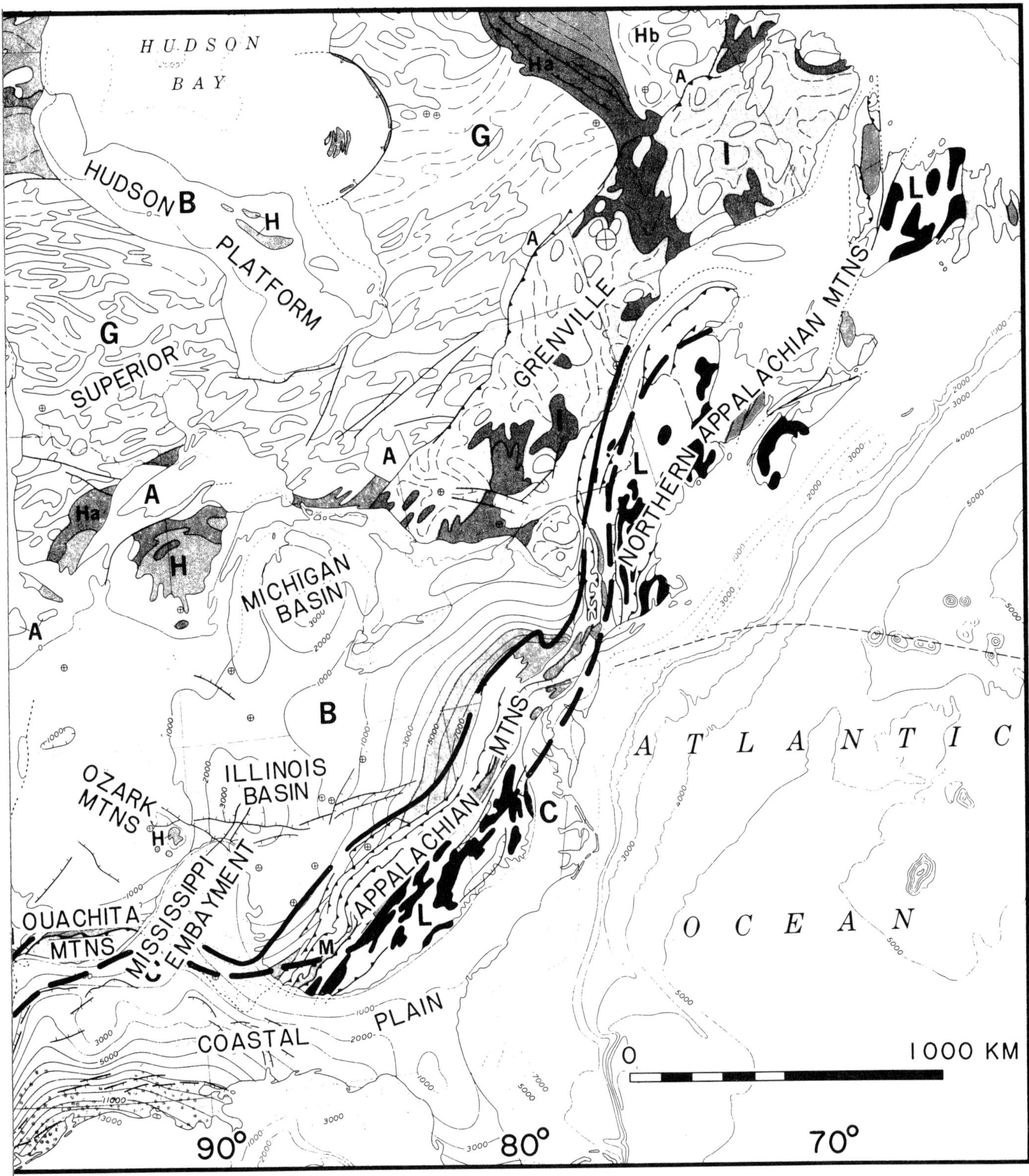
HUDSON BAY
HUDSON PLATFORM
SUPERIOR
GRENVILLE
NORTHERN APPALACHIAN MTNS
MICHIGAN BASIN
ILLINOIS BASIN
OZARK MTNS
MISSISSIPPI EMBAYMENT
OUACHITA MTNS
APPALACHIAN MTNS
COASTAL PLAIN
ATLANTIC OCEAN
0
1000 KM
90°
80°
70°

TABLE 4. SURFACE WAVE DISPERSION MODELS FOR NORTH AMERICA

No.	Province	Reference	No	Province	Reference
S-1	SU	Godlewski and West, 1977	S-49	AP	Oliver and others, 1961
S-2	PB	Panza and Calcagnile, 1974	S-50	BR	Bache and others, 1978
S-3	AP	Dorman and Ewing, 1962	S-51	CP	Bache and others, 1978
S-4	CM	Greensfelder and Kovach, 1982	S-52	BR	Bucher and Smith, 1971
S-5	CM	L. W. Braile, unpublished data	S-53	CP	Bucher and Smith, 1971
S-6	CP	Keller and others, 1979a	S-54	PB	Hadley and Kanamori, 1979
S-7	BR	Keller and others, 1979b	S-55	PB	Hadley and Kanamori, 1979
S-8	GP	Keller and others, 1979b	S-56	BR	Sinno and Keller, 1986
S-9	CR	Losee, 1980	S-57	RM	Daniel and Boore, 1982
S-10	GR	Losee, 1980	S-58	BR	Priestley and others, 1980
S-11	HP	Losee, 1980	S-59	BR	Priestley and Brune, 1982
S-12	MC	McEvilly, 1964	S-60	CM	Priestley and Brune, 1982
S-13	BR	Priestley and Brune, 1978	S-61	GP	Bodnar, 1982
S-14	CO	Austin and Keller, 1982	S-62	GP	Bodnar, 1982
S-15	CH	Brune and Dorman, 1963	S-63	BR	Priestley and others, 1980
S-16	BR	Lowrey, 1977	S-64	BR	Priestley and others, 1980
S-17	BR	Lowrey, 1977	S-65	BR	Priestley and others, 1980
S-18	BR	Lowrey, 1977	S-66	BR	Priestley and others, 1980
S-19	BR	Lowrey, 1977	S-67	BR	Priestley and others, 1980
S-20	BR	Lowrey, 1977	S-68	BR	Priestley and others, 1980
S-21	PB	Mikumo, 1965	S-69	BR	Priestley and others, 1980
S-22	CO	Keller and Shurbet, 1975	S-70	BR	Priestley and others, 1980
S-23	CO	Keller and Shurbet, 1975	S-71	PB	Press, 1956
S-24	CO	Keller and Shurbet, 1975	S-72	PB	Press, 1956
S-25	AP	Long and Mathur, 1971	S-73	PB	Press, 1956
S-26	BR	Thatcher and Brune, 1973	S-74	SN	Press, 1956
S-27	PB	Press, 1957	S-75	HP	Oliver and Ewing, 1957
S-28	MC	Adams, 1975	S-76	BR	Smith, 1962
S-29	GP	Prewitt, 1969	S-77	BR	Smith, 1962
S-30	GP	Stanton, 1972	S-78	RM	Ewing and Press, 1959
S-31	CR	Wickens, 1977	S-79	RM	Smith, 1962
S-32	CR	Wickens, 1977	S-80	GP	Ewing and Press, 1959
S-33	CR	Wickens, 1977	S-81	AP	Ewing and Press, 1959
S-34	CR	Wickens, 1977	S-82	AP	Ewing and Press, 1959
S-35	CR	Wickens, 1977	S-83	AP	Ewing and Press, 1959
S-36	kCR	Wickens, 1977	S-84	BR	Keller and others, 1976
S-37	CR	Wickens, 1977	S-85	CP	Keller and others, 1976
S-38	CR	Wickens, 1977	S-86	RM	Keller and others, 1976
S-39	CR	Wickens, 1977	S-87	CP	Keller and others, 1976
S-40	CR	Wickens, 1977	S-88	BR	Biswas and Knopoff, 1974
S-41	kCR	Wickens, 1977	S-89	kCP	Biswas and Knopoff, 1974
S-42	PB	Thompson and Talwani, 1964	S-90	RM	Biswas and Knopoff, 1974
S-43	SN	Thompson and Talwani, 1964	S-91	MC	Biswas and Knopoff, 1974
S-44	BR	Thompson and Talwani, 1964	S-92	CO	Biswas and Knopoff, 1974
S-45	CR	Wickens, 1971	S-93	MC	Press and Ewing, 1955
S-46	GP	Wickens, 1971	S-94	AP	Taylor and Toksoz, 1982
S-47	HP	Wickens, 1971	S-95	GR	Taylor and Toksoz, 1982
S-48	AP	Wickens, 1971			

COMPARISON OF SEISMIC PROPERTIES WITH REGIONAL GRAVITY AND MAGNETIC ANOMALY MAPS

Regional topography, gravity, and magnetic anomaly maps are shown in Figures 10 through 13 for comparison with the distribution of seismic properties illustrated in the contour maps shown as Figures 3, 4, and 5. Although Simpson and others (1986) stressed the need for caution when using processed maps to make inferences about isostatic adjustment, Woollard (1969) showed that averaged free-air anomaly values provide an approximate measure of the degree to which isostatic balance has been obtained. Thus, the smoothed free-air gravity anomaly map in Figure 11 was prepared with this in mind. Comparison of the free-air map with the contour maps of crustal structure indicates no obvious relation between variations in crustal properties and the isostatic condition of the continental crust. For example, several areas of relatively high average seismic velocity (and therefore presumably high average density) for the crust, such as the Mississippi Embayment and Montana areas, are roughly in isostatic balance, at least partially due to crustal thickening. Thus, the mechanism of isostatic balance for the North American continent appears to correspond to neither purely the Pratt nor purely the Airy hypothesis. Lateral variations of both density (as inferred from average seismic velocity) and crustal thickness contribute to the attainment of isostatic balance. Additionally, the position of the Moho, while certainly affecting isostatic balance, does not provide the only mechanism for compensation. For example, in the Basin and Range province (an area of relatively high elevation), a thin crust and therefore a mantle upwarp is compensated by the combined effects of relatively low-density crust and low-velocity (and presumably low density) upper mantle. Thus, the seismic results and topographic data alone indicate that the mechanism of isostatic compensation for the North American continent must involve lateral variations in density in both the crust and upper mantle as well as differences in crustal thickness.

Regional magnetic anomaly maps of portions of the North American continent are shown in Figures 12 and 13. Figure 12 illustrates a reduced to radial polarization satellite (POGO) magnetic anomaly map corresponding to an elevation of 450 km. Figure 13 shows a smoothed total field magnetic intensity anomaly map utilizing the U.S. Naval Oceanographic Office (NOO) magnetic survey. The wavelength character of the two maps is distinctly different: the satellite map corresponds to more regional features, and the NOO map indicates more local features. The subdued character of the magnetic anomalies west of the Rocky Mountains (as noted by a number of investigators) correlates with areas of thin crust and low upper-mantle seismic velocity and is probably a temperature (Curie point) effect (e.g., Zietz, 1969). The most prominent correlation of the long-wavelength magnetic anomalies, as illustrated by the satellite magnetic map, is with the high average seismic velocity of the crust in the south-central

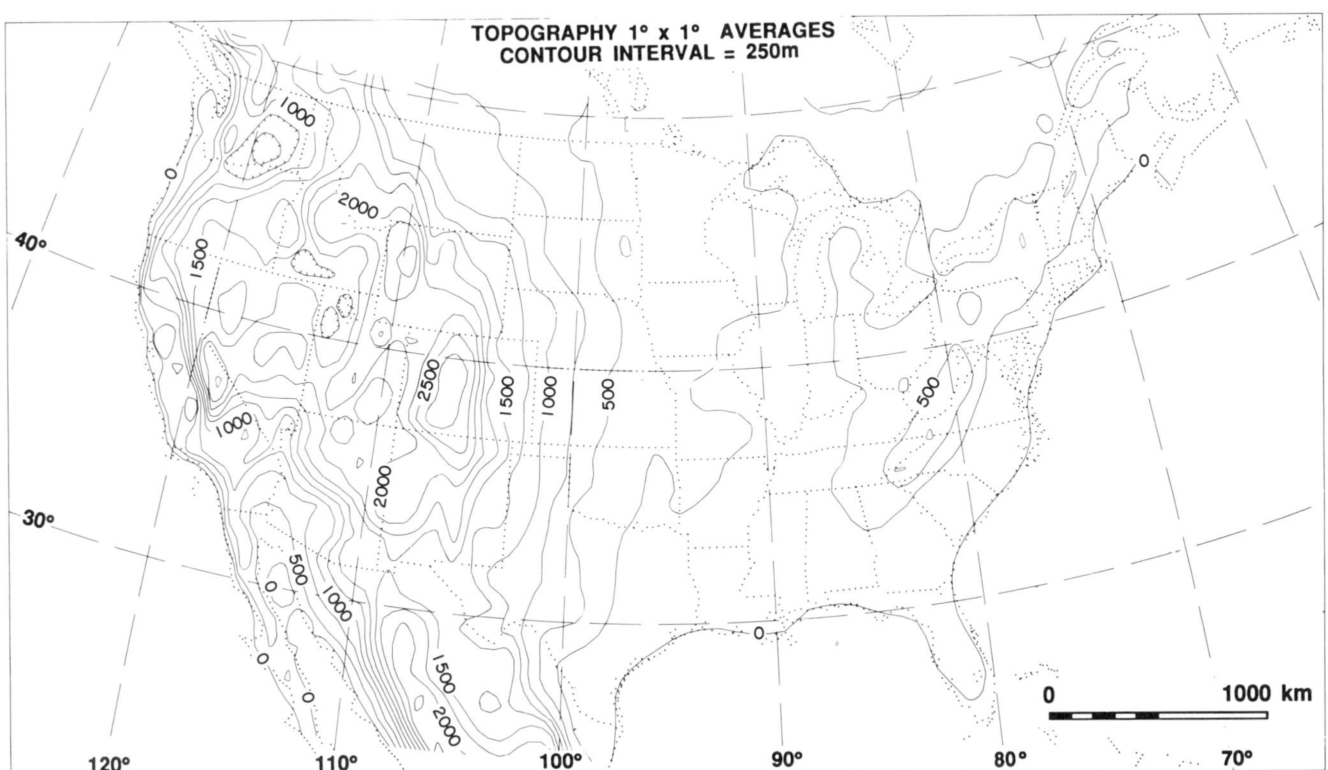

Figure 10. Regional topographic elevation map of a portion of the North American Continent. Elevations are 1° × 1° averages. Contour interval = 250 m.

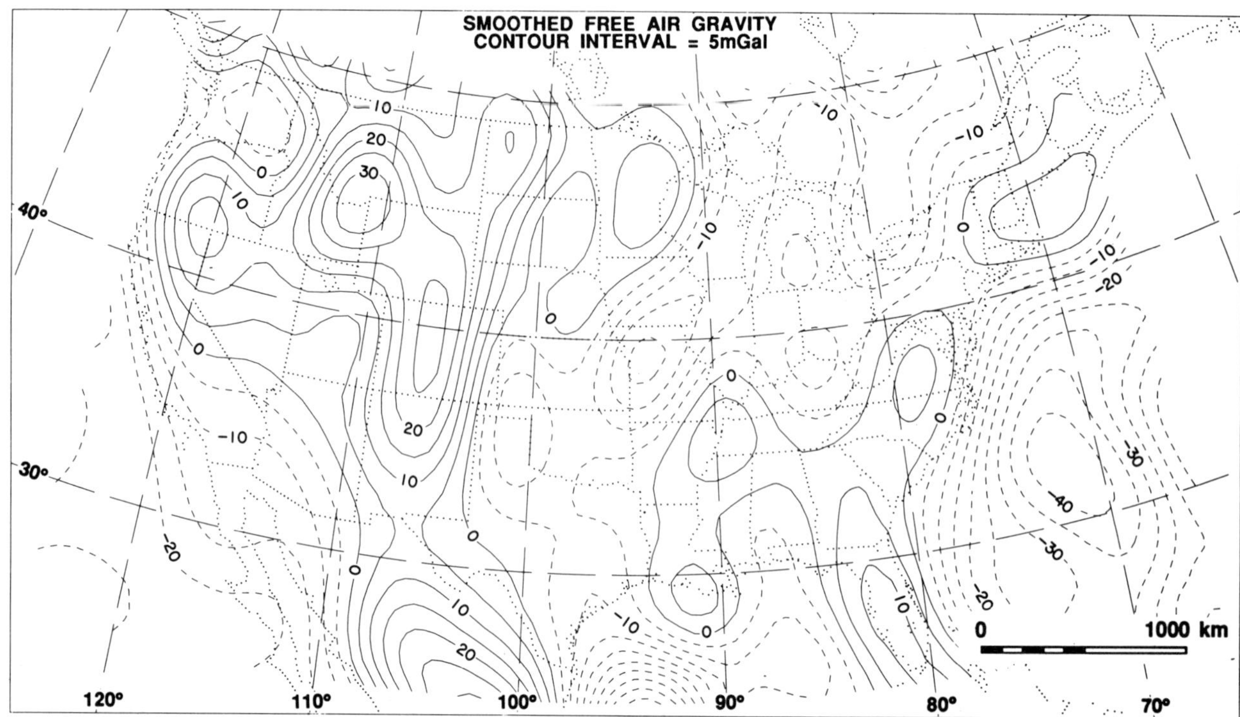

Figure 11. Smoothed free-air gravity anomaly map of North America. Free-air anomaly data have been filtered to remove anomalies with wavelengths smaller than approximately 8°. Modified from von Frese and others (1982). Contour interval = 5 mGal.

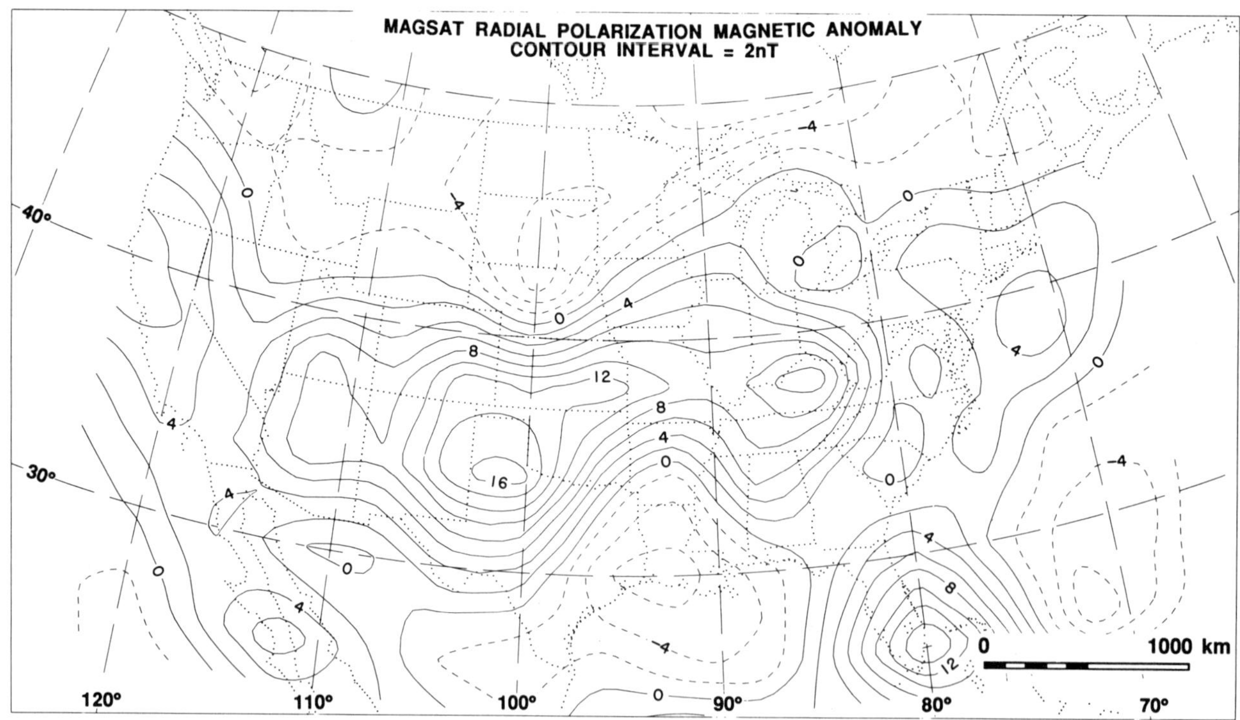

Figure 12. Radial polarization magnetic anomaly map of North America. The MAGSAT satellite magnetometer observations were reduced to a uniform elevation of 450 km and reduced to radial polarization by spherical equivalent source field calculations (von Frese and others, 1982). Contour interval = 2 nT.

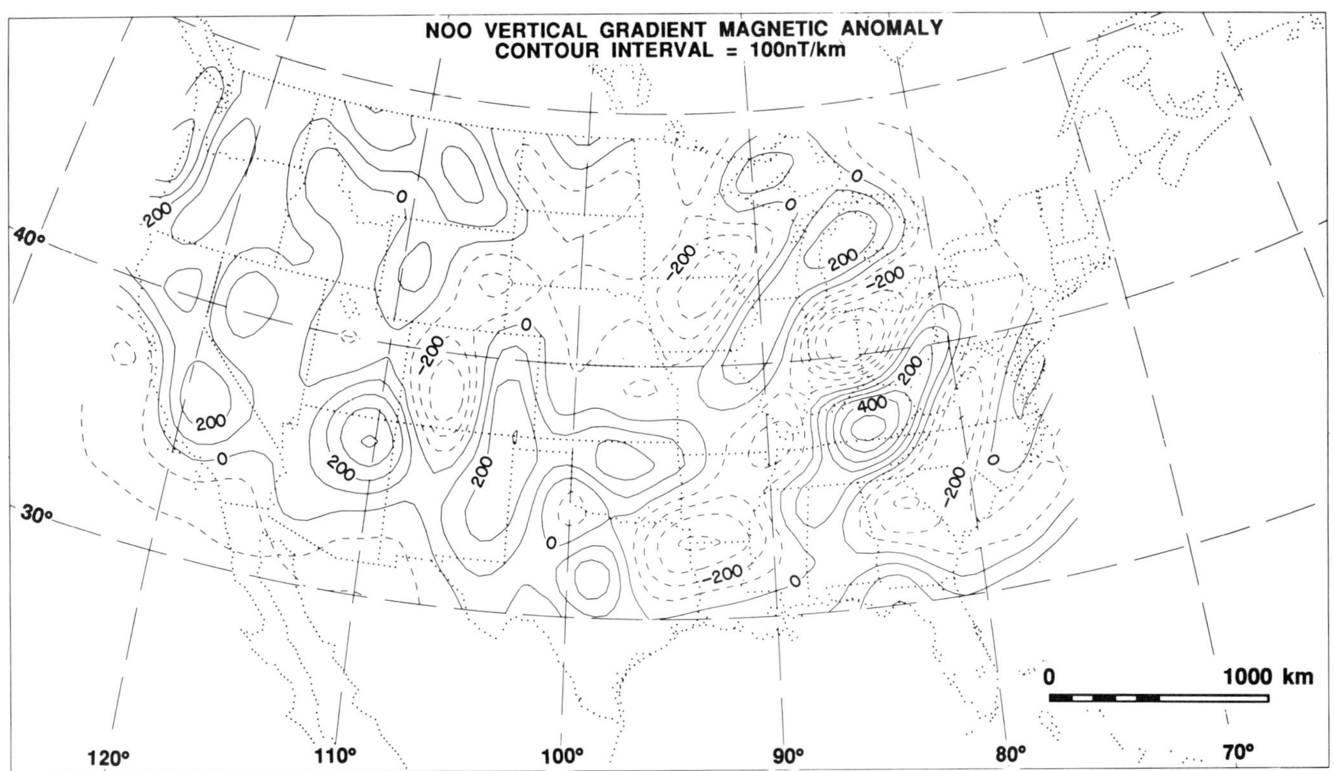

Figure 13. Filtered vertical gradient of magnetic intensity anomaly map of the United States. Data are from the Naval Oceanographic Office (NOO) aeromagnetic survey recorded along north-south tracks and smoothed by applying a high-cut filter that attenuated wavelengths shorter than approximately 200 km. The data were then reduced to the pole (radial polarization) and the vertical gradient calculated by equivalent point source methods. Contour interval is 100 nT/km. Original, smoothed NOO data from Sexton and others (1982).

portion of the United States, which trends roughly east-west and correlates approximately with the prominent east-west magnetic anomaly.

DISCUSSION AND CONCLUSIONS

The compilation of seismic properties of the crust and upper mantle for the North American continent provides important information on the distribution of physical properties of the upper continental lithosphere that can be used to correlate with other geophysical and geologic features. A number of prominent correlations are noted. New information presented here includes an updated and more complete set of data from seismic refraction profiles and studies of surface-wave dispersion, and also includes an analysis of Poisson's ratio in the crust and upper mantle. Contour maps of crustal thickness, average crustal compressional-wave velocity, and P_n velocity are here compared with gravity and magnetic anomaly data.

Although a number of observations and correlations concerning the distribution of seismic properties of the continental crust and relation to potential field data have been made in this chapter, many interesting questions arise from this review of these regional data. For example:

1. What is the cause of the regional variations in seismic properties? Black and Braile (1982) presented evidence to suggest that temperature in the Earth's mantle may be a primary control on the upper-mantle P_n velocity and thus may account for the regional variations observed in Figure 4. Smithson and others (1981) suggested that average seismic velocity of the crust is an indicator of average crustal composition. Compositional variations between geologic provinces, age provinces, or terranes might thus be reflected in the distribution of average crustal velocity. Correlations of $\overline{V}_p$ anomalies with the Snake River Plain, Colorado Plateau, Midcontinent Rift System, the volcanic provinces along the west coast of the United States, and the Mississippi Embayment area demonstrate that this parameter does reflect the presence of some specific geologic features or provinces.

2. The difference in crustal structure associated with the Pacific and Atlantic continental margins suggests fundamental variations due to the differing plate interactions at those margins. For example, the Atlantic continental margin represents a trailing edge (passive margin) of the continental crust, and a gradual thinning and decrease in average seismic velocity of the crust is observed as the oceanic plate is approached. However, the mechanism and timing of the formation of this crustal transition is presently not known. The decrease in average velocity does not

seem to be what one would expect since rifting at the continental margin would probably involve the intrusion of mafic material into the crust.

3. The combination of distribution of seismic properties in the crust and upper mantle, with topographic and long-wavelength free-air gravity anomaly data provides important data for an analysis of the mechanisms of isostatic compensation for continental regions. However, it is clear that this mechanism involves both lateral density changes in the crust and upper mantle, as well as variations in the depth to the Moho discontinuity and the asthenosphere.

4. Prominent lithospheric anomalies such as the Basin and Range province in the western United States require explanation. It is well known that the Basin and Range represents an area of major Cenozoic crustal extension. However, any simple model of extension involving brittle failure (faulting) in the upper crust and ductile flow in the lower crust does not account for the prominent low in average seismic velocity in the crust. Some mechanism of thinning of the lower crust more than the upper crust during stretching, "subcrustal erosion," or addition of low-velocity material is necessary in order to explain the observed pattern of velocities. Alternatively, the low average velocities may be caused by high average temperatures in the Basin and Range, resulting in decreases in velocity of both upper and lower crustal rocks (as compared with the craton and adjacent provinces) due to the negative temperature derivative for seismic velocity and lower metamorphic grade.

5. Although the craton displays relatively stable crustal seismic parameters with a crustal thickness of about 42 km, an upper-mantle P-wave velocity of about 8.1 km/sec, and an average seismic velocity of the crust of about 6.5 km/sec, some local variations in these properties are observed; the most prominent of them tend to be associated with areas of Phanerozoic basin development. At least some of these basins appear to be associated with paleorifts (Keller and others, 1983; Van Schmus and Hinze, 1985; Braile and others, 1986).

In this chapter we have analyzed a large volume of crustal seismic velocity data and qualitatively compared the distribution of seismic parameters in the continental crust to geologic, tectonic, and other geophysical features. These comparisons provide interesting correlations and patterns that aid in our understanding of the nature and development of the continent. However, perhaps the most important contribution of this chapter is the presentation of data that warrant more quantitative analysis and that raise interesting questions concerning the origin and evolution of the continental crust.

ACKNOWLEDGMENTS

We thank Neil Stillman, Mark Sparlin, Paul Black, Kevin Martindale, John McGinnis, Mark Brumbaugh, and William Meyer for aiding in the compilation and plotting of the data presented here. This work was partially supported by National Aeronautics and Space Administration Contract NCC5-21, by Subcontract 9-X60-2133K-1 with the Los Alamos National Laboratory, and by the Earth Sciences Division of the National Science Foundation.

REFERENCES CITED

Adams, H. E., 1975, A crustal study of Kentucky and adjacent areas [M.S. thesis]: Lexington, University of Kentucky, 44 p.

Allenby, R. J., and Schnetzler, C., 1983, United States crustal thickness: Tectonophysics, v. 93, p. 13–31.

Ansorge, J., and 21 others, 1982, Comparative interpretation of explosion seismic data: Journal of Geophysics, v. 51, p. 69–84.

Archambeau, C. B., Flinn, E. A., and Lambert, D. G., 1969, Fine structure of the upper mantle: Journal of Geophysical Research, v. 74, p. 5825–5865.

Austin, C. B., and Keller, G. R., 1982, A crustal structure study of the northern Mississippi Embayment: U.S. Geological Survey Professional Paper 1236, p. 83–93.

Bache, T. C., Rodi, W. L., and Harkrider, D. G., 1978, Crustal structure inferred from Rayleigh-wave signatures of NTS explosions: Seismological Society of America Bulletin, v. 68, p. 1399–1413.

Baldwin, J. L., 1980, A crustal seismic refraction study in southwestern Indiana and southern Illinois [M.S. thesis]: West Lafayette, Indiana, Purdue University, 99 p.

Bamford, D., 1973, Refraction data in western Germany; A time-term interpretation: Zeitschrift fur Geophysik, v. 39, p. 907–927.

Bamford, D., Jentsch, M., and Prodehl, C., 1979, P_n anistropy in northern Britain and the eastern and western United States: Geophysical Journal of the Royal Astronomical Society, v. 57, p. 397–429.

Barr, K. G., 1971, Crustal refraction experiment: Yellowknife: Journal of Geophysical Research, v. 76, p. 1929–1948.

Barrett, D. L., Berry, M., Blanchard, J. E., Keen, M. J., and McAllister, R. E., 1964, Seismic studies on the eastern seaboard of Canada; The Atlantic coast of Nova Scotia: Canadian Journal of Earth Sciences, v. 1, p. 10–22.

Bartelsen, H., and 5 others, 1982, The combined seismic reflection-refraction investigation of the Urach geothermal anomaly, *in* The Urach Project: Stuttgart, Schweizerbart'sche Verlagsbuchhandlung, p. 237–262.

Bateman, P., and Eaton, J., 1967, Sierra Nevada batholith: Science, v. 158, p. 1407–1417.

Bates, A., and Hall, D. H., 1975, Upper mantle structure in southern Saskatchewan and western Manitoba from Project Edzoe: Canadian Journal of Earth Sciences, v. 12, p. 2134–2144.

Bennett, G. T. Clowes, R. M., and Ellis, R. M., 1975, A seismic refraction survey along the southern Rocky Mountain trench: Seismological Society of America Bulletin, v. 65, p. 37–54.

Berg, J. W., Jr., Cook, K. L., Narans, H. D., Jr., and Dolan, W. M., 1960, Seismic investigation of crustal structure in the eastern part of the Basin and Range province: Seismological Society of America Bulletin, v. 50, p. 511–535.

Berg, J. W., Jr., and 10 others, 1966, Crustal refraction profile, Oregon Coast Range: Seismological Society of America Bulletin, v. 56, p. 1357–1362.

Berry, M. J., 1973, Structure of the crust and upper mantle in Canada: Tectonophysics, v. 20, p. 183–201.

Berry, M. J.,a nd Forsyth, D. A., 1975, Structure of the Canadian Cordillera from seismic refraction and other data: Canadian Journal of Earth Sciences, v. 12, p. 182–208.

Berry, M. J., and Fuchs, K., 1973, Crustal structure of the Superior and Grenville provinces of the northeastern Canadian shield: Seismological Society of America Bulletin, v. 63, p. 1393–1432.

Berry, M. J., and West, G. F., 1966, A time-term interpretation of the first-arrival data of the 1963 Lake Superior experiment, *in* Steinhart, J. S., and Smith, T. J., eds., The Earth beneath the continents: American Geophysical Union

Geophysical Monograph 10, p. 166–180.

Biswas, N. N., and Knopoff, L., 1974, The structure of the upper mantle under the United States from the dispersion of Rayleigh wave: Geophysical Journal of the Royal Astronomical Society, v. 36, p. 515–539.

Black, P. R., and Braile, L. W., 1982, P_n velocity and cooling of the continental lithosphere: Journal of Geophysical Research, v. 87, p. 10557–10568.

Blumling, P., and Prodehl, C., 1983, Crustal structure beneath the eastern part of the Coast Ranges (Diablo Range) of central California from explosion seismic and near earthquake data: Physics of the Earth and Planetary Interiors, v. 31, p. 313–326.

Blumling, P., Mooney, W. D., and Lee, W.H.K., 1985, Crustal structure of the southern Calaveras Fault Zone, central California, from seismic refraction investigations: Seismological Society of America Bulletin, v. 75, p. 193–209.

Bodnar, C. A., 1982, Crustal structure of the great plains of North America from Rayleigh wave analysis [M.S. thesis]: El Paso, University of Texas at El Paso, 62 p.

Bolt, B. A., and Gutdeutsch, R., 1982, Reinterpretation by ray tracing of a transverse refraction seismic profile through the California Sierra Nevada, Part 1: Seismological Society of America Bulletin, v. 72, p. 889–900.

Braile, L. W., Smith, R. B., Keller, G. R., Welch, R. M., and Meyer, R. P., 1974, Crustal structure across the Wasatch Front from detailed seismic refraction studies: Journal of Geophysical Research, v. 79, p. 2669–2766.

Braile, L. W., and 9 others, 1982, The Yellowstone–Snake River Plain seismic profiling experiment; Crustal structure of the Eastern Snake River Plain: Journal of Geophysical Research, v. 87, p. 2597–2609.

Braile, L. W., Hinze, W. J., Keller, G. R., Lidiak, E. G., and Sexton, J. L., 1986, Tectonic development of the New Madrid Rift Complex, Mississippi Embayment, North America: Tectonophysics, v. 131, p. 1–21.

Brune, J., and Dorman, J., 1963, Seismic waves and earth structure in the Canadian shield: Seismological Society of America Bulletin, v. 53, p. 167–210.

Bucher, R. L., and Smith, R. B., 1971, Crustal structure of the eastern Basin and Range province and the northern Colorado Plateau from phase velocities of Rayleigh waves, *in* Heacock, J. G., eds., The structure and physical properties of the Earth's crust: American Geophysical Union Geophysical Monograph 14, p. 59–70.

Burdick, L. J., and Langston, C. A., 1977, Modeling crustal structure through the use of converted phases in teleseismic body wave forms: Seismological Society of America Bulletin, v. 67, p. 677–691.

Carder, D. S., 1973, Trans-California seismic profile, Death Valley to Monterey Bay: Seismological Society of America Bulletin, v. 63, p. 571–586.

Carder, D. S., Qamar, A., and McEvilly, T. V., 1970, Trans-California seismic profile Pahute Mesa to San Francisco Bay: Seismological Socioety of America Bulletin, v. 60, p. 1829–1846.

Catchings, R. D., and Mooney, W. D., 1988, Crustal structure of the Columbia Plateau; Evidence for continental rifting: Journal of Geophysical Research, v. 93, p. 459–474.

Catchings, R. D., and 20 others, 1988, The 1986 PASSCAL Basin and Range lithospheric seismic experiment: EOS Transactions of the American Geophysical Union, v. 69, p. 593–598.

Chandra, N. N., and Cumming, G. L., 1972, Seismic refraction studies in western Canada: Canadian Journal of Earth Sciences, v. 9, p. 1099–1109.

Clee, T. E., Barr, K. G., and Berry, M. J., 1974, Fine structure of the crust near Yellowknife: Canadian Journal of Earth Sciences, v. 11, p. 1534–1549.

Clowes, R. M., Spence, G. D., Ellis, R. M., and Waldron, D. A., 1986, Structure of the lithosphere in a young subduction zone; Results from reflection and refraction studies, *in* Barazangi, M., and Brown, L., eds., Reflection seismology; The continental crust: American Geophysical Union Geodynamics Series 14, p. 313–321.

Cohen, T. J., 1966, Explosion seismic studies of the midcontinent gravity high [Ph.D. thesis]: Madison, University of Wisconsin, 329 p.

Cohen, T. J., and Meyer, R. P., 1966, The midcontinent gravity high; Gross crustal structure, *in* Steinhart, J. S., and Smith, T. J., eds., The earth beneath the continents: American Geophysical Union Geophysical Union Geophysical Monograph 10, p. 141–165.

Colburn, R. H., and Mooney, W. D., 1986, Two-dimensional velocity structure along the synclinal axis of the Great Valley, California: Seismological Society of America Bulletin, v. 76, p. 1305–1322.

Cook, F. A., McCullar, D. B., Decker, E. R., and Smithson, S. B., 1979, Crustal structure and evolution of the southern Rio Grande Rift *in* Riecker, R. E., ed., Rio Grande Rift; Tectonics and magmatism: Washington, D.C., American Geophysical Union, p. 195–208.

Cram, I. H., Jr., 1961, Crustal structure refraction survey in south Texas: Geophysics, v. 26, p. 560–573.

Crosson, R. S., 1976, Crustal structure modeling of earthquake data; 1. Simultaneous least squares estimation of hypocenter and velocity parameters; 2, Velocity structure of the Puget Sound region: Journal of Geophysical Research, v. 81, p. 3036–3054.

Cumming, W. B., Clowes, R. M., and Ellis, R. M., 1979, Crustal structure from a seismic refraction profile across southern British Columbia: Canadian Journal of Earth Sciences, v. 16, p. 1024–1040.

Dainty, A. M., Keen, C. E., Keen, M. J., and Blanchard, J. E., 1966, Review of geophysical evidence on crust and upper mantle structure on the eastern seaboad of Canada, *in* Steinhart, J. S., and Smith, T. J., eds., The Earth beneath the continents: American Geophysical Union Monograph 10, p. 349–369.

Daniel, R. G., and Boore, D. M., 1982, Anomalous shear wave delays and surface wave velocities at Yellowstone Caldera, Wyoming: Journal of Geophysical Research, v. 87, p. 2731–2744.

Dehlinger, P., Chilburis, E. F., and Collver, M. M., 1965, Local travel-time curves and their geologic implications for the Pacific northwest states: Seismological Society of America Bulletin, v. 55, p. 587–607.

Delandro, W., and Moon, W., 1982, Seismic structure of Superior-Churchill Precambrian boundary zone: Journal of Geophysical Research, v. 87, p. 6884–6888.

Der, Z., Masse, R., and Landisman, M., 1970, Effects of observational errors on the resolution of surface waves at intermediate distances: Journal of Geophysical Research, v. 75, p. 3399–3409.

Diment, W. H., Stewart, S. W., and Roller, J. C., 1961, Crustal structure from the Nevada test site to Kingman, Arizona, from seismic and gravity observations: Journal of Geophysical Research, v. 66, p. 201–213.

Dorman, J., and Ewing, M., 1962, Numerical inversion of seismic surface wave dispersion data and crust-mantle structure in the New York–Pennsylvania area: Journal of Geophysical Research, v. 67, p. 5227–5241.

Dorman, J., Worzel, J. L., Leyden, R., Cook, T. N., and Hatziemmanuel, M., 1972, Crustal section from seismic refraction measurements near Victoria, Texas: Geophysics, v. 37, p. 225–236.

Dorman, L. M., 1972, Seismic crustal anisotropy in northern Georgia: Seismological Society of America Bulletin, v. 62, p. 39–45.

Douglas, R.J.W., and Price, R. A., 1972, Nature and significance of variations in tectonic styles in Canada, *in* Price, R. A., and Douglas, R.J.W., eds., Variations in tectonic styles in Canada: Geological Society of Canada Special Paper 11, p. 625–688.

Dziewonski, A. M., and Anderson, D. L., 1983, Travel times and station corrections for P waves at teleseismic distances: Journal of Geophysical Research, v. 88, p. 3295–3314.

Dziewonski, A. M., and Hales, A. L., 1972, Numerical analysis of dispersed seismic waves: Methods in Computational Physics, v. 11, p. 39–85.

Eaton, J. P., 1963, Crustal structure from San Francisco, California, to Eureka, Nevada, from seismic-refraction measurements: Journal of Geophysical Research, v. 68, p. 5789–5806.

——, 1966, Crustal structure in northern and central California from seismic evidence *in* Geology of northern California: California Division of Mines Geology Bulletin, v. 190, p. 419–426.

Ellis, R. M., and 15 others, 1983, The Vancouver Island seismic project; A COCRUST onshore-offshore study of a convergent margin: Canadian Journal of Earth Sciences, v. 20, p. 719–741.

Ewing, G. N., Dainty, A. M., Blanchard, J. E., and Keen, J. J., 1966, Seismic studies on the eastern seaboard of Canada; The Appalachian system 1: Canadian Journal of Earth Sciences, v. 3, p. 89–109.

Ewing, M., and Press, F., 1959, Determination of crustal structure from phase velocity of Rayleigh waves; Part 3, The United States: Geological Society of America Bulletin, v. 70, p. 229–244.

Ewing, M., Worzel, J. L., Ericson, D. B., and Heezen, B. C., 1955, Geophysical and geological investigations of the Gulf of Mexico, Part 1: Geophysics, v. 20, p. 1–18.

Fenneman, N. M., 1946, Physical divisions of the United States: U.S. Geological Survey Map, scale 1:7,000,000.

Fenwick, D.K.B., Keen, M. J., Keen, C. E., and Lambert, A., 1968, Geophysical studies of the continental margin northeast of Newfoundland: Canadian Journal of Earth Sciences, v. 5, p. 483–500.

Fernandez, L. M., and Careaga, J., 1968, The thickness of the crust in central United States and La Paz, Bolivia, from the spectrum of longitudinal seismic waves: Seismological Society of America Bulletin, v. 58, p. 711–741.

Forsyth, D. A., Berry, M. J., and Ellis, R. M., 1974, A refraction survey across the Canadian Cordillera at 54°N: Canadian Journal of Earth Sciences, v. 11, p. 533–548.

Fuis, G. S., Mooney, W. D., Healy, J. H., McMechan, G. A., and Lutter, W. J., 1984, A seismic refraction survey of the Imperial Valley region, California: Journal of Geophysical Research, v. 89, p. 1165–1189.

Gibbs, J. F., and Roller, J. C., 1966, Crustal structure determined by seismic-refraction measurements between the Nevada test site and Ludlow, California: U.S. Geological Survey Professional Paper 550-D, p. D-125–D-131.

Ginzburg, A., Mooney, W. D., Walter, A. W., Lutter, W. J., and Healy, J. H., 1983, Deep structure of northern Mississippi Embayment: American Association of Petroleum Geologists Bulletin, v. 67, p. 2031–2046.

Gish, D. M., Keller, G. R., and Sbar, M. L., 1981, A refraction study of deep crustal structure in the Basin and Range; Colorado Plateau of eastern Arizona: Journal of Geophysical Research, v. 86, p. 6029–6038.

Godlewski, M.J.C., and West, G. F., 1977, Rayleigh-wave dispersion over the Canadian Shield: Seismological Society of America Bulletin, v. 67, p. 771–779.

Green, A. G., and 7 others, 1980, Cooperative seismic surveys across the Superior-Churchill boundary zone in southern Canada: Canadian Journal of Earth Sciences, v. 17, p. 617–632.

Green, A. R., 1977, The evolution of the Earth's crust and sedimentary basin development, *in* Heacock, J. G., ed., The Earth's crust; Its nature and physical properties: American Geophysical Union Geophysical Monograph 20, p. 1–17.

Green, R.W.E., and Hales, A. L., 1968, The travel times of P waves to 30° in the central United States and upper mantle structure: Seismological Society of America Bulletin, v. 58, p. 267–289.

Greensfelder, R. W., and Kovach, R. L., 1982, Shear wave velocities and crustal structure of the eastern Snake River Plain, Idaho: Journal of Geophysical Research, v. 87, p. 2643–2653.

Gutenberg, B., 1952, Waves from blasts recorded in southern California: EOS Transactions of the American Geophysical Union, v. 33, p. 427–431.

Hadley, A., and Kanamori, H., 1977, Seismic structure of the Transverse Ranges, California: Geological Society of America Bulletin, v. 88, p. 1469–1478.

Hadley, D., and Kanamori, H., 1979, Regional S-wave structure for southern California from the analysis of teleseismic Rayleigh waves: Geophysical Journal of the Royal Astronomical Society, v. 58, p. 655–666.

Hajnal, Z., and 6 others, 1984, An initial analysis of the earth's crust under the Williston basin; 1979 COCRUST experiment: Journal of Geophysical Research, v. 89, p. 9381–9400.

Hales, A. L., and Nation, J. B., 1973, A seismic refraction survey in the northern Rocky Mountains; More evidence for an intermediate crustal layer: Geophysical Journal of the Royal Astronomical Society, v. 35, p. 381–399.

Hales, A. L., Helsey, C. E., Dowling, J. J., and Nation, J. B., 1968, The east coast onshore-offshore experiment; 1, The first arrival phases: Seismological Society of America Bulletin, v. 58, p. 757–819.

Hales, A. L., Helsey, C. E., and Nation, J. B., 1970, Crustal structure study of Gulf Coast of Texas: American Association of Petroleum Geologists Bulletin, v. 54, p. 2040–2057.

Hall, D. H., and Hajnal, Z., 1969, Crustal structure of northwestern Ontario; Refraction seismology: Canadian Journal of Earth Sciences, v. 6, p. 81–99.

——, 1973, Deep seismic studies in Manitoba: Seismological Society of America Bulletin, v. 63, p. 883–910.

Halls, H. C., 1982, Crustal thickness in the Lake Superior region, *in* Wold, R. J., and Hinze, W. J., eds., Geology and tectonics of the Lake Superior Basin: Geological Society of America Memoir 156, p. 239–243.

Hamilton, R. M., 1970, Time-term analysis of explosion data from the vicinity of the Borrego Mountain, California, earthquake of 9 April 1968: Seismological Society of America Bulletin, v. 60, p. 367–381.

Hamilton, R. M., Ryall, A., and Berg, E., 1964, Crustal structure southwest of the San Andreas fault from quarry blasts: Seismological Society of America Bulletin, v. 54, p. 67–77.

Hartzell, S. H., and Heaton, T. H., 1986, Rupture history of the 1984 Morgan Hill, California, earthquake from the inversion of strong motion records: Seismological Society of America Bulletin, v. 76, p. 649–674.

Hauksson, E., 1987, Seismotectonics of the Newport-Inglewood fault zone in the Los Angeles basin, southern California: Seismological Society of America Bulletin, v. 77, p. 539–561.

Healy, J. H., 1963, Crustal structure along the coast of California from seismic-refraction measurements: Journal of Geophysical Research, v. 68, p. 5777–5787.

Healy, J. H., and Peake, L. G., 1975, Seismic velocity structure along a section of the San Andreas fault near Bear Valley, California: Seismological Society of America Bulletin, v. 65, p. 1177–1197.

Healy, J. H., and Warren, D. H., 1969, Explosion seismic studies in North America, *in* Hart, P. J., ed., The Earth's crust and upper mantle: American Geophysical Union Geophysical Monograph 13, p. 208–220.

Hearn, T. M., 1984, P_n travel times in southern California: Journal of Geophysical Research, v. 89, p. 1843–1855.

Herrin, E., 1969, Regional variations of P-wave velocity in the upper mantle beneath North America, *in* Hart, P. J., ed., The Earth's crust and upper mantle: American Geophysical Union Geophysical Monograph 13, p. 242–246.

Herrin, E., and Taggart, J., 1962, Regional variations in P_n velocity and their effect on the location of epicenters: Seismological Society of America Bulletin, v. 52, p. 1037–1046.

Hersey, J. B., Bunce, E. T., Wyrick, R. F., and Dietz, T. F., 1959, Geophysical investigation of the continental margin between Cape Henry, Virginia, and Jacksonville, Florida: Geological Society of America Bulletin, 70, p. 437–465.

Hill, D. P., 1972, Crustal and upper mantle structure of the Columbia plateau from long-range seismic-refraction measurements: Geological Society of America Bulletin, v. 83, p. 1639–1648.

Hill, D. P., and Pakiser, L. C., 1966, Crustal structure between the Nevada test site and Boise, Idaho, from seismic-refraction measurements, *in* Steinhart, J. S., and Smith, T. J., eds., The Earth beneath the continents: American Geophysical Union Geophysical Monograph 10, p. 391–419.

Hobson, G. D., Overton, A., Clay, D. N., and Thatcher, W., 1967, Crustal structure under Hudson Bay: Canadian Journal of Earth Sciences, v. 4, p. 929–947.

Hodgson, J. H., 1953, A seismic survey of the Canadian Shield; 1, Refraction studies based on rock-bursts at Kirkland Lake, Ontario: Ottawa, Publication of the Dominion Observatory, v. 16, p. 111–163.

Hoffman, L. R., and Mooney, W. D., 1984, A seismic study of Yucca Mountain and vicinity, southern Nevada; Data report and preliminary results: U.S. Geological Survey Open-File Report 83-588, 50 p.

Holbrook, W. S., and Mooney, W. D., 1987, The crustal structure of the axis of the Great Valley, California, from seismic refraction measurements: Tectonophysics, v. 140, p. 49–63.

Horn, J. R., Clowes, R. M., Ellis, R. M., and Bird, D. N., 1984, The seismic structure across an active oceanic/continental transform fault zone: Journal of Geophysical Research, v. 89, p. 3107–3120.

Howell, D. G., ed., 1985, Tectonostratigraphic terranes of the circum-Pacific region: Houston, Texas, Circum-Pacific Council for Energy and Mineral Resources Earth Science Series 1, 581 p.

Huang, H., Spencer, C., and Green, A., 1986, A method for the inversion of refraction and reflection travel times for laterally varying velocity structures: Seismological Society of America Bulletin, v. 76, p. 837–846.

Huestis, S., Molnar, P., and Oliver, J., 1973, Regional S_n velocities and shear velocity in the upper mantle: Seismological Society of America Bulletin, v. 63, p. 469–475.

Hwang, L. J., and Mooney, W. D., 1986, Velocity and Q structure of the Great Valley, California, based on synthetic seismogram modeling of seismic refraction data: Seismological Society of America Bulletin, v. 76, p. 1053–1067.

Ibrahim, A-B.K., and Nuttli, O. W., 1967, Travel-time curves and upper-mantle structure from long-period S waves: Seismological Society of America Bulletin, v. 57, p. 1063–1092.

Jackson, W. H., and Pakiser, L. C., 1965, Seismic study of crustal structure in the southern Rocky Mountains: U.S. Geological Survey Professional Paper 525-D, p. D-85–D-92.

Jackson, W. H., Stewart, S. W., and Pakiser, L. C., 1963, Crustal structure in eastern Colorado from seismic-refraction measurements: Journal of Geophysical Research, v. 68, p. 5767–5776.

Jaksha, L. H., 1982, Reconnaissance seismic refraction-reflection surveys in southwestern New Mexico: Geological Society of America Bulletin, v. 93, p. 1030–1037.

James, D. E., and Steinhart, J. S., 1966, Structure beneath continents; A critical review of explosion studies 1960–1965, *in* Steinhart, J. S., and Smith, T. J., eds., The Earth beneath the continents: American Geophysical Union Monograph 10, p. 293–333.

James, D. E., Smith, T. J., and Steinhart, J. S., 1968, Crustal structure of the Middle Atlantic states: Journal of Geophysical Research, v. 73, p. 1983–2007.

Jardine, W. G., 1988, Seismic study in the Ouachita system of southern Arkansas and the adjacent Gulf Coastal Plain [M.S. thesis]: West Lafayette, Indiana, Purdue University, 112 p.

Johnson, L. R., 1965, Crustal structure between Lake Mead, Nevada, and Mono Lake, California: Journal of Geophysical Research, v. 70, p. 2863–2872.

Johnson, S. H., and Couch, R. W., 1970, Crustal structure in the North Cascade Mountains of Washington and British Columbia from seismic refraction measurements: Seismological Society of America Bulletin, v. 60, p. 1259–1269.

Johnson, S. H., Couch, R. W., Gemperle, M., and Banks, E. R., 1972, Seismic refraction measurements in southeast Alaska and western British Columbia: Canadian Journal of Earth Sciences, v. 9, p. 1756–1765.

Kanamori, H., and Hadley, D., 1975, Crustal structure and temporal velocity change in southern California: Pure and Applied Geophysics, v. 113, p. 257–280.

Kanasewich, E. R., 1966, Deep crustal structure under the plains and Rocky Mountains: Canadian Journal of Earth Sciences, v. 3, p. 937–946.

Kanasewich, E. R., and Chiu, S.K.L., 1985, Least-square inversion of spatial seismic refraction data: Seismological Society of America Bulletin, v. 75, p. 865–880.

Katz, S., 1954, Seismic study of crustal structure in Pennsylvania and New York: Seismological Society of America Bulletin, v. 44, p. 303–325.

Katz, S., and Ewing, M., 1956, Seismic-refraction measurements in the Atlantic Ocean; Part 7, Atlantic Ocean Basin west of Bermuda: Geological Society of America Bulletin, v. 67, p. 475–510.

Kean, E. A., and Long, L. T., 1980, A seismic refraction line along the axis of the southern Piedmont and crustal thickness in the northeastern United States: Earthquake Notes, v. 51, p. 3–13.

Keen, C. E., and Hyndman, R. D., 1979, Geophysical review of the continental margins of eastern and western Canada: Canadian Journal of Earth Sciences, v. 16, p. 712–747.

Keller, G. R., and Shurbet, D. H., 1975, Crustal structure of the Texas Gulf Coastal Plain: Geological Society of America Bulletin, v. 86, p. 807–810.

Keller, G. R., Smith, R. B., and Braile, L. W., 1975, Crustal structure along the Great Basin–Colorado Plateau transition from seismic refraction studies: Journal of Geophysical Research, v. 80, p. 1093–1097.

Keller, G. R., Smith, R. B., Braile, L. W., Heaney, R., and Shurbet, D. H., 1976, Upper crustal structure of the eastern Basin and Range, northern Colorado Plateau, and middle Rocky Mountains from Rayleigh-wave dispersion: Seismological Society of America Bulletin, v. 66, p. 869–876.

Keller, G. R., Braile, L. W., and Morgan, P., 1979a, Crustal structure, geophysical models and contemporary tectonism of the Colorado Plateau: Tectonophysics, v. 61, p. 131–147.

Keller, G. R., Braile, L. W., and Schlue, J. W., 1979b, Regional crustal structure of the Rio Grande Rift from surface wave dispersion measurements, *in* Riecker, R. E., ed., Rio Grande Rift; Tectonics and magmatism: Washington, D.C., American Geophysical Union, p. 115–126.

Keller, G. R., Lidiak, E. G., Hinze, W. J., and Braile, L. W., 1983, The role of rifting in the tectonic development of the Midcontinent, U.S.A.: Tectonophysics, v. 94, p. 391–412.

Keller, G. R., and 6 others, 1989, The Paleozoic continent-ocean transition in the Ouachita Mountains imaged from PASSCAL wide-angle reflection-refraction data: Geology, v. 17, p. 119–122.

Kind, R., 1972, Residuals and velocities of P_n waves recorded by the San Andreas seismograph network: Seismological Society of America Bulletin, v. 62, p. 85–100.

King, P. B., and Edmonston, G. J., 1972, Generalized tectonic map of North America: U.S. Geological Survey Miscellaneous Geologic Investigation Map I-688, scale 1:15,000,000.

Kovach, R. L., 1978, Seismic surface waves and crustal and upper mantle structure: Reviews in Geophysics and Space Physics, v. 16, p. 1–13.

Kurita, T., 1973, Regional variations in the structure of the crust in the central United States from P-wave spectra: Seismological Society of America Bulletin, v. 63, p. 1663–1687.

Langston, C. A., 1977, The effect of planar dipping structure on source and receiver responses for constant ray parameter: Seismological Society of America Bulletin, v. 67, p. 1029–1050.

Langston, C. A., and Helmberger, D. V., 1974, Interpretation of body and Rayleigh waves from NTS to Tucson: Seismological Society of America Bulletin, v. 64, p. 1919–1929.

Langston, C. A., and Isaacs, C. M., 1981, A crustal thickness constraint for central Pennsylvania: Seismological Society of America Bulletin Earthquake Notes, v. 52, p. 13–22.

Leaver, D. S., Mooney, W. D., and Kohler, W. M., 1984, Seismic refraction study of the Oregon Cascades: Journal of Geophysical Research, v. 89, p. 3121–3134.

Lewis, B.T.R., and Meyer, R. P., 1968, A seismic investigation of the upper mantle to the west of Lake Superior: Seismological Society of America Bulletin, v. 58, p. 565–596.

Lomnitz, C., and Bolt, B. A., 1967, Evidence on crustal structure in California from the Chase V explosion and the Chico earthquake of May 24, 1966: Seismological Society of America Bulletin, v. 57, p. 1093–1114.

Long, L. T., and Liow, J., 1986, Crustal thickness, velocity structure, and the isostatic response function in the southern Appalachians, *in* Barazangi, M., and Brown, L., eds., Reflection seismology; The continental crust: American Geophysical Union Geodynamics Series 14, p. 215–222.

Long, L. T., and Mathur, U. P., 1971, Southern Appalachian crustal structure from the dispersion of Rayleigh waves and refraction data: Earthquake Notes, v. 42, p. 31–40.

Losee, B. A., 1980, Rayleigh-wave dispersion applied to lithospheric structure in Canada [M.S. thesis]: West Lafayette, Indiana, Purdue University, 85 p.

Lowrey, W. S., 1977, Crustal structure of the Basin and Range Province using intermediate period surface wave dispersion [M.S. thesis]: West Lafayette, Indiana, Purdue University, 40 p.

Luetgert, J. H., 1985, Depth to Moho and characterization of the northern Appalachians from 1984 Maine-Quebec seismic refraction data: EOS Transactions of the American Geophysical Union, v. 66, p. 1074–1075.

Luetgert, J. H., and Meyer, R. P., 1982, Structure of the western basin of Lake Superior from cross structure refraction profiles, *in* Wold, R. J., and Hinze, W. J., eds., Geology and tectonics of the Lake Superior Basin: Geological Society of America Memoir 156, p. 245–255.

Luetgert, J. H., Mann, C. E., and Klemperer, S. L., 1987, Wide-angle deep crustal reflections in the northern Appalachians: Geophysical Journal of the Royal Astronomical Society, v. 89, p. 183–188.

Lyons, J. S., Forsyth, D. A., and Mair, J. A., 1980, Crustal studies in the La Malbaie Region, Quebec: Canadian Journal of Earth Sciences, v. 17, p. 478–490.

Martin, W. R., 1978, A seismic refraction study of the northeastern Basin and Range and its transition with the eastern Snake River plain [M.S. thesis]: El Paso, University of Texas, 41 p.

McCamy, K., and Meyer, R. P., 1964, A correlation method of apparent velocity measurement: Journal of Geophysical Research, v. 69, p. 691–699.

——, 1966, Crustal results of fixed multiple shots in the Mississippi Embayment, *in* Steinhart, J. S., and Smith, T. J., eds., The Earth beneath the continents: American Geophysical Union Geophysical Monograph 10, p. 166–180.

McCarthy, J., Fuis, G., and Wilson, J., 1987, Crustal structure of the Whipple Mountains, southestern California; A refraction study across a region of large continental extension: Geophysical Journal of the Royal Astronomical Society, v. 89, p. 119–124.

McConnell, R. J., Jr., Gupta, R. N., and Wilson, J. T., 1966, Compilation of deep crustal seismic refraction profiles: Reviews of Geophysics, v. 4, p. 41–55.

McEvilly, T. V., 1964, Central U.S. crust–upper mantle structure from Love and Rayleigh wave phase velocity inversion: Seismological Society of America Bulletin, v. 54, p. 1997–2015.

McMechan, G. A., and Spence, G. D., 1983, P-wave velocity structure of the Earth's crust beneath Vancouver Island: Canadian Journal of Earth Sciences, v. 20, p. 742–752.

Mereu, R. F., and Hunter, J. A., 1969, Crustal and upper mantle under the Canadian Shield from Project Early Rise data: Seismological Society of America Bulletin, v. 59, p. 147–165.

Mereu, R. F., and Jobidon, G., 1971, A seismic investigation of the crust and Moho on a line perpendicular to the Grenville Front: Canadian Journal of Earth Sciences, v. 8, p. 1553–1583.

Mereu, R. F., Majumdar, S. C., and White, R. E., 1976, The structure of the crust and upper mantle under the highest ranges of the Canadian Rockies from a seismic refraction survey: Canadian Journal of Earth Sciences, v. 14, p. 196–208.

Mereu, R., and 11 others, 1986a, The 1982 COCRUST seismic experiment across the Ottawa-Bonnechere graben and Grenville Front in Ottawa and Quebec: Geophysical Journal of the Royal Astronomical Society, v. 84, p. 491–514.

Mereu, R., Wang, D., and Kuhn, O., 1986b, Evidence for an inactive rift in the Precambrian from a wide-range reflection survey across the Ottawa-Bonnechere graben, *in* Barazangi, M., and Brown, L., eds., Reflection seismology; The continental crust: American Geophysical Union Geodynamics Series 14, p. 127–134.

Merkel, R. H., and Alexander, S. S., 1969, Use of correlation analysis to interpret continental margin ECOOE refraction data: Journal of Geophysical Research, v. 74, p. 2683–2697.

Mikumo, T., 1965, Crustal structure in central California in relation to the Sierra Nevada: Seismological Society of America Bulletin, v. 55, p. 65–83.

Mitchell, B. J., and Landisman, M., 1970, Interpretation of crustal section across Oklahoma: Geological Society of America Bulletin, v. 81, p. 2647–2656.

——, 1971, Geophysical measurements in the southern great plains, *in* Heacock, J. G., ed., The structure and physical properties of the Earth's crust: American Geophysical Union Geophysical Monograph 14, p. 77–93.

Mooney, W. D., Andrews, M. C., Ginzburg, A., Peters, D. A., and Hamilton, R. M., 1983, Crustal structure of the northern Mississippi Embayment and a comparison with other continental rift zones: Tectonophysics, v. 94, p. 327–348.

Morel-àL'-Huissier, P., Green, A. G., and Pike, C. J., 1987, Crustal refraction surveys across the Trans-Hudson Orogen–Williston Basin of south-central Canada: Journal of Geophysical Research, v. 92, p. 6403–6420.

Mueller, S., and Landisman, M., 1971, An example of the unified method of interpretation for crustal seismic data: Geophysical Journal of the Royal Astronomical Society, v. 23, p. 365–371.

Müller, G., and Mueller, S., 1979, Travel-time and amplitude interpretation of crustal phases on the refraction profile Delta-W, Utah: Seismological Society of America Bulletin, v. 69, p. 1121–1132.

Nava, F. A., and Brune, J. N., 1982, An earthquake-explosion reversed refraction line in the Peninsular Ranges of southern California and Baja California Norte: Seismological Society of America Bulletin, v. 72, p. 1195–1206.

Oliver, J., and Ewing, M., 1957, Higher modes of continental Rayleigh waves: Seismological Society of America Bulletin, v. 47, p. 187–203.

Oliver, J., Kovach, R., and Dorman, J., 1961, Crustal structure of the New York–Pennsylvania area: Journal of Geophysical Research, v. 66, p. 215–225.

Olsen, K. H., Keller, G. R., and Stewart, J. N., 1979, Crustal structure along the Rio Grande Rift from seismic refraction profiles, *in* Riecker, R. E., ed., Rio Grande Rift; Tectonics and magmatism: Washington, D.C., American Geophysical Union, p. 127–143.

Oppenheimer, D. H., and Eaton, J. P., 1984, Moho orientation beneath central California from regional earthquake travel times: Journal of Geophysical Research, v. 89, p. 10267–10282.

Owens, T. J., Taylor, S. R., and Zandt, G., 1987, Crustal structure at regional seismic test network stations determined from inversion of broadband teleseismic P waveforms: Seismological Society of America Bulletin, v. 77, p. 631–662.

Pakiser, L. C., 1985, Seismic exploration of the crust and upper mantle of the Basin and Range province: Boulder, Colorado, Geological Society of America Centennial Special Volume 1, p. 453–469.

Pakiser, L. C., and Brune, N., 1980, Seismic models of the root of the Sierra Nevada: Science, v. 210, p. 1088–1094.

Pakiser, L. C., and Robinson, R., 1966, Composition of the continental crust as estimated from seismic observations, *in* Steinhart, J. S., and Smith, T. J., eds., The Earth beneath the continents: American Geophysical Union Geophysical Monograph 10, p. 620–626.

——, 1967, Composition and evolution of the continental crust as suggested by seismic observations: Tectonophysics, v. 3, p. 547–557.

Pakiser, L. C., and Steinhart, J. S., 1964, Explosion seismology in the Western Hemisphere, *in* Odishaw, J., ed., Research in geophysics; 2, Solid earth and interface phenomena: Cambridge, Massachusetts Institute of Technology Press, p. 123–142.

Pakiser, L. C., and Zietz, I., 1965, Transcontinental crustal and upper-mantle structure: Reviews in Geophysics, v. 3, p. 505–520.

Panza, G. F., and Calcagnile, G., 1974, Crustal structure along the coast of California from Rayleigh waves: Physics of the Earth and Planetary Interiors, v. 9, p. 137–140.

Phinney, R. A., 1986, A seismic cross section of the New England Appalachians; The orogen exposed, *in* Barazangi, M., and Brown, L., eds., Reflection seismology; The continental crust: American Geophysical Union Geodynamics Series 14, p. 157–172.

Press, F., 1956, Determination of crustal structure from phase velocity of Rayleigh waves; Part 1, Southern California: Geological Society of America Bulletin, v. 67, p. 1647–1656.

——, 1957, Determination of crustal structure from phase velocity of Rayleigh waves; Part 2, San Francisco Bay region: Seismological Society of America Bulletin, v. 47, p. 87–88.

——, 1960, Crustal structure in the California-Nevada region: Journal of Geophysical Research, v. 65, p. 1039–1051.

Press, F., and Ewing, M., 1955, Earthquake surface waves and crustal structure, *in* Poldervaart, A., ed., Crust of the Earth: Geological Society of America Special Paper 62, p. 51–60.

Prewitt, R. H., 1969, Crustal thickness in central Texas as determined by Rayleigh wave dispersion [M.S. thesis]: Lubbock, Texas Tech University, 51 p.

Priestley, K., and Brune, J., 1978, Surface waves and the structure of the Great Basin of Nevada and western Utah: Journal of Geophysical Research, v. 83, p. 2265–2272.

——, 1982, Shear wave structure of the southern volcanic plateau of Oregon and Idaho and the northern Great Basin of Nevada from surface wave dispersion: Journal of Geophysical Research, v. 87, p. 2671–2675.

Priestley, K., Orcutt, J. A., and Brune, J. N., 1980, Higher-mode surface waves and structure of the Great Basin of Nevada and western Utah: Journal of Geophysical Research, v. 85, p. 7166–7174.

Priestley, K. F., Ryall, A. S., and Fezie, G. S., 1982, Crust and upper mantle structure in the northwest Basin and Range province: Seismological Society of America Bulletin, v. 72, p. 911–923.

Prodehl, C., 1979, Crustal structure of the western United States: U.S. Geological Survey Professional Paper 1034, 74 p.

Prodehl, C., and Pakiser, L. C., 1980, Crustal structure of the southern Rocky Mountains from seismic measurements: Geological Society of America Bulletin, v. 91, p. 147–155.

Prodehl, C., Schlittenhardt, J., and Stewart, S. W., 1984, Crustal structure of the Appalachian Highlands in Tennessee: Tectonophysics, v. 109, p. 61–76.

Redmond, R. J., and Kodama, K. P., 1984, Wide-angle reflection study of crustal structure, eastern Pennsylvania–northern New Jersey: Earthquake Notes, v. 55, p. 9–15.

Richards, T. C., and Walker, D. J., 1959, Measurement of the thickness of the Earth's crust in the Albertan plains of western Canada: Geophysics, v. 24, p. 262–284.

Rohay, A., 1982, Crust and mantle structure of the North Cascades Range, Washington [Ph.D. thesis]: Seattle, University of Washington, 163 p.

Roller, J. C., 1964, Crustal structure in the vicinity of Las Vegas, Nevada from seismic and gravity observations: U.S. Geological Survey Professional Paper 475-D, p. D108–D111.

——, 1965, Crustal structure in the eastern Colorado Plateau province from seismic-refraction measurements: Seismological Society of America Bulletin, v. 55, p. 107–119.

Roller, J. C., and Healy, J. H., 1963, Seismic-refraction measurements of crustal structure between Santa Monica Bay and Lake Mead: Journal of Geophysical Research, v. 68, p. 5837–5849.

Roller, J. C., and Jackson, W. H., 1966, Seismic-wave propagation in the upper mantle; Lake Superior, Wisconsin, to Denver, Colorado, *in* Steinhart, J. S., and Smith, T. J., eds., The Earth beneath the continents: American Geophysical Union Geophysical Monograph 10, p. 270–275.

Romanowicz, B. A., 1979, Seismic structure of the upper mantle beneath the United States by three-dimensional inversion of body wave arrival times: Geophysical Journal of the Royal Astronomical Society, v. 57, p. 479–506.

Romney, C., and 5 others, 1962, Travel times and amplitudes of principal body phases recorded from Gnome: Seismological Society of America Bulletin, v. 52, p. 1057–1074.

Ryall, A., and Stuart, D. J., 1963, Travel times and amplitudes from nuclear explosions, Nevada test site to Ordway, Colorado: Journal of Geophysical Research, v. 68, p. 5821–5835.

Sexton, J. L., Hinze, W. J., von Frese, R.R.B., and Braile, L. W., 1982, Long-wavelength aeromagnetic anomaly map of the conterminous United States: Geology, v. 10, p. 364–369.

Shearer, P. M., and Oppenheimer, D. H., 1982, A dipping Moho and crustal low-velocity zone from P_n arrivals at The Geysers–Clear Lake, California: Seismological Society of America Bulletin, v. 72, p. 1551–1566.

Sheriff, S. D., and Stickney, M. C., 1984, Crustal structure of southwestern Montana and east-central Idaho; Results of a reversed seismic refraction line: Geophysical Research Letters, v. 11, p. 299–302.

Shor, G. G., Jr., 1962, Seismic refraction studies off the coast of Alaska; 1956–1957: Seismological Society of America Bulletin, v. 52, p. 37–57.

Shor, G. G., Jr., Dehlinger, P., Kirk, H. D., and French, W. S., 1968, Seismic refraction studies off Oregon and northern California: Journal of Geophysical Research, v. 73, p. 2175–2194.

Sienko, D. A., 1982, Crustal structure of south-central Pennsylvania determined from wide-angle reflections and refractions [M.S. thesis]: University Park, Pennsylvania State University, 127 p.

Simpson, R. W., Jachens, R. C., and Blakely, R. J., 1986, A new isostatic residual gravity map of the conterminous United States with a discussion on the significance of isostatic residual anomalies: Journal of Geophysical Research, v. 91, p. 8348–8372.

Sinno, Y. A., and Keller, G. R., 1986, A Rayleigh wave dispersion study between El Paso, Texas, and Albuquerque, New Mexico: Journal of Geophysical Research, v. 91, p. 6168–6174.

Sinno, Y. A., Keller, G. R., and Sbar, M. L., 1981, A crustal seismic refraction study in west-central Arizona: Journal of Geophysical Research, v. 86, p. 5023–5038.

Sinno, Y. A. Daggett, P. H., Keller, G. R., Morgan, P., and Harder, S. H., 1986, Crustal structure of the southern Rio Grande Rift determined from seismic refraction profiling: Journal of Geophysical Research, v. 91, p. 6143–6156.

Slichter, L. B., 1951, Crustal structure in the Wisconsin area: Arlington, Virginia, Office of Naval Research Report N9, ONR86200.

Smith, R. B., 1978, Seismicity, crustal structure, and intraplate tectonics of the interior of the western Cordillera, *in* Smith, R. B., and Eaton, G. P., eds., Cenozoic tectonics and regional geophysics of the western Cordillera: Geological Society of America Memoir 152, p. 111–144.

Smith, R. B., and 9 others, 1982, The 1978 Yellowstone–eastern Snake River Plain seismic profiling experiment; Crustal structure of the Yellowstone region and experiment design: Journal of Geophysical Research, v. 87, p. 2583–2596.

Smith, S. W., 1962, A reinterpretation of phase velocity data based on the Gnome travel time curves: Seismological Society of America Bulletin, v. 52, p. 1031–1035.

Smith, T. J., Steinhart, J. S., and Aldrich, L. T., 1966, Lake Superior crustal structure: Journal of Geophysical Research, v. 71, p. 1141–1172.

Smithson, S. B., Johnson, R. A., and Wong, Y. K., 1981, Mean crustal velocity; A critical parameter for interpreting crustal structure and crustal growth: Earth and Planetary Science Letters, v. 53, p. 323–332.

Sparlin, M. A., Braile, L. W., and Smith, R. B., 1982, Crustal structure of the eastern Snake River Plain determined from ray-trace modeling of seismic refraction data: Journal of Geophysical Research, v. 87, p. 2619–2633.

Spence, G. D., Clowes, R. M., and Ellis, R. M., 1977, Depth limits on the M discontinuity in the southern Rocky Mountain Trench, Canada: Seismological Society of America Bulletin, v. 67, p. 543–546.

——, 1985, Seismic structure across the active subduction zone of western Canada: Journal of Geophysical Research, v. 90, p. 6754–6772.

Spieth, M. A., Hill, D. P., and Geller, R. J., 1981, Crustal structure in the northwestern foothills of the Sierra Nevada from seismic refraction experiments: Seismological Society of America Bulletin, v. 71, p. 1075–1087.

Stanton, J. C., 1972, Crustal structure of the central High Plains of Texas from Rayleigh wave dispersion [M.S. thesis]: Lubbock, Texas Tech University, 48 p.

Stauber, D. A., 1983, Crustal structure in northern Nevada from seismic refraction data, *in* The role of heat in the development of energy and mineral resources in the northern Basin and Range province: Geothermal Resources Council Special Report 13, p. 319–332.

Stauber, D. A., and Boore, D. M., 1978, Crustal thickness in northern Nevada from seismic refraction studies: Seismological Society of America Bulletin, v. 68, p. 1049–1058.

Steinhart, J. S., and Meyer, R. P., 1961, Explosion studies of continental structure: Carnegie Institution of Washington Publication 622, 409 p.

Steinhart, J. S., Suzuki, Z., Smith, T. J., Aldrich, L. T., and Sacks, I. S., 1964, Explosion seismology: Carnegie Institution of Washington Yearbook, v. 63, p. 311–319.

Steppe, J. A., and Crosson, R. S., 1978, P-velocity models of the southern Diablo Range, California, from inversion of earthquake and explosion arrival times: Seismological Society of America Bulletin, v. 68, p. 357–367.

Stewart, S. W., 1968a, Crustal structure in Missouri by seismic-refraction meth-

ods: Seismological Society of America Bulletin, v. 58, p. 291–323.
—— , 1968b, Preliminary comparison of seismic travel times and inferred crustal structure adjacent to the San Andreas Fault in the Diablo and Gabilan Ranges of central California, *in* Dickinson, W. R., ed., Proceedings of the Geologic Problems of San Andreas Fault System Conference: Stanford, California, Stanford University Publication in the Geological Sciences 11, p. 218–230.
Stierman, D. J., and Ellsworth, W. L., 1976, Aftershocks of the February 21, 1973, Point Mugu, California, earthquake: Seismological Society of America Bulletin, v. 66, p. 1931–1952.
Taber, J. J., Jr., 1983, Crustal structure and seismicity of the Washington continental margin [Ph.D. thesis]: Seattle, University of Washington, 159 p.
Taber, J. J., Jr., and Lewis, B.T.R., 1986, Crustal structure of the Washington continental margin from refraction data: Seismological Society of America Bulletin, v. 76, p. 1011–1024.
Taber, J. J., Jr., and Smith, S. W., 1985, Seismicity and focal mechanisms associated with the subduction of the Juan de Fuca Plate beneath the Olympic Peninsula, Washington: Seismological Society of America Bulletin, v. 75, p. 237–249.
Tatel, H. E., and Tuve, M. A., 1955, Seismic exploration of a continental crust, *in* Poldervaart, A., ed., Crust of the Earth: Geological Society of America Special Paper 62, p. 35–50.
Tatel, H. E., Adams, L. H., and Tuve, M. A., 1953, Studies of the Earth's crust using waves from explosions: Proceedings of the American Philosophical Society, v. 97, p. 658–669.
Taylor, S. R., 1983, Three-dimensional crust and upper mantle structure at the Nevada test site: Journal of Geophysical Research, v. 88, p. 2220–2232.
Taylor, S. R., and Toksoz, M. N., 1979, Three-dimensional crust and upper mantle structure of the northeastern United States: Journal of Geophysical Research, v. 84, p. 7627–7644.
—— , 1982, Crust and upper-mantle velocity structure in the Appalachian orogenic belt; Implications for tectonic evolution: Geological Society of America Bulletin, v. 93, p. 315–329.
Taylor, S. R., Toksoz, M. N., and Chaplin, M. P., 1980, Crustal structure of the northeastern United States; Contrasts between Grenville and Appalachian provinces: Science, v. 208, p. 595–597.
Thatcher, W., and Brune, J. N., 1973, Surface waves and crustal structure in the Gulf of California region: Seismological Society of America Bulletin, v. 63, p. 1689–1698.
Thompson, G. A., and Talwani, M., 1964, Crustal structure from Pacific Basin to central Nevada: Journal of Geophysical Research, v. 69, p. 4813–4837.
Toppozada, T. R., and Sanford, A. R., 1976, Crustal structure in central New Mexico interpreted from the Gasbuggy explosion: Seismological Society of America Bulletin, v. 66, p. 877–886.
Trehu, A. M., and Wheeler, W. H., IV, 1987, Possible evidence in the seismic data of profile SJ-6 for subducted sediments beneath the Coast Ranges of California, U.S.A.: U.S. Geological Survey Open-File Report 87-73, p. 91–104.
Tryggvason, E., and Qualls, B. R., 1967, Seismic refraction measurements of crustal structure in Oklahoma: Journal of Geophysical Research, v. 72, p. 3738–3740.
Tuve, M. A., 1951, The Earth's crust: Year Book of the Carnegie Institution of Washington, v. 50, p. 69–73.
—— , 1953, The Earth's crust: Year Book of the Carnegie Institution of Washington, v. 52, p. 103–108.
—— , 1954, The Earth's crust: Year Book of the Carnegie Institution of Washington, v. 53, p. 51–55.
Tuve, M. A., Tatel, H. E., and Hart, P. J., 1954, Crustal structure from seismic exploration: Journal of Geophysical Research, v. 59, p. 415–422.
Unger, J. D., Stewart, D. B., and Phillips, J. D., 1987, Interpretation of migrated seismic reflection profiles across the northern Appalachians in Maine: Geophysical Journal of the Royal Astronomical Society, v. 89, p. 171–176.
Van der linden, W.J.M., and Srivastava, S. P., 1975, The crustal structure of the continental margin off central Labrador, *in* van der Linden, W.J.M., and Wade, M. A., eds., Offshore geology of eastern Canada; v. 2, Regional geology: Geological Survey of Canada Paper 74-30, p. 233–245.
Van Schmus, W. R., and Hinze, W. J., 1985, The Midcontinent Rift system: Annual Reviews of Earth and Planetary Sciences, v. 13, p. 345–383.
Vetter, U., and Minster, J-B., 1981, Velocity-anisotropy in southern California: Seismological Society of America Bulletin, v. 71, p. 1511–1530.
Von Frese, R.R.B., Hinze, W. J., and Braile, L. W., 1982, Regional North American gravity and magnetic anomaly correlations: Geophysical Journal of the Royal Astronomical Society, v. 69, p. 745–761.
Walter, A. W., and Mooney, W. D., 1982, Crustal structure of the Diablo and Gabilan Ranges, central California; A reinterpretation of existing data: Seismological Society of America Bulletin, v. 72, p. 1567–1590.
Warren, D. H., 1968, Transcontinental geophysical survey (35°-39°N) seismic refraction profiles of the crust and upper mantle: U.S. Geological Survey Maps I-532-D, I-533-D, I-534-D, and I-535-D, scale 1:1,000,000.
—— , 1969, Seismic-refraction survey of crustal structure in central Arizona: Geological Society of America Bulletin, v. 80, p. 257–282.
—— , 1981, Seismic-refraction measurements of crustal structure near Santa Rosa and Ukiah, California: U.S. Geological Survey Professional Paper 1141, p. 167–181.
Warren, D. H., and Healy, J. H., 1973, Structure of the crust in the conterminous United States: Tectonophysics, v. 20, p. 203–213.
Warren, D. H., and Jackson, W. H., 1968, Surface seismic measurements of the project Gasbuggy explosion at intermediate distance ranges: U.S. Geological Survey Open-File Report 1023, 45 p.
Warren, D. H., Healy, J. H., and Jackson, W. H., 1966, Crustal seismic measurements in southern Mississippi: Journal of Geophysical Research, v. 71, p. 3437–3458.
Warren, D. H., Healy, J. H., Bohn, J., and Marshall, P. A., 1972, Crustal calibration of the large aperture seismic array (LASA), Montana: U.S. Geological Survey Open-File Report 72-0444, 163 p.
White, W.R.H., and Savage, J. C., 1965, Seismic refraction and gravity study of the Earth's crust in British Columbia: Seismological Society of America Bulletin, v. 55, p. 463–486.
White, W.R.H., Bone, M. N., and Milne, W. G., 1968, Seismic refraction surveys in British Columbia, 1941–1966; A preliminary interpretation, *in* Knopoff, L., Drake, C. L., and Hart, P. J., eds., The crust and upper mantle of the Pacific area: American Geophysical Union Monograph 12, p. 81–93.
Wickens, A. J., 1971, Variations in lithospheric thickness in Canada: Canadian Journal of Earth Sciences, v. 8, p. 1154–1162.
—— , 1977, The upper mantle of southern British Columbia: Canadian Journal of Earth Sciences, v. 14, p. 1100–1115.
Willden, R., 1965, Seismic-refraction measurements of crustal structure beneath American Falls Reservoir, Idaho, and Flaming Gorge Reservoir, Utah: U.S. Geological Survey Professional Paper 525-C, p. C-44–C-50.
Woollard, G. P., 1969, Regional variations in gravity, *in* Hart, P. J., ed., The Earth's crust and upper mantle: American Geophysical Union Geophysical Monograph 13, p. 329–341.
Young, R. A., Wright, J., and West, G. F., 1986, Seismic crustal structure northwest of Thunder Bay, Ontario, *in* Barazangi, M., and Brown, L., eds., Reflection seismology: The continental crust: American Geophysical Union Geodynamics Series 14, p. 143–155.
Zandt, G., and Owens, T. J., 1986, Comparison of crustal velocity profiles determined by seismic refraction and teleseismic methods: Tectonophysics, v. 128, p. 155–161.
Zietz, I., 1969, Aeromagnetic investigations of the earth's crust in the United States, *in* Hart, P. J., ed., The Earth's crust and upper mantle: American Geophysical Union Geophysical Monograph 13, p. 404–415.
Zucca, J. J., Fuis, G. S., Milkereit, B., Mooney, W. D., and Catchings, R. D., 1986, Crustal structure of northeastern California: Journal of Geophysical Research, v. 91, p. 7359–7382.

Manuscript Accepted by the Sciety October 31, 1988

Printed in U.S.A.

Geological Society of America
Memoir 172
1989

Chapter 29

Upper-mantle velocity structure in the continental U.S. and Canada

H. M. Iyer and T. Hitchcock
U.S. Geological Survey, MS 977, 345 Middlefield Road, Menlo Park, California 94025

ABSTRACT

The available studies on upper-mantle structure in North America can be broadly divided into two categories: delineation of one-dimensional models, that is, the determination of P- and S-velocities as a function of depth; and computation of two- and three-dimensional models to take into account lateral heterogeneities in structure. About 50 one-dimensional models based on traveltimes, synthetic seismograms, and surface-wave velocities are currently available for continental North America. The gross features of these models are sharp velocity increases at depths near 400 and 650 km in the upper mantle beneath the whole continent and the presence of a low-velocity layer in the uppermost part of the upper mantle in the western half of the continent. A few other velocity discontinuities and velocity-gradient changes have also been documented. The most important finding from the available studies is a quantitative confirmation of what was suspected even in the early 1950s, namely, that the upper-mantle structure, particularly the structure related to the low-velocity layer, is drastically different in the tectonically active Cordillera from the stable central and eastern shield of North America. In western North America, in general, the upper-mantle velocities are low, and the low-velocity zone is well developed and occurs at shallow depths. On the other hand, in the central and eastern parts of the continent the upper-mantle velocities are higher than in the west, and a low-velocity layer—if present at all—tends to be deep and to have a smaller velocity contrast than in the west. Available data and modeling techniques are inadequate to unambiguously resolve spatial variation in the depths to the 400-km and 650-km discontinuities and the boundaries of the low-velocity layer in North America. Apart from the broad division of the one-dimensional models into tectonically active and shield structures, any further finer scale quantitative division of the models within each tectonic unit is not warranted by the available data. Qualitatively, however, it is clear that such finer scale differences do exist, particularly in the upper 250 to 300 km of the upper mantle.

The laterally heterogeneous structure of the upper mantle in North America has been studied by three-dimensional modeling using teleseismic P- and S-wave residuals. Three-dimensional inversion of P- and S-wave residual data collected over a substantial part of the North American continent show the existence of long-wavelength heterogeneous structure extending throughout the upper mantle. In addition, short-wavelength lateral heterogeneities are revealed by regional investigations. These include heterogeneous velocity structures associated with: (1) the ongoing subduction of the Juan de Fuca plate beneath Washington and Oregon, (2) the cessation of subduction of the Farallon plate beneath California during early Tertiary time, (3) rifting in Imperial Valley, (4) hot-spot magmatism in the Yellowstone Plateau, (5) large-scale asthenospheric upwelling in the region of the Rio Grande Rift, and (6) the orogenic belts in the northeastern United States.

Iyer, H. M., and Hitchcock, T., 1989, Upper-mantle velocity structure in the continental U.S. and Canada, *in* Pakiser, L. C., and Mooney, W. D., Geophysical framework of the continental United States: Boulder, Colorado, Geological Society of America Memoir 172.

INTRODUCTION

Since seismic waves from man-made sources, such as explosions and the VIBROSEIS instrumentation system have only enough energy to penetrate to the uppermost mantle, even under ideal experimental conditions, medium and large earthquakes are the primary sources of seismic energy used to probe the deep subcrustal structure of the Earth. In addition, large chemical and nuclear explosions with energy equivalent to medium-size earthquakes have provided useful data to model the Earth's upper mantle. This chapter reviews the upper-mantle P- and S-wave velocity models in the continental United States and Canada, with special emphasis on heterogeneous velocity structure. The upper mantle is here defined as the region from the base of the crust down to and including the 670-km discontinuity. The seismological studies discussed herein can be broadly divided into two categories: (1) computation of one-dimensional P- and S-wave velocity models (i.e., models in which velocity is specified as a function of depth) by inversion of traveltime vs. distance data, computation of synthetic seismograms, and modeling of surface-wave dispersion; and (2) derivation of two- and three-dimensional models primarily using teleseismic traveltime residuals.

The emphasis in one-dimensional models is on delineating vertical rather than lateral heterogeneities. The available models can be broadly divided into shield models for the stable platform and shield of central and eastern North America, and nonshield (or tectonic) models for the tectonically active western North America. Shield models exhibit higher P- and S-wave velocities in the upper half of the upper mantle than tectonic models. Also, all tectonic models have pronounced low-velocity zones in the top 200 km of the upper mantle. The other major features of the models are rapid velocity increases near 400-km and 650-km depths. These so-called 400- and 650-km discontinuities seem to be present with varying depths and velocity contrasts both in the shield and tectonic models.

In addition to the shield-type and tectonically active–type of structural differences, studies using teleseismic traveltime residuals even as early as in the 1960s revealed that the velocity structure of the upper mantle in the United States is laterally very heterogeneous on a regional scale. The technique of using teleseismic traveltime residuals has since been used extensively to study this lateral heterogeneity and has successfully modeled anomalies of a few kilometers to several hundred kilometers in size in the crust and upper mantle in the United States. Recently introduced tomographic techniques have considerably enhanced the power of the teleseismic-residual method for modeling velocity anomalies in three dimensions. Typical studies include modeling the upper-mantle structure of the whole continental United States, tectonically diverse regions within the North American continent, volcanic centers, and earthquake source regions.

The bulk of the literature on the seismic structure of the upper mantle in the United States deals with the use of traveltime curves, synthetic seismograms, surface-wave dispersion, and teleseismic residuals, for modeling P- and S-wave velocity variations. However, progress has also been made in modeling velocity anisotropy, mapping discontinuities using reflected phases, and determining 1/Q (attentuation) with depth. Many new techniques have also been developed to simultaneously use arrival times and waveforms of multiple phases to infer structure. However, in this chapter we concern ourselves primarily with one-dimensional velocity structures derived using traveltime data, synthetic seismograms, and surface-wave dispersion, and three-dimensional models derived using teleseismic residuals.

Approximately 25 one-dimensional upper-mantle P-wave velocity models and an almost equal number of S-wave velocity models are available for the North American continent. We find the synthesis of these models to be considerably more ambiguous than in similar synthesis of crustal models. In crustal studies, parameters such as thickness of the upper crust, lower crust, and P_n-velocity, can be specified on local and regional scales and contoured continent-wide from the vast quantity of available data (Braile and others, this volume). On the other hand, we find it difficult to specify with sufficient accuracy for contouring purposes upper-mantle parameters, such as the depth to the top of the low-velocity layer or lithospheric thickness, the magnitude of velocity decrease in the low-velocity layer, the thickness of the low-velocity layer, and depths to other velocity discontinuities in the upper mantle and velocity changes associated with them. The reason is that the available one-dimensional models are averages over broad regions and are of varying accuracy due to the differences in the types and quantity of data and in modeling techniques employed in the studies. However, in spite of these difficulties, as mentioned earlier, it is possible to identify characteristic structural differences between different tectonic provinces in the North American continent. The available continental-scale and regional-scale three-dimensional models derived using P-wave and S-wave residuals provide further evidence of the strong lateral velocity heterogeneities that seem to persist throughout the upper mantle beneath North America. Unfortunately, at present we are unable to combine them with the one-dimensional models to provide a unified picture of the laterally heterogeneous upper mantle structure of the North American continent due to the inadequate resolution of the continental-scale models and the limited number of the regional-scale models. Because of these difficulties, the one-dimensional models and three-dimensional models remain as separate entities in seismological literature and are kept as such in this review.

ONE-DIMENSIONAL VELOCITY MODELS

Historical perspectives

Seismological modeling of the upper mantle began in the United States in the 1930s and 1940s. The conventionally accepted Earth model at that time was that proposed by Jeffreys and Bullen (1940), in which the Earth was divided into seven spherical layers of constant velocity, later designated by Bullen

(1965) as layers A through G. In the Earth's upper mantle (layers B and C), a major velocity jump occurs at a depth of 413 km. There are no other discontinuities in the upper mantle of the Jeffreys-Bullen Earth model, even through layer C has greater velocity gradients at some depths than anywhere in layer B. Early work in the United States was mainly directed toward identifying the differences in deep structure of the North American continent from the global average model of Jeffreys and Bullen (1940).

Gutenberg's low-velocity layer. Even as early as 1926, Gutenberg was convinced that there was a low-velocity layer in the upper mantle starting approximately at a depth of 75 km. He based his inference on theoretical calculations of pressues and temperatures at depth, which led to the presence of partial melt in the upper mantle and amplitude studies of P-waves in the distance range of 200 to 300 km (Gutenberg, 1959). In a series of papers written during the 1940s and 1950s, Gutenberg developed the concept of the low-velocity layer (Gutenberg, 1959). He constructed an average Earth model in which the low-velocity layer extended from immediately below the Moho to a depth of about 170 km for P-waves and 270 km for S-waves. Gutenberg's early insight has proved to be a landmark in seismology and earth sciences. His low-velocity layer has become an integral part of every computed upper-mantle model for western North America and other tectonically active regions of the world including the oceans, and it is now regarded as the viscous asthenosphere over which the brittle lithosphere rides.

The 400- and 650-Km discontinuities. Byerly (1926) was the first to point out, using data from a Montana earthquake, that the slope of the traveltime curve changed abruptly at an epicentral distance of 20°, which could correspond to a velocity discontinuity encountered by the seismic waves at a depth of about 400 km. Jeffreys (1939) showed that the velocity discontinuity responsible for the 20° traveltime discontinuity was at a depth of 413 ± 32 km, the interface between Bullen's B and C layers. This discovery was another landmark in seismology because Jeffreys' 400-km discontinuity is now an integral part of all upper-mantle models of the Earth and is generally interpreted to be due to the phase change of olivine to spinel in the upper mantle. Gutenberg did not believe in the 400-km discontinuity. His upper-mantle models showed a smooth velocity gradient at this depth. The second major upper-mantle discontinuity is near 650 km depth, and is interpreted to be due to a phase change from spinel type of structure to a more compact oxide structure.

Western and eastern models. Even as early as the 1950s and 1960s, seismic studies revealed that significant differences in upper-mantle structure were present between the western and eastern halves of the continental United States. Traveltimes from nuclear explosions (Romney and others, 1962; Carder and others, 1966), P_n-velocities (Herrin and Taggart, 1962; Herrin, 1969), and teleseismic residuals (Cleary and Hales, 1966; Hales and Doyle, 1967), measured across the United States, all showed this difference. Virtually all the upper-mantle models for the United States and Canada that have subsequently been computed using a variety of techniques confirm this difference, namely, the existence of lower velocities in the western Cordillera than in the eastern stable continental mass. Also, the western models invariably show the presence of a marked low-velocity layer at the top of the upper mantle, even though details such as its thickness and the velocity contrast within it vary from model to model. Some of the eastern models also have low-velocity layers, but these are in general of low velocity contrast and/or deeper than in the western models. Finally, the eastern models are, in general, closer to the Jeffreys-Bullen Earth model than the western models.

Observations. A major impetus to study the upper mantle in the United States was provided during the 1960s by the detonations of a large number of nuclear explosions, together with the operation of the Long Range Seismic Measurement (LRSM) stations and large seismic arrays by the Department of Defense, and establishment of the World Wide Standard Seismic Network (WWSSN). These activities resulted in the accumulation of excellent body-wave and surface-wave data; consequent development of several modeling techniques has sustained U.S. research in upper-mantle structure for over two decades. In addition, three major continent-scale seismic experiments were conducted between 1963 and 1966 involving large explosions that were set off in Lake Superior and transcontinental seismic recording along long profiles across the United States and Canada. These experiments, especially the last one with the code name EARLY RISE, remain as the largest controlled-source experiments for deep-structure studies in North America. In the 1966 EARLY RISE experiment, seismic waves generated by 38 large explosions in Lake Superior were recorded along profiles at many azimuths to distances of more than 3,000 km (Fig. 1). The data from this experiment have been analyzed by several investigators to yield average and regional upper-mantle models of North America.

Modeling techniques

Modeling techniques are discussed in detail in the section on Methods (Mooney, this volume) and therefore are mentioned only briefly here. The one-dimensional velocity models discussed in this review are derived using three basic modeling techniques, inversion of traveltime (T) vs. distance (Δ) data, traveltime inversion together with computation of synthetic seismograms, and inversion of surface-wave phase and group-velocity data. Inversion of T-Δ data is carried out using the Herglotz-Wiechert integral (Bullen, 1965) or the τ-p method of Bessonova and others (1974, 1976). Helmberger and Wiggins (1971) introduced the use of synthetic seismograms to provide additional constraints to model the upper-mantle structure using traveltime data. Modeling of surface-wave phase- and group-velocity data is carried out by comparing computed and observed dispersion curves using the Haskell-Thompson matrix formulation, inversion of the dispersion curves, or by expressing surface waves in terms of normal modes of a layered Earth.

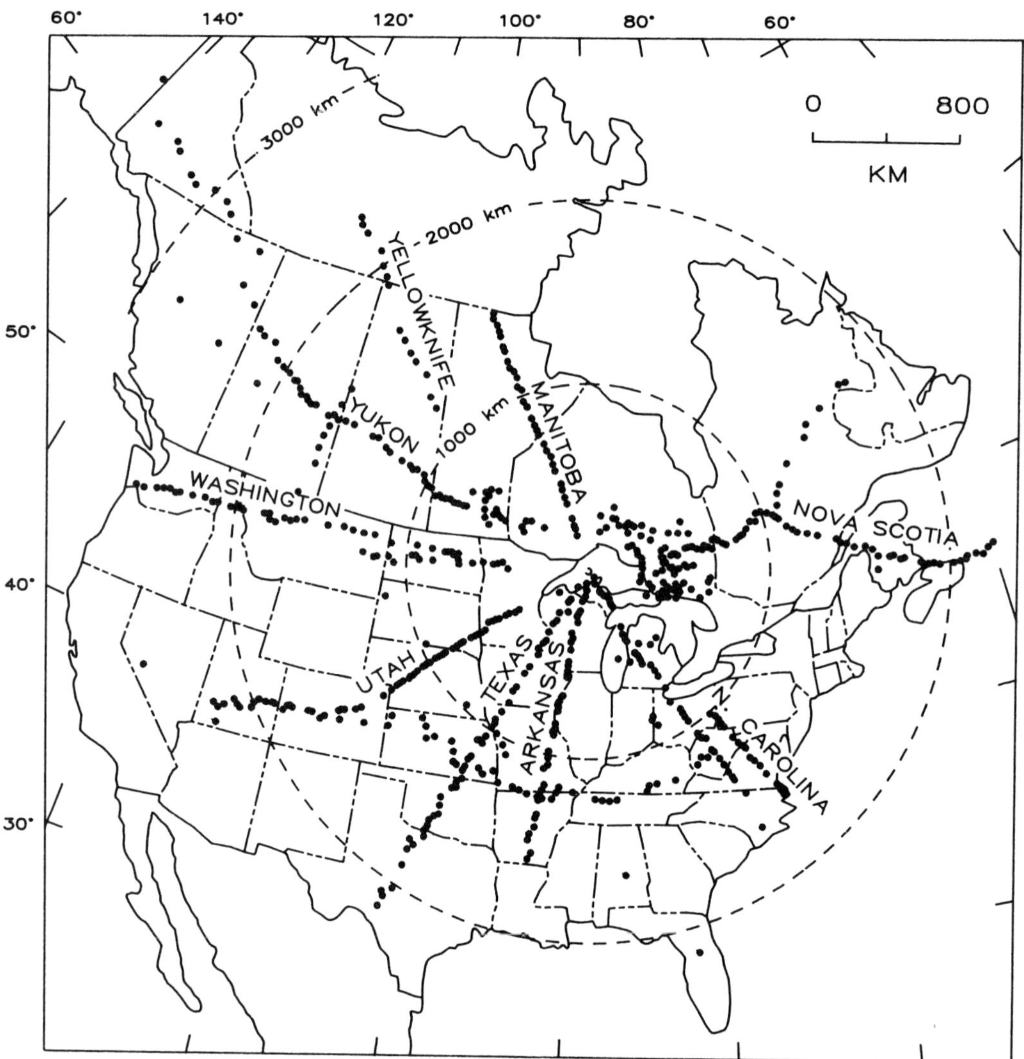

Figure 1. Project EARLY RISE profiles. Dots are locations of recording stations. Geographic names for the profiles are indicated. After Iyer and others (1969).

Discussion of one-dimensional velocity models

In this section we review the available one-dimensional models of upper-mantle structure for North America and synthesize the major findings. A few other reviews of traveltime models (Nuttli, 1963; Anderson, 1966; Hales and Herrin, 1972; and Massé, 1987), and several complete reviews on surface-wave studies (Ewing and others, 1957; Arkhangel'skaya, 1960; Nuttli, 1963; Bolt, 1964; Anderson, 1966; Kovach, 1966; Anderson, 1967; Toksöz and others, 1967; Dorman, 1969; Bucher and Smith, 1971; Kovach, 1978; Knopoff, 1983) are available. The main contribution from the numerous available one-dimensional velocity models to our understanding of the upper-mantle structure of North America is in emphasizing the difference between the stable central and eastern half of the continent and the tectonically active western half of the continent. During the past decade, this difference has been documented with progressively improving modeling techniques. Careful scrutiny of the published velocity models reveals that, although in general the older models may be said to have been superseded by the newer models, each model has its own characteristic uniqueness. Therefore, in this discussion we have chosen to include as many models as possible and to examine what significant new information is provided by each model. However, we must note that models have improved with time as better data sets and modeling techniques have become available. In general, we find that models derived using traveltime data or combined traveltime and wave-form data (i.e., synthetic seismogram modeling) tend to be averages over fairly large regions and do not have the resolution to examine fine-scale variations within a specific tectonic block. Some of the surface-wave models using phase-velocity measurements have provided more localized regional models than the traveltime models. Also, traveltime and synthetic seismogram models, particularly those in which multiple arrivals are used, delineate boundaries of velocity discontinuities much better than surface-wave models.

Models of the stable continent

P-velocity models. Several one-dimensional P-wave velocity models representative of the stable part of the North American continent have been computed using data from the EARLY RISE experiment. Iyer and others (1969) constructed an average P-wave velocity model NCER1 for North America by combining first-arrival time data from several of the EARLY RISE profiles (Fig. 2a). The main features of this model are higher average P-wave velocity in the upper mantle than in the average global model of Jeffreys and Bullen (Bullen, 1965), a smooth increase in velocity with depth, and a rapid increase in velocity gradient at 430 km depth. However, in another average model using first and later P-wave arrivals from combined EARLY RISE profiles, Massé (1973) found velocity discontinuities with rapid velocity increases at depths of 73, 107, 328, and 430 km, and a thin, low-velocity layer at 94 km depth (Fig. 2a). The 73- and 107-km discontinuities in Massé's (1973) model belong to the group of discontinuities found between Moho and about 220 km depth (where the low-velocity zone is usually found in tectonic models) in many models of the upper mantle for the stable continent. These discontinuities vary in number and location from model to model and are not well constrained, nor are there any clear physical explanations for their occurrence. The discontinuity around 328 km depth is also found in a model for the Basin and Range province by Massé and others (1972) and a few other models in addition to or in lieu of the 400-km discontinuity.

Analysis of individual EARLY RISE profiles shows basically the same features found in the average models discussed above, namely, high upper-mantle velocities and thin or nonexistent low-velocity layers. The model NCER4 of Iyer and others (1969) for the north-central United States and south-central Canada using data from the Washington and Yukon profiles (Fig. 1) is essentially the same as the model for the same region computed by Lewis and Meyer (1968) using first and later arrivals except for the series of velocity steps in the latter model near 70 km, 126 km and 450 km and a thin zone of increased velocity gradient near 350 km (Fig. 2b). In another EARLY RISE model using data from the Texas and Arkansas profiles (Fig. 1), Green and Hales (1968) found a velocity jump near 90 km and marginal evidence for a thin low-velocity layer near 150-km depth. Green and Hales (1968) also computed a Nevada model using P-arrivals from nuclear explosions recorded along the arc profile (Fig. 1). This model has, in addition to a discontinuity near 93 km depth, a pronounced low-velocity layer within the upper 160 km and sharp velocity jumps between 362 and 382 km and 623 and 645 km depths. Green and Hales (1968) proposed the upper mantle of the United States to be of the EARLY RISE type to the east and the Nevada type to the west to a depth of 160 km, below which the models for both sections are not significantly different. The validity of this proposal, however, is questionable, since profiles from west to east tend to fold horizontal heterogeneities into vertical structures. Helmberger and others (1985) have shown that the Rocky Mountain region is a zone of rapid transition in lateral velocity structure.

Several models for the Canadian shield have been computed using data from explosions in Lake Superior and in Hudson Bay. For the region from Lake Superior to the Churchill Province,

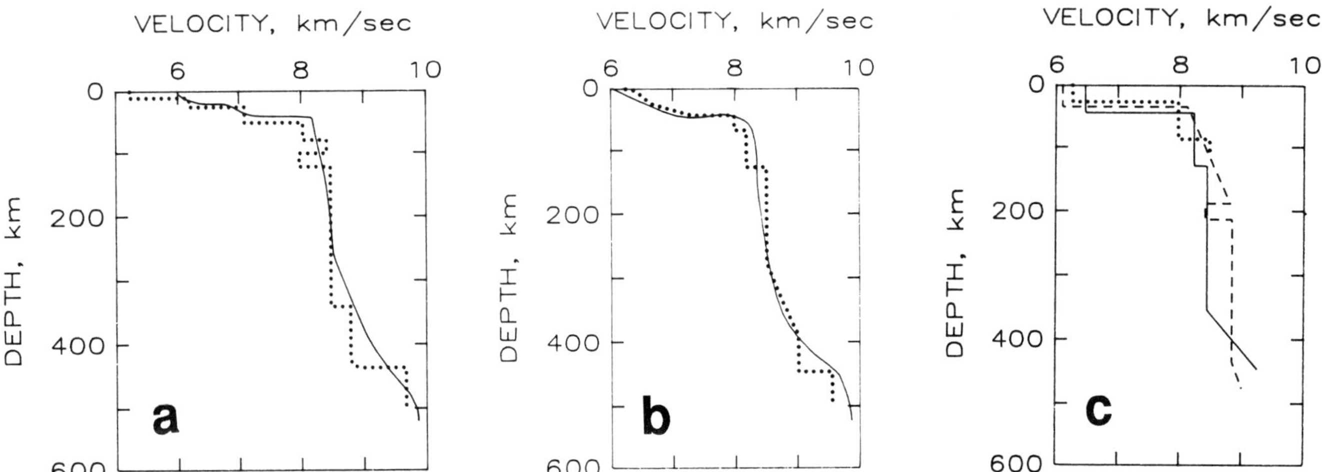

Figure 2. P-velocity models for North America using data from EARLY RISE, and Lake Superior and Hudson Bay explosions. a, Average models using all EARLY RISE data; solid line = NCER 1 (modified from Iyer and others, 1969); dotted line = model of Massé (modified from Masse, 1973). b, Models based on the two western EARLY RISE profiles; solid line = NCER 4 (modified from Iyer and others, 1969) using data from Washington and Yukon profiles; dotted line = the model modified from Lewis and Meyer (1968) from Washington profile. c, Canadian P-velocity models using data from Lake Superior and Hudson Bay explosions: solid line (modified from Barr, 1967); dotted line (modified from Mereu and Hunter, 1969); dashed line (modified from Wiechert, 1968).

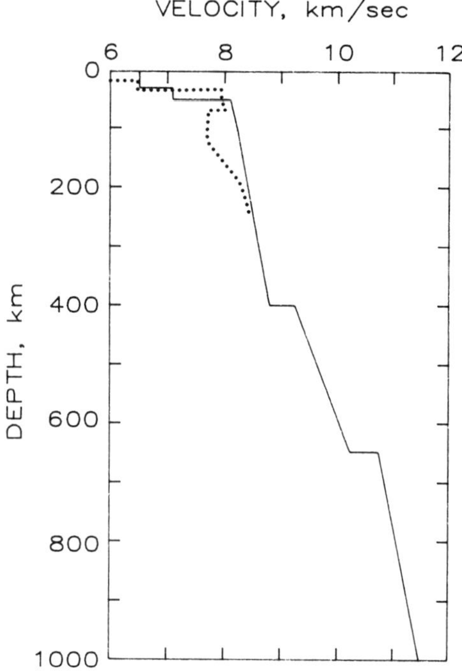

Figure 3. P-velocity model S8 for stable continent (solid line) and for tectonic continent (T9). Modified from Burdick (1981).

Canada, Mereu and Hunter (1969) found a velocity jump near 90 km (Fig. 2c), similar to that found in the model of Green and Hales (1968) for the eastern United States. Barr (1967) found two velocity discontinuities at 126 km and 366 km, respectively, in the eastern Canadian upper mantle (Fig. 2c). Note that these two discontinuities occur at nearly the depths as the top and bottom of the low-velocity layer in the S-wave velocity model of Brune and Dorman (1963) for the Canadian shield, even though Barr's (1967) data did not suggest such a low-velocity layer. Wiechert (1968) found a low-velocity layer at a depth of 200 km in the upper mantle of the Churchill province of the Canadian shield (Fig. 2c), the lower edge of which corresponded with the 200-km discontinuity found invariably in most of Lehman's (1959, 1962, 1964) upper-mantle models.

The most recent P-velocity model for the tectonically stable part of the North American continent is provided by Burdick (1981) using the synthetic seismogram technique. Burdick (1981) showed that his model S8 (Fig. 3) fits the traveltime data from the EARLY RISE shots and nuclear explosions used in some of the earlier studies. It has no low-velocity layer, and below the crust there are only two major discontinuities near 400 km and 650 km, respectively. Between the Moho and the 400-km discontinuity and between the 400-km and 650-km discontinuities, the velocity increases smoothly with depth. The main difference between this model and the western models (discussed in a later section) is the presence of a low-velocity layer in the western models. This is illustrated in Figure 3 by comparison with Burdick's (1981) model T9 for a tectonically active continent.

In summary, the upper-mantle P-wave velocity models for the stable continental mass of North America show, in addition to the 400-km and 650-km discontinuities, one or more velocity discontinuities and no clear indication of a low-velocity layer.

S-Velocity models. An S-wave velocity model for the upper mantle of the continental United States, computed by Ibrahim and Nuttli (1967) using traveltime data, is not very different from Nuttli's (1969) model for the western United States. Both models have low-velocity zones at 150, 350, and 670 km depths, rapid velocity increases at 400 and 700 km, and nearly constant velocity between 220 and 350 km and between 400 and 670 km. The base of the first low-velocity zone is at a depth of 200 km and is reminiscent of Lehman's 215-km discontinuity. Anderson and Julian (1969), using more or less the same data set, computed a model US26 in which the first low-velocity layer is moved up to a depth range of 80 to 125 km. The deeper part of US26 is not significantly different from the model of Ibrahim and Nuttli (1967).

The first and still most widely cited S-velocity model for the North American shield is the Canadian shield model (CANSD) of Brune and Dorman (1963) derived using Rayleigh- and Love-wave dispersion data. In this model, both earthquake sources and station locations were chosen to achieve almost pure-path travel across the Canadian shield. CANSD (Fig. 4a) differs significantly from the model of Ibrahim and Nuttli (1967) and from other "continental-type" models such as the Jeffreys-Bullen and Gutenberg models. It has a thick high-velocity lid, and a deep low-velocity zone with a small velocity contrast. Rapid velocity increases occur at 315 and 395 km. Other S-wave velocity models for the stable part of North America are the models of McEvilly (1964) and Mitchell and Herrmann (1979) for the central-southeast United States using Rayleigh and Love-wave phase velocity data. McEvilly's (1964) model is similar to the CANSD model except for the thinner lid above the low-velocity zone. McEvilly (1964) postulated the need for anisotropy to simultaneously fit both the Rayleigh-wave and Love-wave data. However, Mitchell and Herrmann (1979) found no significant evidence for S-velocity anisotropy in the upper mantle. The mod-

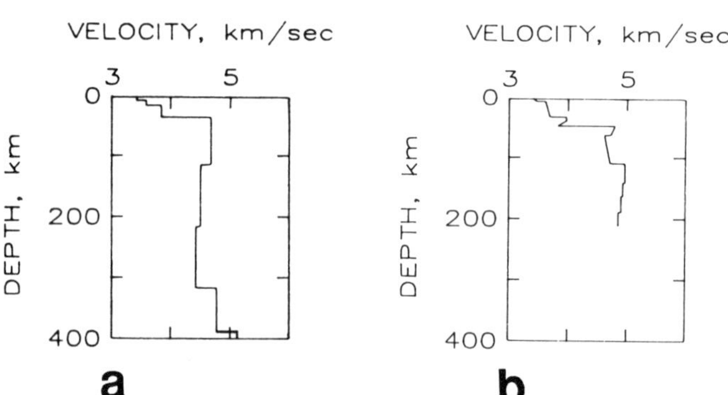

Figure 4. Shield and platform S-velocity models. a, CANSD model modified from Brune and Dorman, 1963. b, Model modified from Mitchell and Herrmann (1979).

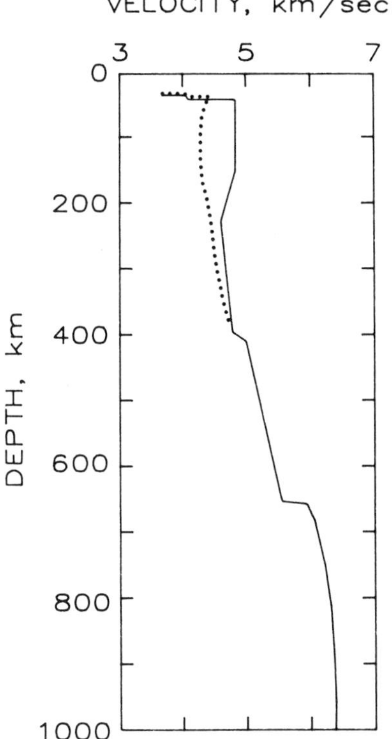

Figure 5. Shield model SNA (solid line) and tectonic model TNA (dotted line). Modified from Grand and Helmberger (1984).

els of Mitchell and Herrmann (1979), computed using Rayleigh and Love waves combined, are shown in Figure 4b. They all have a high-velocity lid with a modest low-velocity layer underneath it. The velocity jump of 0.25 km/sec at 315 km depth at the base of the low-velocity layer found in CANSD corresponds approximately to the location of similar discontinuities in the P-velocity models for stable continent of Massé (1973) at 328 km (Fig. 2a), Lewis and Meyer (1968) at 350 km (Fig. 2b), Green and Hales (1968) at 362 km, and Barr (1967) at 366 km (Fig. 2c). This leads to the speculation that a velocity discontinuity in this depth range may be present in the stable continental models, perhaps as the base of a weak low-velocity zone. However, such a discontinuity is absent in both the P-velocity model S8 (Burdick, 1981) (Fig. 3) and S-velocity model SNA (Grand and Helmberger, 1984) (Fig. 5) for the North American shield.

A good comparison between the upper-mantle S-velocity structures of the tectonic region of North America and for the Canadian shield, respectively, is provided by models TNA and SNA constructed by Grand and Helmberger (1984) using the synthetic seismogram technique (Fig. 5). These models can be considered to be the most current and well-constrained tectonic and shield models of North America. The main difference between the models SNA and TNA is in the top 400 km. The S-velocity in the shield is higher by 10 percent from the base of the crust to 175-km depth and by 5 percent in from 175 to 250 km. The contrast between the two models decreases at greater depths until the two models merge at 405-km depth. The shield model is similar to the CANSD model of Brune and Dorman (1963) in the sense that it has a thick high-velocity lid, beneath which a large, low-contrast low-velocity zone is present. The discontinuity at 315 km found in the CANSD model is absent in SNA, even though the deeper discontinuity (olivine-spinel transition) occurs at almost the same depth in both the models. In comparing the S-wave velocity model pairs SNA and TNA with the pair of P-velocity models T9 and S8 (Fig. 3) of Burdick (1981) for the tectonically active and stable parts of North America, note that the latter two models are different only to a depth of 200 km. If this is true, comparison with the S-velocity models implies that the Poisson's ratio is different between the tectonically active and shield sections of the North American continent in the 200- to 400-km depth range.

In summary, the main feature of the S-wave velocity model for the continental shield is the presence of a low-contrast low-velocity zone in the approximate depth range of 100 to 300 km. Apart from the low-velocity zone, the P-wave and S-wave velocity vs. depth functions are not significantly different from each other.

Helmberger and others (1985) computed a laterally varying upper-mantle S-wave velocity cross section from California to Greenland. In this computation they used TNA and SNA as starting models for the tectonically active and shield portions, respectively, of the North American continent. Using a variation of the synthetic seismogram technique normally used for modeling laterally uniform structure and an extensive data set of multiple-bounce S-waves, they showed that the strongest lateral heterogeneity along the cross section occurs over a distance of about 400 km beneath the Rocky Mountains. In this transition region the low velocity in the upper 300 km of the upper mantle associated with the low-velocity zone of the west is replaced by about 7 percent higher velocity in the east, and the lithospheric thickness increases by 75 km from west to east.

Models of the tectonically active continent

P-velocity models. Gutenberg's (1959) P-wave velocity model, which contains a substantial low-velocity layer in the upper mantle, is representative of the structure of tectonically active western North America. Attempts at a comparison of upper-mantle models for the western and eastern United States using nuclear explosions in Nevada and New Mexico have been made by several investigators (Lehman, 1962, 1964; Dowling and Nuttli, 1964; Archambeau and others, 1969; Massé and others, 1972). Even though many of these models suffer from the complex path effects due to the strong lateral velocity heterogeneity in the transition between the west and east, some obvious differences have been documented. Lehman (1962, 1964) computed two models for the south-central United States and one model for the Northeast. Even though her main emphasis was to demonstrate the presence of a sharp velocity discontinuity at 215 km in these models, the western models had lower velocities than the eastern models in the upper 200 km of the upper mantle. Both the eastern and western models had low-velocity layers of varying thicknesses and locations. Dowling and Nuttli (1964) derived four P-wave velocity models, two for the western United States, and one each for the north-central states and the Southeast. The two western models had lower upper-mantle velocities than the north-central and southeastern models. All the models had low-velocity layers, with the low-velocity layer in the western models being shallower than in the other models. In the north-central and southeastern models, the base of the low-velocity layer occurred at depths close to Lehman's 215-km discontinuity, suggesting that this discontinuity may be a real feature in the upper-mantle models of the stable part of the continental United States. Archambeau and others (1969) computed four models for the western states along northeast and southwest profiles originating from Nevada and encompassing different tectonic provinces. They used additional constraints provided by P_n-velocity and traveltime residual data. Their main conclusions from this study are that strong lateral velocity variations are present in the upper 220 km of the upper mantle and that lateral variations are minimal at greater depths. In the laterally variable zone, the Basin and Range structure differs sharply from the plateau and mountain structure in that the subcrustal high-velocity lid is of negligible thickness or nonexistent. The Basin and Range model of Archambeau and others (1969) is similar to the S-velocity model for rifts proposed by Knopoff (1972) in his classification of surface-wave models into five different categories, each representative of a different tectonic region (see below). It is also similar to the S-velocity model of Daniel and Boore (1982) for the Yellowstone caldera in which extremely low upper-mantle velocities are found, starting from the base of the crust and interpreted to be due to the presence of partial melt. However, other recent studies (summarized in this chapter) show that a subcrustal high-velocity lid is found above the low-velocity zone in the Basin and Range province (Priestley and Brune, 1978; Burdick and Helmberger, 1978; Priestley and others, 1980). These studies place the base of the lithosphere at a depth of about 65 km. In the deeper upper mantle, all the models of Archambeau and others (1969) are similar, with velocity discontinuities occurring at 400-, 650-, and 1,000-km depths.

A typical model for the upper 170 km of the Basin and Range province and the deeper part of the upper mantle of the central United States is that computed by Massé and others (1972) using traveltime data from nuclear explosions recorded by LRSM stations. They determined the P-velocity distributions from record sections "through a combined analysis of first arrivals, large-amplitude retrograde reflections, critical points and undercritical reflections" (Massé and others, 1972, p. 26). The authors assumed that the composite of the Basin and Range structure for the uppermost mantle, and the central U.S. model for the deeper structure, was representative of the total Basin and Range upper-mantle structure (Fig. 6). The new feature of this model is the discontinuity near 305 km depth. This discontinuity is similar to that in Massé's (1973) average model for North America (Fig. 2a).

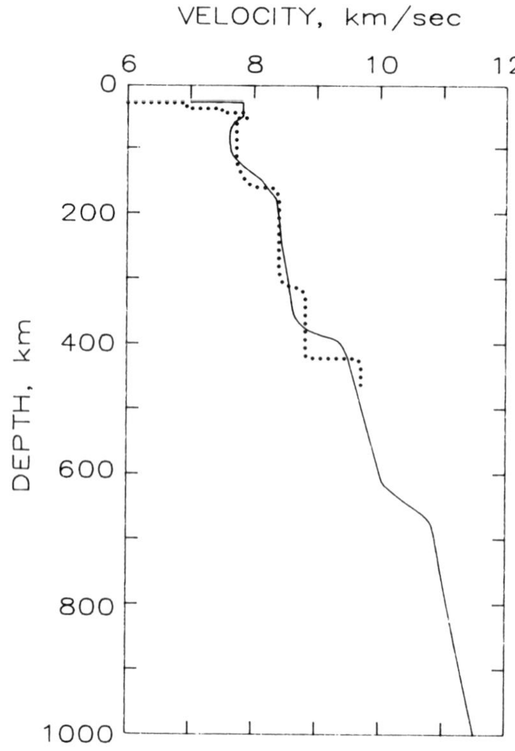

Figure 6. P-velocity models for tectonically active continent. CIT204 modified from Johnson (1967) (solid line); model modified from Massé and others (1972) (dotted line).

The model CIT 204 of Johnson (1967), often cited as a representative model for the western states, is derived by the inversion of $dT/d\Delta - \Delta$ data from the Tonto Forest seismic array in Arizona. Johnson (1967) used first and later arrivals in computing the models. However, the upper 160 km of the final model involving the low-velocity layer was based on trial and error. The characteristics of the final model were a pronounced low-velocity layer in the uppermost part of the upper mantle, and the two velocity gradients near 400 and 650 km depths (Fig. 6). Note that Johnson's (1967) 400-km discontinuity is significantly shallower than that for the average U.S. and eastern models discussed in this chapter. This decreased depth may be real, as the ray paths used in deriving CIT 204 are pure-paths for western North America. Other pure-path models for western North America also show relatively shallow 400-km discontinuities (e.g., Walck, 1984, 1985).

A series of upper-mantle models for the western United States have been derived using the synthetic seismogram technique. Three of these models, HWNE of Helmberger and Wiggins (1971) and models HWA and HWB of Wiggins and Helmberger (1973), were for the region encompassing the Basin and Range and Snake River Plain provinces. The models had a low-velocity zone, major discontinuities at 430 and 660 km depths, and a change in velocity gradient around 550 km depth, a new feature found in most models based on synthetic seismo-

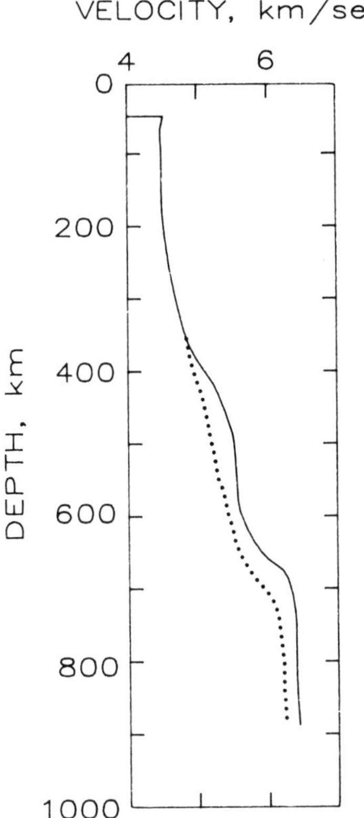

Figure 7. S-velocity models for western U.S. (modified from Kovach and Robinson, 1969): STAN2 (dotted line); STAN3 (solid line).

grams. A new model T7 for the western states, constructed by Burdick and Helmberger (1978), is markedly different from the above models, particularly in the depth and thickness of the low-velocity zone. In the deeper upper mantle, T7 has the first discontinuity at 395 km (as against 425 km in HWA), the second discontinuity at 670 km (as against 625 km in HWA), and no inflection in velocity gradient between the two discontinuities. Burdick and Helmberger (1978) show that T7 fits the traveltime data, $dT/d\Delta$ of multiple arrivals, and long-period synthetic seismograms much better than the earlier models. Burdick (1981) modified T7 to a new model T9 (Fig. 3) in which the 650-km discontinuity was made shallower by 20 km to fit the observations better. Model T9 is a representative P-wave velocity model for the tectonically active part of North America derived using the synthetic seismogram technique and is shown in Figure 3 along with model S8 for the stable shield.

Three models for western Canada were constructed by Dey-Sarkar and Wiggins (1976) using the seismic data collected by the extensive Canadian national seismic network. All three models—WCA, WCB, and WCC—have low-velocity zones and discontinuities near 400 and 600 km depths; models WCA and WCB show a discontinuity in the slope of the velocity-depth curve near 504 km. This slope discontinuity is absent in WCC. Model WCA covers the region west of the Rocky Mountain system and some portions of the oceanic mantle and shows a lithospheric thickness of 92 km. Model WCB is for the eastern Rocky Mountains and shows a lithospheric thickness of 120 km. Model WCC is for Arctic Canada and is similar to WCA and WCB, except that it does not have the transition zone at 504 km. Models WCA and WCB are similar to models HWA and HWB of Wiggins and Helmberger (1973); the only difference is that the 650-km discontinuity is much sharper in WCA and WCB than in HWA and HWB. The major difference between the Canadian models and the Basin and Range models of Johnson (1967) and Archambeau and others (1969) is in the 504-km transition zone.

In summary, the most significant feature of the P-wave velocity model for the tectonically active continent is the pronounced low-velocity layer in the upper 200 km of the upper mantle, in addition to the 400- and 650-km discontinuities.

S-wave velocity models. Wickens and Pec (1968) modeled the S-wave velocity structure of the upper mantle using Love wave-phase velocities in five segments along a great circle path from Mould Bay, Canada, to Tucson, Arizona. They found significant variation in the thickness and depth to the top of the low-velocity zone. From Mould Bay to Yellowknife, the low-velocity zone begins at a depth of about 150 km and is similar to the CANSD model of Brune and Dorman (1963). From Yellowknife to Dugway, Utah, the low-velocity zone is shallower, with its top at a depth of 60 to 75 km, and from Dugway to Tucson it almost reaches the base of the crust.

A representative S-wave velocity model for the western states is that of Kovach and Robinson (1969), constructed using data from a linear seismic array in Arizona for more or less the same region as covered by Johnson's (1967) P-wave velocity model CIT 204. The starting model (STAN2) was obtained by direct inversion of a smooth curve passing through the observed $dT/d\Delta - \Delta$ points. It had a thin lid above a low-velocity zone and an increase in velocity gradient beginning at a depth of 160 km near the base of the low-velocity zone. In the depth range of 350 to 680 km, the velocity gradually increased, followed by a rapid change in velocity gradient between 680 and 740 km depths. In order to be compatible with the model CIT 204 of Johnson (1967) and the models of Archambeau and others (1969), Kovach and Robinson (1969) computed a second model, STAN3, by adding a receding branch from 20° to 14° of the $dT/d\Delta - \Delta$ curve. This *artifact* had the effect of introducing the 400- and 650-km discontinuities. The models STAN2 and STAN3 are shown in Figure 7. Note the close similarity between STAN3 and the S-velocity model TNA of Grand and Helmberger (1984) shown in Figure 5. This agreement between the two models is encouraging and shows that very different techniques can yield similar models if the seismic waves sample the same tectonic region.

Cara (1979) used higher mode surface waves and the generalized inversion theory to model the upper-mantle S-wave velocity structure for the whole of the continental United States and for the two distinct tectonic provinces, one in the central and northeastern states, and the other in the Southwest. The central and

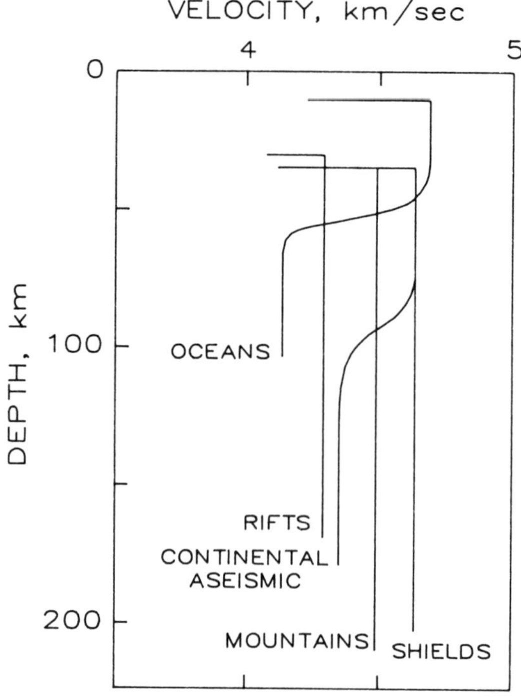

Figure 8. Schematic summary of S-wave velocity models for different tectonic areas. Modified from Knopoff (1972).

northeastern model has a slight low-velocity zone similar to the CANSD model. This low-velocity zone is not needed if only the fundamental mode data are used. The model for the southwestern region has a well-developed low-velocity zone. The average model for the whole of the continental United States has significance only at depths greater than 300 km and reveals two major discontinuities at 360 and 650 km, corresponding to the average depths of occurrence of the 400- and 650-km discontinuities found in other North American models.

Biswas (1971) and Biswas and Knopoff (1974) conducted a systematic study of the regional variation of S-velocity structure of the upper mantle in the continental United States. They analyzed data from WWSSN station pairs to determine Rayleigh-wave dispersion in the 20- to 250-sec period range along pure paths in different tectonic provinces. Suites of models for typical paths were derived using the "Hedgehog" method developed by Keilis-Borok and Yanovskaya (1967) and Press (1968). For each tectonic province, a whole suite of models were derived, providing bounds on velocities and depths. The results show the upper mantle in the north-central states has high S-velocities. The phase-velocity data in this region could be fit with no low-velocity channels, small low-velocity channels, or small positive gradients. Under the south-central states, a high-velocity lid and low-velocity channel were required to fit the data. In the West, very low velocities are found throughout the upper mantle extending to 400 km, the greatest depths reached by the inversions.

Knopoff (1972) supplemented the data of Biswas (1971) with data from other parts of the world and showed that the Rayleigh-wave data and the S-wave velocity models derived using them, can be basically divided into five distinct groups belonging to shields, aseismic platforms, rifts, ocean basins, and mountains (Fig. 8). The shield data are remarkably consistent with those in the Canadian shield and indicate a thick high-velocity lid and poorly developed low-velocity channel. The aseismic platform has a high-velocity lid and low-velocity channel. The oceanic structure has a high-velocity lid and a pronounced low-velocity channel starting at a depth of about 50 km below the sea level. For the rift structure the data fit a model with a high-velocity lid with an extremely well-developed low-velocity channel, or a thin, deep ultra-low-velocity channel, or a low-velocity layer extending from the Moho to great depths. The structures beneath the two mountains, the Andes and the Alps, investigated by Knopoff (1972) were inferred to be in between shield and rift structures.

In a series of papers, Lee and Solomon (1975, 1978, 1979) developed a method to invert surface-wave data for simultaneously estimating S-wave velocity and attenuation (1/Q) as a function of depth. In their most recent study, Lee and Solomon (1979) applied their method to combine Rayleigh and Love wave data to model the western and eastern United States. One important result for the West is the identification of a zone of high attenuation coinciding with the low-velocity zone. The eastern model has no low-velocity zone, but there could be a low-Q zone in the 138- to 250-km depth range. If the low-velocity and low-Q zones of the western states and the low-Q zone of the eastern states correspond to the asthenosphere, then the lithospheric thicknesses are 130 ± 30 km for east-central North America and 80 ± 20 km for western North America.

Regional models

East Pacific Rise and northeast Pacific rim. Walck (1984, 1985) modeled the P-wave velocity structure of the upper mantle beneath the East Pacific Rise and northeast Pacific Ocean rim using data from the dense southern California seismic array. The inversion of the traveltime data was carried out by the Tau method of Bessonova and others (1974, 1976) supplemented by synthetic seismogram computations. The model GCA (Fig. 9) for the East Pacific Rise is similar in shape to the S-wave velocity model TNA of Grand and Helmberger (1984) for western U.S. (Fig. 5) at depths greater than 125 km above which either a smooth crust-mantle velocity transition or a lid with a low-velocity layer can satisfy the data. The model also approximates to the S-wave model of Kovach and Press (1961) for the same region computed using Rayleigh-wave group-velocity data. This model has a thin high-velocity lid overlying a 200-km-thick low-velocity zone.

In another study Walck (1985) modeled the upper-mantle P-wave velocity structure of the Cascade Range–Juan de Fuca region using earthquakes in the northeast Pacific rim recorded by the southern California seismic array. The model CJF for the northeast Pacific rim (Fig. 9) is different from the Gulf of California model GCA (Walck, 1984), mainly in the upper 200 km, in the shape and structure of the low-velocity zone, and in the velocity gradient between the base of the low-velocity zone and the 400-km discontinuity. Note that the model CJF is intermediate between the model TNA of Grand and Helmberger (1984) for tectonically active continent and Walck's (1984) East Pacific Rise model GCA.

Great Basin and the Volcanic Plateau. Priestley and Brune (1978, 1982) modeled fundamental mode Rayleigh and Love waves to yield the upper-mantle S-wave velocity structure of the northern and central Great Basin of Nevada and Utah and the southern Volcanic Plateau of Oregon (Fig. 10). The northern Great Basin model consisted of a 30-km-thick crust with a 30-km-thick high-velocity lid beneath it, and a low-velocity layer similar in thickness but with considerably higher average velocity than in their southern Great Basin model (Priestley and Brune, 1978). The volcanic plateau model consisted of a 40-km-thick crust, a 10-km-thick mantle lid, and a pronounced 130-km-thick low-velocity layer (Priestley and Brune, 1982). Note that the shear velocities in the Great Basin models (Fig. 10) are considerably lower than in other models of the same region (Kovach and Robinson, 1969; Bucher and Smith, 1971; Biswas and Knopoff, 1974) and are comparable with values for East African rifts (Knopoff and Schule, 1972) and the East Pacific Rise (Knopoff and others, 1970). In another study, Priestley and others (1980) used higher mode surface data to confirm the existence of a high-velocity lid above the low-velocity layer in the Great Basin.

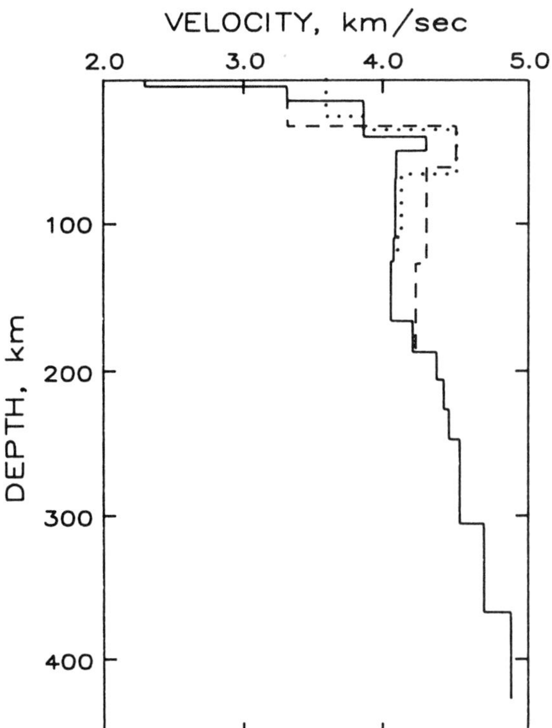

Figure 10. Comparison of S-velocity models for the Great Basin and the southern Volcanic Plateau of Oregon: northern Great Basin (modified from Priestley and Brune, 1978) (dashed line); central Great Basin (modified from Priestley and Brune, 1978) (dotted line); Volcanic Plateau (modified from Priestley and Brune, 1982) (solid line).

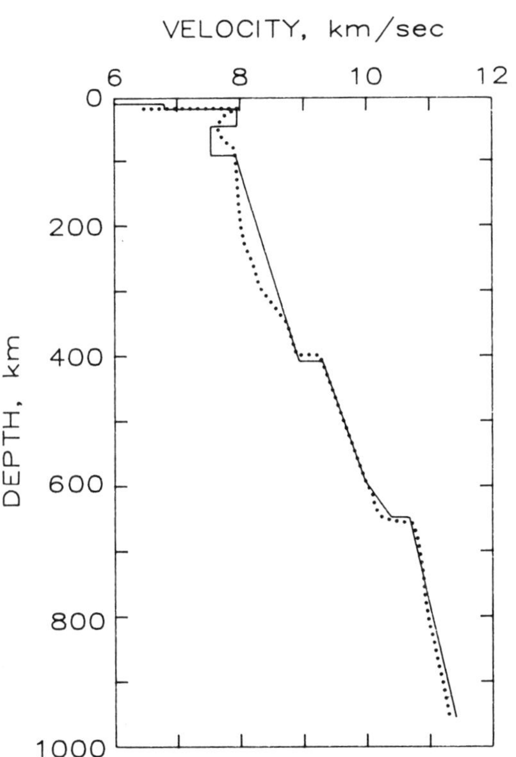

Figure 9. East Pacific Rise Model GCA (dotted line). Modified from Walck, 1984, compared with the Cascades–Juan de Fuca region model CJF (solid line). Modified from Walck (1985).

Hill (1972) used traveltime data from explosions to model the crust and upper mantle along a 600-km-long profile in the Columbia Plateau. He proposed two tentative models for the uppermost mantle of this region. Model CP1 has as 10-km-thick lid at a depth of 100 km, below which is a pronounced low-velocity zone extending to a depth of 145 km. Model CP2 does not have a pronounced lid, but grades directly into a wedge-shaped low-velocity zone beginning at 60 km and ending at 130 km. Recent seismic refraction studies by Catchings and Mooney (1988) in the Columbia Plateau reveal much higher P_n velocity (8.4 km/sec) than that found by Hill (1972). Also, Catchings and Mooney (1988) found that the deep crustal and upper-mantle structure in the Columbia Plateau is suggestive of continental rifting.

Yellowstone and Eastern Snake River Plain. Daniel and Boore (1982) investigated the anomalous S-velocity structure in the Yellowstone Plateau, considered to be a continental hot spot, using intermediate-period (1 to 30 sec) body and surface waves. They found extremely low Rayleigh-wave phase velocities: for example, 2.6 km/sec at 5-sec period, compared with 3.3 km/sec at the same period for the Basin and Range province (Priestley

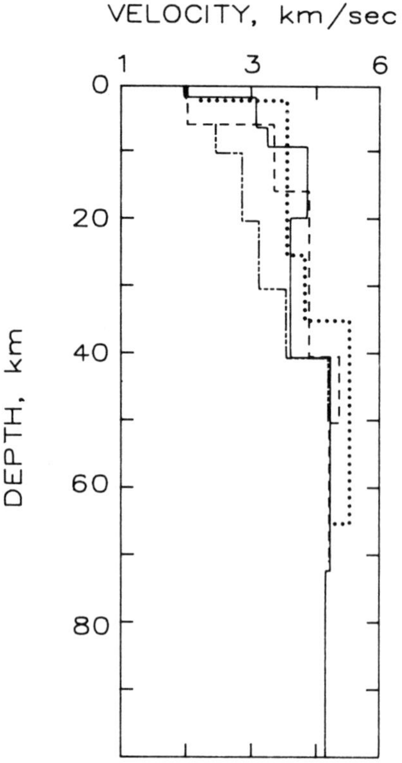

Figure 11. S-velocity models of Daniel and Boore (1982) for the Yellowstone Caldera (dash-dot line), of Greensfelder and Kovach (1982) for the Eastern Snake River Plain (solid line), of Priestley and Brune (1978) for the Great Basin (dotted line), and of Priestley and Brune (1982) for the Volcanic plateau (dashed line). Figure modified from Greensfelder and Kovach (1982).

and Brune, 1978). Using the inversion technique of Backus and Gilbert (1970), they found the crust and upper-mantle velocities to be extremely low beneath the Yellowstone caldera, the scene of very recent volcanism. In another study of the same region, Greensfelder and Kovach (1982) modeled Rayleigh-wave phase and group velocities in the eastern Snake River Plain, immediately to the southwest of the Yellowstone caldera. They found abnormally low velocities in the lower crust and low velocities in the upper mantle to a depth of 70 km, the modeling depth. An upper mantle lid may or may not be present in the model. Figure 11 compares the Yellowstone caldera model of Daniel and Boore (1982), the eastern Snake River Plain model of Greensfelder and Kovach (1982), and the northern Great Basin model of Priestley and Brune (1978).

Montana. Baumgardt and Alexander (1984) modeled the S-velocity structure of the upper mantle in Montana using mode-converted S-phases recorded at the Large Aperture Seismic Array (LASA) in Montana. Mode conversion occurs when S-waves incident at the base of a discontinuity beneath a seismic station are converted into P-phases. Thus S, SKS, and ScS can produce mode-converted phases Sdp, SKSdp, and ScSdp arriving prior to the respective phases. In this technique, starting from a crustal model, discontinuities were placed at appropriate depths to produce Sdp phases that matched shear-wave precursor arrival times with impedance contrasts adjusted to match the wave-form shapes and amplitudes. The computed model LS1M 19 has seven first-order discontinuities, including a low-velocity channel in the depth range of 170 to 226 km.

Synthesis of one-dimensional models

There is no simple way to synthesize the one-dimensional velocity models discussed in the previous sections to accurately reflect regional variations in key parameters of upper-mantle structure of North America. The major problem is that the models, with a few exceptions, are averages for large regions and reveal only differences in structure between major tectonic units such as the western and eastern halves of the continent and do not have the resolution to provide information about lateral and vertical heterogeneous structure within each unit itself. Another problem is that the models are nonunique, due in part to ambiguities in identifying arrivals associated with multiple-branch traveltime curves and in matching actual data with synthetic seismograms. Both of these problems are at least partly responsible for the observed differences between the one-dimensional models computed using traveltime data or traveltime and synthetic seismograms within the same tectonic regime. However, in spite of these difficulties, a broad picture of the heterogeneous structure of the upper mantle in North America, suspected to be present from studies even in the 1950s and 1960s, has emerged. In this section we quantify this heterogeneous structure by tabulating and plotting several of the critical parameters from the numerous P-wave and S-wave velocity models discussed in the previous sections. These parameters are P_n and S_n velocities, lowest velocities within the low-velocity layers, velocities at 100-km and 200-km depths, thicknesses of low-velocity layers, lithospheric thickness, and depths of discontinuities up to and including the 400-km discontinuity. The depths to the 650-km discontinuity are not considered to be accurate enough for inclusion in this discussion.

The models are assembled into four groups: P-wave velocity models for stable continent; S-wave velocity models for stable continent; P-wave velocity models for tectonically active continent, and S-wave velocity models for tectonically active continent. The critical parameters for the various groups are listed in Tables 1 through 4 (p. 704–707). Eight of the above parameters—namely, lowest P-wave and S-wave velocities in the low-velocity layer, P-wave and S-wave velocities at 100 and 200 km depths, and lithospheric thicknesses for P- and S-models—are plotted on maps of North America (Figs. 12a–h). In using or interpreting these maps it should be remembered that the geographic locations at which the estimated values are plotted are only approximate. They are plotted somewhere near the center of the region covered by the sources and stations for traveltime and synthetic seismogram studies and near the middle of the seismic array for surface-wave studies. The low-velocity layers and depths to the various discontinuities are summarized in Figure 13. A discussion of the critical parameters follows.

Figure 12 (this and next 2 pages). Variation of critical parameters (see text) derived from the one-dimensional velocity models for North America. a, Lowest P-wave velocity in the low-velocity layer (see Tables 1, 3). b, Lowest S-wave velocity in the low-velocity layer (see Tables 2, 4). c, P-wave velocity at 100-km depth (see Tables 1, 3). d, S-wave velocity at 100-km depth (see Tables 2, 4). e, P-wave velocity at 200-km depth (see Tables 1, 3). f, S-wave velocity at 200-km depth) see Tables 2, 4). g, Lithosphere thickness (in kilometers from P-models (see Tables 1, 3). h, Lithosphere thickness (in kilometers) from S-models (see Tables 2, 4).

Lowest velocity in the low-velocity layer. Figures 12a and b show the geographic distribution of observed lowest P-wave and S-wave velocities in the low-velocity layer. (The definition of the low-velocity layer is subjective. When a model shows several low-velocity layers, we take into consideration only the first low-velocity layer in the upper mantle.) Both P-wave and S-wave velocities are lower in western North America than in central and eastern North America. Low velocities also occur in the eastern United States, but when taken together with the lithospheric thickness plots (Fig. 12g,h), it can be seen that they occur at relatively greater depths than in western North America.

Velocities at 100 km depth. These values, shown in Figures 12c and d clearly show the partitioning of North America into the tectonically active western structure and shield-type eastern structure. P-wave and S-wave velocities average about 0.5 km/sec lower in the west than in the east.

Velocities at 200 km depth. P-wave and S-wave velocities at 200 km depth displayed in Figures 12e and f show the upper mantle at this depth attaining a reasonable degree of homogeneity compared with velocities at 100 km depth. Lateral heterogeneity, although present at this depth, has dwindled to a variation of about 0.1 to 0.2 km/sec, which is close to the noise in the velocity estimates.

Lithospheric thickness. The seismological definition of lithospheric thickness is ambiguous, particularly for stable continental areas. In this review, lithospheric thickness is defined—for both the tectonically active and the stable regions of North America—as simply the depth to the first low-velocity layer in the upper mantle if such a layer exists or the crustal thickness when no marked low-velocity zone is present but the P_n velocity is abnormally low. This arbitrary definition leads to some confusing values of lithospheric thickness, but a pattern does emerge, as can be seen from Figures 12 g and h, in which lithospheric thicknesses based, respectively, on P-wave and S-wave velocities are depicted. The lithosphere may be as thin as 50 km in western North America and about three times that thick in eastern North America. These values are close to those found by Helmberger and others (1985) in their upper-mantle cross section from California to Greenland. The inferred lithospheric thicknesses from our figures also compare reasonably well with the lithospheric

7.90
8.23
7.97
8.37
8.43
8.30
8.28
7.75
8.20
7.95
7.68
7.94
8.27
7.98
7.87
7.71
8.10
8.20
(G) 7.83
7.70
7.70
7.70
7.82
8.30
7.72
8.30
8.25
8.26
8.34
8.00
7.68
7.90
7.80
7.90
0 800
KM
c
4.72
4.72
4.60
4.10
4.10
4.30
4.43
4.10
4.12
4.38
4.60
4.60
4.50–4.78
4.40
4.38
4.78
4.45–
4.80
4.29
0 800
KM
d
8.21
8.32
8.02
8.50
8.48
8.34
8.46
8.22
8.55
8.06
8.42
8.43
8.43
8.37
7.90
8.50
8.32
7.95
8.31
8.36
8.30
8.25
8.36
8.00
8.34
8.39
7.95
8.38
7.90
8.08
8.20
8.05
0 800
KM
e
4.54
4.38
4.56
4.47
4.50
4.38
4.60
4.60
4.63–4.70
4.38
4.50
4.45–
4.63
4.80
4.39
0 800
KM
f

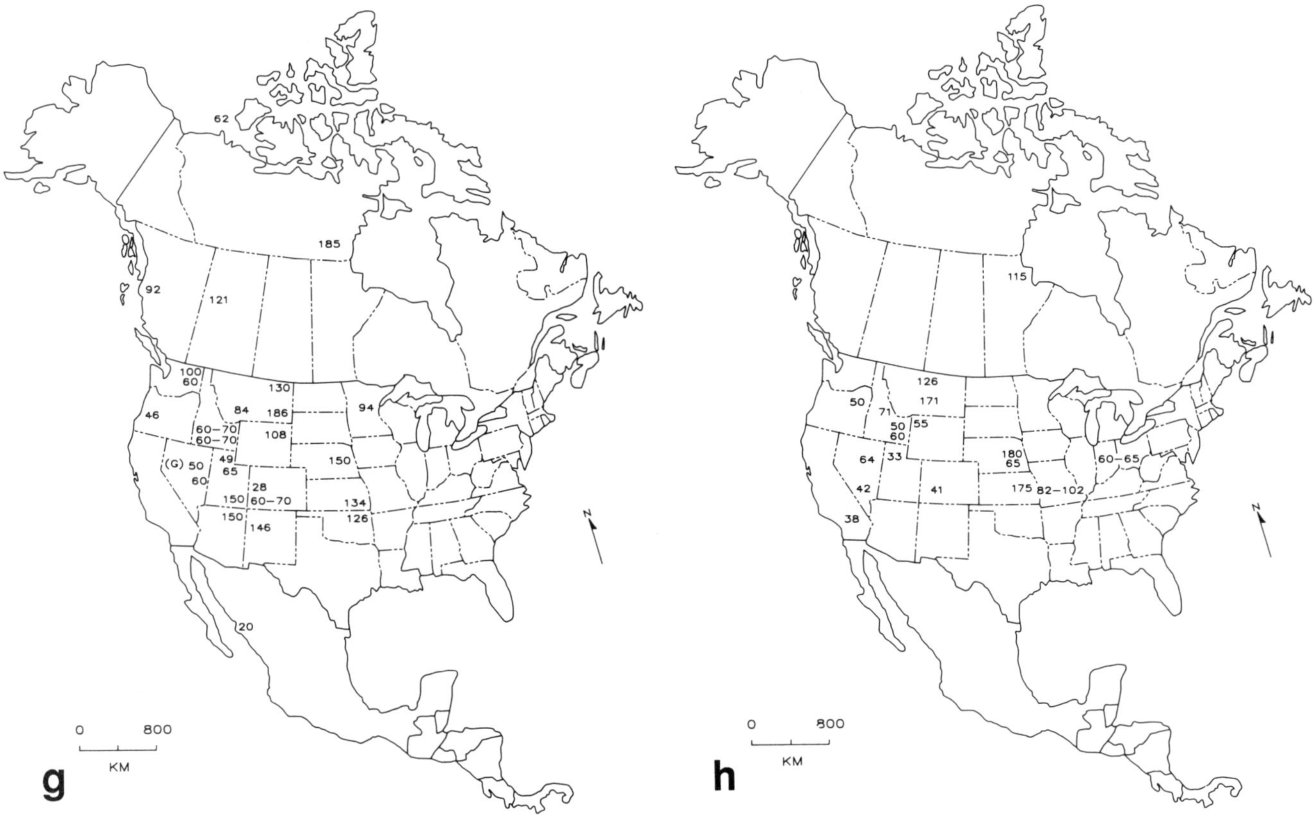

thicknesses of 135 km for the east-central United States and 84 km for the western states, computed by Lee and Solomon (1979) using simultaneous estimation of S-wave velocity and Q-structure of the upper mantle. However, in spite of these favorable comparisons, we urge that considerable caution be exercised in interpreting the lithospheric-thickness maps (Figs. 12g and h) for central and eastern North America, as some of the estimates are based on poorly contrained low-velocity layers.

Depths to discontinuities. In Figure 13 we display depths to the velocity discontinuities revealed by the various models. The low-velocity zones are depicted by dark vertical bars. Their widths are proportional to the velocity decrease within the zones. The following inferences can be made from these figures. The P-wave velocity models for the stable continent, in general, do not have low-velocity zones but instead have one or more velocity jumps between the Moho and the 400-km discontinuity (Fig. 13a). The S-wave velocity models, on the other hand, show low-velocity zones of varying thicknesses, mean depths, and velocity contrasts (Fig. 13a). A consistent picture of low-velocity zones is presented by the P- and S-wave velocity models of the tectonically active continent (Fig. 13b). An observation of some significance is that the shield models show diverse discontinuities in the upper mantle, whereas the tectonic models seldom show discontinuities other than those associated with the low-velocity zone and the 400- and 650-km discontinuities. This leads us to suspect that the extra discontinuities in the shield models may be due to misintepretation of weak low-velocity zones.

HETEROGENEOUS VELOCITY MODELS USING TRAVELTIME RESIDUALS

Introduction

The one-dimensional models presented in the previous sections provide comparison of absolute velocity as a function of depth in different regions, and reveal the existence of strong lateral heterogeneities in the upper mantle in the continental United States. Teleseismic P- and S-wave traveltime residuals provide a complementary way of probing the three-dimensional heterogeneous structure of the crust and upper mantle. An advantage of the traveltime-residual technique over inverting $T - \Delta$ data for one-dimensional models is that the former does not require a continuous distribution of earthquakes as a function of distance from the recording stations. However, both techniques do require a dense array of recording instruments. Also, the instrumentation required for traveltime residual measurements is relatively simple (short-period and/or intermediate-period instruments) compared with the long-period instruments needed for surface-wave studies. Since resolution of the velocity models constructed using teleseismic residuals is a function of instrument spacing in the array, the required resolution can be tailored to the goals. Against these advantages is the disadvantage that the models are, in general, "fuzzy," with uncertain boundaries of velocity discontinuities. Also, the final model is specified in terms of velocity perturbations rather than in terms of absolute velocities. This is not a

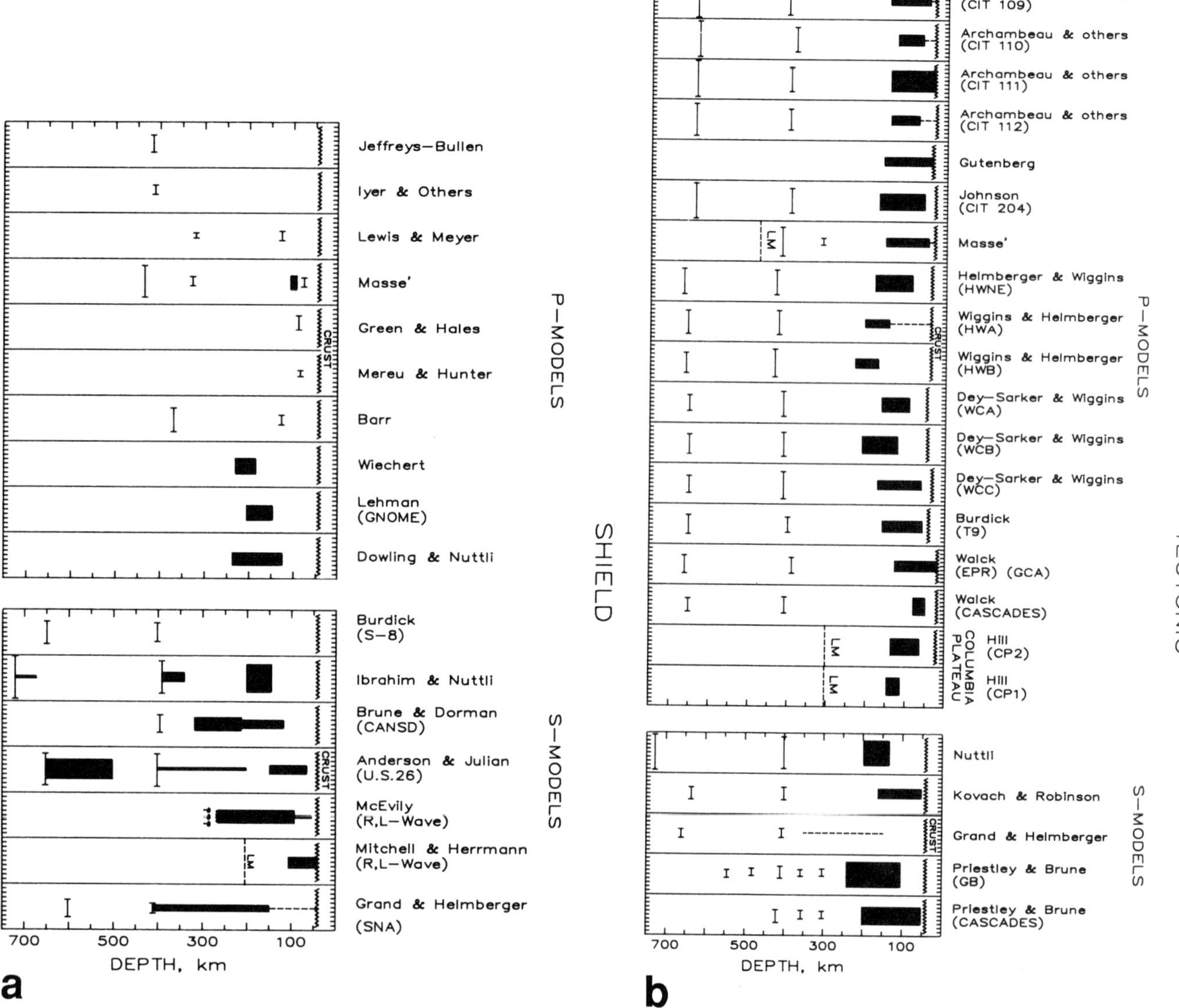

Figure 13. Summary of depths to discontinuities derived from one-dimensional models. Wavy line denotes approximate locations of Moho. Dark strips are low-velocity zones. Bars are depths to other discontinuities. Width of bars and stripes indicate approximate velocity variations. LM = depth of modeling. Question marks = uncertainties. a, Shield and platform models. b, Tectonic models.

serious limitation if an appropriate one-dimensional velocity model is available for the region under investigation. With the advent of three-dimensional inversion or tomographic techniques, the teleseismic-residual technique has become a powerful tool to probe the anomalous structure of the Earth's crust and upper mantle on local, regional, and continental scales.

Modeling techniques

Techniques for deriving three-dimensional seismic velocity models using teleseismic-residual data are discussed in the Methods section (Mooney, this volume) and are therefore discussed here only briefly. Basically the technique consists of computing traveltime residuals at stations in a two-dimensional array of instruments for 3-D models and in a linear array for 2-D models. The computation of residuals, which are simply deviations from specified traveltimes based on standard Earth models, is as routine procedure and is discussed extensively in seismological literature (e.g., Cleary and Hales, 1966; Herrin and Taggart, 1968; Iyer and others, 1981). The two currently available techniques for inversion of teleseismic residual data are the damped least-squares inversion technique of Aki and others (1976, 1977) and the tomographic back-projection technique of Humphreys and others (1984). Of the two techniques, the damped least-squares technique of Aki and others (1976, 1977) has yielded numerous tectonically interpretable models of the crust and upper mantle in the United States (most of which are discussed in this review), and other parts of the world.

Discussion of two- and three-dimensional velocity models

North America. Herrin and Taggart (1962) and Herrin (1969) contoured P_n velocities in the United States, and showed that P_n was systematically lower in the western half than in the eastern half. Romney and others (1962) showed that the traveltimes of seismic P-waves from the GNOME nuclear explosion were as much as 12 sec earlier in the eastern states than at corresponding distances in the western states. A systematic study of variation of traveltime residuals in the continental United States was first started by Cleary and Hales (1966). They computed P-wave residuals with respect to the J-B Traveltime Tables at a large number of WWSSN and LRSM seismic station in the continental United States. They expressed the observed residuals as combinations of source terms, path terms, and station terms, and solved for station terms using a least-squares method. The station terms were, in general, found to be large and positive in the Basin and Range province and large and negative in the central United States. In a similar study using S-wave residuals, Doyle and Hales (1967) found that the station anomalies were about three times as much as P-wave anomalies, in addition to being positive in the West and negative in the East.

Similar spatial patterns and magnitudes of P- and S-wave residuals have also been found by other investigators. Herrin and Taggart (1968), using a large data set of P-wave residuals, represented the station terms by two components, a constant and an azimuthally varying sinusoidal part. The constant part showed a spatial pattern similar to the station terms of Cleary ahd Hales (1966). The azimuthal term also varied by about the same magnitude as the constant term, but did not reveal any systematic spatial pattern. In a similar treatment of a large data set of global traveltime residuals taken from the International Seismological Summary, Dziewonski and Anderson (1983) represented the residuals as a constant plus two cosine terms (cos θ and cos 2θ where θ is the azimuth). The constant (azimuth-independent) term for North America showed arrivals that were systematically late in the Appalachians, earlier in the Great Plains and Canada east of the Rocky Mountains, and late in Nevada and California. The first azimuthal term showed two definite populations, one pointing to almost north and the other in the direction N45°E. Dziewonski and Anderson (1983) proposed that this behavior could be due to the heterogeneous velocity structure of the deep crust and upper mantle. The second azimuthal term showed a great deal of regional consistency. The fast direction was found to be east-west in the western United States and Canada, changing abruptly by about 60° near the coast of southern California. Dziewonski and Anderson (1983) postulated that the second azimuthal term could be related to seismic anisotropy in the mantle, the fast direction coinciding with the higher velocity axis of olivine crystals that are oriented in the direction of mantle flow.

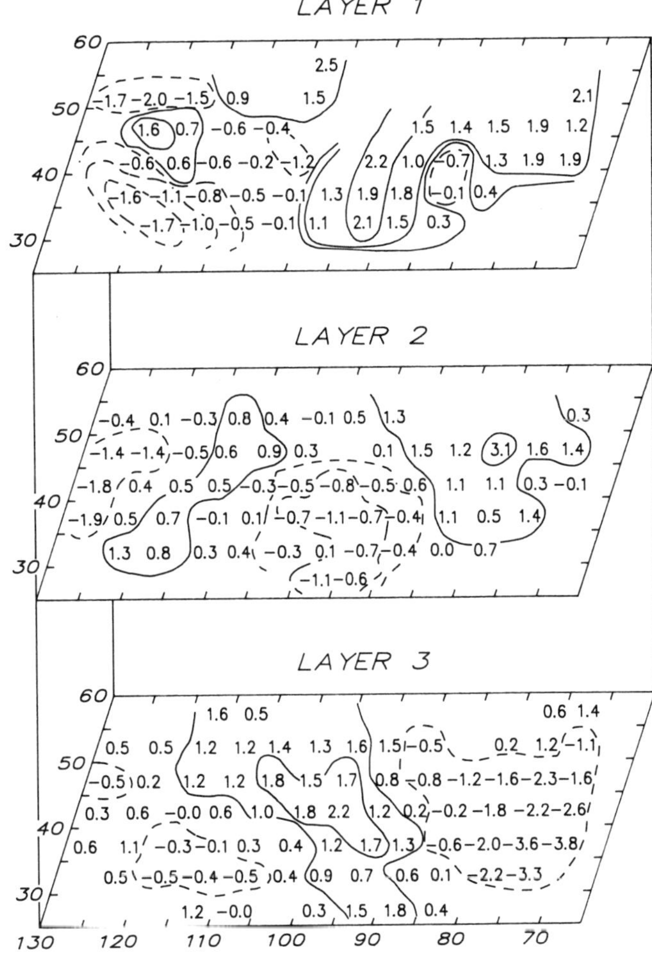

Figure 14. Three-layer inversion of travel-time residuals in U.S. Numbers are relative velocity fluctuations in percent. Solid-line contours outline high-velocity zones and dashed line contours outline low-velocity zones. Layer 1 = 0 to 250 km; layer 2 = 250 to 450 km; layer 3 = 450 to 700 km. Latitudes and longitudes are shown. After Romanowicz (1979).

Romanowicz (1979) computed three-dimensional models of the upper mantle of the continental United States by inversion of the massive numbers of P-wave residuals published in the International Seismological Summary, using the technique of Aki and others (1976, 1977). The final model had a horizontal block size of 5° by 5° and vertical layer thickness of 200 to 250 km and is shown in Figure 14. In the first layer, the P-wave velocity perturbations showed a striking correlation with tectonic features, namely, relatively low velocities in the Basin and Range Province, high velocities in the Colombia Plateau, and in general high velocities in the central and eastern states except for a weak low in part of the Appalachians. Layers 2 and 3 showed three distinct spatial patterns corresponding to the west, central, and eastern United States. A striking feature in layer 3 was a gradual increase in velocity from the west to the east, followed by a sharp drop in velocity in the eastern continental margin. Romanowicz (1980) proposed an interpretation of certain features of the velocity model in terms of the global tectonic evolution of western North America. For example, she postulated that the high-velocity anomaly in the northwestern states found in layer 1 (Fig. 14) could be explained as the subducting Juan de Fuca Plate and the high-velocity region in layer 2 in the central and southern part of the western states could be a remnant of the Farallon Plate, which subducted beneath western California between 10 and 20 Ma. The low-velocity zone in the upper mantle of central California in layer 2 was interpreted to be the void left by the Farallon Plate that disappeared, as proposed by Dickinson and Snyder (1979).

Grand (1987) inverted the S-wave residuals computed at the WWSSN and Canadian Seismic Network using the tomographic back-projection technique to yield a three-dimensional S-wave velocity model for the North American mantle. The most significant finding from this study is the delineation of a high-velocity root to a depth of about 400 km beneath almost all of the shield and platform of North America. Up to this depth, S-wave velocities were relatively low in the western part of the continent. Eastern North America was found to have high velocities to a depth of 140 km and normal to low velocities to 400 km.

Northern and central California. The complex geotectonic setting of northern California, from a subduction-related regime north of the Mendocino triple junction and a transform-fault related regime to the south, is ideally suited for probing using the tomographic technique. But the full potential of the technique is yet to be fully exploited. A few studies, however, hint at the complex heterogeneous upper-mantle structure in northern and central California. Bolt and Nuttli (1966), using a sparse seismic network in northern California, found spectacular azimuthal variation of teleseismic P-wave residuals. They (Nuttli and Bolt, 1969) interpreted these observations in terms of a thinning low-velocity layer from the coast toward the Sierra Nevada. Otsuka (1966a, b) analyzed traveltime data from teleseisms in the Coast Ranges of central California and found systematic cyclic trends in both slowness and azimuth anomalies. He showed that the structural feature responsible for these anomalies could be due to a combination of moderate dip in the Moho plus a steeply dipping deeper interface, probably the top of the low-velocity layer. Moho inclinations in northern and central California have been computed using P_n arrival times recorded by the dense network operated by the U.S. Geological Survey in this region. In the Geysers–Clear Lake region, Shearer and Oppenheimer (1982) found the Moho dip to be 5.3° in the direction N57°E. Oppenheimer and Eaton (1984) found the Moho dips to be 3° to 6° beneath the Coast Ranges and 8° to 17° beneath the Great Valley, and no dip beneath the Sierra Nevada foothills. They showed that the upper mantle had an isotropic P-wave velocity of 7.97 ± 0.05 km/sec beneath all three regions and that there were indications for the presence of deeper heterogeneities in the upper mantle.

Mavko and Thompson (1983) modeled the lithosphere in the Sierra Nevada from Mono Lake to the southern end of the Cascade Range using teleseismic residuals. In their preferred modeling approach they assumed the spatial and azimuthal patterns of the residuals to be due to variations in lithospheric thickness. This model showed a dramatic thinning of the lithosphere, 150 to 100 km in the northwest to 100 to 50 km in the southeast, depending on the assumed velocity contrast between the lithosphere and the asthenosphere (Fig. 15). They proposed that the thinning in the southeast was probably a result of heating from the hot asthenosphere and consequent consumption of the lower lithosphere after the cessation of subduction of the Farallon Plate beneath central California. The thicker lithosphere to the northwest was postulated to be due to the attachment of a remnant of the old plate to the normally thin lithosphere. Mavko and Thompson (1983) showed that, irrespective of the cause for the variation of the thickness of the lithosphere of the region, their model was consistent with isostasy, teleseismic residuals, seismic refraction, and gravity data, as well as the mechanism of Cenozoic uplift of the Sierra Nevada proposed by Crough and Thompson (1977).

San Andreas fault system, central and southern California. The dense seismic network operated by the U.S. Geological Survey in central California has provided excellent data to create a three-dimensional image of the velocity structure beneath the San Andreas fault system; several studies on modeling this

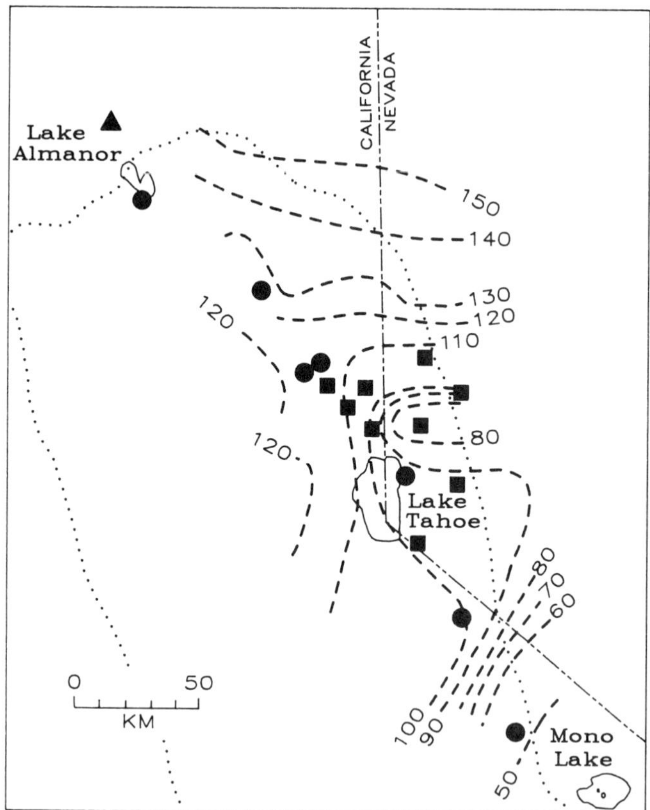

Figure 15. Contoured lithospheric thickness (in kilometers) beneath the Sierra Nevada, after Mavko and Thompson (1983). In this version of the model, a velocity contrast of 0.3 km/sec between the lithosphere and asthenosphere is assumed. Contour interval = 10 km (after Mavko and Thompson, 1983).

region have been presented (Husebye and others, 1976; Cockerham and Ellsworth, 1979; Zandt, 1981). The most significant finding from these studies in central California is the delineation of an inclined low-velocity zone in the upper mantle dipping eastward from a depth of 30 to 60 km beneath the Coast Ranges to at least 225 km farther to the east. Zandt (1981) found only the western edge of this anomaly, which he interpreted to be "the upper mantle track of the migrating pulse of extension" associated with the northward movement of the Mendocino triple junction, and suggested that it could probably be the subcrustal Pacific-American plate boundary. The deeper low-velocity zone delineated by Cockerham and Ellsworth (1979) is about 100 km thick, has 6 percent lower velocity than in the surrounding rock, and is inferred to be the Dickinson and Snyder (1979) window caused by the cessation of subduction in this region as the Mendocino triple junction migrated northward. Aki (1982) has synthesized the results of Cockerham and Ellsworth (1979) for central California and a southern California study (see below) of Raikes (1980). The combined model for the upper mantle beneath California is shown in Figure 16. Aki (1982) has suggested that the low-velocity zone in central California is "probably filled with mobile, ductile, soft material." Note that in southern California an elongated velocity high is found across the San Andreas fault. The transition from low velocity in the north to high velocity in the south, though both the regions are in the Dickinson and Snyder (1979) window, is explained by Aki (1982) as follows; since the migration of the Mendocino triple junction in the south occurred at 12 Ma, the soft material could not persist due to gravitational instability.

The teleseismic residual data from the dense USGS–Caltech seismic network in southern California has been analyzed in detail by several investigators to provide three dimensional models of the crust and upper mantle in this region (Raikes and Hadley, 1979; Raikes, 1980; Humphreys and others, 1984). In a qualitative study of the teleseismic residual pattern in southern California, Raikes and Hadley (1979) found azimuthally varying residuals with amplitudes of about 1 sec, which they interpreted to be due to the presence of strong lateral heterogeneities in the upper mantle. The major heterogeneous structures in the depth range of 50 to 180 km of the upper mantle were delineated in a series of models computed by Raikes (1980). These include marked low velocities to the west of the seismic array in the ocean, interpreted to be due to the ocean-continent transition; a large low-velocity region extending through the Mojave desert southward into the Salton Trough, inferred to be related to the active spreading and consequent high heat flow in this region; and a region of high velocity with a contrast of 2 to 6 percent beneath the Transverse Ranges with well-defined northern and eastern boundaries. The Transverse Ranges anomaly is neither correlated with surface topography nor is there any gravity anomaly at the surface corresponding to it. Hadley and Kanamori (1977) had earlier detected this anomaly. The horizontal extent of this anomaly and its relationship to upper-mantle velocity anomalies in central California are shown in Figure 16 reproduced from Aki

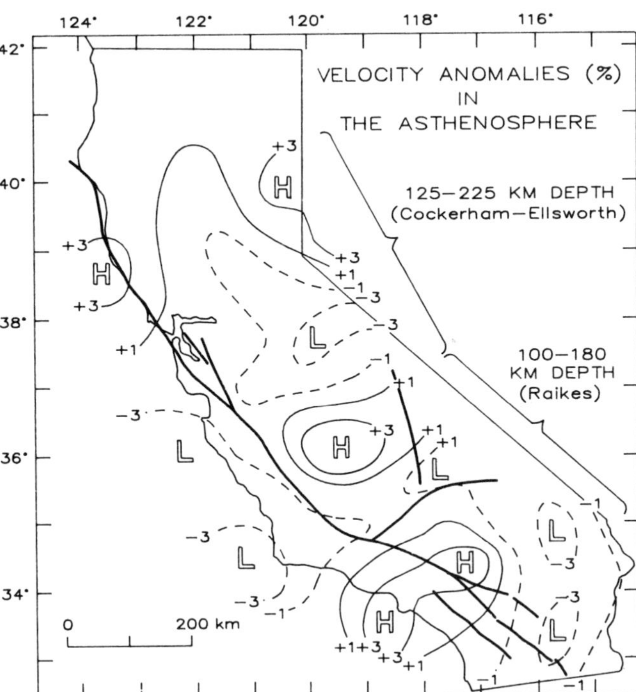

Figure 16. Composite P-velocity variation in the depth range 125 to 225 km in California, constructed by combining the data of Cockerham and Ellsworth (1979) for central California and of Raikes (1980) for southern California. Contour interval = 2 percent (after Aki, 1982).

(1982). In a comprehensive tomographic model of the upper-mantle P-wave velocity structure of southern California, Humphreys and others (1984) confirmed the two major anomalies in the upper mantle found by the earlier investigators, namely, the high-velocity anomaly across the Transverse Ranges and the low-velocity anomaly in the Salton Trough (Fig. 17). The tomographic imaging enabled sharp definition of the anomalies. The Transverse Ranges anomaly was found to be nearly vertical and slab-like with a velocity contrast of about 3 percent from the surrounding mantle. It was about 200 km wide and was deepest on the eastern side, reaching a depth of 250 km.

An intriguing feature of the high-velocity anomaly across the Transverse Ranges in the upper mantle is that it is continuous across the San Andreas fault. Therefore, an interpretation is possible only if it is assumed that the upper mantle below the fault readjusts periodically by flow sustained by temperature and pressure effects, thus compensating for the offset caused by progressive displacement across the fault. Alternatively, in order to accommodate the decoupling of the plate motion of the lithosphere from the asthenosphere, the plate boundary at depth could diverge from the surface expression of the San Andreas fault and could be located to the east of the high-velocity region where the low-velocity material in the upper mantle could accommodate the shear. Hadley and Kanamori (1977) preferred this latter model. Humphreys and others (1984), however, offered two other explanations for the anomaly: first, that it could be a result

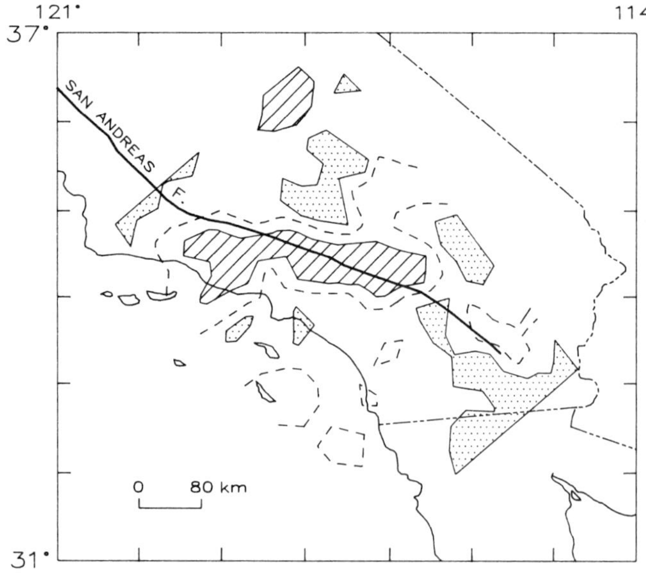

Figure 17. Results of tomographic inversion of P-residuals in southern California (modified from Humphreys and others, 1984). Horizontal section at a depth of 100 km. Contour interval = 1.4 percent, with >−1.5 percent indicated by dotted areas and <−1.5 percent by hatched areas. The zero contour is dashed.

of mini-subduction of the Pacific plate in the void associated with the offset in the San Andreas fault; second, that it could be due to downwelling of the lithosphere caused by a thermal instability.

The Salton Trough low-velocity anomaly (Fig. 17) extends to a depth of 125 km with possibly a thin extension to greater depths and is directly related to rifting. Humphreys and others (1984) suggested that the upwelling in this area probably reinforces the downwelling beneath the Transverse Ranges.

The Cascades. Teleseismic-residual modeling of the heterogeneous structure of the upper mantle in the Cascade Range region has begun to yield useful results on the nature of subduction in this region. Unlike many other convergent margins where the subducting lithosphere can be delineated by mapping hypocenters of deep earthquakes, the subduction of the Juan de Fuca plate beneath the Cascades is aseismic in California and Oregon, and there exists only shallow seismicity in Washington in the presumed subduction zone. Preliminary results from data collected by seismic networks in Oregon by the U.S. Geological Survey show an eastward-dipping high-velocity slab beneath the Oregon Cascades (Iyer and others, 1982; Berge, 1985).

Better information is available from modeling teleseismic residuals in Washington and northern Oregon. McKenzie and Julian (1971) found arrivals from an earthquake in Seattle at a few worldwide stations to be anomalously early and postulated that these could be due to passage of the rays through a 50° east-dipping high-velocity slab. Lin (1973) and Rohay (1982) found teleseismic residuals at stations in the Cascade Range to be negative relative to stations in western Washington. In a detailed study of teleseismic arrivals recorded by a dense seismic network in Washington and northern Oregon, Michaelson and Weaver (1986) modeled the crust and upper mantle in the region to depth of 300 km. The main feature of their three-dimensional upper-mantle tomographic model is a roughly north-south–striking zone of high velocity which shifts progressively northeastward with increasing depth. Michaelson and Weaver (1986) interpreted the high-velocity zone to be the manifestation of the subducting Juan de Fuca plate. In conjunction with forward modeling using ray-tracing methods, they find the slab to be segmented into three sections.

Basin and Range province. Koizumi and others (1973) found the teleseismic residuals at several northern Nevada seismic stations to be negative (early arrivals) with respect to stations in southern Nevada for teleseisms in the southeast azimuth. Based on these observations and the azimuthal variation patterns at some of the other stations, they postulated the presence of "a high-velocity lithospheric plate, striking northeast and dipping southeast under northern Nevada." Iyer and others (1977), in a study of P-wave residuals in the Battle Mountain Heat Flow High region in Nevada, found a similar high-velocity anomaly in the upper mantle, which probably corresponds to the same body found by Koizumi and others (1973). Solomon and Butler (1974) found evidence for a high-velocity body beneath the western edge of the Basin and Range province. It is not clear whether this body is the same as that found by Koizumi and others (1973). However, it would appear that both the high-velocity bodies are related to the broad high-velocity region found by Romanowiscz (1979) in the depth range 250 to 450 km in the western United States (Fig. 14).

The heterogeneity of the upper mantle in the central Basin and Range province is demonstrated by other P-wave residual studies. Monfort and Evans (1982) computed a three-dimensional model of a 150-km-diameter region around the Nevada Test Site and found complex bands of high and low velocities in the upper mantle, which coincided with regional geologic trends. Spence (1974) used observations of P-wave residuals at world-wide seismic stations for different locations of nuclear explosions in the northern section of the Nevada Test Site to delineate a large high-velocity anomaly in the crust and upper mantle beneath the Silent Canyon volcanic center in Nevada. He interpreted the high-velocity body to be the remnant of depleted mantle lavas that had accumulated in the region during volcanic activity between 14 and 1 Ma.

Rio Grande Rift. Parker and others (1984) and Davis and others (1984) modeled the deep structure of the Rio Grande Rift region using teleseismic residuals from a 20-station, 1,000-km-long northwesterly profile of seismographs centered on the Rio Grande Rift. The relative P-wave traveltime residuals across the profile increased toward the center from both ends, reaching a maximum value of about +1.5 sec. The data were modeled in terms of an upwarp of the lithospheric-asthenospheric boundary by projecting the residuals downward and by block inversion (Fig. 18). The downward-projection model showed the litho-

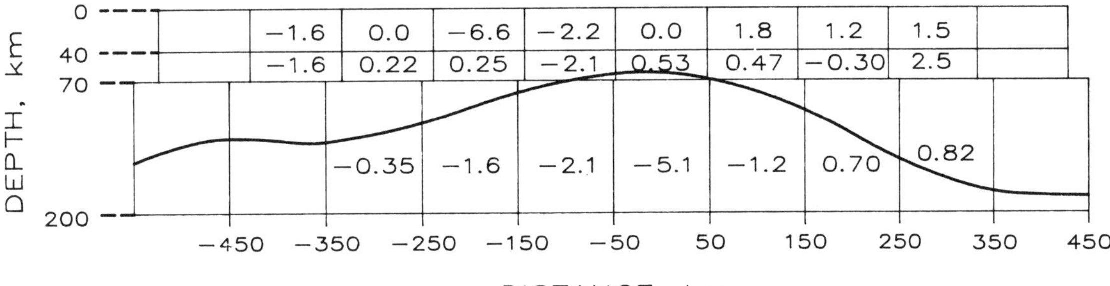

Figure 18. Variation of lithospheric thickness beneath the Rio Grande Rift region derived using P-residuals. The inferred depth to the lithosphere-asthenosphere boundary using forward modeling is shown by the solid line. The surface expression of the rift is approximately symmetrical with the center of the figure. Block inversion showing velocity perturbation in percent is also shown. Modified from Parker and others (1984).

spheric thickness to be about 70 km beneath the Rio Grande Rift, deepening to 150 to 200 km at both ends. The best-fitting velocity contrast between the lithosphere and asthenosphere was about 8 percent. The block inversion model showed a low-velocity anomaly with 5 percent velocity contrast in the depth range of 70 to 200 km beneath the rift. Both models fit the Bouguer gravity anomaly from west to east across the rift, a broad low centered on the rift with a sharp decrease superposed in the region corresponding to the surface expression of the rift. Previous seismic studies have shown a thinning of the crust beneath the Rio Grande Rift and low P_n velocities compared with the Colorado Plateau and the Great Plains (Keller and others, 1979; Olsen and others, 1979). These observations, together with the high observed seismic attenuation, high heat flow, uplift, and the presence of rifting and volcanism indicate that the low-velocity anomaly under the rift, delineated using teleseismic residuals, could be the manifestation of the upwarping of a hot, low-density asthenosphere caused by lithospheric extension or active convection (Davis and others, 1984).

Yellowstone Plateau and Eastern Snake River Plain. The eastern Snake River Plain–Yellowstone Plateau volcanic system stands out as a prominent feature in the geologic, tectonic, and physiographic maps of the western United States, and is believed by many to be the manifestation of a hot spot whose present location is the Yellowstone Plateau. The deep structure of the Yellowstone Plateau has been probed using teleseismic P-wave residuals by Zandt (1978), Iyer and others (1981), and Iyer (1984). Evans (1982) modeled the two-dimensional structure of the crust and upper mantle across the eastern Snake River Plain. The upper-mantle model for the Yellowstone Plateau reveals the presence of a large low-velocity body extending from the crust to at least 190 km depth. The velocity contrast inside the body is 10 to 15 percent in the upper 20 km of the crust directly beneath the caldera and decreases to 2 to 4 percent at 190 km, the maximum depth of modeling reached by the inversion. The P-wave velocity perturbation model for the eastern Snake River Plain also shows a large low-velocity body, but only in the upper mantle (Evans, 1982). The Yellowstone Plateau and eastern Snake River Plain models are compared in Figure 19. Iyer and others (1981) examined in detail various mechanisms of P-wave velocity reduction in crustal and mantle materials and concluded that partial melting is a plausible mechanism to explain the observed low-velocity anomalies. In comparing the Yellowstone Plateau and eastern Snake River Plain models, Evans (1982) explained the absence of significant low-velocity anomalies in the eastern Snake River Plain crust and the similarity between both the models in the upper mantle as the effect of the aging process of the volcanic system. S-wave velocity models for the eastern Snake River Plain computed by Greensfelder and Kovach (1982) using Rayleigh-wave dispersion and body-wave residual data also suggest a healed crust and partially molten mantle beneath the region. If the volcanism associated with the Yellowstone Plateau–eastern Snake River Plain volcanic system is due to a migrating volcanic center whose current location is the Yellowstone Plateau, the inference is that the causative magma body is cooling from the top downward.

A crucial test for the presence of partial melt in a volume of rock is a drastic reduction in S-wave velocity. Daniel and Boore (1982) modeled the S-wave velocity structure beneath the Yellowstone Plateau using Rayleigh-wave phase velocities and S-wave residuals. Their final model shows S-wave velocities beneath the plateau to be lower than in the surroundings by 14 to 29 percent in the upper crust (0 to 20 km), 11 to 22 percent in the lower crust (20 to 40 km), and 12 percent in the upper mantle (Fig. 11).

Eastern and central United States. Fletcher and others (1978) mapped traveltime residuals in parts of the eastern and central United States and southeastern Canada using an extensive seismic network. They found a complex pattern of spatial variation of residuals which could be related to surface geology and tectonics of the region. In a more quantitative modeling of the region, Taylor and Toksöz (1979) inverted teleseismic P-wave residuals from the Northeastern U.S. Seismic Network. In the 35- to 200-km depth range, their model showed a northeast-trending region of low velocities extending through Massachusetts, New Hampshire, and southern Maine, with the velocities increasing to

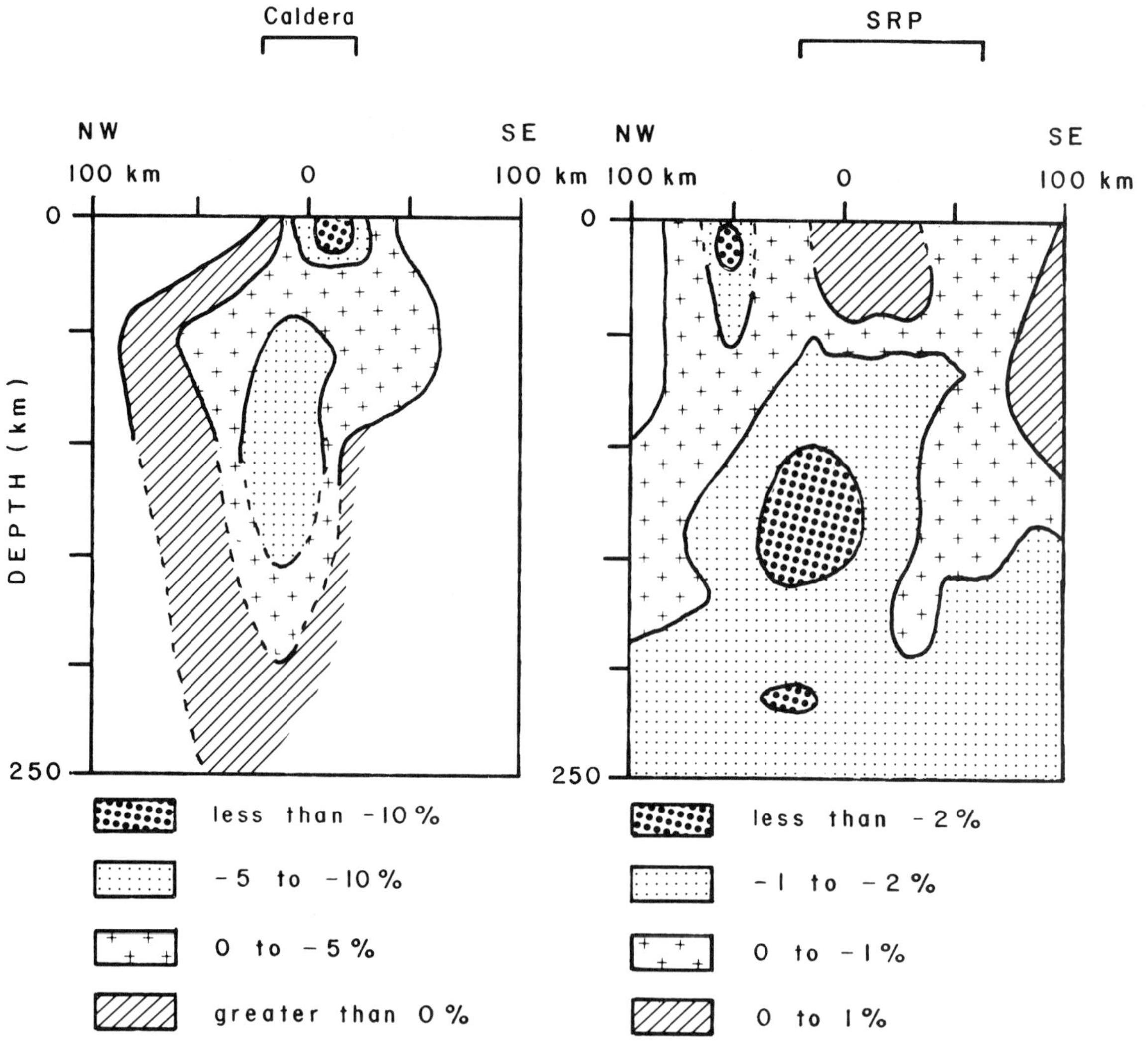

Figure 19. Schematic northwest southeast cross section of the three-dimensional P-wave velocity variation model for the eastern Snake River Plain–Yellowstone Plateau region. Left, Yellowstone Plateau, after Iyer and others, 1981. Right, Eastern Snake River Plain, after Evans (1982).

the west and northwest in the Precambrian Grenville Province (Fig. 20). The inference is that surface structural trends persist into the upper mantle. In the 200- to 350-km depth range, the low-velocity region shifts to the west and northwest and occurs beneath Vermont and central Maine. The important finding from this study is that in the major orogenic belts, although the lateral velocity variations are smaller than in tectonically active regions, there are perturbations that reach well into the upper mantle. Specifically, the high velocities are associated with the Precambrian Grenville Province and low velocities with the Bronson Hill–Boundary Mountains Anticlinorium. Taylor and Toksöz (1979) proposed that the latter may be the result of post-Acadian radioactive heating of the lithosphere that subducted in the region.

The heterogeneous structure of the deep crust and upper mantle around the New Madrid seismic zone located in the northern Mississippi embayment has been delineated by Mitchell and others (1977) and Al-Shukri and Mitchell (1987) using teleseismic residuals. Mitchell and others (1977) constructed a three-dimensional P-wave velocity model of a 300-km by 350-km zone around the New Madrid seismic zone and found a low-velocity zone extending from the upper crust to a depth of 150 km beneath most of the actively seismic area. In a study with higher resolution using a comprehensive data set, Al-Shukri and Mitchell (1987) delineated a band of low velocities in the lower crust and upper mantle to a depth of 156 km. The low-velocity zone correlated well with the position and orientation of the basement rift system. The lowest velocities in the upper mantle coincided with the area of highest seismic activity in the New Madrid seismic zone.

Aki and others (1976) modeled the three-dimensional lithospheric structure in the region of the Large Aperture Seismic

Array (LASA) in Montana by inversion of teleseismic P-wave residual data. The most conspicuous feature of the model is a low-velocity anomaly in the central and northeast part of the array extending from the upper crust into the upper mantle to depths greater than 100 km. Aki and others (1976) interpreted thi anomaly to be due to wrench faulting and shearing, which might have occurred in this region during the Late Jurassic–Early Cretaceous Nevadan orogeny and Upper Cretaceous to Tertiary Laramide orogeny.

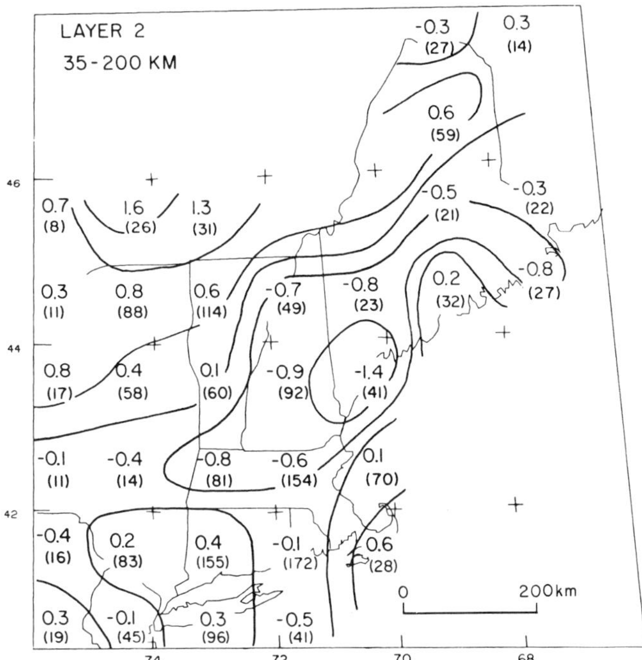

Figure 20. Layer 2 (35 to 200 km) of the three-dimensional P-velocity variation model for northeastern U.S. (after Taylor and Toksöz, 1979). The numbers are velocity perturbations in percent. Numbers in parenthesis indicate the number of rays used for the particular block. Contour interval = 0.5 percent. State boundaries are indicated.

CONCLUSIONS

The one-dimensional P-wave and S-wave velocity models discussed in the preceding pages clearly demonstrate the broad division of the upper mantle of North America into two typical structures, the tectonically active western structure and the stable central and eastern structure. Differences are apparent even within each of the tectonic provinces, but are not well constrained by the one-dimensional models. The two- and three-dimensional models derived using teleseismic residuals, however, show that even within a region having dimensions of a few tens to a few hundreds of kilometers, lateral variations in velocity structure with magnitudes as high as the difference between the western and eastern models may exist in the upper mantle. The survey of seismological literature for the past decade shows that, in spite of rapid improvements in modeling techniques, our advancement in the understanding of the upper-mantle structure of North America has not been progressing at the same rate as in the 1960s. Apart from a broad understanding of structural differences within the continent and heterogeneities associated with specific regions, we have very little new information on the finer scale upper-mantle structure of North America, either horizontally or vertically. A specific instance of this inadequacy is that we are unable even to map variations in thickness of the lithosphere in North America determined by seismology with any degree of confidence. We are unable to contour important parameters associated with the low-velocity layer (lid thickness and velocity, and low-velocity layer thickness and velocity) within the Basin and Range province.

We see the problem as the lack of data collection capabilities to exploit fully the advance in analysis methodologies. Whereas there is considerable national interest and support to study in great detail the crust of North America by seismic reflection, seismic refraction, and deep drilling, upper-mantle studies are not attracting as much academic interest. However, the emergence of new programs like IRIS are quite encouraging. A detailed seismological program using broad-band instruments to model the upper mantle of North America with a horizontal and vertical resolution of about 20 km is a desirable objective to understand the evolution of this continent. Such a program naturally will utilize the power of all modern analysis techniques involving body-wave and surface-wave data.

ACKNOWLEDGMENTS

I am grateful to Walter Mooney, Dave Hill, Bruce Julian, Don Helmberger, and Brian Mitchell for patiently and critically reviewing the earlier, bulkier version of this manuscript. I am indebted to the numerous researchers whose work forms the substance of this review. Finally, I dedicate this work to Dr. Otto Nuttli, whose pioneering work has provided a firm foundation for our understanding of the upper-mantle structure of North America.

**TABLE 1. SUMMARY:
P-WAVE MODELS, AVERAGE, SHIELD, AND STABLE PLATFORM**

Source	P_N Velocity	Low-Velocity Layer Velocity (km/sec)	Low-Velocity Layer Thickness (km)	Velocity at 100 Km (km/sec)	Velocity at 200 Km (km/sec)	Lithospheric Thickness (km)	Depth to 400 Km Discontinuity (km)
Jeffreys–Bullen (Bullen, 1965)	7.78	....	(.......)	7.92-7.95	8.26-8.35	....	....
Iyer and others (1969) (NCER1)	8.10	....	(.......)	8.27	8.42	....	480
Massé (1973)	8.06	7.98	(13)	7.98	8.43	94	425
Roller and Jackson (1966) (not in text)	8.20	....	(.......)	8.20	8.50	....	......
Iyer and others (1969) (NCER4)	8.10	....	(.......)	8.30	8.50	....	440
Lewis and Meyer (1968)	8.08	8.43	(80)	8.20	8.46	130?	449
Green and Hales (1968) (ER2)	8.02	8.25	(134)	8.34	8.39	134	362-382
Mereu and Hunter (1969)	8.05	....	(.......)	8.43	....	....	....
Barr (1967)	8.23	....	(.......)	8.23	....	....	366
Wiechert (1968)	8.65?	8.97	(45)	8.9?	8.9?	185	440
Lehman (1962 (Logan–Blanca)	8.00	7.90	(65)	8.00	7.90	150	....
Lehman (1964) (Gnome-Northeast)	7.95	7.95	(65)	8.30	7.95	150	....
Dowling and Nuttli (1964) (Bilby NE)	7.70	7.90	(116?)	8.10	7.90	180	....
Dowling and Nuttli (1964) (Gnome NE)	8.26	8.00	(126)	8.26	8.00	126	....
Burdick (1981) (S8)	8.10	....	(.......)	8.25	8.34	....	400

TABLE 2. SUMMARY:
S-WAVE MODELS, AVERAGE, SHIELD, AND STABLE PLATFORM

SOURCE	S_N VELOCITY	LOW-VELOCITY LAYER VELOCITY (KM/SEC)	LOW-VELOCITY LAYER THICKNESS (KM)	VELOCITY AT 100 KM (KM/SEC)	VELOCITY AT 200 KM (KM/SEC)	LITHOSPHERIC THICKNESS (KM)	DEPTH TO 400 KM DISCONTINUITY (KM)
Ibrahim and Nuttli (1967)	4.60	4.00	(54?)	4.60	4.60	180	371-391
Anderson and Julian (1969) (US26)	4.59	4.40	(85)	4.60	4.60	65	400
McEvilly (1964) (R and L)	4.47-4.83	4.45-4.80	(180)	4.45-4.80	4.45-4.80	82	...
(Average)	4.75	4.45	(180)	4.61	4.45	102	...
Brune and Dorman (1963) (CANSD)	4.72	4.51	(200)	4.72	4.54	115	...
Lewis and Meyer (1968)	4.72	3.90	(84)	4.72	4.56	126	450
Mitchell and Herrmann (1979) (R and L)	4.80	4.65	(40)	4.70	4.80	65	...
(R)	4.55	4.50	(....)	4.50	4.65	60	...
(L)	4.70	4.65	(45)	4.65	4.70	65	...
Grand and Helmberger (1984) (SNA)	4.80	4.63	(175)	4.78	4.63	175	406

TABLE 3. SUMMARY:
P–WAVE MODELS; TECTONIC MODEL

Source	P_N Velocity	Low-Velocity Layer Velocity km/sec	Low-Velocity Layer Thickness (km)	Velocity at 100 Km (km/sec)	Velocity at 200 Km (km/sec)	Lithospheric Thickness (km)	Depth to 400 Km Discontinuity (km)
Gutenberg (1959)	8.17	7.85	(200)	8.02	7.98	50	....
	8.08	7.83	(140)	7.83	8.12	50	....
Lehman (1964) (NTS-Gnome)	7.95	7.95	(65)	8.30	7.95	150	....
Dowling and Nuttli (1964) (Bilby-SE)	7.95	7.80	(47)	7.80	8.20	40	....
Dowling and Nuttli (1964) (Gnome-NW)	7.90	7.80	(.......)	7.82	8.25	80	....
Archambeau and others (1969) (CIT 111P)	7.72	7.72	(120?)	7.72	8.36	28?	....
(CIT 110P)	7.98	7.66	(70)	7.68	8.38	60-70	....
(CIT 112P)	7.98	7.70	(59)	7.71	8.37	60-70	....
(CIT 109P)	8.05	7.83	(70)	7.87	8.43	60-70	....
Niazi and Anderson (1965) (Z) (not in text)	8.08	7.83	(170)	7.83	8.32	50?	....
Johnson (1967) (CIT 204)	7.88	7.65	(110)	7.70	8.31	60	420
Massé (1972)	7.80	7.70	(116)	7.70	8.36	49	422
Helmberger and Wiggins (1971) (HWNE)	8.01	7.58	(97)	7.68	8.55	84	432
Wiggins and Helmberger (1973) (HWA)	7.87?	7.87?	(90?)	7.90	8.08	146?	427
(HWB)	7.90	7.90?	(186?)	7.94	8.06	186?	437
Burdick and Helmberger (1978) (T7)	7.95	7.70	(91)	7.70	8.30	65	400
(T9)	7.95	7.70	(91)	7.70	8.30	65	400
Dey-Sarker and Wiggins (1976) (WCA)	8.17	7.90	(64)	7.97	8.35	92	410
(WCB)	8.17	8.01	(86)	8.37	8.02	121	409
(WCC)	8.07	7.80	(165)	7.90	8.21	62	409
Hill (1972) (CP1)	7.91	7.50	(32)	8.28	8.48	100	....
(CP2)	7.91	7.66	(72)	7.75	8.34	60	....

TABLE 3. SUMMARY:
P–WAVE MODELS; TECTONIC MODEL (continued)

SOURCE	P_N VELOCITY	LOW-VELOCITY LAYER VELOCITY KM/SEC	LOW-VELOCITY LAYER THICKNESS (KM)	VELOCITY AT 100 KM (KM/SEC)	VELOCITY AT 200 KM (KM/SEC)	LITHOSPHERIC THICKNESS (KM)	DEPTH TO 400 KM DISCONTINUITY (KM)
Walck (1984)	7.90	7.70	(80)	7.90	8.05	20	390
(GCA)							
(GCA')	7.00?	....	....	7.87	8.05	20?	390
Walck (1985)	7.90	7.50	(36)	7.95	8.22	46	410
(CJF)							

TABLE 4. SUMMARY:
S–WAVE MODELS; TECTONIC MODEL

SOURCE	S_N VELOCITY	LOW-VELOCITY LAYER VELOCITY KM/SEC	LOW-VELOCITY LAYER THICKNESS (KM)	VELOCITY AT 100 KM (KM/SEC)	VELOCITY AT 200 KM (KM/SEC)	LITHOSPHERIC THICKNESS (KM)	DEPTH TO 400 KM DISCONTINUITY (KM)
Gutenberg (1959)	4.65	4.25	(250)	4.41	4.38	50	....
Kovach and Robinson (1969)	4.50	4.40	(118)	4.40	4.50	42	425
Baumgardt and Alexander (1984)	4.60	4.47	(59)	4.60	4.47	171	....
Helmberger and Engen (1974)	4.45	4.30	(50?)	4.43	4.50	50	400
Grand and Helmberger (1984) (TNA)	4.40	4.29	(.......)	4.29	4.39	38	406
Kovach and Press (1961)	4.30?	4.30	(173)	4.30	4.30	45	....
Bucher and Smith (1971) (BC3)	4.17	4.17	(45)	4.38	....	33	....
(C7)	4.25	4.25	(45)	4.28	....	41	....
Priestley and Brune (1978) (GB)	4.50	4.05	(116)	4.12	4.38	64	....
(NGB) (1982)	4.50	4.25	(120)	4.30	4.38	60	....
(High Lava Plateau) (1982)	4.30	4.05	(130)	4.10	4.38	50	....
Daniel and Boore (1982)	4.11	4.10	(.......)	4.10?	....	55?	....
Greensfelder and Kovach (1982)	4.07-4.21	4.10	(.......)	4.10+	....	71?	....

REFERENCES CITED

Aki, K., 1982, Three-dimensional seismic inhomogeneities in the lithosphere and asthenosphere; Evidence for decoupling in the lithosphere and flow in the asthenosphere: Reviews of Geophysics and Space Physics, v. 20, p. 161–170.

Aki, K., Christoffersson, A., and Husebye, E. S., 1976, Three-dimensional seismic structure of the lithsphere under Montana LASA: Seismological Society of America Bulletin, v. 66, p. 501–524.

——, 1977, Determination of the three-dimensional seismic structure of the lithosphere: Journal of Geophysical Research, v. 82, p. 277–296.

Al-Shukri, H. J., and Mitchell, B. J., 1987, Three dimensional velocity variations and their relation to the structure and tectonic evolution of the New Madrid seismic zone, Journal of Geophysical Research, v. 97, p. 6377–6390.

Anderson, D. L., 1966, Recent evidence concerning the structure and composition of the earth's mantle, *in* Ahrens, L. H., Press, F., Runcorn, S. K., and Urey, H. C., eds., Physics and chemistry of the earth, v. 6: New York, Pergamon Press, p. 1–131.

——, 1967, Latest information from seismic observations, *in* Gaskell, T. F., ed., The Earth's mantle: New York, Academic Press, p. 355–420.

Anderson, D. L., and Julian, B. R., 1969, Shear velocities and elastic parameters of the mantle: Journal of Geophysical Research, v. 74, p. 3281–3286.

Archambeau, C. B., Flinn, E. A., and Lambert, D. G., 1969, Fine structure of the Upper Mantle: Journal of Geophysical Research, v. 74, p. 5825–5865.

Arkhangel'skaya, J. M., 1960, Dispersion of surface waves and crustal structure: Izvestia Akademy Nauk U.S.S.R., Geophysical Series no. 9, p. 904–927.

Backus, G., and Gilbert, F., 1970, Uniqueness in the inversion of gross earth data: Philosophical Transactions of the Royal Society of London, series A, v. 266, p. 123–192.

Barr, K. G., 1967, Upper mantle structure in Canada from seismic observations using chemical explosions: Canadian Journal of Earth Sciences, v. 4, p. 961–975.

Baumgardt, D. R., and Alexander, S. S., 1984, Structure of the mantle beneath Montana LASA from analysis of long-period, mode-converted phases: Seismological Society of America Bulletin, v. 74, p. 1683–1702.

Berge, P. A., 1985, Teleseismic P-wave traveltime residuals across the Cascade Range in southern Oregon: U.S. Geological Survey Open-File Report 85-622, 104 p.

Bessonova, E. N., Fishman, V. M., Ryaboyi, V. Z., and Sitnikova, G. A., 1974, The tau method for inversion of travel times; 1, Deep seismic sounding data: Royal Astronomical Society Geophysical Journal, v. 36, p. 377–398.

Bessonova, E. N., Fishman, V. M., Shnirman, M. G., Sitnikova, G. A., and Johnson, L. R., 1976, The tau method of inversion of travel times; 2, Earthquake data: Royal Astronomical Society Geophysical Journal, v. 46, p. 87–108.

Biswas, N. N., 1971, The upper mantle structure of the United States from the dispersion of surface waves [Ph.D. thesis]: Los Angeles, University of California, 175 p.

Biswas, N. N., and Knopoff, L., 1974, The structure of the upper mantle under the United States from the dispersion of Rayleigh waves: Royal Astronomical Society Geophysical Journal, v. 36, p. 515–539.

Bolt, B. A., 1964, Recent information on the earth's interior from studies of mantle waves and eigenvibrations, *in* Ahrens, L. H., Press, F., Runcorn, S. K., and Urey, H. C., eds., Physics and chemistry of the earth, v. 5: New York, Pergamon Press, p. 55–119.

Bolt, B. A., and Nuttli, O. W., 1966, P-wave residuals as a function of azimuth: Journal of Geophysical Research, v. 71, p. 5977–5985.

Brune, J., and Dorman, J., 1963, Seismic waves and earth structure in the Canadian shield: Seismological Society of America Bulletin, v. 53, p. 167–210.

Bucher, R. L., and Smith, R. B., 1971, Crustal structure of the eastern Basin and Range province and the northern Colorado Plateau from phase velocities of Rayleigh waves, *in* Heacock, J. G., ed., The structure and physical properties of the Earth's crust; American Geophysical Union Geophysical Monograph 14, p. 59–70.

Bullen, K. E., 1965, An introduction to the theory of seismology: Cambridge, Cambridge University Press, 381 p.

Burdick, L. J., 1981, A comparison of the upper mantle structure beneath North America and Europe: Journal of Geophysical Research, v. 86, p. 5926–5936.

Burdick, L. J., and Helmberger, D. V., 1978, The upper mantle P-velocity structure of the western United States: Journal of Geophysical Research, v. 83, p. 1699–1712.

Byerly, P., 1926, The Montana earthquake of June 28, 1925: Seismological Society of America Bulletin, v. 16, p. 209–265.

Cara, M., 1979, Lateral variations of S-velocity in the upper mantle from higher Rayleigh modes: Royal Astronomical Society Geophysical Journal, v. 57, p. 649–670.

Carder, D. S., Gordon, D. W., and Jordan, J. N., 1966, Analysis of surface-foci travel times: Seismological Society of America Bulletin, v. 56, p. 815–840.

Catchings, R. D., and Mooney, W. D., 1988, Crustal structure of the Columbia Plateau; Evidence for continental rifting: Journal of Geophysical Research, v. 93, p. 459–474.

Cleary, J., and Hales, A. L., 1966, An analysis of the travel times of P waves to North American stations, in the distance range 32° to 100°: Seismological Society of America Bulletin, v. 56, p. 467–489.

Cockerham, R. S., and Ellsworth, W. L., 1979, Three-dimensional large-scale mantle structure in central California [abs.]: EOS Transactions of the American Geophysical Union, v. 60, p. 875.

Crough, S. T., and Thompson, G. A., 1977, Upper mantle origin of Sierra Nevada uplift, Geology, v. 5, p. 396–399.

Daniel, R. G., and Boore, D. M., 1982, Anomalous shear wave delays and surface wave velocities at Yellowstone caldera, Wyoming: Journal of Geophysical Research, v. 87, p. 2731–2744.

Davis, P. M., Parker, E. C., Evans, J. R., Iyer, H. M., and Olsen, K. H., 1984, Teleseismic deep sounding of the velocity structure beneath the Rio Grande Rift, *in* 35th Field Conference Guidebook, Rio Grande Rift, northern New Mexico: New Mexico Geological Society, p. 29–38.

Dey-Sarkar, S. K., and Wiggins, R. A., 1976, Upper mantle structure in western Canada: Journal of Geophysical Research, v. 81, p. 3619–3632.

Dickinson, W. R., and Snyder, W. S., 1979, Geometry of subducted slabs related to the San Andreas transform: Geology, v. 87, p. 609–627.

Dorman, J., 1969, Seismic surface-wave data on the upper mantle, *in* Hart, P. J., ed., The Earth's crust and upper mantle: American Geophysical Union Geophysical Monograph 13, p. 257–265.

Dowling, J., and Nuttli, O., 1964, Travel-time curves for a low-velocity channel in the upper mantle: Seismological Society of America Bulletin, v. 54, p. 1981–1996.

Doyle, H. A., and Hales, A. L., 1967, An analysis of the travel times of S waves to North American stations in the distance range of 28° to 82°: Seismological Society of America Bulletin, v. 57, p. 761–771.

Dziewonski, A. M., and Anderson, D. L., 1983, Travel times and station corrections for P waves at teleseismic distances: Journal of Geophysical Research, v. 88, p. 3295–3314.

Evans, J. R., 1982, Compressional wave velocity structure of the upper 350 km under the eastern Snake River Plain near Rexburg, Idaho: Journal of Geophysical Research, v. 87, p. 2654–2670.

Ewing, M., Jardetzsky, W. S., and Press, F., 1957, Elastic waves in layered media: New York: McGraw-Hill, 380 p.

Fletcher, J. B., Sbar, M. L., and Sykes, L. R., 1978, Seismic trends and travel-time residuals in eastern North America and their tectonic implications: Geological Society of America Bulletin, v. 89, p. 1656–1676.

Grand, S. P., 1987, Tomographic inversion for shear velocity beneath the North American plate: Journal of Geophysical Research, v. 92, p. 14065–14090.

Grand, S. P., and Helmberger, D. V., 1984, Upper mantle shear structure of North America: Royal Astronomical Society Geophysical Journal, v. 76, p. 399–438.

Green, R.W.E., and Hales, A. L., 1968, The travel times of P waves to 30° in the central United States and upper mantle structure: Seismological Society of America Bulletin, v. 58, p. 267–289.

Greensfelder, R. W., and Kovach, R. L., 1982, Shear wave velocities and crustal structure of the eastern Snake River Plain, Idaho: Journal of Geophysical Research, v. 87, p. 2643–2653.

Gutenberg, B., 1959, Physics of the Earth's interior: New York, Academic Press, p. 13–19, 75–99.

Hadley, D., and Kanamori, H., 1977, Seismic structure of the Transverse Ranges, California: Seismological Society of America Bulletin, v. 88, p. 1469–1478.

Hales, A. L., and Doyle, H. A., 1967, P- and S-traveltime anomalies and their interpretation: Royal Astronomical Society Geophysical Journal, v. 13, p. 403–415.

Hales, A. L., and Herrin, E., 1972, Travel times of seismic waves, *in* Robertson, E. C., ed., The nature of the solid Earth; U.S.A.: New York, McGraw-Hill, p. 172–215.

Helmberger, D. V., and Engen, G. R., 1974, Upper mantle shear structure: Journal of Geophysical Research, v. 79, p. 4017–4028.

Helmberger, D., and Wiggins, R. A., 1971, Upper mantle structure in midwestern United States: Journal of Geophysical Research, v. 76, p. 3229–3245.

Helmberger, D. V., Engen, G., and Grand, S., 1985, Upper-mantle cross section from California to Greenland: Journal of Geophysical Research, v. 58, p. 92–100.

Herrin, E., 1969, Regional variations of P-wave velocity in the upper mantle beneath North America: American Geophysical Union Monograph 13, p. 242–246.

Herrin, E., and Taggart, J., 1962, Regional variations in P_n velocity and their effect on the location of epicenters: Seismological Society of America Bulletin, v. 52, p. 1035–1046.

——, 1968, Regional variation in P travel times: Seismological Society of America Bulletin, v. 58, p. 1325–1337.

Hill, D. P., 1972, Crustal and upper mantle structure of the Columbia Plateau from long-range seismic-refraction measurements: Geological Society of America Bulletin, v. 83, p. 1639–1648.

Humphreys, E., Clayton, R. W., and Hager, B. H., 1984, A tomographic image of mantle structure beneath southern California: Geophysical Research Letters, v. 11, p. 625–627.

Husebye, E. S., Christoffersson, A., Aki, K., and Powell, C., 1976, Preliminary results on the 3-dimensional seismic structure of the lithosphere under the USGS central California seismic array: Royal Astronomical Society Geophysical Journal, v. 46, p. 319–340.

Ibrahim, A. B., and Nuttli, O. W., 1967, Travel-time curves and upper-mantle structure from long period S-waves: Seismological Society of America Bulletin, v. 57, p. 1063–1092.

Iyer, H. M., 1984, A review of crust and upper mantle structure studies of the Snake River Plain–Yellowstone volcanic system; A major lithospheric anomaly in western U.S.A.: Tectonophysics, v. 105, p. 291–308.

Iyer, H. M., Pakiser, L. C., Stuart, D. J., and Warren, D. H., 1969, Project Early Rise; Seismic probing of the upper mantle: Journal of Geophysical Research, v. 74, p. 4409–4441.

Iyer, H. M., Roloff, J. N., and Coakley, J. M., 1977, P-wave residual measurements in the Battle Mountain Heat Flow High, Nevada [abs.]: EOS Transactions of the American Geophysical Union, v. 58, p. 1238.

Iyer, H. M., Evans, J. R., Zandt, G., Stewart, R. M., Coakley, J. M., Roloff, J. N., 1981, A deep low-velocity body under the Yellowstone caldera, Wyoming; Delineation using teleseismic P-wave residuals and tectonic interpretation; Part 1, Summary: Geological Society of America Bulletin, v. 92, p. 792–798, part 2, p. 1471–1644.

Iyer, H. M., Rite, A., and Green, S. M., 1982, Search for geothermal heat sources in the Oregon Cascades by means of teleseismic P-residual technique, *in* 52nd Annual International Meeting and Exposition Technical Programs, Abstracts, and Biographies: Tulsa, Oklahoma, Society of Exploration Geophysicists, p. 479–482.

Jeffreys, H., 1939, The times of P, S, and SKS and the velocities of P and S: Monthly Notices of the Royal Astronomical Society, Geophysical Supplement, v. 4, p. 498–533.

Jeffreys, H., and Bullen, K. E., 1940, Seismological tables: British Association, Edinburgh, Gray Milne Trust, 50 p.

Johnson, L. R., 1967, Array measurements of P velocities in the upper mantle: Journal of Geophysical Research, v. 72, p. 6309–6325.

Keilis-Borok, V. I., and Yanovskaya, T. B., 1967, Inverse problems of seismology (structural review): Royal Astronomical Society Geophysical Journal, v. 13, p. 223–234.

Keller, G. R., Braile, L. W., and Schlue, J. W., 1979, Regional crustal structure of the Rio Grande Rift from surface wave dispersion measurements, *in* Riecker, R. E., ed., Rio Grande Rift tectonics and magmatism: American Geophysical Union, p. 115–126.

Knopoff, L., 1972, Observation and inversion of surface wave dispersion: Tectonophysics, v. 13, p. 497–519.

——, 1983, The thickness of the lithosphere from the dispersion of surface waves: Royal Astronomical Society Geophysical Journal,v. 74, p. 55–81.

Knopoff, L., and Schlue, J., 1972, Rayleigh wave phase velocities for the path Addis Ababa–Nairobi: Tectonophysics, v. 15, p. 157–163.

Knopoff, L., Schlue, J. W., and Schwab, F. A., 1970, Phase velocities of Rayleigh waves across the East Pacific Rise: Tectonophysics, v. 10, p. 321–336.

Koizumi, C. J., Ryall, A., and Priestley, K. F., 1973, Evidence for a high-velocity lithospheric plate under northern Nevada: Seismological Society of America Bulletin, v. 63, p. 2135–2144.

Kovach, R. L., 1966, Seismic surface waves; Some observations and recent developments, *in* Ahrens, L. H., Press, F., Runcorn, S. K., and Urey, H. C., eds., Physics and chemistry of the Earth, v. 6: New York, Pergamon Press, p. 251–314.

——, 1978, Seismic surface waves and crustal and upper mantle structure: Reviews of Geophysics and Space Physics, v. 16, p. 1–12.

Kovach, R. L., and Press, F., 1961, Rayleigh wave dispersion and crustal structure in the eastern Pacific and Indian Oceans: Royal Astronomical Society Geophysical Journal, v. 4, p. 202–216.

Kovach, R. L., and Robinson, R., 1969, Upper mantle structure in the Basin and Range provinces, western North America, from the apparent velocities of S waves: Seismological Society of America Bulletin, v. 59, p. 1653–1665.

Lee, W. B., and Solomon, S. C., 1975, Inversion schemes for surface wave attenuation and Q in the crust and the mantle: Royal Astronomical Society Geophysical Journal, v. 43, p. 43–71.

——, 1978, Simultaneous inversion of surface wave phase velocity and attenuation; Love waves in western North America: Journal of Geophysical Research, v. 83, p. 3389–3400.

——, 1979, Simultaneous inversion of surface wave phase velocity and attenuation; Rayleigh and Love waves over continental and oceanic paths: Seismological Society of America Bulletin, v. 69, p. 65–95.

Lehman, I., 1959, Velocities of longitudinal waves in the upper part of the Earth's mantle: Annales de Geophysique, v. 15, no. 1, p. 93–118.

——, 1962, The travel times of the longitudinal waves of the Logan and Blanca atomic explosions and their velocities in the upper mantle: Seismological Society of America Bulletin, v. 52, p. 519–526.

——, 1964, On the travel times of P as determined from nuclear explosions: Seismological Society of America Bulletin, v. 54, p. 123–139.

Lewis, B.T.R., and Meyer, R. P., 1968, A seismic investigation of the upper mantle to the west of Lake Superior: Seismological Society of America Bulletin, v. 58, p. 565–596.

Lin, J., 1973, A study of the upper mantle structure in the Pacific Northwest using P-waves from teleseisms [Ph.D. thesis]: Seattle, University of Washington, 97 p.

Massé, R. P., 1973, Compressional velocity distribution beneath central and eastern North America: Seismological Society of America Bulletin, v. 63, p. 911–935.

——, 1987, Crustal and upper mantle structure of stable continental regions in North America and Europe: Pure and Applied Geophysics, v. 125, no. 2/3, p. 205–239.

Massé, R. P., Landisman, M., and Jenkins, J. B., 1972, An investigation of the upper mantle compressional velocity distribution beneath the Basin and Range province: Royal Astronomical Society Geophysical Journal, v. 30, p. 19–36.

Mavko, B. B., and Thompson, G. A., 1983, Crustal and upper mantle structure of the northern and central Sierra Nevada: Journal of Geophysical Research, v. 88, p. 5874–5892.

McEvilly, T. V., 1964, Central U.S. crust-upper mantle structure from Love and Raleigh wave phase velocity inversion: Seismological Society of America Bulletin, v. 54, p. 1997–2015.

McKenzie, D., and Julian, B., 1971, Puget Sound, Washington, earthquake and the mantle structure beneath the northwestern United States: Geological Society of America Bulletin, v. 82, p. 3519–3524.

Mereu, R. F., and Hunter, J. A., 1969, Crustal and upper mantle structure under the Canadian shield from Project Early Rise data: Seismological Society of America Bulletin, v. 59, p. 147–165.

Michaelson, C. A., and Weaver, C. S., 1986, Upper mantle structure from teleseismic P-wave arrivals in Washington and northern Oregon: Journal of Geophysical Research, v. 91, p. 2077–2094.

Mitchell, B. J., and Herrmann, R. B., 1979, Shear velocity structure in the eastern United States from the inversion of surface-wave group and phase velocities: Seismological Society of America Bulletin, v. 69, p. 1133–1148.

Mitchell, B. J., Cheng, C. C., and Stauder, W., 1977, A three-dimensional velocity model of the lithosphere beneath the New Madrid seismic zone: Seismological Society of America Bulletin, v. 67, p. 1061–1074.

Monfort, M. E., and Evans, J. R., 1982, Three-dimensional modeling of the Nevada test site and vicinity from teleseismic P-wave residuals: U.S. Geological Survey Open-File Report 82-409, p. 66.

Niazi, M., and Anderson, D. L., 1965, Upper mantle structure of western North America from apparent velocities of P waves: Journal of Geophysical Research, v. 70, p. 4633–4640.

Nuttli, O., 1963, Seismological evidence pertaining to the structure of the Earth's upper mantle: Reviews of Geophysics, v. 1, p. 351–400.

—— , 1969, Travel times and amplitudes of S waves from nuclear explosions in Nevada: Seismological Society of America Bulletin, v. 59, p. 385–398.

Nuttli, O. W., and Bolt, B. A., 1969, P wave residuals as a function of azimuth; 2, Undulations of the mantle low-velocity layer as an explanation: Journal of Geophysical Research, v. 74, no. 27, p. 6594–6602.

Olsen, K. H., Keller, G. R., and Stewart, J. N., 1979, Crustal structure along the Rio Grande Rift from seismic refraction profiles, *in* Riecker, R. E., ed., Rio Grande Rift; Tectonics and magmatism: Washington, D.C., American Geophysical Union, p. 127–143.

Oppenheimer, D. H., and Eaton, J. P., 1984, Moho orientation beneath central California from regional earthquake travel times: Journal of Geophysical Research, v. 89, p. 10267–10282.

Otsuka, M., 1966a, Azimuth and slowness anomalies of seismic waves measured on the central California seismographic array; Part 1, Observations: Seismological Society of America Bulletin, v. 56, p. 223–239.

—— , 1966b, Azimuth and slowness of seismic waves measured on the central California seismographic array; Part 2, Interpretation: Seismological Society of America Bulletin, v. 56, p. 655–675.

Parker, E. C., Davis, P. M., Evans, J. R., Iyer, H. M., and Olsen, K. H., 1984, Upwarp of anomalous asthenosphere beneath the Rio Grande Rift: Nature, v. 312, p. 354–356.

Press, F., 1968, Earth models obtained by Monte Carlo inversion: Journal of Geophysical Research, v. 73, p. 5223–5234.

Priestley, K., and Brune, J., 1978, Surface waves and the structure of the Great Basin of Nevada and western Utah: Journal of Geophysical Research, v. 83, p. 2265–2272.

Priestley, K. F., and Brune, J. N., 1982, Shear wave structure of the southern volcanic plateau of Oregon and northern Great Basin of Nevada from surface wave dispersion: Journal of Geophysical Research, v. 87, p. 2671–2675.

Priestley, K., Orcutt, J. A., and Brune, J. N., 1980, Higher mode surface waves and structure of the Great Basin of Nevada and western Utah: Journal of Geophysical Research, v. 85, p. 7166–7174.

Raikes, S. A., 1980, Regional variations in upper mantle structure beneath southern California: Royal Astronomical Society Geophysical Journal, v. 63, p. 187–216.

Raikes, S. A., and Hadley, D. M., 1979, The azimuthal variation of teleseismic P-residuals in southern California; Implications for upper-mantle structure: Tectonophysics, v. 56, p. 89–96.

Rohay, A., 1982, Crust and mantle structure of the North Cascades Range [Ph.D. thesis]: Seattle, University of Washington, 163 p.

Roller, J. C., and Jackson, W. H., 1966, Seismic-wave propagation in the upper mantle; Lake Superior, Wisconsin, to Denver, Colorado, *in* Steinhart, J. S., and Smith, T. J., eds., The Earth beneath the continents: American Geophysical Union Geophysical Monograph 10, p. 270–275.

Romney, C., Brooks, B. G., Mansfield, R. H., Carder, D. S., Jordan, J. N., and Gordon, D. W., 1962, Travel times and amplitudes of principal body phases recorded from GNOME: Seismological Society of America Bulletin, v. 52, p. 1057–1074.

Romanowicz, B. A., 1979, Seismic structure of the upper mantle beneath the United States by three-dimensional inversion of body wave arrival times: Royal Astronomical Society Geophysical Journal, v. 57, p. 479–506.

—— , 1980, Large-scale three-dimensional P-velocity structure beneath the western U.S. and the lost Farallon Plate: Geophysical Research Letters, v. 7, p. 345–348.

Shearer, P. M., and Oppenheimer, D. H., 1982, A dipping Moho and crustal low-velocity zone from P_n arrivals at The Geysers–Clear Lake, California: Seismological Society of America Bulletin, v. 72, p. 1551–1566.

Solomon, S. C., and Butler, R. G., 1974, Prospecting for dead slabs: Earth and Planetary Science Letters, v. 21, p. 421–430.

Spence, W., 1974, P-wave residual differences and inferences on an upper mantle source for the Silent Canyon Volcanic Centre, southern Great Basin, Nevada: Royal Astronomical Society Geophysical Journal, v. 38, p. 505–524.

Taylor, S. R., and Toksöz, M. N., 1979, Three-dimensional crust and upper mantle structure of the northeastern United States: Journal of Geophysical Research, v. 84, p. 7627–7644.

Toksöz, M. N., Chinnery, M. A., and Anderson, D. L., 1967, Inhomogeneities in the Earth's mantle: Royal Astronomical Society Geophysical Journal, v. 13, p. 31–59.

Walck, M. C., 1984, The P-wave upper mantle structure beneath an active spreading centre; The Gulf of California: Royal Astronomical Society Geophysical Journal, v. 76, p. 697–723.

—— , 1985, The upper mantle beneath the north-east Pacific rim; A comparison with Gulf of California: Royal Astronomical Society Geophysical Journal, v. 18, p. 243–276.

Wickens, A. J., and Pec, K., 1968, A crust-mantle profile from Mould Bay, Canada, to Tucson, Arizona: Seismological Society of America Bulletin, v. 58, p. 1821–1831.

Wiechert, D. H., 1968, Upper mantle structure under the Churchill Province of the Canadian shield, east of the Yellowknife array: Journal of Physics of the Earth, v. 16, p. 93–101.

Wiggins, R. A., and Helmberger, D. V., 1973, Upper mantle structure of the western United States: Journal of Geophysical Research, v. 78, p. 1870–1880.

Zandt, G., 1978, Study of the three-dimensional heterogeneity beneath seismic arrays in Central California and Yellowstone, Wyoming [Ph.D. thesis]: Cambridge, Massachusetts Institute of Technology, 490 p.

—— , 1981, Seismic images of the deep structure of the San Andreas fault system, central Coast Ranges, California: Journal of Geophysical Research, v. 86, p. 5039–5052.

Manuscript Accepted by the Society October 31, 1988

Printed in U.S.A.

Geological Society of America
Memoir 172
1989

Chapter 30

Composition of the continental crust and upper mantle; A review

David M. Fountain
Program for Crustal Studies, Department of Geology and Geophysics, University of Wyoming, Laramie, Wyoming 82071
Nikolas I. Christensen
Department of Earth and Atmospheric Sciences, Purdue University, West Lafayette, Indiana 47907

ABSTRACT

At present there are many ambiguities involved in arriving at reasonable models for crustal and upper-mantle compositions. We review several approaches previously employed by geologists and geophysicists to estimate bulk chemical and mineralogic compositions. Recent studies focusing on the petrology of xenolith suites and geologic mapping of high-grade metamorphic massifs, such as the Ivrea zone, support the thesis of heterogeneity in crustal composition. Even though crustal composition varies laterally, there is strong evidence pointing to the importance of metamorphic grade, which generally increases with depth, in controlling crustal petrology. Seismic refraction and reflection methods show the most promise for understanding the lateral variability in petrology. Laboratory studies of the seismic properties of rocks at crustal and upper-mantle pressure and temperature conditions show that seismic data can provide valuable constraints on crustal and upper-mantle composition. Seismic anisotropy is likely an important property of the continental lithosphere at all levels. Field and laboratory experiments carefully designed to investigate this directional dependence of seismic velocities will provide valuable constraints on the fabric and composition of the continental crust and upper mantle. Within the upper crust, physical properties are likely to be strongly influenced by the presence of fractures containing fluids at high pore pressures. A model for the continental crust and upper mantle, emphasizing probable extreme lateral variability, is constructed from information available on exposed deep crustal sections, xenoliths, and laboratory and field seismic studies.

INTRODUCTION

Models depicting the petrology, chemistry, and structure of the continental crust and upper mantle are currently being refined by integration of new geophysical data with the powerful constraints offered by geological investigations. Geological observations that provide important constraints on crustal composition include studies of xenoliths and high-grade metamorphic terrains now exposed on the Earth's surface. Geophysical methods most useful for understanding the nature of the crust and upper mantle are primarily seismic refraction and reflection profiling with important additional insights coming from electrical and magnetic investigations. When coupled with experimental data on seismic properties of continental rocks at pertinent temperatures and pressures, the geological and geophysical data can be used to correlate field measurements of seismic velocities with mineralogic composition at depth.

This chapter reviews the geologic and geophysical evidence pertaining to the chemistry and mineralogy of the continental crust and upper mantle with the goal of developing better constrained models of these horizons of the Earth. We first summarize previous estimates of chemical compositions of the continental crust. Evidence for the mineralogy of the continental lithosphere based on significant observations from xenolith suites

Fountain, D. M., and Christensen, N. I., 1989, Composition of the continental crust and upper mantle; A review, *in* Pakiser, L. C., and Mooney, W. D., Geophysical framework of the continental United States: Boulder, Colorado, Geological Society of America Memoir 172.

is reviewed. The composition of the intermediate and deep levels of the continental crust is discussed in terms of observations of exposed crustal cross sections. Next, limitations that laboratory seismic data place on the estimation of crustal and upper-mantle composition are examined in detail. Finally, seismic constraints on crustal and mantle composition are summarized, and a model for the petrology and structure of the continental crust and upper mantle is proposed.

ESTIMATES OF CRUSTAL CHEMISTRY

Ever since recognition of the Mohorovičić discontinuity in the early part of the century, earth scientists have devised a variety of quantitative estimates of the chemical composition of the continental crust (Table 1 [Tables 1 through 9 are on pages 728 through 736, at the end of this chapter.]). Early estimates were based on the premise that the compositional average of crystalline rocks exposed at the surface of the continents represented the bulk composition of the entire continental crust (Clarke, 1924; Clarke and Washington, 1924; Goldschmidt, 1933, 1954). With the discovery of the Conrad discontinuity in Europe in 1925, however, scientists recognized that the composition of the deeper part of the continental crust may differ significantly from the upper crust. Later estimates, therefore, were predicated on the assumption that crustal composition could be approximated by a mixture of granitic and basaltic compositions (Poldervaart, 1955; Vinogradov, 1962; Taylor, 1964; Pakiser and Robinson, 1966; Ronov and Yaroshevsky, 1967, 1969; Galdin, 1974). Some of these estimates incorporated important geophysical and geochemical data to constrain the bulk composition. Pakiser and Robinson (1966) correlated seismic velocity with rock composition and used the result to estimate composition from seismic refraction data. Taylor (1964) fit the rare-earth element pattern of sediments by a mixture of equal parts of basalt and granite and consequently assumed this reflected the relative abundance of those two components in the crust. Laboratory velocity studies, however, demonstrated compositional models based on igneous rock abundances may be in error because seismic velocities in the upper and lower crust also correspond with a wide range of metamorphic compositions (Christensen, 1965).

More recent estimates were derived either from models of crustal growth or models of crustal structure based on geology of deep terrains. Taylor (1977, 1979) and Taylor and McLennan (1981) modified their models by assuming that 75 percent of the present-day crust was composed of Archean crustal material and the remaining 25 percent was derived by island-arc magmatism. Using whole-rock analyses of xenoliths from the Moses Rock dike, McGetchin and Silver (1972) estimated chemical composition of the crust beneath the Colorado Plateau. Holland and Lambert (1972) assumed that the Lewisian and Dalradian terrains in Scotland constituted a typical crustal section, and their average composition should be representative of the entire crust. Weaver and Tarney (1984a, b) used the upper crustal estimate presented by Taylor (1977) in conjunction with Lewisian terrain data from middle and lower crustal levels to derive their estimate. Smithson (1978) developed a chemical model based on mean crustal velocity that suggested a bulk crustal composition of quartz diorite.

The similarities between the results of crustal composition estimates are surprising, considering the wide variety of methods employed. In fact, the average SiO_2 percentage from all these estimates (60 ± 2.7 percent) is not substantially different from the earliest estimate of Clarke (1924). Certain researchers have obtained a more silicic crust (Smithson, 1978; Weaver and Tarney, 1984a,b), whereas others have estimated a more mafic crust (Pakiser and Robinson, 1966; McGetchin and Silver, 1972; Taylor, 1979; Taylor and McLennan, 1985). It is important to realize, however, that all these estimates are simply models based on a wide variety of simplified, generalized assumptions.

In addition to estimates of bulk crustal composition, some attempts have been made to estimate composition of the crustal seismic layers. Data presented in Table 2 summarize estimates of the composition of the lower continental crust. These values were derived from: (1) analyses of exposed deep-level metamorphic terrains (Lambert and Heier, 1968; Sheraton and others, 1973; Mehnert, 1975; Leyreloup and others, 1977; Smithson, 1978; Weaver and Tarney, 1980, 1984a; Maccarrone and others, 1983); (2) models of lower crustal genesis (Taylor, 1977, 1979; Taylor and McLennan, 1981, 1985); or (3) the assumption that the lower crust is mafic in composition (Poldervaart, 1955; Pakiser and Robinson, 1966). Because these estimates are model-dependent or are based on a variety of deep crustal analogs, it is not surprising that they differ significantly from one another.

Estimates of the composition of the upper crust are based primarily on extensive geochemical data of average chemical compositions of crystalline shield-area surface rocks (Sederholm, 1925; Grout, 1938; Poldervaart, 1955; Eade and others, 1966; Reilly and Shaw, 1967; Shaw and others, 1967; Condie, 1967; Fahrig and Eade, 1968; Ronov and Yaroshevsky, 1969; Eade and Fahrig, 1971; Bowes, 1972; Taylor and McLennan, 1985). These analyses are remarkably similar, yielding an average SiO_2 content of approximately 65 percent, and are well summarized by the upper crustal estimates presented in Table 3. Other workers estimated the petrology of the upper crust from geologic maps of shield areas (Tables 4, 5). This approach is important because it emphasizes the lithologic heterogeneity of the upper crust, an observation obscured in the averaging processes associated with development of geochemical data presented in Tables 1, 2, and 3, but evident in studies of crustal xenoliths and possible exposed crustal sections.

Evidence for Crust and Upper-Mantle Composition from Xenoliths

Xenolith suites provide an excellent catalog of rock types that potentially constitute the crust and upper mantle (e.g., Dawson, 1977; Kay and Kay, 1981; Griffin and O'Reilly, 1987). The geophysical properties of these rocks can be measured or theoret-

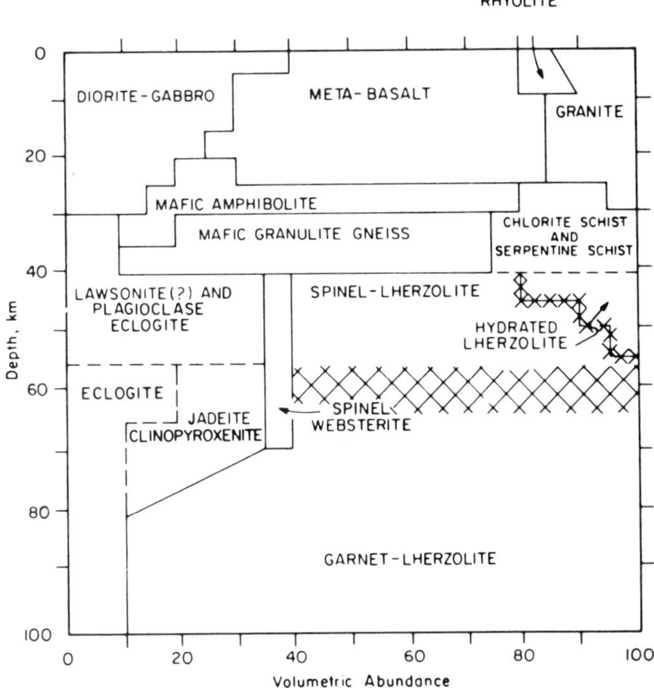

Figure 1. Model of the crust and upper mantle of the Colorado Plateau based on xenoliths (McGetchin and Silver, 1972).

ically calculated so that we can assess their potential importance as lithospheric constituents. The lithologies of xenoliths recovered from the numerous sites in North America are summarized in Table 6. Pressure-temperature (P-T) conditions of metamorphism are listed when reported. No attempt has been made to assess the volumetric abundance of each rock type or to evaluate the quality of the geothermometric and geobarometric data.

Evident in this table is the extent of the lithologic heterogeneity of the lithosphere under many localities. Some xenolith sites (e.g., State Line, central Montana, Navajo field, Moses Rock) exhibit diverse suites that indicate detailed vertical crustal and mantle heterogeneity must exist. In some cases, xenolith researchers have attempted to construct models of the crust and upper mantle by arranging xenoliths into a depth column in an order determined by various parameters such as P-T data, volumetric abundance, and xenolith size.

McGetchin and Silver (1972) presented one of the earliest crustal stratigraphy reconstructions based on xenoliths from the Colorado Plateau (Fig. 1). They envisioned an upper crust dominated by metabasalt, diorite, gabbro, and granite underlain at deep crustal levels by amphibolite, mafic granulite, chlorite schist and serpentine schist. Assignment of specific rock types to various crustal depths was determined by xenolith size. The upper mantle of the plateau was inferred to be lithologically complex, showing significant lateral and vertical heterogeneity. The presence of hydrated rocks at lower crustal levels was questioned by Padovani and others (1982), who, based on petrographic and scanning electron microscope (SEM) analysis, proposed that these rocks were hydrothermally altered during emplacement.

A complex upper-mantle model was presented for the Colorado-Wyoming Front Range (Fig. 2) by McCallum and Eggler (1976), who inferred that the mantle below the Moho was spinel peridotite with zones of garnet websterite. The crustal portion of this section was typified by a wide variety of rocks (Table 6) dominated by granulites and pyroxenites at the base overlain by upper crustal gneisses and granites. Brookins and Meyer (1974) presented a more detailed and complex model for the lithosphere beneath Kansas (Fig. 3). A variety of mafic rocks of generally high metamorphic grade constitute the lower crust and the crust-mantle transition. The upper mantle in this model is dominated by eclogites.

Because portions of the continental lithosphere have likely originated by accretion of island-arc terrains, xenoliths found in island arcs provide information on possible continental crustal composition as well as arc petrology. Compositions of Aleutian lava xenoliths summarized by Kay and Kay (1986) are given in Table 6. Based on xenolith studies and crystal fractionation calculations, Kay and others (1986) have estimated that the upper crust of the Aleutian arc consists of layered volcanic and volcaniclastic sediments intruded by plutons, whereas cumulates from arc rocks, early arc lavas, and original oceanic crust of this model

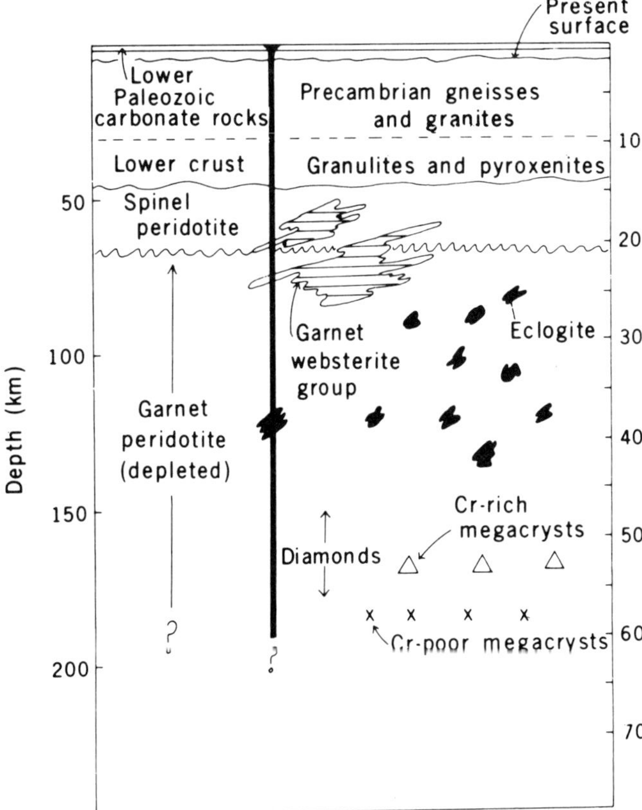

Figure 2. Schematic model of the crust and upper mantle beneath the Colorado-Wyoming Front Range area based on xenoliths (McCallum and Eggler, 1976). Heavy vertical line represents the kimberlite.

vary from basaltic to dacitic in composition and the lower crust is largely basaltic. Arc-related ultramafic cumulates form the uppermost mantle.

Xenoliths have distinct advantages for the study of crustal and upper-mantle composition because they represent samples of the lithosphere between the source of the transporting magma and the site of eruption, they commonly show contact relations between various rock types, and their pressure and temperature conditions of equilibrium can be derived from the compositions of coexisting minerals. They do not, however, show large-scale geometric relationships, they may represent a biased sample set, and they may not show equilibration along an ambient geotherm (e.g., Harte and others, 1981; Griffin and O'Reilly, 1987). In addition, recent isotopic studies show that certain deep crustal xenoliths recovered from Kilbourne Hole, New Mexico (Davis and Grew, 1978), and the Snake River Plain (Leeman and others, 1985) yield Proterozoic and Archean dates of peak metamorphism, respectively, indicating these xenoliths were at deep crustal levels during Precambrian peak metamorphism. Later tectonism may have elevated them to shallower crustal levels, implying these rocks were once in the deep crust or upper mantle but may not necessarily represent contemporary deep crust. The information provided by xenoliths on crustal composition is particularly useful when their lithologies are compared with portions of continental crustal sections now exposed on the Earth's surface.

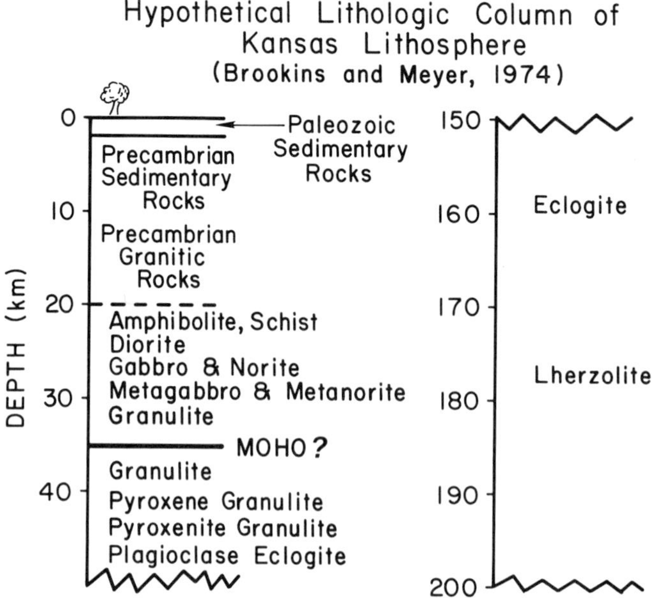

Figure 3. Hypothetical lithologic column for crust and upper mantle of Kansas based on xenoliths (Brookins and Meyer, 1974).

CRUSTAL STRUCTURE AND COMPOSITION BASED ON CRUSTAL CROSS SECTIONS

Several classic papers in the late 1960s (Berckhemer, 1968, 1969; Ansorge, 1968; Giese, 1968) suggested that the deep crust of the Po basin in Italy was thrust to the surface of the Earth and is presently exposed in the Ivrea zone of northern Italy. Evidence for this remarkable structure was found in a large positive Bouguer gravity anomaly and analysis of seismic refraction data showing a high-velocity slab projecting to near-surface depths. Laboratory seismic velocity data for Ivrea-zone rocks (Fountain, 1976) added support to the hypothesis. Later, Fountain and Salisbury (1981) postulated that analogous situations exist elsewhere in the world, and thus, there may be several localities where partial cross sections of the continental crust are exposed for direct observation. In addition to the Ivrea zone, examples of exposed crustal cross sections included the Sachigo-Pikwitonei subprovinces (Manitoba), the Fraser and Musgrave Ranges (Australia), and the Kasila series (West Africa). Several other possible crustal cross sections have recently been identified; these include the Kapuskasing structural zone in Ontario (Percival and Card, 1983), Fiordland in New Zealand (Oliver and Coggon, 1979; Oliver, 1980; Priestley and Davey, 1983), the Tehachapi Mountains in California (Ross, 1985), and Calabria in southern Italy (Schenk and Scheyer, 1978; Moresi and others, 1978-1979). In Table 7 we tabulate the major rock types of the crust based on several of these cross sections.

Two of these examples represent sections through the Archean Superior Province in Canada. Here, the upper crustal levels are dominated by a variety of tonalitic, granodioritic, and granitic rocks, which surround metavolcanic and metasedimentary packages (greenstone belts). In the Pikwitonei example (Table 7), the granite-greenstone terrain passes, with inferred increasing depth, into upper amphibolite to granulite facies tonalitic gneiss with inclusions of remnant greenstone belt lithologies. In contrast, deeper crustal levels in the Kapuskasing structural zone (Table 7) are characterized by various granulite facies rocks (mafic gneiss, tonalite gneiss, metasedimentary gneiss, anorthosite). Percival and Card (1985) have speculated that these rocks are remnants of the pre-greenstone belt crust.

The Ivrea section exhibits different structural and lithologic characteristics than those shown in the Canadian examples. Upper crustal levels are dominated by greenschist to amphibolite facies metasedimentary rocks and quartzo-feldspathic gneisses. Deep crustal levels are composed of upper amphibolite to granulite facies pelitic rocks, marble, amphibolite, gabbroic rocks, and garnet granulite. Mafic granulites dominate along strike of this zone. Ultramafic bodies occur at the base of the exposed section. The transition zone from mafic granulites to the ultramafic bodies has been interpreted as an exposed example of the Moho (Berckhemer, 1969; Hale and Thompson, 1982), but the boundary may be tectonic (e.g., Shervais, 1978-1979).

Geobarometric data for several of these sections indicate that the maximum depths exposed equilibrated at pressures no greater than 700 to 900 MPa. Thus between 3 and 10 km of lower crust is missing if one assumes a mean crustal thickness of 35 km. The crust-mantle transition is generally not observed, so we lack critical information about the nature of the Moho and the uppermost continental mantle. Also, the tops of these sections commonly contain greenschist facies assemblages, suggesting the uppermost crust is missing.

These sections illustrate the great lateral and vertical lithologic heterogeneity that may characterize the crust in one area and over continent-scale distances. This heterogeneity results from the complex, protracted and specific sequence of geologic events experienced by each crustal section. Thus, unlike similar oceanic crustal modeling based on ophiolites, we cannot draw many generalities about overall crustal composition from these sections, but rather we should analyze the crust of each region individually.

SEISMIC CONSTRAINTS ON CRUSTAL AND UPPER-MANTLE PETROLOGY

The models examined to date provide information on crustal and mantle petrology over limited regions. Refraction and reflection methods provide the most promise for understanding lateral variability in the continental lithosphere. Of critical importance in the interpretation of the seismic data are laboratory studies of the compressional (V_p) and shear (V_s) wave velocities of rocks at crustal and upper-mantle P-T conditions. In this section we review some of the more important results from these investigations that are relevant to the interpretation of seismic data, and the constraints that field seismic data place on the estimation of crustal and upper-mantle composition are examined in detail. In this discussion we consider the upper crust to make up the upper third to half of a crustal column. Traditionally, the lower crust is viewed as the portion of the crust below the Conrad discontinuity, but because this discontinuity is not always observed and many velocity profiles show several discontinuities, we regard the lower crust as the bottom third to half of the crust, as defined by seismic methods. In general, V_p is greater than 6.5 km/sec at these depths, and commonly greater than 6.8 km/sec.

Seismic properties of crustal and upper-mantle rocks

Summarized in Table 8 are compressional-wave velocity data for typical continental crystalline rocks that appear to be abundant based on xenolith studies and the petrology of exposed crustal cross sections. The data are tabulated as a function of confining pressure to, in most cases, 1 GPa (10 kbar). Certain samples were classified with reported modal mineralogy according to the IUGS system (Streckeisen, 1976).

For most rocks, the changes of velocity with confining pressure are well known. At low confining pressures, the gradients are large (Table 8), apparently due to the closure of low-aspect ratio microcracks (Birch, 1961; Thill and others, 1969). At confining pressures from 200 to 1,000 MPa, the pressure derivative of compressional wave velocity ($\partial V_p/\partial P$) approaches characteristic single crystal values and at higher pressures reflects the intrinsic elastic behavior of the constituent minerals (Christensen, 1974).

Changes in velocity with pressure for isotropic harzburgites with nearly identical mineralogy from the Twin Sisters Range, Washington, and Kilbourne Hole, New Mexico (xenolith), are shown in Figure 4. The differences in velocities and initial behavior with increasing pressure are explained by the high porosity of the xenolith, which apparently originated during rapid ascent to the Earth's surface. Even at the high pressures simulating those of the lower crust and upper mantle, velocities of the xenolith are low because pressure alone is insufficient to close pore spaces. For this reason, we selected data for Table 8 from rocks collected from crystalline terrains that have not experienced this violent emplacement history.

Compared to measurements at elevated pressure, data on the influence of temperature on velocities are scarce. Recent measurements define the temperature derivatives of compressional wave velocity ($\partial V_p/\partial T$) for a few common crustal and upper-mantle lithologies (Meissner and Fakhimi, 1977; Ramananantoandro and Manghnani, 1978; Christensen, 1979; Kern and Richter, 1981; Kern and Schenk, 1985). In general, these studies show that $\partial V_p/\partial T$ typically ranges from 4 to 6×10^{-4} km sec^{-1} °C^{-1}, but larger and smaller values are reported. Large gradients reported by Christensen (1979) at temperatures greater than 300°C apparently result from crack enlargement associated with anisotropic thermal expansion of minerals at confining pressures of 200 MPa.

To illustrate this effect of temperature, we show the variation of V_p with depth in a granite and a mafic garnet granulite (Fig. 5) calculated on the basis of temperature and pressure derivatives presented in Christensen (1979) and with geothermal gradients given in Pollack and Chapman (1977). These gradients correspond to surface heat-flow values of 40 mW/m^2 (average shield) and 90 mW/m^2 (Basin and Range) and are dashed where extrapolated above 300°C. Also shown, for comparison, are the

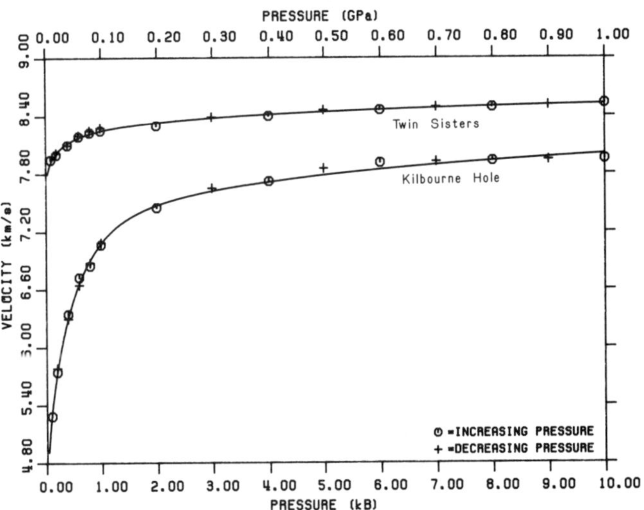

Figure 4. Compressional wave velocity vs. confining pressure for harzburgites from the Twin Sisters Range, Washington, and Kilbourne Hole, New Mexico (N. I. Christensen, unpublished data).

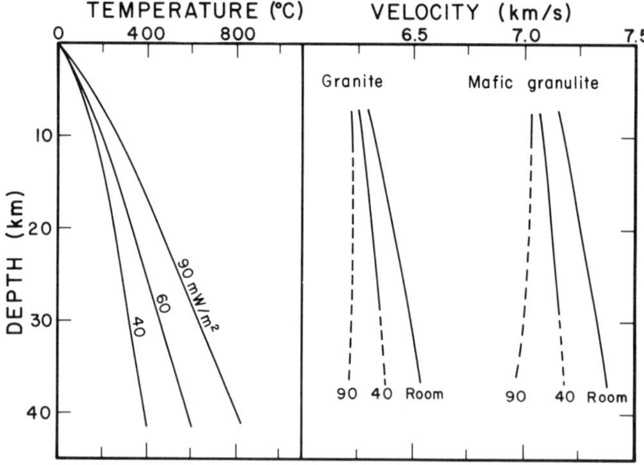

Figure 5. Calculated variation of velocity with depth for a granite and garnet-bearing mafic granulite (right) for room temperature conditions and geothermal gradients corresponding to 40 and 90 mW/m^2 heat-flow provinces (left). Curves are dashed where values are extrapolated above experimental conditions.

room-temperature variations of V_p strictly with pressure. These two rock types represent near end-member lithologies for the continental crust and exhibit similar behavior with temperature. Velocities tend to increase with depth for low geothermal gradients. At depths below approximately 20 km, both rocks show velocity reversals. Thus, decreasing velocity with depth should be common in crustal rcgions whcrc lithology is uniform. In regions characterized by extremely high temperature gradients, partial melting, or the alpha-beta phase transition in quartz will generate low-velocity zones (Gutenberg, 1959; Kern, 1979). Other possibilities include high pore pressure (Christensen, 1986) and intrusions of low-velocity rocks (Mueller, 1977).

It is well known that many rocks exhibit seismic anisotropy. For crustal rocks, this behavior is most notable in metamorphic rocks (Christensen, 1965) including mylonites (Fountain and others, 1984), which exhibit strong fabric elements such as lineation, schistosity, and layering. Anisotropy is reported in Table 8 calculated from $(V_{max}-V_{min})/V_{mean}$ expressed as a percentage. At low pressure, anisotropy can often be related to anisotropy in microcrack orientation (Christensen, 1965; Simmons and others, 1975), whereas at high pressures, preferred orientation of highly anisotropic minerals is primarily responsible for seismic anisotropy. Although most significant in metamorphic rocks, including peridotite tectonites, seismic anisotropy is also detectable in some igneous rocks (Table 8).

The most significant parameter influencing V_p in continental rocks is mineralogy, which is a function of chemical composition and petrologic evolution. Christensen (1965) demonstrated that V_p varies in a systematic manner in coarse-grained igneous rocks (Fig. 6). For instance, there is a consistent increase in V_p between granite and diorite, a change duplicated in the high-grade equivalents, the felsic gneisses (see Table 8). That these changes are related to mineralogy is illustrated by calculations of V_p for combinations of quartz, potassium feldspar, plagioclase, and hornblende following the procedure outlined by Christensen (1966). Instead of presenting simple triangular diagrams, we show a four-component system using four adjacent triangles. These results, calculated for room temperature and midcrustal depths, assuming typical $\partial V_p/\partial P$ values for minerals, are displayed in Figure 7. For the plagioclase composition selected, there is only slight variation of V_p in the quartz-potassium feldspar-plagioclase field. An increase in the amount of the mafic component will produce a significant velocity increase, as will an increase in the anorthite content of plagioclase (Birch, 1961).

Figure 8 shows a similar diagram for a mafic system composed of pyroxene, hornblende, plagioclase (calcic), and garnet. This combination not only describes various gabbroic rocks, but also a variety of metamorphosed mafic rocks, including garnet granulites and amphibolites. The range of V_p for this system is very large, as it is strongly controlled by garnet. This diagram explains the significant increase in V_p associated with the transformation from gabbroic rocks to the higher grade garnet granulites and eclogites. For ultramafic rocks, a triangular diagram presented by Christensen (1966) illustrates the relation between mineralogy and velocity for aggregates of olivine, orthopyroxene, and serpentine (Fig. 9). We emphasize that these diagrams are for isotropic mineral assemblages, but laboratory measurements demonstrate that anisotropy can have greater effects on V_p than significant changes in mineral percentages. For example, average anisotropy appears to be about 7 percent in peridotites. Referring to Figure 9, this corresponds to the total range of velocities between olivine and orthopyroxene.

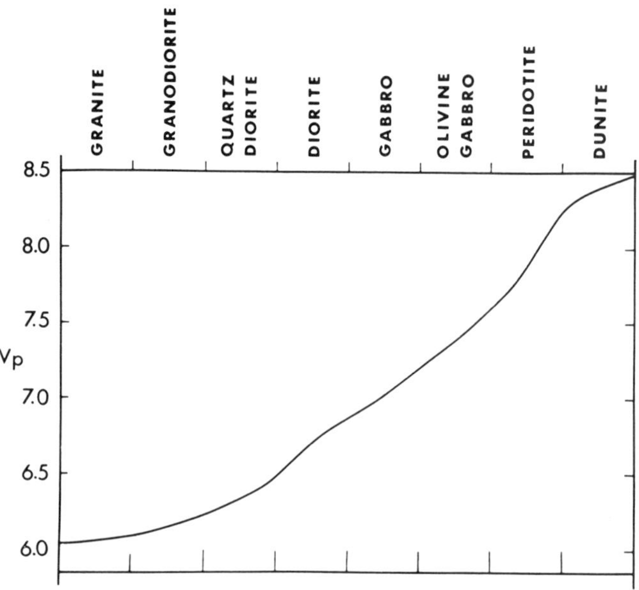

Figure 6. Diagram showing variation of compressional wave velocities at 150 MPa for crustal and upper-mantle rock types (Christensen, 1965).

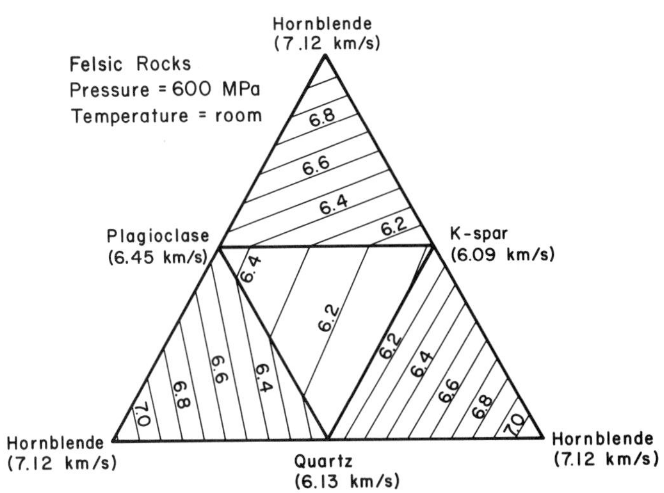

Figure 7. Calculated velocities for the four-component system hornblende-quartz-potassium feldspar-plagioclase (albite). Mineral velocities used in the calculations were obtained from Birch (1961), Alexandrov and Ryzhova (1961, 1962), and McSkimin and others (1965).

Figure 8. Calculated velocities for the four-component system garnet-pyroxene-hornblende-plagioclase (anorthite). Mineral velocities used in the calculations were obtained from Birch (1961), Alexandrov and Ryzhova (1961), Frisillo and Barsch (1972), and Babuska and others (1978).

Seismic constraints on upper-mantle composition

A major source of information on the petrologic nature of the upper continental mantle is P_n velocity. The availability of data for North America varies significantly from region to region. Western North America has been extensively studied, whereas coverage of the eastern and central regions is limited. Several compilations resulted in contour maps of upper-mantle velocity and crustal thickness for North America or portions of North America (e.g., Pakiser and Steinhart, 1964; Warren and Healy, 1973; Smith, 1978; Blair, 1980; Allenby and Schnetzler, 1983; Braile and others, this volume). The P_n contour map of Blair (1980), given in Figure 10, was constructed by computer contouring seismic data at 349 locations from the compilation of Christensen (1982). A histogram of these upper-mantle velocities is presented in Figure 11.

A comparison of the velocities in Figure 11 with laboratory-measured velocities places major constraints on upper-mantle composition. The major rock types with sufficiently high velocities are dunite, peridotite, and some eclogites (Table 8), as shown in the histogram of Figure 12. Some of the eclogite data in Figure 12 were obtained from xenoliths and, as discussed earlier, V_p may be low because of porosity. In addition, the velocities plotted in Figure 12 do not take into account the effects of accessory minerals (including alteration products) and temperatures, both of which lower P_n velocities significantly. A comparison of Figures 11 and 12 shows that P_n data, such as that in Figure 10, does not discriminate a dominantly eclogitic from a peridotitic upper-mantle model. Thus, additional constraints, such as seismic anisotropy, should be considered.

During the past two decades, numerous papers cited evidence for seismic anisotropy in the upper mantle. Azimuthal

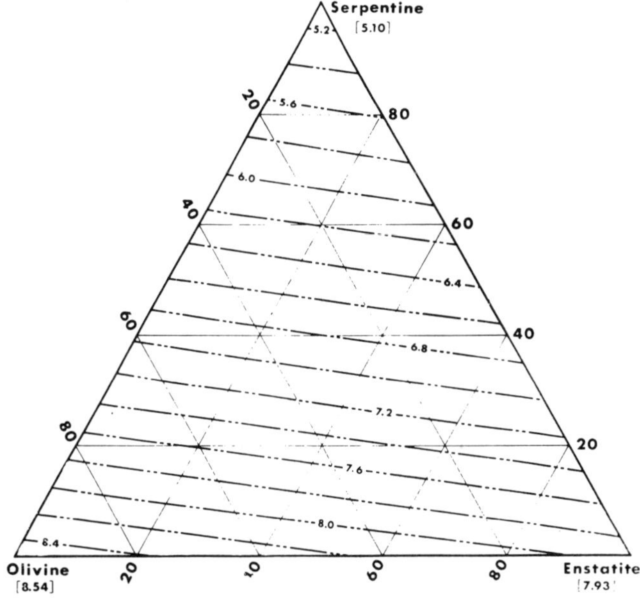

Figure 9. Calculated velocities for the three-component system olivine-enstatite-serpentine at 1 GPa from Christensen (1966).

anisotropy of P_n velocities is a common property of the oceanic upper mantle (e.g., Raitt and others, 1969; Morris and others, 1969; Shimamura and others, 1983) that, based on studies of ultramafic sections of ophiolities, originates from preferred orientation of highly anisotropic olivine crystals (e.g., Christensen and Salisbury, 1979; Christensen, 1984a). Continental upper-mantle anisotropy has been observed in Europe (Bamford, 1973, 1977;

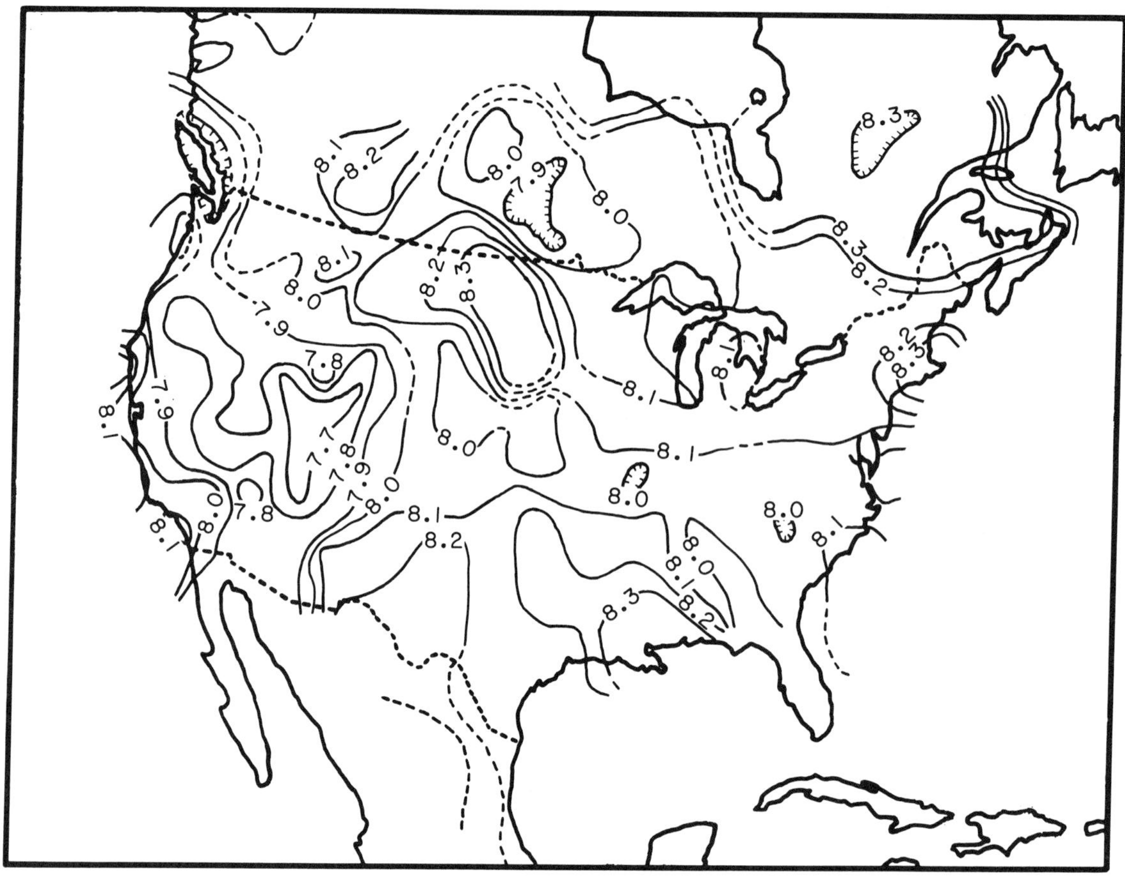

Figure 10. Contour map of P_n velocities for North America from Blair (1980). See Braile and others (this volume) for comparison.

Fuchs, 1975, 1983), the western United States (Bamford and others, 1979; Vetter and Minster, 1981), the USSR (Chesnokov and Nevskig, 1977) and Australia (Leven and others, 1981). Analyses of P_n data have failed to detect anisotropy in the Mojave region of southern California (Vetter and Minster, 1981), the eastern United States (Bamford and others, 1979), and northern Britain (Bamford and others, 1979). It should be emphasized that if upper-mantle anisotropy is transversely isotropic (Christensen and Crosson, 1968) and has a vertical axis of symmetry, P_n anisotropy would not be observed. Significantly, possible upper-mantle peridotite from the Ivrea zone exhibits this type of anisotropy (Fountain, 1976).

Additional observations of the existence of anisotropy to several hundred kilometers depth beneath the continents are based on surface-wave studies (Forsyth, 1975; Mitchell and Yu, 1980; Anderson and Dziewonski, 1982). A global map of 200-sec Rayleigh wave azimuthal variation by Tanimoto and Anderson (1984) was correlated with return flow directions in the asthenosphere.

Anisotropy has been studied in detail in several possible upper-mantle rocks (e.g., Christensen, 1966; Christensen and Ramananantoandro, 1971; Babuska, 1972; Ramananantoandro and Manghnani, 1978). Studies of rock specimens suggest both orthorhombic and axial (transverse isotropy) as possible upper-mantle symmetry. Investigations of anisotropy in ophiolites, believed to represent on-land exposures of oceanic crust and upper mantle, provide strong evidence for an oceanic upper mantle composed of peridotite with olivine *a*-axes producing fast velocities parallel to paleospreading directions (Christensen, 1984a). Our level of understanding of anisotropy beneath the continents is poor because of our lack of knowledge about mantle dynamics and composition. For example, eclogites, which may be an upper continental mantle constituent (Fig. 12), are nearly isotropic or weakly anisotropic. This general low anisotropy of eclogites is illustrated in Figure 13 where anisotropies of possible upper-mantle rocks are plotted against their densities. Examination of this figure shows the strong anisotropy common to most peridotites and dunites and the relatively low anisotropy of many eclogites. Several eclogites, however, do possess significant anisotropy that is likely a consequence of strong preferred orientation of pyroxene. Detailed upper continental mantle seismic studies have the potential to map isotropic versus anisotropic regions and thereby resolve major questions on upper continental mantle composition and dynamics.

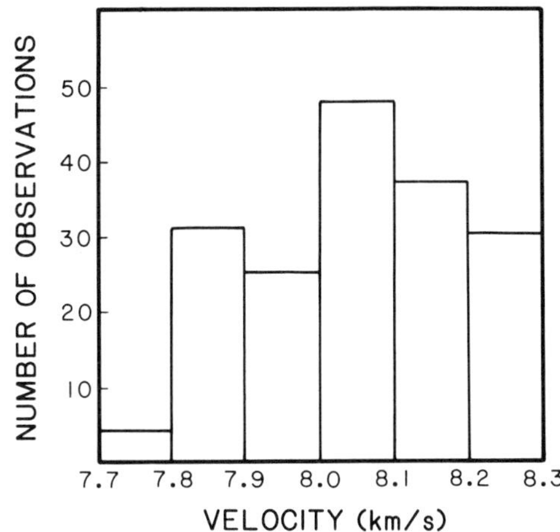

Figure 11. Histogram of P_n velocities for North America.

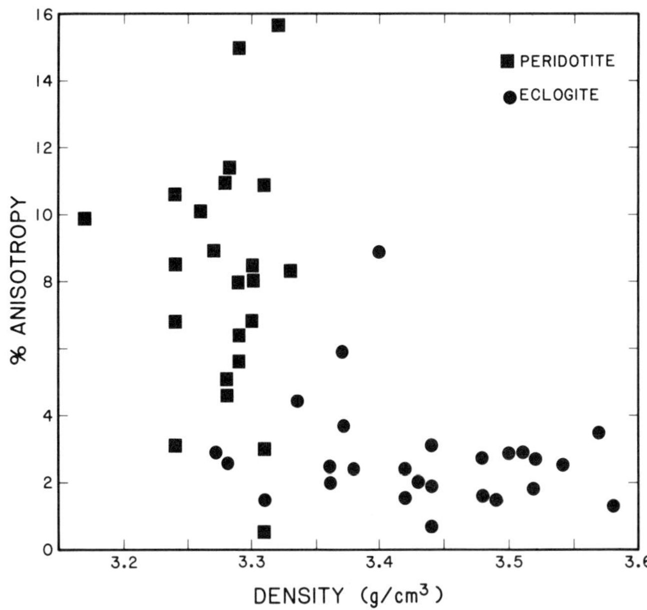

Figure 13. Density versus percent anisotropy for peridotite (squares) and eclogite (circles). Data derived from compilation in Christensen (1982) and Christensen (unpublished data).

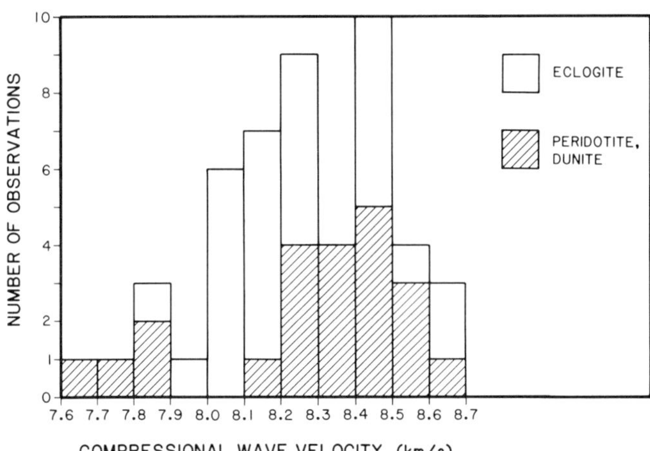

Figure 12. Histogram of compressional wave velocities for eclogite, peridotite, and dunite at 1,000 MPa showing overlap between eclogite and ultramafic rock velocities. Data derived from compilation in Christensen (1982).

Seismic constraints on lower crustal composition

Since Mohorovičić used earthquake arrival-time data to identify the crust-mantle boundary, a variety of seismological methods have been used to investigate the deeper portions of the continental crust. Seismic refraction and reflection profiling are widely used to examine the crust below 10 km depth, and presently provide a more voluminous data set and data of higher resolution than the less widely used methods of magnetic and electrical surveys. In this section we review the seismic evidence bearing on the petrology of middle to lower crustal levels.

Interest in the lower portions of the continental crust has increased lately because of recognition that the deep crust generates significant reflections. One of the earliest examples of deep crustal reflections came from the first Consortium for Continental Reflection Profiling (COCORP) survey in Hardeman County, Texas (Oliver and others, 1976), and is exemplified by recent data from the Basin and Range region (Fig. 14). Since the first COCORP results, reflections from the deep crust have been reported from surveys in Great Britain (e.g., Matthews and Cheadle, 1985), France (e.g., Bois and others, 1985), Germany (Meissner and Lueschen, 1983), Switzerland (Finckh and others, 1985), Australia (Mathur, 1983), and many North American sites (e.g., Oliver and others, 1983). Many seismic reflection workers note that reflective lower crust tends to occur under young extensional terrains such as the Basin and Range (Allmendinger and others, 1983), and the North Sea and the continental shelf of Great Britain (Matthews and Cheadle, 1985), suggesting that deep crustal reflectivity may result from processes associated with crustal extension. Deep crustal reflection events, however, are observed in older terrains with more enigmatic tectonic histories, such as the Adirondacks (Klemperer and others, 1983) and Hardeman County (Oliver and others, 1976). We point out that the cumulative line length over extensional terrains collected by British Institutes Reflection Profiling Syndicate (BIRPS) and COCORP clearly exceeds the length over stable shield areas, that may result in a somewhat biased view of deep crustal reflectivity.

Where reflections are reported, they are generally horizontal to subhorizontal events that exhibit complex, multi-cyclic waveforms with relatively high amplitudes. These characteristics suggest that the deep crust in some regions must be composed of layers of differing acoustic impedances with thicknesses appropriate to cause constructive interference (Fuchs, 1969; Meissner,

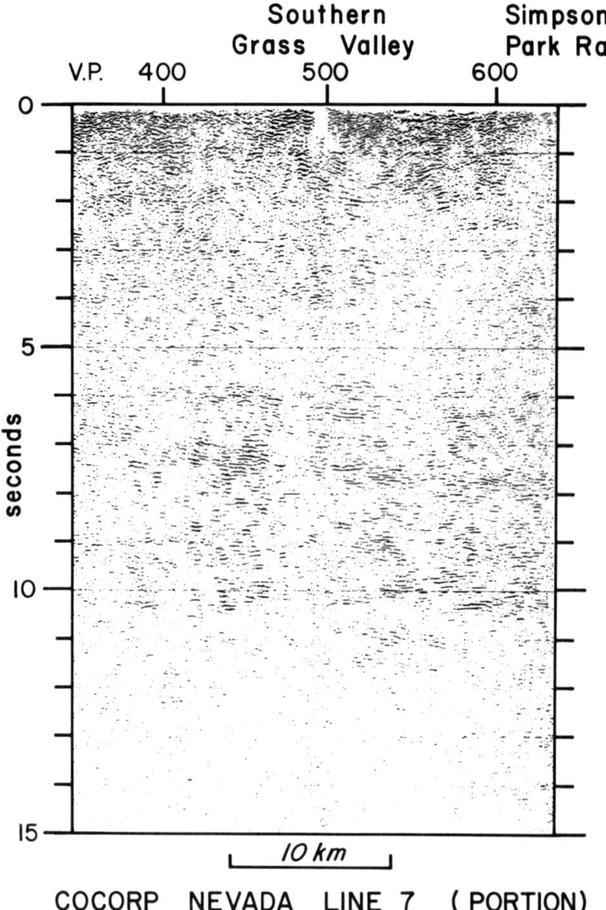

Figure 14. Unmigrated stacked seismic section of the eastern part of COCORP Nevada line 7 (from Potter and others, 1987) from the Basin and Range province showing reflective lower crust between 6- and 10-sec two-way traveltime. Right side of profile is west and left side is east.

1973; Clowes and Kanasewich, 1970; Hurich and others, 1985). This implies that layering on a scale of less than 150 m thick must be prevalent in some regions. Several mechanisms may be responsible for producing this apparent acoustic layering, including variations in petrology, pore pressure, percentages of partial melt, and the degree of anisotropy (e.g., see discussions by Fountain and others, 1984; Matthews and Cheadle, 1985). The relative importance of these mechanisms is poorly understood. Also, to date, there are few assessments of the magnitude of the necessary reflection coefficients (Sandmeier and Wenzel, 1986); thus we have poor constraint on the velocity and density contrasts within the heterogeneous lower crust. Laboratory and modeling studies on Ivrea and Kapuskasing rocks (Hale and Thompson, 1982; Fountain, 1986; Fountain and others, 1987) indicate that reflection coefficients in compositionally layered deep crustal sequences can be very large (absolute value, $\leqslant 0.15$).

In many regions, the middle portion of the crust is reported to be transparent because reflections are either absent or not abundant. This generality was pointed out in BIRPS data (e.g., Smythe and others, 1982) and has been reported elsewhere. Reflections have been observed at this level by groups using shooting or recording parameters different from those used by COCORP (Hurich and others, 1985; Coruh and others, 1987), suggesting that, in some cases, absence of midcrustal reflections may be related to recording and shooting methods. It is important to note that absence of reflections does not imply that the middle crust is lithologically or structurally homogeneous. Inspection of Table 8 shows that there is little significant variation in velocity and density in quartzo-feldspathic lithologies, although there are clear petrologic distinctions. A complex middle crust composed of plutons and various granitic to tonalitic gneisses would not generate large reflections when compared to deep crustal events generated by layers of granulite-facies mafic and tonalitic gneisses (see Table 8).

Whereas reflection methods provide constraints on certain scales of deep crustal geometry, refraction methods and other approaches to velocity inversion can provide velocity data that can, perhaps, better constrain crustal composition. Although numerous refraction surveys have been conducted in the United States, those run in the past 5 to 10 yr are of particular note because of the use of modern recording methods, relatively small station spacing, and new interpretation techniques. Most of these newer surveys were conducted over interesting tectonic features such as rift zones, basins, and recently active magmatic systems. Few have been run over stable shield or platform areas. Thus, the view of the deep crust obtained from these more recent data may be somewhat biased. With that caution in mind, we present some velocity models from these surveys with the rock velocity curves from Figure 5 superimposed for reference.

The Basin and Range and geographically associated Snake River Plain–Yellowstone Plateau (SRP–YP) has the densest refraction coverage of any region in the United States. Figure 15 presents data from the Wasatch Front area (Braile and others, 1974) and the SRP–YP region (Braile and others, 1982; Sparlin and others, 1982; Smith and others, 1982). The deeper crust under the SRP–YP is characterized by a high-velocity (6.8 km/sec) layer that, for a high geothermal gradient, approaches the V_p curve for the mafic garnet granulite gneiss. A variety of mafic lithologies, anorthosites, or even high-grade metapelites could easily match this V_p at appropriate pressure and temperature conditions. Midcrustal levels under these terrains are about 6.5 km/sec, perhaps reflecting a more felsic composition.

Data from the western margin of the United States are shown in Figure 16, which displays velocity profiles for the Coast Range of California (Walter and Mooney, 1982) and the Oregon Cascades (OC) (Leaver and others, 1984). The Cascade model exhibits a thick midcrustal zone characterized by V_p of about 6.5 km/sec, perhaps indicative of felsic rocks. This zone overlies a high V_p (>7 km/sec) lower crust that closely matches the V_p for garnet granulite for the geothermal gradient corresponding to a 90 mW/m^2 heat-flow province. The Diablo and Gabilan Ranges of California exhibit differing V_p profiles and are characterized by

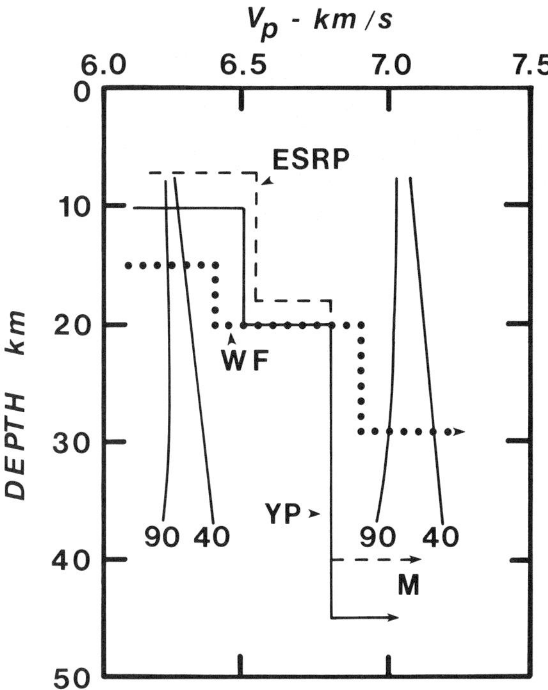

Figure 15. Seismic velocity structure for the eastern Snake River Plain (ERSP), Yellowstone Plateau (YP), and Wasatch Front (WF). M corresponds to Moho. The curves marked 40 and 90 correspond to the variation of velocity with depth for granite (left pair) and mafic garnet granulite (right pair) for heat-flow regimes of 40 and 90 mW/m^2.

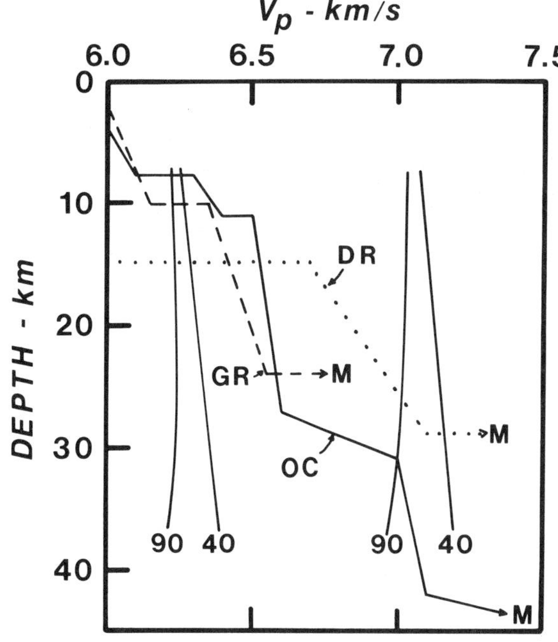

Figure 16. Seismic velocity structure for the Diablo Range (DR) and Gabilan Range (GR) of the California Coast Ranges, and Oregon Cascades (OC) with the same velocity curves for granite and mafic granulite as in Figure 15.

velocity gradients too high to be explained by pressure and temperature changes in a single rock type. The gradients, if real, must reflect significant compositional changes with depth. The Gabilan Range (GR) section is apparently dominated by felsic rocks at depth, whereas felsic midcrustal levels in the Diablo Range (DR) must give way to mafic-dominated lithologies in the lowermost crust. Lin and Wang (1980) and Walter and Mooney (1982) attributed the high velocities under the Diablo Range to a predominantly mafic lower crust, perhaps similar in nature to gabbroic rocks locally found in the dominantly metasedimentary Franciscan complex. Alternative lithologies for the deepest crustal levels might include anorthosite, blueschist, partially serpentinized peridotite, or metapelites.

Recent refraction studies in the midcontinent region are sparse, but two interesting data sets were presented for the Mississippi embayment (Mooney and others, 1983) and the Canadian portion of the Williston basin (Hajnal and others, 1984). The Mississippi embayment was once an active continental rift system (Ervin and McGinnis, 1975). In both sections (Fig. 17), the velocity profile exhibits values between 6.4 and 6.6 km/sec at midcrustal depths, with significant increases to values in excess of 7 km/sec in the lower crust. These high velocities exceed those for garnet granulite, and suggest—assuming isotropic conditions—abundant higher velocity components (e.g., garnet, olivine, pyroxene, hornblende) at this level.

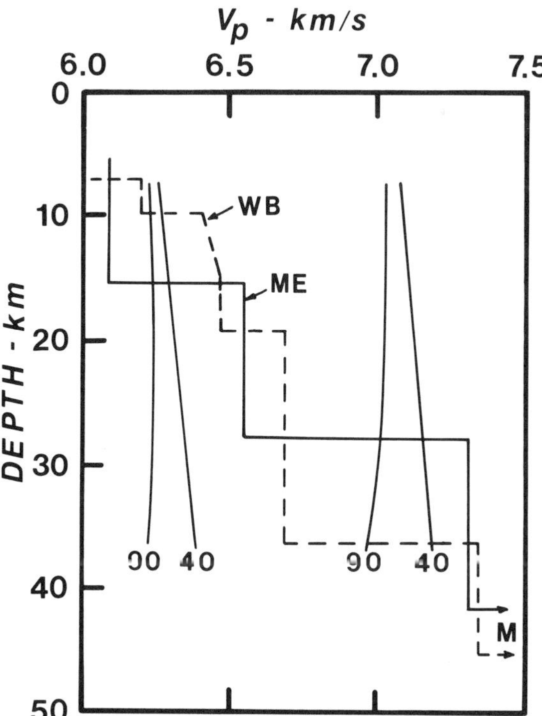

Figure 17. Velocity structure for Mississippi embayment (ME) and Williston basin (WB) with the same velocity curves for granite and mafic granulite as in Figure 15.

Many of these profiles show an intermediate-velocity middle crust and a high-velocity lower crust. These high velocities are generally between 6.8 and 7 km/sec, with a few sections showing higher values. Lower crustal zones dominated by various mafic lithologies, perhaps in upper amphibolite or granulite facies, can explain this range (Table 8). Velocities greater than 7.3 km/sec, however, appear to require the addition of some high-velocity components such as eclogitic or ultramafic rocks if the lower crust is assumed to be isotropic.

Although general compositional information can be gained from interpretation of P-wave velocity data from refraction surveys, the approach has many limitations. First, velocity interpretations based on first arrivals exclusively give the velocity of the refracting horizon and not the entire layer beneath it. Solutions employing later arrivals, amplitudes, and raytracing can give a much more realistic velocity structure. Second, path lengths and wavelengths used in refraction experiments average small-scale heterogeneities so that the velocity determined should not necessarily correspond to a single rock type. Third, direct comparison of V_p from refraction surveys with average V_p for rocks does not take seismic anisotropy into account. Although deep crustal anisotropy has not been reported, laboratory data (Table 8) suggest it may be an important control of wave propagation in the lower crust. Fourth, data in Table 8 indicate that V_p is not a unique function of rock composition, making direct comparisons of field and laboratory data ambiguous. In our previous discussion of the velocity profiles, we elected to correlate V_p with general lithologic categories rather than specific rock types because of this ambiguity. Finally, as stated above, recent refraction studies in the United States have been conducted over areas of tectonic interest, but few have been run over stable shield or platform areas in the midcontinent, resulting in a lack of reference velocity profiles for comparison.

Seismic investigations of the oceanic crust showed that knowledge of shear-wave velocities can provide important constraints on lithologic composition of the crust (Christensen, 1972; Spudich and others, 1978). Shear-wave velocity data for the deeper levels of the continental crust are scarce, generally unreliable, and are derived from a variety of methods including refraction surveys and surface wave dispersion. Available lower crustal shear-wave data for the United States are presented in Table 9. In general, there is considerable variation of V_s in the deep crust, suggesting considerable petrologic differences.

To assess the possibility of constraining deep crustal compositions with shear-wave data, we plot V_p versus V_s for the families of crustal rock types in Table 8 (where shear-wave velocity data are available) at pressures of 600 MPa (Fig. 18). On the same figure we show the V_p and V_s data from Table 9. In some cases there are correspondences between the lower crustal data and the fields of various rock types; in other cases, the field values can be explained by some combination of two or three of the categories.

Shear-wave velocity studies may improve resolution of fine-scale structure of the lower crust. Recently, Owens and others (1984) utilized teleseismic P-waveforms to determine the shear-wave structure beneath the Cumberland Plateau. The resultant models (see Taylor, this volume) show significant lateral heterogeneity in a small geographic region and provide evidence for a thick, laminated crust-mantle transition zone. The P-wave velocity data for the Cumberland Plateau were obtained from a reanalysis of an old U.S. Geological Survey line by Prodehl and others (1984) and are also shown in Taylor (this volume) for comparison with the shear-wave data. In the zones where V_p does not change, we can estimate Poisson's ratio for the southeast profile. For both these zones, Poisson's ratio is close to 0.25, which is low for the fields of crustal rocks presented in Figure 18 but similar to other seismic data. Either the laboratory data are inappropriate for comparison or other factors need to be considered.

It is important to realize that the studies of Owens and others (1984) and Prodehl and others (1984) compare wave energy from different sources traveling in different directions, suggesting that, if the crustal rocks are anisotropic, calculation of Poisson's ratio may be inappropriate. Shear-wave anisotropy in

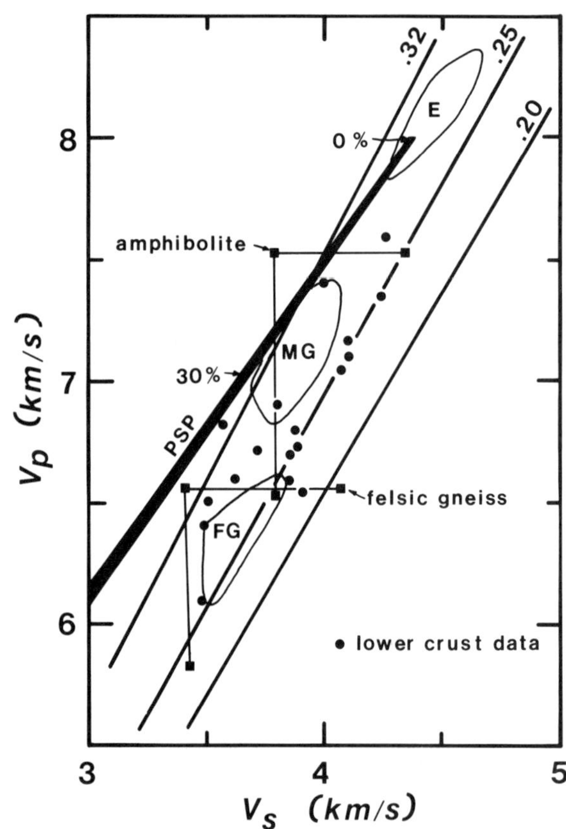

Figure 18. Compressional wave velocity vs. shear-wave velocity showing fields for eclogites (E), mafic gneisses (MG), felsic gneisses (FG), partially serpentinized peridotites with 0 and 30 percent serpentine points marked (PSP = solid area), reported crustal seismic data (circle), and lines of constant Poisson's ratio (0.20, 0.25, and 0.32). Also shown are data for an amphibolite and a felsic gneiss (squares connected by lines) in which shear waves are polarized yielding two shear-wave velocity values for one compressional wave velocity.

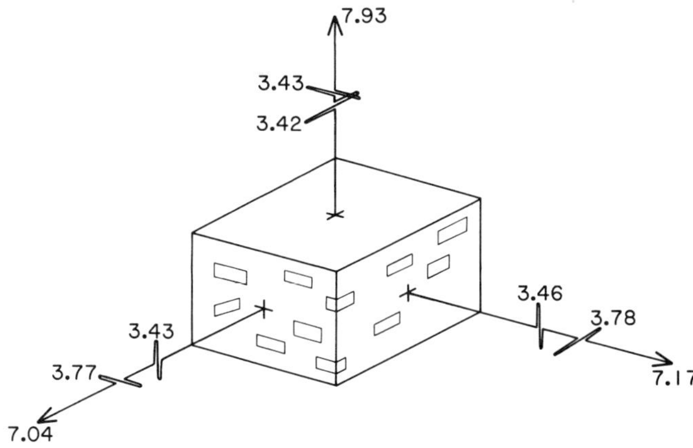

Figure 19. Compressional and shear-wave velocities in different propagation directions for anorthosite (N. I. Christensen, unpublished data). Rectangles schematically represent orientation of plagioclase crystals.

crustal rocks can place severe limitations on the use of V_p-V_s plots such as that presented in Figure 18. In anisotropic rocks, two polarized shear waves with different phase velocities usually propagate in a single direction. Thus, two Poisson's ratios could be calculated for that direction. This effect is illustrated in Figure 19, which shows compressional and shear-wave velocities for several propagation directions for an anorthosite (N. I. Christensen, unpublished data). This behavior introduces considerable complication to the interpretation of V_p-V_s diagrams because the mean V_p and V_s values may not be the appropriate values to use for comparison. This interpretation is also demonstrated in Figure 18, where we show, on a V_p-V_s plot, a single P-wave velocity with its two corresponding polarized shear velocities for several rocks (Christensen, 1971b). Also shown is the slow P-wave velocity normal to foliation and its single corresponding S-wave velocity. These data practically encompass the entire variation of V_p and V_s for the deep crustal data. Furthermore, if shear-wave splitting occurs in large volumes of the deep crust, as it does in some fault zones (e.g., Booth and others, 1985), we may not really know which S-wave was observed, thereby introducing considerable ambiguity in the interpretation of the data. Christensen (1984b) also demonstrated that increasing pore pressure can change V_p/V_s, leading to an increase in Poisson's ratio. This effect raises the possibility that high Poisson's ratios in the lower crust could be related to high pore pressure.

Another issue bearing on the interpretation of velocities in the lowermost crust and upper mantle is the role of the gabbro-eclogite phase transformation, as originally discussed in a series of papers by Ringwood and Green (1964, 1966), Green and Ringwood (1967, 1972), D. Green (1967), T. Green (1967), Cohen and others (1967), and Ito and Kennedy (1968, 1970, 1971). Recently, Furlong and Fountain (1986) estimated the compressional-wave velocities along various geothermal gradients through the gabbro-garnet granulite-eclogite fields, based on new calculations of the phase boundaries (Wood, 1987) and on old estimates of the volumetric abundances of mineral phases through the transformations (Green and Ringwood, 1967). The resulting velocity profiles for an olivine gabbro and quartz tholeiite composition are shown in Figure 20. These curves show no abrupt velocity discontinuities that would correspond to a Moho, but instead show smooth velocity gradients. Of particular importance is the prediction of high velocities (>7 km/sec) at levels roughly equivalent to typical Moho depths. These calculations lend support to the suggestion that high lower crustal velocities may reflect the dominance of garnet-bearing assemblages, as originally suggested by Yoder and Tilley (1962), and in some cases, eclogites in the crust-mantle transition zone. Velocities transitional between crustal and mantle values, such as those predicted in the model, are commonly reported in crust-mantle transition zones in extensional regimes such as the Wasatch Front (Braile and others, 1974).

Constraints on upper crustal composition

Seismological constraints on the composition of the upper portions of the continental crust are primarily derived from interpretation of the P_g phase in refraction studies, upper crustal velocity models from earthquake location algorithms, and seismic reflection data. Traditionally, the upper crust has been considered granitic in composition because of the similarity of P_g velocities to

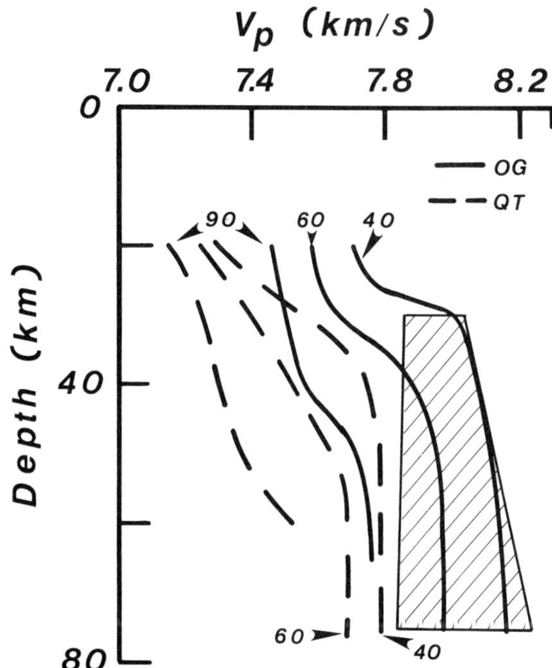

Figure 20. Variation of compressional wave velocity with depth for olivine gabbro (OG) and quartz tholeiite (QT) composition through the gabbro-garnet granulite-eclogite transition based on calculations from Furlong and Fountain (1986). Different curves correspond to calculations for geothermal gradients for 40, 60, 90 (wM/m^2) heat-flow provinces. The shaded area corresponds to velocities for typical ultramafic rocks.

early laboratory compressional wave velocity measurements in granite (Birch, 1958). The surface compositional estimates for shield areas (Table 5), which average all exposed rock types including metavolcanic rocks, suggest, however, an average granodioritic to tonalitic composition. Furthermore, Bott (1982) argued that the average density of the upper crust must be about 2.75 g/cm^3 to account for the significant negative Bouguer anomalies observed over granite and granodiorite plutons. This average is certainly closer to tonalite values than granite values, which is consistent with the conclusion of Woollard (1966) based on density measurements of crystalline rocks. Finally, inspection of the crustal cross sections (Table 7) shows that granites are not dominant, and most likely, that the upper crust is laterally and vertically heterogeneous.

It appears that upper crustal seismic velocities for the frequency range of refraction, reflection, and logging experiments are often lower than expected for the exposed rock types because the upper crust is penetrated by large-scale fractures. Data from the Soviet deep hole in the Kola Peninsula (e.g., Kremenetsky and Ovchinnikov, 1986) suggest that, at least locally, fractures and fluids may be pervasive throughout much of the upper crust, indicating that velocities through this region will tend to be lower than laboratory velocities. Seismic velocities measured in boreholes in crystalline rocks tend to be lower than laboratory-measured values for water-saturated rocks recovered from the hole. Examples of this effect have been reported for a borehole in the Wind River Range (Simmons and Nur, 1968; Smithson and Ebens, 1971), the Matoy well in Oklahoma (Simmons and Nur, 1968), the Stone Canyon well in California (Stierman and Kovach, 1979), and a borehole in the Flin Flon district in Saskatchewan (Hajnal and others, 1983). Cross-hole seismology (Wong and others, 1983) and borehole geophysical tomography (Daily and Ramirez, 1984) identified significant large-scale fractures in borehole environments, an observation that supports the hypothesis that large-scale cracks in the rock mass affect measurements of crustal seismic velocities obtained by field seismic methods. Laboratory velocity measurements represent the maximum velocity of the rock, that is, the value the field measurements would record if the rock mass were free of large-scale fractures (e.g., Hyndman and Drury, 1967; Salisbury and others, 1979; Stierman and Kovach, 1979). Furthermore, rock velocities will be significantly lowered if pore pressure is in excess of hydrostatic pressure (see Christensen, this volume). Thus, estimation of upper crustal composition from velocities requires knowledge of porosity and pore pressure, as has been illustrated by data from the Kola drillhole (see review by Christensen, this volume).

Detailed geometric aspects of the upper crust have been revealed by extensive exploration of the crust with seismic reflection techniques over the past decade. In the United States, these studies revealed important structural detail not evident in seismic refraction data. COCORP and other groups demonstrated that upper crustal faults, fault zones, and faulted terrain geometry can be imaged with reflection techniques. This is well illustrated by data from the Wind River Range (Smithson and others, 1978), the southern Appalachian Mountains (Cook and others, 1979), the Sevier Desert (Allmendinger and others, 1983), southern Oklahoma (Brewer and others, 1983), Kettle dome (Hurich and others, 1985), and the Mojave Desert (Cheadle and others, 1986). Buried basins and associated faults have been identified in Michigan (Brown and others, 1982), Hardeman County in Texas (Brewer and others, 1981), and Kansas (Serpa and others, 1984). The internal seismic structure of fault zones is reviewed by Mooney and Ginzburg (1986). A magma chamber in the upper crust was imaged in the Rio Grande rift (Brown and others, 1980; Brocher, 1981).

There are additional approaches that may yield reliable information about the composition of the upper continental crust. Clearly, surface geology can, in many cases, be reliably extrapolated to depths perhaps as great as 10 km. In some cases such extrapolation can be augmented by shallow borehole information, and future deep drilling will provide a wealth of information on the petrologic nature of the upper continental crust. Because seismic methods can recognize velocity contrasts in the upper crust, high-resolution refraction and reflection techniques can provide key information on the geometry of upper crustal bodies in some cases. In a few situations, exposed cross sections of the crust provide a window through many kilometers of the upper crust. These methods should be used on a case-by-case basis to develop site-specific upper crustal models.

SUMMARY AND CONCLUSIONS

In this chapter, we reviewed several approaches earth scientists have employed to determine the compositions of the crust and upper mantle. Most estimates of bulk chemical composition of the crust center around 60 percent SiO_2, although more silicic and mafic estimates have been proposed. Published estimates of the chemical composition of the lower crust vary from mafic to intermediate, whereas estimates of upper crustal composition, based on averages of surface rocks, tend to be similar (e.g., mean SiO_2 of 65 percent). In contrast, xenolith suites and nearly complete cross sections of the crust, exposed in orogenic belts, demonstrate the lithologic heterogeneity of the crust and upper mantle. Xenolith suites form the basis of several models that exhibit mafic granulites in the lower crust. Upper-mantle models show a complex mixture of various ultramafic and eclogitic rocks. Of significance, most cross sections show the crust is vertically zoned by metamorphic grade, with various composition granulites composing the lower crust.

Constraints offered by seismological methods provide significant information on crustal and upper-mantle composition. Laboratory studies demonstrate that cracks, confining pressure, pore pressure, and temperature, in addition to mineralogy, influence crustal seismic properties. Seismic anisotropy is an important parameter for crustal and upper-mantle rocks. In some cases its effects are greater than compositional changes. Quartzofeldspathic rock types show relatively small variations in compressional wave velocity with composition. In contrast, mafic rocks show a large variation of compressional wave velocity that is primarily related to variable proportions of garnet and plagioclase, or garnet and clinopyroxene. Seismic anisotropy, however, in many rocks can dominate the effect of compositional variations on compressional wave velocity. Constraints offered by seismological methods, combined with laboratory investigations, provide the following significant information on crustal and upper-mantle composition:

1. Upper-mantle velocities vary significantly in North America and correspond to laboratory velocities of dunite, peridotite, and eclogite, or mixtures of these rock types.

2. Seismic anisotropy, when recognized in the continental mantle, provides evidence for a peridotite-dominated upper mantle.

3. In many regions, seismic reflections are observed in the lower crust. Although variations in pore pressure and anisotropy could produce these reflections, we favor an origin due to compositional variation, as is observed in exposures of deep crustal rocks.

4. Midcrustal levels may be seismically transparent because of the dominance of quartzo-feldspathic lithologies.

5. Several recent North American seismic refraction lines, primarily over Phanerozoic tectonic features, suggest that the lower crust is dominated by various mafic rocks (in amphibolite to granulite facies).

6. The predominance of high-velocity components (garnet, olivine, pyroxene) is necessary to explain velocities in the crust-mantle transition zone.

7. Poisson's ratios calculated from compressional and shear-wave velocity data, where available, can be interpreted in terms of specific lithologic composition and variable pore pressure. For both laboratory and field seismic studies, the combined use of shear- and compressional-wave velocity data offers the greatest opportunity in the future to decipher the crustal and upper-mantle composition (e.g., see Christensen and Fountain, 1975).

8. The gabbro-garnet granulite-eclogite transition is likely important in deep, mafic portions of the crust. Garnet granulites can yield velocities intermediate between typical lower crustal values and upper-mantle values.

9. The uppermost crust sampled by P_g phases is likely not granite. Its composition can be approximated by tonalite, but it is lithologically heterogeneous.

10. Upper crustal physical properties, in many regions, are probably strongly controlled by large-scale porosity and fluids. These properties make direct translation of velocity into compositional information difficult.

To this point, our discussion indicates that there are many ambiguities and unknown parameters involved in the interpretation of crustal and upper-mantle composition. We have also discussed many important constraints that available data place on the problem. Despite the mentioned shortcomings of exposed cross sections, these terrains still provide the most accurate view of large portions of the Earth's crust. When the additional constraints of xenolith composition and seismic results are considered, we can formulate some realistic models of the crust and upper mantle in an approach similar to that developed by Smithson and Brown (1977). In Figure 21, we show a crustal and upper-mantle model constructed by interpolating the geologic relations between several exposed cross sections and, by inference, of geologic relations below the sections. The model is intended to be consistent with exposed cross sections, xenoliths, and the seismic nature of the crust and upper mantle. The inset shows which cross sections were used to construct various portions of the model and the variation of metamorphic grade in the model. The greenschist, amphibolite, and granulite boundaries are constrained by their position in the exposed cross sections. We realize that these boundaries may or may not coincide with equilibrium conditions predicted by present geothermal gradients. The granulite-eclogite facies boundary was placed on the basis of calculations from Wood (1987) for low geothermal gradients on the right side of Figure 21 and high gradients on the left side.

There are many characteristics of the model that merit comment. The right side of Figure 21 is dominated by sections derived from Archean and Proterozoic shield areas and can be regarded as examples of ancient crust. The left side of the diagram includes the Ivrea and Fiordland sections, terrains with complex Phanerozoic histories. This portion of the figure presents models for crustal structure and composition in active continental tectonic environments. The model shows significant lateral and vertical changes in lithologic composition. In some regions the crust is dominated by felsic rocks; in others the crust is compositionally zoned, with mafic rocks dominating lower crustal levels. Structural complexity is pervasive throughout, although scale limitations prevent exhibition of the fine-scale layering we expect at many crustal and upper-mantle levels. Metamorphic boundaries vary in depth in the model, and in some cases, high-grade metamorphic assemblages are shown tectonically transported to the surface. The upper mantle in the model shows significant lateral and vertical lithologic heterogeneity, as prescribed by the xenolith models. In some areas the distinction between the lower crust and upper mantle is unclear because mafic rocks are found embedded in peridotite and vice versa.

Despite our attempt to integrate a great deal of data into our model, much of the model is still relatively unconstrained. This will remain the case until seismic and potential field methods are able to provide the enhanced resolution that will allow mapping of small-scale bodies in the crust and upper mantle. Importantly, we need to explore methods to improve our capability to translate field data into compositional information. This will require better understanding of seismic anisotropy, pore pressure, shear-wave velocities, and the effect of rock composition on seismic properties.

In this chapter we have emphasized methods of evaluating the composition of the crust and upper mantle through the combined use of field observations and seismological techniques. An understanding of crustal and upper-mantle composition provides a foundation upon which we can develop insights into the evolu-

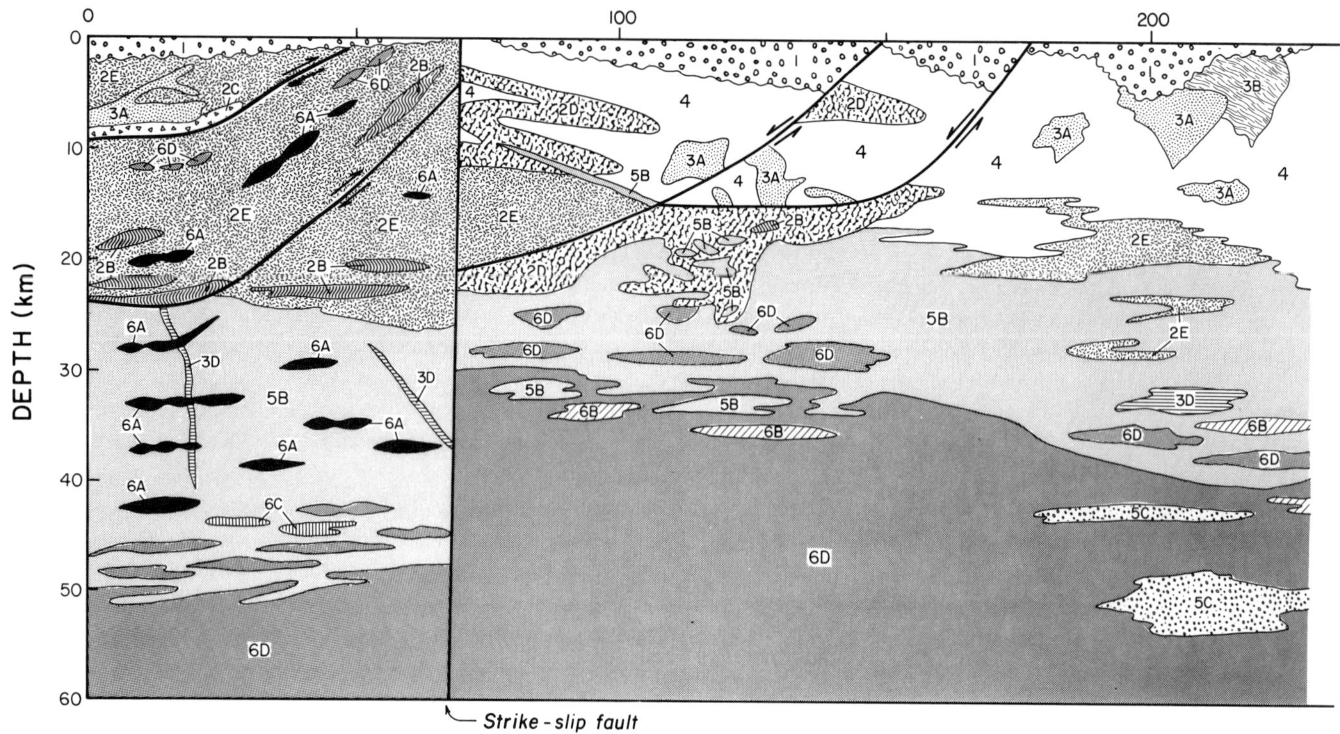

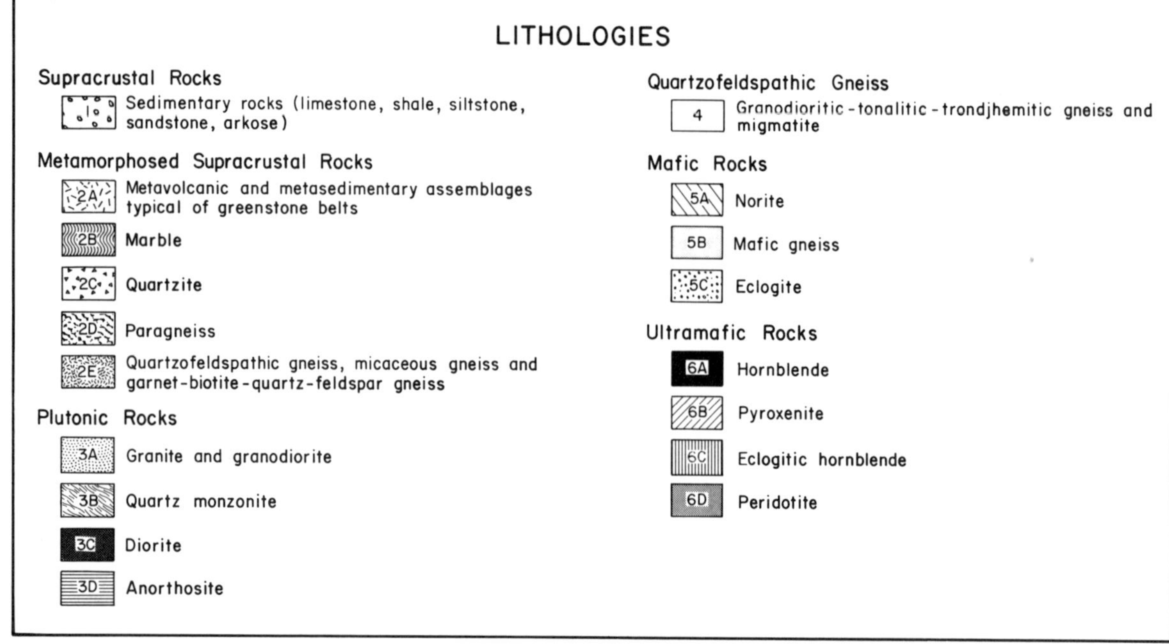

Figure 21. Hypothetical cross section of the continental crust and upper mantle; metamorphic facies and data sources are shown in lower right. Metamorphic facies are unmetamorphosed (U), greenschist (Gr), amphibolite (A), granulite (G), and eclogite (E). Sources for data are (1) Fiordland, New Zealand (Oliver and Coggon, 1979; Oliver, 1980); (2) Ivrea zone, northern Italy (Hunziker and Zingg, 1980; Fountain, 1986); (3) Fraser Range, Australia (Fountain and Salisbury, 1981); (4) Kapuskasing structural zone, Ontario (Percival and Card, 1985); (5) Pikwitonei-Sachigo Provinces, Manitoba (Weber and Scoates, 1978; Manitoba Mineral Resources Division, 1979; Arima and Barrett, 1984); (6) Musgrave Range, Australia (Fountain and Salisbury, 1981). The balance of the model was inferred from data and hypotheses outlined herein.

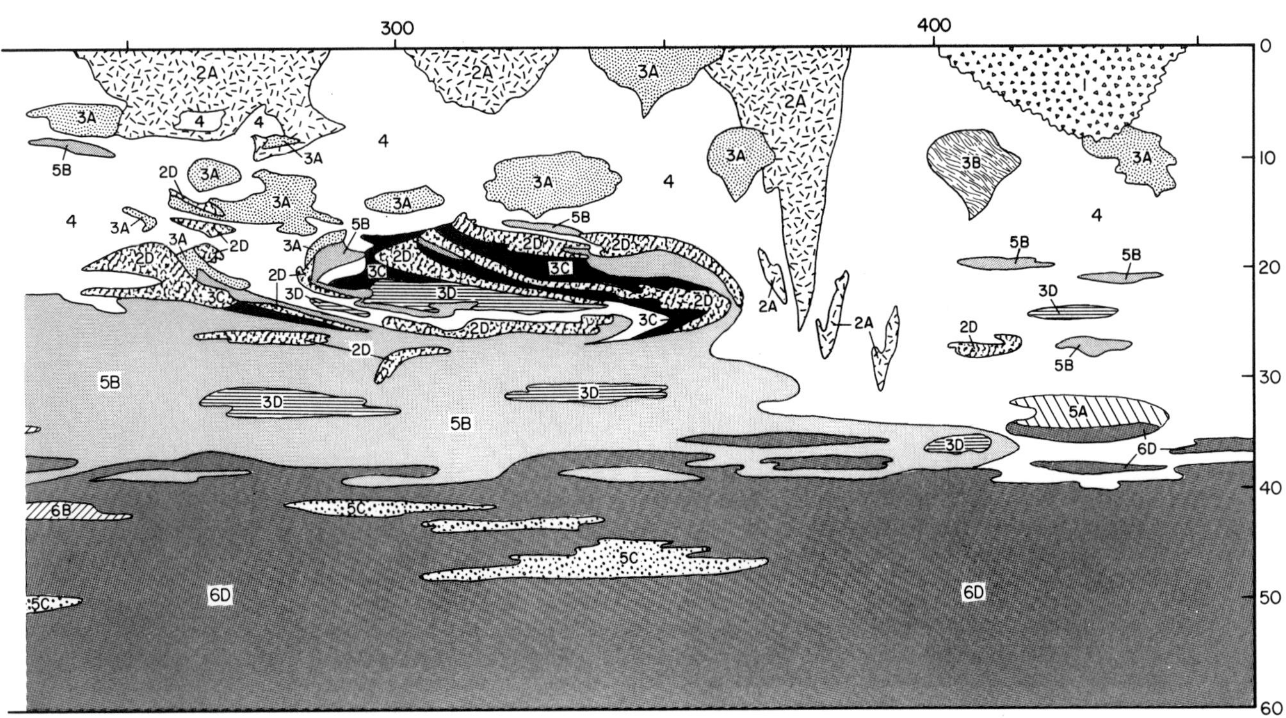

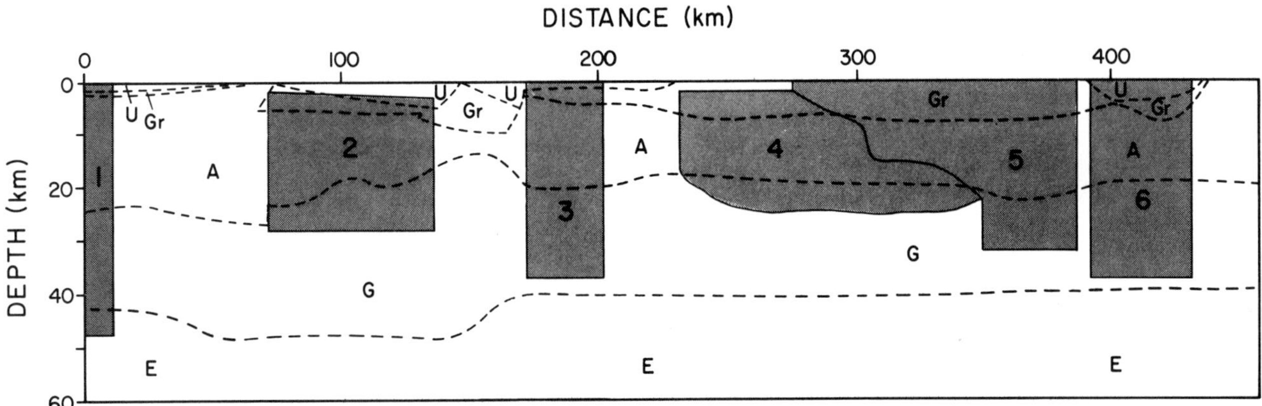

tion of the continental lithosphere. The lithosphere is a complex mosaic of diverse blocks that have evolved through lateral accretion (e.g., Coney and others, 1980), tectonic underplating (e.g., Yorath and others, 1985), magmatic underplating (e.g., Holland and Lambert, 1975; Furlong and Fountain, 1986), Andean margin magmatism (e.g., Taylor, 1977; Hamilton, 1981), and crustal extension (e.g., Wickham and Oxburgh, 1985). Each of these modifies lithospheric structure and composition. In the future, the recognition of deep continental regions that evolved through these processes will require integration of new detailed seismological laboratory data, deep drilling data, and additional geological observations.

ACKNOWLEDGMENTS

We thank R. Kay, W. Mooney, and H. Yoder for their many comments that helped improve the manuscript. This work represents research sponsored by ONR Contract N-00014-84-K-0207 (to N.I.C.), and by National Science Foundation Grants ISP-8111449 (to D.M.F.) and EAR-8418350 (to D.M.F.).

TABLE 1. ESTIMATES OF BULK CHEMICAL COMPOSITION OF CONTINENTAL CRUST

Major Oxide	Author Reference																			
	1	2	3	4	5	6	7	8	9	10	11	12	13	14	15	16	17	18	19	20
SiO_2	59.0	59.1	61.9	59.2	59.8	58.4	63.4	60.3	57.9	60.2	61.9	52.4	62.5	60.6	58.0	63.0	58.0	63.2	63.7	57.3
TiO_2	1.0	1.1	1.1	0.8	1.2	1.1	0.7	1.0	1.2	0.7	0.8	0.7	0.7	0.6	0.8	0.7	0.8	0.6	0.5	0.9
Al_2O_3	15.2	15.3	16.7	15.8	15.5	15.6	15.3	15.6	15.2	15.2	15.6	15.8	15.6	16.3	18.0	15.8	18.0	16.1	15.8	15.9
Fe_2O_3	3.1	3.1	....	3.4	2.1	2.8	2.5	....	2.3	2.5	2.6	2.6	6.1	2.1	....	2.0	....	....	....	....
FeO	3.7	3.8	6.9	3.6	5.1	4.8	3.7	7.2	5.5	3.8	3.9	5.6	....	4.7	7.5	3.4	7.5	4.9	4.7	9.1
MnO	0.1	0.1	....	0.1	0.1	0.2	0.1	....	0.2	0.1	0.1	0.1	0.1	0.1	....	0.1	0.1	0.1	0.1	....
MgO	3.5	3.5	3.4	3.3	4.1	4.3	3.1	3.9	5.3	3.1	3.1	7.7	3.2	4.2	3.5	2.8	3.5	2.8	2.7	5.3
CaO	5.1	5.1	4.2	3.1	6.4	7.2	4.6	5.8	7.1	5.5	5.7	7.7	6.0	5.3	7.5	4.6	7.5	4.7	4.5	7.4
Na_2O	3.7	3.8	3.4	2.1	3.1	3.1	3.4	3.2	3.0	3.0	3.1	1.1	3.4	3.0	3.5	4.0	3.5	4.2	4.3	3.1
K_2O	3.1	3.1	3.0	3.9	2.4	2.2	3.0	2.5	2.1	2.8	2.9	0.2	2.3	1.9	1.5	2.7	1.5	2.1	2.0	1.1
P_2O_5	0.3	0.3	....	0.2	0.2	0.3	0.2	0.2	0.3	0.2	0.3	0.1	0.2	....	....	....	....	0.2	0.2	....
H_2O	1.7	1.6	:.:	3.7	....	....	....	....	....	2.8	....	....	....	....	....	....	....	....	....	....

1, Clarke (1924)
2, Clarke and Washington (1924)
3, Goldschmidt (1933)
4, Goldschmidt (1954)
5, Poldervaart (1955)–shield
6, Poldervaart (1955)–fold belt
7, Vinogradov (1962)
8, Taylor (1964)
9, Pakiser and Robinson (1966)
10, Ronov and Yaroshevsky (1967)
11, Ronov and Yaroshevsky (1969)
12, McGetchin and Silver (1972)
13, Holland and Lambert (1972)
14, Galdin (1974)
15, Taylor (1977, 1979)
16, Smithson (1978)
17, Taylor and McLennan (1981)
18, Weaver and Tarney (1984a)
19, Weaver and Tarney (1984b)
20, Taylor and McLennan (1985)

TABLE 2. ESTIMATES OF BULK CHEMICAL COMPOSITION OF LOWER CONTINENTAL CRUST

Major Oxide	Author Reference 1	2	3	4	5	6	7	8	9	10
SiO_2	48.8	60.6	61.2	57.4	54.0	56.3	59.0	55.4	56.2	59.2
TiO_2	1.8	0.9	0.5	1.4	0.9	1.1	0.8	1.6	1.4	0.9
Al_2O_3	15.6	15.4	15.2	16.00	19.0	17.1	17.1	17.8	18.2	17.2
Fe_2O_3	2.8	....	2.3	2.1	....	3.8	2.0	....	....	....
FeO	8.2	7.3	5.7	7.3	9.0	4.5	3.4	12.0	10.0	6.1
MnO	0.1	0.2	0.2	0.2	....	0.1	0.1	0.2	0.2	0.1
MgO	1.1	3.9	5.6	5.6	4.1	5.0	2.8	5.6	4.8	3.4
CaO	2.6	5.7	7.5	7.1	9.5	5.5	4.6	2.4	3.9	5.9
Na_2O	3.9	2.8	3.0	2.0	3.4	2.1	4.0	2.0	2.0	4.0
K_2O	3.8	2.2	2.0	0.9	0.6	1.4	2.7	2.9	2.5	2.4
P_2O_5	0.2	....	....	....	....	....	....	....	....	0.3

1, Poldervaart (1955); Pakiser and Robinson (1966)
2, Lambert and Heier (1968)
3, Sheraton and others (1973)
4, Mehnert (1975)
5, Taylor (1977, 1979); Taylor and McLennan (1981)–gneisses
6, Leyreloup and others (1977)
7, Smithson (1978)
8, Maccarrone and others (1983–without dioritic and tonalitic gneisses
9, Maccarrone and others (1983)–with dioritic and tonalitic
10, Weaver and Tarney (1984a)

TABLE 3. ESTIMATES OF BULK COMPOSITION OF UPPER CONTINENTAL CRUST

Reference	Major Oxide % SiO_2	TiO_2	Al_2O_3	Fe_2O_3	FeO	MnO	MgO	CaO	Na_2O	K_2O	P_2O_5	H_2O
Galdin (1974)	62.6	0.6	16.5	1.9	3.8	0.1	2.9	4.3	3.3	2.2	....	....
Taylor (1977, 1979)	66.0	0.6	16.0	....	4.5	0.1	2.3	3.5	3.8	3.3	0.2	....
Taylor and McLennan (1985)	66.0	0.5	15.2	....	4.5	....	2.2	4.2	3.9	3.4	....	....

TABLE 4. MINERALOGIC COMPOSITION OF SHIELD AREAS

Mineral	Shaw and others (1967)	Wedehpol (1969)	Nesbitt and Young (1984)*	Nesbitt and Young (1984)†
Plagioclase	39.3	41.0	30.9	34.9
K-Feldspar	8.6	21.0	12.9	11.3
Quartz	24.4	21.0	23.2	20.3
Glass	0.0	0.0	0.0	12.5
Amphibole	0.0	6.0	2.1	1.8
Biotite	11.2	4.0	8.7	7.6
Muscovite	7.6	0.0	5.0	4.4
Chlorite	3.3	0.0	2.2	1.9
Pyroxene	0.0	4.0	1.4	1.2
Olivine	0.0	0.6	0.2	0.2
Oxides	1.4	2.0	1.6	1.4
Other	4.7	0.5	3.0	2.6

*Average mineralogic composition of upper crust.
†Average composition of exposed crust.

TABLE 5. PERCENTAGES OF ROCKS IN SHIELD AREAS*

Lithology	Superior–Slave–Wyoming	Churchill	Central	Grenville
Sedimentary rock	5	18	2	20
Felsic volcanics	0.1	0.5	20	4
Basic volcanics	12	6	3	3
Granitic rock†	76	70	70	66
Diorite and quartz diorite	2	1	0.01	0.01
Peridotite	0.1	Trace	Trace	Trace
Other	4	4	5	5

*From Engel (1963).
†Includes quartz monzonite, granodiorite, quartz porphyry, and gneisses pervasively veined by granite.

TABLE 6. LITHOLOGY OF XENOLITHS FROM NORTH AMERICAN LOCALITIES*

Location	Rock Types	References
1. Rhode Island	Spinel lherzolite (9–25/1,000–1,200), spinel harzburgite, werhlite	Leavy and Hermes (1979)
2. Ithaca, New York	Mafic syenite, garnet clinopyroxenite	Schulze and others (1978); Kay and others (1983)
3. Elliot County, Kentucky	Garnet peridotite	Schulze (1984)
4. Riley County, Kansas	Granite, schist, diorite, gabbro, norite, amphibolite, granulite facies metagabbro, garnet-sillimanite-sapphrine–bearing granulite, pyroxenite, eclogite, plagioclase eclogite	Brookins and Meyer (1974); Meyer and Brookins (1976)
5. State Line, Colorado-Wyoming, and Iron Mountain, Wyoming	Charnockite(?), anorthosite, gabbronorite, hypersthene monzogabbro, two-pyroxene granulite, two-pyroxene garnet granulite, cpx-garnet granulite, hypersthene granulite, garnet-kyanite granulite, augite granulite, kyanite eclogite, eclogite (700-1,300), dunite, garnet-spinel harzburgite (8–20/590–775), spinel lherzolite (900–950), garnet-spinel lherzolite (15–25/650–750), garnet-spinel-olivine websterite d(15–25/650–750), garnet-spinel clinopyroxenite (15–25/650–750)	Eggler and McCallum (1974); McCallum and Maberek (1976); Ater and others (1984); Bradley and McCallum (1984); Kirkley and others (1984)
6. Leucite Hills, Wyoming	Anorthosite, gabbroic granulite (<10/875) granite, quartz diorite, granite gneiss, diorite gneiss, quartz diorite gneiss, schist, dunite	Kay and others (1978); Sperr (1985)
7. Big Belt Mountains, Montana	Cpx-opx-plag-spinel granulite, pyroxenite, harzburgite, wehrlite,lherzolite	Eggler and McCallum (1974)
8. Central Montana	Schist, gneiss, amphibolite, granulite, mafic granulite, mafic amphibolite, spinel peridotite, spinel pyroxenite, dunite, garnet lherzolite (50–50/1,220–1,350), garnet lherzolite (23-42/830-990), garnet pyroxenite (50–60/ 1,220–1,350), garnet harzburgite (50–60/ 1,230–1,350), garnet dunite (50–60/ 1,220–1,350)	Hearn and Boyd (1975); Hearn and McGee (1984)
9. Snake River Plain, Idaho	Rhyolite, pumice, welded tuff, sedimentary rock, biotite-garnet gneiss (730), charnockite, opdalite (hypersthene granodiorite), enderbite, hypersthene diorite, anorthosite, norite (4–8/700–800 for metamorphic rocks)	Leeman (1979); Matty (1984); Leeman and others (1985)
10. Rio Grande Rift (New Mexico)		
a. Black Range	Metagabbro, spinel metagabbro, olivine–spinel metagabbro (11–14/1,100–1,200), spinel lherzolite, spinel clinopyroxenite, clinopyroxenite	Fodor (1978)
b. Kilbourne Hole	Charnockite, anorthosite, sillimanite-bearing garnet granulite, 2-pyroxene granulite (6–10/800–1,000), granet granulite (≤5.4/750–1,000), garnet orthopyroxenite, spinel lherzolite (14-22/900-1100)	Padovani and Carter (1977); Reid and Woods (1978)
c. Engle Basin	Charnockite (7–14/899), pyroxene-plagioclase granulite (5–11/883–934), pyroxenite (8–17/965-1,011), lherzolite (8–20/1,062–1,211)	Warren and others (1979)
d. Elephant Butte	Two-pyroxene granulite (6–13/900—980), clinopyroxenite, spinel lherzolite (14–20/935–1,030)	Baldridge (1979)
e. Abiquiu	Plagioclase-bearing pyroxenite, orthopyroxenite, websterite (877–907)	Baldridge (1979)
f. Cieneguilla	Granulite, harzburgite (943–992)	Baldridge (1979)

TABLE 6. LITHOLOGY OF XENOLITHS FROM NORTH AMERICAN LOCALITIES* (continued)

Location	Rock Types	References
11. Central Colorado Plateau		
a. Navajo Field	Garnetiferous granitic rocks, sillimanite-garnet-biotite gneiss, amphibolite schist, felsic granulite, mafic garnet granulite (>6/555-635), eclogite, clinopyroxenite, websterite, dunite, harzburgite, garnet lherzolite (43/93–1,230)	Ehrenberg (1979); Ehrenberg and Griffin (1979)
b. Moses Rock, Mule's Ear, and Garnet Ridge	Rhyolite porphyry, granite, basalt, chlorite schist, serpentine schist, quartz monzonite, granite gneiss, metabasalt, diorite mafic amphibolite (475–750), garnet amphibolite, garnet-sillimanite schist (750–850) felsic granulite, intermediate granulite, granulite metagabbro gneiss (500–800), eclogite, plagioclase eclogite, garnet pyroxenite (15.5/710), jadeite-clinopyroxenite, spinel websterite, spinel lherzolite	Watson and Morton (1969) McGetchin and Silver (1972); Padovani and others (1982)
c. Buell Park and Green Knobs	Granite, quartz monzonite, granodiorite, felsic volcanic rocks, sandstone, low-grade meta-felsite, biotite schist, amphibolite (530), meta-diabase, two pyroxene granulite (5–8/700–800), amphibole-bearing two-pyroxene granulite, garnet granulite (6–9/640–750), lherzolite-websterite (700–1,100), opx-rich websterite, ultramylonitic and mylonitic peridotite	Smith and Levy (1976); Smith (1979); O'Brien (1983)
12. Colorado Plateau Margin (Arizona)		
a. Chino Valley	Quartzite, schist, two-pyroxene feldspathic granulite, garnet-bearing amphibolite, eclogite (550-700), pyroxenite, garnet pyroxenite, (550–700), garnet websterite (10–20/700–1,000)	Arculus and Smith (1979); Schulze and Helmstaedt (1979; Aoki (1981)
b. Carefree	Plagioclase-bearing amphibolite (assume P = 10/650), garnet-rich amphibolite, garnet-cpx-plagioclase granulite (<8), eclogite (assume P = 10/900), garnet clinopyroxenite	Esperanca and Holloway (1984)
c. San Franciso Volcanic Field	Charnockite, norite, anorthosite, olivine gabbro, two-pyroxene gabbro, two-pyroxene granulite, hypersthene gabbro, pyroxene granulite, plagioclase pyroxenite, wehrlite, olivine clinopyroxenite, clinopyroxenite, websterite, olivine websterite, spinel pyroxenite, dunite	Cummings (1972; Stoesser (1973, 1974)
d. Geronimo volcanic field	Two-pyroxene granulite, harzburgite, websterite, wehrlite, clinopyroxenite, amphibole peridotite, spinel lherzolite	Kempton and others (1984)
13. California		
a. Dish Hill	Granite, two-pyroxene-plagioclase granulite, hornblende clinopyroxenite, clinopyroxenite, spinel lherzolite, spinel wehrlite, garnet clinopyroxenite (15-20/1,100)	Shervais and others (1973)
b. Central Sierra Nevada	Pyroxenite (13), harzburgite, lherzolite, mafic granulites, gabbro	Dodge and others (1986)
14. British Columbia		
a. Kettle River, Lassie Lake, Lightening Peak	Metamorphic lherzolite, dunite, wehrlite (all 9-23/950-1,300)	Ross (1983)
b. Big Timothy Mountain, Jacques Lake	Lherzolite, harzburgite, websterite, dunite, wehrlite (all 10-27/900-1,200), lherzolite tectonite (1,085)	Ross (1983); Littlejohn and Greenwood (1974)

TABLE 6. LITHOLOGY OF XENOLITHS FROM NORTH AMERICAN LOCALITIES* (continued)

Location	Rock Types	References
14. British Columbia (continued)		
c. Castle Rock	Lherzolite and websterite (10-24/1,000-1,300), lherzolite (>1,600)	Ross (1983); Littlejohn and Greenwood (1974)
d. Summit Lake	Wehrlite, spinel lherzolite, clinopyroxenite, olivine websterite, websterite, dunite (18-20/1,080-1,100 or 9-20/950-1,250)	Ross (1983); Brearley and others (1984)
e. Haggens Point, Boss Mountain and Nicola Lake	Lherzolite tectonite (1,085), cumulate lherzolite (840)	Littlejohn and Greenwood (1974)
15. Yukon River and Selkirk Cone, Yukon	Lherzolite (850-1,160), lherzolite, websterite, dunite, wehrlite (all 12-28/950-1,200)	Sinclair and others (1978); Ross (1983)
16. Aleutian arc	Gabbro, diorite, ultramafic cumulate, dunite	Kay and Kay (1981); Kay and others (1986)

TABLE 7. LITHOLOGIES IN EXPOSED CROSS SECTIONS OF THE CONTINENTAL CRUST

Exposed Cross Section	Greenschist Facies and Unmetamorphosed Rocks	Amphibolite Facies	Granulite Facies
Ivera-Verbano and Strona-Ceneri, Italy	Limestone, volcanic rocks, phyllite, orthogneiss, schists	Granite, schist, paragneiss, orthogneiss, amphibolite, marble	Meta-pelite, marble, granetiferous mafic gneiss, mafic gneiss, peridotite, pyroxenite, hornblendite, amphibolite
Sachigo-Pikwitonei, Manitoba, Canada	Metabasalt, metarhyolite, metagraywacke, tonalite, granite, metaconglomerate	Metabasalt, metarhyolite, metagraywacke, tonalite, granite, tonalitic gneiss, granodioritic gneiss	Tonalite gneiss, metagranodiorite, silicate-oxide iron formation, mafic gneiss, ultramafic rocks
Michipicoten-Wawa-Kapuskasing, Ontario, Canada	Metabasalt, metarhyolite, meta-andesite, metagraywacke, granite, tonalite metaconglomerate, iron formation, chert	Granite, granodiorite, tonalite, tonalite-granodiorite gneiss, diorite-monzonite, metavolcanic, rocks, metagraywacke, amphibolite	Anorthosite, tonalite-granodiorite gneiss, mafic gneiss, paragneiss
Fraser Range, Australia	....	Granite, adamellite gneiss, migmatite, quartzo-feldspathic gneiss	Mafic gneiss, metasedimentary gneiss, felsic gneiss, anorthosite, olivine gabbro, norite
Musgrave Range, Australia	Arkose, sandstone	Granite, adamellite, granite gneiss, quartzo-feldspathic gneiss	Maficultramafic plutons, anorthosite, granodioritic gneiss, mafic gneiss, quartzite, calcareous rocks, adamellite
Kasila Series, Western Africa	....	Quartzo-feldspathic gneiss, calc-silicate gneiss, metabasic gneiss	Mafic gneiss, layered gabbro and anorthosite complexes, quartz-magnetite gneiss, quartz-diopside gneiss, sillimanite-bearing gneiss, granite gneiss
Calabria, Italy	Limestone, phyllite, gneiss	Diorite gneiss, tonalite, granite	Metapelite, mafic gneiss, metaquartz monzo-gabbronorite
Fiordland, New Zealand	Granite, granodiorite, diorite, gabbro, sedimentary rocks	Granite, granodiorite, quartzo-feldspathic and micaceous gneiss, schist, metabasite, marble, calc-silicate	Metagabbroic diorite, feldspathic gneiss, ultramafic rocks, anorthosite veins

TABLE 8. COMPRESSIONAL WAVE VELOCITIES AND DENSITIES OF SELECTED CRUSTAL AND MANTLE ROCKS

Rock	Density (g/cm^3)	P(MPa) (km/sec) 10	50	100	200	400	600	1000	Anistropy (%)	Reference*
GRANITE										
Westerly, Rhode Island	2.619	4.1	5.63	5.84	5.97	6.10	6.16	6.23	3.7	1
Barre, Vermont	2.655	5.1	5.86	6.06	6.15	6.25	6.32	6.39	1.3	1
Barriefield, Ontario†	2.672	5.7	6.21	6.29	6.35	6.42	6.46	6.51	1.5	1
TONALITE										
Val Verde, California	2.763	5.1	....	6.33	6.43	6.49	6.54	6.60	1.1	1
San Luis Rey, California	2.798	5.1	....	6.43	6.52	6.60	6.64	6.71	0.7	1
DIORITE										
Dedham, Massachusetts	2.906	5.5	....	6.46	6.53	6.60	6.65	6.70	0.1	1
GABBRO AND NORITE										
Mellen, Wisconsin	2.931	6.8	7.04	7.07	7.09	7.13	7.16	7.21	3.5	1
French Creek, Pennsylvania	3.054	5.8	6.74	6.93	7.02	7.11	7.17	7.23	1.0	1
Transvaal, Africa	2.978	6.6	7.02	7.07	7.11	7.16	7.20	7.28	1.8	1
ANORTHOSITE										
Adirondacks, #1	2.707	5.96	6.40	6.58	6.74	6.83	6.89	6.95	2.3	2
Stillwater Complex, Montana	2.770	6.5	....	6.97	7.01	7.05	7.07	7.10	3.9	1
Bushveld Complex	2.807	5.7	6.92	6.98	7.05	7.13	7.16	7.21	3.1	1
AMPHIBOLITE FACIES GRANITIC GNEISS										
Gneiss 1, Connecticut	2.654	4.5	....	5.85	6.06	6.18	6.24	6.33	2.2	3
Gneiss 2, Connecticut	2.643	4.8	....	5.97	6.12	6.22	6.27	6.35	1.9	3
AMPHIBOLITE FACIES GRANODIORITIC GNEISS										
New Hampshire	2.758	4.4	....	5.95	6.07	6.16	6.21	6.30	4.1	1
AMPHIBOLITE FACIES TONALITIC GNEISS										
Gneiss 3, Connecticut	2.755	5.1	....	6.15	6.32	6.43	6.49	6.57	0.3	3
Gneiss 4, Connecticut	2.824	4.8	....	6.03	6.25	6.40	6.47	6.58	5.2	3
PELITIC SCHIST										
Iverea Zone, Italy (IV-21)	2.700	....	....	5.65	5.92	6.13	6.26	6.37	9.6	4
Thomaston, Connecticut (Garnet Schist)	2.760	5.2	5.95	6.20	6.32	6.43	6.50	6.59	9.9	3
Litchfield, Connecticut	1.750	....	6.06	6.27	6.39	6.52	6.58	6.68	21.0	3
QUARTZITE										
Montana	2.647	5.6	....	6.11	6.15	6.22	6.26	6.35	3.1	1
Clarendon Springs, Vermont	2.630	5.5	5.89	6.05	6.12	6.19	6.24	6.30	0.3	3
GREENSTONE										
Yreka, California	2.910	6.7	6.80	6.84	6.90	6.96	6.99	....	3.3	5
Luray, Virginia	2.930	6.5	6.54	6.58	6.65	6.71	6.75	....	1.0	5
Marin County, California	2.880	5.9	....	6.35	6.46	6.59	6.70	....	2.1	5

TABLE 8. COMPRESSIONAL WAVE VELOCITIES AND DENSITIES OF SELECTED CRUSTAL AND MANTLE ROCKS (continued)

Rock	Density (g/cm^3)	P(MPa) (km/sec) 10	50	100	200	400	600	1000	Anistropy (%)	Reference*
Amphibolite										
Canyon Mountain, Oregon	2.925	....	6.66	6.70	6.76	6.80	6.83	6.86	8.6	6
#2 Bantam, Connecticut	3.030	5.5	6.30	6.63	6.87	7.04	7.10	7.18	14.1	3
Ivrea Zone, Italy (IV-16)	3.044	....	....	6.06	6.60	7.04	7.21	7.32	6.1	4
Hypersthene granodiorite gneiss (charno-enderbite)										
Tichbourne, Ontario	2.712	....	....	6.06	6.17	6.31	6.41	6.48	0.7	7
Pallavaram, India	2.740	6.15	....	6.24	6.30	6.36	6.40	6.46	1.2	1
Hypersthene tonalite gneiss (enderbite)										
#3, Adirondacks	2.703	5.97	6.25	6.34	6.41	6.48	6.53	6.60	3.1	2
New Jersey Highlands	2.681	....	....	6.36	6.44	6.52	6.57	6.63	2.2	7
#2, Adirondacks	2.702	5.78	6.23	6.34	6.42	6.51	6.57	6.64	2.7	2
Hypersthene quartz monzonite (mangerite)										
#5, Adirondacks	2.739	6.04	6.20	6.28	6.35	6.44	6.48	6.54	0.5	2
#8, Adirondacks	2.826	6.20	6.37	6.43	6.53	6.57	6.60	6.67	1.7	2
Saranac Lake, New York	2.830	....	....	6.40	6.54	6.66	6.71	6.77	1.0	7
Granulite facies metapelitic gneiss										
Ivera Zone, Italy (IV-23)	2.954	....	....	6.60	6.74	6.87	6.96	7.05	7.6	4
Ivrea Zone, Italy (IV-7)	3.104	....	....	6.99	7.12	7.27	7.35	7.43	1.0	4
Hornblende-pyroxene granulite (mafic gneiss)										
Santa Lucia, California	2.899	....	....	6.75	6.83	6.91	6.95	7.02	3.3	7
#14,Adirondacks	3.170	5.84	6.80	6.95	7.01	7.11	7.16	7.25	4.5	2
Ivrea Zone, Italy (IV-15)	3.080	....	....	7.12	7.26	7.38	7.45	7.51	5.0	4
Pyroxene-garnet granulite (mafic gneiss)										
Ivrea Zone, Italy (IV-9)	2.942	....	....	6.31	6.58	6.85	6.99	7.09	1.6	4
Ivrea Zone, Italy (IV-17)	2.910	....	....	6.90	7.08	7.21	7.28	7.35	1.2	4
Ivrea Zone, Italy (IV-20)	3.047	....	....	6.69	7.12	7.38	7.48	7.57	0.4	4
Serpentinite										
Mount Boardman, California	2.513	....	4.93	4.96	5.03	5.15	5.25	5.42	1.2	6
Paskenta, California	2.517	....	4.94	5.00	5.07	5.19	5.31	5.49	3.8	6
Canyon Mountain, Oregon	2.535	....	5.28	5.31	5.35	5.42	5.50	5.62	2.1	6
Pyroxenite										
Stillwater Complex, Montana	3.279	7.42	....	7.02	7.66	7.72	7.75	7.83	1.3	1
Sonoma County, California	3.247	6.8	....	7.73	7.79	7.88	7.93	8.01	4.7	1
Bushveld Complex	3.288	....	7.40	7.49	7.60	7.75	7.85	8.02	4.1	1
Dunite										
Addie, North Carolina	3.304	7.7	....	7.99	8.05	8.14	8.20	8.28	7.2	1
#4 Twin Sisters, Washington	3.30	8.3	8.30	8.33	8.39	8.43	8.46	8.51	6.8	8
#11 Twin Sisters, California	3.30	8.1	8.22	8.28	8.35	8.43	8.48	8.52	8.1	8

TABLE 8. COMPRESSIONAL WAVE VELOCITIES AND DENSITIES OF SELECTED CRUSTAL AND MANTLE ROCKS (continued)

Rock	Density (g/cm^3)	P(MPa) (km/sec) 10	50	100	200	400	600	1000	Anistropy (%)	Reference*
ECLOGITE										
#13 Sunmore, Norway	3.539	7.6	7.91	7.98	8.06	8.13	8.17	8.23	2.4	2
#14 Sunmore, Norway	3.577	7.8	8.05	8.14	8.22	8.30	8.35	8.42	3.4	2
#15 Grytinvaag, Norway	3.585	7.7	7.79	8.05	8.16	8.28	8.35	8.43	1.2	2

*References: 1, Birch (1960); 2, Manghnani and others (1974); 3, Christensen (1965); 4, Fountain (1976); 5, Christensen (1970); 6, Christensen (1978); 7, Christensen and Fountain (1975); 8, Christensen (1971a).
†Altered, but typical of many granites.

TABLE 9. POISSON'S RATIO AND COMPRESSIONAL AND SHEAR WAVE VELOCITY DATA FOR THE MIDDLE AND LOWER CRUST OF NORTH AMERICA

Location	Depth to Top of Horizon (km)	V_p (km/sec)	V_s (km/sec)	σ*	Method†	Reference§
New Madrid	25.2	7.17	4.10	0.26	E	1
SW Oklahoma	18.0	6.72	3.88	0.25	R	2
	26.0	7.05	4.07	0.25	R	
Eastern New Mexico	23.0	6.72	3.88	0.25	R	2
	32.0	7.10	4.10	0.25	R	
	41.0	7.35	4.24	0.25	R	
Colorado Plateau	27.0	6.80	3.87	0.26	SWD	3
Eastern Basin and Range	22.5	6.70	3.71	0.28	SWD	3
Colorado Plateau–Basin	14.7	6.5	3.50	0.30	R	4
and Range Transition	24.7	7.4	4.0	0.29	R	
Wasatch Front	19.0	6.90	3.80	0.28	R	5
	28.0	7.60	4.25	0.27	R	
Basin and Range	20.0	6.60	3.85	0.24	SWD	6
Northern Basin and Range	20.0	6.60	3.61	0.29	SWD	6
High Lava Plain, Oregon	15.0	6.70	3.85	0.25	SWD	6
Eastern Snake River Plain	20.0	6.82	3.56	0.31	SWD	7

*Poisson's ratio
†SWD = surface wave dispersion; R = refraction; E = earthquake source.
§References: 1, Mitchell and Hashim (1977); 2, Mitchell and Landisman (1971); 3, Bucher and Smith (1971); 4, Keller and others (1975); 5, Braile and others (1974); 6, Priestley and Brune (1982); 7, Greensfelder and Kovach (1982).

REFERENCES CITED

Alexandrov, K. S., and Ryzhova, T. V., 1961, The elastic properties of rock-forming minerals; 1, Pyroxenes and amphiboles: Bulletin of the Academy of Science U.S.S.R., Geophysics Series, no. 8, p. 871–875.

—— , 1962, Elastic properties of rock-forming minerals; 3, Felspars: Bulletin of the Academy of Science U.S.S.R., Geophysics Series, no. 2, p. 129–131.

Allenby, R. J., and Schnetzler, C. C., 1983, United States crustal thickness: Tectonophysics, v. 83, p. 13–31.

Allmendinger, R., and 7 others, 1983, Cenozoic and Mesozoic structure of the eastern Basin and Range from COCORP seismic reflection data: Geology, v. 11, p. 532–536.

Anderson, D. L., and Dziewonski, A. M., 1982, Upper mantle anisotropy; Evidence from free oscillation: Geophysical Journal of the Royal Astronomical Society, v. 69, p. 383–404.

Ansorge, J., 1968, Die Struktur der Erdkruste an der Westflanke der Zone von Ivrea: Schweizerische Mineralogische und Petrographische Mitteilungen, v. 48, p. 247–254.

Aoki, K., 1981, Chemical composition of mafic-ultramafic xenoliths in Sullivan Buttes: Tohoku, Daigaku, Sendai, Japan Science Reports, Third Series, v. 15, p. 131–135.

Arculus, R. J., and Smith, D., 1979, Eclogite, pyroxenite, and amphibolite in inclusions in the Sullivan Buttes latite, Chino Valley, Yavapai County, Arizona, *in* Boyd, F. R., and Meyer, H.O.A., eds., The mantle sample; Inclusions in kimberlites and other volcanics: Washington, D.C., American Geophysical Union, p. 309–317.

Arima, M., and Barrett, R. L., 1984, Sapphrine-bearing granulites form the Sipiwesk Lake area of the late Archean Pikwitonei granulite terrain, Maintoba, Canada: Contributions to Mineralogy and Petrology, v. 88, p. 102–112.

Ater, P. C., Eggler, D. H., and McCallum, M. E., 1984, Petrology and geochemistry of mantle xenoliths from Colorado-Wyoming kimberlites; Recycled oceanic crust?, *in* Kornprobst, J., ed., Kimberlites; 2, The mantle and crust-mantle relationships: Amsterdam, Elsevier, p. 309–318.

Babuska, V., 1972, Elasticity and anisotropy of dunite and bronzitite: Journal of Geophysical Research, v. 77, p. 6955–6965.

Babuska, V., Fiala, J., Kumazawa, M., Ohno, I., and Sumino, Y., 1978, Elastic properties of garnet solid-solution series: Physics of the Earth and Planetary Interiors, v. 16, p. 157–176.

Baldridge, W. S., 1979, Mafic and ultramafic inclusion suites from the Rio Grande Rift (New Mexico) and their bearing on the composition and thermal state of the lithosphere: Journal of Volcanology and Geothermal Research, v. 6, p. 319–351.

Bamford, D., 1973, Refraction data in western Germany; A time-term interpretation: Journal of Geophysics, v. 39, p. 907–927.

—— , 1977, P_n anisotropy in a continental upper mantle: Geophysical Journal of the Royal Astronomical Society, v. 49, p. 29–48.

Bamford, D., Jeutch, M., and Prodehl, C., 1979, P_n anisotropy studies in northern Britain and the eastern and western United States: Geophysical Journal of the Royal Astronomical Society, v. 57, p. 397–430.

Berckhemer, H., 1968, Topographie des "Ivrea-Korpers" abgeleitet aus seismichen und gravimetrischen Daten: Schweizersiche Mineralogische und Petrographische Mitteilungen, v. 48, p. 235–246.

—— , 1969, Direct evidence for the composition of the lower crust and Moho: Tectonophysics, v. 8, p. 97–105.

Birch, F., 1958, Interpretation of the seismic structure of the crust in light of experimental studies of wave velocities in rocks, *in* Benioff, H., ed., Contributions in geophysics in honor of Beno Gutenberg; New York, Pergamon Press, p. 158–170.

—— , 1960, The velocity of compressional waves in rocks to 10 kilobars, Part 1: Journal of Geophysical Research, v. 65, p. 1085–1102.

—— , 1961, The veolocity of compressional waves to 10 kilobars, Part 2: Journal of Geophysical Research, v. 66, p. 2199–2224.

Blair, S., 1980, Seismic structure of the crust and upper mantle under North America and the Pacific Ocean [M.S. thesis]: Seattle, University of Washington, 64 p.

Bois, C., and 8 others, 1985, Deep seismic profiling of the crust in northern France; The ECORS project, *in* Barazangi, M., and Brown, L., eds., Reflection seismology; A global perspective: American Geophysical Union Geodynamics Series, v. 13, p. 21–29.

Booth, D. C., Crampin, S., Evans, R, and Roberts, G., 1985, Shear-wave polarizations near the North Anatolian Fault; 1. Evidence for anisotropy-induced shear-wave splitting: Geophysical Journal of the Royal Astronomical Society, v. 83, p. 61–73.

Bott, M.H.P., 1982, The interior of the Earth; Its structure, constitution, and evolution, 2nd ed.: London, Elsevier, 403 p.

Bowes, D. R., 1972, Geochemistry of Precambrian crystalline basement rocks, northwest Scotland: 24th International Geological Congress, Section 1, p. 97–103, p. 403.

Bradley, S. D., and McCallum, M. E., 1984, Granulite facies and related xenoliths from Colorado-Wyoming kimberlite, *in* Kornprobst, J., ed., Kimberlites; 2, The mantle and crust–mantle relationships: Amsterdam, Elsevier, p. 205–217.

Braile, L. W., Smith, R. B., Keller, G. R., Welch, R., and Meyer, R. P., 1974, Crustal structure across the Wasatch Front from detailed seismic refraction studies: Journal of Geophysical Research, v. 79, p. 2669–2677.

Braile, L. W., and 9 others, 1982, The Yellowstone–Snake River Plain seismic profiling experiment; Crustal structure of the eastern Snake River Plain: Journal of Geophysical Research, v. 87, p. 2597–2609.

Brearley, M., Scarfe, C. M., and Fujii, T., 1984, The petrology of ultramafic xenoliths from Summit Lake near Prince George, British Columbia: Contributions to Mineralogy and Petrology, v. 88, p. 53–63.

Brewer, J. A., and 5 others, 1981, Proterozoic basin in the southern midcontinent of the U.S. revealed by COCORP deep seismic reflection profiling: Geology, v. 9, p. 569–575.

Brewer, J. A., Good, R., Oliver, J. E., Brown, L. D., and Kaufman, S., 1983, COCORP profiling across the southern Oklahoma aulacogen; Overthrusting of the Wichita Mountains and compression within the Anadarko basin: Geology, v. 11, p. 109–114.

Brocher, T. M., 1981, Geometry and physical properties of the Socorro, New Mexico, magma bodies: Journal of Geophysical Research, v. 86, p. 9420–9432.

Brookins, D. G., and Meyer, H.O.A., 1974, Crustal and upper mantle stratigraphy beneath eastern Kansas: Geophysical Research Letters, v. 1, p. 269–272.

Brown, L. D., Chapin, C. E., Sanford, A. R., Kaufman, S., and Oliver, J. E., 1980, Deep structure of the Rio Grande Rift from seismic-reflection profiling: Journal of Geophysical Research, v. 85, p. 4773–4800.

Brown, L. D., Jensen, L., Oliver, J., Kaufman, S., and Steiner, D., 1982, Rift structure beneath the Michigan basin from COCORP profiling: Geology, v. 10, p. 645–649.

Bucher, R. L., and Smith, R. B., 1971, Crustal structure of the eastern Basin and Range province and the northern Colorado Plateau from phase velocities of Rayleigh waves, *in* Heacock, J. G., ed., The structure and physical properties of the earth's crust: American Geophysical Union Monograph 14, p. 59–70.

Cheadle, M. J., and 6 others, 1986, Geometries of deep crustal faults; Evidence from the COCORP Mojave survey, *in* Barazangi, M., and Brown, L., eds., Reflection seismology; The continental crust: American Geophysical Union Geodynamics Series, v. 14, p. 305–312.

Chesnokov, Y. M., and Nevskig, M. V., 1977, Seismic anisotropy investigations in the USSR: Geophysical Journal of the Royal Astronomical Society, v. 49, p. 115–121.

Christensen, N. I., 1965, Compressional wave velocities in metamorphic rocks at pressures to 10 kilobars: Journal of Geophysical Research, v. 70, p. 6147–6164.

—— , 1966, Elasticity of ultrabasic rocks: Journal of Geophysical Research, v. 712, p. 5921–5931.

—— , 1970, Possible greenschist facies metamorphism of the oceanic crust: Geo-

logical Society of America Bulletin, v. 81, p. 905–908.
—— , 1971a, Fabric, seismic anisotropy, and tectonic history of the Twin Sisters dunite, Washington: Geological Society of America Bulletin, v. 82, p. 1681–1694.
—— , 1971b, Shear wave propagation in rocks: Nature, v. 229, p. 549–550.
—— , 1972, The abundance of serpentinites in the oceanic crust: Journal of Geology, v. 80, p. 709–719.
—— , 1974, Compressional wave velocities in possible mantle rocks to pressures of 30 kilobars: Journal of Geophysical Research, v. 79, p. 407–412.
—— , 1978, Ophiolites, seismic velocities, and oceanic crustal structure: Tectonophysics, v. 47, p. 131–157.
—— , 1979, Compressional wave velocities in rocks at high temperatures and pressures, critical thermal gradients, and low-velocity zones: Journal of Geophysical Research, v. 84, p. 6849–6857.
—— , 1982, Seismic velocities, *in* Carmichael, R. S., ed., Handbook of physical properties of rocks, v. 2: Boca Raton, Florida, CRC Press, p. 1–226.
—— , 1984a, The magnitude, symmetry, and origin of upper mantle anisotropy based on fabric analyses of ultramafic tectonites: Geophysical Journal of the Royal Astronomical Society, v. 76, p. 89–111.
—— , 1984b, Pore pressure and oceanic crustal seismic structure: Geophysical Journal of the Royal Astronomical Society, v. 79, p. 411–424.
—— , 1986, The influence of pore pressure on oceanic crustal seismic velocities, Journal of Geodynamics, v. 5, p. 45–48.
Christensen, N. I., and Crosson, R., 1968, Seismic anisotropy in the upper mantle: Tectonophysics, v. 6, p. 93–107.
Christensen, N. I., and Fountain, D. M., 1975, Constitution of the lower continental crust based on experimental studies of seismic velocities in granulite: Geological Society of America Bulletin, v. 86, p. 227–236.
Christensen, N. I., and Ramananantoandro, R., 1971, Elastic moduli and anisotropy of dunite to 10 kilobars: Journal of Geophysical Research, v. 76, p. 4003–4010.
Christensen, N. I., and Salisbury, M. H., 1979, Seismic anisotropy in the oceanic upper mantle; Evidence from the Bay of Islands ophiolite complex: Journal of Geophysical Research, v. 84, p. 4601–4610.
Clarke, F. W., 1924, The data of geochemistry: U.S. Geological Survey Bulletin 770, 841 p.
Clarke, F. W., and Washington, H. S., 1924, The composition of the Earth's crust: U.S. Geological Survey Professional Paper 127, 117 p.
Clowes, R. M., and Kanasewich, E. R., 1970, Seismic attenuation and the nature of reflecting horizons within the crust: Journal of Geophysical Research, v. 75, p. 6693–6705.
Cohen, L. H., Ito, K., and Kennedy, G. C., 1967, Melting and phase relations in an anhydrous basalt 40 kilobars: American Journal of Science, v. 265, p. 475–518.
Condie, K. C., 1967, Composition of the ancient North American crust: Science, v. 155, p. 1013–1015.
Coney, P. J., Jones, D. L., and Monger, J.W.H., 1980, Cordilleran suspect terranes: Nature, v. 288, p. 329–333.
Cook, F. A., and 5 others, 1979, Thin-skinned tectonics in the crystalline southern Appalachians; COCORP seismic reflection profiling of the Blue Ridge and Piedmont: Geology, v. 7, p. 563–567.
Coruh, C., and 5 others, 1987, Results from regional Vibroseis profiling; Appalachian ultradeep core hole site study: Geophysical Journal of the Royal Astronomical Society, v. 89, p. 147–156.
Cummings, D., 1972, Mafic and ultramafic inclusions, crater 160, San Francisco volcanic field, Arizona: U.S. Geological Survey Professional Paper 800B, p. B95–B104.
Daily, W. D., and Ramirez, A. L., 1984, In situ porosity distribution using geophysical tomography: Geophysical Research Letters, v. 11, p. 614–616.
Davis, G., and Grew, E., 1978, Age of zircon from a crustal xenolith, Kilbourne Hole, New Mexico: Carnegie Institution of Washington Yearbook, v. 77, p. 897–898.
Dawson, J. B., 1977, Sub-cratonic crust and upper mantle models based on xenolith suites in kimberlite and nephelinitic diatremes: Journal of the Geological Society of London, v. 134, p. 173–184.
Dodge, R.C.W., Calk, L. C., and Kistler, R. W., 1986, Lower crustal xenoliths, Chinese Peak lava flow, Central Sierra Nevada: Journal of Petrology, v. 27, p. 1277–1304.
Eade, K. E., and Fahrig, W. F., 1971, Geochemical evolutionary trends of continental plates; A preliminary study of the Canadian shield: Geological Survey of Canada Bulletin 179, 51 p.
Eade, K. E., Fahrig, W. F., and Maxwell, J. A., 1966, Composition of crystalline shield rocks and fractionating effects of regional metamorphism: Nature, v. 211, p. 1245–1249.
Ehrenberg, S. N., 1979, Garnetiferous ultramafic inclusions in minette from the Navajo volcanic field, *in* Boyd, F. R., and Meyer, H.O.A., eds., The mantle sample; Inclusions in kimberlites and others volcanics: Washington, D.C., American Geophsyical Union, p. 330–334.
Ehrenberg, S. N., and Griffin, W. L., 1979, Garnet granulite and associated xenoliths in minette and serpentinite diatremes of the Colorado Plateau: Geology, v. 7, p. 483–487.
Eggler, D. H., and McCallum, M. E., 1974, Preliminary upper mantle–lower crust model of the Colorado-Wyoming Front Range: Carnegie Institution of Washington Yearbook, v. 73, p. 295–300.
Engel, A.E.J., 1963, Geologic evolution of North America: Science, v. 140, p. 143–152.
Ervin, C. P., and McGinnis, L. D., 1975, Reelfoot rift; Reactivated precursor to the Mississippi embayment: Geological Society of America Bulletin, v. 86, p. 1287–1295.
Esperanca, S., and Holloway, J. R., 1984, Lower crustal nodules from the Camp Creek latite, Carefree, Arizona, *in* Kornprobst, J., ed., Kimberlite; 2, The mantle and crust-mantle relationships: Amsterdam, Elsevier, p. 214–227.
Fahrig, W. F., and Eade, K. E., 1968, The chemical evolution of the Canadian shield: Canadian Journal of Earth Science, v. 5, p. 1247–1252.
Finckh, P., and 6 others, 1985, Detailed crustal structure from a seismic reflection survey in northern Switzerland, *in* Barazangi, M., and Brown, L. D., eds., Reflection seismology; A global perspective: American Geophysical Union Geodynamics Series, v. 13, p. 43–54.
Fodor, R. V., 1978, Ultramafic and mafic inclusions and megacrysts in Pliocene basalt, Black Range, New Mexico: Geological Society of America Bulletin, v. 89, p. 451–459.
Forsyth, D. W., 1975, The early structural evolution and anisotropy of the oceanic upper mantle: Geophysical Journal of the Royal Astronomical Society, v. 43, p. 103–162.
Fountain, D. M., 1976, The Ivrea-Verbano and Strona Ceneri zones, northern Italy; A cross section of the continental crust; New evidence from seismic velocities: Tectonophysics, v. 33, p. 145–165.
—— , 1986, Implications of deep crustal evolution for seismic reflection seismology, *in* Barazangi, M., and Brown, L. D., eds., Reflection seismology; The continental crust: American Geophysical Union Geodynamics Series, v. 14, p. 1–7.
Fountain, D. M., and Salisbury, M. H., 1981, Exposed cross sections through the continental crust; Implications for crustal structure, petrology, and evolution: Earth and Planetary Science Letters, v. 56, p. 263–277.
Fountain, D. M., Hurich, C. A., and Smithson, S. B., 1984, Seismic reflectivity of mylonite zones in the crust: Geology, v. 12, p. 195–198.
Fountain, D. M., McDonough, D. T., and Gorham, J. M., 1987, Seismic reflection models of continental crust based on metamorphic terrains: Geophysical Journal of the Royal Astronomical Society, v. 89, p. 61–66.
Frisillo, A. I., and Barsch, G. R., 1972, Measurement of single crystal elastic constants of bronzite as a function of pressure and temperature: Journal of Geophysical Research, v. 77, p. 6360–6384.
Fuchs, K., 1969, On the properties of deep crustal reflectors: Zeitschrift fur Geophysik, v. 35, p. 133–149.
—— , 1975, Seismiche Anisotropie desoberen Erdmantles und Intraplattentektonic: Geologische Rundschau, v. 64, p. 700–716.

——, 1983, Recently formed elastic anisotropy and petrological models for the continental subcrustal lithosphere in southern Germany: Physics of Earth and Planetary Interiors, v. 31, p. 93–118.

Furlong, K. P., and Fountain, D. M., 1986, Continental crustal underplating; Thermal considerations and seismic-petrologic consequences: Journal of Geophysical Research, v. 91, p. 8285–8294.

Galdin, N. E., 1974, Composition of crust in the ancient shields: Geochemistry International, v. 11, p. 270–281.

Giese, P., 1968, Die Struktur der Erdkruste in Bereich der Ivrea–Zone: Schweizerische Mineralogische und Petrographische Mitteilungen, v. 48, p. 261–284.

Goldschmidt, V. M., 1933, Grundlagen der quantitativen geochemie: Fortschritte Mineralogie Kristallographie Petrographie, v. 17, p. 112–156.

——, 1954, Geochemistry: Oxford, Clarendon Press, 730 p.

Green, D. H., 1967, Effects of high pressure on basaltic rock, *in* Hess, H. H., and Poldervaart, A. E., eds., Basalts; The Poldervaart treatise on rocks of basaltic composition: New York, Interscience, p. 401–443.

Green, D. H., and Ringwood, A. E., 1967, An experimental investigation of the gabbro to eclogite transformation and its petrological applications: Geochimica et Cosmochimica Acta, v. 31, p. 767–833.

——, 1972, A comparison of recent experimental data on the gabbro-garnet granulite-eclogite transition: Journal of Geology, v. 80, p. 227–288.

Green, T. H., 1967, An experimental investigation of sub-solidus assemblages formed at high pressure in high-alumina basalt, kyanite eclogite, and grosspydite compositions: Contributions to Mineralogy and Petrology, v. 16, p. 84–114.

Greensfelder, R. W., and Kovach, R. L., 1982, Shear wave velocities and crustal structure of the eastern Snake River Plain, Idaho: Journal of Geophysical Research, v. 87, p. 2643–2653.

Griffin, W. L., and O'Reilly, S. Y., 1987, The composition of the lower crust and the nature of the continental Moho-Xenolith evidence, *in* Nixon, P. H., ed., Mantle xenoliths: London, Wiley, p. 413–430.

Grout, F., 1938, Petrographic and chemical data on the Canadian shield: Journal of Geology, v. 46, p. 486–504.

Gutenberg, B., 1959, Physics of the Earth's interior: New York, Academic Press, 204 p.

Hajnal, Z., and 5 others, 1983, Seismic characteristics of a Precambrian pluton and its adjacent rocks: Geophysics, v. 48, p. 569–581.

Hajnal, Z., and 6 others, 1984, An initial analysis of the Earth's crust under the Williston basin; 1979 COCRUST experiment: Journal of Geophysical Research, v. 89, p. 9381–9400.

Hale, L. D., and Thompson, G. A., 1982, The seismic reflection character of the continental Mohorovičić discontinuity: Journal of Geophysical Research, v. 87, p. 4625–4635.

Hamilton, W., 1981, Crustal evolution by arc magmatism: Philosophical Transactions of the Royal Society of London, Series A, v. 301, p. 279–291.

Harte, B., Jackson, P. M., and Macintyre, R. M., 1981, Age of mineral equilibria in granulite facies nodules from kimberlites: Nature, v. 291, p. 147–148.

Hearn, C., and Boyd, F. R., 1975, Garnet peridotite xenoliths in a Montana, U.S.A., kimberlite: Physics and Chemistry of the Earth, v. 9, p. 247–255.

Hearn, B. C., and McGee, E. S., 1984, Garnet peridotites from Williams kimberlites, north-central Montana, U.S.A., *in* Kornprobst, J., ed., Kimberlites; 2, The mantle and crust-mantle relationships: Amsterdam, Elsevier, p. 57–70.

Holland, J. G., and Lambert, R. St. J., 1972, Major element chemical composition of shields and the continental crust: Geochimica et Cosmochimica Acta, v. 36, p. 673–683.

——, 1975, The chemistry and origin of the Lewisian gneisses of the Scottish mainland; The Scourie and Inver assemblages and sub-crustal accretion: Precambrian Research, v. 2, p. 161–188.

Hunziker, J. C., and Zingg, A., 1980, Lower Paleozoic amphibolite to granulite facies metamorphism in the Ivrea zone (southern Alps, northern Italy): Schweizerische Mineralogische und Petrographische Mitteilungen: v. 60, p. 181–213.

Hurich, C. A., Smithson, S. B., Fountain, D. M., and Humphreys, M. C., 1985, Seismic evidence of mylonite reflectivity and deep structure in the Kettle dome metamorphic core complex, Washington: Geology, v. 13, p. 577–580.

Hyndman, R. D., and Drury, M. J., 1976, The physical properties of oceanic basement rocks from deep drilling on the Mid-Atlantic Ridge: Journal of Geophysical Research, v. 81, p. 4042–4052.

Ito, K., and Kennedy, G. C., 1968, Melting and phase relations in the plane tholeiite-lherzolite-nepheline basanite to 40 kilobars with geological implications: Contributions to Mineralogy and Petrology, v. 19, p. 177–211.

——, 1970, The fine structure of the basalt-eclogite transition: Mineralogical Society of America Special Paper 3, p. 77–83.

——, 1971, An experimental study of the basalt-garnet granulite-eclogite transition, *in* Heacook, J. G., ed., The structure and physical properties of the Earth's crust: American Geophysical Monograph 14, p. 303–314.

Kay, R. W., and Kay, S. M., 1981, The nature of the lower continental crust; Inferences from geophysics, surface geology, and crustal xenoliths: Reviews of Geophysics and Space Physics, v. 19, p. 271–298.

Kay, R. W., Rubenstone, J. L., and Kay, S. M., 1986, Aleutian terranes from Nd isotopes: Nature, v. 322, p. 605–609.

Kay, S. M., Kay, R. W., Hanges, J., and Snedden, T., 1978, Crustal xenoliths from potassic lavas, Leucite Hills, Wyoming: Geological Society of America Abstracts with Programs, v. 10, p. 432.

Kay, S. M., Snedden, W. T., Foster, B. P., and Kay, R. W., 1983, Upper mantle and crustal fragments in the Ithaca kimberlites: Journal of Geology, v. 91, p. 277–290.

Keller, G. R., Smith, R. B., and Braile, L. W., 1975, Crustal structure along the Great Basin–Colorado Plateau transition from seismic refraction studies: Journal of Geophysical Research, v. 80, p. 1093–1098.

Kempton, P. D., Menzies, M. A., and Dungan, M. A., 1984, Petrography, petrology, and geochemistry of xenoliths and megacrysts from the Geronimo volcanic field, southeastern Arizona, *in* Kornprobst, J., ed., Kimberlites; 2, The mantle and crust-mantle relationships: Amsterdam, Elsevier, p. 71–83.

Kern, H., 1979, Effect of high-low quartz transition on compressional and shear wave velocities in rocks under high pressure: Physics and Chemistry of Minerals, v. 4, p. 161–171.

Kern, H., and Richter, A., 1981, Temperature derivatives of compressional and shear wave velocities in crustal and mantle rocks at 6 kbar confining pressure: Journal of Geophysics, v. 49, p. 47–56.

Kern, H., and Schenk, V., 1985, Elastic wave velocities in rocks from a lower crustal section in southern Calabria, Italy: Physics of Earth and Planetary Interiors, v. 40, p. 147–160.

Kirkley, M. B., McCallum, M. E., and Eggler, D. H., 1984, Coexisting garnet and spinel in upper mantle xenoliths from Colorado-Wyoming kimberlites, *in* Kornprobst, J., ed., Kimberlites; 2, The mantle and crust-mantle relationships: Amsterdam, Elsevier, p. 85–96.

Klemperer, S., Brown, L., Oliver, J., Ando, C., and Kaufman, S., 1983, Crustal structure in the Adirondacks, *in* Bally, A. W., ed., Seismic expression of structural styles: American Association of Petroleum Geologists Studies in Geology 15, p. 5–12.

Kremenetsky, A. A., and Ovchinnikov, L. N., 1986, The Precambrian continental crust; Its structure, composition, and evolution as revealed by deep drilling in the U.S.S.R.: Precambrian Research, v. 33, p. 11–43.

Lambert, I. B., and Heier, K. S., 1968, Geochemical investigations of deep-seated rocks in the Australian shield: Lithos, v. 1, p. 30–53.

Leaver, D. S., Mooney, W. D., and Kohler, W. M., 1984, A seismic refraction study of the Oregon Cascades: Journal of Geophysical Research, v. 89, p. 3121–3134.

Leavy, B. D., and Hermes, D. D., 1979, Mantle xenoliths from southeastern New England, *in* Boyd, F. R., and Meyer, H.O.A., eds., The mantle sample; Inclusions in kimberlites and other volcanics: Washington, D. C., American Geophysical Union, p. 374–381.

Leeman, W. P., 1979, Primitive lead in deep crustal xenoliths from the Snake River Plain, Idaho: Nature, v. 281, p. 365–366.

Leeman, W. P., Menzies, M. A., Matty, D. J., and Embree, G. F., 1985, Strontium, neodymium, and lead isotopic compositions of deep crustal xenoliths from the Snake River Plain; Evidence for Archean basement: Earth and

Planetary Science Letters, v. 75, p. 354–368.

Leyreloup, A., Dupuy, C., and Andriambololona, R., 1977, Catazonal xenoliths in French Neogene volcanic rocks; Constitution of the lower crust; 2, Chemical composition and consequences of the evolution of the French Massif Central Precambrian Crust: Contributions to Mineralogy and Petrology, v. 61, p. 283–300.

Leven, J. H., Jackson, I., and Ringwood, A. E., 1981, Upper mantle seismic anisotropy and upper mantle decoupling: Nature, v. 289, p. 234–239.

Lin, W., and Wang, C. Y., 1980, P-wave velocities in rocks at high pressure and temperature and the constitution of the central California crust: Geophysical Journal of the Royal Astronomical Society, v. 61, p. 379–400.

Littlejohn, A. L., and Greenwood, H. J., 1974, Lherzolite nodules in basalts from British Columbia, Canada: Canadian Journal of Earth Science, v. 11, p. 1288–1308.

Maccarrone, E., Paglionico, A., Piccarreta, G., and Rottura, A., 1983, Granulite-amphibolite facies metasediments from Serre, Calabria, southern Italy; Their protoliths and the processes controlling their chemistry: Lithos, v. 16, p. 95–111.

Manghnani, M. H., Ramananantoandro, R., and Clark, S. P., Jr., 1974, Compressional and shear wave velocities in granulite facies rocks and eclogites to 10 kbar: Journal of Geophysical Research, v. 70, p. 5427–5446.

Manitoba Mineral Resources Division, 1979, Geological map of Manitoba: Manitoba Mineral Resources Division Map 79-2, scale 1:100,000.

Mathur, S. P., 1983, Deep reflection probes in eastern Australia reveal differences in the nature of the crust: First Break, v. 1, p. 9–16.

Matthews, D. H., and Cheadle, M. J., 1985, Deep reflections from the Caledonides and Variscides west of Britain and comparison with the Himalayas, *in* Barazangi, M., and Brown, L., eds., Reflection seismology; A global perspective: American Geophysical Union Geodynamics Series, v. 13, p. 5–18.

Matty, D. J., 1984, Petrology of deep crustal xenoliths from the eastern Snake River Plain, Idaho [Ph.D. thesis]: Houston, Texas, Rice University, 203 p.

McCallum, M. E., and Eggler, D. H., 1976, Diamonds in an upper mantle peridotite nodule from kimberlite in southern Wyoming: Science, v. 192, p. 253–256.

McCallum, M. E., and Mabarak, C. D., 1976, Diamond in State-line kimberlite diatremes, Albany County, Wyoming, Larimer County, Colorado: Geological Survey of Wyoming Report of Investigations 12, 36 p.

McGetchin, T. R., and Silver, L. T., 1972, A crustal-upper mantle for the Colorado Plateau based on observation of crystalline rock fragments in the Moses Rock Dike: Journal of Geophysical Research, v. 77, p. 7022–7037.

McSkimin, H. J., Andreatch, P., and Thurston, R. N., 1965, Elastic moduli of quartz versus hydrostatic pressure at 5 and −195.8°C: Journal of Applied Physics, v. 36, p. 1624–1632.

Mehnert, K. R., 1975, The Ivrea Zone; A model of the deep crust: Neues Jahrbuch fuer Mineralogie Abhandlungen, v. 125, p. 156–199.

Meissner, R., 1973, The Moho as a transition zone: Geophysical Surveys, v. 1, p. 195–216.

Meissner, R., and Fakhimi, M., 1977, Seismic anisotropy as measured under high-temperature conditions: Geophysical Journal of the Royal Astronomical Society, v. 49, p. 133–143.

Meissner, R., and Lueschen, E., 1983, Seismic near-vertical reflection studies of the Earth's crust in the Federal Republic of Germany: First Break, v. 1, p. 19–24.

Meyer, H.O.A., and Brookins, D. C., 1976, Sapphrine, sillimanite, and garnet in granulite xenoliths from Stockdale kimberlite, Kansas: American Mineralogist, v. 61, p. 1194–1202.

Mitchell, B. J., and Hashim, B. M., 1977, Seismic velocity determinations in the New Madrid seismic zone; A new method using local earthquakes: Bulletin of the Seismological Society of America, v. 67, p. 413–424.

Mitchell, B. J., and Landisman, M., 1971, Geophysical measurements in the southern Great Plains, *in* Heacock, J. G., ed., The structure and physical properties of the Earth's crust: American Geophysical Union Monograph 14, p. 77–93.

Mitchell, B. J., and Yu, G. K., 1980, Surface wave regionalised models and anisotropy of the Pacific crust and upper mantle: Geophysical Journal of the Royal Astronomical Society, v. 64, p. 497–514.

Mooney, W. D., and Ginzburg, A., 1986, Seismic measurements of the internal properties of fault zones: Pure and Applied Geophysics, v. 124, p. 141–157.

Mooney, W. D., Andrews, M. C., Ginzburg, A., Peters, D. A., and Hamilton, R. M., 1983, Crustal structure of the northern Mississippi embayment and a comparison with other continental rift zones: Tectonophysics, v. 94, p. 327–348.

Moresi, M., Paglionico, A., Piccarreta, G., and Rottura, A., 1978-79, The deep crust in Calabria, Polla-Copanello unit; A comparison with the Ivrea-Verbano Zone: Memorie di Scienze Geologiche gia Memorie degli Istituti di Geologia e Mineralogia dell'Universita di Padova, v. 33, p. 233–242.

Morris, G. B., Raitt, R. W., and Shor, G. G., Jr., 1969, Velocity anisotropy and delay-time maps of the mantle near Hawaii: Journal of Geophysical Research, v. 74, p. 4300–4316.

Mueller, S., 1977, A new model of the continental crust, *in* Heacock, J. G., ed., The Earth's crust: American Geophysical Union Monograph 20, p. 289–317.

Nesbitt, H. W., and Young, G. M., 1984, Prediction of some weathering trends of plutonic and volcanic rocks based on thermodynamic and kinematic considerations: Geochimica et Cosmochimica Acta, v. 48, p. 1523–1534.

O'Brien, T. F., 1983, Evidence for the nature of the lower crust beneath the central Colorado Plateau as derived from xenoliths in the Buell Park–Green Knobs diatremes [Ph.D. thesis]: Ithaca, New York, Cornell University, 203 p.

Oliver, G.J.H., 1980, Geology of the granulite and amphibolite facies gneisses of Doubtful Sound, Fiordland, New Zealand: New Zealand Journal of Geology and Geophysics, v. 23, p. 27–41.

Oliver, G.J.H., and Coggon, J. H., 1979, Crustal structure of Fiordland, New Zealand: Tectonophysics, v. 54, p. 253–292.

Oliver, J., Dobrin, M., Kaufman, S., Meyer, R., and Phinney, R., 1976, Continuous seismic reflection profiling of the deep basement, Hardeman County, Texas: Geological Society of America Bulletin, v. 87, p. 1537–1546.

Oliver, J., Cook, F., and Brown, L., 1983, COCORP and the continental crust: Journal of Geophysical Research, v. 88, p. 3329–3347.

Owens, T. J., Zandt, G., and Taylor, S. R., 1984, Seismic evidence for an ancient rift beneath the Cumberland Plateau, Tennessee; A detailed analysis of broadband teleseismic P waveforms: Journal of Geophysical Research, v. 89, p. 7783–7795.

Padovani, E. R., and Carter, J. L., 1977, Aspects of the deep crustal evolution beneath south-central New Mexico, *in* Heacock, J. G., ed., The Earth's crust; Its nature and physical properties: American Geophysical Union Monograph 20, p. 19–55.

Padovani, E. R., Hall, J., and Simmons, G., 1982, Constraints on crustal hydration below the Colorado Plateau from V_p measurements on crustal xenoliths: Tectonophysics, v. 84, p. 313–328.

Pakiser, L. C., and Robinson, R., 1966, Composition and evolution of the continental crust as suggested by seismic observations: Tectonophysics, v. 3, p. 547–557.

Pakiser, L. C., and Steinhart, J. S., 1964, Explosion seismology in the Western Hemisphere, *in* Odishaw, H., ed., Research in geophysics; 2, Solid Earth and interface phenomena: Cambridge, Massachusetts Institute of Technology Press, p. 123–147.

Percival, J. A., and Card, K. D., 1983, Archean crust as revealed in the Kapuskasing uplift, Superior Province, Canada: Geology, v. 11, p. 323–326.

—— , 1985, Structure and evolution of Archean crust in central Superior province, Canada, *in* Ayres, L. D., Thurston, P. C., Card, K. D., and Weber, W., eds., Evolution of Archean supracrustal sequences: Geological Association of Canada Special Paper 28, p. 179–192.

Poldervaart, A., 1955, Chemistry of the Earth's crust: Geological Society of America Special Paper 62, p. 119–144.

Pollack, H. N., and Chapman, D. S., 1977, On the regional variation of heat flow, geotherms, and lithospheric thickness: Tectonophysics, v. 38, p. 279–296.

Potter, C. J., and 9 others, 1987, Crustal structure of north-central Nevada;

Results from COCORP deep seismic profiling: Geological Society of America Bulletin, v. 98, p. 330–337.

Priestley, K., and Brune, J. N., 1982, Shear wave structure of the southern volcanic plateau of Oregon and Idaho and the northern Great Basin of Nevada from surface wave dispersion: Journal of Geophysical Research, v. 87, p. 2671–2675.

Priestley, K., and Davey, F. J., 1983, Crustal structure of Fiordland, southwestern New Zealand from seismic-refraction measurements: Geology, v. 11, p. 660–663.

Prodehl, C., Schlittenhardt, J., and Stewart, S. W., 1984, Crustal structure of the Appalachian highlands in Tennessee: Tectonophysics, v. 109, p. 61–76.

Raitt, R. W., Shor, G. G., Jr., Francis, T.J.G., and Morris, G. B., 1969, Anisotropy of the Pacific upper mantle: Journal of Geophysical Research, v. 74, p. 3095–3109.

Ramananantoandro, R., and Manghnani, M. H., 1978, Temperature dependence of the compressional wave velocity in anisotropic dunite; Measurements to 500°C at 10 kbar: Tectonophysics, v. 47, p. 73–84.

Reid, J. B., Jr., and Woods, G. A., 1978, Oceanic mantle beneath southwestern U.S.: Earth and Planetary Science Letters, v. 41, p. 303–316.

Reilly, G. A., and Shaw, D. M., 1967, An estimate of the composition of part of the Canadian shield in northwestern Ontario: Canadian Journal of Earth Sciences, v. 4, p. 725–739.

Ringwood, A. E., and Green, D. H., 1964, Experimental investigations bearing on the nature of the Mohorovičić discontinuity: Nature, v. 201, p. 566–567.

——, 1966, An experimental investigation of the gabbro-eclogite transformation and some geophysical implications: Tectonophysics, v. 3, p. 383–427.

Ronov, A. B., and Yaroshevsky, A. A., 1967, Chemical structure of the Earth's crust: Geochemistry International, v. 4, p. 1041–1075.

——, 1969, Chemical composition of the earth's crust, *in* Hart, P. J., ed., The Earth's crust and upper mantle: American Geophysical Union Monograph 13, p. 37–57.

Ross, D. V., 1985, Mafic gneissic complex (batholithic root?) in the southernmost Sierra Nevada, California: Geology, v. 13, p. 288–291.

Ross, J. V., 1983, The nature and rheology of the Cordilleran upper mantle of British Columbia; Inferences from peridotite xenoliths: Tectonophysics, v. 100, p. 321–357.

Salisbury, M. H., and 6 others, 1979, The physical state of the upper levels of Cretaceous oceanic crust from the results of logging, laboratory studies, and the oblique seismic experiment at DSDP sites 417 and 418, *in* Talwani, M., Harrison, C. G., and Hayes, D. E., eds., Deep drilling in the Atlantic Ocean; Ocean crust: American Geophysical Union Maurice Ewing Series 2, p. 113–134.

Sandmeier, K.-J., and Wenzel, F., 1986, Synthetic seismograms for a complex crustal model: Geophysical Research Letters, v. 13, p. 22–25.

Schenk, V., and Scheyer, W., 1978, Granulite-facies metamorphism in the northern Serre, Calabria, southern Italy, *in* Closs, H., Roeder, D., and Schmidt, K., eds., Apennines, Hellenides: Stuttgart, Schweizerbart, p. 341–246.

Schulze, D. J., 1984, Cr-poor megacrysts from the Hamilton Branch kimberlite, Elliot County, Kentucky, *in* Kornprobst, J., ed., Kimberlites; 2, The mantle and crust-mantle relationships: Amsterdam, Elsevier, p. 97–108.

Schulze, D. J., and Helmstaedt, H., 1979, Garnet pyroxenite and eclogite xenoliths from the Sullivan Buttes latite, Chino Valley, Arizona, *in* Boyd, F. R., and Meyer, H.O.A., eds., The mantle sample; Inclusions in kimberlites and other volcanics: Washington, D.C., American Geophysical Union, p. 319–329.

Schulze, D. J., Helmstaedt, H., and Cossie, R. M., 1978, Pyroxene-ilmenite intergrowths in garnet pyroxenite xenoliths from a New York kimberlite and Arizona latite: American Mineralogist, v. 63, p. 258–265.

Sederholm, J. J., 1925, The average composition of the Earth's crust in Finland: Bulletin de la Commission Geologique de la Finland 70, 29 p.

Serpa, L. and 7 others, 1984, Structure of the southern Keeweenawan Rift from COCORP surveys across the Midcontinent Geophysical Anomaly in northeastern Kansas: Tectonics, v. 3, p. 367–384.

Shaw, D. M., Reilly, G. A., Muysson, J. R., Pattenden, G. E., and Campbell, F. E., 1967, An estimate of the chemical composition of the Canadian Precambrian shield: Canadian Journal of Earth Sciences, v. 4, p. 829–853.

Sheraton, J. W., Skinner, A. C., and Tarney, J., 1973, The geochemistry of the Scourian gneisses of the Assynt district, *in* Park, R. G., and Tarney, J., eds., The early Precambrian of Scotland and related rocks of Greenland: Keele, England, University of Keele, p. 13–30.

Shervais, J., 1978–79, Ultramafic and mafic layers in the alpine-type lherzolite massif at Balmuccia, N.W. Italy: Memorie di Scienze Geologiche gia Memorie degli Istituti di Geologia e Mineralogia dell'Universita di Padova, v. 33, p. 135–145.

Shervais, J. W., Wilshire, H. G., and Schwartzner, E. D., 1973, Garnet clinopyroxenite xenoliths from Dish Hill, California: Earth and Planetary Science Letters, v. 19, p. 120–130.

Shimamura, H., Asada, T., Suyehiro, K., Yamada, T., and Inatani, H., 1983, Long-shot experiments to study velocity anisotropy in the oceanic lithosphere to the northwest Pacific: Physics of Earth and Planetary Interiors, v. 31, p. 348–362.

Simmons, G., and Nur, A., 1968, Granites; Relation of properties in situ to laboratory measurements: Science, v. 162, p. 789–791.

Simmons, G., Todd, T., and Baldridge, W. S., 1975, Toward a quantitative relationship between elastic properties and cracks in low porosity rocks: American Journal of Science, v. 275, p. 318–345.

Sinclair, P. D., Templeman-Kluit, D. J., and Medaris, L. D., Jr., 1978, Lherzolite nodules from a Pleistocene cinder cone in central Yukon: Canadian Journal of Earth Sciences, v. 15, p. 220–227.

Smith, D., 1979, Hydrous minerals and carbonates in peridotite inclusions from the Green Knobs and Buell Park kimberlitic diatremes on the Colorado Plateau, *in* Boyd, F. R., and Meyer, H.O.A., eds., The mantle sample; Inclusions in kimberlites and other volcanics: Washington, D.C., American Geophysical Union, p. 345–356.

Smith, D., and Levy, S., 1976, Petrology of the Green Knobs diatreme and implications for the upper mantle beneath the Colorado Plateau: Earth and Planetary Science Letters, v. 29, p. 107–125.

Smith, R. B., 1978, Seismicity, crustal structure, and intraplate tectonics of the interior of the western Cordillera, *in* Smith, R. B., and Eaton, G. P., eds., Cenozoic tectonics and regional geophysics of the western Cordillera: Geological Society of America Memoir 152, p. 111–144.

Smith, R. B., and 9 others, 1982, The 1978 Yellowstone–eastern Snake River Plain seismic profiling experiment; Crustal structure of the Yellowstone region and experiment design: Journal of Geophysical Research, v. 87, p. 2583–2596.

Smithson, S. B., and Brown, S. K., 1977, A model for lower continental crust: Earth and Planetary Science Letters, v. 35, p. 134–144.

Smithson, S. B., and Ebens, R. J., 1971, Interpretation of data from a 3.05-kilometer borehole in Precambrian crystalline rocks, Wind River Mountains, Wyoming: Journal of Geophysical Research, v. 76, p. 7079–7087.

Smithson, S. B., Brewer, J., Kaufman, S., Oliver, J., and Hurich, C., 1978, Nature of the Wind River thrust, Wyoming, from COCORP deep-reflection data and from gravity data: Geology, v. 6, p. 648–652.

Smythe, D. K., and 6 others, 1982, Deep structure of the Scotish Caledonides revealed by the MOIST reflection profile: Nature, v. 299, p. 338–340.

Sparlin, M. A., Braile, L. W., and Smith, R. B., 1982, Crustal structure of the eastern Snake River Plain determined from ray trace modeling of seismic refraction data: Journal of Geophysical Research, v. 87, p. 2619–2633.

Sperr, J. T., 1985, Xenoliths of the Leucite Hills volcanic rocks, Sweetwater County, Wyoming [M.S. thesis]: Laramie, University of Wyoming, 57 p.

Spudich, P.K.P., Salisbury, M. H., and Orcutt, J. A., 1978, Ophiolites found in the oceanic crust?: Geophysical Research Letters, v. 5, p. 341–344.

Stierman, D. J., and Kovach, R. L., 1979, An in situ velocity study; The Stone Canyon well: Journal of Geophysical Research, v. 84, p. 672–678.

Stoesser, D. B., 1973, Mafic and ultramafic xenoliths of cumulus origin, San Francisco volcanic field, Arizona [Ph.D. thesis]: Eugene, University of Oregon, 260 p.

——, 1974, Xenoliths of the San Francisco volcanic field northern Arizona, *in*

Karlstrom, T.N.V., Swan, G. A., and Eastwood, R. E., eds., Geology of northern Arizona; Part 2, Area studies and field guides: Geological Society of America, p. 530–545.

Streckeisen, A., 1976, To each plutonic rock its proper name: Earth Science Reviews, v. 12, p. 1–33.

Tanimoto, T., and Anderson, D. L., 1984, Mapping convection in the mantle: Geophysical Research Letters, v. 11, p. 287–290.

Taylor, S. R., 1964, Abundance of chemical elements in the continental crust; A new table: Geochimica et Cosmochimica Acta, v. 28, p. 1273–1286.

—— , 1977, Island arc models and the composition of the continental crust, *in* Talwani, M., and Pitman, W. C., eds., Island arcs, deep sea trenches, and back arc basins: American Geophysical Union Maurice Ewing Series 1, p. 325–335.

—— , 1979, Chemical composition and evolution of the continental crust; The rare earth element evidence, *in* McElhinny, M. W., ed., The Earth; Its origin, structure, and evolution: London, Academic Press, p. 353–376.

Taylor, S. R., and McLennan, S. M., 1981, The composition and evolution of the continental crust; Rare earth element evidence from sedimentary rocks: Royal Society of London Philosophical Transactions, ser. A, v. 301, p. 381–399.

—— , 1985, The continental crust; Its composition and evolution: Oxford, Blackwell Scientific, 312 p.

Thill, R. E., Willard, R. J., and Bur, T. R., 1969, Correlation of longitudinal velocity variation with rock fabric: Journal of Geophysical Research, v. 74, p. 4897–4909.

Vetter, E., and Minster, J., 1981, P_n velocity anisotropy in southern California: Bulletin of the Seismological Society of America: v. 71, p. 1511–1530.

Vinogradov, A. P., 1962, Average contents of chemical elements in the principal types of igneous rocks of the earth's crust: Geochemistry, v. 7, p. 641–664.

Walter, A. D., and Mooney, W. D., 1982, Crustal structure of the Diablo and Gabilan Ranges, central California: Bulletin of the Seismological Society of America, v. 72, p. 1567–1590.

Warren, D. H., and Healy, J. H., 1973, Structure of the crust in the conterminous United States: Tectonophysics, v. 29, p. 203–213.

Warren, R. G., Kudo, A. M., and Keil, K., 1979, Geochemistry of lithic and single-crystal inclusions in basalt and a characterization of the upper mantle-lower crust in the Engle Basin, Rio Grande Rift, New Mexico, *in* Riecker, R., ed., Rio Grande Rift; Tectonics and magmatism: Washington, D.C., America Geophysical Union, p. 393–415.

Watson, K. D., and Morton, D. M., 1969, Eclogite inclusions in kimberlite pipes at Garnet Ridge, northeastern Arizona: American Mineralogist, v. 54, p. 267–285.

Weaver, B. L., and Tarney, J., 1980, Continental crust composition and nature of the lower crust; Constraints from mantle Nd-Sr isotope correlation: Nature, v. 286, p. 342–346.

—— , 1984a, Major and trace element composition of the continental lithosphere, *in* Pollack, H. N., and Murthy, V. R., eds., Structure and evolution of the continental lithosphere: Oxford, Pergamon, p. 39–68.

—— , 1984b, Empirical approach to estimating the composition of the continental crust: Nature, v. 310, p. 575–577.

Weber, W., and Scoates, R.F.J., 1978, Archean and Proterozoic metamorphism in the northwestern Superior Province and along the Churchill-Superior boundary, Manitobata: Geological Survey of Canada Paper 78-10, p. 5–16.

Wedepohl, K. H., 1969, Composition and abundance of common igneous rocks; The handbook of geochemistry, v. 1: Berlin, Springer-Verlag, p. 227–249.

Wickham, S. M., and Oxburgh, E. R., 1985, Continental rifts as a setting for regional metamorphism: Nature, v. 318, p. 330–333.

Wong, J., Hurley, P., and West, G. F., 1983, Crosshole seismology and seismic imagery in crystalline rocks: Geophysical Research Letters, v. 10, p. 686–689.

Wood, B. J., 1987, Thermodynamics of multicomponent systems containing several solid solutions: Reviews in Mineralogy, v. 17, p. 71–95.

Woollard, G. P., 1966, Regional isostatic relations in the United States, *in* Steinhart, J. S., and Smith, T. J., eds., The earth beneath the continents: American Geophysical Union Monograph 10, p. 557–594.

Yoder, H. S., and Tilley, C. E., 1962, Origin of basalt magmas; An experimental study of natural and synthetic rock systems: Journal of Petrology, v. 3, p. 342–352.

Yorath, C. J., and 7 others, 1985, Lithoprobe, southern Vancouver Island; Seismic reflection sees through Wrangellia to the Juan de Fuca plate: Geology, v. 13, p. 759–762.

MANUSCRIPT ACCEPTED BY THE SOCIETY OCTOBER 31, 1988

Printed in U.S.A.

Geological Society of America
Memoir 172
1989

Chapter 31

Crustal geologic processes of the United States

Warren B. Hamilton
U.S. Geological Survey, Box 25046, Denver Federal Center, Denver, Colorado 80225

ABSTRACT

The evolution of crystalline continental crust probably has been dominated by arc magmatism. Olivine, pyroxene, and garnet are largely crystallized from rising arc melts in the subcontinental mantle. Residual hot, dry gabbroic magmas cross the density filter of the Mohorovičić discontinuity and spread out in the basal crust, heating preexisting crustal rocks and producing widespread secondary melting. The basal crust is dominated by two-pyroxene gabbro, increasingly fractionated and contaminated upward, variably metamorphosed in granulite facies. Migmatites and restites, of granulite or uppermost amphibolite facies, dominate the higher part of the lower crust, and from these rise mixed and secondary melts, of intermediate to felsic compositions, that evolve with complex combinations of crystallization, fractionation, and assimilation. The rising melts heat the middle crust, causing hybridization, migmatization, and secondary melting. The melts absorb much water derived mostly by breakdown of wall-rock micas, in continental crust, and of hornblende in island-arc crust. A hydrated melt cannot rise past the level at which load pressure equals fluid pressure, so hydrous melts expel their volatiles and crystallize in the middle crust, often as two-mica granitic rocks in continental crust. Only melts not hydrated by middle-crust reactions remain hot and dry enough to rise to the upper crust to crytallize as shallow batholiths and to erupt as ash flows.

Rift-related continental magmatism has also been important, although criteria for identifying its products are much less definitive than generally assumed. Volcanic-rift assemblages are in many cases basaltic or bimodal basaltic and rhyolitic, but they can include voluminous rocks of intermediate compositions. Arc assemblages also can be strongly bimodal, on modest scales of space and time. Metaluminous rhyolites are abundant in both rifts and arcs, but peralkaline ones occur primarily in rifts. Ancient rift complexes include upper-crustal layered gabbro-granite-rhyolite complexes. Ancient and modern rift systems display distinctively layered basal crust on reflection profiles, likely recording widespread injection of gabbroic magmas from the mantle.

Foreland thrust belts, wherein preexisting stratal wedges are imbricated cratonward, have formed in both collisional and noncollisional settings. Paleozoic imbrication in Appalachian and Ouachita-Marathon regions was a byproduct of arc-continent and continent-continent collisions; basement overthrusting also affected the southern Appalachians. Cretaceous thrust-belt imbrication in much of the Cordillera occurred in an Andean setting, and gravitational spreading due to magmatic thickening of the crust in the magmatic-arc belt may have been responsible.

The latest Cretaceous and early Paleogene Laramide shortening of the Rocky Mountain sector of the craton represents a clockwise rotation of the Colorado Plateau region of about 4° relative to the continental interior about a New Mexico Euler pole. This, and the synchronous tectonic erosion from beneath of continental crust farther

Hamilton, W. B., 1989, Crustal geologic processes of the United States, *in* Pakiser, L. C., and Mooney, W. D., Geophysical framework of the continental United States: Boulder, Colorado, Geological Society of America Memoir 172.

southwest, may have been byproducts of drag on subducting Pacific lithosphere that did not sink out of the way of the advancing continent.

Early Basin and Range extension occurred in back-arc–spreading mode, behind a trench and subducting margin; late extension has been in oblique mode as the continent has adjusted its shape to that of the evolving San Andreas boundary. The dominant mode of extension has been normal faulting wherein footwalls rise and deform and hingelines migrate between inactivated, undulating sectors of faults and still-active dipping sectors. Basin and Range extension followed widespread magmatic heating of the crust; it has doubled the width of a broad region and has continued for 30 m.y.

Extension starting with cold crust may be geologically more common and may result in extension of much lesser extent and shorter duration before a continent is sundered and oceanic spreading begins. The Midcontinent Rift system is of this type, although extension there stopped before reaching the stage of normal oceanic spreading. The Atlantic continental shelf may be largely prograded over thick Mesozoic basaltic crust, rather than having been built atop severely thinned continental crust.

INTRODUCTION

The syntheses in this volume provide critical data for understanding the structure and composition of the continental crust and upper mantle. Here, I present a personal view of some of the tectonic and petrologic processes and their products, and the vertical variations in processes and products, that bear on geophysical interpretations.

The concepts of plate tectonics have revolutionized our understanding of the evolution of shallow parts of continents. Most Proterozoic and Phanerozoic orogenic belts that are exposed at upper-crustal levels can be interpreted in terms of plate rifting followed by convergence. Rifting of continental masses is followed by the deposition on their margins of sedimentary wedges accumulated as continental shelves and slopes. Subsequent convergence of plates is recorded by accretionary wedges, magmatic arcs, and thrust belts, greatly complicated by continental collisions and by the formation and collision of island arcs. Extension, shortening, and strike-slip faulting are manifestations of plate motions.

Tectonic and magmatic complexes exposed by deep erosion into the continents thus also should have explanations compatible with processes we can now see operating in modern plate-tectonic environments.

MAGMATISM

Variation with depth. The middle and lower continental crust and upper mantle consist largely of igneous and metamorphic rocks crystallized at temperatures far above equilibrium geotherms and then variably retrograded as they cooled. Petrologic variations with depth can be studied in obliquely eroded crustal sections. Within the United States there are widespread exposures of rocks formed in the upper and middle crust, and limited exposures of the upper part of the lower crust. Exposures of the basal continental crust are known in outcrop only outside North America. Studies of xenoliths in basalts and kimberlites further define materials of the lower crust and upper mantle. There are global similarities in the dominant rock assemblages of different levels in continental crust, and there is vertical continuity of magmatic processes displayed in such crustal sections (Fig. 1; Fountain and Salisbury, 1981; Hamilton, 1981, 1988c; Hamilton and Myers, 1967; Hyndman, 1981; Kay and Kay, 1981; Weaver and Tarney, 1982; Wells, 1980; Windley, 1981).

Arc magmatism. The volumetrically dominant volcanism seen in modern continents and island arcs (which are destined to be accreted to continents) is arc magmatism. In fossil continental magmatic systems exposing shallow crustal levels, the magmatism often can be assigned an arc setting. It thus is likely that the evolution of the middle and lower continental crust and upper mantle also has been dominated by arc magmatism.

Rift magmatism. Severe extension of continental crust can be accompanied by mafic, intermediate, silicic, alkalic, or bimodal volcanism; variations are wide, and many provinces display both petrology and setting intermediate between those regarded as typical of arcs or rifts. Plutonic rocks of obliquely eroded upper-crust sections to which rift origins can be most confidently assigned are bimodal granite-gabbro complexes, not unimodal granitic or granitic-and-intermediate ones. Although some petrologists assume that many plutonic igneous rocks from diverse crustal depths represent rift settings, supporting evidence generally is weak.

Bimodality. Because sharply bimodal basalt-and-rhyolite assemblages occur in the modern Snake River Plain and Rio Grande rift, and in parts of the Basin and Range province, many petrologists argue mistakenly that any bimodality in local sections of ancient volcanic rocks is indicative of a rift setting. Many active magmatic arcs are in fact sharply or broadly bimodal, particularly when viewed on a modest scale such as 5,000 km^2 and 0.5 m.y. Bimodality of the volcanic rocks of the Cascade Range is noted in sequel. North Island, New Zealand, exemplifies sharp bimodality: late Quaternary arc volcanism is dominated by silicic rhyolite and much-subordinate basalt (references in Hamilton, 1988c); gabbro likely is much more abundant at depth. On the other hand, nearly unimodal complexes, including abundant intermediate rocks and little or no rhyolite, also occur in the late Neogene rift setting of the western States (Gust and Arculus, 1986; McMillan and Dungan, 1988).

Anorogenic magmatism. Tectonic thickening of silicic or intermediate rocks can depress them to depths at which granitic melts might be generated during a short period after a considerable time lag, if erosion, heat conducted from the mantle, radiogenic heat, tectonic redistributions of hot and cold rocks, and conductivity interact with appropriate concatenations (England and Thompson, 1984). Such conditions might be met in compressively thickened crystalline crust or terrigenous sediments. Mantle diapirs perhaps partially melt basal continental crust (Anderson, 1983). Igneous complexes resulting from such processes could be termed "anorogenic." I know of no province of voluminous Neogene silicic volcanic rocks that clearly is of this type, so actualistic analysis of possible ancient examples is precluded. Some researchers deduce the process to have produced widespread magmatism in the past, whereas I infer that continental anorogenic magmatism generally has been of but minor importance.

Chemical discriminants. Many petrologists have argued that fields or trends of ratios of chemical components, particularly minor elements, discriminate magmatic rocks formed in diverse tectonic settings. Such discriminants commonly are assigned much more resolution than they in fact possess. Many of the proposed discriminants are based on small numbers of modern provinces selected because they provide end-member petrologic characteristics, and are misapplied for diagnostic purposes to other provinces of quite different characteristics. The discriminants are heavily biased for subaerial volcanic rocks (except for spreading-ridge basalts). Discriminants for arc-magmatic rocks are derived primarily from oceanic island arcs, the rocks of which are systematically different from those of continental arcs; and in the oceanic arcs, the discriminants are based only on subaerial rocks and do not incorporate the widespread spilitization of submarine arc rocks. The discriminants commonly cannot cope with processes that have changed through geologic time as the crust and mantle have become more evolved, or with processes that vary with crustal depth. Rigorous studies are needed wherein relatively similar rocks of contrasted known settings are compared.

Facies and equilibration. Facies fields (Fig. 2) apply equally to metamorphic and magmatic rocks. Igneous rocks do not have the compositions of melts, for volatiles are mostly lost before final crystallization, and solids equilibrated elsewhere commonly are contained in melts. Much of igneous petrology represents an attempt to demonstrate equilibria between solids and liquids, and positive results are too often interpreted to indicate that liquid plus solid equal a source rock whose partial melting produced the liquid, or equal a magma whose fractionation produced them both. O'Hara and Mathews (1981) demonstrated the futility of much conventional geochemical modeling even for the simplest systems in which no wall-rock reactions are involved and little variation occurs with depth. Walker (1983) reviewed complexities of magmatic evolution that are receiving increasing attention.

Alumina saturation. Variations in proportions of alumina to K, Na, and Ca in granitic rocks produce conspicuous differences in mineralogy and are expressed by the terms metaluminous, peraluminous, and peralkaline. Metaluminous rocks have molecular $K_2O + Na_2O < Al_2O_3 < K_2O + Na_2O + CaO$; such rocks dominate upper-crustal suites, and typically bear accessory biotite and hornblende. Peraluminous rocks have $Al_2O_3 > K_2O + Na_2O + CaO$, dominate many middle-crust sections, and commonly show evidence for equilibration with micaceous wall rocks; their accessory minerals commonly include biotite plus primary muscovite or garnet. Peralkaline rocks have $Na_2O + K_2O > Al_2O_3$, can be either sodic or potassic, and typically have accessory sodic pyroxene or sodic amphibole.

I-type and S-type granites. A.J.R. White, B. W. Chappell, and their associates (references in White and others, 1986) simplistically assumed granites to be derived by crustal melting of either igneous or sedimentary rocks, and they termed the resulting granites I-type and S-type, respectively. These subjective terms have been widely applied and conflictingly redefined, broadened, narrowed, and argued about by White and Chappel and many others. S-type granites commonly are peraluminous, but just which peraluminous granites qualify is disputed.

In my view, simple one-stage melting is uncommonly a dominant process. Where rising magmas of whatever source and complex prehistory tarry and equilibrate in terrains of micaceous rocks—typically, middle continental crust—they become enriched in the breakdown products of micas; where the micaceous rocks are pelitic metasedimentary rocks, S-type granites result. Such granites thus have environmental significance but not the tectonic and heat-source significance claimed for them by many of their classifiers. Most S-type granites also have depth significance, for most crystallized from richly hydrous melts, which with their mineralogy, commonly requires solidification at depths greater than about 12 km, as discussed in sequel.

EVOLUTION OF CONTINENTAL CRUST BY ARC MAGMATISM

Magmatic arcs now form primarily above those parts of subducted slabs of oceanic lithosphere whose tops are at depths of about 100 km (Fig. 3). Water expelled from subducted oceanic crust, metasedimentary rocks, and serpentinized mantle presumably causes the initial melting at depth, where the proto-arc magmas must be of very mafic compositions. The volcanic rocks erupted at the surface, however, vary systematically in average composition with the character of the crust and mantle through which the magmas have risen. Magmas reaching the surface in a young oceanic island arc are dominantly basaltic; in a mature island arc, andesitic; and in a continental arc, rhyodacitic. Such variations occur not only in the contrasted types of arcs but also as a progression wherever a continuous magmatic arc changes

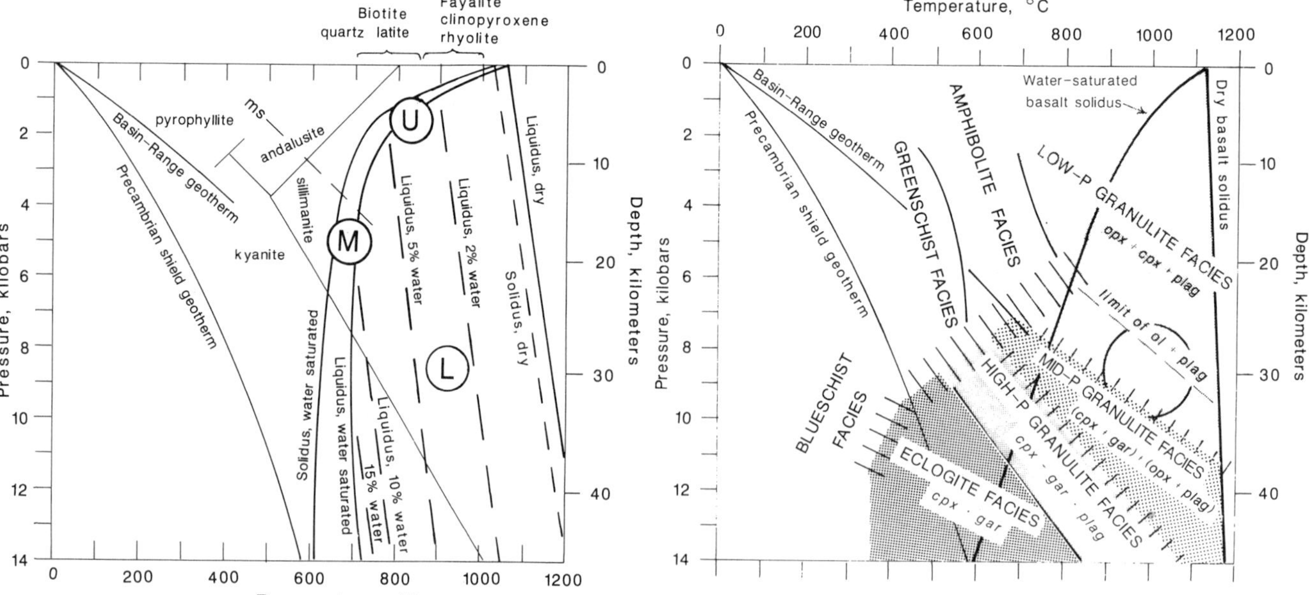

Figure 1. Crystallization relationships for leucogranite, showing the melting interval for water-saturated granite and contours on the liquidus surface for undersaturated magma, after Huang and Wyllie (1981) and Stern and Wyllie (1981). Fields of the aluminum-silicate polymorphs are from Holdaway (1971). Primary muscovite can crystallize in granitic magma only deeper than the intersection of the stability limit of muscovite (ms; after Tracy, 1978; precise position varies with bulk compositions) with the granite solidus. Typical fields are shown by circles for late-stage granitic magmas of the upper (U), middle (M), and lower (L) crust; these could form granites bearing biotite, muscovite, and hypersthene, respectively. Typical eruption temperatures of biotite quartz latite and of fayalite-clinopyroxene rhyolite follow Hildreth (1981) and others. Geotherms from Sclater and others (1980) and Lachenbruch and Sass (1978).

Both wall rocks and magmas are relatively anhydrous in the lower crust. The middle crust, rich in hydrous minerals, is raised toward granite-solidus temperature by transiting magma. Magmas that equilibrate with, or that are produced by partial melting of, micaceous wall rocks, become richly hydrous and commonly peraluminous. Hydrous magmas once formed cannot generally rise to shallow depths, for they cross the water-saturated solidus curve as they are depressurized. Low-water magmas that reach the upper crust without such middle-crust equilibration do not become saturated until crystallization is advanced, at which time either pegmatites or tuffs are expelled.

Figure 2. Generalized pressure-temperature diagram of mineral assemblages relevant to the lower continental crust. Boundaries approximate those for mafic and intermediate rocks but vary with bulk composition; coexisting minerals vary in composition across each facies. Reactions from amphibolite to granulite facies are dominated by dehydration, and reactions between the several granulite facies represent decreasing molar volume—primarily decreasing stability of plagioclase, and increasing stability of garnet and pyroxene—with increasing ratios of pressure to temperature. The boundary between amphibolite and granulite facies at pressures greater than 5 or 6 kbar shifts greatly with availability of H_2O and CO_2, and a garnet-amphibolite facies (not shown) often intervenes within the P-T region of about 5 to 8 kbar and 600° to 750°C (Ghent and Stout, 1986). The assemblage olivine plus plagioclase can be stable to higher pressure in rocks of some compositions. Abbreviations: cpx = clinopyroxene; gar = garnet; ol = olivine; opx = orthopyroxene; plag = plagioclase. Adapted from many published papers, including Hansen (1981), Johnson and others (1983), Newton and Perkins (1982), and papers referred to by each. Geotherms from Sclater and others (1980) and Lachenbruch and Sass (1978).

along trend from oceanic to continental or from young to mature. The magmas that reach the surface even in the most primitive oceanic arcs must be products of serial partial equilibrations and hybridizations in the mantle and crust. Ordering by depth of formation of complexes of known and inferred arc-magmatic origin permits general interpretation of the evolution of upper mantle and crust in terms of magmas that originate as melabasaltic liquids at depths near 100 km.

Continental arc volcanism

Active magmatic arcs developed on mature continental lithosphere above subducting oceanic lithosphere, as in Sumatra (Fig. 3) and the Andes, are characterized by fields of silicic, metaluminous volcanic rocks, dominantly ash-flow sheets erupted from large calderas atop granitic plutons, and by stratovolcanoes of intermediate rocks. Little-eroded fossil magmatic arcs include the vast middle Tertiary ignimbrite fields of the Sierra Madre Occidental–Mogollon Plateau–San Juan Mountains belt in western Mexico and the southwestern United States. In the San Juan Mountains of Colorado, overlapping calderas having diameters to 40 km, and from each of which great ash-flow sheets were erupted, define an underlying shallow, composite batholith (Lipman, 1984). The largest pluton yet mapped in the Cretaceous Sierra Nevada batholith of California is about the same size, 25 × 80 km, as the largest known young caldera in an active magmatic arc, the late Pleistocene Lake Toba caldera of Sumatra. Where erosion has penetrated such volcanic superstructures, the ash flows frequently are seen to cap plutons from which they were

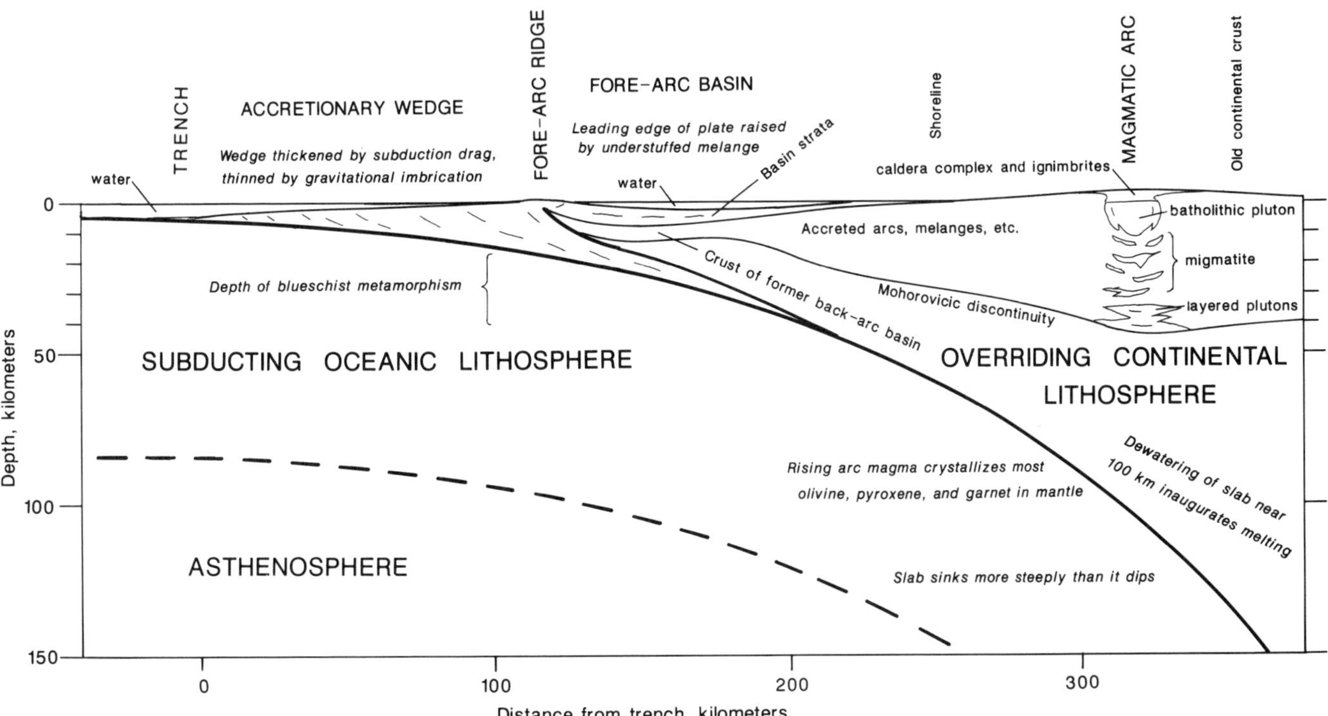

Figure 3. Cross section of a continental-margin subduction system. The dimensions of the model are derived from modern Sumatra (Hamilton, 1979, 1989) but are quantatively like those of the middle Cretaceous components of California. For Cretaceous California, the accretionary wedge is represented by the Franciscan Complex of mélange, the fore-arc basin by the "Valley facies", and the magmatic arc by the Sierra Nevada batholith.

erupted. Most large "epizonal" and "mesozonal" plutons probably breached to erupt voluminous volcanic products. Many plutons form capped only by their own volcanic ejecta.

Silicic magma chambers are zoned thermally and compositionally (Hildreth, 1981). Eruptions are generally from the tops of chambers and drain volatiles from deeper in the chamber. The most highly fractionated magma is erupted first and hence is deposited stratigraphically lowest within ash-flow sheets, which are produced by the radial outflows of successive collapsing magma fountains. Volcanic rocks typically are more silicic and alkalic than are the plutons left beneath. Minerals in some ash flows require that some chambers vent abruptly from depths as great as 10 km, but most source depths are much shallower.

Upper and middle crust

Beneath the volcanic rocks of continental magmatic-arc complexes are upper-crust plutons and composite batholiths of calc-alkalic granitic rocks. Among such complexes are those of the great magmatic arc, formed during most of Cretaceous time, that lies along the Cordillera of North America and that is paired to the fore-arc basin and accretionary wedge farther west (Hamilton, 1978; Figs. 3, 4). The Mesozoic magmatic complexes have been variably uplifted and eroded and show systematic variations with depths of erosion. A common denominator is that typical middle-crust granites formed from magmas both cooler and more hydrous than did upper- or lower-crust granites. Another common denominator is that eastern parts of each batholith display in their isotopes much incorporation of continental crust in their magmas, whereas western parts display little or none.

Sierra Nevada batholith. Most of the medial and eastern Sierra Nevada batholith of California has been eroded only to shallow depths and is a composite of hundreds of plutons, mostly of Cretaceous age, dominantly tonalite and granodiorite in the west and more felsic granodiorite and quartz monzonite in the east (Bateman, 1979; Bateman and others, 1963; Noyes and others, 1983). Petrology of the eastern granitic and metamorphic rocks—the absence of muscovite in the former and the prevalence of andalusite in the latter, for example (Fig. 1)—indicates that they crystallized at depths of only 3 to 10 km. Individual plutons are bounded sharply, and generally steeply, against each other and against the contact-metamorphic rocks that separate many of them. Migmatites occur only as local contact phases. Volcanic rocks comagmatic with the shallow eastern plutons are preserved above them locally, and a Cretaceous caldera complex atop one of them was documented by Fiske and Tobisch (1978). Granitic and metamorphic petrology indicates western Sierran granites to have been eroded to somewhat greater depths.

Cambrian to Permian strata and overlying Triassic and Jurassic volcanic rocks form a broad belt along the east edge of the central part of the batholith. The section is broadly upright and

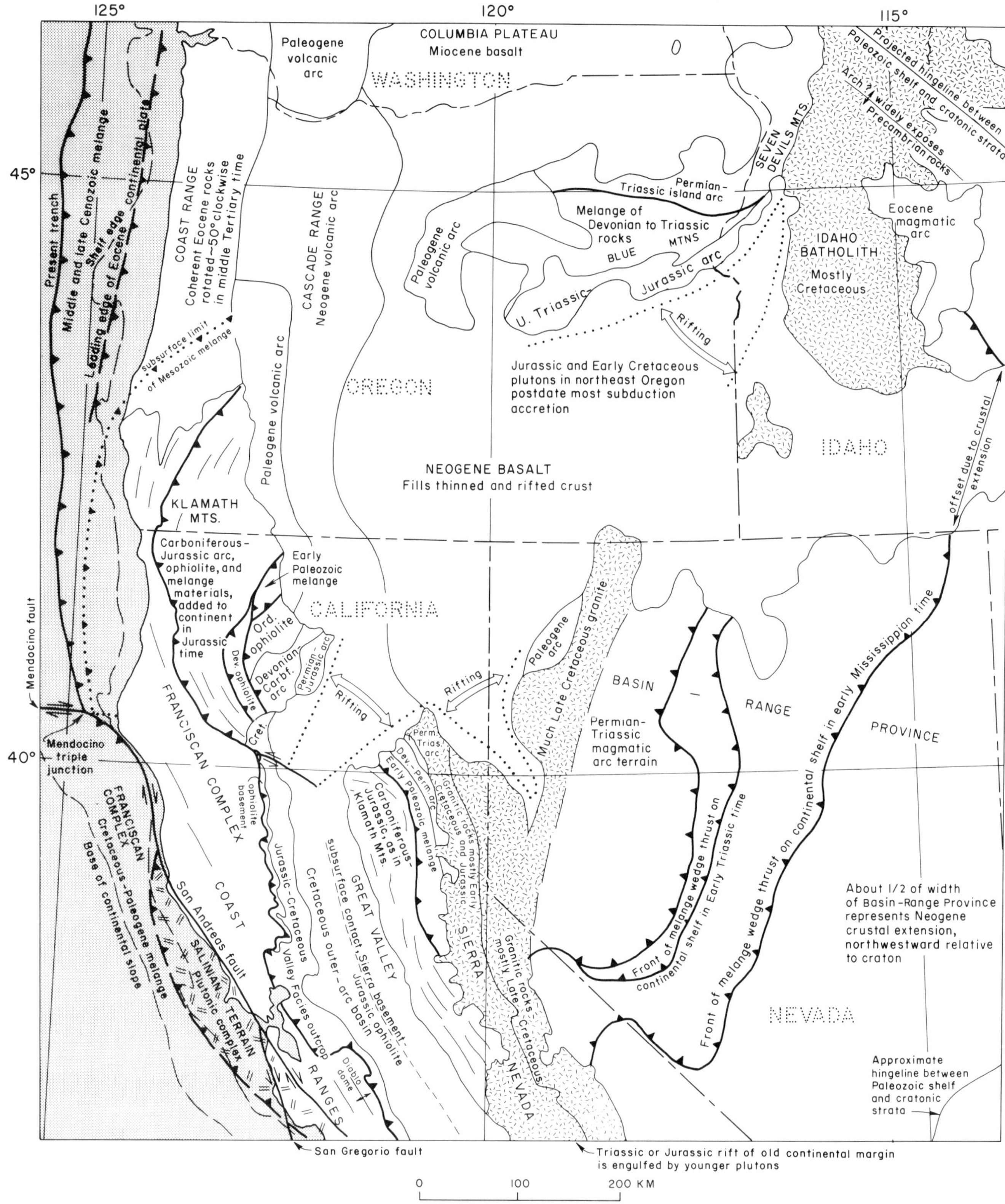

Figure 4. Selected tectonic elements of the west-central United States. Reprinted, with slight corrections, from Hamilton (1978, Fig. 4).

gently dipping in the White and Inyo Mountains, just to the east, but it rolls to steep to vertical west dips in and adjacent to the Sierra Nevada, where it is contact-metamorphosed and highly deformed. Large Late Cretaceous plutons (most conspicuously the Cathedral Peak pluton, and less obviously the Mono Recesses and Mt. Whitney plutons) of the main Sierra Nevada batholith lie with steep contacts against the west edge and stratigraphic top of this section. This geometry indicates these contacts to be the depressed and outward-pushed floors of essentially extrusive, bathtub-shaped plutons, and not the "roof pendants" of popular assumption.

Middle-crust rocks are exposed at the south end of the Sierra Nevada batholith, which was tilted upward and eroded obliquely during latest Cretaceous or early Paleogene time. Depth of formation of rocks exposed increases, over a horizontal distance of 70 km, from less than 10 km to about 30 km (Ross, 1989). The eastern Sierran belt of felsic, metaluminous granodiorite and quartz monzonite trends southward into its substrate of mostly peraluminous granodiorite, which contains widespread primary muscovite and local garnet. Wall rocks contain abundant sillimanite and no andalusite, and are migmatitic or gneissic: a broad terrain was heated almost to granite-solidus temperature at a pressure great enough to permit richly hydrous melts. The medial and western Sierran belts of mostly granodioritic and tonalitic plutons trend southward into a substrate of variably migmatitic hornblende tonalite, and on into tonalitic and more mafic rocks of low-pressure-granulite and garnet-amphibolite facies.

The granulitic character of the coeval deep crust beneath the shallow granites of the central medial Sierra Nevada is indicated also by xenoliths in Neogene trachybasalt (Dodge and others, 1986). Xenoliths of gabbroic granulites of middle Cretaceous age crystallized in the olivine-free part of the low-pressure-granulite facies, thus deeper than about 30 km (Fig. 2). Also present are xenoliths of cumulate pyroxenite, peridotite, and garnet pyroxenite that equilibrated at depths near 40 km.

Granulites formed beneath late Mesozoic and Paleogene granites are exposed also in the California Coast Ranges and in coastal and interior British Columbia.

Peninsular batholith. The southern sector of the composite Cretaceous arc batholith is exposed in the Peninsular Ranges of southwestern California and northern Baja California (Silver and others, 1979; Todd and Shaw, 1979). Like the Sierra, the southern batholith becomes both younger and more silicic and potassic eastward, but unlike the Sierra, it is eroded much more deeply in the east than in the west. Coeval calc-alkalic volcanic rocks are widely preserved atop the shallow, cross-cutting plutons of the western part of the batholith, whereas midcrustal migmatites and gneisses of uppermost amphibolite facies and concordant granites, including many peraluminous ones crystallized from richly hydrous melts, are exposed in the east.

Idaho batholith. The Idaho sector of the Cretaceous batholithic belt is eroded to midcrustal levels (Hyndman, 1983). The migmatitic floor is exposed in a broad arch, trending westward across the center of the batholith, as a zone of intercalated granitic, migmatitic, and gneissic rocks in which dips are more commonly steep or moderate than gentle, and in which the proportion of granitic rocks decreases downward through a vertical range of 2 km as seen in deep canyons (Cater and others, 1973). The part of the batholith north of this arch is rimmed by deep-seated gneisses and migmatites of upper amphibolite facies, the petrology of which indicates a general depth of formation of about 15 km, and is in broad aspect a synformal mass, the remnant of a great batholithic sheet. Granitic rocks are characterized by primary muscovite as well as biotite, hence crystallized from magmas too hydrous to have risen to upper-crustal levels (Hyndman, 1983; Fig. 1).

Arc rocks east of the Sierra Nevada. In latest Cretaceous and early Paleogene time, the North American and Farallon (eastern Pacific) plates probably converged more rapidly than they did during earlier Cretaceous time (Engebretson and others, 1984; Jurdy, 1984; but see also Stock and Molnar, 1988), and the magmatic arc formed to the east of the Sierra Nevada–Peninsular Ranges batholithic belt, in the Basin-Range and other regions. (The northwest United States and Canada were then in contact with the Kula plate, which was spreading relatively northward away from the Farallon plate; very different magmatic and tectonic patterns, including much dextral slip, resulted in the northern Cordillera.) Continental arc rocks of some Jurassic ages also formed inland of the main Cretaceous belt. Extremely variable uplift and erosion has affected the eastern magmatic complexes, which can now be seen at all levels of exposure from ash flows down through batholiths with hydrous-magma deep parts and into the migmatites and concordant granitic sheets beneath those. Much of the erosion was by tectonic denudation: middle-crust rocks are exposed primarily beneath extensional Tertiary detachment faults. Most upper-crust plutons within layered rocks appear to be sills inflated thickly to mushroom shapes. Thus, the small batholiths, mostly Jurassic, of the Inyo Mountains in eastern California in general have steep, outward-dipping contacts that follow limited stratigraphic zones within the thick Paleozoic section for long distances (Ross, 1967). Best known of the batholiths that reached the surface over a broad area is the composite Upper Cretaceous Boulder batholith of Montana, which is 100 km long and 50 km wide; it spread laterally over a floor of all premagmatic rocks, Middle Proterozoic to Upper Cretaceous, rolling them down so that they steepen toward and dip beneath it. The batholith is overlain only by coeval volcanic rocks and is in effect a gigantic extrusive mass (Hamilton and Myers, 1967, 1974).

The exposed middle-crust granites, many as young as middle Tertiary, are commonly peraluminous; they contain muscovite and biotite (and frequently garnet), crystallized from richly hydrous melts, and mostly lie within amphibolite-facies metamorphic terrains. These granites have isotopic compositions indicative of generally much more inclusion of micaceous crustal material in their melts than do the shallow granites.

Northern Appalachians. The transition from upper to middle crust within a magmatic belt mostly of Devonian age is exposed in oblique section in New England. Shallow cross-cutting plutons give way downward to concordant sheets of granite in migmatitic terrains of upper amphibolite, and, where most deeply eroded, low-pressure-granulite facies. The upper-crustal magmatic rocks of this belt appear to be partly of typical arc type and partly of extensional-arc type (formed in an extending region above a subducting slab); I assume them to be byproducts of subduction relatively westward beneath North America from a trench east of the belts now exposed onshore. Most petrologists working with these granites regard them as related instead to rifting or anorogenic processes. One antisubduction argument (Wones, 1980) is that Appalachian peraluminous midcrustal granites are different from Sierra Nevada metaluminous upper-crustal arc granites—but the Appalachian rocks in question are much more like those of the Sierran rocks with which they should be compared, the midcrustal granites of the southern Sierra Nevada. Another dubious antisubduction argument is based on the moderate bimodality of some shallow complexes.

In central and eastern Maine, where erosion has been shallow, plutons are thick, steep-sided, and have contact-metamorphic aureoles (Hodge and others, 1982; Sweeney, 1976). The granites are calc-alkalic biotite and hornblende-biotite granodiorite, quartz monzonite, and granite (Loiselle and Ayuso, 1980)—typical arc-magmatic rocks, crystallized from relatively hot and dry magmas. Volcanic rocks are widely preserved. One batholith retains part of its original cap of coeval granophyre formed directly beneath rhyolite (Griscom, 1976), and I have observed an apparent near-vent ignimbrite breccia intruded by a shallow granite in southeastern Mount Desert Island and nearby small islands.

The magmatic belt is eroded to progressively greater depths as it trends southwestward, then south-southwestward, through New England, obliquely into a north-trending axis of great uplift (compare Carmichael, 1978; Morgan, 1972; Tracy and Robinson, 1980). Within a transition region in southwest Maine, regional metamorphism of wall rocks increases from low to middle grade, and cross-cutting plutons include an increasing proportion of granites bearing primary muscovite and thus eroded deeper than 10 km. The level of exposure passes southwestward through the 13-km depth of the aluminum-silicate triple point (Fig. 1), and migmatite becomes regionally extensive and is injected pervasively by sheets (Hodge and others, 1982; Nielsen and others, 1976) of pegmatites and hydrous-magma granitic rocks dominated by two-mica quartz monzonite (Wones, 1980). Sheets of higher-temperature granodiorite contain abundant large garnets residual from granulite-facies equilibration in the deep crust (Clark and Lyons, 1986). The medial part of the migmatite terrain is on the high-temperature side of the "second sillimanite" isograd, which represents the breakdown of muscovite to K-feldspar, sillimanite, and water; this release of water likely was an important cause of formation of the granitic melts (Grant, 1973; Tracy, 1978). This medial belt was generally heated to the solidus temperature of hydrous granite magma by rising hotter, dryer magmas. The flanking belts, out to the "first" sillimanite isograd, were traversed by less magma but may have been metamorphosed beneath a spreading, aggregating batholith whose magma rose primarily through the medial belt. The magmatic belt is eroded to a depth of 18 to 25 km in southern New England, where it contains migmatites and granites of low-pressure-granulite facies, formed under low-water conditions in which biotite was only partly stable (Tracy, 1978; Tracy and Robinson, 1979). Away from the migmatitic belt, rocks exposed along the axis of maximum uplift were metamorphosed at depths of 20 km but at temperatures near ambient geotherms, largely unperturbed by rising melts.

Lower crust. As seen in outcrop and as defined by xenoliths, the lower continental crust consists largely of rocks in granulite facies. Petrologists disagree as to classification of such rocks; my preferences are summarized in Figure 2.

Mature, lower continental crust is exposed in many parts of the world beneath migmatite terrains such as those noted in the preceding discussion, and basal crust in two areas discussed subsequently. I assume that such lower crust commonly represents the deep parts of magmatic arcs, but only exceptionally can a direct tie be made of lower crustal rocks to upper crustal complexes obviously of arc character. Typical lower crust consists of distinctive igneous and metamorphic rocks equilibrated under low- or middle-pressure-granulite conditions and commonly retrograded, particularly in middle-pressure-granulite facies. Such retrogression, like that to high-pressure granulite and even eclogite exposed in some parts of the world, can be isobaric (Fig. 2). Exposed sections of deep crust, representing depths on the order of 30 to 40 km, generally display differentiates of mafic magmas intruded from the mantle, and also the products of granitic magmas melted from wall rocks by the heat of those intrusions. Associated premagmatic rocks often include metasedimcnts, which have lost most of their initial combined water.

Grenville crustal section. The Grenville province is a broad belt of metamorphism and deformation, which culminated about 1,000 to 1,100 Ma (Easton, 1986; Emslie and Hunt, 1989) developed across a wide variety of Proterozoic and Archean terrains. The exposures are primarily in the southeastern Canadian Shield—southeastern Ontario and southern Quebec and Labrador, with an outlier in the Adirondack Mountains of New York State. At least in Canada, the late Middle Proterozoic deformation represents the northwestward imbrication of continental crust (Davidson, 1986; Green and others, 1988), likely as a response to a collision between continents. Grenville basement lies beneath the Phanerozoic strata of the eastern interior United States. The continental rifting that preceded deposition of the Late Proterozoic and early Paleozoic wedge of continental-shelf strata developed longitudinally within the Grenville, and Grenville basement, variably overprinted by Paleozoic metamorphism and deformation, occurs in uplifts along the northwestern crystalline Appalachians from Newfoundland to Alabama.

The best-known section of middle and lower crustal rocks in the United States is that of the Grenville Province as exposed in the Adirondack Mountains. Middle Proterozoic high-grade metamorphism, magmatism, and deformation affected preexisting magmatic and metamorphic rocks. Studies in the Adirondacks (summary in McLelland, 1986) and nearby southern Ontario (van Breemen and Davidson, 1988; van Breemen and Hanmer, 1986) appear to indicate that most of the magmatism occurred there between 1,100 and 1,300 Ma, not long before the peak of dynamic metamorphism.

Adirondack Grenville metamorphic rocks are exposed at shallowest depths in the far west, where the metamorphic peak represented about 600°C at a depth of 20 km; the deepest rocks, in the east-central region, record about 800°C and 30 km (Bohlen and others, 1985). These high temperatures presumably record heat introduced from the mantle in the younger magmas. Metamorphic facies progress from amphibolite to low-pressure granulite to middle-pressure granulite, in general order of increasing temperature and depth and decreasing water fugacity. Temperature, hence presumably depth, of crystallization of the late pre-Grenville igneous rocks increased in the same sense; many of the igneous rocks are but slightly older than the metamorphism. In the least-eroded region, the granites are of uppermost amphibolite facies. Next come granites and migmatites of low-pressure-granulite facies; contact metamorphism by the largest of these, which was intruded at a temperature of about 1,050°C, occurred at a depth of about 25 km and resulted in partial melting of wall rocks by breakdown of biotite (Powers and Bohlen, 1985). These granites are typically low-quartz, pyroxene-mesoperthite rocks. (Mesoperthite is exsolved from sanidine, high-temperature hypersolvus ternary feldspar.) The deepest rocks in the exposed crustal section are large domes—the tops of megaboudins or culminations of sheets—of anorthosite and subordinate norite that crystallized at about 1,100°C (Bohlen and Essene, 1978; Wiener and others, 1984). Igneous pyroxene in anorthosite crystallized deeper than 30 km (Ollila and others, 1984). The anorthosite contains deep-mantle xenocrysts (Jaffe and Schumacher, 1985) and has ratios of Nd isotopes indicative of derivation from mantle melt (Ashwal and Wooden, 1983). Pyroxene-mesoperthite granite directly overlies much of the anorthosite and forms mixed rocks against it (Wiener and others, 1984); Nd isotopes indicate that the granite is not cogenetic with the mafic rocks (Basu and Pettingill, 1983). Reactions compatible with isobaric cooling from magmatism through metamorphism include the reaction of primary olivine + plagioclase in gabbro to corona assemblages, in order of decreasing temperature, with orthopyroxene, clinopyroxene, and clinopyroxene + garnet (Whitney and McLelland, 1973; Fig. 2).

Grenville metamorphism and deformation in the Adirondacks were severe, and even broad relationships are in dispute. Only 25 to 75 km to the northwest in Ontario, however, metamorphism is low grade; widespread metavolcanic rocks are tholeiitic and calc-alkalic basalt, andesite, dacite, rhyodacite, and rhyolite; they likely record an arc-magmatic system that began as one or several primitive oceanic arcs and evolved to more continental character. The dominant metasupracrustal rocks, now gneisses, of the relatively shallow part of the Adirondack crustal section have dacitic, rhyodacitic, and rhyolitic compositions, indicative to me of derivation from continental magmatic-arc rocks (Popple Hill gneiss of Carl, 1988; quartz-biotite gneiss of Engel and Engel, 1958; biotite-quartz-plagioclase gneiss of Buddington and Leonard, 1962). Metabasalt, now amphibolite, is subordinate; marble and calc-silicate rock are abundant, other metasedimentary rocks much less so.

I infer that the Adirondack plutonic rocks also represent primarily a magmatic-arc setting. A two-stage history is compatible with available dating: the first event produced hot, dry rocks in the deep crust and micaceous, amphibolite facies rocks in the middle crust, whereas magmas of the second event reached a middle crust that had been largely dehydrated during the first event and hence there absorbed little water and remained relatively hot and dry. I see the depression of marble and other metasediments to deep crustal levels as likely recording the sinking and underflow of dense rocks as less dense magmas rose through them and spread out above them. Other geologists have very different interpretations. Thus, McLelland (1986) argued that all Adirondack igneous rocks formed in the upper crust in an extensional or anorogenic setting and were subsequently depressed as the crust was thickened tectonically.

Anorthosite. Most exposed deep-crust anorthosites (as opposed to the anorthosites of such shallow differentiated gabbroic complexes as Wichita and Duluth, discussed subsequently) are of Archean and Proterozoic ages, and it is widely inferred that Precambrian tectonic and magmatic processes were quite different from those operating later and that anorthosites are limited to more or less the Precambrian terrains in which they are now mostly exposed. The best exposures of such anorthosites are those of the fjords of western Norway, where they form stratiform sheets a kilometer or more thick atop norites. It is possible alternatively that anorthosites are common products of deep-crust Phanerozoic magmatism also, the scarcity of Phanerozoic anorthosite correlating with the rarity of exposures of Phanerozoic lower crust. Some terrains of Phanerozoic rock of appropriate depth of exposure do contain numerous anorthosite layers, the largest known to me being the kilometer-thick sheets (Gibson, 1979; other references in Hamilton, 1988c) in the Cretaceous island-arc lower crust of Fiordland, New Zealand; but the Ivrea and Kohistan lower-crust sections, discussed subsequently, contain very little anorthosite. Because two Proterozoic anorthosite masses—Nain in Labrador, Laramie in Wyoming—completed their crystallization in the middle crust, some authorities infer most anorthosites to have formed at shallow levels, but the general restriction of anorthosites to deep-crustal granulite-facies assemblages is strong contrary evidence.

Island-arc magmatism

Island arcs form above subducting plates in oceanic settings but are destined to be accreted tectonically to continents after intervening oceanic lithosphere has been subducted beneath arc, continent, or both. Arcs can change along trend from continental to oceanic above a single subducting plate. Unobstructed island arcs generally migrate and lengthen with time and abandon parts of themselves as back-arc spreading takes place; reversals of subduction polarity and migration direction are common; collisions are progressive along trend when they occur (Hamilton, 1979, 1989). Tectonic and magmatic evolution varies greatly along continuous arcs.

The characteristic subaerial volcanic rocks are tholeiitic basalts in immature arcs of small crustal volume, and high-alumina basalts and andesites and subordinate dacites in mature arcs of large volume. Island-arc volcanic rocks are, however, dominantly submarine—and submarine arc rocks are widely spilitized, their magmatic sodium contents erratically raised, and calcium contents variably reduced, from primary compositions like those of the subaerial rocks, likely by subsolidus exchange with hot seawater brines concentrated by boiling (Hamilton, 1963, 1988c). Lesser reaction affects oceanic-ridge basalts, which form in water deeper than its 2-km critical depth. (Spilitization, which involves large change in bulk composition, is often confused with nearly isochemical greenschist-facies reconstitution of magmatic plagioclase to albite, epidote, and calcite.) Arcs accreted to continents are exposed mostly in submarine facies and so should be compared compositionally with submarine assemblages of young arcs, not, as is almost always done (with the publication of many invalid conclusions as a result), with unaltered subaerial rocks of modern arcs.

Cascade Range. The active, north-trending volcanic arc of the Cascade Range of southern Washington, Oregon, and northern California apparently spans a filled-in oceanic gap between the older continental crust of California and northern Washington. The magmatism is of mature-oceanic-arc type. This arc, in its southern Washington and northern Oregon part, lies outboard of all projections of pre-Tertiary continental crust, and in its southern Oregon and northern California part, is formed within an apparent extensional rift that sundered the pre-Tertiary crust (Hamilton, 1978; Fig. 4). The present crust has a maximum thickness of about 40 to 45 km and has a relatively high-velocity upper part; the upper mantle has low velocity (Leaver and others, 1984). This structure is compatible with an entirely magmatic origin, without any pre-Tertiary continental crust. The volcanic rocks show in their isotopes little or no incorporation of continental crust (Leeman, 1983).

The pattern of Cascade arc magmatism has changed complexly with time in ways that make obvious the shortcomings of many published interpretations of ancient systems in terms of simplistic assumptions regarding modern ones. The high Quaternary stratovolcanoes consist of andesite and high-alumina basalt, plus subordinate silicic rocks in bimodal assemblages with mafic rocks; the big volcanoes are separated by more voluminous correlative high-alumina basalt, erupted concurrently with extensional faulting (Hughes and Taylor, 1986; McBirney and White, 1982). Middle and upper Miocene and Pliocene volcanic rocks define the same general north-trending arc and, although dominantly andesitic, show major temporal and spatial variations (Luedke and Smith, 1982; McBirney and White, 1982). Late Eocene and Oligocene arc rocks also are dominantly of intermediate composition but include voluminous dacite and rhyodacite, and have a disjunct distribution: they occur as north-trending belts along the west side of the Cascades in the north and the south, whereas the part that should be in the center, in northern Oregon, now lies 200 km to the east, in the Blue Mountains (Fig. 4). I infer rifting—clockwise swinging of the Coast and Cascade Ranges about a northern Washington pivot, and lagging behind and clockwise pivoting and strike-slip slicing of the Blue Mountains away from western Idaho—mostly during late Oligocene and early Miocene time, synchronous with major early widening of the Basin and Range Province (Hamilton, 1978; Wells and Heller, 1988; cf. Grommé and others, 1986).

The magmatic arc lay farther east earlier in Eocene time, in northeast Washington and central Idaho. Its jump westward late in the middle Eocene presumably was due to the accretion at that time of a tract, now the Coast Ranges of Oregon and Washington, of Paleocene to middle Eocene oceanic crust and seamounts to what had previously been the western edge of the continent (Heller and others, 1987; Wells and Heller, 1988).

Accreted island arcs. Various oceanic island arcs were accreted to the west edge of North America during Mesozoic time (e.g., Hamilton, 1963, 1978; papers by various authors in Howell, 1985; Fig. 4). Much of the accretion likely was of compound island arcs previously aggregated by offshore collisions, as have been the modern Philippines (Hamilton, 1978, 1979). The supracrustal rocks of the Cordilleran arcs are dominated by variably spilitized submarine basalt, andesite, and dacite. Upper-crust plutons are mostly dioritic to granodioritic. Midcrustal assemblages are typified by amphibolitic, tonalitic, granodioritic, and trondhjemitic gneisses and migmatites. Trondhjemite can be viewed as the island-arc equivalent of muscovite granite (and, indeed, it often contains primary muscovite): it typically represents crystallization of a hydrous, low-temperature melt that was equilibrated with amphibolite and tonalite rather than with micaceous rocks.

The Appalachian Piedmont (LeHuray, 1987; Mittwede, 1988; Pavlides and others, 1982; Whitney and others, 1978) and New England contain accreted Paleozoic oceanic island arcs, the metavolcanic rocks of which are variably spilitic, separated by appropriate accretionary-wedge materials from flanking assemblages.

The Early Proterozoic basement of Colorado consists of aggregated oceanic and continental magmatic arcs (Reed and others, 1987). Grambling and others (1988) suggested plate-tectonic characterizations of the diverse Proterozoic assemblages of New Mexico; Karlstrom and Bowring (1988), those of Arizona; and Sims and others (1987) and Van Schmus and others (1987), those of the largely buried basement of the central United States. Condie (1986) concluded that the Proterozoic metavolcanic terrains of the southwestern states include an oceanic island arc, several continental arcs, and widespread rift-related rocks, but I believe that he has much overestimated the abundance of rift rocks because of his reliance on slight local bimodality as an indicator. Thus, Boardman and Condie (1986) classed as "strongly bimodal" the "rift" metavolcanic rocks of a small area in Colorado in which "mafic rocks" contain as much as 58 percent SiO_2 and "felsic rocks" contain as little as 61 percent SiO_2.

Basal crust and the origin of silicic melts

Although many granulite terrains expose rocks formed in the lower crust, there are only two crustal sections known to me in which the continental upper mantle and basal crust are exposed—the Ivrea Zone of the western Italian Alps, and the Kohistan Arc of northern Pakistan. Both are of Phanerozoic age, both expose magmatic-arc sections from the upper mantle into the upper crust, and both were ramped onto accretionary-wedge complexes during Late Cretaceous or Paleogene collisions. The dominance of arc magmatism in the evolution of these two crustal sections is indicated by the obviously arc-magmatic character of their upper-crust plutonic and volcanic rocks. Kohistan represents relatively primitive island-arc crust, and Ivrea, evolved continental crust. The basal crust of both sections consists of very thick stratiform two-pyroxene gabbro, variably retrograded in granulite facies.

The character of these rocks and their relationships to underlying mantle rocks and to overlying crustal rocks provide in my view the most neglected important evidence available regarding the origin of granitic and rhyolitic rocks and the petrologic evolution of continental crust and so I discuss them here. The crustal-section significance of both complexes is broadly recognized, but neither has yet been described nearly as well as its importance warrants. Only a few of the many Ivrea papers incorporate mapping of lower-crust rocks and structures; many papers consist of descriptions and analyses of small suites of samples whose contexts are poorly constrained. The more limited literature on remote Kohistan consists primarily of descriptions and analyses of sparse samples. Much of my understanding of both complexes comes from extensive field trips in them during 1988. I was guided ably in the Ivrea complex by A. C. Boriani, Luigi Burlini, D. M. Fountain, E. H. Rutter, and Silvano Sinigoi (and, during a prior trip in 1978, by V. J. Dietrich); and in Kohistan, by M. Q. Jan on a trip organized and led by R. D. Lawrence. My comprehension of the Ivrea crustal section was further enhanced by participation in two meetings at which many Ivrea experts were present: the Geological Society of America Penrose Conference on the Geophysics and Structure of Mountain Belts, convened by B. C. Burchfiel and Stephan Müller with National Science Foundation support in Ascona, Switzerland, in 1978; and the Italy–U.S. Workshop on the Nature of the Lower Continental Crust, sponsored by the National Science Foundation and convened by A. C. Boriani and D. M. Fountain in Verbania Pallanza, Italy, in 1988.

The Ivrea and Kohistan complexes display crustal sections like those discussed previously from various American examples but in addition expose the basal crust, which in both cases consists primarily of stratiform two-pyroxene gabbro. This basal crust has no exposed American analog, unless as I infer, the deepest Adirondack rocks, the anorthosites, are equivalent to the upper parts of the basal-gabbro sections. My syntheses here incorporate data from many papers besides those cited and are strongly biased by my application of crustal-zoning and facies concepts.

Kohistan Arc. The Kohistan crustal section is bounded on both north and south by north-dipping collision sutures (Coward and others, 1987). Reconnaissance dating shows the rocks to span late Early Cretaceous, Late Cretaceous, and Paleogene ages; the older rocks are petrologically primitive and formed in an oceanic island arc, whereas the younger are more evolved and may have formed after accretion of the arc to the northern (Eurasian) plate but mostly before collision with the southern (Indian) plate (Petterson and Windley, 1985). The structurally deepest rocks are ramped up to moderate north dips at the south edge of the complex, whereas the shallowest rocks are in the north part of the complex, but major tight folds are present within it. My synthesis here integrates my estimates of initial depths, as deduced from mineral assemblages (compare Fig. 2) described in papers published by others and also as noted by me in the field, with my inferences of megafolds as deduced from the broad symmetry of rock assemblages and attitudes that I saw in the field. My conclusions are in broad, but not detailed, agreement with the structure of the complex as deduced from field data by Petterson and Windley (1985), but are in major disagreement with the variously conflicting interpretations by Bard (1983), Coward and others (1987), and Jan (1988).

Initial thickness of the Kohistan crustal section was about 40 km, of which approximately the lower 30 km are preserved. As much as 5 km of in situ mantle rocks are preserved beneath the crustal section and consist of serpentinized spinel peridotite (mostly harzburgite?) cut by voluminous veins and dikes of spinel-bearing diopsidite; other ultramafic rock types are less abundant (Jan and Howie, 1981; my observations). Above this is a great stratiform sheet of gabbro and metagabbro, which had an initial thickness of perhaps 10 km. The entire gabbro is preserved in the north-dipping southern section (the Jijal complex and the southern third of the Kamila amphibolite, in the terminology of Jan, 1988), and all but the basal part of the gabbro is exposed, also farther north, in a great isocline (the Chilas complex of Jan, 1988, and the granulitic layered lopolith of Bard, 1983, neither of whom recognized the equivalence of this assemblage to the southern section). The basal few kilometers of this sheet have been metamorphosed at high-pressure-granulite and garnet-amphibolite facies. These deep rocks mostly bear plagioclase, but plagioclase-free garnet-clinopyroxene rocks make up the basal 500 m of the section and also abundant thick and thin intercalations higher in the lower few kilometers (Gansser, 1979; Jan and Howie, 1981; my observations). These plagioclase-free rocks formed by the metamorphic destruction of plagioclase-orthopyroxene assemblages in gabbros more mafic than the rest of the sheet (Jan and others, 1984). The gabbro sheet otherwise consists almost entirely of layered two-pyroxene, olivine-free gabbro, of which the lower part is variably metamorphosed at high- to middle-pressure-granulite and garnet-amphibolite facies and the upper part is metamorphosed, less pervasively, at garnet-amphibolite facies (Bard, 1983; Jan, 1988; Jan and Howie, 1981; Khan and others, 1989). The plagioclase-free basal rocks have mantle geophysical properties, so the geophysical Mohorovičić discontinuity (Moho) is about 0.5 km higher in the section than is the base of the magmatic section that otherwise represents the lower crust. (Griffin and O'Reilly, 1987a, deduced that such contrasted positions of geophysical and petrologic discontinuities are typical of the continental Moho.) High in the gabbro, as I read the structure, but low in it as Khan and others (1989) read it, are small complexes of layered cumulates of two-pyroxene gabbro, olivine gabbro, dunite, troctolite (olivine-plagioclase rock) and anorthosite (Khan and others, 1989). Olivine and plagioclase crystallized out together in these small complexes but are everywhere separated by near-solidus reaction rims of pyroxenes (Jan and others, 1984; my observations), so crystallization was at a depth probably near 30 km (Fig. 2). Vertical compositional variations presumably occur in the main gabbro and metagabbro but cannot be defined from published data.

Above the great stratiform gabbro and metagabbro is an upward-varying section, perhaps initially also 10 km thick, of amphibolite and migmatite (the northern two-thirds of the Kamila amphibolite of Jan, 1988, and also the Northern amphibolitic series of Bard, 1983). The lower part of this section is dominated by strongly layered amphibolite, variably garnetiferous and migmatitic, whereas the upper part is a migmatitic complex of amphibolite and diorite and extremely voluminous sheets of tonalite, trondhjemite, granodiorite, and pegmatite (Bard, 1983; my observations). No granitic rocks occur within the stratiform gabbro, so the granitic rocks in the migmatitic amphibolite complex presumably represent partial melts of preexisting rocks, the restites from which are the amphibolites low in the supragabbro section. The Ivrea complex, described next, presents a clearer example of this process, there operating upon metapelitic rocks.

The next 10 km or so of the Kohistan crustal section (the top 10 km is missing) consists of basalt, andesite, rhyodacite, and sedimentary rocks, metamorphosed at lower amphibolite and greenschist facies and intruded by crosscutting plutons of tonalite, granodiorite, and quartz monzonite (Bard, 1983; Jan and Asif, 1983; Petterson and Windley, 1985).

Ivrea Zone. A complex crustal section, dominated by products of Paleozoic arc magmatism, dips subvertically to gently southeastward above high-pressure subduction-complex rocks along the west side of Lake Maggiore in the Italian Alps. The lower part of this section is commonly termed the Ivrea Zone, or Ivrea-Verbano Zone, whereas various terms, among them Strona-Ceneri Zone and Laghi Series, are applied to the upper parts. The most reliable dates—stratigraphic age of rhyolites, lead-uranium dates of felsic rocks, and Sm/Nd dates of mafic and pyroxenitic rocks in the lower crust and upper mantle—indicate the young igneous rocks that dominate the evolution of this crustal section to be of Late Carboniferous and Permian ages and to have intruded a preexisting continental crust with a complex earlier Paleozoic history (Voshage and others, 1987, 1988; Zingg and Hunziker, 1988; and others). Both top and bottom of the late Paleozoic crustal section are preserved—extrusive rhyolite at the top, Moho at the bottom—although much of the middle crust is cut out by a major fault that probably is extensional (Hodges and Fountain, 1984). Zingg (1983) reviewed the geology of the crustal section in terms rather different from mine.

Ultramafic mantle rocks occur in lenses along and near the exposed base of the complex (Schmid, 1967). The best-studied of these, the Balmuccia peridotite (Garuti and others, 1980; Shervais, 1979a, b; Voshage and others, 1988), apparently exposes the undisturbed Moho on the east side of a lens whose west side is separated by a steep thrust from basal-crust mafic rocks. The lens consists of spinel peridotite (mostly bearing two magnesian pyroxenes, rarely garnetiferous, and presumably representing residual mantle of undefined age) injected by dikes and veins of late Paleozoic chrome diopsidite and more aluminous spinel–two-pyroxene pyroxenite.

Geopetally above this mantle lens is a stratiform mass, 6 to 10 km thick, of vertically-varying gabbro and its metamorphosed equivalents (Garuti and others, 1980; Pin and Sills, 1986; Schmid, 1967). Magmatic fabrics and mineralogy are well preserved in some sectors but have been obliterated in others by retrogression, likely isobaric, to high-pressure granulite in deep sections, middle- or low-pressure granulite in shallow ones. The basal 1 km of the great gabbro sheet consists of interlayered cumulate gabbro, pyroxenite, and subordinate peridotite; above this is another kilometer of layered cumulate gabbro. Next higher is 5 km or so of obscurely layered two-pyroxene gabbro, locally anorthositic, which becomes more feldspathic upward and grades into a zone, 2 km or so thick, of leucogabbro, diorite, and hybrid monzonite and quartz-bearing rocks. Isotopes and trace elements show increasing crustal contamination upward in this sequence (Pin and Sills, 1986; Voshage and others, 1987). Much of this upward variation records contamination by felsic melts generated from overlying crustal rocks (Voshage and others, 1987) rather than extreme fractionation of mantle melts or crystallization of discrete melts from diverse sources (as was suggested by Pin and Sills, 1986).

This great stratiform igneous sheet lies beneath metapelites and other premagmatic rocks, and lenses of such rocks also are enclosed high in the sheet. Metapelites within and close to the sheet are restites that are rich in garnet (as much as 30 percent) and often sillimanite but are mica-free. These restites commonly contain only two of the three components quartz, K-feldspar, and plagioclase, and have lost about half of their initial material to hydrous granitic melts that have migrated elsewhere (Schmid, 1979), as is to be anticipated from experimental studies (Vielzeuf and Holloway, 1988). The process of partial melting can be observed in metapelites more distant from the sheet, where similar restites are the melasome component of migmatites extremely rich in leucosomes of pegmatite and leucogranite. Still more distant metapelites are micaceous schists and gneisses that have not been conspicuously degranitized.

The preserved upper crust of the Ivrea section contains

cross-cutting granites with contact-metamorphic aureoles, and extrusive rhyolites. The middle crust has been largely cut out tectonically but presumably included sheets of two-mica granite, as in other sections of continental crust described previously.

Overview. Generalizations can be made from integration of the Kohistan and Ivrea examples with vertical variations, such as those noted in the preceding sections, seen in less complete sections through continental crust. Magmas evolve by exothermically crystallizing refractory minerals and endothermically assimilating, melting, or breaking down fusibile ones, and the reactions involved vary with the changing compositions of melts and solids, the availability of water in wall-rock minerals and in melt, and pressure and temperature. Arc protomelts generated in the mantle by processes related to subduction are much more mafic than basalt. The Moho is a self-perpetuating density filter: melts cannot rise past it until they have crystallized much of their olivine, pyroxene, and garnet components within it and have reached gabbroic compositions. Hot, dry residual gabbroic magmas rise into the dry lower crust. The density of these melts from the mantle inhibits their rising through low-density crust, although they readily rise to the surface through the dense crust of young oceanic island arcs. In mature arcs, either oceanic or continental, the mafic melts crystallize in the deep crust to produce great stratiform complexes of layered two-pyroxene gabbro, or of norite and anorthosite, variably complicated by fractionation and by assimilation and postmagmatic recrystallization at granulite and garnet-amphibolite facies. Such great mafic sheets are exposed completely in the Ivrea and Kohistan complexes, and their upper parts are exposed in many other deeply eroded terrains. Xenoliths in basalts show that such mafic sheets, variably metamorphosed, are typical of the lower continental crust (Griffin and O'Reilly, 1987a, b).

Felsic melts are derived primarily by secondary partial melting of preexisting crustal rocks (which can themselves be entirely of older arc-magmatic origin) by the mafic melts reaching the basal crust and by hybridization of mantle and secondary melts. The resulting melts are mostly tonalitic and mafic granodioritic where formed in petrologically primitive crust (and trondhjemitic where such crust contains abundant hornblende); charnockitic and other dry granitic melts where formed in previously dehydrated continental crust; and hydrous granodiorite to granitic melts where formed in crust containing abundant mica. These melts are variably hybridized with the melts of mantle origin, and the composite melts, which can contain abundant crystals from earlier and deeper crystallization, tend to rise and evolve further. Although little mantle material need be present in melts that reach high crustal levels, mantle heat carried by evolving magma is the primary source of warming of the crust above equilibrium geotherms. Progressive partial equilibrations—crystallization of refractories, melting and assimilation of fusibles—produces progressively more evolved and hydrous melts. Granites and migmatites formed deep in the crust typically form from hotter and dryer melts than do those of the middle crust, a contrast that reflects primarily the availability of water from the dominant wall rocks. The content of combined water in wall rocks tends to increase upward in the crust, and particularly in the middle crust, wall-rock water is abundantly available yet also pressure is high enough to permit the existence of richly hydrous, low-temperature melts. Magmas that equilibrate in the middle crust with micaceous wall rocks, or that are derived from them by secondary melting, are hydrated and cooled, and then are quenched by expulsion of volatiles as they rise to the levels, or crystallize to the points, at which water pressure exceeds load pressure. The resulting granites are commonly peraluminous and typically are in complexes of migmatites and sheets of pegmatites and hydrous-magma granites. Enrichment in water and alumina is greater for magmas in metasedimentary wall rocks than in old plutonic continental crust.

Felsic melts that cross the middle crust without equilibrating there can continue to rise if they remain relatively hot and dry; this likely occurs primarily after the middle crust has been heated and partly dehydrated by earlier magmas stopping there. Voluminous magmas that reach the upper crust spread as shallow batholiths above the deeper migmatites and concordant granites, and erupt as ash-flow sheets and as far-traveling volcanic ash when their own rise and crystallization produce water saturation. Most magmas of upper-crust batholiths probably contain less than 1.5 percent water and reach water saturation only at shallow depth after considerable crystallization, with the resultant expulsion of pegmatites and volcanic materials (Maaloe and Wyllie, 1975), whereas granites in the middle crust mostly crystallize from magmas with 3 to 5 percent water, cross the water-saturated solidus at much greater depth, and solidify there rather than rising to the upper crust.

Contacts between and within granitic and metamorphic rocks typically are steep in shallow complexes and undulating in deep ones. The tendency toward gentle dips in the middle and deep crust is accentuated by the variably pervasive tectonic flattening that accompanies retrograde metamorphism in many regions. Much such metamorphism and deformation probably records extension or shortening, and likely much also records gravitational flow of heated rocks that are displaced outward and downward by rising magmas and then flow beneath shallow spreading batholiths. Depression to deep crustal levels of supracrustal rocks may be due primarily to the repeated injection and eruption of less-dense magma above them.

Archean crust

Archean terrains underlie Phanerozoic strata in much of the northwestern interior of the United States, but are exposed only in northern Minnesota and in uplifts in and near the northern Rocky Mountains (Reed, 1987; Sims and others, 1987); they have been well studied only locally. Archean geology is qualitatively different from that of later Eras, and broad generalizations regarding it can best be made from other countries in which exposures and information are more extensive; see Windley (1984) for a review of Archean tectonics. Archean terrains are typified at shallow exposure levels by elliptical granites, of which many are red or pink and formed from hot, dry potassic magmas, in a network of keel-like greenstone belts. These belts commonly consist of thick sections of petrologically primitive pillow basalt, often including extremely magnesian basalt (komatiite), beneath progressively more evolved intermediate and silicic calc-alkalic volcanic rocks and clastic strata. Archean middle crust commonly consists of migmatites of amphibolite facies, and the upper part of the lower crust consists of granulite-facies rocks that include much anorthosite and pyroxene granite. I know of no exposures of basal Archean crust.

Individual belts of these varied rocks formed in periods of a few tens of millions of years, during which magmatism was more voluminous and, in part, of higher temperature than modern magmatism and recorded higher degrees of melting of less evolved mantle than do younger systems, but postmagmatic geothermal gradients in the resulting continental lithosphere were not demonstrably different from modern ones. Much controversy surrounds tectonic interpretations of Archean terrains. I see them as products of motions of faster and smaller plates than in younger systems, the pillow basalts being products of sea-floor spreading and the other igneous rocks primarily of subduction.

Nonspecialists should be wary of granitic-rock nomenclature as applied by many geologists in Archean terrains. "Granite" can mean merely a pink granitic rock, "trondhjemite" a light gray one, and "tonalite" a dark gray one, regardless of the feldspar compositions and ratios that commonly would be implied by those terms. All three of those properly specific rock names can be misapplied in Archean terrains to rocks that elsewhere would be classified as granodiorite.

CONTINENTAL RIFT MAGMATISM

Magmatism related to extensional rifting of the crust has been of major importance in the Neogene evolution of the western states, and magmatism in similar settings has been widely inferred to have been important in older terrains elsewhere in the United States. The clearly rift-related rocks are extrusive basalts and rhyolites in widely varying proportions, and upper-crust layered complexes of comparably bimodal gabbro and granite. The basal crust in such settings likely consists of other layered intrusive complexes. The nature of the higher part of the lower crust remains problematic.

Discriminating cause and effect in rifting and magmatism is difficult. Rifting commonly is too rapid to permit continuous equilibration by thermal conduction, so geothermal gradients are steepened. A general explanation for continental rift magmatism is that the top of the asthenosphere—that part of the mantle in which the ratio of temperature to pressure is high enough to permit incipient melting—rises because of depressurizing due to extensional thinning of the lithosphere, and of diapirism. Basaltic magma separated in the asthenosphere rises into the crust, where it can crystallize deep or shallow, and can cause secondary melting of crustal rocks.

Silicic non-arc magmatism may be most voluminous where continental crust undergoes particularly rapid extension. The Snake River–Yellowstone province of southern Idaho and northwest Wyoming and the Owens Valley region of eastern California provide active examples. Although arc and extensional magmas are widely assumed to be distinctly different, many of the silicic rocks of the contrasted settings are in fact quite similar. The rift assemblages of volcanic rocks tend to be either basaltic or bimodal basaltic and rhyolitic, in contrast to the broader frequency-distribution peaks typical of arc-volcanic rocks, but intermediate rocks occur in some rift settings and basaltic rocks are all but lacking in some others, some arc assemblages are strikingly bimodal, and many volcanic provinces fall into petrologic and tectonic transitions between the two end-member types. Most arc magmas, both continental and oceanic, probably form in extensional regimes above subducting oceanic plates (references in Hamilton, 1988c).

Basaltic lavas within a subprovince tend to be broadly similar in composition. Such basalts in general are fractionated from more varied and more mafic melts deep in the crust, the subuniform compositions being produced by buffering by variable crystallization of orthopyroxene, clinopyroxene, and plagioclase deeper than about 30 km, and of olivine, clinopyroxene, and plagioclase at shallower depths, with complications added by assimilation (Cox, 1980; Philpotts and Reichenbach, 1985).

As is discussed in subsequent structural sections of this report, rifting of continental crust probably begins in most cases in crust of normal geothermal gradient, and in such cases, magmatism is a byproduct of rifting. In other cases, notably including the Basin and Range province, widespread magmatism precedes most rifting, and the history and petrologic character of such "hot" rifts is quite different from that of the "cold" ones.

Basin and Range province

The Basin and Range province of the western United States has approximately doubled in width during middle and late Cenozoic time, as discussed in a later section. Whereas Oligocene and early Miocene volcanism was dominantly of silicic and intermediate rocks that display hydrid arc and extensional characteristics, the middle Miocene and younger volcanism of most of the province has been of more distinctly rift-related type. The late Neogene magmatism has been erratic in time and space. Basalt, typically moderately potassic olivine basalt, is the most widespread type, but rhyolite was erupted in large volumes in many areas; intermediate rocks, commonly alkaline, are in general much subordinate, and many assemblages are sharply bimodal (Christiansen and Lipman, 1972).

Isotopic compositions show the contrasted rock types to have markedly different sources despite their close association. Apparently the magmatism represents primarily the rise of basaltic magma of largely mantle origin into and through the crust, with important secondary melting of rhyolite magma from lower crustal rocks by the heat of those mantle magmas (Christiansen and McKee, 1978). Both this magmatism and the high heat flow of the province appear to be products, not causes, of the extension (Lachenbruch and Sass, 1978). Nevertheless, the initial middle Tertiary spreading of the Basin and Range province was in an intra-arc or back-arc mode and was accompanied by voluminous magmatism of hybrid extensional-arc type. The inauguration of spreading may have been made possible by softening of the lithosphere due to arc-magmatic transfer of heat.

Rhyolite much outbulks basalt in the areas in which voluminous ash flows were erupted from calderas, beneath some of which granitic plutons must have formed. About 800 km^3 of rhyolite were erupted from and in Long Valley caldera, 15 × 30 km, in eastern California during its climactic middle Pleistocene eruption (Bailey and others, 1976). In southern Nevada, a composite batholith about 50 km in diameter presumably is defined by overlapping middle and late Miocene calderas, each the source of great ash flows (Christiansen and others, 1977). The isostatic-residual-gravity low over this caldera region (Jachens and others, 1985) indicates that granitic plutons probably extend deep into the upper crust. On the other hand, seven large middle Miocene calderas in southeastern Oregon and adjacent Nevada partly overlap within a region of 40 × 100 km (Rytuba and McKee, 1984), yet display slight or no gravity lows.

The dominant rhyolites from most Basin and Range calderas, and all of them from some (e.g., Long Valley: Bailey and others, 1976), are metaluminous and are similar to the rhyolites of continental arc-magmatic settings in both their minor and major elements. These extensional-setting metaluminous rhyolites bear biotite as their common mafic mineral, often with hornblende, and formed mostly from relatively cool and hydrous melts; less commonly they contain ferromagnesian pyroxene or iron-rich olivine, indicative of higher temperature and lower water content. Some of the rhyolites from many calderas (e.g., southern Nevada) and all or most of them from some (e.g., southeastern Oregon: Rytuba and McKee, 1984) are by contrast peralkaline, commonly rich in sodium, and are characterized mineralogically by accessory sodic pyroxene or sodic amphibole. Peralkaline rocks are present in some arc-magmatic assemblages but are far less common in them. Peralkaline rhyolites are erupted commonly with higher temperatures and lower viscosities than metaluminous ones, likely rise rapidly from melting sites in the lower crust, and in important examples do not form upper-crust magma chambers and calderas. Because of their higher eruption temperatures, peralkaline ash flows commonly flow much more after eruption than do metaluminous ones and often are confused for primary rhyolite flows.

Ancient analogs for these complexes could be recognized by several features. Rift rhyolite might be associated with abundant basalt of essentially the same age. Both rhyolite and cogenetic granite likely would include peralkaline rocks with sodic accessory minerals. The rift granites would, in general, be a little less silicic and more mafic than the associated rhyolites, for the most highly fractionated magmas were extruded (Christiansen and others, 1977; Hildreth, 1981). Rift granites representing the common metaluminous complexes would be felsic biotite quartz monzonites of no particular distinction, but granodiorite and tonalite should be but very minor components of the association. Voluminous Neogene mafic igneous rocks probably have formed in the deep Basin and Range crust.

Snake River–Yellowstone province

The eastern Snake River Plain is a northeast-trending depression developed from middle Miocene through Quaternary time directly across Basin and Range fault blocks, which plunge under it and are buried by its volcanic rocks. The depression is about 450 km long and narrows from a width of 100 km in the southwest to 50 km in the northeast. Surface altitudes increase from about 1 km in the southwest to more than 2 km in the Yellowstone Plateau in the northeast, but are low relative to the high flanking ranges. Surface and drilled subsurface rocks of the plain are entirely volcanic and dominantly silicic, and occur mostly as voluminous caldera-related rhyolite ignimbrites and subordinate rhyolite flows, and as flows of basalt (Christiansen, 1984; Leeman, 1982; Morgan and others, 1984). The latest Miocene, Pliocene, and Quaternary complexes have mostly developed in local sequences, each a million years or so long, characterized by early basalt, main-stage voluminous rhyolite, and late voluminous basalt, which overlap in time and space but become younger northeastward in an irregular progression. Older Miocene complexes contain similar rocks but are not exposed completely enough to document details of their evolution. Late Miocene and Pliocene complexes are distributed irregularly about all but the northeast end of the eastern Snake River Plain; middle Miocene complexes are present both in the same region and in a large region to the southwest, west, and northwest; Quaternary basalt is widespread throughout the plain, and minor Quaternary rhyolite occurs in various parts of it. The western Snake River Plain is a bimodal rhyolite and basalt province mostly of middle Miocene age (Leeman, 1982).

Snake River basalts are mostly low-alkali olivine basalts. The rhyolites are mostly high in silica and iron and low in calcium, and are metaluminous rocks, or uncommonly peralkaline ones, mostly erupted at high temperature and bearing accessory iron-rich clinopyroxene and olivine and magnesian orthopyroxene (for example, Christiansen, 1984). Lead and strontium isotopes of Yellowstone rocks accord with an origin of the basalts largely by melting of old continental mantle and of the rhyolites mostly be melting of old, lower continental crust (Doe and others, 1982). Melting of basalt in the mantle and secondary melting of silicic magma in the lower crust by voluminous basalt injected there, with important complications due to magma mixing and contamination, seem indicated.

Intrusive equivalents of the rhyolites would be high-temperature granites and quartz monzonites, or their granophyric counterparts, likely red, bearing hornblende or clinopyroxene, not biotite, as their dominant mafic minerals.

Seismic-refraction and gravity studies define a 40-km crust

of unusual character beneath the eastern Snake River Plain (for example, Sparlin and others, 1982). That crust has a thin, silicic upper part, with $V_p < 6.2$ km/sec; a mafic middle part, V_p near 6.5 km/sec; and a mafic lower part, V_p near 6.8 km/sec. (Normal continental crust, with a thick upper part of $V_p < 6.2$ km/sec, characterizes the flanking regions.) The western part of the Snake River Plain is marked by a high-amplitude linear high in the isostatic-residual-gravity field, the central part by a lower-amplitude linear high, and the eastern part by irregular, low-amplitude highs (Jachens and others, 1985, and this volume). Future cooling of the crust presumably will result in increasingly positive gravity. Voluminous gabbro likely is present at shallow depth in the West, and analogy is inferred with the ancient Wichita and Duluth complexes of the continental interior, discussed subsequently. The latest Miocene to early Quaternary calderas of the eastern Snake River Plain either are associated with minor gravity highs or are without obvious expression; although these calderas must be underlain by granite plutons, the anticipated gravity effect of such plutons apparently is offset by mafic rocks beneath them. There may be little or no pre-Neogene crust remaining within the axial part of most of the Snake River rift.

Seismic-refraction data are interpreted to indicate that there is a hot granite pluton about 10 km thick, with some molten magma remaining in its chamber, beneath the Yellowstone Plateau, where there has been voluminous late Quaternary volcanism (Lehman and others, 1982). The Yellowstone complex is marked by a moderate low in the isostatic-residual-gravity field, which must be due in part to the very high temperature of the crustal column there, but a granite pluton may extend to mid-crustal depths.

"Hot spots"

The erratic progression east-northeastward with time of major magmatism in the eastern Snake River Plain and Yellowstone region has been widely interpreted (e.g., Engebretson and others, 1984) as due to the movement of North America southwestward over a "hot spot" fixed in the mantle beneath the lithosphere. The progression of magmatism is, however, highly irregular: although middle Miocene volcanic rocks are widespread in the far southwest and voluminous late Pliocene and Quaternary silicic rocks occur primarily in the far northeast, the intervening region contains irregularly interspersed late Miocene and Pliocene caldera complexes and also major middle Miocene ignimbrites of uncertain sources. Implied in the hot spot concept is the prediction that the crust has been augmented by magmatism, but actual structure of the crust indicates that, on the contrary, premagmatic crust has been much thinned and may be lacking altogether beneath much of the province. I see the eastern Snake River Plain as a wedge-shaped extensional rift that has propagated eastward with time, and much of its present crust as having been built by mafic magma of mantle origin. Powerful evidence against the notion of a Yellowstone hot spot is provided by the presence in eastern Oregon of volcanic centers by which magmatism progressed 200 km northwestward during the past 10 m.y. (MacLeod and others, 1976), a progression comparable in time to that of the supposed Yellowstone hot spot yet in an altogether different direction.

A vaguely defined zone of Neogene volcanic rocks extending northeastward from Springerville, Arizona, to Raton, New Mexico, has been suggested to be a hot-spot track primarily because it is subparallel to the eastern Snake River Plain. Lipman (1980) demonstrated that there is in fact no age progression within this zone.

Shoulder magmatism

Volcanism affects some regions flanking rift provinces, either accompanying or shortly preceding shoulder uplifts; both such volcanism and the shoulder uplifts may be due primarily to mantle diapirism and other asthenospheric rise consequent on rifting (Bohannon and others, 1989; Buck, 1986; Steckler, 1985). The late Neogene basaltic, trachybasaltic, and rhyolitic volcanism in the southern Sierra Nevada provides an example. Mantle diapirism is the likely cause also of late Neogene volcanism in the southern Colorado Plateau, which is basaltic in many areas but which in central Arizona includes voluminous andesite, dacite, and rhyodacite, as well as basalt (Gust and Arculus, 1986; Wenrich-Verbeek, 1979). The latter assemblage likely would be assigned mistakenly an arc setting if it were met in an ancient terrain.

Old rifts of the interior United States

The central and northern interior of the United States has long been a cratonic region. Its basement rocks and deformational structures mostly lie hidden beneath Phanerozoic strata but are exposed through the erosional window represented by the small part of the Canadian Shield that lies within the United States, and also in Phanerozoic uplifts around the edges of the central stable region. Among the exposed crystalline complexes are two possible upper-crust analogs, the Cambrian Wichita Mountains complex of Oklahoma and the Proterozoic Keweenawan assemblage of the Lake Superior region, for the Neogene Snake River–Yellowstone province of bimodal rift-related volcanism.

Wichita Mountains complex. A Cambrian layered igneous complex is exposed in the Wichita Mountains of southwest Oklahoma and, although largely subsurface, is known from its large positive gravity and magnetic anomalies, and from drilling, to be about 500 km long in a west-northwest direction and to have a mafic core about 50 km wide. Rhyolite, basalt, and diabase extend an additional 50 km over older rocks to the north. Sources of geologic and petrologic data include Flawn (1956), Ham and others (1964), and the reports, and papers cited therein, in Gilbert (1986).

The cap of the Wichita Mountains complex consists of high-temperature rhyolite, probably formed mostly as refluxed ash flows, which has a maximum known preserved thickness of 1.4 km. No calderas have been recognized. Beneath the rhyolite is a layer, a kilometer or so thick, of sheets and irregular plutons of cogenetic high-temperature granite and granophyre. Both rhyolite and granite are high in silica, low in calcium, and rich in iron; they are partly peralkaline, partly metaluminous. The rhyolite is intercalated with subordinate sills and flows of slightly alkalic basalt and diabase with or without olivine, and dikes of similar rocks are numerous in both rhyolite and granite. Beneath the granite in the axial region of the complex are layered gabbros, dominantly augite gabbro with or without olivine or hypersthene, but including also calcic anorthosite, diorite, and troctolite. A drilled section of the mafic rocks, not including either top or bottom, is 2.6 km thick (Ham and others, 1964). Isotopic composition of the gabbroic rocks indicates probable slight crustal contamination of mantle melts (Lambert and others, 1988); analogous data are not available from felsic rocks. Granite intrudes gabbro in their zones of contact, but basalt and diabase widely cut, and are intercalated with, granite and rhyolite.

I early argued (Hamilton, 1956, and other papers) that silicic caprocks and mafic substrate were cogenetic and of the same age; this was long disputed by other students of the complex, who regarded mafic and silicic rocks as unrelated and as separated widely in age, but has recently been confirmed by radiometric data that show all well-dated rocks to be about 530 m.y. old (Lambert and others, 1988). I see the mafic complex as having formed roofed only by its own silicic eruptive cover, which presumably represents primarily secondary melts formed from preexisting crust melted by deep mafic magmas, although it is not yet clear whether mafic and silicic rocks coexisted in the same magma chambers or represent separate magmas that rose from the deep crust or upper mantle.

I infer that the Wichita Mountains complex formed in a rift that largely sundered the continental crust and that was analogous to the Snake River Plain. Although Wichita basalts are too poorly characterized to permit comparisons, Wichita felsic rocks are strikingly similar in composition to Yellowstone rhyolite. The axial crust of the Wichita Mountains complex is thick and dense (Coffman and others, 1986), and its geophysical anomaly resembles that of the old, western part of the Snake River Plain. The young eastern part of the Snake River Plain–Yellowstone Plateau region may develop a similar anomaly after future cooling.

Midcontinent and Lake Superior rifts. En echelon rifts of Middle Proterozoic age are defined by the striking gravity (Jachens and others, 1985, and this volume) and magnetic anomalies of the Midcontinent geophysical anomaly. The rifts trend and widen north-northeastward from Kansas to their exposures, as Keweenawan igneous and sedimentary rocks, in northeastern Minnesota and northern Michigan. The geophysical anomalies are compatible with opening of the rift in response to rigid-plate rotations, the Minnesota side having rotated away from the Wisconsin side by about 2.6° relative to an Euler pole in northern New Mexico, which corresponds to about 90 km of separation in the region of central Lake Superior (Chase and Gilmer, 1973).

The exposed Keweenawan rocks are dominantly bimodal basalt and subordinate rhyolite and their intrusive equivalents, the latter occurring in part in huge layered complexes, and subordinate clastic strata, formed mostly within the brief period from about 1,110 to 1,095 m.y. ago (Green, 1983; Klewin, 1989; Van Schmus and others, 1982; Weiblen, 1982). Basalts are low in alkalis, mostly olivine-bearing but partly olivine-free, and variable but commonly high in alumina; many resemble mid-ocean ridge basalt; giant flows are common (Green, 1982, 1983; Klewin, 1989). Most of the basalt and much of the rhyolite are isotopically primitive and display no evidence for magmatic incorporation of continental crust (Green, 1982, 1983). Most of the rhyolites are metaluminous, high-temperature rocks. Rocks of intermediate silica contents and generally high alkalis and iron are also common. These magmatic rocks show little interaction with continental crustal materials, and I infer that old continental crust is absent in the axial part of the rift system.

Sedimentary rocks, mostly red fluvial clastic strata but partly lacustrine or marine, underlie and are intercalated with early basalts, some of which are pillowed and hence erupted under water (Ojakangas and Morey, 1982a, b). Interbedded with the thick lava section are coarse redbeds, derived mostly from the volcanic rocks but partly from pre-Keweenawan sources outside the rift (Merk and Jirsa, 1982; Morey and Ojakangas, 1982). More voluminous synvolcanic clastic sediments were deposited along the margins of the volcanic terrain, in part synchronously with extensional faulting (Morey and Ojakangas, 1982). After magmatism had ended, the volcanic pile was weathered deeply, then voluminous clastic sediments, derived from outside the rift, were deposited upon it (Daniels, 1982; Kalliokoski, 1982; Morey and Ojakangas, 1982).

I deduce from these relationships that subsidence, not doming, accompanied early stages of rifting, and hence that rifting was not a product of crustal heating, but rather that heating was a product of rifting. The surface of the volcanic terrain rose as volcanism progressed, and subsided after it ended. The rift had high shoulder uplifts, from which sediments were prograded first toward, then across, the volcanic pile. The position of the shoulder uplift on the south side of Lake Superior is defined by biotite closure ages (Peterman and Sims, 1988).

Striking positive gravity anomalies follow the two major exposed complexes of layered gabbro, the huge Duluth complex in Minnesota, 300 km long, and the smaller Mellen Gabbro in Michigan. Anothosite, anorthositic gabbro, troctolite, and olivine gabbro are the dominant exposed fractionates. Granophyre and related silicic rocks are important high in the complexes but are much less voluminous than in the Wichita Mountains complex. The gabbroic complexes lap onto much older sedimentary and crystalline rocks but are overlain only by volcanic and sedimentary rocks that are but slightly older or younger than the gabbros.

Integration of gravity and seismic-refraction data indicates that the crust in the axial part of the rift is uncommonly thick and dense (Chandler and others, 1982; Chase and Gilmer, 1973; Hinze and others, 1982; Luetgert and Meyer, 1982). This crust likely consists almost entirely of igneous rocks derived from mantle melts. We see here sundered, not merely attenuated, continental crust.

The superb GLIMPCE (Great Lakes International Multidisciplinary Program on Crustal Evolution) deep seismic-reflection profiles in the Lake Superior part of the rift show an upper crust, about 30 km thick, of layered rocks, presumably mostly volcanic and sedimentary; a layered basal crust, extending about 6 km above the reflection Moho at a depth of 45 to 55 km, presumably representing layered mafic intrusions; and an intervening lower crust of less distinctive reflection character (Behrendt and others, 1988; Cannon and others, 1989). The upper-crust rocks are broadly synformal; Cannon and others deduced a history of early normal faulting and later reverse faulting, but the synformal geometry leads me to question much of this interpretation. Behrendt and associates regard the nonreflective part of the lower crust as tensionally thinned Archean lower crust, whereas I infer that it, too, consists of Keweenawan mafic intrusive rocks. I deduce from the character of the exposed magmatic rocks and from phase-equilibrium considerations that the shallow igneous rocks are dominantly the much-fractionated products of more magnesian magmas of mantle origin, and that the deep crust consists of rocks fractionated deeper than the stability limit of olivine plus plagioclase (Fig. 2) and that hence are mostly two-pyroxene gabbro; retrograde garnet may now be abundant, giving high densities and velocities.

Overview

Continental arc magmatism and, in its early stages, continental rift and shoulder magmatism, share genetic features, and in many cases their products are more similar than is assumed by most petrologists. Both record the transfer of mantle heat to the crust in the form of voluminous mafic melts. The silicic rocks of both represent in part secondary crustal melts, and in part fractionates of more mafic melts contaminated by crustal materials. As discussed in sequel, arcs can form in rapidly extending environments; there, both tectonic and petrologic contrasts are blurred indeed.

POSSIBLE ANOROGENIC MAGMATISM

Middle Proterozoic granites and quartz monzonites, mostly between 1,350 and 1,500 m.y. old, and associated rhyolites are widespread within the older Proterozoic complexes of the southwestern United States, where they are widely exposed, and in the south-central states, where they are mostly in the subsurface except for small areas of exposure in Oklahoma and Missouri. These granites are relatively silicic, potassic, and ferruginous, are mostly metaluminous but locally peraluminous or peralkaline, and postdate the obvious arc-magmatic and plate-collisional events of their regions by as much as several hundred million years. Anderson (1983) argued that these granites formed in an "anorogenic" setting, perhaps in response to mantle diapirism. Van Schmus and Bickford (1981) suggested that the granites formed by spontaneous melting of the lower crust after tectonic thickening. The lack of obvious modern analogs for such settings was noted previously. Van Schmus and others (1987) argued that the silicic magmatism was instead likely a byproduct of crustal extension. Large dikes and sheets of undated gabbro and diabase cut Middle Proterozoic peralkaline high-temperature ash flows locally in the St. Francis Mountains of Missouri (Anderson, 1970), and a rift origin might be inferred there. The general lack of positive gravity anomalies with these or any of the other Middle Proterozoic granites, and the lack of any geophysical indicators of rift-gabbro complexes of Wichita Mountains or Duluth type, indicate that complete rifting of continental crust has not likely been a factor in any associated crustal extension.

A few younger plutons, dating from about 1040 Ma, in Texas and southern Colorado also have been termed "anorogenic." Barker and others (1975) described one of these, the Pikes Peak batholith of Colorado, which is of moderate-temperature metaluminous granite plus minor peralkaline granite. They speculated that the batholith was the silicic, crustal-melt top of a largely gabbroic complex, but this inference is incompatible with the undistinguished gravity field of the batholith (Jachens and others, 1985).

CONTINENTAL MOHOROVIČIĆ DISCONTINUITY

The Moho, the boundary between crust and mantle, must represent quite different petrologic and tectonic processes in different places. The Moho is commonly defined on the basis of seismic P-wave velocities, and generally represents a change from near 7 to near 8 km/sec. Particularly in regions of extensional tectonism, ancient and modern, the apparent base of the continental crust has been identified on many seismic-reflection profiles, typically as a zone of two or more high-amplitude reflections that separate a crust with a fabric of discontinuous apparent reflectors from a more transparent mantle. This suggests that the boundary is a zone a few kilometers thick of interlayered rocks of markedly different velocities. This is observed to be the case in outcrop in the Ivrea Zone and the Kohistan Arc, where the change from velocities typical of mantle rocks to those typical of crustal rocks occurs within the layered complexes rather than beneath them. Exposures of the lower crust show the high, downward-increasing velocity of lower crust to be due in part to its granulite-facies mineralogy and in part to the mafic composition of the basal crust. Compressional-wave velocities at the usual ambient conditions of the base of continental crust are about 7.9 to 8.3 km/sec for eclogitic and ultramafic rocks, 6.5 to 7.5 km/sec for plagioclase-bearing rocks of high-pressure-granulite facies, and 6.5 to 7.1 km/sec for rocks of low- and middle-pressure-granulite facies (Christensen, 1974; Manghnani and others, 1974). My velocity figures are obtained by subtracting 0.2 km/sec from room-temperature 10- or 12-kbar determinations as an approximate correction for normal lower-crust temperatures; velocities would be an additional 0.1 or 0.2 km/sec slower in the uncommonly hot lower crust of the Basin and Range province.

The Moho is also a rheologic boundary. In crust with normal heat flow, felsic rocks in the deep crust would behave as ductile mylonitic gneisses. In a high-heat-flow region such as the Basin and Range province, deep felsic material would be molten. Mantle rocks would be much more brittle in both regimes.

It has often been assumed that past or present conditions near much of the base of the continental crust in stable settings may be appropriate for the transition between eclogite and plagioclase-bearing rocks, although the pressure-temperature (P-T) position of the boundary (as shown in Fig. 2) is poorly constrained at the relevant low temperatures so that the assumption may be invalid. Analysis is clouded by the likelihood that inversions are slow in both directions so that metastable rocks are widespread. To the extent that phase changes are responsible for major changes in density and seismic velocity, they must be gradations, not interfaces, and in assemblages of varied bulk compositions the gradations could take place over many kilometers.

Mantle rocks

Continental upper-mantle material sampled as inclusions in basalts and kimberlites of the western United States shows that, as is generally the case elsewhere, peridotite is the dominant rock type and consists of olivine, subordinate orthopyroxene, and less clinopyroxene (Menzies and Hawkesworth, 1987; Wilshire and others, 1988). Spinel is important, particularly in uppermost-mantle lherzolite (two-pyroxene peridotite); garnet is important in peridotite formed deeper than about 50 km (Bohlen and Lindsley, 1987); plagioclase occurs in lherzolite recording pressures appropriate for lower crust or uppermost mantle. (Spinel peridotite dominates the uppermost-mantle rocks exposed in place beneath the Ivrea and Kohistan crustal sections discussed previously.) Continental mantle rocks mostly show metamorphic fabrics, but their primary assemblages in general formed in equilibrium with mafic melts or were traversed by such melts (Wilshire and others, 1988). Such mantle rocks are variously residues after partial melting, precipitates of refractory phases from melts, and residues injected by voluminous magmatic materials, mostly pyroxenitic.

Upper oceanic mantle, as observed in ophiolite exposures, largely lacks the components of basalt (which were removed as the melts that produced the basalt, diabase, gabbro, and cumulate ultramafic rocks atop that mantle) and consists mostly of olivine and subordinate magnesian orthopyroxene, in dunite and harzburgite. Continental mantle, though also consisting dominantly of magnesian olivine, contains the equivalent of 10 or 15 percent of basalt within its clinopyroxene, aluminous orthopyroxene, garnet, and spinel (Menzies and Hawkesworth, 1987). Such continental mantle occurs not only beneath old cratons, but also is formed from young oceanic lithosphere in arc-magmatic settings. The basalt component removed from the mantle at spreading centers and behind fast-migrating oceanic arcs is replaced beneath magmatic arcs.

As the arc magmas that reach the lower continental crust have evolved to mafic compositions, the bulk of the olivine and pyroxene components of the protomagmas has crystallized in the mantle. The continental Moho in part represents the shallow limit of crystallization of voluminous rocks dominated by olivine, pyroxenes, and garnet.

Thick crust beneath granulites

Some 30 km of rock have been eroded from above eastern Adirondack granulites, and yet the crust now beneath them is 35 km or so thick. Such a relationship is widely cited as evidence for crust initially 70 km thick, but it is unlikely that plagioclase, and hence geophysical crust, could have been stable another 35 km downward beneath the exposed rocks when those were at 35 km. The deep rocks would likely have been in the eclogite P-T field under equilibrium geotherms, and in or close to it under magmatically heated conditions, if the boundary in Figure 2 is approximately correct. If the present lower crust was indeed beneath the Adirondack granulites when those formed, then the deep crustal rocks may have been produced by phase change from prior garnet-pyroxene rocks. Alternatively, the exposed granulites may have been thrust over other crust when the cover above the granulites was much less than that beneath which they formed.

Magmatic Basin and Range Moho

Middle and late Cenozoic extension has approximately doubled the width of the northern Basin and Range province, yet surface altitudes are now higher than they were early in the extension period, and the Moho is at a depth that varies only between about 9 and 11 sec in reflection time, beneath all of the diverse tectonic assemblages of the upper crust, in the traverse for which the best data are available (Klemperer and others, 1986). Such time constancy is indicative of the operation of isostasy on crust that may vary in thickness and velocity, and not necessarily of horizontality of the surface (Warner, 1987). The Basin and Range Moho cannot represent a phase change—temperatures are too high to permit major inversion of plagioclase to garnet and pyroxene—so a compositional change is required. Presumably the basal crust is being thickened by basaltic magma, representing partial melting of the mantle in response to extension and diapirism, and is being smoothed by gravitational flattening of the deep crust; the subjacent mantle may be thickened by concurrent crystallization of cumulates from magmas parental to those that entered the crust (Klemperer and others, 1986).

MAGMATISM AND DEFORMATION

Deformation and magmatism are in many places related spatially and temporally, so structural and magmatic aspects of evolution of continental crust must be considered together. Until only 15 or 20 years ago, compression was widely viewed as a necessary precursor of most magmatism and metamorphism. Misconceptions from such conjecture linger on. In particular, it is commonly but mistakenly assumed that overriding plates in convergent systems necessarily undergo crustal shortening.

Subduction

Subduction often is visualized as the sliding of oceanic lithosphere down a slot fixed in the mantle beneath an overriding plate; this would, indeed, require shortening of an advancing upper plate if it occurred. The common case, however, is that the hinge rolls back, away from the overriding plate, with a horizontal velocity equal to or greater than the velocity of advance of the overriding plate (Carlson and Melia, 1984; Chase, 1978; Dewey, 1980; Hamilton, 1979, 1988c, 1989; Molnar and Atwater, 1978). Subduction must commonly occur at an angle steeper than the inclination of the Benioff seismic zone, which marks the position, not the trajectory, of a subducting plate.

I know of no young magmatic arcs, oceanic or continental, involving subduction of oceanic lithosphere without associated collision of light crustal masses, across which crustal shortening can be demonstrated. The late Oligocene and early Miocene arc magmatism of the Basin and Range province was synchronous with severe extensional faulting. The Oligocene ash flows that form the great plateau of the Sierra Madre Occidental of Mexico must overlie a batholith dimensionally like the older Sierra Nevada batholith of California, yet are almost undeformed. The lesser Oligocene batholith of the San Juan Mountains of Colorado is similarly capped by subhorizontal volcanic rocks. North Island, New Zealand, is undergoing rapid extension concurrent with its very active arc magmatism (references in Hamilton, 1988c).

Compressional structures

Extreme shear in a compressional regime is shown by the imbrication and mélanging of the materials of accretionary wedges (Fig. 3). These wedges, however, represent surficial debris caught between converging plates, and are not evidence for shortening of lithosphere plates themselves. The leading edges of overriding continental plates typically bear fore-arc basins of little-deformed strata, the presence of which precludes shortening of those thin, tapering leading parts of the overriding plates (Hamilton, 1979, 1988c; Fig. 3). Such undeformed basins are present atop the feather edge even of the Pacific margin of South America, which most proponents of crustal shortening of overriding plates assume to be the site of particularly severe shortening. Collision between plates of thick crust unquestionably produces shortening. So must drag of the bases of continental plates against lithosphere subducted at very gentle dips beneath it, which occurs where absolute velocities of overriding plates are rapid, and where light, young lithosphere is being subducted. Normal subduction of oceanic lithosphere, however, probably does not produce shortening of overriding plates.

Great compressive deformation is recorded within broad fold-and-thrust belts, deformed synchronously with arc magmatism, continentward of some magmatic arcs that are bordered by thick prisms of preexisting strata. Thus, the Cretaceous foreland thrust belt (Sevier Belt) of western North America formed throughout much of the period of arc magmatism recorded in the great batholiths farther west, and compressive structures are continuous between thrust belt and batholith. Such foreland thrusting atop plates of Andean type may have a primary explanation in terms of gravitational spreading of prisms of easily-sheared strata due to inflation of the crust by arc magmatism (Hamilton, 1978; Smith, 1981), although plate shortening can alternatively be inferred (Price, 1981).

The ratio between extension across a magmatic arc and the growth of the crust by the mantle component of magmatism, and decreases in density of crustal and mantle columns through magmatism, diapirism, and perhaps phase changes, will determine the altitude of the top of the magmatic arc, and hence the gravitational potential for driving foreland thrusting. Foreland thrusting is associated with the high part of the modern Andes. Perhaps thrusting in the Cretaceous foreland belt of western North America was a result of similar altitudes of the associated magmatic arc, although I do not know of data with which to evaluate this possibility. No thrusting accompanied formation of the middle Tertiary arc of the Sierra Madre Occidental, the altitude of which perhaps was beneath the threshold needed for major spreading.

If, indeed, magmatic arcs commonly are sites of plate spreading rather than shortening, then structures within their deeper crustal portions must have compatible explanations. Extreme deformation generally is displayed by the contact-metamorphic wall rocks of large, discordant plutons, but this apparently records strains produced by rising and spreading plutons, not strains resulting from externally applied regional stresses. The lower and middle crust and upper mantle also are greatly augmented by magmatism, so they, too, must spread, not narrow, as a result of arc magmatism.

Upper-crust batholiths commonly show little internal deformation, whereas midcrustal granites and migmatites beneath such batholiths typically display severe subsolidus deformation, generally with gently dipping structures. The misleading terms "posttectonic" and "syntectonic" of the older literature are approximately synonymous with upper-crustal and midcrustal, respectively. I see the contrast as due in part to the common case that the middle crust is heated by magma that crystallizes within it before much magma continues to the upper crust, and in part to the flow of dense wall rocks outward from and beneath spreading shallow plutons. The latter process sets up simple-shear flow beneath plutons in the sense of inward underthrusting, which would be read as outward overthrusting in exposed rocks by most structural geologists.

TECTONIC ACCRETION

Much of the width of the Phanerozoic orogenic belts of the United States represents material added tectonically to the continent by processes related to subduction. My initial discussion

(Hamilton, 1969) of such processes, and of discriminants between accreted assemblages of diverse types, has led to a burgeoning literature on "exotic terranes" (e.g., see the many papers in Howell, 1985). Continents, continental fragments, island arcs, and seamount chains have collided with North America as intervening oceanic lithosphere disappeared beneath one or both sides, and ocean-floor materials have been scraped off against the fronts of overriding plates. Some characteristics of subduction systems were discussed in the preceding section, and others are noted here. Elsewhere (Hamilton, 1979, 1988c, 1989), I have documented many features of modern systems.

Behavior of subduction systems

A magmatic arc, continental or oceanic, tracks the depth contours of a subducting plate and hence migrates with regard to the overriding plate. Where such relative migration is extreme, as in some of the modern oceanic island arcs of the western Pacific, a type of sea-floor spreading operates and a continuous sheet of "arc ophiolite" of irregular thickness may be plated out to form a back-arc, or marginal, basin (Hawkins and others, 1984). Many ophiolites now exposed on land may have formed in such settings rather than at spreading mid-ocean ridges and by conventionally visualized back-arc spreading (Hawkins and others, 1984; McCulloch and Cameron, 1983).

Within most oceanic island arcs, extensional structures are widespread, and compressional structures, other than those related to local magmatic spreading, are uncommon.

An arc remains active only as long as oceanic lithosphere is being subducted beneath it, typically at 5 to 10 cm/yr (or 50 to 100 km/10^6 yr). Thousands of kilometers of lithosphere disappear beneath an arc. Most normal oceanic lithosphere, and some of the sediment atop it, is subducted entirely beneath the overriding plate, but most sediment and seamounts on the subducting plate accumulate as dismembered materials in an accretionary wedge at and under the front of the overriding plate. Island arcs and continents are too light to be subducted; hence, they collide and aggregate.

Reversal of subduction polarity commonly follows a collision: convergence of megaplates continues, but a new subduction zone breaks through outside of one of the collided masses, commonly the smaller one. Oceanic island arcs show complex series of collisions and reversals. The southern Philippine Islands contain six distinct Neogene arcs aggregated by these processes, as well as fragments of older arcs (Hamilton, 1979).

Collisions are time-transgressive along strike. Histories of contiguous regions can be very different because arc systems change greatly with both time and space in the small-plate complexes between megaplates. The common cause of inauguration of subduction beneath a continent is polarity reversal following collision with an island arc. On the other hand, where the regime of new subduction is strongly extensional, the magmatic arc may migrate away from the continent, upon which a new stable-margin system can then develop.

As subduction represents sinking, not sliding, it cannot occur beneath opposite sides of an internally rigid plate simultaneously. (Even the hypothetical sliding mode would be impossible beneath opposite sides of a moving plate, which all plates are.) The many published reconstructions invoking two-sided subduction are invalid. Although the sides of North America shared latest Proterozoic rifting and early Paleozoic miogeoclinal development followed by complex sequences of subduction and collision, Cordilleran subduction beneath the continent did not begin until Appalachian subduction had been stopped by closure of the Proto-Atlantic Ocean.

Cordilleran accretion

The Pacific margin of the United States shows a complex history of tectonic accretion and subduction beginning in Mississippian time (Fig. 4). Everything west of a curving line from the central Sierra Nevada through northwest Nevada and western Idaho to northeast Washington has been added to the continent by subduction-related processes, complicated by strike-slip motions. Broad interpretations of this history have been presented by Davis and others (1978), Dickinson (1981), Hamilton (1978), and Saleeby (1983). A number of regional papers and references to many more are in Howell (1985). Here I summarize the tectonic history of central and northern California and Nevada in particular.

Late Paleozoic accretion. Miogeoclinal stable-margin conditions ended in Early Mississippian time with the "Antler orogeny," when continental-rise and -slope materials of early and middle Paleozoic ages were pushed relatively eastward atop the correlative continental-shelf stratal wedge (Dickinson and others, 1983). The resulting complex is preserved from southern California through central Nevada to central Idaho. A possible cause of this was the ramping of deep-water materials ahead of an eastward-advancing island arc that collided with the continent (Dickinson and others, 1983). Although that hypothetical arc did not remain attached to the continent, evidence that it migrated westward away from North America following subduction reversal consequent on the collision is recorded in the western source of volcaniclastic arc debris in deep-water postcollision Mississippian strata (Miller and others, 1984). Much remains to be learned about the geometry of Antler deformation, for geologic mapping of Antler structures largely predated recognition of Cenozoic detachment faults now known to be widely present within the region at issue, and much deformation formerly attributed to Paleozoic thrusting likely is instead the product of Cenozoic extension; further, structures of the Wichita-Uncompahgre family may have been mistaken for Antler ones. (E. M. Moores [oral communication, 1989] has reached the same conclusion regarding the confusion of Antler and detachment structures. Detachment faulting and Wichita-Uncompahgre deformation are discussed subsequently.)

Triassic accretion. Upper Paleozoic strata accumulated on the eroded remnants of this orogenic complex until correlative oceanic volcanic and sedimentary rocks were thrust eastward upon them in Nevada during the Early Triassic "Sonoma orogeny" (Snyder and Brueckner, 1983; Speed, 1979). This deformation apparently was produced by another eastward-migrating arc, which this time remained attached to the continent and is exposed now in northwest Nevada and the eastern Klamath Mountains of northwest California and adjacent Oregon. The later Triassic and earliest Jurassic history is incompletely understood but included offshore island arcs and, within Late Triassic time, subduction beneath the continent, for granites of that age are present in inland California and voluminous silicic volcanic ash was dispersed to the interior Southwest.

Jurassic and younger accretion. Jurassic complexes related to subduction are better known. Subduction beneath the continent, shown by granites and voluminous volcanic rocks, was intermittent, and probably most plate convergence was accommodated by subduction beneath offshore island arcs. Composite Jurassic magmatic arcs and accretionary wedges in the Klamath Mountains of northwestern California and southwestern Oregon are seen by some observers (e.g., Davis and others, 1978; Wright and Fahan, 1988) as having formed by subduction eastward beneath the growing continent. As do Dilek and others (1988) and Ingersoll and Schweickert (1986), however, I (e.g., Hamilton, 1978) see the assemblage as consisting of oceanic island arcs and wedges, the internal indicators of collisions and subduction polarities of which require them to have been formed offshore, and in part also to have been aggregated offshore in Philippine style, and to have collided with the continent during several Middle and Late Jurassic events. The last of these collisions defines the "Nevadan orogeny." This collision was followed by a subduction reversal, and from very early Cretaceous through Oligocene time there was continuous subduction of eastern Pacific oceanic lithosphere beneath mainland California. The new subduction system broke through at the edge of thick arc crust in the Klamath Mountains region, but outboard of a strip of arc(?) ophiolite, 100 km or so wide, that had formed behind the last of the eastward-migrating Jurassic arcs along the Sierra Nevada. This ophiolite strip served as the basement for the fore-arc basin in which accumulated the thick "Valley facies" Cretaceous and Paleogene strata of California, and its upturned west edge served as the ramp beneath and in front of which accumulated the great Franciscan accretionary wedge of mélange (Fig. 3). Ingersoll and Schweickert (1986) made a similar interpretation. The magmatic arc—the Sierra Nevada and other batholiths—paired to the Cretaceous part of this subcontinental subduction was discussed previously.

Accretionary tectonics are less well understood in the northwestern United States. Island arcs, accreted to the continent in Mesozoic time and partly rifted away during Cenozoic time, are exposed in western Idaho and nearby parts of Oregon and Washington (Hamilton, 1978). The minicontinental North Cascades block of northern Washington was not attached to that sector of the continent until middle Cretaceous time, yet a complex Jurassic and Cretaceous assemblage of high-pressure metamorphic rocks, accretionary wedges, and exotic terrains lies west of the block (Frizzell and others, 1987; Gallagher and others, 1988; Garver, 1988; Johnson and others, 1986). The oceanic-crust Coast Ranges part of Oregon and Washington lies oceanward of these tracts and did not arrive until early Tertiary time, and complex rotations have affected the Tertiary complexes (Heller and others, 1987; Wells and Heller, 1988). In the Olympic Mountains of northwestern Washington, the early Tertiary oceanic crust at the leading edge of the North American plate has been domed and deformed oroclinally, exposing the voluminous accretionary-wedge complexes beneath (Tabor and Cady, 1978a, b). Subduction is still going on beneath the Coast Ranges oceanic crust, but the Cenozoic accretionary wedge is entirely offshore along Oregon and most of Washington (EEZ SCAN 84 Scientific Staff, 1986; Snavely, 1987). The subducting plate and related overlying complexes are clearly imaged, down to a depth of about 40 km, on reflection profiles (Clowes and others, 1987).

Appalachian accretion

The crystalline Appalachians east of the Green Mountains in the north and of the Blue Ridge in the south consist largely of diverse tracts added to the continent, or squashed between converging North American, Baltic, and African–South American continental plates, from Middle Ordovician to late Paleozoic time. Understanding lags behind that of the Cordillera because the Appalachian complexes have in general been eroded to greater crustal depths and have been overprinted more severely by postcollision tectonic and magmatic events. Bradley (1983), Hatcher (1987), Robinson and Hall (1980), Thomas (1983), Williams and Hatcher (1982), and Zen (1983) have attempted regional syntheses. Island arcs, mélanges, ophiolites, terrigenous sediments, and continental fragments are variably squashed together, highly metamorphosed, and intruded by granites. The southeastward transition from old continental rocks to accreted oceanic assemblages is approximately followed by a gradient in Bouguer gravity from moderately negative to near-zero values (Thomas, 1983).

CRUSTAL SHORTENING

Crustal shortening occurs where masses of thick crust collide as intervening oceanic lithosphere is subducted beneath one or both of them, and also where a subducting oceanic plate is overridden too rapidly to sink out of the way.

Shortening by lithosphere drag—The Laramide Event

During most of Cretaceous time, convergence of North American and Farallon (eastern Pacific) plates was slow and oblique, subduction was steep, and the continental magmatic arc lay within 300 km or so of the plate boundary. Late in Cretaceous time, the pattern probably changed: convergence between North American and Farallon plates became rapid in the south, and in the north the new oceanic Kula plate moved rapidly northward away from the Farallon plate, so that the northwestern North American margin was the site of highly oblique subduction (Engebretson and others, 1984; Jurdy, 1984; Stock and Molnar, 1988). Convergence between North American and Farallon plates slowed markedly in early Eocene time, between 55 and 50 m.y. ago (Stock and Molnar, 1988). (The slowing was calculated to have taken place late in the Eocene, about 40 m.y. ago, by Engebretson and others, 1984, and Jurdy, 1984, who had less complete magnetic-anomaly data available than did Stock and Molnar.) The very late Cretaceous and Paleocene, and in the continental interior also the Eocene, were times of very different tectonic and magmatic styles than those of earlier Cretaceous time (Dickinson and others, 1988; Engebretson and others, 1984; Hamilton, 1988a). In the Southwest, the belt of arc magmatism migrated rapidly far to the east during very late Cretaceous time as the continent overrode oceanic lithosphere faster than it could sink out of the way. The base of the continental lithosphere was eroded tectonically against the subducting plate, and the western interior of the United States was compressively deformed. In the Northwest, great strike-slip of marginal terrains was produced by the interaction of Kula and North American plates, although I do not discuss that here.

Subcrustal erosion. When North America rapidly overrode subducting Pacific lithosphere in very late Cretaceous and Paleocene time, the continental plate was eroded tectonically from beneath, and oceanic rocks were plated against its truncated base, in the southern California sector. In the Coast Ranges of central California—the fragment of North America bounded on the east by the San Andreas fault, which lay alongside interior southern California until late Neogene time—middle Cretaceous batholithic rocks, formed on strike from those of the medial Sierra Nevada batholith, lie directly upon coeval Franciscan accretionary-wedge materials on the Sur-Nacimiento megathrust (Page, 1982; C. A. Hall, Jr., written communication, 1989, has documented much additional evidence for this megathrusting). The continent-truncating megathrust is exposed in an uplift at the south end of the Sierra Nevada batholith and in a number of uplifts in interior southern California. The subjacent oceanic rocks—the Pelona, Orocopia, and Rand Schists—are mostly terrigenous clastic strata metamorphosed at greenschist and lower amphibolite, or locally blueschist, facies, and include serpentinite, metabasalt, and metachert (Haxel and others, 1986; Jacobson, 1983; Ross, 1989).

Cretaceous, older Mesozoic, and Precambrian midcrust migmatites and granites and lower-crust granulites are truncated downward against these schists. As the Cretaceous granites must have formed 100 km or so above the subducting slab at the time of their formation, about 70 km of the continental mantle and basal crust was subsequently eroded tectonically from beneath. The age of this deformation is bracketed between the middle Cretaceous age of truncated granites and granulites and the middle Eocene age of marine strata that overlie deeply eroded rocks. Although the sense of mylonitic shearing superimposed on the synmetamorphic thrusting of the continent-base structure as exposed in far southeastern California is of northeastward overriding, which led Haxel and others (1986) to infer deformation beneath an eastward-advancing minicontinent sutured against mainland North America, there are North American Paleozoic strata of the same cratonic facies both inboard and outboard of the hypothetical suture. I interpret the late retrograde mylonitization to be a ductile detachment fault, which here cut out the megathrust and formed during the severe middle Tertiary extension that affected the area in question and had a northeastward sense of slip on nearby brittle structures parallel to the ductile fault in question (Hamilton, 1987a, and subsequent field observations). Detachment faulting has also cut out part of the actual thrust surface in the Mojave Desert (Postlethwaite and Jacobson, 1987). Great thrust lenses dipping gently to moderately eastward, outlined by thick zones of amphibolite-facies mylonitic gneiss formed after middle Cretaceous time and uplifted rapidly during Paleocene time (Dokka, 1984), are exposed in the northeastern Peninsular Ranges of south-central California, record westward thrusting (or eastward underthrusting), and may be byproducts of the truncation of the base of the continent.

Broad tracts of the middle crust of southeastern California northeast of the last exposures of subcrustal oceanic schist underwent pervasive regional metamorphism and severe deformation during latest Cretaceous time. Hoisch (1987) showed that the heating was accompanied, and likely was caused, by the passage of vast quantities of water at amphibolite-facies temperature. He proposed that this water was liberated by dehydration metamorphism of the oceanic rocks then being subducted at shallow depth beneath the truncated deeper crust, and that the water also fluxed the hydrous granite melts that formed at the same time.

The distribution of facies of Paleozoic and early Mesozoic strata in southern California disproves speculations that a great Jurassic strike-slip fault, the "Mojave-Sonora megashear," disrupts southeastern California (Hamilton, 1987a), but on the other hand requires latest Paleozoic or earliest Triassic removal of a continental corner that previously projected westward from southern California (Hamilton, 1969; Walker, 1988).

Laramide deformation. The Rocky Mountain region of the continental craton, from southern New Mexico to northern Montana, was distorted by crustal shortening during very late Cretaceous, Paleocene, and early Eocene time, and at least in the south, also during late Eocene and early Oligocene time (Dickinson and others, 1988). The typical products of this deformation are large, asymmetric anticlines of Precambrian rocks bounded on one side by reverse faults or steep monoclines against deep basins that subsided and received sediments concurrent with rise of the uplifts (Fig. 5). Major uplifts and basins have lengths of 200 to 500 km and structural reliefs exceeding 10 km. Although minor concurrent deformation affected the Colorado Plateau and Great Plains, the major structures lie in a belt trending northward and widening through New Mexico into Colorado, where they splay out in successive arcs to the west, west-northwest, northwest, and north. Surface geology, industry drilling and reflection profiling, and limited academic profiling all demonstrate great

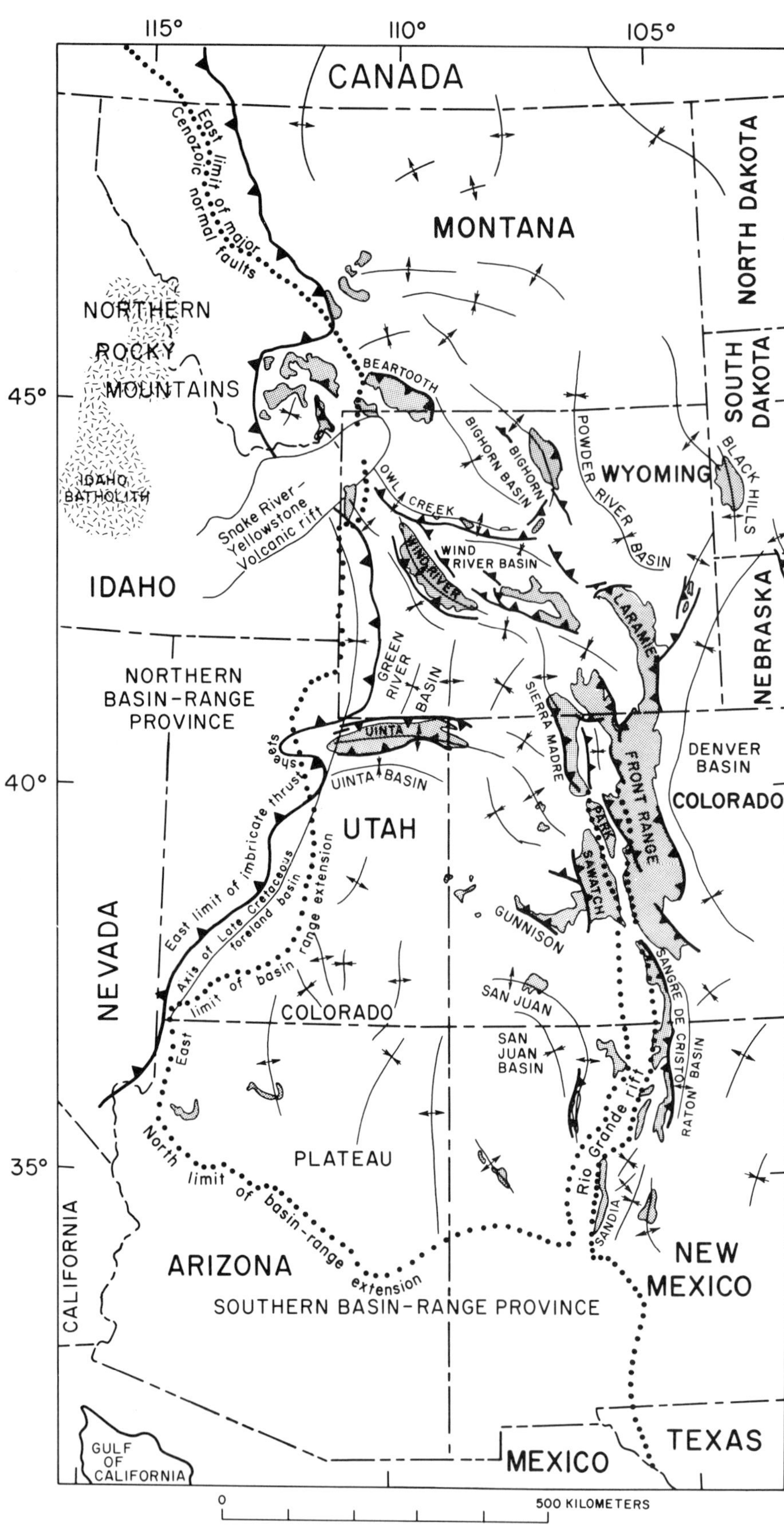

Figure 5. Selected structural elements of the Rocky Mountain region. Most of the uplifts and basins indicated are of Laramide age, very late Cretaceous and early Paleogene. Outcrops of Precambrian rocks east of the foreland fold-and-thrust belt and north of the southern Basin and Range province are shaded.

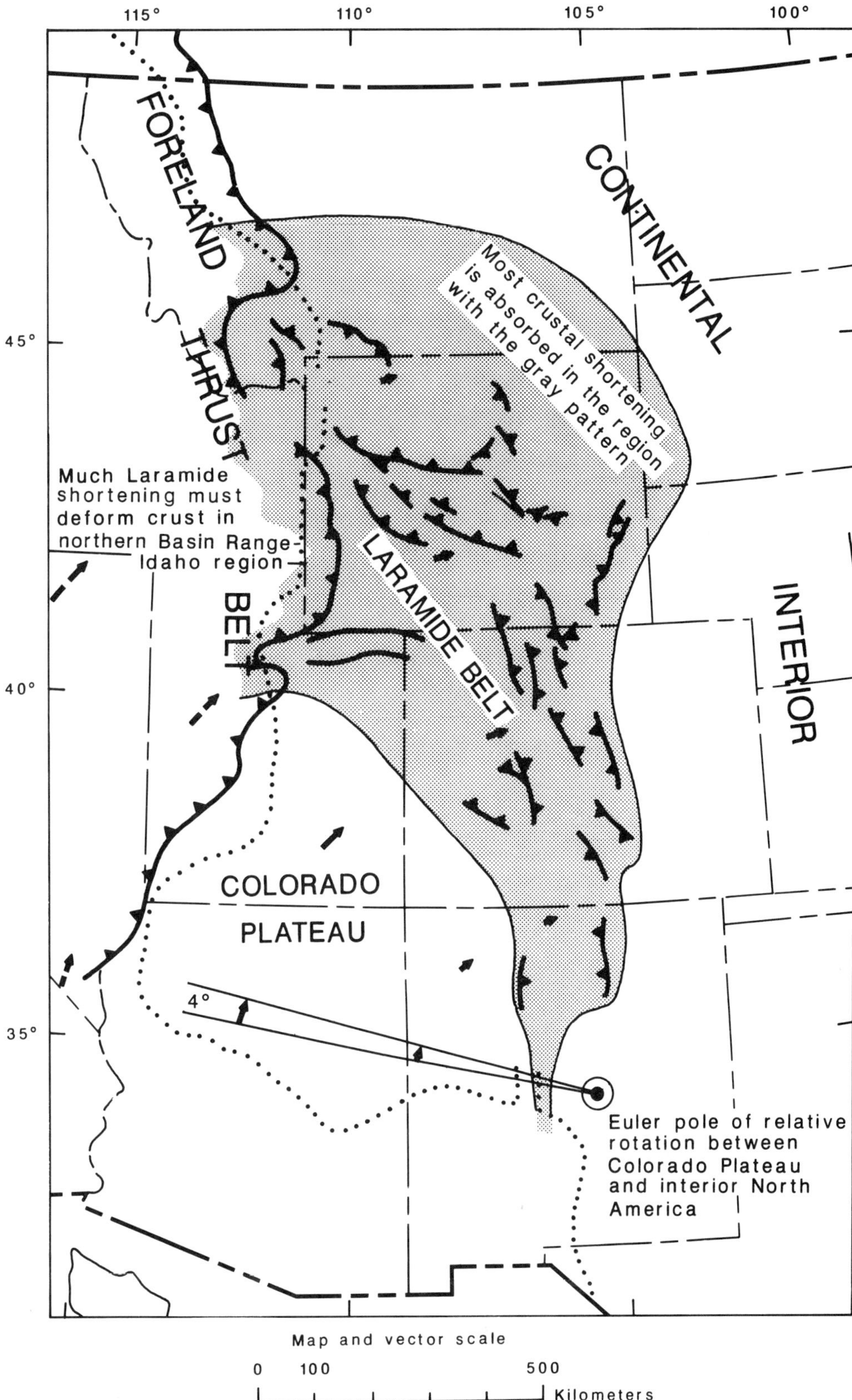

Figure 6. Map illustrating rotation of the Colorado Plateau relative to interior North America as the cause of Laramide deformation of the Rocky Mountain region. The vectors depict, to scale, the relative rotation of the plateau by 4° as though about the inferred Euler pole in central New Mexico. The motion was absorbed by crustal shortening in the Laramide belt.

basement shortening. The amount of total shortening across the belt increases northward and northwestward as both the width of the belt and the number of major structures within it increase. Structures formed with different orientations formed concurrently, disproving speculations regarding sequential regional shortening directions (Dickinson and others, 1988).

The Laramide shortening can be expressed in terms of a rotation of the Colorado Plateau relative to the continental interior, for each behaved as a relatively undeformed plate during this episode of crustal shortening (Hamilton, 1978, 1988a; Fig. 6). The geometry of the compressive structures and of the minor associated strike-slip indicates this rotation to have amounted to about 4° clockwise relative to an Euler pole in or near central New Mexico. A subsequent clockwise rotation of lesser amount, relative to an Euler pole in central Colorado, is required by the middle and late Cenozoic opening of the Rio Grande rift. Steiner (1986) showed that the combined Laramide and Rio Grande rotations accord with the contrasted paleomagnetic orientations of strata of the Colorado Plateau and interior North America. The rotation implied by the Rio Grande extension requires crustal shortening elsewhere; apparently this is accommodated on middle Tertiary compressive structures, in a zone trending westward to northwestward from northern Colorado (Hamilton, 1988a; Steidtmann and others, 1989), which have received little attention in the literature.

Deep reflection profiling (Sharry and others, 1985; Smithson and others, 1979) demonstrates that the basement thrusts flatten downward and permits the inference that they splay into anastomosing ductile shear zones that break the deeper part of the middle crust into lenses. Basement thrusts of Laramide type perhaps generally give way downdip to subhorizontal complexes of lenses separated by thick ductile shear zones. The Proterozoic Grenville deformation in Ontario records thickening of the crust by northwest-verging imbrication of great lenses of deep-crust rocks of granulite and upper amphibolite facies between thick amphibolite-facies mylonite zones (Davidson, 1986) and may be analogous.

Laramide crustal shortening began at the time when convergence between North American and Farallon plates accelerated late in Late Cretaceous time, and the shortening has been widely attributed (e.g., Dickinson and others, 1988; Engebretson and others, 1984; Hamilton, 1988a) to drag of the continental plate against a gently inclined subducting plate beneath it. The shortening in at least the southern part of the province continued, however, long after convergence between these plates slowed in Paleocene time, calling such an explanation into question (Stock and Molnar, 1989). The megathrust truncation of the base of the continent in the southern California sector was certainly complete by early Eocene time, about 15 m.y. before Laramide crustal shortening ended.

Shortening by collision—The Appalachians

Collisions of continents commonly produce extensive thrust sheets of crystalline rocks. Such structures vary greatly along strike because collisions are between irregular crustal masses, proceed along strike with time, and are accompanied by long-continuing shortening and strike-slip adjustments. The magnitude and complexity of motions of the continental masses that finally collided across the Appalachian tectonic system were deduced by Van der Voo (1988) from paleomagnetic data.

Southern Appalachians. The southern Appalachians record a complex and poorly understood Paleozoic history of subduction and tectonic accretion that ended after North America and Africa were firmly squashed together. The Mesozoic rifting that inaugurated opening of the North Atlantic Ocean mostly broke within the broad accretionary tract, but one small bit of Africa remained attached to North America as the subsurface basement of Florida (Opdyke and others, 1987). Byproducts of the continental collision include major overthrusting. A great basement thrust sheet is exposed in the Blue Ridge of western North Carolina and eastern Tennessee (Bryant and Reed, 1970; Hadley and Goldsmith, 1963), where the Middle Proterozoic crystalline bsaement and overlying Late Proterozoic rift-facies rocks and Cambrian miogeoclinal strata of the outer edge of the North American continent were pushed relatively northwestward over inner-shelf rocks and their basement during late Paleozoic time. Basement rocks have been thrust at least 60 km over younger metasedimentary rocks. Late Proterozoic and lower Paleozoic strata beneath the crystalline plate were metamorphosed at upper greenschist facies, and underlying Middle Proterozoic basement rocks were retrograded under the same conditions. The basal kilometer or so of the upper plate was pervasively sheared and retrograded at the same grade. Higher parts of the upper plate preserve the effects of earlier higher-grade metamorphism of miogeoclinal strata, and corresponding higher-grade retrogression of basement rocks, which increases in grade from lower to upper amphibolite facies, and locally to low-pressure-granulite facies, in a metamorphic high in the southern Blue Ridge. I infer from the mineral assemblages that the northwestern exposed subthrust rocks were at a depth near 12 km, and the southeastern ones near 16 km, when they were overridden, whereas the depth of earlier Paleozoic metamorphism of the exposed upper-plate rocks averaged about 20 km.

Northwest of the Blue Ridge is the Valley and Ridge province, a foreland thrust belt in which inner-shelf Paleozoic strata are deformed by thin-skinned, northwest-verging thrust faults and disharmonic folds in front of the basement ramp. Temperatures recorded by alteration of conodonts (Epstein and others, 1977) in Paleozoic strata indicate erosion of only a few kilometers of thrust sheets above the level of present exposures in the northwest but perhaps 10 km in the southeast.

The Blue Ridge is bounded on the southeast by the Brevard fault, which dips steeply to moderately southeastward in exposure and is shown by reflection profiles to sole at depth (Christensen and Szymanski, 1988). The fault has been widely assumed to be a thrust fault, but its straightness and continuity, horizontal stretching lineations and right-slip transport indicators, regional truncation of rock assemblages, and lack of upramping of a tilted crustal section apparently require an origin instead as a strike-slip fault that was pushed over during later deformation (Bryant and Reed, 1970; Vauchez, 1987). Right-slip and oblique compression probably were major factors in the late Paleozoic evolution of the Piedmont province (Gates and others, 1988).

Southeast of the Brevard fault, probable North American Proterozoic basement rocks and overlying miogeoclinal sedimentary rocks, metamorphosed at high-pressure amphibolite facies, are exposed in the Sauratown Mountain window close to the fault in northern North Carolina, and in the large Pine Mountain window in Georgia and Alabama, where they are overlain by amphibolite-facies rocks of oceanic facies that were carried at least 100 km northwestward above them (Schamel and Bauer, 1980).

Although island-arc and other assemblages in the Piedmont, southeast of the Blue Ridge, were accreted tectonically to North America during Paleozoic time, the Blue Ridge overthrust sheet and the rocks in the Piedmont windows appear in both their basement and miogeoclinal assemblages to be of mainland North American ancestry. The Blue Ridge overthrust sheet and the deformation beneath and in front of it record the late Paleozoic telescoping of the early Paleozoic continental shelf strata and their basement and the overriding of those rocks in turn by oceanic materials of the Piedmont sheet. Geologic map patterns require an aggregate shortening of at least 150 km.

Reflection profiles. Much greater thrust offsets in the crystalline terrains have been inferred from seismic-reflection profiles. Discontinuous reflectors in a profile crossing much of Virginia were given a highly speculative interpretation by Pratt and others (1988) as consistent with 200 km of overthrusting of Piedmont and Blue Ridge basement over supracrustal rocks. Profiles across the Blue Ridge through the Grandfather Mountain window show subhorizontal reflectors in the northwest, at a depth near 8 km, which were interpreted by Harris and others (1981) to mark a sole thrust, above which the basement exposed in the window is itself thrust over deeper sedimentary rocks. The inference of a deep thrust is reasonable, but no data require sedimentary rocks, as opposed to shear-layered crystalline rocks, beneath it. The conjecture by Harris and Bayer (1979) that deep sedimentary rocks occur in broad thrust sheets stacked by 300 km of overthrusting is unsupported. In Georgia, the northwest part of the Piedmont is underlain by a zone of reflectors that deepens southeastward, from about 5 to 15 km over a distance of about 75 km from the Blue Ridge, and that may record a thin zone of layered rocks related to the Piedmont thrust fault (Cook and others, 1979; Iverson and Smithson, 1983). Cook and others (1979, 1981) argued that the fault could be identified on reflection profiles for an additional 250 km southeastward through discontinuous domains of variably oriented reflectors. Iverson and Smithson (1983) reprocessed the data and showed that the fault can continue no more than 100 km southeast of the Blue Ridge before it encounters a broad domain of strong reflectors dipping moderately southeast, and they inferred that the fault there splays downward into the lower crust. Thomas (1983) reached a similar conclusion from gravity data.

All rocks beneath the crystalline overthrusts, whatever their original character, must also be crystalline rocks because the rocks exposed in windows were metamorphosed at upper greenschist facies in the Blue Ridge and amphibolite facies in the Piedmont. Ductile transposition in the fault zone of any crystalline rock types initially present can account for the thin zone of layering indicated by the reflection data.

Northern Appalachians. The western Vermont Green Mountains sheet of North American basement rocks and outer-shelf strata was thrust westward over inner-shelf rocks, with the development of a foreland thrust belt in eastern New York State, in Middle Ordovician time, apparently as the result of collision of a west-facing oceanic island arc with the continent (Bradley, 1983). Great complexity was superimposed in Middle and Late Devonian time as a byproduct of continent-continent collision farther east. The Green Mountains sheet appears on reflection profiles to be only about 5 km thick, and to be about 50 km wide as a subhorizontal structure, east of which reflectors dip moderately eastward (Ando and others, 1984). Exposed upper-plate rocks were metamorphosed at middle amphibolite facies, with widespread kyanite and hence likely deeper than 15 km, during the Devonian events. Erosion has removed most of the foreland thrust belt once present to the west. If temperatures of conodont alteration (Epstein and others, 1977), closure of apatite to fission tracks (Johnson, 1986), and slate-grade metamorphism record equilibrium geothermal gradients, then exposed eastern rocks of the foreland thrust belt were buried tectonically 10 to 15 km deep, and the westward-thinning thrust belt extended far to the west. Severe middle and late Paleozoic shortening also affected interior southern New England.

Shortening by collision—Southern orogen

The northern edge—mostly the foreland thrust belt—of a late Paleozoic orogen is exposed in the Ouachita Mountains of Arkansas and Oklahoma and locally in south Texas but is mostly covered by younger strata (Nicholas and Rozendal, 1975). Plate-tectonic aspects of the geology were discussed by Dickinson (1981), Kluth (1986), and Lillie and others (1983). The complexes apparently are byproducts of the collision of the African–South American continent with North America.

Wichita-Uncompahgre event. During Pennsylvanian and Early Permian time, a broad belt of basins and basement uplifts formed with west-northwest trend across the previously cratonic southwestern United States, from southern Oklahoma through Colorado, New Mexico, and Utah (Kluth, 1986). Uplifts and basins are dimensionally similar to the Laramide ones. Drilling and seismic-reflection data demonstrate north-verging thrusting of the Wichita uplift over the Anadarko basin in Oklahoma and southwest-verging thrusting of the Uncompahgre uplift over its complementary basin in Colorado (Kluth, 1986). Thrusting in the eastern part of the belt may have had a left-lateral strike-slip component (Budnik, 1986). Crustal shortening in response to a small rotation of the southwestern United States relative to the continental interior, analogous to the Laramide event, seems indicated.

CRUSTAL EXTENSION

Hot extension—The Basin and Range province

Middle and late Cenozoic extension has approximately doubled the width of the Basin and Range province and related terrains (Hamilton, 1969, 1978; Wernicke and others, 1982, 1988), and the structures on which extension was accommodated in the upper and middle crust are widely exposed.

Plate-tectonic setting. Subduction steepened following Paleogene slowing of the convergence of North American and Farallon (eastern Pacific) lithosphere. It is unclear whether the old, gently inclined slab sank, or became attached to the overriding continental plate and rode with it over the younger subducting plate.

During late Eocene, Oligocene, and early Miocene time, a wave of intermediate, mostly silicic magmatism broadly affected—and was largely limited to—the region that was extended in middle and late Cenozoic time. The thermal softening of the crust that accompanied this magmatism presumably facilitated the uncommon amount and duration of extension. The continental margin evolved during the period of extension from an early one of continuous subduction of Pacific lithosphere to one of strike-slip in the south and subduction along a progressively shortening sector in the north (Atwater, 1970; Dickinson, 1979). Early extension was approximately perpendicular to the continental margin and was synchronous with subduction along most of it, hence was intra-arc or back-arc spreading—severe extension of an overriding plate. Late Neogene extension has been mostly west-northwestward relative to the continental interior, and thus has been directed obliquely between the normal to the continental margin and the northwest strike of the San Andreas plate-boundary direction. Middle Cenozoic extension affected much of the Great Basin; the Basin and Range parts of southern California, Arizona, and New Mexico; the Rio Grande Rift system; and regions north of the eastern Snake River Plain. Late Cenozoic extension has affected most strongly the Great Basin, though with much areal variation; the margin of the Colorado Plateau; the region near the eastern Snake River Plain; and eastern Oregon and adjacent areas. The later extension was in part superimposed on earlier extension and in part affected new regions, and at any one time extension was rapid in some areas and slow or dormant in others.

Crustal setting. Evaluation of paleofloras in quantatitive meteorological terms by Axelrod (1985, and references therein) indicates that regional altitude of the Great Basin, like that of much of the rest of western North America, increased during the long period of extension. Speculation by Coney and Harms (1984) that extension represents gravitational spreading of a plateau that was of Tibetan altitude in late Paleogene time is disproved. The average density of crust and mantle apparently has decreased during extension, despite the thinning of preexisting continental crust, by some combination of thermal expansion of crust and mantle, upward growth of the asthenosphere by diapirism and partial melting, magmatic transfer of material from dense mantle phases to light crustal ones, and phase changes away from lower-crust assemblages rich in garnet and pyroxene. Much of the region likely will be beneath sea level after equilibrium geothermal gradients are reestablished.

Extensional deformation. Undulating extensional detachment faults are widely exposed throughout the extended regions and commonly record both great horizontal offset and omission of thick crustal sections. Upper-plate rocks are rotated to steep to moderate dips and truncated downward against these faults, and the sense of rotation and of slip against the faults is mostly in the same sense, up one side of a dome and down the other, over tracts that may be 100 km square. (Reviews of these features are given by Davis and Lister, 1988; Hamilton, 1987b, Wernicke, 1985; Wernicke and others, 1988.) Lower-plate rocks commonly show great differential uplift, the crustal depth at which the rocks formed increasing in the down-slip direction and commonly reaching the middle crust. The undulating or domiform exposed faults truncate both upper- and lower-plate rocks in the same sense of a plane that was inclined in one direction (Davis and Lister, 1988; Hamilton, 1988b; Howard and John, 1987; Wernicke, 1985; Wernicke and Axen, 1988). Only about half the width of total exposures of upper-plate rocks truncated downward against these faults is of rocks that existed when extension began; the other half consists of synextensional Cenozoic sedimentary and volcanic rocks—this fact alone requires a doubling of the width of upper-plate terrains during extension (Hamilton, 1988b).

Some major undulating detachment faults are known from geologic data (Howard and John, 1987; Wernicke and Axen, 1988) and seismic-reflection profiles (Allmendinger and others, 1983) to headwall into range-front faults, and on both sides of these headwall faults are rocks that were in the upper crust when extension began. Most exposed detachment faults crop out, however, as broadly domiform surfaces, now known in scores of ranges, that typically separate lower-plate rocks that were in the middle crust before extension from upper-plate, upper-crust rocks. Final slip on these structures was brittle under near-surface conditions, but many lower plates display carapaces of sheared rocks that evolved with time from ductile to brittle—from mylonitization, at greenschist or amphibolite facies, through subgreenschist chloritic microbrecciation to gouge—as tectonic denudation progressed and lower-plate rocks rose toward the surface (Davis and Lister, 1988). Preextensional fabrics commonly are preserved beneath the mylonitic carapaces, but in some complexes, mylonitic gneissification is pervasive to the deepest levels seen; some such mylonitic gneisses are of Mesozoic age, but in others the pervasive deformation affects Tertiary granites little older than the final shallow extensional faulting. (Such granites commonly show petrologic evidence for crystallization at midcrust levels.) Some detachment faults are broken by younger steep faults, but more commonly, normal and strike-slip faults that break upper-plate rocks are seen to end downward against the detachment faults.

Howard and others (1982) and Spencer (1984) viewed detachment faults as initially great, gently dipping normal faults whose footwalls rose and warped to undulating configurations as they were unloaded tectonically by slipping down of hanging wall blocks. Buck (1988) evaluated this model in terms of variable flexural rigidities and other quantitative parameters. Wernicke (1985) extended the ramp model to speculate that gently dipping normal faults cut entirely through the crust, a view he has since abandoned but that Davis and Lister (1988) and many others retain. I (Hamilton, 1987b, and earlier papers) proposed that instead the middle and deep crust were extended by flattening, that ramp faults were steep and limited to the upper crust, and that many exposed detachment faults represented the surface exposure of what had been middle-crust zones of flattening.

A composite model has been proposed by Wernicke and Axen (1988), based on the middle Tertiary extensional history of southeast Nevada, and by me (Hamilton, 1988b), based on the late Tertiary and Quaternary history of the Death Valley region of eastern California and adjacent southwest Nevada. Although our concepts differ in detail, we see active normal faulting as taking place on upper-crustal faults that dip moderately or steeply downward to midcrust zones of ductile flattening, and unloaded lower plates as rising to the surface while undergoing much internal, and largely ductile, deformation. The raised part of a fault becomes inactivated when its dip becomes too gentle to permit further slip, and upper-plate blocks are then stranded upon it, whereas slip continues on the still-dipping part of a fault. Stranding progresses with the incremental advances of the hinge between still-dipping and flattened fault.

This process can be seen particularly well in the eastern Death Valley area, where an undulating detachment fault has been inactivated progressively westward from late Miocene to early Quaternary time, and in the far west rolls to moderate west dips to become a normal fault that is now exceedingly active. The most recent upper-plate blocks to be stranded lie atop the western undulations of the detachment fault and are broken on their west sides by rangefront faults that merge with the dipping surface of the master fault to define the modern edge of Death Valley.

A number of conclusions follow from such analyses. The tilted fault-block ranges that characterize upper-crust Basin and Range structure are allochthonous above detachment faults and typically have been separated from initially adjacent similar blocks not only by the widths of the intervening basins but also by the widths of ranges that expose lower-plate rocks (see also Wernicke and others, 1988). The synextensional materials commonly are less resistant to erosion than are older rocks and also dominate subbasin upper-plate materials. The structures widely regarded as typical of the province—tilted panels of bedrock rotated against one another at range-front faults, the depressed part of each panel being covered by basin deposits, as visualized by Zoback and Anderson (1983)—probably are present only where extension has been minor. The ubiquitous rotations of upper-plate blocks record primarily flattening by rotation of the faults that bound them, and not slip on downward-flattening faults. The typical Basin and Range rangefront, sloping 30° or so, is commonly eroded only moderately from the stripped surface of a rotated fault, rather than being a slope eroded far back from an initially steep fault. Symmetrical horsts and grabens are uncommon. Strike-slip faults may be limited to assemblages above detachment faults. Range-sized blocks formed by early extension perpendicular to the continental margin are now being further separated obliquely by slip closer to the San Andreas direction.

Many deep-reflection profiles in the Basin and Range province show various combinations of a middle crust in which zones of dipping reflectors separate acoustically transparent lenses, a lower crust with pervasive subhorizontal fabrics, and an unbroken and highly reflective basal crust (Allmendinger and others, 1987; Frost and Okaya, 1986; Hamilton, 1987b; Hauge and others, 1987; Klemperer and others, 1986; Potter and others, 1987; Valasek and others, 1989). It should be emphasized that acoustic and processing artifacts can contribute much to the apparent high reflectivity of the lower crust. I proposed (Hamilton, 1987b, and earlier papers) that the dominant mode of extension in the middle crust was one of flattening: lenses bounded by gently dipping zones of ductile shear sliding gravitationally apart like a pile of wet halibut. A downward transition from brittle faulting through discontinuous ductile flattening to pervasive flattening is to be expected on rock-mechanic grounds in this region of high heat flow (Fletcher and Hallett, 1983; Sibson, 1983; Zuber and others, 1986).

The transition from inclined upper crustal fault to ductile-flattening structures initially in the middle crust is, I believe, now exposed in a number of ranges. One example is the fairing of a detachment fault into mylonitic gneiss in the Whipple Mountains of southeastern California (which Davis and Lister, 1988, interpreted in different terms).

The possible cause of the modern northwestward extension of the Basin and Range province is discussed following the discussion of the San Andreas fault system.

Neogene rise of the Sierra Nevada. Two-thirds of the rise of the great tilted block—4 km high, 100 km wide and 600 km long—of the Sierra Nevada of California has occurred during the past 9 m.y. (Huber, 1981). As basalts 5 to 6 m.y. old predate most of the block faulting of Owens Valley, which bounds the Sierra Nevada on the east, I infer further that most of the uplift is younger than 6 m.y. and has been synchronous with the crustal extension recorded by Owens Valley and the ranges nearby to the east. The uplift must reflect a thick crustal root or an asthenospheric antiroot, the configuration of which developed in late Neogene time. Seismic-refraction interpretations of lithosphere structure are conflicting, in part probably because of large lateral variations in velocity within crust or mantle. A very thick crustal root was inferred from both longitudinal and transverse refraction data by Pakiser and Brune (1980); Carder (1973) derived an antiroot model from transverse data alone; and Mavko and Thompson (1983) integrated gravity, refraction, and teleseismic data to conclude that both thinning of lithospheric mantle and thickening of the crust occur beneath the uplifted region.

Chase and Wallace (1988) and Kennelly and Chase (1989) assumed the Sierra Nevada to have a very thick crustal root of Mesozoic age, and argued that isostatic compensation of the root was distributed by the great strength of the lithosphere over a broad region of minor uplift until late Neogene extension fragmented the lithosphere and permitted elastic rise and local compensation of the Sierra Nevada.

My own preference is for models that emphasize thermal expansion and rise of asthenosphere due both to physical motion and to heating of previous lithospheric mantle, consequent on severe lithosphere thinning in the adjacent Basin and Range province. The asthenosphere-flow model of Bohannon and others (1989) and the asthenospheric-diapirism, shoulder-uplift model of Buck (1986) and Steckler (1985), both described in the subsequent section on the Red Sea, are such models. The widespread Pliocene alkaline and potassic-basaltic volcanism in the central and southern Sierra Nevada accords with such thermal models. Shallow heat flow in the Sierra is low because conduction has not yet established a thermal gradient in equilibrium with the asthenosphere. This interpretation implies that the Sierra Nevada has a fossil crustal root that extends about 7 km deeper beneath what is now the crest of the range than beneath the western foothills, and that the late Neogene two-thirds of the Sierran uplift is compensated by an antiroot whereby the asthenosphere now rises to near the base of the crust beneath the high part of the range.

Sundered continental crust

Whereas within the Basin and Range province the continental crust has been thinned while remaining continuous, in other rift provinces in the western United States the pre-Cenozoic crust has apparently been much more severely thinned or rifted completely, and the crust has been rebuilt to continental thickness by magmatism dominantly of mantle origin (Fig. 4; Hamilton and Myers, 1966; Hamilton, 1978; Heller and others, 1987; Wells and Heller, 1988). The Cenozoic volcanic terrains of much of Oregon and Washington apparently have been built oceanward of all pre-Cenozoic continental crust. The Cascade Range within southern Oregon and northern California (as discussed above), part of the middle Miocene Columbia River basalt plateau, much of the middle and late Cenozoic volcanic terrain of northeastern California and southeastern Oregon and adjacent parts of Nevada, and the late Neogene Snake River Plain represent new volcanic crust, or severely thinned continental crust greatly augmented by new magmatism. Most of the indicated rifting is of post-Eocene age; patterns have been complex, and deformation is continuing.

The pre-Cenozoic terrains now split and swing around the volcanic terrains; if there has not been more or less complete rifting across those volcanic terrains, then the old rocks that lie beneath them have no counterparts in the exposed terrains. Pre-Cenozoic assemblages move back to reasonable palinspastic fits after removal of these volcanic terrains and reversal of Basin-Range extension (Hamilton, 1978). Paleomagnetic orientations accord with such reconstructions (Beck and others, 1986; Grommé and others, 1986; Wells and Heller, 1988).

Isotopic compositions of the volcanic rocks in these rift terrains mostly display little or no incorporation of continental crust (e.g., Carlson and Hart, 1987). Seismic-refraction studies commonly show marked changes in crustal structure at the margins of the rift terrains, which I regard (as do some seismologists, for example, Catchings and Mooney, 1988, but not others) as best interpreted in terms of severe rifting and crustal rebuilding.

Cold extension—The Red Sea

Extension without preceding magmatism such as that of the Basin and Range province is likely the more general case in the continents, and the continental stretching that precedes the formation of new ocean basins likely starts in most cases without addition of excess heat to the lithosphere. The very oblique rifting of the Salton Trough in southern California is an active example of this process in an early stage, but I here describe briefly instead the rifting of the southern Red Sea because it has reached the stage of normal oceanic spreading and may exemplify the type of direct rifting that commonly precedes the formation of a stable continental margin with its miogeoclinal stratal wedge. The best data come from southwestern Saudi Arabia, and this discussion is based largely on Bohannon (1986, 1989), Bohannon and others (1989), and Gettings and others (1986). Extensive discussions with Bohannon, and an Arabian field trip with him, greatly increased my comprehension.

Active spreading is limited to the axial trough of the Red Sea, where new oceanic crust is forming symmetrically. The trough is bounded not by continental crust but by the carbonate-rich miogeocline of the continental shelf and coastal plain, which has been prograded about 200 km across Neogene oceanic crust. The transition between oceanic and continental crust is exposed in the Arabian pediment region bounding the coastal plain as a zone, about 20 km wide, dominated by detachment faulting in the thin remnant of continental crust above expanded lower crust that consists mostly of basaltic dike-on-dike complexes and gabbroic plutons of very early Miocene age. The next 20 km inland consists of continental crust doubled in width by severe late Oligocene and early Miocene extensional faulting, like that of the Basin and Range province, now recorded by detachment faults that dipped east in some sectors and west in others before they were raised to their present undulating dips. Extension of the same age but much lesser amount continues 40 to 70 km inland, decreasing from about 10 percent to near zero, across the foothills to the Arabian escarpment. That scarp stands more than 2 km high and records uplift that largely postdated extension of continental crust (Bohannon and others, 1989). Around the entire Red Sea, surface altitudes were generally low when extension began; the rifting was not consequent on thermal doming, as has often been inferred, but rather the uplift was a consequence of rifting (Bohannon and others, 1989). Beyond the scarp, the surface of the nonextended Arabian Shield has a slope that decreases exponentially northeastward toward the Persian Gulf. The uplift is attributed to obliquely upward flow of asthenosphere toward the oceanic rift, with thermal conversion of continental lithosphere to asthenosphere by the rising material and thermal expansion of overlying remaining lithosphere, by Bohannon and others (1989), but to mantle diapirism consequent on extension by Buck (1986) and Steckler (1985). Numerous fields of Neogene basalt on both the Red Sea and Arabian Shield sides of the escarpment attest to the operation of asthenosphere diapirism.

Continental crust was thus stretched over only a narrow zone and brief time interval before it was sundered. The geologic future presumably will bring further headward erosion of the scarp with consequent compensating isostatic rise. Subsidence as the thermal uplift decays should ultimately bring the little-eroded part of the shield back to near sea level, whereas both the narrow zone of severely extended continental crust and the broader zone of deep postextensional erosion should subside beneath the sea. Subsidence of young oceanic crust will also continue, as will further progradation and thickening of the miogeoclinal stratal wedge.

Rift-related generation of thick mafic crust

Cold-start rifting and the rapid sundering of continental crust (as in the Red Sea region just discussed, the North American miogeoclines discussed next, and in the Midcontinent and Wichita systems discussed earlier) may be the common mode of continental rifting. Brief and areally limited stretching of the preexisting continental crust in such settings apparently is followed by another period, only 20 m.y. or so long, during which thick new crust is built by mafic magmatism, and only after that does formation of typical oceanic crust progress by sea-floor spreading.

Mutter and others (1988) proposed that the voluminous magmatism of the brief periods of formation of new thick mafic crust in such cold-start rifting episodes is due to "generation of partial melt in which the lateral temperature contrasts evoked by rifting and asthenospheric upwelling drive convection in the upper mantle. Deep, hot mantle material is transported upward by the convection, and hence much larger volumes flow through the region where partial melting occurs than during passive upwelling." The mechanism proposed by Bohannon and others (1989) is similar.

North American miogeoclines

North America was broadly outlined by rifting in very late Proterozoic time, following which uppermost Proterozoic and lower Paleozoic strata were deposited as seaward-thickening miogeoclinal wedges before interruption by collisional and subduction events. The concentric rifting around North America is analogous to that around Africa in Mesozoic time, and presumably the "absolute" motion of North America (the motion of the continent in a zero-sum frame) was similarly slow during this concentric-miogeoclinal stage when plates were migrating away from it on all sides. Collisional events began in Ordovician time in the Appalachian region, and in Devonian and Mississippian time elsewhere. New miogeoclinal wedges were inaugurated following Mesozoic rifting along what are now the Atlantic and Gulf of Mexico margins of the continent.

McKenzie (1978) proposed that miogeoclinal wedges form in response to subsidence accompanying the reestablishment of an equilibrium geothermal gradient by conduction, following rapid extension of continental crust. Broad and others (1985) and Steckler and Watts (1982) showed that histories and geometries of North American miogeoclines can be explained in these terms if there has been extension of crust by factors of two to four over zones with postextension widths of 100 to 500 km.

I infer from the end-member examples of Basin and Range extension of preheated crust and of Red Sea extension of initially cold crust that continental crust cannot be extremely attenuated over a broad region for a prolonged period without being synchronously rebuilt by magmatism from mantle sources, and thus that the McKenzie model lacks broad applicability. Nevertheless, thermal subsidence as a major cause of miogeoclinal deposition seems indicated by the approximate fit of actual subsidence to a time$^{1/2}$ term. Reasoning in substantial part from the Red Sea example, I infer that much of the subsidence of continental crust beneath miogeoclinal wedges follows erosional thinning of thermally uplifted crust, such as the modern Arabian escarpment, and that much crust beneath outer continental shelves is largely basaltic rather than thinned-continental.

Rifted Atlantic margin. The miogeoclinal sedimentary wedge of the Atlantic coastal plain and continental shelf of the United States has been built across the margin of the continent as rifted by the early opening of the Atlantic Ocean in middle Mesozoic time. Although it was long assumed that severely attenuated continental crust extends seaward beneath all or most of the continental shelf, the large-aperture seismic experiment off New Jersey shows that the crust beneath the outer two-thirds of the shelf there is largely or entirely of mafic composition (Diebold and others, 1988; Keen and others, 1986). Most of the miogeoclinal wedge may have formed by progradation across mafic igneous crust, analogous to the progradation now underway along the Red Sea.

Cold-start rifting of Red Sea type perhaps has been the dominant process responsible for miogeoclinal wedges. The McKenzie model may have much less applicability than has been widely assumed. Most reconstructions of continental masses before rifting join them at isobaths well down their continental slopes. The best paleomagnetic-pole data for Late Mississippian, Permian, and Early Triassic North America and Africa produce overlaps of hundreds of kilometers of those continents relative to the generally assumed loose palinspastic fit between them; this has been taken as evidence for great subsequent strike-slip faulting by some paleomagneticists (e.g., Morel and Irving, 1981; Smith and others, 1981). These apparent overlaps can alternatively be taken as evidence that there is much less continental crust beneath many continental shelves, including that of the eastern United States, and that prerift fits were much closer than is generally assumed.

Cratonic basins

The Williston, Michigan, and Illinois Basins of the stable interior of North America each persisted as a slowly subsiding element for a large part of Phanerozoic time. They are broad ellipses, 400 to 700 km long, with a few kilometers of Paleozoic strata in their centers. Intervening regions have been much more stable in altitude, having been neither appreciably eroded nor covered by much sediment during Phanerozoic time except where involved in shortening or extensional events of short duration. Nunn and Sleep (1984) interpreted subsidence histories and geometries of the basins in terms of slow, conductive cooling of heat sources in the lithospheric mantle that have traveled with the continent in its wanderings about the globe. Fowler and Nisbet (1985) showed that the subsidence of the Williston basin did not decrease exponentially with time, as required by this explanation, but instead was almost linear with time. They suggested that the slow conversion of about 5 km of basal-crust gabbro to eclogite could account for the subsidence, but it must again be emphasized that the pressure of the gabbro-to-eclogite transformation has not been calibrated for the temperature conditions relevant to this setting. Klein and Hsui (1987) proposed origins of the basins in localized regional stretching accompanying major continental breakups.

STRIKE-SLIP FAULTING

San Andreas boundary system

Strike-slip faulting is best understood in California, where the San Andreas transform boundary between Pacific and North American plates has lengthened and evolved throughout late Oligocene and Neogene time (Atwater, 1970; Dickinson, 1979). Excellent reviews of California strike-slip tectonics, including oblique extension (transtension) and oblique shortening (transpression), have been published recently by Crowell (1987) and Sylvester (1988).

The very active San Andreas fault presumably extends down to the ductile asthenosphere zone of decoupling, above which the oceanic plate is moving rapidly, the continental more slowly. The San Andreas fault itself, with the associated oblique opening on the oceanic Gulf of California, is a product of only the Pliocene and Quaternary slip along the boundary. The more extensive late Oligocene and Miocene slip, 80 percent of the total, took place mostly on structures oceanward of the San Andreas, and is poorly understood. Miocene deformation was most complex in the region, the offshore southern California borderland, where the Pacific and North American plates first met and where the transition from subduction to strike-slip regime involved young, hot lithosphere (Atwater, 1970). There, complex strike-slip faulting, oblique rifting and shortening, and severe oroclinal rotations affected large blocks torn from the edge of the continent (Hamilton, 1978).

Within even its short period of development, the San Andreas fault itself has undergone marked changes in shape as well as in length and position. The sector through and along the Transverse Ranges of southern California has been rotated from a northwest strike to a west-northwest one; severe associated shortening has raised the Transverse Ranges; and new strike-slip faults, including the San Jacinto fault, have broken through south of the rotated sector.

Basin and Range deformation

Strike-slip faults in the Great Basin are transform faults to upper-crust extensional faults (Hamilton and Myers, 1966; Wernicke and others, 1982). The Furnace Creek right-slip fault of Death Valley, for example, ends in the southeast at the north end of the north-trending Artists' Drive oblique-slip rangefront fault; both have Holocene slip (Hamilton, 1988b). The Artists' Drive fault in turn merges downward with the still-dipping, still-active part of the Amargosa detachment fault, and does not cut lower-plate rocks—so the strike-slip fault also probably cuts only upper-plate rocks.

Extension in the Great Basin is now directed on average west-northwestward relative to the continental interior. This direction is obliquely between the northwestward San Andreas direction on the one hand and the normal to the dominant trend, north to north-northwest, of major young Great Basin fault blocks on the other. This modern extension direction is shown by the pull-apart patterns of active systems, including that of Death Valley, and by the slip-direction indicators on faults of various trends. The obliquity of regional extension was ascribed to distribution of plate-boundary slip inland from the San Andreas fault by Hamilton and Myers (1966) and many others since, but I now perceive a different explanation (Hamilton, 1988b). The present boundary between the North American and Pacific plates along most of California is the northwest-trending, right-slip San Andreas fault system. This plate boundary ends at the northern California triple junction with the north-trending Oregon trench and west-trending Mendocino transform fault. The triple junction is migrating northwestward, relative to the continental interior, with the Pacific plate as that plate slides past North America, yet the triple junction is positioned in a zone of inflection of the trend of the continental margin. Perhaps this is a coincidence, and the margin of North America south of the present triple junction merely happened to lie parallel to the direction of relative motion of the Pacific plate. It appears to me more likely, considering the patterns of crustal extension within the western United States (cf. Hamilton, 1978), that the shape of the continent is changing as needed to maintain contact with the Pacific plate south of the migrating triple junction. The San Andreas fault is the major discontinuity within the broad zone of slippage between the Pacific and North American plates, but presently active slip is now distributed across almost the entire state of California (Ward, 1988). Among the components of this slip may be northwestward drag of components as far east as the Sierra Nevada, and oblique flattening—northwestward spreading—of the deep, hot, ductile, and partly molten crust in the Great Basin, which may lie directly on asthenosphere. Such flattening could be in a pervasive mode in the lower crust and a separating-lens mode in the middle crust. The oblique extension underway at the surface in Death Valley and other regions may be a relatively passive response to the progressively more pervasive flattening at depth.

The brittle upper crust can, in such terms, be viewed as dragged apart by the gravitational spreading of the lower crust. The orientation of upper-crust structures is strongly influenced by preexisting structural grain and lithologic contrasts, but slip directions are controlled by the spreading direction and by discontinuities in the spreading system.

REFERENCES CITED

Allmendinger, R. W., and 6 others, 1983, Cenozoic and Mesozoic structure of the eastern Basin and Range Province, Utah, from COCORP seismic reflection data: Geology, v. 11, p. 532–536.

Allmendinger, R. W., and 7 others, 1987, Overview of the COCORP 40°N transect, western United States; The fabric of an orogenic belt: Geological Society of America Bulletin, v. 98, p. 308–319.

Anderson, J. L., 1983, Proterozoic anorogenic plutonism of North America: Geological Society of America Memoir 161, p. 133–154.

Anderson, R. E., 1970, Ash-flow tuffs of Precambrian age in southeast Missouri: Missouri Geological Survey and Water Resources Report of Investigations 46, 50 p.

Ando, C. J., and 10 others, 1984, Crustal profile of mountain belt; COCORP deep seismic reflection profiling in New England Appalachians and implications for architecture of convergent mountain chains: American Association of Petroleum Geologists Bulletin, v. 68, p. 819–837.

Ashwal, L. D., and Wooden, J. L., 1983, Sr and Nd isotope geochronology, geologic history, and origin of the Adirondack anorthosite: Geochimica et Cosmochimica Acta, v. 47, p. 1875–1885.

Atwater, T., 1970, Implications of plate tectonics for the Cenozoic tectonic evolution of western North America: Geological Society of America Bulletin, v. 81, p. 253–271.

Axelrod, D. I., 1985, Miocene floras from the Middlegate basin, west-central Nevada: University of California Publications in Geological Sciences, v. 129, 279 p.

Bailey, R. A., Dalrymple, G. B., and Lanphere, M. A., 1976, Volcanism, structure, and geochronology of Long Valley caldera, Mono County, California: Journal of Geophysical Research, v. 81, p. 725–744.

Bard, J.-P., 1983, Metamorphic evolution of an obducted island arc; Example of the Kohistan sequence (Pakistan) in the Himalayan collided range: University of Peshawar Geological Bulletin, v. 16, p. 105–184.

Barker, F., Wones, D. R., Sharp, W. N., and Desborough, G. A., 1975, The Pikes Peak batholith, Colorado Front Range, and a model for the origin of the gabbro-anorthosite-syenite-potassic granite suite: Precambrian Research, v. 2, p. 97–160.

Basu, A. R., and Pettingill, H. S., 1983, Origin and age of Adirondack anorthosites; Re-evaluated with Nd isotopes: Geology, v. 11, p. 514–518.

Bateman, P. C., 1979, Cross section of the Sierra Nevada from Madera to the White Mountains, central California: Geological Society of America Map and Chart 28E, scale 1:250,000.

Bateman, P. C., Clark, L. D., Huber, N. K., Moore, J. G., and Rinehart, C. D., 1963, The Sierra Nevada batholith; A synthesis of recent work across the central part: U.S. Geological Survey Professional Paper 414D, 46 p.

Beck, M. E., Jr., Burmester, R. F., Craig, D. E., Grommé, C. S., and Wells, R. E., 1986, Paleomagnetism of middle Tertiary volcanic rocks from the Western Cascade Series, northern California: Journal of Geophysical Research, v. 91, p. 8219–8230.

Behrendt, J. C., and 7 others, 1988, Crustal structure of the Midcontinent rift system; Results from GLIMPCE deep seismic reflection profiles: Geology, v. 16, p. 81–85.

Boardman, S. B., and Condie, K. C., 1986, Early Proterozoic bimodal volcanic rocks in central Colorado, U.S.A.; Part 2, Geochemistry, petrogenesis, and tectonic setting: Precambrian Research, v. 34, p. 37–68.

Bohannon, R. G., 1986, Tectonic configuration of the western Arabian continental margin, southern Red Sea: Tectonics, v. 5, p. 477–499.

—— , 1989, Style of extensional tectonism during rifting, Red Sea and Gulf of Aden: Journal of African Earth Sciences, v. 8 (in press).

Bohannon, R. G., Naeser, C. W., Schmidt, D. L., and Zimmerman, R. A., 1989, The timing of uplift, volcanism, and rifting peripheral to the Red Sea; A case for passive rifting?: Journal of Geophysical Research, v. 94, p. 1683–1701.

Bohlen, S. R., and Essene, E. J., 1978, Igneous pyroxenes from metamorphosed anorthosite massifs: Contributions to Mineralogy and Petrology, v. 65, p. 432–442.

Bohlen, S. R., and Lindsley, D. H., 1987, Thermometry and barometry of igneous and metamorphic rocks: Annual Reviews of Earth and Planetary Sciences, v. 15, p. 397–420.

Bohlen, S. R., Valley, J. W., and Essene, E. J., 1985, Metamorphism in the Adirondacks; 1, Petrology, pressure, and temperature: Journal of Petrology, v. 26, p. 971–992.

Bond, G. C., Christie-Blick, N., Kominz, M. A., and Devlin, W. J., 1985, An early Cambrian rift to post-rift transition in the Cordillera of western North America: Nature, v. 315, p. 742–746.

Bradley, D. C., 1983, Tectonics of the Acadian orogeny in New England and adjacent Canada: Journal of Geology, v. 91, p. 381–400.

Bryant, B., and Reed, J. C., Jr., 1970, Geology of the Grandfather Mountain window and vicinity, North Carolina and Tennessee: U.S. Geological Survey Professional Paper 615, 190 p.

Buck, W. R., 1986, Small-scale convection induced by passive rifting; The cause for uplift of rift shoulders: Earth and Planetary Science Letters, v. 77, p. 362–372.

—— , 1988, Flexural rotation of normal faults: Tectonics, v. 7, p. 959–973.

Buddington, A. F., and Leonard, B. F., 1962, Regional geology of the St. Lawrence County magnetite district, northwest Adirondacks, New York: U.S. Geological Survey Professional Paper 376, 145 p.

Budnik, R. T., 1986, Left-lateral intraplate deformation along the Ancestral Rocky Mountains; Implications for late Paleozoic plate motions: Tectonophysics, v. 132, p. 195–214.

Cannon, W. F., and 11 others, 1989, The North American Midcontinent rift beneath Lake Superior from GLIMPCE seismic reflection profiling: Tectonics, v. 8, p. 305–332.

Carder, D. S., 1973, Trans-California seismic profile, Death Valley to Monterey Bay: Bulletin of the Seismological Society of America, v. 63, p. 571–586.

Carl, J. D., 1988, Popple Hill Gneiss as dacite volcanics; A geochemical study of mesosome and leucosome, northwest Adirondacks, New York: Geological Society of America Bulletin, v. 100, p. 841–849.

Carlson, R. L., and Melia, P. J., 1984, Subduction hinge migration: Tectonophysics, v. 102, p. 399–411.

Carlson, R. W., and Hart, W. K., 1987, Crustal genesis on the Oregon Plateau: Journal of Geophysical Research, v. 92, p. 6191–6206.

Carmichael, D. M., 1978, Metamorphic bathozones and bathograds; A measure of the depth of post-metamorphic uplift and erosion on the regional scale: American Journal of Science, v. 278, p. 769–797.

Catchings, R. D., and Mooney, W. D., 1988, Crustal structure of the Columbia Plateau; Evidence for continental rifting: Journal of Geophysical Research, v. 93, p. 459–474.

Cater, F. W., and 6 others, 1973, Mineral resources of the Idaho Primitive Area and vicinity, Idaho: U.S. Geological Survey Bulletin 1304, 431 p.

Chandler, V. W., Bowman, P. L., Hinze, W. J., and O'Hara, N. W., 1982, Long-wavelength gravity and magnetic anomalies of the Lake Superior region: Geological Society of America Memoir 156, p. 223–237.

Chase, C. G., 1978, Extension behind island arcs and motions relative to hot spots: Journal of Geophysical Research, v. 83, p. 5385–5387.

Chase, C. G., and Gilmer, T. H., 1973, Precambrian plate tectonics; The midcontinent gravity high: Earth and Planetary Science Letters, v. 21, p. 70–78.

Chase, C. G., and Wallace, T. C., 1988, Flexural isostasy and uplift of the Sierra Nevada of California: Journal of Geophysical Research, v. 93, p. 2795–2802.

Christensen, N. I., 1974, Compressional wave velocities in possible mantle rocks to pressures of 30 kilobars: Journal of Geophysical Research, v. 79, p. 407–412.

Christensen, N. I., and Szymanski, D. L., 1988, Origin of reflections from the Brevard fault zone: Journal of Geophysical Research, v. 93, p. 1087–1102.

Christiansen, R. L., 1984, Yellowstone magmatic evolution; Its bearing on understanding large-volume explosive volcanism, *in* Explosive volcanism; Inception, evolution, and hazards: Washington, D.C., National Academy Press, p. 84–95.

Christiansen, R. L., and Lipman, P. W., 1972, Cenozoic volcanism and plate-tectonic evolution of the western United States; 2, Late Cenozoic: Philosophical Transactions of the Royal Society of London, series A, v. 271, p. 249–284.

Christiansen, R. L., and McKee, E. H., 1978, Late Cenozoic volcanic and tectonic evolution of the Great Basin and Columbia Intermontane regions: Geological Society of America Memoir 152, p. 283–311.

Christiansen, R. L., and 5 others, 1977, Timber Mountain–Oasis Valley caldera complex of southern Nevada: Geological Society of America Bulletin, v. 88, p. 943–959.

Clark, R. G., Jr., and Lyons, J. B., 1986, Petrogenesis of the Kinsman Intrusive Suite; Peraluminous granitoids of western New Hampshire: Journal of Petrology, v. 27, p. 1365–1393.

Clowes, R. M., and 7 others, 1987, LITHOPROBE—southern Vancouver Island; Cenozoic subduction complex imaged by deep seismic reflections: Canadian Journal of Earth Sciences, v. 24, p. 31–51.

Coffman, J. D., Gilbert, M. C., and McConnell, D. A., 1986, An interpretation of the crustal structure of the southern Oklahoma aulacogen satisfying gravity data: Oklahoma Geological Survey Guidebook 23, p. 1–10.

Condie, K. C., 1986, Geochemistry and tectonic setting of early Proterozoic supracrustal rocks in the southwestern United States: Journal of Geology, v. 94, p. 845–864.

Coney, P. J., and Harms, T. A., 1984, Cordilleran metamorphic core complexes; Cenozoic extensional relics of Mesozoic compression: Geology, v. 12, p. 550–554.

Cook, F. A., and 5 others, 1979, Thin-skinned tectonics in the crystalline southern Appalachians; COCORP seismic-reflection profiling of the Blue Ridge and Piedmont: Geology, v. 7, p. 563–567.

Cook, F. A., Brown, L. D., Kaufman, S., Oliver, J. E., and Petersen, T. A., 1981, COCORP seismic profiling of the Appalachian orogen beneath the Coastal Plain of Georgia: Geological Society of America Bulletin, v. 92, p. 738–748.

Coward, M. P., Butler, R.W.H., Khan, M. A., and Knipe, R. J., 1987, The tectonic history of Kohistan and its implications for Himalayan structure: Journal of the Geological Society of London, v. 144, p. 377–391.

Cox, K. G., 1980, A model for flood basalt volcanism: Journal of Petrology, v. 21, p. 629–650.

Crowell, J. C., 1987, The tectonically active margin of the western U.S.A.: Episodes, v. 10, p. 278–282.

Daniels, P. A., Jr., 1982, Upper Precambrian sedimentary rocks; Oronto Group, Michigan-Wisconsin: Geological Society of America Memoir 156, p. 107–133.

Davidson, A., 1986, New interpretations in the southwestern Grenville Province: Geological Association of Canada Special Paper 31, p. 61–74.

Davis, G. A., and Lister, G. S., 1988, Detachment faulting in continental extension; Perspectives from the southwestern U.S. Cordillera: Geological Society of America Special Paper 218, p. 133–159.

Davis, G. A., Monger, J.W.H., and Burchfiel, B. C., 1978, Mesozoic construction of the Cordilleran "collage", central British Columbia to central California: Pacific Section Society of Economic Paleontologists and Mineralogists, Pacific Coast Paleogeography Symposium 2, p. 1–32.

Dewey, J. F., 1980, Periodicity, sequence, and style at convergent plate boundaries: Geological Association of Canada Special Paper 20, p. 553–573.

Dickinson, W. R., 1979, Geometry of triple junctions related to San Andreas transform: Journal of Geophysical Research, v. 84, p. 561–572.

——, 1981, Plate tectonic evolution of the southern Cordillera: Arizona Geological Society Digest, v. 14, p. 113–135.

Dickinson, W. R., Harbaugh, D. W., Saller, A. H., Heller, P. L., and Snyder, W. S., 1983, Detrital modes of upper Paleozoic sandstones derived from Antler orogen in Nevada; Implications for nature of Antler orogeny: American Journal of Science, v. 283, p. 481–509.

Dickinson, W. R., and 6 others, 1988, Paleogeographic and paleotectonic setting of Laramide sedimentary basins in the central Rocky Mountain region: Geological Society of America Bulletin, v. 100, p. 1023–1039.

Diebold, J. B., Stoffa, P. L., and LASE Study Group, 1988, A large aperture seismic experiment in the Baltimore Canyon Trough, *in* Sheridan, R. E., and Grow, J. S., eds., The geology of North America, v. I-2, The Atlantic continental margin, U.S.: Geological Society of America, p. 387–398.

Dilek, Y., Moores, E. M., and Erskine, M. C., 1988, Ophiolitic thrust nappes in western Nevada; Implications for the Cordilleran orogen: Geological Society of London Journal, v. 145, p. 969–975.

Dodge, F. C., Calk, L. C., and Kistler, R. W., 1986, Lower crustal xenoliths, Chinese Peak lava flow, central Sierra Nevada: Journal of Petrology, v. 27, p. 1277–1294.

Doe, B. R., Leeman, W. R., Christiansen, R. L., and Hedge, C. E., 1982, Lead and strontium isotopes and related trace elements as genetic tracers in the upper Cenozoic rhyolite-basalt association of the Yellowstone Plateau volcanic field: Journal of Geophysical Research, v. 87, p. 4785–4806.

Dokka, R. K., 1984, Fission-track geochronologic evidence for Late Cretaceous mylonitization and early Paleocene uplift of the northeastern Peninsular Ranges, California: Geophysical Research Letters, v. 11, p. 46–49.

Easton, R. M., 1986, Geochronology of the Grenville Province: Geological Association of Canada Special Paper 31, p. 128–173.

EEZ SCAN 84 Scientific Staff, 1986, Atlas of the Exclusive Economic Zone, western conterminous United States (folio atlas): U.S. Geological Survey Miscellaneous Geological Investigations Series I-1792.

Emslie, R. F., and Hunt, P. A., 1989, The Grenville event; Magmatism and high grade metamorphism: Geological Survey of Canada Paper 89-1C, p. 11–17.

Engebretson, D. C., Cox, A., and Thompson, G. A., 1984, Correlation of plate motions with continental tectonics; Laramide to Basin-Range: Tectonics, v. 3, p. 115–119.

Engebretson, D. C., Cox, A., and Gordon, R. G., 1985, Relative motions between oceanic and continental plates in the Pacific Basin: Geological Society of America Special Paper 206, 59 p.

Engel, A.E.J., and Engel, C. G., 1958, Progressive metamorphism and granitization of the major paragneiss, northwest Adirondack Mountains, New York: Geological Society of America Bulletin, v. 69, p. 1369–1414.

England, P. C., and Thompson, A. B., 1984, Pressure-temperature-time paths of regional metamorphism; 1, Heat transfer during the evolution of regions of thickened continental crust: Journal of Petrology, v. 25, p. 894–928.

Epstein, A. G., Epstein, J. B., and Harris, L. D., 1977, Conodont color alteration; An index to organic metamorphism: U.S. Geological Survey Professional Paper 995, 27 p.

Fiske, R. S., and Tobisch, O. T., 1978, Paleogeographic significance of volcanic rocks of the Ritter Range pendant, central Sierra Nevada, California: Pacific Section, Society of Economic Paleontologists and Mineralogists, Pacific Coast Paleogeography Symposium 2, p. 209–221.

Flawn, P. T., 1956, Basement rocks of Texas and southeast New Mexico: University of Texas Publication 5605, 261 p.

Fletcher, R. C., and Hallett, B., 1983, Unstable extension of the lithosphere; A mechanical model for Basin-and-Range structure: Journal of Geophysical Research, v. 88, p. 7457–7466.

Fountain, D. M., and Salisbury, M. H., 1981, Exposed cross-sections through the continental crust; Implications for crustal structure, petrology, and evolution: Earth and Planetary Science Letters, v. 56, p. 263–277.

Fowler, C.M.R., and Nisbet, E. G., 1985, The subsidence of the Williston Basin: Canadian Journal of Earth Sciences, v. 22, p. 408–415.

Frizzell, V. A., Jr., Tabor, R. W., Zartman, R. E., and Blome, C. D., 1987, Late Mesozoic or early Tertiary mélanges in the western Cascades of Washington: Washington Division of Geology and Earth Resources Bulletin 77, p. 129–148.

Frost, E. G., and Okaya, D. A., 1986, Application of seismic reflection profiles to tectonic analysis in mineral explorations: Arizona Geological Society Digest, v. 16, p. 137–150.

Gallagher, M. P., Brown, E. H., and Walker, N. W., 1988, A new structural and tectonic intepretation of the western part of the Shuksan blueschist terrane, northwestern Washington: Geological Society of America Bulletin, v. 100, p. 1415–1422.

Gansser, A., 1979, The division between Himalaya and Karakorum: University of Peshawar Geological Bulletin, v. 13, p. 9–21.

Garuti, G., Rivalenti, G., Rossi, G., Siena, F., and Sinigoi, S., 1980, The Ivrea-Verbano mafic-ultramafic complex of the Italian western Alps; Discussion of some petrologic problems and a summary: Rendiconti Società Italiana di Mineralogia e Petrologia, v. 36, p. 717–749.

Garver, J., 1988, Stratigraphy, depositional setting, and tectonic significance of the clastic cover to the Fidalgo ophiolite, San Juan Islands, Washington: Canadian Journal of Earth Sciences, v. 25, p. 417–432.

Gates, A. E., Speer, J. A., and Pratt, T. L., 1988, The Alleghanian southern Appalachian Piedmont; A transpressional model: Tectonics, v. 7, p. 1307–1324.

Gettings, M. E., Blank, H. R., Jr., Mooney, W. D., and Healy, J. H., 1986, Crustal structure of southwestern Saudi Arabia: Journal of Geophysical Research, v. 91, p. 6491–6512.

Ghent, E. D., and Stout, M. Z., 1986, Garnet-hornblende geothermometry, $CaMgSi_2O_6$ activity, and the minimum pressure limits of metamorphism for garnet amphibolites: Journal of Geology, v. 94, p. 736–743.

Gibson, G. M., 1979, Margarite in kyanite- and corundum-bearing anorthosite, amphibolite, and hornblendite from central Fiordland, New Zealand: Contributions to Mineralogy and Petrology, v. 68, p. 171–179.

Gilbert, M. C., 1986, ed., Petrology of the Cambrian Wichita Mountains igneous suite: Oklahoma Geological Survey Guidebook 23, 188 p.

Grambling, J. A., Williams, M. L., and Mawer, C. K., 1988, Proterozoic tectonic assembly of New Mexico: Geology, v. 16, p. 724–727.

Grant, J. A., 1973, Phase equilibria in high-grade metamorphism and partial melting of pelitic rocks: Amcrican Journal of Science, v. 273, p. 289–317.

Green, A. G., and 9 others, 1988, Crustal structure of the Grenville front and adjacent teranes: Geology, v. 16, p. 788–792.

Green, J. C., 1982, Geology of Keweenawan extrusive rocks: Geological Society of America Memoir 156, p. 47–55.

—— , 1983, Geologic and geochemical evidence for the nature and development of the Middle Proterozoic (Keweenawan) Midcontinent Rift of North America: Tectonophysics, v. 94, p. 413–437.

Griffin, W. L., and O'Reilly, S. Y., 1987a, Is the continental Moho the crust-mantle boundary?: Geology, v. 15, p. 241–244.

—— , 1987b, The composition of the lower crust and the nature of the continental Moho; Xenolith evidence, *in* Nixon, P. H., ed., Mantle xenoliths: New York, John Wiley and Sons, p. 413–431.

Griscom, A., 1976, Bedrock geology of the Harrington Lake area, Maine [Ph.D. thesis]: Cambridge, Massachusetts, Harvard University, 373 p.

Grommé, C. S., Beck, M. E., Jr., Wells, R. E., and Engebretson, D. C., 1986, Paleomagnetism of the Tertiary Clarno Formation of central Oregon and its significance for the tectonic history of the Pacific Northwest: Journal of Geophysical Research, v. 91, p. 14089–14103.

Gust, D. A., and Arculus, R. J., 1986, Petrogenesis of alkalic and calcalkalic volcanic rocks of Mormon Mountain volcanic field, Arizona: Contributions to Mineralogy and Petrology, v. 94, p. 416–426.

Hadley, J. B., and Goldsmith, R., 1963, Geology of the eastern Great Smoky Mountains, North Carolina and Tennessee: U.S. Geological Survey Professional Paper 349-B, 118 p.

Ham, W. E., Denison, R. E., and Merritt, C. A., 1964, Basement rocks and structural evolution of southern Oklahoma: Oklahoma Geological Survey Bulletin 95, 302 p.

Hamilton, W. B., 1956, Precambrian rocks of Wichita and Arbuckle Mountains, Oklahoma: Geological Society of America Bulletin, v. 67, p. 1319–1330.

—— , 1963, Metamorphism in the Riggins region, western Idaho: U.S. Geological Survey Professional Paper 436, 95 p.

—— , 1969, Mesozoic California and the underflow of Pacific mantle: Geological Society of America Bulletin, v. 80, p. 2409–2430.

—— , 1978, Mesozoic tectonics of the western United States: Pacific Section, Society of Economic Paleontologists and Mineralogists, Pacific Coast Paleogeography Symposium 2, p. 33–70.

—— , 1979, Tectonics of the Indonesian region: U.S. Geological Survey Professional Paper 1078, 345 p.

—— , 1981, Crustal evolution by arc magmatism: Royal Society of London Philosophical Transactions, Section A, v. 302, p. 279–291.

—— , 1987a, Mesozoic geology and tectonics of the Big Maria Mountains region, southeastern California: Arizona Geological Society Digest, v. 18, p. 33–47.

—— , 1987b, Crustal extension in the Basin and Range Province, western United States: Geological Society of London Special Publication 28, p. 155–176.

—— , 1988a, Laramide crustal shortening: Geological Society of America Memoir 171, p. 27–39.

—— , 1988b, Detachment faulting in the Death Valley region, California and Nevada: U.S. Geological Survey Bulletin 1790, p. 51–85.

—— , 1988c, Plate tectonics and island arcs: Geological Society of America Bulletin, v. 100, p. 1503–1527.

—— , 1989, Convergent-plate tectonics viewed from the Indonesian region, *in* Şengör, A.M.C., ed., Tectonic evolution of the Tethyan region: Dordrecht, Netherlands, Kluwer Publishers, p. 655–698.

Hamilton, W. B., and Myers, W. B., 1966, Cenozoic tectonics of the western United States: Reviews of Geophysics, v. 4, p. 509–549.

—— , 1967, The nature of batholiths: U.S. Geological Survey Professional Paper 554-C, 30 p.

—— , 1974, Nature of the Boulder batholith of Montana: Geological Society of America Bulletin, v. 85, p. 365–378, 1958–1960.

Hansen, B., 1981, The transition from pyroxene granulite facies to garnet clinopyroxene granulite facies; Experiments in the system $CaO-MgO-Al_2O_3-SiO_2$: Contributions to Mineralogy and Petrology, v. 76, p. 234–242.

Harris, L. D., and Bayer, K. C., 1979, Sequential development of the Appalachian orogen above a master décollement; A hypothesis: Geology, v. 7, p. 568–572.

Harris, L. D., Harris, A. G., De Witt, W., Jr., and Bayer, K. C., 1981, Evaluation of southern eastern overthrust belt beneath Blue Ridge–Piedmont thrust: American Association of Petroleum Geologists Bulletin, v. 65, p. 2497–2505.

Hatcher, R. D., Jr., 1987, Tectonics of the southern and central Appalachian internides: Annual Reviews of Earth and Planetary Sciences, v. 15, p. 337–362.

Hauge, T. A., and 10 others, 1987, Crustal structure of western Nevada from COCORP deep seismic-reflection data: Geological Society of America Bulletin, v. 98, p. 320–329.

Hawkins, J. W., Bloomer, S. H., Evans, C. A., and Melchior, J. T., 1984, Evolution of intra-oceanic arc-trench systems: Tectonophysics, v. 102, p. 175–205.

Haxel, G. B., Tosdal, R. M., and Dillon, J. T., 1986, Field guide to the Chocolate Mountains thrust and Orocopia Schist, Gavilan Wash area, southeasternmost California: Arizona Geological Society Digest, v. 16, p. 282–293.

Heller, P. L., Tabor, R. W., and Suczek, C. A., 1987, Paleogeographic evolution of the United States Pacific Northwest during Paleogene time: Canadian Journal of Earth Sciences, v. 24, p. 1652–1667.

Hildreth, W., 1981, Gradients in silicic magma chambers; Implications for lithospheric magmatism: Journal of Geophysical Research, v. 86, p. 10153–10192.

Hinze, W. J., Wold, R. J., and O'Hara, N. W., 1982, Gravity and magnetic anomaly studies of Lake Superior: Geological Society of America Memoir 156, p. 203–221.

Hodge, D. S., and 5 others, 1982, Gravity studies of subsurface mass distributions of granitic rocks in Maine and New Hampshire: American Journal of Science, v. 282, p. 1289–1324.

Hodges, K. V., and Fountain, D. M., 1983, Pogallo Line, South Alps, northern Italy; An intermediate crustal level, low-angle normal fault?: Geology, v. 12, p. 151–155.

Hoisch, T. D., 1987, Heat transport by fluids during Late Cretaceous regional metamorphism in the Big Maria Mountains, southeastern California: Geological Society of America Bulletin, v. 98, p. 549–553.

Holdaway, M. J., 1971, Stability of andalusite and the aluminum silicate phase diagram: American Journal of Science, v. 271, p. 97–131.

Howard, K. A., and John, B. E., 1987, Crustal extension along a rooted system of imbricate low-angle faults; Colorado River extensional corridor, California and Arizona: Geological Society of London Special Publication 28, p. 299–311.

Howard, K. A., Goodge, J. W., and John, B. E., 1982, Detached crystalline rocks of the Mohave, Buck, and Bill Williams Mountains, western Arizona, *in* Frost, E. G., and Martin, D. L., eds., Mesozoic-Cenozoic tectonic evolution of the Colorado River region, California, Arizona, and Nevada: San Diego, California, Cordilleran Publishers, p. 377–390.

Howell, D. G., ed., 1985, Tectonostratigraphic terranes of the Circum-Pacific region: Houston, Texas, Circum-Pacific Council for Energy and Mineral Resources Earth Science Series, v. 1, 581 p.

Huang, W. L., and Wyllie, P. J., 1981, Phase relationships of S-type granite with H_2O to 35 kbar; Muscovite granite from Harney Peak, South Dakota: Journal of Geophysical Research, v. 86, p. 10515–10529.

Huber, N. K., 1981, Amount and timing of late Cenozoic uplift and tilt of the central Sierra Nevada, California; Evidence from the upper San Joaquin River basin: U.S. Geological Survey Professional Paper 1197, 28 p.

Hughes, S. S., and Taylor, E. M., 1986, Geochemistry, petrogenesis, and tectonic implications of central High Cascade mafic platform lavas: Geological Society of America Bulletin, v. 97, p. 1024–1036.

Hyndman, D. W., 1981, Controls on source and depth of emplacement of granitic magma: Geology, v. 9, p. 244–248.

—— , 1983, The Idaho batholith and associated plutons, Idaho and western Montana: Geological Society of America Memoir 159, p. 213–240.

Ingersoll, R. V., and Schweickert, R. A., 1986, A plate-tectonic model for Late Jurassic ophiolite genesis, Nevadan orogeny and forearc initiation, northern California: Tectonics, v. 5, p. 901–912.

Iverson, W. P., and Smithson, S. B., 1983, Reprocessing and reinterpretation of COCORP southern Appalachian profiles: Earth and Planetary Science Letters, v. 62, p. 75–90.

Jachens, R. C., Simpson, R. W., Blakely, R. L., and Saltus, R. W., 1985, Isostatic residual gravity map of the United States: U.S. Geological Survey, scale 1:2,500,000.

Jacobson, C. E., 1983, Structural geology of the Pelona Schist and Vincent thrust, San Gabriel Mountains, California: Geological Society of America Bulletin, v. 94, p. 753–767.

Jaffe, H. W., and Schumacher, J. C., 1985, Garnet and plagioclase exsolved from aluminum-rich orthopyroxene in the Marcy anorthosite, northeastern Adirondacks, New York: Canadian Mineralogist, v. 23, p. 457–478.

Jan, M. Q., 1988, Geochemistry of amphibolites from the southern part of the Kohistan arc, N. Pakistan: Mineralogical Magazine, v. 52, p. 147–159.

Jan, M. Q., and Asif, M., 1983, Geochemistry of tonalites and (quartz) diorites of the Kohistan-Ladakh (Transhimalayan) granitic belt in Swat, NW Pakistan, *in* Shams, F. A., ed., Granites of Himalayas, Karakorum and Hindukush: Lahore, Pakistan, Punjab University, p. 355–376.

Jan, M. Q., and Howie, R. A., 1981, The mineralogy and geochemistry of the metamorphosed basic and ultrabasic rocks of the Jijal complex, Kohistan, NW Pakistan: Journal of Petrology, v. 22, p. 85–126.

Jan, M. Q., Parvez, M. K., and Khattak, M.U.K., 1984, Coronites from the Chilas and Jijal-Patan complexes of Kohistan: Peshawar, Pakistan, University of Peshawar Geological Bulletin, v. 17, p. 75–85.

Johnson, C. A., Bohlen, S. R., and Essene, E. J., 1983, An evaluation of garnet-clinopyroxene geothermometry in granulites: Contributions to Mineralogy and Petrology, v. 84, p. 191–198.

Johnson, M. J., 1986, Distribution of maximum burial temperatures across northern Appalachian Basin and implications for Carboniferous sedimentation patterns: Geology, v. 14, p. 384–387.

Johnson, S. Y., Zimmerman, R. A., Naeser, C. W., and Whetten, J. T., 1986, Fission-track dating of the tectonic development of the San Juan Islands, Washington: Canadian Journal of Earth Sciences, v. 23, p. 1318–1330.

Jurdy, D. M., 1984, The subduction of the Farallon plate beneath North America as derived from relative plate motions: Tectonics, v. 3, p. 107–113.

Kalliokoski, J., 1982, Jacobsville Sandstone: Geological Society of America Memoir 156, p. 147–155.

Karlstrom, K. E. and Bowring, S. A., 1988, Early Proterozoic assembly of tectonostratigraphic terranes in southwestern North America: Journal of Geology, v. 96, p. 561–576.

Kay, R. W., and Kay, S. M., 1981, The nature of the lower continental crust; Inferences from geophysics, surface geology, and crustal xenoliths: Reviews of Geophysics and Space Physics, v. 19, p. 271–297.

Keen, C., and 15 others, 1986, Deep structure of the US East Coast passive margin from large aperture seismic experiments (LASE): Marine and Petroleum Geology, v. 3, p. 234–242.

Kennelly, P. J., and Chase, C. G., 1989, Flexure and isostatic residual gravity of the Sierra Nevada: Journal of Geophysical Research, v. 94, p. 1759–1764.

Khan, M. A., Jan, M. Q., Windley, B. F., and Tarney, J., 1989, The Chilas mafic igneous complex; The root of the Kohistan island arc in the Himalayas of north Pakistan: Geological Society of America Special Paper 232 (in press).

Klein, G. D., and Hsui, A. T., 1987, Origin of cratonic basins: Geology, v. 15, p. 1094–1098.

Klemperer, S. L., Hauge, T. A., Hauser, E. C., Oliver, J. E., and Potter, C. J., 1986, The Moho in the northern Basin and Range Province, Nevada, along the COCORP 40°N seismic reflection transect: Geological Society of America Bulletin, v. 97, p. 603–618.

Klewin, K. W., 1989, Polybaric fractionation in an evolving continental rift; Evidence from the Keweenawan Mid-Continent Rift: Journal of Geology, v. 97, p. 65–76.

Kluth, C. F., 1986, Plate tectonics of the Ancestral Rocky Mountains: American Association of Petroleum Geologists Memoir 41, p. 353–369.

Lachenbruch, A. H., and Sass, J. H., 1978, Models of an extending lithosphere and heat flow in the Basin and Range province: Geological Society of America Memoir 152, p. 209–250.

Lambert, D. D., Unruh, D. M., and Gilbert, M. C., 1988, Rb-Sr and Sm-Nd isotopic study of the Glen Mountains layered complex; Initiation of rifting within the southern Oklahoma aulacogen: Geology, v. 16, p. 13–17.

Leaver, D. S., Mooney, W. D., and W. D., and Kohler, W. M., 1984, A seismic refraction study of the Oregon Cascades: Journal of Geophysical Research,

v. 89, p. 3121–3134.
Leeman, W. P., 1982, Development of the Snake River Plain–Yellowstone Plateau province, Idaho and Wyoming; An overview and petrologic model: Idaho Bureau of Mines and Geology Bulletin 26, p. 155–177.
—— , 1983, The influence of crustal structure on compositions of subduction-related magmas: Journal of Volcanology and Geothermal Research, v. 18, p. 561–588.
Lehman, J. A., Smith, R. B., Schilly, M. M., and Braile, L. W., 1982, Upper crustal structure of the Yellowstone caldera from seismic delay time analyses and gravity correlations: Journal of Geophysical Research, v. 87, p. 1723–1730.
LeHuray, A. P., 1987, U-Pb and Th-Pb whole-rock isochrons from metavolcanic rocks of the Carolina slate belt: Geological Society of America Bulletin, v. 99, p. 354–361.
Lillie, R. J., and 7 others, 1983, Crustal structure of Ouachita Mountains, Arkansas; A model based on integration of COCORP reflection profiles and regional geophysical data: American Association of Petroleum Geologists Bulletin, v. 67, p. 907–931.
Lipman, P. W., 1980, Cenozoic volcanism in the western United States; Implications for continental tectonics, *in* Studies in geophysics; Continental tectonics: Washington, D.C., National Academy of Sciences, p. 161–174.
—— , 1984, The roots of ash flow calderas in western North America; Windows into the tops of granitic batholiths: Journal of Geophysical Research, v. 89, p. 8801–8814.
Loiselle, M. C., and Ayuso, R. A., 1980, Geochemical characteristics of granitoids across the Merrimack synclinorium, eastern and central Maine: Blacksburg, Virginia Polytechnic Institute Department of Geological Sciences Memoir 2, p. 117–121.
Luedke, R. G., and Smith, R. L., 1982, Map showing distribution, composition, and age of late Cenozoic volcanic centers in Oregon and Washington: U.S. Geological Survey Map I-1091-D, scale 1:1,000,000.
Luetgert, J. H., and Meyer, R. P., 1982, Structure of the western basin of Lake Superior from cross structure refraction profiles: Geological Society of America Memoir 156, p. 245–255.
Maaloe, S., and Wyllie, P. J., 1975, Water content of a granite magma deduced from the sequence of crystallization determined experimentally with water-undersaturated conditions: Contributions to Mineralogy and Petrology, v. 52, p. 175–191.
MacLeod, N. S., Walker, G. W., and McKee, E. H., 1976, Geothermal significance of eastward increase in age of upper Cenozoic rhyolitic domes in southeastern Oregon: Proceedings Second United Nations Symposium on the Development and Use of Geothermal Resources, v. 1, p. 465–474.
Manghnani, M. H., Ramananantoandro, R., and Clark, S. P., Jr., 1974, Compressional and shear wave velocities in granulite facies rocks and ecologites to 10 kbar: Journal of Geophysical Research, v. 79, p. 5427–5446.
Mavko, B. B., and Thompson, G. A., 1983, Crustal and upper mantle structure of the northern and central Sierra Nevada: Journal of Geophysical Research, v. 88, p. 5874–5892.
McBirney, A. R., and White, C. M., 1982, The Cascade Province, *in* Thorpe, R. S., ed., Andesites: New York, John Wiley and Sons, p. 115–135.
McCulloch, M. T., and Cameron, W. E., 1983, Nd-Sr isotopic study of primitive lavas from the Troodos ophiolite; Evidence for a subduction-related setting: Geology, v. 11, p. 727–731.
McKenzie, D., 1978, Some remarks on the development of sedimentary basins: Earth and Planetary Science Letters, v. 40, p. 25–32.
McLelland, J. M., 1986, Pre-Grenvillian history of the Adirondacks as an anorogenic, bimodal caldera complex of mid-Proterozoic age: Geology, v. 14, p. 229–233.
McMillan, N. J., and Dungan, M. A., 1988, Open system magmatic evolution of the Taos Plateau volcanic field, northern New Mexico; 3, Petrology and geochemistry of andesite and dacite: Journal of Petrology, v. 29, p. 527–557.
Menzies, M. A., and Hawkesworth, C. J., 1987, Upper mantle processes and composition, *in* Nixon, P. H., ed., Mantle xenoliths: New York, John Wiley and Sons, p. 725–738.
Merk, G. P., and Jirsa, M. A., 1982, Provenance and tectonic significance of the Keweenawan interflow sedimentary rocks: Geological Society of America Memoir 156, p. 97–105.
Miller, E. L., Holdsworth, B. K., Whiteford, W. B., and Rodgers, D., 1984, Stratigraphy and structure of the Schoonover sequence, northeastern Nevada; Implications for Paleozoic plate-margin tectonics: Geological Society of America Bulletin, v. 95, p. 1063–1076.
Mittwede, S. K., 1988, Ultramafites, mélanges, and stitching granites as suture markers in the central Piedmont of the southern Appalachians: Journal of Geology, v. 96, p. 693–708.
Molnar, P., and Atwater, T., 1978, Interarc spreading and Cordilleran tectonics as alternates related to the age of subducted oceanic lithosphere: Earth and Planetary Science Letters, v. 41, p. 330–340.
Morel, P., and Irving, E., 1981, Paleomagnetism and the evolution of Pangea: Journal of Geophysical Research, v. 86, p. 1858–1872.
Morey, G. B., and Ojakangas, R. W., 1982, Keweenawan sedimentary rocks of eastern Minnesota and northwestern Wisconsin: Geological Society of America Memoir 156, p. 135–146.
Morgan, B. A., 1972, Metamorphic map of the Appalachians: U.S. Geological Survey Map I-724, scale 1:2,500,000.
Morgan, L. A., Doherty, D. J., and Leeman, W. P., 1984, Ignimbrites of the eastern Snake River Plain; Evidence for major caldera-forming eruptions: Journal of Geophysical Research, v. 89, p. 8665–8678.
Mutter, J. C., Buck, W. R., and Zehnder, C. M., 1988, Convective partial melting; 1, A model for the formation of thick basaltic sequences during the initiation of spreading: Journal of Geophysical Research, v. 93, p. 1031–1048.
Newton, R. C., and Perkins, D., III, 1982, Thermodynamic calibraiton of barometers based on the assemblages garnet-plagioclase-orthopyroxene (clinopyroxene)-quartz: American Mineralogist, v. 67, p. 203–222.
Nicholas, R. L., and Rozendal, R. A., 1975, Subsurface positive elements within Ouachita foldbelt in Texas and their relation to Paleozoic cratonic margin: American Association of Petroleum Geologists Bulletin, v. 39, p. 193–216.
Nielson, D. L., Clark, R. G., Lyons, J. B., Englund, E. J., and Borns, D. J., 1976, Gravity models and mode of emplacement of the New Hampshire Plutonic Series: Geological Society of America Memoir 146, p. 301–318.
Noyes, H. J., Wones, D. R., and Frey, F. A., 1983, A tale of two plutons; Petrographic and mineral constraints on the petrogenesis of the Red Lake and Eagle Peak plutons, central Sierra Nevada, California: Journal of Geology, v. 91, p. 353–379.
Nunn, J. A., and Sleep, N. H., 1984, Thermal contraction and flexure of intracratonal basins; A three-dimensional study of the Michigan basin: Geophysical Journal of the Royal Astronomical Society, v. 76, p. 587–635.
O'Hara, M. J., and Mathews, R. E., 1981, Geochemical evolution in advancing, periodically replenished, continuously fractionated magma chamber: Geological Society of London Journal, v. 138, p. 237–277.
Ojakangas, R. W., and Morey, G. B., 1982a, Keweenawan pre-volcanic quartz sandstones and related rocks of the Lake Superior region: Geological Society of America Memoir 156, p. 85–96.
—— , 1982b, Keweenawan sedimentary rocks of the Lake Superior region; A summary: Geological Society of America Memoir 156, p. 157–164.
Olilla, P. W., Jaffe, H. W., and Jaffe, E. B., 1984, Iron-rich inverted pigeonite; Evidence for the deep emplacement of the Adirondack anorthosite massif: Geological Society of America Abstracts with Programs, v. 16, p. 54.
Opdyke, N. D., and 5 others, 1987, Florida as an exotic terrane; Paleomagnetic and geochronologic investigation of lower Paleozoic rocks from the subsurface of Florida: Geology, v. 15, p. 900–903.
Page, B. M., 1982, Migration of Salinian composite block, California, and disappearance of fragments: American Journal of Science, v. 282, p. 1694–1734.
Pakiser, L. C., and Brune, J. N., 1980, Seismic models of the root of the Sierra Nevada: Science, v. 210, p. 1088–1094.
Pavlides, L., Gair, J. E., and Cranford, S. L., 1982, Massive sulfide deposits of the southern Appalachians: Economic Geology, v. 77, p. 233–272.
Peterman, Z. E., and Sims, P. K., 1988, The Goodman swell; A lithospheric flexure caused by crustal loading along the Midcontinent rift system: Tec-

tonics, v. 7, p. 1077–1099.
Petterson, M. G., and Windley, B. F., 1985, Rb-Sr dating of the Kohistan arc-batholith in the Trans-Himalaya of north Pakistan, and tectonic implications: Earth and Planetary Science Letters, v. 74, p. 45–57.
Philpotts, A. R., and Reichenbach, I., 1985, Differentiation of Mesozoic basalts of the Hartford Basin, Connecticut: Geological Society of America Bulletin, v. 96, p. 1131–1139.
Pin, C., and Sills, J. D., 1986, Petrogenesis of layered gabbros and ultramafic rocks from Val Sesia, the Ivrea Zone, NW Italy; Trace element and isotope geochemistry: Geological Society of London Special Publication 24, p. 231–249.
Postlethwaite, C. E., and Jacobson, C. E., 1987, Early history and reactivation of the Rand thrust, southern California: Journal of Structural Geology, v. 9, p. 195–205.
Potter, C. J., and 9 others, 1987, Crustal structure of north-central Nevada; Results from COCORP deep seismic profiling: Geological Society of America Bulletin, v. 98, p. 330–337.
Powers, R. E., and Bohlen, S. R., 1985, The role of synmetamorphic igneous rocks in the metamorphism and partial melting of metasediments, Northwest Adirondacks: Contributions to Mineralogy and Petrology, v. 90, p. 401–409.
Pratt, T. L., Coruh, C., Costain, J. K., and Glover, L., III, 1988, A geophysical study of the Earth's crust in central Virginia; Implications for Appalachian crustal structure: Journal of Geophysical Research, v. 93, p. 6649–6667.
Price, R. A., 1981, The Cordilleran thrust and fold belt in the southern Canadian Rocky Mountains: Geological Society of London Special Publication 9, p. 427–448.
Reed, J. C., Jr., 1987, Precambrian geology of the U.S.A.: Episodes, v. 10, p. 243–247.
Reed, J. C., Jr., Bickford, M. E., Premo, W. R., Aleinikoff, J. N., and Pallister, J. S., 1987, Evolution of the Early Proterozoic Colorado province; Constraints from U-Pb geochronology: Geology, v. 15, p. 861–865.
Robinson, P., and Hall, L. M., 1980, Tectonic synthesis of southern New England: Blacksburg, Virginia Polytechnic Institute and State University Department of Geological Sciences Memoir 2, p. 73–82.
Ross, D. C., 1967, Generalized geologic map of the Inyo Mountains region, California: U.S. Geological Survey Map I-506, scale 1:125,000.
—— , 1989, Metamorphic and plutonic rocks of the southernmost Sierra Nevada, California, and their tectonic framework: U.S. Geological Survey Professional Paper 1381 (in press).
Rytuba, J. J., and McKee, E. H., 1984, Peralkaline ash flow tuffs and calderas of the McDermitt volcanic field, southeast Oregon and north-central Nevada: Journal of Geophysical Research, v. 89, p. 8616–8628.
Saleeby, J. B., 1983, Accretionary tectonics of the North American Cordillera: Annual Reviews of Earth and Planetary Sciences, v. 15, p. 45–73.
Schamel, S., and Bauer, D., 1980, Remobilized Grenville basement in the Pine Mountain window: Blacksburg, Virginia Polytechnic Institute and State University Department of Geological Sciences Memoir 2, p. 313–316.
Schmid, R., 1967, Zur Petrographie und Struktur der Zone Ivrea–Verbano zwischen Valle d'Ossola und Val Grande (Prov. Novara, Italien): Schweizerische mineralogische und petrographische Mitteilungen, v. 47, p. 935–1117.
—— , 1979, Are the metapelites of the Ivrea–Verbano zone restites?: Università di Padova Memorie di Scienze Geologiche, v. 33, p. 67–69.
Sclater, J. G., Jaupart, C., and Galson, D., 1980, The heat flow through oceanic and continental crust and the heat loss of the Earth: Reviews of Geophysics and Space Physics, v. 18, p. 269–311.
Sharry, J., and 5 others, 1985, Enhanced imaging of the COCORP seismic line, Wind River Mountains: American Geophysical Union Geodynamics Series, v. 13, p. 223–236.
Shervais, J., 1979a, Ultramafic and mafic layers in the Alpine-type lherzolite massif at Balmuccia, N.W. Italy: Università di Padova Memorie di Scienze Geologiche, v. 33, p. 135–145.
—— , 1979b, Thermal emplacement model for the Alpine lherzolite massif at Balmuccia, Italy: Journal of Petrology, v. 20, p. 795–820.
Sibson, R. H., 1983, Continental fault structure and the shallow earthquake source: Geological Society of London Journal, v. 140, p. 741–760.
Silver, L. T., Taylor, H. P., Jr., and Chappell, B., 1979, Some petrological, geochemical and geochronological observations of the Peninsular Ranges batholith near the international border of the U.S.A. and Mexico, *in* Abbott, P. L., and Todd, V. R., eds., Mesozoic crystalline rocks; Geological Society of America 1979 Annual Meeting Guidebook: San Diego, California, San Diego State University, p. 83–110.
Sims, P. K., Kisvarsanyi, E. B., and Morey, G. B., 1987, Geology and metallogeny of Archean and Proterozoic basement terranes in the northern Midcontinent, U.S.A.; An overview: U.S. Geological Survey Bulletin 1815, 51 p.
Smith, A. G., 1981, Subduction and coeval thrust belts, with particular reference to North America: Geological Society of London Special Publication 9, p. 111–124.
Smith, A. G., Hurley, A. M., and Briden, J. C., 1981, Phanerozoic continental world maps: Cambridge, Cambridge University Press, 102 p.
Smithson, S. B., Brewer, J. A., Kaufman, S., Oliver, J. E., and Hurich, C. A., 1979, Structure of the Laramide Wind River uplift, Wyoming, from COCORP deep reflection data and from gravity data: Journal of Geophysical Research, v. 84, p. 5955–5972.
Snavely, P. D., Jr., 1987, Tertiary geologic framework, neotectonics, and petroleum potential of the Oregon–Washington continental margin: Circum-Pacific Council for Energy and Mineral Resources Earth Science Series, v. 6, p. 305–335.
Snyder, W. S., and Brueckner, H. K., 1983, Tectonic evolution of the Golconda allochthon, Nevada; Problems and perspectives, *in* Stevens, C. H., ed., Pre-Jurassic rocks in western North American suspect terranes: Pacific Section, Society of Economic Paleontologists and Mineralogists, p. 103–123.
Sparlin, M. A., Braile, L. W., and Smith, R. B., 1982, Crustal structure of the eastern Snake River Plain determined from ray trace modeling of seismic refraction data: Journal of Geophysical Research, v. 87, p. 2619–2633.
Speed, R. C., 1979, Collided Paleozoic microplate in the western United States: Journal of Geology, v. 87, p. 279–292.
Spencer, J. E., 1984, Role of tectonic denudation in warping and uplift of low-angle normal faults: Geology, v. 12, p. 95–98.
Steckler, M. S., 1985, Uplift and extension at the Gulf of Suez; Indications of induced mantle convection: Nature, v. 317, p. 135–139.
Steckler, M. S., and Watts, A. B., 1982, Subsidence history and tectonic evolution of Atlantic-type continental margins: American Geophysical Union Geodynamics Series, v. 6, p. 184–196.
Steidtmann, J. R., Middleton, L. T., and Shuster, M. W., 1989, Post-Laramide (Oligocene) uplift in the Wind River Range, Wyoming: Geology, v. 17, p. 38–41.
Steiner, M. B., 1986, Rotation of the Colorado Plateau: Tectonics, v. 5, p. 649–660.
Stern, C. R., and Wyllie, P. J., 1981, Phase relationships of I-type granite with H_2O to 35 kilobars; The Dinkey Lakes biotite-granite from the Sierra Nevada batholith: Journal of Geophysical Research, v. 86, p. 10412–10422.
Stock, J., and Molnar, P., 1988, Uncertainties and implications of the Late Cretaceous and Tertiary position of North America relative to the Farallon, Kula, and Pacific plates: Tectonics, v. 7, p. 1339–1384.
Sweeney, J. F., 1976, Subsurface distribution of granitic rocks, south-central Maine: Geological Society of America Bulletin, v. 87, p. 241–249.
Sylvester, A. G., 1988, Strike-slip faults: Geological Society of America Bulletin, v. 100, p. 1666–1703.
Tabor, R. W., and Cady, W. M., 1978a, Geologic map of the Olympic Peninsula, Washington: U.S. Geological Survey Map I-994, scale 1:125,000.
—— , 1978, The structure of the Olympic Mountains, Washington; Analysis of a subduction zone: U.S. Geological Survey Professional Paper 1033, 38 p.
Thomas, M. D., 1983, Tectonic significance of paired gravity anomalies in the southern and central Appalachians: Geological Society of America Memoir 158, p. 113–124.
Todd, V. R., and Shaw, S. E., 1979, Structural, metamorphic and intrusive framework of the Peninsular Ranges batholith in southern San Diego

County, California, *in* Abbott, P. L., and Todd, V. R., eds., Mesozoic crystalline rocks; Geological Society of America Guidebook 1979 Annual Meeting: San Diego, California, San Diego State University, p. 178–231.

Tracy, R. J., 1978, High grade metamorphic reactions and partial melting in pelitic schist, west-central Massachusetts: American Journal of Science, v. 278, p. 150–178.

Tracy, R. J., and Robinson, P., 1980, Evolution of metamorphic belts; Information from detailed petrologic studies: Blacksburg, Virginia Polytechnic Institute Department of Geological Sciences Memoir 2, p. 189–195.

Valasek, P. A., Snoke, A. W., Hurich, C. A., and Smithson, S. B., 1989, Nature and origin of seismic reflection fabric, Ruby–East Humboldt metamorphic core complex, Nevada: Tectonics, v. 8, p. 391–415.

van Breemen, O., and Davidson, A., 1988, U-Pb zircon ages of granites and syenites in the Central Metasedimentary Belt, Grenville Province, Ontario: Geological Survey of Canada Paper 88-2, p. 45–50.

van Breemen, O., and Hanmer, S., 1986, Zircon morphology and U-Pb geochronology in active shear zones; Studies on syntectonic intrusions along the northwest boundary of the Central Metasedimentary Belt, Grenville Province, Ontario: Geological Society of Canada Paper 86-1B, p. 775–784.

Van der Voo, R., 1988, Paleozoic paleogeography of North America, Gondwana, and intervening displaced terranes; Comparisons of paleomagnetism with paleoclimatology and biogeographical patterns: Geological Society of America Bulletin, v. 100, p. 311–324.

Van Schmus, W. R., and Bickford, M. E., 1981, Proterozoic chronology and evolution of the midcontinent region, North America, *in* Kroner, A., ed., Precambrian plate tectonics: Amsterdam, Elsevier, p. 261–296.

Van Schmus, W. R., Green, J. C., and Halls, H. C., 1982, Geochronology of Keweenawan rocks of the Lake Superior region; A summary: Geological Society of America Memoir 156, p. 165–171.

Van Schmus, W. R., Bickford, M. E., and Zietz, I., 1987, Early and Middle Proterozoic provinces in the central United States: American Geophysical Union Geodynamics Series, v. 17, p. 43–68.

Vauchez, A., 1987, Brevard fault zone, southern Appalachians; A medium-angle, dextral, Alleghanian shear zone: Geology, v. 15, p. 669–672.

Vielzeuf, D., and Holloway, J. R., 1988, Experimental determination of the fluid-absent melting relations in the pelitic system: Contributions to Mineralogy and Petrology, v. 98, p. 257–276.

Voshage, H., Hunziker, J. C., Hofmann, A. W., and Zingg, A., 1987, A Nd and Sr isotopic study of the Ivrea zone, southern Alps, N-Italy: Contributions to Mineralogy and Petrology, v. 97, p. 31–42.

Voshage, H., and 5 others, 1988, Isotopic constraints on the origin of ultramafic and mafic dikes in the Balmuccia peridotite (Ivrea Zone): Contributions to Mineralogy and Petrology, v. 100, p. 261–267.

Walker, D., 1983, New developments in magmatic processes: Reviews of Geophysics and Space Physics, v. 21, p. 1372–1384.

Walker, J. D., 1988, Permian and Triassic rocks of the Mojave Desert and their implications for timing and mechanisms of continental truncation: Tectonics, v. 7, p. 685–709.

Ward, S. N., 1988, North America-Pacific plate boundary, an elastic-plastic megashear; Evidence from very long baseline interferometry: Journal of Geophysical Research, v. 93, p. 7716–7728.

Warner, M. R., 1987, Seismic reflections from the Moho; The effect of isostasy: Geophysical Journal of the Royal Astronomical Society, v. 88, p. 425–435.

Weaver, B. L., and Tarney, J., 1982, Andesitic magmatism and continental growth, *in* Thorpe, R. S., ed., Andesites: New York, Wiley and Sons, p. 639–661.

Weiblen, P. W., 1982, Keweenawan intrusive igneous rocks: Geological Society of America Memoir 156, p. 57–82.

Wells, P.R.A., 1980, Thermal models for the magmatic accretion and subsequent metamorphism of continental crust: Earth and Planetary Science Letters, v. 46, p. 253–265.

Wells, R. E., and Heller, P. L., 1988, The relative contribution of accretion, shear, and extension to Cenozoic tectonic rotation in the Pacific Northwest: Geological Society of America Bulletin, v. 100, p. 325–338.

Wenrich-Verbeek, K. J., 1979, The petrogenesis and trace-element geochemistry of intermediate lavas from Humphreys Peak, San Francisco volcanic field, Arizona: Tectonophysics, v. 61, p. 103–129.

Wernicke, B., 1985, Uniform-sense normal simple shear of the continental lithosphere: Canadian Journal of Earth Sciences, v. 22, p. 108–125.

Wernicke, B., and Axen, G. J., 1988, On the role of isostasy in the evolution of normal fault systems: Geology, v. 16, p. 848–851.

Wernicke, B., Spencer, J. E., Burchfiel, B. C., and Guth, P. L., 1982, Magnitude of crustal extension in the southern Great Basin: Geology, v. 10, p. 499–502.

Wernicke, B., Axen, G. J., and Snow, J. K., 1988, Basin and Range extensional tectonics at the latitude of Las Vegas, Nevada: Geological Society of America Bulletin, v. 100, p. 1738–1757.

White, A.J.R., and 5 others, 1986, S-type granites and their probable absence in southwestern North America: Geology, v. 14, p. 115–118.

Whitney, J. A., Paris, T. A., Carpenter, R. H., and Hartley, M. E., III, 1978, Volcanic evolution of the southern Slate Belt of Georgia and South Carolina; A primitive oceanic island arc: Journal of Geology, v. 86, p. 173–192.

Whitney, P. R., and McLelland, J. M., 1973, Origin of coronas in metagabbros of the Adirondack Mts., N.Y.: Contributions to Mineralogy and Petrology, v. 39, p. 81–98.

Wiener, R. W., McLelland, J. M., Isachen, Y. W., and Hall, L. M., 1984, Stratigraphy and structural geology of the Adirondack Mountains, New York; Review and synthesis: Geological Society of America Special Paper 194, p. 1–55.

Williams, H., and Hatcher, R. D., Jr., 1982, Suspect terranes and accretionary history of the Appalachian orogen: Geology, v. 10, p. 530–536.

Wilshire, H. G., and 6 others, 1988, Mafic and ultramafic xenoliths from volcanic rocks of the western United States: U.S. Geological Survey Professional Paper 1443, 179 p.

Windley, B. F., 1981, Phanerozoic granulites: Geological Society of London Journal, v. 138, p. 745–751.

——, 1984, The evolving continents: Chichester, England, John Wiley and Sons, 399 p.

Wones, D. R., 1980, A comparison between granitic plutons of New England, U.S.A., and the Sierra Nevada batholith, California: Blacksburg Virginia Polytechnic Institute and State University Department of Geological Sciences Memoir 2, p. 123–130.

Wright, J. E., and Fahan, M. R., 1988, An expanded view of Jurassic orogenesis in the western United States Cordillera; Middle Jurassic (pre-Nevadan) regional metamorphism and thrust faulting within an active arc environment, Klamath Mountains, California: Geological Society of America Bulletin, v. 100, p. 859–876.

Zen, E-an, 1983, Exotic terranes in the New England Appalachians; Limits, candidates, and ages; A speculative essay: Geological Society of America Memoir 158, p. 55–81.

Zingg, A., 1983, The Ivrea and Strona-Ceneri Zones (southern Alps, Ticino and N-Italy); A review: Schweizerische mineralogische und petrographische Mitteilungen, v. 63, p. 361–392.

Zingg, A., and Hunziker, J. C., 1988, Thermal evolution of the Ivrea Zone; A discussion based on paragneiss data [abs.]: Università degli studi di Milano, Ricerca Scientifica ed Educazione Permanente, Supplemento 65a, p. 78–79.

Zoback, M. L., and Anderson, R. E., 1983, Style of Basin-Range faulting as inferred from seismic reflection data in the Great Basin, Nevada and Utah: Geothermal Resources Council Special Report 13, p. 363–381.

Zuber, M. T., Parmentier, E. M., and Fletcher, R. C., 1986, Extension of continental lithosphere; A model for two scales of Basin and Range deformation: Journal of Geophysical Research, v. 91, p. 4826–4838.

MANUSCRIPT ACCEPTED BY THE SOCIETY OCTOBER 31, 1988

Printed in U.S.A.

Geological Society of America
Memoir 172
1989

Chapter 32

Pore pressure, seismic velocities, and crustal structure

Nikolas I. Christensen
Department of Earth and Atmospheric Sciences, Purdue University, West Lafayette, Indiana 47907

ABSTRACT

The seismic velocity structure of the Earth's crust is examined in relation to the role played by high pore pressures. Compressional and shear-wave velocities measured at carefully controlled confining and pore pressures show significant decreases with increasing pore pressure. This behavior is important in sedimentary rocks and in crystalline rocks likely to occur at deep crustal levels. Velocities are shown to be a function of effective pressure rather than differential pressure. Similar results are found for crystalline rocks in which gas or water is the pore fluid. Thus, pervasive CO_2 at elevated pore pressures will lower crustal velocities. Low-velocity regions originating from high pore pressures have high Poisson's ratios. Within the crust, dehydration accompanying progressive metamorphism can produce high pore pressures and regions of low velocities if pore fluids are contained. A conceptual model is presented for the continental crust in which midcrustal discontinuities such as the Conrad, where present, separate an overpressured upper region of igneous and metamorphic rocks from underlying dry rocks. Fluids released by lower crustal dehydration are trapped above the Conrad discontinuity. In midcrustal regions, where pore pressures are low due to loss of fluids, progressive metamorphism with depth will be detected as a gradual increase in velocity rather than a sharp discontinuity.

INTRODUCTION

It is well known that many physical properties of rocks depend not only on their mineral constituents but also on the nature of contained pore fluids. Of particular significance in geology and geophysics is the pressure of the interstitial pore fluid. For more than half a century, workers have recognized the importance of pore pressure in rock mechanics. For example, wide acceptance of the role of high pore pressure in overthrusting originated from the theory of Hubbert and Rubey (1959), which showed that fracture strength is proportional to the difference between the total normal stress across a fault and the pore fluid pressure within the fault zone. More recently, pore pressure has been shown to have important tectonic implications in the control of earthquakes (e.g., Raleigh, 1971) and in evaluating nuclear waste burial sites (e.g., Trimmer and others, 1980). In refraction seismology, the interpretation of crustal composition relies to a large extent on comparisons of laboratory-measured velocities with velocities determined at various depths. Knowledge of the extent to which pore pressure will influence velocities is critical for such comparisons. Crustal bright spots and deep laminated reflectors (Fuchs, 1969) may also relate to regions of alternating pore pressures.

Compared with standard measurements of velocity as a function of confining pressure, measurements of velocities under controlled pore pressure are complex. A jacket must be constructed that will isolate the rock from the external confining pressure fluid and contain a port so that internal pore pressure can be varied independently. Such experiments have been undertaken for a limited number of rock types, most of which have been sandstones (Wyllie and others, 1958; Banthia and others, 1965; King, 1966; Christensen and Wang, 1985). Nur and Simmons (1969), by comparing jacketed and unjacketed measurements, showed that high pore pressures significantly lower compressional wave velocities in low-porosity rocks such as granite. Other pore-pressure studies of interest include shear-wave velocity measurements on chalk (Banthia and others, 1965), compressional-wave velocity measurements in limestone and granite (Todd and

Christensen, N. I., 1989, Pore pressure, seismic velocities, and crustal structure, *in* Pakiser, L. C., and Mooney, W. D., Geophysical framework of the continental United States: Boulder, Colorado, Geological Society of America Memoir 172.

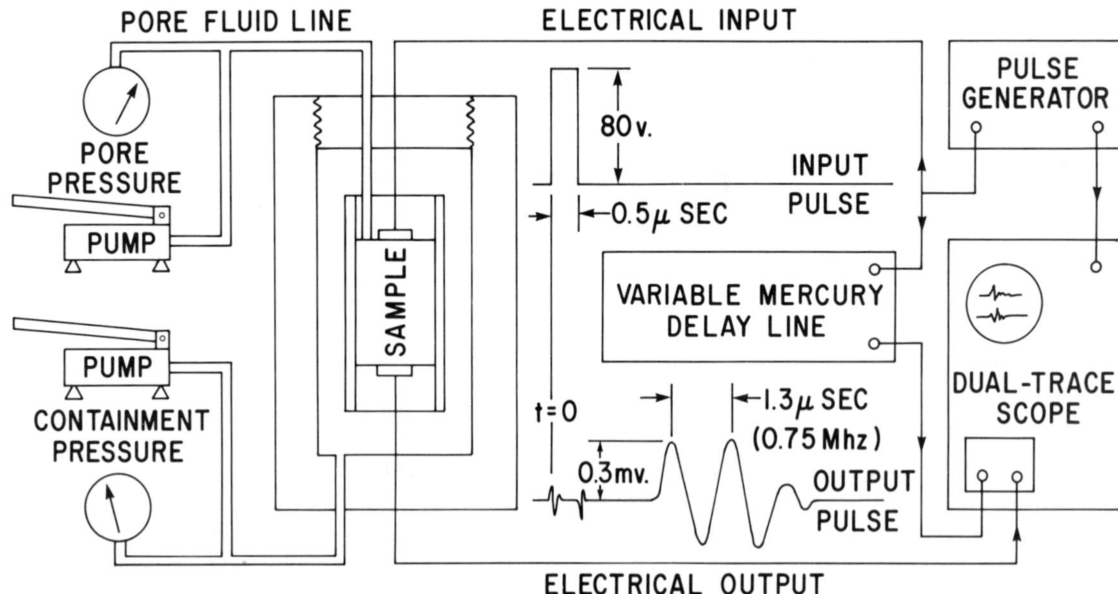

Figure 1. Schematic diagram of electronics and pressure system for velocity measurements at elevated pore and containment pressures.

Simmons, 1972) and both shear- and compressional-wave velocity studies on oceanic basalt and dolerite (Christensen, 1984).

The few papers that have been concerned either directly or indirectly with the influence of pore pressure on seismic velocities and their geophysical consequences differ widely in experimental observations and in interpretation of the data. For example, early velocity measurements by Wyllie and others (1958) for Berea sandstone show that an increase in compressional-wave velocities produced by an increase in confining pressure (P_c) is cancelled by an increase in pore pressure (P_p). Thus velocity was found to be constant for a given differential pressure (defined as $P_c - P_p$). This is in direct conflict with theoretical analyses (Brandt, 1955) and later measurements (King, 1966; Christensen and Wang, 1985), which concluded that an incremental change in pore pressure does not entirely cancel changes in compressional wave velocities produced by a similar change in confining pressure. This and other differences in opinion on this important subject originate, in part, from the lack of detailed experimental data.

An objective of this study has been to obtain detailed velocity data for a wide variety of rocks at carefully controlled confining and pore pressures. Several different rock types, both sedimentary and crystalline, originating from a range of crustal depths, have been chosen for investigation. The measurements provide an understanding of many important aspects of the influence of pore pressure on seismic velocities, thereby demonstrating the significant role played by pore pressure in the seismic velocity structure of the Earth's crust.

EXPERIMENTAL TECHNIQUES

The combined effects of confining and pore pressure on velocities in several principal rock types have been investigated to pressures to 200 MPa (2 kbar). Samples selected were as follows: (1) and (2) relatively pure quartz-rich sandstones with differing porosities, (3) dolomite, (4) a low-porosity basalt, (5) andesite, (6) granite, and (7) lherzolite. In addition, data reported by Christensen (1984) and Christensen and Wang (1985) for (8) a relatively high-porosity oceanic basalt, (9) a greenschist facies meladolerite, and (10) a quartz sandstone with abundant clay clasts and clay cement are included for comparisons with the new measurements. Descriptions of the rocks are given in Table 1.

A schematic diagram illustrating the electrical and hydraulic systems used for the velocity measurements is shown in Figure 1. A 0.5-μsec-wide rectangular pulse of up to 500 V drives the sending transducer on one end of the rock core. Lead-zirconate-titanate (PZT) transducers with 1-MHz resonant frequencies are used for transmitting and receiving compressional waves, whereas similar frequency AC-cut quartz and PZT transducers generate and receive the shear waves, respectively. The elastic wave produced by the sending transducer is transmitted through the sample, detected by the receiving transducer and amplified and displayed on one trace of a dual-trace oscilloscope. The transit times of the elastic waves are measured by superimposing arrivals from the samples and from a calibrated variable-length mercury delay line.

The pore-pressure hydraulic system used distilled water as the pore-pressure medium and operates independent of the confining-pressure hydraulic system. The confining-pressure medium is hydraulic oil. Pressure is generated with hand pumps and measured with Heise bourdon gauges with accuracies of 0.1 percent to 250 MPa (2.5 kbar).

The cylindrical samples are jacketed with aluminum that has longitudinal slots on its inside surface, which are ported to the pore-pressure pumping system, thus exposing the circumferential

TABLE 1. SAMPLE DESCRIPTIONS

Rock	Density (g/cc)	Porosity (%)	Mineralogy (%)	Comments
Mt. Simon Sandstone 1970.2	2.430	9.2	84 quartz, 14 clay, 2 opaque	Obtained from drill core at 1,970.2-ft depth in northern Illinois
Mt. Simon Sandstone 2008.6	2.523	8.0	91 quartz, 8 clay, 1 opaque	Obtained from drill core at 2,008.6-ft depth in northern Illinois
Berea Sandstone	2.253	19.3	70 quartz, 10 clay, 20 lithic fragments, feldspar, and carbonate	Clay-filled pore spaces
Oneota Dolomite	2.755	4.0	93 dolomite, 6 quartz	Obtained from drill core at 539.1-ft depth in northern Illinois
St. Helens Andesite	2.488	3.2	55 plagioclase, 25 amphibole, 3 opaque, 17 groundmass	From massive flow, slightly vesicular
Juan de Fuca Ridge Basalt	2.912	3.1	40 plagioclase, 41 pyroxene, 6 opaque, 13 groundmass	Core of pillow, vesicular
Iceland Basalt	2.788	0.4	55 plagioclase, 37 pyroxene, 5 alteration, 3 opaque	Massive dike, cored at 972.5-ft depth
Oman Dolerite	2.801	1.5	48 plagioclase, 25 orthopyroxene, 5 opaque, 22 alteration	Sheeted dike of Samail ophiolite
Illinois "Granite"	2.648	1.6	35 quartz, 31 K-feldspar, 30 plagioclase, 4 biotite	Coarse grained, from drill core at 4,061-ft depth
Kilborne Hole Lherzolite	3.256	3.9	80 olivine, 12 orthopyroxene, 5 clinopyroxene, 1 opaque, 2 serpentine	Peridotite xenolith, very loose, porous texture typical of many xenoliths

surface of the sample to the controlled pore pressure. Each jacket has a narrow slot parallel to the core axis filled with epoxy that allows for jacket adjustment with changes in confining pressure and assures hydrostatic confining pressure. Electrodes and transducers are placed on the jacketed sample ends and covered with gum rubber tubing to prevent interaction of the confining and pore fluids.

The velocity measurements at elevated temperatures and pressures are made using a two-stage double gas compressor. Argon is the confining pressure medium. Temperatures are produced with a three-zone furnace described by Christensen (1979) and measured with thermocouples located directly on the sample. Temperatures are estimated to be accurate to 1 percent.

VELOCITY DATA

The velocity data as functions of confining and differential pressures are shown in Figure 2. The velocities were measured along paths of increasing and decreasing confining pressure with atmospheric pore pressure, and at constant differential pressure. Data points for each constant differential pressure data set were obtained at random confining pressures. This procedure eliminates possible bias in the determination of the slope of a line of constant differential pressure due to possible hysteresis effects.

The procedure used to obtain velocities along lines of constant differential pressure is illustrated in Figure 3 for the dolomite at $P_d = 5$ MPa (0.05 kbar). After the velocity was measured

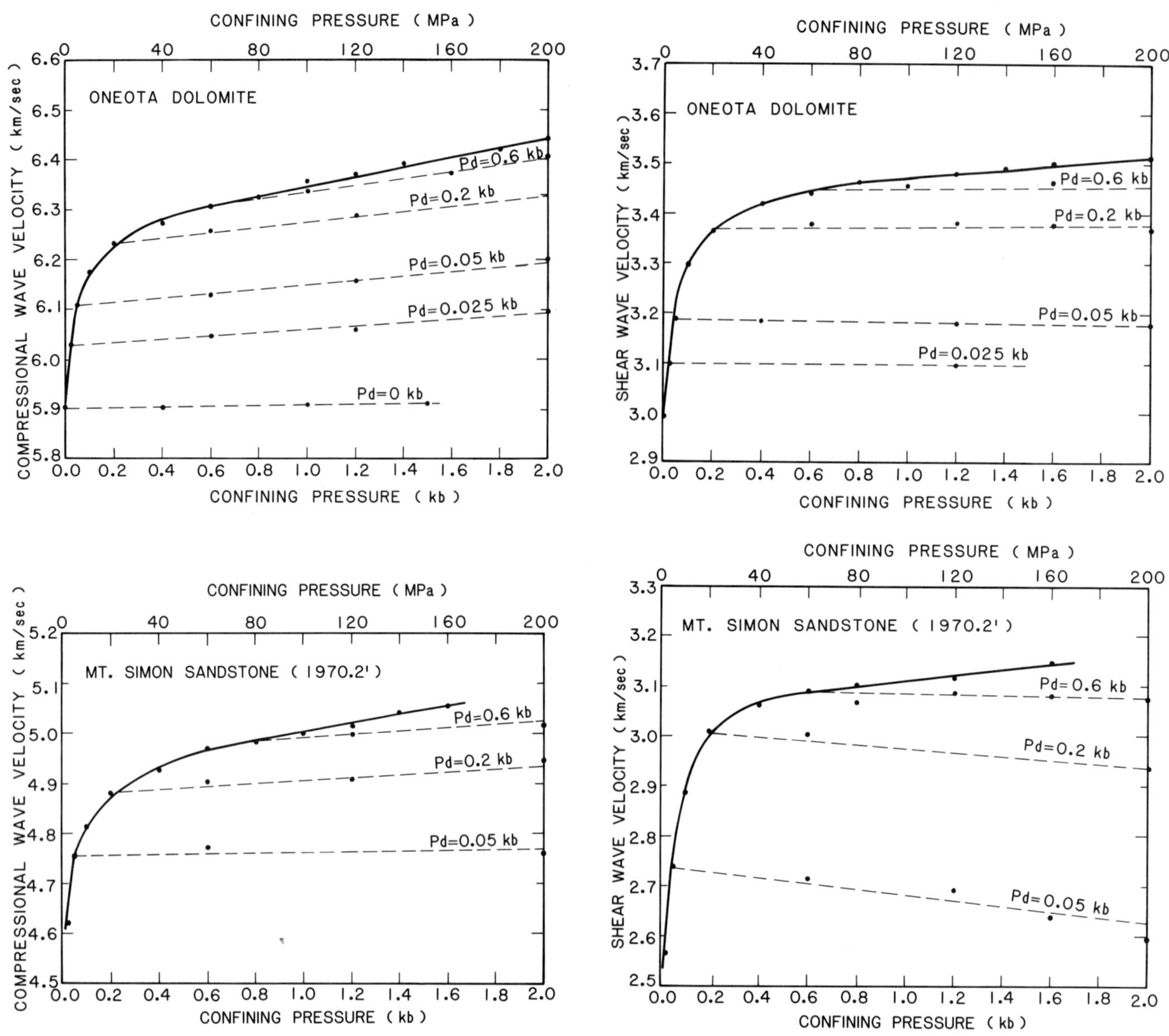

Figure 2a–e. Velocities as a function of confining pressure and differential pressure (P_d).

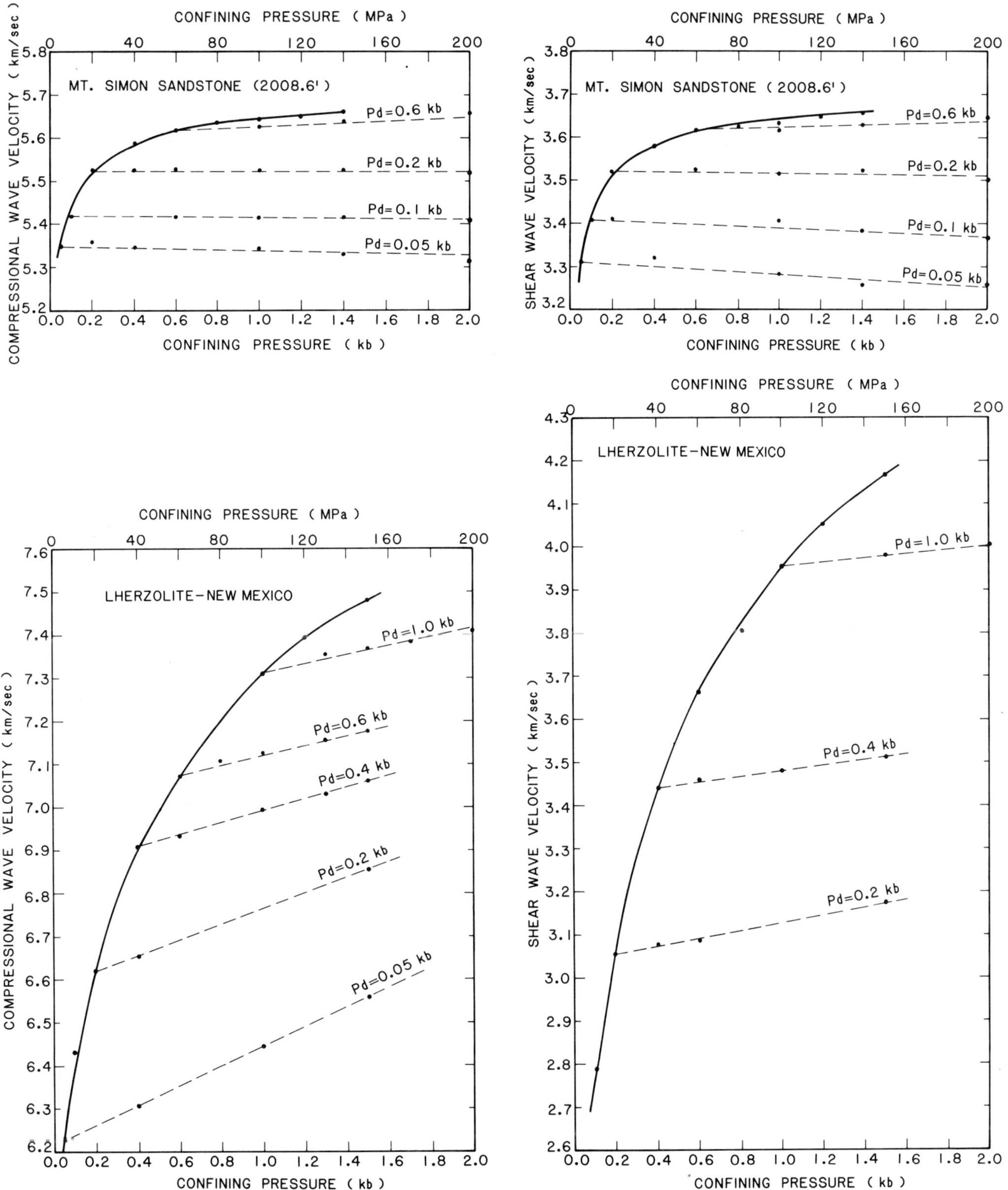
CONFINING PRESSURE (MPa)
MT. SIMON SANDSTONE (2008.6')
Pd=0.6 kb
Pd=0.2 kb
Pd=0.1 kb
Pd=0.05 kb
COMPRESSIONAL WAVE VELOCITY (km/sec)
CONFINING PRESSURE (kb)
SHEAR WAVE VELOCITY (km/sec)
LHERZOLITE-NEW MEXICO
Pd=1.0 kb
Pd=0.4 kb

Figure 2b.

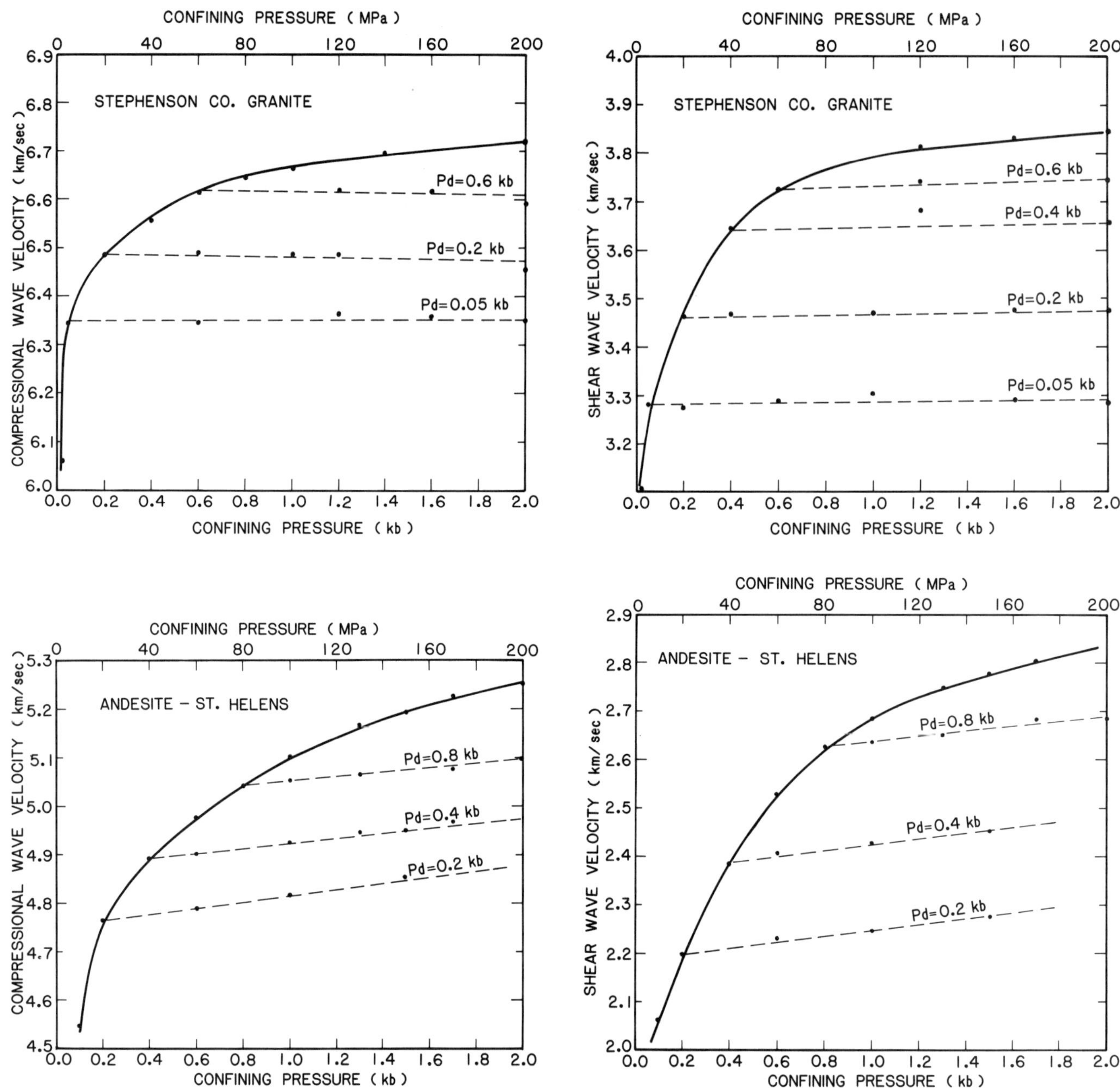
CONFINING PRESSURE (MPa)
STEPHENSON CO. GRANITE
COMPRESSIONAL WAVE VELOCITY (km/sec)
Pd=0.6 kb
Pd=0.2 kb
Pd=0.05 kb
CONFINING PRESSURE (kb)
CONFINING PRESSURE (MPa)
STEPHENSON CO. GRANITE
SHEAR WAVE VELOCITY (km/sec)
Pd=0.6 kb
Pd=0.4 kb
Pd=0.2 kb
Pd=0.05 kb
CONFINING PRESSURE (kb)
CONFINING PRESSURE (MPa)
ANDESITE - ST. HELENS
COMPRESSIONAL WAVE VELOCITY (km/sec)
Pd=0.8 kb
Pd=0.4 kb
Pd=0.2 kb
CONFINING PRESSURE (kb)
CONFINING PRESSURE (MPa)
ANDESITE - ST. HELENS
SHEAR WAVE VELOCITY (km/sec)
Pd=0.8 kb
Pd=0.4 kb
Pd=0.2 kb
CONFINING PRESSURE (kb)

Figure 2c.

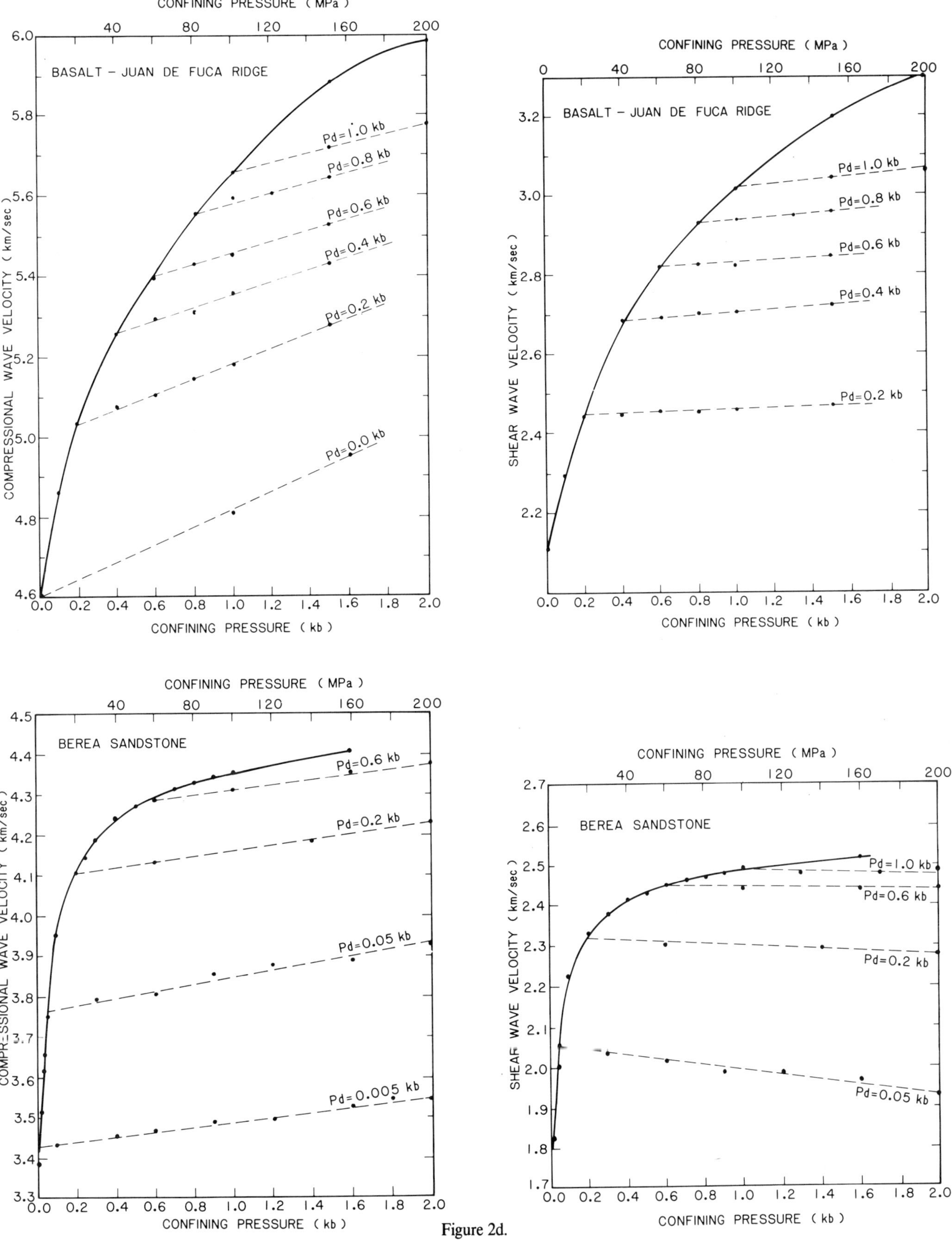

CONFINING PRESSURE (MPa)
BASALT - JUAN DE FUCA RIDGE
COMPRESSIONAL WAVE VELOCITY (km/sec)
Pd=1.0 kb
Pd=0.8 kb
Pd=0.6 kb
Pd=0.4 kb
Pd=0.2 kb
Pd=0.0 kb
CONFINING PRESSURE (kb)
CONFINING PRESSURE (MPa)
BASALT - JUAN DE FUCA RIDGE
SHEAR WAVE VELOCITY (km/sec)
Pd=1.0 kb
Pd=0.8 kb
Pd=0.6 kb
Pd=0.4 kb
Pd=0.2 kb
CONFINING PRESSURE (kb)
CONFINING PRESSURE (MPa)
BEREA SANDSTONE
COMPRESSIONAL WAVE VELOCITY (km/sec)
Pd=0.6 kb
Pd=0.2 kb
Pd=0.05 kb
Pd=0.005 kb
CONFINING PRESSURE (kb)
CONFINING PRESSURE (MPa)
BEREA SANDSTONE
SHEAR WAVE VELOCITY (km/sec)
Pd=1.0 kb
Pd=0.6 kb
Pd=0.2 kb
Pd=0.05 kb
CONFINING PRESSURE (kb)

Figure 2d.

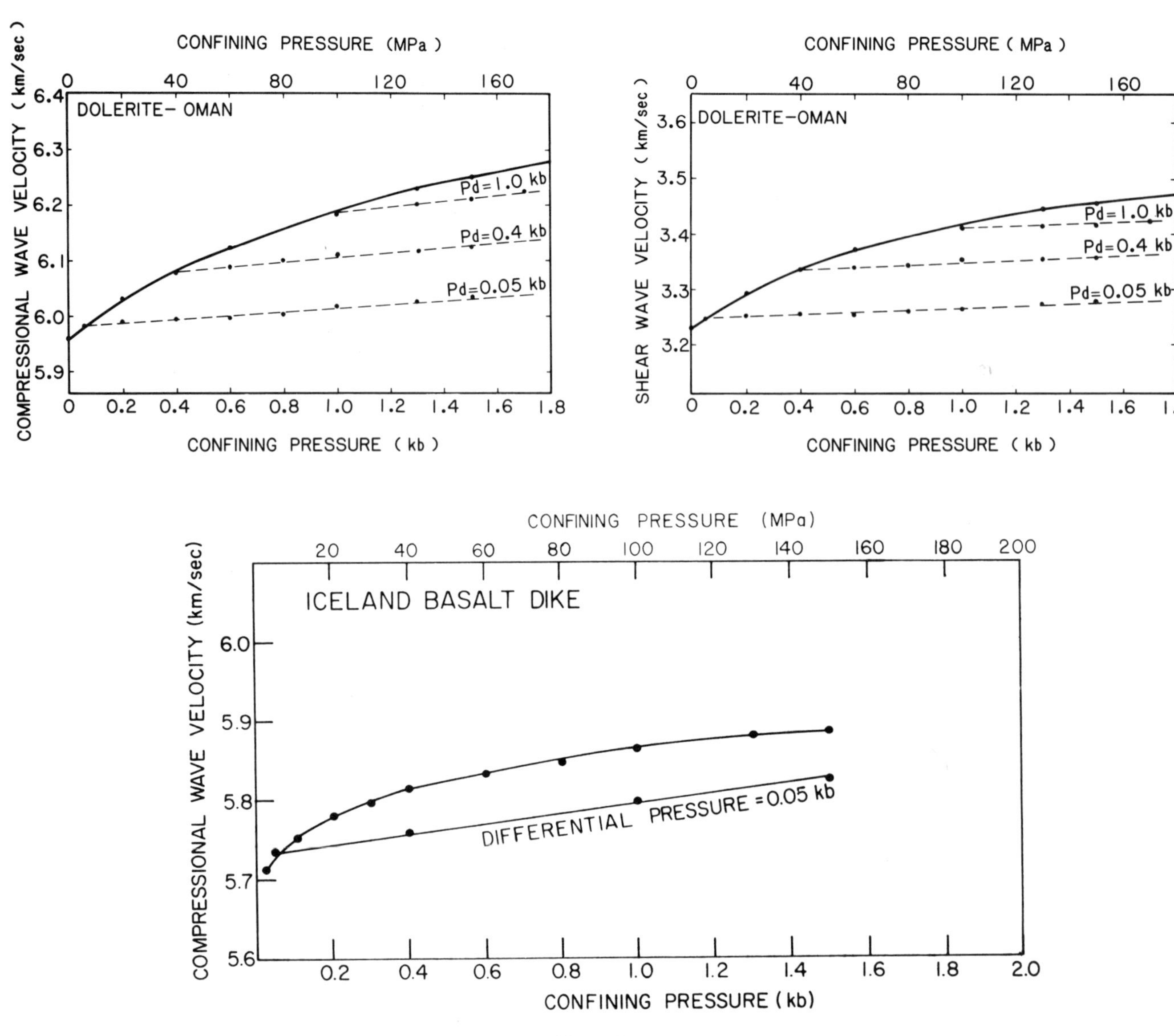

Figure 2e.

at $P_c = 200$ MPa (point 1, Fig. 3), pressures were lowered to $P_c = 100$ MPa and $P_p = 95$ MPa. The initial measured velocity was 5.98 km/s (point 2) and at equilibrium the velocity increased to 6.10 km/sec. Velocity behavior in going to point 5 and back to point 3 is also illustrated.

Compressional and shear-wave velocities were measured on the same samples during separate runs. Runs for a single rock lasted between 20 and 48 days, and pressures were maintained for an average of approximately 24 hr prior to each velocity measurement. The repeatability of a measured velocity is better than 0.5 percent; the accuracy is 1 percent.

VELOCITIES IN TERMS OF EFFECTIVE STRESS

To a first approximation, confining pressure (P_c) and pore pressure (P_p) are expected to have opposite, but roughly equal, effects on velocities. That is, the increase in velocity produced by an increase in confining pressure should be similar to the decrease in velocity produced by an equivalent increase in pore pressure. For this to hold, the lines of constant differential pressure (P_d) in Figure 2 will have zero slopes, i.e., $(\partial V_p/\partial P_c)P_d = 0$.

Several theoretical papers on wave propagation in fluid-saturated rocks (Brandt, 1955; Biot, 1962; Biot and Willis, 1957; Geertsma, 1957; Fatt, 1958) concluded that rather than differential pressure, velocities should be functions of an effective pressure, P_e, defined as $P_e = P_c - nP_p$, where n is an empirical factor usually less than 1. A value of n less than 1 implies that a pore pressure increment does not entirely cancel a confining-pressure increment. Experimental studies usually found the effective pressure law to hold, with values of n less than 1 common for a variety of rocks (Banthia and others, 1965; Todd and Simmons,

TABLE 2. VALUES OF n FOR $P_p = 0$

Rock		Pd (MPa)							
		0.5	5	10	20	40	60	80	100
Berea Sandstone	Vp	0.99	0.98	–	0.95	–	0.70	–	–
	VS	–	1.02	–	1.03	–	0.04	–	1.16
Mt. Simon Sandstone (1920)	Vp	–	1.00	–	0.92	–	0.76	–	–
	Vs	–	1.03	–	1.06	–	1.07	–	–
Mt. Simon Sandstone (2008)	Vp	–	1.02	1.02	1.02	–	1.02	–	–
	Vs	–	1.02	1.02	1.02	–	1.02	–	–
Oneota Dolomite	Vp	1.00	0.98	–	0.87	–	0.68	–	–
	Vs	–	1.00	–	0.99	–	0.88	–	–
Mt. St. Helens Andesite	Vp	–	0.98	–	0.94	0.90	–	0.85	–
	Vs	–	–	–	0.96	0.93	–	0.85	–
Juan de Fuca Basalt	Vp	0.98	–	–	0.86	0.84	0.81	0.81	–
	Vs	–	–	–	0.98	0.96	0.95	0.93	–
Iceland Basalt	Vp	–	0.87	–	–	–	–	–	–
	Vs	–	–	–	–	–	–	–	–
Illinois Granite	Vp	–	1.00	–	0.99	–	0.99	–	–
	Vs	–	1.00	–	0.99	0.99	0.95	–	–
Oman Dolerite	Vp	–	0.91	–	–	0.81	–	–	0.63
	Vsq–	–	0.95	–	–	0.87	–	–	0.81
Kilborne Lherzolite	Vp	–	0.94	–	0.89	0.87	0.86	–	0.79
	Vs	–	–	–	0.96	0.95	–	–	0.91

1972; Dominico, 1977; Christensen, 1984). Christensen and Wang (1985) found that, for equal increments of increased confining and pore pressure in Berea sandstone, V_p increased and V_s decreased. Thus, n is less than 1 for compressional waves but greater than 1 for shear waves. It was postulated that this behavior originated from the presence of high-compressibility clay that lines pores within the quartz framework of the sandstone.

Values of n can be calculated from the expression (Todd and Simmons, 1972; Christensen, 1984):

$$n = 1 - \frac{(\partial V/\partial P_p)_{P_d}}{(\partial V/\partial P_d)_{P_p}}.$$

Values calculated along the curve $P_p = 0$ are given in Table 2. The slopes $(\partial V/\partial P_c)_{P_d}$ of the dashed lines in Figure 2 were used for the $(\partial V/\partial P_p)_{P_d}$ term, since $\partial P_p = \partial P_c$ for constant P_d. The $(\partial V/\partial P_d)_{P_p}$ term was approximated from a curve fit to the $P_p =$ 0 data.

The n values presented in Table 2 show that the effective pressure law holds for the rocks included in this study. For compressional waves, n is usually less than unity and decreases with increasing differential pressure. The three sandstones (Fig. 4), as well as the dolomite, show relatively large decreases in n with increasing confining pressure, which is likely related to the nature of their porosity. Similarly, shear-wave values of n for the crystalline rocks are less than unity and decrease with increasing differential pressure. The three sandstones again are anomalous, with shear-wave n values greater than 1, which remain constant or increase with increasing differential pressure.

APPLICATIONS TO CRUSTAL SEISMIC STRUCTURE

As interest in crustal fluids has increased, a large number of papers have appeared dealing either directly or indirectly with the distribution of water in the crust and the geochemical and geophysical consequences (e.g., O'Neil and Hanks, 1980; Etheridge and others, 1984; Angevine and Turcotte, 1983; Walder and Nur, 1984; Christensen, 1984; Newmark and others, 1985). Several studies dealing primarily with the sedimentary portion of the

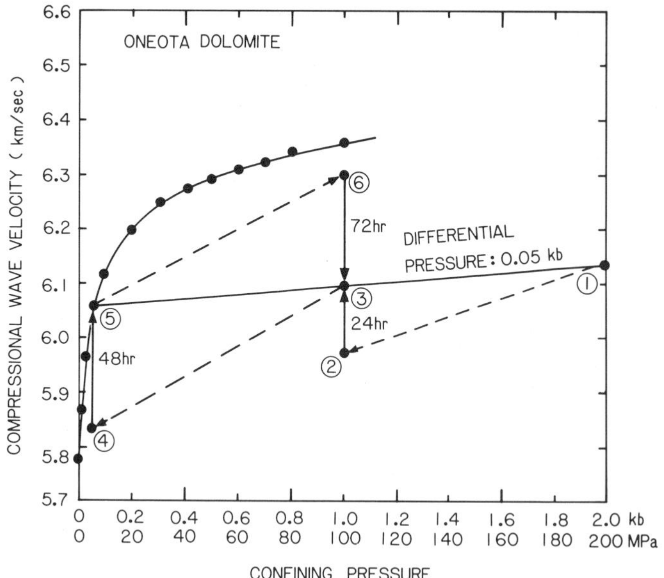

Figure 3. Velocity measurements along a line of constant differential pressure illustrating equilibrium time (see text).

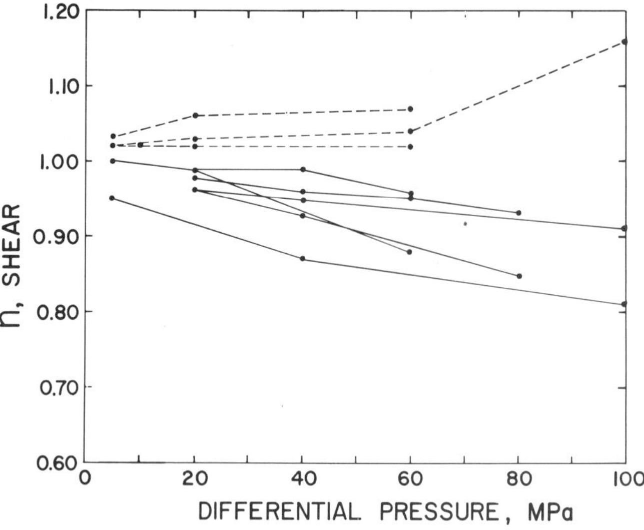

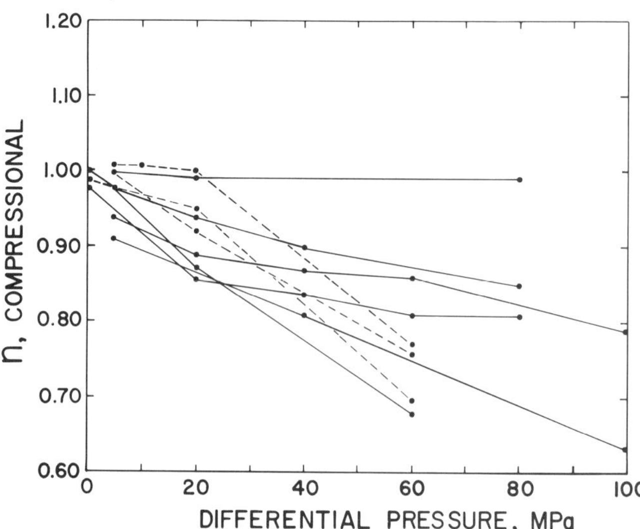

Figure 4. Values of n at various differential pressures (see text for definition of n). Sandstones are shown as dashed lines.

Earth's crust have suggested possible mechanisms that could produce high pore pressures in fluid-saturated rocks. For example, tectonic compression (Hubbert and Rubey, 1959; Fertl, 1976), porosity reduction by precipitation of minerals (Levorsen, 1954), and mineral reactions that release water such as the smectite to illite or gypsum to anhydrite transformations (Burst, 1969; Heard and Rubey, 1966) are all likely to produce excess pore pressures in sedimentary sections. It is apparent that the creation of high pore pressures in sedimentary formations will produce anomalously low seismic velocities. Thus seismic field observations of low velocities or atypically low rates of increase in velocity with depth could easily be attributed to high pore pressure. However, consideration of temperature and changes in mineralogy with depth will complicate this interpretation. If related to pore pressure, the velocities will depend on the magnitudes of the confining and pore pressures based on the effective-pressure law. The response will also be related to porosity: velocities of higher porosity rocks will show larger changes than low-porosity rocks for a given change in pore pressure (Fig. 5).

Many theoretical studies dealing with the influence of porosity on the physical properties of rocks have found that pore aspect ratios are critical parameters (e.g., Walsh, 1965). Crystalline rocks have porosities consisting primarily of microcracks with high aspect ratios, whereas the sandstones and dolomite porosities are in the form of low-aspect ratio pores. Thus it is not surprising that the influence of pore pressure on velocities in igneous and metamorphic rocks is quite different from sedimentary rocks. The results plotted in Figure 5 show that for a given porosity, the velocity response to a change in pore pressure is much greater for crystalline rocks. Even for small porosities, pore pressure markedly affects velocities of the igneous and metamorphic rocks.

As pore pressure is increased to confining pressure, changes in velocity are not linear functions of pore pressure. Initial increases in pore pressure lower velocities only slightly. Maximum decreases in velocity occur as pore pressure approaches confining pressure. This is illustrated in Figure 6 for four crystalline rocks with a range of porosities.

Since increasing pore pressure at constant confining pressure has a greater effect on shear velocities than compressional velocities (Fig. 2), V_p and V_s ratios decrease and Poisson's ratios increase (Fig. 7). Thus, V_p and V_s measurements in overpressured sedimentary strata can give estimates of the magnitudes of the pore pressures, provided reasonable constraints are placed on porosities and mineralogies. For example, the two Mt. Simon sandstones, because of their high quartz contents, have extremely low Poisson's ratios (0.14 and 0.18) at elevated confining pressures and low pore pressures (Fig. 7). However, if pore pressures approach confining pressures, Poisson's ratios increase to values common for many quartz-free rocks at low pore pressures (0.22 and 0.28). Similarly, the 0.29 Poisson's ratio of the dolomite at elevated confining pressure and low pore pressure increases to 0.33 as pore pressures become equivalent to confining pressures.

The applications of these measurements to interpreting seismic velocities in the oceanic crust have been discussed by Christensen (1984). Within the upper few hundred meters of the volcanics that constitute oceanic layer 2, velocities are likely to be strongly influenced by pore pressure. For fractures connected to the surface, pore pressures will be hydrostatic. A major question, still unanswered, is to what depths porosity and water exist within the oceanic crust. Recent deep drilling in the Pacific at DSDP Site 504B found that porosity is extremely low at about 1-km depth below the sea floor (e.g., Salisbury and others, 1985). Deep drilling in Icelandic volcanics, however, recovered cores with large vugs lined with quartz, epidote, and carbonate at depths of 1,900 m (Mehegan and others, 1982).

Several major questions arise concerning pore fluids in the continental crust. What are the vertical and lateral distributions of pore fluid? To what depths are pore fluids no longer connected through crack systems to the surface? What geologic processes are responsible for trapping pore fluids at depth? Are aqueous fluids or CO_2 more important as lower continental crustal pore fluids? The answers to these questions have important implications for the seismic properties of the crust as well as other physical properties.

Researchers concerned with the electrical conductivity structure of the continental crust for many years have been concerned with the abundance and nature of crustal pore fluids. The role of pore water in producing high electrical conductivity is well documented from laboratory studies (e.g., Brace, 1972; Olhoeft, 1981). Thus, aqueous fluids are generally considered to be the principal source of high crustal conductivity (e.g., Connerney and others, 1980; Olhoeft, 1981; Jiracek and others, 1983).

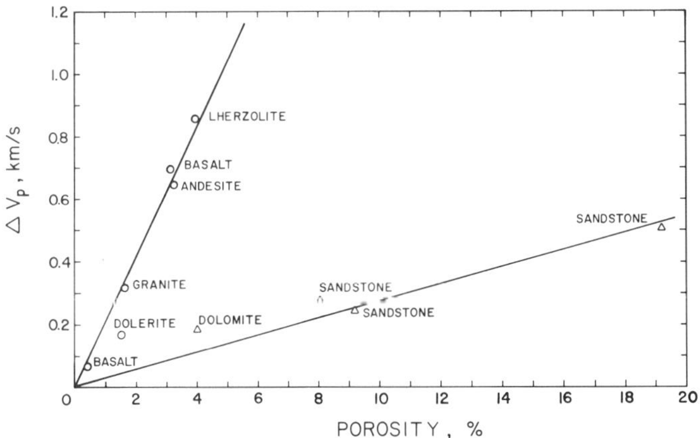

Figure 5. Compressional wave velocity decreases at 100 MPa confining pressure produced by an increase in pore pressure of 95 MPa.

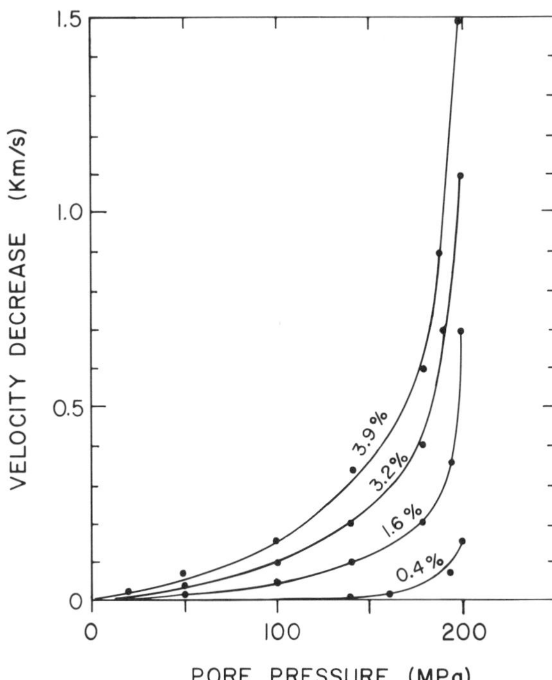

Figure 6. Velocity decreases at 200 MPa confining pressure with increasing pore pressure for basalt (porosity = 0.4 percent), dolerite (porosity = 1.6 percent), andesite (porosity = 3.2 percent), and lherzolite (porosity = 3.9 percent).

Several authors have cited evidence for high electrical conductivity of the lower continental crust (e.g., Hyndman and Hyndman, 1968; Shankland and Ander, 1983; Jones, 1987). Meissner (1986), on the other hand, has cited regions where the lower crust apparently has low electrical conductivity; he emphasized problems in resolution of vertical structure using electromagnetic investigations. Critical to the present study are the observations of Olhoeft (1981), which indicate that hydrous minerals such as amphibole and mica have relatively low conductivities. Thus, lower crustal high-conductivity regions likely contain aqueous fluids that may occur at elevated pore pressures. Fyfe and others (1978) suggested that, under lower crustal conditions, fluid pressures prevail, often approaching lithostatic pressure.

Within the continental crust, progressive metamorphism proceeds by a sequence of reactions involving dehydration (Bowen, 1940; Thompson, 1955; Turner and Weiss, 1963). Common examples include the metamorphism of pelitic rocks, where shale is converted to slate, phyllite, mica schist, and granulite, and the reactions of mafic greenschist through amphibolite to mafic granulite. Reactions involving hydrous silicates such as micas, chlorites, and amphiboles are thus likely to be important in providing H_2O pore fluid at significant crustal depths (e.g., Burst, 1969; Meissner, 1986). In these regions the assumption that the pore pressure of

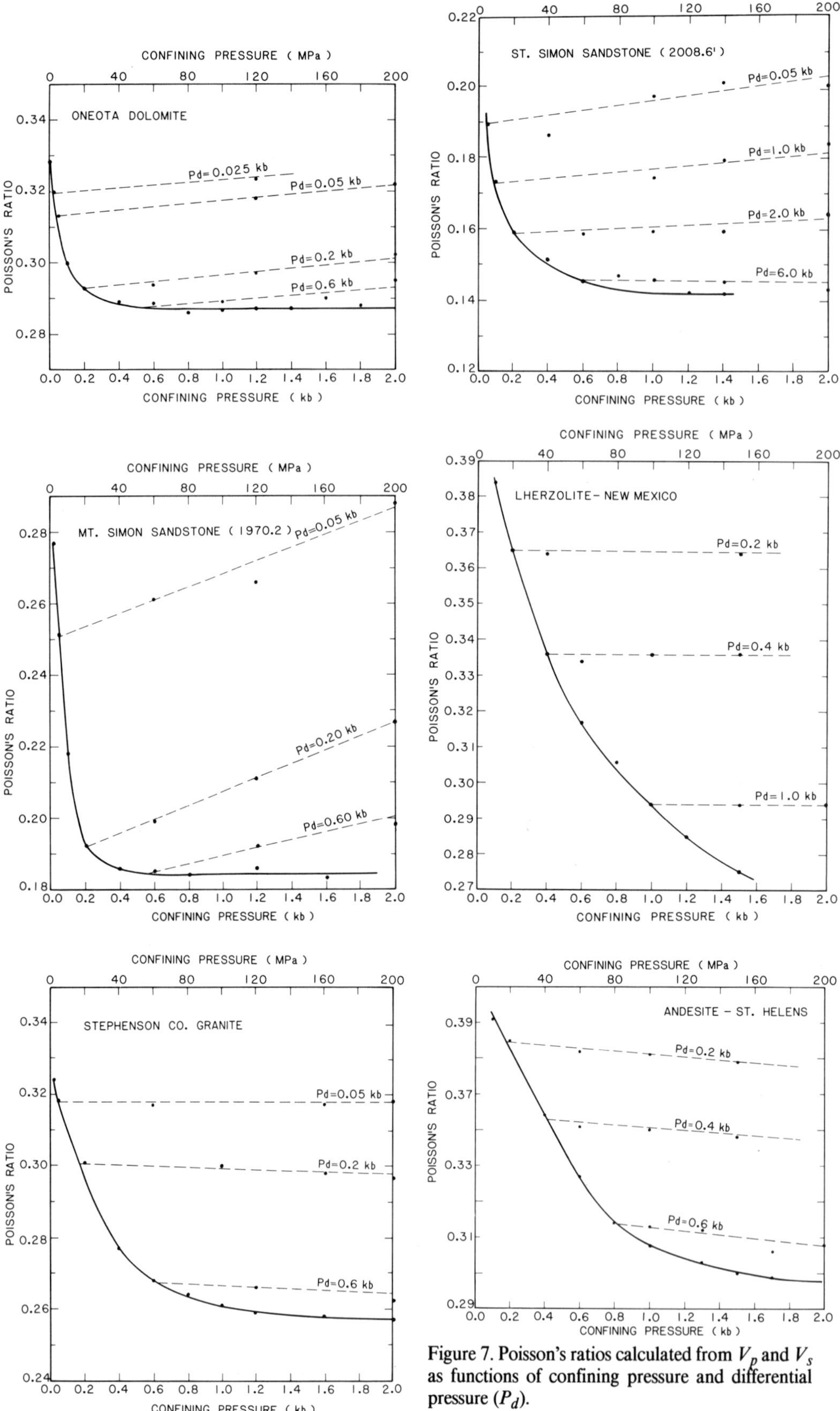

Figure 7. Poisson's ratios calculated from V_p and V_s as functions of confining pressure and differential pressure (P_d).

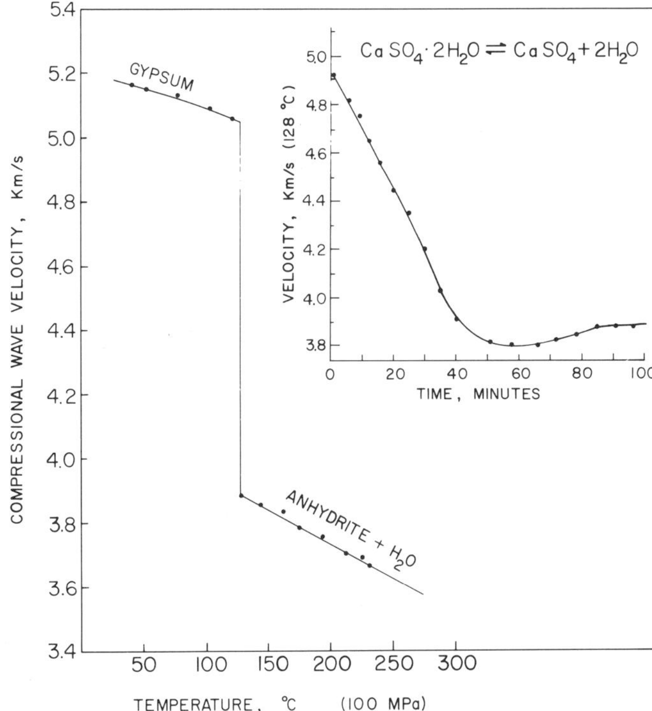

Figure 8. Changes in velocity associated with the gypsum-anhydrite transition. Inset shows velocity as a function of time at 128°C.

H_2O is equivalent to load pressure must, in many instances, be incorrect, since overpressures are likely where permeabilities are low.

The gypsum-anhydrite reaction:

$$\underset{\text{gypsum}}{CaSO_4 \cdot 2H_2O} \rightleftharpoons \underset{\text{anhydrite}}{CaSO_4} + \underset{\text{water}}{2H_2O}$$

illustrates the lowering of seismic velocities that accompanies crustal metamorphic reactions involving pore-pressure buildup during dehydration. The transition temperature is complicated by many factors, including the presence of impurities, the original gypsum porosity, and grain size, pressure, and often the formation of the semihydrate bassanite (Sonnenfeld, 1984). Gypsum has a compressional-wave velocity of approximately 5.2 km/sec at confining pressures high enough to eliminate most crack effects on velocities, whereas anhydrite has velocities of approximately 6.0 km/sec (Birch, 1960). Thus if H_2O is allowed to escape during the dehydration process, velocities will increase rapidly once the rotation proceeds to anhydrite. The anhydrite and water together occupy about 11 percent more volume than the original gypsum. However, if the released pore fluid is contained, much different results are observed due to fluid-pressure buildup (Fig. 8). In this experiment, a monominerallic aggregate of gypsum was completely enclosed in a copper jacket. The sample was placed under a hydrostatic confining pressure of 100 MPa, and velocities were measured with increasing temperature. Velocities first decreased slightly due to the effect of increasing temperature on the gypsum aggregate. At approximately 128°C, velocities decrease significantly due to pore pressure associated with the dehydration process. During the reaction, which occurred over a time interval of 60 min (Fig. 8), the signal from the receiving transducer remained on the oscilloscope screen, and a steady decrease in velocity was observed. After completion of the reaction, a further increase in temperature provided velocity measurements in the assemblage anhydrite + H_2O. A puncture in the copper jacket terminated the experiment. The recovered sample consisted of anhydrite; it is not known if the semihydrate formed during the experiment.

Experiments showing major velocity decreases associated with metamorphic dehydration are particularly important in view of recent findings from deep continental drilling. In 1970 the Soviet Union began a major drilling project in the northeast part of the Baltic shield on the Kola Peninsula above the Arctic Circle (Fig. 9). In 1984 a book published in Russian (Kozlovskiy, 1984) summarized in detail the drilling results to a depth of 11.6 km, nearly one-third of the total continental crustal thickness in this region. The upper 6,847 m of the hole penetrated Proterozoic tholeiitic metabasalt, metapyroxenite, metaperidotite, metatuffs, metasandstones, metaarkoses, and carbonates. Below 6,847 m, the drilling recovered Archean metamorphics consisting of mica-plagioclase gneisses in the upper section and biotite-plagioclase gneisses and amphibolites in the lower section (Fig. 10). Of particular significance was the finding of porous, water-saturated fractured rocks in the lower section of the drillhole that are not connected hydraulically with the overlying zone. At all depths, fractures are coated with various mineral fillings, indicating that they were not formed by the drilling process. In general, porosity was found to *increase* with depth, and logging velocities *decreased* with depth.

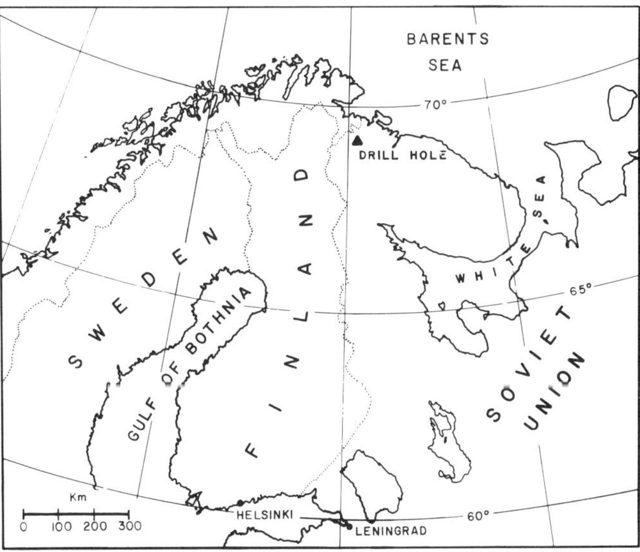

Figure 9. Location of Kola drillhole.

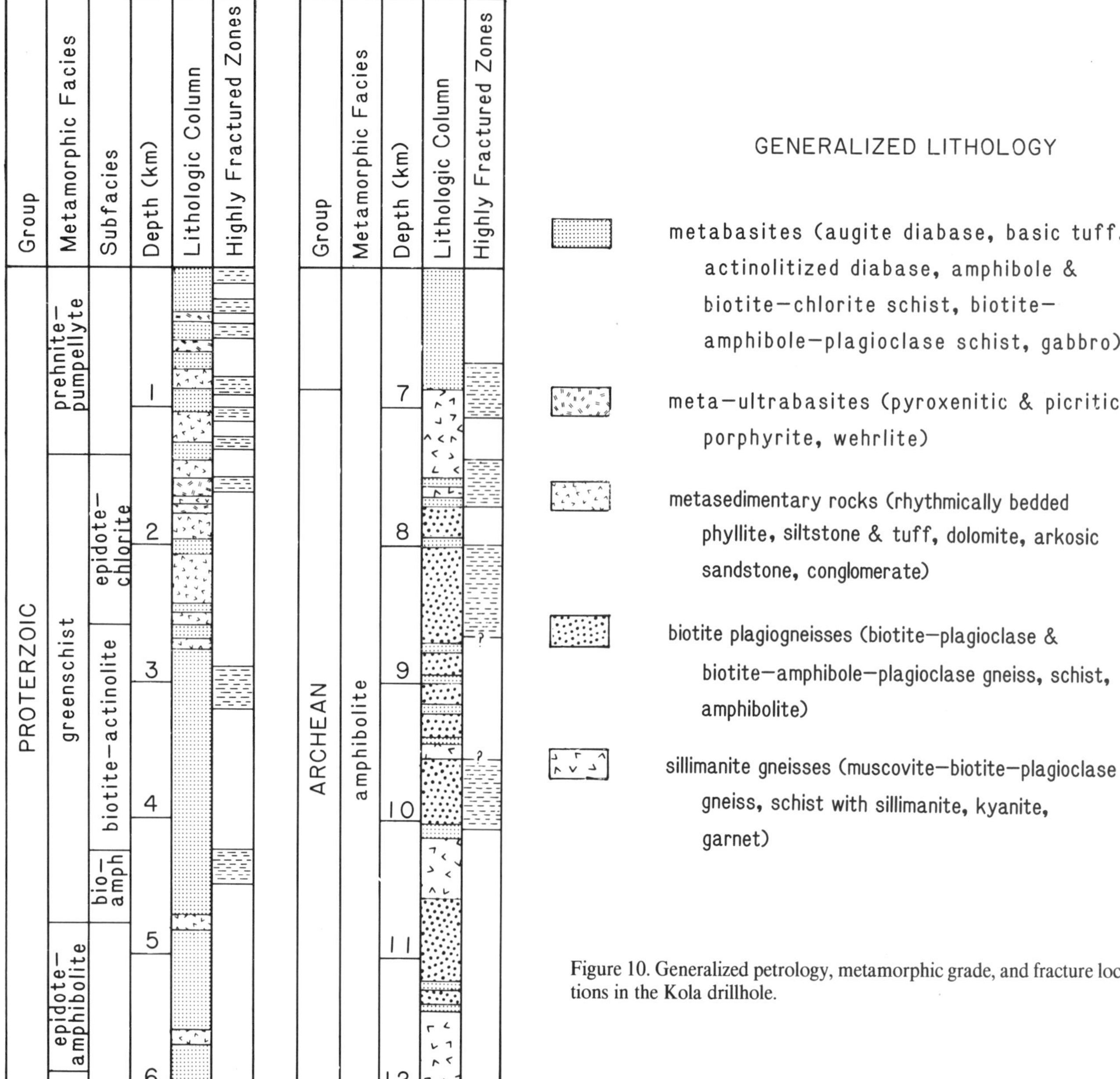

Figure 10. Generalized petrology, metamorphic grade, and fracture locations in the Kola drillhole.

The zones of increased fracturing are nonuniform in depth and separated by nearly impermeable regions. Chemical analyses and textural observations suggest that the origin of water at depths below 4.5 km was mineral dehydration formed during metamorphism. Calculations based on dehydration reactions associated with the greenschist to epidote-amphibolite facies transition show that, in the depth interval of 4.0 to 6.8 km, total water losses were 5.5×10^7 g/m^3, equivalent to 6.7 percent of the original rock volume (Kozlovskiy, 1984). This water loss, if confined, would produce pore pressures of approximately 3 GPa, equivalent to expected mantle pressures at a 100-km depth.

Laboratory experiments on sediments have shown that high pore pressures in rocks containing nonaqueous fluid also lower velocities (Banthia and others, 1965). A preliminary study of compressional-wave velocities in basalt with helium as a pore-pressure saturant shows that crystalline rocks behave in a similar manner (Fig. 11). Note that at equivalent confining and pore pressures, velocities are higher for the water-saturated sample. Also, lines of constant differential pressure for He saturation have lower slopes. Thus, n values are closer to unity with He as a pore-pressure medium. Calculated values of n along the $P_p = 0$ curve range from 0.792 at $P_d = 30$ MPa to 0.940 at $P_d = 80$ MPa.

Studies of mineral equilibria in rocks from exposed lower crustal granulite facies terranes often find evidence for low H_2O activities. One theory to explain these low activities hypothesizes that granulites form from a major influx of CO_2 that dilutes H_2O (Schuiling and Kreulen, 1979; Newton and others, 1980). This is supported by observations of CO_2–rich fluid inclusions in granulites. However, other probable causes of low H_2O activity at lower crustal depths include preferential partitioning of H_2O into partial melt and recrystallization of dry rocks (Valley and O'Neil, 1984). The possibility of the presence of CO_2, at least locally, in the lower continental crust raises speculation as to whether elevated CO_2 pore pressure could play an important role in lowering seismic velocities. Of significance, abundant lower crustal CO_2, because of its nonpolar nature, would not produce a region of high electrical conductivity (Kay and Kay, 1981).

CONCLUSIONS

It is becoming apparent that porosity containing aqueous fluids occurs at considerable depths within the crystalline rocks of the Earth's crust. The interpretation of seismic velocities in these regions requires knowledge of the in-situ pore pressure in addition to confining pressure and temperature. Laboratory studies demonstrate that the application of pore pressure decreases velocities in igneous and metamorphic rocks to values usually attributed to high crustal temperatures, extensive fracturing, or even partial melting.

Velocities in crustal sedimentary sections will be anomalously low if pore pressures and porosities are high. High Poisson's ratios are characteristic of these overpressured regions. Within the oceanic crust, pore pressures within cracks connected to the overlying sea-water column cause lower velocities and may be responsible for some observations of horizontal variability in the velocities of seismic layer 2.

If the findings of the Soviet Union Kola drilling project are typical of continental crustal regions, it is likely that much of the velocity structure of the upper continental crust is directly affected by pore pressure. If connected to the surface, water-saturated regions in the upper few kilometers of the crust are likely to have lowered velocities due to hydrostatic pore pressures. At greater depths, where porous fractured metamorphic rocks are not connected hydraulically with near-surface fractures, pore pressures generated by metamorphic processes and possibly tectonic forces are likely to produce a dramatic lowering of seismic velocities. Dehydration with accompanying high pore pressure resulting from progressive metamorphism is likely to produce hydraulic fracturing. In some deeper crustal regions, high-grade metamorphism to the granulite facies has produced relatively anhydrous coarse-grained rocks, and fluids rich in CO_2 at elevated pore pressure may be important.

Refraction crustal studies have reported a variety of crustal structures at intermediate depths (e.g., Meissner, 1986). Velocity inversions are sometimes observed (e.g., Berry and Fuchs, 1973); a major first-order discontinuity, often referred to as the Conrad (e.g., Steinhart and Meyer, 1961), is common in many regions; and strong positive velocity gradients are frequently reported (e.g., Pavlenkova, 1979). One of many explanations for the origin of low-velocity zones is high pore pressure that arises from dehydration reactions likely involving mica and amphibole. As was discussed earlier, the low-velocity zones will be characterized by high Poisson's ratios and high electrical conductivities. Rather than originating from a major lithologic change, the Conrad discontinuity may in many regions separate an overpressured upper crust from a "dry" lower crustal region. Velocities will increase abruptly across the discontinuity, and electrical conductivity will decrease. In midcrustal regions, where pore pressures are low due to loss of fluids, progressive metamorphism with increasing depth, if isochemical, will be observed as a positive velocity gradient.

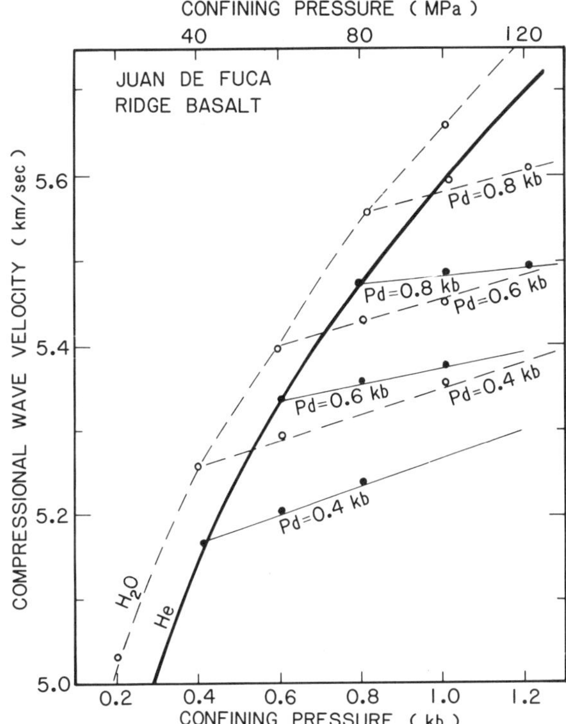

Figure 11. Compressional wave velocities in basalt with water (dashed curves) and helium (solid curves) as pore pressure media.

ACKNOWLEDGMENTS

This study was supported by the Office of Naval Research Contract N00014-84-K-0207.

REFERENCES CITED

Angevine, C. L., and Turcotte, D. L., 1983, Porosity reduction by pressure solution; A theoretical model for quartz arenites: Geological Society of America Bulletin, v. 94, p. 1129–1134.

Banthia, B. S., King, M. S., and Fatt, I., 1965, Ultrasonic shear-wave velocities in rocks subjected to simulated overburden pressure and internal pore pressure: Geophysics, v. 30, p. 117–121.

Berry, M. J., and Fuchs, K., 1973, Crustal structure of the Superior and Grenville provinces of the northeastern Canadian Shield: Seismological Society of America Bulletin, v. 63, p. 1393–1432.

Biot, M. A., 1962, Generalized theory of acoustic propagation in porous media: Acoustical Society of America Journal, v. 34, p. 1254–1271.

Biot, M. A., and Willis, D. G., 1957, The elastic coefficients of the theory of consolidation: Journal of Applied Mechanics, v. 24, p. 594–601.

Birch, F., 1960, The velocity of compressional waves in rocks to 10 kilobars, part 1: Journal of Geophysical Research, v. 65, p. 1083–1102.

Bowen, N. L., 1940, Progressive metamorphism of siliceous limestone and dolomite: Journal of Geology, v. 48, p. 225–274.

Brace, W. F., 1972, Pore pressure in geophysics, *in* Heard, H. C., Borg, I. Y., Carter, N. L., and Raleigh, C. B., eds., Flow and fracture of rocks: American Geophysical Union Monograph 16, p. 265–273.

Brandt, H., 1955, A study of the speed of sound in porous granular media: Transactions of the American Society of Mechanical Engineers, v. 22, p. 479–486.

Burst, J. F., 1969, Diagenesis of Gulf Coast clayey sediments and its possible relation to petroleum migration: American Association of Petroleum Geologists Bulletin, v. 53, p. 73–93.

Christensen, N. I., 1979, Compressional wave velocities in rocks at high temperatures and pressures, critical thermal gradients, and crustal low-velocity zones: Journal of Geophysical Research, v. 84, p. 6849–6857.

—— , 1984, Pore pressure and oceanic crustal seismic structure: Geophysical Journal of the Royal Astronomical Society, v. 79, p. 411–424.

Christensen, N. I., and Wang, H. F., 1985, The influence of pore pressure and confining pressure on dynamic elastic properties of Berea sandstone: Geophysics, v. 50, p. 207–213.

Connerney, J.E.P., Nekut, A., and Kuckes, A. F., 1980, Deep crustal conductivity in the Adirondacks: Journal of Geophysical Research, v. 85, p. 2603–2614.

Domenico, S. N., 1977, Elastic properties of unconsolidated porous sand reservoirs: Geophysics, v. 42, p. 1339–1368.

Etheridge, M. A., Cox, S. F., Wall, V. J., and Vernon, R. H., 1984, High fluid pressures during regional metamorphism and deformation; Implications for mass transport and deformation mechanisms: Journal of Geophysical Research, v. 89, p. 4344–4358.

Fatt, I., 1958, Compressibility of sandstones at low to moderate pressures: American Association of Petroleum Geologists Bulletin, v. 42, p. 1924–1957.

Fertl, W. H., 1976, Abnormal formation pressures: New York, Elsevier, 382 p.

Fuchs, K., 1969, On the properties of deep crustal reflectors: Zeitschrift fur Geophysik, v. 35, p. 133–149.

Fyfe, W. S., Price, N. J., and Thompson, A. B., 1978, Fluids in the Earth's crust: New York, Elsevier, 383 p.

Gerrtsma, J., 1957, The effect of fluid pressure decline on volumetric changes of porous rocks: Society of Petroleum Engineers, v. 210, p. 331–340.

Heard, H. C., and Rubey, W. W., 1966, Tectonic implications of gypsum dehydration: Geological Society of America Bulletin, v. 77, p. 741–760.

Hubbert, M. K., and Rubey, W. W., 1959, Role of fluid pressure in mechanics of overthrust faulting: Geological Society of America Bulletin, v. 70, p. 115–166.

Hyndman, R. D., and Hyndman, D. W., 1968, Water saturation and high electrical conductivity in the lower crust: Earth and Planetary Science Letters, v. 4, p. 427–432.

Jiracek, G. R., Gustafson, E. P., and Mitchell, P. S., 1983, Magnetotelluric results opposing magma origin of crustal conductors in the Rio Grande Rift: Tectonophysics, v. 94, p. 299–326.

Jones, A. G., 1987, MT and reflection; An essential combination: Royal Astronomical Society Geophysical Journal, v. 89, p. 7–18.

Kay, R. W., and Kay, S. M., 1981, The nature of the lower continental crust; Inferences from geophysics, surface geology, and crustal xenoliths: Reviews of Geophysics and Space Physics, v. 19, p. 271–297.

King, M. S., 1966, Wave velocities in rocks as a function of overburden pressure and pore fluid saturants: Geophysics, v. 31, p. 56–73.

Kozlovskiy, Ye. A., 1984, Kol'skaya Sverkhglubokaya: Moscow, Nedra, 490 p.

Levorsen, A. I., 1954, Geology of petroleum: San Francisco, W. H. Freeman, 703 p.

Mehegan, J. M., Robinson, P. T., and Delaney, J. R., 1982, Secondary mineralization and hydrothermal alteration in the Reydarfjordur drill core, eastern Iceland: Journal of Geophysical Research, v. 87, p. 6511–6524.

Meissner, R., 1986, The continental crust; A geophysical approach: Orlando, Florida, Academic Press, 426 p.

Newmark, R. L., Anderson, R. N., Moos, D., and Zoback, M. D., 1985, Structure, porosity, and stress regime of the upper oceanic crust; Sonic and ultrasonic logging of DSDP Hole 504B: Tectonophysics, v. 118, p. 1–42.

Newton, R. C., Smith, J. V., and Windley, B. F., 1980, Carbonic metamorphism, granulites, and crustal growth: Nature, v. 288, p. 45–50.

Nur, A., and Simmons, G., 1969, The effect of saturation on velocity in low porosity rocks: Earth and Planetary Science Letters, v. 7, p. 183–193.

Olhoeft, G. R., 1981, Electrical properties of granite with implications for the lower crust: Journal of Geophysical Research, v. 86, p. 931–936.

O'Neil, J. R., and Hanks, T. C., 1980, Geochemical evidence for water-rock interaction along the San Andreas and Garlock fault zones of California: Journal of Geophysical Research, v. 85, p. 6286–6292.

Pavlenkova, N. I., 1979, Generalized geophysical model and dynamic properties of continental crust: Tectonophysics, v. 59, p. 381–390.

Raleigh, C. B., 1971, Earthquake control at Rangely, Colorado: EOS Transactions of the American Geophysical Union, v. 52, p. 344.

Salisbury, M. H., Christensen, N. I., Becker, K., and Moos, D., 1985, The velocity structure of layer 2 at Deep Sea Drilling Project Site 504 from logging and laboratory experiments, *in* Anderson, R. N., and others, eds., Initial reports of the Deep Sea Drilling Project: Washington, D.C., U.S. Government Printing Office, v. 83, p. 529–539.

Schuiling, R. D., and Kreulen, R., 1979, Are thermal domes heated by CO_2-rich fluids from the mantle?: Earth and Planetary Science Letters, v. 43, p. 298–302.

Shankland, T. J., and Ander, M. E., 1983, Electrical conductivity, temperatures, and fluids in the lower crust: Journal of Geophysical Research, v. 88, p. 9475–9484.

Sonnenfeld, P., 1984, Brines and evaporites: Orlando, Florida, Academic Press, 613 p.

Steinhart, J. S., and Meyer, R. P., 1961, Explosion studies of continental structure: Carnegie Institution of Washington Publication 622, 409 p.

Thompson, J. B., 1955, The thermodynamic basis for the metamorphic facies concept: American Journal of Science, v. 253, p. 65–103.

Todd, T., and Simmons, G., 1972, Effect of pore pressure on the velocity and compressional waves in low-porosity rocks: Journal of Geophysical Research, v. 77, p. 3731–3743.

Trimmer, D., Bonner, B., Heard, H. C., and Duba, A., 1980, Effect of pressure and stress on water transport in intact and fractured gabbro and granite: Journal of Geophysical Research, v. 85, p. 7059–7071.

Turner, F. J., and Weiss, L. E., 1963, Structural analysis of metamorphic tectonites: New York, McGraw-Hill, 545 p.

Valley, J. W., and O'Neil, J. R., 1984, Fluid heterogeneity during granulite facies metamorphism in the Adirondacks; Stable isotope evidence: Contributions to Mineralogy and Petrology, v. 85, p. 158–173.

Walder, J., and Nur, A., 1984, Porosity reduction and crustal pore pressure development: Journal of Geophysical Research, v. 89, p. 11539–11548.

Walsh, J. B., 1965, The effect of cracks on the compressibility of rocks: Journal of Geophysical Research, v. 70, p. 381–390.

Wyllie, M.R.J., Gregory, A. R., and Gardner, G. H., 1958, An experimental investigation of factors affecting elastic wave velocities in porous media: Geophysics, v. 23, p. 459–493.

Manuscript Accepted by the Society October 31, 1988

Printed in U.S.A.

Geological Society of America
Memoir 172
1989

Chapter 33

Geophysical framework of the continental United States; Progress, problems, and opportunities for research

Walter D. Mooney
U.S. Geological Survey, MS 977, 345 Middlefield Road, Menlo Park, California 94025
L. C. Pakiser
U.S. Geological Survey, MS 966, Box 25046, Denver Federal Center, Denver, Colorado 80225

ABSTRACT

Significant progress has been made over the past five decades in determining the geophysical framework of the continental United States. Highlights include detailed maps of gravity and aeromagnetic anomalies, heat flow, crustal thickness, seismicity, state of stress, and paleomagnetic pole positions. Important tectonic insights have come from earthquake studies, and from knowledge of lithospheric structure derived from seismic reflection, refraction/wide-angle reflection, surface-wave, and teleseismic data. Additional major advances in lithospheric geophysics will depend on four key factors: the reduction of uncertainties in the measurement and interpretation of geophysical data, the widespread application of coincident geophysical methods in concert with geological investigations, the collection of a more uniform continent-scale data base for all geophysical measurements, and the investigation of topical geophysical questions regarding the physical state and properties of the lithosphere. The impracticality of repeating most geophysical field measurements introduces poorly known, but likely large, uncertainties. Since most measurements are not repeated, high priority must be given to the reduction and quantification of uncertainties in measurements and interpretations. The most productive future investigations, in terms of resolution and minimum uncertainties in interpretation, will be those that apply different geophysical methods along identical profiles or areas, and that include geological investigations as a vital ingredient. Important gaps remain in our knowledge of the geophysical framework of the United States on a continent-wide scale, including the deep conductivity structure, the nature of the Moho discontinuity, the structure of the subcrustal lithosphere, and the depth of the lithosphere/asthenosphere boundary. Most transition zones separating geologic or physiographic provinces are poorly studied, yet these zones are likely to be the locations of the most profound changes in the physical properties of the lithosphere. Application of coincident geophysical techniques is needed to study these transition zones. Several topical geophysical questions warrant special emphasis in the future. These questions include the rheology of the crust and subcrustal lithosphere; the distribution, composition, and abundance of fluids in the crust; the genesis and evolution of the Moho; the origin of crustal conductivity zones and deep crustal reflections; the evidence for seismic anisotropy; and the short-term prediction of earthquakes.

Mooney, W. D., and Pakiser, L. C., 1989, Geophysical framework of the continental United States; Progress, problems, and opportunities for research, *in* Pakiser, L. C., and Mooney, W. D., Geophysical framework of the continental United States: Boulder, Colorado, Geological Society of America Memoir 172.

INTRODUCTION

Substantial progress has been made in the past half of this century in the study of the geophysical framework of the continental United States; continued research is certain to yield important new insights. In this chapter we summarize the most important advances described in greater detail in other chapters in this volume. We then identify several methodological improvements needed to foster continued progress in geophysical research, and recommend certain tectonic features for high priority in future investigations. Finally, we discuss what we believe are the most important topical geophysical questions to be investigated in the coming years.

PROGRESS IN GEOPHYSICAL STUDIES OF THE CONTINENT AND ITS MARGINS

The 32 other contributions in this volume summarize progress over the past 50 years in studies of the geophysical framework of the continental United States and its margins. The compilation of continent-scale geophysical maps must certainly be included among the highlights of this progress. These maps include: Bouguer and isostatic gravity anomalies, aeromagnetic anomalies, surface heat flow, crustal thickness and average crustal P-wave seismic velocity, uppermost-mantle P-wave seismic velocity, earthquake epicenters and seismic moment release, state of stress, and paleomagnetic pole positions.

When interpreted with geologic maps, these geophysical maps provide a firm basis for the study of continental structure and evolution. Numerous examples of these interpretations are found throughout this volume, ranging from isostatic compensation mechanisms to the correlation of present-day seismicity with faults.

Many advances in geophysics do not lend themselves to presentation in maps, but are no less significant. Seismic reflection and refraction/wide-angle reflection profiles have recently provided detailed cross sections of crustal and uppermost mantle structure throughout the United States (Braile and others, this volume). Seismic reflection profiles in particular have provided unprecedented details on the structure of the crust, and have led to such diverse insights as the confirmation of thin-skin tectonics in the Appalachians, the existence of deeply penetrating detachment faults, and a highly reflective lower crust and a young, reworked Moho in the Basin and Range (Allmendinger and others, 1987; Phinney and Roy-Chowdhury, this volume; Smithson and Johnson, this volume). The application of seismic reflection profiling of the continental lithosphere has spread worldwide and continues to have a profound impact on our understanding of continental structure and tectonics (Fig. 1).

Seismic models of the upper mantle, derived from body and surface waves, have confirmed earlier suggestions that the upper-mantle structure—and particularly the structure of the seismic low-velocity layer—is drastically different beneath the tectonically active western Cordillera, as compared with the stable central and eastern portions of the continent. Models for the deeper mantle have demonstrated the existence of long-wavelength heterogeneous structure to a depth of 700 km (Iyer and Hitchcock, this volume).

Gravity and aeromagnetic data have made sustained contributions to geophysical studies of the continental crust for more than 100 yr (e.g., Airy, 1855; Pratt, 1855; Hayford and Bowie, 1912). Recent progress has come not only from more closely spaced gravity and aeromagnetic measurements, but also from improved data processing, display, and interpretation (Blakely and Connard, this volume). A high point in potential-field studies was the publication, in color, of both the Bouguer gravity anomaly map (Society of Exploration Geophysicists, 1982), and of the composite magnetic anomaly map of the conterminous United States (U.S. Geological Survey and Society of Exploration Geophysicists, 1982). To enhance the display of gravity anomalies caused by shallow bodies of geological significance, an isostatic gravity map was calculated from the Bouguer gravity anomaly map of the United States (Simpson and others, 1986). Filtered gravity maps (Hildenbrand and others, 1982) have revealed long-wavelength anomalies that have been interpreted to reveal density variations in the upper mantle, such as the pronounced thinning of the asthenosphere east of the Rocky Mountains (Kane and Godson, this volume).

Paleomagnetic studies have contributed greatly to understanding the evolution of the continents, particularly by providing evidence for displaced terrains. Recent contributions include recognition of large northward displacements (often accompanied by clockwise rotations) of crustal fragments in the western Cordillera since 125 Ma (Beck, this volume). Displaced terranes have also been identified in older orogens, particularly in the Appalachians. The importance of remagnetization of many formations has been recognized recently, and paleomagnetic data have since been reevaluated, particularly for pre-Cenozoic rocks (Van der Voo, this volume). This reevaluation is providing more accurate models for the Phanerozoic evolution of continents.

Perhaps no research has seen as much recent progress as studies of the thermal regimes of the continental crust (Morgan and Gosnold, this volume; Furlong and Chapman, 1987). More than 2,000 published heat-flow values are now available for the United States, with the highest density of measurements in the western portion of the continent. These data define a single basic thermal regime to the east of the Rocky Mountains, and a highly variable regime in the western United States. Both shallow and deep processes may be inferred from the thermal data. Anomalously low heat-flow measurements are caused in some localities by downward convection of ground water. However, in the Sierra Nevada and northwest Pacific the low heat flow is attributed to downward convection associated with subduction. High heat flow has been correlated with a broad range of parameters, including surface elevation, lithospheric thickness, upper-mantle seismic velocity, a brittle-ductile transition in the crust, electrical conductivity, depth of the Curie isotherm, magmatism, and seis-

GLIMPCE Line F: Lake Superior

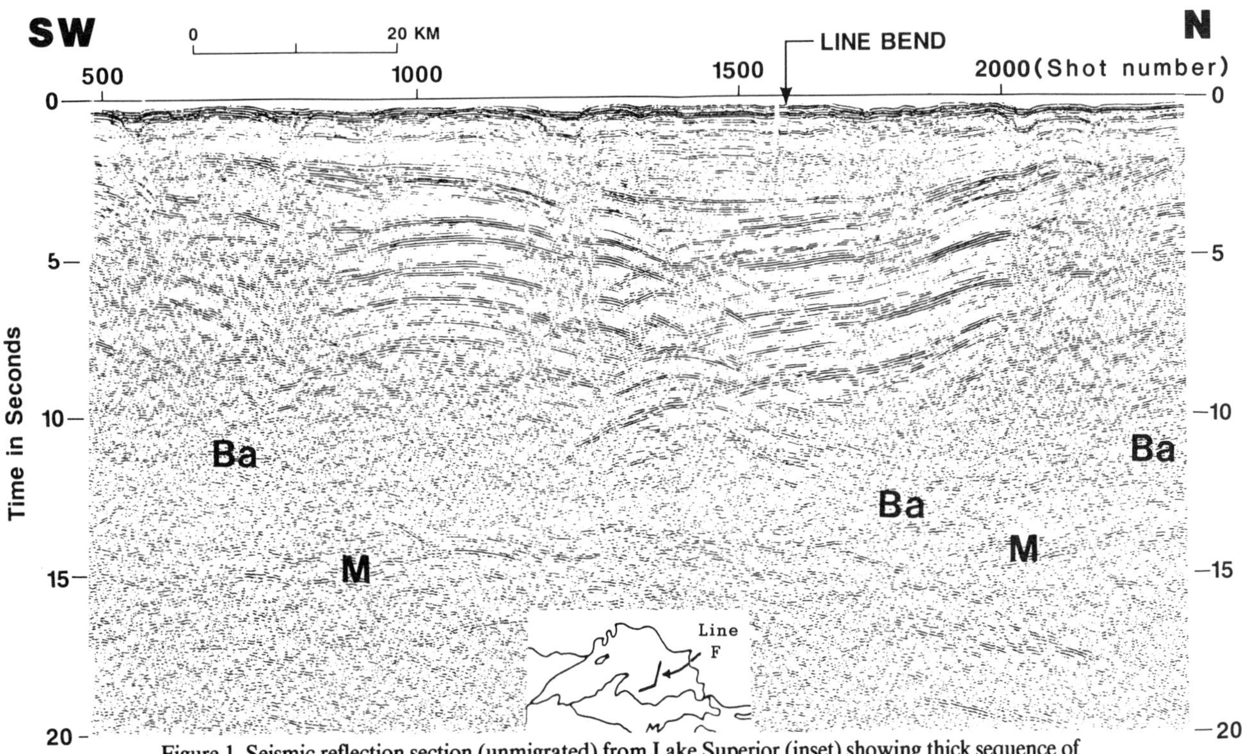

Figure 1. Seismic reflection section (unmigrated) from Lake Superior (inset) showing thick sequence of reflections within the Midcontinent (Keweenawan) rift system. Behrendt and others (1988) interpreted these reflections to arise from volcanic rocks and postvolcanic and interbedded sedimentary rocks that extend to depths as great as 32 km (10.5-sec reflection time). Ba = Archean and/or Proterozoic basement; M = Moho. Data were collected by the Great Lakes International Multidisciplinary Program on Crustal Evolution (GLIMPCE) of the U.S. Geological Survey and the Geological Survey of Canada.

mic reflectivity (Meissner, 1986; Klemperer, 1987; Keller, part 2, this volume; Morgan and Godson, this volume; Blakely, 1988).

A substantial body of information has been obtained regarding the electrical structure of the crust and upper mantle beneath the continental United States. These data show that regions of recent or active extension in the western United States are characterized by high conductivity at shallow depth (Keller, parts 1 and 2, this volume). West of the Basin and Range and east of the Great Plains, depths to the conductive zone in the crust are deeper. The available data for the eastern United States show consistently lower conductivity in the crust and shallow mantle than in the West. Deep electrical surveys indicate that profound lateral changes in the electrical properties of the mantle are present as deep as 350 km. Laboratory data indicate that the most likely explanations for high conductivity are high heat flow causing partial melting, and the presence of significant amounts of water or graphite.

Studies of the continental margin have taken enormous strides since the 1960s when magnetic lineaments, which were critical in developing plate tectonics, were investigated in the Atlantic and Pacific Northwest. Recent progress has come from all aspects of marine geophysics, including multichannel reflection profiles obtained in the 1970s on the Atlantic margin (Trehu and others, this volume). These profiles were the first to reveal the internal structure and stratigraphy of the great marginal basins there. At greater depth, crustal modifications due to stretching and/or igneous intrusions during the final stages of rifting have been identified. These modifications include the development of a thick, seismic high-velocity lower crustal layer in the 30-km-thick crust along the Atlantic coast. By contrast, the crustal thickness along the Pacific margin (and the adjoining 200 km of continent) is almost everywhere 25 km or less. This thin crust may have developed from thinned preexisting crust or from preserved and uplifted former oceanic crust (Couch and Riddihough, this volume).

Great progress has been made in studies of the seismicity of the continental United States. Many (but not all) earthquakes in the western United States can be correlated with faults, and earthquakes in the east can generally be correlated with reactivated pre-Cenozoic faults (Dewey and others, this volume). Variations in the depth of seismicity fit a general model of plate interactions and heat flow, with high heat flow correlating with shallow seismicity. The important concept of a "characteristic earthquake" for a given fault has been developed based on the observation that the segmentation of faults and along-strike variations in strength predispose earthquakes on each section of a fault to have a characteristic rupture length and displacement. Studies of historical seismicity demonstrate that recent inactivity on a fault does not preclude future activity (Dewey and others, this volume).

Recently, seismicity data have been combined with geologic and borehole data to define the tectonic stress field of the continental United States. Four main stress provinces have been defined (Zoback and Zoback, this volume): (1) central U.S. to Atlantic margin, east to northeast compression; (2) uplifted western Cordillera, east-west extension; (3) Pacific Northwest, northeast compression; and (4) San Andreas fault, north-south compression (within a more northeasterly regional compression province). Correlation of stress fields with other geophysical parameters has begun to yield important insights into the forces that deform the crust (Zoback and others, 1987). The recent introduction of the Global Positioning System (GPS) and very long baseline interferometry (VLBI) into studies of crustal deformation will result in improved determinations of the crustal strain and the mechanical properties of the lithosphere (Lichten and Border, 1987).

The integration of new geophysical data with geologic information has substantially refined our understanding of the petrologic and chemical composition of the continental crust and upper mantle (Fountain and Christensen, this volume). Exposed deep crustal sections, xenoliths, and laboratory measurements of the seismic, electrical, and magnetic properties of rocks at elevated pressures and temperatures provide the necessary information to correlate field measurements with composition at depth. Recent results indicate a pronounced vertical zonation by metamorphic grade in the crust, with granulite-facies rocks of various compositions making up the lower crust (Hamilton, this volume). The seismic anisotropy of the continental mantle provides evidence for the dominance of olivine in the upper mantle, and the seismic velocities within the crust-mantle transition zone are best explained by a mixture of metamorphic lower crustal rocks and mantle-derived intrusive rocks. The overall picture to emerge from these studies is one of a more heterogeneous and variable crust compared with the earlier ideas of a granitic upper crust and basaltic lower crust.

MAJOR UNCERTAINTIES IN THE INTERPRETATION OF GEOPHYSICAL DATA

Progress in any field of science depends on steady improvement in the accuracy and resolution of measurements and interpretation. This is particularly true in lithospheric geophysics because it is often costly to duplicate a measurement and is usually impossible to check an interpretation directly by drilling. Achieving the dual goals of improved data and more reliable interpretation within the inevitable logistical and financial limitations requires considerable ingenuity.

Studies of the electrical properties of the lithosphere are a case in point. Of the many electrical methods, magnetotelluric (MT) methods appear to hold the greatest promise for deep crustal and lithospheric measurements (Keller, part 1, this volume). Yet, considerable uncertainty remains regarding the validity or uniqueness of published interpretations of MT data, particularly one- or two-dimensional interpretations that neglect three-dimensional effects. There is an urgent need to reduce uncertainties in electrical studies because the determination of the conductivity structure of the lithosphere is an invaluable constraint on its composition and physical state (Jones, 1987; Christensen, this volume). Improvements will come in the form of significantly increased accuracy and spatial density of field measurements, and from the use of larger computers capable of calculating full three-dimensional conductivity models.

Despite their long-standing and extensive use, there are some major uncertainties in the interpretation of seismic reflection and refraction/wide-angle reflection data (Phinney and Roy-Chowdhury, this volume; Smithson and Johnson, this volume; Mooney, this volume). Uncertainties in seismic-reflection interpretation arise from a variety of sources, including the difficulty in recording high-quality land data owing to poor signal coupling, large near-surface static effects, crooked profile geometry, and signal scattering and attenuation in an absorptive and heterogeneous Earth. As a result, seismic reflectors may appear and disappear in a seismic section even if they are continuous in the subsurface. Furthermore, it is generally difficult to determine the true subsurface geometry and extent of reflectors owing to the problems of migrating deep seismic reflections (Warner, 1987) and the lack of three-dimensional control. In a few cases, the seismic response of very different structures may appear virtually the same, and if the impedance contrasts are insufficient, some important structures may not be imaged with *any* seismic technique. Detachment faults, for example, are generally difficult to follow to great depth. The question of what might be *missing* from our seismic models must be kept in mind.

There are, in addition, major uncertainties regarding the origin and nature of deep crustal reflectors. To address these uncertainties, shear-wave reflection profiles are needed to learn more about the reflecting interfaces, as are truly three-dimensional seismic surveys to determine reflector geometry.

Uncertainties in the interpretation of seismic refraction/wide-angle reflection data are in part due to the limited resolution obtained with the customary large source and receiver spacings, and the limited frequency bandwidth of large-offset arrivals. Likewise, the assumptions of horizontal continuity of refractors and reflectors, and the reliance on seismic ray theory, sometimes lead to oversimplified layered models of crustal structure. Many published crustal models are based on unreversed profiles, and the importance of out-of-plane and multiple arrivals is rarely well understood. Three-dimensional control on structure is generally lacking in seismic refraction surveys, and seismic velocity estimates may be biased by lack of subsurface path reversal or undetected seismic anisotropy. For example, McCarthy and others (1987) reported significant apparent lower crustal anisotropy in a survey of the southern Basin and Range. Future crustal studies

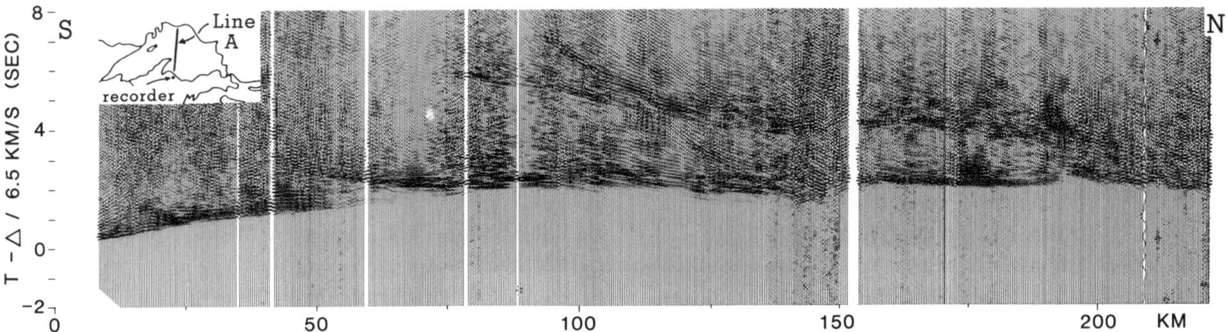

Figure 2. Seismic refraction/wide-angle reflection profile recorded by the University of Wisconsin during the GLIMPCE project in Lake Superior. Seismograms are about 300 m apart and reveal clear (refraction, refracted, and diffracted) phases. Seismic processing procedures normally reserved for seismic reflection data may be applied to these densely sampled data.

should utilize multiple perpendicular profiles with three-component receiver spacings of 200 m or less, and shot-point intervals of 5 km or less. In many areas, the excellent reconnaissance profiles recorded in the 1950s and 1960s should be augmented with multiple, densely sampled lines in order to resolve details of the crustal structure. Moreover, such detailed interpretations should include clear statements that describe which features of the model are well constrained, and which are merely permissible. Alternate acceptable models should be treated as well. Uncertainties in interpretation can be reduced by explicit waveform modeling using synthetic seismogram methods, and many of the data-processing techniques developed for reflection data (such as common midpoint stacking and migration) can be adapted for use in seismic refraction/wide-angle reflection processing. With time, the distinction between seismic reflection and refraction/wide-angle reflection profiling will greatly diminish (Fig. 2).

Teleseismic and geotomographic studies have been widely used in crustal and lithospheric studies in the past decade. As generally applied, these methods use only first-arrival times of compressional waves. In several cases, teleseismic and geotomographic inversions have resulted in velocity models that disagree in detail with earlier inversions or with seismic refraction or earthquake time-term studies (e.g., compare Iyer, 1984 [teleseismic] with Braile and others, 1982 [seismic refraction]; Hearn, 1984 [time-terms] with Hearn and Clayton, 1986 [geotomography]; Steeples and Iyer, 1976 [teleseismic] with Kissling, 1988 [geotomography]). The resolving power of these methods must be improved so that the potential and flexibility of passive seismic methods can be fully exploited. Combining seismic refraction profiles with teleseismic and gravity data would make it possible to strip off near-surface layers, thereby improving resolution of the deeper structure. The uncertainties in these inversions would be reduced dramatically if broad-band data were recorded, and waveform and amplitude information explicitly included in the inversion process, as has recently been developed in single-station methods (e.g., Owens and Zandt, 1985).

Potential-field data provide by far the most extensive and uniform coverage of the continental United States. However, many gravity (or magnetic) models of crustal and/or lithospheric structure are more complex and detailed than the data require. Only simple models are warranted by potential field data unless additional constraints are available from other geophysical or geologic data. Inferred models should be accompanied by discussions of model ambiguities and uncertainties. The separation of regional and residual fields is a particularly difficult problem because the sources of long-wavelength anomalies are not well understood, and their removal, while somewhat arbitrary, can seriously affect modeling results. Research into the origin of these long-wavelength anomalies, and of the magnetization of rocks of the deep crust and upper mantle, is needed.

In addition to density and magnetic properties, laboratory measurements of other physical properties of rocks provide important constraints on the interpretation of field data. However, there are important uncertainties to be considered in the application of laboratory data to the estimation of the in situ properties of the lithosphere. These uncertainties include the question of the influence of scale on measurements of physical properties, the probability that fluid pore pressures are significantly higher at depth than near the surface, and that large connected fracture systems control such properties as permeability and rock strength. As is evident from the initial results from the deep drillhole on the Kola Peninsula in the Soviet Union (Kozlovsky, 1987), ultradeep drilling is required to provide calibration between laboratory data and in situ properties for the upper crust.

A knowledge of the thermal structure of the lithosphere is a necessary prerequisite to determining its rheology and physical state. Thermal-mechanical models have been successful in explaining many aspects of basin formation and convergent plate boundaries, but important details remain unexplained (Furlong and Chapman, 1987). The low heat flow measured on transform faults presents a paradox, since frictional heating would be expected to be generated due to fault motion (Lachenbruch and Sass, 1980). Uncertainties in temperatures at depth appear to be large, with estimates at a depth of 40 km in areas with average heat flow ranging from 450°C to 700°C (Jeanloz and Morris, 1986). Critical information on these problems would be obtained from heat-flow and heat-generation determinations in deep holes over a broad geographic area.

The common reliance on forward modeling in the interpretation of geophysical data makes it difficult to assess whether a model is in any formal sense a unique solution to the interpretational problem. While an investigator may be correct in assuming that other specialists can distinguish free parameters from well-constrained ones, such information should be stated explicitly for nonspecialists. Increased use of inverse methods would begin to provide insight into this "uniqueness problem." The modeling and interpretation through simultaneous inversion of multiple data sets would provide the best constraints on model uniqueness.

A major source of uncertainty that affects all aspects of geophysics is the general impracticality of repeating a field measurement. From the point of view of the other physical sciences, this is a highly atypical situation. A physicist investigating a given phenomenon normally will attempt to reproduce published results before pursuing the phenomenon further, and a chemist will reproduce reported reactions before investigating related ones. In contrast, in geophysics we are accustomed to accepting reported field measurements without serious thought of repeating the experiment. The uncertainties and errors that result from this situation are poorly known, but are likely to be large. In view of this, the raw data of geophysical investigations should be made readily available to other investigators for independent evaluation. Reinterpretation of original data often reveals new insights, and at least partially accomplishes some of the purposes of reproducing results. Resurveying previously studied critical areas using improved methods is an ideal way to confirm and extend earlier interpretations.

APPLICATION OF COINCIDENT GEOPHYSICAL AND GEOLOGICAL METHODS

Any single type of geophysical data, no matter how well determined, is insufficient to define adequately the structure, composition, and physical properties of the lithosphere. The most productive future investigations in terms of increased resolution and reduced uncertainties will be those that apply different geophysical methods over the same area (coincident geophysical investigations), and those that include geological investigations as a vital ingredient.

There have been several recent examples of transects where coincident geophysical methods have been used in concert with broadly based geological investigations. Clowes and others (1987) and Jones (1987) have illustrated the power of coincident magnetotelluric, seismic reflection, and refraction/wide-angle reflection and geologic data to determine crustal structure and composition beneath Vancouver Island, British Columbia. The U.S. Geological Survey Trans-Alaska Crustal Transect (TACT) combines geologic, seismic reflection, seismic refraction/wide-angle reflection, gravity, aeromagnetic, heat-flow, and electrical studies (Page and others, 1986). A similar multidisciplinary effort is underway in the southern Basin and Range–Colorado Plateau transition (CALCRUST: California Consortium for Crustal Studies; PACE: Pacific to Arizona Crustal Experiment; and COCORP; Fig. 3). In the Basin and Range at 40°N., COCORP reflection data (Allmendinger and others, 1987) was recently complemented by coincident seismic refraction/wide-angle reflection and three-component earthquake data collected by PASSCAL (Program for Array Seismic Studies of the Continental Lithosphere; Smith, 1987). Broadly based geophysical investigations with active drilling include the southern Appalachians project (ADCOH, Appalachians Deep Continental Observation Hole) and the DOSEC (Deep Observation and Sampling of the Earth's Continental Crust, Inc.) project at Cajon Pass in southern California.

Coincident seismic reflection and refraction studies combine the structural resolution of the reflection method with the velocity resolution of the refraction/wide-angle reflection method, and therefore reduce the ambiguities in the interpretation of either method alone (Mooney and Brocher, 1987). Such studies should be combined with local earthquake and teleseismic studies that provide information on shear-wave structure and more effectively probe the subcrustal lithosphere.

At present, only the gross shear-wave structure of the crust and subcrustal lithosphere is known, as determined mainly from the study of earthquake traveltimes and amplitudes and from seismic surface waves. However, shear-wave structure needs to be determined with the same resolution as compressional wave structure if Poisson's ratio is to be reliably calculated. Near-vertical incidence shear-wave reflection data would be particularly valuable in determining the origin of deep crustal reflections. Shear waves have the potential for giving higher resolution at the same frequencies as compressional waves and are affected quite differently than compressional waves by the presence of fluids in rocks. Combined compressional and shear-wave seismic reflection sections will provide the most detailed information about seismic anisotropy, fluid content, lithology, density, and physical state in the lithosphere (Fig. 4).

Several chapters in this volume demonstrate the flexibility and versatility of the modern processing of gravity and aeromagnetic data to emphasize lateral changes in density or magnetic properties in either deep or shallow portions of the crust. Filtered long-wavelength gravity maps provide clear evidence for density variations and isostatic compensation in the upper mantle. Future investigations should take advantage of these processing advances. The isostatic gravity map of Jachens and others (this volume) and the filtered gravity maps of Kane and Godson (this volume) define a host of targets for complementary geophysical study.

The reliable interpretation of all field data requires laboratory measurement of the physical properties of rocks, yet field experiment designs to date have not generally included a program of rock sample collection. Future experiments should include laboratory measurements, and where possible, in situ borehole measurements, as an integral part of a coincident geophysical experiment. Laboratory measurements (where necessary, at ele-

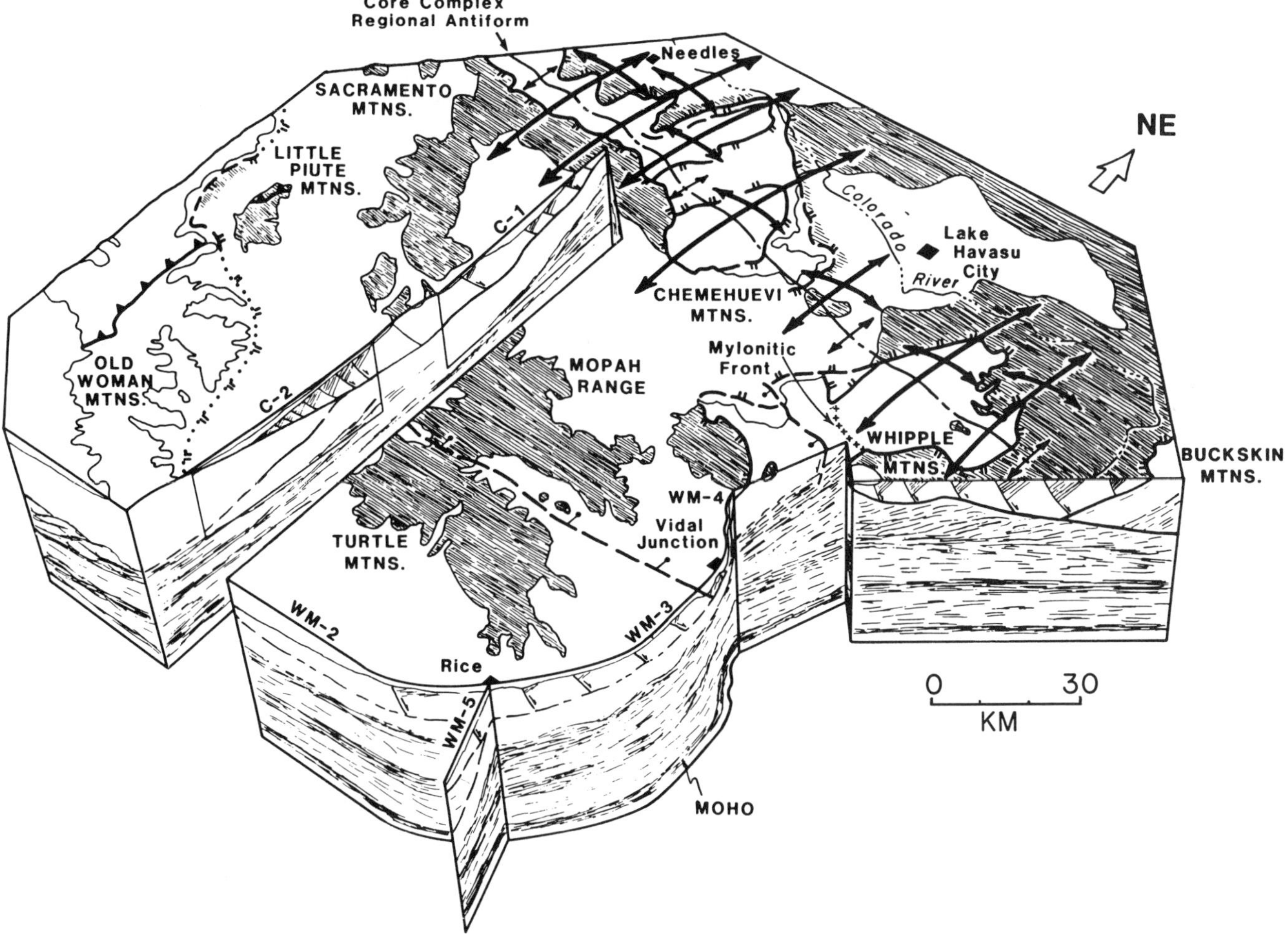

Figure 3. Block diagram (1:1) of the seismic structure of the crust of southeastern California and southwestern Arizona from seismic reflection data collected by CALCRUST (Frost and Okaya, 1989). Doubly plunging anticlines mark core complexes (Hamilton, this volume). Prominent upper crustal reflections are correlated with exposed mylonitic rocks. The Moho is at a depth of about 28 km, and the lower crust is highly reflective. The nature of these reflective lower crustal rocks is poorly understood and will require coincident geophysical measurements for determination.

vated pressures and temperatures) provide basic rock properties such as seismic velocity and attenuation, density, magnetization, heat production, strength, and electrical properties. These studies also yield significant insights into problems of seismic anisotropy and rock pore pressures. These problems are certain to be emphasized in the coming decade (Fountain and Christensen, this volume; Christensen, this volume).

CRITICAL GAPS IN KNOWLEDGE OF THE GEOPHYSICAL FRAMEWORK

At present, a balanced discussion of the geophysical framework of the United States is made difficult by the lack of adequate geophysical data for many areas. In this section we discuss critical gaps in continent-wide surveys and identify important regional studies by tectonic setting.

Gaps in continent-wide surveys

Despite the recent vigorous activity in seismic reflection and refraction/wide-angle reflection profiling in the United States, there remain large areas that are poorly studied or unsurveyed. Although crustal thickness is reasonably well known in nearly all regions, information on the compressional wave velocity structure is inadequate for large portions of the continental United States, particularly in the continental interior (Braile and others, this volume), and the shear-wave structure is essentially unknown. Similarly, seismic reflection studies have been completed only over limited geographic areas, amounting to about 20,000 line-km. Many important research opportunities remain in the study of the seismic reflectivity and seismic velocity structure of the continental United States.

Whereas the depth to the top of the upper mantle (i.e., the

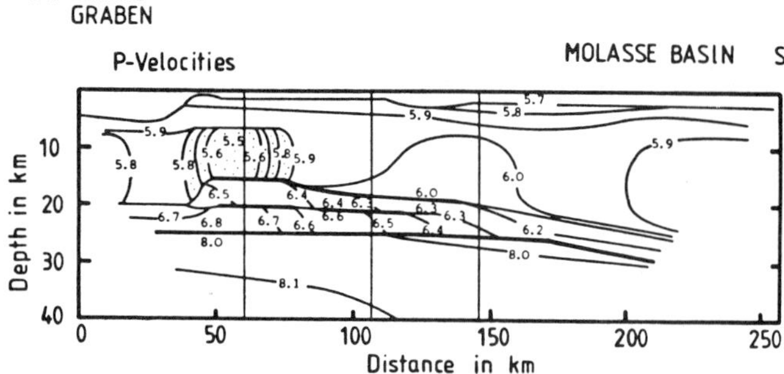

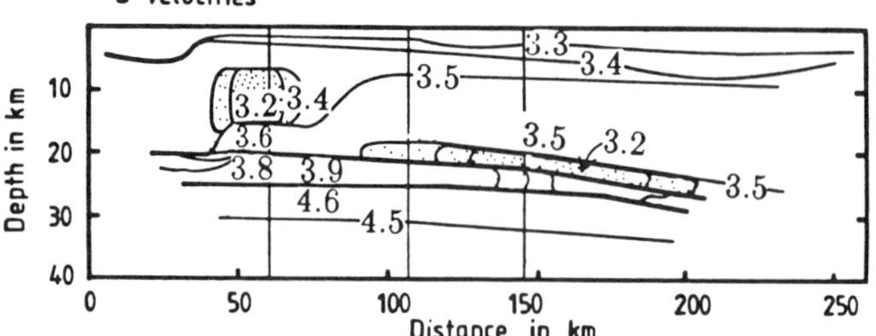

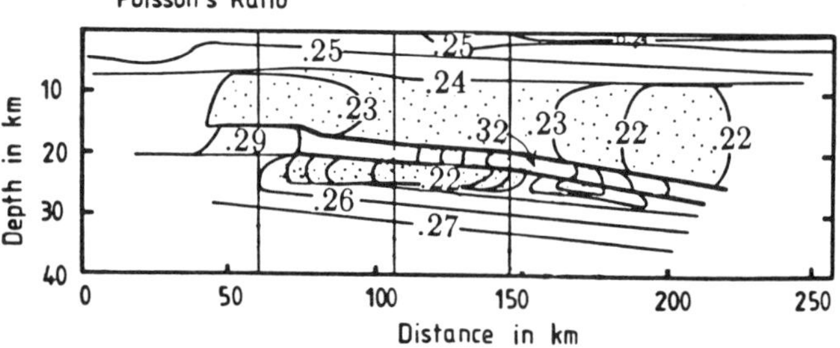

Figure 4. Example of a crustal section for which the seismic velocity of compressional waves (top), shear waves (middle), and Poisson's ratio (bottom) have been determined in two dimensions from seismic refraction/wide-angle reflection data (Holbrook and others, 1989). Crustal section is located in southwest Germany. The combined compressional and shear-wave structure has been used to infer the composition and physical state of the crust and uppermost mantle.

Moho) is generally well known on a continental scale, the summary of Iyer and Hitchcock (this volume) makes it clear that the depth to the lithosphere/asthenosphere boundary is poorly known. Investigations of the structure of the subcrustal lithosphere in the United States lag far behind those of Western and Eastern Europe (Ryaboy, 1977; Prodehl, 1984). We now have the opportunity to make effective and widespread use of seismic and other geophysical methods to measure the seismic velocity structure and other physical properties of the subcrustal lithosphere.

The reduction in uncertainties in the determination of the electrical conductivity of the lithosphere is a high priority for future research. Once uncertainties have been reduced, electrical methods should be widely used. Large portions of the continent, particularly in the eastern two-thirds of the United States, have had few measurements of deep conductivity (Keller, part 2, this volume). In the future, electrical properties will play an increasingly important role in the geophysical characterization of the crust and subcrustal lithosphere because they provide unique insights into the physical state, and particularly to indicate high temperatures, the presence of fluids, partial melts, or other conducting zones.

An important challenge is to exploit aeromagnetic data to estimate the depth of the Curie isotherm throughout the United States (Fig. 5; Blakely and Connard, this volume), and to correlate this depth with other geophysical and geologic parameters, particularly heat flow, conductivity, seismic velocity structure, and the depth to the onset of high reflectivity in the crust. However, for much of the United States, aeromagnetic data are inadequate for this task in terms of quality or density. Systematic regional surveys are needed to ensure reliable determinations of the depth to the Curie isotherm.

The recently developed methods of space geodesy using the Global Positioning System (GPS) and very long baseline interferometry (VLBI) make possible accurate measurements of plate motions and continent-wide crustal strain (Kroger and others, 1987; Lichten and Border, 1987). For example, one of the principal questions regarding plate motions concerns the apparent discrepancy between the estimated relative plate motion across the San Andreas fault determined from geodetic and Holocene geologic data (34 ± 3 mm/yr) as compared with the predictions of global solutions of rigid plate motions (56 ± 3 mm/yr). VLBI data have begun to resolve this well-known "San Andreas discrepancy" by confirming hypothesized plate deformation distributed both east and west of the fault (Minster and Jordan, 1987; Kroger and others, 1987). Space geodetic methods will revolutionize crustal deformation studies by providing a highly mobile, accurate, and flexible method for measuring deformation over large or small areas.

Gaps in regional surveys

The regional summaries of geophysical framework in this volume identify the most important regions in need of further geophysical investigation. Here we discuss briefly these needs by tectonic setting rather than by geographic region.

Craton and platform: The continental interior. With the exception of gravity and magnetic data, this vast region has received by far the least geophysical attention, perhaps because large areas lack exposed crustal rocks to define geological problems and correlate with subsurface information. In addition, important national missions such as nuclear test detection and identification and earthquake prediction have tended to emphasize research in the western states, notwithstanding the major earthquake hazards in the eastern and central United States. Detailed studies of the geophysical framework of the areas that form the nucleus of the continent are needed if questions regarding continental growth and evolution are to be addressed. Furthermore, it has been suggested that there are fundamental differences

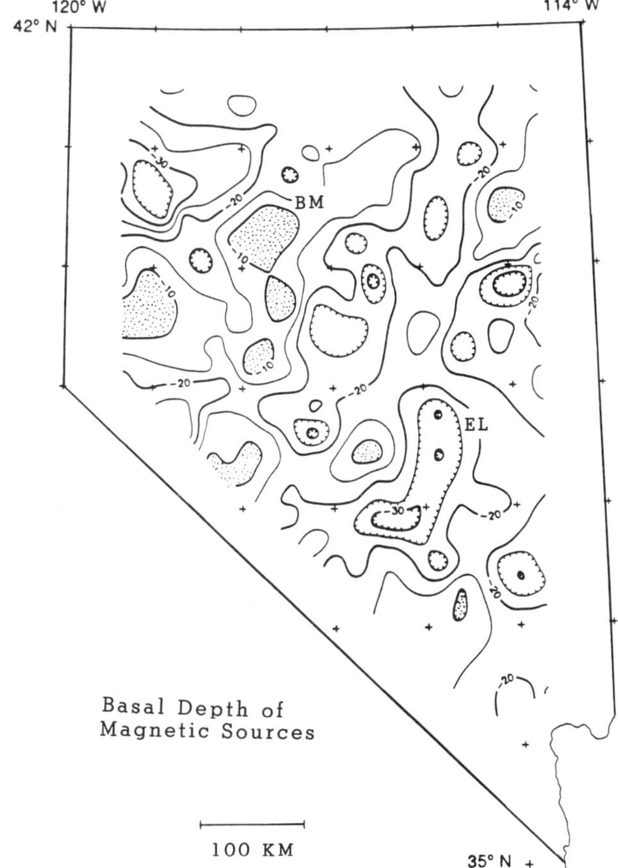

Figure 5. Contour map of depth to the Curie temperature (approximately 580°C) in Nevada derived from aeromagnetic data (Blakely, 1988). Depths less than 10 km shown with stippled pattern. North-south pattern of shallow depths (118° longitude) correlates with the region of historic faulting in Nevada, anomaly labeled B.M. correlates with the Battle Mountain heat-flow high, and anomaly E. L. correlates with the Eureka heat-flow low. The estimation of depth to Curie temperature from magnetic anomalies requires significant geophysical and geological simplifications in the treatment of the properties of the crust. However, the correlation with heat flow, seismic velocities, and reflectivity, and conductivity is potentially very informative regarding the physical state of the crust (Blakely, 1989).

in the seismic structure of old and young crust that reflect a change in geologic processes with time. Continued efforts to investigate this hypothesis are needed. The definition and refinement of age provinces in these areas is an important adjunct to their geophysical investigation.

Rifts and extended crust. Continental rifting constitutes the beginning of the plate-tectonic cycle. In many cases, however, rifting stops at an early stage, leaving an altered and possibly weakened crust that is later reactivated. Rifted and extended regions occur throughout the continental United States. The oldest rifts occur in the continental interior, including the Mid-continent Rift, the Wichita Alacogen, and the Reelfoot Rift (Mississippi Embayment). The Rio Grande Rift is a prominent modern rift system, and the Basin and Range province forms a dramatic extended regime in the western United States. Coincident geophysical measurements are needed to study the deep structure and evolution of rifts and to test hypotheses regarding the mechanisms of extension.

Orogens old and young. Studies of orogenic belts have long had a high priority in geophysics. One of the most important discoveries of the past ten years was the role of thin-skinned tectonics in the Appalachians (c.f., Phinney and Roy-Chowdhury, this volume). Nevertheless, the structure below the Appalachian thrust sheets (Grenville basement?) is poorly known. Similarly, of the many Laramide and later Cenozoic orogens, the deep structure of the Rocky Mountains is among the most poorly studied, despite the spectacular geology of these mountains. Thus, the debate over the deep structure of orogens continues in spite of a 50-yr history of geophysical investigations.

Margins. Critical to an understanding of the evolution of continental lithosphere is the study of continental margins, both rifted (Atlantic and Gulf Coast) and accreted (Pacific). Ancient continental margins recognized in the United States include those of the proto-Atlantic margin beneath the Appalachian thrust sheets, and the Paleozoic margin of the central Basin and Range province. Geophysical studies of the modern margins are summarized by Trehu and others (this volume) and by Couch and Riddihough (this volume); the coverage in these areas is very uneven, with few deep seismic refraction surveys over the Atlantic margin and very limited deep reflection or refraction data over the modern Pacific margin. The entire geophysical data base for the margins must be improved if progress is to be steady in studies of continental margins. Geophysical investigations of continental margins are particularly challenging because of the strong horizontal variations in properties.

Convergence zones, active and recently active. The Pacific Northwest convergence zone offers many important problem-solving opportunities, including the nature and process of the subduction of the Juan de Fuca plate, which lacks a well-defined Benioff zone beneath Oregon. Studies of the seismicity associated with the Cascades volcanoes are important for the prediction of volcanic eruptions. The crustal and upper-mantle structure of the Coast Ranges of northern California and Oregon remain poorly known, although this region is critical to understanding the process of accretion. An asthenospheric window (i.e., a gap in subducted oceanic lithosphere that brings the asthenosphere to the base of the crust) has been postulated south of the onland extension of the Mendocino triple junction based on plate-tectonic considerations, and its existence is supported by heat-flow (Lachenbruch and Sass, 1980) and gravity data (Jachens and Griscom, 1983). This hypothesis should be tested with detailed seismic and geoelectrical studies.

Transition zones. Transition zones separating geologic or physiographic provinces are areas of complex structure, including steeply dipping interfaces and severe lateral changes in velocity, density, and other physical properties. Such complexities involve changes not only in crustal thickness but other changes through-

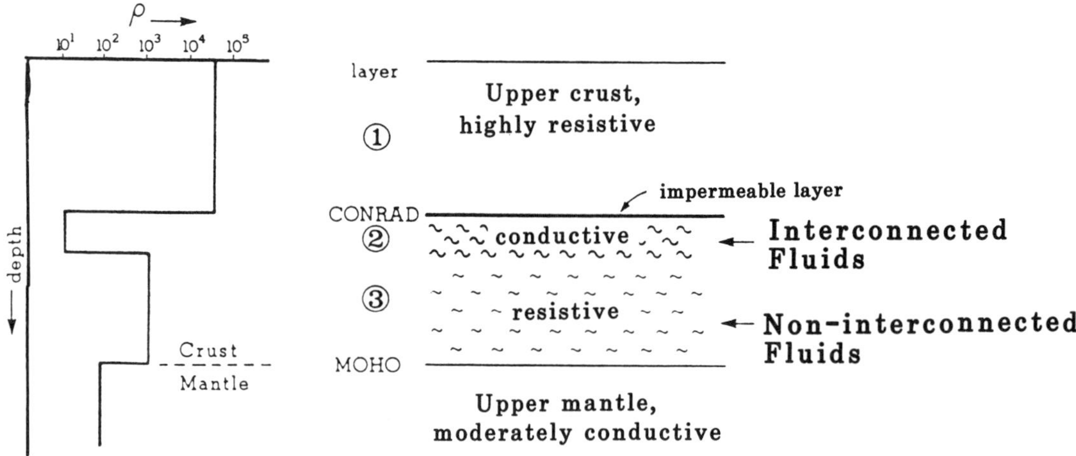

Figure 6. Generalized conductivity–depth structure of the crust and upper mantle with hypothesized fluid-filled sub-Conrad discontinuity in lower crust (modified from Jones, 1987). According to Matthews (1986), fluids in the lower crust also contribute to high seismic reflectivity. Coincident geophysical measurements of seismic velocity, seismic attenuation, conductivity, magnetism, heat production, and seismic reflectivity are needed to test this hypothesis.

out the lithosphere, including the depth to the top of the asthenosphere. These lateral changes in the structure and properties of the lithosphere are evidenced by rather severe variations in seismic wave propagation, gravity and magnetic anomalies, electrical conductivity, heat flow, stress fields, and seismicity. Thus it is especially important to plan geophysical surveys in transition zones with great care, to use the most modern techniques available, and to conduct geophysical studies of more than one type—at the minimum, seismic reflection, refraction/wide-angle reflection, electrical conductivity, and potential-field surveys—to minimize uncertainties and ambiguities in the resulting models.

Some transition zones in the continental United States have been studied intensively, but there is still considerable debate over the nature of the structural changes and variations in rock properties, notably P_n velocities, within them. The transition zones that particularly warrant modern geophysical investigations include those separating the Basin and Range and Colorado Plateau, Basin and Range and Sierra Nevada, Rocky Mountains and Great Plains, and Precambrian crust from accreted crust in the western United States.

TOPICAL GEOPHYSICAL QUESTIONS

Many field experiments are motivated by economic, public safety, or national security interests, or by local or regional geological concerns. Such field projects continue to have a high priority, but there is also a need for coincident geophysical experiments to investigate topical geophysical questions. These questions concern the physical properties, physical state, and geologic processes active within the lithosphere.

The origin and interpretation of deep crustal seismic reflections remains enigmatic despite more than a decade of active seismic reflection profiling. These reflections have been hypothesized as arising from igneous sills, cumulate or metasedimentary layering, open fluid-filled cracks, and ductile flow fabrics in seismically anisotropic rocks (Meissner, 1986; Matthews, 1986; Klemperer, 1987; Mooney and Brocher, 1987). In view of the variations in the character of these reflections (i.e., steeply dipping, subhorizontal, arcuate, and criss-crossing; Smithson, 1986), it appears certain that they have no single cause. However, the identification of the most common causes and their regional and tectonic significance is an important task. An equally interesting problem is the occurrence of seismically transparent zones in the crust, and particularly in the upper mantle. What are the properties required to produce a reflection-free region in the lithosphere? Why does the upper mantle in some places produce prominent (dipping) reflections (Warner and McGeary, 1987) whereas it is generally transparent in presently available data? These and related questions regarding the causes of deep crustal reflections will require the use of coincident geophysical methods.

A question of profound importance to essentially all geologic and geophysical processes in the crust is the distribution and abundance of fluids in the crust (Mase and Smith, 1987; Christensen, this volume). Matthews (1986) and Klemperer (1987) have suggested that deep crustal reflections arise from fluid-filled cracks, and Jones (1987) correlated increased lower crustal conductivity with free water (Fig. 6). Gold and Soter (1985) suggested that deep earthquakes are triggered by fluid ascent through the lithosphere; geologic studies indicate that pore pressures within some fault zones may approach lithostatic values (Parry and Bruhn, 1986). It is critical that well-designed experiments using coincident techniques in a variety of geologic settings be applied to the question of the amount, composition, and distribution of fluids in the crust. Shear-wave seismic reflection profiles and measurements in ultra-deep (up to 15 km) drillholes (Kozlovsky, 1987), as proposed in this country by DOSECC (Raleigh,

1985), hold great potential for contributing to the resolution of this question.

The genesis and evolution of the Moho has emerged as an increasingly challenging geophysical and geological problem. It is uncertain whether the continental Moho is everywhere the same, as is believed for the oceans. While it is well established that the change from lower-crustal velocities to upper-mantle velocities commonly occurs over a transition zone some 1 to 5 km thick, the detailed structure of this transition zone is poorly known. Perhaps more significantly, in areas of active tectonism the Moho appears to be a dynamic rather than a static feature of the lithosphere, as evidenced by the nearly constant two-way traveltime to the seismically transparent upper mantle ("reflection Moho") observed across the Basin and Range province (Klemperer and others, 1986; Allmendinger and others, 1987). How the Moho evolves in extensional and compressional settings is an important question to be studied with coincident geophysical measurements and theoretical modeling.

Seismic anisotropy of the crust and upper mantle has been established experimentally and described theoretically. Seismic anisotropy in the lithosphere is most reliably evidenced by shear-wave splitting, which may result from one of two principal causes. The first is due to deviatoric stress applied to microcracks in otherwise isotropic and homogeneous rocks, and the second to the orientation of intrinsically anisotropic minerals. On entering an anisotropic region, a shear wave splits into two phases, each having the polarization and velocity determined by the anisotropic symmetry of the region. Both mechanisms have tectonic significance. For stress-induced anisotropy the polarization direction and the changes between the slow and fast shear wave over time may reflect stress variations due to stress accumulations and release in seismogenic zones. Mineralogic anisotropy, on the other hand, is likely related to preferred mineral orientation produced by plate motions or extensional processes. Detailed seismic studies are needed to map crustal and lithospheric anisotropy and to relate these observations to lithospheric stress, fracturing, and other geologic processes.

The determination of the rheologic properties of the continental lithosphere is an important research topic. The complex thermal and mechanical properties of the continental lithosphere, as compared with the oceanic lithosphere, present major problems for studies of continental rheology (Kirby and Kronenberg, 1987; McNutt, 1987). Despite these problems, several lines of evidence suggest that the continental lithosphere is stratified into a strong upper crust, weak lower crust, and strong upper mantle (e.g., Meissner and Strehlau, 1982). Kirby and Kronenberg (1987) summarized the evidence in favor of a "crustal asthenosphere" (weak lower crust) and further suggested that the crust-mantle boundary may be a rheologic discontinuity. These concepts have profound implications for the tectonics of continental collision zones, the loading of strike-slip fault zones such as the San Andreas fault, and models of the kinematics of continental rifting.

Geophysical studies have the potential to determine the mechanism for large horizontal transport of displaced terranes. Paleomagnetic measurements have contributed greatly to the study of displaced terranes, and have more recently discovered large block rotations, particularly near continental margins. The mechanisms by which transport and rotation of continental slivers are accommodated is poorly understood (Wells and Coe, 1985; Bruhn, 1987; Hillhouse and McWilliams, 1987), but they may occur over a weak "crustal asthenosphere," as described above. Several of the previously mentioned topics, particularly the measurement of crustal velocities, thermal regimes, and rheologies, are pertinent in these studies.

The emphasis of this volume is on the geophysical framework of the continental United States, rather than on research related to public safety or other direct societal interests. However, the problem of the short-term prediction of earthquakes is closely related to the topics discussed here. It is likely that progress in earthquake prediction, which has been very challenging to date, will occur only as fast as progress is made in the topics discussed in this volume—rheology of the curst, migration of fluids, seismic structure of the crust. It is difficult to imagine that we will understand and predict earthquakes until we understand the physics of the crust. Detailed observations of the earthquake source are important to prediction studies as well, and digital three-component surface and borehole seismic arrays will contribute greatly.

DISCUSSION

The primary physical properties of the lithosphere that we seek to determine are compressional and shear-wave seismic velocity, seismic attenuation (Q), density, conductivity, magnetization, heat production, permeability, strength, and rheology. The primary physical states to be determined are temperature, state of stress, degree of fracturing, and fluid pore pressure. If these physical properties and states can be reliably estimated, lithospheric processes and composition can be reasonably inferred from geologic knowledge and laboratory measurements of rock properties.

We have emphasized four factors that we believe are critical to future significant advances in geophysical research: the reduction of uncertainties in geophysical measurements and interpretations, the widespread application of coincident methods in concert with geological studies, the gathering of a more nearly uniform continental-scale data base, and the vigorous investigation of several topical geophysical questions.

The degree of future progress in geophysics will depend in large part on our ability to work with other specialists in our own fields, and across discipline lines. We are in the world of "big science" with our collective and extensive experiments, just as the physicists and astronomers are, but without their large facilities. The most challenging geophysical problems will require cooperation on a scale equal to any physics experiment. Examples include geophysical transect programs and deep continental drilling.

If we are to work together effectively, we need to understand the strengths, weaknesses, resolution, and uncertainties of our colleagues' methods. It is our hope that this volume will facilitate such understanding, and will foster communication and cooperation across a broad scale.

ACKNOWLEDGMENTS

Summarizing the progress and opportunities in any aspect of science is a hazardous undertaking, and oversights are inevitable. We are grateful for comments, reviews, and suggestions from M. E. Beck, H. M. Benz, T. M. Brocher, C. Craddock, J. W. Dewey, C. L. Drake, E. R. Engdahl, E. R. Flueh, R. C. Jachens, R. A. Johnson, J. McCarthy, R. Meissner, G. R. Olhoeft, K. Roy-Chowdhury, S. Schapper, R. B. Smith, S. B. Smithson, G. A. Thompson, and R. Van der Voo. We accept the responsibility for any errors of omission and commission.

REFERENCES CITED

Airy, G. B., 1855, On the computations of the effect of the attraction of the mountain masses as disturbing the apparent astronomical latitude of stations in geodetic surveys: Philosophical Transactions of the Royal Society of London, v. 145, p. 101–104.

Allmendinger, R. W., and 7 others, 1987, Overview of the COCORP 40°N transect, western United States; The fabric of an orogenic belt: Geological Society of America Bulletin, v. 98, p. 308–319.

Behrendt, J. C., and 7 others, 1988, Crustal structure of the Midcontinental rift system; Results from GLIMPCE deep seismic reflection profiles: Geology, v. 16, p. 81–85.

Blakely, R. J., 1988, Curie-temperature isotherm analysis and tectonic implications of aeromagnetic data from Nevada: Journal of Geophysical Research, v. 93, p. 11817–11832.

Braile, L. W., and 9 others, 1982, The Yellowstone–Snake River Plain seismic profiling experiment; Crustal structure of the eastern Snake River Plain: Journal of Geophysical Research, v. 87, p. 2597–2609.

Bruhn, R. L., 1987, Continental tectonics; Selected topics: Reviews of Geophysics, v. 25, p. 1293–1304.

Clowes, R. M., and 7 others, 1987, LITHOPROBE–southern Vancouver Island; Cenozoic subduction complex imaged by deep seismic reflections: Canadian Journal of Earth Sciences, v. 24, p. 31–51.

Frost, E. G., and Okaya, D. A., 1989, Continuity of exposed mylonitic rocks to middle- and lower-crustal depths within the Whipple Mountains detachment terrane, SE California, and its implications for continental extensional tectonics: Tectonics (in press).

Furlong, K. P., and Chapman, D. S., 1987, Thermal state of the lithosphere: Reviews of Geophysics, v. 25, p. 1255–1264.

Gold, T., and Soter, S., 1985, Fluid ascent through the solid lithosphere and its relation to earthquakes: Pure and Applied Geophysics, v. 122, p. 492–530.

Hayford, J. F., and Bowie, W., 1912, The effect of topography and isostatic compensation upon the intensity of gravity: U.S. Coast and Geodetic Survey Special Publication 10, 132 p.

Hearn, T. M., 1984, P_n travel times in southern California: Journal of Geophysical Research, v. 89, p. 1843–1855.

Hearn, T. M., and Clayton, R. W., 1986, Lateral velocity variations in southern California; 2, Results for the lower crust from P_n-waves: Bulletin of the Seismological Society of America, v. 76, p. 511–520.

Hildenbrand, T. G., Simpson, R. W., Godson, R. H., and Kane, M. F., 1982, Digital colored residual and regional Bouguer gravity maps of the conterminous United States; Wavelength cutoffs of 250 and 1,000 km: U.S. Geological Survey Geophysical Investigations Map GP–953A scale 1:7,500,000.

Hillhouse, J. W., and McWilliams, M. O., 1987, Application of paleomagnetism to accretionary tectonics and structural geology: Reviews of Geophysics, v. 25, p. 951–959.

Holbrook, W. S., Gajewski, D., Krammer, A., and Prodehl, C., 1988, An interpretation of wide-angle compressional and shear-wave data in S.W. Germany; Poisson's ratio and petrological implications: Journal of Geophysical Research, v. 93, p. 12081–12106.

Iyer, H. M., 1984, A review of crust and upper mantle structure studies of the Snake River Plain–Yellowstone volcanic system; A major lithospheric anomaly in western U.S.A.: Tectonophysics, v. 105, p. 291–308.

Jachens, R. C., and Griscom, A., 1983, Three-dimensional geometry of the Gorda plate beneath northern California: Journal of Geophysical Research, v. 88, p. 9375–9392.

Jeanloz, R., and Morris, S., 1986, Temperature distribution in the crust and mantle: Annual Reviews of Earth and Planetary Sciences, v. 14, p. 377–415.

Jones, A. G., 1987, MT and reflection; An essential combination: Geophysical Journal of the Royal Astronomical Society, v. 89, p. 7–18.

Kirby, S. H., and Kronenberg, A. K., 1987, Rheology of the lithosphere; Selected topics: Reviews of Geophysics, v. 25, p. 1219–1244.

Kissling, E., 1988, Geotomography with local earthquake data: Reviews of Geophysics, v. 26, p. 659–698.

Klemperer, S. L., 1987, A relationship between continental heat-flow and the seismic reflectivity of the lower crust: Journal of Geophysics, v. 61, p. 1–11.

Klemperer, S. L., Hauge, T. A., Hauser, E. C., Oliver, J. E., and Potter, C. J., 1986, The Moho in the northern Basin and Range province, Nevada, along the COCORP 40°N seismic-reflection transect: Geological Society of America Bulletin, v. 97, p. 603–618.

Kozlovsky, Ye. A., ed., 1987, The superdeep well of the Kola Peninsula: Berlin, Springer-Verlag, 558 p.

Kroger, P. M., Lysenga, G. A., Wallace, K. S., Davidson, J. M., 1987, Tectonic motion in the western United States inferred from very long baseline interferometry measurements, 1980–1986: Journal of Geophysical Research, v. 92, p. 14151–14161.

Lachenbruch, A. H., and Sass, J. H., 1980, Heat flow and energetics of the San Andreas fault zone: Journal of Geophysical Research, v. 85, p. 6185–6223.

Lichten, S. M., and Border, J. S., 1987, Strategies for high-precision global positioning system orbit determination: Journal of Geophysical Research, v. 92, p. 12751–12762.

Mase, C. W., and Smith, L., 1987, The role of pore fluids in tectonic processes: Reviews of Geophysics, v. 25, p. 1348–1358.

Matthews, D. H., 1986, Seismic reflections from the lower crust around Britain, *in* Dawson, J. B., Carswell, D. A., Hall, J., and Wedepohl, K. H., eds., Nature of the lower continental crust: Geological Society of London Special Publication 24, p. 11–21.

McCarthy, J., Fuis, G. and Wilson, J., 1987, Crustal structure of the Whipple Mountains, southeastern California; A refraction study across a region of large continental extension: Geophysical Journal of the Royal Astronomical Society, v. 89, p. 119–124.

McNutt, M., 1987, Lithospheric stress and deformation: Reviews of Geophysics, v. 25, p. 1245–1253.

Meissner, R., 1986, The continental crust; A geophysical approach: Orlando, Florida, Academic Press, 426 p.

Meissner, R., and Strehlau, J., 1982, Limits of stresses in the continental crust and their relation to the depth-frequency distribution of shallow earthquakes: Tectonics, v. 1, p. 73–89.

Minster, J. B., and Jordan, T. H., 1987, Vector constraints on western U.S. deformation from space geodesy, neotectonics, and plate motions: Journal of Geophysical Research, v. 92, p. 4798–4804.

Mooney, W. D., and Brocher, T. M., 1987, Coincident seismic reflection/refraction studies of the continental lithosphere; A global review: Reviews of

Geophysics, v. 25, p. 723–742.

Owens, T. J., and Zandt, G., 1985, The response of the continental crust-mantle boundary observed on broadband teleseismic receiver functions: Geophysical Research Letters, v. 12, p. 705–708.

Page, R. A., and 6 others, 1986, Accretion and subduction tectonics in the Chugach Mountains and Cooper River Basin, Alaska; Initial results of the Trans-Alaska Crustal Transect: Geology, v. 14, p. 501–505.

Parry, W. T., and Bruhn, R. L., 1986, Pore fluid and seismogenic characteristics of fault rock at depth on the Wasatch fault, Utah: Journal of Geophysical Research, v. 91, p. 730–744.

Pratt, J. H., 1855, On the attraction of the Himalaya Mountains and of the elevated regions beyond upon the plumb-line of India: Philosophical Transactions of trhe Royal Society of London, v. 145, p. 53–100.

Prodehl, C., 1984, Structure of the Earth's crust and upper mantle, *in* Fuchs, K. and Soffel, H., eds., Geophysics of the solid Earth, the Moon, and the planets: Berlin, Springer-Verlag, p. 97–206.

Raleigh, C. B., ed., 1985, Observation of the continental crust through drilling 1: Berlin, Springer-Verlag, 364 p.

Ryaboy, V. Z., 1977, Study of the structure of the lower lithosphere by explosion seismology in the USSR: Journal of Geophysics, v. 43, p. 593–610.

Simpson, R. W., Jachens, R. C., Blakely, R. J., and Saltus, R. W., 1986, A new isostatic gravity map of the conterminous United States with a discussion on the significance of isostatic residual anomalies: Journal of Geophysical Research, v. 91, p. 8348–8372.

Smith, S. W., 1987, IRIS; A university consortium for seismology: Reviews of Geophysics, v. 25, p. 1203–1207.

Smithson, S. B., 1986, A physical model of the lower crust from North America based on seismic reflection data, *in* Dawson, J. B., Carswell, D. A., Hall, J., and Wedepohl, K. H., eds., The nature of the lower continental crust: Geological Society of London Special Publication 24, p. 23–24.

Society of Exploration Geophysicists, 1982, Gravity anomaly map of the United States (exclusive of Alaska and Hawaii): Society of Exploration Geophysicists, scale 1:2,500,000.

Steeples, D. W., and Iyer, H. M., 1976, Low-velocity zone under Long Valley as determined from teleseismic events: Journal of Geophysical Research, v. 81, p. 849–860.

U.S. Geological Survey and Society of Exploration Geophysicists, 1982, Composite magnetic anomaly map of the United States; Part A, Conterminous United States: U.S. Geological Survey Map GP 954-A scale 1:1,250,000, 2 sheets.

Warner, M., 1987, Migration; Why doesn't it work for deep continental data?: Geophysical Journal of the Royal Astronomical Society, v. 89, p. 21–26.

Warner, M., and McGeary, S., 1987, Seismic reflection coefficients from mantle fault zones: Geophysical Journal of the Royal Astronomical Society, v. 89, p. 223–230.

Wells, R. E., and Coe, R. S., 1985, Paleomagnetism and geology of Eocene volcanic rocks of southwest Washington; Implications for mechanisms of tectonic rotation: Journal of Geophysical Research, v. 90, p. 1925–1947.

Zoback, M. D., and 12 others, 1987, New evidence on the state of stress of the San Andreas fault system: Science, v. 238, p. 1105–1111.

Manuscript Accepted by the Society October 31, 1988

Printed in U.S.A.

Index

[Italic page numbers indicate major references]

Typeset by WESType Publishing Services, Inc., Boulder, Colorado
Printed in U.S.A. by Malloy Lithographing, Inc., Ann Arbor, Michigan